*Spencer's
Pathology
of the
Lung*

Spencer's Pathology of the Lung

Fifth Edition

EDITOR

P. S. Hasleton, M.D.

Consultant Histopathologist
Wythenshawe Hospital
Honorary Reader in Histopathology
University of Manchester
Manchester, United Kingdom

McGraw-Hill

Health Professions Division

New York   St. Louis   San Francisco   Auckland   Bogotá   Caracas   Lisbon
London   Madrid   Mexico City   Milan   Montreal   New Delhi   San Juan
Singapore   Sydney   Tokyo   Toronto

McGraw-Hill

A Division of The **McGraw·Hill** *Companies*

1234567890   KGP KGP   9876

ISBN 0-07-105448-0

This book was set in Melior by Better Graphics, Inc.
The editors were Martin J. Wonsiewicz and Steven Melvin;
the production supervisor was Clare Stanley.
Quebecor Printing/Kingsport was printer and binder.

This book is printed on acid-free paper.

Library of Congress Cataloging-in-Publication Data

Spencer's pathology of the lung.—5th ed. / editor, P.S. Hasleton.
    p.   cm.
   Rev. ed. of: Pathology of the lung / H. Spencer. 4th ed. 1985.
   Includes bibliographical references and index.
   ISBN 0-07-105448-0 (alk. paper)
   1. Lungs—Diseases.   I. Spencer, Herbert. Pathology of
the lung. II. Hasleton, P. S.
   [DNLM: 1. Lung Diseases.   WF 600 S746 1996]
RC756.S65   1996
616.2'407—dc20
DNLM/DLC
for Library of Congress   95-19711

Contents

Contributors

Bruce Addis, M.D. [8]
Chairman of Pathology
HCI International Medical Centre
Clydebank
United Kingdom

H. Bachofen, M.D. [23]
Departments of Anatomy and Medicine
University of Berne
Berne
Switzerland

P. W. Bishop, M.D. [10]
Department of Pathology
Wythenshawe Hospital
London
United Kingdom

F. Capron, M.D. [28]
Hôpital Antoine Béclère
Paris
France

Thomas V. Colby, M.D. [25, 27]
Department of Laboratory Medicine and Pathology
Mayo Clinic
Rochester, Minnesota

A. Curry, M.D. [1]
Department of Histiopathology
Withington Hospital
Manchester
United Kingdom

H. M. Doran, M.D. [24]
Department of Pathology
Regional Cardiothoracic Centre
Wythenshawe Hospital
Manchester
United Kingdom

Victor J. Ferrans, M.D. [26]
Chief, Pathology Branch
National Heart, Lung, and Blood Institute
National Institutes of Health
Bethesda, Maryland

Nick Francis, M.D., M.R.C.Path [4]
Charing Cross and Westminster Medical School
London
United Kingdom

A. R. Gibbs, T.D., M.B., Ch.B., F.R.C.Path. [15]
Department of Histopathology and Environmental
* Lung Disease Research Group*
Llandough Hospital NHS Trust
Penarth South Glamorgan
Wales
United Kingdom

Steven J. Gould, M.D. [3]
Department of Pediatric Pathology
Maternity Department
John Radcliffe Hospital
Headington
Oxford
United Kingdom

M. H. Griffiths, M.D. [14]
Department of Histopathology
Middlesex Hospital
London
United Kingdom

M. Harris, M.D. [33]
Department of Pathology
Christie Hospital
Manchester
United Kingdom

P. S. Hasleton [1, 2, 3, 4, 5, 6, 7, 9, 11, 12, 13, 14, 16, 17, 24, 29, 34]
Consultant Histopathology
Wythenshawe Hospital
Honorary Reader in Histopathology
University of Manchester
Manchester
United Kingdom

Donald Heath, M.D. [21]
Southport
Merseyside
United Kingdom

Numbers in brackets refer to chapters written or cowritten by the contributor.

J. Michael Kay, M.D.   [18]
St. Joseph's Hospital
Hamilton Ontario
Canada

Michael N. Koss, M.D.   [26, 32]
Co-Chairman, Department of Pulmonary and
*   Mediastinal Pathology*
Armed Forces Institute of Pathology
Washington, D.C.

D. Lamb, M.D.   [19]
Department of Pathology
Edinburgh University Medical School
Edinburgh
United Kingdom

S. B. Lucas, M.D.   [9]
Department of Pathology
UCL Medical School
Middlesex Hospital
London
United Kindom

W. Mooi, M.D.   [31]
Department of Pathology
The Netherlands Cancer Institute
Amsterdam
The Netherlands

Cesar Moran, M.D.   [32]
Department of Pulmonary and Mediastinal
*   Pathology*
Armed Forces Institute of Pathology
Washington, D.C.

Pamela H. Rabbitts, M.D.   [30]
MRC Clinical Oncology and Radiotherapeutics Unit
Medical Research Council Centre
Cambridge
United Kingdom

William R. Roche, M.D.   [22]
Senior Lecturer in Pathology
University of Southampton
Southampton
United Kingdom

M. Sarno, M.D.   [16]
Department of Respiratory Medicine
City General Hospital
Stoke-on-Trent
Staffordshire
United Kingdom

D.A. Schwartz, M.D., F.C.A.P.   [9]
Associate Professor of Pathology
Adjunct Assistant Professor of Medicine (Infectious
*   Diseases)*
Emory University School of Medicine
Consultant
Centers for Disease Control and Prevention
Atlanta, Georgia

Paul Smith, M.D.   [20]
Department of Pathology
Royal Liverpool Hospital
Liverpool
United Kingdom

M. A. Spiteri, M.D.   [16]
Department of Respiratory Medicine
City General Hospital
Stoke-on-Trent
Staffordshire
United Kingdom

Vasi Sudaresan, M.D.   [30]
Clinician Scientist Fellow
MRC Clinical Oncology and Radiotherapeutics Unit
Cambridge
United Kingdom

William D. Travis, M.D.   [26, 32]
Co-Chairman
Department of Pulmonary and Mediastinal
*   Pathology*
Armed Forces Institute of Pathology
Washington, D.C.

E. R. Weibel, M.D.   [22]
Vice Chairman and Secretary
Maurice E. Miller Foundation
Berne
Switzerland

W. F. Whimster, M.D.   [35]
Department of Pathology
King's College Hospital
London
United Kingdom

Jennifer A. Young, M.D.   [36]
Department of Pathology
Medical School
University of Birmingham
Edgbaston
Birmingham
United Kingdom

Preface

This is the fifth edition of *Pathology of the Lung*. The previous edition came out 11 years ago in 1985. The present book is a complete rewrite. In the current *Spencer's Pathology of the Lung* are sixteen new chapters reflecting the changing nature of cardiothoracic medicine and surgery and our understanding of the mechanisms of lung disease. The remaining chapters have all been completely rewritten. It may appear, with a quick look at the contributors, that the editor has written a large part of this new edition. Professor Spencer was not in favor of multiauthor works and together we started on the fifth edition. When Professor Spencer died it became rapidly apparent that it was not possible for one person to have the breadth of knowledge in all the fields covered in this volume or indeed the time to complete a book of such magnitude. Therefore other colleagues kindly undertook to reread and revamp chapters already written.

It is important at this stage to credit the late Professor Spencer for his contribution to pulmonary pathology. The subject has grown so rapidly that it is a tribute to him that he was able, over a period of some 20 years, to successively produce four editions of *Pathology of the Lung* that were highly regarded throughout the world. Indeed for many years his volume as the only comprehensive treatise on the pathology of lung diseases. Professor Avril A. Liebow, one of the other founding fathers of pulmonary pathology, in the foreword to the first edition said that "never have the lungs been so thoroughly or so well read and the reading so well recorded as in this volume."

The philosophy behind the current volume is different from previous editions. Immunoperoxidase results have been introduced especially to show their value in differential diagnosis. It is hoped that such sections will be of help to doctors not faced with pulmonary problems on a daily basis. For this reason the clinical and radiological features of different diseases have been included. No pathologist exists in a vacuum and in many instances he or she requires full clinical information to come to a sensible diagnosis. There are parts of the book with some repetition, especially in the chapters relating to diseases which cause pulmonary fibrosis. Personally, I find it irritating to have to flick from chapter to chapter to determine pathogenetic mechanisms or histologic features. I have therefore tried, where possible, to save the reader this problem.

I would like to thank numerous people without whom this new edition would not have been possible. First there is my wife, who has propped me up in my frequent hours of despair and forgone many evenings out. My family has also suffered from my surgical attachment to the computer and my lack of availability as a father. To them I offer my thanks for letting me pursue this venture and I want them to know they have not been forgotten.

Apart from family, numerous other people have contributed in very significant ways. Individual authors have been extremely generous in accepting my comments and meeting deadlines. Professors Colby, Travis, and MacKay were extremely generous with illustrations. A good secretary is essential in such a venture as this. Mrs. C. Harris has given unstintingly of her time and has given many of her own hours to make sure that the manuscript was produced on time. Mrs. Judy White, Medical Librarian, chased many references, did unnumerable literature searches, and obtained many hundreds of papers. Mr. Grant Whitehurst and Miss Sue Penny, in Medical Illustration, have over the years taken numerous pictures of the highest quality. My colleagues at Wythenshawe Hospital, Drs. Bishop and Doran, have in addition carried some of my work during the period of writing so that the routine service could be maintained. Their unstinting help in enabling me to get this volume finished is gratefully acknowledged. Helen Doran, in addition, made very constructive comments to the manuscript which were helpful and refreshing. I have, in addition, to thank my clinical colleagues in chest medicine and cardiothoracic surgery. They have unstintingly provided me, over the years, with many interesting cases and much clinical information. In honesty they have provided a milieu in which it is possible to really appreciate and understand pulmonary pathology in its wider aspects. It is this clinical contact that has made my period at Wythenshawe so enjoyable.

Finally I thank Mr. Ed Bolger, Martin J. Wonsiewicz, and Steven Melvin at McGraw-Hill. They pushed, shoved, cajoled, and finally got me to release the book to their safekeeping.

It is my hope that this revised volume, in the face of the considerable change and many advances taking place in pathology, as well as the increased expectations of clinicians, will aid the diagnostic histopathologist in his or her daily work. It still remains as true today as it did over nearly two thousand years ago that "the day is short and the work is great. . . . It is not our duty to complete the work but neither are we free to desist from it." (R. Tarfon. *Avot* 2:15–16).

Chapter 1

Anatomy of the Lung

P.S. Hasleton / A. Curry

MACROSCOPIC ANATOMY

Enclosed within the visceral pleura, the normal adult lungs fill most of the thoracic cavity extending from the root of the neck and the first rib above to the domes of the diaphragm. Medially, the bronchi, pulmonary arteries, veins, and neural connections supply the lung between the layers of the pulmonary ligament. The rest of the medial surface is separated from the mediastinum by two layers of pleura.

The main muscle responsible for breathing is the diaphragm, often neglected by pathologists. It possesses more mitochondria and capillaries than other skeletal muscles, indicating that it has a high oxygen consumption. The diaphragm and the intercostal muscles may be affected by degenerative diseases, with subsequent adverse effects on the lungs. The muscle activity is opposed by the elastic recoil of the lungs, and chest wall, and the resistance to airflow offered by the entire respiratory tract. Diaphragmatic weakness may present as breathlessness.[1] A review article on the human respiratory muscles is available.[2]

The major site for airflow resistance is the larynx, which dilates during inspiration and narrows on expiration, because of the action of the laryngeal muscles.

In the normal adult, each lung weighs approximately 500 g, but this figure is variable. Gough[3] gives a weight of 250 g for each lung in people killed by electrocution. Lung weight tends to diminish slightly in the elderly, probably because of enlargement of the alveoli and a consequent increase of air in the lungs.

The lungs are normally divided into three lobes on the right and two on the left, the former including the upper, middle, and lower lobes and the latter the upper and lower lobes. Each lobe is divided into segments; each segment contains a variable number of lung lobules. In humans, five to 10 acini constitute a lobule, which corresponds to Miller's secondary lobule (Fig. 1-1).[4] The anatomy of the segmental bronchi has been described in detail by Boyden,[5] who delin-

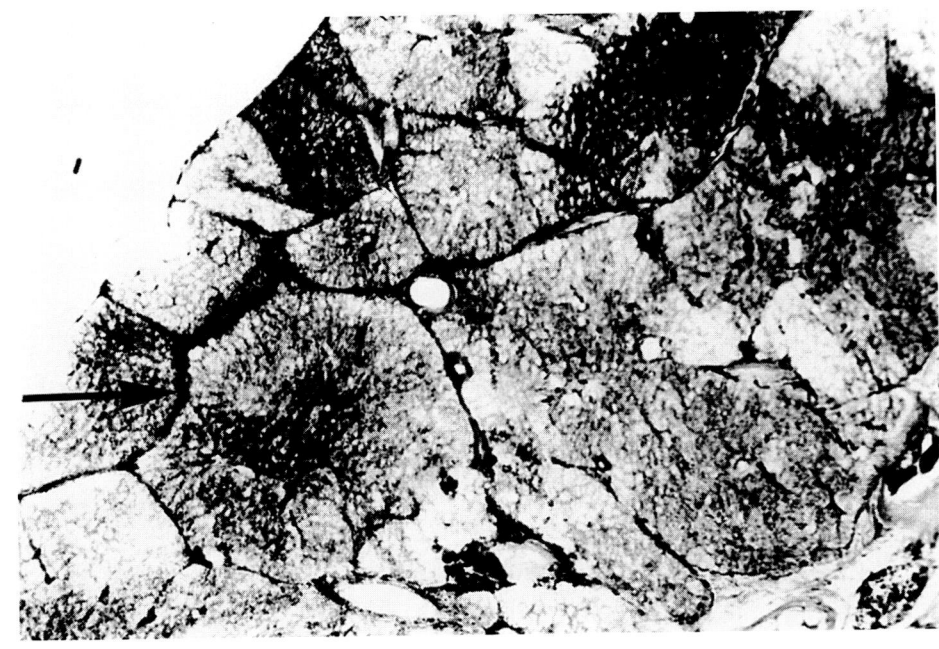

FIGURE 1-1
Part of a lung fixed with formalin vapor, showing well-delineated Miller's secondary lobules (*arrow*). There is much anthracotic pigment.

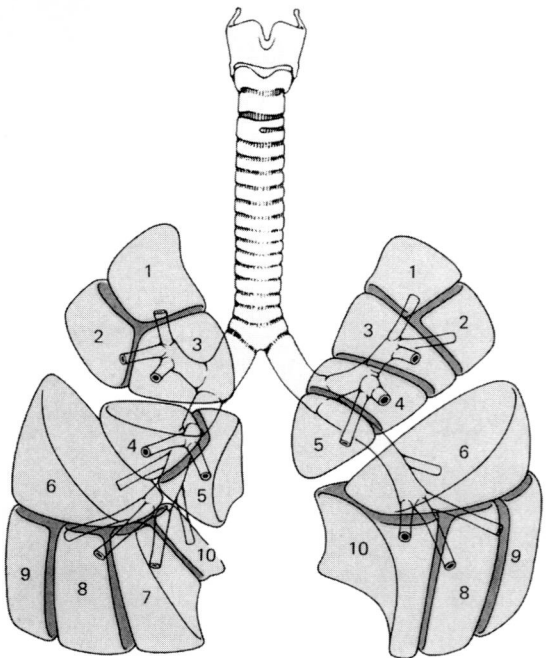

FIGURE 1-2
Bronchopulmonary segments of the human lungs, left and right upper lobes: 1, apical; 2, posterior; 3, anterior; 4, superior lingular and 5, inferior lingular segments; 6, superior (apical); 7, medial-basal; 8, anterior-basal; 9, lateral-basal; and 10, posterior-basal segments. The medial-basal segment (7) is absent in the left lung. (NOTE: The lungs are represented as slightly turned inward in order to display part of the lateral face.)

eates the variations. The normal distribution of the major lobar and segmental bronchi is shown in Fig. 1-2. Alterations in bronchial branching are common: 5 percent or more of left lungs may be three-lobed, and the right may be two-lobed. Up to 25 percent of lungs may show fissures in abnormal locations or incomplete division of lobar fissures. Such structural anomalies have no functional significance but may cause confusion on chest radiographs. A well-known anomaly is the azygous lobe (Fig. 1-3), which is outlined by extrinsic compression of the lung by an aberrant azygous vein in the right upper thorax. It does not reflect any underlying segmental division of the lung.

Normally, the right main bronchus closely follows the general direction of the trachea. The left diverges at a greater angle, especially in the female.

The size of the adult bronchial branches and their angles of bifurcation have been studied by Barnett.[6] In the collapsed lung, the cross-sectional area of the bronchial tree does not increase appreciably until branches with a diameter of 1.5 mm are reached. Below this point, there is a steady increase in the total cross-sectional area. A diagram of the bronchial branching pattern is shown in Fig. 1-4. The diameter of terminal bronchioles increases steadily with advancing age but, relative to the rest of the bronchial system, is greatest in the early years of life. There is a greater angle of divergence of the smaller bronchi as far as the fifth-generation branches. The cross-sectional areas of any two branches are always greater than that of the parent bronchus.

The lung has been divided into three zones (Fig. 1-5).[7] The inner, or hilar, zone consists entirely of lobar bronchi, main vessels, lymph nodes, nerves, and connective tissue but no alveolar tissue. Outside lies the intermediate, or medullary, zone, composed mainly of vessels with their second to fourth generation of branches. Between these structures, wherever space is available, lie lobules of lung tissue. These

FIGURE 1-3
Macroscopic picture of the azygous lobe.

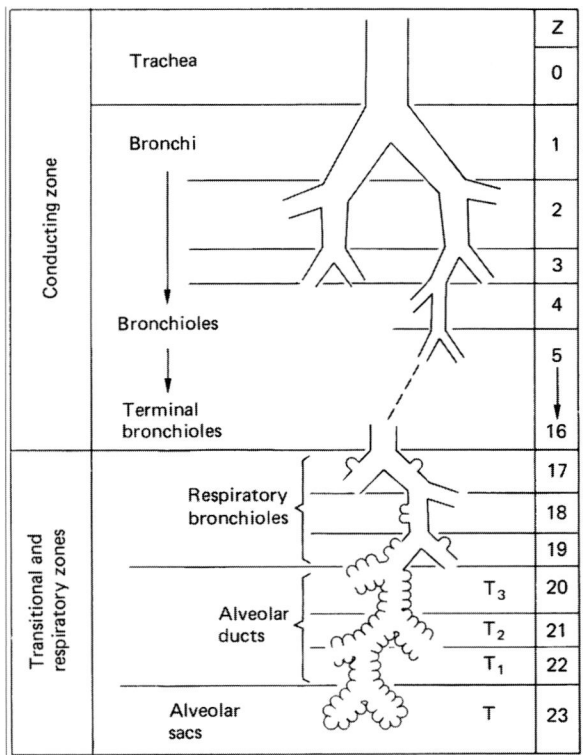

FIGURE 1-4
Airway branching pattern in human lungs by regularized dichotomy from trachea (generation Z = 0) to alveolar ducts and sacs (generations 20 to 23). The first 16 generations are purely conducting; transitional airways lead into the respiratory zone, made of alveoli. (*After Wiebel, 1963.*)

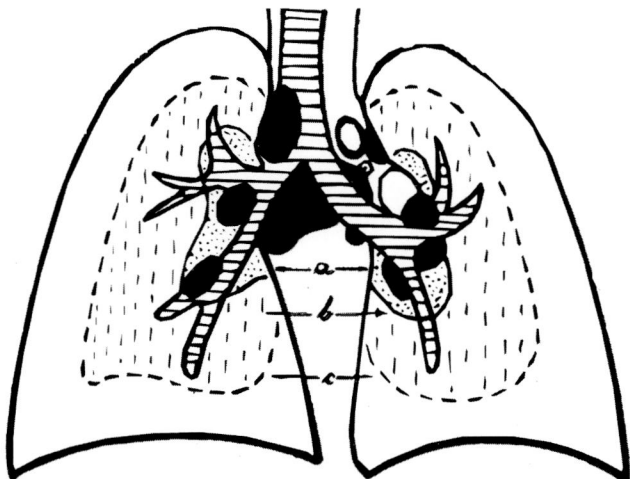

FIGURE 1-5
Diagram of the three zones of the lung (a) the central or hilar zone, (b) the intermediate zone, and (c) the outer or cortical zone. (*Reproduced courtesy of the late Dr. S. Engel, The Child's Lung, Arnold, London.*)

lobules are supplied by small branches arising directly from the large segmental bronchi. The small branches often curve centrally after arising from the parent branch. The outer cortical zone consists entirely of the peripheral branches of the bronchial tree, which end in well-developed lobules. The most superficial subpleural lobules form four-sided pyramids.

The subpleural lobules are separated from each other mainly by fibrous septa, continuous with the pleura. Such septa are better developed in the apical regions and beneath sharp margins of the lung but are almost entirely absent beneath the flat costal and lateral surfaces. Internally the septa are often incomplete, as is necessary if collateral circulation is to have the ability to open.[8] Where septa are well developed and are thus the most mobile parts of the lung, they reduce effective collateral ventilation. In such areas, which include the lung apices and the inferior lingular segment, damage often results in permanent scarring. This is the main reason why biopsies from the inferior lingular segment of the left lung should be at least 3 cm long when the diagnosis of a diffuse interstitial lung disease is suspected.

Beneath the free edges of the lungs, the septa form a network of fibroelastic laminae (Fig. 1-6). Those in the lower lobes become conspicuous (Kerley's B lines) (Fig. 1-7) on radiographs in the presence of pulmonary edema. In the deeper parts of the lung, the septa are poorly developed. They are more conspicuous in the neonatal lung than in the adult lung. The septa normally contain branches of the pulmonary veins and lymphatics and consist of loose fibrous tissue with an elastic membrane continuous with the elastic layer in the visceral pleura. The loose fibrous tissue surrounding bronchi and pulmonary arteries in the deeper parts of the cortex become continuous in places with the inner ends of better-developed septa.

Bronchioles are best defined as conducting airways without cartilage. However, some of these structures contain alveoli and thus have a respiratory function. Bronchioles are approximately 1 mm in diameter.

The terminal bronchiole is the smallest airway completely lined by bronchial epithelial cells. These divide into two to five respiratory bronchioles,[9] which branch at variable angles (Fig. 1-8). They in turn divide into two or three alveolar ducts, each usually ending in four air sacs, which are also known as atria. From the walls of the air sacs arise the alveoli. This is the arrangement in adults, but in infancy and early childhood there are variations in the number of branches during periods of active growth.

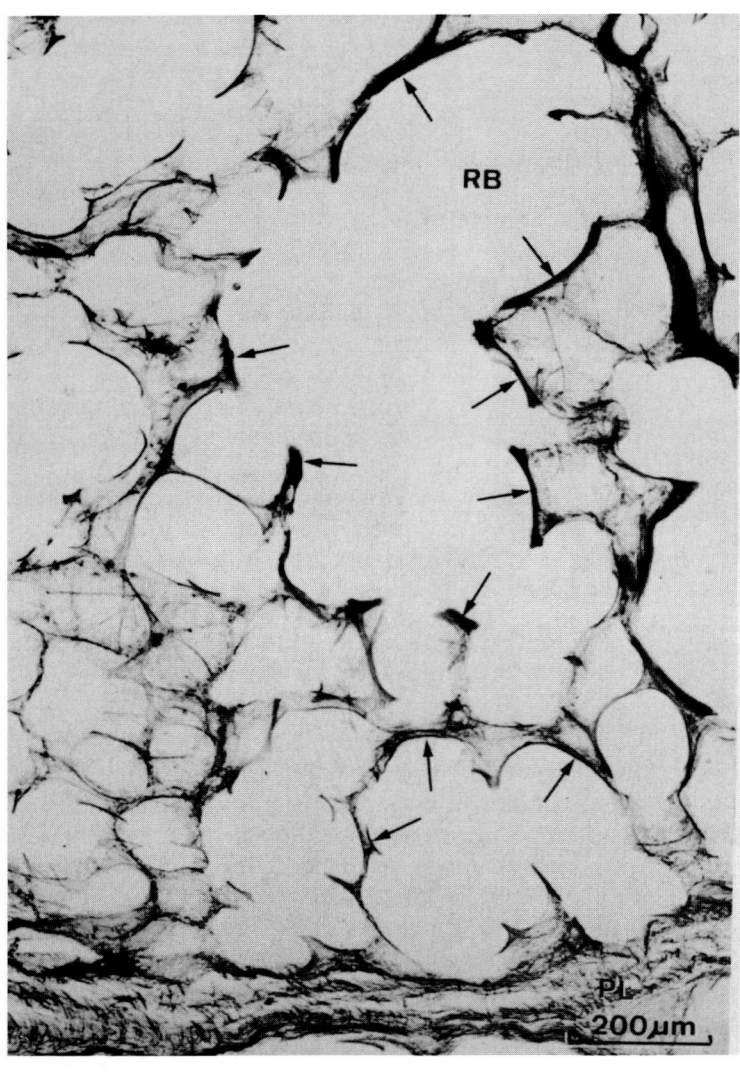

FIGURE 1-6
Connective tissue stain showing the fibroelastic laminae in the lung and pleura. PL, pleura; RB, respiratory bronchiole. (*From Weibel, 1984a.*) (EVG, ×313.)

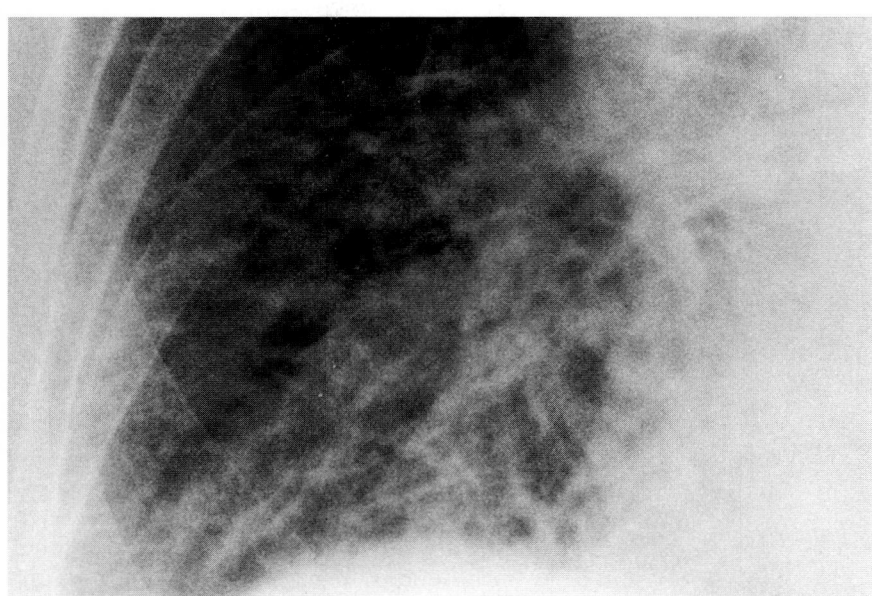

FIGURE 1-7
Chest radiograph of a patient with pulmonary edema, showing prominent Kerley B lines. (*Reproduced by courtesy of Dr. A. Horrocks, Department of Radiology, Wythenshawe Hospital, Manchester, England.*)

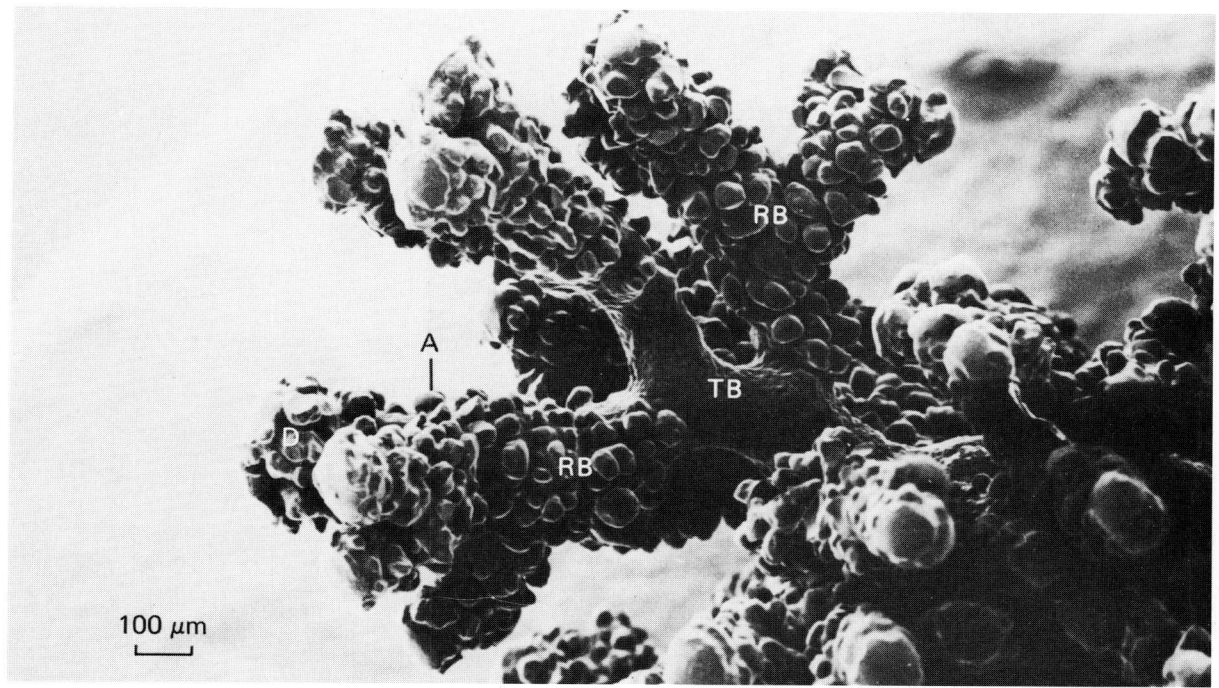

A

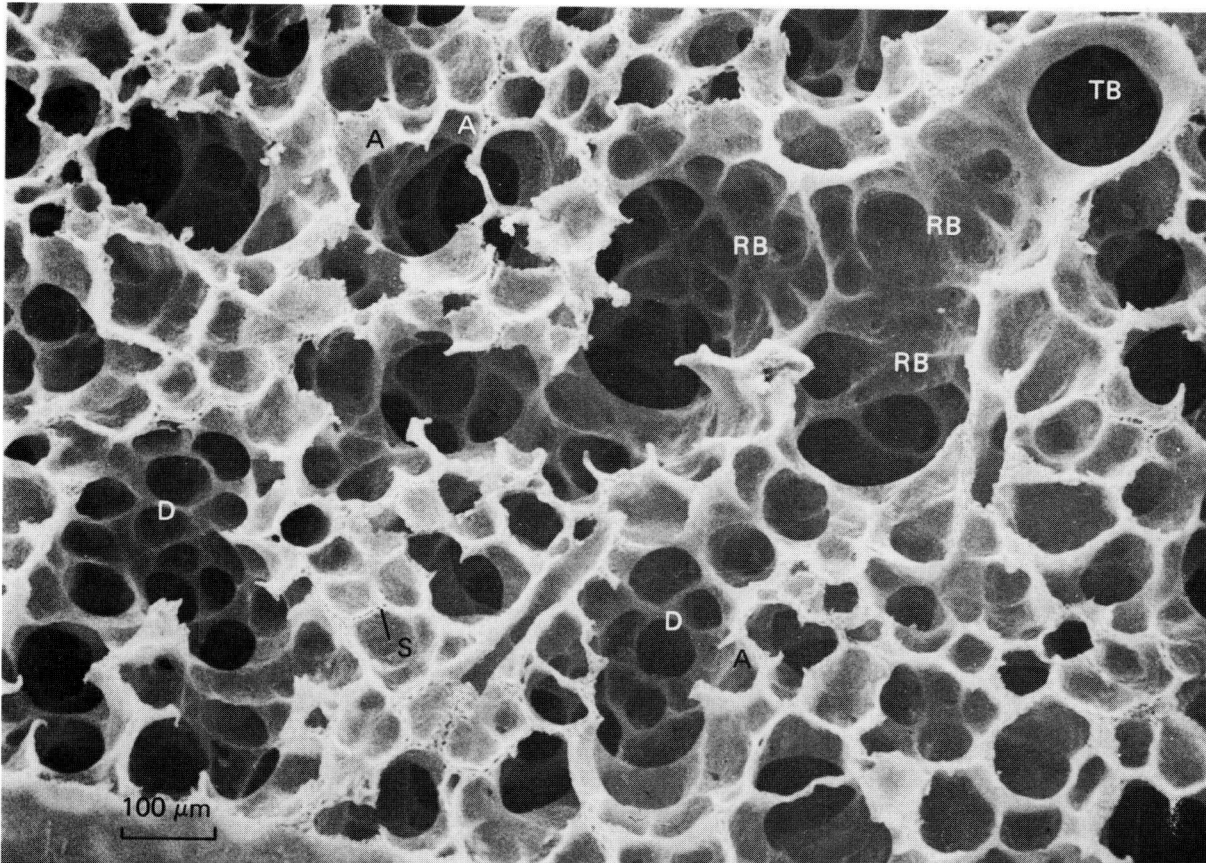

B

FIGURE 1-8
Scanning EM of airway branches peripheral to the terminal bronchiole.   *A.* Silicone-rubber cast of cat lung.   *B.* In whole tissue preparation of air-filled, perfusion-fixed rabbit lung. Note that branching can be followed from the terminal bronchiole to alveolar ducts. A, alveolus; D, alveolar duct; RB, respiratory bronchiole; TB, terminal bronchiole; S, alveolar septum.

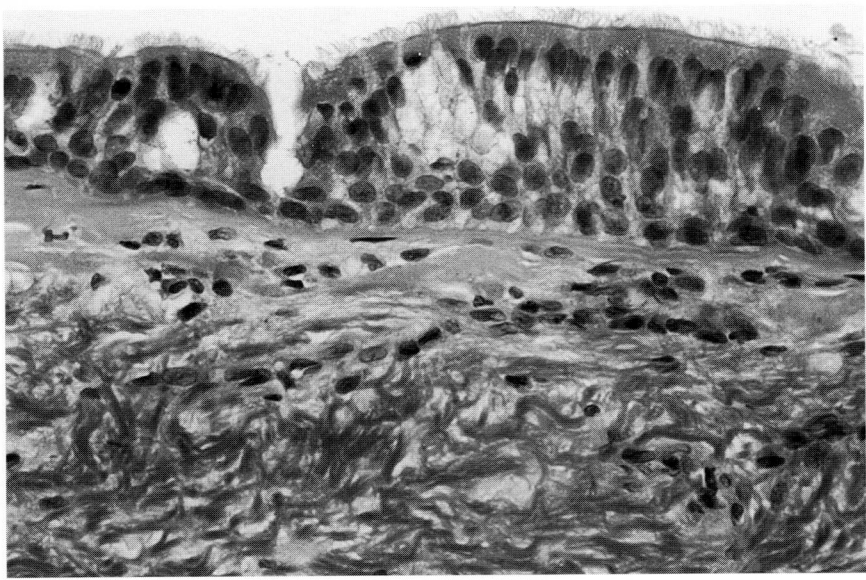

A

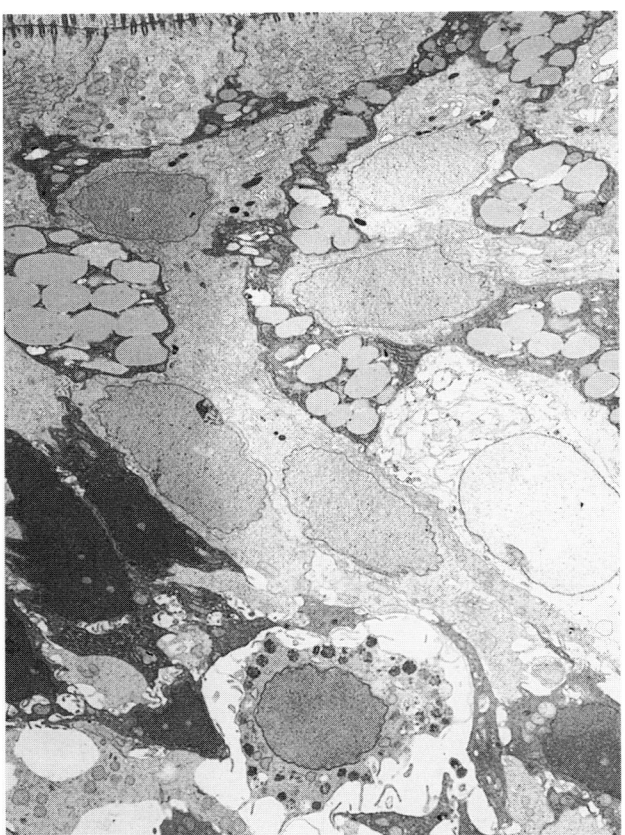

B

FIGURE 1-9
A. Histological section of main bronchus showing ciliated epithelium firmly attached to basement membrane. (H&E, ×350.)
B. Low magnification EM of the wall of the bronchus showing the ciliated epithelial cells and mucus secreting cells which line the lumen. Beneath these lining cells are found intermediate cells and basal cells which overlie the basement membrane. (EM, ×2592.)

MICROSCOPIC ANATOMY OF THE AIRWAYS

It is beyond the scope of this text to give an extremely detailed account of the cells of the airways and all their functions. This information is readily available in the first half of volume I of *The Lung: Scientific Foundations*.[10]

Trachea

The trachea is a straight tube connecting the larynx with the bronchi. It is up to 25 cm long and is supported anteriorly by up to 20 horseshoe-shaped cartilages. Posteriorly, these join the *trachealis muscle*. The histologic structure of the trachea is similar to that of the main bronchi and will be considered below.

Main Bronchi

The trachea and main bronchi have an inner lining of mucus, immunoglobulin, and other substances necessary for pulmonary protection. There is a lining of ciliated, pseudostratified epithelium, which is firmly attached to a basement membrane (Fig. 1-9). The cilia (Fig. 1-10), to be described below, are vital to the mucociliary apparatus and maintain a constant upward flow of respiratory secretions. The surface epithelium contains a variable number of goblet (mucus-secreting) and serous cells.

Also present in the epithelium are brush cells, which, along with the other cells mentioned above, are attached to the basement membrane. Just above the basement membrane, between the deep attachments of the epithelial cells, lie the basal cells, among which are amine precursor uptake decarboxy-

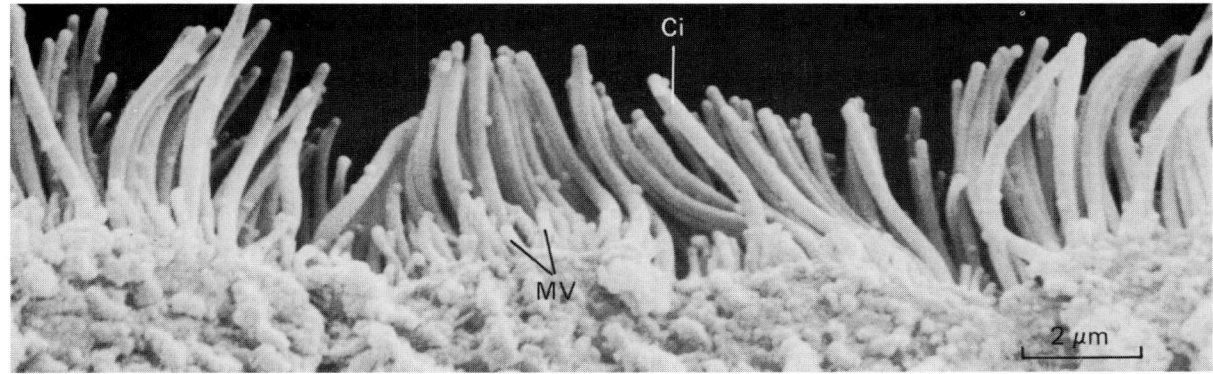

FIGURE 1-10
Cilia (Ci) from human bronchial epithelium seen on sections of epithelial cells in scanning electron micrograph. There are abundant short microvilli (MV) interspersed between cilia.

lation (APUD) cells. These probably act as airway receptors, monitoring the level of oxygen.

Beneath the basement membrane is subepithelial tissue consisting of collagen, elastin, nerves, and lymphatic and vascular channels. Loose collagen surrounds the serous and mucous glands. The smaller the bronchi, the thinner the subepithelial tissue. There is a layer of smooth muscle between the basement membrane and the cartilage, which is seen as C-shaped rings (Fig. 1-11). The outer layer of the bronchus consists of loose connective tissue containing branches of the bronchial arteries and veins, which supply the bronchial wall. Also present in this layer are nerves, derived from the vagus and cervical sympathetic chain (Fig. 1-12). Ganglia may be seen in the loose tissue outside the bronchial wall (Fig. 1-13).

The structure of the large bronchi resembles that of the trachea, with abundant cartilage and muscle fibers posteriorly. In medium-size bronchi, the cartilage is more irregular but the muscle is well developed and the glands are prominent. In the posterior part of large bronchi there are, as well as muscle, dense bands of elastin, running longitudinally, which cause the longitudinal corrugations in the bronchial wall.[11] These elastic fibers extend distally beneath the epithelial lining into the alveolar ducts and alveoli.

CILIATED EPITHELIAL CELLS

Ciliated cells are found from the bronchi to the terminal bronchioles. Their number and cilial length decrease in the more distal airways, implying that mucociliary clearance rates are slower in more peripheral airways.[12] Ciliated cells are roughly columnar, approximately 20 μm long and 7 μm wide,[13]

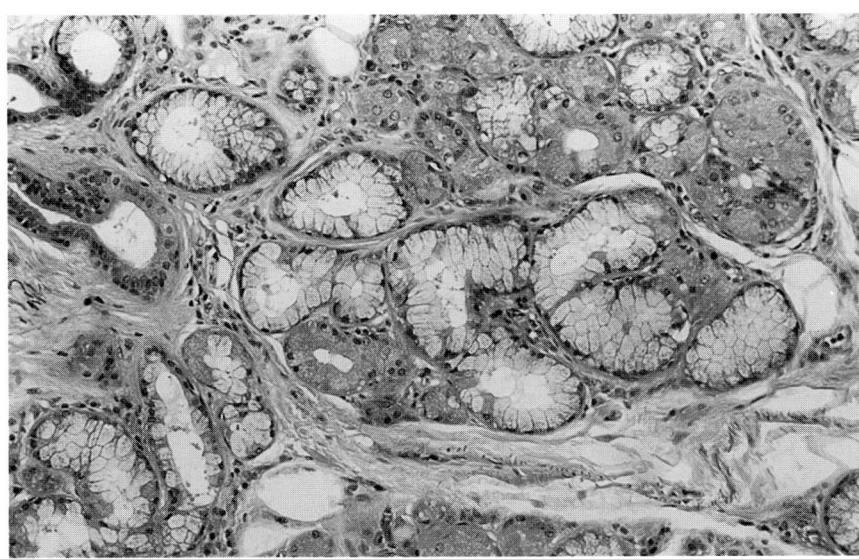

FIGURE 1-11
Mucous and serous glands in the bronchial wall. (H&E, ×219.)

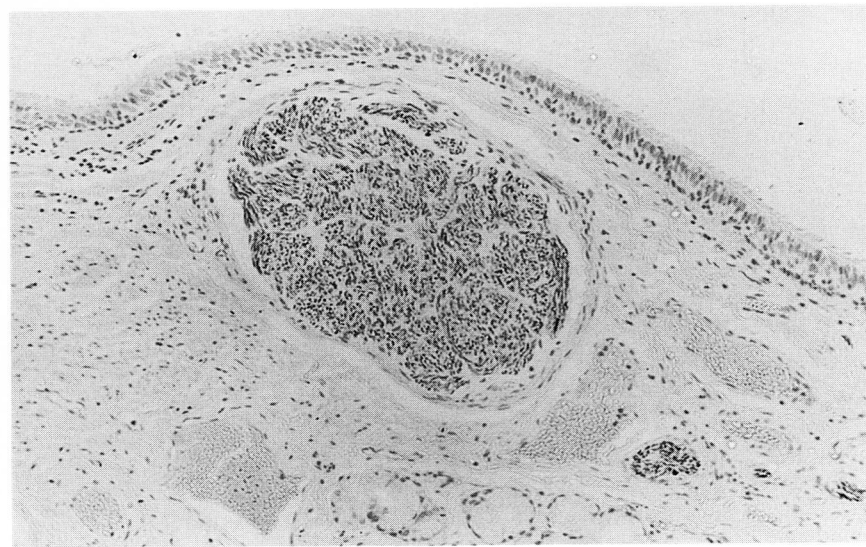

FIGURE 1-12
Prominent nerve in bronchial wall.
(Immunoperoxidase S100, ×87.)

and taper at the base (Fig. 1-14). The lower and lateral cell surfaces are attached to basal and intermediate cells by desmosomes.

Ultrastructurally, the ciliated cell (Figs. 1-15 and 1-16) contains electron-lucent cytoplasm. The nucleus is basal, and above it there is a well-developed Golgi apparatus. The upper part of the cell is rich in mitochondria; these lie just below the basal bodies, which are contiguous with the ciliary shafts.

Approximately 250 cilia are found on the luminal surface of each cell. Interspersed among the cilia are microvilli. Ultrastructural damage is a common finding in biopsies collected by flexible fiberoptic bronchoscopy, and such traumatic change should be taken into account before epithelial changes are attributed to pulmonary disease (see Chap. 3).[14]

The structure of the cilia is similar to that of other ciliated epithelia found in plants and animals. The cilium or ciliary shaft is a cytoplasmic extension from the surface of the cell and is covered by the same plasma membrane. Transverse section of the ciliary shaft reveals an axial filament complex consisting of nine peripheral doublets of microtubules and two central microtubules (Fig. 1-17). Radial spokes extend from the central microtubules to the peripheral circle of microtubular doublets. Such spokes can be difficult to visualize ultrastructurally. Each doublet has inner and outer dynein arms, the site of ATPase activity and essential for ciliary movement.[15] Again, it is sometimes difficult to visualize these dynein arms, and only near-perfect cross-sectional orientation will demonstrate them. There is variation

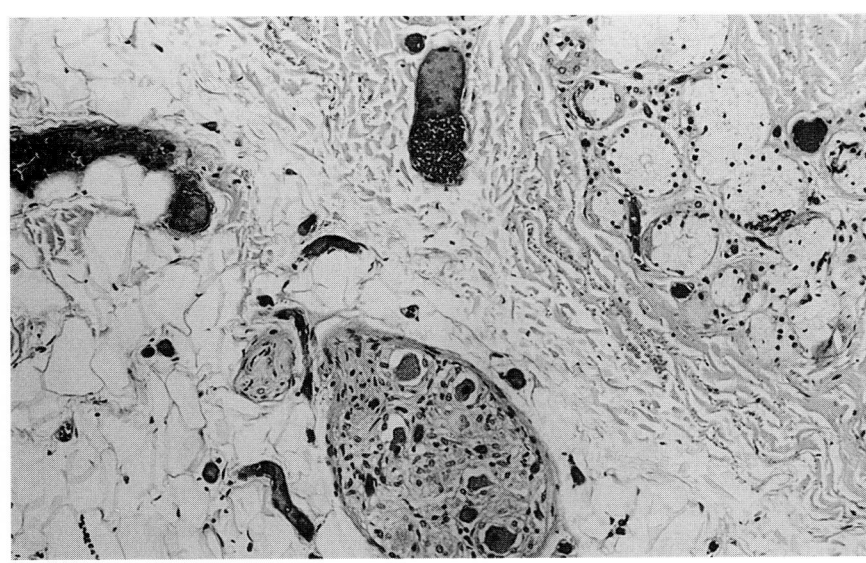

FIGURE 1-13
Ganglion lying outside bronchial wall.
(H&E, ×87.)

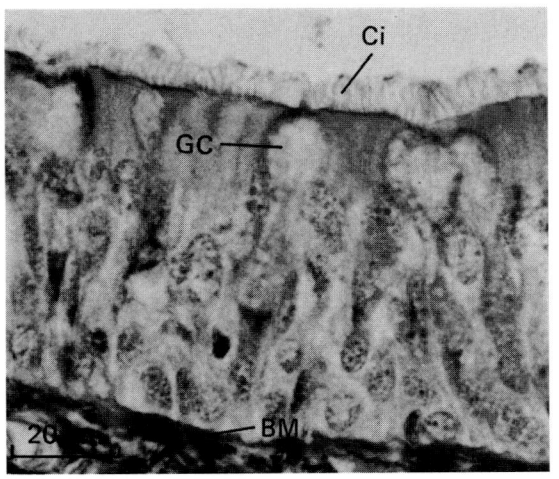

FIGURE 1-14
Light micrograph of bronchial wall showing a pseudostratified ciliated (Ci) epithelium. BM, basement membrane; GC, goblet cell.

in this ciliary structure according to the level of section (Fig. 1-18). Some authors[16] have shown focal ultrastructural abnormalities and have demonstrated variations in different parts of the bronchi. They have failed to show differences between compound cilia or deletion of dynein arms in normal patients compared to those with carcinoma of the lung or chronic infection. Care must therefore be taken in overenthusiastic interpretation of cilial abnormalities on electron microscopy.

Each cilium is contiguous with a basal body, which anchors the axoneme into the cell cytoplasm. Rootlets extend from the basal bodies into the apical part of the cell.[13] There is a transition zone between the ciliary axoneme and the basal body as the cilium enters the cell. The two central microtubules termi-

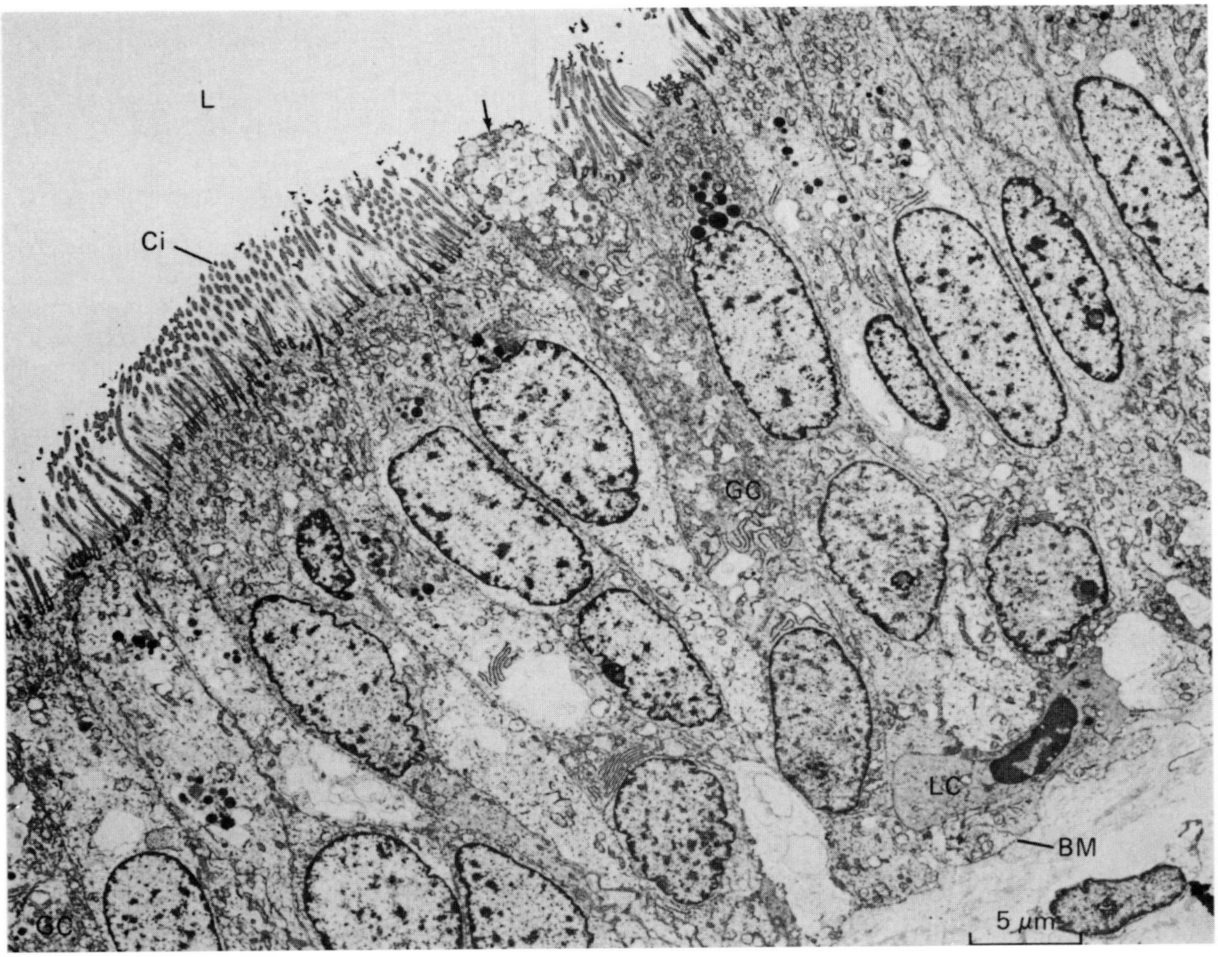

FIGURE 1-15
EM of section across human bronchial epithelium made of high-columnar cells, most of which are ciliated (Ci). A goblet cell (GC) is cut lengthwise; note mucus droplets in the process of accumulating at the cell apex (*arrow*) and leukocyte (LC) caught in the epithelium in the process of diapedesis. L, lumen; BM, basement membrane.

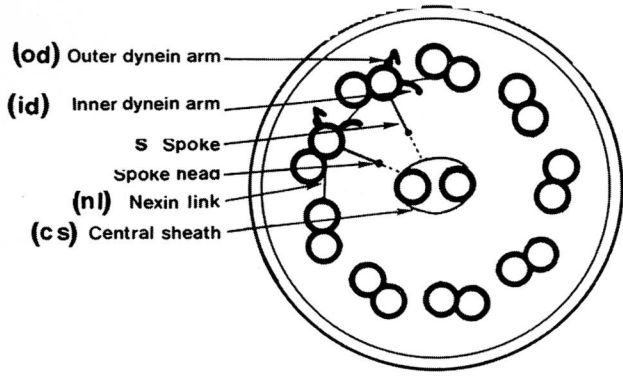

FIGURE 1-16
Diagram of dynein arm.

(od) Outer dynein arm
(id) Inner dynein arm
S Spoke
Spoke head
(nl) Nexin link
(cs) Central sheath

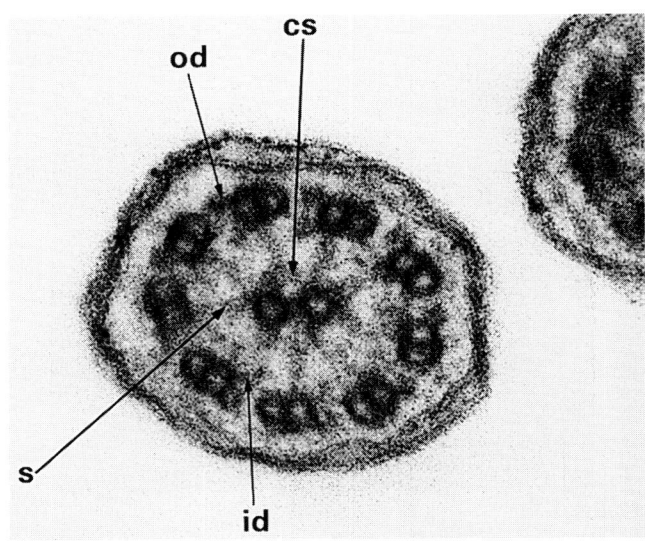

FIGURE 1-17
Electron micrograph of cross section of ciliary shaft showing characteristic "9 + 2" arrangement of microtubules and associated structures. (EM ×189,000.)

nate in this transition zone, and the nine doublets gain another filament to become triplets. Fine filaments link adjacent basal bodies to each other and ultimately to the junctional complexes.[17] At the outer tip of each cilium, the filaments fuse together.

The cilium is thought to bend by the action of microtubules as they slide past each other while anchored distally. This mechanism is mediated by repeated attachments and releases of the dynein arms to adjacent doublets.

Cells with microvillous projections up to 2 μm long and 0.1 μm in diameter are seen on the free border of ciliated cells. Such microvilli are numerous on immature cells undergoing ciliogenesis. The length and number of cilia appear to indicate different stages of maturity.[18]

Mucous and Serous Cells

Interspersed between the ciliated cells are goblet or mucous cells. Such cells stain intensely with diastase/alcian blue pH 2.5/PAS (periodic acid–Schiff), indicating that they contain acid mucopolysaccharides. Ultrastructurally, the mucous gland cells contain distinctive round or oval cytoplasmic vesicles filled with mucin. Their cytoplasm contains abundant endoplasmic reticulum in addition to a well-developed Golgi apparatus. Small microvilli are found on the surface of these cells (Fig. 1-19).

Mucous granules are discharged from the cell, often with intact limiting membranes, as well as some other cell components. These products pass through pores in the luminal surface, which are readily seen ultrastructurally. This indicates apocrine secretion.[19] Mucous cells can proliferate and differentiate into ciliated cells.[20]

Serous cells are rare on the bronchial surface but may be able to transform to mucus cells in response

to adverse conditions. They are more common in the trachea and extrapulmonary bronchi.

Brush Cells

Brush cells, described in the rat bronchial tree and alveoli, account for up to 10 percent of cells present.[21] They are tall cells with prominent microvilli. The cytoplasm contains abundant free ribosomes but little rough endoplasmic reticulum and glycogen. Filaments are prominent and are seen to be continuous with the axial filaments of the microvilli. Brush cells are well documented in animals,[22–24] but their presence in the human bronchial tree has been doubted.[25] Cells described as brush cells are possibly immature ciliated cells. The function of brush cells is unknown, but their resemblance to brush border cells of the gut has led to the supposition that they have an absorptive function.[13]

INTERMEDIATE CELLS

Intermediate cells are spindle-shaped and contain few organelles or filaments (Fig. 1-20). They may be cells undergoing maturation into surface epithelial cells, or they may have an absorptive function or a role in protein transport. Some authors believe the term *intermediate cell* should be avoided, as it refers to several different cell types.[26] In this text it is used to denote cells that are in transition from the basal

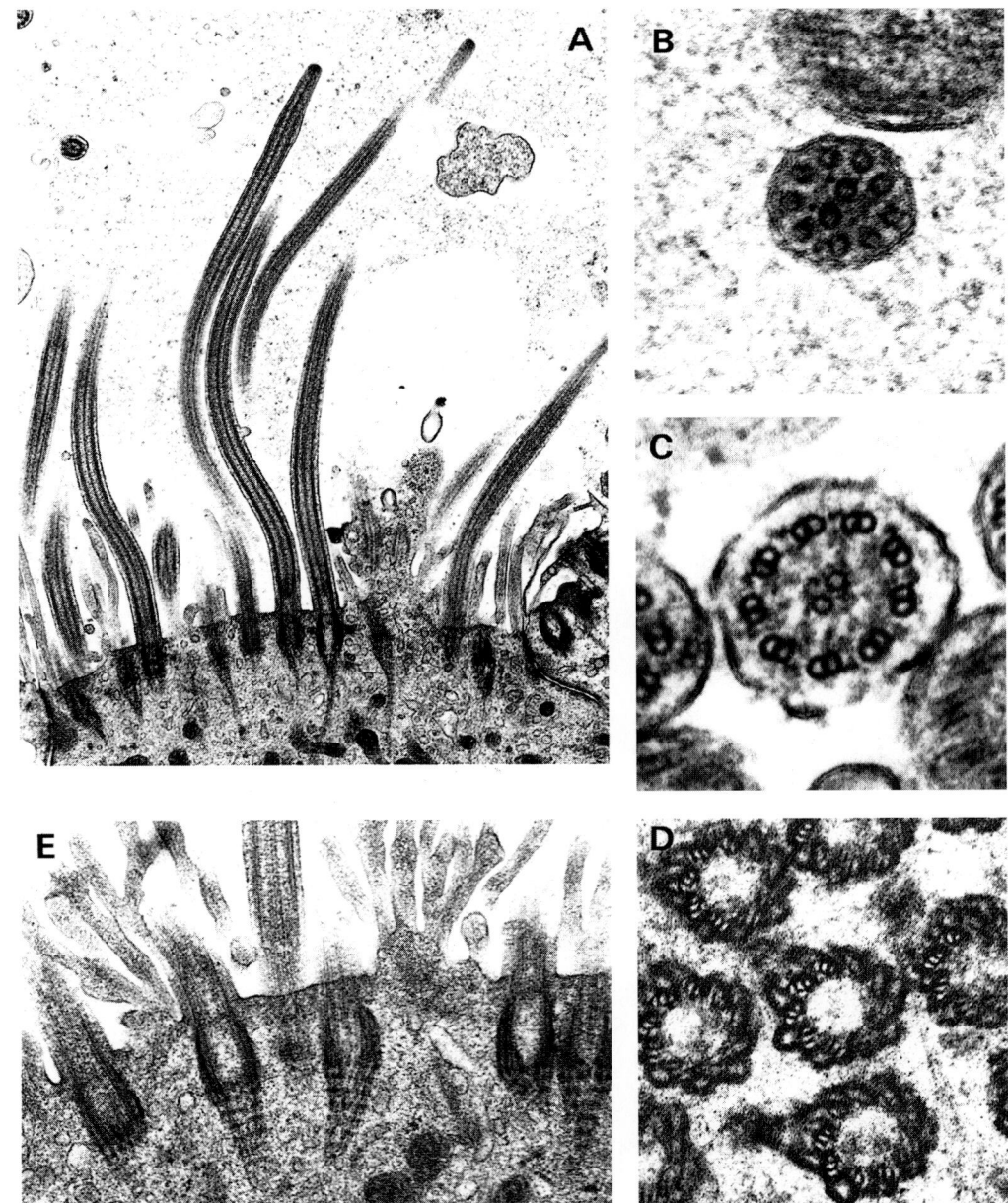

FIGURE 1-18

The ultrastructure of cilia.   *A.* Longitudinal section of several cilia. The external shaft is of uniform thickness except for the tip, which is conical. The entire structure is covered by the plasma membrane, which contains the axoneme, a complex structure composed of microtubules and associated structures. This microtubular complex penetrates into the cell cytoplasm, where it is associated with striated rootlet structures. ×13,500.
B. Transverse section of the conical tip of a ciliary shaft, showing covering plasma membrane and internal microtubules. At this level, only single microtubules are found surrounding the central pair. ×112,000.   *C.* Transverse section of ciliary shaft, showing internal complex of microtubules. There are nine doublets and two central microtubules ("9 + 2" complex). Each doublet has a pair of dynein arms, which originate from the complete microtubule of the doublet and point in the direction of the incomplete microtubule of the succeeding doublet. Spokes also link each doublet to an inner sheath, which surrounds the central pair of microtubules. ×112,000.   *D.* Transverse section of basal bodies from which the ciliary axoneme originates. The basal body is cylindric, with a wall composed of nine triplets of microtubules. ×112,000.   *E.* Longitudinal section of basal bodies, showing distal "9 + 2" ciliary axoneme and proximal striated rootlets. ×31,000.

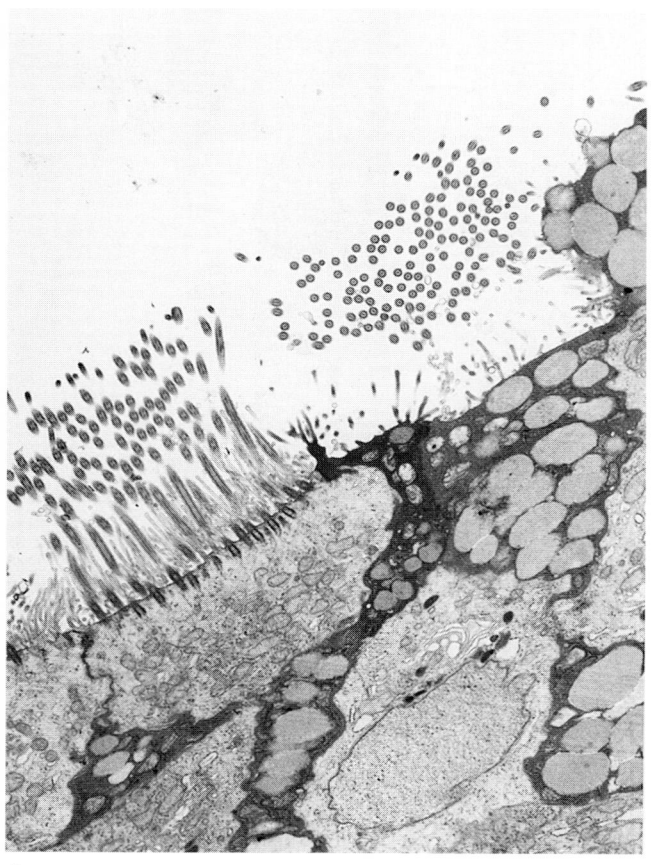

A

layer to more-differentiated cells, but without implying any function.

BASAL CELLS

Basal cells are small pyramidal cells that have the ability to divide into surface epithelial cells and extend as far as the bronchioles, though they are more numerous in the upper airways. They are attached to the basement membrane by hemidesmosomes. The nucleus is prominent, and tonofilaments can be demonstrated (Fig. 1-21). Besides being stem cells, the basal cells have a role in adhesion of columnar epithelium to the basement membrane.[27] This function is mediated by means of their desmosome attachment to columnar cells and their hemidesmosome attachment, as well as cell adhesion molecules, to the basement membrane. Columnar cells do not form hemidesmosome attachments to the basement membrane.

FEYRTER (NEUROENDOCRINE, APUD, KULTSCHITSKY, ARGYROPHIL) CELLS

Pulmonary argyrophil cells have excited much interest because of their endocrine potential and their relationship to carcinoid tumors and small-cell carci-

B

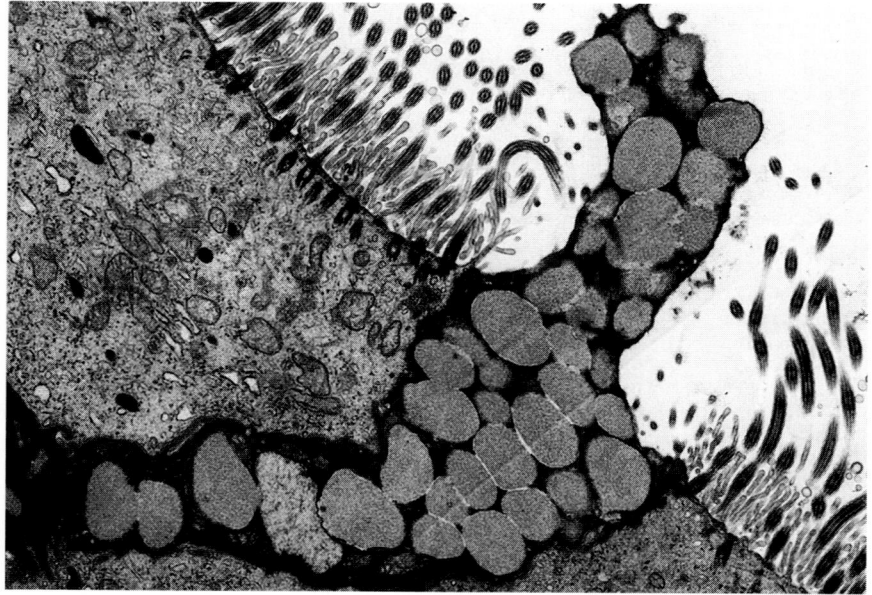

FIGURE 1-19
A. Wall of bronchus. Electron micrograph of a mucus-secreting cell situated between ciliated epithelial cells. (EM, ×4032.)
B. Electron micrograph of mucus-secreting cell discharging mucus droplets into the lumen of the bronchus. (EM, ×6300.)

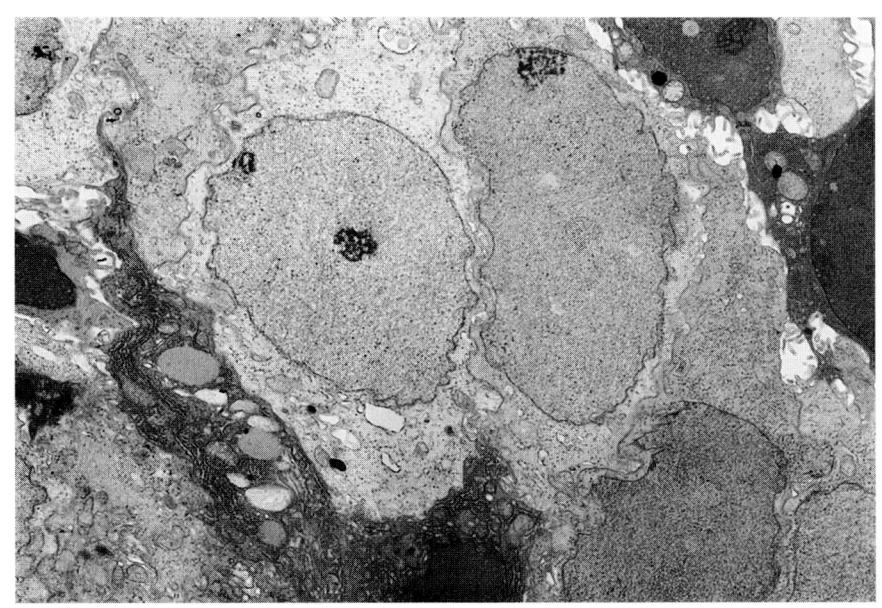

FIGURE 1-20
Electron micrograph of bronchus showing intermediate cells, which lie between the epithelial cells and the basal cells. (EM, ×3920.)

nomas of the bronchus. There are many pseudonyms for this cell type, including Kultschitsky cell, K cell, neurosecretory cell, neuroendocrine cell, Feyrter cell, paracrine cell, and APUD cell.

These cells appear to be basal, but if serial sections are cut, some cells appear columnar and reach the lumen of the airway.[28] By light microscopy, these cells may have a clear cytoplasm, but this is not a uniform feature. They have round to oval nuclei, and the cell tends to be triangular in shape. Argyrophil cells

are more common in the fetus and newborn infant than in adults (Figs. 1-22 and 1-23). They are found at all levels of bronchial and bronchiolar epithelium, as well as in mucous glands, their ducts as well as in alveolar ducts.

APUD cells may occur singly or in clusters of four to 10 cells, each having an oval nucleus and inconspicuous nucleoli (Fig. 1-24). These collections of APUD cells were termed neuroepithelial bodies by Lauweryns and Peuskens.[29] Intraepithelial nerve

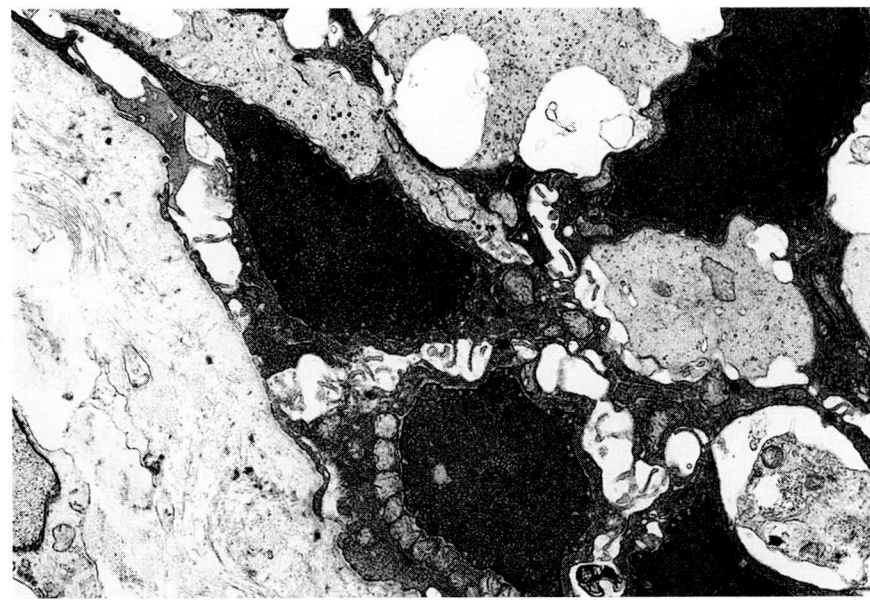

FIGURE 1-21
Electron micrograph of bronchus showing basal cells. (EM, ×6300.)

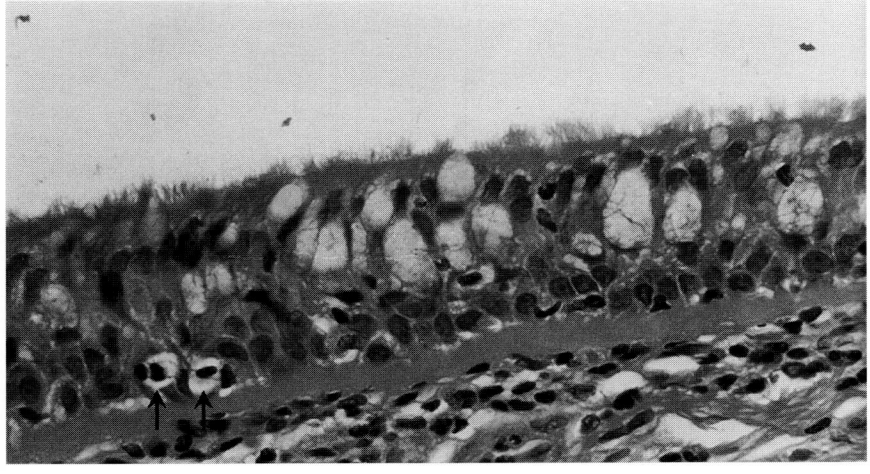

A

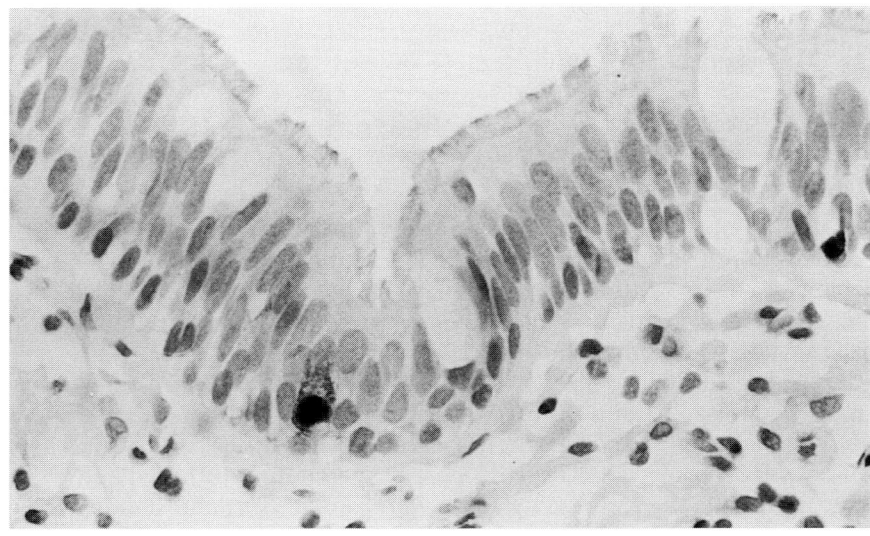

B

FIGURE 1-22
Two neuroendocrine cells (*arrows*) are
seen in the bronchial wall.   *A.* H&E,
×350.   *B.* Immunoperoxidase PGP
9.5, ×350.

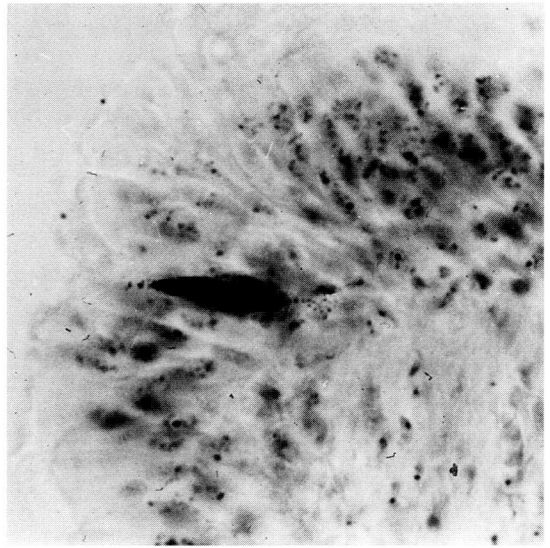

FIGURE 1-23
Argentaffin cell in the bronchial mucosa of a neonate. (Masson's
silver stain, ×400.)

axons have been described by some workers as being
in close association with neuroendocrine cells (Fig.
1-25).[30]

The most common method of demonstrating
these cells used to be Grimelius' stain, which demonstrates fine argyrophilic granules. Such granules predominate in the basal part of the cell. Nowadays the
cells are identified more easily by immunoperoxidase staining. A series of products have been demonstrated in pulmonary neuroendocrine cells, and the
choice of antiserum will rest on the expertise and
availability in a particular laboratory. Antisera to
chromogranin,[31] PGP 9.5,[32] calcitonin, and calcitonin gene–related peptide[33] are commonly used. We
have not found PGP 9.5 a reliable marker for the presence of peptides. Another useful marker is anti–
Leu-7, a monoclonal antibody found in natural killer
cells and myelin-associated glycoprotein as well as
in neuroendocrine cells.[34] In situ hybridization tech-

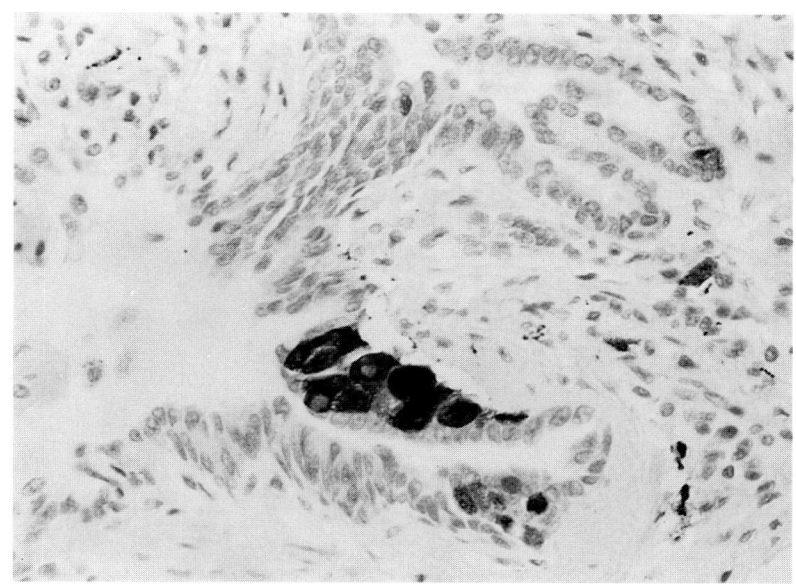

FIGURE 1-24
Neuroepithelial body in the bronchiolar wall, as
well as some scattered neuroendocrine cells.
(Immunoperoxidase PGP 9.5, ×350.)

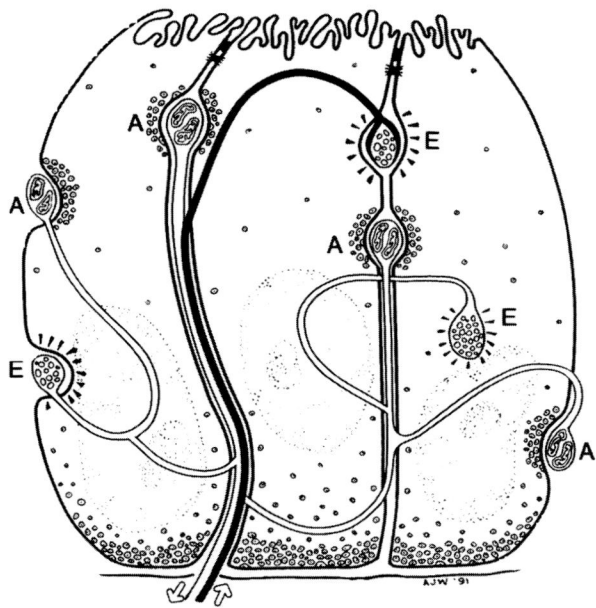

FIGURE 1-25
Diagram of the fine structure and innervation of a typical
mammalian neuroepithelial body, showing vagal nerve endings
ramifying between and making contact with three pulmonary
neuroendocrine cells. As with individual cells, dense core
vesicles are basally concentrated but aggregate also at points of
contact with morphologically afferent nerve endings (A), from
which messages pass centrally to the nodose ganglion. Most
efferent nerve endings (E) are derived from these vagal afferents
and are likely to act to modulate activity of the neuroepithelial
body. Some are in cytoplasmic continuity with them. Others
(black fiber) are terminations, probably of vagal efferent fibers
descending from the brainstem. It is likely that messages relay
also at a lower level than this, possibly within the lung.
(*Reproduced by kind permission of Dr. J. Gosney and Butterworth
Heinemann, Oxford.*)

niques are increasingly being used to demonstrate
such products more accurately. A list of other markers for neuroendocrine cells can be found in the section on bronchial carcinoid tumors (see Chap. 29).
The times at which these cells appear in the fetal respiratory tract can be found in Chap. 2.

Ultrastructurally, the key feature of the neuroendocrine cell is the presence of large numbers of dense
core granules (Fig. 1-26). These range in size from 70
to 170 nm and have a small electron-dense core surrounded by a clear halo with an outer limiting membrane. Neuroendocrine cells also contain abundant
smooth endoplasmic reticulum, a prominent Golgi
apparatus, ribosomes, and bundles of tonofilaments.
Hage[35] has described three types of neuroendocrine
cells in fetal lung, based on the morphologic features
of the granules and differences in ultrastructure, but
there appears to be only one neuroendocrine cell
type in adults.

APUD cells release their granules across the
basal cell membrane into the adjacent peribronchial
connective tissue and therefore near airway smooth
muscle.[36] Some neuroendocrine cells can secrete
hormonal-type products near other epithelial cells by
means of elongated pseudopod connections. Single
cells, unlike neuroepithelial bodies, have no connection with nerves.

The function of the neuroendocrine cells has
still to be fully elucidated. The physiology of the pulmonary neuroendocrine system has been discussed
by Gosney[37] and more fully in a recent conference[38]
that was summarized in an editorial.[36]

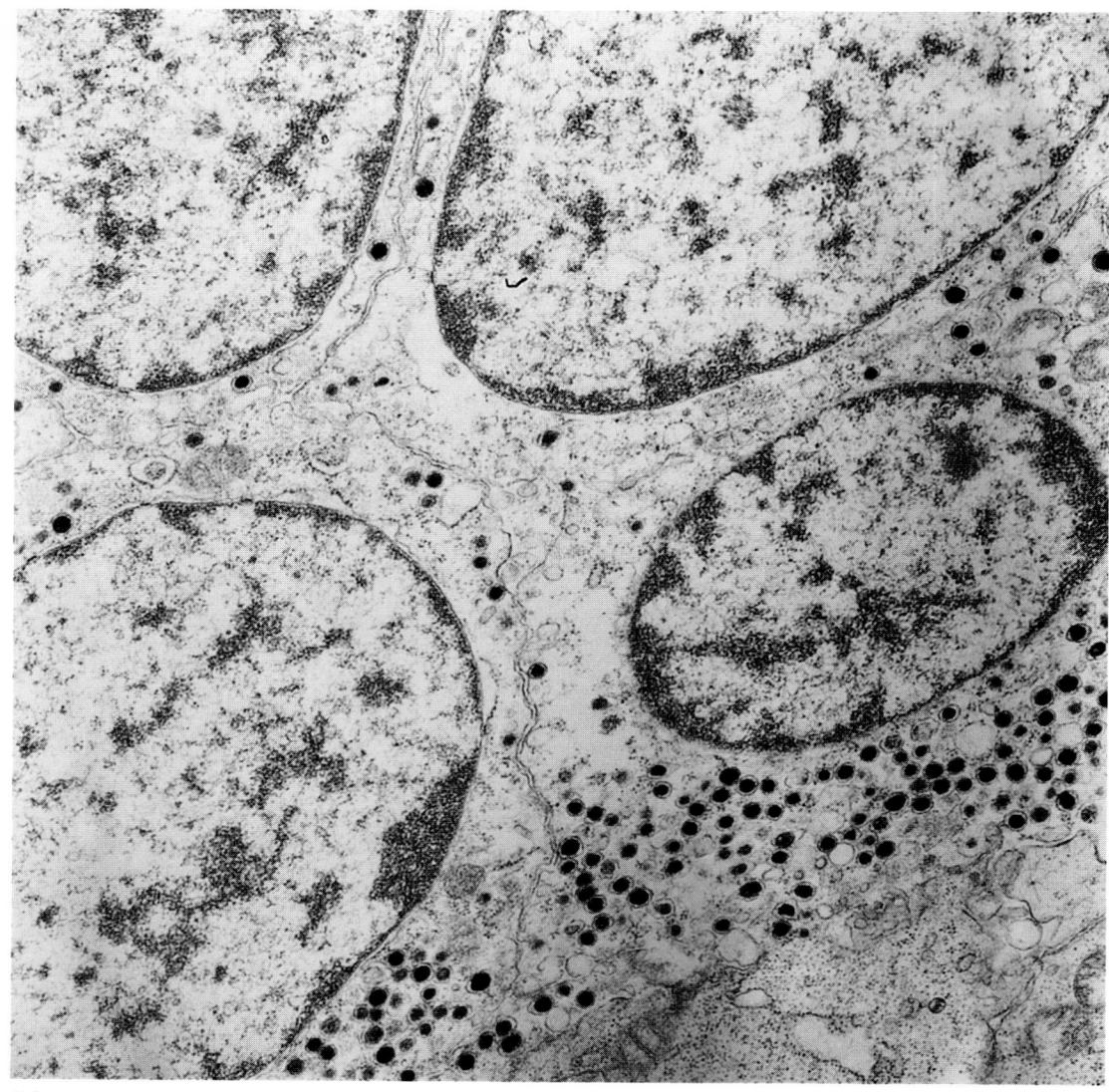

FIGURE 1-26
Electron micrograph of neuroendocrine body with dense core granules. (EM, ×6750.) (*Reproduced by kind permission of Dr. J. Gosney, Department of Pathology, University of Liverpool.*)

APUD cells may act as chemoreceptors. Their location in the airway epithelium has suggested that they may respond to hypoxia or hypercapnia. A number of other functions have been proposed, including bronchoconstriction (bombesin, gastrin-related peptide, and enkaphalin), vasoconstriction, monocyte chemotaxis, and control of the proliferation of epithelial and fibroblastic cells.[39]

The prominence of neuroendocrine cells in the fetus has suggested that these cells contribute to the growth and maturation of the lung. At least one growth factor, gastrin-releasing peptide, with its messenger RNA, has been found in the developing airways.

Pulmonary neuroendocrine cells are increased in number in fibrotic as well hypoxic conditions, suggesting that gastrin-releasing peptide, a growth factor for both epithelial and fibroblastic cells, has a part to play in lung repair. Endothelin is found in neuroendocrine cells. Its early peak in fetal tissue suggests that it may have an action in growth promotion or regulation.[40]

Other peptides in addition to those mentioned above have been identified and may play a role in lung function. These actions are species specific. Calcitonin affects transcellular and intracellular movements of calcium, inhibits synthesis of prostaglandins and thromboxane, increases endothelial generation of prostacyclin, and augments cartilaginous growth. Calcitonin-gene–related peptide functions as a peptidergic neurotransmitter and induces airway proliferation of epithelial cells, endothelial cells, and fibroblasts. Gastrin–releasing peptide causes vasodilation and bronchoconstriction, stimulates mucus secretion, augments surfactant biosynthesis, and promotes the growth of pulmonary neuroendocrine cells as well as other cells of the bronchial epithelium.

BASEMENT MEMBRANE

The epithelial cells rest on a basement membrane, which should be better described as a basal lamina. Pathologists usually comment on this structure only if it is thickened. It is then said to be associated with bronchial asthma. Basement membranes are sheet-like structures that ultrastructurally can be subdivided into electron-dense (lamina densa) and electron-lucent (lamina lucida) zones (Fig. 1-27). The basement membrane provides mechanical support for cells, acts as a semipermeable barrier between tissue compartments, and regulates cellular attachment, migration, and differentiation.[41]

Several structural components can be identified in basement membranes, including collagen type IV, lamin, entactin (nidogen), heparin sulfate, fibronectin, and SPARC (secreted-protein, acidic, rich in cysteine). Interest in basement membranes and their associated proteins is a developing field of interest, particularly for researchers trying to identify factors controlling the spread of tumors. Bronchial epithelial cells in tissue culture produce fibronectin, which stimulates the recruitment of other bronchial epithelial cells. This may be one mechanism by which reepithelialization occurs in the lung.[42] A review of the effect of hormones and growth factors on expression of the fibronectin gene has been published.[43]

SUBEPITHELIAL TISSUES

Beneath the basement membrane is a subepithelial layer, sometimes called a lamina propria. It is doubtful whether the latter term should be used outside the gut. This layer consists of collagen and elastic fibers, capillaries, lymphatics, and nerves. As this layer consists of loose connective tissue, it offers little resistance to the spread of infection or tumors.

Outside the subepithelial layer is an almost circular layer of muscle, interrupted only by collagen and the ducts of bronchial mucous glands. Smooth muscle may be seen as a thin layer when relaxed or as a thicker inner and outer layer suggesting constriction (Fig. 1-28). In pathologic states such as chronic bronchitis or asthma, there is marked hypertrophy of smooth muscle. Despite much investigation, all that is certain is that spasm of bronchial smooth muscle causes asthma.[44] A review of airway smooth muscle including ultrastructure, electrophysiology, mechanical properties, and pathophysiology has been presented by Stephens.[45]

There are probably two types of smooth muscle units in the airway. Multiunit smooth muscle consists of many individual units, each having its own muscle cell with its own nerve. Single-unit smooth muscle has poorer innervation but can exert finer control of the organ it supplies. It is likely that the multiunits supply the main airways and single units the smaller bronchioles.

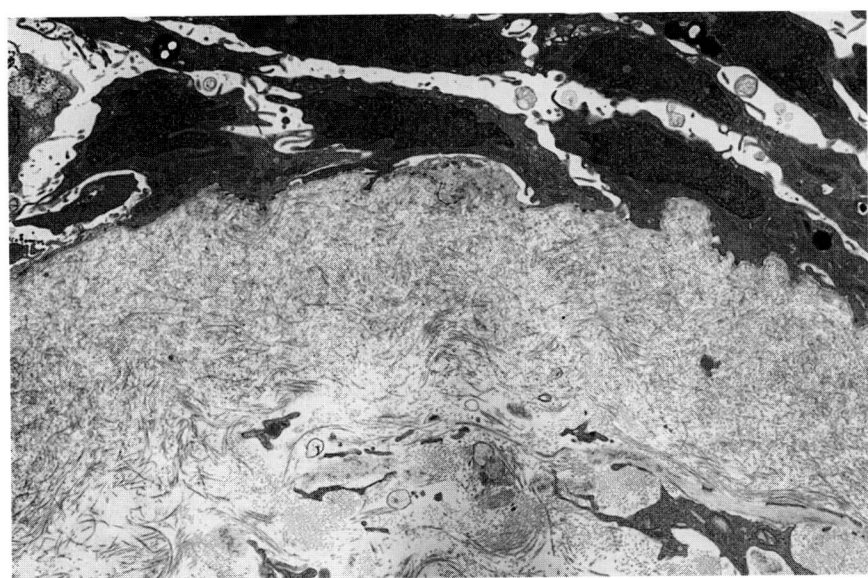

FIGURE 1-27
Electron micrograph of bronchial basement membrane. This extracellular structure is more accurately named "external lamina." (EM, ×4025.)

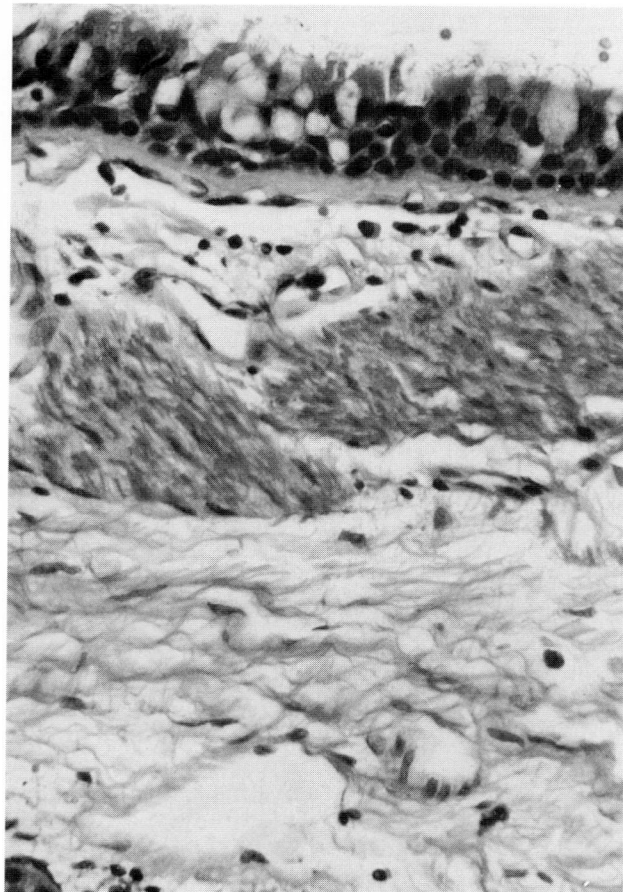

FIGURE 1-28
Bronchial smooth muscle forming a solid band in the subepithelial tissues. (H&E, ×235.)

Various roles have been postulated for bronchial smooth muscle, most relating to regulation of gas flow to maintain the ventilation-perfusion ratio and reflex clearing of sputum or foreign bodies by coughing. It is also probably essential for the maintenance of lung structure.[46] This is because tissues exposed to a constant tension undergo "tissue creep." This term is used to describe the complex viscoelastic properties of soft tissues when exposed to external forces. Such tissues continue to stretch slowly and indefinitely. If this occurs in the lungs, in the absence of counteracting smooth muscle, the bronchi are slowly stretched, causing bronchial dilatation. The anatomic distribution of bronchial smooth muscle is in keeping with this concept of maintenance of lung architecture.

In larger intrapulmonary bronchi, the smooth muscle connects the tips of the cartilage plates together. Thus contraction approximates the cartilage plates, causing a reduction in both diameter and length of the bronchi. In smaller-caliber airways, the muscle forms a separate layer.

Mucous and Serous Glands

Bronchial glands are present to bronchiolar level. Each set of glands, which lie outside the muscle and inside or between the cartilage, has a duct that opens into the bronchial lumen. Serial sections of glands[47] show serous and mucous tubules and collecting and ciliated ducts. Mucous glands are supplied by sympathetic nerve fibers.

Mucous cells make up mucous tubules, which end in serous tubules. However, some serous tubules may arise from the sides of mucous glands. Thus, separate areas in a gland consist of either serous or mucous glands. Myoepithelial cells line part of the mucous glands and are responsible for expelling the contents into the bronchial lumen.

Serous cells contain an irregularly shaped basal nucleus and a variable number of membrane-bound secretory granules, which measure up to 600 nm in diameter (i.e., they are smaller than mucus granules). They have homogeneous electron-dense contents, and are not osmiophilic (Fig. 1-29). A small number of microvilli are seen on the surface.[13] The presence of a population of serous-like cells has recently been identified in human bronchioles as well as small bronchi.[48]

Serous glands are converted into mucous glands in any condition where there is chronic cough. The main function of serous cells is to produce lysozyme[49] as well as possibly helping transport IgA across the glandular epithelium.[50] IgA is formed in plasma cells, found in the region of the bronchial glands.

With increasing age, oncocytes (Fig. 1-30) are seen in the bronchial glands more frequently. These cells have an eosinophilic cytoplasm containing abundant mitochondria. It is unlikely that these cells are degenerate, but their function is unknown.

Clara Cell

The Clara cell (nonciliated bronchiolar secretory cell) is found mainly in terminal and respiratory bronchioles. They are named after Clara[51] and are columnar, protruding above the ciliated cells (Figs. 1-31 to 1-33). At the apex of some cells is a dome-shaped process, which forms an electronlucent cap. The method of secretion is not yet determined; some investigators favor apocrine, others exocrine secretion.[13] The mitochondria in the apical portion of the Clara cell are unusual in that they have no cristae and show an abundant pale matrix. There are ovoid, electron-dense, membrane-bound inclusions as well as a few osmiophilic myelin figures. The electron-dense

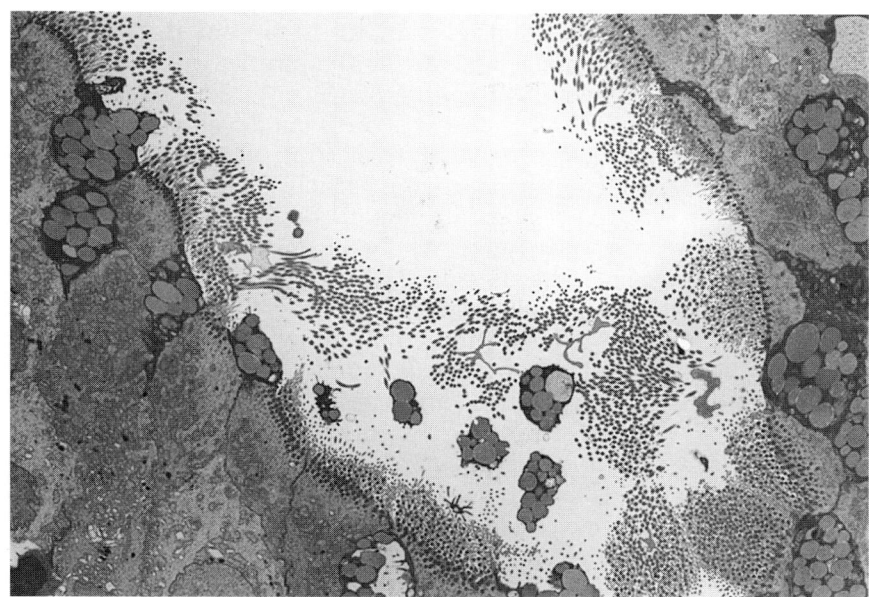

FIGURE 1-29
Low power electron micrograph of
serous gland. The wall of the gland
lumen is lined with ciliated epithelial
cells and some mucus-secreting cells.
(EM, ×1575.)

granules of Clara cells are irregular in most other
species.[52]

The functions of this cell are still not fully elucidated. Surfactant apoprotein A (SP-A) has been
demonstrated in them,[53] as has the carbohydrate
component of surfactant.[54] The cell may produce an
antiprotease[55] and have secretory and ion-absorbing
properties.[56] The Clara cell acts as a reserve cell to
replace both itself and ciliated bronchiolar cells.

Bronchus-Associated Lymphoid Tissue

Bronchus-associated lymphoid tissue (BALT) consists of organized mucosal lymphoid follicles found
between a branch of the pulmonary artery and the
bronchial epithelium (Fig. 1-34).[57] There is considerable variation in BALT in different species. It is most
highly developed in the rat, guinea pig, and rabbit
and not so well developed in humans. It is seen in
only 14 percent of nonsmokers as compared to 82 per-

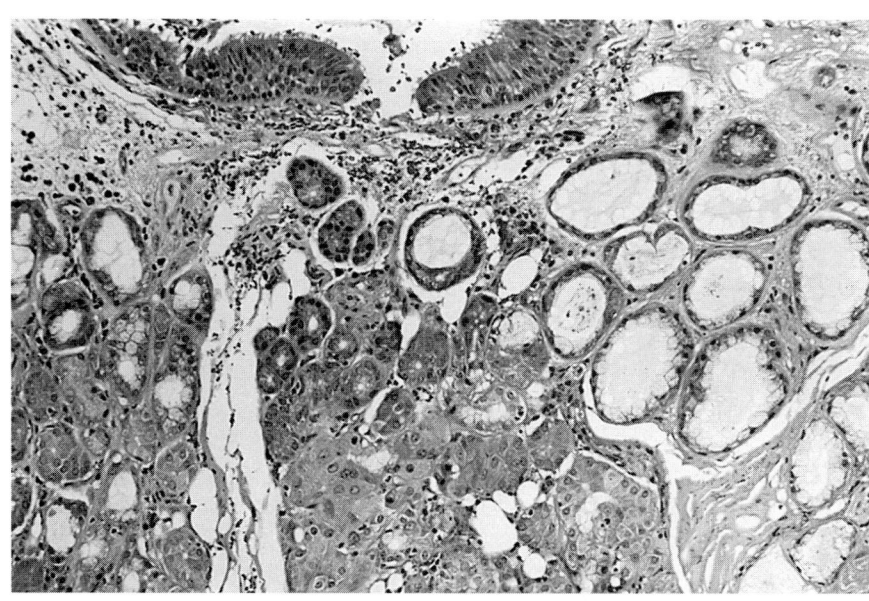

FIGURE 1-30
Part of a bronchial wall with some
mucous and serous glands as well as
oncocytes in the lower half of the
picture. (H&E, ×87.)

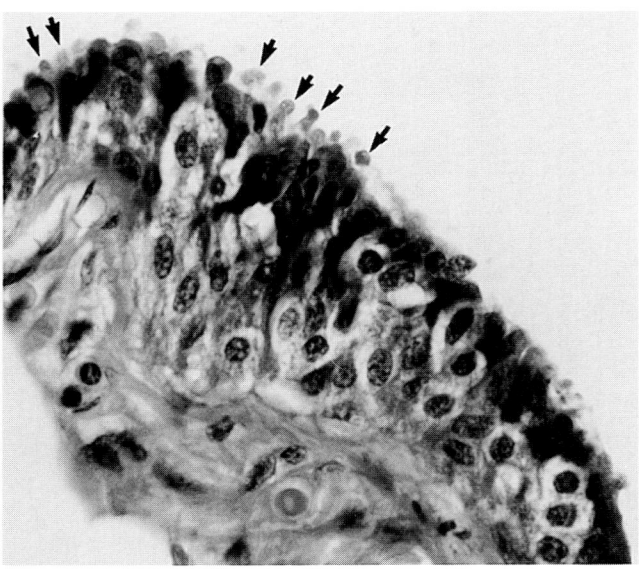

FIGURE 1-31
Part of a bronchiolar wall with many Clara cells (*arrowed*) as well as an increase in neuroendocrine cells. (Immunoperoxidase PGP 9.5, ×375.)

cent of smokers.[58] BALT is present in relation to bronchial glandular epithelium as well as luminal bronchial epithelium.

BALT is absent at birth but develops at points of particle deposition, suggesting that it is produced as a response to airborne particles. BALT is uncommon in the segmental bronchi. Overlying BALT is "lymphoepithelium," where the epithelial cells are attenuated and flattened. No ciliated cells, lymphocytes, or macrophages have been identified in such areas. This altered epithelium allows the passage of both soluble and particulate antigenic material from the airways into the underlying lymphoid follicle, where it can be processed in situ.[59]

BALT in postmortem fetal and infant lungs is associated in the former with chorioamnionitis or in 47 percent of cases with intrauterine pneumonia.[60] The earliest ill-defined lymphoid aggregate had been identified at 16 weeks' gestation, while lymphoep-

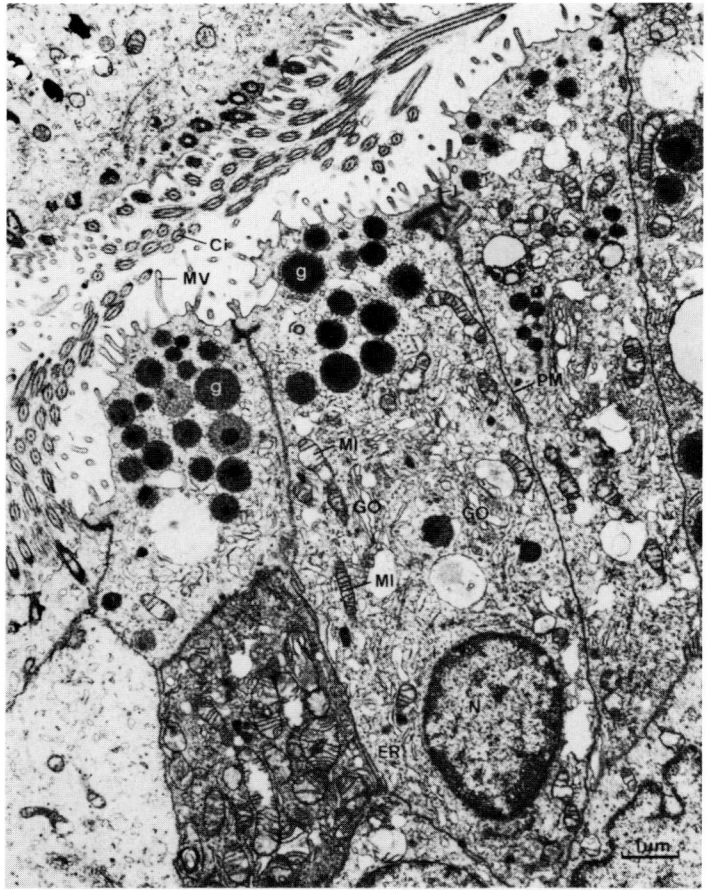

FIGURE 1-32
Clara cells from human bronchiolar epithelium contain dense secretion granules (g) at apex. Note abundant cytoplasmic organelles such as mitochondria (MI). Golgi complex (GO) or endoplasmic reticulum (ER) as well as microvilli (MV) at surface. Cell membranes are closely applied and form tight junctions (J) at apical edge. Ci, cilia; N, nucleus; PM, plasma membrane.

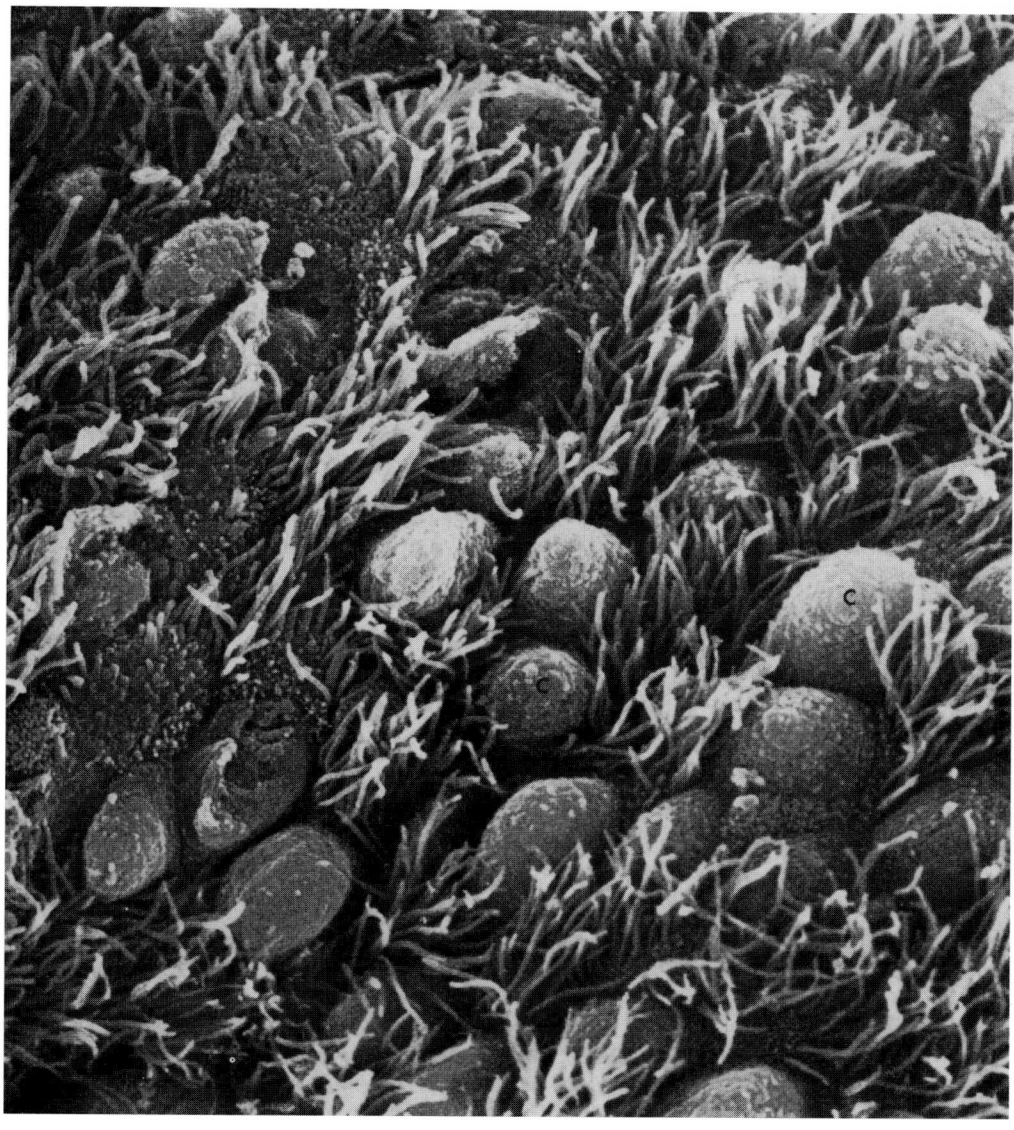

FIGURE 1-33
A scanning electron micrograph of a bronchiole from a pig showing Clara cells (C) projecting among the cilia of neighboring ciliated epithelial cells. (EM, ×3700). (*Reproduced by courtesy of* *the Dr. J.H.L. Watson, Physics & Biophysics Department, Edsel B. Ford Institute for Medical Research, Detroit.*)

ithelium has been seen at 20 weeks. BALT has been identified in 77 percent of infant lungs, whereas a well-developed lymphoepithelium was evident in only four cases.

BALT has been described as being richly innervated, but the nature of the nervous influences on this tissue is unknown. Both B and T lymphocytes are found in BALT, but unlike lymph nodes, it has no T-cell–dependent area. BALT and other mucosal lymphoid tissues are probably the major source of IgA, which is an important part of the local mucosal immune response.

In addition to BALT, the lung contains lymphoreticular aggregates and bronchoalveolar lymphocytes. The former consist of scattered, ill-defined collections of lymphocytes between the bronchial epithelium and the accompanying pulmonary artery. These aggregates occur in bronchi and interlobular septa and pleura, as well as in the alveolar interstitium, at the origin of sites of lymphatic drainage. Such foci are particularly prominent in people with heavy dust exposure.

Bronchoalveolar lymphocytes are approximately 60 percent T cells, 10 percent B cells, and 30 percent "null" cells. Granulated lymphocytes, with the phenotype of natural killer cells, have been demonstrated in human lungs. Bronchoalveolar lymphocytes are thought to originate from the blood.

The mechanism of recruitment and recirculation is a two-step process. There is adherence to endothe-

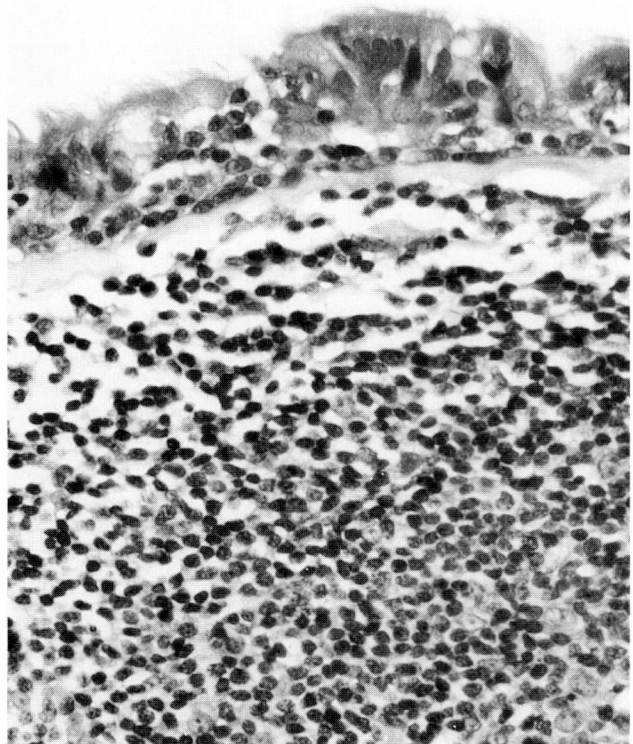

FIGURE 1-34
BALT as shown on light microscopy in a peripheral airway. The lymphocytes have infiltrated the overlying epithelium in one area. (H&E, ×232.)

lia at specific sites, and the induction of a motile response permits the lymphocyte to transit the endothelium and enter the tissue by means of active locomotion. Further details of these processes are described by Berman and colleagues.[59]

CELLS OF THE ALVEOLI

The terminal part of the airways is termed the acinus. This arises distal to the terminal bronchiole, where there are respiratory bronchioles, alveolar ducts, and a terminal cluster of alveoli (Fig. 1-35) called the atrium. The respiratory bronchiole is the first order of bronchiole to bear alveoli. The first respiratory bronchiole has been termed a transitional bronchiole by Rodriguez and coworkers.[61] In functional terms, they divide this bronchiole into two parts: an initial respiratory section beginning with the first alveolus and a purely conductive part. This branching pattern may be oversimplified.

In a study of a single lung, the respiratory bronchioles, while becoming progressively more alveolated, ran for relatively great lengths, with alveolar ducts coming off as side branches. The alveolar duct systems become progressively less complex the

nearer they are to the pleura.[62] There is a significant concentration of collagen (13 percent) and elastin (9 percent) in the alveolar duct walls, consistent with their proposed role as the main load-bearing element of the lung parenchyma.

The alveolar surface area in adults is approximately 70 to 80 m^2, and the alveolar number is about 300 million.[63]

The alveolus is cup-shaped, and by light microscopy normally no lining alveolar cells are seen. Bands of elastin fibers form a continuous ring around each alveolar mouth. Elastin fibers penetrate deep into the alveolar septal walls.[64] The alveolus has an inner diameter of approximately 275 μm, which varies during inspiration and expiration.[65]

The inner surface of the alveolus is covered with surfactant, which consists of a basal layer (hypophase) in contact with the alveolar cells and a surface layer (sialomucin). The air-liquid interface of this duplex layer contains myelin figures (Figs. 1-36 and Fig. 1-37).[66] Surfactant is a phospholipid [dipalmitoylphosphatidylcholine (DPPC)] synthesized by type II pneumocytes and, to a lesser extent, Clara cells. This lowers the surface tension at the air-water interface as well as acting as a "nonwettable" agent. It behaves like a patch of oil on a road surface, repelling water during a shower. This property enables any edema fluid to be converted into droplets and rapidly removed.

The cells of the alveolar wall can be properly visualized only by electron microscopy. However, a discontinuous layer can be seen in the normal lung when sections are stained by the immunoperoxidase technique using antisera against keratins of low molecular weight (AE1/3 or CAM5.2) (Fig. 1-38). Cell-specific markers for the alveolar epithelium have been demonstrated in fetal rat lung.[67]

Between 93 and 96 percent of the alveolar wall is covered by type I or membranous pneumocytes (Fig. 1-39). Type I cells comprise 8 percent of the total alveolar cells.[68] These are specialized cells with no regenerative capacity. They can be likened to peripheral nerve, in that the cytoplasm extends a long way from the nucleus. The cytoplasm is thin, 450 to 600 Å thick, facilitating gas transfer. The cytoplasm contains a few mitochondria, a small amount of smooth endoplasmic reticulum, and an occasional lysosome. Micropinocytotic vesicles, which probably play a major role in transport of solutes through the cell, are seen in the plasmalemma. The edges of adjacent cells are bound together by tight junctions (Fig. 1-40). This tight barrier helps to restrict the movement of ions and water. An intact epithelial barrier is essential for the resolution of intraalveolar

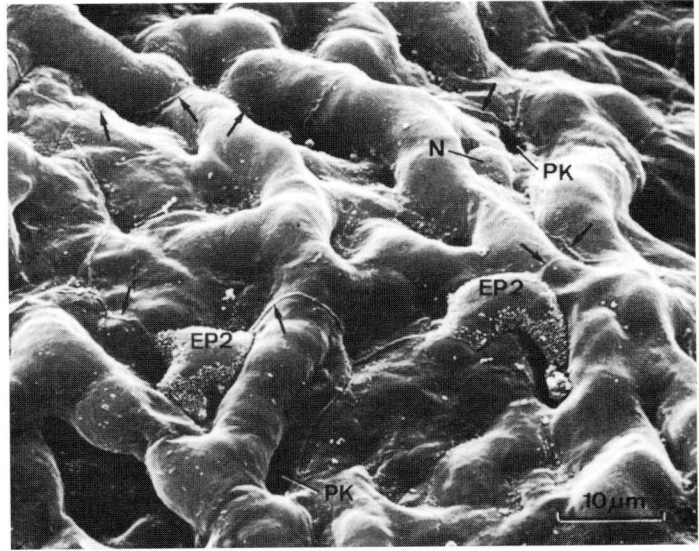

A

FIGURE 1-35
A. Surface of the alveolar wall in the human lung
showing a mosaic of alveolar epithelium consisting of
types I and II (EP2) cells. Arrows indicate boundary of
the cytoplasmic leaflet of the type I cell which extends
over the many capillaries. Note the two interalveolar
pores of Kohn (PK) and nucleus (N) of type I cell. (SEM)
B. Scanning electron micrograph of human lung
parenchyma. Alveolar ducts (D) are surrounded by
alveoli (A), which are separated by thin septa (S). There
are small branches of a pulmonary artery (PA).

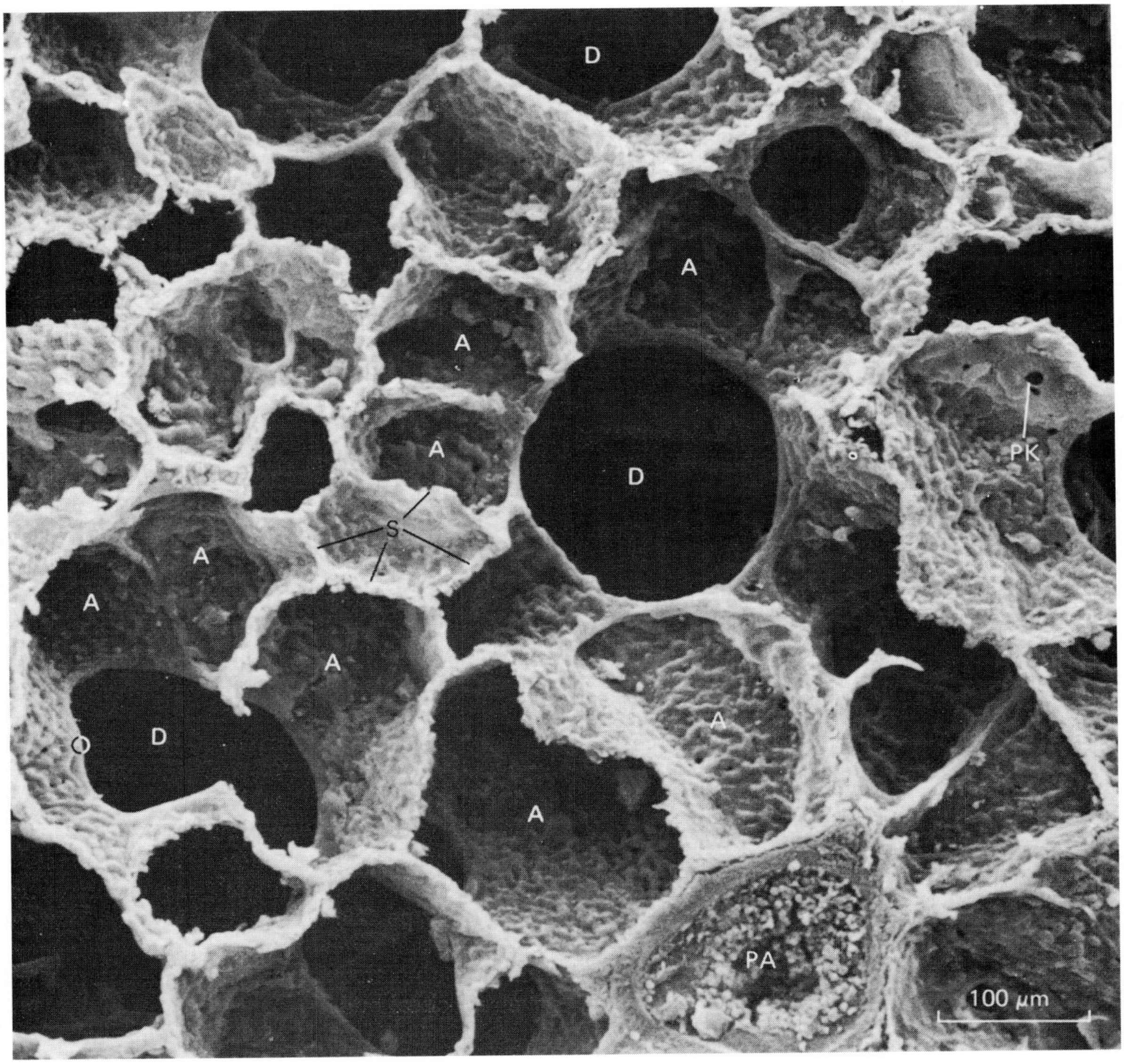

B

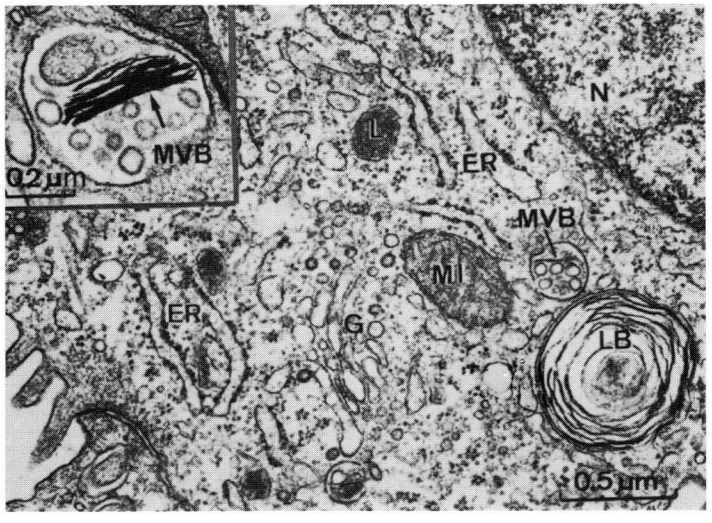

A

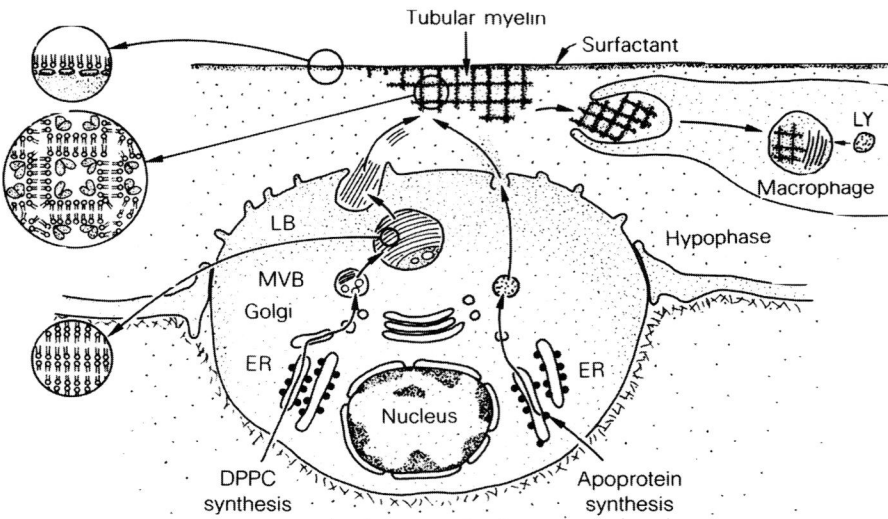

B

FIGURE 1-36
A. Cytoplasmic organelles of the type II cell. There is endoplasmic reticulum (ER), Golgi's complex (G), lysosomes L), multivesicular bodies (MVB), and lamellar bodies (LB). The inset shows a large multivesicular body with a stack of phospholipid lamellae. (*After Wiebel, 1984.*) *B.* Schematic diagram of pathways for synthesis and secretion of surfactant DPPC and apoproteins by a type II cell, and for their removal by macrophages. Note the arrangement of phospholipids and apoproteins in the lamellar bodies, in tubular myelin and in the surface film. (*From Wiebel, 1984.*)

edema. Removal of this fluid is driven primarily by active sodium transport across the tight epithelial barrier.[69]

Weibel[70] demonstrated deep cytoplasmic extensions of type I cells, which pass through the interstitium of the alveolar wall, between the capillaries, to reach and cover parts of the opposing alveolar surface. This anatomic configuration of one cell covering two surfaces, with the nucleus on only one side, makes it impossible for such a cell to undergo mitosis. Type I cells are replaced by proliferation of type II cells.

Type II cells (granular pneumocytes) cover 7 percent of the alveolar surface but account for 16 percent of the total alveolar cells, with a mean volume of half

that of the type I cells.[53] These cells occupy the corners of the alveolar walls and have small microvilli on their free surface. The cytoplasm contains numerous mitochondria, rough endoplasmic reticulum, and many lysosomes. The most characteristic feature is the presence of osmiophilic lamellar bodies (Fig. 1-41), responsible for the production of surfactant (see Chap. 2). This is also produced by Clara cells.

Type II cells posses some phagocytic properties, since experimental evidence shows that thorotrast,[71] hemosiderin, and asbestos fibers[72] are taken up by these cells. If type II cells are truly phagocytic, it would suggest that alveolar epithelium may have a mesodermal rather than an entodermal origin, from the laryngotracheal bud.

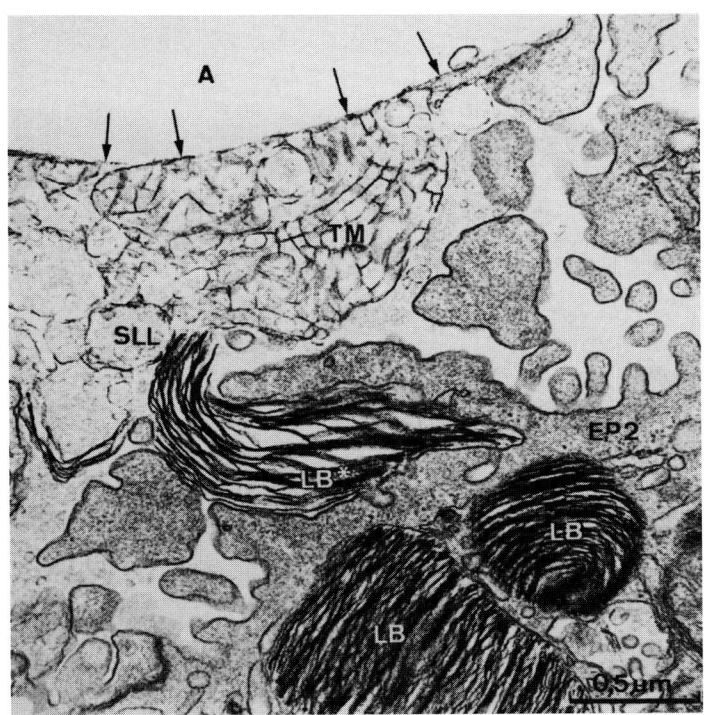

FIGURE 1-37
Myelin figure in acinus. In the apical part of the type II
cell (EP2) with lamellar bodies (LB): one of these (LB*) is
seen in the process of being secreted into the surface
lining layer (SLL). The free surface of the lining layer is
covered by a thin black film of DPPL (dipalmitoylphos-
phatidylcholine) (arrows) which is connected with
tubular myelin (TM) in the hypophase. A, alveolar
lumen. (*From Weibel and Gill, 1977.*)

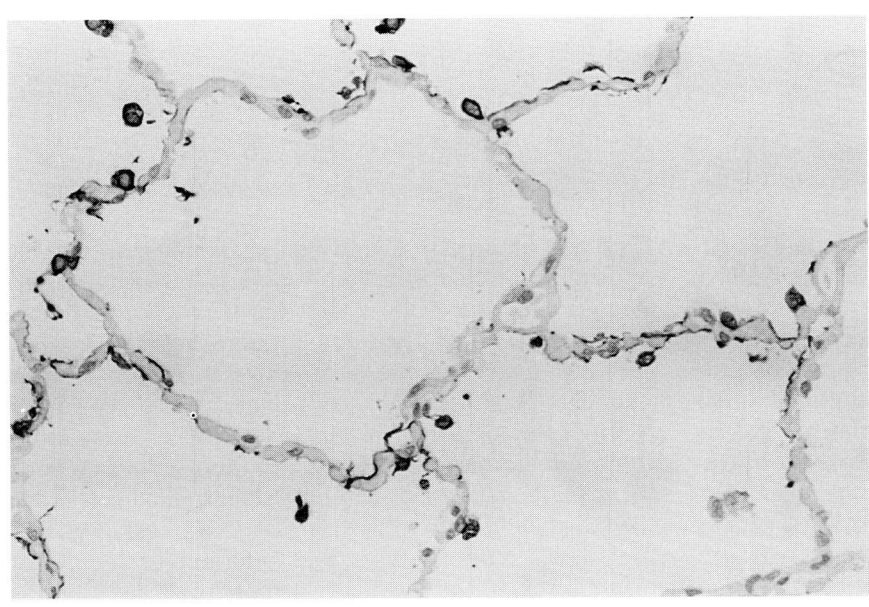

FIGURE 1-38
Cytokeratin (AE1/3) with a discontinuous
layer around the alveolar wall.
(Immunoperoxidase AE1/3, ×350.)

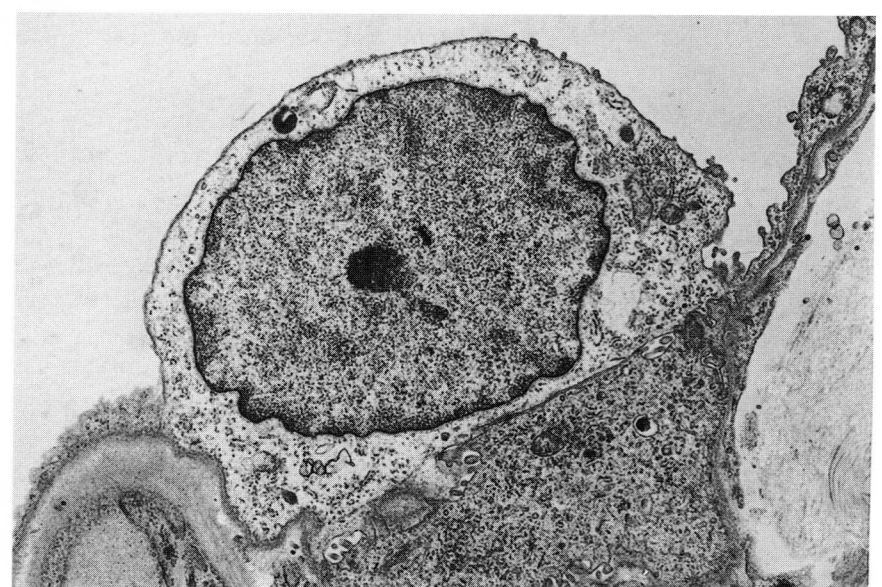

FIGURE 1-39
Type I pneumocyte with a prominent
nucleus. These cells spread a thin layer
of plasma membrane-bound cytoplasm
over the luminal aspect of the alveolar
basement membrane. (EM, ×6300.)

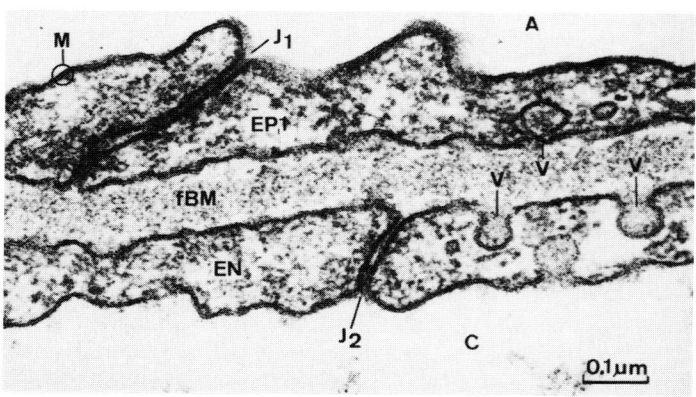

FIGURE 1-40
Barrier between type I epithelial cells (EPI) and
endothelial (EN) cells. In between is the fused basement
membranes (fBM). The cell membranes have a trilaminar
structure (M). A "tight" junction (J1) is formed by close
apposition of the cell membranes over a relatively wide
band. The junction (J2) is relatively less "leaky" because
membranes are only apposed over a narrow strip.

Alveolar Basement Membrane

Below the alveolar epithelial cells lies a basement
membrane (epithelial basement membrane). Where
alveolar epithelium covers capillaries, the alveolar
basement membrane fuses with the basement membrane lying immediately external to the capillary
endothelium (endothelial basement membrane).
Thus, electron microscopically the two basement
membranes form a homogeneous structure. The portion of each membrane subjacent to both the alveolar
and capillary endothelium is less dense and is
termed the lamina lucida.[73] These zones are continuous with a dense zone in each membrane, termed the
lamina densa (Fig. 1-42). This basic architecture is
best seen in the parts of the lung septa lying between
capillary loops or on the opposing surfaces of an
alveolar capillary that is covered only by a thin alveolar epithelial layer.

The basement membrane is produced by both
the epithelial and the endothelial cells.

Interstitial Space

The part of the septal wall lying between the alveolar
epithelial and capillary endothelial basement membranes constitutes the interstitial space. The interstitial space is normally inconspicuous but is distended
in pulmonary edema or in any form of alveolar
damage. Residing in it are macrophages, capillary
pericytes, myofibroblasts, mast cells, occasional
lymphocytes, and a few sensory receptor nerve endings. Collagen and elastin fibers have also been identified.

The myofibroblasts gain attachment by dense
junctional complexes to the opposing septal interstitial surfaces of the alveolar epithelial basement membranes. Contraction of these cells is thought to move

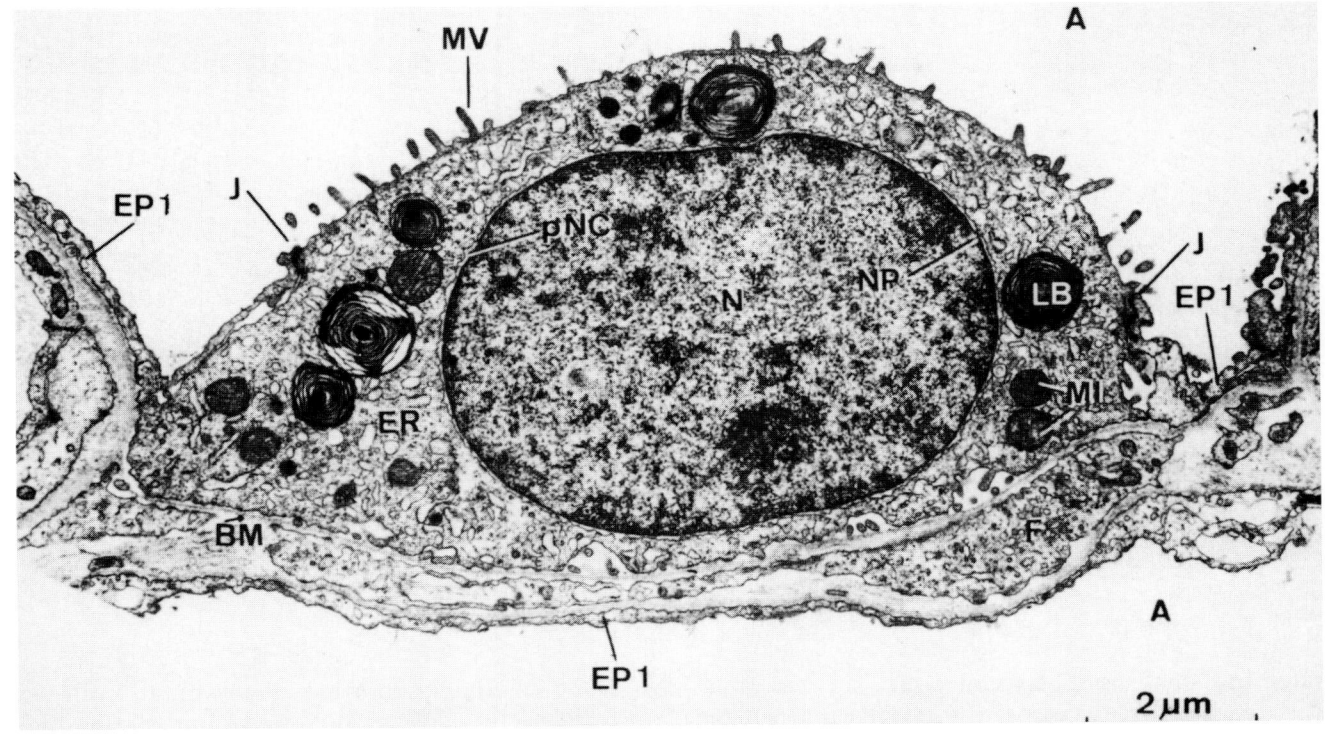

A

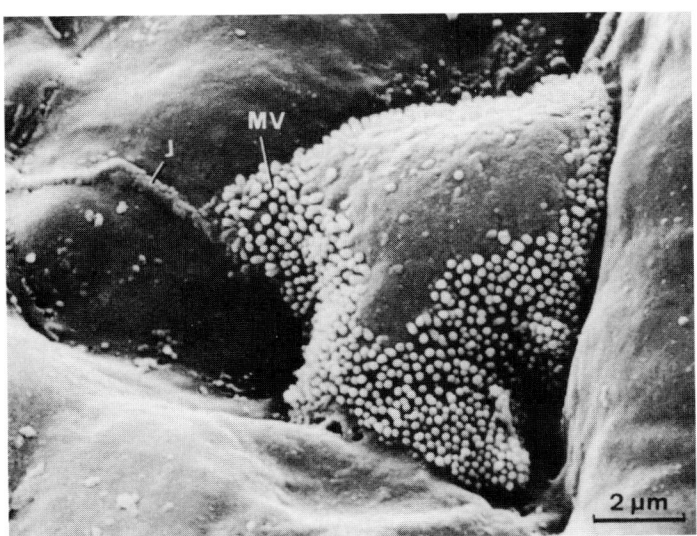

B

FIGURE 1-41

A. Type II pneumocyte from human lung with tight junctions (J) with type I epithelial cells (EPI). The cytoplasm contains lamellar bodies (LB) along with mitochondria (MI) and endoplasmic reticulum (ER). The nucleus (N) is surrounded by a perinuclear cisterna (PNC) which is perforated by nuclear pores (NP). BM, basement membrane; F, fibroblast; MV, microvilli; A, alveolus. *B*. SEM of type II cell with microvilli (MV) and a central "bald patch." The junctions between Types I and II cells are denoted by J.

interstitial fluid toward the peripheral extremity of the pulmonary lymphatic system in the periarterial and perivenous connective tissue sheaths. A fuller discussion of the importance of this space can be found in Fishman and Hecht,[74] Fishman and Renkin,[75] Staub,[76] Yu,[77] and Harris and Heath.[78]

The interstitial space is occupied mainly by a gel, which includes mucopolysaccharides. This gel reduces the flow of water and small molecules.[79] The interstitial gel substance may retain the alveolar and blood vessel configuration by supplying a counter-force to the intraalveolar and intravascular pressures. This function is shared with collagen and elastin.

The pressure in the interstitial liquid is unknown. From several experiments, however, values of between 6 and 9 mmHg less than atmospheric pressure have been recorded, at the level of the heart.[65]

The passage of substances across the epithelium and endothelium is discussed by Harris and Heath.[78]

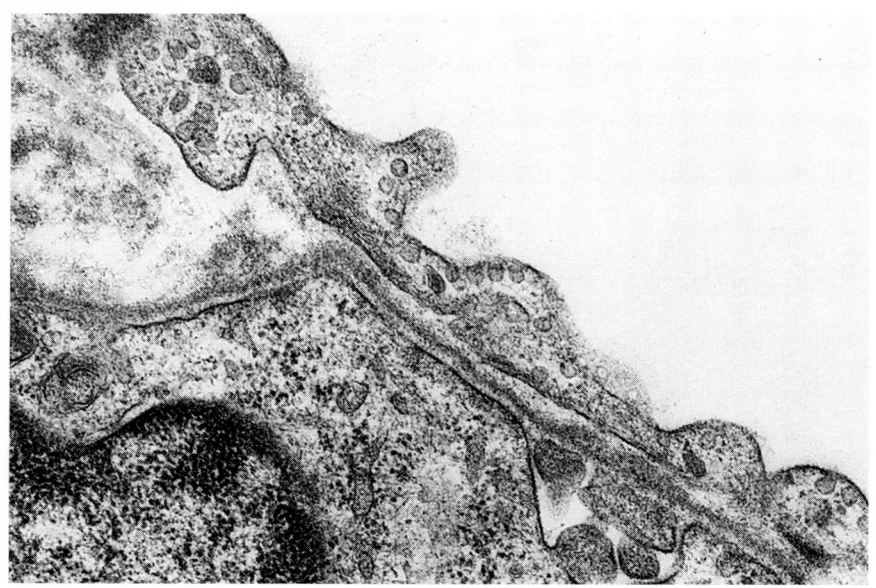

FIGURE 1-42
Alveolar basement membrane showing a central lamina densa and external lamina rare. (EM, ×39,375.)

Water and small nonpolar molecules, such as urea, pass relatively freely across the plasma membrane. Similar free passage is afforded to lipid-soluble substances, such as oxygen. Further modes of transport are through intercellular gaps between endothelial cells, micropinocytotic vesicles, or transcellular channels.[78] Large molecules, such as proteins, must travel across the endothelium, through intercellular gaps or via micropinocytotic vesicles.

In the interstitial space are juxtapulmonary capillary receptors ("J receptors"), which are close to the pulmonary capillaries.[80] These respond to an increased volume in the interstitial space, probably via neural influences. This increases ventilation.

The interstitial connective tissue is continuous with the connective-tissue sheaths surrounding blood vessels and bronchioles in the centers of lung lobules, as well as the pleural and perivenous connective tissue situated at periphery of lung lobules. This continuous connective-tissue framework, extending throughout the lung, is important for the understanding of both fluid transport in the lung as well as the functional disturbances that may follow interstitial disease.

The collagen in normal lung consists mainly of types I and III, but up to 11 other genetic types with specialized adhesive and connecting functions can be found in various lung structures.[81] Adult perivascular and peribronchial connective tissue is composed largely of the strong type I collagen. More type III collagen, which allows greater flexibility, is seen in fetal and neonatal lungs.[82] As chondroblasts are responsible for the formation of type III collagen, this type is normally found only in the bronchial and tracheal cartilages.[83]

The relative proportion of elastin to collagen increases with age.[84] A review documents the biochemistry and turnover of the interstitium.[81]

Pericytes

The pericytes (Rouget cells) are closely applied to the outer walls of the alveolar capillaries and probably have a contractile function (Fig. 1-43). These cells are similar to the myofibroblasts found in arterial walls. The branched cytoplasmic processes of these cells lie partly embedded in the substance of the endothelial basement membrane and come close to the outer sur-

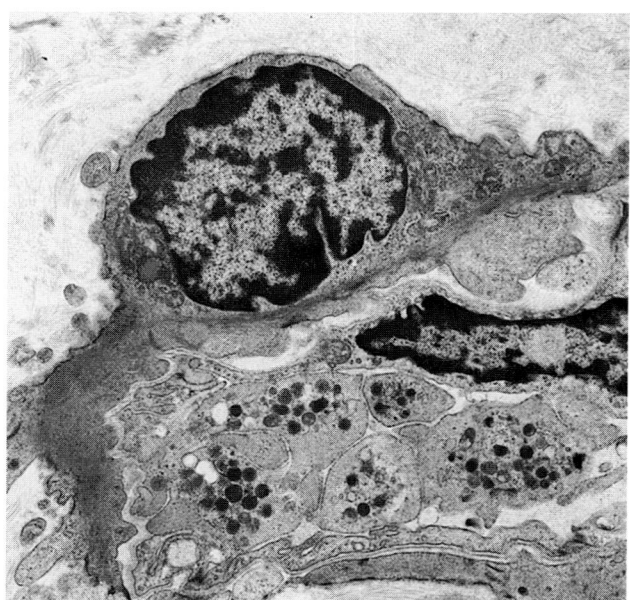

FIGURE 1-43
Alveolar capillary pericyte.

28

face of the endothelial cells. These cytoplasmic processes contain fine filaments similar to those seen in smooth muscle cells. Microtubules, as well as many pinocytotic vesicles, are seen beneath their outer surfaces. Fewer pericytes are found around alveolar capillaries.

Alveolar Macrophages

There has been recent interest in these cells because they can be so readily harvested by bronchoalveolar lavage (Fig. 1-44). Pulmonary macrophages not only have a phagocytic role, keeping the lung sterile, but also are secretory and regulatory cells.

Whether pulmonary intravascular macrophages originate from bone marrow cells (monocytes) or are derived from interstitial or alveolar macrophages and migrate into the capillary lumen is still to be determined.[85] The most commonly held view is that lung macrophages originate from blood monocytes, but alveolar proliferation has been demonstrated. However, interstitial macrophages, as compared with

alveolar ones, have an increased ability to replicate and synthesize DNA.[86] This is probably minimal in the normal individual, owing to the lack of receptors for macrophage colony–stimulating growth factor (M-CSF or CSF-1).[87] The number of tissue macrophages exceeds the circulating macrophages by a factor of 400.[88]

The alveolar macrophage varies in size from 12 to 40 μm in diameter and has a lobulated nucleus, cytoplasm with numerous mitochondria, and electron-dense secondary lysosomes (Fig. 1-45).

Increased numbers of macrophages are found in smokers. They are larger (up to 50 μm and contain pigmented intracytoplasmic inclusions (Fig. 1-46).

Alveolar macrophages have been separated into different subpopulations.[89] Membrane receptor expression and functions, such as phagocytosis and mediator release, vary among these different cell populations. Apart from circulating monocytes and interstitial macrophages, a third and morphologically distinct population of intravascular pulmonary macrophages have been described in humans.[90]

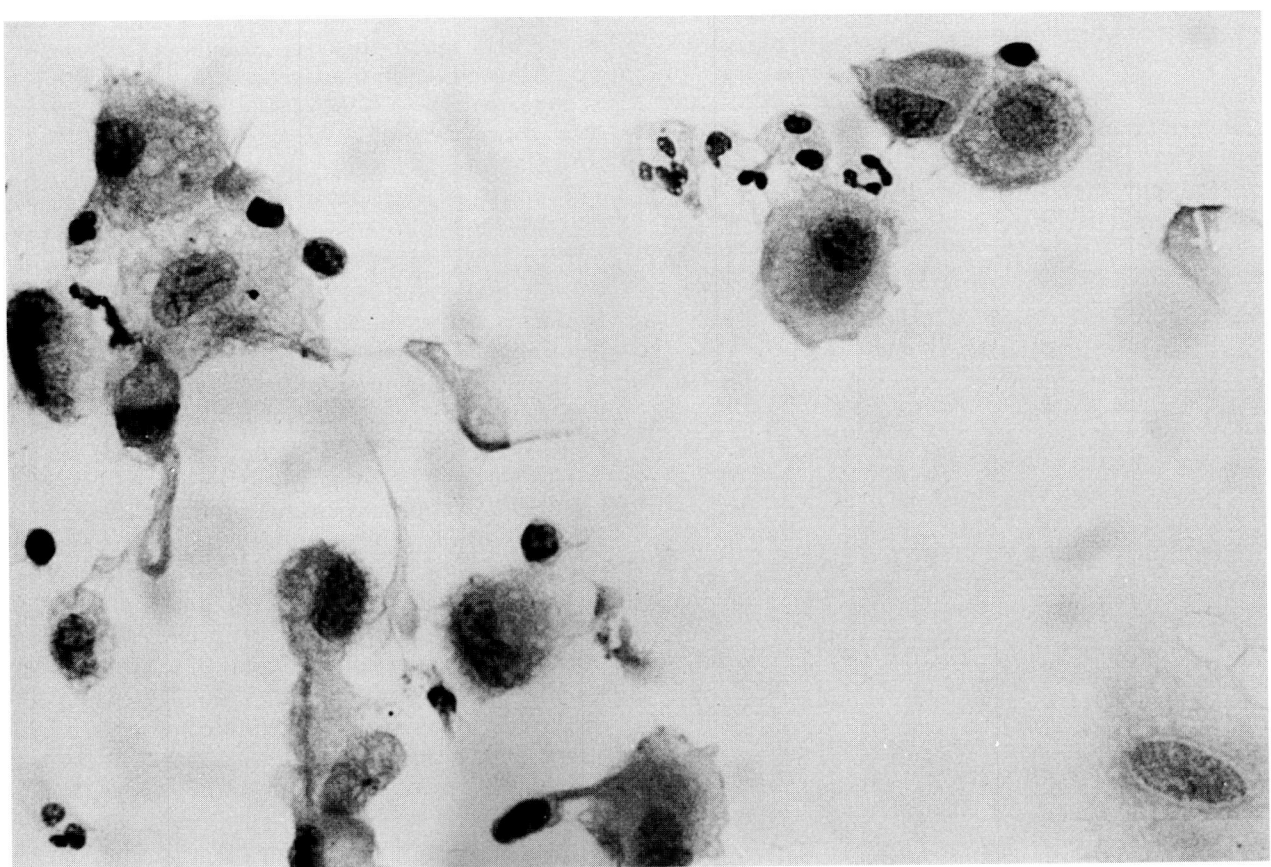

FIGURE 1-44
Alveolar macrophages and a few neutrophil polymorphs in an alveolar lavage. Some scattered lymphocytes are seen. (H&E, ×350.)

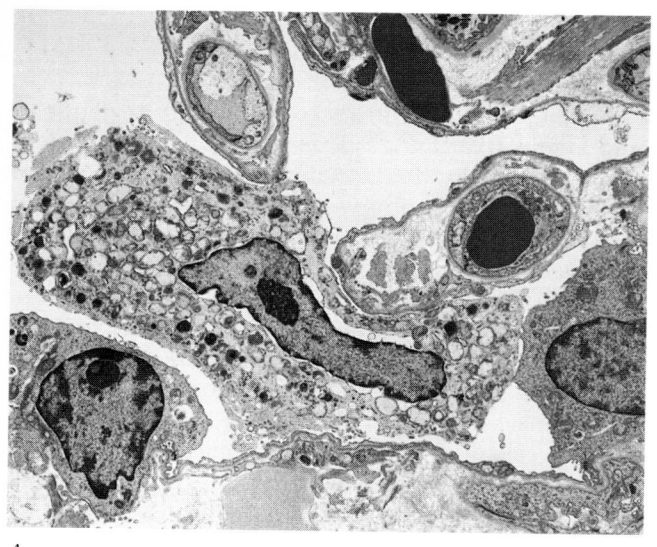
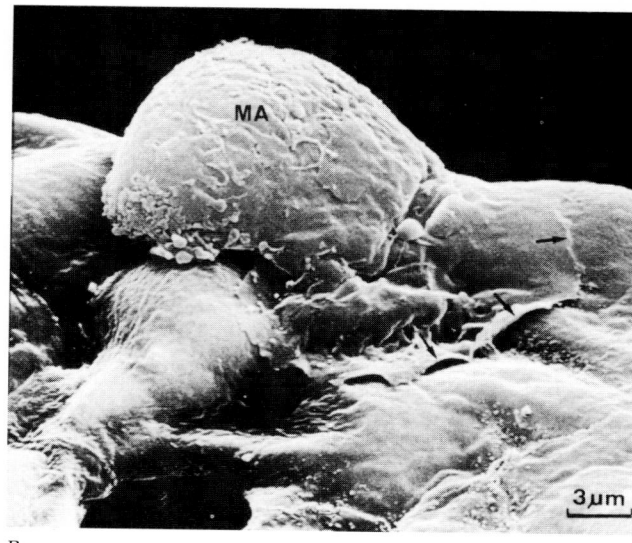

A

B

FIGURE 1-45

A. Alveolar macrophage lying in alveolar lumen. There is an elongated nucleus. (×3920.)
B. Alveolar macrophage (MAA)

Most attention is given to *alveolar* macrophages, but it has been shown that if the lung is fixed by submersion, intravascular fixation, or vapor fixation, many macrophages can be seen in the bronchial tree. Airway fixation displaces most airway macrophages.[91] Some of these cells are present in the mucociliary escalator system, but others are adherent to the bronchial epithelium. One of the least studied macrophage types is the pleural-based macrophage. It is thought that these are similar to peritoneal macrophages, partly reflecting the lower P_{O_2} in

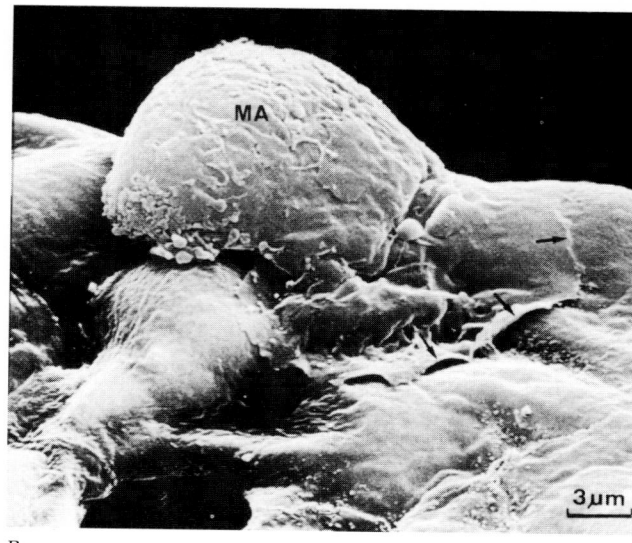

FIGURE 1-46

Needle-like inclusions (*arrow*) in macrophages seen in a cigarette smoker. (EM, ×44,400.)

the pleural and peritoneal cavities than in the alveoli.[92]

There are many functions of pulmonary macrophages.[85] More than 100 substances are secreted by pulmonary macrophages.[93] Some of the most important functions will be described here, but the reader is referred to the articles cited above for fuller descriptions.

Macrophages play an important part in the inflammatory reaction producing tumor necrosis factor and interleukins 1 and 6. Macrophages are important in lymphocyte activation. Some B and T cells cannot recognize free antigen, so a group of macrophages act as antigen-presenting cells for these lymphocytes. They take up antigen, from which they cleave immunogenic fragments. These fragments, together with HLA-DR fragments, stimulate B lymphocytes.[94]

Macrophages are seen at the sites of infection, the prime example being tuberculosis. The cells are attracted to an infected area by both bacterial and polymorph fragments as well as by endotoxins, complement, immune complexes, and collagen fragments.[95] The macrophages remain at the focus of infection because of a migration-inhibition factor generated by B lymphocytes.[96] Besides killing organisms, macrophages play an important role in lysis of tumor cells by releasing substances such as tumor necrosis factor, complement components, and proteases.

Macrophages are present in wounds immediately after injury; they synthesize collagenase and elastase, which clean the wound.[97,98] They release factors regulating bronchial epithelial growth.[99]

Mast Cells

Mast cells are seen throughout the bronchial tree but are most common in the alveoli. They are also found in the interstitium. They measure 10 to 15 μm in diameter, and possess the usual cellular organelles. They are infrequently seen on light microscopy in human formalin-fixed tissues, as they degranulate. Either alcohol or Carnoy's fluid should be used to fix human mast cells, and toluidine blue is commonly used to stain them. A murine monoclonal antibody (G5) has been described that detects human mast cell tryptase.[100] Another mouse monoclonal antibody (AA1), which detects mast cell tryptase in formalin-fixed tissues, has been used on bronchial biopsies.[101] Even with monoclonal antibodies, the numbers of cells identified is higher in Carnoy-fixed material than with formalin-fixed samples.

The specific features used to identify mast cells are

1. the presence of membrane-bound, intracytoplasmic granules 600 to 800 nm in diameter;
2. characteristic intragranular inclusions of various forms; and
3. the presence of rare, long surface filiform microvilli (Fig. 1-47).[102]

These criteria help distinguish mast cells from basophils, which have smaller, finely granular intracytoplasmic inclusions.

Mast cells are derived from hematopoietic precursors and produce two neutral proteases, tryptase and chymase. Tryptase probably activates the Hageman factor as well as cleaving kinin from kininogen. Lysosomal enzymes are also released by pulmonary mast cells: arylsulfatase, which hydrolyzes slow-reacting substance; B-glucuronidase, which splits glucuronide conjugates; superoxide dismutase, which splits oxygen; and myeloperoxidase, which cleaves hydrogen peroxide. In addition, mediators, including histamine, are produced. Histamine causes contraction of airway smooth muscle and pulmonary vasoconstriction and induces bronchiolar leakage, leading to pulmonary edema. Histamine acts on the lung by interacting with two cell membrane-associated receptors, H1 and H2.[103] Also released by mast cells is eosinophil chemotactic factor of anaphylaxis, which causes chemotactic attraction and deactivation of eosinophils and neutrophils,[104] and high-molecular-weight neutrophil chemotactic factor (HMW-NCF), which is present in serum after experimental induction of IgE-dependent bronchospasm. HMW-NCF can also deactivate neutrophils.

An important mediator released by mast cells is the high-molecular-weight proteoglycan, heparin. This can inhibit classical and alternative pathways of complement activation and, because of its ability to interact with antithrombin III, activates many enzymes.

Mast cells act as mediators for IgE-dependent responses to airway allergens. Their ability to alter airway diameter, induce mucus production, and stimulate irritant receptors could provide a mechanism for preventing or removing inhaled toxic materials. Recruitment of plasma cell systems and polymorphs also protects the lung. Mast cells are increased in pulmonary edema and fibrosis.

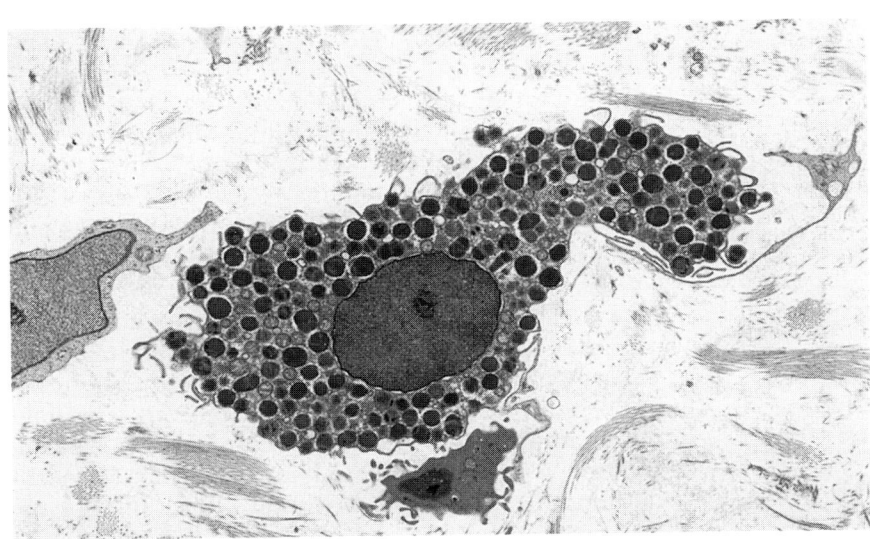

FIGURE 1-47
Electron micrograph of pulmonary mast cell showing typically abundant cytoplasmic granules, the contents of which are granular in nature. Mast cells are often associated with fibroblasts, as in this micrograph. (EM, ×3920.)

Pulmonary Endothelial Cells

Pulmonary endothelial cells are barely detectable by light microscopy, showing as nuclei that ring the vessel lumen. In alveolar capillaries there are long cytoplasmic extensions, which encircle much of the vessel. The endothelial cells rest on the basal lamina, which they secrete. Compared with arteries and veins, the cytoplasm of pulmonary capillary endothelium is poorly endowed with organelles. They contain mitochondria, free ribosomes, rough endoplasmic reticulum, and some Weibel-Palade bodies (Fig. 1-48).[105] Endothelial cells also contain microtubules and microfilaments.

Both epithelial and endothelial cells show junctional complexes between them.[106–108] These are of three main types: tight junctions, adherens junctions, and gap junctions.

Tight junctions are known as *zonula occludens.* These junctions consist of a complex network of interconnected fibers. The endothelium of larger vessels have well-developed tight junctions, so they are relatively impermeable to water and its solutes. Capillary endothelium has loosely organized tight junctions, suggesting that they are more permeable to water.

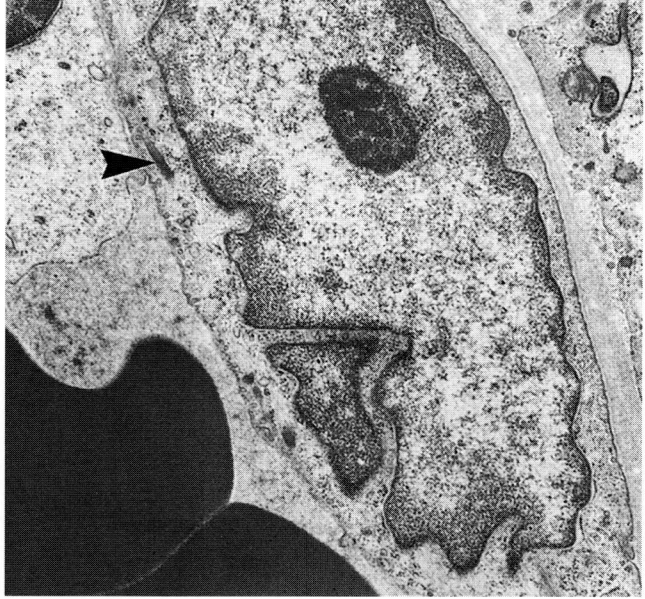

FIGURE 1-48
Electron micrograph of pulmonary endothelial cell, located on the inner, capillary side of the alveolar basement membrane. Numerous micropinocytotic vesicles are seen budding off from the plasma membrane. The cytoplasm of these endothelial cells contain Weibel-Palade bodies (*arrow*). (EM, ×10,800.)

Two molecules have been identified in tight-junctions.[109] They are both on the cytoplasmic side of the junction and are termed ZO-1 and cingulin.

Adherens junctions hold adjacent endothelial and epithelial cells together. On the cytoplasmic side of the junction are plaques, or areas of electron-dense deposits. These contain proteins such as plakoglobulin, vinculin, and alpha-actinin, which probably interact with the actin microfilaments of the cytoskeleton. Actin and myosin play an important part in the regulation of the intercellular cleft width and so control the paracellular pathway of vascular permeability.[110]

For the maintenance of cell-cell contact by the adherens junction, transmembrane molecules, cadherins, have been identified on epithelial cells.[111] An endothelial cadherin, vascular or V cadherin, has been identified.[112]

Gap junctions consist of small pores between adjacent cells. They may allow small molecules to pass through.

Previously, the pulmonary capillaries were considered to be capable only of gaseous exchange. Over the past 20 years, however, other functions have been demonstrated. These include production of angiotensin-converting enzyme, which converts the inactive angiotensin I into the vasoconstrictor, angiotensin II. The same enzyme inactivates bradykinin, a powerful vasodilator.[113] In addition, specific carrier mechanisms operate in the cell membrane to transport serotonin and adenosine into the endothelial cells, where they are metabolized. Besides deactivating chemicals, endothelial cells synthesize substances such as fibronectin, heparan sulfate, interleukin-1, tissue plasminogen activator, various growth-promoting substances, and smaller molecules, such as prostacyclin, endothelium-derived relaxing factor, platelet-activating factor, and endothelin-1.

Endothelin is seen predominantly in the endocrine cells of the developing lung, suggesting that it may play a part in growth regulation. Three family members of the endothelins (ET-1, ET-2, and ET-3) have been demonstrated. The functions of ET-2 and ET-3 are not yet defined. ET-1 is the only endothelin produced by the endothelium. It is a potent vasoconstrictor, bronchoconstrictor, inhibitor of renin release from the juxtaglomerular cells, and modulator of autonomic transmission. It has a positive inotropic effect on the heart. It is unlikely to be an important circulating vasoregulatory hormone, as the concentration is well below that known to cause vasoconstriction.[114] The vasoconstrictor effects of ET-1 are opposed by the concomitant release of prostacyclin, atrial natriuretic peptide, and endothelium-derived relaxing factor. Release of ET-1 immunoreactivity into the plasma may be due to endothelial damage or to direct stimulation by endotoxin. A full description of the normal pulmonary vasculature is given in Chap. 21.

COLLATERAL RESPIRATORY PATHWAYS

Several collateral pathways have been described. These are interalveolar, bronchioalveolar, and inter-bronchiolar channels.[115]

The interalveolar communications (Fig. 1-49) were first described, but not illustrated, over 100 years ago by Kohn.[116] He believed these channels opened only in disease states. There is thus debate as to whether they should be termed pores of Kohn.[117] The average number of pores per alveolus is 13 to 21, with half being located on the bottom walls. The average length of the major axes of the pores is 7 to 19 μm. The distribution, area fraction, and size of the pores are uniform regardless of their location within the alveolus or the size of alveolus.[118] The pores are scarce at birth and increase in number with age. It has been proposed by some that the alveolar pores provide the major pathways for collateral circulation,[119] but their true role has not been elucidated.

Lambert[120] described accessory bronchioalveolar communications extending from respiratory bronchioles to alveolar ducts and sacs subtended by the bronchiole. These have a muscular wall, and so a regional control of airflow seems possible. They range in size from partly closed to 30 μm in diameter.

Martin[121] described respiratory bronchioles connecting terminal bronchioles from adjacent lung segments in dogs. Similar communications have been described in human lungs.[122]

Physiological studies suggest that these pathways are the main route for collateral ventilation. Such ventilation is not important in the healthy individual but becomes significant when the airways are obstructed by disease. Multiple factors may affect

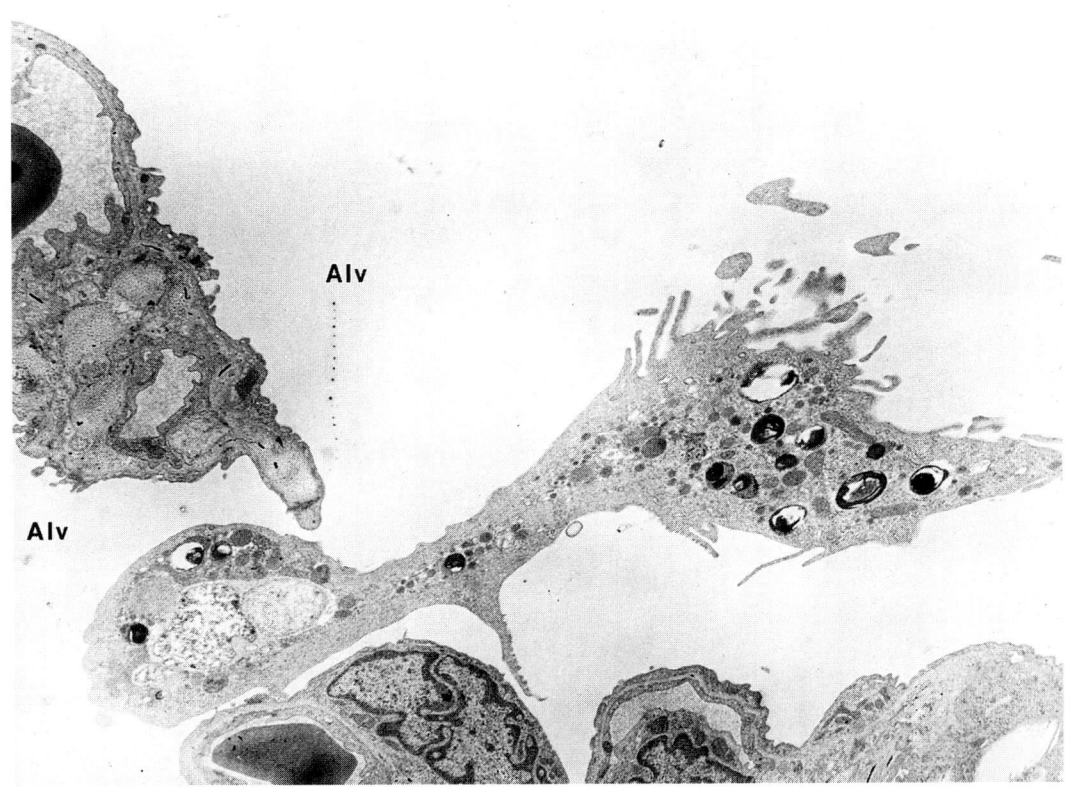

FIGURE 1-49
Pore of Kohn as shown by a macrophage through the alveolar wall, between two alveoli (Alv). Specimen taken from a guinea pig lung. (EM, ×8800.) (*Reproduced by courtesy of Dr. K. G. Bensch and the editor of Pathologic Physiology and Anatomy of the Lung, Williams & Wilkins, Baltimore.*)

the collateral circulation, including an obstructed bronchial segment, an increase in lung volume, an increase in surface tension, hypocapnia, a decrease in pulmonary blood flow, and pulmonary edema as described by Delaunois.[115] The increase in collateral resistance with age or restrictive conditions such as obesity or kyphoscoliosis is one of the mechanisms by which these patients are more prone to develop atelectasis. This condition is especially likely to occur in the middle lobe after infection or obstruction, possibly due the greater ratio of pleural to non-pleural surface.

PROLIFERATION AND DIFFERENTIATION IN AIRWAY EPITHELIUM

There are few studies which determine the length of the cell cycle and its phases for different pulmonary cell types.[20] The results are not uniform as they relate to different animals, and different observers have obtained different results. The time for the cell cycle of the hamster tracheal basal cell is given as 28 h by one group and 159 by another. The figures for mucous cells was 25 or 97 h, respectively.[123]

In differentiation the basal cell acts as a stem cell. It divides into two in the rat, one further basal cell and one superficial cell.[124] There has been the suggestion that the mucous cell plays a major part in the regeneration of conducting airway epithelium.[126] In peripheral bronchioles, where basal cells are absent, the Clara cell acts as the stem cell. In the alveoli, the type II cell is the progenitor cell.

Numerous agents, such as tobacco smoke, ozone, and chemical and infective substances, alter the pattern of airway differentiation. These will be discussed further in the relevant sections.

Pulmonary Lymphatics

The anatomy of the pulmonary lymphatics is important in surgical pathology, not just in relation to the spread of tumors or infection but also as an aid to diagnosis in the distribution of diseases such as sarcoidosis. Miller[127] studied pulmonary lymphatic drainage in both humans and dogs. In adult humans lymphatics surround pulmonary arteries and veins. Two networks are present, a deep plexus accompanying the bronchi and vessels and a superficial one that forms a pleural network (Fig. 1-50). The two communicate below the pleura, where lymph from the deep plexus flows into the superficial channels. Elsewhere, the deep plexus drains centrally toward the hilar nodes.

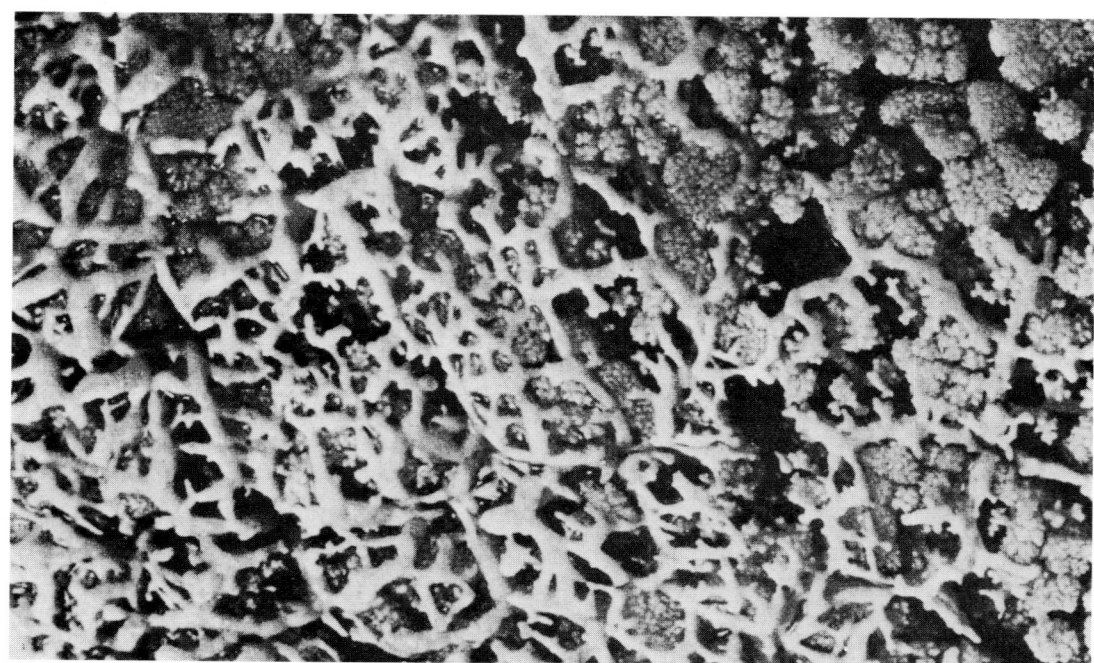

FIGURE 1-50
Corrosion cast of pleural lymphatics plexus in infant lung. The lymphatics form a polyhedral network which roughly outlines the lung lobules. The saccular tissue filling the lung lobules is also injected. (*Reproduced courtesy of Professor J. M. Lauweryus, Louvain, and the publishers of Pathology Annual, 1971, Appleton-Century-Crofts.*)

It has been shown that the subpleural lymphatics can drain directly into the mediastinal nodes in 22 percent of right lung segments studied and in 25 percent of left lung segments. Such communications were found to be more common in the upper lobe. In two cases of carcinoma, the right upper-lobe lymphatics drained directly into the right venous jugular-subclavian junction. Three cases showed basal segment drainage on the right side directly into the thoracic duct. Direct contralateral lymphatics were seen in five cases, four from basal segments of lower lobes.[128] These direct anastomoses into the venous system or thoracic duct, while uncommon, explain the presence of systemic metastases in the absence of affected local nodes. Proper sampling of the mediastinal nodes is a necessity in lung cancer surgery.

A different pattern of drainage may be seen in infants. In a few cases, the lower-lobe lymphatics on both sides drain directly to the celiac nodes through the esophageal hiatus.[129]

The deep system of lymphatics can be traced as far distally as the first generation of respiratory bronchioles. The periarterial, periarteriolar, and peribronchiolar "sumps" in connective-tissue spaces, seen ultrastructurally, receive tissue fluid from the alveolar interstitial spaces (Fig. 1-51). These sumps drain proximally into recognizable lymphatics. The exact means of propulsion of tissue fluid from the alveolar interstitium is unknown, but interstitial myofibroblasts and respiratory movements probably aid the centripetal flow. In addition, the lymphatic wall contains myofibrils.

The lymphatics around small bronchi and bronchioles form a single plexus lying outside the muscle. In large bronchi there is a deep and superficial plexus of lymphatics. The two communicate freely to form the peribronchial plexus.

The superficial pleural lymphatic plexus is best seen over the lower lobes—especially in the mediastinal surface of the lower lobe, where it forms irregular polyhedral rings that outline the pulmonary lobules. Crossing these rings are smaller communicating lymphatics, which in some cases outline acini.

Beneath the free margins of the lungs, especially the lower lobes, the interlobular septa stretch

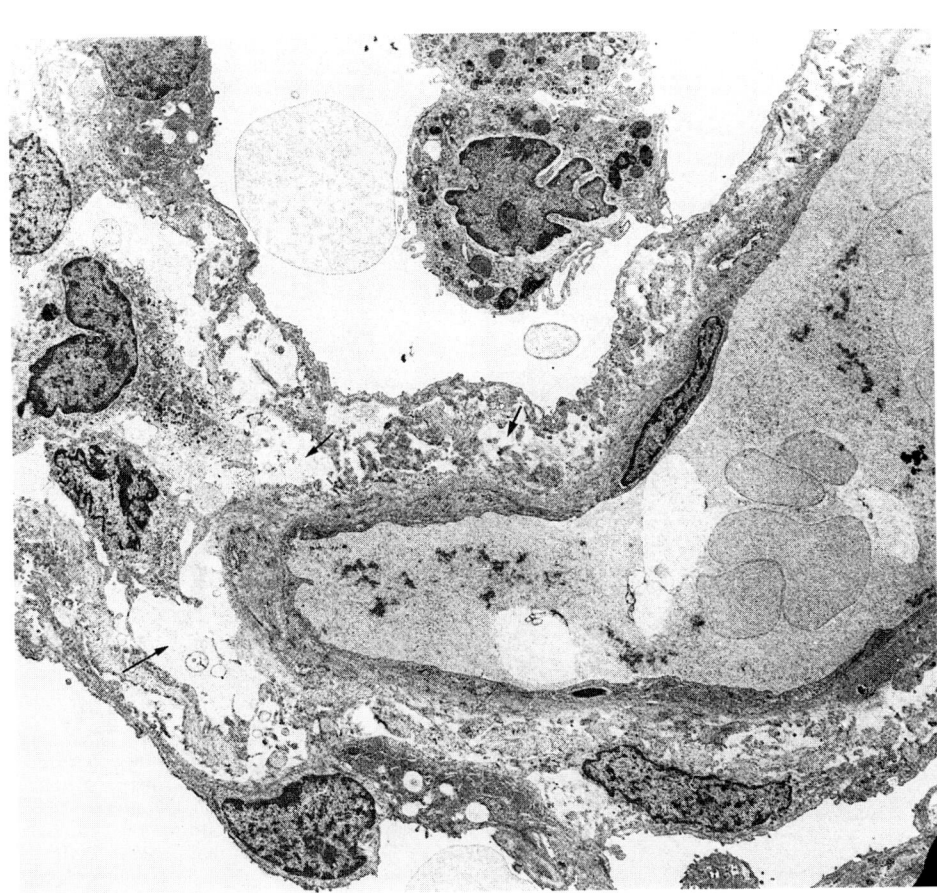

FIGURE 1-51
Lymphatic "sump" of tissue spaces around a pulmonary blood vessel. The spaces are indicated by arrows. (EM, ×14,960.)

between adjacent pleural surfaces. In these loose connective-tissue septa, lymphatics lie next to the paraseptal alveoli. These lymphatics follow the interlobular veins and communicate freely on the surface of the lung with the pleural lymphatics. The deeper portion of the perivenous network receives tributaries from the vein walls and from the peribronchial lymphatics, opposite points of division of the bronchial tree.

Kerley A and B lines are due to septal edema and distention of the septal lymphatics.[130]

The walls of the larger intrapulmonary lymphatics contain both oblique and longitudinally arranged muscle bundles, which do not form a continuous layer. In addition, there are collagen and small elastic fibers, with endothelial cells lining the lumen. A few valves are seen along the course of the lymphatics.

Ultrastructurally, the endothelial cells contain many pinocytotic vesicles as well as myofibrils. There is no continuous basement membrane[131] in lymphatics. This can be demonstrated with antisera to type IV collagen or laminin, which is absent in lymphatics. However, the demonstration of a continuous basement membrane in capillaries by immunohistochemistry to differentiate them from lymphatics is not reliable.

The Nerve Supply of the Lung

Pulmonary nerves are important in the regulation of pulmonary blood flow, secretion from mucous glands, ventilation, and the cough reflex as well as possibly in playing an important role in asthmatic bronchoconstriction (Figs. 1-52 and 1-53). Human lung innervation has been described by several authors.[132–134] The sensory fibers arising in the lung travel in the vagus nerve. The motor supply reaches the lung through both the sympathetic and the parasympathetic (vagus) nervous systems (Fig. 1-54).

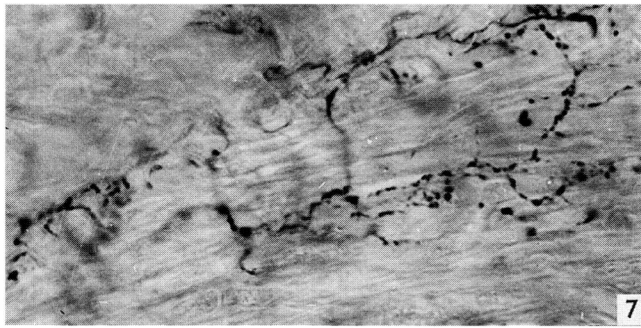

FIGURE 1-52
The motor innervation of the bronchial muscle. The motor nerves form a network of fibers throughout the muscle coat.

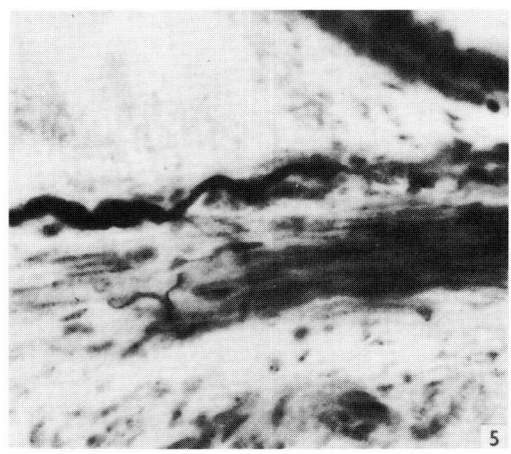

FIGURE 1-53
A sensory nerve giving off a fine twig, which ends in a tendril-like end organ applied near the bronchial muscle. These end organs are thought to be involved in the Hering-Breuer reflex.

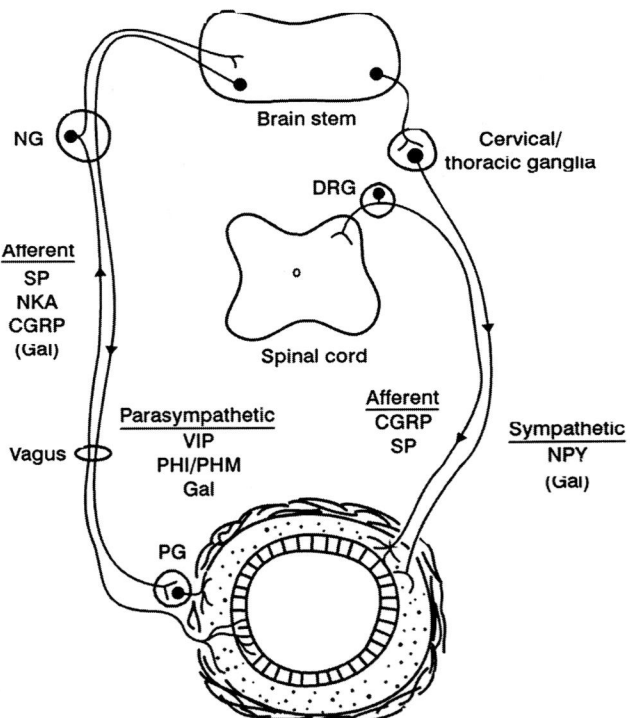

FIGURE 1-54
Innervation of the lower respiratory tract showing neuropeptide colocalization in autonomic nerves. DRG, dorsal root ganglion; NG, nodose jugular ganglion; PG, parasympathetic ganglion; SP, substance P; NKA, neurokinin A; CGRP, calcitonin-gene-related peptide; Gal, galanin; VIP, vasoactive intestinal peptide; PHI, peptide histidine isoleucine; NPY, neuropeptide Y. (*Reproduced by permission of Professor P. Barnes and Am Rev Respir Dis.*)

The preganglionic sympathetic (adrenergic) fibers synapse in the second to fourth thoracic sympathetic ganglia. The preganglionic parasympathetic (cholinergic) as well as the nonadrenergic, noncholinergic fibers relay in the ganglia. These are found in the bronchial arterial wall. Postganglionic fibers arising in the latter ganglia accompany the bronchial arteries. Parasympathetic fibers accompany the large pulmonary veins.

The concept of the classic sympathetic and parasympathetic pathways is now dated. In the 1960s, the existence of nonadrenergic, noncholinergic (NANC) nerves was established.[135] The role of NANC nerves in airways has been reviewed.[136,137] In the airways the existence of purigenic nerves is unlikely, since purine antagonists do not block NANC inhibitory effects.[136] Previously the terms *purigenic* and *NANC* were used interchangeably. In human airways, NANC inhibitory nerves have special importance, since there is no direct adrenergic innervation of airway smooth muscle. NANC may regulate mucus secretion and bronchial blood flow.

It is likely that neuropeptides act as neurotransmitters, and an increasing number have been localized in nerves of the respiratory tract.[138] There is rich vasoactive intestinal peptide (VIP) innervation of human airways. VIP is the most potent known in vitro relaxant of human bronchi, but it has no effect on the smaller peripheral airways.

Nerves have more than one transmitter, as shown by the fact that VIP may coexist with acetylcholine in airway nerves and may have some functionally important interaction with cholinergic control. It is possible that acetylcholine contracts airway smooth muscle, and VIP may act as a bronchodilator. Other examples of cotransmission are the coexistence of neuropeptide Y and norepinephrine in pulmonary adrenergic nerves and galanin with VIP. It is clear that a wide collection of neuropeptides are released from sensory, parasympathetic, and sympathetic neurons in the respiratory tract.

NANC nerves are now envisaged as having neural effects mediated by the release of neurotransmitters from classic autonomic nerves. Cotransmission and coexistence of several peptides occur within the same nerve, and the same peptides occur in different types of nerves. Thus, VIP may be found in sensory, adrenergic, and NANC nerves. The localization and actions of the neuropeptides of the respiratory tract have been reviewed.[139]

Noncholinergic excitatory nerves have been described in animal airways.[140] Substance P, neurokinin A, and calcitonin-gene-related peptide have been found in the human airways and are likely to be neurotransmitters as well as having other actions, such as bronchoconstriction and increased microvascular leakage.

It is important with such a potential source of neurotransmitters that the sensory nerve endings have close contact with the airways. In an ultrastructural study, Laitinen[141] demonstrated axons in the bronchial epithelium. They had two main locations, close to the airway lumen and at the base of the epithelium close to the basement membrane. Nerve fibers penetrating the basement membrane and traversing the epithelium were also identified. Nerves near the lumen were found mainly in the larger airways, with few in the smaller airways. Axon profiles, though showing some differences between airway levels, were constantly identified at every level. Bronchial epithelial nerve fibers are most numerous at points of bronchial bifurcation, reaching their greatest number in lobar and segmental bronchi. The above-mentioned ultrastructural study showed that the axon profiles resembled sensory nerve endings, from which afferent stimuli may originate. These results are supported by the finding of intraepithelial nerves in human airways that are in close contact with neuroepithelial bodies.[30,37]

Afferent impulses from the lung may originate from nerve endings in the bronchial mucosa as far distally as the respiratory bronchioles. These nerve endings form complex candelabra-like arborizations with silver stains, each ending in a fine knob situated between the cells of the bronchial epithelium. The nerve endings are activated by stretch of the bronchial wall. However, as can be seen from the above paragraphs on NANC fibers, epithelial nerve fibers have many other possible functions. Sparse nerve endings have been demonstrated in the alveolar wall by Fox and colleagues[134] and have been described as J receptors.

The bronchial smooth muscle receives its motor nerve supply through afferent fibers connecting with both sympathetic and parasympathetic nerves. Nerve fibers also supply the elastic pulmonary arteries, as well as the vasa vasorum. These fibers arborize in the outer media. The muscular pulmonary arteries are surrounded by an anastomosing plexus of nerve fibers, which link with the peribronchial plexus. Most mammalian species have substantial adrenergic and a less pronounced cholinergic innervation.

Similarly, pulmonary veins as well as their vasa vasorum have a large sensory nerve supply; the nerve fibers reach just below the endothelium, with an extensive intramural nerve plexus. The medium and small pulmonary veins are surrounded by a plexus of fine nerve fibers, as well as a spirally disposed nerve.

These reach the distal branches. The bronchial arteries are also surrounded by a nerve plexus derived from the central peribronchial ganglia, which lie adjacent to the arteries.

Sympathetic and parasympathetic influences on the pulmonary circulation are well established. DeBurgh Daly and von Euler[142] collected references from as early as 1822 to show that electrical stimulation of the vagi caused vasoconstriction or weak vasodilatation. The parasympathetic stimuli cause vasodilatation. However, it is not known whether neural control of the pulmonary circulation is important in either normal hemostasis or disease states.[143] Small pulmonary arterioles are probably the site for the pulmonary vascular resistance, but as can be seen from the innervation described above, the venous side may well be affected. In dogs, sympathetic stimulation constricts pulmonary veins as well as arteries,[144] and PGE$_1$ constricts pulmonary veins.[145]

The extent and distribution of sensory nerve fibers are almost unknown. Nerve fibers end in the walls of subpleural alveoli, and a few accompany pulmonary veins in the interlobular septa and end in the subseptal alveolar walls.

The Pleura

The pleural space is approximately 20 μm wide[146] and contains a small amount of clear, colorless fluid with a protein concentration of less than 1.5 g/dl. The pleural cavity is formed by the parietal and visceral pleura. The parietal pleura covers the inner surface of the thoracic cage, mediastinum, and diaphragm. The visceral pleura covers the lung surfaces, including the interlobar fissures. The visceral pleura is reflected at the hilum, where it becomes the parietal pleura. Both visceral and parietal pleurae consist of a single layer of mesothelial cells, basement membrane, and layers of collagen and elastic tissue (Fig. 1-55).

In the caudal part of the mediastinal pleura, Kampmeier's foci (Fig. 1-56) have been described.[147] These consist of accumulations of macrophages, along with pleuripotential mesenchymal, lymphoid, and plasma cells. These surround a central lymphatic channel or vessel, which is covered by cuboidal mesothelial cells. It is possible that these structures protect the pleura by impeding absorption of harmful agents into the mediastinum, although their role is not firmly established.

The pleura also contains small blood vessels, lymphatics, nerves, and some smooth muscle bundles. This last element can be traced from the walls of blood vessels. In common with other serous cavities, numerous mast cells are present, but they are unrelated to blood vessels.

The main blood supply of the visceral pleura is from the bronchial arteries, but the pulmonary circulation supplies the inner part of the pleura. The parietal pleura is supplied by branches of the intercostal and internal mammary arteries. The mediastinal pleura is supplied by branches of the bronchial, upper diaphragmatic, internal mammary, and mediastinal arteries. The apical pleura is supplied by branches of the subclavian arteries or its collaterals.

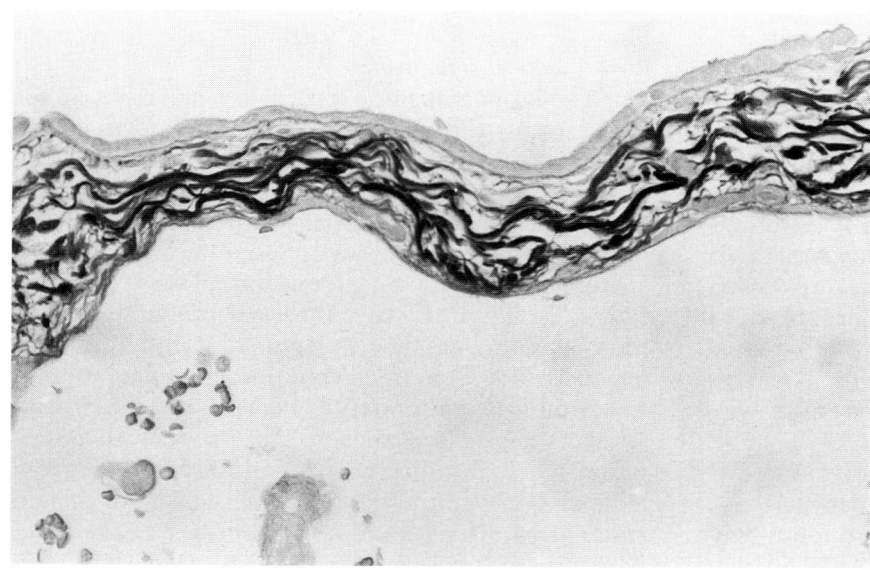

FIGURE 1-55
Pleura showing bands of elastic tissue and some fibrous tissue. (EVG, ×219.)

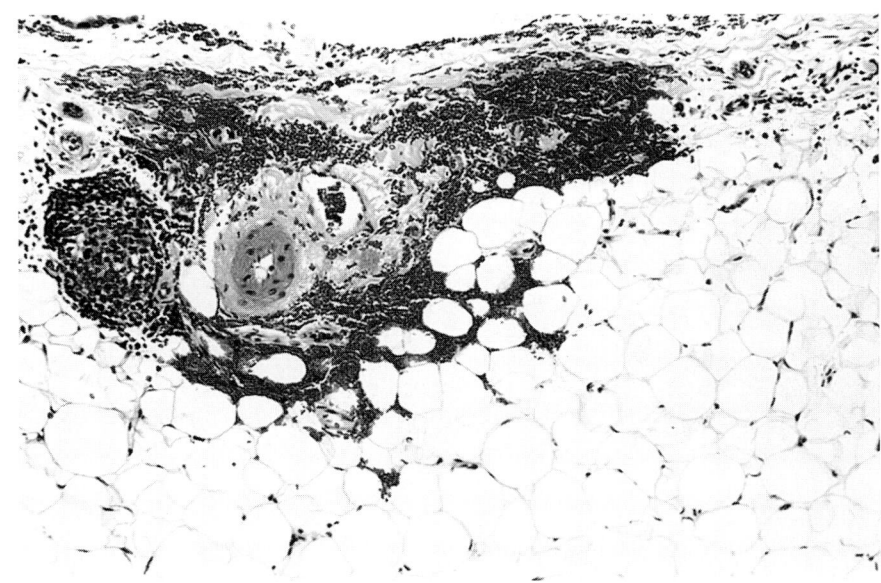

FIGURE 1-56
Kampmeier's focus in the pleura, with a collection of lymphoid cells, in one area showing a primitive germinal center and a central artery and vein. There is focal crush artefact in this surgical biopsy. (H&E, ×87.)

The mesothelial cell is variable in shape—flat, cuboidal, or columnar (Figs. 1-57 and 1-58). Consequently, the nucleus is either elongated or round. In the normal state the cells remain as a single layer.

Ultrastructurally, mesothelial cells are characterized by long, bushy microvilli (Fig. 1-59), which can fork like a large cactus plant. There is a higher density of microvilli on the visceral than on the parietal pleura. Mesothelial cytoplasm is rich in mitochondria as well as many intermediate filaments. These filaments stain with antisera to low-molecular-weight keratins—useful in identification of the cell. The microvilli help to trap hyaluronic acid, which decreases the friction between the lung and chest wall.

Only in the parietal pleura are there stomata, ranging from 2 to 12 μm in diameter (Fig. 1-60). These stomata connect directly with the lymphatics and are thus the usual exit point for pleural fluid and cells. Recently, using scanning electron microscopy, researchers cast doubt on the presence of stomata in man.[148] These authors found that the pleural surface was homogeneous, and the presence of both stomata and microvilli in the same or neighboring cells suggested secretion and absorption at the same level. Discontinuities and clefts were seen at cell junctions, and it was postulated that absorption of high-molecular-weight proteins may be accomplished at this level. Further confirmatory studies are needed to exclude the presence of a significant number of stomata.

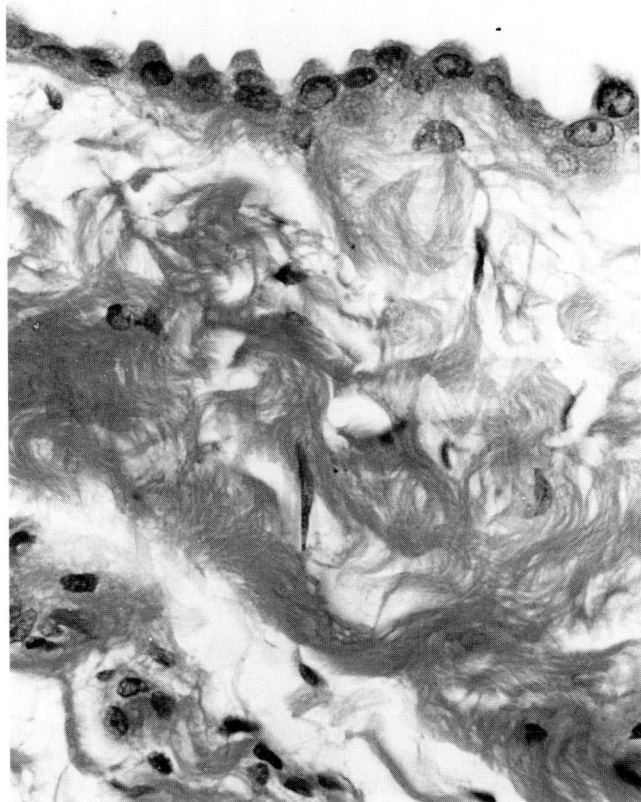

FIGURE 1-57
Pleura showing a single layer of surface mesothelial cells with regular nuclei, beneath which there is loose, fibrous tissue. (H&E, ×375.)

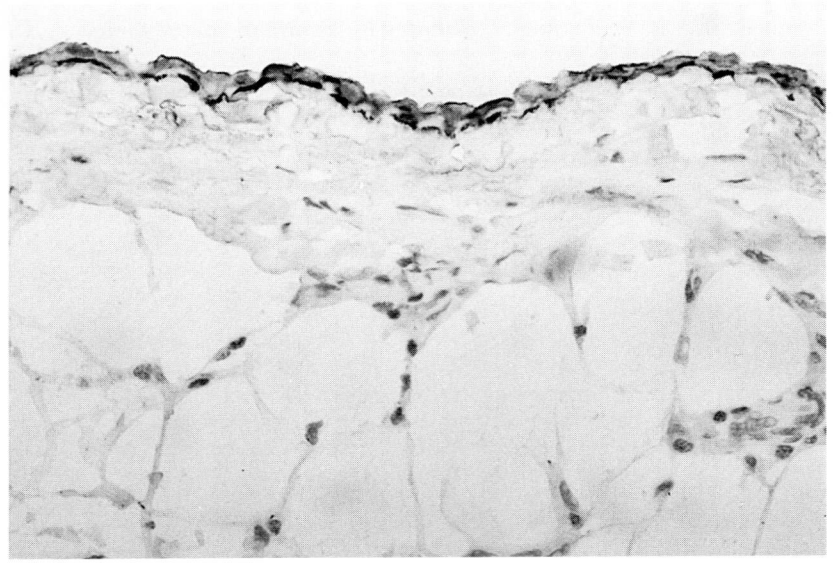

FIGURE 1-58
Pleura showing mesothelial cells staining with CAM 5.2 (cytokeratin). (Immunoperoxidase ×219.)

FIGURE 1-59
SEM of pleura showing many microvilli and red blood corpuscles. (×5000.)

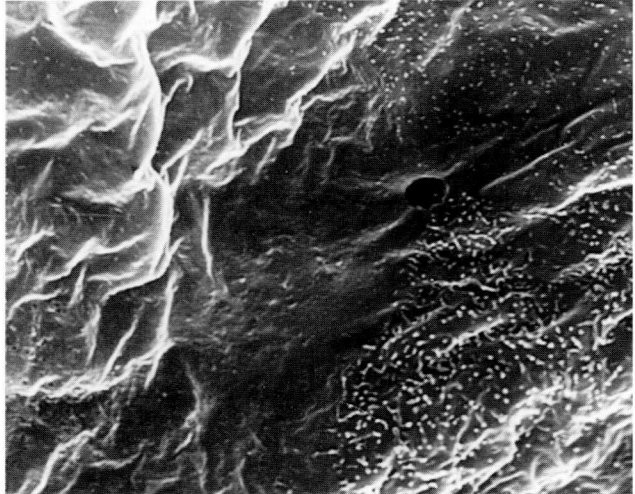

FIGURE 1-60
SEM with pleural surface and microvilli. A stoma is clearly shown to the right of center. (×1162.) (*Reproduced by courtesy of Dr. M-J Peng and editor of Chest.*)

There are multipotential subserosal cells, which are distinct from any other connective-tissue cells in their ability to coexpress vimentin and keratin. This type of cell serves as the replicative cell, which can differentiate into surface mesothelium.[149]

The physiology of pleural fluid formation is complex. It was thought that the role of lymphatics is predominant in fluid exchange. Sahn[146] has suggested that the parietal pleura, with its systemic blood supply, is the most important factor in pleural fluid formation. Pleural fluid is essentially the excess interstitial fluid of the parietal pleura. The subatmospheric pressure in the pleural space and the pressure in the systemic circulation supplying the parietal pleura cause its formation through a pressure gradient. The contribution of the visceral pleura to pleural liquid and protein formation is probably minimal.

REFERENCES

1. Royal Brompton Grand Rounds: Idiopathic diaphragmatic weakness. *Br Med J* 304:492–494, 1992.

2. Muzino M: Human respiratory muscles: fiber morphlogy and capillary supply. *Eur Respir J* 4:587–601, 1991.

3. Gough J: In discussion of paper by Comroe JH Jr: Physiological and biochemical effects of pulmonary artery occlusion. *CIBA Foundation Symposium on Pulmonary Structure and Function.* London, J. and A. Churchill, 1962, p 259.

4. Miller WS: *The Lung,* 2d ed. Springfield, Ill., Charles C Thomas, 1947.

5. Boyden EA: *Segmental Anatomy of the Lungs: A Study of the Patterns of the Segmental Bronchi and Related Pulmonary Vessels.* New York, McGraw-Hill, 1955.

6. Barnett CH: A note on the dimensions of the bronchial tree. *Thorax* 12:175–176, 1957.

7. Herrnheiser G, Hinson KFW: An anatomical explanation of the formation of butterfly shadows. *Thorax* 9:198–210, 1954.

8. Reid L: The connective tissue septa in the adult human lung. *Thorax* 14:138–145, 1959.

9. Heppleston AG: The pathological anatomy of simple pneumoconiosis in coal workers. *J Pathol Bacteriol* 66:235–246, 1953.

10. Crystal RG, West JB, Barnes PJ, Cherniak NS, Weibel ER: *The Lung: Scientific Foundations.* New York, Raven Press, 1991.

11. Monkhouse WS, Whimster WF: An account of the longitudinal mucosal corrugations of the human tracheo-bronchial tree, with observations on those of animals. *J Anat* 122:681–695, 1976.

12. Sturgess JM: Mucous secretions of the respiratory tract. *Pediatr Clin North Am* 26:481–501, 1979.

13. Breeze RG, Wheeldon EB: The cells of the pulmonary airways. *Am Rev Respir Dis* 116:705–777, 1977.

14. Soderberg M, Hellstrom S, Sandstrom T, Lundgren R, Bergh A: Structural characterization of bronchial mucosal biopsies from healthy volunteers: A light and electron microscopical study. *Eur Respir J* 3:261–266, 1990.

15. Warner FD, Mitchell DR, Perkins CR: Structural conformation of the ciliary ATPase dynein. *J Mol Biol* 114:367–384, 1977.

16. Fox B, Bull TB, Makay AR, Rawbone NR: The significance of ultrastructural abnormalities of human cilia. *Chest* 80(Suppl):796–799, 1981.

17. Rhodin JAG: The ciliated cell: Ultrastructure and function of the human tracheal mucosa. *Am Rev Respir Dis* 93(Suppl):1–15, 1966.

18. Frasca JM, Auerbach O, Parks VR, Jamieson JD: Electron microscopic observations of the bronchial epithelium of dogs: 1. Control dogs. *Exp Mol Pathol* 9:363–379, 1968.

19. Freeman JA: Fine structure of the goblet cell mucous secretory process. *Anat Rec* 144:341–357, 1962.

20. Ayers MM, Jeffery PK: Proliferation and differentiation in mammalian airway epithelium. *Eur Respir J* 1:58–80, 1988.

21. Meyrick B, Reid L: The alveolar brush cell: A third pneumocyte. *J Ultrastuct Res* 23:71–80, 1968.

22. Hama K, Nagata F: A stereoscope observation of tracheal epithelium of mouse by means of the high voltage electron microscope. *J Cell Biol* 45:654–659, 1970.

23. Inoue S, Hogg JC: Intercellular junctions of the tracheal epithelium in guinea pigs. *Lab Invest* 31:68–74, 1974.

24. Baskerville A: Ultrastructural studies of the normal pulmonary tissue of the pig. *Res Vet Sci* 11:150–155, 1970.

25. Jeffery PK, Reid L: New observations of rat airway epithelium: A quantitative and electron microscopic study. *J Anat* 120:295–320, 1975.

26. McDowell EM, Combs JW, Newkirk C: A quantitative light and electron microscopic study of hamster tracheal epithelium with special attention to so-called "intermediate cells." *Exp Lung Res* 4:205–226, 1983.

27. Evans, MJ, Plopper CG: The role of basal cells in adhesion of columnar epithelium to airway basement membrane. *Am Rev Respir Dis* 138:481–483, 1988.

28. McDowell EM, Barrett LA, Trump BF: Observations on small granule cells in adult human bronchial epithelium and in carcinoid and oat cell tumors. *Lab Invest* 34:202–206, 1976.

29. Lauweryns JM, Peuskens JC: Neuroepithelial bodies (neuroreceptor or secretory organs?) in human infant bronchial and bronchiolar epithelium. *Anat Rec* 172:471–482, 1972.

30. Jeffery P, Reid L: Intraepithelial nerves in normal rat airways: A quantitative electron microscopic study *J Anat* 114:35–45, 1973.

31. Lauweryns JM, Van Ranst LV, Lloyd R, O'Connor DT: Chromogranin in bronchopulmonary neuroendocrine cells: Immunocytochemical detection in human, monkey, and pig respiratory mucosa. *J Histochem Cytochem* 35:113–118, 1987.

32. Bhatnager M, Springall DR, Ghatei MA, et al.: Localisation of mRNA and co-expression and molecular forms of GRP gene products in endocrine cells of fetal human lung. *Histochemistry* 90:299–307, 1988.

33. Tsutsumi Y: Immunohistochemical analysis of calcitonin and calcitonin gene-related peptide in human lung. *Hum Pathol* 20:896–902, 1989.

34. Lauweryns JM, Van Ranst L: Leu-7 immunoreactivity in human, monkey and pig bronchopulmonary neuroepithelial bodies and neuroendocrine cells. *J Histochem Cytochem* 35:687–691, 1987.

35. Hage E: Electron microscopic identification of several types of endocrine cells in the bronchial epithelium of human foetuses. *Z Zellforsch Mikrosk Anat* 141:401–412, 1973.

36. Becker KL: The coming of age of a bronchial epithelial cell. *Am Rev Respir Dis* 148:1166–1168, 1993.

37. Gosney JR: *Pulmonary Endocrine Pathology: Endocrine Cells and Endocrine Tumours of the Lung.* Oxford, Butterworth-Heinemann, 1992.

38. Sorokin SP, Hoyt RF, Jr (eds): Workshop on pulmonary neuroendocrine cells in health and disease. *Anat Rec* 236:1–256, 1993.

39. Miller YE: The pulmonary neuroendocrine cell: A role in adult lung disease? *Am Rev Respir Dis* 140:283–284, 1989.

40. Giad A, Polak J, Gaitonde V, et al.: Distribution of endothelin-like immunoreactivity and mRNA in the developing lung. *Am J Respir Cell Mol Biol* 4:50–58, 1991.

41. Schittny JC, Yurchenco PD: Basement membranes: Molecular organization and function in development and disease. *Curr Opin Cell Biol* 1:983–988, 1989.

42. Shoji S, Ertl RP, Linder J, Romberger DJ, Rennard SI: Bronchial epithelial cells produce chemotactic activity for bronchial epithelial cells: Possible role for fibronectin in airway repair. *Am J Respir Dis* 141:218–225, 1990.

43. Dean DC: Expression of the fibronectin gene. *Am J Respir Cell Mol Biol* 1:5–10, 1989.

44. Otis AB: A perspective of respiratory mechanics. *J Appl Physiol* 54:1183–1187, 1983.

45. Stephens NL: State of art: Airway smooth muscle. *Am Rev Respir Dis* 135:960–975, 1987.

46. Buckley CJ: The role of bronchial musculature. *Lancet* 3:836–837, 1989.

47. Meyrick B, Sturgess JM, Reid L: A reconstruction of the duct system and secretory tubules of the human bronchial submucosal gland. *Thorax* 24:729–736, 1969.

48. Rogers AV, Dewar A, Corrin B, Jeffery PK: Identification of serous-like cells in the surface epithelium of human bronchioles. *Eur Respir J* 6:498–504, 1993.

49. Bowes D, Corrin B: Ultrastructural and immunocytochemical localisation of lysozyme in human bronchial glands. *Thorax* 32:163–170, 1977.

50. Brandtzaeg P: Mucosal and glandular distribution of immunoglobulin components: Differential localisation of free and bound SC in secretory epithelial cells. *J Immunol* 112:1553–1559, 1974.

51. Clara M: Zür Histobiologie des Bronchialepithels. *Z Mikrosk Anat Forsch* 41:321–347, 1937.

52. Jeffery PK, Gaillard D, Moret S: Human airway secretory cells during development and in mature airway epithelium *Eur Respir J* 5:93–104, 1992.

53. Auten RL, Watkins RH, Shapiro DL, Horowitz S: Surfactant apoprotein-A (SP-A) is synthesized in airway cells. *Am J Respir Cell Mol Biol* 3:491–496, 1990.

54. Gil J, Weibel E: Extracellular lining of bronchioles after perfusion-fixation of rat lungs for electron microscopy. *Anat Rec* 169:185–200, 1971.

55. Kramps JA, Franken C, Meijer CJ, Dijkman JH: Localization of low molecular weight protease inhibitor in serous secretory cells of the respiratory tract. *J Histochem Cytochem* 29:712–719, 1981.

56. Van Scott MR, Hester S, Boucher RC: Ion transport by rabbit nonciliated bronchiolar epithelial cells (Clara cells) in culture. *Proc Natl Acad Sci USA* 84:5496–5500, 1987.

57. Bienenstock J: Bronchus-associated lymphoid tissue. *Int Arch Allergy Appl Immunol* 76(Suppl 1):62–69, 1985.

58. Richmond I, Pritchard GE, Ashcroft T, Avery A, Corris PA, Walters EH: Bronchus associated lymphoid tissue (BALT) in human lung: Its distribution in smokers and non-smokers. *Thorax* 48:1130–1134, 1993.

59. Berman JS, Beer DJ, Theodore AC, Kornfeld H, Bernardo J, Center DM: Lymphocyte recruitment to the lung. *Am Rev Respir Dis* 142:238–257, 1990.

60. Gould SJ, Isaacson PJ: Bronchus-associated lymphoid tissue (BALT) in human fetal and infant lung. *J Pathol* 169:229–234, 1993.

61. Rodriguez MS, Bur A, Favre A, Weibel ER: The pulmonary acinus: Geometry and morphometry of the peripheral airway system in rat and rabbit. *Am J Anat* 180:143–155, 1987.

62. Berend N, Rynell AC, Ward HE: Structure of the human pulmonary acinus. *Thorax* 46:117–121, 1991.

63. Weibel ER: *Morphometry of the Human Lung*. Berlin, Springer-Verlag, 1963, p. 69.

64. Mercer RR, Crapo JD: Spatial distribution of collagen and elastin fibers in the lungs. *J Appl Physiol* 69:756–765, 1990.

65. Rhodin JH: *Histology*. London, Oxford University Press, 1974, p. 632.

66. Weibel ER, Gil J: Electron microscopic demonstration of an extracellular duplex lining layer of alveoli. *Respir Physiol* 4:42–57, 1968.

67. Williams MC, Dobbs LG: Expression of cell-specific markers for alveolar epithelium in fetal rat lung. *Am J Respir Cell Mol Biol* 2:533–542, 1990.

68. Crapo J, Barry BA, Gehr P, Bachofen M, Weibel ER: Cell number and cell characteristics of the normal human lung. *Am Rev Respir Dis* 125:332–337, 1982.

69. Mathay MA, Wiener-Kronish JP: Intact epithelial barrier function is critical for the resolution of alveolar edema in humans. *Am Rev Respir Dis* 142:1250–1257, 1990.

70. Weibel ER: The mystery of "non-nucleated plates" in the alveolar epithelium of the lung explained. *Acta Anat* 78:425–443, 1971.

71. Corrin B: Phagocytic potential of pulmonary alveolar epithelium with particular reference to surfactant metabolism. *Thorax* 25:110–115, 1970.

72. Suzuki, Y, Churg J, Ono T: Phagocytic activity of the alveolar epithelial cells in pulmonary asbestosis. *Am J Pathol* 69:373–379, 1972.

73. Low FN: The extracellullar portion of the human blood-air barrier and its relation to tissue space. *Anat Rec* 139:105–123, 1961.

74. Fishman AP, Hecht HH (eds): *The Pulmonary Circulation and Interstitial Space*. Chicago, University of Chicago Press, 1969.

75. Fishman AP, Renkin EM (eds): *Pulmonary Edema*. Bethesda, Md., American Physiological Society, 1979.

76. Staub NC: *Lung Water and Solute Exchange*. New York, Marcel Dekker, 1978.

77. Yu PN: Pulmonary edema: Key references. *Circulation* 63:724–728, 1981.

78. Harris P, Heath D: *The Human Pulmonary Circulaton: Its Form and Function in Health and Disease*, 3d ed. Edinburgh, Churchill Livingstone, 1986.

79. Ogston AG, Sherman TF: Effects of hyaluronic acid upon diffusion of solutes and flow of solvent. *J Physiol* 156:67–74, 1961.

80. Paintal AS: The mechanism of stimulation of type J pulmonary receptors. *J Physiol (Lond)* 203:511–532, 1969.

81. Davidson JM: Biochemistry and turnover of lung interstitium. *Eur Respir J* 3:1048–1068, 1990.

82. Bateman E, Turner-Warwick M, Adelmann-Grill BC: Immunohistochemical study of collagen types in human foetal lung and fibrotic lung disease. *Thorax* 36:645–653, 1981.

83. Hance AJ, Crystal RG: The connective tissue of lung. *Am Rev Respir Dis* 112:657–711, 1975.

84. Pierce JA, Ebert RV: Fibrous network of the lung and its change with age. *Thorax* 20:469–476, 1965.

85. Sibille Y, Reynolds HY: Macrophages and polymorphonuclear neutrophils in lung disease and injury. *Am Rev Respir Dis* 141:471–501, 1990.

86. Lenhert BE, Valdez YE, Holland LM: Pulmonary macrophages: Alveolar and interstitial populatons. *Exp Lung Res* 9:177–190, 1985.

87. Hogg N: Factor-induced differentiation and activation of macrophages. *Immunol Today* 7:65–66, 1986.

88. Whitelaw DM: Observations on human monocyte kinetics after pulse labelling. *Cell Tissue Kinet* 5:311–317, 1972.

89. Sandron D, Reynolds HY, Laval AM, Venet A, Israel-Beit D, Chretien J: Human alveolar macrophage subpopulations isolated on discontinuous albumin gradients: Cytological data in normals and sarcoid patients. *Eur J Respir Dis* 68:177–185, 1986.

90. Dehring DJ, Wismar BL: Intravascular macrophages in pulmonary capillaries of humans. *Am Rev Respir Dis* 139:1027–1029, 1989.

91. Brain JD, Gehr P, Kavet R: Airway macrophages: The importance of the fixation method. *Am Rev Respir Dis* 129:823–826, 1984.

92. Brain JD: Lung macrophages: How many kinds are there? What do they do? *Am Rev Respir Dis* 137:507–509, 1988.

93. Auger MJ: Mononuclear phagocytes. *Br Med J* 298:546–548, 1989.
94. Unanue ER, Allen PM: The basis for the immuno-regulatory role of macrophages and other accessory cells. *Science* 236:551–557, 1987.
95. Lasser A: The mononuclear phagocyte system: A review. *Hum Pathol* 14:108–126, 1983.
96. Rocklin RE, Benzden K, Greineder D: Mediators of immunity: Lymphokines and monokines. *Adv Immunol* 29:55–136, 1980.
97. Werb Z, Gordon S: Secretion of a specific collagenase by stimulated macrophages. *J Exp Med* 142:346–360, 1975.
98. Werb Z, Gordon S: Elastase secretion by stimulated macrophages: Characterization and regulation. *J Exp Med* 142:361–377, 1975.
99. Takizawa H, Beckmann JD, Shoji S, et al: Pulmonary macrophages can stimulate cell growth of bovine bronchial epithelial cells. *Am J Respir Cell Mol Biol* 2:245–255, 1990.
100. Craig SS, DeBlois G, Schwartz LB: Mast cells in human keloid, small intestine and lung by an immunoperoxidase technique using a murine monoclonal antibody against tryptase. *Am J Pathol* 124:427–435, 1986.
101. Djukanovic R, Wilson JW, Britten KM, et al: Quantitation of mast cells and eosinophils in the bronchial mucosa of symptomatic atopic asthmatics and healthy control subjects using immunohistochemistry. *Am Rev Respir Dis* 142:863–871, 1990.
102. Fox B, Bull TB, Guz A: Mast cells in the human alveolar wall: An electronmicroscopic study. *J Clin Pathol* 34:1333–1342, 1981.
103. Tucker A, Weir EK, Reeves JT, Grover RF: Histamine H1 and H2-receptors in pulmonary and systemic vasculature of the dog. *Am J Physiol* 229:1008–1013, 1975.
104. Wasserman SI: The human lung mast cell. *Environ Health Perspect* 55:259–269, 1984.
105. Weibel ER, Palade GE: New cytoplasmic components in arterial endothelia. *J Cell Biol* 23:101–112, 1964.
106. Franke WW, Cowin P, Grund C, Kapprell H-P: The endothelial junction: The plaque and its components, in Simonescu N, Simonescu M (eds), *Endothelial Cell Biology in Health and Disease*. New York, Plenum Press, 1988, pp. 147–165.
107. Palade GE: The microvascular circulation revisited, in Simonescu N, Simonescu M (eds), *Endothelial Cell Biology in Health and Disease*. New York, Plenum Press, 1988, pp. 3–22.
108. Schneeberger HJ, Lynch RD: Tight junctions: Their structure, composition, and function. *Circ Res* 5:723–733, 1984.
109. Stevenson BR, Heintzelman MB, Anderson JM, Citi S, Mooseker MS: ZO-1 and cingulin: Tight junction proteins with distinct identities and localizations. *Am J Physiol* 257 [*Cell Physiol* 26]:C621–C628, 1989.
110. Schnittler HJ, Wilde A, Gress T, Suttorp N, Drenckhahn D: Role of actin and myosin in the control of paracellular permeability in pig, rat, and human vascular endothelium. *J Physiol* 431:379–401, 1990.
111. Takeichi M: Cadherins: A molecular family important in selective cell-cell adhesion. *Annu Rev Biochem* 59:237–252, 1990.
112. Heimark RL, Degner M, Schwartz SM: Identification of a Ca²-dependent cell-cell adhesion molecule in endothelial cells. *J Cell Biol* 110:1745–1756, 1990.
113. Vane JR, Anggard EK, Botting RM: Regulatory functions of the vascular endothelium. *N Engl J Med* 323:27–36, 1990.
114. Endothelins [editorial]. *Lancet* 337:79–83, 1991.
115. Delaunois L: Anatomy and physiology of collateral respiratory pathways. *Eur Respir J* 2:893–904, 1989.
116. Kohn HN: Zür Histologie der indurirenden fibrosen Pneumonie. *Munch Med Wochenschr* 8:42–45, 1893.
117. Mitzner W: Collateral ventilation, in Crystal RG, West JB, Barnes PJ, Cherniack NS, Weibel ER (eds), *The Lung: Scientific Foundations*. New York, Raven Press, 1991, chap 5.1.2.7, pp 1053–1063.
118. Kawakami M, Takizawa T: Distribution of pores within alveoli in the normal human lung. *J Appl Physiol* 63:1866–1870, 1987.
119. Menkes H, Gardiner A, Gamsu G, Limpert J, Macklem PT: The influence of surface forces on collateral ventilation. *J Appl Physiol* 31:544–549, 1971.
120. Lambert MW: Accessory bronchiole-alveolar communications. *J Pathol Bacteriol* 70:311–314, 1955.
121. Martin HB: Respiratory bronchioles as the pathway for collateral ventilation. *J Appl Physiol* 21:1443–1447, 1966.
122. Anderson JP, Jesperson N: Demonstration of intersegmental respiratory bronchioles in normal human lungs. *Eur J Respir Dis* 61:337–341, 1980.
123. Boren HG, Paradise LJ: Cell proliferation and carinogenesis. Carcinogenesis Program. Fourth Annual Collaborative Conference. Washington: US Department of Health, 1976, p. 73.
124. Boren HG, Paradise LJ: Cytokinetics of lung, in Harris, CC (ed), *Pathogenesis and Therapy of Lung Cancer* New York Marcel Dekker, 1978, pp. 369–418.
125. Blenkinsopp WK: Proliferation of respiratory tract epithelium in the rat. *Exp Cell Res* 46:144–154, 1967.
126. McDowell EM, Keenan KP, Huang M: Restoration of the mucociliary tracheal epithelium following deprivation of vitamin A: A quantitative morphologic study. *Virchows Arch* [B] 45:221–240, 1984.
127. Miller WS: Studies on tuberculous infection: III. The lymphatics and lymph flow in the human lung. *Am Rev Tuberc* 3:193–209, 1919.
128. Riquet M, Hidden G, Debesse B: Direct lymphatic drainage of lung segments to the mediastinal nodes: An anatomic study on 200 adults. *J Thorac Cardiovasc Surg* 97:623–632, 1989.
129. Meyer KK: Direct lymphatic connections from the lower lobes of the lung to the abdomen. *J Thorac Surg* 35:726–733, 1958.
130. Trapnell DH: in McClaren JW (ed), *Modern Trends in Diagnostic Radiology*. London, Butterworth, 1970, p. 39.
131. Lauweryns JM, Baert A, Boussauw L: The pulmonary lymphatics: Macroscopic and microscopic studies. *Am Rev Respir Dis* 101:448–450, 1970.
132. Spencer H, Leof D: The innervation of the human lung. *J Anat* 98:599–609, 1964.
133. Richardson JB, Beland J: Nonadrenergic inhibitory nervous system in human airways. *J Appl Physiol* 41:764–771, 1976.
134. Fox B, Bull TB, Guz A: Innervation of alveolar walls in the human lung: An electron microscopic study. *J Anat* 131:683–692, 1980.
135. Burnstock G, Campbell G, Bennett M, Holman ME: Inhibition of the smooth muscle of *Taenia coli*. *Nature* 200:581–582, 1963.
136. NANC nerves in airways [editorial]. *Lancet* 2:1253–1254, 1986.
137. Barnes PJ: The third nervous system of the lung: Physiology and clinical perspectives. *Thorax* 39:561–567, 1984.
138. Polak J, Bloom SR: Regulatory peptides of the gastrointesti-

nal and respiratory tracts. *Int Arch Pharmacol* 280(Suppl):16–49, 1986.

139. Barnes PJ, Baraniuk JN, Belvisi MG: Neuropeptides in the respiratory tract. Parts 1 and 2. *Am Rev Respir Dis* 144:1187–1198, 1391–1399, 1991.

140. Andersson RGG, Grundstrom N: The excitatory non-cholinergic, non-adrenergic nervous system of the guinea-pig airways. *Eur J Respir Dis* 64:141–157, 1983.

141. Laitinen A: Ultrastructural organisation of intraepithelial nerves in the human airway tract. *Thorax* 40:488–492, 1985.

142. De Burgh Daly I, von Euler V: The functional activity of the vasomotor nerves to the lungs in the dog. *Proc R Soc Lond* [*Biol*] 110:92–111, 1932.

143. Long WA, Brown DL: Central neural regulation of the pulmonary circulation, in Fishman AP (ed), *The Pulmonary Circulation: Normal and Abnormal.* Philadelphia, University of Pennsylvania Press, 1990, p. 132.

144. Kadowitz PJ, Joiner PD, Hyman AL: Influence of sympathetic stimulation and vasoactive substances on the canine pulmonary veins. *J Clin Invest* 56:354–365, 1975.

145. Altura BM, Chand N: Differential effects of prostaglandins on canine intrapulmonary arteries and veins. *Br J Pharmacol* 73:819–827, 1981.

146. Sahn SA: The pleura. *Am Rev Respir Dis* 138:184–234, 1988.

147. Wang NS: Mesothelial cells in situ, in Chretien J, Hirsch A (eds), *Diseases of the Pleura.* New York, Marcel Dekker, 1986, pp. 23–24.

148. Gaudio E, Rendina EA, Pannarale L, Ricci C, Marinozzi G: Surface morphology of the human pleura: a scanning electron microscopic study. *Chest* 92:149–153, 1988.

149. Bolen JW, Hammar SP, McNutt MA: Reactive and neoplastic serosal tissue: A light-microscopic, ultrastructural and immunocytochemical study. *Am J Surg Pathol* 10:34–47, 1986.

Chapter 2

Embryology and Development of the Lung

P. S. Hasleton

A knowledge of pulmonary development is essential to the understanding of the many congenital anomalies that occur in the lung. The key events of pulmonary embryogenesis and postnatal development are given in this chapter, but for a more detailed account the reader is referred to the monograph edited by Hodson.[1]

EARLY INTRAUTERINE DEVELOPMENT: AIRWAYS

The laryngotracheal groove or respiratory anlage is a central diverticulum that grows from the foregut, appearing 22 to 26 days after fertilization. With development, the laryngotracheal groove grows caudally, separated initially by a spur of entoderm and then by mesenchyme that surrounds the foregut. During the fourth week of gestation, the caudal end of the trachea divides into two bronchial buds, each proceeding to form main bronchi (Fig. 2-1). Further bronchial growth is asymmetric, the right being larger. Each main bronchus ends in flask-shaped swellings. As these buds grow (Fig. 2-2), they are covered with mesenchyme, which differentiates to form cartilage, muscle, vessels, and other connective tissues. The right bronchial bud divides to give two branches on the left and three on the right. By 32 to 35 days, the buds of lobar bronchi have formed, and up to 10 days later the segmental and subsegmental bronchial buds have grown.[1] Seventy percent of the bronchial tree is formed in the period 10 to 16 weeks (Fig. 2-3).[2] This growth spurt is greater in the basal segments of the lower lobes, as the general growth of both lungs proceeds more in a caudal than a cephalic direction, resulting in a greater number of lower-lobe segmental bronchi.

After the 20th week, the number of segmental bronchi appears to decrease, as some of the most distal airways lose their epithelium after capillary ingrowth. There is conversion of the tubes into partly epithelialized and partly alveolated respiratory bronchioles. At this stage, the most distal airways cannulate to form immature alveoli. After 24 weeks, terminal bronchioles develop up to four orders of nonalveolated bronchioles destined to become respiratory bronchioles. Each generation of respiratory bronchioles gives rise to two or three branches. The most distal respiratory bronchioles end in two clusters of thin-walled saccules, which are the precursors of immature alveoli (Fig. 2-4).[3] These immature respiratory saccules suffice for gas exchange in premature babies. The alveolar saccules subdivide at birth into four lobules, with depressions forming primitive alveoli (Fig. 2-5). The development of alveoli is

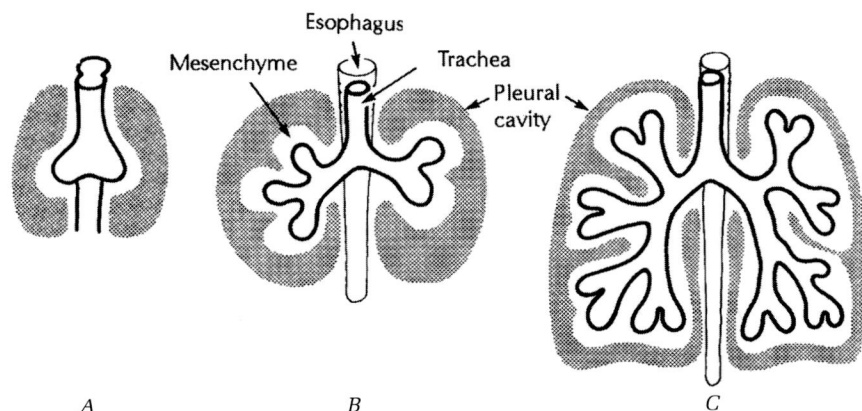

A   *B*   *C*

FIGURE 2-1
Diagram showing the growth of the trachea and esophagus from primitive buds (*A*) into the main bronchi surrounded by mesenchyme (*B*) and further differentiation of the bronchi (*C*). *(Reproduced with permission of Professor E. Weibel and Harvard University Press.)*

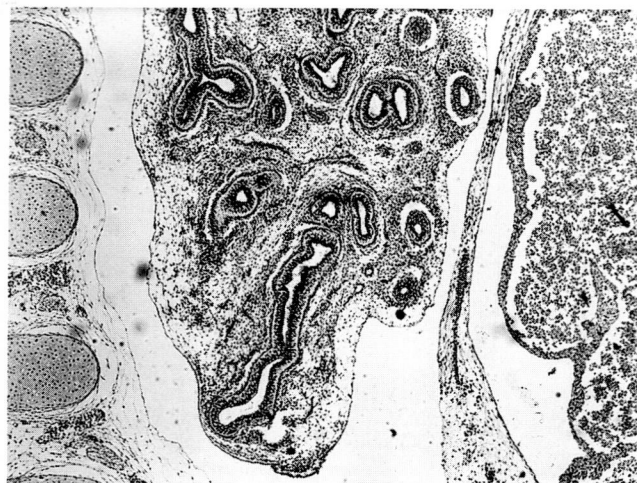

FIGURE 2-2
Developing lung, 20-mm stage (approximately eight weeks), showing bronchial tubes but no attempt at formation of alveolar tissue. H&E, ×30.

accompanied by vascular proliferation so that alveoli are invaded by capillaries, which appear to disrupt the developing alveolar epithelium.

Further intrauterine lung growth leads to further attenuation of the alveolar epithelium until at full term only a few alveolar cell nuclei are visible. After birth, respiratory movements lead to still further thinning of the alveolar epithelium, which assumes its inconspicuous postnatal appearance.

The mode of growth of alveoli both in utero and after birth is uncertain. Loosli and Potter[4] claimed that they grow because of septal division within existing saccules. Macklin[5] regarded alveoli as analogous to interstitial emphysematous spaces in which

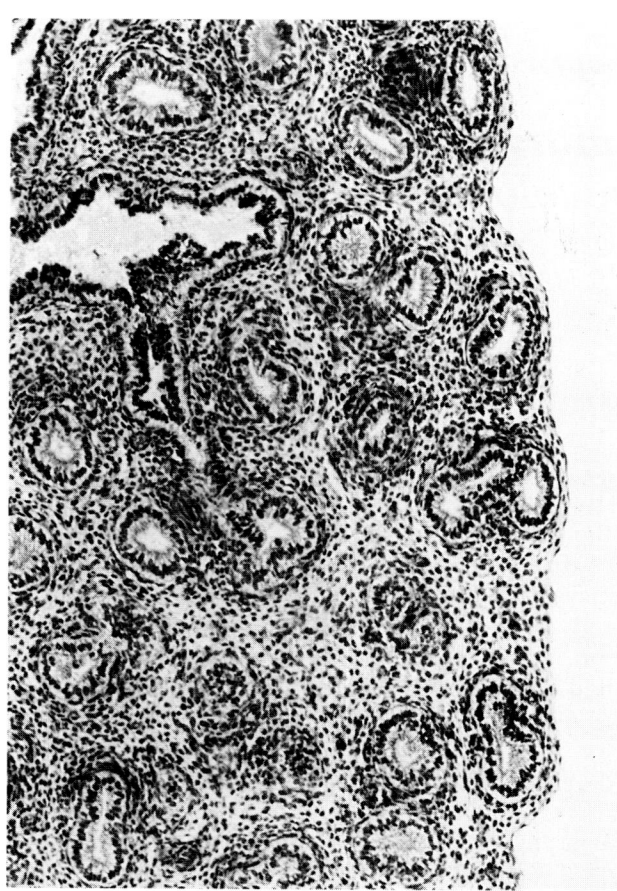

FIGURE 2-4
Fetal lung between 22 and 25 weeks, with the development of rudimentary alveolar structures. H&E, ×100.

the lung mesenchyme splits and allows the capillaries to come into intimate contact with fluid, and later air, in the alveolar spaces.

Along with the first appearance of alveoli, lung septa form.[6] Septa are best developed beneath the sharp margins and apices but are incomplete and do

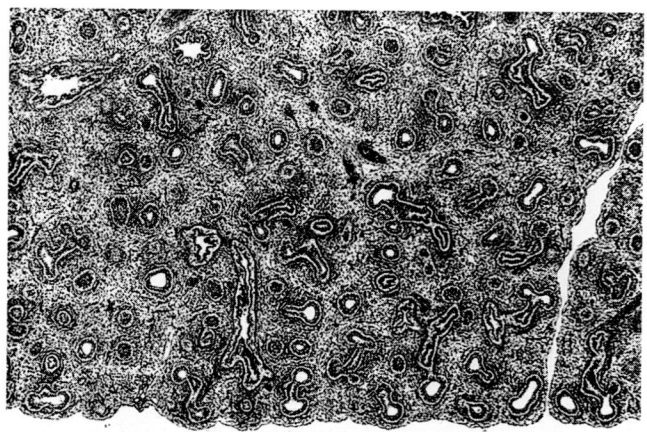

FIGURE 2-3
Later stage of development than Fig. 2-2 in a 60-mm stage fetus (just over 12 weeks), with more extensive branching of the bronchial tubes but still no attempt to form alveoli. H&E, ×30.

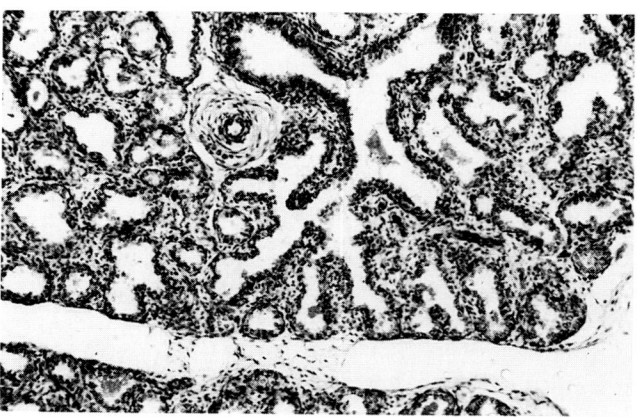

FIGURE 2-5
Fetal lung at 30 weeks, with a greater degree of alveolar development and prominent alveolar epithelium. H&E, ×77.

not isolate one portion of lung from another. As Boyden[7] and Reid[8] demonstrated, the pleural fissures between lobes are often incomplete, failing to extend to the hilum.

In the adult lung, the number of segmental bronchial branches along axial bronchi may increase, compared with that of the full-term infant. Bucher and Reid[2] thought this was due to conversion of respiratory bronchioles into fully epithelialized terminal bronchioles during childhood. Boyden[9] believed in the reverse process, i.e., that terminal bronchioles were converted into respiratory bronchioles. The further course of postnatal lung development is considered below. Bronchial division probably occurs from the growing tip (Fig. 2-6). The failure of bronchial growth causes absence of lobes, segments, or smaller lung volumes. Cessation at an early stage can be responsible for congenital bronchiectasis.

The evolution of the asymmetric adult pattern is probably caused by levorotation of the heart and persistence of the left dorsal aorta and fourth left aortic arch to form the aortic arch. Bronchial growth is dependent on the available celomic space. The cessation of branching along axial segmental lower-lobe bronchi is most probably due to the fact that the lungs do not fill the pleural (coelomic) cavity during the earlier weeks of fetal development, and growth continues caudally for a longer time in the lower lobes. Fetal peribronchial mesoderm is a vital factor acting as an inducing agent for bronchial growth.[10] Similarly, mesenchyme is important in promoting alveolar growth and the development of alveoli. Thus, there is a dual mode of lung development: conducting airways derived by the repeated division of entodermal bronchial buds and mesenchymal development of distal lung structures. The occurrence of pulmonary blastomas and the study of certain congenital pulmonary abnormalities support a dual mode of development of the lung. Organs with a dual arterial circulation elsewhere in the body often reflect an origin from two primary embryonic layers.

PHASES OF LUNG DEVELOPMENT

There are three well-defined periods of lung development.[11] These are (1) the *glandular* period, when the bronchial divisions are established; (2) a *canalicular* phase, which extends from the beginning of the seventh month to term (in this stage, the fetus is viable and the lung becomes delineated and vascularized); and (3) the *alveolar* period, which occurs in the first few weeks after birth, when alveoli are differentiated.

Pleural Cavity Development

The mesenchyme on the outer lung surface forms the medial wall of the pleuroperitoneal canal. This thins to form a layer of connective tissue, which is covered by mesothelial cells to become the visceral pleura. The growth of the pleuroperitoneal cavity occurs at the expense of the mesenchyme forming the future chest wall, which is pushed outward. Simultaneously, the heart grows both caudally and ventrally, causing the lung and its pleura to ensheath the pericardial sac. The pericardial sac becomes sealed off from the celomic cavity. Occasionally, herniation of the lung into the pericardial sac may be found after birth.

At four weeks, the lesser sac of the peritoneum sends an upward prolongation, the infracardiac bursa, toward the back of the root of the lungs. This normally becomes sealed off but may persist, forming a thin-walled pneumoenteric cyst in adult life. These cysts are found behind the root of the lungs.

Bronchial Wall Development

Bronchial cartilage develops at 10 weeks from chondroblasts with an eosinophilic cytoplasm, bounded by a distinct PAS-staining membrane. At first there is no intercellular ground substance, but at 12 to 14 weeks acid mucopolysaccharide appears along with a fine network of reticulin fibers. During the latter half of fetal life, the ground substance changes to a neutral polysaccharide. Bronchial cartilage growth lags behind bronchial development. Sinclair-Smith[12]

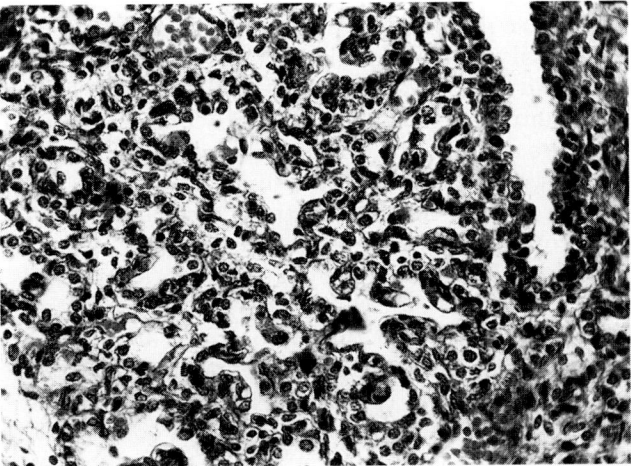

FIGURE 2-6
Developing fetal lung at 10-cm stage showing the growing bronchial tip. There is clustering of the mesenchyme at the tip. H&E, ×160.

noted that new cartilages continue to appear until 48 weeks but thereafter are increased only in area with the sitting height of the child.

Bronchial mucous glands appear at about week 14, growing as solid buds from the lining epithelium and extending outward between the cartilaginous plates. They rapidly canalize, form acini, and are functional by 24 weeks. Serous glands appear at 26 weeks. Larger mucous glands are seen in cystic fibrosis. The density of mucous glands in relation to each unit of bronchial surface is greatest at about 14 weeks in the region of the central bronchial bifurcations. The growth of bronchial mucous glands proceeds at a slower rate in the more distal bronchi, reaching the fourth generation of bronchi at birth. In the distal bronchi, some of the glands may be immature and uncanalized.

Initially, the bronchial buds are lined by pseudostratified columnar epithelium. Cilia are well developed by week 13 and extend to the smallest bronchi during later fetal life. Goblet cells are seen in the largest bronchi at the same time and extend as far as the most distal cartilage-containing bronchi by birth. At 10 weeks, there is a discontinuous bronchial epithelial basement membrane, which becomes uniform in the largest bronchi two weeks later.

Elastic tissue is ill-developed at birth, being found only in the walls of the larger bronchi. The stimulus provided by respiratory movements leads to its early appearance, and by the third postnatal month it is seen in alveolar walls, in pleura, and in the walls of blood vessels.

Development of the Nerve Supply

Bronchial motor and sensory nerves are present by the 15th week.[13] The nerves can be traced to most distal air passages. Immature peri- and intrabronchial parasympathetic ganglion cells are seen supplying postganglionic fibers to developing bronchial muscle in the larger bronchi. The pulmonary arterial nerve plexus develops between 18 weeks and full term.

Development of Pulmonary Vessels

Congdon[14] showed that the initial development sequence in humans is similar to that in albino rats. The heart tube, which develops at the end of the third week, is connected to the dorsal aortae by the vessels of the first branchial arch (Fig. 2-7).[15] Six pairs of aortic arches are formed. The first two disappear, the fifth is vestigial, the third forms the common carotid arteries, and the fourth the aortic arch. The sixth, or pulmonary, arch appears at about 32 days from the

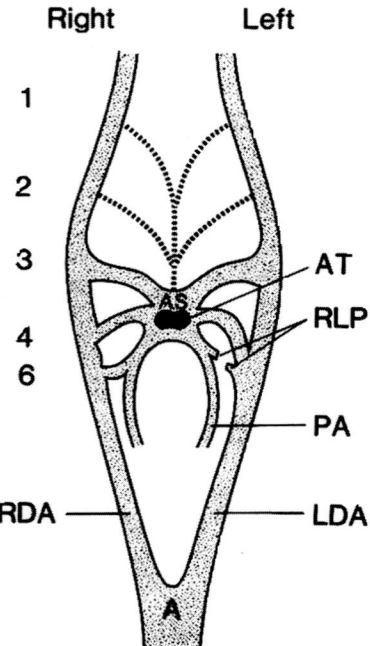

FIGURE 2-7
Brachial arch arteries connecting the ventral aortic sac (AS) with the right and left dorsal aortae (LDA and RDA) in a 5-mm embryo. On the left the dorsal and ventral sprouts of the sixth (pulmonary) arch have nearly met (RLP), and on the right side the arch is complete. From the ventral sprouts plexiform vessels (PA) pass to the lung bud. *(Reproduced with permission of Professor L. Reid, and Heinemann.)*

ventral end of the fourth aortic arch and grows caudally and dorsally, with its central end becoming incorporated into the aortic bulb (truncus arteriosus). By 37 days the aortic sac is divided, so blood from the right ventricle flows into the sixth arch and lungs. The left sixth arch develops into the main pulmonary artery, while the right starts to atrophy. The sixth arch supplies branches to the existing lung bud. The lung bud is also supplied by paired branches arising from the dorsal aorta. Normally such communications are lost, and bronchial arteries develop between the ninth and 12th weeks.[16] If the early branches arising from the aorta persist, this causes one type of anomalous systemic supply to the lung.

By 50 days the right sixth arch has disappeared, and from the left arise the main, left, and right pulmonary arteries. Most of the pulmonary arteries arise from the caudally directed *ventral buds* of the sixth arches and not from the sixth aortic arches themselves. Normally, the dorsal portion of the left sixth arch develops into the ductus arteriosus. The aortic bud twists on itself, and this rotation, coupled with the development of the internal bulbar septum, leads to the separation of the two major arteries. The root of the aorta and stem of the pulmonary artery then assume their normal adult configuration.

The intrapulmonary arteries develop as a separate vascular network from the pulmonary systemic vascular tree, which connects with the bronchial and other aortic arterial branches. Anastomoses form between the two systems and persist into adult life. The paired aortic segmental arteries supplying the developing lungs usually disappear, except the first pair, which become the adult bronchial arteries. The diversion of much pulmonary blood flow through the ductus in intrauterine life is brought about by the vasoconstriction of the muscular pulmonary arteries, and this is reflected by their structure (see Chap. 21).

Initially the pulmonary capillaries lie in the supporting tissues of the lung, but they later grow into the walls of developing terminal air passages. By 20 weeks, the capillaries insinuate themselves between the lining alveolar cells.

The structure and development of the smaller pulmonary arteries have been described by Reid.[17] Normal neonatal pulmonary arteriograms fill the preacinar vessels but not the intraacinar vascular tree. The preacinar pulmonary arteries grow *pari passu* with bronchial development, but intraacinar vessels—i.e., those distal to the terminal bronchioles—await subsequent alveolar development. Supernumerary preacinar arterial development occurs, and from such vessels later grow collateral channels to supply developing alveoli. Thus, alveoli receive not only the main axial intraacinar pulmonary arterial supply but also a collateral arterial supply from the supernumerary arteries.

Postnatal growth of pulmonary arteries[18] occurs mainly in the first three months. During this time the larger pulmonary arteries increase three to four times in diameter—with considerable growth in the number of small arteries (below 200 μm), which parallels the rapid alveolar growth. A recognizable arterial branching pattern can be seen at 14 weeks,[19] and by 20 weeks the arterial and venous branching pattern is the same as in the adult.

The development of both the pulmonary and the bronchial veins is closely related, and they are considered together. Pulmonary vein development was studied by Butler[20] and that of the bronchial veins by Shaner.[21] The main pulmonary veins develop later than the arteries. At about 30 days, the bronchial buds are covered by mesenchyme and drained by the splanchnic venous plexus. This venular plexus drains and surrounds the hind end of the pharynx, esophagus, and stomach together with the laryngotracheal bud. The plexus drains especially into the anterior and posterior cardinal veins. The central part of the common pulmonary venous trunk arises from the sinus venosus, which fuses with the developing pulmonary veins. The sinus venosus lies behind and opens into the common atrium. Part of the right horn of the sinus venosus is incorporated to form part of the common atrium, and its entrance into the atrium later forms the ostium of the coronary sinus.

At about 35 days, an outgrowth arises from the left side of the atrial chamber: the primitive pulmonary vein,[22] which grows posteriorly to contact the lung buds. After connections are made with the buds, the first two branches of the primitive pulmonary veins fuse and are resorbed as a common stem into the atrial chamber to form much of the left atrium. As most of the proximal branches have been resorbed by the 11th week, four separate venous channels enter the left atrium. In the adult, connections remain between esophageal and pulmonary veins at the lung roots.[20] The subsequent growth of the interatrial septum leaves the pulmonary veins draining into the left atrium.

During early bronchial growth, there are free anastomoses in the posterior mediastinum between the developing venous plexuses of both lungs and the rest of the splanchnic venous plexus, surrounding the foregut and vagus nerves. Shaner[21] showed that in the 12-week fetus the venous plexus surrounding the vagus nerves and their recurrent branches condenses to form venae comitantes. These drain at several points into cardinal veins. The largest connection drains into the anterior cardinal vein. Also formed is a persistent venous channel, connecting the common pulmonary vein with the esophageal and gastric venous plexus. This connection communicates with the ductus venosus of the liver. The left bronchial vein is formed from the veins accompanying the recurrent branch of the left vagus nerve. These veins serve to connect the tracheal and bronchial venous plexus with the anterior cardinal vein, into which they drain. Later, partial atrophy and obliteration of the left cardinal system of veins result in the left bronchial veins draining into the accessory hemiazygos vein or the left superior intercostal vein. The right bronchial vein is formed from other venous channels and drains into the upper part of the azygos vein.

Persistence of fetal connections between the pulmonary, bronchial, and cardinal venous systems is responsible for most of the congenital venous abnormalities. Persistence of the connections between the pulmonary veins and the vitelloumbilical system of veins may become important in cirrhosis in adults and allows portosystemic anastomoses, which bypass the pulmonary circulation.

Lymphatic Development

The lymphatics[21] first appear at 60 days at the hilum. By 70 days, lymphatic vessels are seen in the lung itself, and by the 100th day, valves appear in the hilar lymphatics. At birth there is an adult arrangement of lymphatic channels. Small collections of lymphocytes, without germinal centers, are seen in peribronchial tissue at 18 weeks and are prominent at 30 weeks. Further collections of lymphocytes grow around terminal bronchioles up to two years postnatally.

Ultrastructural Features of the Fetal Lung

The electron microscopic features of the developing lung have been described by Campiche and coworkers.[23] Similar patterns of cell growth and differentiation are seen in the rat, except for differences in the time scale.[24]

In the early glandular phase, the primitive lung buds constitute a highly irregular columnar epithelium with frequent mitoses and prominent nucleoli. An abundance of intracytoplasmic glycogen is present in the epithelium up to the fourth month, after which it progressively diminishes. With branching the epithelium gradually becomes cuboidal, with regular spherical nuclei. The cytoplasm contains small mitochondria, scarce cisternae of endoplasmic reticulum, much glycogen, and some fat. The bronchial epithelial cells are joined near their luminal surfaces by terminal bars, forming tight junctions. Small scattered desmosomes are also present. At about the fifth month, the epithelium merges distally with the attenuated flattened epithelium lining the developing saccules. Underlying the entire bronchial and future alveolar epithelium is a membrane 30 μm thick, which eventually forms the adult epithelial basement membrane. The endothelium of the future alveolar capillaries is attached on its outer surface to a similar basement membrane, destined to become the adult endothelial basement membrane. The interstitial space lying between the two basement membranes contains fibrils that initially are thin and lack the periodic banding of collagen. Near full term, the fibrils acquire the characteristics of collagen fibers.

The attenuation of the epithelium is characterized by the appearance of osmiophilic lamellar bodies. This occurs at about six months. In the early stages the lamellar bodies probably develop from multivesicular bodies.[25]

In animals preterm, the alveolar spaces contain tubular myelin figures as well as osmiophilic lamellae. Tubular myelin is a highly surface-active fraction, and its appearance in the developing alveoli may presage the viability of a fetus.[26] Surfactant will be discussed in more detail below.

The type I pneumocyte, vital for gaseous exchange, is recognizable at about six months. In the canalicular phase, the epithelial cells begin to move downward with a reduction in height. This process occurs most often in cells with few lamellar bodies and sparce endoplasmic reticulum. In the saccular phase, the lining forms cytoplasmic extensions but the secretory cells are unchanged.[24]

Neuroendocrine cells as well as neuroepithelial bodies are seen in the conducting airways in the 10-week-old fetus. Immunoreactive bombesin appears first, and the cells increase in number, primarily in bronchioles as well as intrapulmonary bronchi, as gestation progresses. Immunoreactive bombesin-staining cells were not seen in terminal budding airways.[27] Calcitonin gene-related peptide (CGRP) is also demonstrated at 10 weeks and increases in the second trimester.[28] CGRP may influence arterial tone. Serotonin-containing cells are also seen in the first trimester, whereas calcitonin is demonstrated late in the second trimester. Leuenkephalin is not seen in fetal lung. Bombesin may respond to local airway distortion by releasing a locally acting constrictor, but the role of calcitonin is unknown. It may regulate local serotonin action on vascular smooth muscle. Growing axons associated with both neuroendocrine cells and bodies are seen as early as 10 weeks. Innervation to some of these structures occurs throughout fetal life.[29]

Pulmonary Surfactant

Surfactant is responsible for the maintenance of the mechanical stability of the alveoli as well as possibly being involved in some defense mechanisms.[30] Some excellent reviews on surfactant have been published.[31–33] In this section only a little detail of the biochemical synthesis of surfactant will be discussed.

As a monolayer at the air-tissue interface, surfactant prevents the alveoli from closing at low lung volumes by reducing the surface tension to near 0 dyn/cm. The liberation of surfactant coincides with the ability of the newborn to begin breathing.

There are thought to be two main pulmonary surfactant pools. One is the intraalveolar pool that can be identified in bronchoalveolar lavage fluid. The second is the lamellar body, which is the intracellular component. This is a simplistic view,[33] as electron-microscopic studies show that there are several morphologically distinct structures in the extracellular surfactant. These include tubular myelin, newly secreted lamellar body contents, the monolayer at the air-tissue interface, and multi- and unilamellar vesicles. These structures probably represent different stages in the life cycle of surfactant.

The biochemical composition of surfactant has been established for humans.[34]

Surfactant purified from bronchoalveolar lavage consists of approximately 90 percent lipid, 5 to 10 percent protein, and small amounts of carbohydrate. Glycolipids appear in surfactant in lung injury. Eighty to 90 percent of surfactant lipid is phospholipid, of which phosphatidylcholine accounts for 70 to 80 percent. Cholesterol is the major neutral lipid. The phospholipid patterns seen in surfactants from mammals are generally similar.[35]

An enzyme called choline phosphate cytidyltransferase plays a key role in the regulation of surfactant phosphatidylcholine synthesis. This enzyme is an important target for several hormones that promote surfactant synthesis.

Chevalier and Collet[36] demonstrated the intracellular routing and storage of surfactant using radioactively labeled choline and followed its pathway in mice. Soon after injection, the label appears in the endoplasmic reticulum. It then travels to the Golgi apparatus and a few hours later accumulates in the lamellar bodies. Thus, small lamellar bodies may carry the surfactant lipids from the Golgi apparatus to larger lamellar bodies for storage. Multivesicular bodies transport the surfactant proteins from the Golgi apparatus to the growing lamellar bodies, where they meet the proteins for the final storage form of surfactant.

Lamellar bodies leave type II cells by exocytosis. There is fusion of the limiting membrane of the organelle and the cell membrane of the type II cell. At this stage, the lamellar body contents are extruded into the alveoli. Part of these lamellae become tubular myelin[37] with a characteristic lattice-like structure, consisting of a system of densely packed, rectangular tubules.

Clearance of surfactant occurs mainly by uptake by type II pneumocytes.[38] The mechanisms controlling the recycling process are poorly understood. At least some of the phosphatidylcholines taken up by the type II cells are degraded. The subsequent products are probably transferred to the endoplasmic reticulum and then used for resynthesis.[38] The life cycle of surfactant is complex and the recycling and biosynthesis have been reviewed.[39,39a]

ASSESSMENT OF FETAL LUNG MATURITY

A number of indices based on the estimation of some surfactant constituents have been suggested. As noted above, amniotic fluid contains phospholipids such as phosphatidylcholine (lecithin), sphingomyelin, phosphatidylinositol, and phosphatidylglycerol. The most commonly accepted method of assessing fetal lung maturation is the lecithin/sphingomyelin (L/S) ratio in amniotic fluid (normally greater than 2.7 at 37 weeks in normal pregnancies).[40] This ratio can be used to predict whether respiratory distress syndrome (RDS) will develop in the infant born within 24 to 28 h of amniocentesis. It has been estimated that use of fetal lung maturity tests has decreased the incidence of RDS by 30 percent.[41] However, false mature values, ranging from 1 to 15 percent, occur, especially in pregnancies complicated by diabetes mellitus, and an immature L/S ratio may predict RDS in only about 50 percent of cases.[42]

Various factors may affect the results. One is the site of amniocentesis, higher phospholipid concentrations being found near the fetal mouth than in a sample close to the breech. The phospholipid concentration also varies inversely with amniotic fluid volume, and this volume varies greatly, especially after midgestation.[42] Contamination of the sample with blood, meconium, or antiseptics causes error.

Phosphatidylglycerol (PG) is the second most abundant phospholipid in the surfactant system in the mature lung. Immature pulmonary tissue contains phosphatidylinositol (PI) rather than PG. Lung maturation appears to be accomplished by a decline in PI and a consequent increase in PG.[42] Although it is likely that PG is essential for proper surfactant function, the full physiologic role of this phospholipid has to be determined. Preterm babies delivered before the appearance of surfactant rich in PG are at an increased risk of RDS.[43] Conversely, when PG is present in amniotic fluid, RDS does not occur, even when the L/S ratio is less than 2.[43]

There may be acceleration of fetal pulmonary maturation in women with preeclampsia or severe Rh-isoimmunization and in some diabetics.[44]

REGULATION OF FETAL LUNG MATURATION

Many factors play a part in fetal lung maturation. Following are the factors that are now known.

Mesenchymal-Epithelial Interactions
The close relationship between epithelium and mesenchyme has been shown above. If the epithelial anlages are separated from mesenchyme, the branching stops.[10]

Hormonal Factors
A number of hormones accelerate the maturation of the pulmonary surfactant system but do not initiate it. There have been several reviews of hormonal control of fetal surfactant.[33,45] The identification of the

surfactant proteins and their genes has helped the understanding of the mechanisms regulating fetal lung maturation. There are several surfactant proteins.[46]

Surfactant-associated protein A (SP-A), whose gene is located on chromosome 10, is thought to play a role in regulating the flow of surfactant in and out of the type II cell.[47] SP-B and SP-C both play an important role by promoting the spreading of the surfactant layer. SP-A and SP-B are important for the formation of tubular myelin. SP-B is encoded by a gene on chromosome 2 and SP-C by two genes on chromosome 8. SP-D is a collagenous surfactant-associated glycoprotein.[45]

Glucocorticoids are the most studied hormones that accelerate lung differentiation. They enhance surfactant synthesis, and this response appears to be cytoplasmic receptor–mediated.[48] SP-A, -B, and -C are all increased. Glucocorticoids increase the content of a fibroblast-pneumocyte factor, which is produced by fetal lung fibroblasts and enhances differentiation of type II cells with respect to surfactant production.[49] Glucocorticoids also increase fetal lung compliance. They may be given to the mother 24 h before delivery, but they are not effective after 33 to 34 weeks. If given along with thyrotropin-releasing hormone, they do not decrease the incidence of RDS, but there is less bronchopulmonary dysplasia.[50]

Thyroid hormone. Like that of glucocorticoids, the action of T3 and T4 is receptor-mediated. They enhance the synthesis of surfactant phospholipids, especially lecithin, but not the surfactant proteins.

Thyrotropin-releasing hormone (TRH). This readily crosses the placenta and thus increases fetal T3. TRH increases alveolar lecithin and markedly increases lung distensibility when combined with glucocorticoids.[51]

Cyclic adenosine monophosphate (cAMP) causes increased synthesis of lecithin and PG as well as being an important regulator of surfactant protein synthesis.[47]

Growth factors include epidermal growth factor, insulin-like growth factor, transforming growth factor–β, and peptide neurotransmitters. These act on either epithelial or mesenchymal elements.

Prostaglandins affect fetal breathing movements and PGE_2 stops breathing in sheep.[52] Prostaglandins may act on the respiratory center in the central nervous system and thus modulate growth through modulation of breathing activity.[53] Prostaglandins are synthesized within lung tissue and could affect cell proliferation directly through cAMP.

Gamma-interferon works experimentally and appears to increase SP-A, but the significance of this response is not known.[54]

Prolactin, estrogen, and sex of the fetus. Prolactin probably has an additive effect to glucocorticoids and T3.[47] It is not known if estrogen plays a part in lung maturation. Male fetuses lag one to two weeks behind females in their lung development. Part of this delay may be due to androgen-dihydrotestosterone decreasing lecithin synthesis.[55]

Diabetes mellitus. Babies of diabetic mothers with raised levels of insulin, glucose, and alpha-butyric acid have an increased incidence of RDS if premature.[47] Both insulin and alpha-butyric acid decrease SP-A synthesis. Experimentally, insulin antagonizes glucocorticoid-induced enhancement of phospholipid synthesis.[48]

OTHER FACTORS INFLUENCING FETAL LUNG GROWTH

Much space has been devoted above to factors regulating surfactant synthesis, as this is important in the transformation of the fetus into its extrauterine environment. Surfactant has little influence on fetal lung growth. There are other factors that regulate lung growth, and these have been summarized by Pringle[56] and Kitterman.[57] Lung growth is determined by the following physical factors:

1. *Adequate intrathoracic space.* Any lesion that decreases the size of the intrathoracic space limits lung growth. These include fetal pleural effusions, intrathoracic neoplasms, congenital diaphragmatic hernia, large abdominal masses, and ascites.

2. *An adequate amount of amniotic fluid.* Oligohydramnios retards lung growth by pulmonary compression, by altering fetal lung movements, or by affecting the fluid volume within the potential airways.

3. *Fetal breathing movements of normal frequency and amplitude.* When there is absence of fetal breathing movements throughout pregnancy, the baby is born with hypoplastic lungs.

4. *Normal balance of volumes and pressures in the potential airways and air spaces.* In the fetus the tracheal pressure > amniotic pressure > pleural pressure. This differential produces a distending pressure that may promote lung growth.

5. *Obstruction to outflow of lung fluid.* In fetal lambs, tracheal ligation causes overexpanded lungs with very thin alveolar walls and scarce type II cells.[58] The human equivalent of this anomaly is probably congenital lobar emphysema.

6. *Pulmonary artery blood flow.* There has to be a severe or total obstruction to pulmonary artery

flow to cause impaired lung growth. Bronchial blood supply alone is insufficient to supply the nutritional needs of the growing fetal lung.

7. *Smoking.* Smoking appears to affect fetal lung development.[59]

POSTNATAL LUNG GROWTH

Birth, while a critical event, is not a specific point in lung development. Lung growth must be considered in relation to the entire infant. The growth rate is most rapid postnatally, especially during the first year of life, declining gradually apart from a brief reversal at the time of puberty.[60]

There have been many morphometric studies on postnatal lung development, but different methods and terminology have been used, making comparison of data difficult. It is generally accepted that there is new alveolar formation after birth. At birth there are, on average, 50 million alveoli, with a wide variation.[61] The adult figure is thought to be 300 million.[62] This figure is reached between the ages of 5 and 8 years. However, Angus and Thurlbeck[63] found that the final number of alveoli depended on body length and could vary between 212 million and 605 million. In a study of fetuses at 29 weeks' gestation and infants to 18 weeks' postnatal age, the alveolar number showed a linear relationship with body weight.[64] At this age, an increase in body weight is likely to be synonymous with an increase in body height.

The alveolar surface area (the air-tissue interface) increases in the growing lung. Dunnill[62] calculated the surface area was 2.8 m² at birth, 12 m² at one year, almost doubling to 22 m² at four years, and 32 m² at eight years. In the adult, the alveolar surface area ranges between 40 and 120 m².[62,65–67]

Using this growth in alveolar surface area, Burri and Weibel[24] described three distinct phases of postnatal lung growth: (1) lung expansion, whereby the increase in lung volume is due to increased air-space volume only; (2) tissue proliferation, whereby the alveolar surface area per unit volume of lung tissue is increased (in experimental animals this occurs mainly by septation of the primitive saccules, leading to the formation of alveoli); and (3) proportionate growth, which is an ill-defined phase where growth consists mainly of enlargement of existing structures.

The mean thickness of the air-blood barrier decreases from 4 μm at birth to 1.7 μm three weeks after birth.[24]

The Aging Lung

With increasing age, changes in lung function include an increase in functional residual capacity and residual volume and decreases in diffusing capacity, forced vital capacity, and arterial oxygen tension. Lung function tests are potentially good markers of aging.[68] However, it is difficult to distinguish true aging from a cumulative effect of the environment. There are three main factors causing the changes in lung function: a decrease in lung elasticity, an increase in stiffness of the chest wall, and a decrease in respiratory muscle strength.[69,70]

The increase in chest wall stiffness is due to osteoporosis in the ribs and vertebrae and calcification of the costal cartilages. In addition, there is evidence of kyphoscoliosis and decreased height.[67] The changes in muscle strength are due to a reduced diaphragmatic mass.[71] In the lung itself, there is a decrease in elastin content, which can be demonstrated immunocytochemically.[72] There is an increase in type III collagen in the alveoli, and in some cases there is an increase in type IV collagen and laminin in the alveolar basement membrane. These changes cause dilatation of alveolar ducts and respiratory bronchioles, though the mechanism is not fully elucidated. The decrease in gas transfer is accounted for by the increased type IV collagen and laminin. Thurlbeck and Angus[73] speculated that the decreased volumetric density of alveoli could be due to loss of alveolar septa.

The elastic pulmonary arteries develop atherosclerotic plaques with age, and there is increased collagen in muscular pulmonary arteries. Intimal fibrosis is seen in muscular pulmonary arteries and veins.

The aging lung also has reduced airway defenses. There are decreased mucociliary clearance rates, as compared with those in the young. In addition, diseases such as stroke may affect respiration by predisposing to inhalation. A review of the immunology of the aging lung has recently been presented.[74]

Postpneumonectomy Lung

Relatively little is known about the effect of pneumonectomy on the contralateral lung. The remaining lung increases in volume by 30 to 40 percent. One clinical case report suggested that alveolar multiplication occurred after an extensive resection for bronchiectasis.[75] In another pathologic study, the residual lung was overinflated but there was no suggestion of alveolar multiplication.[76]

REFERENCES

1. Hodson WA (ed): *Development of the Lung*, Vol 6, *Lung Biology in Health and Disease.* New York, Marcel Dekker, 1977.

2. Bucher, U, Reid L: Development of the intrasegmental bronchial tree: The pattern of branching and development of cartilage at various stages of intra-uterine life. *Thorax* 16:207–218, 1961.

3. Boyden EA: Development of the Human Lung, in Kelly VC (ed), *Brennemann's Ann's Practice of Pediatrics*, vol 4. Hagers-town, Md., Harper & Row, 1972, Chap. 64, pp. 1–12.

4. Loosli CG, Potter EL: The prenatal development of the human lung. *Anat Rec* 109:32, 1951.

5. Macklin CC: Alveolar pores and their significance in the human lung. *Arch Pathol* 21:202–216, 1936.

6. Reid L, Rubino M: The connective tissue septa in the foetal human lung. *Thorax* 14:3–13, 1959.

7. Boyden EA: Developmental anomalies of the lungs. *Am J Surg* 89:79–89, 1955.

8. Reid LM: Pathology of chronic bronchitis. *Lancet* 1:275–278, 1954.

9. Boyden EA: The terminal air sacs and their blood supply in a 37-day-old infant. *Am J Anat* 116:413–428, 1965.

10. Wessells NK: Mammalian lung development: Interactions in formation and morphogenesis of tracheal buds. *J Exp Zool* 175:455–466, 1970.

11. Dubreuil G, La Coste A, Raymond R: Observations sur developpement du poumon humaine. *Bull Histol Appl Physiol* 13:235–245, 1936.

12. Sinclair-Smith CC, Emery JL, Gadson D, Dinsdale F, Baddeley J: Cartilage in children's lungs: A quantitative assessment using the right middle lobe. *Thorax* 31:40–43, 1976.

13. Spencer H, Leof D: The innervation of the human lung. *J Anat* (*Lond*) 98:599–609, 1964.

14. Congdon ED: Transformation of the aortic-arch system during the development of the human embryo. *Contrib Embryol* [*Carnegie Inst*] 14:47–110, 1922.

15. Hislop A, Reid LA: Formation of the Pulmonary Vasculature in Hodson WA (ed), *Development of the Lung*. New York, Marcel Dekker, 1977, pp. 37–83.

16. Boyden EA: The developing bronchial arteries in a fetus of the twelfth week. *Am J Anat* 129:357–368, 1970.

17. Reid LM: The pulmonary circulation: Remodeling in growth and disease. *Am Rev Respir Dis* 119:531–546, 1979.

18. Davies GM, Reid L: Growth of the alveoli and pulmonary arteries in childhood. *Thorax* 25:669–681, 1970.

19. Hislop A, Reid L: Intra-pulmonary arterial development during fetal life: Branching pattern and structure. *J Anat* 113:35–48, 1972.

20. Butler H: Some derivatives of the foregut venous plexus of the albino rat, with reference to man. *J Anat* (*Lond*) 86:95–108, 1952.

21. Shaner RF: The development of the bronchial veins with special reference to anomalies of the pulmonary veins. *Anat Rec* 140:159–165, 1961.

22. Anderson H, Ashley GT: Growth and development of the cardiovascular system: Anatomical development, in Davis JA, Dobbing J (eds), *Scientific Foundations of Paediatrics*. London, Heinemann, 1974.

23. Campiche M, Prodhom S, Gautier A: Étude du microscopa électronique du poumon de prématurés morts en détresse respiratoire. *Ann Pediatr* (*Paris*) 96:81–95, 1961.

24. Burri PH, Weibel ER: Ultrastructure and morphometry of the developing lung, in Hodson WA (ed), *Development of the Lung*. New York, Marcel Dekker, pp. 215–231, 1977.

25. Soronkin SPA: Morphologic and cytochemical study on the great alveolar cell. *J Histochem Cytochem* 14:884–897, 1967.

26. Shelly SA, Tagaki LR, Balis JU: Assessment of surfactant activity in amniotic fluid for evaluation of fetal lung maturity. *Am J Obstet Gynecol* 16:369–376, 1973.

27. Stahlman MT, Kasselberg AG, Orth TH, Gray ME: Otogeny of neuroendocrine cells in human fetal lung: II. An immunohistochemical study. *Lab Invest* 52:52–60, 1985.

28. Johnson MD, Gray ME, Stahlman MT: Calcitonin generelated peptide in human fetal lung and in neonatal lung disease. *J Histochem Cytochem* 36:199–204, 1988.

29. Stahlman MT, Gray ME: Otogeny of neuroendocrine cells in human fetal lung: I. An electron microscopic study. *Lab Invest* 51:449–463, 1984.

30. Hoffman RM, Claypool WD, Katyal SL, Singh G, Rogers RM, Dauber JH: Augmentation of rat alveolar macrophage migration by surfactant protein. *Am Rev Respir Dis* 135:1358–1362, 1987.

31. Possmayer F: Biochemistry of pulmonary surfactant during fetal development and in the perinatal period, in Robertson B, Van Golde LMG, Battenburg JJ (eds), *Pulmonary Surfactant*. Amsterdam, Elselvier, 1984, pp. 295–355.

32. Rooney SA: Biochemical development of the lung, in Warshaw JB (ed), *The Biological Basis of Reproductive and Developmental Medicine*. New York, Elselvier Biomedical, pp. 239–287.

33. Post M, van Golde LMG: Metabolic and developmental aspects of the pulmonary surfactant system. *Biochim Biophys Acta* 947:249–286, 1988.

34. Hallman M, Spragg R, Harrell JH, Moser KM, Gluck LJ: Evidence of lung surfactant abnormality in respiratory failure: Study of bronchoalveolar lavage phospholipids, surface activity, phospholipase activity and plasma myoinositol. *J Clin Invest* 70:673–683, 1982.

35. Possmayer F, Shou-Hwa Y, Weber JM, Harding PGR: Pulmonary surfactant. *Can J Biochem Cell Biol* 62:1121–1133, 1984.

36. Chevalier G, Collet AJ: In vivo incorporation of choline-3H, leucine-3H and galactose-3H in alveolar type II pneumocytes in relation to surfactant synthesis: A quantitative radiographic study in mouse by electron microscopy. *Anat Rec* 174:289–310, 1972.

37. Williams MC: Conversion of lamellar body membranes into tubular myelin in alveoli of fetal rat lungs. *J Cell Biol* 72:260–277, 1977.

38. Chander A, Claypool WD Jr, Strauss, JF III, Fisher AB: Uptake of liposomal phosphatidylcholine by granular pneumocytes in primary culture. *Am J Physiol* 245:C397–C404, 1983.

39. Wright JR, Clements JA: Metabolism and turnover of lung surfactant. *Am Rev Respir Dis* 135:426–444, 1987.

39a. Jobe A, Ikegami M: Surfactant for the treatment of respiratory distress syndrome. *Am Rev Respir Dis* 136:1256–1275, 1987

40. Wright JR, Wager RE, Hamilton RL, Huang M, Clements JA: Uptake of lung surfactant subfractions into lamellar bodies of adult rabbit lungs. *J Appl Physiol* 60:817–825, 1986.

41. Hack M, Faranoff AA, Klaus MH, Mendelawiz BB, Merkatz R: Neonatal distress following elective delivery: A preventable disease. *Am J Obstet Gynecol* 126:3447–3451, 1976.

42. Cosmi EV, Di Renzo GC: Assessment of foetal lung maturity. *Eur Respir J* 2(Suppl 3):40s–49s, 1989.

43. Feijen HWH, Di Renzo GC, Nederstigt J, Houx PCW, Eskes TKAB: Evaluation of the fetal lung profile including the two dimensional L/S ratio for the establishment of fetal lung maturation. *Gynecol Obstet Invest* 14:142–150, 1982.

44. Zapata A, Hernandez-Garcia JM, Grande C, et al: Pulmonary

phospholipids in amniotic fluid of pathologic pregnancies: Relationship with clinical status of the newborn. *Scand J Clin Lab Invest* 49:351–357, 1989.

45. Mendelson CR, Boggarar V: Hormonal and developmental regulation of pulmonary surfactant synthesis in fetal lung. *Baillieres Clin Endocrinol Metab* 4:351–378, 1990.

46. Hawgood S: Pulmonary surfactant apoproteins: A review of protein and genomic structure. *Am J Physiol* 257 [*Lung Cell Mol Physiol* 1]:L13–L22, 1989.

47. Gross I: Regulation of fetal lung maturation. *Am J Physiol* 259 [*Lung Cell Mol Physiol* 3]:L337–L344, 1990.

48. Gross I, Ballard PL, Ballard RA, Jones CT, Wilson CM: Corticosteroid stimulation of phosphatidylcholine synthesis in cultured fetal rabbit lung. Evidence for *de novo* protein synthesis mediated by glucocorticoid receptors. *Endocrinology* 112:829–837, 1983.

49. Smith BT, Post M: Fibroblast-pneumocyte factors. *Am J Physiol* 257 [*Lung Cell Mol Physiol* 1]:L174–L178, 1989.

50. Ballard RA, Ballard PL, Creasy R, Gross I, Padbury JF, TRH Study Group: Prenatal thyrotropin releasing hormone plus corticosteroid decreases chronic lung disease in very low birth weight infants [abstract]. *Clin Res* 38:192A, 1990.

51. Liggins GC, Schellenberg J, Manzai J, Kitterman JA, Lee CH: Synergism of cortisol and thyrotropin-releasing hormone on lung maturation in fetal sheep. *J Appl Physiol* 65:1880–1884, 1988.

52. Guarra FA, Savich RD, Wallen KD, et al: Prostaglandin E2 causes hyperventilation and apnea in newborn lambs. *J Appl Physiol* 64:2160–2166, 1988.

53. Ballard PL: Hormonal control of lung maturation. *Baillieres Clin Endocrinol Metab* 3:723–753, 1989.

54. Ballard PL, Lilley HG, Gonzales MW, et al: Interferon-gamma and synthesis of surfactant components by cultured human fetal lung. *Am J Respir Cell Mol Biol* 2:137–143, 1990.

55. Torday JS: Androgens delay human fetal lung maturation in vitro. *Endocrinology* 126:3240–3244, 1990.

56. Pringle KC: Human fetal lung development and related animal models. *Clin Obstet Gynecol* 29:502–513, 1986.

57. Kitterman JA: Physiological factors in fetal lung growth. *Can J Physiol Pharmacol* 66:1122–1128, 1988.

58. Alcorn D, Adamson TM, Lambert TF, Maloney JE, Ritchie BC, Robinson PM: Morphological effects of chronic tracheal ligation and drainage in the foetal lamb lung. *J Anat* 123:649–660, 1977.

59. Collins MH, Moessinger AC, Kleinerman J, et al: Fetal lung hypoplasia associated with maternal smoking: A morphometric analysis. *Pediatr Res* 19:408–412, 1985.

60. Tanner JM: Growing up. *Sci Am* 229:35–43, 1973.

61. Langston C, Kida K, Reed M, Thurlbeck WM: Human lung growth in late gestation and in the neonate. *Am Rev Respir Dis* 129:607–613, 1984.

62. Dunnill MS: Postnatal growth of the lung. *Thorax* 17:329–333, 1962.

63. Angus GE, Thurlbeck WM: Number of alveoli in the human lung. *J Appl Physiol* 32:483–485, 1972.

64. Hislop AA, Wigglesworth JS, Desai R: Alveolar development in the human fetus and infant. *Early Hum Dev* 13:1–11, 1986.

65. Weibel ER: *Morphometry of the Human Lung*. Heidelberg, Springer-Verlag, 1963.

66. Thurlbeck WM: Internal surface area and other measurements in emphysema. *Thorax* 22:483–496, 1967.

67. Hasleton PS: The internal surface area of the adult human lung. *J Anat* 112:391–400, 1972.

68. Kauffman F, Frette C: The aging lung: An epidemiological perspective. *Respir Med* 87:5–7, 1993.

69. Mahler DA, Rosiello RA, Loke J: The aging lung. *Gerontol Clin North Am* 2:215–225, 1986.

70. Tolep K, Kelsen SG: Effect of aging on respiratory skeletal muscles. *Clin Chest Med* 14:363–378, 1993.

71. Arora NS, Rochester DF: Effect of body weight and muscularity on human diaphragm muscle mass, thickness and area. *J Appl Physiol* 52:64–70, 1982.

72. D'Errico A, Scarani P, Colosimo E, Spina M, Grigoni WF, Mancini AM: Changes in the alveolar connective tissue of the aging lung. *Virchows Archiv* [A] 415:137–144, 1989.

73. Thurlbeck WM, Angus GE: Growth and aging of the normal human lung. *Chest* 67[suppl 2]:3S–6S, 1975.

74. Gyetko MR, Toews GB: Immunology of the aging lung. *Clin Chest Med* 14:379–391, 1993.

75. Bates DV, Macklem PT, Christie RV: *Respiratory Function in Disease. An Introduction to the Integrated Study of the Lung*. Philadelphia, W. B. Saunders Co., 1971.

76. Dunnill MS: Quantitative observations on the anatomy of chronic nonspecific lung disease. *Med Thorac* 22:261–274, 1965.

Chapter 3

Congenital Abnormalities

Stephen J. Gould / P. S. Hasleton

This chapter discusses congenital malformations of the lung, including the trachea and bronchi, together with those inheritable disorders that have significant pulmonary complications such as cystic fibrosis and α-1- antitrypsin deficiency. In addition it will also consider lung pathology acquired in utero, and those conditions associated with the transition from an intrauterine to an extrauterine existence, particularly when this occurs preterm.

Not surprisingly, many of the conditions discussed here are far more likely to be seen in the course of pediatric pathologic work, at least with any frequency. Collectively, however, congenital, perinatal, and pediatric conditions of the lung are sufficiently common that they will form part of the practice of most general pathologists. Further, the advances in management of these conditions has progressed in recent years to significantly influence the pattern of lung pathology.

Many bronchopulmonary malformations will occur as incidental findings in isolation or as part of a syndrome in spontaneous abortions or stillbirths. Some will present at or shortly after birth and may be amenable to corrective surgery. The outcome of surgery is sometimes less dependent on the technical difficulties of the operation than on the extent to which the abnormality has interfered with the development of adjacent lung. The earlier any compression is removed, the better the prospect of adequate lung growth.

The rapid improvements in antenatal obstetric ultrasound have allowed a more "controlled" approach to many of these abnormalities and malformations such as diaphragmatic hernia or adenomatoid malformations, which are increasingly diagnosed antenatally, often in mid-gestation. Although termination may be an option with early diagnosis, it also allows the prospect of earlier intervention. Prenatal diagnosis holds out the prospect of in utero intervention or surgery in selected cases and some success has already been achieved with diaphragmatic hernia repairs and lung cysts.[1–4]

In perinatal pathology, the lung is frequently the major determinant of survival.[5] Although anatomic immaturity is significant, the immaturity of the biochemical systems is equally important in determining the outcome. The development of artificial surfactant has played an important part in reducing mortality in the very preterm and the incidence and severity of hyaline membrane disease has fallen. Partly as a result of this, and the increased survival of the very preterm infant, there is evidence of an increase in the levels of chronic lung disease.

In inheritable lung disease lies perhaps one of the most exciting success stories and prospects for the future. Progress in cystic fibrosis, one of the most common serious single gene defects recognized, has been rapid and life expectancy is now estimated at approximately 40 years.[6] Thus, it is now a condition that is very much in the realm of the adult as well as the pediatric pathologist. With the cloning of the cystic fibrosis gene, there is a real possibility of a cure. By inserting the gene into the affected cells it should be possible to correct the defect in intracellular transport caused by the abnormal or absent production of the protein, cystic fibrosis transmembrane conductance regulator. The success already achieved in correcting the defect in animal models indicates this is more than just a theoretical possibility.[7]

CONGENITAL MALFORMATIONS

Congenital malformations are morphologic defects in organs or parts of organs due to an intrinsically abnormal developmental process.[8] A congenital abnormality may also result when there is interference with an intrinsically normal developmental process by some external agent such as anencephaly secondary to amniotic bands. Strictly, this is a disruption rather than a malformation. Malformations may be a result of genetic defects ranging from whole chromosomal anomalies to single gene mutations or deletions. Some malformations are presumed multifactorial due to interaction between genetic factors and environmental agents; yet others are purely environmental in etiology. In practice, it may be difficult to be sure where in this etiologic spectrum some malformations should be placed. To understand defects in lung development it is essential to remember the embryology outlined in Chap. 1.

Bronchopulmonary malformations are uncommon and indeed many are rare so few centers will obtain much experience. Further, the spectrum and frequency with which any abnormality is encountered will vary depending on the nature of the clinical practice. Consequently, it is extremely difficult to determine the incidence of many abnormalities with any degree of precision. An additional problem is that prenatal diagnostic ultrasound provides evidence that lesions, such as adenomatoid malformation, may "disappear" in the postnatal period.[9] Probably the most common malformation, although not strictly bronchopulmonary, is diaphragmatic hernia with a incidence of 1/2000 live births, of which approximately 40 to 50 percent will die in the early neonatal period.[10,11] Recently Bailey and colleagues [12] reviewed their experience over an 18-year period of infants coming to surgery for congenital bronchopulmonary malformation, excluding isolated diaphragmatic hernia. They identified 45 patients from birth to 13 years of age. Thirty-seven had solitary lesions, the most common of which were bronchogenic cyst, cystic adenomatoid malformation, congenital lobar emphysema, and pulmonary sequestration. Eight patients had two simultaneous anomalies and three had congenital diaphragmatic hernias associated with extralobar sequestration and adenomatoid malformation. Thoracotomy with excision of the lesion by lobectomy or pneumonectomy resulted in a 93 percent survival rate.

In perinatal or fetal pathology, malformations encountered are likely to be dominated by diaphragmatic hernias, tracheoesophageal fistulae, and adenomatoid malformations. As a group, various bronchial branching and lobation anomalies are also frequent, if only as incidental findings. Bronchogenic cysts and lobar emphysema are rarely encountered. Unlike the pediatric surgical experience quoted above, the lung abnormalities encountered in fetal and perinatal work are usually associated with malformations elsewhere.

Trachea

Tracheal abnormalities are not easily classified, in part due to the rarity of many of the anomalies. Further, many abnormalities involve defects in the cartilaginous skeleton of the trachea and detailed examination is difficult without the use of clearing techniques. This is not a routine practice and subtle abnormalities may well be missed. The normal tra-

chea has 22 transversely arranged, C-shaped cartilagenous rings, open at the back. They extend from the lower border of the larynx to the carina. Even in normal tracheas, however, there may be a significant number of incomplete or Y-shaped rings.[13,14]

AGENESIS

Tracheal agenesis is a rare, fatal anomaly thought to be caused by ventral or dorsal displacement of the tracheoesophageal septum.[15] It has been classified into as many as seven variants (Fig. 3-1)[16] based on

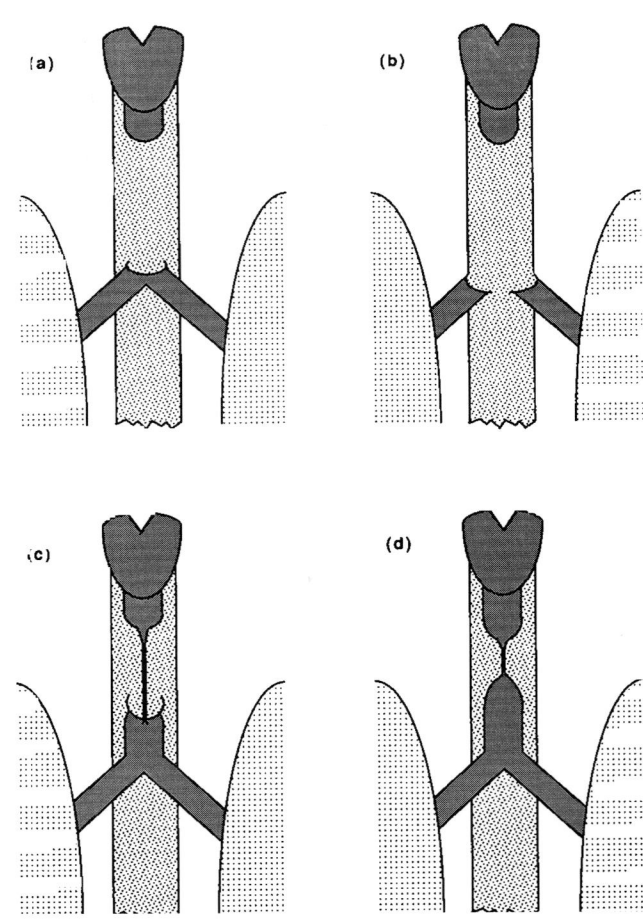

FIGURE 3-1
Tracheal agenesis. Diagram illustrates the main subtypes of tracheal agenesis. A. Long segment tracheal agenesis with broncheoesophageal fistula between the fused main bronchi and the esophagus (56 percent of cases). B. Long segment tracheal agenesis with direct fistulous connection between each bronchus and esophagus (10 percent) C. Bronchi fuse to form the lower trachea and an atretic band joins the proximal with distal trachea (10 percent). A fistula may not be present (5 percent) nor the atretic band (5 percent), D. Short segment tracheal agenesis linked by an atretic fibrous band (5 percent). Total absence of the trachea, bronchi, and lungs accounts for a further 8 percent of cases.

the length of the agenetic segment and the presence or absence of an esophageal fistula.

An atretic strand of fibrous tissue replaces the agenetic segment or there is an esophageal fistula to individual bronchi, fused bronchi, or the lower trachea. The most common variant, occuring in over 50 percent of cases, is complete absence of the trachea below the larynx and a fistulous connection between fused bronchi and the esophagus (Fig. 3-2). The other variants constitute between 5 and 10 percent of cases.[16] The lungs are frequently normally developed but, in the absence of a fistula, pulmonary histology may resemble an extralobar sequestration.[16] Rarely, the lungs are also agenetic.

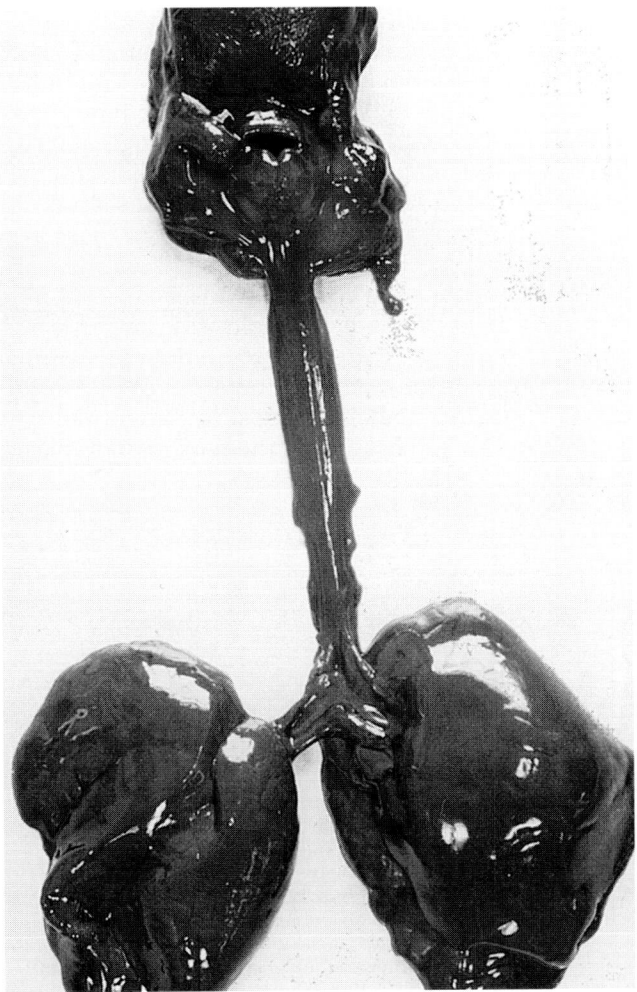

FIGURE 3-2
Complete absence of the trachea below the larynx. The fused bronchi, which have been opened, connect directly with the esophagus. *(Reproduced with permission from Keeling J (ed), The Respiratory System in Fetal and Neonatal Pathology, Springer-Verlag, 1993.)*

Extrapulmonary associations include cardiac and genitourinary system abnormalities.[15,17,18] The abnormalities seen are very similar to those observed in the VACTERL association (see below).[19,20]

ABNORMALITIES OF TRACHEAL LENGTH

In the presence of a congenitally short neck, a reduction in the number of tracheal cartilages is a common finding. The most common cause of brevicollis, in which a reduction in tracheal ring number is described, is the Klippel-Feil syndrome. This consists of a short neck associated with a low occipital hairline and decreased neck mobility, usually with cervical vertebral fusion.[21] Other conditions include various skeletal dysplasias, hypoplastic left heart syndrome, some chromosomal disorders, and neural tube defects.[22] Too many tracheal rings may be a component of some variants of tracheal stenosis (see below).

TRACHEAL STENOSIS

Tracheal stenosis typically presents as neonatal stridor or respiratory distress, but in the older infant, recurrent pneumonia may be a more prominent feature.[23] Tracheal stenosis can be divided into intrinsic abnormalities and stenosis due to extrinsic pressure (Table 3-1). Acquired stenosis secondary to the trauma of endotracheal intubation will not be considered.

Intrinsic Stenosis

Diffuse stenosis This can be subdivided into two broad subtypes. If there is a diffuse abnormality, the trachea is narrowed for most of or its entire length (Fig. 3-3 and 3-4). The luminal diameter in the newborn is often as small as 1 to 3 mm. The abnormality commences immediately below the cricoid and is limited to the trachea. The major bronchi are not involved. The major underlying pathology is posterior fusion of the cartilage ring.[24] Abnormalities that have been associated with diffuse stenosis include

TABLE 3-1
Classification of Tracheal Stenosis[12,15]

Intrinsic
Diffuse generalized hypoplasia (30%)
Diffuse, posterior cartilage ring fusion
Solid cartilagenous sleeve
Funnel-like tapering stenosis (20%)
Segmental stenosis (50%)
Extrinsic

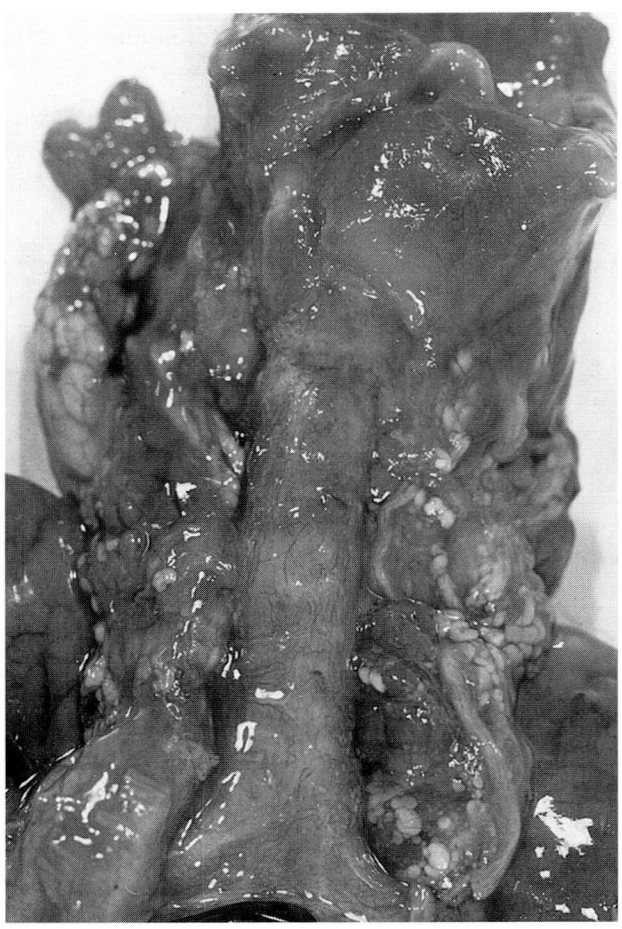

FIGURE 3-3
Posterior view of trachea with diffuse tracheal stenosis although the presence of a reduced tracheal lumen is not obvious from this view. No pars membranacea is visible (for comparison see Fig. 3-7*A*) and stenosis is associated with complete tracheal rings.

Fallot's tetralogy, agenesis of one lung, and diaphragmatic hernia.[25]

In the second variant of diffuse tracheal stenosis, the trachea is typically described as a solid or complete cartilagenous sleeve, although Davis and associates[26] show that this is a slight oversimplification. Rather than a solid tube, they argue that a rudimentary pars membranacea may be present posteriorly and that the trachea may also show other abnormalities and deformity. This can involve the bronchial tree and include an unusually long right main bronchus or a short left main bronchus together with pulmonary pseudoismerism.[21] It is not clear if the functional deficits in ventilation associated with this abnormality result from the tracheal stenosis or are due to the other features of the trachea such as an abnormal length, rigidity, and angulation.

Almost invariably, the diffuse cartilagenous sleeve variant of tracheal stenosis is reported in children with a craniosynostosis syndrome, including Crouzon's disease and Apert and Pfeiffer syndromes. This has led David and coworkers[26] to suggest it represents a fundamental mesenchymal defect in which normally discrete structures fuse. It is thus similar to the other mesenchymal defects present in these syndromes.

Funnel-shaped tracheal stenosis In this subtype the trachea immediately below the larynx is normal, but there is a gradual reduction in luminal diameter to a minimum immediately above the bifurcation.[24] Again, the main tracheal ring abnormality is a reduction in, or absence of, the pars membranacea.

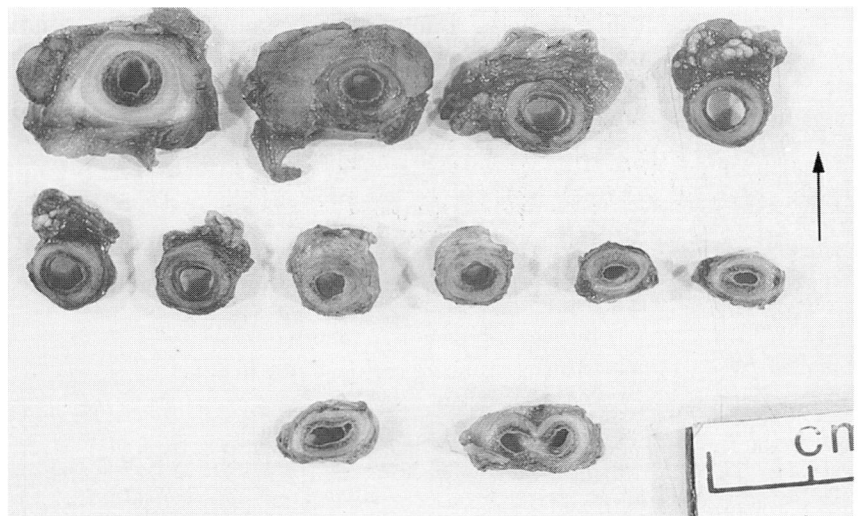

FIGURE 3-4
Same trachea illustrated in Fig. 3-3, transversely sectioned from cricoid to carina. In transverse section the trachea has more resemblances to "funnel-shaped" diffuse tracheal stenosis. The mucosa of the upper trachea is severely ulcerated from attempted intubation. A complete ring is clearly visible in the upper right hand corner and other blocks show cartilage posteriorly. The lumen is reduced to 2 mm diameter at its narrowest. All blocks are orientated in the same direction (arrow points anteriorly).

Absence leads to complete fusion of the tracheal cartilages. Funnel-shaped tracheal stenosis is particularly associated with a "sling" left pulmonary artery. In this condition the left pulmonary artery arises from the right pulmonary artery and reaches the hilum of the left lung by passing over the right main bronchus and between the trachea and the esophagus (Fig. 3-5). The lower trachea is typically stenotic with a complete tracheal ring.[21,27,28] Although it might be thought that the stenosis is therefore due to extrinsic compression, there does appear to be a primary abnormality of the trachea. This particular anomaly has been studied and subclassified in more detail by Wells and colleagues.[29] Funnel-shaped stenosis is rarely present in the absence of anomalies elsewhere; unilateral pulmonary agenesis is a particularly well described association.[25,30]

Segmental stenosis This is the most common form of tracheal stenosis and takes the form of complete cartilagenous rings ("napkin ring" cartilages)[13] or an hour-glass area of tracheal narrowing. In some instances, the stenosis is formed by a focus of muscular hypertrophy.[31] The abnormal segments are reported to be evenly distributed in the subcricoid, mid-trachea and supracarinal areas.[21] Association with a wide variety of other extrabronchopulmonary abnormalities is described and, indeed, is usual. Overlap of many of these associations with those seen in some of the less localized forms of stenosis suggests that although morbidity and mortality are heavily influenced by the length of stenosis, these subtypes do not form discrete entities.[25,30]

Extrinsic Stenosis

In neonates and infants, the most common cause of extrinsic tracheal compression is an abnormally situated or large blood vessel (Fig. 3-6). Clinical presentation is similar to intrinsic stenosis and includes stridor, recurrent chest infection, cough, respiratory distress during feeding, episodic respiratory distress, and cyanosis. Landing and Dixon[21] provide relative frequencies of these anomalies: vascular ring due to a double aortic arch (47 percent); vascular ring due to a right aortic arch with a left ligamentum arteriosum (20 percent); aberrant (retroesophageal) right subclavian artery (14 percent); anomalous innominate artery arising further to the left than usual (11 percent); anomalous left carotid artery, arising further to the right on the arch than normal (4 percent); sling (retrotracheal) left pulmonary artery (3 percent); and right aortic arch with aberrant left subclavian artery 1 percent).

TRACHEOMALACIA

Tracheomalacia implies an inadequacy or softness of the tracheal cartilages. This leads to collapse of the trachea at some point during the respiratory cycle, usually expiration. The condition may be segmental or diffuse. There is overlap with the cartilagenous abnormalities discussed above under tracheal stenosis and it could be argued that true "malacia" only occurs in association with some of the skeletal dysplasia syndromes.[13] Sometimes, as a component of the tracheal stenosis due to traumatic damage, tra-

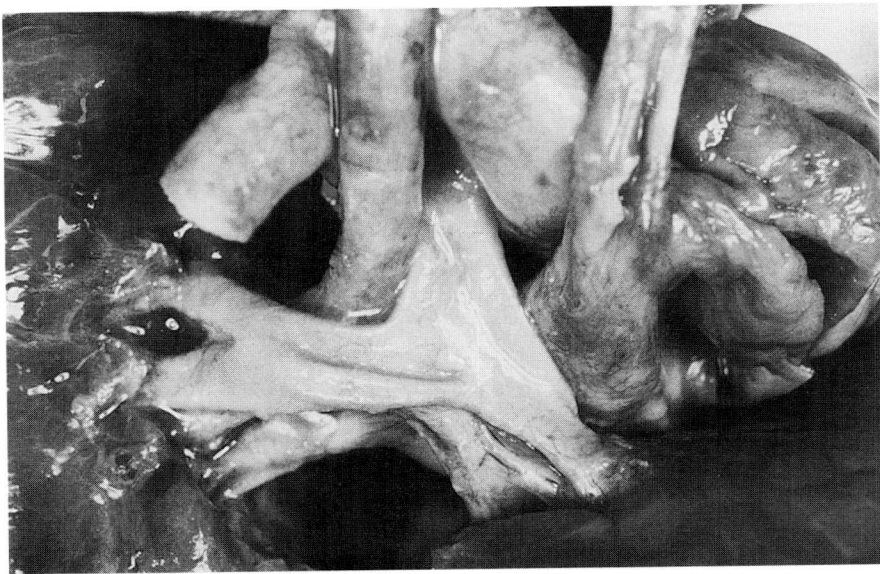

FIGURE 3-5
A "sling" left pulmonary arising from the right pulmonary artery and passing posterior to the right main bronchus and trachea to reach the left pulmonary hilum. The lower part of a diffuse tracheal stenosis with fusion of the tracheal rings posteriorly is present. *(Illustration kindly provided by Dr. J. Keeling, Royal Hospital for Sick Children, Edinburgh.)*

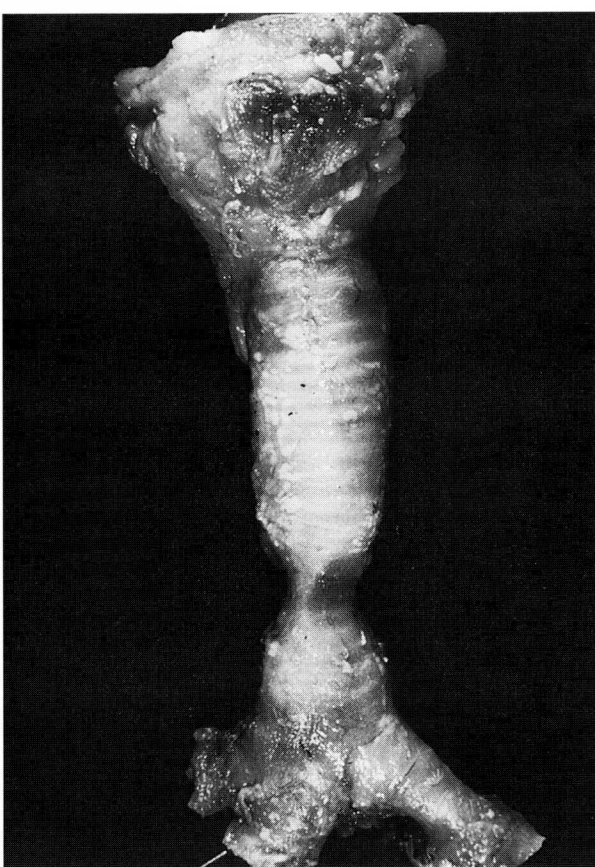

FIGURE 3-6
Tracheal stenosis due to extrinsic arterial compression. This baby had congenital heart disease with aortic arch anomalies. *(Reproduced with permission from Keeling J (ed),* The Respiratory System in Fetal and Neonatal Pathology, *Springer-Verlag, 1993.*

ratio of cartilage to soft tissue, in some instances as low as 2:1 (Fig. 3-7).[33,34]

Thus, there is a primary reduction in the tracheal cartilage length. The relative increase in the posterior myoelastic soft tissue allows an anteroposterior collapse of the airway during expiration.[33,34] The concept of tracheomalacia being a form of "immaturity" is not entirely inappropriate, however. In the lower trachea of a normal infant, at the trigone, the posterior membrane is relatively large and the tracheal profile may be flattened in the anteroposterior diameter (see Fig. 7A).[37] This feature is not present in the adult. The significance of this profile in recurrent respiratory symptoms is uncertain. Secondary tracheomalacia due to abnormalities of tracheal form and cartilagenous deficiency is well documented in association with tracheoesophageal fistulae.[38] Tracheomalacia is also seen in some examples of

cheomalacia can be "acquired" as a complication of mechanical ventilation.[32] Collapse of a structurally normal trachea may be associated with excessive expiratory effort as in asthma. Thus a primary tracheomalacia should only be considered after external compression and localized intratracheal airway obstruction has been excluded.[21]

Primary tracheomalacia, as delineated by Benjamin,[33] is an isolated tracheal anomaly. It presents usually with expiratory but occasionally with inspiratory stridor, cough, neck hyperextension, recurrent respiratory infection, and, rarely, "reflex" apnea.[34] Spontaneous improvement in some of the more mild variants over an 18- to 24-month period has suggested the pathology in tracheomalacia is an abnormally thin or floppy cartilage and represents a form of "tracheal immaturity."[35] However, the abnormality is more than just thin cartilage. For most of its length, the normal tracheal cartilage-to-soft tissue ratio is approximately 4.5:1, a ratio that remains fairly constant throughout childhood.[33,36] In tracheomalacia, there is a demonstrable reduction in the

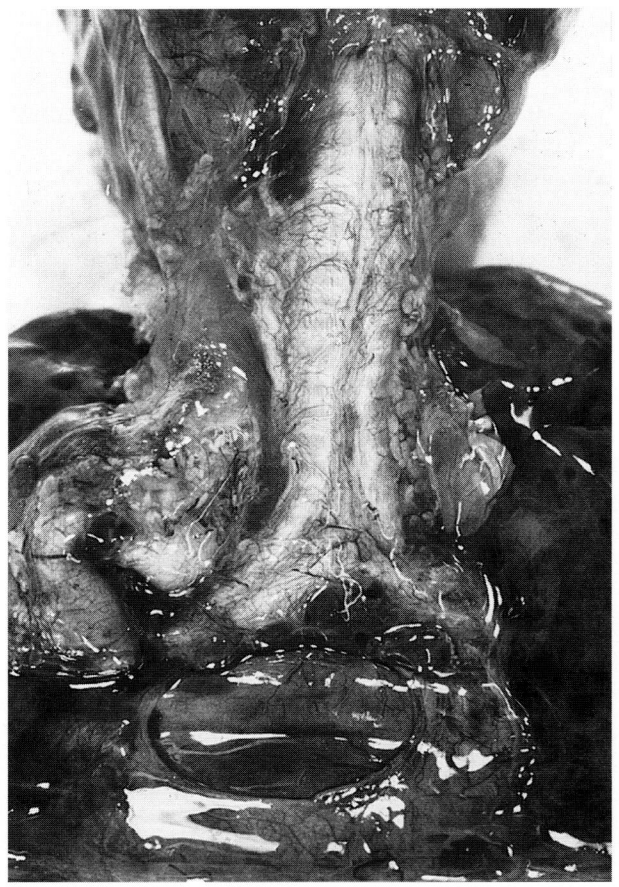

A
FIGURE 3-7
Posterior view of normal trachea (*A*) compared with trachea with tracheomalacia (*B*) The relative increase in the pars membranacea in the malacic trachea is clearly apparent especially distally.

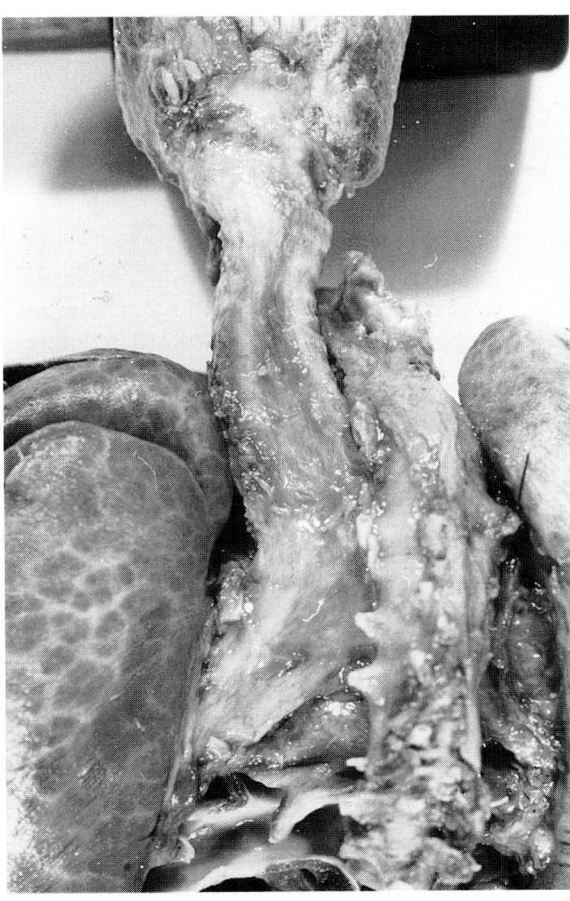

B

FIGURE 3-7 (*Continued*)

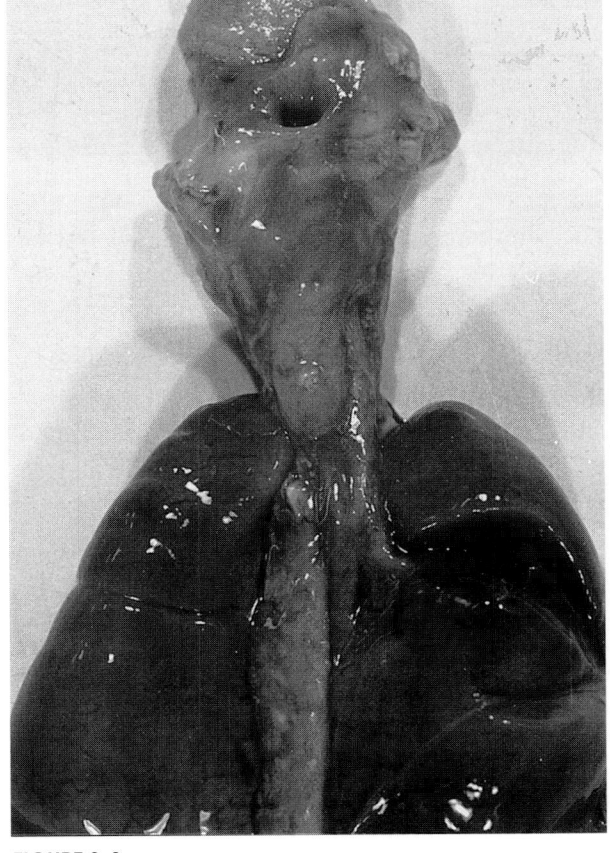

FIGURE 3-8
Posterior view of the most common variant of tracheoesophageal fistula. The upper esophagus terminates in a blind-ending sac. The lower esophagus forms a fistulous connection with the trachea, slightly to the left of the midline.

extrinsic compression by anomalous major arteries.[25]

TRACHEOESOPHAGEAL FISTULA

Tracheoesophageal fistula with esophageal atresia is one of the more commonly encountered abnormalities affecting the trachea. Incidences range from 1/800 to 1/5000 births.[39] It may cause maternal polyhydramnios or, in the neonate, excessive secretions or choking, usually in association with feeding.

Classification is into five main subtypes:

1. A short atretic esophagus with the lower esophagus arising from the lower trachea.[40] This is the most common variant, accounting for over 80 percent of cases (Fig. 3-8).
2. Esophageal atresia with no fistula
3. Tracheoesophageal fistula with no atresia
4. Esophageal atresia with fistula between a proximal atretic esophagus and trachea
5. Esophageal atresia with tracheal fistulae between both proximal and distal segments of esophagus

In a particularly rare variant there is a bronchoesophageal fistula. Tracheomalacia is a frequent accompaniment of the variants with a fistula. If there is esophageal atresia in the absence of a fistula, tracheal development is usually normal (see above).[41]

Tracheoesophageal fistula is a sporadic malformation with a very high rate of associated malformations. The figure is up to 70 percent in low birth weight infants.[42] It is frequently these other abnormalities that determine the outcome. In term infants, survival from isolated tracheoesophageal fistula is almost 100 percent.[43] Low birth weight carries a high risk of a poor outcome. The main association of tracheoesophageal fistula is seen in the VACTERL syndrome, a nonrandom association of **V**ertebral anomaly, **A**nal atresia, **C**ardiac, **T**racheo**E**sophageal fistula with **A**tresia, **R**enal anomalies, **L**imb abnormality (often radial). Bifid or hemivertebrae are the most common features of the vertebral column and the limb abnormality is usually a radial hypo- or aplasia. A ventricular septal defect is the most common lesion seen in the heart.

The embryologic basis of tracheoesophageal fistula and esophageal fistula is uncertain. The traditional concepts of an abnormality of tracheo-esophageal septum development have been recently questioned.[42,43] Further, it is suggested that there may be a fundamental difference in etiology between atresia variants with tracheoesophageal fistulae and those without.

TRACHEOBRONCHIOMEGALY (MOUNIER-KUHN SYNDROME)

Tracheobronchiomegaly is characterized by a trachea or main bronchi of more than three standard deviations greater than the normal diameter.[44,45] Investigation may demonstrate saccular bulging of the intercartilaginous membranes amounting to tracheal diverticulae, usually on the posterior wall. Presentation tends not to be until the fourth or fifth decade although childhood cases have been reported. Clinically there is a chronic cough with ineffective production of secretions and consequent repeated respiratory tract infections. Diagnosis is primarily radiographic or bronchoscopic. The underlying pathogenesis is unclear. An association with inheritable conditions such as Ehlers-Danlos syndrome, cutis laxa, and skeletal dysplasia[45–47] indicates that there may be an increased compliance in the trachea due to a primary defect in elastic tissue. The occurrence in siblings suggests the condition is an autosomal recessive.

OTHER CAUSES OF TRACHEAL OBSTRUCTION

A variety of masses and intraluminal obstructions have been recorded particularly in the newborn, including mucous retention cysts[48] and fibrous webs.[49] Perhaps not strictly a congenital lesion, tracheobronchopathia osteoplastica represents an overgrowth of tracheal cartilage possibly with ossification into the lumen. Although often an incidental finding in later life or autopsy, it can cause obstructive problems particularly in the smaller airways.[50]

Bronchus

BRONCHIAL ATRESIA

Atresia of the bronchus is a rare anomaly but has been the subject of a number of reviews.[51–53] It typically affects the sexes equally and presents most commonly in the second decade.[51] Diagnosis may be incidental following radiology or there may be a history of recurrent chest infection or dyspnea.[51] Clinically the usual sign is reduced breath sounds over the affected segment or lobe. The most common

affected lobe is the left upper, but atresia of the right upper and right lower lobes is also well recorded.[51] The segmental bronchus is the more usual level of atresia, but subsegmental and lobar bronchi can also be affected.

The distal lung is said to show emphysema, free of anthracotic pigment, and as a result of its anatomic isolation, is hyperinflated. The mechanism of this change is uncertain. Pleural thickening and fibrosis are a result of recurrent infection. There is a deficiency of bronchi and vessels in the affected lobe as well as an absence of segmentation and interlobular septae. Histologically, the alveoli are enlarged in the involved parenchyma but there is both a relative and absolute decrease in number. This alveolar hypoplasia is thought to be due to decreased ventilation and blood flow. A similar explanation is given for the hypoplastic compressed lung in nonatretic segments. In a few cases there may be multiple segments involved with coexistent sequestration. Collateral ventilation occurs by the channels described in Chap. 2. It is likely that the interbronchiolar channels are the most important.

The pathology of the atresia itself is variable. The bronchus may be obstructed by circumferential or eccentric luminal fibrosis possibly secondary to inflammation; the timing of the inflammatory reaction might be in utero or postnatal.[16] Thus, in some instances, bronchial atresia might not strictly represent a congenital lesion. In some cases, there is an atretic fibrous cord at the site of obstruction[51] or the lumen may be occluded by relatively normal-appearing bronchial mucosa (Fig. 3-9). The distal atretic bronchus may appear as a series of mucous-filled cystic dilatations amounting to a mucocele, which should not be confused with bronchiectasis. The fact that alveolar development has occurred at all suggests that the origin of many atresias may be around 15 weeks' gestation. The atresia may possibly be a vascular accident. A case of bronchial atresia associated with a bronchogenic cyst has been used to provide evidence of a much earlier developmental event,[54] perhaps as early as 5 weeks' gestation. Atresia of the main bronchus may be associated with cystic adenomatoid malformation type 3 (see below).

BRONCHIAL STENOSIS

As with tracheal stenosis, bronchial stenosis can be intrinsic or due to extrinsic compression. Acquired stenosis due to posttraumatic fibrosis[55] may also occur. Compared with extrinsic bronchial stenosis, congenital intrinsic stenosis is extremely rare and amounts to a very occasional case report.[56] In this

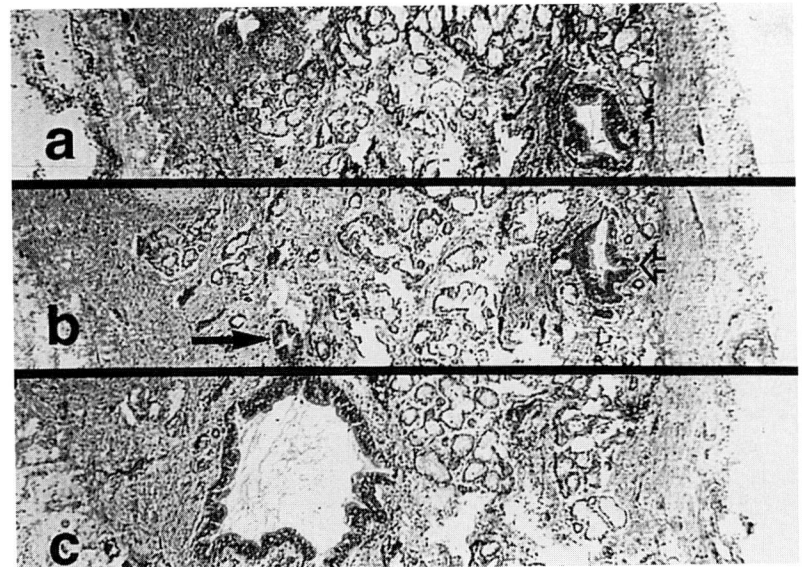

FIGURE 3-9
Montage of serial sections progressing from the distal (*a*) to proximal (*c*) part of a right main bronchial atresia. Sections show lumen of distal bronchus in *a*, both distal (open arrow) and proximal (solid arrow) lumen in *b* and only proximal bronchial lumen in *c*. The lumen is occluded by normal-appearing bronchial mucosa and the proximal and distal bronchial lumens are no longer in continuity. The right lung in this case was an example of adenomatoid malformation type 3.

instance, the stenosis was associated with anomalous cartilage segmentation and secondary squamous metaplasia.

Extrinsic stenosis is most commonly associated with congenital heart disease (Fig. 3-10). Compression occurs mainly when pulmonary arteries, enlarged secondary to pulmonary hypertension, compress the superior aspect of the left upper lobe bronchus. The left main bronchus may also be compressed in conditions associated with an enlarged left atrium.[12] Other causes include bronchogenic cyst and teratoma.[16,57]

BRONCHOMALACIA

Similar to tracheomalacia, bronchomalacia can be considered a cause of bronchial stenosis; it is due to an abnormality of bronchial cartilage leading to airway collapse during respiration. The most common cause is acquired damage to the cartilage from mechanical ventilation. In congenital disease, there may be a widespread cartilagenous deficiency that may lead to childhood bronchiectasis,[58,59] but segmental bronchomalacia is also described.[60,61] The pathology of bronchomalacia is poorly described. Gupta and coworkers attribute an intrapulmonary segment of bronchomalacia to separated and flattened cartilage plates.[60] MacMahon illustrates an abnormal main bronchus with cartilage formed of isolated islands rather than a well formed plate (see Fig. 3-10B)[61]; the extrapulmonary main bronchus should have a horseshoe-shaped cartilage similar to the trachea. Bronchomalacia is not infrequently associated with other anomalies[60,62] and various skeletal dysplasias.[21] Familial isolated bronchomalacia is

also recorded.[63] The lung distal to the focus of malacia may demonstrate congenital lobar emphysema.

ABNORMAL BRONCHIAL ORIGIN AND BRANCHING

Abnormal bronchial origin and branching are not uncommon occurrences and usually incidental findings (Fig. 3-11). Atwell[64] described a variety of abnormalities based on 1200 bronchograms. Additional major bronchi were confined to the right upper lobe exclusively. The anomalies included double right upper lobe bronchi and an origin of the right upper lobe bronchus from the trachea. Fewer cases with a reduction in major bronchi were found and again these exclusively involved the right upper lobes. Minor abnormal branching patterns were more common in this study, notably trifurcation of the left upper lobe bronchus accounting for 10 percent of cases. Branching abnormalities are frequently associated with other anomalies such as the association of a "bridging" bronchus with a sling left pulmonary artery[29] (see Fig. 3-5) and tracheal stenosis. The "bridge" is formed by the bronchus supplying the right lower or middle lobes crossing the mediastinum from its origin on the left main bronchus. Although frequently incidental findings, abnormal origins or branching patterns such as tracheal or accessory cardiac bronchi can be associated with recurrent pneumonia.[65,66]

Bronchial Isomerism Syndromes

Isomerism might be regarded as a curiosity in which the abnormal bronchial branching leads to two similar lungs, that is, two bilobed "left" lungs or two

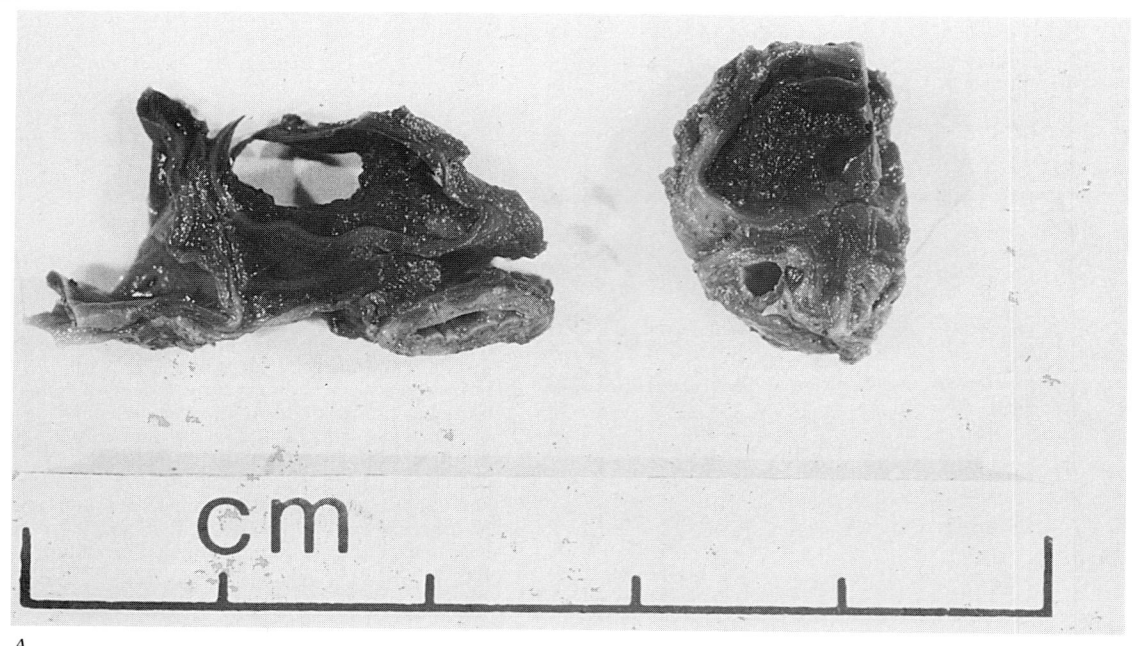

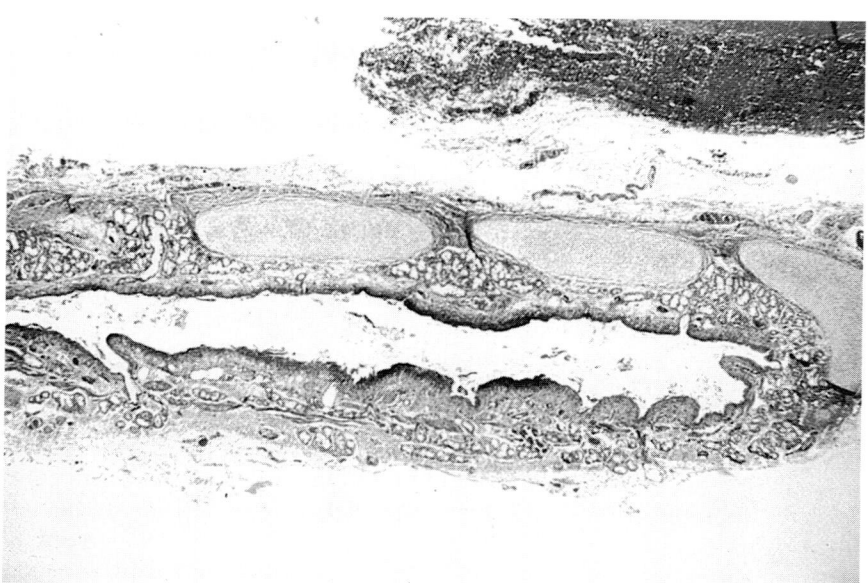

FIGURE 3-10
A. Main bronchi from a case of severe
congenital heart disease with absent
pulmonary valve and ventricular septal
defect. Both pulmonary arteries are
grossly enlarged. There is severe
compression of the right main bronchus
with flattening of the bronchial profile.
The left main bronchus is stenotic but
retains a round profile. *B.* The histology
of the right main bronchus shows
bronchomalacia with only isolated
islands of bronchial cartilage.

trilobed "right" lungs. It should be recognized that
isomerism does represent a fundamental developmental anomaly. In most cases, pulmonary isomerism is associated with multiple other defects, most
notably heart and spleen, and the lung abnormality is
of relatively little significance. These syndromes
appear to reflect an inability of the developing
embryo to establish early left-right asymmetry. This
process is beginning to be linked to a variety of chromosomal loci.[67,68]

A large number of malformation complexes

involving heterotaxy have been described in the literature,[69] but five main groupings have been delineated by Landing and colleagues.[21,70]

Ivemark asplenia syndrome. Occurring mainly in males, this consists of intestinal malrotation,
symmetrical liver, asplenia, and, usually, bilateral
three-lobed ("right") lungs with epiarterial bronchi
for both lungs. Major cardiac anomalies are typical
including transposition and anomalous pulmonary
venous return. The heart will have atria both of
which are "right" in type.

66

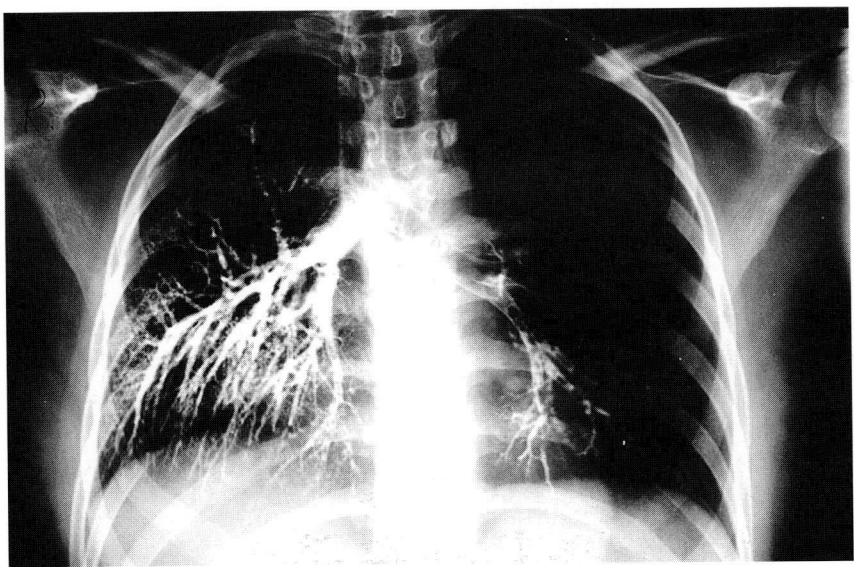

FIGURE 3-11
Bronchogram showing abnormal
bronchial tree.

M-anisosplenia (presence of one or more larger or smaller spleens). Infants have a relatively normal visceral status with congenital heart disease and bilateral three-lobed lungs. The "M" denotes the majority of patients are male.

F-anisosplenia. The female (F) counterpart, but with bilateral two-lobed ("left") lung pattern. The associated congenital heart disease is usually a double outlet right ventricle.

Polysplenia syndrome (up to 14 uniform small spleens). This consists of a bilateral left lobe pattern with intestinal malrotation, a symmetrical liver, and congenital cardiac disease. The heart will typically demonstrate left-sided symmetry. The sex incidence is equal.

O-anisosplenia. Multiple spleens are present with bilateral two-lobed lungs, intestinal malrotation, and congenital heart disease. This is typically a common atrioventricular valve or double outlet right ventricle. There is an equal sex ratio.

In general, most of these syndromes are of sporadic inheritance but autosomal recessive, dominant, and sex-linked patterns of inheritance have been recorded. Overlap in syndromes occurs.[71]

BRONCHOBILIARY FISTULA

This is a rare anomaly in which the right mainstem bronchus and left hepatic duct are connected by a fistula that passes with the esophagus through the diaphragm.[72] The embryologic basis of this condition is uncertain, but it may represent a form of upper gastrointestinal tract duplication.[73] There is production of bile-stained sputum with recurrent chest infec-

tions. The clinical picture may suggest cystic fibrosis, but chest infections are uncommon in this condition before the age of 3 months.

Lung

Pulmonary agenesis has been the subject of a variety of classifications with modification but can be classified as below.

1. Bilateral complete agenesis (aplasia)
2. Unilateral agenesis
   a. Complete absence of bronchi, alveolar tissues, and blood supply
   b. Rudimentary bronchus but no pulmonary tissue
   c. Rudimentary bronchus invested by ill-developed pulmonary tissue
3. Lobar agenesis and other abnormalities

BILATERAL COMPLETE AGENESIS (APLASIA)

In this very rare anomaly the laryngotracheal bud either develops as far as the larynx[74] or the trachea.[75,76] In the latter case the pulmonary artery joins the aorta by way of the ductus arteriosus. Bronchial arteries and veins are usually absent. In some instances the pulmonary and tracheal agenesis are associated (see above). There may be asplenia as well as other anomalies affecting the esophagus and face.[77,78]

Incomplete agenesis occurs when bronchial development is present but the most distal air passages and alveoli fail to grow (Fig. 3-12).

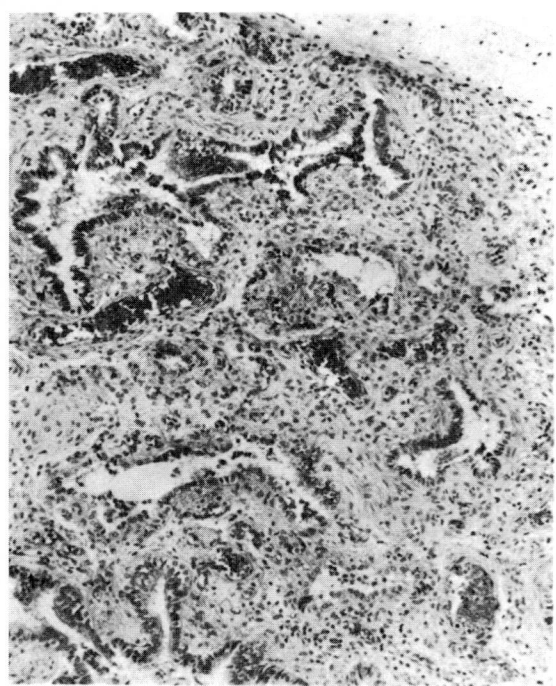

FIGURE 3-12
Pulmonary agenesis showing the pleural surface in the top right-hand corner. Beneath the pleural surface are small ill-developed and scanty bronchi but no respiratory bronchiolar or alveolar tissue. (Medium magnification.)

UNILATERAL PULMONARY AGENESIS (APLASIA)

This anomaly is commoner than bilateral agenesis but is still rare.[79–81] It appears to be a little more common in females and the left and right sides are affected equally. It is not invariably fatal, particularly in the absence of other anomalies, and adult examples of unilateral agenesis are recorded.[82] The trachea may pass smoothly into the remaining main bronchus and lung (group a) or a rudimentary bronchus to the absent lung may still be present (group b). Although the literature does not use the distinction consistently, the former is sometimes labeled as agenesis and the latter, aplasia.[21,81] This division has formed the basis of the classification proposed by Schneider[83] outlined above. It is argued that the group (c) cases in Schneider's classification, in which rudimentary lung is present, do not strictly constitute pulmonary agenesis but hypoplasia.[79] These cases are particularly rare and only 8 examples were present in the series of over 100 cases collected by Maltz and Nadas.[79]

The pulmonary artery and vein on the side of the agenetic lung are usually absent or hypoplastic.[84] Rarely, unilateral agenesis is associated with tracheal agenesis and bronchoesophageal fistula. Unilateral agenesis may be associated with generalized fetal hydrops[85] or with a wide variety of other anomalies.[21] These include cardiovascular abnormalities such as septal defects and anomalous pulmonary venous return.[86] Where there is an absent left lung, vascular compression of the trachea or bronchus between a posterior pulmonary artery and the anterior aorta may occur. Skeletal abnormalities, including vertebral defects, are common and it has been suggested that an absent lung may be considered a part of the VACTERL association (see above).[87]

The majority of cases have been discovered at autopsy within the first year of life, but unilateral pulmonary agenesis may be asymptomatic or cause tachypnea.[21] There is an increased incidence of patent ductus arteriosus due to persistence of the sixth left arch.[80] Where there is a rudimentary bronchus, especially in childhood, the main problem relates to a solitary lung with poor resistance to infection.[88] An important factor in these repeated infections is the blind bronchial stump giving rise to a "spillage" pneumonia.[21] In the remaining lung, morphometry indicates that the alveolar number is increased and "compensates" for the alveolar loss from the agenetic lung.[81]

LOBAR AGENESIS AND OTHER LESSER CONGENITAL PULMONARY ANOMALIES

Cases with absence of the right middle lobe[21] or left lower lobe associated with stenosis of the left main bronchus have been described. Anomalous bronchial distribution usually causes no disabilities and falls into the province of anatomic variation.

Congenital pulmonary anomalies of pulmonary lobes are varied (Fig. 3-13) but include common abnormalities such as the azygos lobe, cardiac lobe, double-lobed upper lobe of left lung in association with Fallot's tetralogy, and isomerism of the right lung.[89] An azygous lobe is a portion of an upper lobe of the right lung growing medial to the right posterior and common cardinal veins. It may rarely be seen on the left side of the chest. A cardiac lobe occurs when the anterior basal segment of the lower lobe of the right lung is separated from the rest of the lobe.

A horseshoe lung is an uncommon congenital abnormality caused by the partial fusion of the bases of both lungs behind the pericardial sac (Fig. 3-14). In the absence of other abnormalities it is usually asymptomatic, but it is described in association with wide variation in pulmonary vascular distribution and a few other anomalies.[90–93] Most cases have been associated with right lung hypoplasia and the scimitar syndrome, but a recent case of horseshoe lung and left lung hypoplasia has been recorded.[94]

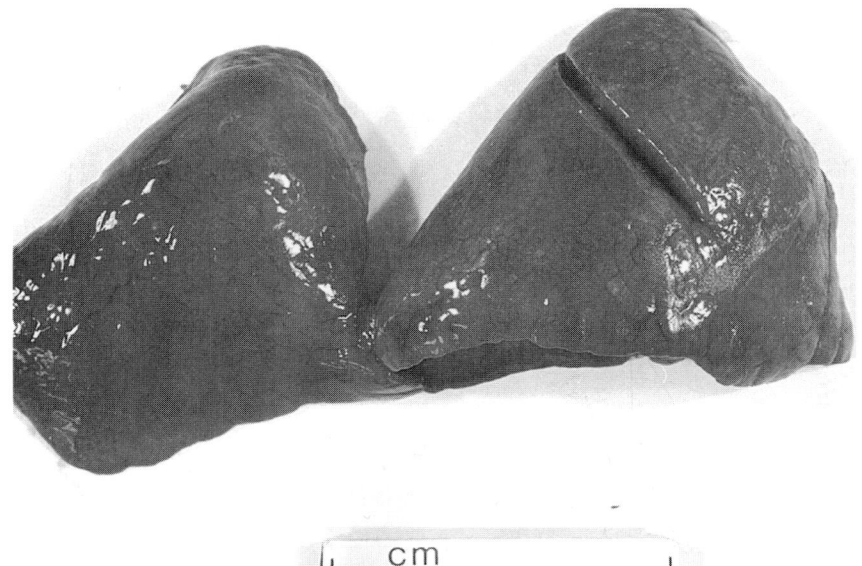

FIGURE 3-13
Abnormal pulmonary lobation with a single left lobe and partial formation of two lobes on the right. Although of no functional significance and sometimes an incidental finding, not infrequently they are associated with abnormalities elsewhere.

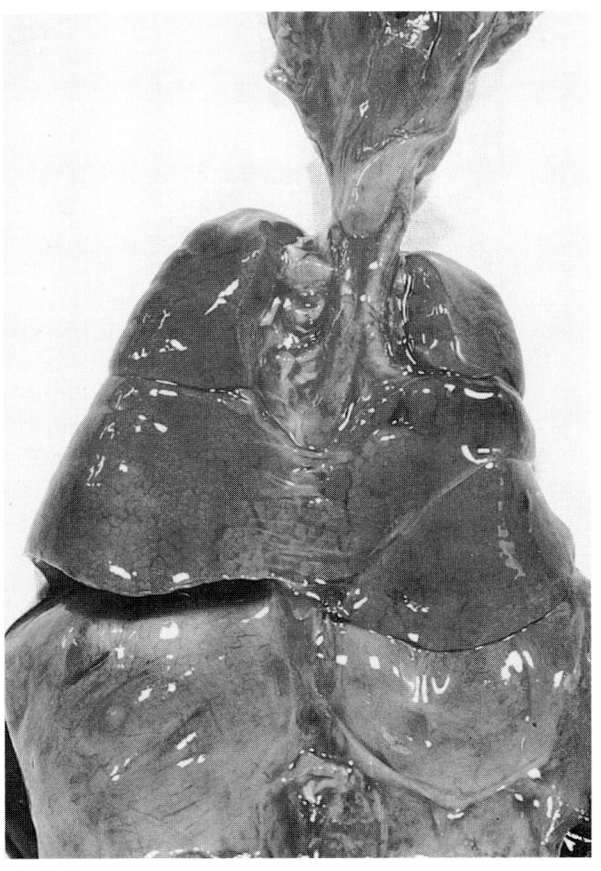

FIGURE 3-14
Horseshoe lung with fusion of lower lobes behind the esophagus. In this case, the lung anomaly was associated with a tracheoesophageal fistula.

Herniation of the lungs into the neck (ectopia) has been described in iniencephalus, Klippel-Feil syndrome, and the "cri du chat" (5-p) syndrome.[95–98] The cause of the pulmonary herniation is unknown although there does appear to be a reduction or absence of fascia in this area. An unusual herniation of the lung into an extrathoracic location, "bagpipe" lung, has also been described in a 16-week gestation fetus by Bridger and colleagues.[99] Again, the underlying mechanism is unknown but is probably secondary to an early intrauterine event.

PULMONARY CYSTIC DISEASE

Classification of cystic lung disease is far from ideal, but at least the potential to present as a form of pulmonary cyst or pulmonary cystic disease is a feature by which many of these conditions can be approached. Although acquired cystic disease is included in Table 3-2, consideration of these lesions will not be included in this chapter. A recent comprehensive review of bronchogenic cysts in adults has been published.[99a]

Bronchogenic Cysts

Bronchogenic cysts are congenital cysts most commonly found in the anterior mediastinum or around the hilum (Fig. 3-15).[100] Presentation is often as an incidental radiologic finding although the cysts may cause bronchial obstruction especially in the very young (Fig. 3-16). In adults symptoms may be related to infection of the cyst. The sex incidence is equal at all ages, and the majority of cysts present in the first decade. They are more common in the left lung. They have been described in association with other pul-

TABLE 3-2
Classification of Lung Cysts

Congenital Cysts
    Bronchopulmonary malformation
        Bronchogenic cysts; (peripheral cysts)
        Pulmonary sequestration; bronchopulmonary foregut malformations
        Congenital cystic adenomatoid malformation; polyalveolar lobe
    Congenital lobar emphysema
    Lymphangiomatous cyst/lymphagiomatosis
    Other cysts, e.g., enterogenous cysts
Aquired Cysts
    Tension cysts in infancy due to
        Infection (see Chap. 7)
        Hypoplastic bronchus
        Compression of bronchus by an enlarged pulmonary artery
        Foreign body (see Chap. 10)
    Emphysematous bullae and giant air cysts of adult life.
    Healed abscesses (see Chap. 10)
    Parasitic cysts
        Hydatid (see Chap. 9)
        Paragonimiasis (see Chap. 9)

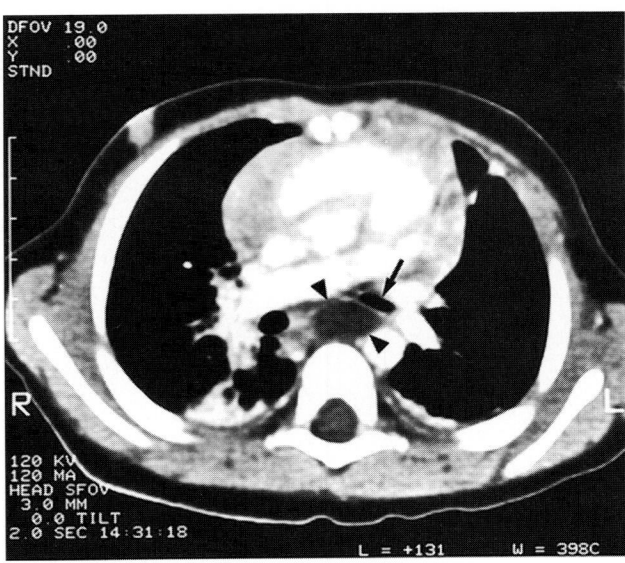

FIGURE 3-16
Computed tomography scan of bronchogenic cyst (*arrowheads*) demonstrating compression of the left main bronchus (*arrow*). *(We thank Dr. K. McHugh, Department of Radiology, Oxford Radcliffe Hospital Trust, for providing the scan for this figure.)*

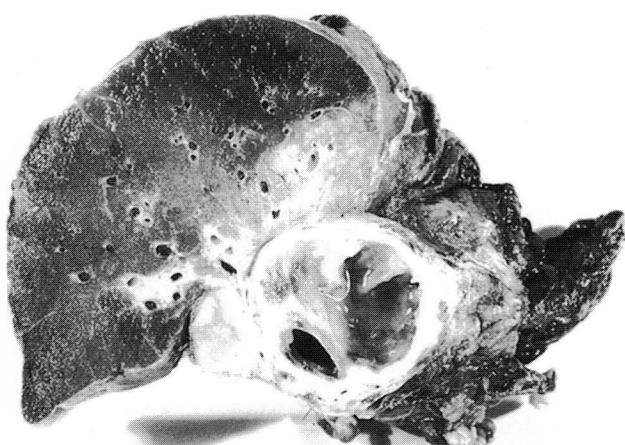

FIGURE 3-15
Bronchogenic cyst with a thick fibrous wall and containing inspissated mucus.

monary abnormalities, often coexisting with other congenital pulmonary defects such as bronchial atresia.[54,101] There is no association with congenital cystic disease in other organs.

Besides the more usual location, bronchogenic cysts are well recorded in a paratracheal location in the neck,[102] esophageal wall,[103] pericardium,[104] and a number of locations below the diaphragm.[105,106] Unusual locations include in the base of the tongue[107] and subcutaneous tissues.[108]

Peripheral[109] or intrapulmonary bronchogenic cysts have also been reported and may be important surgically because of secondary infection. A clear distinction from abscesses or other acquired pathology may be difficult. They are lined by respiratory-type epithelium and small cartilage plates may be present in the wall. Communication with the bronchial tree, it is argued, may predispose to infection.[109] Stocker[16] believes many of these lesions represent examples of type 1, congenital cystic adenomatoid malformations (see below).

Clear diagnostic criteria are required especially for diagnosis of bronchogenic cysts in unusual locations. The cysts are unilocular and, at least in older patients, may be up to 10 cm in diameter and are not in communication with the tracheobronchial tree. Attachment, but not communication, with the tracheobronchial tree should be found. If a communication is noted, particularly when multiple, then other pathology such as an abscess should be suspected. The cyst fluid may be clear or, if infected, contain pus

FIGURE 3-17
Histologically the wall of the bronchogenic cyst should contain cartilage and mucous glands and is lined by respiratory epithelium. Loss of epithelium is common in association with infection and the cartilage may need to be searched for.
(Medium magnification.)

and hemorrhage in a thickened fibrous wall. The lining of the cyst is respiratory in character, sometimes with focal squamous metaplasia. This is not specific to bronchogenic cysts because esophageal cysts may also demonstrate these epithelia. Perhaps the most reliable criterion is the presence of cartilage in the wall,[110] but this may require serial sections for detection (Fig. 3-17). Seromucinous glands and fibromuscular connective tissue are also present in the wall.

Bronchogenic cysts are thought to arise as abnormal "late" buds from the primitive foregut but to explain some of the more unusual locations, the concept of "migrating bronchogenic cysts" is invoked.[107] The failure of separation of primitive foregut and notocord may explain the occurrence of intradural bronchogenic cysts.[111]

PULMONARY SEQUESTRATION AND BRONCHOPULMONARY FOREGUT MALFORMATION

The classification and definition of bronchopulmonary malformations such as sequestrations might be considered a relatively simple task, but a review of the literature over the past 20 years or so, quickly dispels this notion. Recognition that many lesions may coexist, suggesting overlap, or do not fit into simple concepts of embryogenesis has led various approaches to these malformations to be recommended.

Classification and Embryogenesis

It is beyond this chapter to consider the various theories on the embryogenesis of bronchopulmonary abnormalities in detail; this section will attempt to give no more than a brief outline of the problems and the relationship to classification. Some of the more recent papers that review this subject in more detail, and that sometimes present their own solution to the confusion, are given in the references.[112–121]

Much of the early literature uses the term "accessory lung" or "lobe" to label pulmonary tissue that is separate from the main pulmonary mass but that occurs within the lung or below the diaphragm. Many of these were "true" accessory lungs and connected to parts of the gut, including the esophagus or stomach, by an accessory bronchus. The term "sequestration" was originally introduced by Pryce[122] in 1946 to label a disconnected mass with an anomalous pulmonary supply and has gradually replaced the notion of accessory lobes or lungs. Sequestrations were divided into intralobar (ILS) and extralobar (ELS). The many case reports that followed detailed a wide variety of arterial, venous, and bronchial connections. This led to difficulty in incorporating them into the simple concept of ILS or ELS with or without gastrointestinal tract communication. Gerle and colleagues[112] suggested the term bronchopulmonary foregut malformation to describe those lesions with gastrointestinal tract fistulae. This was used initially to include all lesions of sequestration type but currently it describes those lesions with only gastrointestinal tract communication.

An implication in the use of such similar terms as ILS and ELS is that there is inevitably a close developmental relationship between them. There is little doubt the latter is a malformation and various theories have been used to explain its origin, such as the concept of an outpouching that develops separately and later than the primitive lung bud (accessory lung bud). Alternatively, it might represent a separation from the developing lung that becomes invested by its own pleura.[115] ILSs, however, are not always clearly malformations and good evidence suggests that many may result from repeated infection (see below). A developmental origin, however, is suggested by the occurrence of foregut fistulae with both ILS and ELS, that is, a form of overlap. This persuaded Hruban and coworkers[121] that there must be a common pathogenesis to all three lesions.

Clements and Warner[120] proposed the term "malinosculation" (Latin: *mal*-abnormal; *in*-in; *osculum*-mouth) as the establishment of (abnormal) communications by means of small openings or anastomoses. They applied this term especially to the establishment of such communications between existing blood vessels or other tubular structures that came into contact. The term malinosculation describes a congenitally abnormal connection or opening of one or more components of the bronchopulmonary-vascular complex.

The classification has a three-step construction. The first describes the abnormality of the bronchopulmonary airway or arterial blood supply or both. Thus, the abnormality may be either in the bronchopulmonary airway but with a normal pulmonary arterial supply. This is bronchial or bronchopulmonary malinosculation. At the other end of the spectrum is an anomalous arterial supply to an area of lung with a normal bronchopulmonary airway. This is arterial pulmonary malinosculation. There are grades in between that are termed bronchoarterial malinosculations. The second step in the classification defines the associated anomalies of venous drainage. The final stage is to then describe any associated abnormality of the lung parenchyma.

The major advantage of this approach is that it enables logical description of an abnormality however complex or unusual and will allow inclusion of lesions such as tracheal agenesis. Further, it can be used as a basis for understanding the embryogenesis of the malformations depending on timing and severity of the insult to the developing lung bud. To date, despite the logic of this approach, this concept and nomenclature are not widely used. This is perhaps because it is potentially all too embracing, and because the term is somewhat ungainly.

In this chapter, a more traditional approach is used in which the terms ILS and ELS are used as conventionally defined. Bronchopulmonary foregut malformation will refer specifically to those lesions in which there is a communication with the foregut.

Sequestration

Pulmonary sequestration is defined as a mass of abnormal lung tissue that does not communicate with the tracheobronchial tree through a normally connected bronchus and receives its blood supply from a systemic artery.[21] It is subdivided into two main subtypes: ELS and ILS.

Extralobar sequestration. ELS is the rarer of the two types and is recognized as an isolated mass of lung invested in its own pleura. In many of the earlier reports it has been described under the term accessory lung or Rokitansky lobe.[21]

In an analysis of 52 examples, Stocker,[16] has described the more frequent characteristics of ELS: age at diagnosis under 1 month; male-female ratio of 3:1; left side twice as commonly affected as right, with the most common site between lower lobe and diaphragm; arterial supply—thoracic aorta followed by abdominal aorta; and venous return variable. At surgery, any combination of arterial or venous drainage needs to be considered when operating on sequestrations.[117]

Presentation is usually in infancy but, increasingly, pulmonary lesions later identified as ELS are being identified in utero by prenatal diagnostic ultrasound. Antenatal evidence of ELS includes fetal hydrops and maternal polyhydramnios. Clinically, ELS may be suspected in children, usually under the age of 1 year, who present with respiratory distress. The lesion may be an incidental radiologic finding in older children or even adults, where there is a triangular-shaped density or densities at the costophrenic angle.

Macroscopically an ELS is an isolated mass of lung tissue, often cystic, lying anywhere from the neck to the diaphragm (Fig. 3-18). Occasionally they may be intradiaphragmatic, below the diaphragm, or even lie in the pericardial cavity.[124] By definition, there is always a complete pleural investment. Surface lymphatics may be prominent.

Histologically there are small but often dilated and cartilage-bound bronchi, together with focal development of alveolar tissue. In some cases the dilated bronchi may extend out to the pleura with little alveolar development. The bronchioles and alveolar ducts are abnormally shaped and branch at right angles rather than dichotomously.[125] The bronchi do not connect with the tracheobronchial tree. The lymphatics are ectatic and there may be lymphangiectasia resembling congenital lymphangiectasis.[126] These ectatic lymphatics result from disruption of the normal efferent lymphatic pathways. Arteries supplying an extralobar sequestration are systemic in nature although thin-walled pulmonary arteries may also be found. Because the bronchi fail to drain normally, secretions are retained, with consequent recurrent infection and production of a cystic and

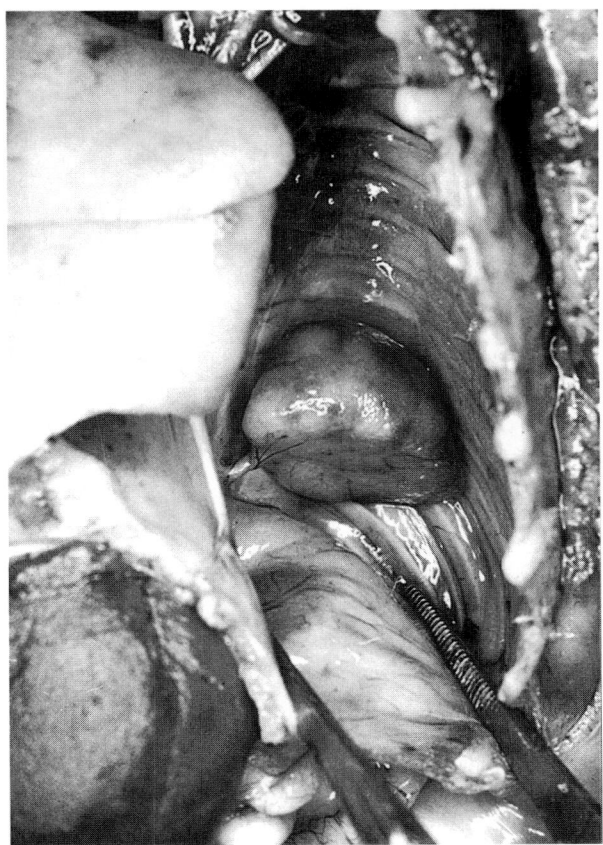

FIGURE 3-18
Extralobar sequestration in the left pleural cavity with arterial communication with the descending aorta. *(We thank Professor P. J. Berry, St. Michael's Hospital, Bristol for providing this figure.)*

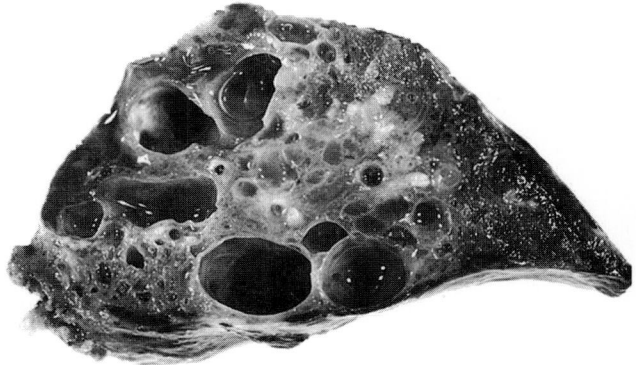

FIGURE 3-19
Extralobar sequestration with marked cystic change on section. Histologically, there were changes of adenomatoid malformation type II.

fibrotic lesion. In nearly half the cases there is a well formed bronchus at one edge of the resected specimen.[126] Congenital cystic adenomatoid malformation (CCAM) and ELS may coexist (Fig. 3-19),[112] the CCAM being of the type 2 or 3 (see below). It is suggested that this typical association results from a common point in embryologic development before 6 weeks' gestation.[113]

Associated abnormalities are present in approximately 65 percent of cases,[16] the more common ones being CCAM (considered below), diaphragmatic hernia,[114,115] and pulmonary hypoplasia.[127] Other bronchopulmonary abnormalities are also recorded[123,128] as are cardiovascular abnormalities[115] and pectus excavatum.

Intralobar sequestration. ILS lies within the visceral pleura and is usually seen in the posterior basal segment of the lower lobe and receives its blood supply from the thoracic or upper abdominal aorta or from an artery arising from one of these portions of the aorta (Fig. 3-20). The venous drainage is via the pulmonary veins of the affected segment or may occasionally join the azygos veins.[129] Fifty-five per-

cent of cases are seen on the left side[126] and bilateral involvement has been reported. Rarely ILS is seen in the upper lobes[129] or in the lingula.[130] ILS is rarely associated with other anomalies.

Clinically, patients often present over the age of 20 with cough, sputum, and recurrent chest infections. There is a slight male predominance. In about 15 percent of cases, ILS is asymptomatic and many cases may be detected on routine chest radiographs as a cystic area or a discrete mass. Bronchography shows ectatic bronchi near the lesion but none entering it.[116] Computed tomography (CT) scanning may confirm the cystic nature of the mass as well as demonstrating the systemic feeding vessel. An aortogram is the method of choice to make the preoperative diagnosis of sequestration as well aiding the surgeon in establishing the origin of the supplying arteries.

Macroscopically the abnormal segment presents as an atelectatic mass of lung or may be replaced entirely by a collection of cysts, which contain thin to viscid yellow-white fluid or sometimes gelatinous material. There is pleural fibrosis and adhesions to adjacent structures. The blood supply is often by a large elastic artery. Because of repeated infection the cystic cavities may appear bronchiectatic but bronchiectasis does not have a systemic blood supply and is rarely as localized as an ELS. The older the patient the greater the degree of fibrosis and cystic change because of the repeated infections. These are caused by the accumulation of secretions in the noncommunicating bronchi.

Histology[126] shows the pulmonary parenchyma replaced by the results of chronic inflammation. The cysts are lined by cuboidal, columnar, or rarely squamous epithelium and contain amorphous eosinophilic material as well as foamy macrophages.

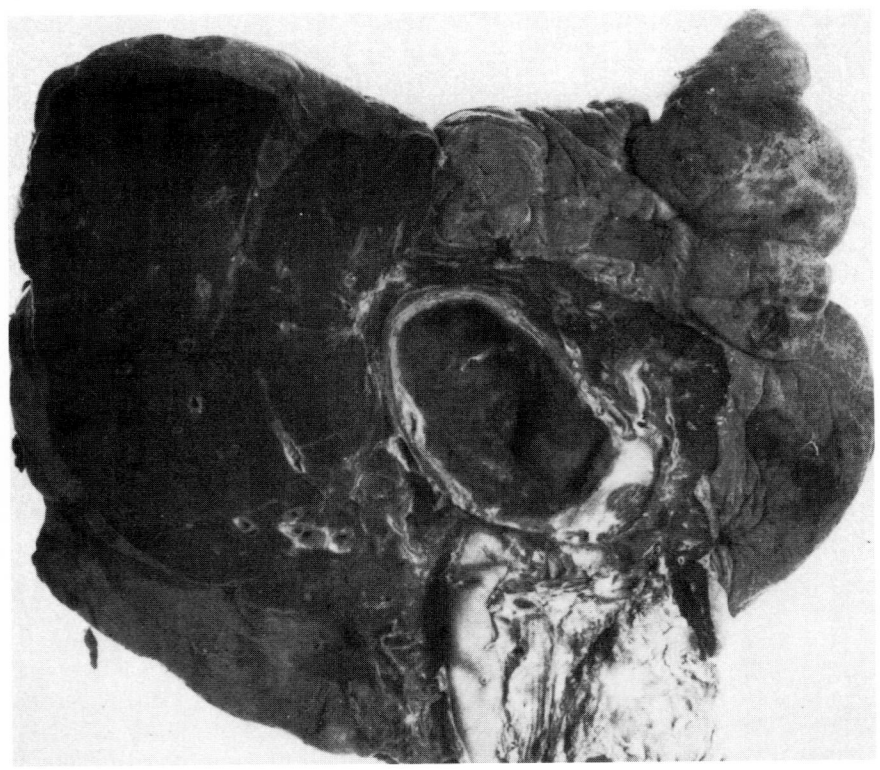

FIGURE 3-20
Intralobar sequestrated lung showing the cystic change in the sequestrated lung and the disproportionately large elastic aberrant artery that had arisen from the abdominal aorta.

Residual bronchi and bronchioles are surrounded by chronic inflammation. There is cuboidalization of alveoli, surrounded by marked chronic inflammation and thick-walled systemic vessels are seen in the interstitium. The edge of the lesion may be sharply demarcated from the adjacent normal lung or blend diffusely with it. Acute bronchopneumonia may be seen. Squamous cell carcinoma has been described.[131]

Although classified as a congenital malformation, good evidence indicates that probably only 5 percent of ILS are developmental in origin. The majority are acquired secondary to infection.[115] ILS is a very rare finding in neonatal autopsies and Stocker and Kagan-Hallet[115] despite finding 15 examples of ELS do not report finding one case in over 47,000 autopsies. The reputed ILS-to-ELS ratio is 6:1[132] and this ratio should be maintained if ILS is a congenital anomaly. Associated malformation is rare, partial duodenal obstruction being the only feature of note in one series.[114] Lastly, ILS usually presents relatively late, with a history of recurrent infection and pneumonia. The fact that the vasculature to the sequestered segment is systemic in origin may be due to hypertrophy of normal pulmonary ligament arteries.[115] A summary of the differences between ELS and ILS is presented in Table 3-3.[114]

Bronchopulmonary Foregut Malformations
The term has sometimes been used to bring together a number of bronchopulmonary lesions under one umbrella,[117] but here it will be used in a more limited way and applied to sequestrations that retain a communication with the gastrointestinal tract.[112,118]

Before the term bronchopulmonary foregut malformation was coined,[112] lesions were described, similar to ELS, under the term accessory lung.[133,134] The abnormal pulmonary segment arises either from the esophagus (usually the lower part) or from the stomach; the former is by far the more common.

TABLE 3-3
Criteria for Distinguishing Extralobar Sequestration (ELS) from Intralobar Sequestration (ILS)

	ELS	ILS
Relation to lower lobe of lung	Separate from lung	Within lower lobe (typically posterior-basal segment)
Age of diagnosis	59% < 1 year	50% > 20 yr
Association with other congenital anomaly	Frequent	Uncommon
Side affected	Left > 90%	Left approx. 55–65% Bilateral, rare
Sex ratio	M:F = 4:1	M:F = 1:1
Venous drainage	Systemic or portal	Pulmonary
Presence of diaphragmatic defect	About 60%	Rare

The sequestration is approximately equally divided between ILS and ELS but unlike sequestrations with no foregut communication, it is right sided in 70 to 80 percent.[118] The sex incidence is equal and the majority of cases present in the first year of life although adult presentation is recorded. Presentation may be due to chronic cough, recurrent pneumonia, or respiratory distress. Associated anomalies are common, overlapping with ELS, and include diaphragmatic hernia and abnormalities of the VACTERL type.[135,136]

The macroscopic and microscopic appearance of the abnormal lung segment will broadly correspond to whether the segment is extra- or intralobar. Thus, a foregut malformation that is extralobar will have the appearance of an ELS;[119] an intralobar bronchopulmonary foregut malformation will have the appearance of an ILS. The main histologic difference is that the secondary change of lung infection is more likely because of the open communication with the gastrointestinal tract.

CONGENITAL CYSTIC ADENOMATOID MALFORMATION

Adenomatoid malformation is a hamartomatous lesion of the lungs in which the main abnormality is an "adenomatoid" increase in the terminal respiratory structures.

Classification usually follows that proposed by Stocker and associates,[119] who described three subtypes distinguishable on the basis of size of the cysts, extent of the lung involved, and histologic features. This classification will be used here, because its use is widespread, particularly in clinical circles, and it recognizes the widely different appearances of CCAM. However, it is worth noting that some authors do not accept entirely the value of such a classification. They argue that because of overlap between the subtypes and the blurred association of some forms with other malformations, it is only of value to describe the abnormality in terms of whether it is cystic or solid or a mixture of both.[137,138] Undoubtedly, a difficulty in subclassifying a pulmonary lesion into one of the three main subtypes as described by Stocker should not inhibit a diagnosis or recognition of a form of "adenomatoid malformation."[139]

The classification below is based on original descriptions from infants.[140] Consequently, the dimensions need appropriate interpretation when applied to fetuses.

Type 1 CCAM. Multiple large cysts, usually 3 to 7 cm. in diameter, often occupy only one lobe but occasionally more are seen (Fig. 3-21). Smaller cysts may merge with normal lung. Communication with the normal bronchial tree does seem to be present in

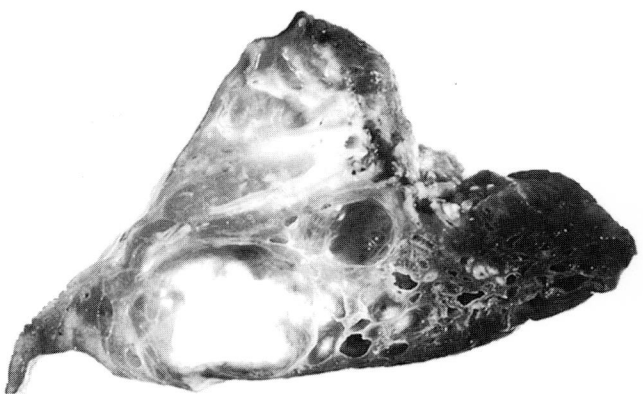

FIGURE 3-21
Resected segment of lung showing CCAM type I.

that distention of cysts with mediastinal shift is a potential complication. Histologically, the cysts are lined by respiratory or sometimes mucigenic epithelium that may be compressed. In the cyst wall, a thin fibromuscular layer is present together with occasional small islands of cartilage. Between the cysts, lung tissue may appear entirely normal and inflammation is not a feature (Fig. 3-22).

FIGURE 3-22
Histology of CCAM type I illustrated in Fig. 3-21 showing large cysts with relatively normal adjacent lung. Cysts are lined by respiratory-type epithelium on a thin fibroelastic wall. (Low magnification.)

Type II CCAM.   This is composed of multiple, evenly spaced cysts, less than 1.2 cm. in entire diameter, usually occupying a whole lobe of lung (Fig. 3-23A) or occasionally an entire lung. The cyst lining resembles normal bronchiolar epithelium and the main distinguishing feature is their excess (Fig. 3-23B). A fibromuscular layer with excess elastic tissue is present in the wall but cartilage is absent. A particular feature, that does appear unique to type II CCAM, is the occasional presence of striated muscle fibers within the interstitium between cysts (Fig. 3-24). Distention of cysts occurs after birth. Although this might argue for communication with the normal bronchial tree, Moerman and colleagues[141] suggest that detailed study of the bronchial tree shows adenomatoid malformation is associated with an underlying bronchial atresia and aeration of the adenomatoid segment is via collateral communication. Recently it has been suggested that CCAM type II may be causally related to polyalveolar lobe,[142] a condition characterized by an increase in alveolar number but with a normal bronchial tree and pulmonary vasculature.

Type III CCAM.   This is usually the more solid variant and the abnormality affects an entire lobe or lung (Fig. 3-25). In infants the cut surface shows few cysts larger than 0.5 cm. in diameter. Occasionally, atresia of a major bronchus is present.

Histologically, the lobe is replaced by evenly spaced cysts lined by ciliated cuboidal epithelium (Fig. 3-26). Around some cysts a thin layer of fibromuscular tissue is present and elastin increased. Cartilage is absent. In some respects, the lung resembles very immature canalicular stage lung. Ultrastructure[143] of type II and III lesions shows columnar or cuboidal cells with varying numbers of microvilli and differing amounts of mucinous differentiation. The cytoplasm in one case showed numerous tonofilaments. In another case myelin figures were detected in ciliated bronchiolar-type cells. These features were thought to be characteristic of normal fetal lung. The authors felt there was no correlation between the ultrastructural changes and the type of CCAM.

CCAM, especially type III, affects males more than females, and lower lobes are affected more com-

A

FIGURE 3-23

A. CCAM type II identified at autopsy in part of the left lower lobe of a fetus with multiple congenital abnormalities, mainly renal agenesis and sirenomelia (mermaid fetus). The upper lobe shows the radiating pattern of the bronchial cut surface. This is absent from the area of adenomatoid malformation where small

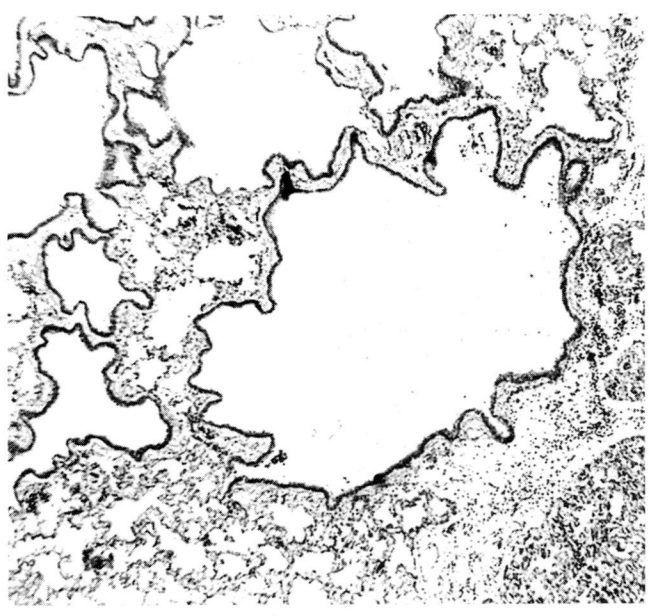

B

parenchymal cysts are just visible. *B.* CCAM type II. On cursory inspection, the lung may not appear abnormal. Bronchiolar type structures are, however, present to excess and is the more typical appearance of type-II CCAM.

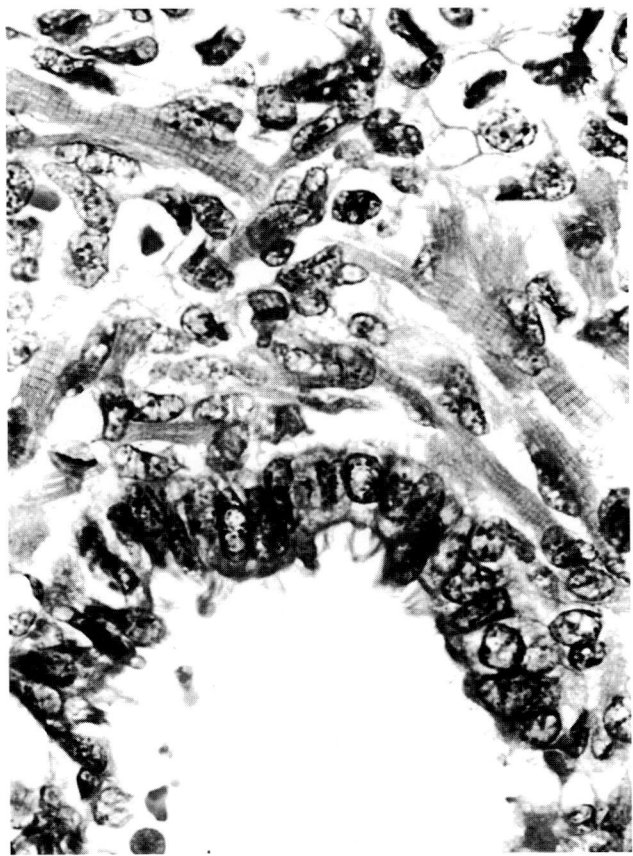

FIGURE 3-24
CCAM type II. Rarely, CCAM type II contains striated muscle in the interstitium between the cysts.

FIGURE 3-25
Section across solid CCAM type III lung from an infant of 30 weeks' gestation. The other lung was hypoplastic and the baby died at 4 hours of age.

monly than upper. In one ultrasound study, the left lung was involved in 51 percent of cases, the right in 35 percent and both in 14 percent.[144] Type II is particularly associated with other malformations including hydrocephalus, diaphragmatic hernia, jejunal atresia, bilateral renal agenesis, tetralogy of Fallot, ventricular septal defect, tracheoesophageal fistula, and agenesis of the rest of the urinary tract, large bowel, uterus, cervix, and vagina. All subtypes appear to have an increased association with renal tract abnormalities.[138]

Broadly, the presentation of CCAM is related to the amount of lung affected by the abnormality than other features, and problems are due to it acting as a "space-occupying lesion." Consequently, type III, where there is usually more lung involvement, tends to present earlier and ultimately carry a worse prognosis. As with other bronchopulmonary abnormalities, the use of prenatal ultrasound has led to more intrauterine diagnosis. There may be difficulties in distinguishing CCAM from other lesions such as ELS or even diaphragmatic hernia. CCAM type I will not infrequently be an incidental finding, whereas type

III may present as maternal polyhydramnios and fetal hydrops, the latter thought to be due to venous obstruction. There is evidence that CCAM, presumably type I, detected by prenatal ultrasound can disappear.[10] In general, the prognosis and outcome of prenatally diagnosed CCAM is still under study.[145,146] Maternal hydramnios and fetal hydrops are poor prognostic indicators.[144]

Postnatally, type I CCAM may be asymptomatic or may give rise to increasing respiratory distress associated with an expanding lesion producing mediastinal shift. Symptoms are usually present in the first week of life but may be delayed up to 4 weeks. Vomiting may be the initial presentation. Radiologically, the air-filled cysts may be difficult to detect, but a multicystic pattern with surrounding nonhomogenous lung corresponding to the less distended cysts may be seen (Fig. 3-27). Type II will not infrequently present due to the associated abnormalities rather than the lung lesion itself. Otherwise presentation may be similar to type I. Type III is rarely silent and produces rapidly progressive respiratory distress. Death from inadequate ventilation and

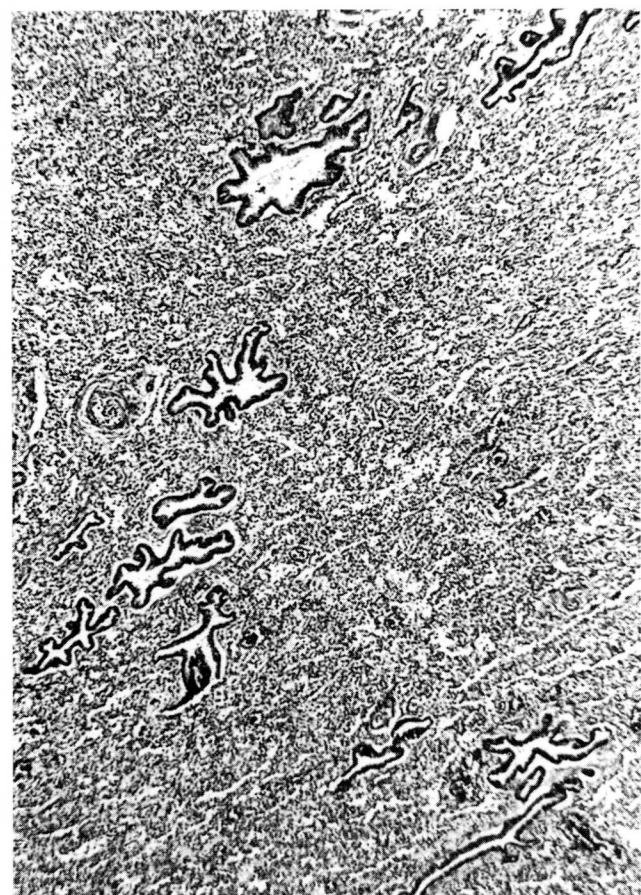

FIGURE 3-26
CCAM type III in an infant of 28 weeks' gestation. Bronchiolar-type structures are present to excess. The structure of the "alveoli" is abnormal and the interstitium contains collagen.

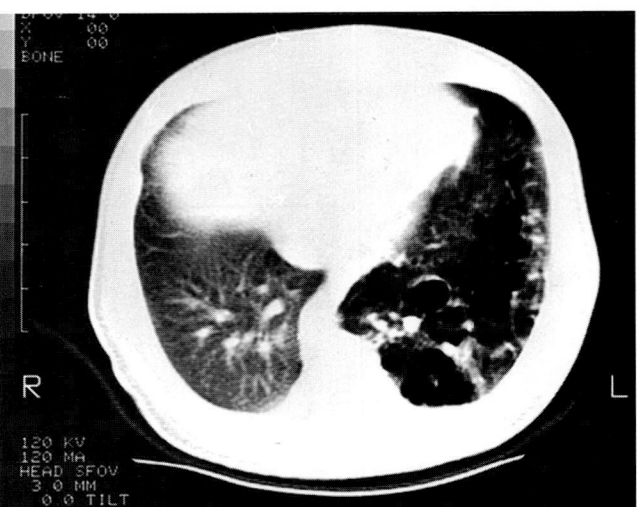

FIGURE 3-27
CT scan of CCAM type I affecting the left lung. *(We thank Dr. K. McHugh, Department of Radiology, Oxford Radcliffe Hospital Trust, for providing the CT scan for this figure.)*

mediastinal shift is usual within 1 to 5 hours after birth.

Pneumothorax is a rare clinical presentation in the newborn.[147] This case has been suggested as a new variant because the right middle lobectomy showed cystic air spaces lined by type II pneumocytes as well as vascular proliferation of the interstitium. CCAM has been associated with bronchioloalveolar carcinoma in two young adults.[148,149] It was suggested that the proliferation of mucous cells was a premalignant condition.

Because not all lobes are involved, CCAM is thought to be due to an intrauterine insult after the fifth week. However, associated renal abnormalities place the injury in type II lesions as prior to 31 days. Type I lesions could occur as late as 49 days.

Congenital bronchogenic cysts and CCAM can only be explained by postulating a dual mode of lung development. The bronchial system develops from the laryngotracheal bud and air spaces distal to the terminal bronchioles develop from mesoderm. Normally the two systems connect. These two condi-tions represent varying degrees of failure of the bronchial bud growth together with complete or partial failure (agenesis) of the distal respiratory system. Cyst formation results from expansion of atretic bronchi or distention of abnormally developed respiratory lung structures in response to respiratory movements in the presence of a small and imperfectly developed lung. CCAM of all types results from failure of the bronchi to join up with distal alveolated tissue in a normal manner.

POLYALVEOLAR LOBE

First described by Hislop and Reid,[150] polyalveolar lobe is characterized by a significant increase in alveoli in the presence of a normal bronchial tree and vascular pattern. It primarily affects upper lobes, particularly on the left side and causes complications due to compression of adjacent structures and mediastinal shift. The underlying cause of the abnormality is unclear although it presumably results from a relatively late gestational event because the bronchial tree is established at 16 weeks of gestation. The relationship to airway obstruction, adenomatoid malformation, and the hyperplastic lung seen in laryngeal atresia is not certain.[142,151]

CONGENITAL LOBAR EMPHYSEMA

This is a localized area of emphysema that usually presents in the neonate or early infancy due to respiratory distress that can be life-threatening. The upper lobes are more frequently affected than the lower and males slightly more than females (Fig. 3-28).

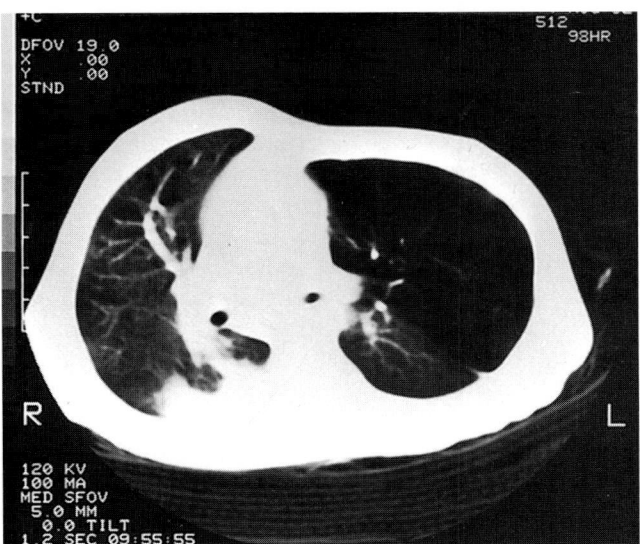

FIGURE 3-28
CT scan of congenital lobar emphysema. Much of the anterior part of the left lower lobe is overinflated. A small segment, normally inflated, is present posteriorly. There is asymmetry between the left and right hemithoraces. The reason for this is unknown. *(We thank Dr. K. McHugh, Department of Radiology, Oxford Radcliffe Hospital Trust, for providing the CT scan for this figure.)*

Pathologically, there are few distinguishing features within the lobe itself, the parenchyma showing nonspecific distention (Fig. 3-29) and sometimes rupture of the alveoli. The underlying cause is primarily due to an abnormality of the supplying bronchus in which there is bronchial compression or collapse due to an intrinsic process[152] such as bron-

chomalacia (Fig. 3-30). Because it may result from external bronchial compression, other congenital abnormalities are a common accompaniment, particularly congenital heart disease,[153] such as ventricular septal defects or tetralogy of Fallot.

Acquired lobar emphysema may also be due to secondary bronchial damage in early childhood. Bronchial damage may follow ventilation associated with chronic lung disease and is reported in neonatal survivors.[154] A form of lobar emphysema or emphysema affecting a whole lung may occur following bronchial damage during childhood viral bronchiolitis (Macleod's syndrome).[155–157]

LYMPHANGIOMATOUS CYSTS/CONGENITAL PULMONARY LYMPHANGIECTASIS

Lymphagiomatous cysts are a group of cysts lined by endothelial cells, filled with lymph, or if they communicate with a bronchus, air and lymph. They are probably best regarded as being part of congenital lymphangiectasis.

Congenital lymphangiectasis is a rare condition and has been reviewed by Stocker and colleagues.[119] It is more common in males than females by a ratio of 2:1 and frequently associated with other congenital anomalies such as asplenia and congenital heart disease. These include total anomalous venous drainage, common atrioventricular valve defects, ostium secundum, pulmonary stenosis, ventricular septal defect, mitral atresia, hypoplastic left heart, and cor triatrium. Accessory lobes and renal abnormalities such as a hydroureter have been docu-

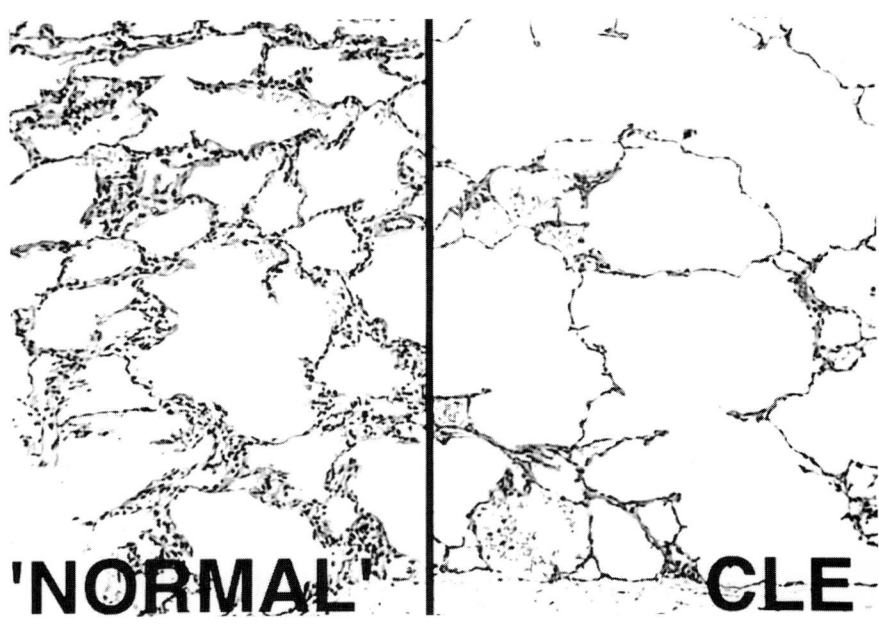

FIGURE 3-29
Normal lung compared with lung from right upper lobectomy specimen resected for congenital lobar emphysema at 1 month of age.

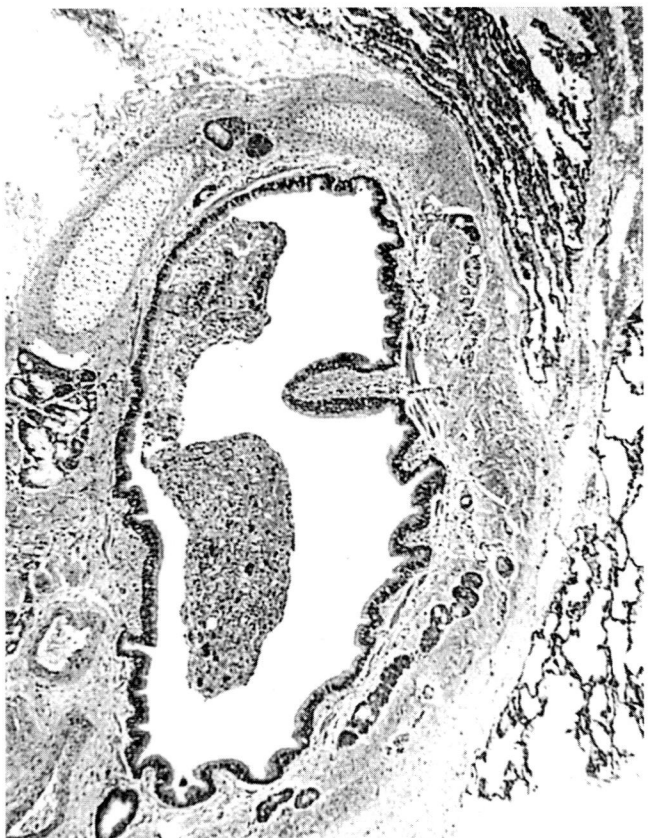

FIGURE 3-30
Hilar bronchus taken from the same case as in Figure 3-26. There is a striking deficiency in the amount of cartilage, allowing bronchial collapse.

mented.[158] The condition has also been reported in association with Noonan's syndrome,[159] congenital icthyosis,[160] and so-called vanishing bones.[161] It may be familial.[162]

Wagenaar and associates[163] proposed the following classification, which takes into account the late cases:

1. Primary
   a. Limited to the lungs
   b. With pulmonary and mediastinal involvement
   c. Generalized (including intestinal, hepatic and osseous involvement)
2. Secondary to obstruction of the pulmonary venous outflow[158]

Pulmonary lymphangiomatosis with venous obstruction is the most common presentation. The vascular anomaly usually is total anomalous venous return or primary pulmonary venous stenosis. A similar but frequently used classification is that of Noonan and coworkers.[164] Either is acceptable although perhaps

that of Wagenaar and colleagues emphasizes more the distinction between a primary abnormality of the lymphatics and secondary lymphatic dilatation due to a congenital anomaly.

It occasionally presents in stillbirths, but more commonly is seen in the newborn within hours of birth, where it gives rise to respiratory distress. A rapidly fatal outcome is usual but longer-term survival has been reported particularly where the anomaly only involves one or two lobes.[163]

Macroscopically the lungs are larger than normal, inelastic, lobulated, and firm. Multiple small cysts, fluctuant on palpation, and up to 1.5 cm in diameter, are seen through the pleura and may be mistaken for air cysts due to interstitial emphysema. The cysts may be present around bronchi and interlobular septa. Subpleural lymphatics are easily seen and there is a honeycomb effect with small cysts containing clear fluid increasing in size toward the hilum. Laurence[165] has shown these lymphatics form a network of channels rather than isolated cysts.

Histologically there are prominent lung septa and dilated lymphatics with thin-walled endothelial-lined spaces varying from 15 μ to 1.5 cm in diameter with sporadic smooth muscle (Fig. 3-31). These dilated lymphatics stain with factor VIII or *Ulex europeus* lectin to confirm the endothelial origin. The intrapleural and interlobular lymphatics are usually more prominent than the others but they are all devoid of valves. The ectatic lymphatics are seen in the early stages of the disease close to pleura and next to pulmonary veins. The vessels are infiltrated by lymphocytes, sometimes with germinal follicle formation. The lumina contain eosinophilic material. There is no endothelial overgrowth to suggest a lymphangioma. No hyperplastic muscle is seen in the wall and there is no chylothorax. Both these are features are seen with acquired lymphatic obstruction in adult life.

ENTERIC (ENTEROGENOUS) CYSTS

These are a form of unilocular gastrointestinal duplication cysts, which may occur anywhere along the gastrointestinal tract. In the thorax they typically occur in the posterior mediastinum on the right side attached to the esophagus with a bronchial connection present only very rarely. Histologically, squamous epithelium is typical but gastric fundal or small intestinal epithelium can also be present; the wall may contain muscle but not cartilage. Reported are pancreatic islets in the wall as well as occasional intrapulmonary cysts.[166] The gastric epithelium can give rise to ulceration and cyst rupture.[167] These

FIGURE 3-31
Multiple dilated lymphatics in the interlobular septae in congenital lymphagiectasia.

cysts may lie in the mediastinum and can be associated with anomalies of the lower cervical and upper thoracic vertebrae.

DIFFERENTIAL DIAGNOSIS OF LUNG CYSTS

The list below provides a brief synopsis of the more important diagnostic points of cystic congenital bronchopulmonary lesions and the more immediate acquired lesions that may enter into the differential diagnosis.

Lung abscess. This may be difficult to distinguish from a bronchogenic cyst or any other type of lung cyst because the latter often become infected. However, an abscess often has multiple bronchial communications.

Extra- and intralobar sequestrations. As has been noted above, ELS and CCAM may coexist. Typically sequestrations have an anomalous systemic arterial supply. ELSs are well demarcated and most are seen in the posterior costophrenic angle adjacent to the lower esophagus. ILSs are found usually in the posterior basal segment of the lower lobe.

Bronchopulmonary foregut malformations. These should be regarded as sequestrations with foregut communications.

Bronchogenic cysts. These are typically lined by respiratory epithelium with islands of cartilage. Multiple sections may be needed to demonstrate cartilage. In the region of the esophagus, bronchogenic cysts should not be confused with esophageal cysts, which are lined by squamous epithelium.

Enteric cysts. Lying in the posterior mediastinum, these are lined by gastric epithelium.

Lymphangiectasis. May be confused with an ELS because lymphangiectasia is seen frequently in the latter condition. No valves are seen in lymphangiectasis and in this condition there is an increase in size of the cyst toward the hilum. The cysts are lined by a flattened endothelium, which stains positively with factor VIII and *Ulex europeus* lectin.

Anterior mediastinal cysts. Cysts in the anterior mediastinum may be confused macroscopically with cystic teratomas or thymomas although the latter are uncommon in childhood.

Pulmonary interstitial emphysema. This may be confused with lymphangiectasis because of the subpleural blebs. A history of hyaline membrane disease, meconium aspiration, or pulmonary hypoplasia should point to emphysema. The presence of giant cells around a cyst with a fibrous wall is an indication of encysted pulmonary interstitial emphysema.

Pseudosequestration. This may clinically mimic an ILS but is in fact a right lower lobe abscess due to an inhaled foreign body.

Hernia. In pediatric surgical practice a differential diagnosis from a congenital Bochdalek's hernia should be considered.

There appears to be a long-term risk of neoplasia in some bronchopulmonary cysts.[131,168,169]

ALVEOLAR CAPILLARY DYSPLASIA

Also known as misalignment of lung vessels,[170] there is a failure of normal vascularization of lung parenchyma at the alveolar level with few capillaries in alveolar wall (Fig. 3-32).[171] In addition, the pulmonary veins are present in association with the pulmonary arteries as an abnormal vascular bundle (Fig. 3-33). Ventilation is required almost from birth and pulmonary hypertension may also be a feature.[172] It may present in the differential diagnosis of persistent pulmonary hypertension of the newborn in the presence of a structurally normal heart.

ABNORMAL AND ECTOPIC TISSUES IN THE LUNGS

Rhabdomyomatous dysplasia with striated muscle fibers in bronchioles and interlobular septa are well documented. It often presents as an incidental finding in association with other abnormalities, particu-

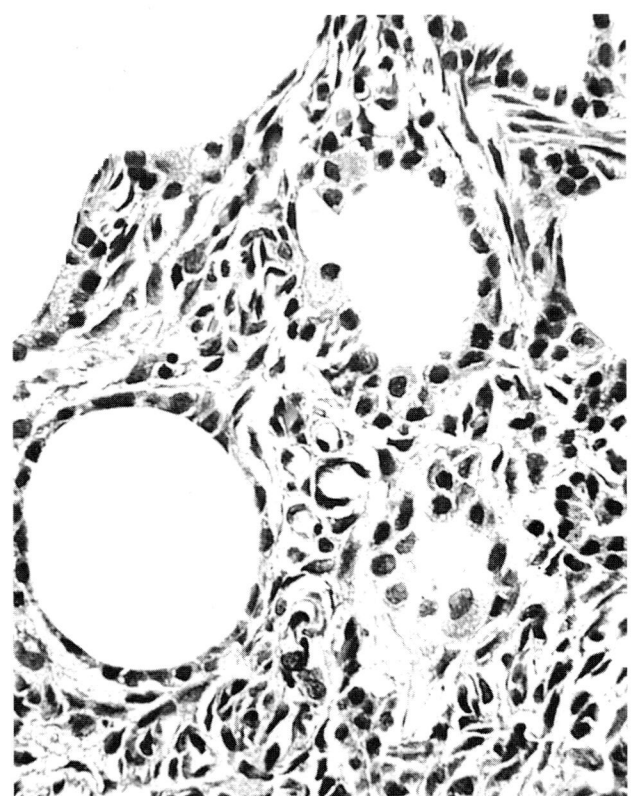

FIGURE 3-32
Alveolar capillary dysplasia. Interstitium has a deficit in the number of capillaries pushing into alveoli.

larly congenital heart disease.[173–176] Ectopic adrenal cortical tissue[177] has also been described (Fig. 3-34). It is suggested that adrenal cortical tissue develops from early fetal celomic cells lying lateral to the root of the mesentery. These cells may become transposed above the growing septum transversum and incorporated into the pleura. Aberrant pancreas has been noted in the presence of an ILS.[178] Pulmonary glial nodules are associated with anecephaly and this may be related to the cerebral disruption and transfer of tissue to the lung, either via blood vessels or ingestion.[179] Intravascular cerebral tissue, usually cerebellum, may rarely occur as an embolic phenomenon following traumatic delivery[180] and should not be mistaken for hamartomatous or heterotopic tissue. Ectopic liver tissue, including bile duct epithelium, may be found in the right lower lobe, invested by the pleura.

PULMONARY HYPOPLASIA

Pulmonary hypoplasia occupies an uneasy place within a section on congenital malformation because the majority of examples are probably misclassified. Nonetheless, it is one of the most common conditions encountered in perinatal pathology and may be a cause or contributory cause of death in up to 10 to

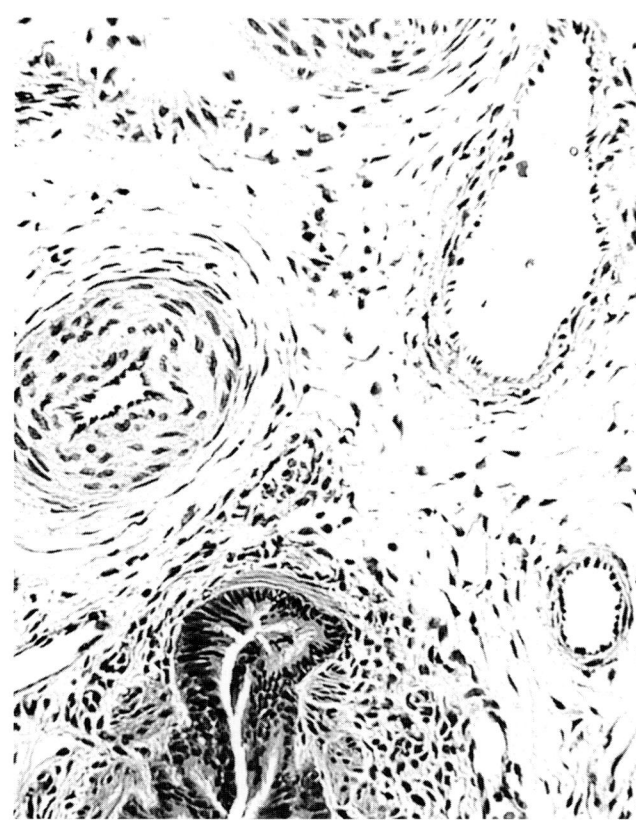

FIGURE 3-33
Capillary alveolar dysplasia. Associated with the bronchus is both a pulmonary artery with grossly thickened wall and a pulmonary vein. Section is from an infant presenting with pulmonary hypertension but with no structural abnormality of the heart.

15 percent of neonatal necropsies.[5,181] In the majority of cases, it is secondary to pathology elsewhere.

Prenatal diagnosis of pulmonary hypoplasia is not very reliable and so presentation is usually at or very shortly after birth. The infant may be born apparently normally but deteriorates rapidly with severe respiratory distress that responds poorly to resuscitative measures. With ventilation, chest movements may be poor and, even if some response is obtained initially, increasing pressures are required with diminishing results. Death may occur in minutes or hours. Occasionally, less severe cases may permit successful ventilation but a marked respiratory distress syndrome develops often requiring relatively high pressures to maintain adequate oxygenation. Outcomes may vary. Some infants may still die in the neonatal period and others may progress to chronic lung disease (see below). In these "clinical" cases, the criteria for diagnosis are often presumptive.

In fatal cases, diagnosis is usually straightforward at autopsy although there is good evidence it is infrequently recognized.[182] In stillbirths and early neonatal deaths, the lungs appear small within the thorax (Fig. 3-35) or after removal the lungs are clearly small relative to the heart (Fig. 3-36). If the infant is resuscitated successfully for a time, this feature may not be so striking. Careful inspection of the surface may reveal subpleural pulmonary interstitial

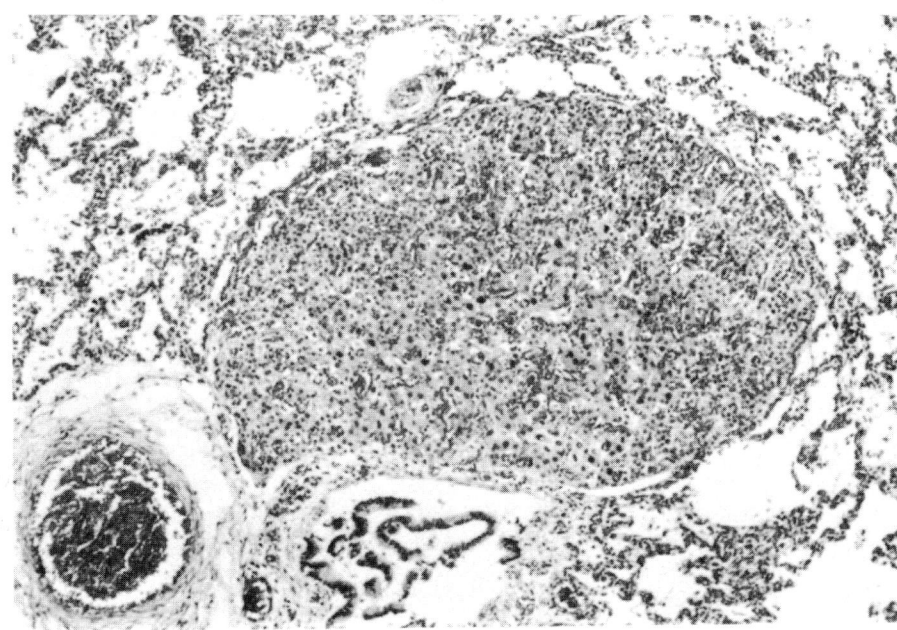

FIGURE 3-34
Ectopic adrenal cortical tissue in neonatal lung lying in an interlobular septum. (*Reproduced by courtesy of Dr. C. Bozic, Lausanne*)

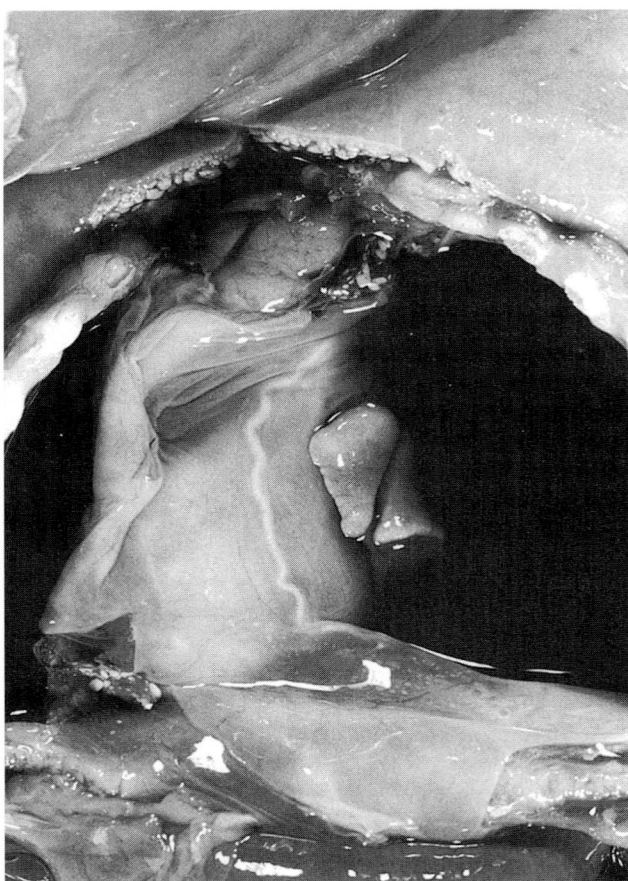

FIGURE 3-35
Bilateral pulmonary hypoplasia in an infant dying with chronic hydrops and pulmonary effusions.

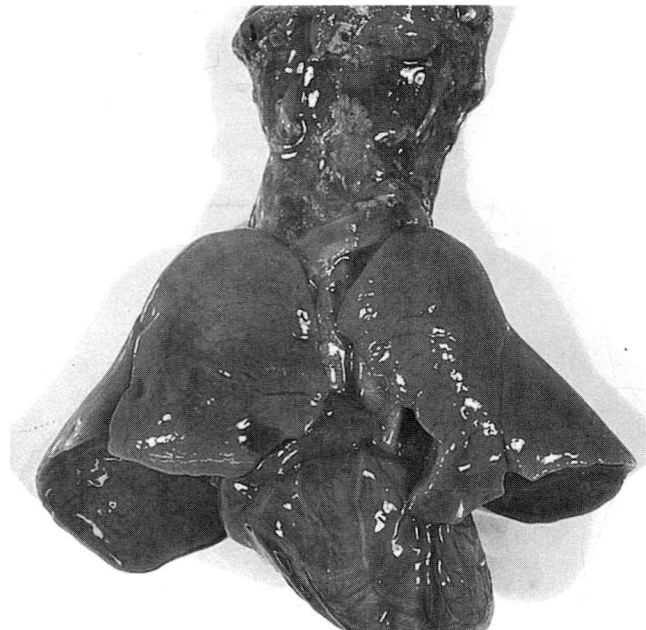

FIGURE 3-36
Hypoplastic lungs relative to the size of the heart. The lungs should approximately reach the level of the cardiac apex.

emphysema. A more objective measure can be used by determining the lung-body weight ratios. Ratios of less than 0.015 in fetuses and neonates less than 28 weeks' gestation or less than 0.012 in fetuses and neonates at 28 weeks or more of gestation are indicative of pulmonary hypoplasia.[183,184] Interpretation of ratios will need caution in the presence of pneumonia or significant neonatal survival that will tend to increase lung weights.

Wigglesworth has defined two main groups of pulmonary hypoplasia—those with poor growth but apparently normal maturation of the epithelial component and those lungs with impairment of both lung growth and maturation (Fig. 3-37). This latter form is seen in pulmonary hypoplasia associated with oligohydramnios (see below). Special stains will show that elastin is absent from septal crests especially in this latter group.[185–188] Morphometric study indicates that hypoplastic lungs from any cause have a reduced radial alveolar count.[183,189] The radial count measures the number of alveoli between respiratory bronchioles (or terminal bronchioles when respira-

tory bronchioles are absent) and the periphery of the acinus. It is therefore a measure of the complexity of the acinus. A reduction in the radial count indicates intraacinar hypoplasia. The normal radial alveolar count is 5.3 with a low limit of normal of 4.1.

Other methods of assessment, although clearly not for routine use, include measurement of lung DNA as an index of a low lung-body ratio and indicates a low total lung cell population.[184] Type IV collagen is less prominent immunohistochemically in hypoplastic lungs than controls.[190] Cooney and Thurlbeck[191] suggested that the most satisfactory evaluation of lung growth was obtained by combining the radial count and the fixed lung volume. They found, in a small number of cases studied, that radial count was a poor predictor of lung maturity in hypoplastic lungs. The ratio of lung to body weight was only of use if severely depressed.

There is a long potential list of causes and associations of pulmonary hypoplasia. More recent understanding of the factors important in normal lung growth (see Chap. 2) allows a logical classification and approach (Table 3-4). Indeed, it has been the study of pulmonary hypoplasia that has contributed to much of this progress. Although many of these associations are established, the precise reasons as to how lung growth is impaired is not always so clear. It is also probably true that more than one of the mechanisms outlined below may be acting in any one case.

Inadequate thoracic volume The most obvious explanation of poor lung growth is that simple phys-

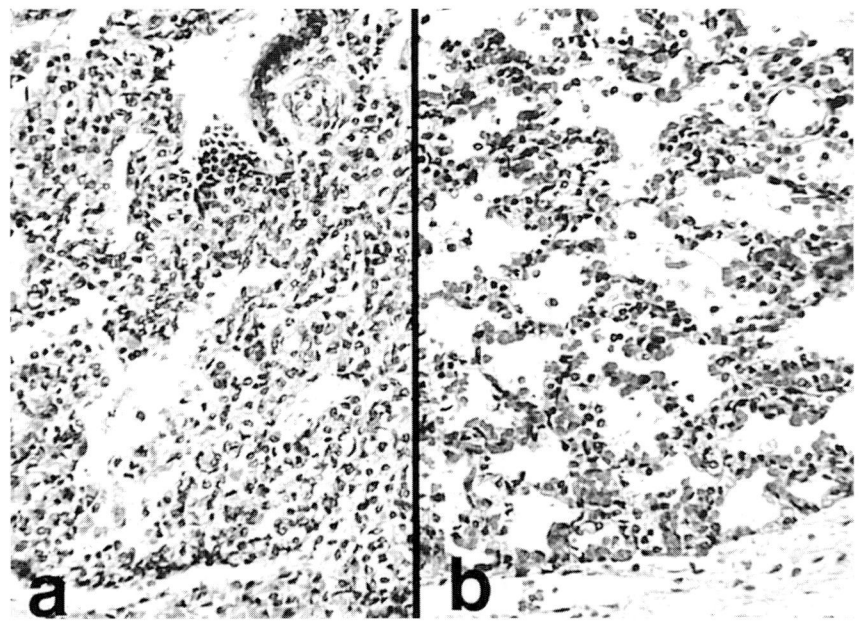

FIGURE 3-37
Hypoplastic lung (*A*) compared with normal lung (*B*) at the same gestation of 32 weeks. The hypoplastic lungs still show relatively thick interstitium with poor maturation of the pneumocytes.

TABLE 3-4
Classification of Pulmonary Hypoplasia

Mechanism	Subgroup	Examples
Primary or idiopathic	Chromosomal abnormality	Trisomy 21, 18, 13
	Familial	
Inadequate thoracic volume	Skeletal	Skeletal dysplasia
	Diaphragmatic hernia	
	Effusions	Rhesus disease
		Fetal hydrops
Impairment of fetal breathing	Cerebral	Hypoxia/ischemia
		Malformation Anencephaly with brain stem involvement
	Skeletal Muscle	Congenital dystrophy
	Diaphragm	Amyoplasia ? Exomphalos
Oligohydramnios	Inadequate production	Urinary tract anomalies Renal agenesis Cystic dysplasia Urethral obstruction
	Excessive loss	Chronic leakage of liquor amnii

It could be argued that some of the specific examples might fit into more than one group, e.g., diaphragmatic hernia might cause pulmonary hypoplasia due to the presence of herniated bowel in the thorax but also the absence of normal breathing movements.

ical restriction of the thoracic space will cause hypoplasia and accounts for perhaps 30 to 40 percent of cases. Specific causes will include many of the skeletal dysplasias where there is poor rib growth and a small thoracic volume. Displaced abdominal organs (e.g., in diaphragmatic hernia) is one of the most common causes, particularly of unilateral pulmonary hypoplasia (Fig. 3-38). Pleural effusions form a further important subgroup, for instance in association with fetal hydrops.

Impairment of fetal breathing This is a difficult group to define and may only account for 5 to 10 percent of cases where it is acting in isolation. There is good experimental evidence that the low-volume, high-frequency breathing normally performed by the fetus in utero is important in the development of the lung.[191–195] Interference in fetal breathing may be central as in severe anencephaly when there is loss of brain stem. It may be acquired as in cerebral damage from hypoxia or ischemia affecting respiratory centers. Poor muscular activity such as might occur in some congenital muscular dystrophies may produce a similar result.

Oligohydramnios With the category of inadequate thoracic volume, this is one of the main associations with hypoplasia and accounts for some 30 to 40 percent of cases. Oligohydramnios may result from reduced production of fetal urine due to renal tract abnormalities (e.g., renal agenesis, cystic dysplasia, or urinary tract obstruction) or chronic loss of amniotic fluid from premature rupture of membranes. From the pathogenetic viewpoint, this is one of the more problematic groups. Initial concepts sug-

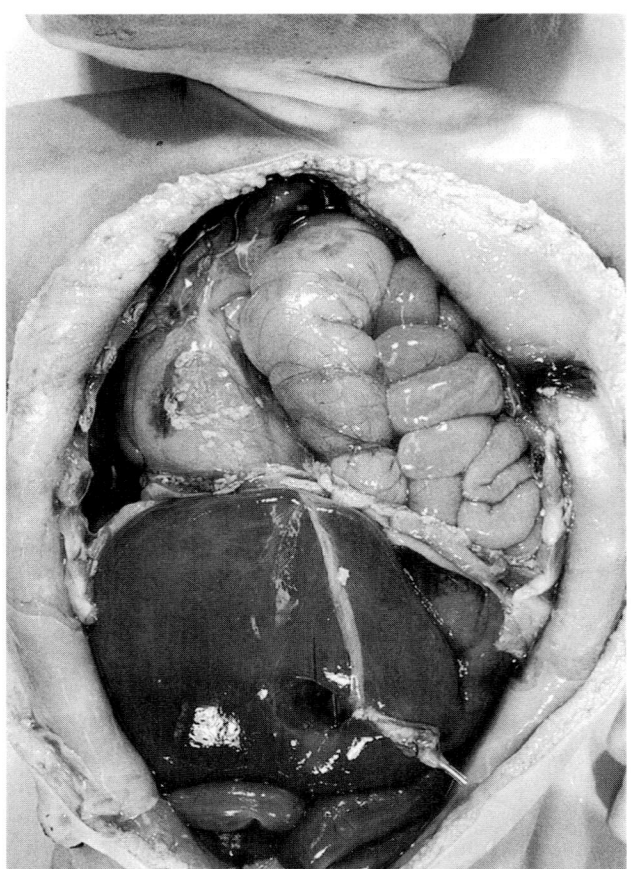

A

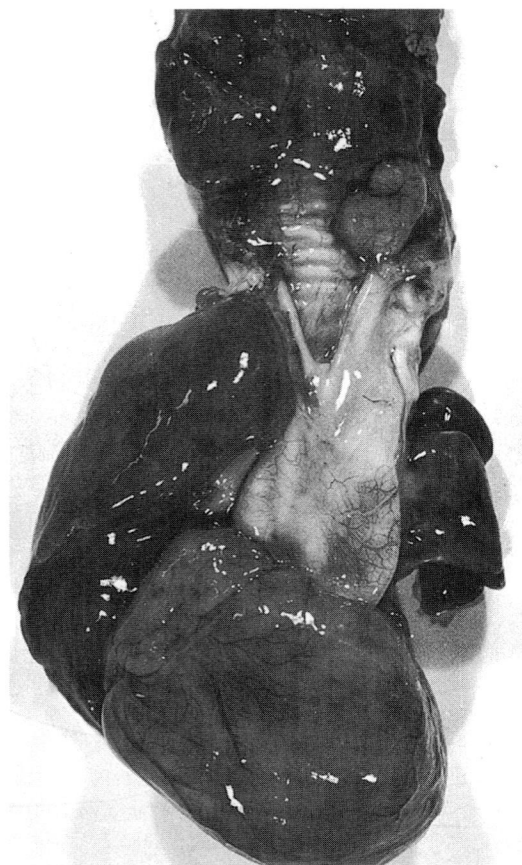

B

FIGURE 3-38

A. Diaphragmatic hernia with gut and liver in the left hemithorax causing marked mediastinal shift. After removal the severe hypoplasia of the left lung in particular is apparent. *B.* The right lung is also hypoplastic in this case due to the severity of the mediastinal shift. *(Figure 3-38B, reproduced with permission from Pulmonary hypoplasia, hyaline membrane disease and chronic lung disease. S. J. Gould, in Fetus and Neonate, vol 2 Breathing. Cambridge University Press, 1994.)*

gested compression of the developing thorax was the significant factor, but later theories have concentrated on the production and loss of lung liquid. It is suggested that for normal lung development, lung liquid needs to be "retained" within the lung under pressure. Low-pressure oligohydramnios may allow too rapid lung liquid efflux.[196–198]

Primary or idiopathic This is a diagnosis of exclusion but there are situations where lung hypoplasia cannot be attributed to any of the above causes.[199,200] Sometimes, in a severely growth-retarded infant, the lungs appear to be particularly affected. Genetic factors are important in some cases and hypoplasia can be found in association with trisomy 13, 18, or 21.[191] The occurrence of hypoplasia in twins[201] or familial cases[202] are examples of primary hypoplasia. Some rare cases of severe acinar maldevelopment have also been reported where there is almost total failure of airway development beyond the small terminal bronchioles.[203] This may represent an extreme form of primary pulmonary hypoplasia.

PULMONARY HYPERPLASIA

Similar to pulmonary hypoplasia, pulmonary hyperplasia is not strictly a malformation but rather a phenomenon secondary to malformation elsewhere. At autopsy, by far the most common cause of enlarged lungs is adenomatoid malformation (see above), and this is usually unilateral.

True pulmonary hyperplasia is bilateral and is due to obstruction of the upper respiratory tract. To date, in reported cases the obstruction has always been laryngeal atresia sometimes in association with Fraser's syndrome. This is an autosomal recessive condition with features that may include cryptophthalmos, ear anomalies, syndactyly, renal abnormalities, and cryptorchidism. It may prove lethal if the renal anomalies are severe or laryngeal atresia is present. Macroscopically, in gross examples, the diaphragm is compressed by the lungs and is horizontal at a low level. The lungs may show an impression of the ribs.[204] Lung-to-body weight ratios are high. The most striking feature histologically is that the lungs appear more mature than might be expected for gestation (Fig. 3-39). Morphometry may demonstrate an increased alveolar surface area for body weight and sometimes for gestation. Hyperplastic lungs in the setting of laryngeal atresia is due to the obstruction of normal lung liquid outflow into the amniotic fluid.[205]

DIAPHRAGMATIC AND CHEST WALL ANOMALIES

Diaphragmatic Eventration
Eventration of the diaphragm describes a membrane-covered hernial sac, representing one half of the diaphragm, which protrudes into the thorax.[8] It is thought that this is due to denervation atrophy with failure of muscle development.[186] There may rarely be pulmonary hypoplasia[206] or extralobar sequestration, especially if the eventration is on the left side. Eventration is often not distinguished from a muscu-

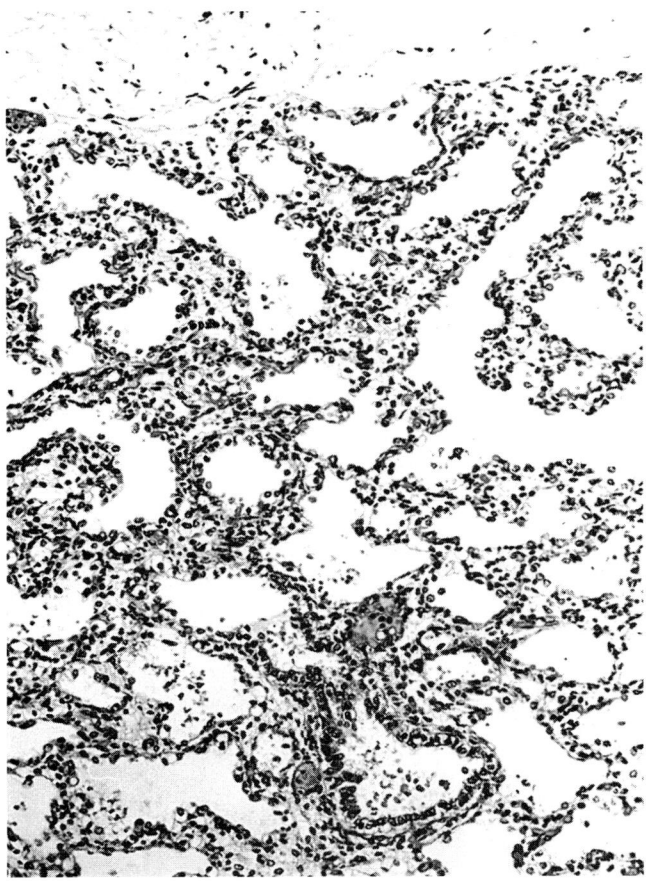

FIGURE 3-39
Hyperplastic lung from a stillborn infant with laryngeal atresia associated with Fraser's syndrome and dying at 27 weeks of gestation. The lung was grossly overdistended and uniformly expanded as illustrated.

lar deficiency with enlargement of the central tendinous portion of the diaphragm. This is termed a hernia of the foramina of Morgagni. It is formed by protrusion through the muscular defect at the site of perforation of the diaphragm by the internal mammary artery. These hernias are usually small and covered by both pleura and peritoneum. They may be associated with trisomy 18.

Diaphragmatic hernia Congenital diaphragmatic hernias may also be central, left, or right sided. Left-sided defects are most common. In such cases stomach, small bowel, spleen and even part of the left lobe of the liver may be found in the thorax (see Fig. 3-38). It may be associated with pulmonary hypo-plasia on the left side. In the past these hernias have carried the eponym of Bochdalek. This is arguably a misnomer because it is not a hernia through Bochdalek's canal. In many instances the pathology is less that of a hernia through the diaphragm than diaphragmatic aplasia or agenesis.

Diaphragmatic hernias are associated with other congenital abnormalities, particularly those of the central nervous system, gastrointestinal tract, genitourinary tract, and heart.[207] Often these other anomalies determine the outcome. In most instances, diaphragmatic hernias are sporadic, but there is a high incidence of associated chromosomal abnormalities[208] particularly in those cases detected early by fetal ultrasound.[208] Familial examples are well recorded.[209] Other described syndrome associations includes Cantrell's pentad: left ventricular diverticulum, absence of the lower sternum, diastasis of the rectus muscles, and defects in the pericardium.[210,211] This is probably sporadic. Fryns syndrome[212] is an autosomal recessive syndrome in which diaphragmatic hernia is usually a major feature. Other major elements include craniofacial (including neurologic) abnormalities, distal limb and nail hypoplasia, and genitourinary and cardiac anomalies (particularly ventricular septal defect).[213]

Primary Ciliary Dyskinesia

This has taken its place alongside cystic fibrosis and α1-antitrypsin deficiency as one of the "big three" genetic causes of chronic lung disease. Investigation of such cases is not simple, as already alluded to in Chap. 2.

The story was initially clear in that a group of men with situs inversus, bronchiectasis and chronic rhinosinusitus, and absent frontal sinuses were described by Kartagener in 1933.[214] Subsequently some of these men were found to have immotile cilia. The syndrome has been described in women.

Kartagener's syndrome has been described with motile spermatozoa[215] in a normal child but having recurrent upper and lower respiratory infections. This case suggested a discordance between the bronchial cilia, which showed axonemal abnormalities and the sperm, which were ultrastructurally normal. Conversely normal cilia have been demonstrated in the turbinate and the bronchial tree in an infertile man with Kartagener's syndrome.[216] Another cause of repeated sinopulmonary infections is Young's syndrome,[217] where there is azoospermia

due to obstruction of the middle third of the epididymis with amorphous, yellow material. In a recent study of Young's syndrome there was a relative disorientation of the distal ciliary axoneme, which was thought to be due to abnormal mucus rather than a structural defect.[218]

The initial defect in Kartagener's syndrome was a lack of dynein arms in the ciliary axoneme (Fig. 3-40).[219] Since this paper it has been realized that there are many different abnormalities in primary ciliary dyskinesia (PCD) or the immotile cilia syndrome.

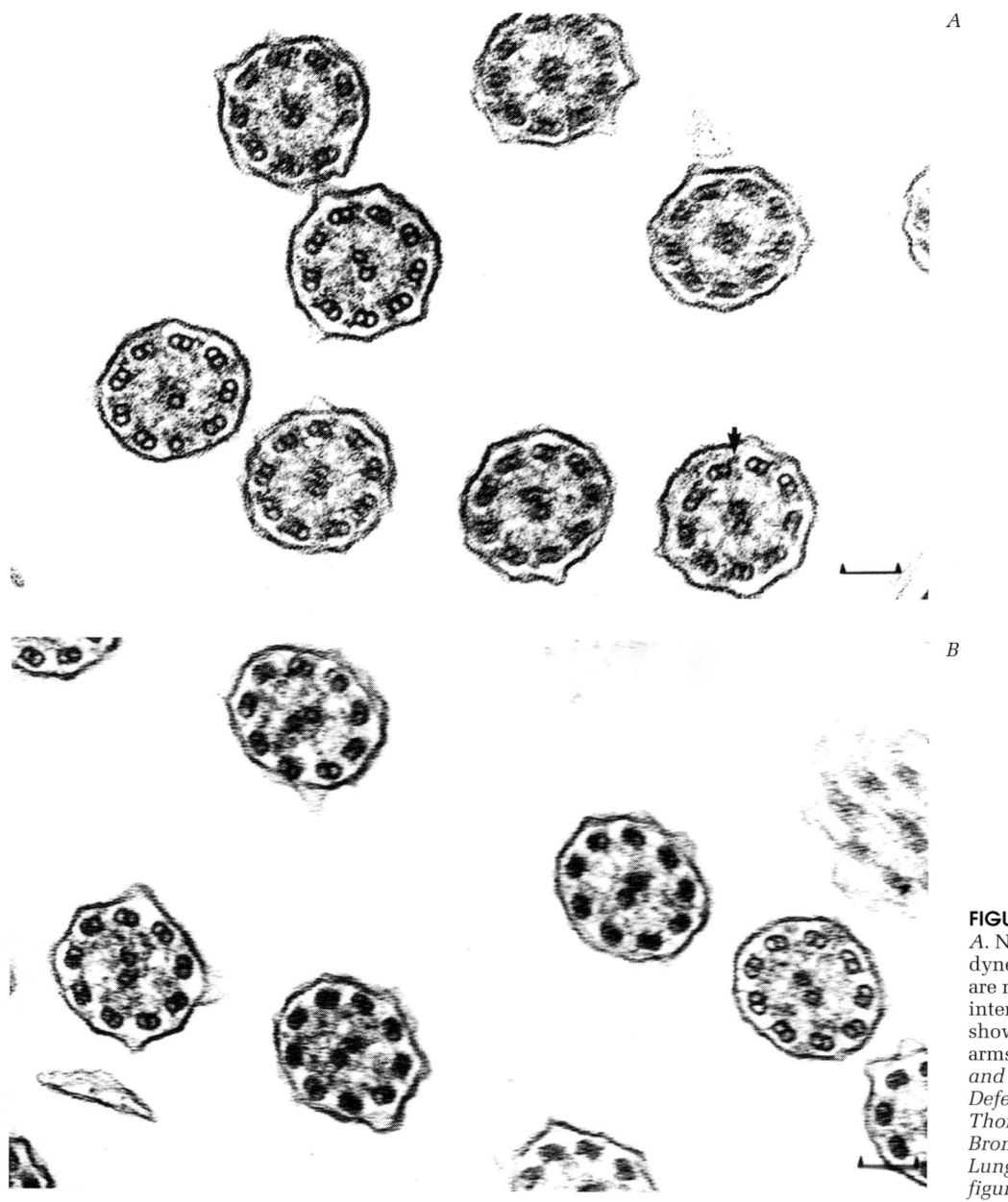

A

B

FIGURE 3-40
A. Normal cilia. The external dynein arms (example arrowed) are more readily seen than the internal.   *B.* By comparison, cilia showing an absence of dynein arms. *(We thank Drs. R. Wilson and A. Rutman of the Host Defence Unit, Department of Thoracic Medicine, Royal Brompton National Heart and Lung Institute for providing these figures.)*

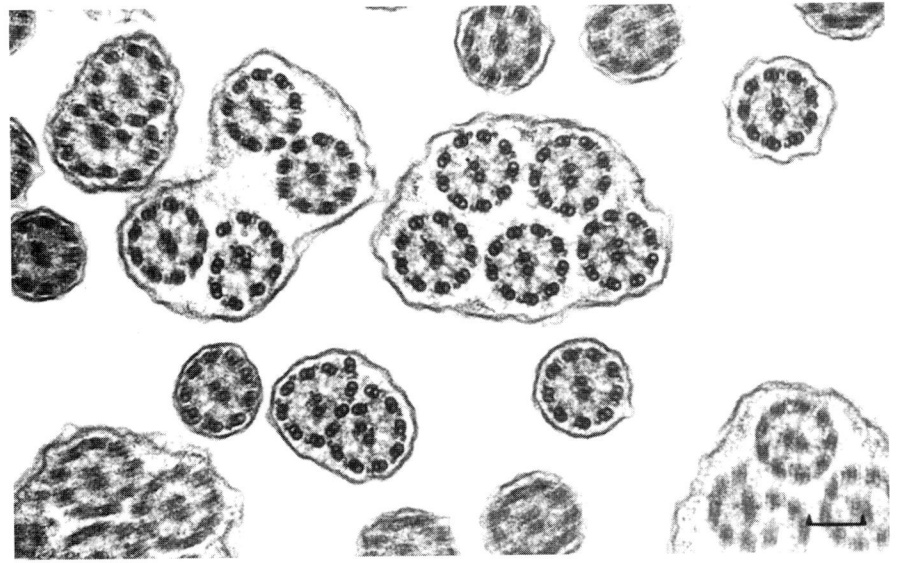

FIGURE 3-41
Complex cilia. Assessment of cilia in which the abnormalities are primarily tubular requires caution. Despite their striking appearance, they can be secondary to infection. *(We thank Drs. R. Wilson and A. Rutman of the Host Defence Unit, Department of Thoracic Medicine, Royal Brompton National Heart and Lung Institute for providing this figure.)*

Before making a diagnosis of PCD, it is important to exclude aquired defects (Fig. 3-41). Some of the ultrastructural abnormalities seen in PCD have been identified in the general population and thus the current "gold standard" of electron microscopy of individual cases is not specific. Respiratory tract viruses, air pollutants, and other chronic respiratory diseases can cause abnormal cilia. Heavy smokers may show three types of abnormality. These include compound cilia with 2 to 27 axial filament complexes, swollen cilia with an increase in matrix but only a single axial filament, and cilia with no central complexes. Such cilia were considered functionally incompetent.[220] Further abnormalities noted in patients apparently without the immotile cilia syndrome were intracytoplasmic microtubular doublets and cilia within periciliary sheaths.[221] Patients with retinitis pigmentosa have abnormal cilia with irregularly arranged outer doublets and a variation in the number from 2 to 10. Accessory single microtubules may be seen in the outer ring and there was a higher incidence of compound cilia than in controls with nasal abnormalities.[222] To qualify for the syndrome (PCD) patients should have one or more of the following criteria:

1. Kartagener's syndrome
2. Men without situs inversus but with chronic bronchitis or bronchiectasis and rhinitis since childhood and with immotile or poorly motile spermatozoa (though the latter criterion is not necessary in every case[223])
3. Patients with bronchitis or bronchiectasis and rhinitis having a sibling with the criteria given in 1 or 2
4. Patients with a typical history with specific ultrastructural defects seen in the syndrome[224]

Most important is a history of chronic productive cough that can be traced back to childhood, as well as chronic rhinitis. Most patients lead a fairly normal life. It is possible to assess nasal respiratory ciliary motility and ultrastructure 16 hours after death and possibly longer.[225]

It would appear that electron microscopy is no longer sufficient for diagnosis of PCD. In addition to a saccharin test, a measurement of beat frequency is required.[226] This is a noninvasive test that measures on a microscope slide the beat frequency of a sample of mucosa taken from the inferior turbinate and lateral nasal wall by a cytology brush. The ciliary beat frequency is measured photometrically and the strips of epithelium are easily seen at a magnification of ×320 by bright-field illumination. As with any test there are some pitfalls and it is necessary to quantitate not only ciliary beat frequency but also the percentage of motile cilia. Ciliary beat measurements must be made in random areas rather than in sections where the cilia are beating fastest. In PCD the beat frequency is significantly reduced and the ciliary wave form is grossly abnormal.[227]

The saccharin test involves a small particle of saccharin, approximately 1 mm,^{3} being placed on the inferior turbinate. The normal patient usually tastes the strong sweet taste in less than 30 min and this is taken as the nasal mucociliary clearance time.[228] This test may be inappropriate for children under the age of about 6 because of compliance problems. The dynein arms are best shown in biopsies fixed in a modified Karnovsky fixative.[229]

A variety of defects may be seen in PCD. These include loss (Fig 3-40*B*) or abnormally short dynein arms, defective radial spokes, transposition of microtubules, including eccentrically located central tubules and transposed peripheral tubules, absence of axonomal structure within the ciliary shafts, defective basal bodies of the so-called half centriole type, abnormal basal bodies with an unusual dense granule in their central part as well as abnormal accessory structures such as giant striated roots and double basal feet arising from single basal bodies.[230] A case has been described with random ciliary orientation, where the ciliary ultrastructure was otherwise normal. The direction of effective ciliary stroke in this patient was random so that effective mucociliary clearance did not occur. There was a reduction in ciliary beat frequency.[231]

Separating acquired from congenital defects may be difficult. However, PCD represents a generalized ciliary disorder, whereas acquired injury to the ciliated epithelium may be restricted to one site in the body. Compound cilia, supernumery microtubules in the axoneme, and loss of cilia or ciliated cells are nonspecific changes found in either congenital or acquired disease.[232] As a guide, structural abnormalities of varying type and found in less than 5 percent of cells are unlikely to indicate a primary cilial disorder.[226] Some abnormalities in PCD, such as loss of dynein arms, absence of both spokes and a central sheath, transposition of one of the central microtubules, and short central microtubules have been taken as more specific features. Where nasal biopsy is inconclusive, bronchial biopsy may be of benefit. The nasal mucosa may show secondary changes more frequently, possibly because it is a more susceptible to infection and other insults.[233]

Cystic Fibrosis

Cystic fibrosis, probably more than any other pulmonary disease of childhood, has been given much attention in recent years. This is largely due to identification of the cystic fibrosis gene and the advent of heart/lung transplantation, the former holding out the possibility of a fundamental development in therapy and perhaps cure.

GENETICS

Cystic fibrosis is an autosomally recessively inherited multisystem disease that affects approximately 1/2500 infants and some 350 affected children will be born each year in the United Kingdom. It is due to a defect in the gene on the long arm of chromosome 7 that codes for the cystic fibrosis transmembrane conductance regulator (CFTR),[234] a protein of 1480 amino acids. Of the 250 or so mutations described to date, the most common in Great Britain leads to dele-

tion (Δ) of phenyalanine (F) at position 508 (ΔF-508) in this protein and accounts for approximately 75 to 80 percent of cases. However, there are a number of other gene abnormalities some leading to absence of CFTR. Others produce an abnormal gene product, the gene frequency varying in different populations around the world.[235] It is probably not necessary for the abnormal gene to be the same on each chromosome. This may account for the wide variety of clinical manifestations of the disease.[236,237] Abnormality of one of the more common mutations permits first trimester diagnosis of cystic fibrosis.[238] In general clinical practice, the key to diagnosis is an elevated sweat sodium and chloride(> 60 mEq/L).

BIOCHEMISTRY

The major defect in cystic fibrosis is a reduction in chloride permeability across epithelial membrane. CFTR is a cAMP-regulated membrane protein whose main demonstrated function is to act as a low conductance, chloride channel across epithelium. Although normal chloride transport seems important for normal secretion in some tissues (e.g., the pancreatic duct), in other sites there is evidence for an associated or additional defect in sodium reabsorption. Thus, in addition to defective chloride transport, airway epithelia may have increased capacity for sodium resorption leading to drying of the surface. In the lung, CFTR mRNA can be detected in abundance in the distal airways by the early second trimester although, interestingly, there is a relative lack of expression in adult lungs suggesting a degree of developmental regulation.[239]

PATHOGENESIS

The clinical manifestations are protean and we shall restrict ourselves to the respiratory effects, which are the most significant element in patient morbidity and mortality.[240] With the block in transport of chloride into the bronchial lumen and the excessive resorption of sodium, bronchial mucus is poorly hydrated. Mucociliary transport of bacteria from the lungs is impaired due to the combined effects of increased mucous viscosity and the presence of abnormal, sulfated rather than sialated mucins. Lung damage is caused by the resultant chronic infection and inflammatory reaction. The main damage to the lung is from the inflammatory reaction to bacteria, particularly the neutrophils,[241] and recent strategies have been directed toward reducing the inflammatory response rather than the bacterial colonizing load. This has had a measure of success.[242,243]

PATHOLOGY

Bronchitis, bronchiectasis, and bronchiolitis usually begin in childhood (Fig. 3-42) but very rarely respiratory distress occurs in the neonatal period. In this situation, presentation is with respiratory distress and a picture not unlike bronchopulmonary dysplasia (see below) may develop. The lung pathology may be similar. Whether abnormal mucus is a significant factor is not clear.

Several series have documented cases presenting in adolescence or in young adults.[244,246] Usually these patients had respiratory symptoms in childhood but a few were asymptomatic until adult life. Cough and sputum are present in all patients, but an important feature is the presence of particular organisms that characterize this disease. These are *Staphylococcus pyogenes, Haemophilus influenzae,* and *Pseudomonas aeruginosa.* The latter organism is often hospital acquired and the mucoid strain implies a poor prognosis. *Pseudomonas* infection is seldom eradicated and is often found in the nasopharynx, oropharynx, middle auditory meatus, external nose, and inferior turbinates. The same strain is found in these sites as in the lungs and it is thus difficult to eradicate.[246] Recently a new strain of *Pseudomonas, P.cepacia,* has been demonstrated in

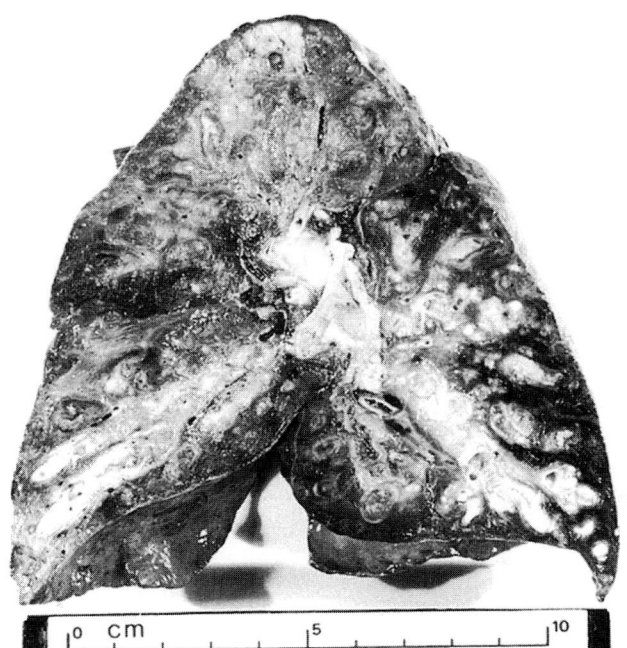

FIGURE 3-42
Pneumonia and bronchiectasis in a right lung of a 5-year-old child with cystic fibrosis.

cystic fibrosis patients and has been found to be rapidly lethal. The defect permitting *Pseudomonas* to flourish is limited to the airways. Cystic fibrosis patients are not more susceptible to infections outside the respiratory tract and septicemia virtually never occurs. The organism is acquired by person-to-person contact.[247]

Fungal infection can be a problem in cystic fibrosis and *Aspergillus species, Candida species,* and *H. capsulatum* may all be found. Although these organisms were found in a postmortem study, there are relatively few reports of fungal infection posing a significant clinical problem.[248] Fungal infection appears to be closely associated with aggressive therapeutic intervention. Rarely a bronchocentric granulomatosis picture has been described.[249] The fungal infection may become disseminated. Pneumothoraces may be a complication. Death is from respiratory failure and cor pulmonale.

One of the most comprehensive studies of the lung in cystic fibrosis is that of Bedrossian and coworkers[249] who demonstrated bronchitis and bronchopneumonia in children of all ages. Patients with fungal infection had either acute inflammation or in the case of *H. capsulatum* there were granulomata, often extrapulmonary.

With increasing age, squamous metaplasia and bronchiectasis developed. Emphysema was not seen under 2 years of age but was detected in 11 percent of patients aged 2 to 6 years and 40 percent of patients over 6 years. The emphysema never reached severe proportions. An exact figure for the upper limit of the percentage of emphysema is not readily available but figures are in the region of 10 percent. The Reid index was used to quantitate the mucous gland hyperplasia and this index was increased in all age groups, the presence of squamous metaplasia further compromising mucociliary clearance. Pulmonary hemorrhage was an occasional finding and is probably due to the bronchopulmonary anastomoses, which complicate the bronchiectasis. Anastomoses between pulmonary veins and the bronchial arteries were detected. Subpleural air cysts especially in the upper lobe are found and explain the pneumothoraces that may occur in cystic fibrosis. They have been categorized into three types: bronchiectatic cysts, interstitial cysts, and emphysematous bullae.[250] A small series of cases with a necrotizing vasculitis associated with intraluminal thrombi has been reported. It may be found in vessels involving skin, lung, and gastrointestinal tract. Interestingly, in 40 percent of patients there were positive antineutrophil cytoplasmic antibodies, which were not detected in patients without vasculitis.[251] The etiology of the vasculitis is obscure.

FUTURE DEVELOPMENT

Two new approaches result from the identification of the cystic fibrosis gene. Antenatal diagnosis can specifically identify the presence of cystic fibrosis in a fetus of known carriers. Further, screening programs can now be developed. Due to the large number of different cystic fibrosis mutations, however, and the resource only to examine for a relatively limited number of the more common mutations, current programs are only likely to detect approximately 85 percent of cases. The other major potential advance is in the development of gene therapy. Insertion of the cystic fibrosis gene into the defective cell should result in normal CFTR production. In vitro, gene transfer has been successful using vectors such as adenovirus and liposomes. Transgenic mice have also been successfully transfected. The significant differences between the manifestation of cystic fibrosis in murine models and their pulmonary anatomy, however, still leaves a major jump from animal model to human therapeutic success.[252,253]

α₁-Antitrypsin deficiency

α_1-Antitrypsin (AAT), a 52-kDal glycoprotein, is a major serum protease inhibitor and is produced in the hepatocytes as well as in macrophages. A deficiency in AAT results from mutations in a gene located on chromosome 14q31-32.3.[254] There is extensive genetic heterogeneity at the AAT locus with approximately 75 alleles identified[255]; the discovery of AAT deficiency has been documented by Eriksson.[256]

Classically the phenotype (Pi type) is determined by isoelectric focusing of serum. Over 30 phenotypes have been described and the nomenclature is based on letters corresponding to the position of migration on isoelectric focusing of the serum protein. The most common phenotype in Great Britain is PiM. Low serum AAT levels are seen in PiZ and PiS. In addition there are rare Pi 'null' forms associated with no detectable serum AAT.[257] The M and Z variants differ by a single base substitution in exon V of the normal gene, causing a glutamic acid-lysine substitution at position 342 in the molecule. The epidemiology of AAT deficiency has been reviewed by Hutchinson.[258]

α_1-Antitrypsin deficiency is an autosomally recessive condition seen in 1/3000 caucasians but the incidence varies geographically. The typical pulmonary manifestation is basal panacinar emphysema although in some, the emphysema may be centrilobular. In some cases a clinical picture more typical of chronic bronchitis or bronchiectasis may develop.[259] Although it is well recognized that the lack of AAT leaves proteolytic enzymes released by neutrophil polymorphs and macrophages free to cause damage to the alveolar walls, the cause of the basal distribution of the disease is unknown. It is unlikely that the basal distribution is due solely to chronic bronchitis and it is postulated that the proteolytic enzymes are preferentially distributed to the lung bases. In the supine posture, the apices of the lung are better perfused than the bases and should thus be at least at equal or greater risk from blood-borne pathogens during any illness confining the patient to bed. Similar changes can be caused by installation of papain into the trachea of experimental animals. In addition to emphysema, certain AAT phenotypes are also associated with other pulmonary diseases. There is a significant increase in PiZ phenotype in patients with fibrosing alveolitis, both with and without rheumatoid disease. It is postulated that low levels of AAT associated with non-M phenotypes may predispose to tissue damage with subsequent fibrosis.[260]

Patients with the PiZ phenotype are at risk from emphysema whether or not they smoke. In a recent study by Eriksson and colleagues, 60 percent of the patients with PiZ phenotype developed emphysema. These patients are also at risk of developing cirrhosis[261] and AAT bodies may be demonstrated on D/PAS or immunoperoxidase staining. Such bodies are not conclusive evidence of an abnormal AAT phenotype.[262]

Smoking will accelerate a decline in lung function in patients homozygous for AAT deficiency (PiZ).[263] There is still disagreement as to whether the heterozygous deficiency state for AAT (PiMZ) will lead to emphysema. There is a belief that smoking and the heterozygous state will cause lung disease at an earlier age in smokers. It appears that soon an artificial AAT derived by recombinant DNA will be developed to replace the missing inhibitor and this will be used in treating these patients.

Other Pulmonary Disorders

The lung is often the target organ when a child suffers a systemic disease such as chronic granulomatous disease, where there is defective neutrophil function or in hypogammaglobulinaemia. Such disorders are outside the scope of this book, except to note that in chronic granulomatous disease the histologic picture may be confused with diseases such as tuberculosis.

PERINATAL PATHOLOGY

At birth, the lung turns from being largely redundant into a fully functioning respiratory organ. Perinatal pathology is dominated by the problems related to this transition. This is especially so in the very preterm infant where the lung is immature both structurally and biochemically. The archetypal condition in this category is hyaline membrane disease (HMD). Near term, however, the problems of transition for the infant are less the result of immaturity than uteroplacental dysfunction. This may occur shortly before or more frequently during labor leading to asphyxia. "Acquired" pulmonary pathology manifests itself primarily as meconium aspiration or infection.

Hyaline Membrane Disease

Hyaline membrane disease and respiratory distress syndrome (RDS) are frequently used synonymously. RDS describes a clinical condition of cyanosis, intercostal and sternal retraction, and a grunting respiration starting within an hour or so of birth. The chest radiograph shows a ground-glass appearance in the lung fields with an air bronchogram. Blood gases show hypoxia and hypercarbia. The incidence varies sharply with gestation and RDS will be seen in some 2 percent of babies; up to 15 percent of infants less than 2500 g will suffer respiratory distress as will 60 percent of very preterm infants under 1500 g in weight.[264] These figures are now modified by the early use of artificial surfactant that has reduced both the incidence and severity of respiratory distress in the preterm. This clinical syndrome will almost but not invariably be associated with the pathologic change in the lung of HMD.[265]

Macroscopically, the lungs of babies dying with acute RDS are collapsed, are deep red-purple in color, and have the texture of liver. Histologically, the earliest change is necrosis of respiratory epithelium lining the distal bronchi and bronchioles.[266] This is often a particularly striking feature in the very preterm infants (approximately 24/25 weeks) surviving only a few hours. Some of this epithelium may dislodge and block some of the more distal airways. Early hyaline membrane formation will be found in parts of the lung, occasionally within half an hour of birth, in the very preterm. In most cases usually they take more than 2 hours to form. The membrane lines the distal airways, especially the terminal bronchioles and developing alveolar ducts (Fig. 3-43). The most peripheral part of the lung, the terminal sacs, is collapsed and is rarely lined by membrane.

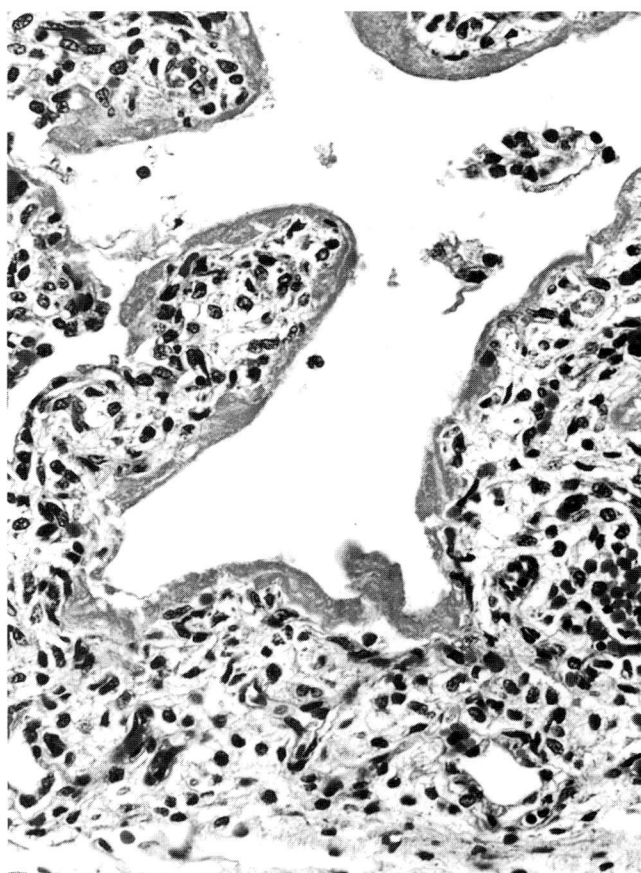

FIGURE 3-43
Hyaline membrane in a 25-week gestation infant at 6 hours of age. Note that the hyaline membranes do not involve the terminal sacs.

The eosinophilic membranes are composed mainly of proteinaceous exudate derived from plasma together with some nuclear debris from necrotic epithelial lining cells but scant if any fibrin.[267] Fibrin thrombi may be present in pulmonary arterioles. Occasionally, the membrane may be stained yellow, particularly if the bilirubin level is raised. The presence of the "yellow HMD" has no further significance. Some membranes will show yellow staining on the luminal surface but the more typical eosinophilic appearance in the deeper layers.

Reaction to the injury and presence of hyaline membrane proceeds rapidly from about 24 hours after birth. By 48 hours macrophages (membranophages) are a prominent feature ingesting the membrane. Macrophages are also present in the surrounding interstitium beneath the membrane. Neutrophils are relatively inconspicuous histologically although they usually are a prominent component of bronchoalveolar lavage specimens taken during the acute phase of HMD.[268] With time there may be removal of the membranes as well as regeneration of bronchiolar epithelium. Focal edema and fibrosis are present around respiratory bronchioles. When the infant has not required ventilation or high levels of inspired oxygen there may be resolution with reepithelialization. Only residual peripheral balls of membranes are likely to be found by 5 days after onset of the disease.[269] This uncomplicated progression of HMD should not, of itself, cause death in modern neonatal practice.

PATHOGENESIS AND ETIOLOGY

Hyaline membrane disease is an indication of acute lung injury where the barrier between the airspace and vascular lumina is damaged. The failure of epithelial-endothelial integrity allows efflux of plasma components and, following partial resorption, deposition of eosinophilic blood proteins onto the wall of the bronchiole or alveolus, that is, the hyaline membrane.

The most common situation in which HMD occurs neonatally is with prematurity. The lack of surfactant, which leads to collapse of terminal air sacs at end expiration, has long been recognized as having a fundamental role in the pathogenesis of the hyaline membranes. In an attempt to expand air sacs, either by the infant's own respiratory efforts or the ventilatory pressures required to prevent lung collapse, uneven transpulmonary pressures are probably generated.[270] Shear forces caused by the expansion of terminal and respiratory bronchioles give rise to damage to the epithelial-endothelial barrier and passage of proteins into the airway. These proteins may compound the problem by further inhibiting surfactant activity.[271]

Ischemia may play a role in the HMD of prematurity with variation in blood flow to different parts of the airway caused by uneven lung expansion.[272] After vasoconstriction, hyaline membranes could form in a period of reflow and the surfactant deficiency result from ischemic injury to the type II pneumocytes. The importance of ischemia and vascular damage receives support from evidence of increased pulmonary intravascular coagulative activity.[273] In addition the pulmonary vessels in the neonate are very reactive, muscularized, and capable of vasoconstriction.

Hyaline membranes in the neonate are not exclusively associated with surfactant deficiency and prematurity. They may be caused by a variety of insults to epithelial-endothelial integrity. Group B streptococcal infection may mimic the RDS of prematurity clinically.[274] Pathologically the hyaline membranes may show a blue tinge due to the overwhelming

numbers of bacteria. The epithelial-endothelial damage has been shown to be associated with induction of tumor necrosis factor.[275]

Some neonates present with HMD at term, usually in association with birth asphyxia. In this subgroup, the clinical and pathologic profile is more similar to that seen in adult HMD than the typical preterm disease.[276–278] In this instance, epithelial-endothelial injury may be primarily ischemic in origin with toxic radical production, microthrombi, and leukocyte accumulation in the microvasculature amplifying the initial damage. Injury to type II pneumocytes producing a secondary surfactant deficiency may further contribute to hyaline membrane formation.

Bronchopulmonary Dysplasia

The term bronchopulmonary dysplasia (BPD) was originally used to describe a constellation of clinical, radiologic, and pathologic features following ventilatory therapy for HMD.[279] BPD has been defined in slightly different ways but, at a clinical level, this must be:

Acute lung injury (usually HMD)
Some form of continuing ventilatory support
Pulmonary radiographic abnormalities, such as persistent increased densities[280,281]

Importantly, the chronic changes must be present beyond a specified time, for example, 28 days. Although this may be a reasonable approach to a clinical diagnosis, there is little doubt the onset of BPD can be recognized earlier both clinically and pathologically.[282]

A further difficulty in defining BPD is that the character of the disease has changed over the years with improvements in ventilatory technique and with the introduction of new therapies, such as surfactant replacement. Broadly, BPD is now rarely a severe lung injury. However, some infants still require some form of long-term ventilatory support. Increasingly, infants are described as having "chronic lung disease." Unfortunately, this "diagnosis" has not been defined and it is not always clear whether it is to be considered equivalent to BPD. One perceived advantage with the term "chronic lung disease" is that while BPD often implies a particular etiology (see below), chronic lung disease does not. Also some infants may develop a chronic lung disease without fulfilling all criteria for BPD, particularly the requirement for an early acute lung disease process.[283,284]

INCIDENCE

Partly because of the difficulties in definition, it is not easy to arrive at precise figures for the incidence of BPD. Kraybill and colleagues[285] found chronic lung disease in up to 54 percent of very low birth weight survivors at 30 days of age but only 5 percent at 6 months. Wide geographic variations were a feature in this study. Shaw and coworkers[286] note that the incidence is inversely related to gestational age and the incidence may be increasing probably due to improved survival of the very immature infants.

PATHOLOGY

This change in clinical severity is also matched by an alteration in the spectrum of lung pathology seen at autopsy. The pathology of BPD has been described in terms of three stages,[138,287] an exudative, early reparative stage; a subacute fibroproliferative stage; and a chronic fibroproliferative stage.

Early, Exudative Phase
The pathology is that of acute lung injury together with a reaction to the often aggressive ventilatory support. This phase lasts from approximately 3 to 9 days.

Bronchial and bronchiolar damage may be particularly prominent. There will be squamous metaplasia, submucosal muscular hyperplasia, and obliterative bronchiolitis (Fig. 3-44). Many airways will also show residual hyaline membranes some of which become incorporated into the fibroproliferative process. Unlike uncomplicated HMD described above, hyaline membranes may continue to form beyond the first day. The membranes may contain more fibrin and there may be involvement of the peripheral alveoli. The interstitium is edematous with florid interstitial fibrosis, smooth muscle proliferation, and type II pneumocyte metaplasia. In many affected infants the immediate cause of death may well be due to respiratory failure.

Subacute, Fibroproliferative Stage
This stage is more a transitional one from the early acute lung disease to the chronic phase described below. It is seen from the end of the second week up to the end of the first month. Obliterative bronchiolitis is still present and interstitial fibrosis becomes more pronounced. Reepithelialization progresses with type II pneumocytes.

Chronic BPD
Macroscopically, the lung is firm, has a rather nodular appearance on the pleural surface (Fig. 3-45). Examination of the pleural or cut surface may reveal emphysematous air spaces or a fairly uniform firm

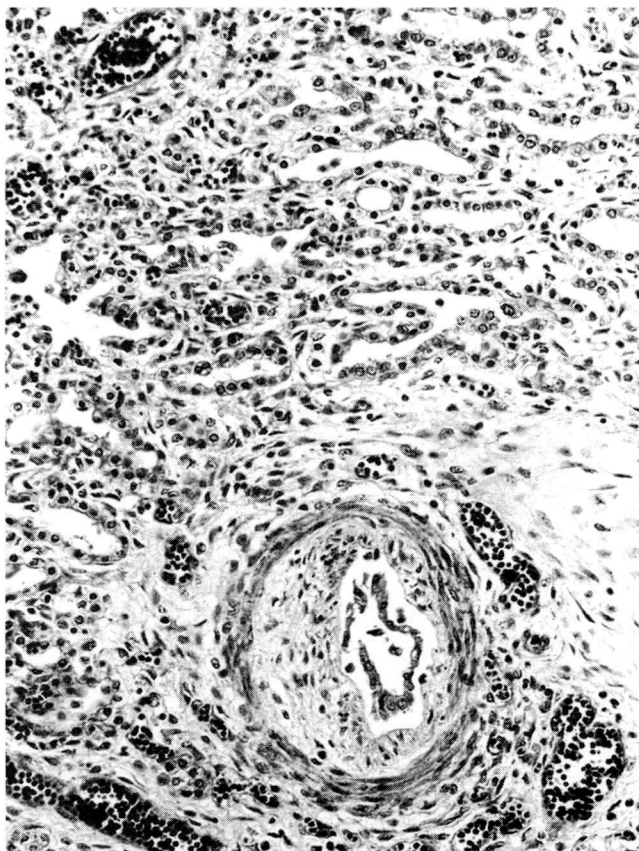

FIGURE 3-44
Obliterative bronchiolitis in acute bronchopulmonary dysplasia.

parenchyma. Histologically, the low power appearance can be remarkably variable. In some cases, large emphysematous airspaces may be adjacent to more normal-sized spaces (Fig. 3-46). This geometric distortion of the airways probably contributes significantly to the degree of respiratory failure and may reduce gas exchange capacity to 25 percent of normal.[37] Such variation may not be present, but the lung will show more uniformly increased alveolar spaces with reduced alveolar surface area.[288,289] Special stains will reveal thickened elastin plaques at alveolar crests. Airspaces are frequently lined by type II pneumocytes. More proximal airways will be relatively normal although there may be some persisting metaplastic squamous epithelium and smooth muscle hypertrophy. Some 25 to 50 percent of cases may show glandular hyperplasia and patchy chronic inflammation.[289,290]

The vasculature in this late stage may be one of the most clinically significant elements leading to cor pulmonale. The more common features are hypertrophy of the medial smooth muscle of the smaller muscular arteries and an increase in adventitial fibrous tissue.[291,292] Endothelial cells may remain plump

95

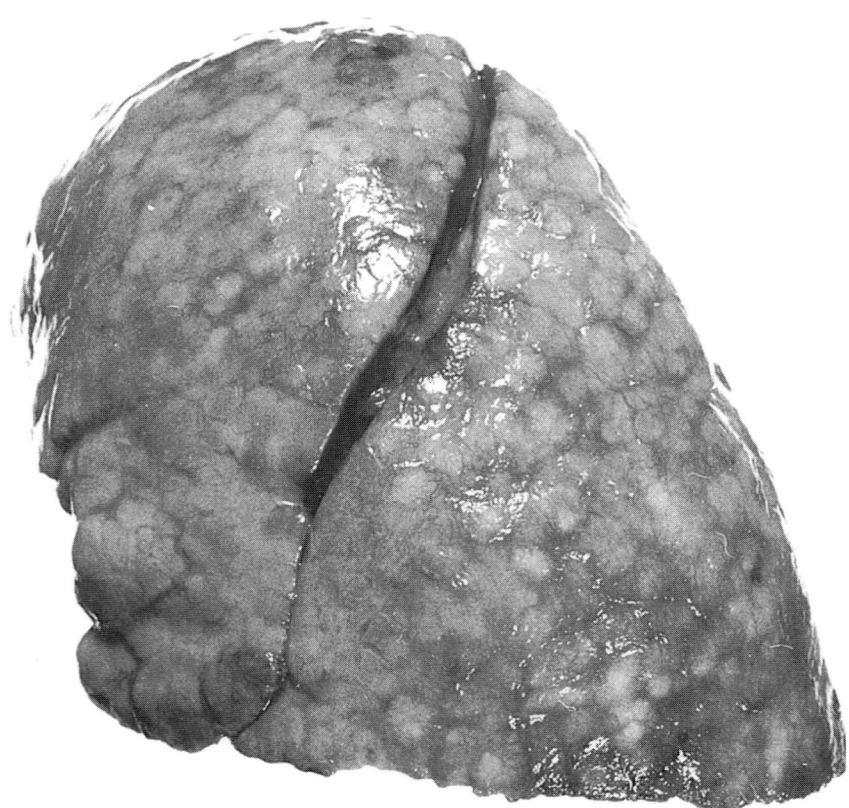

FIGURE 3-45
Nodular surface of lung with
bronchopulmonary dysplasia.

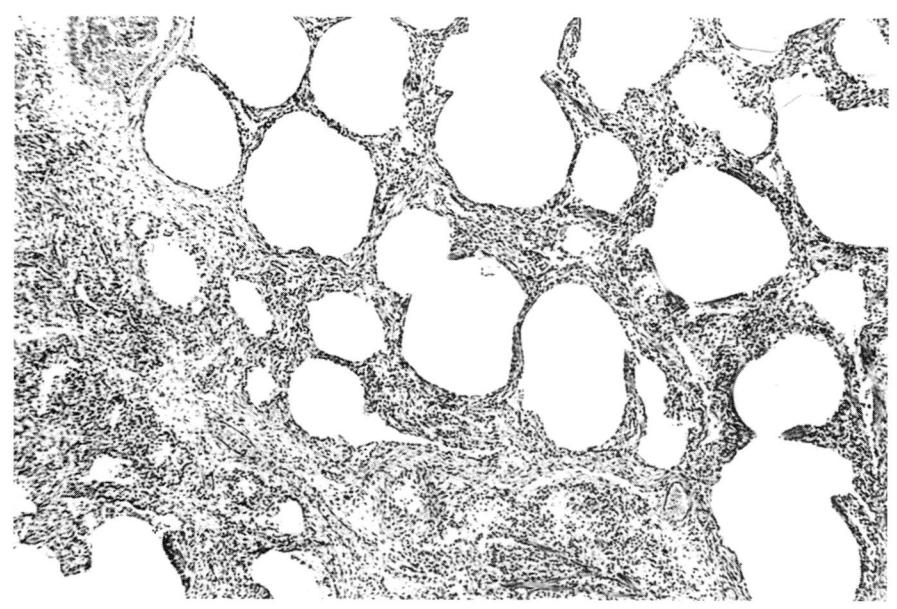

FIGURE 3-46
Lung with chronic bronchopulmonary
dysplasia and geometric distortion.
These emphysematous areas adjacent to
more collapsed fibrotic lung may be
established relatively early in the disease
process and have a significant effect on
ventilatory efficiency.

and contribute to the vascular obstruction. A reduction in normal peripheral arterial recruitment in the immediate neonatal period may lead to a reduction in peripheral arterial numbers.

In some cases, where the neonate has survived many months, the pathologic changes within the lung are surprisingly unimpressive. This is despite a clear history of, and indeed death due to, chronic lung disease clinically. Architecturally, many of the alveoli are larger than normal and there has been significant interference with normal development. Interstitial thickening is often minimal. However, increased reticulin can be detected and thickened elastin is present at septal crests. Collagen may be demonstrated between capillary walls and the alveolar airspace. Bronchial and bronchiolar changes are unimpressive. Much of this appearance will reflect a remodeling process that follows the earlier more fibroproliferative phases.

The vascular changes should not be ignored. Overt evidence of vascular pathology may be absent and arterial medial thickening unimpressive. It is likely, however, that the lack of alveolar development is associated with a significantly reduced capillary vascular bed. This may be compounded, as noted above, by a reduced peripheral arterial supply. The long-term functional effects of chronic lung disease on survivors is as yet unknown.[293]

WHAT CONSTITUTES BPD?

There is no clear definition, with minimum criteria, that defines BPD pathologically. Further, it is probable that most infants who develop and perhaps die from "chronic lung disease," have barely passed through the acute and subacute phases of BPD outlined above. The above description of the early phases of "acute" BPD during which infants die undoubtedly can be seen by pathologists, but it is increasingly rare.[294] Indeed, there have been no cases at the John Radcliffe Hospital in the past 3 years coming to autopsy, in which acute BPD was the cause of death in the absence of extreme prematurity (under 26 weeks' gestation) or a significant congenital anomaly. The lungs of these infants may well show changes but they are less florid and simply represent the usual repair and regenerative processes associated with HMD, usually with an element of reaction to ventilatory support. Bronchial changes such as obliterative bronchiolitis and even squamous metaplasia are rarely a feature. It is doubtful that this sort of appearance would justify the term "broncho-pulmonary dysplasia," acute or otherwise. In some cir-

cumstances it is possible to attribute the lung changes to a particular etiology (e.g., repair, "ventilator lung", "oxygen toxicity," etc.)[37,295] without becoming excessively concerned as to whether or not the label BPD should be attached to the lung.

PATHOGENESIS

The emergence of BPD as a disease process was closely linked to the respiratory support needed to treat neonatal HMD. The underlying mechanism in chronic lung disease is probably damage to the epithelial-endothelial barrier with chronic leakage of fluid and protein into the interstitium, stimulating fibrosis. Although its role in adult respiratory distress syndrome has been questioned,[296] in the neonate, probably the most important factor in damaging the endothelium-epithelium is the high oxygen concentrations. Notably, the preterm lung may be susceptible to relatively normal inspired oxygen concentrations. These levels would still be much higher than the in utero environment to which the lung is adapted. Highly reactive toxic radicals generated either by normal cellular metabolism or from recruited inflammatory cells[297,298] may cause endothelial-epithelial damage. This would lead to protein leakage from vessels and interstitial fibrosis. Rapid activation of normally low antioxidant mechanisms in the preterm infant may be an important defense mechanism.[299–301] Enzymes and other chemical mediators from inflammatory cells may also contribute to increased vascular permeability or tissue destruction.[297,302–304]

An important additonal factor in the more acute forms of BPD is the high ventilatory pressure sometimes needed to maintain adequate oxygenation. It is suggested that this is especially associated with the airway damage and obliterative bronchiolitis seen in acute BPD.[305] The pressure and volume effects of ventilation are complex and they may be also important in causing damage at the alveolar level, for example, by reducing capillary perfusion.[306]

Wilson-Mikity Syndrome

In 1960, Wilson and Mikity described a new form of lung disease.[307] They reported data on five preterm infants, all weighing between 1200 and 1300 g, dying from a chronic lung disorder after a few months. One case perished after only 27 days. The pathology was similar to chronic BPD. Subsequently, only isolated cases have been reported. An important aspect of the syndrome that has persuaded authors to retain it as a separate disorder is the context in which the condition developed. All infants had either no or only a very mild initial lung disease. All had a period of often a month with no outward sign of pulmonary disease.

The etiology of the chronic lung disease in this circumstance has never been satisfactorily explained although the interruption of normal development probably plays an important role. In recent years, the advent of surfactant therapy has seen a rise in the number of cases where the severity of the initial acute lung disorder is significantly reduced, or in some cases removed but chronic lung disease still occurs. It might be inappropriate to label these cases Wilson-Mikity syndrome. However, it highlights the fact that acute lung disease with its associated therapy, which feature prominently in discussions on the etiology of BPD, is not a requirement for the development of neonatal chronic lung disease.

Pulmonary Interstitial Emphysema

Air leaks are a complication of RDS and its ventilatory therapy. The relatively high pressures damage the delicate loose connective tissue of the lung. Another cause of damage is the increase in alveolar pressure associated with aspiration of foreign material such as meconium that occludes or has a ball-valve effect on the bronchi or bronchioles. An air leak at the pleural surface will precipitate a pneumothorax that can cause acute collapse and require rapid drainage. An air leak confined to the interstitium gives rise to pulmonary interstitial emphysema. This can track along tissue planes to create a pneumopericardium. If severe, compression of surrounding alveoli may significantly embarass respiration and may even cause mediastinal shift. Unilateral pulmonary interstitial emphysema may allow selective intubation and ventilation of the nonaffected lung while the other recovers. The persistence of the disease is probably related to the continuous and prolonged use of positive pressure ventilation.[308]

Macrosopically the lungs show air- or air and fluid-filled cysts localized to the interlobular septa, the cysts extending radially from the hilum to the subpleura. These are irregularly shaped, are lined by a glistening smooth membrane, and tend to isolate or partially surround bronchi and vessels. Small blebs may be visible on the subpleural surface.

Microscopically, acute interstitial emphysema in a very immature lung simply appears as large circular or oval empty spaces surrounded by compressed lung. In many instances the emphysema is within distended lymphatics of the interlobular septae. In some of the more immature infants, pulmonary hemorrhage into the emphysema is a terminal event. If the interstitial emphysema is persistent, it stimulates fibrosis and a foreign body giant cell reaction (Fig. 3-47). Chronic pulmonary interstitial emphysema may sometimes feature as a differential diagnosis of

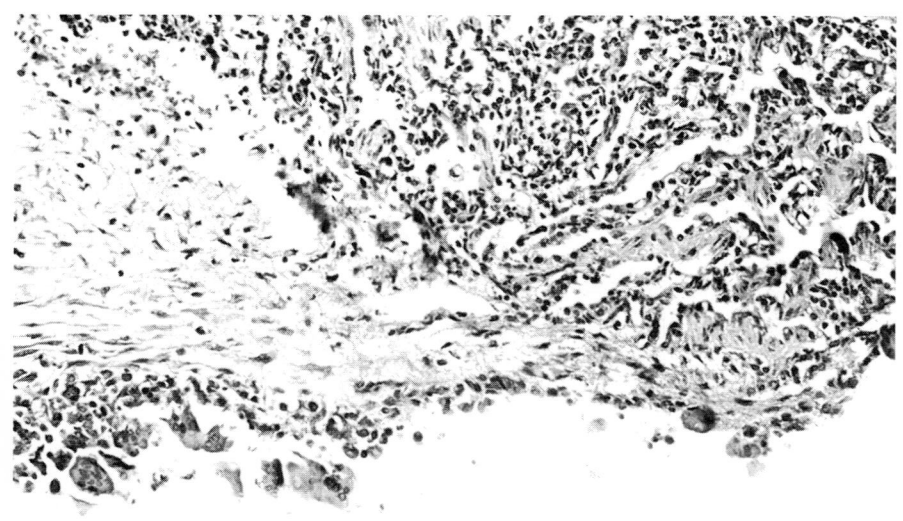

FIGURE 3-47
Cyst wall in chronic pulmonary interstitial emphysema. The fibrous wall is lined in part by macrophages including giant cells.

cystic lung disease, but cysts lined by fibrous tissue and giant cells are a useful diagnostic feature.

Pulmonary Hemorrhage

Focal hemorrhage into the alveoli or interstitium is a common finding in neonates, usually in association with HMD. Occasionally, however, massive hemorrhage may be found with fresh blood involving large confluent areas of lung. Often there is a clinical history of collapse and aspiration of a considerable amount of heavily blood-stained material up the endotracheal tube during attempted resuscitation. Preterm infants are especially affected. There may be a history of birth asphyxia but the underlying cause is unclear.[309,310] It has been suggested that it just represents an extremely hemorrhagic acute pulmonary edema and reflects terminal heart failure.[311]

Surfactant Replacement Therapy

A major feature of respiratory distress is surfactant deficiency and appropriate replacement or prophylactic therapy has long been a goal of neonatologists. A number of preparations are now available some natural, others artificial. Studies are continuing as to the best formulation and therapy regimen. There is little doubt that artificial surfactant is effective in reducing the incidence, severity, and mortality of HMD. Evidence indicates that other complications including air leaks and bronchopulmonary dysplasia[284] are also being reduced. These results are not confirmed in all studies.[312–315] At present the only complication that is usually noted is an increase in pulmonary hemorrhage.

To date, published descriptions of the lung pathology after surfactant therapy are limited.[301,302] Our own experience is that the lungs show little gross difference from other babies dying with HMD who have not been treated with surfactant. Pinar and coworkers[316] describe lungs in infants dying having received a natural surfactant from bovine lungs. No qualitative differences were found when compared with lungs that did not receive surfactant. However, surfactant-treated lungs showed a reduced number of hyaline membranes, less interstitial emphysema, and epithelial necrosis. No significant difference in the level of pulmonary or intracranial hemorrhage was noted. Whether these conclusions are applicable to all surfactants is not yet resolved. Thus with some surfactants, giant cells may be more of a feature. It is usually impossible to detect the artificial surfactant itself although, if fixation occurs in unbuffered formalin, pigment can precipitate out in association with the lipid in alveoli.

Intralipid Embolism

Lipid can sometimes be found in the lung capillaries of infants. Although this probably results mainly from lipid given as a part of parenteral nutrition[318] some argue lipid infusion is not a prerequisite.[319] Histologically, it is most easily identified by lipid stains on fresh tissue but is also usually apparent even after normal processing. Capillaries are round and are devoid of red blood cells (Fig. 3-48). At vessel margins, slightly refractile material can usually be seen.

The effect of lipid on lung function is difficult to assess[320] and evidence of acute clinical deterioration due to vascular obstruction is lacking. There is some evidence of a reduction in blood oxygenation in infants receiving lipid infusion.[321] Further, some studies suggest that infants receiving lipid are at an increased risk of chronic lung disease particularly if it is given too early in the neonatal period.[322] The mechanism of action of lipid damage is unclear, but various metabolites released from the emulsions may increase pulmonary vascular tone.[320]

Meconium Aspiration

Passage of meconium by the fetus in utero or intrapartum may result in aspiration and the meconium aspiration syndrome. Fetal discharge of meconium appears to be a reflex response to acute hypoxia, which both increases peristalsis and causes reflex anal dilatation. These reflexes are gestation dependent and meconium release is unusual before 38 weeks[323] and rare before 34 weeks. Infants present with respiratory distress and chest radiographs are abnormal with evidence of atelectasis. Clinically, some infants may appear surprisingly normal at birth with good Apgar scores before subsequent deterioration.[324]

Macroscopically, lung surfaces may not show any gross abnormality in the acute phase, except the nonspecific feature of petechial hemorrhages due to asphyxia (Fig. 3-49). The major airways may contain residual sticky dark green meconium, but this may have been washed out during the resuscitation process. With compression, the cut surface may exude meconium from the smaller airways. The lung pathology of meconium aspiration is due initially to obstruction of smaller airways with the tenacious meconium causing distal collapse; in other parts of the lung, emphysema may develop due to a ball-

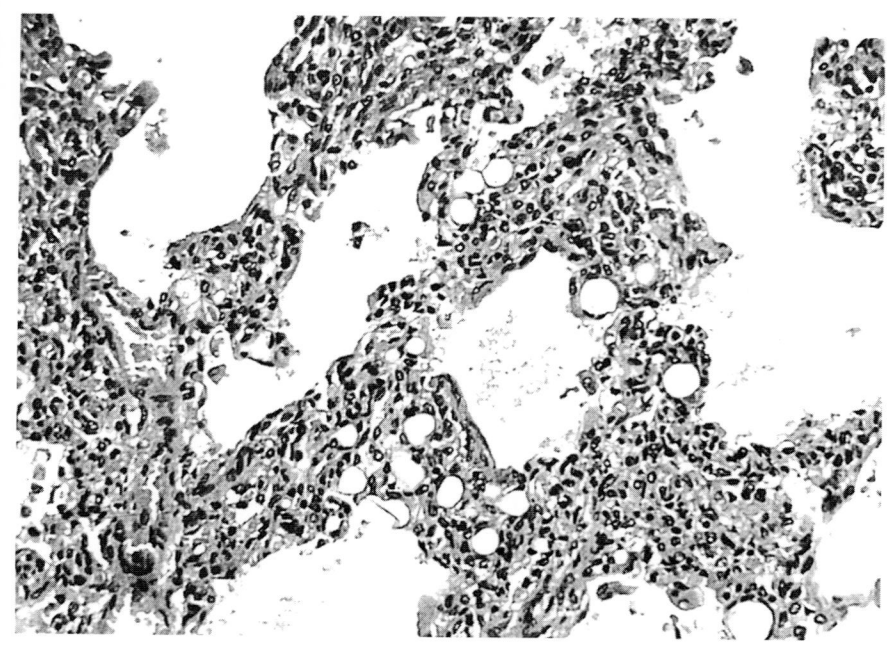

FIGURE 3-48
Intralipid in pulmonary capillaries. Capillaries are round, empty and contain slightly refractile material. (H&E.)

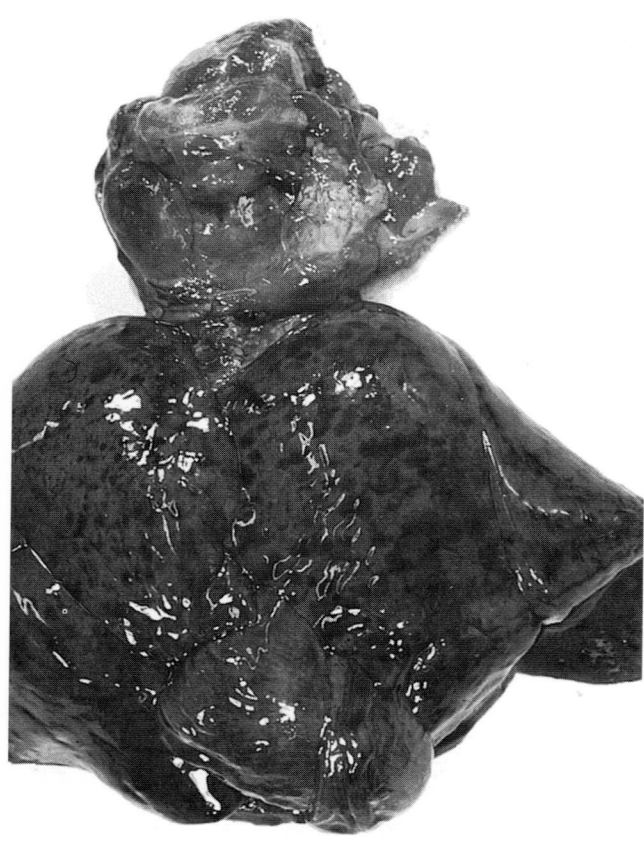

FIGURE 3-49
Surface of lungs in infant dying from meconium aspiration. The petechial hemorrhages are an indication of acute asphyxia not meconium aspiration specifically.

valve effect and this overdistention can precipitate a pneumothorax (see above). An acute inflammatory response to the presence of the meconium can occur within hours. Some of this may be chemical in origin and there is evidence that bile salts have a direct toxic effect on pneumocytes.[325] However, some experimental evidence suggests the pneumonia occurs due to bacteria inhaled with the meconium that stimulates reaction.[326]

Histologically, the meconium appears as a mixture of mucus with granular eosinophilic material often containing small yellow meconium bodies. Squames are a prominent feature of the aspirated material (Fig. 3-50). In addition to the inflammatory infiltrate, hyaline membranes may be present reflecting the acute hypoxic insult to the lung that usually accompanies the meconium aspiration.

A feature that often dominates the clinical course of meconium aspiration is persistent pulmonary hypertension or persistent fetal circulation. The ventilation-perfusion mismatch associated with meconium aspiration produces an acidosis and hypoxia causing vasoconstrictive pulmonary hypertension. Histologically, muscularization of pulmonary arterioles may be found, usually taking at least 2 days to become established. Some authors argue that the excessive muscularization of preacinar and intra-acinar arterioles found in these infants has already been established in utero. As such it is not a complication of the meconium aspiration but reflects a common etiology of chronic intrauterine stress and hypoxia.[327,328] Others have suggested that method-

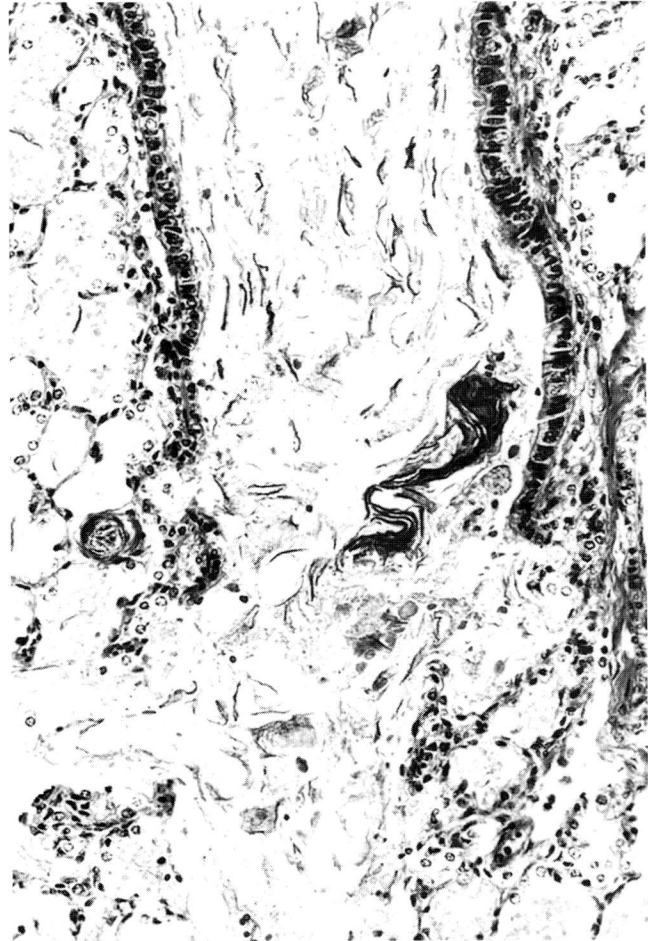

FIGURE 3-50
Meconium in a respiratory bronchiole and alveoli. Aspirated material contains mucus, squames, and amorphous granular eosinophilic material. *(Reproduced with permission from J. Keeling (ed), The Respiratory System in Fetal and Neonatal Pathology, Springer-Verlag, 1993.)*

ologic defects have led to an underestimation of the normal muscularization of preacinar arterioles in normal infants[329] and that the vascular changes are of postnatal origin.

Persistent Pulmonary Hypertension of the Newborn

Persistent pulmonary hypertension of the newborn (PPHN) is almost invariably related to increased pulmonary vascular resistance rather than increased flow. There is a left-to-right intrapulmonary shunt and shunting will also occur across the foramen ovale and ductus arteriosus if these structures remain patent. Thus PPHN is often called persistent fetal circulation.[329] It may occur in association with congenital heart disease or congenital alveolar dysplasia (congenital misalignment of the pulmonary veins; see above). More common associations include meconium aspiration syndrome, sepsis including

pneumonia, and pulmonary hypoplasia. Very rarely, there will be no recognized underlying condition and it may be appropriate to consider the pulmonary hypertension primary. Some authors doubt the existence of idiopathic PPHN.[330]

Normally, there is a postnatal fall in pulmonary vascular resistance but in PPHN this fails to occur or be maintained. Structurally there is hyperplasia of the media in the preacinar and intra-acinar pulmonary arteries and arterioles with extension of smooth muscle to involve the precapillary vessels. In some instances, muscularization of these vessels is so complete that luminal obliteration almost occurs.[331] Whether idiopathic PPHN or that associated with meconium aspiration syndrome is due to an intrauterine structural abnormality of the vessels is a matter of some debate.[327,330,331]

Other Therapeutic Complications

NECROTIZING TRACHEOBRONCHITIS

Necrotizing tracheobronchitis[332,333] is primarily a complication of high-frequency ventilation (HVF), a ventilatory strategy not widely available or used in the United Kingdom at present. HVF is a technique aimed at reducing the high tidal volumes used in more conventional approaches with a view to reducing lung damage.[334] It may be present in up to 90 percent of infants dying after HFV.[335]

The major histologic characteristics are ulceration, with mucosal hemorrhage, inflammation, and accumulation of luminal mucus. Some of the changes may be seen where the endotracheal tube is in contact with the trachea, the occurrence of the lesions distal to the tip of the tube, and indeed extending into the main bronchi, indicates that the ulcerations are not simply contact trauma-related. The pathogenesis is not entirely clear but may be related to excessive drying of tissues or the "jet-hammer" effect on tissues subject to HFV.[336,337]

EXTRACORPOREAL MEMBRANE OXYGENATION

Extracorporeal membrane oxygenation (ECMO) is an increasingly used ventilatory modality in respiratory failure where conventional therapy is failing to maintain adequate ventilation.[338] In neonates its main use is in the treatment of conditions such as meconium aspiration syndrome, diaphragmatic hernia, and sepsis. ECMO may benefit the lung partly by protecting it from the effects of the underlying disease process, improving tissue oxygenation, and improving capillary perfusion. It reduces the severity of hypoxia-induced vasoconstrictive pulmonary hypertension. Although some data suggest there may be an increase in pleural effusions and hemothoraces, there are few direct pathologic data on the effect of ECMO on the lung. The presumption is that the pulmonary pathology is dominated by the underlying disease process for which ECMO is used. Chou and colleagues[339,340] have reported on lungs examined at autopsy following ECMO therapy. Although it is difficult to disentangle the effects of the underlying disease process from the effects of ECMO, some changes did appear to be directly attributable to the therapy itself.

In the early stages the major findings were interstitial and intra-alveolar hemorrhages together with hyaline membranes. Type II hyperplasia was an early feature and by day 7 of ECMO therapy interstitial fibroplasia was prominent. This was often accompanied by other changes affecting the bronchi including epithelial hyperplasia, squamous metaplasia, and smooth muscle hyperplasia. Intraalveolar calcification was present after 7 to 15 days. In three cases where ECMO had been given for between 2 and 3 weeks, mucinous metaplasia affecting terminal bronchioles was noted. These latter two features appear characteristic of ECMO lungs and are not seen secondary to meconium aspiration or conventional ventilation.[339,340] These authors argue that ECMO is important in the generation of these histologic features because their development is more related to the period of ECMO therapy than the nature of the underlying primary condition or length of preceding conventional ventilation.

Infection

Pulmonary infections are largely dealt with elsewhere. This section will discuss briefly the problem of infection only in the context of the perinate.

ASCENDING INFECTION

In utero, fetal infection is due mainly either to ascending infection through the maternal os or transplacentally, the former being the more common. Amniotic fluid infection is common before 22 weeks of gestation and evidence may be found in 20 to 30 percent of spontaneously aborted fetuses. Fetal lungs may also show evidence of the infection with polymorphs in the airways and developing terminal sacs. Contrary to former opinion, studies on male fetuses have shown the inflammatory cells visible in the airways contain the Y chromosome and are therefore fetal cells, not aspirated maternal cells from amniotic fluid.[341] In response to infection, lymphoid aggregates may develop adjacent to bronchial epithelium; these are part of the mucosal-associated lymphoid system and have parallels to the Peyer's patches of

the gut.[342] Infection may be associated with spontaneous abortion, stillbirth, or early onset neonatal pneumonia (see below). Organisms that may be isolated are often bacterial vaginal commensals such as group B streptococci but fungal infection with *Candida* may also occur. In the abortuses, despite a florid histologic pneumonia, culture may be negative. The involvement of organisms such as *Mycoplasma* or *Ureaplasma* has been suggested for a proportion of these cases.[343,344]

NEONATAL BACTERIAL PNEUMONIA

Neonatal pneumonia can be divided into early and late onset. The former is acquired shortly before or during birth and results from aspirated infected or colonized material such as amniotic fluid, blood, or vaginal material. Infecting organisms will include group B streptococci, enteric bacteria, or staphylococci. Infection may be overwhelming, particularly with group B streptococci. Death may occur so rapidly after birth that diagnosis will only be made at autopsy. The pathology is nonspecific. At autopsy, the lungs will be heavy and often appear red-purple in color similar to HMD. Occasionally a blood-stained effusion is present. Histologically, the lung will show a nonspecific pneumonia but, particularly with group B streptococci, hyaline membranes with a blue tinge due to the overwhelming numbers of bacteria may be found (Fig. 3-51).[345]

Late-onset neonatal pneumonia occurs toward the end of the first week of life. It may be identical to early onset disease in terms of causative organism, mode of infection, and pathology. Organisms such as *Pseudomonas* but occasionally *Proteus* may feature, often with characteristic involvement of vessel walls, thrombosis and a hemorrhagic exudate.[346,347] Sometimes it is only at autopsy that organisms such as *Pseudomonas* will be recognized as causing an infection, as opposed to colonization. In septicemia, the lung may be a site of infection, possibly secondary to an indwelling catheter. Candida is by far the more common fungal infection (Fig. 3-52) but aspergillosis is reported and both may be associated with hemorrhagic pneumonias due to vessel involvement.[348,349]

More uncommon pneumonias may be due to organisms such as *Listeria monocytogenes*. Often part of a disseminated infection, it leads to small abscess formation with a gradually increasing proportion of macrophages with granuloma formation. Syphilis acquired transplacentally leads to disseminated infection with white patches identifiable on the cut surface of the lung. Histologically, there are

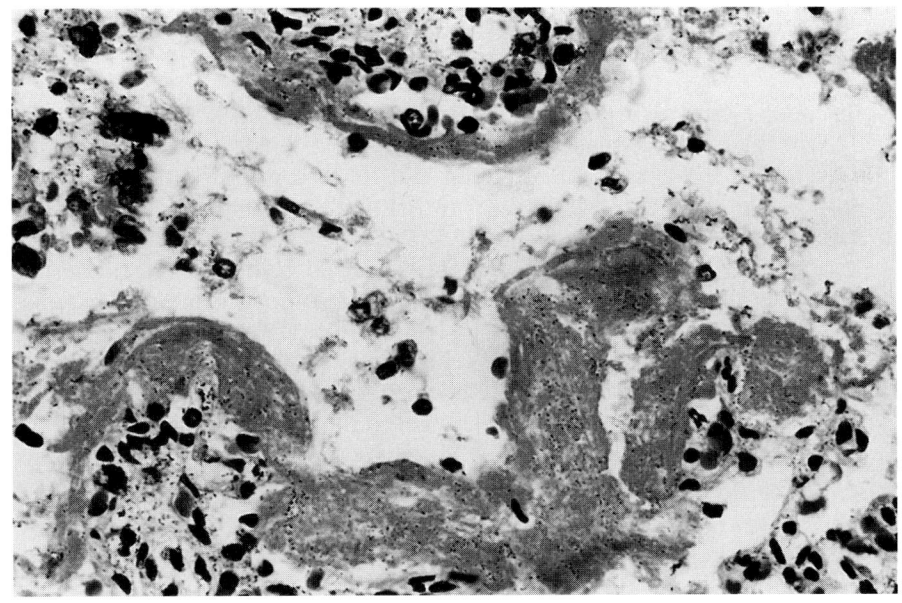

FIGURE 3-51
Hyaline membrane in association with group B streptococcal infection. Bacteria are readily visible in the membrane. Frequently, the bacterial growth is far more overwhelming than illustrated here.

areas of fibrosis and a mononuclear interstitial infiltrate. *Chlamydia trachomatis* may cause a mild pneumonic illness with a mixed alveolar infiltrate of macrophages, lymphocytes, plasma cells, eosinophils, and neutrophils. Rarely, lung collapse associated with a bronchiolitis may occur.[350,351]

VIRAL INFECTION

Cytomegalovirus
The most serious cytomegalovirus (CMV) infections occur after early maternal primary infection with transplacental transmission to the fetus. In abortuses, the infection is often unsuspected although it may well present with hydrops, occasionally associated with pulmonary hypoplasia.[352] At birth, up to 90 percent of neonates with congenital disease may be asymptomatic but extrapulmonary problems are more likely to predominate, with jaundice, hepatosplenomegaly, and a purpuric rash.[353] Neonatal disease is usually acquired from the mother either during birth or via breast milk, but pulmonary disease in this instance is primarily confined to a self-limiting pneumonitis. A more severe pneumonitis can occur if the neonate is transfused with infected blood giving rise to the "gray baby" syndrome.

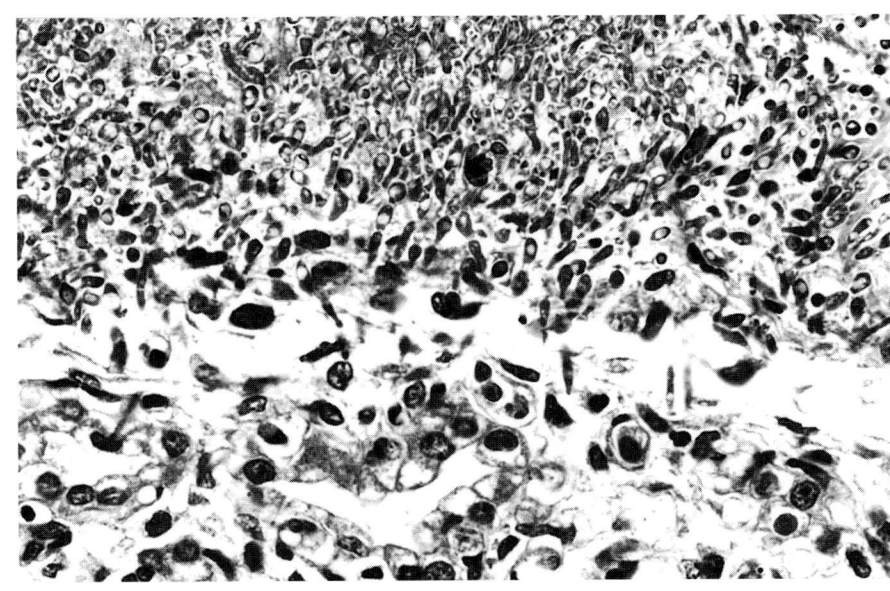

FIGURE 3-52
Candida within a small pulmonary artery. Typically, this mode of infection is associated with an infected catheter.

Nosocomially acquired disease also can occur with prolonged stays on neonatal units due to problems such as BPD.

Macroscopically, the lung pathology is nonspecific and may be relatively normal or show focal consolidation. Histologically, the typical CMV inclusions are detectable in endothelial cells and macrophages (Fig. 3-53) although, even in the presence of known CMV infection diagnosed serologically, the inclusions may be scant. Interstitial fibrosis may accompany the lung infection.

Herpes Simplex Virus

Herpes simplex virus (HSV), predominantly type 2, involves the lung as part of a disseminated disease. The virus is acquired almost invariably during the birthing process from maternal secretions. Mothers with primary, rather than secondary disease, generate the greatest risk. Presentation with disseminated viral infection is usually around days 9 to 11 and up to 50 percent of infants die.

The lung may appear normal or consolidated with small white necrotic foci on the cut surface. Histologically, the necrotic foci are bland and inflammation may be remarkable for its absence. Hyaline membranes or a pneumonitis may be present and inclusions can usually be detected around the margins of the necrotic foci. Where there is any doubt, anti-herpes antibodies are now readily available for immunohistologic confirmation.

Respiratory Syncytial Virus

Most infection with respiratory syncytial virus (RSV) occurs in older infants but may be seen in the perinatal period especially in association with chronic lung disease or congenital heart defects.[354] Thirty to 70 percent of babies will have a pneumonia with their first infection and 1 to 3 percent of hospitalized infants will die.

Two broad patterns of histology can be found but these probably represent opposite ends of a spectrum.[355–357] The bronchiolar pattern shows plugging of bronchioles by mucus and inflammatory cells associated with epithelial desquamation. Air trapping can occur distal to the affected airway. Alveolar spaces may be lined by giant cells with squamous metaplasia more proximally. The distal pattern of RSV infection shows more alveolar plugging by eosinophilic debris and more giant cells (Fig. 3-54), and inclusions may be seen that are paranuclear in location, globular but varying in size.[357] Giant cells are not specific and may be found in other viral diseases such as measles and parainfluenza. RSV culture from specimens such as nasal wash, aspirate, or throat swab may produce a cytopathic effect in 3 to 7 days. Direct immunoflourescence allows a rapid diagnosis although specimens, such as nasopharyngeal swabs, must contain epithelial cells.

Other Viruses

A variety of other viruses have been reported as causing neonatal lung disease but particularly enterovirus, rhinovirus, adenovirus, and parainfluenza.[358] Neonatal enterovirus, including coxsackie A, B, and echovirus,[358–360] have been reported as causing neonatal pneumonia and can be acquired through the aspiration or ingestion of infected secre-

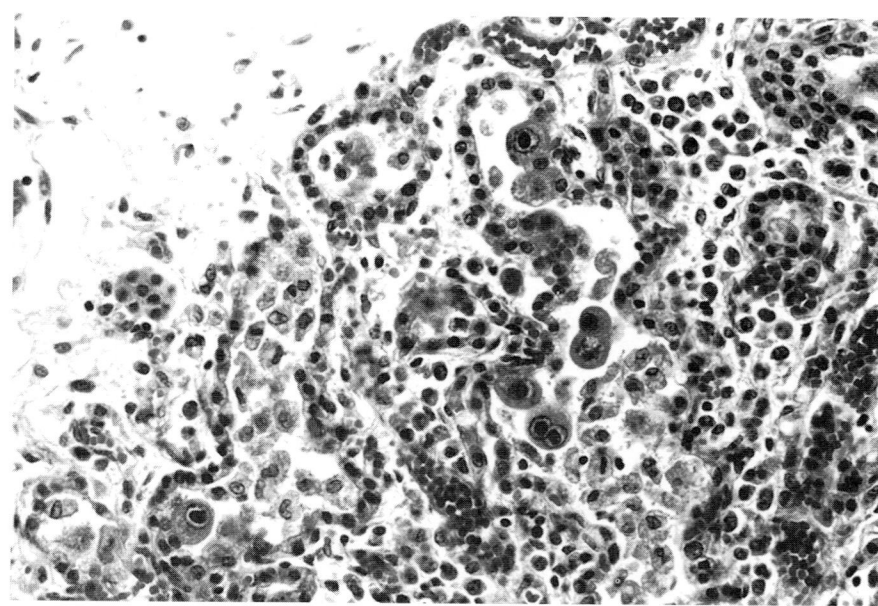

FIGURE 3-53
Cytomegalovirus pneumonitis. Other than the presence of the inclusions, pulmonary histology is largely nonspecific with interstitial fibrosis and regenerating pneumocytes.

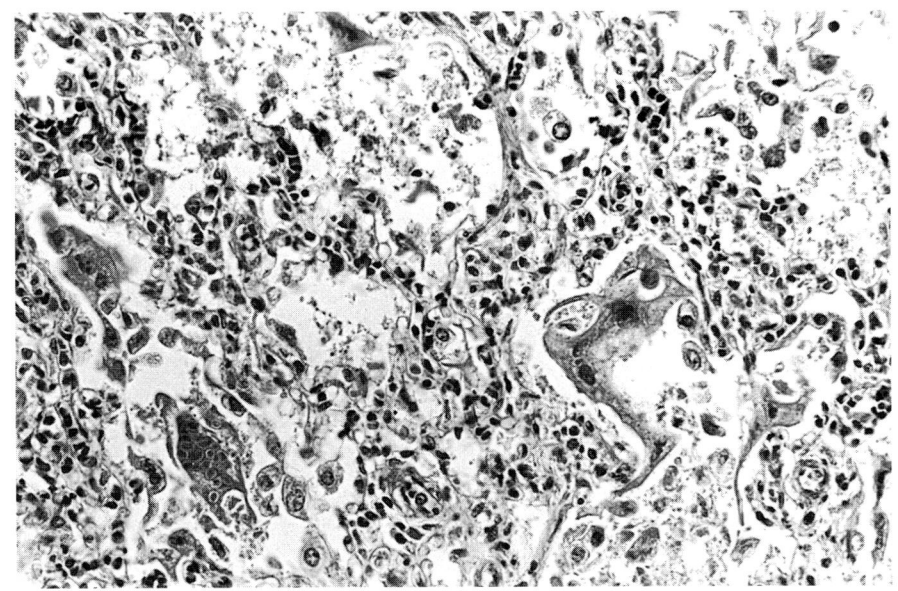

FIGURE 3-54
Viral giant cell pneumonia from which respiratory syncytial virus was isolated. Alveoli contain macrophages, many giant cells, and eosinophilic debris. In other parts of this lung, patchy hyaline membrane disease was also present.

tions during birth. Tranplacental infection may occur but is less common. In general, vertically acquired congenital pneumonias are more serious and can be fatal[359,361]; echovirus 11 infections may resemble acute bacterial infections. Horizontally acquired disease generally gives rise to a mild, self-limiting condition.[362] Adenovirus may give rise to a very severe acute pneumonia when congenitally acquired.[363]

The pathology of neonatal viral pneumonia is relatively nonspecific and most of these viruses will give a broadly similar picture. Macroscopically, the lungs will be firm with patchy consolidation and hemorrhage. Like RSV, the microscopic appearance is a spectrum from a primarily airway disease with bronchiolar necrosis, epithelial plugging, and air trapping. The more distal pattern involves an interstitial pneumonitis. Hyaline membranes may occur focally and some of the changes may be attributable to disseminated intravascular coagulation.[364] In adenovirus, desquamated bronchiolar epithelial cells may contain inclusions.[365]

REFERENCES

1. Clark SL, Vitale DJ, Minton SD, Stoddard RA, Sabey PL: Successful fetal therapy for cystic adenomatoid malformation associated with second–trimester hydrops. *Am J Obstet Gynecol* 157:294–296, 1987.
2. Blott M. Nicolaides KH, Greenough A: Postnatal respiratory function after chronic drainage of fetal pulmonary cyst. *Am J Obstet Gynecol* 159:858–859, 1988.
3. Harrison MR, Adzick NS, Lonaker MT, et al: Successful repair in utero of fetal diaphragmatic hernia after removal of herniated viscera from the left thorax. *N Engl J Med* 322:1582–1584, 1990.
4. Harrison MR, Langer JC, Adzick NS, et al: Correction of congenital diaphragmatic hernia in utero, V: Initial clinical experience. *J Pediatr Surg* 25:47–57, 1990.
5. Wigglesworth JS, Desai R: Is fetal respiratory function a major determinant of perinatal survival? *Lancet* i:264–267, 1982.
6. Santis G, Geddes D: Recent advances in cystic fibrosis. *Postgrad Med J* 70:247–251, 1994.
7. Hyde SC, Gill DR, Higgens CF: Correction of the ion transport defect in CF transgenic mice by gene therapy. *Nature* 362:250–255, 1993.
8. Spranger J, Benirschke K, Hall JG, et al: Errors or morphogenesis: Concepts and terms. Recommendations of an International Working Group. *J Pediatr* 100:160–165, 1982.
9. Young ID: Incidence and genetics of congenital malformation, in Brock DJH, Rodeck CH, Ferguson–Smith MA (eds), *Prenatal Diagnosis and Screening.* London, Churchill Livingstone, 1992, pp 000.
10. Meagher SE, Fisk NM, Harvey JG, Watson GF, Boogert A: Disappearing lung echogenicity in fetal bronchopulmonary malformations: A reassuring sign? *Prenat Diagn* 13:495–501, 1993.
11. Steinhorn RH, Kriesmer PJ, Green TP, McKay CJ, Payne NR: Congenital diaphragmatic hernia in Minnesota. Impact of antenatal diagnosis on survival *Arch Pediatr Adolesc Med* 148:626–631, 1994.
12. Bailey PV, Tracy T, Connors RH, deMello D, Lewis JE, Weber TR: Congenital bronchopulmonary malformations: Diagnostic and therapeutic considerations. *J Thorac, Cardiovasc Surg* 99:597–603, 1990.
13. Landing BH, Wells TR: Tracheobronchial anomalies in children. *Perspect Pediatr Pathol* 1:1–32, 1973.
14. Fearon B, Whalen JS: Tracheal dimensions in the living infant (preliminary report). *Ann Otol Rhinol Laryngol* 76:964–974, 1967.
15. Diaz EM, Adams JM, Hawkins HK, Smith RJH: Tracheal agenesis. A case report and literature review. *Arch Otolaryngol Head Neck Surg* 115:741–745, 1989.
16. Stocker JT: The respiratory tract, in Stocker JT, Dehner LP (eds), *Pediatric Pathology*, Philadelphia, JB Lippincott, 1992, pp 505–573.
17. McNie DJM, Pryse–Davies J: Tracheal agenesis. *Arch Dis Child* 45:143–144, 1970.

18. Koltai PJ, Quiney R: Tracheal agenesis. *Ann Otol Rhinol Laryngol* 101:560–566, 1992.

19. Milstein JM, Lau M, Bickers RG: Tracheal agenesis in infants with VATER association. *Am J Dis Child* 139:77, 1985.

20. Anderson CH, Ahmed G, Caldwell CC: Tracheal atresia associated with congenital absence of the radii. *Clin Pediatr* 20:478–480, 1981.

21. Landing BH, Dixon LG: Congenital malformations and genetic disorders of the respiratory tract (larynx, trachea, bronchi and lungs). *Am Rev Respir Dis* 120:151–185, 1979.

22. Wells TR, Wells AL, Galvis DA, Senac MO, Landing BH, Vachon LA: Diagnostic aspects and syndromal associations of short trachea with bronchial intubation. *Am J Dis Child* 144:1369–1371, 1990.

23. Loeff DS, Filler RM, Vinograd I, et al: Congenital tracheal stenosis: A review of 22 patients from 1965–1987, *J Pediatr Surg* 25:492–495, 1990.

24. Cantrell JR, Guild HG: Congenital stenosis of the trachea. *Am J Surg Pathol* 108:297–303, 1964.

25. DeLorimer AA, Harrison MR, Hardy K, Howell LJ, Adzick NS: Tracheobronchial obstructions in infants and children. *Ann Surg* 212:277–289, 1990.

26. Davis S, Bove KE, Wells TR, Hartsell B, Weinberg A, Gilbert E: Tracheal cartilagenous sleeve. *Pediatr Pathol* 12:349–364, 1992.

27. Cohen SR, Landing BH: Tracheostenosis and bronchial abnormalities associated with pulmonary arterial sling. *Ann Otol Rhinol Laryngol* 85:1–9, 1976.

28. Sailer R, Zimmermann T, Bowing B, Scharf J, Zeilinger G, Stehr K: Pulmonary artery sling associated with tracheobronchial malformations. *Arch Otolaryngol Head Neck Surg* 118:864–867, 1992.

29. Wells TR, Gwinn JL, Landing BH, Stanley P: Reconsideration of the anatomy of sling left pulmonary artery: The association of one form with bridging bronchus and imperforate anus. Anatomic and diagnostic aspects. *J Pediatr Surg* 23:892–898, 1988.

30. Weber TR, Connors RH, Tracy TF Jr: Congenital tracheal stenosis with unilateral pulmonary agenesis. *Ann Surg* 213:70–74 1991.

31. Benisch BM, Wood WG, Kroeger GB, Breitenbach EE, Cohen JJ: Focal muscular hyperplasia of the trachea. *Arch Otolaryngol* 99:226–227, 1974.

32. Sotomayor JJ, Godinez RI, Borden S, Wilmott RW: Large airway collapse due to acquired tracheobronchomalacia in infancy. *Am J Dis Child* 140:367–371, 1986.

33. Benjamin B: Tracheomalacia in infants and children. *Ann Otol Rhinol Laryngol* 93:430–437, 1984.

34. Chen JC, Holinger LD: Congenital tracheal anomalies: Pathology study using serial macrosections and review of the literature. *Pediatr Pathol* 14:513–537, 1994.

35. Cogbill TH, Moore FA, Accurso FJ, Lilly JR: Primary tracheomalacia. *Ann Thorac Surg* 35:538–541, 1983.

36. Wailoo MP, Emery JL: Normal growth and development of the trachea. *Thorax* 37:584–587, 1982.

37. Fagan DG, Emery JL: A review and restatement of some problems in histological interpretation of the infant lung. *Semin Diagn Pathol* 9:13–23,1992.

38. Wailoo MP, Emery JL: The trachea in children with tracheo–oesophageal fistula. *Histopathology* 3:329–338, 1979.

39. Romero R, Ghidini A, Gabrielli S, Jeanty P: Gastrointestinal tract and abdominal wall defects, in Brock DJH, Rodeck CH, Ferguson-Smith MA (eds): *Prenatal Diagnosis and Screening.* London, Churchill Livingstone, 1992, pp 227–255.

40. Waterston DJ, Bonham Carter RE, Aberdeen E: Oesophageal atresia: Tracheo–oesophageal fistula. A study of survival in 218 infants. *Lancet* 1:819–822, 1962.

41. Rideout DT, Hayashi AH, Gillis DA, Giacomantonio JM, Lau HY: The absence of clinically significant tracheomalacia in patients having esophageal atresia without tracheoesophageal fistula. *J Pediatr Surg* 26:1303–1305, 1991.

42. German JC, Mahour GH, Woolley MM: Esophageal atresia and associated anomalies. *J Pediatr Surg* 11: 299–306, 1976.

43. Kluth D, Steding G, Seidl W: The embryology of foregut malformations. *J Pediatr Surg* 22:389–393, 1987.

44. Aaby GV, Blake HA: Tracheobronchiomegaly. *Ann Thorac Surg* 2:64–70, 1966.

45. El–Mallah Z, Quantock OP: Tracheobronchiomegaly. *Thorax* 23:320–324, 1968.

46. Wanderer AH, Ellis EF, Goltz RW, Cotton EK: Tracheobronchomegaly and acquired cutis laxa in a child. *Pediatrics* 44:709–710, 1969.

47. Sane AC, Effmann EL, Brown SD: Tracheobronchiomegaly. The Mounier–Kuhn syndrome in a patient with the Kenny–Caffey syndrome. *Chest* 102:618–619, 1992.

48. Denson DE, Taussig LM, Pond GD: Intraluminal tracheal cyst producing airway obstruction in the newborn infant. *J Pediatr* 88:521–522, 1976.

49. Kushner DC, Clifton Harris GB: Obstructing lesions of the larynx and trachea in infants and children. *Radiol Clin North Am* 16:181–193, 1978.

50. Martin CJ: Tracheobronchopathia osteoplastica. *Arch Otolaryngol* 100:290–293, 1974.

51. Meng RL, Jensik RJ, Faber LP, Matthew GR, Kittle CF: Bronchial atresia. *Ann Thorac Surg* 25:184–192, 1978.

52. Jederlinic PJ, Sicilian LS, Baigelman W, Gaensler EA: Congenital bronchial atresia. A report of 4 cases and a review of the literature. *Medicine (Baltimore)* 66:73–83, 1987.

53. Mori M, Kidogawa H, Moritaka T, Ueda N, Furuya K, Shigematsu S: Bronchial atresia: Report of a case and review of the literature. *Surg Today* 23:449–455, 1993.

54. Kuhn C, Kuhn JP: Coexistence of bronchial atresia and bronchogenic cyst: Diagnostic criteria and embryologic considerations. *Pediatr Radiol* 22:568–570, 1992.

55. Nagaraj HS, Shott R, Fellows R, Yacoab U: Recurrent lobar atelectasis due to acquired bronchial stenosis in neonates. *J Pediatr Surg* 15:411, 1980.

56. Chang N, Hertzler JH, Gregg RH, Lofti MW, Brough AJ: Congenital stenosis of the right mainstem bronchus. A case report. *Pediatrics* 41:739–742, 1968.

57. Weicher RF, Lindsey ES, Pearce CW, Waring WW: Bronchogenic cyst with unilateral obstructive emphysema. *J Thorac Cardiovasc Surg* 59:287–291, 1970.

58. Mitchell RE, Bury RG: Congenital bronchiectasis due to deficiency of bronchial cartilage. *J Pediatr* 87:230–232, 1975.

59. Wayne KS, Taussig LM: Probable familial congenital bronchiectasis due to cartilage deficiency (Williams–Campbell syndrome). *Am Rev Respir Dis* 114:15–22, 1976.

60. Gupta CM TG, Goldberg SJ, Lewis E, Fonkalsrud EW: Congenital bronchomalacia. *Am J Dis Child* 115:88–90, 1968.

61. MacMahon HE, Ruggieri J: Congenital segmental bronchomalacia. *Am J Dis Child* 118:923–926, 1969.

62. Van Benthem LHBM, Driessen O, Haneveld GT, Rietema HP: Cryptorchidism, chest deformities and other congenital

anomalies in three brothers. *Arch Dis Child* 45:143–144, 1970.

63. Agosti E, Filippi G, Chiussi F: Generalized familial bronchomalacia. *Acta Paediatr Scand* 63:616–618, 1974.

64. Atwell SW: Major anomalies of the tracheobronchial tree with a list of minor anomalies. *Dis Chest* 52:611, 1967.

65. Atwell SW: An aberrant bronchus. *Ann Thorac Surg* 2:438–441, 1966.

66. McLaughlin FJ, Stieder DJ, Harris GBC, Vawter GP, Eraklis AJ: Tracheal bronchus: Association with respiratory morbidity in childhood. *J Pediatr* 106:751, 1985.

67. Horwich A, Brueckner M: Left, right and without a cue. *Nature Genetics* 5:321–322, 1993.

68. Casey B, Devoto M, Jones KL, Ballabio A: Mapping a gene for familial situs abnormalities to human chromosome Xq24–q27.1. *Nature Genetics* 5:403–407, 1993.

69. Wainwright H, Nelson M: Polysplenia syndrome and congenital short pancreas. *Am J Med Genet* 47:318–320, 1993.

70. Landing BH: Five syndromes (malformation complexes) of pulmonary symmetry, congenital heart disease, and multiple spleens. *Pediatr Pathol* 2:148, 1984.

71. Winter RM, Baraitser M: *Ivemark Syndrome—Asplenia, Polysplenia*. Multiple Congenital Anomalies. A Diagnostic Compendium. London, Chapman and Hall Medical, 1991, pp 309–310.

72. Save SM, Sieber WK, Girdany BR: Congenital bronchobiliary fistula. *Surgery* 69:599–608, 1971.

73. Chang CC, Giulian BB: Congenital bronchobiliary fistula. *Radiology* 156:82, 1985.

74. Schmit H: Ein fal von Vollstandiger agenisie bei der Lungen. *Virchows Arch Pathol Anat* 134:25–32, 1893.

75. Claireaux AE, Ferreira HP: Bilateral pulmonary agenesis *Arch Dis Child* 33:364, 1958.

76. Ostor AG, Stillwell R, Fortune DW: Bilateral pulmonary agenesis. *Pathology* 10:243–248, 1978.

77. Tuynman RE, Gardner LW: Bilateral aplasia of the lung. *Arch Pathol* 54:306–313, 1952.

78. DeBuse PJ, Morris G: Bilateral pulmonary agenesis, oesophageal atresia and the first arch syndrome. *Thorax* 28:526–528, 1973.

79. Maltz DL, Nadas AS: Agenesis of the lung: Presentation of eight new cases and review of the literature. *Pediatrics* 42:175–188, 1968.

80. Booth JB, Berry CL: Unilateral pulmonary agenesis. *Arch Dis Child* 42:361–374, 1967.

81. Ryland D, Reid L: Pulmonary aplasia: A quantitative analysis of the development of a single lung. *Thorax* 26: 602–609, 1971.

82. Oyamada A, Gasul RM, Holinger PH: Agenesis of the lung: Report of a case with a review of all previously reported cases. *Am J Dis Child* 85:182–201, 1953.

83. Schneider P: in Schwalbe E (ed), Der Morphologie der Missbildungen des Menschen und der Tiere, Jena, G Fischer, 1912, vol 3, p 772.

84. Sbokos CG, McMillan IKR: Agenesis of the lung. *Br J Dis Chest* 71:183, 1988.

85. Engellener W, Kaplan C, Van de Vegte GL: Pulmonary agenesis association with non–immune hydrops. *Pediatr Pathol* 9:725–730, 1989.

86. Boyer RA, Hayes CJ, Hordof AJ, Mellins RB: Agenesis of the left lung and total anomalous pulmonary venous connection. *Chest* 74:106, 1978.

87. Knowles S, Thomas RM, Lindenbaum RH, Keeling JW, Winter RM: Pulmonary agenesis as part of the VACTERL sequence. *Arch Dis Child* 63: 723–726, 1988.

88. Rosenberg DML: Pulmonary agenesis. *Dis Chest* 42: 68–73, 1962.

89. Brandt HM, Liebow AA: Right pulmonary isomerism associated with venous, splenic, and other anomalies. *Lab Invest* 7:469–503, 1958.

90. Figa FH, Yoo SJ, Burrows PE, Turner–Gomes S, Freedom RM: Horseshoe lung—a case report with unusual bronchial and pleural anomalies and a proposed new classification. *Pediatr Radiol* 23:44–47, 1993.

91. Cerruti MM, Marmolejos F, Cacciarelli T: Bilateral intralobar pulmonary sequestration with horseshoe lung. *Ann Thorac Surg* 55:509–510, 1993.

92. Obregon MG, Giannotti A, Digilio MC, Barbuti D, Mingarelli R, Dallapiccola B: Horseshoe lung: An additional component of the Vater association. *Pediatr Radiol* 22:158, 1992.

93. Cerruti MM, Marmolejos F, Cacciarelli T: Bilateral intralobar pulmonary sequestration with horseshoe lung. *Ann Thorac Surg* 55:509–510, 1993.

94. Ersoz A, Soncul H, Gokgoz L, et al: Horseshoe lung with left lung hypoplasia. *Thorax* 47:205–206, 1992.

95. Chaurasia BD, Waugh KV: Iniencephalus with ectopic lungs. *Anat Anz* 136:447–452, 1974.

96. Chaurasia BD, Singh MP: Ectopic lungs in a human foetus with Klippel–Feil syndrome. *Anat Anz* 142:205–208, 1977.

97. Cunningham MD, Peter ER: Cervical hernia of the lung associated with cri du chat syndrome. *Am J Dis Child* 118:769–771, 1969.

98. Munnell ER: Herniation of the lung. *Ann Thorac Surg* 5:204–212, 1968.

99. Bridger JE, Kreczy A, Wigglesworth JS: Transthoracic herniation of the fetal lung: bagpipe lung. *Pediatr Pathol* 12:417–423, 1992.

99a. Patel SR, Meeker DP, Biscotti CV, Kirby TJ, Rice TW: Presentation and management of bronchiogenic cysts in adults. *Chest* 106:79–85, 1994.

100. Ramenofsky ML, Leape LL, McCauley RGK: Bronchogenic cyst. *J Pediatr Surg* 14:219–224, 1979.

101. Grewal RG, Yip CK: Intralobar pulmonary sequestration and mediastinal bronchogenic cyst. *Thorax* 49:615–616, 1994.

102. Maier HC: Bronchogenic cysts of the mediastinum. *Ann Surg* 127:476–502, 1948.

103. Wenig BL, Abramson AL: Tracheal bronchogenic cyst: a new clinical entity? *Ann Otol Rhinol Laryngol.* 96:58–60, 1987.

104. Gomes MN, Hufnagel CA: Intrapericardial bronchogenic cysts. *Am J Cardiol* 36:817–822, 1975.

105. Sumiyoshi K, Shimizu S, Enjoji M, Iwashita A, Kawakami K: Bronchogenic cyst in the abdomen. *Virchows Arch Pathol Anat* 408:93–98, 1985.

106. Keohane ME, Schwartz I, Freed J, Dische R: Subdiaphragmatic bronchogenic cyst with communication to the stomach: A case report. *Hum Pathol* 19:868–871, 1988.

107. Boue DR, Smith GA, Krous HF: Lingual bronchogenic cyst in a child: An unusual site of presentation. *Pediatr Pathol* 14:201–205, 1994.

108. Tresser NJ, Dahms B, Berner JJ: Cutaneous bronchogenic cyst of the back: A case report and review of the literature. *Pediatr Pathol* 14:207–212, 1994.

109. Shamji FM, Sachs HJ, Perkins DG: Cystic disease of the lungs. *Surg Clin North Am* 68:581–620, 1986.

110. Salyer DC, Salyer WR, Eggleston JC: Benign developmental cysts of the mediastinum. *Arch Pathol Lab Med* 101:136–139, 1977.

111. Wilkinson N, Reid H, Hughes D: Intradural bronchogenic cysts. *J Pediatr Surg* 23:993–995, 1988.

112. Gerle RD, Jaretzki A III, Ashley CA, Berne AS: Congenital bronchopulmonary foregut malformation: Pulmonary sequestration communicating with the gastrointestinal tract. *N Engl J Med* 278: 1413–1419, 1968.

113. Zangwill BC, Stocker JT: Congenital cystic adenomatoid malformation within an extralobar pulmonary sequestration. *Pediatr Pathol* 13:309–315, 1992.

114. DeParades CG, Pierce WS, Johnson DG, Waldhausen JA: Pulmonary sequestration in infants and children: A 20 year experience and review of the literature. *J Pediatr Surg* 5:136–147, 1970.

115. Stocker JT, Kagan–Hallet K: Extralobar pulmonary sequestration. *Am J Clin Pathol* 72:917–925, 1979.

116. Flye MW, Conley M, Silver D: Spectrum of pulmonary sequestration. *Ann Thorac Surg* 22:478–482, 1976.

117. Heithoff KB, Sane SM, Williams JW, et al: Bronchopulmonary foregut malformations. A unifying concept. *AJR (Am J Roentgenol)* 126:46–55, 1976.

118. Srikanth MS, Ford EG, Stanley P, Hossein Mahour G: Communicating bronchopulmonary foregut malformations: Classification and embryogenesis. *J Pediatr Surg* 27:732–736, 1992.

119. Stocker JT, Drake RM, Madewell JE: Cystic and congenital lung diseases in the newborn. *Perspect Pediatr Pathol* 4:93–154, 1978.

120. Clements BS, Warner JO: Pulmonary sequestration and related congenital bronchopulmonary–vascular malformations: nomenclature and classification based on anatomical and embryological considerations. *Thorax* 42:401–408, 1987.

121. Hruban RH, Shumway SJ, Orel SB, Dumler JS, Baker RR, Hutchins GM: Congenital bronchopulmonary foregut malformations: Intralobar and extralobar pulmonary sequestrations communicating with the foregut. *Am J Clin Pathol* 91:403–409, 1989.

122. Pryce DM: Lower accessory pulmonary artery with intralobular sequestration of lung: A report of seven cases. *J Pathol* 58: 457–467, 1946.

123. Clements BS, Warner JO, Shinebourne EA: Congenital bronchopulmonary vascular malformations: Clinical application of a simple anatomical approach in 25 cases. *Thorax* 42:409–416, 1987.

124. Hayashi AH, McLean DR, Peliowski A, Tierney AJ, Finer NN: A rare intrapericardial mass in a neonate. *J Pediatr Surg* 27:1361–1363, 1992.

125. Case records of the Massachusetts General Hospital. *N Engl J Med* 324:980–986, 1991.

126. Stocker JT: Sequestrations of the lung. *Semin Diagn Pathol* 3:106–121, 1986.

127. Savic B, Birtel FJ, Tholen W, Funke HD, Knoche R: Lung sequestration: Report of seven cases and review of 540 published cases. *Thorax* 34:96–101, 1979.

128. Buntain WL, Isaacs H Jr, Payne VC, Lindesmith GG, Rosenkrantz JG: Lobar emphysema, cystic adenomatoid malformation, pulmonary sequestration, and bronchogenic cyst in infancy and childhood: A clinical group. *J Pediatr Surg* 9:85, 1974.

129. Witten DA, Clagett OT Woolner LB: Intralobar bronchopulmonary sequestration involving the upper lobes. *J Thorac Cardiovasc Surg* 43:523–529, 1962.

130. Parke WW: Intralobar sequestration of the lingula pulmonalis. *Dis Chest* 41:378–383, 1962.

131. Bell–Thompson J, Missier P, Sommers SC: Lung carcinoma arising in bronchopulmonary sequestration. *Cancer* 44:334–339, 1979.

132. Carter R: Pulmonary sequestration. *Ann Thorac Surg* 7:68–88, 1969.

133. Gans SL, Potts WJJ: Anomalous lobe of lung arising from esophagus. *Thorac Surg* 21:313–318, 1951.

134. Boyden EA, Bill AH Jr, Creighton SA: Presumptive origin of a left lower accessory lung from an esophageal diverticulum. *Surgery* 52:323–329, 1962.

135. Davies RP, Kazlowski K, Wood BP: Right upper lobe esophageal bronchus (with VATER anomalies). *Am J Dis Child* 252:1989, 1989.

136. Tsang TM, Tam PKH: Arrest of foregut development in a congenital bronchopulmonary foregut malformation. *Pediatr Surg Int* 9:401–402, 1994.

137. Bale PM: Congenital cystic malformation of the lung. A form of congenital bronchiolar (adenomatoid) malformation. *Am J Clin Pathol* 71:411–420, 1979.

138. Askin F: Respiratory tract disorders in the fetus and neonate, in Wigglesworth JS, Singer DB (eds), *Textbook of Fetal and Perinatal Pathology*, Oxford, Blackwell Scientific Publications, 1991, pp 643–688.

139. Fisher JE, Nelson SJ, Allen JE, Holzman RS: Congenital cystic adenomatoid malformation of the lung. A unique variant. *Am J Dis Child* 136:1071–1074, 1982.

140. Stocker JT, Madewell JE, Drake RM: Congenital cystic adenomatoid malformation of the lung: classification and morphologic spectrum. *Hum Pathol* 8:155–171, 1977.

141. Moerman P, Fryns JP, Vandenberghe K, Devlieger H, Lauweryns JM: Pathogenesis of congenital cystic adenomatoid malformation of the lung. *Histopathology* 21:315–321, 1992.

142. Wagenvoort CA, Zonervan PE: Polyalveolar lobe and congenital cystic adenomatoid malformation type II: Are they related? *Pediatr Pathol* 11:311–320, 1991.

143. Alt B, Shikes FH, Stanford RE, Silverberg SG: Ultrastructure of congenital cystic adenomatoid malformation of the lung. *Ultrastruct Pathol* 3:217–228, 1982.

144. Thorpe Beeston JG, Nicolaides KH: Cystic adenomatoid malformation of the lung: Prenatal diagnosis and outcome. *Prenatal Diagn* 14:677–688, 1994.

145. Mendoza A, Wolf P, Edwards DK, Leopold GR, Voland JR, Benirscheke K: Prenatal ultrasonographic diagnosis of congenital adenomatoid malformation of the lung: Correlation with pathology and implications for pregnancy management. *Arch Pathol Lab Med* 110:402–404, 1986.

146. Saltzman DH, Adzick NS, Benacerraf BR: Fetal cystic adenomatoid malformation of the lung: Apparent improvement in utero. *Obstet Gynecol.* 71:1000–1002, 1988.

147. Bentur L, Canny G, Thoner P, Superina R, Babyn P, Levison H: Spontaneous pneumothorax in cystic adenomatoid malformation: Unusual clinical and histologic features. *Chest* 99:1292–1293, 1991.

148. Sheffield EA, Addis BJ, Corrin B, McCabe MM: Epithelial hyperplasia and malignant change in congenital lung. *J Clin Pathol* 40:612–614, 1987.

149. Benjamin DR, Cahill JL: Bronchioloalveolar carcinoma of the lung and congenital cystic adenomatoid malformation. *Am J Clin Pathol* 95:889–892, 1991.

150. Hislop A, Reid L: New pathological findings in emphysema of childhood: 1. Polyalveolar lobe with emphysema. *Thorax* 25:682–690, 1970.

151. Silver MM, Thurston WA, Patrick JE: Perinatal pulmonary hyperplasia due to laryngeal atresia. *Hum Pathol* 19:110–113, 1988.

152. Warner JO, Rubin S, Heard BE: Congenital lobar emphysema: A case with bronchial atresia and abnormal bronchial cartilage. *Br J Dis Chest* 76:177, 1982.

153. Jones JC, Almond CH, Snyder HM, Meyer BW, Patrick JR: Lobar emphysema and congenital heart disease in infancy. *J Thorac Cardiovasc Surg* 49:1–7, 1965.

154. Azizkhan RG, Grimmer DL, Askin FB, Lacey SR, Merten DF, Wood RE: Acquired lobar emphysema (overinflation): Clinical and pathological evaluation of infants requiring lobectomy. *J Pediatr Surg* 27:1145–1151, 1992.

155. Macleod WM: Abnormal transradiency of one lung. *Thorax* 9:147–153, 1954.

156. Hislop A, Reid L: New pathological findings in emphysema of childhood: 2. Overinflation of a normal lobe. *Thorax* 26:190–194, 1971.

157. Cumming GR, MacPherson RI, Cherniak V: Unilateral hyperlucent lung syndrome in children. *J Pediatr* 250:78, 1971.

158. France NE, Brown RJK: Congenital pulmonary lymphangiectasis: Report of 11 examples with special reference to cardiovascular findings. *Arch Dis Child* 46:528–532, 1971.

159. Hernandez RJ, Stern AM, Rosenthal A: Pulmonary lymphangiectasis in Noonan syndrome. *AJR Am J Radiol* 1 34:75–80, 1980.

160. Rhatigan RM, Hobin FP: Congenital pulmonary lymphangiectasis and icthyosis congenita: A case report. *Am J Clin Pathol* 53:95–99, 1970.

161. Bhatti MAK, Ferrante JW, Gielchincshy I, Norman JC: Pleuropulmonary and skeletal lymphangiomatosis with chylothorax and chylopericardium. *Ann Thorac Surg* 40:398–401, 1985.

162. Scott Emukapor AB, Warren ST, Kapur S, Quiachon EB, Higgins JV: Familial occurrence of congenital pulmonary lymphangiectasis: Genetic implications. *Am J Dis Child* 135:532–534, 1981.

163. Wagenaar SjSc, Swiergena J, Wagenvoort CA: Late presentation of primary pulmonary lymphangiectasis. *Thorax* 33:791–795, 1978.

164. Noonan JA, Walters LR, Reeves JT: Congenital pulmonary lymphangiectasis. *Am J Dis Child* 120:314–319, 1970.

165. Laurence KM: Congenital pulmonary lymphangiectasis. *J Pathol Bacteriol* 70:325–333, 1955.

166. Ward IM, Krahl JB: Enterogenous pulmonary cyst. *Am J Dis Child* 63:924–933, 1942.

167. Case records of the Massachusetts General Hospital. *N Engl J Med* 310:36–41, 1984.

168. Ueda K, Gruppo R, Unger F, Martin L, Bove K: Rhabdomyosarcoma of the lung arising in a congenital adenomatoid malformation. *Cancer* 40:383–388, 1977.

169. Domitzio P, Liesner RJ, Dicks–Mireaux C, Risdon RA: Malignant mesenchymoma associated with a congenital lung cyst in a child. Case report and review of the literature. *Pediatr Pathol* 10:785–797, 1990.

170. Wagenvoort CA: Misalignment of lung vessels: A syndrome causing persistent neonatal pulmonary hypertension. *Hum Pathol* 17:727–730, 1986.

171. Janny CG, Askin FB, Huhn CK: Congenital alveolar dysplasia—an unusual cause of respiratory distress in the newborn. *Am J Clin Pathol* 76:722–727, 1981.

172. Khorsand J, Tennant R, Giliies C, Phillipps AF: Congenital alveolar capillary dysplasia: A developmental vascular anomaly causing persistent pulmonary hypertension of the newborn. *Pediatr Pathol* 3:299, 1985.

173. Remberger K, Hubner G: Rhabdomyomatous dysplasia of the lung. *Virchows Arch Pathol Anat* 363:363–369, 1974.

174. Chi JG, Shong YK: Diffuse striated muscle heteroplasia of the lung. An autopsy case. *Arch Pathol Lab Med* 106:641–644, 1982.

175. Chellam VG: Rhabdomyomatous dysplasia of the lung: A case report with review of the literature. *Pediatr Pathol* 8:391–394, 1988.

176. Chen MF, Onerheim R, Wang NS, Huttner I: Rhabdomyomatosis of the newborn lung. *Pediatr Pathol* 11:123–129, 1991.

177. Bozic C: Ectopic fetal adrenal cortex in the lung of a newborn. *Virchows Arch Pathol Anat* 363:371–374, 1974.

178. Beskin CA: Intralobar enteric sequestration of the lung containing aberrant pancreas. *J Thorac Cardiovasc Surg* 41:314, 1961.

179. Wigglesworth JS: Pathology of intrapartum and early neonatal death in the normally formed infant in Wigglesworth JS, Singer DB (eds). *Textbook of Fetal and Perinatal pathology*. Oxford, Blackwell Scientific Publications, 1991, pp 285–306.

180. Campo E, Bombi JA: Central nervous system heterotopia in the lung of a fetus with cranial malformation. *Virchows Arch Pathol Anat* 391:117–122, 1981.

181. Page DV, Stocker JT: Anomalies associated with pulmonary hypoplasia. *Am Rev Respir Dis* 125:216, 1982.

182. Rushton DI: West Midlands perinatal mortality survey, 1987. An audit of 300 perinatal autopsies. *Br J Obstet Gynecol* 98:624–627, 1991.

183. Askenazi SS, Perlman M: Pulmonary hypoplasia: Lung weight and radial alveolar count as criteria of diagnosis. *Arch Dis Child* 54:614–618, 1979.

184. Wigglesworth JS, Desai R: Use of DNA estimation for growth assessment in normal and hypoplastic fetal lungs. *Arch Dis Child* 56:601–605, 1981.

185. Wigglesworth JS, Desai R, Guerrini P: Fetal lung hypoplasia: Biochemical and structural variations and their possible significance. *Arch Dis Child* 56:606–615, 1981.

186. Wigglesworth JS: Pathology of the lung in the fetus and neonate, with special reference to problems of growth and maturation. *Histopathology* 11:671–689, 1987.

187. Haidar A, Ryder TA, Wigglesworth JS: Failure of elastin development in hypoplastic lungs associated with oligohydramnios; an EM study. *Histopathology* 18:471–447, 1991.

188. Wigglesworth JS, Hislop AA, Desai R: Biochemical and morphometric analyses in hypoplastic lungs. *Pediatr Pathol* 11:537–549, 1991.

189. Emery JL, Mithal A: The number of alveoli in the terminal respiratory unit of man during late intrauterine life and childhood. *Arch Dis Child* 35:544–547, 1960.

190. Haidar A, Wigglesworth JS, Krausz T: Type IV collagen in developing human lung: A comparison between normal and hypoplastic fetal lungs. *Early Hum Develop* 21:175–180, 1990.

191. Cooney TP, Thurlbeck WM: Lung growth and development in anencephaly and hydranencephaly. *Am Rev Respir Dis* 132:596–601, 1985.

192. Wigglesworth JS, Desai R: Effects on lung growth of cervical cord section in the rabbit fetus. *Early Hum Develop* 3:51–65, 1979.

193. Liggins GC, Vilos GA, Campos GA, Kitterman JA, Lee CH: The effect of bilateral thoracoplasty in lung development in fetal sheep. *J Develop Physiol* 3:275–282, 1981.

194. Kitterman JA: Fetal lung development. *J Develop Physiol* 6:67–82, 1984.

195. Harding R, Hooper SB, Han VK: Abolition of fetal breathing movements by spinal cord transection leads to reductions in

fetal lung liquid volume, lung growth, and IGF–II gene expression. *Pediatr Res* 34:148–153, 1993.

196. Nicolini U, Fisk NM, Rodeck CH, Talbert DG, Wigglesworth JS: Low aminotic pressure in oligohydramnios—Is this the cause of pulmonary hypoplasia? *Am J Obstet Gynecol* 161:1098–1101, 1989.

197. Harding R, Hooper SB, Dickson KA: A mechanism leading to reduced lung expansion and lung hypoplasia in fetal sheep during oligohydramnios. *Am J Obstet Gynecol* 163:1904–1913, 1990.

198. Dickson KA, Harding R: Fetal breathing and pressures in the trachea and amniotic sac during oligohydramnios in sheep. *J Appl Physiol* 70:293–299, 1991.

199. Mendelsohn G, Hutchins GM: Primary pulmonary hypoplasia. *Am J Dis Child* 131:1220–1223, 1977.

200. Swischuk LE, Richardson CJ, Nichols MM, Ingman MJ: Primary pulmonary hypoplasia in the neonate. *J Pediatr* 95:714–718, 1979.

201. Marechal M, Gillerot Y, Chef R: L'hypoplasie pulmonaire. A propos d'une observation chez des jumeaux. *J Gynecol, Obstet, Biol Reprod* 143:391–400, 1984.

202. Boylan P, Howe A, Gearty J, O'Brien NG: Familial pulmonary hypoplasia. *Irish J Med Sci* 146:179–180, 1977.

203. Chambers HM: Congenital acinar aplasia: An extreme form of pulmonary maldevelopment. *Pathology* 23:69–71, 1991.

204. Silver MM, Thurston WA, Patrick JE: Perinatal pulmonary hyperplasia due to laryngeal atresia *Hum Pathol* 19:110–113, 1988.

205. Wigglesworth JS, Desai R, Hislop AA: Fetal lung growth in congenital laryngeal atresia. *Pediatr Pathol* 7:515–525, 1987.

206. Wayne ER, Campbell JB, Burrington JD, Davis WS: Eventration of the diaphragm. *J Pediatr Surg* 9:643–651, 1974.

207. David T, Illingworth C: Diaphragmatic hernia in the southwest of England. *J Med Genet* 13:253–262, 1976.

208. Benacerraf B, Adzick N: Fetal diaphragmatic hernia: Ultrasound diagnosis and clinical outcome in 19 cases. *Am J Obstet Gynecol* 156:573–576, 1987.

209. Norio R, Kaariainen H, Rapola J, Herva R, Kekomaki M: Familial congenital diaphragmatic defect. *Am J Med Genet* 17:471–483, 1984.

210. Cantrell J, Haller J, Ravitch M: A syndrome of congenital defects involving the abdominal wall, sternum, diaphragm, pericardium and heart. *Surg Gynecol Obstet* 107:602–614, 1958.

211. Symbas PM, Ware RE: A syndrome of defects of the thoraco–abdominal wall, diaphragm, pericardium and heart. One stage surgical repair and analysis of the syndrome. *J Thorac Cardiovasc Surg* 65:914–919, 1973.

212. Fryns J, Moerman F, Goddeeris P, Bossuyt C, Van den Bergher H: A new lethal syndrome with cloudy cornea, diaphragmatic defects and distal limb deformities. *Hum Genet* 50:65–70, 1979.

213. Pinar H, Carpenter MW, Abuelo D, Singer DB: Fryns syndrome: A new definition. *Pediatr Pathol* 14:467–478, 1994.

214. Kartagener M: Zur Pathologie der Bronchieckasien: Bronchiectasien bei Situs viscerum inversus. *Beitr Klin Tubrk* 83:489–501, 1933.

215. Ailsby RL, Ghadially FN: Atypical cilia in human bronchial mucosa. *J Pathol* 109:75–77, 1973.

216. Escudier E, Escalier D, Homasson JP, Pinchon MC, Bernaudin JF: Unexpectedly normal cilia and spermatozoa in an infertile man with Kartagener's syndrome. *Eur J Resp Dis* 70:180–186, 1987.

217. Handelsman DJ, Conway AJ, Boylan LM, Turtle JR: Young's syndrome: Obstructive azoospermia and chronic sinopulmonary infections. *N Engl J Med* 310:3–9, 1984.

218. de Longh R, Ing A, Rutland J: Mucociliary function, ciliary ultrastructure and ciliary orientation in Young's syndrome. *Thorax* 47:184–187, 1992.

219. Afzelius BA: A human syndrome caused by immotile cilia. *Science* 193:317–319, 1976.

220. McDowell E, Barrett LA, Harris CC, Trump BF: Abnormal cilia in human bronchial epithelium. *Arch Pathol Lab Med* 100:429–436, 1976.

221. Fox B, Bull TB, Arden GB: Variations in the ultrastructure of human nasal cilia including abnormalities found in retinitis pigmentosa. *J Clin Pathol* 33:327–333, 1980.

222. Camner P, Mossberg B, Afzelius BA: Evidence for congenitally nonfunctioning cilia in the tracheobronchial tract in two subjects. *Am Rev Respir Dis* 112:807–809, 1975.

223. Mossberg B: The immotile cilia syndrome: Ultrastructurally heterogeneous and clinically homogeneous. *Eur J Respir Dis* 66:161–163, 1985.

224. Lee RMKW, Rossman CM, O'Brodovich H: Assessment of postmortem respiratory ciliary motility and ultrastructure. *Am Rev Respir Dis* 136:445–447, 1987.

225. Rutland J, Cole PJ: Non–invasive sampling of nasal cilia for measurement of beat frequency and study of ultrastructure. *Lancet* 2:564–565, 1980.

226. Rossman CM, Lee RMKW, Forrest JB, Newhouse MT: Nasal ciliary ultrastructure and function in patients with primary ciliary dyskinesia compared with that in normal subjects with various respiratory diseases. *Am Rev Respir Dis* 129:161–167, 1984.

227. Rutland J, Cole PJ: Nasal mucociliary clearance and ciliary beat frequency in cystic fibrosis compared with sinusitis and bronchiectasis. *Thorax* 36:654–658, 1981.

228. Fox B, Bull TB, Arden GB: Variations in the ultrastructure of human nasal cilia including abnormalities found in retinitis pigmentosa. *J Clin Pathol* 33:327–335, 1980.

229. Lungarella G, De Santi MM, Palatresi R, Tozi P: Ultrastructural observations on basal apparatus of respiratory cilia in immotile cilia syndrome. *Eur J Respir Dis* 66:165–172, 1985.

230. Afzelius BA: Genetic disorders of cilia, in Schweiger HG (ed), *International Cell Biology*, 1980–1. New York, Springer, 1981, pp 440–447.

231. Rutland J, de Longh RU: Random ciliary orientation: A cause of respiratory tract disease. *N Engl J Med* 323:1681–1684, 1990.

232. Afzelius BA, Camner P, Mossberg B: Acquired ciliary defects compared to those seen in the immotile–cilia syndrome. *Eur J Respir Dis* 64 (suppl 127): 5–10, 1983.

233. Verra F, Fleury–Feith J, Boucherat M, Pinchon MC, Bignon J, Escudier E: So nasal ciliary changes reflect bronchial changes? An ultrastructural study. *Am Rev Respir Dis* 147:908–913, 1993.

234. Knowlton RG, Cohen–Haguenauer O, Van Cong N, et al: A polymorphic DNA marker linked to cystic fibrosis is located on chromosome 7. *Nature* 318:380–382, 1985.

235. Cystic Fibrosis Genetic Analysis Consortium. *Am J Hum Genetics* 47:354–359, 1990.

236. Cutting G, Kasch LM, Rosenstein BJ, et at: Two patients with cystic fibrosis, nonsense mutations in each cystic fibrosis gene and mild pulmonary disease *N Engl J Med* 323:1685–1689, 1990.

237. Champbell PE, Phillips JA, Krishnamani MRS, Maness KJ,

Hazinki TA: Cystic fibrosis: Relationship between clinical status and ΔF508 deletion. *J Pediatr* 118:239–241, 1991.

238. Farrall M, Rodeck CH, Stainer P, et al: First–trimester prenatal diagnosis of cystic fibrosis with linked DNA probes. *Lancet* 1:1402–1404, 1986.

239. Trezise AEO, Chambers JA, Wardle CJ, Gould S, Harris A: Expression of the cystic fibrosis gene in human foetal tissues. *Hum Mol Genetics* 2:213–218, 1993.

240. Elborn JS, Shale DJ: Pathogenesis of lung injury in cystic fibrosis. *Thorax* 45:970–973, 1990.

241. Santis G, Geddes D: Recent advances in cystic fibrosis. *Postgrad Med J* 70:247–251, 1994.

242. McElvaney NG, Hubbard RC, Birrer P, et al: Aerosol alpha–1–antitrypsin treatment for cystic fibrosis. *Lancet* 337:392–394, 1991.

243. Laurence R, Sorrell T: Eicosapentoanoic acid in cystic fibrosis: Evidence of a pathogenetic role for leukotriene B4. *Lancet* 342:465–469, 1993.

244. Mitchell–Heggs P, Mearns M, Batten JC: Cystic fibrosis in adolescents and adults. *Q J Med* (179): 479–504, 1976.

245. Di Sant'Agnese PA, Davis PB: Cystic fibrosis in adults: 75 cases and a review of 232 cases in the literature. *Am J Med* 66:121–132, 1979.

246. Taylor RFH, Morgan DW, Nicholson PS, MacKay IS, Hodson ME, Pitt TL: Extrapulmonary sites of *Pseudomonas aeruginosa* in adults with cystic fibrosis. *Thorax* 47:426–428, 1992.

247. Lipuma JJ, Dasen SE, Nielson DW, Stern RC, Stull TL: Person–to–person transmission of *Pseudomonas cepacia* between patients with cystic fibrosis. *Lancet* 336:1094–1096, 1990.

248. Bhargava V, Tomashefski JF Jr, Stern RC, Abramowsky CR: The pathology of fungal infection and colonization in patients with cystic fibrosis. *Hum Pathol* 20:977–986, 1989.

249. Bedrossian CWM, Greenberg SD, Singer DB, Hansen JJ, Rosenberg HS: The lung in cystic fibrosis: A quantitative study including prevalence of pathologic findings among different age groups. *Hum Pathol* 7:195–204, 1976.

250. Tomashefski JF Jr, Bruce M, Stern RC, Dearborn DG, Dahms B: Pulmonary air cysts in cystic fibrosis: Relation of pathologic features to radiologic findings and history of pneumothorax. *Hum Pathol* 16:253–261, 1985.

251. Finnegan MJ, Hinchcliffe J, Russell Jones D, et al: Vasculitis complicating cystic fibrosis. *Q J Med* (NS) 72:609–621, 1989.

252. Hyde SC, Gill DR, Higgens CF: Correction of the ion transport defect in CF transgenic mice by gene therapy. *Nature* 362:250–255, 1993.

253. Alton EWFW, Middleton PG, Caplen NJ, et al: Non–invasive liposome mediated gene delivery can correct the ion transport defect in CF mutant mice. *Nature Genet* 5:135–142, 1993.

254. Schroeder WT, Miller MF, Woo SLC, Saunders GF: Chromosomal localization of the human alpha–1–antitrypsin gene to 14q31–32. *Am J Hum Genet* 37:868–872, 1985.

255. Brantly M, Nukiwa T, Crystal PG: Molecular basis of alpha–1–antitrypsin deficiency. *Am J Med* 84 (suppl 6A): 13–31, 1988.

256. Eriksson S: Discovery of alpha–1–antitrypsin deficiency. *Lung* (suppl): 523–529, 1990.

257. Brind AM, Bassendine MF: Molecular genetics of chronic liver diseases. *Clin Gastroenterol* 4:233–253, 1990.

258. Hutchinson DCS: The epidemiology of alpha–1–antitrypsin deficiency. *Lung* (suppl): 535–542, 1990.

259. Hutchinson DCS: Alpha–1–antitrypsin deficiency and pulmonary emphysema: The role of proteolytic enzymes and their inhibitors. *Br J Dis Chest* 67:171–196, 1973.

260. Geddes DM, Webly M, Brewerton DA, et al: Alpha–1–antitrypsin phenotypes in fibrosing alveolitis and rheumatoid arthritis. *Lancet* Z: 1049–1051, 1977.

261. Eriksson S, Carlson J, Velez R: Risk of cirrhosis and primary liver cancer in alpha–1–antitrypsin deficiency. *N Engl J Med* 14:736–739, 1986.

262. Bradfield JWB, Blenkinsopp WK: Alpha–1–antitrypsin globules in the liver and PiM phenotype. *J Clin Pathol* 30:464–466, 1977.

263. Janus ED, Philips NT, Carrell RW: Smoking, lung function and alpha–1–antitrypsin deficiency. *Lancet* 1:152–154, 1985.

264. Morley CJ: The respiratory distress syndrome, in Roberton NRC (ed), *Textbook of Neonatology*. London, Churchill Livingstone, 1986, pp 274–239.

265. Wigglesworth JS: Pathology of neonatal respiratory distress. *Proc R Soc Med* 70:861–862, 1977.

266. de la Monte S, Hutchins GM, Moore GW: Respiratory epithelial cell necrosis is the earliest lesion of hyaline membrane disease of the newborn. *Am J Pathol* 123:155–160, 1986.

267. Lauweryns JM: "Hyaline membrane disease" in newborn infants. Macroscopic, radiographic and light and electron microscopic studies. *Hum Pathol* 1:175–204, 1970.

268. Ogden BE, Murphy SA, Saunders GC, Pathak D, Johnson JD: Neonatal lung neutrophils and elastase/proteinase inhibitor imbalance. *Am Rev Respir Dis* 130:817–821, 1984.

269. Wigglesworth JS: *Perinatal Pathology*. Philadelphia, WB Saunders, 1984, pp 186–194.

270. Robertson B: The origin of neonatal lung injury (editorial). *Pediatr Pathol* 11:iii–vi, 1991.

271. Fuchimukai T, Fujiwara T, Takahashi A, Enhorning G: Artificial surfactant inhibited by proteins. *J Appl Physiol* 62:429–437, 1987.

272. de la Monte S, Hutchins GM, Moore GW: Respiratory epithelial cell necrosis is the earliest lesion of hyaline membrane disease of the newborn. *Am J Pathol* 123:155–160, 1986.

273. Schmidt B, Vegh P, Weitz J, Johnson M, Caco C, Roberts R: Thrombin/antithrombin III complex formation in the neonatal respiratory distress syndrome. *Am Rev Respir Dis* 145:767–770, 1992.

274. Ablow RC, Driscoll SG, Effman EL, et al: A comparison of early onset group B streptococcal neonatal infection and the respiratory distress syndrome of the newborn *N Engl J Med* 294:65–70, 1976.

275. Gibson RL, Redding GJ, Henderson WR, Truog WE: Group B streptococcus induces tumor necrosis factor in neonatal piglets. *Am Rev Respir Dis* 143:598–604, 1991.

276. Royall JA, Levin DL: Adult respiratory distress syndrome in pediatric patients. 1. Clinical aspects, pathophysiology, pathology and mechanisms of lung injury. *J Pediatr* 112:169–180, 1988.

277. Faix RG, Viscardi RM, DiPietro MA, Nicks JJ: Adult respiratory distress syndrome in full–term newborns. *Pediatrics* 83:971–976, 1989.

278. Pfeffinger J, Tschaeppler H, Wagner BP, Weber J, Zimmerman A: The paradox of adult respiratory distress syndrome in neonates. *Pediatr Pulmonol* 10:18–24, 1991.

279. Northway WH, Rosan RC, Porter DY: Pulmonary disease following respiratory therapy for hyaline membrane disease. *N Engl J Med* 276:357–368, 1967.

280. O'Brodovich HM, Mellins RB: Bronchopulmonary dysplasia. *Am Rev Respir Dis* 132:694–709, 1985.

281. Greenough A: Bronchopulmonary dysplasia: Early diagno-

sis, prophylaxis, and treatment. *Arch Dis Child* 65:1082–2088, 1990.

282. Abman SH, Groothius JR: Pathophysiology and treatment of bronchopulmonary dysplasia: current issues. *Pediatr Clin North Am* 41:277–316, 1994.

283. Krauss AN, Klain DB, Auld PAM: Chronic pulmonary insufficiency of prematurity (CPIP). *Pediatrics* 55:55–58, 1975.

284. Hudak BB, Egan EA: Impact of lung surfactant therapy on chronic lung diseases in premature infants. *Clin Perinatol* 19:591–602, 1992.

285. Kraybill EN, Bose CL, D'Ercole J: Chronic lung disease in infants with very low birth weight. A population based study. *Am J Dis Child* 141:784–788, 1987.

286. Shaw N, Gill B, Weindling M, Cooke R: The changing incidence of chronic lung disease. *Health Trends* 25:50–53, 1993.

287. Anderson WR, Engel RR: Cardiopulmonary sequelae of reparative stages of bronchopulmonary dysplasia. *Arch Pathol Lab Med* 107:603–608, 1983.

288. Hislop AA, Wigglesworth JS, Desai R, Aber V: The effects of preterm delivery and mechanical ventilation on human lung growth. *Early Hum Develop* 15:147–164, 1987.

289. Margraf LR, Tomashefski JF Jr, Bruce MC, Dahms BB: Morphometric analysis of the lung in bronchopulmonary dysplasia. *Am Rev Respir Dis:* 143:391–400, 1991.

290. Stocker JT: Pathologic features of long–standing "healed" bronchopulmonary dysplasia: A study of 28, 3–40 month old infants. *Hum Pathol* 17:943–961, 1986.

291. Hislop AA, Haworth SG: Pulmonary vascular damage and the development of cor pulmonale following hyaline membrane disease. *Pediatr Pulmonol* 9:152–161, 1990.

292. Gorenflo M, Vogel M, Obladen M: Pulmonary vascular changes in BPD. *Pediatr Pathol* 11:851–866, 1991.

293. Shaheen SO, Barker DJP: Early lung growth and chronic airflow obstruction. *Thorax* 49:533–536, 1994.

294. Van Lierde S, Cornelis A, Devlieger H, Moerman P, Lauweryns J, Eggermont E: Different patterns of pulmonary sequelae after hyaline membrane disease: Heterogeneity of bronchopulmonary dysplasia? A clinicopathologic study. *Biol Neonate* 60:152–162, 1991.

295. Fagen DG: Recent advances in neonatal lung pathology in Scadding JG, Cumming G, Thurlbeck WM (eds), *Scientific Foundations of Respiratory Medicine*, London Heinemann, 1981, pp 573–592.

296. Hasleton PS: Adult respiratory distress syndrome—a review. *Histopathology* 7:307–332, 1983.

297. Barry BE, Crapo JD: Patterns of accumulation of platelets and neutrophils in rat lungs during exposure to 100% and 85% oxygen. *Am Rev Respir Dis* 132:548–555, 1985.

298. Clement A, Chadelat K, Sardet A, Grimfeld A, Tournier G: Alveolar macrophage status in bronchopulmonary dysplasia. *Pediatr Res* 23:470–473, 1988.

299. Frank L, Sosenko IRS: Prenatal development of lung antioxidant enzymes in four species. *J Pediatr* 110:106–110, 1987.

300. Frank L, Sosenko IRS: Development of lung antioxidant enzyme system in late gestation: Possible implications for the prematurely born infant. *J Pediatr* 110:9–14, 1987.

301. Frank L, Sosenko IRS: Failure of premature rabbits to increase antioxidant enzymes during hyperoxic exposure: Increased susceptibilty to pulmonary oxygen toxicity compared with term rabbits. *Pediatr Res* 29:292–296, 1991.

302. Gerdes JS, Harris MC, Polin RA: Effects of dexamethasone and indomethacin on elastase, alpha 1–proteinase inhibitor, and fibronectin in bronchoalveolar lavage fluid from neonates. *J Pediatr* 113:727–731, 1988.

303. Yoder MC Jr, Chua R, Tepper R: Effect of dexamethasone on

pulmonary inflammation and pulmonary function of ventilator–dependent infants with bronchopulmonary dysplasia. *Am Rev Respir Dis* 143:1044–1048, 1991.

304. Bruce MC, Schuyler M, Martin RJ, Starcher BC, Tomashefski JF Jr, Wedig KE: Risk factors for the degradation of lung elastic fibers in the ventilated neonate. Implications for impaired lung development in bronchopulmonary dysplasia. *Am Rev Respir Dis* 146:204–212, 1992.

305. Taghizadeh A, Reynolds EOR: Pathogenesis of bronchopulmonary dysplasia following hyaline membrane disease. *Am J Pathol* 82:241–264, 1976.

306. Parker JC, Hernandez LA, Peevy KJ: Mechanisms of ventilator–induced lung injury. *Crit Care Med* 21:131–143, 1993.

307. Wilson MG, Mikity VG: A new form of respiratory disease in premature infants. *Am J Dis Child* 99:119–129, 1960.

308. Stocker JT, Madewell JE: Persistent interstitial pulmonary emphysema: Another complication of the respiratory distress syndrome. *Pediatrics* 59:847–857, 1977.

309. Fredrick J, Butler NR: Certain causes of neonatal death: Massive pulmonary haemorrhage. *Biol Neonate* 18:243–262, 1971.

310. Trompter R, Yu VYH, Aynsley–Green A, Roberton NRC: Massive pulmonary haemorrhage in the newborn infant. *Arch Dis Child* 50:123–127, 1975.

311. Cole VA, Normand ICS, Reynolds EOR, Rivers RPA: Pathogenesis of hemorrhagic pulmonary edema and massive pulmonary haemorrhage in the newborn. *Pediatrics* 51:175–186, 1973.

312. Robertson B: Neonatal respiratory distress syndrome and surfactant therapy; a brief review. *Eur Respir J* (suppl 3): 73s–76s, 1989.

313. Surfactant treatment for premature babies—a review of clinical trials. *Arch Dis Child* 66:445–450, 1991.

314. Long W, Corbet A, Cotton R, et al: A controlled trial of synthetic surfactant in infants weighting 1250 g or more with respiratory distress syndrome. *N Engl J Med* 325:1696–1703, 1991.

315. Pramanik AK, Holtzman RB, Merritt TA: Surfactant replacement therapy for pulmonary disease. *Pediatr Clin North Am* 40:913–936, 1993.

316. Pinar H, Makarova N, Rubin LP, Singer DB: Pathology of the lung in surfactant–treated neonates. *Pediatr Pathol* 14:627–636, 1994.

317. Hagstrom N, Waters BL: Morphometric analysis of exogenous surfactant's effect on neonatal lung: A retrospective, case controlled postmortem study. (abstract). *Pediatr Pathol* 13:112, 1993.

318. Barson AJ, Chiswick ML, Doig CM: Fat embolism in infancy after intravenous fat infusions. *Arch Dis Child* 53:218–223, 1978.

319. Hertel J, et al: Intravascular fat accumulation after intralipid infusion in very low birth weight infant. *J Pediatr* 100:975, 1982.

320. Stahl GE, Spear ML, Hamosh M: Lipid infusions and pulmonary function abnormalities, in Polin RA, Fox WW (eds), *Fetal Neonatal Physiology*, London, WB Saunders, 1992, pp 346–352.

321. Pereira GR, et al: Decreased oxygenation and hyperlipaemia during intravenous fat infusions in premature infants. *Pediatrics* 66:26, 1980.

322. Cooke RWI: Factors associated with chronic lung disease in preterm infants. *Arch Dis Child* 66:776–779, 1991.

323. Ostrea EM, Naqvi M: The influence of gestational age on the ability of the fetus to pass meconium in utero. *Obstet Gynecol Scand* 61:275–277, 1982.

324. Wiswell TE, Tuggle JM, Turner BS: Meconium aspiration

syndrome: Have we made a difference? *Pediatrics* 85:715–721, 1990.

325. Oelberg DG, Downey SA, Flynn MM: Bile salt induced intracellular Ca accumulation in type 2 pneumocytes. *Lung* 168:297, 1990.

326. Lauweryns J, Bernat R, Lerut A, Detournay G: Intrauterine pneumonia. An experimental study. *Biol Neonate* 22:301–318, 1973.

327. Murphy JD, Vawter GF, Reid LM: Pulmonary vascular disease in fetal meconium aspiration. *J Pediatr* 104:758–762, 1984.

328. Katz VL, Bowes WA: Meconium aspiration syndrome: Reflections on a murky subject. *Am J Obstet Gynecol* 166:171–183, 1992.

329. Gersony WM: Persistence of the fetal circulation. A commentary. *J Pediatr* 82:1103–1106, 1973.

330. Perlman EJ, Moore GW, Hutchins GM: The pulmonary vasculature in meconium aspiration. *Hum Pathol* 20:701–706, 1989.

331. Raine J, Hislop AA, Redington AN, Haworth SG, Shinebourne EA: Fatal persistent pulmonary hypertension presenting late in the neonatal period. *Arch Dis Child* 66:398–402, 1991.

332. Boros SJ, Mammel MC, Lewallen PK, Coleman JM, Gordon MJ, Opheoven J: Necrotizing tracheobronchitis: A complication of high–frequency ventilation. *J Pediatr* 109:95–100, 1986.

333. Metlay LA, Mackherson TA, Doshi N, Milley JR: A new iatrogenic lesion in newborns requiring assisted ventilation. *N Engl J Med* 109:111–112, 1983.

334. Clark RH: High–frequency ventilation. *J Pediatr* 124:661–670, 1994.

335. Mammel NC, Boros SJ: Airway damage and mechanical ventilation: A review and commentary. *Pediatr Pulmonol* 98:915–921, 1987.

336. Metlay LA, Macpherson TA, Doshi N, Milley JR: Necrotizing tracheobronchitis in intubated newborns: A complication of assisted ventilation. *Pediatr Pathol* 7:575–584, 1987.

337. Nordin ULF, Klain M, Kessler H: Electron microscopic studies of tracheal mucosa after high frequency jet ventilation. *Crit Care Med* 10:211, 1982.

338. Kein MD, Whittlesey GC: Extra–corporeal membrane oxygenation. *Pediatr Clin North Am* 41:365–384, 1994.

339. Chou P, Shen–Schwartz S, Blei ED, Crusse FG, Reynolds M: Pulmonary epithelial changes following ECMO therapy. Analysis of 17 autopsy cases. *Lab Invest* 64:2p, 1991.

340. Chou P, Blei ED, Shen–Schwarz S, Gonzalez–Crussi F, Reynolds M: Pulmonary changes following extracorporeal membrane oxygenation: Autopsy study of 23 cases. *Hum Pathol* 24:405–412, 1993.

341. Grigg J, Arnon S, Chase A, Silverman M: Inflammatory cells in the lungs of premature infants on the first day of life: Perinatal risk factors and origin of cells. *Arch Dis Child* 69:40–43, 1993.

342. Gould SJ, Isaacson PG: Bronchus–associated lymphoid tissue (BALT) in human fetal and infant lung. *J Pathol* 169:229–234, 1993.

343. Gravett MG, Nelson P, DeRouen T, Chrtchlow C, Eschenbach DA, Holmes KK: Independent associations of bacterial vaginosis and *Chlamydia trachomatis* infection with adverse pregnancy outcome. *JAMA* 256:1899–1903, 1986.

344. Cassell GH, Waites KB, Watson HL, Crouse DT, Harasawa R: *Ureaplasma urealyticum* intrauterine infection: Role in prematurity and disease in newborns. *Clin Microbiol Rev* 6:69–87, 1993.

345. Jeffery H, Mitchison R, Wigglesworth JS, Davies PA: Early neonatal bacteremia: Comparison of group B streptococcal, other gram–positive, and gram–negative infections. *Arch Dis Child* 52:683–686, 1977.

346. Teplitz C: Pathogenesis of *Pseudomonas* vasculitis and septic lesions. *Arch Pathol* 80:297–307, 1965.

347. Singer DB: Infections of fetuses and neonates, in Wigglesworth JS, Singer DB (eds), *Textbook of Fetal and Perinatal pathology.* Oxford, Blackwell Scientific Publications, 1991, pp 525–591.

348. Knox WF, Hooton VN, Barson AJ: Pulmonary vascular candidiasis and use of central venous catheters in neonates. *J Clin Pathol* 40:559–565, 1987.

349. Rhine WD, Arvin AM, Stevenson DK: Neonatal aspergillosis. A case report and review of the literature. *Clin Pediatr* 25:400–403, 1986.

350. Frommel GT, Bruhn FW, Schwartzman JD: Isolation of *Chlamydia trachomatis* from infant lung tissue. *N Engl J Med* 296:1150–1152, 1977.

351. Griffin M, Pushpanathan C, Andrews W: *Chlamydia trachomatis* pneumonitis: A case study and literature review. *Pediatr Pathol* 10:843–852, 1990.

352. Stocker JT: Congenital cytomegalovirus infection presenting as massive ascites with secondary pulmonary hypoplasia. *Hum Pathol* 16:1173, 1985.

353. Yow MD, Williamson DW, Leeds LJ, et al: Epidemiologic characteristics of cytomegalovirus infections in mothers and their infants. *Am J Obstet Gynecol* 158:1189–1195, 1988.

354. Hall CB, Kopelman AE, Douglas RG Jr, Geiman JM, Meagher MP: Neonatal respiratory syncytial virus infection. *N Engl J Med* 300:393–396, 1979.

355. Aherne W, Bird T, Court SDM, Gardner PS, McQuillin J: Pathological changes in virus infections of the lower respiratory tract in children. *J Clin Pathol* 23:7–18, 1970.

356. Delage G, Brochu P, Robillard L, Jasmin G, Joncas JH, Lapointe N: Giant–cell pneumonia due to respiratory syncytial virus. *Arch Pathol Lab Med* 108:623–625, 1984.

357. Neilson KA, Yunis EJ: Demonstration of respiratory syncytial virus in an autopsy series. *Pediatr Pathol* 10:491–502, 1990.

358. Abzug MJ, Beam AC, Gyorkos EA, Myron J: Viral pneumonia in the first month of life. *Pediatr Infect Dis J* 9:882–885, 1990.

359. Kaplan MH, Klein SW, McPhee J, Harper RG: Group B coxsackie virus infections in infants younger than 3 months of age: a serious childhood illness. *Rev Infect Dis* 5:1019–1032, 1983.

360. Modlin JF: Perinatal echovirus infection: Insights from a literature review of 61 cases of serious infection and 16 outbreaks in nurseries. *Rev Infect Dis* 8:918–926, 1986.

361. Toce SS, Keenan WJ: Congenital echovirus 11 pneumonia in association with pulmonary hypertension. *Pediatr Infect Dis J* 7:360–362, 1988.

362. Isaacs D, Dobson SRM, Wilkinson AR, Hope PL, Eglin R, Moxon ER: Conservative management of an echovirus II outbreak in a neonatal unit. *Lancet* 1:543–545, 1989.

363. Abzug MJ, Levin MJ: Neonatal adenovirus infection: Four patients and review of the literature. *Pediatrics* 87:890–896, 1991.

364. Berry PJ, Nagington J: Fatal infection with echovirus II. *Arch Dis Child* 57:22–29, 1982.

365. Abbondanzo SL, English CK, Kagan E, McPherson RA: Fatal adenovirus pneumonia in a newborn identified by electron microscopy and in-situ hybridization. *Arch Pathol Lab Med* 113:1349–1353, 1989.

Chapter 4

The Pathology of the Lung in Human Immunodeficiency Virus Infection and AIDS

Nick Francis / P.S. Hasleton

In almost all cases, respiratory disease develops in patients infected with human immunodeficiency virus (HIV) at some stage during the progression of disease and, indeed, this was one of the first manifestations that led to the recognition of AIDS in 1981. Since that time, pulmonary problems of a variety of causes continue to be of major importance in morbidity and mortality.

The immunodeficiency viruses (HIV-1 and HIV-2) belong to the retroviruses (subfamily Lentivirinae), which replicate in a unique manner. They have a reverse transcriptase enzyme that allows the copying of a single RNA strand into multiple copies of DNA, which are then incorporated into the host-cell genome. The virologic and immunologic details of the organisms are well reviewed elsewhere.[1,2]

The exact way in which the virus produces immunodeficiency is not known but is in part due to cytopathic effects on the CD4+ subset of human T lymphocytes and the formation of syncytia between infected and noninfected CD4 lymphocytes. The precise mechanism (or mechanisms) of this toxic effect on the lymphocytes is controversial. There are other possible mechanisms for the immunosuppression, including a defective humoral immune component and infection of pulmonary macrophages by the virus, impairing their function.[3,4]

While HIV-1 is mainly responsible for the global pandemic of acquired immunodeficiency syndrome (AIDS), HIV-2, identified more recently, appears at the moment to be largely limited to countries in West Africa and to Portugal. The viruses are spread by sexual contact, exposure to infected blood or blood products, and perinatal transmission from mother to child.

The magnitude of the AIDS problem makes it one of the most important medical challenges for the present and future. Globally, the World Health Organization (WHO) estimates that 13 million people are currently infected with HIV-1 and by the year 2000 30 to 40 million will be infected.[5] Though the incidence of AIDS in the homosexual population has received much media attention, the majority of cases worldwide (60 percent) are acquired heterosexually.[1,2,5]

Changes in the Lung in HIV Infection

Bronchoalveolar lavage (BAL) shows an increase in both the total number and the percentage of cells that are lymphocytes,[6] and the majority of these cells are the suppressor-cytotoxic subset. The percentage of helper-inducer lymphocytes in BAL is decreased. These findings are important in the face of decreased numbers of peripheral circulating helper-inducer and suppressor-cytotoxic lymphocytes, and especially in view of the fact that HIV-1 selectively depletes helper-inducer cells. It is possible that both types of lymphocytes are sequestered in the lung. A small subset of patients with AIDS have a neutrophilia in their BAL, but there is no consistent association with a particular pathogen. These cellular abnormalities may not be the only pulmonary defect in AIDS. The presence of increased quantities of albumin in BAL fluid suggests that there is an abnormality in the alveolar wall in these patients. There are significant increases in IgG and IgA but not IgM.

General Considerations

It is important that all specimens from HIV-positive patients are labeled as posing a danger of infection. The local safety codes require specific identification of certain categories, such as HIV. This is best done by means of the associated paperwork with a specimen. All staff working in laboratories and postmortem rooms must be fully trained in safety procedures, and the number of people involved in an autopsy should be limited to three, consisting of a pathologist, an assistant, and a circulator.[7] The last-named person should perform duties such as photography, removal of specimens in containers, communication, recording, and so on. Precise guidelines are laid down by the Health and Safety Commission. Guidance for surgeons and their teams dealing with HIV infection have been laid down by the Department of Health in England and the Royal College of Pathologists Working Party report (July 1995).[8]

The defect in cell-mediated immunity that occurs with AIDS predisposes to a variety of pulmonary pathogens and opportunistic infections as well as reactive or neoplastic processes, which are listed in Table 4-1. The commonest of these is *Pneumocystis carinii*. It is likely that this list is not exhaustive and that with time, further, more obscure infections and complications will be described. It must be emphasized that more than one pathogen or process may be present in the respiratory systems of patients with HIV and AIDS and a high level of suspicion for multiple pathologies must be maintained in the examination of tissue specimens. It must also be borne in mind that because of the immune suppression, both the clinical presentation and pathologic response are frequently atypical. Therefore special stains for pathogens must be routinely used in the assessment of specimens derived from these patients. Indeed, almost any of the conditions affecting the lungs described in the following sections may present with a clinical or radiologic pattern that can mimic *Pneumocystis carinii* pneumonia.

THE INFECTIONS

Fungi

Pneumocystis carinii PNEUMONIA

Pneumocystis carinii, originally described as a protozoan, is now thought to be a fungus on the basis of ultrastructural, biochemical, and RNA ribosomal probe analysis. It is a widespread and latent infection in the general population. As a cause of disease, it was first recognized in orphans during World War II and since has been widely recognized as a cause of lung disease in patients immunosuppressed for a variety of reasons (therapeutic, transplantation, and malignancies). Most evidence suggests that disease is due to reactivation of latent infection. In patients with HIV, disease is uncommon unless the CD4 count has fallen below 200.

Pneumocystis carinii pneumonia (PCP) is the most common serious respiratory complication occurring in AIDS patients. Nearly 75 percent will have at least one episode of PCP during their lifetime.[9] The usual pathology of PCP is described below, but there are several features of this infection in AIDS patients that should be mentioned here, as

TABLE 4-1
Pulmonary Complications of HIV Infection

Infectious
Viruses
  Cytomegalovirus
  Herpes simplex virus
  Varicella zoster virus
  Epstein-Barr virus
  Human immunodeficiency virus?
Bacteria
  Pyogenic organisms (especially *Streptococcus pneumoniae*, *Haemophilus influenzae*)
  Mycobacterium tuberculosis
  Mycobacterium avium complex
  Other nontuberculous mycobacteria
  Bacillary angiomatosis
Fungi
  Pneumocystis carinii
  Histoplasma capsulatum
  Coccidioides immitis
  Cryptococcus neoformans
  Candida species
  Aspergillus species
Parasites
  Toxoplasma gondii
  Cryptosporidium
  Microsporida (*Encephalitozoon hellum*)
  Strongyloides stercoralis
Noninfectious
  Kaposi's sarcoma
Interstitial pneumonias
  Lymphocytic interstitial pneumonitis
  Nonspecific interstitial pneumonitis
  Drug-induced reactions
Other
  Adult respiratory distress syndrome
  Secondary alveolar proteinosis
  Pulmonary hypertension
Malignancies
  Non-Hodgkin's lymphoma
  ?Carcinoma of lung (especially adenocarcinoma)

the disease in AIDS patients is clinically different from that seen in others.[10] There may be an insidious, slowly progressive presentation, lasting weeks to months, and without the typical cough, fever, and dyspnea. In the early stages, the patients may appear surprisingly healthy, and physical findings are variable. Although typically presenting as bilateral shadowing on a chest radiograph, disease may be confined to the upper lobes. Rapidly progressive disease may occur in some patients, and adverse drug reactions are commoner in AIDS than non-AIDS patients.

Both animal and human studies show that specific antimicrobial therapy does not completely eradicate the organism from the lungs, even though clinical cure rates of up to 75 percent can be achieved.[11] Despite clinical improvement, follow-up lung biopsies showed that in 38 percent of episodes, there was positivity for *Pneumocystis* 25 days after the initial diagnosis and after treatment had been instituted.[12] However, the clinical response of AIDS patients with residual cysts was no different from that of those without the organisms, while other biopsy-negative cases proceeded to second episodes of the disease. A more recent study has shown a correlation between the number of *Pneumocystis carinii* clusters and early relapse.[13] The technique for cluster counting of the organisms is simple and reproducible. This study had only a small number of patients, with 18 responders, 6 relapsing within 6 months. Of 5 patients who had more than 20 clusters of *Pneumocystis* per 500 nucleated cells, all died within 21 days of their initial lavage. This group had significantly more organisms than the other groups studied. In the responders and nonresponders without relapse, there was a significant decline in the number of organisms on repeat lavage. In responders without relapse, the number of organisms decreased by more than 50 percent in all but one case. In responders with relapse, no patient had a reduction

in numbers of *P. carinii* of more than 50 percent.[13] However, the issue is complicated by a small study suggesting that *Pneumocystis* detected by silver or immunofluorescent staining in BAL more than 1 month after treatment may represent a new infection rather than relapse of prior disease.[14] Another small study, examining cyst density in induced sputum and BAL, showed no correlation with prognosis but suggested that a low cyst count may indicate a greater proportion of trophozoites.[15] A high neutrophil lavage count may be an indicator of more morbidity and respiratory difficulty, suggesting that this subgroup of patients have a capillary leak syndrome, as they also have a higher lavage protein concentration than controls with AIDS but no *Pneumocystis*.[16]

Pathologic Changes Associated with *Pneumocystis*

The characteristic histologic appearances are of an intraalveolar, pinkish, foamy, honeycomb exudate that contains the cysts and trophozoites of *P. carinii* (Fig. 4-1). These are often visible on a good hematoxylin and eosin (H&E) stain but should be confirmed by any of the methods indicated below (Fig. 4-2). The associated inflammation of the alveolar wall is very variable and ranges from almost no reaction to marked chronic interstitial inflammation and eventually to alveolar wall fibrosis. These changes are partly dependent on the duration of the infection and on whether the patient has had previous episodes. Patients with significant respiratory

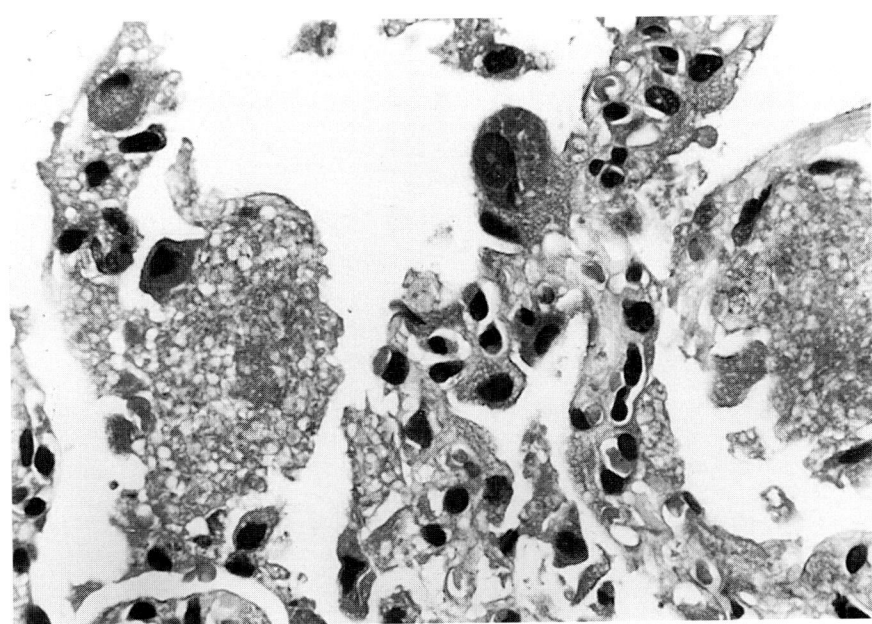

FIGURE 4-1
Lung alveoli containing granular material typical of PCP, with alveolar walls containing a mild inflammatory infiltrate. Note also large cell typical of CMV infection. (H&E, high magnification.)

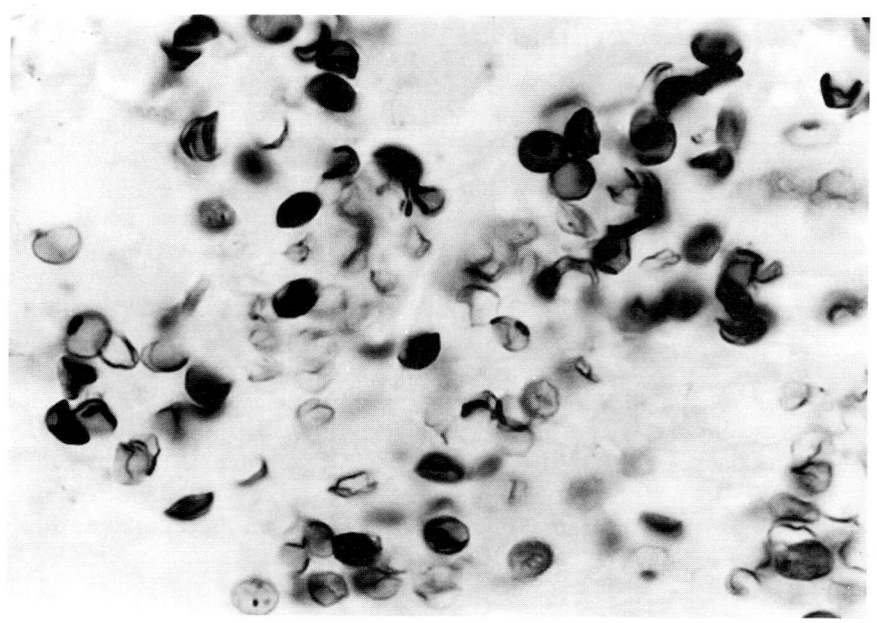

FIGURE 4-2
Grocott silver stain of PCP from same case as in Fig. 4-1. (High magnification.)

TABLE 4-2
Patterns of Pneumocystis in the Lung

Minimal
Typical
Nodular
Necrotizing
Granulomatous
Bullous with pneumothorax or hemothorax
Calcification
Vasculitic
Diffuse alveolar damage
BOOP
Lymphoplasmacytic interstitial pneumonitis pattern
Interstitial fibrosis
Extrapulmonary

impairment may have little exudate and few organisms.

In addition to this spectrum of variants of the usual pattern of involvement, there are an increasing number of cases where the histologic pattern falls into a number of other recognizable categories (Table 4-2). Some of these types may relate to improved survival and the influence of therapeutic regimens.

Granulomatous PCP may be of the sarcoid or caseating type and be associated with cavitation and necrosis. It may also be associated with minor or extensive calcification (Figs. 4-3, 4-4, and 4-5). The organisms are usually found within the necrotic areas, but clearly the overall pattern may raise the

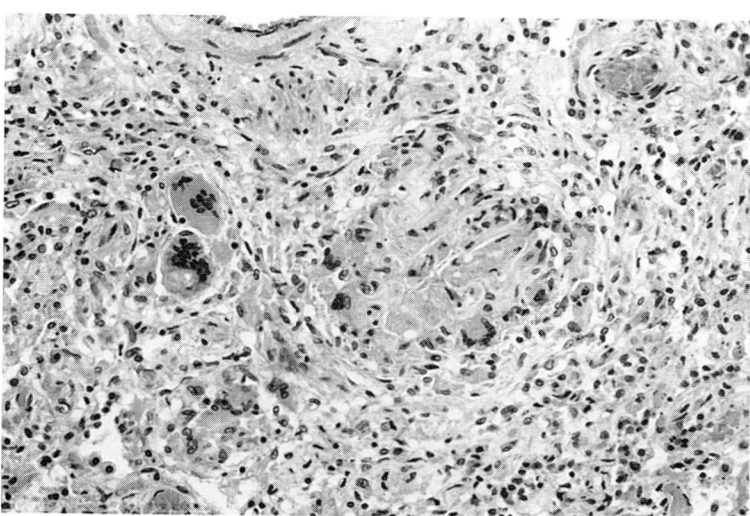

FIGURE 4-3
Granulomatous PCP. (Medium magnification.)

118

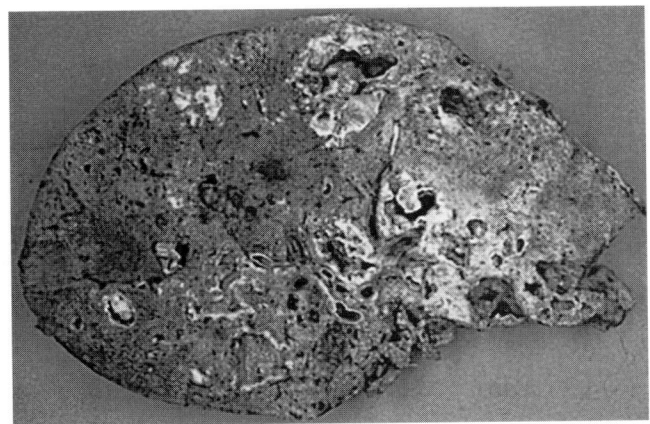

FIGURE 4-4
Macrophotograph of cavitating PCP.

possibility of a tuberculous infection. Cavitating lesions may be associated with pulmonary fistula formation and pneumothorax. In these unusual and complicated cases, organisms may be found in the alveolar and peribronchial vasculature—a feature that may account for the increasing reports of disseminated extrapulmonary *Pneumocystis*. Severe and rapid respiratory failure may be associated with a pattern of diffuse alveolar damage, with a proteinaceous exudate, hyaline membrane formation, loss of and damage to type I pneumocytes and alveolar wall fibrosis.

A pattern of lymphocytic interstitial pneumonitis may also be encountered when the infiltrate is of a lymphoplasmacytic composition. This pattern is more common in children and African adults.

These atypical patterns are well represented in the summary of the following three reports. The National Institutes of Health (United States) showed atypical pathologic findings in 123 lung biopsies from 76 patients with AIDS and *Pneumocystis* infection. These authors found interstitial and intraluminal fibrosis, an absence of alveolar exudate, numerous alveolar macrophages, granulomatous inflammation (Fig. 4-3), hyaline membranes, marked interstitial pneumonitis, cavitation (Fig. 4-5), interstitial microcalcification (Fig. 4-4), minimal histologic reaction, vasculitis, and cyst formation.[17] The cysts, which were small intrapulmonary or subpleural bullous blebs ranging from 0.5 to 1 cm in size, gave rise to pneumothoraces in two out of three patients. In these two, cysts were associated with noncaseating granulomas. Bronchiolitis obliterans may be seen.[18] In two cases, the *Pneumocystis* was strikingly subpleural in its distribution. Calcification may be "bubbly," platelike, elongated, or conchoidal. All foci of calcification contain cysts, as demonstrated by methenamine silver. In some cases, calcification may be the only histologic manifestation of previous *Pneumocystis* infection.[19] In the granulomas, which were sarcoidlike in some cases, *P. carinii* organisms were found in the majority. In one case, the interstitial inflammation had a pattern resembling lymphocytic interstitial pneumonia (LIP) (see Fig. 4-27). This diagnosis was associated with a low index of suspicion for *Pneumocystis*. Acute interstitial inflammation was seen in four of the *Pneumocystis*-positive biopsy specimens. In most cases, the interstitial infiltrate was scanty[10] and there was minimal interstitial edema or fibrosis. *Pneumocystis* may present as an endobronchial mass without any parenchymal involvement.[20] In the case of pleural *Pneumocystis*, infection has been described following a pneumothorax.[21] Pneumocystis pneumonia may be complicated by pneumothoraces in 9 percent of cases.[22] The picture of adult respiratory distress syndrome may be a strong predictor of mortality and is common at autopsy.[10]

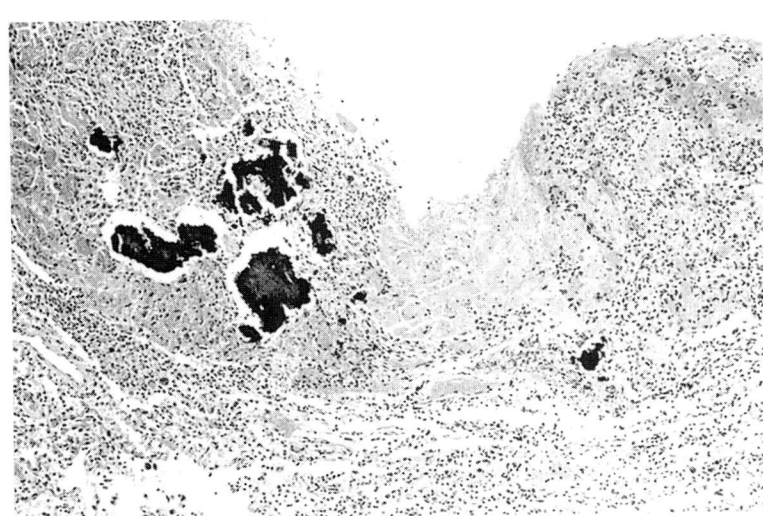

FIGURE 4-5
Edge of cavity due to PCP showing PCP amorphous debris in pale area and dark-staining areas of calcification. (Low magnification.)

The pathology of treated *Pneumocystis* has been studied in biopsy-proven disease.[23] The cases were classified into three groups. The first group died in the first week of the disease and one-third of the alveolar volume was filled with the characteristic foamy exudate, accounting for the respiratory failure. The second group of patients died from 8 days to 2 months after diagnosis and respiratory failure was due to intraalveolar fibrosis as well as pulmonary arterial thromboemboli. The last group appeared cured of *Pneumocystis* but died months to years later of other manifestations of AIDS.

In one autopsy study, respiratory failure was the commonest immediate cause of death in AIDS, and the organism most frequently responsible was *Pneumocystis*.[24] Cytomegalovirus (CMV) pneumonia was said to be the next most common cause of respiratory death, but—as noted below—this virus may be only a passenger.

Extrapulmonary *Pneumocystis* infection is seen increasingly in AIDS patients in a variety of sites but particularly in regional lymph nodes, spleen, liver, bone marrow, and more unusual sites such as the gut, thyroid, heart, and skin. In these locations, the same characteristic clusters of organisms and exudate are usually seen. This increase is thought to be associated with pentamidine treatment, and prophylaxis.[25]

Diagnosis of *Pneumocystis* Infection

Early in the history of AIDS, the two main sampling methods were sputum examination and endoscopic transbronchial biopsy with or without bronchoalveolar lavage. However, due to the complication rate of biopsy, the investigative protocol has been refined to induced sputum and empirical treatment in clinically suspected cases. Resort to biopsy is only when there are atypical clinical and/or radiologic features, identifiable mass lesions, or a failure to respond to 5 days of empirical therapy against *P. carinii*.

The methods that have been advocated for the diagnosis of PCP are discussed below. One series examined 91 paired biopsy and cytology specimens from 72 AIDS patients.[26] There was no significant difference between biopsy and cytology in diagnosing *P. carinii*. Biopsy was more useful than cytology in the diagnosis of other infections, which were present in some of these patients. Bronchoalveolar lavage gives a good diagnostic yield but may not be clinically appropriate in all cases.[27] The need for bronchoscopy is reduced if mucolyzed induced-sputum examination is used.[28] Sputum is induced by inhalation of 3% saline mist, mucolyzed, concentrated by centrifugation, and stained with modified

Giemsa or methenamine silver. Using this method, PCP was diagnosed in 55 percent of 404 episodes of suspected *Pneumocystis*. In 180 episodes where the sputum was negative for *Pneumocystis,* bronchoscopy with transbronchial biopsy and/or BAL demonstrated the organism in 50 cases (42 percent). This underlines the importance of proceeding to a more invasive method if induced sputum gives a negative result.

The standard stain used for the detection of *Pneumocystis* cysts is methenamine silver (Fig. 4-2). This has suffered the drawback of taking 1 to 3 h, but development of a rapid (10-min) stain technique has overcome this problem.[29] To the uninitiated, mucus, carbon pigment, red blood corpuscles (in overdifferentiated sections) and *Candida* spores, *Histoplasma capsulatum, Cryptococcus neoformans*, and *Torulopsis glabrata* may resemble *Pneumocystis*; identification of the typical moon-shaped cysts with intracystic bodies is very useful in making a diagnosis. Wright's stain can be used in the rapid diagnosis of paraffin sections of *Pneumocystis* and correlates well with methenamine silver staining.[30] Giemsa, Gram-Weigart, a modified Gridley, and Diff-Quik (a differential stain similar to Wright-Giemsa),[31] all show trophozoites[32]; Papanicolaou and toludine blue have also been used for identification of the organism. It is for the histologist to choose the method that gives the best results. We still prefer the methenamine silver backed by the immunoperoxidase or the immunofluorescence stain (see below). A positive control must always be examined.

In treated patients, silver staining shows either ballooned cysts or failure to stain because of a degenerated cyst wall.[27] Trophozoites and intracystic dots are rarely seen in treated cases. Immunostaining usually outlines the cyst walls, even after therapy.

The immunofluorescence technique for detection of the organism has been shown to give reliable results as compared with methenamine silver staining. Lavage fluid must have the HIV inactivated by 1% paraformaldehyde or 50% ethanol for 30 s. The immunofluorescent stain was positive in 93 percent of cases, as compared with 70 percent with methenamine silver.[33] Results were better with lavage specimens than with those obtained at bronchial wash. These results have been confirmed in other studies[34–36]; in two of these,[35,36] the technique was used on induced sputum with good results.

In general, immunoperoxidase staining does not detect any more cases of *Pneumocystis* than the methenamine silver.[37] The latter stain was thought to be superior, as it also detected other fungi. Immunoperoxidase has the advantage that the organisms may be detected at a lower magnification (×100) than with methenamine silver (×400). Theoretically, immunoperoxidase staining should detect trophozoites, but in practice they are easily confused with background debris. Thus, from the literature to date with induced sputa and lavage specimens, immunofluorescent methods are more sensitive than silver stains. This advantage is not maintained in paraffin sections. Not all studies have used the same antibody, and thus comparison of data is difficult. Most have used the monoclonal 3F6, but one study utilized a polyclonal rat antibody,[38] and it is possible that there is a different strain of *Pneumocystis* in this animal. We have found peroxidase useful when few cysts are present in small transbronchial biopsies. It has been estimated that immunoperoxidase is at least 50 times more expensive than methenamine silver.[37]

Material embedded in 0.9-μm methacrylate sections stained with silver methenamine borate and counterstained with toluidine blue demonstrates both cyst walls and intracystic sporozoites.[39] The technique is very labor-intensive and lengthy, requiring 8 h for embedding alone. In a laboratory with rapid processing facilities and immunoperoxidase experience, paraffin sections will give a quicker reliable answer.

DNA amplification has been used to detect *Pneumocystis*.[40] In this study, DNA amplification was superior to silver staining in both lavage and induced sputum, but immunoperoxidase techniques were not used. The appropriate oligonucleotide primers and probes are necessary. The sensitivity appears to be related to the time after starting treatment, with a reduction of positive results after a few days of treatment; therefore specimens should be collected early for DNA analysis.[41]

It is possible to detect antibody to *P. carinii* in serum using indirect immunofluorescence, but negative serology does not exclude active infection.[42] The detection rate using this method in histologically confirmed or strongly suspected PCP was only 53.8 percent, which is low compared to other methods given above.

OTHER FUNGAL INFECTIONS

Generally other fungal pneumonias are relatively rare in HIV patients, accounting for about 2 percent of pneumonias. However, as many of them may be associated with disseminated infection and they respond to treatment, accurate recognition and diagnosis are important.

Cryptococcosis

Infection by *C. neoformans* is probably the commonest fungal pulmonary infection in AIDS. Though the most life-threatening manifestation is cryptococcal meningitis, the infection may be limited to the lung in up to 10 percent of cases at the time of presentation.[43] It may also be associated with other concurrent lung infections. The presentation may vary widely from diffuse interstitial inflammation or aggregates of pale macrophages (Fig. 4-6) to pulmonary granulomas and extensive involvement of

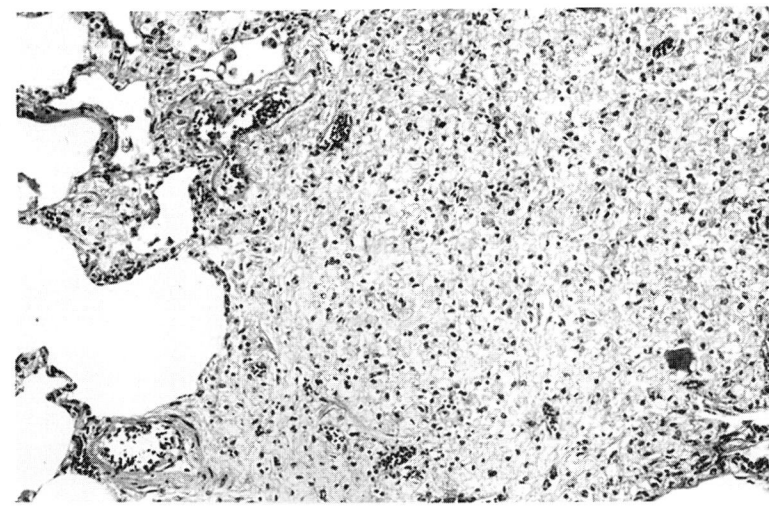

FIGURE 4-6
Pale cell interstitial infiltrate of pulmonary cryptococcosis. (Medium magnification.)

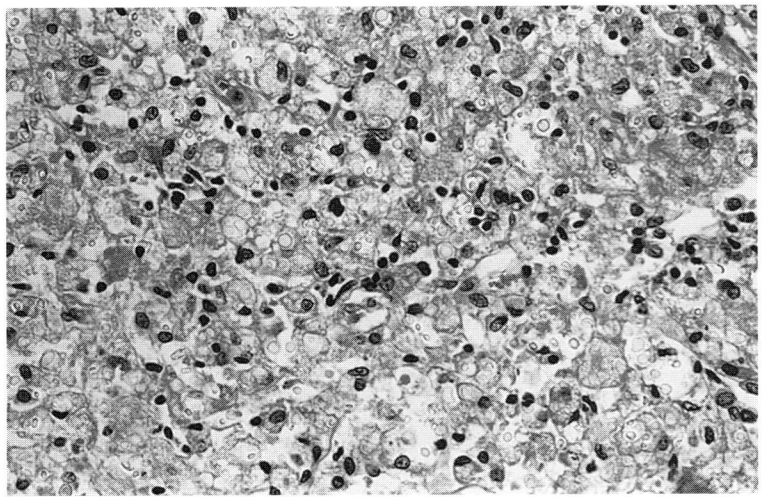

FIGURE 4-7
High-power H&E showing semirefractile fungi of
Cryptococcus.

mediastinal lymphoid tissue. In most patients the pattern is that of a rather insidious interstitial involvement. Once again, the need to perform special stains is emphasized, as organisms may easily be missed, since they do not stain with H&E (Fig. 4-7). They are 4 to 7 μm in size, oval, budding yeasts (Fig. 4-8) whose capsules may be clearly seen in mucicarmine stains. The response may be gelatinous or granulomatous.

In AIDS patients, it has been suggested that unencapsulated (less pathogenic) forms of the yeast may be present in significant numbers, and this is thought to be related to the level of immunosuppression.[44] The pulmonary abnormalities described at postmortem are peripheral granulomata, granulomatous pneumonia, interstitial pneumonia, and massive pulmonary disease. Among these, the interstitial

pattern is the most common (Fig. 4-6). Cavitation, lymph node enlargement, and pleural effusion are infrequent.[45]

Histoplasmosis

Histoplasmosis is endemic in some areas, such as the Ohio and Mississippi river valleys. Latent infection rates are therefore high, and reactivation is associated with disseminated histoplasmosis in up to 47 percent of AIDS patients from such areas. Histoplasmosis may be seen in AIDS patients from nonendemic areas.

Patients present with nonspecific features, including fever, and weight loss, but a large number may have cough or dyspnea. Radiologically, there are diffuse reticulonodular infiltrates as well as intersti-

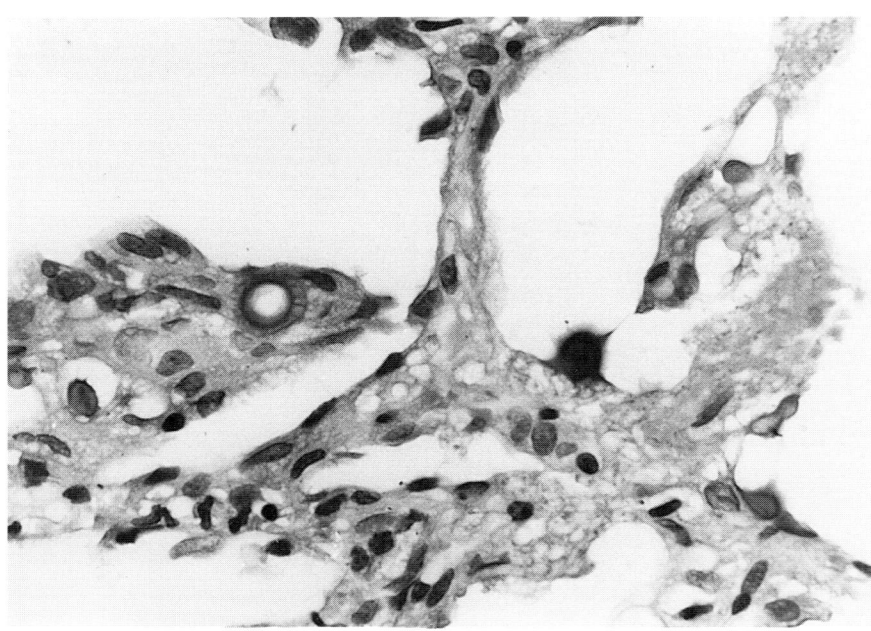

FIGURE 4-8
High-power view of *Cryptococcus* in
alveolar exudate. (Alcian blue stain.)

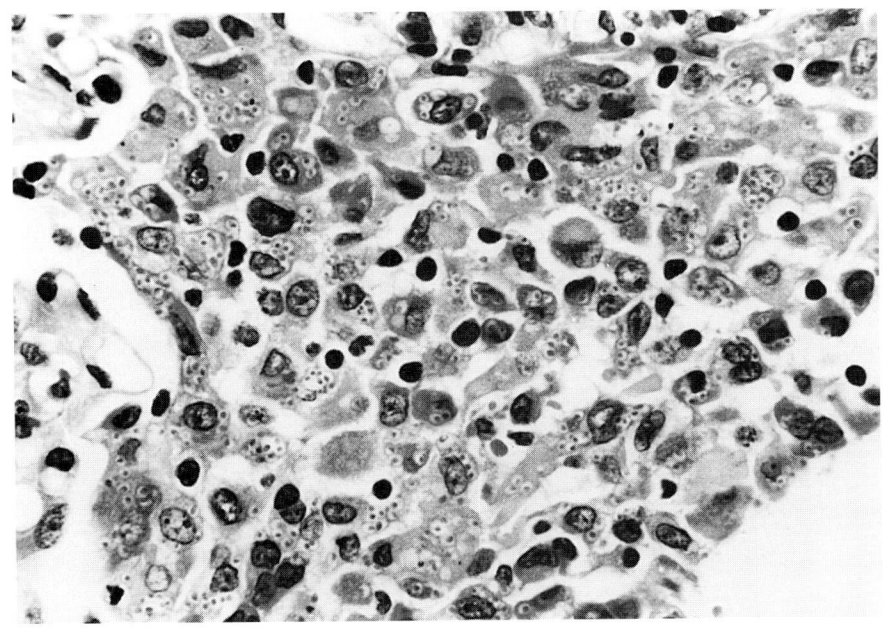

FIGURE 4-9
Alveolar space filled with a histiocytic infiltrate and lymphoid cells, the former containing numerous *Histoplasma* organisms. (High magnification.)

tial or intraalveolar patterns. However, at least one-fifth of the patients may have no radiologic changes.[46] A syndrome that includes adult respiratory distress syndrome, disseminated intravascular coagulation, encephalopathy, and acute renal failure has also been described in AIDS.[47] In disseminated disease, the organism may be found in the peripheral blood.[48]

The histology associated with histoplasmosis may be nongranulomatous or may consist of sheets of histiocytes forming noncaseating granulomas with abundant organisms (Fig. 4-9). The fungi are recognized as 2- to 5-μm ovoid yeasts (Fig. 4-10), which may be difficult to distinguish from *Pneumocystis* or *Leishmania*. The latter organism does not stain with fungal stains. The organisms are usually found within macrophages and have a characteristic "fried egg" appearance in H&E or Giemsa (Fig. 4-11). In Africa *Histoplasma duboisii* is the prevalent form, which is recognized by a much larger 10- to 15-μm thick-walled yeast. It may also cause disseminated disease.

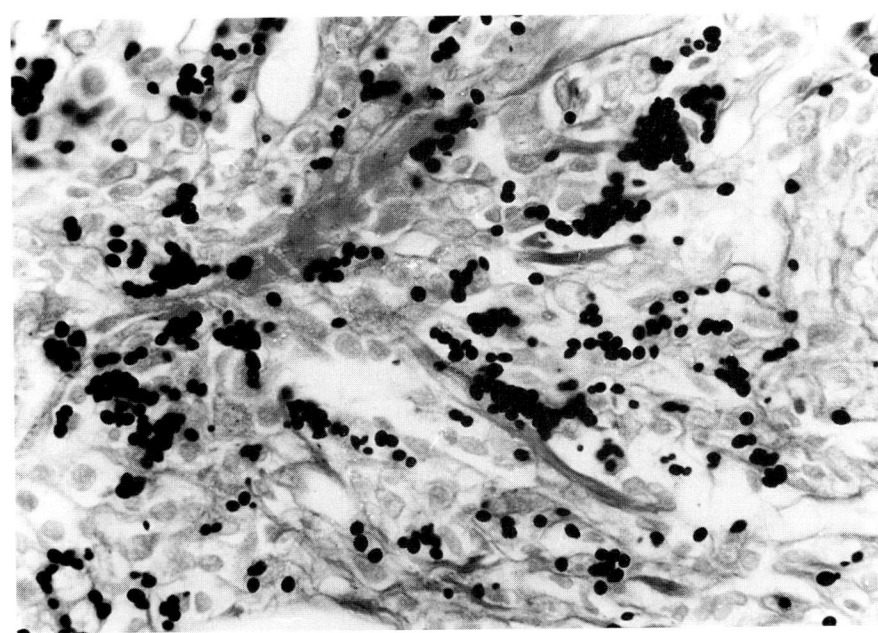

FIGURE 4-10
Grocott stain of same case as in Fig. 4-9. (High magnification.)

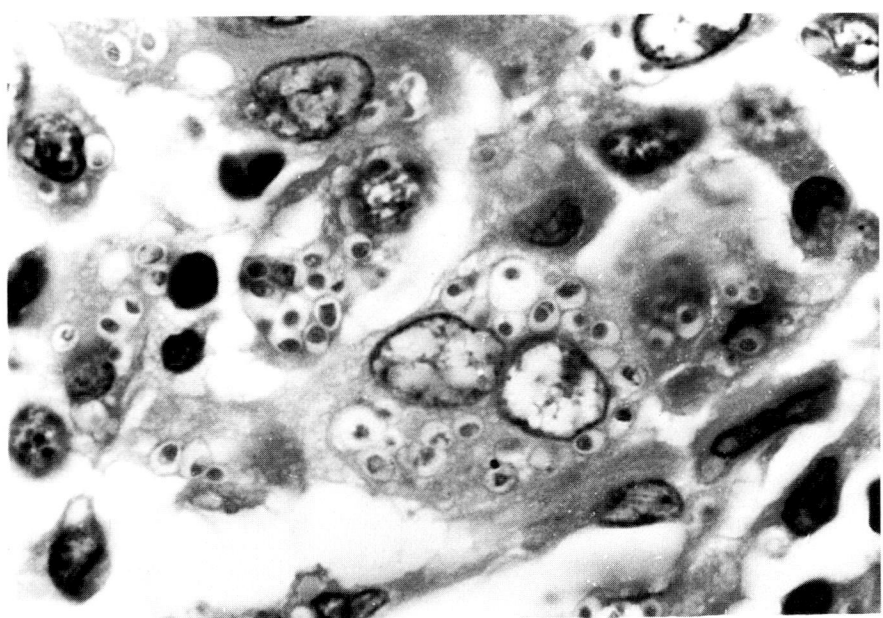

FIGURE 4-11
High-power view of "fried egg" appearance of *Histoplasma*.

Coccidiomycosis

In coccidiomycosis, the chest radiographs show diffuse bilateral reticulonodular or nodular infiltrates. Less commonly, there may be a single nodular lesion, hilar or mediastinal lymphadenopathy, cavitation, or even miliary lesions. Invasion of blood vessels or lymphatics may accompany the acute pulmonary infection, and then disseminated lesions are seen in the meninges, skin, bone, and lymph nodes. The organisms of *Coccidioides immitis* are large spherules ranging in size from 5 to 100 μm, depending on their maturity. The endospores, seen in abundance in mature organisms, are usually found in the inflammatory exudate. Associated tissue inflammation may range from suppurative to granulomatous.

Penicilliosis

Penicillium marneffei has recently been described as causing disseminated disease in the United Kingdom[49] and is also reported as the third most common infection in HIV patients in Thailand. Given that Southeast Asia is one of the areas of the world where HIV infection is rising particularly rapidly, it can be anticipated that this and additional unusual infections will be increasingly identified in the context of HIV disease.

Candidiasis

Mucocutaneous candidiasis is one of the most common opportunistic infections that occurs in AIDS. In AIDS patients as in others who are immunocompromised, pulmonary candidiasis is rare and may affect just the trachea or the lungs themselves. To diagnose pulmonary involvement, histologic tissue invasion must be seen. The appearances of the organism are well described elsewhere.

Aspergillosis

This is relatively uncommon in AIDS patients, being diagnosed in 7 of 449 patients in one series.[50] The disease may range from necrotizing pseudomembranous bronchoaspergillosis,[51] aspergilloma developing in the preexisting cavity,[52] or mutilple cavities with invasive disease and dissemination.[53] Diagnosis in histologic samples depends upon the identification of the characteristic septate hyphae, which branch at 45° (Fig. 4-12). Depending on the duration of the infection, the hyphae also show irregular swelling, producing balloon-like, clear, enlarged areas of the hyphae.

VIRAL INFECTIONS

Cytomegalovirus

Cytomegalovirus (CMV) may be incriminated histologically as a cause of respiratory tract disease in HIV-positive patients and may cause a necrotizing bronchiloitis.[54] It may be part of a systemic viral infection and give rise to the typical inclusions. The virus is common in BAL, being found in up to 58 percent of all cases,[55] but its presence does not affect survival.[56] The incidence is higher in patients coming to postmortem, with the lungs being the commonest organ involved.[57]

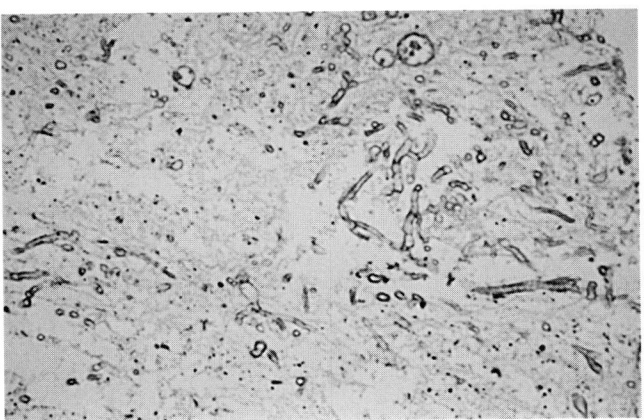

FIGURE 4-12
Aspergillus hyphae in necrotic debris from cavitating lesion in lung. (PAS stain, medium magnification.)

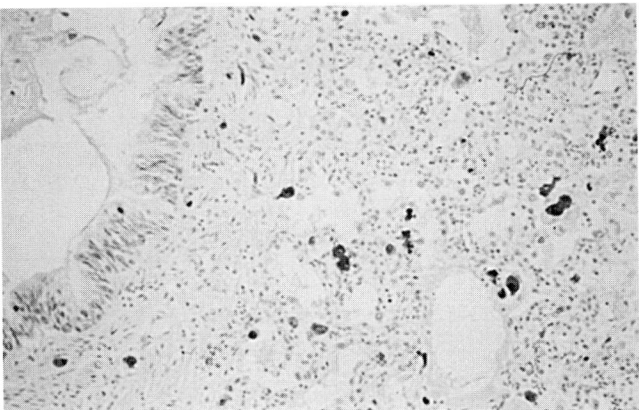

FIGURE 4-13
Immunocytochemistry stain of diffuse interstitial pattern of cytomegaloviral infection. (Medium magnification.)

The role of CMV in pulmonary disease in AIDS patients is controversial, and it is believed that the virus may have only a trivial effect on the patient's clinical course.[58] Patients presenting with pulmonary disease are likely to have another pathogen in their lungs causing the symptoms, apart from CMV.

The criteria for CMV pneumonitis in an AIDS patient include (1) CMV recovered on culture, (2) CMV inclusions demonstrated on BAL cells or preferably transbronchial biopsy, (3) no other pathogen being recovered or visualized, (4) progressive pneumonitis with worsening radiographic or pulmonary function tests, and (5) a response to an antiviral agent against CMV.[59] Using these strict criteria, only 1 percent of all patients with AIDS have clinically significant pulmonary CMV infection.

In another study, CMV was found in only 6 of 166 cases (4 percent).[60] Identification of the infection may be easy when classic Cowdry A inclusion cells are present (see Figs. 4-1 and 4-21), but it may be difficult when atypical cells are the only form present. These may have minimally enlarged nuclei with variably increased amounts of smudgy cytoplasm, within which small basophilic clumps of virions may be seen. Immunocytochemical staining may be very helpful in the diagnosis of CMV in such cases (Fig. 4-13).

Herpes Simplex Virus

While it is a recognized cause of pneumonia in other immunosuppressed patients, this virus has only rarely been reported in patients with AIDS.[61] Similarly, varicella zoster pneumonia is a rare complication in AIDS patients. A recent paper described 11 cases with disseminated viral infection with this agent.[62] Ulcerative lesions of the mucocutaneous margins are much more common, and rare cases of

herpetic tracheobronchial ulceration have been reported.

Epstein-Barr Virus

Epstein-Barr virus (EBV) may act in concert with HIV, since it has been shown in vitro that EBV-infected B cells are unusually susceptible to infection by HIV.[63] Epstein-Barr virus has been implicated in the lymphoid interstitial infiltrates in AIDS (see below and Fig. 4-28), and recently in smooth muscle tumors in HIV patients.[64]

Adenovirus

This virus is rarely associated with pulmonary disease in AIDS; it probably represents reactivation of latent infection. Types 4 and 7 are those usually associated with lung infections and may also be associated with other concurrent infections.

BACTERIAL INFECTIONS

These occur with greater frequency in HIV-infected individuals (Fig. 4-14). The commonest organisms isolated are *Streptococcus pneumoniae* and *Haemophilus influenzae*. However, *Branhamella catarrhalis* infection is also seen in severe pneumonia, and *Pseudomonas* infection can occur, causing either acute pneumonia in patients with neutropenia or a relapsing insidious pneumonia in patients with very low CD4 counts (<25). *Straphylococcus aureus* and gram-negative infections can occur.[65] Rare cavitating infections with *Rhodococcus* have also been reported in AIDS. Bacterial infection is seen more commonly than PCP in drug abusers with HIV infection.[66] Infection with the pneumococcus and *H. influenzae* may frequently occur when the patient

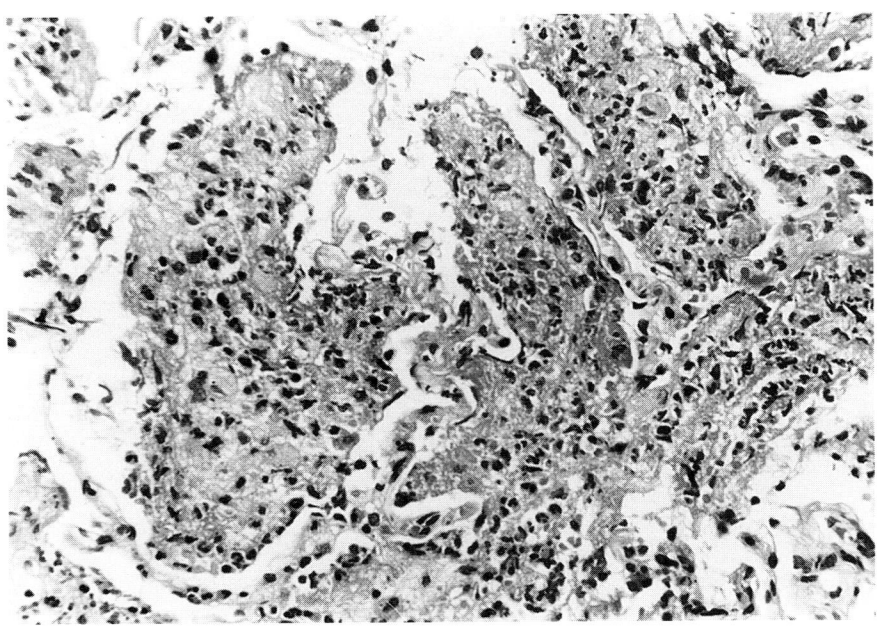

FIGURE 4-14
Alveoli filled with acute inflammatory exudate of bacterial pneumonia. (High magnification.)

has only lymphadenopathy or AIDS-related complex.[67]

Tuberculosis

Tuberculosis (TB) may occur at any stage of HIV infection. Although only extrapulmonary TB[68,69] was in the original Centers for Disease Control (CDC) case definition criteria (1987) for AIDS, this has now been revised to include patients with pulmonary TB. Late in the clinical course of the HIV disease, TB presents with nonspecific features including fatigue, weight loss, and lymphadenopathy. There may be cough or no pulmonary symptoms. The common sites of extrapulmonary disease are bone, pericardium, peritoneum, lymph nodes, central nervous system, bone marrow, and liver. Miliary disease is also common and has a high mortality. The results of tuberculin testing will depend on the individual and residual cell-mediated immunity. If this is poor, false-negative reactions are likely to occur.

Tuberculosis is not at the moment a major problem in relation to HIV in the United Kingdom. However, this state of affairs may change if the United Kingdom follows trends in the United States, where there has been an increase in the number of cases of tuberculosis, mainly due to HIV-related disease.

Tuberculosis had been declining up to 1984, but this trend was reversed and the cases rose by 3 percent in 1986 and 5 percent in 1989. The provisional rise is put at 6 percent in 1990.[68] However, the incidence varies in different ethnic groups. In the United States, 60 percent of Haitian patients with AIDS also had TB.[69] This figure decreased to 28 percent in drug

absusers in New York.[70] In developing countries, HIV-related TB is commoner, producing profound health problems; from 17 to 55 percent of all Africans with tuberculosis are HIV-positive.[71] Tuberculosis is now the commonest infection in 44 percent of African patients dying with HIV infection, and it is of similar prevalence in those with "slim disease." More importantly, it is the major cause of death in 36 percent and is more rapidly progressive. This problem is going to increase as HIV infection in Africa is associated with increased susceptibility to reactivation of previous TB or with the appearance of newly acquired infection.[71–73]

Histologic examination of material from AIDS patients with TB often reveals numerous acid-fast bacilli. The cellular reaction present will depend on the stage of the disease and level of immunosuppression. With advanced stages, there is often a nonreactive macroscopic, (Fig. 4-15) and microscopic pattern, poor granuloma formation (Fig. 4-16), and extensive necrosis with numerous bacilli (Figs. 4-17 and 4-18). We have also seen patients with no cellular reaction and normal alveolar walls on transbronchial biopsy but with distinct clumps of tubercle bacilli. It is therefore imperative that any transbronchial biopsy or lavage from a patient with known or suspected AIDS be stained for TB.

Atypical Mycobacterial Infection

Mycobacterium avium-intracellulare (MAI) is found in the majority of AIDS patients at autopsy. However, up to 50 percent are undiagnosed during life, partly because the infection occurs toward the end of the natural history of AIDS, when CD4 counts are below

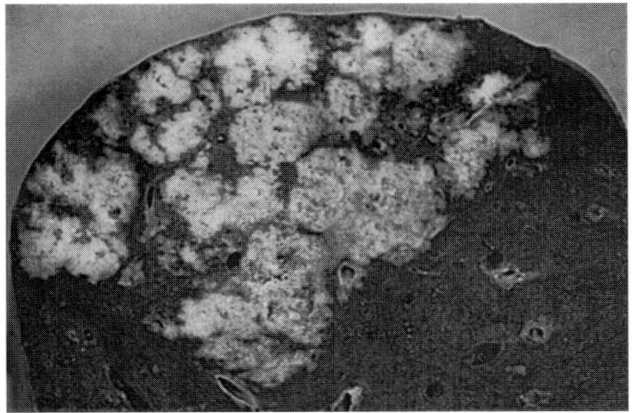

FIGURE 4-15
Macrophotograph of nonreactive tuberculosis; multiple semiconfluent whitish necrotic areas in upper zone.

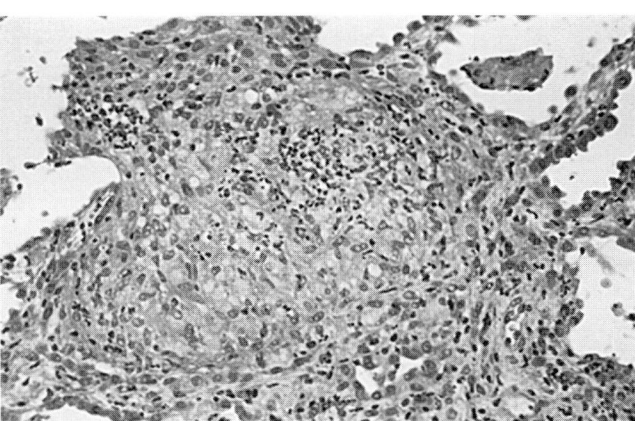

FIGURE 4-16
Field showing poorly developed granuloma with focal central necrosis in a case of tuberculosis. (Medium magnification.)

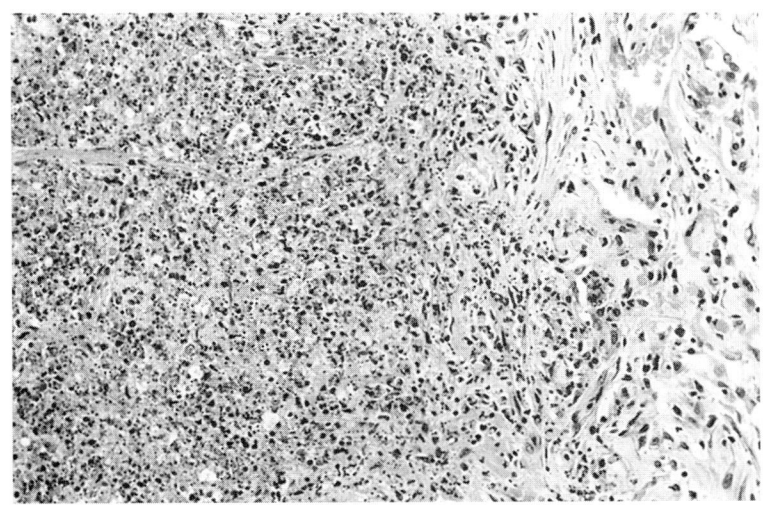

FIGURE 4-17
Medium-power view of edge of area from Fig. 4-15 showing necrotic inflammatory appearance.

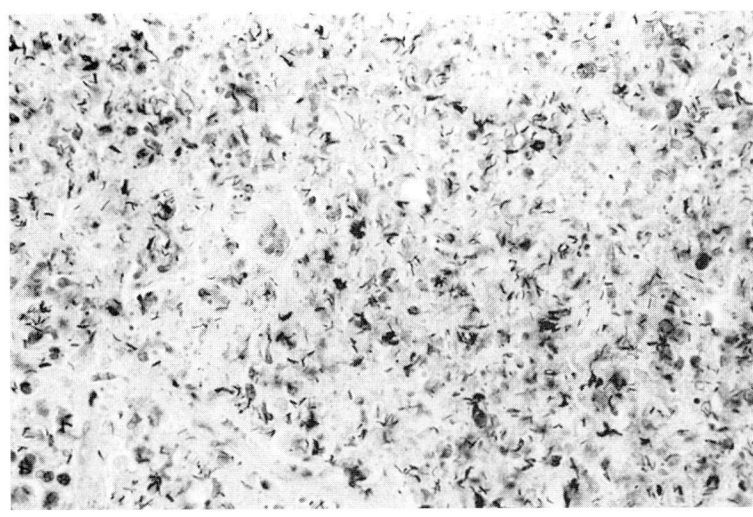

FIGURE 4-18
Ziehl-Neelsen stain of same area as in Fig. 4-17 showing diffuse and abundant organisms. (High magnification.)

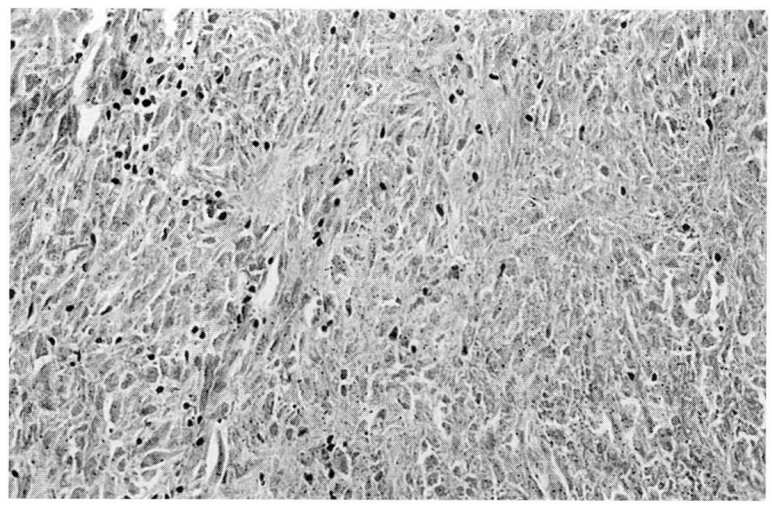

FIGURE 4-19
Hilar lymph node in a case of diffuse atypical mycobacteriosis, with necrosis in background. (Medium magnification.)

100 and usually below 50.[74,75] In contrast to TB, MAI is relatively rarely seen in developing countries. Infection occurs as a result of primary acquisition of the pathogen. The organism is present in soil, water, and food. It enters the body via the lungs or gastrointestinal tract and is found in lymph nodes (Figs. 4-19 and 4-20), liver, spleen, and bone marrow. It can be cultured from BAL fluid. Patients present with the same nonspecific features as noted above with TB, but anemia and abdominal features are common, with hepatosplenomegaly, diarrhea, malabsorption, and abdominal pain. Focal pneumonia or endobronchial lesions without pneumonia may be present.[76]

Large numbers of acid-fast bacilli are usually visible on aspiration of biopsy material from lymph nodes, liver, spleen, or bone marrow. Bronchoalveolar lavage is a source of the bacilli for culture. The typical histologic picture is of numerous acid-fast bacilli within groups or sheets of pale macrophages, with a minimal tissue reaction. A spindle-cell pseudotumor (Fig. 4-21)[77] may rarely be caused by MAI; it can involve lymph nodes or bone marrow and cause confusion with mesenchymal tumors or Kaposi's sarcoma. However the routine use of a Ziehl-Neelsen stain should avoid this problem. Electron microscopic examination shows that the spindle cells contain many lysosomes (compatible with histiocytes) as well as the bacteria. Apparently new mycobacteria may appear in patients with AIDS, as shown by a recent case report.[78] Other species of mycobacteria causing disseminated nontuberculous mycobacterial infection are uncommon.[79] They include *M. kansasii*, *M. gordonae*, *M. fortuitum*, and *M. chelonei*.

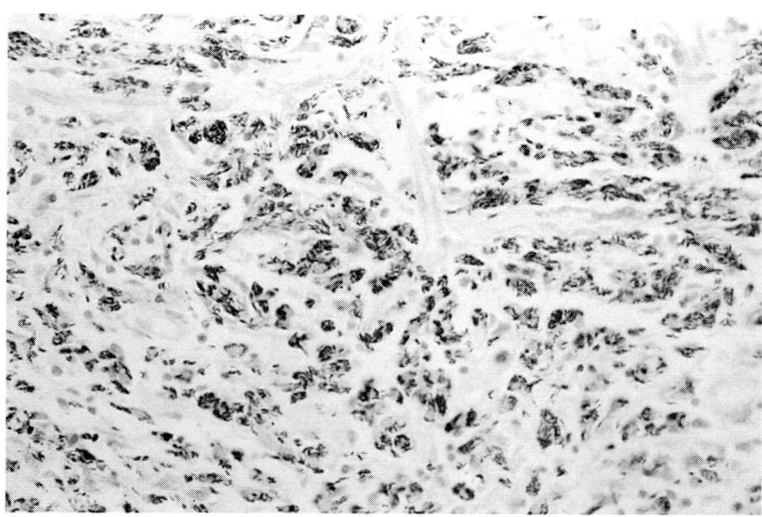

FIGURE 4-20
Ziehl-Neelsen stain showing vast numbers of bacilli within macrophages. (Medium magnification.)

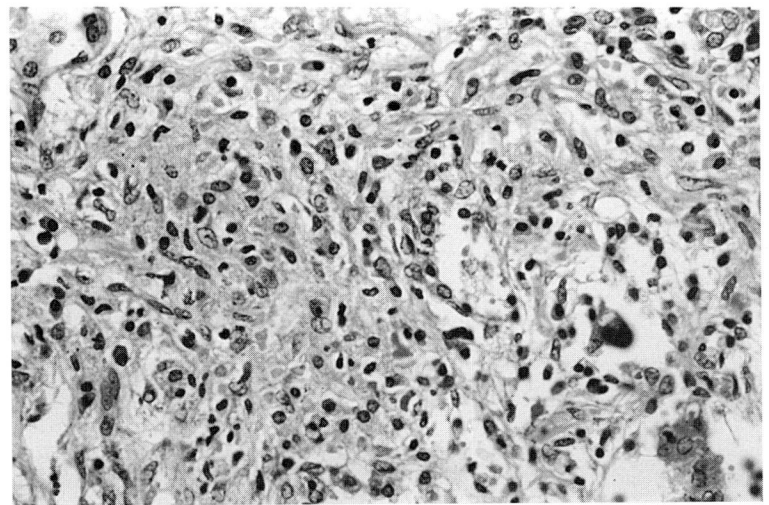

FIGURE 4-21
Spindle-cell variant of atypical mycobacterial infection. Note also cytomegalovirus cell in right side of field within an alveolar space. (This section was taken from close to Kaposi's sarcoma in the lung in Fig. 4-29.) (Medium magnification.)

Nocardial Infections

To date nearly 80 cases of nocardial infections have been reported in AIDS patients, the majority being *Nocardia asteroides*, affecting the lung.[80] Although it is therefore an uncommon infection, the associated macroscopic and microscopic changes and the nature of the bacteria mean that it should be included in the differential diagnosis of pneumonias. The patterns of involvement have been diffuse bronchopneumonia with predominantly upper-zone cavitation (Fig. 4-22) and/or miliary lesions. Granulomatous lesions are also occasionally seen. The bacteria are fine, filamentous, beaded, and branching. They are gram-positive and positive with Wade-Fite or Grocott stains (Fig. 4-23). A number of the reported cases have had disseminated infection to hilar nodes, brain, bone marrow, kidney, and liver. In a significant number of patients, the nocardial element may be obscured by other pathologies, notably PCP. However, the mortality is high and *Nocardia* has been implicated as the cause of death in more than 70 percent of those infected.

Bacillary Angiomatosis

This is a relatively uncommon condition that principally presents as hemorrhagic skin lesions, sometimes resembling pyogenic granuloma and sometimes flatter purplish lesions. It therefore enters the clinical differential diagnosis of Kaposi's sarcoma. Since it was described in 1983 in association with HIV,[81] it has been recognized as being associated with peliosis in the liver and hemorrhagic foci in the spleen, lymph nodes, and bone. It may possibly appear in the brain and be seen in other visceral lesions.[82–84] It is now known to be caused by infection with *Rochalimaea henselae*.[82] A few cases of pulmonary involvement have been reported, produc-

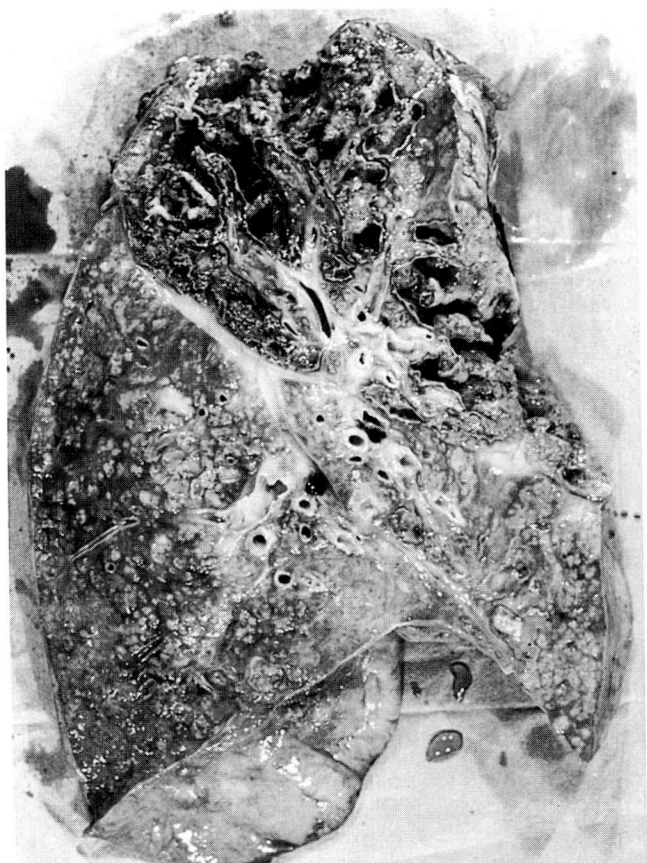

FIGURE 4-22
Nocardia, showing upper-lobe cavitation and diffuse lower lobe bronchopneumonic consolidation.

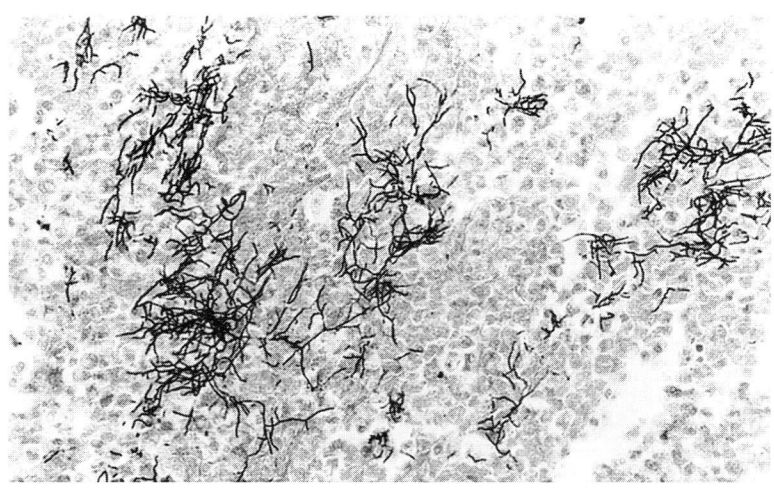

FIGURE 4-23
Grocott stain showing fine filamentous *Nocardia*.
(Medium magnification.)

ing lesions not unlike those in the skin. The characteristic histology is of vascular (capillary-sized) proliferation with acute inflammation and scattered necrosis and karyhorectic debris within the edematous connective tissue. Smudgy clumps of slightly basophilic/purplish material are found within the tissue and often surround the small vessels (Fig. 4-24). This purplish material contains numerous clumps of the bacilli measuring 0.3 by 1 to 3 μm, which are positive with the Warthin-Starry silver stain (Fig. 4-25).[83,85] Response to treatment with erythromycin is usually very good, emphasizing the need for accurate diagnosis.

PARASITIC INFECTIONS

Parasitic infections include *Toxoplasma gondii* (Fig. 4-26), *Cryptosporidium*, and *Strongyloides stercoralis*. It should be noted that, in the case of *Cryptosporidium*, it is usually part of a multiple infection, though recently four cases have been described where it was the sole pulmonary pathogen as diagnosed on BAL.[86] The patients had cough, dyspnea, and fever. The oocysts were stained by a modified Ziehl-Neelsen technique.[87] This does not stain the cysts positively in tissue sections. A more sensitive immunofluorescence technique can also be used for diagnosis.

Toxoplasmosis in AIDS is usually associated with necrotizing encephalitis. However, cases with pulmonary involvement have shown infection of bronchial mucosa and alveolitis with vasculitis and, more typically, parenchymal necrotizing lesions. Tachyzoites are identified as oval and measuring 2 to 6 μm, while the cysts are filled with numerous bradyzoites (Fig. 4-26). Immunocytochemistry may be helpful in diagnosis.[88]

Strongyloidiasis of the respiratory tract results from migration of the larvae and may be diagnosed in

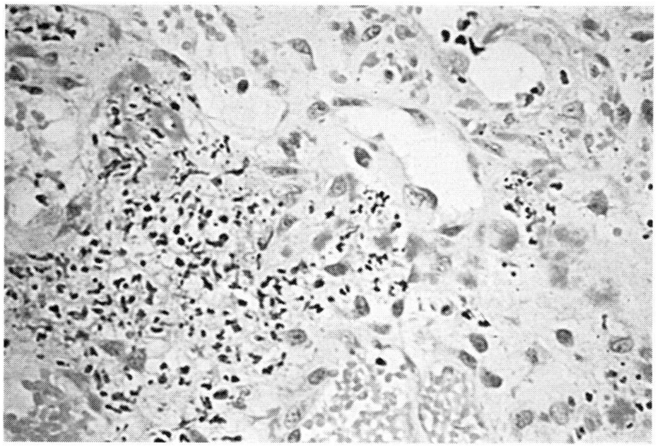

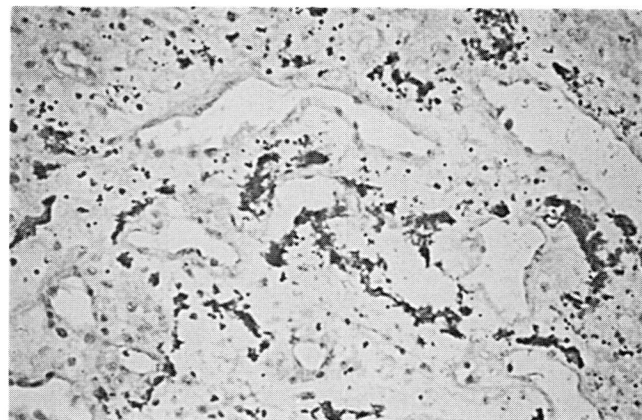

FIGURE 4-24
Bacillary angiomatosis showing necrosis, endothelial-cell hyperplasia, inflammatory-cell infiltrate, and smudgy purplish deposits. (Medium magnification.)

FIGURE 4-25
Warthin-Starry stain showing clumps of bacteria around vessels.
(Medium magnification.)

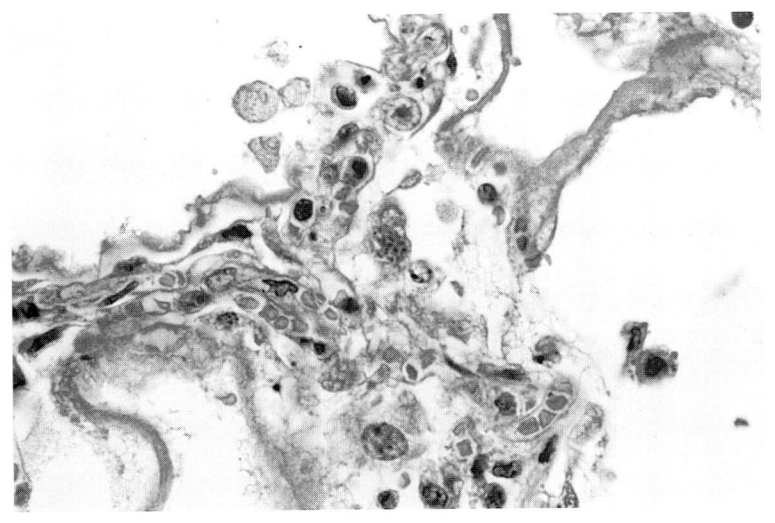

FIGURE 4-26
High-power view of toxoplasmosis showing cell in center with intracytoplasmic aggregate of zoites. Lung showed small areas of hemorrhagic necrosis and intralveolar exudate, with some hyaline membrane formation.

sputum and BAL samples. Microsporidia of the genus *Encephalitozoon* (*E. hellum*) have a propensity to infect epithelia of the respiratory passages, including the sinuses, pharynx, and tracheobronchial tree.[89] The organisms are identified by the presence of developing meront and sporont stages and clusters of spores. These are usually located in a supranuclear position of the luminal side of the epithelium and in the macrophages of the subjacent stroma. The spores stain with Brown-Brenn-Gram's and may show a PAS-positive polar dot; occasionally, they may be positive with the Ziehl-Neelsen stain.

INTERSTITIAL PNEUMONITIS AND LYMPHOID INTERSTITIAL PNEUMONIA

Nonspecific interstitial pneumonitis is diagnosed as either the pattern of diffuse alveolar damage or interstitial fibrosis without any specific cause.[90] It should

be noted that some of these patients had previous PCP and this could have accounted for some of the lung damage. Nonspecific interstitial fibrosis is commoner than lymphoid interstitial pneumonia (LIP), occurring in 46 of 50 lung biopsy specimens from adult HIV-infected patients.[91] The remaining four cases in that study had LIP. It may be that HIV plays a significant role in the pathogenesis of both LIP and nonspecific interstitial of pneumonitis, as it is found in the areas of inflammation. Epstein-Barr virus has also been suggested as a factor in LIP (Figs. 4-27 and 4-28),[92] but it cannot be demonstrated in the serum or tissue with any regularity. Lymphoid interstitial pneumonia occurs mostly in children and Haitians, with smaller numbers reported in African patients.

In LIP, the inflammation is predominantly peribronchial, with a dense proliferation of lymphocytes and plasma cells. With time, the infiltrate extends to the interstitium and parenchyma, but it does not

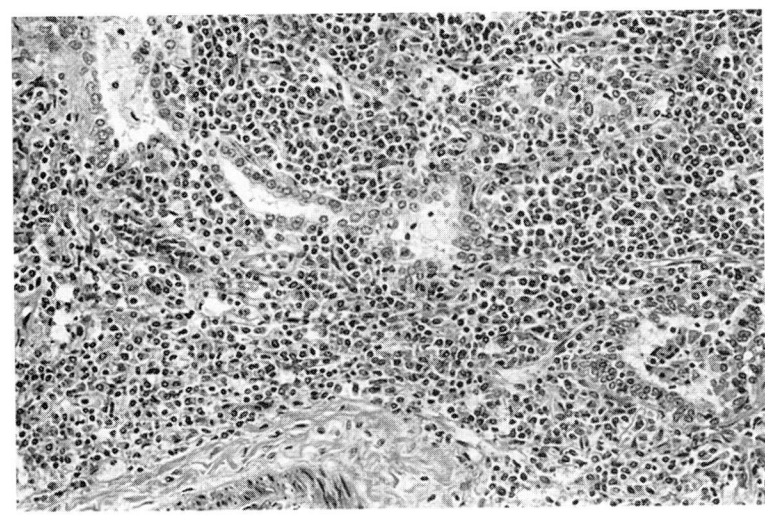

FIGURE 4-27
Child's lung section showing diffuse lymphoid interstitial pneumonitis. (H&E, medium magnification.)

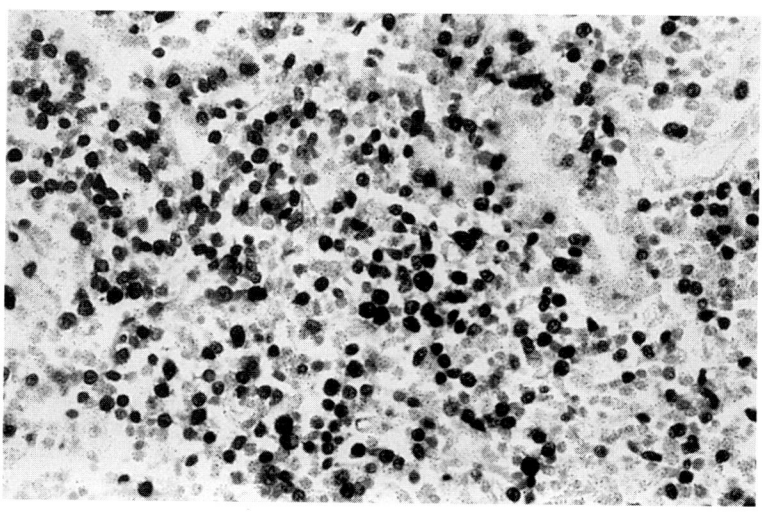

FIGURE 4-28
Epstein-Barr virus (EBV) (EBER) stain showing strong nuclear positive staining in many cells of the infiltrate seen in Fig. 4-27. (Medium magnification.)

appear to progress to interstitial fibrosis. Interstitial pneumonitis is seen predominantly in adults and may have a history of a variety of infections, as indicated above. Histologically, apart from the inflammation, there may be hyaline membranes and a varying degree of type II hyperplasia.

OTHER PULMONARY INVOLVEMENT

Drug-Related Reactions

These include the pulmonary complications of intravenous drug abuse, such as septic emboli, bacterial pneumonia, subsequent pulmonary complications, and foreign-body granulomas in relation to some of the fillers used for the drugs (see Chap. 26). In some cases, this is talc. Pulmonary drug reactions are only rarely described in patients with HIV infections.

Adult respiratory distress syndrome has been described in AIDS.[93] In these cases, there was sepsis due to *Clostridia*, *Salmonella*, or *Klebsiella* pneumonia. This complication is a rare manifestation of HIV infection.

In pulmonary hypertension, the commonest lesion is plexogenic pulmonary arteriopathy. In one patient there was an associated lymphocytic interstitial pneumonia.[94] These patients present with dyspnea and have pulmonary artery systolic pressures ranging from 50 to 100 mmHg, but only a few had marginally elevated pulmonary capillary wedge pressures. Another report describes plexogenic arteriopathy and venoocclusive disease as well as a mild, patchy pulmonary interstitial lymphoid infiltrate.[95] The vascular changes may be due to damage from a specific immune response to HIV infection. Pulmonary hypertensive changes may, of course, be seen in association with the lung diseases of intravenous drug users.

There are some reports suggesting a vastly increased risk of lung cancer in HIV patients. Many of these studies are based on relatively small numbers, but what has emerged suggests that there is usually a strong association with smoking and that the tumors present earlier and are more extensive than in the non-HIV population. Adenocarcinoma in particular seem to be the predominant type.[96]

Kaposi's Sarcoma

Kaposi's sarcoma is much more common in patients with AIDS than the general population. Pulmonary and thoracic involvement is variable, being found at autopsy in 38 to 75 percent of patients who have cutaneous disease.[97] Conversely, of those with pulmonary Kaposi's sarcoma, about 20 percent may have no evidence of cutaneous disease.[98,99]

Patients present with radiographic abnormalities or shortness of breath, nonproductive cough, and fever. Hemoptysis and chest pain have also been reported. Radiologically there may be a predominance of linear densities following septal lines. The infiltrates may have a nonhomogeneous pattern with focal areas of consolidation. In other cases there is evidence of ill-defined nodular infiltrates.[97] The yield from endobronchial biopsy is low, since the lesion is seen deep in the submucosa. It may be unwise to do a biopsy, as the procedure is associated with bleeding and has a low diagnostic yield. It has a distinctive appearance endobronchially, with bright-red flat or raised lesions, measuring up to 1 cm in diameter. They are found predominantly at points of branching in the airways. Parenchymal or even pleural disease may be noted. Some centers use the endoscopic features as diagnostic without considering it necessary to take a biopsy. In the absence of endobronchial lesions, an open lung biopsy may be

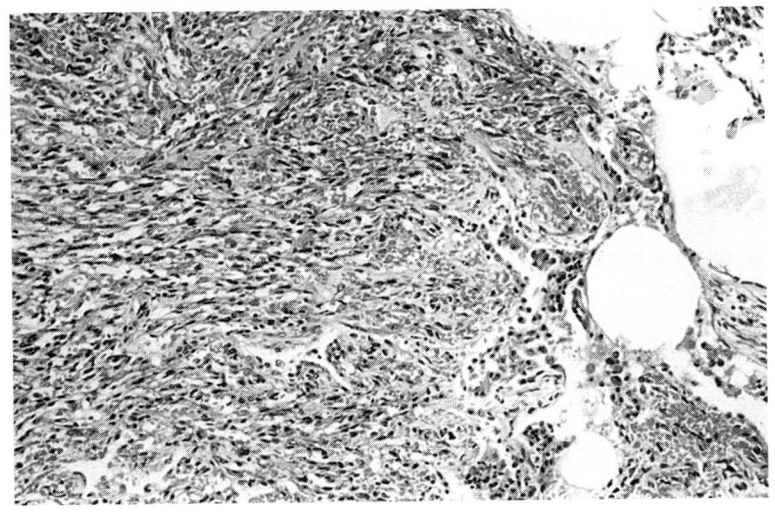

FIGURE 4-29
Kaposi's sarcoma, interstitial type, but a large area is becoming nodular. (Taken from same case and close to spindle-cell pattern of MAI illustrated in Fig. 4-21.) (H&E, medium magnification.)

necessary. Bronchoalveolar lavage and cytology are unhelpful, though the presence of many hemosiderin-laden macrophages may be suggestive of the disease. Transbronchial biopsies do not give a high yield and may be associated with excessive bleeding.[100] However, false-negative open lung biopsies may be found in a significant number of patients. In one study, 4 of 9 patients were negative and were subsequently found to have pulmonary Kaposi's sarcoma.[101]

The histology of the lesions may be straightforward (Fig. 4-29), especially in endobronchial lesions (Fig. 4-30), but where the distribution is paraseptal or of diffuse type in the parenchyma, the diagnosis can be difficult.

Lymphomas

Lymphomas are more common in HIV patients and frequently present at extranodal sites. The tumors are predominantly of B-cell high-grade type. They either have a lymphoblastic (Burkitt-like and non-Burkitt's) or immunoblastic pattern. In the latter type, there may be a striking plasmacytoid appearance. Lung involvement is perhaps surprisingly uncommon, with reports of 1 to 3 percent of those with lymphoma presenting with pulmonary manifestations.[102,103] Presentation in these cases is usually with pleural effusion or interstitial lung disease.

Other unusual lymphoproliferative lung diseases and lymphoid malignancies have included

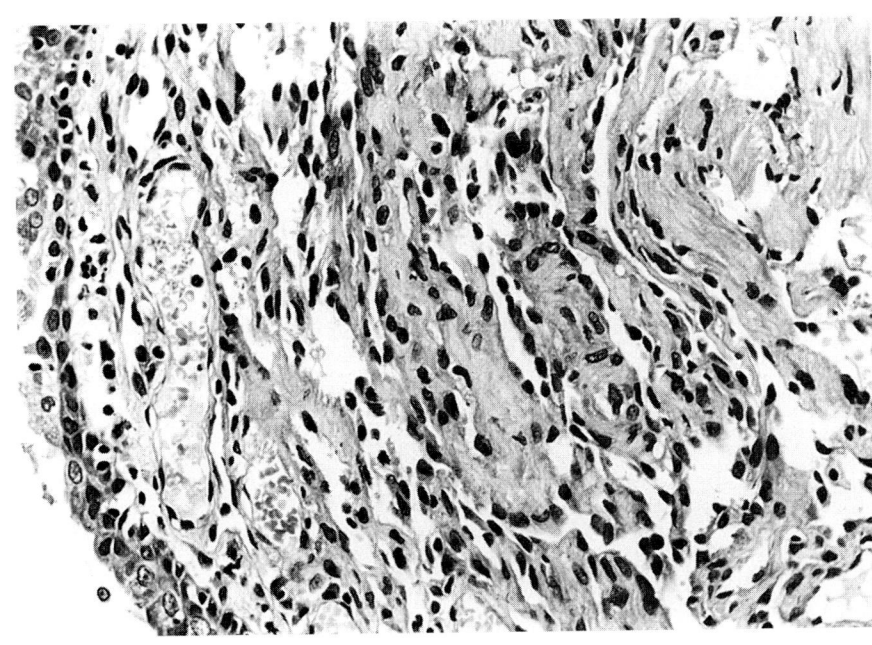

FIGURE 4-30
Bronchial biopsy showing stromal involvement of the bronchial submucosa by Kaposi's sarcoma. There are arborizing vascular channels with enlarged endothelial cells. (H&E, high magnification).

angioimmunoblastic lymphadenopathy-like lesions; T-cell (immunoblastic type) lymphomas; angiocentric lymphomatoid granulomatosis-like lesions, and some cases of Hodgkin's disease.[104]

The relationship of EBV and HIV to these conditions is unclear. Certainly EBV and HIV have been identified in some of the LIP lesions in children and Haitians, but there is no convincing evidence to date that these lesions progress to lung lymphoma. Epstein-Barr virus has also been identified in extranodal lymphomas at sites other than the lung, notably brain and gut. However, to date, although EBV has been identified in tumor cells from lung lymphomas, it is not invariably present.

DIAGNOSIS OF LUNG DISEASE IN ACQUIRED IMMUNODEFICIENCY SYNDROME

As mentioned above, the use of mucolyzed sputum may be useful in diagnosis. The commonest and probably most diagnostically useful investigation is BAL, and the pathologist should routinely look at H&E, Giemsa, or Wright's stain, depending on preference, as well as methenamine silver for *Pneumocystis* and other fungi, a Ziehl-Neelsen for tubercle bacilli, and Gram's stain for other bacteria. Cytomegalovirus or other viruses should be detected on the H&E. It may be difficult to obtain good immunoperoxidase staining on BAL material since the preparations do not stay on the slide.

If no diagnosis can be reached and the patient's condition is stable, closed or even open biopsy should be arranged. Biopsy was found to be more useful in Kaposi's sarcoma, but note should be made of the poor yield and dangers of biopsy in this tumor. The total diagnostic yield of biopsy and cytology together was 78.3 percent. Fine needle aspiration of the lung carries risks in patients with AIDS. There may be syncopal attacks and even fatalities. It is suggested that HIV infection is associated with autonomic neuropathy, which exposes these patients to this risk after invasive procedures.[105] Necropsy may play a part in detecting diagnoses that were unsuspected during life in AIDS patients.[106] The commonest unsuspected AIDS-related disease was cytomegalovirus infection (49 percent of all cases). This appears to have been disseminated in most cases. In keeping with the findings described above, CMV infection produced few pathologic lesions except in six, where there was morphologically significant injury of the lung. Some 20 percent of cases were found to have systemic fungal infection; 14 percent, systemic Kaposi's sarcoma; 11 percent, MAI;

and 9 percent systemic herpes infection. Cryptococcosis, CMV retinitis, PCP were almost always diagnosed antemortem. Four cases of central nervous system lymphoma were diagnosed only at necropsy and had been treated initially for toxoplasmosis. Bacterial pneumonias contributed considerably to mortality in 30 percent of the patients. Of the total, 51 percent had a lymphocytic myocarditis, which was sometimes associated with toxoplasmosis. The AIDS necropsy rate in this study was 41 percent. Though necropsy is a valuable tool for detecting diagnoses that may have been missed in life, it is important to adhere to the guidelines laid down by the Health and Safety Commission, and only a senior pathologist should undertake such examinations.

ACKNOWLEDGMENT

The author wishes to thank Professor Sebastian Lucas for providing Figs. 4-4, 4-15, 4-22, and 4-23. Professor Lucas also provided the cases/slides from which photographs for the following figures were taken: 4-3, 4-5 through 4-7, 4-17 through 4-21, and 4-26 through 4-29.

REFERENCES

1. Topley WWC, Wilson G: *Principles of Bacteriology, Virology and Immunity.* Volume 4: *Virology,* 7th ed. Edward Arnold, 1990, p 633.
2. Green WC: The molecular biology of human immunodeficiency virus type I infection. *N Engl J Med* 324:308–317, 1991.
3. Lane HC, Masur H, Edgar LC, Whalen G, Rook A, Fauci A: Abnormalities of B-cell activation and immunoregulation in patients with the acquired immunodeficiency syndrome. *Ann Intern Med* 103:522–533, 1985.
4. Salahuddin SZ, Rose R, Groopman JE, Markham PD, Gallo RC: Human T lymphotropic virus type III infection of human alveolar macrophages. *Blood* 68:281–284, 1986.
5. Stoneburger RL, Sato P, Burton A, Mertens T: The global HIV pandemic. *Aeta Paediatrica* 400(s) 1–4, 1994.
6. Rankin JA, Collman R, Danielle RP: Acquired immune deficiency syndrome and the lung. *Chest* 94:155–164, 1988.
7. *Safe Working and Prevention of Infection in the Mortuary and Postmortem Room.* Health Services Advisory Committee, 1991. DHSS publication.
8. *Acquired Immune Deficiency Syndrome: AIDS Booklet 3.* London, England, Department of Health and Social Security, 1986.
9. Covak SJA: Diagnosis, treatment and prevention of *Pneumocystis carinii* and pneumonia in HIV infected patients. *AIDS Update* 2(March/April):1–12, 1989.
10. Gal AA, Koss MN, Strigle S, Angritt P: Pneumocystis carinii infection in the acquired immune deficiency syndrome. *Semin Diagn Pathol* 6:287–299, 1989.
11. Hughes W: Persistence of *Pneumocystis. Chest* 88:4–5, 1985.

12. DeLorenzo LJ, Maguire GP, Wormser GP, Davidian MM, Stone DJ: Persistence of *Pneumocystis carinii* pneumonia in the acquired immune deficiency syndrome: Evaluation of therapy by follow-up transbronchial lung biopsy. *Chest* 88:79–83, 1985.

13. Colangelo G, Baughman RP, Dohn MN, Frame PT: Follow-up bronchoalveolar lavage in AIDS patients with *Pneumocystis carinii* pneumonia: *Pneumocystis carinii* burden predicts early relapse. *Am Rev Respir Dis* 143:1067–1071, 1991.

14. Epstein LJ, Meyer RD, Antonson S, Stigle SM, Mohsenifar Z: Persistence of *Pneumocystis carinii* in partients with AIDS receiving chemoprophylaxis. *Am J Respir Crit Care Med* 150:1456–1459, 1994.

15. Blumenfeld W, Miller CN, Chew KL, Mayall BH, Griffiss JM: Correlation of *Pneumocystis carinii* cyst density with mortality in patients with acquired immunodeficiency syndrome and *Pneumocystis* pneumonia. *Hum Pathol* 23:612–618, 1992.

16. Sadaghar H, Huang Z-B, Eden E: Correlation of bronchoalveolar lavage findings to severity of *Pneumocystis carinii* pneumonia in AIDS: Evidence for the development of high-permeability pulmonary edema. *Chest* 102:63–69, 1992.

17. Travis WD, Pittaluga S, Lipschik GY, et al: Atypical pathologic manifestations of *Pneumocystis carinii* pneumonia in the acquired immune deficiency syndrome: Review of 123 lung abscesses from 76 patients with emphasis on cysts, vascular invasion and vasculitis and granulomas. *Am J Surg Pathol* 14:615–625, 1990.

18. Saldana M, Mones JM: Cavitation and other atypical manifestations of *Pneumocystis carinii* pneumonia. *Semin Diagn Pathol* 6:273–286, 1989.

19. Lee MM, Schinella RA: Pulmonary calcification caused by *Pneumocystis carinii* pneumonia: A clinicopathologic report of three cases and review of the literature. *Chest* 98:266–270, 1990.

20. Gagliardi AJ, Stover DE, Zaman MK: Endobronchial *Pneumocystis carinii* infection in a patient with the acquired immune deficiency syndrome. *Chest* 91:463–464, 1987.

21. Mariuz P, Raviglione C, Gould IA, Mullen MP: Pleural *Pneumocystis carinii* infection. *Chest* 7:74–76, 1991.

22. McClellan MD, Miller SB, Parsons PE, Cohn DL: Pneumothorax with patients with *Pneumocystis carinii* pneumonia in AIDS. *Chest* 100:1224–1228, 1991.

23. Saldana M, Mones JM, Martinez GR: The pathology of treated *Pneumocystis carinii* pneumonia. *Semin Diagn Pathol* 6:300–312, 1989.

24. Moskowitz L, Hensley GT, Chan JC, Adams K: Immediate causes of death in acquired immunodeficiency syndrome. *Arch Pathol Lab Med* 109:735–738, 1985.

25. Coker RJ, Clark D, Clayton EL, Ainsworth JG, Lucas SB, Miller RF: Disseminated *Pneumocystis* infection in AIDS. *J Clin Pathol* 44:820–823, 1991.

26. Francis ND, Goldin RD, Forster SM, et al: Diagnosis of lung disease in acquired immune deficiency syndrome: Biopsy or cytology and implications for management. *J Clin Pathol* 40:1269–1273, 1987.

27. Bedrossian CWM, Mason MR, Gupta PK: Rapid cytologic diagnosis of *Pneumocystis*: A comparison of effective techniques. *Semin Diagn Pathol* 6:245–261, 1989.

28. Ng VL, Gartner I, Weymouth LA, Goodman CD, Hopewell TC, Hadley K: The use of mucolised induced sputum for the identification of pulmonary pathogens associated with human immunodeficiency virus infection. *Arch Pathol Lab Med* 113:488–493, 1989.

29. Musto L, Flanigan M, Elbadawi A: Ten-minute silver stain for *Pneumocystis carinii* and fungi in tissue sections. *Arch Pathol Lab Med* 106:292–294, 1982.

30. Domingo J, Waskal HW: Wright's stain in rapid diagnosis of *Pneumocystis carinii*. *Am J Clin Pathol* 81:511–514, 1984.

31. Tollerud DJ, Wesseler TA, Kim CK, Baughman RP: Use of a rapid differential stain for identifying *Pneumocystis carinii* in bronchoalveolar lavage fluid: Diagnostic efficiency in patients with AIDS. *Chest* 95:494–497, 1989.

32. Lindley RP, Mooney PA: Rapid stain for *Pneumocystis*. *J Clin Pathol* 40:811–812, 1987.

33. Baughman RP, Strohofer SS, Clinton BA, et al: The use of an indirect fluorescent antibody test for detecting *Pneumocystis carinii*. *Arch Pathol Lab Med* 113:1062–1065, 1989.

34. Magee JG, McDade KJ, Cunningham J, Harrison V: *Pneumocystis carinii* pneumonia: Detection of parasites by immunofluorescence based on a monoclonal antibody. *Med Lab Sci* 48:235–237, 1991.

35. Kovacs JA, Ng VL, Masur H, et al: Diagnosis of *Pneumocystis carinii* pneumonia: Improved detection in sputum with use of monoclonal antibodies. *N Engl J Med* 318:589–593, 1988.

36. Blumenfield W, Kovacs JA: Use of monoclonal antibody to detect. *Pneumocystis carinii* in induced sputum and bronchiolar lavage fluid by immunoperoxidase staining. *Arch Pathol Lab Med* 112:1233–1236, 1988.

37. Amin MB, Mezger E, Zarbo RJ: Detection of *Pneumocystis carinii*: Comparative study of monoclonal antibody and silver staining. *Am J Clin Pathol* 98:13–18, 1992.

38. Lim SK, Eveland WC, Porter RJ: Direct fluorescent antibody method for the diagnosis of *Pneumocystis carinii* pneumonitis from sputa or tracheal aspirates from humans. *Appl Microbiol* 27:144–149, 1974.

39. Schwartz DA, Munger RG, Katz SM: Plastic embedding evaluation of *Pneumocystis carinii* pneumonia in AIDS: Simultaneous demonstration of cyst and sporozoite forms. *Am J Surg Pathol* 11:304–309, 1987.

40. Wakefield AE, Guiver L, Miller RF, Hopkin JM: DNA amplification on induced sputum samples for diagnosis of *Pneumocystis carinii* pneumonia. *Lancet* 337:1378–1379, 1991.

41. Leigh TR, Gazard BG, Rowbottam A, Collins JV: Quantitative and qualitative comparison of DNA amplification by polymerase chain reaction with immunocytochemical staining for diagnosis of *Pneumocystis carinii* pneumonia. *J Clin Pathol* 46:140–144, 1993.

42. Chatterton JMW, Joss AWL, Williams H, Ho-Yen DO: *Pneumocystis carinii* antibody testing. *J Clin Pathol* 42:865–868, 1989.

43. Fels AOS: Bacterial and fungal pneumonias. *Clin Chest Med* 9:449–457, 1988.

44. Bartone EJ, Toma M, Johansson SE: Poorly encapsulated *Cryptococcus neoformans* from patients with AIDS. *AIDS Res* 2:211–218, 1986.

45. Grant IH, Armstrong D: Fungal infections in AIDS: Cryptococcosis. *Infect Dis Clin North Am* 2:457–464, 1988.

46. Johnson TC, Howell RJ, Sarosi GA: Clinical review: Progressive disseminated histoplasmosis in the AIDS patient. *Semin Respir Infect* 4:139–146, 1989.

47. Johnson PC, Khardor IN. Najjar AF, Butt S, Mansell PWA, Sarosi J: Progressive disseminated histoplasmosis with acquired immunodeficiency syndrome. *Am J Med* 85:152–158, 1988.

48. Tomita T, Chiga M: Disseminated histoplasmosis in acquired

immunodeficiency syndrome: Light and electron microscopic observations. *Hum Pathol* 19:438–444, 1988.

49. Pitkin AD, Grant AD, Foley NM, Miller RF. Changing patterns of respiratory disease in HIV positive patients in a referral centre in the UK between 1986–7 and 1990–1. *Thorax* 48:204–207, 1993.

50. Blaser MJ, Cohn DL: Opportunistic infections in patients with AIDS: Clues to the epidemiology of AIDS and the relative virulence of pathogens. *Rev Infect Dis* 8:21–30, 1986.

51. Pervez NK, Kleinerman J, Kattan M, et al: Pseudomembranous necrotising bronchoaspergillosis: Variant of invasive aspergillosis in a patient with haemophilia and acquired immunodeficiency syndrome. *Am Rev Respir Dis* 131:961–963, 1985.

52. Lombardo GT, Anandarao N, Lin C-S, Abbate A, Becker WH: Fatal hemoptysis in a patient with AIDS-related complex and pulmonary aspergilloma. *NY State J Med* 87:306–308, 1987.

53. Asnis DS, Chitkara RK, Jakobson M, Goldstein J: Invasive aspergillosis: An unusual manifestation of AIDS. *NY State J Med* 88:653–655, 1988.

54. Vasudevan VP, Mascarenhas DAN, Klapper P, Lombardias S: Cytomegalovirus necrotising bronchiolitis with HIV infection. *Chest* 97:483–484, 1990.

55. Miles PR, Boughman RP, Linneman CC: Cytomegalovirus in bronchiolar lavage fluid in patients with AIDS. *Chest* 97:1072–1076, 1990.

56. Broaddus C, Dake MD, Stulbarg MS, et al: Bronchiolar lavage and transbronchial biopsy for the diagnosis of pulmonary infection in the acquired immunodeficiency syndrome. *Ann Intern Med* 102:747–752, 1985.

57. Hui AN, Koss MN, Meyer PR: Necropsy findings in acquired immune deficiency syndrome. *Hum Pathol* 15:70–76, 1984.

58. Jacobson MA, Mills J: Cytomegalovirus infection. *Clin Chest Med* 9:443–448, 1988.

59. Murray JF, Mills J: Pulmonary infectious complications of human immunodeficiency virus infection: Part 1. *Am Rev Respir Dis* 141:1356–1372, 1990.

60. Miller AB, Paton G, Miller RS, et al: Cytomegalovirus in the lungs of patients with AIDS: Respiratory pathogen or passenger? *Am Rev Respir Dis* 141:1474–1477, 1990.

61. Suster B, Ackerman M, Orenstein M, Wax R: Pulmonary manifestations of AIDS: Review of 106 episodes. *Radiology* 61:87–93, 1986.

62. Cohen PR, Beltrani VP, Grossman ME: Disseminated herpes zoster in patients with human immunodeficiency virus infection. *Am J Med* 84:1076–1080, 1988.

63. McClain KL, Leach CT, Jenson HB, et al: Association of Epstein-Barr virus with leiomyosarcomas in young people with AIDS. *N Engl J Med* 332:12–18, 1995.

64. Montagnier L, Guest J, Charmar TS, Dauguet C, Axler C, Guetar DD: Adaptation of lymphadenopathy associated virus (LAV) to replication in (EBV) transformed B lymphoblastoid cell lines. *Science* 63–66, 1984.

65. Mitchell DM, Miller RB: Recent developments in the management and pulmonary complications of HIV disease. *Thorax* 47:381–390, 1992.

66. Magnenat JL, Nicod LP, Aeuckenthaler R, et al: Mode of presentation and diagnosis of bacterial pneumonia in human immunodeficiency virus infected patients. *Am Rev Respir Dis* 144:917–922, 1991.

67. Witt DG, Craven DE, McCabe WR: Bacterial infections in adult patients with the acquired immunodeficiency syndrome (AIDS) and AIDS related complex. *Am J Med* 82:900–906, 1987.

68. Barnes PS, Bloch AB, Davidson PT, Snider DE: Tuberculosis in patients with human immunodeficiency virus infection. *N Engl J Med* 324:1644–1650, 1991.

69. Pitchenik AE, Cole C, Russel BW, et al: Tuberculosis, atypical mycobacteriosis and the acquired immunodeficiency syndrome among Haitian and non-Haitian patients in South Florida. *Ann Intern Med* 101:641–645, 1984.

70. Sunderan G, McDonald RJ, Maniatis T, et al: Tuberculosis and the manifestations of the acquired immunodeficiency syndrome (AIDS). *JAMA* 256:362–366, 1986.

71. Harries AD: Tuberculosis from human immunodeficiency virus infection in developing countries. *Lancet* 335:387–389, 1990.

72. Lucas SB, Hounnou A, Peacock C, et al: The mortality and pathology of HIV infection in a West African city. *AIDS* 7:1569–1579, 1993.

73. Lucas SB, DeCock KM, Hounnou A, et al: Contribution of tuberculosis to slim disease in Africa. *Br Med J* 308:1531–1533, 1994.

74. Hawkins CE, Gold JWM, Whimbey E, et al: *Mycobacterium avium* complex infections in patients with acquired immunodeficiency syndrome. *Ann Intern Med* 105:184–188, 1986.

75. Wallis JM, Hannah JB: *Myocobacterium avium* complex infection in patients with the acquired immunodeficiency syndrome. *Chest* 93:926–932, 1988.

76. Horsburgh CR: *Mycobacterium avium* complex infection in the acquired immunodeficiency syndrome. *N Engl J Med* 324:1332–1337, 1991.

77. Umlas J, Federman M, Crawford C, O'Hara C, Fitzgibbon JS, Modeste A: Spindle cell pseudotumor due to *Mycobacterium avium-intracellulare* in patients with acquired immunodeficiency syndrome (AIDS): Positive staining of mycobacteria for cytoskeleton filaments. *Am J Surg Pathol* 15:1181–1187, 1991.

78. Hirschel B, Chang HR, Mach N, et al: Fatal infection with a novel unidentified mycobacterium in a man with the acquired immunodeficiency syndrome. *N Engl J Med* 323:109–113, 1990.

79. Horsburgh CR, Selik RM: The epidemiology of disseminated nontuberculous mycobacterial infection in the acquired immunodeficiency syndrome (AIDS). *Am Rev Respir Dis* 139:4–7, 1989.

80. Lucas SB, Hounnou A, Peacock C, Beaumel A, Kadio A, De Cock KM: Nocardiosis in HIV positive patients: An autopsy study in West Africa. *Tuber Lung Dis* 75:301–307, 1994.

81. Stoler MH, Bonfiglio TA, Steigbigel RJ, Pereira M: An atypical subcutaneous infection associated with acquired immunodeficiency syndrome. *Am J Clin Pathol* 80:714–717, 1983.

82. Slater LN, Welch DF, Min K-W: *Rochalimae henselae* cause bacillary angiomatosis and peliosis hepatis. *Arch Int Med* 152:602–606, 1992.

83. Koehler JE, Quin FD, Berger TG, LeBoit PE, Tappero JW: Isolation of *Rochalimaea* species from cutaneus and osseous lesions of bacillary angiomatosis. *N Engl J Med* 327:1625–1631, 1992.

84. Hadfield TL, Warren R, Kass M, et al: Endocarditis caused by *Rochalimaea henselae*. *Hum Pathol* 24:1140–1141, 1993.

85. Cockrell CJ, Whitlow MA, Webster GF, Friedman-Kein AE: Epithelioid angiomatosis: A distinct vascular disorder in

patients with acquired immunodeficiency syndrome or AIDS-related complex. *Lancet* 2:654–656, 1987.

86. Hojlyng N, Jensen BN: Respiratory cryptosporidiosis in HIV-positive patients. *Lancet* 1:590–591, 1988.

87. Henriksen SA, Pohlenz JFH: Staining of *Cryptosporidium* by a modified Ziehl-Neelsen technique. *Acta Vet Scand* 22:594–596, 1981.

88. Nash G, Kerschmann RL, Herndier B, Dubey JP: The pathological manifestations of pulmonary toxoplasmosis in the acquired immunodeficiency syndrome. *Hum Pathol* 25:652–658, 1994.

89. Schwartz DA, Visvesvara GS, Leitch GJ, et al: Pathology of symptomatic microsporidial (*Encephalitozoon hellum*) bronchiolitis in the acquired immunodeficiency syndrome. *Hum Pathol* 24:937–943, 1993.

90. Ramaswamy G, Jagadha V, Tchertkoff V: Diffuse alveolar damage and interstitial fibrosis in acquired immunodeficiency syndrome patients without concurrent pulmonary infection. *Arch Pathol Lab Med* 109:408–412, 1985.

91. Travis WM, Fox CH, Devaney KO, et al: Lymphoid interstitial pneumonia in 50 adult patients infected with the human immunodeficiency virus: Lymphocytic interstitial pneumonitis versus nonspecific interstitial pneumonitis. *Hum Pathol* 23:529–541, 1992.

92. Purtilo DT, Kipscomb H, Volsky DJ, et al: Role of Epstein-Barr virus in acquired immunodeficiency syndrome. *Adv Exp Med Biol* 187:93–96, 1985.

93. Stover DE, White DA, Romano PA, Gellene RA, Robeson WA: Spectrum of pulmonary diseases associated with the acquired immunodeficiency syndrome. *Am J Med* 78:429–437, 1985.

94. Tollis PG, Wolfe D, Harley RA, Strange C, Sar N: Pulmonary hypertension and human immunodeficiency virus infection: Two reports and a review of the literature. *Chest* 101:474–478, 1992.

95. Jacques C, Richmond G, Tierney L, Curtis JL, Kerrow J, Warnock ML: Primary pulmonary hypertension of human immunodeficiency virus infection in a non-hemophiliac. *Hum Pathol* 23:191–194, 1992.

96. Fraire AE, Awe RJ: Lung cancer in association with human immunodeficiency virus infection. *Cancer* 70:432–436, 1992.

97. White DA, Matthay RA: Non-infectious pulmonary complications of infection with the human immunodeficiency virus. *Am Rev Respir Dis* 140:1764–1787, 1989.

98. Ognibene F, Steis R, Macher A, et al: Kaposi's sarcoma causing pulmonary infiltrates and respiratory failure in the acquired immunodeficiency syndrome. *Ann Intern Med* 102:471–475, 1985.

99. Miller R. Tomlinson M, Cottrill C, Donald J, Spittle M, Semple S: Bronchopulmonary Kaposi's sarcoma patients with AIDS. *Thorax* 47:721–725, 1992.

100. Stover DE, White DA, Romano PA, Gellene RA: Diagnosis of pulmonary disease in acquired immune deficiency syndrome (AIDS): Role of bronchoscopy and bronchoalveolar lavage. *Am Rev Respir Dis* 130:659–662, 1984.

101. Garay SM, Belenko M, Fazzini E, Schinella R: Pulmonary manifestations of Kaposi's sarcoma. *Chest* 91:39–43, 1987.

102. Raphael M, Gentilhomme O, Tulliez M, Byron PA, Diebold J: French study group of pathology for human immunodeficiency virus associated tumors: Histopathological features of high grade non-Hodgkins lymphomas in acquired immunodeficiency syndrome. *Arch Pathol Lab Med* 115:15, 1991.

103. Knowles DM, Chamuluck GA, Subar M, et al: Lymphoid neoplasia associated with the acquired immunodeficiency syndrome (AIDS). *Ann Intern Med* 108:744, 1988.

104. Sheib IG, Siegel RS: Atypical Hodgkin's disease and the acquired immunodeficiency syndrome. *Ann Intern Med* 102:554, 1985.

105. Craddock C, Pascol G, Ball R, Trotheroe A, Hopkin J: Cardiorespiratory arrest and autonomic neuropathy in AIDS. *Lancet* 2:16–18, 1987.

106. Wilkes MS, Fortin A, Felix JC, Godwin TA, Thompson WG: Value of necropsy in acquired immunodeficiency syndrome. *Lancet* 2:85–88, 1988.

Chapter 5

Viral Infections

P. S. Hasleton

PULMONARY DEFENSE MECHANISMS

The individual cells of the normal respiratory tract have already been described (Chap. 1). The aim of this section is to describe some of the pulmonary defense mechanisms, which protect the lung from infection. Host defense mechanisms against pulmonary infection have recently been reviewed.[1]

Most pathogenic agents are inhaled but bacteria, viruses, and other parasites can reach the lungs via the pulmonary circulation. In some cases organisms reach the lung by direct spread from an adjacent infected focus, such as the liver. This section will deal with the protection of the respiratory tract against inhaled particles, the most important of which are microorganisms.

In the description of individual organisms given below, there will be minimal attention paid to general factors that affect the lung and its response to infection.

Nasal Physiology

The nose plays an important part in protecting the upper and lower respiratory tracts. It is a useful site for studying the inflammatory response of the tracheobronchial tree, but not the terminal airways and parenchyma.[2] Humans prefer to breathe through their noses, despite the fact that nasal breathing doubles the total respiratory resistance to airflow. This is especially so in the infant, where nasal breathing is the norm, except when crying. The nose accounts for approximately one half of the total respiratory resistance to airflow.[3] The smallest cross-sectional area of the conducting airways is at the internal nasal ostium (0.3 cm.[2] on each side).[4] The main contribution to resistance is made by the turbinate bones with their folds and the nasal septum.

The nose has many arteriovenous anastomoses and these can increase the temperature of inspired air by as much as 25°C between the external nares and the nasopharynx. As well as providing warm, moist air for the lower respiratory tract, it also guards against large particles and soluble, noxious gases, such as formaldehyde and sulphur dioxide.[5,6] There is also a suggestion that the nose and pharynx may possess receptors that affect ventilation, but more research is needed in this area.[7]

Sneezing is a very effective mechanism of ridding the upper respiratory tract of noxious agents. Particles greater than 10 μm are filtered in the nasopharynx (Fig. 5-1).

The mucosa lining these air passages consists of ciliated epithelium, beneath which are mainly mucous and serous glands. Mucociliary function is

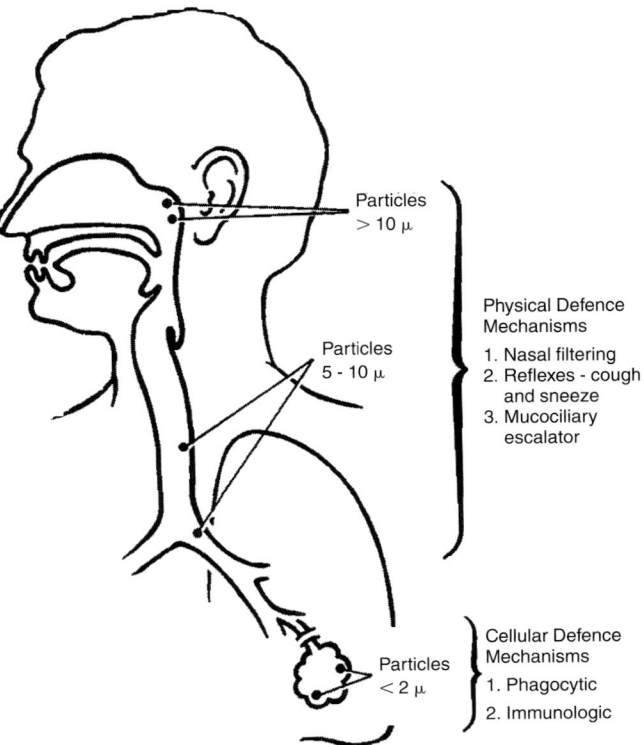

FIGURE 5-1
Diagram showing the sites at which particles may be deposited in the respiratory tract from the nose to the terminal airways. (Modified from Harada and Repine, reproduced courtesy of the authors and *Chest*.)

important in the upper airways and any deficiency in dynein arms will prevent cilia propelling airway secretions upwards. Cilia beat more slowly in the distal than the proximal airways. They function within a watery (sol) layer, the origin and composition of which is still unknown. The thickness of this sol is thought to be of the same magnitude as the length of the cilia. This enables their tips to touch the underside of the overlying secretion layer (epiphase) and propel it caudally. Claws can be identified at the tips of the cilia. It is possible that their role is to hook to the underside of the epiphase and help its upward propulsion.[8] The mucociliary escalator with its cells and secretions is an effective first line of defense in removing microorganisms.

Nasal airway secretions and their constituent proteins play an important part of the defense of the respiratory tract. The serous cell is probably the resident antimicrobial cell in mucous membranes.[9] Respiratory secretions consist of a mixture of mucus glycoproteins, glandular products, and plasma proteins.

Human nasal secretions have several functions. Some of these are protective, including humidification and lubrication of the respiratory tract and providing a proper medium for ciliary action. The secretions have barrier functions as well as host defense mechanisms, which include IgA/IgG and antimicrobial functions. The latter are provided for by substances such as lysosyme, lactoferrin, and glandulin. The last substance is a small molecule which is bacteriocidal to multiple bacteria especially gram-negative organisms such as *Pseudomonas*. Lysosyme and lactoferrin are also antibacterial and are produced by serous glands. The latter substance is an oxygen free radical scavenger.

The main specific mediators of host defense are immunoglobulins, especially IgA and IgG. These act in different ways. IgG is a plasma protein that is distrubuted in the nasal mucosa by microvascular permeability. Only 25 percent of the plasma cells in the nasal mucosa produce IgG. IgA is largely produced by plasma cells located close to the submucous glands. Locally produced IgA is dimeric and is then joined by a J chain before secretion. Dimeric IgA binds to secretory components produced by serous cells and forms secretory IgA (S-IgA). This is transported transcellularly through the serous cells into glandular secretions. S-IgA acts primarily by binding microorganisms in the airway lumen and planting attachments of these pathogens to the mucosa. IgA acts primarily on the mucosa itself to limit invasion by microorganisms that reach the epithelium.

In addition, binding of IgG or even IgM to the surfaces of microorganisms initiates activation of the classic pathway of complement. Complement components are also seen in the lower respiratory tract[10] and they may be activated by the alternate pathway. Both pathways of complement activation bind C3b to the infectious agent. This enables it to be recognized by complement receptors on neutrophil polymorphs and macrophages. Cleavage of C5 into C5b yields a highly reactive antimicrobial component.

Tracheobronchial Secretions

The angulation of the tracheobronchial tree, like the nose, effectively filters larger particles from the air.[11]

The main sources of tracheobronchial secretions are the submucosal glands, goblet or mucous cells, Clara cells, and type II pneumocytes. The control of airway mucus secretion is poorly understood. There are a number of mechanisms which may be responsible for increasing secretion. These include reflexes initiated by nervous receptors in the airways with the efferent limbs in the autonomic nerves. Cigarette smoke short-circuits this reflex pathway and nicotine absorbed from the smoke stimulates ganglion cells directly, therefore driving secretion. Secretory and motor nerves which control the output from submucosal glands include cholinergic, adrenergic, and NANC (non-adrenergic, non-cholinergic) fibers. Prostaglandins and leukotrienes released by inflammation increase secretion. Cytokines are very important in the response to infection.[12] Other stimuli are also important, for example, dust inhalation, which stimulates cough receptors.[13]

Sleep and aging may result in a decrease in mechanical and mucociliary clearance. There is an age-related decline of mucociliary clearance, after the age of 60 years.[14] There is also a depression of mucociliary clearance in patients with chronic bronchitis, bronchiectasis, and asthma. Respiratory defenses are less effective in the elderly due to loss of elastic recoil, structural alteration of the chest wall, and a decreased strength of cough.[15] Diseases, more common with aging, such as stroke, disordered breathing, or drugs adversely affect airway protection.

In addition to mechanical problems with aging, there are also changes in the cell populations. One of the most commonly observed deficits with aging is a decrease in antigen-driven lymphocyte proliferation.[16] A possible explanation for this could be a decline in the absolute number of lymphocytes in the aged, or alternatively, a decrease in a critical T lymphocyte subpopulation. T-cell proliferation is the component of cell-mediated immunity most consistently showing a defect associated with aging. The effect of aging on the macrophage population remains controversial. Some studies show no change in macrophage chemotaxis, adherence or phagocytosis.[17,18] An increase in the intraalveolar macrophage population in nonsmokers with aging has been documented.[19]

Acute viral respiratory infections affect mucociliary transport by a direct cytotoxic effect on the ciliated cells.[20] Cilia obtained from sites of purulent secretions beat more slowly than those from normal controls, when measured in vitro on nasal brushings. Some organisms, such as *H. influenzae*, produce a factor called cilotoxin, which retards ciliary beat frequency.

Pharyngeal contents enter the lower respiratory tract during sleep in normal individuals. Postmortem cultures of the lung from previously anesthetized dogs yielded small numbers of bacteria, similar to those cultured from the pharynx.[21]

Cough is a reserve clearance mechanism for the tracheobronchial tree. It may be brought into effect in both chronic lung disease and emergency situations such as inhalation of a foreign body. In the case of cough, air is accelerated through the airways and transfers its momentum into the lining mucus, resulting in its shearing and expulsion from the lungs. Cough is probably effective distal to the seventh generation of airways.[22] Under conditions of excess mucus production, the effect of cough can extend to the level of respiratory bronchioles, i.e., the 17th airway generation.

The secretions of the nasal passages, trachea, cough reflex, are by-passed in patients on mechanical ventilators or with tracheostomies. Such patients, who are usually cared for on intensive care units, have an increased risk of nosocomial pneumonia (pneumonia developing after 48 h in hospital or an intensive care unit). This complication has an estimated incidence on intensive care units of between 21 and 26 percent.[23] This figure may be an overestimate because of lack of specific clinical criteria in the critically ill. An endotracheal tube can produce acute inflammation and consequent purulent sputum. The diagnosis of a pneumonia may be difficult in life, since such patients may already have pulmonary infiltrates, due to conditions such as acute respiratory distress syndrome (ARDS).

The gold standard for the diagnosis of nosocomial pneumonia is histological examination and quantitative culture of lung biopsy specimens.[23] The main organisms are gram-negative. In seriously ill patients, the oropharynx is colonized by these organisms because of aspiration.[24] This process is aided by a supine position and the presence of a nasogastric tube, which impairs gastroesophageal sphincter function.[23] There is a change in the gastrointestinal tract flora as gram-negative bacteria colonize the stomach and small bowel. This colonization is aided by antacids and H2 antagonists. Little is known about the pathogenesis of lower airway colonization. Nosocomial pneumonia may be due to bacteria crossing from the intestinal lumen to the blood, due to mucosal damage. This process has been termed gut-bacterial translocation.[25]

Mucus

For several reasons mucosal fluids from the trachea and largest intrapulmonary conducting airways are probably least precisely characterized. First, methods of obtaining lavage fluid cannot control the possible contamination of the lower respiratory tract with material from the tip of the bronchoscope. In addition the mucus blanket is not spread homogeneously over the entire epithelial surface.[26] As one proceeds distally down the respiratory tract equal amounts of IgG and IgA can be retrieved from the major bronchi in middle-aged adults.[13]

Mucus is a mixture of glycoproteins, proteoglycans, lipids, other proteins, and sometimes DNA.[27] The last component probably originates in dead neutrophil polymorphs.

Mucus glycoproteins or mucins represent the main components of respiratory mucus. They are secreted by goblet cells of the respiratory mucosal mucous glands. These substances are a diverse group with molecular masses ranging from a few hundred thousand to more than 1 million daltons.[28] In human respiratory mucins, glycosylation generates a remarkable diversity of O-glycosidically linked carbohydrate chains. These chains can vary in length from 1 to 20 sugars and may be neutral, sialated, or sulphated. This diversity of chains has many interactions with microorganisms and might be an important factor in maintaining the sterility of the distal respiratory tree. In diseases, such as cystic fibrosis, there may be an alteration in this interaction with *Pseudomonas sp.* Mucus also contains lysozyme and lactoferrin.

A list of the airway secretions and their origin is given by Jeffrey.[29]

Protection at an Alveolar Level

Particles may reach the alveoli via the blood stream or the airways. In the former case there is likely to be alveolar wall damage causing ARDS. Particles, whether microbial or inorganic, such as asbestos, deposited in alveoli have to be transported to the terminal bronchioles before they can contact cilia. These enable them to be carried upwards on the mucociliary transport mechanism. Foreign materials may also enter lymphatics or perivascular interstitial tissues.

Cellular Components of Lung Defense

At the alveolar level, cells are one of the most important protective mechanisms. Studies using bronchoalveolar lavage have shown that in normal adults there are about 93 percent macrophages, 6 percent lymphocytes, and 1 percent neutrophils. IgG and IgA are also found in bronchiolar lavage fluid, but as noted above it is difficult to determine the exact contribution from either the alveoli or the smaller airways.

The cells important to the defense of the lower respiratory tract are neutrophil polymorphs, alveolar macrophages, and T and B lymphocytes. Neutrophils are not a normal feature of the alveolar lumen but with the appropriate message they are rapidly recruited to the lower respiratory tract. Polymorph adherence is partly mediated by expression of neutrophil surface glycoproteins of the integrin family.[30] Neutrophils are recruited across the alveolar wall by C5a and interleukin-8. Polymorphs secrete elastase and collagenase, which aid their passage across tissue planes.

These enzymes may also play a part in the antimicrobial activity of polymorphs, which also contain positively charged proteins called defensins.[31] These proteins probably are attached to negatively charged regions on the organism's surface. This leads to alteration in cell membrane permeability and death. Other antimicrobial factors in polymorphs include lactoferrin, myeloperoxidase, and lysozyme.

Macrophages are important in the defense of the lower pulmonary airways. These cells kill microbes by endocytosis. Once inside the cell, the microorganism is destroyed by generation of oxygen free radicals and digestion by lysosomes. In addition there is secretion of tumor necrosis factor (TNF) alpha and transferrin. Macrophages also inhibit growth of organisms by antibody-dependent cytotoxity.

Some organisms have evolved mechanisms to avoid being killed by macrophages. *M. tuberculosis* is able to stop fusion of the primary phagosome with lysosomes.[32] The most famous organism which evades destruction by macrophages is human immunodeficiency virus (HIV). One of its strategies is to live in intracytoplasmic vacuoles.

T lymphocytes are important for their direct cytotoxicity and their ability to produce macrophages, activating cytokines, and stimulate lymphocyte-mediated cytotoxicity. B lymphocytes produce IgA, IgG, and IgM.

The surfactant-containing fraction of lavage fluid is bactericidal for *Pneumococci sp.* and some other gram-positive bacteria, excluding *Staphylococci sp.* The antibacterial factors are probably long-chain free fatty acids (FFA). Polyunsaturated FFA appear to be particularly active.[33] Significant species variability exists in the efficiency and clearance of oropharyngeal organisms. *Streptococci sp.* are cleared promptly and *Branhamella catarrhalis* slowly. Non-typeable *Haemophilus influenzae* multiply before they are killed. Thus several factors are responsible for destroying organisms in the lower respiratory tract.[34]

The Clara cells are a source of a low-molecular-weight protease inhibitor as well as mucosubstances and lipids. All three compounds help to protect terminal bronchioles and alveoli.

Cell Adhesion Molecules

Fibronectin, a dymeric matrix protein, which acts as a substrate for cell adhesion and migration helps to protect the bronchial tree.[35] This is an important substance both in the alveolar wall and the upper respiratory tract. Gram-positive bacteria adhere to fibronectin through binding sites. Some gram-negative organisms are prevented from adherence to the fibronectin. Interaction between cells and fibronectin is important in a number of biologic processes, including those mentioned above. Fibronectin is destroyed in ill patients by proteolytic enzymes from saliva or neutrophils.

The leukocyte integrins, selectins, and members of the immunoglobulin supergene family, help to mediate adhesive interactions between polymorphs and endothelial cells.[36] ICAM-1 and ICAM-2 are counterreceptors on endothelial cells. Endothelial cells also express selectins that recognize carbohydrate ligands on the surface of the polymorph. Vascular cell adhesion molecule (VCAM-1) and granule-associated membrane protein 140 (GMP 140), as well as endothelial-leucocyte adhesion molecule (ELAM-1), are expressed during inflammation and are important in white blood cell adhesion.[37]

Deposition of Inhaled Particles in the Lung

The distribution of inhaled particles in the lung and other factors concerning pulmonary defense have been reviewed by Dunnill.[38] The site of deposition of inhaled particles depends on their physical properties as well as the depth and frequency of respiration. Particles larger than 10 μm in diameter remain in the upper air passages. Only 20 to 30 percent of particles between 0.5 and 2 μm. reach the alveoli. The aerodynamic factors involved in particle deposition are summarized by Dunnill.[38]

Ciliary action is important in providing pulmonary defense. Any factor impairing the host defense will predispose to infection. Thus atmospheric pollution or cigarette smoke alter cilial or pulmonary macrophage function. Cytotoxic agents may depress polymorph, lymphocyte, or macrophage function. Alcohol has a profound effect on the lung. Apart from depression of the central nervous system, there is a reduction in neutrophil function and delayed removal of organisms by macrophages.[39] Another effect of alcohol is to predispose to aspiration which greatly lower the lung's natural resistance to infection because of the acid in the gastric juices.

VIRAL INFECTIONS

Viruses are an important cause of pulmonary damage. This role is compounded by their mutagenic capacity, which results in a constant evolution of new strains from already recognized pathogens. This enables new and sometimes highly virulent variants of familiar viruses, such as influenza, to cause pandemics of serious respiratory disease. New viruses will be isolated from the respiratory tract and cause severe infections. A recent viral cause of respiratory failure is the Hantavirus, described in May 1993.[40]

Many of the viruses reach the respiratory tract by droplet infection and subsequent inhalation. Some viruses, such as influenza and parainfluenza, can invade and damage an immunologically competent host at any age. Others, such as *Cytomegalovirus* and *Varicella* virus, only affect the respiratory tract when immunity is depressed and are best regarded as opportunistic infections. Their portal of entry does not necessarily have to be through the airways. Many respiratory tract viruses predispose to secondary bacterial infection and it is often the synergy between the two that causes death. This is demonstrated in influenza, where secondary staphyloccal pneumonia usually kills the patient.

Diagnosis of Viral Infection

The histological diagnosis of viral infection is not easy unless there are characteristic intracytoplasmic or intranuclear inclusion bodies. These can be identified either on H&E sections or by immunocytochemistry. If one suspects a viral infection, it is always prudent to collect some blood. A previous blood sample may have been taken earlier in the illness and a rise in titre may be demonstrated. Similarly one should always put some tissue into transport medium for virological culture from any case of pneumonia. Those most commonly used are Eagle's minimal essential medium (MEM), Parker's 199, and RPMI 1640. The sample must be taken in a sterile fashion.

Electron microscopy (EM), widely used in the diagnosis of viral infections, poses problems. It is difficult to process a large number of specimens and expertise is needed to operate the instrument and interpret the results. In addition it is subject to sampling errors. However, the virus can be concentrated using ultrafiltration or by the use of immune EM. In the latter technique the specimen is first coated with specific antiviral antibody. This has the effect of concentrating the virus onto the grid and enabling more specific identification or typing of viruses.

Other methods used in virus identication are:

1. Virus isolation, e.g., inoculating a clinical specimen into laboratory animals. For example Coxsackie viruses can be demonstrated by intracerebral cultivation in newborn mice.
2. Cell culture.
3. Cytopathic effects. These can be either intranuclear or intracytoplasmic. The former is well demonstrated in *Herpes* infection.
4. Direct demonstration of viruses has recently been revolutionized by the introduction of monoclonal antibodies and newer DNA techniques. The latter include in-situ hybridization and the polymerase chain reaction. The polymerase chain reaction is potentially capable of detecting a single viral genome in a cell suspension or tissue section. The problem that faces the histopathologist is the significance of finding such a small quantity of viral particles and attributing it as the cause of the disease. The result must be correlated with the

histopathological findings. The techniques are labor intensive and often take 1 to 2 days. For quicker results, it is possible using rapid tissue processing to obtain H&E sections within 4 h.

In any case in which direct demonstration of viruses needs to be established the following samples, where appropriate, should be obtained: swabs or washings taken from the nose and throat; if a thoracotomy has been done a piece of sterile lung tissue should be obtained from an immunocompromised patient.

Paired serum samples are the commonest way of detecting viral disease. These demonstrate an antibody response. Neutralization, complement-fixation, hemagglutination inhibition tests, radioimmunoassay using enzyme-linked immunosorbent assay, (ELISA), and specific IgM antibody tests are all used in detecting viral infection.

Influenza

ORGANISM

The influenza viruses belong to the *Orthomyxoviridae*. They are pleomorphic, spherical, or frequently filamentous particles. These particles contain RNA in the nucleocapsid and are surrounded by a lipid-containing envelope.

There are three types of orthomyxovirus: A, B, and C. They are classified on the basis of the antigenic structures of the nucleoprotein and matrix antigens.[41]

The major structural proteins of influenza virus are hemagglutinin, nucleoprotein, major membrane or matrix protein, and neuraminidase. One of the most important and most extensively investigated is hemagglutinin, which is a functional protein involved in the absorption of influenza virus into the cell. Hemagglutinin possesses important antigenic sites, which are recognized by neutralizing antibody.

An important property of the influenza A virus is its ability to undergo antigenic drift and shift. This occurs because of changes in both the hemagglutinin and the neuraminidase antigens. The result is that, over several years, strains evolve that are considerably different from the original pandemic virus. In antigenic *drift* only certain epitopes of the hemagglutinin and neuraminidase antigens change. It is antigenic shift which is probably responsible for pandemic influenza.

The virus is cytopathic to respiratory epithelium. It appears to enter the cell by endocytosis.[42] The hemagglutinin spike glycoprotein of the influenza virus catalyzes a low pH–induced membrane fusion event, which releases the viral genome into the host cell cytoplasm.[43] Influenza virus also impairs chemotaxis in polymorphs as well as macrophages. There is also impaired bacterial antibody response to *Pneumococci* in experimental animals.[44]

The natural hosts of influenza virus include humans, pigs, horses, seals, whales, and birds.

CLINICAL FEATURES

Outbreaks of influenza are seen annually. They usually occur in temperate climates over a 6- to 10-week period during the winter months.[45]

The clinical features of influenza are well known. The virus has an incubation period of 1 to 4 days.[46] There is often headache, intense shivering, nonproductive cough, and high temperature. Myalgia is a prominent feature. The symptoms last between 2 and 5 days and may leave a marked postinfluenzal depression. Influenza has a more devastating course in patients with preexisting renal or cardiac disease or in diabetics.

Diagnosis may be difficult and is often made with the knowledge of an outbreak of the disease in the community. Viral titers in respiratory secretions are highest on days 1 to 3 of the illness.

Pandemics are usually caused by influenza A. Influenza B and C cause less severe infections and result in epidemics less frequently. Spread of the virus is facilitated by close contact and crowding.

PATHOLOGY

Macroscopically the appearances in the lungs vary depending on the severity and extent of the disease. In rare but rapidly fatal cases, the lungs are heavy, bulky, plum-colored and often show numerous subpleural petechial hemorrhages. The changes are most prominent in the lower lobes. The bronchi and cut surfaces exude blood-stained, frothy, edema fluid. The parenchyma shows widespread dark areas of hemorrhage. The septa throughout the lung are edematous and there is hilar node enlargement.

Secondary infection is frequent and the form taken depends on the agent responsible. Thus with a staphylococcal pneumonia there will be multiple lung abscesses. In all cases there is tracheal ulceration.

The distribution of the histopathologic lesions in 148 virologically confirmed fatal cases of influenza studied through the Asian influenza outbreak of 1957 is shown in Fig. 5-2.[47] The initial changes are in bronchial epithelium, where the surface cells disso-

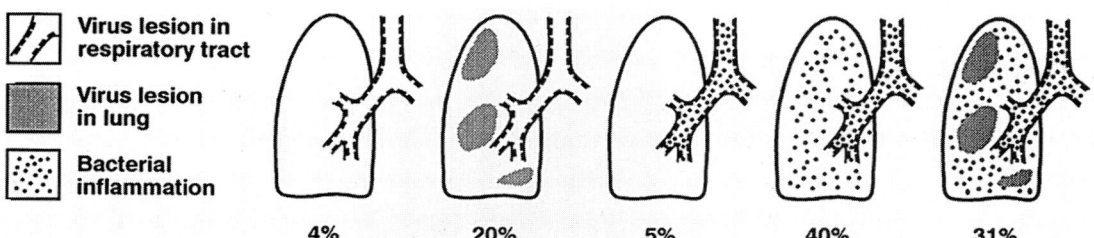

Virus lesion in respiratory tract

Virus lesion in lung

Bacterial inflammation

4%   20%   5%   40%   31%

FIGURE 5-2
Distribution of viral lesions in the respiratory tract in influenza as well as the distribution of the superimposed bacterial infection in the main bronchi and lung. (After Hers et al. Reproduced by permission of the *Lancet*.)

ciate, and undergo necrosis, and some are shed. Much of the bronchial mucous gland epithelium is lost. In severe cases all the bronchial epithelial cells, except the basal layer, undergo necrosis. There is edema of bronchial and bronchiolar walls. They are diffusely infiltrated with lymphocytes and very occasional neutrophil polymorphs (Fig. 5-3).

Aggregates of virus in dying cells can be demonstrated by immunofluorescent microscopy.

A recent study of lung biopsies taken from the Mayo Clinic files showed a spectrum of acute lung injury and repair (Table 5-1).[48] The acute changes varied from patchy fibrinous alveolar exudates and hyaline membranes (Fig. 5-4) with interstitial edema to severe alveolar damage and necrosis of bronchiolar mucosa. Capillaritis was seen in one case with neutrophil polymorphs in both the alveolar wall and lumen. There was atelectasis and a decrease in pulmonary surfactant.

Capillary thrombosis and intraalveolar hemorrhage have been described.[49] Focal alveolar wall necrosis may occur in addition to fibrin thrombi in the acute phase of primary influenzal virus pneumonia. There is an associated mononuclear cell infiltrate in the affected lung in 50 percent of cases. These alveolar changes are at their height at the fourth to fifth day.

Repair is by type II cell proliferation and there are mild chronic interstitial infiltrates. In addition there is a bronchiolitis obliterans organizing pneumonia pattern. In the ensuing days there is epithelial regeneration which may develop into squamous metaplasia (Fig. 5-5) and tumorlet formation. This reepithelialization occurs first in the proximal ciliated airways, as early as 1 to 2 weeks after infection[50] (Fig. 5-6).

Epithelial recovery may take weeks to months.[51] Lourdes et al.[51] studied sequential lung biopsies. The bronchial, bronchiolar and alveolar epithelium showed widespread bronchial and interstitial alveolar lymphocytic infiltration, followed by interstitial fibrosis. There were lymphoid follicles with germi-

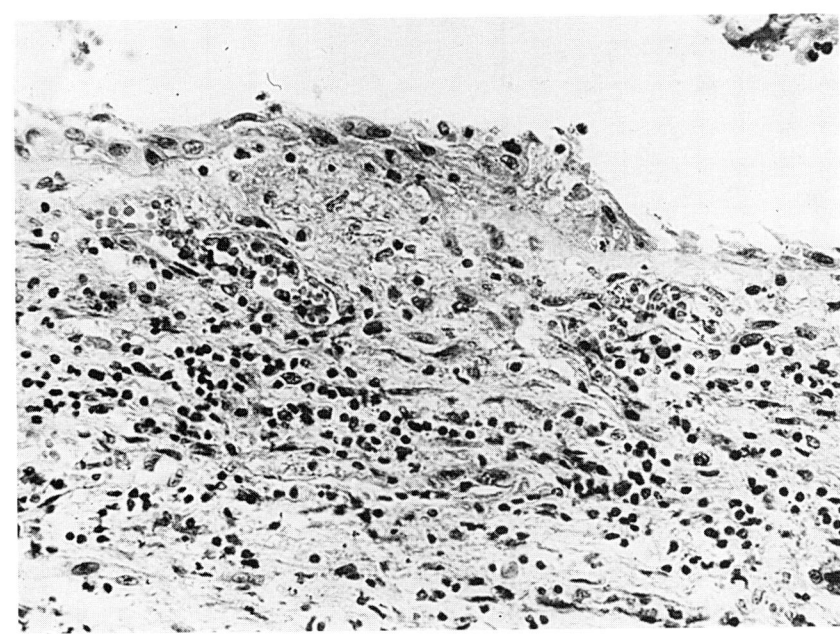

FIGURE 5-3
Influenza A. Section of a large bronchus with loss of most of the epithelial cells. There are dilated capillaries, oedema and chronic inflammation in the wall. (H&E, ×213.)

145

TABLE 5-1
Pathologic Features of Influenza Pneumonia

Fatal Cases

Early Cases:

Necrotizing tracheobronchitis, diffuse alveolar damage

Later:

Organized diffuse alveolar damage with variable degree of fibrosis. Peribronchiolar squamous metaplasia

A secondary bacterial pneumonia may be superimposed on either of these patterns.

Nonfatal Cases

Mild alveolar damage manifesting primarily as fibrinous exudates, alveolar septal oedema, type II cell metaplasia and small foci of early organization; some neutrophils may be present

Later stages showing patchy organization, including bronchiolitis obliterans with organizing pneumonia (BOOP pattern)

SOURCE: From Yeldani and Colby,[48] with permission.

nal centers. One patient developed bronchiectasis and bronchiolitis obliterans.

Experimentally Jakab et al.[52] gave mice a sublethal dose of influenza virus and followed them for up to a year. They showed an ongoing inflammatory response with patchy mononuclear interstitial pneumonia, deposition of collagen in the affected areas, and marked hyperplasia of bronchus-associated lymphoid tissue. Infective virus could not be recovered after day 9, but viral antigen persisted in high concentrations in the lung. The authors concluded that influenza produced a long-term alveolitis and raised the possibility that it may play a role in interstitial lung disease. Superinfection with *Klebsiella pneumoniae*, but not *S. aureus*, appears also to play a role in the development of fibrosis, as shown by increased lung hydroxyproline levels.[53]

Some of the complications of influenza have been noted above. A recent report described a 19-year-old man who developed postinfluenza toxic shock syndrome due to staphylococcal pneumonia.[54] Patients with a toxic shock syndrome have fever, hypotension, and a rash that shows desquamation. It is associated with an exotoxin-producing *Staphylococcus aureus*. Desquamation of the skin is predominantly over the palms and soles.

Parainfluenza

Parainfluenza belongs to the family *Paramyxovirdae* and to a genus *Paramyxovirus*. There are four strains of parainfluenza. Type 1 is called Sendai virus, type 2 is occasionally known as Newcastle disease virus, and type 3 occasionally has the number TI3. Types 1 and 2 occur in epidemics whereas Type 3, where it is endemic, occurs in all seasons. Parainfluenzae Type 4 has been associated with mild, sporadic, upper respiratory tract disease.

Parainfluenza appears to share a common progenitor with mumps virus.[55]

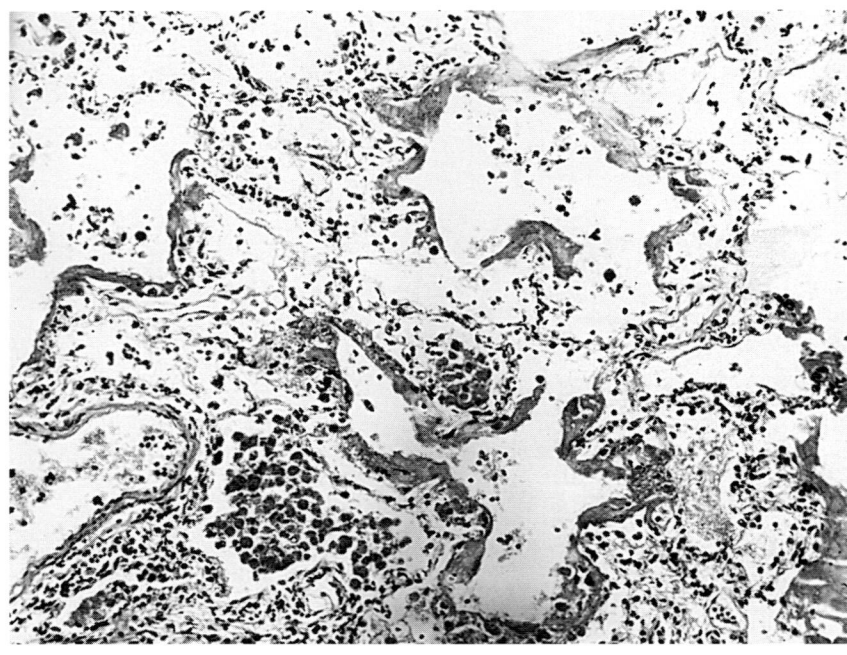

FIGURE 5-4
Hyaline membranes in influenza pneumonitis (H&E, ×104.)

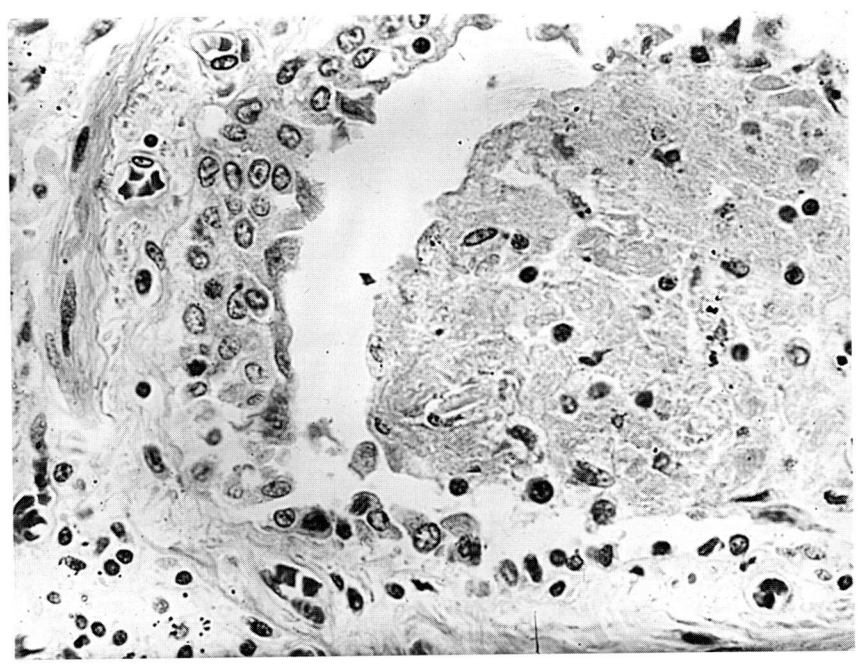

FIGURE 5-5
Influenza A in the healing phase. A bronchiole with squamous metaplasia and prominent nuclei are seen. (H&E, ×280.)

Most parainfluenza viruses cause upper respiratory tract infections, especially in young children. The infection presents as a laryngotracheobronchitis and croup. In children there may be lower respiratory disease.[56] In severe forms there is wheezing, croup, bronchitis, bronchiolitis, and pneumonia. Immunocompromised patients may have severe pneumonia, especially after marrow transplantation.[57]

Relatively little is known of the pathogenesis or pathology of parainfluenza infections. It would appear that in their severe forms they have an effect on the bronchoepithelium similar to influenza virus. There may be giant cell pneumonia along with viral inclusions and an interstitial pneumonia (Fig. 5-7).

The infection can be diagnosed by viral culture from nasopharyngeal specimens, BAL, or hemagglutination inhibition tests on paired acute and convalescent sera. Rapid diagnosis may be achieved using immunofluorescent staining of cells aspirated from the respiratory tract.[58]

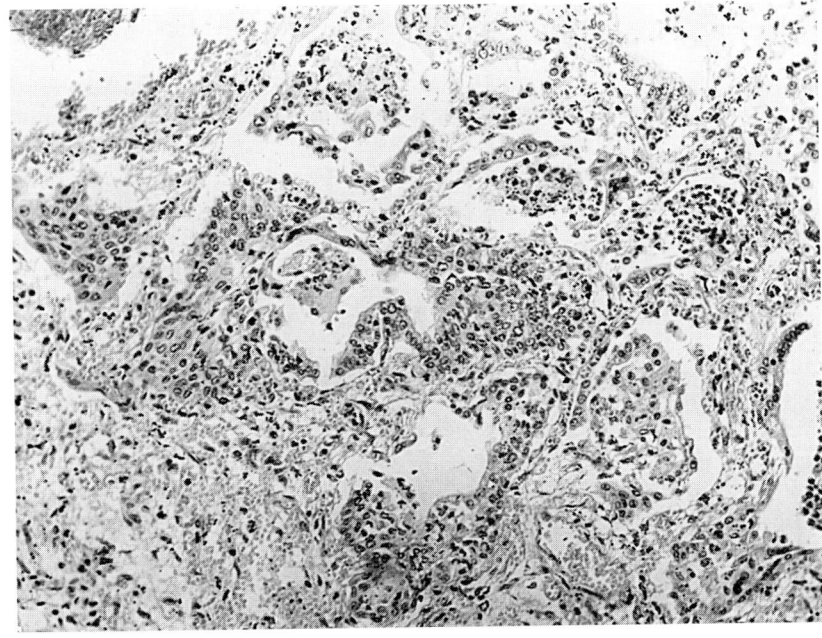

FIGURE 5-6
Hyperplastic epithelium lining the alveolar spaces with squamous change. This was during the recovery phase of influenza pneumonia. (H&E, ×140.)

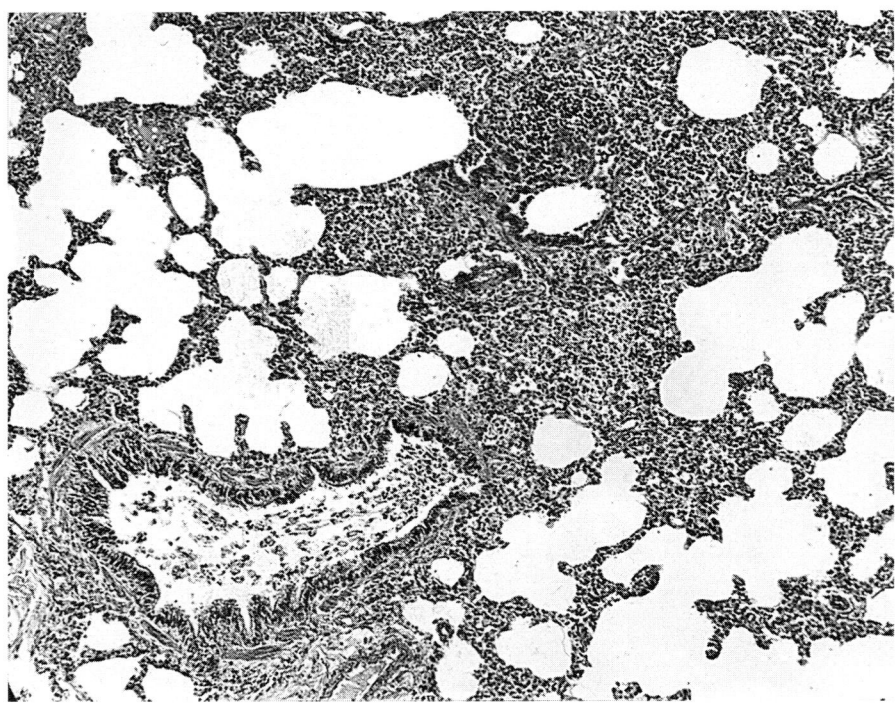

FIGURE 5-7
Parainfluenza pneumonia with an
interstitial component. (H&E, ×100.)

Respiratory Syncytial Virus (RSV)

This virus belongs to the family *Parviridae* but is
included under the genus *Pneumonvirus*. Serologically there are at least three different strains. The
virus measures 120 to 300 nm. They are nonenveloped, double-stranded, DNA viruses.

CLINICAL FEATURES

It causes more severe disease in children than
immunocompetent adults. It is estimated that each
year in the United States 95,000 people are hospitalized with lower respiratory tract disease, especially
bronchiolitis, due to RSV. More than 4 of 500 infected
children die due to the virus,[50] or 1 percent of cases
admitted to hospital.[50a]

It produces epidemics that in temperate climates
peak in the winter and early spring (Fig. 5-8). The
season lasts 4 months, peaking in February or
March.[60] The rate of infection is highest in the first
four years of life, but especially in the first year[61]
(Fig. 5-9).

Transmission is by droplets, via either the nose
or the eyes. The lower respiratory tract is involved in
25 to 33 percent of cases. Fever, cough, coryza, and
rhinorrhoea may be typical presenting symptoms
(Fig. 5-10). RSV may also be complicated by otitis
media and sinusitis. Up to 50 percent of children
hospitalized with RSV infection are premature or
have cardiac, pulmonary, or other congenital disease.
Twenty percent of infants hospitalized with RSV
have severe episodes of apnea.

Older children and healthy adults generally
have a mild upper respiratory tract infection and usually this precedes the lower respiratory infection.

In immunosuppressed patients, RSV infection is
more severe with pneumonia occurring at all ages
and having a higher mortality rate. Children receiving long-term steroid therapy do not appear to have
more severe clinical manifestations than normal children but show significantly greater viral shedding.[62]
It has been thought that infants hospitalized with a
diagnosis of RSV infection were more likely than
controls to have a diagnosis at the age of 3 years of
asthma. Such children may not have persistent
asthma.[62a]

The enigma of the disease is that the most severe
disease occurs in the first months of life, when maternal antibodies are abundant. Affected infants may
have a defect in both suppressor cell numbers and
function, as shown by overproduction of IgE and an
exaggerated lymphocyte response.[63] There is the suggestion that RSV immunity may be mediated by specific cytotoxic lymphocytes.[64] An infection evokes
an antibody response but reinfection may occur
within months.[65]

PATHOLOGY

The virus predominantly affects the epithelium of
small bronchi and bronchioles (Fig. 5-11). As mentioned above, however, it can cause low-grade infection in lymphoid tissue. It replicates in the epithelial
cytoplasm and after maturation buds off from the cell
membrane. The epithelial cells swell, ultimately

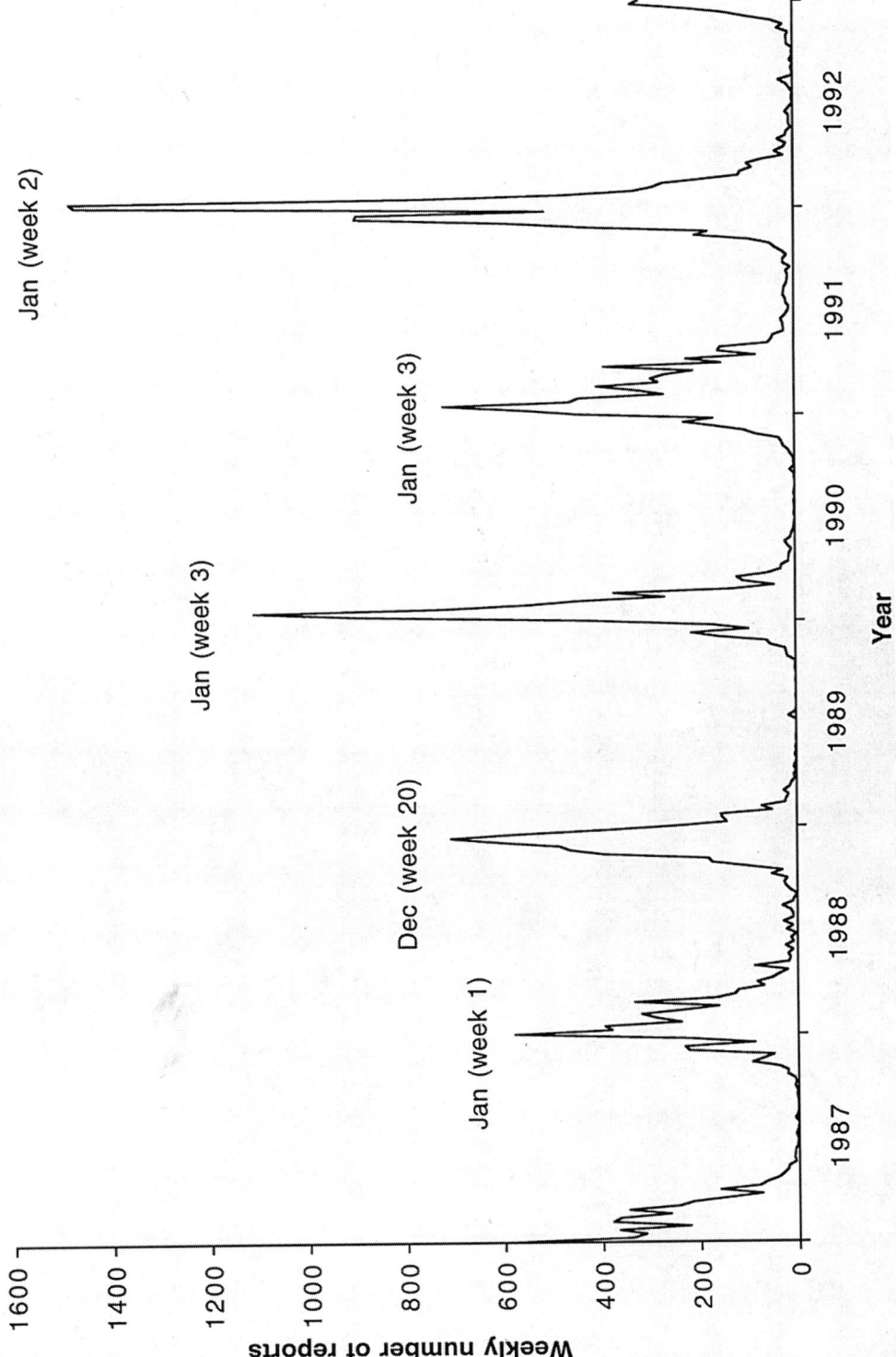

FIGURE 5-8
Reports of infections due to RSV in children under 5 years of age, by week, England and Wales, 1987 to 1992. (Reproduced by permission of the Lung and Asthma Information Agency, Department of Public Health Services, St. George's Hospital, London.)

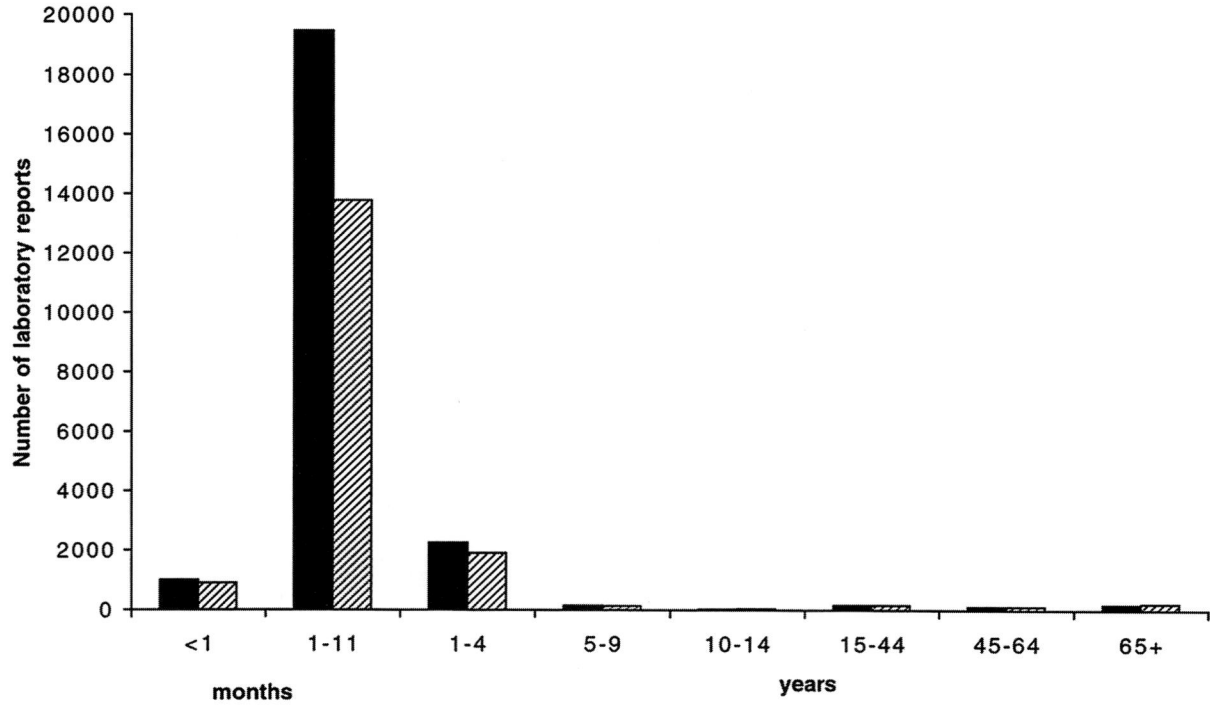

FIGURE 5-9
The age and sex distribution of RSV infection, England and Wales, 1987 to 1992. ■, Male; ▨, female. (Reproduced by permission of the Lung and Asthma Information Agency, Department of Public Health Services, St. George's Hospital, London.)

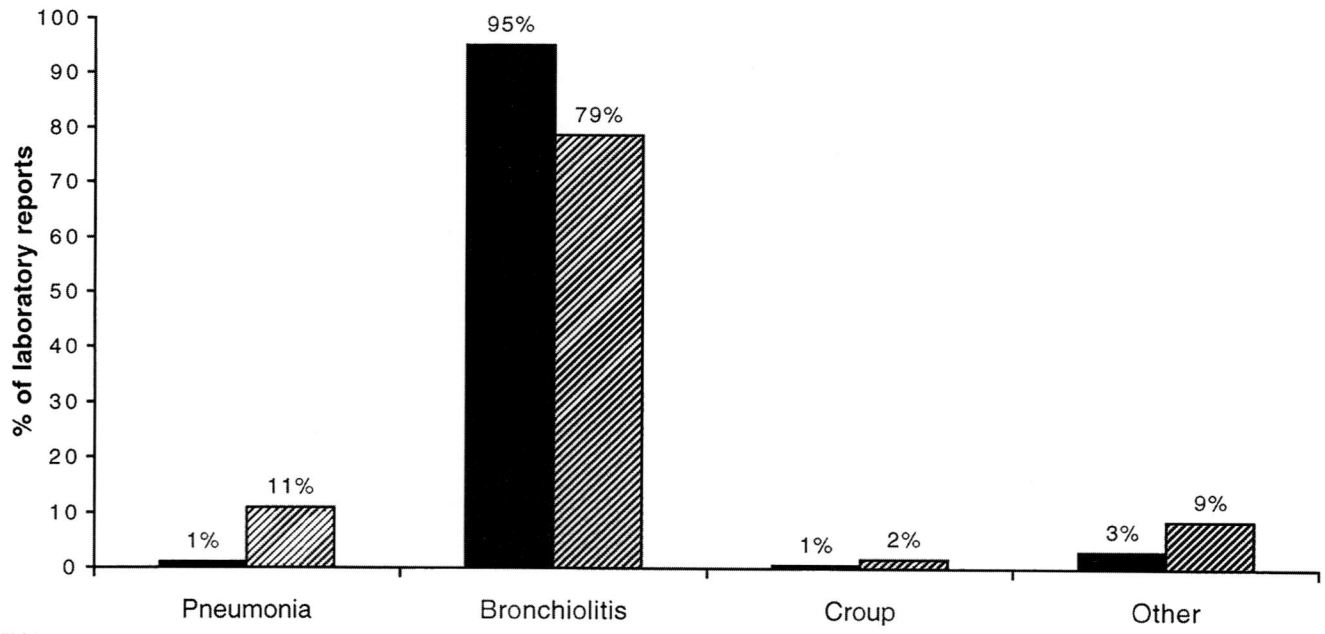

FIGURE 5-10
Clinical features reported with RSV infection in children, England and Wales, 1987 to 1992. ■, < 1 year; ▨, 1–4 years. (Reproduced by permission of the Lung and Asthma Information Agency, Department of Public Health Services, St. George's Hospital, London.)

150

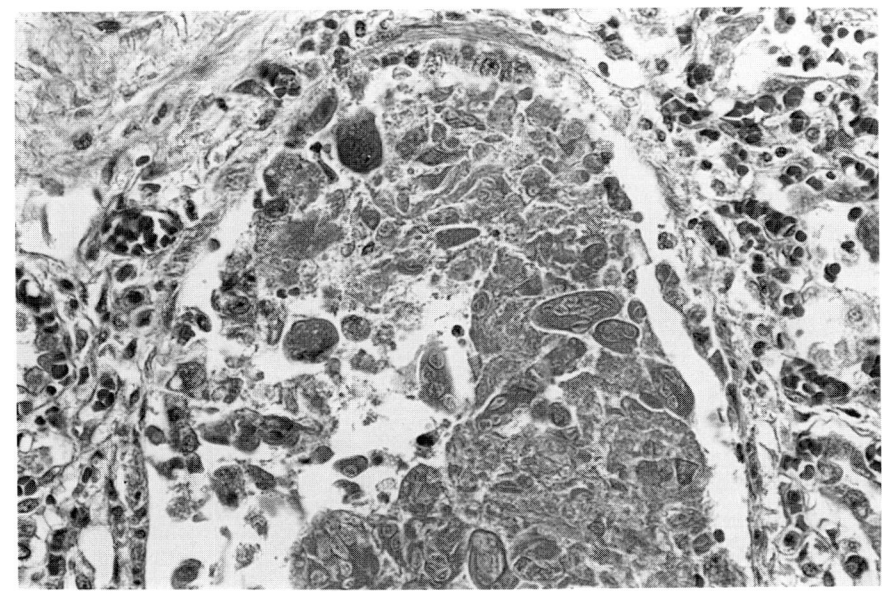

FIGURE 5-11
RSV-infected bronchiolar epithelial cells, which have become necrotic and sloughed into the lumen. (H&E, ×219.) (Reproduced courtesy of Dr. Gould, Oxford.)

desquamate and die. These form plugs of mucus along with cellular debris causing partial or complete airway obstruction. There is an associated lymphocytic, plasma cell, histiocytic and neutrophilic infiltrate the bronchiolar walls. During recovery there is proliferation of cuboidal lining epithelium, which may block the lumina of air passages. Histological sections show multinucleated giant cells (Figs. 5-12 and 5-13) suggesting a giant cell pneumonia. The multinucleated giant cells are usually found lining the alveolar spaces. They are irregular in shape and do not appear foamy.

The virus may be detected on bronchoalveolar lavage. Thus, as a routine in immunosuppressed patients, some of the specimen should be innoculated onto a monolayer of tissue culture cells.

The inclusions contain oval to round intracytoplasmic, acidophilic material measuring 3 to 17 μm (Fig. 5-14). No intranuclear inclusions are found. Ultrastructurally they consist of membrane-limited aggregates and dense amorphous particles (Fig. 5-15), surrounded by smaller inclusions containing tubuloannular structures. The giant cells show numerous surface microvilli. No lamellar bodies were identified in one publication and the primary infected cell was thought to be the Type 1 pneumocyte.[66] Clearly bronchiolar cells may be affected.

Giant cell pneumonia may be found in measles, RSV infection, and parainfluenza type 3 viral infections, especially in immunodeficiency states.

Diagnosis is made by the syncytial cytopathic effect but this takes 2 to 10 days on HELA cells. Direct

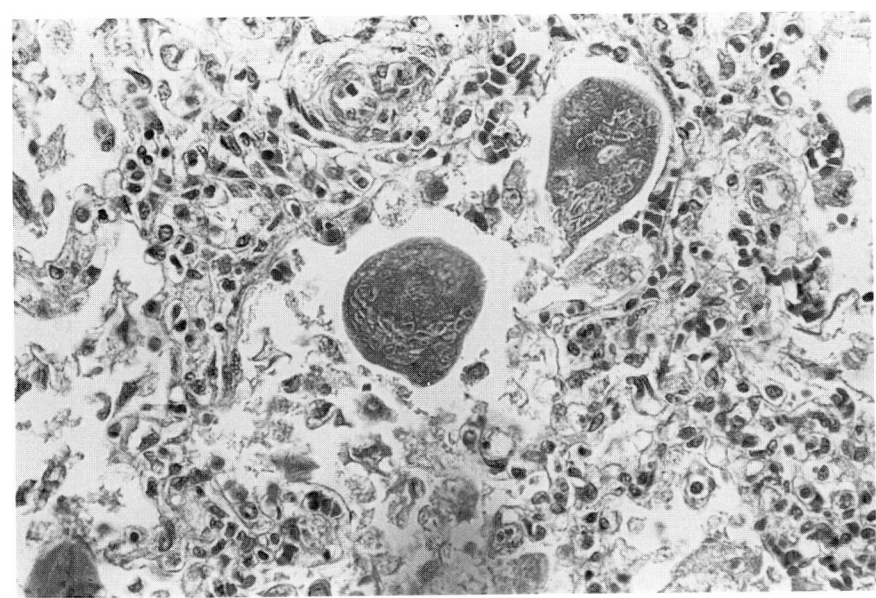

FIGURE 5-12
Giant cells, due to RSV, lying in alveolar ducts. There is a mild interstitial chronic inflammatory infiltrate. (H&E, ×219.) (Reproduced courtesy of Dr. Gould, Oxford.)

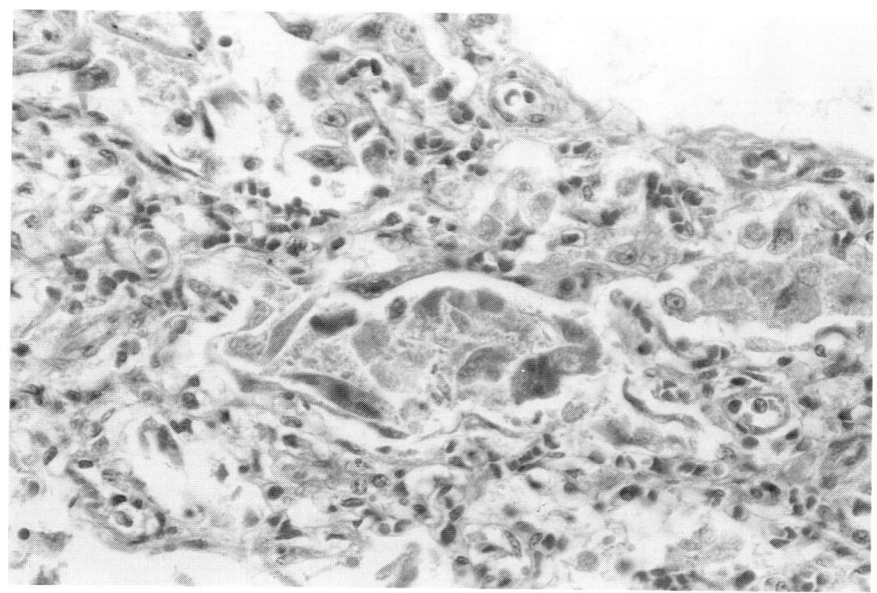

FIGURE 5-13
Giant cells lining an alveolus in RSV
pneumonia. (H&E, ×219.) (Reproduced
courtesy of Dr. Gould, Oxford.)

immunofluorescent staining of cells in respiratory
secretions is a quicker way of reaching a diagnosis.[67]
Such staining may persist for weeks after viral cultures have become negative.[68] Diagnosis can also be
made by complement fixation tests on paired acute
and convalescent sera. Rapid detection of RSV antigen is now possible by commercially available ELISA
kits. Infected infants shed abundant virus in their respiratory secretions. The infected respiratory epithelial cells, obtained by nasopharyngeal aspiration or
washings, can be stained directly with fluorescein-
labeled specific antibody. The slides are ready for
interpretation within a few hours.

RSV infection can be detected on BAL in pulmonary macrophages and epithelial cells in a small
group of marrow, heart/lung, or renal transplant
patients by in-situ hydridization.[69]

Measles Pneumonia

Measles, which is related to canine distemper virus,
belongs to the genus *Morbillivirus* and is in the family *Paramyxoviridae*.

CLINICAL FEATURES

The virus causes disease in children, but may persist
in adults if immunity fails. It can also cause subacute
sclerosing encephalitis, a disease which strikes in
early life and whose pathogenesis is poorly understood. Measles pneumonitis is rare in immunosup-

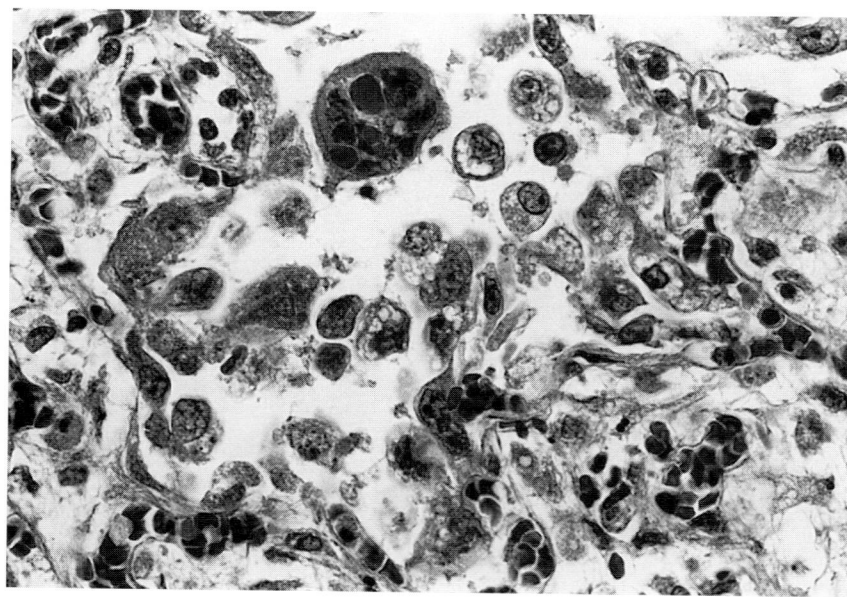

FIGURE 5-14
RSV pneumonia showing a large
inclusion. (H&E, ×219.) (Reproduced
courtesy of Dr. Gould, Oxford.)

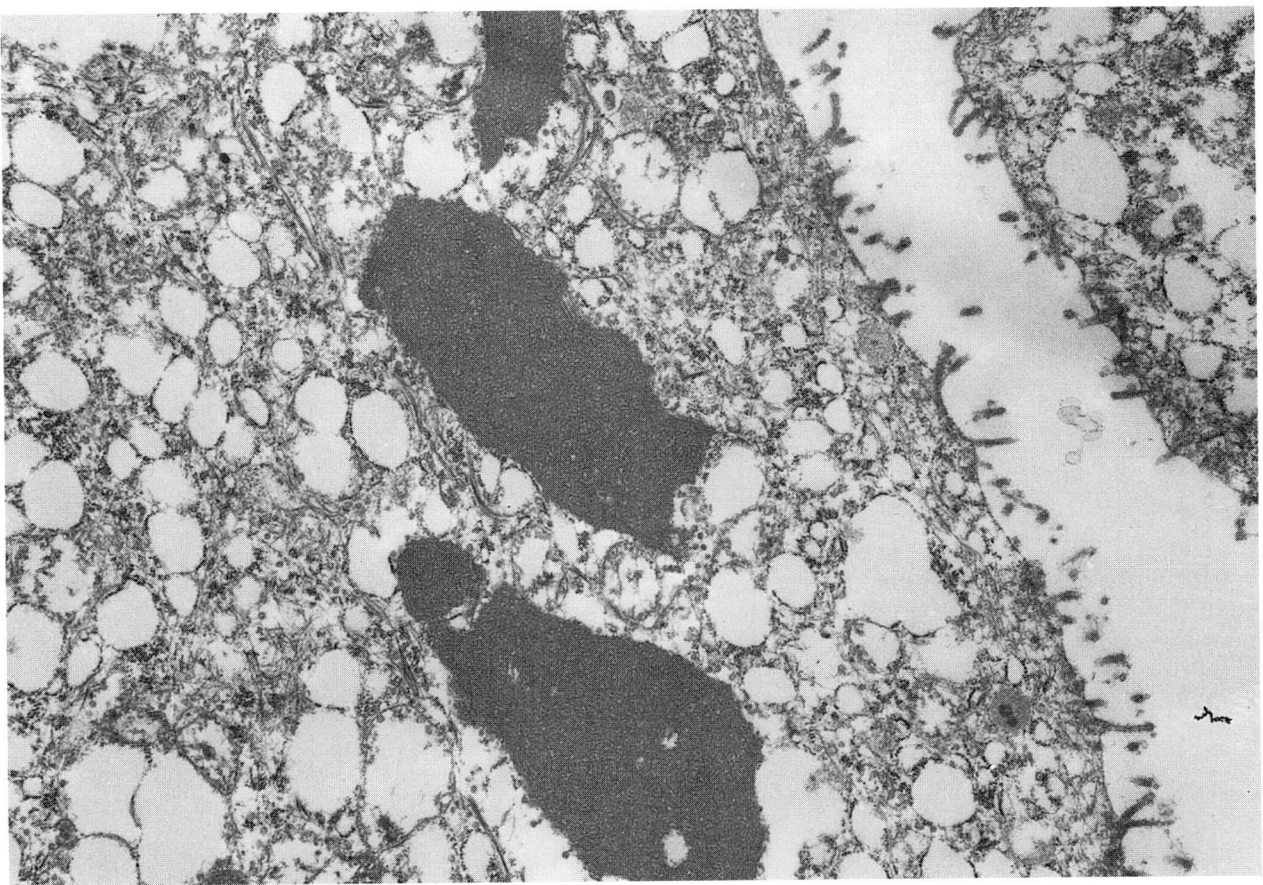

FIGURE 5-15
Electron micrograph of RSV particles. These are discrete cytoplasmic inclusions with crystalloid arrays of packed tubular virions. (Reproduced by permission of Dr. D. Parham, and *The American Journal of Clinical Pathology.*)

pressed adults, who most probably have acquired immunity via previous infection. However, there are known instances where a child has transmitted fatal infection to an immunocompromised adult.

Measles is a serious disease in third world countries, where children have kwashiorkor and malnutrition. Case fatality rates in Africa are up to 21 percent.[70] Denton[71] described 11 fatal cases following an epidemic in Panama in 1923. In three cases death occurred very early in the course of the disease.

The disease tends to occur in the winter and is thought to be spread in aerosol form. Measles vaccination has altered the timing of the epidemic, which used to be biannual.

The virus probably enters the respiratory tract, replicates in the lymphoid tissue, and then spreads systemically to the skin and other organs. The incubation period is 10 to 14 days. It involves mainly the skin, brain, and lungs.

Clinically there is initially fever, conjunctivitis, cough, and rhinorrhoea. Koplik spots, which are areas of erythema with central grey foci, are seen in the mouth. The rash starts on the head and neck,

spreading over the rest of the body during the next couple of days.

PATHOLOGY

Macroscopically there are no characteristic appearances except for grayish-white tissue lining the smaller bronchi.

Histologically the bronchial and bronchiolar epithelium shows vacuolar degeneration. This may also affect alveolar epithelial cells. In time the whole epithelium may separate leaving a bare basement membrane. There is also hyperplastic change in the lining epithelium which replicates to form six or more layers of cells. (Fig. 5-16) Some of these cells may show mitoses. There are amphophilic intranuclear inclusions due to clumping of the nuclei into large masses (Fig. 5-17) as well as intracytoplasmic inclusion bodies. These latter inclusions may be found in individual hyperplastic bronchiolar epithelial cells or even in alveolar lining cells. The inclusions vary in diameter from 1 to 10 μm in diameter. They are identified with picric acid or the appropri-

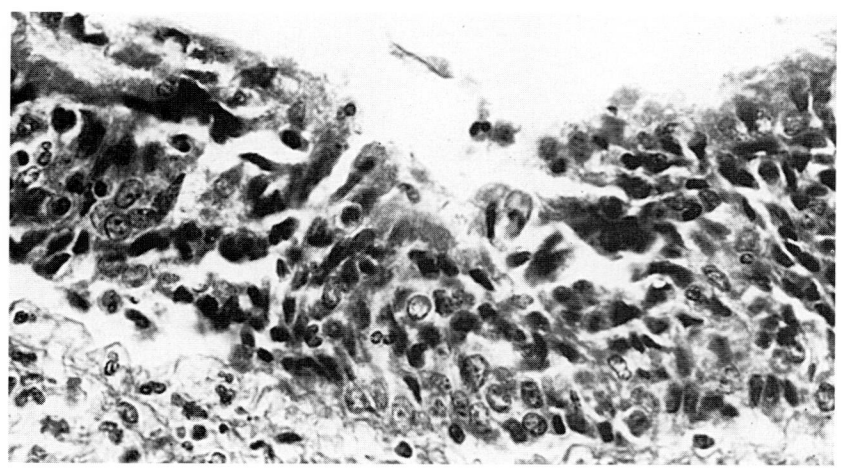

FIGURE 5-16
Hyperplastic epithelium in a bronchiole in a case of measles. (H&E, ×375.) (Reproduced by permission of Dr. G. B. S. Roberts.)

ate immunoperoxidase stain (Fig. 5-18). These giant cells are distinct from those described by Warthin[72] and Finkeldy[73] in that Warthin-Finkeldy cells occur in the germinal centers of lymphoid tissue.

In addition to the epithelial changes there is a peribronchial mononuclear infiltrate (Fig. 5-19) as well as edema which extends into the adjacent alveolar septa. The infiltrate is predominantly lymphocytic and plasmacytoid but a few polymorphs may also be found. There is an increase in BALT (bronchial associated lymphoid tissue). The epithelial changes described above last only several days. Secondary bacterial pneumonia usually supervenes.

The disease may heal without any scarring but some cases may progress to develop type II hyperplasia as well as interstitial fibrosis (Fig. 5-20).

Squamous metaplasia is seen in bronchi and bronchioles.

Other Causes of Giant Cell Pneumonia

The concept that giant cell pneumonia results solely from measles is now no longer tenable. As noted above it may occur in viral disease with RSV and parainfluenza type 3.

DIAGNOSIS

Complement fixation tests on acute and convalescent sera can be used. The presence of a specific IgM antibody indicates recent infection. In immunosuppressed patients, where rapid diagnosis is required, the demonstration of giant cells in the respiratory tract secretions should be used. Viral antigens can be

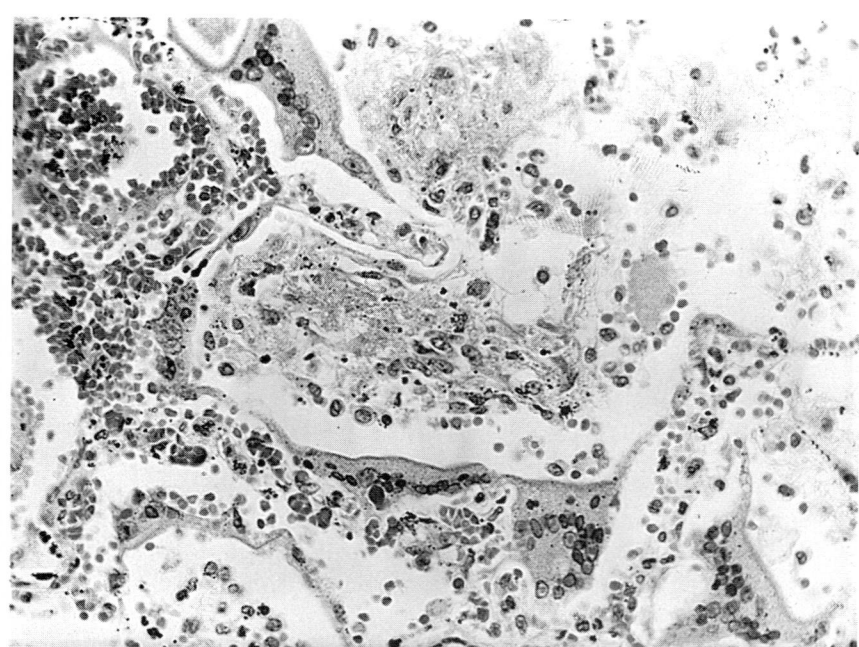

FIGURE 5-17
Multinucleated cells lining alveoli as seen in measles. (H&E, ×148.)

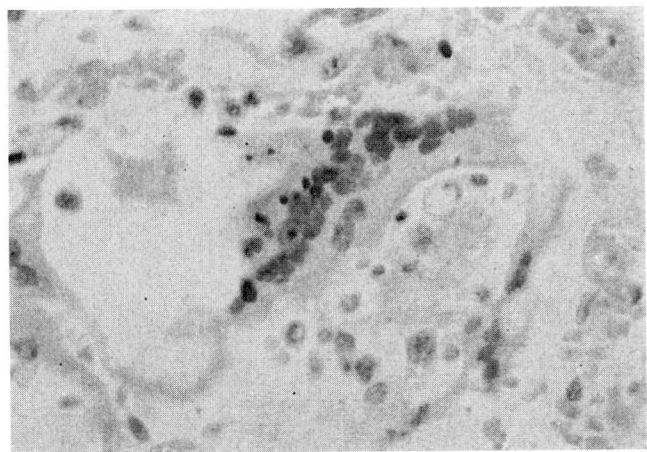

FIGURE 5-18
Measles inclusion body as shown by immunoperoxidase. (×500.)

demonstrated in such specimens by immunofluorescence with monoclonal antibodies.

Adenoviral Infections of the Lung

Adenoviruses obtain their name from the fact that the first isolate was made from human adenoid tissue. This tissue shows degenerative changes after culture with the virus for several weeks.[74] They infect epithelial cells but also cause low-grade infection in lymphoid tissue.

Adenoviruses belong to the family *Adenoviridae* and are characterized by nonenveloped icosahedral virions containing double-stranded DNA that grows in the nucleus of the infected cell.[75] There are, at the time of writing, 41 adenovirus serotypes which can be divided into six genera, according to the hemagglutination of monkey and rat erythrocytes.[76]

CLINICAL FEATURES

The virus can affect many body systems, including the bowel, producing diarrhea, the meninges, and the eyes. The commonest site for the infection is in the lung. There is often a severe bronchiolitis and cough. In some patients there may be a whooping cough-like syndrome. The virus may play a role in the pathology of asthma.[77] It may cause a severe and fatal pneumonia, especially in children. The disease may occur in the newborn and viral acquisition may occur via the birth canal. Neonatal adenovirus infection is frequently disseminated and often fatal.[78]

Predominent serotypes causing respiratory diseases are 1, 2, 3, 4, 5, 6, 7, 14, and 21. Recently type 35 has been incriminated in a fatal neonatal pneumonia.[79]

These viruses may play a part in pulmonary complications in AIDS or other immunocompromised patients. Adenovirus infection may also be a problem in liver transplant patients, occurring in 20/484 (4 percent) of recipients in one series and leading to death in nine.[80] Altogether 49 patients had 53 episodes of adenovirus infection. The serotypes involved in these cases were 1, 2, and 5.

PATHOLOGY

The beagle, with experimentally induced adenoviral infection, serves as a good experimental model for

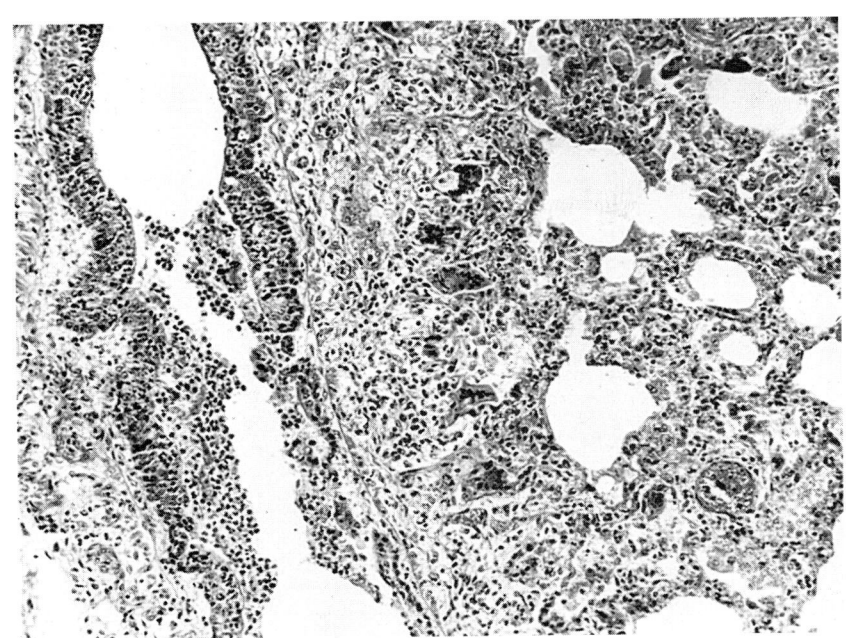

FIGURE 5-19
Measles pneumonia showing a peribronchiolar infiltrate as well as giant cells. (H&E, ×108) (Reproduced by permission of Dr. G. B. S. Roberts.)

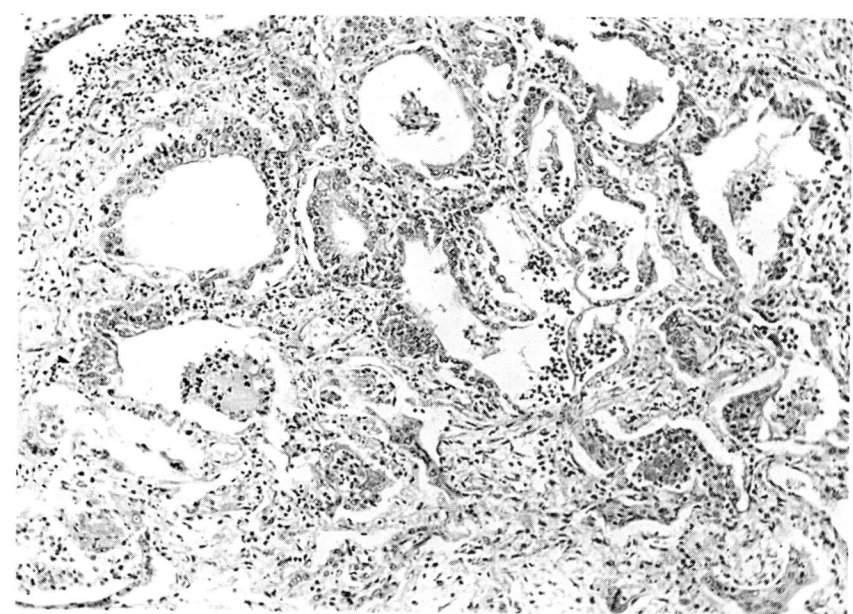

FIGURE 5-20
Post-measles pneumonia with prominent epithelial lining cells and interstitial fibrosis. (H&E, ×75.)

pulmonary damage shown in the human during early life. Castleman[81] inoculated young beagles with canine adenovirus type 2. The virus was recovered from the lungs at 2, 3, 5, and 8 days after inoculation. He demonstrated virions and viral antigen by ultrastructural and immunoperoxidase studies. They were found in nonciliated bronchiolar epithelial cells and mucous cells in bronchioles, bronchi, and the trachea, as well as the bronchial and tracheal submucosal epithelial cells. The viral replication in airways was associated with severe and proliferative bronchitis and bronchiolitis. He also demonstrated viral antigen in type II pneumocytes. There was evidence of bronchiolitis obliterans as well as interstitial pneumonia. Viral particles can be shed from

normal tonsils and adenoids for long periods after infection.

Macroscopically the lungs are hyperemic with multiple small dark, red-purple, collapsed areas, mainly in the lower and posterior parts of the lungs. The bronchi are ulcerated, red, and filled with mucoid material. There is coexistent enlargement of the hilar lymph nodes. Focal interstitial emphysema is seen.

Histologically the most prominent feature occurs in the bronchi and bronchioles, where the epithelium may be completely shed. The ulcerated surface is covered by PAS-positive, eosinophilic material (Fig. 5-21). The lumina of small bronchioles are filled with either dense plugs of fibrin and cellu-

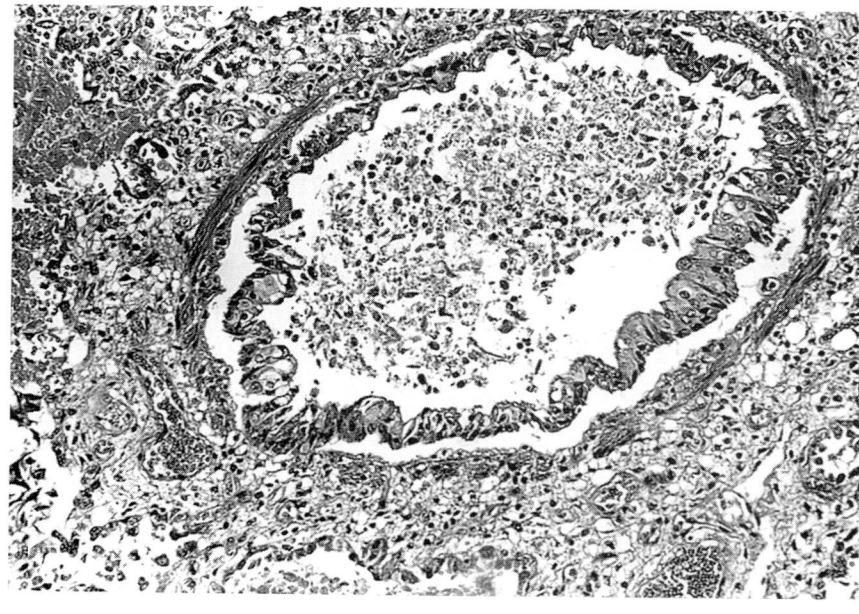

FIGURE 5-21
Adenovirus. Bronchiolar epithelium covered by PAS positive eosinophillic material. (H&E, ×125.)

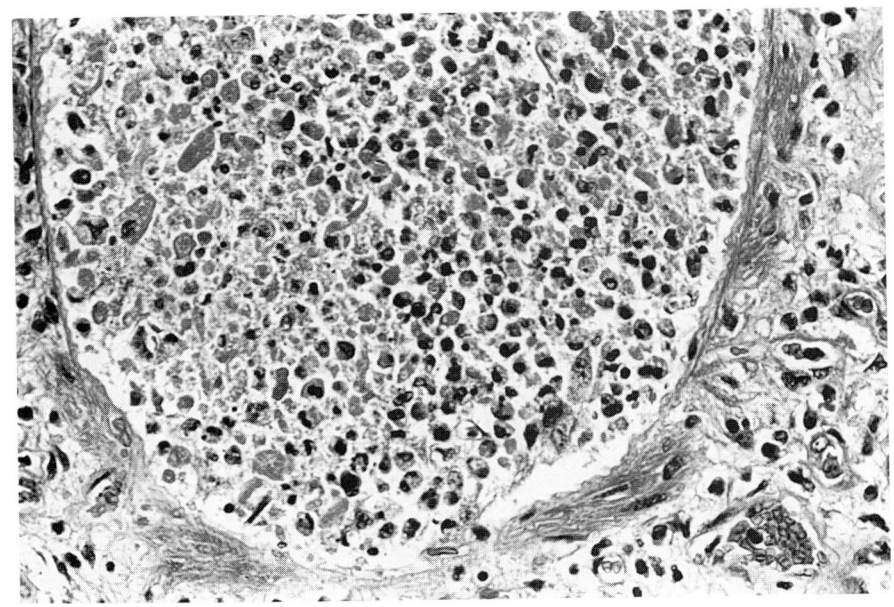

FIGURE 5-22
Adenovirus. Degeneration of epithelial cell cytoplasm with a full bronchial lumen (H&E, ×250.).

lar debris or, later in the process, organizing fibrous tissue. There is a nuclear hyperchromasia and pyknosis with degeneration of epithelial cell cytoplasm, often without diagnostic intranuclear inclusions (Fig. 5-22). A few nuclei may contain brightly eosinophilic, rectangular inclusions, which may have a clear halo. These inclusions are Feulgen-negative and positive with picric acid. They measure up to 5 μm in diameter. As the inclusion bodies mature they coalesce and stain basophilically with H & E. It is when they disrupt, that the "smudge" cells are seen (Figs. 5-23 and 5-24). The combination of smudged nuclei and bricklike, intranuclear inclu-

sions in pulmonary epithelial cells has been reported on several occasions in adenoviral pneumonia.[82,83]

Consequent to adenoviral infection, there is interstitial pneumonia and bronchiolitis obliterans.[84] Bronchiolitis obliterans is not specific for adenovirus infection. It may also be found secondary to *RSV*, *M. pneumoniae*, *H. influenzae*, *Mycobacterium tuberculosis*, post heart/lung transplantation, and in a variety of other conditions.

The virus may be demonstrated in paraffin sections by use of the DNA probes and in-situ hybridization.[85] Adenoviral DNA may be detected by in-situ hybridization and PCR in some cases of bronchiecta-

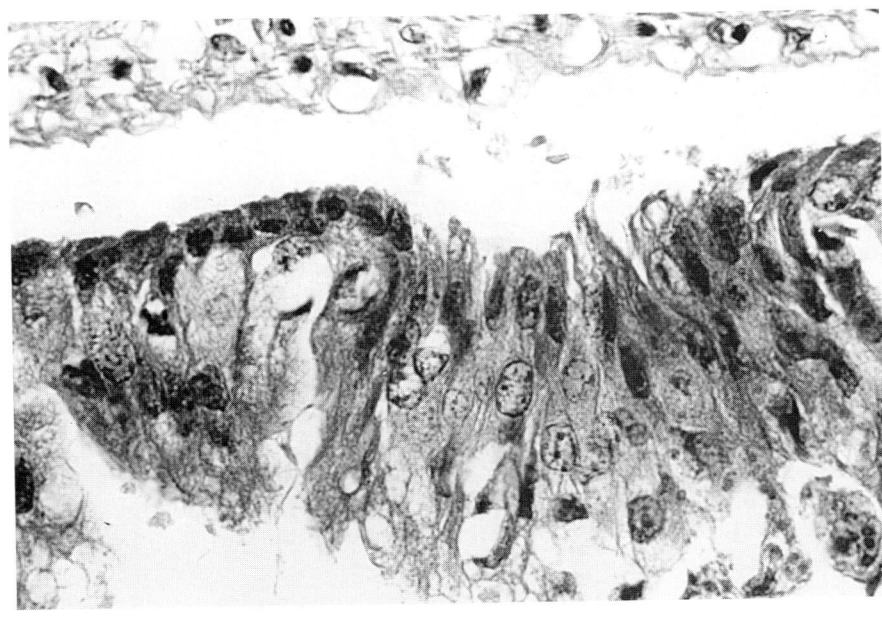

FIGURE 5-23
Adenovirus. Smudge cells in desquamated bronchiolar epithelium. (H&E, ×375.).

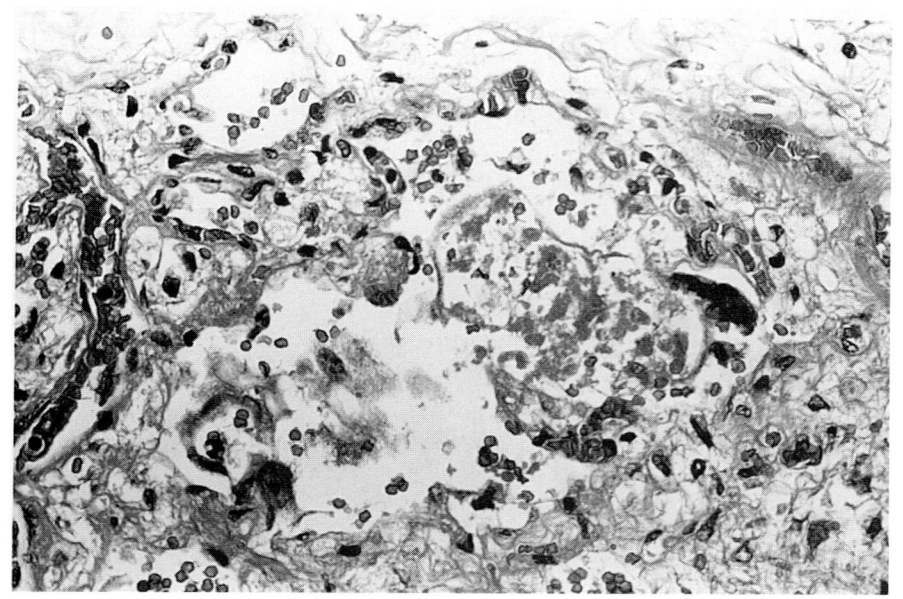

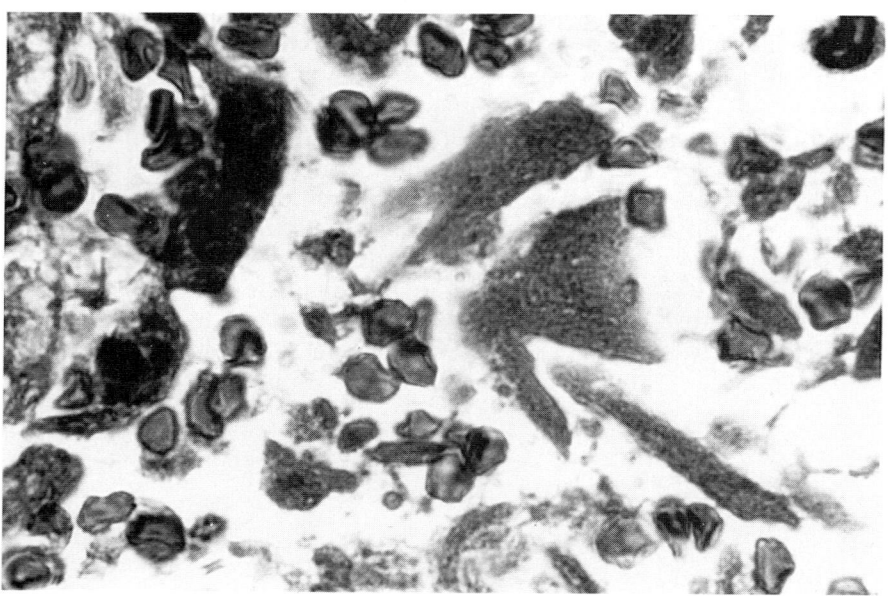

FIGURE 5-24
Adenovirus. *A*. Loss of alveolar
surface epithelium with some
smudge cells. (H&E, ×250.) *B*.
Prominent smudge cells are seen
on the left. (H&E, ×375.)

sis. There was a lower prevalence of latent adenovirus than that found in smokers with normal lung
function. There was no causal link between adenovirus infections and bronchiectasis.[85a] The virus
may be isolated from the respiratory tract, urine, or
feces. It is best to take a nasal or throat swab and put
it into transport medium immediately after collection. It can be grown on a monolayer of Hela cells and
viral effects will appear in 1 to 4 weeks. Complement
fixation or ELISA are most commonly used to identify the virus. Definitive diagnosis requires viral
identification (Fig. 5-25) or serologic changes, as disease produced by the virus must be distinguished
from asymptomatic viral shedding.

Varicella (Chickenpox) Pneumonitis

CLINICAL FEATURES

Varicella zoster is one of the herpes group of viruses.
Over the last 10 years there have been major
advances in our understanding of the varicella-zoster
virus. Varicella is usually a benign manifestion of primary infection and zoster is a result of reactivation of
latent virus. Varicella or chickenpox is a common
childhood illness, though it may affect adults, especially those who are immunosuppressed. Zoster
affects the skin giving rise to painful vesicles within
a single dermatome. It appears that *Herpes zoster*,
like *H. simplex* can reside in ganglia.

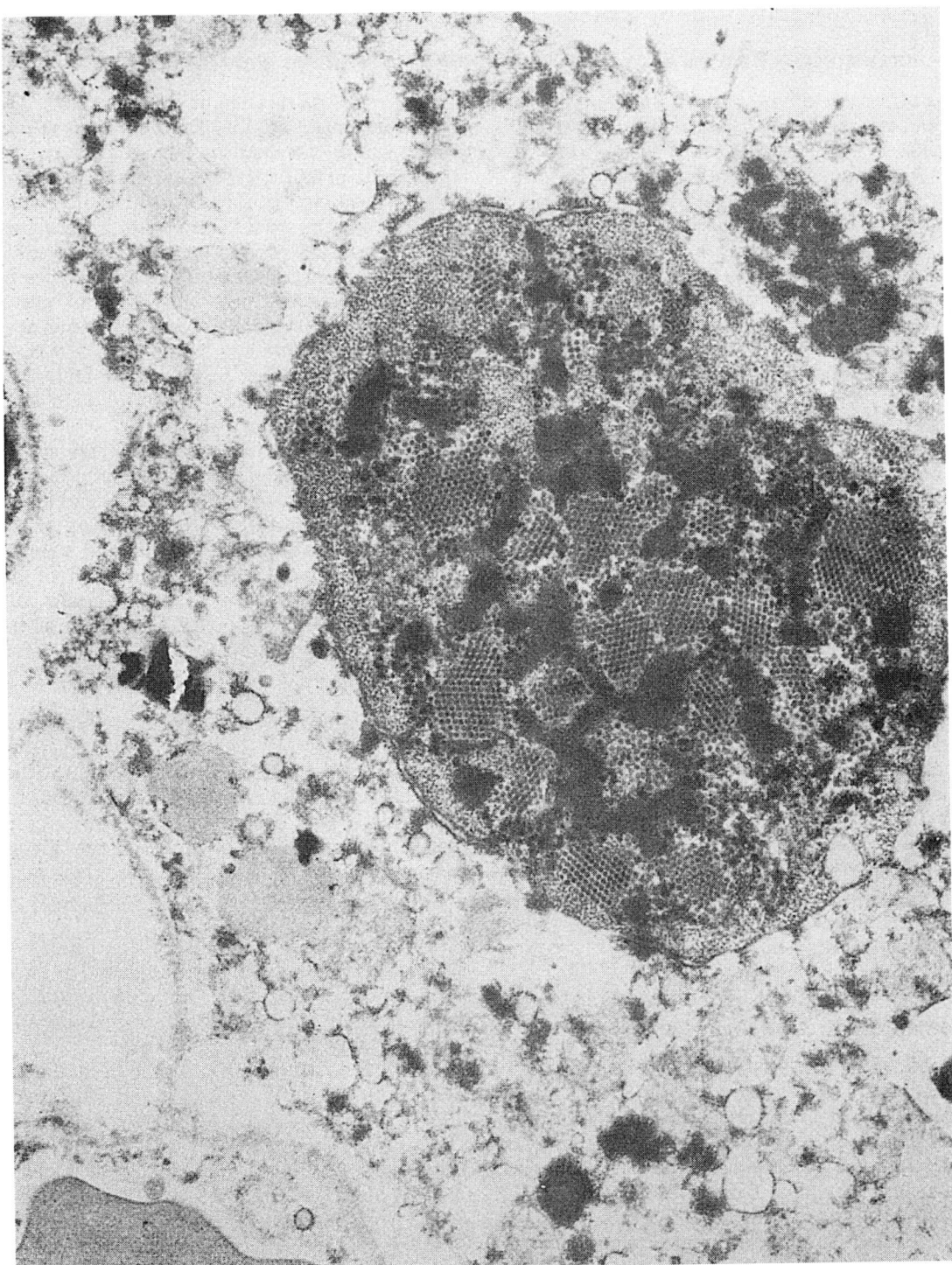

FIGURE 5-25
Adenovirus showing packed viral bodies. The intranuclear inclusions consist of aggregations of whole virus particles. (EM, ×16,000.) (Reproduced by courtesy of Dr. M. Schonland, Durban.)

Pulmonary complications are most common in adults, where there is dyspnea, cough, and pneumonia, sometimes ending in respiratory failure. There is a significant mortality in adults, rising from 13.5 to 45 percent in pregnant women.

Radiologically there are nodular or reticular densities, which may resolve to leave calcific pneumonia.[86]

PATHOLOGY

Macroscopically the lungs are heavy, edematous, and plum-colored. There are foci of pale tissue, mea-

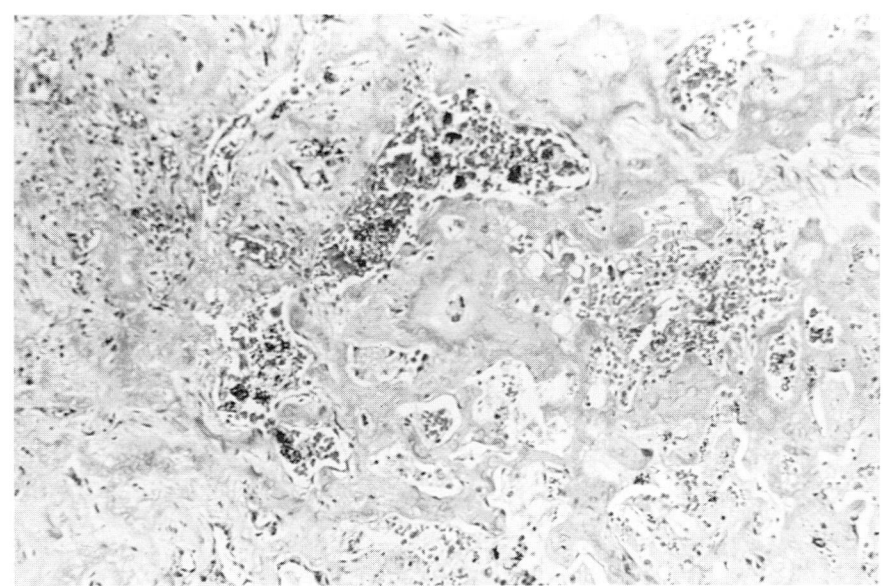

FIGURE 5-26
Varicella pneumonia with necrosis of the alveolar walls. (H&E, ×125.) (Reproduced by permission of Dr. Gould, Oxford.)

suring up to several millimeters in diameter. The bronchi contain blood-stained mucus.

Histologically there is a mononuclear, interstitial infiltrate with intraalveolar edema, hemorrhage, fibrin, and hyaline membranes. Focal areas of necrosis may be seen in the alveolar walls (Fig. 5-26) and the adjacent blood vessels, as well as the bronchioles. Intranuclear inclusions are seen in the lining epithelium (Fig. 5-27), small bronchi, and desquamated alveolar epithelial cells.

The inclusions are acidophilic and occupy the nuclei which become vesicular. They may be demonstrated by in-situ hybridization or electron microscopy of fresh tissue.

Usually there is no permanent damage but occasionally calcified nodules may appear up to 2 years after the disease. The nodules are uniformly distributed throughout the lungs and vary in size, measuring up to 6 mm in diameter (Fig. 5-28). Only the centers of the nodules are calcified and they are surrounded by collagen and lymphocytes with occasional giant cells and hemosiderin-laden macrophages (Figs. 5-29 and 5-30). The postmortem appearance of late calcified nodules has been described by Raider.[87]

As with other viral infections, immunofluorescence and the ELISA test can be used for detecting *Varicella zoster*. There is a specific IgM. In histological material, in-situ hybridization or electron microscopy may be of use.

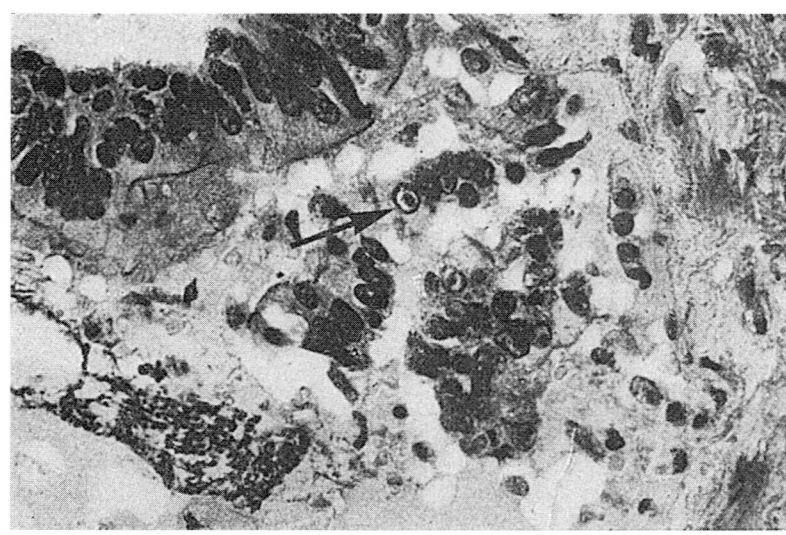

FIGURE 5-27
Varicella pneumonia with an intranuclear inclusion (*arrowed*). (H&E, ×400.)

FIGURE 5-28
Old, healed calcified nodules of Varicella.

Herpes Simplex Pneumonia

CLINICAL FEATURES

There are two types of *Herpes simplex*. Type I (HSV-1) which classically affects the head and neck, i.e., above the waist. Type II (HSV-2) affects the anogenital regions. HSV-1 infection is common and apparently spread by kissing. HSV-2 is spread by sexual intercourse. Most of the viral infections in this group are probably HSV-1, due to reactivation.

Infection occurs on a background of lung damage in patients who have suffered burns, ARDS, or heart/lung transplantation.[88–90] *Herpes simplex* pneumonia, localized or generalized, may occur in any immunocompromised patient. Recently a group of patients with no underlying disease presented with bronchospasm and were found to have herpetic tracheobronchitis.[91]

In a study of 42 consecutive patients with respiratory tract cultures positive for HSV-1, 64 percent were immunocompetent.[92] These patients had a higher incidence of tobacco smoking and had endotracheal intubation prior to HSV isolation. Their mortality was greater.

Mucocutaneous lesions precede or coincide with pulmonary disease in all patients.[93]

Disseminated HSV-2 can rarely occur in neonates[94] and undergo calcification.[95] No note was

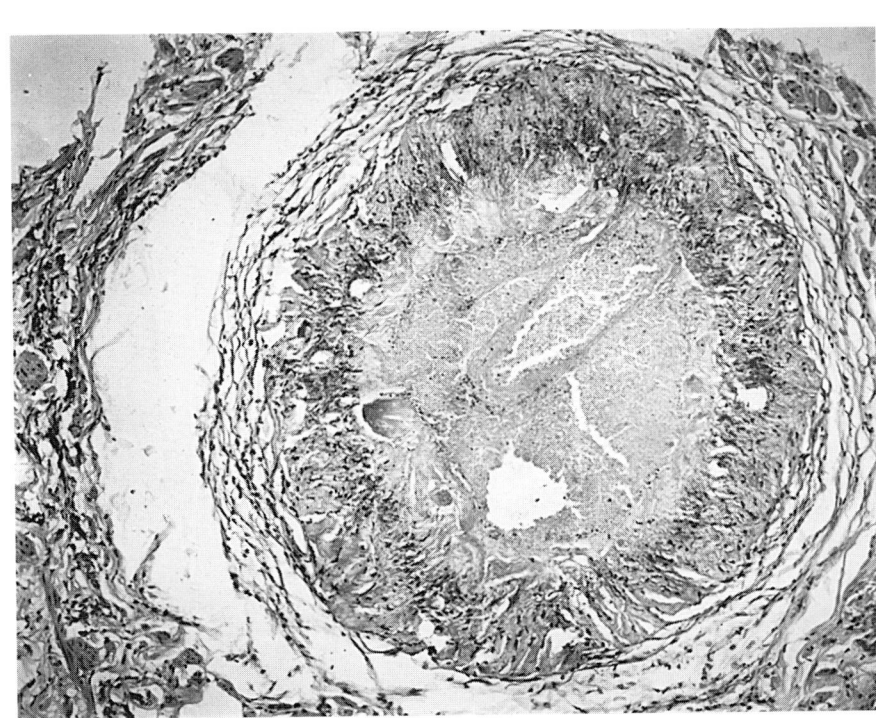

FIGURE 5-29
Calcified Varicella with surrounding giant cells. This was 9 years after a severe attack of the disease in an adult. (H&E, ×71.) (Reproduced by courtesy of Dr. E. H. Bailey.)

161

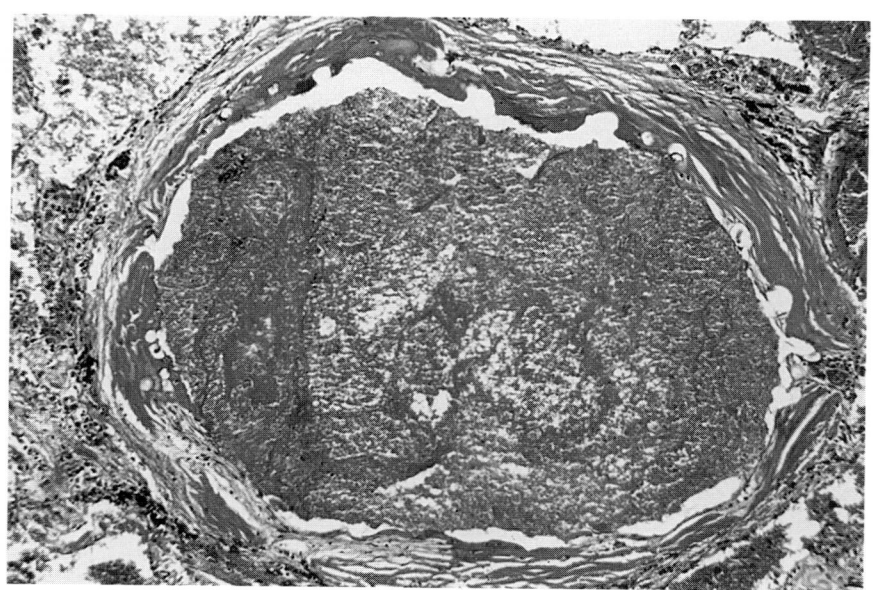

FIGURE 5-30
Calcified Varicella nodule from the case
shown in Fig. 5-28. There is bone and
fibrosis but no giant cells. (H&E, ×56.)

made in this case as to whether the lesions resembled
chickenpox.

PATHOLOGY

Macroscopically the lungs show pale areas (Fig. 5-31), a few millimeters in diameter.

Histologically there is often necrosis of the respiratory tract. This may extend from the trachea out to the alveoli (Fig. 5-32). There is a surrounding interstitial infiltrate of lymphocytes and plasma cells. The bronchial epithelium shows subtle changes with Cowdrey type A intranuclear inclusions, which have a ground glass appearance (Fig. 5-33). The inclusions may have an eosinophilic appearance. Viral particles can be detected by EM inside nuclei as round, discrete organisms with a central dot (Fig. 5-34). There may be syncytial cells and ballooning degeneration of epithelial cells.

Specific HSV-1 and HSV-2 antibodies are present and these can be used along with DNA probes in tissue sections. Both immunocytochemistry and in-situ hybridization will detect the virus but the former technique is better since the in-situ stains are difficult to prepare and in some cases difficult to interpret. Interestingly both methods correlated better with viral inclusions than culture methods.[96]

Recently *Herpes simplex* virus DNA has been demonstrated in lung tissue by the polymerase chain reaction in an immunocompetent patient dying from ARDS. No known infective agent had been found antemortem. Cultures were negative for the virus.[97]

As with adenovirus, EBV, and *Cytomegalovirus*, care must be taken that one is distinguishing between

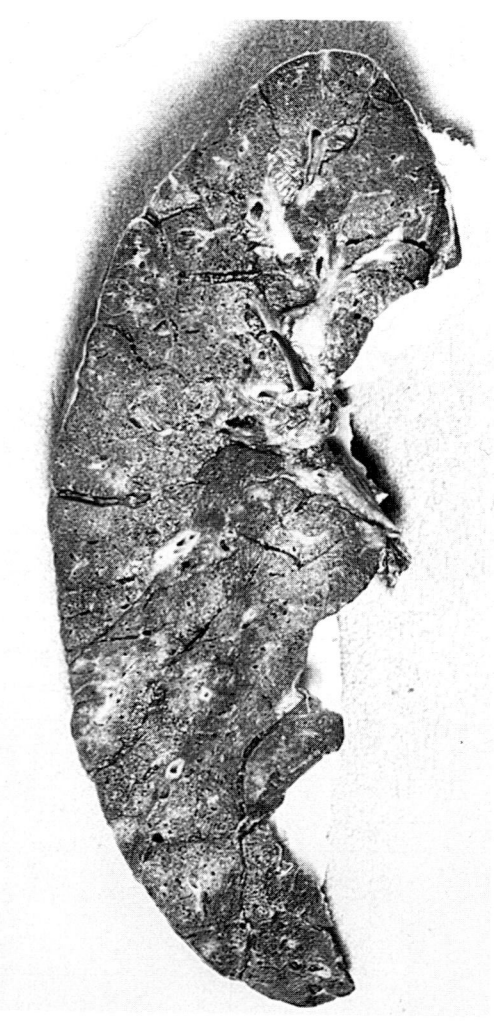

FIGURE 5-31
Herpes pneumonia in a neonatal lung showing pale consolidated areas. (×0.5.)

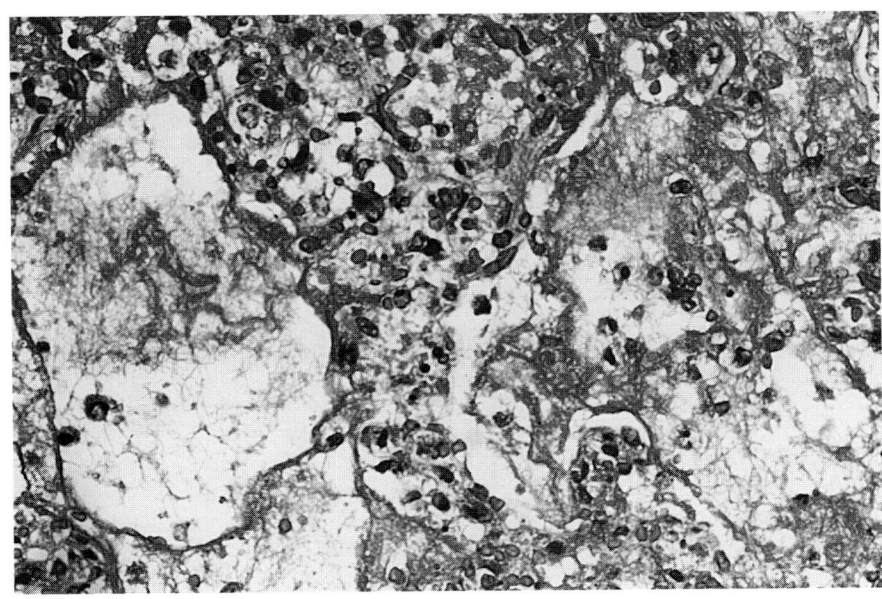

FIGURE 5-32
Herpes pneumonia with foci of necrosis.
(H&E, ×125.)

asymptomatic shedding of virus and disease caused by the virus. Neutralization, complement-fixation, and hemagglutination inhibition are not as accurate in detecting disease as some of the methods given below. Immunofluorescence, ELISA, or radioimmunoassay can also be used for diagnosis. Tracheal aspirates are superior to sputum samples, since the organism is common at this site.

Demonstration of Cowdrey bodies from BAL, the appropriate immunoperoxidase stain, or the polymerase chain reaction performed on a bronchial biopsy from an ulcerated area are the best methods of diagnosis.

Cytomegalovirus (CMV)

The incidence of CMV infection is low in Europe, Australia, and parts of North America but is higher in southeast Asia.[98]

There may be primary or secondary infections and the distinction between the two is still ill-defined. Primary infection occurs in the sero-negative patient whereas a secondary infection represents reactivation of a latent infection or reinfection from an outside source.

CMV has come into prominence due to solid organ transplantation, as well as being a passenger in patients with AIDS.

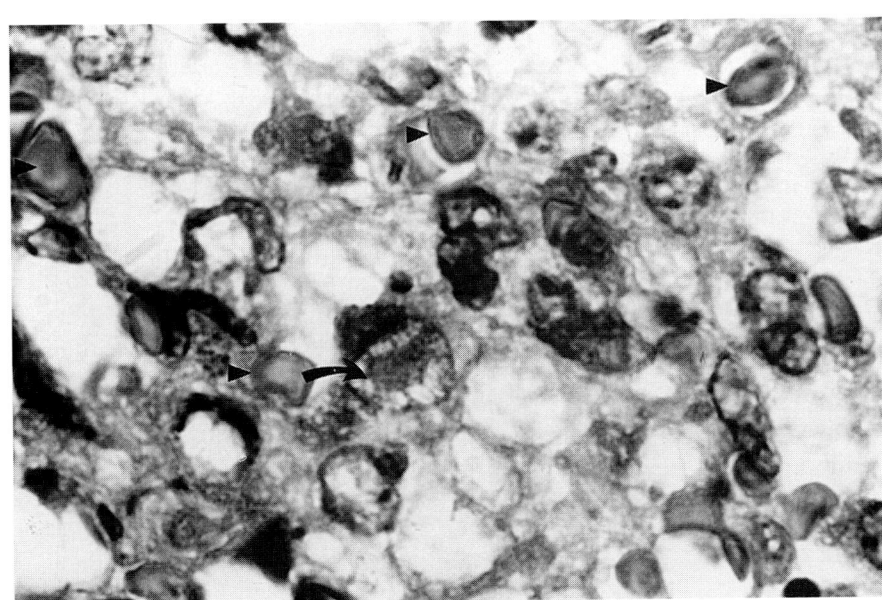

FIGURE 5-33
Ground-glass intranuclear inclusions (*arrowheads*) in Herpes pneumonia. There are also eosinophilic inclusions (*curved arrow*). (H&E, ×400.)

163

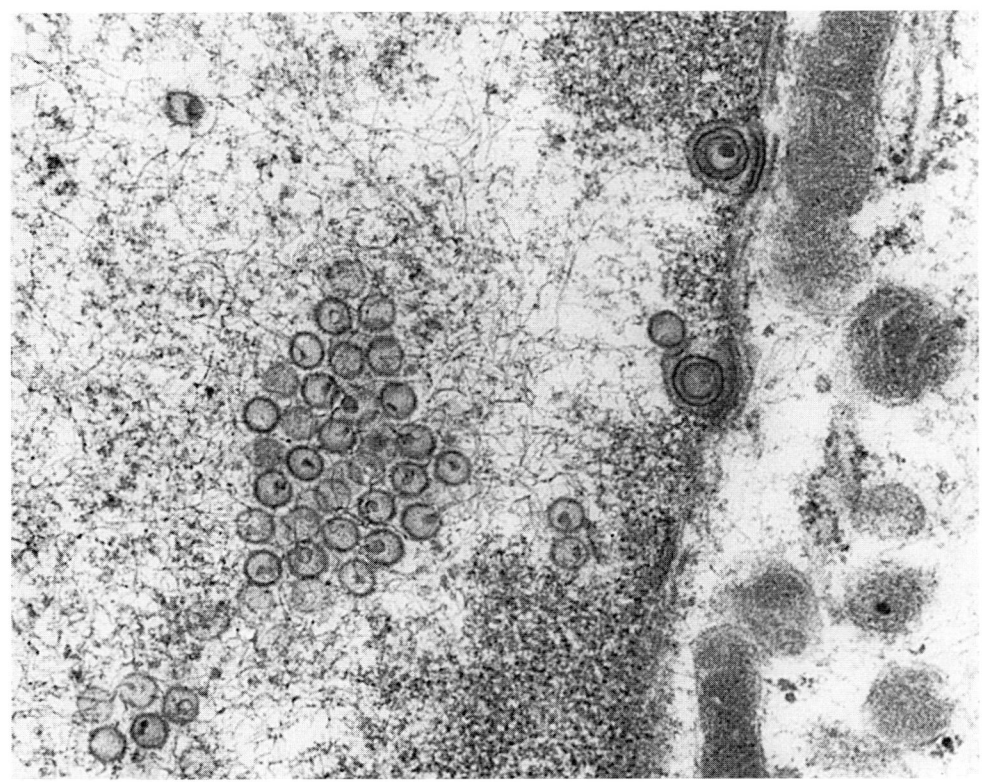

FIGURE 5-34
Herpes virus lying in a nucleus. As the viral particles are expelled from the nucleus they are invested with a capsular layer formed from the nuclear membrane. (EM, ×3392.) (Reproduced courtesy of Professor K. Porter.)

Congenital or intrauterine infection apparently occurs in the babies of mothers who are thought to have had a primary infection during pregnancy. This is one of the least frequent modes of transmission of the virus and the factors determining congenital infection are poorly understood.[98] Infection may be contracted at birth due to CMV cervical infection. Neonatal transmission may occur from human breast milk[99] or from other infected children in nurseries or day care centers.[100] The virus is transmitted sexually. This is especially striking in male homosexuals, where more than 90 percent are seropositive. They often asymptomatically secrete virus in the urine and semen.

Virus may be transmitted during solid organ transplantation or blood transfusion, or between drug abusers. The risk of acquiring CMV infection from a blood transfusion appears proportional to the number of units transfused and is estimated as 5 to 12 percent per unit.[101] The importance of CMV-negative blood and donor organs has been stressed by the Papworth Group (Cambridge, England). They recommended that CMV antibody-negative recipients should receive organs only from antibody-negative donors. Of eight patients who received organs from CMV antibody-positive donors, five developed pneumonitis, which was fatal in three. One patient, who survived, developed primary CMV infection of the gastrointestinal tract.[102]

Patients with malignancy, especially hematological, are prone to CMV infection but the incidence is not as high as in solid organ transplantation.

CMV ENCODED PROTEINS

Electron microscopically CMV belongs to the *Herpes* group of viruses. It is icosahedral in shape with 162 capsomeres and contains DNA.

Convalescent human serum contains antibodies that will react to at least 20 different CMV proteins.[103] These can be classed as capsid, tegument, nonstructural, and envelope. The envelope glycoproteins are important in producing immunogenicity. High levels of neutralizing antibodies are produced that combine with the surfaces of intact virions and CMV-infected cells.[104] The most immunogenic of the CMV envelope glycoproteins is GP 55-116, which has been called the integral membrane protein. CMV tegument proteins do not induce the production of monoclonal antibodies with neutralizing activity or have the capacity to bind to virus or virus-infected cells.

Key regulatory events occur during infection and these appear immediately following penetration and uncoating of the virus.[105] These are essentially transcriptional events that lead to the activation of the first set of genes expressed in viral infection, the alpha or intermediate early genes. These alpha gene products regulate subsequent phases of viral gene expression. These events are important, not just in viral growth but also in processes such as latency. It is thought that regulation of or by these gene products controls latency. While the immune system obviously plays an important part in controlling virus replication, the virus gene products also play an important, but not fully understood, role.

PEDIATRIC CMV INFECTION

Congenital infection, occurring in utero, must be distinguished from perinatal disease, seen after the birth of the child. In the former group, 90 percent show no clinical signs of infection. The remaining 10 percent have some of the signs or symptoms listed below. Congenital disease produces microcephalus, intellectual or developmental impairment, neuromuscular disorders, hearing loss, and a host of other abnormalities, including chorioretinitis.[106]

In congenital CMV infection, pneumonia is uncommon and has the same clinical signs as *Pneumocystis carinii* or *Chlamydia trachomatis* infection.[107] These are paroxysmal respiration, cough, tachypnea, intercostal retraction, and bilateral pulmonary infiltrates, along with air trapping.

CMV IN ADULTS

In adults there may also be a granulomatous hepatitis, encephalitis, myocarditis, thrombocytopenia, and a hemolytic anemia. Lung disease is more common in immunosuppressed patients. It may be complicated by gram-negative bacteria, *Candida sp.*, and *Pneumocystis carinii*.

There is fever, nonproductive cough, dyspnea, and hypoxia.[108] Cytomegalovirus can give a mononucleosis that may be difficult to distinguish from EBV-induced mononucleosis. In both diseases there is a relative or absolute lymphocytosis with abundant atypical lymphocytes. The diseases are similar in that they occur sporadically without any traceable source.

The pneumonitis is usually interstitial but there may rarely be nodules or cavities and atypical lymphocytes. CMV pneumonia may range from symptomatic viral shedding to a rapidly fatal pneumonia. Interstitial CMV pulmonary infection is rare in immunocompetent patients.

The changes affect both bronchiolar and alveolar epithelium: The cells enlarge and the first change is an intranuclear acidophilic dot surrounded by a clear zone, beyond which lies the remainder of the nuclear membrane. The intranuclear inclusion is seen at high magnification to consist of a morula of granules. As it

enlarges, condensations of chromatin appear in the nuclear membrane. When fully developed the inclusion and orbital bodies[109] cause the whole nucleus to resemble a bird's or owl's eye.

While the nuclear changes are progressing, cytoplasmic inclusions (Fig. 5-35) appear beneath the free margin of the cell. These are up to 4 μm in diameter and may form two or more rows, separated from the nucleus by a clear zone of cytoplasm. Eventually the cells detach into the alveolar lumen.

Several pathologic patterns may be seen in CMV histologically. In most cases no recognizable macroscopic lesions are found and there may be cytomegalic inclusion bodies in the alveolar epithelium with normal underlying alveolar walls. In overwhelming infections there are small miliary lesions up to 4 cm in diameter. There is central hemorrhage and the alveolar walls are necrotic, with a sharply demarcated periphery. CMV inclusions are seen within these lesions.[110] In a third group of cases there may be a diffuse interstitial alveolar infiltrate with interstitial fibrosis and an underlying lymphoplasmacytic infiltrate. There is a chronic inflammatory reaction in the interstitium and intraalveolar fibrin. In addition hyaline membranes and cytomegalic inclusions (Fig. 5-36) may be seen throughout the lung.

In immunosuppressed patients it is important to look for other pathogens, such as *Pneumocystis carinii*, since such organisms could be the cause of the pulmonary changes, especially in AIDS patients.

CMV IN SOLID ORGAN TRANSPLANTATION

Pulmonary dysfunction is common during CMV infection especially after renal or lung transplantation. This occurs even in asymptomatic patients and may be related to complement activation.[111] It is thought that CMV infection occurs in up to 50 percent of patients who receive major organ transplants.[98] In cardiac transplants CMV infection is a major cause of mortality, not only from severe CMV disease but also from its association with graft rejection and increased susceptibility to bacterial and fungal infections.[112] Cytomegalovirus is also shown to increase the risk of opportunistic infection and there is an increased rate and extent of atherosclerosis in a transplanted heart.[113] Among renal transplant recipients there are two large studies from individual centers that document decreased allograft survival among patients with primary and reactivated CMV infection as opposed to those without experience of the virus.[114,115] The effect of CMV in heart and heart/lung transplantation is described in Chap. 24.

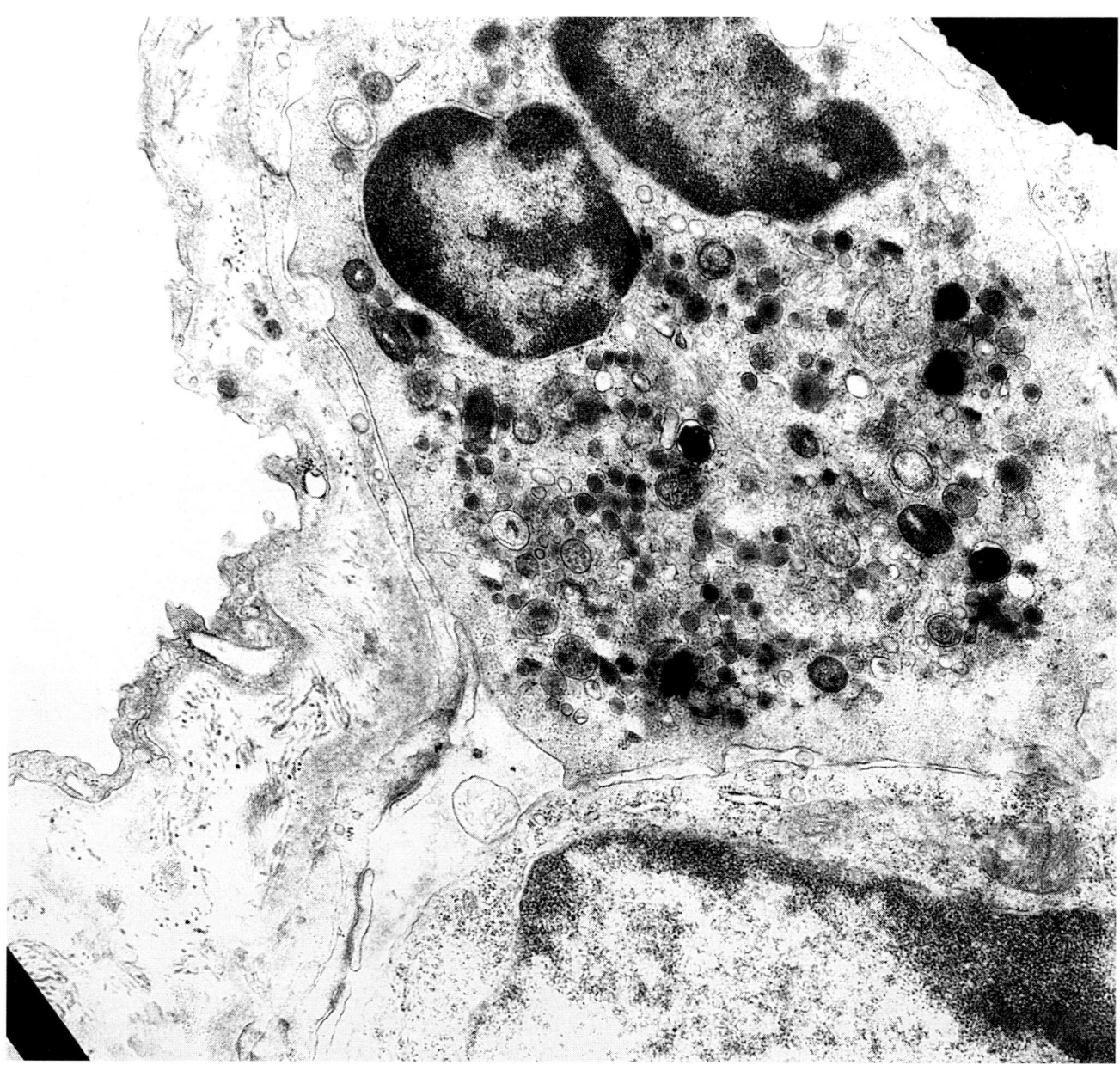

A

FIGURE 5-35
A. Cytomegaloviral particles filling the cytoplasm in a pneumonitis. (EM ×1125.)

CMV IN ATHEROMA

There is increasing evidence that CMV is involved in atherogenesis.[116] This is especially important in heart and heart-lung transplantation. Studies have shown accelerated coronary atherosclerosis in transplanted human hearts and in some heart-lung transplants. A recent study has described four patients with acute heart failure, 90 days after transplantation. In these cases there was no significant rejection or the concentric intimal thickening with dense collagen typical of chronic vascular rejection. The patients described had a prominent lymphocytic infiltrate and the loosely organized intimal thickening, opposed to smooth muscle cells, as well as extensive endothelial injury. CMV disease was present in six of seven patients with vascular rejection, of whom 43 percent were CMV-negative recipients of

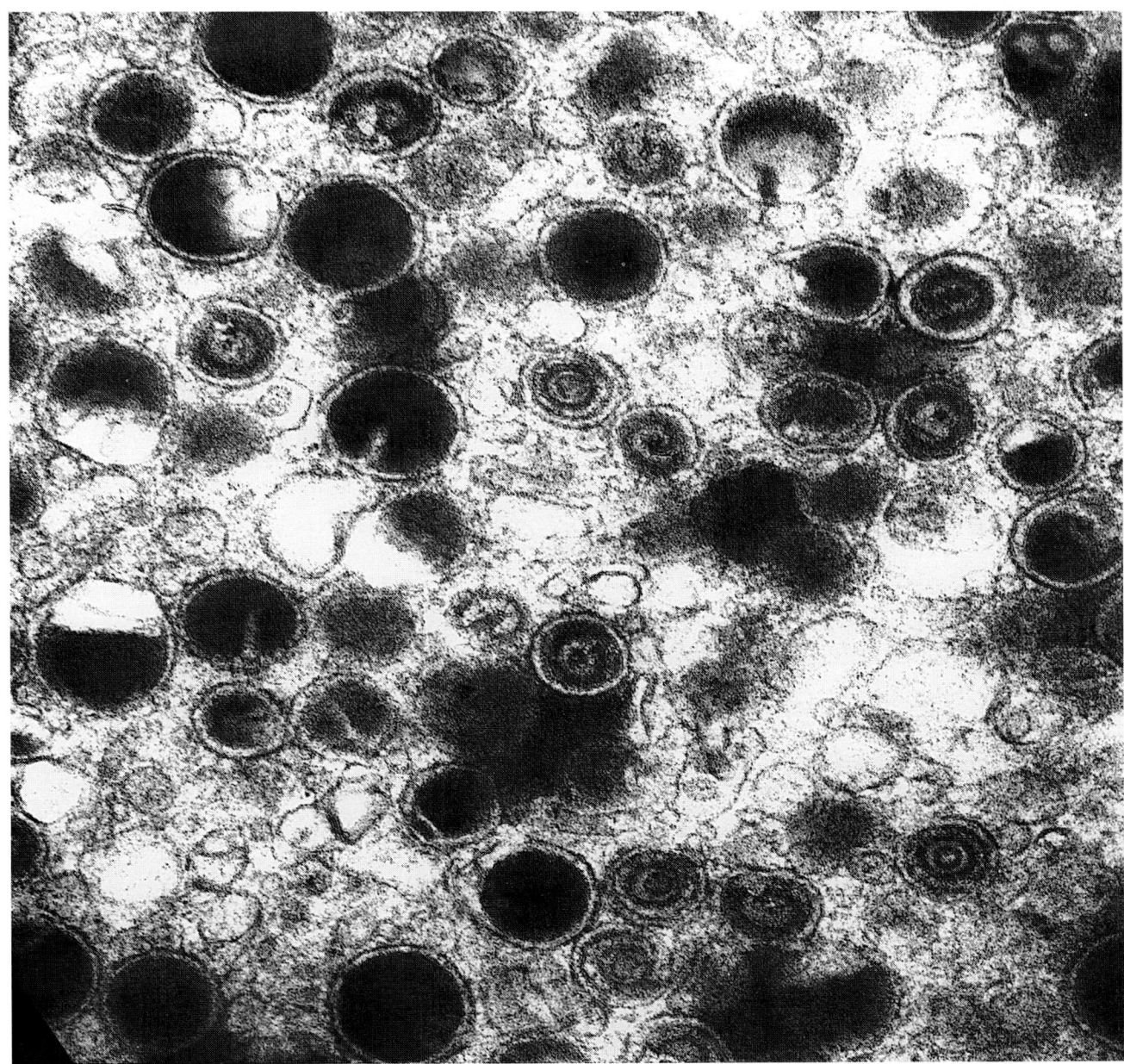

B

FIGURE 5-35

(Continued) B. Higher power of CMV viruses, some show an outer membrane that has been lost by fusion with the plasma membrane (exocytosis). Also in the cytoplasm and scattered among the virions are membrane-bound dense core granules, the origin of which is not known. (EM, ×72,000.)

hearts from CMV-positive donors.[117] The mechanism by which CMV or other Herpes viruses might contribute to atherosclerosis is still a matter for speculation.

DIAGNOSIS OF CMV PNEUMONITIS

The quickest way of diagnosing pulmonary CMV is by bronchoalveolar lavage and identifying the typical inclusion bodies. Cytologic BAL preparations are only positive in two thirds of patients with CMV pneumonia.[118] Spin amplification of lavage fluid followed by staining with a monoclonal antibody to the early nuclear antigen (EA-assay) was found to be positive in all patients with proven or probable CMV pneumonia. However, it was also seen in 92 percent of patients without such documented pneumonia.

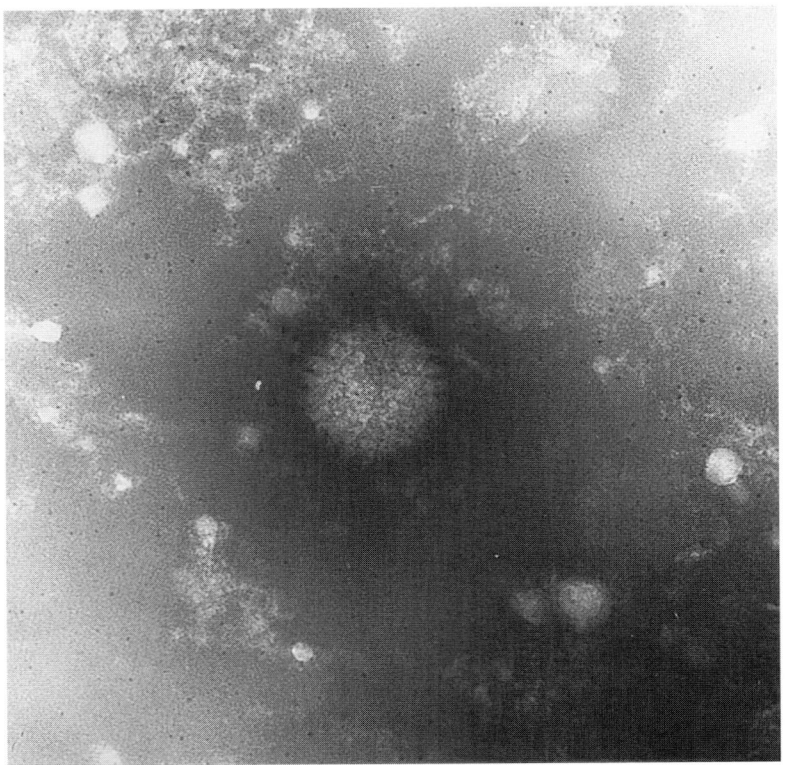

C

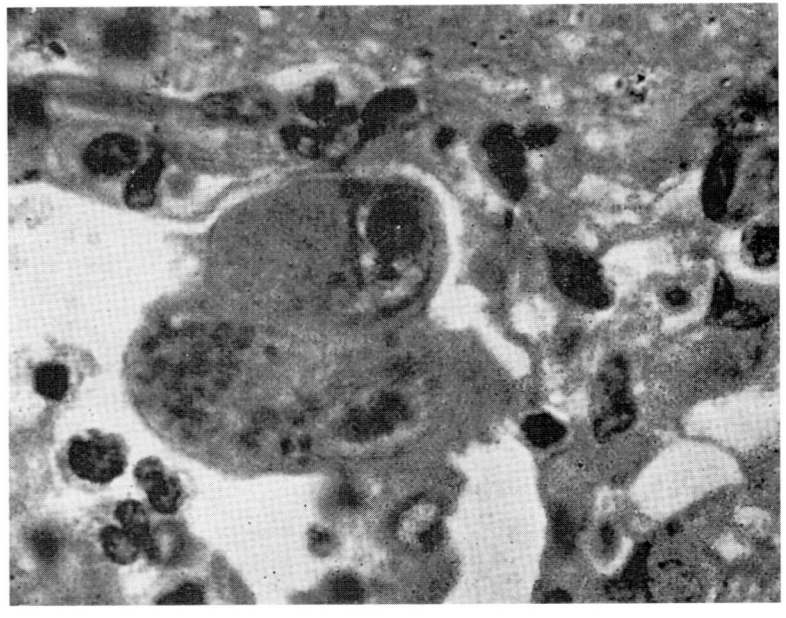

D

FIGURE 5-35
(Continued) C. Negatively stained CMV
virus as shown by electron microscopy.
The outer membrane envelope is missing.
D. Intracytoplasmic and intranuclear
CMV inclusions. (H&E, ×1000.)

The lymphocyte count in patients with CMV pneumonia was decreased (6.2 ± 3.6) as opposed to controls (11.8 ± 1.3). Similarly the total number of lymphocytes recovered from patients with CMV pneumonia (2.5 ± 1.6 × 10^6) was significantly less than the total number of lymphocytes recovered from those without CMV pneumonia (5.3 ± 2.1 × 10^6). These authors concluded that the EA-assay along with an alveolar leukopenia should be thought of as highly suggestive of disease.

A problem sometimes confronting the pathologist is the significance of a single inclusion body in a

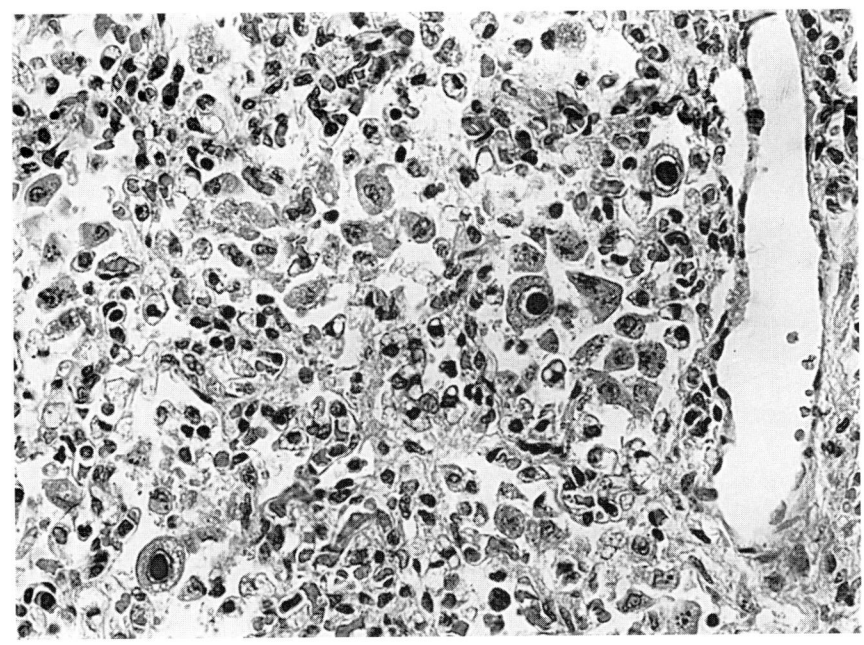

FIGURE 5-36
Pneumonia due to *Cytomegalovirus* with many intranuclear inclusions, as well as a mild lymphoplasmacytic reaction. (H&E, ×400.)

lavage. It is important to ascertain if the patient has pneumonia, is HIV positive, or, in the case of a transplant patient, has clinical or radiological evidence of pneumonia. It is only by correlating the lavage findings, the transbronchial biopsy, and the clinical features, that a diagnosis of CMV pneumonitis can be made.

Some authors[119] found Wright-Giemsa touch imprints, as compared with H & E, a superior way of detecting the virus. Only a small number of specimens were examined.

In-situ hybridization has been used for detection of CMV. This technique is still in its early stages and not every paper is using the same technique or identical probes. One study[120] used an IgG1 monoclonal antibody, obtained by immunization with a lysate of cells infected with CMV and screened for selective reactivity of infected fibroblasts. In-situ hybridization was less sensitive than immunocytochemistry using the same antibody. Using the probes, both nuclear and cytoplasmic staining were recognized. There was a similar pattern of staining using immunocytochemistry.

Wolber and Lloyd,[121] using sequential in-situ hybridization and immunocytochemistry on standard histologic sections with a biotonylated DNA probe, showed that CMV was localized predominantly within cell nuclei. However, with immunostaining, CMV antigen was predominantly cytoplasmic. The hybridization technique provided more intense staining, detected greater numbers of inclusions, and had less background staining than the immunoperoxidase technique. 17 tissues from 12 patients were studied.

Another study with a small number of cases was that of Jiwa et al.[122] who used in-situ hybridization and the polymerase chain reaction for CMV-DNA detection and immunocytochemistry for the detection of CMV immediate early antigens. Five patients had CMV-related interstitial pneumonia. All three techniques yielded positive results, whereas only 1 of 10 patients with idiopathic interstitial pneumonia was positive. No patients with treated CMV interstitial pneumonia showed any evidence of positivity.

It is clear that while in-situ hybridization and the polymerase chain reaction are useful tools in the diagnosis of CMV infection, larger number of cases need to be studied, all using the same probes and with standardized methods to achieve the most accurate method of diagnosis of this infection.

Epstein-Barr Virus (EBV)

The exact role of this virus in causing pulmonary disease is not known but several papers have described the virus in lung tissue using sensitive techniques. Jiwa et al.[122] found the virus in two patients with interstitial pneumonia and Wallace[124] demonstrated that EBV or HIV or both have a role in the pneumonitis in this disease. EBV has recently been described in extrapulmonary inflammatory pseudotumor[124] and fibrosing alveolitis.[124a,124b] Further work is needed on the role of EBV in lung disease.

Congenital Rubella Pneumonia

Rubella viruses have been placed in a genus *Rubivirus*.

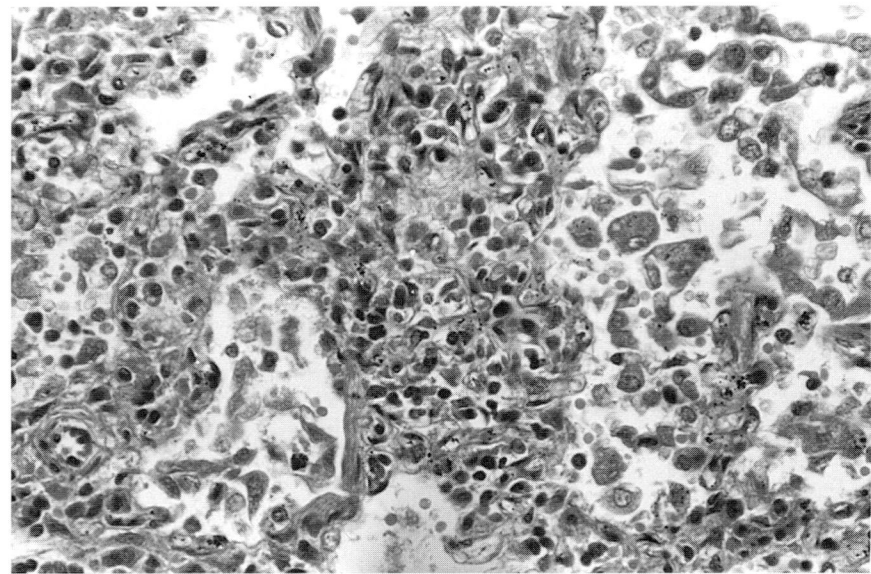

A rare form of viral pneumonia may affect the fetus after rubella infection in pregnancy. Phelan and Campbell[125] described seven cases, of which six had died. The infants usually survived for up to six months after birth but suffered from increasing respiratory disability. The pneumonitis has recently been thought of as late-onset disease but may respond to corticosteroids.[126]

The rubella syndrome includes persistent diarrhea and a chronic rubelliform rash, thought to represent an immunopathological phenomenon. There are circulating immune complexes which contain rubella antigen.[127,128]

Histologically there is a severe, interstitial, alveolar, lymphocytic, plasma cell and macrophage infiltration (Fig. 5-37). Hyaline membranes may be seen (Fig. 5-38). This is often complicated by other opportunistic infections, such as *Pneumocystis carinii*. In time interstitial pulmonary fibrosis develops and the lung is indistinguishable from idiopathic or cryptogenic interstitial pulmonary fibrosis. At that time only culture of the virus could characterize the disease but now lavage can be used.

Clinically most laboratories would only use radioimmunoassay or hemabsorption inhibition tests, which employ red blood corpuscles for the detection of bound rubella antigen. Another method available is an enzyme immunosorbant assay, which has a strong positivity and can give results in a day. Passive hemagluttination has the same degree of

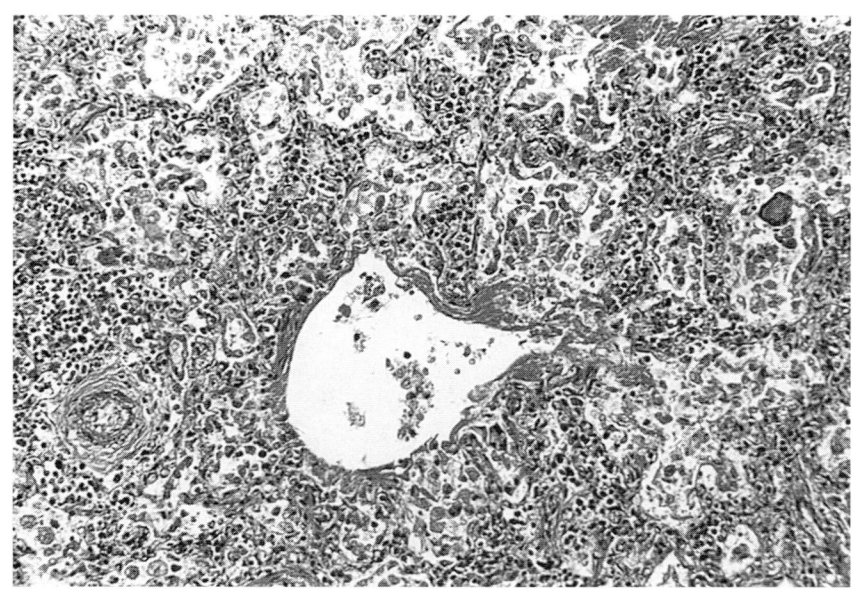

FIGURE 5-38
Rubella. A lower magnification of the case seen in Fig. 5-37. A small, central bronchiole shows a hyaline membrane and there is diffuse interstitial inflammation. (H & E, ×125.)

sensitivity and will give a result in up to 3 h. Immunofluorescence is less sensitive, but will also give results in 2 to 3 h.

Coxsackie Virus Pneumonia

Coxsackie viruses belong to the genera *Picoranoviridae*. The commonest types causing disease are A4, A6, and A10. They can cause pleurodynia (Bornholme disease), where there is severe chest pain which may be on either side or substernal and is sometimes intensified by movement. It is a self-limited disease and may be confused with dissecting aortic aneurysm or more commonly myocardial infarction.

Coxsackie virus pneumonia has been described by Lerner et al.[129] It is usually caused by group A or B virus, occurs as a sequel to a coxsackie viremia and is found with an aseptic meningitis, acute myocarditis, and petechial skin rashes.

Macroscopically the lungs show numerous petechial hemorrhages on the pleural surface and widespread foci of collapse. Histologically there is extensive interstitial lymphocytic, histiocytic, and some polymorph infiltration throughout the lungs but there are no distinctive features of this pneumonia. Confirmation of the diagnosis requires isolation of the virus from the blood and body tissues and demonstration of antibodies in the serum.

Hantavirus

The genus Hantavirus belongs to the *Bunyaviradae* family, which includes the causative agents of a group of febrile nephropathies known collectively as hemorrhagic fever with a renal syndrome.[130] This syndrome occurs throughout Europe and East Asia. Respiratory symptoms are generally not pronounced and pulmonary involvement has not been a prominent feature of the known hantaviral syndromes.[131]

CLINICAL FEATURES

In May 1993, an outbreak of severe respiratory illness occurred in the southwestern United States.[132] The patients were mostly young, with an age range of 30 to 64 years. Seventy-two percent were native American and 22 percent white.

The outbreak was first identified on May 19, 1993 and by August 9 Nichol and his colleagues had submitted a report to Science[133] containing the key elements of the outbreak. The agent was identified as Hantavirus using molecular biology. Samples were sent to the Centers for Disease Control and Prevention (CDC) in Atlanta. Together with the U.S. Army Medical Research Institute for Infectious Diseases they had developed accurate and sensitive enzyme-linked immunosorbent assays for serological diagnosis of the viral hemorrhagic fever. The Hantavirus group gave a positive signal. Nichol knew the sequences to draw out of the gene bank to design the primers for the polymerase chain reaction. Zoologists were quickly recruited because all previously known Hantaviruses had been isolated from rodents. Amplified DNA products were obtained by June 8 and sequenced the next day.[134]

Rodents are the primary resovoir for all Hantaviruses, shedding virus from saliva, urine, and feces.[135] Humans acquire infection by inhalation of rodent excreta. In a case of a hantavirus pulmonary syndrome, the deermouse (*Peromyscus maniculatus*) have been closely linked to infection. Since the initial reports from the southwestern United States, cases have been described in other parts of United States. A recent summary[136] dated December 3, 1993 showed infection had been confirmed in 48 patients. Half of these were native American Indians.

The common prodromal symptoms were fever and myalgia, cough or dyspnea, and headache. Gastrointestinal symptoms were common and included nausea, vomiting, and diarrhea. In addition some cases had abdominal as well as chest pain.

The most common physical findings were tachypnea, tachycardia, and hypotension. Rapidly progressive, acute pulmonary edema developed in 88 percent of patients, and 13 patients, all of whom had profound hypotension, died giving a case fatality rate of 66 percent.

There was a leukocytosis with a mean peak white cell count of 20,000/mm^3. Myeloid precursers were often present. There was an increased haematocrit, thrombocytopenia (median lowest platelet count 64,000/mm^3), and prolonged prothrombin and partial thromboplastin times. In addition there were elevated serum lactate dehydrogenase, decreased serum protein and proteinuria.

The chest radiograph showed diffuse interstitial and alveolar infiltrates.

PATHOLOGY

Pathology consistently showed large serous pleural effusions with severe pulmonary edema.

Histology of the lung showed intraalveolar edema with scant to moderate numbers of hyaline membranes and scant to moderate numbers of interstitial mononuclear cells resembling immunoblasts (Fig. 5-39). These cells were T cells, as well as cells of

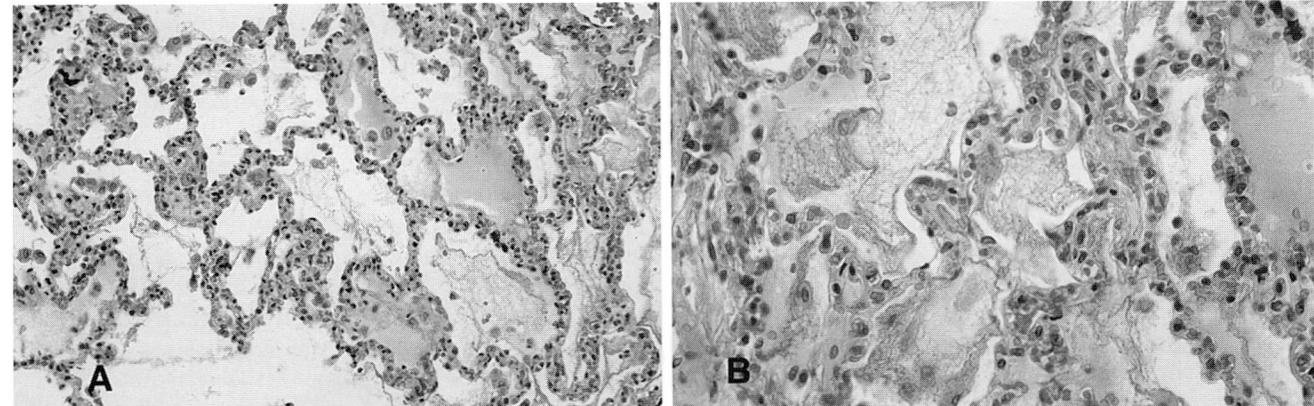

FIGURE 5-39
A. Typical case of Hantavirus pulmonary syndrome with interstitial pneumonitis and intraalveolar edema. (H&E, ×50.) *B.* Higher magnification showing intraalveolar fibrin and a mononuclear cell infiltrate. (H&E, ×100.) (Reproduced with permission of Dr. S. Zaki, Centers for Disease Control and Prevention, Atlanta, Georgia and *American Journal of Pathology*.)

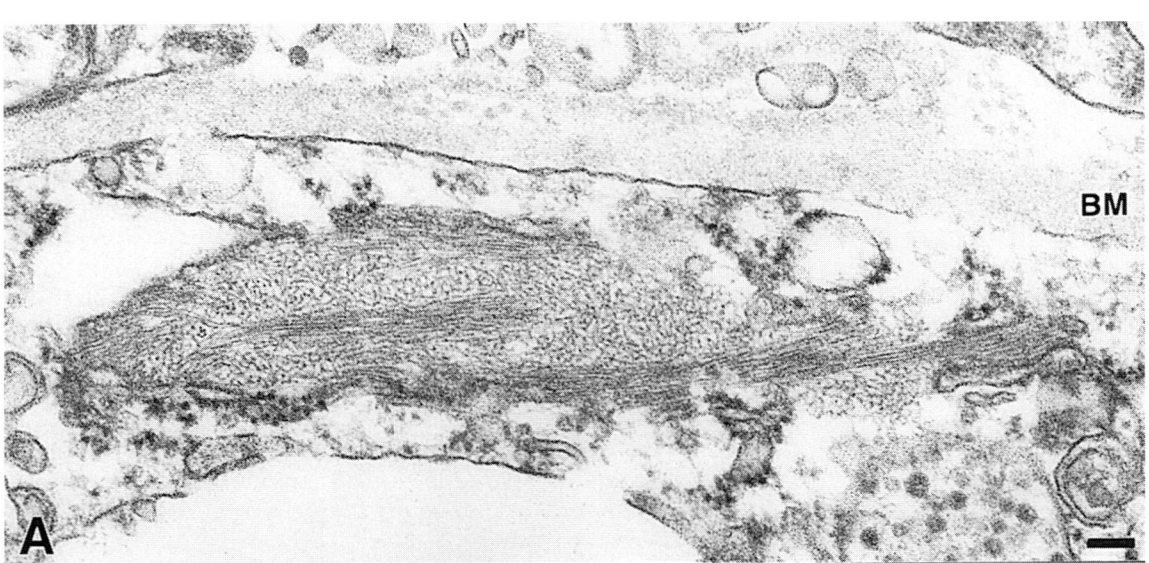

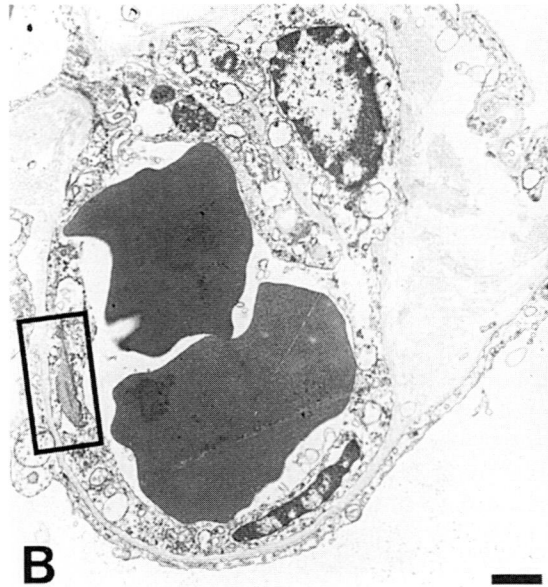

FIGURE 5-40
A. Typical hantaviral inclusions as seen within the pulmonary vasculature. *A.* High power magnification of boxed area in *B* showing a granulofilamentous inclusion within capillary basement membrane. (BM, basement membrane; Scale bars: *A*, 100 μm; *B*, 1 μm.) (Reproduced with permission of Dr. S. Zaki, Centers for Disease Control and Prevention, Atlanta, Georgia and *American Journal of Pathology*.)

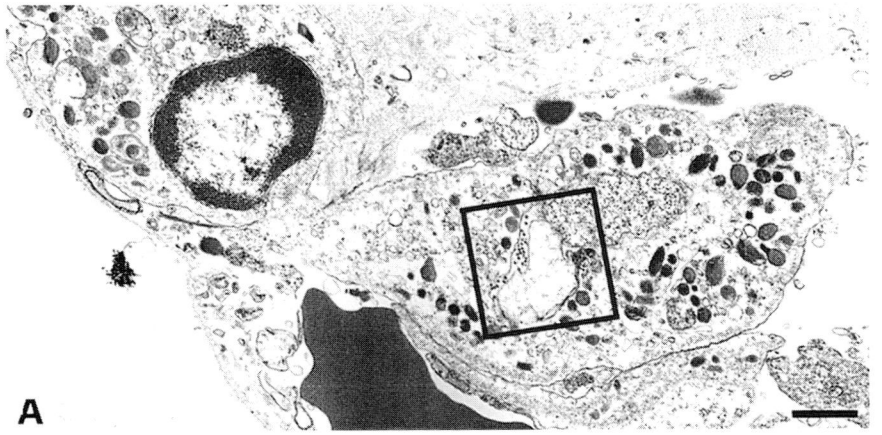

FIGURE 5-41
A collection of virus-like particles in a pulmonary macrophage. *A.* Interstitial macrophage showing virus-like particles in association with a phagolysosome containing fragments of cellular debris. *B.* Higher magnification of boxed area showing an accumulation of several virus-like particles including one budding particle (*arrow*). (Scale bars: *A*, 1 μm; *B*, 100 μm.) (Reproduced with permission of Dr. S. Zaki, Centers for Disease Control and Prevention, Atlanta, Georgia and *American Journal of Pathology*.)

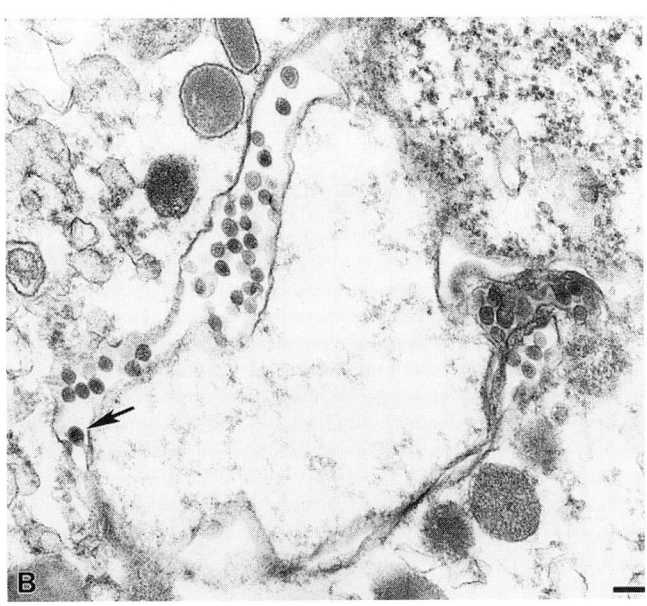

firmed using immunogold electron microsopy.[138] Collections of whole viral particles were identified in interstitial pulmonary macrophages (Fig. 5-41).

In areas the continuity of the endothelial cells, as well as the epithelial cells along the basement membrane, was interrupted. This was thought to be due to postmortem change.

The pulmonary changes initially resemble ARDS. However, the minimal number of neutrophils in alveolar or interstitial spaces and the presence of an intact alveolar wall, as well as the relatively limited hyaline membrane formation, suggest a different pathogenesis for the pulmonary edema. The mechanism may be due to florid capillary leakage, as a result of functional derangement of the vascular endothelium.[138]

monocyte/macrophage lineage. No Type II pneumocyte proliferation was identified in the majority of cases.[137,138] Neutrophils were noticeably scarce. There was no evidence of a viral cytopathological effect or any viral inclusions.

A recent elegant study on the pathology of this disease has been published.[138] These authors showed by immunocytochemistry that there was widespread presence of hantaviral antigens in endothelial cells of the microvasculature. They were also demonstrated in follicular dendritic cells, macrophages, and lymphocytes.

Electron microscopy confirmed the presence of hyaline membranes and there were spherical hantavirus-like particles, 90 to 110 nm in diameter in the endothelial cells (Fig. 5-40). The inclusions were seen in the perinuclear region of the endothelial cells. The viral nature of the inclusions was con-

REFERENCES

1. Skerrett SJ: Host defenses against respiratory infection. *Med Clin North Am* 78:941–966, 1994.
2. Persson CGA, Svennson C, Greiff L, et al: The use of the nose to study the inflammatory response of the respiratory tract. *Thorax* 47:993–1000, 1992.
3. Proctor DF: The upper airways (1) Nasal physiology and defense of the lungs. *Am Rev Respir Dis* 115:97–129, 1977.
4. Godfrey R: Editorial: The nose and the lower airways. *Lancet* 343:991–992, 1994.
5. Proctor DF, Anderson I (eds): *The Nose: Upper Airway Physiology and the Atmospheric Environment.* Amsterdam, Elsevier, 1982.
6. Andersen I, Lundqvist GR, Jensen PL, Proctor DF: Human response to controlled levels of sulphur dioxide. *Arch Environ Health* 28:31–39, 1974.
7. Editorial. The nose and the respiratory system. *Lancet* 339:1511–1512, 1992.
8. Pavia D, Agnew JE, Lopez-Vidriero MT, Clarke SW: General review of tracheo-bronchial clearance. *Eur J Respir Dis* 71(suppl 153):123–129, 1987.
9. Kaliner MA: Human nasal respiratory secretions and host defense. *Am Rev Respir Dis* 44:S52–S56, 1991.

10. Giclas PC, King TE, Baker SL, et al: Complement activity in normal rabbit bronchoalveolar fluid. Description of an inhibitor of C3 activation. *Am Rev Respir Dis* 135:403–411, 1987.

11. Kalteider HB: Expression of immune mechanisms in the lung. *Am Rev Respir Dis* 113:347–379, 1976.

12. Van Deuren M, Dofferhoff ASM, Van Der Meer JW: Cytokines and the response to infection. *J Pathol* 168:349–356, 1992.

13. Richardson PS, Peatfield AC: The control of airway mucus secretion. *Eur J Respir Dis* 71(Suppl 153):43–51, 1987.

14. Incalzi RA, Maini CL, Fuso L, Giordano A, Carbonin PU, Galli G: Effects of aging on mucociliary clearance. *Compr Gerontol A* 3(Suppl):65–68, 1989.

15. Dhar S, Shasti SR, Lenora RAK: Aging and the respiratory system. *Med Clin N Am* 60:1121–1139, 1976.

16. Gyetko MR, Toews GB: Immunology of the aging lung. *Clin Chest Med* 14:379–390, 1993.

17. Ganz T, Selsted ME, Lehrer RI: Antimicrobial activity of phagocyte granule proteins. *Semin Respir Infect* 1:107–117, 1986.

18. Sondell K, Athlin L, Bjermer L, et al: The role of sex and age in yeast cell phagocytosis by monocytes from healthy blood donors. *Mech Ageing Dev* 51:55–61, 1990.

19. Wallace WAH, Gillooly M, Lamb D: Age related increase in the intra-alveolar macrophage population of non-smokers. *Thorax* 48:668–669, 1993.

20. Pavia D: Acute respiratory infections and mucociliary clearance. *Eur J Respir Dis* 71:219–226, 1987.

21. Lindsey JO, Pierce AK: An examination of the microbiologic flora of the normal lung of the dog. *Am Rev Respir Dis* 117:501–505, 1978.

22. Scherer PW: Mucus transport by cough. *Chest* 80(suppl):830–833, 1981.

23. A'Court C, Garrard S: 1-Nosocomial pneumonia in the intensive care unit: Mechanisms and significance. *Thorax* 47:465–473, 1992.

24. Johanson WG, Pierce AK, Sanford JP: Changing pharyngeal bacterial flora of hospitalized patients. *N Engl J Med* 281:1137–1140, 1969.

25. Fiddian-Green RG, Baker S: Nosocomial pneumonia in the critically ill: Product of aspiration or translocation? *Crit Care Med* 19:763–769, 1991.

26. Reynolds HY: Identification and role of immunoglobulins in respiratory secretions. *Eur J Respir Dis* 71(suppl 153):103–116, 1987.

27. Coles SJ, Bhaskar KR, O'Sullivan DD, et al: Airway mucus: Composition and regulation of its secretion by neuropeptides in vitro. *Ciba Found Symp* 109:40, 1984.

28. Roussel P, Lamblin G, Lhermitt M, et al: The complexity of mucins. *Biochimie* 70:1471–1482, 1988.

29. Jeffery PK: The origins of secretions in the lower respiratory tract. *Eur J Respir Dis* 71(Suppl 153):34–42, 1987.

30. Zimmerman GA, Prescott SM, McIntyre TM: Endothelial cell interactions with granulocytes: Tethering and signaling molecules. *Immunol Today* 13:93–100, 1992.

31. Ganz T, Selsted ME, Lehrer RI: Defensins. *Eur J Haematol* 44:1–8, 1990.

32. Goren MB, D'Arcy Hart P, Young MR, et al: Prevention of phagosome-lysosome fusion in cultured macrophages by sulfatides of *Mycobacterium tuberculosis*. *Proc Natl Acad Sci USA* 73:2510–2514, 1976.

33. Coonrod JD: Role of surfactant free fatty acids in antimicrobial defenses. *Eur J Respir Dis* 71(suppl 153):209–214, 1987.

34. Toews GB, Hansen EJ, Strieter RM: Pulmonary host defenses and oropharyngeal pathogens. *Am J Med* 88(Suppl 5A):20S–24S, 1990.

35. Kornblihtt AR, Gutman A: Molecular biology of the extracellular matrix proteins. *Biol Rev Cam Philos Soc* 63:465–507, 1988.

36. Pilewski JM, Albelda SM: Adhesion molecules in the lung: An overview. *Am Rev Respir Dis* 148(suppl):S31–S37, 1993.

37. Albelda SM: Endothelial and epithelial cell adhesion molecules. *Am J Respir Cell Biol* 4:195–203, 1991.

38. Dunnill M: Some aspects of pulmonary defence. *J Pathol* 128:221–236, 1979.

39. Green GM, Kass EH: The role of the alveolar macrophage in the clearance of bacteria from the lung. *J Exp Med* 119:167–176, 1964.

40. Centers for Disease Control and Prevention: Outbreak of acute illness—southwestern United States, 1993. *MMWR* 42:421–424, 1993.

41. Oxford JS, Schild GC: The *orthomyxoviridae* and influenza. In Parker TM, Collier LH (eds): *Topley and Wilson's Principles of Bacteriology, Virology and Immunity*, 8th ed, Vol 4. London, Edward Arnold, 1990, pp 291–322.

42. Marsh M: The entry of enveloped viruses into cells by endocytosis. *Biochem J* 218:1–10, 1984.

43. Doms RW, Helenius A, White J: Membrane fusion activity of the influenza virus hemagglutinin: The low pH-induced conformational change. *J Biol Chem* 260:2973–2981, 1985.

44. Giebink GS, Berzins IK, Marker SC, Schiffman G: Experimental otitis media after nasal inoculation of *Streptococcus pneumoniae* and *Influenza A* virus in chinchillas. *Infect Immunol* 30:445–450, 1980.

45. Kendal AP: Epidemiologic implications of changes in the influenza virus genome. *Am J Med* 82(6A):4–14, 1987.

46. Cate TR: Viral pneumonia in immunocompetent adults. In Niederman MS, Sarosi GA, Glassroth J (eds): *Respiratory Infections: A Scientific Basis for Management*, Chap 10. Philadelphia, W.B. Saunders, 1994, pp 471–479.

47. Hers JFAPh, Masurel N, Mulder J: Bacteriology and histopathology of the respiratory tract and lungs in fatal Asian influenza. *Lancet* 2:1141–1143, 1958.

48. Yeldandi AV, Colby TV: Pathologic features of lung biopsy specimens from influenza pneumonia cases. *Hum Pathol* 25:47–53, 1994.

49. Stinson SF, Ryan DP, Hertweck S, Hardy JD, Sui-yee-hwang-kow MS, Loosli CG: Epithelial and surfactant changes in influenzal pulmonary lesions. *Arch Pathol Lab Med* 100:147–153, 1976.

50. Walsh JJ, Dietlin LF, Low FN, Burch GE, Mogabgab WJ: Tracheobronchial response in human influenza. *Arch Intern Med* 108:376–388, 1961.

50a. Editorial: Bronchiolitis. *Br Med J* 310:4–5, 1995.

51. Lourdes R, Laraya-Cuasay RL, Deforest A, Palmer J, Huff DS, Lischner HW, Huang NN: Chronic pulmonary complications of early influenza virus infection. *Am Rev Respir Dis* 109:703, 1974.

52. Jakab GJ, Astry CL, Warr GA: Alveolitis induced by influenza virus. *Am Rev Respir Dis* 128:730–739, 1983.

53. Jakab GJ: Sequential virus infections, bacterial superinfections and fibrogenesis. *Am Rev Respir Dis* 142:374–379, 1990.

54. Prechter GC, Gerhard AK: Post-influenza toxic shock syndrome. *Chest* 95: 1153–1154, 1989.

55. Pringle CR, Heath RB: *Paramyxoviridae*. In Parker MT, Collier LH (eds): *Topley and Wilson's Principles of*

Bacteriology, Virology and Immunity, 8th ed, vol 4 London, Edward Arnold, 1990, pp. 273–289.

56. Chanock RM, Vargoso A, Luckey A, et al: Association of hemadsorption viruses with respiratory illness in childhood. *JAMA* 169:548–553, 1959.

57. Wendt CH, Weisdorf DJ, Jordan MC, Balfour HH Jr, Hertz MI: Parainfluenza virus respiratory infection after bone marrow transplantation. *N Engl J Med* 326:921–926, 1992.

58. Gardner PS, McQuillin J, McGuckin R, Ditchburn RK: Observations on clinical and immunofluorescent diagnosis of parainfluenza virus infections. *BMJ* 2:7–12, 1971.

59. Hall CB, McBride JT: Respiratory syncytial virus—from chimps with colds to conundrums and cures. *N Engl J Med* 325:57–58, 1991.

60. Editorial: Nosocomial infection with respiratory syncytial virus. *Lancet* 340:1071–1073, 1992.

61. Glezen WWP, Taber LH, Frank AL, Kasel JA: Risk of primary infection and reinfection with respiratory syncytial virus. *Am J Dis Child* 140:543–546, 1986.

62. Hall CB, Powell KR, MacDonald NE, et al: Respiratory syncytial viral infection in children with compromised immune function. *N Engl J Med* 315:77–81, 1986.

62a. Editorial: RSV and chronic asthma. *Lancet* 34:789–790, 1995.

63. Baker JC: Human and bovine respiratory syncytial virus: Immunopathologic mechanisms. *Vet Quart* 13:47–59, 1991.

64. La Via WV, Marks MI, Stutman HR: Respiratory syncytial virus puzzle: Clinical features, pathophysiology, treatment and prevention. *J Pediatr* 121:503–510, 1992.

65. Hall CB, Walsh EE, Long CE, Schnabel KC: Immunity to and frequency of reinfection with respiratory syncytial virus. *J Infect Dis* 163:693–698, 1991.

66. Delage G, Brochu P, Robillard L, Jasman G, Joncas JH, Lapointe M: Giant cell pneumonia due to respiratory syncytial virus. Occurrence in severe combined immunodeficiency syndrome. *Arch Pathol Lab Med* 108:623–625, 1984.

67. Gardner PS, McQuillin J: Application of immunofluorescent antibody technique in rapid diagnosis of respiratory syncytial virus infection. *BMJ* 3:340–343, 1968.

68. Gardner PS, McQuillin J, McGuckin R: The late detection of respiratory syncytial virus in cells of respiratory tract by immunofluorescence. *J Hyg* 68:575–580, 1970.

69. Panuska JR, Hertz MI, Taraf H, Villani A, Cirino NM: Respiratory syncytial virus infection of alveolar macrophages in adult transplant patients. *Am Rev Respir Dis* 145:934–939, 1992.

70. Morley D: Severe measles in the tropics. *BMJ* 1:297–300, 1969.

71. Denton J: The pathology of fatal measles. *Am J Med Sci* 169:531–543, 1925.

72. Warthin AS: Occurrence of numerous large giant cells in the tonsils and pharyngeal mucosa in the prodromal stage of measles. *Arch Pathol* 11:864–874, 1931.

73. Finkeldy W: Uber riesenzellbefunde in den gaumenmandeln zugleich ein beitrag zur histopathologie der mandelveranderungen im maserninkubationsstadium. *Virchows Arch Path* 281:323–329, 1931.

74. Rowe WP, Huebner RJ, Gilmore LK, Parrott RH, Ward TG: Isolation of a cytopathogenic agent from human adeoids undergoing spontaneous degeneration in tissue culture. *Proc Soc Exp Biol Med* 84:570–573, 1953.

75. Norrby E, Bartha A, Boulanger P, et al: Adenoviridae. *Intervirology* 7:117–125, 1976.

76. Stott EJ, Garwes DJ: *Ranoviruses, Adenoviruses* and *Corona viruses*. In Parker TM, Collier LH (eds): *Topley and Wilson's Principles of Bacteriology, Virology and Immunity*, 8th ed, vol 4. London, Edward Arnold, 1990, p 253.

77. Hogg JC: Persistent and latent viral infections in the pathology of asthma. *Am Rev Respir Dis* 145:S7–S9, 1992.

78. Abzug MJ, Levin MJ: Neonatal adenovirus infection: Four patients and review of the literature. *Paediatrics* 87:890–896, 1991.

79. Pinto A, Beck R, Jadavji T: Fatal neonatal pneumonia caused by adenovirus type 35. Report of one case and review of the literature. *Arch Pathol Lab Med* 116:95–99, 1992.

80. Michaels MG, Green M, Wald ER, Starzl TE: Adenovirus infection in paediatric liver transplant receipients. *J Infect Dis* 165:170–174, 1992.

81. Castleman WL: Bronchiolitis obliterans and pneumonia induced in young dogs by experimental adenovirus infection. *Am J Pathol* 119:495–504, 1985.

82. Becroft DNO: Histopathology of fatal adenovirus infection of the respiratory tract in young children. *J Clin Pathol* 20:561–569, 1967.

83. Strano J: Light microscopy of selected viral diseases (morphology of viral inclusion bodies). *Pathol Annu* 11:53–75, 1976.

84. Aherne W, Bird T, Court SDM, et al: Pathological changes in virus infections of the lower respiratory tract in children. *J Clin Pathol* 23:7–18, 1970.

85. Abbondanzo SL, English CK, Kagan E, McPherson RA: Fatal adenovirus pneumonia in a newborn identified by electron microscopy and in situ hybridization. *Arch Pathol Lab Med* 113:1349–1353, 1989.

85a. Bateman ED, Hayshi S, Kuwano K, Wilke TA, Hogg JC: Latent adenoviral infection in follicullar bronchiectasis. *Am J Respir Crit Care Med* 151:170–176, 1995.

86. Ruben FL, Nguyen MLT: Viral pneumonitis. *Clin Chest Med* 12:223–235, 1991.

87. Raider L: Calcification in chickenpox pneumonia. *Chest* 60:504–507, 1971.

88. Nash G, Foley FD: Herpetic infection of the middle and lower respiratory tract. *Am J Clin Pathol* 54:857–863, 1970.

89. Tuxen DV, Cade JF, McDonald MI, Buchanan RC, Clark RJ, Pain MCF: *Herpes simplex* virus of the lower respiratory tract in adult respiratory distress syndrome. *Am Rev Respir Dis* 126:416–419, 1982.

90. Smyth RL, Higenbottam TW, Scott JP, et al: *Herpes simplex* virus infection in heart-lung transplant recipients. *Transplantation* 49:735–739, 1990.

91. Sherry MK, Klainer AS, Wolff M, Gerhard H: Herpetic tracheobronchitis. *Ann Intern Med* 109:229–233, 1988.

92. Schuller D, Spessert C, Fraser VJ, Goodenberger D: *Herpes simplex* virus from respiratory tract secretions: Epidemiology, clinical characteristics, and outcome in immunocompromised and nonimmunocompromised hosts. *Am J Med* 94:29–33, 1993.

93. Ramsey PG, Fife KH, Hackman RC, et al: *Herpes simplex* virus pneumonia: Clinical, virologic and pathologic features in 20 patients. *Ann Intern Med* 97:813–820, 1982.

94. Nakamura Y, Yamamoto S, Tanaka S, et al: *Herpes simplex* viral infection in human neonates: An immunohistochemical and electron microscopic study. *Hum Pathol* 16:1091–1097, 1985.

95. Mannhardt W, Schumacher R: Progressive calcifications of lung and liver in neonatal *Herpes simplex* virus infection. *Pediatr Radiol* 21:236–237, 1991.

96. Strickler JG, Manivel JC, Copenhaver CM, Kubic VL:

Comparison of in situ hybridization and immunohisto-chemistry for detection of *Cytomegalovirus* and *Herpes simplex* virus. *Hum Pathol* 21:443–448, 1990.

97. Geradts J, Warnock M, Yen TS: Use of the polymerase chain reaction in the diagnosis of unsuspected Herpes simplex viral pneumonia: Report of a case. *Hum Pathol* 21:118–121, 1990.

98. Ho M: *Cytomegalovirus: Virology and Infection.* New York, Bingham Medical Book Company, 1982, p 309.

99. Stagno S, Reynolds DW, Pass RF, Alford CA, et al: Breast milk and the risk of cytomegalovirus infection. *N Engl J Med* 302:1073–1076, 1980.

100. Pass RF, August AM, Dworsky M, et al: Cytomegalovirus infection in a day care centre. *N Engl J Med* 307:477, 1982.

101. Henle W, Henle G, Scriba M, et al: Antibody responses to the Epstein Barr virus and cytomegaloviruses after open-heart and other surgery. *N Engl J Med* 282:1068–1074, 1970.

102. Hutter JA, Scott J, Wreghitt T, Higenbottam T, Wallwork J: The importance of cytomegalovirus in heart-lung transplant recipients. *Chest* 95:627–631, 1989.

103. Britt WJ: Recent advances in the identification of significant human cytomegalovirus-encoded proteins. *Transpl Proc* 23(suppl 3):64–69, 1991.

104. Britt WJ, Auger D: Identification of a 65,000 dalton virion envelope protein of human cytomegalovirus. *Virus Res* 4:31–36, 1985.

105. Mocarski ES Jr: Initial events involved in cytomegalovirus–cell interactions. *Transpl Proc* 23(suppl 3):43–47, 1991.

106. Britt WJ, Pass RF, Stagno S, Alford CA: Pediatric cytomegalovirus infection. *Transpl Proc* 23(suppl 3):115–117, 1991.

107. Stagno S, Brasfield DM, Brown MB, et al: Infant pneumonitis associated with *Cytomegalovirus, Chlamydia, Pneumocystis* and *Ureaplasma.* A prospective study. *Paediatrics* 68:322–329, 1981.

108. Abdallah PS, Mark JBD, Merigan TC: Diagnosis of cytomegalovirus pneumonia in compromised hosts. *Am J Med* 61:326–332, 1976.

109. Cappell DS, McFarlane LM: Inclusion bodies (protozoon-like cells) in the organs of infants. *J Pathol Bacteriol* 59:385–398, 1947.

110. Beschorner WE, Hutchins JM, Burns WH, Saral AR, Tutschka PJ, Santos WG: Cytomegalovirus pneumonia in bone marrow transplant recipents: Miliary and diffuse patterns. *Am Rev Respir Dis* 122:107–114, 1980.

111. Van Son WJ, Tegzess AM, Hauwthe T, et al: Pulmonary dysfunction is common during a cytomegalovirus infection after renal transplantation, even in asymptomatic patients. Possible relationship with complement activation. *Am Rev Respir Dis* 136:580–585, 1987.

112. Gorensek MJ, Stewart RW, Keys TF, McHenry MC, Babiak T, Goormastic M: Symptomatic cytomegalovirus as a significant risk factor for major infection after cardiac transplantation. *J Infect Dis* 158:884–887, 1988.

113. Grattan MT, Morenocabral CE, Starnes VA, Oyer PE, Stinson EB, Shumway NE: Cytomegalovirus infections is associated with cardiac allograft rejection and atherosclerosis. *JAMA* 261:3561–3566, 1989.

114. Fryd DS, Peterson PK, Ferguson RM, et al: Cytomegalovirus as a risk factor in renal transplantation. *Transplantation* 30:436–439, 1980.

115. Smiley ML, Wlodaver CG, Grossman RA, et al: The role of pre-transplant immunity in protection from cytomegalovirus disease following renal transplantation. *Transplantation* 40:157–161, 1985.

116. Adam E. Melnick JL, Probtsfield JL, et al: High levels of cytomegalovirus antibody in patients requiring vascular surgery for atherosclerosis. *Lancet* 2:291–293, 1987.

117. Normann SJ, Salomon DR, Leelachaikul P, et al: Acute vascular rejection of the coronary arteries in human heart transplantation: Pathology and correlations with immunosuppression and cytomegalovirus infection. *J Heart Lung Transplant* 10:674–687, 1991.

118. Woods GL, Thompson AB, Rennard SL, Linder J: Detection of cytomegalovirus in bronchoalveolar lavage specimens: Spin amplification and staining with a monoclonal antibody to the early nuclear antigen for diagnosis of cytomegalovirus pneumonia. *Chest* 98:568–575, 1990.

119. Shulman HM, Hackman RC, Sale GE, Meyers JD: Rapid cytologic diagnosis of cytomegalovirus interstitial pneumonia on touch imprints from open lung biopsy. *Am J Clin Pathol* 70:90–94, 1982.

120. Niedobitek G, Finn T, Herbst H, et al: Detection of *Cytomegalovirus* by insitu hybridization and immunohisto-chemistry using new monoclonal antibodies CCH2: A comparison of methods. *J Clin Pathol* 41:1005–1009, 1988.

121. Wolber RA, Lloyd RV: Cytomegalovirus detection by non-isotopic in situ DNA hybridization and viral antigen immuno-staining using a 2-color technique. *Hum Pathol* 19:36–41, 1988.

122. Jiwa M. Steenbergen RD, Zwaan FE, et al: Three sensitive methods for the detection of cytomegalovirus in lung tissue of patients with interstitial pneumonitis. *Am J Clin Pathol* 93:491–494, 1990.

123. Wallace JM: Pulmonary infection in human immunodeficiency disease: Viral pulmonary infections. *Semin Respir Infect* 4:147–154, 1989.

124. Arber DA, Kamel OW, van de Rijn M, et al: Frequent presence of the Epstein-Barr virus in inflammatory pseudo-tumor. *Hum Pathol* 26:1093–1098, 1995.

124a. Egan JJ, Stewart JP, Hasleton PS, Arrand JR, Carroll KB, Woodcock AA: Epstein-Barr virus replication within pulmonary epithelial cells in cryptogenie fibrosing alveolitis. *Thorax* 50:1234–1239, 1995.

124b. Hogg JC: Epstein-Barr virus and cryptogenic fibrosing alveolitis. *Thorax* 50:1232, 1995.

125. Phelan P, Campbell P: Pulmonary complications of rubella embryopathy. *Paediatrics* 75:202–212, 1969.

126. Marshall WC: Intrauterine infections, Ciba Foundation Symposium No. 10. Amsterdam, Associated Scientific Publishers, 1973, p 3.

127. Coyle PK, Wolinsky JS, Buomovici-Klein E, et al: Rubella-specific immune complexes after congenital infection and vaccination. *Infect Immun* 36:498, 1982.

128. Tardieu M, Grospierre B, Durandy A, Griscelli C, et al: Circulating immune complexes containing rubella antigens in late onset rubella syndrome. *J Paediat* 97:370, 1980.

129. Lerner AM, Klein JO, Levin HS, Finland M: Infections due to Coxsackie virus group A, type 9 in Boston, 1959, with specific reference to xanthems and pneumonia. *N Engl J Med* 263:1265–1272, 1960.

130. Haemorrhagic fever with a renal syndrome. Memorandum from a WHO Meeting. *Bull World Health Organ* 61:269–275, 1993.

131. Giles RB, Sheedy JA, Ekman CM, et al: The sequelae of hemorrhagic fever with a note on causes of death. *Am J Med* 16:629–638, 1954.

132. Duchin JS, Koster FT, Peters CJ, et al: Hantavirus, pulmonary syndrome: A clinical description of 17 patients with a newly recognized disease. *N Engl J Med* 330:949–955, 1994.

133. Nichol S, Spiropoulou C, Morzunov S, et al: Genetic identification of Hantavirus associated with an outbreak of acute respiratory illness. *Science* 262:914–917, 1993.

134. Le Guenno B: Identifying hantavirus associated with acute respiratory illness: A PCR victory? *Lancet* 342:1438–1439, 1993.

135. Lee HW, Lee PW, Johnson KM: Isolation of the etiologic agent of Korean Hemorrhagic Fever. *J Infect Dis* 137:298–308, 1978.

136. Levy H, Simpson SQ: Hantavirus plumonary syndrome. *Am J Respir Crit Care Med* 149:1710–1713, 1994.

137. Nolte KB, Feddersen RM, Foucar K, et al: Hantavirus pulmonary syndrome in the United States. A pathological description of a disease caused by a new agent. *Hum Pathol* 26:110–120, 1995.

138. Zaki SR, Greer PW, Coffield LM, et al: Hantavirus pulmonary syndrome. *Am J Pathol* 146:552–579, 1995.

Chapter *6*

Atypical Pneumonias

P. S. Hasleton

The atypical pneumonias do not usually pose a problem for the histopathologist, since they are diagnosed clinically. They are included here first because it is useful for the pathologist to understand the pathology and second because they may be encountered in a biopsy or postmortem. They are put in a separate chapter, since the organisms do not belong with bacteria or viruses.

CHLAMYDIAL PNEUMONIA

Chlamydia have been defined as spherical or ovoid gram-negative bacteria that undergo a well-defined life cycle in the host cell cytoplasm. They cannot be grown on artificial culture media. The infective form is an elementary body 200 to 300 nm in diameter, which becomes transformed in the host cell cytoplasm to the vegetative or reticulate body. This is 600 to 1000 nm in diameter.[1] The organisms belong to the order *Chlamydiales* and three species have been described: *C. trachomatis*, *C. psittaci*, and *C. pneumoniae*. In the old literature these agents were often referred to as *Bedsonia* or *Psittacosis-Lymphogranuloma-Trachoma* (PLT) group. A recent study has suggested the possibility of a second strain of *C. pneumoniae*.[2] Ultrastructurally *C. pneumoniae* has a pear-shaped elementary body and a large periplasmic space. This is in contrast to *C. trachomatis* and *C. psittaci*, which contain circular elementary bodies and narrow periplasmic spaces.[3]

The chlamydial growth cycle distinguishes this group of organisms from other microorganisms.[4] The elementary body (EB) (Fig. 6-1) is the infectious particle which attaches to susceptible host cells and is ingested by what appears to be receptor-mediated endocytosis. The organism enters the cell in a phagosome, where it remains throughout the cell cycle. A few hours after entering the cell the infectious EB changes into a larger noninfectious reticulate body. This divides by binary fission for 16 to 24 h and then condenses into EBs. The length of the growth cycle is dependent on the host cell and the specific chlamydial strain but is typically 48 to 72 h. Towards the end of the growth cycle, EBs predominate. Ultimately the entire cell cytoplasm is displaced by the growing elementary body, which is full of chlamydial particles. The cell bursts and releases the infectious EBs.

Chlamydia pneumoniae

The main serotype of *C. pneumoniae* has been named TWAR (the TW from "TWAIN" and AR denoting "acute respiratory").[5] The first isolate was obtained from the eye of a primary school child in Twain in October 1965.[6]

It is a common cause of pneumonia, ranking as the fourth or fifth among recognized causes of pneumonia in outpatients.[7] Humans are the natural host for this organism.

In the "Maresme" region (Barcelona, Spain) *C. pneumoniae* infection is one of the commonest causes of community-acquired pneumonia, being detected in 16 of 54 cases. This is probably an underestimate, since chlamydial antibodies were only sought in 64 of 105 patients with community-acquired pneumonia.[8] More than 50 percent of adults worldwide have antibodies to this organism.[8a]

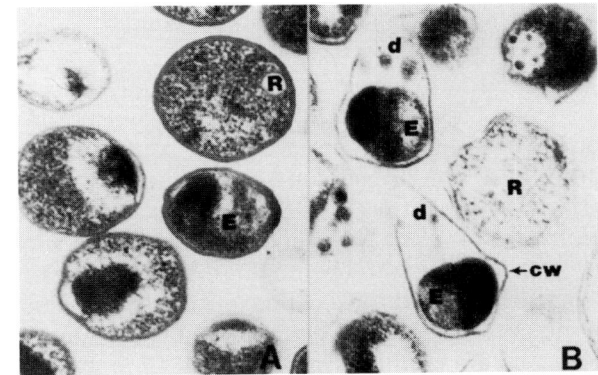

FIGURE 6-1
Illustration of two ultrastructural morphologies in elementary bodies (E). *A*. Round (*C. trachomatis* and *C. psittaci*). *B*. Pear-shaped (*C. pneumoniae*). Note no or little periplasmic space in *A* versus large periplasmic space and presence of electron-dense minature bodies (d) in *B*. Reticulate bodies are all round. cw = cell wall. (×80,000.)

The disease only occurs in man[9] and is probably spread person to person. A symptomless carrier state can last for up to one year.[10]

Clinically acute *C. pneumoniae* infection can range from being asymptomatic to being a life-threatening pneumonia. Ninety percent of cases are at the mild end of the spectrum.[11] The disease is uncommon in children of less than 5 years of age and antibody prevalence studies in the United States and Denmark suggest that frequent infections occur between the ages of 8 and 15 years. The disease is reported more frequently in patients hospitalized with pneumonia, but such individuals usually have underlying chronic disease. This gives rise to a second peak which starts at the age of 70.[12] The pneumonia may be endemic or epidemic and can spread within families.[5] High antibody levels persist into old age, suggesting antibody boosts from multiple reinfections.[13] A recent review of the clinical spectrum has been published.[13a]

In young adults the disease is mild. It may be biphasic, the patient presenting with pharyngitis and then developing pneumonia 1 to 3 weeks later. It may cause bronchitis, laryngitis, and pharyngitis, as well as sinusitis. The patient is often hoarse and afebrile. It may be associated with wheezing.[14] The pneumonia is usually mild, with fever, cough, and crackles.[15] The infection is more severe in older adults.

The organism was isolated in 10 percent of lavage specimens from HIV-infected patients with pneumonia.[16] As with other pneumonic episodes in these patients, other pathogenic organisms were found. In one case however the infection mimicked *Pneumocystis carinii*. A recent outbreak of the infection has been documented in a community of former intravenous drug abusers, half being HIV-positive. The pneumonia was more common in this latter group.[17]

A fascinating association has been described between *C. pneumoniae* antibody and coronary artery disease.[18] These workers continued this study[19] and thought that the mechanism for this association was as follows. The lipopolysaccharide in the *Chlamydia* binds to low-density lipoproteins, modifying them. They then become immunogenic or toxic to endothelial cells. Modified antibody-associated, low-density lipoprotein causes foam cell formation. Such cells are common in atheroma. *Chlamydia* are also strong inducers of TNF, which inhibits lipoprotein lipase, leading to altered lipid metabolism and the accumulation of triglycerides in the blood. Recently the association with *C. pneumoniae* and coronary risk factors has included *Helicobacter pylori*.[19a]

Radiologically there may be single subsegmental lesions or more diffuse changes with consolidation and even pleural effusions.

There are few reports of the pathology of this disease but one case is described.[20] The patient had underlying treated Wegener's granulomatosis. There were discrete nodules, measuring between 2 and 5 mm in diameter, with inflammation in bronchioles and alveoli. Lymphocytes and histiocytes were the predominant cells and there was organizing fibrin filling some of the alveoli at the edge of the inflammatory nodules. Acute and organizing bronchiolitis characterized by focal mucosal necrosis and a filling of the lumen by neutrophils and polymorphs was found in the center of some nodules. The intervening lung was relatively normal. No organisms or viral inclusions were present. Diagnosis in this case was made on the presence of intracytoplasmic elementary bodies in McCoy cell culture.

Laboratory diagnosis of *C. pneumoniae* is based on isolation of the organism, serologic evidence of acute infection, or both. Recently there has been development of a TWAR-specific monoclonal antibody conjugated with fluorescein. This enhances identification of the few inclusions that may be seen in cell culture. TWAR-specific monoclonal antibodies are available from the Washington Research Foundation (Seattle). The usual criteria for infection are a fourfold or greater rise in either IgM or the IgG antibody or IgM antibody at a titer $\geq 1/16$ and/or an IgG antibody titer $\geq 1/512$.[5] Inclusions may be seen on cell culture and are well visualized by iodine staining (Fig. 6-2).

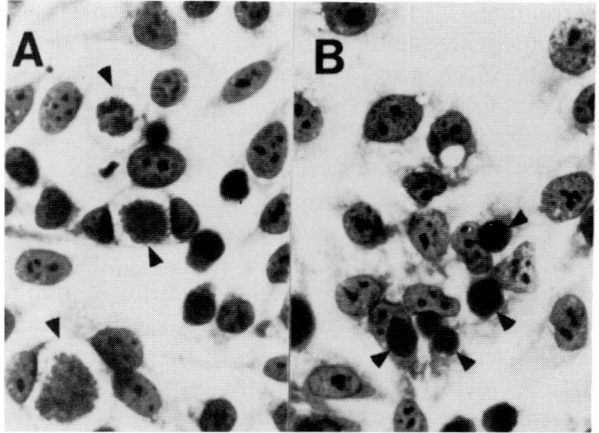

FIGURE 6-2
Illustrations of two kinds of inclusions (*arrowheads*) in Giemsa stain of Hela cell culture. *A.* Vacuolar and glycogen positive by iodine stain (*C. trachomatis*). *B.* Dense and glycogen negative (*C. psittaci* and *C. pneumoniae*). (×820.)

There is still much work to be done on this organism since the mode of transmission of *C. pneumoniae* is unknown, as is the incubation period. Several mechanisms of transmission, including infected aerosols, may be important in the transfer of infection from person to person.[13a]

Chlamydia psittaci

This organism causes psittacosis/ornithosis. The incidence of the disease appears to be increasing, though it is less common than *C. pneumoniae*.[21] A recent review of the disease has been published by Crosse.[22] In the Yorkshire region of England 219 cases were identified between January 1965 and September 1989. There was a wide age range, with most cases occurring between 20 and 70 years. Un-like *C. pneumoniae*, cases are usually sporadic. The organism is highly infectious and is thought to be transmitted by aerosol. Person-to-person spread is rare.

The disease is contracted by aerosol or direct handling of infected bird tissues.[23] Avian contact preceeded infection in 62 percent of cases. In 41 percent of cases the birds had either been ill or died and in 15 percent they had been newly acquired. All common pet birds, especially budgerigars and pigeons and many species of wild birds can carry *C. psittaci*. They often remain healthy while shedding the organism in their secretions or excretions.[24] The incubation period for *C. psittaci* in humans is between 6 and 15 days, though periods of up to 30 days have been reported.

The predominant clinical features are cough, sputum, chest pain, dyspnea, hemoptysis, fever, and malaise. There are neurological and gastrointestinal features including headache, confusion, meningism, photophobia, vomiting, and diarrhea. Myalgia may be present. In most cases the infection had resolved within 6 days. Where the illness was prolonged it was because the diagnosis had not been considered and the appropriate antibiotics, erythromycin and tetracycline, were not given. The disease may cause complications at other sites in the body (Table 6-1).

In patients admitted to hospital the diagnosis of pneumonia was made in 29 percent. In only 11 percent was a diagnosis of psittacosis considered.

Radiology shows a variable degree of consolidation and some cases have lobar consolidation.

Macroscopically the lungs are bulky and show a variable degree of hemorrhagic consolidation, confined mainly to the lower lobes. A fibrinous pleurisy is usual and may be accompanied by subpleural petechial hemorrhages. On cut surface there is a lobular contour to the consolidation but these areas fuse to give rise to a lobar pattern of infection. The con-

TABLE 6-1

Extrapulmonary Complications of Chlamydia Psittaci Infection

System	Manifestation
Hematologic	Severe anemia, hemolytic anemia, disseminated intravascular coagulation
Neurologic	Rarely, meningoencephalitis, lymphocytic meningitis, fits, focal neurologic signs
Cardiovascular	Myopericarditis, culture-negative endocarditis
Rheumatologic	Reactive arthritis
Gastrointestinal	Granulomas causing hepatitis, pancreatitis, nausea, and diarrhea
Renal	Rhabdomyolysis causing renal failure
Dermatologic	Erythema nodosum
Other	Tonsillitis, splenomegaly, thyroiditis.

SOURCE: Adapted from Nash TW, Murray HW: The atypical pneumonias, in Fishman AP (ed): *Pulmonary Diseases and Disorders*, 2d ed, vol 2. New York: McGraw-Hill, 1988, p 1621, with permission.

solidated lung is friable, swollen, and dark purple. The bronchi are congested and there is a regional lymphadenopathy.

Histologically there is an intraalveolar inflammatory response and a lesser degree of interstitial inflammation. Early changes are not usually seen, since the disease is not fatal at this stage. McGavan and coworkers[25] were able to study the early stages of the disease in monkeys. Three days after inhalation of an aerosol containing the organism, there is aggregation of mononuclear cells and some polymorphs in the walls of respiratory bronchioles. The inflammatory changes then spread outwards to involve peribronchiolar alveoli and later extend to the entire lobule.

In human infection, the initial changes are in the bronchial and bronchiolar epithelium, which show a varying degree of necrobiosis and desquamation. There is congestion of the alveolar capillaries and alveoli contain red blood corpuscles, fibrin, and mononuclear cells (Fig. 6-3). The alveolar walls are lined by what appear to be proliferating type II pneumocytes, which desquamate. There is a lymphocytic interstitial infiltrate, which may be perivascular and extends into peribronchial tissues (Fig. 6-4). Many of the alveolar lining cells and those lying free in the alveoli contain Levinthal intracytoplasmic inclusions (Fig. 6-5). In older lesions there is evidence of

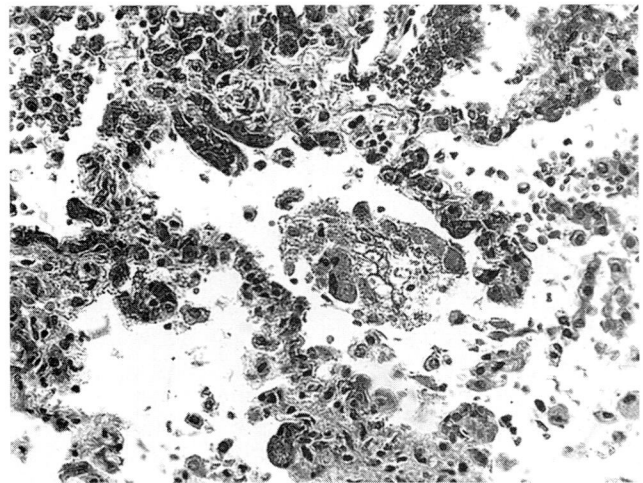

FIGURE 6-3
Psittacosis with congested alveolar walls and a mild, interstitial chronic inflammatory infiltrate. (H&E, ×210.)

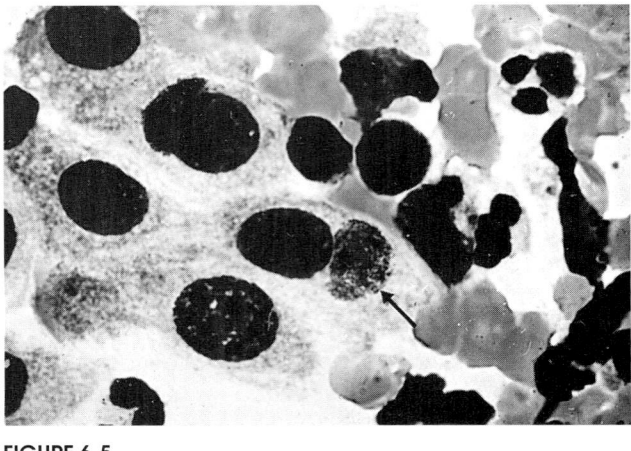

FIGURE 6-5
Psittacosis. A Levinthal body in a cell of a smear preparation. (Approx. ×1340.) *(Reproduced by courtesy of Professor J. D. Williams.)*

capillary thrombosis (Fig. 6-6) and this may extend to involve pulmonary arterioles. Less specific changes may be found in the spleen and liver due to a generalized bacteremia. Inclusions can be demonstrated by prolonged Giemsa staining, where the reticulate bodies are basophilic, and Castaneda's stain, where the organisms stain blue as small coccoid bodies, just visible under the microscope and vary in size from 0.25 to 0.45 μm in diameter.

LABORATORY DIAGNOSIS

Because of the danger of aerosol transmission of infection to laboratory personnel, full precautions should be taken in the handling of any specimens. A direct immunoperoxidase method for demonstrating

C. psittaci in tissue sections has been described.[26] Immunofluorescence and ELISA (enzyme linked immunoabsorbent assay) have been developed for *Chlamydia* antigen detection. The latter test is based on a specific monoclonal antibody, directed against the lipopolysaccharide present in the cell envelope of *C. trachomatis* and *C. psittaci*. Isolation of *C. psittaci* is possible from sputum, and pleural and pericardial fluid. It may also be recovered from autopsy specimens such as lung, spleen, and liver. The most commonly used test is complement fixation on paired sera in the acute and convalescent phases collected 2 to 3 weeks apart. A fourfold rise in titer is considered positive for infection. A single or stable titer of 1 to 64 or greater suggests recent infection.

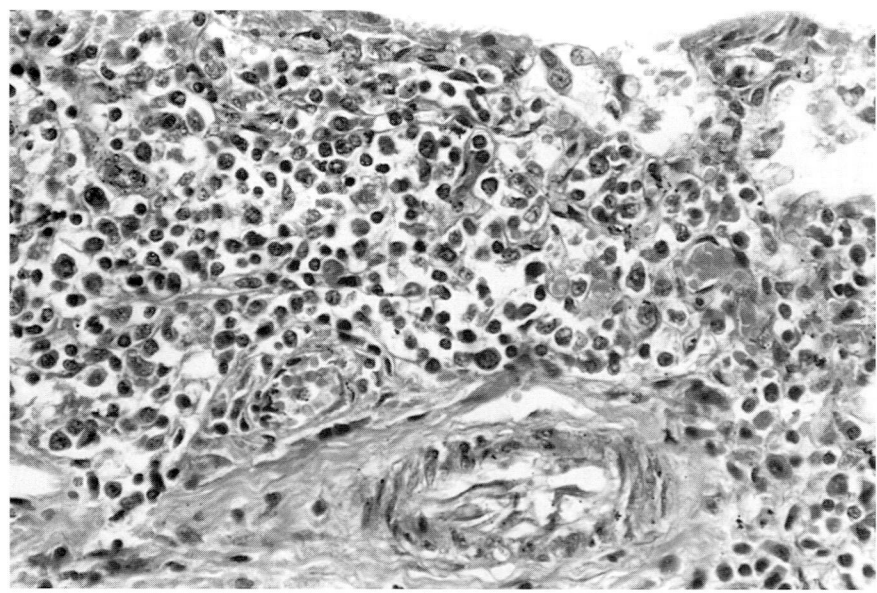

FIGURE 6-4
Psittacosis. Peribronchiolar tissue with chronic inflammation and thrombi in a few small vessels. (H&E, ×219.)

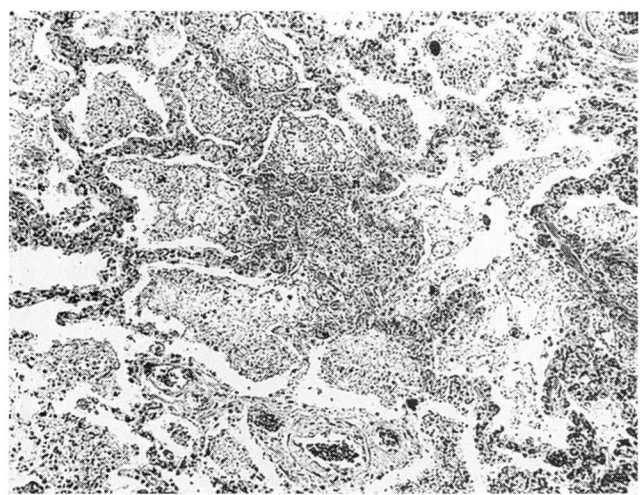

FIGURE 6-6
Psittacosis with intraalveolar fibrin and congested capillary walls, some of which contain capillary thrombi. (H&E, ×106.) *(Reproduced by kind permission of Professor D. S. Russell.)*

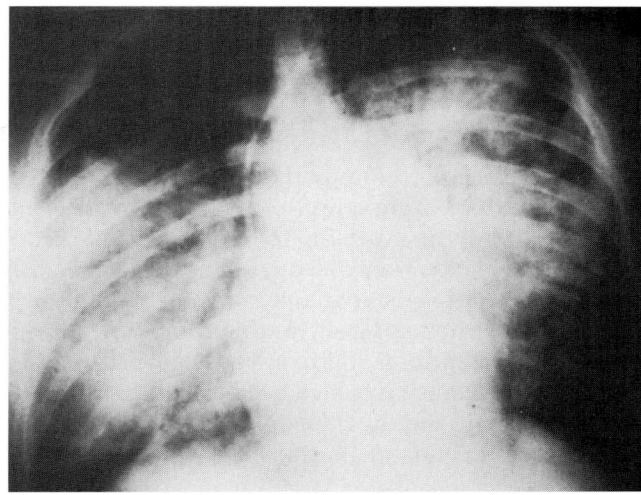

FIGURE 6-7
Psittacosis. This chest radiograph was from a 31-year-old woman 24 h after admission with fulminant disease.

As with all the chlamydial infections, DNA amplification using PCR has been proposed[27] but awaits sensitivity and utility testing.

C. Trachomatis

C. trachomatis causes oculogenital infections but has recently been described as a cause of pneumonia in the neonatal period. There has been a recent case study and a literature review.[28] Infants born to mothers with cervical chlamydial infections may either develop a congenital conjunctivitis or pneumonia. Most of the latter cases were seen at under six months of age.[29] Transmission was thought to occur at birth, presumably through contact with vaginal secretions. However, *C. trachomatis* infections have been reported in infants delivered by caesarian section[30] and apparent intrauterine infections associated with prematurity have been described.[31] Rare cases have been reported in adults and it has been suggested that sexual transmission through the oral-genital route causes such cases.[29] The two other possible mechanisms are self-inoculation from an unrecognized genital infection or reactivation of latent infection.

Clinically the pneumonia usually presents between 4 and 10 weeks of age. Conjunctivitis and rhinorrhea are initial manifestations, followed 1 to 2 weeks later by pneumonia. The main symptom is a paroxysmal staccato cough, which becomes progressively worse. There is tachypnea with periods of respiratory distress, fever is uncommon and there may be little to hear on auscultation. There is an eosinophilia. The condition is usually self-limiting and most cases appear to resolve slowly, even without treatment. There may be abnormalities in lung function subsequent to the disease.[32] *C. trachomatis* pneumonia may occur in immunocompromised patients.

Radiologically there is bilateral hyperinflation with diffuse interstitial infiltrates (Fig. 6-7).

A wide range of histological features has been found including interstitial pneumonia and diffuse interstitial and intraalveolar mononuclear infiltrate with eosinophils and focal aggregates of neutrophils. In some cases intranuclear inclusion bodies are seen. Chronic interstitial inflammation with granulomatous features and hyperplasia of type II pneumocytes has been described. Plasma cells may be seen in the alveolar lumina joining macrophages, lymphocytes, eosinophils, and neutrophils. No germinal centers are present. There may be necrosis of the bronchiolar epithelium with plugging of bronchioles by desquamated necrotic cells, as well as mucoid debris.

The inclusions may be difficult to see and are best identified on Epon-embedded, 0.5 μm sections, stained with toluidine blue. The intracytoplasmic inclusions are surrounded by a clear halo. Mature inclusions may appear as cytoplasmic vacuoles which are usually single, often displacing the nucleus, containing many eosinophilic elementary bodies.

Such inclusions stain blue with a Castaneda's method, and with Giemsa early inclusions are basophilic. Mature inclusions may appear as cytoplasmic vacuoles which are usually single, often displacing the nucleus, containing many eosinophilic elementary bodies.

LABORATORY DIAGNOSIS

Culture of *C. trachomatis* is possible from specimens obtained from the conjunctivae, nasopharynx, rectum, vagina, cervix, and urethra. A rapid immunofluorescence assay is available for detecting inclusions from smears made from swab specimens.[33] An ELISA for *C. trachomatis* is also available. One of the best methods for detection is a microimmunofluorescence test, which is both specific and sensitive. Complement fixation is also used and a fourfold rise in titer, a single titer of 1:16, or the presence of IgM is considered diagnostic of infection.

RICKETTSIAL DISEASE

The genus *Rickettsia* comprises ten or more species, which form five biological groups. These are typhus, spotted fever, scrub typhus, trench fever, and Q fever. The main five organisms causing human pulmonary disease are *R. rickettsii*, which causes Rocky Mountain spotted fever. It is transmitted to man by tick bite. *R. tsutsugamushi*, which causes scrub typhus, is found in Asia and transmitted by mites and rodents. Q fever is caused by *Coxiella burnetii*. The important organism is *Coxiella burnetii*, which has a worldwide distribution.[34] The organisms are coccobacilli, with a gram-negative cell wall. They are obligate intracellular parasites. *C. burnetii* is resistant to dessication and is highly infectious, in that one organism appears the infective dose for human disease.[35]

C. burnetii occurs worldwide and affects a wide variety of domestic animals, especially sheep, goats, cattle, and ticks. More recently it had been associated with infected parturient cats, even in urban settings.[36] Person-to-person transmission is rare.[37] Infection occurs by inhalation of aerosol particles that contain the organisms. Rarely it can be transmitted by raw milk or blood transfusions.[23]

The average incubation period for *C. burnetii* is 20 days with a range of 14 to 39 days.[38]

A recent study of 164 community acquired cases in the Basque region of Spain[39] showed the most common clinical symptoms were high fever, cough, severe headache, and myalgia. The cases were often sporadic. Nearly half the patients had no respiratory symptoms, although 34 percent of cases reported pleural pain. Hemoptysis was present in just over 10 percent of cases. There are thought to be three distinct pulmonary manifestations: atypical pneumonia, rapidly progressive pneumonia, and most commonly an asymptomatic infiltrate in a febrile

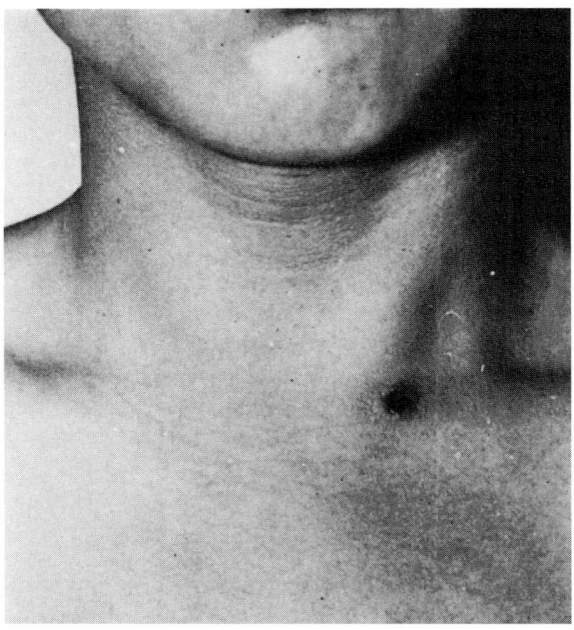

FIGURE 6-8
Classic eschar of scrub typhus (tsutsugamushi disease) at the site of a mite attachment.

patient.[40] Rarely bacterial endocarditis can occur, sometimes many years after the acute infection.

Chest radiographs show single or multiple, rounded, segmental densities, usually in the lower lobes.

Scrub typhus has a characteristic primary skin lesion (eschar) (Fig. 6-8), generalized lymphadenopathy, rash, and nonspecific symptoms, such as headache, myalgia, and cough. Rocky Mountain spotted fever is caused by *R. rickettsii* and is caused by tick bites.

The histopathologic lesions of rickettsial infections are similar.[41] The main target is the endothelium of capillaries and venules. In Rocky Mountain spotted fever, arteriolar lesions are conspicuous, due to invasion into vascular smooth muscle cells by the organism. There is endothelial swelling followed by perivascular and intramural infiltration by lymphocytes, macrophages, and many neutrophil polymorphs (Fig. 6-9). Thrombi and microinfarcts occur only in a tiny minority of infected blood vessels.[42] The basic pathology in established disease is disseminated intravascular coagulation. Thus there is thrombocytopenia in a significant number of cases.

Macroscopically the lungs show gray foci of consolidation, confined mainly to the lower lobes of one or both lungs. The residual lung tissue is edematous with inflamed main bronchi. A case has been described where Q fever resulted in an inflammatory pseudotumor of the lung. *C. burnetii* was demonstrated by electron microscopy.[43] The spleen is

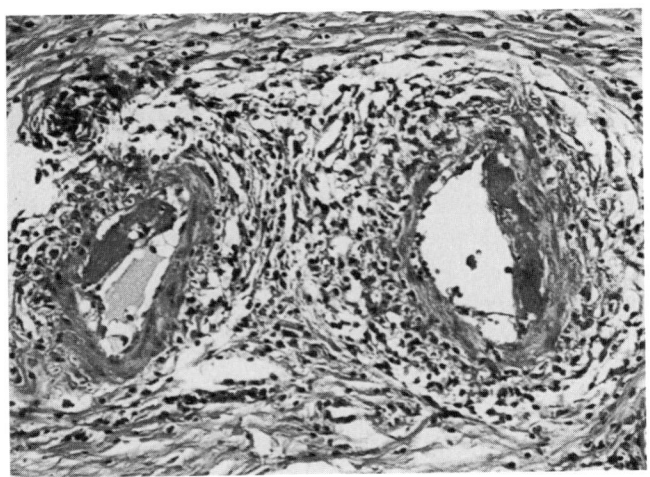

FIGURE 6-9
Rocky Mountain spotted fever. Skin biopsy at day 10 of the illness, with two arterioles showing perivascular inflammation and mural thrombus. (*See also Color Plate 1.*)

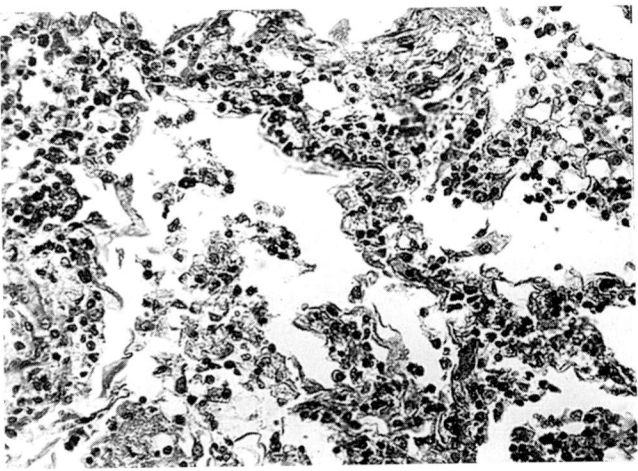

FIGURE 6-10
Q fever. There is interstitial inflammation and edema of the alveolar wall. (H&E, ×207.) (*Reproduced by kind permission of Professor N. Gowing.*)

enlarged, soft, and red but other organs show no distinctive changes.

Histologically the lung shows a diffuse interstitial pneumonia (Fig. 6-10) with an intraalveolar exudate consisting of edema fluid. This progresses to fibrin, containing many mononuclear cells, lymphocytes, a few polymorphs, and red blood corpuscles (Fig. 6-11). The alveoli are lined partly by prominent type II pneumocytes and the walls, and subpleural tissues are infiltrated by lymphocytes, plasma cells, and histiocytes. These cells are also seen around bronchioles and smaller branches of the pulmonary artery, where there is lymphocytic cuffing. There is marked edema of all the interstitial tissues. The intraalveolar exudate gives a cribriform appearance, caused by the retraction of the fibrin network. The small bronchioles contain a similar exudate and epithelium is frequently shed. Alveolar septal necrosis is rare, though alveolar epithelial cells are shed into the lumen. The organisms may be detected intracellularly in some cases, but the pulmonary changes are nonspecific in their absence.

In extrapulmonary sites such as marrow and liver there is granuloma formation and in some cases this has a classic "doughnut" appearance.[44]

Isolation of *C. burnetti* is difficult and hazardous. Care should be taken if a postmortem is to be carried out in the case of Q fever. One should prevent visceral blood from drying and the splashing of body contents should be avoided. Blood should also be taken for estimation of complement-fixing antibodies. There should be a four-fold rise in the serologic titer. The most common methods of measurement are complement fixation and indirect immunofluores-

cent assay. Immunofluorescence can distinguish IgM from IgG antibodies. *C. burnetii* produces phase 1 and 2 antibodies, the significance of which is not fully understood. Levels of phase 2 antibodies rise in acute self-limiting infection. Phase 1 antibodies are predominant in a more chronic form of Q fever such as endocarditis.

Electron microscopy[45] shows the organism is pleomorphic and forms terminal bodies that resemble endospores of gram-positive bacteria. In the case of *C. burnetii* the organisms tend to be found by a phagolysosomal membrane.

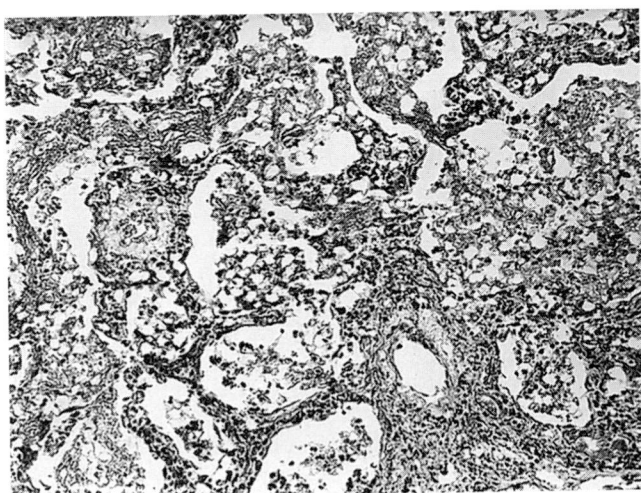

FIGURE 6-11
Q fever. A more severe stage of the disease than in Fig. 6-10. There is marked interstitial inflammation and chronic inflammatory cells in the alveolar lumina. (H&E, ×105.) (*Reproduced by kind permission of Professor N. Gowing*)

MYCOPLASMA PNEUMONIA

Mycoplasma belongs to the group *Mycoplasmatales*. They are pleomorphic organisms varying in shape from coccoid to filamentous, with a triple-layered limiting membrane but no rigid cell wall and thus are deformable. *Mycoplasma pneumoniae* most commonly exists as a filamentous form. The organism is 0.3 μm in diameter.[46]

They pass through filters that tend to exclude bacteria and this property led to their being classed as viruses initially. They are gram-negative and stain poorly with other bacterial stains, but they are readily identifiable with Giemsa.

Mycoplasma pneumonia is an important cause of community acquired pneumonia, especially during epidemics which are seen every 4 years or so.[47] In a British Thoracic Society/Public Health Laboratory Survey Study of community acquired pneumonia in adults during 1982 to 1983 the organism was responsible for 18 percent of cases. It was the most commonly identified cause of pneumonia after the *Streptococcus pneumoniae*.[48] Spread of the infection is usually slow and it is seen throughout the year. Transmission is by droplets. Infection rates are often higher among family members and the disease especially affects children. The incubation period ranges from 2 to 3 weeks.

The range of illness in Mycoplasma pneumonia extends from mild upper respiratory tract symptoms to frank pneumonia. It is estimated that 25 percent of *M. pneumoniae* infections are symptomless.[49] Some believe that asymptomatic infection is uncommon, accounting for only 1.5 to 4 percent of cases and that 90 percent of the patients have a history of cough and tracheobronchitis.[50] There may be neurological manifestions including meningitis, encephalitis, and Guillain-Barre syndrome. The neurological damage is thought to be mediated by autoantibodies. In severe cases there may be hemolytic anemia and disseminated intravascular coagulation, hepatitis, myo- and pericarditis, arthritis, arthropathy, and erythema multiforme including the Stevens Johnson syndrome.[51]

Chest radiographs show consolidation which is either uni- or multilobar and maybe bilateral. Lower lobe involvement is the most common.[52]

Some cases may have large pleural effusions. There are persistent radiological abnormalities including pleural shadowing, loss of volume in a lobe or streaky intrapulmonary shadowing.

Macroscopically the lungs are heavy and plum-colored and there may be subpleural hemorrhages. The cut surface may show only generalized pulmonary edema and a variable amount of intrapulmonary hemorrhage. In some cases the small bronchial walls are thickened, filled with mucus or mucopus, while the larger bronchi contain frothy fluid. There may be diffuse alveolar damage.[53]

Histologically the findings, given the macroscopic picture, are predictable, with many of the alveoli being filled with edema fluid together with red blood corpuscles and macrophages. There may be hyaline membranes and the lung may demonstrate diffuse alveolar damage. Focal chronic inflammation is confined to the edematous walls of alveolar ducts and the alveoli (Figs. 6-12 and 6-13). Most of the inflammatory cells are lymphocytes and plasma cells. The affected alveoli are almost bloodless and show elastic fiber destruction. There are lymphocytic infiltrates in the walls of bronchioles, proliferation of the epithelium, and partial desquamation of the bronchiolar epithelial cells. During recovery damaged alveoli show fibrosis. In a case reported in the *New England Journal of Medicine* there were hemorrhagic infarcts, pulmonary fibrosis, and thrombi in pulmonary arteries and veins, as well as in bronchial arteries.[53] This was thought to be due to disseminated intravascular coagulation and not embolism. Importantly no organisms could be stained or cultured. The best way histologically of detecting the organism is by the *M. pneumoniae* DNA probe test.

In the laboratory, complement fixation tests are used and a rise in antibody titer of more than fourfold in paired samples is thought to be evidence of recent infection. *M. pneumoniae* IgM is a reliable indicator

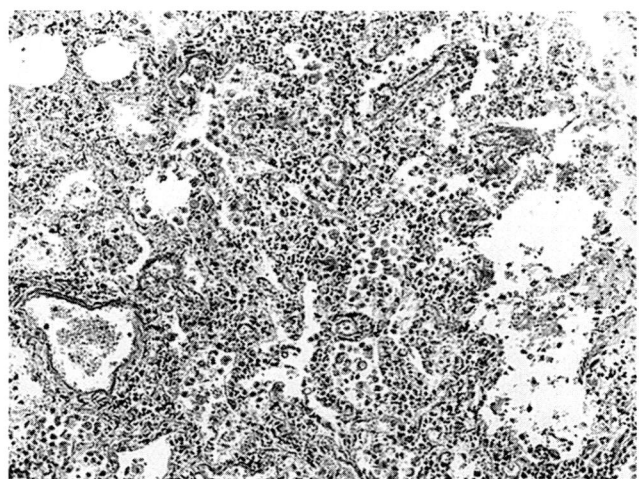

FIGURE 6-12
Mycoplasma pneumonia with diffuse interstitial inflammation. The chronic inflammatory cells also surround bronchioles and blood vessels. (H&E, ×109.)

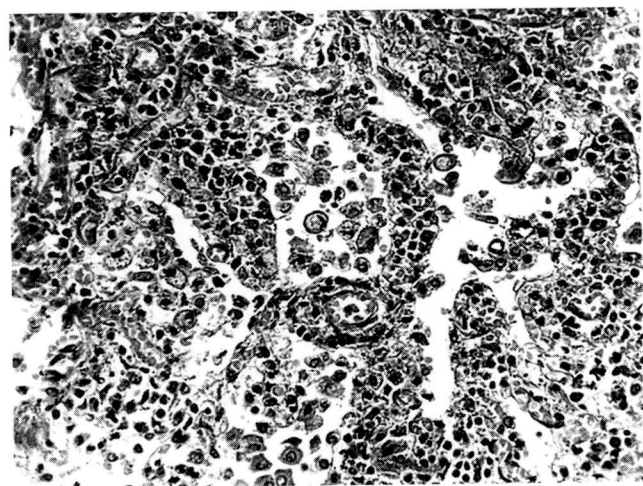

FIGURE 6-13
Mycoplasma pneumonia at a higher magnification and showing the chronic interstitial inflammation. (H&E, ×216.)

of recent infection but since this antibody is produced less frequently during reinfection, a negative result does not exclude a recent infection, especially in patients over the age of 45. Newer antibody tests, such as ELISA, probably provide a more reliable indication of recent infection.

The *M. pneumoniae* antigen ELISA has a sensitivity of 90 percent. In a pediatric study, IgM Anti-P1 immunoblotting was found to have an 84.2 percent sensitivity and a 95.5 percent specificity for *Mycoplasma*, though the test cannot differentiate primary disease from a superimposed infection. Five to eight days of symptoms are required in immunocompetent patients before the test is positive.[54]

REFERENCES

1. Collier LH: Chlamydia, in Parker MT, Collier LH (eds): *Topley and Wilson's Principals of Bacteriology, Virology and Immunity*, 8th ed. London: Edward Arnold, 1990, p 630.
2. Black CM, Sarshy CE, Brown TM: Antigenic variations in *Chlamydia pneumoniae*, in *Proceedings of the 7th International Symposium on Human Chlamydial Infections*. Har-rison Hot Springs, Canada, 1990, p 410.
3. Chi EY, Kuo CC, Grayston JT: Unique ultrastructure in the elementary body of Chlamydia SP strain. *TWAR J Bacteriol* 169:37–57, 1987.
4. Schachter J: Pathogenesis of Chlamydial infections. *Pathol Immunopathol Res* 8:206–220, 1989.
5. Marrie TJ: *Clamydia pneumoniae*. *Thorax* 48:1–4, 1993.
6. Kuo C-C, Chen HH, Wang S-P, Grayston JT: Characterisation of TWAR strains, a new group of *Clamydia psittaci*. In Oriel D, Ridgway G, Schachter J, Taylor-Robinson D, Ward M (eds): *Chlamydial Infections. Proceedings of the Sixth International Symposium on Human Chlamydial Infections.* Cambridge: Cambridge University Press, 1987, pp 321–324.
7. Grayston JT, Thom DH: The Chlamydial pneumonias. *Curr Clin Top Infect Dis* 11:1–18, 1991.

8. Almirall J, Morato I, Riera F, et al: Incidence of community-acquired pneumonia and *Chlamydia pneumoniae* infection: A prospective multicentre study. *Eur Respir J* 6:14–18, 1993.
8a. Yaakov MB, Lazarovich Z, Beer S, Levin T, Shoham I, Boldur I: Prevalence of *Chlamydia pneumoniae* antibodies in patients with acute respiratory infections in Israel. *J Clin Pathol* 47:232–235, 1994.
9. Grayston JT, Wang SP, Kuo C-C, Campbell LA: Current knowledge on *Chlamydia pneumoniae*, strain TWAR, an important cause of pneumonia and other respiratory diseases. *Eur J Clin Microbiol Infect Dis* 8:191–202, 1989.
10. Chirgwin K, Roblin PM, Gelling M, Hammerschlag MR, Schachter J: Infection with *Chlamydia pneumoniae* in Brooklyn. *J Infect Dis* 163:757–761, 1991.
11. Grayston JT, Campbell LA, Kuo CC, et al: A new respiratory tract pathogen: *Chlamydia pneuomoniae*, strain TWAR. *J Infect Dis* 161:618–625, 1990.
12. Grayston JT: *Chlamydia pneumoniae*, strain TWAR pneumonia. *Annu Med* 43:317–323, 1992.
13. Marrie TJ, Grayston JT, Wang ST, et al: Pneumonia associated with the (TWAR) strain of Chlamydia. *Ann Intern Med* 106:507, 1987.
13a. Bourke SJ, Lightfoot NF: *Chlamydia pneumoniae*: Defining the clinical spectrum requires precise laboratory diagnosis. *Thorax* 50:543–548, 1995.
14. Hahn DL, Dodge RW, Golubjatnikov R: Association of *Chlamydia pneumoniae* (strain TWAR) infection with wheezing, asthmatic bronchitis, and adult-onset asthma. *JAMA* 266:225–230, 1991.
15. Martin RE, Bates JH: Atypical pneumonia. *Infect Dis Clin N Am* 5:585–601, 1991.
16. Augenbraun MH, Roblin PM, Chirgwin K, Landman D, Hammerschlag MR: Isolation of *Chlamydia pneumoniae* from the lungs of patients infected with the human immunodeficiency virus. *J Clin Microbiol* 29:401–402, 1991.
17. Blasi F, Boschini A, Cosentini R, et al: Outbreak of Chlamydia pneumonia infection in former injection-drug users. *Chest* 105:812–815, 1994.
18. Saikku P, Leinonen M, Mattila K, et al: Serological evidence of an association of a novel *Chlamydia* TWAR with chronic coronary artery disease and acute myocardial infarction. *Lancet* 2:983–986, 1988.
19. Saikku P, Leinonen M, Tenkanen L, et al: Chronic *Chlamydia pneumoniae* infection as a risk factor for coronary artery disease in the Helsinki Heart Study. *Ann Intern Med* 116:273–278, 1992.
19a. Patel P, Mendall MA, Carrington D, et al: Association of *Helicobacter pylori* and *Chlamydia pneumoniae* infections with coronary heart disease and cardiovascular risk factors. *Br Med J* 311:711–714, 1995.
20. Case records of the Massachusetts General Hospital: *N Engl J Med* 323: 1546–1555, 1990.
21. Public Health Laboratory Service Communicable Disease Surveillance Centre. Trends in human ornithosis/psittacosis. 1975–80. *Br Med J* 282:1411, 1981.
22. Crosse BA: Psittacosis: A clinical review. *J Infect* 21:251–259, 1990.
23. Greenberg SB, Atmar RL: Atypical pneumonia, in Niederman MS, Sarosi GA, Glassroth J (eds): *Respiratory Infections: A Scientific Basis for Management*. Philadelphia, W.B. Saunders, 1994, pp 331–343.
24. Schachter J, Dawson CR: *Human Chlamydial Infection*. PSG Publishing Company, Littleton, MA, 1978.

25. McGavran MH, Beard CW, Berendt RS, Nakamura LM: The pathogenesis of Psittacosis. *Am J Pathol* 40:653–661, 1962.

26. Finlayson J, Buxton D, Anderson IE, Donald KM: Direct immunoperoxidase method for demonstrating *C.psittaci* in tissue sections. *J Clin Pathol* 38: 712–714, 1985.

27. Holland SM, Gaydos CA, Quinn TC: Detection and differentiation of *Chlamydia trachomatis, Chlamydia psittaci,* and *Chlamydia pneumoniae* by DNA amplification. *J Infect Dis* 162:984–987, 1990.

28. Griffin M, Pushpanathan C, Andrews W: Chlamydia trachomatis pneumonitis: A case study and literature review. *Pediatr Pathol* 10:843–852, 1990.

29. Grayston JT: The chlamydial pneumonias. *Curr Clin Top Infect Dis* 11:1–18, 1991.

30. Lascolea LJ, Paroski JS, Burzyk IL, Fadden HS: *Chlamydia trachomatis* infection in infants delivered by caesarian section. *Clin Paediat* 23:118–120, 1984.

31. Mard PA, Johansson PG, Svenningsen N: Intrauterine lung infection with *Chlamydia trachomatis* in a premature infant. *Acta Paediatr Scand* 73:569–572, 1984.

32. Harrison HR, Tassig AM, Fulgitit AF: *Chlamydia trachomatis* in chronic respiratory disease in childhood. *Paediat Infect Dis* 1:29–33, 1982.

33. Lisby SM, Nahata MC: Recognition and treatment of chlamydia infections. *Clin Pharm* 6:25–36, 1987.

34. Marmion BP: *The Rickettsiae,* in Parker MT, Collier LH (eds): *Topley and Wilson's Principles of Bacteriology, Virology and Immunity,* 8th ed, vol 2. London, Edward Arnold, 1990, p 648.

35. Bacca OG, Paretsky D: Q fever and Coxiella burnetii: A model for host-parasite interactions. *Microbiol Rev* 127–149, 1983.

36. Kosatsky T: Household outbreak of Q fever pneumonia related to a parturient cat. *Lancet* 2:144–147, 1984.

37. Mann JS, Douglas JG, English JM, et al: Q. fever: Person to person transmission within family. *Thorax* 41:974, 1986.

38. Sawyer LA, Fishbein DB, McDade JE: Q fever: Current concepts. *Rev Infect Dis* 9:935, 1987.

39. Sobradillo V, Ansola P, Baranda S, Corral C: Q fever pneumonia: A review of 164 community acquired cases in the Basque country. *Eur Respir J* 2:263–266, 1989.

40. Edgmartin R, Bates JH: Atypical pneumonia. *Infect Dis Clin N Am* 5:585–601, 1991.

41. Walker DH: Diagnosis of rickettsial diseases. *Pathol Annu* 23:69–96, 1988.

42. Lilley RD, Dyer RE: Brain reaction in guinea pigs infected with endemic typhus, epidemic (European typhus) and Rocky mountain spotted fever, Eastern and Western types. *Public Health Rep* 51:1295, 1936.

43. Janigan DT, Marrie TJ: An inflammatory pseudotumor of the lung in Q fever pneumonia. *N Engl J Med* 308:86–88, 1983.

44. Travis LB, Travis WD, Li C-Y, Pierre RV: Q fever: A clinicopathologic study of 5 cases. *Arch Pathol Lab Med* 110:1017–1020, 1986.

45. Silverman DJ: Some contributions of electron microscopy to the study of the Rickettsiae. *Eur J Epidemiol* 7:200–206, 1991.

46. Taylor-Robinson D: *Mycoplasma* in Parker MT, Collier LH (eds): *Topley and Wilson's Principles of Bacteriology, Virology and Immunity,* 8th ed, vol 2. London, Edward Arnold, 1990, p 663.

47. Editorial: Mycoplasma pneumoniae. *Lancet* 337:651–652, 1991.

48. Research committee of the British Thoracic Society and Public Health Laboratory Service: Community acquired pneumonia in adults in British hospitals, 1982-3, a survey of aetiology, mortality, prognostic factors and outcome. *QJ Med* 62:195–220, 1987.

49. Surveillance of Mycoplasma infections, 1980-90. *Commun Dis Rep* 46:3–4, 1990.

50. Sillis M: Mycoplasma pneumonia. *Lancet* 37:1101, 1991.

51. Ali MJ, Sillis M, Andrews BE, Jenkins PF, Harrison BDW: The clinical spectrum and diagnosis of mycoplasma pneumoniae infection. *QJ Med* 227:241–251, 1986.

52. Mansell JK, Rosenow EC, Smith TF, Martin JW: Mycoplasma pneumonia. *Chest* 95:639–646, 1989.

53. Case records Massachusetts General Hospital. *N Engl J Med* 326:324–326, 1992.

54. Cimolai N, Cheong ACH: IgM Immunoblotting: A standard for the rapid serologic diagnosis of *Mycoplasma pneumoniae* infection in pediatric care. *Chest* 102:477–481, 1992.

Chapter 7

Pulmonary Bacterial Infection

P. S. Hasleton

COMMON BACTERIAL PNEUMONIAS

Throughout this text the terms *pneumonia* and *pneumonitis* are used synonymously. Following the introduction of chemotherapeutic and antibiotic drugs, the mortality from many bacterial diseases, including many of the acute bacterial pneumonias, has been reduced dramatically. The decline in clinical importance of this group has led to a lessening of the interest in their pathology. However, complacency was shaken in 1976 when a new organism, *Legionella pneumophila,* struck down 147 people attending an American Legion convention in Philadelphia; 29 of these patients died. Subsequently there have been outbreaks of this type of pneumonia in either hospitals or resorts, hitting the national headlines.

Pneumonia remains an important disease in the immunosuppressed patient, and cases may tax the pathologist to differentiate between infection or rejection in heart-lung transplantation (Chap. 24). With the increased success of chemotherapy for tumors, pneumonia is a common complication and has to be differentiated from a drug reaction or recurrent tumor. In such cases, open lung biopsy may be necessary to provide a definitive diagnosis and aid rational treatment.

Some types of bacterial pneumonia, such as plague and anthrax, are largely diseases of history. However, there was an outbreak of so-called pneumonic plague[1] on the Indian subcontinent in September 1994, and anthrax was documented in the Soviet Union. It is therefore obvious that these illnesses cannot be forgotten.

The decline of some forms of acute bacterial pneumonia antedated the introduction of chemotherapeutic agents. This was in part attributable to a decline in western countries in virulence and pathogenicity of some of the causative bacteria, especially beta-hemolytic *Streptococcus pyogenes.* In addition, because of better nutrition and housing, there is increased host immunity. Microorganisms sometimes appear to undergo cyclic phases, often spanning a century or more, in their virulence and pathogenicity.

In the past it was customary to divide pneumonias into lobar, bronchogenic, hypostatic, and aspiration forms, depending on the etiology and anatomic type of reaction caused in the lung. It is more realistic to classify them according to the organism responsible. A correlation between the pathology of some of the common bacterial pneumonias and the causative organisms is given in Table 7-1. This table is not comprehensive, however, and should be studied in conjunction with the text. In many cases the bacteriology from an open lung biopsy, lavage, or material taken at postmortem will help pinpoint the exact organism causing the disease.

The macroscopic diagnosis of pneumonia in the postmortem room is very inaccurate. The collapsed lung, filled with blood and edema fluid, masks many disease processes. In all cases where pneumonia is a clinical or pathological diagnostic possibility, histology is required as well as bacteriology. In addition histology may not be representative if small tissue blocks are taken.

Pakko and colleagues[3a] showed postmortem radiology of the air-inflated lung was both a sensitive and accurate method for studying the prevalence of any pulmonary lesions. Morphological evidence of acute bacterial infection was seen in 42.2 percent of lungs from hospital necropsies. Only one non-hospital necropsy showed incipient bronchopneumonia. These authors showed an association between the occurrence of gram-negative bacteria in the bronchial mucus at necropsy and the presence of pulmonary lesions of any kind. Pulmonary lesions did not always produce apparent clinical disease. There was, however, a highly significant association between the presence of gram-negative rods in the bronchial mucus and signs of acute bacterial infection in lung tissue.

With the advent of molecular biology, the diagnosis of infectious diseases is at the beginning of a remarkable phase. Pathologists in some centers have

TABLE 7-1

Patterns of Pulmonary Infection with Acute Bacterial Pneumonia

Organism	Patterns of Disease
Staphylococcus	Tracheitis (neonates)
	Lobar or lobular consolidation; abscesses
	Acute hemorrhagic pulmonary edema
β-Hemolytic streptococcus	Bronchopneumonia
S. pneumoniae	Lobar pneumonia
	Bronchopneumonia
B. anthracis	Pulmonary edema and hemorrhage
H. influenzae	Mucopurulent bronchitis
	Lobar or bronchopneumonia
Klebsiella	Lobular or lobar consolidation
	Cavitation
B. pertussis	Bronchitis, bronchiolitis
Pseudomonas	Infarct-like areas
	Lobular consolidation
	Bronchopneumonia
B. proteus	as for Pseudomonas
Escherichia coli	as for Pseudomonas
Leptospira ictero-haemorrhagiae	Intrapulmonary hemorrhage
Yersinia pestis	Hemorrhagic pulmonary edema
	Lobular consolidation
Franeisella tularensis	Small gray-white nodules
	Lobar consolidation (abscess)
Brucella	Rare—few pathologic descriptions
Neisseria meningitidis	Lobular, rarely lobar consolidation
Salmonella	Bronchitis, bronchopneumonia
Pasteurella multocida	Bronchitis, bronchopneumonia
Neisseiria meningitides	Little pathologic data
Moraxella catarrhalis	Tracheitis
	Lobar consolidation
Legionella	Bronchopneumonia
	Lobar pneumonia

highly sensitive, rapid, and specific molecular methods for identifying microorganisms by direct detection of DNA or RNA sequences unique to a particular organism. These techniques are useful in the diagnosis of slowly growing organisms, such as the tubercle bacillus, and when too few organisms are present for detection by other means. Thus the polymerase chain reaction (PCR) may be used for the detection of *Borrelia burgdorferi*, the causative agent of Lyme disease.[2,3]

These investigative methods are still in their early stages and continue to be evaluated. The extreme sensitivity of PCR is one its drawbacks, as there are false-positive results. Small amounts of amplified target sequence, of which up to 10^9 copies may be present in a single PCR solution, can contaminate reagents or laboratory equipment. Contamination can be caused by airborne organisms. There is also the problem of the relevance of detection of a small number of organisms in samples from clinically unaffected patients.

In the present section little reference will be made to these methods, as they are still experimental and are not available in most laboratories.

Despite an increasing number of antibiotics and improvement in social conditions, the death rate from pneumonia continues to rise in patients aged 85 or over (Table 7-2 and Fig. 7-1). Since 1984, mortality attributed to pneumonia has remained relatively constant at around 3.5 deaths per thousand patients aged 65 or older (Fig. 7-2). Mortality statistics do not reflect the full impact of pneumonia in the elderly. It is classified as the underlying cause of death in only 6 percent of cases in the 65-plus age group. It is mentioned as an immediate or contributory cause of death in approximately one death in four in this age group. Pathologists do not have to be reminded of the inaccuracy of many death certificates.

It is well known that there is more morbidity and mortality in the winter. Two causes have been highlighted by Curwen: low temperatures and influenza epidemics, the former being the more important.[4]

TABLE 7-2

Deaths from Selected Respiratory Causes, Age 65+, England and Wales, 1990

	No. of Deaths
Cancer of trachea, bronchus, and lung	25,659
Pneumonia	25,473
Chronic obstructive pulmonary disease	24,489
Influenza	735
Upper respiratory disorders	523
Pneumoconioses	488
Pulmonary tuberculosis	283
All causes	456,916

SOURCE: OPCS. Reproduced with the permission of the Lung and Asthma Information Agency, Department of Public Health Services, St. George's Hospital Medical School, London, England.

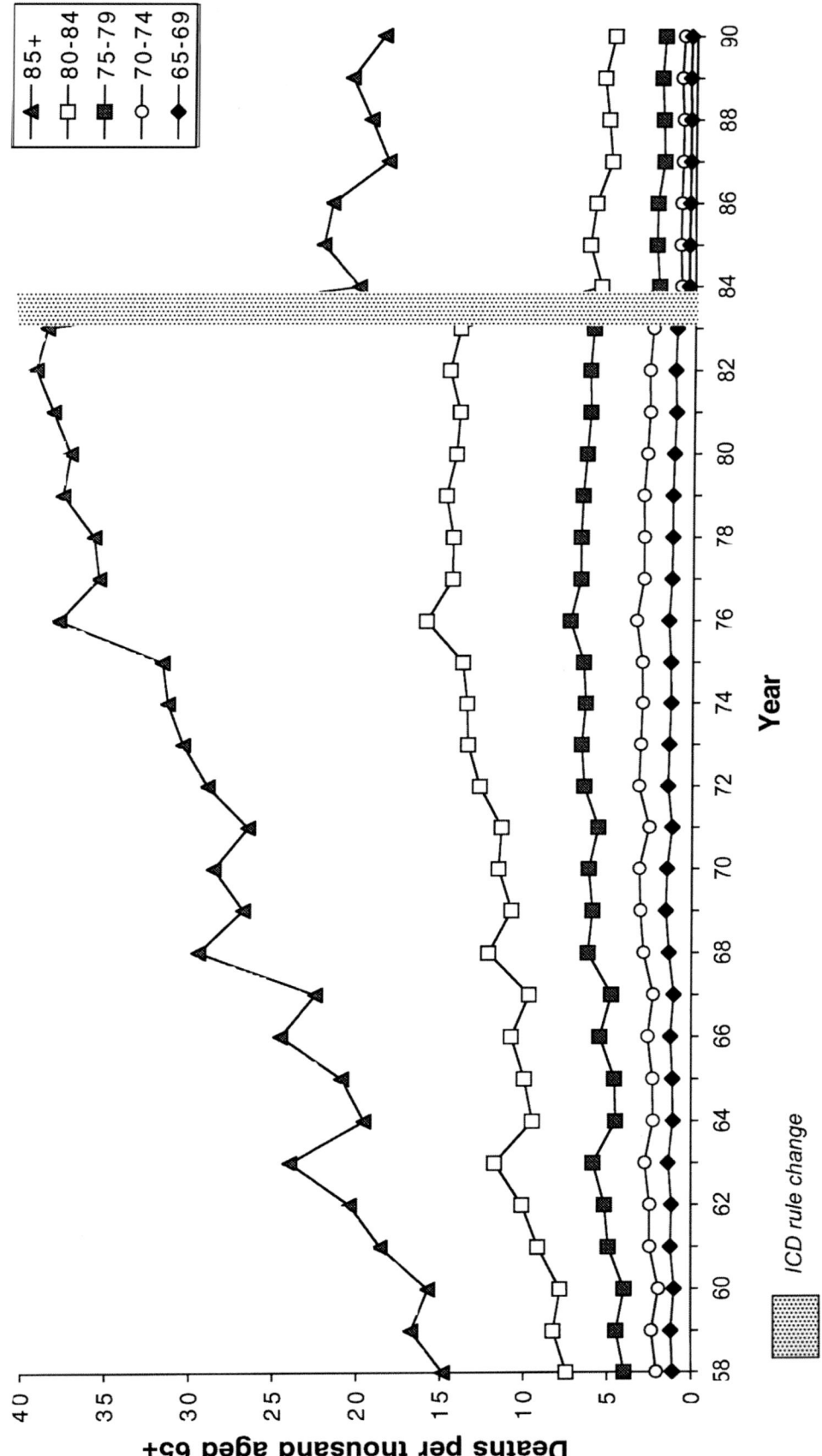

FIGURE 7-1
Graph showing the incidence of pneumonia in various age groups; males and females combined, England and Wales, 1958–1990. *(Reproduced with permission of the Lung & Asthma Information Agency, Department of Public Health Services, St. George's Hospital, London, England.)*

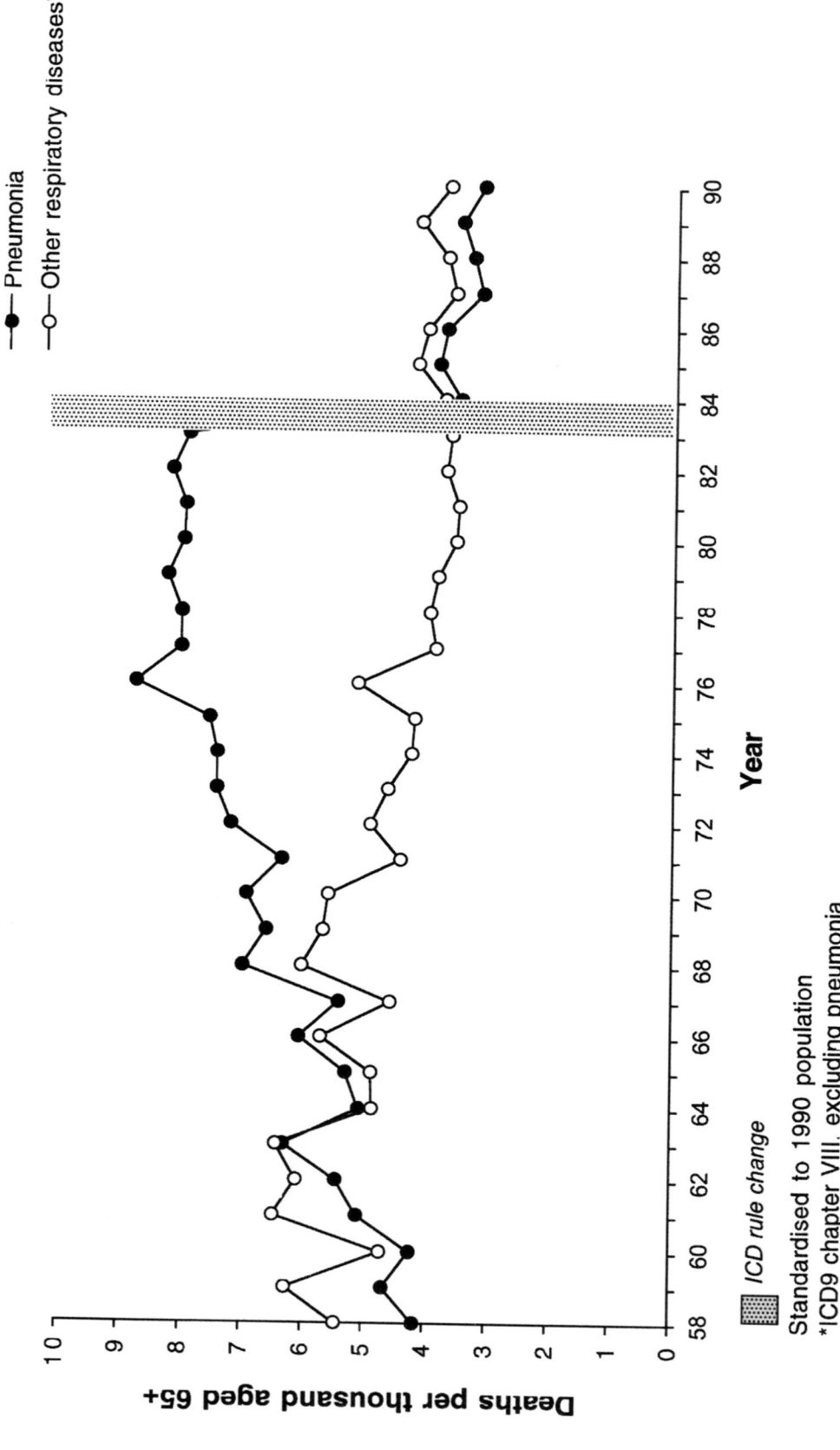

FIGURE 7-2
Age- and sex-standardized mortality rates for pneumonia and other respiratory diseases, age 65+, England and Wales, 1958–1990. *(Reproduced with permission of the Lung & Asthma Information Agency, Department of Public Health Services, St. George's Hospital, London, England.)*

What is unexplained is why Britain appears to be lagging behind other countries in overcoming adverse winter conditions that tend to be more severe in countries such as the United States, Canada, and Norway.

GRAM-POSITIVE PNEUMONIAS

Staphylococcus

ORGANISM AND EPIDEMIOLOGY

Staphylococci are gram-positive, nonmotile organisms that characteristically divide in more than one plane to form irregular clusters. They grow under aerobic and anaerobic conditions.[5] Staphylococci are naturally found on the surface of sebaceous glands of the skin and in mucous membranes of warm-blooded animals. The main habitats of the organism in humans are the anterior nares and the skin.

Hospital patients have higher carrier rates, which increase during prolonged hospital stay.[6] Most of these cases result from inhalation of ward dust, contaminated with virulent strains of staphylococci. These are derived from surgical wounds, rather than from the nursing and medical personnel. Staphylococcal pneumonia was found at necropsy in 6 percent of a large British series of patients dying in hospital.[7]

Staphylococci used to be an important cause of nosocomial infection, now largely replaced by gram-negative bacteria. It was a common organism in patients with previous influenza or cystic fibrosis and in the immunocompromised. It may complicate wounds or soft-tissue infections.[8] In young adults, intravenous drug abuse should be considered when there are recurrent staphylococcal infections.

The clinical features[9] are variable, depending on the age and previous health of the patient.

ADULT STAPHYLOCOCCAL PNEUMONIA

Fulminating

Hers and colleagues[10] described staphylococcal pneumonia in up to two-thirds of fatal cases after an influenza epidemic in the Netherlands.

A fulminating pneumonia can complicate postoperative staphylococcal septicemia.[11] Because of the ability of the staphylococci to enter blood vessels, there may be endocarditis, cerebral abscesses, or foci of infection elsewhere in the body. A fulminating type of staphylococcal pneumonia may complicate postoperative staphylococcal septicemia. Such cases are apyrexial with purulent bloody sputum with gram-positive cocci and a negative chest radiograph.

This type of infection is rare and is not associated with any antecedent viral infection.

In these cases the lungs show little distinctive change apart from a few scattered hemorrhages and a variable amount of edema, but microscopically there is generalized slight interstitial infiltration with polymorphonuclear leukocytes and mononuclear phagocytic cells both in the alveolar walls and in septal tissues. Unless the possibility of fulminating pneumonia is considered, the disease will remain undetected because the diagnosis depends on the culture of the organism from tissues and blood.

Acute

In the common type of staphylococcal pneumonia, the staphylococci invade the lungs very soon after the onset of the viral disease and can often be found within 36 h. Some cases cannot be distinguished macroscopically from the rare cases of primary influenzal viral pneumonia. The lungs are plum-colored, bulky, and heavy with edema fluid, which may replace up to four-fifths of the aerating tissue. On section the bronchi are filled with fluid that drains, leaving an inflamed bronchial mucosa often partly covered in the larger bronchi with a pseudomembrane of fibrin, resembling diphtheritic bronchitis. The cut surface exudes hemorrhagic fluid on gentle pressure, and widespread bleeding occurs. In brief, the picture is one of overwhelming hemorrhagic pulmonary edema, often fibrinous, which may show little or no evidence of actual pneumonic consolidation. Staphylococci are grown readily from the lung, the edema fluid, or the bronchial contents. Apart from the presence of a serohemorrhagic effusion, the pleural reaction may be negligible.

The disease starts initially as acute bronchitis and bronchiolitis. The epithelium in the larger bronchi is extensively destroyed, and the underlying walls are heavily infiltrated with polymorphs. In many places the ulcerated patches are covered with a fibrinopurulent membrane containing the organisms, which are readily demonstrable by the use of Gram's stain. In the smaller bronchi the rapidly developing necrotic bronchitis leads to thrombosis of the adjacent lobular branches of the pulmonary arteries and release of septic emboli. The alveoli are filled with proteinaceous edema fluid and often contain hyaline membranes, much extravasated blood, a few phagocytic cells, and a variable number of polymorphonuclear leukocytes. The microorganisms are readily found in the edema fluid.

In the less acute cases of staphylococcal bronchopneumonia, the initial reaction consists of focal patches of grayish-yellow peribronchial consolida-

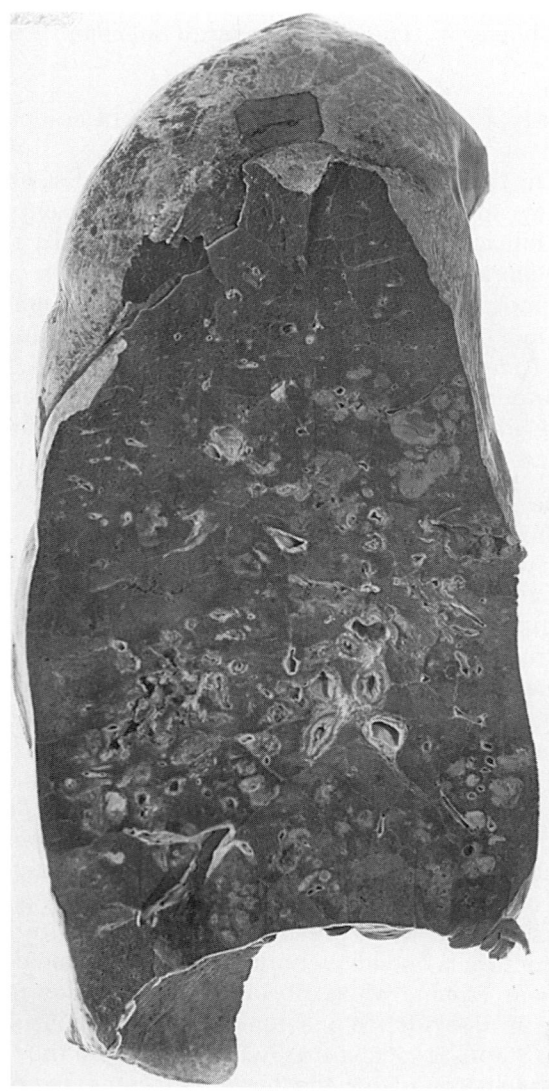

FIGURE 7-3
Acute staphylococcal pneumonia with multiple acute lung abscesses arising from a background of consolidation. (One half natural size.)

FIGURE 7-4
Chronic staphylococcal pneumonia with multiple chronic abscesses affecting the entire lung. (One half natural size.)

tion, situated mainly in the posterior segments of the lower lobes. These later break down to form small, ill-defined abscess cavities connecting with small bronchi, which are filled with yellowish sticky pus (Fig. 7-3).

The affected lung is filled with pus cells, and the infection spreads outward through the walls of the inflamed and ulcerated bronchioles. In the first instance the infection is a purulent bronchiolitis and bronchitis, and destruction of the bronchiolar wall allows free escape of infection outward as well as distally into the related air passages and alveoli. The abscesses enlarge rapidly and drain into the bronchi; if the patient does not die from septicemia, the abscesses spread quickly to invade much of a lobe and may even reduce the lung to a honeycombed

mass of chronic abscesses. Such abscesses may later become lined by granulation and fibrous tissue, and any intervening lung tissue becomes extensively damaged and fibrosed (Figs. 7-4 and 7-5). Many of the pulmonary vessels are obliterated by endarteritis. In long-standing cases an extensive chronic inflammatory cell reaction is found in the abscess walls. There may be an associated chronic empyema or dense pleural adhesions.

STAPHYLOCOCCAL PNEUMONIA IN NEONATES AND INFANTS

Bacterial tracheitis

This disease, previously referred to as nondiphtheritic laryngitis with marked exudate, was commonly described in pediatric textbooks before 1940. It apparently disappeared but has been recorded with increasing frequency since 1979. Donnelly and coworkers[12] described eight new cases and reviewed 110 previously described. Sabiston[13] noted a 10-fold

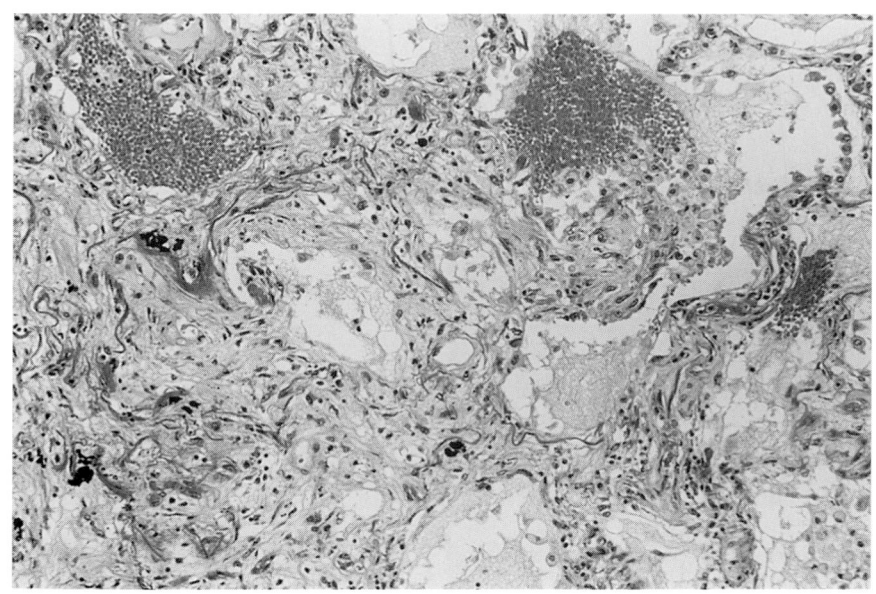

FIGURE 7-5
Intervening lung parenchyma showing organizing intraalveolar and interstitial fibrosis in a case of staphylococcal bronchopneumonia. (H&E, ×81.)

increase in incidence over two periods, 1951 to 1952 and 1955 to 1958.

In neonates, staphylococcal pneumonia occurs in small epidemics, usually associated with a high nasal carrier rate both in those who attend the infants, especially in hospitals, and in the infants themselves. Such infections may occur in the absence of any previous viral pneumonia.[14] In the first few weeks of life, staphylococcal pneumonia has a fulminating course, causing death in less than 48 h from onset. In children aged less than 1 year, staphylococcal pneumonia is almost always related to an antecedent viral infection, such as influenza or measles.

Clinically, there is a prodromal upper respiratory illness with stridor, fever, and a variable degree of respiratory distress. Bronchial obstruction can occur.[15] Most patients require endotracheal intubation or even a tracheostomy. There may be a preceding viral infection, such as parainfluenza. *Staphylococcus aureus* and *Hemophilus influenzae* are the predominant causes of bacterial tracheitis. Recurrent attacks of staphylococcal pneumonia in a child raises the possibility of chronic granulomatous disease.

Pathology
Macroscopically, there are purulent tracheal secretions, subglottic edema, and narrowing of the airways. There is either lobular or lobar hemorrhagic consolidation, which affects most of the lung. Areas of dark-red collapsed lung are seen below the surrounding aerated tissue (Fig. 7-6). There is little pleural reaction.

The affected lung shows alveoli filled with blood, edema, occasional macrophages, and a few

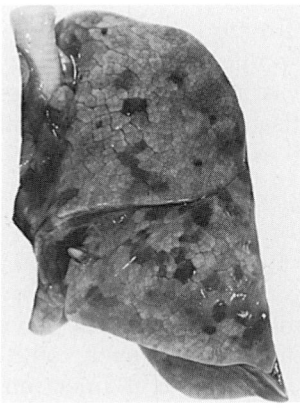

FIGURE 7-6
Neonatal staphylococcal pneumonia showing scattered hemorrhagic zones of consolidation. This appearance is typical of the early stages of staphylococcal pneumonia in infants. (Natural size.)

polymorphs. Many organisms are present. The bronchi contain pus and variable mucosal damage. Some cases pursue a less severe course; in these the pleural surfaces may be grayish-yellow with consolidation. There is a fibrinous pleurisy, and empyema is common in children under 4 months of age. On section, many of the zones of consolidation have broken down to form abscess cavities, which increase in number the longer the child survives. The smaller bronchi in these zones are filled with pus.

Abscess formation may lead to formation of tension pneumatoceles (Fig. 7-7). Air reaches the necrotic foci through the bronchi and, as a result of valvular obstruction, leads to difficulty in expiration. Because of the consolidation surrounding the pneumatocele, collateral ventilation is unable to remove the excess air. The cavities may then become greatly distended, causing severe respiratory embarrassment and interference with venous return. These air sacs may rupture, causing pneumothoraces and later a pyopneumothorax. Watkins and Hering[16] proved the

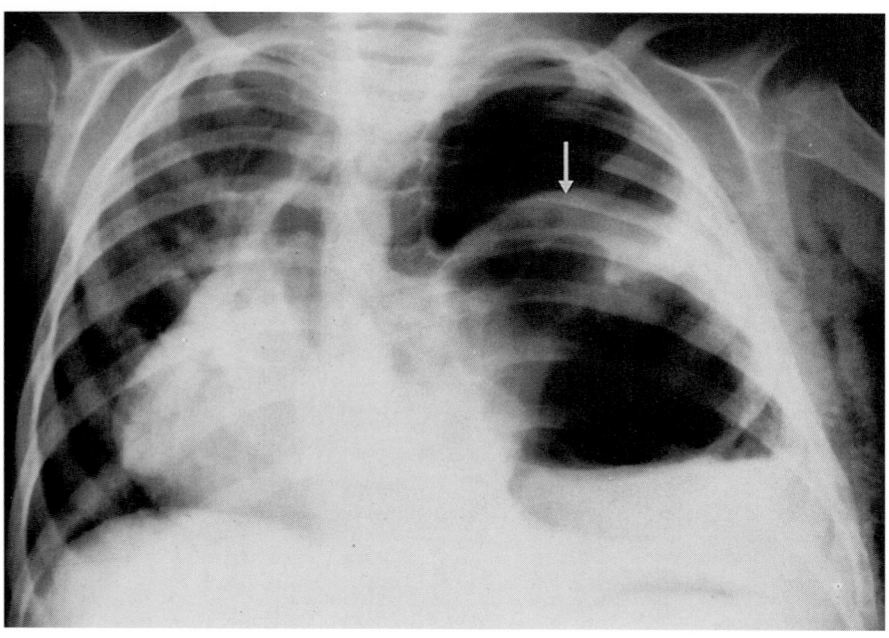

FIGURE 7-7
Radiograph of a tension pneumatocele (*arrow*) due to pneumonia in an infant. There is a pneumothorax in the upper part of the left pleural cavity.

valve-like nature of the obstruction. Pneumatoceles of this type may disappear spontaneously if the surrounding pneumonia resolves, allowing the excess intracystic pressure to be dissipated by the natural process of collateral ventilation.

Toxic Shock Syndrome

Toxic shock syndrome was first described by Todd and colleagues.[17] This is a multisystem disease occurring in women of reproductive age who have used tampons. There is sudden onset of fever, vomiting, diarrhea, hypotension, conjunctival injection, strawberry tongue, and a skin rash with subsequent desquamation. Pneumonia does not tend to be a feature of this disease. The pulmonary picture is that of the adult respiratory distress syndrome.

Streptococcus pyogenes (Beta-Hemolytic Streptococcal Pneumonia)

ORGANISM

Streptococci are gram-positive nonsporing organisms. Most will grow in air but some require the addition of CO_2 for growth.[17a] They are spherical in shape and are arranged in pairs or chains.

This infection resembles staphylococcal pneumonia, since it often follows an antecedent viral infection.

Many cases were noted in the 1918–1919 influenza pandemic. In the 1958 influenza pandemic, the incidence of secondary streptococcal pneumonia fell to less than 1 percent.[10] There is still a high incidence of the disease, as well as a high mortality in the first few days of hospitalization.[18–20]

The description that follows is based on material collected before the introduction of sulfonamides and antibiotics.

Streptococcal pneumonia is characterized by fever, rigors, cyanosis, pharyngitis, delerium, dyspnea, cough, blood-streaked sputum, and pleuritic chest pain. Death often occurs within 36 h of onset. Radiologically there rarely is bronchopneumonia.

Bronchopneumonia

ACUTE CASES

Macroscopically the lungs are bulky, dark, and edematous. Serosanguinous pleural effusions cause lower lobe collapse. Blood-stained, sometimes purulent fluid fills the bronchi and much of the parenchyma, where there is extensive hemorrhage. This leaves little functioning aerated tissue.

Histologically the bronchial and bronchiolar walls are infiltrated by polymorphs and mononuclear cells. The epithelium is shed, filling the lumina with debris, pus cells, and edema, containing large numbers of streptococci (Fig. 7-8). The alveoli are filled with edema, red blood corpuscles, and a few polymorphs and are often lined by hyaline membranes. If the patient lives 4 to 5 days, there is alveolar wall necrosis, followed by abscess formation caused by alveolar capillary thrombosis. The infection spreads rapidly because of liberation of fibrinolysins and hyaluronidase.

SUBACUTE CASES

Subacute cases are seen after a viral infection, and the patients live longer than a week. The lungs are

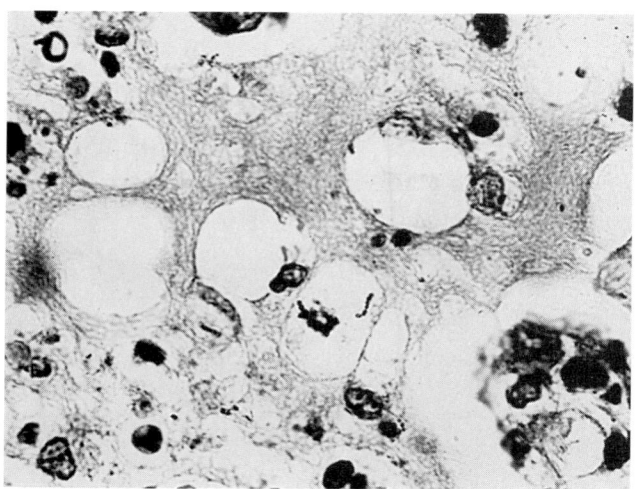

FIGURE 7-8
Hemolytic streptococci with readily identifiable organisms inside macrophages. (H&E, ×720.)

less edematous, and the smaller bronchi and bronchioles are surrounded by firm yellow tissue. Intervening parenchyma shows ill-defined foci of consolidation. There is a pleurisy and often empyema.

Hisologically there is bronchopneumonia, with severe damage to the bronchial and bronchiolar walls and loss of the lining epithelium. In addition there is associated edema, and an infiltrate of polymorphs and lymphocytes causing macroscopic thickening of smaller airways. The inflammation spreads interstitially into the walls of adjacent alveoli (Fig. 7-9), which show type II cell hyperplasia. Distal spread to related alveoli occurs. The larger bronchi are covered by a fibrinopurulent pseudomembrane. The bronchiolar lumina are filled with debris, pus, and large numbers of organisms. In time there is fibrosis of

peribronchiolar tissues and interlobular septa, which causes edema due to lymphatic obstruction. There may be lung abscesses, bronchiectasis, and widespread destruction and fibrosis.

Streptococcal pnemonia combined with *H. influenzae* infection, as occurred in the 1918–1919 pandemic, causes severe lung damage.

Pneumococcal Pneumonia (Lobar Pneumonia)

ORGANISM AND EPIDEMIOLOGY

Classic pneumococcal pneumonia, as taught to medical students, was the archetypal example of acute inflammation and the course it may follow. It might seem to be a disease on the wane in the Western world. In the 1930s, *S. pneumoniae* accounted for more than 80 percent of all community-acquired pneumonia[20a] but now accounts for 8 to 10 percent of all cases.[20b] This figure may be an underestimate since 60 percent of cases of bacteremic pneumonia are due to *S. pneumoniae*. Certainly epidemics of pneumococcal infections have rarely been reported in the antibiotic era. In the past decade, however, shelters for the homeless have emerged as a new site for institutional outbreaks.

In 1980, DeMaria and coworkers[21] reported a group of cases of pneumonia caused by capsular type I *Streptococcus pneumoniae* in a men's shelter in Boston. Nearly all the patients were alcoholics, and this association has been confirmed.[22] Many of the patients also smoked and had chronic bronchitis. These factors, as well as overcrowding, accounted for the high prevalence of the disease in these institu-

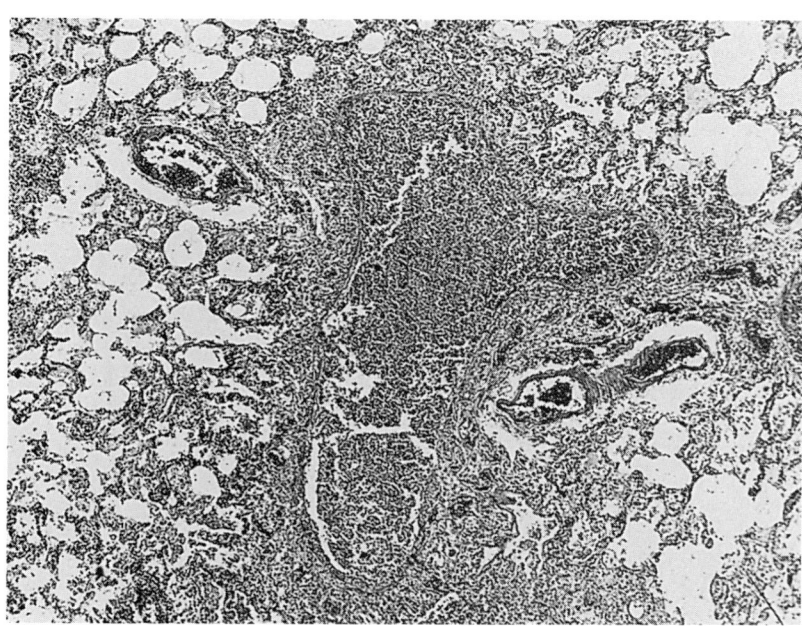

FIGURE 7-9
β-Hemolytic streptococcal pneumonia in a child showing peribronchial extension. (H&E, ×32.)

197

tions. In addition, shelter residents have a relatively high incidence of congestive cardiac failure as well as dementia and seizure disorders, which may predispose to pneumococcal infection. HIV infection, common in some urban shelters, is associated with an increased risk of pneumococcal pneumonia.[23]

The disease is uncommon in healthy young adults, with an estimated incidence of 0.2 case/year/1000 persons. It is more common in the elderly (incidence: 2.8 cases/year/1000 persons >70 years) and infants and children (incidence: 6.4 cases/year/1000 persons).[24]

S. pneumoniae is a gram-positive diplococcus that is capable of causing disease only if it is not ingested by macrophages. It can be a normal pharyngeal commensal. The organism measures 0.5 to 1.25 μm in diameter and is lancet-shaped. The composition of the polysaccharide capsule varies among the various serotypes. *S. pneumoniae* organisms characteristically associate in pairs and can be differentiated from other alpha-hemolytic streptococci because of their susceptibility to Optochin and their solubility in bile.

Capsular polysaccharide and a constituent of the cell wall, called techoic acid, appear to participate in the pathogenesis of the disease.[25] Pneumococcal antisera confer type-specific immunity in mice. The role of cytokines in the development of infection and the place of immunoprophylaxis in this infection have recently been summarized.[26]

Classic lobar pneumonia is still a common disease in many areas of the world. There is a high mortality in tropical regions. It is now seldom seen at autopsy in the West, as the organism is highly responsive to antibiotics. However, *S. pneumoniae* is developing resistance to pencillin and other antibiotics.[20b,26a] Major deficiencies still exist in our knowledge of interactions between the host and *S. pneumoniae*, especially in children in developing countries. The defects in host mechanisms that predispose to pneumococcal infection are given in Table 7-3.[27]

Surface IgA and IgG antibodies probably play some role in protecting against mucosal penetration of pneumococci, but it is difficult to understand why normal people should develop this disease. Antibody, complement, and phagocytosis must be important. In both adults and children, infection occurs after recent acquisition of a particular serotype rather than prolonged carriage of the organism. Rates of pneumococcal disease are highest in infants who have low antibody levels and whose production of antibody, in response to pneumococcal nasopharyngeal carriage, is relatively poor. Infants may develop

TABLE 7-3

Defects in Host Defense Mechanisms That Predispose to Pneumococcal Infection

Defects in nonimmunologic systems
  Skull fracture
  Obstruction of the eustachian tube
  Ethanol and narcotic intoxication
  Decreased vascular perfusion (sickle cell disease causing infarction, edema due to congestive failure or nephrotic syndrome)
Defects of phagocyte function
  Neutropenia
  Hyposplenia
Defects of humoral systems
  Hypogammaglobulinemia
  Specific antibody deficiency
  C2, C3, or factor I deficiency

SOURCE: After Johnston RB Jr: Pathogenesis of pneumococcal pneumonia. *Rev Infect Dis* 13(suppl 6):S509–S517, 1991.

pneumococcal septicemia due to colonization from the mother's genital tract.[28] Increase in human contact also plays a role, as does cold air or cold-water immersion.

Antecedent viral infections may predispose to the disease. Fekety and colleagues[29] found parainfluenza to be the most common preceding virus; 32 percent of their patients with pneumococcal pneumonia gave a history of an upper respiratory infection preceding the onset of the pneumonia.

CLINICAL FEATURES

The clinical course may be dramatic, with an abrupt onset of a severe shaking chill that lasts for approximately 1 h. Thereafter the patient develops a fever of up to 106°F, tachypnea, dry cough, and severe pleuritic chest pain. Nausea and vomiting are present in about a third of patients. By day 2 to 3 the patient is severely ill and markedly weak with prominent respiratory distress. The cough becomes productive of rusty thick tenaceous sputum, and there are coarse inspiratory rales, increased vocal fremitus, and bronchial breath sounds as well as a localized pleural friction rub. These are the classic signs of a pneumococcal pneumonia, but in the antibiotic era the signs are likely to be different. Even common manifestations, such as fever and sputum production, are not invariably present.

Radiologically, it may be difficult to distinguish this disease from other infections, such as those caused by *Legionella, Mycoplasma, H. influenzae,* and *Klebsiella pneumoniae.* Typically there is lobar consolidation. However, more than 50 percent of patients have only a small infiltrate, involving less than one segment.[30] Many patients have multifocal or patchy infiltrates.

In untreated cases the patient is ill for up to 10 days, after which the infection resolves promptly by crisis or more gradually by lysis. Penicillin will bring about an improvement within 36 h.

Engel[31] studied the early stages of lobar pneumonia in children below the age of 4. He found that in most cases the disease was initially localized to the posterior (basal) segment of the right upper lobe and the apical segments of the lower lobes. This distribution argues in favor of inhalational infection. For some unknown reason, in children the lesion remains more localized than it does in adults.

The changes described are those present in untreated adult cases. Following the use of sulfonamides and antibiotics, the macroscopic picture is altered and will be discussed separately.

PATHOLOGY

Macroscopically, lobar pneumonia may progress through (1) spreading inflammatory edema, (2) red hepatization, (3) gray hepatization, and (4) resolution.

1. The early spreading stage of the disease is seldom seen and is unlikely to be recognized in the absence of later stages elsewhere in the same lung. It resembles any other form of pulmonary edema, but the fluid contains the causative organisms.
2. In red hepatization the affected lobe is firm, airless, and brick-red. There are petechiae beneath the pleura, which shows a thin coat of fibrin. The cut surface is intensely congested and airless, and anthracotic pigment is conspicuous in city dwellers. Smaller bronchi in the affected lobes are plugged with fibrin.
3. In the stage of gray-hepatization (Fig. 7-10), the lung is firm and noncrepitant and covered by a thick film of fibrin. On section it is gray-yellow, and the anthracotic pigment gives it an appearance similar to that of gray granite.
4. The stage of resolution is marked by a return to the normal red color, although the affected lung may remain airless for a further short period, during which fibrin is absorbed.

Histologically, during the initial stage of invasion the affected lung is filled with edematous fluid containing a few polymorphs and an abundance of

S. pneumoniae. This stage is seldom recognized and is most likely to be found in patients who die after illness lasting only a few hours.

The first stage that is usually recognized is red hepatization (Fig. 7-11), when the alveoli are filled with red blood corpuscles and fibrin but few polymorphs. At this stage the alveolar capillaries are very congested, and it is thought that the bronchial arteries become blocked proximal to the affected lung.[32] The flow through the bronchial arteries is resumed during gray hepatization.

At the stage of gray hepatization (Fig. 7-12), the alveoli are filled with fibrin, large numbers of polymorphs, and few red blood corpuscles. Alveolar capillaries are inconspicuous, and pulmonary arterioles may become thrombosed. When recovery begins, alveolar macrophages appear in increasing numbers, replacing the polymorphs. Although polymorphs engulf *S. pneumoniae,* they do not destroy them. In the absence of a macrophage reaction, resolution of the exudate fails and the organisms persist.

Incomplete resolution occurs when the fibrinous exudate organizes instead of becoming absorbed in the normal manner. The alveoli become filled with a proliferating mass of fibroblasts, which extend through the pores of Kohn (Fig. 7-13). The remains of the alveoli frequently epithelialize, and the whole process is referred to as *carnification.*

Symmers and Hoffman[33] found incomplete resolution progressing to fibrosis in 3.2 percent of 125 postmortem cases of lobar pneumonia. In 1949, Gleichman and coworkers[34] showed delayed resolution in 26 percent of their cases. Three years later, Auerbach and colleagues[35] drew attention to the increasing number of cases of all forms of pneumonia that were failing to resolve completely as compared with cases in the period 1930–1940. This was attributed to the introduction of antibiotics and the increase in incidence of viral pneumonia. The fibrosis occurs mainly in the peribronchial and subpleural regions and may be related to the paucity of the polymorph response in antibiotic-treated cases.

Short-term complications are uncommon. There may be pleural effusion, which is small and self-limited. Empyema may develop, and because the organism can spread rapidly. Pneumococcal pericarditis is another complication. There may be pneumococcal meningitis, infective endocarditis, septic arthritis, and rhabdomyolysis secondary to the bacteremia.[36] However in men pneumonia in early childhood (organism unknown) is related to the development of chronic obstructive pulmonary disease in later adult life. Measles and whooping cough did not affect lung function.[36a]

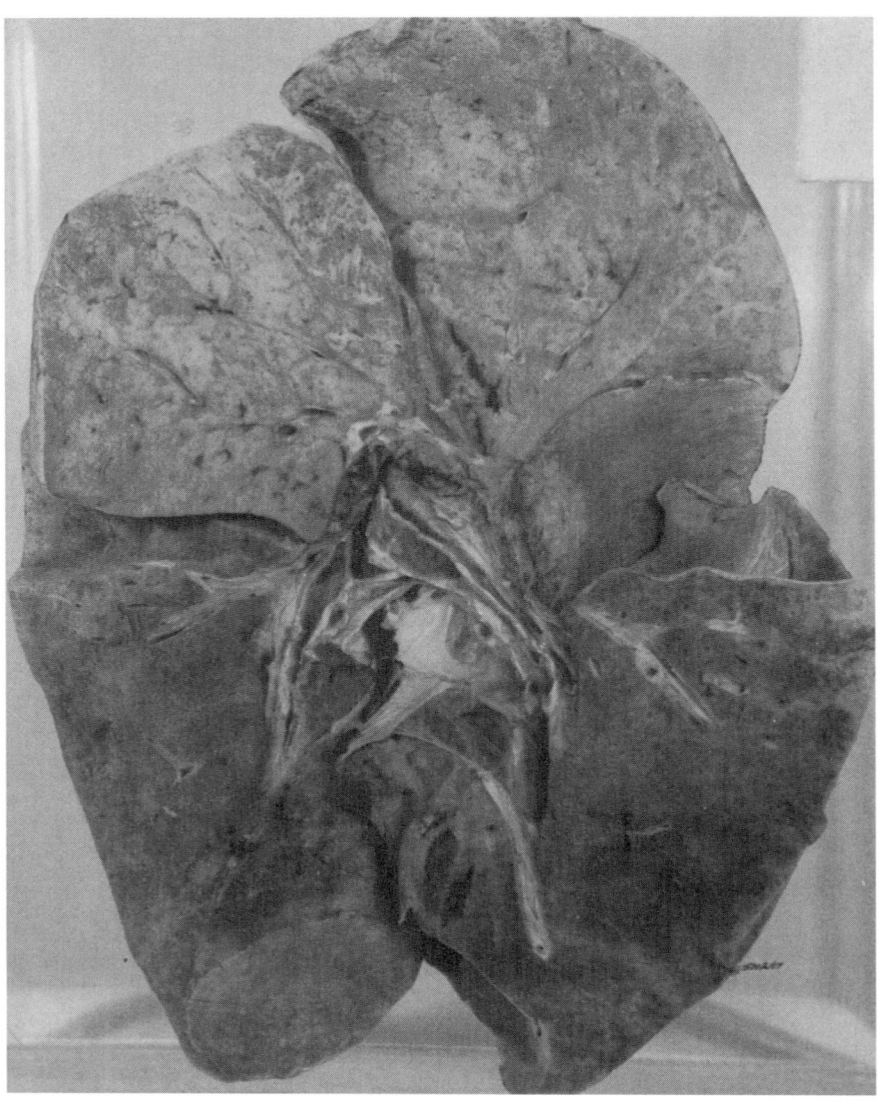

FIGURE 7-10
Acute pneumococcal lobar pneumonia in the upper lobe at the stage of gray hepatization and in the lower lobe at the stage of red hepatization. (Three-fourths natural size.)

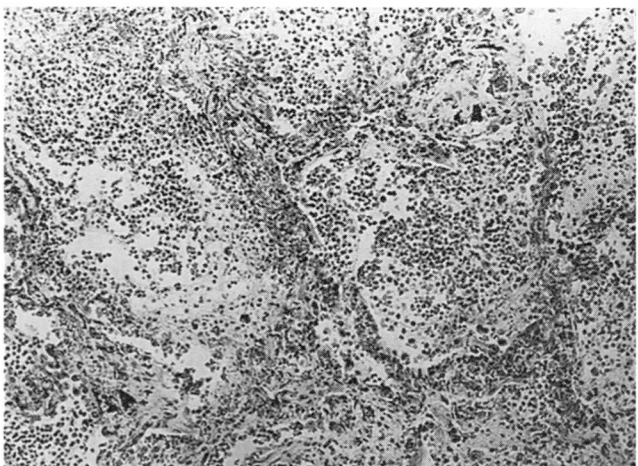

FIGURE 7-11
Pneumococcal lobar pneumonia at the stage of red hepatization showing dilated and congested alveolar capillaries. There are also numerous red blood corpuscles in the alveolar lumina. (H&E, ×112.)

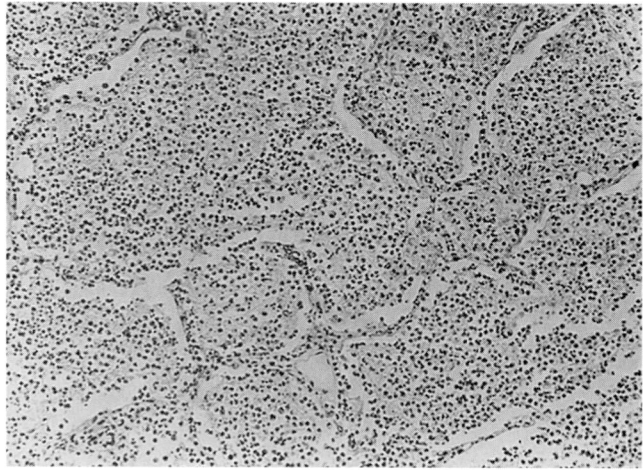

FIGURE 7-12
Pneumococcal lobar pneumonia. Gray hepatization with avascular alveolar capillaries and alveoli full of neutrophil polymorphs. (H&E, ×112.)

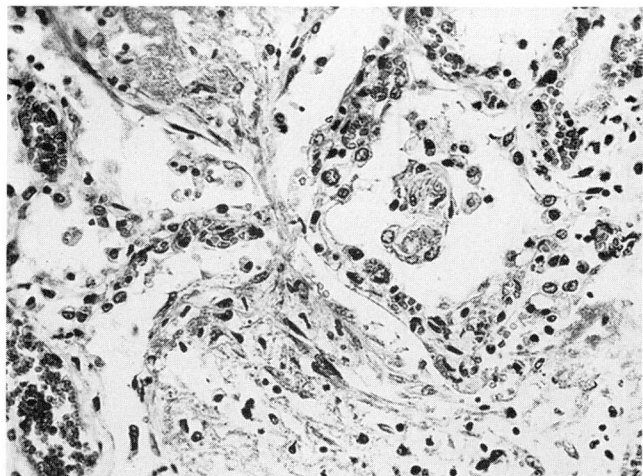

FIGURE 7-13
Organizing pneumococcal lobar pneumonia with intraalveolar fibrosis extending between alveoli through a pore of Kohn. (H&E, ×216.)

DIAGNOSIS

The diagnosis is usually made from examination of the sputum and isolation of *S. pneumoniae*, which may be obtained from blood, pleural fluid, or even lung tissue. The expectorated sputum must contain numerous polymorphonuclear leukocytes, unless the patient is neutropenic, and a few bronchial epithelial cells or alveolar macrophages to ensure that one is not examining oropharyngeal contents. In debilitated or elderly patients, sputum specimens may be difficult to obtain and invasive techniques, such as bronchoscopy and transtracheal aspiration, are used.

It has been shown that pneumococcal C and capsular polysaccharide antigens can be detected in various body fluids.[37] Most studies have used countercurrent immunoelectrophoresis to detect these antigens (Omniserum, Statens Serum Institut, Copenhagen). They are found most frequently in the sputum, followed by urine, and may persist for many weeks after admission; detection is unaffected by the prior use of antibiotics. The assays are rapid, giving a same-day result. For sputum assay the sensitivity is 68 to 94 percent, and the specificity for pneumococcal pneumonia is good. The widespread routine use of antigen detection has to await further research to identify the best antigen to assay, which specimens to collect, and when. Positive blood cultures in lobar pneumonia were found in 65 percent of cases before the introduction of modern chemotherapy. It is rare to detect circulating microorganisms in childhood pneumococcal pneumonia.[38]

Alternative Manifestations of Pneumococcal Infection

Some serotypes of *S. pneumoniae* seem able to cause a lobular type of bronchopneumonia and may complicate a wide variety of diseases at all ages (Fig. 7-14) p. 196. Pneumonia in these cases occurs most commonly in early childhood and infancy as well as a terminal illness in adult life. *S. pneumoniae* may cause adult respiratory distress syndrome.[39] Another unusual occurrence is severe pleuritis in the apparent absence of bronchopneumonia. These patients developed persistent high fever and intense pleuritic pain after severe pharyngitis.[40]

Anthrax Pneumonia (Wool-Sorters' Disease, Rag-Pickers' Disease, Pulmonary Anthrax)

ORGANISM AND EPIDEMIOLOGY

Bacillus anthracis is an aerobic, endospore-forming bacillus that is gram-positive, at least in its early stages of formation. The cells are rod-shaped, and most are motile with peritrichous flagella. Endospores, one per cell, are formed and are resistant to adverse conditions.[41]

B. anthracis produces three toxins,[41] referred to as factor I, the edema factor; factor II, the protective antigen (PA); and factor III, the lethal factor (LF). The edema toxin possibly damages endothelial cells, rendering them more permeable and resulting in intravascular thrombosis.

Anthrax, a disease mainly of cattle, sheep, and goats, has slowly declined in incidence as these animals have been vaccinated. Occasionally cutaneous and septicemic forms of the disease occur in humans. A disease known as anthrax pneumonia first appeared in the United Kingdom in 1837. This followed the importation of alpaca wool from Peru and mohair from eastern Anatolia for use in the mills in Bradford, Yorkshire. A dusty atmosphere was created in the sheds where bales of wool were first opened. The sorters were liable to inhale anthrax spores carried in the raw material.

Several years ago, there was an outbreak of mainly cutaneous anthrax in Switzerland that affected 25 workers in one textile factory.[42] Now only occasional cutaneous lesions are seen. Sporadic cases of pulmonary anthrax have been seen in North Africa in those engaged in handling sacks of dry bones, used in the preparation of bone meal. Some of the animals probably died from anthrax. Anthrax now is mainly confined to the arid subtropical regions, especially in the Middle East (Iran), north

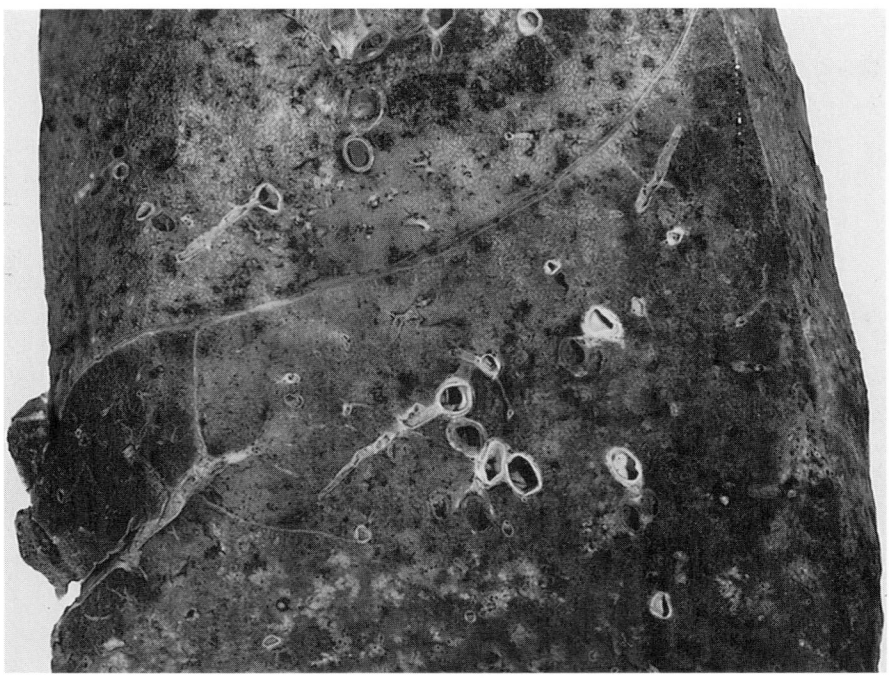

FIGURE 7-14
Pneumococcal bronchopneumonia with pale zones of consolidation, some related to small bronchi. In many areas the consolidation has become confluent. (×1.6.)

Africa, South West Africa, Ethiopia, Somalia, and the Indian subcontinent. All these are regions where goats, sheep, and cattle are reared. A small outbreak of pulmonary anthrax occurred in New Hampshire in 1957, and the pulmonary pathology was subsequently described by Albrink and colleagues.[43]

The organism has been manufactured for biologic warfare. Recently the British government handed back a remote Scottish island, Guinard Island, which had been requisitioned 50 years earlier for biologic warfare experiments[44]; the monitoring and detection of the organism on the island have been described.[45] In 1979, 96 deaths were ascribed to anthrax poisoning in the Sverdlovsk region of Russia, 850 miles from Moscow, in the Urals (Fig. 7-15). The cases had probably arisen from the release of anthrax spores into the air from a military establishment situated upwind.[46]

CLINICAL FEATURES

The typical dermatologic features in the Swiss epidemic were initially a pruritic, insectlike pimple and then a painless ulcer surrounded by serous hemorrhagic, often rapidly confluent vesicles and pitting edema. In the region of the ulcer, a black area of necrosis developed that was never colliquative but was transformed into the typical pitch-black, firmly adherent eschar. Lymphadenitis and lymphangitis were important manifestations.[43]

FIGURE 7-15
Map from *The Independent* showing the Sverdlovsk region.

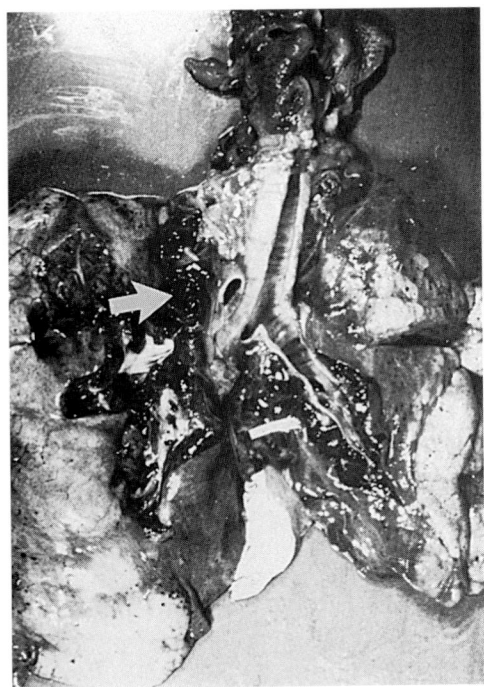

FIGURE 7-16
Anthrax showing hemorrhagic enlargement of the hilar nodes (*arrow*). (*Courtesy of Dr. Haghighi.*)

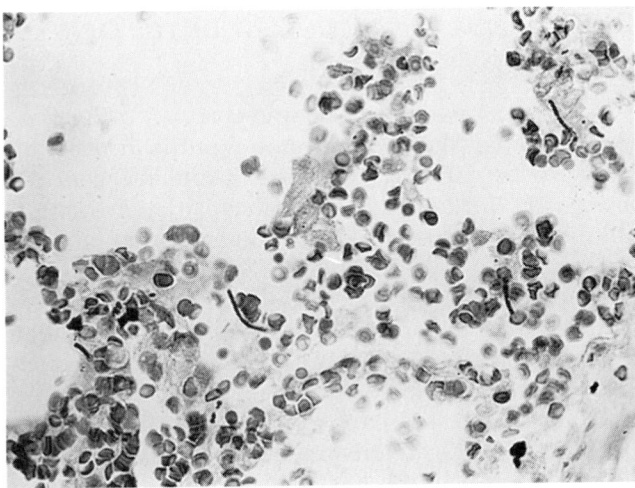

FIGURE 7-17
Anthrax pneumonia with little inflammation. Large anthrax bacilli lie amid intraalveolar hemorrhagic exudate.

Initially, pulmonary anthrax gives rise to nonspecific flulike symptoms with a low-grade fever, malaise, myalgias, and nonproductive cough. After several days of apparent improvement the patient's condition deteriorates, with marked dyspnea and occasional stridor due to tracheal compression from enlarged mediastinal lymph nodes.

The spores of *B. anthracis* are first taken up by pulmonary macrophages, are carried to the lymphatics, and reach the hilar lymph nodes.[47] The vegetative forms of the organism develop within the macrophages and are mostly liberated in the sinuses of the regional nodes. From there they spread via the lymphatics to the bloodstream, causing septicemia.

PATHOLOGY

Macroscopically, there is extensive hemorrhagic edema surrounding the hilar nodes (Fig. 7-16). This extends widely into the mediastinum, resembling the appearance of a ruptured thoracic aneurysm. The lymph nodes are enlarged, hemorrhagic, and edematous. Pulmonary arteries may become dissected by an extension of the mediastinal hemorrhage. The pleural cavities contain much blood-stained fluid.

The lungs are bulky, purple, and filled with brown-stained fluid together with areas of hemorrhage. The larger bronchi contain blood and mucus.

In addition to the pulmonary changes, there may be anthrax meningitis and ileitis. Death results from shock due to loss of blood and fluid into the extravascular compartment.

Histologic examination shows few neutrophil polymorphs, unless there is a terminal pneumonia caused by secondary pyogenic infection. The bronchial epithelium may be partly lost and the lumen filled with edema, epithelial debris, red blood corpuscles, and bacilli. The alveoli are filled with red blood corpuscles and edematous fluid, swarming with the large gram-positive organisms (Fig. 7-17). Hemosiderin-laden macrophages are found if the patient survives beyond a few hours. Longer survival may be followed by necrosis of the lung tissue.

In the majority of sporadic cases, diagnosis may be made only after postmortem examination is completed. In all cases in which the clinical diagnosis is made during life, the subsequent postmortem examination should *not* be undertaken for fear of disseminating the spores in spilled blood.

The diagnosis may be made retrospectively using a Russian allergen, "anthraxin," an intracutaneous test unknown until recently in the West, and also by enzyme-linked immunoassay. It is important to look for cutaneous skin lesions.

GRAM-NEGATIVE PNEUMONIAS

Haemophilus influenzae Pneumonia

ORGANISM AND EPIDEMIOLOGY

H. influenzae are small to medium-sized coccobacilli or rods. There are often pleomorphic, sometimes filamentous, gram-negative, nonmotile, nonsporing organisms.[48] They possess capsular antigens, enabling them to be divided into groups A to E.

Up to 80 percent of people carry *H. influenzae*. In the United States, children 6 months or older are likely to be carriers.[49,50] Most nasopharyngeal isolates of *H. influenzae* are unencapsulated (nontypable). Most episodes of infection in young children are caused by type B strains.[51] Spread of *H. influenzae* type B in families occurs over weeks or months.[52] This suggests that affected children may disseminate organisms very efficiently. The only known reservoirs for *H. influenzae* in humans are the respiratory tract and, to a lesser extent, the conjunctival and genital surfaces. The organism is associated with chronic bronchitis and in the East with diffuse panbronchiolitis (p. 247). Patients with AIDS are at increased risk of developing *H. influenzae* bacteremia.[53]

In comparison to those in children, adult infections due to *H. influenzae* type B are uncommon. There have been hospital outbreaks of nontypable strains of *H. influenzae*.[54] These outbreaks occurred in respiratory, general medical, and geriatric units. Usually there is underlying respiratory pathology.

There may be intraamniotic infection with *H. influenzae*; most cases occur with preterm rupture of membranes or with preterm labor and intact membranes. The infant described in the case reported by Mazor's team[55] developed respiratory distress syndrome and *H. influenzae* septicemia.

Macroscopically and histologically, the disease may resemble *Streptococcus pneumoniae* infection. In patients with chronic bronchitis, the organism appears to live saprophytically in the bronchial mucus between episodes of purulent bronchitis and bronchopneumonia. In most cases there is a surface mucopurulent bronchitis, often causing partial separation of the lining epithelium from the basement membrane. Bronchiolar mural and perimural lymphocytic infiltration accompanies these surface changes. Abscess formation may occur.

Klebsiella pneumoniae (Friedlander Pneumonia)

ORGANISM

K. pneumoniae is a nonmotile, usually capsulate gram-negative bacillus found in the bowel and respiratory tract of humans and animals as well as in soil and water.[56] The capsular polysaccharide enables different serologic types to be defined. There are now more than 80 different types, and the most common isolates in British series are K2, K3, and K21.

CLINICAL FEATURES

The classic clinical presentation of *K. pneumoniae* is rapid onset of acute pulmonary symptoms with rigors, fever, and cough productive of a thick, purulent yellow sputum in an alcoholic man who is also a chronic smoker. The patient is extremely ill, and frank hemoptysis may occur. The chest radiograph may show either lobar or lobular consolidation or a chronic stage with abscess formation and suppuration.[57] A bulging interlobar fissure used to be a common feature, but it is now uncommon.[58]

K. pneumoniae was believed to be a frequent cause of community-acquired pneumonia, especially in alcoholics. One recent study has suggested, however, that the disease is changing, in that cavitation is uncommon and immunosuppression is one of the primary risk factors.[58] Further studies will be needed to confirm these findings.

Another study[59] confirmed the positive association with alcoholism noted above. All of the hospitals from the series reviewed were public municipal hospitals in the United States, emphasizing the patient population at risk for community-acquired *Klebsiella* pneumonia. The incidence was far lower in private hospitals. From 1967 on, four series showed no more than four cases in any one year. This author questioned the use of sputum bacteriology to diagnose *Klebsiella* pulmonary infections unless precise criteria were followed—i.e., only sputum specimens with a predominance of polymorphonuclear cells were accepted for Gram's stain and culture.

PATHOLOGY

Primary *Klebsiella* pneumonia usually occurs in the right lung, especially in the posterior segment of the right upper lobe or in the apical segments of the lower lobe—a distribution that is characteristic of inhalation. Initially there is a lobular pattern to the pneumonia (Fig. 7-18), but this soon spreads into the surrounding lung, eventually involving one or more lobes (Fig. 7-19). Initially the consolidated areas are gray and exude a characteristic sticky exudate, which on direct smear shows numerous gram-negative, large, encapsulated bacilli. In a few days there is abscess formation, which often becomes extensive.

Histologically in the early stages there is an outpouring of viscid edematous fluid and mononuclear cells in which there are large numbers of organisms. Alveolar destruction follows, together with a polymorphonuclear leukocyte reaction, and later there is granulation tissue. In the early stages there is alveolar edema and distended perivascular and peribronchial lymphatics as well as interlobular septal edema.

In the chronic stage, seen more often in the preantibiotic era, there are multiple abscesses, with walls formed of granulation tissue. Secondary bacterial infection is common. There may be empyema.

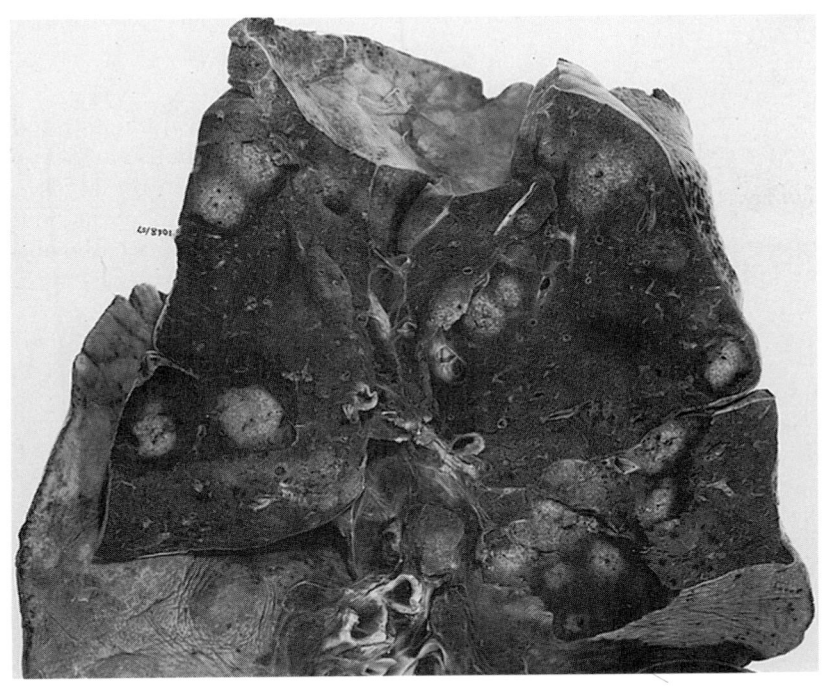

FIGURE 7-18
Discrete lobular zones of consolidation in early *Klebsiella* pneumonia. (Two fifths natural size, approx.)

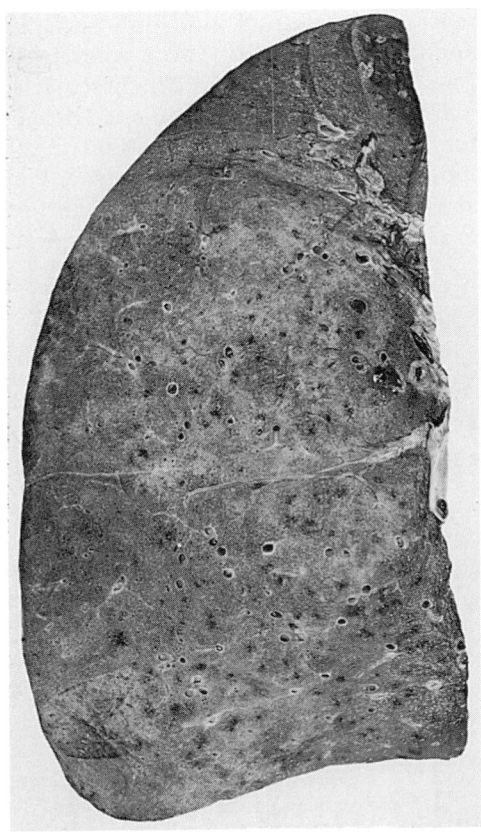

FIGURE 7-19
A later stage of *Klebsiella* pneumonia with lobar consolidation. (One third natural size, approx.)

Chronic *Klebsiella* pneumonia may simulate tuberculosis or chronic staphylococcal pneumonia. Without a clear history of the initial infection, it may be impossible to distinguish from the latter two diseases. The development of acute lung abscesses after acute *K. pneumoniae* infection is significantly higher than after any other common cause of community-acquired pneumonia, with the possible exception of anaerobic organisms.

There is a problem with the diagnosis of chronic pneumonia due to *Klebsiella* because some of the studies published on this subject have not looked for anaerobic organisms. Some documented cases[60] may have displayed chronic illness due to superinfection with another organism.

PULMONARY GANGRENE

The most dramatic complication of *Klebsiella* infection is massive pulmonary gangrene, which brings rapid total destruction of part of the lung due to vascular compromise. The main features separating pulmonary gangrene from necrotizing pneumonia and lung abscess are the extent of the necrosis and the presence of thrombosis in large vessels. Most cases are seen in men.

In their review of 24 cases, Penner and coworkers[61] found no identifiable host risk factor common to all or most cases. There were, however, the known risk factors for community-acquired pneumonia, such as alcoholism, chronic respiratory disease, diabetes mellitus, and nutritional deficiency.

Most cases are due to *Klebsiella*, but *S. pneumoniae, H. influenzae, Escherichia coli, S. aureus,* and *Bacteroides* have been implicated.[61]

The pathology is characterized by four main findings. These are the predominance of upper lobe involvement, especially of the right upper lobe, a prominent vasculitis in both bronchial and pulmonary arteries. In addition there is extensive vascular thrombosis in large and small pulmonary vessels and an intact architecture of the sloughed lung. Vascular thromosis is seen in lobar pneumonia but is limited to small vessels. The bronchial arteries dilate. The hallmark of pulmonary gangrene is infarction of an entire segment or lobe.

SECONDARY *KLEBSIELLA* PNEUMONIA

Secondary infection by *K. pneumoniae* is discovered only if the lungs are examined bacteriologically at postmortem. Kneeland and Price[62] recovered *K. pneumoniae* as often from normal lungs as from those that were pneumonic. This presumably is a result of terminal aspiration as well as colonization of the oropharynx with gram-negative rods, common in postoperative patients or those in intensive care units.

KLEBSIELLA RHINOSCLEROMATIS

This may be a rare cause of pulmonary infection. It has been documented in a pancytopenic patient who had respiratory failure and hypotension. Classically there are large vacuolated histiocytes, known as Mikulicz's cells (Fig. 7-20), with intracellular bacteria. The *Klebsiella* organisms are best demonstrated by Giemsa and silver impregnation methods.[63] The lungs show areas of hemorrhage within the alveoli, some disruption of alveolar walls, and occasional hyaline membranes, lymphocytes, plasma cells, and Mickulicz's cells in the alveoli. Polymorphs are rare.

WHOOPING COUGH (PERTUSSIS PNEUMONIA)

Organism

Whooping cough is caused by *Bordetella pertussis*, a minute gram-negative, non–acid-fast, nonsporing coccobacillus. *B. parapertussis* has given rise to cases of mild whooping cough, and *B. bronchiseptica* has been identified from the respiratory tracts of dogs with distemper and mistakenly identified as the cause of that disease. A fourth species, *B. avium*, causes respiratory tract disease in turkeys.[64]

Despite the availability of pertussis vaccine, whooping cough remains an important cause of morbidity and mortality. The yearly incidence in children under one year is 35 cases per 100,000, with a case fatality rate of 0.5 to 1 percent in children under the age of 6 months.[65]

Clinical Features

The disease is transmitted by droplet infection and has an incubation period of 1 to 2 weeks. There are then three stages of the disease. The first is a catarrhal phase, which cannot be distinguished from a viral respiratory tract infection. The patient may have a low-grade fever with sneezing, injected conjunctivae, and cough. In the paroxysmal stage the cough becomes more severe, with a series of characteristic

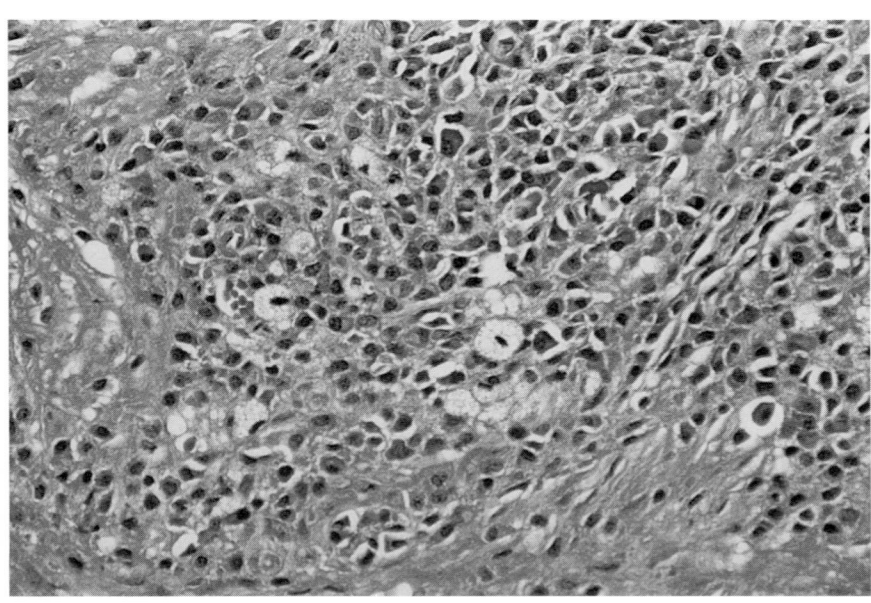

FIGURE 7-20
Klebsiella rhinoscleromatosis with large pale Mikulicz's cells as well as chronic inflammation and fibrosis. (H&E, ×219.) *(Courtesy of Prof. T. V. Colby, Mayo Clinic, Scottsdale, Ariz.)*

whoops. These are short coughs of increasing severity, followed by a large inspiration through a partially closed glottis. There is often vomiting at the end of the paroxysm, and there may be apneic attacks. The last stage is secondary bacterial pneumonia or otitis media. Convulsions occur in these patients, and they may require mechanical ventilation.

Infants have the highest death rate, and epidemiologic evidence[66] suggests that some cases of sudden infant death syndrome may be due to unrecognized pertussis infection. There may be very high leukocyte counts, sometimes reaching up to 100,000 cells per mm³, 80 percent of which are lymphocytes.

In adults the infection is less severe and may even be subclinical.[67,68] They usually do not show the leukocytosis or lymphocytosis seen in children. The major manifestation in adults may be rhinorrhea or sore throat, and the cough is rarely typical, with a whoop limited to only a small proportion of cases. Adults may act as a major reservoir of infection for children.

Pathology

Opportunities to study the pulmonary changes pathologically are rare, as the disease is seldom fatal. The most extensive account of the changes was provided by Feyrter,[69] based on 225 postmortems from whooping cough cases with histology in 100. Other fatal cases have been described by Goodpasture and colleagues.[70]

There is bronchitis and bronchiolitis, which extend to cause inflammation around the bronchi, and an interstitial pneumonia. Small bronchi and bronchioles are filled with mucopurulent material. In severe infections the epithelium shows patchy destruction and is shed into the lumen, leaving shallow ulcers. The bronchiolar walls are infiltrated by lymphocytes and plasma cells.

Goodpasture and colleagues[70] found intranuclear acidophilic inclusion bodies on the surface epithelium and in the mucous gland epithelial cells. There is a clinical similarity of some cases of adenovirus to classic whooping cough, and these acidophilic inclusions may have been viral.

A rapid diagnosis of whooping cough is now available with counterimmunoelectrophoresis for antigen detection, using a monoclonal antibody that has 85 percent sensitivity and 94 percent specificity in serum.[71] The lymphocytosis that occurs with whooping cough in infants and young children is characterized by morphologically abnormal cells that are predominantly CD4-positive and appear to represent an expansion of an immunophenotypically normal lymphocyte population.[72]

Gram-Negative Pneumonia

Gram-negative pneumonias and subsequent septicemia are now major causes of hospital deaths. This term does not usually include *H influenzae.* They occur in patients who have been severely ill, have been on the intensive care unit, or have taken broad-spectrum antibiotics, steroids, or immunosuppressive drugs. Very often the organisms may be multiple, and they include *Pseudomonas aeruginosa, Proteus* sp., *E. Coli,* and, less commonly, *Enterobacter* sp. and *Serratia* sp. As has been noted in the section on *Klebsiella,* gram-negative rods colonize the oropharyngeal cells because of a decreased amount of fibronectin, which usually coats sugar receptors and prevents attachments of such organisms to these cells. The fibronectin is destroyed by elastase elaborated by polymorphs. *Klebsiella* infection is one of the other hospital-acquired infections, but it is relatively uncommon.

Clinically the diagnosis of the pneumonia may be difficult, and the presenting feature may be septicemic shock. There is fever, lymphocytosis, and mucopurulent sputum. The presence of gram-negative bacilli in the sputum is insufficient to prompt a diagnosis of gram-negative pneumonia, since such organisms can occur in any hospitalized patient.

Raised endotoxin levels in bronchoalveolar lavage fluid have been demonstrated in patients with gram-negative pneumonia. An endotoxin level equal to or greater than 6 EU/ml distinguished patients with gram-negative pneumonia from colonized patients as well as from those with pneumonia due to gram-positive cocci.[73]

Pseudomonas Pneumonia

ORGANISM AND EPIDEMIOLOGY

Pseudomonas are straight or slightly curved gram-negative rods. They are motile by means of one or more polar flagella, although some strains also have lateral flagella of different lengths.[74]

This organism can be isolated in 24 percent of cases coming to postmortem.[62] In a fifth of the cases there is no pneumonia, but 93 percent of the patients have received antibiotic therapy. Lung infection, when present, is rapid in onset. Other factors that encourage pseudomonal infection are tracheostomy and mechanical ventilation. It may be seen on premature baby units after mechanical ventilation, and humidifying aerosols are usually the major source of contamination.[75,76] The organism is a common pathogen on burns units. It is also found as a secondary pathogen in diffuse panbronchiolitis.

Pseudomonas is common in cystic fibrosis, especially the mucoid strain. *Ps. (Burkholderia) cepacia* has recently been a notable cause of death in these patients.[76a] This organism may be disseminated in the air from affected individuals[76b] though others remain unconvinced of the transmittability of *Ps. cepacia*.[76a]

The morbidity and mortality resulting from nosocomial pneumonia caused by *Pseudomonas* species are thought to be due to an ineffective humoral response to the organism. IgG2 antibodies in response to pseudomonal lipopolysaccharide are poorly opsonic. IgG humoral response to *Ps. aeruginosa* is made ineffective by another unique mechanism. It is now known that *Pseudomonas* can produce an elastase that cleaves IgG and is thus an important virulence factor for patients with these infections.[77]

PATHOLOGY

The pathology of *Pseudomonas* pneumonia was described by Teplitz.[78] Macroscopically, it may present as numerous multifocal, indurated, hemorrhagic areas resembling infarcts. There are serosanguinous pleural effusions. Alternatively, the lungs contain scattered firm, yellow, irregular, lobulated areas of consolidated tissue (Fig. 7-21) that project above the cut surface and are usually surrounded by a hemorrhagic border. The centers of some foci become soft, gelatinous, and, in severe cases, necrotic (Fig. 7-22). The bronchial mucosa in the large airways is red and velvety, and the lumen is filled with frothy fluid. The lungs have a characteristic odor, associated with cultures of *Ps. aeruginosa*.

Histologically, the bronchopneumonia is usually associated with a polymorph response (Fig. 7-23), but around the pneumonic area there is usually a zone of hemorrhage and a fibrinous exudate. The alveolar walls become necrotic and show a hyaline structure. There is thrombosis of alveolar capillaries, which extends centripetally to involve the pulmonary arteries; the organisms invade and destroy the walls of pulmonary arteries and veins, spreading inwards from the adventitia causing an arteritis (Fig. 7-24). This may lead to necrosis of the arterial and venous walls and subsequent hemorrhage. Numerous gram-negative bacteria can be demonstrated in the outer vascular wall. When the vascular destruction is less severe, there is intimal fibrosis. The vascular changes are most common when there is leukopenia and few polymorphs in the lung.

As the pneumonic changes progress, the nuclei of polymorphs, lymphocytes, and mononuclear cells

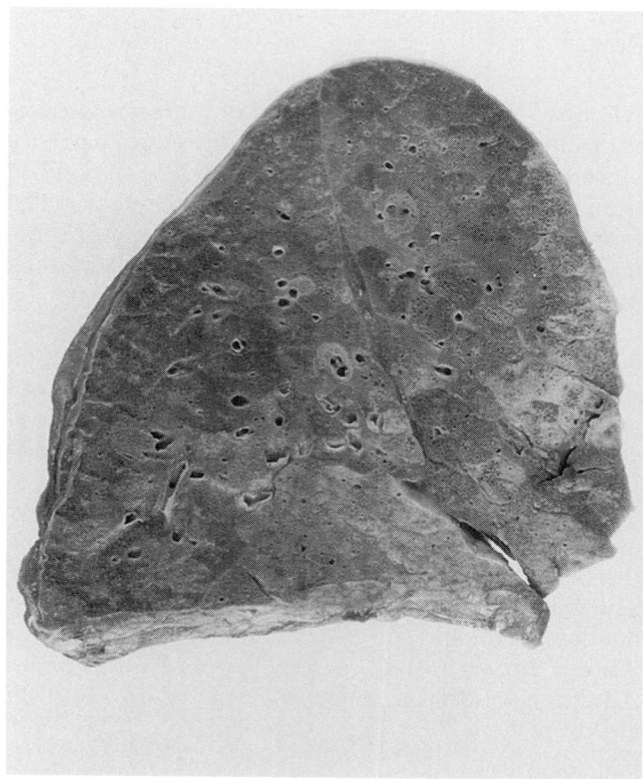

FIGURE 7-21
Lobular consolidation in *Pseudomonas* pneumonia. (×1.2, approx.)

undergo necrosis. Nuclear debris and fibrin fill the centers of the lesions along with necrotic lung tissue. The contents of the resulting abscesses consist of sticky material; the abscesses may be single or multiple. All forms of *Pseudomonas* pneumonia are associated with varying degrees of hemorrhage into the surrounding tissues, which is reflected in the macroscopic appearance. The smaller bronchi and bronchioles may be caught up in the spreading necrotic lesions, and in the larger bronchi the lining epithelium often undergoes metaplasia to a multilayered type of epithelium.

The organisms are localized in cases of cystic fibrosis predominantly around small airways (less than 1 mm) and associated with obliterative bronchiolitic changes.

Bacillus proteus Pneumonia

The same etiological background as for *Pseudomonas* pneumonia holds true for *Bacillus proteus*. However, the bronchopneumonia is less fulminating and presents as a lobular areas of consolidation. Diagnosis is made from bacteriologic culture.

E. coli Pneumonia

E. coli is probably a secondary contaminant in most cases of respiratory infection in adults. In the first 3

FIGURE 7-22
Confluent pseudomonal pneumonia with early abscess formation.

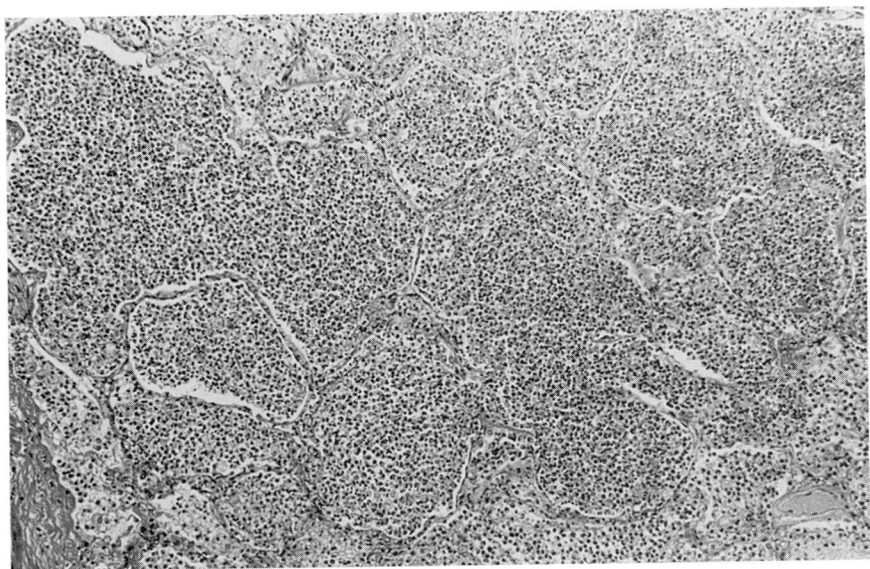

FIGURE 7-23
Pseudomonas bronchopneumonia with confluent consolidation. (H&E, ×4.4.)

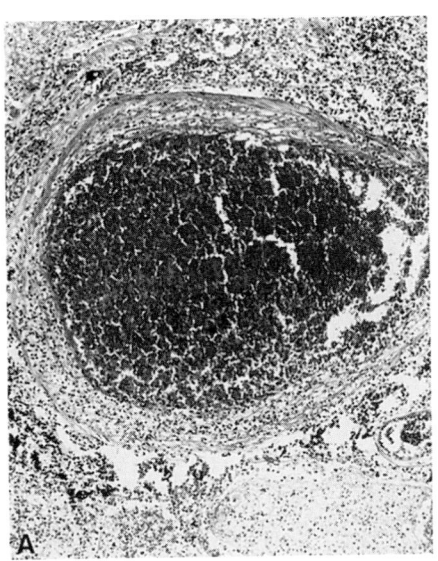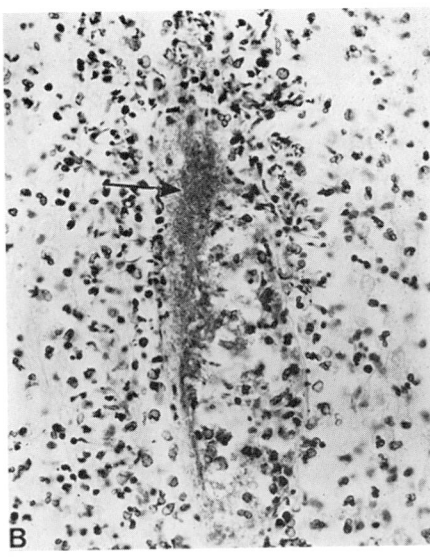

FIGURE 7-24
A. Pseudomonas pneumonia with arteritis. H&E, ×53. *B.* Another vessel with a colony of *Pseudomas* species in one area (*arrow*) of the wall. (Gram stain, ×210.)

weeks of life, however, *E. coli* may be pathogenic and cause pneumonia. *E. coli* has been cultured from amniotic fluid and may be responsible for neonatal pneumonia after aspiration in utero.[79]

Malakoplakia

This uncommon inflammatory disorder has been linked with *E. coli* infection, although other gram-negative and gram-positive bacteria (including mycobacteria) and fungi have been implicated.[80]

Rare cases have been documented in the lungs. It may be seen with emaciation,[81] alcohol abuse,[82] renal[83] or heart transplantation,[84] or AIDS; it may be due to *Rhodococcus equi*[85,86] or may be a complication of endobronchial Hodgkin's disease.[84]

CLINICAL FEATURES

Patients present with multiple pulmonary nodules, which appear bronchial-based, or cavitating nodular lesions, which contain yellow friable material or yellow plaques. The multiple nodules may suggest a radiologic diagnosis of malignancy.[82] In AIDS patients there is often cavitating upper lobe pneumonia with an associated empyema.

PATHOLOGY

Macroscopically there is firm gray to yellow-tan necrotic tissue that forms nodules up to 2.5 cm diameter; these are often subpleural and may extend across the fissures. Cavitation may be seen.

Histologically there is obliteration of the underlying architecture by macrophages and by smaller numbers of lymphocytes, plasma cells, and polymorphs (Fig. 7-25). Multiple abscesses are present, and there is granulation tissue at the periphery of the lesion.

The macrophages are the typical von Hansemann's histiocytes, with abundant eosinophilic, coarsely granular, vacuolated cytoplasm (Fig. 7-26). The cytoplasmic granules vary markedly in size and shape and are PAS-positive and diastase-resistant. Many cells contain one or more round or targetlike calcific inclusions (Michaelis-Gutmann bodies) measuring up to 20 μm in diameter. Sections stained with Gram or the appropriate stain may show the appropriate organisms, but they may be scanty.

Ultrastructure of the Michaelis-Gutmann bodies shows a wide variety of electron-dense membrane- and non–membrane-bound cytoplasmic inclusions (Fig. 7-27). Many have concentric laminations of varying electron density. Some bodies have laminae of radially oriented crystalline material, alternating with layers of a less dense amorphous to granular substance. Other bodies are composed predominantly of granular material.

The pathogenesis of malakoplakia is unknown, but it is thought to represent an abnormal response to an infection, mediated by an acquired defect in macrophage function.[87–89] The underlying defect may be a deficiency of 3′,5′-guanosine monophosphate dehydrogenase, causing diminished lysosomal breakdown and decreased bacterial killing.[87,88]

The disease may resemble endogenous lipid pneumonia, granular cell myoblastoma, Whipple's disease, tuberculosis, or a storage disease. The pres-

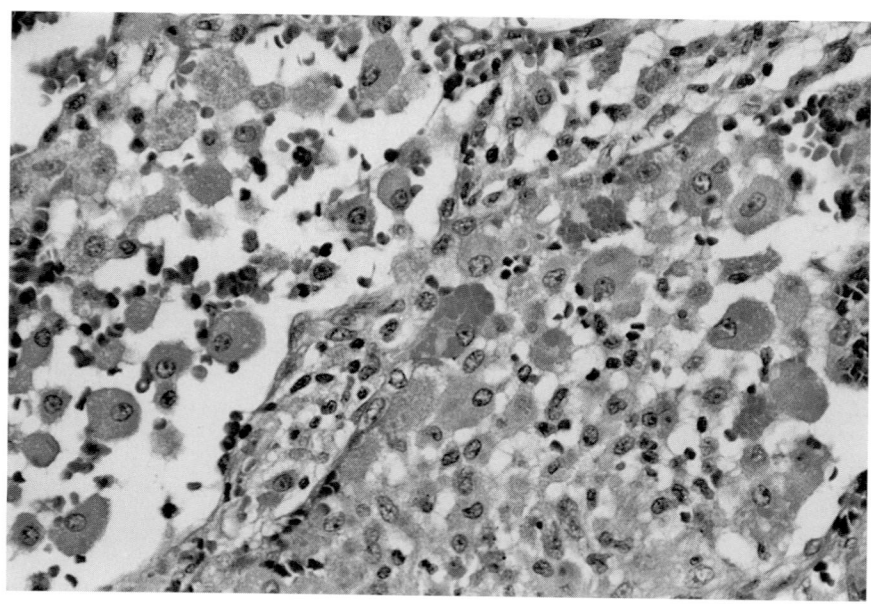

FIGURE 7-25
Malakoplakia with large macrophages and scattered chronic inflammatory cells. (H&E, ×219.) *(Courtesy of Prof. T. V. Colby, Mayo Clinic, Scottsdale, Ariz.)*

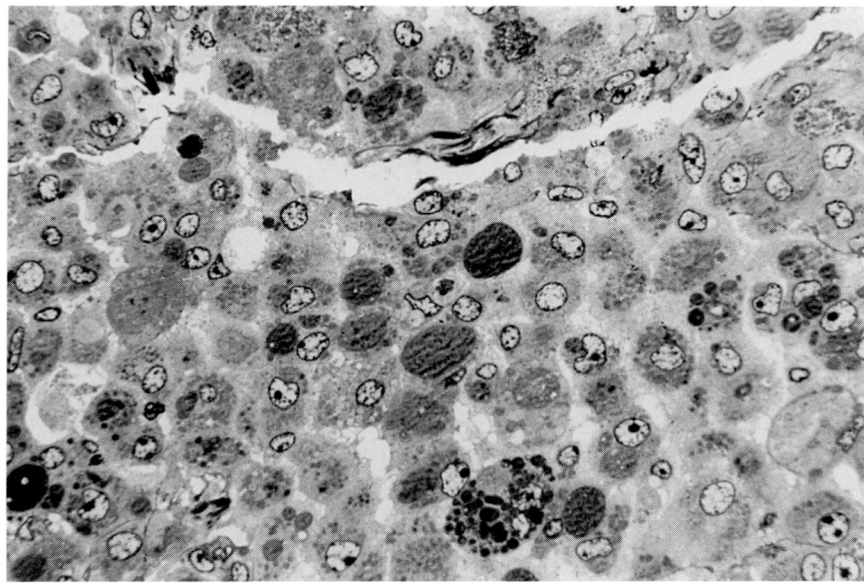

FIGURE 7-26
Malakoplakia. Ultrathin section with large coarsely granular histiocytes. Note Michaelis-Gutmann bodies. (Toluidine blue, ×350.) *(Courtesy of Prof. T. V. Colby, Mayo Clinic, Scottsdale, Ariz.)*

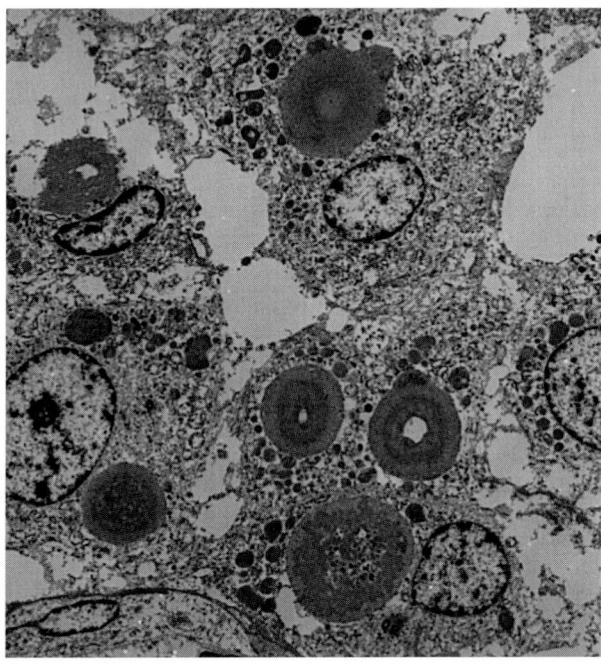

FIGURE 7-27
Electron micrograph of Michaelis-Gutmann bodies with concentric laminations. *(Courtesy of Prof. T. V. Colby, Mayo Clinic, Scottsdale, Ariz.)*

ence of Michaelis-Gutmann bodies is specific to malakoplakia.

Leptospirosis

ORGANISM AND EPIDEMIOLOGY

The genus *Leptospira* comprises three species: *L. biflexa*, which includes saprophytic serovars (serotypes), *L. interrogens*, which contains 202 serovars in 23 serogroups, and *L. parva*.[89a,90] *L. ictero-*

hemorrhagiae is one of the most common causes of human disease.[91] Leptospira are flexible, helical motile orgamisms that differ from other spirochaetes in having closely set primary coils that are regular and permanent and in their characteristic rotating and bending movements. They stain poorly with aniline dyes and can be demonstrated by the use of fluorescent antibody or silver deposition techniques. Since the organisms are narrow (0.1 μm) leptospires are best visualized by dark-field phase-contrast or electron microscopy. Organisms vary from 3 to 20 μm in length and have one or both ends characteristically hooked or bent. It is also one of the most widespread zoonoses in the world. The organisms have an animal reservoir; rodents are the most common, although dogs, cattle, and other mammals may be incriminated in the passage of the bacterium.

CLINICAL FEATURES

A full description of leptospirosis was presented by Feigin and Anderson.[92] The disease may be seen in adults, usually presenting with cough. Hemoptysis and respiratory failure are rare, as is lung abscess. The latter may be due to secondary infection.[93] Pulmonary symptoms occur in approximately 11 percent of patients, usually in the septicemic phase of the illness.[94] The organisms are rarely seen in lung tissue, and they may produce a toxin that directly affects the pulmonary vascular endothelium.[95]

PATHOLOGY

The pathology has been described in seven fatal cases.[95] The pulmonary changes are those of hemorrhage rather than acute inflammation. Petechial hemorrhages are seen in the tracheobronchial tree, lung, and pleura. There are scattered confluent areas of consolidation, as well as sub- and intrapulmonary hemorrhage. Pleural exudates were seen in four of the seven cases. Feigin and Anderson[92] reviewed the pulmonary pathologic findings and, in addition to those mentioned above, described acute hemorrhagic lobar pneumonia with massive hemoptysis.

Histologically there was intraalveolar edema, hemorrhage, fibrin, hyaline membranes, necrosis of the alveolar wall, and a neutrophilic cellular infiltrate in the alveolar spaces as well as in the interstitium. Some necrosis of alveolar walls and intra-alveolar fibrosis were present.

The heart may display interstitial myocarditis. There is no relationship between the myocarditis and the severity of the lung disease.[92]

Because of the difficulty in isolating the organism, the diagnosis of leptospirosis is made serologically.

PLAGUE PNEUMONIA

Organism and Epidemiology

Plague pneumonia is caused by *Yersinia pestis (Pasteurella pestis)*, a gram-negative coccobacillus 0.5 μm in length. The organism is conveyed from rodent to rodent or from rodent to humans by fleas, mainly *Zenopsylla cheopis*. House cats are a growing source of human plague in 13 southwestern states of the United States.[96] Bacilli are ingested in blood sucked from the infected host and multiply in the proventriculus of the flea. They are then regurgitated with the next feed either into another rodent or, when these become scarce due to the ravages of the disease, into humans. If rubbed into an abraded surface, flea feces can cause the disease. Cases have been recorded in hunters or trappers exposed to infected carcasses of deer, antelope, badger, coyote, and bobcat.[97] Much remains unknown about the enzootic cycle.

In England the last outbreak was in 1666 (the Great Plague), and the last case was documented in 1918. The Black Death, an earlier epidemic of plague affecting the skin, swept Western Europe in the Middle Ages. The Black Death obtained its name from the septicemic occlusion of small vessels, producing gangrenous skin lesions. The disappearance of the disease followed the introduction of strict quarantine regulations and public health measures to control the rodent population. The natural virulence of the disease has also decreased during the past three centuries.

Human plague is almost always preceded by an outbreak of plague in the local rodent population. The World Health Organization has received official notification of 1000 to 1500 cases a year from North and South America, Africa, and Southeast Asia.[98]

The disease recently made the headlines with an outbreak in Surat, a city in Gujarat state, in Western India (Fig. 7-28). This outbreak claimed at least 58 lives. There has been doubt cast on the diagnosis of plague, since attempts to culture the organism failed and the direct demonstration of bipolar bacilli from samples was not confirmed with fluorescent antibody staining for *Y. pestis* fraction 1 (F1) antigen.[1] It was suggested that the outbreak in India could have been due to hantavirus pulmonary syndrome (Chap. 5); others have suggested that it might have been tularemia.[99]

Clinical Features

Clinically there may be one of four clinical presentations; bubonic plague, septicemic plague, pneumonic plague, and meningitis. The early stages of most cases are of bubonic type. Initially, 2 to 7 days after exposure, there is a febrile illness. The flea bite is usually on the legs and is rapidly followed by lymphatic spread of the infection to the regional lymph nodes, usually the groin or axillae. The nodes enlarge to form characteristic painful swellings called "buboes" and hence the name *bubonic plague*. The lungs become involved in the general dissemination of the infection, giving rise to secondary pneumonic plague. The disease is contagious, and the bacilli are readily inhaled by those nursing the sick, causing primary pneumonic plague. The pulmonary changes are identical in primary and secondary pneumonic plague. Primary septicemia can occur without progression from bubonic plague. There is a systemic polymorphonuclear leukocytosis, up to 20,000 per mm^3.

Pathology

Macroscopically, plague starts as bronchitis, bronchiolitis, and alveolitis causing lobular consolidation (Fig. 7-29). Adjacent consolidated lobules fuse, causing lobar pneumonia with an associated fibrinous pleurisy. The consolidated zones are grayer in the center, while the periphery and much of the intervening tissue are congested, hemorrhagic, and edematous. Gray-white fluid, devoid of fibrin, can be scraped from the cut surface of the consolidated zones and is full of the causative organisms. The

FIGURE 7-28
People fleeing Surat in September 1994 and wearing face masks to avoid contracting "plague." *(Courtesy of Associated Press.)*

regional lymph nodes are large, edematous, and hemorrhagic.

Histologically, bacilli are present in macrophages in large numbers. They are seen in small bronchi, bronchioles, and adjacent alveoli (Fig. 7-30). Spread occurs into peribronchial and subpleural tissue as well as into the intralobular septa. The organisms pass into blood vessels and lymphatics to become disseminated throughout the body. The presence of large numbers of the organisms in the alveoli causes severe hemorrhagic pulmonary edema and hyaline membranes. Later, polymorphonuclear leukocytes and macrophages are found in increasing numbers in the alveolar exudate (Fig. 7-31).

If the patient survives sufficiently long, there is destruction of alveolar septa, and smaller branches of the pulmonary artery become filled with antemortem thrombus.

Confirmation of the diagnosis is obtained by either aspirating a bubo or growing the organism from blood and sputum while the patient is alive, or from the lung and blood after death. In Gram-stained smears of sputum, the bipolar staining gram-negative bacteria resemble safety pins. Rapid identification of the organism can be made at reference laboratories by means of a fluorescent antibody test. It is clear that postmortem examinations should be done with the greatest care, *if at all*, since the organism is released as a spray and any workers in the postmortem room would be at risk.

TULAREMIC PNEUMONIA

Organism and Epidemiology

The organism *Francisella tularensis* receives its name from the researcher (Edward Francis) who carried out many of the early bacteriologic and epidemiologic studies of the disease and from Tulare County, California, where the disease was first described in squirrels.

The organism is widespread in rodents in many parts of the Northern Hemisphere, including the United States, Canada, Norway, Sweden, Japan and Turkey. Most human infections have been reported in the United States, where rodent infections have been detected in almost every state.

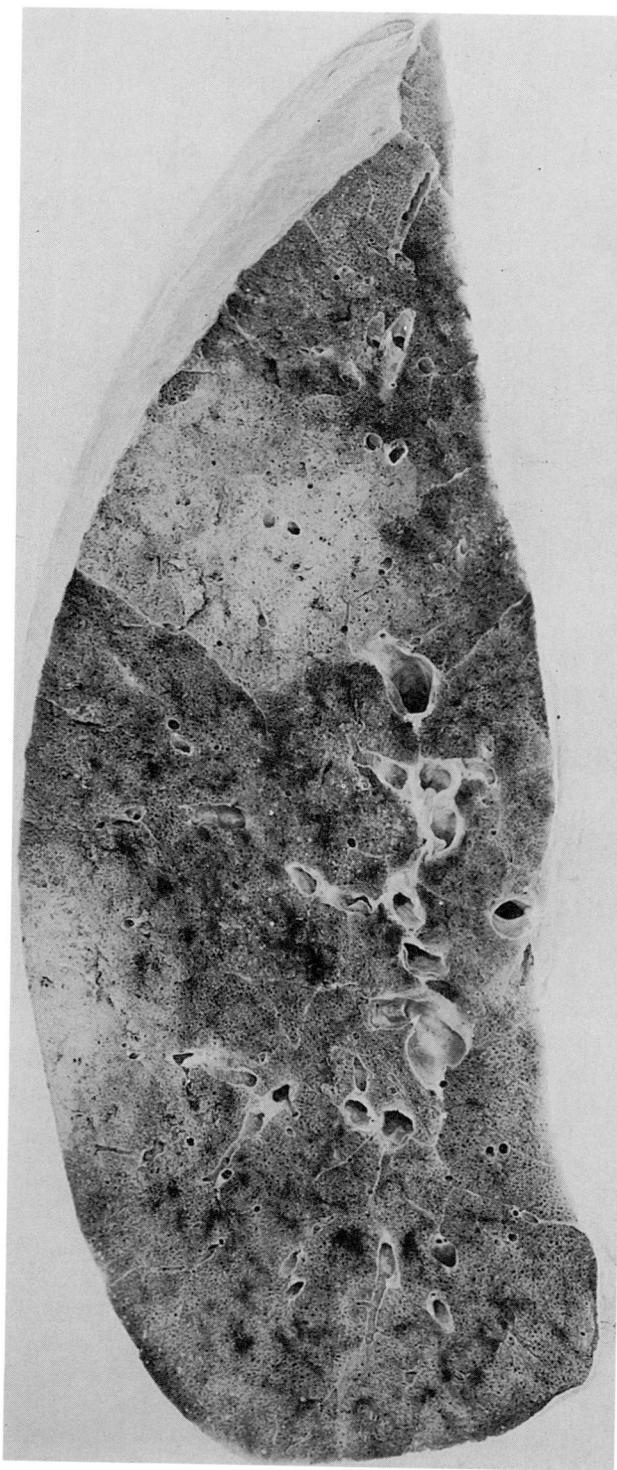

FIGURE 7-29
Plague pneumonia with lobular consolidation and foci of hemorrhage. *(Courtesy of Dr. L. Iverson, AFIP, Washington, DC.)*

Tularemia is primarily a disease of wild rabbits, hares, squirrels, possums, and foxes, although it has been seen in common rodents. Human infection is likely to occur among those who handle the skins and carcasses of wild rabbits, hares, and muskrats, but it may follow the bite of the deer fly (*Chrysops discalis*).[97] As in other tic-borne infections, the causative organisms may be transmitted from a parent tic to its offspring. Human infection may also follow the consumption of insufficiently cooked infected rabbit meat, and several cases have been reported in laboratory staff handling cultures of the organism. Two other modes of transmission are inhalation of infected aerosols and contamination of ocular or oral mucous membranes.

The responsible organism, *F. tularensis* (formally called *Pasteurella tularensis*), is a small gram-negative coccobacillary organism (0.2 to 0.7 μm) found in an encapsulated form. This capsule appears as a clear area. The organism is often pleomorphic and nonmotile.[100]

Several distinct types are recognized and can be distinguished by their virulence, biochemical characteristics, epidemiology, and the spectrum of clinical illness. The North American strain, designated type 1 or A strain, is highly virulent: as few as 10 organisms can produce clinical illness. The less virulent American, European, and Asian strain is referred to as type 2 or B strain. This requires 100,000 or more organisms to produce infection and is often associated with rodents and water. The organism is quite resistant to drying or cold.

After introduction of the organism in the susceptible host, multiplication occurs locally, with spread to regional lymph nodes by 96 h. There is a cutaneous ulcer or soft-tissue focus, which in time shows necrosis and granuloma formation. Septicemia may result. The organisms are ingested by macrophages that, as they become activated, are eventually capable of killing them. Containment, as well as long-term immunity, results from cell-mediated immunity.

Clinical Features

Clinical manifestations involve the skin, which is the most common site of presentation in ulceroglandular tularemia. The cutaneous ulcer (chancriform) lesion has a crater with elevated margins and evolves from the macular and papular lesions. The ulcers are usually seen in the hands and forearms in hunters but tic bites may occur on the abdomen, groin, or axilla.[97] The adjacent nodes are enlarged and tender.

Ocular, glandular, and pharyngeal tularemia may follow contact with contaminated hands or the splattering with infected tissue fluids from animals or insects, as when tics are cut in sheep shearing.

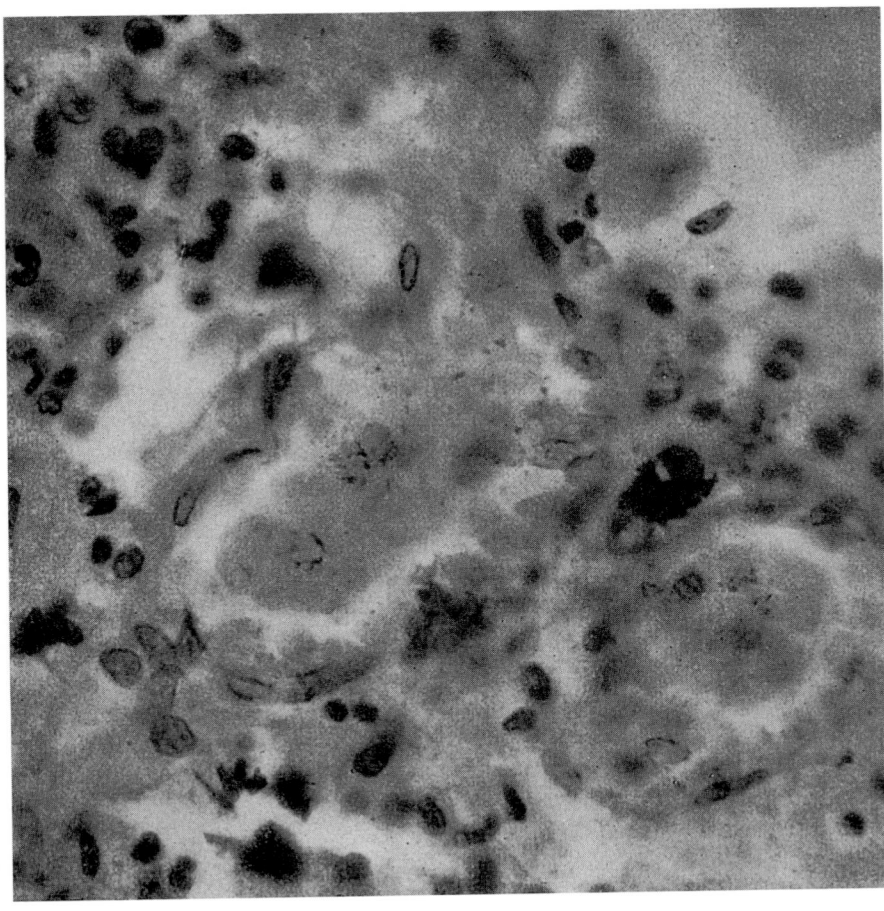

FIGURE 7-30
Y. pestis in macrophages. *(Courtesy of Dr. L. Iverson, AFIP, Washington, D.C.)*

Pulmonary tularemia is a severe atypical pneumonia that has a high fatality rate when untreated. It may be primary after inhalation of infected aerosols. Secondary pneumonia occurs during hematogenous dissemination of the organism. Regardless of the route of inoculation, all forms of tularemia are associated with high fever, and half of the patients complain of headache and rigors. There may be meningitis, rhabdomyolysis, hypotension, and intravascular coagulation.

Pathology

Stewart and Pullen[101] reviewed 268 cases from North America. Macroscopically there is a fibrinopurulent pleurisy and pleural effusions. The cut surface showed either small gray-white solid nodules or abscess formation. These often become confluent, resulting in lobar consolidation, and the condition of the lung resembles the gray hepatization of pneumococcal pneumonia. The consolidated lobes in these cases are grayish pink and often break down to form abscesses filled with caseous material. Consolidation is prominent in the peribronchial tissues and in the region of interlobular septa, which become very prominent. The regional nodes show similar abscesses and are markedly enlarged. Occasionally the pulmonary abscesses ulcerate through a bronchus and the contents are expectorated, leaving a cavity.

Microscopically the early lesion is an acute ulcerative bronchiolitis, followed by acute fibrinous bronchopneumonia that rapidly undergoes necrosis. The necrotic areas resemble infarcts, with ghost outlines of the destroyed tissues and edges bounded by a zone of degenerate nuclei. Beyond this is a layer composed predominantly of epithelioid cells and lymphocytes. Outside the cellular zone, the surrounding alveoli are filled with a fibrinous exudate containing mononuclear cells and some polymorphs (Fig. 7-32). Alveoli are lined by swollen type II cells. The alveolar walls and peribronchial and septal connective tissues may all show a moderate mononuclear infiltrate. Necrotic changes spread throughout the lung, destroying medium and small blood vessels along with smaller bronchi and bronchioles. Most arteries show extensive endarteritis and finally thrombosis (Fig. 7-33). Tularemia is marked by an absence of giant cells. It is difficult to demonstrate the causative organisms in human lesions, but Foulger and

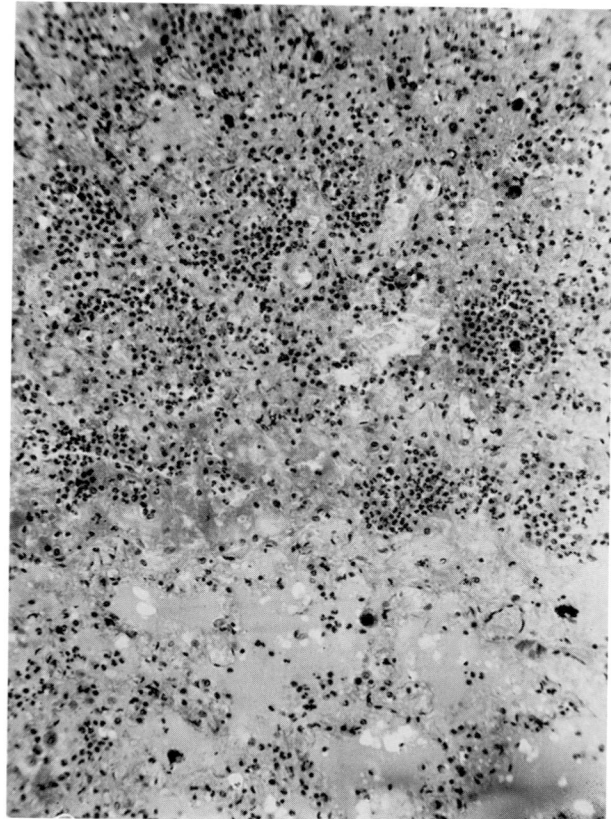

FIGURE 7-31
Plague pneumonia with hemorrhagic and edematous exudate.
(Courtesy of Dr. L. Iverson, AFIP, Washington, D.C.)

coworkers[102] recommended a saturated solution of Nile blue sulfate for this purpose.

Diagnosis is now made by serum agglutination tests. Specific agglutinins appear 8 to 10 days after the onset of illness, and the maximum titer occurs at about 4 weeks. *Brucella* agglutinins may cross-react with tularemia antigens.[103] The organism is fastidious in its growth and requires cysteine; it will be seen in 48 to 72 h on glucose-cysteine blood agar.

BRUCELLA INFECTION

The previous edition of this book cast considerable doubt on whether the *Brucella* organisms (*B. melitensis*, *B. abortis*, and *B. suis*) were responsible for causing pneumonia. It would appear it does occur but is rare. Perhaps the most convincing examples are reported by Weed and colleagues.[104] They described three cases of brucellosis with circumscribed large caseous nodules found in surgically excised specimens of lung tissue. *B. suis* was isolated in all the cases. Inhalation of infectious aerosols has been noted as an important vehicle in spread of the disease.[105]

The rarity of pneumonia in *Brucella* infection has been underlined by three reports. Lulu's team[106] described 400 cases of brucellosis in Kuwait but noted that respiratory complications were only seen in 1 percent. Three patients had pneumonitis and one a pleural effusion. It is not clear whether these patients had *Brucella* infection of the lung or some superimposed infection. A case report from Saudi Arabia[107] describes multisystem disease in an 8-year-old girl who had a 4-month history of fever, sweating, cough, dyspnea, malaise, bilateral rhonchi, and crepitations. Chest radiographs showed bilateral hilar lymphadenopathy, miliary mottling, and scattered parenchymal nodules. The patient made an uneventful recovery on oral tetracycline and intra-

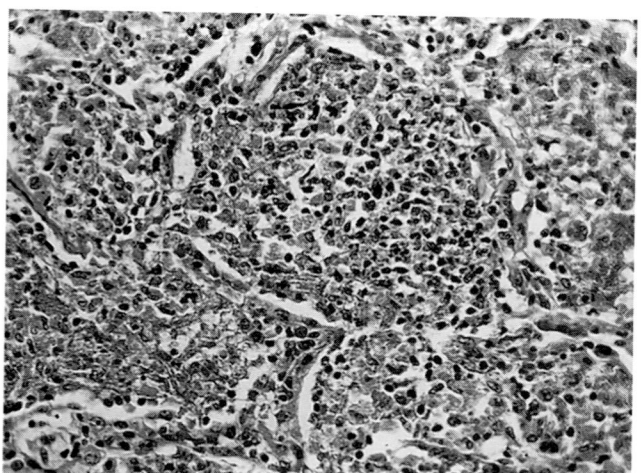

FIGURE 7-32
Tularemia pneumonia with intact alveolar walls and intraalveolar macrophages, fibrin, and lymphocytes. (H&E, ×218.) *(Courtesy of Prof. R. D. Baker.)*

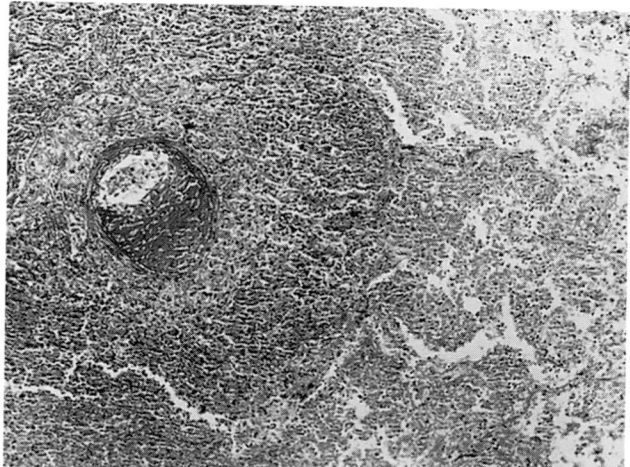

FIGURE 7-33
Tularemic pneumonia with thrombosis in a vessel and large areas of necrosis. (H&E, ×108.) *(Courtesy of Prof. R. D. Baker.)*

muscular streptomycin. The last case was of *Brucella* empyema in a Spanish agricultural worker from an area hyperendemic for the disease.[108] The last authors, in a review, noted the rarity of respiratory symptoms. Many of the cases quoted were from the 1930s and 1940s, and in many countries the incidence of *Brucella* infection has fallen.

MENINGOCOCCAL PNEUMONIA

Meningococcal pneumonia was first recognized, after the influenzal pneumonias swept Europe in the 1918–1919 pandemic, by Davison and colleagues.[109] Among 78 patients who died of pneumonia, postmortem cultures revealed *Neisseria meningitidis* in 23 cases; in seven it was found in pure culture. A study in a 1000-bed Danish general hospital over a 2-year period found 19 patients with infections from whom *N. meningitidis* was isolated from lower respiratory tract secretions and in pure culture in 17. It was a potential pathogen in 0.2 percent of all lower respiratory specimens. All but three patients had chronic pulmonary disease, primarily chronic bronchitis.[110] There was also a relationship with viral infection, 42 percent of cases showing serologic evidence of influenza A or parainfluenza type I virus. Meningitis may precede or follow the development of the pneumonia. The organism may be found alongside other bacteria such as *H. influenzae* and *M. tuberculosis*. Rarely it may be a nosocomial infection.[121,122]

Pathology

Macroscopically, the pneumonia is lobular and rarely lobar. The pleura is seldom involved.[111]

Histologically, the infected bronchioles and alveoli are filled with neutrophil polymorphs, which contain gram-negative intracellular diplococci.

SALMONELLA PULMONARY INFECTIONS

The *Salmonella* group of organisms, both *S. typhi* and nontyphoid strains, are occasionally responsible for pneumonia and lung abscesses. A rare case in which typhoid pneumonia and lung abscess occurred several decades after the initial infection was reported by De Matteis and Armani.[112] In this case there was active pulmonary arteritis with intramural plasma cells and lymphocytic and fibroblastic proliferation. The center of the lesion had undergone suppuration, part of which was attributable to infarction. *S. typhi* is known to cause endothelial damage.

Eleven cases of nontyphoid *Salmonella* pleuropulmonary disease with a 63 percent mortality were seen in a Madrid general hospital over a 26-year period.[113] All patients with pneumonia had positive blood cultures. Pneumonia was found in eight patients, lung abscess in two, and empyema in one. Eight of the 11 patients were 60 years or older, and all had one or more serious underlying diseases. A gastrointestinal source of infection was thought unlikely, since only two patients had positive stool cultures.

An older study, dating back to 1946, showed that bronchitis occurred in 85 percent of patients, followed by bronchopneumonia.[114] However, in a review of the world literature, it was estimated that pulmonary manifestations occur in only 1 percent of patients with typhoid fever.[115]

The pathogenesis of this infection is unknown. Sputum culture is often negative. A *Salmonella* lung abscess has been described complicating Wegener's granulomatosis[116] and AIDS.[117]

Histologically there are no distinctive features of *Salmonella* pneumonia.

Pasteurellosis

Pasteurella multocida may be found in cats and dogs.[118,119] The organism appears to be opportunistic and colonizes the respiratory tract in patients with chronic lung disease. Occasionally there is purulent bronchitis, bronchopneumonia, or a suppurative pleural effusion, possibly forming secondary to aspiration pneumonia.[120]

Moraxella (Branhamella) catarrhalis

This is a large, kidney-bean shaped, gram-negative diplococcus that until the 1970s was considered a respiratory tract commensal. Recently it has been described as a cause of upper and lower respiratory tract infections in both adults and children, with a peak distribution in the winter months. It often occurs in patients with lower respiratory tract disease and causes pneumonia. It is also found in patients who are immunocompromised. It is frequently part of a mixed bacterial pattern in the sputum, coexisting with *H. influenzae* and *S. pneumoniae*.

Chest radiographs show consolidation of a lobar or segmental nature in nearly half of the patients. In children there may be a purulent tracheitis. Good reviews of the clinical aspects of this disease and the epidemiology have been presented by Wright and Wallace[123] and Verghase and Berk.[124]

Legionnaires' Disease

ORGANISM AND EPIDEMIOLOGY

In July 1976, the American Legion held its annual convention in Philadelphia, during and after which there was a large outbreak of what came to be known as Legionnaires' disease. This was centered on the Philadelphia hotel where the conventioneers gathered. The patients had a respiratory infection characterized by pyrexia, dry cough, pleuritic chest pain, dyspnea, mental confusion, and gastrointestinal symptoms. Many of the victims were already suffering from other serious disorders, including malignant disease, chronic bronchitis, emphysema, and chronic heart disease. In this outbreak, 29 of 186 patients (15 percent) died.[125] Cases were soon reported from other parts of the world, and it was apparent that previously healthy patients could be struck down by the disease.[126]

A second large outbreak, killing 29 persons, occurred in Stafford, England, in 1985 and was judged worrying enough to warrant an official inquiry. The outbreak was traced to a hospital's water-cooled air-conditioning system.

Other reports have located this ubiquitous organism in warm-water sprays, showers, and vacation resorts. A gap in the cooling tower, above the air intake level, is the probable route for the bacterium to enter a water-cooling system.[127] A further outbreak occurred in a renal transplant unit in Oxford.[128]

The organisms are short, occasionally filamentous, gram-negative rods that stain poorly with most aniline dyes. They are nonsporing and are usually motile by means of one or more polar flagella (Fig. 7-34). In infected tissue they are short rods or coccobacilli, measuring approximately 1.3×0.5 μm.[129] The organisms are saprophytic and need water for their growth. Water temperature has a crucial influence on the growth of the organism.[130] *Legionella pneumophila* can infect and multiply within amoebae and ciliated protozoa.[131] Both *Legionella* and amoebae grow well in warm water. The amoebae may provide a shelter for the organism.

There are at least 34 *Legionella* species and more than 52 serogroups.[132] About half of the species are pathogenic for humans.[133] The most common causes of Legionellosis are *L. pneumophila*, serogroups 1, 4, and 6. There is probably a variable degree of virulence in different *Legionella* strains.[135] The cell wall lipopolysaccharide is the main constituent that determines serogroup-specific antigen.[134]

The disease is transmitted by inhalation, and aerosol-generating systems play an important role in the production of the disease. Colonization of the

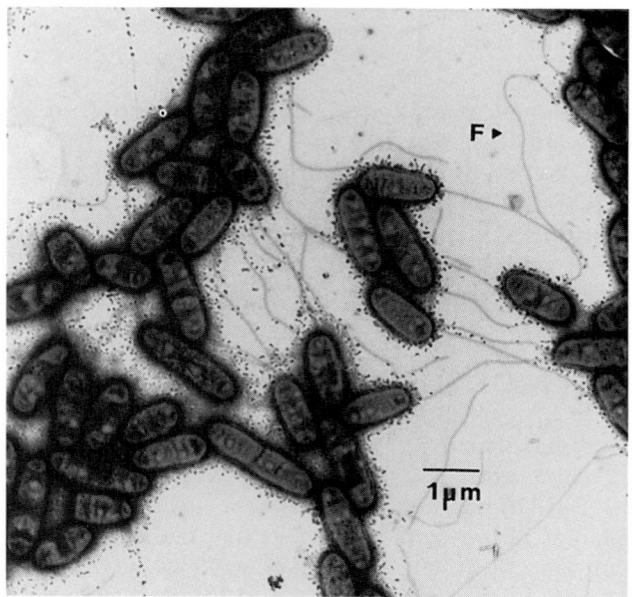

FIGURE 7-34
Morphology of *Legionella pneumophila* serogroup 1 grown in liquid medium and showing flagellae (F). (*By kind permission of Drs. Dennis and Lee.*)

water supply is probably the most consistent epidemiologic finding in nosocomial outbreaks.[136] Person-to-person transmission of infection has never been shown.[133] However, pathologists are at risk, and a case has been reported in a pathologist who carried out a postmortem on an infected case and then developed the disease himself.[137] Physicians often use tap water to rinse respiratory apparatus and tubing for use on mechanical ventilators; if this water is contaminated, the organism can be instilled directly into the lung.[138]

There are, however, risk factors for development of the disease. These include cytotoxic chemotherapy, use of corticosteroids,[139] chronic lung disease, diabetes, cigarette smoking, alcohol abuse, and advanced age.[133] The incidence of Legionnaires' disease depends on the degree of contamination by the organism in the aquatic reservoir, the susceptibility of the patients exposed to that water, and the intensity of exposure.[140] Transplant recipients, of all types, are at the highest risk.

Clinical Features

There appear to be two different manifestations of infection. The first is Pontiac fever, which has been caused by at least five species of *Legionella*, the most recently described being *L. anisa*.[141] There is an acute upper respiratory illness and in some cases a flulike illness with high fever, myalgia, headache, and occasionally malaise. Pneumonia does not occur.

In the second group of patients there is a full-blown pneumonia as described above, with an incubation period of 2 to 10 days. However, the severity of the symptoms is variable, from mild respiratory illness to a fulminating course. Dry cough is present in almost 90 percent of patients, and there is pleuritic chest pain in up to a third. Fifty to 75 percent of patients ultimately develop purulent sputum and hemoptysis. There may be gastrointestinal symptoms with abdominal discomfort, diarrhea, and loose or watery but nonbloody stools. Change in mental status is observed in 20 to 30 percent of patients. However, the presence of gastrointestinal or neurologic signs or symptoms does not help differentiate Legionnaires' disease from other pneumonias.

Hyponatremia is more frequent in the initial stages of Legionnaires' disease than in other pneumonias.[142] In some cases there is a rise in serum creatinine phosphokinase.

There are no definite differential diagnostic features based on radiographic features that will distinguish Legionnaires' disease from other nosocomial pneumonias.[143]

Pathology

The pathology of the disease has been described by several authors.[144–146] Macroscopically, the consistent finding is confluent bronchopneumonia; less often a lobar pattern is seen (Fig. 7-35). No lobe is preferentially affected, and consolidation is usually bilateral. When there is a lobar pattern, it is in the stage of red or gray hepatization. The cut surface of the consolidated areas is reddish-gray, friable, and granular. Abscesses are relatively uncommon. In rare cases the disease may present with white miliary nodules that resemble tuberculosis.[147] Histologically, these small foci are microabscesses. In some patients, tubercle and *Legionella* may coexist.[148] There is often a straw-colored or serosanguinous pleural effusion and fibrinous pleurisy.

A unique case occurred in a previously healthy 46-year-old man in whom the disease had disseminated to the kidneys, bone marrow, spleen, and multiple peripheral lymph nodes at autopsy. It was suggested that the spread occurred via hematogenous and lymphatic pathways.[146]

Histologically, in the acute phase the alveoli are filled with macrophages and neutrophils in varying proportions (Fig. 7-36). There is a fibrinous exudate and variable numbers of red blood corpuscles. In some cases the exudate is predominantly fibrinous with few inflammatory cells (Fig. 7-37). The lung parenchyma shows areas of coagulative necrosis and septal edema. There is a mild interstitial inflammation with lymphocytes and neutrophils in some cases; others may show a hemorrhagic component.

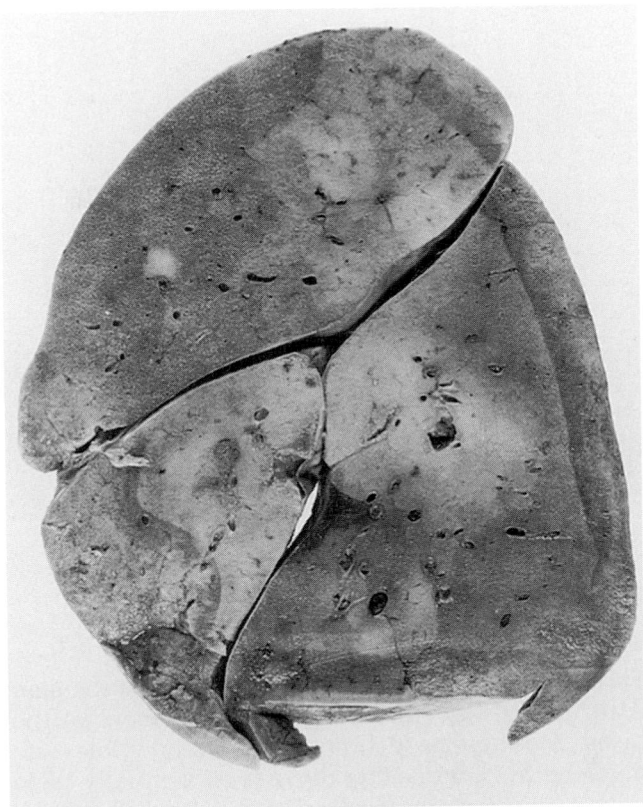

FIGURE 7-35
Lobular and diffuse consolidation in *Legionella* pneumonia.

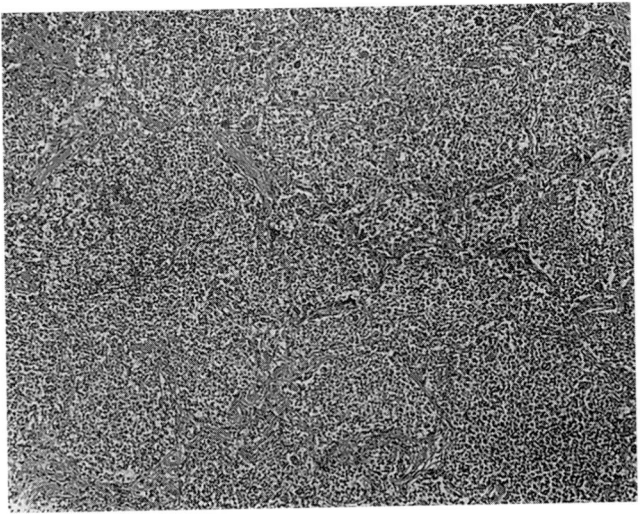

FIGURE 7-36
Legionnaires' disease with alveoli filled with neutrophil polymorphs. The picture resembles gray hepatization in lobar pneumonia. (H&E, ×73.)

Microabscesses may occur; in one case a nonspecific vasculitis was present, with infiltration by neutrophils and bacteria of the vessel walls. Bronchiolar epithelium appears necrotic in some areas. Diffuse alveolar damage is present, and there is reactive type II cell hyperplasia as well as hyaline membranes.

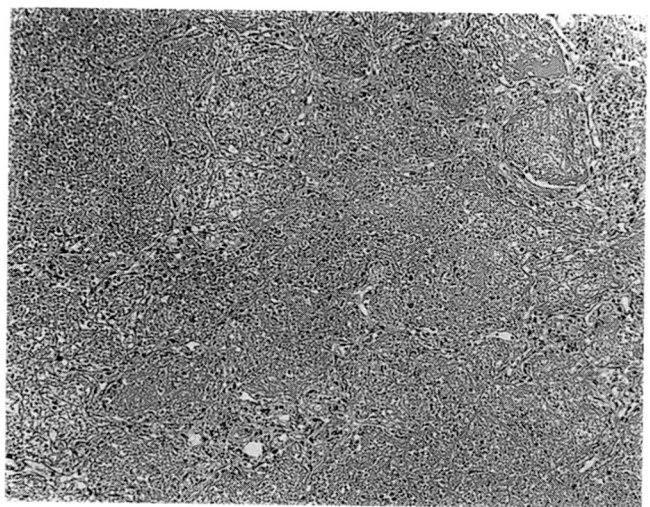

FIGURE 7-37
Legionnaires' disease with a fibrinous exudate inside alveoli and relatively few cells. (H&E, ×73.)

Thrombosis of small pulmonary vessels associated with infarct-like areas occurs in 25 percent of the cases. Medium-sized bronchi and bronchioles are involved in 75 percent of the cases. Eventually there is intraalveolar organization of the exudate, which progresses to diffuse fibrosis, and obliteration of the alveolar septal framework (Fig. 7-38).[149] Pulmonary cavitation leading to fibrosis is a rare complication.[150]

Diagnosis

Diagnosis of Legionnaires' disease is established by visualization of the organisms in clinical samples.

They are small, faintly staining, pleomorphic gram-negative rods. There is a rise in antibody titer in the patient's serum. However, positive anti-*Legionella* antibodies are a frequent occurrence among outpatient subjects, 36 percent having positive antibody titers equal to or greater than 1:2 to 1:8. This has important implications for the interpretation of single or static antibody titers from patients who are acutely ill.[151]

In lung tissue the organism can be demonstrated by the Dieterle silver stain,[152] where the bacilli, which are both intra- and extracellular, appear intensely brown to black against a pale-yellow background (Fig. 7-39).

Immunohistochemistry using a panel of monoclonal antibodies on formalin-fixed paraffin-embedded sections enables a diagnosis of *L. pneumophila* infection to be made and the appropriate serogroup given.[153] The antibody was raised as a rabbit polyclonal anti–*L. pneumophila* serogroup I, which is used at a dilution of 1 in 200. A direct immunofluorescence antibody technique may allow detection of the organism (Fig. 7-40). In situ hybridization has also been used to detect the organism; the probe used consisted of synthetic oligodeoxynucleotides complementary to the ribosomal RNA of all clinically urban *Legionella* species labeled with biotinolated de-oxyuridine triphosphate (dUTP) at their 3′ ends. No cross-hybridization was observed with other common causes of pneumonia, such as *S. pneumoniae, Ps. aeruginosa,* or *K. pneumoniae.*[154]

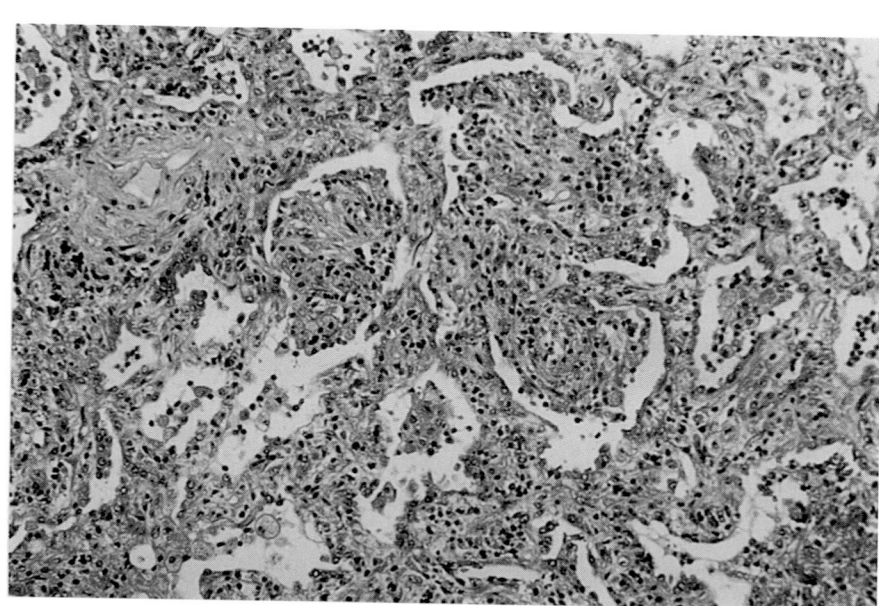

FIGURE 7-38
Legionella. Interstitial and intraalveolar fibrosis. (H&E, ×88.)

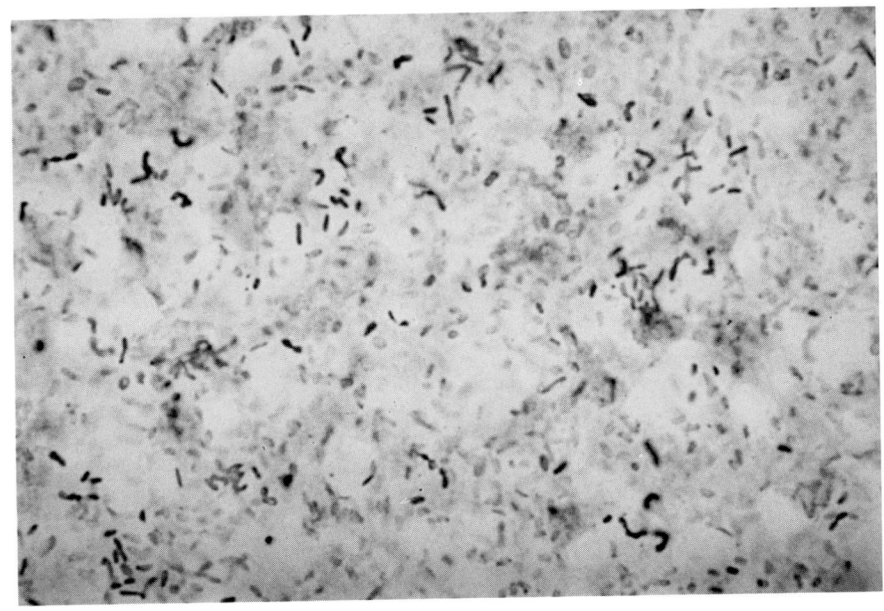

FIGURE 7-39
Legionella as demonstrated by the Dieterle stain. *(Reproduced with permission of Dr. D. Schurman, Department of Infectious Diseases, Rudolf Virchow University Hospital, Berlin, Germany.)*

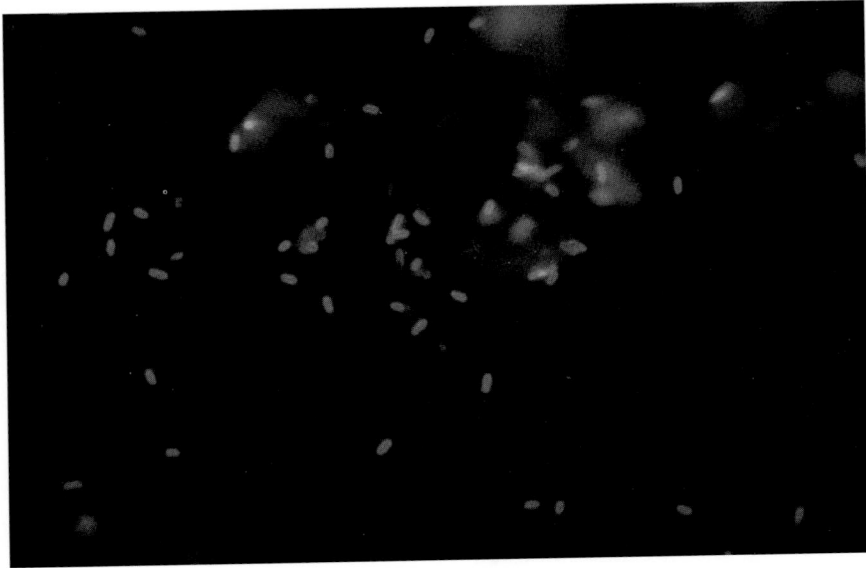

FIGURE 7-40
Legionella shown by direct immunofluorescence. *(Reproduced with permission of Dr. D. Schurman, Department of Infectious Diseases, Rudolf Virchow University Hospital, Berlin, Germany.)*

Culture is commonly used for diagnosis and sensitivity but the value varies according to the source and quality of the specimen. It is possible to use bronchial washings to obtain a sample, and this method was used to detect disease in eight patients.[155] The organism can also be demonstrated rapidly on transbronchial biopsy by the use of fluorescent antibody stains.[156] A specific anti-*Legionella* antibody obtained from immunized animals and labeled with fluorescein can be used in identification of the organism in clinical specimens, which takes only 2 to 4 h. Sensitivity can be variable, however, because positive results demand the presence of large numbers of organisms. Therefore, a negative result does not rule out disease.

CHRONIC INFECTIVE PNEUMONIAS

Before we consider in detail the changes in the different chronic infective pneumonias, it is important to outline the manner in which the lungs respond to injury in infection.

In most cases of bacterial infection, all the alveolar lining cells are damaged to a variable extent. The alveoli themselves become filled with fibrinous exudate, which leaks from the damaged alveolar wall. In most cases the exudate lyses because of fibrinolytic enzymes, and the alveolar wall heals normally. If lysis fails, however, after a short time alveolar epithelial cells spread from the intact areas of the alveoli to cover the surface of the contracting intraluminal fi-

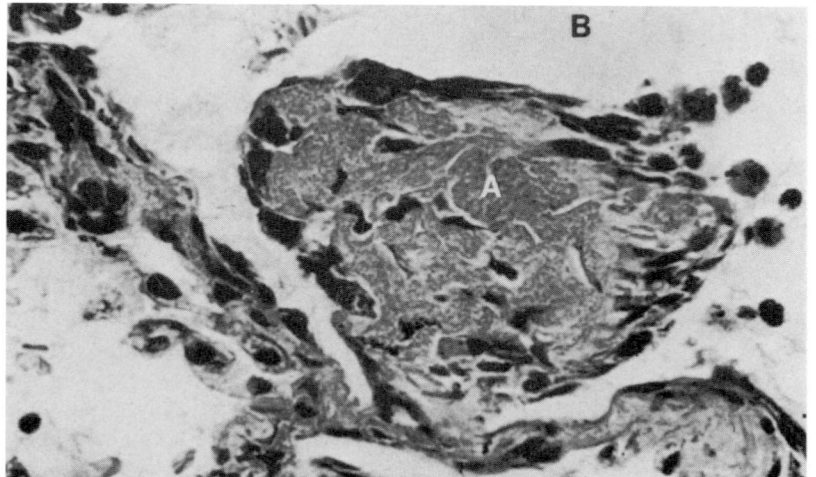

FIGURE 7-41
Intraalveolar fibrinous material (A) that has contracted and is in process of being covered on its surface by type II pneumocytes. Contraction of the fibrin mass has caused clefts to appear into which cells have grown. Eventually such a fibrinous mass will form a minute patch of interstitially situated fibrous tissue. B is the alveolar space in which A is lying.

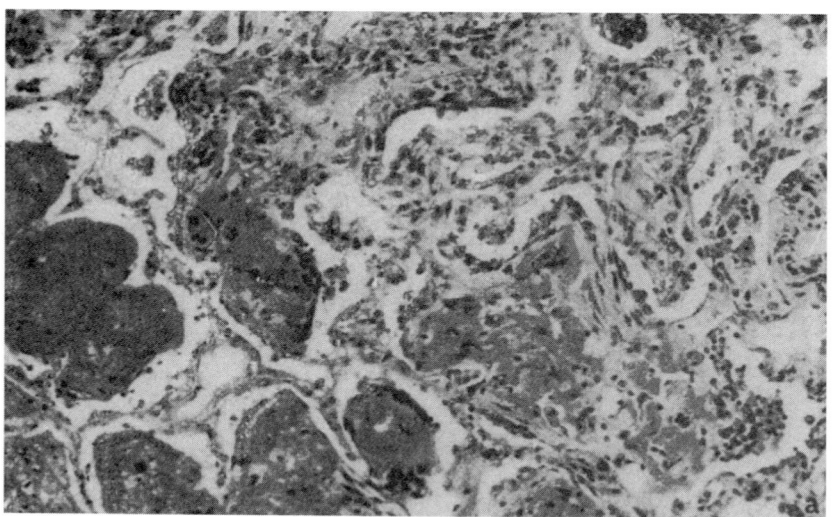

A

FIGURE 7-42
Lung showing widespread fibrinous intraalveolar exudate in various stages of organization. *A.* The fibrin masses are becoming surrounded on the surface by proliferating alveolar epithelial cells. In the more mature parts, fibrosis has commenced on the surface of the fibrin mass. H&E, ×100. *B.* Intraalveolar fibrin masses show the earliest reticulin fiber formation near the surface of the fibrin masses. (Modified Foot's reticulin stain, ×400.)

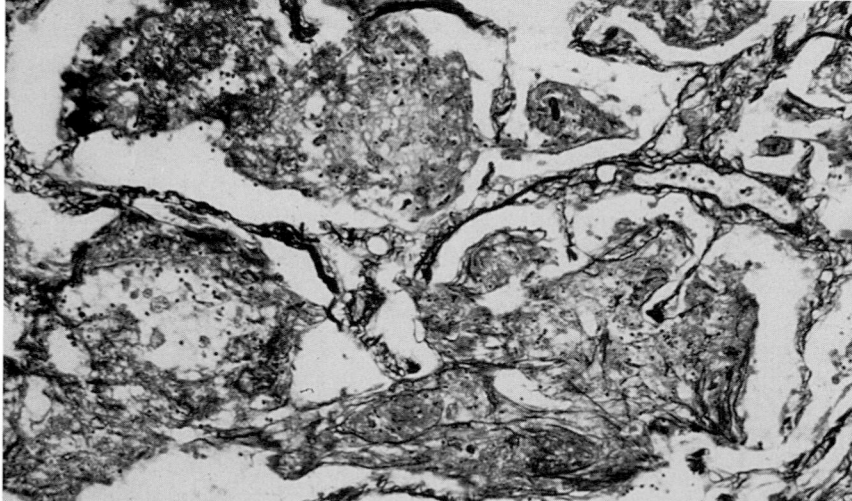

B

brinous mass (Fig. 7-41). The process resembles the incorporation of a mural thrombus in a blood vessel by the ingrowth of neighboring endothelial cells. While this process is occurring, fibroblasts and histiocytes derived from the alveolar interstitial cells grow into the fibrin and form young reticulin fibrils (Fig. 7-42).

The newly formed reticulin can be distinguished from similar but thicker fibers that form part of the normal alveolar wall. The former are fine, more sharply defined, and branched. In time they are converted to type III collagen. All that remains is an interstitial plaque of collagen covered by a visible alveolar epithelium, which probably consists of type II pneumocytes. Concurrently with the interstitial fibrosis there is intraalveolar fibrous tissue deposition (Fig. 7-43). It is common to find intraalveolar as well as interstitial pulmonary fibrosis in end-stage chronic infections or even in interstitial lung disease.

In a diagram showing the stages of organization of intraalveolar fibrin (Fig. 7-44), it can be seen (C) that there is intraalveolar as well as interstitial fibrosis. A fuller description of the possible early stages of interstitial lung disease (fibrosing alveolitis) will be given in Chap. 12.

The above model, however, assumes that all the inflammatory insult occurs at the alveolar level. There are descriptions (p. 242, p. 243) of constrictive bronchiolitis and bronchiolitis obliterans–organizing pneumonia. Both of these processes may be the end result of an infectious disease such as *Legionella* or *Nocardia*.

PULMONARY TUBERCULOSIS

A brief history of pulmonary tuberculosis, or the "white plague," was given by Murray in the J. Burns Amberson Lecture in 1989.[157] He presented archeologic findings to support the belief that the disease had been present at the time of the ancient Greek and Roman civilizations. Hippocrates described the disease, "as one following ulceration and suppuration of the lungs [and] this concept is particularly evident in the search of etiological causes in pulmonary tuberculosis."[158] Tuberculosis (TB) probably increased in incidence during the Middle Ages and was prominent during the Renaissance period in Europe. In the 16th and 17th centuries, TB is said to have caused nearly 20 percent of all deaths in England. The decline began during the late 1700s and was well under way by the time that Koch, in 1882, discovered the causative organism. This decline in mortality is thought to be due to better living standards and the realization that it was an infectious disease and consequent introduction of isolation practices. Between 1920 and 1940 the rate of decline in TB mortality increased, probably because of the lethal effects of influenza on such patients.[159] The final decline in mortality up to 1985 was due to chemotherapy.

The incidence of tuberculosis leveled off in the United States in 1985 partly as a result of HIV-positive cases (Fig. 7-45).

In an update of this data Professor Murray has shown that the excess cases are not all HIV-related.

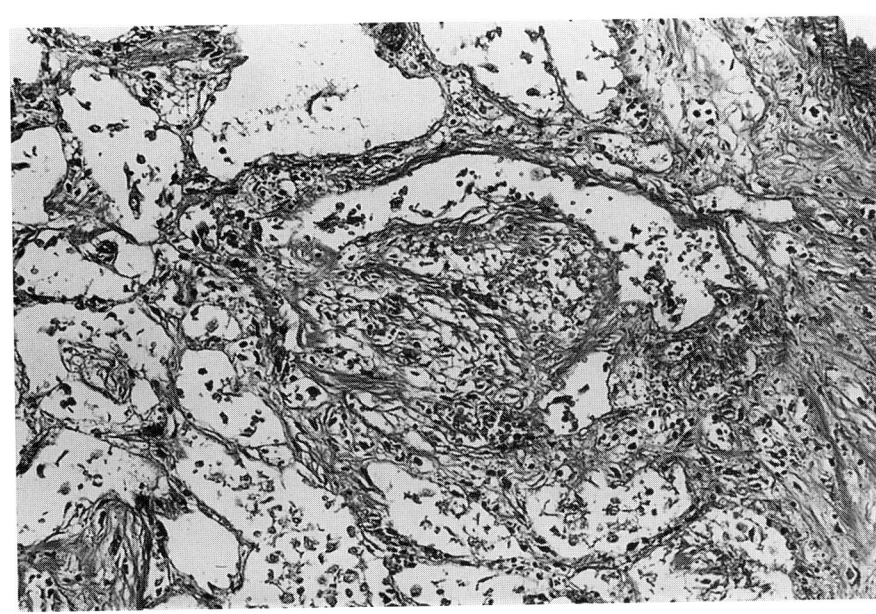

FIGURE 7-43
Intraalveolar fibrosis filling alveolar lumina in resolving pneumonia. Interstitial fibrosis is also present. (H&E, ×313.)

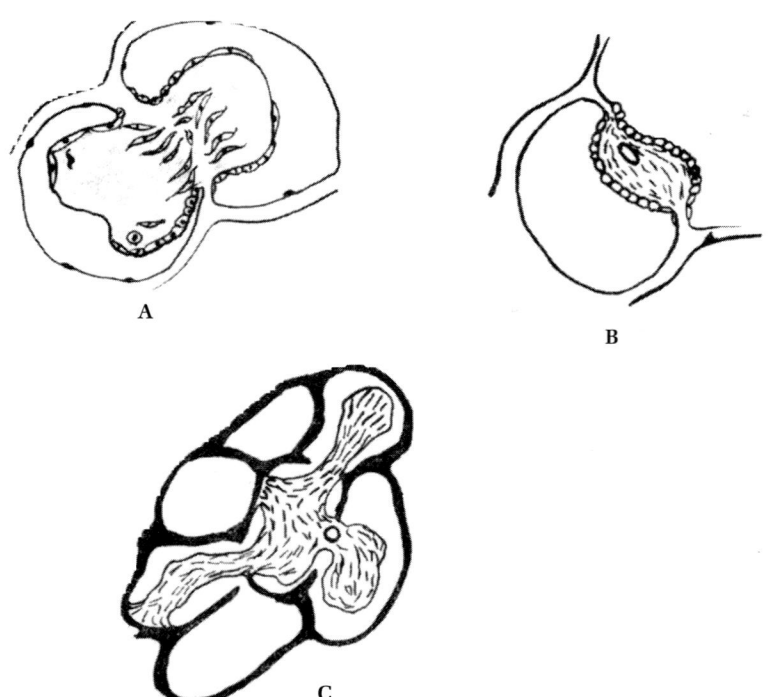

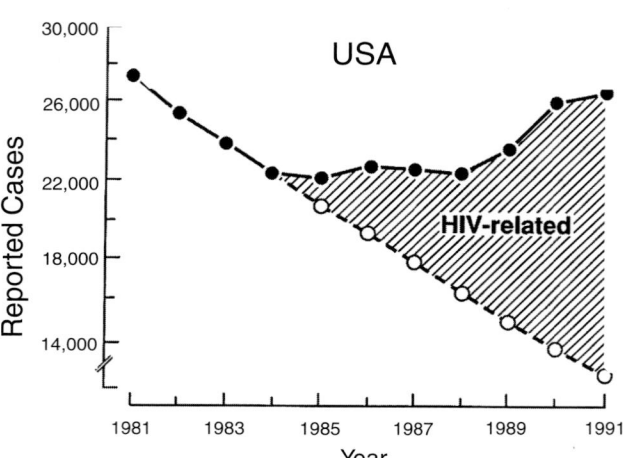

FIGURE 7-44
Diagram showing stages in the organization of an intraalveolar fibrinous exudate. *A.* Fibrin (gray) is being covered by overgrowth of alveolar epithelium and is being invaded by young mural fibroblasts. *B.* Organization has resulted in mural fibrous plaques. *C.* Simultaneous organization in many alveoli has resulted in partial alveolar dissolution.

FIGURE 7-45
Fall in the number of reported cases of TB in the United States. *(Courtesy of Dr. J. Murray and the* American Review of Respiratory Diseases.*)* (See text.)

Relapses of tuberculosis in foreign-born immigrants and transmission of new infections in "high-risk" environments such as hospitals and prisons also contribute.[159a]

However beginning in 1986 the number of reported cases of tuberculosis increased for the first time in 33 years. The number of reported cases has increased nearly every year until 1993. Similar increases have been observed in other countries such as Denmark, Italy, the Netherlands, Spain, Switzerland, France, and the United Kingdom.[159b] However almost 90 percent of the "extra" cases live in the impoverished countries of Asia, Africa, and South America.[159c]

Without acceptable autopsy and audit rates it is possible that cases are being missed. This is exemplified in a study from the Bronx, New York,[160] which reported 21 cases of TB over a 22-year period; the disease was undiagnosed until necropsy. This may seem a small number of cases, but the author does not give the autopsy rate or the the total number of deaths in the institutions where the study was carried out. A study in Scotland showed that almost 40 percent of patients with a convincing combined clinical and pathologic history of TB were not notified.[161] Diagnosis of this disease is of great social importance, since it is potentially curable and diagnosis at autopsy enables surveillance of contacts.

An increasing problem in the United States and throughout the Western world is the emergence of drug-resistant TB, occurring in both the homeless and the immigrant population. Death rates in recent outbreaks are as high as 89 percent, and patients may transfer drug-resistant TB to others.[162]

Organism and Epidemiology

The genus *Mycobacterium* consists of more than 55 well-defined species, including the causative agents of TB, leprosy, and chronic hypertrophic enteritis (Johne's disease) of cattle. The organisms are either straight or slightly curved rods, but coccobacillary, filamentous, and branched forms may also occur.

The organisms are gram-positive, although staining by this method is difficult. They are typically acid-fast, nonmotile, and nonsporing.[163] The organisms usually measure 1 to 4 μm by 0.3 to 0.6 μm. The morphology varies from species to species. Thus, *M. kansasii* is often elongated with a distinct banded or beaded appearance, while *M. avium* is almost coccoid. Mycobacterial diseases other than TB are dealt with below.

Certain groups of people are at increased risk for developing TB. Patients who are HIV-positive or malnourished, the homeless, and those who smoke and drink heavily are the most susceptible. Workers in the mining, quarrying, sandblasting, and tool-grinding industries have a higher incidence of TB because of the lung damage caused by exposure to the consequent dust.[164] There is a risk to health workers and especially pathology laboratory staff, but this risk has been minimized by health and safety procedures. Immunosuppressed patients, other than those with AIDS, are at increased risk, but TB is relatively uncommon as an infective complication in transplantation.

The clinical features of the disease depend on the virulence of the organism and the host immune defenses.[165] No single determinate of virulence has been found, but it appears to be related to the organism's ability to survive within macrophages. The risk of infection is related to the number of tubercle bacilli in the air. Riley's team[166] showed that, on average in TB wards, one infected particle is found in every 15,000 to 20,000 cubic feet of air.

Pathology

Only a brief outline of the pathology of TB will be given, and the reader interested in this disease is referred to the very detailed account by Rich.[167] The pathology of TB is conveniently divided into primary and secondary tuberculosis.

Primary Pulmonary Tuberculosis

Inhalation of *M. tuberculosis* by a nonimmune person leads to the formation of a caseous and granulomatous focal lesion known as the Ghon focus. This may be situated subpleurally at any level in either lung. Mycobacteria spread rapidly via the intrapulmonary lymphatics to the regional hilar lymph nodes, where the organisms are ingested by the reticuloendothelial cells. Caseous and granulomatous lymphadenitis rapidly ensues, causing considerable hilar lymph node enlargement.

The primary lung lesion (Ghon focus) and the regional lymph node lesions are together referred to

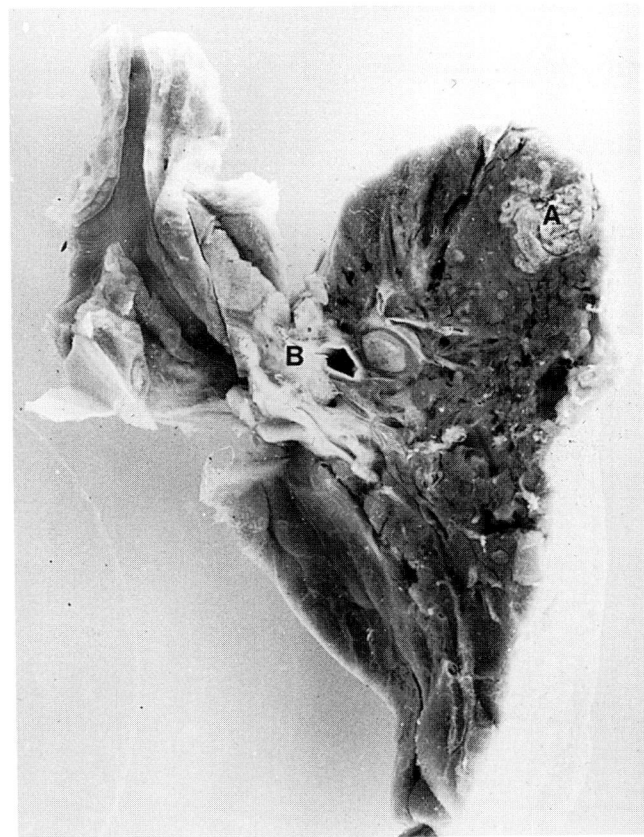

FIGURE 7-46
Primary pulmonary complex in a young child's lung. *A.* Primary subpleural Ghon focus. *B.* Tuberculous hilar nodes lying outside but around a major bronchus.

as the primary complex (Fig. 7-46). The further course depends on the immune response of the individual, and this depends to a great extent on age. In the very young (less than 2 years of age), further rapid spread of the primary complex may result in erosion of a bronchus or a blood vessel, resulting in tuberculous bronchopneumonia or even miliary TB. In most older people the primary infection is clinically silent and undetectable, eventually heals to leave fibrosis and calcification in the Ghon focus and affected lymph nodes (Fig. 7-47).

Primary lesions in some adults, particularly Asians, may be associated with large pleural effusions.

Secondary Pulmonary Tuberculosis

This almost invariably follows a further inhalation of *M. tuberculosis* and is not usually the consequence of reactivation of a dormant primary lesion. Reactivation of a primary lesion would result in the secondary lesion being located at the site of the original Ghon focus—i.e., anywhere in the subpleural tissue of an upper lobe.

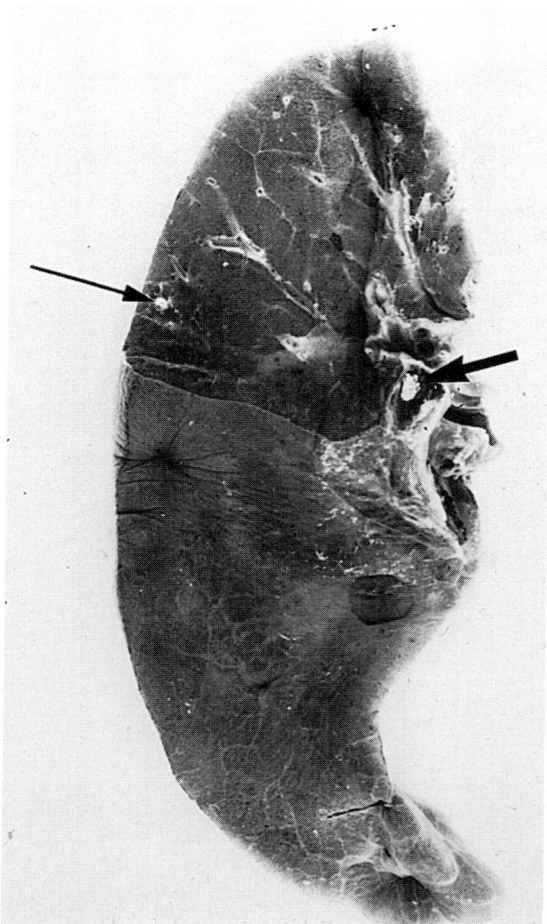

FIGURE 7-47
Healed primary pulmonary tuberculosis. The thin black arrow indicates healed subpleural Ghon focus. The thick black arrow points to healed and calcified lesions in a hilar node.

The immediate tissue response to the inhaled second dose of *M. tuberculosis* is to cause a hypersensitivity reaction characterized by tissue necrosis indistinguishable from caseation. Lymphocytes, epithelioid (histiocytic) cells, plasma cells, and Langhan's giant cells rapidly collect around the zone of necrosis. They are followed by the formation of a wall of fibrous tissue, which effectively seals the area from the surrounding lung.

Lymphatic spread of the disease outside the confines of the main lesion is minimal, but may result in a few extramural satellite tuberculous granulomas. The hilar lymph nodes are not usually involved in the spread of secondary infections, and in most cases the secondary lesions tend to heal slowly; they shrink as fibrosis and calcification take place.

Some cases present as either hilar or mediastinal lymphadenopathy. In these cases an extrinsic mass may impinge on the tracheal or bronchial walls. Occasionally, after initial shrinkage, as the healing proceeds, the secondary lesions may remain static or even increase in size. Coexistent sarcoidosis has to be excluded, although such dual diagnoses should be made with great caution.[168]

Such secondary lesions are often referred to as a tuberculomas. Their continued presence and possible further enlargement constitute a danger to the patient (Fig. 7-48). Enlargement of a tuberculoma may cause erosion of an adjacent bronchial wall and discharge of its caseous contents into the lumen. When this occurs, the center of the tuberculoma often forms a cavity that opens directly into the bronchus, allowing the continued discharge of the infected caseous contents (Fig. 7-49). The patient then has open cavitated TB and is a danger to other persons, as tubercle bacilli continue to be expectorated in the sputum.

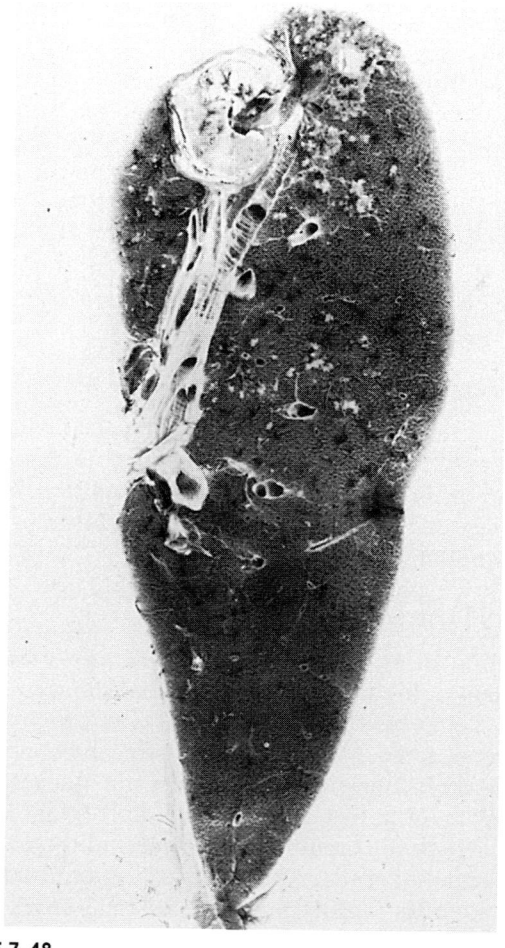

FIGURE 7-48
An apical "tuberculoma." This is close to a bronchus, and some miliary tubercles are seen.

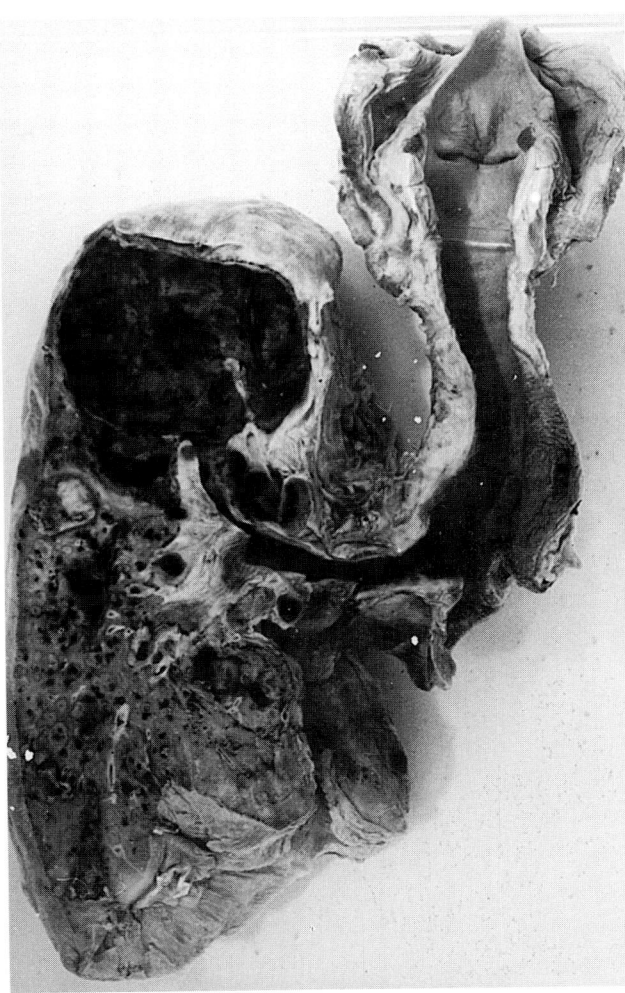

FIGURE 7-49
Chronic cavitated pulmonary TB. The large cavity communicates with a bronchus.

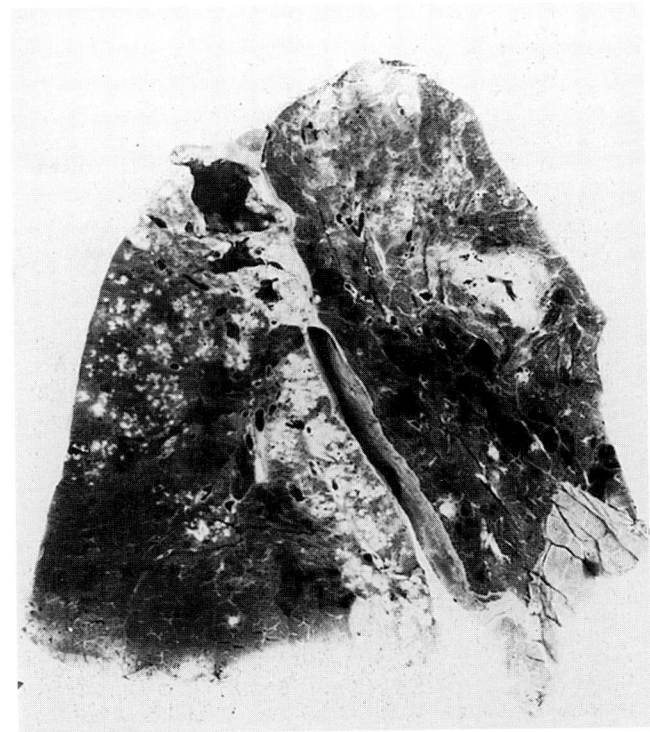

FIGURE 7-50
Chronic secondary tuberculosis. A cavitated lesion lies ruptured into a bronchus, causing tuberculous bronchopneumonia.

The contents of the tuberculous cavity may also be inhaled into other parts of the patient's own bronchial tree and cause tuberculous bronchopneumonia (Fig. 7-50). The nature and extent of the pneumonia depend on the quantity of infected material aspirated and on the patient's resistance. The resulting lung lesions may vary from small focal caseating lesions to a rapidly spreading and almost lobar consolidative change, with little or no cellular or fibrotic reaction.

When there is granulomatous pneumonia in terminal bronchiolar units, it is termed *acinar-nodose pneumonia*. Bronchiolar walls are thickened by granulomatous inflammation, with loss of mucosa, which may progress to fibrous obliteration. At any time, a secondary tuberculous lesion or any of the complications can develop into miliary TB, after the rupture of a tubercle into the lumen of a vein (Fig. 7-51). Granulomatous inflammation may be seen in the walls of arteries and veins and, when extensive, is associated with thrombosis and infarction.[169]

The disease may present as an endobronchial tumor, as inflammation (Fig. 7-52), or with bronchial stenosis.[170] It is very important that TB be considered in any necrotic endobronchial biopsy. Patients with endobronchial disease have a high incidence of bronchial stenosis.

Another presentation is as a spindle cell pseudotumor. The case reported by Sekosan's team[171] presented at postmortem with numerous firm round gray nodules ranging from 0.2 to 2.5 cm in diameter. Some were centered around small bronchi. Histologically there was a proliferation of spindle cells arranged in fascicles with ill-defined borders. There were no granulomas and only a few scattered lymphocytes and plasma cells. A Ziehl-Neelsen stain showed numerous acid-fast bacilli.

There may be pleural involvement. This may be in the form of an effusion. There is pleural fibrosis, with a dense infiltrate of lymphocytes and granulomas. Small numbers of neutrophils and eosinophils may be present, but acid-fast bacilli may be difficult to detect. Needle biopsies of the pleura are positive in up to 80 percent of cases, and in these the culture is usually positive. Mycobacteria can reach the pleura as part of miliary disease, from rupture of an intrapulmonary focus, from rupture of a caseous intrathoracic lymph node, or in rare cases, from a focus in the thoracic spine.

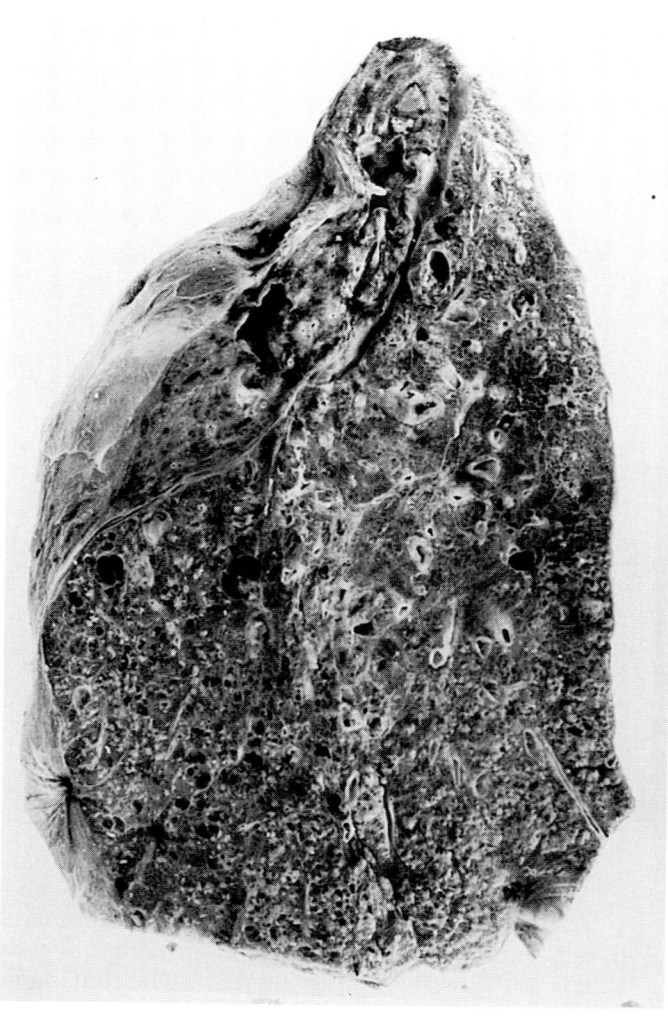

FIGURE 7-51
Secondary pulmonary TB with apical fibrosis and cavitation and miliary TB.

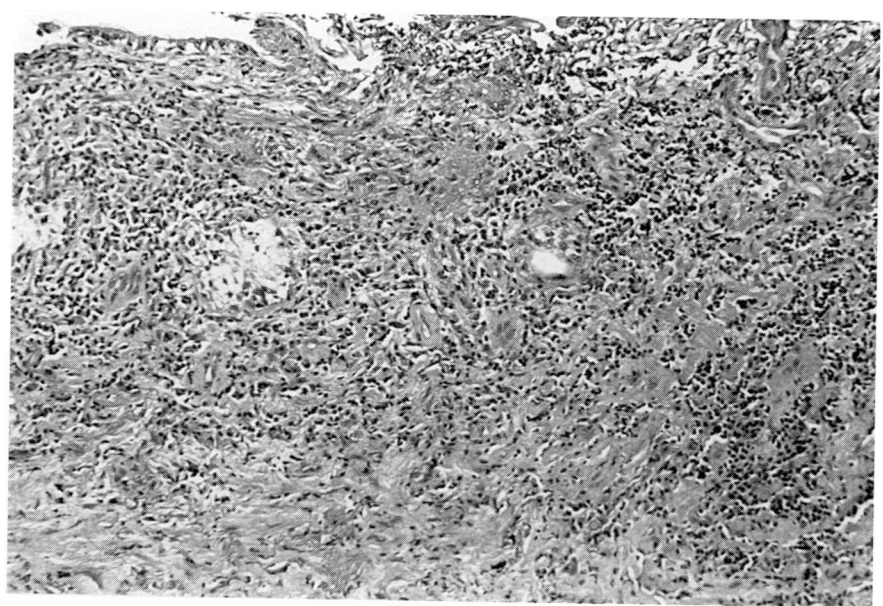

FIGURE 7-52
Bronchial wall with marked chronic inflammation, fibrosis, and a few ill-defined granulomas. Tubercle bacilli were demonstrated. (H&E, ×88.)

The microscopy of the pulmonary lesions, as of tuberculous lesions anywhere in the body, largely mirrors the immune status of the host. The initial tissue reaction to a primary encounter with *M. tuberculosis* is a local mobilization of neutrophil polymorphs at the site of implantation—a change not normally seen in human lesions by the time they can be examined. Tissue necrosis (caseation) rapidly ensues, provoking a variable lymphocytic, histiocytic (epithelioid) cell and giant cell reaction. This is usually followed by mural fibrosis. Giant cells (Langhans' cells) are formed by fusion of histiocytes. In lymph nodes these may be the main tissue reaction, simulating sarcoidosis (Fig. 7-53).

Epituberculosis is a disease, usually seen in young children, in which a lobe of lung becomes airless, grayish white, and consolidated. Careful examination usually reveals enlarged tuberculous hilar lymph nodes. It is thought that seepage of their caseous contents into a lobar bronchus excites inflammatory and edematous changes in the affected lobe in a person with an unusually highly developed hypersensitivity reaction rather than a tuberculous bronchopneumonia.

Tuberculous bronchopneumonia in a person with reduced or destroyed immunity to infection results in widespread caseous consolidation of alveoli, with little or no cellular response.

A review of a cellular and molecular aspects of granulomatous inflammation was presented by Kunkel and coworkers,[172] who described cytokines that are potentially active during the initiation and maintainance of chronic inflammation, including tumor necrosis factor (TNF), interleukin 1, and a novel class of chemotactic cytokines. This last group of mediators belongs to a supergene family of immune signals that play a key role in the selective recruitment of inflammatory cells to an area of inflammation. It appears that the coordinative synthesis of these cytokines is important to the development of the granulomatous response. The participation of molecular signals by fibroblasts and epithelial cells serves an important role in inducing pulmonary granulomatous inflammation.

Bronchoalveolar lavage in patients with slow disease regression in active pulmonary tuberculosis showed a decreased CD4/CD8 ratio and an increase in CD8 cells in the alveolar spaces. These patients did not have HIV infection. This suggests that the balance of T lymphocyte subsets may play an important role in the modulation of host defences against tuberculosis.[172a]

Mycobacterial Diseases Other Than Tuberculosis

As noted above, the incidence of TB in the Western world has declined steadily up to 1986. However, during the same period the frequency of disease attributable to other mycobacteria has increased both in actual numbers and in proportion to the total burden of mycobacterioses.[173] Chronic pulmonary disease, lymphadenitis in children, soft-tissue and skin involvement, and skeletal system involvement are predominant. The principal etiologic agents are *M.*

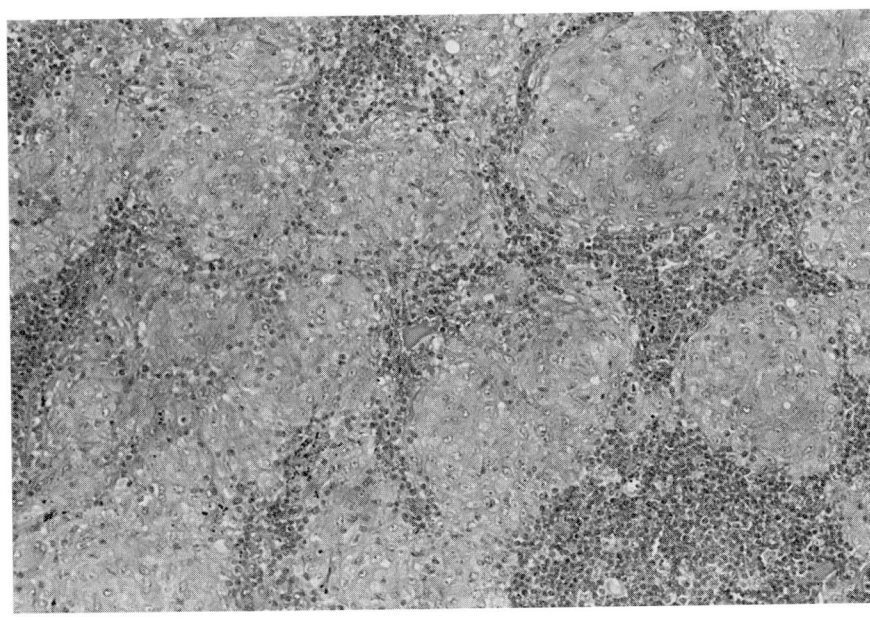

FIGURE 7-53
Lymph node replaced by noncaseating granulomas and resembling sarcoidosis. (H&E, ×88.)

avium–*M. intracellulare* complex,[174,175] *M. kansasii*,[176] *M. marinum, M. fortuitum–M. chelonae* complex and *M. scrofulaceum*. These diseases have become more common in patients with AIDS. The increase has also been attributed to the growing number of people who are immunocompromised because of malignancy, aggressive treatment of rheumatic diseases, or organ transplantation.

Person-to-person transmission of nontuberculous mycobacterial disease rarely occurs, and the environment is the prime source of the free-living strains.

There are geographic differences in the frequency of the disease. Discounting AIDS patients, there is a high incidence of *M. kansasii* infection in Chicago and New Orleans and in some coalmining regions of England and Wales. *M. xenopi* is a common pathogen in southeastern England and Ontario, Canada.

Pulmonary disease has a wide spectrum of clinical presentations. The disease varies in severity from infected bullae, which may heal spontaneously, to a progressive destructive lung disease. Contamination of specimens or transient colonization may pose a problem, since the organisms mentioned above are often free-living.

There are four main pulmonary patterns.[177] In decreasing order of frequency these are solitary nodules, chronic bronchitis or bronchiectasis, tuberculous-like infiltrates (in older white women), and diffuse infiltrates (in AIDS patients). Lymphadenitis occurs almost exclusively in children between 16 months and 10 years of age. The infected nodes are in the submandibular or preauricular areas.

This atypical mycobacterial disease may be difficult to differentiate from TB. Certain features can provide helpful clues, such as thin-walled cavities with relatively little surrounding parenchymal infiltration. Bronchogenic spread to the same or the opposite lobe is rare, and thickening of the pleura over the lesion is common, but a basal pleurisy is not often evident.

Atypical mycobacterial infection is rarely accompanied by pleural involvement. A rare case of *M. avium–intracellulare* pleuritis with a massive pleural effusion has been reported.[177a]

Predisposing conditions include pneumoconiosis, especially that associated with silicosis and coalmining and any chronic lung disease. Nontuberculous mycobacterial infection has been described in cystic fibrosis.[178] Cultures for mycobacteria in patients with cystic fibrosis were frequently overgrown with other bacteria. Seventeen of 87 patients had at least one positive culture for nontu-

berculous mycobacteria, and the most common organism was *M. avium–intracellulare*.

PATHOLOGY

The pathology of nontuberculous mycobacterial infection has been described by Marchevsky and colleagues,[179] who recognized three distinct clinicopathologic groups of patients. The first had solitary pulmonary nodules resected. These showed granulomas with varying degrees of necrosis. Mediastinal nodes had no granulomas except in one case. Focal pleural granulomas overlay eight of 18 solitary nodules. Granulomatous bronchitis was present in 11 cases, and granulomatous vasculitis was seen in two.

The second group had radiologic evidence of bilateral diffuse interstitial infiltration, and the most commonest organisms were *M. avium–intracellulare* or *M. gordonae*. There was nonspecific chronic inflammation with organizing pneumonia. One case had a focal aggregate of noncaseating epithelioid granulomas. *Pneumocystis* was seen in two cases and herpes-like intranuclear inclusions in one.

The third group had multiple discrete radiologic infiltrates, and *M. avium–intracellulare* was isolated from the lung tissue. There were histiocytic infiltrations or necrotizing or hyalinized granulomas. Some of these patients had preexisting lung disease, such as Wegener's granulomatosis or Hodgkin's disease.

DIAGNOSIS OF TUBERCULOSIS

Detection of the tubercle bacilli by either direct smear or culture remains the only available way to confirm the diagnosis with certainty. It is often difficult to demonstrate tubercle bacilli in histologic sections. In such cases it is necessary to rely on multiple sections or to confer with the bacteriology department to see if they have been able to isolate the organism. In any specimen in which TB is suspected, material should be sent for bacteriology, not only so that the diagnosis can be confirmed but also to enable sensitivities to be worked out. It is the rule rather than the exception to be able to demonstrate tubercle in lymph nodes. Nearly half of pulmonary tuberculous infections are smear-negative.

Measurement of an *M. tuberculosis*–specific antibody has proved helpful in diagnosis. There is a specificity of 98 percent with a single monoclonal antibody to *M. tuberculosis*–specific antigen. Seventy percent of smear-negative pulmonary TB cases were positive with this technique.[180]

Another technique that has been used in diagnosis is the polymerase chain reaction (PCR).[181,182] The target DNA of a 123-based pair segment of IS6110—

which is repeated in the *M. tuberculosis* chromosome and is specific for *M. tuberculosis* complex—gives the best results. This primer (IS6110 DNA) is highly specific. It was identified in all six patients with active TB, from 15 of 18 patients with past TB, in five of nine contacts of patients with TB, and from nine of 54 patients with lung disease unrelated to TB.[183] A study in a chest hospital in Hong Kong using PCR on both the organisms and sputa also gave reliable results but used an amplified product of 239 base pairs. The PCR remained positive after 4 weeks of antituberculous therapy, when 16 of 29 patients had become culture-negative.[184]

There have been two reviews on the various molecular amplification techniques used in the diagnosis of tuberculosis.[184a,184b] PCR may result in false positives in some laboratories. In addition there may be variation in the levels of sensitivity between different workers. False-negative results may be caused by inhibitors of the DNA polymerase.*

Recently a commercial PCR amplification kit for the detection and identification of *M. tuberculosis* complex bacteria has become available (Amplicor; F. Hoffman-La Roche, Basel, Switzerland). The target for the PCR is the 16S rRNA sequence. The detection system is based on hybridization with a *M. tuberculosis* complex-specific capture probe in a microplate format.

PCR has also been used on histologic sections. Popper and colleagues[185] compared conventional staining techniques and PCR for mycobacterial DNA sequences in 24 selected tissue samples from patients with TB. In all samples, either positive or negative with the acid-fast stain, mycobacterial DNA fragments were detected. However, the validity of this investigation may be in some doubt, since two of 15 cases of clinically proven sarcoidosis had strong signals for mycobacterial DNA. Other studies with PCR have demonstrated the presence of mycobacterial DNA in sarcoidosis.[186,187] It has been suggested that the disease is an abnormal response to the tubercle bacillus.

Bronchoalveolar lavage (BAL) fluid from patients with pulmonary TB may yield the bacillus. Differential counts show that neutrophils and lymphocytes are increased in BAL fluid from patients with miliary TB. The number of alveolar macrophages is decreased. The number of these cells was unchanged in BAL fluid from unaffected regions of the lungs of patients with active or inactive pulmonary TB.[188]

DIFFERENTIAL DIAGNOSIS

For a definitive diagnosis of TB, one must demonstrate tubercle bacilli histologically, by culture, or perhaps by PCR. If there are no organisms the following differential diagnoses should be entertained.

Sarcoidosis

This is a common problem in differential diagnosis. Clinically in both diseases one might observe bilateral hilar lyphadenopathy. The presence of a raised serum angiotensin-converting enzyme will favor a diagnosis of sarcoid. Histologically the granulomas in sarcoid tend to follow the lymphatics and often show little caseation. However, in necrotizing sarcoid granulomatosis there is evidence of caseation as well as bronchiolar and vascular involvement. Traditionally the reticulin stain has been said to be useful in differentiating sarcoid from TB. In sarcoid the reticulin pattern is preserved in the small necrotic foci, whereas in TB it is lost. The author has found this of little value. Cases have been described in which sarcoid and TB may have coexisted, but these were in the early literature.[189] It is wise to do a Ziehl-Neelsen stain in every case of suspected sarcoid.

Fungal Infection

Fungal infection, including cases of *Pneumocystis carinii* pneumonia in AIDS patients, may produce granulomas. It is essential in any granulomatous infection to do a methenamine silver stain to demonstrate the causative organism.

Extrinsic Allergic Alveolitis

Extrinsic allergic alveolitis (EAA) is misdiagnosed by the unwary as TB. In EAA there is no caseation, and the granulomas are ill defined and always interstitial. The infiltrates consist of plasma cells, lymphocytes, and some eosinophils, which expand the interstitium. In EAA the process is usually bronchocentric.

Rheumatoid Nodule

This may present before the onset of rheumatoid disease. There are pallisaded histiocytes and usually large areas of central necrosis with or without vasculitis. If there is no clinical history, it is important to look at the special stains, such as a Ziehl-Neelsen and a fungal stain before making a diagnosis of a rheumatoid nodule.

---

* A recent Dutch study using culture and clinical history as the gold standard showed a sensitivity and specificity for PCR of 92.1 percent and 99.8 percent respectively. This paper suggests that with reliable precautions PCR is a suitable and reliable method for the detection of *M. tuberculosis*.[184c]

Bronchocentric Granulomatosis

This disease usually occurs in asthmatics and is a form of allergic bronchopulmonary aspergillosis. It is worthwhile cutting serial sections to look for fungi. Myers and Kazenstein[190] described four cases of granulomatous infection with histologic features indistinguishable from those of bronchocentric granulomatosis. No patient had a history of asthma, and eosinophils were not numerous in the biopsy specimens. Organisms were rare, but tubercle bacilli were found in two cases and histoplasma in one. These authors noted that extensive peribronchial localization of granulomas can occur in infection and thus cause confusion with bronchocentric granulomatosis.

Wegener's Granulomatosis

It is important in this disease to identify necrosis of the vessel wall. In many of the diseases listed above it is possible to see a degree of transmural inflammation, but that feature is not sufficient for a diagnosis of Wegener's granulomatosis. A positive ANCA test may help clinically. Elastic–van Gieson and Martius Scarlet Blue (MSB) stains are essential.

Pulmonary Syphilis

Cases of congenital pulmonary syphilis will show numerous spirochetes. In adults gummas may occasionally cause confusion. These are usually multiple but may fuse to form a large solitary lesion. There is associated interstitial fibrosis. Ghost outlines or vessels and alveoli are well seen in EVG-stained sections. Giant cells are scanty and small. Plasma cells, not lymphocytes, predominate.

Pulmonary Syphilis

Pulmonary syphilis is still a rare disease, but there has been an increase in primary and secondary syphilis over the past several years in the United States.[191] Pulmonary syphilis is rare in countries where medical treatment of the primary lesion is readily available; the best reports come from earlier papers by Karshner and Karshner,[192] Carrera,[193] Howard,[194] and McIntyre.[195]

CONGENITAL PULMONARY SYPHILIS (PNEUMONIA ALBA)

Most infants with congenital pulmonary syphilis are stillborn or rarely survive more than a few hours. The other stigmata of the disease are present. The lungs tend to be larger than normal, pale, and firm. If the child has breathed, islands of pale indurated tissue remain depressed below the general surface of the more vascular aerated lung.

Histologically, the lungs show diffuse interstitial thickening caused by proliferating fibroblasts. These may be obvious in alveolar walls and in peribronchial and perivascular connective tissue with the aid of a reticulin stain. There is also a diffuse lymphoplasmacytic infiltrate (Fig. 7-54). In some areas, small necrotic foci are surrounded by chronic lymphocytes cells along with neutrophil polymorphs, forming microscopic gummas. These may fuse to form larger, macroscopic gummas. The alveoli become lined by cuboidal epithelium and resemble acini of fetal life. There may be obliterative endarteritis. Spirochetes are usually present in large numbers and can be shown by the Levaditi stain (Fig. 7-55).

Pneumonia due to congenital syphilis must be differentiated from other causes of pneumonitis occurring in this age group, such as *P. carinii* and *Cytomegalovirus*.

ACQUIRED PULMONARY SYPHILIS

Descriptions of pulmonary syphilis usually refer to the tertiary disease. Karshner and Karshner remarked that secondary pulmonary syphilis can be "dismissed in a word; in a few of the cases mild bronchitic symptoms date from the period of the exanthem."[192] Bronchitis and pleurisy may occur in the secondary stages, but there is no evidence of any interstitial disease.

Four varieties of pulmonary lesions may be seen in the tertiary stage: gummas, which may be multiple and small or, if they coalesce, may form a large

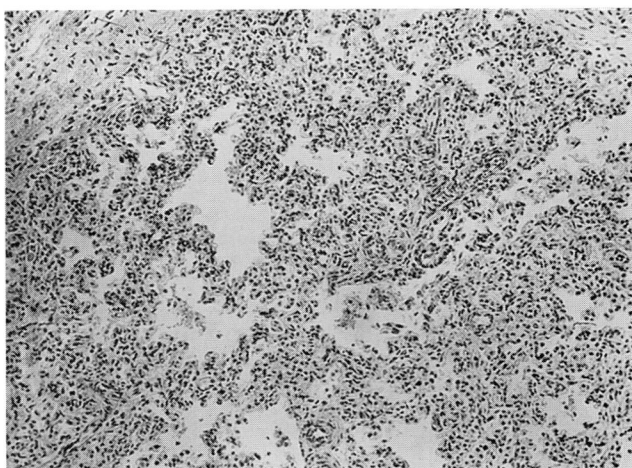

FIGURE 7-54
Congenital syphlis. There is a generalized chronic interstitial inflammatory cell infiltrate and thickening of the alveolar walls. (H&E, ×104.)

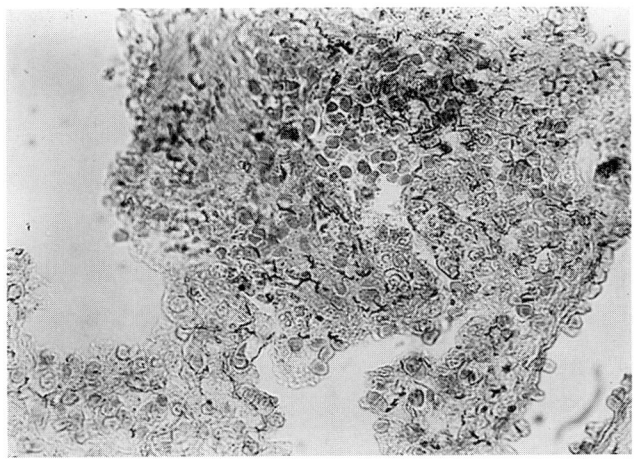

FIGURE 7-55
Congenital syphlis with many *Treponema pallidum* organisms. (Levaditi, ×370.)

gumma; syphilitic pulmonary fibrosis of a nonspecific nature; syphilitic chronic interstitial fibrosis; and gummatous ulceration of the trachea and bronchi.

There may be a bronchiectatic form—described by Karshner and Karshner[192]—but this is not specific

FIGURE 7-56
Pulmonary gumma presenting as a fairly well demarcated lesion and a fibrous wall around much of it. (×1.8.)

for syphilis. In the absence of a gumma, it may be impossible to be certain of the syphilitic nature of the less specific changes noted in the first two varieties mentioned above.

Gummas are usually multiple and small in the early stages but later fuse to form a large solitary lesion (Fig. 7-56), which is usually associated with diffuse interstitial fibrosis. The two lesions together are referred to as a sclerogummatous lung. In pulmonary gummas there is a central necrotic area, which is sharply differentiated from and bounded by a thick layer of gray-white scar tissue. Bands of fibrous tissue radiate into the surrounding lung.

Histologically, the necrotic central area (Fig. 7-57) shows ghost outlines of dead tissues, including vessels and septa, which can be demonstrated by elastic stains. Plasma cells predominate, and this helps to differentiate syphilis from TB. In addition, giant cells are scanty and small (Fig. 7-58). The arteries in the vicinity of a gumma show occlusion by intimal fibrosis and, in some cases, chronic inflammation (Fig. 7-59). *Treponema pallidum* is not normally found.

Diffuse interstitial fibrosis causes derangement of the alveolar wall along with epithelialization, fibrosis, and an infiltrate of lymphocytes and plasma cells. There is also thickening of the interlobular, subpleural, and perivascular connective tissue. The lung may lose all semblance of an alveolar pattern, being reduced to a mass of fibrous tissue. Bronchiectasis may result from the pulmonary fibrosis. Gummatous ulcers in the main bronchi or trachea are uncommon. If they do occur, they leave deep penetrating ulcers and perforations of these structures.

Before a diagnosis of syphilis is made, it is important to exclude pulmonary TB, a fungal disease, or, less likely, Wegener's granulomatosis.

SYPHLITIC PULMONARY ARTERITIS

While syphilitic aortitis used to be common, pulmonary arteritis due to this disease was rare. It caused aneurysmal stretching of the walls of the large pulmonary arteries. Warthin[196] reported a case of a leaking syphilitic aneurysm in the left upper lobe. Macroscopically, the disease closely resembles syphilitic aortitis. "Spider web" intimal scarring and fusiform or saccular aneurysms are the two most distinctive features. The aneurysms are more likely to fill with thrombus than their aortic counterparts, and this leads to occlusion of their major branches.[197]

The histologic appearances are similar to those found in aortic disease, being characterized by peri-

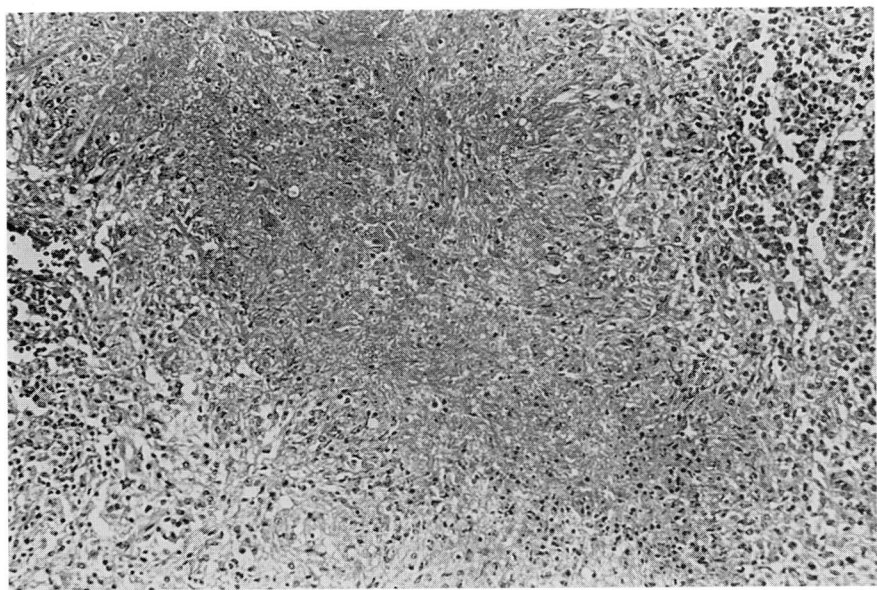

FIGURE 7-57
Central necrotic area (gumma) surrounded by fibrosis and chronic inflammation. (H&E, ×88.)

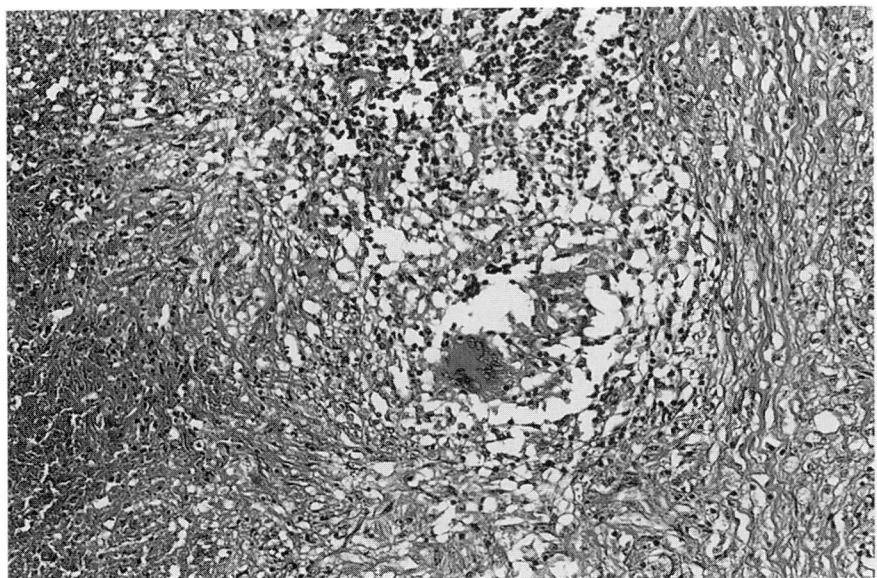

FIGURE 7-58
Scattered small giant cells at the margin of a gumma. (H&E, ×88.)

adventitial and perivascular lymphocytic and plasma cell infiltrates. The internal elastic lamina may show greater destruction than is usually seen in the aorta. The small pulmonary arteries are rarely affected by syphilitic changes.

Pulmonary Actinomycosis

ORGANISM AND EPIDEMIOLOGY

Actinomyces, Arachnia, and *Nocardia* belong to the order Actinomycetales. These are gram-positive branching bacteria. The most common organism causing actinomycosis in humans is *Actinomyces israeli,* although some of the *Arachnia* species may cause a clinical disease resembling actinomycosis.

Many strains are microaerophilic, but the organism grows best anaerobically; some strains grow aerobically.

The term *actinomycosis* was first used by Bollinger[198] to describe a disease of cattle, where they developed a hard and indurated swelling in the tongue and jaw. Israel[199] was the first to identify the organism in human postmortem sections. For 60 years the organism was known as *Actinomyces bovis,* but in 1949 the name was changed to *A. israeli.*

A. israeli is found around carious teeth and in the tonsillar crypts of healthy people. The organisms can grow on the surface of normal teeth, however, and its natural habitat is the oropharyngeal area.

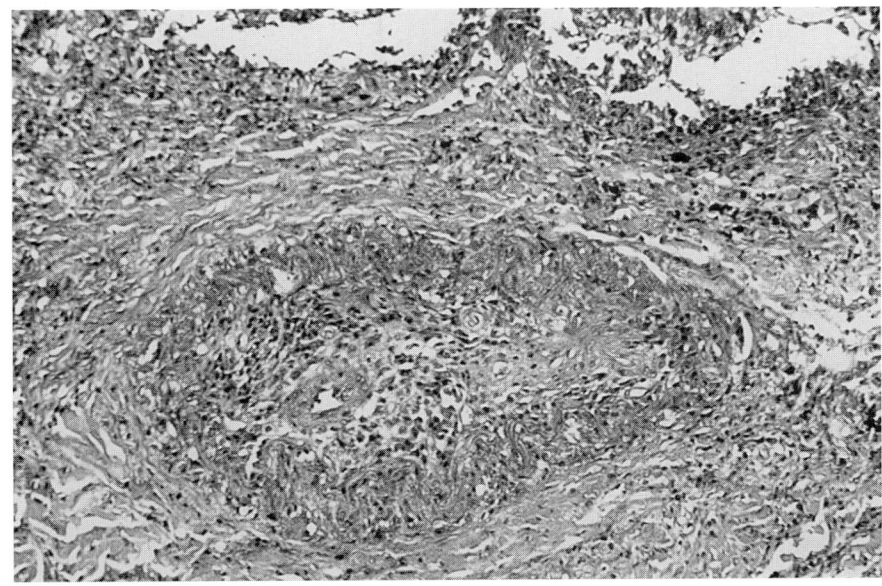

FIGURE 7-59
Occlusion of a muscular pulmonary artery along with some inflammation in pulmonary syphilis. There is a new vessel amid the fibrosis. (H&E, ×88.)

The disease is three times more common in men. Although it may occur at any age, according to Bates and Cruikshank[200] there are two peak periods of incidence. The first is between the ages of 11 and 20 years and the second between 35 and 50. These two periods probably coincide with infection being commonly present in the tonsils and carious teeth, respectively.

The disease is becoming rarer, and a recent study over a 15-year period identified only 19 cases, of which 15 were in males.[201] The incidence has been given as 1 in 2.8 million in Cologne, Germany,[202] or as five cases per year in the United Kingdom.[203]

Thoracic actinomycosis accounts for 15 to 50 percent of all cases of actinomycosis.[204–206] It has many similarities to lung abscesses and other anaerobic bacterial infections. Most cases of actinomycosis gain entrance by aspiration of oral contents and initially produce a pulmonary parenchymal infection in dependent segments. A cervicofacial focus of actinomyces infection is uncommon. Secondary infection resulting from spread of the disease below the diaphragm—particularly the liver or, in rare instances, in the neck from a cervicofacial focus—may occur. Occasionally, pulmonary actinomycosis may complicate pulmonary tuberculosis or carcinoma of the lung.[200,207]

CLINICAL FEATURES

Patients rarely present at an early stage of the disease. The pulmonary lesions may be confused with TB when there is cavity formation or, more commonly, with a bronchial carcinoma if there is a mass. Actinomycosis may be diagnosed clinically as a nonspecific pneumonia, lung abscess, or empyema.[208] It is unusual for the diagnosis to be made clinically and is often made on a frozen section or a resection specimen. Patients have a productive cough, dyspnea, chest pain, hemoptysis, and sometimes superior vena caval obstruction with mediastinal involvement. There is fever and low-grade leukocytosis.

Five cases of endobronchial actinomycosis simulating bronchogenic carcinoma have been described. The diagnosis was correctly made from bronchial biopsy.[209] A further such case was found in a patient with AIDS.[210]

PATHOLOGY

As mentioned above, primary pulmonary actinomycosis occurs most commonly in the lower lobes, although it is unusual for the disease to remain localized by the time the affected lung is removed or the patient dies. Usually more than one lobe is involved, and classically the disease presents as a dense fibrotic lesion containing a honeycomb of small abscess cavities (Fig. 7-60). Occasionally a single large actinomycotic abscess occurs. Primary pulmonary actinomycosis is mainly a peripheral lung lesion and usually involves the pleura early. Pleural involvement results in either an actinomycotic empyema or dense pleural adhesions. The infection may spread directly to invade the chest wall and cause destruction of the ribs and sternum. This may result in discharging sinuses. However, these are now rare, probably because of the use of antibiotics early in the disease process. Thus, the diagnosis is made from colonies of organisms in sputum rather than in discharging sinuses.

The disease may directly invade adjacent structures, such as the esophagus and pericardium, and

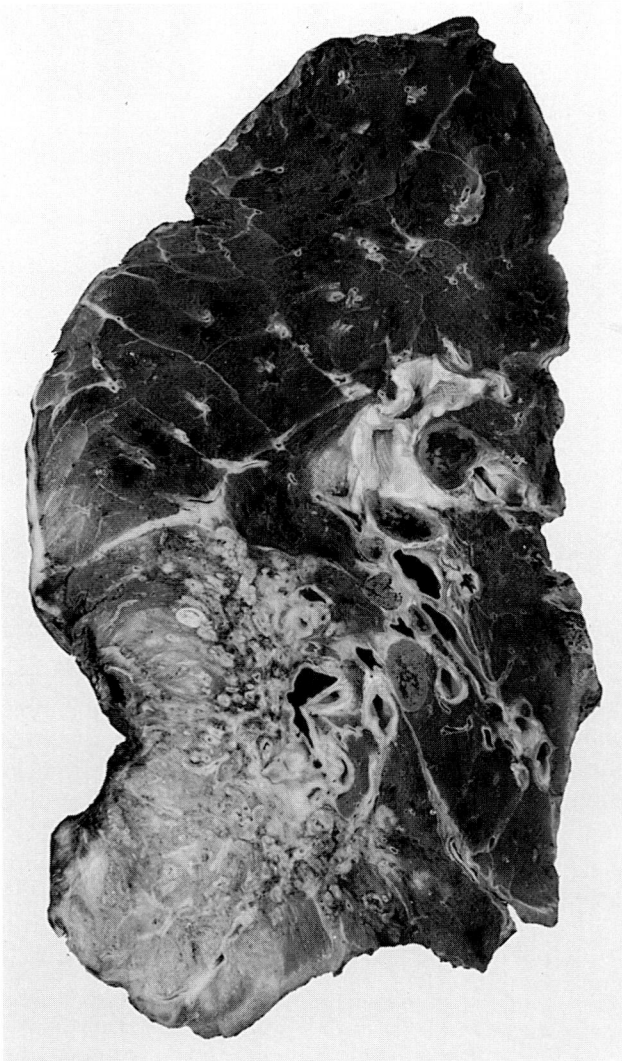

FIGURE 7-60
Pulmonary actinomycosis with a honeycomb of small abscesses amid fibrosis in the lower lobe. (One half natural size.)

lung before a colony of *Actinomyces* is found. In 26 percent of cases the diagnosis of actinomycosis is based on the finding of a single sulfur granule in a lesion.[205]

The actinomycotic colonies occupy the centers of some abscess cavities. These consist of neutrophil polymorphs with a few eosinophils and are surrounded by chronic inflammation, occasional giant cells, and a fibroblastic reaction. (Fig. 7-61 and 7-62).

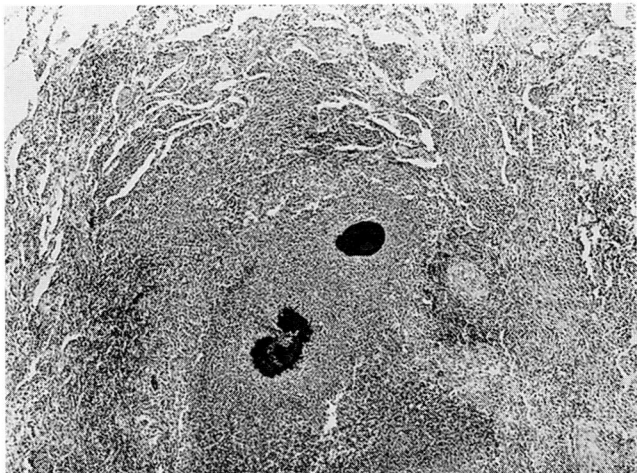

FIGURE 7-61
Low-power view of pulmonary actinomycosis. The dark-staining colonies of *A. israeli* are seen in the center of the abscess cavity. Outside this there is a zone of chronic inflammation. (H&E, ×30.)

there may be cerebral abscesses if there is bloodstream spread. Pleuropulmonary actinomycosis may be complicated by a bronchopleural fistula and amyloid.

Actinomycosis is a chronic granulomatous disease resulting in abscess cavities filled with pus. In the center of some of the abscesses are colonies of *A. israeli*, formed of a felted network of filaments staining intensely with hematoxylin. At the edge of the colony, the individual hyphae may be surrounded with eosinophilic "clubbing" material (Splendore-Hoeppli reaction), giving the whole colony an irregular narrow eosinophilic margin. In some places, "clubbing" material may be absent. The diagnosis should be considered when acute or chronic inflammation is seen causing damage to the lung parenchyma and no other cause is seen. It may be necessary to take several sections from the affected

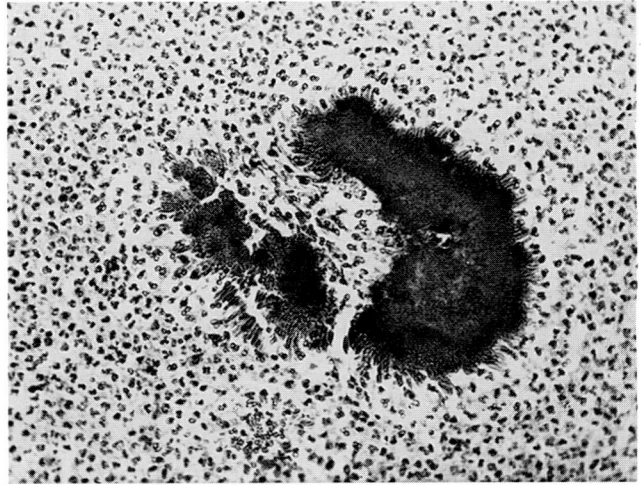

FIGURE 7-62
Higher-power view of *A. israeli* showing a surrounding polymorph reaction. (H&E, ×216.)

An actinomycosis granule is called a "sulfur" granule. This Splendore-Hoeppli phenomenon may be seen around some fungi; it is not specific for actinomycosis and may be seen with *Streptomyces*, staphylococci, streptococci, *E. coli, Pseudomonas* species, and *B. proteus.*

The characteristic sulfur granule ranges from 25 to 3,700 μm in diameter and may be visible on gross examination if a hand lens is used.

A. israeli within the colonies is best shown by the use of a methenamine silver or Gram's stain. The organisms differ from *Nocardia* only in their staining properties. *Nocardia* is acid-fast with the use of 1% H_2SO_4 for 1 min, whereas *Actinomyces* is generally not. However, affinity of the organisms for acid-fast stains varies with the strength of the decolorizing agent. With the Putt modification of the acid-fast stain, which uses a weak decolorizing agent, the organisms are uniformly stained, as is *Nocardia.*[211] Fluorescein species-specific antiserum is available for staining material obtained when sulfur granules are not seen.[212] *Actinomyces* also stains with PAS, and although the organisms are usually gram-positive, it is possible to see gram-negative forms.

Actinomyces consists of long branching filamentous rods, whereas granules may be seen in botromycosis.

Pulmonary Botromycosis (Bacterial Pseudomyocosis, Discomycosis)

Botromycosis is the term applied to an unusual tissue response to a variety of bacterial infections as well as some fungi. It differs from ordinary acute suppurative inflammation inasmuch as the causative organisms become surrounded by eosinophilic "clubbing" material (Splendore-Hoeppli reaction), resembling *A. israeli.* There may be a foreign-body giant cell reaction, surrounded by lymphocytes and plasma cells. Several varieties of bacteria may produce botromycotic lesions, including *S. aureus,* streptococcus species, *E. coli, Pseudomonas* species, *B. proteus,* and *Streptomyces.* Fungi and the eggs of helminths can also give this eosinophilic "clubbing" effect, which occurs when a delicate balance is struck in the host between the numbers and invasiveness of the pathogen and the ability of the host's immunologic and cellular response to contain the invading organism.

The eosinophilic "clubbing" material is an antigen-antibody complex. The colonies of microorganisms thus surrounded are often referred to as *actinomycetoid forms.* Recently primary pulmonary botromycosis has been noted as a manifestation of granulomatous disease[213]; it may occur as part of

AIDS[214] and is common in cystic fibrosis.[215] It can rarely present in normal hosts.[215a]

CLINICAL FEATURES

The skin and subcutaneous tissue are most commonly involved. The pulmonary symptoms are nonspecific and include dyspnea, pleuritic chest pain, cough, and hemoptysis. The upper lobe is more commonly involved with a diffuse infiltrate, a discrete mass, or a cavitating lesion.[215a,216] The disease may spread to hilar nodes, pleura, ribs, and vertebrae. Diagnosis is usually made either at autopsy or on biopsy by histologic and bacteriologic examination.

PATHOLOGY

In the lungs a single large or multiple small abscesses present as hard greenish-yellow nodules or chronic suppurative bronchiectasis.

Histologically, the abscesses contain the usual purulent material, but in the pus are colonies of either bacillary or coccal bacteria surrounded by eosinophilic hyaline "clubbing" material identical to that seen in actinomycosis. A variable number of chronic inflammatory cells and fibrous tissue form the outer part of the abscess walls. Gram stain usually shows the morphology of the bacterial colony and stains the "clubbing" material in a variable fashion. Silver impregnation and periodic acid Schiff (PAS) stain usually show no fungi, unless these are the causative organisms. The inner part of the "clubbing" material surrounding the microorganisms is stained positively by PAS, which can be abolished by previous digestion with 10% diastase. The "clubbing" material also stains metachromatically with toluidine blue.

Pulmonary Nocardiosis

ORGANISM AND EPIDEMIOLOGY

Nocardiosis is a disease that may occur in humans and cattle. Nocard first described the causative organisms in bovine farcy in Guadeloupe.[217] The causative nocardial organisms belong to the order Actinomycetales and comprise *Nocardia asteroides, N. brasilensis,* and *N. caviae.* Most human pulmonary infections are caused by the first two species.

N. asteroides has a worldwide distribution, but *N. brasilensis* is confined to Central and South America. Cases have been reported from Zimbabwe, and a common underlying cause was alcoholism. Nocardiosis may be more common in developing countries than is generally recognized.[218]

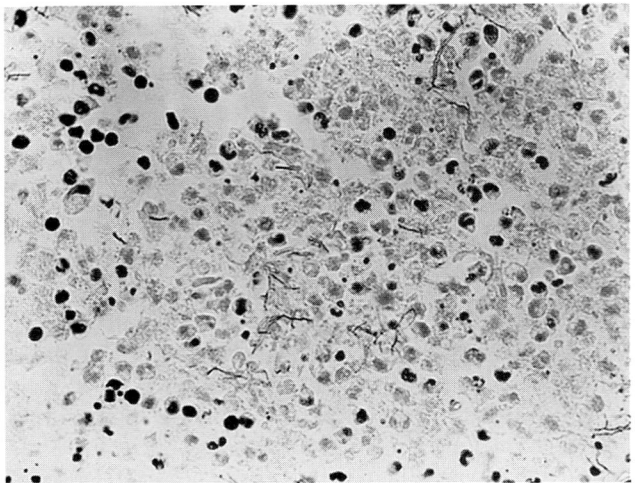

FIGURE 7-63
Fine filamentous branching *Nocardia* organisms. (Gram, ×390.)

The organism is airborne and can also be found in grasses, straw, and grains. It is an aerobic, gram-positive, delicately branching, beaded, filamentous organism 0.5 to 1.0 μm in diameter. No spores are formed. It is weakly acid-fast, which is useful in distinguishing the organism from *Actinomyces*. In ordinary hematoxylin and eosin sections the causative organisms are difficult to visualize but are readily identified when Gram, weak Ziehl-Neelsen, or methenamine silver stain is used (Fig. 7-63). Usually only an occasional filament or part of a filament is seen if the organism is acid-fast in sections stained by Ziehl-Neelsen. They decolorize in 5% aqueous sulfuric acid, although strongly acid-fast strains closely resembling *M. tuberculosis* occur.

Nocardiosis is a rare disease, and the lungs may be involved either primarily or secondary to a bloodstream infection. The disease is more common in immunocompromised patients.[219] With the increase in transplantation and the rising incidence of AIDS,[220] the frequency of nocardiosis is increasing. Diabetics and alcoholics have an increased incidence.[221] The disease may also be associated with pulmonary alveolar proteinosis.[222] This association is not unique, and an alveolar proteinosis picture may be found with other infections, such as *cytomegalovirus*, *Mycobacteria*, *Aspergillus*, Cryptococcus, *Mucor*, *Histoplasma*, and *Pneumocystis*.[223]

CLINICAL FEATURES

The most common clinical presenting features are fever and productive cough with a small amount of mucoid or mucopurulent sputum. There may be hemoptysis, dyspnea, and pleuritic chest pain. The initial diagnosis is usually acute bacterial pneumonia. The real diagnosis is often delayed, and in its chronic form the disease may mimic TB.[224] There is

no clinical syndrome characteristic of pulmonary nocardiosis.[225]

The radiographic findings are not specific and include nodular infiltrates, cavitation, and pleural effusion, and it is possible for the chest radiograph to be normal even though sputum culture may be positive.

PATHOLOGY

In acute nocardiosis the lungs, in common with other tissues, contain numerous yellowish-white abscesses resembling miliary tubercles (Fig. 7-64). In the center of the abscesses are neutrophil polymorphs with a few plasma cells, lymphocytes, and histiocytes. Both the abscess and the surrounding alveoli show a fibrinous exudate.

Nocardia grow within macrophages, and their virulence has been related to inhibition of phagosome and lysosome fusion, reduction in intracellular levels of lysosomal acid phosphatase, and the production of superoxide dismutase and catalase.[225] Sulfur granules and "clubbing" material are not found with *Nocardia*.

Chronic nocardial infection has been de-

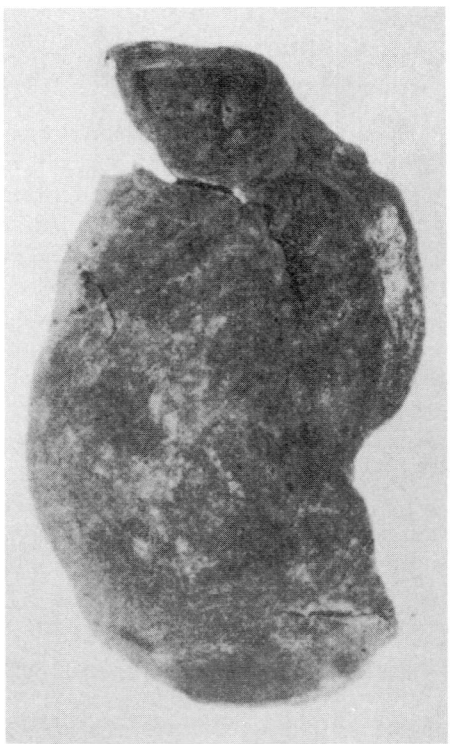

FIGURE 7-64
Pulmonary nocardia showing multiple miliary nodules resembling TB. (Three fourths natural size, approx.) *(Courtesy of Dr. D. R. L. Wilcox and the Editor of Guy's Hospital Reports.)*

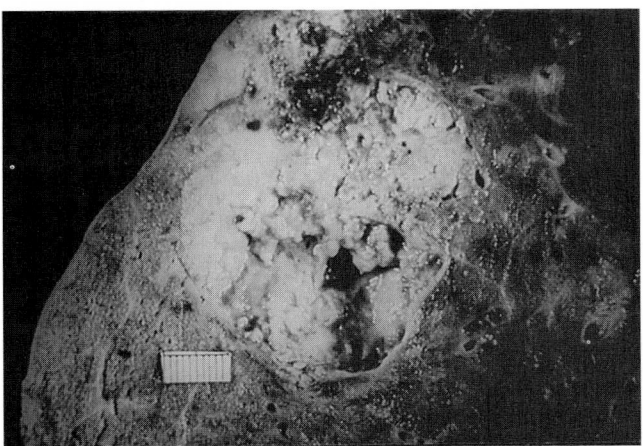

FIGURE 7-65
Chronic lung abscess due to *N. asteroides. (Courtesy of Dr. A. A. Liebow.)*

Chronic nocardial infection has been described.[226] This results in a chronic pneumonia in much of the lobe with single or multiple lung abscesses (Fig. 7-65).

Microscopically the lungs show extensive fibrosis, with lymphocytic, histiocytic, and giant cell infiltration along with masses of foamy histiocytes (Fig. 7-66). There are multiple small lung abscesses, and both macroscopically and histologically the appearance may resemble that of pulmonary actinomycosis. There are no sulfur granules, however, and unless the sections are treated with Gram's stain, the nature of the disorder will probably remain undetected.

Nocardia may occasionally cause a pulmonary mycetoma, a chronic fibrotic cavitating lesion usually found in the upper lobe of a lung containing a

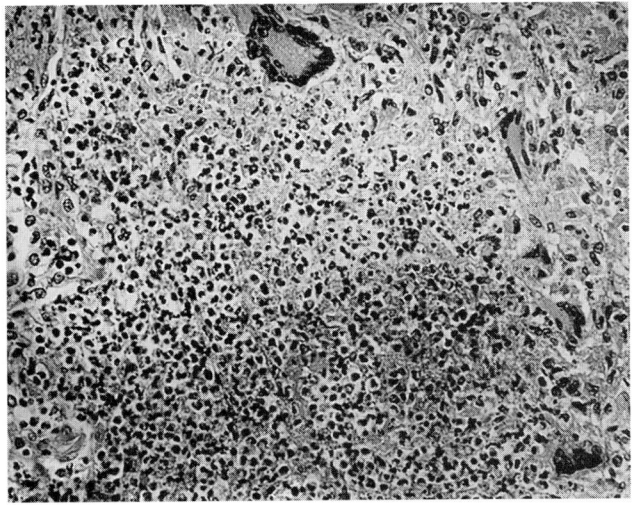

FIGURE 7-66
Chronic nocardial infection with a granulomatous response. (H&E, ×90.)

ball of necrotic debris with a tangled mycelium of the causative organism.

In both acute and chronic forms of pulmonary disease, direct extension of infection may involve the blood vessels, pleura, and chest wall. Nocardial empyema and bronchopleural fistula may ensue. Away from the damaged lung there may be diffuse alveolar damage with hyaline membranes. Necrotic extensions may grow from the abscess cavity toward the pleura. Parenchyma away from the abscess, apart from diffuse alveolar damage, also shows a chronic organizing pneumonia with many lipid-laden macrophages. As mentioned above, there may be evidence of coexistent pulmonary alveolar proteinosis in adjacent lung tissue.

The diagnosis of nocardiosis is made by direct microscopic examination of tissue sections, sputum, or purulent material. A presumptive diagnosis is made if there are delicately branching gram-positive beaded filaments, 0.5 to 1 μm in diameter. Cultures of *Nocardia* from routine clinical specimens require 5 to 21 days for growth. Since most routine specimens are discarded at 3 days, the laboratory should be instructed to incubate the specimens for a longer period if *Nocardia* is suspected.

Glanders and Melioidosis

ORGANISMS AND EPIDEMIOLOGY

Human infection with glanders or meliodiosis occurs as either a localized or a generalized form. Local glanders follows infection of an abrasion or mucous membrane by *Ps. mallei*. This causes a chronic ulcerative lesion with a copious foul discharge. Local lesions are found in the nose or mouth as well as on the skin. There is lymphatic spread, but pyemia often intervenes, causing generalized glanders, which is rapidly fatal.

Melioidosis is a disease caused by *Ps. pseudomallei*. The bacillus was originally isolated in 1885 from a horse dying of the disease and was named *B. mallei* by Zopth.[227] Subsequently an organism resembling it in many respects was isolated by Whitmore and Krishnaswani[228] from a glanders-like disease in humans. There was subsequent confusion over these two organisms' taxonomic position, and today *Ps. pseudomallei* (Whitmore's bacillus) and *Ps. mallei* (the glanders bacillus) are both included in the genus *Pseudomonas*. The two organisms are serologically related. The former organism has the ability to lie dormant in the body and emerge after some years, at a time of stress.

Ps. pseudomallei is usually found in soil and stagnant water in endemic areas, such as Southeast

Asia, the Philippines, northern Australia, parts of Africa, and Central and South America.[229] The disease was brought back to the United States by soldiers during the Vietnam war. A damp climate and a flood terrain appear to favor the organism's growth. It was described after inhalation of infected dust following a blast injury in Kuala Lumpur, Malaysia.[230]

Melioidosis is common in animals in endemic areas. Those affected include sheep, dogs, pigs, horses, and birds. As in human subjects, the clinical manifestations may be varied.[230]

CLINICAL FEATURES

In humans the disease has protean manifestations and affects many organ systems. Skin and soft tissue involvement is characterized by abscess formation with draining sinuses. There may be superinfection of these lesions with other bacteria. In the acute septicemic form, rigors, high temperatures, and shock are seen.

In acute cases, respiratory symptoms range in severity from mild bronchitis to overwhelming pneumonia. There is cough, dyspnea, pleuritic chest pain, and sometimes hemoptysis. In chronic pulmonary disease there is a history of fever, cough, and weight loss, closely resembling that of TB, especially in immigrants and refugees.[231] It should even be considered in patients who have traveled to endemic areas up to 18 years previously.[232] It has been invoked as a cause of unexplained death in immigrant Thai construction workers.[233]

PATHOLOGY

The disease is often septicemic[234] in acute cases. The other organs most frequently affected are liver, spleen, lymph nodes, and every other organ system except the gut. The lung shows patchy areas of consolidation, which become confluent. There is then a suppurative acute bronchopneumonia, with multiple abscesses up to 3 mm in diameter (Fig. 7-67), as well as pleural nodules and effusions. There may be obliteration of the pleura due to fibrosis. In chronic pulmonary disease the lesions are usually found in the upper lobe, and there are thin-walled cavities.

The cellular response is predominantly polymorphonuclear, with some lymphocytes and macrophages. As the abscesses become confluent, bronchitis becomes more severe. Lymph nodes show either abscess formation or reactive changes. Gram's stain shows many slender, slightly curved gram-negative rods.

In chronic disease granulomas are present with epithelioid cells and giant cells, often surrounded by a zone of fibrosis. The giant cells may be of Langhans' or foreign-body type, with central necrosis that may resemble TB. It is difficult to demonstrate the organism in tissue sections.

Diagnosis may be difficult, and the indirect hemoglutination and complement fixation tests may provide evidence of *Ps. pseudomallei* infection. The former test is more sensitive and specific. Immunofluorescence microscopy has been used to give a rapid diagnosis from sputum, pus, or urine.[235] Antibodies become detectable in the second week of illness at a titer of at least 1 in 40.[236] Isolation of the organism from tissue, body fluids, or secretions is the most rapid and definitive means of diagnosis available. Confirmatory identification by reference laboratory, including the use of fluorescent antisera, should be made because of possible confusion with other species, such as *Ps. (Burkholderia) cepacia*.

Whipple's Disease

The rare Whipple's disease affects primarily the gastrointestinal tract and lymph nodes. There may, however, be chronic cough.[237] Organisms may be identified in pulmonary emboli.[238] Schiff-positive bacilli have been identified in the media of pulmonary arteries from two patients dying from this disease.[239] There may be involvement of the alveolar septae by bacilli-laden macrophages (Fig. 7-68).[240] In some areas, cystic medial degeneration has been noted, but there was no inflammation apart from a few scattered adventitial lymphocytes. In one case, macrophages with Schiff-positive bacilli were also demonstrated in the coronary arteries.

BRONCHIOLITIS

There are many causes of bronchiolitis and bronchiolitis obliterans–organizing pneumonia as will be discussed below. The inclusion of this topic in the bacterial chapter was made with some difficulty. It could equally have justified a section in viral, collagen-vascular, drug-induced diseases or in transplantation-related disease. However it may be present in some bacterial infections and as these are probably numerically the commonest cause, bronchiolitis has been described in this section.

Inflammatory airways diseases that are not distinctly viral include a broad range of pathologically and clinically disparate lesions. These are divided into constrictive bronchiolitis, cryptogenic organizing pneumonia (idiopathic bronchiolitis obliterans–organizing pneumonia), and diffuse panbronchio-

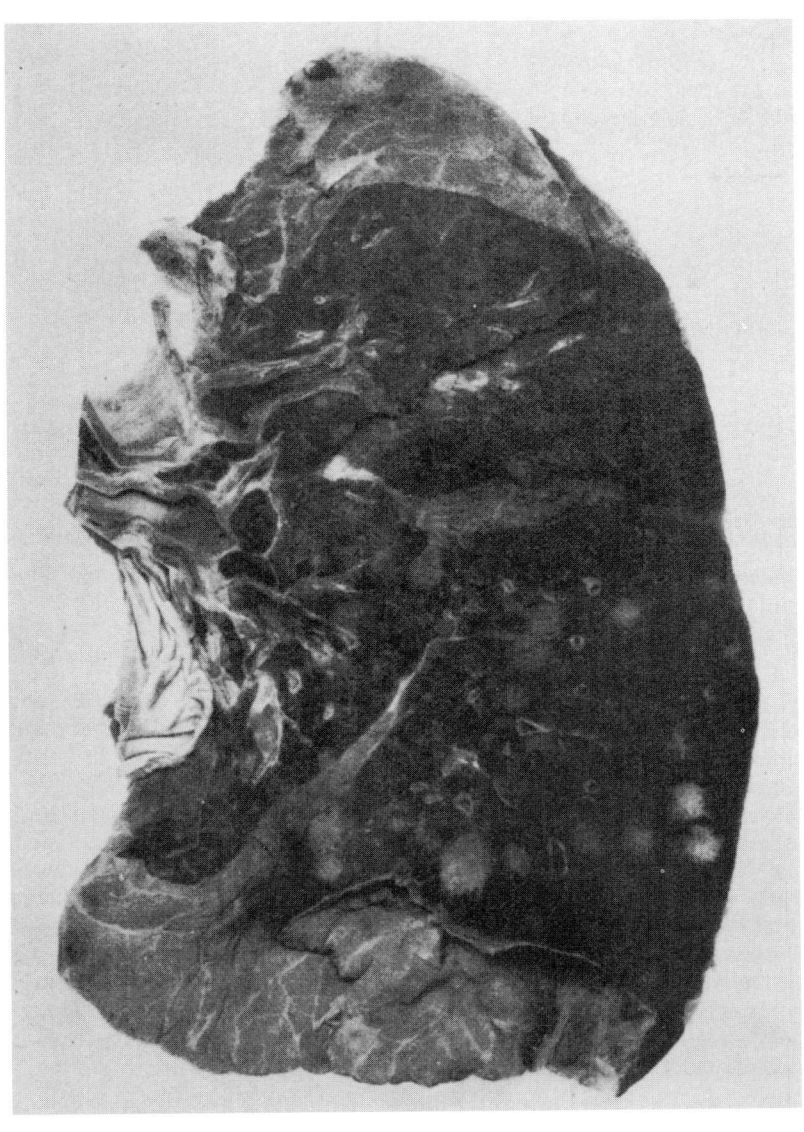

FIGURE 7-67
Meloidiosis of the lung with multiple abscesses. *(Courtesy of the curator of the Museum of the Royal College of Surgeons.)*

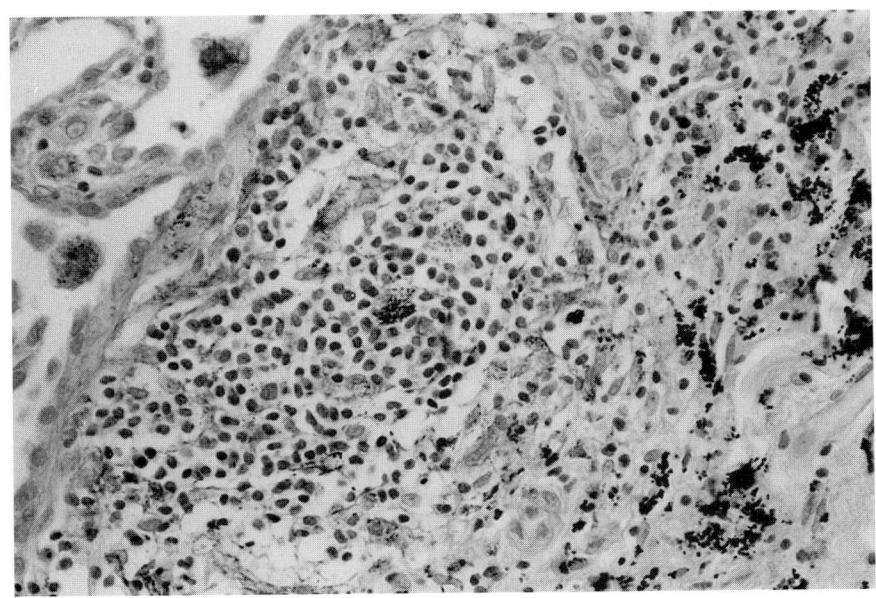

FIGURE 7-68
Whipple's disease. PAS-positive bacilli in intraalveolar macrophages. (PAS, ×219.) *(Courtesy of Prof. T. V. Colby, Mayo Clinic, Scottsdale, Ariz.)*

TABLE 7-4
Histologic Types of Bronchiolitis

Histologic lesion/syndrome	Synonyms
Acute bronchiolitis	Viral/infectious
Small-airway disease	Adult bronchiolitis, Constrictive bronchiolitis
Respiratory bronchiolitis	Smoking or associated with fibrosing alveolitis
Mixed dust fibrosis	Early pneumoconiosis
Follicular bronchiolitis	Lymphoid hyperplasia
Constrictive bronchiolitis	Obliterative bronchiolitis
BOOP	Cryptogenic organizing pneumonia, bronchiolitis obliterans–organizing pneumonia–like process, idiopathic intraluminal pneumonia, proliferative organizing pneumonia
Diffuse panbronchiolitis	

SOURCE: After Myers JL, Colby TV: Pathologic manifestations of bronchiolitis, cryptogenic organizing pneumonia and diffuse panbronchiolitis. *Clin Chest Med* 14:611–622, 1993.

litis. Further causes of bronchiolitis, not described in this section, are given in Table 7-4.

These diseases are rare, and their precise incidence is unknown. In a study of the incidence in 1941, Ladue found only one case of bronchiolitis obliterans in more than 42,000 autopsies.[241] Hardy and coworkers[242] found only seven cases in almost 3000 pediatric autopsies. However, these authors suggested that the incidence was higher.

The first description of bronchiolitis obliterans was that of Lange,[243] although Gosink and colleagues refer to a report by Renault in 1835 as the first description of the disease.[244] Two papers in the early 1980s by Davison and coworkers[245] and Grinblatt and colleagues[246] drew attention to patients with intraalveolar fibrosis of unknown cause. In the study by Epler's team,[247] the pattern was accepted as a distinct clinicopathologic entity. It can be called either cryptogenic organizing pneumonia or cryptogenic bronchiolitis obliterans–organizing pneumonia.

The classification of bronchiolitis has been given in Table 7-4. This section will deal with constrictive bronchiolitis, cryptogenic organizing pneumonia, and diffuse panbronchiolitis.

Lymphocytic bronchiolitis is not described here. It is not a common process. It is seen with interstitial lung disease and also is described with hypersensitivity pneumonitis and immunodeficiency. It in unlikely to be a pure process and is part of either cryptogenic fibrosing alveolitis or some immune deficiency.

Constrictive Bronchiolitis

Myers and Colby[248] used the term *constrictive bronchiolitis* in a specific sense to refer to a syndrome of airflow limitation due to bronchiolar and peribronchiolar inflammation and fibrosis (Fig. 7-69). The terms *obliterative bronchiolitis* and *pure bronchiolitis obliterans* have also been used for this entity.[249]

The term *constrictive bronchiolitis* was coined by Gosink and coworkers.[244] The causes of constric-

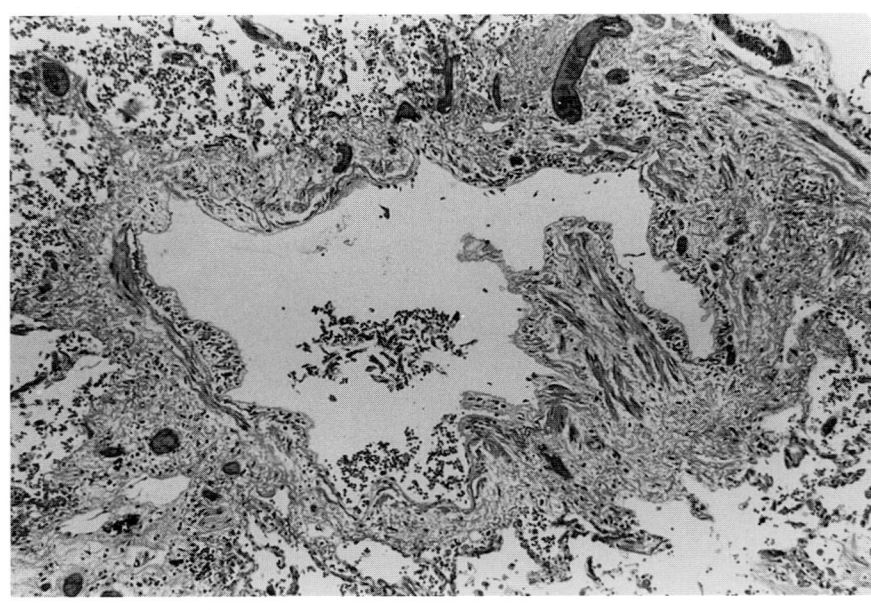

FIGURE 7-69
Constrictive bronchiolitis with an increase in fibrosis in the bronchiolar wall. There is little inflammation and only mild luminal narrowing. (H&E, ×88.)

TABLE 7-5
Clinical Syndromes Associated with Constrictive Bronchiolitis

Allograft recipients
    Bone marrow transplant recipients (Chap. 24)
    Heart-lung or lung transplant recipients (Chap. 24)
Collagen vascular diseases (Chap. 21)
    Rheumatoid arthritis
    Eosinophilic fasciitis
    Systemic lupus erythematosus
Postinfectious[250,251]
    Viral (respiratory syncytial virus, adenovirus, influenza, parainfluenza)
    Mycoplasma
Inhaled toxins[250,252]
    Nitrogen dioxide
    Sulfur dioxide
    Ammonia
    Chlorine
    Phosgene
Drugs (Chap. 18)
    Penicillamine
    Lomustine
Ulcerative colitis[253]
Idiopathic

tive bronchiolitis are given in Table 7-5. The disease is rare and is seen mainly in patients with collagen diseases, especially rheumatoid disease (Chap. 26).

CLINICAL FEATURES

Patients typically present with breathlessness, and radiographs of the chest either are normal or show evidence of hyperinflation. Pulmonary function studies show airflow limitation, often associated with air trapping. The course of the disease is relentless, ending in respiratory failure.

PATHOLOGY

The fibrosis is mainly submucosal and peribronchiolar in distribution.[248] It results in extrinsic narrowing and obliteration of the bronchiolar lumen. There is a spectrum of change, ranging from bronchiolar inflammation to peribronchiolar fibrosis.[248,250,255] There is ultimately complete scarring of the bronchiolar lumen. Even at postmortem the changes may be patchy, and the diagnosis can be missed if the lung is inadequately sampled.

In the initial stages there is intraluminal mucosal, submucosal, and peribronchiolar inflam-

mation, seen mainly in membranous and respiratory bronchioles, and there is associated epithelial necrosis. The mixed inflammatory infiltrate includes variable numbers of neutrophil polymorphs, lymphocytes, and plasma cells. Neutrophils tend to predominate within the lumen, whereas lymphoplasmacytic cells are seen within the peribronchiolar interstitium.

The inflammatory process is followed by peribronchiolar fibrosis that encroaches into part of the bronchial lumen in an ectatic fashion. As the fibrosis progresses, the lumen is reduced to a slitlike space and is eventually obliterated. An elastic–van Gieson stain is often helpful in identifying bronchioles.

CRYPTOGENIC BRONCHIOLITIS OBLITERANS–ORGANIZING PNEUMONIA

Cryptogenic bronchiolitis obliterans–organizing pneumonia (BOOP) affects the sexes equally and has a peak commonly in middle age, especially in the 60s.

Clinical Features

Most patients have cough and shortness of breath of several weeks' duration, frequently accompanied by pyrexia. The symptoms are often related to an acute upper respiratory infection, which is presumed to be viral. Hemoptysis and wheeze are rare, and in approximately 75 percent of the patients the duration is less than 2 months. On examination in the cases reported by Epler and colleagues.[247] there was no finger clubbing, and there were fine end-inspiratory crackles in two-thirds of patients. The erythrocyte sedimentation rate is often raised in such patients.

Chest radiographs show bilateral patchy infiltrates and bilateral linear or nodular opacities in 20 percent of patients. Occasionally, pleural effusions are identified.

Pulmonary function studies show an abnormally low vital capacity in one-half of the patients. Airflow obstruction is seen only in smokers. All patients have a normal diffusion capacity.

Several subgroups appear to be emerging; one such group has recently been described by Cohen and colleagues.[256] Their study reported a series of 10 patients with rapidly progressive BOOP characterized by severe respiratory failure. The main clinical manifestations were as noted above and included dyspnea, cough, fever, crackles on chest examination, and hypoxemia at rest. Underlying conditions or exposures were similar to those listed in Table 7-6 and included connective tissue disease, exposure to

TABLE 7-6
Clinical Syndromes Associated with Organizing Pneumonia

Collagen vascular diseases (Chap. 21)
    Rheumatoid arthritis
    Systemic lupus erythematosus
    Dermatomyositis
    Mixed connective tissue disease
Toxins
    Inhalants (nitrogen dioxide: silo filler's disease)[257]
    Systemic (drugs: bleomycin, amiodorone)[258]
Chronic ("organizing") infection[252]
    Bacteria (*Legionella, Nocardia*)[259,260]
    Viruses (influenza, cytomegalovirus)
    Mycoplasma
    Pneumocystis carinii
    HIV/AIDS[260a]
Idiopathic (cryptogenic organizing pneumonia)

birds, and chronic nitrofurantoin therapy. There is overlap between some of these causes and those documented in Table 7-5 as etiological factors in constrictive bronchiolitis.

At autopsy the predominant pattern is that of alveolar septal inflammation and fibrotic honeycombing. BOOP may progress to end-stage pulmonary fibrosis with honeycombing.

Another unusual variant has been named *seasonal cryptogenic organizing pneumonia*. The group of patients studied by Spiteri and coworkers[261] had dyspnea, cough, pleuritic chest pain, and widespread shadows on the chest radiographs. Liver function tests showed a raised alkaline phosphatase, varying between 340 and 1029 IU/L. In all 12 patients the symptoms recurred in late February and early March each year and tended to increase in severity in successive years. The disease resolved between June and January.

There were many neutrophils within medium-sized airways, and some terminal bronchioles showed obstruction by granulation tissue; many bronchioles were patent. Buds of intraalveolar granulation tissue contained fibroblasts, lymphocytes, plasma cells, and a few neutrophils but no multinucleated giant cells. Areas of adjacent lung had a mononuclear infiltrate, but the lung interstitium was not affected.

Liver biopsy in two patients showed small numbers of nonnecrotizing granulomas within hepatic parenchyma, unrelated to bile ducts. There was no microscopic evidence of cholestasis or damage to the bile ducts. Lavage showed an increased percentage of neutrophils (10.4 percent) and lymphocytes (1 percent).

It was suggested that there was some inhaled source for the disease, but the exact agent was unknown.

BAL in cases of BOOP shows that neutrophils account for more than 50 percent of the cells recovered.[262] In some cases the neutrophil lavage count exceeded 90 percent. This BAL neutrophilia is thought to reflect an inflammatory process at bronchiolar level. BAL also shows an increase in lymphocytes, especially in patients who respond to steroid therapy.[263] There may be an increase in eosinophils and a significant decrease in the ratio of CD4+ to CD8+ lymphocytes compared with the ratio in healthy subjects.[264] Lavage cell findings return to normal when chest radiographs show an improvement. However, there may be a significant increase in lymphocytes and neutrophil polymorphs in lavage even when the chest radiographs are normal.

The lymphocyte count and the good response to steroids suggest a type of hypersensitivity pneumonitis. These lavage findings are nonspecific but were not found in patients with usual interstitial pneumonia.

Pulmonary physiology shows a predominantly restrictive defect associated with a diminishing diffusion capacity.

The most common radiologic finding is bilateral patchy opacities that may have a ground-glass appearance.

PATHOLOGY

The histology of BOOP consists of plugs of granulation tissue within the lumina of small airways (Fig. 7-70). These extend distally into alveolar ducts and alveoli (Figs. 7-71 and 7-72). There may be proliferation of connective tissue from the intraluminal polyps. This has been termed *proliferative bronchiolitis obliterans*. There are also fibrinous exudates; alveolar accumulations of foamy macrophages due to obstruction; inflammation of the alveolar walls with lymphocytes, plasma cells, and a few neutrophils and eosinophils; varying degrees of interstitial thickening and epithelial hyperplasia; and evenly spaced round cores of myxomatous connective tissue. Occasional multinucleated giant cells can be seen within loose granulomatous tissue.

Focal acute intraalveolar inflammation is also present. There may be type II pneumocyte hyperplasia and squamous metaplasia along with chronic interstitial pneumonia. Most often the interstitial

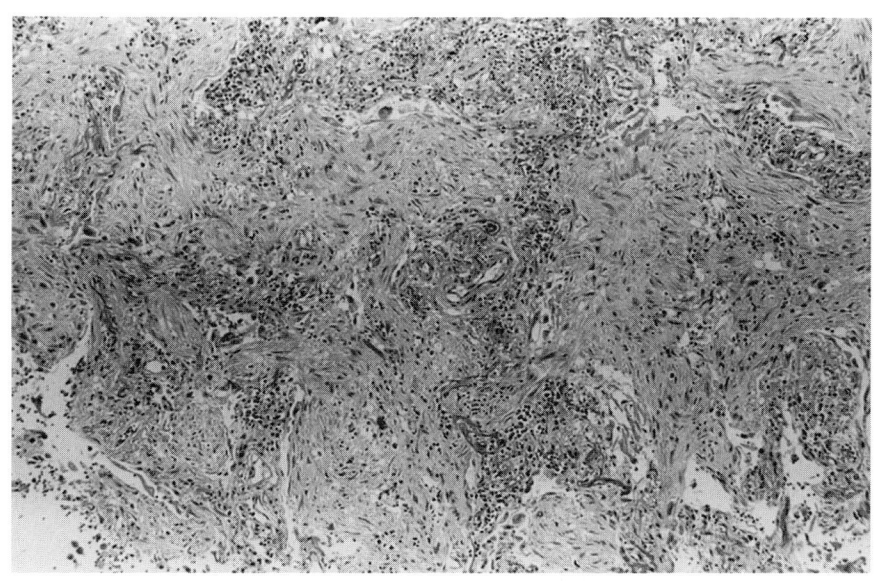

FIGURE 7-70
BOOP with fibroblastic tissue in alveolar ducts and terminal bronchioles. (H&E, ×55.)

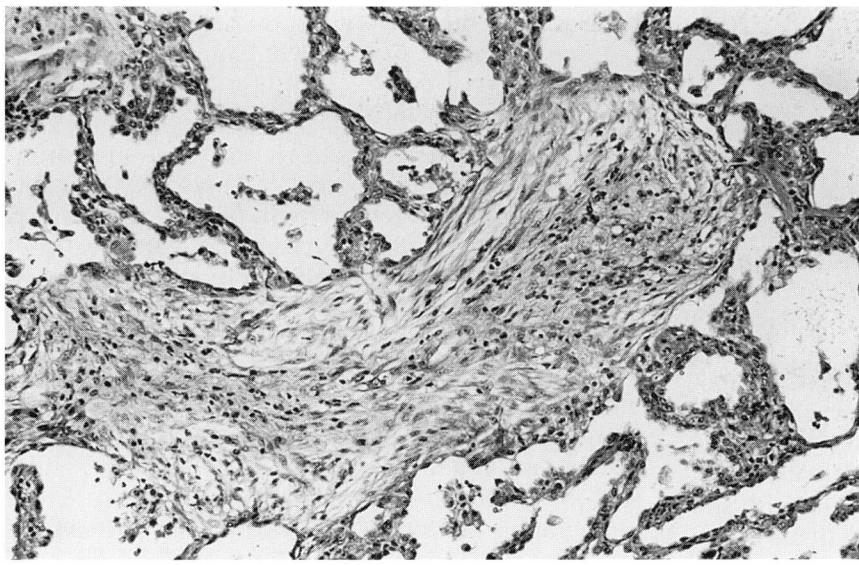

FIGURE 7-71
BOOP. Loose myxoid tissue filling an alveolar duct. There is some chronic inflammation. (H&E, ×219.)

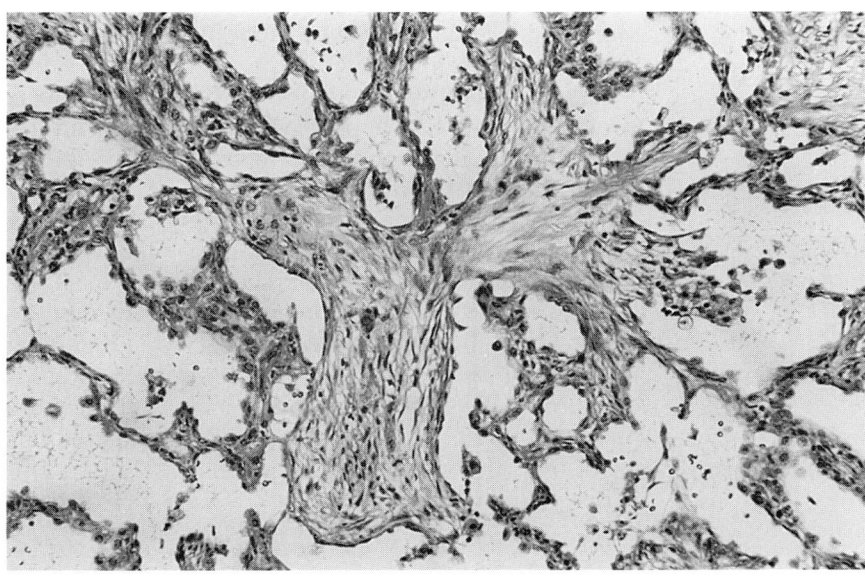

FIGURE 7-72
BOOP. Plugs of fibroblastic tissue in alveoli and alveolar ducts. (H&E, ×219.)

pneumonia is focal and confined to areas of bronchiolitis obliterans.

Ultrastructural studies[265] have shown that BOOP resembles other forms of acute lung injury. There is necrosis of the epithelium of the bronchioles, alveolar ducts, and alveoli. This leaves a bare basement membrane. The fibroblasts and myofibroblasts migrate from the interstitial compartment into the airways and peribronchiolar spaces, through the damaged basement membranes. These form the intraluminal fibroblastic plugs. Proliferating bronchiolar cells and type II pneumocytes eventually incorporate the fibroblastic plugs into the interstitial compartment.

Transbronchial biopsy was sufficient to give a diagnosis in seven of 11 cases.[266] One case that on transbronchial biopsy had the features of BOOP lacked them in open lung biopsy. The authors maintained that the open lung biopsy, while the gold standard for interstitial lung disease, may produce negative results because of sampling error in BOOP.[266]

Constrictive bronchiolitis and BOOP have been discussed as two separate diseases. However, their relationship is not well defined. Myers and Colby[248] stated that in most cases the diseases appear to be distinct entities with different clinical, radiographic, and morphologic features.

Constrictive bronchiolitis is relatively unresponsive to steroids and often associated with progressive airflow limitation. BOOP is steroid-responsive and readily reversible. However, in some patients—such as heart-lung transplant recipients, patients with silo filler's disease, and occasionally patients with rheumatoid disease—there is evidence to suggest that the intraluminal fibrosis, which is localized to membranous bronchioles, can cause obstructive airways disease. This may progress to lesions indistinguishable from those in the late stage of constrictive bronchiolitis.[257,267]

Pathogenesis

The sequence of events leading to BOOP is unclear. It is postulated that the inflammatory process is confined to bronchioles. Once the airway damage has occurred, there is release of the inflammatory mediators and subsequent repair, causing the characteristic obliterative bronchiolar lesion.[268]

The effect of neutrophil polymorphs causing damage to lung tissue has been described in the chapter on adult respiratory distress syndrome (Chap. 11).

The fact that neutrophil myloperoxidase and collagenase have been recovered from the airways of patients with BOOP in significantly greater quantities than in normal volunteers[269] gives additional support to the role of the neutrophil polymorph in the pathogenesis of this disease. Clearly, from Tables 7-5 and 7-6 a vast range of processes and diseases that cause neotrophila can be seen.

Transplant bronchiolitis is a specific entity described in Chap. 24.

The differential diagnosis of BOOP includes a wide spectrum of infiltrative lung diseases. Some of these have already been listed in the tables, but it is worthwhile discussing a few in more detail.[270]

INFECTIOUS PULMONARY DISEASES

Viruses such as adenovirus, *M. pneumoniae*, and *Legionella* may all give a BOOP pattern.

LOCALIZED ORGANIZING PNEUMONIA

This is a focal lesion that shows organization in bronchioles and distal air spaces.[271] There may be a loss of background architecture with fibrosis and marked fibrotic and inflammatory lesions in the walls of bronchioles, causing luminal stenosis.

USUAL INTERSTITIAL PNEUMONIA

The differential diagnosis for usual interstitial pneumonia (UIP) was considered in detail by Katzenstein and Askin.[249] Bronchiolitis obliterans has a more acute onset and is often associated with fever, while UIP is insidious, with dyspnea and cough. Chest radiographs uniformly show bilateral interstitial opacities in patients with UIP; these are more variable in patients with bronchiolitis obliterans. Mortality is much lower in bronchiolitis obliterans (12.5 percent) than in UIP (62 percent).

Pathologically, bronchiolitis obliterans affects air spaces and has a peribronchiolar distribution, while the changes in UIP are mainly interstitial, diffuse, and randomly distributed. The fibrosis in bronchiolitis obliterans consists of proliferating fibroblasts, as compared with mature collagen deposition in UIP. Honeycombing does not occur in BOOP but is seen UIP and other forms of interstitial pneumonia. Katzenstein and Askin[249] have shown that the distinction between UIP and bronchiolitis obliterans may be difficult. Foamy macrophages are unusual in UIP but common in bronchiolitis obliterans. The differences in prognosis make it important to distinguish these two disease entities.

DIFFUSE ALVEOLAR DAMAGE

Resolving diffuse alveolar damage, without any hyaline membranes, may be difficult to differentiate from BOOP.

CHRONIC EOSINOPHILIC PNEUMONIA

This will show bronchiolitis obliterans, but BOOP does not show eosinophilic microabscesses with intraalveolar necrosis or sarcoid-like granulomas. In idiopathic BOOP eosinophils are sparse in tissue sections.

EXTRINSIC ALLERGIC ALVEOLITIS

This typically shows interstitial granulomas, which are not seen in BOOP. Bronchiolitis obliterans is seen as part of extrinsic allergic alveolitis in some cases.

COLLAGEN DISEASES

There is an overlap in these diseases with the picture described for BOOP as shown by Davison and coworkers,[245] whose study included two patients, one with a digital vasculitis and the other with scleroderma.

PULMONARY LESIONS DUE TO DRUGS

Some drugs may cause a BOOP-like pattern (Chap. 18). These include gold, cephalosporine, sulfasalazine, sulfametopyrazine, amiodorone, and sulindac as well acebutolol.[272]

DIFFUSE PANBRONCHIOLITIS

Diffuse panbronchiolitis (DPB) has been recognized in Japan since the late 1960s. It has only recently impinged on Western consciousness and has been the subject of three recent papers.[273-275]

Clinical Features

The disease was first described in 1969 by Yamanaka and colleagues[276] and later reviewed by Homma and coworkers.[277] The latter authors gave the following clinical diagnostic criteria:

1. Symptoms of chronic cough, sputum, and dyspnea on exertion.
2. Physical signs of rales and rhonchi.
3. A chest radiograph showing diffusely disseminated fine nodular shadows, mainly in the lower lung fields with hyperinflation.

4. Lung function studies showing at least three of the four following abnormalities: FEV_1/FVC less than 70 percent, VC less than 80 percent of the predicted value, residual volume greater than 150 percent of the predicted value or a P_{O2} less than 80 mmHg.

Additional features are commonly present, such as current or previous chronic paranasal sinusitis. This may be the initial presentation in the second or third decade of life. Patients often require sinus surgery. The respiratory symptoms are seen several years after the onset of sinus disease, which usually occurs in the second to fifth decade (average age: 39.5 years). Two-thirds of the patients are nonsmokers. There is no sex predominance.[273]

Leukocytosis is common, rheumatoid factor is positive, and serum IgG and IgA are increased. The most characteristic feature is a persistent elevation of cold agglutinins. The titer may be increased up to 16-fold. These findings are similar to those found in infections caused by *Mycoplasma*.

The CD4+:CD8+ ratio in the peripheral blood is increased. In addition, there is a significant increase in frequency of HLA Bw54 in patients with DPB. Familial cases of DPB have been reported in Japan.[278, 279] The locus HLA Bw54 or its related haplotype is limited to Orientals, such as Japanese, Chinese, and Koreans, and to some Ashkenazi Jews.[280]

Radiology

The typical radiologic findings are diffusely disseminated small nodular shadows up to 2 mm in diameter with ill-defined margins. They are most prominent in the lung bases. In some cases there is associated lung hyperinflation. Mild bronchiectasis usually develops in the lower lobe and the lingula. With progression, some cases have cystic change or diffuse bronchiectasis.

CT scans show diffuse small rounded and linear opacities, dilatation of small bronchi and bronchioles, and thickening of the bronchial walls.

Pathology

The most distinctive pathologic feature of DPB is chronic inflammation, with an accumulation of foam cells in the walls of respiratory bronchioles, adjacent alveolar ducts, and alveoli (Figs. 7-73 and 7-74). This has been referred to as the PB unit lesion.[274] Terminal and respiratory bronchioles are narrowed by chronic

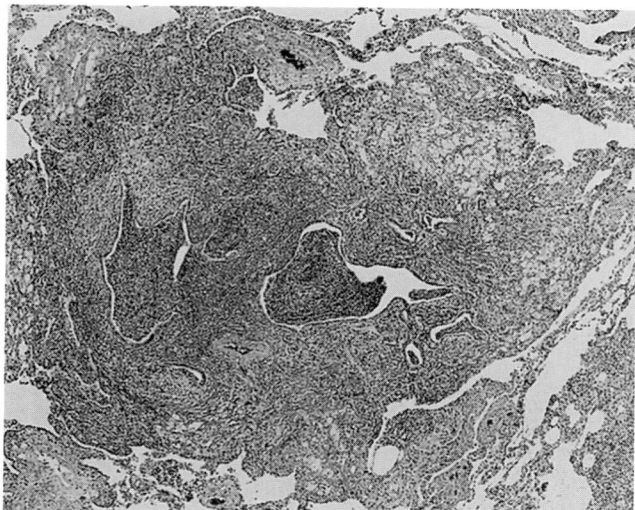

FIGURE 7-73
Diffuse panbronchiolitis with chronic inflammation in the walls of respiratory brochioles and an accumulation of foam cells. (H&E, ×39.) *(Courtesy of Prof. T. V. Colby, Mayo Clinic, Scottsdale, Ariz.)*

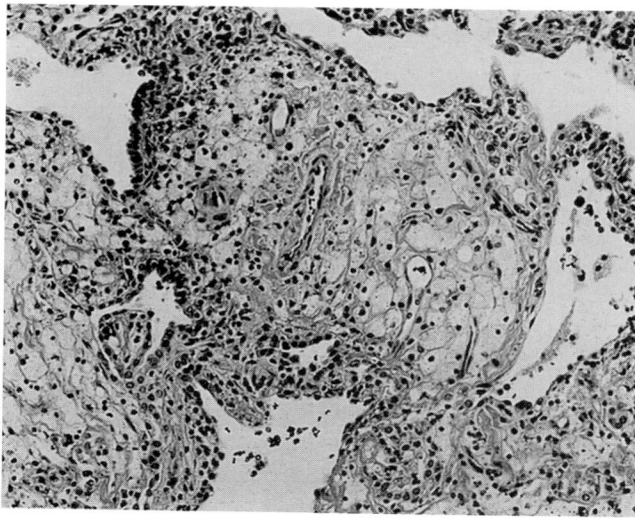

FIGURE 7-74
Diffuse panbronchiolitis with a focal accumulation of foam cells. (H&E, ×154.) *(Courtesy of Prof. T. V. Colby, Mayo Clinic, Scottsdale, Ariz.)*

inflammation. The proximal membranous bronchiole is slightly dilated. There is an accumulation of foam cells, accompanied by an infiltration of lymphocytes and plasma cells in the walls of respiratory bronchioles and alveolar ducts. These inflammatory changes extend into the peribronchiolar tissues. In advanced stages there is narrowing and constriction of the respiratory bronchioles by infiltration of these cells, proliferation of lymphoid follicles, accumulation of foamy cells within the walls of neighboring areas and ectasia of the proximal terminal bronchioles.[277]

Etiology

The cause of DPB is unknown. As mentioned above, it is associated with the HLA Bw54 locus or its related haplotype. An increase in HLA Bw54 is also noted in patients with juvenile-onset diabetes mellitus[281] and rheumatoid disease.[280]

H. influenzae infections occur repeatedly or continually over the course of the disease. These cause damage to the respiratory bronchioles, which become obstructed as nearby bronchioles become dilated. Superinfection with *Ps. aeruginosa* occurs, and the disease becomes intractable.[282]

The pathogenesis of DPB has been related to immunologic events.[283] In some cases there is a relationship to human T-cell lymphotropic virus type I. This has been noted because DPB has been known to precede the development of adult T-cell leukemia.[284] However, most patients with DPB do not have anti–human T-cell lymphotropic virus type I antibody.[285]

The survival rate has been quoted as 62.1 percent at 5 years and 33.2 percent at 10 years.[252] With long-term erythromycin therapy, the 10-year survival rate is now more than 90 percent.[282]

Differential Diagnosis

The diffuse panbronchiolitis picture can be seen in other chronic airways diseases. These can be divided into two groups.[274] The first includes constrictive bronchiolitis, follicular bronchiolitis, cystic fibrosis, and bronchiolectasis. In this group the chronic inflammation stems from proximal bronchioles to involve respiratory bronchioles secondarily. This is the main difference between the unique lesion in classic DPB and the PB-like lesions of the group mentioned above. This is unlikely to cause problems at postmortem, but it may cause difficulty in biopsy specimens.

The second group includes such diseases as extrinsic allergic alveolitis, aspiration pneumonia, Wegener's granulomatosis, bronchocentric granulomatosis, and lymphoma. In this group the PB-like lesion is focal and minor, and the other features common to these diseases predominate.

REFERENCES

1. Dar L, Thakur R, Dar VS: India: Is it plague? *Lancet* 344:1359, 1994.
2. Naber SP: Molecular pathology—diagnosis of infectious disease. *N Engl J Med* 331:1212–1215, 1994.
3. Murphy TF: Modern molecular biology and respiratory bacterial infections: A revolution on the horizon. *Thorax* 45:552–559, 1990.

3a. Pääkko P, Särkioja T, Hirvonen J, Nurmi T, Lahti R, Suttinen S: Postmortem radiographic, histological and bacteriological studies in terminal respiratory infections and other pulmonary lesions in hospital and non-hospital necropsies. *J Clin Pathol* 37:1282–1288, 1984.

4. Curwen M: *Trends in Respiratory Mortality, 1951–75, England and Wales.* London, HMSO, 1981.

5. Easmon CSF, Goodfellow M: *Staphylococcus* and *Micrococcus*, in Parker MT, Collier LH (eds), *Topley and Wilson's Principles of Bacteriology, Virology and Immunity*, vol 2, *Systematic Bacteriology*, 8th ed. London, Edward Arnold, 1990, p 162.

6. Zierdt CH: Long-term *Staphylococcus aureus* carrier state in hospital patients. *J Clin Microbiol* 16:517–20, 1982.

7. Public Health Laboratory Service. Necropsy survey of staphylococcal infection in patients dying in hospitals. *Br Med J* 1:313–319, 1966.

8. Tsao T C-Y, Tsai Y-H, Shieh W-B, Lee C-H: Pulmonary manifestations of *Staphylococcus aureus* septicemia. *Chest* 101:574–576, 1992.

9. McKinsey DS, Bisno AL: Pneumonias caused by gram-positive bacteria, in Fishman AP (ed), *Pulmonary Disease and Disorders*, 2nd ed. New York, McGraw-Hill, 1988, p 1484.

10. Hers JF, Masurel N, Mulder J: Bacteriology and histopathology of the respiratory tract and lungs in fatal Asian influenza. *Lancet* 2:1141–1143, 1958.

11. Morgan EH, Lawrence LJ, Lancaster L, Pearsall H: Staphylococcal pneumonitis in the post-operative patient. *Dis Chest* 38:1–12, 1958.

12. Donnelly BW, McMillan JA, Weiner LB: Bacterial tracheitis: Report of 8 new cases and review. *Rev Infect Dis* 12:729–735, 1990.

13. Sabiston DC: The management of staphylococcic tension pneumatoceles by intracavitary suction tube drainage (discussion). *J Thorac Surg* 36: 642–655, 1958.

14. MacGregor AR: Pneumonia in the newborn. II. *Arch Dis Child* 14:336–351, 1939.

15. Cazzola M, Matera MG, Rossi F: Bronchial hyper-responsiveness and bacterial respiratory infections. *Clin Ther* 13:157–171, 1991.

16. Watkins E, Hering AC: The management of staphyloccocic tension pneumatoceles by intracavitary suction tube drainage. *J Thorac Surg* 36:642–653, 1958.

17. Todd J, Fisaut M, Capral S, Welch T: Toxic-shock syndrome associated with phage–group-1 staphylococci. *Lancet* 2:1116–1118, 1978.

17a. Colman G: Streptococcus and Lactobacillus, in Parker MT, Collier LH (eds): *Topley and Wilson's Principles of Bacteriology, Virology, and Immunity, Systematic Bacteriology*, vol 2, 8th ed. London, Edward Arnold, 1990, pp 120–148.

18. British Thoracic Society Research Committee: Community-acquired pneumonia in adults in British hospitals in 1982–1983: A Survey of aetiology, mortality, prognostic factors and outcome. *Q J Med* 62:195–220, 1987.

19. Austrian R: Some observations on the pneumococcus and on the current status of pneumococcal disease and its prevention. *Rev Infect Dis* 3:S1–S17, 1981.

20. Austrian R, Gold J: Pneumococcal bacteremia with special reference to bacteremic pneumococcal pneumonia. *Ann Intern Med* 60:759–760, 1964.

20a. Heffron R: *Pneumonia: With Special Reference to Pneumococcal Lobar Pneumonia*. Cambridge, Harvard University Press, 1939 (2d printing, 1979).

20b. Marrie TJ: New aspects of old pathogens of pneumonia. *Med Clin North Am* 78:987–995, 1994.

21. DeMaria A Jr, Browne K, Berk SL, Sherwood EJ, McCabe WR: An outbreak of type 1 pneumococcal pneumonia in a men's shelter. *JAMA* 244:1446–1449, 1980.

22. Lipsky BA: Pneumococcal pneumonia: Predispositions and prevention. *Chest* 99:2–3, 1991.

23. Polsky B, Gold JWM, Whimbey E: Bacterial pneumonia in patients with the acquired immunodeficiency syndrome. *Ann Intern Med* 104:38–41, 1986.

24. Musher DM: Infections caused by *Streptococcus pneumoniae:* Clinical spectrum, pathogenesis, immunity and treatment. *Clin Infect Dis* 14:801–809, 1992.

25. McDaniel LS, Waltman WD, Gray B, Briles DE: A protective monoclonal antibody that reacts with a novel antigen of pneumococcal techoic acid. *Microb Pathog* 3:249–260, 1987.

26. Perry FE, Catterall JR: The pneumococcus: Host-organism interactions and their implications for immunotherapy and immunoprophylaxis. *Thorax* 49:946–950, 1994.

26a. Mandigers CMPW, Diepersloot RJA, Dessens M, Mol SJM, van Klingerman B: A hospital outbreak of penicillin-resistant pneumococci in the Netherlands. *Eur Respir J* 7:1635–1639, 1994.

27. Johnston RB Jr: Pathogenesis of pneumococcal pneumonia. *Rev Infect Dis* 13(suppl 6):S509–S517, 1991.

28. Geelen SPM, Gerards LJ, Fleer A: Pneumococcal septicemia in the newborn: A report on seven cases and a review of the literature. *J Perinat Med* 18:125–129, 1990.

29. Fekety FR Jr, Caldwell J, Gump D, et al: Bacteria, viruses and mycoplasmas in acute pneumonia in adults. *Am Rev Respir Dis* 104:499–507, 1971.

30. Ort S, Ryan JL, Barden G, D'Esopo N: Pneumococcal pneumonia in hospitalized patients: Clinical and radiographical presentations. *JAMA* 249:214–218, 1983.

31. Engel S: *The Child's Lung*. London, Edward Arnold, 1947.

32. Cudkowicz B: Some observations of the bronchial arteries in lobar pneumonia and pulmonary infarction. *J Tuberc* 46:99–102, 1952.

33. Symmers D, Hoffman AM: The increased incidence of organizing pneumonia: Preliminary communication. *JAMA* 81:297–298, 1923.

34. Gleichman TK et al: Major etiological factors producing delayed resolution in pneumonia. *Am Med Sci* 218:369–373, 1949.

35. Auerbach SH, Mims OH, Goodpasture EW: Pulmonary fibrosis secondary to pneumonia. *Am J Pathol* 28:69–87, 1952.

36. Hronchich ME, Rudinger AN: Rhabdomyolysis with pneumococcal pneumonia: A report of two cases. *Am J Med* 86:467–468, 1989.

36a. Shaheen SO, Barker DJP, Shiell AW, Crocker FJ, Wield GA, Holgate ST: The relationship between pneumonia in early childhood and impaired lung function in late adult life. *Am J Respir Crit Care Med* 149:616–619, 1994.

37. Venkatesan P, Macfarlane JT: Role of pneumococcal antigen in the diagnosis of pneumococcal pneumonia. *Thorax* 47:329–331, 1992.

38. Austrian R: Current status of bacterial pneumonia with special reference to pneumococcal infection. *J Clin Pathol* 21(suppl 2 R Coll Pathol):93–97, 1968.

39. Fruchtman SM, Gombert ME, Lyons HA: Adult respiratory distress syndrome as a cause of death in pneumococcal pneumonia. *Chest* 83:598–601, 1983.

40. Braman SS, Donat E: Explosive pleuritis: Manifestation of group A beta-hemolytic streptococcal infection. *Am J Med* 81:723–726, 1988.

41. Turnbull P, Kramer J, Melling J: Bacillus, in Parker MT, Collier LH (eds), *Topley and Wilson's Principles of Bacteriology, Virology, Immunity. Systematic Bacteriology*, vol 2, 8th ed. London, Edward Arnold, 1990, p 188.

42. Pfisterer RM: Eine Milzbrandepidemie in der Schweiz. Klinische, diagnostische und epidemiologische Aspekte einer weitgehend vergessenen Krankheit. *Schweiz Med Wochenschr* 121:813–825, 1991.

43. Albrink WS, Brooks SM, Biron RE, Kopel M: Human inhalation anthrax: A report of three fatal cases. *Am J Pathol* 36:457–471, 1960.

44. Aldhous P: Gruinard Island handed back. *Nature* 344:801, 1990.

45. Titball RW, Turnbull PCB, Hutson RA: The monitoring and detection of *Bacillus anthracis* in the environment. *J Appl Bacteriol* (suppl)70:9S–18S, 1991.

46. Pringle P: Anthrax answer was blowing in the wind. *The Independent* (London, England), Nov 18, 1994, p 13.

47. Ross JM: The pathogenesis of anthrax following the administration of spores by the respiratory route. *J Pathol Bacteriol* 73:485–494, 1957.

48. Slack MPE: *Haemophilus*, in Parker MT, Collier LH (eds), *Topley and Wilson's Principles of Bacteriology, Virology and Immunity*, vol 2, 8th ed. *Systematic Bacteriology*, London, Edward Arnold, 1990, p 356.

49. Moxen E R: The carrier state: *Haemophilus influenzae. J Antimicrob Chemother* 18(suppl A):17–24, 1986.

50. Michaels RH, Poziviak CS, Stonebraker EE, Norden CW: Factors affecting pharyngeal *Haemophilus influenzae* Type B: Colonization rates in children. *J Clin Microbiol* 4:413–417, 1976.

51. Moxom ER, Wilson R: The role of *Haemophilus influenzae* in the pathogenesis of pneumonia. *Rev Infect Dis* 13(suppl 6):S518–S527, 1991.

52. Turk DC: Nasopharyngeal carriage of *Haemophilus influenzae* type B. *J Hyg (Camb)* 61:247–256, 1963.

53. Casadevall A, Dobroszyki J, Small C: *Haemophilus influenzae* type B bacteremia in adults with AIDS and at risk for AIDS. *Am J Med* 92:587–590, 1992.

54. Howard AJ: Nosocomial spread of *Haemophilus influenzae. J Hosp Infect* 19:1–3, 1991.

55. Mazor M, Chaim W, Maymon E: Intra-amniotic infection with *Haemophilus influenzae*: Report of a case and review of the literature. *Arch Gynecol Obstet* 249:47–50, 1991.

56. Holmes B, Gross RJ: Coliform bacteria: various other members of the Enterobacteriachae, in Parker MT, Collier LH (eds), *Topley and Wilson's Principles of Bacteriology, Virology and Immunity, vol 2, Systematic Bacteriology*, 8th ed. London, Edward Arnold, 1990, p 416.

57. Ritvo M, Martin F: The clinical and roentgen manifestations of pneumonia due to bacillus mucosus capsulatus (primary Friedlander pneumonia). *Am J Roentgenol* 62:211–222, 1949.

58. Korvick JA, Hackett AK, Yu VL, Muder RR: *Klebsiella* pneumonia in the modern era: Clinicoradiographic correlations. *South Med J* 84:200–204, 1991.

59. Carpenter JL: Klebsiella pulmonary infections: Occurrence at one medical center and review. *Rev Infect Dis* 12:672–682, 1990.

60. Soloman S: Chronic Friedlander infections of the lung. *JAMA* 115:1527–1536, 1940.

61. Penner C, Maycher B, Long R: Pulmonary gangrene: A complication of bacterial pneumonia. *Chest* 105:567–573, 1994.

62. Kneeland Y Jr, Price KM: Antibiotics and terminal pneumonia. *Am J Med* 29:967–979, 1960.

63. Porto R, Heavia O, Hensley GT, Mayer PR: Disseminated *Klebsiella rhinoscleromatis* infection. *Arch Pathol Lab Med* 113:1381–1383, 1989.

64. Wardlaw AC: Bordetella, in Parker MT, Collier LH (eds), *Topley and Wilson's Principles of Bacteriology, Virology and Immunity*, vol 2, *Systematic Bacteriology*, 8th ed. London, Edward Arnold, 1990, p 322.

65. Woods M II: Pertussis: An old disease in a new era. *Semin Respir Infect* 6:37–43, 1991.

66. Nicholl A, Gardner A: Whooping cough and unrecognized postperinatal mortality. *Arch Dis Child* 63:41–47, 1988.

67. Hewlett EL: Pertussis in adults: Significance for disease transmission and immunisation therapy. *J Med Microbiol* 36:141–142, 1992.

68. Aoyama T, Takeuchi Y, Goto A, Iwai H, Murase Y, Iwata T: Pertussis in adults. *Am J Dis Child* 146:163–166, 1992.

69. Feyrter F: Ueber die pathologische Anatomie der Lungenveränderungen beim Keuchhusten. *Frankfurt Ztschr Path* 35:213–255, 1927.

70. Goodpasture EW, Auerbach SH, Swanson HS, Cotter AF: Virus pneumonia of infants secondary to epidemic infections. *Am J Dis Child* 57:997–1011, 1939.

71. Boreland PC, Gillespie SH, Ashworth LAE: Rapid diagnosis of whooping cough using monoclonal antibody. *J Clin Pathol* 41:573–575, 1988.

72. Kubic CL, Kubic PT, Brunning RD: The morphologic and immunophenotypic assessment of the lymphocytosis accompanying *Bordetella pertussis* infection. *Am J Clin Pathol* 95:809–815, 1991.

73. Pugin J, Auckenthaler R, Delaspre O, van Gessel E, Suter PM: Rapid diagnosis of gram-negative pneumonia by assay of endotoxin in bronchoalveolar lavage fluid. *Thorax* 47:547–549, 1992.

74. Pitt L: Pseudomonas, in Parker MT, Collier LH (eds), *Topley and Wilson's Principles of Bacteriology, Virology and Immunity*, vol 2, *Systematic Bacteriology*, 8th ed. London, Edward Arnold, 1990, p 256.

75. Barson AJ: *Pseudomonas aeruginosa* pneumonia in a children's hospital. *Arch Dis Child* 46:55–60, 1971.

76. Grieble HG, Colton FR, Bird TJ, et al: Fine-particle humidifiers: Source of *Pseudomonas aeruginosa* infections in a respiratory-disease unit. *N Engl J Med* 182:531–535, 1970.

76a. Stableforth DE, Smith DL: *Pseudomonas cepacia* in cystic fibrosis (editorial). *Thorax* 49:629–630, 1994.

76b. Humphreys H, Peckham D, Patel P, Knox A: Airborne dissemination of *Burkholderia (Pseudomonas) cepacia* from adult patients with cystic fibrosis. *Thorax* 49:1157–1159, 1994.

77. Fick RB, Magel GP, Squire SU, Wood RE, Gee JB, Reynolds HY: Proteins of the cystic fibrosis respiratory tract: Fragmented IgG opsonic antibody causing defective opsonophagocytosis. *J Clin Invest* 74:236–248, 1984.

78. Teplitz C: Pathogenesis of *Pseudomonas* vasculitis and septic lesions. *Arch Pathol* 80:297–307, 1965.

79. MacGregor AR: Scottish Scientific Advisory Committee: *Neonatal Deaths Due to Infection: Report of a Subcommittee*. Edinburgh, HMSO, 1947.

80. Damjanov I, Katz SM: Malakoplakia. *Pathol Annu* 16:103–126, 1981.

81. Gupta RK, Shuster RA, Christian WD: Autopsy findings in a unique case of malakoplakia. *Arch Pathol* 93:42–48, 1972.

82. Crouch E, White V, Wright J, Churg A: Malakoplakia mimicking carcinoma metastatic to the lung. *Am J Surg Pathol* 8:151–156, 1984.

83. Hodder RV, St George-Hyslop P, Chalvardjian A, Bear RA, Thomas P: Pulmonary malakoplakia. *Thorax* 39:70–71, 1984.

84. Colby TV, Hunt S, Pelzmann K, Carrington CB: Malakoplakia of the lung: Report of two cases. *Respiration* 39:295–299, 1980.

85. Schwartz DA, Ogden PO, Blumberg HM, Honig E: Pulmonary malakoplakia in a patient with the acquired immunodeficiency syndrome: Differential diagnostic considerations. *Arch Pathol Lab Med* 114:1267–1272, 1990.

86. Scannell KA, Portoni EJ, Finkle HI, Rice M: Pulmonary malakoplakia and *Rhodococcus equi* infection in a patient with AIDS. *Chest* 97:1000–1001, 1990.

87. Biggar WD, Keating A, Brear RA: Malakoplakia: Evidence for an acquired disease secondary to immunosuppression. *Transplantation* 31:109–112, 1981.

88. Shabtai M, Anaise D, Frei L, et al: Malakoplakia in renal transplantation: An expression of altered tissue reactivity under immunosuppression. *Transplant Proc* 21:3725–3727, 1989.

89. Schreiber AG, Maderazo EG: Leukocytic function in malakoplakia. *Arch Pathol Lab Med* 102:534–537, 1978.

89a. Penn C, Pritchard D: The spirochaetes, in Parker MT, Collier LH (eds): *Topley and Wilson's Principles of Bacteriology Virology and Immunity, Systematic Bacteriology*, vol 2, 8th ed. London, Edward Arnold, 1990, pp 617–618.

90. International Union of Microbiological Societies Subcommittee on the Taxonomy of Leptospira. *Revised List of Leptospira Serovars*. Groningen University Press, 1988.

91. Leptospirosis update (Editorial). *BMJ* 302:128–129, 1991.

92. Feigin RD, Anderson DC: Human leptospirosis. *Crit Rev Clin Lab Sci* 5:413–468, 1975.

93. Winter RJD, Richardson A, Lehner MJ, Hoffbrand BI: Lung abscess and reactive arthritis: Rare complications of leptospirosis. *Br Med J* 288:448–449, 1984.

94. Heath CW, Alexander AD, Galton MM: Leptospirosis in the United States: Analysis of 483 cases in man. *N Engl J Med* 273:857–864, 1965.

95. Ramachandran S, Perera MVF: Cardiac and pulmonary involvement in leptospirosis. *Trans R Soc Med Hyg* 71:56–59, 1977.

96. Charatan FB: House cats transmit human plague in US. *BMJ* 308:1060, 1994.

97. Craven RB, Barnes AM: Plague and tularemia. *Infect Clin North Am* 5:165–172, 1991.

98. Public Health Link. From the Chief Medical Officer of Health (U.K.), September 28, 1994.

99. John TJ: India: Is it plague? *Lancet* 344:1359–1360, 1994.

100. Pearson AD: *Franciscella*, in Parker MT, Collier LH (eds), *Topley and Wilson's Principles of Bacteriology, Virology and Immunity*, vol 2, *Systematic Bacteriology*, 8th ed. London, Edward Arnold, 1990, p 595.

101. Stewart BM. Pullen RL: Tularemic pneumonia: Review of American literature and report of 15 additional cases. *Am J Med Sci* 210:223–236, 1945.

102. Foulger M, Glazer AM, Foshay L: Tularemia: Report of a case with postmortem observations and a note on the staining of *Bacterium tularense* in tissue sections. *JAMA* 98:951–954, 1932.

103. Francis E: A summary of present knowledge of tularaemia. *Medicine* 7:411–432, 1928.

104. Weed LA, Sloss PT, Clagett OT: Chronic localized brucellosis. *JAMA* 161:1044–1047, 1956.

105. Kaufman AF, Fox MD, Boyce JM, et al: Airborne spread of brucellosis. *Ann NY Acad Sci* 353:105–144, 1980.

106. Lulu AR, Araj GF, Chateeb MI, Mustafa MY, Yusuf AR, Fenech FF: Human brucellosis in Kuwait: A prospective study of 400 cases. *Q J Med* 249(NS 66):39–54, 1988.

107. al-Eissa Y, al-Zamil F, al-Mugeiren M, al-Rasheed S, al-Sanie A, al-Mazyad A: Childhood brucellosis: A deceptive infectious disease. *Scand J Infect Dis* 23:129–133, 1991.

108. García-Rodríguez JA, García-Sánchez JE, Muñoz Bellido JL, Ortiz de la Tabla V, Bellido Barbero J: Review of pulmonary brucellosis: A case report on brucellar pulmonary empyema. *Diagn Microbiol Infect Dis* 11:53–60, 1988.

109. Davison WC, Holme ML, Emmons VB: Meningcoccus pneumonia: II. The epidemiology of post-influenzal pneumonia in which diplococcus intracellularis meningitidis was isolated. From observations at Camp Coetquidon A.E.F., France. *Bull Johns Hopkins Hosp* 30:329–331, 1919.

110. Christensen JJ, Gadeberg O, Bruun B: *Neisseria meningitidis*: Occurrence in non-pneumonic pulmonary infections. *APMIS* 96:218–222, 1988.

111. Reymann AK: *The Pneumonias*. Philadelphia, W.B. Saunders, 1938, p 381.

112. De Matteis A, Armani G: *Salmonella typhi* pneumonia without intestinal lesions. *J Pathol Bacteriol* 94:464–467, 1967.

113. Aguado JM, Obeso G, Cabanillas JJ, Fernández-Guerrero M, Alés J: Pleuropulmonary infections due to nontyphoid strains of *Salmonella*. *Arch Intern Med* 150:54–56, 1990.

114. Stuart B, Roscoe P: Typhoid: Clinical analysis of 360 cases. *Arch Intern Med* 78:629–661, 1946.

115. Sharma AM, Sharma OP: Pulmonary manifestations of typhoid fever: Two cases and a review of the literature. *Chest* 101:1144–1146, 1992.

116. Chan JC, Raffin TA: Salmonella lung abscess complicating Wegener's granulomatosis. *Respir Med* 85:339–341, 1991.

117. Ankobiah WA, Salehi F: Salmonella lung abscess in a patient with acquired immunodeficiency syndrome [letter]. *Chest* 100:591, 1991.

118. Owen CR, Bucker EO, Bee JF, et al: *Pasteurella multocida* in animals' mouths. *Rocky Mount Med J* 65:45–46, 1968.

119. Saphir DA, Carter GR: Gingival flora in the dog with special reference to bacteria associated with bites. *J Clin Microbiol* 3:344–349, 1976.

120. Weber DJ, Wolfson JS, Swartz MN, et al: *Pasteurella multocida* infections: Report of 34 cases and review of the literature. *Medicine (Baltimore)* 63:133–154, 1984.

121. Christensen JJ, Gadeberg O, Bruun B: *Neisseria meningitidis*: Occurrence in non-pneumonic pulmonary infections. *APMIS* 96:218–222, 1988.

122. Yagyu Y, Sawaki M, Mikasa K, et al: A clinical study on 5 cases of respiratory infections caused by *Neisseria meningitidis*. *Kansenshogaku Zasshi* 64:822–829, 1990.

123. Wright PW, Wallace RJ: Pneumonia due to *Moraxella (Branhamella) catarrhalis*. *Semin Respir Infect* 4:40–46, 1989.

124. Verghase A, Berk SL: *Moraxella (Branhamella) catarrhalis*. *Infect Dis Clin North Am* 5:523–538, 1991.

125. Fraser DW, Tsai TR, Orenstein W, et al: Legionnaires' disease: Description of an epidemic of pneumonia. *N Engl J Med* 297:1189–1203, 1977.

126. The spectrum of Legionnaires' disease (Editorial). *Lancet* 2:976, 1978.

127. Lessons from Stafford (Editorial). *Lancet* 1:1363–1364, 1986.

128. Tobin JO'H, Beare J, Dunnill MS, et al: Legionnaire's disease in a transplant unit: Isolation of the causative agent from shower baths. *Lancet* 2:118–121, 1980.

129. Fallon RJ: Legionella, in Parker MT, Collier LH (eds), *Topley and Wilson's Principles of Bacteriology, Virology and Immunity*, vol 2, *Systematic Bacteriology*, 8th ed. London, Edward Arnold, 1990, p 276.

130. Stout JE, Yu VL, Yee YC, Vaccarello S, Divan W, Lee TC: *Legionella pneumophila* in residential water supplies: Environmental surveillance with clinical assessment for Legionnaires' disease. *Epidemiol Infect* 109:49–57, 1992.

131. Fields BS, Sanden G, Barbaree J, et al: Intracellular multiplication of *Legionella pneumophila* in amoebae isolated from hospital hot water tanks. *Curr Microbiol* 18:131–137, 1989.

132. Roig J, Domingo C, Morera J: Legionnaires' disease. *Chest* 105:1817–1825, 1994.

133. Thacker WL, Dyke JW, Benson RF, et al: *Legionnella lansingensis* sp. nov. isolated from a patient with pneumonia and underlying chronic lymphocytic leukemia. *J Clin Microbiol* 30:2398–2401, 1992.

134. Ciesielski CA, Blaser MJ, Wang WLL: Serogroup specificity of *Legionella pneumophila* is related to lipopolysaccharide characteristics. *Infect Immun* 51:397–404, 1986.

135. Bollin GE, Plouffe JF, Para MF, Prior RB: Difference in virulence of environmental isolates of *Legionella pneumophila*. *J Clin Microbiol* 21:674–677, 1985.

136. Hoge CW, Brieman RF: Advances in the epidemiology and control of Legionella infections. *Epidemiol Rev* 13:329–340, 1991.

137. van Den Bergen HA, Meenhorst PL, Ruiter DJ, Mauw BJ, Meijer CJLM: Legionnaires' disease: case report with special emphasis on electron microscopy and potential risk of infection at autopsy. *Histopathology* 3:523–530, 1979.

138. Kaan JA, Simoons-Smith AM, MacLaren DM: Another source of aerosol causing nosocomial Legionaires' disease. *J Infect* 11:145–148, 1985.

139. Carratala J, Gudiol F, Pallares R, et al: Risk factors for nosocomial *Legionella pneumophila* pneumonia. *Am J Respir Care Med* 149:625–629, 1994.

140. Nguyen MH, Stout JE, Yu VL: Legionellosis. *Infect Dis Clin North Am* 5:561–584, 1991.

141. Fenstersheib MD, Miller M, Biggins C, et al: Outbreak of Pontiac fever due to *Legionella anisa*. *Lancet* 336:35–37, 1990.

142. Yu VL, Kroboth FJ, Shonnard J, Brown A, McDearman S, Magnussen M: Legionnaires' disease: New clinical perspective from a prospective pneumonia sudy. *Am J Med* 73:357–361, 1982.

143. Roig J, Agular X, Ruiz J, et al: Comparative study of *Legionella pneumophila* and other nosocomial acquired pneumonias. *Chest* 99:344–350, 1991.

144. Winn WC, Glavin SL, Perl DP, et al: The pathology of Legionnaires' disease: fourteen fatal cases from the 1977 outbreak in Vermont. *Arch Pathol Lab Med* 102:344–350, 1978.

145. Blackmon JA, Hicklin MD, Chandler FW, Special Expert Pathology Panel: Legionnaires disease: Pathological and historical aspects of "a new" disease. *Arch Pathol Arch Med* 102:337–343, 1978.

146. Weisenburger DD, Rappaport H, Ahluwalia MS, Melvini R, Renner ED: Legionnaires' disease. *Am J Med* 69:476–482, 1980.

147. Cluroe AD: Legionnaires' disease mimicking pulmonary miliary tuberculosis. *Histopathology* 22:73–75, 1993.

148. Case Records of the Massachusetts General Hospital. *N Engl J Med* 330:557–564, 1994.

149. Chastre J, Raghu G, Soler P, Brun P, Basset F, Gibert C: Pulmonary fibrosis following pneumonia due acute Legionnaires' disease: Clinical, ultrastructural and immunofluorescence study. *Chest* 91:57–62, 1987.

150. Hughes JA, Anderson PB: Pulmonary cavitation, fibrosis and Legionnaires' disease. *Eur J Respir Dis* 66:59–61, 1985.

151. Nichol KL, Parenti CM, Johnson JE: High prevalence of positive antibodies to *Legionella pneumophila* among outpatients. *Chest* 100:663–666, 1991.

152. Schurmann D, Ruf B, Ferenbach FJ, Jautzke G, Pohle HD: Fatal Legionnaires' pneumonia: Frequency of legionellosis in autopsied patients with pneumonia from 1969 to 1985. *J Pathol* 155:35–39, 1988.

153. Theaker JM, Tobin JO'H, Jones SEC, Kirkpatrick P, Vina MI, Fleming KA: Immunohistological detection of *Legionella pneumophila* in lung sections. *J Clin Pathol* 40:143–146, 1987.

154. Fain JS, Bryan RN, Cheng L, Lewin KJ, Porter DD, Grody WW: Rapid diagnosis of Legionella infection by a nonisotopic *in situ* hybridization method. *Am J Clin Pathol* 95:719–724, 1991.

155. Cohorst W, Sheonfeld SA, Macklin JE, et al: Rapid diagnosis of Legionnaires' disease by bronchiolar lavage. *Chest* 84:186–190, 1983.

156. Chiodini P, Williams A, Barker J, et al: Bronchial lavage and transbronchial lung biopsy in the diagnosis of Legionnaires' disease. *Thorax* 40:154–155, 1985.

157. Murray JF: The white plague: Down and out, or up and coming? J Burns Amberson Lecture. *Am Rev Respir Dis* 140:1788–1795, 1989.

158. Waksman SA: *The Conquest of Tuberculosis*. Berkeley, University of California Press, 1964, p 8.

159. Hinshaw AC, Murray JF: *Diseases of the Chest*. Philadelphia, W.B. Saunders, 1980, pp 1–1044.

159a. Murray JF: Personal communication, December 1992.

159b. Murray JF: The International Union against tuberculosis and lung disease: Its contribution to world lung health. *Am J Respir Crit Care Med* 151:1697–1699, 1995.

159c. Sudre P, ten Dam G, Kochi A: Tuberculosis: a global overview of the situation today. *Bull World Health Organ* 70:149–159, 1992.

160. Bobrowitz ID: Active tuberculosis undiagnosed until autopsy. *Am J Med* 72:650–658, 1982.

161. Bradley BL, Kerr KM, Leitch AG, Lamb D: Notification of tuberculosis: Can the pathologist help? *BMJ* 297:595, 1988.

162. Sibbison JB: USA: Action plan against multiresistant TB. *Lancet* 339:1161, 1992.

163. Grange JM: The mycobacteria, in Parker MT, Collier LH (eds), *Topley and Wilson's Principles of Bacteriology, Virology and Immunity*, vol 2, *Systematic Bacteriology*, 8th ed. London, Edward Arnold, 1990, p 74.

164. Hart PDA, Haslett EA: Special report series. Medical Research Council, London, No. 243, 1942.

165. Grange JM: Tuberculosis, in Parker MT, Collier LH (eds), *Topley and Wilson's Principles of Bacteriology, Virology and Immunity*, vol 3, *Bacterial Disease*, 8th ed. London, Edward Arnold, 1990, p 98.

166. Riley RL, Mills CC, O'Grady F, Sutton LU, Wittstadt F, Shipuri DN: Infectiousness of air from a tuberculosis ward. *Am Rev Respir Dis* 85:511–525, 1962.

167. Rich AR: *The Pathogenesis of Tuberculosis*, 2d ed. Springfield, Ill., Charles C Thomas, 1951.

168. Case Records of the Massachusetts General Hospital. *N Engl J Med* 322:1728–1738, 1990.

169. Khan FA, Rehman M, Marcus P, Azueta V: Pulmonary gangrene occurring as a complication of pulmonary tuberculosis. *Chest* 77:76–80, 1980.

170. Hoheisel G, Chan BKM, Chan CHS, Chan KS, Teschler H, Costabel U: Endobronchial tuberculosis: Diagnostic features and therapeutic outcome. *Respir Med* 88:593–597, 1994.

171. Sekosan M, Cleto M, Senseng C, Farolan M, Sekosan J: Spindle cell pseudotumors in the lungs due to *Mycobaterium tuberculosis* in a transplant patient. *Am J Surg Pathol* 18:1065–1068, 1994.

172. Kunkel SL, Chensue SW, Striete R, Lynch JP, Remick S: Cellular and molecular aspects of granulomatous infection. *Am J Respir Cell Mol Biol* 1:439–447, 1989.

172a. Yu C-T, Wang C-H, Huang T-J, Lin H-C, Kuo H-P: Regulation of bronchoalveolar lavage T lymphocyte subpopulations to rate of regression of active pulmonary tuberculosis. *Thorax* 50:869–874, 1995.

173. Wolinsky E: Mycobacterial disease other than tuberculosis. *Clin Infect Dis* 15:1–12, 1992.

174. Wallace RJ Jr: *Mycobacterium avium* complex lung disease and women: Now an equal opportunity disease. *Chest* 105:6–7, 1994.

175. Maesaki S, Kohno S, Koga H, Miyazaki Y, Kaku M: A clinical comparison between *Mycobacterium avium* and *Mycobacterium intracellulare* infections. *Chest* 104:1408–1411, 1993.

176. Davies PDO: Infection with *Mycobacterium kansasii*. *Thorax* 49:435–436, 1994.

177. Teirstein AS, Damsker B, Kirschner PA, Krellenstein DJ, Robinson B, Chuang MT: Pulmonary infection MAI: Diagnosis, clinical patterns, treatment. *Mt Sinai J Med* 57:209–215, 1990.

177a. Okada Y, Ichinose Y, Yamaguchi K, Kamazawa M, Yamasawa F, Kawashiro T: *Mycobacterium avium-intracellulare* pleuritis with massive plural effusion *Eur Respir J* 8:1428–1429, 1995.

178. Kilby JM, Gilligan PH, Yankaskas JR, Highsmith WE Jr, Edwards LJ, Knowles MR: Nontuberculous mycobacteria in adult patients with cystic fibrosis. *Chest* 102:70–75, 1992.

179. Marchevsky A, Damsker B, Gribetz A, Tepper S, Geller SA: The spectrum of pathology of nontuberculous mycobacterial infections in open-lung biopsy specimens. *Am J Clin Pathol* 78:695–700, 1982.

180. Bothamley GH, Rudd R, Festenstein F, Ivanyy I: Clinical value of the measurement of the *Mycobacterium tuberculosis* specific antibody in pulmonary tuberculosis. *Thorax* 47:270–275, 1992.

181. de Lassence A, Lecossier D, Pierre C, Cadranel J, Stern M, Hance AJ: Detection of mycobacterial and DNA in pleural fluid from patients with tuberculous pleurisy by means of the polymerase chain reaction: Comparison of two protocols. *Thorax* 47:265–269, 1992.

182. Eisentch KD, Sifford MD, Cave MD, Bates JH, Crawford JT: Detection of *Mycobacterium tuberculosis* in sputum samples using a polymerase chain reaction. *Am Rev Respir Dis* 114:1160–1163, 1991.

183. Walker DA, Taylor IK, Mitchell DM, Shaw RJ: Comparison of polymerase chain reaction amplification of two mycobacterial DNA sequences, IS110 and the 65kDa antigen gene, in the diagnosis of tuberculosis. *Thorax* 47:690–694, 1992.

184. Yuen KY, Chan KS, Chan CM, Ho BSW, Dai LK, Chau PY, Ng MH: Use of PCR in routine diagnosis of treated and untreated pulmonary tuberculosis. *J Clin Pathol* 46:318–322, 1993.

184a. Godfrey-Fausselt P: Molecular diagnosis of tuberculosis: The need for new diagnostic tools. *Thorax* 50:709–711, 1995.

184b. Kox LFF: Tests for detection and amplification of mycobacteria. How should they be used? *Respir Med* 89:399–408, 1995.

184c. Noordhock GT, Kaan JA, Mulder S, Wilke H, Kilk AHJ: Routine application of the polymerase chain reaction for detection of *Mycobacterium tuberculosis* in clinical samples. *J Clin Pathol* 48:810–814, 1995.

185. Popper HH, Winter E, Hofler G: DNA of *Mycobacterium tuberculosis* in formalin-fixed, paraffin-embedded tissue in tuberculosis and sarcoidosis detected by polymerase chain reaction. *Am J Clin Pathol* 101:738–741, 1994.

186. Saboor SA, Johnson NMcI, McFadden J: Detection of mycobacterial DNA in sarcoidosis and tuberculosis with polymerase chain reaction. *Lancet* 339:1012–1015, 1992.

187. Fidler HM, Rook GA, Johnson NMcI, McFadden J: *Mycobacterium tuberculosis* DNA in tissue affected by sarcoidosis. *BMJ* 306:546–548, 1993.

188. Ozaki T, Nakahira S, Tani K, Ogushi F, Yasuoka S, Ogura T: Differential cell analysis in bronchoalveolar lavage fluid from pulmonary lesions of patients with tuberculosis. *Chest* 102:54–59, 1992.

189. Kent DC, Houk VN, Elliot RC, Sokolowoski JW Jr, Sorensen K: The definitive evaluation of sarcoidosis. *Am Rev Respir Dis* 101:721–727, 1970.

190. Myers JL, Katzenstein LA: Granulomatous infection mimicking bronchocentric granulomatosis. *Am J Surg Pathol* 10:317–322, 1986.

191. Centers for Disease Control: Syphilis trends in the US. *MMWR* 30:441–449, 1981.

192. Karshner RG, Karshner CF: Syphilis of the lung. *Ann Med* 1:371–401, 1920.

193. Carrera JL: A pathologic study of the lungs in 152 autopsy cases of syphilis. *Am J Syph* 4:1–33, 1920.

194. Howard CP: Pulmonary syphilis. *Am J Syph* 8:1–33, 1924.

195. McIntyre MC: Pulmonary syphilis: Its frequency, pathology and roentgenologic appearance. *Arch Pathol* 11:258–280, 1931.

196. Warthin AF: Studies of the pulmonary artery; syphilitic aneurysm of the left upper division; demonstration of spirochaete pallida in wall of artery and aneurysmal sac. *Am J Syph* 1:693–711, 1917.

197. Karsner HT: Post-cicatricial syphilitic disease of the pulmonary artery. *Arch Intern Med* 51:367–386, 1933.

198. Bollinger O: Uber eine neue Pilzkrankheit beim Rinde. *Zbl Med Viss* 15:481–485, 1877.

199. Israel J: Neue Beobachtungen auf dem Gebiete der Mykosen des Menschen. *Virchows Arch Path Anat* 74:15–53, 1878.

200. Bates M, Cruikshank G: Thoracic actinomycosis. *Thorax* 12:99–124, 1957.

201. Kinnear WJM, MacFarlane JT: A survey of thoracic actinomycosis. *Resp Med* 84:57–59, 1990.

202. Schaal K, Beaman BL: Clinical significance of the actinomycetes, in Goodfellow M, Modarski M, Williams ST (eds), *The Biology of the Actinomycetes.* London, Academic Press, 1984, pp 389–424.

203. Wright E, Holmberg K, Houston J, Morrison IM, Roberts C: Pulmonary actinomycosis simulating a bronchial neoplasm. *J Infect* 6:179–181, 1983.

204. McQuarrie EG, Hall WH: Actinomycosis of the lung and chest wall. *Surgery* 64:905–911, 1968.

205. Brown JR: Human actinomycosis: A study of 101 subjects. *Hum Pathol* 4:319–330, 1973.

206. Weese WC, Smith IM: A study of 57 cases of actinomycosis over a 36-year period: A diagnostic "failure" with good prognosis after treatment. *Arch Intern Med* 135:1562–1568, 1975.

207. Slade PR, Slesser BV, Southgate J: Thoracic actinomycosis. *Thorax* 28:73–85, 1973.

208. Hseih M-J, Liu H-P, Chang J-P, Chang C-H: Thoracic actinomycosis. *Chest* 104:366–370, 1993.

209. Ariel H, Brewer R, Kamal S, Bendov I, Mogle P, Rosenmann E: Endobronchial actinomycosis simulating bronchogenic carcinoma: Diagnosis by bronchial biopsy. *Chest* 99:493–495, 1991.

210. Cenden I, Klapholz A, Talavera W: Pulmonary actinomycosis: A cause of endobronchial disease in a patient with AIDS. *Chest* 103:1886–1887, 1993.

211. Robboy SJ, Vickery AL Jr: Tinctorial and morphologic properties distinguishing actinomycosis and nocardiosis. *N Engl J Med* 282:593–596, 1970.

212. Lerner PR: Actinomyces and Arachnia species, in Mandell GL, Douglas RG, Bennett JE (Eds), *Principles and Practice of Infectious Disease.* New York, John Wiley, 1979, pp 1969–1978.

213. Paz HL, Little BJ, Ball WC Jr, Winkelstein JA: Primary pulmonary botryomycosis: A manifestation of chronic granulomatous disease. *Chest* 101:1160–1162, 1992.

214. Toth IR, Kazal HC: Botryomycosis in acquired immunodeficiency syndrome. *Arch Pathol Lab Med* 111:246–249, 1987.

215. Katznelson D, Vawter GF, Foley GE, Shwachman H: Botromycosis: A complication of cystic fibrosis. *J Pediatr* 65:525–539, 1964.

215a. Multz AS, Cohen R, Azenta V: Bacterial pseudomycosis: a rare cause of haemoptysis *Eur Respir J* 7:1712–1713, 1994.

216. Speir WA, Mitchener JW, Galloway RF: Primary pulmonary botryomycosis. *Chest* 60:92–93, 1971.

217. Stanford JL: A simple view of nocardial taxonomy. *J Hyg (London)* 91:369–376, 1983.

218. Baily GG, Neill P, Robertson VJ: Nocardiosis: A neglected chronic lung disease in Africa. *Thorax* 43:905–910, 1988.

219. Heffner JE: Pleuro-pulmonary manifestations of actinomycosis and nocardiosis. *Semin Respir Infect* 3:352–361, 1988.

220. Kim J, Minamoto GY, Griecho MH: Nocardial infection and the complication of AIDS: A report of six cases and review. *Rev Infect Dis* 13:624–629, 1991.

221. Curry WA: Human nocardiosis: A clinical review with selected case reports. *Arch Intern Med* 140:818–826, 1980.

222. Carlsen ET, Hill RB Jr, Rowlands DT Jr: Nocardiosis and pulmonary alveolar proteinosis. *Ann Intern Med* 60:275–281, 1964.

223. Case Records of the Massachusetts General Hospital. *N Engl J Med* 308:1147–1156, 1983.

224. Wongthim S, Udompanich V: Pulmonary nocardiosis in Chulalongkorn Hospital. *J Med Assoc Thai* 74:271–277, 1991.

225. Chapman SW, Wilson JP: Nocardiosis in transplant patients. *Semin Respir Dis* 5:74–79, 1990.

226. Connar RG, Ferguson TB, Sealy WC, Conant NF: Nocardiosis: Report of a single case with recovery. *J Thorac Surg* 22:424–433, 1951.

227. Buchanan RE, Holt JC, Lessel EF: *Index Bergeyana.* Baltimore, Williams & Wilkins, 1966.

228. Whitmore A, Krishnaswani CS: An account of the discovery of a hitherto undescribed disease among the population of Rangoon. *Indian Med Gazette* 47:262–267, 1912.

229. Smith CJ, Alan SC, Noor Embi M, Othman O, Razak N, Ismail G: Human melioidosis: An emerging medical problem *MIRCEN* 3:343–366, 1987.

230. Wang CY, Yap BH, Delilkan AE: Melioidosis pneumonia and blast injury. *Chest* 103:1897–1899, 1993.

231. Iralu JV, Maguire JH: Pulmonary infections in immigrants and refugees. *Semin Respir Infect* 6:235–246, 1991.

232. Koponen MA, Zlock D, Palmer DL, Merlin TL: Melioidosis: Forgotten but not gone! *Arch Intern Med* 151:605–608, 1991.

233. Yap EH, Chan YC, Goh KT, et al: Sudden unexplained death syndrome: A new manifestation in melioidosis? *Epidemiol Infect* 107:577–584, 1991.

234. Piggott JA, Hochholzer L: Human melioidosis: A histopathologic study of acute and chronic melioidosis. *Arch Pathol* 90:101–111, 1970.

235. Walsh AL, Smith MD, Wuthiekanun V, et al: Immunofluorescence microscopy for the rapid diagnosis of melioidosis. *J Clin Pathol* 47:377–379, 1994.

236. Alexander AD, Huxsoll DL, Warner AR Jr, Shepler V, Dorsey A: Serological diagnosis of human melioidosis with indirect hemagglutination and complement fixation tests. *Appl Microbiol* 20:825–833, 1970.

237. Winberg CD, Rose ME; Rappaport H: Whipple's disease of the lung. *Am J Med* 65:873–880, 1978.

238. Spain DM, Kliot DA: PAS and Sudan positive pulmonary emboli in Whipple's disease. *Gastroenterol* 43:202–205, 1962.

239. James JN, Bulkley BH: Whipple bacilli within the tunica media of pulmonary arteries. *Chest* 86:455–458, 1984.

240. Sieracki JC, Fine G: Whipple's disease—observations on systemic involvement: II. Gross and histologic observations. *Arch Pathol* 67:81–93, 1959.

241. Ladue JS: Bronchiolitis obliterans fibrosa. *Arch Intern Med* 68:663–673, 1941.

242. Hardy KA, Schidlow DV, Zaeri N: Obliterative bronchiolitis in children. *Chest* 93:460–466, 1988.

243. Lange W: Ueber eine eigenthümliche Erkrankung der kleinen Bronchien und Bronchiolen. *Deutsche Arch Klin Med* 70:342–364, 1901.

244. Gosink BB, Friedman PJ, Liebow AA: Bronchiolitis obliterans: reontgenologic-pathologic correlation. *Am J Roentgenol* 117:816–832, 1973.

245. Davison AG, Heard BE, McAllister WAC, Turner-Warwick MEH: Cryptogenic organising pneumonitis. *Q J Med* 52:382–394, 1983.

246. Grinblatt J, Mechlis S, Lewitus Z: Organizing pneumonia–like process: An unusual observation of steroid responsive cases with features of chronic interstitial pneumonia. *Chest* 80:259–263, 1981.

247. Epler GR, Colby TV, McCloud TC, Carrington CB, Gaensler EA: Bronchiolitis obliterans organizing pneumonia. *N Engl J Med* 312:152–158, 1985.

248. Myers JL, Colby TV: Pathologic manifestations of bronchiolitis, constrictive bronchiolitis, cryptogenic organizing pneumonia and diffuse panbronchiolitis. *Clin Chest Med* 14:611–622, 1993.

249. Katzenstein A-LA, Askin F: *Surgical Pathology of Nonneoplastic Lung Disease*, 2nd ed. Philadelphia, W.B. Saunders, 1990, pp 40–57.

250. Colby T, Myers J: Clinical and histologic spectrum of bronchiolitis obliterans including bronchiolitis obliterans organizing pneumonia. *Semin Respir Med* 13:119–132, 1992.

251. King T Jr: Bronchiolitis obliterans. *Lung* 167:69–93, 1989.

252. Homma H: Annual report on the study of diffuse disseminated lung disease. Grant-in-AID from the Ministry of Health and Welfare, Japan 1982.

253. Wright JL, Cagle P, Churg A, Colby TV, Myers J: Diseases of the small airways. *Am Rev Respir Dis* 146:240–262, 1992.

254. Colby T, Myers J: Clinical and histologic spectrum of bronchiolitis obliterans including bronchiolitis obliterans organising pneumonia. *Semin Respir Med* 13:119–132, 1992.

255. Epler G, Sneider G, Gaensler E, et al: Bronchiolitis and bronchitis in connective tissue disease: Possible relationship to the use of penicillamine. *JAMA* 242:528–532, 1979.

256. Cohen AJ, King TE Jr, Downey GP: Rapidly progressive bronchiolitis obliterans with organizing pneumonia. *Am J Respir Crit Care Med* 149:1670–1675, 1994.

257. Douglas WW, Hepper NG, Colby TV: Silo-filler's disease. *Mayo Clin Proc* 64:291–304, 1989.

258. Myers J: Pathology of drug-induced lung disease, in Katzenstein A-LA, Askin F (Eds), *Surgical Pathology of Non-neoplastic Lung Disease*, 2nd ed. Philadelphia, W. B. Saunders, 1990, pp 97–127.

259. Camp M, Mehta J, Whitson M: Bronchiolitis obliterans and *Nocardia asteroides* infection of the lung. *Chest* 92:1107–1108, 1987.

260. Sato P, Madtes D, Thorning D, et al: Bronchiolitis obliterans caused by *Legionella pneumophila*. *Chest* 87:840–842, 1985.

260a. Sanito NJ, Morley TF, Condoluci DV: Bronchiolitis obliterans organising pneumonia in an AIDS patient. *Eur Respir J* 8:1021–1024, 1995.

261. Spiteri M, Klemerman P, Sheppard M, et al: Seasonal cryptogenic organising pneumonia with biochemical cholestasis: a new entity. *Lancet* 340:281–284, 1992.

262. Dorinski PM, Davis WB, Lucas JG, et al: Adult bronchiolitis: Evaluation by broncho-alveolar lavage and response to prednisone therapy. *Chest* 88:58–63, 1985.

263. Cordier J-F, Loire R, Brune J: Idiopathic bronchiolitis obliterans organizing pneumonia: Definition of characteristic clinical profiles in a series of 16 patients. *Chest* 96:999–1004, 1989.

264. Nagai S, Aung H, Tanaka S, et al: Bronchoalveolar lavage cell findings in patients with BOOP and related diseases. *Chest* (suppl) 102:32S–37S, 1992.

265. Myers JL, Katzenstein A-LA: Ultrastructural evidence of alveolar epithelial injury in idiopathic bronchiolitis obliterans–organizing pneumonia. *Am J Pathol* 132:102–109, 1988.

266. Dina R, Sheppard MN: The histological diagnosis of clinically documented cases of cryptogenic organizing pneumonia: Diagnostic features in transbronchial biopsies. *Histopathology* 23:541–545, 1993.

267. Abernathy E, Hruban R, Baumgartner W, et al: The two forms of bronchiolitis obliterans in heart/lung transplant recipients. *Hum Pathol* 22:1102–1110, 1991.

268. St John RC, Dorinsky PM: Cryptogenic bronchiolitis. *Clin Chest Med* 14:667–675, 1993.

269. Kindt CG, Weiland JE, Davies WB, et al: Bronchiolitis in adults. *Am Rev Respir Dis* 140:483–492, 1989.

270. Kitaichi M: Differential diagnosis of bronchiolitis obliterans organizing pneumonia. Chest (suppl)102:44S–49S, 1992.

271. Epler GR, Colby TV: The spectrum of bronchiolitis obliterans. *Chest* 83:161–162, 1983.

272. Costabel U, Teschler H, Schoenfeld B, et al: BOOP in Europe. *Chest* (suppl)102:14S–20S, 1992.

273. Sugiyama Y: Diffuse panbronchiolitis. *Clin Chest Med* 14:765–772, 1993.

274. Iwata M, Colby TV, Kitaichi M: Diffuse panbronchiolitis: Diagnosis and distinction from various pulmonary diseases with centrilobular interstitial foam cell accumulations. *Hum Pathol* 25:357–363, 1994.

275. Hoiby N: Diffuse panbronchiolitis and cystic fibrosis: East meets West. *Thorax* 49:531–532, 1994.

276. Yamanaka M, Saiki S, Tamura S, et al: The problems of chronic obstructive pulmonary disease, especially concerning diffuse panbronchiolitis. *Naika* 23:442–451, 1969.

277. Homma H, Yamanaka A, Tanimoto S, et al: Diffuse panbronchiolitis: A disease of the transitional zone of the lung. *Chest* 83:63–69, 1983.

278. Suzuki M, Usui K, Tamura S, et al: Familial cases of diffuse panbronchiolitis. *Nippon Kyobu Shikkan Gakkai Zasshi* 19:645–651, 1981.

279. Danbara T, Matsuoka R, Nukiwa T, et al: Familial occurrence of diffuse panbronchiolitis accompanied by elevation of cold agglutinin titer in the father and his two daughters. *Nippon Kyobu Shikkan Gakkai Zasshi* 20:597–603, 1982.

280. Sugiyama Y, Kudoh S, Maeda H, et al: Analysis of HLA antigens in patients with diffuse panbronchiolitis. *Am Rev Respir Dis* 141:1459–1462, 1990.

281. Wakisaka A, Iizawa M, Matsuura N, et al: HLA in juvenile diabetes mellitus in the Japanese. *Lancet* 2:970, 1976.

282. Tanmoto H: Review of the recent progress in treatment of patients with diffuse panbronchiolitis associated with *Pseudomonas aeruginosa* infection in Japan, in Homma JY, Yanimoto H, Holder IA, Hoiby N, Doring G (eds), *Pseudomonas aeruginosa* in human diseases. *Antibiot Chemother* 44:94–98, 1991.

283. Hirata T, Nishikawa S, Izumi T: An immunological study on diffuse panbronchiolitis. *Jpn J Chest Dis* 38:90–95, 1979.

284. Ono K, Shimamoto Y, Matsuzaki M, et al: Diffuse panbronchiolitis as a pulmonary complication in patients with adult T-cell leukemia. *Am J Hematol* 30:86–90, 1989.

285. Kimura I: Anti-ATLA antibody in diffuse panbronchiolitis. *Igaku no Ayami* 147:30–33, 1988.

Chapter *8*

Pulmonary Mycotic Disease

Bruce Addis

Until nearly fifty years ago, many of the pulmonary mycotic diseases were little known and fungi were considered to be unimportant as respiratory pathogens. The pulmonary mycoses have now assumed greater importance as opportunistic infections in debilitated patients during the terminal stages of many diseases or during the course of active infection with human immunodeficiency virus (HIV). Fungi take advantage of alterations in the normal bacterial ecology due to antibiotic therapy and compromised defense mechanisms due to steroid therapy, cytotoxic drugs and other immunosuppressants, organ transplantation, and HIV infection.

Although innumerable species of fungi are known and the spores of some are ubiquitous, very few affect humans. Of the potentially pathogenic species, many are almost entirely saprophytic. Following the sustained use of antibiotics, the normal human bacterial ecology may become greatly disordered, allowing the growth of normally non-pathogenic fungi, which multiply and sometimes invade the body tissues. Immunosuppression leads to decreased cellular and humoral immunity, which may already have been depressed as a result of disease involving the lymphoreticular system. Furthermore, debilitating and metabolic diseases such as pulmonary tuberculosis and diabetes mellitus also depress the immune mechanism, providing a more suitable milieu for fungi to survive and invade the body.

Some of the fungi capable of causing lung disease are almost purely saprophytic; these include the genera *Monilia, Aspergillus, Allescheria*, and *Mucor*, all of which may be found growing on the surface of cavities or within necrotic tissue in the lung. Pulmonary mycotic diseases due to these fungi occur throughout the world. Penetration of tissue by the invading organism can result in generalized spread of infection. Other fungal diseases, including those due to fungi of the genera *Blastomyces, Coccidioides, Paracoccidioides*, and *Histoplasma*, are seen in well-

defined geographic areas, in which the fungi exist in spore form in the soil. Exposure to these truly parasitic organisms can cause primary invasive infections in previously healthy people in the absence of any predisposing factors.

The effects produced by fungi in the lungs vary from tissue necrosis, resulting in lung destruction as seen in coccidioidomycosis and histoplasmosis, to allergic manifestations such as asthma, which follows proliferation of saprophytic fungi such as *Aspergillus* species on the surfaces of bronchi and in lung cavities. The allergic changes result from absorption and sensitization to fungal products, as shown by the occurrence of positive skin reactions to fungal allergens, the presence of circulating precipitins, and evidence of obstructive changes due to bronchospasm following the inhalation of certain fungal extracts in aerosol form. In this situation, fungal disease is often unsuspected and fungal elements are sparse.

The diagnosis of fungal disease is based on culture, histologic identification, and evidence of altered immunity. Whereas culture of the parasitic fungi—from biopsy material, sputum, or lung washings—provides convincing evidence of infection, this does not necessarily apply to the saprophytic fungi, when histology of biopsy or autopsy material may be required. The most useful stains for the identification of fungi in sections are the periodic acid-Schiff stain, or Gridley's modification of this, and the Grocott methenamine-silver stain. No classification of the pathogenic fungi is entirely satisfactory, as future research may demonstrate further changes in the life cycle of many of the Fungi imperfecti and necessitate their reclassification.

Normally, saprophytic fungi that cause lung disease in immunocompromised patients include: *Aspergillus* spp., Mucorales, *Candida* spp., and *Pneumocystis carinii*.

Fungi that cause lung disease in normal and immunocompromised hosts include: *Cryptococcus* neoformans, Histoplasma capsulatum, Coccidioides immitis, Blastomyces dermatitidis, Paracoccidioides brasiliensis, and Sporothrix schenckii. The main fungi causing pulmonary disease in humans are shown in Table 8-1.

PULMONARY ASPERGILLOSIS

Epidemiology

Aspergillus is an important cause of pulmonary disease in debilitated patients. The fungus is widespread in the environment and several hundred species are known, being found in the soil and in decaying organic matter such as manures, compost, dead leaves, hay, straw, spoiled grains, and rotting wood. Their spores are ubiquitous and readily disseminated by air currents. The lungs are constantly exposed.

The fungus receives its name from the shape of the fruiting head, which resembles the brush used for sprinkling holy water (aspergillum). Normally the vegetative form is found in human lesions, but conidiophores develop if there is sufficient oxygen tension in the tissue. The septate hyphae branch at an acute angle and swell at intervals to form foot cells, from which conidiophores arise as stalks. The ends of the conidiophores are swollen and form the vesicle. From the vesicle arise phialides, and these, in turn, bear the conidia, or spores, which vary in color in different species. The structure of the fruiting organs also varies in different species. Pulmonary disease is most commonly caused by A. fumigatus, A. flavus, and A. niger.

Human pulmonary aspergillosis was, in the past, divided into primary and secondary forms of the disease. Primary infections followed inhalation of spores with a previously normal lung and secondary

TABLE 8-1
Fungi Causing Pulmonary Disease in Humans

Organism	Disease
Aspergillus sp.*	Spectrum of pulmonary aspergillosis
Mucorales	Zygomycosis
Rhizopus	
Absidia	
Mucor	
Conidiobolus	
Cunninghamella	
Rhizomucor	
Candida sp.*	Candidiasis
Torulopsis glabrata	Pulmonary torulopsis
Cryptococcus neoformans	Cryptococcosis
Histoplasma capsulatum	N. American histoplasmosis
Histoplasma duboisii	African histoplasmosis
Coccidioides immitis*	Coccidioidomycosis
Blastomyces dermatitidis	N. American blastomycosis
Paracoccidioides brasiliensis	Paracoccidioidomycosis (S. American blastomycosis)
Sporothrix schenkii*	Sporotrichosis
Pseudallescheria boydii*	Fungus ball (pseudallescherioma)
Penicillin decumbens*	Fungus ball
Emmensia parva	Adiaspiromycosis
Geotrichum candidum	Geotrichosis
Malassezia furfur	Malassezia infection of vessels
Pneumocystis carinii	Pneumocystis pneumonia

*Fungi known to cause fungus ball.
NOTE: Bold type indicates important causes of pulmonary disease.

infections were defined as those occurring in a previously damaged lung or in an immunologically compromised or hypersensitive host. This division is now regarded as unhelpful, as it is recognized that the host immunologic response or failure of response to the infecting spores determines the outcome and resultant tissue response seen in the lung. A great increase in the number of pulmonary *Aspergillus* infections has followed the widespread introduction of immunodepressant, steroid, cytotoxic, and antibiotic drugs. These drugs artificially depress both humoral and cellular immunity and, often coupled with the natural immunosuppressive effects of the primary disease, predispose the host to severe forms of *Aspergillus* infection. Gowing and Hamlin have reviewed the historical aspects, mycology, and pathology of the condition.[1] Serologic tests are of value in diagnosis. These are based on demonstration of precipitins, which denotes infection, colonization of a cavity, or hypersensitivity to *Aspergillus* antigens.

Pulmonary disease due to *Aspergillus* can take a number of different forms, depending on the immune status of the individual, preexisting changes in the underlying lung architecture, and the degree of tissue invasion (Table 8-2).[2,3]

Allergic Forms of Pulmonary Aspergillosis

Allergic forms of disease are discussed in Chaps. 13, 14, 22, and 27. They include asthma (Chap. 22), eosinophilic pneumonia (Chap. 14), mucous impaction of the bronchi (Chap. 22), allergic bronchopulmonary aspergillosis (Chap. 27), bronchocentric granulomatosis (Chap. 27), and extrinsic alveolitis (see Chap. 13).

Most often, these separate manifestations are combined to produce a clinicopathologic spectrum of disease. A variable peripheral eosinophilia is present and patients are frequently atopic. Patients with cystic fibrosis are particularly prone to develop hypersensitivity to *Aspergillus.*

Proximal bronchi may become occluded by plugs of thick mucus, which may be visible on chest radiographs. They are frequently expectorated and appear as bronchial casts up to 2 cm in diameter. Airways distended by impacted mucus show a variable inflammatory infiltrate but are able to return to normal if the plug is removed. In fully developed lesions of allergic bronchopulmonary aspergillosis, damage to the bronchial wall becomes irreversible; there is loss of bronchial epithelium, squamous metaplasia, and replacement by inflammatory granulation tissue. Eosinophils are numerous and histiocytes are conspicuous, with scattered histiocytic giant cells. Severely damaged airways appear dilated with replacement by granulation tissue and fibrosis, a stage known as bronchocentric granulomatosis. Since this is a hypersensitivity response to *Aspergillus*, the fungus is not present in the tissues. Occasional fungi can be identified in bronchial secretions by fungal stains.

Aspergilloma

The terms *mycetoma, aspergilloma,* and *fungus ball* are sometimes used interchangeably, producing confusion. *Mycetoma* is usually defined as a clinical syndrome of localized, indolent, deforming, swollen lesions and sinuses involving cutaneous and subcutaneous tissue, fascia, and bone.[4,5] *Aspergilloma* is a more precise term, which refers to colonization of a cavity by *Aspergillus*. The first pulmonary aspergilloma was described by Virchow in 1856.[36] Typically, this develops in a preexisting cavity, formed during the course of such diseases as sarcoidosis, tuberculosis, bronchiectasis, lung abscess, cavitated neoplasms, and even cavities resulting

TABLE 8-2
Spectrum of Pulmonary Aspergillosis

Clinical Manifestation	Immune Status	Underlying Lung Architecture Away from Lesion	Tissue Invasion
Asthma/eosinophilic pneumonia	Increased	Normal	None
Allergic bronchopulmonary aspergillosis	Increased	Normal	None
Extrinsic allergic alveolitis	Increased	Normal	None
Saprophytic colonization (aspergilloma)	Normal	Preexisting cavitation	None
Invasive pulmonary aspergillosis	Decreased	Normal	+++
Tracheobronchial aspergillosis	Decreased	Normal	+

SOURCE: From Albida and Talbot,[2] with permission.

from other forms of fungal disease, notably *histoplasmosis*.[35] It is probably best to use the term of *fungus ball* until the organism has been identified, since macroscopically it is impossible to be certain if one is dealing with *Aspergillus* or another organism. Other fungi capable of producing fungus balls include *Candida, Nocardia, Sporothrix,* and *penicillium* species such as *P. decumbens,*[6] and *Pseudallescheria boydii.*[7]

Pulmonary aspergillomas most commonly occur in the upper lobes and may sometimes be multiple. The most consistent clinical manifestation is hemoptysis, which may be severe enough to justify embolization or resection. A high mortality rate is reported. Chest radiographs show a rounded opacity with a radiolucent crescent of air outlining the upper border (Monod sign).

Aspergillus colonizes the necrotic debris lining cavities, and saprophytic surface growth leads to the accumulation of layers of fungus, cellular debris, fibrin, and inflammatory cells to form a brownish-yellow mass filling the cavity (Fig. 8-1). Expansion of the cavity can occur as the fungus ball grows. This is accompanied by increasing fibrosis with dense pleural adhesions. The intracavitary mycotic membrane may calcify, producing a broncholith, part of which may be expectorated if bronchial obstruction is incomplete. Rarely, the fungus invades the cavity wall and underlying lung, in which case the aspergilloma may rupture into the pleural cavity, forming a mycotic empyema.

Zonation is characteristic of aspergillomas. Hyphae from the center of the mass appear degenerate and dilated. They stain poorly and resemble the zygomycetes. At the periphery, hyphae are more typical, with regular septa and dichotomous branching at 45° angles (Fig. 8-2). Conidiospores are formed if the cavity communicates with a bronchus (Fig. 8-3). Hoeppli-Splendore material is frequently deposited around the margins. This is brightly eosinophilic amorphous material, which has been shown to contain immunoglobulin and complement. It outlines the periphery of colonies and can be seen around the tips of individual hyphae (Fig. 8-4).[8]

Extensive local deposition of calcium oxalate occasionally occurs due to oxalic acid produced by the fungus. The crystals are birefringent under polarized light (Fig. 8-5). Although first described in association with *A. niger,* this cannot be used to distinguish between different species. Nime and Hutchins[41] found 11 such cases among 68 cases of paranasal or pulmonary aspergillosis; in one case, death was due to renal failure secondary to oxalosis.

FIGURE 8-1
Pulmonary aspergilloma. The cavity contains a mass of fungus, with layers of dead and degenerate organisms.

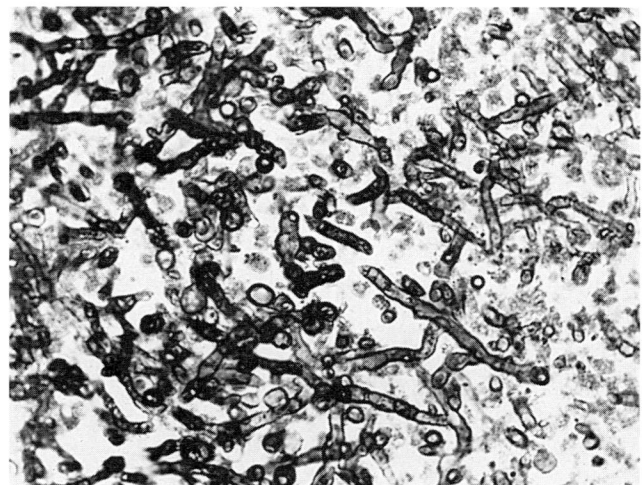

FIGURE 8-2
Hyphae of *Aspergillus fumigatus*: They branch at 45° angles and regular septa are present.

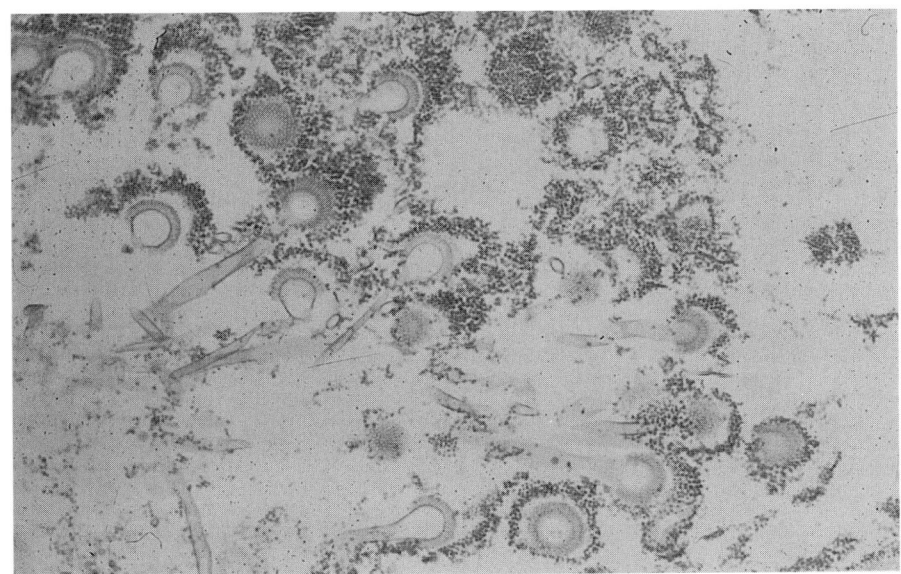

FIGURE 8-3
Aspergillus sporing heads. Different
species can be identified by differences
in the fruiting heads.

Invasive Aspergillosis (Nonsuppurative Necrotizing Pulmonary Aspergillosis)

This is a severe, fulminating type of pulmonary infection in which there is little tissue reaction to the invading fungus. It may be associated terminally with bloodstream dissemination of the infection. A primary invasive form has been described in individuals exposed repeatedly to overwhelming numbers of spores.[9] More frequently, however, infection occurs in immunosuppressed patients, in both those with acquired immunodeficiency syndrome (AIDS) and groups that are not infected with AIDS.

Aspergillus fumigatus and *A. flavus* are the most commonly involved species. Patients are usually chronically ill and debilitated and develop cough, pyrexia, and dyspnea.

Macroscopically, the affected lung is dark red and hemorrhagic, resembling a red infarct. In addition, it is edematous, and areas of necrosis may appear paler. In AIDS patients, upper lobe cavitation has been described, with diffuse infiltrates and pleural-based lesions in the lower lobes.[10] Histologically there are areas of hemorrhagic coagulative necrosis containing some nuclear debris. The original structure of the lung is no longer recognizable or

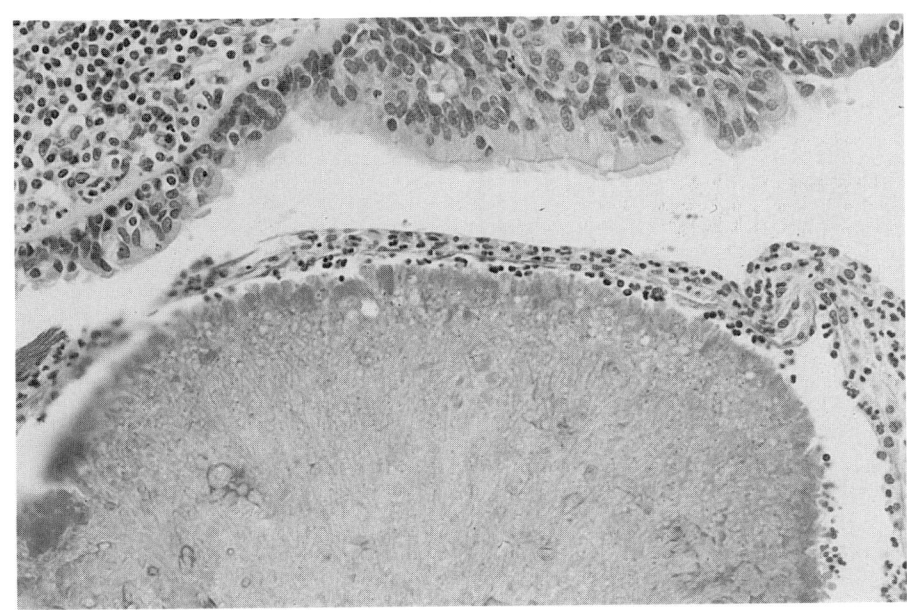

FIGURE 8-4
Pulmonary aspergilloma: Hoeppli-
Splendore material is present around
the edges of the fungus ball. The cavity
is lined by respiratory epithelium.

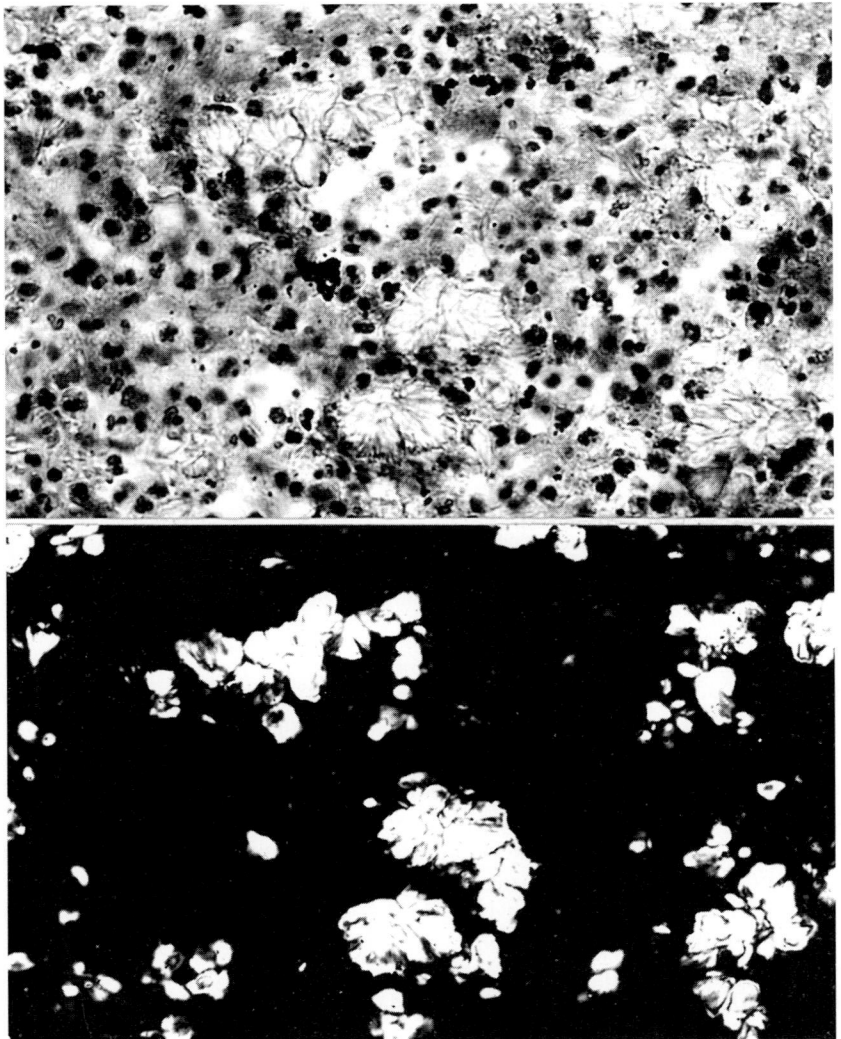

Birefringent calcium oxalate crystals in the vicinity of a pulmonary *Aspergillus* infection. The same field is viewed by ordinary light and polarized light.

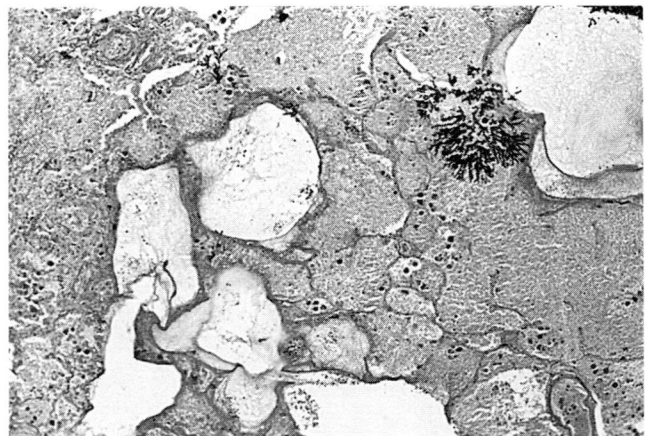

FIGURE 8-6
Invasive pulmonary aspergillosis. The alveolar tissue is necrotic with a colony of *Aspergillus* and virtual absence of a cellular response.

there may be a few stainable remains of elastic and reticulin fibers. Inflammatory cells are inconspicuous, with scattered neutrophils and lymphocytes. Hyphae may be scattered through the necrotic tissue and may be difficult to identify without fungal stains. If they are more numerous, they may form small colonies attached to alveolar walls (Fig. 8-6). These are best seen at the edges of the necrotic zone, where it is bounded by epithelium-lined alveoli filled with fibrinous material.

The bronchi at the margins and within the lesion contain friable brown membranous material containing a tangled mycelium. This was regarded by Gowing and Hamlin as the primary lesion and portal of entry of fungus into the lung.[1]

Fungal proliferation within and around arteries (angioinvasive aspergillosis) produces infarction with a segmental or lobar distribution (Fig. 8-7). This

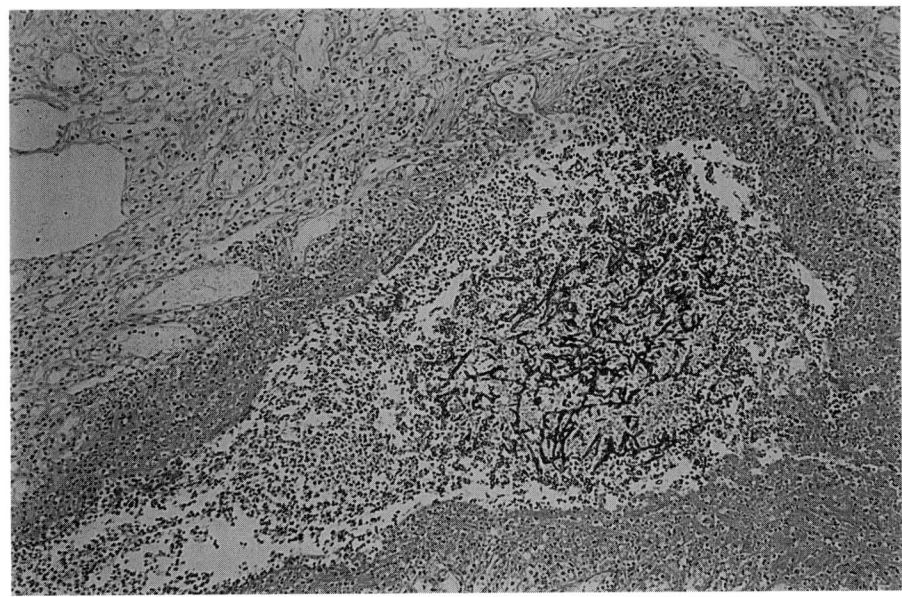

FIGURE 8-7
Angioinvasive aspergillosis. *Aspergillus* is growing within the lumen of a pulmonary artery with inflammation and partial necrosis of the vessel wall (PAS stain).

leads to rapid hematogenous dissemination to other organs, particularly the brain and endocardium. Fulminating necrotizing pulmonary aspergillosis has been associated with bacterial septicemia due to *Pseudomonas pyocyaneus*.

Granulomatous (Tuberculoid) Pulmonary Aspergillosis

In this unusual variant of invasive pulmonary aspergillosis, the cellular response is similar to that seen in tuberculosis. Tubercle-like granulomas consist of a central area of necrosis surrounded by histiocytic giant cells with lymphocytes and neutrophils (Fig. 8-8). Fungal hyphae are present both in the necrotic tissue and in the cytoplasm of the giant cells (Fig. 8-9). They are best demonstrated by the Grocott methenamine silver stain. Occasionally, actinomycetoid forms of *Aspergillus* may be present in this type of lesion.[39] A dense fibrotic granulomatous type of lesion leading to stenosis of major bronchi has also been described.[40]

Tracheobronchial Aspergillosis

This form of pulmonary aspergillosis is being seen increasingly commonly in immunosuppressed patients. They present with bronchial obstruction and the diagnosis is made by bronchoscopy, when a felt-like membrane is seen to spread throughout the tra-

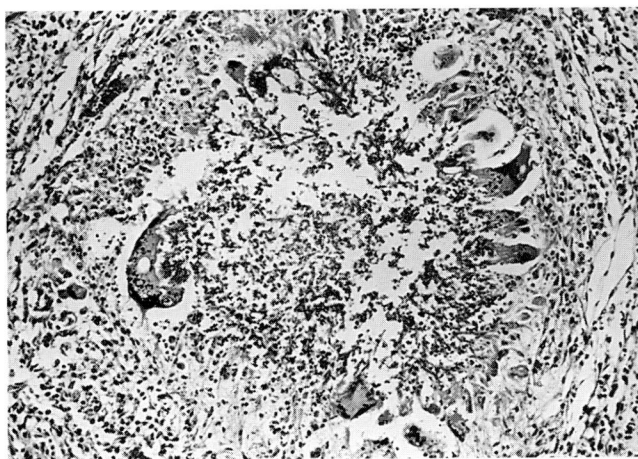

FIGURE 8-8
Granulomatous aspergillosis. Giant cells surround a small collection of neutrophils and fungal hyphae.

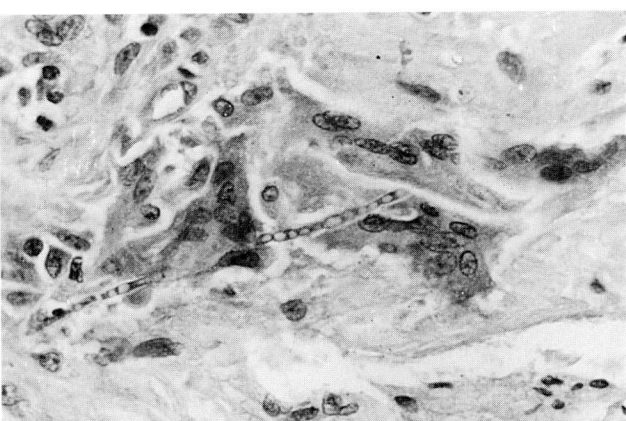

FIGURE 8-9
Granulomatous aspergillosis. A fungal hypha is contained within two giant cells.

263

FIGURE 8-10
Tracheobronchial aspergillosis. The mucosa of the larynx and trachea is partially coated by a membrane.

cheo-bronchial tree and occluded airways (Figs. 8-10 and 8-11). This bleeds readily on contact and histologically a carpet of fungal hyphae can be seen replacing the epithelium and invading the walls of airways (Fig. 8-12).

Other Forms of Pulmonary Aspergillosis

Diffuse pneumonic and suppurative aspergillosis was described by Bech[11] as an acute pneumonia with a marked polymorph response. Macroscopically, there are multiple small abscesses filled with yellow pus and surrounded by hemorrhagic borders. Histologically, the abscess cavities are filled with pus cells and fungal hyphae; they are encircled by chronic inflammation, including foreign-body giant cells. This is presumed to be due to inhalation of fungal spores into terminal airways and is usually seen in immunosuppressed or debilitated patients. In most cases the fungal infection is secondary and follows a bacterial pneumonia, often staphylococcal, with abscess formation.

In 1969 Young and colleagues reported a group of patients with a necrotizing hemorrhagic form of lobar pneumonia due to *Aspergillus*.[12] Another form of saprophytic *Aspergillus* infection occurs when necrotic lung tissue is colonized by the fungus. This may be seen after pulmonary embolism and infarction when cavitation is seen within the infarcted lung.[13]

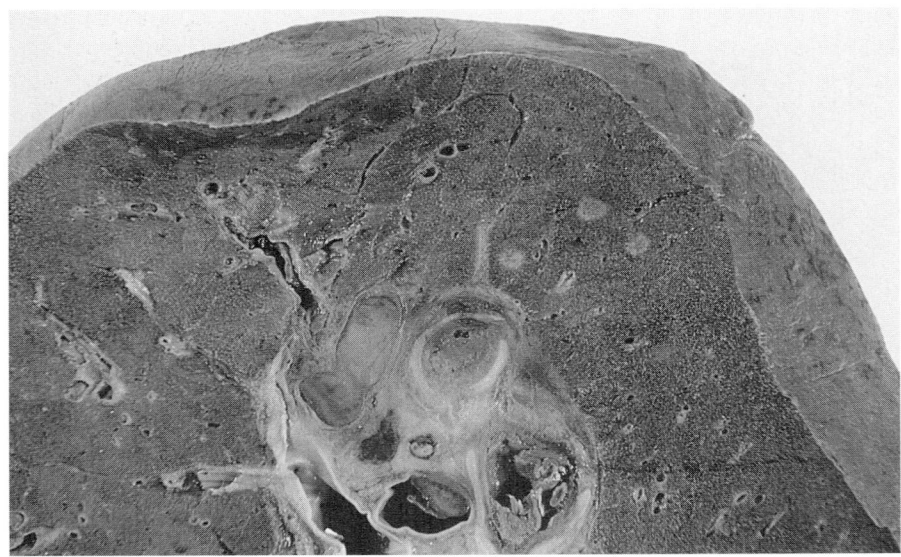

FIGURE 8-11
Tracheobronchial aspergillosis. Main bronchi are partially occluded by a mass of fungus.

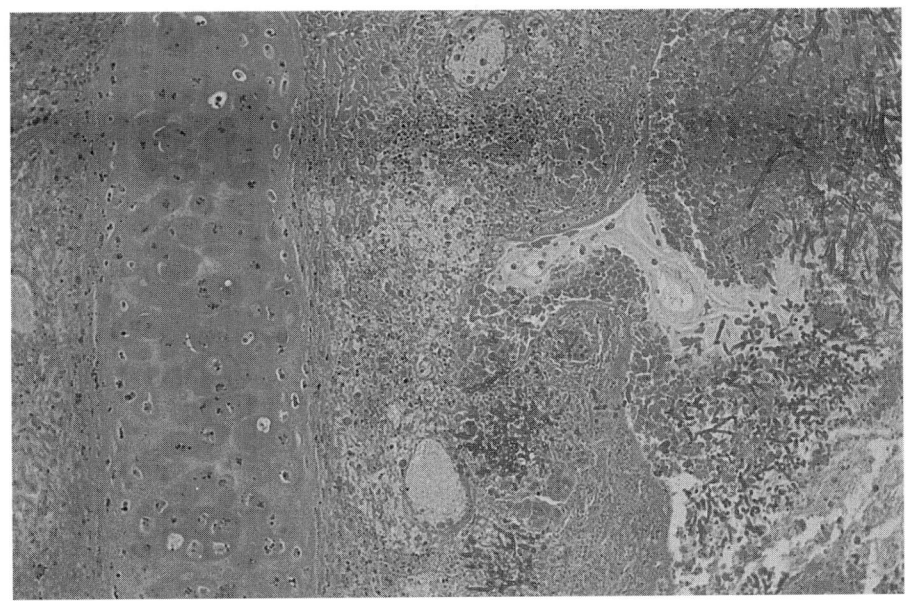

FIGURE 8-12
Tracheobronchial aspergillosis. The mucosa of the bronchus is invaded and replaced by *Aspergillus* hyphae.

Diagnosis of Aspergillus Infection

The presence of *Aspergillus* in cultures must be interpreted with caution, as these organisms may be present in material from the respiratory tract as saprophytes. Careful consideration should be given to the clinical situation, and a positive sputum culture from a patient with severe neutropenia is suggestive of the invasive form of the disease. Nasal cultures may also be useful in this situation.[14]

Many normal individuals have precipitating antibodies of IgG and IgE classes to *A. fumigatus*. Serum precipitins are present in 90 percent of cases of aspergilloma and 70 percent of patients with the allergic forms of bronchopulmonary aspergillosis. Skin tests with *A. fumigatus* antigen show an immediate (type I) reaction in asthmatic patients and an additional delayed (type III) reaction in patients with allergic bronchopulmonary aspergillosis. Patients with invasive disease often become anergic.

Several major antigenic and allergenic components of *Aspergillus* have been identified.[46] Three allergens have been characterized: Ag3, Ag7, and Ag13, which is identical to antigen C and possesses chymotryptic activity.[15] Monospecific antisera enable immunochemical characterization and measurement of antigen-specific IgG by means of an enzyme-linked immunosorbent assay (ELISA) technique.

PULMONARY ZYGOMYCOSIS (MUCORMYCOSIS)

Zygomycosis and *mucormycosis* are terms now used to describe the disease, previously known as phy-comycosis, caused by fungi of the order Mucorales. These are ubiquitous saprophytic fungi found in soil, fruit, decaying vegetable matter, and moldy bread. Human disease is most frequently caused by the genera *Rhizopus arrhizus*[16] and *Absidia corymbifera*. Less common causes include *Mucor, Conidiobolus, Cunninghamella*, and *Rhizomucor*. Mucorales have characteristic broad, nonseptate hyphae, branching randomly at right angles. They occur worldwide and may manifest themselves in several clinical forms. These include infections of the skin, lungs, gastrointestinal tract, paranasal sinuses, and brain as well as disseminated multiorgan disease.

Pathogenesis and Clinical Features

There are many predisposing factors to *zygomycosis*, but the disease is rarely seen in normal people. It more commonly occurs in patients with diabetes mellitus, leukemia, or lymphoma. Patients undergoing immunosuppressive or antibiotic therapy are susceptible, particularly transplant patients, and those with burns or severe neutropenia. The organism may reach the lungs by inhalation or via the bloodstream. Zygomycosis may present as pneumonia, abscess formation, pleural infection, or endobronchial mycosis.

Patients present with cough and fever, often with pleuritic chest pain. Sputum production is variable; when present, the sputum may be white, yellow, blood-stained, or grossly bloody. Patients are often profoundly ill with marked gas exchange abnormalities and rapidly advancing respiratory failure. Spread of the infection is hematogenous, and in such cases it is rapidly fatal.[17]

Pathology

The lung pathology of *zygomycosis* bears a close resemblance to pulmonary aspergillosis. In most cases, the fungus invades the bronchial wall, producing a surface film of exudate and hyphae with occasional sporing structures. The organism then penetrates the bronchial wall to invade pulmonary veins and arteries (Fig. 8-13). This, in turn, leads to thrombosis and infarction, producing the typical appearance of patchy hemorrhagic consolidation. Abscess formation and cavitation may follow, and erosion into a vessel may cause fatal hemorrhage. A pyogenic inflammatory response is usually elicited, with necrosis and abscess formation. If the tissue is infarcted, the response will be reduced. In more chronic lesions, granulomatous inflammation with multinucleate giant cells is a feature. An unusual form of zygomycosis is described by Mamlock and coworkers, in which the fungal hyphae are surrounded by eosinophilic sheaths; a form of Hoeppli-Splendore phenomenon.[18]

Endobronchial zygomycosis is the rarest form of the disease and usually presents in diabetics. Friable yellow tissue lines a main bronchus and protrudes into the lumen as a gelatinous plug. The patient may then develop subacute obstructive pneumonia distal to this. Mediastinal disease with superior vena caval obstruction is also recorded.[19]

The fungi may complicate bacterial inflammatory lesions to produce a mycetoma. Typically, this has a yellowish-white, concentrically layered appearance within a fibrous walled cavity.

Diagnosis

The presence of the fungus in cultures from the respiratory tract of an immunocompromised patient with clinical evidence of pulmonary infection is highly suggestive but not diagnostic of invasive zygomycosis.[20] Identification of fungi in biopsy material is based on the presence of the characteristic broad (5- to 50-μm) nonseptate hyphae with random right-angled branching (Fig. 8-14). Recently, spherical or ovoid chlamydoconidia have been described in pulmonary mucormycosis.[21] The chlamydoconidia are thick-walled, 15 to 30 μm in diameter, and attached to the sometimes septate hyphae. The same hyphae may have terminal bulbous swellings and lateral oval to spherical structures.

Unlike most filamentous fungi, the zygomycetes are best seen in tissue sections stained with hematoxylin and eosin (H&E). Iron hematoxylin demonstrates them particularly well. The usual fungal

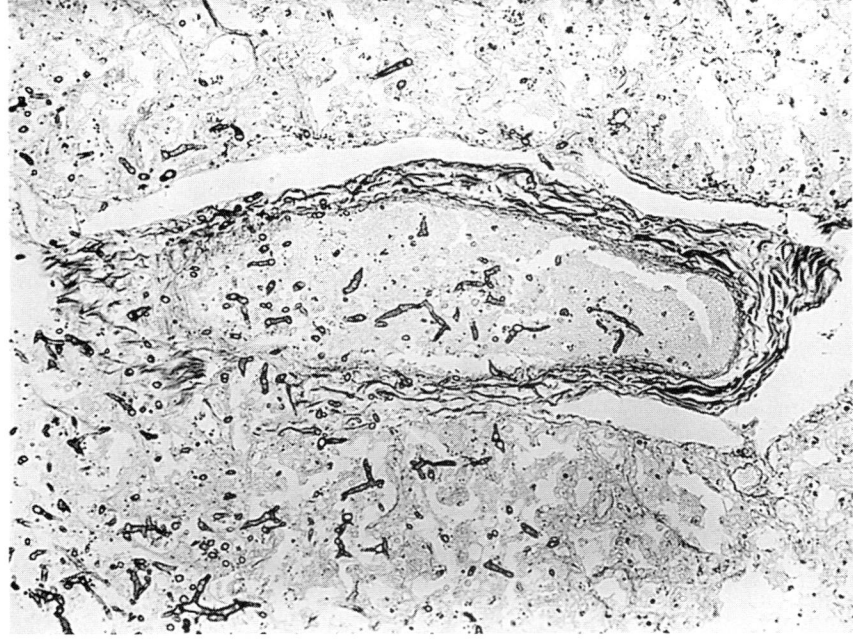

FIGURE 8-13
Zygomycosis. Fungal hyphae are invading the wall of a pulmonary vein (Grocott's stain).

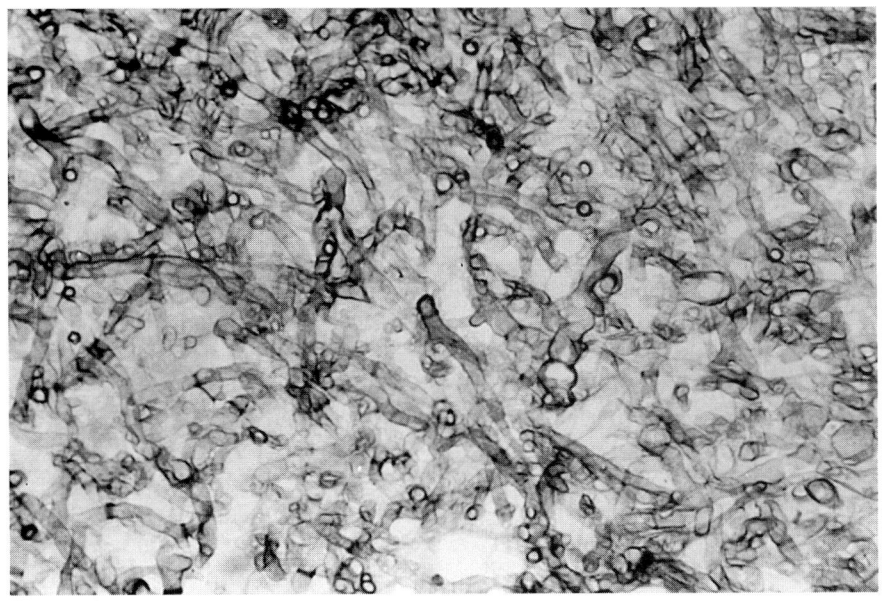

FIGURE 8-14
Zygomycosis. Hyphae vary in diameter, lack septa, and branch at right angles.

stains—Gomori's methenamine-silver (GMS) or periodic acid-Schiff (PAS)—are confirmatory but produce rather variable staining. Antizygomycete antibodies are available, with fluorescein or peroxidase conjugates for specific labeling of the fungi in the tissues. Alternatively, fluorescein-conjugated lectins can be used.[22]

The principal differential diagnosis is between zygomycosis and aspergillosis. The presence of septa does not rule out zygomycosis entirely, because these fungi occasionally produce septa. However, *Aspergillus* hyphae are typically narrower, with parallel walls, frequent septa, and dichotomous branching.

PULMONARY CANDIDIASIS

Candida, which are yeasts, are normal commensals of humans, being found on the skin, in the mouth and gastrointestinal tract, and in the female genital tract. More than 150 species exist in nature but only about 10 are regarded as important human pathogens, including *C. albicans, C. tropicalis, C. pseudotropicalis, C. guilliermondi, C. krusei, C. parapsilosis, C. stellatoidea*, and *C.* (now *Torulopsis*) *glabrata* (considered separately). The first two account for 80 percent of yeasts isolated from patients with invasive candidiasis. *Candida tropicalis* is found more commonly in leukemic patients.[23] *Candida parapsilosis* is usually associated with intravascular lines. *Candida albicans* can be found in 8.5 percent of all swabs and 10 percent of throat swabs taken from healthy

adults as well as 62 percent of sputum specimens taken from patients with pulmonary tuberculosis. Only repeated sputum cultures yielding heavy growth of these fungi suggest a true saprophytic or invasive lung lesion.

The natural habitat of this group of fungi is dead wood, leaves, decaying fruit, and unpasteurized milk. They are found in hospitals on the floors, in food, in air conditioning systems, and in respirators.[24]

Epidemiology and Pathogenesis

Intact skin and mucosal surfaces provide the first lines of defense against candidal infection. The normal defense mechanisms are extremely complex. Neutrophils and macrophages are able to ingest and kill the organisms, a mechanism that is dependent on myeloperoxidase, hydrogen peroxide, and the superoxide anion system. Delayed-type hypersensitivity to *Candida* antigens is present in 70 to 80 percent of healthy people, indicating a role for T lymphocytes. Humoral factors, IgG and complement, appear to have a role in opsonizing the organism.[25] Interruption of these normal defense mechanisms is necessary for *Candida* to become a pathogen. A survey carried out between 1980 and 1990 showed that nearly 80 percent of nosocomial pathogens were fungi, and *Candida* species accounted for 79 percent of these.[26] Thus *Candida* has become nearly as common a cause of hospital-acquired sepsis as the familiar bacterial pathogens.

Fungemia is associated with increased length of hospital stay.[27] Risk factors have been identified and include the presence of a central venous line or bladder catheter, use of multiple antibiotics, uremia, transfer between hospitals, diarrhea, and candiduria.[28] Other factors associated with candidal invasion are hemodialysis, prosthetic joints, laryngeal prostheses, vascular grafts, burns, intravenous drug abuse, and especially, among neonates, low birthweight.[29] The use of central venous catheters for prolonged parenteral feeding of preterm infants is thought to have accounted for an increase in fatalities among infants with systemic candidiasis. General factors include diabetes mellitus and immunosuppression of any cause.[24] Candidemia is common in AIDS patients and organ transplant recipients, particularly those undergoing marrow transplantation.

Candida albicans occurs in lung lesions in three forms: yeast forms or blastospores, pseudohyphae, and hyphae. All may coexist in the same lesion. The blastospores are oval and 3 to 5 μm in size. Hyphae represent the virulent phase of the organism and are associated with tissue invasion.[30] The pseudohyphal form consists of elongated oval cells joined together end to end, while the hyphae are slender, uniform in thickness, septate, branching, and may form blastospores.

Candida reach the lung either by aspiration or via the bloodstream. Pulmonary candidiasis can be divided into the bronchial type and candidal pneumonia.

Clinical and Pathologic Features

CANDIDA BRONCHITIS

This usually occurs in the very young and is relatively rare as compared with candidiasis in the pharynx and esophagus. Bronchial *Candida* are seen mainly in association with leukemia and chronic pulmonary infections, especially in cystic fibrosis. The larger bronchi may be covered with a gray-yellow membrane composed of fibrin, hyphae, and spores.

CANDIDAL PNEUMONIA

The clinical picture of candidal pneumonia is usually overshadowed by the underlying systemic disease. There may be evidence of septic shock at one end of the spectrum or a low-grade fever at the other end. In systemic disease, there may be skin lesions and endophthalmitis.[24] The fungus reaches the terminal airways either by inhalation or from spread through the walls of larger airways. The fungal hyphae may provoke little in the way of an inflammatory response, but there may be a polymorphic, histiocytic, or even giant cell reaction to blastospores in the small terminal airways. Blastospores can be identified in giant cells. There is edema and focal necrosis of alveolar walls, accompanied by hemorrhage and a variable amount of collapse. In some cases hyphae, pseudohyphae, and blastospores grow in the bronchioles and alveoli, causing edema and alveolar wall necrosis with minimal cellular reaction (Fig. 8-15). The fungus may also be seen within ves-

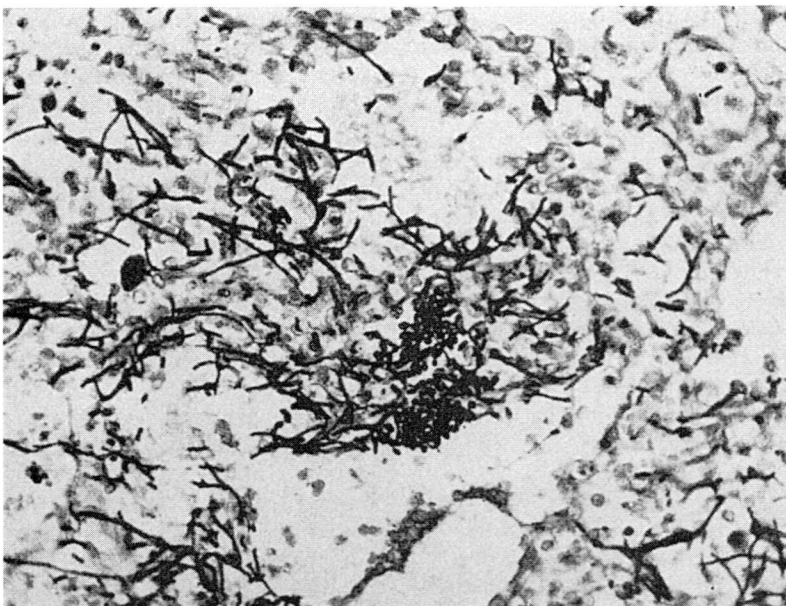

FIGURE 8-15
Candida pneumonia. Neonatal lung heavily colonized by *C. albicans* (PAS stain). A mixture of yeast forms and hyphal forms is present with little inflammatory response.

sels, infiltrating their walls (Fig. 8-16). Widespread dissemination of the fungus in the lung, whether or not a cellular response follows, indicates that infection occurred before death. In infants, the development of oral or pharyngeal thrush during the first 10 days of life is not uncommon, and this is the probable source of pulmonary infection, rather than inhalation of *Candida* from the vagina. A rare case of true congenital candidiasis has been described in a child delivered by cesarean section. The infection appears to have developed within apparently intact fetal membranes prior to delivery.[31] *Candida* may appear in adults in areas of pulmonary fibrosis. Such lesions are usually detected only by means of histology.

CANDIDAL FUNGEMIA

Dissemination of the fungus may occur as part of a fungal pneumonia. Systemic candidiasis has been reported complicating acute hepatic failure in patients treated with cimetidine, and it was suggested that in these cases the drug suppressed gastric acid secretion, allowing fungal colonization of the upper gastrointestinal tract.[32] A potential source of candidal pneumonia is embolization from an intravenous catheter.[33]

Diagnosis

Serologic tests have a role in diagnosis and in monitoring therapy for the disease. These include immunodiffusion, agglutination, and indirect fluorescent antibody tests. Since many noninfected individuals may have antibodies, the results should be interpreted with caution. A test for candidal cytoplasmic antigen enolase in serum has been described.[34] This appears to be specific, but the sensitivity for deep tissue infections was only 52 percent. In the future, ELISA may be useful.

In tissue sections, the presence of blastospores mixed with pseudohyphae or hyphae is characteristic. Cultural studies are required to identify the species. *Candida* are poorly stained in H&E sections, but they are intensely stained by PAS or silver stains. Blastospores alone may be confused with *H. capsulatum*, *B. dermatitides*, or poorly encapsulated *C. neoformans*.

Pulmonary Torulopsis

Torulopsis (Candida) glabrata is an opportunist, and pulmonary infection usually follows aspiration of infected gastric contents. This organism, unlike other *Candida* species, does not usually show mycelial forms. In H&E stained sections, the yeasts show variable staining and tend to be amphophilic. They have thin, refractile capsules. They are best seen in Grocott silver-stained preparations as black, round to oval spores, 1 to 3 μm in diameter, occurring in small clusters (Fig. 8-17). Inflammation may be minimal, or there may be an acute inflammatory response or a granulomatous reaction.[35] Pulmonary blood vessels may be invaded. *Torulopsis glabrata* can be distinguished from *H. capsulatum* by its larger size, more frequent budding, and predominantly extracellular situation.

CRYPTOCOCCOSIS (TORULOSIS, EUROPEAN BLASTOMYCOSIS)

Pulmonary cryptococcosis is caused by the encapsulated basidiomycetous yeast *Cryptococcus neoformans*. There are four serotypes (A, B, C, D), and most infections are caused by serotypes A and D, which are commonly found in bird excreta and contaminate soil throughout the world. Pigeons carry *C. neoformans* in their crops in large numbers and the yeast is most abundant in pigeon habitats. Serotypes B and C (*var. gatti*) have an association with *Eucalyptus camaldulensis* in Australia and elsewhere.[36] It has been suggested that dispersal of basidiospores occurs with the flowering of the plants in late spring or early summer, when airborne basidiospores are infectious. After dissemination, basidiospores synthesize capsular material and undergo transformation into encapsulated yeast cells.

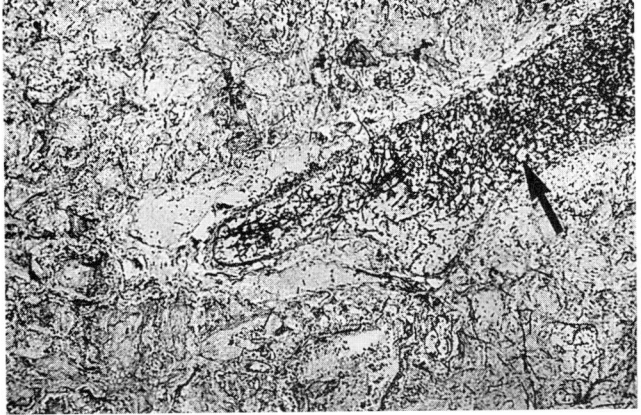

FIGURE 8-16
Candida pneumonia. A small pulmonary artery (*arrow*) contains thrombus with a mass of hyphae. Hyphae are invading into the surrounding infarcted lung (Grocott's stain).

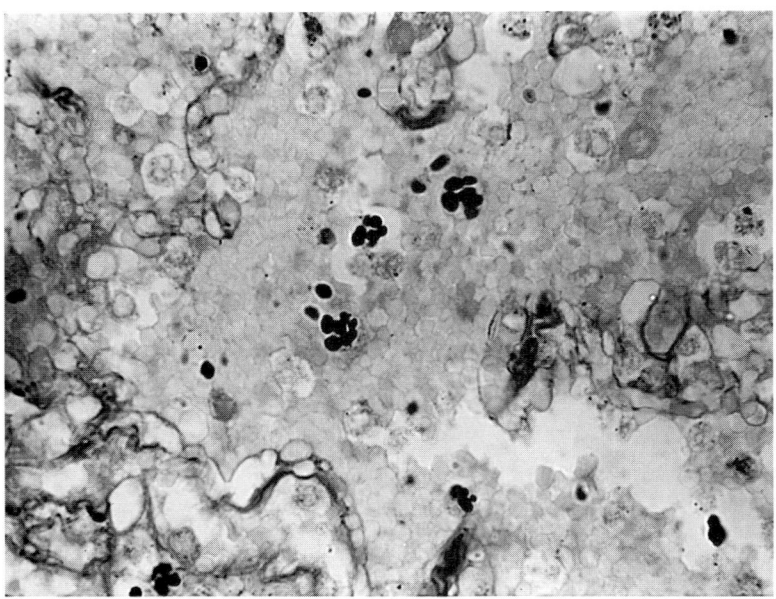

FIGURE 8-17
Pulmonary torulopsis. Clusters of *C. glabrata*
spores in the lung following aspiration pneumonia
(Grocott's stain).

Pathogenesis

The polysaccharide capsule plays an important part in determining the virulence of the fungus. Regulation of the encapsulation process may be adaptive for survival in the saprophytic environment, where encapsulation is suppressed in favor of growth. In the host, encapsulation allows the organism to resist phagocytosis by interfering with attachment of phagocytic cells. The capsular polysaccharide induces T suppressor cells, impairing both cell-mediated and antibody responses. Cell-mediated immunity is the major defense against cryptococcal infection.

Infection is worldwide and some cases are asymptomatic. Sporadic cases with no predisposing condition do occur, but in recent years the incidence of the disease has risen dramatically, reflecting increased numbers of immunocompromised patients, especially those with AIDS. Other high-risk groups include patients with leukemia, connective tissue disease, lymphoma, and sarcoidosis as well as transplant recipients and those on steroid therapy.

The primary portal of entry is the lung and the initial infection is invariably pulmonary, with subsequent dissemination. Meningitis is often the first evidence of infection, and other manifestations of the disease include cutaneous, mucocutaneous, osseous, and visceral forms. The predilection of the organism for the central nervous system remains unexplained but cerebrospinal fluid (CSF) acts as a good medium for growth of the organism and lacks immunoglobulins and complement.[37]

Clinical Features

Pulmonary disease often remains asymptomatic even when marked radiographic changes are present. There may be fever, chest discomfort, weight loss, dyspnea, and night sweats. Hemoptysis is rare, but a cough is present in nearly half the cases. Dissemination of the infection leads to central nervous system manifestations, such as headache, lethargy, and personality changes. Among 48 patients with AIDS and disseminated cryptococcosis, the organism was first isolated from the lungs in 12 patients.[38]

Radiologic findings include normal x-rays, nodular circumscribed infiltrates, pleural effusions, and lobar consolidation. Interstitial infiltrates usually indicate the presence of another opportunistic lung infection. The bronchial tree usually appears normal, but endobronchial abnormalities include circumferential narrowing, white thrushlike plaques, mucosal granularity. and red plaques.

Pathology

Primary complex disease resembles primary tuberculosis and produces small subpleural cryptococcal scars usually less than 1 cm in diameter.[39] These occur in any lobe and may contain minute caseous foci surrounded by dense fibrosis. The lesion is often subclinical and discovered as an incidental finding at postmortem. It represents healed or healing primary cryptococcal infection and may be associated with cryptococcal granulomatous infection in regional hilar lymph nodes, thus constituting a primary cryptococcal complex. The subpleural scars tend not to calcify. In fatal cases of cryptococcal meningitis, a careful search at postmortem will often reveal a dormant and apparently healed cryptococcal lesion. The tissue reaction to cryptococcus varies widely from a minimal inflammatory response to a granulomatous reaction with necrosis (Fig. 8-18). When the tissue reaction is least, large numbers of yeast cells tend to be present; this is particularly true of patients with AIDS. The yeasts displace the normal tissues and compress alveoli. In other cases, there is a mixed purulent and granulomatous reaction and the cryptococci tend to be less well encapsulated. Multinucleate foreign-body giant cells may contain large numbers of yeasts and are mixed with plasma cells and lymphocytes. Neutrophils tend not to be numerous. Necrosis may sometimes be conspicuous in fibrocaseous nodules, where the organisms may be difficult to identify.

CRYPTOCOCCOMA

Localized granulomatous lesions, known as cryptococcomas or torulomas, are seen radiographically as circumscribed, rounded shadows and are commonly mistaken for malignancy. They may be several centimeters in diameter and are usually subpleural, solitary, and nonencapsulated. When multiple, they may involve one lobe or both lungs. Central necrosis and cavitation may occur, but calcification is rare. The lesions are firm and well defined, with a white or gray cut surface that may appear gelatinous (Fig. 8-19).

Progression of the disease may lead to a form of pneumonia, with segmental or lobar consolidation and a gelatinous appearance. This occurred in over half the patients reviewed at Johns Hopkins Hospital in Baltimore.[40] The lesions heal with fibrosis and scarring (Fig. 8-20).

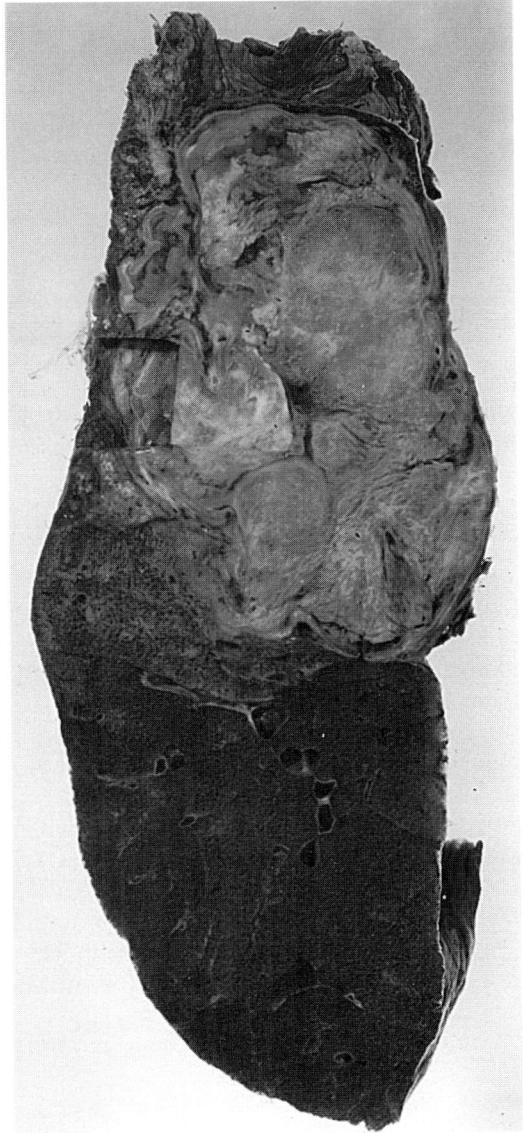

FIGURE 8-19
Pulmonary cryptococcoma. The irregular, well-defined mass consists of nodules of white tissue.

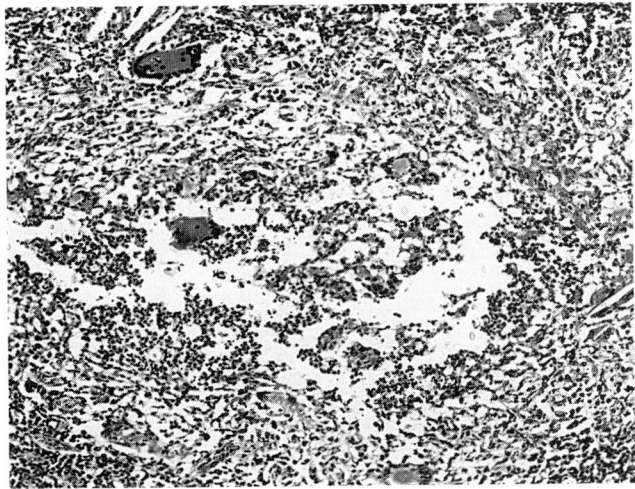

FIGURE 8-18.
Pulmonary cryptococcosis. A mixed reaction is present with central necrosis and surrounding granulomatous inflammation.

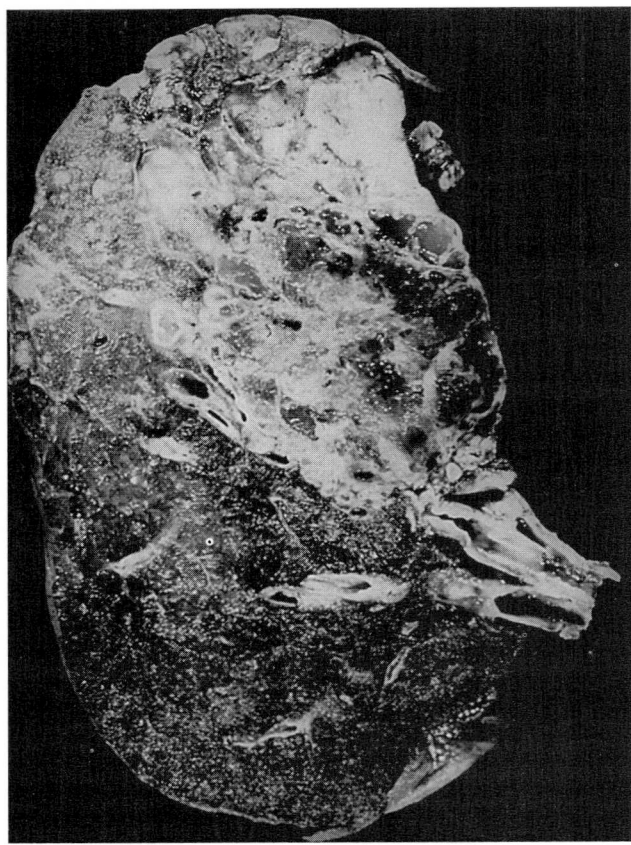

FIGURE 8-20
Pulmonary cryptococcosis. An extensive irregular area of fibrosis has resulted from cryptococcal pneumonia consolidation.

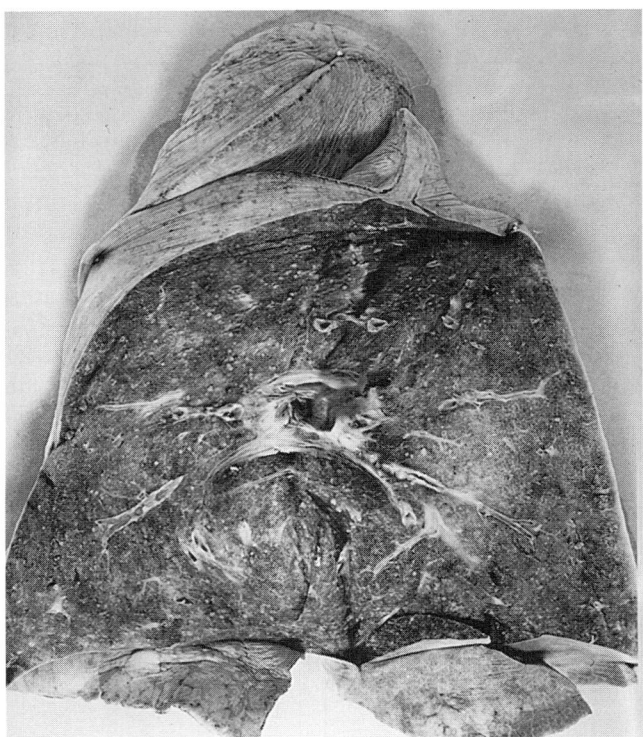

FIGURE 8-21
Disseminated pulmonary cryptococcosis. Multiple miliary granulomas resemble miliary tubercles.

DISSEMINATED PULMONARY CRYPTOCOCCOSIS

Disseminated pulmonary cryptococcosis is usually recognized after death in profoundly immunodeficient patients. It results in diffuse edema, necrosis, and hemorrhage, which fills both lungs and resembles lobar pneumonia in the stage of red hepatization with a gelatinous, slimy consistency. Numerous cryptococci are found throughout small airways, alveoli, and pulmonary blood vessels. Blood-borne infection can result in multiple translucent miliary nodules resembling miliary tubercles (Figs. 8-21 and 8-22). In addition, a larger, more chronic lesion is present, from which the infection originated. In a third variant of blood-borne infection, no lesions are visible macroscopically but general dissemination of cryptococci is identified in small pulmonary blood vessels on histology.

Diagnosis

The diagnosis of pulmonary cryptococcosis depends on demonstrating the organism in sputum or biopsy material. The identity of *Cryptococcus* should be confirmed by culture and specific fluorescent anti-

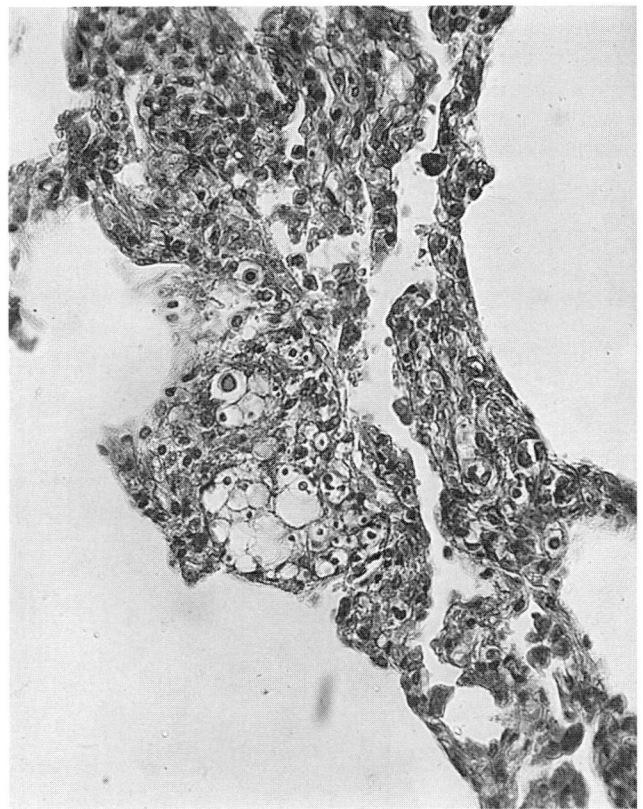

FIGURE 8-22
Disseminated pulmonary cryptococcosis. Yeasts are present within the interstitium of the lung (PAS stain).

body conjugates.[41] In H&E stained sections, the yeasts stain pinkish-blue with a diameter ranging from 2 to 20 μm. The shape varies from round or oval to crescent-shaped. The capsule appears as an unstained halo around the cell, 2 to 10 μm wide (Fig. 8-23). Mayer's mucicarmine stains the capsule bright red. Fungal stains, such as PAS and GMS, give greater detail and show single or multiple budding of the yeasts in active lesions. The diagnosis can be attempted on bronchial lavage material or brushings, but biopsy may be necessary.

In most larger mucoid solitary lesions, the cryptococci are encapsulated and readily identified. In small solid primary lesions, the organisms may be intracellular small, and poorly encapsulated ("dry" variants). They may be difficult to differentiate from other yeasts, notably *Histoplasma capsulatum*. Serologic tests are of diagnostic value, the most useful being the latex agglutination test for cryptococcal polysaccharide antigen.

HISTOPLASMOSIS

Two forms of histoplasmosis are recognized, the North American form, caused by *H. capsulatum*, an important cause of lung disease, and the African form, caused by *H. duboisii*, in which lung manifestations are unusual.

Histoplasma capsulatum is one of the dimorphic fungi, existing in its native habitat, the soil, in a mycelial phase but converting to a yeast phase at body temperature. Hyphae bear spores 2 to 6 μm in diameter, and these form the airborne infectious agent, reaching the alveoli or small airways when inhaled.

Histoplasmosis was first described in 1905 by Samuel Darling, who found parasites in the spleen, lung, and bone marrow of a Negro, an inhabitant of Martinique, who died following an undiagnosed febrile illness.[42] He named the organism *Histoplasma capsulatum*, believing it to be an encapsulated plasmodium. In 1934, DeMonbreun showed that the organism was a dimorphic fungus rather than a protozoan.[43]

Epidemiology

The fungus is soil-inhabiting and requires organic nitrates for growth. These are provided by bird and bat feces, and the disease may be seen in relation to enclosed spaces such as chicken coops, cellars, attics, and caves. Excavation of earth at construction sites is a common cause of the disease in urban dwellers; this may have been the cause of a recent epidemic that infected more than 100,000 people in Indianapolis.[44] Skin testing with histoplasmin has been used to map endemic areas in the United States and elsewhere. The largest endemic area is in the United States in the Mississippi and higher river valleys, where over 50 percent of adults show a positive skin test. In parts of the Mississippi Valley, virtually the entire population has been infected, but most people remain asymptomatic. Rates of 34 to 80 per-

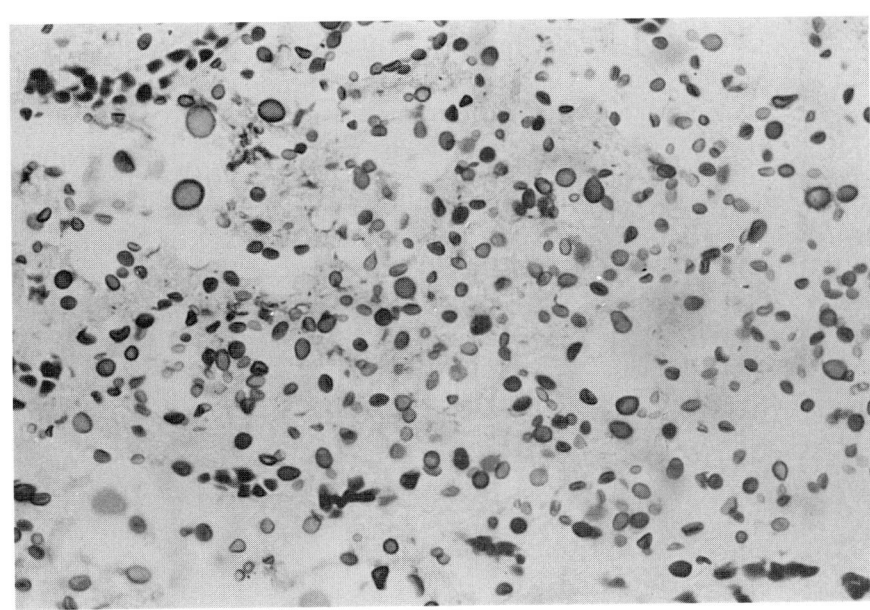

FIGURE 8-23
Cryptococcosis. The yeasts are round to oval and vary considerably in size. Some form a single bud. The capsule is seen as a clear space with the Grocott's stain.

cent have been recorded in parts of Burma, Vietnam, Indonesia, and Japan.[45,46] The disease is also found along the Amazon, Orinoco, and Paraguay river basins of South America; in Central and East Africa; in the Ganges River basin in India; in parts of Southeast Asia, especially along the major river valleys; and in Malaysia. Very few cases occur in Europe, but occasional cases have been reported from the United Kingdom. The localization of the disease to fertile river basins and lakes has been ascribed to periodic flooding, which disseminates the infective microconidia (chlamydospores) in the floodwater.

Pathogenesis

Inhalation of large numbers of spores of *H. capsulatum* can cause symptomatic disease in normal people; the fungus behaves as an opportunistic organism in immunocompromised patients.

Animal studies have been carried out in mice using intranasal inoculation of yeast forms.[47] After the first week, the focus of inflammation consists of neutrophils, lymphocytes, and macrophages, infiltrating small airways and adjacent alveoli. In nonimmune animals the organisms proliferate within macrophages, eventually killing them (Fig. 8-24). Infected macrophages transport the organisms to lymph nodes, spleen, and liver with dissemination of the disease. Between 9 and 15 days, cellular immunity develops, mediated by T lymphocytes. This results in tissue necrosis and caseation with granuloma formation. Intracellular fungi are killed by macrophages and healing occurs, with subsequent fibrosis, encapsulation, and calcification. Yeasts trapped in the fibrotic process remain as a source of antigen. In the immune animal, there is a more rapid cell-mediated host response with less tissue reaction. Acquired immunity and skin reactivity to histoplasmin both diminish with time.

Pulmonary Histoplasmosis

The clinical manifestations of histoplasmosis depend on the size of the inoculum and the immune status of the host. They have been classified by Goodwin and colleagues[48] as shown in Table 8-3.

Clinical Features

MILD EXPOSURE

The majority of people exposed to *H. capsulatum* are clinically unaffected. If the number of spores inhaled is small, the patient remains asymptomatic and fre-

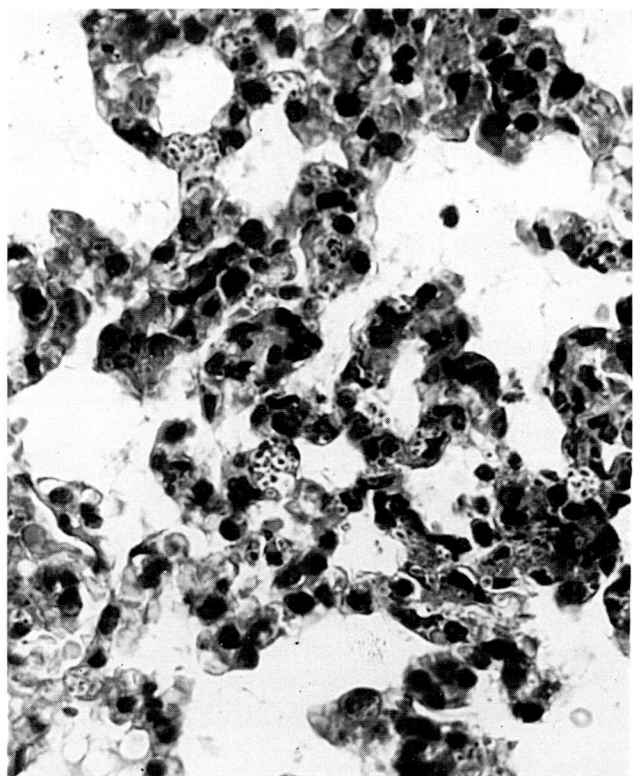

FIGURE 8-24
Histoplasmosis. An early stage of infection in an experimental animal shows the fungi within pulmonary macrophages.

quently the chest radiograph also remains normal, although the histoplasmin stain test becomes positive. However, focal pulmonary infiltrates that may be present will later calcify.

MODERATE TO HEAVY EXPOSURE

In nonimmune individuals, moderate exposure will cause symptomatic disease, with incubation periods between 10 and 16 days. Immune individuals usually require heavier exposure, have shorter incubation periods (3 to 5 days), and develop less severe disease. Symptoms are very variable but usually consist of an influenza-like illness with fever, myalgia, cough, and chest pain. Dyspnea may be a feature after heavy inhalation, and acute respiratory failure may rarely occur, with adult respiratory distress syndrome. Symptoms usually last from 5 to 10 days.

During the active phase of the disease, pulmonary infiltrates may be visible on chest radiograph, frequently with associated enlargement of hilar and mediastinal lymph nodes. Eventually both pulmonary lesions and lymph nodes calcify. If exposure has been heavy, infiltrates are more widespread and larger. They later calcify or ossify, producing multiple rounded, calcified nodules (Fig. 8-25).

TABLE 8-3
Clinical Manifestations of Histoplasmosis

Normal host	
a. Mild exposure (usual)	Asymptomatic primary
	Reinfection
b. Heavy exposure (unusual)	Acute histoplasmosis
c. Healed 1° infection with excessive fibrosis	
Lung	Histoplasmoma
Extrapulmonary	Mediastinal fibrosis
Abnormal host—opportunistic infection	
a. Immune defect	Disseminated histoplasmosis
b. Structural defect	Chronic pulmonary histoplasmosis

SOURCE: From Goodwin et al.,[48] with permission.

Pathologic Findings

The primary transient infection causes a combination of pulmonary and nodal changes that resemble the Ghon focus of pulmonary tuberculosis. The yeasts are taken up by the cells of the mononuclear phagocyte system, with a local tissue destruction

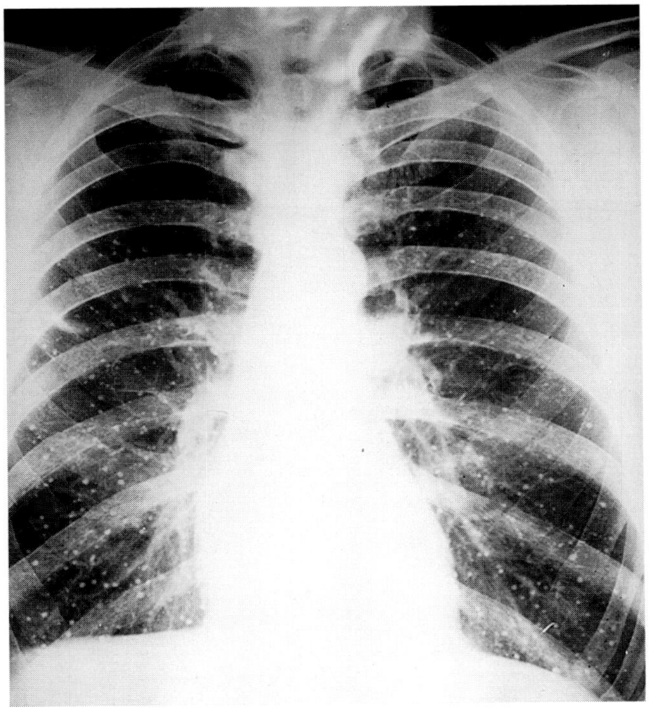

FIGURE 8-25
Histoplasmosis. Chest radiograph showing multiple healed calcified lesions due to previous infection.

similar to caseation. Primary lesions heal rapidly by fibrosis, with subsequent calcification. A similar sequence occurs in the regional lymph nodes, with enlargement, necrosis, and calcification. Enlarged hilar nodes may compress the right middle lobe bronchus, causing a middle lobe syndrome. Occasionally a focus of calcification, either in the lung or lymph node, may erode through the wall of a bronchus, and the softened contents are extruded into the lumen to form a broncholith.

EXCESSIVE FIBROSIS

Histoplasmoma
Exuberant fibrosis is sometimes a feature of healing lesions in the lung, lymph nodes, and mediastinum. Delayed healing of a primary focus of infection in the lung tends to occur in older patients, with greater tissue destruction. Continuing fibrosis over a number of years leads to concentric layers of fibrous tissue, forming a nodule several centimeters in diameter—a histoplasmoma (Fig. 8-26). Degenerate yeasts can usually be demonstrated in the necrotic areas (Fig. 8-27).

Mediastinal Fibrosis
Involvement of mediastinal lymph nodes by histoplasmosis may lead to necrosis and fusion into a single encapsulated mass, which causes pressure symptoms in adjacent structures. However, even around small infected nodes, exuberant fibrosis can occur and extend into tissues. Symptoms are due to compression of the superior vena cava, bronchi, or esophagus. A mass may be visible on chest radi-

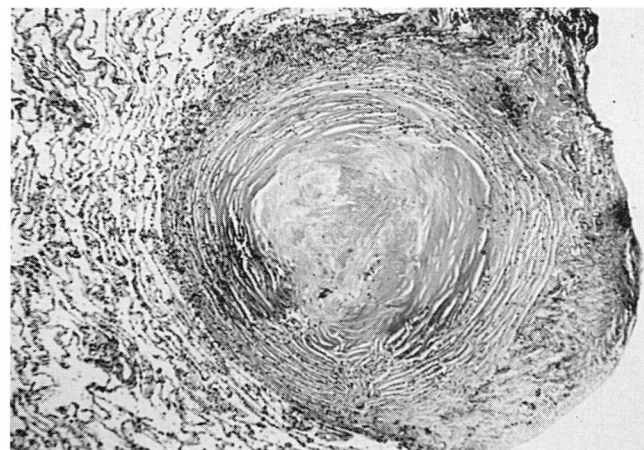

FIGURE 8-26
Pulmonary histoplasmoma. This is a slowly healing primary lesion with surrounding fibrosis, readily confused with a tuberculous lesion.

ographs, but CT scans are often required to localize the process. Biopsy may reveal only dense fibrous tissue. Necrosis and multinucleate giant cells are diagnostic.[49,50]

OPPORTUNISTIC INFECTION

With Abnormal Lung Structure

In patients with normal immune function, *H. capsulatum* may take advantage of abnormal lung structure to cause chronic disease. Inhalation of spores into emphysematous bullae causes chronic apical and subapical disease that may be mistaken for tuberculosis. The walls of the bullae become thickened and fibrotic and the pleural space becomes obliterated. Large numbers of organisms are identified in the walls of the cavities. Smaller rounded lesions occurring in the central zones are produced when the fungus colonizes centrilobular emphysematous spaces. They are fluid-filled and the organisms are sparse.

Infected spaces are associated with pneumonic lesions. These are segmental areas with foci of necrosis. Rarely, a primary lesion spreads to cause a granulomatous pneumonia, leading to widespread consolidation (Figs. 8-28 and 8-29). Large areas of the lung undergo necrosis with fibrosis and granulomatous inflammation where *H. capsulatum* is found (Figs. 8-30 and 8-31).

Disseminated Histoplasmosis

Disseminated disease occurs almost invariably in immunocompromised patients. From the time of pri-

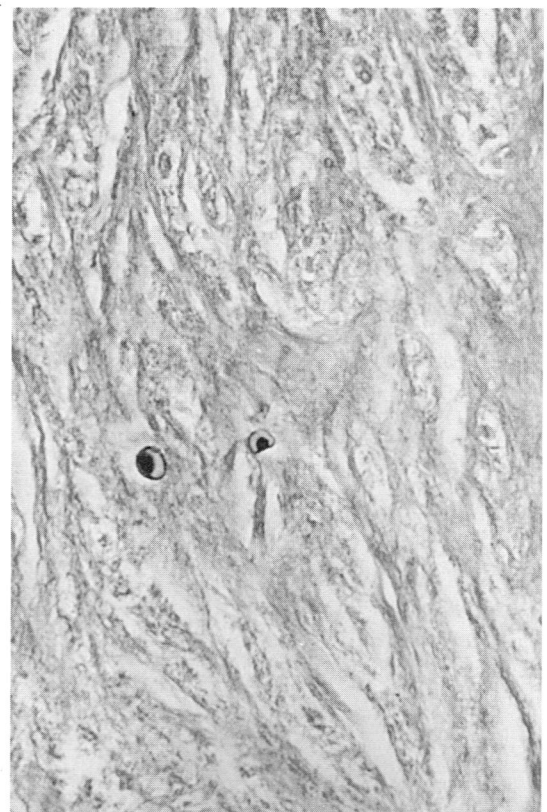

FIGURE 8-27
Pulmonary histoplasmoma. *Histoplasma capsulatum* is still recognizable in the lesion.

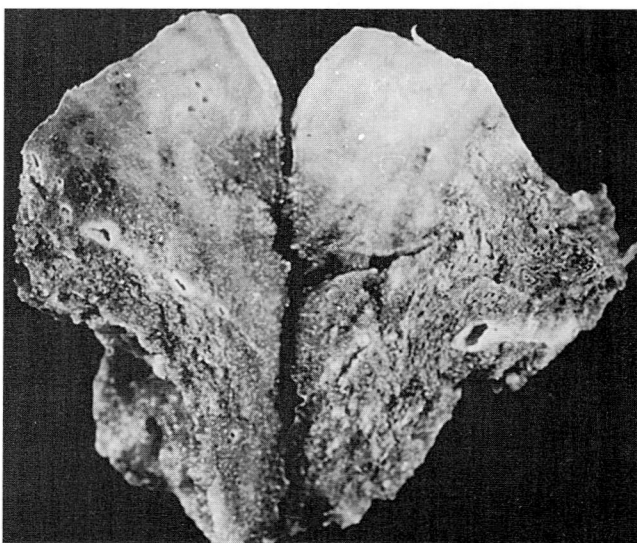

FIGURE 8-28
Chronic pulmonary histoplasmosis. Spread of a primary lesion has resulted in widespread consolidation.

276

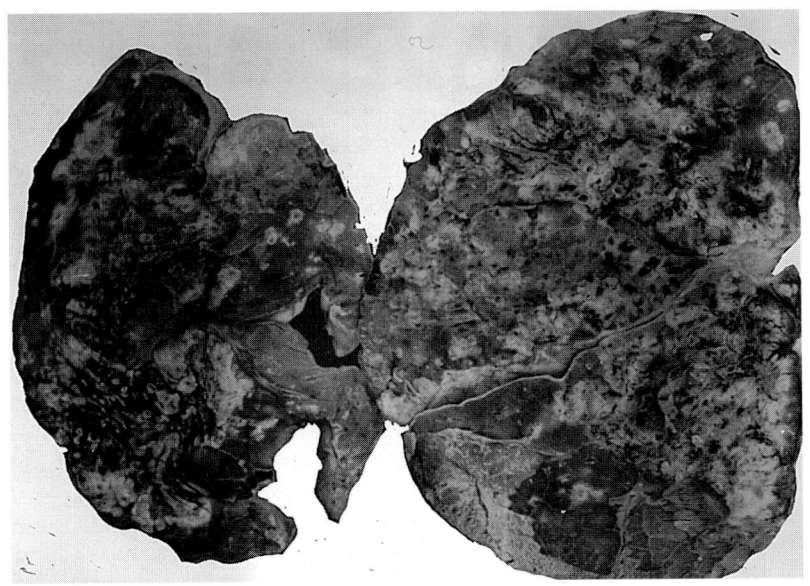

FIGURE 8-29
Pulmonary histoplasmosis. Lung showing a widespread bronchopneumonic form of the disease.

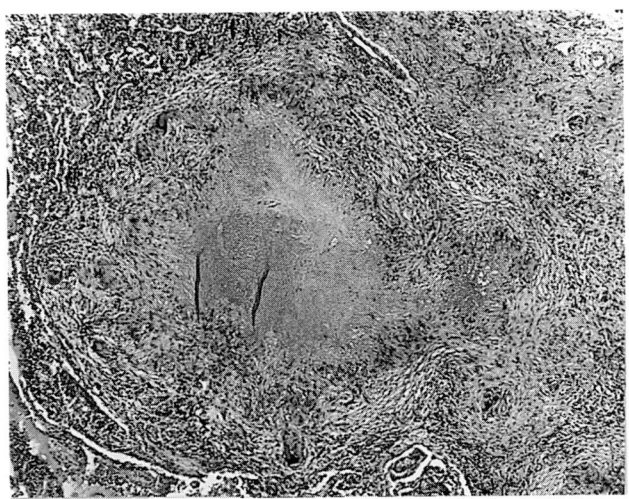

FIGURE 8-30
Histoplasmosis. This chronic granulomatous lesion with central necrosis resembles a tuberculous lesion.

mary infection, *H. capsulatum* remains dormant in cells of the mononuclear phagocyte system and is able to spread if cellular immunity is depressed. The syndrome is most often seen today in patients with AIDS, but previously it was a disease of young children, patients receiving corticosteroids or chemotherapy, or patients with malignant lymphoma. In addition to pulmonary involvement, the disease attacks the liver, spleen, central nervous system, gastrointestinal tract, adrenal glands, and other organs. Chest radiographs show reticulonodular or diffuse interstitial infiltrates.[51,52] The diagnosis is frequently established by bone marrow biopsy or culture. Lung biopsy or culture is positive in a minority of patients.[53]

Diagnosis

Skin testing with histoplasmin is helpful in determining previous exposure but not disease activity.

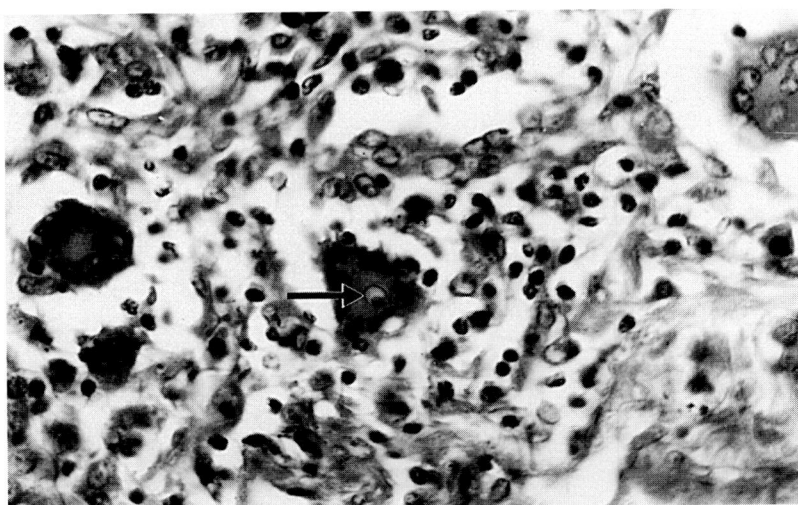

FIGURE 8-31
Histoplasmosis. In this granulomatous lesion, the fungus is present within the giant cell (PAS stain).

Direct examination of sputum is unlikely to reveal the organisms, but cultures are often positive. Complement fixation tests become positive 3 to 4 weeks after the onset of infection. Active disease is suggested by a fourfold rise in titers in convalescent-phase serum or in titers greater than 1 in 32. Immunodiffusion testing for antibodies to the H and M antigens (mycelial and yeast antigens) is specific but is negative in 50 percent of patients with acute disease. The test becomes positive 4 to 6 weeks after exposure; the M band stays positive for several years, whereas the H band disappears in 6 months. A radio-immunoassay has been developed. This becomes positive before the other tests and gives positive results in 80 percent of patients.[54,55] *Histoplasma capsulatum* polysaccharide antigen can be identified in bronchoalveolar lavage fluid, urine, and serum. Nuclear DNA probes to polymorphic restriction fragments from *H. capsulatum* may also provide a sensitive means of detecting the fungus.[56]

On histology, free organisms can be found in necrotic material. Others are seen within the cytoplasm of macrophages, which have abundant cytoplasm that appears bubbly due to the presence of the yeasts. These are round or oval and between 2 and 4 μm in diameter with an eccentric nucleus. They stain poorly with H&E and are best seen with GMS or PAS staining. Each is surrounded by a clear zone and they form single buds with narrow bases.

In immunocompetent patients, compact necrotizing epithelioid and giant cell granulomas are formed, similar to those seen in tuberculosis. In immunosuppressed patients, the granulomas are less well formed, consisting of loose aggregates of histiocytes and lymphocytes.

Histoplasma duboisii

Histoplasma duboisii is found only in those parts of Africa where the annual rainfall ranges from 40 to 80 in. Lesions are found predominantly in skin and bone and less commonly in the lung, gastrointestinal tract, lymph nodes, liver, and spleen. Descriptions of pulmonary disease are unusual. Clark and Greenwood described two patients and cited 10 cases from the literature.[57] The lesions consist of caseating granulomas and the yeasts are present in large numbers within the cytoplasm of histiocytes and giant cells (Fig. 8-32). *Histoplasma duboisii* yeast forms are round to oval and 8 to 15 μm in diameter. They have double-contoured walls and form buds with narrow bases. The distinction from *Blastomyces dermatitidis* may be difficult. *Blastomyces dermatitidis* cells are smaller, with broader-based buds, and are found within macrophages rather than multinucleate giant cells.

COCCIDIOIDOMYCOSIS

Coccidioidomycosis is caused by the dimorphic fungus *Coccidiodes immitis*. This exists in soil in the mycelial phase. The hyphae fragment into structures

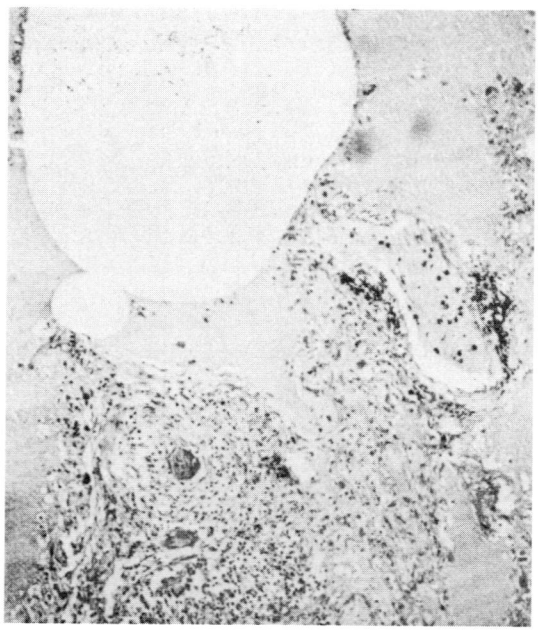

A

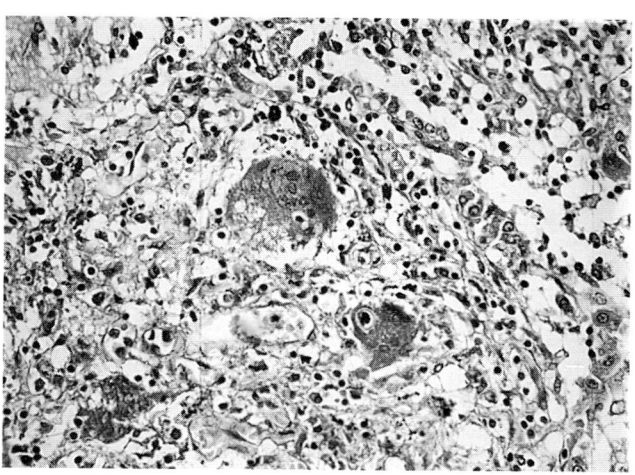

B

FIGURE 8-32

Histoplasma duboisii: *A.* The low-power view show a granulomatous lung lesion. *B. Histoplasma duboisii is* present within giant cells and histiocytes (PAS stain).

called arthroconidia, which become airborne, either to start the cycle again in the soil or to be inhaled by an animal host. In the host, the spores swell, become spherical, and develop a thick wall, forming a spherule. Cleavage occurs within the spherule to form as many as several hundred endospores. When the spherule bursts, endospores are released, and each can develop into a new spherule (the parasitic spherule-endospore cycle). If endospores are returned to the soil, they develop into hyphae and the cycle begins again.

Epidemiology

Coccidioidomycosis is endemic in certain areas of North, Central, and South America.[58] The endemic area stretches from the San Joaquin Valley in central California south to Mexico and east to central Texas. Climatic conditions are semiarid, with a short, intense rainy season, ideal for the mycelial growth phase. With the onset of dry weather, the arthroconidia break off and become airborne if the soil is disturbed. Outbreaks occur in the vicinity of construction sites and archeological digs. It is estimated that 100,000 people are infected annually in the United States. Sporadic cases are being increasingly recognized outside the endemic areas. These arise in travelers who have visited an endemic area or are exposed from fomites such as fruit and cotton. Infection is widespread in wild rodents, dogs, cattle, and sheep, but there is no evidence that animals spread the disease directly to humans. Accidental laboratory infections have been recorded, and great care should be exercised in the handling of cultures. Interhuman infection occurs only in very exceptional circumstances. Drainage of pus from a coccidioidal osteomyelitis sinus beneath a plaster cast has led to mycelial growth beneath the cast and shedding of arthrospores when the cast was opened.[59]

Primary Pulmonary Infection

CLINICAL FEATURES

About 60 percent of infections are asymptomatic. Symptoms of pulmonary disease usually develop between 1 and 3 weeks after inhalation of the arthroconidia. These symptoms include cough, fever, pleuritic chest pain, and myalgia. About half the patients have a fine erythematous rash and some develop erythema nodosum or erythema multiforme. The majority of patients suffer little or no inconvenience but—rarely—symptoms are severe, with acute respiratory insufficiency.[60] Chest radiographs show single or multiple patchy infiltrates that may coalesce, and hilar nodes may be prominent.

PATHOLOGIC FINDINGS

As the inhaled arthrospores develop into spherules, the initial host response involves the mobilization of macrophages and neutrophils, which is accompanied by edema.[61] The complement sequence is activated and T cells are stimulated, but the fungus, particularly the spherules, remains resistant to killing.[62] If lymph nodes are involved, the primary complex resembles that seen in tuberculosis or histoplasmosis. Caseation may occur, and most primary lesions heal at this early stage, leaving a small peripheral scar that may later calcify.

Chronic Pulmonary Infection

Coccidioidal spherules may lie dormant for many years in apparently healed primary lung lesions, and positive cultures from autopsy tissue have been obtained. Reactivation can occur years after the patient has left the endemic area. Reactivation also occasionally occurs in old age, particularly among American Indians, as immunity wanes. About 5 percent of those infected develop chronic lung disease, a pulmonary nodule, or a cavity.[63,64]

COCCIDIOIDOMA

Chronic granulomatous consolidation may result in one or more localized lesions known as coccidioidomas. These become rounded and usually measure between 0.5 and 3.5 cm in diameter. They undergo central caseous necrosis and sometimes calcify (Fig. 8-33). If a bronchus is eroded, cavitation results. Coccidioidomas are sharply demarcated from the surrounding lung by fibrous tissue, containing numerous chronic inflammatory cells, histiocytes, and giant cells. Microscopic satellite lesions are often present around the main lesion. Coccidioidomycosis is sometimes diagnosed only when a nodule is resected to exclude malignancy. Spherules are often difficult to find in chronic lesions, and when there is free access of air to the lining of a cavity, arthrospores may occasionally be present. Many cavities and granulomas contain unusual degenerate and mycelial forms of the fungus, which can be revealed only by fungal stains.

PROGRESSIVE PULMONARY DISEASE

In the other form of chronic pulmonary coccidioidomycosis, the primary infection persists or is reactivated. Diabetics and immunocompromised patients are particularly prone to develop progressive disease. The chest radiographs show persisting extensive infiltrates, sometimes associated with

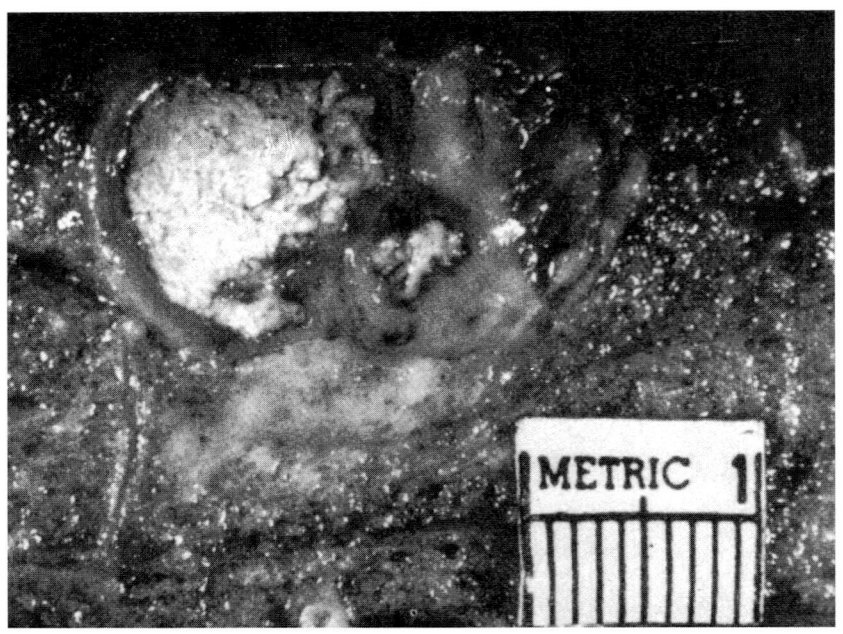

FIGURE 8-33
Coccidioidomycosis. A localized chronic coccidioidal granuloma (coccidioidoma) is present.

pleural effusions. Some patients develop acute pneumonia, and this may be fatal or progress to chronic disease. Acute pneumonia is associated with a polymorphonuclear leukocyte response with much necrosis and edema (Fig. 8-34). A blood eosinophilia greater than 20 percent and the presence of eosinophilic microabscesses carries a poor diagnosis.[65] Macroscopically, the lung resembles a spreading tuberculous bronchopneumonia with irregular spreading areas of gelatinous consolidation (Fig. 8-35). In surviving patients, this process results in extensive fibrosis, chronic granulomatous bronchi-

tis, bronchiectasis, and the development of abscess cavities, which persist after the pneumonia has subsided. Winn considered that the cavities resulted from necrosis of the pulmonary parenchyma and later distension with air followed bronchial obstruction due to chronic bronchial damage.[66] After resolution of the pneumonia, the cavities remain as permanent thin-walled cystic spaces lined by a fine layer of scar tissue (Fig. 8-36). Cavities, usually solitary, occur in 0.5 to 1.5 percent of patients. Unlike tuberculous cavities, they provide no source of infection, though hemorrhage may occur from their walls,

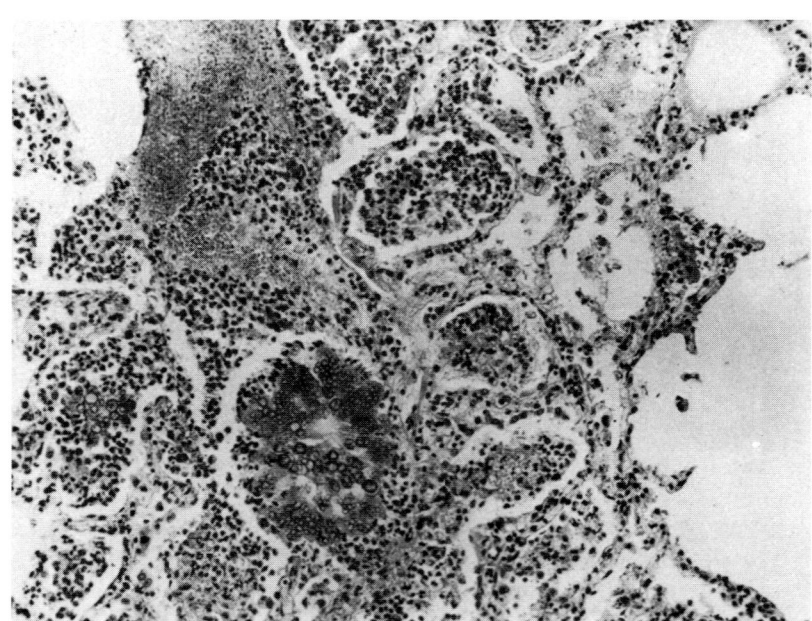

FIGURE 8-34
Acute coccidioidal bronchopneumonia. In the center, a spherule has burst, liberating numerous endospores. This has elicited an acute inflammatory reaction with fibrinous edema.

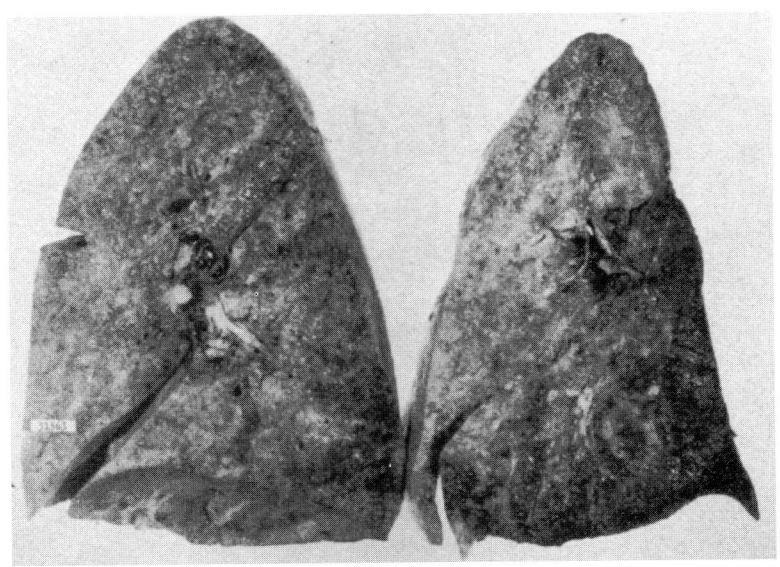

FIGURE 8-35
Pulmonary coccidioidomycosis. Diffuse broncho-pneumonic coccidioidal infection with extensive consolidation.

causing hemoptysis, and rupture into the pleural space may lead to pneumothorax and bronchopleural fistula. The cavities fluctuate in size due to air trapping and, very rarely, a fungus ball containing the mycelial form of *C. immitis* may form within the cavity.[67]

Disseminated Coccidioidomycosis

About 0.5 percent of patients develop disseminated disease with extrapulmonary spread to involve almost any organ of the body. Rarely, widespread rapid dissemination is fatal. Dissemination is more common in men, pregnant women, immunocompromised patients, and very young or elderly hosts. Dark-skinned races are more susceptible, particularly Filipinos and Mexicans.

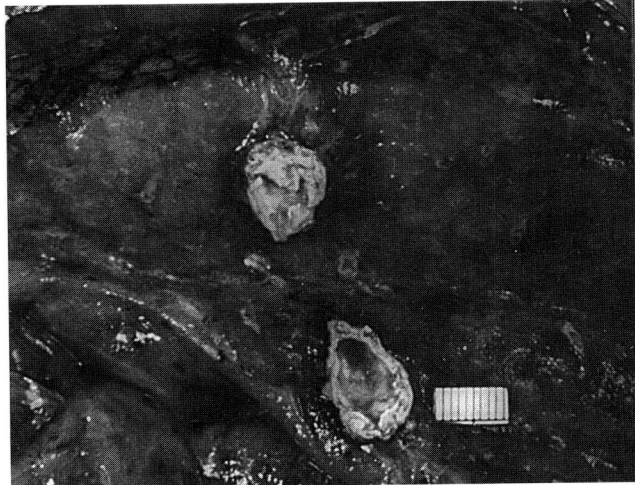

FIGURE 8-36
Chronic coccidioidomycosis. A chronic cavitated lesion has been divided and the two halves are shown.

Diagnosis

Coccidioides immitis can be identified in sputum or other fluids by examination of wet-mount preparations with addition of 10% potassium hydroxide. The spherules are relatively large and their walls appear doubly retractile if the light is reduced.[68] Papanicolaou-stained smears have been shown to be superior to potassium hydroxide (KOH) preparations.[69] Culture of the fungus from sputum and specimens obtained by bronchoscopy is more likely to be helpful in the evaluation of pulmonary infiltrates than of nodules. It should be borne in mind, however, that cultures are hazardous for laboratory staff.

Skin testing with coccidioidin or spherulin is specific and is useful for epidemiologic investigations or to determine the patient's immune status. Skin tests become positive within 4 weeks of the primary infection and usually remain so for life. While conversion is helpful in diagnosing recent infection, disease activity cannot be assessed.

Complement-fixation tests measure IgG and are useful in diagnosis, becoming positive in 90 percent of patients by the eighth week of the illness and lasting for 6 to 8 months. High or steadily rising titers indicate dissemination of the disease. Serum IgM precipitins can be demonstrated by tube precipitation, latex agglutination, or immunodiffusion methods. They appear within 3 weeks of primary infection and disappear within 4 months. Persistence may indicate progression or reactivation.

The histologic recognition of coccidioidal infection rests on the identification of the spherule stage in biopsy or autopsy material. Spherules are between 20 and 200 μm in diameter and have a thick, doubly refractile wall. They may be present as small, immature and maturing spherules, mature spherules containing endospores, or collapsed spherules without

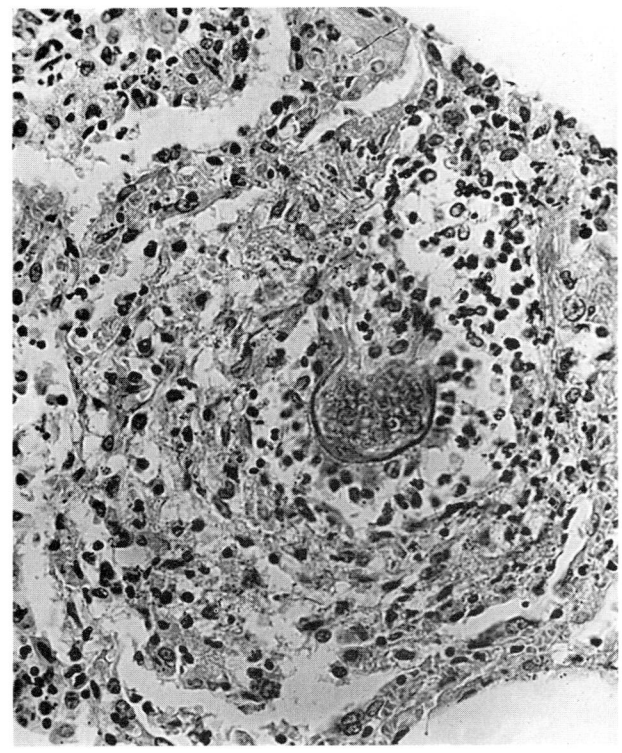

FIGURE 8-37
Coccidioidomycosis. A ruptured spherule of *C. immitis* is seen centrally with an acute inflammatory response.

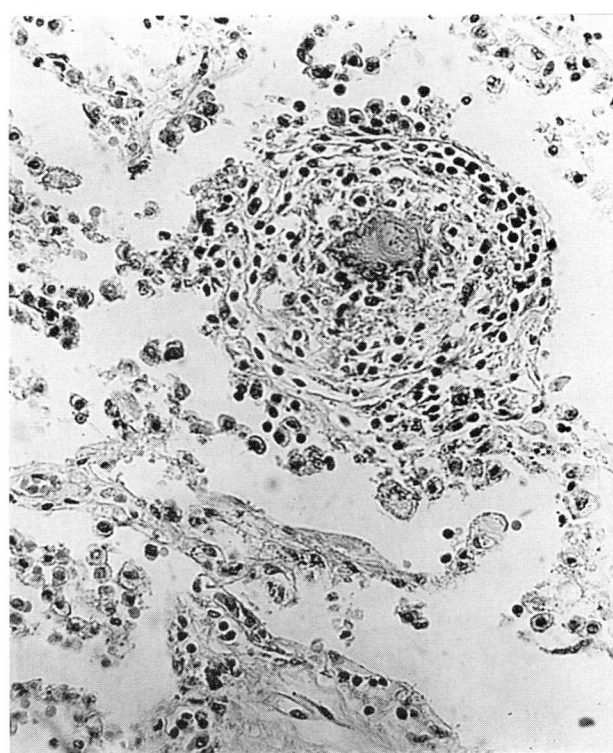

FIGURE 8-38
Coccidioidomycosis. A miliary granuloma is present with a young spherule in a giant cell.

endospores (Fig. 8-37). As the spherule enlarges and forms endospores, the wall becomes thinner. Routine H&E staining is often adequate, as spherules and endospores are basophilic. Endospores and spherules are stained by silver techniques, whereas the PAS stain stains endospores but not the walls of spherules. The inflammatory response is typically mixed and pyogenic. A localized acute inflammatory response is seen when spherules rupture and release endospores. As spherules mature, they are typically surrounded by epithelioid cells and Langhans' cells or foreign-body giant cells and may lie within the cytoplasm of the giant cells (Fig. 8-38). Occasionally spherules may become coated with eosinophilic clubbing material and form an asteroid or actinomycetoid form (Fig. 8-39). Empty spherules of *C.*

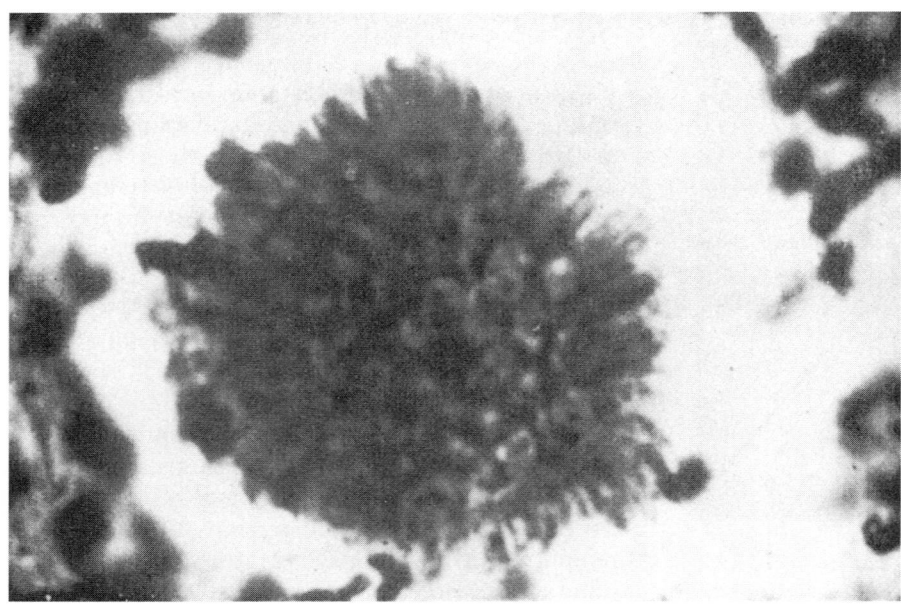

FIGURE 8-39
Coccidioidomycosis. The asteroid form of *C. immitis* resembles an *Actinomyces* colony.

immitis may be mistaken for *B. dermatitidis*; it is best to examine multiple sections to demonstrate both spherules and endospores. Free endospores may also be confused with *H. capsulatum* or *C. neoformans*.

PULMONARY BLASTOMYCOSIS (NORTH AMERICAN BLASTOMYCOSIS)

Like *H. capsulatum* and *C. immitis, Blastomyces dermatitidis* is a dimorphic soil-dwelling fungus, growing as a mycelial form at room temperature and converting to a yeast form at 37°C. Conidiophores arise from the hyphae and produce single terminal conidia, which are thought to be infectious for humans and animals, particularly dogs, when the mycelia are disturbed.

Epidemiology

Lack of a specific skin test and the difficulty in isolating the fungus in the environment has made epidemiologic studies difficult. The disease is found predominantly in North America in the states surrounding the Mississippi and Ohio rivers, with sporadic cases in humans and dogs as well as clusters and epidemics. It also occurs in midwestern states, the Canadian provinces bordering on the Great Lakes, and along the St Lawrence River.[70,71] Cases are also reported from Venezuela, Mexico, Australia, Israel, and the Middle East.[72] A number of cases have been reported from Africa, and it has been suggested that the disease originated there and was carried by the slave trade to America.[73]

Most attempts to trace the sources of epidemics have been unsuccessful. However, Klein and associates were able to isolate the organism from soil in the vicinity of a beaver lodge, suggesting that it requires moist, acid soil with decaying vegetable material and animal manure.[74]

Clinical Features

The lung is the probable portal of entry for blastomycosis, even in those cases where it manifests itself elsewhere. Primary cutaneous infections have been described in immunosuppressed patients,[75] or resulting from laboratory or autopsy accidents and dog bites. The incubation period is uncertain. In one outbreak, a mean of 45 days was established, but other estimates range from 21 to 106 days.

Infection may be inapparent or there may be an acute systemic illness with pyrexia, cough, arthralgia, and myalgia. The cough is nonproductive at first, but mucopurulent sputum follows, sometimes with hemoptysis. The chest radiograph is very variable, with appearances ranging from one or more pleural based, rounded opacities to lobar consolidation. Mass lesions that mimic carcinoma are not uncommon. Cavitation and calcification are less common than in histoplasmosis or tuberculosis. Pleural thickening, pleural effusions, and pneumothorax may occur. Hematogenous spread may lead to miliary disease, with acute respiratory failure.[76] Other patients may develop diffuse consolidation due to endobronchial spread of the infection.[77,78]

In some patients, the disease is self-limiting. In others, resolution of the pulmonary symptoms is followed by extrapulmonary disease. Skin disease occurs in 40 to 80 percent of cases. Other sites include bones, joints, the genitourinary tract, and the central nervous system.

Pathology Of Pulmonary Lesions

Following inhalation of conidiophores, which lodge predominantly in the lower lobes, budding yeast forms develop. From the alveoli, these pass into the interstitium, causing an acute inflammatory response. This is followed by necrosis and an influx of eosinophils, giant cells, lymphocytes, and fibroblasts. The granulomatous response is rarely as conspicuous as in histoplasmosis or tuberculosis. The characteristic feature of the disease is the mixture of pyogenic and granulomatous inflammation forming a suppurating granuloma (Fig. 8-40).[79] The yeasts may be found in clusters in severe infections but usually occur singly and may be contained within giant cells. They may be very scanty in chronic lesions. Extensive tracheal involvement may be seen with numerous white, streaky, nodular, and vesicular lesions. Bronchial lesions are common and cause mucosal destruction with spread into the surrounding lung. Bronchial stenosis may occur.[80] Pulmonary lesions vary in extent from small granulomatous nodules to extensive necrotic areas containing multiple abscesses and involving much of the lung. Small lesions may eventually heal and calcify. In chronic infections, cavity formation may occur; in severe cases, a dense fibrocaseous mass may replace parts or the whole of a lobe (Fig. 8-41).

Blastomycotic infection of the regional lymph nodes seldom results in the same degree of caseous destruction as is seen in tuberculosis. Hilar nodes are therefore not often involved and primary complex lesions are rare. In widespread disease, the pleura is invaded, resulting in extension of the disease to the chest wall. Amyloid has been reported as a late com-

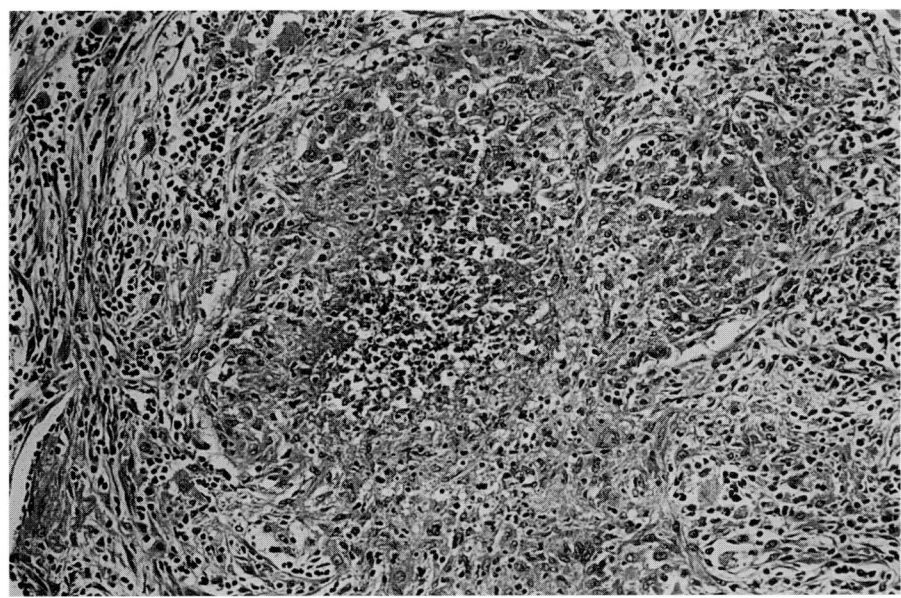

FIGURE 8-40
Blastomycosis. The characteristic lesion is a combination of granulomatous inflammation and an acute inflammatory response, producing a suppurative granuloma.

plication. Arrest and healing of the primary infection may be followed by later recrudescence.

Diagnosis

Blastomyces dermatitidis can be cultured from appropriate material such as sputum, induced sputum, or specimens obtained by bronchoscopy. When skin lesions are present, they should be cultured; prostatic fluid also provides suitable material.

Examination of freshly obtained material after digestion with 10% KOH also has a high yield and produces rapid results in experienced hands. Material stained by the Papanicolaou technique may also provide the diagnosis.[70] Skin testing using blastomycin is unreliable because of variations in sensitivity and cross-reactivity with histoplasmin and coccidioidin. Complement-fixation tests are also insensitive and lack specificity. Immunodiffusion tests have become more specific since A and B antigens have been identified for blastomycosis.[81] Radioimmunoassay and enzyme immunoassay have both been used to provide rapid screening tests and show good sensitivity,[82] but they are no more specific than complement fixation testing. Lymphocyte transformation in the presence of a blastomyces alkaline and water-soluble antigen (B-ASWS) is useful as an indicator of cell-mediated immunity, being positive in 81 percent of cases tested. The specificity of this test is undetermined and it is not helpful in the very early stages of the disease.

In tissue sections, *B. dermatitidis* appears as round to oval yeasts with thick, sharply defined, refractile walls (Fig. 8-42). The cells are usually between 6 and 15 μm in diameter, but they may show

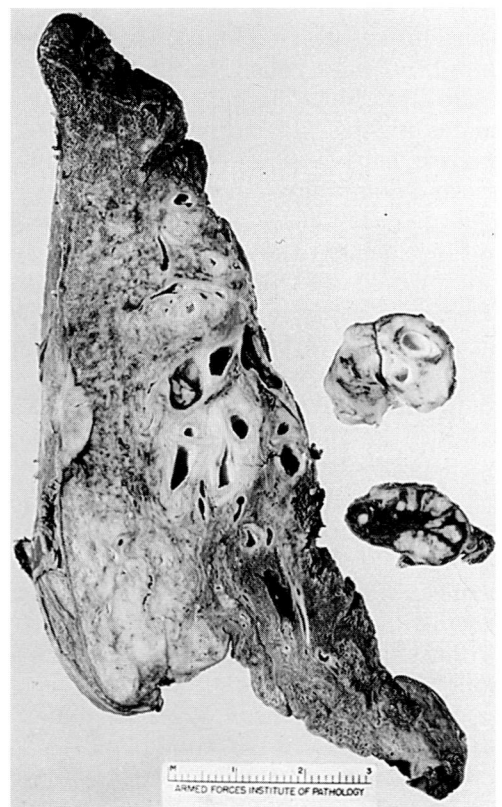

FIGURE 8-41
Blastomycosis. The lung shows widespread fibrocaseous changes. Infection has spread to hilar nodes, causing enlargement.

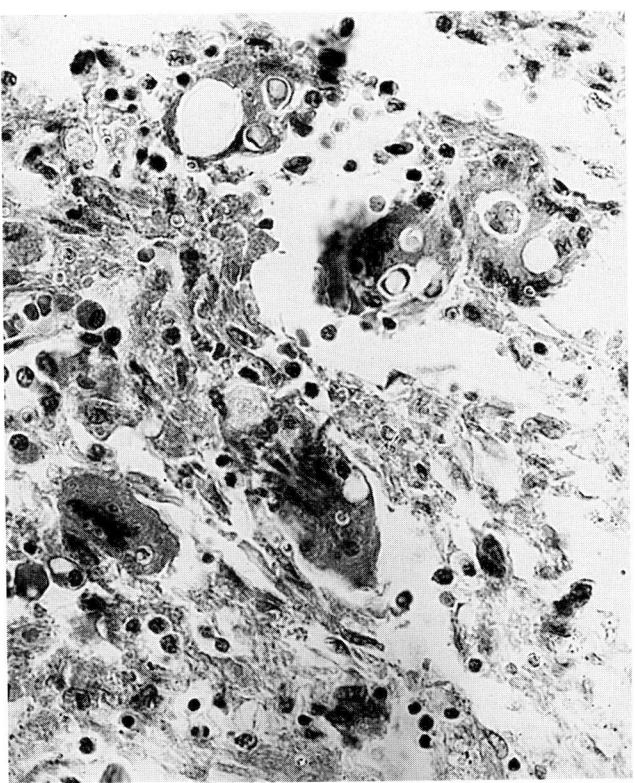

FIGURE 8-42
Blastomycosis. Chronic granulomatous inflammation with giant cells containing *B. dermatitidis*. Small forms are lying free outside the cells.

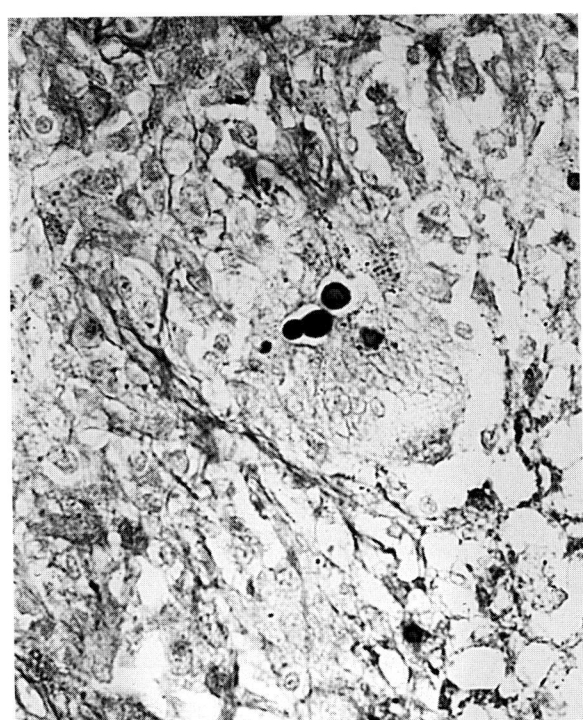

FIGURE 8-43
Blastomycosis. The yeast forms single broad-based buds, sometimes as large as the parent cell (Grocott's stain).

much greater variation, from small forms 2 to 4 μm diameter to giant forms 40 μm in diameter.[83] In H&E stains, the protoplasm is basophilic and separated from the wall by a clear space. Several nuclei may be visible. The yeasts reproduce by single budding and the bud is characteristically attached by a broad base (Fig. 8-43). Fungal stains are essential to demonstrate the organisms if they are sparse. Mucicarmine may be useful to distinguish *B. dermatitidis* from lightly encapsulated forms of *C. neoformans*. Empty single cells of *B. dermatitidis* may be mistaken for the empty spherules of *C. immitis*. *Paracoccidioides brasiliensis* may be mistaken for *B. dermatitidis* if only single buds are seen. Small intracellular forms of *B. dermatitidis* may be confused with *H. capsulatum*, but the two fungi can be distinguished by the morphology of budding forms.

PARACOCCIDIOIDOMYCOSIS (SOUTH AMERICAN BLASTOMYCOSIS)

South American blastomycosis was first described by Lutz in 1908, who also recognized that the causative organism differed from *C. immitis*.[84] *Paracoccidioides brasiliensis* is a dimorphic fungus. At 37°C in

tissues and in cultures, it forms oval to round yeasts with multiple buds. At lower temperatures, it grows as a mycelium and forms arthroconidia.

Epidemiology

Paracoccidioidomycosis has a very restricted geographic distribution, being confined to Central and South America. Brazil is the center of the endemic area, which is characterized by forests with high temperatures and humidity. Sporadic cases have been reported in North America, Europe, and elsewhere in patients who had previously resided in endemic areas. The natural habitat of *P. brasiliensis*, which is difficult to isolate, is thought to be the soil. The route of infection has not been determined, although inhalation seems more likely than direct inoculation. No animal reservoir has been identified, but the disease has been reported in armadillos.

The disease is rare in children and young people; most patients are over 30 years of age. It affects males far more often than females except in prepubertal children, and this may be related to an inhibitory action by estrogens on the transition from mycelium to yeast forms.[85] Agricultural workers are particularly vulnerable to the disease.

Clinical Features

Symptoms of paracoccidioidomycosis often develop many years after exposure. This long latent period may be due to the fungus lying dormant in lymph nodes.[86]

Generally, the severity of the clinical disease is related to host immunity. In people with a normal immune system, it may be asymptomatic. If cell-mediated immunity becomes depressed, the disease manifests itself. Immunosuppressed patients, including those with AIDS, may develop severe disease.[87]

The lungs are regarded as the primary site of infection, and most patients have respiratory symptoms. However, a small number of infections probably remain subclinical. Calcified lung nodules containing *P. brasiliensis* have been described as an incidental finding at autopsy.[88]

PROGRESSIVE PULMONARY DISEASE

In patients with progressive lung disease, the symptoms are dyspnea, cough, hemoptysis, and pyrexia. Chest radiographs show patchy or confluent nodular infiltrates, often bilateral and symmetrical, in the mid- or basal lung fields.

PATHOLOGY OF PULMONARY LESIONS

The pulmonary pathology has been described by Salfelder and associates.[89] Macroscopically the lesions resemble blastomycosis and tuberculosis,

with which the disease may coexist. Lung scarring leaves a coarse hobnail appearance (Fig. 8-44). Cavities may result from the breakdown of caseous lesions, but calcification is said to be unusual. Extensive fibrosis may result in honeycomb change with cor pulmonale (Figs. 8-45 and 8-46). Small airways become extensively damaged and macroscopically resemble areas of tuberculous caseation. Involvement of larger airways causes ulceration, with resulting bronchiectasis. Localized caseating lesions with surrounding fibrosis are called *paracoccidioidomas*. Pleural involvement is common and results in dense fibrous adhesions, and less commonly, pleural effusions. Hilar lymph nodes are involved in 50 percent of cases. Other pulmonary complications include pneumothorax and pulmonary arterial thrombosis.

A mixed pyogenic and granulomatous inflammatory response, similar to that seen in blastomycosis, is typical of the disease. Neutrophils may predominate in some areas, whereas others contain poorly defined epithelioid cells and giant cell granulomas (Fig. 8-47). Numerous organisms are usually present within giant cells, in the abscesses, or scattered throughout the tissue. In progressively enlarging lesions, necrosis occurs, with large numbers of free organisms, and this is accompanied by fibrosis (Fig. 8-48).

A more rapidly spreading form of lung infection results in a widespread confluent granulomatous and

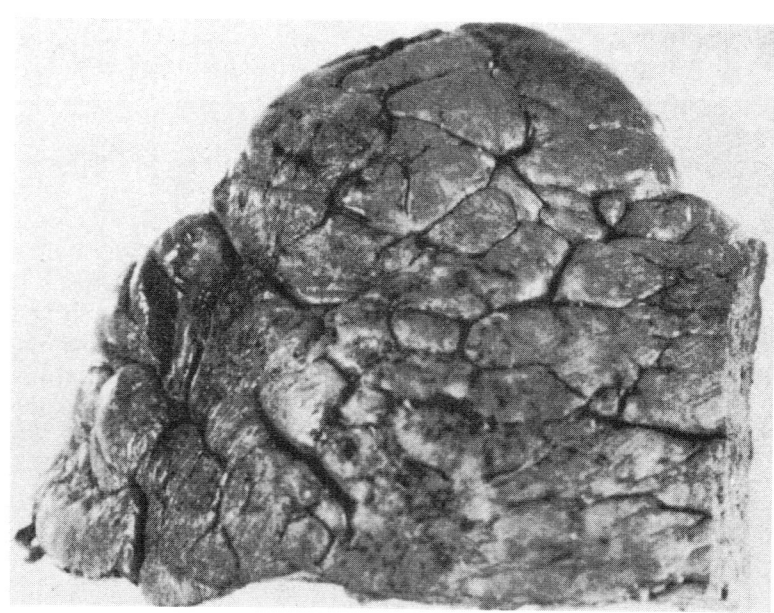

FIGURE 8-44
Paracoccidioidomycosis. Chronic infection may lead to extensive scarring, with nodularity of the pleural surface.

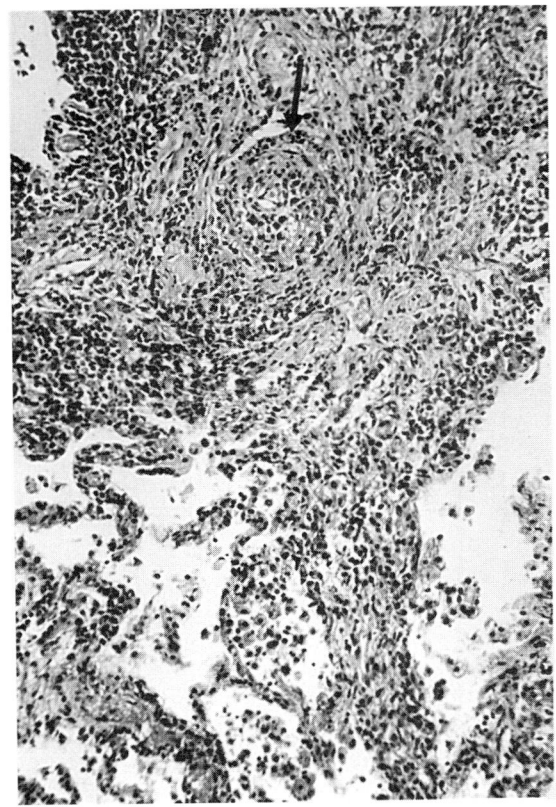

FIGURE 8-45
Paracoccidioidomycosis. Chronic granulomatous inflammation has resulted in almost total obliteration of a small pulmonary artery (*arrow*). Such changes contribute to pulmonary hypertension.

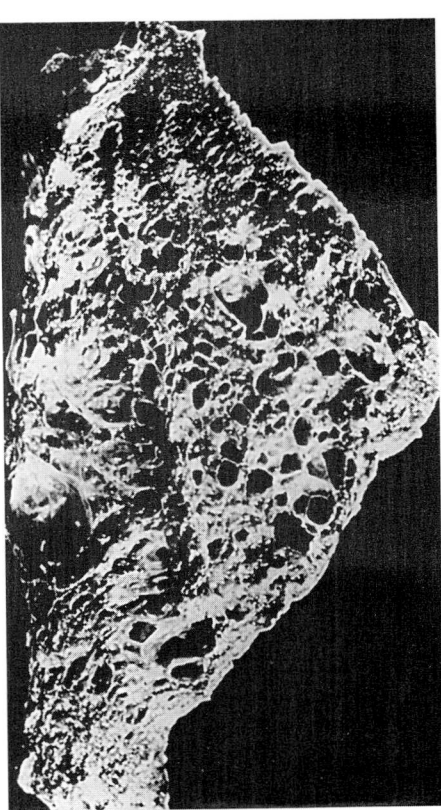

FIGURE 8-46
Paracoccidioidomycosis. Chronic disease has resulted in fibrosis and remodeling, resulting in a honeycomb pattern.

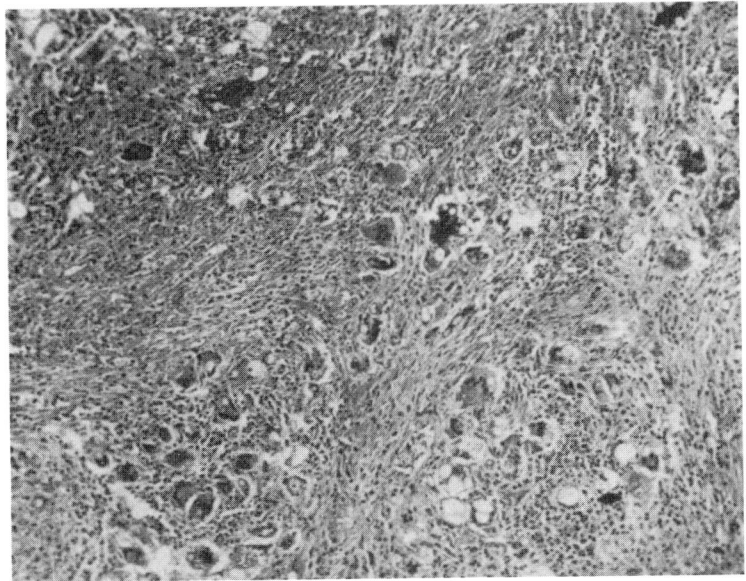

FIGURE 8-47
Paracoccidioidomycosis. An area of chronic inflammation in the lung with confluent, poorly defined giant cell granulomas.

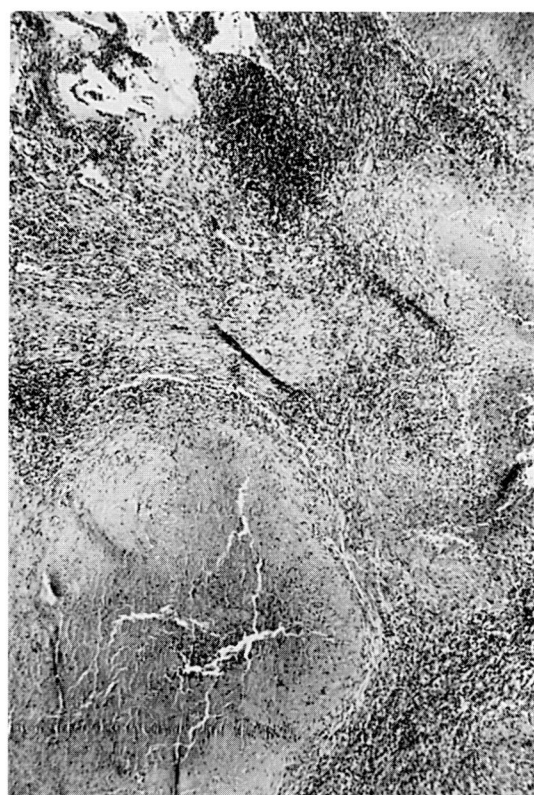

FIGURE 8-48
Paracoccodioidomycosis. Necrotizing granulomatous inflammation. The fungi are often found in the outer layers of the caseous material.

fibrinous intraalveolar exudate with microabscesses called *paracoccidioidal pneumonia*. When the lungs are involved in a blood-borne infection, there are numerous miliary granulomas, which contain fungi.

Extrapulmonary Manifestations

MUCOCUTANEOUS-LYMPHANGITIC PARACOCCIDIOIDOMYCOSIS

It has been suggested that the mucocutaneous form of the disease might be the result of direct inoculation. The present view is that it is more likely to be the result of dissemination from a pulmonary infection. Oral, nasal, and anal mucosa is most often involved and spread to regional nodes causes lymphadenopathy.

SYSTEMIC DISEASE

Dissemination from a primary lung infection may involve multiple organs including the liver, spleen, gastrointestinal tract, adrenals, bones, and central nervous system.

Diagnosis

Skin testing with paracoccidioidin is not reliable, as some patients with active infection are nonreactive.

Crossreaction with histoplasmin and blastomycin also occurs.

Circulating antibodies can be detected in 95 percent of cases by agar gel immunodiffusion; the presence of precipitation bands 1 and 2 is specific. The complement fixation test is helpful in judging the activity of the disease. An ELISA technique is now also available.

Culture is not straightforward but should be attempted in suspected cases. The diagnosis is most readily made by demonstrating the budding yeast forms in sputum or lavage material.[90] If this is warmed with 5% KOH, the yeasts appear as doubly refractile structures. Cell-block preparations or smears stained with GMS are also sensitive techniques.

In tissue sections, the organisms are round to oval and 5 to 60 μm in diameter. They have thin, refractile walls. The contents may be basophilic or amphophilic with H&E stains and tend to retract from the wall. They reproduce by multiple buds, attached to the parent by narrow necks. Buds may be roughly equal or may vary in size (Fig. 8-49). Hyphae

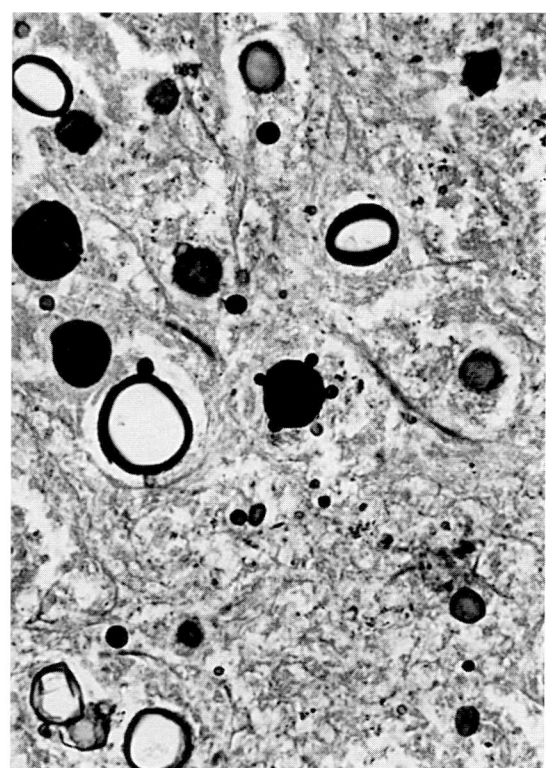

FIGURE 8-49
Paracoccidioidomycosis. Grocott's stain shows the cell wall of the fungus. Multiple buds are formed from the parent cell (gemmation).

are rarely formed in tissues. In long-standing fibro-granulomatous lesions, the yeasts may be fragmented, distorted, and unevenly stained by fungal stains. They may be calcified and appear as empty shells or rings, when they may be confused with the empty spherules of *C. immitis*. *Blastomyces dermatitides* occasionally forms multiple buds and must therefore be distinguished from *P. brasiliensis*.

PULMONARY SPOROTRICHOSIS

Sporothrix schenckii is a dimorphic fungus, growing at room temperature as a mycelial form and developing at 37°C in the tissues into a yeast form. The organism is found in the soil, wild and domesticated plants, timber, moss, straw, and other organic matter. Skin infection usually occurs by accidental inoculation of the fungus. This produces a nodular or ulcerative lesion at the site of trauma, usually in the distal extremity, associated with regional lymphadenitis. Systemic *sporotrichosis* may be localized to a single organ system such as the joints or lower respiratory tract, or it may disseminate to involve multiple organs, in some cases from a primary pulmonary focus.[91] The vast majority of pulmonary cases are due to primary infection of the lung resulting from inhalation of the conidia of *S. schenckii*.[92]

Clinical Features

The disease occurs mainly in middle-aged male alcoholics. There is often preexisting pulmonary disease, such as emphysema or bronchiectasis, and the presenting complaints are nonspecific. They include hemoptysis, fever, chills, cough, chest pain, dyspnea, malaise, and weight loss. The radiologic appearances are nonspecific. There may be chronic cavitary disease, resembling tuberculosis and histoplasmosis. Other patients develop solitary peripheral pulmonary nodules, and one case is described where reticulonodular infiltrates developed and then cleared before death.[93] Fluid levels within the cavities are infrequent and fungus balls have not been documented.

Pathologic Features

Macroscopic findings in the series of England and Hochholzer are based on tissue from two wedge resections, three lobectomies, and three autopsies.[92] There were cystic, cavitating lesions, ranging from 0.5 to 6 cm in diameter. The larger cysts had thin walls containing nodular gray to yellow tissue peripherally and old and hemorrhagic necrotic tissue

in the center of the cavity. The cavities communicated with bronchi. Two cases had well circumscribed, solitary, peripheral, encapsulated nodules that were yellow-tan with necrotic centers and had no airway connections. The same authors described a case with bilateral, diffuse granulomatous nodules.[93]

Cavities are lined by chronic inflammatory granulation tissue with giant cells of foreign-body type (Fig. 8-50). Hilar lymph nodes may be sufficiently enlarged to cause bronchial obstruction. Satellite lesions surround small pulmonary vessels and consist of macrophages, lymphocytes, plasma cells, and a few eosinophils without central necrosis. Less frequently, the pulmonary architecture is effaced by numerous, often confluent, large granulomas containing central areas of infarct-like or liquefactive necrosis. The margins of the granulomas are demarcated by palisaded epithelioid cells intermixed with varying numbers of multinucleate Langhans' giant cells. Eosinophils are infrequent at the periphery and there is a lymphoplasmacytic infiltrate. Smaller granulomas, both necrotizing and nonnecrotizing, are seen around bronchi and bronchioles. A similar histology is seen in some of the regional lymph nodes. The yeast forms of *S. schenckii* may be difficult to find in pulmonary lesions. They are round, oval or elongated, "cigar-shaped," single or budding cells with a diameter of 2 to 6 μm. Budding yeasts are relatively uncommon and budding may be multiple. Branching hyphae may occasionally be found.[94]

H&E staining rarely demonstrates the fungi and the PAS stain is the method of choice, showing brilliant red-purple yeasts. The organisms are seen within granulomas, giant cells, and macrophages, but free organisms are also present in areas of necrosis. Their morphology may not be sufficiently distinctive to allow differentiation from other yeasts such as *Candida*, capsule-deficient variants of *C. neoformans*, or *Histoplasma*. The organism may be detected on sputum cytology, where macrophages contain small intracellular eosinophilic yeasts with a faint halo. Extracellular yeast forms may also be found.[95] By electron microscopy, England and Hochholzer demonstrated that a capsule was present, and others have noted similar features.[93,96] The organisms may be surrounded by eosinophilic Hoeppli-Splendore material as a stellate, radial corona, the so-called asteroid body (Fig. 8-51).[93]

PSEUDALLESCHERIA BOYDII

This fungus is surrounded by a confusion of names, also being known as *Allescheria boydii* or *Petriel-*

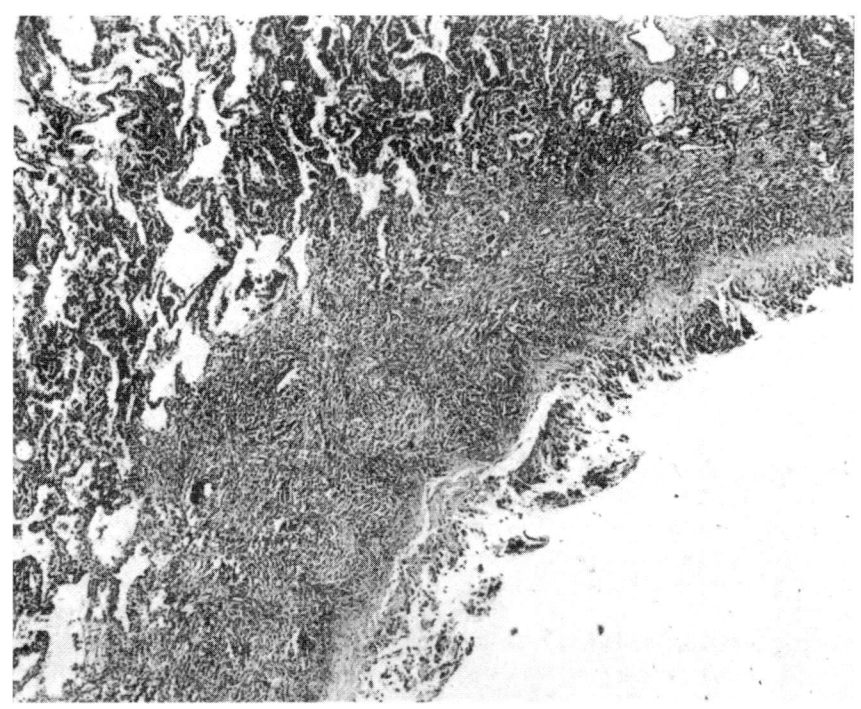

FIGURE 8-50
Sporotrichosis. Section of chronic abscess cavity in the lung. *Sporothrix schenckii* may be very difficult to identify in the tissue.

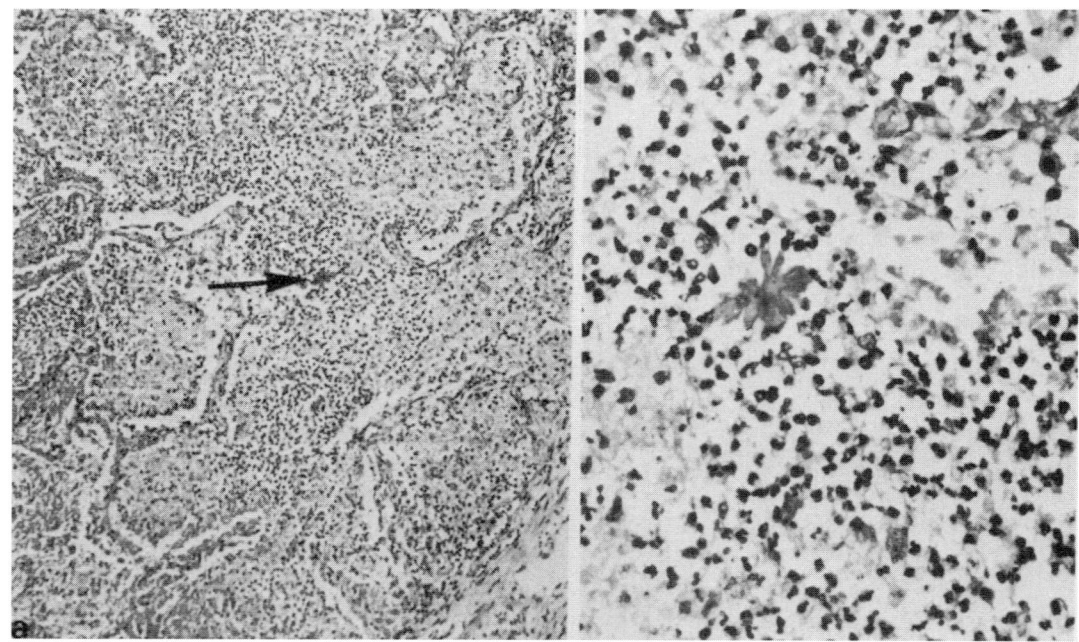

FIGURE 8-51
Sporotrichosis. An acute terminal infection with a widespread acute inflammatory response and abscess containing occasional asteroid forms.

lidium boydii in the perfect state and *Monosporium apiosporium* or *Sedosporium apiosporium* in the imperfect state. *Pseudallescheria boydii* is readily cultured to form a white mold. It has septate hyphae with oval or club-shaped conidia. In nature, it is found in soil, decaying organic matter, and polluted water and sewage. The most frequent form of infection is due to penetrating injury with subsequent mycetoma formation.

Pulmonary Infection

Most patients with pulmonary *pseudallescheriosis* have some form of preexisting chronic lung disease with cysts, cavities, or ectatic airways. Saprophytic colonization by *P. boydii* results in fungus ball formation, similar to that caused by *Aspergillus*, and infection is due to inhalation of spores. Invasive disease occurs in immunosuppressed patients, and this may result in abscess formation in the lung.[97–99]

Macroscopically there are fibrosed cavitating lesions, usually in the upper lobes and sometimes bilateral. These may contain fungus balls, which are yellow-black in color, and the whole colony of the fungus appears to the naked eye as a yellow grain.

Histologically, the chronically inflamed cavity walls may be partly epithelialized or consist of granulation tissue. The cavity contains a tangled mass of branching, sparsely septate hyphae with numerous chlamydospores situated near the periphery (Figs. 8-52 and 8-53). These should be distinguished from *Aspergillus* species. There is a considerable polymorphonuclear leukocyte response and the infection can spread, giving a honeycomb appearance to the lung. Systemic spread has been noted to involve the brain, thyroid, and kidneys.[99] The fungi are well stained with PAS or methenamine silver and show branching septate hyphae with terminal or lateral conidia. The conidia measure 3 to 7 by 6 to 10 μm in size and usually occur singly. Hyphae are 1 to 5 μm in width.

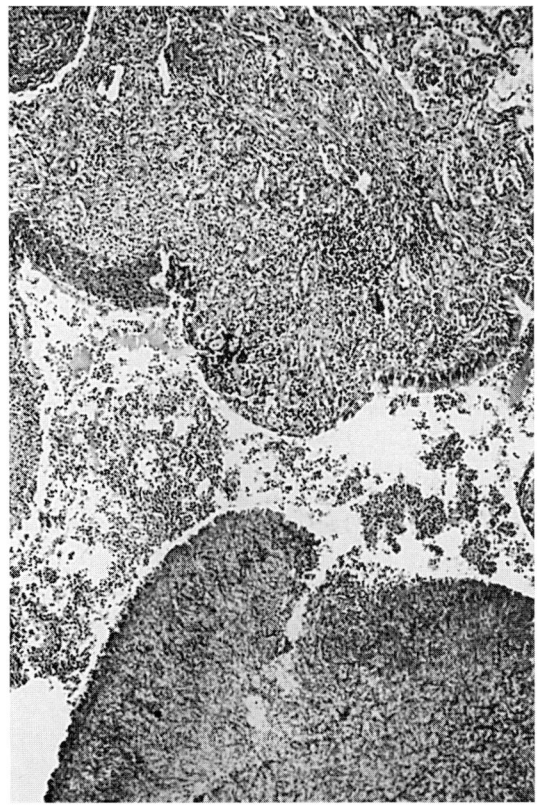

FIGURE 8-52
Pseudallescheria boydii. A colony of the fungus is seen lying in a chronic lung abscess cavity.

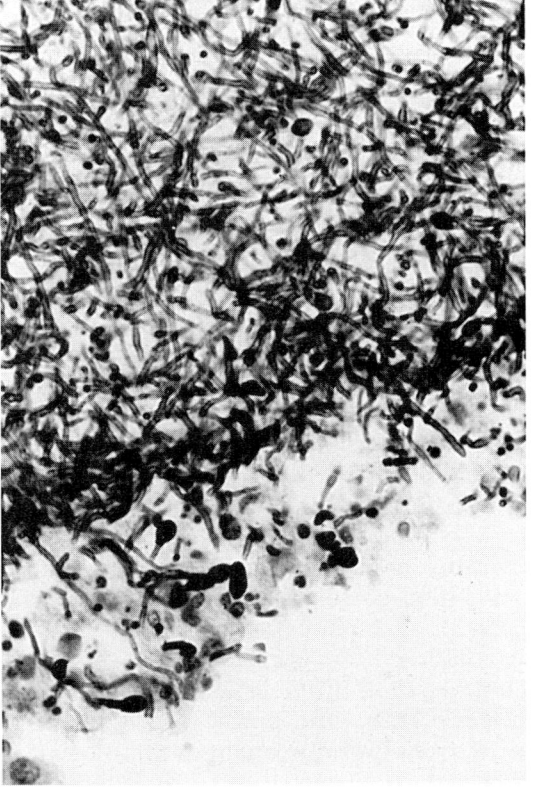

FIGURE 8-53
Pseudallescheria boydii. A high-power view of a colony showing sparsely septate hyphae and chlamydospores (Grocott's stain).

RARE FUNGAL INFECTIONS

Adiaspiromycosis

Adiaspiromycosis (syn.: adiaspirosis, haplomycosis) is an uncommon disease that is still in need of further investigation. It is ubiquitous in rodents and small wild mammals throughout the world but rarely affects humans. The first established human case was diagnosed in France in 1964.[100] Further cases have been described in eastern Europe and more recently in Brazil.[101-103]

The disease is caused by the fungus *Chrysosporium parvum* var *crescens* (*Emmonsia parva*), which was first described in Arizona rodents. The term *adiaspiromycosis* is derived from the conidia of this fungus, the adiaconidia of which show the unique property of progressive enlargement with replication.[104] The inhaled conidium, which is 2 to 4 μm in diameter, can grow to 200 to 400 μm or more. Proliferation or replication of adiaconidia does not occur in human tissues. The disease remains confined to the lungs.

CLINICAL FEATURES

Clinically the disease is usually self-limiting, benign, and localized, with few if any symptoms. Some infections have been discovered during the course of an autopsy or examination of the lung for other causes, such as bronchiectasis.[105] A few cases can end in respiratory failure.[104]

PATHOLOGY

Macroscopically there are multiple lesions or occasionally a single lesion in the lung, forming gray-white, firm nodules up to 2 cm in diameter. The centers of these nodules are glassy or gelatinous and the whole lesion resembles frog's spawn.

Histologically there are multiple nodular granulomas distributed throughout the lung, resembling miliary tuberculosis. These granulomas have a predominantly interstitial localization, compressing adjacent blood vessels and small bronchi. There is considerable necrosis and caseation. A solitary, round adioconidium occupies the center of some granulomas and multinucleated giant cells surround the surface of these adioconidia (Fig. 8-54). A zone of epithelioid cells usually encloses giant cells, which are surrounded by thick, concentric layers of fibrous connective tissue containing a great number of eosinophils and a variable number of lymphocytes, plasma cells, macrophages, and fibroblasts. The conidia vary greatly between 50 and 200 μm in diameter, with the cell walls also varying from 5 to 20 μm in thickness. The inside of the conidium is usually empty except for a little eosinophilic material. The membrane stains positively with PAS and methenamine silver. The organism might be diagnosed as a parasitic infection if the pathologists were unaware of it.

Geotrichosis

The importance of *Geotrichum candidum*, a saprophytic fungus, is difficult to estimate. The spores are found in the mouths and intestinal contents of normal persons and the fungus occurs as a saprophyte in the soil, especially in Brazil. It has frequently been isolated from throat swabs and sputum when radiographic changes have been apparent in the lungs. The lung shadows subsequently cleared following treatment with potassium iodide and nystatin. In the absence of fatal infections, the pathology of this infection remains unknown. Unfortunately such cases lack histologic proof. The fungus must be distinguished from *B. dermatitidis* and *Aspergillus* species.

Malassezia Furfur

This is an incompletely understood fungus known as the causative organism of taenia versicolor. It has recently been described in seven patients, who had all received intralipid infusion through central venous catheters. The clinical presentation was characterized by pulmonary infiltrates, fever, and, in infants, thrombocytopenia. Neonates with cardiopulmonary disease and adults with severe gastrointestinal disease and immunosuppression appear at greatest risk for *Malassezia* infection. The pulmonary arterial deposits laid by the intralipid appear to be colonized by the organisms, suggesting that these lipophilic and lipid-dependent organisms are introduced into the bloodstream from venous catheters and need high lipid concentrations to proliferate in tissue.[106] Chest radiographs in these patients show either pulmonary infiltrates, lower lobe consolidation, or even a normal appearance.

Histologically the lungs show the underlying disease, either graft-versus-host disease or bronchopulmonary dysplasia, and evidence of a pulmonary vasculitis in small and medium sized arteries. Small pulmonary emboli were seen in one case; in another, there was diffuse alveolar damage. The fungi show as globose to elliptical cells, 2 to 4 μm in diameter, budding in a unipolar fashion on a more or less broad base. They occasionally show distinctive scar collarettes from past buddings. The

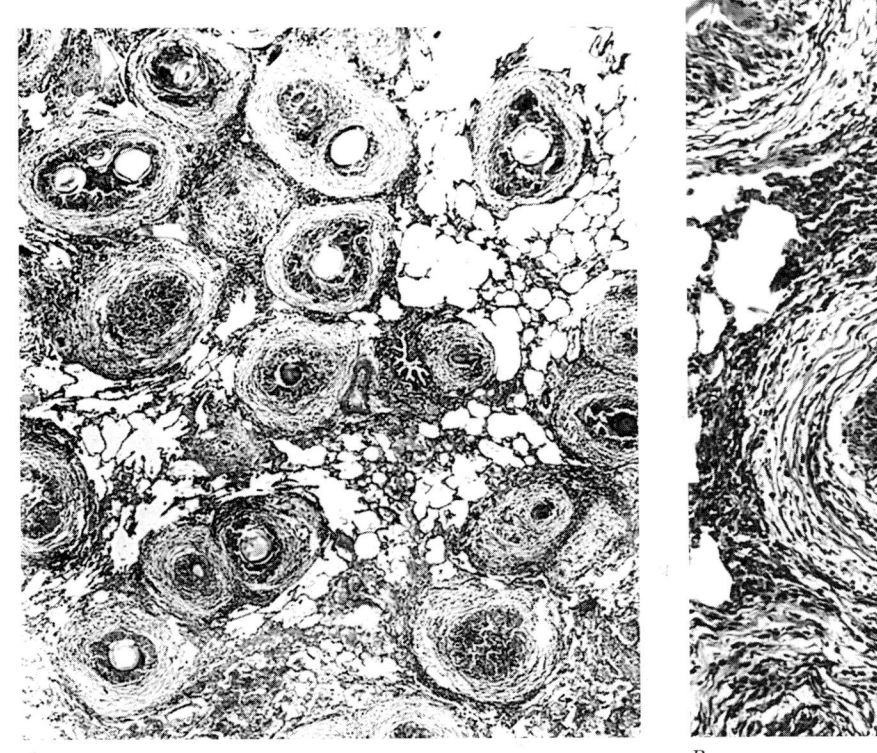

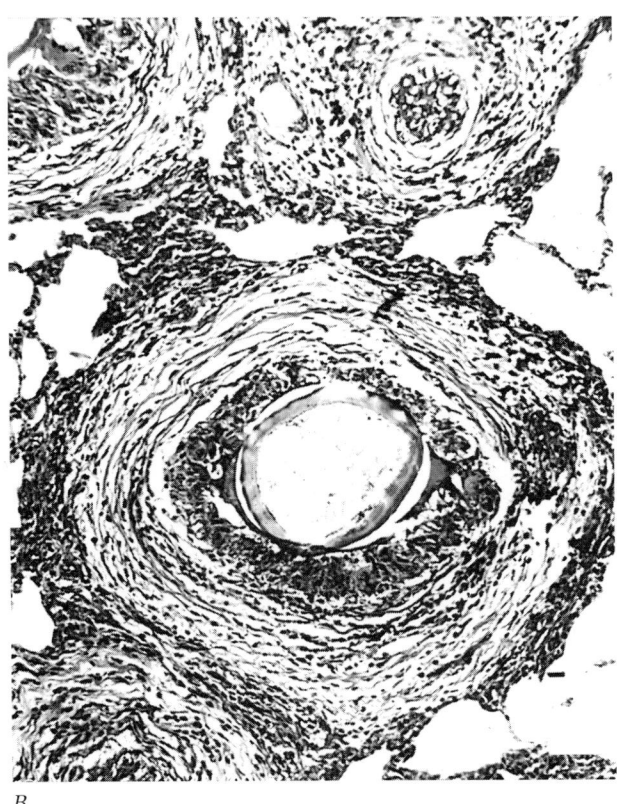

A   *B*

FIGURE 8-54
Adiaspiromycosis. Adiaspores are present in the lung, with a giant cell and granulomatous reaction.

organisms stain with methenamine silver and very faintly with H&E. Pulmonary arterial wall lipid deposits containing the organisms were found in two cases.

Phaeohyphomycosis

Phaeohyphomycoses are cutaneous and systemic infections characterized by darkly pigmented, septate hyphae. They are rarely described in the lung, but recently a case has been demonstrated by bronchial lavage from a patient who had presumed bronchiectasis.[107] The organisms have septate hyphae, which are long and branch far less frequently than *Aspergillus*. On culture, the conidiospores have areas of narrowing and produce typical conidia.

Cunninghamella bertholletiae

This infection is extremely rare, though the fungus is ubiquitous. It has been reported mainly in immunosuppressed or immunoincompetent patients.[108,109] Diagnosis is usually made by isolation of the fungus on Sabouraud's dextrose medium at room temperature.

Histologically, the organism shows large hyphae with rare septa, rhizoids, and branched and unbranched stalklike sporangiophores with single spores or sporangia. These are best demonstrated on methenamine-silver staining.

Pneumocystis Carinii INFECTIONS

Pneumocystis carinii was discovered by Chagas in 1909 but bears the name of Carini, another early worker. Its taxonomy has been controversial. It was once considered to be a protozoan but is now thought to be more closely related to the fungi. This is partly based on its ultrastructural features,[110] but there is now other confirmatory evidence: analysis of ribosomal RNA sequences suggests that *Pneumocystis* is most closely related to the *Saccharomyces*, and Pixley and coworkers demonstrated that the mitochondrial gene sequences of *P. carinii* show fungal homology.[111,112]

Three developmental stages have been described (Fig. 8-55). Trophozoites are 1 to 5 μm in size. Cysts are larger (5 to 8 μm) and contain up to eight sporozoites. When the cyst reaches maturity, sporozoites are liberated and mature into haploid trophozoites.

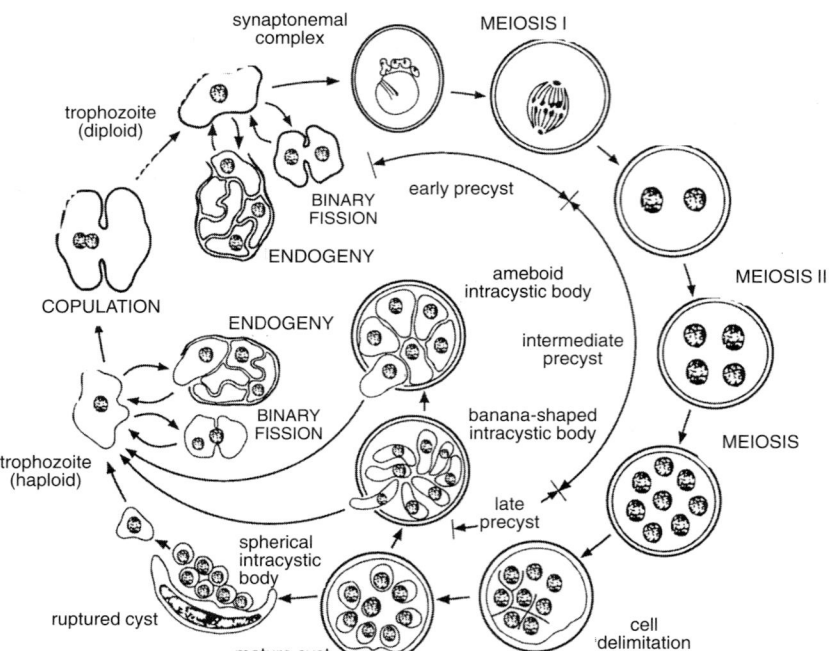

FIGURE 8-55
Life cycle of *Pneumocystis carinii*, showing sexual and asexual stages. *(Reproduced by kind permission of Dr. Hopkin, Oxford, England and Oxford Medical Publications.)*

Trophozoites multiply asexually, possibly by binary fission, but they appear to develop into cysts by a sexual cycle.

Epidemiology and Pathogenesis

Pneumocystis carinii is thought to exist normally within the lungs of humans and other mammals.[113] Antibodies to *Pneumocystis* are virtually absent before 1 year of age but increase rapidly thereafter, so that, by the age of 4, more than 80 percent of children show evidence of latent infection, which persists throughout adult life.[114]

Transmission is thought to be aerogenous, either as small aerosol particles or droplet nuclei. The only evidence for this route of transmission is from animals. Hughes showed that when infected animals were placed is close proximity to germ-free immunosuppressed rats, there was transmission of the infection.[115] However, there are differences in structure between *Pneumocystis* occurring in rats and humans. An interval oligonucleotide PAZ 102/LI differs in the two species. The exact mode of transmission, therefore, remains to be clarified, and an environmental reservoir, such as the soil, has not yet been identified.

Reactivation of a latent infection has been regarded as the main mechanism of infection in immunosuppressed patients.[116] In patients with HIV infection, *P. carinii pneumonia* (PCP) develops when the CD4 count falls below 200/mm^3.[117] However, there is now evidence that reactivation is not always the explanation. Postmortem studies using a monoclonal antibody in both immunocompetent and immunosuppressed patients failed to identify *Pneumocystis*.[118] This study was extended using the polymerase chain reaction (PCR) to amplify *Pneumocystis* DNA in postmortem lung from immunosuppressed patients. No *Pneumocystis*-specific DNA amplification product was identified.[119] A further study using lavage fluid from immunosuppressed and immunocompetent patients undergoing bronchoscopy detected *Pneumocystis* DNA only in those patients who had PCP diagnosed by conventional methenamine silver staining.[120] Reactivation of pre-existing latent infection is not, therefore, a completely satisfactory explanation for PCP. It seems likely that reinfection is a more likely cause, at least in some cases, and the origin and method of spread of the organism requires further clarification.

Pneumocystis carinii pneumonia occurs almost exclusively in immunocompromised hosts. During and after the Second World War, in the period between 1940 and 1960, there were outbreaks of pneumonia in malnourished children in European orphanages; this was initially described as "plasma cell pneumonia." *Pneumocystis* was identified as the causative agent in 1952 by Vanek and Jirovec. The first documented case in Great Britain was described by Barr[121] in a full-term infant girl whose mother suffered from recurrent respiratory infections. Although no evidence of immunodeficiency was identified in either mother or baby using the techniques then available, it is possible that both were immunocompromised and that the infection was transferred from mother to baby. The importance of malnutrition as a key factor in reducing immunity in babies with PCP is highlighted by a report from southern Iran, describing endemic PCP in babies with diarrhea and resulting gross malnutrition.[122]

Pneumocystis carinii pneumonia was first recognized as a major complication in immunosuppressed patients during the 1960s and 1970s, when chemotherapy for cancer, particularly leukemias and lymphomas, was developed. Chemotherapeutic agents, including corticosteroids, profoundly affect immunity. T-cell immunity defects are recognized as the major risk factor, but hypogammaglobulinemia *per se* is a risk.

More recently, PCP has most frequently been seen in patients with AIDS; and affects up to 80 percent of individuals with HIV infection.[123] Symptoms are very variable, but death occurs in up to 20 percent of patients and recurrence follows treatment in up to 65 percent. Children have a worse prognosis than adults and the initial infection is frequently fatal, although their CD4 counts may be higher. Interestingly, PCP is infrequently seen in Africa, particularly in Uganda, where the incidence of AIDS is the highest in the world, suggesting an absence of the organism in the environment.[124]

In transplant patients, the risk of PCP is greatest when T-cell function is lowest. However, infection may occur 2 or 3 years after transplantation in cardiac transplant patients if there have been multiple episodes of rejection.

Recently a group of patients without any obvious underlying immunosuppression have been described as suffering from PCP. The patients were elderly and suffered from a variety of chronic diseases. They showed an absence of CD25 (interleukin-2 alpha-chain receptor CD25) and there was poor response to stimulation with mitogen and tetanus toxoid.[125] This study can be questioned, as only a small number of organisms were identified in each case and the stain employed was the Gram-Weigert, which is not routinely the stain of choice.

Clinical Features

Most patients with PCP present with dyspnea, cough, and pyrexia. Sputum production and chest pain are relatively uncommon. Chest radiographs may be normal or may show diffuse interstitial, nodular, or reticulonodular infiltrates with focal areas of consolidation.[126–128] There are, however, significant differences between AIDS patients and non-AIDS patients.[127,129] In AIDS patients, the onset of the disease tends to be more insidious, with a more prolonged course. It responds less well to therapy but has a high recurrence rate. Atypical presentations

may also occur. These include localized disease, usually confined to the upper lobes, cyst formation or cavitation, and spontaneous pneumothorax.

Pathology

Typically, PCP causes diffuse disease, with consolidation. Both lungs appear bulky and contain firm areas of yellow or pink tissue, with intervening normal lung (Fig. 8-56). The pale, consolidated areas may resemble lobular pancreatic tissue. The septal tissues are prominent and thickened and the pleura is usually normal apart from the presence in some cases of interstitial emphysema. In very acute and fatal infantile infections, the lungs may resemble liver tissue. In typical cases of PCP, many alveoli and alveolar ducts are filled with characteristic amorphous, foamy eosinophilic material (Fig. 8-57). Careful examination may reveal small dots within the foamy spaces, but the organisms are not easily visible on routine H&E stained material. There is an associated interstitial pneumonia with an infiltrate that consists mainly of lymphocytes and plasma cells. Eosinophils are occasionally a feature. Alveolar walls may appear thickened due to the cellular infiltrate and edema, but fibrosis is not a feature in the early stages of the disease. The interstitial infiltrate is very variable and may even be minimal. Alveolar

FIGURE 8-56
Pneumocystis carinii pneumonia. The lung shows almost total consolidation with firm, granular areas.

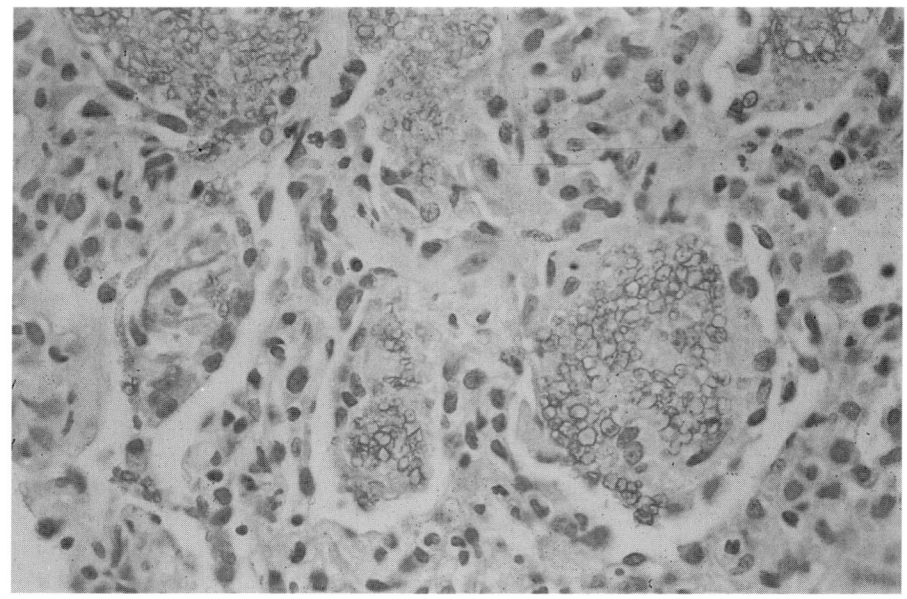

epithelial cells show evidence of damage with regeneration of type II pneumocytes. The appearances may be confused with intraalveolar oedema or alveolar proteinosis. The amount of intraalveolar exudate varies, and it may be only focal.[130] A high index of suspicion may therefore be required in immunosuppressed patients.

Diagnosis

Pneumocystis carinii pneumonia can be diagnosed via material obtained by transbronchial biopsy with bronchial brushing, percutaneous lung biopsy, open lung biopsy, and bronchoalveolar lavage. A noninvasive technique that is gaining in popularity is saline-induced sputum analysis. In AIDS patients, this procedure has a sensitivity of about 55 percent using routine stains to detect the organisms but the sensitivity can be increased by immunostaining.[131]

In tissue sections of biopsy material, the diagnosis of PCP is best made with silver stains such as GMS. The walls of the cysts stain black and they appear as round, oval, or crescent-shaped bodies about 5 to 7 μm in diameter. A very typical feature is a focal, darkly staining area of thickening of the capsule 1 to 2 μm in diameter (Fig. 8-58). In patients who have been treated, degenerative changes occur with fragmentation and blurring of the cyst walls.[132] In cytological preparations, such as lavage material or touch imprints from biopsy material, the Giemsa stain can be used to demonstrate the trophozoites, but this is less useful in tissue sections.

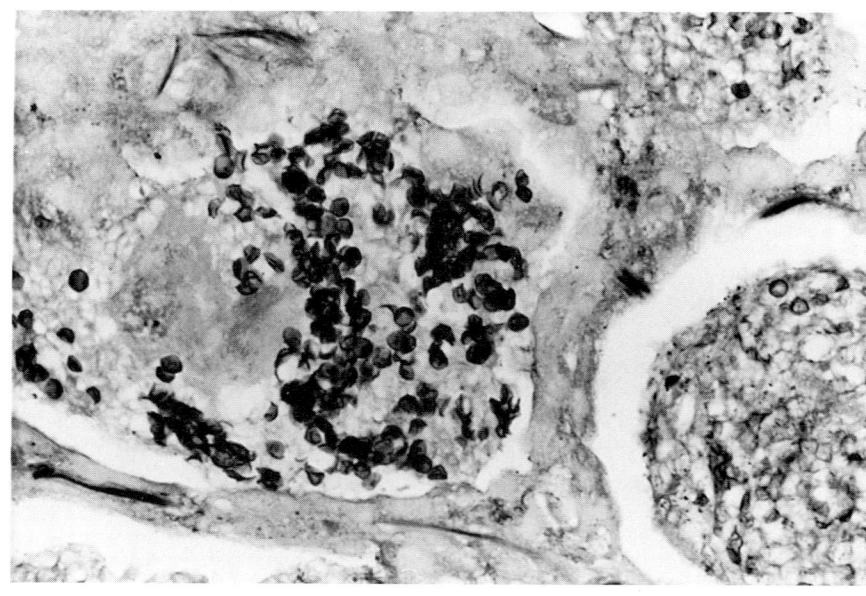

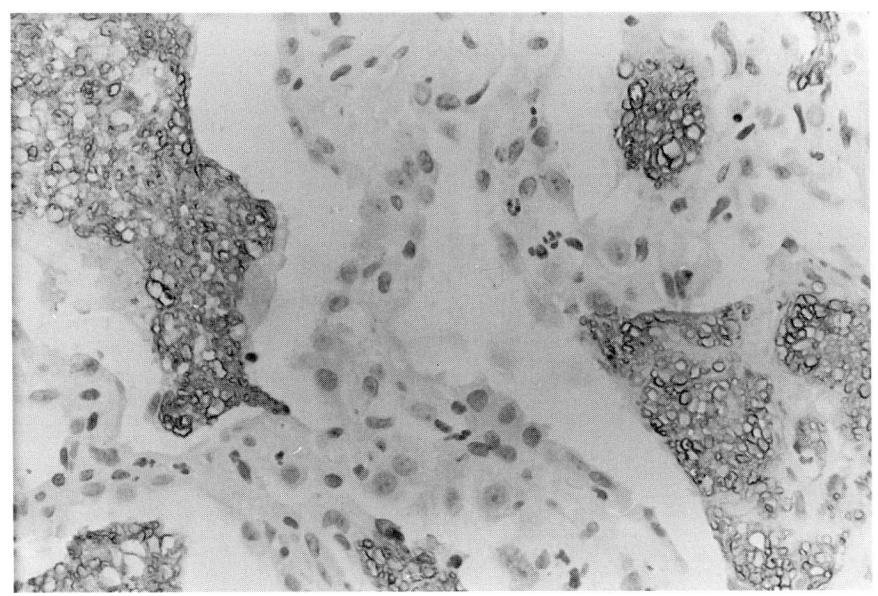

FIGURE 8-59
Pneumocystis carinii pneumonia. Immune staining using antibodies to the cyst walls increases the sensitivity (immunoperoxide method).

Monoclonal antibodies that react specifically with either trophozoites or cyst walls are now available commercially and are more sensitive than the traditional silver stains in both tissue sections and cytological material (Fig. 8-59).[133] The use of *in situ* hybridization using DNA probes, or the polymerase chain reaction increases sensitivity and specificity further.[134,135]

Electron Microscopy of *Pneumocystis carinii* Pneumonia

The ultrastructural features of PCP are well described and have helped elucidate the life cycle of the organism (Fig. 8-55).[136–138] Two basic forms are identified. The trophozoites range from 1.5 to 5 μm in diameter with a single nucleus and a clear nuclear membrane. There are glycogenlike particles in the cytoplasm and rough endoplasmic reticulum as well as some mitochondria. Nuclear division appears to occur by fission (Fig. 8-60). An important feature of the trophozoites is the prominent cytoplasmic processes that may be important in oxygen transport systems for the organism (Fig. 8-60). With increasing size, the endoplasmic reticulum recedes. The second form is cystic, measures 5 to 8 μm in diameter, and has a wall 0.3 to 0.4 μm thick, comprising an electron-lucent middle layer sandwiched between two slimmer electron dense bands. It is this wall that is seen on silver staining. Within the cysts there are variable numbers of intracystic bodies, each comprising a nucleus 1 μm in diameter and a cytoplasm with a small amount of rough endoplasmic reticulum (Fig. 8-60). Some cysts with ruptured walls may be identified. Crescent-shaped collapsed cysts are a common feature ultrastructurally and are similar to those seen on silver staining. Some organisms may be seen within polymorphonuclear leukocytes. A prominent feature in the early stages of disease is the lack of proliferation of type II cells or a marked prominence of intraalveolar macrophages.

Scanning electron microscopy shows the cell wall of the cysts and slight roughening with a few processes, as opposed to the extremely smooth surface of erythrocytes and the markedly hairy surface of type II pneumocytes.[110] The cysts show focal eccentric thickening. The surface of the trophic-stage organisms is ruffled and beaded. Cytoplasmic processes of type II pneumocytes frequently touch the conglomerate mass of organisms. Explant cultures of *P. carinii* in parasitized rabbit lung tissue have been grown and it has been shown that the organisms have close contact with host cells. The attachment of the organism to the alveolar epithelium seems to be a critical step in development. When the attachment of the trophozoites to feeder cells is examined by serial sections using transmission electron microscopy, the development of filopodia, which penetrate deep into invaginations of the feeder cell membrane, is seen. Then the apical tips of the filopodia become bulged, anchoring the parasite to the feeder alveolar cell.[139]

Atypical Features of *Pneumocystis carinii* Pneumonia

INTERSTITIAL LUNG DISEASE

Although some studies suggest that residual pulmonary fibrosis is unusual in PCP,[140] there is no doubt that interstitial disease is an important complication in some cases.[141] Several investigators have described diffuse alveolar damage with rapidly pro-

297

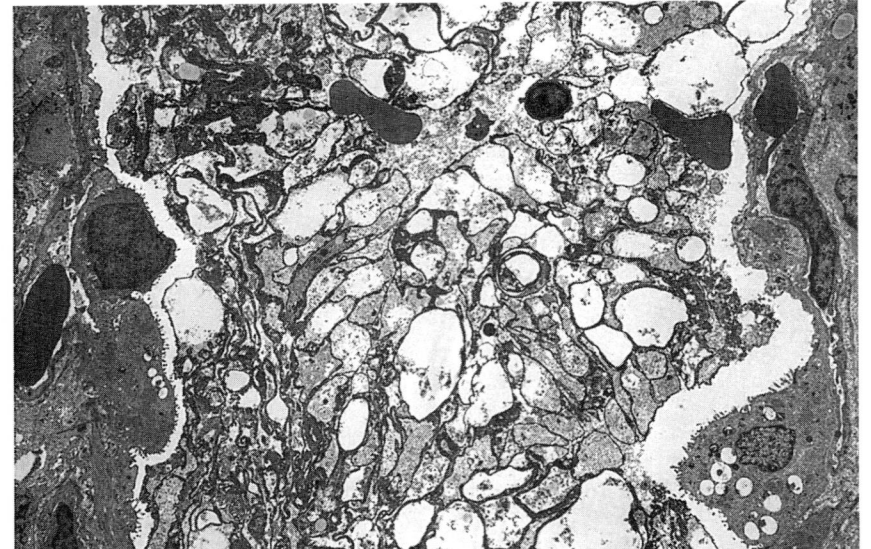

A

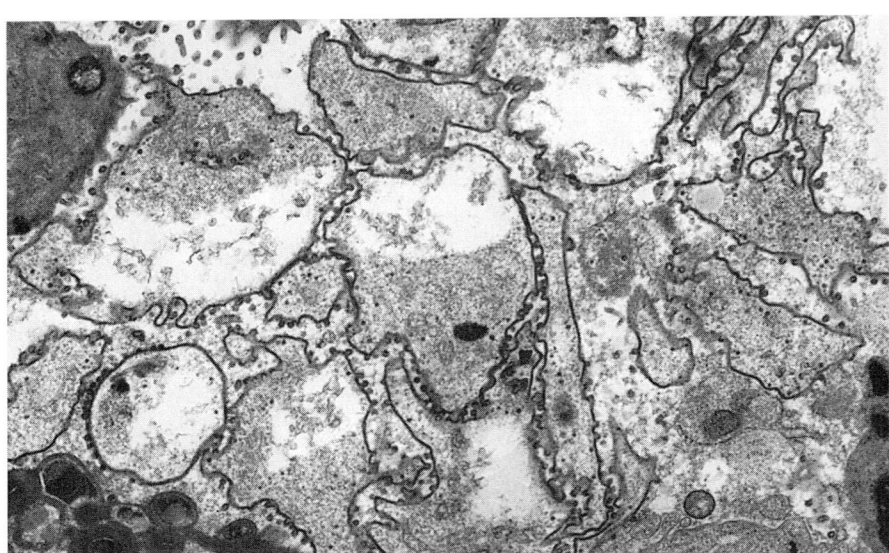

B

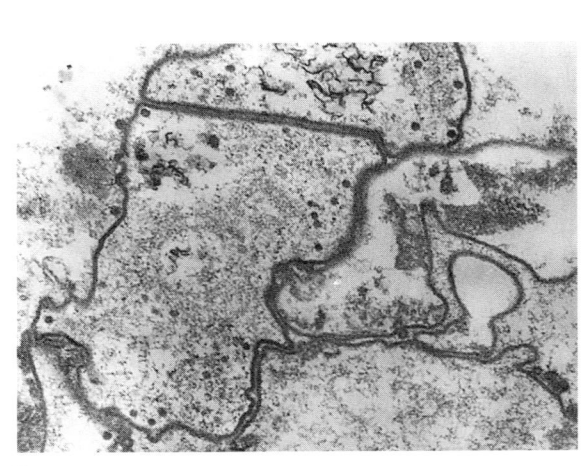

C

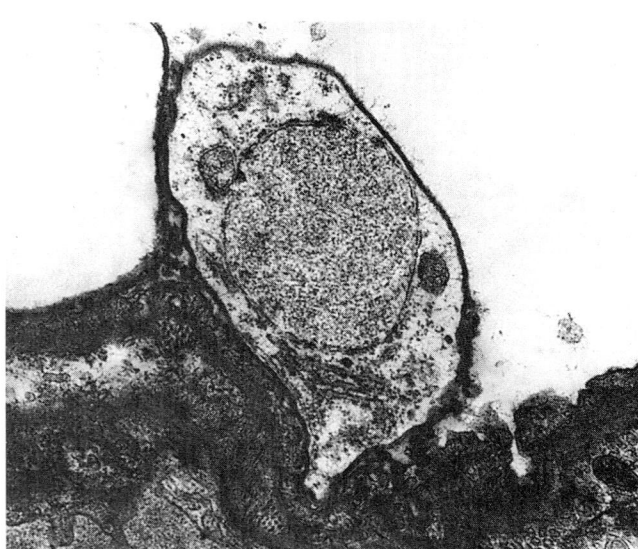

D

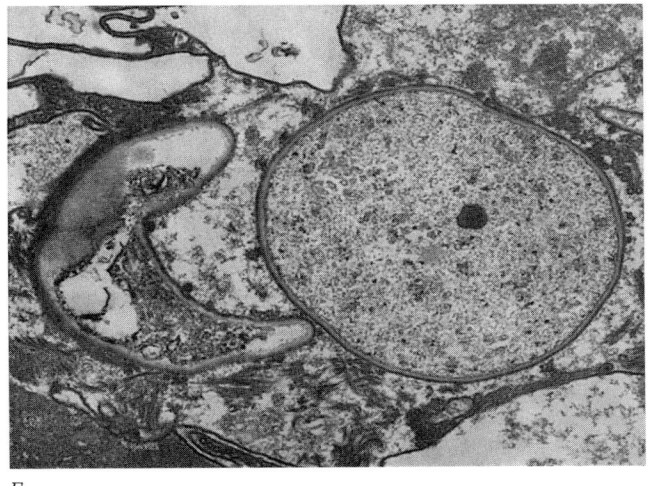

E

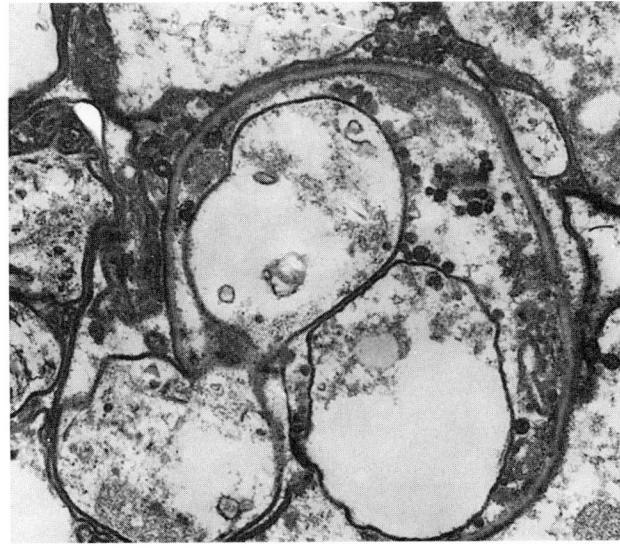

F

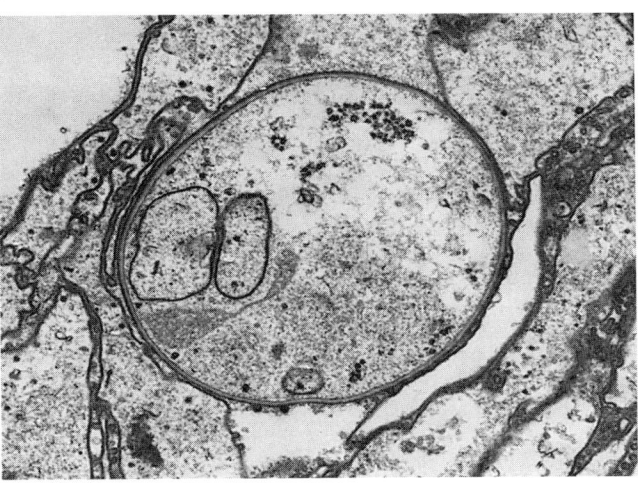

G

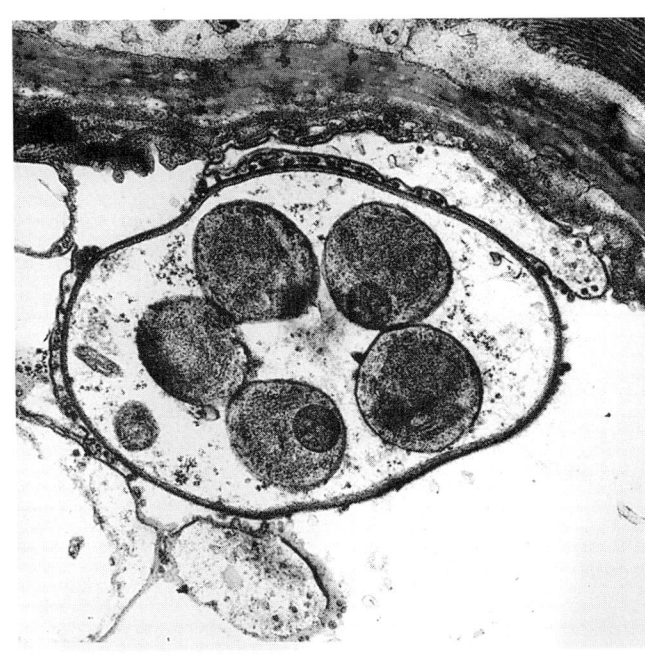

H

gressive respiratory failure,[142] and some patients with treated PCP show organization of intraalveolar exudates.[132,143] It is difficult in such patients—who have often received oxygen therapy and show evidence of coexisting cytomegalovirus infection—to determine the role of *Pneumocystis*. Damage to alveolar epithelium may occur when the trophozoites attach to type I pneumocytes. The degree of fibrosis is generally mild, but occasionally it is more severe, with remodeling. Rarely, it is associated with bronchiolitis obliterans and organising pneumonia.[144] *Pneumocystis carinii* pneumonia may be associated with a diffuse lymphoplasmacytic interstitial infiltrate, which is more prominent in children with the endemic form of the disease. In AIDS patients, parenchymal dystrophic calcification may occur, usually associated with cavitation.

CYSTS, NODULAR LESIONS, AND GRANULOMATOUS LESIONS

Cysts or centrally cavitating nodular lesions are found most frequently in the upper lobes and are frequently multiple and bilateral.[143,144] (see Chap 4). Cysts vary in size from microscopic lesions to grossly visible cysts and may be intrapulmonary or subpleural. Localized nodular PCP may be unilateral and therefore may simulate carcinoma or tuberculosis.[145] Granulomatous inflammation is seen in about 5 percent of cases of PCP. This varies from isolated giant cells of foreign-body type related to the alveolar exudate to well-formed epithelioid and giant cell granulomas of sarcoid type. Central necrosis may be a feature and the necrotic material may calcify. About

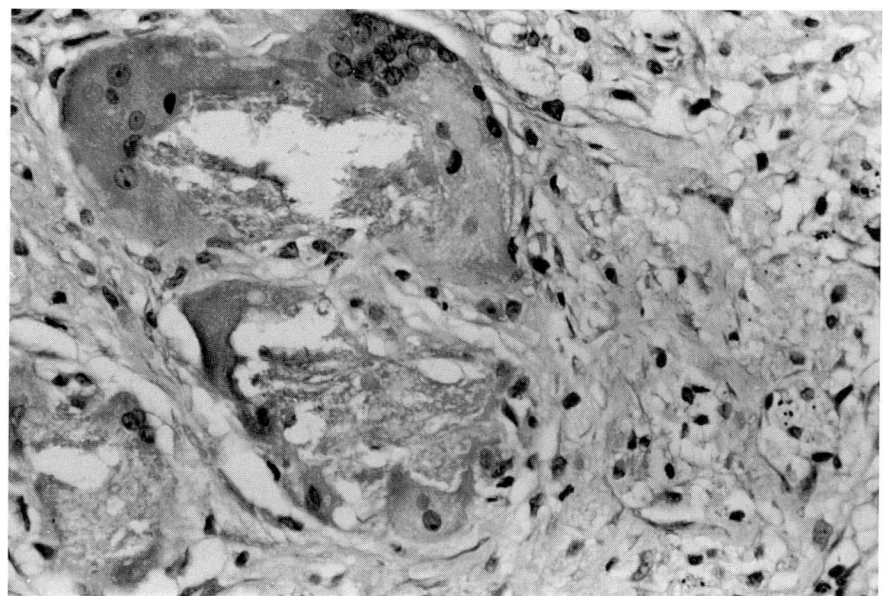

FIGURE 8-61
Pneumocystis carinii. Pleural involvement with a granulomatous and giant cell response to the organisms.

half of the patients with nodular PCP show histologic features of granuloma formation, and necrosis with cavitation may occur within the nodules.[143,144] Granulomatous inflammation may also occur in the overlying pleura (Fig. 8-61).

VASCULAR INFILTRATION AND VASCULITIS

Occasionally the cysts of *P. carinii* are seen infiltrating vessel walls, including arteries, veins, and capillaries. This is associated with a transmural chronic inflammatory infiltrate and vascular necrosis.[146] Vascular spread and vasculitis may be followed by subsequent extrapulmonary dissemination of the organism.[144]

EXTRAPULMONARY *PNEUMOCYSTIS CARINII* INFECTION

Dissemination of *Pneumocystis* infection to extrapulmonary sites has been reported in both AIDS and non-AIDS patients but is more commonly seen in the first group.[143] Regional lymph nodes are most often affected, followed by liver, spleen, and bone marrow. However, any organ may be involved, and cases are reported of spread to other sites such as the brain, pericardium, thymus, and palate.[147] Transplacental spread has also been recorded.[147,148]

REFERENCES

1. Gowing NFC, Hamlin IME: Tissue reactions to aspergillus in cases of Hodgkin's disease and leukaemia. *J Clin Pathol* 13:396–413, 1960.
2. Albilda SM, Talbot GH: Pulmonary aspergillosis. In Fishman AP (ed): *Pulmonary diseases and disorders*, vol. 2, 2nd ed. New York, McGraw-Hill, 1988.
3. Bardana EJ (Jr): The clinical spectrum of aspergillosis—Part II: Classification and description of saprophytic, allergic and invasive variants of human disease. *Crit Rev Clin Lab Sci* 13:85–159, 1981.
4. Rippon JW: Mycetoma, in *Medical Mycology. Pathogenic Fungi and the Pathogenic Actinomycetes*, 3d ed. Philadelphia, Saunders, 1988 p 80.
5. McGinnis MR, Fader RC: Mycetoma: A contempory concept. *Infect Dis Clin North Am* 2:939–954, 1988.
6. Yoshida K, Hiraoka T, Ando M, Uchida K, Mohsenin V: *Penicillium decumbens.* A new cause of fungus ball. *Chest* 101:1152–1153, 1992.
7. Schwartz DA: Organ-specific variation in the morphology of the fungomas (fungal balls) of *Pseudallescheria boydii:* Development within necrotic host tissue. *Arch Pathol Lab Med* 113:476–480, 1989.
8. Nime FA, Hutchins GM: Oxalosis caused by aspergillus infection. *John Hopkins Med J* 133:183–194, 1973.
9. Cook DJ, Achong MR, King DE: Disseminated aspergillosis in an apparently healthy patient. *Am J Med* 88:74–76, 1990.
10. Denning DW, Follansbee SE, Scolaro M, Norris S, Edelstein H, Stevens DA: Pulmonary aspergillosis in the acquired immunodeficiency syndrome. *N Eng J Med* 324:654–662, 1991.
11. Bech AO: Diffuse bronchopulmonary aspergillosis. *Thorax* 16:144–152, 1961.
12. Young RC, Vogel CL, de Vila VT: Aspergillus lobar pneumonia. *JAMA* 208:1156–1162, 1969.
13. Buchanan DR, Lamb D: Saprophytic invasion of infarcted pulmonary tissue by aspergillus species. *Thorax* 37:693–698, 1982.
14. Aisner J, Murillo J, Schimpff SC, Steere AC: Invasive aspergillosis in acute leukaemia: Correlation with nose cultures and antibiotic use. *Ann Int Med* 90:4–7, 1979.
15. Longbottom JL: Allergens of Aspergillus fumigatus, in Latge JP (ed): *Proceedings of International Symposium of Fungal Antigens*. New York, Plenum, 1988, pp 223–236.

16. Rinaldi MG: Zygomycosis (review). *Infect Dis Clin North Am* 3:19–41, 1989.

17. Bigby TD, Serota ML, Tierney LM Jr, Matthay MA: Clinical spectrum of pulmonary mucormycosis. *Chest* 89:435–439, 1986.

18. Mamlock V, Cowan WT Jr, Schnadig V: Unusual histopathology of mucormycosis in acute myelongenous leukaemia. *Am J Clin Pathol* 88:117–120, 1987.

19. Marwaha RK, Banerjee AK, Thapa BR, Agrawal SM: Mediastinal zygomycosis. *Postgrad J Med* 61:733–735, 1985.

20. Carbone KM, Pennington LR, Gimenez LF, Burrow CR, Watson SM: Mucormycosis in renal transplant patients: A report of two cases and review of the literature. *Q J Med* 57:825–831, 1985.

21. Chandler FW, Watts JC, Kaplan W, Hendry AT, McGinnis MR, Ajello L: Zygomycosis: Report of four cases with formation of chlamydoconidia in tissue. *Am J Clin Pathol* 84:99–103, 1985.

22. Benbown EW, Delamore IW, Stoddart RW, Reid H: Disseminated zygomycosis associated with erythroleukaemia. Confirmation by lectin stains. *J Clin Pathol* 38:1039–1044, 1985.

23. Komshian SV, Uwaydah AK, Sobel JD, Crane LR: Fungemia caused by candida species and *Torulopsis glabrata* in the hospitalised patient: Frequency, characteristics and evaluation in factors influencing outcome. *Rev Infect Dis* 11:379–390, 1989.

24. Sobel JD, Vazquez J: Candidemia and systemic candidiasis (Review). *Semin Respir Infect* 5:123–137, 1990.

25. Lehrer RI, Cline MJ: Leukocyte candidacidal activity and resistance to systemic candidiasis in patients with cancer. *Cancer* 27:1211–1217, 1971.

26. Jarvis WR: The National Nosocomial Infections Surveillance Systems. Hospital infections programme centres for disease control. Nosocomial fungal infection in the United States: National Nosocomial Infections Surveillance System (NNIS). January 1980–April 1990, presented at the 3d International Conference on Nosocomial Infections. Atlanta, July 31–August 1, 1990.

27. Wey SB, Mori M, Pfaller MA, Woolson RF, Wenzel RP: Hospital-acquired candidemia. The attributable mortality and excess length of stay. *Arch Intern Med* 148:2642–2645, 1988.

28. Bross J, Talbot GH, Maislin G, Hurwitz S, Strom BL: Risk factors for nosocomial candidemia: A case study in adults without leukaemia. *Am J Med* 87:614–620, 1989.

29. Edwards JE Jr: Invasive candida infections: Evolution of a fungal pathogen. *N Eng J Med* 324:1060–1062, 1991.

30. Rogers TJ, Balish E: Immunity to Candida albicans. *Microbiol Rev* 44:660–682, 1980.

31. Dvorak AM, Gavaller B: Congenital systemic candidiasis. Report of a case. *N Eng J Med* 274:540–543, 1966.

32. Triger DR, Goepel JR, Slater DN, Underwood JC: Systemic candidiasis complicating acute hepatic failure in patients treated with Cimetidine. *Lancet* 2:837–838, 1981.

33. Knox WF, Hooton VN, Barson AJ: Pulmonary vascular candidiasis and use of central venous catheters in neonates. *J Clin Pathol* 40:559–565, 1987.

34. Walsh TJ, Hathorn JW, Sobel JD, et al: Detection of circulating Candida enolase by immunoassay in patients with cancer and invasive candidiasis. *N Eng J Med* 324:1026–1031, 1991.

35. Chandler W, Kaplan W, Ajello L: Histopathology of mycotic diseases. London, Wolfe Medical Publications, 1980, p 45.

36. Ellis DH, Pfeiffer TJ: Natural habitat of *Cryptococcus neoformans var gatti*. *J Clin Microbiol* 28:1642–1644, 1990.

37. Graybill JR, Ahrens J: Immunisation and complement interaction in host defence against murine cryptococcosis. *J Reticuloendothel Soc* 30:347–357, 1981.

38. Chechani V, Kamholz SL: Pulmonary manifestations of disseminated cryptococcosis in patients with AIDS. *Chest* 98:1060–1066, 1990.

39. Baker RD: The primary pulmonary lymph node complex of cryptococcosis. *Am J Clin Pathol* 65:83–92, 1976.

40. McDonnell JM, Hutchins GM: Pulmonary cryptococcosis. *Hum Pathol* 16:121–128, 1985.

41. Kaplan W: Direct fluorescent antibody tests for the diagnosis of mycotic diseases. *Ann Clin Lab Sci* 3:25–29, 1973.

42. Darling ST: A protozoon general infection producing pseudotubercles in the lungs and focal necroses in the liver, spleen and lymph nodes. *JAMA* 46:1283–1285, 1906.

43. DeMonbreun WA: The cultivation and cultured characteristics of Darling's *H. capsulatum*. *Am J Trop Med* 14:93–125, 1934.

44. Wheat LJ, Slama TG, Eitzen HE, Kohler RB, French ML, Biesecker JL: A large urban outbreak of histoplasmosis: Clinical features. *Ann Intern Med* 94:331–337, 1981.

45. Randhawa HS: Occurrence of histoplasmosis in Asia. *Mycopathologia et Mycologia Applicata* 41:75–89, 1970.

46. Siddiqi SH, Stauffer JC: Prevalence of histoplasmin sensitivity in Pakistan. *Am J Trop Med Hyg* 29:109–111, 1980.

47. Baughman RP, Kim CK, Vinegar A, Hendricks DE, Schmidt DJ, Bullock WE: The pathogenesis of experimental pulmonary histoplasmosis. Correlative studies of histopathology, bronchoalveolar lavage and respiratory function. *Am Rev Respir Dis* 134:771–776, 1986.

48. Goodwin RA Jr, Des Prez RM: State of the art: Histoplasmosis. *Am Rev Respir Dis* 117:929–956, 1978.

49. Goodwin RA, Nickell JA, Des Prez RM: Mediastinal fibrosis complicating healed primary histoplasmosis and tuberculosis. *Medicine* (Baltimore) 51:227–246, 1972.

50. Lloyd JE, Tillman BF, Atkinson JB, et al: Mediastinal fibrosis complicating histoplasmosis. *Medicine* (Baltimore) 67:295–310, 1988.

51. Goodwin RA Jr, Shapiro JL, Thurman GH, Thurman SS, Des Prez RM: Disseminated histoplasmosis: Clinical and pathologic correlations. *Medicine* (Baltimore) 59:1–33, 1980.

52. Martin RA, Williams T, Monalto N: AIDS with disseminated histoplasmosis. *J Fam Pract* 29:628–632, 1989.

53. Johnson PC, Khardori N, Najjar AF, Butt F, Mansell PW, Sarosi GA: Progressive disseminated histoplasmosis in patients with the acquired immunodeficiency syndrome. *Am J Med* 85:152–158, 1988.

54. Sarosi GA, Davis SF: *Histoplasmosis.* In Fishman AP (ed): *Pulmonary Diseases and Disorders*, vol 2, 2d ed. New York, McGraw-Hill, 1988.

55. Wheat LJ, Connelly-Springfield P, Williams B, et al: Diagnosis of histoplasmosis in patients with the acquired immunodeficiency syndrome by detection of Histoplasma capsulatum polysaccharide antigen in bronchoalveolar lavage fluid. *Am Rev Respir Dis* 145:1421–1424, 1992.

56. Keath EJ, Spitzer ED, Painter AA, Travis SJ, Kobayashi GS, Medoff G: DNA probe for the identification of *Histoplasma capsulatum*. *J Clin Microbiol* 27:2369–2372, 1989.

57. Clark BM, Greenwood BM: Pulmonary lesions in African histoplasmosis. *J Trop Med Hyg* 71:4–10, 1968.

58. Bronnimann DA, Galgiani JN: Coccidioidomycosis. *Eur J Clin Microbiol Infect Dis* 8:466–473, 1989.

59. Eckmann BH, Schaefer GL, Huppert M: Bedside interhuman transmission of coccidioidomycosis via growth on fomites. *Am Rev Respir Dis* 89:175–185, 1964.

60. Larsen RA, Jacobson JA, Morris AH, Benowitz BA: Acute respiratory failure caused by primary pulmonary coccidioidomycosis: Two case reports and a review of the literature. *Am Rev Respir Dis* 131:797–799, 1985.

61. Frey CL, Drutz DJ: Influence of fungal surface components on the interaction of *Coccidioides immitis* with polymorphonuclear neutrophils. *J Infect Dis* 153:933–943, 1986.

62. Galgiani JN, Yam P, Petz LD, Williams PL, Stevens DA: Complement activation by Coccidioides immitis. *Infect Immun* 28:944–949, 1980.

63. Deppisch LM, Donowho EM: Pulmonary coccidioidomycosis. *Am J Clin Pathol* 58:489–500, 1972.

64. Huntington RW: Pathology of coccidioidomycosis. In Stevens DA (ed): *Coccidioidomycosis: A Text*; New York, Plenum, 1980, p 125.

65. Echols RM, Palmer DL, Long GW: Tissue eosinophilia in human coccidioidomycosis. *Rev Infect Dis* 4:656–664, 1982.

66. Winn WA: A long term study of 300 patients with cavitary-abscess lesions of the lung of coccidioidal origin. An analytical study with special reference to treatment. *Dis Chest* 54(suppl 1):268, 1968.

67. Drutz DJ, Catanzaro A: Coccidioidomycosis. Part II (Review). *Am Rev Respir Dis* 117:727–771, 1978.

68. Galginei JN, Wack EE: Coccidioidomycosis. In Fishman AP (ed): *Pulmonary Disease and Disorders*, vol. 2, New York, McGraw-Hill, 1988.

69. Warlick MA, Quan SF, Sobonya RE: Rapid diagnosis of pulmonary coccidioidomycosis. Cytologic vs potassium hydroxide preparations. *Arch Intern Med* 143:723–725, 1983.

70. Sarosi GA, Davies SF: Blastomycosis (Review). *Am Rev Respir Dis* 120:911–938, 1979.

71. Chick EW: The epidemiology of blastomycosis, in Al Dory Y. (ed): *The epidemiology of Human Mycotic Disease*. Springfield, IL, Charles C Thomas, 1975, p 103.

72. Londero AT, Ramos CD: Paracoccidioidomycosis: A clinical and mycologic study of 41 cases observed in Santa Maria, RS, Brazil. *Am J Med* 52:771–775, 1972.

73. Davidson L, Gelfand M: North American blastomycosis in Rhodesia. *Cent Afr J Med* 18:129–136, 1972.

74. Klein BS, Vergeront JM, Week RJ, et al: Isolation of *Blastomyces dermatitidis* in soil associated with a large outbreak of blastomycosis in Wisconsin. *N Engl J Med* 314:529–534, 1986.

75. Butka BJ, Bennett SR, Johnson AC: Disseminated inoculation blastomycosis in a renal transplant recipient. *Am Rev Respir Dis* 130:1180–1183, 1984.

76. Skillrud DM, Douglas WW: Survival in adult respiratory distress syndrome caused by blastomycosis infection. *Mayo Clin Proc* 60:266–269, 1985.

77. Griffith JE, Campbell GD: Acute miliary blastomycosis presenting as fulminating respiratory failure. *Chest* 75:630–632, 1979.

78. Evans ME, Haynes JB, Atkinson JB, Devaux TC Jr, Kaiser AB: Blastomycosis dermatitidis and the adult respiratory distress syndrome. Case report and review of the literature. *Am Rev Respir Dis* 126:1099–1102, 1982.

79. Davis SF, Sarosi GA: Blastomycosis (review). *Eur J Clin Microbiol Infect Dis* 8:474–479, 1989.

80. Kaufman J: Tracheal blastomycosis. *Chest* 93:424–425, 1988.

81. Kaufman L, McLaughlin DW, Clark MJ, Blumer S: Specific immunodiffusion tests for blastomycosis. *Appl Microbiol* 26:244–247, 1973.

82. Lo CY, Notenboom RH: A new enzyme immunoassay specific for blastomycosis. *Am Rev Respir Dis* 141:84–88, 1990.

83. Watts JC, Chandler FW, Mihalov ML, Kammeyer PL, Armin AR: Giant forms of *Blastomyces dermatitidis* in the pulmonary lesions of blastomycosis. Potential confusion with *Coccidioides immitis*. *Am J Clin Pathol* 93:575–578, 1990.

84. Ajello L: Paracoccidioidomycosis: A historical review, in *Paracoccidioidomycosis*. Washington DC, Pan American Health Organisation, Scientific Publication No 254, 1972, pp 3–10.

85. Restrepo A, Salazar ME, Cano LE, Stover EP, Feldman D, Stevens DA: Oestrogens inhibit mycelium to yeast transformation in the fungus *P. brasiliensis*: Implications for resistence of females to paracoccidioidomycosis. *Infect Immun* 46:346–353, 1984.

86. Restrepo A, de Bedout C, Cano LE, Arango MD, Bedoya V: Recovery of *P. brasiliensis* from a partially calcified lymph node lesion by microaerophilic incubation in liquid media. *Sabouraudia* 19:295–300, 1981.

87. Sugar AM, Restrepo A, Stevens DA: Paracoccidioidomycosis in the immunosuppressed host. *Am Rev Respir Dis* 129:349–352, 1984.

88. Angulo-Ortega A: Calcification in paracoccidioidomycosis: Are they morphological manifestations of subclinical infections?, in *Paracoccidioidomycosis*. Washington DC, Pan American Health Organisation, Scientific American Publication No 254, 1972, pp 129–133.

89. Salfelder K, Doehnert G, Doehnert HR: Paracoccidioidomycosis. Anatomic study with complete autopsies. *Virchows Arch Pathol* 348:51–76, 1969.

90. de Mattos MC, Menes RP, Marcondes-Machado J, et al: Sputum cytology in the diagnosis of pulmonary paracoccidioidomyces. *Mycopathologia* 114:187–191, 1991.

91. Watts JC, Chandler FW: Primary pulmonary sporotrichosis. *Arch Pathol Lab Med* 111:215–217, 1987.

92. England DM, Hochholzer L: Primary pulmonary sporotrichosis. Report of 8 cases with clinicopathologic review. *Am J Surg Pathol* 9:193–204, 1985.

93. England DM, Hochholzer L: Sporothrix infection of the lung without cutaneous disease. Primary pulmonary sporotrichosis. *Arch Pathol Lab Med* 111:298–300, 1987.

94. Berson SD, Brandt FA: Primary pulmonary sporotrichosis with unusual fungal morphology. *Thorax* 32: 505–508, 1977.

95. Farley ML, Fagan MF, Mabry LC, Wallace RJ Jr: Presentation of *Sporothrix schenckii* in pulmonary pathology cytology specimens. *Acta Cytol* 35:389–395, 1991.

96. Garrison RG, Mirikitani FK: Electron cytochemical demonstration of the capsule of yeast-like *Sporothrix schenckii*. *Sabouraudia* 21:167–170, 1983.

97. Seale JP, Hudson JA: Successful medical treatment of pulmonary petriellodosis. *South Med J* 78:473–476, 1985.

98. Reddy PC, Christianson CS, Gorelick DF, Larsh HW: Pulmonary monosporosis: An uncommon pulmonary mycotic infection. *Thorax* 24:722–728, 1969.

99. Shih LY, Lee N: Disseminated petriellidiosis (allescheriasis) in a patient with refractory acute lymphoblastic leukaemia. *J Clin Pathol* 37:78–82, 1984.

100. Doby-Dubois M, Chevrel ML, Doby JM, Louvet M: Premier cas humain d'adiaspiromycosis, par emmonsia crescens, emmons et jellison. *Bull Soc Pathol Exot* 57:240–44, 1964.

101. Kodousek R, Vortel V, Fingerland A, et al: Pulmonary adiaspiromycosis in man caused by *Emmonsia crescens*: Report of an unique case. *Am J Clin Pathol* 56, 294–399, 1971.

102. Watts JC, Callaway CS, Chandler FW, Kaplan W: Human pulmonary adiospiromycosis. *Arch Pathol* 99:11–15, 1975.

103. Barbas Filho JV, Amato MB, Deheinzelin D, Saldiva PH, de Carvalho CR: Respiratory failure caused by adiaspiromycosis. *Chest* 97:1171–1175, 1990.

104. Watts JC, Chandler FW: Adiaspiromycosis. An uncommon disease caused by an unusual pathogen. *Chest* 97:1030–1031, 1990.

105. Vermeil C, Gordeeff A, Geffriaud M: Human pulmonary adiaspiromycosis: Comment on a new case in Brittany. *Mycopathologia* 56:109–111, 1975.

106. Redline RW, Redline SS, Boxerbaum B, Dahms BB: Systemic Malassezia furfur infections in patients receiving intralipid therapy. *Human Pathol* 16:815–822, 1985.

107. Barenfanger J, Ramirez F, Tewari RP, Eagleton L: Pulmonary phaeohyphomycosis in a patient with haemoptysis. *Chest* 95:1158–1160, 1989.

108. Ventura GJ, Kantarjian HM, Anaissie E, Hopfer RL, Fainstein V: Pneumonia with Cunninghamella species in patients with haematologic malignancies. A case report and review of the literature. *Cancer* 58:1534–1536, 1986.

109. Zeilander S, Drenning D, Glauser FL, Bechard D: Fatal Cunninghamella bertholletiae infection in an immunocompetent patient. *Chest* 97:1482–1483, 1990.

110. ul Haque A, Plattner SB, Cook RT, Hart MN: *Pneumocystis carinii*. Taxonomy as viewed by electron microscopy. *Am J Clin Pathol* 87:504–510, 1987.

111. Edman JC, Kovacs JA, Masur H, Santi DV, Elwood HJ, Sogin ML: Ribosomal RNA sequence shows *Pneumocystis carinii* to be a member of the fungi. *Nature* 334:519–522, 1988.

112. Pixley FJ, Wakefield AE, Banerji S, Hopkin JM: Mitochondrial gene sequences show fungal homology for *Pneumocystis carinii*. *Mol Microbiol* 5:1347–1351, 1991.

113. Gutierrez Y: The biology of *Pneumocystis carinii*. *Semin Diagn Pathol* 6:203–211, 1989.

114. Pifer LL, Hughes WT, Stagno S, Woods D: Pneumocystis carinii infection: Evidence of a high prevalence in normal and immunosuppressed children. *Pediatrics* 61:35–41, 1978.

115. Hughes WT: Natural mode of acquisition for de novo infection with *Pneumocystis carinii*. *J Infect Dis* 145:842–848, 1982.

116. Hughes WT, Feldman S, Aur RJ, Verzosa MS, Hustu HO, Simone JV: Intensity of immunosuppressive therapy and incidence of Pneumocystis carinii pneumonitis. *Cancer* 36:2004–2009, 1975.

117. Said JW. Pathogenesis of HIV infection, in Nash G, Said J (eds): *Pathology of AIDS and HIV infections*. Philadelphia, Saunders, 1992, p 15.

118. Millard PR, Heryet AR: Observations favouring Pneumocystis carinii pneumonia as a primary infection: A monoclonal antibody study on paraffin sections. *J Pathol* 154:365–370, 1988.

119. Peters SE, Wakefield AE, Sinclair K, Millard PR, Hopkin JM: A search for Pneumocystis carinii in postmortem lungs by DNA amplification. *J Pathol* 166:195–198, 1992.

120. Wakefield AE, Pixley FJ, Banerji S, et al: Detection of *Pneumocystis carinii* with DNA amplification. *Lancet* 336:451–453, 1990.

121. Barr HS: Interstitial plasmacellular pneumonia due to *Pneumocystis carinii*. *J Clin Pathol* 8:19–24, 1955.

122. Post C, Dutz W, Nasarian I: Endemic Pneumocystis carinii pneumonia in South Iran. *Arch Dis Child* 39:35–40, 1964.

123. Marchevsky AN, Rosen MJ, Crystal G, Kleinerman J: Pulmonary complications of the acquired immunodeficiency syndrome: A clinicopathologic study of 70 cases. *Hum Pathol* 16:659–670, 1985.

124. Goodgame RW: AIDS in Uganda, clinical and social features. *N Engl J Med* 323:383–389, 1990.

125. Jacobs JL, Libby DM, Winters RA, et al: A cluster of Pneumocystis carinii pneumonia in adults without predisposing illness. *N Engl J Med* 324:246–250, 1991.

126. Walzer PD, Perl DP, Krogstad DJ, Rawson PG, Schultz MG: Pneumocystis carinii pneumonia in the United States. Epidemiologic, diagnostic and clinical features. *Ann Intern Med* 80:83–93, 1974.

127. Kovacs JA, Hiemenz JW, Macher EM, et al: Pneumocystis carinii pneumonia: A comparison between patients with the acquired immunodeficiency syndrome and patients with other immunodeficiencies. *Ann Intern Med* 100:663–671, 1984.

128. Peters SG, Parkash UB: Pneumocystis carinii pneumonia: Review of 53 cases. *Am J Med* 82:73–78, 1987.

129. Sterling RP, Bradley BB, Khalil KG, Kerman RH, Conklin RH: Comparison of biopsy proven Pneumocystis carinii pneumonia in acquired immunodeficiency syndrome patients and renal allograft recipients. *Ann Thorac Surg* 38:494–499, 1984.

130. Luna MA, Cleary KR: Spectrum of pathologic manifestations of Pneumocystis carinii pneumonia in patients with neoplastic diseases. *Semin Diagn Pathol* 6:262–272, 1989.

131. Blumenfeld W, Kovacs JA: Use of a monoclonal antibody to detect *Pneumocystis carinii* in induced sputum and broncho-alveolar lavage fluid by immunoperoxidase staining. *Arch Pathol Lab Med* 112:1233–1236, 1988.

132. Saldana MJ, Mones JM, Martinez JR: The pathology of treated Pneumocystis carinii pneumonia. *Semin Diagn Pathol* 6:300–312, 1989.

133. Radio SJ, Hansen S, Goldsmith J, Linder J: Immunohistochemistry of *Pneumocystis carinii*. *Mod Pathol* 3:462–469, 1990.

134. Gal AA, Koss MN, Strigle S, Angritt P: Pneumocystis carinii infection in the acquired immunodeficiency syndrome. *Semin Diagn Pathol* 6:287–299, 1989.

135. Blumenfeld W, McCook O, Holodniy M, Katzenstein DA: Correlation of morphological diagnosis with the presence of Pneumocystis DNA amplified by the polymerase chain reaction. *Mod Pathol* 5:103–106, 1992.

136. Pleiss G, Siefert K: Elecktronemotische untersunchungen bei experimenteller Pneumocystose. *Beitr Patholigische Anatomie* 120:399, 1959.

137. Campbell WG Jr: Ultrastructure of Pneumocystis in human lung: Life cycle in human Pneumocystosis. *Arch Pathol* 93:312–324, 1972.

138. Hasleton PS, Curry A, Rankin EM: Pneumocystis carinii pneumonia: A light microscopical and ultrastructural study. *J Clin Pathol* 34:1138–1146, 1981.

139. Dei Cas E, Jackson H, Palluault F, et al: Ultrastructure observations on the attachment of Pneumocystis carinii in vitro. *J Protozool* 38:205S–207S, 1991.

140. Sanyal SK, Mariencheck WC, Hughes WT, Parvey LS, Tsiatis AA, Mackert PW: Course of pulmonary dysfunction in chil-

dren surviving Pneumocystis carinii pneumonitis: A prospective study. *Am Rev Respir Dis* 124:161–166, 1981.

141. Whitcomb ME, Schwarz MI, Charles MA, Larson PH: Interstitial fibrosis after Pneumocystis carinii pneumonia. *Ann Intern Med* 73:761–765, 1970.

142. Askin FB, Katzenstein AL: Pneumocystis infection masquerading as diffuse alveolar damage. A potential source of diagnostic error. *Chest* 79:420–422, 1981.

143. Saldana MJ, Mones JM: Cavitation and other atypical manifestations of Pneumocystis carinii pneumonia. *Semin Diagn Pathol* 6:273–286, 1989.

144. Travis WD, Pittaluga S, Lipschik GY, et al: Atypical pathologic manifestation of Pneumocystis carinii pneumonia in the acquired immune deficiency syndrome. Review of 123 lung biopsies from 76 patients with emphasis on cysts, vas-

cular invasion, vasculitis and granulomas. *Am J Surg Pathol* 14:615–625, 1990.

145. Cross AS, Steigbigel RT: Pneumocystis carinii pneumonia. *Chest* 69:422–423, 1976.

146. Liu YC, Tomashefski JF Jr, Tomford JW, Green H: Necrotising Pneumocystis carinii vasculitis associated with lung necrosis and cavitation in a patient with acquired immunodeficiency syndrome. *Arch Pathol Lab Med* 113:494–497, 1989.

147. Awen CF, Baltzan MA: Systemic dissemination of Pneumocystis carinii pneumonia. *Can Med Assoc J* 104:809–812, 1971.

148. Pavlica F: The first observation of congenital Pneumocystis pneumonia in a fully developed stillborn child. *Ann Paediatr* 198:177–184, 1962.

Chapter *9*

Parasitic Lung Disease

S. B. Lucas / D. A. Schwartz /
P. S. Hasleton

A significant proportion of the parasitic infections of humans involve the lung as the final location of infection or during the migration of parasites in the body. The pathogenetic mechanisms of disease are diverse. Some parasites cause pulmonary disease by forming a space-occupying lesion (e.g., hydatid cyst); others block pulmonary arteries and induce infarction (e.g., dirofilariasis), or they may stimulate a marked host inflammatory response with necrotic, even cavitating lesions (e.g., paragonimiasis). Malaria is associated with a pulmonary shock syndrome. The most important parasitic infections, numerically and clinically, are malaria, hydatid cyst, schistosomiasis, paragonimiasis, and filarial tropical pulmonary eosinophilia.

The recent pandemic of human immunodeficiency virus (HIV) infection and acquired immunodeficiency syndrome (AIDS) has altered the epidemiology of only a few parasitic infections that involve the lung. Because of immunosuppression, pulmonary *Toxoplasma* and *Leishmania* infections are more frequently seen than hitherto. Pulmonary cryptosporidiosis and microsporidiosis are new diseases, induced by HIV disease.

In this chapter, life cycles are not described in detail. These are discussed in standard parasitology texts. A full description of all parasitic infections in humans is provided in the monograph by Gutierrez.[1]

PROTOZOA

Several genera of protozoa cause respiratory tract disease, affecting the airway mucosal surfaces, the lung parenchyma, the blood vessels, or the interstitium. Globally, the most frequent parasitic infection is *falciparum* malaria, although significant pulmonary disease is less common. Infection with HIV has introduced several new protozoal lung diseases, such as cryptosporidiosis and microsporidiosis. It has increased the frequency of previously known entities, toxoplasmosis being one example.

Pulmonary Amebiasis

EPIDEMIOLOGY AND PATHOGENESIS

Amebic dysentery is caused by the protozoan *Entamoeba histolytica*. The disease is common in third world countries but is neither tropical nor exotic and can occur anywhere.[2] Approximately 10 percent of the world's population harbor *E. histolytica* parasites. Intestinal disease is the commonest manifestation of amebiasis, and the incidence of amebic lung abscess in any locality reflects the level of amebic dysentery in the population. A recent paper from California[3] describes 30 patients with pleuropulmonary manifestations of hepatic amebiasis; they ranged in age from 2 to 64 years, with a mean of 31 years. Nearly all patients were male immigrants from third world countries, including Mexico, Vietnam, and India. At least 73 percent of the cases occurred in persons who had been in the United States for less than 2 years. Amebic liver abscesses, from which pulmonary lesions develop, are found in about 5 percent of patients with amebic colitis.

The genus *Entamoeba* is now recognized to comprise two strains that are morphologically identical.[4] The most frequent infection, sometimes termed *E. dispar*, is nonpathogenic and causes no significant disease; it maintains a vegetative cycle within the colon, with excretion of cysts. The other, sometimes termed *E. dysenteriae* but probably best labeled *E. histolytica*, is potentially invasive and pathogenic. Trophozoites in the colon may invade the mucosa and then spread to liver, lung, and other sites. The distinction of the species is based on isoenzyme electrophoresis of cultured trophozoites and, more recently, on the identification of specific DNA.

The pathogenesis of invasive amebiasis is still controversial.[5] It is likely that the trophozoites, after binding to target host cells, secrete an ionophore ("amebaphore") that lyses the host cell; the ameba then phagocytoses the cell. A focus of liquefactive necrosis ensues, multiple lesions coalesce, and the amebas continue multiplying by binary fission within the tissues. Both species infect humans by the same route; oral ingestion of food contaminated by amebic cysts from human feces.

The amebas cause colonic ulceration, and *E. histolytica* trophozoites invade branches of the portal vein and thus reach the liver. There, they cause destruction of the parenchyma and produce a liver abscess. Abscesses of this type, occurring mainly in the right lobe, may reach a massive size, occasionally rupturing into the subdiaphragmatic space. At this stage it is common for a sterile effusion, usually right-sided, to occur in the corresponding pleural sac, but this does not necessarily indicate extension of amebic infection through the diaphragm (Fig. 9-1A and B). If inadequately treated or untreated, a proportion of liver abscesses will burst through the diaphragm into the pleural cavity or the lung, causing an amebic empyema or amebic lung abscess. There may be an accompanying subdiaphragmatic abscess in some cases.

Entamoeba histolytica can also reach the pleura via diaphragmatic lymphatics and, rarely, by hematogenous spread.[6,7] These abscesses were found in the left upper lobe in the absence of amebic liver disease. Parasites may reach systemic veins from a small, clinically silent amebic lesion in the liver or from the invasion of the inferior hemorrhoidal veins by amebas, situated in low rectal ulcers. Pulmonary amebiasis has been described by Webster.[8,9]

CLINICAL FEATURES

The patients present with right-sided chest pain, cough, right shoulder pain, dyspnea, and some signs at the right base. As thoracic amebiasis is usually a complication of liver involvement, many patients have elevation of the right diaphragm and right basal atelectasis. As mentioned above, liver involvement may also be the cause of a pleural effusion. This may be either serous, serosanguinous, bloody, or frankly purulent. There may be rupture of an abscess into an airway, causing a hepatopleurobronchial fistula. Rarely an abscess may rupture into the pericardial space, with consequent high mortality.[10] Extension of a liver abscess may lead to amebic empyema with no pulmonary involvement.

PATHOLOGY

The diagnosis of an amebic lung abscess can be made with certainty if *E. histolytica* is discovered in the sputum during life or in the wall of the abscess after death. Serology (see below) is used to support a diagnosis or exclude amebiasis. With careful dissection,

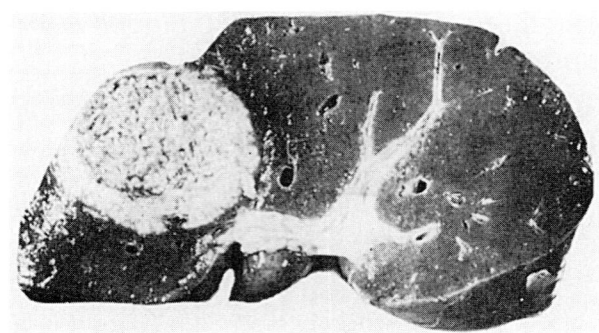

A

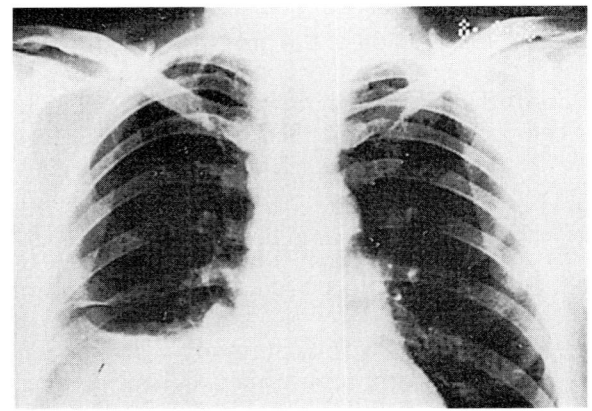

B

FIGURE 9-1
A. Liver with a large amebic abscess. (*From Warren and Mahmoud,*[188] *with permission.*) *B.* Right-sided pleural effusion secondary to an hepatic abscess. There is elevation of the right diaphragm. (*From Warren and Mahmoud,*[188] *with permission.*)

it is often possible to find an abscess tract through the diaphragm (Fig. 9-2).

The affected lobe is swollen, with the pleural surface adherent to the chest wall. On section, there is an abscess filled with glairy, reddish-brown liquid contents, likened to anchovy paste or chocolate sauce. The latter term is more applicable to the contents of liver abscess. Lung abscesses usually contain more hemorrhagic and sticky contents. There is a shaggy, reddish-brown, vascular inner lining to the abscess cavity with a fibrous wall. Outside the wall, the lung is partly collapsed, hemorrhagic and edematous. Blood-borne amebic abscesses can occur in any part of the lung (Fig. 9-3).

Histologically, most amebic lung abscesses are preceded by several foci of colligative necrosis. The alveolar pattern initially remains intact, though multiple hemorrhages and areas of necrosis occur later (Fig. 9-4). The early changes may be accompanied by a polymorph leukocytic exudate, but later this disap-

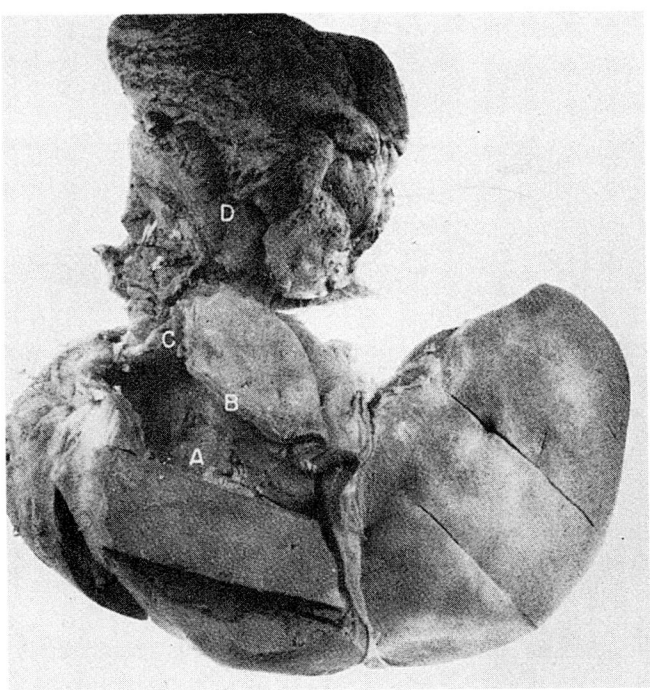

FIGURE 9-2
An amebic abscess in the liver, which has penetrated the diaphragm and burst into the lung. Here it has given rise to a further abscess cavity (*A*). The liver abscess cavity (*B*) and the cut edge of the diaphragm (*C*) are illustrated. There is communication between the subdiaphragmatic cavity and lung abscess (*D*). Approximately one-sixth natural size. (*Reproduced by courtesy of the Curator of the Wellcome Museum of Medical Science, London, England. From Spencer,*[189] *with permission.*)

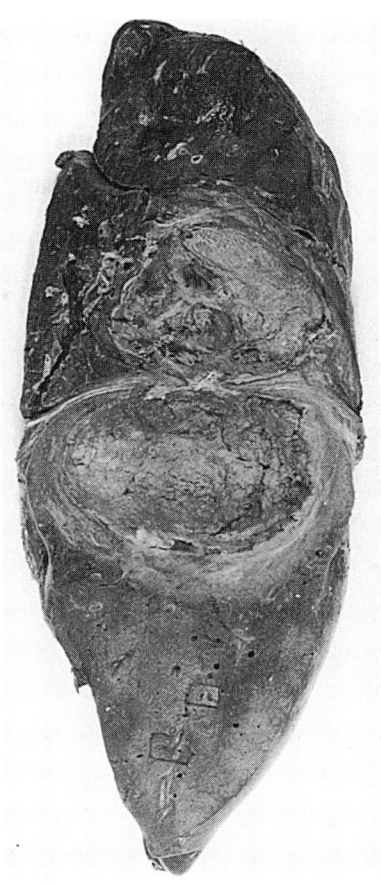

FIGURE 9-3
Chronic pulmonary amebic abscess. The situation in the central region of the lung is indicative of blood-borne rather than transdiaphragmatic spread from the liver. Approximately one-quarter natural size. (*Reproduced by courtesy of Professor J. Wainwright, Durban, South Africa. From Spencer,*[189] *with permission.*)

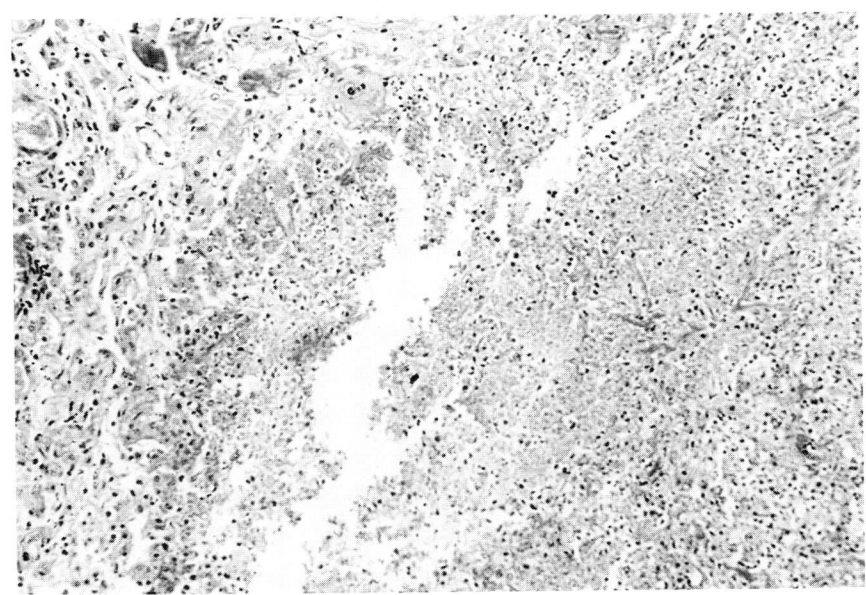

FIGURE 9-4
Entamoeba histolytica with abscess formation. (H&E, ×87.5).

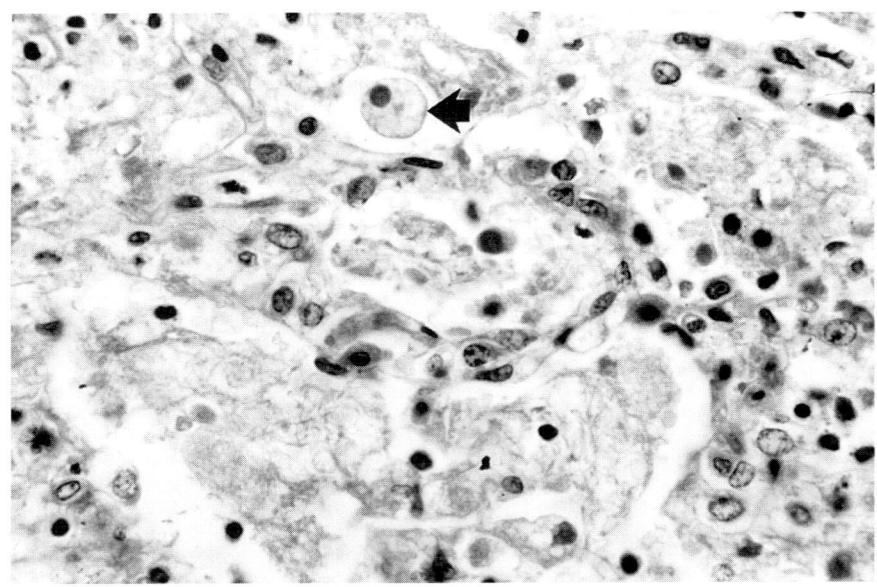

FIGURE 9-5
Entamoeba histolytica. Edge of an abscess with a typical trophozoite (*arrow*). (H&E, ×350.)

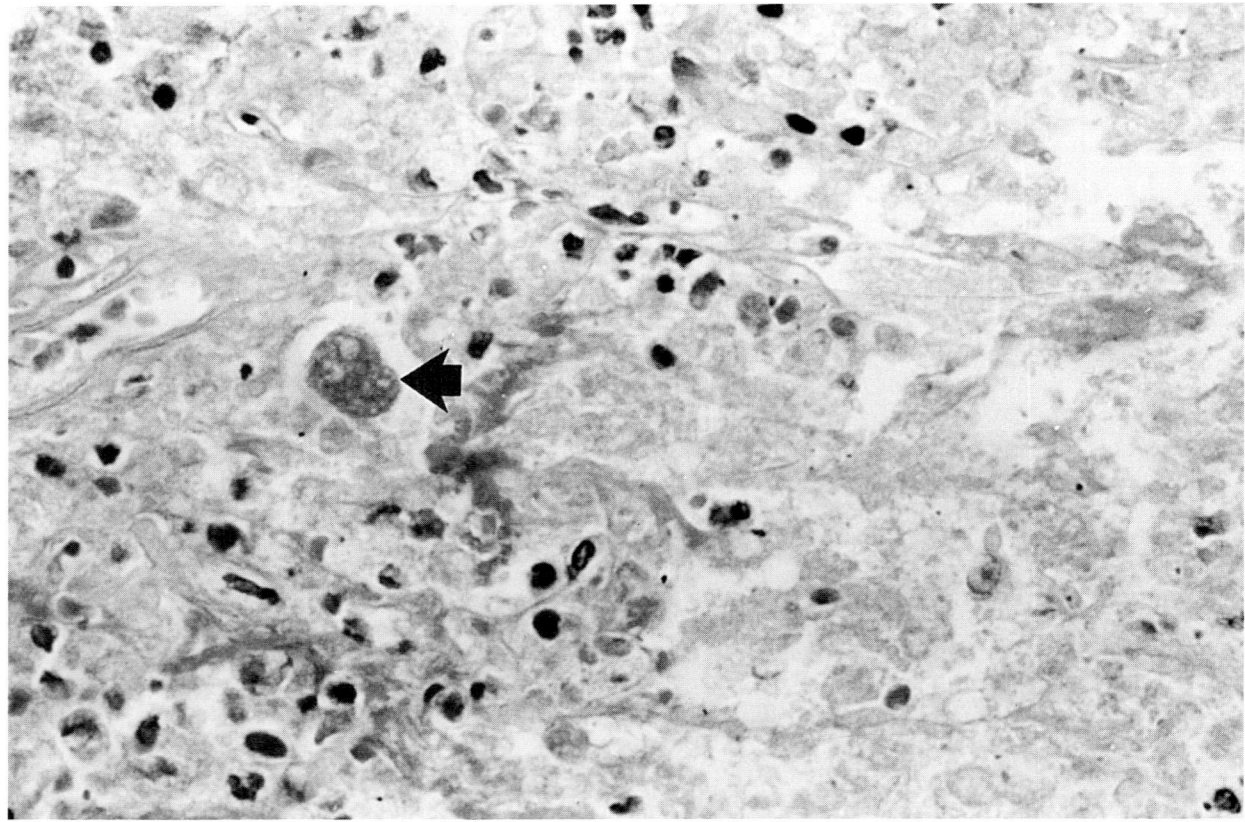

FIGURE 9-6
Edge of an amebic abscess with trophozoite enhanced by immunocytochemistry (*arrowed*), using serum from an infected patient as the primary antibody. (Immunoperoxidase, ×350.)

pears. In fully established abscess formation, the contents consist of necrotic tissue, neutrophils, lymphocytes, macrophages, and *Entamoeba* trophozoites. The wall of the abscess consists of granulation tissue and fibrosis. After the abscess has ruptured into a bronchus, the contents are expectorated, but the residual abscess cavity often becomes secondarily infected.

Entamoeba histolytica organisms are usually found in the boundary zone formed by a necrotic center and the encircling wall of chronic inflammatory tissue. Trophozoites are 30 to 50 μm in diameter (Fig. 9-5). The cytoplasm is characteristically purple and granular on being stained with hematoxylin and eosin (H&E). Erythrophagocytosis is typical but is not seen in every ameba. The nucleus is round, with a thick nuclear membrane; it is deep purple and contains a single nucleolus. Periodic acid-Schiff (PAS) stain highlights trophozoites, since they contain much glycogen. Immunocytochemistry with monoclonal antibodies or human serum from a seropositive patient demonstrates the parasites well (Fig. 9-6).

Patients with invasive amebiasis produce antibodies to *Entamoeba*, enabling efficient serodiagnosis using enzyme-linked immunosorbent assay (ELISA) or an immunofluorescence assay.[11] Seropositivity persists after therapy.

Visceral Leishmaniasis (Kala-Azar)

EPIDEMIOLOGY AND PATHOGENESIS

This disease is caused by the hemoflagellate protozoan *Leishmania donovani*. The disease is endemic in parts of Sudan, Ethiopia, Kenya, the Middle East, India, China, and Brazil. The organism is conveyed to man by various species of sand flies (*Phlebotomus* sp.). It does not produce a primary skin lesion but replicates in macrophages and monocytes and may spread throughout the lymphoreticular system. The majority of infections are subclinical. The parasite exists in macrophages in the leishmanial amastigote form in humans—that is, a protozoan 2 to 3 μm in diameter with a hematoxyphilic nucleus, a rod-shaped hematoxyphilic kinetoplast, and clear refractile cytoplasm (Fig. 9-7).

CLINICAL FEATURES

The disease usually presents with anemia, lymphadenopathy, and hepatosplenomegaly. However, any organ with an interstitium where macrophages are found, such as the lung, may be affected. Patients with visceral leishmaniasis often have a persistent

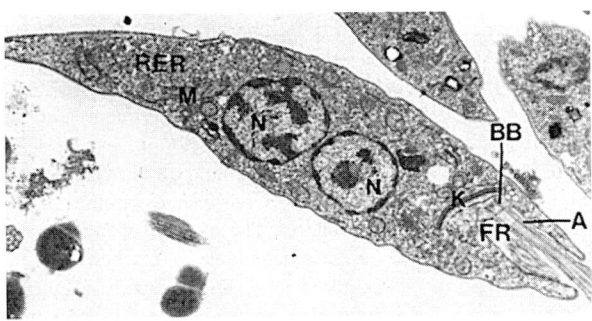
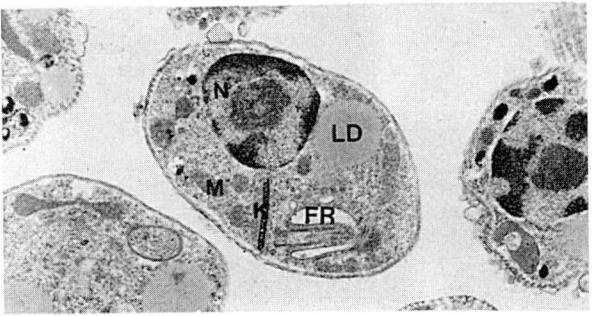

FIGURE 9-7

Electron micrograph showing *Leishmania* ultrastructure. Promastigote (*top* ×13,000). This is the flagellated form found in the sandfly vector. The amastigote is seen in the bottom half of the picture. N=nucleus; K=kinetoplast; M=mitochondria; RER=rough endoplasmic reticulum; A=axoneme; FR=flagellar reservoir; BB=basal body; LD=lipid droplet. (*Courtesy of Dr. Paulo Pimenta. From Warren and Mahmoud,*[188] *with permission.*)

dry cough that appears early, consistent with an interstitial pneumonitis.

PATHOLOGY

An autopsy series in Brazil of patients with visceral leishmaniasis found evidence of lung infection in 10 of 13 cases.[12] They showed irregularly distributed interalveolar septal thickening due to infiltration by macrophages, lymphocytes, and plasma cells. Only 3 of 10 had visible amastigotes (Figs. 9-8 and 9-9), the others showing positive interstitial immunocytochemical staining using specific anti-*L. donovani* serum. Fine septal fibrosis was seen, but types I and II pneumocytes and endothelial cells were unaffected. Since terminal bronchopneumonia is a common event in visceral leishmaniasis, acute inflammation is also seen in a proportion of patients.

The histologic differential diagnosis when parasites are visible includes histoplasmosis, toxoplasmosis, and Chagas' disease. *Histoplasma capsulatum* is larger, with a more eosinophilic nucleus and, critically, stains as a fungus with silver stains or PAS; *Leishmania* do not take up these stains. *Toxoplasma gondii* tachyzoite parasites are of similar size to *Leishmania* amastigotes but lack the kinetoplast; and toxoplasmic bradycysts are often also present. The

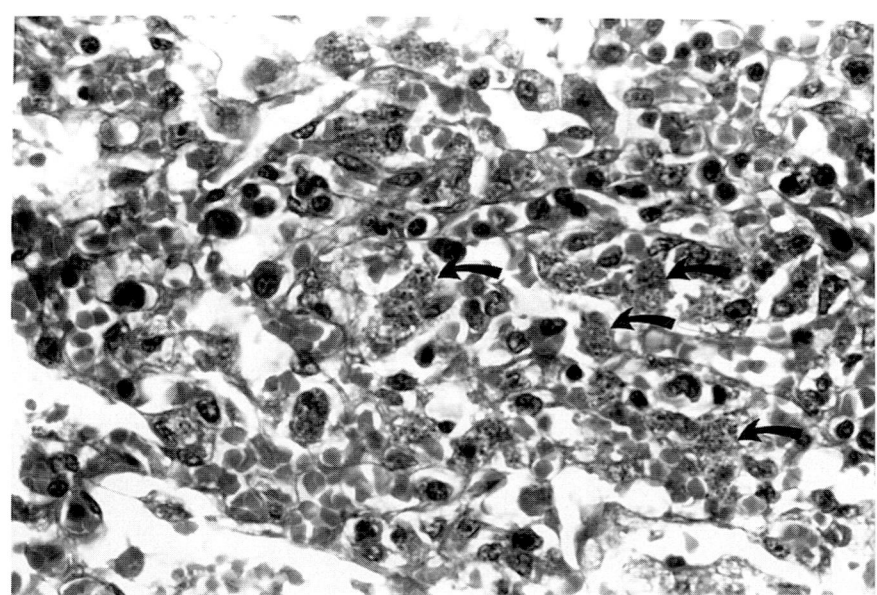

FIGURE 9-8
Leishmaniasis with parasites inside
macrophages in the lung interstitium
(*arrows*). (×350.)

tissue form of *Trypanosoma cruzi*, the agent of Chagas' disease, is morphologically similar to *Leishmania*. The *T. cruzi* amastigotes, however, also parasitize other epithelial and endothelial cells. Unlike *Leishmania*, they may produce multiparasitic cyst forms in tissues. In the unlikely event of clinicopathologic confusion, clinical data, serology, and immunocytochemical confirmation will determine the nature of the infection.

Falciparum Malaria

Of the four species of *Plasmodium* that cause malaria in humans, *Plasmodium falciparum* is the most

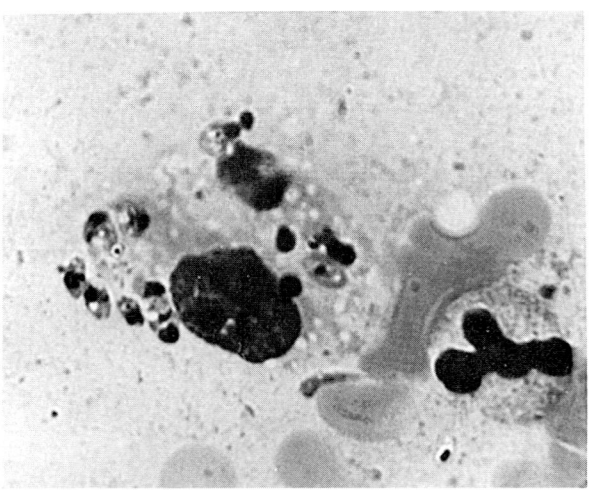

FIGURE 9-9
Stained impression smear from the edge of a biopsy from cutaneous leishmaniasis. Note the intracellular amastigotes in or adjacent to the cell; some show characteristic kinetoplast. (×1000.) (*From Warren and Mahmoud,*[188] *with permission.*)

important, since it may be fatal. The infection is transmitted by female *Anopheles* mosquitoes and is endemic throughout the tropics.

CLINICAL FEATURES

Clinically, anemia and cerebral malaria are the major features of childhood disease, while in adults, cerebral malaria, anemia, renal failure, pulmonary complications, and hypoglycemia are important.[13]

The main lung complications are pulmonary edema and shock lung. A mild interstitial pneumonitis is frequent,[14] particularly in children. Shock lung with hyaline membranes, conversely, is rarely seen in children but is common in adults. This may reflect differences in therapy rather than primary pathogenesis. The etiology of the edema is, broadly, iatrogenic fluid overload or hemodynamic abnormalities due to the systemic infection of malaria itself. Hypoalbuminemia may contribute in some patients.[12,15]

PATHOLOGY

Necessarily, pathologic observations of the lung in malaria derive from autopsy material, with the inherent complications of agonal and other changes superimposed on the manifestations of malaria. Thus, patients maintained on intensive care before they die will usually have secondary bronchopneumonia as well as changes secondary to mechanical ventilation.

At autopsy, the lungs are congested and heavy and may be dark in color from the accumulated hemozoin pigment, the products of the parasitic metabolism of red cell hemoglobin. Microscopically, large and small vessels are distended by red cells. The characteristic feature in untreated and recently treated (i.e., less than 2 to 3 days) patients is ring-

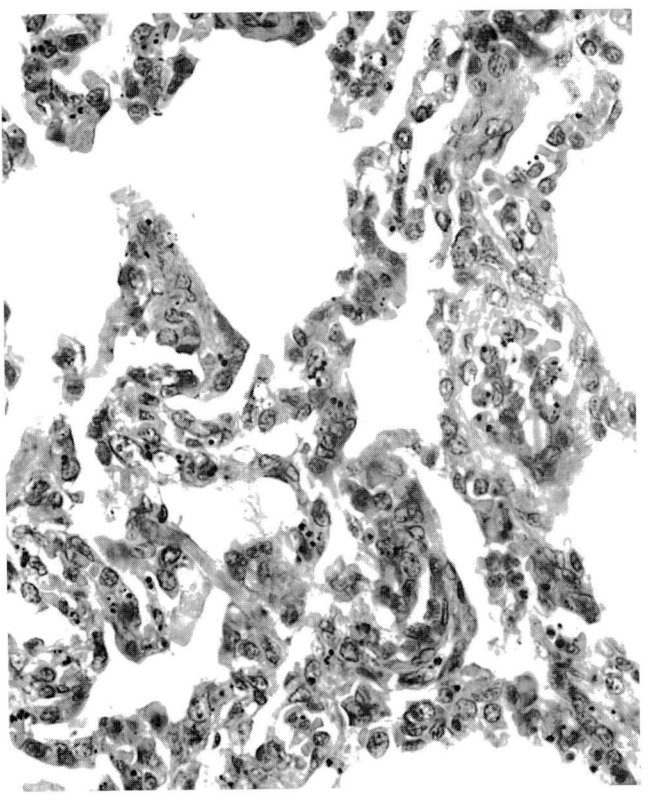

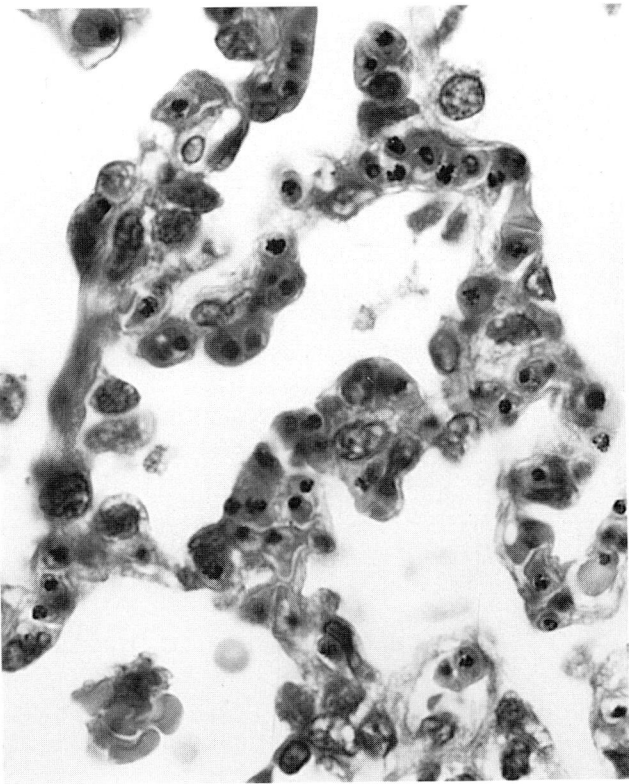

FIGURE 9-10
A. Lung tissue obtained from an autopsy of an African patient with *Plasmodium falciparum* infection. There is thickening of alveolar septa and capillary congestion. Macrophages are present in the alveolar spaces, which contain malarial pigment.

Trophozoites are also present in the erythrocytes in the alveolar capillaries, although they are difficult to see at these magnifications. (*A:* ×170; *B:* ×340.)

form trophozoites of *P. falciparum* within the erythrocytes in the alveolar capillaries (Fig. 9-10*A* and *B*). Although the peripheral blood parasitemia may approach 50 percent in severe cases of malaria, the proportion of red cells in pulmonary small vessels that are parasitized is usually < 15 percent[16] (Fig. 9-11*A* to *D*). Septal macrophages are seen with dark-brown granules of hemozoin pigment. A mild chronic lymphocytic interstitial pneumonitis may be seen.

The degree of pulmonary edema is variable, as there is shock lung. Evidence of disseminated intravascular coagulation is also variable, being more frequent in intensively treated patients as opposed to the untreated.

Electron microscopic observations show capillary endothelial edema and intravascular monocytes.[17] Parasitized erythrocytes present an abnormal surface with numerous "knobs" (Fig. 9-12*A* and *B*). These are accumulations of parasitic proteins and red cell structural filaments causing the red cell to adhere to endothelial surfaces and to other red cells. Hence the cytoadherence that may reduce and obstruct blood flow at the microvascular level.[12,18]

As well as cytoadherence, there have been intensive investigations of the complementary roles of humoral cytokine activation and the activation of intercellular adhesion molecules in the pathogenesis of *falciparum* malaria. Cytokines may play a direct role in the genesis of shock lung.[19]

Toxoplasmosis

EPIDEMIOLOGY

Toxoplasmosis results from infection with the facultative intracellular sporozoan parasite *Toxoplasma gondii*. This coccidian is the most common cause of parasitic infection in the United States, where approximately one-half of the population has serologic evidence of previous infection. Despite this high figure, clinical illness resulting from *T. gondii* in immunologically intact persons is uncommon. Most symptomatic infections occur in the immunocompromised host, especially patients with AIDS, or in neonates as a result of congenital transmission.

Toxoplasma gondii is a worldwide zoonosis. Members of the cat family are the definitive hosts. Following sporogony in the intestinal epithelium of

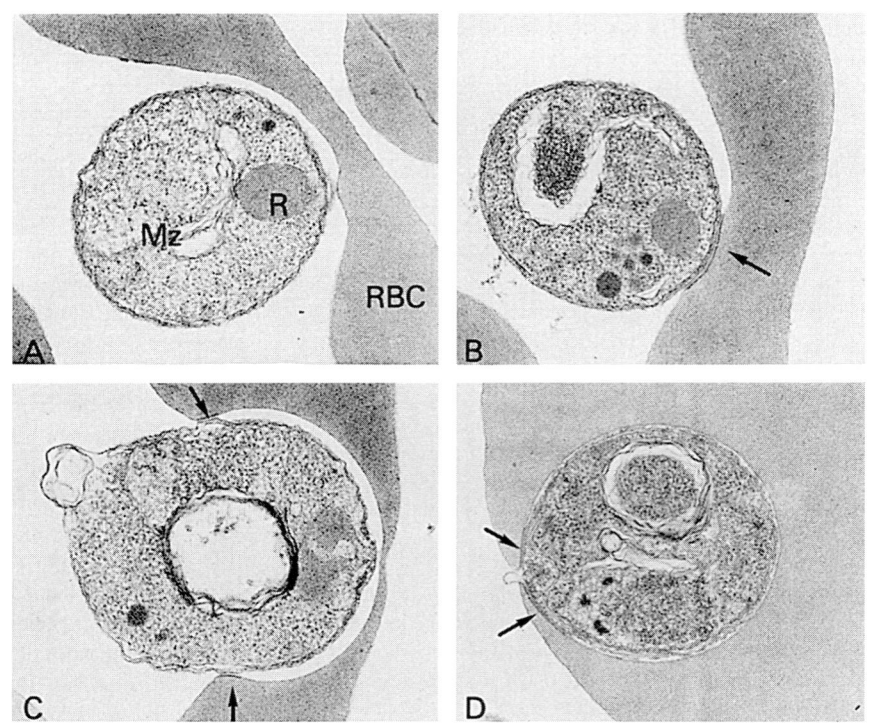

FIGURE 9-11
Merozoite (MZ) invading a red blood corpuscle (RBC).   *A*. Attachment.   *B*. Junction formation (*arrow*);   *C* and *D*. Movement of junction around the merozoite, bringing the merozoite into a vacuole in the RBC. (*From Warren and Mahmoud,*[188] *with permission.*)

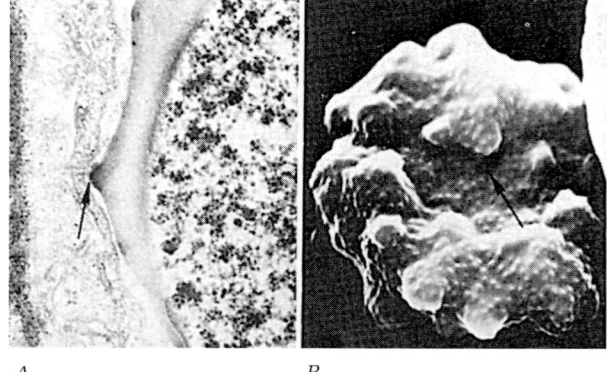

A   *B*

FIGURE 9-12
A. Scanning electron micrograph showing knobs (*arrow*) over the *P. falciparum*–infected erythrocyte surface. *B*. Electron micrograph showing adhesions (*arrow*) between electron-dense knobs on infected erythrocytes and capillary endothelial cells. (*From Warren and Mahmoud,*[188] *with permission.*)

cats, oocysts containing sporoblasts are shed in the feces. Following a 3- to 4-day period of sporulation, infective sporozoites are formed within the oocysts. Humans can acquire the infection in four ways. Ingestion of oocysts by contact with cat feces or ingestion of tissue cysts by eating undercooked or raw animal meat are considered the most common mechanism for contracting a primary infection. Other modes of infection include the inadvertent transfer of cysts from infected allografts and transplacental transmission from a mother with primary infection to a susceptible fetus.

Following infection of a human host, cysts containing hundreds of viable organisms are formed within a variety of tissues. In the acute form of the disease, cysts are termed pseudocysts and contain motile tachyzoites or trophozoites. During chronic infections, the organisms multiply more slowly, forming cysts that contain bradyzoites. Cysts originate in intracellular vacuoles. They gradually enlarge beyond the size of the normal host cell, displacing the nucleus to the side and sometimes causing it to degenerate. Upon lysis of the host cell, motile forms are liberated from the host to re-enter a new cell and create a new generation of cysts. Tachyzoites of *T. gondii* contain an apical complex comprising a complex system of organelles that enable them to enter virtually any host cell, whether or not it is phagocytic.[20]

CLINICAL FEATURES

In the immunocompetent adult host, only a small number of *Toxoplasma* infections are symptomatic, usually producing a benign and self-limited illness similar to mononucleosis or cytomegalovirus infection. The clinical findings consist of lymphadenopathy (usually cervical), fever, malaise, sore throat, hepatosplenomegaly, and night sweats. Chorioretinitis may occur in cases of acute acquired toxoplasmosis. There have been rare reports of immunocompetent patients with acute acquired toxoplasmosis who developed potentially fatal disseminated disease.[21]

The most common consequence of primary infection is latent or asymptomatic infection. These persons harbor dormant cysts containing potentially infective bradyzoites within various organs.

Congenital toxoplasmosis is a member of the TORCH complex and results from the transplacental transmission of organisms to the fetus from a mother who acquires primary infection during pregnancy. The clinical manifestations to the fetus are variable. There may be no sequelae, or sequelae may develop at varying times after birth. The signs and symptoms of congenital toxoplasmosis are nonspecific and include anemia, jaundice, rash, petechiae, pneumonitis, encephalitis, microcephaly, hydrocephalus, intracranial calcification, chorioretinitis, blindness, strabismus, epilepsy, and psychomotor or mental retardation. In neonates with severe congenital toxoplasmosis containing numerous cysts and tachyzoites.[22]

In patients with AIDS and other immunocompromised individuals, toxoplasmosis is considered to be the result of a reactivated latent infection, but primary infections may rarely occur. Toxoplasmosis in the immunocompromised adult usually presents as encephalitis.[23] It is the commonest cause of an intracerebral mass in a patient with AIDS.[24] Although the brain is the most frequent organ involved, disease may occur in the heart, skeletal muscle, lungs and eye. Organisms have been found in almost every organ of the body in disseminated cases. There has been a recent increase in the numbers of AIDS patients developing pulmonary toxoplasmosis, both with and without encephalitis. This raises the possibility that it is more common than had previously been thought.[25–31]

Pulmonary toxoplasmosis in any patient is indicative of a systemic infection and should raise the possibility of central nervous system involvement, with subsequent risk of death from meningoencephalitis or bronchopneumonia.

PULMONARY PATHOLOGY

The lungs of autopsied patients with pulmonary toxoplasmosis are heavy, often having a combined weight of greater than 2000 g. Cut surface shows congestion, consolidation, cavitation, petechial hemorrhages, or minute (3- to 5-mm) yellow-white or brown nodules representing foci of necrotizing bronchopneumonia.[21,25]

The characteristic microscopic patterns of pulmonary toxoplasmosis include an interstitial pneumonitis/diffuse alveolar damage (DAD)[25,32] and focal or diffuse necrotizing pneumonia[25,33] similar to that seen with herpesvirus infections (Figs. 9-13 and 9-14). In one study of pulmonary toxoplasmosis in patients with AIDS, the pattern of interstitial pneumonitis/DAD accompanied by a fibrinous alveolar exudate appeared to predate necrotizing pneumonitis, with large areas of parenchymal necrosis. Diagnostic, predominantly intracellular tachyzoites were less frequent in lungs with interstitial pneumonitis or DAD. In contrast, cases of necrotizing pneumonitis were associated with numerous tachy-

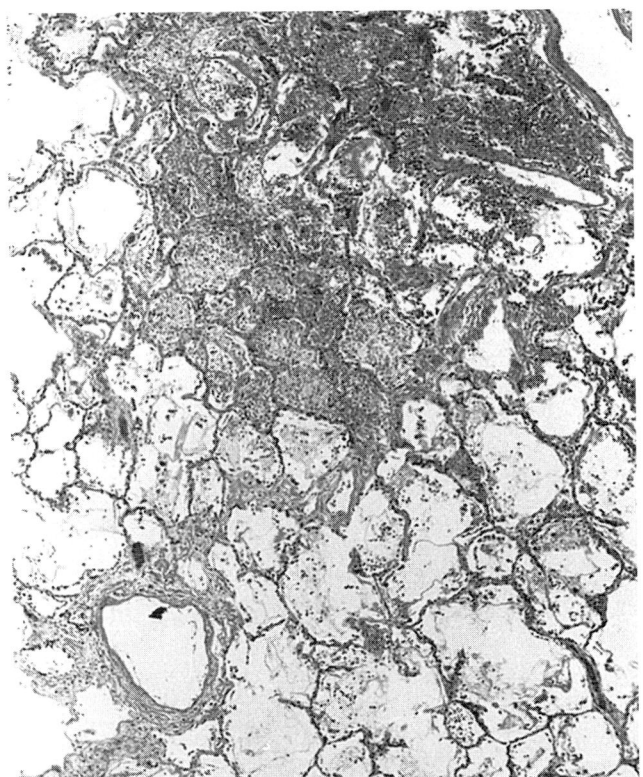

FIGURE 9-13
Area of necrotizing bronchopneumonia from a patient developing pulmonary toxoplasmosis following cardiac transplantation. (H&E, ×33.)

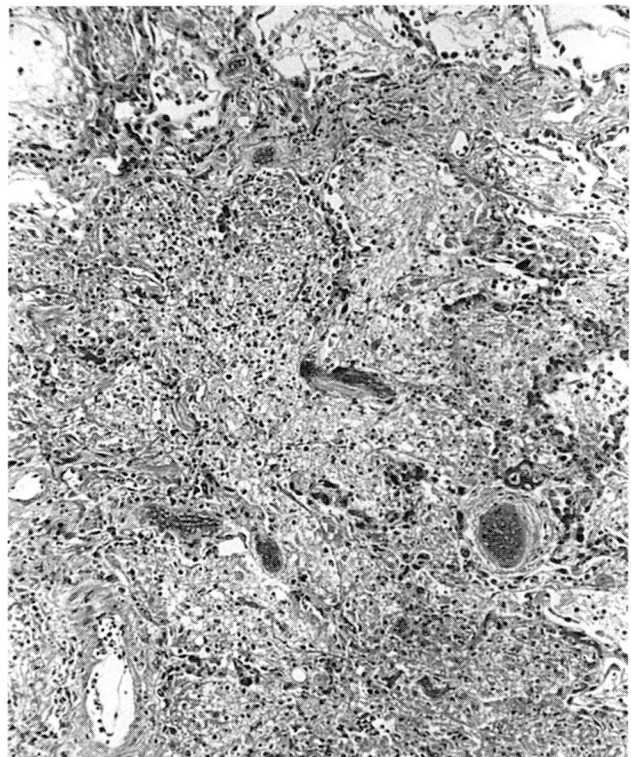

FIGURE 9-14
Pulmonary toxoplasmosis showing focus of necrotizing pneumonia associated with intraalveolar exudate, fibrin, and neutrophils. (H&E, ×83.)

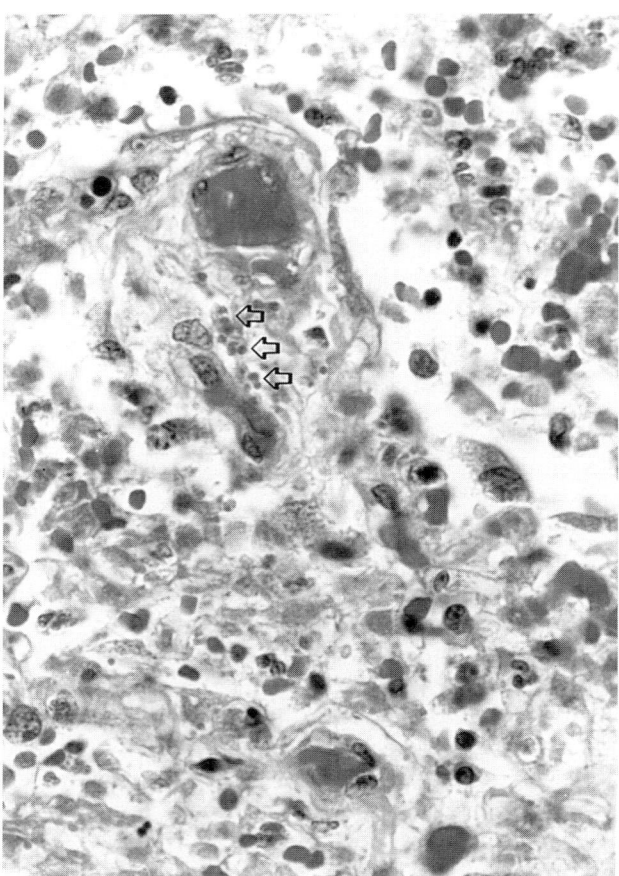

FIGURE 9-15
Clusters of extracellular trophozoites of *Toxoplasma* (*arrows*) can easily be mistaken for cellular debris in this area of alveolar necrosis. (H&E, ×190.)

zoites; these were located in foci of necrosis, where they were both intra- and extracellular.[25] Areas of intraalveolar fibrinous exudate containing nuclear debris, ghost cells, neutrophils and macrophages were focal or diffuse (Fig. 9-14). Alveoli may contain regenerating type II pneumocytes. In necrotic areas, tachyzoites can easily be confused with necrotic nuclear and cellular debris (Fig. 9-15). Organisms can also be present within the cytoplasm of histiocytes or alveolar macrophages, endothelial cells, alveolar lining cells, and smooth muscle cells lining the airways and blood vessels (Figs. 9-16 and 9-17). In the acute pulmonary infection, cyst forms of the organism may be difficult or impossible to find.[25]

A single tissue cyst of *T. gondii* may range from 10 to 200 μm, contain up to 3000 organisms, and stains well with routine H&E. The cyst wall is argyrophilic and stains weakly with PAS stain. However, the organisms within the cyst are strongly PAS-positive. The tachyzoites of *T. gondii* are crescentic and measure approximately 2 to 3 by 4 to 6 μm. They stain well with H&E, revealing a basophilic nucleus and eosinophilic or amphophilic cytoplasm. Tachyzoites also stain well using PAS, Giemsa, methylene blue, and toluidine blue stains.

DIAGNOSIS

The patterns of pulmonary tissue injury caused by *Toxoplasma* are nonspecific, and thus diagnosis of the etiologic agent is dependent on finding the cysts or tachyzoites. Immunohistochemical staining of formalin-fixed biopsy or autopsy tissues using commercially available anti–*T. gondii* monoclonal antisera are useful in the identification of organisms. Some authors found greatly increased numbers of organisms by using immunohistochemical staining procedures.[25] If there is sufficient tissue, touch impressions of lung tissue stained with Giemsa, PAS, methylene blue, or anti–*T. gondii* antisera is a rapid method for the demonstration of organisms. Electron microscopy of infected lung tissue can be a valuable adjunct in cases where it is difficult to differentiate *Toxoplasma* from similar infectious agents.

Cytologic diagnosis of pulmonary toxoplasmosis using bronchoalveolar lavage (BAL) is useful. The best staining results using BAL fluid have been reported with Giemsa, May-Grunewald-Giemsa, Wright-Giemsa, and H&E stains.[34] Although BAL has

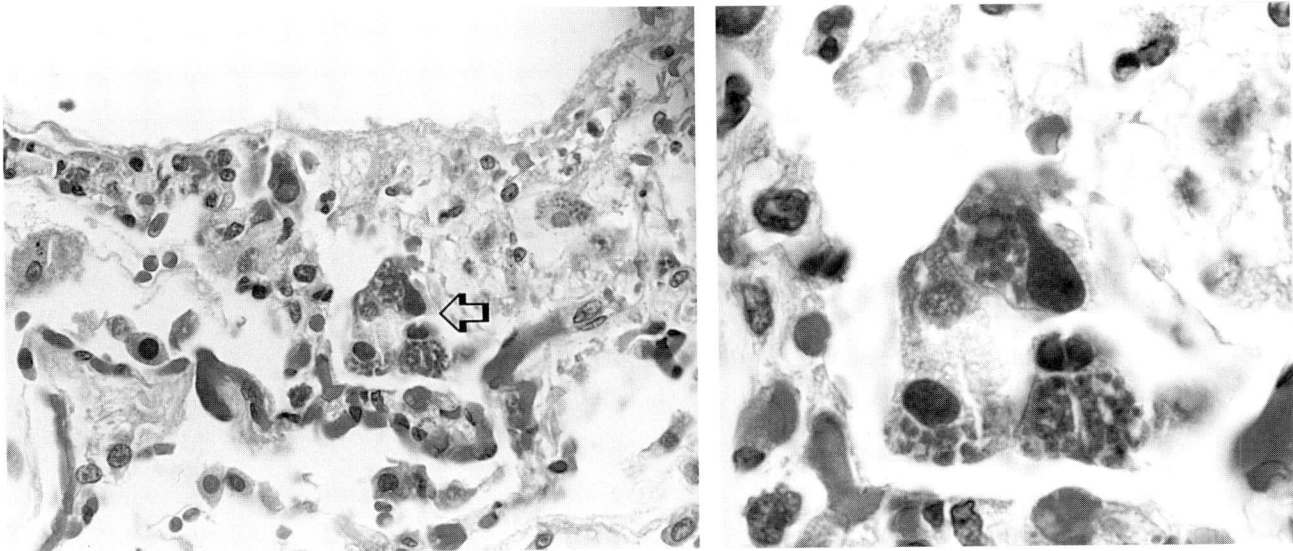

FIGURE 9-16
Toxoplasma pseudocysts (*arrow*) present within the cytoplasm of three alveolar macrophages (*left*) adjacent to an area of necrotizing pneumonia. Higher magnification reveals the presence of nuclei within each trophozoite (*right*). (H&E; ×212.5 left, ×425 right.)

been useful for the diagnosis of pulmonary infection in recipients of kidney[35,36] and cardiac transplants,[34] AIDS patients,[27,37,38] and patients with hematologic malignancies,[39] the sensitivity of this technique remains unknown.

DIFFERENTIAL DIAGNOSIS

Toxoplasma gondii should be differentiated from intracellular *Histoplasma capsulatum*, *Pneumocystis carinii*, amastigote forms of *Trypanosoma cruzi* and *Leishmania* spp., microsporidia, cytoplasmic inclusions of cytomegalovirus (CMV), and foreign material ingested by macrophages. *Leishmania* and *Trypanosoma* are distinguished from *Toxoplasma* by the presence in the former of a kinetoplast. The intracytoplasmic granules of CMV do not contain a nuclear structure, and intranuclear inclusions indicate a viral infection. A methenamine silver stain is useful for the recognition of *Pneumocystis* and *Histoplasma*, including budding organisms of the latter. Microsporidia often stain gram-positive, have a beltlike equatorial stripe, are slightly smaller (1 to 2.5 μm) than *Toxoplasma*, and are birefringent using polarized light.

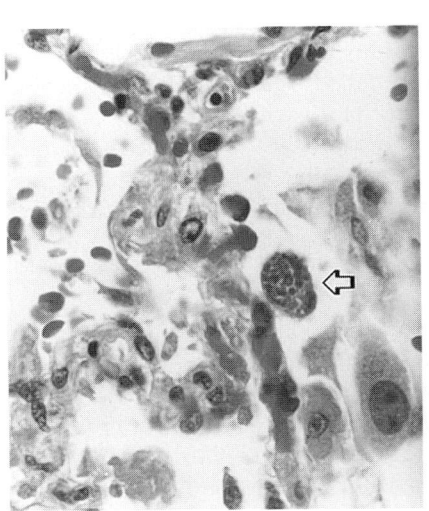
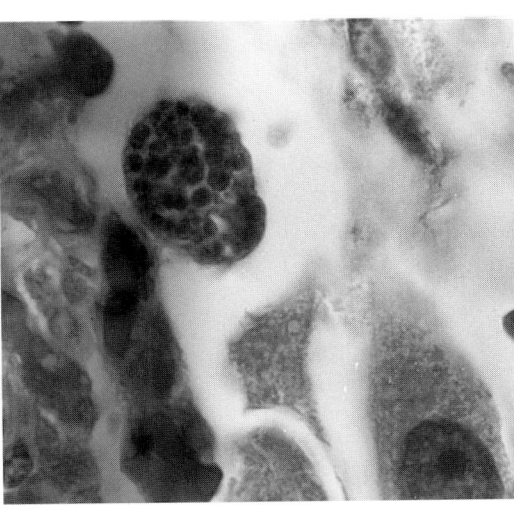

FIGURE 9-17
This pseudocyst of *Toxoplasma* has completely filled the cytoplasm of the host cell (*arrow*) within an alveolar space. (H&E; ×225 left, ×750 right.)

MICROSPORIDIOSIS

Epidemiology

Microsporidia are obligate intracellular protozoa belonging to the phylum Microspora. They infect mammals, other vertebrates, and intervertebrates. Four genera of microsporidia—*Enterocytozoon*, *Encephalitozoon* (including *Septata intestinalis*), *Pleistophora*, and *Vittaforma*—infect humans.[40] Within a host cell, microsporidia have a unique life cycle involving a proliferative merogonic stage followed by a sporogonic stage. The latter stage results in the production of distinctive and resistant infective spores. The mature spores of all microsporidia are characterized by a unique coiled polar tubule and accompanying tubule extrusion apparatus. Under the appropriate environmental conditions, the polar tubule is forcibly extruded from the spore and injects infective sporoplasm into a suitable host cell. Little is known of the source(s) or mechanism(s) of transmission of these agents to humans.

CLINICAL FEATURES

Microsporidial infections in human were rarely reported until the advent of AIDS. Microsporidiosis is, along with the unrelated cryptosporidiosis, one of the most important and rapidly emerging parasitic infections. The most frequently reported microsporidian species infecting AIDS patients is *Enterocytozoon bieneusi*.[41] This organism has been identified in a bronchial biopsy and BAL from one AIDS patient with pulmonary symptoms.[42] However, lung involvement in patients infected with *Enterocytozoon* is probably very rare.

Respiratory tract infection due to microsporidia is associated almost exclusively with disseminated disease produced by members of the genus *Encephalitozoon*.[43,44] Three organisms in this genus—*E. hellem*, *E. cuniculi*, and *E. intestinalis*—all infect the respiratory tract. Schwartz and colleagues identified the first patient with pulmonary microsporidiosis following the autopsy of a patient with AIDS and disseminated *E. hellem* infection involving the eyes, upper and lower urinary tract, and tracheobronchial mucosa.[43] The pattern of microsporidian colonization of the superficial tracheobronchial mucosa was suggestive of respiratory acquisition. Further cases with pulmonary involvement with microsporidia have been described.[40,45,46] Patients with AIDS have been described with disseminated *E. hellem*, microsporidian spores present in sputum, and no respiratory symptoms.[47,48] *Encephalitozoon intestinalis* (formerly *Septata in-testinalis*) has a propensity for infecting the gastrointestinal tract but may infect the respiratory tract. It is associated with a clinical presentation similar to that of other members of this genus.[49]

Microsporidian infections may be found in patients with organ transplants (Josiah Rich, M.D., personal communication), in non–HIV-infected immunocompetent adults,[50] and children[51] having traveler's diarrhea. It remains to be seen whether microsporidiosis is associated with pulmonary infections in these and other non–HIV-infected patients.

PULMONARY PATHOLOGY

The spectrum of pathologic changes in the lungs of persons with microsporidiosis has not been fully defined because of their recent identification as pulmonary pathogens and the small number of cases examined. Because *Encephalitozoon cuniculi* and *E. hellem* preferentially parasitize epithelial cells, they are concentrated within the tracheobronchial lining epithelium. In the sole patient undergoing autopsy with disseminated *E. hellem* infection, massive numbers of gram-positive microsporidian spores were present diffusely throughout the length of the tracheobronchial tree, extending into terminal bronchioles. In some areas they were associated with an erosive tracheitis, bronchitis of major airways, and bronchiolitis (Figs. 9-18 through 9-23). Biopsies of patients with pulmonary *Encephalitozoon* infections have all demonstrated acute bronchiolitis (Figs. 9-21 and 9-22) with or without pneumonia. Large numbers of gram-positive ovoid spores, having an average length of 2 μm and a characteristic equatorial beltlike stripe, were present in epithelial cells lining bronchi and bronchioles. They were also seen in neutrophils within the bronchiolar wall, cells lining the alveoli, and neutrophils in alveolar spaces.[44,46] In well-oriented sections of tracheobronchial epithelium, spores are concentrated in the supranuclear or subapical regions of host cells (Figs. 9-20 and 9-23).

The spores do not stain well with H&E or cytologic (PAP) stains and are best visualized in tissues using the Gram-stain method.[52,53]

There are scant data available on the pulmonary pathologic findings of *E. intestinalis* infection; however, spores have been reported in the bronchial epithelial cells.[54]

Electron microscopic examination is useful for confirmation of a light microscopic diagnosis of microsporidial infection and permits the differentiation between *Enterocytozoon bieneusi*, *Encephalitozoon intestinalis*, and the *E. hellem/E. cuniculi* complex.[40] The parasitophorous vacuole of *E. intesti-*

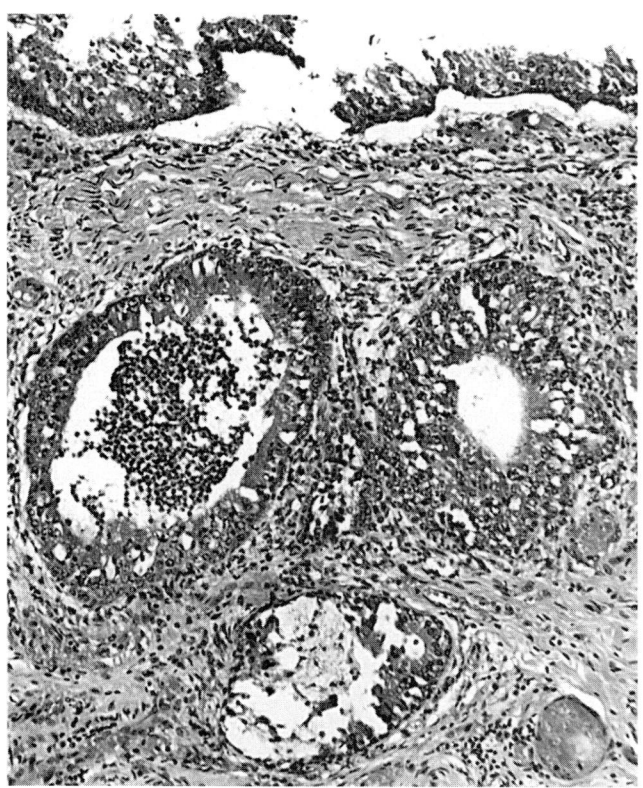

FIGURE 9-18
Bronchus obtained from autopsy of a patient with disseminated
E. hellem infection, showing acute inflammation involving the
lining epithelium and glands. (H&E, ×95.)

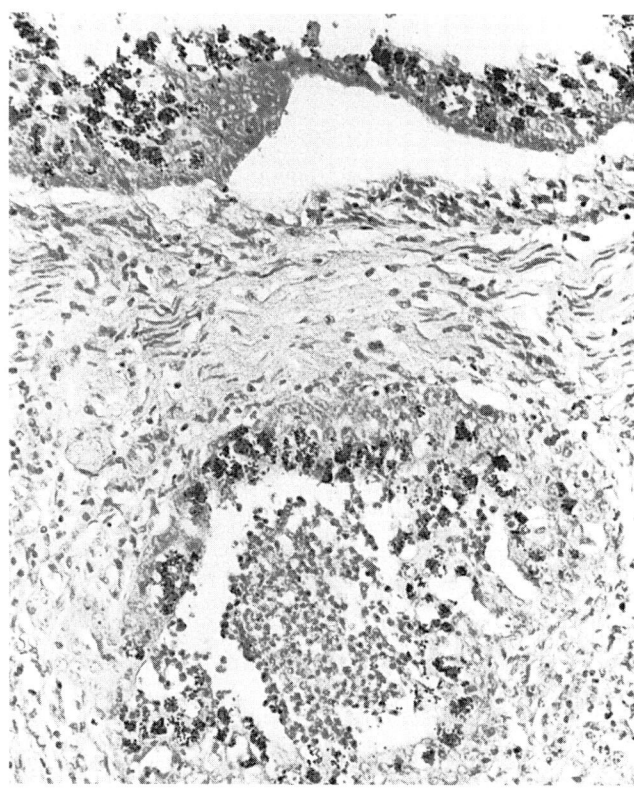

FIGURE 9-19
Gram stain of bronchus from same patient reveals numerous
gram-positive microsporidian spores in lining epithelium and
bronchial glands. The spores were poorly visualized using H&E.
(Tissue Gram stain, Brown and Hopps method, ×105.)

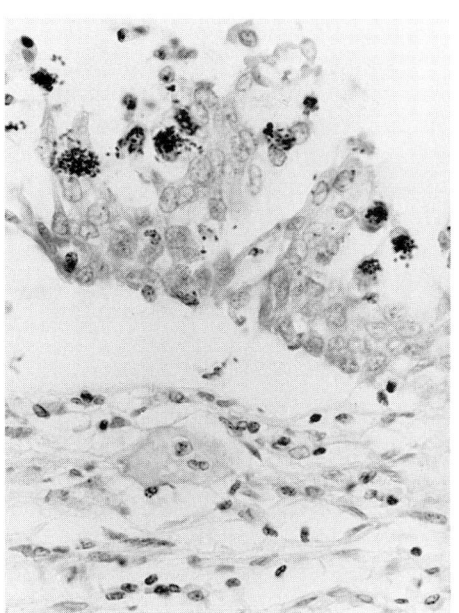
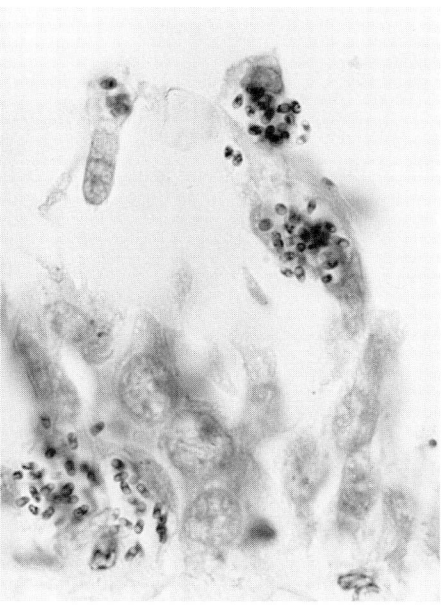

FIGURE 9-20
Bronchial epithelium infected with
E. hellem. The characteristic morphologic features of the spores,
including the equatorial beltlike
stripe, are easily seen using Gram
stain. (Tissue Gram stain, Brown
and Hopps method; ×140 left,
×350 right.)

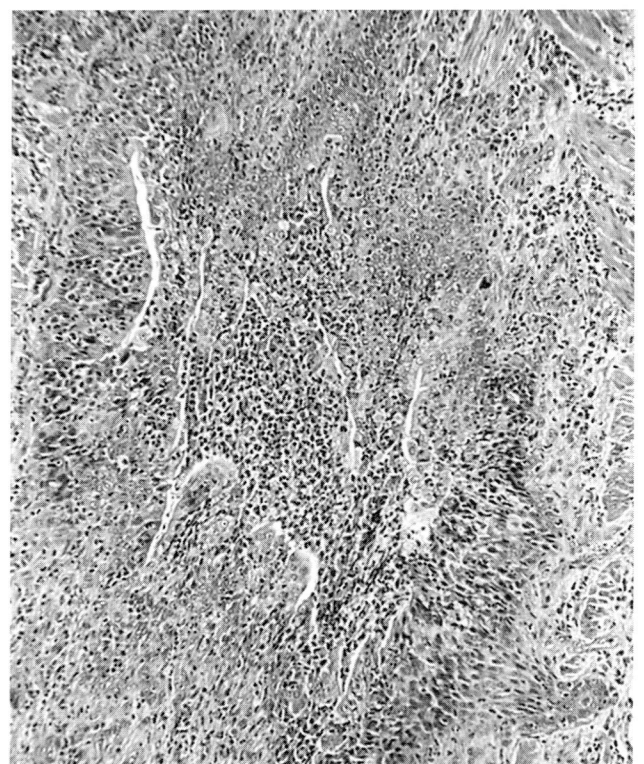

FIGURE 9-21
Open lung biopsy from a patient with *E. cuniculi* infection, acute bronchiolitis, and bronchopneumonia. Purulent exudate containing microsporidian spores fills the lumen of this inflamed airway. Sputum from patients with pulmonary infection is almost always positive for microsporidian spores. (H&E, ×83.) (*Case courtesy of Dr. Michael L. Wilson.*)

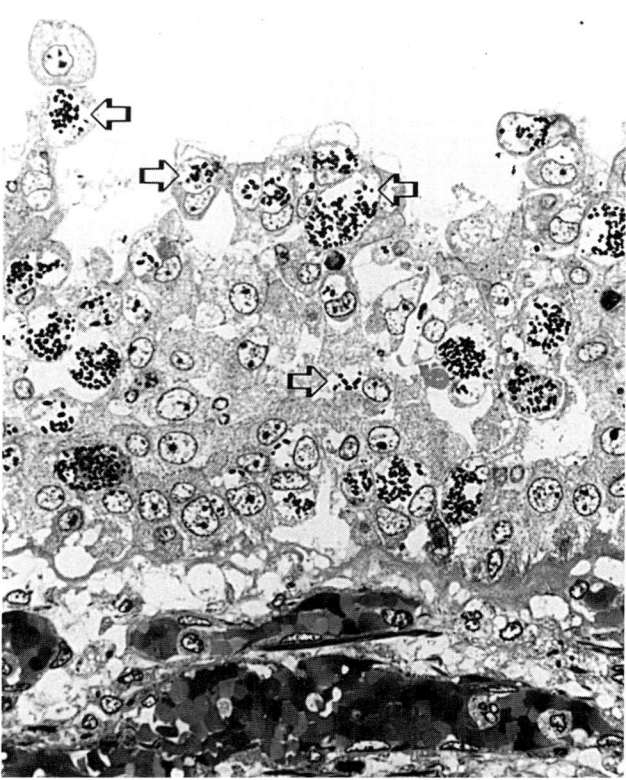

FIGURE 9-23
In this plastic-embedded semithin section of tracheal mucosa infected with *E. hellem*, the parasitophorous vacuole is seen as an intracytoplasmic clear zone (*arrows*) surrounding the spores. (Toluidine blue, ×380.)

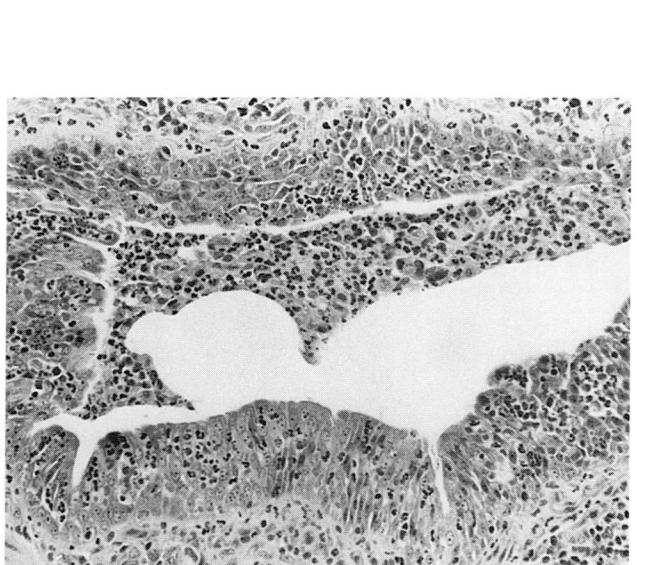

FIGURE 9-22
High magnification of microsporidial bronchiolitis due to *Encephalitozoon*. The etiology of this process may not be suspected unless a Gram stain is performed. (H&E, ×134.)

nalis (Fig. 9-24) is distinguished ultrastructurally from those of *E. hellem* and *E. cuniculi* (Fig. 9-25) by the formation of septations between developing spores. Species identification of *E. hellem* or *E. cuniculi* can only be made using antibody– or nucleic acid–based methods.[55] Although these immunohistochemical and nucleic acid methods are an increasingly important tool for the precise diagnosis of microsporidian infection in both fresh and formalin-fixed cytologic and biopsy specimens,[45] they are currently performed in only a few research laboratories. For clinical purposes, it is useful to identify a microsporidian agent only to the level of genus for the institution of appropriate medical treatment.

DIAGNOSIS

Microsporidian spores in tissues obtained from autopsy, transbronchial biopsy, or open lung biopsy are best seen using the tissue Gram's stain (either Brown and Brenn or Brown and Hopps modifications). However, other staining methods are favored by some investigators, including the chromotrope,

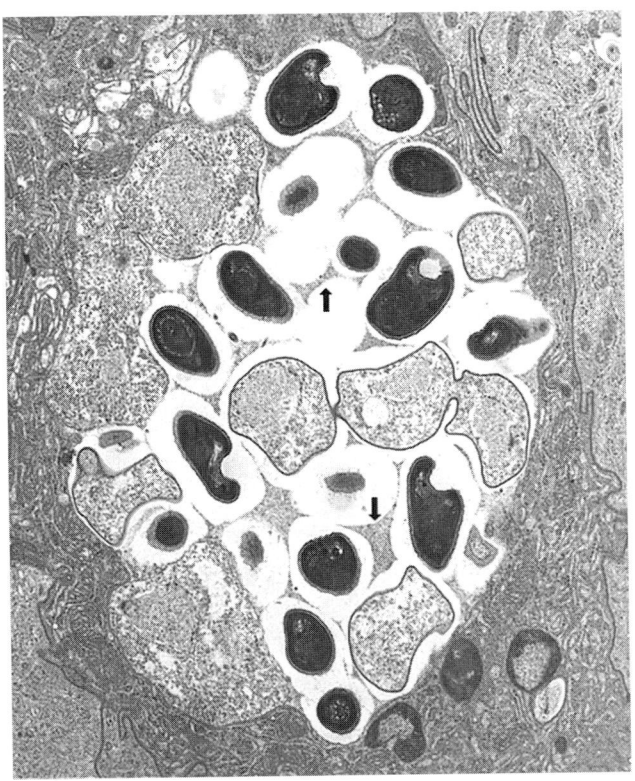

FIGURE 9-24
Electron micrograph of the parasitophorous vacuole of
Encephalitozoon (formerly *Septata) intestinalis*. Both mature and
immature spores are present, and the characteristic matrix
(*arrows*) is present between spores, forming a "honeycomb"
appearance. (×6142.)

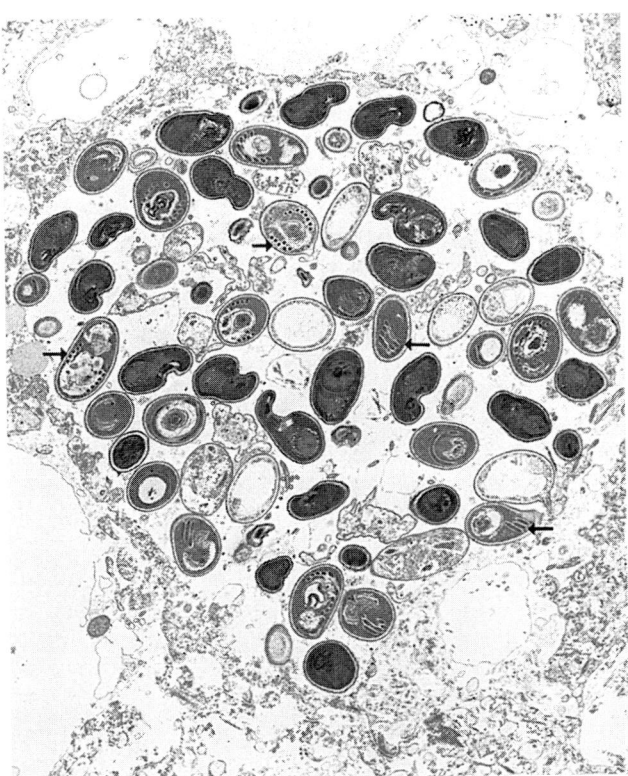

FIGURE 9-25
In contrast to *E. intestinalis*, this parasitophorous vacuole of *E.
hellem* lacks matrix between spores. The coiled polar tubules that
characterize all microsporidia are seen in various planes of section (*arrows*). (×4335.)

Giemsa, and Warthin-Starry stains and chemofluo-
rescent techniques (Uvitex 2B).[45] In many but not all
cases, spores of *Encephalitozoon* may appear birefringent with polarized light.

DIFFERENTIAL DIAGNOSIS

Because of the propensity for systemic dissemination
of *Encephalitozoon* infections, all patients diagnosed
as having *E. hellem*, *E. cuniculi*, or *E. intestinalis*
should be examined for respiratory tract involvement. Sputum cytology is a safe, rapid, non-invasive,
and apparently sensitive method for screening
patients for respiratory tract microsporidiosis.[45]
Bronchoalveolar lavage may be useful in persons
with radiographic evidence of disease and those with
suspected lower respiratory tract microsporidiosis,
those with radiographic evidence of disease, and
those with negative repeated sputum examinations
in whom there is a strong suspicion of microsporidial
bronchiolitis. Upper respiratory infection has been
diagnosed using smears of nasal and sinus exudate as
well as mucosal biopsy. Cytologic preparations
should be stained with either Gram's stain or Weber's

modified chromotrope stain to identify characteristic
spores. Patients who are treated for pulmonary
microsporidiosis can be followed by repeated sputum examinations to assess the efficacy of therapy.

Because of their small size and oval shape,
microsporidian spores must be differentiated from
other similar-sized agents including *Toxoplasma
gondii*, *Trypanosoma cruzi*, amastigotes of *Leishmania* spp., *Histoplasma capsulatum*, bacteria, and
intracellular and neuroendocrine granules. However,
if a diagnosis of microsporidiosis is restricted to
those organisms that show the characteristic
morphology—including appropriate size, shape,
beltlike equatorial stripe, and no evidence of kinetoplast or budding—the diagnosis can be made with
confidence.

Cryptosporidiosis

This disease is caused by the protozoan *Cryptosporidium parvum*, a member of the phylum
Apicomplexa. It is widespread and infects the gastrointestinal tract of a variety of hosts, including fish,
birds, reptiles, and mammals.[56] The main clinical

features produced by this parasite are gastrointestinal. There is acute watery diarrhea, weight loss, crampy abdominal pain, anorexia, and malaise lasting from 10 to 14 days. Respiratory tract involvement by *Cryptosporidium* is rare and has never been reported in immunologically intact individuals, only in patients who are severely immunocompromised. Most of these individuals had concomitant gastrointestinal tract cryptosporidiosis. It is therefore difficult to ascertain if the organism is a contaminant, a colonizer or a true respiratory tract pathogen. In six HIV-positive patients with respiratory symptoms including cough and dyspnea, *Cryptosporidium* was the only pathogen isolated on BAL and brush biopsy.[57]

There are few pathologic reports of cryptosporidial lung disease. One described the organisms on the surface of the bronchial and tracheal epithelium, associated with some metaplasia and involving the glands but no inflammation.[58–64] In one case the organisms were found in the alveolar inflammatory exudate of an AIDS patient's open lung biopsy.[65] This patient also had CMV and *Mycobacterium avium-intracellulare.*

The organisms are easily identified on H&E sections as small, 2- to 5-μm blue-pink staining organisms on the apical border of the epithelial cells (Figs. 9-26 and 9-27). Some organisms have a membranous zone formed by parallel folds of the cryptosporidial double-unit membrane or pellicle, which develops adjacent to the attachment site with the host cell membrane (Figs. 9-27 and 9-28). The organisms can be confused with cellular debris or mucus because of their small size and indistinct structure. A modified acid-fast stain may help in diagnosis,[62,64] as may electron microscopy.

Pulmonary Trichomoniasis

Trichomonads have a worldwide distribution and are occasionally associated with respiratory tract disease. Pulmonary trichomoniasis is usually caused by aspirated *Trichomonas tenax* from the oral cavity, but it can rarely be caused by the genitourinary *T. vaginalis* or the intestinal *T. hominis*. *Trichomonas vaginalis* was isolated from the respiratory tract of infants with respiratory disease.[66] It is not yet clear whether this organism is the primary cause of pulmonary infection or a secondary invader. Most cases of infection with this organism have underlying lung disease, such as carcinoma of the lung, lung abscess, or bronchiectasis.[67] Aspirated *T. tenax* causes infection due to poor oral hygiene and may also be an opportunist.

The organism is a pear-shaped flagellate protozoan with an undulating membrane. It has a worldwide distribution. The trophozoite of *T. tenax* is up to 12 μm long, smaller than *T. vaginalis*, and has five flagellae.

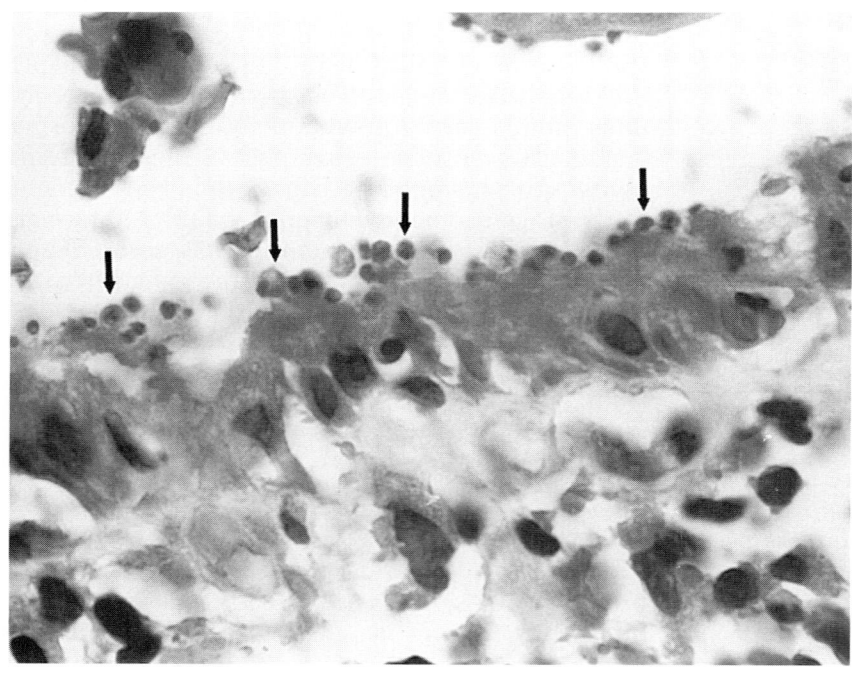

FIGURE 9-26
Numerous cryptosporidia (*arrows*) on the apical surface of epithelial cells from a patient with AIDS. (H&E, ×900.)

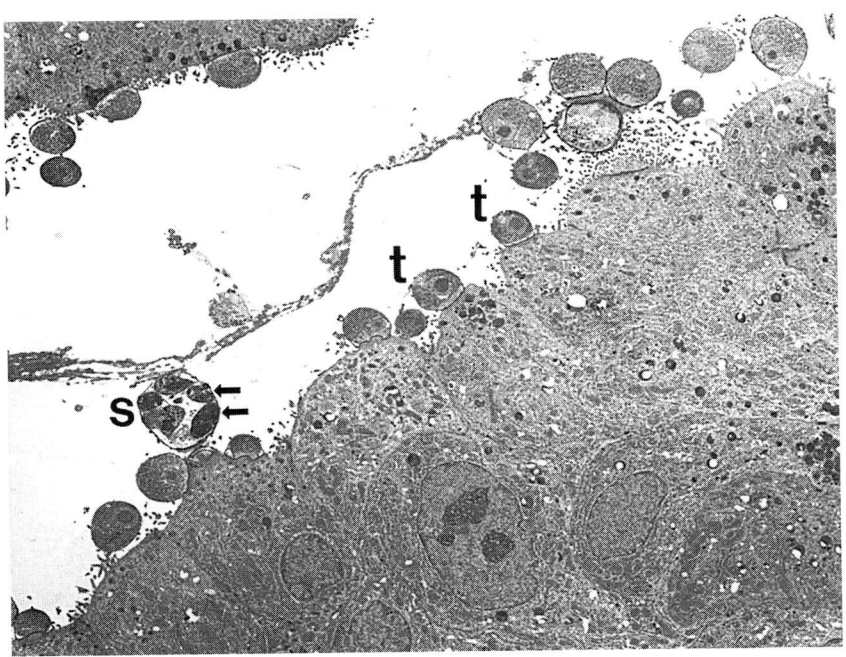

FIGURE 9-27
Electron micrograph showing trophozoites (t) and a schizont (s) containing merozoites (*arrows*) in a case of cryptosporidiosis. (×1260.)

Trypanosoma cruzi (Chagas' Disease)

EPIDEMIOLOGY

This is an insect-transmitted protozoan. It is transmitted to over 150 species of wild and domestic mammals by various species of hematophagous triatomid or reduviid bugs. The insect vectors become infected following a blood meal from an infected mammal. The peripheral blood contains the free-swimming trypomastigote, infective stage of the parasite. Following a period of development in the midgut of the insect, the organisms transform into infective metacyclic trypomastigotes. The infected reduviid bug then takes another blood meal and defecates after feeding. The trypomastigotes are discharged in the insect's feces onto skin, mucous membrane, or conjunctiva. When the host inadvertently rubs the feces into the bite wound, trypomastigotes enter the broken skin. They invade and multiply in a variety of host cells. In humans, *T. cruzi* can infect any cell type but prefers mesenchymal cells, especially macrophages, cardiac and skeletal muscle, neuroglial and adipose cells, cells of the thyroid, and adrenal cortex.

Trypanosoma cruzi is endemic throughout Central and most of South America. It is an enzootic infection in parts of the southern United States. Some 10 percent of the rural population in Brazil have serologic evidence of infection.[68] It is common in rural areas, especially where there is substandard accommodation. Domesticated and farm animals serve as reservoirs for infection. Congenital infection via the placenta is not uncommon.[69]

CLINICAL FEATURES

There is an acute stage of the disease, seen most commonly in children, but as the lung is not involved in this, it is not discussed further here. Chronic Chagas' disease is diagnosed more frequently and is com-

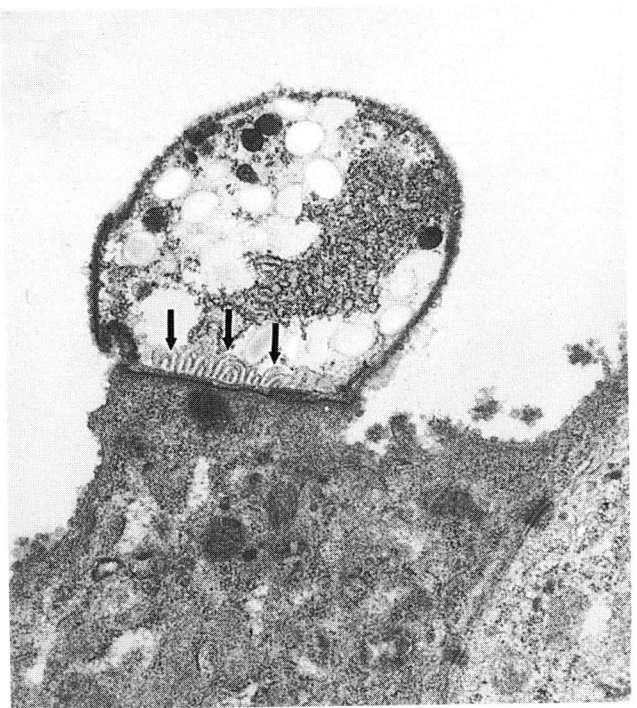

FIGURE 9-28
The attachment of this cryptosporidian to the host cell membrane is clearly seen, including the highly folded parasite pellicle (*arrows*). (×8500.)

moner in adults. The heart is most often involved with abnormalities of conduction, myocarditis, and aneurysm formation. The colon and esophagus may show dilatation and muscular hypertrophy due to destruction of their autonomic ganglia. The megesophagus can cause pulmonary problems secondary to aspiration.

Pulmonary disease due to *T. cruzi* has been reported only in infants with congenital infection.[70,71] Diagnosis was made in almost all cases in stillbirths or children dying in the first few days of life. Two children surviving 41 days with pulmonary disease had hyperinfection and bronchopneumonia.[72]

PATHOLOGY

There is interstitial lung disease. The septa show edema and there is endothelial swelling. Many macrophages and a few neutrophils are seen. Some histiocytes are swollen with hundreds of amastigotes, causing the cells to protrude into the alveolar lumina. The amastigotes are round to ovoid and measure 3 to 5 μm in diameter. They can be distinguished from other organisms such as *Histoplasma*, *Toxoplasma*, and microsporidia by the bar-shaped kinetoplast (Fig. 9-29). The kinetoplast of *Leishmania* spp. is smaller, and this organism parasitizes different tissues. Electron microscopy is helpful in diagnosis (Fig. 9-30).

Acanthamoebiasis

Acanthamoebiasis results from human infection with members of the genus *Acanthamoeba*. These are a group of free-living amebas that do not need a parasitic relationship with an animal host for survival. These amebas have been isolated from soil, water, and air throughout the world. There are no described human carriers or insect vectors. The *Acanthamoeba* species pathogenic for humans include *A. castellani*, *A. polyphaga*, *A. culbertsoni*, and *A. rhysodes*.[73]

CLINICAL FEATURES

Illness due to this organism is uncommon. When present, it takes two main forms, keratitis and granulomatous amebic encephalitis. The latter is an oppurtunistic infection that spreads hematogenously to involve the central nervous system, skin, and lungs. The main presenting symptoms relate to neurologic deficits. Underlying conditions include AIDS, renal transplantation, liver disease, skin ulcers, diabetes mellitus, and immunosuppression with glucocorticoids or other drugs. The skin and

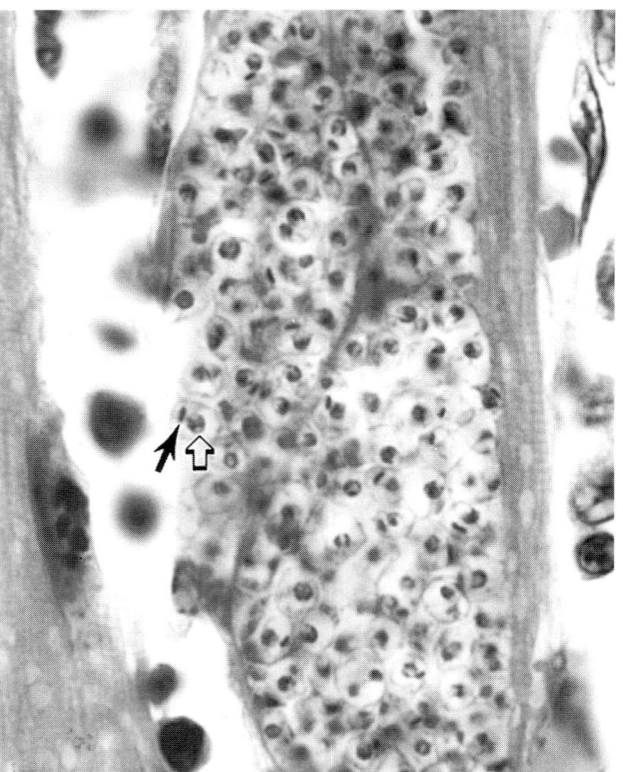

FIGURE 9-29
Intracellular amastigotes of *T. cruzi* in an infected cell. Each organism contains a nucleus (*open arrow*) and bar-shaped kinetoplast (*solid arrow*). (H&E, ×1050.)

lung are the major extraneural organs involved.[74,75] The frequency of pulmonary involvement in granulomatous amebic encephalitis is unknown, but pneumonitis was reported in 2 of 13 patients with central nervous system disease.[73]

PATHOLOGY

There are nodular areas from 1 mm to several centimeters in diameter of necrotizing pneumonia, resembling the target lesions of disseminated candidiasis. There are central areas of necrosis with intraalveolar fibrosis at the periphery. Granulomatous lesions have been described.

On H&E sections, both cysts and trophozoites can be seen. The latter vary in size from 15 to 45 μm. They may initially be difficult to distinguish from necrotic tissue cells and histiocytes, but they have a large, dense, centrally situated intranuclear structure, termed a karyosome (Figs. 9-31 and 9-32). Trophozoites have abundant spongy cytoplasm. *Acanthamoeba* cysts are smaller than the trophozoites, varying from 15 to 20 μm in diameter. They have a stellate or spherical cell wall, which often appears wrinkled or collapsed in tissue sections. This is secondary to shrinkage. The number of organisms seen in infected tissue is inversely proportional

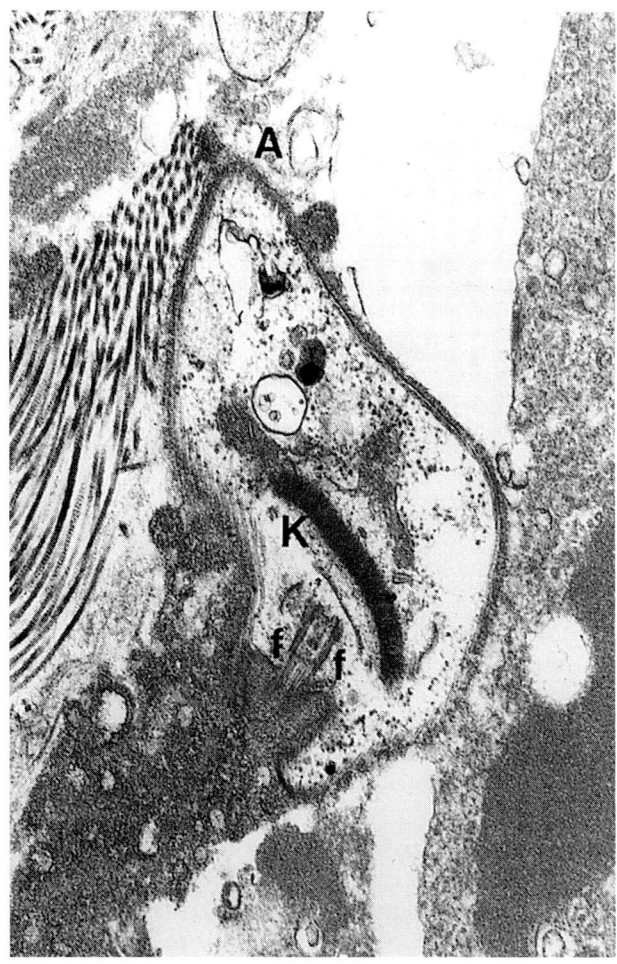

FIGURE 9-30
Electron micrograph of *T. cruzi* showing the anterior tip (A), kinetoplast (K), and flagellar pocket (f). (EM, ×28,500.)

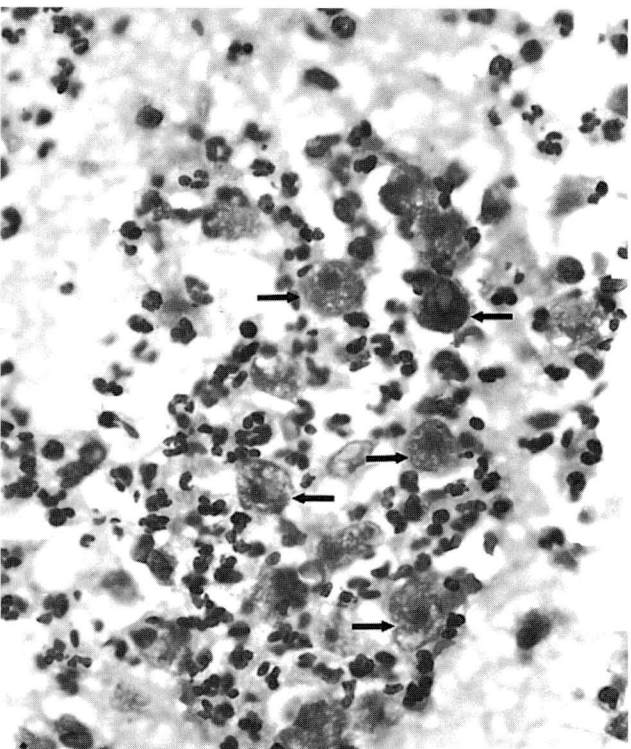

FIGURE 9-31
A cluster of *Acanthamoeba* trophozoites (*arrows*) within an area of necrosis and inflammation. The organisms may be confused with macrophages and other amebas. (H&E, ×348.)

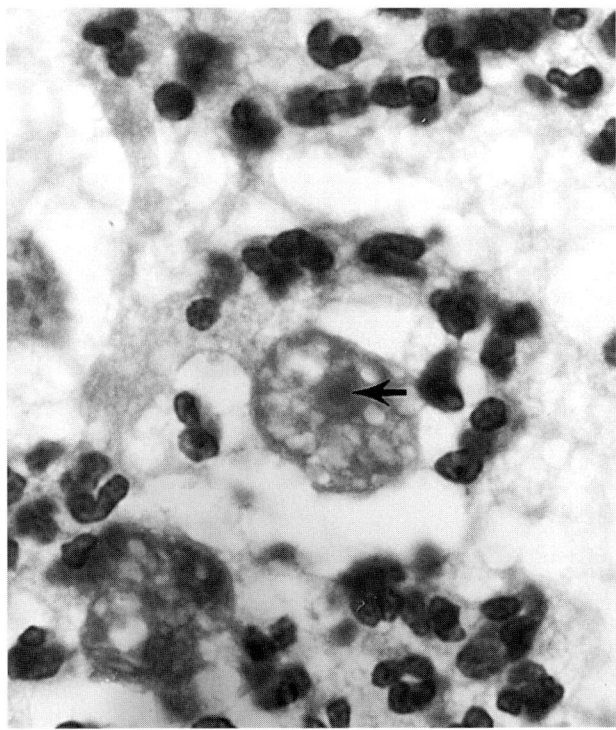

FIGURE 9-32
The diagnostic central nuclear karyosome (*arrow*) and vacuolated cytoplasm is present in this *Acanthamoeba* trophozoite. (H&E, ×1087.)

to the severity of the inflammatory reaction. It is not usually difficult to find diagnostic amebas in lesions.

The trophozoites of *Naegleria fowleri* may show some similarities to *Acanthamoeba*, but the former is smaller and has a poorly developed karyosome. In addition, the cyst is not seen in tissues. Trophozoites of *Acanthamoeba* do not as frequently demonstrate erythrophagocytosis as those of *Entamoeba histolytica*. Species differentiation in the genus *Acanthamoeba* can be reliably performed only by using species-specific antisera.

TREMATODES

Of the trematodes parasitic in humans, the most important are the *Schistosoma* species, but they are not all major causes of pulmonary disease. *Paragonimus* species are significant causes of lung lesions that clinically mimic pulmonary tuberculosis. Rare cases of ectopic *Fasciola* infection in the lung and of infection with *Alaria* larvae are recorded.

All trematode infections have a similar basic life cycle. The egg is released from parasitized persons via feces, urine, or sputum and hatches in water into a miracidium. This enters an aquatic snail, in which cercariae then develop and are released into the water. In schistosomiasis, cercariae directly infect humans, penetrating the skin after water contact. In paragonimiasis, the cercariae enter crayfish or crabs and develop in metacercariae. Ingestion of these uncooked foods leads to infection in humans. *Schistosoma* has separate male and female adult worms; the other species are hermaphroditic. All have two suckers for attachment to host tissues.

Paragonimiasis

THE PARASITE

This disease is caused by infection by numerous *Paragonimus* species. It is endemic in South America, sub-Saharan Africa (including South Africa), India, Southeast Asia, and China. *Paragonimus westermani*, *P. miyazakii*, or *P. heterotremus* account for most infections in Asia, and *P. africanus* and *P. uterobiliateralis* for most African infections.[76] In South and Central America, *P. mexicanus* and *P. ecuadoriensis* are responsible.[77] In North America, the disease is more prevalent among immigrants and travelers, but a few local cases due to *P. kellicotti* have been described.[78] The speciation is currently determined by the morphology of the adult worm; the eggs are similar for all species. All species produce a similar clinicopathologic disease.

The adult flukes found in human lung are reddish-brown and fleshy, measuring 0.8 to 1.4 cm in length, 0.6 cm in width and 0.5 cm in thickness. They are slightly flattened on the ventral surface and have oral and ventral suckers. The cuticle is covered with wedge-shaped spines. The eggs are golden-brown and operculate (Fig. 9-33*A* and *B*).

LIFE CYCLE

Humans acquire the fluke by ingesting larvae in raw or incompletely cooked crustaceans, mainly crayfish and crabs. Preservation of crustacea in brine or rice wine fails to kill the cysts; only boiling or deep-fat frying will effectively destroy them. These metacarcariae hatch in the human duodenum and pass through the bowel wall into the peritoneal cavity and then through the diaphragm. They burrow into the pleural cavity and finaly enter the lung parenchyma. The parasites mature in about 70 days and lodge

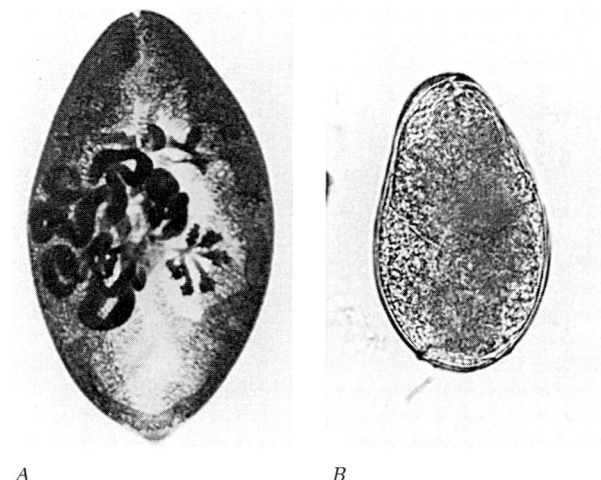

A   *B*

FIGURE 9-33
Adult worm (*A*) and egg (*B*) of *P. westermani*. (*From Warren and Mahmoud,*[188] *with permission.*)

near the bronchioles and bronchi, where they deposit their ova.

The host mounts an inflammatory reaction and immature ova may be discharged into the bronchial lumen, along with blood and inflammatory debris. Some of these ova will be expectorated and others will be swallowed and passed in the feces. They require moist, warm conditions (between 25 and 30°C), so the eggs can develop into miracidia that infect freshwater snails, particularly of the genus *Melania*. Because of the temperature needed, the growth cycle is restricted to the warmer summer months in the more northerly countries of endemic areas such as Korea and Japan. In endemic zones, reinfection is likely, since effective immunity does not develop in paragonimiasis.

CLINICAL ASPECTS

Paragonimiasis causes chronic hemoptysis, slight dyspnea and fever, and, in severe cases, anorexia and weight loss. Because hemoptysis is such a common symptom of patients in endemic areas, the disease is also called "endemic hemoptysis." Chest pain often presents and there might be night sweats in the early stages. Physical examination may be normal, and there may be a pleural effusion.

In a recent review of the radiologic findings,[79] there was patchy air-space consolidation with or without cystic change; there were also ring shadows and peripheral linear opacities that were more prominent in patients with pleural effusions (Fig. 9-34). Pneumothorax was seen in some patients. On computed tomography (CT), round, low-attenuation cystic lesions up to 15 mm filled either with fluid or gas were characteristically present in the consolidated areas. Peripheral linear shadows were also seen on CT and were thought to be suggestive of

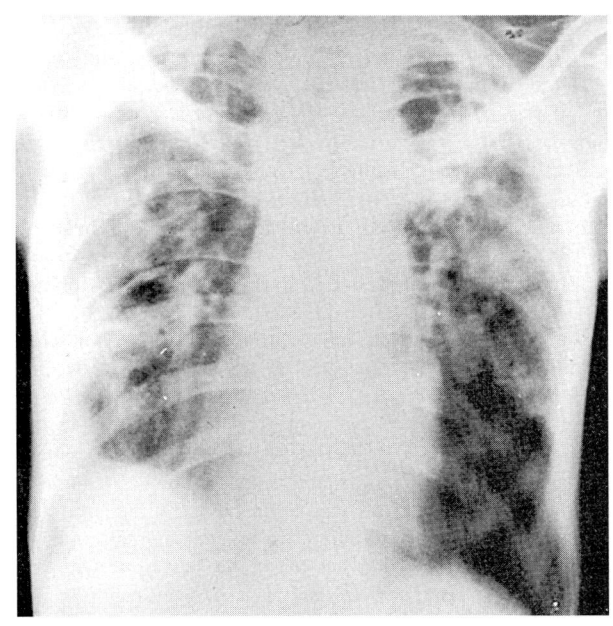

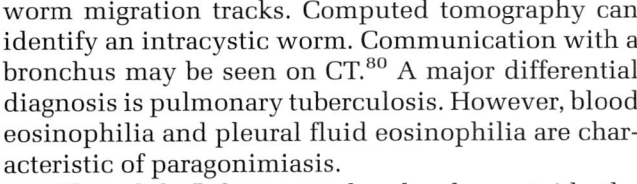

FIGURE 9-34
Chest radiograph showing generalized fibrosis, calcification, and cavitation. This patient was 79 years old, weighed 26 kg, and had had hemoptysis for 50 years. (*From Warren and Mahmoud,*[188] *with permission.*)

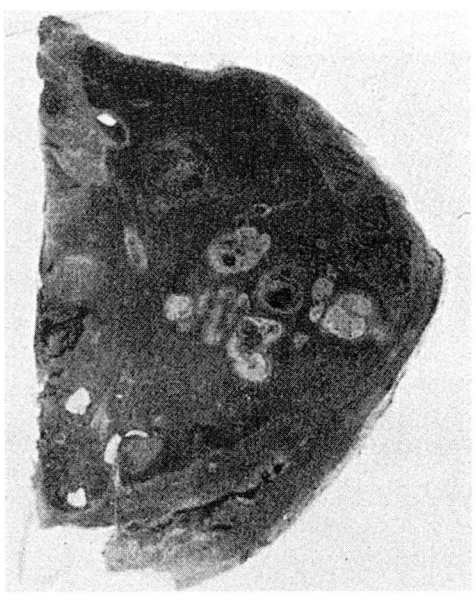

FIGURE 9-35
Part of a lung from a case of paragonomiasis with cavities, some containing caseous material. (*Reproduced by courtesy of the Curator of the Wellcome Museum of Medical Science, London, England. From Spencer,*[189] *with permission.*)

worm migration tracks. Computed tomography can identify an intracystic worm. Communication with a bronchus may be seen on CT.[80] A major differential diagnosis is pulmonary tuberculosis. However, blood eosinophilia and pleural fluid eosinophilia are characteristic of paragonimiasis.

The adult flukes may also develop outside the lung and cause hepatic and splenic lesions, lymphadenopathy, and cerebral space-occupying lesions. The infection may resolve spontaneously with radiologic healing.[81] The life span of the fluke is in the range of 6 to 7 years.

PATHOLOGY

The benign chronic nature of the disease gives few opportunities for the study of the histology.[82–84] As well as the initial invasion of the lung with consequent hemorrhage, there is tissue destruction and chronic abscess formation (Fig. 9-35). Such abscesses may measure up to 5 cm in diameter. These are gray-white in color in the early stages and contain adult worms, usually three in number. The inflammation may involve vessels causing infarction. In addition, the inflammation may erode into the airways, causing subsequent bronchiectasis and distal atelectasia. The inflammation is induced both by the adult parasite itself, with its spiny cuticle and possibly toxic secretions, and by the granulomatous response to eggs deposited around the worm.[85,86] The upper lung zones are affected more than the lower.

Histologically, early lesions contain worms surrounded by granulation tissue with macrophages, polymorphs, and eosinophils. The parasites may be seen in bronchi or bronchioles (Fig. 9-36). This causes either chronic inflammation (Fig. 9-37) or squamous metaplasia (Fig. 9-38). In later lesions, the worms disintegrate, leaving an eosinophilic amorphous debris or an empty fluid-filled cystic cavity. Adjacent lesions can fuse to form a multicystic cavity. Charcot-Leyden crystals are frequently present. Peripheral fibrosis develops and, in this layer, there is progressive deposition of eggs, many of which cause a granulomatous reaction (Fig. 9-39). The eggshells are acid-fast with Ziehl-Neelsen stain. The peripheral lung is collapsed and the overlying pleura thickened. Subpleural lesions may lead to a prominent effusion. Old lesions may calcify, and healing with focal scar formation is frequent.

Ova of *Paragonimus* may be seen in tissue sections and are typically oval with one side slightly flattened; they have a single operculum and the shell at the opposite end is thicker (Fig. 9-40). They are 48 to 60 by 80 to 118 μm in size.[87] Occasionally the ova may be covered with eosinophilic material and may resemble corpora amylacea. The worms are bean-shaped and measure up to 1.4 cm long, 0.6 cm wide, and 0.5 cm thick.[88]

The disease is diagnosed definitively by identification of the ova either in sputum, feces, or fine needle aspiration samples.[89,90] Surgical resection

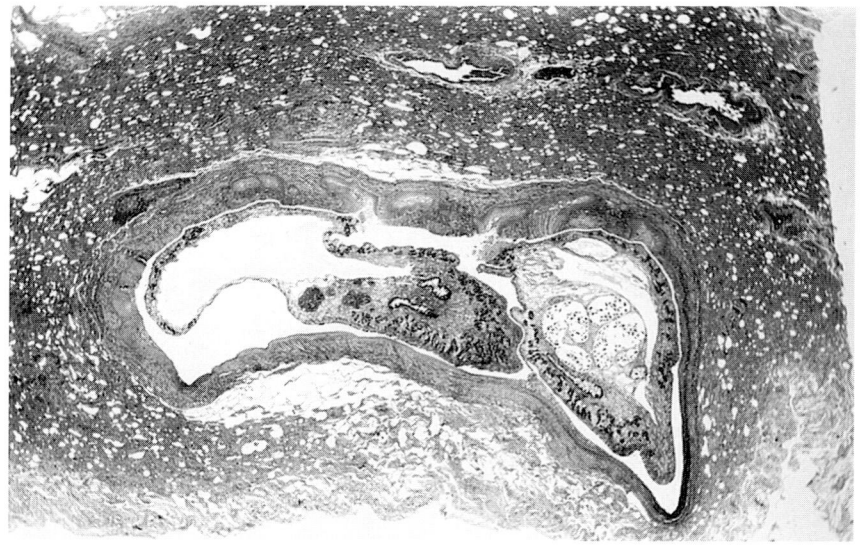

FIGURE 9-36
Low-power view of *Paragonimus*, showing a pair of adults lying in a bronchus.

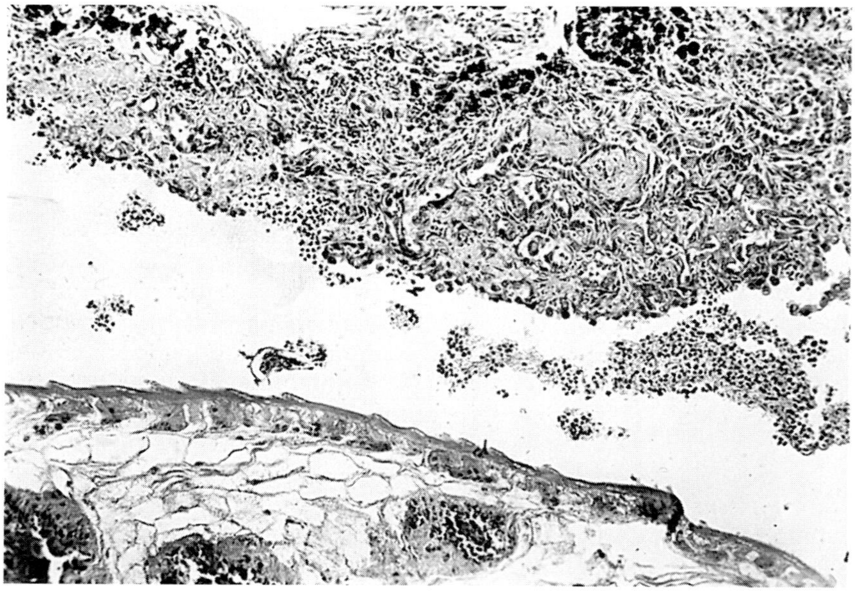

FIGURE 9-37
The spiculated cuticle of the *Paragonimus* is seen in the lower half of the picture. In the upper half there is granulation tissue and chronic inflammation. (Hematoxylin and Bebrich scarlet, ×140.) (*Reproduced by courtesy of Colonel LRS McFarlane, Royal Army Medical College, London, England.*)

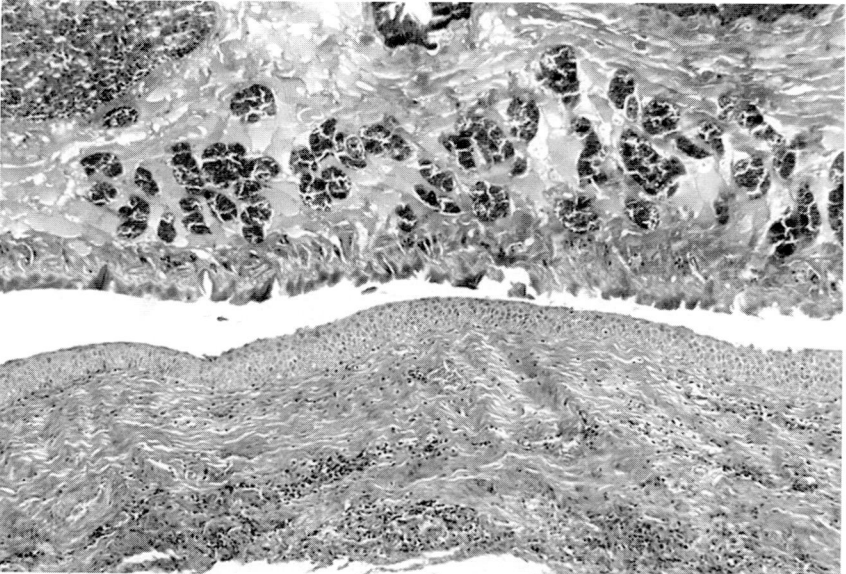

FIGURE 9-38
Paragonimus adult lying in a bronchus with squamous metaplasia. It shows a spiny cuticle. (H&E, ×56.)

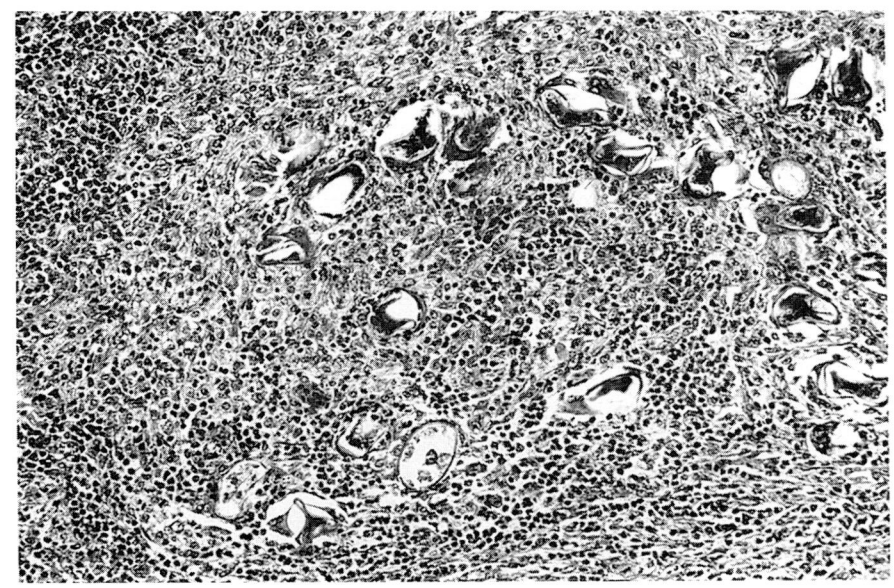

FIGURE 9-39
Paragonomiasis. Numerous *Paragonimus* eggs with an associated granulomatous reaction. (H&E, ×219.)

specimens are uncommon as diagnostic material. Recent advances in serodiagnosis permit sensitive and specific diagnosis of pulmonary paragonimiasis from blood and pleural fluid.[91,92]

The treatment of paragonimiasis is with antihelminthic agents and local resection if necessary.

Schistosomiasis

Among parasitic diseases, schistosomiasis rates second only to malaria in global importance. It occurs mainly in rural and agricultural areas of tropical countries where there is poverty and poor hygiene.

Four epidemiologically distinct parasites cause the disease: *Schistosoma haematobium*, *S. japonicum*, *S. mansoni*, and *S. intercalatum*. The first causes predominantly bladder disease, the others hepato-splenic and intestinal disease.[93] Lung disease is relatively common in *S. mansoni* infection, uncommon in *S. japonicum* and *S. haematobium* infections, and as yet unrecorded in human *S. intercalatum* infection.

Schistosoma haematobium occurs through the Arabian Peninsula, Sudan, Egypt, countries of the North African littoral, and sub-Saharan Africa. *Schistosoma mansoni* occurs in Eygpt, throughout

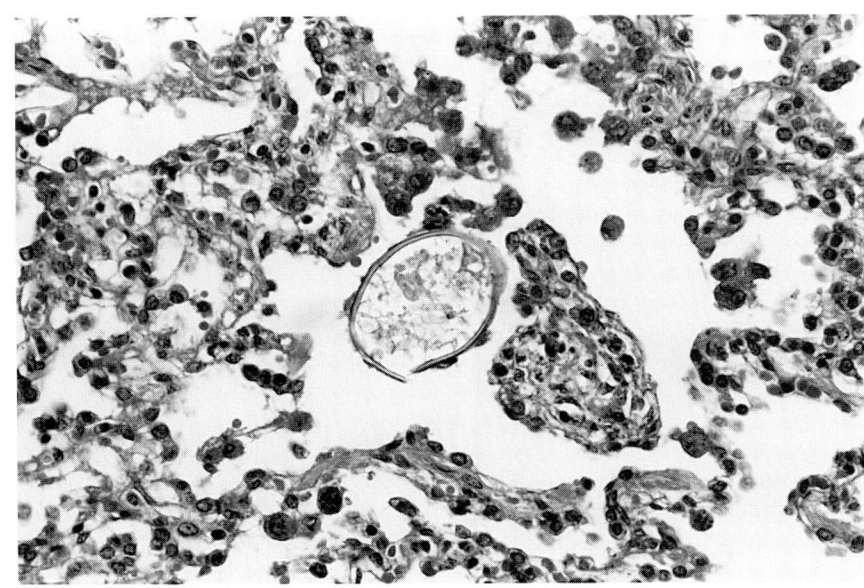

FIGURE 9-40
Paragonimus egg, open at the operculum. (H&E, ×219.)

sub-Saharan Africa, and also in the New World: Brazil, Venezuela, and certain Caribbean islands. *Schistosoma japonicum* has a wide geographic distribution ranging from China to the Philippines to Indonesia. *Schistosoma intercalatum* occurs in a few areas of Central Africa.

For a detailed account of the life history of these parasites, the reader is referred to works on parasitology and tropical pathology.[94] Humans are infected after contact with water containing cercariae. They penetrate the skin, transform into immature schistosomes, migrate to the lung for a maturation phase, and then enter the portal venous system. Thereafter, *S. haematobium* worms are mainly located in perivesical and colorectal veins; the other species reside in mesenteric and portal veins. Adult males and females pair in veins and deposit eggs from 6 to 8 weeks after infection. Schistosome worms do not replicate in humans, but infection intensity builds up from repeated reinfections. Most worms live for 3 to 5 years, with exceptional longevities of two decades or more. Although worms may end up in ectopic sites (e.g., veins of the spinal cord, brain, and skin) only 1 percent of worms are found in pulmonary vessels.

There are two patterns of schistosomal disease, acute and chronic. Acute schistosomiasis (sometimes known as Katayama fever) occurs in the first weeks after infection and is a multisystem disorder (see below). It is usually seen in adults from nonschistosomal endemic areas who become infected for the first time (e.g., tourists in Africa), and it may be caused by one of the schistosome species. Acute schistosomiasis has some features of an immune complex disease and involves reactions to maturing adult worms and/or the phase of first egg deposition.[95]

Chronic schistosomiasis is the more important disease and results from the host cellular inflammatory reaction to deposited eggs. Approximately 50 percent of eggs are excreted via the bladder and large bowel mucosa with *S. haematobium* and via small and large bowel mucosa with *S. mansoni* and *S. japonicum*. Retained eggs and eggs being extruded cause disease. Their secreted products are antigenic and induce a mixed eosinophilic and granulomatous reaction through a T cell–mediated immune response. Accumulating granulomas induce fibrosis, as in other granulomatous diseases.

With heavy chronic infections from *S. mansoni* and *S. japonicum*, there may be extensive noncirrhotic portal fibrosis of the liver; portal hypertension, and splenomegaly. As a consequence of portal hypertension, eggs deposited in mesenteric and portal veins may pass via collateral veins to the lung. With

S. haematobium infection, eggs may pass directly to the lung in systemic veins. Schistosome eggs are ovoid and range in size from 60 by 80 μm (*S. japonicum*, with a minimal spine on the eggshell) to 60 by 150 μm *S. mansoni* (lateral spine) and *S. haematobium* (terminal spine) (Fig. 9-41). The eggs impact in the pulmonary arterial system.

Chronic pulmonary schistosomiasis is therefore secondary to small vessel lesions and causes pulmonary hypertension and cor pulmonale. However, the prevalence of clinical disease varies with the species of schistosome. *Schistosoma haematobium* eggs are frequently found in lung tissue, but actual secondary disease is rare. In an Egyptian study, about one-third of all patients with schistosomiasis (predominantly *S. haematobium*) showed some degree of lung involvement. In only 2 percent of these cases was pulmonary vascular disease responsible for death.[96] Similarly in *S. japonicum* infection with portal hypertension, cor pulmonale is uncommon.[97]

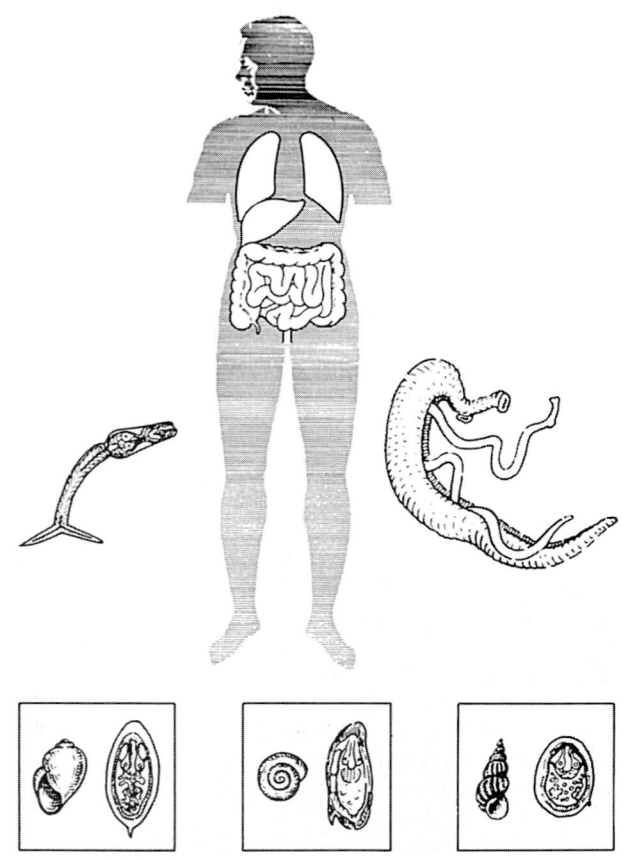

FIGURE 9-41
Characteristic shapes of *S. hematobium, S. mansoni,* and *S. japonicum. (From Warren and Mahmoud,*[188] *with permission.)*

It is among patients with *S. mansoni* infection that cor pulmonale most frequently occurs. In one series from Egypt, right heart failure was seen in 7.5 percent of 682 patients with schistosomal hepatic fibrosis.[98] These species differences are not yet explained, although it has been proposed that intense oviposition over a short period (as with *S. mansoni*) is more likely to induce cor pulmonale than a gradual buildup of eggload (as with *S. haematobium*).[99]

CLINICAL FEATURES

In acute schistosomiasis there is fever, rigors, headache, anorexia, weight loss, abdominal pain, myalgia, diarrhea, blood eosinophilia, and a dry cough.[100] Radiologically, these patients have miliary mottling resembling miliary tuberculosis or condensed basilar and midzone infiltrates. The etiology of this pulmonary syndrome is unknown and the pathology has not been evaluated. It may be a form of allergic alveolitis secondary to high levels of immune complexes found in the serum.[101,102] The syndrome usually subsides over a period of 2 months. Therapy is with steroids in addition to antischistosomal chemotherapy.

Chronic pulmonary schistosomiasis passes through sequential stages. Initially patients are symptomatic, but radiology shows lower-zone widened proximal arteries, thicker peripheral arteries, and focal nodules greater than 1 mm in diameter near peripheral vessels. Later, symptoms are breathlessness and fatigue, cyanosis, and then right ventricular failure. Radiologic findings are right ventricular enlargement, aneurysmal dilatation of the pulmonary artery and its main branches, and fine mottling throughout the lung fields.

Other Forms of Pulmonary Schistosomiasis

New lung involvement can occur after treatment for schistosomiasis, mainly among children. Patients may develop eosinophilia and new pulmonary symptoms such as cough and wheeze.[103,104] Other patients may show atypical features and present with a parenchymatous infiltrate, a cavity, and even hilar lymphadenopathy suggesting pulmonary tuberculosis.[105] Generalized enlargement of mediastinal nodes may raise the possibility of lymphoma.[106] Intense focal egg deposition in the lung may cause coin lesions on lung radiographs, mimicking a tumor; these represent bilharziomas (inflammatory schistosomal pseudotumors).

PATHOLOGY

In chronic schistosomiasis, eggs reach the pulmonary arteries and cause a pulmonary arteritis.[107] The eggs are localized in pulmonary arteries (Figs. 9-42 and 9-43) and there is a surrounding granuloma formation with macrophages and eosinophils. Grossly, the lung may appear normal or have numerous 0.5-mm white granuloma nodules on the cut surface (Fig. 9-44). The pulmonary artery may be atheromatous from pulmonary hypertension.

The significant pathology is that of granulomatous lesions in arteries and mainly arterioles.[106,108,109] The majority of eggs lodge in

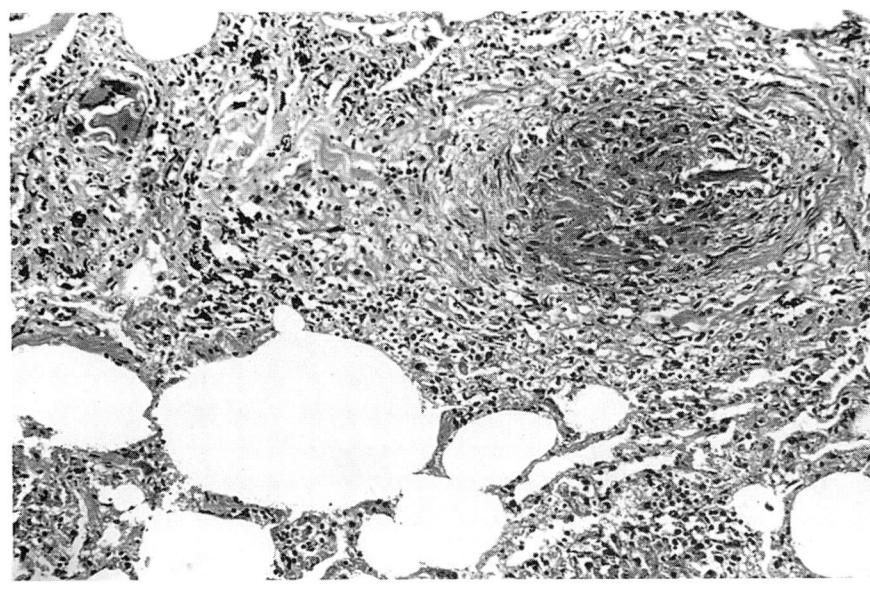

FIGURE 9-42
Schistosomiasis. Active obliterative arteritis with adjacent egg and giant cell reaction.

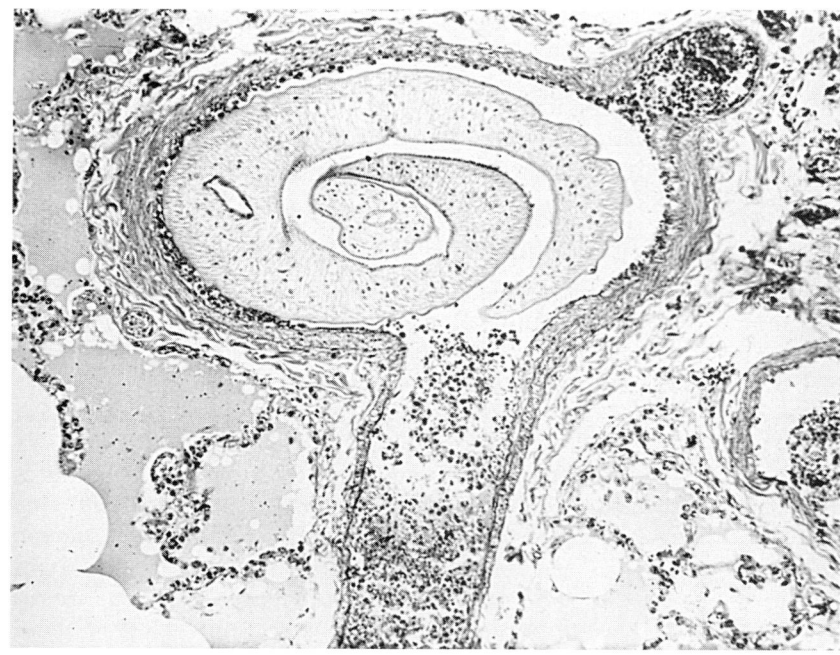

FIGURE 9-43
Pulmonary artery containing adult male and female *S. mansoni*. The adult flukes are alive, hence the lack of any vascular reaction. (H&E, ×140.)

pulmonary arterial vessels ranging from 50 to 100 μm in diameter. The endothelial cells are hyperplastic and there is fibrin deposition.[110] The eggs cause intravascular (Fig. 9-45) and perivascular granulomatous reactions and the vessels are surrounded by histiocytes, giant cells, eosinophils, and lymphocytes, causing a schistosomal granuloma. This eventually destroys the whole or part of the vessel wall and there is intimal fibrosis (Figs. 9-46 and 9-47). An egg can be seen in the middle of the vessel. Other ova may excite little extravascular reaction after eroding through the vascular wall and are ejected into an alveolus or peribronchial tissues. Eggs may impact in larger, 100- to 250-μm arteries and induce an acute necrotizing arteritis along with an eosinophilic granulomatous reaction and lumenal thrombosis. In severe cases,

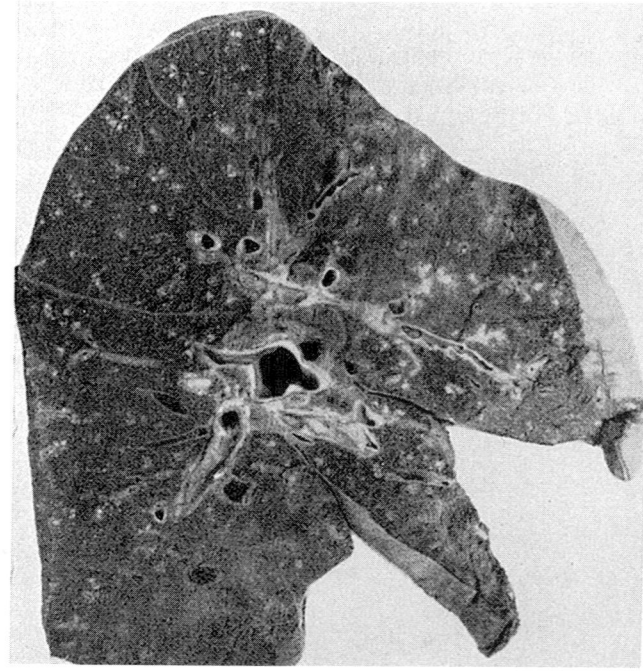

FIGURE 9-44
Pulmonary schistosomiasis (*S. mansoni*) with numerous pale granulomatous areas. (*Reproduced by courtesy of Dr J. Lopez de Faria.*)

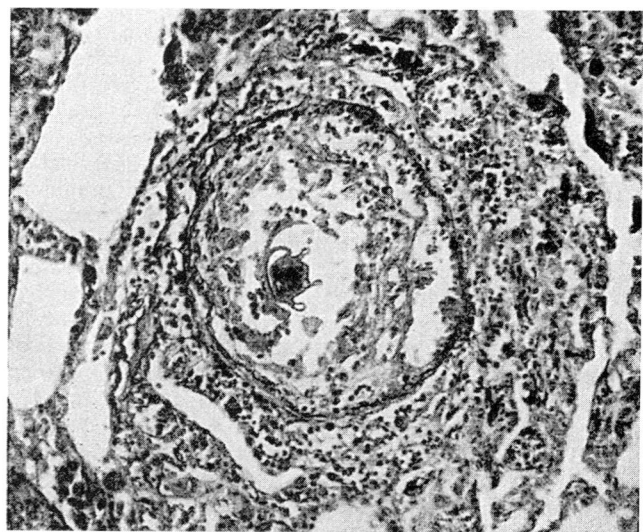

FIGURE 9-45
Small pulmonary artery containing an egg of *S. mansoni* and an intravascular granuloma, which has practically obliterated the lumen. (EVG, ×260.) (*Reproduced by courtesy of Dr J. Lopez de Faria.*)

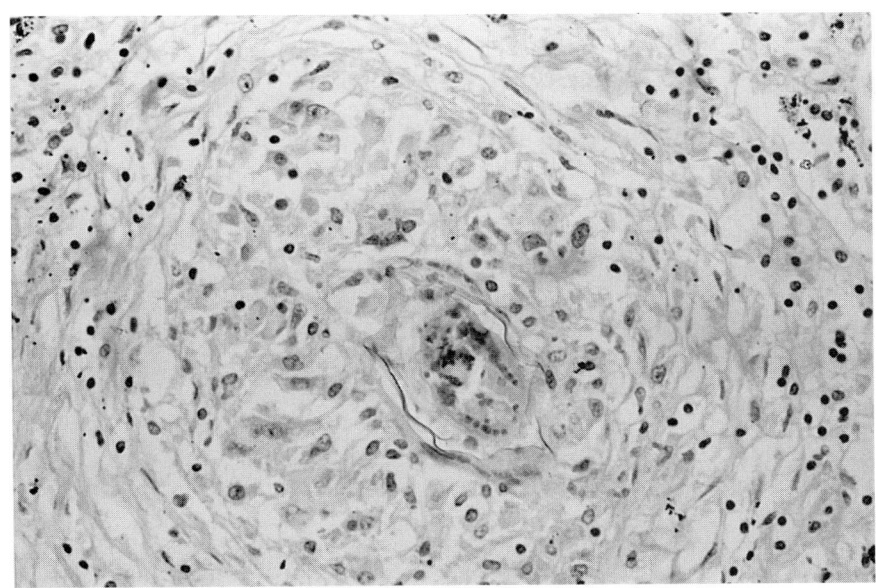

FIGURE 9-46
Schistosome egg with a tuberculoid granulomatous reaction. (H&E, ×350.)

plexiform (Fig. 9-48) lesions result. These are similar to those found in pulmonary hypertension of other etiologies, and there is general medial hypertrophy of arterioles and arteries (Fig. 9-49).

Plastic vascular casting of the lungs of autopsied patients with schistosomal cor pulmonale showed plexiform lesions that included tortuous new vessels, ending up as capillaries in the alveolar walls. The plexiform lesions were glomeruloid in appearance and purely arterial in origin; no vascular shunts were seen.[110]

The *S. mansoni* egg has a lateral spine and the shell is acid-fast with Ziehl-Neelsen stains (Figs. 9-50 and 9-51). The eggs live for about 3 weeks and then degenerate, with macrophages and giant cells phago-

cytosing the contents. Calcified fragments of eggs may remain in tissues for years.

Extensive schistosomal lung lesions may ultimately cause widespread interstitial fibrosis, in which calcified eggshells may be seen (Figs. 9-52 and 9-53).

Bronchiectasis due to *S. haematobium* lesions of the lung has been described,[111] perhaps caused by bronchial damage due to schistosomal granulomas together with ischemic damage from obliterative endarteritis in the vessels of the bronchial wall. The pathology of a pulmonary biharzioma is of confluent schistosomal granulomas with much fibrosis.

The pathology of the lung in acute schistosomiasis is unclear. In the posttreatment pulmonary syn-

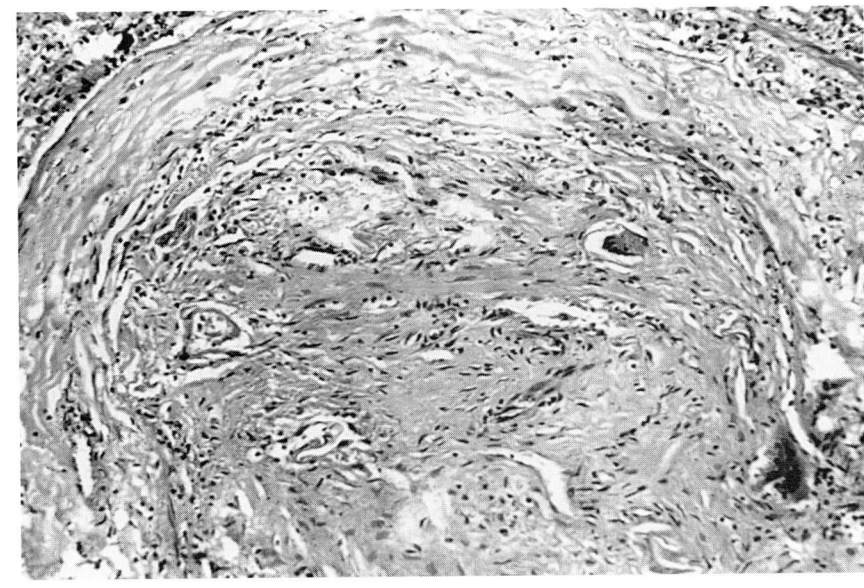

FIGURE 9-47
Schistosomiasis. Pulmonary activity with occlusive intimal fibrosis and recanalization. (H&E, ×219.)

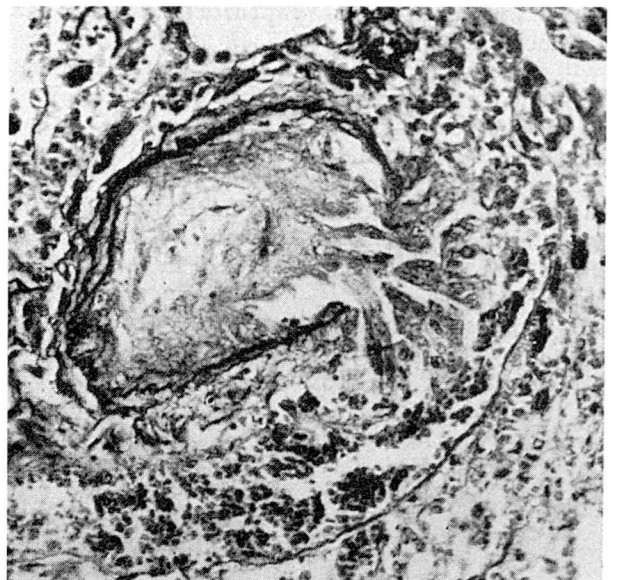

FIGURE 9-48
Plexiform lesion of a branch of pulmonary artery with occlusive intimal fibrosis and a series of sinuous vascular channels opening from the vessel. (EVG, ×300.) (*Reproduced by courtesy of Dr J. Lopez de Faria.*)

drome (see above), lavage shows many eosinophils; on transbronchial biopsy there is interstitial infiltration with macrophages and eosinophils and a patchy alveolar exudate. No schistosomes or eggs are seen and the syndrome is thought to be self-limiting, possibly affecting the immunologic response to a new release of antigens.[103,104] This form of disease is therefore analogous to acute schistosomiasis.

With *S. haematobium* and *S. japonicum* infec-

tions, the vascular lesions are less severe than with *S. mansoni*, but eggs and granulomas are seen related to arterioles and in the alveolar walls.

DIAGNOSIS AND TREATMENT

Definitive diagnosis of schistosomiasis depends on demonstration of the ova. Examination of stool and urine detects most cases of infection, but lung biopsy may be needed to prove that a pulmonary syndrome is directly due to schistosomiasis in an infected person.

Immunodiagnostic techniques such as ELISA, radioimmunoassay, and complement-fixation tests are useful in supporting the diagnosis of schistosomiasis. A disadvantage is that they indicate past exposure without specifying the duration or quantum of infection and remain positive after definitive chemotherapy.

The most used chemotherapy for schistosomiasis is praziquantel, which kills the adult worms. It has no effect on eggs already deposited; thus chronic established disease persists.

Other Trematode Infections in the Lung

Fasciola hepatica is a fluke with worldwide distribution. It is acquired from eating contaminated watercress and the adult resides in the liver. Occasional cases of ectopic infection in the lung are reported.[112] The flukes do not achieve maturity in such sites; they become incorporated into a granuloma and often calcify.

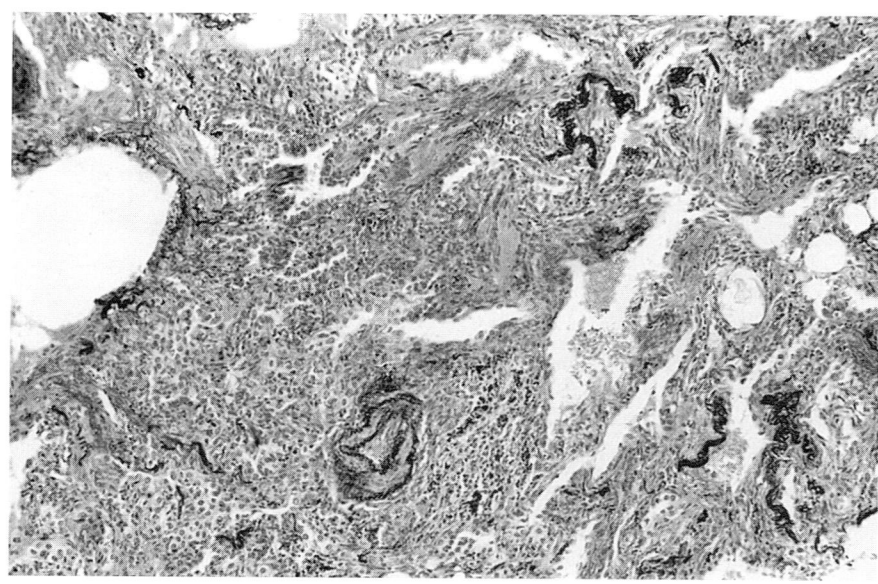

FIGURE 9-49
Schistosomiasis. Hypertrophy of a muscular pulmonary artery amid pulmonary fibrosis.

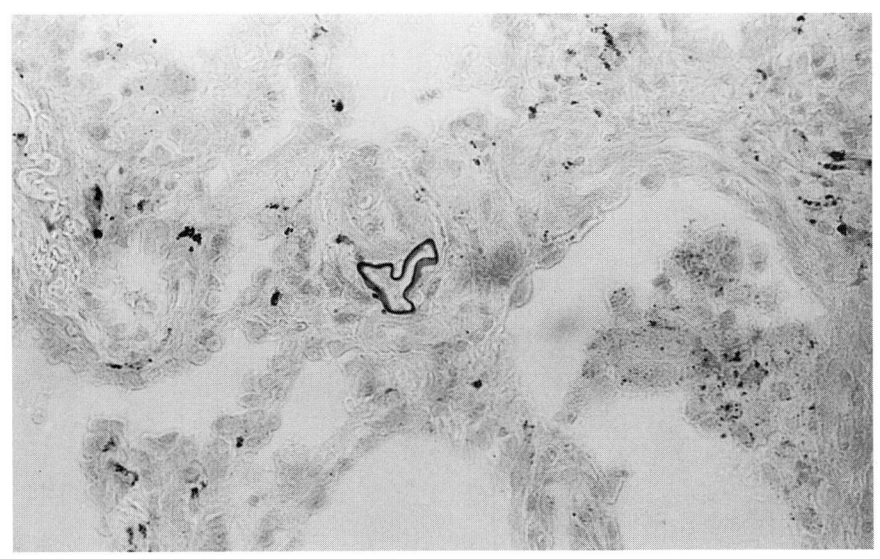

FIGURE 9-50
Schistosoma mansoni egg with acid fast shell. (Ziehl-Neelsen, ×350.)

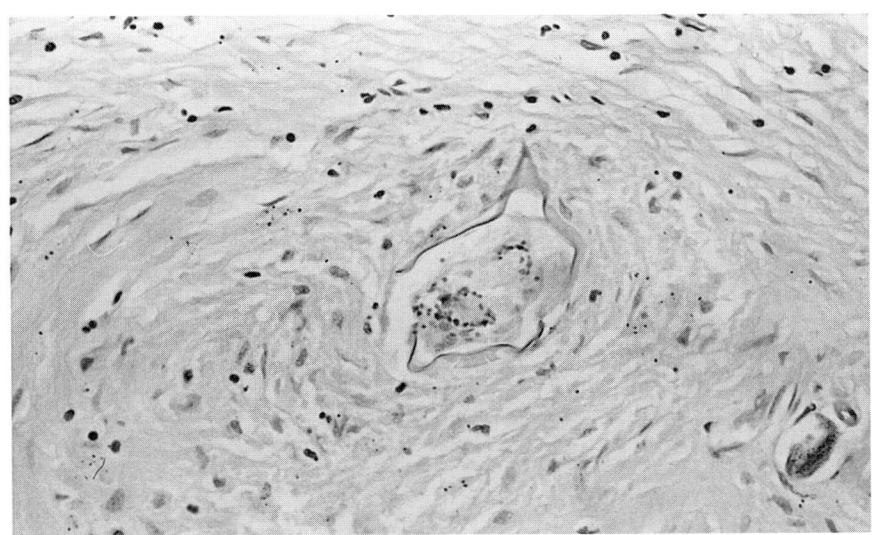

FIGURE 9-51
Schistosoma mansoni egg with characteristic lateral spine. (H&E, ×350.)

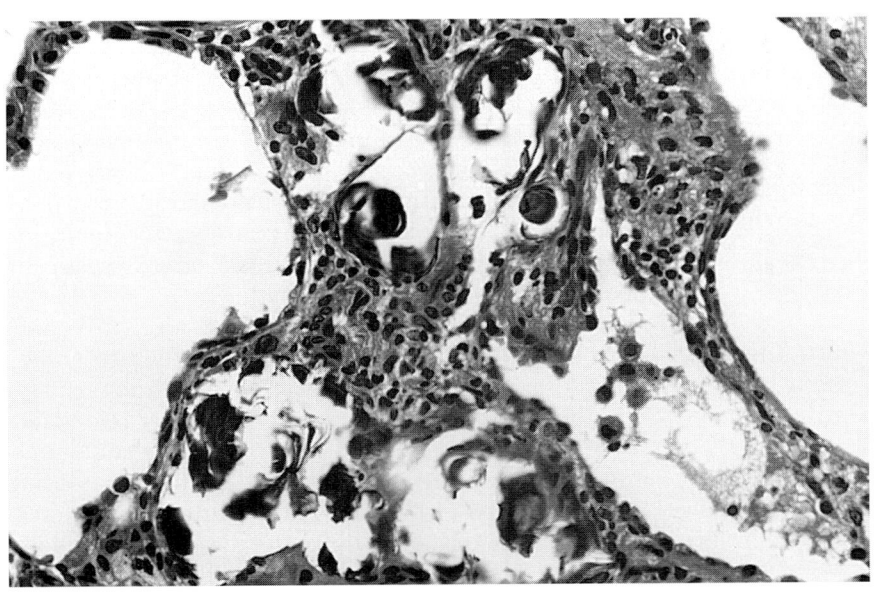

FIGURE 9-52
Cluster of dead calcific fragmented schistosome eggs. (H&E, ×219.)

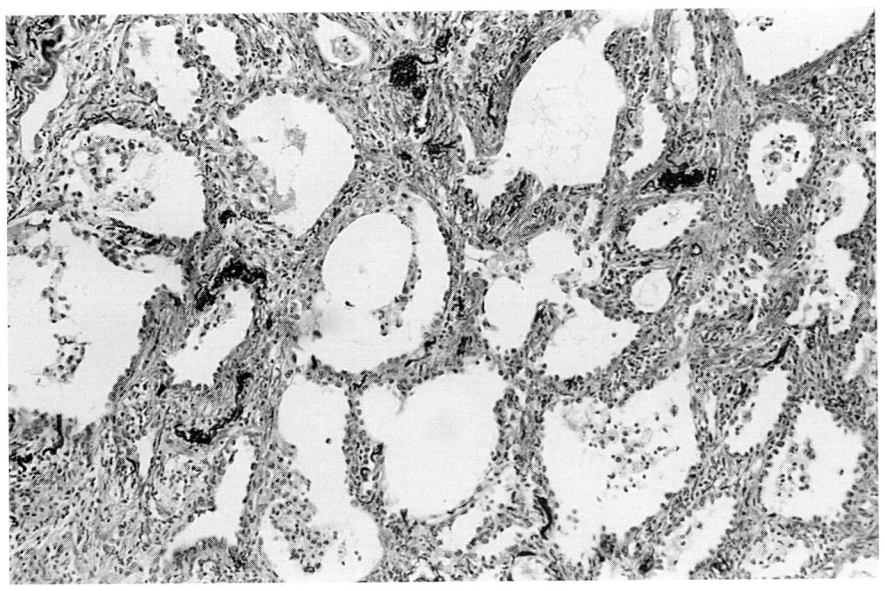

FIGURE 9-53
Schistosomiasis. Diffuse interstitial fibrosis. (H&E, ×87.5.)

The larvae of *Alaria* species rarely cause disease in humans in North America. As part of disseminated nonmaturing infection, the lung may be involved. Focal parenchymal hemorrhage is seen around the larvae.[113]

NEMATODES

These worms are the commonest parasites to infect humans, the adult forms residing most often in the gut. However, during migration and maturation, many gut worms pass through the tracheopulmonary system and may cause lung disease. Additionally, the lung is primary locus of infection of certain nematodes. Many of these inflammatory lung lesions present as coin lesions on a chest radiograph; they are often asymptomatic and mimic carcinoma or tuberculosis.

Filarial Lung Disease

The filarial worms include those causing lymphatic filariasis, onchocerciasis, loiasis, and dirofilariasis. A variety of pulmonary syndromes are associated with several of these infections.

Pulmonary Dirofilariasis

This disease is caused by *Dirofilaria immitis*, the dog heart worm, and is a zoonosis. It occurs in most states of the United States[114–116] and transmission has occasionally been reported in Europe.[117]

The adult worm of *D. immitis* reside in the right ventricle and pulmonary arteries of the dog, but other hosts include cats, wolves, coyotes, and foxes. Larval microfilariae circulate in the blood and are picked up by biting mosquitoes, in which larval maturation takes place prior to further infection of a dog. In humans, larvae introduced by an infected mosquito cannot develop to a sexually mature form, and the human is therefore a "dead-end" host.

CLINICAL FEATURES

As in the dog, the worm in human dirofilariasis inhabits the pulmonary artery. The clinical features are those of pulmonary infarction.[118] The male: female ratio is 2:1, with a median age in the fifties. Patients have chest discomfort, malaise, low-grade fever, cough, and occasional hemoptysis. However, about 60 percent of those infected are asymptomatic and the lesions are detected on routine chest radiograph. The radiologic presenting feature is a solitary pulmonary nodule; in 5 percent of cases, there may be multiple nodules present at the periphery of the lung. The initial radiologic change may be that of a pleural effusion and associated pulmonary infiltrates. In the case diagnosed in Europe,[117] there was a transitory coin lesion and it was presumptively diagnosed by ELISA. Following conservative management the nodules disappeared. Blood eosinophilia occurs in up to 15 percent of patients.

Dirofilariasis may be confused clinically and radiologically with primary or metastatic carcinoma, tuberculoma, and thromboembolic pulmonary infarction.[119] In one case report that was associated with *D. immitis*, well-differentiated squamous epithelial cells in an infiltrative pattern was seen within a well-circumscribed nodule of infarcted necrotic lung tissue. Initially the diagnosis of a well-differentiated squamous carcinoma was considered, but on subsequent examination, the typical worm of *D. immitis* was identified. The changes were thought to be squamous metaplasia in response to the organism.[120] There have been case reports of human dirofilariasis associated with malignancies.[121,122]

PATHOLOGY

Macroscopically, the dirofilarial granuloma is essentially an embolic infarct. Unlike a classic thromboembolic infarct, it is a spherical nodule. This has been attributed to the fact that there may be a radial effusion of antigen from the degenerating worm into the surrounding parenchyma, with a consequent granulomatous reaction.[123] The lesions are well circumscribed, grayish-yellow nodules, 1 to 4 cm in diameter, surrounded by normal parenchyma. They are usually subpleural.[124]

Histologically, there is central coagulative necrosis surrounded by fibrosis. At the periphery there are histiocytes, lymphocytes, eosinophils, and occasional multinucleated foreign-body giant cells. Charcot-Leyden crystals are frequent. Within the necrotic tissue, a pulmonary artery with intimal proliferation contains a dirofilarial parasite (Fig. 9-54 *A* and *B*). The association of a spherical infarct, eosinophilic pneumonitis, and a proliferative endarteritis should alert one to search for the parasite. Fine needle biopsy of a dirofilarial lung lesion has been diagnostic.[125]

The parasite is immature (i.e., the genital tract contains no larval microfilariae) and is often degenerate. It is 100 to 300 μm in diameter and has a thick cuticle (up to 25 μm). Characteristically, there are fine longitudinal ridges. Other nematodes that may lodge in pulmonary arteries to cause local necro-inflammatory lesions are *Angiostrongylus*, *Brugia*, and *Wuchereria*. Distinction of these lesions involves both a geographic history and morphology.

Serologic diagnostic tests for dirofilariasis still lack sensitivity and specificity.[126,127] After the death of the worm, following impaction in the pulmonary artery, antibody titers decline.

The Lymphatic Filariae—*Wuchereria* and *Brugia*

Lymphatic filariasis is a very common disease throughout the tropics, with about 90 million people infected. It is caused by *Wuchereria bancrofti* in Africa, India, and South America and *Brugia malayi* in Southeast Asia. Humans are infected by mosquito, and the adult worms normally reside in lymphatics and lymph node sinuses. The microfilaria (larvae) circulate in the blood with nocturnal periodicity.

CLINICAL FEATURES

The adult cause bouts of fever and lymphangitis, which in some patients proceed to elephantiasis (lymphedema) of the limbs, external genitalia, or breast.

Pulmonary disease from these worms takes in two distinct forms: rare obstructive pulmonary arterial lesions (caused by ectopic adult mosquitoes) and tropical pulmonary eosinophilia, caused by microfilariae.

PATHOLOGY

The arterial lesion is a thrombus around an immature adult worm that has impacted in the vessel. It produces an inflammatory pulmonary nodule with local infarction. *Wuchereria* female worms are 250 μm in diameter (Fig. 9-55), and *Brugia* are smaller at 100 μm (Fig. 9-56). Within the thrombus, there is an eosinophilic infiltrate around the worms and a Hoeppli-Splendore reaction.[128] The parasite has also been observed in pleural fluid.[129]

Tropical Pulmonary Eosinophilia

Tropical pulmonary eosinophilia (TPE) is the lung component of the tropical eosinophilia syndrome (Weingarten's syndrome). This is a multiorgan disease caused by an abnormal host response to microfilariae. It can occur with various parasites (see Table 14-2) and is most frequent in India and Southeast Asia.[130,131]

CLINICAL FEATURES

Clinically, patients with tropical eosinophilia have an increased male:female ratio (4:1) and are usually in their twenties and thirties. The disease may involve the liver, spleen, and nodes. There are few localizing signs or symptoms in the early stages, but there is a high blood eosinophilia, over 3000/mm³. With lung involvement, there are bouts of cough, which is often nocturnal; dyspnea; wheezing; chest discomfort, often accompanied by low grade fever; weight loss; and fatigue.

Radiology shows increased bronchovascular markings and diffuse interstitial lesions, 1 to 3 mm in size, or mottled opacities involving mainly the middle and basal regions. Rarely, the disease may be confined to one lung.[132] Pulmonary function tests show restrictive defects, diminished residual volume and

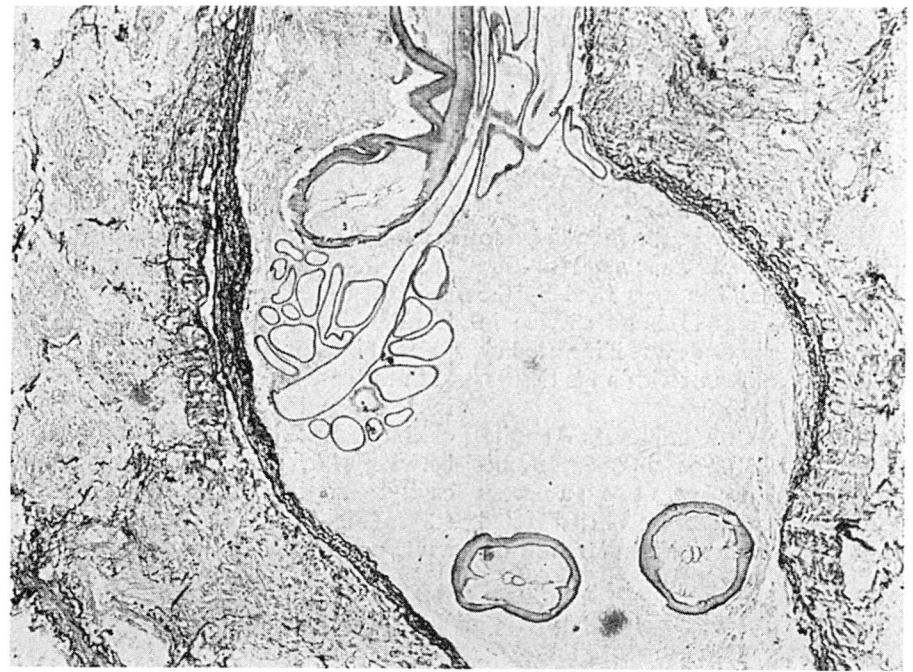

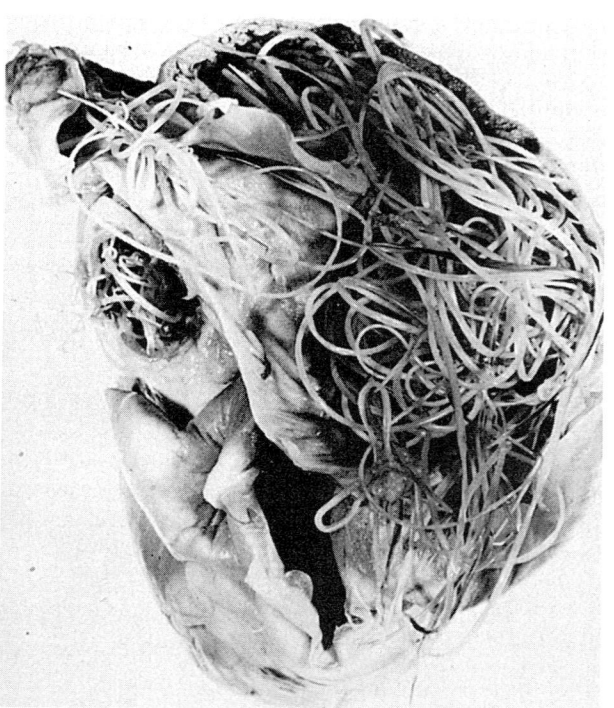

A

B

FIGURE 9-54

A. Dead adult *Dirofilaria immitis* in a pulmonary artery. The surrounding lung shows infarction. (H&E, ×75.) (*Reproduced by courtesy of Dr M Millard. From Spencer,[189] with permission.*) *B. Dirofilaria* organisms filling the right atrium of a dog and protruding through the pulmonary artery. (*From Fishman,[190] with permission.*)

total lung capacity, as well as abnormal diffusion capacity.

PATHOLOGY

Though the disease causes much morbidity, there is little mortality, and pathologic descriptions are based on surgically excised lung lesions.[133–135] Macroscopically, the pleura is often thickened and adherent over both lungs. Red-brown hemorrhagic areas are seen beneath the pleural surface. On cut section, pale gray-white nodules up to 1 cm in diameter along with small hemorrhages are found scattered throughout both lungs. The smaller bronchi are filled with blood-stained mucopus containing many eosinophils.

Histologically in the initial stages, there is an eosinophilic alveolitis.[136] Later, in some areas, interstitial and intraalveolar eosinophilic infiltrates with lymphocytes, plasma cells, and histiocytes are present, as well as granulomas, in the centers of which eosinophilic, structureless debris occurs, surrounded by foreign-type giant cells (Fig. 9-57). The central necrotic debris probably represents the products of dead microfilariae and stains positively by the PAS method. The eosinophilic debris is sometimes termed a Myers-Kouwenaar body and is analogous to a Hoeppli-Splendore reaction.[137] Rarely, by study of large numbers of serial sections, fragments microfilariae, 3 to 7 μm diameter, are observed within these

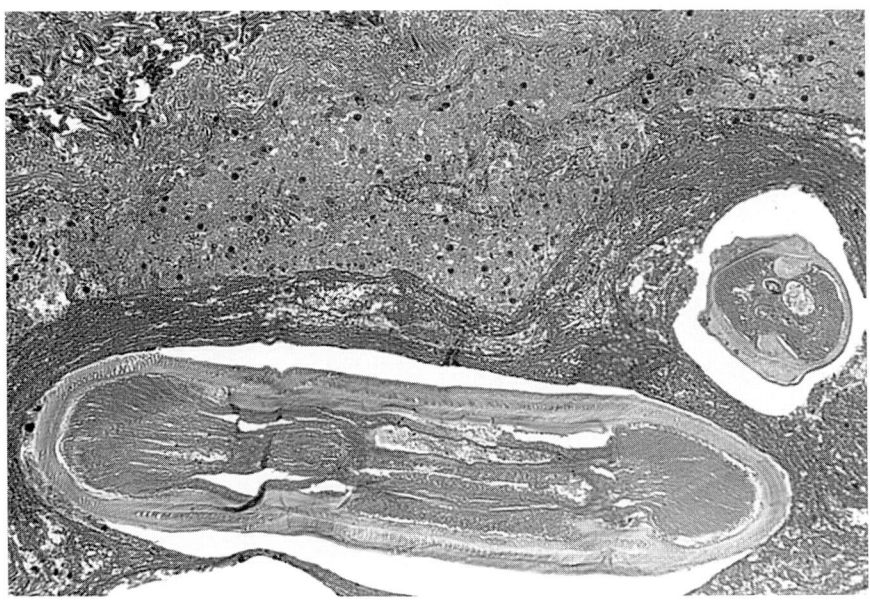

C

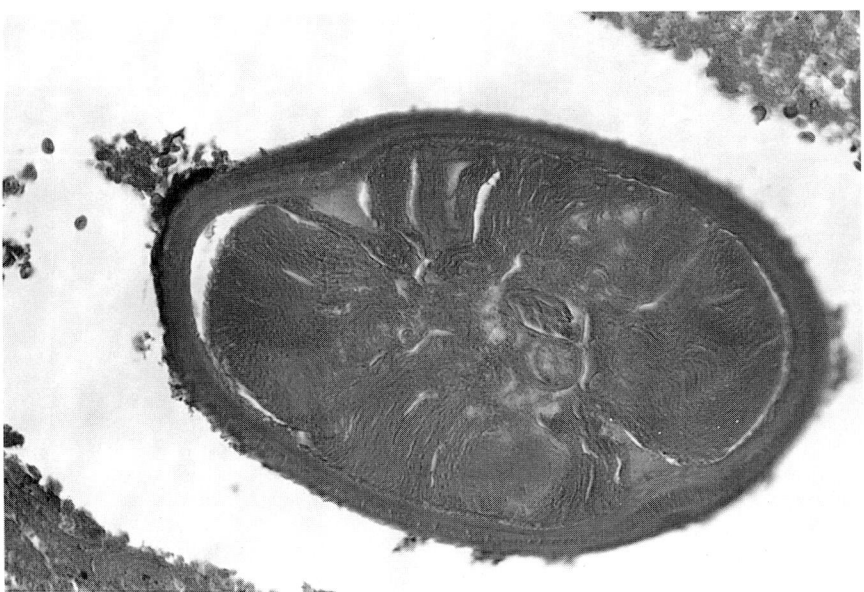

D

FIGURE 9-54
(*Continued*) *C. Dirofilaria* organisms, which are dead, lying in a pulmonary artery. There is infarcted lung nearby. (H&E, ×87.5.) *D.* Higher magnification of necrotic *Dirofilaria*. (H&E, ×219.)

lesions. In other areas, a slowly progressive interstitial fibrosis occurs, and the alveoli in these areas are lined by cuboidalized epithelium. They contain hemosiderin-laden macrophages along with eosinophilic bronchopneumonia.

Two years after onset, the eosinophils become sparse, the infiltrate is histiocytic, and interstitial fibrosis is prominent, though usually patchy. There may be eosinophilic infiltration of the bronchioles, with disruption of the mucosa and muscle.[136]

In TPE, adult lymphatic filariae are not evident, although they are present somewhere in the system.

Microfilariae are not seen in the peripheral circulation. Similar pathology, with occasional microfilariae, is also seen in the lymph nodes, liver, and spleen.[138]

PATHOGENESIS

In TPE, microfilariae are released into the circulation from adult worms dwelling in lymphatics. They give rise to hypersensitivity reactions, usually types I, III, and IV, and are rapidly opsonized with antifilarial antibody and cleared in the pulmonary vascula-

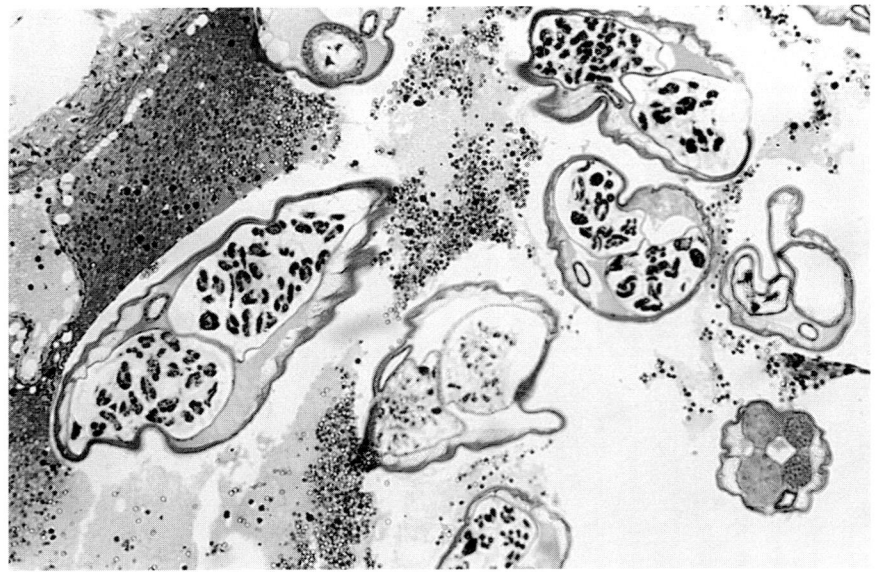

FIGURE 9-55
Adult *Wuchereria bancrofti* worm in multiple cross section in a vessel. There are numerous microfilarial larvae in the uterus. (H&E, ×87.5.)

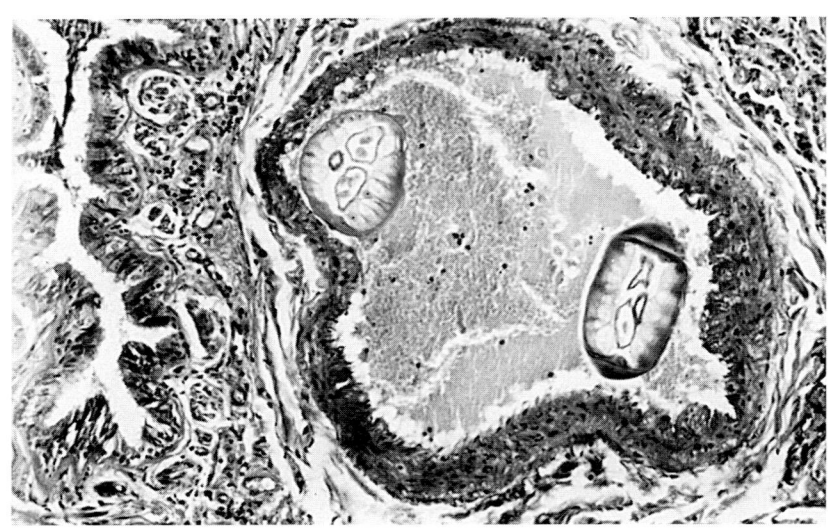

FIGURE 9-56
A sexually mature female, unidentified *Brugia sp*, lying in a branch of a child's pulmonary artery. (Methylene blue, ×100.) (*Reproduced by courtesy of Dr G.H. Smith.*)

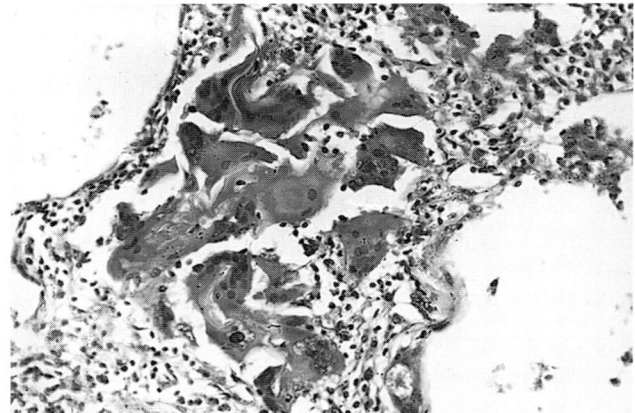

FIGURE 9-57
Granulomatous focus in tropical eosinophilic lung. There are also lymphocytes and eosinophils. Some of the giant cells contain refractile threads of an unknown etiology. (H&E, ×207.) (*Reproduced by courtesy of Dr. T.J. Danaraj.*)

ture.[139] Individuals with tropical pulmonary eosinophilia have markedly elevated levels of parasite-specific IgE, IgM, and IgG antibody.[140] The eosinophils that accumulate in the lung will cause damage to alveolar lining cells through the generation of oxidizing radicals and elaboration of cationic eosinophil granule proteins such as major basic protein, eosinophil cationic protein, and eosinophilic-derived neurotoxin. The eosinophils also destroy the microfilariae. The process is an idiosyncratic host immune response to a common infection.

DIFFERENTIAL DIAGNOSIS

The differential diagnosis is that of pulmonary eosinophilia and includes Loeffler's syndrome, allergic bronchopulmonary aspergillosis, allergic granulomatosis with angiitis (Churg-Strauss syndrome), the systemic vasculitides (polyarteritis nodosa), Wegener's granulomatosis, chronic eosinophilic pneumonia, and idiopathic hypereosinophilic syndrome. Features that distinguish TPE are the geographic exposure to filariasis, nocturnal wheezing, the high levels of antifilarial antibodies in the serum, and, critically, a rapid therapeutic response to the antifilarial drug diethylcarbamazine.

Pulmonary Manifestations of Hookworm Infections and Ascariasis (Loeffler's Syndrome)

The human hookworms *Ancylostoma duodenale* and *Necator americanus* have a wide geographic distribution throughout the tropics and warmer temperate zones (including also Europe and the United States). The 1-cm-long adults reside in human small bowel; eggs deposited in the feces develop in the soil into larvae, which then infect humans by penetrating the skin. After hematogenous passage to the lungs, the larvae penetrate into the alveolar spaces, ascend the tracheobronchial tree, and descend the esophagus to locate in the intestine.

Ascaris lumbricoides is a larger gut nematode (up to 10 cm long). It is also widely distributed throughout the tropics and subtropics. Its eggs are ingested directly, the larvae hatch in the small bowel, and then they invade veins and lymphatics and reach the lungs. Finally, they pass out through the lungs to the gut, in a similar fashion to hookworms.

In these pulmonary migration phases during which the larvae molt, the host inflammatory reaction to the larval antigens may induce a transient pneumonitis, Loeffler's syndrome.[141] The syndrome may also be caused by *Strongyloides* during its migration and by various zoonotic hookworm and ascarid larvae that pass to the lungs but do not finally mature in humans: *Ancylostoma caninum, A. braziliense,* and *Ascaris suum* (an ascarid of pigs).[142,143] In terms of clinical severity, *A. lumbricoides* and *A. suum* cause the more severe pulmonary disease, typically when there is a heavy synchronous infection.[143]

Clinically, Loeffler's syndrome is a self-limiting, mild disease lasting 10 to 14 days. There is cough, bronchitis, and asthma and blood eosinophilia up to about 8000/mm^3. A chest radiograph shows transient pulmonary infiltrates.[144]

There have been few opportunities to study the pulmonary pathology due to these organisms.[145,146] Hookworm larvae are about 300 by 20 μm in size, while the larvae of *A. lumbricoides* are 1 to 2 mm in length by 75 μm in diameter, with lateral alae on the cuticle. The relatively large size of ascarid larvae results in considerable alveolar capillary damage, causing hemorrhage when the parasite migrates into the alveolus from the capillary. Associated with this migration there is pulmonary edema, an allergic eosinophilic pneumonitis, and desquamation of the alveolar lining cells. Also, within the alveoli there are neutrophil polymorphs, eosinophils, fibrin, and Charcot-Leyden crystals. The lower airways are also inflamed. The larvae are seen in the interstitium, as well as the alveoli and bronchioles (Figs. 9-58 to 9-60). If the larvae die during pulmonary migration, a granulomatous reaction with eosinophils is induced around the degenerative fragments.

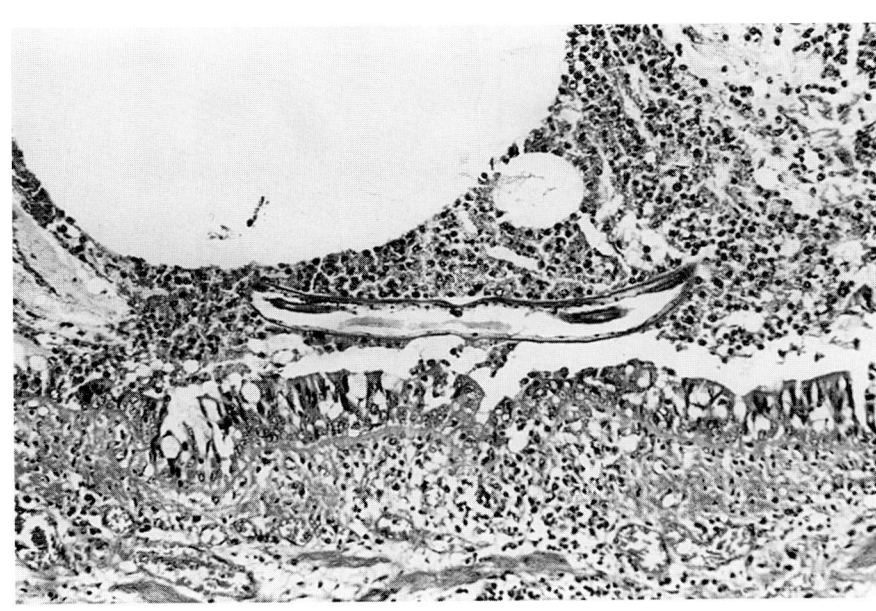

FIGURE 9-58
Ascaris lumbricoides larvae migrating in bronchi with associated inflammation. (H&E, ×87.5.)

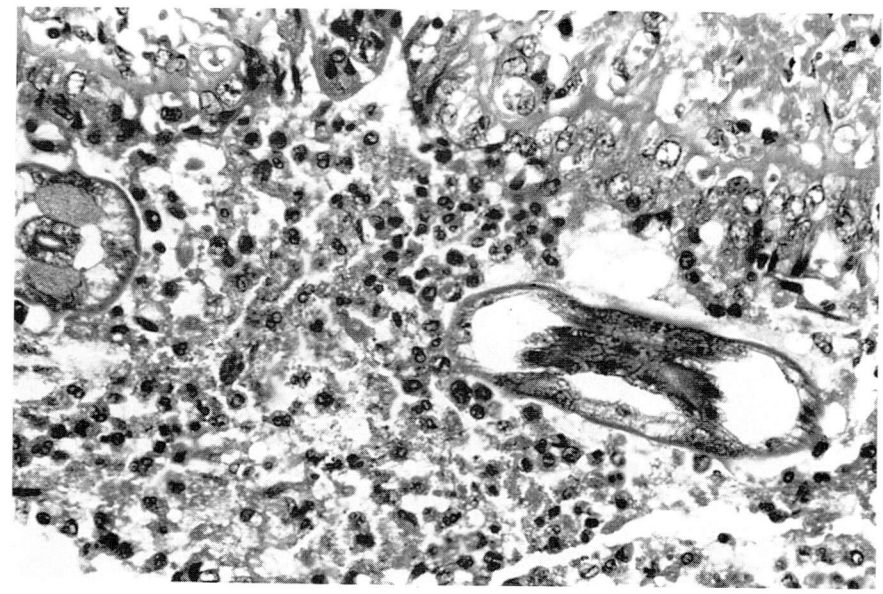

FIGURE 9-59
Higher power of same slide as Fig. 9-54 showing *A. lumbricoides* in both longitudinal and tangential sections. (H&E, ×87.5.)

Mature ascarid worms in the bowel can damage the lungs indirectly. They are prone to migrate, and fatal asphyxia from obstruction to the trachea is reported.[147] Ascarids may obstruct the bile ducts and induce a liver abscess, which can penetrate the diaphragm, resulting in an empyema.[148]

Visceral Larva Migrans (Toxocariasis)

The syndrome of visceral larva migrans (VLM) indicates disease caused by larvae migrating through human tissues. Many species do this, including *Toxocara* spp., *Ancylostoma* spp., *Strongyloides ster-*coralis, *Capillaria* spp., *Baylisascaris* spp., *Gnathostoma* spp., and *Spirometra* spp. However, the term *VLM* is usually restricted to those infections where the larvae do not develop and mature in humans (i.e., paratenic infections), and the major pathogens are *Toxocara canis* and *T. cati*. These are ascarid gut parasites of dogs and cats respectively. Humans, usually children who eat dirt, are infected by eggs in the feces of these animals.

In symptomatic children, up to 80 percent of those with the highest antitoxocaral antibody titers have cough, with wheeze and chest radiographs showing peribronchial or perihilar infiltrates.[149]

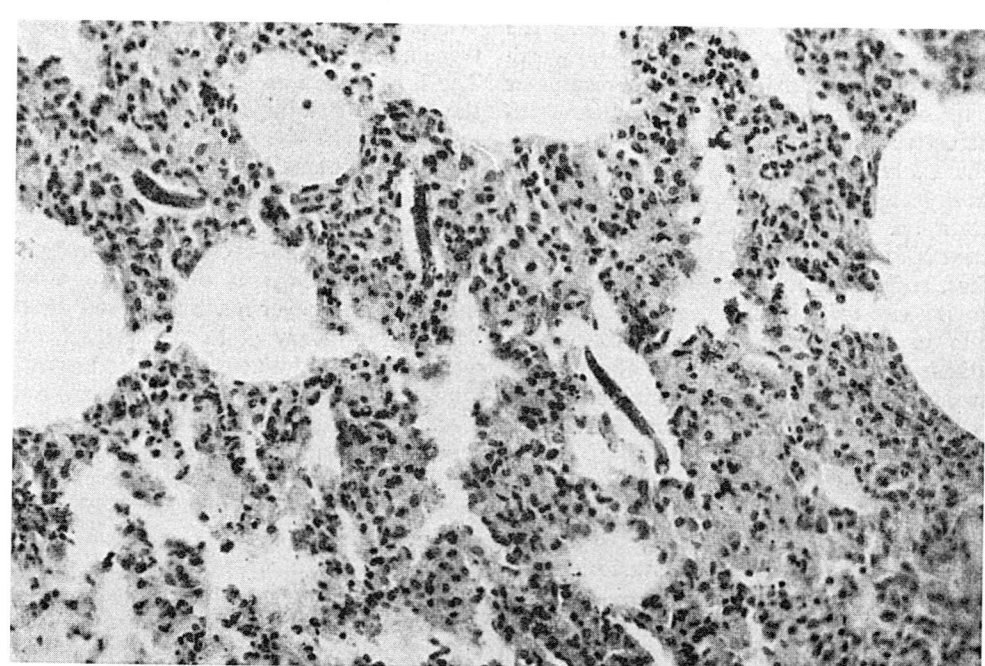

FIGURE 9-60
Migrating intrapulmonary larvae of *A. lumbricoides*. (*From Muller,*[191] *with permission.*)

340

Blood eosinophilia of up to 50,000/mm³ is frequent. The clinical diagnosis is supported by positive toxocaral serology.

The pulmonary histopathology is of giant cell granulomas (Fig. 9-61) in which larvae may be seen.[150] The larvae are PAS-positive, 500 μm long, and 20 μm in diameter, they have lateral alae (longitudinal ridges) on the cuticle. A peripheral eosinophilic infiltrate is usual.

Mixed infections of *A. lumbricoides* and *T. canis* are described.[151] [152] There was a widespread hemorrhagic bronchopneumonia and many of the small bronchi were partly filled with necrotic debris, eosinophils, and nematode larvae.

Uncommon Nematode Lung Infections

There are many nematode infections, mostly zoonoses, where single or a few cases of human pulmonary infection are documented. The detailed criteria are not given here, but morphology, geography of infection, and serology all contribute to making the diagnosis.

Dirofilariasis repens

In southern Europe, the more frequent dirofilarial infection is *D. repens* (*not D. immitis*, see above), a mosquito-borne zoonosis that usually produces a cold abscess in the subcutis. An uncommon example of *D. repens* migrating to the lung is reported.[153] A coin-shaped infarct resulted. This was caused by thrombosis of a pulmonary artery, in which a nongravid dirofilarial worm with characteristic dimensions was seen.

Capillariasis

Capillaria aerophila is a parasite of carnivores that burrows under the airway mucosa. In a rare biopsied human infection in Iran, the small worm invaded the bronchial mucosa and elicited a mixed granulomatous response with eosinophilia.[154]

Loiasis

The filarial parasite *Loa loa* is characterized by migratory cutaneous angioedema (Fig. 9-62) and conjunctivitis associated with the passage of an adult worm. The microfilariae circulate in the blood, and most infections are asymptomatic. The infection is endemic in West and Central Africa.

A case is described in a 40-year-old man who had traveled extensively in West Africa. He presented with right-sided chest pain, cough with blood-tinged sputum, and shortness of breath. There was a loculated right-sided pleural effusion that contained numerous eosinophils and *Loa* microfilariae. Similar microfilariae were identified in a daytime blood sample.[155] There was a posttreatment rise in blood eosinophils as well as an antifilarial antibody titer, which is commonly seen in filarial infection and probably represents a reaction to the liberation of antigens in dying parasites.[156]

Anisakiasis

Human infection with *Anisakis* nematodes from eating fish is common where fish is not cooked or frozen properly to kill the larvae. The result is an eosinophilic lesion of the gastrointestinal mucosa, but occasionally the worm may migrate in the abdominal cavity and even into the subcutis. A patient with eosinophilic pleural effusion is reported who had an appropriate history of eating raw fish.

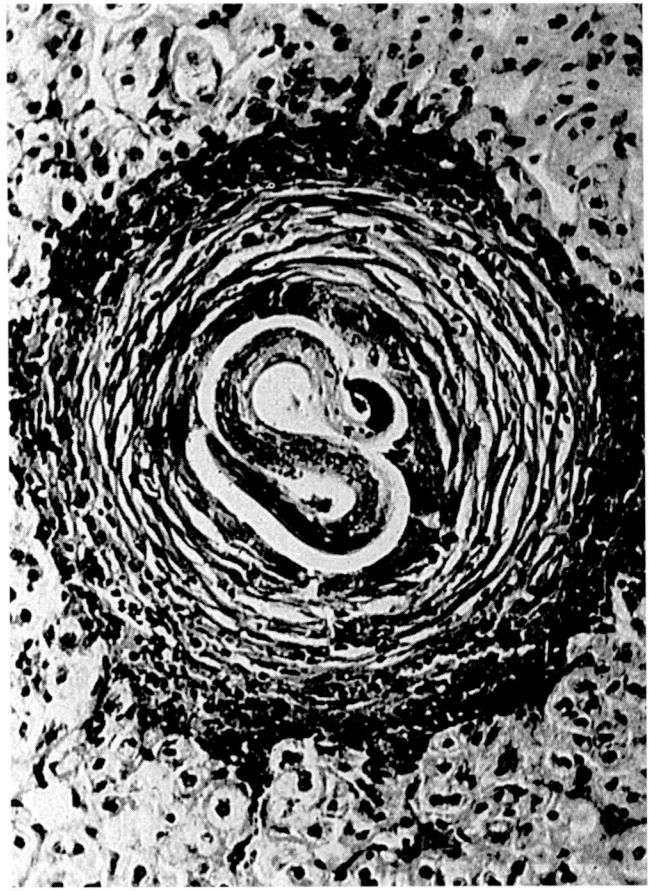

FIGURE 9-61
Toxacara canis. Granulomatous response to a larva in the liver. (*Courtesy of the American Society of Pathologists. From Fishman,*[190] *with permission.*)

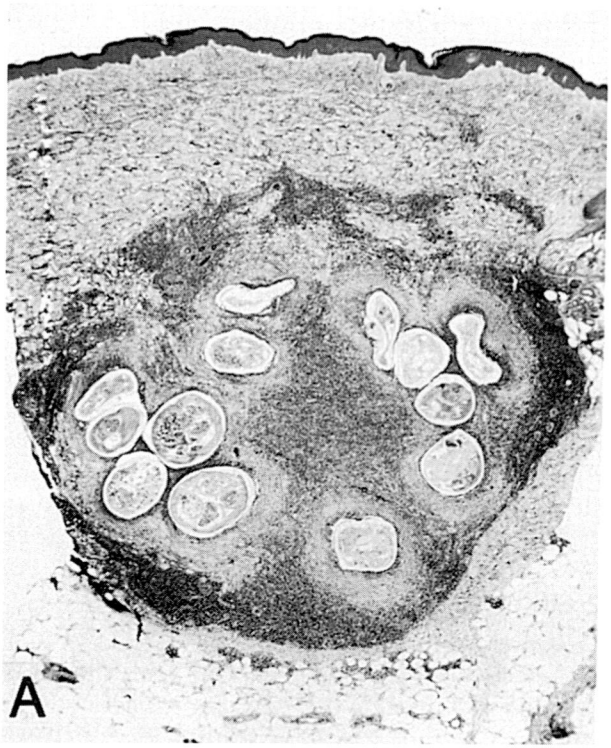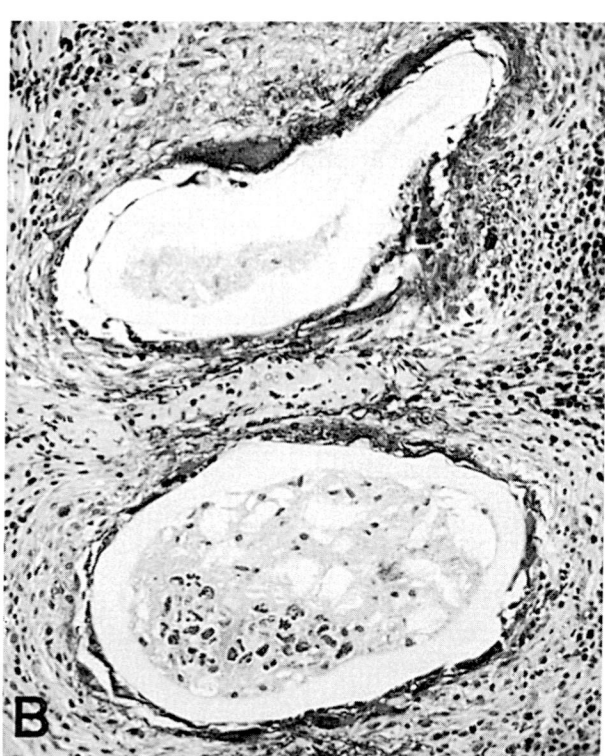

FIGURE 9-62

A. Cross section of a deep inflammatory lesion containing a coiled, partially degenerated gravid adult female *Loa loa.* The surrounding inflammatory reaction is made up of granulomatous and scar tissue infiltrated by histiocytes, epithelioid cells, giant cells, plasma cells, eosinophils, and lymphocytes. (Movat stain, ×25.)   *B.* A higher magnification of tangential sections of the parasites in (*A*). Irregularly spaced cuticular bosses characteristic of *L. loa* are seen in the lower organism. The cuticle is intact but the internal structures are degenerate. Nuclear fragments (*lower left*) are recognizable as degenerating microfilariae. (Movat stain, ×160.) (*Reproduced by courtesy of Dr. D.H. Connor, AFIP. From Warren and Mahmoud,*[188] *with permission.*)

Serum antibodies precipitated only *Anisakis* antigens, indicating the probable etiology.[157]

Enterobiasis

The pinworm *Enterobius vermicularis* is extremely common as a colorectal luminal infection among children in temperate zones. Ectopic enterobiasis follows aberrant migration and death of worms, usually within the abdominal cavity or genital tract in females. A patient is reported who had a 18-mm coin lesion on routine chest x-ray which was then excised. On histology, within the necrotic eosinophilic inflammatory mass was a degenerate female *E. vermicularis* worm. Characteristic lateral alae and eggs were visible. The route of infection was not clear, but direct entry via mouth or nose and subsequent migration through the airways appear probable.[158]

Mammomonogamiasis

Mammomonogamiasis laryngeus is the tapeworm that resides in the upper airways of bovines, mainly in South America. Human infection causes a dry cough, and there may be blood eosinophilia. Physical examination shows a reddened bronchial mucosa to which are attached a pair of white worms (up to 20 by 0.5 mm).[159] Eggs may also be found in the sputum or feces.

Angiostrongyliasis

Angiostrongylus cantonensis is the lungworm of rats in the tropics, particularly the Far East. In rats, it is a parasite in the pulmonary arteries, having migrated through the central nervous system. As a zoonosis in humans, the larvae are usually arrested in the brain, causing eosinophilic meningoencephalitis. Rarely, worms are also located in the pulmonary artery (Figs. 9-63 and 9-64).[160]

Gnathostomiasis

Gnathostoma spinigerum is a parasite of felines, seen mainly in Asia. The worm burrows into the bowel wall and may migrate in humans to the lungs. It then causes pleural effusion, cough, and sometimes a coin lesion on x-ray. Histologically, the worm is about 1 mm in diameter, surrounded by hemorrhage and a mixed granulomatous reaction with eosinophilia.[161]

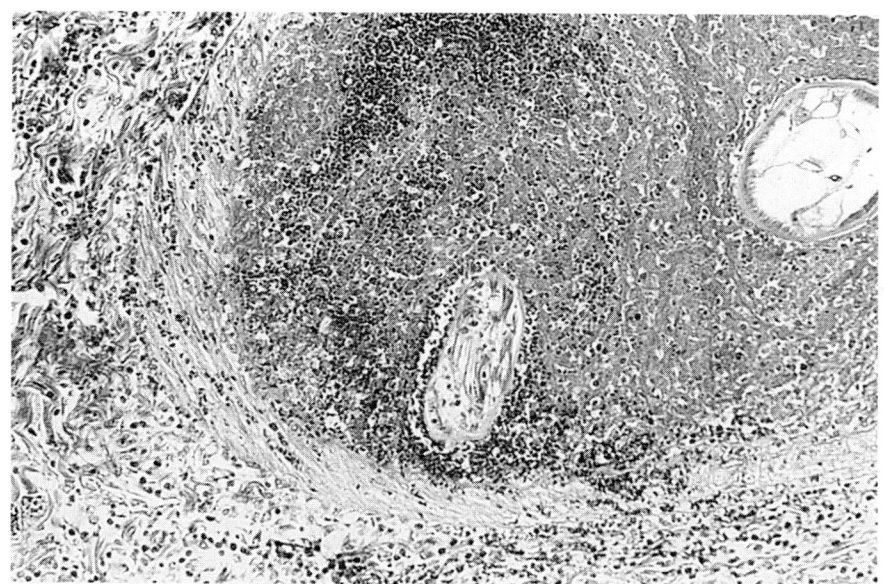

Strongyloidiasis

EPIDEMIOLOGY

The nematode *Strongyloides stercoralis* or threadworm is a member of the order Rhabdita. Adult females produce embryonated ova which transform into noninfective rhabditiform larvae in the human small intestine. These are passed in the stool, where, in moist soil, they may either molt into free-living adult worms or transform into infective filariform larvae. These larvae infect humans by penetrating intact skin to enter subcutaneous capillaries. They are then carried to the lungs, where they invade the alveoli, migrate to the pharynx, and are coughed up and swallowed. They penetrate the duodenal and jejunal mucosa to become mature females and produce a new generation of larvae. Male worms are not identified in humans.

Strongyloides is unique in having an internal or autoinfective stage of development. This takes place in the intestines or anus.

Humans are an important definitive host of *Strongyloides*, but natural infections occur in cats and dogs. The disease is common in warm climates

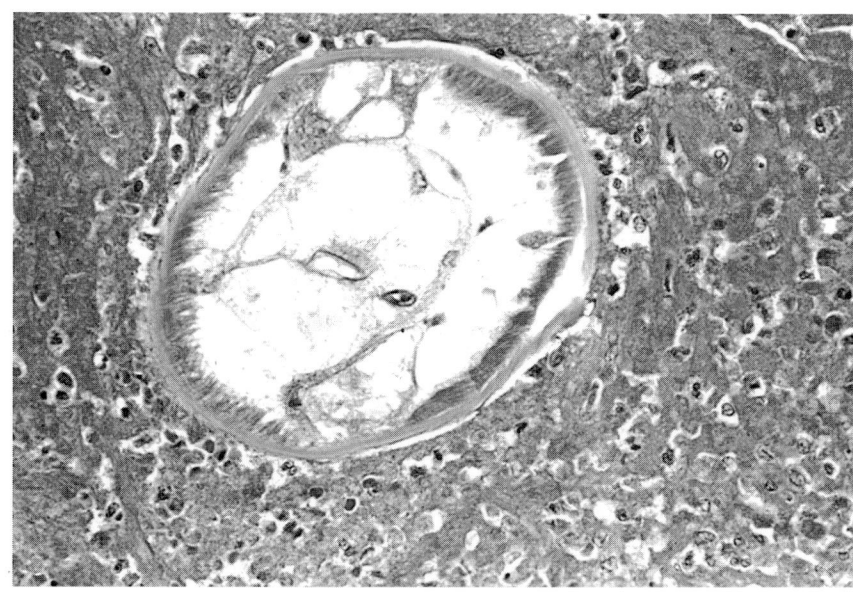

FIGURE 9-64
Same case as Fig. 9-59, showing a cross section of *A. cantonensis* within thrombus. (H&E, ×350.)

and in persons who are institutionalized. In the United States, the incidence is estimated as 0.4 to 4 percent in some southern states.[77]

CLINICAL FEATURES

The infection may manifest as cutaneous, pulmonary, and intestinal disease. Pulmonary infection is a normal component of the life cycle of *Strongyloides* in human hosts. It results when filariform larvae in the peripheral circulation undergo an obligatory migration to the lungs. In patients sensitized to the larvae, pneumonia can occur. Symptoms include cough, hemoptysis, dyspnea, cyanosis, and wheezing. Bronchospasm associated with pulmonary strongyloidiasis can often be mistaken for asthma. The administration of glucocorticoids to these patients can be disastrous.[162] Eosinophilia is a common component of the disease.

Radiologically, the changes are nonspecific. There may be diffuse or segmental alveolar infiltrates with or without cavitation, diffuse interstitial infiltrates, nodular opacities, or abscess formation.

Massive and life-threatening larval invasion of the lungs and other organs can occur as a result of autoinfection in immunocompromised patients.[163,164] Only a few patients with AIDS have been reported to have strongyloidiasis.[162]

PATHOLOGY

The pulmonary changes produced by *Strongyloides* are dependent on the number of larvae produced and, as a consequence, are correlated with the host's immune status. During the period of larval migration through the lungs, there are petechial hemorrhages and infiltration of the alveolar septa by neutrophils and monocytes. A Loeffler-like syndrome with mild pneumonitis and eosinophilia can occur in immunocompetent hosts.

Autoinfection can produce massive larval invasion of the lungs, causing severe intraalveolar hemorrhage and hemoptysis. Cavities and abscesses have been reported in the lung in strongyloidiasis.[165,166] Granulomatous inflammation with multinucleated giant cells, eosinophils, histiocytes, and plasma cells can be associated with larvae in the center of lesions.[167] Interlobular septal fibrosis has also been reported in pulmonary strongyloidiasis. Both larval and adult *Strongyloides* can be seen in the bronchial epithelium. The presence of rhabditiform larvae, ova, or adult worms in the lung is indicative of hyperinfection.

Adult parasitic females can be identified on H&E sections. They are 1.5 by 2.5 mm in length and 30 to 40 μm in width (Figs. 9-65 and 9-66). Females usually contain numerous ovoid thin-walled eggs, measuring 50 to 58 by 30 to 34 μm in paired uteri. Rhabditiform larvae are 380 μm in length and 20 μm wide. Filariform larvae are longer and measure 600 μm in length and 16 μm in width (Fig. 9-67).

Strongyloides may be a difficult parasite to detect, since often the parasite load is low and the larval output in the stools is irregular and minimal. Pulmonary disease is diagnosed by the detection of larvae, ova, or less commonly adult worms in the sputum.[165,168,169] The parasite may be detected in bronchial washings.[170] In open lung or transbronchial biopsy, the diagnosis is dependent on identification of characteristic larval or adult worms in the alveolar septa, lumina, or bronchial epithelium. Larvae must be differentiated from *Ascaris lumbricoides*, hookworm, *Bayliascaris procyonis,* and *Toxacara* spp.

CESTODES

Cestode worms (tapeworms) consist of a chain of hermaphrodite segments and a head segment for attachment. Disease in humans is from such adult parasites, usually in the gut, or their larvae in many tissues, including the lung. The most important pulmonary infection is hydatid disease caused by *Echinococcus* spp. Cysticercosis and sparganosis may involve the chest wall and, less often, the lung parenchyma.

Cysticercosis

This is the larval phase of infection with *Taenia solium*, the pork tapeworm. The normal life cycle has the human as definitive host, with the tapeworm in the bowel. But if ingestion of another human's fecal eggs occurs, the ingestor becomes the intermediate host and develops cysticercosis.

Pulmonary cysticerci are rare and patients are usually asymptomatic.[171] The larval cysts are white, about 10 mm in diameter, and contain milky fluid. The usual host reaction is a fibrous capsule until the parasite dies spontaneously or because of chemotherapy. The eosinophilic reaction is minimal.

Echinococcus

Three species of *Echinococcus* affect humans: *E. granulosus, E. multilocularis,* and *E. vogeli, E. granulosus* causes the classic unilocular hydatid cyst and is by far the most frequent. *Echinococcus multilocu-*

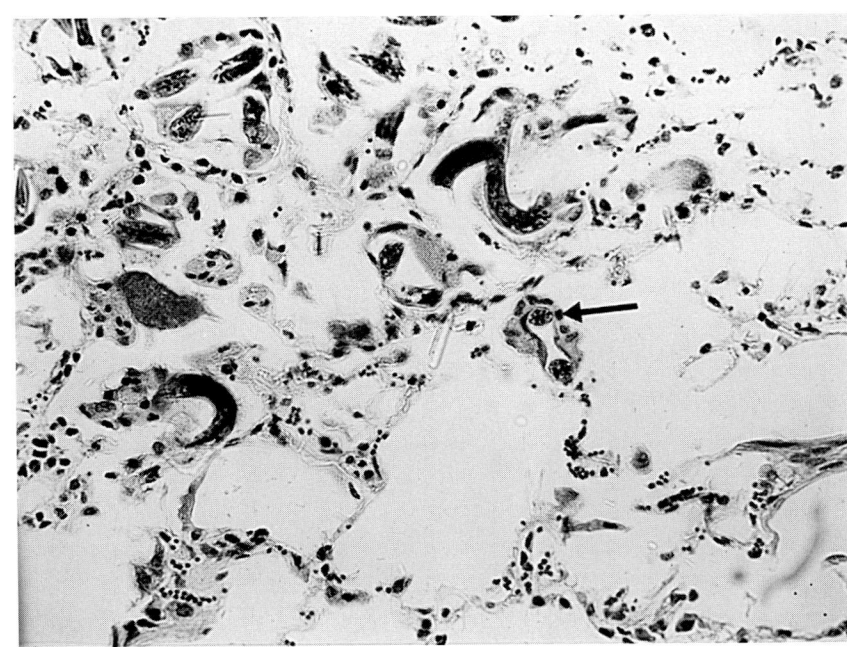

FIGURE 9-65
Larvae of *Strongyloides stercoralis* lying within alveolar lumina and alveolar capillaries (*arrow*). (H&E, ×280.)

laris may cause proliferative spreading lesions known as alveolar hydatid disease, which behaves more like a neoplasm. Lesions are less cystic and tend to undergo necrosis. Pulmonary involvement is less common with *E. multilocularis.* The third organism is *E. vogeli*, which is mainly confined to South and Central America.[172] *Echinococcus vogeli* mostly affects the liver; only 3 of 15 known cases have involved the lung or pleura; pathologically, it is intermediate between the other two species.

The life cycles of all *Echinococcus* species involve canines or foxes as definitive hosts with adult worms in the bowel (Fig. 9-68); excreted eggs are ingested by intermediate hosts: horses, cattle, sheep, and camels for *E. granulosus*, voles and other small mammals for *E. multilocularis.* In them, the eggs develop into the cystic larvae, which are eventually consumed as offal by the definitive hosts. Thus the human is an accidental intermediate host in hydatid disease.

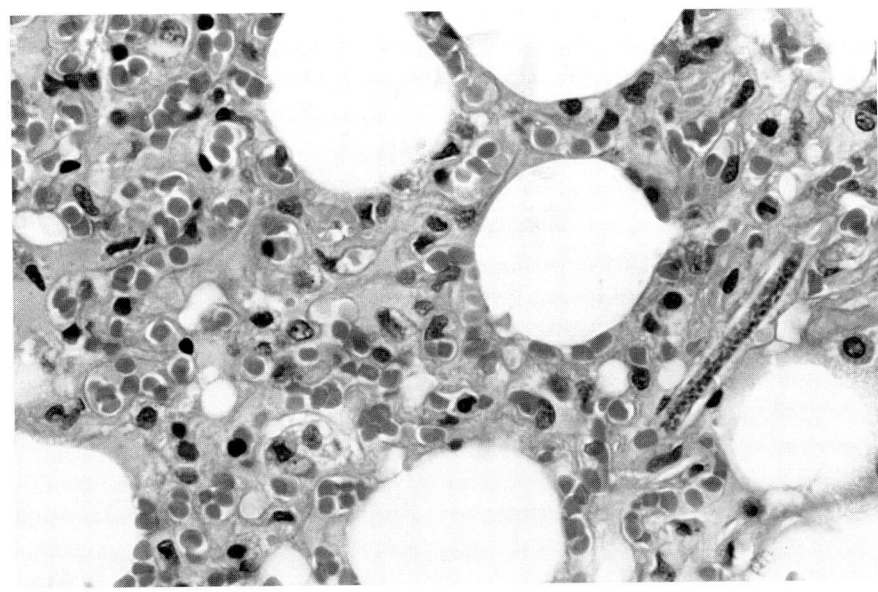

FIGURE 9-66
Strongyloides stercoralis. A filariform larva in congested, edematous lung. (H&E, ×219.)

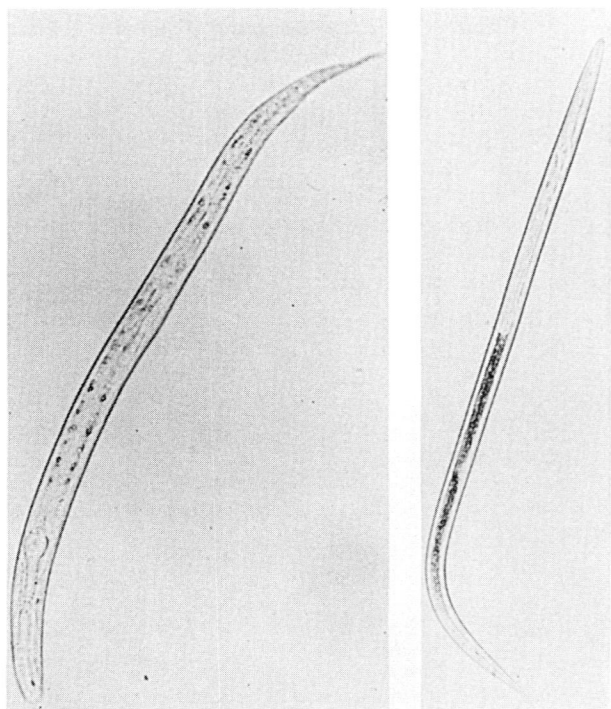

FIGURE 9-67
Rhabditiform larva measuring approximately 250 μm in length (*left*) and filariform larva approximately 600 μm in length (*right*) (*From Warren and Mahmoud,*[188] *with permission.*)

Echinococcus Granulosus Hydatid

EPIDEMIOLOGY

Pulmonary hydatid disease is still a common problem, as evidenced by two recent papers from Chile[173] and Tunisia,[174] which document a total of 717 cases. Hydatid disease is a global problem, and there are few countries where the disease has not been recorded.[175] The highest incidence of hydatidosis in the world is found among the Turkana people living in Kenya, where, after all the slaughtering is done at home, any hydatid cysts are fed to the dogs. This produces a high prevalence of infection in the canine population.[176] There are programs for controlling the disease, the first of which was introduced in Iceland 120 years ago, where an estimated 1 in 6 people had the disease.[177] Other control programs have been introduced in New Zealand, Tasmania, and—more recently—in Cyprus and the Falkland Islands. The infection is also endemic in the United Kingdom. Humans often become infected in childhood through fondling and handling infected dogs, which carry dried fecal dust in their coats and paws.

Men are more frequently affected than women in Europe but not in Northern Africa, where there is a female predominance, due to sex-related behavioral differences with respect to exposure to dogs and dog feces.[174] More than half the patients are between 10 and 30 years of age. In the study by Jerray and

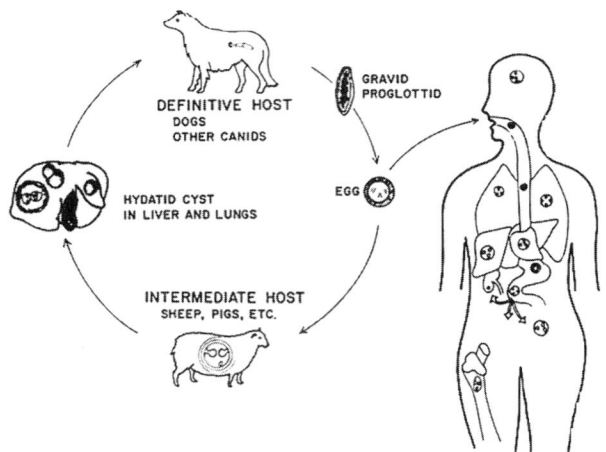

FIGURE 9-68
Life cycle of *Ecchinococcus granulosus* and the mode of infection for humans. (*From Warren and Mahmoud,*[188] *with permission.*)

coworkers, over 20 percent of patients for whom occupational data were available, 34 percent were agricultural workers and the remainder pursued indoor occupations, but half owned dogs.

Because the ingested eggs hatch in the small bowel and pass to the liver, where they are filtered from the circulation, some 75 percent of hydatid cysts are hepatic. The remaining involve in descending order of frequency, the lungs, muscle, spleen, kidney, brain, and bones. Cysts in multiple sites are frequent, partly from the original multisite dissemination and from metastasis or rupture of established cysts. This can also occur from inadequate surgical intervention.

CLINICAL FEATURES

Within the thoracic cavity, hydatid cysts present most commonly in the lung parenchyma, mediastinum, pleural cavity, or chest wall. Rupture from the liver across the diaphragm into the pleura may occur. Essentially, hydatid cysts behave as space-occupying lesions. The majority of pulmonary cysts are solitary.

The clinical presentation is hemoptysis, chest pain, and cough. In 20 percent of cases, there was a sudden cough with expectoration of cyst fluid, membranes, and scolices. Fever and purulent sputum are often present. Dyspnea and rashes are unusual and just over 2 percent of cases are asymptomatic. Eosinophilia (i.e., $>500/pmm^3$) is noted in approximately 20 percent of patients. Some patients, where the cyst has ruptured from the liver into the lung, will also cough up bile.

The radiologic findings vary. There may be well-defined rounded cysts, which are the commonest feature, but hydatid disease may appear as a lung

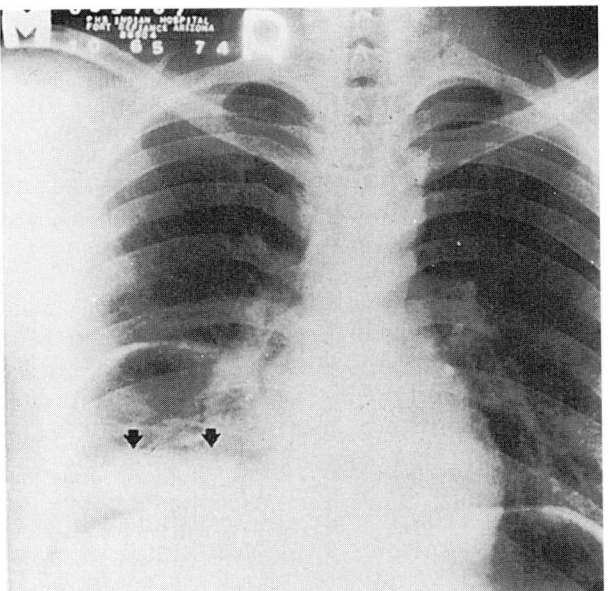

FIGURE 9-69
Chest radiograph of a 19-year-old man with a ruptured hydatid cyst. Note the fluid levels. (*From Warren and Mahmoud,*[188] *with permission.*)

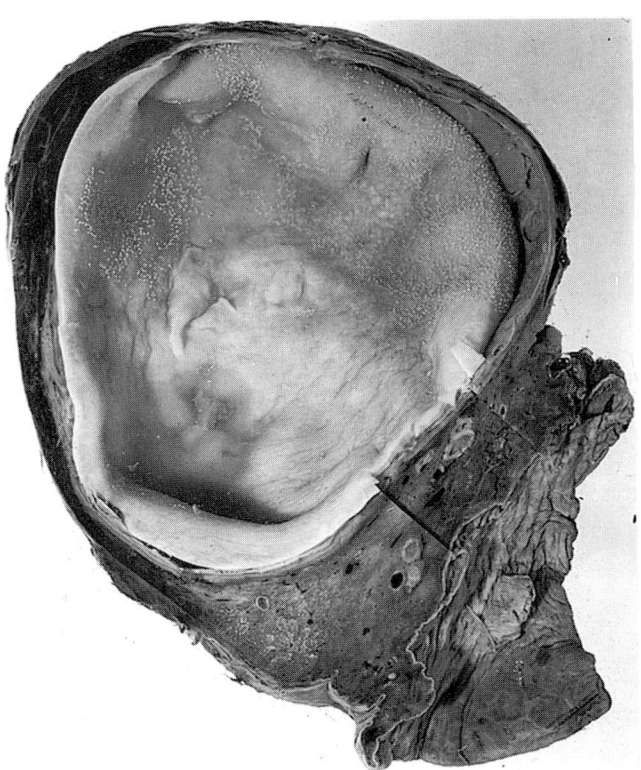

A

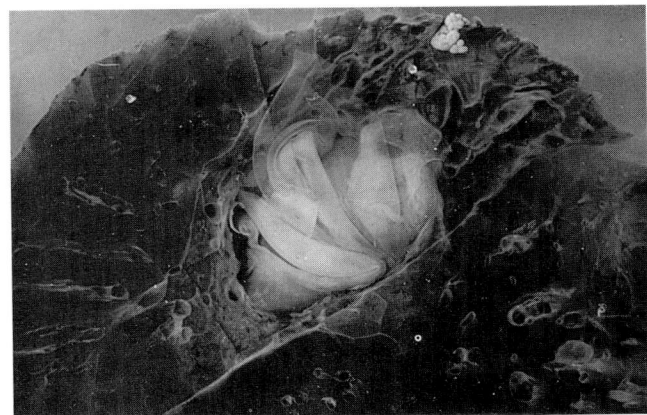

B

FIGURE 9-70
A. Hydatid cyst of the lung with partial separation of the laminated coat from the adventitia in a few places. The inner surface shows multiple granules, scolices, attached to the germinal layer. *B.* Smaller hydatid cyst in the upper lobe of the lung. A superficial glance may suggest this is a carcinoma.

abscess (Fig. 9-69), show a "water-lily" or crescent sign or may present in just over 10 percent of patients as a pleural effusion. Less than 1 percent calcify and 0.2 percent show a hydropneumothorax. Lung cysts were solitary in the majority of cases (74 percent). There is a predilection for the right lung (57 percent) and especially for the lower lobes (59 percent). Smaller air passages may rupture in the adventitial plane and escaping air may then seep between the laminated layer of the cyst and host tissue. This causes a stripping up of the laminated layer and the space created fills with air. As seen on chest radiographs, this is known as "perivascular pneuma."

PATHOLOGY

The cyst is lined on its inner surface by a syncytial layer one cell thick known as a germinal layer (endocyst) (Fig. 9-70). Outside this layer lies the thick laminated membrane (ectocyst), derived from the germinal layer. It is structureless and elastic, with the feel of soft satin. Outside the laminated layer is a layer of fibrous tissue and some granulation tissue formed by the host, known as the pericyst. Provided that rupture (Fig. 9-71) of the cyst does not occur, there is only a mild chronic inflammatory reaction in this adventitia. Eosinophilia is not normally marked. However, if cyst extravasation occurs, there is a considerable granulomatous reaction, with eosinophilia persisting in the adventitia for a long period.[178]

The parent cyst brood capsules develop from the germinal layer. These brood capsules are small vesic-

ular structures which, in turn, bud to form the scolices (Figs. 9-72 to 9-74). These are the future heads of adult tapeworms. They form by the proliferation and invagination of the brood capsule wall and are liberated as the capsules burst. As the parent cyst grows, daughter cysts develop within the original cyst, both from the brood capsules and directly from the germinal center. Daughter cysts are thin-walled structures,

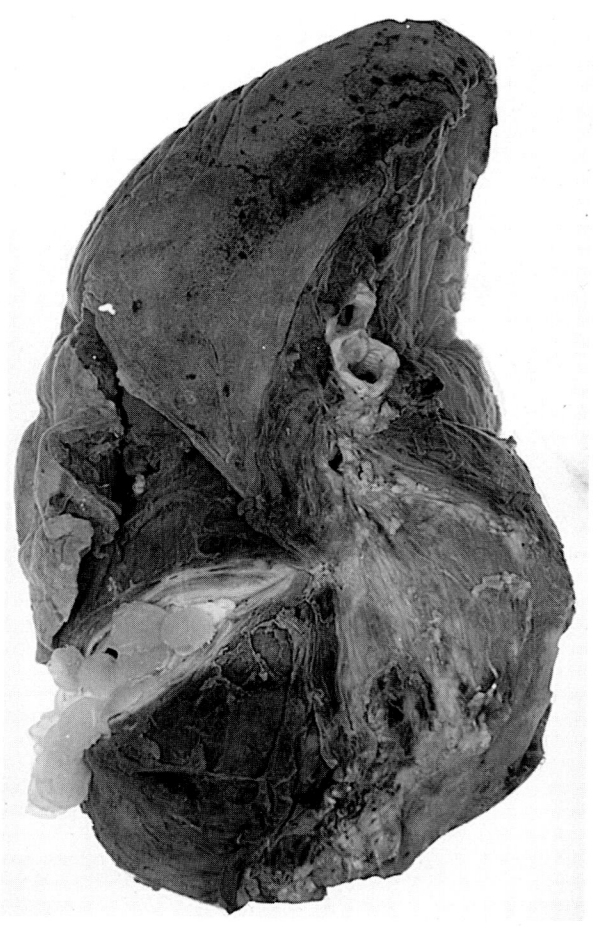

FIGURE 9-71
Hydatid cyst of the lung, which has ruptured and shows daughter cysts escaping. (*Reproduced by courtesy of Dr. K.F.W. Hinson.*)

and those found in *E. echinococcus* are lined by germinal cells and surrounded by poorly developed laminated layers. The daughter cysts lie free in the cavity of the parent cyst (Fig. 9-72), but should there be a break in the adventitia due to trauma or infection, they may grow outward.

Growth of hydatid cysts has usually been regarded as very slow. In most organs, they take several years to reach a diameter of 5 cm. However, in richly vascularized organs such as the lung, growth may be faster. Growth by a pulmonary hydatid cyst of 7 cm within a year is recorded.[179]

Old cysts and those treated with chemotherapy often lack identifiable germinal layers and scolices. The laminated membrane persists, intact or fragmented, and the cyst contents are amorphous. Hooklets may be found, their identification aided by a Ziehl-Neelsen stain, since they are acid-fast. The pericyst in old cysts may include giant cell granulomas, the giant cells containing bits of laminated membrane.

DIAGNOSIS

A diagnosis is usually made on plain chest radiographs. Calcification may be detected on straight radiograph but is more likely to be found in small amounts on CT scanning. It may be possible to make the diagnosis on finding scolices, hooklets, or pieces of chest wall in material that is coughed up. There are dangers in using invasive diagnostic techniques for

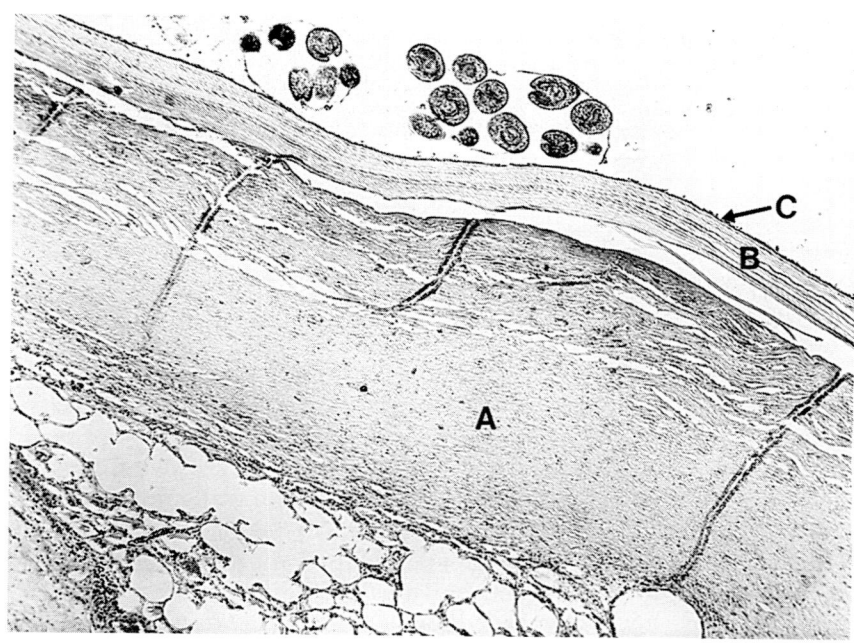

FIGURE 9-72
Wall of a hydatid cyst showing the adventitial fibrous layer formed by the host (*A*), the laminated layer forming the chitinous layer of the cyst and part of the parasite (*B*), and the very thin unicellular germinal layer from which brood capsules containing scolices are arising (*C*). (H&E, ×40.) (*Reproduced by courtesy of Colonel LRS MacFarlane, Royal Army Medical College, London.*)

FIGURE 9-73
The fragment of liver on the right is lined by an *Ecchinococcus* membrane. The brood capsules are on the left. *(Reproduced by courtesy of Dr. D.H. Connor, AFIP. From Fishman,*[190] *with permission.)*

diagnosis. These are the implantation of material, such as scolices or germinal membranes, that could lead to metastatic cysts and release of hydatid antigen, possibly causing anaphylactic shock. Aspiration of cysts used not to be recommended. However, fine needle aspiration with ultrasound guidance is now a reliable diagnostic technique, and a recent series found no complications.[180]

Immunoelectrophoresis in the Tunisian series was found to be the most sensitive (75 percent positivity) compared with 54 percent for indirect hemagglutination. Enzyme-linked immunosorbent assay and latex fixation have also been used. Although these assays are sensitive, false-positive reactions

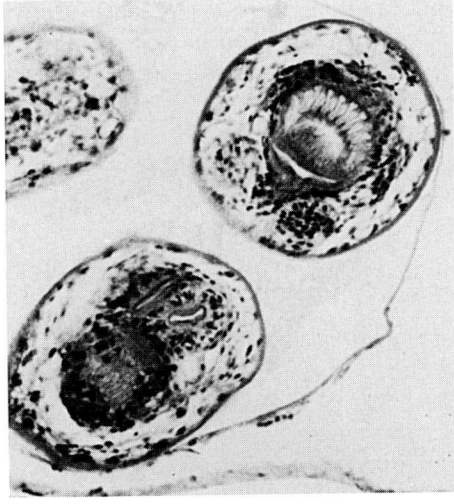

FIGURE 9-74
Three scolices of *E. granulosus*. The upper right scolex shows the hooklets of the organism. *(Reproduced by courtesy of Dr. D.H. Connor, AFIP. From Fishman,*[190] *with permission.)*

can occur, most commonly in patients with other helminthic infections. Negative serology does not exclude the diagnosis of echinococcosis.[181] Computed tomography scanning is highly sensitive and can discover additional small intrathoracic lesions.

Hydatid cysts may rupture into the pleural cavity, a process that may result from injury due to needling in an attempt to establish a diagnosis. There may be superimposed infection, which usually results in the death of the parasite. However, daughter cysts developing in the cavity following removal of parent cysts survive despite the presence of intracavitary infection.

Pulmonary hydatid cysts may be resectable. Surgical removal of a hydatid cyst can cause release of daughter cysts or brood capsules into the cavity created. Both these structures may subsequently develop directly into a fresh hydatid cyst without undergoing the full cycle of development. This is known as retrograde metamorphosis. The commonest treatment in the surgical series described from Chile was cystectomy in 312 of 508 (61 percent) cases. During cystectomy, there was careful isolation of the cystic area, puncture, suction of the hydatid liquid, extraction of the hydatid membranes, and complete extirpation of the endocyst. If cysts are not resectable, chemotherapy is used, and on radiology, they regress. However, complete cure of pulmonary hydatid disease without surgery is infrequent.

Echinococcus Multilocularis Hydatid Cyst

The alveolar hydatid is endemic in temperate eastern Europe, Russia, China, and North America, particularly Alaska.

Most lesions are in the liver, with lung spread in less than 10 percent of patients. The multilocular type of hydatid disease is characterized by the growth of a larval stage in which the laminated (ectocyst) layer and germinal (endocyst) layer are attentuated. The alveolar arrangement of the metacestode is attained by extensions of the germinal membrane, which invades adjacent host tissues. It then gradually acquires a lining of laminated membrane. This process forms small, continuous chambers containing brood capsules (Fig. 9-75). Instead of a rounded cyst, the parasite material is more trabecular and fragmented. It appears acellular and proliferates and spreads like an infiltrating neoplasm in the invaded tissues, provoking a necrotic, fibrotic, and granulomatous reaction. Scolices are not seen in human infection. However, if transplanted into the appropriate intermediate host, parasite fragments will develop into normal viable alveolar cysts.[182]

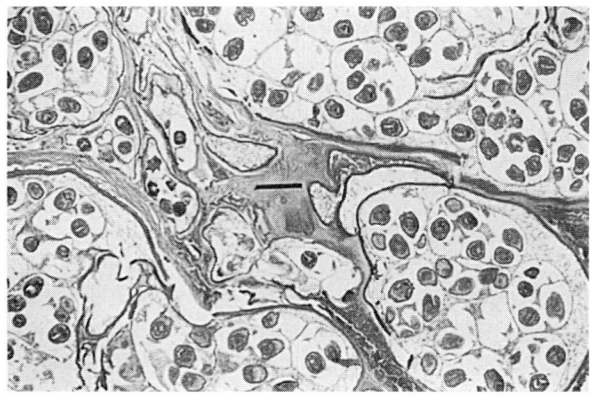

FIGURE 9-75
Larval *E. multilocularis* in the liver of an infected rodent. There are well-marked alveolarlike microvesicles with proscolices. (H&E, bar=100 μm.) (*From Warren and Mahmoud,*[188] *with permission.*)

Although many of the cysts die, others continue to grow and invade tissues. Some infected individuals are asymptomatic, since the cyst has spontaneously died.[183]

Clinical presentation is usually from cancer-like liver symptoms and systemic wasting. There may be blood eosinophilia.

Grossly, an alveolar hydatid is solid, gray-white in color and centrally necrotic. The histology is characteristic (Fig. 9-75, top, and Fig. 9-76). The laminated membrane is well stained by the PAS method. Outside the membrane, there is variable eosinophilic tissue necrosis, fibrosis, and granuloma formation.

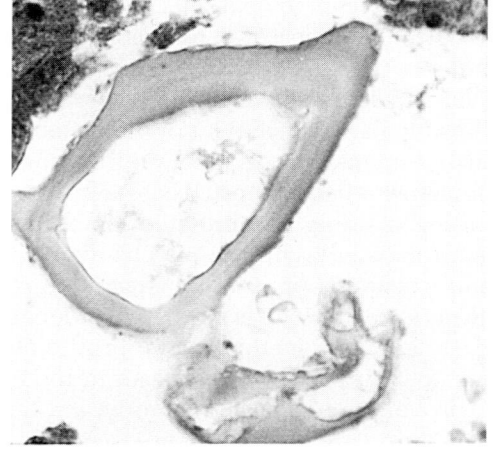

FIGURE 9-76
Alveolar hydatid. Fragments of parasite membrane. (H&E, ×313.)

ARTHROPODS

Pentastomiasis (Porocephalosis)

Pentastomes are a phylum of arthropodlike and annelidlike parasites of reptiles and mammals, distributed worldwide.[184–186] They have complicated life cycles, often involving several intermediate hosts. The human is an accidental zoonotic host. Two genera of pentastomes cause the majority of human disease. These are *Linguatula serrata* and *Armillifer armillatus* and *A. moniliformis*; they produce different respiratory tract lesions.

Armillifer Pentastomiasis

The parasite lives in the respiratory tree of snakes and lizards. Humans are infected by eating the uncooked meat of an intermediate host that contains larvae. Another mode of infection is by direct contact if such meat is used as a poultice. The larvae migrate in the body but do not mature; they die and usually end up as a fibrotic calcified nodule. During migration, they may cause significant inflammatory lesions if they are present in large numbers. Most cases are reported from Africa and Southeast Asia. An autopsy study in Malaysia documented a 45 percent prevalence of pentastomiasis among aborigines, the liver and lung being the most affected organs.[187]

In the lung, migrating larvae can cause collapse of a lobe and pneumonitis.[186] Usually old fibrocalcified nodules are observed, the infection having been asymptomatic. In these lesions, the parasite has degenerated but retained its characteristic shape. Histologically viable *Armillifer* nymphs are 1 cm long and C-shaped; they show acidophilic glands, an intestinal tract, striated muscle, and a chitinous cuticle with fine openings (Figs. 9-77 to 9-79). The normal host tissue is separated from the encysted nymphs by a thin, delicate fibrous capsule. There may be a variable acute or chronic inflammatory reaction and granules around degenerating nymphs.[185]

Linguatulosis

Linguatula serrata adults inhabit the nasal cavities of dogs, cats, and foxes. They are most common in the Middle East. When these infected animals sneeze or defecate, they contaminate humans via their eggs. The larvae may then migrate in the body, as with *Armillifer*, but they rarely cause disease. Nasopharyngeal linguatulosis ("halzoun" syndrome) may develop if undercooked visceral organs containing

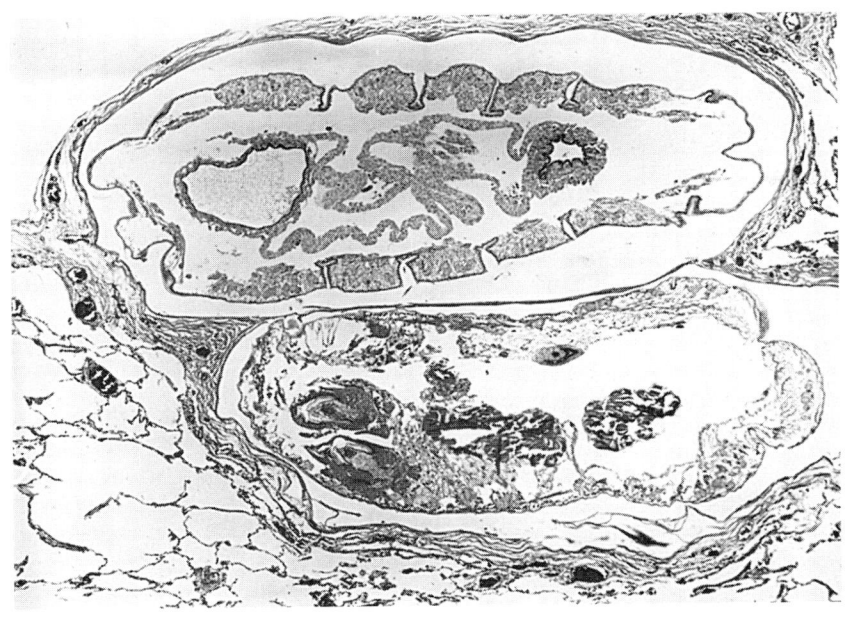

FIGURE 9-77
Pulmonary pentastomiasis due to *Armillifer monoliformis*. The parasite is situated in the lung, beneath the pleura. It is cut in two places because of its crescentic shape. (H&E, ×30.)

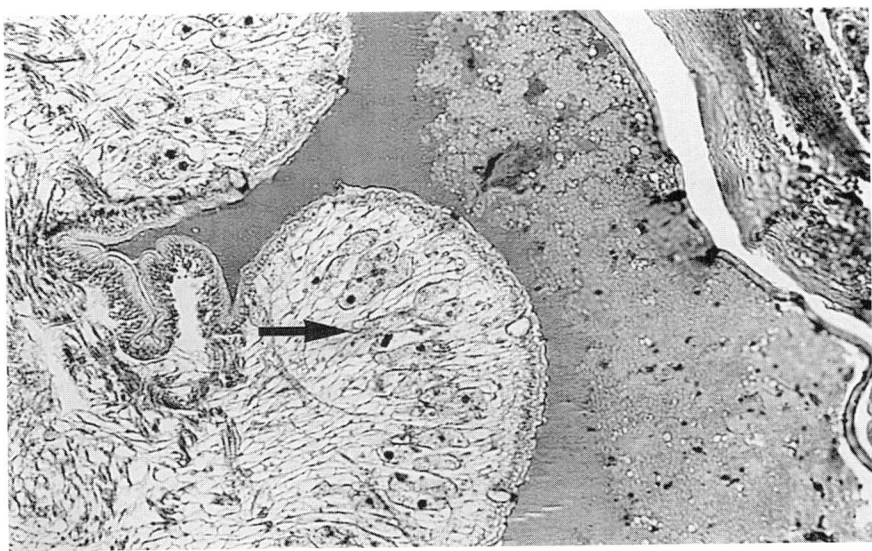

FIGURE 9-78
Pentastomiasis. (*A. monoliformis*). A viable parasite in a capsule. The parasite has typical large subcuticular glands (*arrow*). (H&E, ×87.5.)

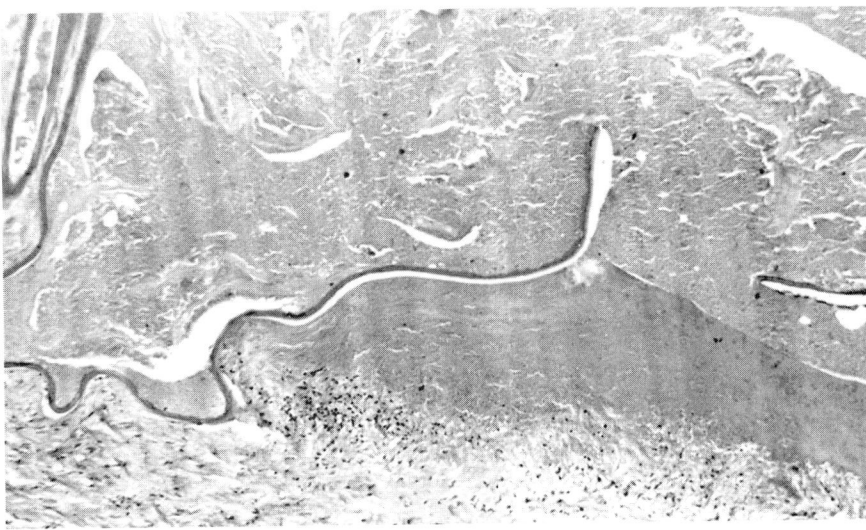

FIGURE 9-79
Pentastomiasis. Dead parasite with a residual cuticle (H&E, ×87.5).

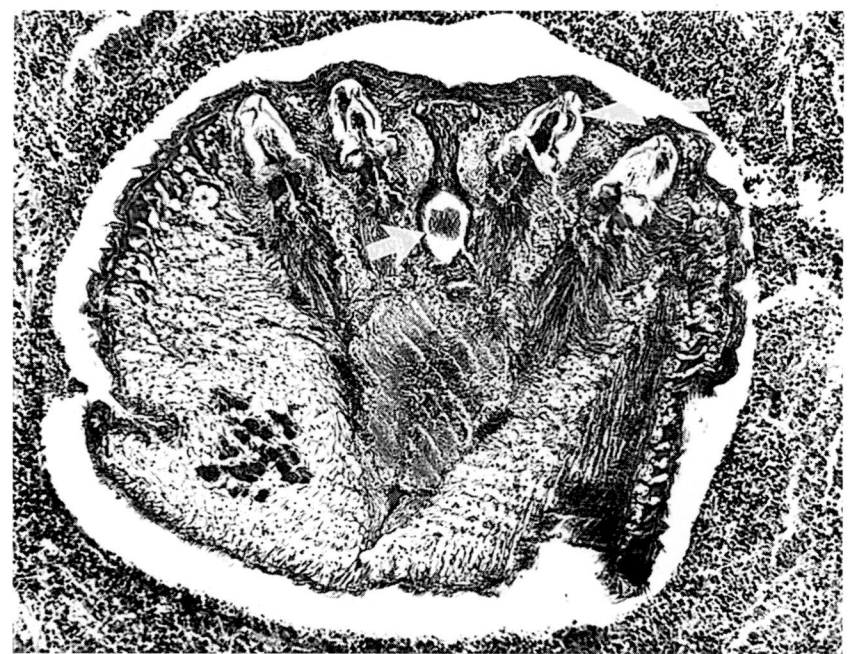

FIGURE 9-80
Lingulata serrata in human lung. The long arrow points to one of the four hooklets and the short arrow indicates the buccal cavity, leading ventrally to the mouth. There are spines projecting from the edge of the larva. (H&E, ×100.)

larvae are ingested. These usually come from the livers of sheep, goats, or camels, who act as the intermediate hosts. There is initial discomfort in the throat, followed by congestion of the nasopharyngeal mucosa and upper airways as the larval nymphs attach to the mucosa. Itching, discomfort, and sneezing result, often with expulsion of the parasites. These are white and motile, measuring 5 to 10 mm in length (Fig. 9-80).

REFERENCES

1. Gutierrez Y: *Diagnostic Pathology of Parasitic Infections with Clinical Correlations* Philadelphia, Lea & Febiger, 1990.
2. Patterson M, Schoppe LE: The presentation of amoebiasis. *Med Clin North Am* 66:689–705, 1982.
3. Lyche KD, et al: Pleuropulmonary manifestations of hepatic amebiasis. *West J Med* 153:275–278, 1990.
4. Clark CG, Diamond LS: Pathogenicity, virulence and *Entamoeba histolytica. Parasitology Today* 10:46–47, 1994.
5. Spice WM, et al: *Entamoeba histolytica*, in: *Molecular and Cell Biology of Opportunistic Infections in AIDS.* Myint S, and Cann A (eds): London, Chapman & Hall, 1993, pp 95–137.
6. Takaro T, Bond WM: *Int Abstr Surg* 107:209,1958.
7. Daniels AC, Childress ME: *Dis Chest* 156:360, 1956.
8. Webster BH: Pleuropulmonary amebiasis: A review with an analysis of 10 cases. *Am Rev Respir Dis* 81:683–688, 1960.
9. Le Roux BT: Pleuro-pulmonary amebiasis. *Thorax* 24:91–101, 1969.
10. Ibarra-Perez C: Thoracic complications of amebic abscess of the liver: Report of 501 cases. *Chest* 79:672–677, 1981.
11. Healy GR: Immunologic tools in the diagnosis of amebiasis. *Rev Infect Dis* 8:239–246, 1986.
12. Duarte MI, et al: Interstitial pneumonitis in human visceral leishmaniasis. *Trans R Soc Trop Med Hyg* 83:73–76, 1989.
13. World Health Organisation, Division of Control of Tropical Diseases: *Severe and Complicated Malaria*, 2d ed. *Trans R Soc Trop Med Hyg* 84:(Suppl 2):1–65 1990.
14. Cayea PD, et al: Atypical pulmonary malaria. *AJR* 137:51–55, 1981.
15. Charoenpan P, et al: Pulmonary edema in severe falciparum malaria: Hemodynamic study and clinico-physiologic correlation. *Chest* 97:1190–1197, 1990.
16. MacPherson G G, et al: Human cerebral malaria: A quantitative ultrastructural analysis of parasitized erythrocyte sequestration. *Am J Pathol* 119:385–401, 1985.
17. Duarte MIS, et al: Ultrastructure of the lung in falciparum malaria. *Am J Trop Med Hyg* 34:31–35, 1985.
18. Corbett CEP, et al: Cytoadherence in human falciparum malaria as a cause of respiratory distress. *J Trop Med Hyg* 92:112–120, 1989.
19. Miller LH, et al: Malaria pathogenesis. *Science* 264:1878–1883, 1994.
20. McCabe RE, Remington JS: *Toxoplasma gondii*, in Mandell GL et al(eds): *Principles and Practice of Infectious Diseases*, 3d ed., New York, Churchill Livingstone, 1990, pp 2090–2103.
21. Frenkel JK: Toxoplasmosis, in Binford CH, et al (eds): *Pathology of Tropical and Extraordinary Disease. An Atlas.* Washington DC, Armed Forces Institute of Pathology, 1976, pp 284–300.
22. von Litchtenberg F: *Pathology of Infectious Diseases.* New York, Raven Press, 1991, pp 287–295.
23. Ruskin J, Remington JS: Toxoplasmosis in the compromised host. *Ann Intern Med* 84:193–199, 1976.
24. Holliman RE: Toxoplasmosis and the acquired immune deficiency syndrome. *J Infect Dis* 16:121–128, 1988.
25. Nash G, et al: The pathological manifestations of pulmonary toxoplasmosis in the acquired immunodeficiency syndrome. *Hum Pathol* 25:652–658, 1994.
26. Schnapp LM, et al: *Toxoplasma gondii* pneumonitis in patients infected with the human immunodeficiency virus. *Arch Intern Med.* 152:1073–1077, 1992.
27. Derouin F, et al: Laboratory diagnosis of pulmonary toxo-

plasmosis in patients with acquired immunodeficiency syndrome. *J Clin Microbiol* 27:1661–1663, 1989.

28. Oksenhendler E, et al: *Toxoplasma gondii* pneumonia in patients with the acquired immunodeficiency syndrome. *Am J Med* 88:18N–21N, 1990.

29. Tawney S, et al: Pulmonary toxoplasmosis: An unusual nodular radiographic pattern in a patient with AIDS. *Mt Sinai Med J* 53:683–685, 1986.

30. Garcia LW, et al: Acquired immunodeficiency syndrome with disseminated toxoplasmosis presenting as an acute pulmonary and gastrointestinal illness. *Arch Pathol Lab Med* 115:459–463, 1991.

31. Mendleson MH, et al: Pulmonary toxoplasmosis in AIDS. *Scand J Infect Dis* 19:703–706, 1987.

32. Pomeroy C. Filice GA: Pulmonary toxoplasmosis: A review. *Clin Infect Dis* 14:863–870, 1992.

33. Jautzke G, et al: Extracerebral toxoplasmosis in AIDS: Histological and immunohistological findings based on 80 autopsy cases. *Pathol Res Pract* 189:428–436, 1993.

34. Gordon SM, et al: Diagnosis of pulmonary toxoplasmosis by bronchoalveolar lavage in cardiac transplant recipients. *Diagn. Cytopathol* 9:650–654, 1993.

35. Jacobs F, et al: Role of bronchoalveolar lavage in diagnosis of disseminated toxoplasmosis. *Rev Infect Dis* 13:637–641, 1991.

36. Renoult E, et al: Generalised toxoplasmosis in two renal transplant recipients who received a kidney from the same donor. *Rev Infect Dis* 13:180–181, 1991.

37. Bottone EJ: Diagnosis of acute pulmonary toxoplasmosis by visualisation of invasive and intracellular tachyzoites in Giemsa-stained smears of broncho-alveolar lavage fluid. *J Clin Microbiol* 29:2626–2627, 1991.

38. Knani L, et al: Toxoplasmosis pulmonaire au cours du SIDA. *Ann Med Interne (Paris)* 141:469–471, 1990.

39. Bendelac A, et al: Decourverte d'une localization pulmonary de *Toxoplasma gondii* chez un malade immunodeprime. *Presse Med* 13:1213–1214, 1984.

40. Weber R, et al: Human microsporidial infections. *Clin Microbiol Rev* 7:426–461, 1994.

41. Desportes I, et al: Occurrence of a new microsporidian: *Enterocytozoon bieneusi*, n.g., n.sp., in the enterocytes of a human patient with AIDS. *J Protozool* 32:250–254, 1985.

42. Weber R, et al: Pulmonary and intestinal microsporidiosis in a patient with the acquired immunodeficiency syndrome. *Am Rev Respir Dis* 146:1603–1605, 1992.

43. Schwartz DA, et al: Disseminated microsporidiosis (*Encephalitozoon hellem*) and aquired immune deficiency syndrome: Autopsy evidence for respiratory acquisition. *Arch Pathol Lab Med* 116:660–668, 1992.

44. Schwartz DA, et al: Pathology of symptomatic microsporidial (*Encephalitozoon hellem*) bronchiolitis in aquired immune deficiency syndrome: A new respiratory pathogen diagnosed from lung biopsy, bronchoalveolar lavage, sputum and tissue culture. *Hum Pathol* 24:937–943, 1993.

45. Schwartz DA, et al: A nasal microsporidian with unusual features from a patient with AIDS. 9th International Conference on AIDS, 1993, p 384, abstr.-B-10-1495.

46. De Groote MA, et al: Polymerase chain reaction and culture confirmation and disseminated *Encephalitozoon cuniculi* in a patient with AIDS: Successful therapy with albendazole. *J Infect Dis* 171:1375–1378, 1995.

47. Weber R, et al: Disseminated microsporidiosis due to *Encephalitozoon hellem*: pulmonary colonization, micro-

hematuria and mild conjunctivitis in a patient with AIDS. *Clin Infect Dis* 17:415–419, 1993.

48. Schwartz DA, et al: Pathologic features and immunofluorescent antibody demonstration of ocular microsporidiosis (*Encephalitozoon hellem*) from seven patients with acquired immune deficiency syndrome. *Am J Ophthalmol* 115:285–292, 1993.

49. Molina JM, et al: Disseminated microsporidiosis due to *Septata intestinalis* in patients with AIDS: Clinical features and response to albendazole therapy. *J Infect Dis* 172:245–249, 1995.

50. Sandfort J, et al: *Enterocytozoon bieneusi* infection in an immunocompetent patient who had acute diarrhoea and who was not infected with the human immunodeficiency virus. *Clin Infect Dis* 19:514–516, 1994.

51. Sobottka I, et al: Self-limited travellers diarrhoea due to a dual infection with *Enterocytozoon bieneusi* and *Cryptosporidium parvum* in an immunocompetent HIV-negative child. *Eur J Clin Microbiol Infect Dis* in press.

52. Schwartz DA, et al: Diagnostic approaches for *Encephalitozoon* infections in patients with AIDS. *J Eukaryot Microbiol* 41:59S–60S, 1994.

53. Schwartz DA, et al: Microsporidiosis in HIV-positive patients: current methods for diagnosis using biopsy, cytologic, ultrastructural, immunological and tissue culture techniques. *Folia Parasitol Praha* 41:101–109, 1994.

54. Orenstein JM, et al: A microsporidian previously undescribed in humans, infecting enterocytes and macrophages, and associated with diarrhea in an acquired immunodeficiency syndrome patient. *Hum Pathol* 23:722–727, 1992.

55. Visvesvara GS, et al: Polyclonal and monoclonal antibody and PCR-amplified small-subunit rRNA identification of a microsporidian, *Encephalitozoon hellem*, isolated from an AIDS patient with disseminated infection. *J Clin Microbiol* 32:2760–2768, 1994.

56. Current AL, et al: Human cryptosporidiosis in immunocompetent and immunodeficient persons. *N Engl J Med* 308:1252–1257, 1983.

57. Hoyling N, Jensen BN: Respiratory cryptosporidiosis in HIV-positive patients. *Lancet* 1:590–591, 1988.

58. Forgacs P, et al: Intestinal and bronchial cryptosporidiosis in an immunodeficient homosexual man. *Ann Intern Med* 99:793–794, 1983.

59. Booth CC, et al: Immunodeficiency and cryptosporidiosis. *Br Med J* 281:1123–1127, 1980.

60. Gross TL, et al: AIDS and systemic involvement with *Cryptosporidium*. *Am J Gastroenterol* 81:456–458, 1986.

61. Harari MD, et al: *Cryptosporidium* as a cause of laryngotracheitis in an infant. *Lancet* 1:590–591, 1988.

62. Kibbler CC, et al: Pulmonary cryptosporidiosis occurring in a bone-marrow transplant recipient. *Scand J Infect Dis* 19:581–584, 1987.

63. Kocoshis SA, et al: Intestinal and pulmonary cryptosporidiosis in an infant with severe combined immune deficiency. *J Pediatr Gastroenterol Nutr* 3:49, 1984.

64. Miller RA, et al: Detection of *Cryptosporidium* oocysts in sputum during screening for mycobacteria. *J Clin Microbiol* 20:1192–1193, 1984.

65. Brady EM, et al: Human cryptosporidiosis in acquired immune deficiency syndrome. *JAMA* 252:89–90, 1984.

66. McLaren LC, et al: Isolation of *Trichomonas vaginalis* from the respiratory tract of infants with respiratory disease. *Pediatrics* 71:888–890, 1983.

67. Hersch SM: Pulmonary trichomoniasis and *Trichomonas tenax*. *J Med Microbiol* 20:1–10, 1985.

68. Camargo ME, et al: Inquerito sorologico da prevalencia da infeccao chagasica no Brasil 1975/1980. *Rev Inst Trop Sao Paulo* 26:192–204, 1984.

69. Lora J, et al: Placental pathology of congenital Chagas' disease in Cochabama, Bolivia. *Proceedings of the Annual Meeting of the American Society of Tropical Medicine and Hygiene*, Atlanta, 1993, p 271.

70. Bittencourt AL: Congenital Chagas' disease: A review. *Am J Dis Child* 130:97–103, 1976.

71. Bittencourt AL, et al: Pneumonitis in congenital Chagas' disease: A study of 10 cases. *Am J Trop Med Hyg* 30:38–42, 1981.

72. Szarfman A, et al: Immunologic and immunopathologic studies in congenital Chagas' disease. *Clin Immunol Immunopathol* 4: 489–499, 1975.

73. Petri WA, Ravdin JI: Free-living amebae, in Mandell GL et al (eds): *Principles and Practice of Infectious Diseases*. 3d ed. New York, Churchill Livingstone, 1990, pp 2049–2056.

74. Martinez AJ, et al: Experimental pneumonitis and encephalitis caused by *Acanthameba* in mice: Pathogenesis and ultastructural features. *J infect Dis* 131:692–699, 1975.

75. Myerowitz RL: *The Pathology of Opportunistic Infections, with Pathogenetic, Diagnostic and Clinical Correlations.* New York, Raven Press, 1983, pp 235–242.

76. Sun T (ed): *Paragonimiasis in Pathology and Clinical Features of Parasitic Disease.* New York, Masson, 1982, pp 253–260.

77. Beaver PC, et al: Amphistomate and distomate flukes, in Beaver PC et al (eds): *Clinical Parasitology,* 9th ed. Philadelphia, Lea & Febiger, 1984, pp 464–471.

78. Mariano EG, et al: Human infection with *Paragonimus kellicotti* (lung fluke) in the United States. *Am J Clin Pathol* 84:685–687, 1986.

79. Im J-G, et al: Pleuropulmonary paragonimiasis: Radiologic findings in 71 patients. *AJR* 159:39–43, 1992.

80. Moon WK, et al: Pulmonary paragonimiasis simulating lung abscess in a 9 year old: CT findings. *Pediatr Radiol* 23:626–627, 1993.

81. Kimura H, et al: A case of spontaneous remission of paragonimiasis miyazakii. *Nippon Kyobu Shikkan Gakkai Zasshi* 31:1151–1156, 1993.

82. Bercovitz A: Clinical studies on human lung fluke disease (endemic hemoptysis caused by *Paragonimus westermani* infestation) *Am J Trop Med* 17:101–122, 1937.

83. Meyers WM, Neafie RN: Paragonomiasis, in Binford CH, Connor DH (eds): *Pathology of Tropical and Extraordinary Diseases: An Atlas* Washington DC, Armed Forces Institute of Pathology, 1976, pp 517–523.

84. Gutierrez Y: Paragonimus-paragonimiasis in Gutierrez Y. (ed): *Diagnostic pathology of parasitic infections with clinical correlations.* Philadelphia, Lea & Febiger, 1990, p 375.

85. Diaconita GH, Goldis GH: Investigations on pathomorphology and pathogenesis of pulmonary paragonimiasis. *Acta Tuberc Scand* 44:51–75, 1964.

86. Yokogawa M: *Paragonimus* and paragonimiasis. *Adv Parasitol* 3:99–158, 1965.

87. Yamaguchi T (ed): *Color Atlas of Clinical Parasitology.* Philadelphia, Lea & Febiger, 1981, pp 8–21.

88. Shim YS, et al: Pulmonary paragonimiasis: A Korean perspective. *Semin Respir Med.* 12:35–45, 1991.

89. Sharma OP: The man who loved drunken crabs: A case of pulmonary paragonimiasis. *Chest* 95:670–672, 1989

90. Rangdaeng S, et al: Pulmonary paragonimiasis: report of a case with diagnosis by fine needle aspiration cytology. *Acta Cytol* 36:31–36, 1992.

91. Zhang Z, et al: Diagnosis of active *Paragonimus westermani* infections with a monoclonal antibody-based antigen detection assay. *Am J Trop Med Hyg* 49:329–334, 1993.

92. Ikeda T, et al: Parasite-specific IgE and IgG levels in the serum and pleural effusion of paragonimiasis patients. *Am J Trop Med Hyg* 47:104–107, 1992.

93. Davis A: Recent advances in schistosomiasis. *Q J Med* 226:(NS 58): 95–110, 1986.

94. Jordan P, et al: *Human Schistosomiasis.* Wallingford, CT, CAB International, 1993.

95. Hiatt RA, et al: Factors in the pathogenesis of acute schistosomiasis mansoni. *J Infect Dis* 139:659–666, 1979.

96. Shaw FB, Gareeb AA: The pathogenesis of pulmonary schistosomiasis in Egypt with special reference to Ayerza's disease. *J Pathol Bacteriol* 46:401–429, 1938.

97. Watt G, et al: Cardiopulmonary involvement is rare in severe *Schistosoma japonicum* infection. *Trop Geogr Med* 38:233–239, 1986

98. Chen MG, Mott KE: Progress in assessment of morbidity due to *Schistosoma mansoni* infection. *Trop Dis Bull* 85:(No 10): R2—R56, 1988.

99. Cheever AW, Kamel IA, Elwi AM, Mosimann JE, Danner R, Sippel JE: *Schistosoma mansoni* and *S. haematobium* infections in Egypt. III. Extrahepatic pathology. *Am J Trop Med Hyg* 27:55–75, 1978.

100. Nash TB, Cheever AW, Ottesen EA, Cook JA: Schistosome infections in humans: Perspectives and recent findings. *Ann Intern Med* 97:740–754, 1982.

101. Barrett-Connor E: Parasitic pulmonary disease. *Am Rev Respir Dis* 126:558–563, 1982.

102. Hiatt RA, Ottesen EA: Sera obervations of circulating immune complexes in patients with acute schistosomiasis. *J Infect Dis* 142:665–666, 1980.

103. Ottesen AE, Weller PF: Eosinophilia following treatment of patients with schistosomiasis mansoni and Bancroft's filariasis. *J Infect Dis* 139:343–347, 1979.

104. Davidson BL, El-Kassimi F, Uz-Zaman A, Pillai DE: The "lung shift" in treated schistosomiasis. Broncho-alveolar lavage evidence of eosinophilic pneumonia. *Chest* 89:455–457, 1986.

105. Schaberg T, Rahn W, Racz P. Lode: Pulmonary schistosomiasis resembling acute pulmonary tuberculosis. *Eur Respir J* 4:1023–1026, 1991.

106. Al-Fawaz TM, Al-Shaheed SA, Al-Najed SA, Ashour M: Schistosomiasis associated with a mediastinal mass: case report and review of the literature. *Ann Trop Paediatr* 10:293–297, 1990.

107. Andrade ZA, Andrade SG: Pathogenesis of schistosomal pulmonary arteritis. *Am J Trop Med Hyg* 19:305–310, 1970.

108. Marchand P, Gilroy JC, Wilson VH: An anatomical study of the bronchial vascular system and its variations in disease. *Thorax* 5:207–221, 1957.

109. Lopez De Faria J, Czapski J, Ribeiro Leite MO, et al: Cyanosis in Manson's schistosomiasis. *Am Heart J* 54:196–204, 1957.

110. Sadigursky M, Andrade ZA: Pulmonary changes in schistosomal cor pulmonale. *Am J Trop Med Hyg* 31:779–784, 1982

111. Jawahri and Shamma *Dis Chest* 38:569, 1960.

112. Chen MG, Mott KE: Progress in assessment of morbidity due to *Fasciola hepatica* infection: A review of the recent literature. *Trop Dis Bull* 87:R1–R38, 1990.

113. Freeman RS, Stuart PF, Cullen JB, et al: Fatal human infec-

tion with mesocercariae of the trematode, *Alaria americana.* *Am J Trop Med Hyg* 25:803–807, 1976.

114. Asimacopoulos PJ, Katras A. Christie B: Pulmonary dirofilariasis. The largest single-hospital experience. *Chest* 102:851–855, 1992.

115. Kochar AS: Human pulmonary dirofilariasis: Report of three cases and brief review of the literature. *Am J Clin Pathol* 84:19–23, 1985.

116. Risher WH, Crocker EN, Beckman EM, Blalok JV, Ochsner JL: Pulmonary dirofilariasis, the largest single institution experience. *J Thorac Cardiovasc Surg* 97:303–308, 1989.

117. Cordero M, Munoz MR, Muro A. Simon F: Transient solitary pulmonary nodule caused by *Dirofilaria immitis. Eur Respir J* 3:1070–1071, 1990.

118. Ciferri F: Human pulmonary dirofilariasis in the United States: A critical review. *Am J Trop Med Hyg* 31:302–308, 1982.

119. Ro JY, et al: Pulmonary dirofilariasis: The great imitator of primary or metastatic lung tumour: A clinico-pathologic analysis of 7 cases and a review of the literature. *Hum Pathol* 20:69–76, 1989.

120. Mizrachi HM, et al: Pulmonary dirofilariasis: Mimicry of well differentiated squamous cell carcinoma. *Hum Pathol* 20:818–819, 1989.

121. Hardman B, Rudders RA: Benign pulmonary nodule and small cell carcinoma. *Ann Intern Med* 97:140–141, 1982.

122. Khan FW, et al: Pulmonary dirofilariasis and transitional cell carcinoma: Benign lung nodules mimicking metastatic malignant neoplasm. *Arch Intern Med* 143:159–160, 1983.

123. Dayal Y, Neafie RC: Human pulmonary dirofilariasis: A case report and review of the literature. *Am Rev Respir Dis* 112:437–443, 1975.

124. Case Records of the Massachusetts General Hospital *N Engl J Med* 300:723–729, 1979.

125. Hankins AG, et al: Pulmonary dirofilariasis diagnosed by fine needle aspiration biopsy. A case report. *Acta Cytol* 29:19–22, 1985.

126. Glickman LT, et al: Serologic diagnosis of zoonotic pulmonary dirofiloriasis. *Am J Med* 60:161–164, 1986.

127. Akao N, et al: Immunoblot analysis of *Dirofilaria immitis* recognized by infected humans. *Ann Trop Med Parasitol* 85:455–460, 1991.

128. Beaver PC, Cran IR: *Wuchereria*-like filaria in an artery associated with pulmonary infarction. *Am J Trop Med Hyg* 23:869–876, 1974.

129. Neva FA, Ottesen EA: Tropical (filarial) eosinophilia. *N Eng J Med* 298:1129–1131, 1978.

130. Ottesen EA, Nutman TV: Tropical pulmonary eosinophilia. *Annu Rev Med* 43:417–424, 1992.

131. Weingarten RJ: Tropical eosinophilia. *Lancet* 1:103–105, 1943.

132. Arora VK, Varma VR: Pneumonia left lung: an unusual presentation of tropical pulmonary eosinophilia. *Ind J Chest Dis Allied Sci* 34:29–32, 1992.

133. Webb KG, et al: Tropical eosinophilia: demonstration of microfilariae in lung, liver and lymph nodes. *Lancet* 1:835–842, 1960.

134. Udwadia FE: Tropical eosinophilia—a correlation of clinical histopathologic and lung function studies. *Dis Chest* 52:531–538, 1967.

135. Danaraj TJ, et al: The etiology and pathology of eosinophilic lung (tropical eosinophilia). *Am J Trop Med Hyg* 15:183–189, 1966.

136. Uwadia FE: Tropical eosinophilia: A review. *Respir Med* 87:17–21, 1993.

137. Crandall RB, et al: The ferret (*Mustela purius furo*) as an experimental host for *Brugia malayi* and *B. pathangi. Am J Trop Med Hyg* 31:752–759, 1982.

138. Beaver PC: Filariasis without microfilaremia. *Am J Trop Med Hyg* 19:181–189, 1970.

139. Udwadia FE.: Tropical eosinophilia in Hertzog H (ed): *Pulmonary Eosinophilia: Progress in Respiration Research.* Basel, S Karger, 7:35–155, 1975.

140. Nutman TB, et al: Tropical pulmonary eosinophilia: Analysis of antifilarial antibody localized to the lung. *J Infect Dis* 160:1042–1050, 1989.

141. Loeffler W: Transient lung infiltration with blood eosinophilia. *Arch Allergy* 8:54–59, 1956.

142. Muhleisen JP: Demonstration of pulmonary migration of the causitive organism of creeping eruption. *Ann Intern Med* 38:595–600, 1953.

143. Phills JA, et al: Pulmonary infiltrates, asthma and eosinophilia due to *Ascaris suum* infestation in man. *N Eng J Med* 286:965, 1972.

144. Gelpi AP, Mustafa A: *Ascaris* pneumonia. *Am J Med* 44:377–389, 1968.

145. Beaver PC, Danaraj TJ: Pulmonary ascariasis resembling eosinophilic lung. Autopsy report with description of larvae in the bronchioles. *J Trop Med Hyg* 7:810–805, 1958.

146. Piggott J, et al: Human ascariasis. *Am J Clin Pathol* 53:223–234, 1970.

147. Mittal VK, et al: Fatal respiratory obstruction due to round worm. *Med J Aust* 2:210–212, 1976.

148. Beaver PC, et al: *Clinical Parasitology* 9th ed. Philadelphia, Lea & Febiger, 1984, p 316.

149. Taylor MRH, et al: The expanded spectrum of toxocaral disease. *Lancet* 1:692–695, 1988.

150. Mahmoud AAF: Helminthic diseases of the lungs, in Fishman AP (ed): *Pulmonary Disease and Disorders,* 2d ed. New York, McGraw Hill, 1988, p 1724.

151. Proffitt RD, Walton BC: *Ascaris* pneumonia in a 2-year-old girl: Diagnosis by gastric aspirate. *N Engl J Med* 266:931–934, 1962.

152. Simson I W. Heinz H J: Pulmonary ascariasis. *Leech* 30:140–144, 1960.

153. Pampiglione S, Rivasi F, Trotti GC: Human pulmonary difilariasis in Italy. *Lancet* 1:333, 1984.

154. Aftandelians R, et al: Pulmonary capillariasis in a child in Iran. *Am J Trop Med Hyg* 26:64–71, 1977.

155. Klion AD, et al: Pulmonary involvement in loiasis. *Am Rev Respir Dis* 145:961–963, 1992.

156. Francis H, et al: The Mazzotti reaction following treatment of onchocerciasis with diethylcarbamazine: Clinical severity as a function of infection intensity. *Am J Trop Med Hyg* 34:529–536, 1985.

157. Kobayashi A, et al: Probable pulmonary anisakiasis accompanying pleural effusions. *Am J Trop Med Hyg* 34:310–313, 1985.

158. Beaver PC. et al: Pulmonary nodule caused by *Enterobius vermicularis. Am J Trop Med Hyg* 22:711–713, 1973.

159. Timmons RF, et al: Infection of the respiratory tract with *Mammomanogamus laryngeus:* A new case in Largo, Florida. *Am Rev. Respir Dis* 128:566–569, 1983.

160. Yii C-J, et al: Human angiostrongliasis involving the lungs. *Chin J Microbiol* 1:148–150, 1968.

161. Nagler A, et al: Human pleuropulmonary gnathostomiasis:

A case report from Israel. *Isr J Med Sci* 19:834–837, 1983.

162. Liu LX, Weller PF: Strongyloidiasis and other intestinal nematode infections. *Infect Dis Clin North Am* 7:655–682, 1993.

163. Higenbotham TW, Heard BE: Opportunistic pulmonary strongyloidiasis complicating asthma treated with steroids. *Thorax* 31:226–233, 1976.

164. Venizelos PC, et al: Respiratory failure due to *Strongyloides stercoralis* in a patient with a renal transplant. *Chest* 78:104–106, 1980.

165. Harris RA, et al: Disseminated strongyloidiasis. Diagnosis made by sputum examination. *JAMA* 244:65–66, 1980.

166. Ford J, et al: Pulmonary strongyloidiasis and lung abscess. *Chest* 79:239–240, 1981.

167. Lin AL, et al: Restrictive pulmonary disease due to interlobular septal fibrosis associated with disseminated infection by *Strongyloides stercoralis*. *Am J Respir Crit Care Med* 151:205–209, 1995.

168. Humphreys K, Heiger LR: *Strongyloides stercoralis* in routine Papinicolaou-stained sputum smears. *Acta Cytol* 23:471–476, 1979.

169. Chu E, et al: Pulmonary hyperinfection syndrome with *Strongyloides stercoralis*. *Chest* 97:1475–1477, 1990.

170. Rassiga AL, et al: Diffuse pulmonary infection due to *Strongyloides stercoralis*. *JAMA* 230:426–427, 1974.

171. Choi JH, et al: A case of pulmonary cysticercosis. *Korean J Intern Med* 6:38–43, 1991.

172. D'Alessandro A, et al: First observation of *Echinococcus vogeli* in man, with a review of polycystic hydatid disease in Colombia and neighboring countries. *Am J Trop Med Hyg* 28:303–317, 1979

173. Burgos L, et al: Experience in 331 patients with pulmonary hydatidosis. *J Thorac Cardiovasc Surg* 102:427–430, 1991.

174. Jerray M, et al: Hydatid disease of the lungs: Study of 386 cases. *Am Rev Respir Dis* 146:185–189, 1992.

175. Editorial: Man, dogs and hydatid disease. *Lancet* 1:21–22, 1987.

176. Macpherson CNL, et al: Hydatid disease in the Turkana district of Kenya, IV. The prevalence of *Echinococcus granulosus* infection in dogs and observations on the role of the dog in the lifestyle of the Turkana. *Am Trop Med Parasitol* 79:51–61, 1985.

177. Schwabe CW: Veterinary medicine and human health. 3d ed. Baltimore, Williams & Wilkins, 1984.

178. Dutz W, et al: The cyst wall of *Echinococcus granulosus* in ruptured and non-ruptured pulmonary cysts in man. *Z Tropenmed Parasitol* 22:191–195, 1971.

179. Sterman MM, Brown HW: *Echinococcus* in man and dog in the same household in New York City. *JAMA* 169:938–940, 1957.

180. Hira PR, et al: Diagnosis of cystic hydatid disease: Role of aspiration cytology. *Lancet* 2:655–657, 1988.

181. Case Records of the Massachusetts General Hospital. *N Engl J Med* 317:219–218, 1987.

182. Rausch RL, Wilson JF: Rearing of the adult *Echinococcus multilocularis* Leuckart, 1863, from sterile larvae from man. *Am J Trop Med Hyg* 22:357–360 1973

183. Rausch RL, et al: Spontaneous death of *Echinococcus multilocularis*: Cases diagnosed serologically (by Em2 ELISA) and clinical significance. *Am J Trop Med Hyg* 36:576–585, 1987.

184. Guardia S, et al: Pentasomiasis in Canada. *Arch Pathol Lab Med* 115:515–517, 1991.

185. Self JT, et al: Pentastomiasis in Africans. *Trop Geogr Med* 27:1–13, 1975.

186. Drabick JJ: Pentastomiasis. *Rev Infect Dis* 9:1087–1094, 1987.

187. Prathap K, et al: Pentastomiasis: A common finding at autopsy among Malaysian aborigines. *Am J Trop Med Hyg* 18:20–27, 1969.

188. Warren KS, Mahmoud AAF (eds): *Tropical and Geographical Medicine*, 2d ed. New York, McGraw-Hill, 1990.

189. Spencer H: *Pathology of the Lung*, 4th ed. Oxford, Pergamon, 1985.

190. Mahmoud AAF: Helminthic diseases of the lung, in Fishman AP (ed): *Pulmonary Diseases and Disorders*, vol 2, 2d ed. New York, McGraw-Hill, 1988.

191. Muller R: *Worms and Disease*. New York, Heineman.

Chapter 10

Aspiration Pneumonia, Lung Abscess, Bronchiectasis, and Atelectasis

P. W. Bishop

ASPIRATION PNEUMONIA

Introduction

Aspiration pneumonia follows the inhalation of gasric contents or infected material from the oropharyngeal region. There are five major syndromes associated with aspiration.[1] The first is airways obstruction caused by aspiration of particulate matter, foreign bodies, or large volumes of fluid. The second is chemical pneumonitis, the main cause of which is aspiration of liquid gastric contents. Other causes are inhalation of exogenous chemicals, lipid pneumonitis, and smoke inhalation. The third clinical picture is infectious pneumonitis. Infection can follow aspiration in the first two categories but may also be related to inhalation of oropharyngeal secretions. The fourth is aspiration after near-drowning, and finally there is evidence of chronic aspiration in heart-lung transplant recipients.[2] The clinical syndromes developing with aspiration vary and depend on the frequency of aspiration, the quantity and nature of the aspirated material, and the underlying host response.

The factors associated with increased risk of gastric aspiration are given in Table 10-1.

Fifty percent of healthy persons and up to 70 percent of those with impaired consciousness aspirate during their sleep. This has been demonstrated by placing contrast in the oropharynx of sleeping patients. Most of these patients remained asymptomatic, although most had contrast in the lower airways as shown on chest radiographs the next day. The incidence of aspiration in patients having upper gastrointestinal endoscopy has been reported as high as 25 percent. Any mechanism bypassing normal swallowing, such as a tracheostomy or the presence of a nasogastric tube, will increase the dangers of aspiration. Disease in the upper air passages, especially the mouth and gums, acts as a source of infection by anaerobic organisms and may cause pleuropulmonary infections.

TABLE 10-1

Factors Associated with Increased Risk of Gastric Aspiration[3-5]

1. Impaired level of laryngeal sensation (e.g., decreased consciousness either in normal people during sleep or in patients with head injuries
2. Impaired pharyngeal peristalsis/function (neuromuscular disease, myopathies, or head and neck surgery)
3. Impaired laryngeal closure (laryngeal nerve dysfunction, endotracheal or tracheostomy tube, etc.)
4. Cricopharyngeal spasm (e.g., gastroesophageal reflux)
5. Increased gastric volume (e.g., delayed gastric emptying, recent ingestion, obesity)
6. Increased intragastric pressure (nausea and vomiting, upper abdominal surgery, pregnancy, ascites)
7. Lower esophageal sphincter pressure (e.g., increased gastric acidity, exogenous agents such as ethanol and tobacco, and mechanical factors such as nasogastric tubes)
8. Decreased antegrade esophageal propulsion (e.g., Trendelenburg's positon, abnormal esophageal motility such as scleroderma, achalasia, and esophageal strictures)

It is important not to confuse anaerobic and aspiration pneumonia. Aspiration is one of the most common causes of anaerobic pneumonia, but anaerobic bacteria can reach the lung from subphrenic collections and hematogenous spread from septic thrombophlebitis.

The classic description of aspiration of gastric contents and the role of hydrochloric acid in the pathophysiology of the associated lung disease came from Mendelson,[6] who described aspiration in 0.15 percent of cases of obstetric anesthesia between 1932 and 1945. Another important cause of aspiration is the so-called café coronary, in which a large volume of liquid or particulate material has been aspirated (Fig. 10-1). Aspiration seldom occurs during normal swallowing. In a study of 500 normal people by radi-

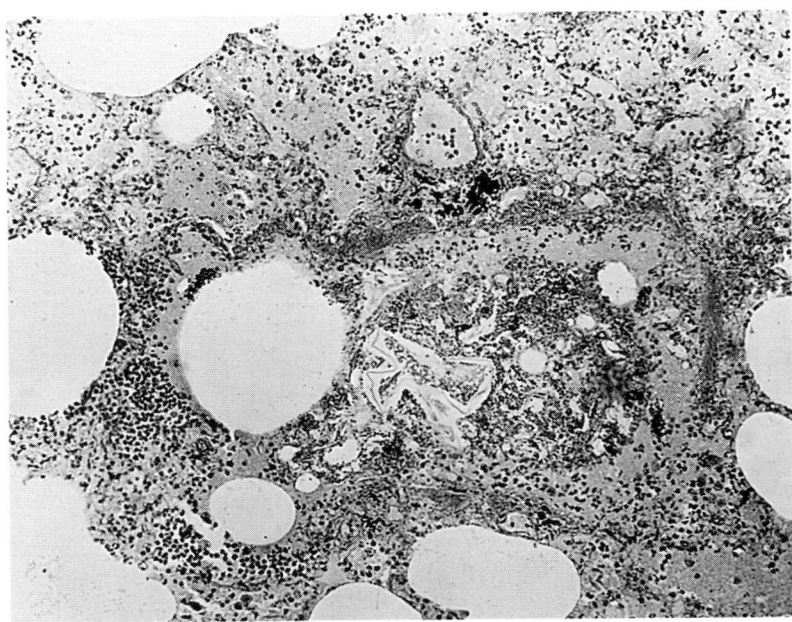

FIGURE 10-1
Acute aspiration pneumonia showing large quantities of inhaled vegetable matter. The foreign material is surrounded by a severe polymorph infiltrate.

ographic cinematography, in no case did barium get through the glottis.[7] When aspiration of food does occur in healthy people, it is usually the result of incautious inspiration, laughing, or sneezing while the mouth is full. However, massive aspiration occurs in alcoholics or during anesthesia for emergency surgery and can lead to rapid death. In gastric contents of semidigested or undigested food, the amount of acid may be low but the bacterial content is often much greater. Inhaled food particles cause a foreign body reaction in the lung (Fig. 10-2). In the unconscious patient, it may be very difficult to obtain

a history of aspiration. Aspiration pneumonia may be caused by *Vibrio parahemolyticus* after near-drowning accidents in parts of the Mediterranean, especially on the eastern coast of Spain. The infection probably follows inhalation of the organism.

Thus, many factors are associated with the development of aspiration pneumonia. If, in addition, there is underlying lung disease, such as chronic bronchitis and emphysema, the incidence of aspiration pneumonia is increased. The pH of the aspirate, its volume, the presence of foreign bodies such as food, highly irritant bile, fecal contamina-

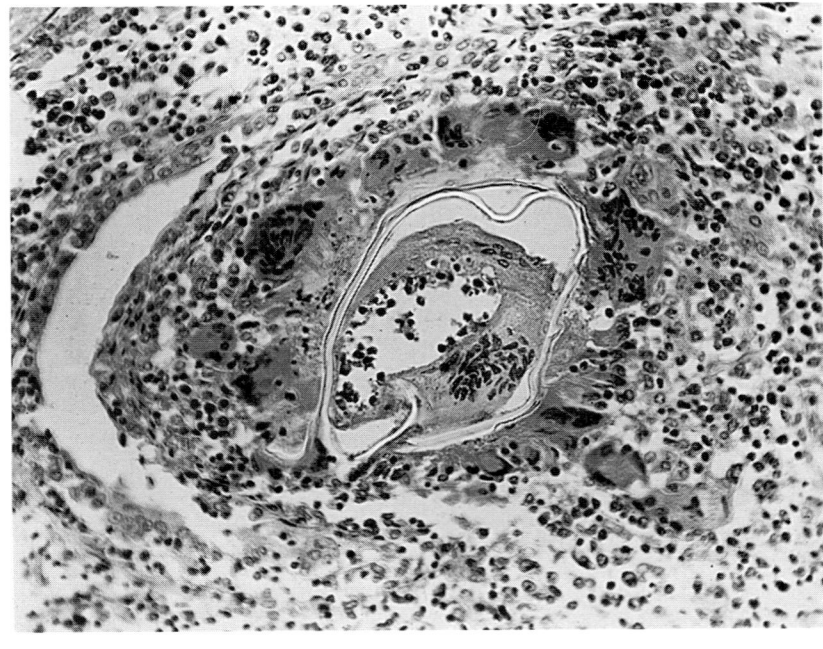

FIGURE 10-2
An inhaled vegetable body showing surrounding giant cell reaction and inflammatory cell infiltration.

tion, as well as the bacterial concentration, are all important determinants in the lung injury after aspiration.[8] The clinical features of aspiration have a wide range. With the typical Mendelson syndrome, there is respiratory arrest with apnea or dyspnea, cough, and wheezing; at the other extreme there may be chronic indolent infections, mistakenly diagnosed as tuberculosis or even bronchial carcinoma. The right side is involved more commonly, because of the more direct path of the right main-stem bronchus. The superior segments of the lower lobes and posterior segment of the upper lobes are more commonly involved, since these segments are dependent when the patient is recumbent.

Bacteriology

A wide range of organisms may be found in aspiration pneumonia, including not only anaerobes but also aerobes (Table 10-2).[9] There may be mixed infections.

Macroscopically, there is a putrefactive pneumonia with ill-defined yellow-brown areas and areas of central softening in the lung. The centers of these foci are filled with necrotic contents with a characteristic malodor.

Histology depends on the material aspirated. If there is food, this will be reflected by inhaled vegetable bodies surrounded by a granulomatous reaction. There is abscess formation, as well as many histiocytes and neutrophil polymorphs.

In chronic aspiration, such as with achalasia of the cardia, there is extensive fibrosis in the lower lobes, and the changes are those of an inhalation lipid pneumonia. For the pathologist this diagnosis is easy if vegetable fibers or a definite history is present. If it supervenes on some other disease process, such as carcinoma of the esophagus, it may be misdiagnosed as a concurrent pneumonia. It is likely that many cases are unrecorded as aspiration pneumonia, since it is often a preterminal event in an unconscious patient or one with swallowing difficulties.

A major problem is the diagnosis of aspiration and anaerobic pleuropulmonary infection in life.[11] Aspiration of lung abscess to obtain bacteriology has been described, and it is necessary to obtain bacteriology in such patients, especially those with necrotizing pneumonia. The source of the infecting organism in these patients is always the upper airways. The specimen must be collected to avoid indigenous and colonizing organisms. To culture delicate anaerobes, the specimen must be promptly placed under anaerobic conditions. Empyema fluid provides reliable specimens in sufficient volume so it can be placed in an anaerobic transport vial or tube. Transtracheal aspiration, when not contraindicated, may be the choice if no empyema fluid is available.

TABLE 10-2
Bacteriology of Aspiration Pneumonia

Anaerobes	Aerobes
Peptostreptoccus	Microaerophilic streptococci
Fusobacterium species	*Eikenella corrodens*
Bacteroides species	*Pseudomonas aeruginosa*[a]
Pigmented bacteroides (*B. melaninogenicus*) (*B. intermedius*)	*Staphylococcus aureus*[a]
B. fragilis group	Enterobacteriaceae[a]
B. ureolyticus group	*Streptococcus pyogenes*
B. oralis	*Klebsiella pneumoniae*
Ensibacterium sp.	*Enterobacter*
Clostridium	*Proteus*
Bifidobacterium[a]	
Actinomyces israelii	
Arachnia sp.	

[a] Isolates found in anaerobic infection.

SOURCE: From Hill and Sanders.[10]

LUNG ABSCESS

Introduction

Lung abscesses may result from many causes, chief among which are inhalation of foreign bodies and infected material, conditions secondary to bronchial obstruction by neoplasm, complicating bacterial pneumonia, pyemic and septic infarcts, trauma to the lung, transdiaphragmatic spread of infection, infected parasitic cysts, miscellaneous, and cryptogenic.

Inhalational Lung Abscess

Inhalation of infected material has long been recognized as a cause of lung abscesses, and its importance was emphasized more than 70 years ago.[12] Later, Touroff and Moolten[13] found that one-fourth of their cases of lung abscess followed tonsillectomy and less commonly tooth extraction. Neuhof and Touroff[14] analyzed 45 consecutive surgical cases of lung abscess. In 19 they were able to find a definite source of inhaled infected material. The conditions responsible included tonsillectomy, dental extraction,

FIGURE 10-3
A small lung abscess with adjacent bronchiectasis. The patient was known to be epileptic.

opening of peritonsillar abscesses, acute tonsillitis, and inhalation of foreign bodies during epileptic attacks (Fig. 10-3). Others[15,16] have drawn attention to the occurrence of lung abscesses after electroconvulsive therapy and after alcoholic intoxication. Some cases, without a source of gross inhaled material, have severe gingivodental infection.[13,17–19]

Flick and coworkers[20] and Brock and coworkers[21] showed that most solitary abscesses were in the upper lobes. These authors analyzed their series of lung abscesses requiring surgical treatment and found that of 184 monolobar abscesses, 105 (57 percent) occurred in the upper lobes. In 116 cases the abscess was situated in the right lung. The reason for the right-sided predominance of inhalational

abscesses is the more direct path pursued by the right bronchus after it arises from the trachea.

When inhaled material reaches the lung, it passes to the terminal air passages, where it causes a severe purulent reaction. The center of the abscess liquefies, and as the lesion grows, bronchi are eroded and provide partial drainage. This is often accompanied by foul-smelling sputum. In most untreated cases the drainage is inadequate, and continuing enlargement of the abscess erodes blood vessels. Severe hemorrhage occurs in 10 to 15 percent of cases.[22] The abscess may spread across interlobar fissures to invade adjacent lobes and proceed out into the pleural cavity, resulting in an empyema. Since the introduction of antibiotics, it has become possible to sterilize the abscess cavity. Such a cavity is lined first by granulation and fibrous tissue and then is covered by a newly formed squamous or even ciliated columnar epithelium. This may eventually provide a diagnostic problem, but the distinguishing features are the multiple bronchial communications and the trabeculae, which form folds lining the cavity. If the inhaled material is bulky and becomes impacted in a large airway, the result is suppurative bronchitis, followed by bronchiectasis and destruction of lung parenchyma; the end result of healing and epithelialization is a saccular bronchiectasis.

Lung Abscesses Due to Bronchial Obstruction by Neoplasia

Between 7 and 17 percent of all lung abscesses are related to bronchogenic carcinoma. The proportion in the elderly is much higher—up to 40 percent. Most abscesses are in proximity to the primary tumor. They may be distant to the tumor because of spillover of infection from a necrotic growth. Necrosis, cavitation, and abscess formation are most common in squamous cell carcinomas. This disorder needs to be differentiated from the rare event of a cancer arising in the wall of a chronic bronchiectatic or abscess cavity. The presence of mural fibrosis and chronic inflammatory change is indicative of such a secondary neoplasm. Abscess formation may occur in neoplasms of various histologic types after chemotherapy and tumor necrosis.[23]

Lung Abscess Complicating Bacterial Pneumonia

Bacterial pneumonia due to a range of organisms may be complicated by lung abscess. *Staphylococcus aureus* may result in multiple abscess formation, causing very extensive destruction and scarring. *Klebsiella pneumoniae* may similarly produce pneumonia followed by multiple abscesses, often confined to an upper lobe, which is replaced by a honeycomb of thin fibrous-walled cavities. Pneu-mococcal pneumonia only rarely results in abscess formation and when it occurs is usually caused by serotype III pneumococci. It is seen more often in young children and the elderly. It is preceded by alveolar capillary thrombosis. This may be related to the reduction of bronchial arterial blood flow,[24] which is seen during the stage of red hepatization. *Pseudomonas* pneumonia may cause both acute and chronic lung abscesses. Lung abscess as a complication of bacterial pneumonia is common in children in developing countries.[25] It may rarely occur as a late complication of adult respiratory distress syndrome.[26]

Pyemic Lung Abscess and Septic Infarcts

Embolic abscesses may occur in septic thrombophlebitis. Septic emboli may be liberated from any peripheral vein, but they most commonly arise within the veins of the lower limbs and pelvis or may follow septic thrombosis due to an indwelling catheter in the superior vena cava. The source of sepsis may be a peripheral venous catheter. The abscesses are usually small and predominantly subpleural. They often yield a pure culture of *S. aureus*. In rare instances, lung abscesses occur in association with right-sided endocarditis in drug addicts.[27]

Traumatic Lung Abscesses

These are very rare but may result from open or closed lung injury, and may follow either the implantation of infected material or infection of a traumatic intrapulmonary hematoma. A case has been reported 24 years after injuries sustained in the Vietnam war.[28]

Transdiaphragmatic Spread of Infection

Spread upward through the diaphragm may occur from a number of sites of abdominal sepsis, including amebic and *Klebsiella* liver abscess,[29] a pancreaticobronchial fistula,[30] and a nephrobronchial fistula infected with *Proteus mirabilis*.[31]

Lung Abscesses in Infected Parasitic Cysts

Hydatid cysts may become infected and give rise to abscesses.[32,33] Similar abscess formation has been reported with paragonimiasis.[34]

Other Secondary Lung Abscesses

Congenital immunodeficiency resulting in recurrent bacterial infection predisposes to lung abscess. Such conditions include *chronic granulomatous disease*[35] and Job's syndrome.[36] Acquired constitutional or immunologic deficiency (alcoholism, malnutrition, drug abuse, and AIDS) predispose adults to abscess formation.[37] Eosinophilic lung disease,[38] fulminant sarcoidosis,[39] and Wegener's granulomatosis[40] are all rare causes of lung abscess. Stevens and colleagues[41] reported on 600 middle-aged patients with celiac disease of whom seven developed lung abscesses or cavities[41]; next to malignancy, lung abscesses were the most common cause of death in these patients.

Cryptogenic Lung Abscesses

It has been stated that up to 25 percent of lung abscesses have no identifiable cause. This is probably a considerable overestimate, but there remain a minority of cases in which, despite thorough investigation, no underlying cause of the lung abscess can be identified.

Bacteriology

Identifying the pathogenic organism requires samples uncontaminated by oropharyngeal flora. These may be obtained by tracheal aspiration[42] or percutaneous lung aspiration.[43] Quantitative culture of bronchoalveolar lavage fluid has also been advocated.[44] Appropriate culture conditions for anaerobic and microaerophilic organisms are required. The organisms discussed below are particularly those found in inhalational lung abscesses. In many cases, a mixture of aerobic, microaerophilic, and anaerobic organisms is found.

The causative organisms include all of those listed in Table 10-1 as causes of aspiration pneumonia. In addition, anaerobic causes of lung abscess include *Streptococcus pneumoniae*[45,46] and β-hemolytic streptococcus,[47,48] *Pseudomonas cepacia*,[49] *Hemophilus influenzae*,[45] *Brucella melitensis*,[50] *Salmonella*,[40] *Lactobacilli*,[51] *Legionella*,[52,53] and *Pasteurella*.[54] *Mycobacterium tuberculosis*[55,56] and other mycobacteria[57] have been found, sometimes in association with anaerobes. *Rhodococcus* (formerly *Corynebacterium*) *equi* has been found,[58–61] is acid-fast, and needs to be distinguished from mycobacteria. Typhoid abscesses may occur years after the initial infection.[62] Fungi identified include *Candida*[63] and *Zygomyces*.[64] Protozoan amebic abscesses occur.[65,66]

Complications of Lung Abscesses

Infection may spread from a lung abscess to the pleura, giving rise to an empyema. Horner's syndrome is rarely a reported complication.[67] Drainage into a bronchus may be followed by a persistent bronchopulmonary fistula. Metastatic abscesses occur, most seriously to the brain, or spread of infection that gives rise to meningitis. Chronic lung abscesses may occasionally result in secondary amyloidosis.

BRONCHIECTASIS

Bronchiectasis was first described in 1819 by Laennec.[68] He later attributed the disorder to retention of bronchial secretions, which he believed led to damage to the bronchial walls and their subsequent weakening and dilatation.[69] The term *bronchiectasis* is used pathologically to refer to irreversible dilation of bronchi. The term is also used more loosely, often by radiologists, to include acute reversible bronchial dilatation. The most useful subdivision is by etiology into obstructive and nonobstructive types.

Obstructive Bronchiectasis

Obstructive bronchiectasis may follow obstruction from any cause. It is localized to the region distal to the obstruction. Particular causes of obstruction tend to be more common at particular sites. Epithelial tumors more commonly affect the upper lobes. The clinical presentation of a tumor as bronchiectasis is more likely with the rarer, more slowly growing tumors, such as carcinoid, chondroma, papilloma, lipoma,[70] and granular cell tumor. Hilar lymphadenopathy may also obstruct bronchi; classically, tuberculous lymphadenopathy at the hilum leads to bronchiectasis and collapse in the right middle and part of the right lower lobe.[71] This predisposition of the right middle lobe bronchus to either extrinsic compression or obstruction of its relatively narrow lumen has been called the *middle lobe syndrome*. The middle lobe symdrome may be caused by a wide variety of neoplastic and nonneoplastic disorders. These include bronchial asthma, mediastinal lymphadenopathy, traction diverticula of the esophagus, tracheal foreign bodies, cystic fibrosis, *M. avium-intracellulare* infection, allergic bronchopulmonary aspergillosis, and cardiovascular anomalies. The pathology of this condition has recently been reviewed.[71a]

Aspiration of foreign bodies is commonly confined to the right lower lobe or the posterior segment of the right upper lobe. Most patients are under 10 years of age,[72] and half are between 1 and 4 years old.[73] Aspiration is followed by bronchiectasis in 3.5 to 16 percent of cases.[74] Even with delayed diagnosis and marked dilatation of distal airways radio-

logically, there may be complete resolution after removal of the foreign body.[75] The materials inhaled by children include seeds (peanuts), unchewed food (popcorn), and beads. Grass seeds with rows of angled spikes tend to travel deep down the bronchial tree and may reach the pleura, causing empyema, osteomyelitis of ribs, and even a pneumocutaneous fistula.[76] Some plant materials not only cause mechanical obstruction but also contain chemical irritants. The factors predisposing to aspiration are discussed above in relation to aspiration pneumonia. Heroin and other drugs of abuse, taken parenterally, may cause bronchial damage with or without an episode of aspiration.

Obstructive bronchiectasis may also occur distal to inspissated mucus in allergic bronchopulmonary aspergillosis and distal to broncholiths. Allergic aspergillosis preferentially affects the upper lobes. Cystic fibrosis as a cause of bronchiectasis is discussed in Chap. 3.

In contrast to nonobstructive bronchiectasis, the pathogenesis is not considered problematic. At first, bronchial secretions accumulate distal to the obstruction, and these were considered by Laennec[69] to be the cause of bronchiectasis. Although uninfected mucoceles of the bronchi do occur, they are usually soon followed by infection in the obstructed portion of lung. Inflammatory damage to the bronchial wall ensues and weakens it.

Nonobstructive Bronchiectasis

Various classifications have been proposed for this disorder. The older classifications were based on the shape of the dilated bronchi as judged by bronchographic examination. Thus, Reid[77] recognized saccular (cystic), cylindric (fusiform or tubular), and varicose forms. Although bronchography is no longer performed, computed tomography allows the anatomy of the abnormal bronchi to be visualized, and the terms may be of some descriptive value. Whitwell[78] introduced a new division into follicular, saccular, and collapsed (atelectatic). This was modified by Spencer[79] into postpneumonic infective, collapsed (atelectatic), and congenital (including saccular). Such classifications are based on an unsatisfactory combination of anatomic, histologic, and imperfectly associated etiologic criteria. The existence of congenital bronchiectasis is disputed: of course, a distinction must be made between bronchiectasis present at birth and hereditary conditions predisposing to acquired bronchiectasis. Few studies in the past 10 years have regarded such subtyping as of importance.

MORPHOLOGY

Nonobstructive bronchiectasis is usually more widespread within the lung. Most frequently it affects the basal segments of the lower lobes. The left lower is affected twice as often as the right. Tuberculosis and cystic fibrosis may involve only the upper lobes. One-third of cases are bilateral. Spread from diseased to normal bronchi has been blamed on "spillage" of infected mucus, but that is rare.[80] Apparent extension probably results from progression of disease within bronchi damaged by the initial insult. Bronchiectasis typically involves the medium-sized bronchi (fourth and nineth generations). The shape of the dilated bronchi is variable (Fig. 10-4). Cylindric bronchiectasis consists of uniform dilation along the length of a bronchus. In saccular bronchiectasis, the dilation is greatest in the distal portions of the dilated bronchi. Varicose bronchiectasis describes focal dilatation along the length of a

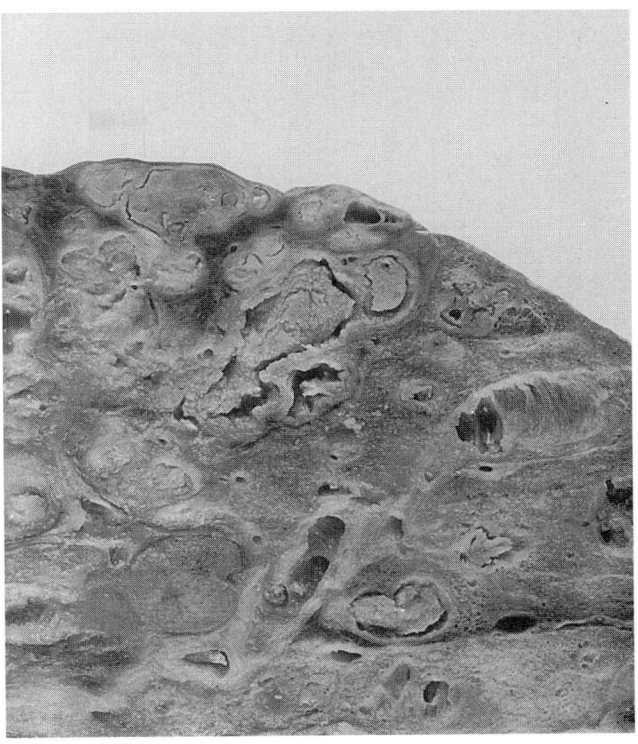

FIGURE 10-5
In this case of cystic fibrosis, the ectatic bronchi are filled with mucopurulent material. This is typical of "wet" bronchiectasis.

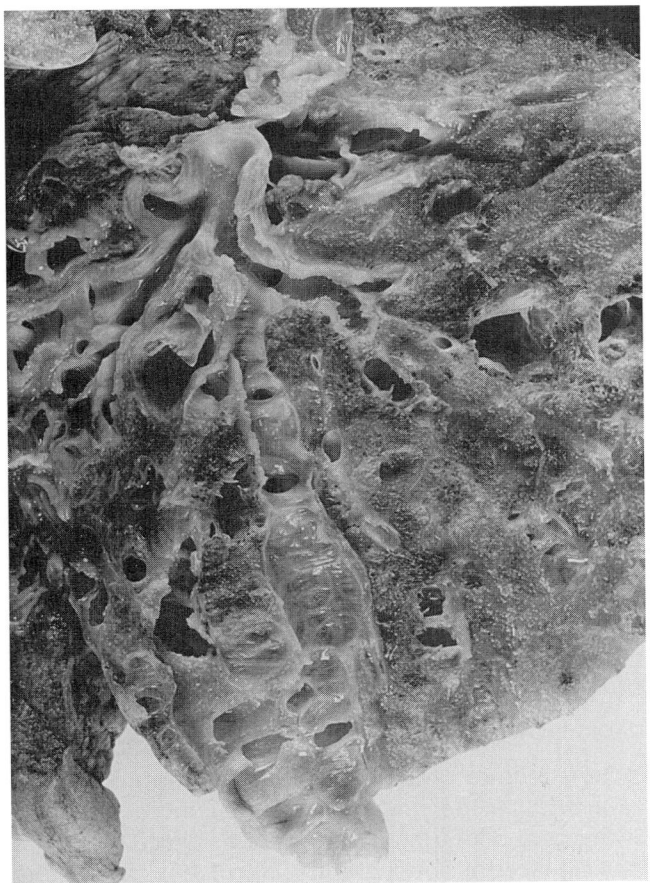

FIGURE 10-4
Prominent ectatic bronchi from an area of bronchiectasis. The abnormal bronchi appear crowded and extend more peripherally than normal.

bronchus. Bronchiectasis may cause abscess formation extending into the surrounding lung parenchyma.

A distinction has been made between *wet* and *dry* forms of bronchiectasis, according to the degree of sputum production; in "wet" bronchiectasis, the bronchial lumen contains mucus with neutrophil polymorphs, macrophages, and desquamated epithelial cells (Fig. 10-5). The epithelium shows reserve cell hyperplasia (Fig. 10-6), squamous metaplasia, and inflammation and may be ulcerated. Within the bronchial wall is an inflammatory cell infiltrate, usually chronic inflammatory cells, including large numbers of T cells.[81] The lymphocytic infiltrate may form prominent aggregates—hence the term *follicular* bronchiectasis. The submucosal glands are usually decreased in number. The bronchial cartilage may be eroded (Fig. 10-7). In some cases there are granulomas. These may be simply the response to inspissated material, but if they are well formed, aspiration or infection by mycobacteria or fungi should be considered. Chronic infection often causes extensive destruction of the alveoli, with obliteration of many small bronchi and bronchioles distal to the bronchiectasis.

Secondary hyperplasia of neuroendocrine cells is common. This is part of any infective process.

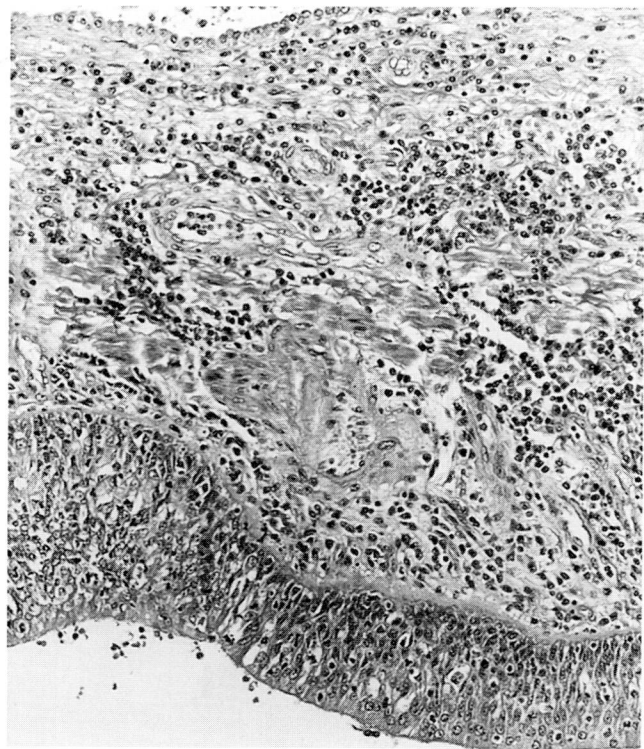

FIGURE 10-6
The wall of an ectatic bronchus showing a predominantly mononucluer inflammatory cell infiltrate and reserve cell hyperplasia.

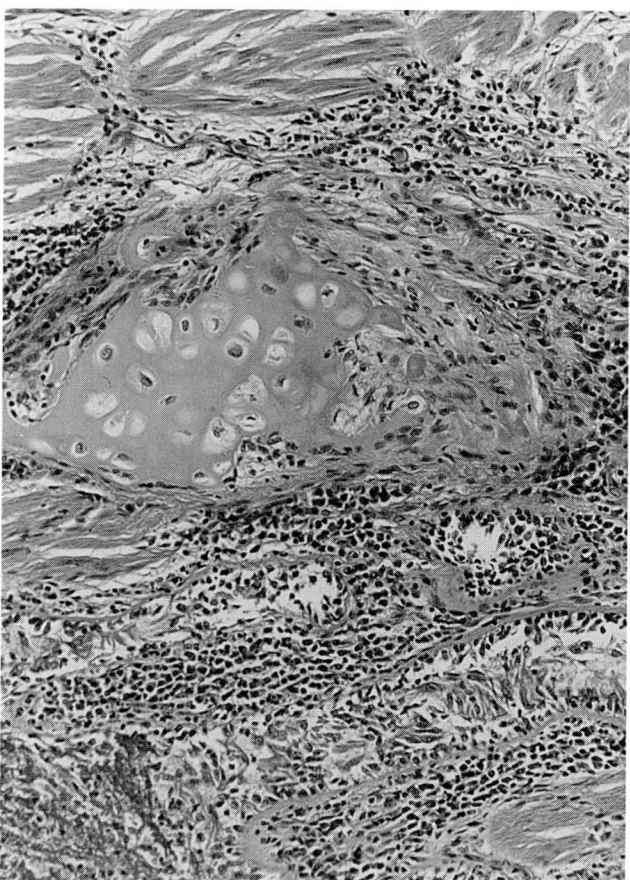

FIGURE 10-7
The wall of an ectatic bronchus showing a mononuclear inflammatory cell infiltrate and erosion of bronchial cartilage.

These focal proliferations give rise to tumorlets. Vascular changes occur, with the formation of large bronchopulmonary anastomoses and a greatly increased bronchial arterial blood flow.[82-85] It is unclear whether these anastomoses are enlarged preexisting vasa vasorum or new vessels forming within areas of inflammation and granulation tissue. Other bronchopulmonary anastomoses result from an expansion of the normally minute, inconspicuous precapillary anastomoses known to exist between the two circulations. There is destruction of the capillary bed as alveolar tissue is destroyed by inflammation. The combination of increased resistance from obliteration of the capillary bed and the formation of anastomoses raises the pulmonary arterial pressure and gives rise to cor pulmonale. Ulceration of these vessels results in hemoptysis, which may be clinically significant.

MICROBIOLOGY

In Western countries, the most common isolates in bronchiectasis not due to cystic fibrosis are *H. influenzae, S. pneumoniae,* and *S. aureus.* A study from Hong Kong reported a different pattern of isolates, with *Pseudomonas aeruginosa* and *P. fluorescens* being relatively common.[86] In cystic fibrosis, *P. aeruginosa* and *P. cepacia* are important infecting organisms. *P. aeruginosa* infection is associated with radiologically severe bronchiectasis.[87] Fungi may superinfect bronchiectatic cavities, and in allergic bronchopulmonary aspergillosis, numerous noninvasive fungal hyphae are seen within mucus plugs in association with numerous eosinophils. Tuberculosis is a well-recognized cause of bronchiectasis; it may produce lower lobe bronchiectasis, and there may be little clinical suspicion of the nature of the infection.[88] Bronchiectasis may be complicated by atypical mycobacterial infections. There are individual case reports of rare complicating infections such as those from *Neisseria sicca*[89] or *P. multicida*[90] and phaeohyphomycosis.[91]

RADIOLOGY

After standard chest radiography, computed tomography (CT) is the imaging method of choice for examining patients with suspected bronchiectasis[92-94]; it has replaced bronchography (Fig. 10-8). In conjunction with the appropriate clinical history, CT findings are sufficiently specific to allow a confident diagnosis of bronchiectasis in most cases (Fig. 10-9).

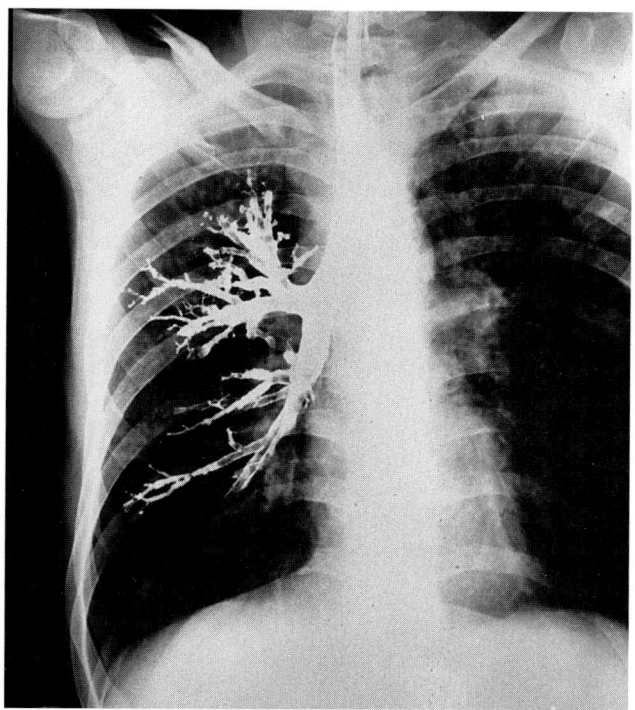

FIGURE 10-8
A bronchogram of a case of bronchiectasis showing dilated and truncated large bronchi. The distal bronchi fail to fill.

The abnormal bronchi apppear as "tramlines" or "signet rings," depending on their orientation; visible airways extend into the peripheral lung fields. The measurements of bronchial circumference and the scoring of the extent of bronchiectasis are sufficiently reproducible to correlate with the degree of physiologic impairment as measured by FEV_1[95] and may be useful in monitoring progression of disease.[96] When bronchiectasis coexists with multiple small lung nodules, this is commonly due to infection by *Mycobacterium avium-intracellulare*.[97]

Etiology and Pathogenesis

Patients with bronchiectasis often give a history of childhood respiratory infection. This has usually occurred before the age of 2 years.[98,99] This may be a viral infection, commonly measles or adenovirus, primary bacterial infection *(Bordetella pertussis)*, or bacterial superinfection complicating a viral infection. In developed countries, the prevalence of bronchiectasis is falling. The relative contributions of improved living conditions, immunization, and antibiotic treatment of bacterial infections are difficult to disentangle. An increasing number of under-

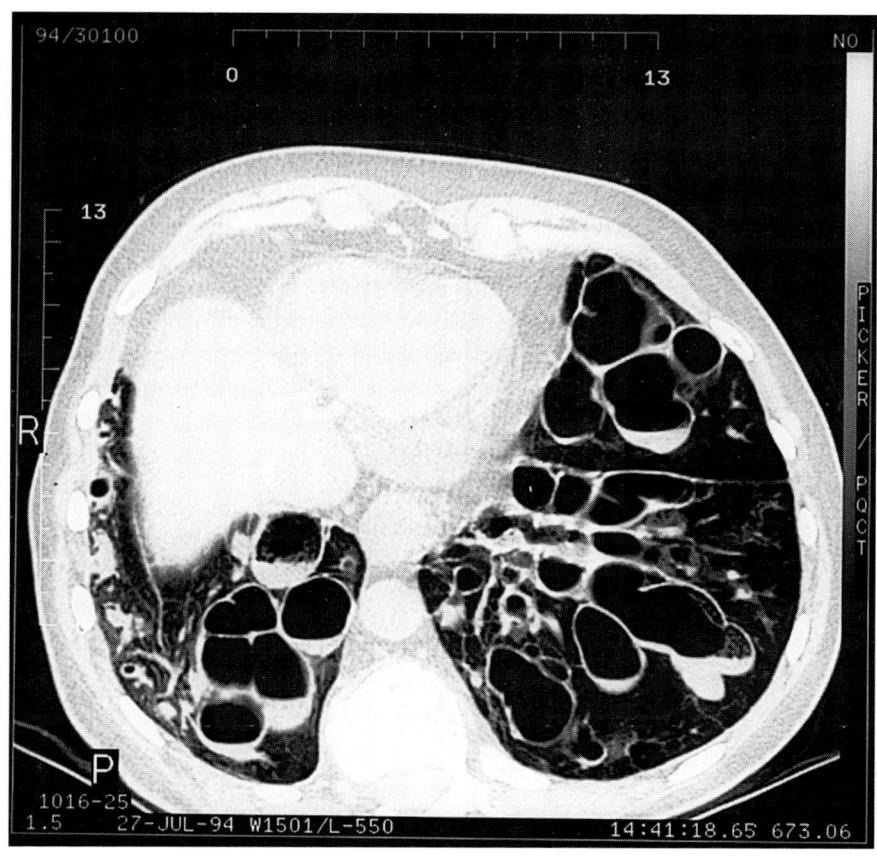

FIGURE 10-9
A CT scan of lung showing severe bronchiectasis. In this view the ectatic bronchi are seen in transverse section.

TABLE 10-3
Underlying Conditions Implicated as Causing Bronchiectasis

Congenital malformations of lung
    Tracheobronchomegaly[101]
    Williams-Campbell syndrome[102,103]
Cystic fibrosis
Ciliary abnormalities[104–108]
Young's syndrome[109]
Woake's syndrome[110]
Endobronchial sarcoid[111]
α_1-antitrypsin deficiency[112–116]
Primary immune defects[117,118]
    Neutrophil abnormalities
    Panhypogammaglobulinemia
    Hyper-IgE syndrome/Job's syndrome[36]
    IgA deficiency[119]
    IgG subclass deficiencies[120,121]
    Ataxia-telangiectasia
    Bare lymphocyte syndrome[122]
    Yellow nail syndrome[123,124]
Secondary immune defects
    AIDS[125–127]
    Paragononimiasis[128]
    Coccidioidomycosis[129]
Inhalation injury
    Smoke[130,131]
    Acrolein[132]
    Mustard gas[133]
    Ammonia[134]
Inflammatory bowel disease[135–140]
Rheumatoid arthritis[141–145]
Nephronophthisis[146]
Dyskeratosis congenita[147]
Klinefelter's syndrome[148,149]

lying, sometimes subtle, factors predisposing to infection have been reported. Nikolaizik and coworkers[100] were able to identify predisposing factors in 63 percent of 41 children with chronic suppurative lung disease. Conditions causing or associated with bronchiectasis are listed in Table 10-3; some of these conditions are discussed below. Cystic fibrosis, ciliary abnormalities, and congenital malformations are discussed in Chap. 3, along with other pediatric pulmonary diseases.

The permanent dilatation of bronchi has been attributed to a combination of damage to bronchial walls, with resultant weakening and traction from the surrounding lung due to collapse and fibrosis.

Hedblom[150] and Lander and Davidson[151] thought that collapse of the lung was more important than infection in causing bronchiectasis. However, the importance of pulmonary inflammation as a cause of nonobstructive bronchiectasis has been repeatedly stressed.[78,152–155] More recently, increasing emphasis has been placed on the mechanisms by which the bronchial wall is weakened. Inflammation appears to be pivotal in the pathogenesis of bronchiectasis and has been described as a "two-edged sword."[156] The hypothesis is that instead of transient inflammation followed by resolution, a "vicious circle" of inflammatory damage is initiated. Infection is the most common initiator of the inflammatory process. Once damage to the bronchi impairs mucociliary clearance,[157] persistent infection continues to induce inflammation. A number of microorganisms directly damage the mucociliary apparatus.

Damage by the Host
Most studies of damage by host inflammatory cells have been directed at neutrophil proteases. The elastase of human neutrophils both causes epithelial disruption and reduces ciliary beat frequency.[158] It has a secretagogue action on mucous glands.[159] Neutrophil elastase can degrade elastin, collagen, and proteoglycan,[160] damaging the bronchial wall. The effects of neutrophil elastase are abolished by protease inhibitors, such as alpha-1-antitrypsin.[158] However, the protease inhibitors, in turn, can be inhibited both by oxidants from neutrophil polymorphs and by tobacco smoke.[160] Other neutrophil proteases that may be of importance are collagenase and cathepsin-G, which, in synergy with oxidants, may mediate epithelial cell detachment.[161] Eosinophils may also induce disruption of epithelium and reduce ciliary beat frequency, effects mediated by eosinophil major basic protein.[162] These cells are not usually numerous in bronchiectasis, except in association with asthma and allergic bronchopulmonary aspergillosis. The role of mononuclear inflammatory cells in damaging both epithelium and the bronchial wall is less clear. A rat model of bronchiectasis mimics the prominent T-cell infiltrate and may give greater insight into the role of a cell-mediated immune response.[163]

Damage by Microorganisms
Viral infection probably only has a transient role in initiating damage to the bronchial tree. Attempts to show persistence of adenoviral infection in bronchiectasis have given negative results.[164] In children, however, infection by several common respira-

tory viruses produces ciliary abnormalities, at least in the upper respiratory tract, lasting 2 to 10 weeks.[165] There is evidence that both *H. influenzae* and *S. pneumoniae* can inhibit ciliary action. Although these are the most common infecting organisms in non–cystic fibrosis bronchiectasis, most of the experimental study of the mechanisms of bacterial action have been directed to *P. aeruginosa*. *Pseudomonas* produces phenazine pigments (pyocyanin and 1-hydroxyphenazine) that inhibit ciliary action.[158,166,167] This effect is mediated by reduced intracellular cyclic adenosine monophosphate and adenosine triphosphate.[168] *Pseudomonas* pyocyanin also induces epithelial disruption.[168] *Pseudomonas* rhamnolipids are toxins with ciliostatic and secretagogue actions.[169] *Pseudomonas* is the only microorganism known to produce an elastase,[160] which causes epithelial disruption and, in the presence of divalent metal ion, also impairs ciliary beat frequency.[158]

SPECIFIC CONDITIONS ASSOCIATED WITH BRONCHIECTASIS

Bronchial asthma (not due to *Aspergillus*) has been cited in textbooks as a cause of bronchiectasis, although studies of the association are scanty. Murphy and coworkers[170] found no significant increase in the prevalence of atopy in 23 patients with bronchiectasis. Pang's team[171] studied 36 patients with bronchiectasis and a similar number of well-matched controls. There was increased bronchial reactivity in the patients with bronchiectasis, but there was no difference in the prevalence of asthma or atopy in the patients or their families. Ip and colleagues[172] studied 85 patients with bronchiectasis, of whom 23 had asthma: in 13 the asthma preceded the onset of bronchiectasis; in seven it developed after a long duration of bronchiectasis; in three the temporal relationship was uncertain.[172]

α_1 Antitrypsin deficiency is usually associated with emphysema, but there are a few reports of its association with bronchiectasis. The published reports were considered by Barker,[115] who pointed out that most patients had some other factor in their history, such as childhood pertussis or an abnormal sweat test. The report of three cases by Jones and coworkers[116] appears the most satisfactory. The subject is also complicated by the reactive rise in α_1 antitrypsin in patients with bronchiectasis.[112] There is evidence that some patients with either bronchiectasis or emphysema may have subtle abnormalities of their α_1 antitrypsin at a molecular level, without an overt defect in the enzyme activity.[113]

Inflammatory bowel disease is associated with a variety of pulmonary conditions; in some cases the pulmonary disease is caused by drug therapy. These associations have been reviewed by Camus and colleagues.[135] Of their 33 patients with inflammatory bowel disease, six had bronchiectasis. The association is specifically with ulcerative colitis;[136,137] Neilly and coworkers[140] found no cases of bronchiectasis in 29 patients with Crohn's disease. Bronchiectasis may develop after colectomy has been performed for ulcerative colitis.[173]

Rheumatoid arthritis may also give rise to a range of pulmonary conditions, of which bronchiectasis is probably the least common. Shadick and colleagues[142] identified 23 patients with the association; in 18 the arthritis preceded the lung disease. By contrast, McMahon and coworkers[145] found 32 patients with the association; in 30, bronchiectasis preceded the arthritis, and there was an additional association with xerophthalmia. They discussed the explanations proposed for the association, including the possibility that an exocrine dysfunction might predispose to both bronchiectasis and rheumatoid arthritis. Solanki and Neville[141] approached the problem from the opposite direction, looking for rheumatoid arthritis in patients with bronchiectasis, and confirmed the association. Attempts to find a three-way association between bronchiectasis, rheumatoid arthritis, and α_1 antitrypsin have yielded negative results.[143,144]

A variety of pulmonary diseases, including bronchiectasis, are associated with Klinefelter's syndrome.[148,149] Dyskeratosis congenita is an unusual familial disorder primarily affecting the skin and its appendages[147]; there have been a few case reports of bronchial sepsis, probably due to the associated immunodeficiency seen in this syndrome. Woake's syndrome consists of nasal polyps with necrotic ethmoiditis, frontal sinus aplasia, bronchiectasis, and highly viscous mucus.[110] Nephronophthisis–medullary cystic disease is a complex of hereditary disorders characterized by urinary concentrating defects and progressive renal failure in association with retinal degeneration, hepatic fibrosis, hyperuricemia, skeletal abnormalities, cerebellar ataxia, mental retardation, and seizures. There is one case report of associated bronchiectasis.[146] Yellow nail syndrome is a combination of yellow nails, lymphedema, chronic sinusitis, and bronchiectasis; the underlying pathologic condition is thought to be lymphatic hypoplasia.[123,124] Bare lymphocyte syndrome is a lack of HLA class I antigen expression by lymphocytes, with resulting immunodeficiency.[122] Young's syndrome is impaired mucociliary clearance with resulting chronic sinusitis and bronchiectasis, possibly due to mercury intoxication in infancy.[109] Williams-Campbell syndrome is a developmental deficiency of bronchial cartilage.[103]

COMPLICATIONS

Bronchiectasis may give rise to a number of complications. Right ventricular hypertrophy and failure have already been mentioned. Metastatic abscesses occur, most devastatingly to the brain. The association with AA Type amyloidosis is well recognized, although with improving management it is decreasing in incidence. Vasculitis is due to the formation of immune complexes within the areas of bronchiectasis. Rose and Spencer[174] drew attention to the high incidence of bronchiectasis in patients with pulmonary polyarteritis. The bronchial sepsis may long predate the development of microscopic polyarteritis or Wegener's granulomatosis.[175–177]

ATELECTASIS OF COLLAPSE OF THE LUNG

The term atelectasis is now increasingly used to describe alveolar collapse. Atelectasis in the older sense of macroscopic collapse of the lung may be due to external compression or to bronchial obstruction. It is in this older connotation that atelectasis will be described below.

External Compression

External compression of the lung may be caused by pleural effusion, hemothorax, pyothorax, or pneumothorax; rarely, it results from surgical procedures designed to collapse the lung. The pulmonary changes consequent on this type of collapse differ from those which follow absorption collapse, inasmuch as the absence of bronchial obstruction leaves the bronchial and lung secretions free to drain up the bronchial tree. Such changes as eventually occur in the lung parenchyma do not result from infection but from the hemodynamic alterations and subsequent vascular changes.

Unlike small areas of absorption collapse, which may re-aerate because of collateral ventilation, there is no possibility of this occurring in compressed lung. Schlaepfer[178] investigated the structural changes that followed pneumothorax. The collapsed lung had edema and there was an increased lymph flow from the lung. The persistent chronic edema of the alveolar wall caused by chronic collapse results in the alveolar walls developing an excess of reticulin and later collagen fibers. There is an increase of the perivascular and peribronchial connective tissue together with pleural fibrous thickening. As a result of these changes, the lungs are often unable to reexpand.

The changes in blood flow in pressure collapsed lungs were studied in man[179] and in the rabbit.[180] There was no appreciable alteration in the pulmonary arterial blood flow 1 h after the lung was collapsed, though this partly depended on the extent of the collapse and the degree of mechanical vascular obstruction. During the first month following collapse, the flow became gradually reduced until it was negligible. If the cause of the collapse was removed and the lung allowed to reexpand, the blood-flow was almost wholly restored. After the lungs were collapsed for a year, only a negligible amount of blood circulated through the pulmonary arteries and the main arteries become contracted to half their normal size. The smaller branches of the pulmonary arterial tree were also reduced in size. The main obstruction appeared to be situated in the capillary bed; this was explained in the light of the structural changes noted above.

The bronchial arteries are slightly enlarged in chronic collapse. The presence of pulmonary arterial blood with a higher oxygen saturation than in the pulmonary veins indicates bronchopulmonary anastomoses causing a slight reversal of blood-flow in the affected pulmonary artery. Cor pulmonale is rare but has been described following collapse of a lung by thoracoplasty. In such cases there is usually considerable tuberculous damage of the contralateral lung.

Absorption Collapse and Acute Bronchial Obstruction Syndrome

Absorption collapse is more common than pressure collapse and results from a rapid complete obstruction of a large bronchus. The size of the blocked bronchus and the freedom of the adjacent lung from edema, hemorrhage, and pneumonic consolidation determine whether collateral ventilation will be successful in keeping the obstructed segment of the lung filled with air.

In both absorption and pressure collapse the lung is smaller than normal, reddish-purple and has a wrinkled pleural surface. The cut surface is airless and the lung may sink in water.

The mechanisms of bronchial obstruction are discussed in the section on bronchiectasis.

Spain[181] blocked the main lower lobe bronchi in dogs with cotton wool and observed the changes at intervals up to 36 h later. After 6 h no macroscopic changes were observed but microscopically the alveolar capillaries were congested in the obstructed lobe. At 12 h the changes were more exaggerated with alveolar edema. After 24 h the obstructed lobe was smaller but all the alveoli contained some edema fluid. Finally, after 36 h the lobe, though small, showed very congested alveolar capillaries, and the alveoli were filled with both edema fluid and polymorph leukocytes.

Spain[181] also quoted an example of human right main bronchial obstruction caused by a foreign body which resulted in death 8 h later. The lung was very congested and edematous, but was only slightly reduced in size. The lack of ventilatory movements in the obstructed lung impeding lymph clearance is a factor causing pulmonary edema. Jackson[182] described the same changes seen in the human lungs both at bronchoscopy and after death.

The experiments quoted above, and observation of human cases of bronchial obstruction, lend little support to the widely accepted view that when air beyond an acute obstruction is rapidly absorbed, it leads to severe collapse of the lung. As the air disappears its place is to a great extent replaced by edema fluid and such a reduction in lung size as does occur is not great.

Massive Collapse of the Lungs

Massive collapse of the lungs is a rare condition usually caused by closed or open wounds and injuries of the chest wall; it may follow the use of lipiodol for bronchography and rarely may complicate laryngeal paralysis in diphtheria.

Of all types of lung collapse, massive collapse produces the greatest shrinkage of the lung and until recent years the cause of the condition was completely unknown. Lilenthal[183] described a case he had seen during the First World War, which followed an extreme degree of bronchospasm. He also described a case of massive collapse by bronchospasm, observed during an open chest operation.

Burford and Burbank[184] showed that the first reaction of the lung to chest injury, closed or open, was the production of acute pulmonary edema. Rodbard[185] emphasized the importance of bronchospasm as a means of raising the intraalveolar pressure to combat the exudation of edema fluid. Almost all cases of massive collapse have been preceded by acute and severe pulmonary edema. The edema may result either from injury, the irritant (hypersensitivity) reaction to a contrast medium used for bronchography, or the accumulation of fluid within the lung due to the inability of the patient to cough and rid him- or herself of bronchial and pulmonary edema fluid, as may occur in laryngeal paralysis caused by diphtheria.

Rounded atelectasis is described in Chap. 34.

REFERENCES

1. Wynne JW, Modell JH: Respiratory aspiration of stomach contents. *Ann Intern Med* 87:466–474, 1977.
2. Reid KR, McKenzie FN, Menkis AH, et al: Importance of chronic aspiration in recipients of heart/lung transplants *Lancet* 336:206–208, 1990.
3. Gardner AMN: Aspiration of food and vomit. *QJM* 106:227–242, 1957.
4. Arms RA, Dines DE, Tinstman TC: Aspiration pneumonia. *Chest* 65:136–139, 1974.
5. Ribaudo CA, Grace WJ: Pulmonary aspiration. *Am J Med* 50:510–520, 1971.
6. Mendelson CL: The aspiration of stomach contents into the lungs during obstetric anesthesia. *Am J Obstet Gynecol* 52:191–205, 1946.
7. Ardran GM, Kemp FH: The protection of the laryngeal airway during swallowing. *Br J Radiol* 25:406–416, 1952.
8. DePaso WJ: Aspiration pneumonia. *Clin Chest Med* 12:269–284, 1991.
9. Lode H: Microbiological and clinical aspects of aspiration pneumonia. *J Antimicrob Chemother* 21(suppl C):83–90, 1988.
10. Hill MK, Sanders CV: Anaerobic disease of the lung. *Infect Dis Clin North Am* 5:453–466, 1991.
11. Bartlett JG: Anaerobic bacterial infections of the lung and pleural space. *Clin Infect Dis* 16:S248–255, 1993.
12. Lockwood AL: Abscess of the lung. *Surg Gynecol Obstet* 35:461–492, 1922.
13. Touroff ASW, Moolten SE: The symptomatology of putrid abscess of the lung. *J Thorac Surg* 4:558–572, 1934.
14. Neuhof H, Touroff ASW: Acute putrid abscess of the lung: Analysis of 45 consecutive operative cases. *Surg Gynecol Obstet* 66:836–857, 1938.
15. Wayl P, Rakower J: Lung abscess as a complication of shock therapies. *Thorax* 9:216–221, 1954.
16. Duffy TJ, Chofnas I: Primary lung abscess. *Am J Med Sci* 243:269–278, 1962.
17. Stern L: Etiological factors in the pathogenesis of putrid abscess of the lung. *J Thorac Surg* 6:202–211, 1936.
18. Ban B: Lung abscess in India. *J Thorac Surg* 32:254–267, 1956.
19. Bartlett JG, Gorbach SL, Finegold SM: The bacteriology of aspiration pneumonia. *Am J Med* 56:202–207, 1974.
20. Flick JB, Clerf LH, Fink EH, Farrell JT: Pulmonary abscess: An analysis of 172 cases. *Arch Surg* 19:1292–1312, 1929.
21. Brock RC, Hodgkiss F, Jones HO: Bronchial embolism and posture in relation to lung abscess. *Guy's Hosp Rep* 91:131–139, 1942.
22. Philpott NJ, Woodhead MA, Wilson AG, Millard FJ: Lung abscess: A neglected cause of life threatening haemoptysis. *Thorax* 48:674–675, 1993.
23. Senan IS, Kaye SB: Lung abscess: A fatal complication of treatment for testicular teratoma. *Clin Oncol (R Coll Radiol)* 3:345–347, 1991.

24. Cudkowicz L: Some observations of the bronchial arteries in lobar pneumonia and pulmonary infarction. *Br J Tuberc* 46:99–102, 1952.

25. Gonzago NC, Navarro EE, Lucero MG, Queipo SC, Schroeder I, Tupasi TE: Etiology of infection and morphological changes in the lungs of Filipino children who die of pneumonia. *Rev Infect Dis* 12(suppl 8):S1055–S1064, 1990.

26. Knapp MJ, Bunn WB, Stave GM: Adult respiratory distress syndrome from sulfuric acid fume inhalation. *South Med J* 84:1031–1033, 1991.

27. Dickens P, Chan AC, Lam KY: Isolated right-sided endocarditis in Hong Kong Chinese. *Am J Cardiovasc Pathol* 4:367–370, 1993.

28. Kollef MH: A persistent left lower lobe infiltrate and chronic cough following chest wounds sustained in Vietnam twenty-four years earlier. *Milit Med* 158:499–500, 1993.

29. Cheng DL, Liu YC, Yen MY, Liu CY, Wang RS: Septic metastatic lesions of pyogenic liver abscess. Their association with *Klebsiella pneumoniae* bacteremia in diabetic patients. *Arch Intern Med* 151:1557–1559, 1991.

30. Misare BD, Gagner M, Braasch JW, Shahian DM: Pancreaticobronchial fistula causing lung abscess: Case report and brief discussion of the literature. *Surgery* 110:549–551, 1991.

31. Kyriakopoulos M, Stathopoulos P, Kourti A, Pandis B: Nephrobronchial fistula. Case report. *Scand J Urol Nephrol* 25:245–246, 1991.

32. Lamy AL, Cameron BH, LeBlanc JC, Culham JA, Blair GK, Taylor GP: Giant hydatid lung cysts in the Canadian northwest: Outcome of conservative treatment in three children. *J Pediatr Surg* 28:1140–1143, 1993.

33. Woolf DC: Presentation of echinococcus infection as lung abscess. *Trop Geogr Med* 43:297–299, 1991.

34. Moon WK, Kim WS, IM JG, Kim IO, Yeon KM, Han MC: Pulmonary paragonimiasis simulating lung abscess in a 9-year-old: CT findings. *Pediatr Radiol* 23:626–627, 1993.

35. Paz HL, Little BJ, Ball WC, Winkelstein JA: Primary pulmonary botryomycosis. A manifestation of chronic granulomatous disease. *Chest* 101:1160–1162, 1992.

36. Lui RC, Inculet RI: Job's syndrome: A rare cause of recurrent lung abscess in childhood. *Ann Thorac Surg* 50:992–994, 1990.

37. Cordice JW, Chitkara RK: The role of surgery in treating pleuropulmonary suppurative disease—review of 77 cases managed at Queens Hospital Center between 1986 and 1989. *J Natl Med Assoc* 84:145–150, 1992.

38. Kondo T, Suzuki H, Hirokawa Y, Ohta Y, Yamabayashi H: Chronic eosinophilic pneumonia with small abscesses in the tracheo-bronchial mucosa and lung parenchyma. *Intern Med* 31:391–393, 1992.

39. Biem J, Hoffstein V: Aggressive cavitary pulmonary sarcoidosis. *Am Rev Respir Dis* 143:428–430, 1991.

40. Chan JC, Raffin TA: Salmonella lung abscess complicating Wegener's granulomatosis. *Respir Med* 85:339–341, 1991.

41. Stevens FM, Connolly CE, Murray JP, McCarthy CF: Lung cavities in patients with coeliac disease. *Digestion* 46:72–80, 1990.

42. Yamashita Y, Kohno S, Tanaka K, et al: Anaerobic respiratory infection—evaluation of methods of obtaining specimens. *Kansenshogaku Zasshi* 68:631–638, 1994.

43. Ohnishi Y, Sawaki M, Mikasa K, et al: Clinical study of anaerobic respiratory infection. *Kansenshogaku Zasshi* 67:336–341, 1993.

44. Henriquez AH, Mendoza J, Gonzalez PC: Quantitative culture of bronchoalveolar lavage from patients with anaerobic lung abscesses. *J Infect Dis* 164:414–417, 1991.

45. Germaud P, Poirier J, Jacqueme P, et al: Monotherapy using amoxicillin/clavulanic acid as treatment of first choice in community-acquired lung abscess: Apropos of 57 cases. *Rev Pneumol Clin* 49:137–141, 1993.

46. Gesner M, Desiderio D, Kim M, et al: *Streptococcus pneumoniae* in human immunodeficiency virus type 1–infected children. *Pediatr Infect Dis J* 13:697–703, 1994.

47. Tumwine JK: Lung abscess in children in Harare, Zimbabwe. *East Afr Med J* 69:547–549, 1992.

48. Frieden TR, Biebuyck J, Hierholzer WJ Jr: Lung abscess with group A beta-hemolytic Streptococcus. Case report and review. *Arch Intern Med* 151:1655–1657, 1991.

49. Snell GI, de Hoyos A, Krajden M, Winton T, Maurer JR: *Pseudomonas cepacia* in lung transplant recipients with cystic fibrosis. *Chest* 103:466–471, 1993.

50. Papiris SA, Maniati MA, Haritou A, Constantopoullous SH: Brucella haemorrhagic pleural effusion. *Eur Respir J* 7:1369–1370, 1994.

51. Manresa F, Dorca J, Prats E: Fatal lung abscess due to *Lactobacillus casei* ss rhamnosus (letter). *Thorax* 47:992, 1992.

52. Nechwatal R, Ehret W, Klatte OJ, Zeissler HJ, Prull A, Lutz H: Nosocomial outbreak of legionellosis in a rehabilitation center: Demonstration of potable water as a source. *Infection* 21:235–240, 1993.

53. Halberstam M, Isenberg HD, Hilton E: Abscess and empyema caused by *Legionella micdadei*. *J Clin Microbiol* 30:512–513, 1992.

54. Boitout A, Lion C, Chabot F, Delorme N, Burdin JC, Polu JM: Respiratory pasteurellosis: Apropos of 32 cases. *Rev Pneumol Clin* 47:208–213, 1991.

55. Ceyhan B, Celikel T, Korten V: Lung abscess due to coexisting *Mycobacterium tuberculosis* and anaerobic bacteria in a cavitary bronchogenic carcinoma (letter). *Tuberc Lung Dis* 74:407, 1993.

56. Makanjuola D, Adeyemo AO: Lower lung field tuberculosis in a rural African population. *West Afr J Med* 10:412–419, 1991.

57. Vadakekalam J, Ward MJ: *Mycobacterium fortuitum* lung abscess treated with ciprofloxacin. *Thorax* 46:737–738, 1991.

58. Takasugi JE, Godwin JD: Lung abscess caused by *Rhodococcus equi*. *J Thorac Imaging* 6:72–74, 1991.

59. Pialoux G, Dupont B: Lung abscess caused by *Rhodococcus equi* in HIV infection: Two cases (letter). *Presse Med* 21:1086, 1992.

60. Pialoux G, Fournier S, Dupont B, et al: Lung abscess caused by *Rhodococcus (Corynebacterium) equi* in HIV infection. Two cases. *Presse Med* 21:417–421, 1992.

61. Heudier P, Taillan B, Garnier G, et al: *Rhodococcus equi* infection in AIDS: A case with pulmonary abscess. Review of the literature. *Rev Med Interne* 15:268–72, 1994.

62. De Matteis A, Armani G: *Salmonella typhi* pneumonia without intestinal lesions. *J Pathol Bacteriol* 94:464–467, 1967.

63. Mori T, Ebe T, Takahashi M, Isonuma H, Ikemoto H, Oguri T: Lung abscess: Analysis of 66 cases from 1979 to 1991. *Intern Med* 32:278–284, 1993.

64. Rubin SA, Chaljub G, Winer-Muram HT, Flicker S:

Pulmonary zygomycosis: A radiographic and clinical spectrum. *J Thorac Imaging* 7:85–90, 1992.

65. Ghosh AK, Paul K, Panja M, De PK, Mitra J: Primary amoebic lung abscess. *J Assoc Physicians India.* 40:200–201, 1992.

66. García Uribe JA, Padua Gabriel A, Quintana O, García Vázquez A, de la Escosura G: Three cases of hematogenous lung abscess of amebic origin. *Rev Invest Clin* 43:264–268, 1991.

67. Dugdale DC, Ritter KJ, Wilhyde DE: Lung abscess causing Horner's syndrome. *West J Med* 153:196–197, 1990.

68. Laennec RTH: *De l'auscultation mediate ou traite du diagnostic des maladies des poumons et du coeur, fonde, principalement sur ce noveau moyen d'exploration.* Paris, Brosson et Claude, 1819.

69. Laennec RTH: *A Treatise on the Diseases of the Chest and on Mediate Auscultation* (transl. J. Forbes), 1st ed. London, T. G. Underwood, 1829.

70. Box K, Kerr KM, Jeffrey RR, Douglas JG: Endobronchial lipoma associated with lobar bronchiectasis. *Respir Med* 85:71–72, 1991.

71. Brock RC: Post-tuberculous broncho-stenosis and bronchiectasis of the middle lobe. *Thorax* 5:5–39, 1950.

71a. Kwon KY, Myers JL, Swensen SJ, Colby TV: Middle lobe syndrome: A clinicopathological study of 21 patients. *Hum Pathol* 26:302–307, 1995.

72. Weissberg D, Schwartz I: Foreign bodies in the tracheobronchial tree. *Chest* 91:730–733, 1987.

73. Limper AH, Prakash UB: Tracheobronchial foreign bodies in adults. *Ann Intern Med* 112:604–609, 1990.

74. Barker AF, Bardana EJ Jr: Bronchiectasis: Update of an orphan disease. *Am Rev Respir Dis* 137:969–978, 1988.

75. Ernst KD, Mahmud F: Reversible cystic dilatation of distal airways due to foreign body. *South Med J* 87:404–406, 1994.

76. Maayan C, Avital A, Elpeleg ON, Springer C, Katz S, Godfrey S: Complications following oat head aspiration. *Pediatr Pulmonol* 15:52–54, 1993.

77. Reid L McA: Reduction in bronchial subdivision in bronchiectasis. *Thorax* 5:233–247, 1950.

78. Whitwell F: A study of the pathology and pathogenesis of bronchiectasis. *Thorax* 7:213–239, 1952.

79. Spencer H: *Pathology of the Lung*, 4th ed. Oxford, Pergamon Press, 1984.

80. Perry KMA, King DS: Bronchiectasis: A study of prognosis based on follow-up of 400 patients. *Am Rev Tuberc* 41:531–548, 1940.

81. Lapa e Silva JR, Jones JAH, Cole PJ, Poulter LW: The immunological component of the cellular inflammatory infiltrate in bronchiectasis. *Thorax* 44:668–673, 1989.

82. Liebow AA, Hales MR, Harrison W, Bloomer W, Lindskog GE: The genesis and functional implications of collateral circulation of the lungs. *Yale J Biol Med* 22:637–660, 1949.

83. Liebow AA, Hales MR, Lindskog GE: Enlargement of the bronchial arteries, and their anastomoses with pulmonary arteries in bronchiectasis. *Am J Pathol* 25:211–231, 1949.

84. Marchand P, Gilroy JC, Wilson VH: An anatomical study of the bronchial vascular system and its variation in disease. *Thorax* 5:207–221, 1950.

85. Cockett FB, Vass CCN: The collateral circulation to the lungs. *Br J Surg* 38:97–103, 1950.

86. Pang JA, Cheng A, Chan HS, Poon D, French G: The bacteriology of bronchiectasis in Hong Kong investigated by pro-

tected catheter brush and bronchoalveolar lavage. *Am Rev Respir Dis* 139:14–17, 1989.

87. Nagaki M, Shimura S, Tanno Y, Ishibashi T, Sasaki H, Takishima T: Role of chronic *Pseudomonas aeruginosa* infection in the development of bronchiectasis. *Chest* 102:1464–1469, 1992.

88. Chan CH, Ho AK, Chan RC, Cheung H, Cheng AF: Mycobacteria as a cause of infective exacerbation in bronchiectasis. *Postgrad Med J* 68:896–899, 1992.

89. Gris P, Vincke G, Delmez JP, Dierckx JP: *Neisseria sicca* pneumonia and bronchiectasis. *Eur Respir J* 2:685–687, 1989.

90. Inoue Y, Fujii T, Ohtsubo T, et al: Three cases of *Pasteurella multocida* infection in the respiratory tract. *Kansenshogaku Zasshi* 68:242–248, 1994.

91. Barenfanger J, Ramirez F, Tewari RP, Eagleton L: Pulmonary phaeohyphomycosis in a patient with hemoptysis. *Chest* 95:1158–1160, 1989.

92. McGuinness G, Naidich DP, Leitman BS, McCauley DI: Bronchiectasis: CT evaluation. *AJR: Am J Roentgenol* 160:253–259, 1993.

93. Grenier P, Cordeau MP, Beigelman C: High-resolution computed tomography of the airways. *J Thorac Imaging* 8:213–229, 1993.

94. Padley SP, Adler B, Muller NL: High-resolution computed tomography of the chest: Current indications. *J Thorac Imaging* 8:189–199, 1993.

95. Wong-You-Cheong JJ, Leahy BC, Taylor PM, Church SE: Airways obstruction and bronchiectasis: Correlation with duration of symptoms and extent of bronchiectasis on computed tomography. *Clin Radiol* 45:256–259, 1992.

96. Desai SR, Wells AU, Cheah FK, Cole PJ, Hansell DM: The reproducibility of bronchial circumference measurements using computed tomography. *Br J Radiol* 67:257–262, 1994.

97. Swensen SJ, Hartman TE, Williams DE: Computed tomographic diagnosis of *Mycobacterium avium-intracellulare* complex in patients with bronchiectasis. *Chest* 105:49–52, 1994.

98. Boyd G: Bronchiectasis in children. *Can Med Assoc J* 25:174–182, 1931.

99. Warner WP: Factors causing bronchiectasis. *JAMA* 105:1666–1670, 1935.

100. Nikolaizik WH, Warner JO: Aetiology of chronic suppurative lung disease. *Arch Dis Child* 70:141–142, 1994.

101. Van Schoor J, Joos G, Pauwels R: Tracheobronchomegaly—the Mounier-Kuhn syndrome: Report of two cases and review of the literature. *Eur Respir J* 4:1303–1306, 1991.

102. Lee P, Bush A, Warner JO: Left bronchial isomerism associated with bronchomalacia, presenting with intractable wheeze. *Thorax* 46:459–461, 1991.

103. Jones VF, Eid NS, Franco SM, Badgett JT, Buchino JJ: Familial congenital bronchiectasis: Williams-Campbell syndrome. *Pediatr Pulmonol* 16:263–267, 1993.

104. De Iongh R, Rutland J: Orientation of respiratory tract cilia in patients with primary ciliary dyskinesia, bronchiectasis, and in normal subject; *J Clin Pathol* 42:613–619, 1989.

105. Verra F, Fleury-Feith J, Boucherat M, Pinchon M-C, Bignon J, Escudie E: Do nasal ciliary changes reflect bronchial changes? An ultrastructural study. *Am Rev Respir Dis* 147:908–913, 1993.

106. Eavey RD, Nadol JB Jr, Holmes LB, et al: Kartagener's syndrome: A blinded, controlled study of cilia ultrastructure. *Arch Otolaryngol Head Neck Surg* 112:646–650, 1986.

107. De Santi MM, Magni A, Valletta EA, Gardi C, Lungarella G: Hydrocephalus, bronchiectasis, and ciliary aplasia. *Arch Dis Child* 65:543–544, 1990.

108. Verra F, Escudier E, Bignon J, et al: Inherited factors in diffuse bronchiectasis in the adult: A prospective study. *Eur Respir J* 4:937–944, 1991.

109. Hendry WF, A'Hern RP, Cole PJ: Was Young's syndrome caused by exposure to mercury in childhood? *BMJ* 307:1579–1582, 1993.

110. Abbud-Neme F, Reynoso VM, Deutsch Reiss E: Woake's syndrome: A case report in a teenager. *Int J Pediatr Otorhinolaryngol* 12:327–333, 1987.

111. Udwadia ZF, Pilling JR, Jenkins PF, Harrison BD: Bronchoscopic and bronchographic findings in 12 patients with sarcoidosis and severe or progressive airways obstruction. *Thorax* 45:272–275, 1990.

112. el-Kassimi FA, Warsy AS, Uz-Zaman A, Pillai DK: Alpha 1-antitrypsin serum levels in widespread bronchiectasis. *Respir Med* 83:119–121, 1989.

113. Kalsheker NA, Hodgson IJ, Watkins GL, White JP, Morrison HM, Stockley RA: Deoxyribonucleic acid (DNA) polymorphism of the α_1-antitrypsin gene in chronic lung disease. *Br Med J* 294:1511–1514, 1987.

114. Shin MS, Ho KJ: Bronchiectasis in patients with alpha 1-antitrypsin deficiency. A rare occurrence? *Chest* 104:1384–1386, 1993.

115. Barker AF: Alpha-1-antitrypsin deficiency presenting as bronchiectasis (letter). *Br J Dis Chest* 80:97, 1986.

116. Jones DK, Godden D, Cavanagh P: Alpha-1-antitrypsin deficiency presenting as bronchiectasis. *Br J Dis Chest* 79:301–304, 1985.

117. Misbah SA, Spickett GP, Zeman A, et al: Progressive multifocal leucoencephalopathy, sclerosing cholangitis, bronchiectasis and disseminated warts in a patient with primary combined immune deficiency. *J Clin Pathol* 45:624–627, 1992.

118. Barker AF, Craig S, Bardana EJ: Humoral immunity in bronchiectasis. *Ann Allergy* 59:179–182, 1987.

119. Björkander J, Bake B, Oxelius V-A, Hanson LÀ: Impaired lung function in patients with IgA deficiency and low levels of IgG2 or IgG3. *N Engl J Med* 313:720–724, 1985.

120. Feldman C, Weltman M, Wadee A, Sussman G, Smith C, Zwi S: A study of immunoglobulin G subclass levels in black and white patients with various forms of obstructive lung disease. *S Afr Med J* 83:9–12, 1993.

121. Tezcan I, Ersoy F, Sanal O, Gocmen A, Yeniay I: IgG subclass deficiency in children with recurrent infections. *Turk J Pediatr* 33:163–166, 1991.

122. Sugiyama Y, Maeda H, Okumura K, Takaku F: Progressive sinobronchiectasis associated with the "bare lymphocyte syndrome" in an adult. *Chest* 89:398–401, 1986.

123. Parry CM, Powell RJ, Johnston IDA: Yellow nails, bronchiectasis and low circulating B cells. *Respir Med* 88:475–476, 1994.

124. Camilleri AE: Chronic sinusitis and the yellow nail syndrome. *J Laryngol Otol* 104:811–813, 1990.

125. Holmes AH, Trotman-Dickenson B, Edwards A, Peto T, Luzzi GA: Bronchiectasis in HIV disease. *QJM* 85:875–882, 1992.

126. McGuinness G, Naidich DP, Garay S, Leitman BS, McCauley DI: AIDS associated bronchiectasis:CT features. *J Comput Assist Tomogr* 17:260–266, 1993.

127. Amorosa JK, Miller RW, Laraya-Cuasay L, et al: Bronchiectasis in children with lymphocytic interstitial pneu-

128. Im JG, Kong Y, Shin YM, et al: Pulmonary paragonimiasis:clinical and experimental studies. *Radiographics* 13:575–576, 1993.

129. Batra P: Pulmonary coccidioidomycosis. *J Thorac Imaging* 7:29–38, 1992.

130. Narita H, Kikuchi I, Ogata K, Inoue S, Uehara K, Takehara Y: Smoke inhalation injury from newer synthetic building materials—a patient who survived 205 days. *Burns Incl Therm Inj* 13:147–152, 1987.

131. Slutzker AD, Kinn R, Said SI: Bronchiectasis and progressive respiratory failure following smoke inhalation. *Chest* 95:1349–1350, 1989.

132. Mahut B, Delacourt C, de Blic J, Mani TM, Scheinmann P: Bronchiectasis in a child after acrolein inhalation. *Chest* 104:1286–1287, 1993.

133. Freitag L, Firusian N, Stamatis G, Greschuchna D: The role of bronchoscopy in pulmonary complications due to mustard gas inhalation. *Chest* 100:1436–1441, 1991.

134. Leduc D, Gris P, Lheureux P, Gevenois PA, De Vuyst P, Yernault JC: Acute and long term respiratory damage following inhalation of ammonia. *Thorax* 47:755–757, 1992.

135. Camus P, Piard F, Ashcroft T, Gal AA, Colby TV: The lung in inflammatory bowel disease *Medicine (Baltimore)* 72:151–183, 1993.

136. Gibb WR, Dhillon DP, Zilkha KJ, Cole PJ: Bronchiectasis with ulcerative colitis and myelopathy. *Thorax* 42:155–156, 1987.

137. Moles KW, Varghese G, Hayes JR: Pulmonary involvement in ulcerative colitis. *Br J Dis Chest* 82:79–83, 1988.

138. Gabazza EC, Taguchi O, Yamakami T, et al: Bronchopulmonary disease in ulcerative colitis. *Intern Med* 31:1155–1159, 1992.

139. Gionchetti P, Schiavina M, Campieri M, et al: Bronchopulmonary involvement in ulcerative colitis. *J Clin Gastroenterol* 12:647–650, 1990.

140. Neilly JB, Main AN, McSharry C, Murray J, Russell RI, Moran F: Pulmonary abnormalities in Crohn's disease. *Respir Med* 83:487–491, 1989.

141. Solanki T, Neville E: Bronchiectasis and rheumatoid disease: Is there an association? *Br J Rheumatol* 31:691–693, 1992.

142. Shadick NA, Fanta CH, Weinblatt ME, O'Donnell W, Coblyn JS: Bronchiectasis. A late feature of severe rheumatoid arthritis. *Medicine (Baltimore)* 73:161–170, 1994.

143. Steers G, McMahon MJ, Grennan DM, Hillarby MC: Lack of association of the alpha-1-antitrypsin PIZ allele with rheumatoid arthritis or with its extra articular complications. *Dis Markers* 10:151–157, 1992.

144. Bate AS, Sidebottom D, Cooper RG, Loftus M, Chattopadhyay C, Grennan DM: DNA variants of alpha-1-antitrypsin in rheumatoid arthritis with and without pulmonary complications. *Dis Markers* 8:317–321, 1990.

145. McMahon MJ, Swinson DR, Shettar S, et al: Bronchiectasis and rheumatoid arthritis: A clinical study. *Ann Rheum Dis* 52:776–779, 1993.

146. Bagga A, Vasudev A, Kabra SK, Mukhopadhyay S, Bhuyan UN, Srivastava R: Nephronophthisis with bronchiectasis. *Child Nephrol Urol* 10:211–213, 1990.

147. Verra F, Kouzan S, Saiag P, Bignon J, de Cremoux H: Bronchoalveolar disease in dyskeratosis congenita. *Eur Respir J* 5:497–499, 1992.

monia and acquired immune deficiency syndrome: Plain film and CT observations. *Pediatr Radiol* 22:603–606, 1992.

148. Castleman B, McNeely BU: Case Records of the Massachusetts General Hospital. Case 44-1965. *N Engl J Med* 273:816–823, 1965.

149. Perlemuter L, Quevauvilliers J, Manigand G, Deparis M, Hazard J, Fraisse B: Klinefelter's syndrome associated with bronchiectasis and aplasia of the pulmonary artery. *Nouv Presse Med* 1:1507–1510, 1972.

150. Hedblom CA: Pathogenesis, diagnosis and treatment of bronchiectasis. *Surg Gynecol Obstet* 52:406–417, 1931.

151. Lander FPL, Davidson M: The aetiology of bronchiectasis (with special reference to pulmonary atelectasis). *Br J Radiol* 11:65–89, 1938.

152. Erb IH: Pathology of bronchiectasis. *Arch Pathol* 15:356–386, 1933.

153. Robinson WL: Bronchiectasis: A study of the pathology of 16 surgical lobectomies for bronchiectasis. *Br J Surg* 21:302–312, 1933.

154. Ogilivie AG: The natural history of bronchiectasis. *Arch Intern Med* 68:395–465, 1941.

155. Lisa JR, Rosenblatt MB: *Bronchiectasis*. Oxford, Oxford University Press, 1943.

156. Cole PJ: Inflammation: A two-edged sword—the model of bronchiectasis. *Eur J Respir Dis* 147(Suppl):6–15, 1986.

157. Currie DC, Pavia D, Agnew JE, et al: Impaired tracheobronchial clearance in bronchiectasis. *Thorax* 42:126–130, 1987.

158. Amitani R, Wilson R, Rutman A, et al: Effects of human neutrophil elastase and *Pseudomonas aeruginosa* proteinases on human respiratory epithelium. *Am J Respir Cell Mol Biol* 4:26–32, 1991.

159. Fahy JV, Schuster A, Ueki I, Boushey HA, Nadel JA: Mucus hypersecretion in bronchiectasis: The role of neutrophil proteases. *Am Rev Respir Dis* 146:1430–1433, 1992.

160. Hutchison DC: The role of proteases and antiproteases in bronchial secretions. *Eur J Respir Dis* 153(Suppl):78–85, 1987.

161. Mendis AH, Venaille TJ, Robinson BW: Study of human epithelial cell detachment and damage: Effects of proteases and oxidants. *Immunol Cell Biol* 68(Pt 2):95–105, 1990.

162. Hastie AT, Loegering DA, Gleich GJ, Kueppers F: The effect of purified human eosinophil major basic protein on mammalian ciliary activity. *Am Rev Respir Dis* 135:848–853, 1987.

163. Lapa e Silva JR, Guerreiro D, Noble B, Poulter LW, Cole PJ: Immunopathology of experimental bronchiectasis. *Am J Respir Cell Mol Biol* 1:297–304, 1989.

164. Bateman ED, Hayashi S, Kuwano K, Wilkie TA, Hogg JC: Latent adenoviral infection in follicular bronchiectasis. *Am J Respir Crit Care Med* 151:170–176, 1995.

165. Carson JL, Collier AM, Hu S-C: Acquired ciliary defects in nasal epithelium of children with acute viral upper respiratory infections. *N Engl J Med* 312:463–468, 1985.

166. Wilson R, Sykes DA, Watson D, Rutman A, Taylor GW, Cole PJ: Measurement of *Pseudomonas aeruginosa* phenazine pigments in sputum and assessment of their contribution to sputum sol toxicity for respiratory epithelium. *Infect Immun* 56:2515–2517, 1988.

167. Seybold ZV, Abraham WM, Gazeroglu H, Wanner A: Impairment of airway mucociliary transport by *Pseudomonas aeruginosa* products. *Am Rev Respir Dis* 146:1173–1176, 1992.

168. Kanthakumar K, Taylor G, Tsang KW, et al: Mechanisms of action of *Pseudomonas aeruginosa* pyocyanin on human ciliary beat in vitro. *Infect Immun* 61:2848–2853, 1993.

169. Rendell NB, Taylor GW, Somerville M, Todd H, Wilson R, Cole PJ: Characterisation of Pseudomonas rhamnolipids. *Biochim Biophys Acta* 1045:189–193, 1990.

170. Murphy MB, Reen DJ, Fitzgerald MX: Atopy, immunological changes and respiratory function in bronchiectasis. *Thorax* 34:179–184, 1984.

171. Pang J, Chan HS, Sung JY: Prevalence of asthma, atopy, and bronchial hyperreactivity in bronchiectasis: A controlled study. *Thorax* 44:948–951, 1989.

172. Ip MS, So SY, Lam WK, Yam L, Liong E: High prevalence of asthma in patients with bronchiectasis in Hong Kong. *Eur Respir J* 5:418–423, 1992.

173. Zenone T, Heyraud JD, Gontier C: Bronchiectasis following colectomy for hemorrhagic rectocolitis. *Rev Med Interne* 14:326–327, 1993.

174. Rose GA, Spencer H: Polyarteritis nodosa. *QJM* 26:43–81, 1957.

175. Sitara D, Hoffbrand BI: Chronic bronchial suppuration and antineutrophil cytoplasmic antibody (ANCA) positive systemic vasculitis. *Postgrad Med J* 66:669–671, 1990.

176. Pinching AJ, Lockwood CM, Pussell BA, et al: Wegener's granulomatosis: Observations on 18 patients with severe renal disease. *QJM* 52:435–460, 1983.

177. Tanaka E, Tada K, Amitani R, Kuze F: Systemic hypersensitivity vasculitis associated with bronchiectasis. *Chest* 102:647–649, 1992.

178. Schlaepfer K: Ligation of the pulmonary artery of one lung with and without resection of the phrenic nerve: An experimental study. *Arch Surg* 9:25–94, 1924.

179. Bjork VO, Salen EF: Circulation through an atelectatic lung in man. *J Thoracic Surg* 26:533–543.

180. Ahmed FS, Harrison CV: Morphological effects of serotonin on the pulmonary artery, an experimental study in rabbits. *J Pathol Bacteriol* 87:325–332, 1963.

181. Spain DM: Acute non-aeration of the lung: Pulmonary oedema versus atelectasis. *Dis Chest* 25:550–558, 1954.

182. Jackson CL: Bronchial obstruction. *Dis Chest* 17:125–150, 1950.

183. Lilenthal H: Pneumonectomy for sarcoma of the lung in a tuberculosis patient. *J Thorac Surg* 2:600–611, 1932–1933.

184. Burford TM, Burbank B: Traumatic wet lung obstruction on certain physiological fundamentals of thoracic trauma. *J Thorac Surg* 14:483–486, 1945.

185. Rodbard S: Bronchomotor tone. *Am J Med* 15:356–367, 1953.

Chapter 11

Adult Respiratory Distress Syndrome

P. S. Hasleton

INTRODUCTION

Adult respiratory distress syndrome was first described by Ashbaugh and colleagues,[1] who described 12 patients with various factors contributing to respiratory distress; these ranged from multiple trauma, blunt chest injury, and acute pancreatitis to possible viral pneumonia and aspiration. The patients had cyanosis refractory to oxygen and diffuse alveolar infiltrations radiologically. Shortly afterward, Petty and Ashbaugh[2] coined the term *adult respiratory distress syndrome* (ARDS). Since the early 1970s, this has been a growth area in chest medicine and surgery. There have been numerous articles, conferences, and books devoted to the syndrome. Despite all this work, many problems remain. Disagreement persists about the nature of ARDS, its causes, and the optimum treatment, since mortality is high, being in the range of 40 to 50 percent.[3]

Some investigators believe that ARDS is not a disease entity.[4,5] These authors think that the same clinical picture can be caused by other diseases, such as myocardial infarction, pulmonary emboli, pneumonia (especially of viral origin), and pulmonary edema secondary to overenthusiastic rehydration.

The important point is that while ARDS has many causes, the pathologic picture is the same irrespective of etiology. This should cause no surprise, since the lung, like most other organs, can respond to an insult only in a limited number of ways. Thus, fibrosing alveolitis has more than 100 causes,[6] but in many patients the histology is similar and only a clinicopathologic approach can help to elucidate the etiology. Similarly, with ARDS a combined approach is needed. A summary of the causes of ARDS is given in Table 11-1.

The pathologist should not consider this disorder as being manifest only in the postmortem room. Increasingly in patients on chemotherapy with bilateral pulmonary infiltrates, open lung biopsy is needed to determine whether there is an infective cause, an ARDS-type picture possibly due to drugs, or recurrent disease.

One series examined the role of open lung biopsy in patients with an ARDS-type picture.[7] This procedure provided a specific etiologic diagnosis in 66 percent of patients and influenced therapy in 70 percent. Only 30 percent survived to hospital discharge. It is interesting to note that survival rates did not depend on the availability of a specific diagnosis, changes in diagnosis, or changes in therapy. In this group of patients, it would appear that open lung biopsy could well be contraindicated. This view is fortified by the fact that nearly 20 percent of the patients suffered complications, possibly related to the biopsy procedure.

MULTIPLE ORGAN FAILURE

The term *adult respiratory distress syndrome* is probably a misnomer, since it is part of multiple organ system failure. This hypothesis was first advanced by Bell and coworkers.[8] These authors studied 141 patients with ARDS, of whom 37 (26.2 percent) survived. Necropsies were carried out on 47 of 104 nonsurvivors, and the analysis was based on findings in the 37 survivors and the postmortem cases.

They showed that disseminated intravascular coagulation, which is part of ARDS, and failure of the endocrine and central nervous systems, gastrointestinal tract, and kidneys were common in all patients. The high incidence of multiorgan dysfunction has recently been noted by Bone and colleagues.[9]

INFECTION AND ARDS

Infection is common in ARDS, both in survivors and in nonsurvivors. All patients with bacteremia, without a clinically identified site of infection, died.[8] Montgomery and coworkers[10] studied the "sepsis syndrome," giving laboratory and other criteria to cover seven organ systems. They listed in their Table 1 the criteria for evidence of damage to the cardiovascular, respiratory, central nervous, hematologic, hepatic, renal, and gastrointestinal systems.

375

TABLE 11-1
Causes of ARDS

Shock of Any Origin
Infection

   Gram-negative sepsis

   Pneumonia (viral, bacterial, or fungal)

   Pneumocystis carinii

Trauma

   Fat emboli

   Lung contusion

   Nonthoracic trauma

   Head injury (neurogenic pulmonary edema)

Aspiration

   Gastric juice

   Near-drowning

   Hydrocarbon fluids

Drug Overdose or Sensitivity

   Heroin

   Methadone

   Propoxyphene

   Barbiturates

   Bleomycin, etc.

Inhaled Ingested Toxins

   Oxygen (?)

   Smoke (HCN, CO, etc.)

   Corrosive chemicals (NO_2, Cl_2, NH_3, phosgene, cadmium)

   Paraquat

Hematologic Disorders

   Disseminated intravascular coagulation

   Massive blood transfusion

   Postcardiopulmonary bypass

Metabolic Disorders

   Pancreatitis

   Uremia

Miscellaneous

   Heart/lung transplantation

   Lymphangitis carcinomatosa

   Eclampsia

   Postcardioversion

   Radiation pneumonitis

   Amniotic and air emboli

   High altitude

SOURCE: Adapted from Hopewell and Murray.[132]

They identified a sixfold increase in the "sepsis syndrome" after ARDS, compared with the control group. When this syndrome preceded ARDS, the abdomen was the usual site of infection, but when sepsis syndrome occurred after ARDS, there was usually a pulmonary source.

There has been a recent attempt to redefine *sepsis*.[11] This author, as one of a group at a consensus conference, restricted the term *sepsis* to settings in which infection was documented. A new term, *systemic inflammatory response syndrome* (SIRS) was coined. This was defined as the systemic inflammatory response to a variety of severe clinical insults, including infection, pancreatitis, ischemia, multiple trauma and tissue injury, hemorrhagic shock, immune-mediated injury, and exogenous administration of inflammatory mediators, such as tumor necrosis factor or other cytokines. All of these are potent causes of ARDS. SIRS is manifested (but not limited to) two or more of the following conditions:

- Temperature $> 38°$ or $< 36°$ C
- Heart rate > 90 beats/min
- Respiratory rate > 20 breaths/min or arterial oxygen tension < 32 mmHg
- White blood cell count $> 12,000$ cells/mm^3 or > 10 percent immature (band) forms.

This new term was used to underline the fact that the response was not necessarily to infection, but to an underlying stimulus. There was widespread inflammation affecting vascular endothelium, which explained the multiple organ dysfunction.

Seidenfeld and coworkers[12] studied infections in patients with ARDS and showed that gram-negative bacilli caused 57 percent of cases and gram-positive cocci, 36 percent. Only 7 percent of infections were caused by other organisms. Gram-negative organisms were more common in the lung, abdomen, and pleura. Bacteremia arising from abdominal infections was more common than from infections from other sites. Interestingly more patients (59 percent) with abdominal infections survived compared with those with lung infections (13 percent). They retrospectively reviewed the in vivo organism susceptibility and the antibiotics administered. Patients receiving adequate antibiotic therapy did not have a higher survival rate. This finding is not unexpected since trauma causes immune-system paralysis.

CLINICAL FEATURES

As described by Ashbaugh and coworkers,[1] the syndrome is characterized by severe dyspnea, tachypnea, and marked hypoxemia with cyanosis; it is refractory to oxygen, and there is a decrease in lung compliance. Radiologically, there are diffuse bilateral pulmonary infiltrates. The patient with ARDS has usually been healthy before the current illness.

There is a latent period, varying from several hours to a few days after hospitalization, during which the clinical features are those of the underlying illness, such as trauma or a perforated viscus, and respiratory abnormalities are minimal or absent. A recent review of the acute respiratory distress syndrome has been documented.[12a] Some of the clinical risks of the development of ARDS have recently been presented.[12b]

The term *ARDS* is a misnomer, since the syndrome may be seen in children.[13] It is thought to account for up to 1 percent of all admissions to intensive care units of children.[14] This figure is probably an underestimate, given the predilection of this age group for trauma as well as the increasing number of children undergoing transplantation, chemotherapy, and other conditions associated with this syndrome.

The basic pathology, irrespective of the causes, is given in Table 11-1. The pathologic conditions all affect the alveolar wall. The ultrastructural features are first described, followed by a review of the light microscopic features, and then consideration is given to the complications and pathogenetic mechanisms of ARDS.

Ultrastructural Features

The early changes of ARDS in humans, for obvious reasons, are not well described.[15,16] We have examined a series of patients with coronary artery bypass grafts who had an open lung biopsy before and after cardiopulmonary bypass—a potential cause of ARDS. The changes tended to be more severe after bypass and consisted of damage to type I pneumocytes, with formation of papillary processes (Fig. 11-1), enlargement of the cell, increase in mitochondrial

size, and separation from the basement membrane. There was damage to type II pneumocytes, edema of the basement membrane (Fig. 11-2), and focal endothelial damage (Fig. 11-3). One patient with sickle cell trait developed ARDS.[17] A review of the pulmonary disease occurring in sickle hemoglobinopathies has recently been published.[18]

Epithelial Changes

Bachofen and Weibel[19] studied lung tissue from nine patients dying from ARDS due to septicemia. In the acute phase, up to four days after the onset of respiratory symptoms, type I pneumocytes showed a range of degenerative changes, especially over the thinnest parts, exposing a bare basement membrane (Fig. 11-4). The changes included cytoplasmic swelling and membrane fragmentation. Some type I cells showed prominent endoplasmic reticulum, suggesting increased metabolic activity, and fibrin appeared to leak into some cells. No fibrin thrombi or megakaryocytes were identified.

Twenty-four hours after onset, type II pneumocytes were proliferating within the alveolar lumina. Hyaline membranes,[20] comprising a mixture of plasma proteins, fibrin strands, and cell debris, and degenerative changes in elastic fibers were also noted (Fig. 11-5).

Pietra and colleagues[21] studied a series of patients dying from trauma at periods ranging from 7 hours to 38 days. They noted interstitial edema, which was most prominent in the thick regions of the alveolar septa. Many patients had cerebral damage, and thus neurogenic pulmonary edema may have complicated the picture.

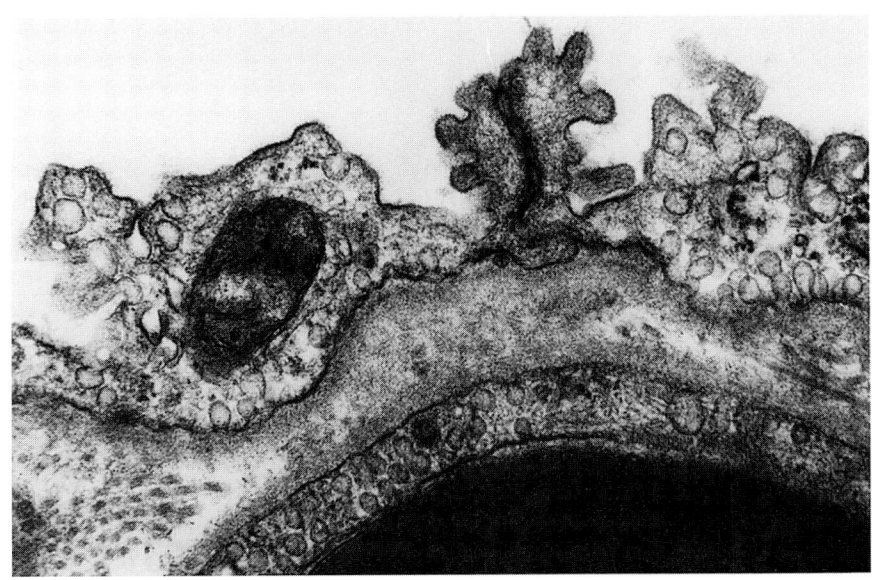

FIGURE 11-1
Hyperplastic type I cells with a prominent nucleus and well-formed papillary processes as well as a cell junction. (Uranyl acetate and lead citrate, ×45,500.)

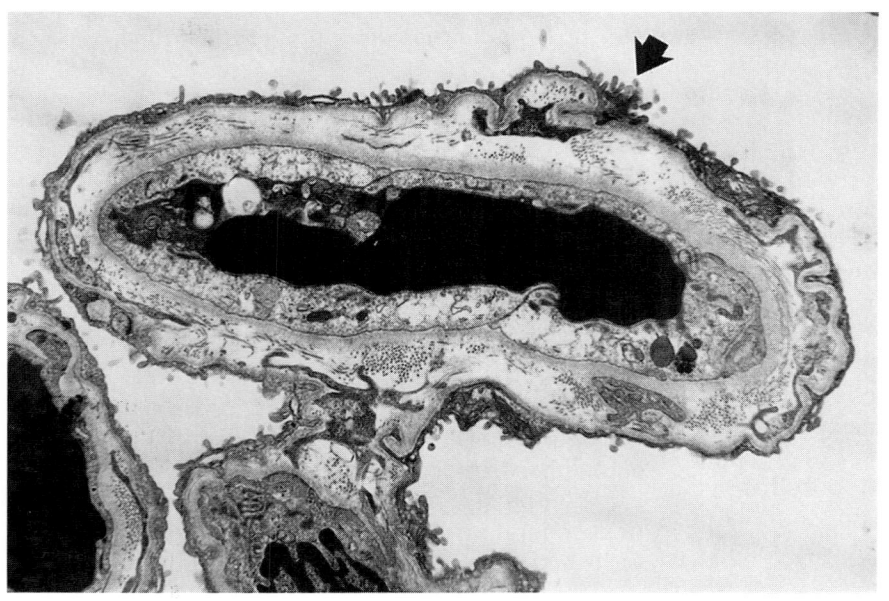

FIGURE 11-2
Postbypass lung with focal papillary
processes (*arrow*) and marked edema
of the basement membrane as well as
electron lucency of endothelial cells.
(Uranyl acetate and lead citrate, ×7350.)

Endothelial Changes

Pietra and coworkers[21] noted that all but one of their
acute cases (up to 5 days) had fat emboli. The
endothelium in relation to these showed no evidence
of injury. Also present was interstitial edema, which
was most prominent in the thick regions of the alveolar septa. Other authors[16,20,22] have also emphasized
the lack of severe endothelial changes in the acute
phase of ARDS.

Teplitz,[23] using scanning electron microscopy,
demonstrated alveolar capillary proliferation. He
believed that the endothelial cell junctions widen in
ARDS. No such feature has been noted on transmission electron microscopy. There is irregular endothelial cell cytoplasmic swelling or vacuole formation.
However, the endothelial cell junctions stay closed,
and there is no widespread destruction of the
endothelial cells.

A discussion of epithelial and endothelial cell
junctions is given in Chap. 2. Proteolytic enzymes,
from cells such as neutrophil polymorphs, will
break down these junctions, allowing fluid to flow
freely into the alveolar lumina. In addition, these

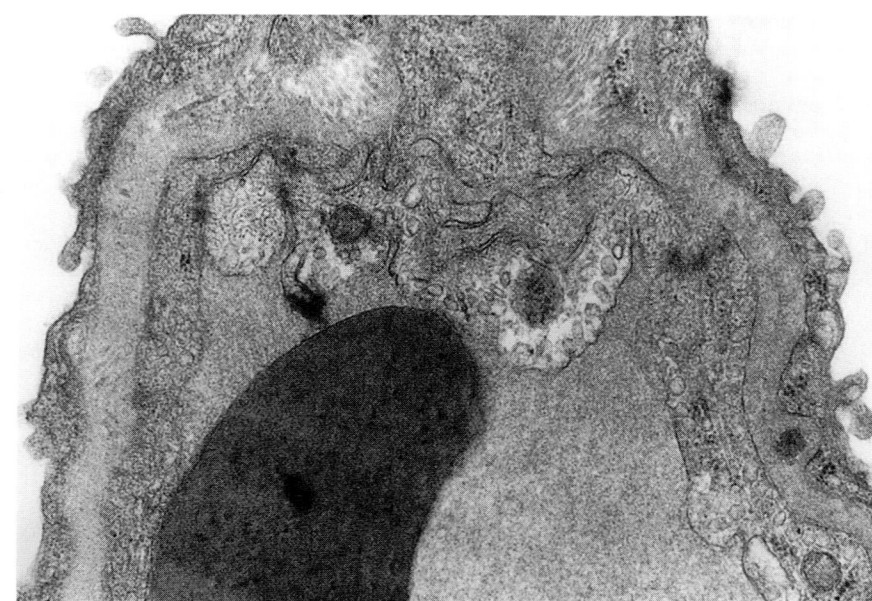

FIGURE 11-3
Postbypass lung with focal electron
lucency of the endothelial cells as well
as prominent mitochondria. (Uranyl
acetate and lead citrate, × 31,500.)

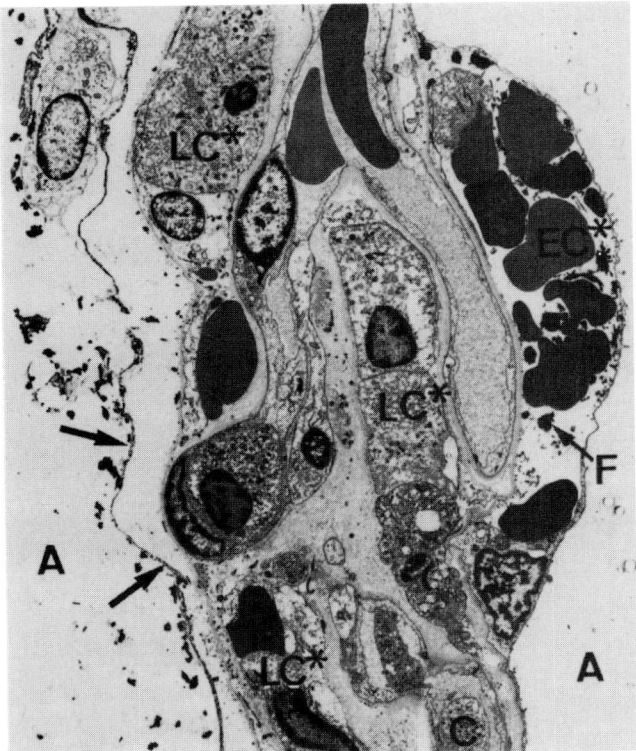

FIGURE 11-4
Alveolar septum in acute ARDS. Note early epithelial lesions.
Squamous epithelial lining is completely lifted up from basement
membrane on one side (*arrows*) and forms a huge blister filled
with red blood cells (EC*) and fibrin scraps (F) on the other side.
In the interstitial space there is accumulation of extravasated
leukocytes (LC*). [*From Bachofen M, Weibel ER: Sequential morphologic changes in the adult respiratory distress syndrome, in
Fishman AP (ed): Pulmonary Diseases and Disorders, vol 3, 2d
ed. New York, McGraw-Hill, 1988, p 2216.*]

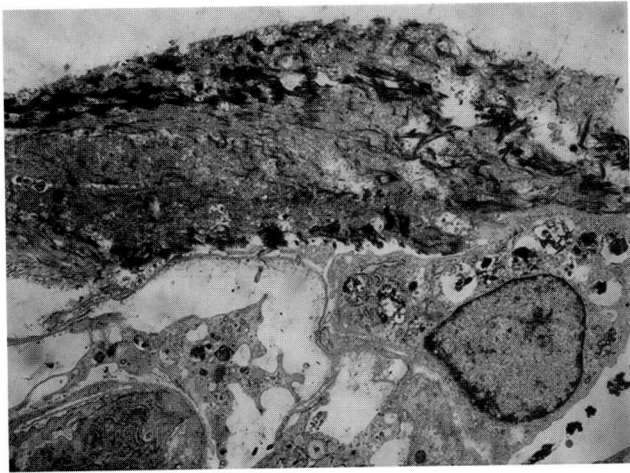

FIGURE 11-5
Hyaline membrane with fibrin lying on a bare basement membrane. (Uranyl acetate and lead citrate, ×4538.)

enzymes will also disrupt the peripheral actin fibers,
causing cell retraction and the formation of wider
gaps.

Bachofen and Weibel[24] described focal cytoplasmic swelling, vacuole formation, and occasional
endothelial cell destruction. The last feature is associated with microthrombi formation.[15] Fibrin in the
subendothelial space is probably due to the fibrinogen leaking from the alveolar capillaries. Endothelial
cell extensions quickly repair leaks in this layer.

Early in the process, there is interstitial or
intraalveolar pulmonary edema, and Hurley[25] has
reviewed the mechanisms of its production. Type I
pneumocytes, with their tight junctions, rather than
endothelial cells, control the passage of water across
the alveolar wall. Intraalveolar edema is a terminal
event of "shock lung," since excess fluid in the interstitial space is transported by the pulmonary lymphatics. Edema accumulates when the lymphatic
capacity is exceeded. The edema spreads to invade
the tissue adjacent to a thick part of the alveolar capillaries. The thin part of the alveolar wall, where
gaseous exchange occurs, does not become water-
logged until much later, usually when fluid is present
in the alveoli.[26]

The relative volumes of the alveolar septa at different periods during ARDS are shown in Fig. 11-6.

Not all patients exhibit the sequence of diffuse
alveolar damage progressing to interstitial and
intraalveolar fibrosis. Some recover within a few
days; in others it takes several months before the lung
returns to normal. A third group eventually develops
progressive fibrosing alveolitis, which will be discussed later. This cohort has erroneously been
regarded by pathologists as a typical end result of
ARDS.

REPAIR

The epithelial side of the alveolar wall is repaired
by cuboidal type II pneumocytes with prominent
microvilli and poorly developed lamellar bodies
(Fig. 11-7). The cells may be more than one layer
thick.[19] Type I cells are seen more frequently and
show large nucleoli, prominent organelles, and endoplasmic reticulum. Partly fragmented red blood corpuscles are present inside the cytoplasm of such
cells. In some cases, there is complete transformation
of the lining of the alveolar wall to type II cells within
two weeks. At the same time as the alveolar cell proliferation, there is interstitial edema and fibrosis.[19,21,22,27]

There is also an influx of plasma cells, histio-

379

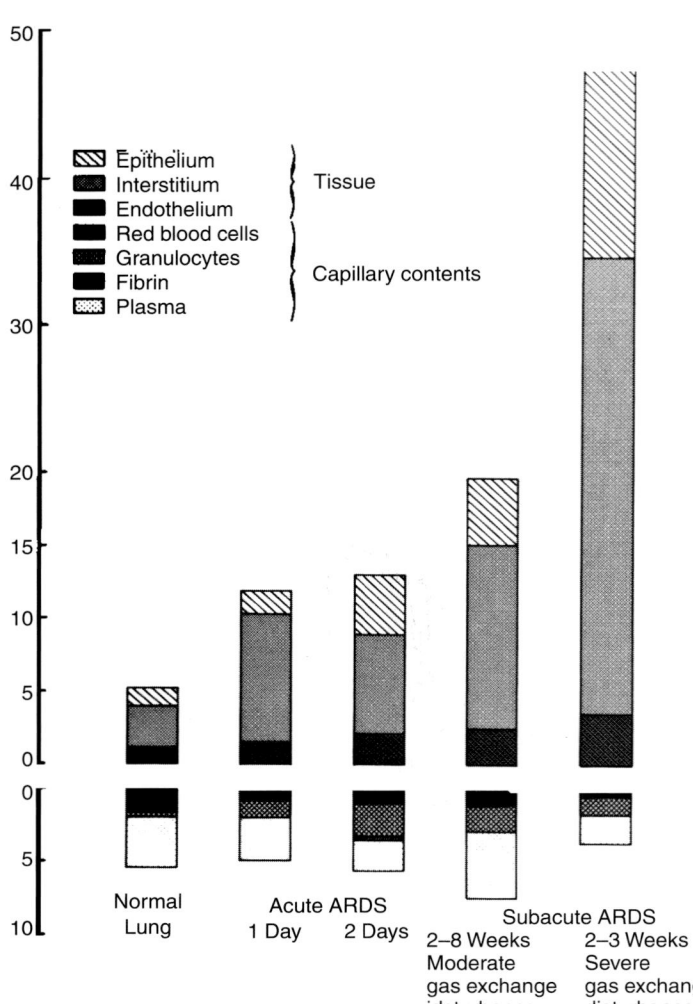

FIGURE 11-6
Relative volumes of constituents of alveolar septa in normal lungs and in ARDS at different stages of the disease. *[From Bachofen M, Weibel ER: Sequential morphologic changes in the adult respiratory distress syndrome, in Fishman AP (ed): Pulmonary Diseases and Disorders, vol 3, 2d ed. New York, McGraw-Hill, 1988, p 2217.]*

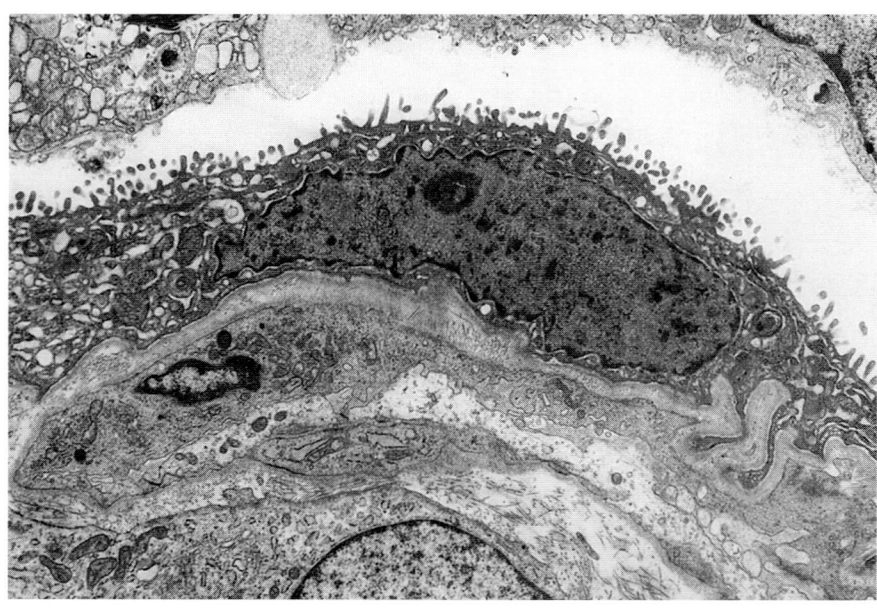

FIGURE 11-7
Repair of alveolar wall with prominent type II cells having microvilli but poorly developed lamellar bodies. (Uranyl acetate and lead citrate, ×4900.)

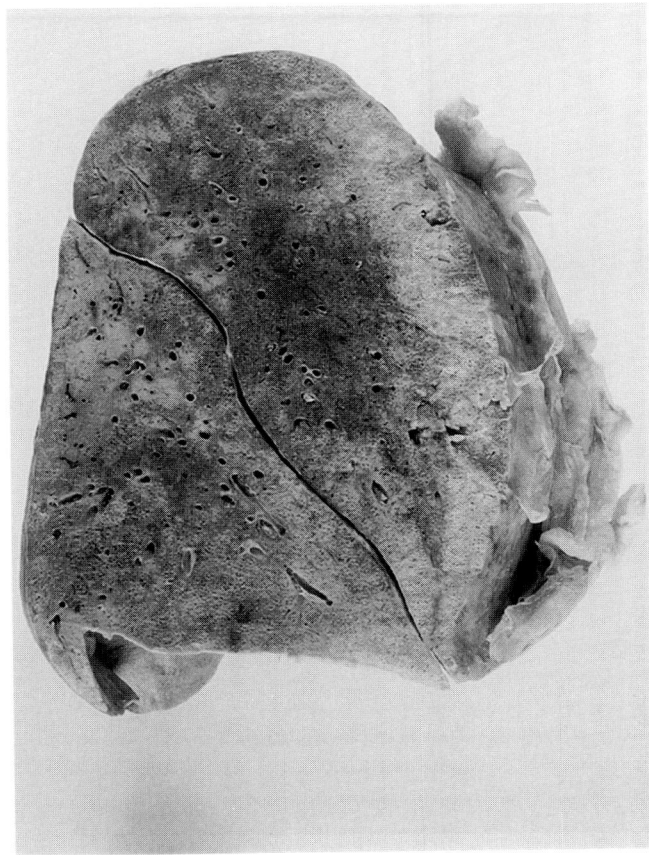

FIGURE 11-8
Whole lung slice from a case of ARDS, with fibrosis becoming confluent in the apex of the left lower lobe.

cytes, and lymphocytes. Proliferative activity in pericytes has been noted.[28]

Finally there is interstitial and intraalveolar fibrosis, with disorganization of the alveolar architecture. This is best appreciated with light microscopy.

MACROSCOPIC AND MICROSCOPIC FEATURES

Exudative Phase

In the early stages, the lungs are dusky, reddish blue, and heavy and sometimes have a combined weight of over 2000 g. The cut surface is hemorrhagic with edema. After some weeks the lungs become solid, fleshy, and reddish gray (Fig. 11-8). They are often misdiagnosed macroscopically as displaying pneumonia.

The sequence of pulmonary changes in patients suffering from burns, a cause of ARDS, has been demonstrated.[29] In the immediate period after injury, there is interstitial, intraalveolar, and septal edema. Congestion of the alveolar walls and capillary proliferation[30] are striking features (Figs. 11-9 and 11-10). This capillary proliferation is due to new vessel formation as demonstrated by reticulin. Pratt[31] attributed such new vessel formation to oxygen therapy, but this has been challenged.[15,30]

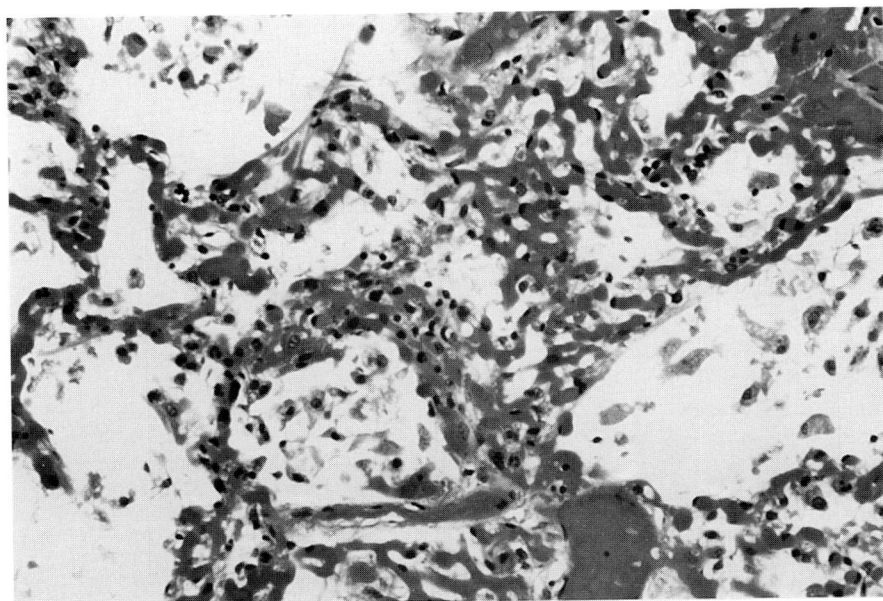

FIGURE 11-9
Capillary proliferation in a burn case. (H&E, ×88.)

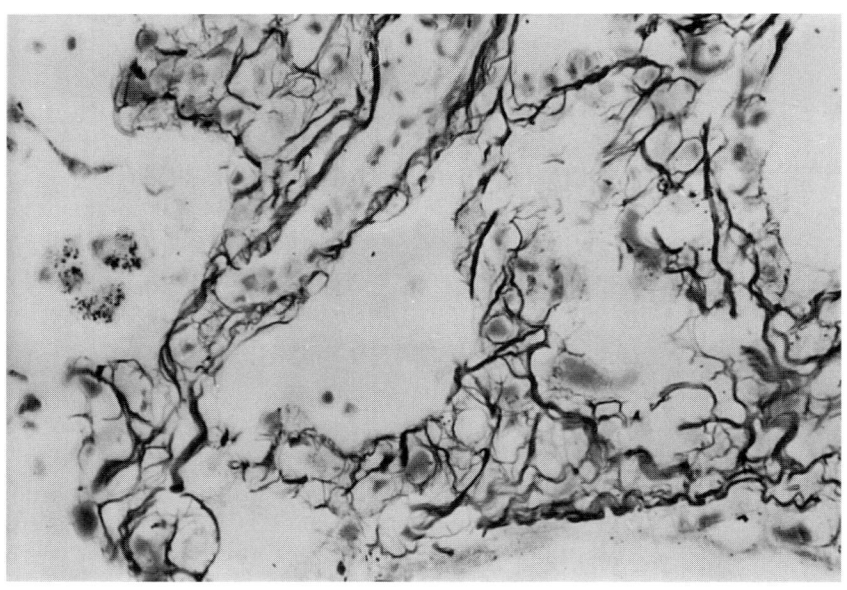

FIGURE 11-10
Capillary proliferation. (Reticulin, ×219.)
*(Reprinted with permission from
Histopathology 7:307–332, 1983.)*

In patients dying within 48 h, there are pulmonary megakaryocytes and microthrombi in pulmonary arterioles and capillaries (Figs. 11-11 and 11-12). There is a significant association between the numbers of megakaryocytes and microthrombi (Fig. 11-13), supporting a relationship between disseminated intravascular coagulation and the number of pulmonary megakaryocytes. No correlation has been found between antemortem platelet counts and either intravascular thrombi or megakaryocytes.[32]

Megakaryocytes travel to the lung, where they receive a constant buffeting, which transforms them into platelets.[33,34] Megakaryocytes have also been identified in other vital organs, such as kidney.[29] Megakaryocytes and platelets as well as platelet microthrombi may be recognized by a monoclonal antibody, Y2/51, which detects platelet glycoprotein IIIa.[35]

In cases with septicemia, there is intravascular basophilic staining (Fig. 11-14). Vasculitis may occasionally be seen in bacterial, viral, or fungal infection. In the intermediate and late stages of ARDS,[36] many small arteries and veins show obstruction by eccentric and concentric intimal fibrosis as a result of

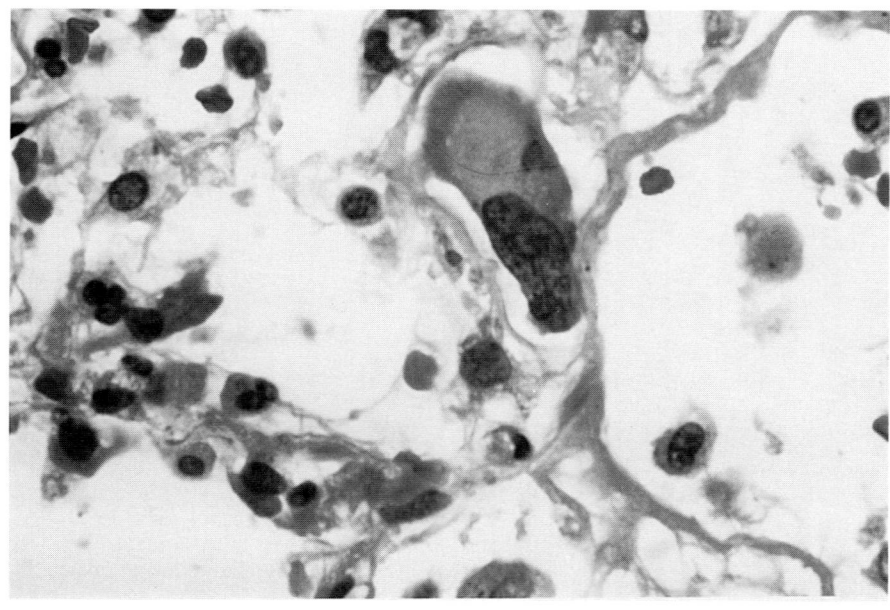

FIGURE 11-11
Pulmonary megakaryocyte with
ample eosinophilic cytoplasm and a
large nucleus. Often all that remains
of these cells is a bare nucleus. (H&E,
×350.) *(Reprinted with permission
from Histopathology 7:307–332,
1983.)*

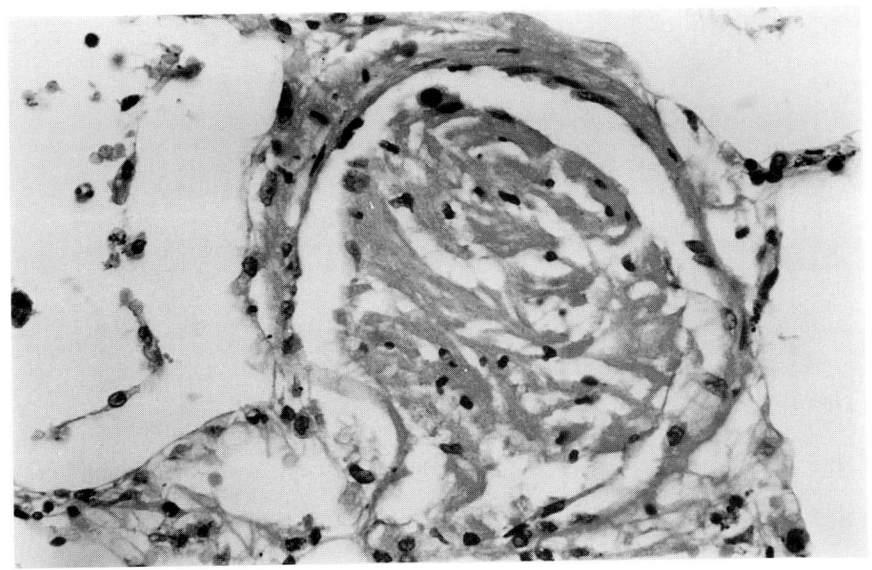

FIGURE 11-12
Intravascular thrombus in a pulmonary artery. (H&E, ×219.) *(Reprinted with permission from J Clin Pathol 34:1147–1154, 1981.)*

intravascular fibrin or platelet thrombi. Large pulmonary emboli are seen in patients dying in the early stages of the disease. Infarction may be present. The pulmonary lymphatics show prominent dilatation, and in some patients there is focal narrowing of the interlobular and subpleural lymphatic lumina by sparsely cellular, loose connective tissue.

As the duration of the ARDS increases, there is a corresponding rise in the medial thickness of muscu-

lar pulmonary arteries and muscularization of arterioles. Both the acute phase changes with fibrin or platelet thrombi and the later changes with intimal fibrosis in pulmonary arteries and veins contribute to pulmonary arterial hypertension.

While this work suggests that pulmonary capillary thrombi are probably related to an excess of megakaryocytes in the lung, Eeles and Sevitt,[37] who studied burn and trauma patients, suggested that

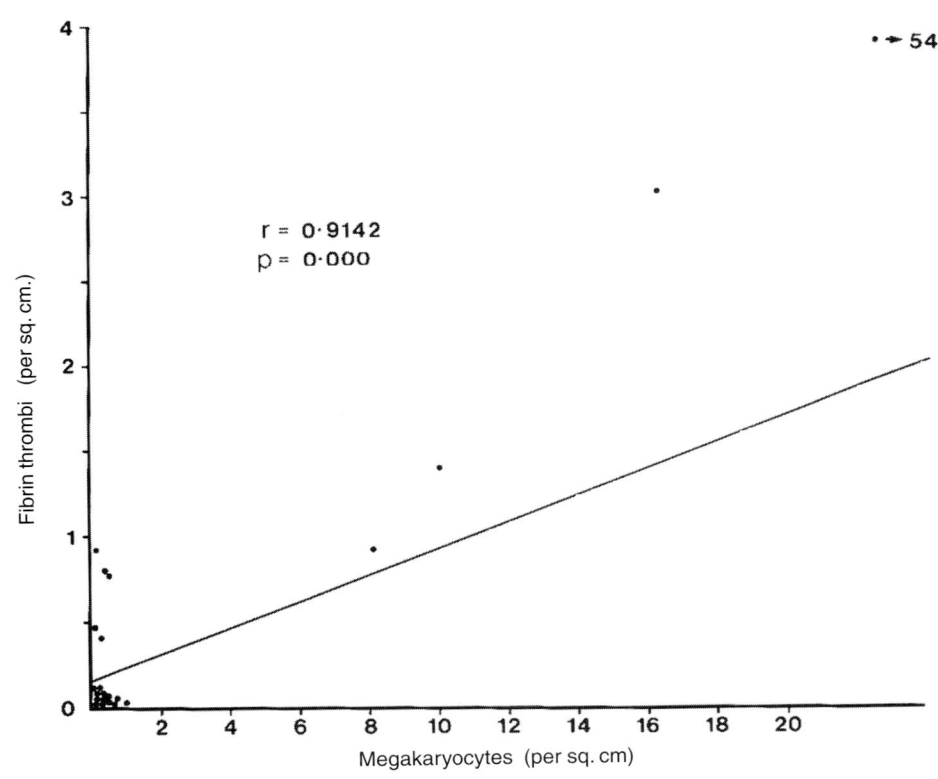

r = 0·9142
p = 0·000

• ← 546

FIGURE 11-13
Graph showing relationship between megakaryocytes and fibrin thrombi ($r=0.9142$; $p=.000$.) *(Reprinted with permission from Histopathology 8:517–527, 1984.)*

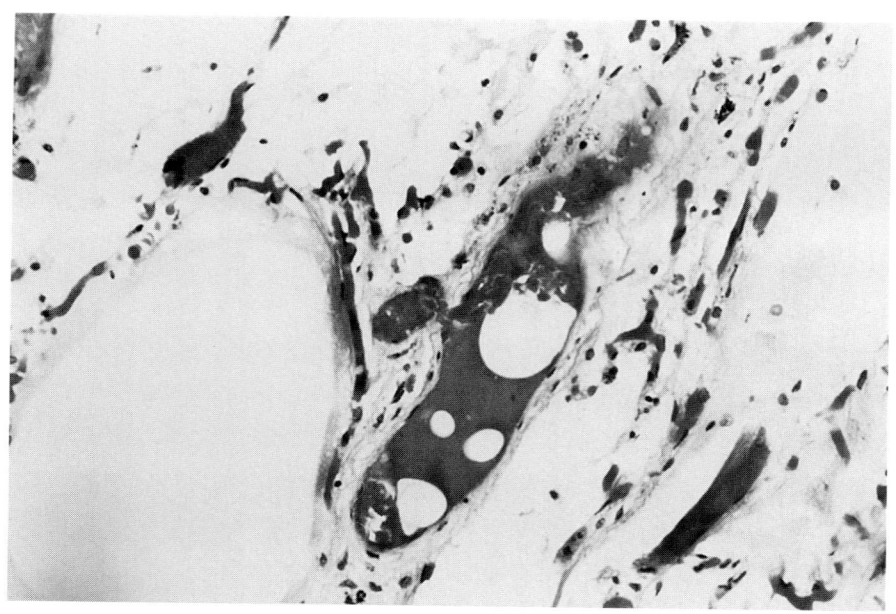

FIGURE 11-14
Basophilic staining of blood vessel lumina. (H&E, ×219.)

both macro- and microthrombi originate as systemic venous thrombi related to hypercoagulation after severe injury. They found deep venous thrombosis in the extremities in most patients with pulmonary emboli.

Hyaline membranes, which line alveoli and the walls of alveolar ducts (Fig. 11-15), are usually seen after 48 h and there may be necrosis of the underlying epithelium. Sevitt[38] used the presence of hyaline membranes as one of his criteria for pulmonary oxygen toxicity. These structures were detected in 15 of 21 of his cases.

Immunohistochemical and immunofluorescent staining of hyaline membranes demonstrates fibrinogen and, to a lesser extent, complement. Fibronectin is minimally layered on the surface of the membrane.[39] Alveolar ducts are dilated in contrast to the adjacent congested, edematous, and partly collapsed alveoli in the acute exudative phase.

Hyaline membranes are not specific for oxygen toxicity, and we have identified them in patients not receiving large amounts of oxygen. Such membranes may be found in other disease processes or conditions, such as fat embolism, drowning, viral pneumo-

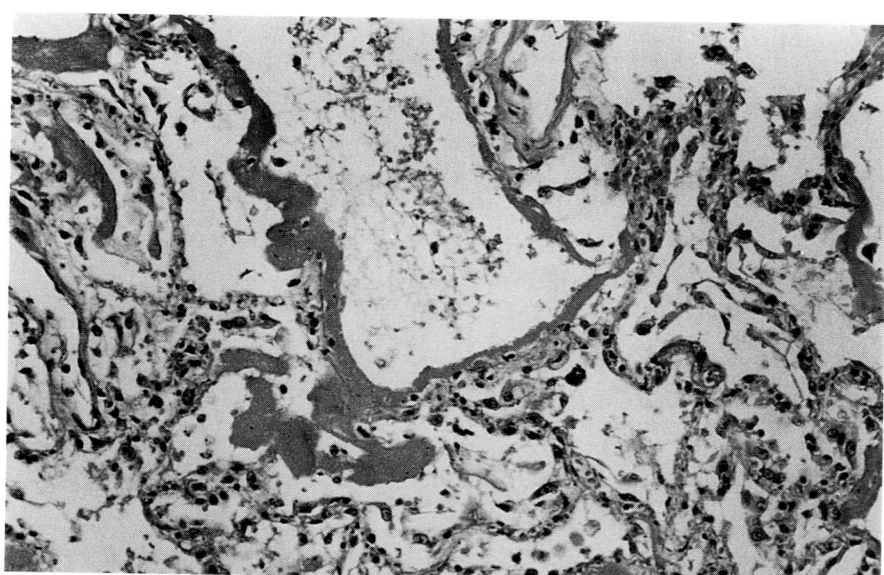

FIGURE 11-15
Hyaline membranes lining alveolar ducts and extending into alveoli. (H&E, ×219.)

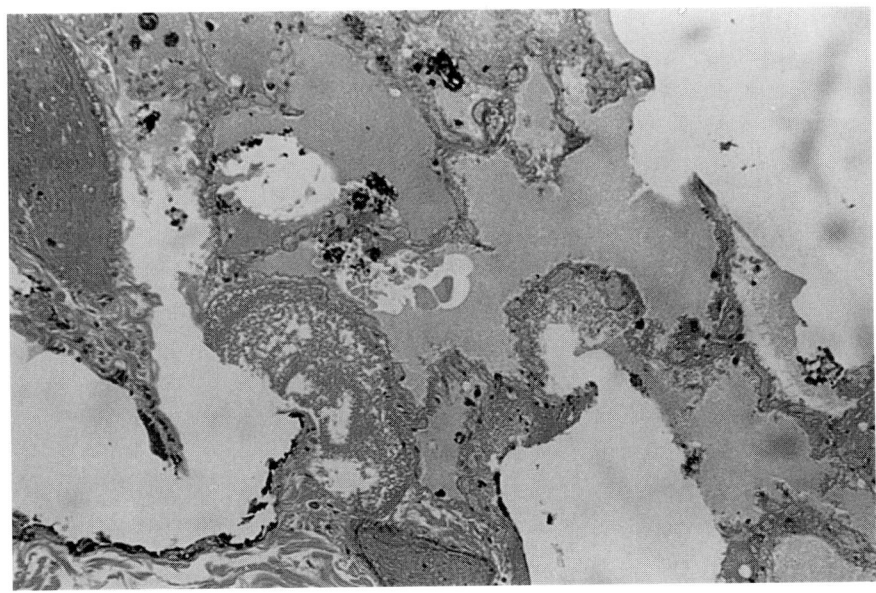

FIGURE 11-16
Bile appearing as dark material lining some of the alveolar walls in ARDS. (H&E, ×88.)

nia, sepsis, and paraquat poisoning, or in any other cause of ARDS.

It is also important to identify bile (Fig. 11-16) or ingested food particles as indicators of aspiration.

Large airways may be affected by shock, especially in burn victims. There may be evidence of soot deposition in the main airways, and the bronchial mucosa may be necrotic (Fig. 11-17) or replaced by acute inflammation. Laryngeal edema is also a feature of respiratory tract burns. A review of the complications of burns has been provided by Sevitt.[40] Airway lesions, including squamous metaplasia, are said to reflect the length of survival in relation to

patients with burns.[41] The type of airway lesion, including squamous metaplasia, is a reflection of the length of survival after the burn episode.

Proliferative Phase

In the proliferative phase of ARDS, there is both intraalveolar and interstitial fibrosis (Fig. 11-18) and cuboidalization of epithelium with proliferation of type II pneumocytes. Loose, myxoid interstitial fibrosis may be identified as early as 48 h. It is important to distinguish this new fibrosis from any old fibrosis related to preexisting lung disease.

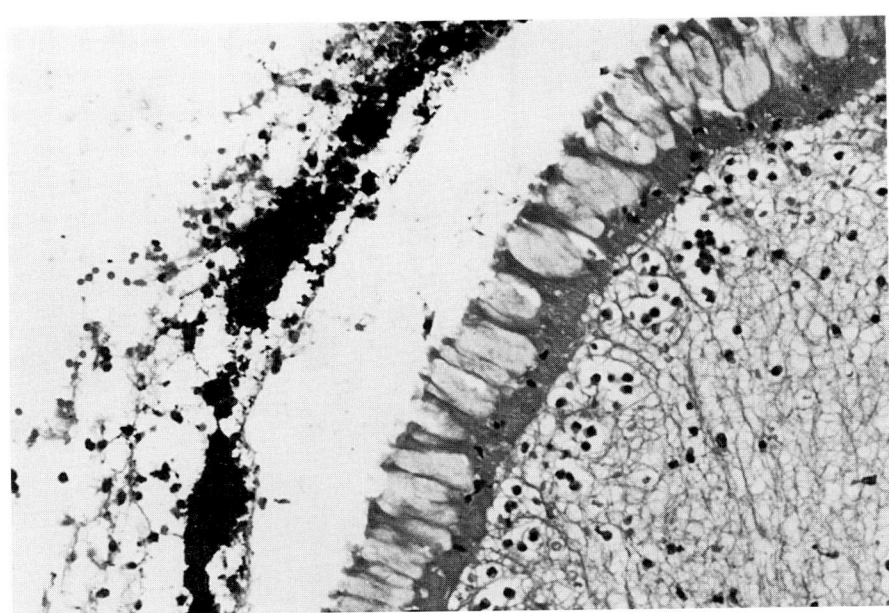

FIGURE 11-17
Necrosis of bronchial mucosa with subepithelial edema and a layer of carbon from smoke inhalation. (H&E, ×219.) *(Reproduced with permission from Histopathology 7:307–332, 1983.)*

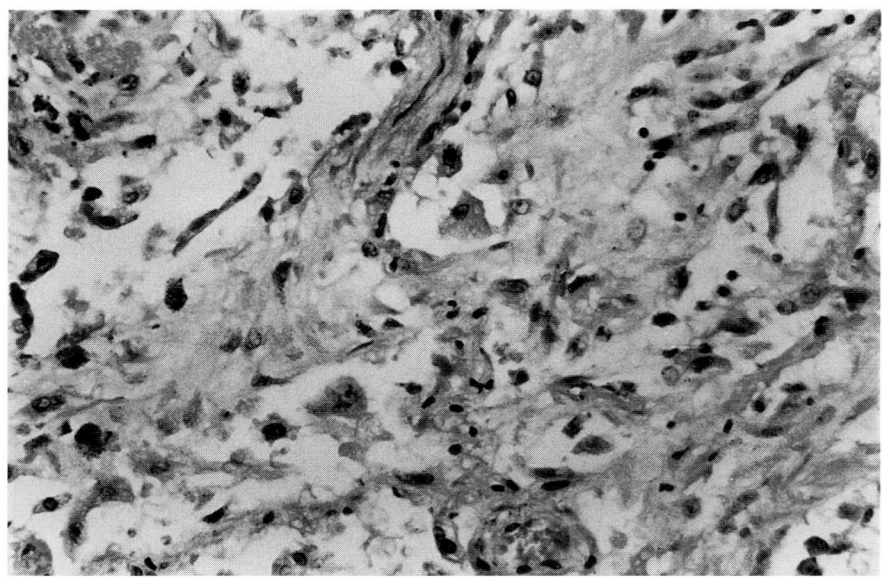

FIGURE 11-18
Intraalveolar and interstitial fibrosis
occurring 10 days after blast injury. (H&E,
×219.) *(Reproduced with permission
from J Clin Pathol 34:1147–1154, 1981.)*

The reactive hyperplasia of type II pneumocytes can cause diagnostic problems on cytology. The danger is misinterpreting these cells as arising from a well-differentiated bronchoalveolar carcinoma.

Cytologically, hyperplastic type II cells occur in cohesive groups or singly. They are often large and easily located on low magnification. The cyanophilic cytoplasm ranges from finely vacuolated to widely distended by one or more large vacuoles. Nonvacuolated areas are finely granular. The nuclear-cytoplasmic ratio varies widely but is generally high in cells not distended by vacuoles. Nuclei are either single or multiple, and in some cells nuclear shapes show angulation or lobulation. Most cells show one or more macronuclei. The chromatin varies from uniformly distributed to clumped with areas of clearing. Group-to-group variability was the prominent finding in cases with many hyperplastic type cells.[42] The only cytologic finding that appears significantly more characteristic of type II pneumocyte hyperplasia than of adenocarcinoma is the greater group-to-group variability in the benign lesions.

Classically, the cell balls or sheets in bronchoalveolar carcinoma are composed of monotonous cells, which are usually fairly small and uniform in size and shape, with round to oval nuclei and a finely granular chromatin. Nuclei may be even, very small, or even absent. The occurrence of pale and hyperchromatic nuclei and cells with prominent but few nucleoli together in the same preparation would be unusual in a carcinoma. However, a definite distinction between bronchoalveolar carcinoma and hyperplastic type II cells may be impossible, and it is vital that the cytopathologist knows the full clinical history; even then it may be wise to take repeat lavage specimens and match these with the evolving chest radiograph before making a definitive diagnosis of carcinoma on purely cytologic grounds.

Usually there is a neutrophilia along with these hyperplastic-type cells and this may account for up to 60 percent of cells if lavage is done in the early stages of ARDS. In a recent bronchoalveolar lavage (BAL) study, all patients had increased leukocytes on day 3 after the onset of ARDS. In patients with ARDS, the leukocyte count was higher on day 7 and especially on day 14 in patients who died. There was a constant trend of a higher polymorph concentration on all days in patients who died than in patients who lived. In patients after trauma and other risks, BAL polymorph counts did not distinguish survivors from patients who died. Alveolar macrophages increased in survivors of ARDS, both in absolute numbers and as a percentage of total cells. This pattern was most pronounced in sepsis patients.[43]

REGIONAL ALVEOLAR DAMAGE

ARDS has also been termed *diffuse alveolar damage*.[44] A localized form of this process, regional alveolar damage, has been described in 10 percent of adult autopsies.[45] These patients had chronic systemic diseases, but presented with life-threatening illnesses. The probable causes of regional alveolar damage in these patients were multifactorial and included hypotension, septicemia, pneumonia, hypoxia, and pancreatitis. All patients developed respiratory failure, requiring oxygen and in some cases ventilation. Chest radiographs showed alveolar or combined alveolar and interstitial infiltrates corresponding to the postmortem lesions.

Total lung weights were increased, averaging 1485 g. Most had discrete tan-yellow or hemorrhagic, indurated foci with fine airspace enlargement (Fig. 11-19). The lesions could be found in any part of the lungs but were more common on the right side. In some patients, multiple lobes were involved. The extent of the involvement ranged from 5 to 75 percent of the cut lung surface.

Histologically, all lesions showed ARDS in either the exudative or the proliferative (fibrotic) phase. Microemboli were sparse. Four patients had bronchopneumonia, but in three it was "slight."

Complications of ARDS

The functional and physiologic sequelae of ARDS have been summarized in review articles and are outside the scope of this text.[46–48] However, in one study of 25 patients, 21 (84 percent) examined between 4 months and 7.5 years after the event had symptoms.[49] The most common complaints were cough, sputum production, and dyspnea. Fewer than half of those with pulmonary complaints had abnormal

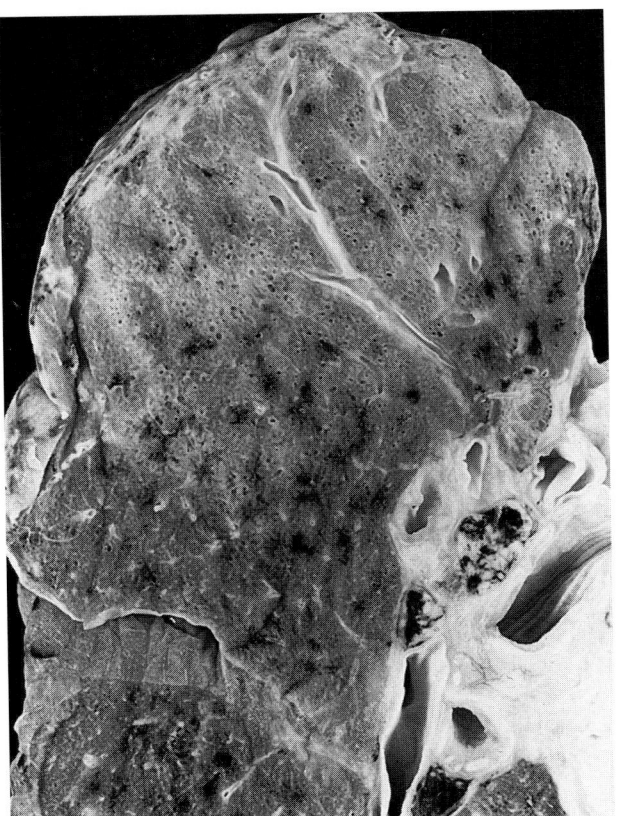

FIGURE 11-19
Right upper lobe of a patient with regional ARDS showing a consolidated area and enlarged airspaces. There is secondary tumor in a node. *(Courtesy of Dr. Tomashefski; reprinted with permission from Am J Clin Pathol 92:10–15, 1989.)*

results on pulmonary function tests. Conversely, those with abnormal pulmonary function were no more likely to be symptomatic than those with normal pulmonary function.

PULMONARY FIBROSIS

Interstitial pulmonary fibrosis and proliferation of type II pneumocytes are integral parts of the proliferative phase of ARDS (Fig. 11-20). The macroscopic picture may end as diffuse cystic change, as seen in infants with bronchopulmonary dysplasia, or honeycomb lung.[50] The pneumocytes may have cytoplasmic hyaline material, Mallory's alcoholic hyaline, which is a nonspecific marker of cell injury. In addition, they may show squamous metaplasia. There may be tumorlet formation. The fibrotic process is similar to that occurring in any pneumonia. Pratt and colleagues[51] believed the fibrosis was centered around the alveolar duct. This is not always the case, and there may be myxoid intraalveolar fibrosis (Masson balls) as well as interstitial fibrosis. Bronchiolitis obliterans is not a prominent feature in ARDS.[52]

Other features contribute to the fibrosis in ARDS.[53] The accretion that occurs is probably due to reorganization of the hyaline membranes and any other material leaking into the alveolar lumina. There is also interstitial edema, which, if continuous, along with mast cell proliferation is a powerful stimulus for fibrosis. A further mechanism for fibrosis is "collapse induration," occurring at the same time as fibrosis of the alveolar walls. It is thought that after alveolar wall injury the alveoli collapse, either partly or completely, and their "mouths" are sealed by organizing fibrin, hyperplastic epithelium, and any other material lying in the alveolar space.[54] This process is not specific for ARDS and may occur in any form of interstitial pulmonary fibrosis. Lack of surfactant, due to loss of type II cells, will also contribute to alveolar wall collapse.

Myofibroblasts play an active part in lung remodeling in ARDS. They migrate from the interstitium through gaps in the epithelial basement membrane and pass into the air spaces. These cells attach to the luminal side of the epithelial basement membrane, replicate, and secrete matrix products.[39]

Factors stimulating these mesenchymal cells have been the subject of much recent research. Growth factors such as platelet-derived growth factor and fibronectin act early in the cell cycle (G0-G1 phase).[55] Epidermal growth factor or transforming growth factor appears later (late in the G1 phase), just before the action of insulin-like growth factors, which signal entry into the S phase of the cell cycle.[56]

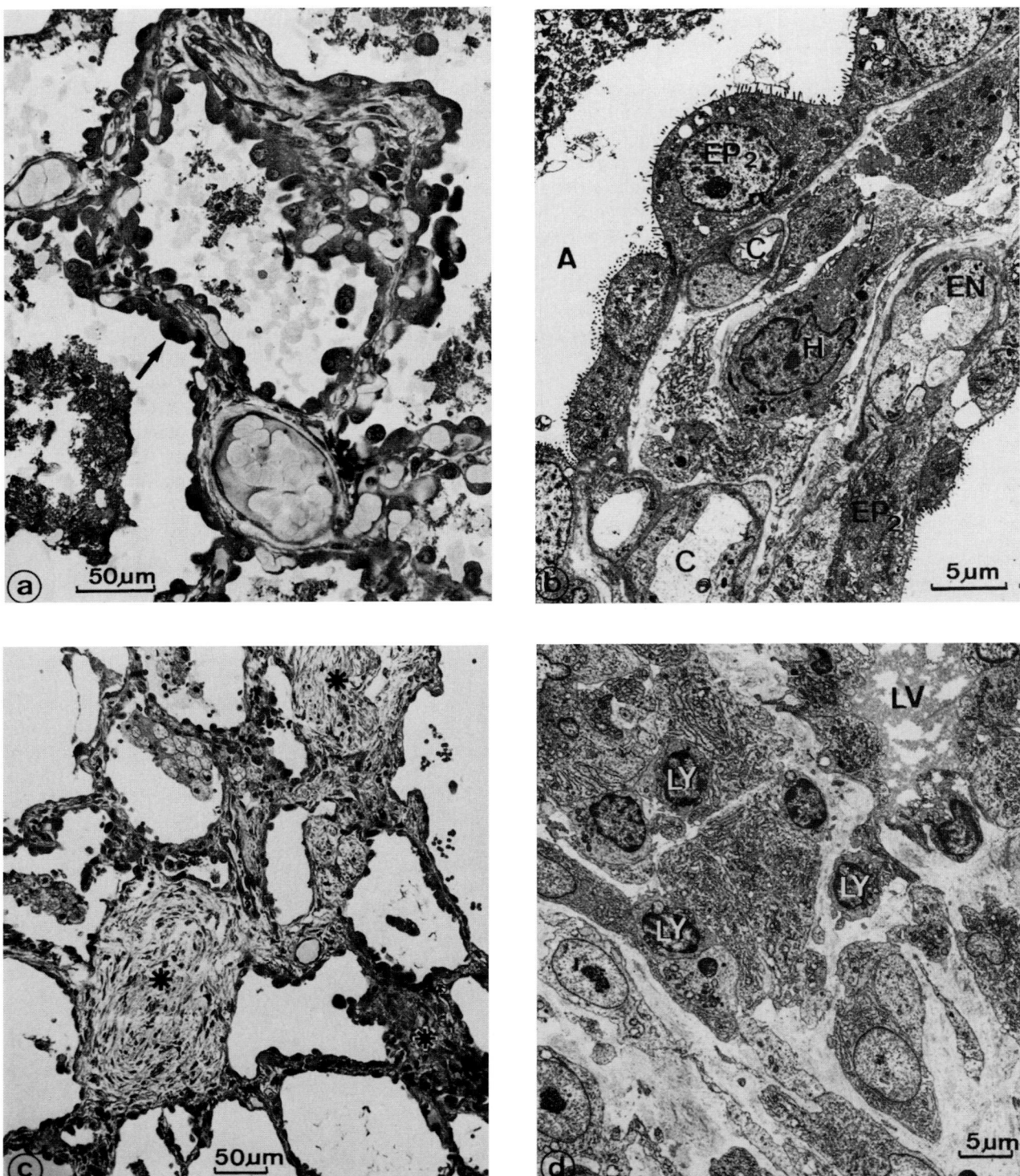

FIGURE 11-20

Subacute ARDS lasting 3½ weeks. *A.* Light microscope view. Note cuboidal epithelial cell proliferation with cell mitosis (*arrow*), septal enlargement, and intraalveolar content. *B.* Septum of same lung. Complete epithelial transformation (EP₂), interstitial cell accumulation, and reduced capillary lumen by endothelial swelling are main features. *C.* Structural heterogeneity of pulmonary parenchyma in the subacute stage of ARDS. Fibrous tissue plates (*) adjoin moderately widened septa. *D.* Interstitial cell composition near small lymphatic vessel (LV). A, alveolar space; C, capillary; EN, endothelium; H, histiocyte; LY, lymphocyte. (*Reproduced with permission from Bachofen M, Weibel ER: Sequential morphologic changes in the adult respiratory distress syndrome, in Fishman AP (ed): Pulmonary Diseases and Disorders, vol 3, 2d ed. New York, McGraw-Hill, 1988, p 2219.*)

Human alveolar macrophages are a rich source of growth factor for lung mesenchymal cells.[57] Prostaglandins, especially PGE, serve to limit the fibroproliferative process and inhibit mesenchymal cell proliferation.[58]

Fibronectin is probably pivotal in controlling collagen formation. It combines to denature collagens,[59] as well as serving as a nidus for mesenchymal cell attachment and deposition of connective tissue matrix protein.

An index of damage to the alveolar wall has been shown by detection of 7S collagen in BAL fluid in patients with ARDS.[60] 7S collagen is a small, disulfide-rich, helical domain at the end of the amino terminals of the four individual molecules that have linked together in type IV collagens. It is responsible for the characteristic arrangement of the type IV molecules. 7S collagen is fairly resistant to enzymatic degradation and can be obtained from any tissue containing considerable amounts of basement membrane by use of various digest procedures. Therefore, detection of 7S collagen is useful in evaluation of basement membrane disruption. In one study 12 of 14 patients with ARDS had 7S collagen in BAL fluid, but none was detected in controls.[60] The level of 7S collagen correlated strongly with the neutrophil counts and elastase complex.

Type III as well as type I collagenase activity has also been detected in patients with ARDS.[61] Type III collagenase peaked before type I. The *N*-terminal peptide of type III procollagen, as well as types I and III collagenase and galactosylhydroxylysyl glucosyltransferase—an intracellular enzyme responsible for the specific glycosylation of free pro-alpha chains before their incorporation into the triple helix of procollagen—are all raised in lavage material from patients with ARDS. The level of type III procollagen was higher in patients with histologically proven intraalveolar fibrosis.[62]

PULMONARY HYPERTENSION

A threefold rise in pulmonary arterial pressure was seen in 30 cases of pulmonary hypertension due to ARDS.[63] Chronic hypoxia causes muscularization of pulmonary arterioles,[64] but hypoxia was corrected in these patients. The pulmonary vascular changes that occur after ARDS have been described above. Early on there is pulmonary vasoconstriction, thromboembolism, and interstitial edema, all of which are reversible. The density of the precapillary arterioles is reduced by 50 percent by thrombi as well as intimal fibrosis.[65] It appears that mesenchymal cells, besides growing into the alveolar lumina, can occupy the vascular lumina and cause focal microvascular obstruction.[36]

There is marked remodeling of the pulmonary vascular bed in the late proliferative and fibrotic phases of ARDS. Small muscular pulmonary arteries are stretched around thin, fine-walled cysts and dilated air spaces. In patients dying in the late fibrotic stages, the arteries appear serpentine and show intimal fibrosis. There is muscularization of pulmonary arterioles and hypertrophy of muscular pulmonary arteries. With prolonged duration of ARDS, the medial thickness increases.[36]

LARGE PULMONARY EMBOLI, PNEUMOTHORAX, OR PNEUMOMEDIASTINUM

Pneumothorax and pneumomediastinum are due to alveolar rupture. Swan-Ganz catheters used in patients with ARDS can cause pulmonary hemorrhage due to a rupture of a pulmonary artery[66] and pulmonary infarction. Some of the infarcts may cavitate. The other complication is pulmonary embolism.

PNEUMONIA

Pneumonia is a common complication and in some series accounts for up to three-fourths of the cases of sepsis occurring after the onset of ARDS.[10] In view of the airway damage, it is not surprising that there is colonization and infection of the respiratory tract, especially with hospital-acquired pathogens. Loss of the normal ciliated epithelium and surfactant as well as an influx of neutrophil polymorphs contributes to epithelial damage and consequent infection.

The early stages of ARDS are associated with a pulmonary vascular and alveolar neutrophilia (Fig. 11-21). Neutrophils harvested from lavage fluid are abnormal.[67] Their microbicidal activity for *Staphylococcus aureus* and production of superoxide anion and hydrogen peroxide are significantly impaired. In addition, neutrophil migration is reduced in response to a variety of stimuli. However, neutrophils from patients' pulmonary arteries have normal chemotactic and microbicidal responses. This lack of activity of lung neutrophils may contribute to the increased incidence of pulmonary infection in ARDS patients.

One team of investigators has postulated several reasons for the "defective" alveolar neutrophil function. Oxidants in the alveoli may alter neutrophil function, as well as altering proteins.[68] Autooxidation of neutrophils has been proposed as a mechanism impairing neutrophil function.[69] In addition, when neutrophils migrate into the lung, they lose

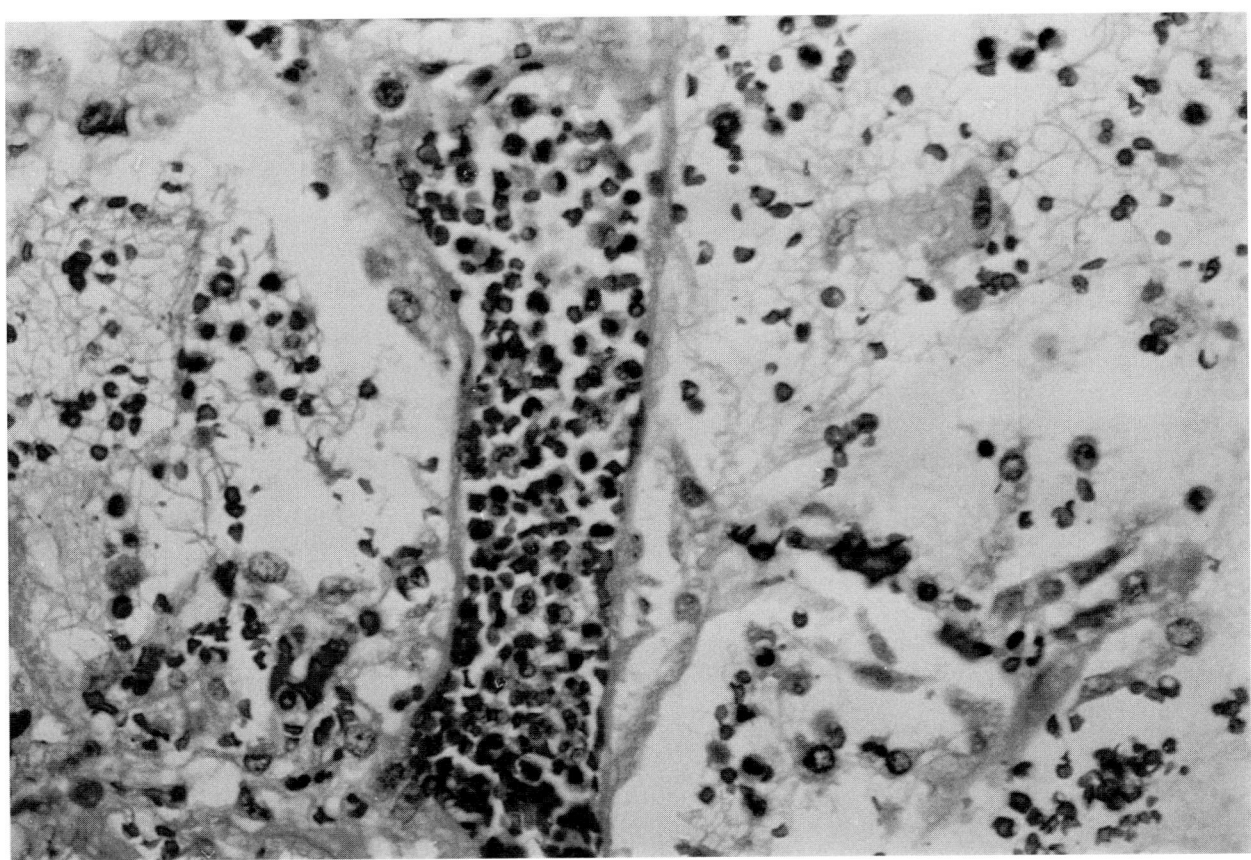

FIGURE 11-21
Pulmonary neutrophilia, showing a small pulmonary vessel filled with neutrophil polymorphs. There is intraalveolar edema, fibrin, and some inflammatory cells in the adjacent alveoli. (H&E, ×219.)

some of the smaller specific granules, but the number of primary granules containing myeloperoxidase and proteolytic enzymes remains the same.[70] The overall effect on the lung is harmful, since the proteolytic enzymes are free to damage the alveoli, but there is reduced defense against organisms.

There is a decrease in immunity after severe trauma, impairing resistance to infection. Selective decontamination of the gastrointestinal tract and the oropharynx in ventilated patients, as well as improvement of hygiene by medical staff and measures to reduce bacterial carriage by intensive care staff, may help to reduce the incidence of infection.[71] Our studies have shown that in blast injury, organisms may be carried directly into the lungs from the surrounding air or via the burn injury.[30]

An ARDS-type picture may occur as a direct cause of pneumonia, especially viral pneumonia. Only by detection of viral inclusions or virologic studies can the difference in etiology be elucidated. Staphylococcal, pneumococcal, and gram-negative

pneumonias are all important in causing ARDS, but other organisms such as *Pneumocystis*, *Mycoplasma*, and *F. tularensis* cause this histologic picture. Many early studies did not make a distinction between pneumonia as a cause of ARDS and pneumonia complicating ARDS.

Recently we found an increased incidence of herpesvirus (type I) in the lung in burn cases.[72] The presence of the virus correlated with ARDS and pneumonia. Viral inclusions were identified only by means of the immunoperoxidase technique (Fig. 11-22) and could well be missed in the presence of a pneumonia in routine H&E sections. In susceptible subjects the virus may cause cytolysis of alveolar, bronchiolar, and bronchial epithelium, leaving a bare basement membrane free for infection.

The factors predisposing to nosocomial pneumonia in ARDS are given in Table 11-2.[73] With a knowledge of the ultrastructural changes in ARDS, it is easy to understand how a nutrient-rich medium in the alveoli and loss of types I and II pneumocytes

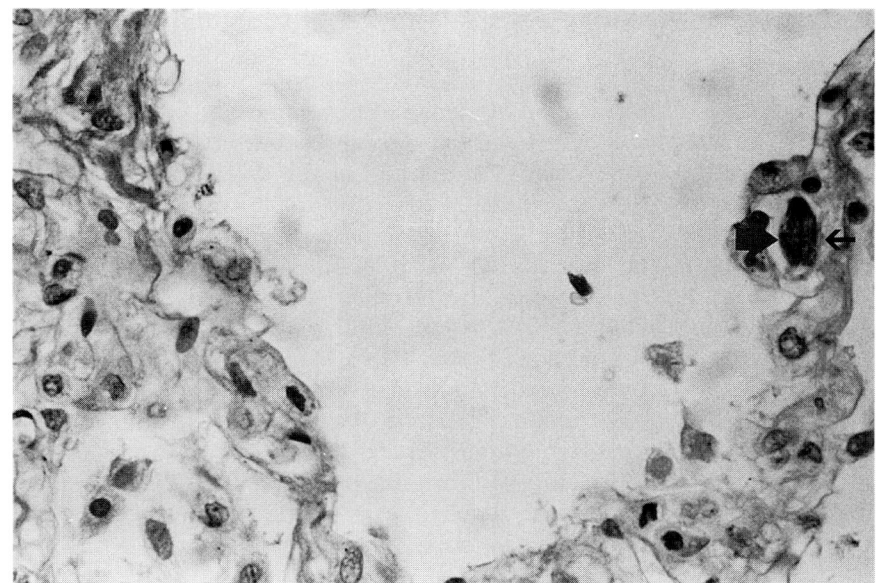

FIGURE 11-22
Herpes virus as shown by immuno-peroxidase (*arrow*) in early ARDS. (Immunoperoxidase HSV I, ×350.)

TABLE 11-2
Factors Predisposing to Nosocomial Pneumonia in ARDS

Acute lung injury
Predisposing factors
   Systemic sepsis
   Extrapulmonary infection
   Hemorrhagic shock
   Head injury
Coexisting illnesses
   Extremes of age
   Diabetes mellitus
   Cardiopulmonary disease
   Malignancy
   Obesity
   Chronic liver or renal disease
Therapeutic interventions
   Endotracheal intubation and tracheostomy
   Nasogastric tubes
   Steroid therapy
   Intestinal bleeding prophylaxis with acid
     neutralization
   Antibiotic therapy
   Sedatives
   Oxygen (?)
Malnutrition

SOURCE: Adapted from Hopewell and Murray.[132]

predispose patients to development of this complication.

PULMONARY OXYGEN TOXICITY

Sevitt[38] suggested the criteria of hyaline membranes and proliferative pneumonitis as specific features of oxygen toxicity. He demonstrated intraalveolar fibrosis as early as 54 h after oxygen therapy, though it appeared well-established by 5 days. Pratt[31] attributed pulmonary capillary proliferation to oxygen.

There have been numerous reviews on the effect of oxygen on the lung,[74–77] as well as a chapter in the companion volume to this book.[78] The theoretical basis of the changes is that oxygen diffuses across the alveolar membranes into the capillaries, where it is taken up by hemoglobin to be distributed to other tissues. Oxygen is utilized by the tissues in a variety of biochemical processes with different end-products,[76] which can be expressed as follows:

$$O_2 + 4H^+ + 4e^- \longrightarrow 2\,H_2O \qquad (1)$$
$$O_2 + 2H^+ + 2e^- \longrightarrow H_2O_2 \qquad (2)$$
$$O_2 + e^- \longrightarrow O_2^- \qquad (3)$$
$$O_2^- + H_2O_2 \;\; fe \longrightarrow 2 \bullet OH + O_2 \qquad (4)$$

Though from equation 2 the formation of hydrogen peroxide appears biologically frightening, it is tolerated in moderate concentrations by most aerobic organisms. Superoxide radical (O_2^-) is more reactive than hydrogen peroxide and may react with hydrogen peroxide to form a hydroxyl radical (OH, equation 4), which is highly reactive with most biologic materials. Hydrogen peroxide, superoxide radical, hydroxyl radical, and singlet oxygen are all considered possible agents of hyperoxic tissue damage. Nitrogen oxides have been shown to react with superoxides, and the reaction products may play a part in the development of ARDS.[79,80]

Enzymes such as xanthine and urate oxidases can all generate reduced oxygen products. Macrophages in the respiratory burst give rise to superoxide generation. The ultrastructural level of free radical formation during hyperoxia is unknown. It is possibly at the level of mitochondria or endoplasmic reticulum.

Nearly all cell components react with oxygen free radicals, and the metabolic damage varies between cell types.[78] The most susceptible cells are those containing lipids with unsaturated fatty acids. However, complex carbohydrates and metabolic cofactors such as NADH (nucleonic acids) are also susceptible. Because much of the cell membrane is composed of lipids, there is a potential for cell damage in hyperoxia.

There are natural defenses against these products in the form of superoxide dismutase, which immobilizes superoxide, and catalase and peroxidases that control hydrogen peroxide.

Little work has been carried out on the possible mechanisms of oxygen toxicity in humans and large mammals.[76] This is important, since the changes of oxygen toxicity are nonspecific and have been described in ARDS.[44] The pathology consists of atelectasis, edema, alveolar hemorrhage, inflammation, fibrin deposition, and thickening and hyalinization of alveolar membranes. There is early damage to capillary endothelium in animals, and plasma leaks into the interstitial and alveolar spaces.

Using bronchoalveolar lavage, Crystal's group[81] showed that after 16 h of breathing more than 95 percent oxygen, there was significant alveolar capillary leak as detected by increased plasma albumin and transferrin in lavage fluid. These changes were reversible, as shown by repeat lavage two weeks later. There was no alteration in the total number of inflammatory cells in the lung or the immune effector cells. Alveolar macrophages from hyperoxic patients released increased amounts of fibronectin and alveolar macrophage-derived growth factor for fibroblasts. These mediators were thought to modulate fibroblast recruitment and proliferation in the alveolar wall. None of the patients had any ill effects, and there was

no clinical evidence of pulmonary edema. The main side effect was substernal discomfort.

Airway examination at bronchoscopy after oxygen exposure revealed mild erythema in six of 14 subjects but no increase in secretions. There was no accumulation of neutrophil polymorphs in the hyperoxic patients and no evidence that macrophages were induced to release neutrophil chemotactic factor. This last factor was not identified in the supernatants of alveolar macrophages from the experimental or control groups. The patients were exposed to high levels of oxygen for less than 24 h.

Some authors have demonstrated permanent structural damage, including parenchymal cell injury and alveolar wall fibrosis, after two to three days of oxygen.[82,83]

Because of the lack of specific histology, it is necessary to decide whether the changes of oxygen toxicity are related to the underlying disease giving rise to the need for this gas, or whether the histologic picture is brought about by oxygen. With humans this is a difficult question to answer, since most patients with ARDS will have received oxygen. Animal models have been used, and the findings have been summarized.[44] Small animals, such as mice and rats, are more sensitive to oxygen than larger ones, but in different animals there are different sites of oxygen injury. Thus, in rats the initial injury is to the endothelial cells, whereas in dogs the primary injury is to type I pneumocytes, the endothelial cells being injured later.

Kaplan and coworkers[84] and Kapanci and coworkers[85] studied the effect of oxygen in monkeys (*Macca mulatta*), mammals whose susceptibility to oxygen toxicity most closely resembles that of humans. They exposed the monkeys to 100 percent oxygen at 750 mmHg for periods of up to 12 days. Ultrastructurally, type I pneumocytes were lost first and replaced by type II cells. There was interstitial edema and swelling of the endothelial cells. These latter cells showed changes at two days, and by seven days there was a variable degree of endothelial destruction. Hyaline membranes were not a prominent feature on light microscopy. The absence of hyaline membranes and the conspicuous endothelial cell damage are thus in contrast to the findings in human subjects.

Clinical studies have not elucidated the problem of oxygen toxicity. Caldwell and Weibel[86] believe that pulmonary oxygen toxicity occurs in any person breathing oxygen (P_{O_2} 700 torr) for longer than 6 h, but they were careful to point out that it was difficult to assess the contribution oxygen makes in respiratory failure. Bachofen and Weibel[19] found that there was no correlation between the ultrastructural alterations and the administration of high concentrations of oxygen in their patients with septic shock. Some of their cases showed changes that could not be explained by oxygen toxicity.

In a study of patients with irreversible brain damage,[87] no hyaline membranes were found in a group receiving pure oxygen for a mean range of 52 h (range: 31 to 76 h).

We have investigated the incidence of hyaline membranes and pulmonary fibrosis in patients exposed to high levels of oxygen after severe burns.[29,30] In our first report,[30] we noted hyaline membranes in only three of 10 cases. In one patient, who died at 13 days, there was focal interstitial and intraalveolar fibrosis. This patient had received 70 percent oxygen for more than 10 days. However, his oxygen requirement fell the day before death, associated with treatment of his renal failure by hemodialysis and probable reduction in his pulmonary edema. If there had been progressive fibrosis due to oxygen toxicity, his oxygen requirements would have been expected to increase.

In the second study, hyaline membranes were present in four of 64 cases. In one of these, aspiration pneumonia may have been a contributory factor in hyaline membrane formation. In two of the three remaining cases, the amount of oxygen given ranged from 35 percent for two days to 60 percent for four days. Significantly, in cases without hyaline membranes, the amount of oxygen administered was greater—ranging from 100 percent for seven days to 60 percent for two months in one patient. Similarly, the presence of fibrosis showed no consistent relationship to the amount of oxygen given.

Thus, while it is possible that oxygen per se has a toxic effect on the lung, the case against it—especially when there are other causes of lung damage, as in ARDS—is not fully substantiated.

RIGHT VENTRICULAR FUNCTION IN ARDS

A review[88] showed increased right ventricular diastolic and systolic volumes and decreased ejection fractions. In septic shock or viral myocarditis, which could be part of a viral pneumonia, there is depression of right ventricular contractility, which can influence the clinical course of ARDS. The functional effect of fibrin thrombi affecting the myocardial microvasculature, as part of multiorgan involvement in ARDS, must also be considered.

PATHOGENETIC MECHANISMS

As can be seen from the list of causative factors, ARDS is a heterogeneous disorder. In some cases, the damage is initiated on the epithelial side of the alveolar wall as with *Pneumocystis carinii*.[89] In fat embolism,[90] ARDS is characterized by tachypnea, hypoxemia, and "snowstorm" infiltrates on the chest radiograph. The respiratory insufficiency is thought to be due to a coagulation disorder, and therefore this type of ARDS is initiated on the endothelial cell side of the alveolar wall. There remain, however, a large group of disorders causing ARDS whose pathogenesis is poorly understood.

The current situation is well stated by Cochrane: "No longer is there any illusion that any specific molecule mediates ARDS. Rather it is clear that many of yesterday's putative 'mediators' of ARDS actually are enhancers, amplifiers, modulators and regulators that fine-tune the evolution and resolution of inflammatory sequences rather than cause them."[91] These pathogenetic mechanisms will now be briefly reviewed.

Mechanisms and Mediators in ARDS

NEUTROPHIL POLYMORPHS

There have been many articles indicating that the neutrophil polymorph is central to lung injury in ARDS.[92–96] A neutrophil polymorph stimulation index, a reflection of the degree of hyperresponsiveness, correlated with elevated levels of tumor necrosis factor–alpha (TNF-alpha) in plasma.[97] TNF-alpha has also been shown to induce an increase in airway responsiveness in normal subjects and is associated with a neutrophil infiltration. It is thought to be an important cytokine for the induction of airway inflammation and hyperresponsiveness.[97a] Neutrophil-derived collagenase has been detected in lavage fluid from patients with ARDS.[98] Polymorphs are also capable of releasing elastase and lysosomal cationic protein, thought to cause increased capillary permeability.[99]

There is a recent review of pulmonary adhesion molecules, thought to be important in neutrophil-endothelium reactions.[100] There are three main classes of cell adhesion molecules: selectins, integrins, and immunoglobulins. In experimental animals, neutrophils kill endothelial cells.[101] This polymorphonuclear leukocyte–mediated injury entails interaction between oxygen products and proteases. Neutrophils cross into the interstitium after adherence to endothelial cells by expressing cell surface adhesion molecules, especially the integrin CD 18/CD 11b, which acts as a ligand for the intercellular adhesion molecule. Endothelium also produces endothelial leukocyte adhesion molecules, which bind neutrophils. This reaction is enhanced by gamma-interferon, TNF-alpha, and interleukin 1 (IL-1).[102]

A role for selectins in ARDS has been described.[103,103a] These are single-chain surface glycoproteins with a common *N*-terminal lectin domain.[104] Selectins are a family of adhesion molecules implicated in leukocyte-endothelial adhesion. Their receptors exist in a cleaved, soluble form. Among patients at risk of ARDS, initial plasma-soluble L-selectin (sL-selectin) levels were significantly lower in patients who developed ARDS than in those who did not. A significant correlation was found between low values of sL-selectin and indices of subsequent lung injury, including requirement for ventilation and the degree of respiratory failure. A significant correlation was found between low values of sL-selectin and patient mortality. Reduced sL-selectin may represent a peripheral blood marker of widespread panendothelial activation. This is an important event in the pathogenesis of ARDS and multiorgan failure.

However, Glauser and Fairman have expressed doubts about the central role of the polymorph in acute lung damage.[105] Much of their review is based on experimental work in animals, and the authors show that other factors, including endotoxin production and microembolism, may cause neutrophil trapping. Reasonably, they conclude that the neutrophil may not be the primary mediator in ARDS, but just one of many. Possibilities include combinations of other formed elements, such as platelets, fibrin, fibrindegradation products, eicosanoids, bradykinin, histamine, alveolar macrophages, and certain substances such as oleic acid.[106]

Two reports have described ARDS in patients with neutropenia and no pulmonary neutrophil sequestration.[107,108]

Despite some of the above evidence against neutrophils playing an active part in ARDS, it is still considered by most authorities an important cell in initiating injury in the early stages of the syndrome.

COMPLEMENT

The role of complement in ARDS was first suspected in 1977. It was then observed that dialysis-induced neutropenia could be reproduced in animals and caused activation of complement via the alternative pathway.[109] Neutrophils were aggregated in the small pulmonary vessels. Further studies have confirmed that complement activation occurs in ARDS. C3 and C5a and properdin factor B activation occurs in the epithelial lining fluid of most patients with ARDS.[110] C5a could account for the neutrophil influx seen in this disease. Elevated C3a levels in the first few hours after injury were associated with the later development of ARDS.[111] A more sensitive indicator than C3a was the C3a:C3 ratio, which discriminated ARDS from non-ARDS patients. A second rise of C3a levels and the C3a:C3 ratio from day 4 onward paralleled the course of extravascular lung water. These authors demonstrated that in the first 48 h complement activation occurs via the alternative pathway only and is later followed by additional activation via the classical pathway. Complement acts with circulating endotoxin to produce the disease.[112]

Fibroblasts and type II pneumocytes in tissue culture fluid have the ability to synthesize and secrete a number of complement components, including C3 and C5.[113] This finding has parallels in work, considered below, in which TNF-alpha has been demonstrated in the same cells.

MACROPHAGES

Macrophages produce many mediators, including neutrophil and eosinophil chemotactic factors and cytokines, such as TNF-alpha and IL-1. These have recently been reviewed.[114] Many of these factors alter the permeability of endothelial cells and adhesion molecule expression. Platelet-derived growth factor (PDGF), insulin-like growth factor–1, and fibronectin are released from macrophages and act as factors that stimulate fibroblast proliferation. Macrophages release proteases as well as substances such as thromboxane and leukotriene B_4. Macrophage action could account for the lung damage that occurs in neutropenic patients.

PLATELETS

Platelets aggregate together in ARDS to form microthrombi. During this process they release cytokines such as PDGF, thromboxane A, epidermal growth factor, and transforming growth factor. These substances produce vasoconstriction, which contributes to pulmonary arterial hypertension, and bronchoconstriction by the action of thromboxane A2 on bronchial smooth muscle. The platelet products probably do not directly damage the alveolar capillary membranes, but the vasoconstriction will augment lung edema.[115] Platelets may also activate complement and amplify neutrophil-induced injury.

LEUKOTRIENES

Leukotrienes (LT) are a family of biomolecules derived from arachidonic acid by oxidative metabolism through the lipoxygenase pathway. The release of arachidonic acid from cell membranes may be stimulated by a variety of biologic signals, including specific receptor activation, physical stimuli such as cold or altered ionic environments, antigen-antibody reactions, and cell death. High concentrations of 5-lipoxygenase products, LTB 4, LTD 4, and 11-trans LTC 4, along with elevated numbers of neutrophils, have been identified in lavage fluid from four patients with ARDS and in the injured lungs of six patients with unilateral lung disease after chest trauma or aspiration.[116] The values for LTB 4 and LTD 4 in this study were high; at such a concentration, leukotrienes could cause not only capillary damage but also airway smooth muscle constriction. The neutrophils detected could injure the pulmonary microvasculature by producing free radicals or releasing mediators such as LTB 4 and LTC 4. These, in turn, can be metabolized to LTD 4 and LTE 4. Recently, LTE 4 has been found to be increased in urine of patients with ARDS and has been associated with abnormalities of gas exchange and lung compliance.[117]

CYTOKINES

One of the main cytokines to have been incriminated in ARDS and septic shock is TNF-alpha (Fig. 11-23). This is produced predominantly by macrophages[118] and neutrophil polymorphs and, along with IL-1, is responsible for the production and release of polymorphs and probably fever.[119] TNF-alpha is thought to be central to septic shock[120] as well as ARDS.[121] It increases collagen synthesis and is chemotactic for fibroblasts.[122]

Immunostaining of autopsy lung tissue from 13 patients who died from ARDS and 10 control tissues showed positive staining with TNF-alpha in 12 of 13 ARDS cases and 0/10 controls. TNF-alpha was

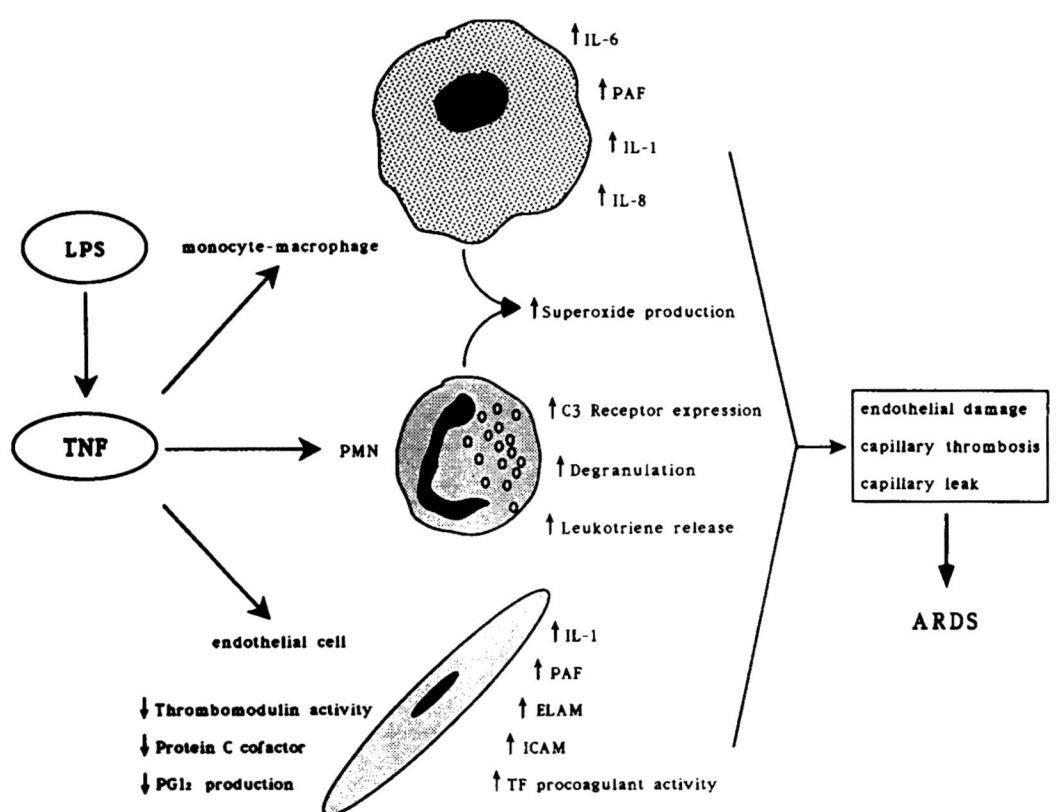

FIGURE 11-23
Role of the macrophage, polymorph, and TNF-alpha. (*Reprinted with permission from Fishman.[121]*)

located in the proliferating type II pneumocytes, suggesting that this cytokine is synthesized or possibly absorbed by these cells and may be involved in the production or recovery from ARDS.[123] Nash recently described TNF-alpha in patients with fibrosing alveolitis.[124]

Two new cytokines have also been described that may have a role in ARDS. These are macrophage inflammatory protein (MIP) and neutrophil activating peptide-1/interleukin 8 (NAP-1/IL-8).[125]

The reader is referred to the review by for more information on pulmonary cytokines.[114]

SURFACTANT

Respiratory distress syndrome in infants is due to immaturity of the lung and can be corrected by surfactant replacement. Since the pathophysiologies of that disease and ARDS show similarities, there are reasons to be optimistic that surfactant replacement could be valuable in the treatment of the latter disease has been demonstrated in experimental animals.[126] In addition to loss of surfactant caused by type II cell damage, proteins released through the alveolar capillary wall have an inhibiting effect on pulmonary surfactant.[127,128]

Surfactant has an increased surface compressibility in patients with ARDS.[129] Lavage samples, from these patients have abnormal surface properties and measurements of surfactant function have correlated with the severity of disease in adult patients.[130] The role of surfactant in ARDS has recently been reviewed.[131] Early clinical trials of exogenous surfactant therapy in ARDS have given encouraging results, and a prospective, randomized clinical trial evaluating its efficacy is in progress. Early results show a trend toward decreased mortality and improvements in gas transfer.

REFERENCES

1. Ashbaugh DG, Bigelow DB, Petty TL, Levine BE: Acute respiratory distress in adults. *Lancet* 2:319–323, 1967.
2. Petty TL, Ashbaugh BG: The adult respiratory distress syndrome: Clinical features, the factors influencing prognosis and principles of management. *Chest* 60:223–239, 1971.
3. du Bois RM: Respiratory medicine. *BMJ* 310:1594–1597, 1995.
4. Fishman AP: Shock lung: A distinctive non-entity. *Circulation* 47:921–923, 1973.
5. Murray JS: Respiratory distress syndrome, in Flenley DC (ed), *Recent Advances in Respiratory Medicine.* Edinburgh, Churchill-Livingstone, 1980, pp 69–90.
6. Crystal RG: Alveolitis: The key to interstitial lung disorders. *Thorax* 37:1–10, 1982.
7. Warner DO, Warner MA, Divertie MB: Open lung biopsy in patients with diffuse pulmonary infiltrates and acute respiratory failure. *Am Rev Respir Dis* 137:90–94, 1988.
8. Bell RC, Coalson JJ, Smith JD, Johanson WG Jr: Multiple organ system failure and infection in adult respiratory distress syndrome. *Ann Intern Med* 99:293–298, 1983.
9. Bone RC, Balk R, Slotman G, et al: Adult respiratory distress syndrome: Sequence and importance of development of multiple organ failure. *Chest* 101:320–326, 1992.
10. Montgomery AB, Stager MA, Carrico CJ, Hudson LD: Causes of mortality in patients with the adult respiratory distress syndrome. *Am Rev Respir Dis* 132:485–489, 1985.
11. Bone RC: Why new definitions of sepsis and organ failure are needed. *Am J Med* 95:348–350, 1993.
12. Seidenfeld JJ, Pohl DF, Bell RC, Harris GD, Johanson WG Jr: Incidence, site and outcome of infections in patients with the adult respiratory distress syndrome. *Am Rev Respir Dis* 134:12–16, 1986.
12a. Kollef MH, Schuster DP: The acute respiratory distress syndrome. *N Engl J Med* 332:27–37, 1995.
12b. Hudson LD, Milberg JA, Anardi D, Mounder RJ: Clinical risks for development of the acute respiratory distress syndrome. *Am J Respir Crit Care Med* 151:293–301, 1995.
13. Sarnaik AP, Leih-Lai M: Adult respiratory distress syndrome in children. *Pediatr Clin North Am* 41:337–363, 1994.
14. Lyrene RK, Truog WE: Adult respiratory distress syndrome in a pediatric intensive care unit: Predisposing conditions, clinical course, and outcome. *Pediatrics* 67:790–795, 1981.
15. Hasleton PS: Adult respiratory distress syndrome: A review. *Histopathology* 7:307–332, 1983.
16. Hasleton PS, McWilliam L, Curry A, Lawson RAM: Ultrastructural changes pre- and post-cardiopulmonary bypass. In preparation.
17. Hasleton PS, Orr K, Webster A, Lawson RAM: Evolution of acute chest syndrome in sickle cell trait: An ultrastructural and light microscopic study. *Thorax* 44:1057–1058, 1989.
18. Weil JV, Castro O, Malik AB, Rodgers G, Bonds DR, Jacobs TP: Pathogenesis of lung disease in sickle hemoglobinopathies. *Am Rev Respir Dis* 148:249–256, 1993.
19. Bachofen M, Weibel ER: Alterations in the gas exchange apparatus in adult respiratory insufficiency associated with septicemia. *Am Rev Respir Dis* 116:589–615, 1977.
20. Nash G, Foley FD, Langlinlas PC: Pulmonary interstitial edema and hyaline membranes in adult burn patients: Electron microscopic observations. *Hum Pathol* 5:149–160, 1974.
21. Pietra GG, Ruttner JR, Wust W, Glinz W: The lung after trauma and shock: Fine structure of the alveolar-capillary barrier in 23 autopsies. *J Trauma* 21:454–462, 1981.
22. Schnells G, Voight WH, Riedl H, Schlag G, Glatzl A: Electron microscopic investigation of lung biopsies in patients with post-traumatic respiratory insufficiency. *Acta Chir Scand* (Suppl) 499:9–20, 1980.
23. Teplitz C: The core pathobiology and integrated medical science of adult acute respiratory insufficiency. *Surg Clin North Am* 56:1091–1133, 1976.
24. Bachofen M, Weibel ER: Structural alterations of lung parenchyma in the adult respiratory distress syndrome. *Clin Chest Med* 3:35–56, 1982.
25. Hurley JV: Current views of the mechanisms of pulmonary oedema. *J Pathol* 125:59–79, 1978.
26. Cunningham AL, Hurley JV: Alpha-naphthyl-thiourea induced pulmonary oedema in the rat: A topographic and electron microscope study. *J Pathol* 106:25–35, 1972.

27. Riede UN, Mittermayer CH, Friedburg H, Wybitul K, Sandritter W: Morphological development of human shock lung. *Pathol Res Pract* 165:269–286, 1979.

28. Bachofen M, Weibel ER: Basic pattern of tissue repair in human lungs following unspecific injury. *Chest* (Suppl) 65:14S–19S, 1974.

29. Hasleton PS, McWilliam L, Haboubi NY: The lung parenchyma in burns. *Histopathology* 7:333–347, 1983.

30. Hasleton PS, Penna P, Torry J: Effect of oxygen on the lungs after blast injury and burns. *J Clin Pathol* 34:1147–1154, 1981.

31. Pratt PC: Pulmonary capillary proliferation induced by oxygen inhalation. *Am J Pathol* 34:1033–1049, 1958.

32. Wells S, Sissons M, Hasleton PS: Quantitation of pulmonary megakaryocytes and fibrin thrombi in patients dying from burns. *Histopathology* 8:517–527, 1984.

33. Martin JF, Slater DN, Trowbridge EA: Abnormal intrapulmonary platelet production: A possible cause of vascular and lung disease. *Lancet* 1:793–796, 1983.

34. Slater D, Martin J, Trowbridge A: Circulating megakaryocytes [letter]. *Histopathology* 7:136–140, 1983.

35. Gatter KC, Cordell JL, Turley H, et al: The immunohistological detection of platelets, megakaryocytes and thrombi in routinely processed specimens. *Histopathology* 13:257–267, 1988.

36. Tomashefski JF Jr, Davies P, Boggis C, Greene R, Zapol WM, Reid LM: The pulmonary vascular lesions of the adult respiratory distress syndrome. *Am J Pathol* 112:112–126, 1983.

37. Eeles GH, Sevitt S: Microthrombosis in injured and burned patients. *J Pathol Bacteriol* 93:275–293, 1967.

38. Sevitt S: Diffuse and focal oxygen pneumonitis: A preliminary study on the threshold of pulmonary oxygen toxicity in man. *J Clin Pathol* 74:21–30, 1974.

39. Fukuda Y, Ishizaki M, Masuday et al: The role of intra-alveolar fibrosis in the process of pulmonary structural remodelling in patients with diffuse alveolar damage. *Am J Pathol* 126:171–182, 1987.

40. Sevitt S: A review of the complications of burns, their origin and importance for illness and death. *J Trauma* 19:358–369, 1979.

41. Toor AH, Tomashefski JF Jr, Kleinerman J: Respiratory tract pathology in patients with severe burns. *Hum Pathol* 21:1212–1220, 1990.

42. Stanley MW, Henry-Stanley MJ, Cmiac MS, Gajl-Peczalska KJ, Bitterman PM: Hyperplasia of type II pneumocytes in acute lung injury: Cytologic findings in sequential bronchoalveolar lavage. *Am J Clin Pathol* 97:669–677, 1992.

43. Steinberg KP, Milberg JA, Martin TR, Maunder RJ, Cockrill BA, Hudson LD: Evolution of bronchalveolar cell populations in the adult respiratory distress syndrome. *Am J Respir Crit Care Med* 150:113–122, 1994.

44. Katzenstein A, Bloor CM, Liebow AA: Diffuse alveolar damage, the role of oxygen, shock related factors. *Am J Pathol* 210–222, 1976.

45. Yazdy AM, Tomashefski JF Jr, Yagan R, Kleinerman J: Regional alveolar damage (RAD): A localized counterpart of diffuse alveolar damage. *Am J Clin Pathol* 92:10–15, 1989.

46. Hert R, Albert RK: Sequelae of the adult respiratory distress syndrome. *Thorax* 49:8–13, 1994.

47. Suchyata MR, Clemmer TP, Elliot CG, Orme JF Jr, Weaver LK: The adult respiratory distress syndrome: A report of survival and modifying factors. *Chest* 101:1074–1079, 1992.

48. McHugh LG, Milberg JA, Whitcomb ME, Schoene RB, Maunder RJ, Hudson LD: Recovery of function in survivors of the acute respiratory distress syndrome. *Am J Respir Crit Care Med* 150:90–94, 1994.

49. Ghio A, Elliot C, Crapo R, Berlin S, Jensen R: Impairment after adult respiratory distress syndrome: An evaluation based on American Thoracic Society recommendations. *Am Rev Respir Dis* 139:1158–1162, 1989.

50. Churg A, Golden J, Fligiel S, Hogg J: Case reports: Bronchopulmonary dysplasia in the adult. *Am Rev Respir Dis* 127:117–120, 1983.

51. Pratt PC, Vollmer RT, Shelburne JD: Pulmonary morphology in a multihospital collaborative extra-corporeal membrane oxygenation project: 1. light microscopy. *Am J Pathol* 95:191–205, 1979.

52. Eppler GR, Colby TV: A spectrum of bronchiolitis obliterans. *Chest* 83:161–162, 1983.

53. Tomashefski JF Jr: Pulmonary pathology in the adult respiratory distress syndrome. *Clin Chest Med* 11:593–619, 1990.

54. Burkhardt A: Alveolitis and collapse in the pathogenesis of pulmonary fibrosis. *Am Rev Respir Dis* 140:513–524, 1989.

55. Harrington MA, Pledger WJ: Characterization of growth factor modulated events regulating cellular proliferation. *Methods Enzymol* 147:400–407, 1987.

56. Derynck R: Transforming growth factor alpha: Structure and biological activities. *J Cell Biochem* 32:293–304, 1986.

57. Bitterman PB, Adelberg S, Crystal RG: Mechanisms of pulmonary fibrosis: Spontaneous release of the alveolar macrophage derived growth factor in the interstitial lung disorders. *J Clin Invest* 72:1801–1813, 1983.

58. Bitterman PB, Wewers MD, Rennard FI, Adelberg S, Crystal RG: Modulation of alveolar macrophage-driven fibroblast proliferation by alternative macrophage modulators. *J Clin Invest* 77:700–708, 1986.

59. Engvall E, Ruoslaht IE: Binding of soluble form of fibroblast surface protein, fibronectin, to collagen. *Int J Cancer* 20:1–5, 1977.

60. Kodoh Y, Taniguchi H, Taki F, Takagi K, Satake T: 7S collagen in bronchoalveolar lavage fluid in patients with adult respiratory distress syndrome. *Chest* 101:1091–1094, 1992.

61. Christner P, Fein A, Goldberg S, Lippman M, Abrams W, Weinbaum G: Collagenase in the lower respiratory tract of patients with adult respiratory distress syndrome. *Am Rev Respir Dis* 131:690–695, 1985.

62. Farjanel J, Hartmann DJ, Guidet B, Luquel L, Offenstadt G: Four markers of collagen metabolism as possible indicators of disease in the adult respiratory distress syndrome. *Am Rev Respir Dis* 147:1091–1099, 1993.

63. Zapol WM, Snider MT: Pulmonary hypertension in severe respiratory failure. *N Engl J Med* 296:476–480, 1977.

64. Hasleton PS, Heath D, Brewer DB: Hypertensive pulmonary vascular disease in states of chronic hypoxia. *J Pathol Bacteriol* 95:431–440, 1968.

65. Snow RL, Davies P, Pontoppidan H, Zapol WM, Reid L: Pulmonary vascular remodeling in adult respiratory distress syndrome. *Am Rev Respir Dis* 126:887–892, 1982.

66. Pace NL: A critique of flow directed pulmonary arterial catheterization. *Anesthesiology* 47:455–465, 1977.

67. Martin TR, Pistorise BP, Hudson LD, Maunder RJ: The function of lung and blood neutrophils in patients with the adult respiratory distress syndrome: Implications for the pathogenesis of lung infections. *Am Rev Respir Dis* 144:254–262, 1991.

68. Cochrane CG, Spragg R, Revak SD: Pathogenesis of the adult

respiratory distress syndrome: Evidence of oxidant activity in bronchoalveolar lavage fluid. *J Clin Invest* 71:754–761, 1983.

69. Baehner RL, Boxer LA, Allen JM, Davis J: Auto-oxidation as a basis for altered function by polymorphonuclear leucocytes. *Blood* 50:327–335, 1977.

70. Martin TR, Pistorese BP, Chi EY, Goodman RB, Matthay MA: Effects of leukotriene B₄ in the human lung: Recruitment of neutrophils into the alveolar spaces without a change in protein permeability. *J Clin Invest* 84:1609–1619, 1989.

71. ARDS: A clinical view [editorial]. *Lancet* 2:439, 1986.

72. Byers RJ, Hasleton PS, Quigley A, Dennett C, Klapper PE, Cleator GM: Pulmonary *Herpes simplex* infection in the lungs of burns patients [abstract]. *J Pathol* 172:159A, 1994.

73. Niederman MS, Fein AM: Sepsis syndrome, the adult respiratory distress syndrome and nosocomial pneumonia: A common clinical sequence. *Clin Chest Med* 11:633–656, 1990.

74. Claireaux AE: Drugs in the lung: The effect of oxygen on the lung. *J Clin Pathol* 28 (Suppl) 9:75–80, 1975.

75. Lee F, Massaro D: The lung and oxygen toxicity. *Arch Intern Med* 139:347–350, 1979.

76. Deneke SM, Fanburg BL: Normobaric oxygen toxicity and the lung. *N Engl J Med* 303:76–86, 1980.

77. Fisher AB, Forman HJ, Glass M: Review mechanisms of pulmonary oxygen toxicity. *Lung* 162:255–259, 1984.

78. Fisher AB: Pulmonary oxygen toxicity, in Fishman AP (ed), *Pulmonary Diseases and Disorders,* vol 3, 2d ed. New York, McGraw-Hill, 1988, pp 2331–2338.

79. Gaston B, Drazen JM, Loscalzo J, Stamler JS: State of the art: The biology of nitrogen oxides in the airways. *Am J Crit Care Med* 149:538–551, 1994.

80. Ischiropoulos H, Mendiguen I, Fisher D, Fisher AB, Thom SR: Role of neutrophils and nitric oxide in lung alveolar injury from smoke inhalation. *Am J Respir Crit Care Med* 150:337–341, 1994.

81. Davis WB, Rennard SI, Bitterman PB, Crystal RG: Pulmonary oxygen toxicity: Early reversible changes in human alveolar structures induced by hyperoxia. *N Engl J Med* 309:878–883, 1983.

82. Nash G, Blennerhassett JB, Pontoppidan H: Pulmonary lesions associated with oxygen therapy and artificial ventilation. *N Engl J Med* 276:368–374, 1977.

83. Gould VE, Tosko R, Whellis RF, Gould NS, Kapanci Y: Oxygen pneumonitis in man: Ultrastructural observations in the development of alveolar lesions. *Lab Invest* 26:499–508, 1972.

84. Kaplan HP, Robinson FR, Kapanci Y, Weibel ER: Pathogenesis and reversibility of the pulmonary lesions of oxygen toxicity in monkeys, clinical and light microscopic studies. *Lab Invest* 20:94–100, 1969.

85. Kapanci Y, Weibel ER, Kaplan HP, Robinson FR: Pathogenesis and reversibility of the pulmonary lesions of oxygen toxicity in monkeys: II. Ultrastructural and morphometric studies. *Lab Invest* 20:101–118, 1969.

86. Caldwell PRB, Weibel ER: Pulmonary oxygen toxicity, in Fishman AP (ed), *Pulmonary Diseases and Disorders,* 1st ed. New York, McGraw-Hill, 1980, pp 800–805.

87. Barber RE, Lee J, Hamilton WD: Oxygen toxicity in man: A prospective study in patients with irreversible brain damage. *N Engl J Med* 283:1478–1484, 1970.

88. Dhainaut J-F, Brunet F: Right ventricular performance in adult respiratory distress syndrome. *Eur Respir J* 3 (Suppl 11):490S–495S, 1990.

89. Hasleton PS, Curry A, Rankin EM: Pneunocystis carinii pneumonia: A light microscopical and ultrastructural study. *J Clin Pathol* 34:1138–1146, 1981.

90. Alho A: Fat embolism syndrome: Etiology, pathogenesis and treatment. *Acta Chir Scand* 499 (Suppl):75–85, 1980.

91. Cochrane CG: The enhancement of inflammatory injury [editorial]. *Am Rev Respir Dis* 136:1–2, 1987.

92. Corrin B: Lung pathology in septic shock. *J Clin Pathol* 33:891–894, 1980.

93. Staub NC: Pulmonary edema due to increased microvasculature permeability. *Am Rev Med* 32:291–312, 1981.

94. Tate RM, Repine JE; Neutrophils in the adult respiratory distress syndrome. *Am Rev Respir Dis* 125:552–559, 1983.

95. Repine JE, Beehler CJ: Neutrophils and adult respiratory distress syndrome: Two interlocking perspectives in 1991. *Am Rev Respir Dis* 144:251–252, 1991.

96. Tetley TD: Proteinase imbalance: Its role in lung disease. *Thorax* 48:560–565, 1993.

97. Chollet-Martin S, Montravers P, Jibert C, et al: Subpopulation of hyperresponsive polymorphonuclear neutrophils in patients with adult respiratory distress syndrome: Role of cytokine production. *Am Rev Respir Dis* 146:990–996, 1992.

97a. Thomas PS, Yates DH, Barnes PJ: Tumour necrosis factor-alpha increases airway responsiveness and sputum neutrophilie in normal subjects. *Am J Respir Crit Care Med* 152:76–80, 1995.

98. Christner P, Fein A, Goldberg S, Lippman M, Abrams W, Weinbaum G: Collagenase in the lower respiratory tract of patients with adult respiratory distress syndrome. *Am Rev Respir Dis* 131:690–695, 1985.

99. Chang S-W, Voelkel NF: Charge-related lung microvascular injury. *Am Rev Respir Dis* 139:534–535, 1989.

100. MacNee W, Selby C: Neutrophil traffic in the lungs: Role of haemodynamics, cell adhesion, and deformability. *Thorax* 48:79–88, 1993.

101. Varani J, Ginsburg I, Schuger L, et al: Endothelial cell killing by neutrophils: Synergistic interaction as oxygen products and proteases. *Am J Pathol* 135:435–438, 1989.

102. Pober JS: Cytokine-mediated activation of vascular endothelium. *Am J Pathol* 133:426–433, 1988.

103a. Albert RK: Mechanisms of the adult respiratory distress syndrome: Slectins. *Thorax* 50 (suppl 1):S49–S52, 1995.

103. Donnelly SC, Haslett C, Dransfield I, et al: Role of selectins in development of adult respiratory distress syndrome. *Lancet* 344:215–219, 1994.

104. Keelan E, Haskard DO: CAMs and anti-CAMS: The clinical potential of cell adhesion molecules. *J R Coll Phys Lond* 26:17–24, 1992.

105. Glauser FL, Fairman RP: The uncertain role of the neutrophil in increased permeability pulmonary edema. *Chest* 88:601–607, 1985.

106. Schuster DP: ARDS: Clinical lessons from the oleic acid model of acute lung injury. *Am Rev Respir Crit Care Med* 149:245–260, 1994.

107. Laufe MD, Simon RH, Flint A, Keller JB: Adult respiratory distress syndrome in neutropenic patients. *Am J Med* 80:1022–1025, 1986.

108. Ognibene FP, Martin SE, Parker MM, et al: Adult respiratory distress syndrome in patients with severe neutropenia. *N Engl J Med* 315:547–551, 1986.

109. Craddock PR, Fehr J, Brigham KL, et al: Complement and leukocyte-mediated dysfunction in hemodialysis. *N Engl J Med* 296:769–774, 1977.

110. Robbins RA, Russ WD, Rasmussen JK, Clayton MM:

Activation of the complement system in the adult respiratory distress syndrome. *Am Rev Respir Dis* 135:651–658, 1987.

111. Zilow G, Sturm JA, Rother U, Kirschfink M: Complement activation and the prognostic value of C3a in patients at risk of adult respiratory distress syndrome. *Clin Exp Immunol* 79:151–157, 1990.

112. Parsons PE, Worthen GS, Moore EE, et al: The association of circulating endotoxin with the development of the adult respiratory distress syndrome. *Am Rev Respir Dis* 140:294–301, 1989.

113. Rothman BL, Merrow M, Bamba M, Kennedy T, Kreutzur DL: Biosynthesis of the third and fifth components of bi-isolated human lung cells. *Am Rev Respir Dis* 139:212–220, 1989.

114. Nicod LP: Cytokines: Overview. *Thorax* 48:660–667, 1993.

115. Heffner JE, Sahn, SA, Repine JE: The role of platelets in the adult respiratory distress syndrome: Culprits or bystanders? *Am Rev Respir Dis* 135:482–492, 1987.

116. Antonelli M, Bufi M, De Blasi RA, et al: Detection of leukotrienes B4, C4 and of their isomers in arterial, mixed venous blood and bronchoalveolar lavage fluid from ARDS patients. *Intensive Care Med* 15:296–301, 1989.

117. Bernard GR, Korley V, Chee P, Swindell B, Ford-Hutchinson AW, Tagari P: Persistent generation of peptido-leukotrienes in patients with the adult respiratory distress syndrome. *Am Rev Respir Dis* 144:263–267, 1991.

118. Van Nhieu JT, Misset B, Lebargy F, Carlet J, Bernaudin J-F: Expression of tumor necrosis factor–alpha gene in alveolar macrophages from patients with adult respiratory distress syndrome. *Am Rev Respir Dis* 147:1585–1589, 1993.

119. Tracey KJ, Cerami A: Metabolic responses to cachectin/TNF: A brief review. *Ann NY Acad Sci* 587:325–331, 1990.

120. Marks JD, Marks CB, Luse JM, et al: Plasma tumor necrosis factor in patients with septic shock: Mortality rate, incidence of adult respiratory distress syndrome and effects of methylprednisolone administration. *Am Rev Respir Dis* 141:94–97, 1990.

121. Fahey TJ III, Tracey KJ, Cerami A: Tumor necrosis factor (cachectin) and the adult respiratory distress syndrome, in Fishman AP (ed), *Update: Pulmonary Diseases and Disorders*. New York, McGraw-Hill, 1992, pp 175–183.

122. Elias AA, Feundlich B, Adams S, Rosenbloom J: Regulation of human lung fibroblast collagen proliferation by recombinant interleukin 1, tumor necrosis factor, and interferon gamma. *Ann NY Acad Sci* 580:233–244, 1990.

123. Nash JRG, McLaughlin PJ, Hoyle C, Roberts D: Immunolocalisation of tumour necrosis factor alpha in lung tissue in patients dying with adult respiratory distress syndrome. *Histopathology* 19:395–402, 1991.

124. Nash JRG, McLaughlin PJ, Butcher D, Corrin B: Expression of tumour necrosis factor–alpha in cryptogenic fibrosing alveolitis. *Histopathology* 22:343–347, 1993.

125. Rinaldo JE, Christman JW: Mechanisms and mediators of the adult respiratory distress syndrome. *Clin Chest Med* 11:621–632, 1990.

126. Matalon S, Holm BA, Notter RH: Mitigation of pulmonary hyperoxic injury by administration of exogenous surfactant. *J Appl Physiol* 62:756–761, 1987.

127. Seeger W, Stohr G, Wolf HRD, Neuhof H: Alteration of surfactant function due to protein leakage: Special interaction with fibrin monomer. *J Appl Physiol* 58:326–338, 1985.

128. Holm BA, Notter R, Finkelstein JN: Surface property changes from interactions of albumin with a natural lung surfactant and extracted lung lipids. *Chem Phys Lipids* 38:287–298, 1985.

129. Petty TL, Silvers GW, Paul GW, Stanford RE: Abnormalities in lung elastic properties and surfactant function in adult respiratory distress syndrome. *Chest* 75:571–574, 1979.

130. Pison U, Seeger W, Buchhorn R, et al: Surfactant abnormalities in patients with respiratory failure after multiple trauma. *Am Rev Respir Dis* 140:1033–1039, 1989.

131. Lewis JF, Jobe AH: State of the art: Surfactant and the adult respiratory distress syndrome. *Am Rev Respir Dis* 147:218–233, 1993.

132. Hopewell PE, Murray JF: The adult respiratory distress syndrome. *Am Rev Med* 27:343–356, 1976.

Chapter *12*

Fibrosing Alveolitis

P. S. Hasleton

INTRODUCTION

Fibrosing alveolitis is also termed *interstitial pulmonary fibrosis, cryptogenic pulmonary fibrosis* (if there is no known cause), or *honeycomb lung.* Progress in understanding the disease has been hampered by many factors,[1] including the perceived relative rarity and the spectrum of clinical activity. The disease may have a rapidly progressive course, or lasting many years. Different treatment methods may have been used, making comparison of groups difficult. In many cases, the disease presents as end-stage fibrosis, making identification of etiologic factors almost impossible. There is no ideal animal model for the disease, making research into the early phases difficult.

The incidence of the disease is estimated at one case per 3000 to 4000 of the population in Britain,[2] and every year it accounts for 1500 deaths.[3] This figure is rising, and in a prospective study in New Mexico, the estimated incidence was 31.5 per 100,000 in males and 26.1 per 100,000 in females.[4] This incidence may be higher than usual, as mining has been a major industry in New Mexico. It is a salutory fact that despite the increased number of cases, few patients are investigated to confirm the diagnosis.[5]

ETIOLOGIC FACTORS

The key to understanding interstitial lung disease of unknown cause is the concept of alveolitis.[6] As a result of an initial injury, both acute and chronic inflammatory cells accumulate in alveoli, the lumina of small bronchioles, and the walls of small pulmonary arteries and veins. This alveolitis causes injury and subsequent alveolar wall fibrosis.

The histopathologist is often faced with end-stage lung disease. In such cases there is diffuse interstitial and sometimes intraalveolar fibrosis, and determination of a cause is often impossible. This should come as no surprise, since in chronic hepatic or renal disease, cirrhosis or end-stage kidney is not an infrequent pathologic diagnosis. Without a history, it may be impossible to determine a cause in these conditions.

There are more than 130 disorders that can cause interstitial pulmonary fibrosis. Some, such as asbestosis, silicosis, sarcoidosis, or histiocytosis X (Langerhans cell histiocytosis), may be easy to diagnose in the early stages. In many cases the initial insult is unknown, and the term *idiopathic* or *cryptogenic pulmonary fibrosis* is applied.

There have been attempts to separate desquamative interstitial pneumonia[7] as well as Hamman-Rich syndrome from fibrosing alveolitis. Hamman-Rich syndrome is a rapidly progressive interstitial pneumonitis. A recent review has questioned the concept of splitting these fibrosing lung diseases and proposed that they be lumped together under the term *pulmonary fibrosis of unknown origin.*[8]

Some of the causes given in Table 12-1 are taken from Dunnill's article.[8] A fuller list is available in the classic chapter by Davis and Crystal.[9]

Apart from systemic diseases, many etiologic factors cause interstitial fibrosis. Adult respiratory distress syndrome (ARDS) may progress to pulmonary fibrosis. This is important because many of the pathogenetic mechanisms are similar to those in the lungs of patients with ARDS. Diseases such as extrinsic allergic alveolitis, sarcoidosis, histiocytosis X (Langerhans cell histiocytosis), and lymphangioleiomyomatosis, while strictly classified as interstitial lung diseases, will be discussed in their appropriate sections.

CLASSIFICATION

Because there are so many causes of fibrosing alveolitis, any rational classification is almost impossible, since the pathologist is looking at end-stage disease. Liebow gave the first classification of idiopathic pulmonary fibrosis, putting the diseases into

TABLE 12-1

Some Causes of Diffuse Pulmonary Damage Proceeding to Interstitial Fibrosis[a]

Systemic diseases	*Noncytotoxic drugs*
Rheumatoid arthritis	Antibacterial agents
Systemic lupus erythematosus	Nitrofurantoin
Systemic sclerosis	Sulfasalazine
Dermatomyositis	Anticonvulsants
Polymyositis	Hydantoin
Sjögren's syndrome	Antiarrhythmics
Chronic active hepatitis	Amiodarone
Ulcerative colitis	Lidocaine
Celiac disease	Antirheumatics
Renal tubular acidosis	Penicillamine
Atopy[128]	Gold
	Colchicine
Dusts	Narcotics and opiates
Organic	Heroin
(e.g., those associated with extrinsic allergic alveolitis)	Hexamethonium
Inorganic	Others
Coal, asbestos, silica, aluminum, etc.	Procarbazine
	Vinca alkaloids
Inhalants	
Oxygen	*Miscellaneous*
Nitrogen dioxide	Infection
Cadmium	Viral pneumonias
Smoke	Mycoplasma
Mercury vapor	Opportunistic (in immunosuppressed patients)
Sulfur dioxide	Ingested agents
Beryllium fumes	Paraquat
	Verosene
Cytotoxic drugs	Ionizing radiation
Antibiotics	Late-stage ARDS
Bleomycin	Uremia
Mitomycin	Sarcoidosis
Zimostatin	Histiocytosis X (Langerhans cell histiocytosis)
Nitrosoureas	Lymphangioleiomyomatosis
Alkylating agents	Familial
Busulfan	Idiopathic
Cyclophosphamide	Vascular
Chlorambucil	Venoocclusive disease
Melphalan	
Antimetabolites	
Methotrexate	
Azathioprine	
Cytosine arabinoside	

[a] In most patients with diffuse pulmonary fibrosis, there is no established cause.

TABLE 12-2
Classification of Cryptogenic Fibrosing Alveolitis[11]

Acute interstitial pneumonia (Hamman-Rich syndrome)
Usual interstitial pneumonia
Desquamative interstitial pneumonia
Chronic interstitial pneumonia
Giant cell interstitial pneumonia

SOURCE: From Katzenstein and Askin,[11] with permission.

five distinct groups: interstitial pneumonia, desquamative interstitial pneumonia, bronchiolitis obliterans interstitial pneumonia, lymphoid interstitial pneumonia, and giant-cell interstitial pneumonia.[10] This classification is now outdated. Lymphoid interstitial pneumonia is now regarded as a lymphoma, and bronchiolitis obliterans organizing pneumonia as well as respiratory bronchiolitis is considered in the section on bacterial infections.

Katzenstein and Askin's classification[11] (Table 12-2) is now probably the most often used, although, as noted, there is reservation about acute interstitial pneumonia appearing as a separate disease. In addition, granulomatous interstitial pneumonia, excluding sarcoidosis, is not listed, since many of these cases may have the picture of extrinsic allergic alveolitis. Cases of hypersensitivity pneumonitis have been described without an identified allergen.[12,13] Finally, chronic interstitial pneumonia is regarded by most histopathologists as a nonspecific feature and is part of fibrosing alveolitis. It is doubtful that it merits a separate category.

CLINICAL FEATURES

The disease may be seen in children[14,15] from birth to 16 years. Children have the same clinical findings as adults, and the long-term prognosis is poor, with only four of 14 children in one series[14] having normal vital capacity, on follow-up, ranging from 2 to 7 years.

In adults the disease often presents in the late 50s and early 60s, although it has been diagnosed in persons over the age of 80.[16] The sex ratio varies between series, but there is a tendency to a male predominance, sometimes in a ratio of 2:1.

Patients have an insidious onset of exertional dyspnea, which may be overlooked in older, sedentary patients. A dry, nonproductive cough is common.

In the early stages, auscultation reveals end-inspiratory crackles posteriorly over the lower lobes. Clubbing may be identified at presentation or as the disease progresses and is seen in 70 percent of patients. This has been related to the development of bronchopulmonary anastomoses, as well the development of smooth muscle.[17] Central cyanosis and respiratory failure are late manifestations, along with pulmonary hypertension. In cases of fibrosing alveolitis without a connective tissue origin, polyarthralgia may be present.[16]

Physiologically there is a restrictive ventilatory defect, with reduced lung volumes and decreased compliance.

Radiologically there is an infiltrative pattern, which may be confined to the lower lobes or in some cases is diffuse. There is a loss of lung volume (Fig. 12-1). The nature of the radiologic pattern differs, probably according to the disease stage. In some cases there may be a fine "ground glass" appearance. In more advanced cases there is a honeycomb pattern. Chest radiographs may be entirely normal in up to 10 percent of patients with biopsy-proven diffuse lung disease.[18]

CT scanning is more accurate than conventional radiography, since attempts to discriminate between alveolar space and interstitial disease on plain chest radiographs have been discredited.[19] Another advantage of CT scanning is that it may detect other diseases, such as carcinoma (Fig. 12-2), which may arise on a background of interstitial pulmonary fibrosis. There is good correlation between high-resolution CT findings and the degree of disease activity in

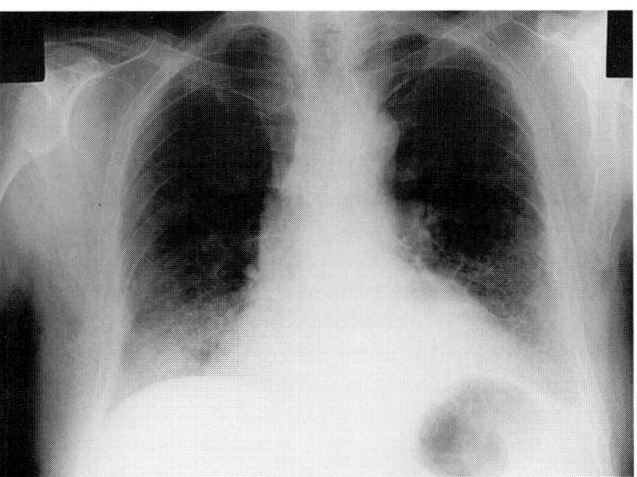

FIGURE 12-1
Chest radiograph with decreased lung volumes and lower zone infiltrates. *(Reproduced by courtesy Dr. A. Horrocks, Wythenshawe Hospital, Manchester.)*

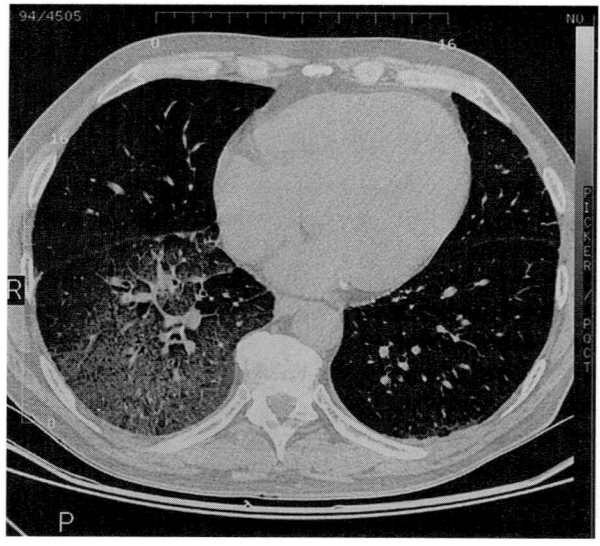

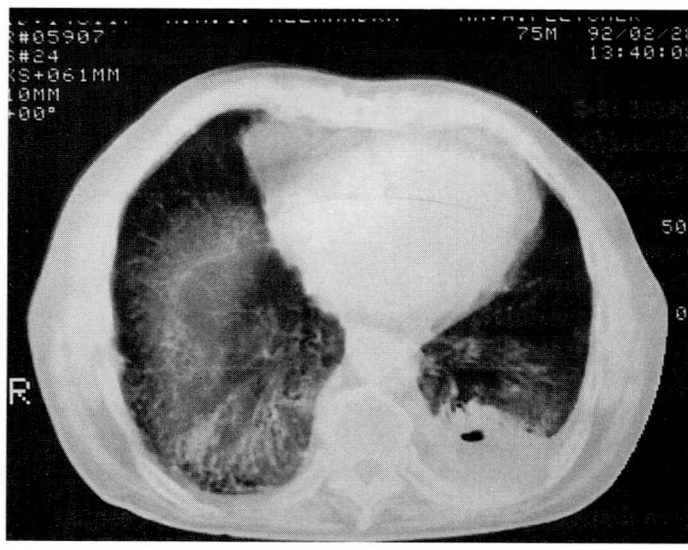

A

B

FIGURE 12-2

A. CT scan of pulmonary fibrosis and prominent fibrosis on the right side. There is some fibrosis posteriorly on the other side. It is not uncommon radiologically to see a difference in the shadowing on CT between the right and left sides on different "cuts."

B. CT scan showing an increase in cellularity in both hemithoraces. However, the left side shows a homogeneous mass, which was a carcinoma. This is eroding a rib. *(Reproduced by courtesy of Dr. A. Horrocks, Wythenshawe Hospital, Manchester.)*

fibrosing alveolitis. An amorphous parenchymal picture represents cellularity in both the air spaces and the interstitium, whereas a reticular pattern is the result of fibrosis.[20,21] Emphysema can be differentiated from fibrosing alveolitis with CT.[22]

The disease tends to progress relentlessly in adults, and the mean duration from first symptom to death is 4 to 5 years. The rate of progression is variable, and about 25 percent of patients survive for more than 10 years. This tendency to progression, despite treatment, is in contrast to other chronic inflammatory diseases of the lung, including asbestosis, organizing pneumonia, and silicosis. In those diseases, after initial lung destruction the condition remains stable over many years.

Approximately 60 percent of patients with fibrosing alveolitis die as a direct consequence of their pulmonary disease. Some die from terminal infection, others from respiratory failure. Nearly one-third of patients develop pulmonary hypertension and right-sided heart failure. There is an increased incidence of lung cancer, with all three main cell types identified. As many patients are seen in later life, some die from unassociated disease.

Fibrosing alveolitis is seen in a wide range of collagen diseases, which will be considered in Chap. 26. In any case of pulmonary fibrosis, it is wise to take an occupational history to exclude such disorders as end-stage extrinsic allergic alveolitis and asbestosis.

MACROSCOPIC FEATURES

The description of the histopathology and electron microscopy given below relates to fibrosing alveolitis

(idiopathic pulmonary fibrosis, cryptogenic pulmonary fibrosis, usual interstitial pneumonia).

Macroscopically the lungs are usually seen only in end-stage disease. The outer surface is sometimes labeled "cirrhotic" and has a hobnail appearance (Fig. 12-3). On the cut surface there are microcysts with intervening fibrosis (Figs. 12-4 to 12-6). These changes are most marked in the lower lobes, but very often the fibrosis spreads in a subpleural manner into the upper lobes. It is important to carefully sample lungs to detect any coexistent carcinoma.

The nodules seen externally correspond to dilated bronchioles and alveolar ducts surrounded by fibrosis. These cause the microcysts. With time the cyst formation extends into the upper lobes. Since patients with cryptogenic fibrosing alveolitis are often in their 60s and some have been smokers, it is not unusual to see a mixture of fibrosis and pulmonary emphysema (Fig. 12-7). In the pathologic setting the differentiation of the two diseases usually causes few problems, since emphysema is not surrounded by fibrosis and does not have microcysts.

In end-stage fibrosing alveolitis there is right ventricular hypertrophy.[23] The mean right ventricular weight, using the method of Fulton and colleagues,[24] is 88.8 g (upper limit of normal: 65 g).

HISTOLOGIC FEATURES

There are several histologic forms of idiopathic pulmonary fibrosis. They consist of a desquamative pattern as well as acute or exudative, proliferative, and fibrotic phases.

FIGURE 12-3
Pleural surface of a case of cryptogenic fibrosing alveolitis showing a nodular ("cirrhotic") pleural surface.

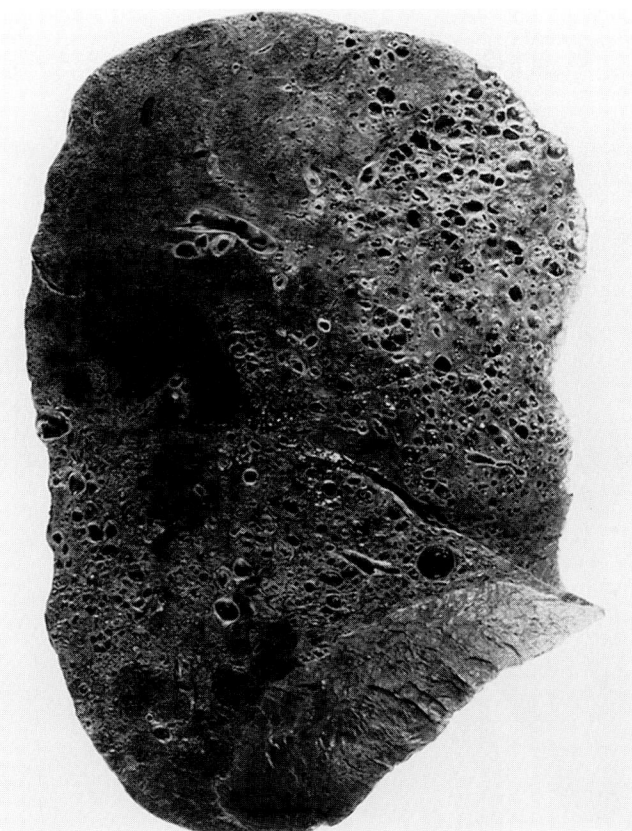

FIGURE 12-4
Cut surface of lung with fibrosing alveolitis. On the anterior border of the upper lobe as well as in the lower lobe, there is evidence of well-marked cystic change (see Fig. 12-5).

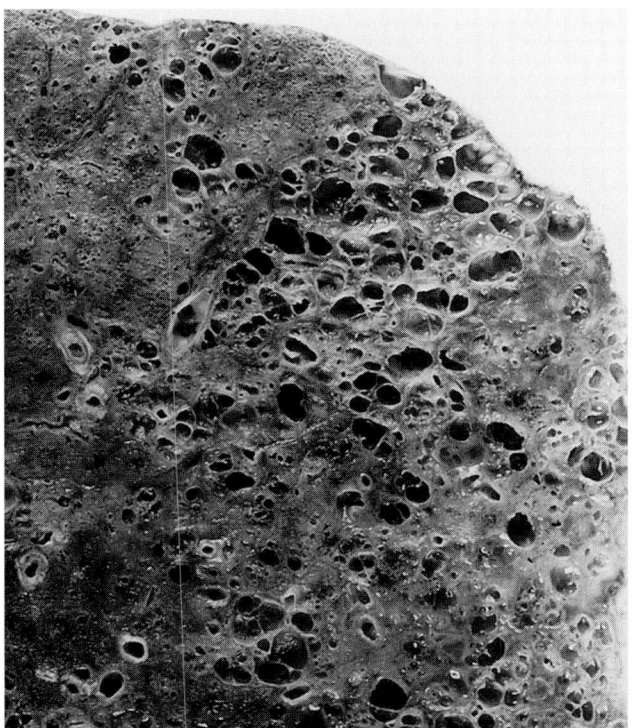

FIGURE 12-5
Part of the upper lobe of a case of fibrosing alveolitis with discrete cyst formation. Well-marked pale fibrous tissue is seen along one border.

Desquamative Pattern

The desquamative pattern of interstitial pneumonia is thought to be an early form of cryptogenic fibrosing alveolitis.[25–28] However, Carrington and coworkers[29] and Katzenstein and Askin[30] believe that this is a different entity from usual interstitial pneumonia or cryptogenic fibrosing alveolitis. This desquamative pattern may occur with asbestos[31] as well as in the region of pulmonary nodules.[32]

In the desquamative phase of the disease, macrophages fill alveoli (Figs. 12-8 to 12-10). These cells have bland, oval nuclei and abundant cytoplasm with small brown granules. The granules are negative with Perls' stain but PAS-positive. There is an increase in thickness in alveolar septa as demonstrated by reticulin, along with a lymphocytic and mast cell infiltrate. Little fibrosis, either within the airways or in the interstitium, is present. No hyaline membranes are present at this stage.

Ultrastructurally the cells are predominantly macrophages.[33] Some have suggested that most of

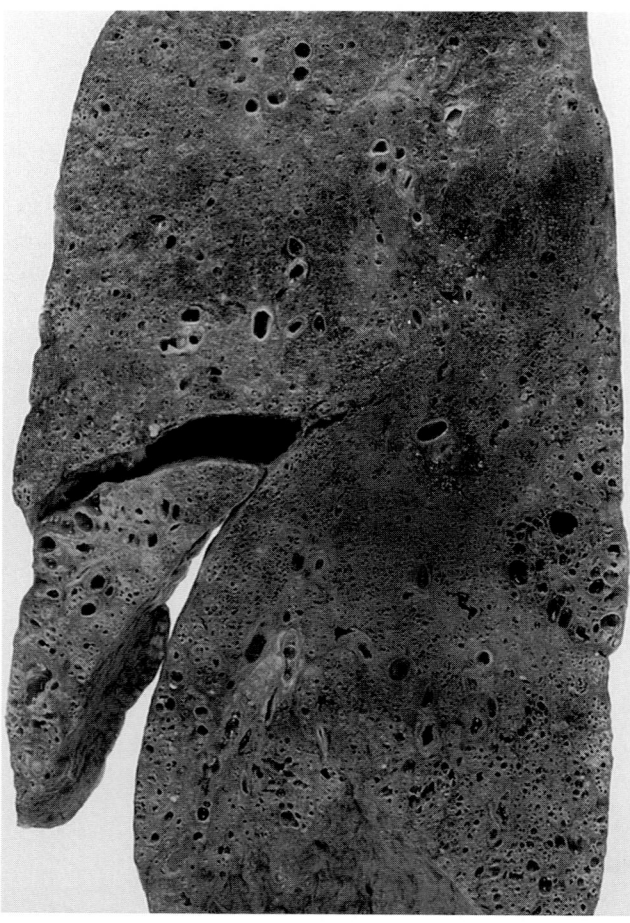

FIGURE 12-6
Part of a right lung affected by interstitial pulmonary fibrosis. Foci of confluent fibrosis are seen in the right middle lobe as well as the upper lobe.

these cells are type II pneumocytes,[34] but this view is no longer held by most authors.

Desquamative interstitial pneumonia (DIP) may be associated with other inorganic particles.[35] Katzenstein and Askin[30] separate DIP from a DIP-like reaction. In the former the macrophages do not form tightly clustered nests and tend not to distend the alveolar lumina.

Prominent type II pneumocytes line the alveolar walls. Gardener and Uff[36] described blue bodies (see Chap. 25), which are usually intracellular, inside macrophages or surrounded by them. They are pale blue or gray, laminated, and Perls' positive. These bodies are nonspecific and may be found in any disorder in which alveolar macrophages aggregate.

Evolution of Pulmonary Fibrosis

The evolution of pulmonary fibrosis through the acute or exudative to the final fibrotic stage has been charted by Fukuda and colleagues.[37,38]

THE ACUTE PHASE

The acute phase closely resembles the exudative phase of ARDS. There is intraalveolar edema and hyaline membranes. Ultrastructurally there is type I pneumocyte necrosis, leaving a bare basement membrane. It is unusual to find cases in this early stage of the process.

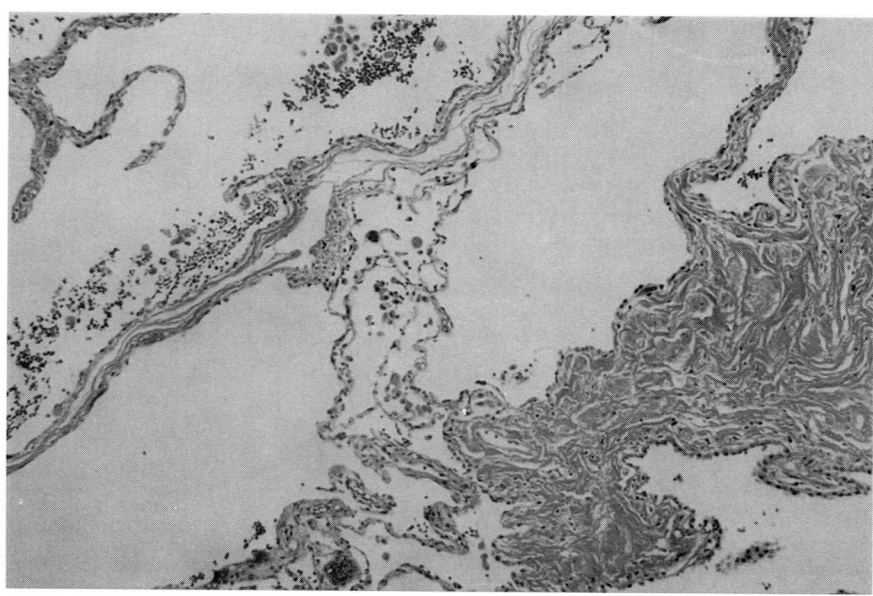

FIGURE 12-7
Histologic picture of an area of emphysema and to the right a focus of interstitial pulmonary fibrosis lined with cuboidal epithelium. (H&E, ×55.)

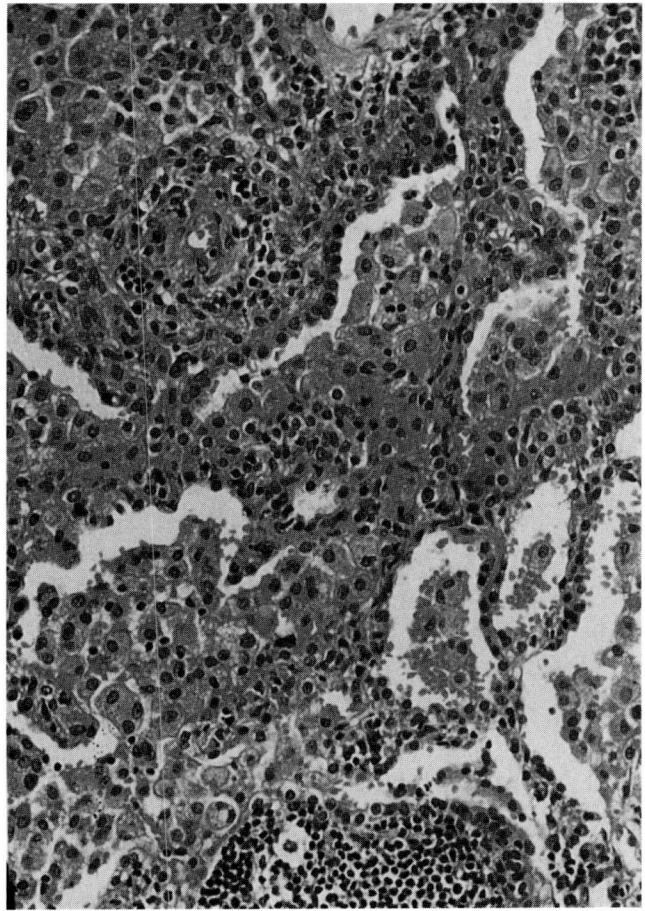

FIGURE 12-8
Desquamitive interstitial pneumonia with many intraalveolar histiocytes and only mild thickening of the alveolar wall. (H&E, ×approx 313.)

PROLIFERATIVE STAGE

In this phase there is fibroblastic proliferation, which is both intraalveolar and interstitial (Fig. 12-11). The fibrosis not only is centered around alveoli but also involves alveolar ducts and respiratory bronchioles (Fig. 12-12). It is often impossible to separate interstitial from intraalveolar fibrosis in some areas of lung tissue.

There is cuboidalization of the epithelium (Fig. 12-13) due to proliferation of type II pneumocytes, and hyaline membranes (Fig. 12-14) are still prominent. Foci of squamous metaplasia are seen (Fig. 12-15). Other metaplastic epithelium may be mucinous (Fig. 12-16) or ciliated (Fig. 12-17).

In a few cases there is metaplastic bone (Fig. 12-18), which has presumably arisen as a metaplastic change in the fibrous tissue.

Immunocytochemically, tumor necrosis factor may be identified in these type II cells.[39] Transforming growth factor-β 1 can be demonstrated immunocytochemically in alveolar macrophages, bronchial epithelium, and hyperplastic type II cells.[40]

In the early stages of pulmonary fibrosis, the endocrine cells are normal in appearance but few in number. No neuroendocrine cells have been identified in honeycomb lung even with intact, well-preserved epithelium. The loss of pulmonary endocrine cells is thought to reflect the generalized epithelial

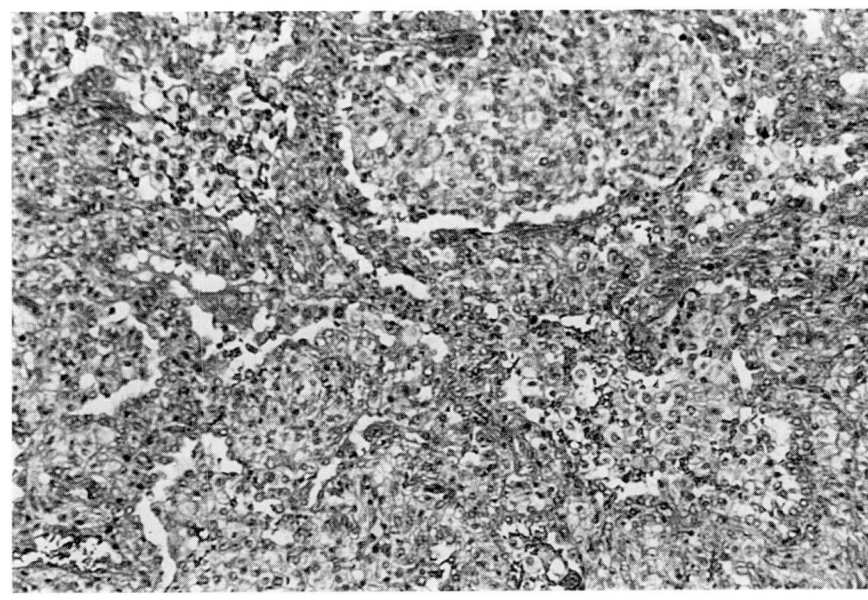

FIGURE 12-9
Desquamative interstitial pneumonia (DIP): A low-power picture showing alveoli distended with macrophages. It is difficult to discern the alveolar walls. (H&E, ×88.)

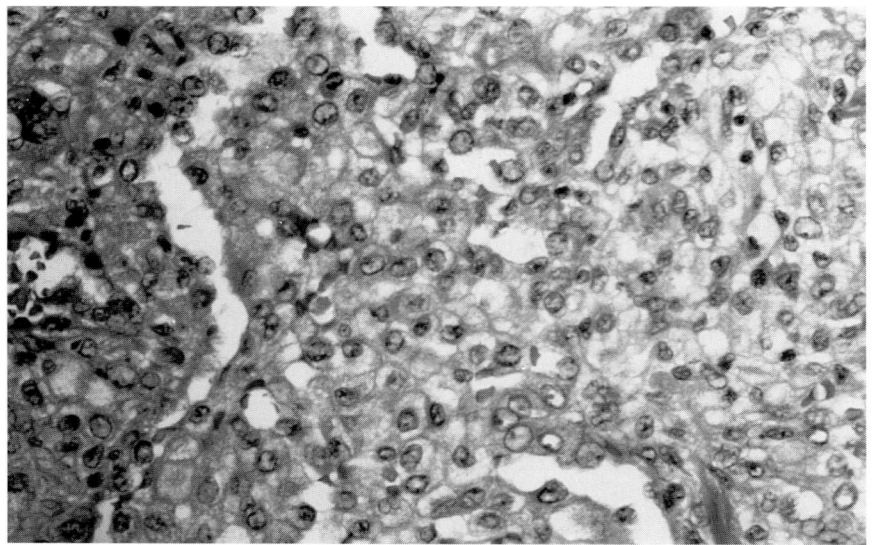

FIGURE 12-10
High-power view of DIP showing many macrophages with bland nuclei along with some cells having clear cytoplasm. (H&E, ×219.)

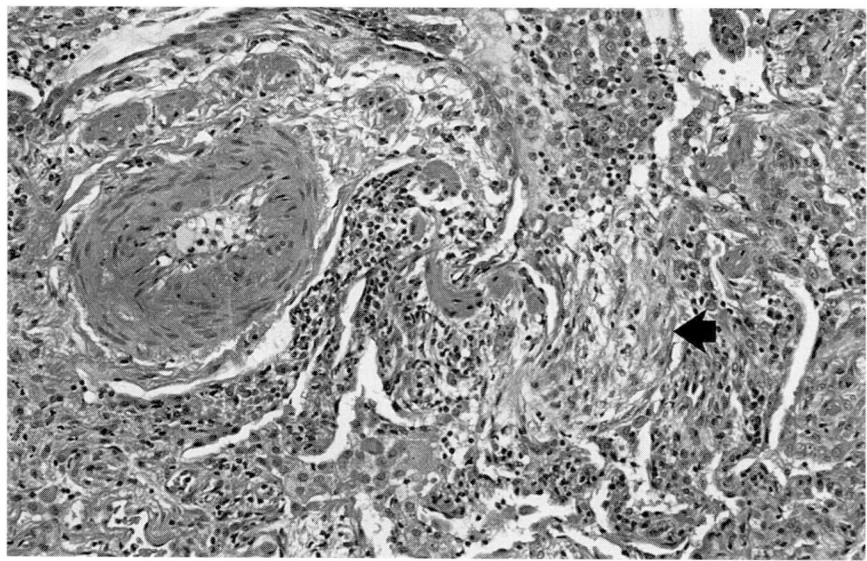

FIGURE 12-11
Cryptogenic fibrosing alveolitis showing a thick-walled pulmonary artery. There is evidence of intraalveolar fibrosis (*arrow*) as well as interstitial fibrosis. (H&E, ×88.)

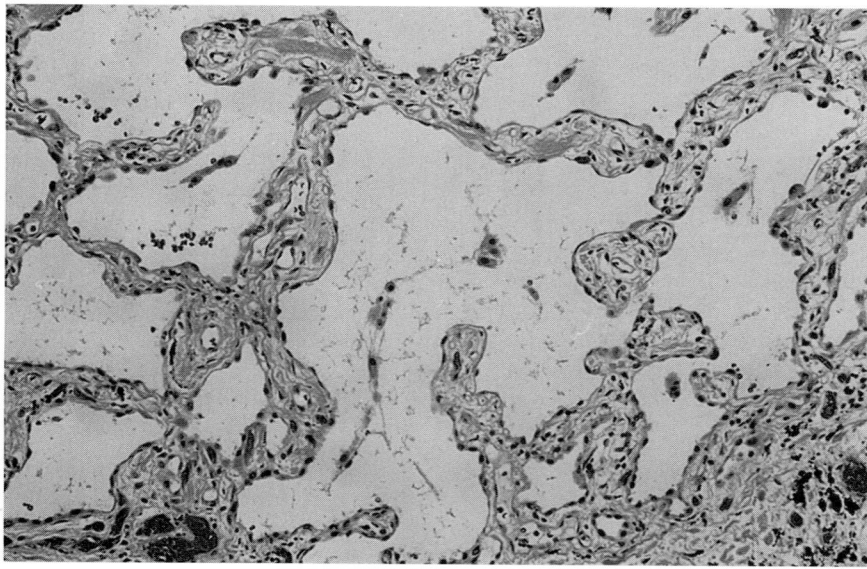

FIGURE 12-12
In this case of fibrosing alveolitis, there is some preservation of the alveolar walls and alveolar ducts. These are thickened by relatively acellular fibrosis, along with some early smooth muscle proliferation. (H&E, ×88.)

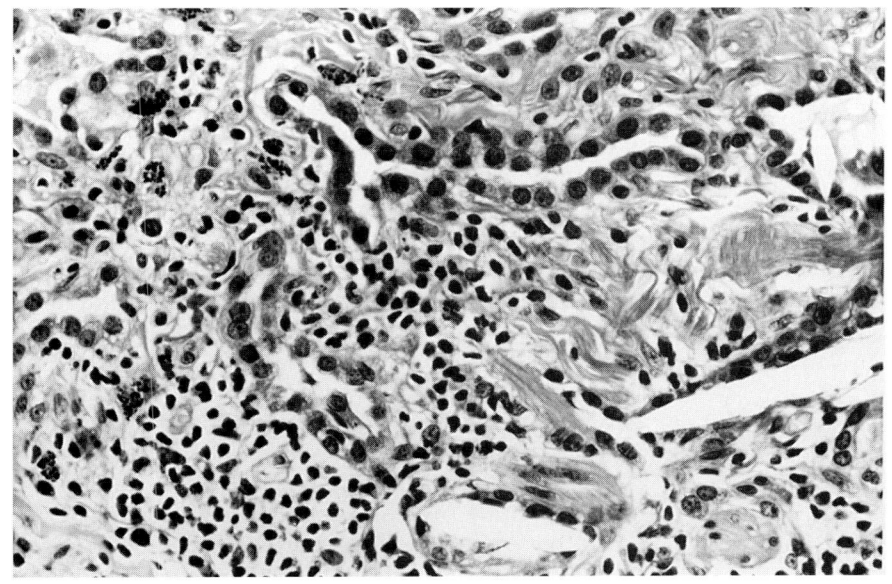

FIGURE 12-13
Cuboidalized-lining epithelium, with prominent nuclei with well-marked interstitial fibrosis. (H&E, ×219.)

damage.[41] However, some cases show tumorlet formation (Fig. 12-19).

Fukuda and colleagues[38] showed that the important cell at this stage of the disease is the myofibroblast. This cell, normally located in the interstitium (Fig. 12-20), migrates through gaps in the epithelial basement membranes into the alveolar lumina.

In the interstitium there is a prominent cellular infiltrate composed of neutrophils, lymphocytes, plasma cells, eosinophils, and macrophages (Fig. 12-21).

STAGE OF FIBROSIS AND REMODELING

There has to be considerable architectural destruction of the lung parenchyma and reorganization before the final fibrosis and remodeling occur. The airspaces are of irregular shape, with thickened, edematous, fibrotic walls (Fig. 12-22). There is much intraalveolar fibrosis as well as intrabronchiolar scar tissue.

In addition to fibrosis, there is proliferation of smooth muscle in the interstitium (Fig. 12-23). The incidence of clubbing is significant in patients with lesser degrees of honeycombing and higher grades of smooth muscle proliferation. This smooth muscle proliferation is also correlated with the duration of symptoms and the extent of radiologic pulmonary infiltrates. Patients with most marked smooth muscle proliferation have the longest duration of symptoms.[42]

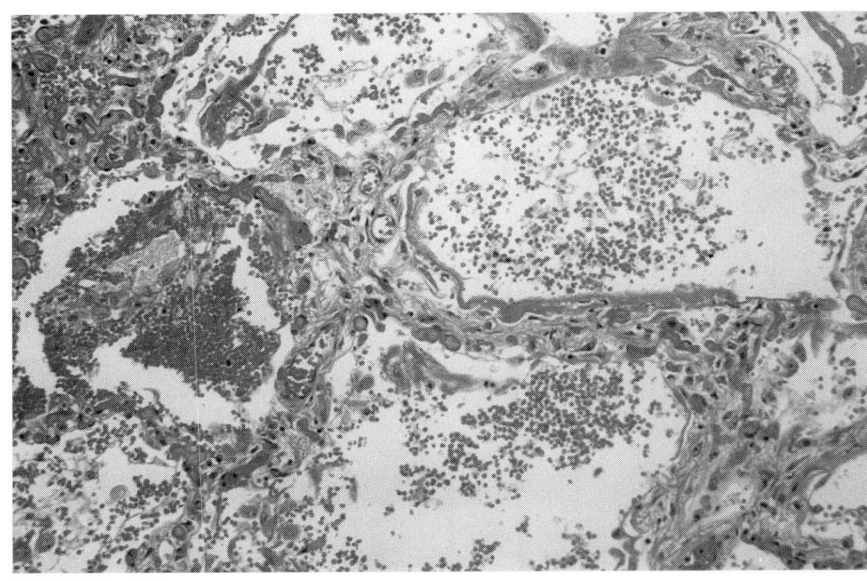

FIGURE 12-14
Cryptogenic pulmonary fibrosis with some prominent hyaline membranes. (H&E, ×219.)

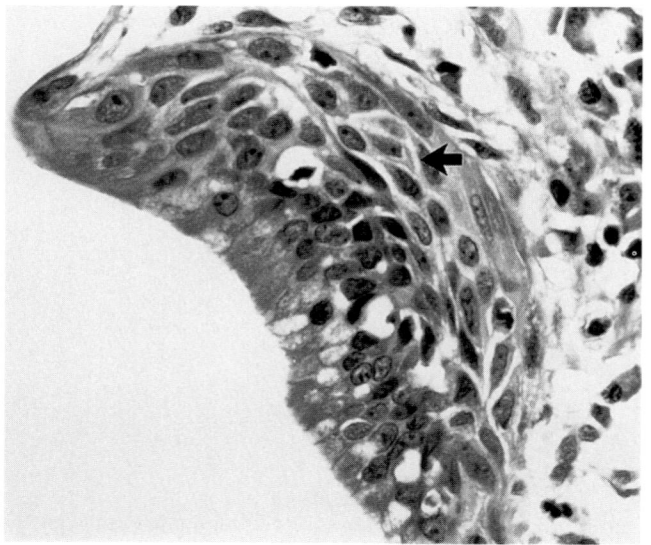

FIGURE 12-15
A focus of squamous metaplasia with intercellular bridges (*arrow*) and some pseudostratified epithelium in fibrosing alveolitis. (H&E, ×340.)

Basset and coworkers[43] described three patterns of alveolar fibrosis. The first consisted of intraluminal buds, partly filling alveoli, alveolar ducts, and respiratory bronchioles. The second pattern showed obliterative change, with total occlusion of airspaces by connective tissue. In the last pattern, exudate and connective tissue were identified inside alveoli, became epithelialized on their surface, and then were incorporated into the wall. Despite describing these three patterns, the authors noted that there was a common pathogenesis and that these different pictures did not imply different causes.

The initial concept that idiopathic pulmonary fibrosis is due to alveolitis should not be construed to mean that only alveoli are involved. In 67 percent of a small series (18 patients), there was narrowing of small bronchioles as well as some bronchiolitis.[44]

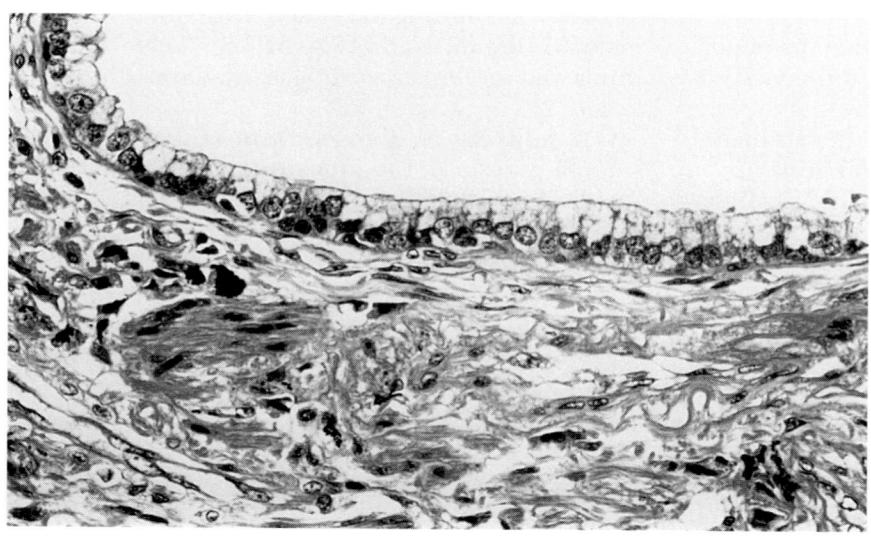

FIGURE 12-16
A focus of mucinous metaplasia with underlying fibrosis and smooth muscle proliferation in fibrosing alveolitis. (H&E, ×219.)

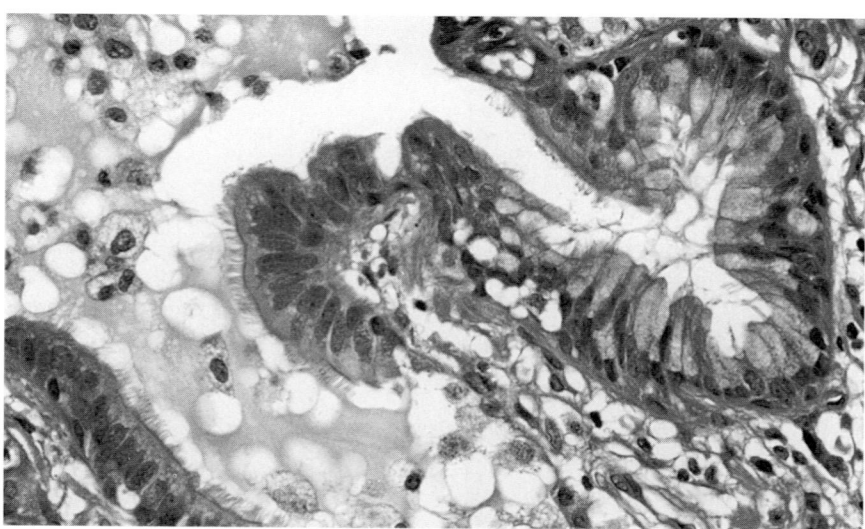

FIGURE 12-17
A well-defined area of ciliated epithelium with some underlying interstitial fibrosis. There is intraalveolar edema and a foamy histiocytes in this case of cryptogenic fibrosing alveolitis. (H&E, ×219.)

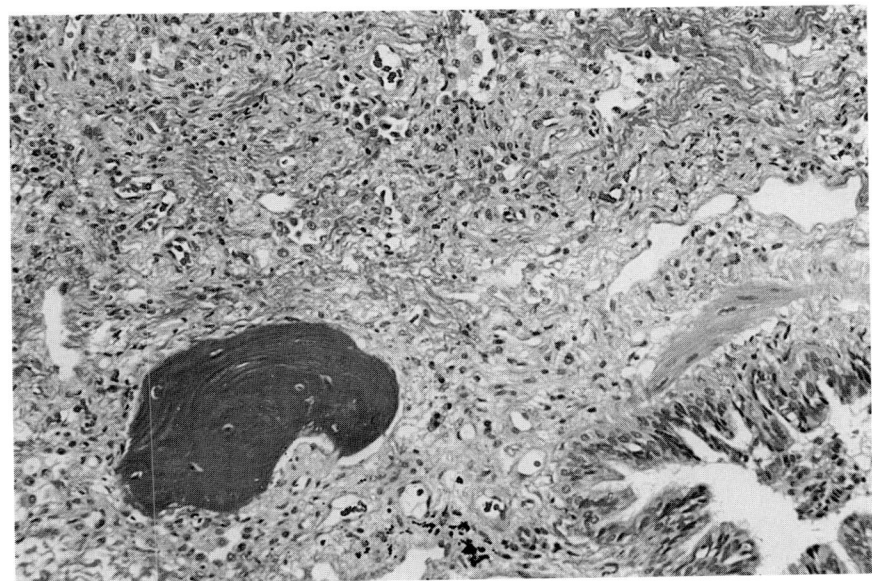

FIGURE 12-18
A focus of osseous metaplasia set in a background of interstitial and almost confluent fibrosis. (H&E, ×88.)

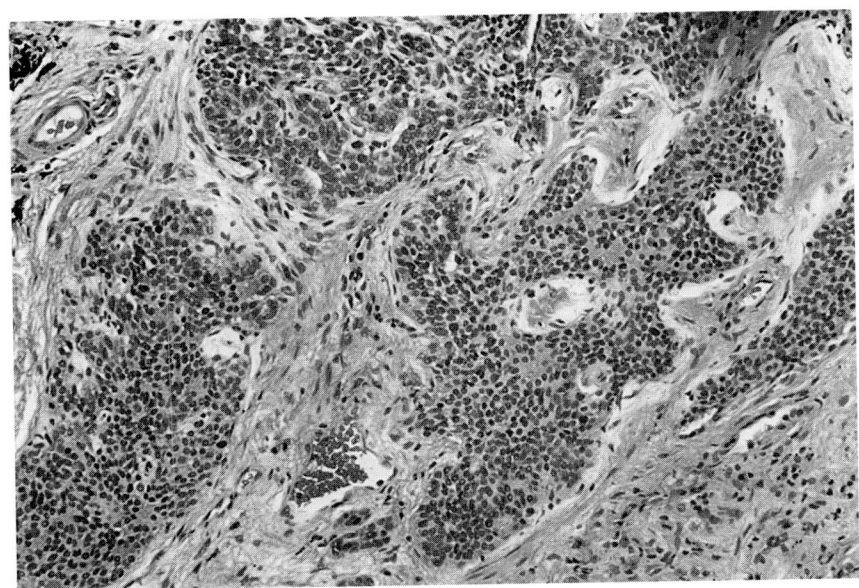

FIGURE 12-19
Prominent tumorlet consisting of small hyperchromatic cells set amidst a background of fibrosis. (H&E, ×88.)

Comparisons between the morphologic and physiologic data showed a significant correlation between the results of dynamic compliance and the overall estimate of small-airways diameter. The fibrosis was predominantly peribronchiolar; bronchiolitis was seen in four cases and peribronchiolar inflammation in another eight cases.

Mast cells are markedly increased in lung biopsies from patients with idiopathic pulmonary fibrosis. These cells are usually difficult to define on tissue that has undergone formalin fixation because of loss of their metachromatic staining. They are present directly adjacent to the lumina of small airways.[45] Mast cells in cases of idiopathic pulmonary fibrosis have an altered appearance, with irregularity of the plasma membrane and release of extracellular tryptase. The presence of mast cells is consistent with their association with fibrosis.

VASCULAR CHANGES

The pulmonary blood vessels react to the changes in the surrounding lung parenchyma—muscularization

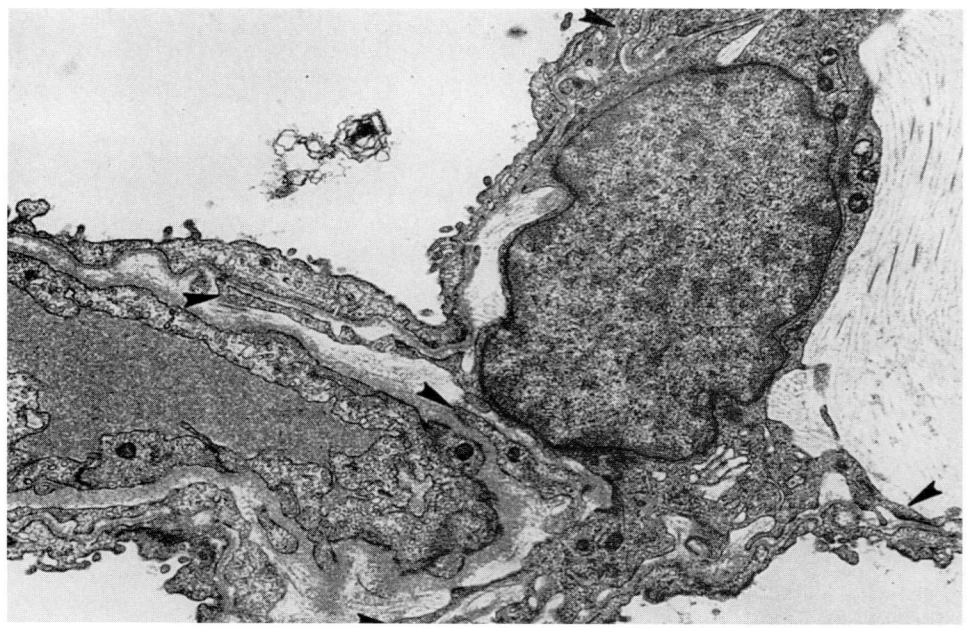

FIGURE 12-20
Electron micrograph of a myofibroblast in aleovar interstitium. The myofibroblast has cytoplasmic processes which extend through the alveolar septum and around a capillary (*arrowheads*). (EM, ×5,920.)

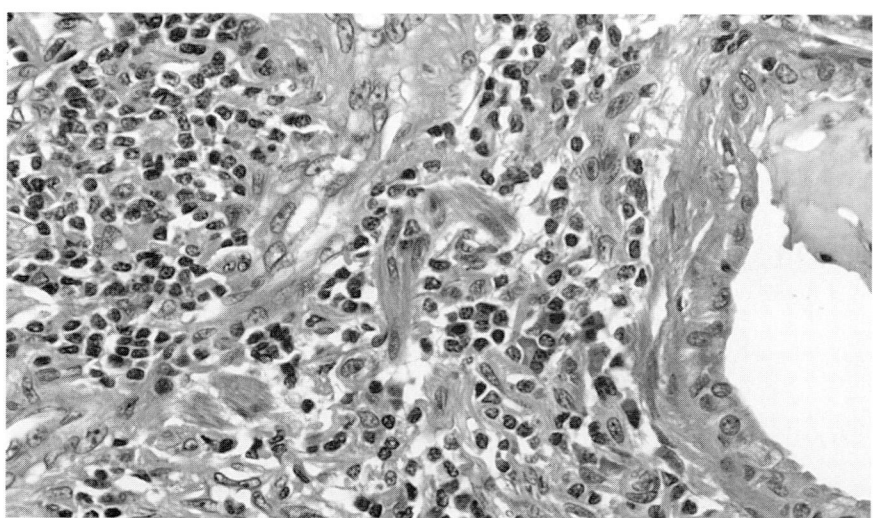

FIGURE 12-21
Marked chronic interstitial infiltrate consisting of lymphocytes and plasma cells set amid some smooth muscle and fibrosis. Cuboidal lining epithelium is seen to the right of the picture. (H&E, ×219.)

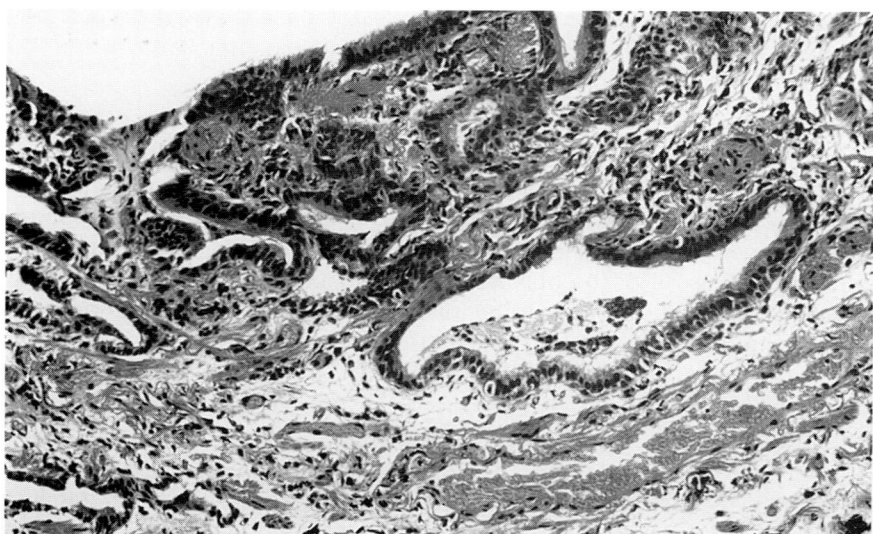

FIGURE 12-22
End-stage fibrosis with a lining cuboidal and columnar epithelium with underlying fibrosis as well as edema. There is little chronic inflammation in the thickened alveolar walls. There is distortion of the air spaces. (H&E, ×88.)

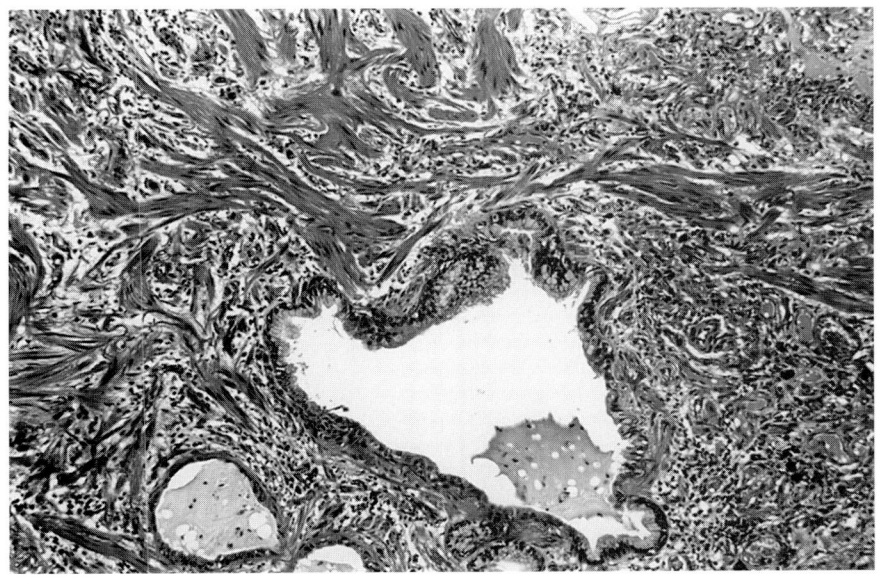

FIGURE 12-23
Marked proliferation of smooth muscle amid some chronic inflammation and fibrosis in cryptogenic fibrosing alveolitis. (H&E, ×88.)

of pulmonary arterioles (Fig. 12-24) and marked intimal fibrosis in pulmonary arterioles and muscular pulmonary arteries. In some cases there is total fibrous obliteration of vascular lumina as well as recanalization. The latter vessels show evidence of medial hypertrophy (Fig. 12-25), but there are no dilatation lesions or evidence of marked pulmonary hemosiderosis, as would be seen in congenital cardiac shunts. Intimal longitudinal muscle is identified in pulmonary arterioles and muscular pulmonary arteries.

Pulmonary veins show marked vascular occlusion by intimal fibrosis (Fig. 12-26). In some cases there is evidence of arteriolization of the vein wall

(Fig. 12-27).[46] There may be thin-walled vessels, possibly representing bronchopulmonary anastomoses (Fig. 12-28).

Quantitative Assessment of the Degree of Pulmonary Fibrosis

To quantify the cellular and fibrotic changes, reproducible quantitative parameters should be used. Ashcroft and coworkers[47] used a scoring system for fibrosis from 0 to 8. This was found to be easily applicable and reproducible, but inflammation and other pathologic changes were excluded from this system.

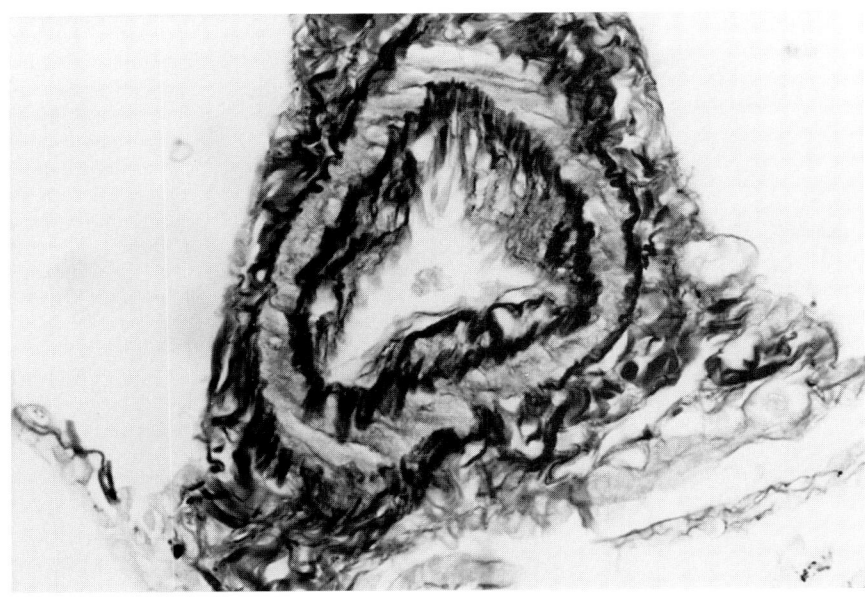

FIGURE 12-24
Muscularized pulmonary arteriole showing well-marked internal and external elastic lamina as well as some intimal fibrosis and intimal longitudinal muscle. (EVG, ×350.)

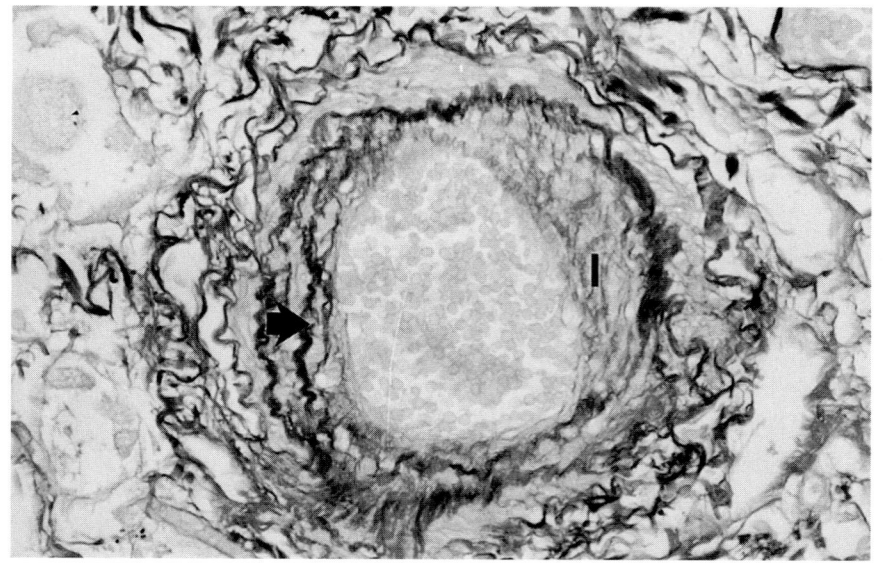

FIGURE 12-25
Medial hypertrophy in a muscular pulmonary artery showing evidence of some intimal longitudinal muscle (*arrow*) as well as intimal fibrosis (I). (EVG, ×219.)

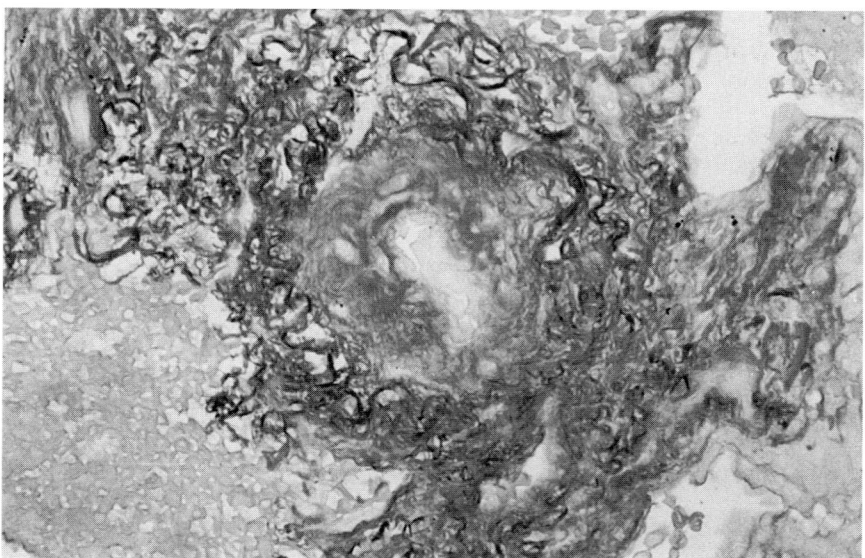

FIGURE 12-26
Pulmonary vein shows severe occlusion by old intimal fibrosis. There is formation of a small vascular lumen, which appears to be lined by smooth muscle. (EVG, ×219.)

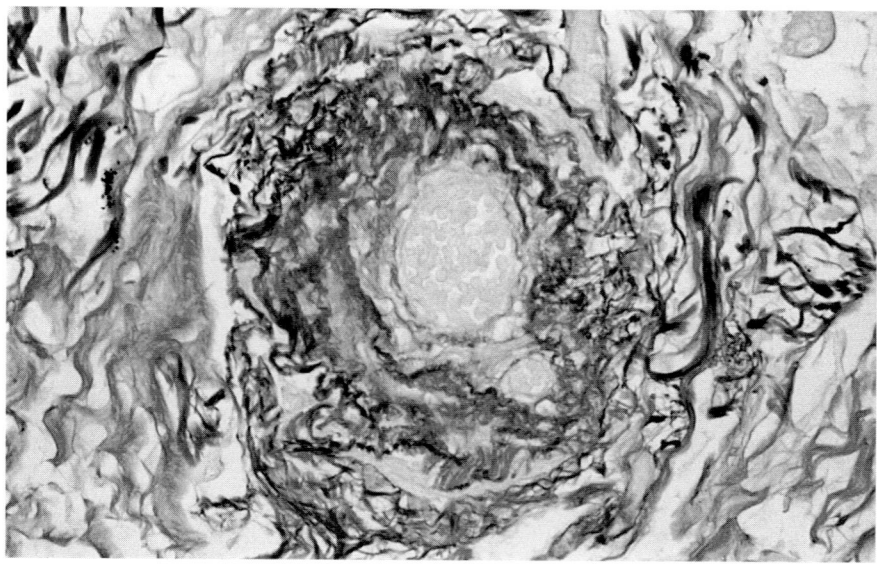

FIGURE 12-27
Early muscularization of a pulmonary vein along with intimal fibrosis. (EVG, ×219.)

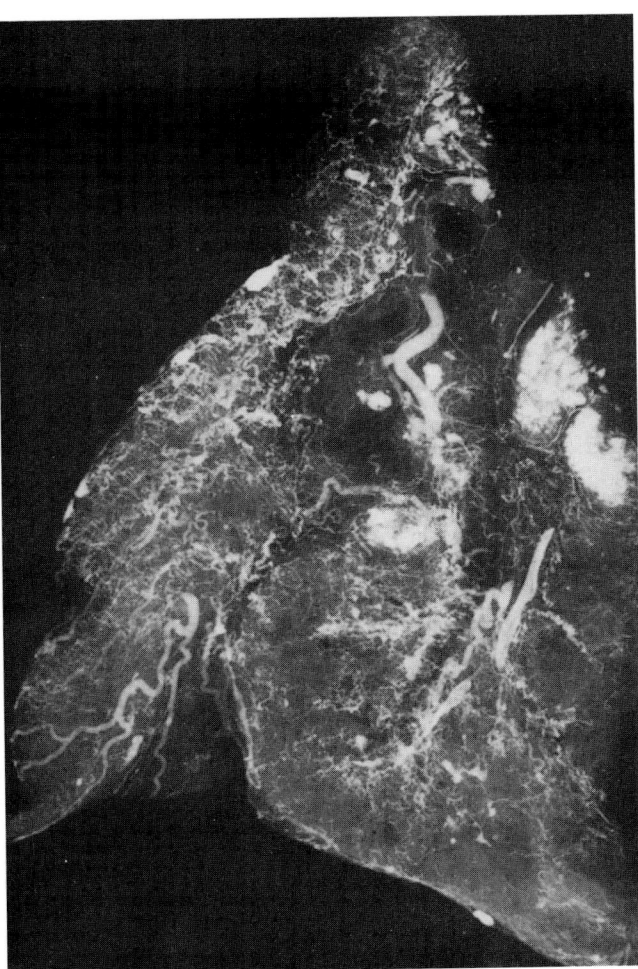

FIGURE 12-28
Bronchopulmonary anastomoses.

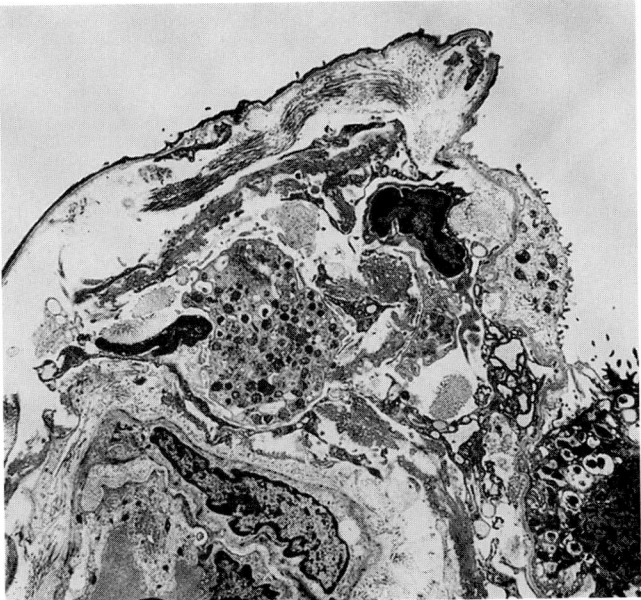

FIGURE 12-29
Electron micrograph showing loss of types I and II cells, with a thin layer of fibrin covering part of the alveolar wall. A type II cell is seen on the right of the picture. There is edema of the wall as well as a mast cell. (Lead citrate and uranyl acetate, ×2960.)

Electron Microscopy

The earliest lesion is probably interstitial edema. There is then loss of types I and II cells (Fig. 12-29), focal to extensive denudation of alveolar epithelial basement membranes (Fig. 12-30), cytoplasmic edema, sloughing, and edema of capillary endothe-

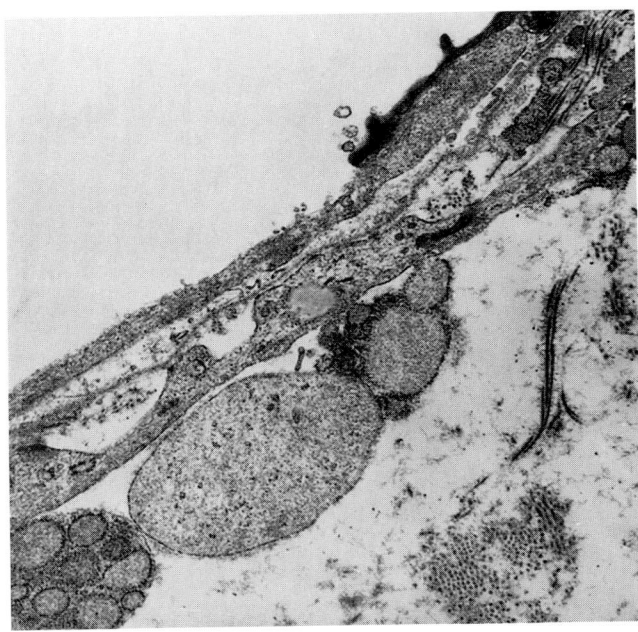

FIGURE 12-30
Higher-power view of the alveolar wall with loss of epithelial cells, fibrin on the right, and much subepithelial edema. (EM, ×11,200.)

More recently, Cherniack's team[48] described a semiquantitative assessment by four pathologists of open lung biopsy tissue that had been gently inflated through the pleura with formalin with the aid of a syringe and needle. Fifteen features were examined, and an overall assessment was based on cellularity (inflammatory cells) and fibrosis. Also noted were the inflammatory or exudative changes, the total number of lymphoid aggregates, cellularity, and the presence of granulation tissue in alveolar spaces, alveolar ducts, and respiratory bronchioles. The presence or absence of respiratory and terminal bronchioles was noted, and the degree of mural inflammation and luminal granulation tissue was also quantitated. Mural fibrosis in the airways was also noted. There was good agreement between pathologists. The correlations were as follows:

0.6 for fibrosis

0.55 for alveolar wall cellularity

0.51 for alveolar space cellularity

0.48 for connective/granulation tissue.

415

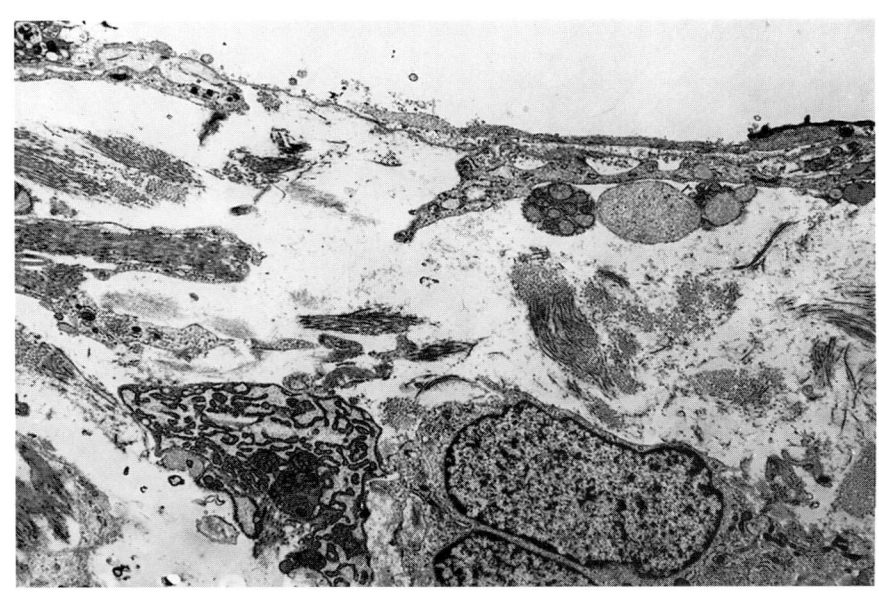

FIGURE 12-31
Fibroblasts, collagen, but no epithelial cells lining the basement membrane. (EM, ×4200.)

lial cells. Cell debris mixed with fibrin, red blood corpuscles, and surfactant-like materials is often layered along the alveolar surfaces; it is most prominent in cases with hyaline membranes. Scattered macrophages, lymphocytes, and plasma cells are present within alveolar lumina. Other inflammatory cells, including lymphocytes, plasma cells, macrophages, fibroblasts, and primitive mesenchymal cells, are seen in the interstitium (Fig. 12-31).[49]

In addition, infolding of portions of alveolar septa or collapse of entire alveoli and permanent apposition of their walls has been described.[49] This process occurs in areas that have lost alveolar lining epithelium. Such a change would seem logical, since surfactant is produced by type II cells and helps to keep alveoli open.

Granular pneumocytes attempt to reepithelialize the denuded basal lamina (Fig. 12-32). They tend to proliferate over the surface of apposed septa, forming a single thickened septum. Intraalveolar exudates are incorporated into the interstitium. Eosinophils and neutrophil polymorphs are less common than the other inflammatory cells.

With increasing duration there is an increased

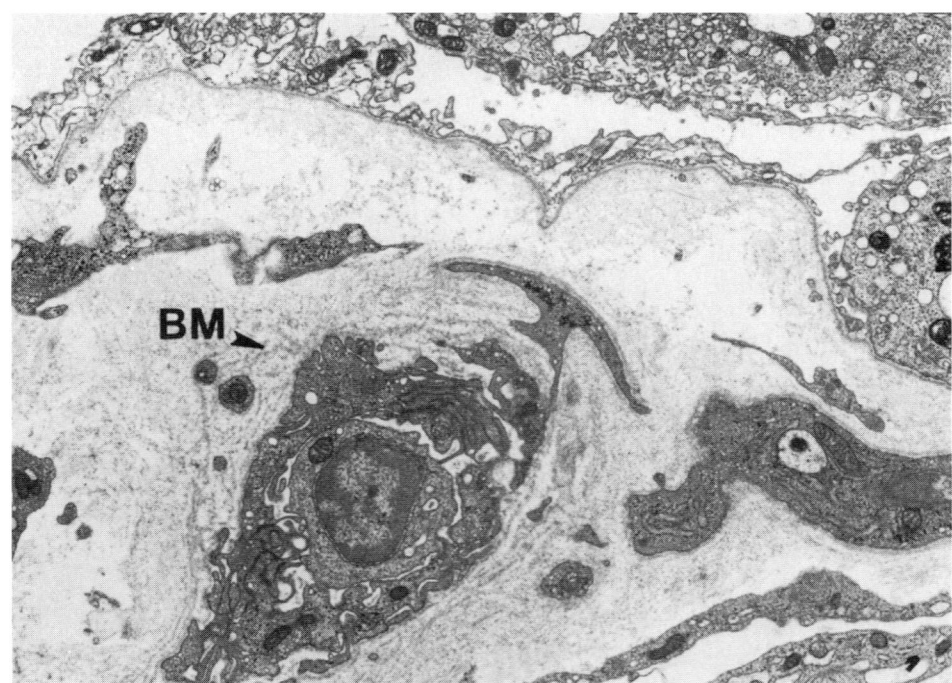

FIGURE 12-32
Damaged type I cell and type II cell with lamellar bodies line the surface of a fibrotic, edematous alveolus in a case of fibrosing alveolitis. Several myofibroblasts are present in the interstitium, close to a capillary with a reduplicated basement membrane (BM). (EM, ×6660.)

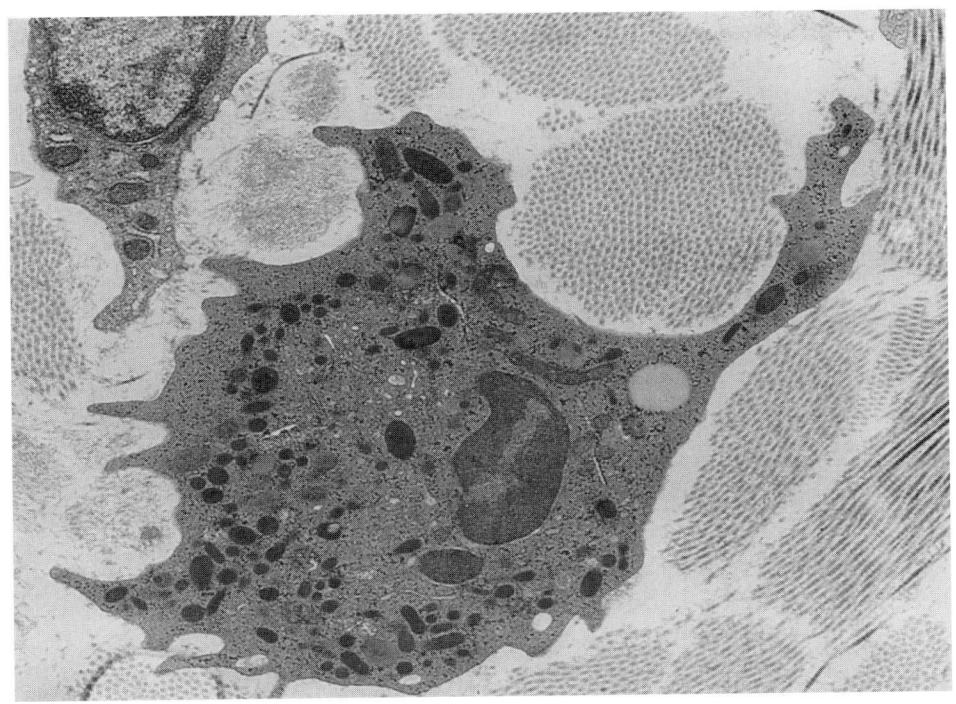

FIGURE 12-33
Polymorph adjacent to collagen fibrils in the interstitium of a patient with cryptogenic fibrosing alveolitis. (EM, ×13,300.)

number of fibroblasts together with collagen fibers and elastin. There is prominent apposition of fibroblasts to mast cells and lymphocytes as well as polymorphs (Fig. 12-33).[50]

Collagen fibers are located in the thick part of the alveolar wall and separate the basement membranes of the capillary and alveolar epithelium in the thinner parts (Fig. 12-34). The intraalveolar accumulation of macrophages, lymphocytes, and some neutrophils or eosinophils, as seen with light microscopy, is confirmed ultrastructurally. There is proliferation of blood vessels that arise from the bronchial circulation.[17]

Bronchoalveolar Lavage

One of the first studies using bronchoalveolar lavage (BAL) employed a rigid bronchoscope.[51] Reynolds and Newball[52] adapted a fiberoptic bronchoscope for use with this technique, which is discussed in

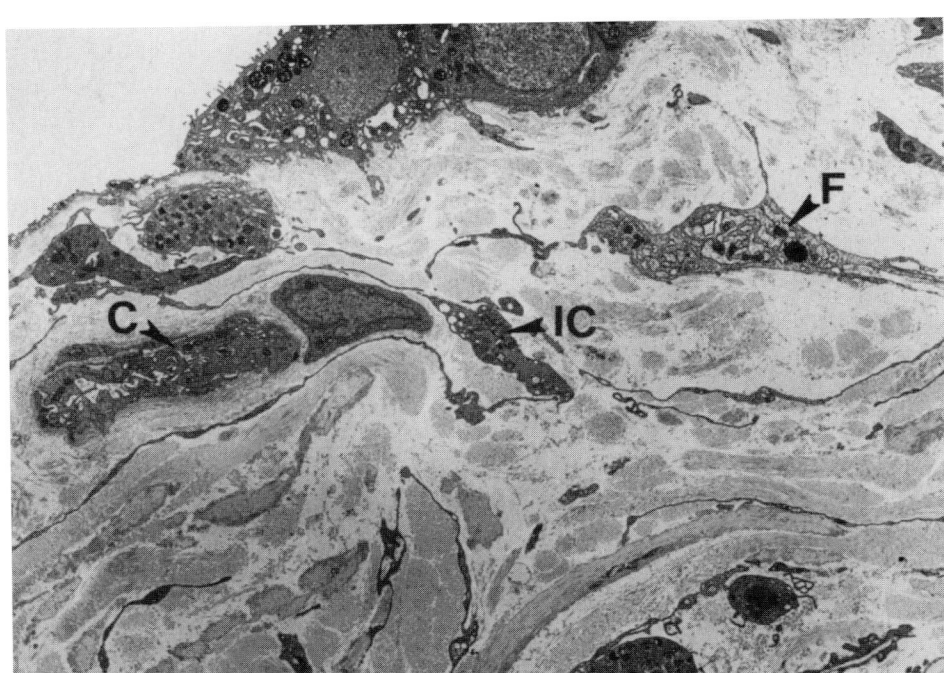

FIGURE 12-34
Thickened alveolar septum in fibrosing alveolitis with much collagen. (C, capillary; IC, interstitial cells; F, fibroblasts.) (EM, ×4100.)

Chap. 16. The basic premise of lavage is that sampling of alveolar space contents accurately reflects the inflammatory and immune mechanisms throughout the lung. The validity of this concept has been partly established by studies comparing both the types and state of activation of the cells recovered in lavage fluid with those seen in open lung biopsies in the same subjects.[53]

There are discrepancies between the relative distribution of immune effector cells when histologic and lavage samples are compared. Plasma cells are common in the interstitium but are usually found only in small numbers in lavage samples or in the intraalveolar spaces. Lymphocytes are predominant in the interstitium but generally found in small numbers in fibrosing alveolitis lavage samples. It is also well recognized that the increased yield of macrophages in lavage samples from patients with fibrosing alveolitis as compared with those from normal controls is far less than might be expected from the histologic appearance. This suggests that many resident macrophages are not sampled by lavage in fibrosing alveolitis.

The increase in neutrophil polymorphs in BAL in fibrosing alveolitis is initially more surprising. They are less common than the other inflammatory cells in fibrosing alveolitis (Fig. 12-35). This discrepancy may be because lavage samples intraalveolar cell populations. Though the number of neutrophils and eosinophils is small in histologic sections, BAL potentially samples a much greater area of lung. The

correlation between neutrophils and eosinophils in histologic and BAL samples is poor for individual patients but good in group analysis.

Chest physicians should still strive to obtain an open lung biopsy for histologic confirmation of the diagnosis and assessment of the stage of the disease. Biopsy confirmation is important because treatment is often with cytotoxic drugs, and response to these agents may be only partial and difficult to assess.

The methods used for BAL differ between laboratories and in the number of lobes lavaged, ranging from one to three. Consequently, the total volume retrieved ranges from 100 to 250 ml.[9]

In Crystal's unit at the National Institutes of Health, the tip of the bronchoscope is wedged into a third- or fourth-order bronchus in the lingula, left lower lobe, or right middle lobe. Sterile isotonic saline is infused in five 20-ml aliquots, and each aliquot is immediately sucked back. For clinical purposes, lavage of one lobe (100-ml total) is adequate. Usually, 40 to 60 percent of the fluid should be recovered. The recovered lavage fluid is then pooled and the total volume measured.

A small aliquot of fluid is taken for a cell count by hemocytometer and for a differential cell count using a Wright-Giemsa stained cytocentrifuge preparation.

The bulk of the lavage is then centrifuged at 600 g for 15 min to separate the cellular and noncellular components. The supernatant can be stored at −70°C for subsequent use. The cells can then be

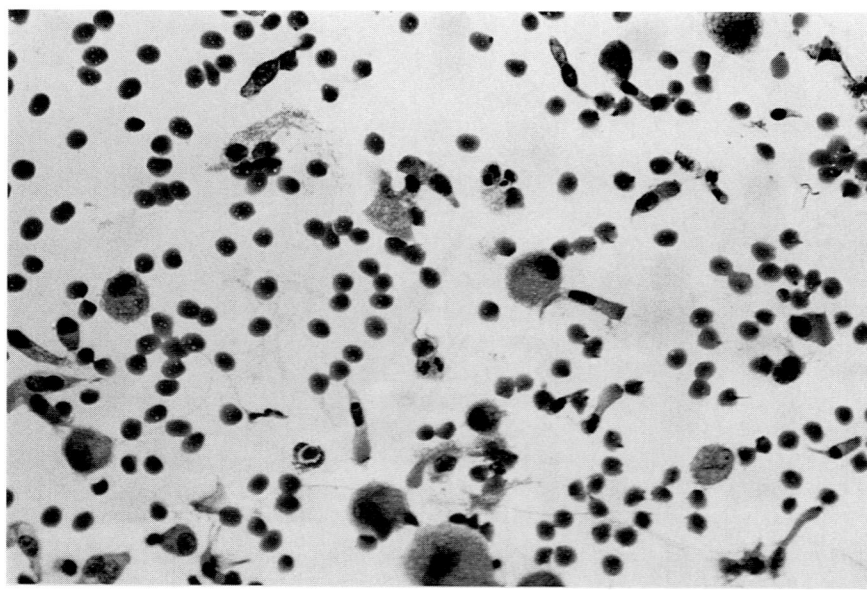

FIGURE 12-35
Lavage in cryptogenic fibrosing alveolitis showing macrophages, detached, ciliated epithelial cells, a few polymorphs, and an increase in lymphocytes. (H&E, ×219.)

suspended in Hank's balanced salt solution without calcium or magnesium for use in various assays.

In nonsmoking adults, the normal lavage count is approximately 10 to 15 × 10⁶ cells per milliliter. The most common cells are alveolar macrophages, which account for more than 90 percent of the population; lymphocytes, which account for 10 percent or less; and neutrophil polymorphs and eosinophils, which constitute less than 1 percent of the recovered cells.

Lower volumes of lavage fluid are recovered from smokers with cryptogenic fibrosing alveolitis, and the fluid contains lower percentages of macrophages but increased percentages of eosinophils and neutrophils. Similar changes are seen in ex-smokers and nonsmokers. There is also an increase in the percentage of lymphocytes when cryptogenic fibrosing alveolitis patients are taken as a group and compared with controls.[53] The only feature distinguishing smokers from nonsmokers with cryptogenic fibrosing alveolitis is the presence of pigmented cytoplasmic inclusions in the macrophages of smokers.

In cryptogenic pulmonary fibrosis, lavage results are not diagnostic but give a guide to prognosis. Patients with eosinophil counts of 3 percent or greater and neutrophil counts of over 4 percent or both these counts along with a lymphocyte count of less than 11 percent have a poor clinical response to corticosteroids.[53] A lymphocyte count greater than 11 percent is associated with a better prognosis.

In systemic sclerosis, which is histologically identical to cryptogenic fibrosing alveolitis, a BAL neutrophilia is associated with extensive pulmonary fibrosis, whereas BAL eosinophilia is often seen in less advanced disease.[54]

The Brompton Hospital group[55] compared BAL inflammatory cell percentage counts with defined clinical categories and related the results to prognosis. At the initial lavage, 97 percent of patients had an increased percentage of at least one cell type—neutrophils in 81 percent, eosinophils in 66 percent, and lymphocytes in 25 percent. There is an inverse correlation between lymphocytes and eosinophils. On follow-up, 12 patients showed a definite and sustained clinical improvement, and their serial lavage counts tended to return to normal. A fall in neutrophils is significant in patients responding to prednisolone. Decreases in eosinophils are significant in patients showing improvement on cyclophosphamide.

In contrast, neutrophil and eosinophil counts remain high in patients who fail to improve. Several nonresponders apparently had stable disease, showing that persisting granulocytes are not necessarily paralleled by clinical deterioration.

BAL SUPERNATANT

Histopathologists tend to concentrate on the cellular component of lavage fluid, and the supernatant is often discarded. This component is rich in factors causing both fibroblast activation and lung injury. Some of these reflect neutrophil, eosinophil, and macrophage damage. Thus TNF-alpha, interleukin-1, proteases, and collagenases are all related to neutrophil damage. Raised concentrations of eosinophil cationic protein are seen in lavage fluids from patients with fibrosing alveolitis.[56]

Mast cells release histamine and tryptase, which have been observed in BAL from patients with fibrosing alveolitis.[57,58] Mast cells also produce some leukotrienes—especially TG4 and PGD 2, which promote vascular leakage.[59]

Besides releasing TNF-alpha and IL-1, macrophages may also release fibronectin, PDGF, and insulin-like growth factor–1.

Platelets may produce TGF-β 1, which stimulates the synthesis of procollagens, fibronectins, and glysoaminoglycans in lung fibroblast cell lines.[60–63]

Albumin is useful in the assessment of interstitial lung disease. Patients with fibrosing alveolitis have increased albumin, which has been strongly correlated with an increase in the lavage lymphocyte count and is unrelated to any age or smoking effect.[64]

PATHOGENETIC MECHANISMS

Many cells compete to be considered as the prime effector cell in fibrosing alveolitis. These will be reviewed. There is some overlap between the descriptions given below and that presented above for ARDS (Chap. 11). However, the references relate mainly to fibrosing alveolitis and thus are given for completeness.

Neutrophil Polymorphs

As noted above, lavage in the early stages shows a neutrophilia, though fewer polymorphs are seen in open lung biopsy tissue. The role of this cell in causing pulmonary damage has been described in the section on ARDS (Chap. 11). Neutrophil polymorphs are capable of releasing enzymes such as collagenase and elastase and generating free oxygen radicals. These enzymes have been detected in lavage fluid in fibrosing alveolitis.[65] Loss of normal fibronectin architecture follows neutrophil degranulation and is more rapid and extensive if endothelium is pretreated with IL-1. Degraded fibronectin is stimulatory for neutrophils and is likely to induce further fibronectin breakdown. It is evident that this sequence has a potential to set up an amplified inflammatory loop with neutrophil-mediated loss of vascular hemostasis.[66]

The mismatch of neutrophils in lavage and lung biopsy specimens may be due to the increased traffic from the capillaries into the airspaces.

Macrophages

The role of macrophages in fibrosing alveolitis is shown by the increased numbers found in both the alveolar spaces and the lung interstitium. These cells recruit neutrophils by means of different neutrophil chemotactic factors[67] as well as IL-8.[68] Monocyte chemoattractant protein–1 has been localized in the cuboidal and flattened metaplastic epithelial cells, alveolar macrophages, and endothelial cells.[69] The mechanism for the action of these cytokines in causing lung damage has been described in ARDS (Chap. 11).

Both macrophages and T-lymphocytes produce interferon-gamma. This cytokine is released in the lungs of a number of patients with fibrosing alveolitis, suggesting that it may contribute to the disease process.[70]

Eosinophils

An increase in eosinophils and neutrophil polymorphs suggests a poor prognosis. Eosinophils release eosinophilic cationic protein, major basic protein, and eosinophilic peroxidase, all of which damage tissues.[71]

Mast Cells

Mast cells are associated with fibrosis and consequently raised levels of histamine, and tryptase may be seen in BAL fluid in fibrosing alveolitis. Heparin derived from mast cells is mitogenic for fibroblasts.[72] Mast cells produce prostaglandins such as PGD2 and LTC4, which promote vascular leakage.[59]

Lymphocytes

Most series on the etiology of fibrosing alveolitis suggest some form of autoimmunity as a causative factor. This was initially based on the fact that many of the collagen diseases present with fibrosing alveolitis. In some cases there may be a lymphocytic and mononuclear interstitial infiltration accompanied by fibrosis. A local B-cell response, sometimes organized into follicles, is regarded as the main reaction.[73] Other authors have shown that the lymphocytic infiltrate in the alveolar walls is composed predominantly of T8 cytotoxic/suppressor cells.[74] These authors also detected class II MHC antigens on alveolar epithelium and demonstrated that both type I and type II pneumocytes stain strongly with anti-1A (HLA-DR). Such staining was absent in control cases. Their findings suggested that there was a "local presentation" of cell antigens and an ensuing local immune response. The nature of the possible autoantigens is discussed below.

Markers of T-cell activation, such as IL-2 and soluble CD8, have been identified in the sera of patients with fibrosing alveolitis.[75] Activated T cells produce interferon-gamma, which is present in lavage fluid from patients with fibrosing alveolitis.[70] Suppressor T cells may have a role in regulating pulmonary fibrosis, as shown by the fact that an increased number of lymphocytes in lavage fluid appears to be associated with good prognosis.

Lavage fluid from patients with fibrosing alveolitis is known to secrete more B-cell growth factor than that from controls.[76] A recent workshop recommended that if a T-cell process was implicated early in the pathogenesis of fibrosing alveolitis, there would need to be a reevaluation of the possible association of this disease with specific HLA classes I and II alleles. The number of cases studied to date is relatively few.[77]

The most likely cellular model is thought to be T lymphocytes that drive the inflammatory process. Macrophages and neutrophils promote the subsequent injury and fibrosis, with eosinophils and mast cells contributing, but to a lesser extent.[77]

Fibroblasts

There are several mechanisms for pulmonary fibroblast activation. These are dealt with by Sheppard and Harrison.[75] They fall into four categories:

1. An increase in fibroblast number, occurring as a result of either fibroblast proliferation or recruitment of fibroblasts from mesenchymal cells. A further mechanism is chemoattraction of cells from adjacent areas.

2. Increased collagen biosynthesis.

3. Diminished collagen regulation, either through a reduction in the rate at which newly synthesized procollagens are degraded or by reduced enzymatic breakdown of collagen in the extracellular matrix.

4. Small clones of fast-growing fibroblasts. The conditions under which such clones emerge and become dominant are uncertain.[78] Alveolar macrophages in patients with fibrosing alveolitis

release fibroblast mitogens when cultured in vitro.[79] The significance of this cell in the fibrotic process, given the ubiquitous nature of pulmonary macrophages, is uncertain.

There are probably two types of signal for cell division: competence and progression factors (Fig. 12-36). These signals are required for mitosis and are thought to occur at different stages of the cell cycle.

Other factors play a role in pulmonary fibrosis. Fibronectin acts as a chemoattractant for fibroblasts and also as a competence factor. Platelet-derived growth factor acts as competence factor for fibroblast proliferation as well as a chemoattractant for mesenchymal cells. IGF-1 stimulates collagen production by pulmonary fibroblasts, as well as acting as a progression factor for fibroblasts.[75]

Laminin is an extracellular matrix glycoprotein active in cellular movement, growth, and differentiation. Laminin fragment P1 is detectable in lavage fluid and is increased in all interstitial lung diseases. There is a positive correlation with lymphocyte counts, BAL albumin, and laminin fragment P1.[80]

Fibroblasts are effector cells in their own right.[81] They actively produce collagenase, gelatinase, and stromelysin—enzymes that cleave collagen or its breakdown products.[82] In addition, they produce fibronectin and an IGF–1-like molecule. They also secrete PGE-2, which can inhibit fibroblast proliferation.[83] They produce IL-1 in response to synergystic actions of IL-1 and TNF-alpha.[84] Also produced is IL-6, which stimulates lymphocyte proliferation.[85] Human lung fibroblasts may regulate the inflammatory reaction by release of granulocytes and macrophage colony–stimulating factor. This enhances in vitro survival of eosinophils.[86] Lung fibroblast replication responds to primary growth-promoting signals released by alveolar macrophages. They are inhibited by PGE-2 and modestly augmented by IL-1.[87]

It has been suggested that expression of viral genes could enhance the promotor activity of genes that play an important role in the initiation or progression of pulmonary fibrosis.[88] Retroviruses and adenoviruses appear able to produce proteins that modulate the expression of genes encoding extracellular matrix proteins, providing a possible mechanism for their participation in the pathogenesis of lung fibrosis. One adenoviral gene[89] can induce type II cells to produce an abnormal type I collagen—not type IV, as is normally manufactured by this cell. Pulmonary fibrosis may be seen in some patients with HIV infection. Recent work suggests that the Epstein-Barr virus may play a role in pulmonary fibrosis.[90,91]

Immunohistochemical studies of the pulmonary fibrosis have previously given inconsistent results, perhaps because biopsies were taken at different stages of the disease. Collagen types I, II, III, and V constitute the vast majority of pulmonary collagen. In interstitial fibrosis there is an increase in the ratio of types I, II, and III collagen.[92] It is now possible to determine procollagen types, and procollagen type III and amino terminal peptide-related antigens have been shown in increased amounts in lavage fluid from patients with fibrosing alveolitis.[93] BAL contains a heterogeneous mixture of type III procollagen *N*-terminal peptides, which are greater in patients with cryptogenic fibrosing alveolitis than in controls.[94] It was concluded that processes other than type III collagen production affect the total concentration of type III procollagen *N*-terminal peptides. The confounding effects of possible abnormal collagen production, as noted above,[89] are now being investigated.

There are also changes in collagen cross-linking in pulmonary fibrosis, with increases in hydroxypyridinium, a reaction product of a difunctional collagen cross-link created by condensation of two hydroxylysine residues. This increase in hydroxypyridinium is probably due to an increase in lysyl hydroxylase activity. This is a key intracellular enzyme responsible for specific posttranslational modification of collagen.[95] The regulation of collagen production in the lung may occur at sites of posttranslational modification to procollagen molecules at the level of procollagen gene transcription or translation of procollagen mRNA. These mechanisms are outside the range of this text.

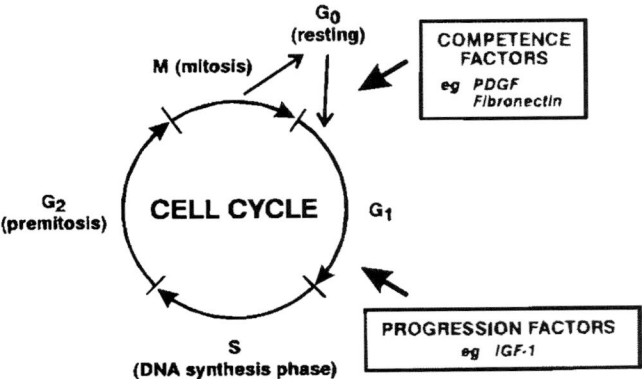

FIGURE 12-36
Cell cycle showing competence and progression factors influencing fibroblast mitosis. (PGDF: platelet-derived growth factor; IGF-1: insulin growth factor–1.) *(Reprinted with permission from MN Sheppard and NK Harrison: Thorax 47:1064–1074, 1992.)*

Other Possible Pathogenetic Mechanisms

DUST EXPOSURE

Patients with fibrosing alveolitis are more likely to report occupational exposure to metal or wood dust, to have worked with cattle, or to have lived in a house heated by a wood fire than are their matched controls. Environmental exposure to dust may be an important factor in the etiology of pulmonary fibrosis. One study used clinical and radiographic features for the definition of fibrosing alveolitis. No pathology was included in the study, and thus it is possible that not all the cases were cryptogenic fibrosing alveolitis; other diseases, such as extrinsic allergic alveolitis, may have been included.[96] Cherniack and colleagues[76] considered that inorganic dusts or drugs could "incite" fibrosing alveolitis in a person with the appropriate genetic susceptibility.

FAMILIAL IDIOPATHIC PULMONARY FIBROSIS

Familial idiopathic pulmonary fibrosis is rare, only approximately 41 families having been recorded in the literature up to 1992.[97–104]

In one study,[104] 17 clinically unaffected members from three families with an autosomal dominant form of idiopathic pulmonary fibrosis were examined for evidence of alveolar inflammation. Just under half of these patients had evidence of alveolar inflammation based on increased neutrophils in the lavage as well as activated macrophages; four of these eight patients had a positive gallium scan. During a follow-up period of up to four years, seven of eight patients with evidence of inflammation had no clinical stigmata of pulmonary fibrosis. It was concluded that alveolar inflammation occurred in almost 50 percent of clinically unaffected family members at risk of inheriting autosomal dominant idiopathic pulmonary fibrosis. Only longer-term follow-up would show whether these patients may develop fibrosing alveolitis. Further research is needed in this area so that genes or markers of susceptibility can be evaluated.

AUTOIMMUNE DISORDER

Chronic inflammation may develop as a response to injury due to viruses or drugs. There is suspicion of a viral etiology in fibrosing alveolitis. Circulating antibodies to plasma IgG autoantibodies to protein(s) associated with both types of alveolar lining cells have been demonstrated in cryptogenic fibrosing alveolitis, but their significance is not yet clear.[105]

METHOD OF DIAGNOSIS

Transbronchial biopsy and lavage are frequent methods of diagnosis in fibrosing alveolitis. Since with established disease there is much fibrous tissue, it is often impossible to obtain an adequate sample of the diseased lung by transbronchial biopsy. In view of possible treatment with long-term cytotoxics or immunosuppressive drugs, it is vital that such patients have open lung biopsies in the hands of an experienced surgeon. Some of the biopsy should be sent immediately for bacteriology, as *Mycoplasma, Legionella, Mycobacteria, Chlamydia,* and some fungal infections may give a radiologic and clinical picture similar to that of fibrosing alveolitis. If the tissue is received fresh in the laboratory, some material should be placed immediately into gluteraldehyde. This will enable the identification of pentalaminar inclusions in histiocytosis X. The rest of the tissue should then be gently inflated by injection of 10% formol-saline via a fine needle into the pulmonary parenchyma. If the biopsy is large enough, some of the tissue should be left in formalin, since it may be necessary to proceed to an asbestos fiber count.

OTHER INTERSTITIAL PNEUMONIAS

Lymphoid Interstitial Pneumonia

The term *lymphoid interstitial pneumonia* is a misnomer, since it represents a low-grade lymphoma of the lung. It will be discussed in Chap. 33.

Granulomatous Interstitial Pneumonia

Coleman and Colby[13] described a cellular interstitial pneumonia of unknown origin, with prominent epithelial histiocytes forming loose, nonnecrotizing granulomas. Interstitial histiocytes were accentuated around bronchioles and the picture differed from sarcoid in that granulomas were loose and there was a predominance of the interstitial infiltrate.

Clearly, there is a histologic spectrum between granulomatous interstitial pneumonia and extrinsic allergic alveolitis. It may be impossible to distinguish the two.

Giant Cell Interstitial Pneumonia

Giant cell interstitial pneumonia is a rare disease[106] in which the alveoli contain giant cells, often showing phagocytosis. There is interstitial fibrosis. Some of the cases are probably a manifestation of exposure to hard metals. This will be discussed in Chap. 15.[107]

Hamman-Rich Syndrome

As noted above, it is doubted that Hamman-Rich syndrome is a discrete disease entity. Katzenstein and coworkers[49,108] and more recently Olsen and colleagues[109] described a total of 37 cases of this syndrome. Only 13 patients survived.

The histologic picture was organizing acute alveolar damage with no predisposing cause. The most prominent feature was the presence of extensive interstitial fibrosis, along with accentuation of fibrosis around respiratory bronchioles and alveolar ducts. Lymphocytes, plasma cells, and, less commonly, neutrophils were seen in the interstitium. Hyaline membranes were more common in patients who died. Type II hyperplasia was a prominent feature. In some cases there was fibroblastic proliferation in alveolar spaces.

The patients were younger than those suffering from fibrosing alveolitis, with a mean age of 39 years (range: 13 to 77 years). In at least 17 patients, there was a prodromal viral upper respiratory tract illness, followed by increasing dyspnea and nonproductive cough. No specific infectious agent was identified. In Katzenstein's series, two patients had renal failure, one having been treated by transplantation. Another had preeclampsia. Such diseases might be a cause of pulmonary edema and transplantation may cause an ARDS picture.

The changes were those of ARDS, without a predisposing cause. Diffuse alveolar damage is a common reaction and can be due to a wide range of noxious agents. Hamman-Rich syndrome[110] is not a chronic interstitial pulmonary disease. At the present state of our knowledge, it would be wisest to regard this "syndrome" as a manifestation of acute alveolar damage of unknown etiology.

Chronic Interstitial Pneumonia

The term *chronic interstitial pneumonia* (CIP) is used by Katzenstein and Askin[111] to encompass the cases of interstitial fibrosis that do not fall into the diagnostic categories given in Table 12-2. As mentioned above, this term is not used by many histopathologists, as the microscopic picture is nonspecific.

There are various causes of chronic interstitial pneumonia, including collagen diseases, drugs, *Mycoplasma pneumoniae* infection, extrinsic allergic alveolitis, bone marrow transplantation, asbestosis, and severe pulmonary hypertension.[111]

There is chronic interstitial inflammation with plasma cells and lymphocytes (Fig. 12-37). There is interstitial fibrosis, but there are relatively few macrophages.

MARKERS OF ACTIVITY

Histology and Lavage

Patients with a predominantly cellular lavage or biopsy, especially with lymphocytes, have the most

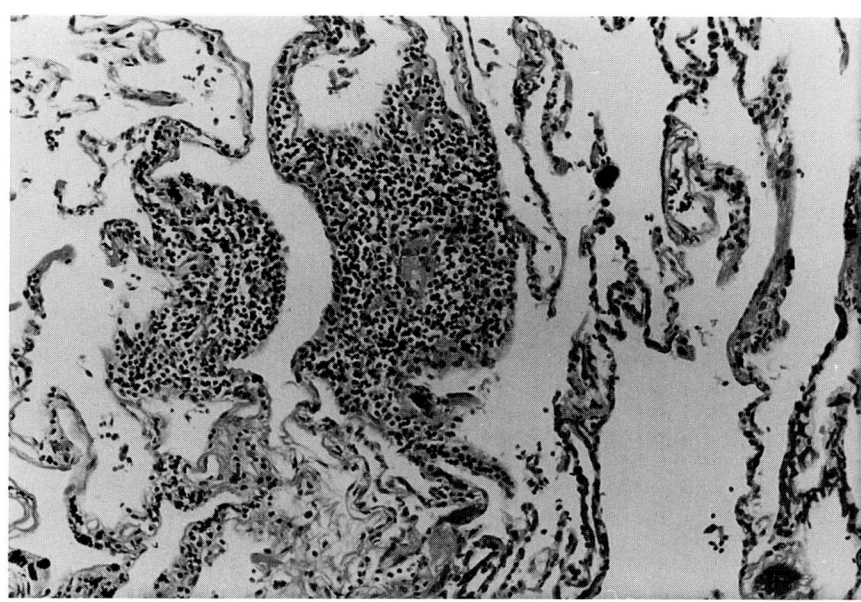

FIGURE 12-37
Chronic interstitial pneumonia in a case of scleroderma. (H&E, ×55.)

favorable prognosis. The prognostic value of lavage has been dealt with above. A recent series demonstrated that biopsies with pure inflammation and no fibrosis signified good prognosis.[112] The inflammation was mainly lymphocytic, and thus the findings are consistent with good-prognosis lavage patients. The series was atypical of cryptogenic fibrosing alveolitis in that one-third of patients had either connective-tissue diseases or possible causes of extrinsic allergic alveolitis.

A small series showed no relationship between IL-2 receptor, soluble CD8, and tumor necrosis factor and clinical measures of outcome.[113]

Lung Function

Abnormalities of elastic recoil and pulmonary gas exchange usually correlate with parenchymal fibrosis in patients with cryptogenic pulmonary fibrosis. Lung function measurements are relatively insensitive in assessing advanced disease.[77]

Immunologic Markers

Serum and immune complexes do not appear reliable indices of disease activity because the association of C1Q binding and clinical disease activity is qualitative.

CT Scanning

CT scanning offers promise as a noninvasive, sensitive indicator of altered lung function. Scans may be able to distinguish between advanced fibrotic disease and early cellular changes, such as edema and cellular infiltration.

ALVEOLAR/CAPILLARY BLOCK

From the above account of the histopathology of fibrosing alveolitis, it may be thought that the reduced diffusing capacity is due to the thickened alveolar wall. The term *alveolar/capillary block* was coined by Austrian and colleagues[114] to describe a syndrome characterized by reduced lung volume, reasonably normal ventilatory capacity, hyperventilation, and normal arterial P_{O2} at rest, but with desaturation on exercise.

From the description of the histology given above, it may seem logical that the cause of the alveolar/capillary block is the interstitial fibrosis. This is probably too simplistic an answer. Ventilation/perfusion scans (Fig. 12-38) from cases of interstitial pulmonary fibrosis may strongly resemble those seen in pulmonary infarction.

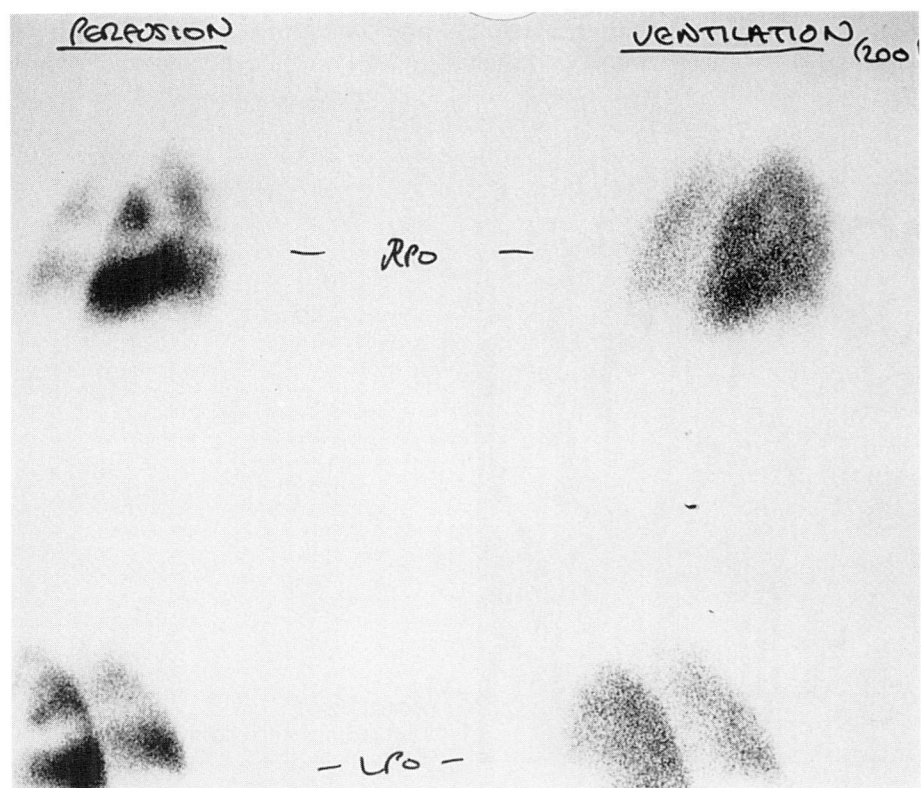

FIGURE 12-38
V/Q scan of the chest, showing a marked mismatch of ventilation and perfusion. This is thought to be due to the vascular changes described in the text and not to pulmonary emboli. *(Courtesy of Dr. Testa, Manchester Royal Infirmary and Dr. K. B. Carroll, Wythenshawe Hospital.)*

Arndt and colleagues[115] studied 10 patients with alveolar capillary block due to a variety of causes. Two patients with alveolar cell carcinoma and cryptogenic interstitial pulmonary fibrosis had hypoxemia at rest due to large shunts. In the next four patients, who had sarcoid, DIP, and eosinophilic granuloma, the hypoxemia was explained as follows. The "slow alveolus," which is a functional unit with high airway resistance or a high compliance,[116] has a low ventilation-perfusion ratio but still has sufficient perfusion to explain the hypoxemia.

This left four patients with interstitial pulmonary edema, sarcoidosis, and systemic sclerosis, with no appreciable shunt and with identical ventilation-perfusion ratios in "slow" and "fast" alveolar compartments. A "fast" compartment is a functional airway unit with a low airway resistance or a low compliance. In these cases the oxygen saturation was normal at rest, and there was no reason for believing that a maldistribution affected the oxygenation of the arterial blood. Only the reduction in the diffusing capacity in these last four cases caused the hypoxemia on exercise. Thus the functional changes seen in fibrosing alveolitis are not easily explained on histologic grounds.

DIFFERENTIAL DIAGNOSES

Connective Tissue Diseases

It may be impossible to distinguish the histology of a connective tissue, such as rheumatoid disease or systemic sclerosis, from cryptogenic fibrosing alveolitis or in some cases chronic interstitial pneumonia.[111] In any case of fibrosing alveolitis, a full clinical history as well as the appropriate serologic studies must be taken to exclude a collagen disease. In polymyositis and dermatomyositis there may be a histologic pattern of fibrosing alveolitis, diffuse alveolar damage, or bronchiolitis obliterans organizing pneumonia. This last group of patients have a good prognosis.[117] Fibrosing alveolitis associated with systemic sclerosis has a better prognosis than "lone" cryptogenic fibrosing alveolitis.[118] The two conditions are indistinguishable histologically and radiologically. Patients with systemic sclerosis should fulfill the diagnostic criteria of the American Rheumatism Association.[119]

Extrinsic Allergic Alveolitis

Extrinsic allergic alveolitis shows loose-knit interstitial granulomas, usually patchily distributed around respiratory bronchioles. The giant cells, when present, may contain inclusions. Intervening areas of lung are normal. Besides the interstitial granulomas, there is minimal fibrosis, and the predominant cells are lymphocytes. Plasma cells and histiocytes are also seen. Eosinophils and neutrophils are not prominent. Bronchiolitis obliterans may be seen, as may hyaline membranes. As the disease progresses, if the allergen is not withdrawn, there may be end-stage pulmonary fibrosis. This is uncommon, and separation from fibrosing alveolitis is usually possible. Some of the patients in the series reported by Katzenstein[112] had a history of exposure to allergens and thus may have had a form of extrinsic allergic alveolitis.

Sarcoidosis

In end-stage sarcoidosis, honeycomb lung may be found with relatively few granulomas. In most cases there are definite granulomas, which tend to follow the lines of the pulmonary lymphatics.

Acute Interstitial Pneumonia (Hamman-Rich Syndrome)

The histology of this disease is the same as that of cryptogenic fibrosing alveolitis and ARDS. Acute interstitial pneumonia has been discussed above.

Histiocytosis X

This disease may cause confusion in a few cases. Besides a centrilobular distribution of the stellate scars, which contain the typical Langerhans (or histiocytosis-X) cells, there are eosinophils, plasma cells, lymphocytes, and fibrosis. In most cases the intervening lung parenchyma is normal or may show emphysema, as well as honeycomb change. In histiocytosis the lesions are usually of varying ages. The Langerhans cells stain positively for S100 protein. Langerhans cells are not confined to histiocytosis, and they may be seen in fibrosing lung disease—but as scattered cells and not in large clusters.

Raised Left Atrial Pressure

Increased left atrial pressure causes interstitial fibrosis, but in nearly all cases there are also many hemosiderin-laden macrophages. In practice this differential diagnosis is usually not difficult, since clinicians have detected mitral stenosis or regurgitation or severe left ventricular failure. Pulmonary venoocclusive disease has *focal* areas of fibrosis, which may cause an erroneous diagnosis of fibrosing alveolitis if full account is not taken of the clinical history and radiology. There may be a CIP picture, but usually the inflammation is focal.

Immunosuppression

Interstitial pulmonary fibrosis may occur as a complication of infection with HIV or as a result of pulmonary damage in heart-lung transplantation, single lung transplantation, or allogeneic bone marrow transplantation. In most cases of pulmonary transplantation, there is also bronchiolitis obliterans.

Infection

Pneumocystis carinii infection can present with prominent interstitial fibrosis but with few organisms. In any immunosuppressed patient, a battery of stains for organisms, including fungi, should be done. *Mycoplasma* may also cause interstitial fibrosis, sometimes giving a CIP picture. This organism is usually detected serologically.

Drug Reactions

Drugs such as BCNU, bleomycin, and busulfan can cause interstitial lung disease, resembling CIP in some cases. In practice, most patients will have a history of treatment with the appropriate drug. In these cases it is important not to miss an associated infective process. With either a transbronchial or, better, an open lung biopsy from such a patient, all the appropriate bacteriologic stains should be done.

A helpful feature in some cases may be marked nuclear pleomorphism in the proliferating type II cells. These initially may be taken as evidence of malignant change. This feature is not specific for drug-induced lung disease.

ARDS (Diffuse Alveolar Damage)

The picture in cryptogenic fibrosing alveolitis in the early stages is histologically inseparable from that seen in healing ARDS. The presence of hyaline membranes does not help to distinguish the two disorders. In practice, there is usually a history of a preceding cause, such as trauma, in patients with ARDS.

Asbestosis

In the presence of a few asbestos bodies, it may be difficult to separate asbestosis from cryptogenic fibrosing alveolitis. With very few asbestos bodies and a low asbestos fiber count, consideration should be given to a dual diagnosis of cryptogenic fibrosing alveolitis and mild asbestosis. The exact significance of the asbestos in such cases may be difficult for the pathologist to quantify. A desquamative pattern of interstitial pneumonia may be associated with asbestos[31] or CIP.[111]

Other Causes of Desquamative Interstitial Pneumonia

Talc and silica may also give a desquamative interstitial pneumonia (DIP) type of pattern. A DIP reaction has been reported in pulmonary alveolar proteinosis.[120]

Respiratory Bronchiolitis

Respiratory bronchiolitis is a distinct pathologic lesion of cigarette smokers, characterized by the accumulation of pigmented macrophages within respiratory bronchioles and adjacent alveoli. There is associated mild thickening of the peribronchiolar interstitium.[121] Most of these lesions have been seen in autopsy specimens from victims of sudden, nonrespiratory deaths and in lungs excised for unrelated lesions.

Myers and coworkers[122] described six patients with chronic interstitial lung disease in whom open lung biopsy showed respiratory bronchiolitis. The patients were young (age range: 28 to 46 years), all were heavy cigarette smokers, and their main symptoms were cough and dyspnea. Radiology showed diffuse interstitial infiltrates, and pulmonary function tests showed decreased diffusing capacity in four cases. Two patients had histories of occupational asbestos exposure. One worked as a brake shoe grinder and the other had noted an acute onset of his respiratory complaints after exposure to an aluminum cleaner of unknown composition.

The main abnormality was the presence of pigmented macrophages in respiratory bronchioles as well as in neighboring alveolar ducts and alveoli (Figs. 12-39 and 12-40). Typically, the macrophages had abundant cytoplasm with a finely granular golden-brown pigment. Alveolar septae in these areas often showed nonspecific interstitial fibrosis and a mild chronic inflammatory infiltrate, with hyperplastic pneumocytes lining the alveolar spaces. The peribronchiolar interstitium was focally expanded by a mild infiltrate of mononuclear cells consisting primarily of lymphocytes and histiocytes.

Three of these cases were initially interpreted as variants of idiopathic pulmonary fibrosis, because of the frequent thickening of the peribronchiolar alveolar septa and associated type II hyperplasia. The correct diagnosis of respiratory bronchiolitis was based on the patchy, strictly peribronchiolar distribution of the interstitial changes. In the cryptogenic fibrosing alveolitis, unlike respiratory bronchiolitis, the changes are microscopically heterogeneous, with alternating areas of inflammation, fibrosis, and hon-

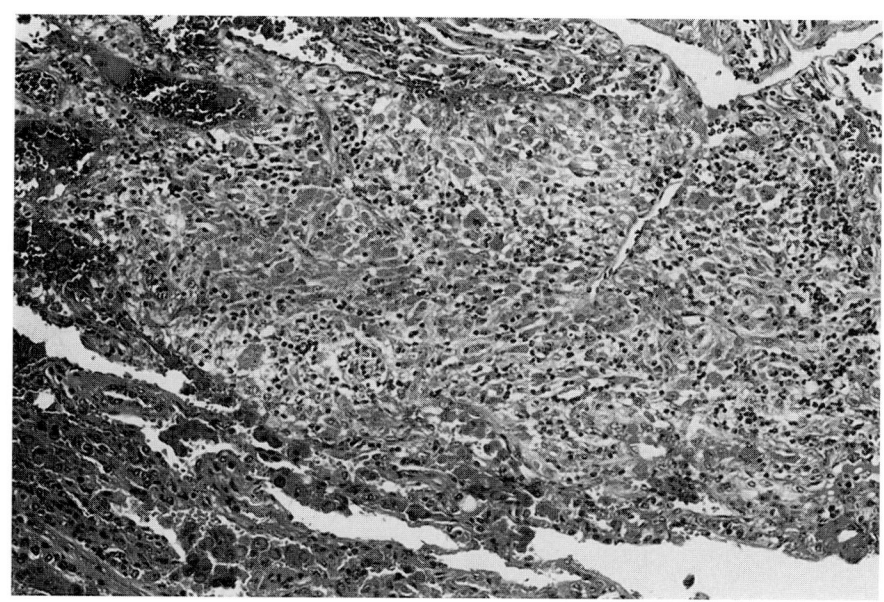

FIGURE 12-39
Respiratory bronchiolitis with a large mass of dark histiocytes filling alveolar ducts and alveoli. (H&E, ×88.)

eycombing. Fibrosis is minimal in respiratory bronchiolitis, and honeycomb change is not present.

There is a superficial resemblance between respiratory bronchiolitis and DIP. In respiratory bronchiolitis the process is patchy, whereas in DIP it is diffuse.

The macrophages in respiratory bronchiolitis contain lysosomes and phagolysosomes with many elongated needle-shaped inclusions (Fig. 12-41), which are thought to originate in cigarette smoke.

Bronchiolitis Obliterans–Organizing Pneumonia

Bronchiolitis obliterans–organizing pneumonia (BOOP) may be associated with an infection, drugs such as amiodarone, or collagen vascular diseases. Thus it shares a common etiology with interstitial pulmonary fibrosis. BOOP is commonly misdiagnosed as the latter disease.

Characteristically, it is a patchy process with areas of normal parenchyma alternating with fibrosis, consisting of myofibroblasts with a loose myxoid pattern and forming plugs or polyps inside terminal and respiratory bronchioles, alveolar ducts, and alveoli. Besides these plugs of fibrous tissue there may be interstitial fibrosis, type II pneumocyte hyperplasia, and a focal chronic inflammatory infiltrate. Foamy histiocytes are a common feature. The interstitial disease is confined to the areas with fibrosis and does

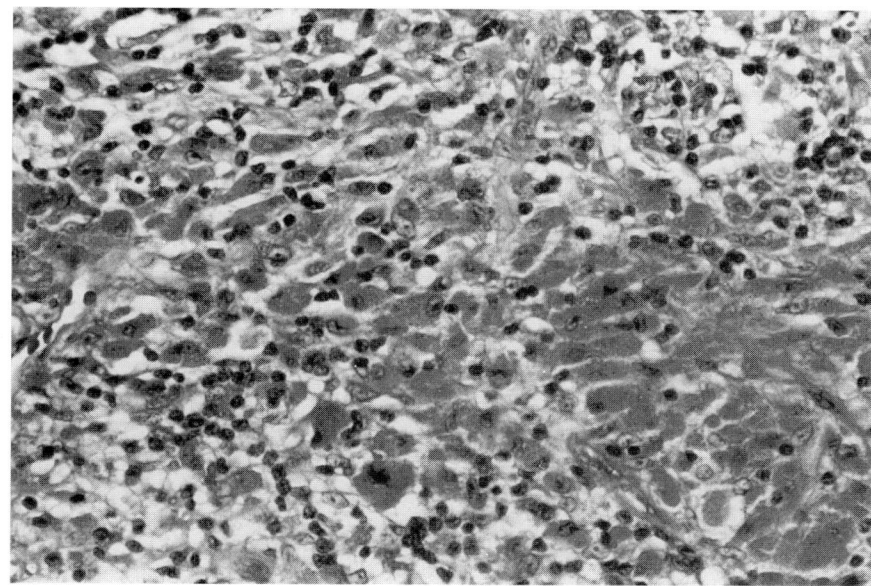

FIGURE 12-40
Higher magnification of Fig. 12-39, with many dark macrophages. (H&E, ×219.)

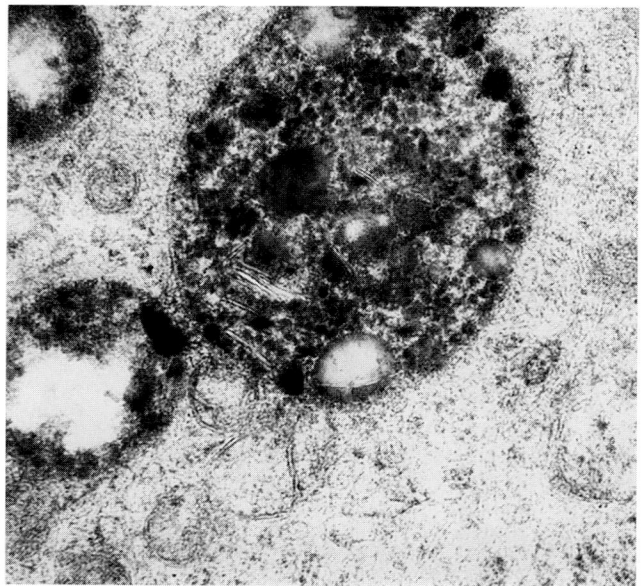

FIGURE 12-41
Phagolysosomes with typical needle-shaped crystals, due to cigarette smoke. This patient had received a lung transplant and was found to be smoking. (EM, ×44,400.)

not extend to the unaffected lung. In BOOP the disease is not diffuse as in fibrosing alveolitis, in which there is no bronchiolitis obliterans or prominent intraalveolar reaction.

Diagnostic problems may occur when biopsies are not well distended.

BOOP is essentially a low-power diagnosis, which is aided by an elastic van Gieson stain. This confirms the predominantly intrabronchiolar and intraalveolar nature of the fibroblastic reaction.

Other Associations with Interstitial Lung Disease

Cryptogenic fibrosing alveolitis has been associated with chronic active hepatitis[123] as well as primary biliary cirrhosis.[124] It has been argued that the pulmonary disease in primary biliary cirrhosis is due to associated Sjögren's disease and not to underlying liver disease.[125]

Interstitial Lung Disease Associated with Bowel Disease

Pulmonary fibrosis is a rare manifestation of inflammatory bowel disease. There have been reports of reduced diffusing capacity, presumably due to the interstitial lung disease, in patients with ulcerative colitis and Crohn's disease.[126] Patients with celiac disease show partial fibrous obliteration of small airways and dilatation of larger airways, but no evidence of any granulomatous or primary interstitial lung disease. In one study, 18 of 73 patients with gluten-sensitive enteropathy had an abnormality of

gas transfer; in 13 of the 18 patients, the respiratory problem was of sufficient severity to justify transbronchial biopsy. One patient came to open lung biopsy.

In two patients the biopsies were normal; in others there were varying degrees of interstitial, peribronchiolar, and perivascular fibrosis. The adjacent lung parenchyma showed patchy thickening of alveolar septa and occasional prominence of type II pneumocytes. The open lung biopsy showed fibrosis and thickening of alveolar walls, with an excess of intraalveolar macrophages. Bronchioles were surrounded by cuffs of fibrous tissue with an infiltrate of lymphocytes and plasma cells, but the extensive peribronchiolar fibrosis seen in an autopsy case was not present. The interstitial, perivascular, and peribronchiolar fibrosis in the biopsy material was thought to be nonspecific.

Most of the patients were cigarette smokers or ex-smokers, and it was concluded that these findings are common in patients with emphysema. The nonspecific chronic pneumonitis present in the open lung biopsy was thought to be obstructive and caused by airway damage.

Seven patients with celiac disease and lung cavities have been described by Stevens and colleagues.[127] These authors, from the west of Ireland, had a large experience of celiac disease. The researchers found that, after malignancy, lung abscess was the most common cause of death in these patients, who were 42 to 71 years of age. There were predisposing factors in four patients, but in three no definite cause of the lung abscess could be found.

REFERENCES

1. King A, Talmadge E Jr: Interstitial lung disease, in Bone RC, Petty TL (eds), *Year Book of Pulmonary Disease, 1992.* St. Louis, Mosby–Year Book, 1992, pp 281–287.
2. Office of Population Censuses and Surveys: Mortality Statistics Cause, 1990. London, Her Majesty's Stationery Office, 1991 (DH2 No. 17).
3. Johnston I, Britton J, Kinnear W, Logan R: Rising mortality from cryptogenic fibrosing alveolitis. *BMJ* 301:1017–1021, 1990.
4. Coultas DB, Zumwalt RE, Black WC, Sobonya RE: The epidemiology of interstitial lung diseases. *Am J Respir Crit Care* 150:967–972, 1994.
5. Johnston IDA, Gomm SA, Kalra S, Woodcock AA, Evans CC, Hind CRK: The management of cryptogenic fibrosing alveolitis in three regions of the United Kingdom. *Eur Respir J* 6:891–893, 1993.
6. Keogh BA, Crystal RG: Alveolitis: The key to the interstitial lung disorders. *Thorax* 37:1–10, 1982.
7. Liebow AA, Steer A, Billingsley JG: Desquamative interstitial pneumonia. *Am J Med* 39:369–404, 1991.

8. Dunnill MS: Pulmonary fibrosis. *Histopathology* 16:321–329, 1990.

9. Davis WB, Crystal RG: Chronic interstitial lung disease, in Simons D (ed), *Current Pulmonology,* vol. V. New York, Wiley Medical, 1984, pp 347–473.

10. Liebow AA: Definition and classification of interstitial pneumonias in human pathology. *Prog Respir Res* 8:1–33, 1975.

11. Katzenstein A-LA, Askin FB: *Surgical Pathology of Nonneoplastic Disease,* 2d ed. Philadelphia, WB Saunders, 1990, pp 58–59.

12. Klennerman P, Woodhead MA, Spiteri M, et al: Extrinsic allergic alveolitis without an identified causative antigen. *Thorax* 46:313P, 1990

13. Coleman A, Colby TV: Histologic diagnosis of extrinsic allergic alveolitis. *Am J Surg Pathol* 12:514–518, 1988.

14. Steinkamp G, Müller KM, Schirg E, von der Hardt H: Fibrosing alveolitis in childhood: A long-term follow-up. *Acta Paediatr Scand* 79:823–831, 1990.

15. Schroeder SA, Shannon DC, Mark EJ: Cellular interstitial pneumonitis in infants: A clinicopathological study. *Chest* 101:1065–1069, 1992.

16. Turner-Warwick M, Burrows B, Johnson A: Cryptogenic fibrosing alveolitis: Clinical features and their influence on survival.*Thorax* 35:171–180, 1980.

17. Turner-Warwick M: Systemic arterial patterns in the lung and clubbing of the fingers. *Thorax* 18:238–250, 1963.

18. Hansell DM, Kerr IH: The role of high resolution computative tomography in the diagnosis of interstitial lung disease. *Thorax* 46:77–84, 1991.

19. McCloud TC, Carrington CB, Gaensler EA: Diffuse infiltrative lung disease: A new scheme for description. *Radiology* 149:353–363, 1991.

20. Muller NL, Staples CA, Miller RR, Vedal F, Thurlbeck WM, Ostrow DM: Disease activity in idiopathic pulmonary fibrosis: CT and pathologic correlation. *Radiology* 165:721–724, 1991.

21. Hansell DM, Wells AU, du Bois R, Corrin B: Disease activity in fibrosing alveolitis: Assessment by high resolution CT and histological correlation [abstract]. *Clin Radiol* 42:375, 1990.

22. Wiggins J, Strickland B, Turner-Warwick M: Combined cryptogenic fibrosing alveolitis and emphysema: The value of high resolution computed tomography in assessment. *Respir Med* 84:365–369, 1990.

23. Packe GE, Cayton RM, Edwards CW: Comparison of right ventricular weight at necropsy in interstitial pulmonary fibrosis and in chronic bronchitis and emphysema. *J Clin Pathol* 39:594–595, 1986.

24. Fulton RM, Hutchinson EC, Morgan Jones A: Ventricular weight in cardiac hypertrophy. *Br Heart J* 14:413–420, 1952.

25. Crystal RG, Gadek JE, Ferrans VJ, Fulmer JD, Line BR, Hunninghake GW: Interstitial lung disease: Current concepts of pathogenesis, staging and therapy. *Am J Med* 70:542–568, 1981.

26. Scadding JG, Hinson KW: Diffuse fibrosing alveolitis (diffuse interstitial fibrosis of the lungs): Correlation of histology at biopsy with prognosis. *Thorax* 22:291–304, 1967.

27. Tubbs RR, Benjamin SP, Reich NE, et al: Desquamative interstitial pneumonitis: Cellular phase of fibrosing alveolitis. *Chest* 72:159–165, 1977.

28. Case records of the Massachusetts General Hospital. *N Engl J Med* 308:511–519, 1983.

29. Carrington CB, Gaensler EA, Coutu RE, Fitzgerald MX, Gupta RG: Natural history and treated cause of usual and desquamative interstitial pneumonia. *N Engl J Med* 298:801–809, 1978.

30. Katzenstein A-LA, Askin SB: *Surgical Pathology of Non-Neoplastic Lung Disease,* 2d ed. Philadelphia, WB Saunders, 1990, p 75.

31. Corrin B, Price AB: Electron microscopic studies in desquamative interstitial pneumonia associated with asbestosis. *Thorax* 27:324–331, 1972.

32. Bedrossian GWM, Kuhn C III, Luna MA, Conklin RH, Byrd RB, Kaplan ED: Desquamative interstitial pneumonia–like reaction accompanying pulmonary lesions. *Chest* 72:166–169, 1977.

33. Brewer DB, Heath D, Asquith P: Electron microscopy of desquamative interstitial pneumonia. *J Pathol* 97:317–323, 1969.

34. Shortland JR, Dark CS, Crane WAJ: Electron microscopy of desquamative interstitial pneumonia. *Thorax* 24:192–208, 1969.

35. Abraham JL, Hertzberg MA: Inorganic particles associated with desquamative interstitial pneumonia. *Chest* 80:67S–69S, 1981.

36. Gardiner IT, Uff JS: "Blue bodies" in a case of cryptogenic fibrosing alveolitis (desquamative type): An ultrastructural study. *Thorax* 33:806–813, 1978.

37. Fukuda Y, Ferrans VJ, Schoenberger CI, Rennard SI, Crystal R: Patterns of pulmonary structural remodelling after experimental paraquat toxicity: The morphogenesis of intra-alveolar fibrosis. *Am J Pathol* 118:452–475, 1985.

38. Fukuda Y, Ishizaki M, Masuda Y, Kimura G, Kawanami O, Masugi Y: The role of intraalveolar fibrosis in the process of pulmonary structural remodeling in patients with diffuse alveolar damage. *Am J Pathol* 126:171–182, 1987.

39. Nash JRG, McLaughlin PJ, Butcher D, Corrin B: Expression of tumour necrosis factor–alpha in cryptogenic fibrosing alveolitis. *Histopathology* 22:343–347, 1993.

40. Corrin B, Butcher D, McNaulty BJ, et al: Immunohistochemical localisation of transforming growth factor–β 1 in the lungs of patients with systemic sclerosis, cryptogenic fibrosing alveolitis and other lung disorders. *Histopathology* 24:145–150, 1994.

41. Wilson NJE, Gosney JR, Mayall F: Endocrine cells in diffuse pulmonary fibrosis. *Thorax* 48:1252–1256, 1993.

42. Kanematsu T, Nishimura K, Nagai S, Izumi T: Clubbing of the fingers and smooth muscle proliferation in fibrotic changes in patients with idiopathic pulmonary fibrosis. *Chest* 105:339–342, 1994.

43. Basset F, Ferrans VS, Soler P, Takemura T, Fukuda Y, Crystal RG: Intraluminal fibrosis in interstitial lung disorders. *Am J Pathol* 122:448–461, 1986.

44. Fulmer JD, Roberts WC, von Gal ER, Crystal RG: Small airways in idiopathic pulmonary fibrosis: Comparison of morphologic and physiologic observations. *J Clin Invest* 60:595–610, 1977.

45. Hunt LW, Colby TV, Weiler DA, Sanjiv Sur MS, Butterfield JH: Immunofluorescent staining for mast cells in idiopathic pulmonary fibrosis: Quantification and evidence for extracellular release of mast cell tryptase. *Mayo Clin Proc* 67:941–948, 1992.

46. Heath D, Gillund TD, Kay JM, Hawkins CF: Pulmonary vascular disease in honeycomb lung. *J Pathol Bacteriol* 95:423–430, 1968.

47. Ashcroft T, Simpson JM, Tumbrell V: Simple method of estimating severity of pulmonary fibrosis on a numerical scale *J Clin Pathol* 41:467–470, 1988.

48. Cherniack M, Colby TV, Flint A, et al: BAL Cooperative Group Steering Committee: Quantitative assessment of lung pathology in idiopathic pulmonary fibrosis. *Am Rev Respir Dis* 144:892–900, 1991.

49. Katzenstein A-LA: Pathogenesis of "fibrosis" in the interstitial pneumonias: An electron microscopic study. *Hum Pathol* 16:1015–1024, 1985.

50. Heard BE, Dewar A, Corrin B: Apposition of fibroblasts to mast cells and lymphocytes in a normal human lung and in cryptogenic fibrosing alveolitis: Ultrastructure and cell parameter measurements. *J Pathol* 166:303–310, 1992.

51. Falk GA, Okinaka AJ, Siskind GW: Immunoglobulins in the bronchial washings of patients with chronic obstructive pulmonary disease. *Am Rev Respir Dis* 105:14–21, 1972.

52. Reynolds HY, Newball HH: Analysis of proteins and respiratory cells obtained from human lungs by bronchial lavage. *J Clin Lab Med* 84:559–573, 1974.

53. Haslam PL, Turton CWG, Herd B, et al: Bronchoalveolar lavage in pulmonary fibrosis: Comparison of cells obtained with lung biopsy and clinical features. *Thorax* 35:9–18, 1980.

54. Wells AU, Hansell DM, Rubens MB, et al: Fibrosing alveolitis in systemic sclerosis: Bronchoalveolar lavage findings in relation to computed tomographic appearance. *Am J Respir Crit Care Med* 150:462–468, 1994.

55. Turner-Warwick M, Haslam PL: The value of serial bronchoalveolar lavages in assessing clinical prognosis in patients with a cryptogenic fibrosing alveolitis. *Am Rev Respir Dis* 135:26–34, 1987.

56. Haslam PL, Dewar A, Turner-Warwick M: Lavage eosinophils and histamine, in Cumming G, Bonsignor G (eds), *Cellular Biology of the Lung*. New York, Plenum, 1981, pp 77–94.

57. Haslam PL, Cromwell O, Dewar A, Turner-Warwick M: Evidence of increased histamine levels in lung lavage fluids in patients with cryptogenic fibrosing alveolitis. *Clin Exp Immunol* 44:587–593, 1981.

58. Jordana M, Befus AD, Newhouse MT, Bienenstock J, Gauldi J: Effect of histamine on proliferation of normal human adult lung fibroblasts. *Thorax* 43:552–558, 1988.

59. Djukanovic R, Roche WR, Wilson JW, et al: Mucosal inflammation in asthma. *Am Rev Respir Dis* 142:434–457, 1990.

60. Fine A, Goldstein RH: The effect of transforming growth factor–beta on cell proliferation and collagen formation by lung fibroblasts. *J Biol Chem* 262:3897–3902, 1987.

61. Raghu G, Mastra S, Meyers D, Narayan AS: Collagen synthesis by normal and fibrotic human lung fibroblasts and the effect of transforming growth factor–β. *Am Rev Respir Dis* 140:95–100, 1989.

62. Dubaybo BA, Thet LA: Effect of transforming growth factor beta on synthesis of glycosaminoglycans by lung fibroblasts. *Exp Lung Res* 16:389–403, 1990.

63. McAnulty RJ, Campa JS, Cambrey AD, Laurent GJ: The effect of transforming growth factor beta on rates of procollagen synthesis and degradation in vitro. *Biochim Biophys Acta* 1091:231–235, 1991.

64. Roberts CM, Cairns D, Bryant DH, et al: Changes in epithelial lining of fluid albumin associated with smoking and interstitial lung diseases. *Eur Respir J* 6:110–115, 1993.

65. Gadek JE, Kelman JA, Fells G: Collagenase in the lower respiratory tract of patients with idiopathic pulmonary fibrosis. *N Engl J Med* 301:737–742, 1979.

66. Forsyth KD, Levinsky RJ: Fibronectin degradation: An invitro model of neutrophil mediated endothelial cell damage. *J Pathol* 161:313–319, 1990.

67. Ozaki T, Hayashi H, Tani K, Ogushi F, Yosoka S, Ogura T: Neutrophil chemotactic factors in the respiratory tract of patients with chronic airway disease or idiopathic pulmonary fibrosis. *Am Rev Respir Dis* 145:85–91, 1992.

68. Lynch JP III, Standiford TJ, Rolfe MW, Kunkel SL, Strieter RM: Neutrophilic alveolitis in idiopathic pulmonary fibrosis: Role of interleukin-8. *Am Rev Respir Dis* 145:1433–1439, 1992.

69. Iyonaga K, Takeya M, Saita N, et al: Monocyte chemoattractant protein–1 in idiopathic pulmonary fibrosis and other interstitial diseases. *Hum Pathol* 25:455–463, 1994.

70. Robinson BWS, Rose AH: Pulmonary gamma interferon production in patients with fibrosing alveolitis. *Thorax* 45:105–108, 1990.

71. Gleach GJ, Adolphson CR: The eosinophil leucocyte: Structure and function. *Adv Immunol* 39:177–253, 1986.

72. Roche WR: Mast cells and tumors: The specific enhancement of tumor proliferation *in vitro*. *Am J Pathol* 119:57–64, 1985.

73. Campell DA, Poulter LW, Janossi G, du Bois RM: Immunohistological analysis of lung tissue from patients with cryptogenic fibrosing alveolitis suggesting local expression of immune hypersensitivity. *Thorax* 40:405–411, 1985.

74. Kallenberg CGM, Schilizzi BM, Beaumont F, Poppema S, De Leij L, The TH: Expression of class II MHC antigens on alveolar epithelium in fibrosing alveolitis. *Clin Exp Immunol* 67:182–190, 1987.

75. Sheppard MN, Harrison NK: Lung injury, inflammatory mediators and fibroblast activation in fibrosing alveolitis. *Thorax* 47:1064–1074, 1992.

76. Emura M, Nagai S, Takeuchi M, Kitaichi M, Izumi T: In vitro production of B-cell growth factor and B-cell differentiation factor by peripheral blood mononuclear cells and bronchoalveolar lavage T lymphocytes from patients with idiopathic pulmonary fibrosis. *Clin Exp Immunol* 82:133–139, 1990.

77. Cherniack RM, Crystal RG, Kalica AR: NHLBI workshop summary: Current concepts in idiopathic pulmonary fibrosis: A road map for the future. *Am Rev Respir Dis* 143:680–683, 1991.

78. Jordana M, Schulman J, McSharry C, et al: Heterogeneous proliferative characteristics of human adult lung fibroblast lines and clonally derived fibroblasts from control and fibrotic tissue. *Am Rev Respir Dis* 137:379–384, 1988.

79. Bitterman PB, Rennard SI, Hunninghake GW, Crystal RG: Human alveolar macrophage growth factor for fibroblasts: Regulation and partial characterization. *J Clin Invest* 70:806–822, 1982.

80. Pérez-Arellano J-L, Pedraz M-J, Fuertes A, de la Cruz J-L, González de Buitrago J-M, Jiménez A: Laminin fragment P1 is increased in the lower respiratory tract of patients with diffuse interstitial lung diseases. *Chest* 104:1163–1169, 1993.

81. Raghu G, Kavanagh T: The human lung fibroblast: A multi-faceted target and effector cell, in Selman Lama M, Barrios R (eds): *Interstitial Pulmonary Diseases: Selected Topics*. Boca Raton, Fla., CRC Press, 1991, pp 1–34.

82. Sakamoto S, Sakamoto M: Degradative processes of connective tissue proteins with special emphasis on collagenolysis bone resorption. *Mol Aspects Med* 10:299–428, 1988.

83. Goldstein RH, Polgar P: The effect and interaction of bradykinin and prostaglandins on protein and collagen production by lung fibroblasts. *J Biol Chem* 257:8630–8633, 1982.

84. Elias JA, Reynolds MM, Kotloff RM, Kern JA: Fibroblast interleukin-1beta: Synergistic stimulation by recombinant interleukin 1 and tumor necrosis factor and posttranscriptional regulation. *Proc Natl Acad Sci* (USA) 86:6171–6175, 1989.

85. Gauldie J, Richards C, Harnish D, Lansdorp P, Baumann H: Interferon beta$_2$/B cell stimulatory factor type 2 shares identity with monocyte-derived hepatocyte-stimulating factor and regulates the major acute phase protein response in liver cells. *Proc Natl Acad Sci USA* 84:7251–7255, 1987.

86. Vancheri C, Geldie J, Bienstock J, et al: Human lung fibroblast–derived granulocyte-macrophage colony stimulating factor (GM-CSF) mediates eosinophil survival in vitro. *Am J Respir Cell Mol Biol* 1:289–295, 1989.

87. Bitterman PB, Wewers MD, Rennard SI, Adelberg S, Crystal RG: Modulation of alveolar macrophage–driven fibroblast proliferation by alternative macrophage mediators. *J Clin Invest* 77:700–708, 1986.

88. Jiminez SA: New insights into the pathogenesis of interstitial pulmonary fibrosis. *Thorax* 49:193–195, 1994.

89. Matsui R, Goldstein RH, Mihal K, Brody JS, Steele MP, Fine A: Type I collagen formation in rat type II alveolar cells immortalised by viral gene products. *Thorax* 49:201–206, 1994.

90. Egan J, Stewart JP, Hasleton PS, et al: Epstein-Barr virus–associated pulmonary fibrosis following heart/lung transplantation. *Eur Respir J* (submitted).

91. Egan J, Stewart J, Hasleton PS, Woodcock AA, Carrol KB: Epstein-Barr virus is identified in type II pneumocytes in fibrosing alveolitis. *Thorax* (in press).

92. Madri JA, Furthmayr H: Collagen polymorphism in the lung: An immunohistochemical study of pulmonary fibrosis. *Hum Pathol* 11:355–365, 1980.

93. Cantin AM, Boileau R, Begin R: Increased procollagen III amino terminal peptide-related antigens and the fibroblast growth signals in the lungs of patients with idiopathic pulmonary fibrosis. *Am Rev Respir Dis* 137:572–578, 1988.

94. Harrison NK, McAnulty RJ, Kimpton WG, Fraser JRE, Laurent TC, Laurent GJ: Heterogeneity of type III procollagen N-terminal peptides in BAL fluid from normal and fibrotic lung. *Eur Respir J* 6:1443–1448, 1993.

95. Last JA, King TE Jr, Nerlich AG, Reiser KM: Collagen cross-linking in adult patients with acute and chronic fibrotic lung disease: Molecular markers for fibrotic collagen. *Am Rev Respir Dis* 141:307–313, 1990.

96. Scott JM, Johnston I, Britton J: What causes cryptogenic fibrosing alveolitis? A case-control study of environmental exposure to dust. *BMJ* 301:1015–1021, 1990.

97. King TE: Idiopathic pulmonary fibrosis, in Schwartz MI, King TE (eds), *Interstitial Lung Disease.* New York, Marcel Dekker, 1988, pp 139–169.

98. Rosenberg DM: Inherited forms of interstitial lung disease. *Clin Chest Med* 3:635–641, 1982.

99. Watters LC: Genetic aspects of idiopathic pulmonary fibrosis and hypersensitivity pneumonitis. *Semin Respir Med* 7:317–325, 1986.

100. Driessen P, Demedts M: Inherited forms of interstitial lung disease. *Bull Fijnvl Longafw* 22:59–63, 1990.

101. McKusick VA: Autosomal dominant phenotypes, in McKusick VA (ed), *Mendelian Inheritance in Man.* Baltimore, Johns Hopkins University Press, 1990, pp 820–821.

102. Buchino JJ, Keenan WJ, Algren JT, Bove KE: Familial desquamative interstitial pneumonitis occurring in infants. *Am J Med Gen Suppl* 3:285–291, 1987.

103. Farrell PM, Gilbert EF, Zimmerman JJ, Warner TF, Saari TN: Familial lung disease associated with proliferation and desquamation of type II pneumocytes. *Am J Dis Child* 140:262–266, 1986.

104. Bitterman PB, Rennard SI, Keogh B, Wewers MD, Adelberg S, Crystal RG: Familial idiopathic pulmonary fibrosis: Evidence of lung inflammation in unaffected family members. *N Engl J Med* 314:1343–1347, 1986.

105. Wallace WAH, Roberts SN, Caldwell H, et al: Circulating antibodies to lung protein(s) in patients with cryptogenic fibrosing alveolitis. *Thorax* 49:218–224, 1994.

106. Sokolowski JW, Carday DR, Cantow EF, Elliot RC, Seal RB: Giant cell interstitial pneumonia: Report of a case. *Am Rev Respir Dis* 105:417–420, 1972.

107. Ohori MP, Sciurba FC, Owens GR, Hodgson MJ, Yousem SA: Giant-cell interstitial pneumonia and hard metal pneumoconiosis: A clinicopathologic study of four cases and review of the literature. *Am J Surg Pathol* 13:581–587, 1989.

108. Katzenstein AL-A, Myers JL, Mazur MT: Acute interstitial pneumonia: A clinicopathologic, ultrastructural, and cell kinetic study. *Am J Surg Pathol 10:256–267, 1986.*

109. Olsen J, Colby TV, Elliot CG: Hamman-Rich syndrome revisited. *Mayo Clin Proc* 65:1538–1548, 1990.

110. Hamman L, Rich AR: Acute diffuse interstitial fibrosis of the lungs. *Bull Johns Hopkins Hosp* 74:177–212, 1944.

111. Katzenstein A-LA, Askin FB: *Surgical Pathology of Nonneoplastic Disease,* 2d ed. Philadelphia, WB Saunders, 1990, pp. 85–88.

112. Katzenstein A-LA, Fiorelli RF: Nonspecific interstitial pneumonia/fibrosis: Histologic features and clinical significance. *Am J Surg Pathol* 18:136–147, 1994.

113. Meliconi R, Lalli E, Borzi RM, et al: Idiopathic pulmonary fibrosis: Can cell-mediated immunity markers predict clinical outcome? *Thorax* 45:536–540, 1990.

114. Austrian R, McClement JH, Renzetti AD Jr, Donald KW, Riley RL, Cournand A: Clinical and physiologic features of some types of pulmonary disease with impairment of alveolar-capillary diffusion. *Am J Med* 11:667–685, 1951.

115. Arndt H, King TKC, Briscoe WA: Diffusing capacities and ventilation:perfusion ratios in patients with the clinical syndrome of alveolar capillary block. *J Clin Invest* 49:408–422, 1970.

116. Nunn JF: *Applied Respiratory Physiology* 3d ed. London, Butterworth, 1987, p. 31.

117. Tazelaar HD, Viggiano RW, Pickersgill J, Colby TV: Interstitial lung disease in polymyositis and dermatomyositis: Clinical features and prognsis as correlated with histologic findings. *Am Rev Respir Dis* 141:727–733, 1990.

118. Wells AU, Cullinan P, Hansell DM, et al: Fibrosing alveolitis associated with systemic sclerosis has a better prognosis than lone cryptogenic fibrosing alveolitis. *Am J Respir Crit Care Med* 149:1583–1590, 1994.

119. Subcommittee for Scleroderma Criteria of the American Rheumatism Association Diagnostic and Therapeutic Criteria Committee: Preliminary criteria for the classification of systemic sclerosis (scleroderma). *Arthritis Rheum* 23:581–590, 1980.

120. Bhagwat AG, Wentworth P, Conen PE: Observations on the relationship of desquamative interstitial pneumonia and pulmonary alveolar proteinosis in childhood: A pathologic and experimental study. *Chest* 58:326–332, 1970.

121. Niewoehner D, Kleinerman J, Rice D: Pathologic changes in the peripheral airways of young cigarette smokers. *N Engl J Med* 291:755–758, 1974.

122. Myers JL, Veal CF Jr, Shin MS, Katzenstein A-LA:

Respiratory bronchiolitis causing interstitial lung disease: A clinicopathologic study of six cases. *Am Rev Respir Dis* 135:880–884, 1987.

123. Turner-Warwick M: Fibrosing alveolitis and chronic liver disease. *Q J Med* 145:133–149, 1968.

124. Golding PL, Smith M, Williams R: Multisystem involvement in chronic liver disease. *Am J Med* 55:772–782, 1973.

125. Rodriguez-Roisin R, Pares A, Bruguera M, et al: Pulmonary involvement in primary biliary cirrhosis. *Thorax* 36:208–212, 1981.

126. Eade OE, Smith CL, Alexander JR, et al: Pulmonary function in patients with inflammatory bowel disease. *Am J Gastroenterol* 73:154–156, 1980.

127. Stevens FM, Connolly CE, Murray JP, McCarthy CF: Lung cavities in patients with coeliac disease. *Digestion* 46:72–80, 1990.

128. Marsh P, Johnston I, Britton J: Atopy as a risk factor for cryptogenic fibrosing alveolitis. *Respir Med* 88:369–371, 1994

Chapter *13*

Hypersensitivity Pneumonitis (Extrinsic Allergic Alveolitis)

P. S. Hasleton

Farmer's lung, following exposure to moldy hay, was first described in 1932 by Campbell.[1] An increasing number of different occupational and environmental antigens have since been associated with hypersensitivity pneumonitis. Despite increasing recognition of the disease, its epidemiology, risk factors, pathogenesis, and prognosis remain poorly understood.

EPIDEMIOLOGY

There have been few population-based studies of hypersensitivity pneumonitis.[2] When a strict case definition was followed, the prevalence of farmer's lung in a population of Wisconsin dairy farmers was 42 out of 100,000.[3] More recently, outbreaks of hypersensitivity pneumonitis secondary to bacterial contamination in office or industrial populations, with "attack rates" of up to 70 percent in exposed subjects, have been described.[4]

Antigens

There are numerous antigens capable of inducing hypersensitivity pneumonitis. A list is given in Table 13-1, but further putative antigens are continually being added. The majority of antigens, such as thermophilic spores, are small enough to enter alveoli. Others, such as spores of *Alternaria,* causing "woodworker's lung" in handlers of wood pulp, deposit in the airways and then become solubilized. Chemicals are inhaled as fine mists or dust particles with absorbed antigen or antigenic determinants.

Most of the organisms associated with hypersensitivity pneumonitis are thermophilic, thriving best at 56 to 60°C. *Actinomycetes* require a moisture content of at least 30 percent to thrive. They are abundant in wet hay and dispersed into the air when the hay pile is disturbed (Fig. 13-1). Organisms thrive in ventilation systems containing some stagnant water and at a temperature of around 60°C. Bird breeder's lung, caused by avian proteins, may affect not only the person closely in contact with the pigeons, geese, etc., but also his or her family, exposed to antigens on clothing.[5] Drugs, including amiodarone and bleomycin, and chemicals as well as yet unidentified antigens have all been incriminated in hypersensitivity pneumonitis.

One group of chemicals, the isocyanates, deserve special mention because they may cause occupational asthma and, less commonly, hypersensitivity pneumonitis. They are used for the large-scale production of polyurethane polymers. Their chemical structure, industrial uses, and pathogenesis have been reviewed.[6]

The disease has been diagnosed on open lung biopsies when no causative antigen has been identified.[7]

CLINICAL PRESENTATION

Hypersensitivity pneumonitis has been divided into three syndromes—acute, subacute, and chronic.[8] In practice, only the acute and chronic phases are discernible. This rough classification has stood the test of time but should not be taken to imply that the disease progresses from acute through subacute to chronic. This sequence is uncommon, and patients with acute hypersensivity pneumonitis do not usually progress to chronic disease. Conversely, patients with chronic disease may not have had an acute exposure.[9] Long-term low-level exposure can lead to insidious chronic disease.[10] The disease is usually seen in adults but has been described in infants.[11]

In an acute attack, there is a history of exposure to the causative allergen 4 to 8 h before the onset of symptoms. This lag period often makes it difficult for patients to identify the offending antigen.

The patient complains of cough, dyspnea, chest tightness, chills, and fever. The cough is nonproductive, and hemoptysis is rare. The symptoms last from 12 to 60 h but recur with repeated exposure to the antigen. Fine crepitant rales are heard on auscultation.

TABLE 13-1
Some Etiologic Agents in Hypersensitivity Pneumonitis

Major antigens	Exposure or source	Disease
Thermophilic bacteria		
Micropolyspora faeni	Moldy hay	Farmer's lung
Thermoactinomyces faeni	Moldy grain	Grain handler's lung, Thresher's lung
M. faeni and *T. vulgaris*	Mushroom compost	Mushroom worker's lung
Thermoactinomyces sacchari	Moldy sugar cane (bagasse)	Bagassosis
T. Vulgaris, M. faeni, and *Aureobasidium pullulans*	Heated water reservoirs	Humidifier or air-conditioner lung
T. viridis and *P. frequetens*	Cork tree bark	Suberosis
Other bacteria		
Bacillus subtilis	Water	Detergent worker's lung
Bacillus sereus	Water reservoir	Humidifier lung
True fungi		
Cryptostroma corticale	Moldy bark	Maple bark stripper's lung
Aspergillus clavatus	Moldy malt, barley	Malt worker's lung
Aureobasidium pullulans and *Graphium* spp.	Moldy redwood dust	Sequoiosis
Mucor solonifer	Moldy paprika pods	Paprika splitter's lung
Penicillum caseii	Cheese mold	Cheese worker's lung
Penicillum frequentans	Moldy cork dust	Suberosis
Aspergillus spores	Water reservoir	Aspergillosis
Trichosporon cutaneum	House dust, bird droppings	Japanese summer-type hypersensitivity pneumonitis[56]
Aspergillus spp.	Moldy hay	Farmer's lung
Asp. versicolor	Moldy straw	Dog house disease
Animal proteins		
Avian proteins (feathers, serum, and excreta)	Pigeons, parakeets	Bird breeder's lung
Chicken feathers and serum	Chickens	Chicken plucker's lung
Turkey feathers and serum	Turkeys	Turkey handler's lung
Duck feathers	Ducks	Duck fever
Rat urine and serum	Rats	Rodent handler's disease
Porcine and bovine protein	Pituitary snuff	Pituitary snuff-taker's lung
Fish meal dust	Fish meal	Fish meal worker's lung
Animal fur dust	Animal pelts	Furrier's lung
Silkworm antigens	Silk products	Sericulturist's lung[57]
Amoeba		
Acanthamobea castellani	Water	Humidifier lung
Naegleria gruberi	Water	Humidifier lung
Bacterial products		
Lipopolysaccharide	Cotton brac	Byssinosis
Streptomyces verticillus glycopeptides	Bleomycin	Bleomycin hypersensitivity (in contrast to fibrosis)
Grass (*Stipatenacissama*)	Grass	Hypersensitivity pneumonitis[58]
Other proteins		
Insect products		
Sitophilus granarius	Contaminated grain	Miller's lung (Wheat weevil disease)
Chemicals		
Trimellitic anhydride	Plastics	Chemical worker's lung
Toluene diisocyanate and methylene diisocyanate	Polyurethane foam or rubber manufacture	Hypersensitivity pneumonitis
Zirconium	Ceramic tile work	Hypersensitivity pneumonitis

FIGURE 13-1
Spores being disturbed in a haystack. *(Reproduced with permission in Dr. A. Gibbs Penarth, Wales.) (See also Color Plate 2.)*

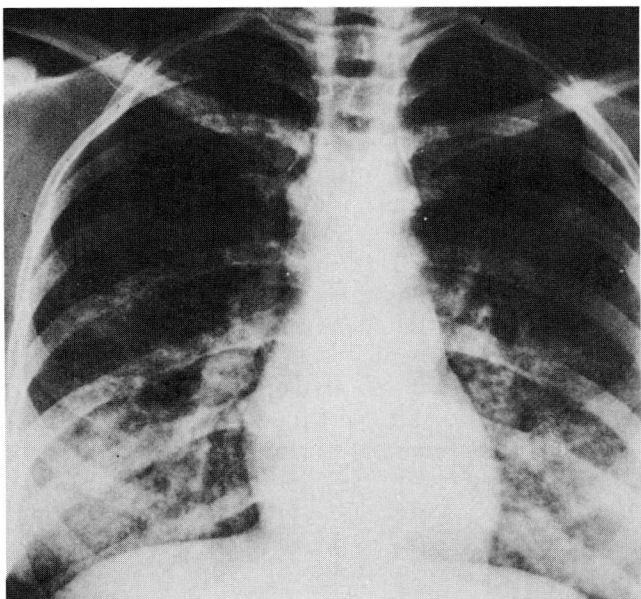

FIGURE 13-2
Chest radiograph of acute hypersensitivity pneumonitis with lower lobe disease.

Chest radiographs show bilateral ground-glass haziness, fine nodular shadows 1 to 4 mm in diameter, and linear striations. The disease spares the apices (Fig. 13-2).

In subacute disease the symptoms are similar but less severe and last over a 10- to 14-day period.

In chronic disease there is dyspnea, chronic cough, fatigue, anorexia, and weight loss. There may be respiratory failure, with cyanosis and clubbing. Often a clinical diagnosis of underlying malignancy, sarcoidosis,[12] or tuberculosis is made.

Chest radiographs show coarse reticulonodular upper and middle zone infiltrates progressing to lung shrinkage and honeycomb lung. The sparing of the lower zones militates against a diagnosis of cryptogenic fibrosing alveolitis. Some cases have hilar lymphadenopathy.

Mortality of patients with chronic disease is 9 percent,[13] although a recent Finnish study of mortality due to farmer's lung gave a mortality of 0.7 percent.[14] On average, death occurred eight years after the diagnosis of farmer's lung.

Braun and coworkers[13] found that the most important factor causing chronic disease was a history of five or more acute symptomatic attacks.

The factors determining prognosis in the various forms of hypersensitivity pneumonitis are poorly understood. Age at diagnosis, duration of antigen exposure after onset of symptoms, and the total number of years of exposure before diagnosis have a predictive value for recovery in pigeon breeder's and farmer's lung.[15,16]

Hypersensitivity pneumonitis is not a single disease process but a complex syndrome. More studies are needed in the assessment of prognosis.

IMMUNOPATHOGENESIS

The immunology of hypersensitivity pneumonitis is poorly understood. The disease shows characteristics of both a type III antigen-antibody complex reaction and a type IV cell-mediated hypersensitivity response. There was once much diagnostic reliance on serum-precipitating antibodies. These are usually present in high titers against the causative antigen. IgA and IgG antibodies against *Actinomycetes fumigatus* correctly identify 94 percent of cases of farmer's lung.[17] A positive skin test may occur 4 to 8 h after injection of the causal antigen, suggesting that immune complexes are formed and may mediate the disease.[18]

The role of precipitins in hypersensitivity pneumonitis has been queried,[19] and some have concluded that they have no pathogenetic role. They can be used to identify agents in the environment to which the patient has been exposed. Precipitating antibodies are thought by some to reflect exposure, but not necessarily disease.[20]

The presence of a pulmonary lymphocytic infiltrate and ill-defined granulomas, as seen in hypersensitivity pneumonitis, are consistent with a cell-mediated immune response. Costabel[21] suggested that the initial response, between 4 and 48 h after antigen inhalation, is immune-complex mediated. This is characterized by an early increase in neutrophils in the bronchoalveolar lavage (BAL) fluid. Histopathologically there is intraalveolar edema, neutrophilic alveolitis, and, less commonly, vasculitis.

After 12 h to several days, the immune response probably shifts to a cell-mediated reaction. The alveolar infiltrate consists of cytotoxic-effector cells as well as the suppressor cells required to modulate the B-cell response of antibody production by plasma cells. Lymphocytes are prominent in BAL fluid in this phase.

After weeks to months there is a delayed hypersensitivity reaction, with an increase in lymphocytes and natural killer (NK) cells. Histology shows a mononuclear infiltrate with granulomas. Finally, with repeated immune-mediated alveolar wall injury and consequent release of proteolytic enzymes and fibroblastic growth factors, there may be end-stage pulmonary fibrosis. This may also cause an increase in neutrophils in the lavage.

LYMPHOCYTE RESPONSE

Most of the lymphocyte subsets show T-cell markers.[22] CD8+ T lymphocytes are the predominant cells, with resulting inversion of the CD4:CD8 ratio. In patients who are repeatedly exposed to specific antigens, there is a persistent increase of CD8+ cells and a reversal of the CD4:CD8 ratio.[23] Other patients, who continued to live in agricultural environments but were not exposed further to specific antigens at work, showed a recovery of CD4+ cells and a decrease of CD8+ cells. There was a consequent increase in the CD4:CD8 ratio to normal levels within 6 months. A persistent increase in the NK cells has also been demonstrated in patients with hypersensitivity pneumonitis.[23]

CD8+ T cells may have a protective effect on the lung, and high levels are associated with an absence of fibrosis. These cells probably have an important role in the pathogenesis of pulmonary fibrosis of hypersensitivity pneumonitis in the chronic phase.[24]

In addition to the increase in CD8+ lymphocyte numbers there is an increased number of activation markers in symptomatic antigen-exposed patients.[25] These include interleukin-2 receptor, very late activation antigen, and HLA-DR antigen.[26] A significant increase in cytotoxicity has been shown in patients with hypersensitivity pneumonitis.[27] Lavage lymphocytes from exposed but asymptomatic farmers show in vitro cytotoxicity similar to that of controls.[28] Lung T cells from patients with hypersensitivity pneumonitis have also shown in vitro suppressor activity.

Yamasaki and colleagues[29] showed decreased proliferative responsiveness to mitogens in T cells from the lower respiratory tract, compared with peripheral blood lymphocytes from the same patients. These BAL T lymphocytes had a high level of CD18 antigen. This antigen is one of the surface markers strongly expressed on "memory" T cells compared with "naive" T cells. "Memory" T cells show reduced responses to mitogens compared to "naive" T cells.[30] The presence of different T-cell subsets is important in the regulation of granuloma formation, probably by the release of lymphokines. Helper T cells are associated with active granuloma formation, whereas suppressor/cytotoxic T cells and NK cells are associated with regression of granulomas.[22] The suppressor/cytotoxic population of T cells found in the lungs of patients with hypersensitivity pneumonitis could be responsible for inhibiting granuloma development.

MAST CELLS

Mast cells, often ignored because they quickly degranulate in aqueous fixatives, are identified in hypersensivity pneumonitis. Mast cell numbers relate to the presence of edema and interstitial fibrosis. They are often seen in young connective tissue, in either the interstitium or the alveoli, and have few granules.[31]

Mast cell numbers may be helpful as a discriminant in the diagnosis of hypersensitivity pneumonitis.[32] Counts rarely exceed 0.5 percent in sarcoidosis, cryptogenic fibrosing alveolitis, asbestosis, or controls. There is an associated increased concentration of histamine in the lavage. The cells resemble "mucosal" rather than "connective tissue" mast cells.

OTHER FACTORS

There is an increased turnover of collagens and proteoglycans in patients with hypersensitivity pneumonitis of recent onset.[33] Fibroblast growth factors are increased.[34] The BAL concentration of procollagen-III-peptide correlated with the amount of vitronectin, fibronectin, and mast cells, but this has not been confirmed in all studies.[35] Vitronectin and fibronectin are components of the extracellular matrix, produced by alveolar macrophages, and mediate fibroblast attachment. Similar results are seen in sarcoidosis and cryptogenic fibrosing alveolitis.[36,37] Transforming growth factor–beta is produced in the course of experimental mouse hypersensitivity pneumonitis and contributes significantly to collagen synthesis.[38]

Hypersensitivity pneumonitis is less common in smokers.[39] Smoking increases pulmonary macrophages, which has an effect on the T-cell subsets responsible for the maintenance of such granulomas.

Host factors may contribute to pathogenesis of the disease. Only a relatively small percentage of antigen-exposed individuals develop hypersensitivity pneumonitis, suggesting that ill-defined genetic factors play a role in this disease. There is an increased incidence of HLA-B8 in patients with farmer's lung.[40] Other studies have failed to confirm any link between histocompatibility loci with disease activity in hypersensitivity pneumonitis.[41]

DISTRIBUTION OF PARTICLES IN THE LUNG

Airway morphology is important in relation to the lodging of spores in terminal parts of the lung. Gravity plays an important role in the distribution of particles because of its effect on lung ventilation. Ventilation is approximately three times greater at the lung bases than in the apices.[42] Most inhaled organic antigen is deposited in the dependent regions of the lung. With chronic exposure, clearance

mechanisms may be impaired. Particulate substances are removed from the distal air spaces via the lymphatics, but this function is not uniform throughout the lung.[43] Lymphatic clearance of particulate matter is slower from the upper than from the lower zones. Regionally, antigen concentration is most likely to depend on the balance between deposition (which is predominantly in the lower zones) and clearance (lymphatic clearance is also predominantly lower zonal). Therefore the upper lobes are more severely affected in chronic disease, owing to retention of a greater number of particles in this area. This disease distribution is also seen in other diseases, such as coal worker's pneumoconiosis and silicosis.

BRONCHOALVEOLAR LAVAGE

Patients with acute disease have a raised total cell count ($93.1 \pm 55.7 \times 10^6$/mL, compared with normal control values of $10.9 \pm 6.2 \times 10^6$/mL). The predominant cells are lymphocytes (66.2 percent), macrophages (23.3 percent), and neutrophils (10.1 percent) (controls: macrophages 88.6 percent, lymphocytes 9.6 percent, and neutrophils 1.7 percent) (Fig. 13-3).[44] The lymphocyte count alone is insufficient to differentiate sarcoid from hypersensitivity pneumonitis. A neutrophilia is often present in cryptogenic fibrosing alveolitis.

Plasma cells may be identified in lavage fluid from patients with hypersensitivity pneumonitis, who have increased absolute numbers of lympho-

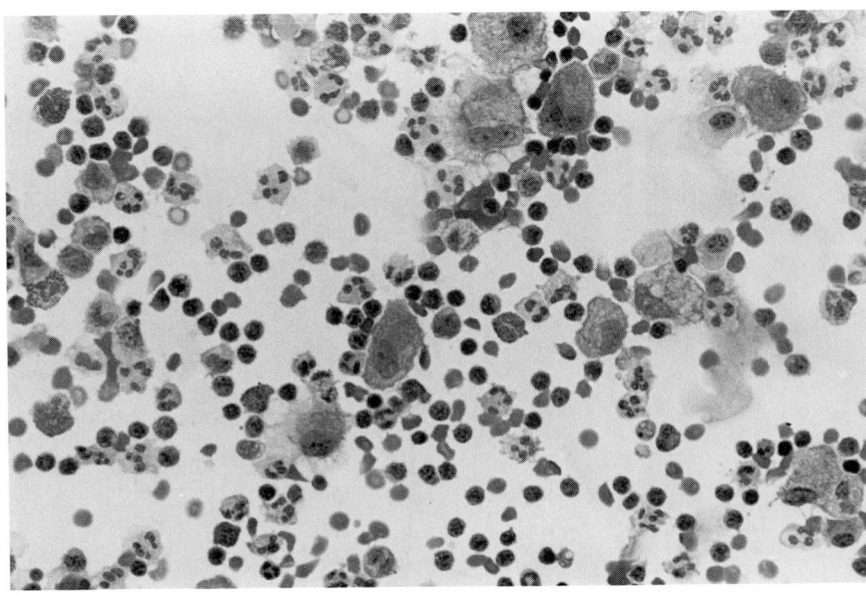

FIGURE 13-3
Lavage in hypersensitivity pneumonitis showing a lymphocytosis.

cytes, eosinophils, and mast cells.[45] In addition, they have lower levels of alveolar macrophages than do patients with hypersensitivity pneumonitis but with no plasma cells in their lavage. Plasma cells are associated with a more intense alveolitis. Mast cells are also increased (Fig. 13-4).[32]

Patients who had previous farmer's lung disease and were still in daily contact with hay had lower lavage counts (total cell count: 51.7×10^6/mL, macrophages 49.6 percent, lymphocytes 49.9 percent, and neutrophils 2.8 percent). These counts were still elevated as compared with those of controls.[44]

Patients with serum precipitins to *Micropolyspora faeni* showed a raised total cell count (24.8×10^6/mL) but a large percentage of macrophages 81.4 percent and fewer lymphocytes (15.6 percent) and only 3 percent neutrophils. The percentage of lavage lymphocytes correlated with the radiographic changes.[46] The lung disease tends to be poorly responsive to treatment if the lymphocyte and neutrophil counts remain high.[47]

Surfactant protein-A (SP-A) is increased in alveolar macrophages in hypersensitivity pneumonitis.[48] SP-A is important in the maintenance of the alveolar surfactant monolayer. It may modulate the activity of immunoreactive cells and increase migration and phagocytotic and bactericidal activity of macrophages. The increase in the number of SP-A–containing macrophages could be due to activation of type II cells or a decrease in reuptake of SP-A by these cells, with a secondary increase in BAL.

BAL hyaluronic acid was raised in patients with farmer's lung but not in farmers with asymptomatic alveolitis.[49] The latter group had no pulmonary symptoms but raised total cell, lymphocyte, and neutrophil counts were detected in BAL.

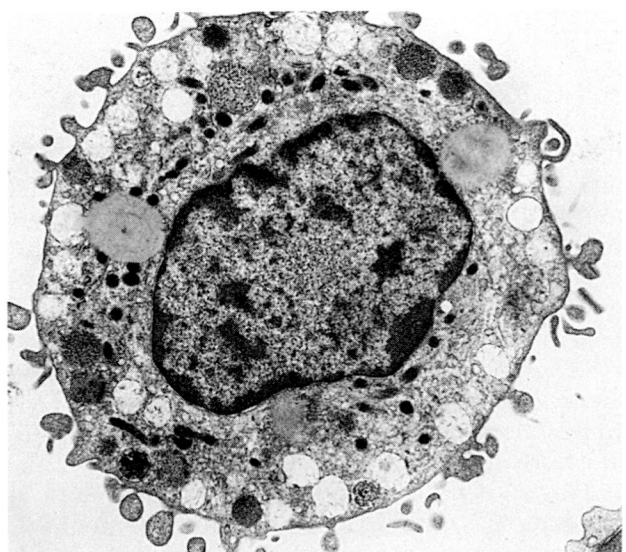

A

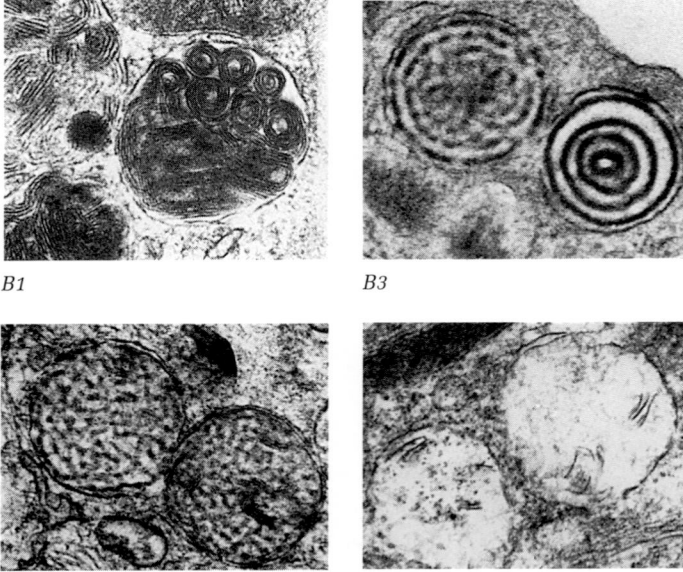

B1   *B3*

B2   *B4*

FIGURE 13-4
A. Mast cell recovered from BAL in a case of hypersensitivity pneumonitis. There are surface projections, lipid inclusions, and a variety of granules. (Uranyl acetate and lead citrate, ×8500.)
B. Mast cell granules. 1: Typical scroll form. 2: Particulate form. 3: Coil pattern. 4: Pattern with no granules. (Uranyl acetate and lead citrate, ×8500.)
(Reprinted with permission from P. Haslam, A. Dewar, P. Butchers, Z. S. Primett, A. Newman-Taylor, M. Turner-Warwick: Am Rev Repir Dis 135:35–47, 1987.)

PATHOLOGY

The gross features seen in hypersensitivity pneumonitis are variable. In the acute phase, the lung may appear normal or show areas of consolidation with varying degrees of nodularity. It is rare to see lung tissue in the acute stage.

Seal and colleagues[50] noted that the lung at thoracotomy looked yellow and indurated on palpation, with the lower lobe more affected than the upper. The yellow appearance is probably due to areas of obstruction, with lipid-laden histiocytes filling alveolar lumina. The lung was difficult to inflate in the acute stage. Some cases showed pleural adhesions.

In the chronic phase there is marked interstitial fibrosis in the lower and upper lobes, with superimposed cyst formation. The major bronchi are not ectatic.

Other cases show yellow consolidated foci and fine interstitial fibrosis. The fibrosis, when early, tends to invade the lateral portions of the lungs. There are focal areas of fibrosis around bronchi and pulmonary blood vessels.

HISTOLOGY

The largest series describing the abnormalities in the acute stage is that documented by Reyes and coworkers.[51] The most common feature they noted was an interstitial infiltrate consisting primarily of lymphocytes and plasma cells with occasional neutrophils (Figs. 13-5 and 13-6) and eosinophils. An early neu-

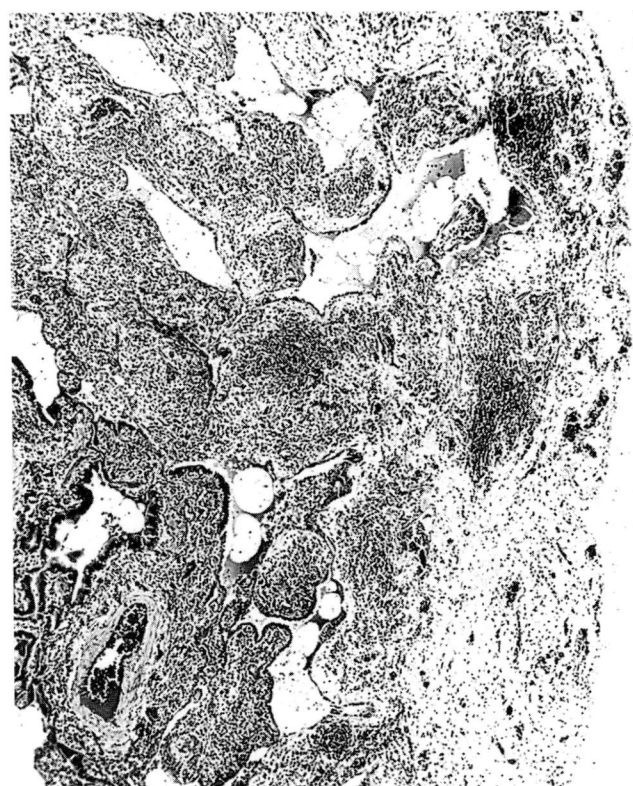

FIGURE 13-5
Low-power histology of farmer's lung with prominent interstitial chronic inflammation. (H&E, ×42.)

trophilic alveolitis due to activation of the alternative pathway of complement has been described.[52]

The intervening lung parenchyma is unimpaired. In some areas the alveolar epithelium is destroyed. Edema of the interstitial spaces and interstitial fibrosis (Fig. 13-7) are common, as is bronchi-

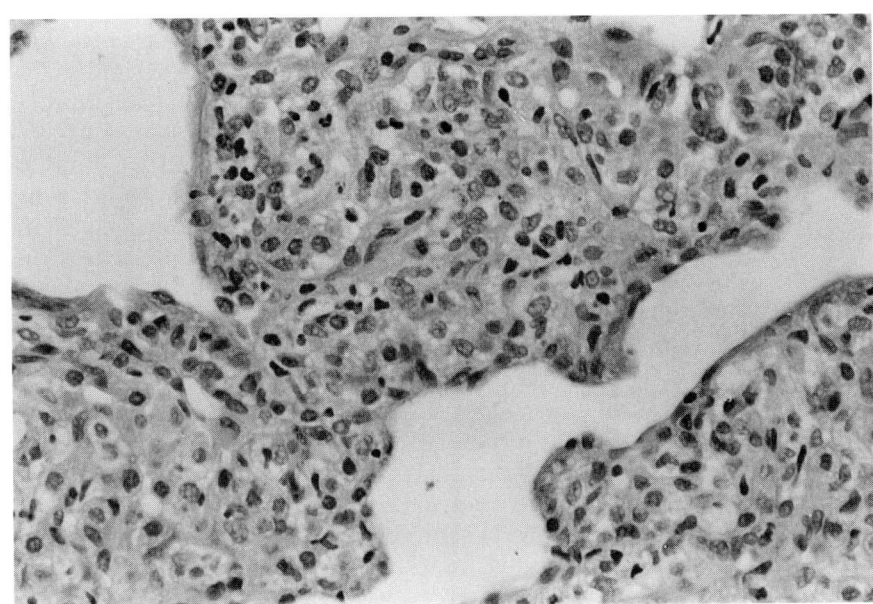

FIGURE 13-6
Thickening of the interstitium in hypersensitivity pneumonitis with an infiltrate of plasma cells, lymphocytes, and neutrophils. Eosinophils are not easily identified in this picture. (H&E, ×219.)

439

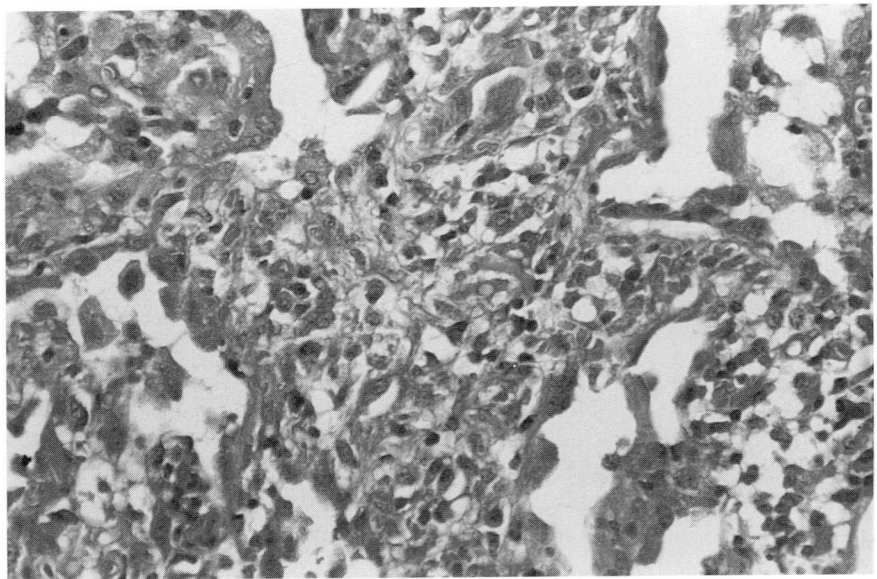

FIGURE 13-7
Interstitial fibrosis in a case of hypersensitivity pneumonitis. There is a mild interstitial inflammatory infiltrate and some prominent alveolar lining cells. (H&E, ×219.)

olitis obliterans (Fig. 13-8). Foam cells—large histiocytic cells with abundant foamy cytoplasm—are found in clusters within the alveolar spaces (Fig. 13-9). This finding has been associated with pigeon breeder's disease but is not specific for that entity.

An unresolved pneumonia, characterized by organizing fibrinous exudate and large numbers of neutrophils within the alveoli, is also seen. Patchy intraalveolar edema was also seen.

Granulomas are seen in 70 percent of the biopsies (Fig. 13-10). They are characterized by epithelioid cells, which are often loose-knit. The giant cells are either Langerhan's or of the foreign-body variety. Both could be found in the same case. Foreign-body

material, usually within the cytoplasm of giant cells (Fig. 13-11), is found in 60 percent of the biopsies. In some cases it is birefringent and ranges from 8 to 12 μm in length and 3 to 4 μm in width.

The giant cells are mainly clustered around small bronchi and bronchioles or within the interstitium, though occasionally they can be seen within an alveolar space. The granulomas are nonnecrotizing, and in some cases giant cells are associated with Schaumann bodies, calcium oxalate crystals, cholesterol clefts, and asteroid body inclusions.

There is an increase in reticulin in the interstitium and around the granulomas (Fig. 13-12). In the acute stage, collagen formation is minimal. Masson bodies (intraalveolar buds of collagen) are seen in some cases in the acute phase (Fig. 13-13). In occasional cases, hyaline membranes are present. Some of the alveoli contain large mononuclear cells with PAS-positive granules. In one case described by Seal and coworkers,[50] there was fine interstitial fibrosis and dense lymphocytic infiltration.

The chronic phase shows interstitial fibrosis, with collagenous thickening of interalveolar septa (Fig. 13-14) and a mild focal lymphoplasmacytic infiltrate, which is perivascular in some areas. Confluent foci of fibrosis are also seen.

Pulmonary arteries show changes of pulmonary hypertension, with intimal fibrosis in muscular pulmonary arteries. Granulomas are difficult to identify. Vasculitis, a feature of immune-complex-mediated injury, is not usually present in biopsies from hypersensitivity pneumonitis.

As the disease progresses, there is diffuse interstitial fibrosis with formation of cystic spaces, many lymphocytes but few plasma cells. There is also cuboidalization of the epithelium. Fibrosis is most

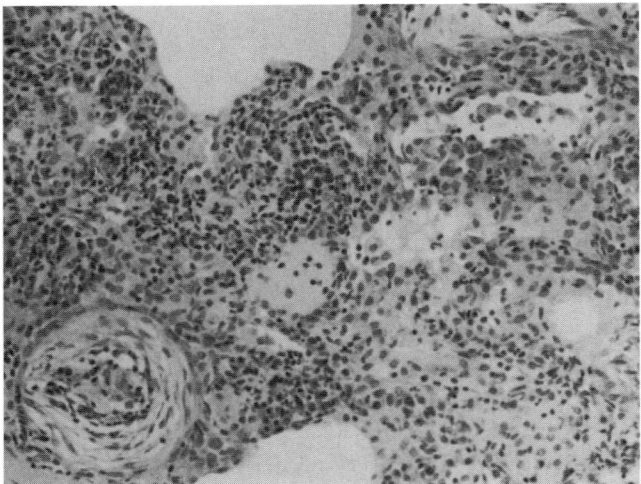

FIGURE 13-8
Bronchiolitis obliterans in hypersensitivity pneumonitis. (H&E, ×125 approx.)

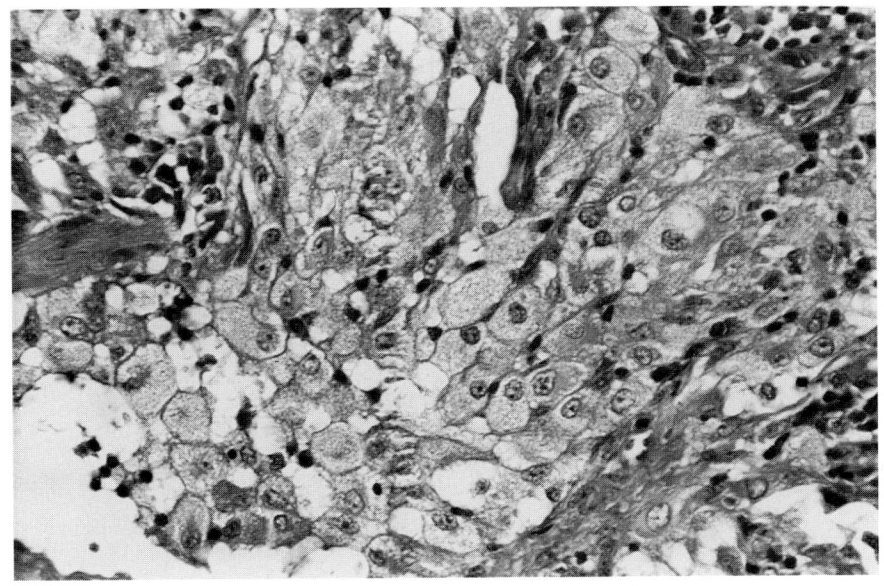

FIGURE 13-9
Foamy macrophages in peripheral air space in hypersensitivity pneumonitis. There is some focal interstitial fibrosis. (H&E, ×219.)

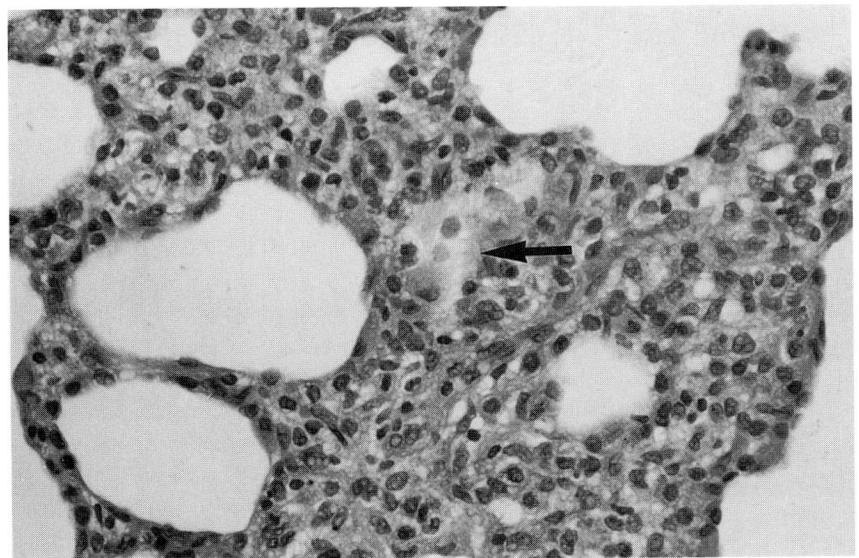

FIGURE 13-10
Loose-knit granulomas (*arrow*). (H&E, ×219.)

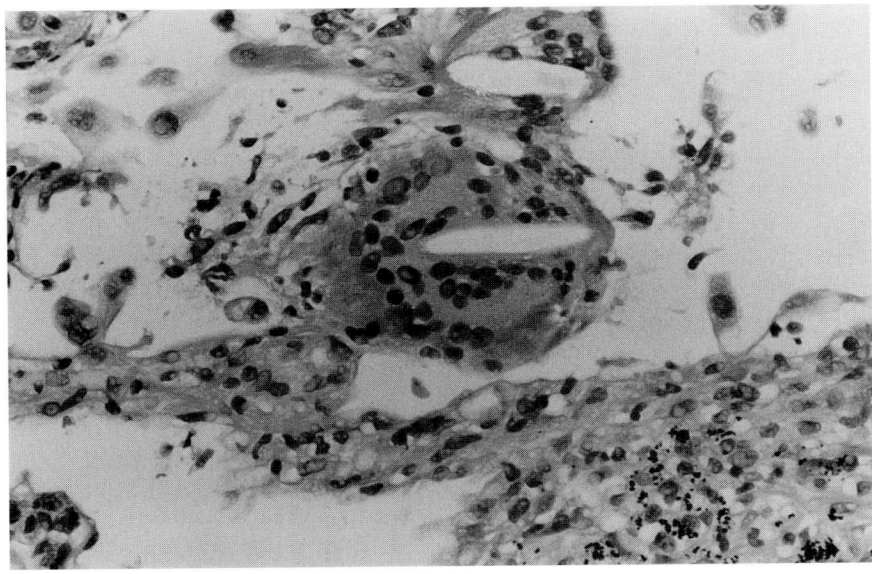

FIGURE 13-11
Hypersensitivity pneumonitis. Foreign-body giant cells associated with cholesterol clefts and some adjacent interstitial fibrosis. (H&E, ×219.)

441

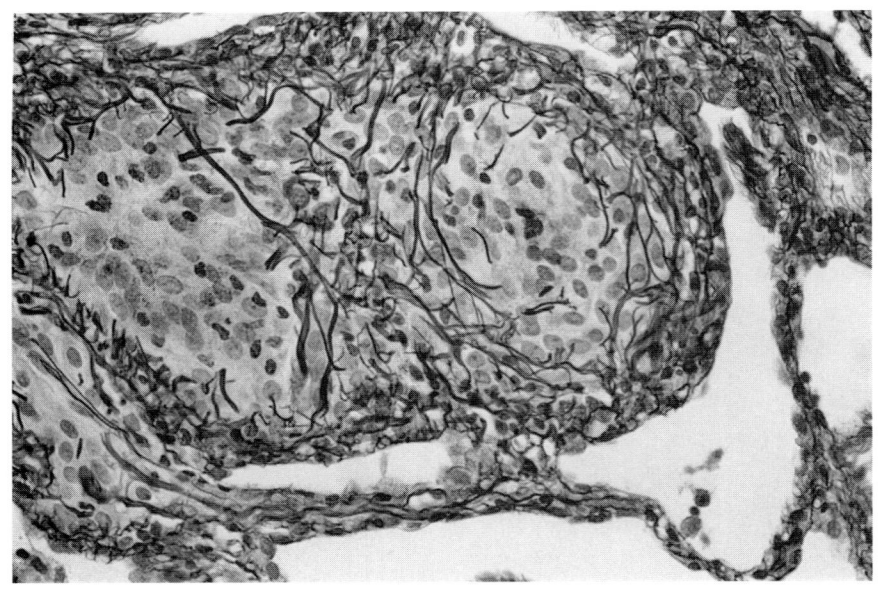

FIGURE 13-12
Increased reticulin in the interstitium in hypersensitivity pneumonitis. (Reticulin, ×219.)

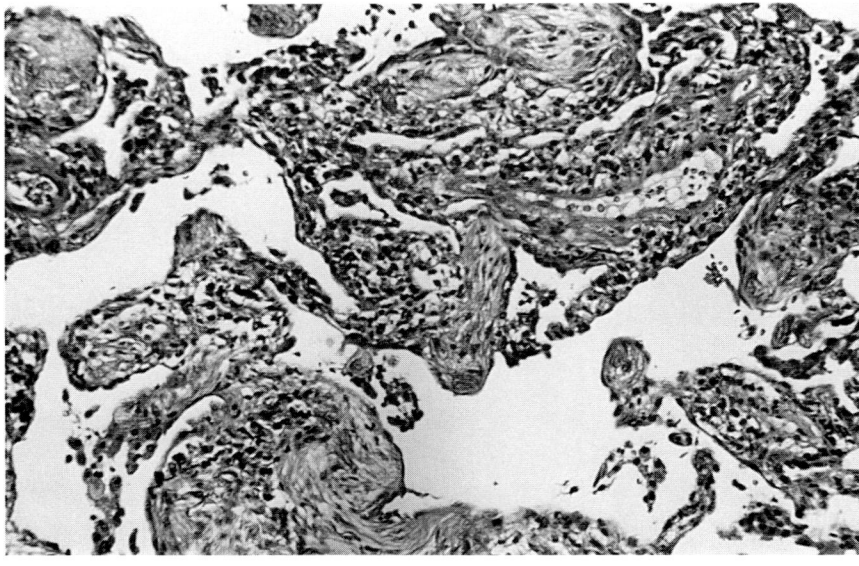

FIGURE 13-13
Masson bodies in hypersensitivity pneumonitis.

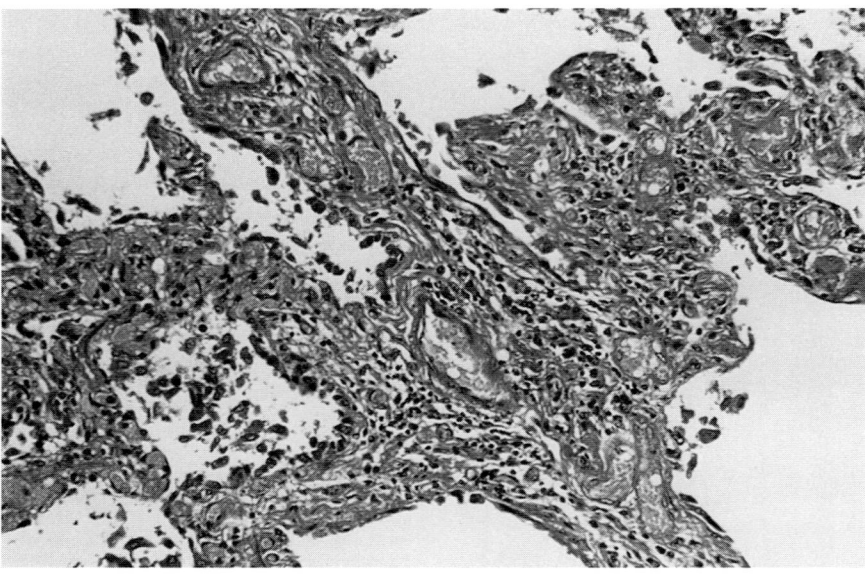

FIGURE 13-14
Chronic phase of hypersensitivity pneumonitis with marked interstitial fibrosis. This man kept pigeons in his loft for many years. (H&E, ×88.)

prominent around respiratory bronchioles and pulmonary vessels. The larger areas of fibrosis contain small bacillary-sized, doubly refractile particles. Hilar nodes show collagenous foci with spindle-shaped cells containing refractile particulate matter.

The major bronchial mucous glands are increased in number.

ELECTRON MICROSCOPY

Several studies of the ultrastructure of this disease have been documented.[53–55] The study by Kawanami and colleagues[53] examined 18 cases of chronic disease. Most of the patients were allergic to bird products. Histologically there was both intraalveolar and interstitial lung disease. In the second study,[54] most of the patients had been exposed to moldy hay but none had fibrosis; therefore, this study is more of an indicator of the early changes.

The early changes were loss of microvilli in the ciliated bronchiolar epithelium and many lymphocytes, macrophages, and giant cells in the bronchiolar lumina. In the bronchioles there were granulomas, detachment of basal cells from one another, and disintegration of the basement membrane. Type I pneumocytes separated from the alveolar wall, leaving a bare basement membrane. In the alveoli there was hyperplasia of type II pneumocytes, detached from the basement membrane in areas (Fig. 13-15). Lymphocytes and mast and plasma cells predominated in the interstitium (Fig. 13-16) Interstitial macrophages were seen in seven of 13 patients. The

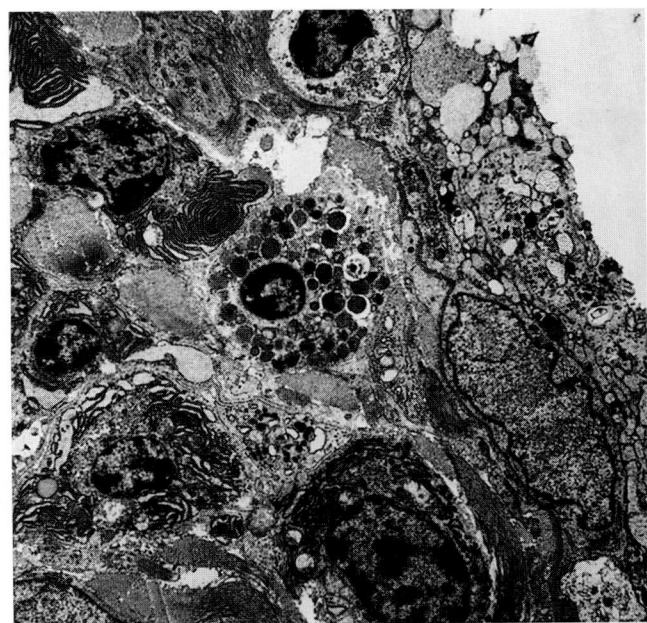

FIGURE 13-15
Hypersensitivity pneumonitis. There is degeneration of the alveolar lining cell, probably a type II pneumocyte. The interstitium is thickened with edema, plasma, and mast cells. Some small foci of collagen are identified. (Uranyl acetate and lead citrate, ×3330.)

lymphocytes were round or oval with thin surface projections or philopodia. The nuclei were occasionally bilobed, eccentric, or horseshoe-shaped. Granulomas were also present (Fig. 13-17). In some cases these contained foreign material, resembling hyphal fragments. Intraalveolar edema was seen. Neutrophils were frequently identified in the alveolar lumina but not in the interstitium.

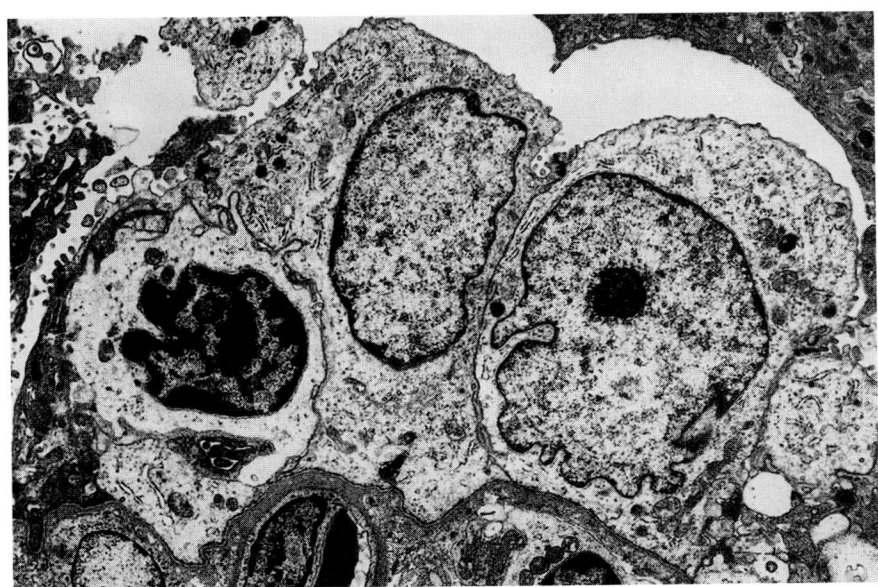

FIGURE 13-16
Hypersensitivity pneumonitis. Prominent alveolar lining cells with large nuclei but very few identifiable lamellar bodies. (Uranyl acetate and lead citrate, ×4900.)

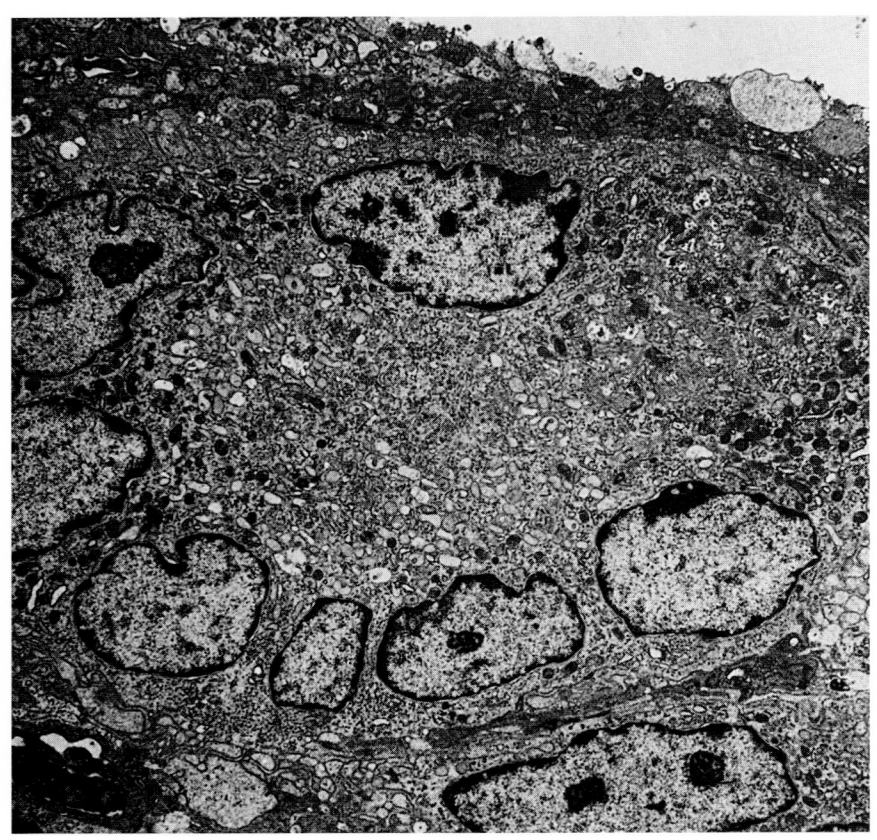

FIGURE 13-17
Granuloma lying beneath the epithelium in a case of hypersensitivity pneumonitis. (Uranyl acetate and lead citrate, ×2886.)

Intraalveolar macrophages and giant cells containing lamellar bodies as well as electron-dense inclusions, probably representing ingested foreign material, were noted. Eosinophils were rare. Even in early-stage disease there was an increase in colla-gen and elastic fibers. With increasing duration of the disease, fibroblasts become more prominent (Fig. 13-18).

Similar findings were found in the other study[53]—although, as would be expected in an exam-

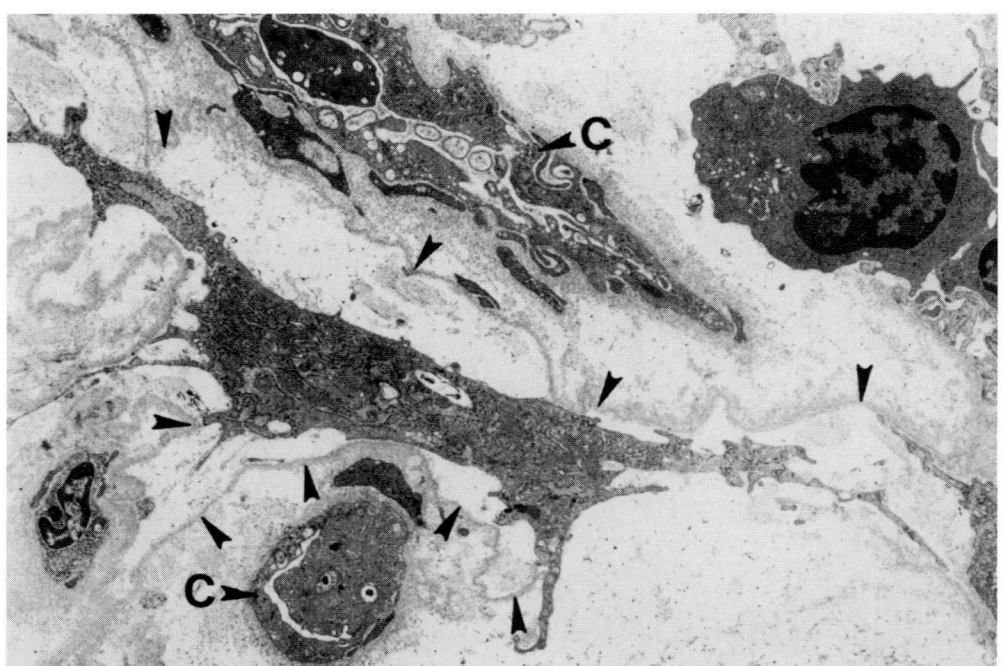

FIGURE 13-18
Fibroblast with elongated cytoplasmic processes appears to be passing through a defect in the denuded alveolar epithelial basement membrane from a patient with hypersensitivity pneumonitis. The fibroblast is adjacent to the luminal sides of the two basement membranes (arrowheads), and a capillary (C) is seen on the septal side of each basement membrane. (Uranyl acetate and lead citrate, ×9000.)

ination of later stages of the disease, there was more fibrosis. Polymorphs were uncommon, but eosinophils were readily identified.

The third study[55] identified changes ultrastructurally at the level of the carina of the right upper lobe. Ciliated goblet cells and in some cases even basal cells had disappeared. The pathogenesis of the epithelial damage was obscure.

DIFFERENTIAL DIAGNOSIS

Granulomatous Lung Disease

The granulomas in extrinsic allergic alveolitis are loose-knit and usually interstitial. If there is necrosis and prominent acute inflammation, a search for an infective cause should be made. The most common diseases causing confusion are tuberculosis and disorders caused by fungi such as *Aspergillus,* bacteria such as *Nocardia* and *Actinomyces,* and in rare instances, parasites. Miliary infections with herpesvirus or cytomegalovirus in HIV-positive patients may be associated with granulomatous necrotizing lesions. *Pneumocystis* cause a granulomatous reaction in some HIV-positive patients. In sarcoidosis and berylliosis, the granulomas are better defined and there is less of an interstitial infiltrate or cellular bronchiolitis. The granulomas in sarcoid follow the lymphatics, bronchi, and blood vessels. Despite this anatomic location, some cases of sarcoid may be difficult to differentiate from hypersensitivity pneumonitis. Intravenous drug abuse may present with granulomas and some chronic inflammation, but polarization will show intravascular foreign material.

A granulomatous reaction is part of localized pulmonary interstitial emphysema. In this disease, large cysts destroy the lung and the granulomas are identified around air-filled spaces and not in the interstitium, as in hypersensitivity pneumonitis.

Lymphoid Interstitial Pneumonia

In this disease, unlike hypersensitivity pneumonitis, the infiltrate is diffuse and the lymphocytes are atypical. There is no peribronchiolar distribution. The interstitial infiltrate is thicker in lymphoid interstitial pneumonia than in hypersensitivity pneumonitis.

Fibrosing Alveolitis (Cryptogenic Interstitial Pulmonary Fibrosis)

Fibrosing alveolitis may be very difficult to differentiate from hypersensitivity pneumonitis in end-stage disease. The pathologist must rely heavily on the clinical history, which is not always helpful. The interstitial infiltrate tends to be less cellular than in hypersensivity pneumonitis and does not show a peribronchiolar distribution.

Bronchiolitis Obliterans–Organizing Pneumonia

This shows well-marked Masson balls in alveoli as well as fibrosis in bronchioles. There is a minimal cellular reaction and no granulomas. The interstitium is not as thickened in bronchiolitis obliterans–organizing pneumonia as in hypersensitivity pneumonitis. Bronchiolitis obliterans may be seen in some cases of hypersensitivity pneumonitis.

Adult Respiratory Distress Syndrome

In the early stages of ARDS there are hyaline membranes, intraalveolar fibrosis and edema, interstitial edema, and some inflammation. This is compared with the more scanty interstitial infiltrate in hypersensitivity pneumonitis. Granulomas are uncommon in ARDS.

Drug Reactions

Hypersensitivity reactions occur to some drugs, such as gold, procarbazine, and amiodarone, and should be considered an integral part of hypersensitivity pneumonitis.

Diffuse Panbronchiolitis

This disease, rare in the West, is seen in Japan. Hypersensitivity pneumonitis can show areas with a diffuse panbronchiolitis picture, but it is only a focal component; other features, such as granulomas or interstitial thickening, predominate.

REFERENCES

1. Campbell JM: Acute symptoms following work with hay. *Br Med J* 2:1143–1166, 1932.
2. Rose C, King TE Jr: Controversies in hypersensitivity pneumonitis. *Am Rev Respir Dis* 145:1–2, 1992.
3. Gruchow HW, Hoffman RG, Marx JJ, Emanuel DA, Rimm AA: Precipitating antibodies to farmer's lung antigens in a Wisconsin farming population. *Am Rev Respir Dis* 124:411–415, 1981.
4. Kreiss K, Hodgson MJ: Building-associated epidemics, in Walsh PJ, Dudney CS, Copenhaver ED (eds), *Indoor Air Quality.* Boca Raton, Fla., CRC Press, 1984, pp 87–108.
5. Riley DJ, Saldana M: Pigeon breeder's lung: Subacute course and the importance of indirect exposure. *Am Rev Respir Dis* 107:456–460, 1973.
6. Vandenplas O, Malo J-L, Saetta M, Mapp CE, Fabbri LM: Occupational asthma and extrinsic allergic alveolitis: Current status and perspectives. *Br J Ind Med* 50:213–228, 1993.

7. Klenerman P, Woodhead MA, Spiteri M, et al: Extrinsic allergic alveolitis without an identified causative antigen. *Thorax* 46:313P, 1991.

8. Fuller CH: Farmer's lung: A review of the present knowledge. *Thorax* 8:59–64, 1953.

9. Bourke SJ, Banham SW, Carter R, Lynch P, Boyd G: Longitudinal course of extrinsic alveolitis in pigeon breeders. *Thorax* 44:415–418, 1989.

10. Schleuter DP: Infiltrative lung disease hypersensitivity pneumonitis. *J Allergy Clin Immunol* 70:50–55, 1982.

11. Eisenberg JD, Montanero A, Lee RG: Hypersensitivity pneumonitis in an infant. *Pediatr Pulmonol* 12:186–190, 1992.

12. Forst LS, Abraham J: Hypersensitivity pneumonitis presenting as sarcoidosis. *Br J Ind Med* 50:497–500, 1993.

13. Braun SR, doPico GA, Tsiatis A, et al: Farmer's lung disease: Long-term clinical and physiologic outcome. *Am Rev Respir Dis* 119:185–191, 1979.

14. Kokkarinen J, Tukiainen H, Terho EO: Mortality due to farmer's lung in Finland. *Chest* 106:509–512, 1994.

15. deGracia J, Morell S, Bofill JM, et al: Time of exposure as a prognostic factor in avian hypersensitivity pneumonitis. *Respir Med* 83:139–143, 1989.

16. Allen DH, Williams GV, Woolcock AJ: Bird breeder's hypersensitivity pneumonitis: Progress studies of lung function after cessation exposure to the provoking antigen. *Am Rev Respir Dis* 114:555–566, 1976.

17. Ojanen T: Class specific antibodies in serodiagnosis of farmer's lung. *Br J Ind Med* 49:332–336, 1992.

18. Pepys J: Hypersensitivity diseases of the lungs due to fungi and organic dusts. *Monogr Allergy* 4:1–199, 1969.

19. Burrell R, Rylander R: A critical review of the role of precipitins in hypersensitivity pneumonitis. *Eur J Respir Dis* 62:332–343, 1981.

20. Cormier Y, Bélanger J: The fluctuant nature of precipitating antibodies in dairy farmers. *Thorax* 44:469–473, 1989.

21. Costabel U: The alveolitis of hypersensitivity pneumonitis. *Eur Respir J* 1:5–9, 1988.

22. Semenzato G: Immunology of interstitial lung diseases: Cellular events taking place in the lung of sarcoidosis, hypersensitivity pneumonitis and HIV infection. *Eur Respir J* 4:94–102, 1991.

23. Trentin L, Marcer G, Chilosi M, et al: Longitudinal study of alveolitis in hypersensitivity pneumonitis patients: An immunological evaluation. *J Allergy Clin Immunol* 82:577–585, 1988.

24. Murayama J-i, Yoshizawa Y, Ohtsuka M, Hasegawa S: Lung fibrosis in hypersensitivity pneumonitis: Association with CD4+ but not CD8+ cell dominant alveolitis and insidious onset. *Chest* 104:38–43, 1993.

25. Millburn HJ: Lymphocyte subsets in hypersensitivity pneumonitis. *Eur Respir J* 5:5–7, 1992.

26. Trentin L, Migone M, Zambello R, et al: Mechanisms accounting for lymphocytic alveolitis in hypersensitivity pneumonitis. *J Immunol* 145:2147–2154, 1991.

27. Semenzato G, Trentin L, Zambello R, Agostini C, Cipriani A, Marcer G: Different types of cytotoxic lymphocytes recovered from the lungs of patients with hypersensitivity pneumonitis. *Am Rev Respir Dis* 137:70–74, 1988.

28. Semenzato G, Agostini C, Zambello R, et al: Lung T cells in hypersensitivity pneumonitis: Phenotypic and functional analyses. *J Immunol* 137:1164–1172, 1986.

29. Yamasaki H, Kinoshita T, Ohmura T, et al: Lowered responsiveness of bronchoalveolar lavage T-lymphocytes in hypersensitivity pneumonitis. *Am J Respir Cell Mol Biol* 4:417–425, 1991.

30. Sanders ME, Makjoba MW, Shaw S: Human naive and memory T cells: Reinterpretation of helper-inducer and suppressor-inducer subsets. *Immunol Today* 9:195–199, 1988.

31. Pesi A, Bertorelli G, Gabrielli M, Olivieri G: Mast cells in fibrotic lung disorders. *Chest* 103:989–996, 1993.

32. Haslam P, Dewar A, Butchers P, Primett ZS, Newman-Taylor A, Turner-Warwick M: Mast cells, atypical lymphocytes and neutrophils in bronchoalveolar lavage in extrinsic allergic alveolitis: Comparison with other interstitial lung diseases. *Am Rev Respir Dis* 135:35–47, 1987.

33. Teschler H, Thompson AB, Pohl WR, Konietzko N, Rennard SI, Costabel U: Bronchoalveolar lavage procollagen-III-peptide in recent onset hypersensitivity pneumonitis: Correlation with extracellular matrix components. *Eur Respir J* 6:709–714, 1993.

34. Cormier Y, Laviolette M, Cantin A, Tremblay GM, Begin R: Fibrogenic activities in bronchoalveolar lavage fluid of farmer's lung. *Chest* 104:1038–1042, 1993.

35. Lalancette M, Carrier G, Laviolette M et al: Farmer's lung: Long-term outcome and lack of predictive value of bronchoalveolar lavage fibrosing factors. *Am Rev Respir Dis* 148:216–221, 1993.

36. Eklund AG, Sigurdardottir O, Ohrn M: Vitronectin and its relationship to other extracellular matrix components in bronchoalveolar lavage fluid in sarcoidosis. *Am Rev Respir Dis* 145:646–650, 1992.

37. Pohl WR, Conlan MG, Thompson AB, et al: Vitronectin in bronchoalveolar lavage fluid is increased in patients with interstitial lung disease. *Am Rev Respir Dis* 143:1369–1375, 1991.

38. Denis M, Ghadirian E: Transforming growth factor-beta is generated in the course of hypersensitivity pneumonitis: Contribution to collagen synthesis. *Am J Respir Cell Mol Biol* 7:156–160, 1992.

39. Warren CPW: Extrinsic allergic alveolitis: A disease commoner in non-smokers. *Thorax* 32:567–569, 1977.

40. Flaherty DK, Iha T, Chmelik F, Dickie H, Reed CE: HLA-8 and farmer's lung. *Lancet* 2:507, 1975.

41. Stankus RP, Morgan JE, Salvaggio JE: Immunology of hypersensitivity pneumonitis. *Crit Rev Toxicol* 11:15–32, 1982.

42. West JB: Regional differences in gas exchange in the lung of erect man. *J Appl Physiol* 17:893–898, 1962.

43. Gurney JW, Schroeder BA: Upper lobe disease: Physiologic correlates. *Radiology* 167:359–366, 1988.

44. Cormier Y, Bélanger J, Leblanc P, Laviolette M: Bronchoalveolar lavage in farmers' lung disease: Diagnostic and physiological significance. *Br J Ind Med* 43:401–405, 1986.

45. Drent M, Wagenaar SS, van Velzen-Blad H, Mulder PGH, Hoogsteden HC, van den Bosch JMM: Relationship between plasma cell levels and profile of bronchoalveolar lavage fluid in patients with subacute extrinsic allergic alveolitis. *Thorax* 48:835–839, 1993.

46. Cormier Y, Bélanger J, Tardif A, Leblanc P, Laviolette M: Relationships between radiographic change, pulmonary function and bronchiolar lavage fluid lymphocytes in farmer's lung disease. *Thorax* 41:28–33, 1986.

47. Hunninghake GW, Bedell N: Interstitial lung disease: Concepts of pathogenesis. *Semin Respir Med* 6:31–39, 1984.

48. Guzman J, Wang YM, Kalaycioglu OY, et al: Increased surfactant protein-A content in human alveolar macrophages in hypersensitivity pneumonitis. *Acta Cytol* 36:668–673, 1992.

49. Larsson K, Eklund A, Malmberg P, Bjermer L, Lundgren R, Belin L: Hyaluronic acid (hyaluronan) in BAL fluid distinguishes farmers with allergic alveolitis from farmers with asymptomatic alveolitis. *Chest* 101:109–114, 1992.

50. Seal ME, Hapke EJ, Thomas GO, Meek JC, Heyes M: The pathology of the acute and chronic stages of farmer's lung. *Thorax* 23:469–489, 1968.

51. Reyes CN, Wenzel FJ, Lawton BR, Emanuel DA: The pulmonary pathology of farmer's lung disease. *Chest* 81:142–146, 1982.

52. Edwards JH, Baker JT, Davies BH: Precipitin test negative farmer's lung: Activation of the alternative pathway of complement by mouldy hay dust. Clin Allergy 4:379–384, 1974.

53. Kawanami O, Basset F, Barrios R, Lacronique JG, Crystal RG: Hypersensitivity pneumonitis in man: Light and electron-microscopic studies of 18 lung biopsies. *Am J Pathol* 110:275–289, 1983.

54. Reijula K, Sutinen S: Ultrastructure of extrinsic allergic bronchiolo-alveolitis. *Pathol Res Pract* 181:418–429, 1986.

55. Heino M, Monkare S, Haahtela T, Laitinen LA: An electron-microscopic study of the airways in patients with farmer's lung. *Eur J Respir Dis* 63:51–61, 1982.

56. Ando M, Arima K, Yoneda R, Tamura M: Japanese summer-type hypersensitivity pneumonitis: Georgraphic distribution, home environment, and clinical characteristics of 621 cases. *Am Rev Respir Dis* 144:755–769, 1991.

57. Nakazawa T, Yumegae Y: Sericulturist's lung disease: Hypersensitivity pneumonitis related to silk production. *Thorax* 45:233–234, 1990.

58. Gamboa PM, de las Marinas MD, Antepara I, Jauregui I, Sanz MML: Extrinsic allergic alveolitis caused by esparto (*Stipa tenacissima*). *Allergol et Immunopathol* 18:331–334, 1990.

Chapter 14

Eosinophilic Pneumonia

M. H. Griffiths / P. S. Hasleton

Eosinophilic pneumonia is usually taken to refer to disease in which the lung parenchyma is involved with an inflammatory process consisting largely of eosinophils and histiocytes. The term *eosinophilic pneumonia* is a misnomer, since the infiltrates contain other inflammatory elements, such as lymphocytes, neutrophil polymorphs, plasma cells, and mast cells.

In common usage, the term *chronic eosinophilic pneumonia* may well be used synonymously with *eosinophilic pneumonia*. However, the term *eosinophilic pneumonia* is often loosely applied to eosinophilic pulmonary infiltrates that may accompany bronchial asthma, polyarteritis, and other hypersensitivity diseases. The pathologic finding of eosinophilic pneumonia, either on transbronchial biopsy or at postmortem, is not specific. This is because a variety of factors may produce the same histologic picture.

Frequently eosinophilia of the blood and sputum accompanies the lung infiltrates.[1] The introduction of bronchoalveolar lavage (BAL) has expanded the list of eosinophilic lung diseases by identifying disorders characterized by an increase of intraalveolar eosinophils but not necessarily hematogenous eosinophils.

The understanding of the biology of the eosinophil has increased remarkably over the past decade. These cells play a key role in the pathogenesis of asthma (Chap. 22). Four different cationic proteins are recognized in the eosinophilic granules: major basic protein, eosinophil-derived neurotoxin, or protein X, eosinophil cationic protein, and eosinophil peroxidase. These cells have numerous membrane and surface markers, including those for adhesion molecules, complement proteins, lipid mediators, and steroid mediators. The function of the eosinophil has recently been summarized in several review articles.[2–6] It is beyond the limitations of this text to discuss eosinophil biology, and the reader is referred to the review articles.

The eosinophilic lung diseases are a heterogeneous group of disorders that bear little clinical relationship to one another. Some are mainly airways-based, some are parenchymal, and others are a mixture of both.[7] There are three ways in which the eosinophilic lung diseases may be diagnosed. The first is by demonstration of peripheral blood eosinophilia and radiologic pulmonary infiltrates. This is the least satisfactory means of diagnosis, since there are many causes of blood eosinophilia. A generalized eosinophilia does not necessarily mean that there is eosinophilic pneumonia. The second method is by open lung biopsy, but this is rarely used because of the invasive nature of the procedure and the fact that most clinicians diagnose pulmonary eosinophilia on the radiologic and clinical findings.

Probably the most frequently employed method of diagnosis is BAL, often combined with a transbronchial biopsy. Lavage especially has the advantage of being a safe procedure that can also be used to monitor both the progress of the disease and treatment. The limitation is that it is not possible to determine whether the eosinophils have an alveolar or a bronchial origin. An increased percentage of eosinophils in lavage fluid correlates with the presence of tissue eosinophils on lung biopsy.[8] The normal component of eosinophils in lavage fluid is less than 1 percent.[9]

The presence of eosinophils does not necessarily imply an eosinophilic pneumonia. One study[10] noted counts of greater than 5 percent eosinophils in lavage fluid in cases with interstitial lung disease, drug-induced lung disease, and *Pneumocystis carinii* pneumonia associated with HIV infection.

CLASSIFICATION

There are numerous classifications, demonstrating that many patients are difficult to fit into defined categories. Reeder and Goodrich[11] introduced the term *PIE* (pulmonary infiltrates with eosinophilia). In the same year, Crofton and coworkers[12] divided the eosinophilic lung diseases into five groups: simple pulmonary eosinophilia (Löffler's syndrome), prolonged pulmonary eosinophilia, tropical eosinophilia, pulmonary eosinophilia with asthma, and

TABLE 14-1
Classification of Eosinophilic Pneumonia

Simple eosinophilia

Chronic eosinophilia

Acute eosinophilic syndrome

Churg-Strauss syndrome (Allergic granulomatosis and angiitis)

Idiopathic hypereosinophilic syndrome

Bronchial asthma

Allergic bronchopulmonary aspergillosis

Bronchocentric granulomatosis

Some parasitic infections

Some drug reactions

SOURCE: From Allen and Davis.[7]

polyarteritis nodosa. This classification was revised in 1969 to include chronic eosinophilic pneumonia.[13] A recent classification has been given in an excellent review article on the eosinophilic pneumonias[7] and is reproduced in Table 14-1.

The list in this table is limited to diseases in which eosinophils are thought to be an integral part of the lung inflammation. Diseases such as Churg-Strauss syndrome (allergic granulomatosis and angiitis), bronchial asthma, allergic bronchopulmonary aspergillosis, and bronchocentric granulomatosis are discussed elsewhere in this volume.

Simple Pulmonary Eosinophilia

In 1932, working in Zurich, Löffler described four patients who had presented with minimal respiratory symptoms, fever, and transient radiographic infiltrates, which resolved spontaneously in 6 to 12 days. Two cases had elevated eosinophil counts.[14]

The most common symptom in this condition is cough, which is usually dry, although some patients produce mucoid sputum containing eosinophils. Chest tightness and pain, malaise, headaches, night sweats, and a low-grade fever may also be features of the disease.

The chest radiographs show unilateral or bilateral transient, migratory, nonsegmental densities of varying sizes. These are often peripheral and sometimes appear pleural-based. The histologic changes have been reported in only a few cases, but they are the same as those described for chronic eosinophilic pneumonia.

Many cases of Löffler's syndrome are caused by an infestation by parasites, such as *Ascaris lumbricoides*. The first pathologic demonstration of pulmonary parasites in simple eosinophilic pneumonia

or Löffler's syndrome was by von Meyenberg.[15] Two of his four patients had intestinal infection with *Ascaris*, but no larvae were detected in the lungs. It has been suggested that the accumulation of eosinophils could result from filtering of these cells by the lung capillaries while the actual cause resides elsewhere—intestinal parasites, for example.

Sprent[16] sensitized mice to various *Ascaris* extracts and induced pulmonary changes, including an eosinophilic infiltrate. This suggests that a sensitivity reaction rather than larval migration per se plays an important role in the production of the pulmonary changes. Most patients initially thought to have simple pulmonary eosinophilia will ultimately be found to have a parasitic infection or a drug reaction. However, no etiologic agent can be found in one-third of patients.[17] There may be a seasonal variation in simple pulmonary eosinophilia, suggesting that environmental antigens may be responsible in some areas.[18]

The prognosis for patients with acute transient eosinophilic pneumonia is excellent. The infiltrates resolve spontaneously, and treatment with corticosteroids is rarely required. A careful search for parasitic infection or a drug reaction should be undertaken so that recurrence of the problem may be averted.

Chronic Eosinophilic Pneumonia

This disorder was originally described in 1960.[19] The first large series was described by Carrington and colleagues.[20] The disease has been reported in every age group, but the peak incidence is in the fourth decade. There is a female predominance, with a ratio of 2:1.[21,22]

The onset is insidious, symptoms having been present for an average of 7.7 months at the time of diagnosis.[22] The symptoms usually consist of cough, fever, dyspnea, and weight loss. Sometimes there is sputum production, malaise, wheezing, and night sweats. Asthma is noted in 50 percent of patients and will usually have been present for 5 years.[21] Peripheral blood eosinophilia, usually defined as greater than 6 percent of the total blood count, has been found in 88 percent of patients.[22]

Eosinophils are detected in the sputum in approximately half of the patients. Serum IgE levels are increased in two-thirds of the patients, and rheumatoid factors or immune complexes may be identified.[23,24]

Pulmonary function tests may be normal in mild cases but usually show restrictive defects with a reduced diffusing capacity. Obstructive defects may

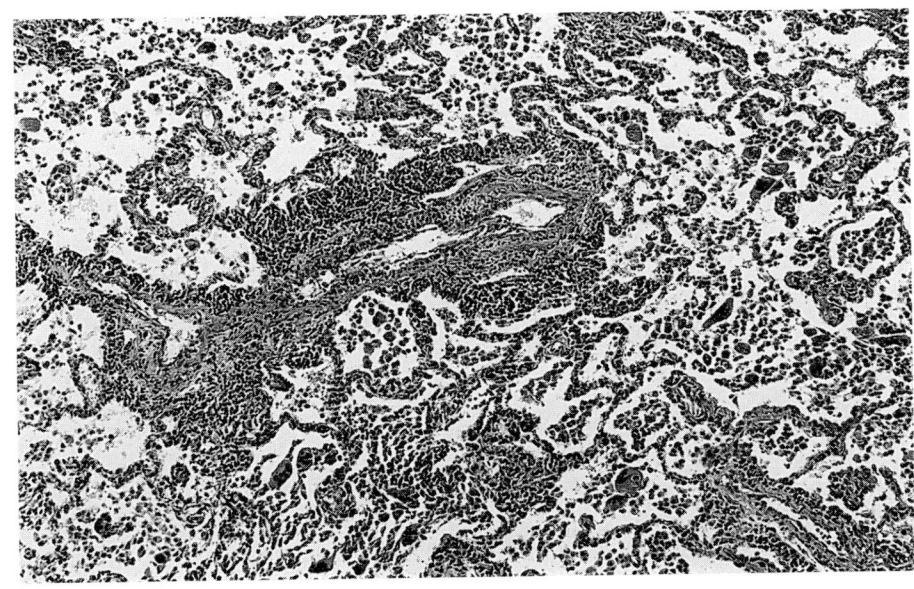

FIGURE 14-1
An alveolar infiltrate of eosinophils and histiocytes is associated with an interstitial infiltrate that includes lymphocytes and plasma cells. (H&E, ×18.)

be noticed. Their presence probably reflects concurrent asthma, rather than chronic eosinophilic pneumonia.[22]

Chest radiographs show peripheral infiltrates in approximately two-thirds of patients.[22] The so-called negative image of pulmonary edema is said to be virtually diagnostic of chronic eosinophilic pneumonia.[25] This finding is not specific and may be found in only 25 percent of patients.[22] It may also be seen in bronchiolitis obliterans–organizing pneumonia (BOOP)[26] and sarcoidosis.[27] CT scans show peripheral air-space disease, which may be identified even when the chest radiograph fails to show the periph-

eral nature of the infiltrates.[22] In addition, 50 percent of patients have mediastinal lymph node enlargement on CT scans.[28]

Histologically, there is an interstitial infiltrate of lymphocytes, plasma cells, histiocytes, and eosinophils. Eosinophils may be present in the interlobular septa (Figs. 14-1 and 14-2). The underlying lung architecture remains intact. There may be a component of BOOP. Multinucleated giant cells may be prominent (Fig. 14-3), along with a smaller number of other inflammatory cells. There may be filling of the alveolar spaces with eosinophils and histiocytes (Fig. 14-4). Alveoli may contain Charcot-Leyden

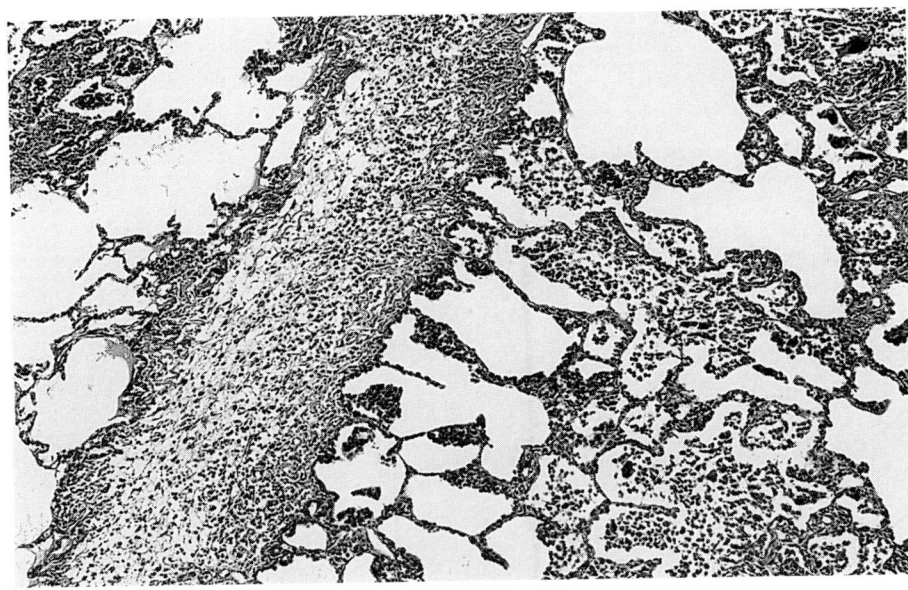

FIGURE 14-2
There may be eosinophils in interlobular septa. (H&E, ×75.)

451

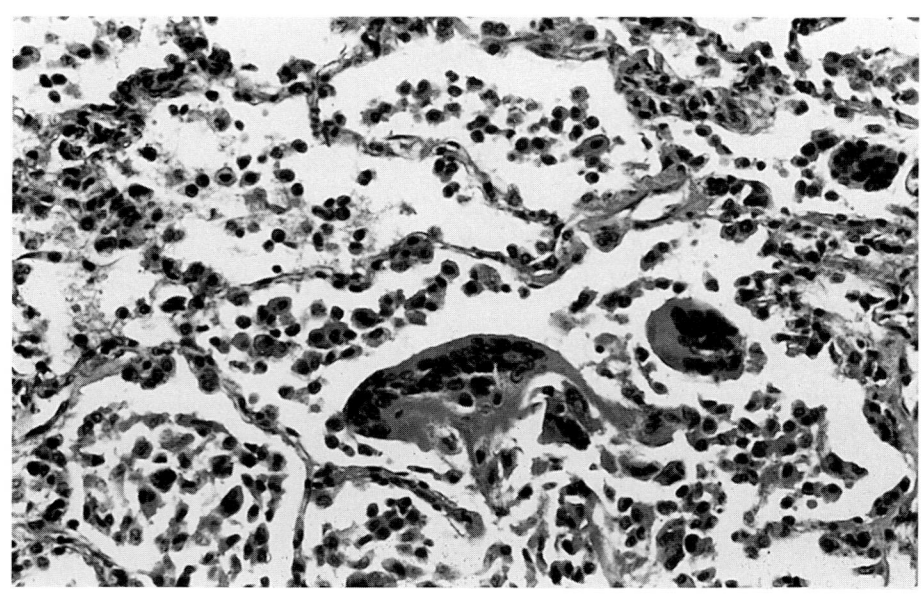

FIGURE 14-3
Multinucleated giant cells may be prominent in the infiltrate. (H&E, ×187.)

crystals (products of eosinophil granule degradation) (Fig. 14-5). Eosinophilic microabscesses may be seen in alveoli, histiocytes may palisade around eosinophilic debris, and sarcoid-like granulomas are sometimes identified. The granulomas are noncaseating and are seen in only minority of patients.[22] A low-grade vasculitis may be seen, but associated vascular granulomas are absent.[29] The eosinophils may be degranulated and difficult to find on routine staining, making histologic diagnosis difficult.[21]

BAL shows large numbers of eosinophils, usually more than 25 percent. After treatment with corticosteroids, the amount returns to normal.[30] The BAL lymphocyte and neutrophil counts may be either normal or slightly elevated.[31]

Fewer than 10 percent of patients experience spontaneous resolution of symptoms. Respiratory failure has been recorded in accelerated cases and when the diagnosis has been delayed.[10] Treatment with corticosteroids will produce a dramatic resolution of symptoms, and the chest radiographs usually clear within 10 days. If steroids are discontinued, relapses will occur within the next 6 months. If the dose is reduced gradually, most patients will be able to discontinue steroids completely. A few patients have to remain on the drug indefinitely.

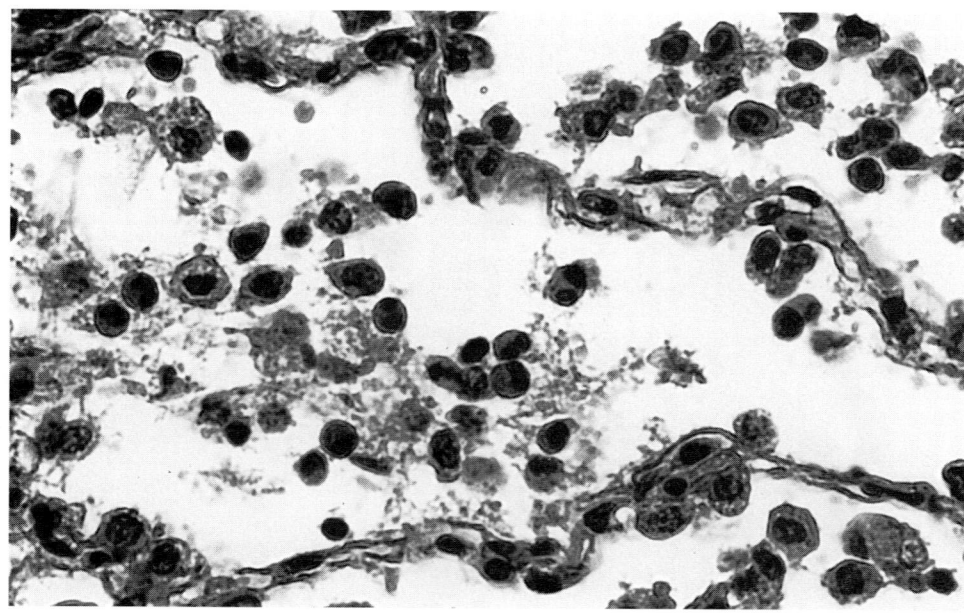

FIGURE 14-4
The alveoli contain a mixture of eosinophils and histiocytes. (H&E, ×400.)

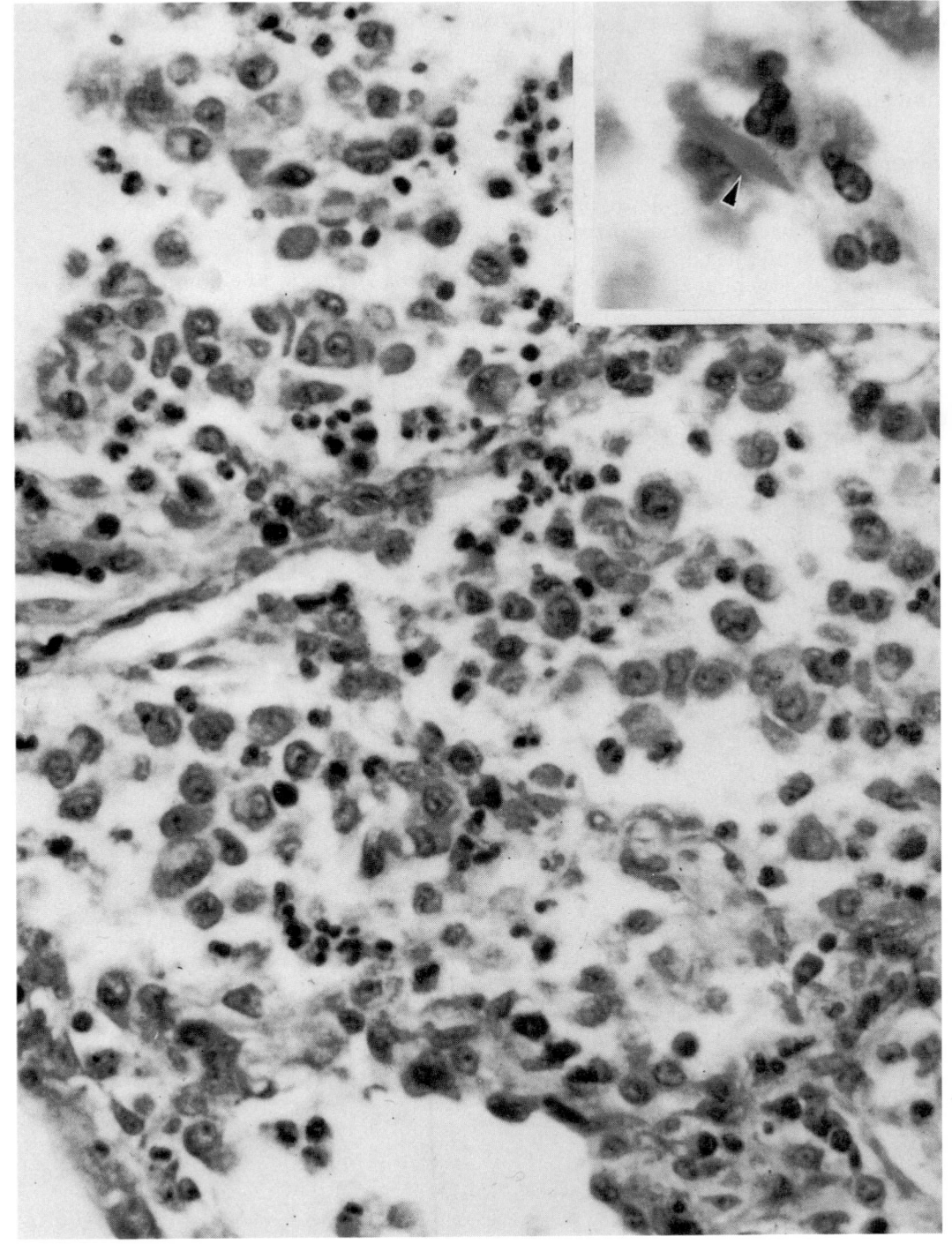

FIGURE 14-5
Alveoli filled with eosinophils and macrophages. Inset shows a Charcot-Leyden crystal (*arrowhead*) within a macrophage. (Luna stain, ×485, inset ×650.) (*See also Color Plate 3.*)

Differential Diagnosis

BRONCHOPULMONARY ASPERGILLOSIS

This may have a similar appearance to chronic eosinophilic pneumonia. Therefore, in any case in which chronic eosinophilic pneumonia is suggested, a search should be made for *Aspergillus* precipitins, as well as for the fungi.

EOSINOPHILIC GRANULOMA

The infiltrate, as in chronic eosinophilic pneumonia, is composed of a mixture of eosinophils and histiocytes, but in eosinophilic granuloma the lesions are well defined and there are intervening zones of normal-appearing pulmonary parenchyma. There are also histiocytosis cells, which stain positively with S-100, and Birbeck granules (pentalaminar bodies) are identified on electron microscopy. In addition, eosinophilic granuloma shows a finely reticular radiologic pattern, in contrast to the more homogeneous areas of consolidation in eosinophilic pneumonia.

DESQUAMATIVE INTERSTITIAL PNEUMONIA

The characteristic feature of desquamative interstitial pneumonia is the massive proliferation and accumulation of type II pneumocytes and/or macrophages in the alveolar spaces. The interstitial infiltrate is usually sparse, and eosinophils and plasma cells are present in relatively small numbers, mainly in the interstitium. There is no necrosis of intraalveolar or perivascular infiltrates.

CHURG-STRAUSS SYNDROME (ALLERGIC ANGIITIS AND GRANULOMATOSIS)

In this desease there is peripheral eosinophilia, a history of asthma and allergic rhinitis, followed by systemic vasculitis. Histologically there is a necrotizing, granulomatous vasculitis in the small arteries and veins. In addition, there are areas of eosinophilic pneumonia. Extravascular granulomas may be identified, with palisading histiocytes around central necrotic eosinophils.

POLYARTERITIS NODOSA

This may be difficult to distinguish from allergic angiitis and granulomatosis. The vasculitis is confined to arteries, and extravascular granulomas are not seen. Pulmonary involvement is rare in polyarteritis.

WEGENER'S GRANULOMATOSIS

This may occasionally present on a background of bronchial asthma and eosinophilia. Glomerulonephritis is common. The antineutrophil cytoplasmic antibody is usually positive. In Wegener's granulomatosis, however, there are areas of necrosis, fibrinoid necrosis of muscular pulmonary arteries and veins, along with multinucleated cells. These areas are surrounded by lymphocytes, plasma cells, histiocytes, and fibroblasts. Eosinophils are not usually prominent.

Etiology

Chronic eosinophilic pneumonia appears to be the result of the introduction of foreign material into the body via a variety of routes, including ingestion, injection, inhalation, skin contact, and vaginal absorption.

Drug reactions and parasitic manifestations will be mentioned later in this chapter.

While the incidence due to parasitic causes is falling in the West, drugs are playing an increasing role in eosinophilic pneumonia.

Acute Progressive Eosinophilic Pneumonia

This form of eosinophilic pneumonia was first described in 1989.[32,33] It appears to be a distinct entity from the other eosinophilic lung diseases.

Patients with this syndrome may present at any age, and there is equal sex distribution. The youngest patient was 13 years of age.[34]

The diagnostic criteria for acute progressive eosinophilic pneumonia have been given by Allen and Davis.[33] These are acute febrile illness of less than 5 days' duration, hypoxemic respiratory failure, diffuse alveolar or mixed alveolar radiologic infiltrates, a bronchoalveolar eosinophilia of greater than 25 percent, and absence of parasitic or fungal infection, and a prompt or complete response to corticosteroids. There is also a failure to relapse after discontinuation of corticosteroids. Patients have myalgias and pleuritic chest pain. None of the original patients had a history of either previous or subsequent asthma or other atopic illnesses. However, several reports have described patients with a history of allergic rhinitis.[35]

On examination there is fever, respiratory distress, and basal or diffuse crackles. There is no evidence of infection or drug reaction.

The earliest radiologic findings are those of a subtle interstitial infiltrate, often with Kerley's B lines. This is followed within several hours to 2 days by extensive alveolar and interstitial infiltrates, which involve all lobes. Unlike chronic eosinophilic pneumonia, there are rarely peripheral-based infiltrates. CT scans show diffuse parenchymal alveolar infiltrates, pleural effusions, and prominent septa. The lymph nodes are normal.

Lung biopsies show intraalveolar eosinophils and edema of the alveolar walls. There is also some degree of interstitial edema. No vasculitis or fungi are identified.[34,36,37]

Patients respond rapidly to high doses of steroids.

ETIOLOGY

The cause or causes of this syndrome are unknown. It has been postulated that it may result from an acute hypersensitivity phenomenon due to an unidentified inhaled antigen.[32,37]

The differential diagnosis is similar to that given for chronic interstitial pneumonia. It is also important, apart from excluding *Aspergillus*, to search for *Coccidioides immitis*, which has been noted to produce a syndrome resembling acute progressive eosinophilic pneumonia.[38]

Idiopathic Hypereosinophilic Syndrome

The idiopathic hypereosinophilic syndrome is a rare but often fatal disorder. It is characterized by a blood eosinophilia of greater than $1500/\mu l$ for more than 6 months and an absence of parasitic or other cause of secondary eosinophilia. There is increasing end-organ damage related to the tissue eosinophilia.[39]

The cause of the disease is unknown. It is possible that the profound eosinophilia is lymphocyte-dependent, since an abnormal clonal proliferation of T-helper lymphocytes has been identified in some of these patients.[40,41]

Patients present in the third or fourth decade, although patients as old as 70 have been reported.[10] There is a 7:1 male predominance.[42] Patients present with night sweats, anorexia, weight loss, cough, fever, and itching.

There is a profound peripheral eosinophilia, with up to 70 percent of the total white count being eosinophils.[43] Eosinophil metamyelocytes and myelocytes may be found in the peripheral blood. In addition to being found in the lung, eosinophils are abundant in the bone marrow.

Pulmonary involvement occurs in up to 40 percent of patients and typically presents as a cough,

which is worse at night.[39] In long-standing disease there may be pulmonary fibrosis.

The chest radiograph shows nonlobar interstitial infiltrates, and approximately 50 percent of patients have pleural effusions.

BAL shows a high percentage of eosinophils, accounting for up to 73 percent of the total number of cells recovered.[44]

One of the most important complications is cardiac involvement. This includes endocardial fibrosis, a restrictive type of cardiomyopathy, damage to the valves, and mural thrombus formation.[45] Arterial thromboembolism is seen in two-thirds of patients.[10] Splinter hemorrhages are the most common manifestation, but renal or splenic infarcts and retinal arteriolar embolism may be seen. In addition, there may be cerebral hemorrhages, diffuse small-vessel cerebrovascular occlusions, and deep venous thrombosis. In advanced disease, peripheral neuropathy is present.[39] Other organs infiltrated by eosinophils are the gastrointestinal tract, kidneys, joints, skin, and muscle.[46]

Fifty percent of patients have a good clinical response to oral corticosteroids. These patients typically have high IgE levels, angioedema, and a dramatic reduction of blood eosinophil numbers after the first dose of corticosteroids.[47]

Tropical Eosinophilia

Many parasites can cause pulmonary infiltrates with blood or alveolar eosinophilia. The causative parasites are given in Table 14-2.[7]

TABLE 14-2
Parasites Causing Eosinophilic Lung Disease

Ancylostoma sp.
Ascaris sp.
Brugia malayi
Clonorchis sinensis
Dirofilaria immitis
Echinococcus sp.
Opisthorchis sp.
Paragonimus westermani
Schistosoma sp.
Strongyloides stercoralis
Toxacara sp.
Trichinella spiralis
Wuchereria bancrofti

SOURCE: From Allen and Davis.[7]

Strongyloides, Ascaris, Toxocara, and *Ancylostoma* are the most frequent parasitic causes of eosinophilia in the United States. However, the prevalence of individual parasitic infection varies from one region to another. Patients must be examined fully, since infection with one parasite indicates an increased risk for other parasites.[7]

Tropical pulmonary eosinophilia may be caused by the filarial worms *Wuchereria bancrofti* and *Brugia malayi.* Most cases have been reported from India, Africa, South America, and Southwest Asia. Microfilaria are transported to humans by mosquitoes. The adult worms eventually reside in the lymphatics; there they release microfilaria, which travel to the lung and create an intense inflammatory reaction. Patients have a nocturnal cough, dyspnea, wheezing, fever, weight loss, and malaise. BAL eosinophils are 50 percent.[48] The peripheral blood eosinophil count generally exceeds 3,000/μL. Another parasite that causes eosinophilia is *Schistosoma. Clonorchis sinensis* and *Opisthorchis* have also been noted in Asia as frequent causes of pulmonary eosinophilia.[49]

Drug Reactions

Drug reactions are considered in Chap. 18. For the sake of completeness, however, the drugs causing eosinophilic lung disease are listed in Table 14-3.[7]

Other Lung Diseases Associated with an Eosinophilia

IDIOPATHIC PULMONARY FIBROSIS

This disease is in some instances associated with an increase in lavage eosinophils and a peripheral blood eosinophilia.[87] It has been suggested that eosinophilic infiltration or deposition of major basic protein can be found in up to three-fourths of all patients with idiopathic pulmonary fibrosis.[88] A BAL eosinophilia may presage a poor response to treatment in patients with idiopathic pulmonary fibrosis.[89]

HYPERSENSITIVITY PNEUMONITIS

There is an increased percentage of eosinophils by lavage in up to 15 percent of patients with this disease.[90] Histology in these cases, however, does not show an intraalveolar eosinophilic infiltrate, as with chronic eosinophilic pneumonia. Ill-defined granulomas are identified.

TABLE 14-3
Drugs Causing Eosinophilic Lung Disease

Ampicillin[50]	Methylphenidate (Ritalin)[52]
Beclomethasone	Minocycline[54]
dipropionate (inhaled)[51]	Naproxen[56]
Bleomycin[53]	Nickel[58]
Carbamazepine[55]	Nitrofurantoin[50]
Chlorpromazine[57]	Paraaminosalicylic acid[61]
Clofibrate[59]	Penicillin[63]
Cocaine (inhaled)[60]	Pentamidine (inhaled)[65]
Cromolyn (inhaled)[62]	Phenytoin[67]
Desipramine[64]	Pyrimethamine[69]
Diclofenac[66]	Rapeseed oil[71]
Febarbamate[68]	Sulfadimethoxine[73]
Fenbufen[70]	Sulfadoxine[75]
Glafenine[72]	Sulfasalazine[76]
GM-CSF[74]	Sulindac[78]
Ibuprofen[56]	Tamoxifen[50]
Interleukin 2[77]	Tetracycline[81]
Interleukin 3[79]	Tolazamide[83]
Iodinated contrast dye[80]	Tolfenamic acid[85]
L-Tryptophan[82]	Vaginal sulfonamide
Mephenesin carbamate[84]	cream[86]
Methotrexate[50]	

SOURCE: From Allen and Davis.[7]

MALIGNANCY

There may be an increase in lavage eosinophils in non–small-cell lung cancer.[87] The mechanism of eosinophilia in bronchial carcinoma is probably related to tumor-derived eosinophilic factors. High levels of granulocyte-macrophage colony-stimulating factor have been identified in two subjects.[91] Eosinophils are also a feature of Hodgkin's disease, but diagnostically they should not cause a problem. Non-Hodgkin's lymphoma and lymphocytic leukemia may be associated with pulmonary infiltrates and a peripheral blood eosinophilia.[92]

Eosinophilic leukemia can occur in a few patients with myeloblastic leukemia.

FUNGAL INFECTIONS

Pulmonary eosinophilic infiltrates may be noted on lavage and lung biopsy in primary coccidiomycosis.[93] *P. carinii* is associated with an increase in lavage eosinophils in cases of AIDS.

TRANSPLANTATION

A tissue eosinophilia may be noted in transplantation, and this may be due to rejection but may be secondary to infection with a virus, such as Coxsackie (Chap. 24).

OTHER DISORDERS

A lavage eosinophilia of 20 percent may be seen in BOOP.[94] It may be difficult to distinguish BOOP from cases of chronic eosinophilic pneumonia. This problem is made more acute because some patients with chronic eosinophilic pneumonia have evidence of bronchiolitis obliterans. Several cases of *Mycobacterium tuberculosis* pneumonia have been associated with 40 to 50 percent BAL eosinophils.[25]

Other conditions associated with pulmonary infiltrates and either blood or alveolar eosinophilia are infection with *Mycobacterium simiae*,[96] hereditary ataxia telangiectasia,[97] Sjögren's syndrome,[98] postradiation fibrosis, graft-versus-host disease,[98] scorpion sting,[99] and human *Corynebacterium pseudotuberculosis* infection.[100]

It can therefore be seen that pulmonary eosinophilia is not specific to entities such as chronic eosinophilic pneumonia. As in all lung disease, there needs to be a careful workup of the cases with a full clinical history and in some instances even open lung biopsy. It is important that the pathologist is aware of all the potential causes of pulmonary eosinophilia and is not tempted to make a diagnosis of chronic eosinophilic pneumonia based solely on a lavage.

REFERENCES

1. Mayock RL, Iozzo RV: The eosinophilic pneumonias, in Fishman AP (ed), *Pulmonary Diseases and Disorders*, 2d ed, vol 1. New York, McGraw-Hill, 1988, pp 683–698.
2. Kroegel C, Virchow J-C Jr, Luttmann W, Walker C, Warner JA: Pulmonary immune cells in health and disease: The eosinophil leucocyte (part I). *Eur Respir J* 7:519–543, 1994.
3. Kroegel C, Warner JA, Virchow J-C Jr, Matthis S: Pulmonary immune cells in health and disease: The eosinophil leucocyte (part II) *Eur Respir J* 7:743–760, 1994.
4. Weller PF: The immunobiology of eosinophils. *N Engl J Med* 324:1110–1118, 1991.
5. Spry CJS: *Eosinophils: A Comprehensive Review and Guide to the Scientific and Medical Literature*. New York, Oxford University Press, 1988.
6. Gleich GJ, Adolphson CR: The eosinophilic leucocyte: Structure and function. *Adv Immunol* 39:177–253, 1986.
7. Allen JN, Davis WB: Eosinophilic lung diseases. *Am J Respir Crit Care Med* 150:1423–1438, 1994.
8. Umeki S: Reevaluation of eosinophilic pneumonia and its diagnostic criteria. *Arch Intern Med* 152:1913–1919,1992.
9. The BAL Cooperative Group Steering Committee: Bronchoalveolar lavage constituents in healthy individuals, idiopathic pulmonary fibrosis, and selected comparison groups. *Am Rev Respir Dis* 141:S169–S202, 1990.
10. Allen JN, Davis WB, Pacht ER: Diagnostic significance of increased bronchoalveolar lavage fluid eosinophils. *Am Rev Respir Dis* 142:642–647,1990.
11. Reeder WH, Goodrich BE: Pulmonary infiltration with eosinophilia (PIE syndrome). *Ann Intern Med* 36:1217–1240, 1952.
12. Crofton JW, Livingstone JL, Oswald NC, Roberts ATM: Pulmonary eosinophilia. *Thorax* 7:1–35, 1952.
13. Liebow AA, Carrington CV: The eosinophilic pneumonias. *Medicine* 48:251–285, 1969.
14. Löffler W: Zur Differential-Diagnose der Lungeninfiltrierungen: Uber flüchtige Succedan-Infiltrate (mit Eosinophilie). *Beitr Klin Tuberk* 79:368–382, 1932.
15. von Meyenberg H: Das eosinophilie Lungeninfiltrat: Pathologische Anatomie und Pathogenese. *Schweiz Med Wochenschr* 23:809–811, 1942.
16. Sprent JFA: On the toxic and allergic manifestations produced by the tissue and fluids of ascaris: Effect on different tissues. *J Infect Dis* 84:221–229, 1949.
17. Ford RM: Transient pulmonary eosinophilia and asthma: A review of 20 cases occurring in 5,702 asthma sufferers. *Am Rev Respir Dis* 93:797–803, 1966.
18. Zifroni A, Rosenberg M: Transitory lung infiltrates with blood eosinophilia. *Geriatrics* 27:162–179, 1972.
19. Christoforides HA, Molnar W: Eosinophilic pneumonia: Report of two cases with pulmonary biopsy. *JAMA* 173:157–161, 1960.
20. Carrington CB, Addington WW, Goff AM, et al: Chronic eosinophilic pneumonia. *N Engl J Med* 280:787–798, 1969.
21. Fox B, Seed WA: Chronic eosinophilic pneumonia. *Thorax* 35:570–580, 1980.
22. Jederlinic PJ, Sicilian L, Gaensler EA: Chronic eosinophilic pneumonia: A report of 19 cases and review of the literature. *Medicine* 67:154–162, 1988.
23. Naughton M, Fahy J, FitzGerald MX: Chronic eosinophilic pneumonia: A long-term follow-up of 12 patients. *Chest* 103:162–165, 1993.
24. Demedts M, De Man F: Circulating immune complexes in chronic eosinophilic pneumonia. *Acta Clin Belg* 46:75–81, 1991.
25. Gaensler EA, Carrington CB: Peripheral opacities in chronic eosinophilic pneumonia: The photographic negative of pulmonary edema. *AJR* 128:1–13, 1977.
26. Bartter T, Irwin RS, Nash G, Balikian JP, Hollingsworth HH: Idiopathic bronchiolitis obliterans organizing pneumonia with peripheral infiltrates on chest roentgenogram. *Arch Intern Med* 149:273–279, 1989.
27. Scherpenisse J, van der Valk PD, van den Bosch JM, van Hees PA, Nadorp JH: Olsalazine as an alternative therapy in a patient with sulfasalazine-induced eosinophilic pneumonia. *J Clin Gastroenterol* 10:218–220, 1988.
28. Mayo JR, Müller NL, Road J, Sisler J, Lillington G: Chronic eosinophilic pneumonia: CT findings in six cases. *AJR* 153:727–730, 1989.
29. Rogers RM, Christiansen JR, Coulson JJ, Patterson CD: Eosinophilic pneumonia: Physiologic response to steroid therapy and observations on light and electron microscopic findings. *Chest* 68:665–671, 1975.

30. Degaejher P, Demedts M: Bronchoalveolar lavage in eosinophilic pneumonia before and during corticosteroid therapy. *Am Rev Respit Dis* 129:631–632, 1984.

31. Costabel U, Teschler H, Guzman J: Bronchiolitis obliterans organizing pneumonia (BOOP): The cytological and immunocytological profile of bronchoalveolar lavage. *Eur Respir J* 5:791–797, 1992.

32. Badesch DV, King TE, Schwarz MI: Acute eosinophilic pneumonia: A hypersensitivity phenomenon. *Am Rev Respir Dis* 139:249–252, 1989.

33. Allen JN, Pacht ER, Gadek JE, Davis WB: Acute eosinophilic pneuomonia as a reversible cause of non-infectious respiratory failure. *N Engl J Med* 321:569–574, 1989.

34. Buchheit AJ, Eid N, Rodgers G Jr, Feger T, Yakoub O: Acute eosinophilic pneumonia with respiratory failure: A new syndrome? *Am Rev Respir Dis* 145:716–718, 1992.

35. Hayakawa H, Sato A, Toyoshima M, Imokawa S. Taniguchi M: A clinical study of idiopathic eosinophilic pneumonia. *Chest* 105:1462–1466, 1994.

36. Ogawa H, Fugimura M, Matsuda T, Nakamura H, Kumabashiri I, Kitagawa S: Transient wheeze: Eosinophilic bronchobronchiolitis in acute eosinophilic pneumonia. *Chest* 104:493–496, 1993.

37. Davis WB, Wilson HE, Wall RL: Eosinophilic alveolitis and acute respiratory failure: A clinical marker for a non-infectious etiology. *Chest* 90:7–10, 1986.

38. Meeker DP, Gephardt GN, Cordasco EM, Wideman HE: Hypersensitivity pneumonitis versus invasive pulmonary aspergillosis: Two cases with unusual pathologic findings and review of the literature. *Am Rev Respir Dis* 143:431–436, 1991.

39. Fauci AS, Harley JB, Roberts WC, Ferrans VJ, Gralnick HR, Bjornson BH: The idiopathic hypereosinophilic syndrome: Clinical, pathophysiologic, and therapeutic considerations. *Ann Intern Med* 97:78-92, 1982.

40. Raghavachar A, Fleischer S, Frickhofen N, Heimpel H, Fleischer B: T lymphocyte control of human eosinophil granulopoiesis: Clonal analysis in an idiopathic hypereosinophilic syndrome. *J Immunol* 139:3753–3758, 1987.

41. Cogan E, Schandené L, Crusiaux A, Cochaux P, Velu T, Goldman M: Brief report: Clonal proliferation of type 2 helper T cells in a man with a hypereosinophilic syndrome. *N Engl J Med* 330:535–538, 1994.

42. Spry CJF, Davies J, Tai PC, Olsen EGJ, Oakley CM, Goodwin JF: Clinical features of fifteen patients with a hypereosinophilic syndrome. *Q J Med* 205:1–22, 1983.

43. Chusid MJ, Dale DC, West BC, Wolff SM: The hypereosinophilic syndrome: Analysis of 14 cases with review of the literature. *Medicine* 54:1–27, 1975.

44. Wynn RE, Kollef MH, Meyer JL: Pulmonary involvement in the hypereosinophilic syndrome. *Chest* 105:656–660, 1994.

45. Parrillo JE, Borer JS, Henery WL, Wolff SM, Fauci AF: The cardiovascular manifestions of the hypereosinophilic syndrome: Prospective study of 26 patients with review of the literature. *Am J Med* 67:572–582, 1979.

46. Spry CJF: The hypereosinophillic syndrome: Clinical features, laboratory findings and treatment. *Allergy* 37:539–551, 1982.

47. Parrillo JE, Fauci AS, Wolff SM: Therapy of the hypereosinophilic syndrome. *Ann Intern Med* 89:162–172, 1978.

48. Pinkerton P, Vijayan AK, Nutman TV, et al: Acute tropical eosinophilia; characterization of the lower respiratory tract inflammation and its response to therapy. *J Clin Invest* 82:160–225, 1987.

49. Ly LI, Zue CS, Luo WC: Clinical analysis 90 cases of eosinophilia and pulmonary infiltrates with peripheral eosinophilia. *Chung Hua Nei Ko Tsa Chih* 29:659–662, 1992.

50. Poe RH, Condemi JJ, Weinstein SS, Schuster RJ: Adult respiratory distress syndrome related to ampicillin sensitivity. *Chest* 77:449–451, 1990.

51. Mollura JL, Bernstein R, Fine SR, Vevaina J: Pulmonary eosinophilia in a patient receiving beclomethasone dipropionate. *Ann Allergy* 42:326–329, 1979.

52. Wolf J, Fein A, Fehrenbacher L: Eosinophilic syndrome with methylphenidate abuse. *Ann Intern Med* 89:224–245, 1978.

53. White DA, Kris MG, Stover DE: Bronchoalveolar lavage cell populations in bleomycin lung toxicity. *Thorax* 42:551–552, 1987.

54. Sitbon O, Bidel N, Dussopt C, Azarian R, Piard F, Camus P: Minocycline-induced eosinophilic pneumonitis: An analysis in 5 patients (abstract). *Am Rev Respir Dis* 147:A76, 1993.

55. Culinan SA, Bower GC: Acute pulmonary hypersensitivity to carbamazepine. *Chest* 68:580–581, 1975.

56. Goodwin SD, Glenny RW: Nonsteroidal anti-inflammatory drug–associated pulmonary infiltrates with eosinophilia: Review of the literature and Food and Drug Administration Adverse Drug Reaction reports. *Arch Intern Med* 152:1521–1524, 1992.

57. Shear MK: Chlorpromazine-induced PIE syndrome. *Am J Psychiatry* 135:492–493, 1978.

58. Gray K: Löffler's syndrome following ingestion of a coin. *Can Med Assoc J* 127:999–1000, 1982.

59. Hendrickson RM, Simpson F: Clofibrate and eosinophilic pneumonia (letter). *JAMA* 247:3082, 1982.

60. Nadeem S, Nasir N, Israel RI: Löffler's syndrome secondary to crack cocaine. *Chest* 105:1599–1600, 1994.

61. Cuthbert RJ: Löffler's syndrome occurring during streptomycin and PAS therapy. *Br Med J* 2:298–299, 1954.

62. Repo UK, Nieminen P: Pulmonary infiltrates with eosinophilia and urinary symptoms during disodium cromoglycate treatment: A case report. *Scand J Respir Dis* 57:1–4, 1976.

63. Riechlin S, Loveless MH, Kane EG: Löffler's syndrome following penicillin therapy. *Ann Intern Med* 38:113–120, 1953.

64. Panuska JR, King TR, Korenblat PE, Wedner HJ: Hypersensitivity reaction to desipramine. *J Allergy Clin Immunol* 80:18–23, 1987.

65. Dupon M, Malou M, Rogues AM, Lacut JY: Acute eosinophilic pneumonia induced by inhaled pentamidine isothionate. *BMJ* 306:109, 1993.

66. Khalil H, Molinary E, Stoller JK: Diclofenac (Voltaren)–induced eosinophilic pneumonitis: Case report and review of the literature. *Arch Intern Med* 153:1649–1652, 1993.

67. Mahatma M, Haponik EF, Nelson S, Lopez A, Summer WR: Phenytoin-induced acute respiratory failure with pulmonary eosinophilia. *Am J Med* 87:93–94, 1989.

68. Gali JM, Vilanova JL, Mayos M, Cornudella R, de las Heras P, Rodriguez Arias JM: Febarbamate-induced pulmonary eosinophilia: A case report. *Respiration* 49:231–234, 1986.

69. Davidson AC, Bateman C, Shovlin C, Marrinan M, Burton GH, Cameron IR: Pulmonary toxicity of malaria prophylaxis. *BMJ* 297:1240–1241, 1988.

70. Burton GH: Rash and pulmonary eosinophilia associated with fenbufen. *BMJ* 300:82–83, 1990.

71. Kilbourne EM, Rigau-Perez JG, Heath CW Jr, et al: Clinical epidemiology of toxic-oil syndrome: Manifestations of a new illness. *N Engl J Med* 309:1408–1414, 1983.

72. Gheysens B, Van Mieghem W: Pulmonary infiltrates with eosinophilia due to glafenine. *Eur J Respir Dis* 65:456–459, 1984.

73. Fiegenberg DS, Weiss H, Kirshman H: Migratory pneumonia and eosinophilia associated with sulfonamide administration. *Arch Intern Med* 120:85–89, 1967.

74. Gonzales-Chambers R, Rosenfeld C, Winkelstein A, Dameshek L: Eosinophilia resulting from administration of recombinant granulocyte-macrophage colony-stimulating factor (rhGM-CSF) in a patient with T-gamma lymphoproliferative disease. *Am J Hematol* 36:157–159, 1991.

75. Daniel PT, Holzschuh J, Berg PA: Sulfadoxine specific lymphocyte transformation in a patient with eosinophilic pneumonia induced by sulfadoxine-pyrimethamine (Fansidar). *Thorax* 44:307–309, 1989.

76. Sullivan SN: Sulfasalazine lung: Desensitization to sulfasalazine and treatment with acrylic coated 5-ASA and azodisalicylate. *J Clin Gastroenterol* 9:461–463, 1987.

77. van Haelst Pisani C, Kovach JS, Kita H, et al: Administration of interleukin-2 (IL-2) results in cancer. *Blood* 78:1538–1544, 1991.

78. Fein M: Sulindac and pneumonitis (letter). *Ann Intern Med* 95:245, 1981.

79. Lindermann A, Ganser A, Herrmann F, Frisch J, Seipelt G, Schulz G: Interleukin-3 in vivo. *J Clin Oncol* 9:2120–2127, 1991.

80. Jennings CA, Deveikis J, Azumi N, Yeager H: Eosinophilic pneumonia associated with reaction to radiographic contrast medium. *South Med J* 84:92–95, 1991.

81. Ho D, Tashkin DP, Bein ME, Sharma O: Pulmonary infiltrates with eosinophilia associated with tetracycline. *Chest* 76:33–36, 1979.

82. Kaufman LD, Seidman RJ, Gruber BL: L-Tryptophan–associated eosinophilic perimyositis, neuritis, and fasciitis: A clinicopathologic and laboratory study of 25 patients. *Medicine (Baltimore)* 69:187–199, 1990.

83. Bondi E, Slater S: Tolazamide-induced chronic eosinophilic pneumonia (letter). *Chest* 80:652, 1981.

84. Rodman T, Fraimow W, Myerson RM: Löffler's syndrome: Report of a case associated with administration of mephenesin carbamate (Tolseram). *Ann Intern Med* 48:668–674, 1958.

85. Stromberg C, Palva E, Alhava E, Aranko K, Idänpään-Heikkilä J: Pulmonary infiltrations induced by tolfenamic acid (letter). *Lancet* 2:685, 1987.

86. Donlan CJ, Scutero JV: Transient eosinophilic pneumonia secondary to use of a vaginal cream. *Chest* 67:232–233, 1975.

87. Allen JN, Davis WB, Pacht ER: Diagnostic significance of increased bronchoalveolar lavage fluid eosinophils. *Am Rev Respir Dis* 142:642–647, 1990.

88. Noguchi H, Kephart GN, Colby T, Gleich GJ: Tissue eosinophil degranulation in syndromes associated with fibrosis. *Am J Pathol* 140:521–528, 1992.

89. Rudd RM, Haslam PL, Turner-Warwick M: Cryptogenic fibrosing alveolitis: Relationships of pulmonary physiology and bronchoalveolar lavage to response to treatment and prognosis. *Am Rev Respir Dis* 124:1–16, 1981.

90. Davis WB, Fells GA, Sun XH, Gadek JE, Venet A, Crystal RG: Eosinophil-mediated injury to lung parenchymal cells and interstitial matrix: A possible role for eosinophils in chronic inflammatory disorders of the lower respiratory tract. *J Clin Invest* 74:269–278, 1984.

91. Sawyers CL, Gold DW, Quan S, Niner ST: Production of granulocyte-macrophage colony-stimulating factor in two patients with lung cancer, leukocytosis, and eosinophilia. *Cancer* 69:1342–1346, 1992.

92. Tan AM, Downie PJ, Ekert H: Hypereosinophilia syndrome with pneumonia in acute lymphoblastic leukaemia. *Aust Paediatr J* 23:359–361, 1987.

93. Lombard CM, Tazelaar MD, Krasen DL: Pulmonary eosinophilia in coccidioidal infections. *Chest* 91:734–736, 1987.

94. Izumi T, Kitaichi M, Nishimura K, Nagai S: Bronchiolitis obliterans organising pneumonia: Clinical features and differential diagnosis. *Chest* 102:715–719, 1992.

95. Vijyam V-K, Reetha A-M, Jawahar MS, Sankaran K, Prabhakar R: Pulmonary eosinophilia in pulmonary tuberculosis. *Chest* 101:1708–1709, 1992.

96. Wright JL, Pare PD, Hammond M, Donevan RE: Eosinophilic pneumonia and atypical mycobacterial infection. *Am Rev Respir Dis* 127:497–499, 1983.

97. Zagami AS, Colebatch HJH, Wakefield D: Chronic Eosinophilic pneumonia in a patient with ataxia telangiectasia. *Aust NZ J Med* 17:592–595, 1987.

98. Janin A, Torpier G, Courtin P, et al: Segregation of eosinophil proteins in alveolar macrophage compartments in chronic eosinophilic pneumonia. *Thorax* 48:57–62, 1993.

99. Shah PKD, Lakhotia M, Chittora M, Mehta S, Purohit A: Pulmonary infiltration with blood eosinophilia after scorpion sting. *Chest* 95:691–692, 1989.

100. Keslin MH, McCoy EL, McCusker JJ, Lutch JS: *Corynebacterium pseudotuberculosis*: A new cause of infectious and eosinophilic pneumonia *Am J Med* 67:228–231, 1979.

Chapter 15

Occupational Lung Disease

A. R. Gibbs

Zenker first coined the term *pneumoconiosis* for lung disease caused by inhalation of dust. Today many different forms of pneumoconioses are known, each caused by different dusts, fumes, smokes, vapors, or gases. The definition of pneumoconiosis varies, some authors restricting it to the nonneoplastic reaction of the lungs to inhaled minerals or organic dusts excluding asthma, bronchitis, and emphysema,[1] whereas others, including myself, use the term more widely to encompass any pulmonary reaction to inhaled particles from an industrial source. The exposure may not necessarily be by direct exposure to a particular mineral within an industry but can be from neighborhood or domestic (paraoccupational) exposures.

The damage caused by inhalation of different types of mineral particles varies according to the cumulative exposure, size, shape, surface properties, and durability of the particles. It should be realized that minerals given the same name but originating from different locations and processes may vary in physicochemical properties and thus biologic activity. This is particularly important with regard to asbestos and talc.

Immunologic processes are also involved to a variable extent in the response of the lungs to mineral exposures. They are important, for example, in chronic berylliosis and extrinsic allergic alveolitis. Also individual variability exists in deposition and retention of particles and the presence of other diseases (e.g., rheumatoid disease) can modify the response to the deposited particles.

Clearance mechanisms are important. It has been estimated that a nonsmoking city dweller inhales about 1 g of particulate matter each year, about half of which is deposited in the respiratory tract. Without efficient clearance mechanisms the lungs of an individual would "silt up" at a young age. Smokers inhale even greater quantities of particulate material—about 150 g/y from 20 cigarettes per day.[2]

DEPOSITION AND CLEARANCE OF PARTICLES

Airborne particles possess a range of shapes and may exist individually or as aggregates and this will affect their deposition in the human respiratory tract. Particles that are near to a sphere are designated compact and those that have a length-to-breadth ratio of 3:1 or greater as fibers. Some authorities require a minimum length of $5\mu m$ to designate a particle fibrous. Particles are deposited in the respiratory tract according to their size, density, shape and aerodynamic properties and by the volume of each inspiration.[3–5] Those in the size range of 1 to $5\mu m$ aerodynamic equivalent diameter (AED) are the most likely to deposit in the lung parenchyma. Particles of extreme shape such as fibers and plates and particles of very low density show a greater tendency to deposit in the airway walls. The likelihood of fiber deposition within the lung depends mainly on the diameter because those with diameters below $3\mu m$ align axially within the air stream.[6] The majority of fibers that deposit in the lung have diameters less than $3\mu m$ but may exceed $200\mu m$ in length. Short fibers ($<5\mu m$ in length) reach the lung parenchyma more readily than long fibers ($>5\mu m$ in length) because the latter exhibit greater aerodynamic drag, which increases impaction and sedimentation in the larger airways. Aggregation of particles effectively increases their AED, which will reduce deposition within the respiratory tract. Particle size also influences the toxicity of a mineral.

The pattern and rate of breathing also influence particle deposition. Because the internal volume of the airways increases with successive generations, the net effect of normal breathing at the level of the respiratory bronchioles is that the forward air flow virtually ceases. However, with heavy exertion the flow velocity increases and this increases the deposition of particles—in the upper airways by impaction and within the smaller airways by sedimentation and diffusion.[7] The nose is more efficient than the mouth at removing particles from the air. All particles greater than 5μm in diameter are removed from the air by impaction in the nose. Mouth breathing increases the number of particles that can reach the lower respiratory tract.

Clearance of particles from the lung depends on a number of factors including the anatomic site of deposition, particle solubility, particle size and shape, the efficiency of the phagocytic system, the presence of fibrosis, and cigarette smoking. Soluble particles dissolve in the lung secretions, but insoluble particles within the acinus either lie free or within macrophages and may then move to the ciliated epithelium whereby they can be cleared by the mucociliary apparatus. Some particles within the acinus, however, may penetrate epithelial membranes and enter the interstitum,[8–10] which is more likely to lead to fibrosis. This is particularly likely to happen when the macrophage system becomes overloaded by heavy dust burdens. The particles located within the interstitium may gain access to the vascular and lymphatic systems and may then be carried to other organs, the visceral pleura, or regional lymph nodes.[11] This is reflected in the black outlining of subpleural lymphatic pathways observed in the lungs of urban dwellers and coal workers. There is a greater retention of large compared to small compact and fibrous particles within the lungs. Also the presence of fibrosis within a lung interferes with compact and fibrous particle clearance. Tobacco smoking impedes particle clearance by interfering with the mucociliary action and possibly macrophage function.[12,13] Substantial evidence indicates that for many lung diseases caused by mineral particle exposure, it is the retained mineral particle burden that is important in pathogenesis.

IN VITRO AND IN VIVO STUDIES

Experimental studies provide useful information concerning the effects of mineral exposures on the lung and pleura.[14] However, when considering the value of in vitro and in vivo experiments with mineral particles in relation to the likelihood of developing human disease, certain limitations and criticisms should be kept in mind.[15] Experimental procedures use artificial "pure" and heavy exposures to a particular mineral, often causing overload of the macrophage system in animals. It should be realized that in such heavy doses even relatively inert dusts can

cause pulmonary changes. Also the experiments often utilize relatively short high exposures that do not reflect the lower long-term exposures that occur in human beings.

Intratracheal exposures result in a very uneven deposition of mineral particles quite unlike that occurring in the normal state. Intrapleural and intraperitoneal procedures bypass the normal defense mechanisms and deliver considerably higher doses of a mineral to the mesothelium than would occur in life. Mesotheliomas might occur in serosal membranes when a fibrous particle is directly implanted, but this is irrelevant if the particle cannot normally penetrate the membrane.

Inhalation studies are the most appropriate, but great care has to be taken in conducting these; the mineral used should be well characterized and sized, dosage should be appropriate and simulate long-term human exposures, and appropriate controls should be used. Unfortunately, well-conducted inhalation studies are extremely expensive and technically demanding.

Interpretation of in vitro and in vivo data on the pathogenicity of a specific mineral should be done in conjunction with the human epidemiologic, pathologic, and pulmonary mineral burden evidence. It is important that unjustified conclusions should not be made about human consequences of exposure to a mineral only from in vitro and in vivo studies. They should be regarded as screening procedures.

PATTERNS OF DISEASE

Mineral exposures can produce almost the whole gamut of pulmonary pathology. Table 15-1 gives examples.[16] A pathologist confronted by almost any form of pulmonary disease should consider mineral exposures in the list of etiologies. Of course, some types of pulmonary pathology are more closely associated with mineral exposures than others. When a subject develops a pulmonary disease a good and complete occupational history from commencement of work is essential because the pathologic sequelae may not develop until several decades after exposure (latency). The occupational histories of relatives may also need to be ascertained because some diseases (e.g., berylliosis and mesothelioma) may result from indirect paraoccupational exposures. Also it may be necessary to ask about hobbies and pastimes because these can result sometimes in exposure to hazardous minerals.

TABLE 15-1
Types of Pathologic Reaction That May Occur with Various Minerals

Compartment	Type of Reaction	Examples
Airways	Asthma	Isocyanates, metals
	Bronchiolitis	nitrogen dioxide
Airways—parenchyma	Centrilobular dust accumulations	
	Slight fibrosis	Coal, kaolin, talc, mica
	Moderate stellate fibrosis	Silica plus silicates or iron
	Marked circumscribed fibrosis	Silica
	Diffuse interstitial fibrosis	Asbestos, kaolin, talc
	Extrinsic allergic alveolitis	Farmer's lung
	Sarcoid-like	Beryllium
Parenchyma	Diffuse alveolar damage	Toxic fumes
	DIP	Asbestos, silica, silicates
	GIP	Hard metal
	Pulmonary alveolar proteinosis	Silica
	Emphysema	Coal, cadmium
Diffuse—random	PMF	Coal, kaolin, talc
	Carcinoma	Asbestos, nickel, arsenic
	Infection	Tubercle in silicosis
Pleura	Benign effusion	Asbestos
	Plaques	Asbestos, talc, wollastonite
	Diffuse fibrosis	Asbestos
	Mesothelioma	Asbestos

DIP = desquamative interstitial pneumonia; GIP = giant cell interstitial pneumonia; PMF = progressive massive fibrosis

GENERAL APPROACH TO INTERPRETATION OF PATHOLOGIC SPECIMENS

Pneumoconioses have important connotations medicolegally and a careful macroscopic examination and recording of lesions from specimens of lung and pleura is essential if the pathologist wishes to avoid considerable difficulties and potential embarassments later on. Often correlations between pathology and radiology and lung function tests will be necessary to correctly diagnose and attribute disability or death to a mineral exposure. It is wise to retain lung tissue from such cases for a considerable period of time, if possible, because sometimes issues are raised that were not initially considered.

Adequate sampling for histologic examination should be performed. It is important to take samples from many areas of the lung and to include background lung as well as tumor when considering an occupational cause for tumors. It is necessary to sample from areas away from the tumor and preferably from the other lung to evaluate fibrosis, when tumor is present. At least one block should be taken from each lobe to include the most abnormal, normal, and intermediate areas and at least one block from the major airways, pleura, and lymph nodes. Table 15-2 is a suggested schema for evaluating and recording macroscopic findings of the lungs and pleura.[17]

The microscopic assessment follows the usual procedure for any pulmonary disease, namely, first of all a low-power examination to decide which anatomic compartment the disease is in followed by higher-power examination to characterize the nature of the cellular and any particulate component.

TABLE 15-2
Macroscopic Recording of Lung Diseases Associated with Occupational Exposures

Primary dust foci
Average size (grade 0–3)
    1 = 1–3 mm; 2 = 4–5 mm; 3 = > 5 mm
Profusion (grade 0–3)
    Proportion of lobules involved:
      1 = <33%; 2 = 33–66%; 3 = >66%
Secondary dust foci
Number of stellate foci according to size
Number of round foci according to size
Size of PMF lesion(s)
Emphysema
Type
    Centriacinar, panacinar, irregular
Severity (grade 0–3)
    1 = 1–3 mm; 2 = 4–5 mm; 3 = >5 mm
Profusion (grade 0–3
    Proportion of lobules involved (similar to dust foci)
Interstitial fibrosis
Severity (grade 0–4)—microscopic
Extent (grade 0–3)
    1 = up to 10% of the lung involved
    2 = involvement of 10–25%
    3 = involvement of >25%

PMF = progressive massive fibrosis

SOURCE: Modified from Gibbs and Seal,[17] courtesy of the editor of *Journal of Clinical Pathology.*

In some cases the pathologic examination will not provide the complete answer to diagnosis and attribution and mineralogic examination of lung tissue will be advantageous.[18] This has been used most extensively in asbestos-related diseases but it can also provide useful information in assessing silicate- and metal-induced diseases.

SILICA

The word silica is derived from the Latin noun *silex,* a flint, a stone that contains silica (silicon dioxide). Pneumoconiosis caused by the inhalation of dusts containing a large proportion of silica is called silicosis and it is a disease as old as the history of man himself. Until the discovery of the tubercle bacillus by Koch in 1882, silicosis was confused with tuberculosis, a disease to which it predisposes and with which it is often associated. Because silica and silicates are the most common constituents of the earth's crust, dusts containing them constitute an important occupational hazard in mining and innumerable industrial processes. Historical evidence of pneumoconiosis, including silicosis, has been described in Egyptian mummies,[19] Bohemian miners in the sixteenth century,[20] and stone cutters in the eighteenth century.[21] With the coming of the Industrial Revolution many new trades developed in the United Kingdom; the danger of occupational dust inhalation was not appreciated and the incidence of silicosis rose. It was during this period that occupational diseases first began to be studied seriously. In 1831, Thackrah began a study of longevity of British workmen in dusty trades and found that sandstone workers died on average before the age of 40 years.[22] The first use of the term *silicosis* has been credited to Visconti in 1870.[23] The South African authorities and investigators provided several important reports on the prevalence of disease in gold miners and inaugurated compensation schemes and improved ventilation systems at the mines in the first 20 years of the twentieth century.[24] Just after the commencement of the twentieth century sandblasting with silica was introduced, which resulted in many cases of a particularly fulminating form of silicosis.[25] In the United States at Gauley Bridge, in the early 1930s, silicosis affected a large number of tunnelers.[26] These cases led to the establishment of occupational dust standards. Regulations have become increasingly tighter in developed countries over the past 50 years. Cases of silicosis are still observed although they are uncommon. In underdeveloped countries the disease is still a significant problem.

The effect of silica inhalation on the lung depends on the dose and duration of exposure, the size of the particles, the mineral form (polymorph), the presence of associated minerals, individual variability, and presence of mycobacterial infection. Relatively high exposures to finely particulate silica dust, such as has occurred in sandblasters and silica flour workers, can lead to the acute form of silicosis (silicoproteinosis), whereas prolonged exposure to lower doses can lead to the typical chronic form of silicosis. Silica exists naturally in crystalline and amorphous forms. Amorphous forms include opal and flint, the former hydrated form being found in diatomaceous earth (kieselguhr). These are relatively nonfibrogenic unless heated to high temperatures when the crystalline forms, cristobalite and tridymite, are produced. There are several natural crystalline forms, which include *a* quartz, the most common form and which is usually meant by the terms silica and quartz. The others are tridymite, cristobalite, stishovite, and coesite. The latter two forms are very rare. Tridymite and cristobalite are formed when quartz is subjected to high temperatures, for example, in refractories; experimentally it appears to be more fibrogenic than quartz. Virtually all human exposures to silica are associated with varying amounts of other minerals, particularly silicates, and these will modify the response to silica depending on their concentration and their own mineralogic properties. When the effects of silica and other minerals are considered to be roughly in balance, the response is referred to as *mixed dust fibrosis*. When the effect of silica is predominant, the response is referred to as *silicosis*. There is also individual variation in response to these exposures, most noticeable in those with the rheumatoid diathesis (Caplan's syndrome). Each form of silicosis predisposes to mycobacterial infection and the risk is increased also in workers exposed to silica but without radiographic abnormalities. In vitro and in vivo studies have shown that the presence of silica enhances the growth and virulence of mycobacteria.[27,28]

Sources of Exposure

There are many potential occupational sources of exposure to dusts containing a high proportion of silica and those that are given in Table 15-3 are by no means comprehensive.

Clinical Features

Generally chronic (classical or nodular) silicosis develops after more than 20 years of exposure; uncommonly cases develop within 5 to 10 years of exposure when the term *accelerated silicosis* is used. The pathology is similar in both forms. Symptoms or signs are not usually present in the simple form but dyspnea may occur in the complicated form, which progresses with the extent of the radiologic opacities. Radiographic and functional abnormalities correlate

TABLE 15-3
Industries with Possible Silica Exposure

Mining and milling of nearly all metals and nonmetals
Quarrying of granite, sandstone, and slate
Granite and stone industries
Iron and steel foundries
Manufacture of glass, pottery, and ceramics
Manufacture of refractories
Manufacture of abrasive powders
Enameling
Boiler scaling
Steel manufacture
Rubber industry
Processing of diatomaceous earth

with cumulative exposure to respirable silica.[29,30] Lesions may progress after cessation of exposure and the subject may present clinically several years after exposure has ceased. *Acute silicosis* (silico lipoproteinosis) develops within 3 years of heavy exposure, is usually rapidly fatal, and shows a different pathology to the other forms.

Chronic (Nodular, Classical) Silicosis

Macroscopically, the lungs in chronic silicosis reveal firm, grayish black, sometimes calcified, well-circumscribed, whorled nodules that are most prevalent in the upper zones. They vary from a few millimeters in diameter to large massive [progressive massive fibrosis (PMF)] lesions that can occupy large areas of the lung and can extend across fissures into the adjacent lobe (Fig. 15-1). The PMF lesions are most frequently observed in the upper lobes and when present are usually bilateral. The centers of the smaller nodules and the massive lesions can cavitate and may contain grayish black fluid. This cavitation has usually resulted from ischemic necrosis but sometimes it has been caused by mycobacterial infection. In the latter case, the cavities often contain gray friable material. If the massive lesions are carefully examined, they are made up of apparently fused, individual hard nodules, so-called conglomeration. This contrasts with the massive lesions seen in the majority of cases of coal worker's pneumoconiosis and silicatoses, which are diffuse and soft. In contrast to coal worker's pneumoconiosis, focal emphysema is not a feature of silicosis. The pleural surfaces frequently show firm fibrous adhesions to the chest wall and there may be areas of pleural thickening. In the earliest stages of the disease the nodules will be observed in the subpleural tissues and as the disease progresses they will be seen in the remainder of the parenchyma. The hilar lymph nodes typically are enlarged and contain whorled, hard, often calcified nodules (Fig. 15-2) which is responsible for the "eggshell" calcification often present in chest radiographs. The changes in the lymph nodes develop before those in the lung and if nodules are absent from the lung the diagnosis of silicosis should not be made.

Microscopically, the well-developed silicotic nodule, which is the hallmark of exposure to dusts with a high percentage of silica, consists of concentric layers of hyaline fibrous tissue at its center with a peripheral zone of dust-laden macrophages, fibroblasts, and chronic inflammatory cells of variable width, which extends out irregularly into the adjacent interstitial tissues[31] (Figs. 15-3 and 15-4). The center of the nodule can exhibit necrosis and calcifi-

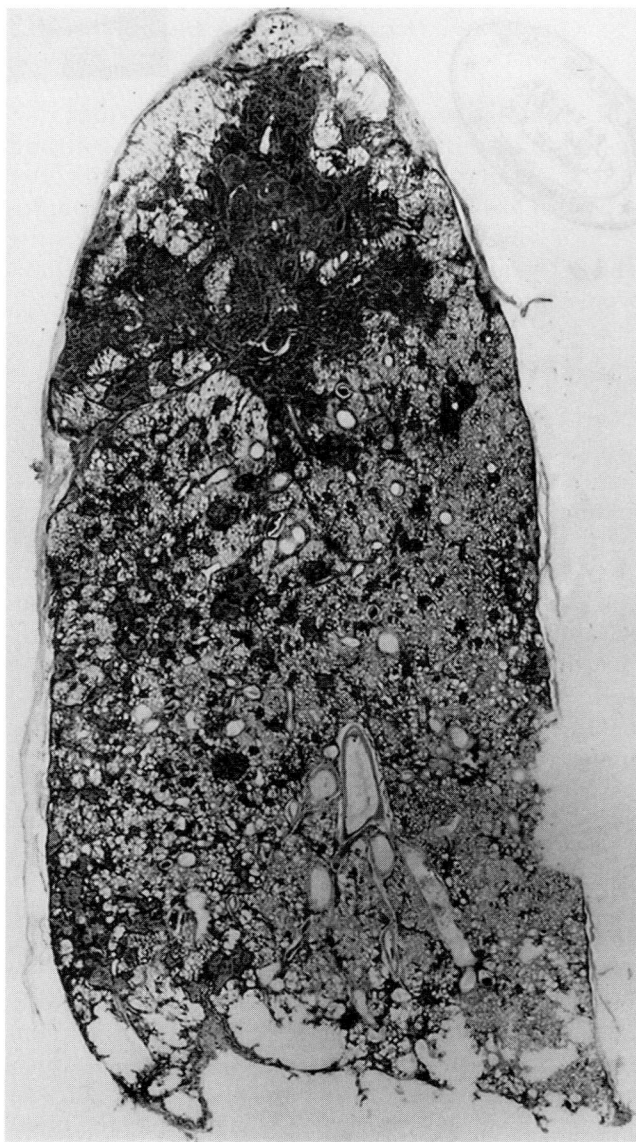

FIGURE 15-1
Gough-Wentworth whole lung section from a North Wales slate worker shows numerous silicotic nodules with an area of progressive massive fibrosis formed by conglomeration of nodules in the upper lobe.

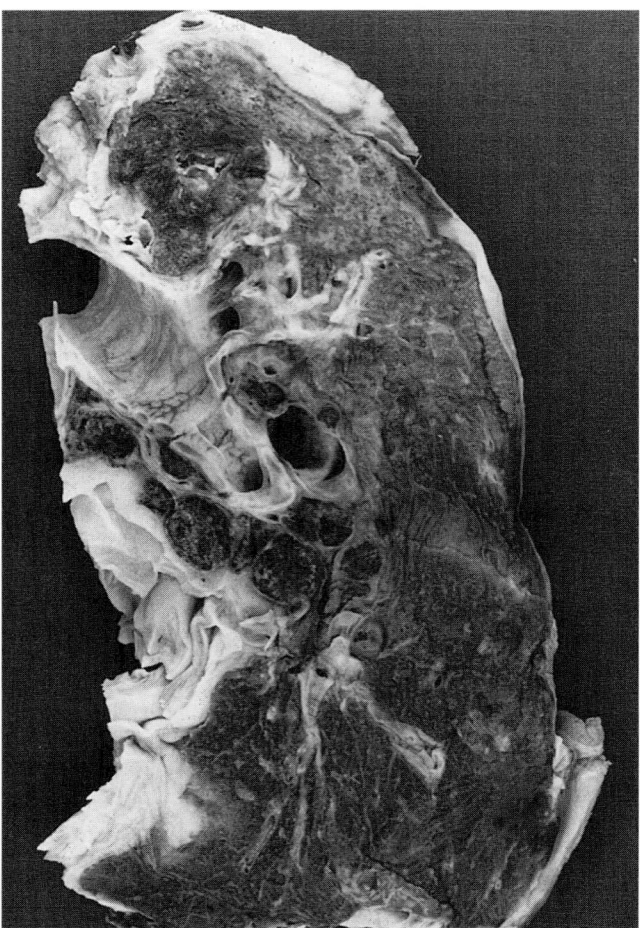

FIGURE 15-2
The lung of a stonemason shows enlarged hilar nodes with gray whorled nodules.

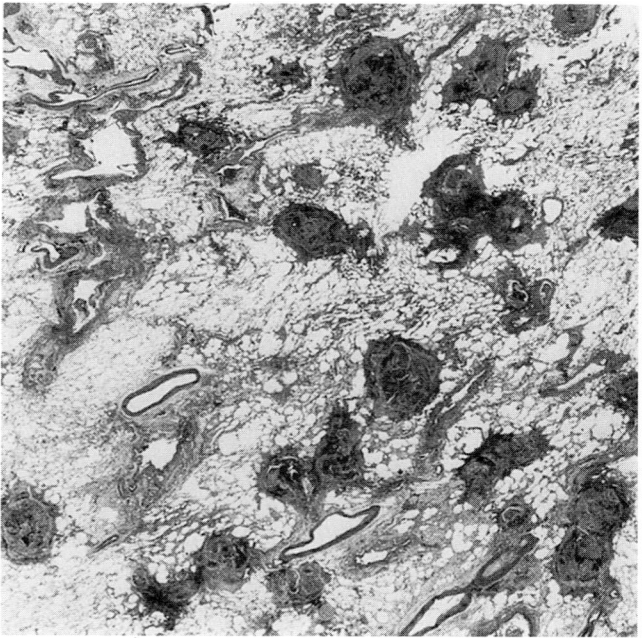

FIGURE 15-3
Numerous rounded, well-demarcated nodules in the lung of a North Wales slate worker. (Low power.)

cation. The necrosis is usually a consequence of ischemia from the formation of nodules obliterating pulmonary arterioles and perivascular lymphatics. Even so, whenever conspicuous necrosis is seen, even in the absence of granulomas, direct staining and culture for acid-fast bacilli should be performed to exclude mycobacterial infection.

The evolution of these silicotic nodules has been traced in animals and man. The earliest lesion consists of collections of free dust particles and dust-containing macrophages situated in the alveoli and interstitial tissues around the respiratory bronchioles, alveolar ducts, septa, and beneath the pleura. This is followed by the development of reticulin and collagen fibers, at first irregularly, and then at the

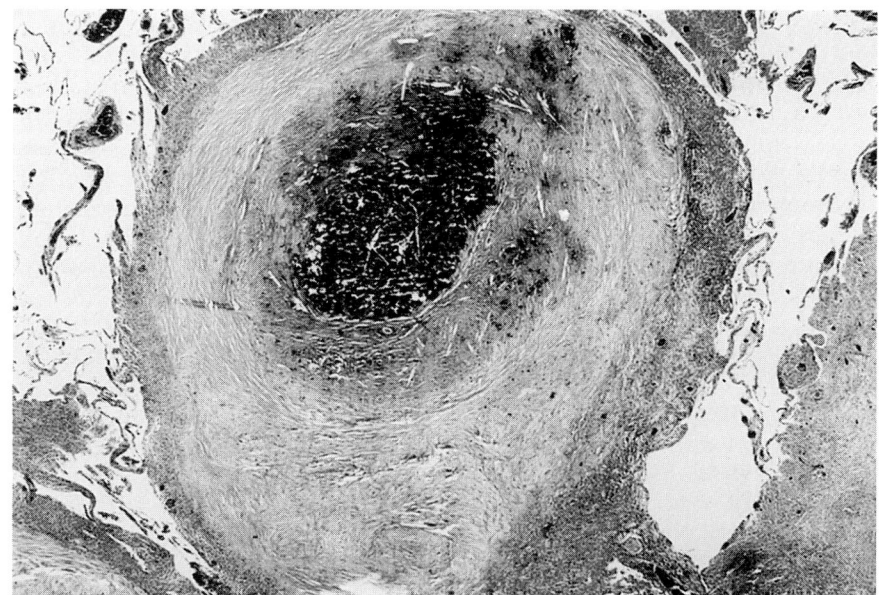

FIGURE 15-4
A silicotic nodule shows central necrosis with calcification, an intermediate zone of circumferentially arranged collagen, and a peripheral zone of dust-laden macrophages intermingled with inflammatory cells. (Low power.)

center of the lesions (Fig. 15-5). With progression the central zone takes on a concentric laminar arrangement and the dust-containing macrophages orient peripherally. This leads to obliteration of the bronchiolar and vascular architecture. Often nodules at different stages can be visualized in the same lung. Under polarized light the mineral particles at the centers of the nodules exhibit dull birefringence (quartz), whereas those in the peripheral zone exhibit strong birefringence (silicates and quartz). From the above description it will be appreciated that silicotic lesions are much more fibrous than those seen in coal worker's pneumoconiosis.

INTERSTITIAL FIBROSIS

On microscopic examination of lungs of subjects exposed to silica and silicates it is not uncommon to find foci of interstitial fibrosis in addition to the typical silicotic nodules. Occasionally the interstitial fibrosis is sufficiently severe to be associated with honeycombing. The precise contributions of silica and silicates to these lesions have not been clarified.

Rheumatoid Pneumoconiosis

In subjects with the rheumatoid diathesis the nodules may develop more quickly and become large and

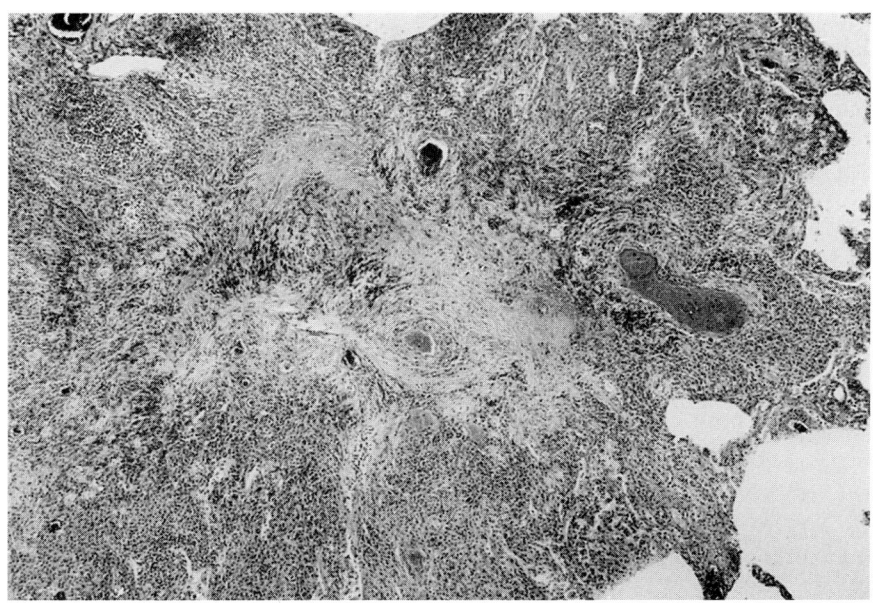

FIGURE 15-5
An "early" silicotic nodule in a North Wales slate worker shows a large collection of dust-laden macrophages and inflammatory cells with fibrosis at the center. (Low power.)

set within a background of relatively sparse small nodules, as they do in the Caplan form of coal worker's pneumoconiosis.[32] This has been described in association with silica exposures on rare occasions.

Interestingly, it has been claimed that there is an increased incidence and prevalence of rheumatoid arthritis, systemic lupus erythematosus, and scleroderma among workers exposed to silica, but not necessarily in the presence of silicosis, possibly due to the immunologic abnormalities set off by the exposure.[33,34] Others have not found this to be so.

Pathogenesis

In vitro and in vivo studies have shown that the toxicity of silica and its polymorphs depends on particle size and surface properties.[35] As the particle size distribution diminishes the cytotoxicity increases. The crystallinity of the particle's surface reflects the silanol and ionized silanol groups; the former determine mineral–cell attachment and the latter determines the interaction. The presence of trace metals, such as iron and aluminium, within the silica particle can modify the surface interaction by blocking the ionized silanol groups, which results in a lower fibrotic response. A key event is the uptake and interaction between the silica particles and macrophages, which causes damage to lysosomal membranes. This causes release of chemical attractants, which increase macrophage production and release. Death of some of the macrophages causes release of the silica particles that are then rephagocytosed by more macrophages. Some of the macrophages, however, will survive and carry the particles to other parts of the lung. There is evidence that polymorph neutrophils are also involved in the inflammatory process and that there is direct damage to type 1 pneumocytes in the centrilobular location in the early phase.[36] Macrophages containing quartz are found in the bronchoalveolar lavage fluids of silicotics and the percentage increases with duration of exposure.[37] Macrophages and neutrophils damaged by silica can release oxidants, which can result in degradation of connective tissue components of the lung and epithelial damage.[38] The macrophages can release factors, such as interleukin-1, which stimulates helper thymus-derived lymphocytes and fibroblast growth factors, which may lead to fibrosis.[39] The specific roles of these various factors and how they interact are still being elucidated.

SILICOSIS AND LUNG CANCER

There has been considerable debate for several years about the carcinogenic potential of crystalline silica.[40,41] Experimental studies have shown carcinogenicity in the rat but not the hamster or mouse[42] and this has been linked to fibrogenesis. Human studies have provided conflicting results and have been difficult to interpret because of confounding factors such as tobacco usage, radon exposure, and other industrial carcinogens that have not been adequately considered. A recent review of the subject concluded that lung cancer should not be classified as an occupational disease linked to silica exposure.[43]

Acute Silicosis (Silicoproteinosis)

This pattern of reaction occurs in association with high-dose exposures to finely particulate crystalline silica and has been described in sandblasters, tunnelers, and silica flour workers. In contrast to the other forms of silica reaction it can develop relatively rapidly over a few years. Clinically there is breathlessness and the chest radiograph resembles that of edema.

It is characterized by filling of the alveolar spaces with granular, eosinophilic material that stains positively with the periodic acid–Schiff (PAS) reaction and is also rich in lipid (Fig. 15-6). There is hyperplasia and metaplasia of the type II pneumocytes and an excess of these cells has been found on bronchoalveolar lavage.[44] It is similar to idiopathic lipoproteinosis but tends to show more interstitial inflammation and fibrosis, which is said to be a useful differential diagnostic feature. Well-developed silicotic nodules are usually absent although there may be focal collections of dust-laden macrophages with slight amounts of central fibrosis present. Polarization often, but not always, shows large numbers of birefringent particles. Mineral analysis of the lung tissue can be helpful in determining the etiology of this type of pathologic reaction.[45]

A similar condition has been produced in animals by inhalation of large amounts of quartz, tridymite, and cristobalite particles over short periods of time,[46,47] which damages and activates the type II pneumocytes causing an excess release of laminated bodies containing phospholipid, which accumulates in the alveolar spaces. This is followed by type II pneumocyte hyperplasia and further release of phospholipid. There is also evidence of impaired phagocytosis and clearance of the lipid material and particulate matter.[48]

Mixed Dust Fibrosis

This term was coined to imply concomitant exposure to silica in combination with less fibrogenic dusts such as iron oxides, carbon, kaolin, and mica.[49,50]

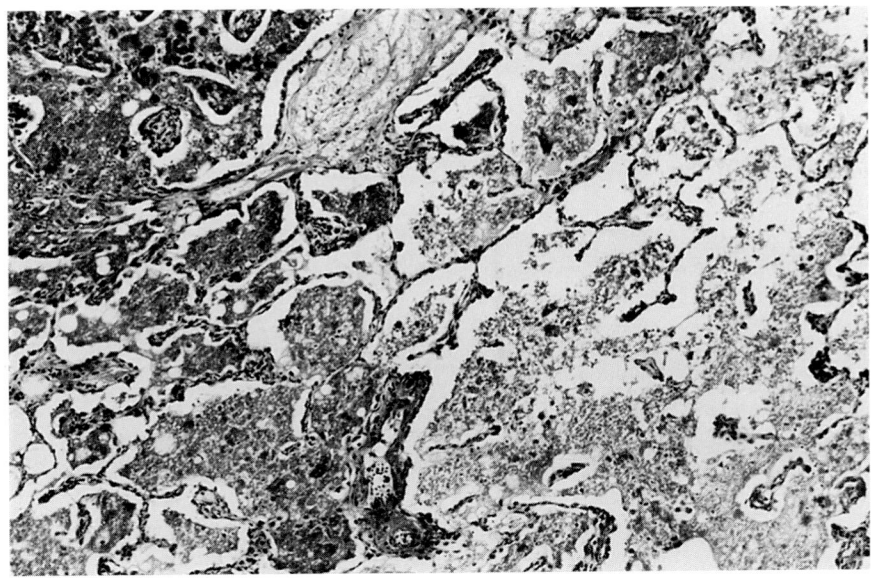

FIGURE 15-6
Acute silicoproteinosis shows a granular acellular intraalveolar exudate. (Low power.)

Pathologic descriptions have emphasized the presence of the stellate nodule with a "medusa-head" configuration and on chest radiographs typically there are irregular opacities. These lesions usually predominate when the silica content is less than 10 percent. In some ways it is an artificial distinction because these lesions are frequently found alongside typical silicotic nodules and virtually all exposures to silica are associated with exposures to other minerals. Typical occupations where these lesions have been described include hematite workers, foundry workers, arc welders, china stone workers, and a small proportion of coal workers.

PATHOLOGY

Macroscopically the lesions are typically stellate, firm, and their color varies according to the dusts present—reddish black in hematite miners, grayish white in china stone workers, and grayish black in coal workers. The lesions are most prevalent in the upper zones. Microscopically the typical mixed dust fibrotic nodule contains a central hyalinized collagen zone with a peripheral zone composed of radially and linearly arranged collagen and reticulin fibers intermingled with dust-laden macrophages (Fig. 15-7).

SILICATES

The nonfibrous silicates are a large group of minerals that are widespread and have an extensive range of industrial uses. Pulmonary disease has been associated with exposure to these minerals in a number of different industries.[50] They are less fibrogenic than silica and asbestos and pulmonary disease has been observed only after heavy prolonged exposures. The clinicopathologic features of these pneumoconioses show many similarities.

Kaolin Pneumoconiosis

Kaolin or china clay was discovered in a hill of extruded granite in Cornwall in Great Britain in 1750 by William Cuckworthy. It has been used in the United Kingdom and abroad for the manufacture of porcelain, as a filler and coating for paper, and in the rubber, plastics, paint, and fertilizer industries. It has also been exploited from deposits in Devon and from sedimentary deposits in Georgia and South Carolina

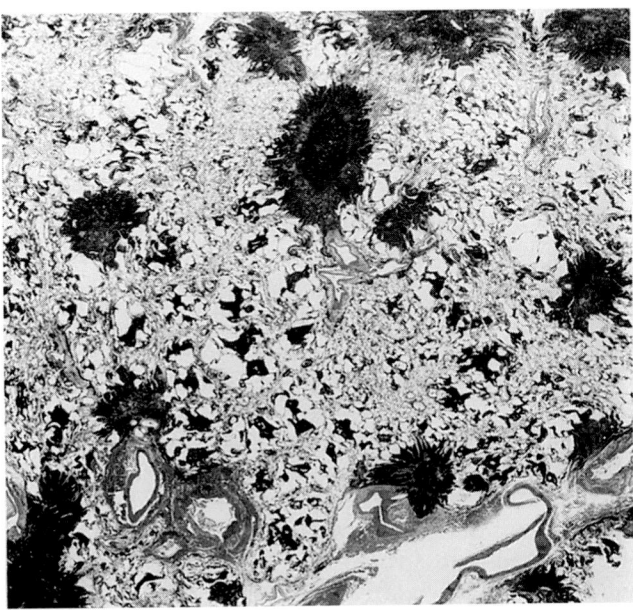

FIGURE 15-7
Mixed dust fibrotic nodules with stellate margins and central collagenization superimposed on simple macular lesions of coal worker's pneumoconiosis associated with emphysema. (Low power.)

in the United States. To a lesser extent it has also been produced in Australia, Brazil, France, Spain, and Germany.

During the commercial preparation and purification of china clay the granite, mica, and quartz are allowed to sediment out, the lighter kaolin particles being carried away in suspension. Nowadays, the kaolin is then allowed to settle and dry out using filter presses and mechanized drying plants. Formerly open kilns and the transfer of the dry material into storage linhays, wagons, or bags was accomplished by hand. After drying, the process was dusty but since 1970 the dust levels have been dramatically reduced.[51] After refining, the main constituent of the material is kaolinite and there may be up to 10 percent mica but quartz is less than 1 percent.

Several radiologic surveys of workers from this industry in Cornwall[52–54] and Georgia[55,56] have shown a pneumoconiosis generally characterized by small rounded opacities but occasionally PMF. Pneumoconiosis is generally seen in the millers, baggers, and loaders, where the dust is dry. Exposures after 1971 are not considered likely to lead to pneumoconiosis.

In some of the previously reported cases of kaolin pneumoconiosis it has been suggested that quartz was responsible. In the past some of the china clay workers had also worked with china stone and in these subjects pathologic examination of the lungs revealed silicotic and mixed dust fibrotic nodules. However, there are cases of pneumoconiosis that develop from exposure to kaolin without quartz—the typical china clay workers' exposure—and in these the pathology shows differences.

PATHOLOGY

On gross examination the lungs show gray, soft stellate lesions between 2 and 7 mm in diameter. Rarely massive lesions may be seen and these are diffuse soft and easily cut (Fig. 15-8). These massive lesions may cavitate. Sometimes fine honeycombing impregnated by gray dust is observed.[57] Microscopically, masses of dust-laden particles are first observed around the respiratory bronchioles but with increasing size they extend along the alveolar ducts and the interstitium of the adjacent alveoli (Figs. 15-9 and 15-10). Lesions from adjacent lobules may link up to give diffuse interstitial collections and fine fibrosis. The particles are strongly birefringent on polarization. Ferruginous bodies, formed on the platy particles, may be present but usually are not conspicuous. The massive lesions consist of masses of dust-laden particles mixed with randomly arranged collagen fibers. Mineral analysis of the lung tissue from sub-

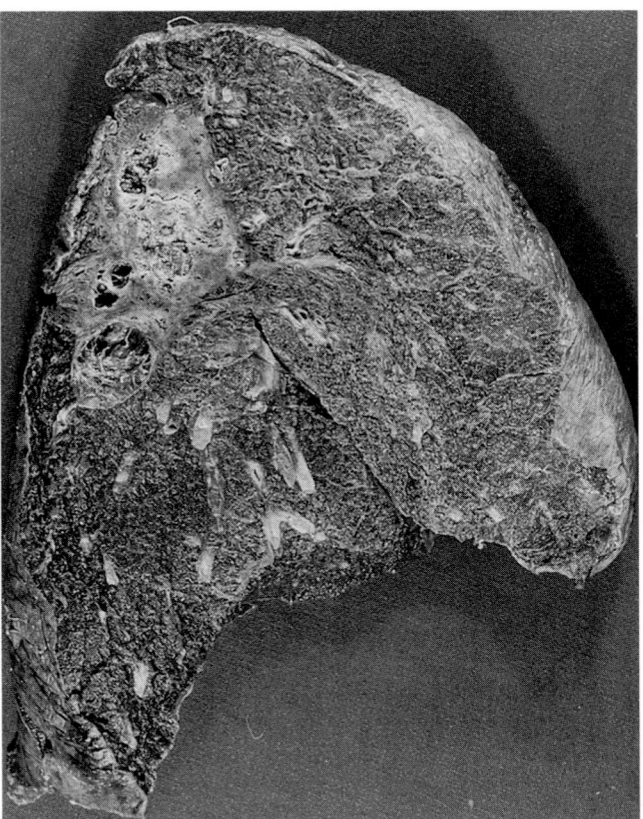

FIGURE 15-8
The lung of a Cornish china clay worker shows a cavitated grayish white progressive massive fibrosis lesion in the upper part of the lower lobe and background stellate simple lesions.

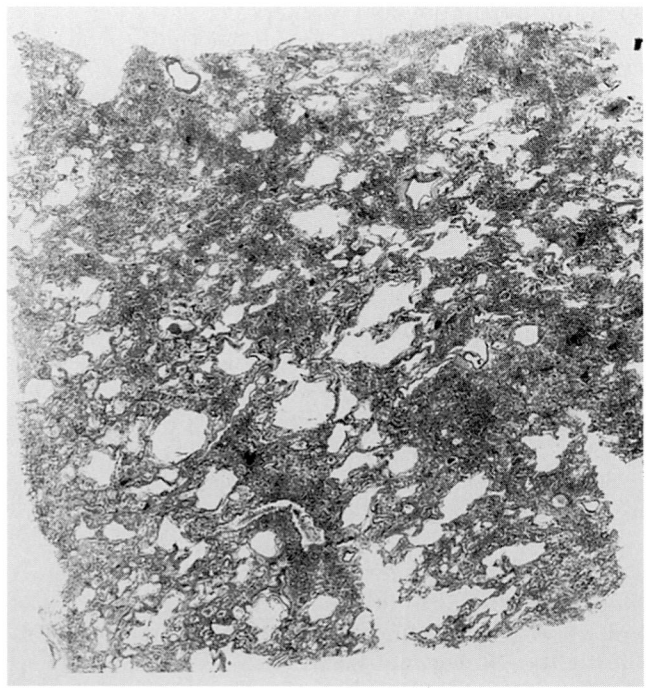

FIGURE 15-9
Extensive interstitial collections of dust-laden macrophages in the lung of a Cornish china clay worker. (Low power.)

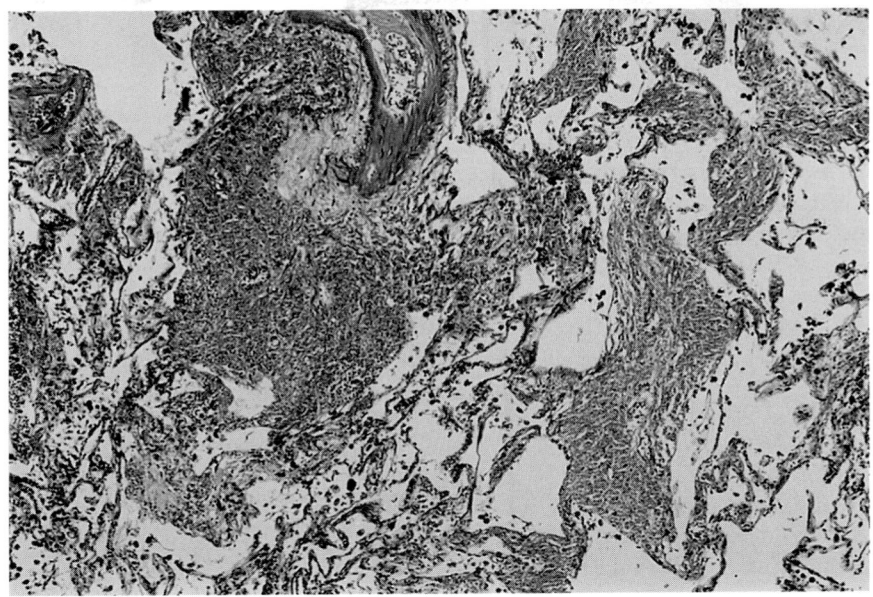

FIGURE 15-10
Interstitial collections of dust-laden macrophages associated with slight fibrosis in the lung of a Cornish china clay worker. (Low power.)

jects exposed to china clay show a similar composition to the original dust[58]: >90 percent kaolinite, < 1 percent quartz, < 1 percent feldspar, 1 to 10 percent mica.

In summary, it is possible to develop a pneumoconiosis from exposure to kaolin in the absence of quartz when exposures are heavy. However, with the improvement in the hygiene conditions of the industry since the early 1970s this problem is likely to disappear.

Talc Pneumoconiosis

Talc or hydrated magnesium silicate ($Mg_3Si_4O_{10}(OH)_2$) is formed during the breakdown and weathering of serpentine, tremolite, anthophyllite, and rocks of similar nature. It is found in and has been exploited from the United States, Canada, France, Norway, India, Australia, Italy, Spain, and Austria. Commercial talcs can contain a number of impurities that will depend on the geologic source of the material. In some products the percentage of nontalc mineral will be greater than that of the talc; for example, the "talc" from New York State contains approximately 30 to 55 percent tremolite and 12 to 50 percent talc. Approximately 50 percent of the world's tonnage of talc contains predominantly talc (95 percent or more); in the other 50 percent talc is associated with variable percentages (10 to 50 percent) of carbonates or chlorites.[59] A further complication is that some products sold as "talcs" may not even contain the mineral talc. Therefore, one should be cautious in interpreting studies of pulmonary disease in one cohort of talc workers and extrapolating them to other cohorts exposed to different varieties of talcs. Particular care has to be taken in attributing pul-

monary disease to talc exposure in workers employed in secondary industries because the mineral to which they were exposed may not be talc; for example, in the tire plants, mica and kaolin were often used instead of talc for various applications.

Talc has been used in many industries (Table 15-4). Pulmonary fibrosis has been associated with talc exposure in secondary industries and also in the mining and milling of talc. Rare cases of fibrosis have been described with tertiary exposure to cosmetic talc. The question arises as to whether the lung changes were caused by the talc or nontalc components or whether the exposure was to a different mineral completely. Often the only way to settle this question in an individual case is by mineral analysis of the lung tissues, which is difficult and requires a range of sophisticated techniques. A recent study of a series of purported cases of talc pneumoconiosis, in which these techniques were performed, showed in several of them significant quantities of minerals, such as kaolin and mica, in addition to talc.[60] In one case the predominant mineral was mica. Nevertheless there appear to be genuine cases of pulmonary fibrosis that have followed heavy exposures to talc uncontaminated by significant amounts of other minerals.[60–63] Deaths shortly after extremely

TABLE 15-4
Industries with Possible Talc Exposure

Ceramics	Tires
Paper	Rubber
Plastics	Cosmetics and pharmaceuticals
Building materials	Animal feeds
Paints	Fertilizers

high exposures to talc have been reported rarely, but these have been due to suffocation rather than the toxic effects on the lung. A range of radiographic lesions has been reported in association with talc, uncontaminated by amphiboles, including fine reticular infiltrates, honeycombing, PMF, and pleural thickening.

PATHOLOGY

On gross examination of the cut surfaces of the lungs there are stellate gray macules, which in severe cases link up and are associated with honeycombing (Fig. 15-11). Less commonly present are palpable gray nodules. PMF lesions characterized by a diffuse soft gray appearance, which may occupy most of an upper lobe, sometimes cavitated, have been described occasionally.[60,63,64] Microscopically, the most common lesions are stellate interstitial collections of dust-laden macrophages associated with mild fibrosis situated around the respiratory bronchioles (Figs. 15-12 and 15-13). Progression of these lesions results in linking up of adjacent lesions that if widespread will lead to greater fibrosis and honeycombing. Severe pathologic changes are usually associated with very high mineral burdens. Less constantly present are rounded and irregular nodules composed of dust-laden macrophages with a greater amount of collagen. Mixed dust fibrotic nodules may be present when there is associated quartz exposure. PMF lesions are characterized by large collections of dust and dust-laden macrophages mixed with irregu-

larly arranged collagen. Other features include ferruginous bodies formed on talc plates, which can be mistaken for asbestos bodies, foreign body granulomas containing large numbers of mineral particles (Fig. 15-14), and rarely sarcoid-like granulomas. It is uncertain at the present time whether the latter are a genuine feature of talcosis or a coincidental finding in a few subjects with undiagnosed sarcoidosis. Polarized light will reveal large numbers of strongly birefringent platy and acicular crystals (Fig. 15-15).

CARCINOMA OF THE LUNG

Some reports have claimed an increased risk of malignancy when exposure occurs to talcs contaminated by amphibole minerals. This has mainly centred around the New York State talc workers who were exposed to a high proportion of amphibole minerals in addition to talc. Several investigators have argued that the amphibole component is not asbestiform and therefore not hazardous.[65–69] Early cohort studies of this group found an excess mortality from lung cancer, but this was in short-term workers with less than a year's work; there was no relation to duration of exposure and the effect of smoking was not evaluated. More recent studies of the New York State talc workers and comparisons with the Vermont talc workers, who were exposed to talc free of amphiboles, indicated that there was no causal connection between lung cancer and amphiboles in the New York State workers.[70,71]

FIGURE 15-11
Lung from a worker in the rubber industry showing interstitial fibrosis with fine honeycombing. Analysis of the lung tissue shows significant quantities of talc, kaolin, and mica. *(Courtesy of the editor of Human Pathology.[60])*

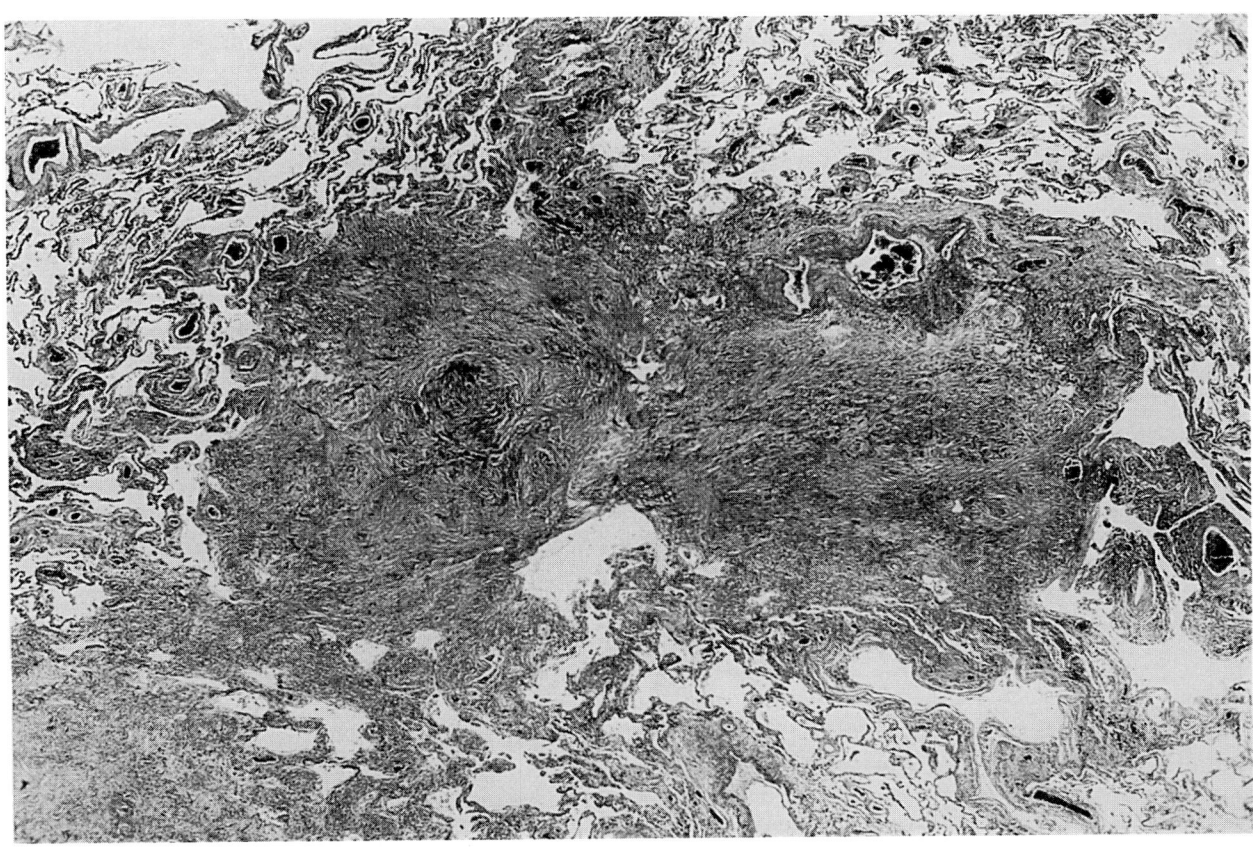

FIGURE 15-12
Lung from a talc worker shows interstitial collections of dust-laden cells associated with a mild degree of fibrosis. (Low power.)

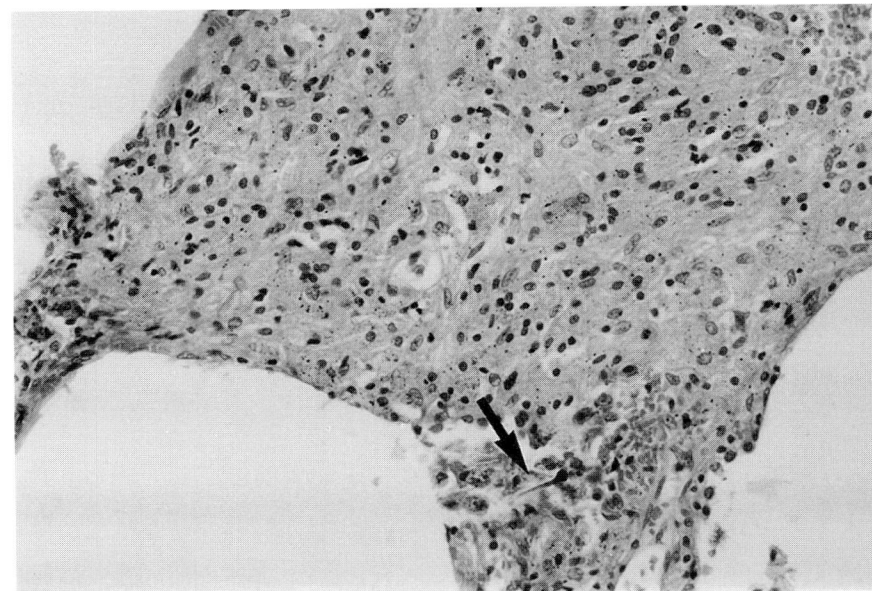

FIGURE 15-13
Lung from a talc worker shows accumulation of fine particulate material and inflammatory cells and a ferruginous body (*arrow*). (Medium power.)

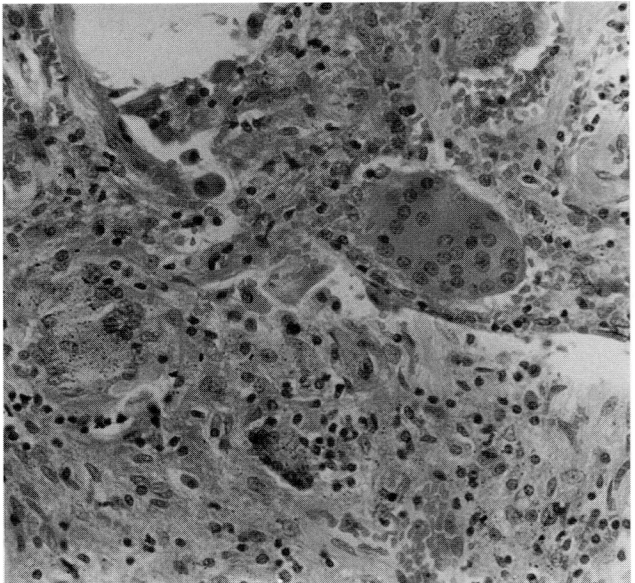

Mica Pneumoconiosis

The term mica, for practical purposes, encompasses the minerals muscovite, phlogopite, and vermiculite. Micas are produced in India, Brazil, Madagascar, the United States, France, Spain, Finland, and China. These minerals have been used in paints, plastics, fire protection boards, pigments, oilwell drilling muds, the electrical industry, and textured coatings. Exposures to relatively pure micas have occurred mainly in the grinding and packing of the minerals.

Occasional cases of pneumoconiosis and pleural thickening have been described in association with prolonged exposures.[72,73]

PATHOLOGY

On gross examination the lung shows gray, soft stellate lesions that measure a few millimeters in diameter. Rarely massive lesions have been reported (Fig. 15-16). The small lesions may link up to give a fine interstitial pattern. Microscopic examination shows initially dust-laden macrophages situated around respiratory bronchioles, which later extend out into the walls of the adjacent alveoli and alveolar ducts. In severe cases there is a diffuse interstitial collection of dust-laden macrophages associated with fine fibrosis. The dust particles show strong birefringence under polarized light. Ferruginous bodies formed on the yellowish brown mica plates and foreign body giant cells may be observed[57,74] (Fig. 15-17).

Fuller's Earth Pneumoconiosis

There is some confusion as to what is understood by the term fuller's earth because it does not mean the same mineral(s) in every country. In the United States it includes an attapulgite/calcium range of products, whereas in the United Kingdom it refers to calcium montmorillonite only. It has strong absorptive properties and is used in animal feeds, pet litter, pharmaceuticals, herbicides, and the processing of woolen garments. Some fuller's earths contain significant quantities of quartz, but there have been rare cases of pneumoconiosis described in association

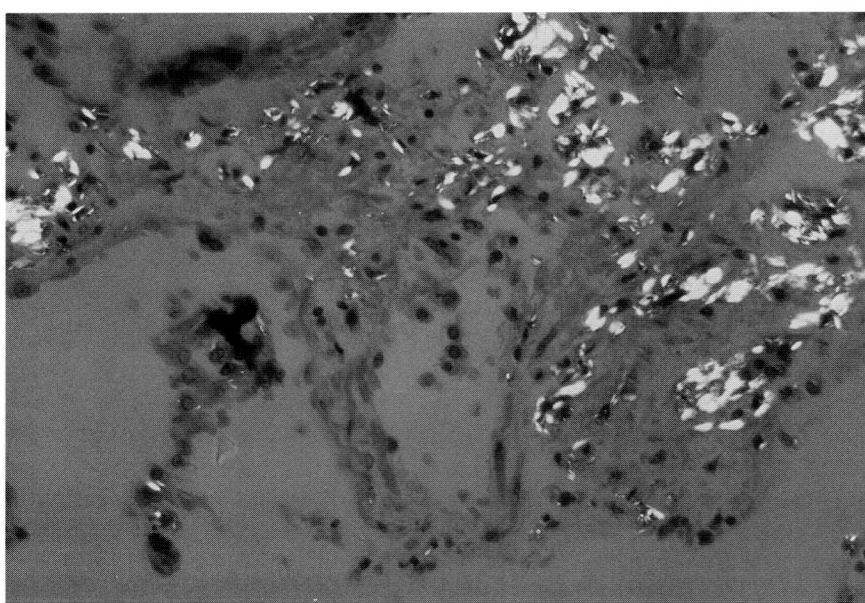

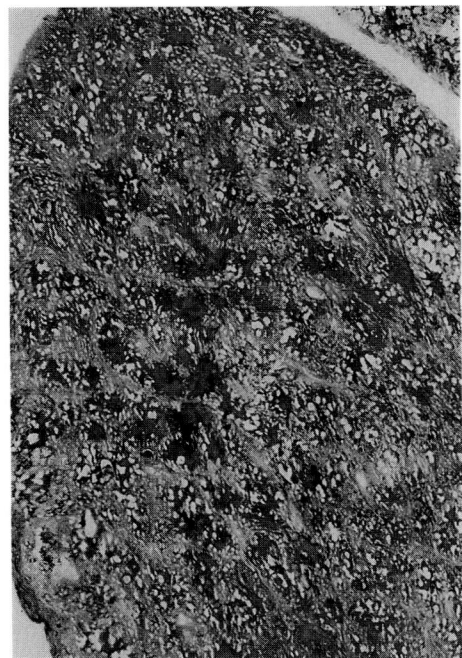

FIGURE 15-16
Mica grinder—numerous stellate gray lesions.

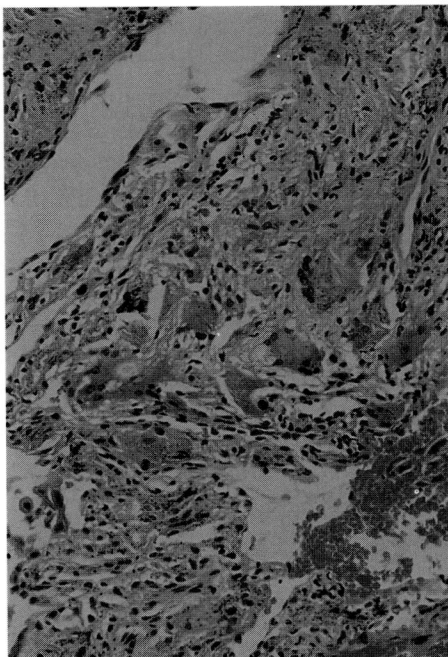

FIGURE 15-17
Mica bagger. The lung shows finely particulate gray-brown material in macrophages and foreign body giant cells in association with ferruginous bodies.

with heavy prolonged exposures to fuller's earth uncontaminated by quartz, for example, in workers at the Redhill and Combe Down deposits in the United Kingdom.[75,76] It has resulted in little clinical disability and the radiologic changes have usually taken several decades to develop.

PATHOLOGY

In those exposed to fuller's earth without quartz the macroscopic changes have consisted of gray-black stellate nodular lesions measuring a few millimeters in diameter. Microscopic examination shows peribronchiolar and interstitial collections of dust-laden macrophages associated with a slight degree of fibrosis (Fig. 15-18). The particles are strongly birefringent under polarized light.

COAL

Coal has been mined for over a thousand years in Great Britain, being first mentioned in the Saxon Chronicle of Peterborough in 852. Laennec described coal worker's pneumoconiosis in 1819.[77] With the coming of the Industrial Revolution in the first half of the nineteenth century the coal mining industry assumed major importance and it was not long before the many occupational hazards of coal mining became recognized in Scotland.[78] In 1838, a young Scots doctor (Stratton) coined the term anthracosis to describe the black discolored lungs of Fifeshire miners.[79]

In 1936, a Medical Research Council (MRC) survey was set up of coal miner's pneumoconiosis in South Wales because the incidence was higher in this than in any other British coal field. This survey contributed to much of the knowledge of this form of pneumoconiosis with regard to the clinical, radiologic, histopathologic, and experimental aspects. The survey also included an extensive geologic and petrologic study of the mines and identification of the different aspects of mine dust. The reader seeking detailed information on this subject is referred to the Medical Research Special Reports of 1942, 1943, and 1945.

The term coal worker's pneumoconiosis (CWP) is now used to cover a variety of pulmonary radiologic and pathologic changes resulting from the inhalation of coal dust and its impurities. In addition to causing pneumoconiotic changes in various categories of underground workers, Gough[80] showed that coal trimmers employed in loading coal into ships' holds might also develop similar pneumoconiotic lesions. This and the previous MRC studies settled

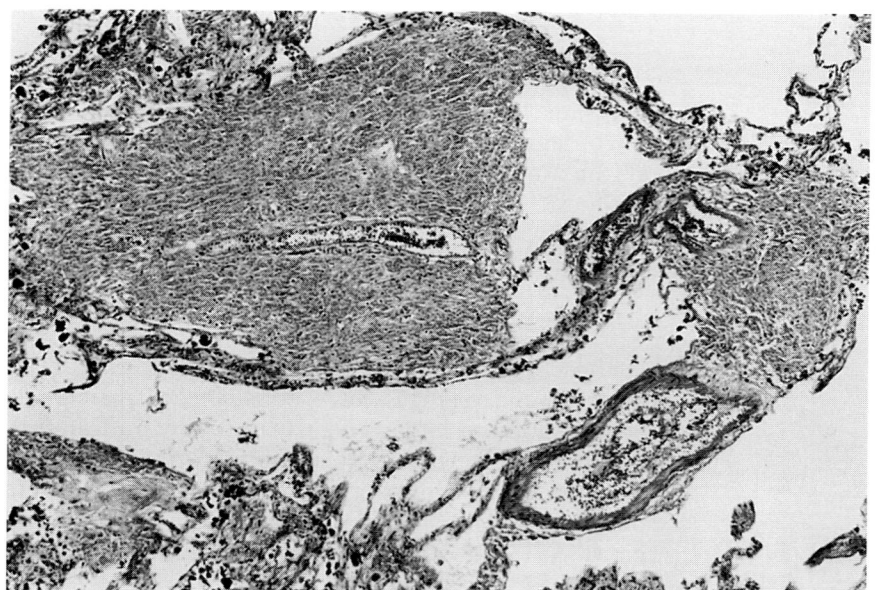

FIGURE 15-18
Fuller's earth pneumoconiosis shows interstitial accumulation of dust-laden macrophages associated with slight fibrosis. (Low power.)

the dispute as to whether CWP was caused by coal or silica exposure, showing that it could occur without significant silica exposure. In Great Britain CWP has been recognized as a disease separate from silicosis for the purposes of industrial benefits since 1943.

Further studies by the MRC in South Wales and the National Coal Board longitudinal study of 30,000 miners from 25 British collieries has taken the research further and allowed realistic standards of respirable dust exposure to be formulated.[81,82] The reduction in dust levels in the mines, decline in numbers of the workforce, early retirement, and the gradual closure of uneconomic mines has led to a decline in incidence of CWP in Great Britain, Europe, and the United States. CWP rarely develops in coal workers before the age of 50 years.

The prevalence and types of pathologic lesions observed in CWP depend on several factors (Table 15-5), which may explain the varying incidence in different coal fields and even between mines in the same coal fields. For example, in Great Britain it has occurred most frequently among miners in the South Wales coal field and especially in the miners employed in the deep anthracite mines in West Glamorgan. It is also more common among miners working in pits with small, deeply situated coal seams.

Coals are rich in carbon and are complex mineralogically, the composition varying from mine to mine. Coals are ranked according to calorific value into low, medium, and high. The higher ranked coals are the oldest and contain the least amounts of

TABLE 15-5
Determining Factors in Coal Worker's Pneumoconiosis

Cumulative exposure—dose and duration
Particle size distribution
Coal rank
Concentration and nature of silica
Associate minerals—muscovite, illite, and kaolin
Individual responses
Infection

volatile matter, whereas the younger, lower ranked coals contain greater amounts of volatile matter. Inorganic material is present in coals, for the most part clays—muscovite, illite, and kaolin—which constitutes a higher proportion of lower than higher ranked coals. Coal may also contain a number of trace elements including arsenic, lead, manganese, titanium, and beryllium. Uranium is present in certain low ranked coals and lignites (up to 0.1 percent). Exposure to respirable particles will depend on the type of coal being mined and adjacent rock strata. Therefore, the job of the underground worker plays a very important part in determining the amount and composition of the respirable dust to which the worker is exposed and consequently the type and severity of pulmonary lesions the worker may develop. Miners employed as hard headers and rock workers in the construction of communicating shafts between adjacent seams or bridging geologic faults work almost entirely in hard siliceous rocks and are exposed to high silica concentrations, which may result in silicosis rather than CWP. Other groups of coal workers exposed to high concentrations of silica include firemen and screen workers.

The incidence and progression of CWP, as one would expect, is related to cumulative exposure to respirable dust,[81] but this is not the whole story. There is also a relationship to coal rank. Several studies have shown that for equivalent exposures the incidence and severity of CWP increases as the rank increases.[81-83] For unknown reasons miners of high ranked coals show significantly more retention of dust in the lungs than those exposed to low rank coals. The histologic appearances of the pneumoconiotic lesions also varies with coal rank.[84,85] Another variable factor is quartz content, which affects the prevalence of CWP unpredictably for several reasons. Associated clay minerals may inhibit the toxic effects of quartz on the lung and this has also been demonstrated experimentally.[86-89] This inhibition may not be permanent and may vary with geologic source.[90] There is evidence that cytotoxicity of coal mine dusts increases as particle size decreases which in turn correlates with high mineral and quartz content.[91] Some miners exposed to dust with a high quartz content have exhibited unusually rapid radiologic changes of pneumoconiosis.[92]

In life the diagnosis and assessment of severity of CWP depends on the prevalence and size of opacities observed on chest radiographs. It is customarily divided into two forms—simple and complicated. The criterion for separation is size of the dust foci and this has varied with time and organization. In the International Labor Organization system of radiographic assessment of pneumoconiotic lesions, if an opacity is present which is greater than 1 cm in diameter then the condition is regarded as complicated CWP or PMF. Generally, simple CWP is not associated with symptoms or clinical signs, significant reduction in lung function or reduction in longevity.[86] Complicated CWP, particularly when lesions are very large, is associated with productive cough, breathlessness, and significant impairment of lung function and it can result in premature death. Complicated CWP is more likely to develop on a background of higher grades of simple CWP.[85] It can develop several years after leaving the industry and in my experience of South Wales coal miners over the last decade it has affected elderly retired miners who had sustained exposures several decades ago.

Pathology[93,94]

The pathology of CWP is most appropriately considered under five subdivisions: (1) simple CWP, (2) complicated CWP, (3) Caplan lesions, (4) emphysema, and (5) interstitial fibrosis.

SIMPLE CWP

Macroscopically, this is characterized by multiple bilateral, soft, stellate, impalpable, black (macular) lesions scattered through the lungs but which are most prevalent in the upper two-thirds (Fig. 15-19). At the earliest stage the lesions are only slightly larger than the soft dust foci associated with urban living. The lesions vary from about 1 mm to several millimeters in size but by definition do not exceed 10

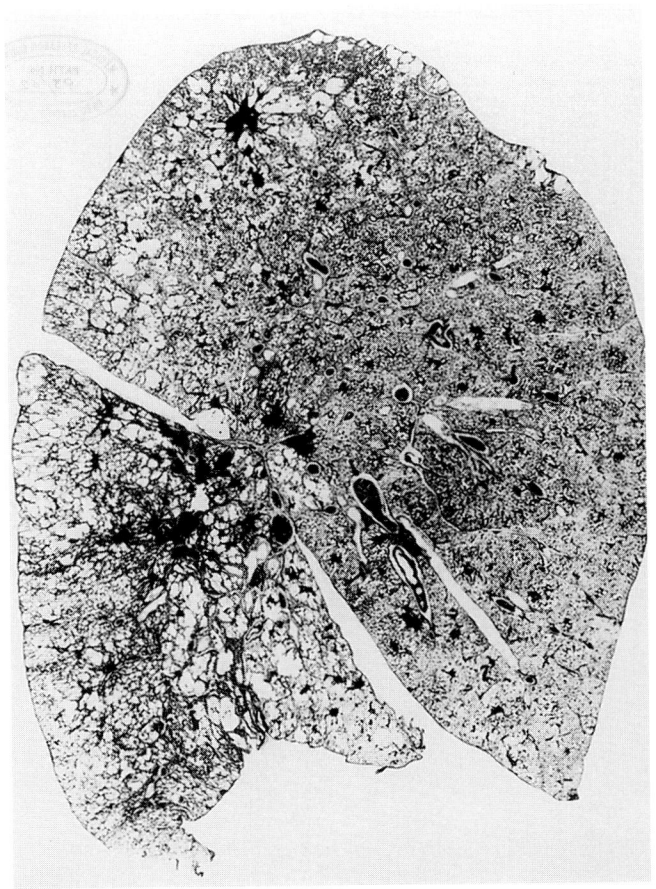

FIGURE 15-19
Gough-Wentworth whole lung section from a coal worker of many years shows numerous black macules affecting most of the lung lobules, a small number of larger stellate black lesions, and centrilobular emphysema affecting about two-thirds of the lung lobules.

mm. With progression they may come to occupy the majority of the lung lobules. This pathologic picture can be seen without there necessarily being any radiologically detectable pneumoconiosis. The number of lesions seen radiologically always appears to be less than is actually demonstrated pathologically.[95] Associated with these macular lesions are frequently variable degrees of emphysema situated at the centers of the acini (focal emphysema). Therefore, the black lesions appear to be highlighted by the surrounding pale emphysematous areas. The radiologic changes of simple CWP are due to the dust and the small amount of collagen present and do not reflect the emphysema.[96] In addition to these impalpable macular lesions there may be firm, palpable, black lesions that may be stellate or rounded in shape (fibrotic nodules). These nodules are usually much less in number and often larger than the macules (Fig. 15-20). Uncommonly they occur on their own but usually they are superimposed on a background of macules. Other features include dust impregnation of interlobular septa and the pleura; the latter is usually focal but occasionally may be diffuse. The peribronchial, hilar, and mediastinal lymph nodes are often enlarged, densely black and may be soft or firm. Sometimes they contain whorled silicotic-like nodules, usually in the absence of similar lesions in the lung. They are caused by the preferential clearance of silica from the lung to the lymph nodes.[97]

Microscopically the earliest change consists of the accumulation of dust-laden macrophages around the respiratory bronchioles and adjacent alveoli, blood vessels, septa, and beneath the pleura. Free dust particles may also be seen. The macules consist of uniformly distributed dust particles free or within macrophages intermingled with reticulin and small amounts of collagen, which contrasts with the considerable formation of collagen in silicotic nodules (Figs. 15-21 to 15-23). The dust packing of these lesions increases with decreasing ash content and increasing coal rank.[84] The nodules show much more collagen than the macules and it is arranged in bundles that radiate in all directions. Nodules are associated with the mining of lower ranked coals and increasing ash content. Nodules may show cavitation, calcification, and cholesterol clefts. If the coal worker has been exposed to a high silica concentration, mixed dust fibrotic or silicotic-type nodules may be observed, in which case the collagen fibers will either be arranged in a linear or circumferential pattern. Examination of macules or nodules under polarized light may reveal small doubly refractile

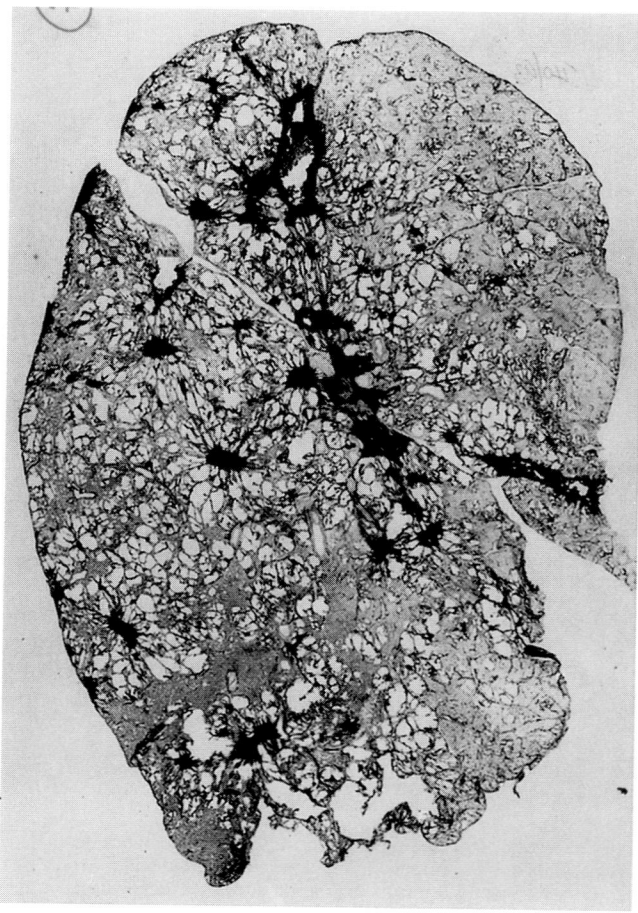

FIGURE 15-20
Gough-Wentworth whole lung section from a coal worker shows several stellate black lesions, which were palpable, and centrilobular emphysema.

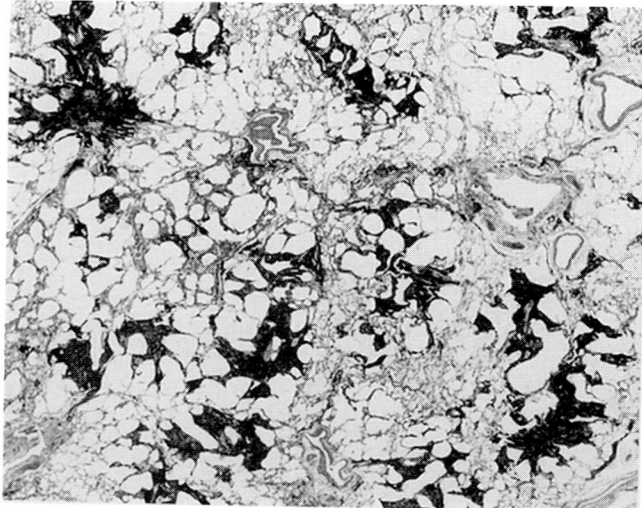

FIGURE 15-21
Simple coal worker's pneumoconiosis shows numerous stellate macular lesions associated with emphysema. (Low power.)

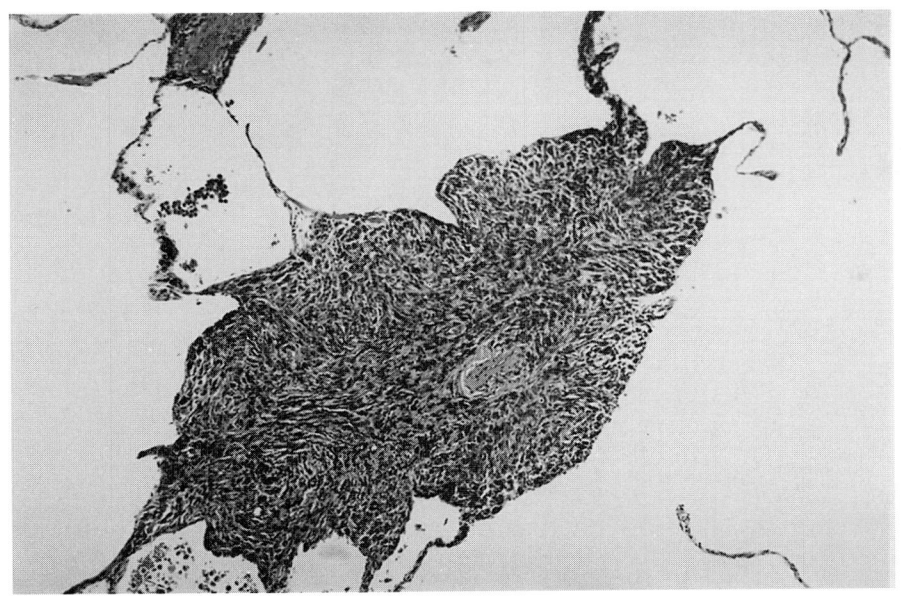

FIGURE 15-22
Simple lesion in a coal worker showing dust packing and relatively mild fibrosis. (Low power.)

particles, which are the silicates present within the coal dust.

COMPILICATED CWP OR PMF

The PMF lesions, which by definition are greater than 1 cm in diameter, nearly always occur on the background of severe simple pneumoconiosis. The PMF lesions most commonly occur in one or both upper lobes but can occur in any lobe. They may reach several centimeters in size and can destroy the

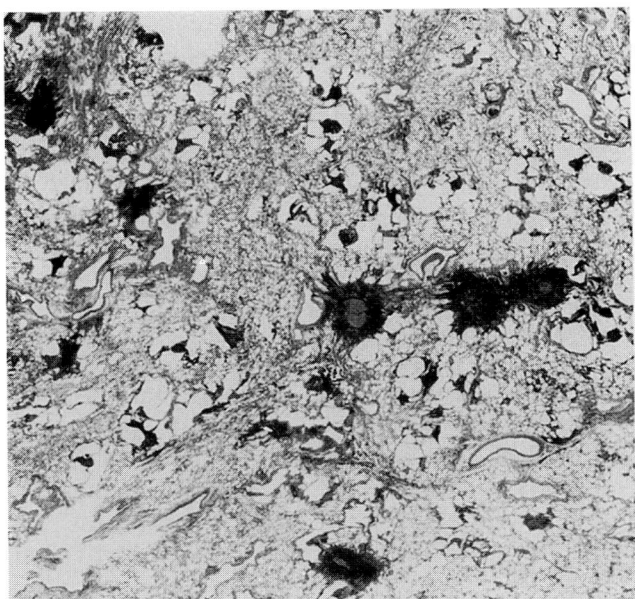

FIGURE 15-23
Coal worker. In addition to stellate macular lesions, some nodules show central collagenization. (Low power.)

whole of a lobe or sometimes several lobes. The lesion is typically soft, jet black, homogeneous, and well delineated and may show varying degrees of cavitation (Fig. 15-24). The cavities may contain fluid resembling India ink. The affected lobes may be adherent to the chest wall. Microscopically, the lesions are composed of dust particles, both free and within macrophages mixed with randomly arranged collagen and reticulin fibers (Fig. 15-25). The centers of the lesions may evince necrosis and cholesterol clefts. There are frequently ghost outlines of vessels and airways within the centers and at the periphery small vessels cuffed by lymphocytes and plasma cells. Necrosis within PMF lesions is usually a result of ischemia rather than infection. Often severe emphysema surrounds PMF lesions.

A second type of PMF lesion may occur. This is characterized by hardness and a conglomerate pattern of nodules (Fig. 15-26). Mineralogic studies of lung mineral content have indicated that the first type of PMF is associated with exposure to high rank coals and the second type to lower ranked coals.[84] Obviously these PMF lesions represent ends of a spectrum and one often encounters some with intermediate features that are difficult to classify (Fig. 15-27).

CAPLAN'S SYNDROME (RHEUMATOID PNEUMOCONIOSIS)

In 1953, Caplan[98] described unusual radiologic changes in coal miners suffering from rheumatoid arthritis, and subsequently radiologists in Germany, France, Belgium, and Holland reported similar findings. The chest radiographs showed large, rounded, and well-defined shadows mainly in the peripheral

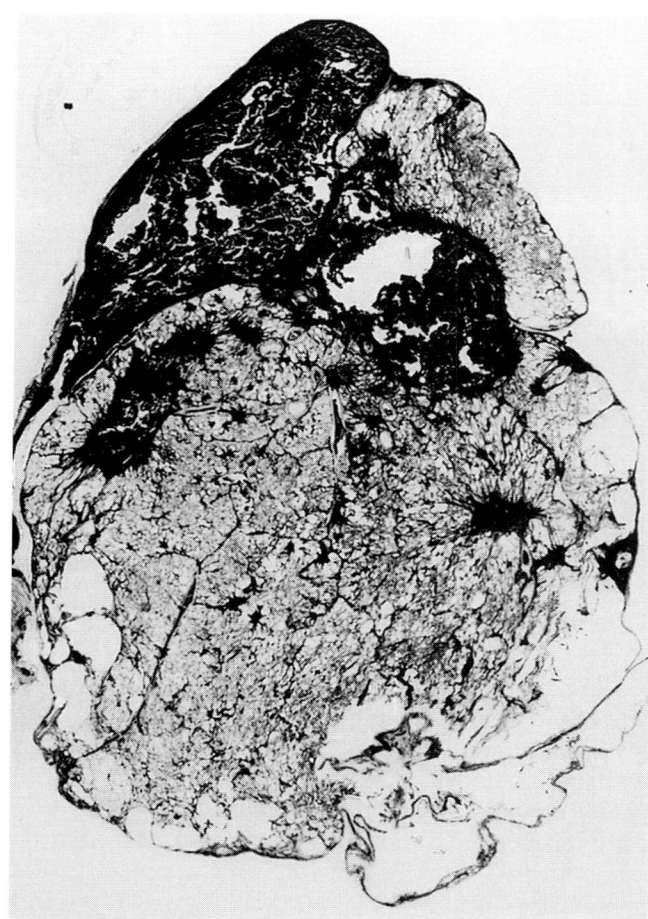

FIGURE 15-24
Gough-Wentworth whole lung section from a coal worker shows diffusely black progressive massive fibrosis mainly in the upper lobe.

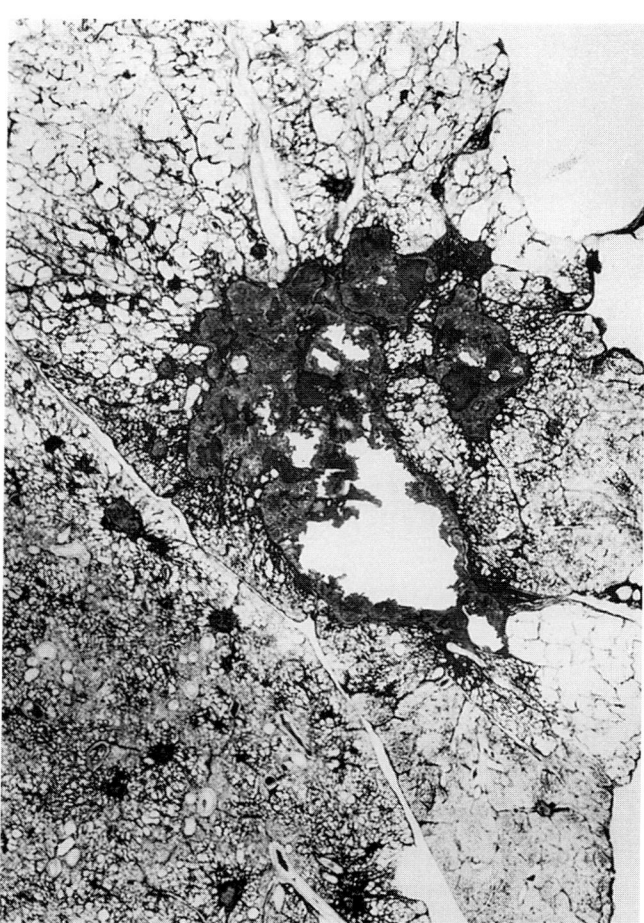

FIGURE 15-26
Coal worker. An area of progressive massive fibrosis shows conglomeration of nodules and an area of cavitation.

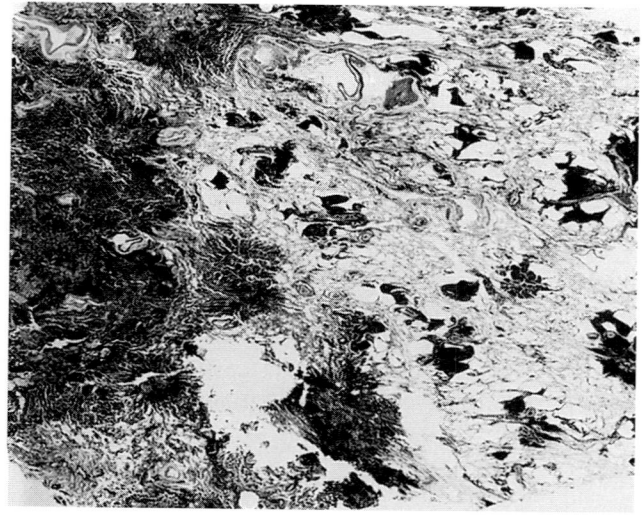

FIGURE 15-25
Coal worker. Part of a progressive massive fibrosis lesion shows diffuse black area with adjacent multiple black macules. (Low power.)

parts of the lung fields, but unlike the shadows caused by PMF they were not restricted to the upper lobes and tended to enlarge more rapidly. The appearance of the pulmonary lesions sometimes preceded the onset of arthritic lesions, as can ordinary pulmonary rheumatoid nodules. In 1962, Caplan and coworkers[99] found that 65 percent of such patients showed positive serologic tests for rheumatoid disease.

In 1955, Gough and colleagues[100] described the pathologic appearances of these lesions. They are rounded or oval, well-demarcated firm nodules that may be discrete or confluent (Fig. 15-28). They exhibit a laminated appearance with light and dark layers. They vary from a few millimeters to several centimeters in size and often are superimposed on a background of relatively slight pneumoconiosis. The nodules may show areas of liquefaction, cavitation, and calcification.

Microscopically, the central zone of the nodule consists of necrotic collagen, coal dust, and debris. Dust often lies in rings. At the edge of the necrotic

FIGURE 15-27
Progressive massive fibrosis lesion from a coal worker shows an intermediate pattern between the diffuse black form and the conglomerate form of the condition. (Low power.)

nuclear debris (Figs. 15-29 and 15-30). Outside the fibroblastic zone there is circumferentially arranged collagen and numerous lymphocytes and plasma cells, which sometimes form germinal centers. The small vessels around the periphery of the nodule frequently show endarteritis with infiltrates of the walls composed of lymphocytes and plasma cells.

The major differential diagnosis includes tuberculous and silicotic nodules. The clue to the rheumatoid nature of these lesions is the zone of neutrophils and pallisaded fibroblasts. Direct staining for acid-fast bacilli (Ziehl-Nielsen) will be negative in Caplan lesions. In some instances, however, the external capsule may be entirely collagenous—so-called burnt out lesions—which may be extremely difficult to distinguish from encapsulated tuberculous lesions.

In subjects with rheumatoid arthritis, Caplan-like lesions also have been associated with exposure to other dusts including silica[101] and asbestos.[102,103] In those associated with asbestosis concentric dust rings are not observed but instead asbestos bodies are seen.

EMPHYSEMA AND CWP

The issue of coal dust exposure and the development of emphysema has been controversial.[104] It is important because the degree of functional disability in CWP correlates better with the degree of emphysema

zone a cleft separates it from the adjacent layer of spindle-shaped fibroblasts, which show variable degrees of pallisading. Between the cellular and necrotic zones is a narrow belt of neutrophils and

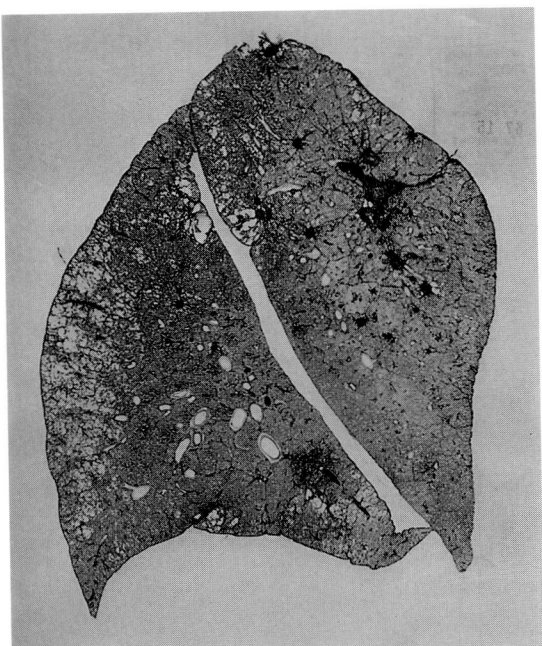

FIGURE 15-28
Caplan's syndrome. Well-demarcated whorled nodules superimposed on a background of very slight pneumoconiosis.

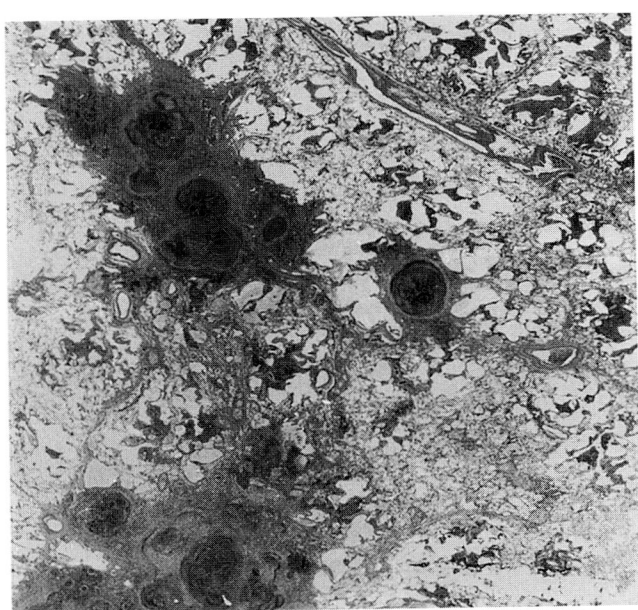

FIGURE 15-29
Rounded, well-demarcated confluent nodules in a coal worker with rheumatoid arthritis—Caplan's syndrome. (Low power.)

481

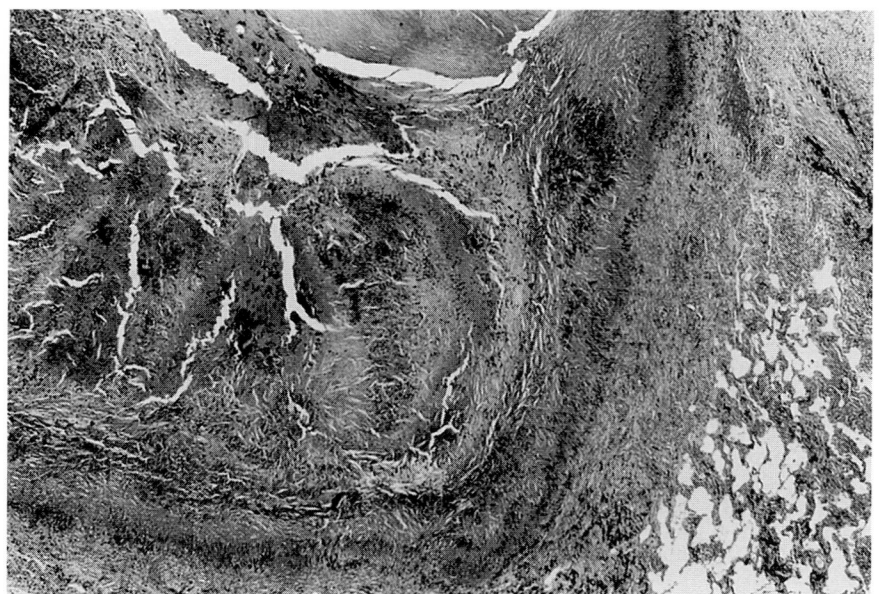

FIGURE 15-30
Caplan lesion shows central necrosis and a peripheral zone of inflammatory cells with pallisading. (Low power.)

present than with the size and number of simple dust foci.[105,106] The major confounding factor is cigarette smoking. A considerable body of evidence indicates that chronic bronchitis and emphysema in coal workers is directly related to tobacco usage and cumulative exposure to respirable dust during life.[107,108] Postmortem studies have also shown that there is a significant excess of emphysema in coal workers with pneumoconiosis when compared to a nonmining control population.[108–110] This is specific to centriacinar emphysema and unrelated to panacinar emphysema. Its occurrence is associated with increasing amounts of dust retained in the lungs and palpable lesions.[107]

When considering emphysema in relation to coal dust exposure two forms should be distinguished: focal and centriacinar. In many studies this has not been done and it has to be admitted that there is no sharp boundary where one ceases and the other begins.

Focal emphysema was the term coined by Heppleston[111] for the dilatation of the respiratory bronchioles that clustered around the stellate dust foci and that was associated with little architectural destruction. It is mild, the air spaces not usually exceeding 2 mm in diameter, and not associated with any functional abnormality (see Figs. 15-19 and 15-23). Duguid and Lambert[112] thought that it was a form of compensatory emphysema caused by increasing fibrosis and contraction of the dust focus over many years. This implies that the contracted dust focus becomes relatively less and the emphysematous part more conspicuous over time. The emphysematous changes are located at the center of the acinus and secondary lobule because of the location of the dust focus. This is where the changes of centriacinar emphysema are also located. It should be noted that similar lesions may be observed around the carbonaceous dust foci of town dwellers.

Centriacinar emphysema, in contrast to focal emphysema, was considered to be accompanied by architectural destruction of the acinus from its earliest stage and inflammation within bronchi and bronchioles.[113] Apart from the presence of the dust foci the pathologic appearances are identical to those seen in noncoal workers with centriacinar emphysema (see Figs. 15-19 to 15-21). If extensive it is likely to be accompanied by impairment of lung function of an obstructive type. It has a strong association with tobacco usage but as stated above it is related also to cumulative respirable coal dust exposure.

In the United Kingdom, the government has accepted the report of the Industrial Injuries Advisory Council's recommendation that chronic bronchitis and emphysema in *underground* coal workers should be added to the present prescribed diseases for Industrial Injury Disablement benefit. The conditions for compensation were:

1. Underground coal mining for a minimum aggregate of 20 years
2. A reduction of lung capacity at least 1 liter below expected level
3. A radiologic pneumoconiosis category of 1/1 or greater

On occasion at autopsy a pathologist will be faced with the problem of determining in a coal miner, who has died from obstructive airways disease and who

482

has centriacinar emphysema, whether coal dust exposure was contributory. If the subject has not had lung function tests and a chest radiograph in the last few years, this might be difficult. In this situation the pathologist will need to make a judgment of the likely appearances of the chest x-ray based on the appearance of the dust foci and emphysema at autopsy. In practice for category 1/1 there would be extensive simple coal dust lesions present.

INTERSTITIAL FIBROSIS AND CWP

There appears to be an increased prevalence of interstitial fibrosis in the lungs of coal workers. This has been observed in both South Wales and U.S. coal miners and is approximately 16 percent.[114,115] Sometimes this can be observed macroscopically but in many cases it is only seen by microscopy (Fig. 15-31). A radiologic-pathologic study of irregular opacities in CWP has indicated that they represent a combination of emphysema and interstitial fibrosis; the interstitial fibrosis can show various degrees of dust deposition.[116] The precise connection between the interstitial fibrosis and coal dust or work underground remains to be determined. The disease appears to run a slower course than cryptogenic fibrosing alveolitis.[114]

GRAPHITE

This is a natural or synthetic crystalline form of carbon. Many commercial forms of natural graphite (plumbago) contain other minerals such as silica and various clays. It is used in refractory ceramics and crucibles, foundry facings, steel and iron manufacture, pencils, lubricants, and brake linings. Synthetic graphite is used in graphite electrodes, atomic reactors, and lubricants.[1] A number of cases of graphite pneumoconiosis have been described but in many exposure to silica or silicates may have played a part in their causation.[117]

METALS

Pulmonary responses to metals have been inadequately studied because they tend to occur in small industries with poorly characterized exposures. There have been several epidemiologic studies of various industries but there have been few systematic pathology studies. Most of the pathologic information comes from isolated case reports, with poorly characterized exposures, no mineralogic analysis of the lung tissues, and no proper controls for comparison. This account is highly selective and discusses only those metals with fairly well-documented effects on the lung. Agents that will not be considered are those that predominantly cause responses in organs outside the respiratory system even though they are inhaled.

It is rare for human exposures to occur to the pure metallic forms but it is usually to the oxides or other compounds. Not all compounds of a toxic metal are necessarily harmful to the lung. A wide variety of pulmonary responses can result from exposure to metals, which can be in the form of fumes or dusts. These responses can be acute, chronic, or neoplastic. The acute response may not develop immediately after exposure but may be delayed for some hours. Death or complete recovery may follow the acute response and rarely chronic sequelae such as fibrosis. Table 15-6 gives a broad classification of the pathologic reactions associated with exposures to metals together with examples. As with exposure to other minerals, the responses will vary with dose, physicochemical forms, associated minerals, and host factors.[118,119]

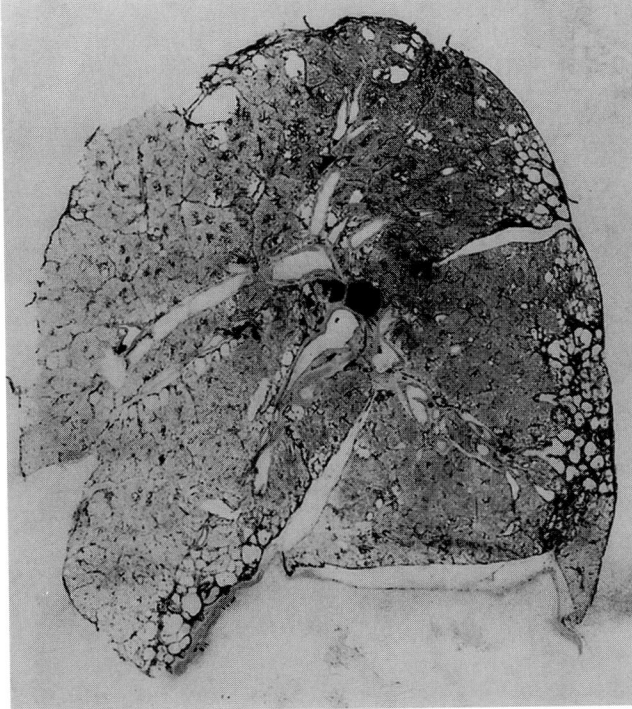

FIGURE 15-31
Gough-Wentworth whole lung section from a coal worker shows dust-impregnated honeycombing in the posterior and inferior zones.

TABLE 15-6

Pulmonary Reactions Associated with Exposures to Metals

Reaction	Examples
Diffuse alveolar damage or acute bronchitis	Beryllium, cadmium, cobalt, manganese, mercury, nickel
Industrial asthma	Chromium, cobalt, nickel, platinum
Emphysema	Cadmium
Centrilobular dust accumulation with minimal fibrosis	Barium, iron, tin
Interstitial fibrosis	Aluminum, cobalt
Desquamative interstitial pneumonia	Aluminum
Giant cell interstitial pneumonia	Cobalt (hard metal)
Alveolar proteinosis	Aluminum
Granulomatous disease	Aluminum, beryllium, titanium
Carcinoma of lung	Arsenic, chromium, nickel, uranium

Aluminum

There are numerous uses for aluminum metal and its compounds and a considerable number of workers have been exposed to them. Sources of exposure include the mining and crushing of bauxite ore, smelting of bauxite (potroom workers), the manufacture of abrasives or explosives containing aluminum, and the grinding, polishing, and welding of aluminum. Various forms of lung disease, the earliest being fibrosis, have been linked to exposure to aluminum in manufacturing or processing factories but they have been rare considering the large numbers of workers potentially at risk. Some reports have been difficult to assess because of concomitant exposures to other minerals. For example, aluminum potroom workers are exposed to carbon monoxide, coal tar pitch volatiles, fluorides, sulfur dioxide, and alumina.[120] Bauxite lung (also known as Shaver's disease, corundum smelter's lung) was first recognized during World War II among a small group of workers engaged in the manufacture of aluminum abrasives (corundum, Al_2O_3) by Shaver[121] and then by Riddell.[122] Reports from Germany during the same period described severe lung damage occurring in association with aluminum metal powder. In 1959, Mitchell[123] and McClaughlin and coworkers[124] in 1962 described further cases in Britain.

Several forms of nonmalignant lung disease have been associated with exposure to aluminum and its compounds.[125–133] These are listed together with the associated exposures in Table 15-7. None of these reactions are specific to aluminum and the association cannot be made on the histopathologic appearances alone (Fig. 15-32). However, the aluminum-associated diffuse interstitial fibrosis pattern is usually more severe in the upper zones of the lung, whereas the cryptogenic form and asbestosis are usually more severe in the lower zones. Mineral analysis of the lung tissue is essential in identifying the aluminum and excluding other mineral exposures.

TABLE 15-7

Aluminum-Associated Pulmonary Reactions

Reaction	Exposure	References
Diffuse interstitial fibrosis	Bauxite smelting	Abramson et al.[120]; Gilks and Churg[126]
	Fireworks and explosives	Mitchell[123]
	Abrasives	Shaver[121]; Riddell[122]; Jederlinic et al.[125]
	Welding	Vallyathan et al.[131]
	Polishing	DeVuyst et al.[132]
DIP	Welding	Herbert et al.[130]
GIP	Grinding	Ferryman et al.[133]
Alveolar proteinosis	Grinding	Miller et al.[127]
Granulomatous disease	Welding	Chen et al.[129]
	Chemist	DeVuyst et al.[128]

DIP = desquamative interstitial pneumonia; GIP = giant cell interstitial pneumonia

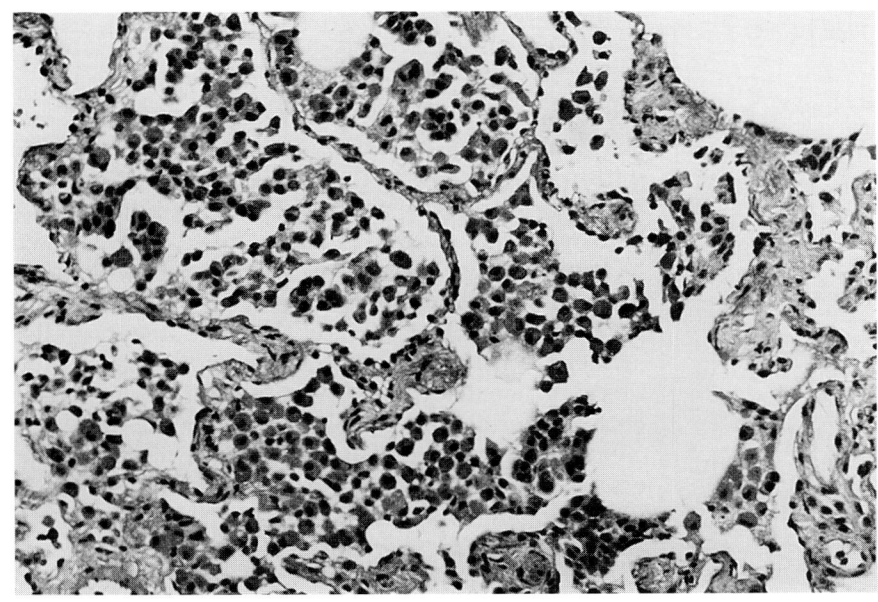

FIGURE 15-32
Diffuse interstitial fibrosis-like reaction characterized by intraalveolar collections of macrophages. (Medium power.)

Asthma-like symptoms have been reported in several groups of workers exposed to aluminum compounds,[120,134] but it is not certain whether the exposure initiates asthma or exacerbates symptoms in subjects with subclinical asthma.

An excess of lung cancer has been described in aluminum production workers but it is probably linked to coal tar pitch volatiles.[135]

Barium

Barium lung (baritosis) can follow the inhalation of dust particles from inspired air or by aspiration of barium suspensions during radiographic procedures. In the former circumstance, such as in barytes miners and grinders, the particles will tend to be uniformly and bilaterally distributed, whereas in the latter circumstance the particles will accumulate in localized areas. The radiologic changes are dense but there is usually little functional disability because there is little associated fibrosis. Microscopically, the lungs show alveolar collections of particle-laden macrophages with few other inflammatory cells. The particles are refractile and strongly birefringent on polarization.

Beryllium

Berylliosis is a disease caused by exposure to beryllium metal or its salts. The term was first used in 1935 by Fabroni[136] with reference to the experimentally produced acute form of beryllium poisoning, although an acute disease following beryllium exposure was described in Germany in 1933.[137] Hardy and Tabershaw were the first to draw attention to a chronic form of berylliosis.[138] Beryllium most frequently enters the body through the respiratory tract and most conspicuously affects the lungs. However, it is a multisystem disorder[139] and changes may occur in the skin, liver, kidney, spleen, and lymph nodes.[140] Dermatitis or chronic skin ulcers may occur as a local disease or in conjunction with systemic disease.

Acute berylliosis follows relatively high exposures to dusts, fumes, or mists, whereas the chronic form may follow limited or prolonged exposure to low doses. Individual hypersensitivity appears to be involved, which is mediated by a specific delayed-type hypersensitivity reaction to beyllium in which there is proliferation and accumulation of CD4+ (helper/inducer) T cells in the lungs.[141] The disease is uncommon even though the use of beryllium in industry has increased over the last few decades. Exposure may occur in those engaged in the extraction of the metal from its ores, most commonly beryl, and in those who handle beryllium compounds. The latter are used in refractory ceramics, the electronics industry, space and atomic engineering, laboratories, and beryllium metal and alloy production. Exposure may also occur during the reclamation of scrap metals.[139] The use of beryllium in flourescent light tubes ceased in the late 1940s and early 1950s; this industry was the cause of 46 percent of the chronic berylliosis cases listed in the U.S. Berylliosis Case Registry up to 1966.[142] Cases of berylliosis claimed to have been caused by paraoccupational and neighborhood exposure have been reported.[143–145]

Some studies have claimed that exposure to beryllium compounds causes an increased risk of lung cancer; however, the evidence for this is unconvincing.

ACUTE BERYLLIOSIS

This condition is rare nowadays with modern industrial conditions. Respiratory symptoms of breathlessness, cyanosis, productive cough, and nasal discharge manifest within 72 h following a severe exposure, but if exposure is lower the onset may take several weeks. The majority resolve without long-term sequelae, but approximately one-sixth have progressed to chronic disease. Death has occurred in about 10 percent of the cases reported.[139]

The gross and microscopic appearances are those of diffuse alveolar damage (see Chap. 11) and granulomas are absent. Diagnosis rests on a history of exposure and demonstration of beryllium within the lung tissues.

CHRONIC BERYLLIOSIS

This is an insidious disease that can follow the last exposure to beryllium with a latent period varying from 1 month to more than 40 years. It closely resembles sarcoidosis and can lead to death from pulmonary fibrosis.

Macroscopically, the lungs appear contracted and have bosselated pleural surfaces, sometimes with thickening and adhesions. The cut surface shows variable degrees of honeycombing, which tends to be more conspicuous in the upper zones.

Microscopically, the hallmark of the disease is the presence of numerous noncaseating compact epithelioid sarcoid-like granulomas located in the interstitium, septa, bronchovascular bundles, and subpleural areas. Occasionally there is necrosis within the centers of the granulomas.[146] Fusion of granulomas may occur to give rise to nodules that are 2 to 3 cm in diameter. These may show marked hyalinization. Schaumann bodies are frequently present; in the chronic fibrotic stage the granulomas may disappear leaving only the Schaumann bodies as their "calling card." In addition to the granulomas there is usually a pronounced diffuse interstitial chronic inflammatory infiltrate, which is much greater than that usually observed in cases of sarcoidosis (Fig. 15-33). There is a variable degree of diffuse interstitial fibrosis present, which may be accompanied by bronchiolectasis (honeycombing) and bronchioloalveolar epithelial metaplasia.

The following criteria for diagnosis of chronic berylliosis have been put forward.[146,147]

1. History of beryllium exposure
2. Consistent clinico/physiologic/pathologic features
3. Presence of granulomas in tissue
4. Detection of beryllium in tissues
5. Evidence of beryllium hypersensitivity
6. Exclusion of other granulomatous diseases

Occupational histories are often difficult to assess because exposure frequently occurred several years previously, it is often mixed, and other granu-

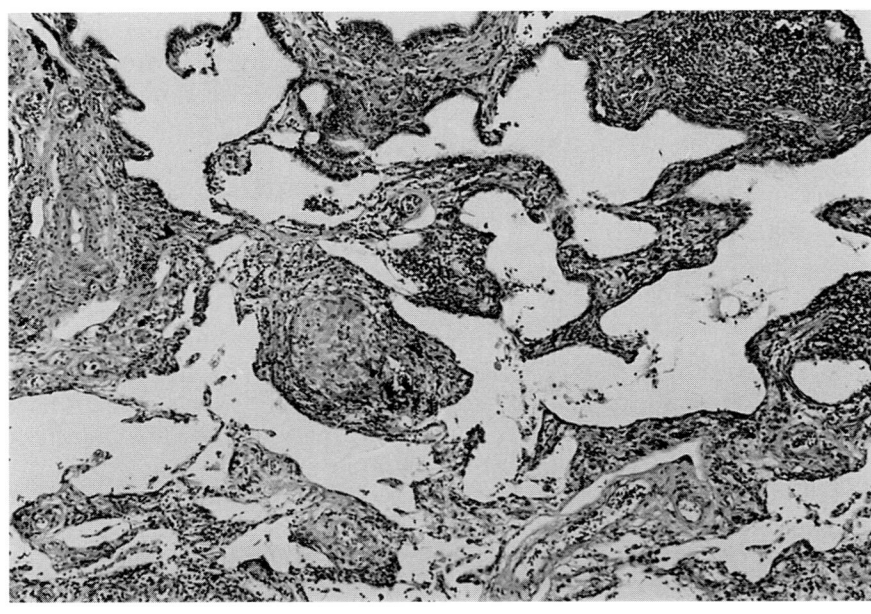

FIGURE 15-33
Diffuse interstitial chronic inflammation with a granuloma in a worker exposed to beryllium. (Low power.)

lomatous diseases can occur in workers with claimed exposure to beryllium. The detection of beryllium in the granulomas is particularly valuable but unfortunately, because of its low atomic weight, it is one of the more problematic minerals to detect. The laser microprobe mass spectometry technique can be applied to conventional tissue sections and can detect elements in low concentrations. A recent study using the technique demonstrated beryllium in the granulomas of all 33 cases of chronic beryllium disease and in none of 30 cases of sarcoidosis.[148] It should be noted that it demonstrated beryllium in the dust lesions of coal workers. Its disadvantages are that the technique is available in few centers and it provides qualitative and not quantitative data. Other techniques for the detection of beryllium in tissues have used electron microscopy and have included ion microprobe mass spectometry and electron energy mass spectometry.

It has been postulated that chronic berylliosis is caused by immunopathologic mechanisms because of an absence of a dose–response relationship and the presence of manifestations of delayed (type 4) hypersensitivity. The latter can be useful in diagnosis. Beryllium patch testing on the skin of subjects with suspected berylliosis has been discontinued because of its unreliability and possible exacerbation of the disease. However, in vitro transformation tests to beryllium on peripheral blood and bronchoalveolar lavage lymphocytes has been positive in nearly all subjects with chronic berylliosis tested.[149–151] It can also be used for screening of workers but there is still debate as to whether a positive beryllium lymphocyte test in an asymptomatic worker represents latent disease.

The major conditions from which chronic berylliosis needs to be distinguished include sarcoidosis, chronic extrinsic allergic alveolitis, and cryptogenic fibrosing alveolitis. Table 15-8 gives helpful features in differential diagnosis.

Cadmium

Cadmium is used in the manufacture of alloys, nickel–cadmium batteries, and in electroplating and plating of automobile parts and musical instruments. Additional sources of exposure include various welding processes and the smelting of zinc, lead, and copper ores that contain cadmium. Cigarettes may contain appreciable quantities of cadmium. Three forms of lung disease have been associated with cadmium exposure: diffuse alveolar damage (acute cadmium pneumonitis), emphysema, and cancer.

Inhalation of large amounts of cadmium fume (usually the oxide) can cause diffuse alveolar damage, sometimes with death of the subject.[152] The histopathologic features are similar to diffuse alveolar damage caused by other agents. Diagnosis depends on the history of exposure and mineral analysis of the lung tissues.

The link between chronic cadmium exposure and the development of centrilobular emphysema was first noted in the 1950s, but the results of subsequent epidemiologic surveys have been controversial. This is because the majority did not consider various confounding factors particularly cigarette smoking. However, a recent well-conducted study of a cohort of workers from a copper–cadmium alloy factory reported decrements of lung function and radiologic changes consistent with emphysema.[153]

TABLE 15-8

Comparison of Chronic Berylliosis (CB) with Sarcoidosis (SAR), Chronic Extrinsic Allergic Alveolitis (EAA), and Cryptogenic Fibrosing Alveolitis (CFA)

Feature	CB	SAR	EAA	CFA
Exposure to beryllium	Yes	No	No	No
Kveim test	Neg	Pos	Neg	Neg
Serum precipitins	Neg	Neg	Pos	Neg
Predominant distribution of fibrosis	Upper	Upper	Upper	Lower
Compact granulomas	++	++	+	0
Interstitial inflammation	++	+	++	+
Hilar lymphadenopathy	Rare	++	0	0

NOTE: +, mild/moderate; ++, prominent.

Furthermore, the decrements of lung function increased with cumulative exposure to cadmium. Inhalation of cadmium chloride by rats has led to emphysema.[154]

Several epidemiologic studies of cadmium-exposed workers have reported increased numbers of lung cancers. However, recent information has cast doubt on its role in causing lung cancer because two major confounders—smoking and arsenic exposure—were not controlled for.[119]

Chromium

Alloys and compounds of chromium have been in commercial use for over a century. Chromium is a gray metal that can combine with other elements in a number of different ratios, which are termed valencies. There are divalent (e.g., chromous chloride $CrCl_2$), trivalent (e.g., chromic chloride $CrCl_3$), and hexavalent compounds (e.g., potassium chromate K_2CrO_4) compounds. Experimental and human studies indicate that the hexavalent compounds are pathogenic, particularly the relatively insoluble forms such as calcium, lead, strontium, and zinc chromates, but also soluble forms.

Occupational exposure to chromium compounds can occur in a number of situations (Table 15-9) and can cause perforation of the nasal septum and an increased risk of nasal and lung cancer.[155] These have been observed in chromium plating, chromate production, and chromate pigment production and the risk increases with duration and severity of exposure.[156–158] The occurrence of nasal perforation in a cohort of workers does not necessarily indicate that they have been exposed to types and amounts of chromium compounds capable of producing lung cancers.

Cobalt—Hard Metal

Hard metal (tungsten carbide) is a synthetic tungsten alloy used in machine tools. Pneumoconiosis has been associated with the manufacture and grinding

TABLE 15-9
Exposures to Chromium Compounds

Production of chromates and pigments

Chromium plating

Production, welding, cutting, and grinding of chromium alloys

Leather tanning

Spray painting

Steel smelting and welding

of tungsten carbide.[159–161] The finished product contains between 5 and 25 percent of cobalt, which is thought to be the agent responsible for the disease.

Machine tools requiring very hard textured metal are made by blending tungsten and carbon and heating them to form tungsten carbide. Cobalt and other metals, such as titanium, tantalum, chromium, molybdenum, vanadium, niobium, and nickel, are then added, according to the desired properties, and then sintered into the hard metal.

Exposure to hard metal has been associated with diffuse interstitial fibrosis, desquamative interstitial pneumonia (DIP), giant cell interstitial pneumonia (GIP), hypersensitivity pneumonitis, and industrial asthma.[159–164] GIP is a rare disease and when it is encountered it is nearly always caused by exposure to hard metal (Fig. 15-34). In this disorder the giant cells may be found in the bronchoalveolar lavage fluid.[163] When an interstitial disorder is caused by hard metal exposure, analysis of the lung tissues consistently reveals tungsten but not always cobalt, because of its solubility. Experimental studies have indicated that cobalt is responsible for the interstitial diseases. A similar disease has been reported in diamond polishers exposed to cobalt but not to hard metal.[165] In hard metal asthma evidence indicates that sensitivity to both cobalt and nickel may be involved.[164]

Iron

Exposure to dusts or fumes composed predominantly of iron oxide causes little damage to the lung because the particles are weakly fibrogenic. The chest radiographs may show opacities but there is usually little functional impairment. In cases where exposure has been heavy and deposits of iron particles marked, the lung will appear reddish brown or brick red on gross examination (Fig. 15-35). Ferruginous bodies containing black cores are frequently present, which should not be mistaken for asbestos bodies. Microscopically, there will be collections of dark brown, dust-laden macrophages with relatively slight fibrosis (siderosis). This may occur in silver polishers, ocher workers, and arc welders. The latter may experience exposure to a variety of other mineral particles and may show a variety of pathologies (vide infra).

In some situations exposure to significant amounts of silica also will occur,[166] for example, in hematite miners, foundry workers, and boiler scalers. The lung will reflect the effects of both silica and iron—so-called silicosiderosis. Simple and complicated pneumoconiosis can develop in these workers,

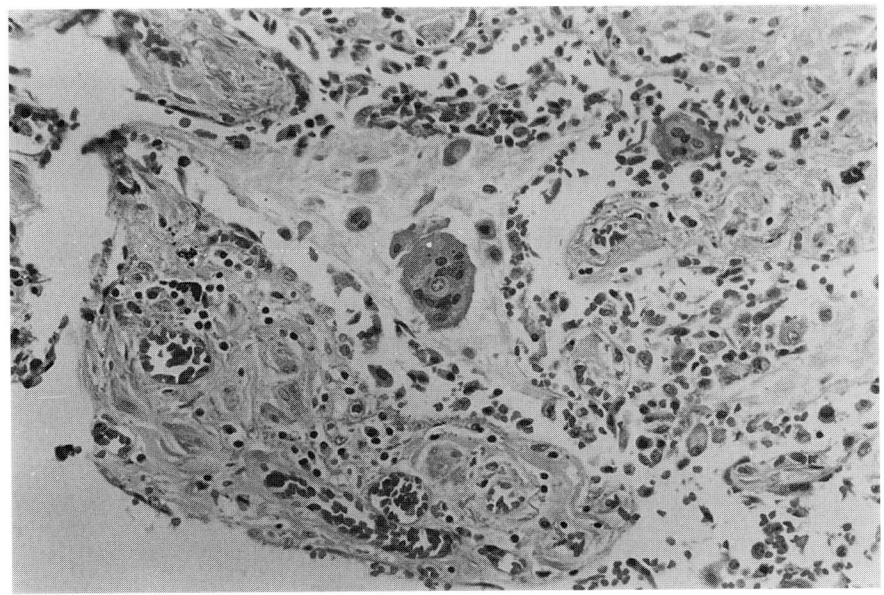

FIGURE 15-34
Chronic interstitial fibrosis with giant cells in a worker exposed to hard metal. (Low power.)

which will be similar to silicosis or mixed dust fibrosis depending on the amount of silica present.[167]

A high prevalence of lung cancer has been reported in the hematite miners from Cumberland, which is probably due to high levels of radon in the mines.

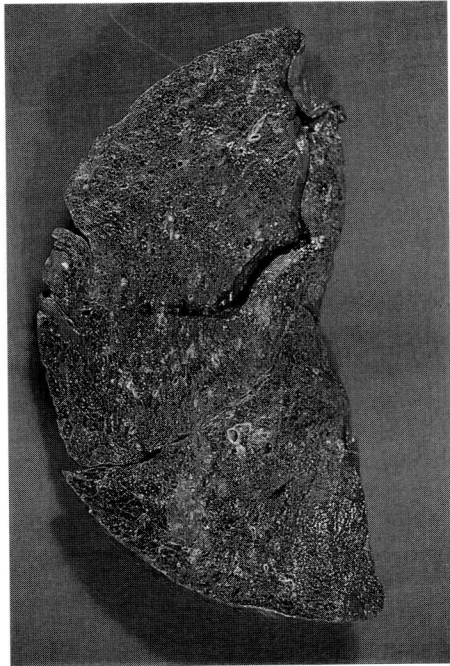

FIGURE 15-35
The lung from a subject who had been exposed to ochers for many years showing a diffuse reddish brown appearance and multiple stellate gray-brown lesions. (*See also Color Plate 4.*)

Welding is the process of joining two materials by fusion or coalescence under conditions of pressure and temperature. There are many different types of welding processes and workers will be exposed to a wide variety of airborne particulates depending on the process and other operations in the facility. The welding fumes will be contributed to by: (1) vaporization of the wire, rod, or metallic/alloying coatings; (2) decomposition and vaporization of the flux materials; (3) spatter from the arc region and weld pool and fumes therefrom; and (4) vaporization of the molten weld metal.[158] Welding fumes can comprise metal oxides such as iron, titanium, chromium, nickel, manganese, copper, zinc, and aluminum, organic materials, and crystalline materials such as silica and silicates. Welders can be exposed to asbestos. The lung's reaction to welding will therefore depend on the composition of the fumes and the cumulative exposure. Welders' lungs typically show large collections of dark brown spherical particles (iron oxide) and ferruginous bodies but little fibrosis (Figs. 15-36 and 15-37). However, a variety of pathologic responses have been described in the lungs of welders depending on the type of exposure and it should be realized that there is no such entity as welder's lung.

Nickel

Alloys and compounds of nickel have been in commercial use for over a century. Exposure to metallic nickel compounds in the forms of fumes and dusts can occur in nickel mining, refining, calcining, smelting, and electrolysis and in the production of nickel alloys. It has been known for a long time that a

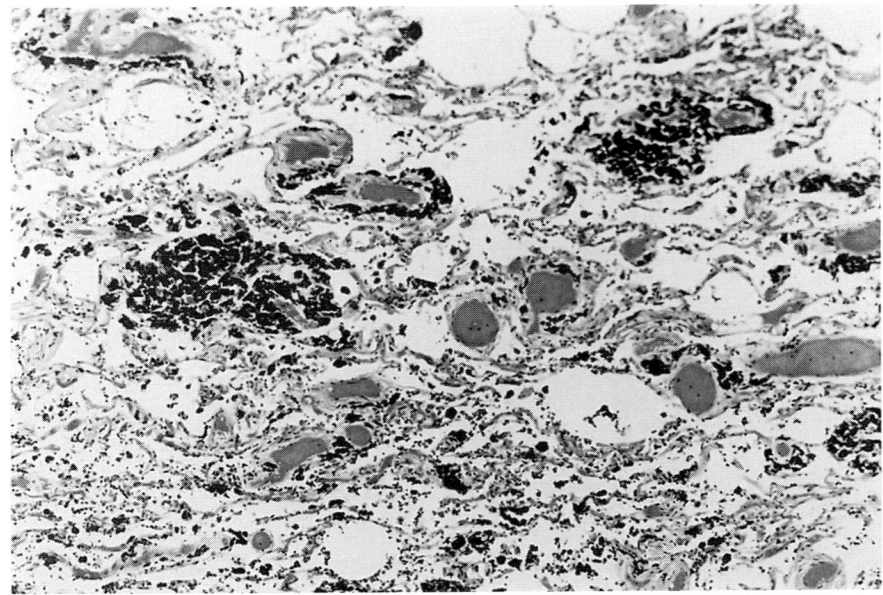

FIGURE 15-36
Lung from an arc welder shows collections of dust-laden macrophages with little fibrosis. (Low power.)

high incidence of nasal and lung cancer has been associated with exposures to nickel during refining and these have become prescribed diseases. Workers involved in nickel mining and smelting have not exhibited the same levels of lung and nasal cancer risk as those in refineries. No excess risk of these cancers has been found among men working in nickel alloy production. However, because the majority of epidemiologic studies have lacked details of amounts and types of nickel compounds to which the workers have been exposed, it is not known precisely which nickel forms are carcinogenic.[168] No particular histologic type is specific to nickel exposure although squamous are the most frequent lung carcinomas.

Acute heavy exposures to nickel have caused diffuse alveolar damage. Exposure to nickel may cause occupational asthma.

Tin

A pneumoconiosis can result from the inhalation of tin dioxide (stannosis). This occurs principally among those who bag and sample the cassiterite or who smelt and refine the ores and metal. Tin dioxide ores (cassiterite) are mined in Cornwall, Great

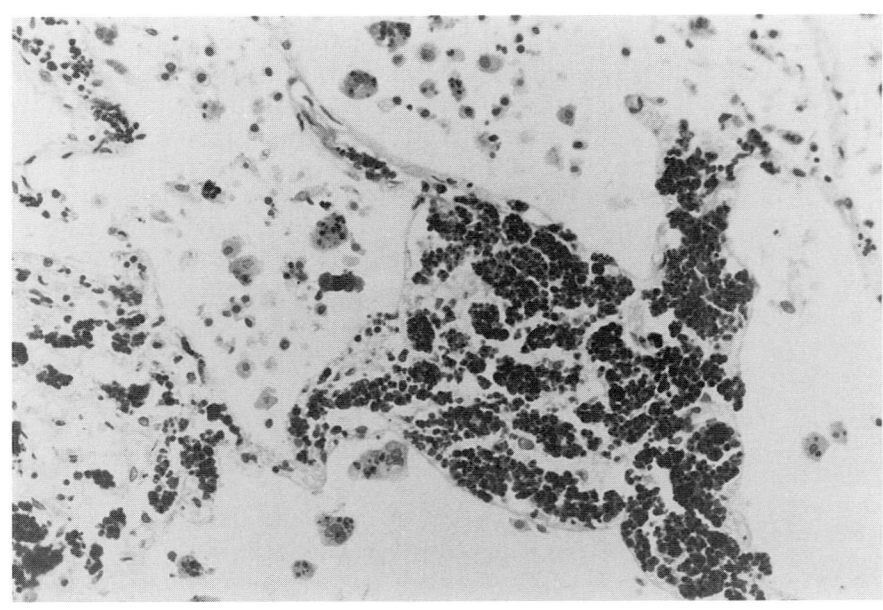

FIGURE 15-37
The same case as Fig. 15-36 shows the dust-laden macrophages. (Medium power.)

490

Britain, and Bolivia and those who mine the ore are also exposed to substantial amounts of silica. Therefore, in tin miners the major problem is silicosis and silicotuberculosis although in Cornwall the incidence has declined considerably since 1950.[169]

Stannosis, which is caused by exposure to tin per se, is characterized by striking radiologic opacities because tin has a high atomic number. However, the subject is usually well and shows little impairment in lung function.

Macroscopically, the lungs contain numerous gray-black macules up to about 5 mm in diameter.

Microscopically, the macules are characterized by numerous dust-laden macrophages associated with little if any fibrosis.[170] PMF does not occur. The particles are strongly birefringent. The dust content in individual lungs can reach 3 g tin dioxide. Microprobe analysis of the tissues will reveal the tin.

An inreased risk of lung cancer has been reported in Cornish tin miners but this was probably due to radon exposure.[171]

Titanium

Titanium dioxide is a white powder used in the dye industry. During the grinding of titanium dioxide, dust particles within the range 0.5 to 1.0 μm are released and can be inhaled into the lung. It is a very inert dust and in animals it is sometimes used as a negative control for inhalation studies. Elo and associates[171] studied the lung tissues from three factory workers employed in processing titanium dioxide pigments and found collections of macrophages containing black particles. These were situated around the bronchovascular bundles and in subpleural areas. The particles were birefringent when viewed with polarized light. There was little fibrosis.

A granulomatous pneumonitis has been described rarely in association with titanium exposure.[172]

ASBESTOS

Asbestos minerals are a group of naturally occurring fibrous silicates that possess good thermal, acoustic, and electrical insulating properties; they have flexibility and high tensile strength and are stable in alkaline and acid environments. Their use has been widespread and they have been incorporated into several thousand different products, including textiles, insulation products, cement, reinforced plastics, friction materials, filters, floor tiles, and paints. In fact, their use goes back several thousand years. They were used in the manufacture of Finnish pottery in 2500 B.C. The Romans knew of its fire-resistant properties. Asbestos has been referred to by Marco Polo in the thirteenth century and by Benjamin Franklin in the early eighteenth century.[173] It has been present ubiquitously in the environment for greater than 10,000 years since it has been demonstrated in Antarctic ice samples.[174] However, the major industrial exploitation of asbestos began in the 1870s when large deposits were found in Quebec (chrysotile) followed by Russia (chrysotile) and South Africa (crocidolite and amosite). The production and uses of asbestos throughout the world have increased from about 50 tons year in 1877 to a peak of nearly 6 million tons in the mid-1970s followed by a reduction to about 4 million tons in the 1980s. It should be realized that everyone is exposed to a low background level of asbestos and the normal lung contains a low level of asbestos fibers.

Asbestos minerals comprise two separate mineralogic groups: serpentine, the only member of which is chrysotile (white) asbestos, and the amphiboles, which comprise amosite, crocidolite, anthophyllite, tremolite, and actinolite. The major commercial forms have been chrysotile (90 to 95 percent) and crocidolite and amosite (5 percent).

Individual chrysotile fibers are pliable and curly with diameters averaging 0.06 μm.[175] Amphibole fibers, on the other hand, are straight, rigid and more brittle structures. The finest diameter amphibole fibers on average are crocidolite followed by amosite and the thickest are anthophyllite. Tremolite shows the greatest variation being relatively thick in some locations (e.g., California and Greenland) and thin in others (e.g., South Korea).[176] The physical forms of the fibers are important in their relative pathogenicity. The difference in chemistries between the fibers allows them to be readily identified in airborne and tissue samples by modern analytical transmission electron microscopy.

Exposure to asbestos can occur in a variety of ways, broadly classified by Zielhuis[177] into direct, indirect (bystander), paraoccupational (e.g., housewives washing their husbands' contaminated work clothes), neighborhood (living in the vicinity of asbestos mines, dumps, or factories), and ambient (in buildings where there are asbestos-containing materials). Table 15-10 gives a list of sources of exposure.

The level and size distribution of airborne fibers at these industrial sites will vary according to the amount of asbestos used, the manipulation processes (e.g., sawing, grinding, milling), and fiber type and this will determine the risk of disease. It is generally accepted that exposure to amphiboles is far more dangerous than chrysotile. This is because there is a

TABLE 15-10

Exposures to Asbestos Materials

Mining and milling of asbestos

Insulation work

Production of cement and cement products

Friction products

Shipbuilding

Textiles

Filters

Electrical work

Construction work

Railway work

Power stations

Iron and steel industry

Gas mask production

Paraoccupational (family) exposure

Neighborhood exposure—living near asbestos factories or dumps

differential rate of clearance from the lungs between chrysotile and amphiboles and it appears that it is the retained fiber burden within the lungs that is an important determinant of disease. Rapid clearance of chrysotile from the lung has been shown both in animals and human beings.[178–182] Even with continuing high exposure to chrysotile a steady state is reached after about 3 months and the level within the lung remains constant, whereas amphiboles are less rapidly removed and tend to accumulate over time if there is continuing exposure and this increased level persists after cessation. The half-life of amphiboles in the human lung is measured in decades. The clearance rate of fibers is also influenced by their length; short fibers are cleared efficiently whereas long fibers ($> 10 \ \mu m$) are not. Therefore, with increasing time following cessation of exposure the relative proportion of long fibers retained increases.[183,184] Animal inhalation studies using short-fiber preparations of asbestos and erionite produced little or no disease, whereas long-fiber preparations produced fibrosis and tumors.[185,186] Fiber clearance may also be inhibited by the concomitant exposure to other dusts such as quartz and titanium[187] and cigarette smoke.[188] No studies of exposure to asbestos in human beings have examined the possible potentiating role of other dusts in the disease-causing potential.

Exposure to asbestos may cause lung fibrosis (asbestosis), lung cancer, rounded atelectasis, parietal pleural plaques, diffuse pleural fibrosis, benign pleural effusion, and mesothelioma. The pleural lesions will be described in Chap. 34. However, it is worth making one very important point—exposures necessary to produce the pleural complications are far less than those required to produce asbestosis and lung cancer, probably one to two orders of magnitude difference.

Asbestosis (Including Airways Disease)

Asbestosis is defined as diffuse interstitial pulmonary fibrosis caused by asbestos exposure and the term should never be used for any pleural changes. The first histologically confirmed case was described in Great Britain by Murray in 1907 but it was not until 1924 that Cooke proposed the term "asbestosis" and published a detailed description of the disease. Following the report by Merewether and Price in 1930[189] asbestosis was officially recognized in the United Kingdom and included within the provisions of the Workmen's Compensation Acts and the Asbestos Industry Regulations were formulated. Over the decades these regulations have become progressively more stringent so that exposures to asbestos have decreased from well in excess of 100 f/mL before the 1930s to about 1 f/mL in 1980.[173]

The Merewether and Price study was of heavily exposed asbestos textile workers who showed a very high incidence of lung fibrosis with symptoms developing as early as 7 years after starting work and death occurring within 12 years. With reduction in airborne fiber levels in the workplace the disease has become less frequent and less severe. Asbestosis is usually seen after prolonged substantial direct exposure and the latency from first exposure to clinical pneumoconiosis usually exceeds 20 years. The prevalence of asbestosis is associated with age, cumulative exposure, fiber type, and smoking.[190–192] Clinically asbestosis may be associated with dyspnea, nonproductive cough, and basal crackles on auscultation. In advanced disease finger clubbing, cyanosis, and tachypnea may be present. However, many patients are asymptomatic.

Small irregular opacities are the radiologic hallmark of asbestosis although they are not specific. In particular the milder grades can be caused by smoking per se.[193,194] Pathologic examination of the lungs may reveal asbestosis when it is radiologically undetectable. Severe degrees of asbestosis may result in a honeycomb appearance although nowadays this is uncommon. Pleural abnormalities such as plaques or diffuse thickening may also be present. It has been claimed that computed tomography is more sensitive and specific for the recognition of asbestosis,[195] but this has not been validated by histopathologic studies of large series of cases.

The alterations in lung function in asbestosis are those found in restrictive forms of lung disease—a reduced vital capacity and reduced diffusion capacity for carbon dioxide. With progression the total lung capacity decreases.

MACROSCOPIC

When asbestosis is present there is frequently fibrotic thickening of the parietal and visceral pleura. In severe cases the lungs may be encased in a thick and often calcified covering of dense hyaline fibrous tissue. Parietal pleural plaques may be present. However, there is no correlation between the presence or extent of pleural change and the presence or extent of parenchymal abnormalities. In severe cases of asbestosis, if there is little or no pleural thickening, the pleural surfaces will appear bosselated. The lungs are reduced in size and show honeycomb changes, which are most marked in the lower and subpleural bases. In lesser degrees of asbestosis the lungs may just feel indurated or normal, but on microscopic examinations considerable fibrosis will be observed. At the earliest stage the fibrosis is observed as fine gray irregular streaks but with progression they become thicker and are associated with fibrocystic spaces (Fig. 15-38). Typically the changes are greatest in the lower zones but occasionally this is reversed. Because many of these subjects smoke, emphysema may also be present. This disease will counteract the tendency for shrinkage of the lungs and will be associated with an obstructive profile on lung function. It may be extremely difficult to assess the precise implications of asbestosis and emphysema to impairment of lung function and clinical disability and will require extensive sampling of the lung for microscopic evaluation.

MICROSCOPIC

Careful sampling of the lungs is necessary to diagnose and assess the severity of asbestosis. If a tumor is present the assessment should be on tissue away from the tumor and preferably on the contralateral side because tumors can result in local fibrosis. Tissue samples should be taken from the apex of the upper lobe, apex of lower lobe, and basal segments and major bronchus to include nodes,[17] but some authors have recommended as many as 15 blocks from each lung in cases suspected of asbestos-related disease.[196] The earliest changes are seen in and around the respiratory bronchioles with fibrosis of their walls. The next stage is extension proximally in the walls of the terminal bronchioles and distally in the walls of the alveolar ducts and adjacent alveolar

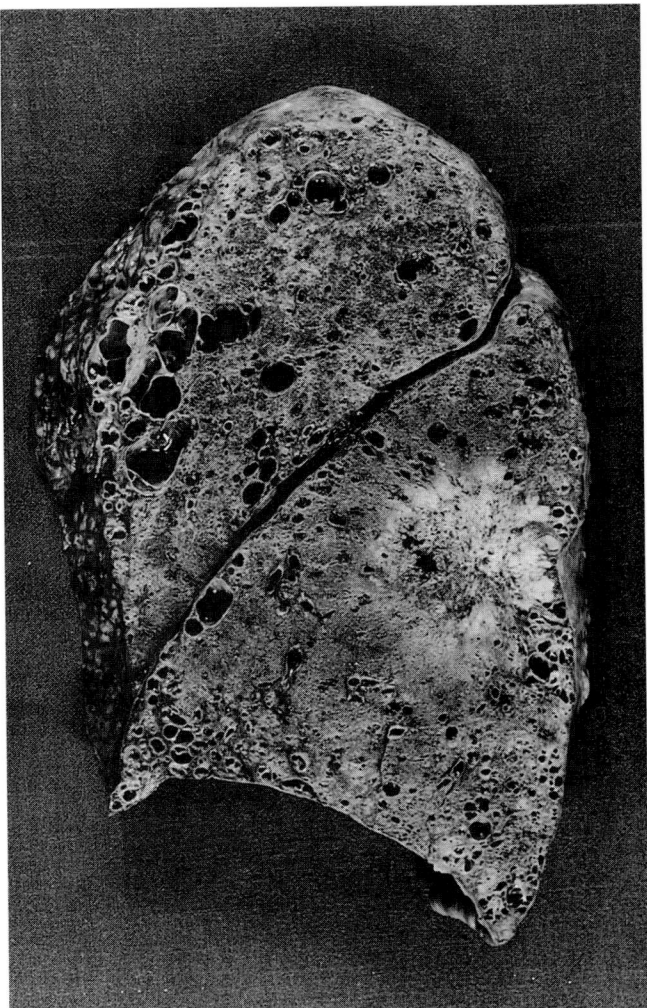

FIGURE 15-38
Lung shows extensive honeycomb change with a peripheral carcinoma in the upper part of the lower lobe.

septa. With further progression there is linking up by fibrosis of alveolar walls of adjacent respiratory bronchioles (Figs. 15-39 to 15-42). Obliteration of alveoli, alveolar ducts, and respiratory bronchioles by fibrosis then occurs, which leads to dilatation of the residual structures—so-called honeycomb change. The residual cystic spaces, which may measure up to 1.5 cm, may become lined on their internal surfaces by hyperplastic, cuboidal, or columnar epithelium, sometimes containing mucous cells. Squamous metaplasia may also occur. Cytoplasmic eosinophilic hyaline bodies, similar to the Mallory's hyaline observed in liver cells, has been observed in approximately 7 percent of cases of asbestosis.[197] The fibrosis is not uniformly distributed through the lung, the most extensive changes being seen in the lower lobes and subpleural area. However, with progression the microscopic changes will also be observed in the

493

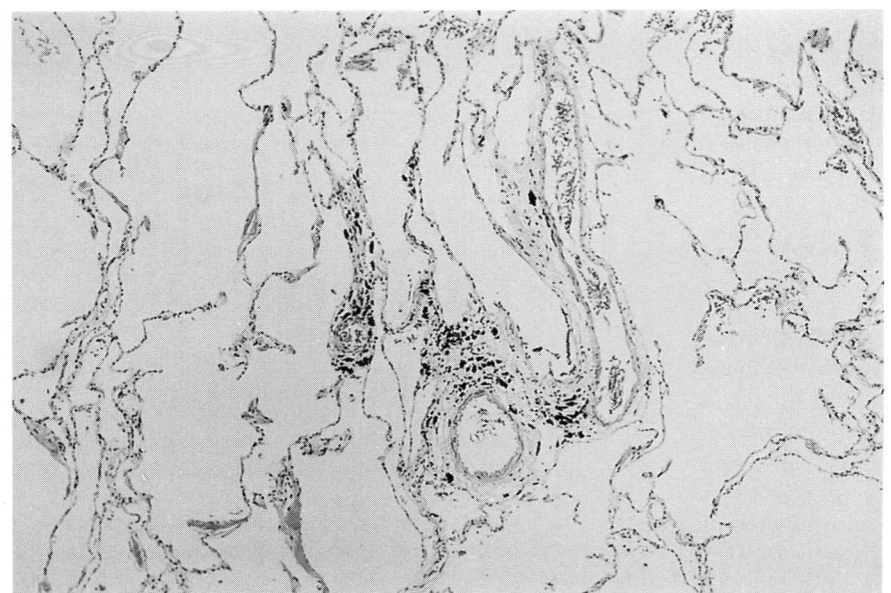

upper lobes. The fibrosis will be accompanied by variable numbers of chronic inflammatory cells.

Several grading systems have been applied to these stages of asbestosis.[17,196,198] The following is a simple scheme based on microscopic examination.

Grade 1 Slight increase of reticulin and collagen around the respiratory bronchioles

Grade 2 Fibrosis around respiratory bronchioles, which extends into adjacent alveolar ducts, atria, and alveoli but does not join up with fibrosis extending from other respiratory bronchioles

Grade 3 Fibrosis that links up adjacent respiratory bronchioles and with little architectural distortion

Grade 4 Widespread fibrosis with or without honeycombing

Asbestos bodies are the hallmark of pulmonary fibrosis due to asbestos exposure. Because they form on asbestos fibers—nearly always amphiboles—they

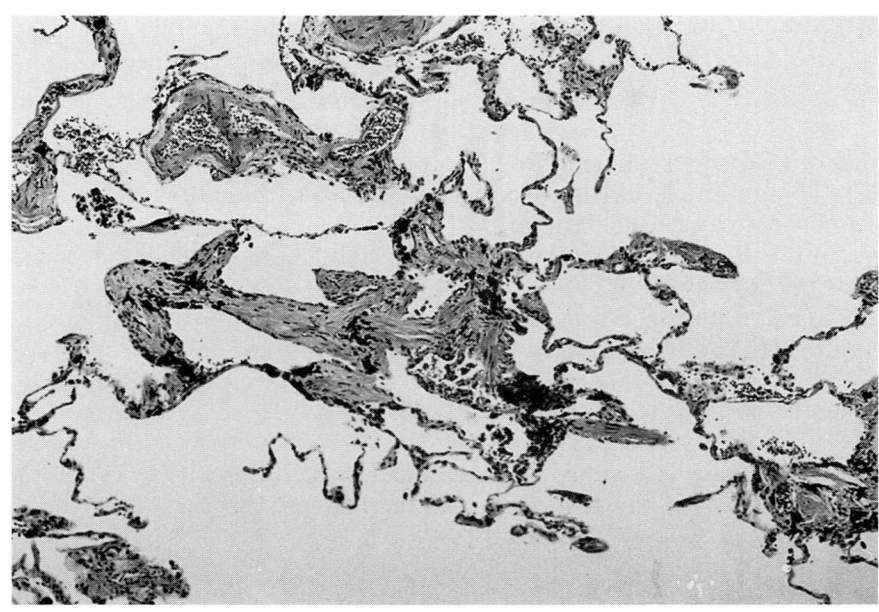

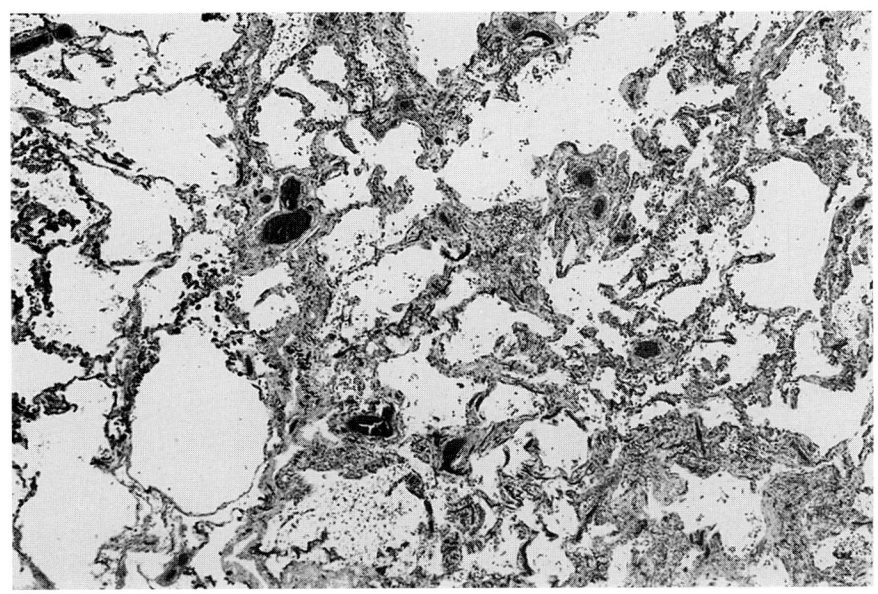

FIGURE 15-41
Interstitial fibrosis linking up adjacent lobules but with little architectural distortion—grade 3. (Low power.)

possess a clear fibrous core. They are coated with an iron–protein–mucopolysaccharide substance and appear as golden brown, beaded, dumbbell or drumstick structures (Figs. 15-43 to 15-45). They show positivity with Prussian blue stain because of the iron present. This can be used as a screening procedure for histologic sections. Asbestos bodies average 2 to 5 μm in diameter and usually exceed 20 μm in length. Great care needs to be taken in identifying them; it is the core that is important and it is characterized by a transparent fibrous appearance. There are other forms of ferruginous bodies that can be mistaken for them,[199,200] but their cores appear different.

They can be located within the alveolar spaces or interstitium and in cases of moderate to severe asbestosis are usually easily found. When they are not easily found it is usual that the fibrosis has not been caused by asbestos exposure; it is uncommon in my experience to demonstrate a very high asbestos fiber lung burden by mineral analysis in such cases.

In the majority of instances only a minority of fibers become coated to form bodies (Fig. 15-46) and this depends on fiber length, fiber width, fiber type, and the presence of the dusts in the lungs.[201,202] In general longer thicker fibers tend to become more easily coated than shorter thinner ones and amphi-

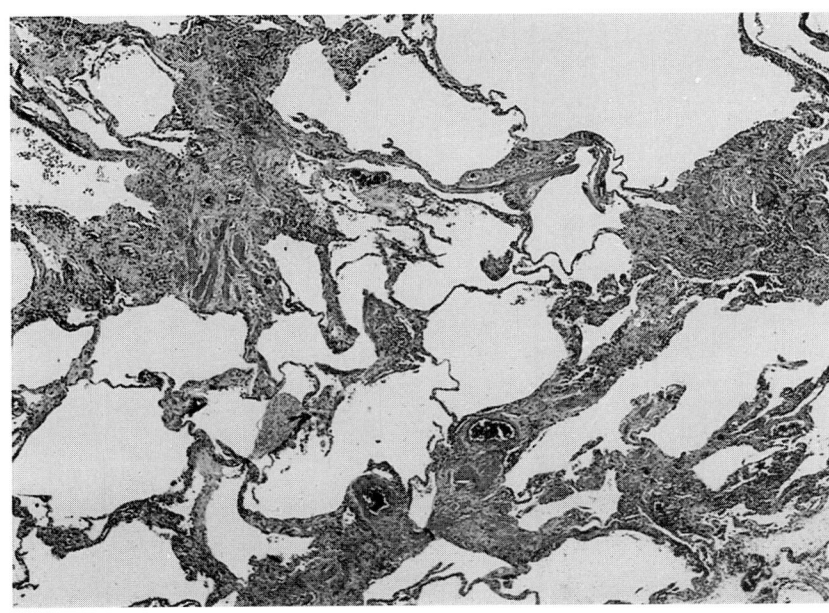

FIGURE 15-42
Severe interstitial fibrosis with architectural distortion—grade 4 lesion. (Low power.)

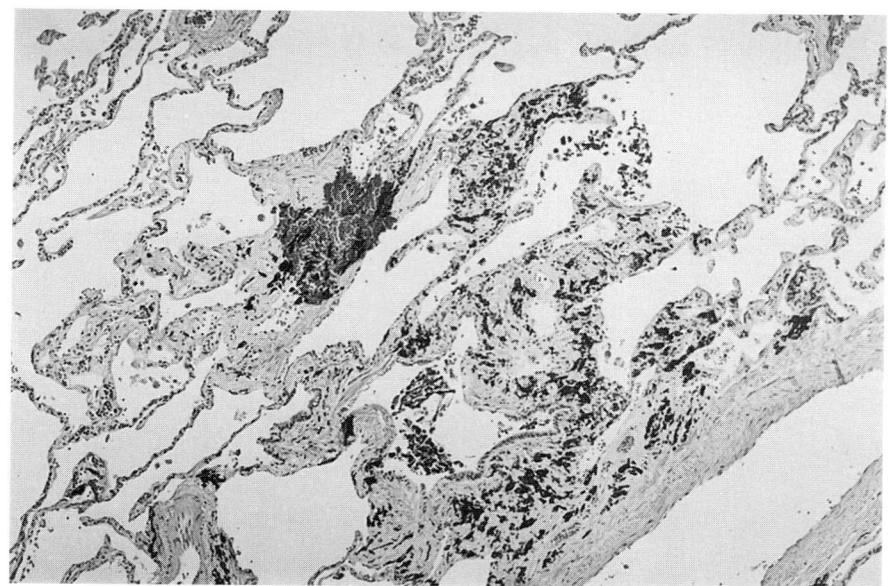

FIGURE 15-43
Grade 3/4 interstitial fibrosis associated with numerous asbestos bodies. (Low power.)

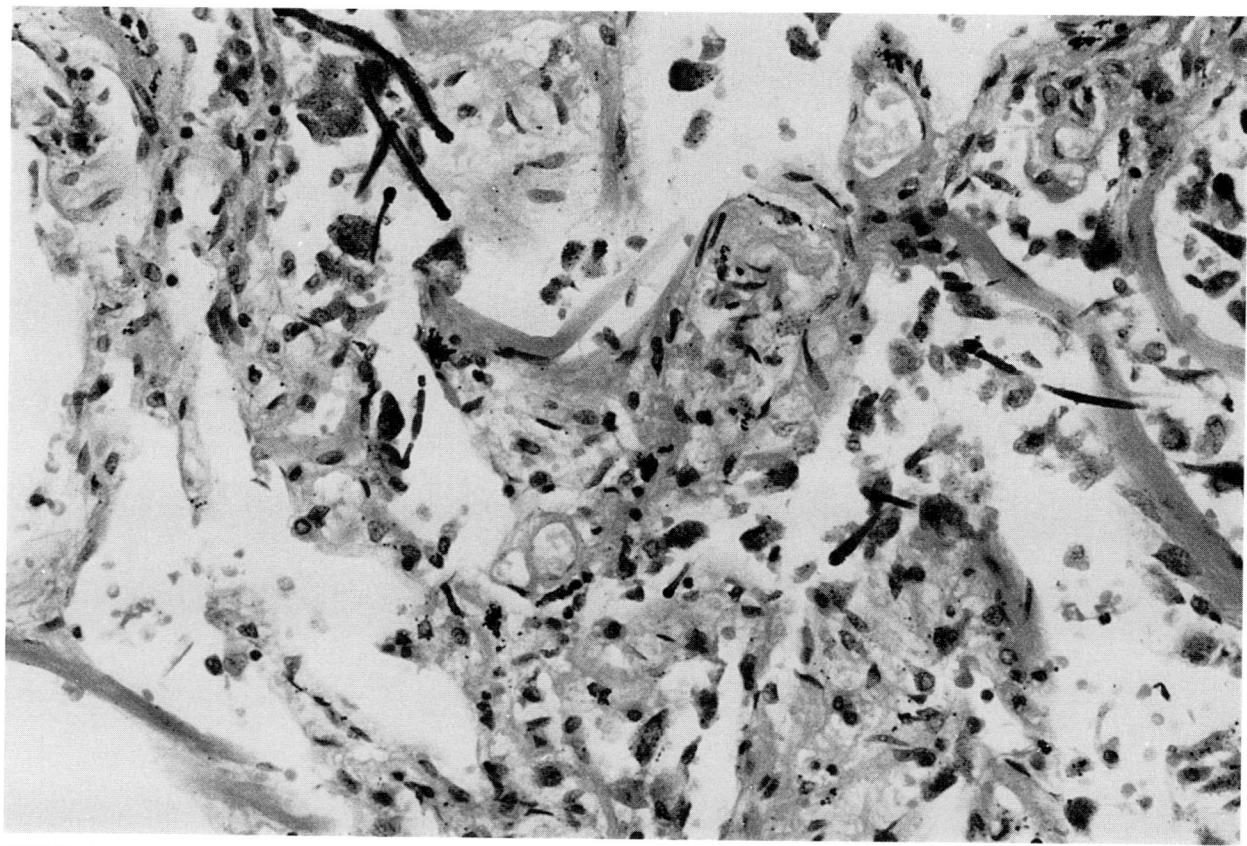

FIGURE 15-44
Same case as in Fig. 15-43 showing numerous asbestos bodies. (Medium power.)

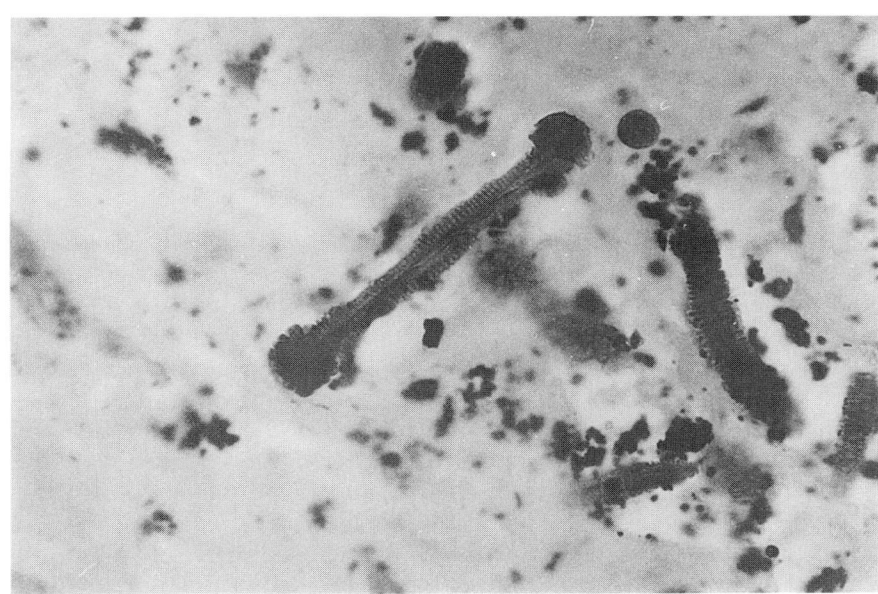

FIGURE 15-45
Asbestos bodies with clear fibrous cores.
(High power.)

boles more easily than chrysotile. Subjects with a high iron oxide content in their lungs, for example, welders and steel workers, tend to form ferruginous bodies more easily both on asbestos fibers and other mineral particles.[199] In these subjects great care has to be taken in evaluating lung fibrosis in relation to ferruginous body content because many of the ferruginous bodies will not be formed on asbestos. One can also use a grading scheme for asbestos bodies.[203] This shows a good correlation with asbestos fiber level determined microanalytically and extent of fibrosis. Other features that may be seen in cases of asbestosis are foreign body giant cells, often located around the asbestos bodies, osseous metaplasia

![Electron microscopic appearance of naked amphibole fibers and asbestos bodies.]

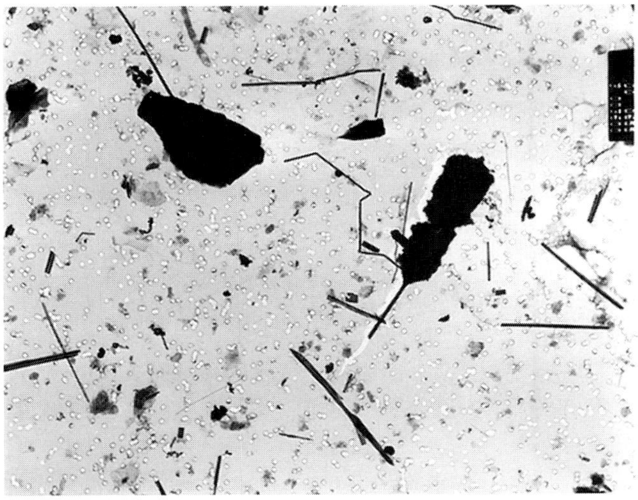

FIGURE 15-46
Electron microscopic appearance of naked amphibole fibers and asbestos bodies.

within the fibrous tissue[204] and pulmonary blue bodies.

Some authors have agreed that peribronchiolar fibrosis associated with asbestos bodies (i.e., grade 1 asbestos) should not be called asbestosis but "mineral dust mining disease."[205] This is largely a matter of semantics. However, it should be realized that peribronchiolar fibrosis is frequent in the lungs of routinely autopsied subjects in whom no exposure to asbestos has occurred and in most cases is related to cigarette smoking. Secondly, a variety of other mixed dusts including silica, coal dust, iron oxide, and aluminum oxide can be associated with similar changes. Mineral analysis is required in this situation to assess the relationship to asbestos. However, these changes on their own are unlikely to be of great functional significance. Also they are unlikely to be progressive unless the asbestos fiber load within the lung is great enough to cause such changes.

DIFFERENTIAL DIAGNOSIS

The macroscopic and microscopic changes within the lungs are similar in both idiopathic pulmonary fibrosis and asbestosis except for the presence of numerous asbestos bodies indicating a heavy asbestos fiber load. Pleural thickening and plaques are more common in cases of asbestosis. However, pleural changes result from much lower levels of exposure than asbestosis. Cases of idiopathic pulmonary fibrosis can occur in subjects with asbestos exposure,[206] and they may also show pleural changes but examination of the lung tissues will reveal scant or absent asbestos bodies. Mineralogic analysis of these cases is useful in demonstrating a relatively low fiber level.

The diagnosis of asbestosis should not be made on small biopsies because they are frequently unrepresentative. Also if a tumor is present in the lung, fibrotic changes in the vicinity should not be overinterpreted because tumors frequently result in fibrotic changes per se. If one is dealing with a postmortem case the assessment of fibrosis is best performed on tissue specimens obtained from the nontumorous lung.

Sarcoidosis, extrinsic allergic alveolitis, and histiocytosis X also have to be distinguished from asbestosis. They tend to show a greater upper zonal distribution of fibrosis although there are exceptions to this. The demonstration of granulomas or foci of histiocytosis X will exclude asbestosis. If there are any doubts mineralogic analysis will be helpful.

Other mineral dusts including talc, mica, kaolin, and hard metals can result in interstitial fibrosis. Lung fibrosis induced by nonfibrous silicates is associated with extremely heavy mineral burdens,[203] and there should be a good history of exposure and numerous easily identified mineral particles within the lung. Mineral analysis can be helpful in this situation.

Hammar has reported focal ossification in about one-third of cases of grade 4 (severe) asbestosis[207] and has implied that it is helpful in distinguishing asbestosis from interstitial pulmonary fibrosis (Fig. 15-47). However, this feature may be seen in interstitial pulmonary fibrosis and the most reliable diagnostic indicator is a high level of asbestos within the lungs whether determined by asbestos body content or mineralogically.

Asbestos-Related Lung Cancer

The link between lung cancer and asbestos exposure was first established by the systematic epidemiologic studies of Doll[208] and Breslow[209] published in 1955. In 1979, Hammond and colleagues published a study examining the effects of smoking and asbestos exposure and the risk of the development of lung cancer.[210] They found that the effect of smoking and asbestos exposure on the risk of lung cancer was multiplicative. It should be realized, however, that this risk was applicable to a heavily exposed group of insulation workers, nearly all of whom had asbestosis[211] and should not be applied to all other groups of asbestos-exposed workers. The excess lung cancer risk was much less in other cohorts and varied with industrial process, being relatively high in textile workers and very low in friction products and asbestos cement manufacturing.[212] In some groups the risk was between additive and multiplicative.[213] It is rare to observe a lung cancer in a nonsmoking individual exposed to asbestos.

The risk of lung cancer also varies with fiber type; consistently lower rates occur in occupational cohorts exposed to chrysotile relative to similar cohorts exposed to amphiboles. For example, the standard mortality rate for lung cancer for chrysotile mining and milling is 125 whereas for amphibole mining and milling it is 264.[213]

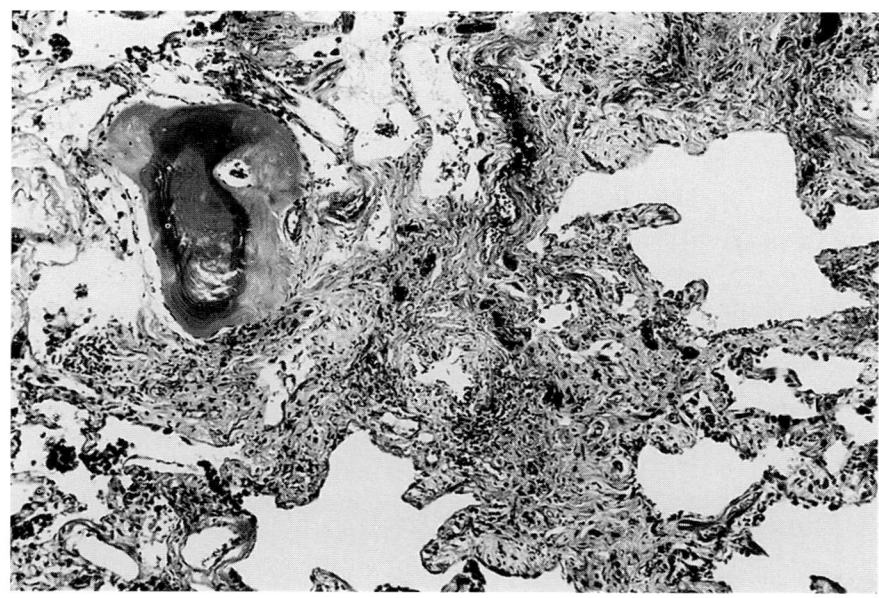

FIGURE 15-47
Severe interstitial fibrosis with an area of ossification. (Low power.)

The latent period from the first exposure to asbestos and the development of an asbestos-related lung cancer is usually in excess of 20 years and usually requires prolonged heavy direct exposure. Churg reviewed the literature pertaining to lung cancer type and asbestos exposure and found that all the major histologic types occurred.[215] However, there are several limitations to this kind of analysis. In many series there has been no attempt to separate within the cohorts those tumors that are related to asbestos exposure and those that are not. Similarly little attempt was made to control for confounders such as sex and year of diagnosis. A study by Johansson and associates,[216] which controlled for these factors, found that there was an association between adenocarcinoma of the lung and a peripheral localization with asbestos exposure. However, the data are still not sufficient to use the histologic type of carcinoma in an asbestos-exposed worker for ascription or not of lung cancer to asbestos exposure.

There is now reasonable evidence to use the presence or absence of asbestosis as the criterion for ascribing a lung cancer to asbestos (see Fig. 15-38). Epidemiologic and histopathologic studies have indicated that lung cancer risk from asbestos exposure is increased in those industries where asbestosis occurs but not in those where it does not and that it increases with severity of asbestosis.[217,218] Furthermore, the increase appears to develop when there is radiologic evidence of fibrosis,[219] and this corresponds to histopathologic grades of 2 or 3 or greater. Animal studies have shown a strong link between the development of lung fibrosis and tumor formation.[220]

Nonasbestos Fibers

These include natural and synthetic fibers (Table 15-11), a proportion of which are respirable and therefore potentially pathogenic. Information concerning their health effects is limited but continued vigilance for possible health effects from exposure to these fibers should be maintained because of the long latent periods between first exposure to a fiber and the development of pleural and pulmonary disease. Their use as asbestos substitutes, particularly the synthetic fibers, has been increasing over recent years.

The natural group includes sepiolite, attapulgite (palygorskite), wollastonite, and zeolite. The information on these fibers is very limited. Attapulgite exposure has been linked to lung cancer[221] and wollastonite has been associated with pleural plaques and lung fibrosis.[222] Further studies of these fibers

TABLE 15-11
Commonly Used Inorganic Nonasbestos Mineral Fibers

Natural	Synthetic
Sepiolite	Mullite
Attapulgite/palygorskite	Glass
Wollastonite	Wool
Zeolite	Fiber
	Continuous filament
	Mineral wools
	Rock
	Slag
	Ceramic

are needed. Environmental exposures to erionite (a fibrous zeolite) have been associated with very high rates of mesothelioma and lung cancer in three villages in Turkey.[223] In animal inhalation studies erionite has produced the highest rate of mesothelioma from any known fiber.

The synthetic group includes a variety of different fibers and both workers and the general public have been exposed to them. Mullite fibers are frequently found in the lungs of normal individuals because the fibers are formed from combustion of fossil fuels; increased numbers are found in the lungs of coal-fired power station workers but they do not appear to have any pathogenetic significance.

The other fibers are rarely encountered on mineral analysis of human lungs. This is probably because the majority are not durable in lung tissues. The majority of epidemiologic studies of glass wool, rock/slag wool, and continuous filament production workers have not shown an increased risk of lung cancer or mesothelioma. A few studies have shown an excess of lung cancer in rockwool and slagwool production workers but this has been in the workers in the early production phase and the studies have not been well controlled for confounders such as smoking and asbestos. In summary, the excess lung cancer mortality found in some production workers may not necessarily be caused by exposure to airborne manmade mineral fibers.[224,225]

There is some concern about refractory ceramic fibers because these are durable in lung tissues. Animal inhalation studies with these fibers have produced lung cancers and mesotheliomas.[226] They have been produced in limited quantities and only represent approximately 1 to 2 percent of world production. They have been used in paper, boards, blankets, and heat-resistant linings of furnaces. No mesotheliomas or lung cancers in humans have been linked to refractory ceramic fiber exposure so far.

ORGANIC DUST DISEASE

A simple classification of organic dust diseases is shown in Table 15-12.

Occupational Asthma

"Occupational asthma is asthma which is due, in whole or part, to agents encountered at work."[227] Largely due to better recognition, but also due to an increased number of chemicals in the workplace, the prevalence of occupational asthma has increased over the last two decades. More than 200 chemical agents have been implicated in its causation. Some of the most frequent are isocyanates, anhydrides, colophony, metals, animal proteins, wood dusts, and grain and flour dusts. Occupational asthma is now the most prevalent occupational lung disease in developed countries.[228]

There is a lag period varying from months to years after starting a job before symptoms commence. Initially, symptoms occur at work and improve when the subject is away from the workplace but as the condition progresses the symptoms may become persistent and the typical work related pattern lost.

Agents can cause occupational asthma by several mechanisms.[229]

1. *Immunologic*—through an immunoglobulin (IgE)-dependent mechanism where specific IgE antibodies can be identified in the subject. Positive skin tests to the antigen may be present, although these do not always coincide with symptoms and may therefore be a marker of exposure rather than disease. This is usually induced by high molecular weight compounds such as animal and plant proteins.

Through non–IgE-dependent mechanisms but probably involving cell-mediated immunity. Specific IgE antibodies to offending agents are not usually identified in the subjects. The majority of low molecular weight compounds that cause occupational asthma act through this pathway, for example, isocyanates, colophony, and wood dusts.

2. *Nonimmunologic*—In this type reexposure to small amounts of the offending agent does not cause the symptoms. Some may be due to direct irritant or pharmacologic reactions.

The pathologic changes for each type of occupational asthma have not been well studied but appear to be similar to those of nonoccupational asthma.

Byssinosis

Byssinosis, derived from the Greek work meaning flax, is the term applied to respiratory disease caused by the inhalation of cotton dust and now also caused by flax and hemp.

For over a hundred years it had been recognized that some of the workers engaged in the Lancashire cotton mills developed a chest disease similar to chronic bronchitis. It has also been reported among cotton mill operatives in North and South Carolina. A similar illness has been described among flax workers in Northern Ireland,[230] and hemp workers in Italy and France.[231] In the cotton industry it occurs particularly among those employed in blowing and carding the raw cotton, formerly a very dusty process and the prevalence is increased in cigarette smokers.[232]

Clinically, the condition presents as a tightness of the chest, cough, and asthmatic wheezing, which is most noticeable when the operative returns to work on Monday after being away from the mill for the weekend. It does not usually develop with less than 20 years of exposure. Roach and Schilling[233] divided the severity of byssinosis into several grades.

Grade 0	Symptomless on Mondays
Grade 1/2	Occasional chest tightness on the first working day of the week or mild symptoms such as irritation of the respiratory tract
Grade 1	Chest tightness or breathlessness on the first day of the working week only
Grade 2	Chest tightness or breathlessness on the first and other days
Grade 3	Chest tightness or breathlessness accompanied by permanent respiratory incapacity from reduced ventilatory capacity

Edwards and coworkers[234] studied the lungs from 43 byssinotics and found that the proportions of

TABLE 15-12
Classification of Organic Dust Diseases

Allergic	Occupational asthma
	Humidifier fever
	Byssinosis
	Hypersensitivity pneumonititis
Toxic	Acute organic dust toxic syndrome (mycotoxicosis)

mucous gland, muscle, and cartilage were increased in the named bronchi. Emphysema does not appear to be increased.[235] So-called byssinosis bodies have been described but these are probably ferruginous bodies formed on carbonaceous material and are neither specific for nor found in every case of byssinosis. It will be realized from this description that the pathologic changes are nonspecific and the diagnosis has to be made on clinical grounds.

Extrinsic Allergic Alveolitis (Hypersensitivity Pneumonitis)

See Chap. 13.

Humidifier Fever

Subjects exposed to "conditioned" air contaminated by microorganisms such as bacteria and fungi can develop a condition called humidifier fever, characterized by mild malaise, fever, sweats, cough, breathlessness, and generalized aches. It usually develops on the first day of work after a weekend or holiday but resolves spontaneously within a day or two even though there is continued exposure. The pathology and pathogenesis of the condition are unknown. The clinical and laboratory changes are similar to those of acute extrinsic allergic alveolitis except radiologic changes are absent and it does not progress to lung fibrosis.

Acute Organic Dust Toxic Syndrome (Mycotoxicosis)

This syndrome follows exposure to very high levels of organic dust containing fungal spores and hyphae and it is thought to act through a toxic rather than a hypersensitivity mechanism.[236,237] It has been described in farm workers who have unloaded storage silos, cotton workers ("mill fever"), and storage elevator workers ("grain fever"). After exposure the subject develops an acute influenza-like illness and crackles are frequently heard on auscultation. Chest radiographs show diffuse infiltrates. The disease is usually self-limiting although it may persist for several days. It shares many features with acute farmer's lung and silo-filler's disease, which is due to nitrogen dioxide toxicity.

Few cases have been studied pathologically. A neutrophilic infiltrate affects bronchioles, alveoli, and interstitium and fungal spores may be present. Granulomas are not seen (cf extrinsic allergic alveolitis). Bronchoalveolar lavage fluid taken within a few days of the onset shows an increased number of neutrophils and frequently fungal spores.[238]

REFERENCES

1. Parkes WR: *Occupational Lung Disorders*, 3rd ed. Oxford, Butterworth-Heinemann, 1994.
2. Harley RA Jr: Tobacco, in Dail DH, Hammar SP (eds), *Pulmonary Pathology*, 2d ed. New York, Springer-Verlag, 1994, pp 831–846.
3. Brain JD, Valberg PA: Deposition of aerosol in the respiratory tract. *Am Rev Respir Dis* 120:1325–1373, 1979.
4. Timbrell V: Deposition and retention of fibres in the human lung. *Ann Occup Hyg* 26:347–369, 1982.
5. Gerrity TR, Garrard CS, Yeates DB: A mathematical model of particle retention in the air spaces of human lungs. *Br J Ind Med* 40:121–130, 1983.
6. Timbrell V, Pooley FD, Wagner JC: Characteristics of respirable asbestos fibres, in Shapiro HP (ed), *Pneumoconioses: Proceedings of the International Conference, Johannesburg.* London, Oxford University Press, 1970, pp 120–125.
7. Dennis WL: The effect of breathing rate on the deposition of particles in the human respiratory system, in Walton WH (ed), *Inhaled Particles III, vol. 1.* Old Woking, Unwin Brothers, 1971, pp 91–102.
8. Adamson IYR, Bowden DH: Effects of irradiation on macrophagic response and transport of particles across the alveolar epithelium. *Am J Pathol* 106:40–46, 1982.
9. Lee KP. Lung response to particulates with emphasis on asbestos and other fibrous dusts. *Crit Rev Toxicol* 14:33–36, 1985.
10. Pinkerton KE, Pratt PC, Brody AR, Crapo JD: Fiber localization and its relationship to lung reaction in rats after chronic inhalation of chrysotile asbestos. *Am J Pathol* 117:484–498, 1984.
11. Vincent JH, Jones AD, Johnston AM, McMillan C, Bolton RE, Cowie H: Accumulation of inhaled mineral dust in the lung and associated lymph nodes: Implications for exposure and dose in occupational lung disease. *Ann Occup Hyg* 31:375–393, 1987.
12. Cohen D, Arai SF, Brain JD: Smoking impairs long term dust clearance from the lung. *Science* 204: 514–517, 1979.
13. McFadden D, Wright JL, Wiggs B, Churg A: Smoking inhibits asbestos clearance. *Am Rev Respir Dis* 133:372–374, 1986.
14. Mossman BT, Begin RO: *Effects of Mineral Dusts on Cells.* Berlin, Springer-Verlag, 1989, NATO ASI Series H, Vol 30.
15. Wagner JC: Overview of experimental pneumoconiosis, in Gil J (ed), *Models of Lung Disease.* New York, Marcel Dekker, 1990, pp 753–760.
16. Gibbs AR: Pathological reactions of the lung to dust, in Morgan WKC, Seaton A (eds), *Occupational Lung Disease,* 3d ed. Philadelphia, WB Saunders, 1995, pp 127–157.
17. Gibbs AR, Seal RME: Examination of lung specimens. *J Clin Pathol* 43:68–72, 1990.
18. Gibbs AR, Pooley FD: Mineral analysis in lung disorders. *Thorax* (in press)
19. Collis EL: Industrial pneumoconiosis with special reference to dust phthisis. *Pub Health Lond* 28:252–264, 1915.
20. Hoover HC, Hoover L: Agricola G. *De Re Metallica* (trans). New York, Dover, 1950, p 214.
21. Ramazzini B: *Diseases of Workers.* New York, Hefner Publishing Co., 1964, p 250.
22. Thackrah CT: *The Effects of Arts, Trades and Professions,* 2d ed. Leeds, Baines and Newson, 1832.
23. Visconti reported by Rovida CL: Un casoi di silicosi del pulmone con analisi chimica. *Pilli Annali Chimica* 1871.
24. Lanza AJ: *Silicosis and Asbestosis.* London, Oxford University Press, 1938.

25. Middleton EL: Industrial pulmonary disease due to the inhalation of dust. *Lancet* 2: 59, 1936.

26. Cherniack M: *The Hawk's Nest Incident: America's Worst Industrial Disaster.* New Haven, CN, Yale University Press, 1986.

27. Gardner LU: The reactivation of healing primary tubercles in the lung by inhalation of quartz, granite and carborundum dusts. *Am Rev Tuberc* 20:833–875, 1929.

28. Allison AC, Hart PD: Potentiation by silica of the growth of mycobacterium tuberculosis in macrophage cultures. *Br J Exp Pathol* 49:465–476, 1968.

29. Westerholm P. Silicosis: Observations on a case register. *Scand Work Environ Health* 6 (suppl 2):1–86, 1980

30. Banks DE, Morring KL, Boehlecke BA, et al: Silicosis in the 1980's. *Am Ind Hyg Assoc J* 42:77–79, 1981.

31. Gibbs AR, Wagner JC: Diseases due to silica, in Churg A, Green FHY (eds), *Pathology of Occupational Lung Disease.* New York, Igaku Shoin, 1988, pp 155–176.

32. Sluis-Cremer GK, Hessel PA, Hnizdo E, Churchill AR: Relationship between silicosis and rheumatoid arthritis. *Thorax* 41:596–601, 1986.

33. Klockars M, Koskela R-S, Jarvinen E, Kolari PJ, Rossi A: Silica exposure and rheumatoid arthritis: A follow up study of granite workers 1940–81. *Br Med J* 294:997–1000, 1987.

34. Sluis-Cremer GK, Hessel PA, Hnizdo EH, Churchill AR, Zeiss EA: Silica, silicosis and progressive systemic sclerosis. *Br J Ind Med* 43:838, 1985.

35. Langer AM, Nolan RP: Physicochemical properties of minerals relevant to biological properties: State of the art, in Beck EG, Bignon J (eds), *In Vitro Effects of Mineral Dusts.* Berlin, Springer-Verlag, 1985, pp 9–24.

36. Bowden DH, Adamson IYR: The role of cell injury and the continuing inflammatory response in the generation of silicotic pulmonary fibrosis. *J Pathol* 144:149–161, 1984.

37. Christman JW, Emerson RJ, Graham WGB, Davis GS: Mineral dust and cell recovery from the bronchiolar lavage of healthy Vermont granite workers. *Am Rev Respir Dis* 132:393–399, 1985.

38. Brown GM, Donaldson K: Degradation of connective tissue components by lung derived leucocytes in vitro: Role of proteases and oxidants. *Thorax* 43:132–139, 1988.

39. Brown GP, Monick M, Hunninghake GW: Fibroblast proliferation induced by silica-exposed human alveolar macrophages. *Am Rev Respir Dis* 138:85–89, 1988.

40. Goldsmith DF, Guidotti TL, Johnston DR: Does occupational exposure to silica cause lung cancer? *Am J Ind Med* 3:423–440, 1982.

41. Heppleston AG: Silica, pneumoconiosis and carcinoma of the lung. *Am J Ind Med* 7:285–294, 1985.

42. International Agency for Research into Cancer: Silica and some silicates. *IARC Monogr Eval Carcino Risks Hum (Lyon)* 42:39–143, 1987.

43. Pairon JC, Brochard P, Jaurand MC, Bignon J: Silica and lung cancer: A controversial issue. *Eur Respir J* 4:730–744, 1991.

44. Schuyler M, Gauner HR, Stankus RP, Kaimal V, Hoffman E, deSalvaggio J: Bronchoalveolar lavage in silicosis. *Lung* 157:95–102, 1980.

45. McEuen DD, Abraham JL: Particulate concentrations in pulmonary alveolar proteinosis. *Environ Res* 17:334–339, 1978.

46. Prakash UBS, Barham SS, Carpenter HA, Dines DE, Marsh HM: Pulmonary alveolar phospholipoproteinosis: Experience with 34 cases and a review. *Mayo Clin Proc* 62:499–518, 1987.

47. Gross P, DeTreville RTP: Alveolar proteinosis: Its experimental production in rodents. *Arch Pathol* 86:255, 1968.

48. Heppleston AG, Wright NA, Stewart JA: Experimental alveolar lipo-proteinosis following the inhalation of silica. *J Pathol* 101: 293–307, 1970.

49. McLaughlin AIG: Pneumoconiosis in foundry workers. *Tuberculosis* 51: 297–309, 1957.

50. Craighead JE, Kleinerman J, Abraham J, et al: Diseases associated with exposure to silica and nonfibrous silicate minerals. *Arch Pathol Lab Med* 112: 673–720, 1988.

51. Sheers G: The china clay industry—lessons for the future of occupational health. *Respir Med* 83: 173–175, 1989.

52. Sheers G: Prevalence of pneumoconiosis in Cornish kaolin workers. *Br J Ind Med* 21: 218–225, 1964.

53. Oldham PD: Pneumoconiosis in Cornish china clay workers. *Br J Ind Med* 40:131–137, 1983.

54. Ogle CJ, Rundle EM, Sugar ET: A survey of china clay workers in the South West of England. *Br J Ind Med* 46:261–270, 1989.

55. Kennedy T, Rawlings W, Baser M, Tockman M: Pneumoconiosis in Georgia kaolin workers. *Am Rev Respir Dis* 127: 215–220, 1983.

56. Morgan WKC, Donner A, Higgins ITT, Pearson MG, Rawlings W: The effects of kaolin on the lung. *Am Rev Respir Dis* 138: 813–820, 1988.

57. Gibbs AR: Human pathology of kaolin and mica pneumoconiosis, in Bignon J (ed), *Human Related Effects of Phyllosilicates.* Berlin, Springer-Verlag, NATO ASI Series Vol G21, 1990, pp 217–226.

58. Wagner JC, Pooley FD, Gibbs A, Lyons J, Sheers G, Moncrieff CB: Inhalation of china stone and china clay dusts: Relationship between the mineralogy of dust retained in the lungs and pathological changes. *Thorax* 41: 190–196, 1986.

59. Ferret J, Moreau P: Mineralogy of talc deposits, in Bignon J (ed), *Human Related Effects of Phyllosilicates.* Berlin, Springer-Verlag, NATO ASI Series Vol G21, 1990, pp 147–158.

60. Gibbs AR, Pooley FD, Griffiths DM, Mitha R, Craighead JE, Ruttner JR: Talc pneumoconiosis: A pathologic and mineralogic study. *Hum Pathol* 23: 1344–1354, 1992.

61. Wegman DH, Peters JM, Boundy MG, Smith TJ: Evaluation of respiratory effects in miners and millers exposed to talc free of asbestos and silica. *Br J Ind Med* 39: 233–238, 1982.

62. Gamble J, Greife A, Hancock J: An epidemiological–industrial hygiene study of talc workers. *Ann Occup Hyg* 26: 841–859, 1982.

63. Vallyathan NV, Craighead JE: Pulmonary pathology in workers exposed to nonasbestiform talc. *Hum Pathol* 12: 28–35, 1981.

64. Leophonte P, Didier A: French talc pneumoconiosis, in Bignon J (ed), *Human Related Effects of Phyllosilicates.* Berlin, Springer-Verlag, NATO ASI Series Vol G21, 1990, pp 203–210.

65. Kleinfeld M, Messite J, Kooyman O, Zaki MH: Mortality experiences among talc miners and millers in New York State. *Arch Environ Health* 14: 663–667, 1967.

66. Kleinfeld M, Messite J, Zaki MH: Mortality experiences among talc workers: A follow up study. *J Occup Med* 16: 345–349, 1974.

67. Stille WT, Tabershaw IR: The mortality experience of upstate New York talc workers. *J Occup Med* 24: 480–484, 1982.

68. Reger RB, Morgan WKC: On talc, tremolite and tergiversation. *Br J Ind Med* 47: 505, 1990.

69. Lamm SH, Levine MS, Starr JA, Tirey SL: Analysis of excess lung cancer risk in short-term employees. *Am J Epidemiol* 127: 1202–1209, 1988.

70. Lamm SH, Starr JA: Similarities in lung cancer and respiratory disease mortality of Vermont and New York State talc workers. *Proceedings of VIIth International Pneumoconioses Conference.* DHHS (NIOSH) Publication No. 90-108, 1990, pp 1576–1581.

71. Gamble JF: A nested case control study of lung cancer among New York talc workers. *Int Arch Occup Environ Health* 64: 449–456, 1993.

72. Cullinan P, McDonald JC: Respiratory disease from occupational exposure to non-fibrous phyllosilicates, in Bignon J (ed), *Human Related Effects of Phyllosilicates.* Berlin, Springer-Verlag, NATO ASI Series Vol G21, 1990, 161–178.

73. Skulberg KR, Gylseth B, Skaug V, Hanoa R: Mica—pneumoconiosis—a literature review. *Scand J Work Environ Health* 11: 65–74, 1985.

74. Davies D, Cotton R: Mica pneumoconiosis. *Br J Ind Med* 40: 22–27, 1983.

75. Sakula A: Pneumoconiosis due to fuller's earth. *Thorax* 16: 176–179, 1961.

76. Gibbs AR, Pooley FD. Fuller's earth (montmorillonite) pneumoconiosis. *Occup Environ Med* 51: 644–646, 1994.

77. Laennec RTH: Traite de l'auscultation mediate, 1st ed. Paris, 1819.

78. Gregory JC: Case of peculiar black infiltration of the whole lungs, resembling melanosis. *Edinburgh Med Surg* 36: 389–394, 1831.

79. Stratton TML: Case of anthracosis or black infiltration of the whole lungs. *Edin Med Surg* 49: 490–491, 1838.

80. Gough J: Pneumoconiosis in coal trimmers. *J Pathol Bacteriol* 51: 277, 1940.

81. Jacobsen M, Rae S, Walton WH, Rogan JM: The relation between pneumoconiosis and dust exposure in British coalmines, in Walton WH (ed), *Inhaled Particles III.* Old Woking, Unwin Brothers, 1971, pp 903–917.

82. Hurley JF, Burns J, Copland L, Dodgson J, Jacobsen M: Coalworkers' simple pneumoconiosis and exposure to dust at 10 British coal mines. *Br J Ind Med* 39: 120–127, 1982.

83. Bennett JG, Dick JA, Kaplan YS, et al: The relationship between coal rank and the prevalence of pneumoconiosis. *Br J Ind Med* 36: 206–210, 1979.

84. Douglas AN, Robertson A, Chapman JS, Ruckley VA: Dust exposure, dust recovered from the lung, and associated pathology in a group of British coalminers. *Br J Ind Med* 43: 795–801, 1986.

85. Davis JMG, Chapman J, Collings P, Fernie J, Lamb D, Ruckley VA: Variations in the histological patterns of the lesions of coal workers' pneumoconiosis in Britain and their relationship to lung dust content. *Am Rev Respir Dis* 128: 118–124, 1983.

86. Reisner MTR, Robock K: Results of epidemiological, mineralogical and cytological studies on the pathogenicity of coalmine dusts, in Walton WH (ed), *Inhaled Particles IV.* Oxford, Pergamon Press, 1977, pp 703–715.

87. Gormley IP, Collings P, Davis JMG, Ottery J: An investigation into the cytotoxicity of respirable dusts from British collieries. *Br J Exp Pathol* 60: 523–536, 1979.

88. Le Bouffant L, Daniel H, Martin JC: The therapeutic action of aluminum compounds on the development of experimental lesions produced by pure quartz or mixed dust, in Walton WH (ed), *Inhaled Particles IV.* Oxford, Pergamon Press, 1977, pp 389–400.

89. Davis JMG, Addison J, Brown GM, et al: *Further Studies on the Importance of Quartz in the Development of Coalworkers' Pneumoconiosis.* Edinburgh, Institute of Occupational Medicine. (IOM Report TM/91/05), 1991.

90. Heppleston AG: Minerals, fibrosis, and the lung. *Environ Health Perspect* 94: 149–168, 1991.

91. Seemayer NH: Biological effects of coal mine dusts on macrophages in vitro: Importance of grain size and mineral content. *VIth International Pneumoconiosis Conference, 1983.* Bochum ILO, 1984, pp 513–528.

92. Jacobsen M, Maclaren WM: Unusual pulmonary observations and exposure to coal mine dust: A case control study, in Walton WH (ed), *Inhaled Particles V.* Oxford, Pergamon Press, 1982, pp 735–765.

93. Kleinerman J, Green F, Laquer W, et al: Pathology standards for coal workers' pneumoconiosis. *Arch Pathol Lab Med* 103: 375–385, 1979.

94. Heppleston AG: The essential lesion of pneumoconiosis in Welsh coal miners. *J Pathol Bacteriol* 67: 51–63, 1954.

95. Theron GP, Walters LG, Webster J: The international classification of radiographs in the pneumoconiosis. *Med Proc* 10:352–354, 1964.

96. Gough J, James WRL, Wentworth JE: A comparison of the radiological and pathological changes in coalworkers' pneumoconiosis. *J Faculty Radiol* 1: 28–39, 1949.

97. Chapman JS, Ruckley VA: Microanalyses of lesions and lymph nodes from coalminers' lungs. *Br J Ind Med* 42: 551–555, 1985.

98. Caplan A: Certain unusual radiological appearances in the chest of coal miners suffering from rheumatoid arthritis. *Thorax* 8: 29–37, 1953.

99. Caplan A, Payne RB, Withey JL: A broader concept of Caplan's syndrome related to rheumatoid factors. *Thorax* 17: 205–212, 1962.

100. Gough J, Rivers D, Seal RME: Pathological studies of modified pneumoconiosis in coal miners with rheumatoid arthritis (Caplan's syndrome). *Thorax* 10: 9–18, 1955.

101. Chatgidakis CB, Theron CP: Rheumatoid pneumoconiosis (Caplan's syndrome). *Arch Environ Health* 2: 397–408, 1961.

102. Rickards AG, Barrett GM: Rheumatoid lung changes associated with asbestosis. *Thorax* 13: 185–193, 1958.

103. Greaves IA: Rheumatoid "pneumoconiosis" (Caplan's syndrome) in an asbestos worker: A 17 year follow up. *Thorax* 34: 404–405, 1979.

104. Seaton A: Coalmining, emphysema and compensation. *Br J Ind Med* 47: 433–435, 1990.

105. Ryder R, Lyons JP, Campbell H, Gough J: Emphysema in coal workers' pneumoconiosis. *Br Med J* 3: 481–487, 1970.

106. Glick M, Outhred KG, McKenzie HJ: Pneumoconiosis and respiratory disorders of coal mine workers in New South Wales, Australia. *Ann NY Acad Sci* 200: 316–334, 1972.

107. Miller BG, Jacobsen M: Dust exposure, pneumoconiosis, and mortality of coalminers. *Br J Ind Med* 42: 723–733, 1985.

108. Ruckley VA, Gauld SA, Chapman JS, et al: Emphysema and dust exposure in a group of coal workers. *Am Rev Respir Dis* 129: 528–532, 1984.

109. Lamb D: A survey of emphysema in coal workers and the general population. *Proc R Soc Med* 69: 14, 1976.

110. Cockcroft A, Seal RME, Wagner JC, Lyons JP, Ryder R, Anderson N: Postmortem study of emphysema in coal workers and noncoal workers. *Lancet* 2: 600–603, 1982.

111. Heppleston AG: The pathological anatomy of simple pneumoconiosis in coal workers. *J Pathol Bacteriol* 66: 235–246, 1953.

112. Duguid JB, Lambert MW: Pathogenesis of coalminer's pneumoconiosis. *J Pathol Bacteriol* 88: 389–403, 1964.

113. Leopold JG, Gough J: The centrilobular form of hypertrophic emphysema and its relation to chronic bronchitis. *Thorax* 12: 219–235, 1957.

114. McConnochie K, Green FHY, Vallyathan V, Wagner JC, Seal RME, Lyons JP: Interstitial fibrosis in coal miners—experience in Wales and West Virginia. *Ann Occup Hyg* 3 (Suppl. 1): 553–560, 1988.

115. Green FHY: Coal workers' pneumoconiosis and pneumoconiosis due to other carnoaceous dusts, in Churg A, Green FHY (eds), *Pathology of Occupational Lung Disease.* New York, Igaku Shoin, 1988, pp 89–154.

116. Cockcroft AE, Wagner JC, Seal RME, Lyons JP, Campbell MJ: Irregular opacities in coalworkers' pneumoconiosis—correlation with pulmonary function and pathology. *Ann Occup Hyg* 26: 767–787, 1982.

117. Hanoa R: Graphite pneumoconiosis. A review of etiologic and epidemiologic aspects. *Scand J Work Environ Health* 9: 303–314, 1983.

118. Nemery B: Metal toxicity and the respiratory tract. *Eur Respir J* 3: 202–219, 1990.

119. Magos L: Epidemiological and experimental aspects of metal carcinogenesis: physicochemical properties, kinetics, and the active species. *Environ Health Perspect* 95: 157–189, 1991.

120. Abramson MJ, Wlodarczyk JH, Saunders NA, Hensley MJ: Does aluminum smelting cause lung disease? *Am Rev Respir Dis* 139: 1042–1057, 1989.

121. Shaver CG: Pulmonary changes encountered in employees engaged in the manufacture of alumina abrasives. *J Med* 5: 718–728, 1948.

122. Riddell RR: Pulmonary changes encountered in the manufacture of alumina abrasives. *J Med* 5: 710–717, 1948.

123. Mitchell J: Pulmonary fibrosis in an aluminum worker. *Br J Ind Med* 16: 123–125, 1959.

124. McLaughlin A, Karantzis G, King E, Teare D, Porter R, Owen R: Pulmonary fibrosis and encephalopathy associated with the inhalation of aluminum dust. *Br J Ind Med* 19: 253–263, 1962.

125. Jederlinic PJ, Abraham JL, Churg A, et al: Pulmonary fibrosis in aluminum oxide workers. *Am Rev Respir Dis* 142: 1179–1184, 1990.

126. Gilks B, Churg A: Aluminum induced pulmonary fibrosis: Do fibers play a role? *Am Rev Respir Dis* 136: 176–179, 1987.

127. Miller RR, Churg A, Hutcheon M, et al: Pulmonary alveolar proteinosis and aluminum dust exposure. *Am Rev Respir Dis* 130: 312–315, 1984.

128. DeVuyst P, Dumortier P, Schandene L, et al: Sarcoidlike lung granulomatosis induced by aluminum dusts. *Am Rev Respir Dis* 135: 493–497, 1987.

129. Chen WJ, Monnat RJ, Chen M, Mottet NK: Aluminum induced pulmonary granulomatosis. *Hum Pathol* 9: 705–711, 1978.

130. Herbert A, Sterling G, Abraham J, et al: Desquamative interstitial pneumonia in an aluminum welder. *Hum Pathol* 13: 694–699, 1982.

131. Vallyathan V, Bergeron W, Robichaux P, Craighead JE: Pulmonary fibrosis in an aluminum arc welder. *Chest* 81: 372–374, 1982.

132. De Vuyst P, Dumortier P, Rickaert F, Vande Wayer R, Lenclud C, Yernault JC: Occupational lung fibrosis in an aluminum polisher. *Eur J Respir Dis* 68: 131–140, 1986.

133. Ferryman SR, Morrison HM, Gibbs AR, Pooley FD, Byrne P: Giant-cell interstitial pneumonia: An unusual reaction to aluminum? (in preparation)

134. Kilburn KH, Warshaw RH: Irregular opacities in the lung, occupational asthma, and airways dysfunction in aluminum workers. *Am J Ind Med* 21: 845–853, 1992.

135. Wright JL, Churg A: Diseases caused by metals and related compounds, fumes and gases, in Churg A, Green FHY, eds, *Pathology of Occupational Lung Disease.* New York, Igaku Shoin 1988, pp 31–71.

136. Fabroni SM: Pulmonary disease due to beryllium. *Med D Lavoro* 26: 297–303, 1935.

137. Weber HH, Englehardt WE: Uber eine Apparat zur Erzeugung niedriger Staubkonzentratonen von grosse Konstanzt und eine Methode zur microgravinctrischen Staubbestimmung. Andwendung bei der Untersuchung von Stauben aus der Beryllium Gewinnung. *Zentralblatt fur Gewerbehygiene und Unfallverhutung* 10: 41, 1933.

138. Hardy HL, Tabershaw JR: Delayed chemical pneumonitis occurring in workers to beryllium compounds. *J Ind Hyg Toxicol* 28: 197–211, 1946.

139. Jones-Williams W: Beryllium disease; in Parkes WR (ed), *Occupational Lung Disease,* 3d ed. Oxford, Butterworth and Heinemann, 1994.

140. Hardy HL: Epidemiology, clinical character and treatment of beryllium poisoning. *Arch Ind Health* 11: 273–279, 1955.

141. Saltini C, Winestock K, Kirby M, Pinkston P, Crystal RG: Maintenance of alveolitis in patients with chronic beryllium disease by beryllium specific helper T cells. *N Engl J Med* 320: 1103–1109, 1989.

142. Hardy HL, Rabe EW, Lorch S: United States beryllium case registry (1952–1966): Review of its methods and utility. *J Occup Med* 9: 271–276, 1967.

143. Lieben J, Metzner F: Epidemiological findings associated with beryllium excretion. *Am Ind Hyg Assoc J* 20: 494–499, 1959.

144. Newman LS, Kreiss K: Non-occupational beryllium disease masquerading as sarcoidosis: Identification by blood lymphocyte proliferative responses to beryllium. *Am Rev Respir Dis* 145: 1212–1214, 1992.

145. Kniskhowy B, Baker EL: Transmission of occupational disease to family contacts. *Am J Ind Med* 9: 543–550, 1986.

146. Jones-Williams W: Beryllium disease–pathology and diagnosis. *J Soc Occup Med* 27: 93–96, 1977.

147. Kriebel D, Brain JD, Sprince NL, Kazemi H: The pulmonary toxicity of beryllium. *Am Rev Respir Dis* 137: 464–473, 1988.

148. Williams WJ, Wallach ER: Laser microprobe mass spectometry (LAMMS) analysis of beryllium, sarcoidosis and other granulomatous diseases. *Sarcoidosis* 6: 111–117, 1989.

149. Jones-Williams W, Williams WR: Value of beryllium lymphocyte transformation tests in chronic beryllium disease and potentially exposed workers. *Thorax* 38: 41–44, 1983.

150. Newman LS, Keiss K, King TE, Seay S, Campbell PA: Pathologic and immunologic alterations in early stages of beryllium disease. *Am Rev Respir Dis* 139: 1479–1486, 1989.

151. Mroz MM, Kreiss K, Lezotte DC, Campbell PA, Newman LS. Re-examination of the blood lymphocyte transformation test in the diagnosis of chronic beryllium disease. *J Allergy Clin Immunol* 88: 54–60, 1991.

152. Beton DC, Andrews GS, Davies HJ, Howells L, Smith GF: Acute cadmium fume poisoning. *Br J Ind Med* 23: 292–301, 1966.

153. Davison AG, Newman Taylor AJ, Darbyshire J, et al: Cadmium fume inhalation and emphysema. *Lancet* 1: 663–667, 1988.

154. Snider GL, Hayes JA, Korthy AL, Lewis GP: Centrilobular emphysema experimentally induced by cadmium chloride aerosol. *Am Rev Respir Dis* 108: 40–48, 1973.

155. Langard S. One hundred years of chromium and cancer. *Am J Ind Med* 17: 189–215, 1990.

156. Davies JM, Easton DF, Bidstrup PL: Mortality from respiratory cancer and other causes in United Kingdom chromate production workers: *Br J Ind Med* 48: 299–313, 1991.

157. Sorahan T, Burge DC, Waterhouse JA: A mortality study of nickel/chromium platers. *Br J Ind Med* 44: 250–258, 1987.

158. International Agency for Research on Cancer: Chromium, nickel and welding. Lyon, *IARC Monogr Eval Carcinog Risks Hum* 49, 1990.

159. Coates EO, Watson JHL: Diffuse interstitial lung disease in tungsten carbide workers. *Ann Intern Med* 75: 709–716, 1971.

160. Ohori NP, Sciurba FC, Owens GR, Hodgson MJ, Yousem SA: Giant-cell interstitial pneumonia and hard-metal pneumoconiosis. *Am J Surg Pathol* 13: 581–587, 1989.

161. Sprince NL, Chamberlain RI, Hales CA, Weber AL, Kazemi H: Respiratory disease in tungsten carbide production workers. *Chest* 86: 549–557, 1984.

162. Cugell DW, Morgan WKC, Perkins A, Rubin A: The respiratory effects of cobalt. *Arch Intern Med* 150: 177–183, 1990.

163. Davison AG, Haslam PL, Corrin B, et al: Interstitial lung disease and asthma in hard-metal workers: Bronchoalveolar lavage, ultrastructural, and analytical findings and results of bronchial provocation tests. *Thorax* 38: 119–128, 1983.

164. Shirakawa T, Kusaka Y, Fujimura N, Kato M, Heki S, Morimoto K: Hard metal asthma: Cross immunological and respiratory reactivity between cobalt and nickel. *Thorax* 45: 267–271, 1990.

165. Demedts M, Gheysens B, Nagels J, et al: Cobalt lung in diamond polishers. *Am Rev Respir Dis* 139: 130–135, 1984.

166. Stewart MJ, Faulds JS: The pulmonary fibrosis of haematite miners. *J Pathol Bacteriol* 39: 233–253, 1934.

167. Faulds JS, Nagelschmidt GS: The dust in the lungs of haematite miners from Cumberland. *Ann Occup Hyg* 4: 255–263, 1962.

168. Report of the International Committee on Nickel Carcinogenesis in Man. *Scand J Work Environ Health* 16: 1–82, 1990.

169. Hodgson JT, Jones RD: Mortality of a cohort of tin miners 1941–86. *Br J Ind Med* 47: 665–676, 1990.

170. Robertson AJ, Rivers D, Nagelschmidt G, Duncumb P: Stannosis: Pneumoconiosis due to tin oxide. *Lancet* 1: 1089–1095, 1961.

171. Elo R, Maatta K, Uksila E, Arstila AU: Pulmonary deposits of titanium dioxide in man. *Arch Pathol* 94: 417–424, 1972.

172. Redline S, Barna BP, Tomashefski J, Abraham JL: Granulomatous disease associated with pulmonary deposition of titanium. *Br J Ind Med* 43: 652–656, 1986.

173. Liddell D, Miller K: *Mineral Fibers and Health.* Boca Raton, FL, CRC Press, 1991.

174. Kohyama N: Airborne asbestos fibre levels in non-occupational environments in Japan, in Bignon J, Peto J, Saracci R (eds), *Non-occupational Exposure to Mineral Fibres.* Lyon, IARC Scientific Publications No. 90, 1989, pp 262–276.

175. Pooley FD, Mitha R: Fibre types, concentrations, and characteristics found in lung tissues of chrysotile-exposed cases and controls, in Wagner JC (ed), *The Biological Effects of Chrysotile.* Philadelphia, JB Lippincott, 1986, pp 1–11.

176. Wagner JC, Chamberlain M, Brown RC, et al: Biological effects of tremolite. *Br J Cancer* 45: 352–360, 1982.

177. Zielhuis RL: *Public Health Risks of Asbestosis.* Oxford, Pergamon Press, 1977.

178. Wagner JC, Berry G, Skidmore JW, Timbrell V: The effects of the inhalation of asbestos in rats. *Br J Cancer* 29: 252–269, 1974.

179. Davis JMG, Beckett ST, Bolton RE, Collings P, Middleton AP: Mass and number of fibres on the pathogenesis of asbestos-related lung disease in rats. *Br J Cancer* 37: 673–688, 1978.

180. Pooley FD: An examination of the fibrous mineral content of asbestos lung tissue from the Canadian chrysotile mining industry. *Environ Res* 12: 281–298, 1976.

181. Rowlands N, Gibbs GW, McDonald AD: Asbestos fibres in the lungs of chrysotile miners and millers: A preliminary report. *Ann Occup Hyg* 26: 411–415, 1982.

182. Churg A: Fiber size and number in workers exposed to processed chrysotile asbestos, and members of the general population. *Am J Ind Med* 9: 143–152, 1986.

183. Morgan A, Talbot RJ, Holmes A: Significance of fibre length in the clearance of asbestos fibres from the lung. *Br J Ind Med* 35: 146–153, 1978.

184. Davis JMG: Mineral fibre carcinogenesis: Experimental data relating to the importance of fibre type, size, deposition, dissolution and migration, in Bignon J, Peto J, Saracci R (eds), *Non-occupational Exposure to Mineral Fibres.* Lyon, IARC Scientific Publications No. 90, 1989, pp 33–45.

185. Davis JMG, Addison J, Bolton RE, Donaldson K, Jones AD, Smith T: The pathogenicity of long versus short fibre samples of amosite asbestos administered to rats by inhalation and intraperitoneal injection. *Br J Exp Pathol* 67: 415–430, 1986.

186. Wagner JC, Skidmore JW, Hill RT: Griffiths DM: Erionite exposure and mesotheliomas in rats. *Br J Cancer* 51: 727, 1985.

187. Davis JMG, Jones AD, Parker I: Experimental studies in rats on the effects of asbestos inhalation coupled with the inhalation of titanium dioxide, in *Proceedings of the VIIth International Pneumoconioses Conference 1988,* DHHS (NIOSH) Publication No. 90-108, 1990, pp 159–162.

188. McFadden D, Wright JL, Wiggs B, Churg A: Smoking inhibits asbestos clearance. *Am Rev Respir Dis* 133: 372–374, 1986.

189. Merewether ERA, Price CW: Report on effects of asbestos dusts in the lungs and dust suppression in the asbestos industry. London, HMSO, 1930.

190. McMillan GHG, Pethybridge RJ, Sheers G: Effect of smoking on attack rates of pulmonary and pleural lesions related to exposure to asbestos dust. *Br J Ind Med* 37: 268–272, 1980.

191. Kilburn KH, Lilis R, Anderson HA, Miller A, Warshaw RH: Interaction of asbestos, age, and cigarette smoking in producing radiographic evidence of diffuse pulmonary fibrosis. *Am J Ind Med* 80: 377–381, 1986.

192. Berry G, Gilson JC, Holmes S, Lewinsohn HC, Roach SA: Asbestosis: A study of dose-response relationships in an asbestos textile factory. *Br J Ind Med* 36: 98–112, 1979.

193. Hnizdo E, Sluis-Kremer GK: Effect of tobacco smoking on the presence of asbestosis at post mortem and the reading of irregular opacities on roentgenograms in asbestos-exposed workers. *Am Rev Respir Dis* 138: 1207–1212, 1988.

194. Weiss W: Cigarette smoking, asbestos and small irregular opacities. *Am Rev Respir Dis* 130: 293–301, 1984.

195. Friedman AC, Fiel SB, Fisher MS, Rudecki PD, Lev-Touff AS, Caroline DF: Asbestos related pleural disease and asbestosis: A comparison of CT and chest radiography. *Am J Roentgenol* 150: 269–275, 1988.

196. Craighead JE, Abraham JL, Churg A, et al: The pathology of asbestos associated diseases of the lungs and pleural cavities: Diagnostic criteria and proposed grading schema. *Arch Pathol Lab Med* 106: 544–596, 1982.

197. Roggli VL: Pathology of human asbestosis: A critical review, in Fenoglio-Preiser CM (ed), *Advances in Pathology,* vol 2. Chicago, Year Book Medical Publishers, 1989, pp 31–60.

198. Hinson KFW, Otto H, Webster I, Rossiter CE: Criteria for the diagnosis and grading of asbestosis, in Bogovski P, Gilson JC, Timbrell V, Wagner JC (eds), *Biological Effects of Asbestos.* Lyon, IARC Scientific Publications No. 8, 1973, pp 54–57.

199. Roggli VL: Asbestos bodies and nonasbestos ferruginous bodies, in Roggli VL, Greenberg SD, Pratt PC (ed), *Pathology of Asbestos-Associated Diseases.* Boston, Little, Brown, 1992, pp 39–76.

200. Crouch E, Churg A: Ferruginous bodies and the histologic evaluation of dust exposure. *Am J Surg Pathol* 8: 109–116, 1984.

201. Morgan A, Holmes A: The enigmatic asbestos body: Its formation and significance in asbestos-related disease. *Environ Res* 38: 283–292, 1985.

202. Pooley FD, Ransome DL: Comparison of the results of asbestos fibre dust counts in lung tissue obtained by analytical electron microscopy and light microscopy. *J Clin Pathol* 39: 313–317, 1986.

203. Gibbs AR, Pooley FD, Griffiths DM, Mitha R: Silica and silicate pneumoconioses—A pathological and mineralogical study. *Ann Occup Hyg* 38(suppl 1): 851, 1994.

204. Joines RW, Roggli VL: Dendriform pulmonary ossification: Report of two cases with unique findings. *Am J Clin Pathol* 91: 398–402, 1989.

205. Churg A, Wright JL: Small airway lesions in patients exposed to nonasbestos mineral dusts. *Hum Pathol* 14: 688–693, 1983.

206. Gaensler EA, Jederlinic PJ, Churg A: Idiopathic pulmonary fibrosis in asbestos-exposed workers. *Am Rev Respir Dis* 144: 689–696, 1991.

207. Hammar SP: Controversies and uncertainties concerning the pathologic features and pathologic diagnosis of asbestosis. *Semin Diag Pathol* 9: 102–109, 1992.

208. Doll R: Mortality from lung cancer in asbestos workers. *Br J Ind Med* 12: 81–86, 1955.

209. Breslow L: Industrial aspects of bronchogenic neoplasms. *Dis Chest* 28: 421–430, 1955.

210. Hammond EC, Selikoff IJ, Seidman IJ: Asbestos exposure, cigarette smoking and death rates. *Ann NY Acad Sci* 330: 473–490, 1979.

211. Kipen HM, Lilis R, Suzuki Y, Valciukas JA, Selikoff IJ: Pulmonary fibrosis in asbestos insulation workers with lung cancer: A radiological and histopathological evaluation. *Br J Ind Med* 44: 96–100, 1987.

212. McDonald JC, McDonald AD: Epidemiology of asbestos related lung cancer, in Antman K, Aisner J (eds), *Asbestos Related Malignancy.* Orlando, FL, Grune and Stratton, 1987, pp 57–79.

213. McDonald, JC, Liddell FDK, Gibbs GW, Eyssen GE, McDonald AD: Dust exposure and mortality in chrysotile mining, 1910–75. *Br J Ind Med* 37: 11–24, 1980.

214. Hughes JM: Epidemiology of lung cancer in relation to asbestos exposure, in Liddell D, Miller K (eds), *Mineral Fibers and Health.* Boca Raton, FL, CRC Press, 1991, pp 135–145.

215. Churg A: Lung cancer cell type and asbestos exposure. *JAMA* 253: 2984–2988, 1985.

216. Johansson L, Albin M, Hakobsson K, Mikoczy Z: Histological type of lung carcinoma in asbestos cement workers and matched controls. *Br J Ind Med* 49: 626–630, 1992.

217. Newhouse ML, Sullivan KR: A mortality study of workers manufacturing friction materials: 1941–86. *Br J Ind Med* 46: 176–179, 1989.

218. Sluis-Cremer GK, Bezuidenhout BN: Relation between asbestosis and bronchial cancer in amphibole asbestos miners. *Br J Ind Med* 46: 537–540, 1989.

219. Hughes JM, Weill H: Asbestosis as precursor of asbestos related lung cancer: Results of a prospective mortality study. *Br J Ind Med* 48: 229–233, 1991.

220. Davis JMG, Cowie HA: The relationship between fibrosis and cancer in experimental animals exposed to asbestos and other fibres. *Environ Health Perspect* 88: 305–309, 1990.

221. Waxweiler RJ, Zumvalde RD, Ness GO, Brown DP: A retrospective cohort mortality of males mining and milling attapulgite clay. *Am J Ind Med* 13: 305, 1988.

222. Huuskonen MS, Tossvainen A, Koskinen H, et al: Wollastonite exposure and lung fibrosis. *Environ Res* 30: 291–304, 1983.

223. Baris YI, Simonato L, Artvinli M, et al: Epidemiological and environmental evidence of the health effects of exposure to erionite fibres: A four years study in the Cappadocian region of Turkey. *Int J Cancer* 39: 10–17, 1987.

224. Wagner JC, Gibbs AR: Diseases due to synthetic mineral fibres, in Churg A, Green FHY (eds), *Pathology of Occupational Lung Disease.* New York, Igaku Shoin, 1988, pp 327–330.

225. Meek ME: Lung cancer and mesothelioma related to man-made mineral fibres: The epidemiological evidence, in Liddell D, Miller K (eds), *Mineral Fibers and Health.* Boca Raton, FL, CRC Press, 1991, pp 175–186.

226. Hesterberg TW, Mast R, McConnell EE, et al: Chronic inhalation toxicity of refractory ceramic fibres in Syrian hamsters, in Brown RC, Hoskins JA, Johnson NF (eds), *Mechanisms of Fiber Carcinogenesis.* New York, Plenum Press, 1990, pp 531–538.

227. Sherwood Burge P: Occupational asthma, in Brewis RAL, Gibson GJ, Geddes DM (eds), *Respiratory Medicine.* London, Bailliere Tindall, 1990, pp 704–721.

228. Chan-Yeung M, Malo JL: Aetiological agents in occupational asthma. *Eur Respir J* 7: 346–371, 1994.

229. Mapp CE, Saetta M, Maestrelli P, et al: Mechanisms and pathology of occupational asthma. *Eur Respir J* 7: 544–554, 1994.

230. Elwood PC, Pemberton J, Merrett JD, Carey GCR, McAuley IR: Byssinosis and other respiratory symptoms in flax workers in Northern Ireland. *Br J Ind Med* 22: 27–37, 1965.

231. Cinkotai FF, Emo P, Gibbs ACC, Caillard J-F, Jouany J-M: Low prevalence of byssinotic symtoms in 12 flax scutching mills in Normandy, France. *Br J Ind Med* 45: 325–328, 1988.

232. Berry G, Molyneux MKG, Tombleson JBL: Relationships between dust levels and byssinosis and bronchitis in Lancashire cotton mills. *Br J Ind Med* 31: 18–27, 1974.

233. Roach SA, Schilling RSF: A clinical and environmental study of byssinosis in the Lancashire cotton industry. *Br J Ind Med* 17: 1–9, 1960.

234. Edwards C, Carlile A, Rooke G: The larger bronchi in byssinosis: A morphometric study. *J Clin Pathol* 37: 20–22, 1984.

235. Pratt PC, Vollmer RT, Millar JA: epidemiology of pulmonary lesions in nontextile and cotton textile workers: A retrospective analysis. *Arch Environ Health* 35: 133–138, 1980.

236. Emanuel DA, Wenzel FJ, Lawton BR: Pulmonary mycotoxicosis. *Chest* 67: 293–297, 1975.

237. doPico GA: Health effects on organic dusts in the farm environment, Report on diseases. *Am J Ind Med* 10: 261–265, 1986.

238. Lecours R, Laviolette M, Cormier Y: Bronchoalveolar lavage in pulmonary mycotoxicosis (organic dust toxic syndrome). *Thorax* 41: 924–926, 1986.

Chapter *16*

Sarcoidosis

M. Sarno / P. S. Hasleton / M. A. Spiteri

What is sarcoidosis?—"It is a riddle wrapped in a mystery, inside an enigma."

W. S. Churchill, 1951

Sarcoidosis is a multisystem granulomatous disorder whose etiology remains an enigma. This disease commonly affects young adults and frequently presents with bilateral hilar lymphadenopathy with or without pulmonary infiltration or ocular or cutaneous lesions. The diagnosis is firmly established when well-recognized clinical and radiographic findings are supported by histologic evidence of discrete noncaseating epithelioid cell granulomata in one or more organs. In addition, a positive Kveim-Siltzbach skin test and cutaneous anergy to tuberculoprotein are often present. Immunologic features suggest aberrant cell-mediated reactions at the site of inflammation, in the presence of hypergammaglobulinemia.[1] The disease is usually self-limiting with spontaneous resolution; however, in a few patients a progressive downhill course culminates in irreversible fibrosis and severe impairment of organ function.

This description of sarcoidosis* has been known for over 100 years. Early reports on sarcoidosis focused on a number of independently described manifestations occurring in various tissues and organs. By the beginning of this century, it became apparent that sarcoidosis was a distinct entity, with a mysterious etiology, giving rise to a wide spectrum of immunologic and clinical expressions that shared a common histologic pattern and that could coexist in the same patient.[2]

To amalgamate the observations of various sarcoid research groups, the first world conference was held in 1958 at the Brompton Hospital, London.[3] Such meetings have continued to the present day. In particular, these conferences have witnessed the promotion of the use of immunologic techniques in diagnosing and monitoring the disease; the recognition of the importance of the cellular and noncellular components of bronchoalveolar lavage (BAL) fluid in the pathogenesis, as well as the advent of monoclonal antibody probes and molecular technology to investigate immune effector cells in BAL and tissue biopsies.

These innovations have made a tremendous impact on the current understanding of the underlying immunopathologic mechanisms operative in sarcoidosis. For although the cause of sarcoidosis remains unknown, they have provided fresh insights into the disease. It is now clear that sarcoidosis is *not* a disorder mediated by a generalized impaired cellular immunity as early reports tended to suggest. Its protean clinical expressions are a reflection of localized aberrant immunologic responses (involving both lymphoid and nonlymphoid cellular elements), to an as yet unknown stimulus or stimuli. These give rise to a granulomatous inflammation in one or more organs.

POSSIBLE ETIOLOGIC FACTORS

Despite extensive research there is no identifiable etiologic agent to account for the granulomata that characterize sarcoidosis. Infectious agents, chemicals and drugs, allergy, autoimmunity, and genetic factors have all been explored as potential causes.

Most studies have focused on an infectious etiology. Many workers have tried to show that sarcoidosis is an aberrant form of tuberculosis.[4,5] In the most complete summary available to date on the evidence for and against tuberculosis as an etiologic factor in sarcoidosis, Siltzbach concluded that tuberculosis did not appear to give rise to or protect against sarcoidosis.[6] Attempts to fulfill Koch's postulates by isolating the mycobacteria from sarcoid tissue samples,[7] or by inducing clinical tuberculosis with injection of sarcoid tissue into laboratory animal hosts[8] have failed. A search for circulating antibodies to *Mycobacterium tuberculosis* and to other mycobacteria in sarcoid patients has also proved inconclusive.[9]

One study, on a limited number of patients with systemic sarcoidosis, detected acid-fast coccobacillary forms within the biopsy material.[10] It was postu-

---

*Derived from the Greek *sarkos* meaning "flesh," hence the word "sarcoid" for "resembling flesh."

lated that the cell wall-deficient bacteria, possibly related to mycobacteria or corynebacteria, may be the causative agents in some sarcoid patients. The recent finding of mycobacterial nucleic acids, albeit in variable proportions of patients with sarcoidosis, suggests that mycobacterial antigens might play a role in the pathogenesis of sarcoid disease in this small group of patients.[11–13]

Although nondiphtheroid corynebacteria may produce a recognizable clinical syndrome resembling sarcoidosis,[14] such case reports remain only interesting incidental observations because they have not been, in general, successfully related to the etiology of sarcoidosis.

Mycobacterial and viral infections may interact to produce sarcoidosis.[15] The T- and B-lymphocyte disturbance in sarcoid patients might be due to the effects of a viral infection depressing T-cell function, and of simultaneous mycobacterial infection stimulating B-cell function. Other investigators have hypothesized that sarcoidosis could follow bacille Calmette-Guérin (BCG) vaccination or a tuberculous infection in a person with a virus-induced T-cell defect.[16,17] Sporadic reports have been made of virus isolation, such as mumps, influenza, parainfluenza, Newcastle agent, and measles virus particles in patients with sarcoidosis.[3,16] These have, however, been subsequently dismissed as possible laboratory contaminants.[18] High antibody titers to a variety of viruses (e.g., Epstein-Barr virus) have also been found in sarcoid patients.[19] On the other hand, it has been suggested that organisms found intermittently in sarcoid tissues are able to survive there because of the altered immunity in such patients.

Despite the above random claims, tissue cultures and electron microscopy have failed to uncover any specific infectious agent in sarcoid patients. Examination of the suspended sarcoidal tissue used in the Kveim test has so far not revealed the nature of the responsible agent.[20]

In view of the immunologic features found in sarcoidosis, it has been hypothesized that sarcoidosis may represent a form of hypersensitivity to inhalation of environmental organic antigens. Inhalation of pine pollen and peanut dust,[21] clay soil,[22] talc,[23] and secondary oxalosis[24] have all been incriminated as contributory regional factors in different areas; their role in the pathogenesis of sarcoidosis remains unclear.

Exhaustive skin testing with metals and other inorganic substances in sarcoid patients and controls have not revealed any peculiar hypersensitivity to chemicals. Beryllium[25] and zirconium[26] produce granulomata in the sensitized individual. Although such granulomata are found at the injection site 1 month after inoculation, and are histologically similar to those produced in the Kveim response in sarcoid patients, each skin test is specific for its own disorder, and there is no overlap. The common denominator to such dissimilar stimuli appears to be *susceptible* individuals, with *heightened* local cellular-mediated immune response to known (or unknown) antigen or antigens.

The occurrence of sarcoidosis in members of the same family has suggested that genetic factors might be involved,[27,28] but no firm relationship has been demonstrated. Some workers have suggested an autosomal recessive inheritance.[29] Sarcoidosis has been reported to be more common in monozygotic rather than heterozygotic twins.[30] Various features of sarcoidosis may be associated with specific antigens of major histocompatibility loci. HLA-B8 has been associated with erythema nodosum[31] and arthritis,[32] whereas HLA-B27 is found in a high percentage of patients with uveitis.[33] Persson and colleagues reported a statistically significant increase of HLA-B7 in 47 patients with negative reactivity to tuberculin.[34] In contrast, the frequency of this allele was zero in the group with positive response to tuberculin.

The most interesting HLA association in sarcoidosis has been that of B8, DR3 in patients with acute sarcoidosis.[35] Kremer suggested that B8, DR3 phenotype identifies a group of patients who are more likely to develop acute sarcoid arthritis and hilar adenopathy, progressing to chronic disease.[36] Others have shown an association of HLA-B8 with early resolution of the disease.[37] Patients expressing HLA-B13 are more prone to have persistent disease.[38] Despite such findings there is as yet no definitive evidence for the "linkage" of the HLA-associated alleles with disease susceptibility gene(s) in sarcoidosis.

In conclusion, the etiology of sarcoidosis remains elusive. It appears that genetic, environmental, nutritional, and socioeconomic factors could play a critical governing role in the development of sarcoidosis in an immunologically susceptible individual. Additional support for this view may be gained by the observation that nothing similar to sarcoidosis is seen in the veterinary world! Nonspecific local sarcoid reactions are seen in some animals, but these appear to be different from the generalized multisystem disease seen in human beings.[39]

CLINICAL PRESENTATION

Sarcoidosis is a relatively common disease, occurring worldwide with varying incidence and prevalence. In Europe, its frequency ranges from 3 to 500 cases per 100,000 population, usually affecting the 20- to 40-year age group.[40] There is no sex predominance although in a few studies more women are affected.[41] Although certain geographic locations have a high prevalence of the disease, there is no identifiable or etiologically significant pattern.[42] Sarcoidosis is more prevalent and probably more severe in certain populations. The prevalence in the caucasian population in the United States is about 5/100,000 in contrast to about 40/100,000 in the African American population.[43] The disease also seems to be more severe, chronic, and disabling in the African American population. In support of these observations is the clinical finding that caucasian patients with sarcoidosis are more likely to present with acute symptoms, which are often associated with a good response.[44]

Approximately 20 to 40 percent of patients are symptom free and their disease is usually discovered by routine chest radiograph. Another 25 percent of patients seek medical advice because of cough and dyspnea, while an additional 25 percent present with eye, skin, or nasal complaints. Constitutional symptoms such as fever, fatigue, malaise, anorexia, and weight loss are usually absent or mild; yet in 20 to 30 percent of patients with sarcoidosis these systemic features are striking. In most patients, all manifestations of the disease disappear within a few months or years. In 10 to 15 percent of sarcoid patients the disease is progressive, with ensuing major chronic disability. In this latter course, extrathoracic features are prominent. These may range from skin lesions to peripheral lymphadenopathy, parotid enlargement, central nervous system involvement, cardiac syndromes, hepatosplenomegaly, arthralgia, hypercalcemia, and nephrocalcinosis. These features appear to be more severe in elderly patients.[45,46] The proportions of patients presenting with various manifestations are, however, likely to be affected by the way the epidemiologic data are collected. The British Thoracic and Tuberculosis Association survey of four areas in Great Britain (1969) gives the most reliable estimate of relative frequency of the various ways in which patients with sarcoidosis first come to medical notice.[47] The protean clinical manifestations and course of sarcoidosis lead to considerable inherent variability within the patient population as a whole, as well as in the same individual under evaluation.

PULMONARY MANIFESTATIONS

The lung is the organ most commonly involved in sarcoidosis. At least 90 percent of patients with sarcoidosis exhibit abnormalities on their chest radiograph during the course of their disease.[48] Twenty to 25 percent of these patients have a permanent loss of lung function; 5 to 10 percent die from complications of the disease.[49] Patients are commonly asymptomatic or their respiratory symptoms may start insidiously with dry cough, progressive dyspnea, exercise intolerance, and chest pain.[50] Lung function typically shows a decrease in diffusion capacity, with or without a loss in lung volume.[49]

The type and degree of the physiologic abnormality relates to the location and the extent of the pathology.[51,52]

The lung is probably one of the first sites of the body involved. The inflammatory process seems to extend through the lymphatics to the hilar and mediastinal nodes (over 85 percent of patients with pulmonary sarcoidosis have radiographically apparent mediastinal and hilar lymph node involvement). As in miliary tuberculosis, lymphohematogenous spread may then occur throughout the lung as well as to other commonly affected organs, such as the liver.[45]

The clinical course of the pulmonary sarcoidosis may be related, at least in part, to the radiologic appearance of the disease.[53–55] Patients with pulmonary sarcoidosis have been divided into clinical groups according to the appearance of the chest film, ranging from the more common, usually asymptomatic, bilateral hilar lymphadenopathy without parenchymal involvement[56] (Fig. 16-1) to diffuse dense progressive and irreversible parenchymal fibrosis[42–48] (Fig. 16-2) with respiratory failure.[57] Using this radiologic staging, Siltzbach noted a clear relationship between the prognosis of sarcoidosis and the initial radiologic appearance as well as associated extrathoracic features.[58]

STAGING OF DISEASE ACTIVITY

Prognostic signs and indicators of disease activity are necessary to determine the need for treatment. Such measures have included clinical, radiographic, pul-

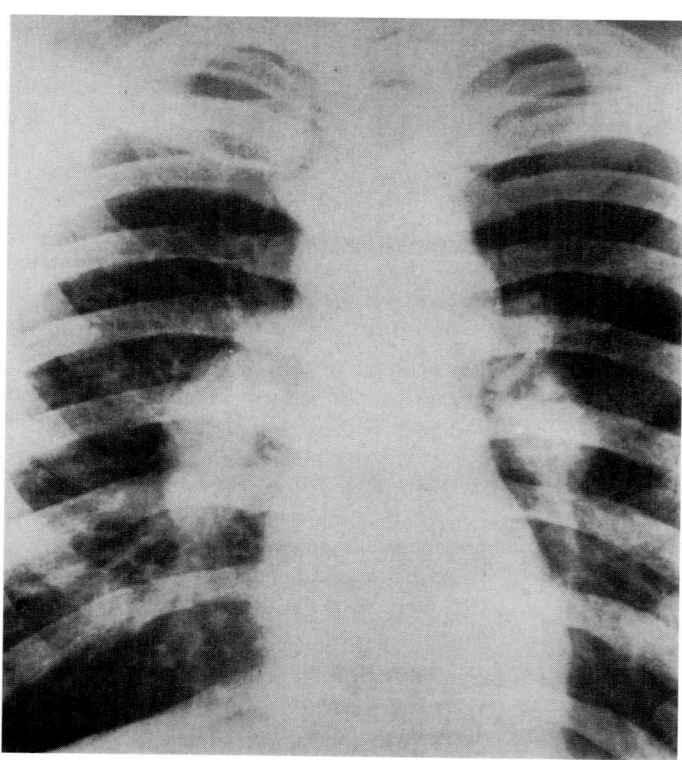

FIGURE 16-1
Chest radiograph shows bilateral hilar lymphadenopathy as well as enlargement of nodes in both paratracheal regions. The enlarged hilar nodes characteristically stand away from the cardiac borders. The lung fields are virtually clear.

monary, and physiologic parameters. Unfortunately, although general guidelines have been established, these measurements are often unreliable in predicting the course of individual patients.

Some reports have mistakenly interpreted activity in terms of the stage of granuloma formation (recent or old) or the extent of the disease.[59] The term "activity" has also been applied to reflect the evolution and outcome of the disease. However, in practice in particular groups of affected patients, although the disease is often graded as intense by certain indices, the majority of the patients recover spontaneously within a short time and without treatment.[60] Another aspect to be considered is the multisystem involvement of sarcoidosis. In practice, patients who have intense granulomatous inflammation in one organ do not necessarily have serious systemic involvement. Studies have shown that the old concept of a single or multiple extrathoracic organ involvement, as an indicator of poor prognosis, remains largely unconfirmed.[61]

Recent guidelines on the management of sarcoidosis suggest that routine tests to stage disease activity can be limited to clinical indices including worsening respiratory symptoms, deterioration of lung function or chest radiography.[62] Optionally, in a select clinical setting, the following may also be useful: biochemical markers such as serum angiotensin-converting enzyme (ACE), gallium 67 scanning, high-resolution computed tomography, BAL cell populations, and CD4/CD8 ratio.[62]

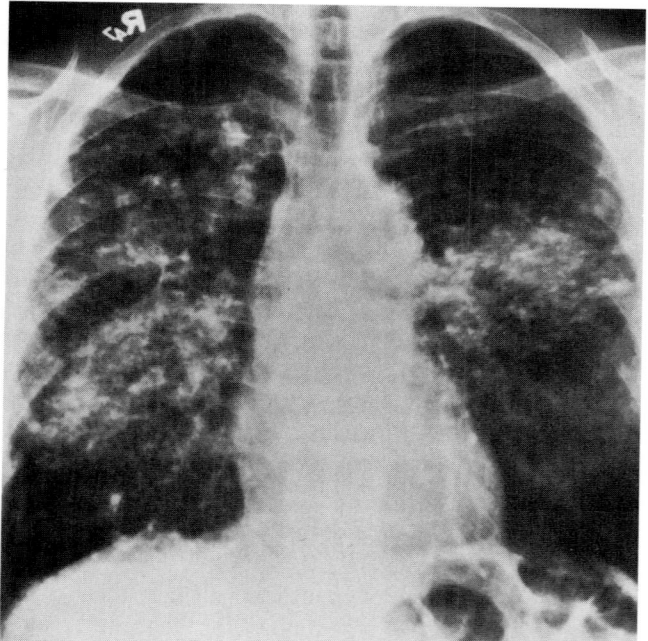

FIGURE 16-2
Chest radiograph shows diffuse parenchymal infiltrates.

510

BRONCHOALVEOLAR LAVAGE

Bronchoalveolar lavage is a simple extension of routine fiberoptic bronchoscopy. It permits a repeatable, safe, and quantitative evaluation of the inflammatory and immune processes in the alveolar structures.[63] Analysis of the cellular constituents and biochemical supernatant measurements in BAL fluid can provide a dynamic and kinetic impression of events in the lower respiratory tract in health and disease.[64] A fuller discussion of BAL is given in Chap. 12.

Validity of Assessing Pulmonary Sarcoidosis by BAL

Before the widespread use of BAL, an understanding of the inflammatory processes in this disease was based on a paucity of pathologic material. Now several lines of evidence support the view that the cells and soluble products recovered by segmental lavage may be used to assess the inflammatory events occurring in the lung parenchyma in this disorder.

First, the types and numbers of the cells retrieved by BAL are similar to those found in lung biopsies from the same sarcoid patients.[65-67] They parallel the severity of the acute illness resolving as the disease abates or responds to treatment.[68] This does not discount the possibility that the inflammatory response in the alveolar air space may not be in phase with that in the interstitium.

Secondly, the cellular constituents obtained by BAL from sarcoid patients are "immunocompetent"

in terms of proliferation to mitogenic and antigenic stimulation, as well as in the production of immunologically active mediators. Therefore, these cells appear to be part of the local immune apparatus of the lung.

Finally, the lung in sarcoidosis appears to be the site of a compartmentalized inflammatory response, not reflected in the peripheral blood.[69] Critics of the lavage technique maintain that because lavage cannot sample all of the lung tissue, segmental BAL findings (at best) can only be regarded as a partial representation of the actual disease pattern.[70]

A diagram of the proportions as well as the total numbers of cells seen in normal and sarcoid patients is given in Fig. 16-3.

BAL Analysis Correlated to Other Indices of Sarcoid Activity

A number of investigations have shown BAL may be superior to any clinical, radiologic, and physiologic parameters in assessing the present status of the alveolitis and subsequent events in any individual patient.[71-76] These studies suggested that the degree of lymphocytosis in the lavage fluid was predictive of disease activity and progression. These observations have not been confirmed[77-79] and it is now less widely accepted that a high initial lymphocyte count in BAL reflects a poor prognosis.

Although Bjermer and colleagues agree with this latter view, they have also observed that a BAL lymphocytosis present on two successive investigations

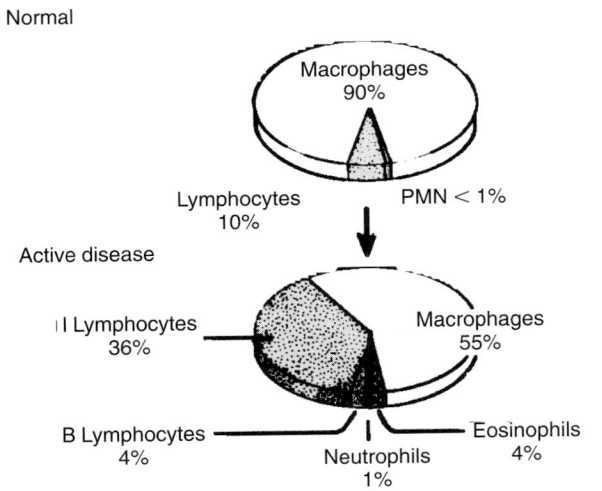

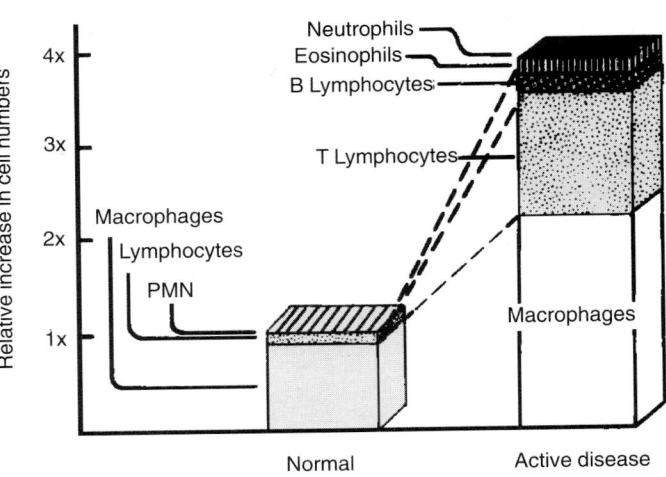

A   *B*

FIGURE 16-3
A. Diagram of proportion of effector cells, which are different in sarcoid from those normally observed.   *B.* This diagram shows the increase in the total number of effector cells but to a different extent for each cell type.

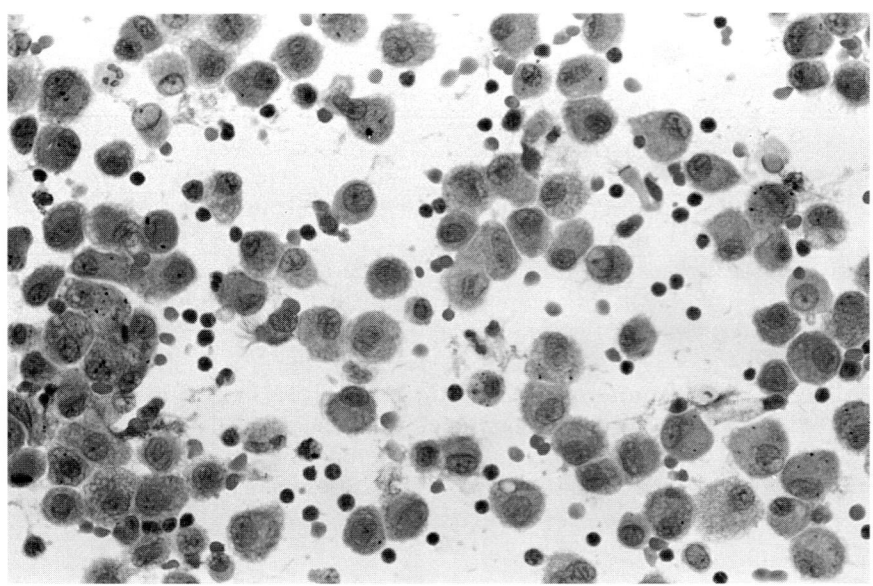

FIGURE 16-4
BAL cytospin in sarcoid shows a lymphocytosis.

tends to correlate well with a deterioration of lung function.[78] They suggest serial BAL measurements are a more reliable estimate of disease activity. Ward and associates have proposed the observed disparity between studies could be partly due to genetic and racial differences, as well as the heterogeneity within study populations.[79] In addition, they found that an acute onset of sarcoidosis with a high lymphocytosis in BAL offered a better prognosis (Fig. 16-4) than disease of more insidious onset with low lavage lymphocyte counts. Foley and coworkers concluded that the presence of a high BAL lymphocytosis could indicate a favorable prognosis for the outcome of lung function in pulmonary sarcoidosis even in patients with chronic disease.[80] This suggests that the absence of lymphocytosis in lavage does not necessarily indicate burnt out disease. In patients with more extensive long-standing radiographic shadows, BAL neutrophils may be as important as lymphocytes in the assessment of disease activity, perhaps serving as an indicator to therapy.[81]

Inconsistent correlations are found in the literature relating to the proportion of lavage T cells as recovered by BAL and the amount of ^{67}Ga uptake by the lung parenchyma.[82–84] This inconsistency may be due to the possibility that, although BAL fluid analysis mirrors the alveolitis, gallium uptake reflects granuloma load. Thus, the two parameters measure different factors. Of these two investigations, BAL is now believed to be more sensitive and to yield more information, at least in the initial stages of the disease.[85,86] Several studies have shown good agreement between the cellular findings in BAL fluid

and those in lung biopsy specimens of the same patients.[87–89] Data also suggest that the delineation of specific lavage lymphocyte and macrophage subpopulations could serve as better parameters in assessing the current degree of lung inflammation and in predicting prognosis in a given patient. A close relationship between the raised ratio of helper CD4+/suppressor CD8+ T lymphocytes and more progressive disease has been reported.[73] A correlation between phenotypically distinct macrophage subsets and radiographic staging has been identified.[90]

PATHOGENESIS

Many studies have aimed at determining how the immune aberrations in sarcoid fit the current model of immune stimulation. Such immune phenomena have been observed both systemically and locally in the lung.

Lung Cellular Immune Events

Early disease in sarcoid patients is characterized by a mononuclear cell infiltrate comprised of macrophages and T lymphocytes.[91,92] This lymphocytic response distinguishes sarcoidosis from other interstitial lung diseases such as fibrosing alveolitis, where polymorphonuclear cells usually predominate in early disease.[91] Furthermore, the presence of a mononuclear cell infiltrate is an important clue that immune mechanisms are involved.

LYMPHOCYTES

The lymphocytes associated with granulomata are larger than usual and have the morphologic features of *activated* cells. The accumulation of lymphocytes around the granulomata is more prominent within fresh, active lesions and tends to diminish as the granulomata mature. These lymphocytes are mainly of the helper T-cell type, which express CD4+ surface antigens. As the lesions become less active these cells decrease, and suppressor CD8+ T cells predominate.[67] A similar switch in immunoregulatory T cells has also been seen in animal models of granulomatous lung disease. This process may modulate granuloma formation.[93,94]

Such changes within the lung interstitium are accompanied by an increase in the proportion and total number of lymphocytes in BAL fluid. This increase consists mainly of T-helper cells. The helper/suppressor T-lymphocyte ratio in active sarcoid lavage is 4 to 10 times greater than in normal BAL,[95] suggesting an immunoregulatory imbalance. Functional studies using these two lymphocyte subsets agree with their phenotypic expression.

The lavage T lymphocytes obtained from patients with active sarcoidosis also appear to be *activated* both morphologically and functionally. These cells show an augmented response to external stimuli such as lectins and antigens.[96] They exhibit active proliferation and spontaneously release various biologically active substances called lymphokines. These influence the migration and function of macrophages as well as other target cells. These T cells also possess markers of cell activation including the capacity to form stable E rosettes at 37°C, the expression of class II major histocompatibility complex (MHC) antigens on their surface, and the acquisition of the receptors from interleukin-2 (IL-2 or Tac antigen).[97] Activation of such T cells is further supported by flow cytometry analysis of DNA content, which shows a subpopulation of T lymphocytes in the proliferative phase of their cell cycle (S/G2).[98]

Separate in vitro studies[99,100] have demonstrated that lung T lymphocytes from active sarcoid patients spontaneously secrete IL-2, the glycoprotein that stimulates proliferation in responsive T cells. IL-2 is a central mediator, which has been postulated to be involved in at least three interrelated events in the pathogenesis of sarcoidosis: (1) the stimulation of T-cell proliferation and the expansion of the number of T cells at the sites of inflammation, (2) the differentiation of T lymphocytes into effector cells that produce additional lymphokines, and (3) the recruitment of additional helper T cells from the peripheral blood. In support, T cells within inflamed tissue in sarcoidosis replicate faster than normal. In addition, raised levels of IL-2 have been found in active sarcoid BAL and are chemotactic for T cells.[98,101,102] Such findings may partly explain the expansion in number of *activated* T cells observed within the lung parenchyma and BAL in active sarcoidosis. It is postulated that these cells are deployed to amplify the immune inflammatory response within the lung interstitium.

The T lymphocytes recovered from active sarcoid BAL spontaneously release monocyte chemotactic factor (MCF) and monocyte migration inhibition factor (MIF).[103] Pulmonary T cells release approximately 25 times more MCF on a per cell basis than blood T cells from the same patients. This observation, together with findings that there are increased numbers of pulmonary T cells and decreased number of T cells in peripheral blood, suggests that a gradient of monocyte chemotactic activity exists between the lung and blood, the purpose of which is to attract monocytes to sites of disease. Such a recruitment of monocytes to the pulmonary interstitium may be an initial step in the assembly of granulomata. Other distinct lymphokines secreted by the same or a different subpopulation of T cells may then be responsible for the activation and differentiation of recruited monocytes into *activated* macrophages, giant cells, and epithelioid cells, all constituents of sarcoid granulomata. One of these mediators, interferon γ (IFNγ), is a potent activator of macrophages and has been shown to be liberated by lung T lymphocytes and alveolar macrophages in patients with active sarcoidosis.[104]

The as "yet unknown antigen(s)" in sarcoidosis is probably recognized through the T-cell antigen receptor (TCR) expressed on the surface of T lymphocytes. TCR is a heterodimer composed of two chains in which the combination of the constant and variable regions may be responsible for the different antigen specificity of T lymphocytes. In blood samples from healthy volunteers, the TCR is composed of an α and β chain (αβ T cells) with less than 10 percent composed to a γ and δ chain (γδ T cells).[105–107] In sarcoid BAL, the vast majority of T lymphocytes exhibit increased levels of mRNA transcripts for the β chain of TCR, with significant correlation to the enhanced expression of accessory molecules (such as HLA-DR and CD29) on alveolar macrophages.[108,109] These observations support the hypothesis that sarcoidosis is caused by persistent, exogenous, or self-specific antigens leading to cell-mediated immune responses. However, recent work has shown that although γδ T cells are increased in the peripheral blood of active long-term sarcoidosis, they are scarcely seen in BAL and in biopsy specimens, suggesting that these cells may not be directly related to granuloma formation.[110]

ALVEOLAR MACROPHAGES

Macrophages are necessary to initiate and maintain these T cell-mediated immune reactions that lead to granuloma formation. BAL samples, as well as lung biopsies, have recorded an increase in the total number of macrophages present in active sarcoid.[96] Studies suggest that these alveolar macrophages (AM) might serve as antigen-presenting cells and initiate the alveolitis seen in pulmonary sarcoidosis by presenting some as yet unidentified antigen(s) to locally resident and recruited T lymphocytes.[111] In support, two independent studies have shown that AM from active sarcoid patients have enhanced capacity to present recall antigen to T cells in vitro.[112,113] Furthermore, sarcoid AM have an enhanced expression of HLA-D region molecules, as well as increased HLA-DR membrane determinants.[114,115]

Sarcoid AM spontaneously secrete IL-1, which further stimulates the local T-cell population.[116] In addition, AM from active sarcoid patients have increased phagocytosis, as well as increased Fc receptor expression.[117,118] Increased IL-2 expression on sarcoid AM has been observed, and it is postulated that this could be involved in macrophage activation.[119]

Consistent phenotypic and functional aberrations exist within the lavage macrophage population in sarcoid patients.[120] Recent studies have shown that the presence of sarcoid inflammation in the lung induces specific changes within the local heterogeneous AM population, with the particular emergence of a specific AM subset whose suppression of T-cell proliferation is enhanced in sarcoidosis.[121,122] These macrophages also exhibit sarcoid-related differences in surface receptor expression and physiologic capabilities such as an increased expression of a separate antigen that identifies epithelioid cells in tissue and enhanced fibronectin content. Interestingly the increased emergence of this specific AM subset occurs in direct proportion to the degree of local lymphocytosis. It has been postulated that this marked rise in functionally specific "suppressor" AM in

active sarcoidosis is aimed at controlling the T-lymphocyte responses from potentially precipitating a cascade of immune events within the lung. In support Ainslie and associates noticed reduced levels of CD7 (a marker of blast formation in T cells) and HLA-DR expression of CD4+ T cells, at the same time as proportions of the "suppressor" type AM were increasing in sarcoid BAL.[90] It would follow, therefore, that such AM could arise as part of a secondary response to stimuli in the immediate milieu, to contain events arising from the initial macrophage–T cell interaction. With such a sophisticated inbuilt macrophage-mediated regulation of local T-cell responses, it would be expected that the sarcoid reaction would be self-limiting.[120,123] Indeed, this is seen clinically in the majority of patients with pulmonary sarcoidosis. In the relatively small number of patients who progress to fibrosis, reduced numbers of lymphocytes and increased neutrophils are found in the BAL. In addition there are increased levels of fibronectin, AM-derived growth factors, and enhanced expression of membrane fibronectin receptors on macrophages. The role of suppressor AM in these patients remains speculative.

Prior and associates have shown an increased and concerted release of cytokines [IL-1β, IL-6, tumor necrosis factor (TNF) α and granulocyte–monocyte stimulating factor (GMCSF)] with presumed monocyte/macrophage origin in sarcoid BAL. The primary drive for the release of these proinflammatory factors is unknown but appears not to be related to lymphocyte-derived cytokines like IFNγ or IL-2. In peripheral blood there is a deficit in the spontaneous production of IFNγ and neopterin. The release of IL-1β, IL-6 and GMCSF by sarcoid peripheral blood mononuclear cells correlates with neutrophil influx into the bronchoalveolar compartment. There appears to be no correlation between local or systemic cytokine release and the severity of the disease as assessed by radiographic staging and detailed lung function testing.[124]

Systemic Cellular Immune Events

Although lung T cells appear *activated* in sarcoidosis, the opposite is true in the peripheral blood where there is an absolute decrease in the number of circulating T cells. Functionally the response of peripheral lymphocytes to mitogen and recall antigen is partially impaired, and B-cell function (in vitro) is consequently depressed. Such findings in the peripheral blood are also accompanied by a depression in certain cutaneous delayed hypersensitivity responses.[96]

Many active sarcoid patients seem to have a partial to complete anergy to tuberculin purified protein derivative. This was one of the earliest immunologic abnormalities to be observed in sarcoidosis. Such skin anergy extended to certain antigens such as *Trichophyton,* mumps virus, streptokinase/streptodornase, *Candida,* as well as dinitrochlorobenzene. The incidence of anergy to tuberculoprotein and other antigens in sarcoid patients varies from 30 to 70 percent depending on the number of antigens used.[49,53,125] This anergy also appears to parallel disease severity, being especially marked when the sarcoidosis is clinically active and disappearing months to years after resolution of the disease. For a time, therefore, there appeared a puzzling paradox: as such, cutaneous anergy is usually associated with depressed cellular immunity. Lung studies have shown the presence of an active, if not overactive, immune response. The nature of such cellular anergy, like the etiology of the disease itself, remains an enigma.

To add to this paradox, although sarcoid patients do not respond to the above cutaneous antigens, they are uniquely capable of mounting a cutaneous response to a Kveim preparation.[89] These observations have raised the question of whether the proposed defect in cellular mediated immunity in sarcoidosis is in fact *real* or *apparent*.

First, the partial skin anergy observed in sarcoid patients is frequently associated with a reduction in the number and function of circulating T lymphocytes.[88] It has been suggested that the reduction of delayed-type skin test reactivity could reflect a depletion of the population of immunoreactive cells being sampled from the skin. In addition, studies of animal models of granulomatous disease suggest that there is an altered circulating pattern for T lymphocytes. In particular, T-helper cells become sequestrated in central lymphoid organs and sites of inflammation, thereby leaving a proportionately increased number of suppressor T cells in the peripheral blood.[20] In sarcoid patients, the ratio of helper/suppressor T cells in the blood may be normal or slightly decreased.[20] Other studies have also noted serum inhibitory factors that are partly derived from monocytes in the blood of active sarcoid patients. These factors have been shown to inhibit the in vitro response of normal lymphocytes to mitogens and antigens. Such circumstantial evidence has also been postulated to account for the observed suppressed function of sarcoid peripheral blood lymphocytes in vitro.[126,127]

Secondly, the mechanics of the Kveim response are not entirely clear. Although Kveim test lesions exhibit histopathologic similarities to sarcoid granulomata, there are marked differences between the Kveim reaction and the classic delayed-type hypersensitivity skin reaction. The most obvious difference is in the kinetics of the two reactions. The delayed-type hypersensitivity reaction begins at 8 h, peaks at 24 to 48 h, and resolves by 96 to 120 h. By contrast, the Kveim reaction takes 4 to 6 weeks to develop. Furthermore, no specific antigen in the Kveim preparation has been identified as the inducing agent in this response, and no convincing data have shown reproducibly that sarcoid lymphocytes are actively sensitized to the soluble or insoluble products in the Kveim material. Attempts to induce blood lymphocytes from sarcoid patients to proliferate or secrete lymphokines, after in vitro exposure to the Kveim material, have been unrewarding.[128] Thus, the nature of the stimulus and the reaction itself in the Kveim test remain the subjects of much controversy.

Finally, this apparent impairment of cellular immune function (in blood and skin) in sarcoid patients was noted to be similar but less profound than that in Hodgkin's disease or non-Hodgkin's lymphomas. However, unlike with the latter diseases, patients with sarcoidosis do not seem to be specifically predisposed to opportunistic infections or to develop cancer.

Humoral Events

There appears to be a hyperreactivity of the humoral immune system in sarcoidosis.[129] This is expressed mainly as an increased response to exogenous antigens and the development of autoantibodies to host antigens (e.g., rheumatoid factor, antinuclear antibody, autoantibodies to T cells).[130–132] One of the first observed humoral immune abnormalities was a polyclonal elevation in γ globulins. This is present in the blood of 70 percent of patients with active disease.[53] These abnormalities are often associated with an exaggerated response to certain common antigens (e.g., mycoplasma and respiratory viruses)[133] and the presence of immune complexes. A polyclonal hypergammaglobulinemia is also found in active sarcoid BAL.[134] Early reports suggested that the high titers of antibodies directed against specific antigen or antigens might have initiated the inflammatory process.[131,132] There is no convincing evidence to support these views. Available data suggest that immunoglobulin formation in sarcoid patients is nonspecific and merely a by-product of the presence of large numbers of activated T cells at the site of granuloma formation. Lung CD4+ T cells, but not blood T lymphocytes, from patients with active sarcoidosis are capable of polyclonally activating normal B cells (without added antigen or mitogen) to differentiate into immunoglobulin-secreting cells.[129] In contrast, lung T lymphocytes from normal subjects or patients with idiopathic pulmonary fibrosis do not exhibit such functional features. Consistent with such in vitro observations, in vivo studies have demonstrated that immunoglobulins are produced primarily at sites of granuloma formation in sarcoid patients,[135] and that the numbers of immunoglobulin-secreting cells at these sites are directly correlated with the numbers of activated T-helper cells.[129] However, no significant numbers of B cells are found in sarcoid BAL.[87] Although it is clear that lung B lymphocytes are actively producing immunoglobulin in sarcoidosis, the relevance to pathogenesis of this disease is unknown. The hypergammaglobulinemia probably reflects an epiphenomenon, unrelated to the morphologic derangements of this disease. Suggestions on the pathogenesis of the humoral abnormalities in sarcoidosis are incomplete and largely speculative. The elevated levels of immunoglobulin found in BAL and serum of sarcoid patients could reflect a defect in T cell regulation of B cell function, with consequent immunoglobulin synthesis by lung B cells at sites of inflammation and subsequent diffusion into the blood.[136]

The prevalence of immune complexes ranges from 23 to 70 percent of sarcoid patients.[137] It has been postulated that such variability in incidence could either signify the degree of disease severity or the fact that different techniques are used by different laboratories to identify the immune complexes. The latter, when present, appear to persist in active disease but disappear with resolution. A high prevalence of immune complexes is noted in sarcoid patients present with erythema nodosum and arthritis. It is unclear what role they play in the evolution of the disease because there is no apparent association between the presence of immune complexes and stage of disease or presence of other extrapulmonary features.[138,139] In addition, some of the serious clinical effects of immune complexes such as renal disease seen in other diseases (e.g., systemic lupus erythematosus) do not occur in sarcoidosis.

Firm evidence now supports the notion that sarcoidosis is primarily an immunologic disorder. Recent data suggest that sarcoidosis does not represent a generalized depression of the immune system, but a *heightened* inflammatory reaction at sites of disease activity. This has led to further speculation on the possible histopathologic sequence of events occurring in the lung interstitium. In this regard, BAL has been instrumental in showing that the initial lesion in the lungs of sarcoid patients arises as a result of an exaggerated local cellular-mediated immune response to unknown stimulus or stimuli. Such a reaction involves the accumulation and compartmentalization of mononuclear cells in the lung parenchyma, thereby providing the appropriate setting for granuloma formation.

Available evidence suggests an inverse relationship between the number of granulomata and the total number of inflammatory cells in sarcoid lung biopsies.[140] In patients with early disease there is a tendency toward greater numbers of inflammatory cells and fewer granulomata and a less prominent alveolitis. Such findings are consistent with experimental models of granulomatous lung disease. These demonstrate that an accumulation of mononuclear cells in the lung precedes granuloma formation.[141,142] Such observations support the hypothesis that the presence of an initial cellular inflammatory process is essential to the development of the granulomatous lesion.

HISTOPATHOLOGY

Macroscopic Pathology

Although it is now generally accepted that the initial lung lesion in sarcoidosis is a lymphocytic alveolitis,[143–145] macroscopic descriptions at this early stage of the disease are unknown because cases coming to autopsy are extremely rare. The definition of sarcoidosis is based on the formation of granulomata, even though these appear after the lymphocytic alveolitis. The recommended pathologic definition for sarcoidosis is a disease characterized by the formation in all or several affected organs or tissues of discrete noncaseating epithelioid cell tubercles. Fibrinoid necrosis may be present at the centers of the granulomata, proceeding either to resolution or conversion into hyaline fibrous tissue.

At postmortem the organs most frequently affected are lymph nodes (78 percent), lung (77 percent), liver (67 percent), spleen (50 percent), heart (20 percent), skin (16 percent), central nervous system (8 percent), kidney (7 percent), eyes and parotid glands (6 percent), thyroid (4 percent), intestine (3 percent), stomach (3 percent), and pituitary (3 percent).[147] Also affected are the eye, lachrymal glands, minor salivary glands, bone and joints, muscle and hematopoietic system, causing thrombocytopenia and autoimmune hemolytic anemia.[148] Histopathologists are less likely to see this disease than clini-

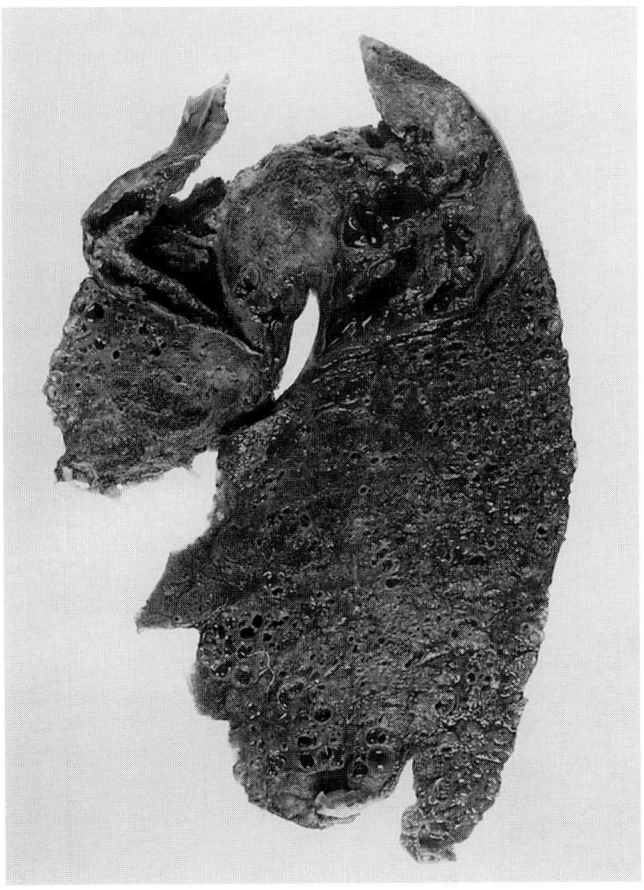

FIGURE 16-5
Lung slice from a case of sarcoidosis, with marked upper lobe fibrosis.

FIGURE 16-6
Macroscopic picture of lung in sarcoid. There is both upper and lower lobe fibrosis as well as early cyst formation.

cians because sites such as skin or eyes are infrequently biopsied.

Involvement of bronchi and bronchioles by sarcoid has been noted previously.[149] Distal to the bronchostenosis there is a lipid pneumonia.[150] The consequence of this bronchiolar narrowing and occlusion is pulmonary fibrosis, usually most marked in the upper lobes, which are contributed and largely replaced by irregular fibrous scars (Figs. 16-5 and 16-6). Honeycombing may be present. In addition there is enlargement of the adjacent subcarinal and peribronchial nodes. Upper lobe bronchiectasis is seen. In some cases there may be cavity formation (Fig. 16-7) and *Aspergillus* tends to invade such relatively avascular spaces, causing hemoptysis.

Apart from severe end-stage fibrosis involving both parenchyma and the pleura, there may be emphysema. A rare complication is systemic amyloidosis.[151,152]

The heart is involved in the sarcoidosis and this may cause sudden death. The common sites of involvement[153] are the free wall of the left ventricle, the basal aspect of the ventricular septum, the free wall of the right ventricle, the sinoatrial node, atrioventricular node, and Purkinji system. Uncommon locations of cardiovascular involvement are the atrial walls, pericardium, valves, aorta, the main pulmonary arteries and veins, and superior vena cava.

Histology

Granulomata are characteristic of this disease. All the lesions are approximately the same age, which is typical of sarcoidosis.[154,155] They may involve the walls of bronchi or bronchioles, follow the lines of the pulmonary lymphatics in the interstitium (Fig. 16-8) as well as lymph nodes, especially those in peribronchial, hilar, and mediastinal locations. The diagnostic yield of granulomata in transbronchial lung biopsies is increased by step sectioning.[156] It in-

517

FIGURE 16-7
Apical *Aspergillus* cavity set on a background of pulmonary fibrosis secondary to sarcoid.

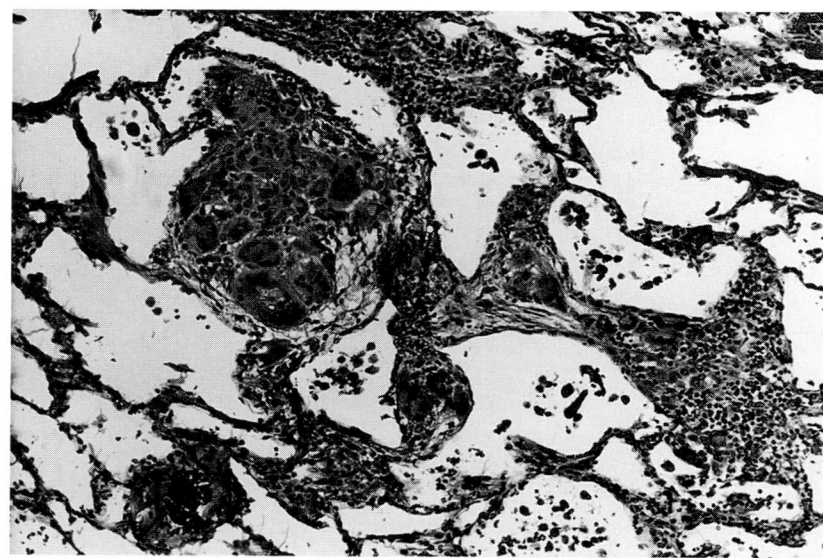

FIGURE 16-8
Multiple sarcoid granulomata, seen in the interstitium. There is also a chronic inflammatory reaction. (H&E, ×219.)

creased from 38 to 47 percent in stage I patients and from 57 to 82 percent in stage II. All the newly detected granulomata were identified between the third and seventh sections.

Descriptions of the sarcoid granulomata have been given by various authors.[157,158] They consist of focal collections of macrophages, a few multinucleate giant cells and scanty surrounding lymphocytes (Fig. 16-9). Various inclusion bodies may be present. The epithelioid cells are large mononuclear cells approximately 20 μm in diameter with a round or oval nucleus, with loose chromatin and prominent nucleoli. The cytoplasm is pale, granular, and often vacuolated. There is a lack of cytoplasmic demarcation (Fig. 16-10).

Electron Microscopy[159,160]

A typical well-developed granuloma ultrastructurally shows two characteristic zones. The central zone is composed of closely opposed epithelial cells, a few giant cells, lymphocytes, histiocytes, and intermediate cells. The intermediate cells resemble both histiocytes and epithelioid cells. Cytoplasmic organelles are profuse with many mitochondria, much rough endoplasmic reticulum, and some mem-

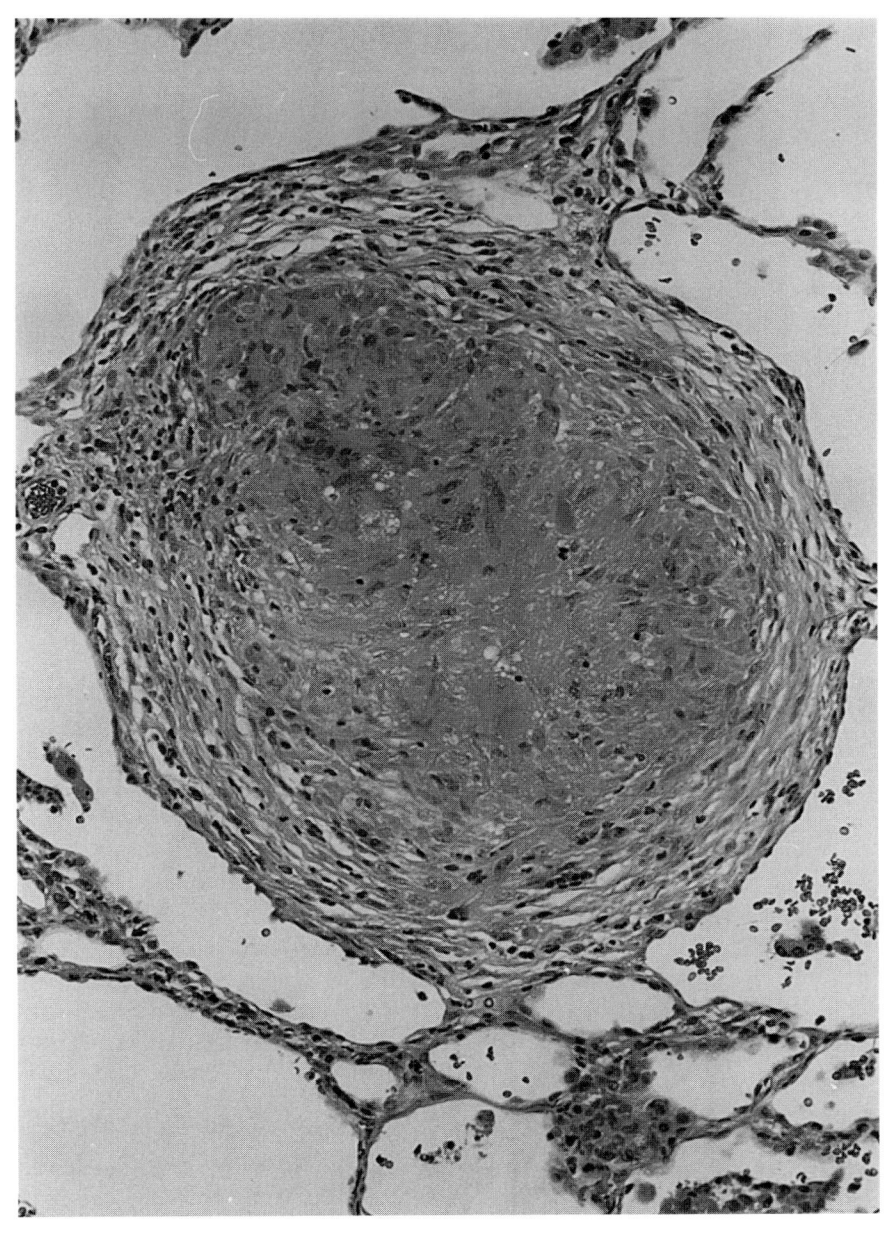

FIGURE 16-9
Large sarcoid granuloma, with well
defined giant cells, surrounded by fibrosis. There is little inflammation. (H&E,
×88.)

brane-lined vesicles. There is a well-developed Golgi
apparatus but only a few lysosomes. The peripheral
zone contains loosely arranged cellular elements,
including fibroblasts, myofibroblasts, occasional
epithelioid cells, collagen, fibrils, and a mixture of
chronic inflammatory cells including lymphocytes,
plasma cells, and eosinophils.

The giant cells are predominantly of the
Langhan's type. They may be up to 300 μm in diameter and have up to 30 peripherally arranged nuclei,
derived from the fused macrophages and are themselves phagocytic (Fig. 16-11). Foreign body-type
giant cells are also identified most frequently around
Schaumann bodies. These are larger, more irregular
in shape and contain more nuclei.

In the lymph nodes there is replacement by dis-
creet noncaseating epithelioid cell tubercles (Fig.
16-12).

Necrosis may be identified in up to 39 percent of
cases (Fig. 16-13).[161,162] Necrosis is usually spotty
and relatively inconspicuous, unlike the necrosis
seen in tuberculosis, which is usually much more
advanced. In sarcoid the necrotic areas are said to
show maintenance of fine reticulin and collagen
fibers ultrastructurally. However, the use of a reticulin to differentiate sarcoid from tuberculosis has
never been successful in our hands.

Rosen and colleagues[158] suggest the term caseous necrosis should not be used because it is difficult
to define and its description is imprecise. These

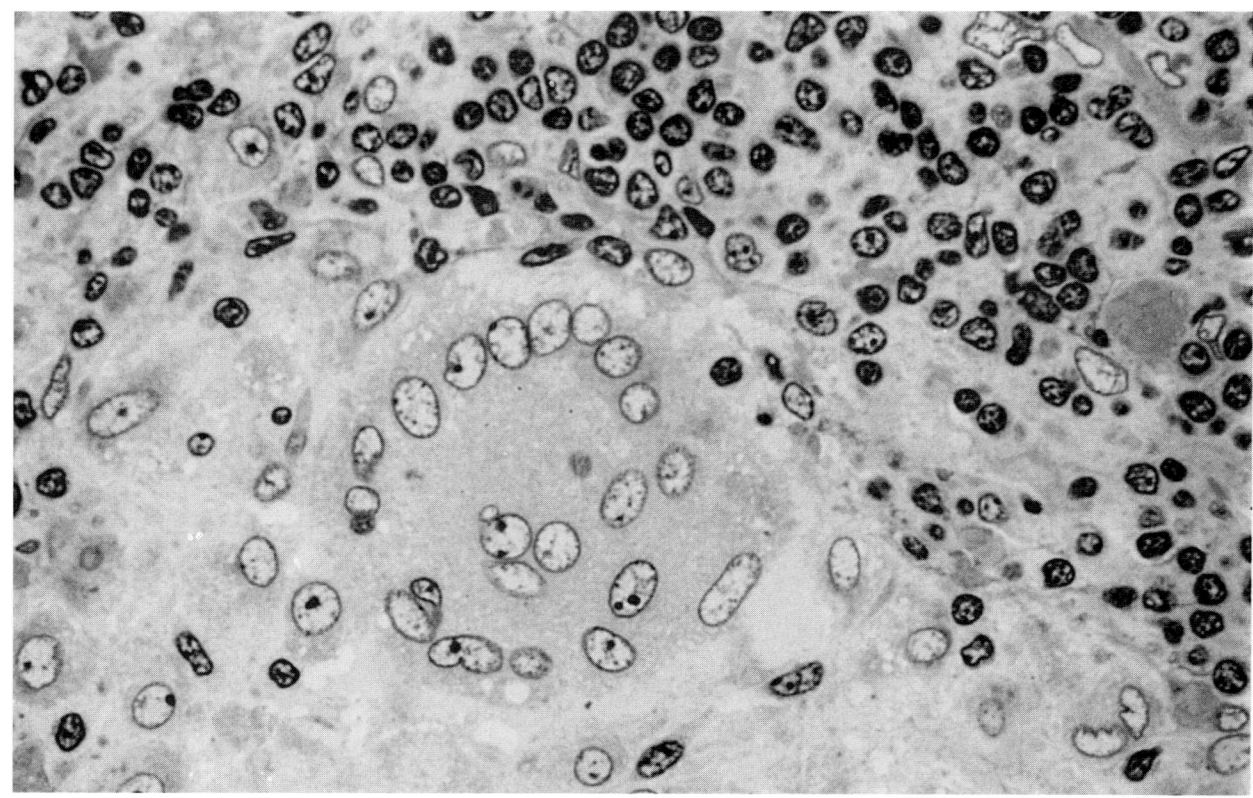

FIGURE 16-10
Semithin section with giant cell and ill-defined cell borders. There is evidence of chronic inflammation. (Toluidine blue, ×350.)
(Reproduced with permission of Dr. A. Curry, Department of Pathology, Withington Hospital, Manchester, England)

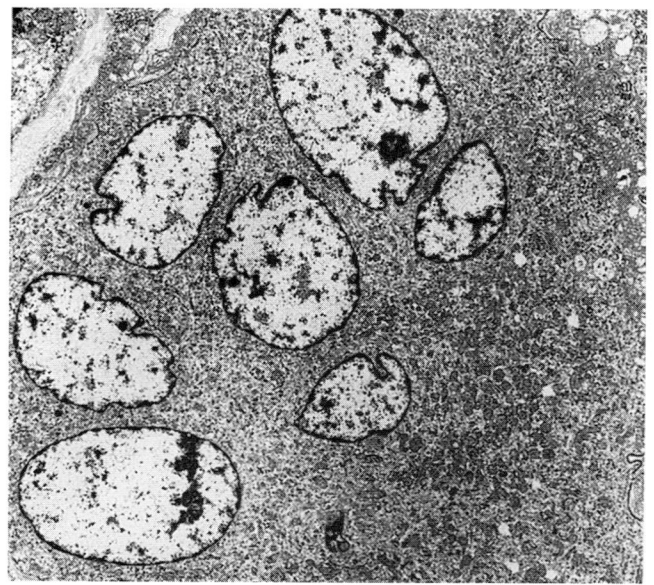

FIGURE 16-11
Electron microscopic picture of a giant cell, with no cytoplasmic membranes between the giant cells. The cells show many organelles. (Uranyl acetate-lead citrate, ×2775.)

authors propose that granulomas should either be described as necrotizing, nonnecrotizing, or showing minimal necrosis. In the presence of a granuloma it is necessary to exclude an infective condition before a diagnosis of sarcoid can be firmly established, especially if necrosis is present.

Pleural involvement is common in sarcoid but clinically causes few symptoms. Rosen and coworkers identified pleural granulomas in 36 percent of open lung biopsy specimens. In one series with clinically identified pleural sarcoid, 20 of 37 cases had pleural effusions.[162]

A vascular component of sarcoid may be seen with extensive disease.[164] There are noncaseating granulomas in the intima (Fig. 16-14), media and adventitia of vessels. There is no fibrinoid necrosis of the vessel walls but rarely pulmonary hypertension may occur secondary to this occlusion.[165,166] Rarely pulmonary veins may be affected, simulating veno-occlusive disease, without pulmonary arterial involvement.[167]

Cellular Inclusions

SCHAUMANN (CONCHOIDAL) BODIES

These are large (25 to 200 μm), concentrically lamellated structures present within the giant cells of sar-

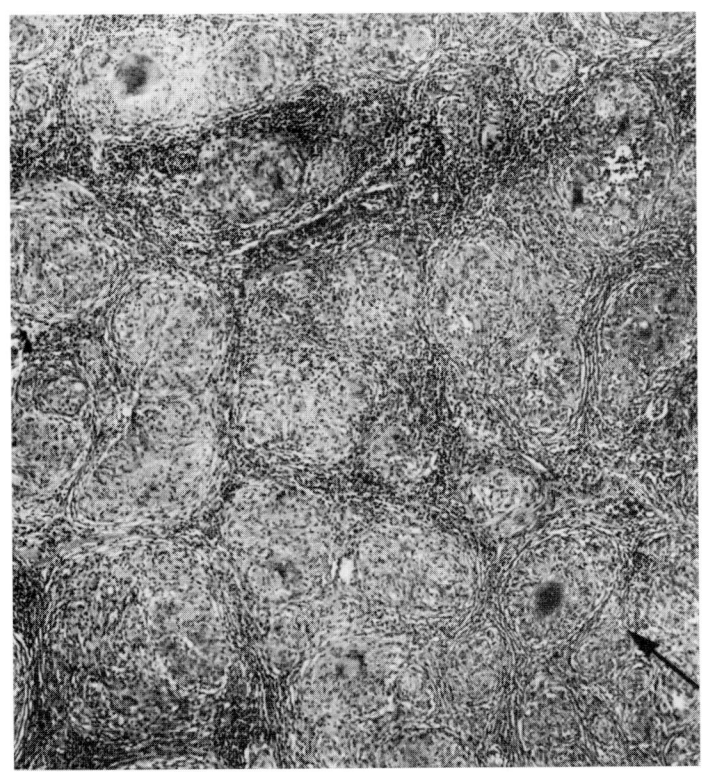

FIGURE 16-12
Lymph node in sarcoid showing no normal residual architecture. The node has been replaced by discrete and confluent granulomata (*arrow*). There is no necrosis. (H&E, ×88.)

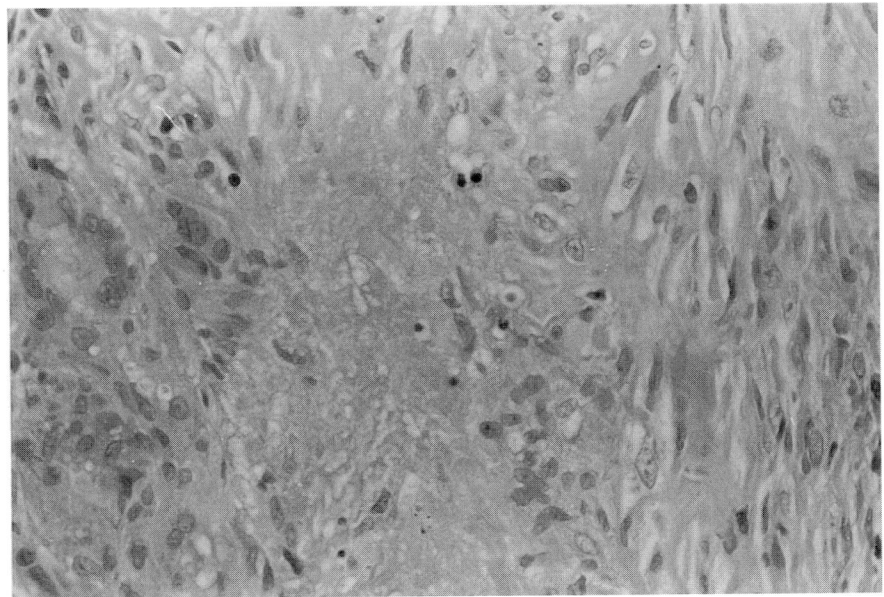

FIGURE 16-13
Microscopic area of necrosis in a case of sarcoid. No infectious cause for this feature was identified. (H&E, ×219.)

coid (Figs. 16-15 and 16-16). They are not specific to this condition and may be found, as are the other inclusions, in other granulomatous diseases. Jones-Williams described them in chronic berylliosis (62 percent), Crohn's disease (10 percent), and tuberculosis (6 percent).[168] They are common in sarcoid (88 percent). Schaumann bodies comprise aggregated spherules and 70 percent contain brilliantly birefrin-

gent, crystalline material. They stain positively for iron and calcium. Electron diffraction studies suggest they are largely composed of calcium and phosphorous with smaller peaks of aluminum and iron (Fig. 16-17).[169] These structures have also been described in extrinsic allergic alveolitis.

Schaumann bodies probably originate in epithelioid cells as small (1 to 2 μm) sharply angulated,

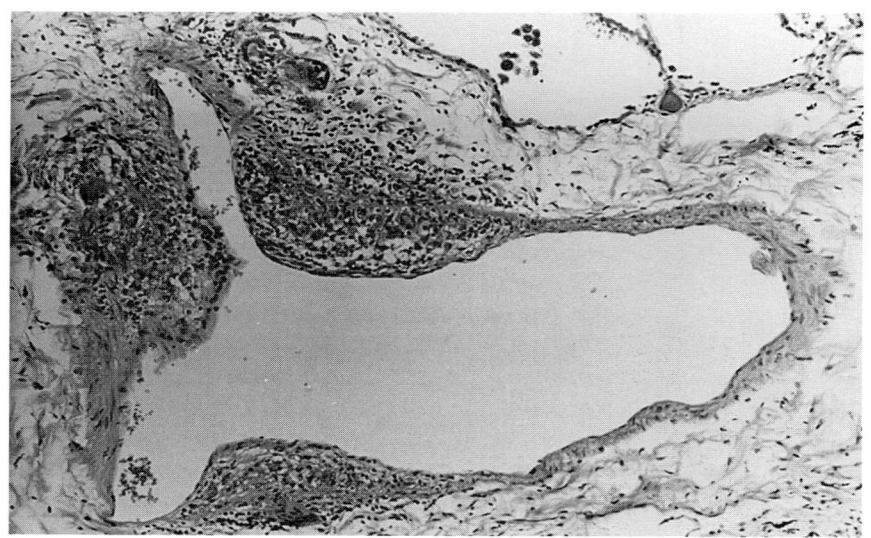

FIGURE 16-14
Vascular involvement by sarcoidal granu-lomatous inflammation. (H&E, ×88.)

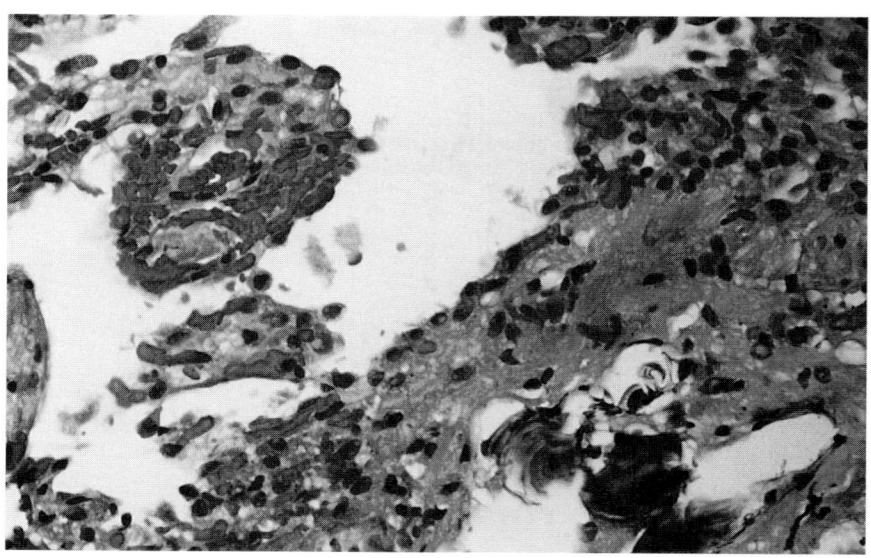

FIGURE 16-15
Schaumann body in the lower right of the picture. These are laminated, calcified bodies. The adjacent lung shows intersti-tial fibrosis. (H&E, ×219.)

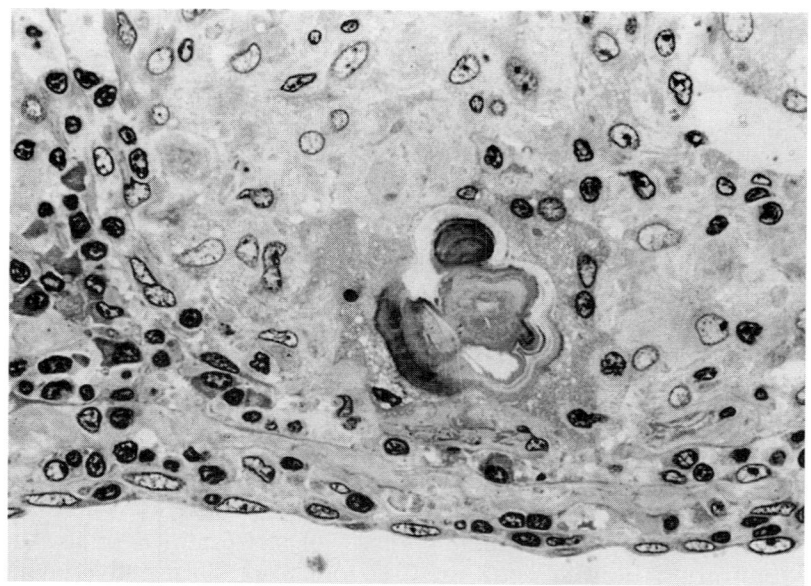

FIGURE 16-16
Semithin section with a well formed Schaumann body having a laminated appearance. (Toluidine blue, ×350.) *(Reproduced with permission of Dr. A. Curry, Department of Pathology, Withington Hospital, Manchester, England.)*

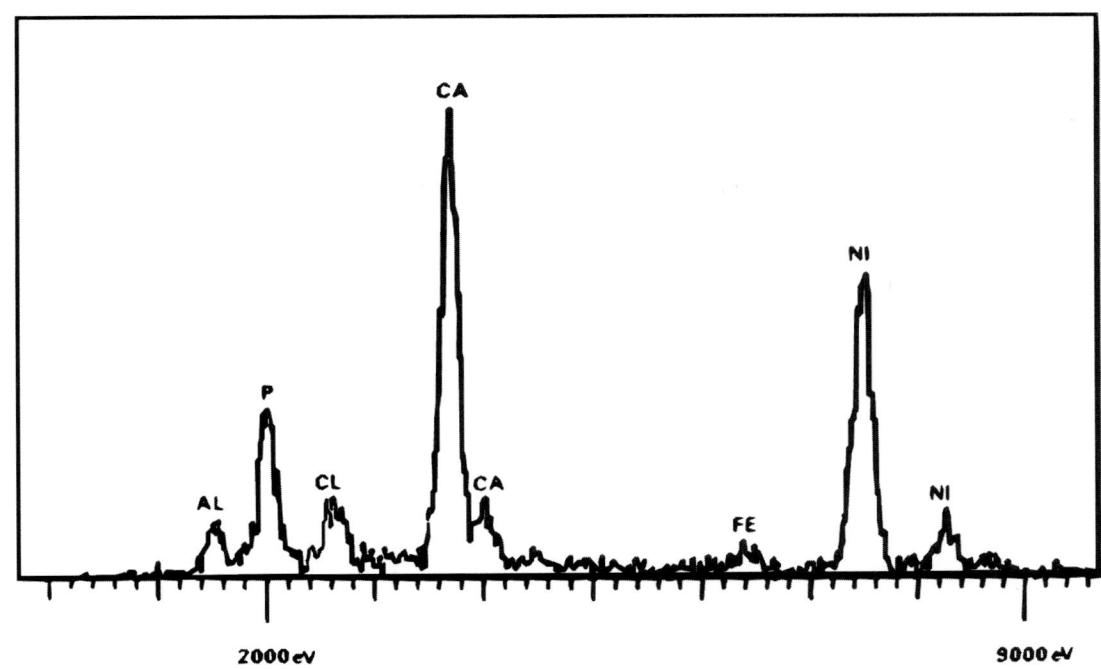

2000 eV   9000 eV

FIGURE 16-17
Elemental spectrum of a Schaumann body in nonosmicated tissue. There is much calcium (CA), aluminum (AL), and iron (FE). The peaks for chlorine (CL) and nickel (NI) are from the supporting grid. *(Reproduced with permission of Professor J. Kirkpatrick, Drs. A. Curry and D. L. Bissett, Department of Pathology, Withington Hospital, Manchester, England and the editor of Ultrastuctural Pathology.)*

brilliantly birefringent crystals. Further deposition of crystals leads to aggregates up to 20 μm forming a nidus for the development of conchoidal bodies.[168] Conchoidal bodies consist of a complex mixture of acidic and nonacidic mucoproteins, lipoproteins, iron, and often calcium.[157,168] The Schaumann body may derive from residual bodies, resulting from activated lysosomes.[170]

More recently it has been shown that the nidus for mineral deposition in Schaumann bodies is the intracytoplasmic myelinoid membranous inclusions[169] (Figs. 16-18 and 16-19).

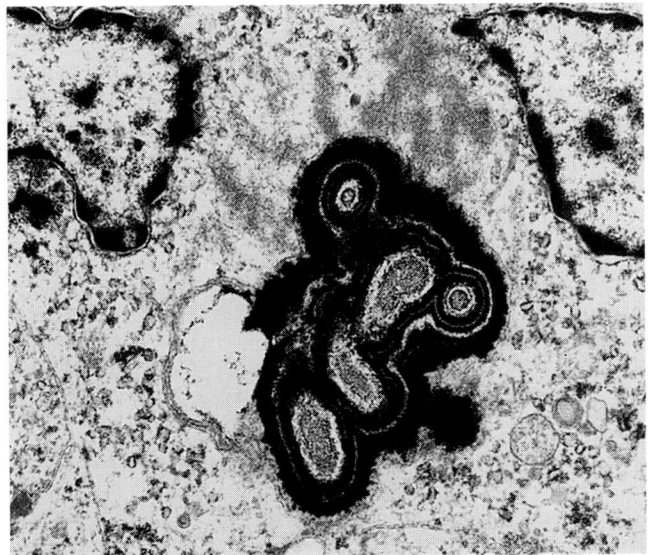

FIGURE 16-18
Altered myelin figure bound by a multilamellar membrane. There is a finely granular matrix and core. (Uranyl acetate-lead citrate, ×11,100) *(Reproduced with permission of Professor J. Kirkpatrick, Dr. A. Curry, Department of Pathology, Withington Hospital, Manchester, England and the editor of Ultrastructural Pathology.)*

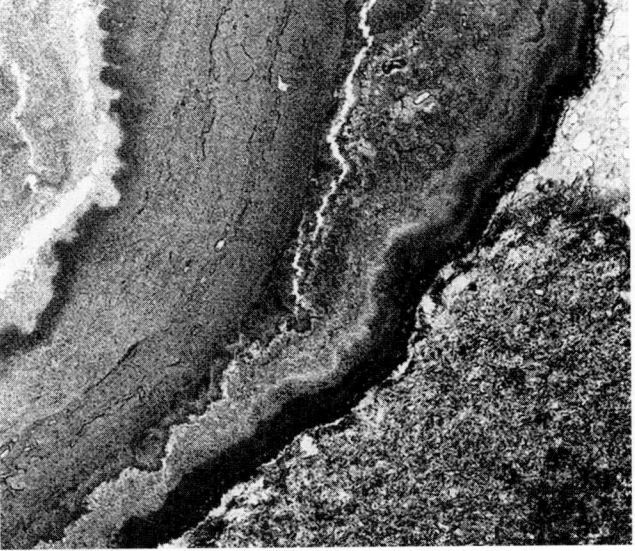

FIGURE 16-19
Periphery of part of a Schaumann body with concentric rings. (Uranyl acetate-lead citrate, ×6993) *(Reproduced with permission of Professor J. Kirkpatrick, Dr. A. Curry, Department of Pathology, Withington Hospital, Manchester, England and the editor of Ultrastructural Pathology.)*

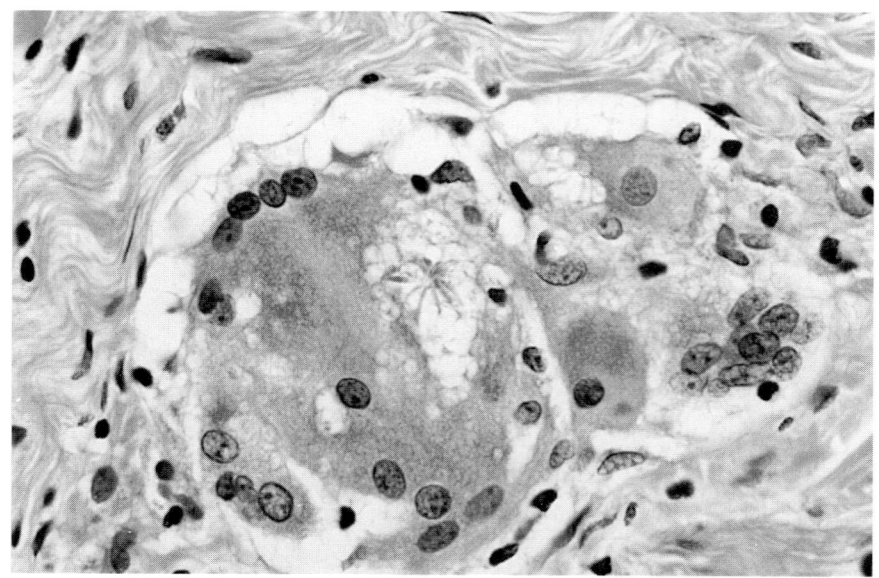

ASTEROID BODIES

These consist of a star-shaped body with a central
core approximately 1 μm in diameter with radiating,
slightly curved spines (Fig. 16-20). They are found in
epithelioid cells but are more common in giant cells.
Rosen and colleagues detected asteroid bodies in 11
(9 percent) of 127 open lung biopsies in sarcoid
patients.[158] They are not specific for this disease,
being found in tuberculosis, leprosy, histoplasmosis,
schistosomiasis, and lipoidal granulomas.[161]

Ultrastructurally asteroid bodies are composed
of filamentous arms radiating from a central granular

core (Fig. 16-21). Closely associated with them were
myelinoid membranous components. These are
closely related to the asteroid arm. These myelinoid
membranous elements took several forms.[169] There
were either single or a few electron-dense laminae
intimately related to the radiating arms of the asteroid body. In addition there may be complex, multilamellar membranous structures, some of which had
the appearance of classical myelin figures, located
between the "spokes" of the asteroid body as well as
probable intermediate forms. The latter consisted of

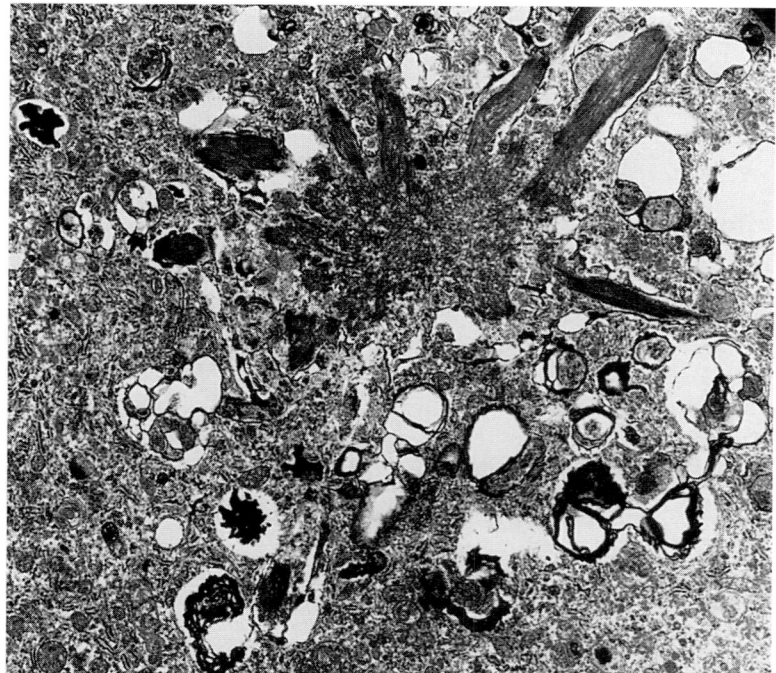

multilamellar myelinoid membranes that, although in places were contiguous with the radiating arms, also projected into the surrounding cytoplasm.

Cain and Kraus[171] showed asteroid bodies were composed of a noncollagenous, filamentous and some microtubular material, possibly derived from a cytosphere. Kirkpatrick and colleagues[169] failed on serial sections through asteroid bodies to demonstrate either centrioles or microtubular structures in close proximity to the asteroid body. The assembly of the asteroid body is thus probably independent of these organelles. It is speculated that the organization of the vimentin filaments to form the asteroid body is mediated by the extensive lipid component, of which the asteroid body forms the hub.

Phospholipids, which are the skeleton of these myelinoid membranes, are avid binders of calcium.[172] Elemental analysis of an asteroid body shows a small calcium peak and a possible phosphorus peak.[169]

HAMAZAKI-WESENBERG BODIES (FIG. 16-22)

As with the other inclusions, these are nonspecific for sarcoid. They are oval or spindle shaped, pleomorphic and vary from 0.5 to 8 μm in diameter. They are yellow-brown on hematoxylin and eosin sections thus the term yellow bodies[173] or yellow-brown bodies.[174] They are both intra- and extracellular and are seen predominantly at or near the peripheral sinuses of lymph nodes. Hamazaki-Wesenberg bodies have been identified in up to 68 percent of sarcoid lymph nodes.[174] Rosen identified these bodies using methenamine silver in up to 41 percent of mediastinal lymph nodes and 16 percent of open lung biopsy specimens.[164] These structures may be mistaken for fungi. They are giant intracellular and extracellular lysosomes and residual bodies.

SLITLIKE CRYSTALS (FIG. 16-23)

Elongated, acicular, birefringent crystals resembling cholesterol are seen in giant cells in sarcoid as well as other granulomatous diseases.

Immunocytochemistry

Angiotensin-conversing enzyme (ACE) has been detected in the cytoplasm of epithelioid granulomas in 38 of 39 patients with clinical sarcoidosis but not in 37 nonsarcoid granulomas.[175] This mode of diagnosis is not routinely used.

Natural History of the Disease

A large series of cases was followed for over 2 years in the 1950s. Seventy-one percent with bilateral hilar lymphadenopathy (BHL) showed spontaneous radiographic improvement within a year. Fifty percent of

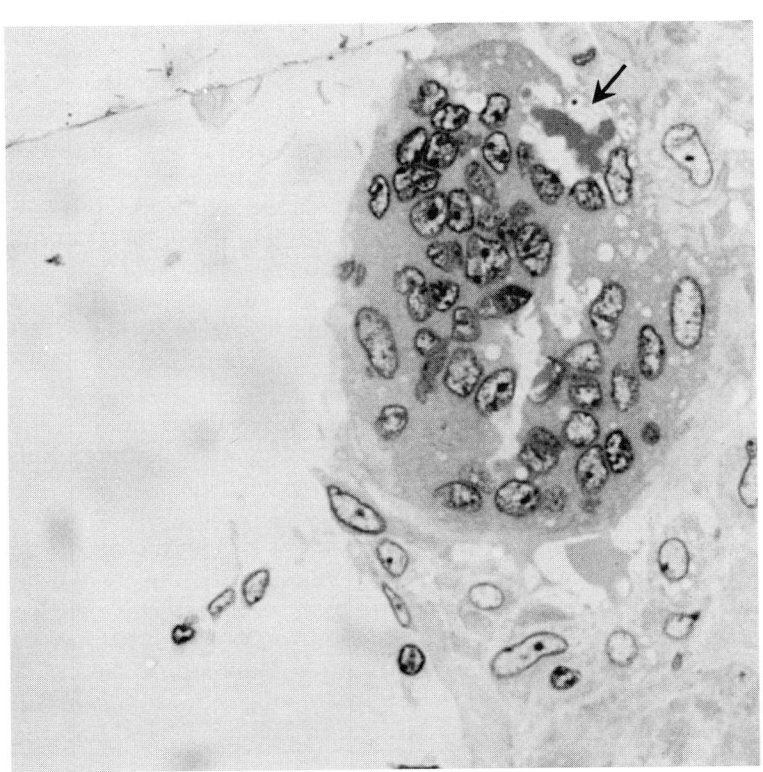

FIGURE 16-22
Semithin section showing a Hamazaki-Wesenberg body (*arrow*) amidst a granuloma. (Toluidine blue ×450.)

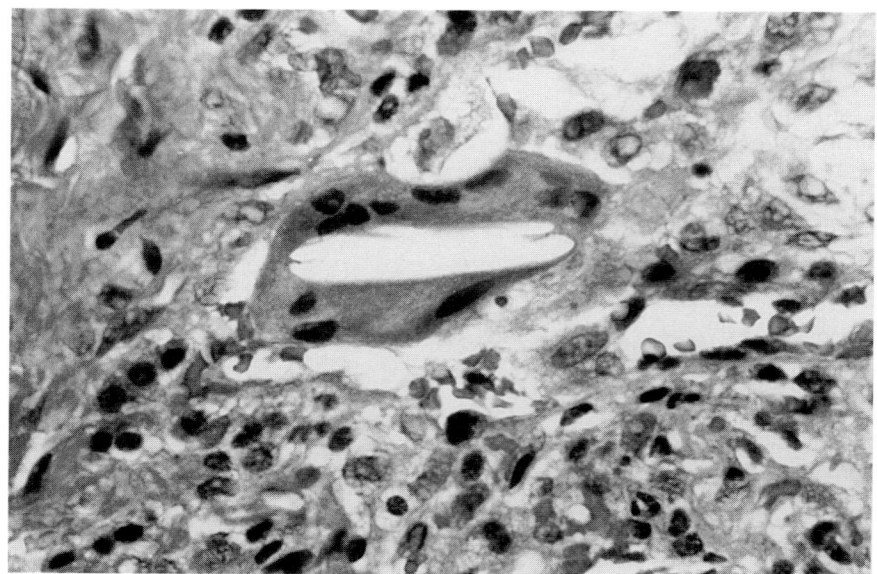

FIGURE 16-23
Slitlike crystals with an associated giant cell reaction. As with the other cellular inclusions, these crystals are not specific for sarcoidosis. (H&E, ×350.)

patients with pulmonary infiltrates improved radio-logically. There was a better outlook for patients first diagnosed as erythema nodosum or BHL and normal lung fields. None of the patients with a normal radiograph relapsed.[176] Pulmonary infiltrates persisting for more than 2 years were unlikely to remit spontaneously. In this group two fifths showed evidence of pulmonary fibrosis by 5 years, with dyspnea as their main symptom. Two fifths of these patients became respiratory cripples progressing to massive fibrosis, secondary infection, bronchiectasis, occasional hemoptysis, or spontaneous hemothoraces. Cor pulmonale developed from between 10 and 20 years from the initial diagnosis and was the most common terminal event. All studies prior to 1955[161,177–179] showed half the deaths were unrelated to sarcoidosis.

Subsequent studies indicate that death is more frequently related to the disease.[180–183] A North American study between 1960 and 1977[184] detected 41 deaths in 1090 patients with a clinical diagnosis of sarcoidosis. Thirteen patients died of causes unrelated to sarcoid and death was related to the disease in 68 percent (28 patients). Advanced pulmonary fibrosis, seen in 22 of 28 patients, was the most common cause of death. The average age at death was 39 years. The patients died of cardiorespiratory failure, gram-negative pneumonias, and fungal infections. Autopsy in 25 patients showed systemic granulomas in all cases, including asymptomatic patients as well as those presenting with clinical manifestations limited to the central nervous and cardiovascular systems. The overall mortality rate in this study was 3.8 percent, which compared with 4.1 percent in an earlier, larger series.[182]

These findings seem to indicate there had been a *mild,* worldwide decrease in the mortality rate of sarcoid. Cardiac sarcoidosis accounted for between 13[183] and 25 percent[185] of deaths. Not infrequently the diagnosis was unsuspected in these cases during life. Such cases had pulmonary as well as cardiac lesions.

POSSIBLE CONCOMITANT OR ASSOCIATED DISEASES

These have been discussed by Scadding and Mitchell.[146]

Malignant Disease
These authors came to the conclusion that there was no evidence of increased frequency of malignant tumors, with the *possible* exception of lymphoma. There is 11 times the expected number of cases of malignant lymphoma in patients with previous sarcoidosis.[186] It should be remembered that sarcoid has a marked chronic inflammatory reaction and it is possible that some of the cases of sarcoid were misdiagnosed as lymphoma. In our experience this malignancy is rarely seen as a complication of sarcoidosis.

Other tumors associated with sarcoid include ovarian cancer, testicular malignancy, either seminoma or teratoma,[187] and breast cancer.[188] A sarcoid-like reaction may be seen in carcinoma of the lung. Some cases of carcinoma of the lung have sarcoid granulomas in the lung away from the tumor or in the draining nodes, suggesting that there is a systemic reaction to the tumor. It may be difficult to determine if this is concomitant sarcoid or a sarcoid-like reaction in the lung. A similar sarcoid-like reaction may be seen in axillary nodes draining carcinoma of the breast or in retroperitoneal nodes associated with testicular tumors.

Hairy cell leukemia,[189] myeloma,[190] macrocryoglobulinemia,[191] and IgA deficiency[192] have been associated with sarcoid. Other associations mentioned by Scadding and Mitchell[146] were systemic lupus erythematosis, rheumatoid disease, ankylosing spondylitis, gout, psoriasis, and polyarteritis nodosa. These were all thought to be more than just chance associations.

Celiac Disease

Douglas and coworkers[193] showed an association between sarcoidosis and celiac disease in five patients. In three cases the gastrointestinal symptoms of celiac disease preceded those of sarcoid, and in the other two symptoms of both diseases appeared at the same time. The likelihood of a chance association between these two diseases was thought to be remote.

Sarcoid has also been associated with membranous glomerulonephritis.[194] There is an association between sarcoid and thyroiditis, Addison's disease, and Sjögren's syndrome (TASS).[195]

Primary Biliary Cirrhosis[196–198]

Sarcoidosis may share overlapping features with other multisystem, granulomatous disorders. In some patients the distinction between sarcoidosis and primary biliary cirrhosis may be impossible. These similarities were further emphasized by the finding of a subclinical, mononuclear inflammatory pulmonary infiltrate (termed "alveolitis") in both sarcoid and primary biliary cirrhosis patients, who were clinically asymptomatic and had a normal chest radiograph.[199] These findings strongly suggested that both diseases were promoted as an immunologic response but also raised the possibility of an identical underlying granuloma-producing mechanism.[200]

The mitochondrial antibody test is the hallmark of primary biliary cirrhosis and is positive in 99 percent of patients but usually negative in sarcoid. Maddrey concluded after a literature review that there was still a doubt as to whether sarcoid and primary biliary cirrhosis were related.[201]

Differential Diagnosis

INFECTION

The optimum any pathologist can expect in any case of granulomatous disease is some fresh tissue. A portion should be dispatched to the bacteriology laboratory and the rest processed for hematoxylin-eosin sections and stained for fungi and tubercle bacilli.

There has been a rare association of *M. tuberculosis* infection with sarcoid.[146] Diagnosis was either made by the detection of acid-fast bacilli or patients having close contact with tuberculosis. *Histoplasma capsulatum, Coccidioides immitis, Cryptococcus neoformans, Blastomyces dermatidis,* and *Sporothrix schenkii* may all cause pulmonary granulomas.

OCCUPATIONAL LUNG DISEASE

The best known is berylliosis, which causes noncaseating granulomas. Erythema nodosum and uveitis are absent. The Kveim test is negative and a beryllium patch test and lymphocyte transformation test are both positive. Serum ACE levels are normal and laser microprobe mass spectrometry confirms the presence of the metal.

Talc granulomatosis will be diagnosed under the polarizing microscope. Talc granulomas may occur in intravenous drug abuse and the vascular and interstitial component of this disease may suggest sarcoid.

HYPERSENSITIVITY PNEUMONITIS

This is frequently misdiagnosed as sarcoid but the granulomas are less well defined. They are interstitial and do not follow the lymphatics as in sarcoid.

CARCINOMA

Carcinoma of the bronchus may cause a sarcoid-like reaction in pulmonary lymph nodes as well as in the lung parenchyma away from the tumor. In a few cases this may be difficult to differentiate from true sarcoidosis.

PULMONARY FIBROSIS

Other causes of pulmonary fibrosis, such as fibrosing alveolitis or bronchiectasis may be confused with sarcoid. Usually there are some residual granulomas in sarcoid.

NECROTIZING SARCOID GRANULOMATOSIS

This is a rare condition, originally described by Liebow in 1973.[202] There have been three subsequent large series[203–205] describing a total of 86 cases.

The etiology of the condition is unknown and there is doubt whether it should be even classified along with sarcoid. Churg argued that necrotizing sarcoid granulomatosis (NSG) was in fact a form of sarcoid, based on the frequency of hilar lymphadenopathy, the occasional ocular abnormalities, especially uveitis, and liver damage.[204]

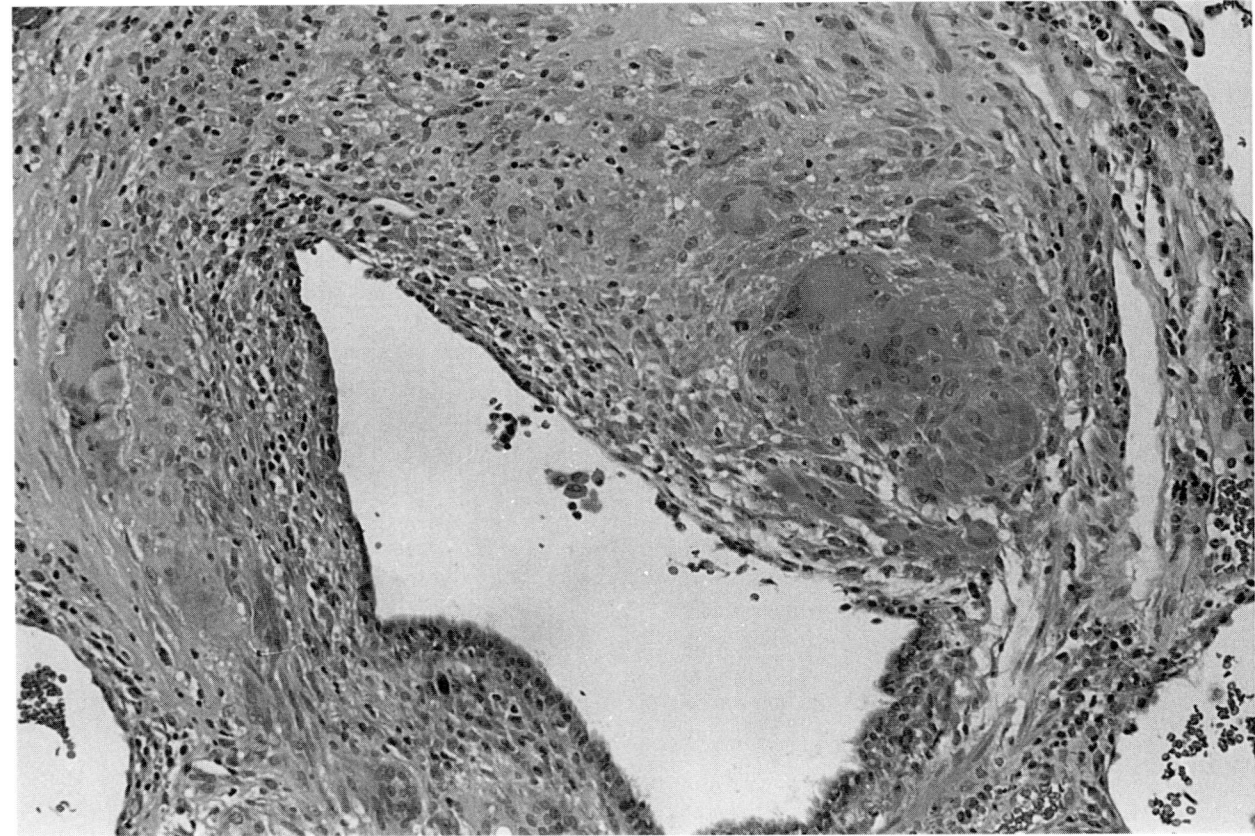

FIGURE 16-24
Necrotizing sarcoid granulomatosis—bronchiolar involvement. (H&E, ×88.)

Saldana[203] believes the disease is unrelated to classic sarcoid for the following reasons:

1. The radiologic findings of NSG are unusual for sarcoid, including the frequent absence of hilar lymphadenopathy.
2. Granulomatous vasculitis is seldom so extensive and severe in sarcoid as in NSG.
3. The extensive necrosis of the lesions in NSG is most unusual for sarcoidosis.
4. Negative ACE levels in both serum and tissue have been noted in one patient with NSG. Three patients with negative Kveim tests have been identified.[206]
5. Systemic vasculitis has been noted in two patients with NSG but does not occur in sarcoidosis.

Clinical Features

This disease is rare in children and adolescents and is most common in the fifth decade of life. It affects women more than men by a ratio of 4:1.[204] There are chest symptoms, including chest pain, dyspnea, and nonproductive cough, but with extensive disease there may be fever, malaise, and weight loss.[206] Only

a small percentage of patients have extrapulmonary signs and symptoms.[204] There may be diabetes insipidus due to hypothalamic deficiency, uveitis, iritis[205] and leg weakness due to spinal cord involvement.[207,208] Some may be asymptomatic and are diagnosed on routine chest radiographs.[204]

Radiology

The lesions tend to involve the lower lobes and the incidence of cavitation ranges from 0[204] to 23 percent.[205] Multiple nodular infiltrates can measure up to 5 cm in diameter. BHL is rare, being present in less than 10 percent of cases.[202,203,205]

Macroscopic Pathology

There are solid masses or irregular infiltrates measuring up to 8 cm in diameter. They are yellow-white, firm, and less necrotic than in Wegner's granulomatosis.

Histology

There are sarcoid-like granulomas with variable necrosis and vasculitis. Between the granulomas

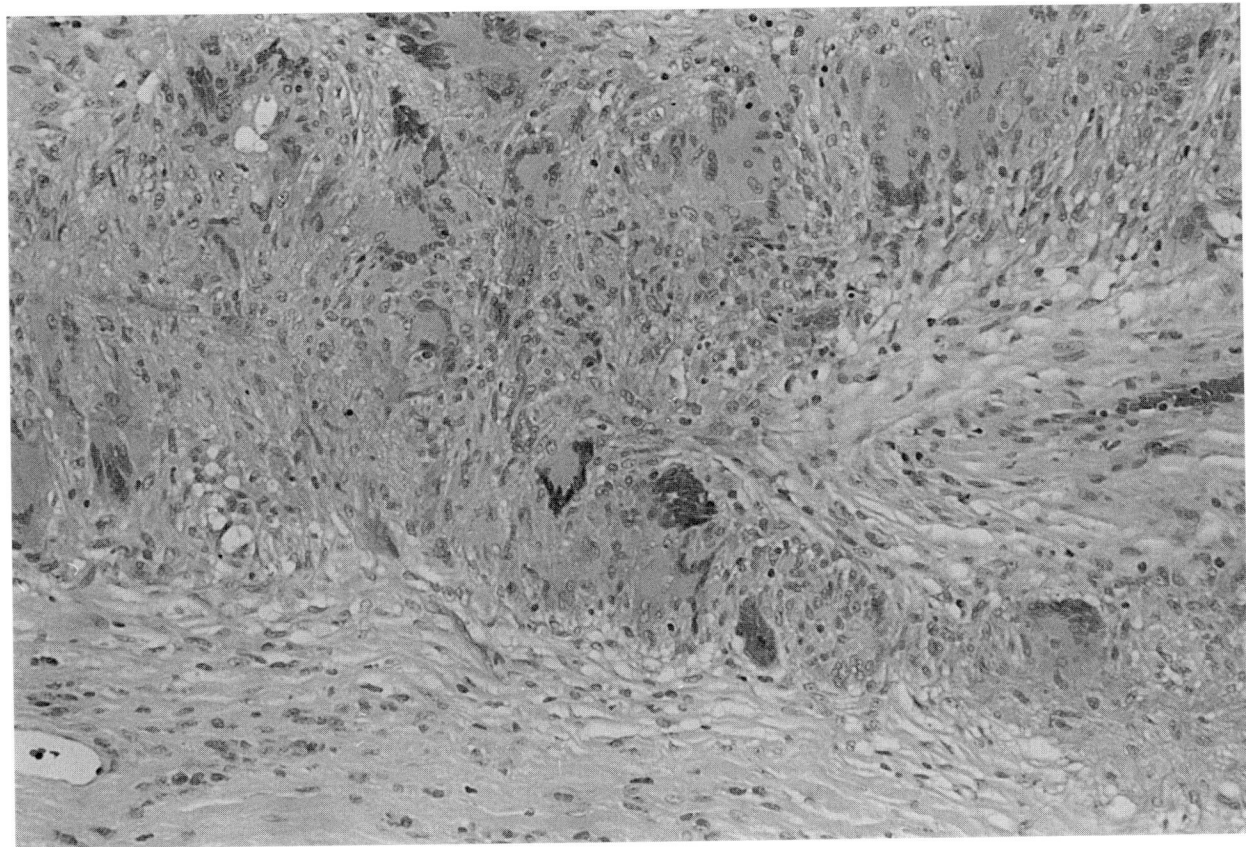

FIGURE 16-25
Necrotizing sarcoid granulomatosis—vascular involvement. There is extensive granulomatous infiltration of the media. (H&E, ×88.)

there are large numbers of chronic inflammatory cells with lymphocytes, plasma cells as well as desquamated alveolar cells and macrophages. The granulomas are sited as in typical sarcoid (i.e., in the bronchovascular bundles and interlobular septa). Bronchial and bronchiolar (Fig. 16-24) involvement is common. The granulomas become confluent.

The necrosis may affect small central areas but Churg[204] did not require this feature for the diagnosis of NSG. The necrosis is coagulative and located in the center of the largest granuloma aggregation. The necrosis may have a stellate outline with prolongations between individual granulomas. Rare cases have been described, where the necrosis was suppurative, resembling Wegner's or bronchocentric granulomatosis.[209] The necrosis is replaced by acellular hyalinized tissue.

Vascular involvement is prominent (Fig. 16-25) and affects both arteries and veins. The granulomas may extend through arterial walls, causing thrombosis and obliteration. Similar lesions are seen in veins. More rarely a giant cell arteritis is noted in large pulmonary arteries up to elastic type vessels.

An association between primary biliary cirrhosis and granulomatous colitis has been noted.[202]

Course of the Disease

The prognosis of NSG is good and patients with bilateral infiltrates respond well to corticosteroids. Relapse may occur in up to 25 percent of patients.[202,205,210]

Differential Diagnosis

INFECTION

This must be excluded either by culture or the appropriate stains.

SARCOIDOSIS

Vascular involvement in sarcoidosis occurs in half the cases diagnosed on transbronchial biopsy.[211] A postmortem study by the same Japanese group,[212] showed extensive vascular involvement, ranging from lymphatics, bronchial arteries, interlobular veins, venules, arterioles to muscular and elastic pulmonary arteries. The frequency of vascular involvement in sarcoid was much greater than in tuberculosis. ACE is usually positive in sarcoid.

GIANT CELL ARTERITIS

This has been described by Doyle and coworkers.[213] It is rare in the lung and follows giant cell arteritis in the head and neck.

WEGNER'S GRANULOMATOSIS

This disease must have necrosis of vessel walls. In NSG there are granulomatous infiltrates but no fibrinoid necrosis. There are many more nonnecrotizing granulomas in NSG.

BRONCHOCENTRIC GRANULOMATOSIS

This disease is confined to bronchiolar walls, with complete replacement by necrotizing granulomatous inflammation. Remains of fungal hyphae may be identified in levels of any impacted mucoid material. There are numerous eosinophils in bronchocentric granulomatosis and peripheral blood eosinophilia is common. Admixed with the eosinophils are plasma cells and other chronic inflammatory cells. In bronchocentric granulomatosis there are areas resembling eosinophilic pneumonia, a feature not identified in NSG. Patients have a history of chronic asthma with severe wheezing.

ALLERGIC GRANULOMATOSIS AND ANGIITIS (CHURG-STRAUSS SYNDROME)

This is characterized by granulomatous inflammation or giant cell infiltration of vessel walls with a transmural eosinophilic and histiocytic infiltrate as well as fibrinoid necrosis. Eosinophils and the fibrinoid necrosis are absent in NSG. The chronic asthmatic change in bronchioles with the thickened basement membranes may be of help in diagnosis. Unlike NSG the radiologic infiltrates are often transient. The patients have chronic asthma and peripheral blood eosinophilia, features which distinguish it from NSG.

REFERENCES

1. James DG, Turiaf J, Hosoda Y, et al: Description of sarcoidosis: Report of the Subcommittee on Classification and Definition. *Ann NY Acad Sci* 278: 742, 1976.
2. Scadding JG, Mitchell DN: *Sarcoidosis.* London, Chapman and Hall, 1985.
3. James DG, Jones-Williams W: In *Sarcoidosis and Other Granulomatous Disorders.* Philadelphia, WB Saunders, 1985, pp 10–11.
4. Buck AA, McKeesick VA: Epidemiologic investigations of sarcoidosis III. *Am J Hyg* 74: 174–188, 1961.
5. Udderfeld TM, Stjetrnberg N, Jundgren K: Sarcoidosis or tuberculosis—a case report. *Tubercle* 63: 221–223, 1968.
6. Siltzbach LE: Sarcoidosis and mycobacteria. *Am Rev Respir Dis* 97: 1–4, 1968.
7. Nethersott SE, Strawbridge WG: Identification of bacterial residues in sarcoid lesions. *Lancet* 2: 1132–1134, 1956.
8. Bowman BU, Amos WT, Geer JC: Failure to produce experimental sarcoidosis in guinea pigs with *M. tuberculosis* and mycobacteriophage DS6A. *Am Rev Resp Dis* 105: 85–94, 1972.
9. Chapman JS, Speight M: Further studies of mycobacterial antibodies in the sera of sarcoidosis patients. *Acta Med Scand* (suppl) 425: 61–67, 1964.
10. Cantwell AR: Histologic observations of variably acid-fast pleomorphic bacteria in systemic sarcoidosis; a report of three cases. *Growth* 46: 113–121, 1982.
11. Saboor SA, Johnson NMcI, McFadden J: Detection of mycobacterial DNA in sarcoidosis and tuberculosis with polymerase chain reaction. *Lancet* 339: 1012–1015, 1992.
12. Mitchell IC, Turk JL, Mitchell DN: Detection of mycobacterial rRNa in sarcoidosis with liquid-phase hybridisation. *Lancet* 339: 1015–1017, 1992.
13. Bocart D, Lecossier D, De Lassence A, Valeyre D, Battesti JP, Hance AJ: A search for mycobacterial DNA in granulomatous tissues from patients with sarcoidosis using the polymerase chain reaction. *Am Rev Respir Dis* 145: 1142–1148, 1992.
14. Lipsky BA, Goldberger AC, Tomkins LS, Plorde JJ: Infections caused by non-diphtheria corynebacteria. *Rev Infect Dis* 4:1220–1235, 1982.
15. Hanngren A, Biberfeldt G, Carlens E: Is sarcoidosis due to an infectious interaction between virus and mycobacterium? In Iwai K, Hosoday (eds), *Proceedings of the VIth International Conference on Sarcoidosis, Tokyo,* University of Tokyo Press, 8–11, 1974.
16. Lofgren S, Lundback H: Isolation of a virus from six cases of sarcoidosis; preliminary report. *Acta Med Scand* 138: 71–75, 1950.
17. Lantrop K, Wahren B, Hanngren A: Infectious mononucleosis and depression of cellular immunity. *Br Med J* 4: 668–669, 1972.
18. Lundback H, Lofgren S: Attempts at isolation of virus strains from cases of sarcoidosis and malignant lymphoma. *Acta Med Scand* 143: 98–109, 1952.
19. Hirshault Y, Glade P, Viera LO: Sarcoidosis, another disease associated with serological evidence for herpes-like virus infection. *N Engl J Med* 283: 502–506, 1970.
20. Rocklin RE: In Fanburg BL (ed), *Sarcoidosis and Other Granulomatous Diseases of the Lung.* New York, Marcel Dekker, 1983, pp 203–224.
21. Konig G, Baur X, Fruhmann G: Sarcoidosis or extrinsic allergic alveolitis? *Respiration* 42: 150–154, 1981.
22. Buck AA, Sartwell PE: Epidemiologic investigations of sarcoidosis II. *Am J Hyg* 74: 152–173, 1961.
23. Farber HW, Fairman RP, Glauser FL: Talc granulomatosis: laboratory findings similar to sarcoidosis. *Am Rev Respir Dis* 125: 258–261, 1982.
24. Fayemi AO, Ali M: Sarcoid-like granulomas in secondary oxalosis—a case report. *Mt Sinai J Med (NY)* 47: 255–257, 1980.
25. William WJ, Williams R: The value of beryllium lymphocyte transformation tests in chronic beryllium disease and in potentially exposed workers. *Thorax* 38: 41–44, 1983.

26. Shelley WB, Hurley HJ: The allergic origin of zirconium deodorant granuloma. *Br J Dermatol* 70: 75–77, 1958.

27. Prendiville J, Robinson A, Young M: Familial sarcoidosis. *Isr J Med Sci* 151: 258–260, 1982.

28. Priestly S, Delaney JC: Familial sarcoidosis presenting with stridor. *Thorax* 36: 636–637, 1981.

29. James DG, Neville E, Piyasena KH, Walker AN, Hamlyn AN: Possible genetic influences in familial sarcoidosis. *Postgrad Med J* 50: 664–670, 1974.

30. Sharma OP, Neville E, Walker AN, James DG: Familial sarcoidosis: A possible genetic influence. *Ann NY Acad Sci* 278: 335–346, 1976.

31. Guyatt GH, Bensen WG, Stolman LP, Fagnilli L, Singal DP: HLA-B8 and erythema nodosum. *Can Med Assoc J* 127: 1005–1006, 1982.

32. James DG, Neville E: Pathology of sarcoidosis. *Pathobiol Ann* 7: 31–36, 1977.

33. Scharf Y, Zonis S: Histocompatibility antigens (HLA) and uveitis. *Surv Ophthalmol* 24: 220–228, 1977.

34. Persson IB, Ryder LP, Nielson SL, Svejgaard A: The HLA-B histocompatibility antigen in sarcoidosis in relation to tuberculin sensitivity. *Tissue Antigens* 6: 50–53, 1975.

35. Hedfors E, Lindstorm F: HLA-B8/DR3 in sarcoidosis: Correlation to acute onset disease with arthritis. *Tissue Antigens* 22: 200–203, 1983.

36. Kremer JM: Histologic findings in siblings with acute sarcoid arthritis: Association with the B8, DR3 phenotype. *J Rheumatol* 13: 593–597, 1986.

37. Smith MJ, Turton CW, Mitchell DN, Turner-Warwick M, Morris LM, Lawler SD: Association of HLA-B8 with spontaneous resolution in sarcoidosis. *Thorax* 36: 296–298, 1981.

38. Neville E, James DG, Brewerton DA: HLA antigens and features of sarcoidosis, in Jones-Williams W, Davies BH (eds), *Sarcoidosis*, Cardiff, Alpha and Omega Press, 1980, pp 201–205.

39. Hall G, Naish P, Sharma OP, Doe W, James DG: The epidemiology of sarcoidosis. *Postgrad Med J* 45: 241–250, 1969.

40. Levinsky L, Cummiskey J, Romer FK, et al: Sarcoidosis in Europe: A co-operative study. *Ann NY Acad Sci* 278: 335–434, 1976.

41. Neville E, Walker AN, James DG: Prognostic factors predicting the outcome of sarcoidosis. An analysis of 818 patients. *Q J Med* 208: 525–533, 1983.

42. Sartwell PE: Racial differences in sarcoidosis. *Ann NY Acad Sci* 278: 368–370, 1976.

43. Honeybourne D: Ethnic differences in the clinical features of sarcoidosis in southeast London. *Br J Dis Chest* 74: 63–69, 1980.

44. Johns CJ, MacGregor MI, Zachary JB, Ball WC: Extended experience in the long-term corticosteroid treatment of pulmonary sarcoidosis. *Ann NY Acad Sci* 278: 722–731, 1976.

45. Katz S: Clinical presentation and natural history of sarcoidosis, in Tanburg BL (ed), *Sarcoidosis and Other Granulomatous Diseases of the Lung*. New York, Marcus Dekker, 1983, pp 3–36.

46. James DG, Neville E, Siltzbach LE: A worldwide review of sarcoidosis. *Ann NY Acad Sci* 278: 321–334, 1976.

47. British Thoracic and Tuberculosis Association: Geographical variation in the incidence of sarcoidosis in Great Britain: A comparative study in four areas. *Tubercle* 50: 211–232, 1969.

48. Dunbar RD: Sarcoidosis and its radiologic manifestations. *Crit Rev Diagn Imaging* December: 185–221, 1978.

49. Mitchell DN, Scadding JG: Sarcoidosis. *Am Rev Respir Dis* 110: 774–802, 1974.

50. Israel HL: Prognosis of sarcoidosis. *Ann Intern Med* 73: 1038–1039, 1970.

51. Westcott JL, Noehren TH: Bronchial stenosis in chronic sarcoidosis. *Chest* 63: 893–897, 1973.

52. Dines DE, Stubbs SE, McDougall JC: Obstructive disease of the airways associated with stage I sarcoidosis. *Mayo Clin Proc* 53: 788–791, 1978.

53. Siltzbach LE, James DG, Neville E, et al: Course and prognosis of sarcoidosis around the world. *Am J Med* 57: 847–852, 1974.

54. Wurm K, Rosner R: Prognosis of chronic sarcoidosis. *Ann NY Acad Sci* 278: 732–735, 1976.

55. DeRemee RA: The roentgenographic staging of sarcoidosis: Historic and contemporary perspective. *Chest* 83: 128–132, 1983.

56. Kirks DR, McCormick VD, Greenspan RH: Pulmonary sarcoidosis. Roentgenologic analysis of 150 patients. *Am J Radiol* 117: 777–786, 1973.

57. Freundlich IM, Libshitzh I, Glassman LM, Israel HL: Sarcoidosis, typical and atypical thoracic manifestations and complications. *Clin Radiol* 21: 376–383, 1970.

58. Siltzbach LE: Sarcoidosis: Clinical features and management. *Med Clin North Am* 51: 483–502, 1967.

59. De Labarthe B, Chretien J: Les indices d'active de la sarcoidose: Signification et valeur. *Ann Med Interne* 132: 221–224, 1981.

60. Valeyre D, Saumon G, Georges R, et al: The relationship between disease duration and non-invasive pulmonary exploration in sarcoidosis with erythema nodosum. *Am Rev Respir Dis* 129: 938–943, 1984.

61. Chretien J, Venet A, Israel-Biet D, Clavel F, Sandron D: Summary statement on Disease Activity Assessment. *Ann NY Acad Sci* 465: 479–481, 1986.

62. Panel of the World Association of Sarcoidosis and Other Granulomatous Diseases. Consensus conference: Activity of sarcoidosis. *Eur Respir J* 7: 624–627, 1994.

63. Reynolds HY: Bronchoalveolar lavage. *Am Rev Respir Dis* 135: 250–263, 1987.

64. Reynolds HY: Bronchoalveolar lavage has extended the usefulness of bronchoscopy. *Eur Respir Rev* 28: 48–53, 1992.

65. Campbell D, Poulter LW, duBois RM: Immunocompetent cells in bronchoalveolar lavage reflect the cell populations in transbronchial biopsies in pulmonary sarcoidosis. *Am Rev Respir Dis* 132: 1300–1306, 1985.

66. Semenzato G, Chilosi M, Ossi E, et al: Bronchoalveolar lavage and lung histology: Comparative analysis of inflammatory and immunocompetent cells in patients with sarcoidosis and hypersensitivity pneumonitis. *Am Rev Respir Dis* 132: 400–404, 1985.

67. Paradis IL, Dauber JH, Rabin BS: Lymphocyte phenotypes in bronchoalveolar lavage and lung tissue in sarcoidosis and idiopathic pulmonary fibrosis. *Am Rev Respir Dis* 133: 858–860, 1986.

68. Spiteri MA, Poulter LW: Autologous mixed lymphocyte reactions probe macrophage function in sarcoidosis, in Grassi C, Rizzato G, Pozzi E (eds), *Sarcoidosis and Other Granulomatous Disorders,* Amsterdam, Elsevier, 1988, pp 173–176.

69. Hudspith BN, Flint KC, Geraint James D, Brostoff J, Johnson NM: Lack of immune deficiency in sarcoidosis: Compartmentalization of the immune response. *Thorax* 42: 250–255, 1987.

70. Reynolds HY: Bronchoalveolar lavage, in Murray JF, Nadel JA (eds), *Textbook of Respiratory Medicine*. Philadelphia, WB Saunders, 1988, pp 597–610.

71. Keogh BA, Crystal RG: Alveolitis—the key to the interstitial lung disorders. *Thorax* 37: 1–10, 1982.

72. Hollinger WM, Staton GW, Fajman WA, Gilman MJ, Pine JR, Check IJ: Prediction of therapeutic response in steroid-therapy pulmonary sarcoidosis. Evaluation of clinical parameters, bronchoalveolar lavage, gallium-67 scanning and serum angiotensin-converting enzyme levels. *Am Rev Respir Dis* 132: 65–69, 1985.

73. Costabel U, Bross KJ, Guzman J, Nilles A, Ruhle KH, Matthys H: Predictive value of bronchoalveolar T-cell subsets for the course of pulmonary sarcoidosis. *Ann NY Acad Sci* 465: 418–426, 1985.

74. Israel-Biet D, Venet A, Chretien J: Persistent high alveolar lymphocytosis as a predictive criterion to chronic pulmonary sarcoidosis. *Ann NY Acad Sci* 465: 395–406, 1986.

75. Buchalter S, Aap W, Jackson WAL, Chandler D, Jackson R, Fulmer J: Bronchoalveolar lavage cell analysis in sarcoidosis, a comparison of lymphocyte counts and clinical course. *Ann NY Acad Sci* 465: 678–684, 1986.

76. Delval P, Pencole C, Bourguet P: Predictive value of serum angiotensin-converting enzyme, bronchoalveolar lavage T-lymphocyte subsets and gallium-67 lung scan in pulmonary sarcoidosis. *Sarcoidosis* 3: 177, 1986.

77. Turner-Warwick M, McAllister W, Lawrence R, Britten A, Haslam PL: Corticosteroid therapy in pulmonary sarcoidosis: Do serial lavage lymphycyte counts, serum angiotensin converting enzyme measurements, and gallium-67 scans help management? *Thorax* 4: 903–913, 1986.

78. Bjermer L, Rosenhall L, Angstrom T, Hallgren R: Predictive value of bronchoalveolar lavage cell analysis in sarcoidosis. *Thorax* 43: 284–288, 1988.

79. Ward K, O'Connor C, Odlum C, Fitzgerald MX: Prognostic value of bronchoalveolar lavage in sarcoidosis: The critical influence of disease presentation. *Thorax* 44: 732–738, 1989.

80. Foley NM, Coral AP, Tung K, Hudspith BN, James DG, Johnson NM: Bronchoalveolar lavage cell counts as a predictor of short-term outcome in pulmonary sarcoidosis. *Thorax* 44: 732–738, 1989.

81. Lin YH, Haslam PL, Turner-Warwick M: Chronic pulmonary sarcoidosis—relationships between lung lavage cell counts, chest radiograph and results of standard lung function tests. *Thorax* 40: 501–507, 1985.

82. Gupta RG, Berkerman C, Catchatourian R: Relationship between bronchoalveolar cellular analysis and 67 gallium-citrate uptake in active sarcoidosis. *Clin Res* 27: 662, 1979.

83. Line BR, Hunninghake GW, Keogh BA, Jones AE, Johnston GS: Crystal RG: Gallium-67 scanning to stage the alveolitis of sarcoidosis: Correlation with clinical studies, pulmonary function tests and bronchoalveolar lavage. *Am Rev Respir Dis* 123: 440–446, 1981.

84. Beaumont D, Herry JY, Sapene M, Bourguet P, Larzul JJ, De Labarthe B: Gallium-67 in the evaluation of sarcoidosis: Correlation with serum angiotensin-converting enzyme and bronchoalveolar lavage. *Thorax* 37: 11–18, 1982.

85. Keogh BA, Crystal RG: Pulmonary function testing in interstitial pulmonary disease: What does it tell us? *Chest* 78: 856–865, 1980.

86. Hunninghake GW, Keogh BA, Line BR: Pulmonary sarcoidosis: Pathogenesis and therapy, in Boros D, Yoshida T (eds), *Basis and Clinical Aspect of Granulomatous Diseases,* New York, Elsevier 1980, pp 274–290.

87. Davis GS, Giancola MS, Costanza MC, Low RB: Analysis of sequential bronchoalveolar lavage samples for healthy human volunteers. *Am Rev Respir Dis* 126: 611–616, 1982.

88. Daniele RP, Elias JA, Epstein PE, Rossman ME: Bronchoalveolar lavage: Role in pathogenesis, diagnosis and management of interstitial lung diseases. *Ann Intern Med* 102: 93–108, 1985.

89. Pingleton SK, Harison GR, Stechschulte DJ, Wesselius LJ, Kerby GR, Ruth WE: Effect of location, pH and temperature of instillate in bronchoalveolar lavage in normals. *Am Rev Respir Dis* 128: 1035–1037, 1983.

90. Ainslie G, duBois RM, Poulter LW: Relationship between immunocytological features of bronchoalveolar lavage and clinical indices in sarcoidosis. *Thorax* 44: 501–509, 1989.

91. Hunninghake GW, Gadek JE, Kawanami O, Ferrans VJ, Crystal RG: Inflammatory and immune processes in the human lung in health and disease. Evaluation by bronchoalveolar lavage. *Am J Pathol* 97: 149–206, 1979.

92. Hunninghake GW, Kawanami O, Ferrans VJ, Young RG Jr, Roberts WC, Crystal RG: Characterization of the inflammatory and immune effector cells in the lung parenchyma of patients with interstitial lung disease. *Am Rev Respir Dis* 123: 407–412, 1981.

93. Chensue SW, Boros DL, David CS: Regulation of granulomatous inflammation in murine schistosomiasis: In vitro characterization of T-lymphocyte subsets involved in the production and suppression of migration inhibition factor. *J Exp Med* 151: 1398–1412, 1980.

94. Chensue SW, Wellhausen SR, Boros DL: Modulation of granulomatous hypersensitivity II. Participation of Ly1+ and Ly2+ lymphocytes in the suppression of granuloma formation and lymphokine production in *Schistosoma mansoni*-infected mice. *J Immunol* 127: 363–367, 1981.

95. Hunninghake GW, Crystal RG: Pulmonary sarcoidosis—a disorder mediated by excess helper T-lymphocyte activity at sites of disease activity. *N Engl J Med* 305: 429–434, 1981.

96. Daniele RP, Dauber JH, Rossman MD: Immunologic abnormalities in sarcoidosis. *Ann Intern Med* 92: 406–416, 1980.

97. Thomas PD, Hunninghake GW: Current concepts of pathogenesis of sarcoidosis. *Am Rev Respir Dis* 135: 747–760, 1987.

98. Mornex JF, Cordier G, Pages J, et al: Pulmonary sarcoidosis—flow cytometry measurement of lung T-cell activation. *J Lab Clin Med* 105: 70–76, 1985.

99. Hunninghake GW, Bedell GN, Zevala DC, Monick M, Brady M: Role of interleukin-2 release by lung T-cells in active pulmonary sarcoidosis. *Am Rev Respir Dis* 128: 634–638, 1983.

100. Pinkston P, Bitterman PB, Crystal RG: Spontaneous release of interleukin-2 by lung T-lymphocytes in active pulmonary sarcoidosis. *N Engl J Med* 308:793–800, 1983.

101. Alvarez JM, Silva A, de Landzuri MO: Human T-cell growth factor. Optimal condition for its production. *J Immunol* 123: 977–983, 1979.

102. Watson J, Mochizuki D: Interleukin-2: A class of T-cell growth factors. *Immunol Rev* 51: 257–278, 1980.

103. Hunninghake GW, Gadek JE, Young RC, Kawanami O, Ferrans VS, Crystal RG: Maintenance of granuloma formation in pulmonary sarcoidosis by T-lymphocytes within the lung. *N Engl J Med* 302: 594–598, 1980.

104. Robinson BWS, McLemore TL, Crystal RG: Gamma interferon is spontaneously released by alveolar macrophages and lung T-lymphocytes in patients with pulmonary sarcoidosis. *J Clin Invest* 75: 1488–1495, 1985.

105. Balbi B, Valle MT, Oddera S, et al: T-lymphocytes with γ δ + γ δ 2+ antigen receptors are present in increased proportions in a fraction of patients with tuberculosis or with sarcoidosis. *Am Rev Respir Dis* 148: 1685–1690, 1993.

106. Raulet DH: The structure, function and molecular genetics of the γδ+T cell receptor. *Annu Rev Immunol* 7: 175–205, 1989.

107. Bluestone JA, Matis LA: TCR γδ+ cells; minor redundant T-cell subset or specialized immune system component? *J Immunol* 142: 1785–1788, 1989.

108. Agostini C, Chilosi M, Zambello R, Trentin L, Semenzato G: Pulmonary immune cells in health and disease—lymphocytes. *Eur Respir J* 6: 1378–1401, 1993.

109. duBois RM, Kirby M, Balbi B, Saltini C, Crystal RG: T-lymphocytes accumulation in the lung in sarcoidosis have evidence of recent stimulation of the T-cell antigen receptor. *Am Rev Respir Dis* 45: 1205–1211, 1992.

110. Nakata K, Sugie T, Nakano H, Sakai T, Aoki M: Gamma-delta T-cells in sarcoidosis. Correlation with clinical features. *Am J Respir Crit Care Med* 149: 981–988, 1994.

111. Daniele RP, Rossman MD, Kern JA, Elias JA: Pathogenesis of sarcoidosis—state of the art. *Chest* 89: 174–177, 1986.

112. Venet A, Hance AJ, Saltini C, Robinson BW, Crystal RG: Enhanced alveolar macrophage-mediated antigen-induced T-lymphocyte proliferation in sarcoidosis. *J Clin Invest* 75: 293–301, 1985.

113. Toews GB, Lem VM, Weissler JC, et al: Antigen presentation by alveolar macrophages in patients with sarcoidosis. *Ann NY Acad Sci* 465: 74–81, 1986.

114. Razma AG, Lynch JP, Wilson BS, Ward PA, Kunkel SL: Expression of Ia-like (DR) antigens on human alveolar macrophages isolated by bronchoalveolar lavage. *Am Rev Respir Dis* 129: 419–424, 1984.

115. Campbell DA, duBois RM, Butcher RG, Poulter LW: The density of HLA-DR antigen expression on alveolar macrophages is increased in pulmonary sarcoidosis. *Clin Exp Immunol* 65: 165–171, 1986.

116. Hunninghake GW: Release of interleukin-1 by alveolar macrophages of patients with pulmonary sarcoidosis. *Am Rev Respir Dis* 129: 569–572, 1984.

117. Schyler MR, Steinberg D: Activated alveolar macrophages: IgG and complement receptor. *J Lab Clin Med* 100: 932–942, 1982.

118. Saltini C, Brugni N, Magnani C, et al: Chemiluminescence measurement of phagocytic activity of pulmonary macrophages in sarcoid alveolitis. *Respiration* 4: 291–295, 1984.

119. Hancock WW, Kobzik L, Colby AJ, O'Hara CJ, Cooper AG, Gooleski JJ: Detection of lymphokines and lymphokine receptors in pulmonary sarcoidosis. Immunohistologic evidence that inflammatory macrophages express IL-2 receptors. *Am J Pathol* 123: 1–8, 1986.

120. Spiteri MA, Clarke SW, Poulter LW: Phenotypic and functional changes in alveolar macrophages contribute to the pathogenesis of pulmonary sarcoidosis. *Eur Respir J* 74: 359–364, 1989.

121. Spiteri MA, Poulter LW: Characterization of immune inducer and suppressor macrophages from the normal human lung. *Clin Exp Immunol* 83: 157–162, 1991.

122. Spiteri MA, Clarke SW, Poulter LW: Alveolar macrophages that suppress T-cell responsiveness may be crucial to the pathogenic outcome of pulmonary sarcoidosis. *Eur Respir J* 5: 394–403, 1992.

123. Spiteri MA, Poulter LW, James DG: The macrophage in sarcoid granuloma formation. *Sarcoidosis* 6: 12–14, 1989.

124. Prior C, Knight RA, Spiteri MA: Cytokine networks in pulmonary sarcoidosis. *Eur Respir J* (in press), 1995.

125. Broom BC, Maclaurin BP: Sarcoidosis—correlation of delayed hypersensitivity, MLC reactivity and lymphocytox-icity with disease activity. *Clin Exp Immunol* 15: 355–364, 1973.

126. Mangi RJ, Dwyer JM, Kantor FS: The effect of plasma upon lymphocyte response in vitro—demonstration of a humoral inhibitor in patients with sarcoidosis. *Clin Exp Immunol* 18: 519–528, 1974.

127. Goodwin JS, DeHoratius R, Israel H, Peake GT, Messner RP: Suppressor cell function in sarcoidosis. *Ann Intern Med* 90: 169–173, 1979.

128. Daniele RP: Immunology of sarcoidosis, in *Immunology and Immunological Disease of the Lung*. Boston, Blackwell Scientific Publications, 1988, pp 335–349.

129. Hunninghake GW, Crystal RG: Mechanisms of hypergamma-globulinaemia in pulmonary sarcoidosis: Site of increased antibody production and role of T-lymphocytes. *J Clin Invest* 67: 86–92, 1981.

130. Oreskes I, Siltzbach LE: Changes in rheumatoid factor activity during the course of sarcoidosis. *Am J Med* 44: 60–67, 1968.

131. Veien NK, Hardt F, Bendixen G, et al: Humoral and cellular immunity in sarcoidosis. *Acta Med Scand* 203: 321–326, 1978.

132. Lobop I, Suratt PM: Studies on the autoantibody to lymphocytes in sarcoidosis. *J Clin Lab Immunol* 1: 283–288, 1979.

133. Byrne EB, Evans AS, Fouts DW: Serological hyper-reactivity to Epstein-Barr virus and other viral antigens in sarcoidosis; in Iwai K, Hosoda Y (eds), *Proceedings of the VIth International Conference on Sarcoidosis*. Baltimore, University Park Press, 1974, pp 218–225.

134. Rankin JA, Olchowski J, Naegel GP, Merrill WW, Reynolds C: Immunoglobulin G subclasses in sarcoidosis. *Ann NY Acad Sci* 465: 122–129, 1986.

135. Lawrence EC, Martin RR, Blaese RM, et al: Increased bronchoalveolar IgG secreting cells in interstitial lung disease. *N Engl J Med* 302: 1186–1188, 1980.

136. Daniele RP: Abnormalities of the humoral immune system in sarcoidosis, in Fanburg BL (ed) *Sarcoidosis and Other Granulomatous Disease of the Lung*. New York, Marcell Dekker, 1983, pp 225–242.

137. Hedfors E, Norberg R: Evidence for circulating immune complexes in sarcoidosis. *Clin Exp Immunol* 16: 493–496, 1974.

138. Gupta RC, Kueppers F, De Remee RA, Huston KA, McDuffie FC: Pulmonary and extrapulmonary sarcoidosis in relation to circulating immune complexes. A quantification of immune complexes by two radioimmunoassays. *Am Rev Respir Dis* 116: 261–266, 1977.

139. Daniele RP, McMillan LJ, Dauber JH, Rossman MJ: Immune complexes in sarcoidosis. A correlation with activity and duration of disease. *Chest* 74: 261–264, 1978.

140. Garrett KC, Richerson HB, Hunninghake GW: Pathogenesis of the granulomatous lung disease II—mechanisms of granuloma formation. *Am Rev Respir Dis* 130: 477–483, 1984.

141. Adams DO: The granulomatous inflammatory process: A review. *Am J Pathol* 84: 164–191, 1976.

142. Boros DL: Granulomatous inflammation. *Prog Allergy* 24: 183–267, 1978.

143. Crystal RG, Roberts WC, Hunninghake GW, Gadek JE, Fulmer JD, Line BR: Pulmonary sarcoidosis: A disease characterized and perpetuated by activated T-lymphocytes. *Ann Intern Med* 94: 73–94, 1981.

144. Rosen Y, Athanassiades TJ, Mood S, Lyons HA: Non-granulomatous interstitial pneumonitis in sarcoidosis: Relationship to the development of epithelioid granulomas. *Chest* 74: 122–125, 1978.

145. Bernaudin JF, LaCronique J, Soler P, Lance F, Kawanami O, Basset F: Alveolitis and granulomas: sequential onset and evolution in pulmonary sarcoidosis. *Bull Eur Physiopathol Respir* 17: 27–64, 1981.

146. Scadding JG, Mitchell DN: *Sarcoidosis,* 2d ed. London, Chapman and Hall, 1985.

147. Branson JH, Park JH: Sarcoidosis: Hepatic involvement. *Ann Intern Med* 40: 111–145, 1954.

148. Schonfeld SA, Johns CJ: Sarcoidosis, in, *Recent Advances in Respiratory Medicine.* Edinburgh, Churchill Livingstone, 1986, pp 109–130.

149. Case records of the Massachusetts General Hospital. *N Engl J Med* 310: 1245–1252, 1984.

150. McCann BG, Harrison BDW: Bronchiolar narrowing and occlusion in sarcoidosis—correlation of pathology with physiology. *Respir Med* 85: 65–67, 1991.

151. Swanton RH, Peters DK, Burn JI: Sarcoidosis and amyloidosis. *Proc Soc Med* 64: 1002–1003, 1971.

152. Fresko D, Lazarus SS: Reactive systemic amyloidosis complicating longstanding sarcoidosis. *NY State J Med* 82: 232–234, 1982.

153. Sharma OP: Sarcoidosis. *Disease-a-Month* 36:474–535, 1990.

154. Uehlinger EA: Sarcoid tissue reaction. The origin and significance of inclusion bodies. Differential delineation from tuberculosis. *Acta Med Scand* (suppl) 425: 7–13, 1964.

155. Teilum G: Morphogenesis and development of sarcoid lesions: Similarities to a group of collagenoses. *Acta Med Scand* (suppl) 425: 14–18, 1964.

156. Takayama K, Nobuhiko N, Miyagawa J, Hirano H, Shigematsu N: The usefulness of step-sectioning of transbronchial lung biopsy specimen in diagnosing sarcoidosis. *Chest* 102: 1441–1443, 1992.

157. Jones-Williams W: Pathology of sarcoidosis. *Hosp Med* 2: 21–27, 1967.

158. Rosen Y, Vuletin JC, Pertschuk LP, Silverstein E: Sarcoidosis from the pathologist's vantage point. *Pathol Annu* 14: 405–439, 1979.

159. Spector WG, Epithelioid cells, giant cells and sarcoidosis, in Siltzbach LE (ed), Proceedings VIIth International Conference on Sarcoid, New York, *Ann NY Acad Sci* 278, 3–6, 1976.

160. Carr I, Norris P. The fine structure of human macrophage granules in sarcoidosis. *J Pathol* 122: 29–32, 1977.

161. Ricker W, Clark M: Sarcoidosis: A clinicopathologic review of three hundred cases, including twenty two autopsies. *Am J Clin Pathol* 19: 725–749, 1949.

162. Carrington CB, Giensler EA, Mikus JP, et al: Structure and function in sarcoidosis, in Siltzbach LE (ed) Proceedings VIIth International Conference on Sarcoidosis. *Ann NY Acad Sci* 278: 265–283, 1976.

163. Beekman JF, Zimmet SM, Chun BK, Miranda AA, Katz S: Spectrum of pleural involvement in sarcoidosis. *Arch Intern Med* 136: 323–330, 1976.

164. Rosen Y, Moon S, Huang CT, et al: Granulomatous pulmonary angiitis in sarcoidosis. *Arch Pathol Lab Med* 101: 170–174, 1977.

165. Levine B, Saldana M, Hunter A: Pulmonary hypertension in sarcoidosis. A case report of a rare but potentially treatable cause. *Am Rev Respir Dis* 103: 413–417, 1971.

166. Smith L, Lawrence JB, Katzenstein A: Vascular sarcoidosis: A rare cause of pulmonary hypertension. *Am J Med Sci* 285: 38–44, 1983.

167. Hoffstein V, Ranganathan N, Mullen JMB: Sarcoidosis simulating pulmonary veno-occlusive disease. *Am Rev Respir Dis* 134: 809–811, 1986.

168. Jones-Williams W: The nature and origin of Schaumann bodies. *J Pathol Bacteriol* 79: 193–201, 1960.

169. Kirkpatrick CJ, Curry A, Bisset DL: Light and electron-microscopic studies on multinucleated giant cells in sarcoid granuloma: New aspects of asteroid and schaumann bodies. *Ultrastruct Pathol* 12: 581–597, 1988.

170. Jones-Williams W, Williams D: The properties and development of conchoidal bodies in sarcoid and sarcoid-like granulomas. *J Pathol Bacteriol* 96: 491–496, 1968.

171. Cain H, Kraus B: Asteroid bodies: Derivatives of the cytosphere. An EM contribution to the pathology of the cytocentre. *Virchows Arch (Cell Pathol)* 26: 119–132, 1977.

172. Boskey AL: The role of calcium-phosphlolipid-phosphate complexes in tissue materialisation. *Metab Bone Dis Relat Res* 1: 137–142, 1978.

173. Boyd JF, Valentine JC: Unclassified yellow bodies in human lymph nodes. *J Pathol* 102: 58–60, 1970.

174. Doyle WF, Brahman HD, Burgess JH: The nature of yellow-brown bodies in peritoneal lymph nodes. Histochemical and electron microscopic evaluation of these bodies in a case of suspected sarcoidosis. *Arch Pathol Lab Med* 96: 320–326, 1973.

175. Pertsehuk LP, Silvestein E, Friedland J: Immunohistologic diagnosis of sarcoidosis: Detection of angiotensin converting enzyme in sarcoid granulomas. *Am J Clin Pathol* 75: 350–354, 1981.

176. Smellie H, Hoyle C: The natural history of pulmonary sarcoidosis. *Q J Med* 116: (new series) 539–558, 1960.

177. Pinner M: Non-caseating tuberculosis. *Am Rev Tuberc* 37: 690–728, 1938.

178. Riley ER: Boeck's sarcoid: A review based upon a clinical study of 52 cases. *Am Rev Tuberc* 52: 231–285, 1950.

179. Scadding JG: Discussion on sarcoidosis. *Proc R Soc Med* 49: 799–802, 1956.

180. Bashout FA, McConnell T, Skinner W, Hanson M: Myocardial sarcoidosis. *Dis Chest* 53: 413–420, 1968.

181. Reisner D: Observations on the cause and prognosis of sarcoidosis with special consideration of intra-thoracic manifestations. *Am Rev Respir Dis* 96: 361–380, 1967.

182. Siltzbach LE, James DG, Turiaf J, et al: Course and prognosis of sarcoidosis around the world. *Am J Med* 57: 847–852, 1974.

183. Sones M, Israel HL: Course and prognosis of sarcoidosis. *Am J Med* 29: 84–93, 1960.

184. Huang CT, Heurich AE, Sutton AL, Lyons HA: Mortality in sarcoidosis: A changing pattern of the causes of death. *Eur J Respir Dis* 62: 231–238, 1981.

185. Gendel BR, Young JM, Greiner DJ: Sarcoidosis. A review with 24 additional cases. *Am J Med* 12: 205–218, 1952.

186. Brincker H, Wilbek E: The incidence of malignant tumours in patients with respiratory sarcoidosis. *Br J Cancer* 29: 247–251, 1974.

187. Toner GC, Bosl GJ: Sarcoidosis, "Sarcoid-like lymphadenopathy" and testicular germ cell tumours. *Am J Med* 89: 651–656, 1990.

188. Suen JS, Forse MS, Hyland RH, Chan CK: The malignancy–sarcoidosis syndrome. *Chest* 98: 1300–1302, 1990.

189. Myers JJ, Granville MB, Witter BA: Hairy cell leukemia and sarcoidosis. *Cancer* 43: 1777–1781, 1979.

190. Falini B, Tabilio A, Velardi A, Cenetti C, Aversa F, Martelli

MF: Multiple myeloma with sarcoidosis-like reaction. *Scand J Haematol* 29: 211–216, 1982.

191. Turkington RW, Buckley CE: Macrocryoglobulinaemia and sarcoidosis. *Am J Med* 40: 156–164, 1966.

192. Goldstein RA, Israel HL, Rawnsley HM: Effect of race and stage of disease on the serum immunoglobulins in sarcoidosis. *JAMA* 208: 1153–1155, 1969.

193. Douglas JG, Gillon J, Logan RSA, Grant IWB, Crompton GK: Sarcoidosis and coeliac disease: An association? *Lancet* 2: 13–15, 1984.

194. Taylor RG, Fisher C, Hoffbrand BI: Sarcoidosis and membranous glomerulonephritis. Significant association. *Br Med J* 284: 1297–1298, 1982.

195. Seinfield ED, Sharma OP: The TASS syndrome. *J R Soc Med* 76: 883–885, 1983.

196. Stanley NN, Fox RA, Whimster WF, Sherlock F, James DG: Primary biliary cirrhosis or sarcoidosis or both? *N Engl J Med* 287: 1282–1284, 1972.

197. Fagan EA, Moore-Gillon JC, Turner-Warwick M: Multi-organ granulomas and mitochondrial antibodies. *N Engl J Med* 308: 572–575, 1983.

198. Bass NM, Burroughs AK, Scheuer PJ, et al: Chronic intrahepatic cholestasis due to sarcoidosis. *Gut* 23: 417–421, 1982.

199. Wallert B, Bonniere P, Prin L, et al: Primary biliary cirrhosis. Subclinical inflammatory alveolitis in patients with normal chest roentgenograms. *Chest* 90: 842–828, 1986.

200. Spiteri MA, Johnson MA, Epstein O, et al: Immunological features of lung lavage cells from patients with primary biliary cirrhosis may reflect those seen in pulmonary sarcoidosis. *Gut* 31: 208–212, 1990.

201. Maddrey WC: Sarcoidosis and the primary biliary cirrhosis-associated disorders? *N Engl J Med* 308: 588–590, 1983.

202. Leibow AA: The J Burns Amberson Lecture. Pulmonary angiitis and granulomatosis. *Am Rev Respir Dis* 108: 1–18, 1973.

203. Saldana MJ: Necrotising sarcoid granulomatosis: Clinicopathologic observations in 24 patients (abstract). *Lab Invest* 38: 364, 1978.

204. Churg A: Pulmonary angiitis and granulomatosis revisited. *Hum Pathol* 14: 868–883, 1983.

205. Koss MN, Hochholzer L, Faigan S, Garancis TC, Ward PA: Necrotising sarcoid-like granulomatosis: Clinical, pathologic and immunopathologic findings. *Hum Pathol* 11: 510–519, 1980.

206. Saldana MJ, Israel HL: Necrotising sarcoid granulomatosis, benign lymphocytic angiitis and granulomatosis: Do they exist? *Semin Respir Med* 10: 182–188, 1989.

207. Singh HN, Cole S, Krause PJ, Conway M, Garcia L: Necrotizing sarcoid granulomatosis with extrapulmonary involvement. Clinical, pathologic, ultrastructural and immunologic features. *Am Rev Respir Dis* 124: 189–192, 1981.

208. Beach RC, Corrin B, Scopes JW, Graham E: Necrotizing sarcoid granulomatosis with neurologic lesion in a child. *J Paediatr* 97: 950–953, 1980.

209. Rolfes DB, Weiss MA, Sanders MA: Necrotising sarcoid granulomatosis with suppurative features. *Am J Clin Pathol* 82: 602–607, 1984.

210. Spiteri MA, Gledhill A, Campbell D, Clark SW: Necrotising sarcoid granulomatosis. *Br J Dis Chest* 81: 70–75, 1987.

211. Takemura T, Matsui Y, Oritsu M, et al: Pulmonary vascular involvement in sarcoidosis: Granulomatous angiitis and microangiopathy in transbronchial lung biopsies. *Virchows Archiv (A) (Pathol Anat)* 418: 361–368, 1991.

212. Takemura T, Matsui Y, Saiki S, Mikami R: Pulmonary vascular involvement in sarcoidosis: A report of 40 autopsy cases. *Hum Pathol* 23: 1216–1223, 1992.

213. Doyle LD, McWilliam L, Hasleton PS: Giant cell arteritis with pulmonary involvement. *Br J Dis Chest* 82: 88, 1992.

Chapter *17*

Lung Injury Due to Radiation

P. S. Hasleton

Pulmonary radiation injury is seen because this treatment is used not only for carcinoma of the bronchus but also for Hodgkin's disease, carcinoma of the esophagus and breast, and for other tumors. Recently intraluminal radiotherapy has been used for the palliation of lung cancer with a high-dose rate micro-Selectron.[1] Although this treatment is effective in the treatment of hemoptysis, dyspnea, and cough and may limit the spread of radiation to the surrounding lung, the long-term effect of such a concentrated therapy has not yet been elucidated.

Further problems may be encountered by combined treatment with radiotherapy and chemotherapy, often used for solid tumors and lymphomas. These combined treatments increase the risks of pulmonary toxicity.[2]

MECHANISMS OF RADIATION INJURY

These have been reviewed.[3] Irradiation produces breaks in DNA. Single-strand breaks are probably of little consequence because the cell can efficiently repair them. Without intact templates for their mutual repair, double-stranded breaks will be misrepaired and disrupt the integrity of the chromosome. Chromosomal aberrations do not normally affect the survival or function of cells between the time they are irradiated and attempt to replicate themselves. Cell death usually occurs at the first or subsequent mitotic divisions. Therefore, normal tissues and tumors show radiation responses at a rate proportional to their proliferative turnover rate. Respiratory epithelium, which is actively proliferative, develops a detectable reaction within 2 to 3 weeks of first exposure. The rate of appearance of injury depends also on the life span of the differentiating progeny of these cells. Platelets and white cells may decrease quickly after irradiation but anemia is uncommon because of the slow turnover of mature erythrocytes and the ability of the surviving precursor cells to compensate for injury before the effects of radiation injury become obvious.

Slowly proliferating tissues, such as lung, respond tardily to irradiation. Signs of damage appear only months or years after exposure. Damaged DNA is repaired within a few hours, but the extent of repair is unequal in all tissues. In general slow-responding tissues, including lung, are capable of greater repair than malignant tissues. However, lung lesions secondary to radiation are common.[4]

In addition to molecular changes, there is a distinct inflammatory syndrome.[5] This may be initiated by the release of free radicals,[6] in a similar fashion to that seen in adult respiratory distress syndrome (ARDS). In acute radiation injury involving the thorax, hyaline membranes are seen. The highly charged, often OH molecules, directly damage DNA. Another source of tissue injury in radiation is the influx into the interstitium and alveolar spaces of neutrophils, macrophages, and lymphocytes.[7,8]

Repopulation by surviving cells of the proliferating normal tissue occurs as a homeostatic response to injury. This allows slowly proliferating tissues (e.g., respiratory epithelium), to tolerate an increased dose given to a surrounding tumor. Radiation injury of small vessels such as capillaries and arterioles is commonly seen in different tissues.[9] Damage to large arteries is less common but has been documented in several instances.[10]

Pulmonary endothelial cells have a doubling time of approximately 8 weeks,[11] which is reduced to 4 weeks after injuries such as radiation. Type I pneumocytes are fixed postmitotic cells and are replaced in radiation damage by type II pneumocytes, with a turnover time of 28 to 35 days in experimental animals. Cells of the bronchial epithelium have a short turnover time, between 1 and 3 weeks.[12]

The pathogenesis of the fibrosis is unknown. It is possible that radiation may directly injure the fibroblasts or their matrix. The mechanisms of collagen production are probably similar to that seen in ARDS.

CLINICAL SYNDROME OF RADIATION PNEUMONITIS AND FIBROSIS

There are two clinical phases—acute pneumonitis and postradiation fibrosis.[13] Because of a slow turnover rate of pulmonary cells, there is considerable delay in the evolution of any radiation tissue injury in the lung. Often no pulmonary physical signs are seen, but there may be consolidation in the region corresponding to the radiation.[14]

The onset of pneumonitis is insidious and depends on the amount of lung tissue irradiated. Radiation pneumonitis usually begins 2 to 6 months after cessation of the radiation therapy, although this interval may be as short as two weeks.[7] There is initially a troublesome, nonproductive cough, with subsequent expectoration of thick white sputum. Hemoptyis is unusual. Dyspnea is usually seen on exertion but there may be progression to severe respiratory distress. In addition, fever and nonspecific respiratory symptoms such as chest pain occur. Rarely spontaneous pneumothoraces may be found.[14] A pleural effusion or a friction rub may be encountered. Radiologic abnormalities are seen in half the patients receiving irradiation but only 5 to 15 percent develop the clinical syndrome of radiation pneumonitis.[15]

In the early stages, the chest radiograph may be normal but develops into alveolar infiltrates, which coalesce to define the treatment portals in a "straightedge effect"[16] (Fig. 17-1). Changes may occur outside the radiation field. The reason for changes in a contralateral lung are unknown.

Adult respiratory distress syndrome may rarely develop after limited thoracic radiotherapy.[17] The

three patients described had received no drugs known to cause lung injury and cardiogenic pulmonary edema and infection were excluded. Reference should be made to the section on ARDS where a form of localized ARDS is described.[18] It is possible that such patients have lung damage due to the irradiation, then for some, as yet unknown reason, proceed to the full-blown picture of ARDS. Such ARDS follows days or weeks after radiotherapy.[19] ARDS may follow whole body irradiation in preparation for bone marrow transplantation.

Several factors influence the severity of the pneumonitis. These are a high radiation dose, the volume of lung treated, and the greater the dose per fraction.[20]

Most patients are asymptomatic because there is a great functional pulmonary reserve. The pulmonary changes may resolve spontaneously. Patients who have had severe pneumonitis may progress to chronic respiratory failure with dyspnea, cor pulmonale, cyanosis, and even finger clubbing. If there is unilateral irradiation, mediastinal shift may be seen.

PATHOLOGY

Much work has been carried out in animals, but problems exist with extrapolation of the experimental data to human beings.

Most mammals, especially the rat and rabbit, do not develop hyaline membranes.[9] Sequential ultrastructural studies in rats and mice[21,22] show the initial lesion is in the endothelial cell and occurs within 5 days. These cells swell and detach from a basement membrane and platelet thrombi form, which subsequently obstruct the capillary lumina. Alveolar epithelial cells are also damaged but not as early as the endothelial cells. Patchy areas of atelectasis are common in the first few months after radiation exposure.[23] Fibrosis begins at 1 month and progresses thereafter (Fig. 17-2). These studies are confounded by the fact that there is either whole body irradiation or radiation that covers the entire thorax and therefore is not equivalent to most lesions seen in man.

The sequence of injury is shown in Fig. 17-3. This shows capillary injury with formation of blebs 8 weeks after irradiation at the stage where acute and subacute periods merge. Twenty weeks after irradiation (see Fig. 17-3C) there are damaged atypical type II cells, regenerating capillaries and infiltrates of neutrophil polymorphs, and mast cells as well as fibrosis. Six months after irradiation there is an increase in collagen in the alveolar septum as well as an

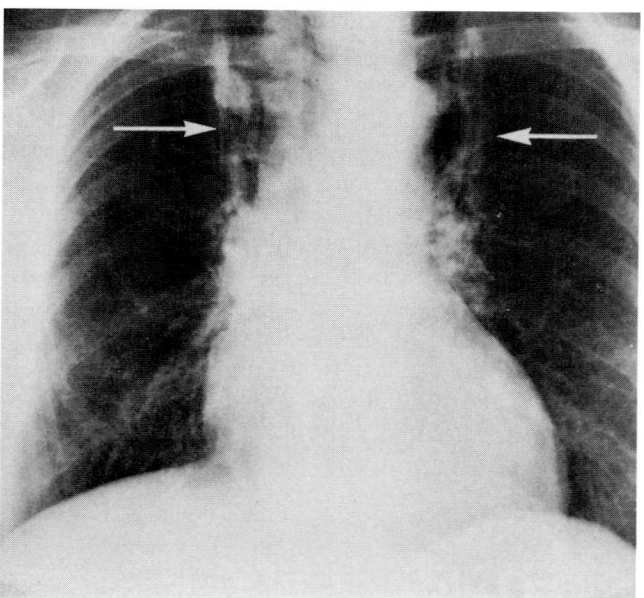

FIGURE 17-1
Straight-edge effect of radiation therapy on chest radiograph. Patient had carcinoma of the upper esophagus at the thoracic inlet. Onset of acute radiation pneumonitis in the paramediastinal areas (*arrows*) 3 months after 6000 cGy in 35 fractions.

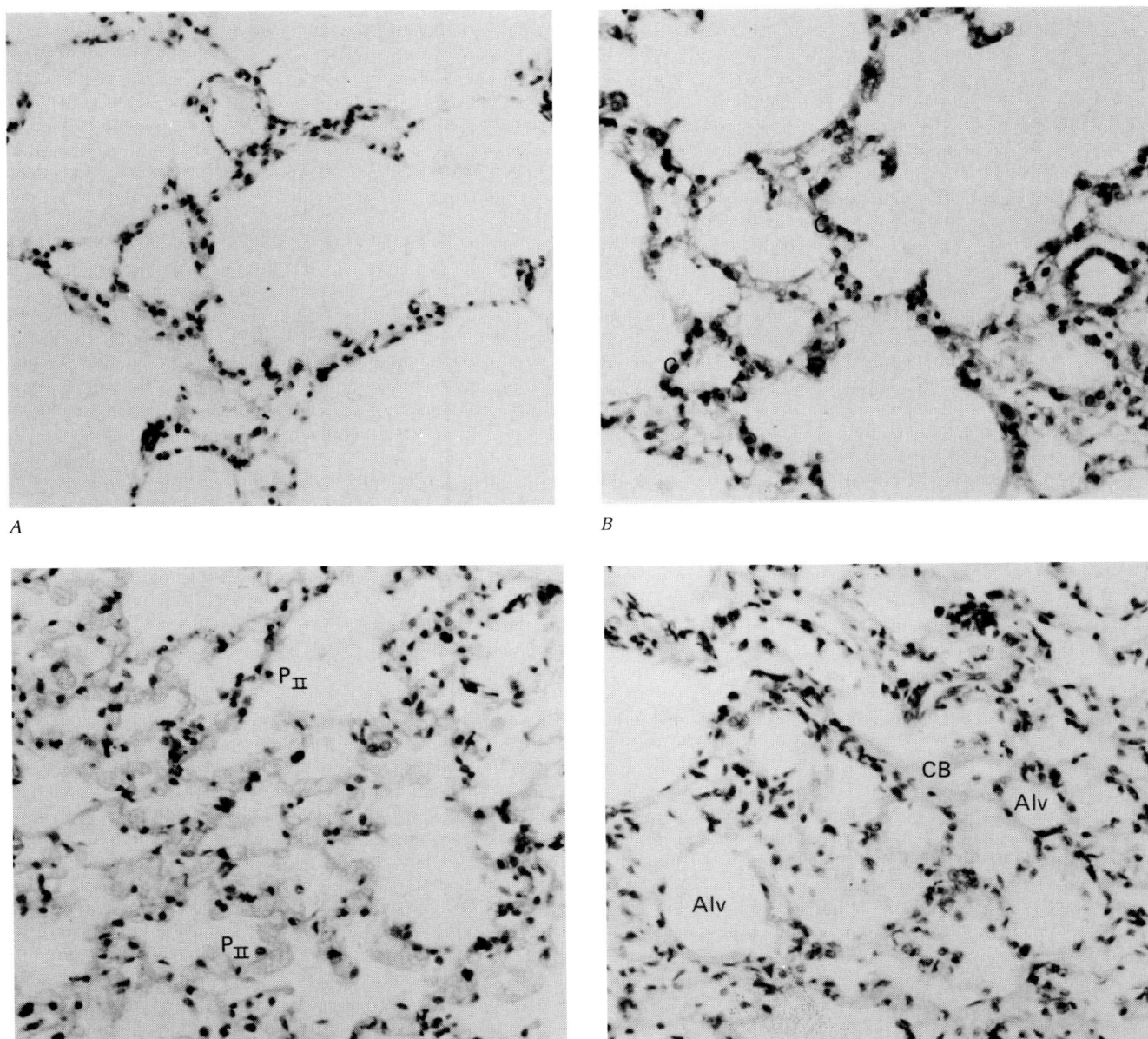

A

B

C

D

FIGURE 17-2
Histologic changes in rat lung (SPF Sprague-Dawley) irradiated at 3 months of age. *A.* Control unirradiated lung from a 5-month-old rat. *B.* Irradiated rat lung 4 months after 2000 cGy in one fraction to the left hemithorax. The capillaries (C) are engorged. Lacking are inflammatory cells or macrophage collections. (×74.) *C.* Rat lung 6 months after irradiation with 2000 cGy. Alveolar volume is decreased; capillaries are engorged and obstructed. The number of atypical type II pneumocytes (P$_{II}$) in the epithelium is increased. (×74.) *D.* Rat lung 12 months after 2000 cGy. The reaction has now progressed to fibrosis. Many alveoli have been obliterated, and many of the residual alveoli (Alv) are small. The septa are thickened by collagen and deposits of basement membrane material (CB). The number of epithelial cells is not increased, but some atypical type II cells remain; interstitial infiltrates persist. (×74.)

increased number of mast cells (see Figure 17-3*E* and *F*).

Twelve months after irradiation some of the alveolar septa contain capillaries located within a thickened basement membrane containing prominent collagen, elastin, and reticulin fibers as well as enlarged endothelial cells.

If lethal doses of irradiation are given there is congestion, intraalveolar edema with an increase in lymphocytes, polymorphs, and macrophages. The injury extends from alveoli to alveolar ducts and bronchioles but usually there is little change in the bronchi.[24]

Histologically, acute radiation injury has been divided into (1) a latent period, (2) acute pneumonitis, and (3) a late phase.

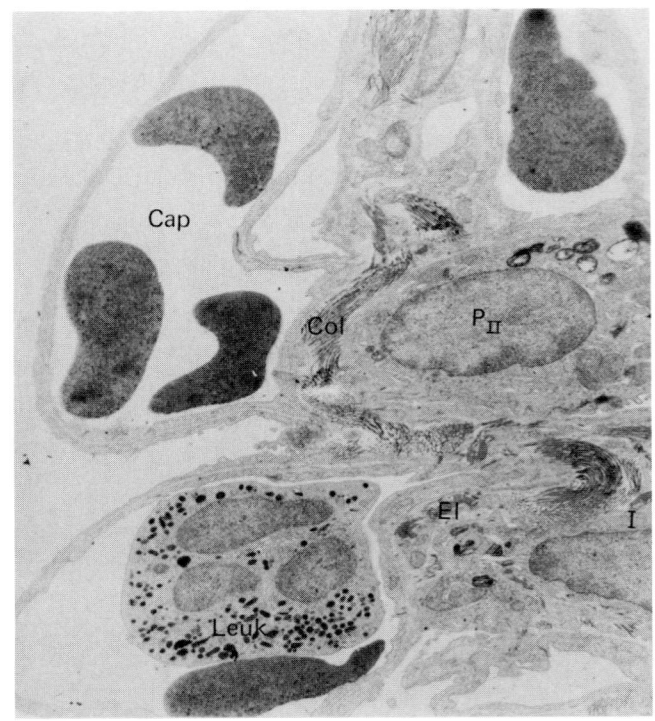

A

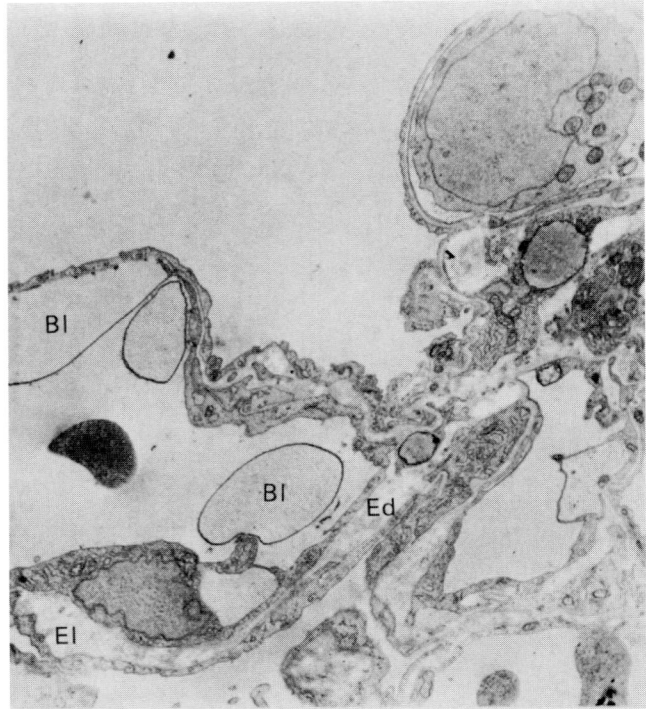

B

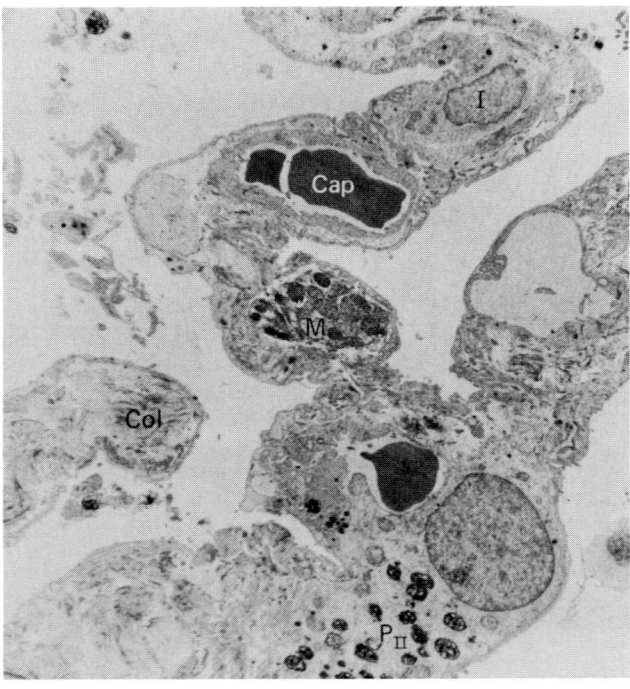

C

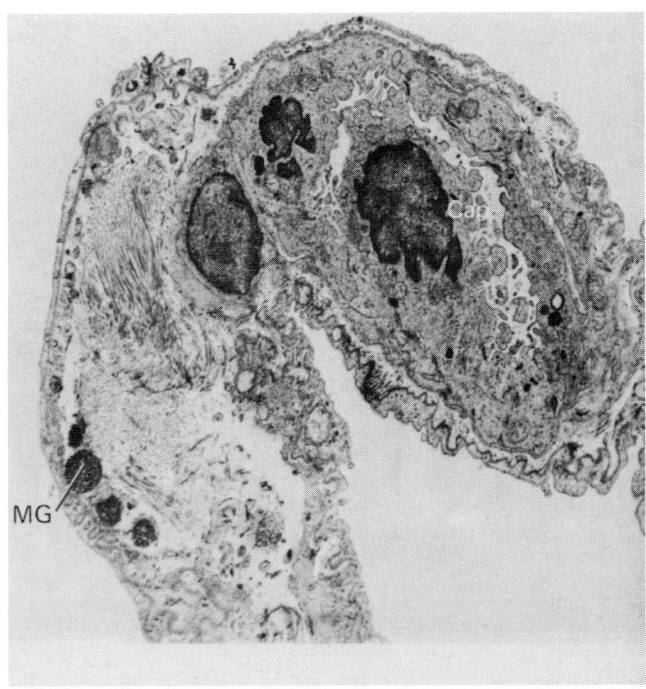

D

FIGURE 17-3

Electron micrographs of rat lung taken from the same group of animals as Fig. 17-2. Fixed in osmium tetroxide, embedded in araldite. *A.* Control unirradiated lung. The capillary lumen (Cap) contains red cells and a polymorphonuclear leukocyte (Leuk). Collagen (Col), elastic fibers (El), an interstitial cell (I), and a type II pneumocyte (P$_{II}$) are also present. (×4950.) *B.* Alveolar septum 8 weeks after 2000 cGy. The endothelium is abnormal and contains blebs (Bl). The pericapillary space is edematous (Ed). (×4950.) *C.* Several alveolar septa 20 weeks after 2000 cGy. A new capillary (Cap), mast cell (M), type II pneumocyte (P$_{II}$), collagen (Col), and hyperplastic interstitial cell (I). (×2475.) *D.* Alveolar septum partly replaced by collagen and mast cell granules (MG) and by an abnormal capillary (Cap), 6 months after irradiation. (×5050.)

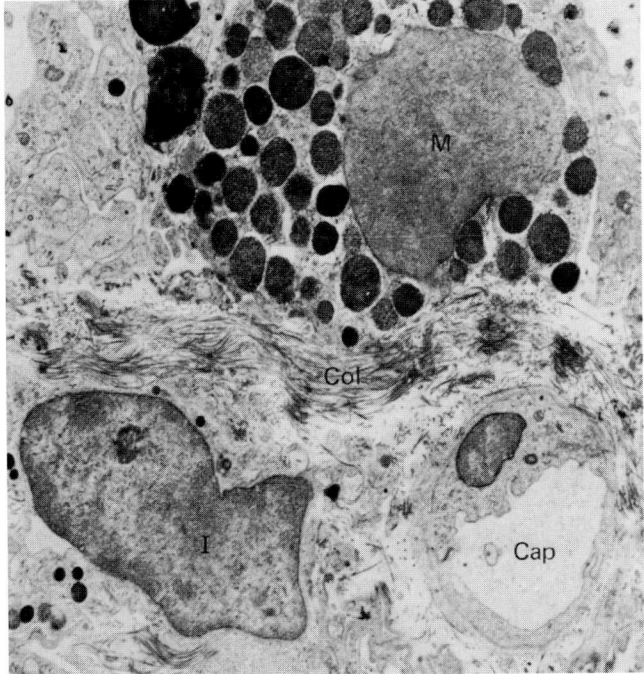

E

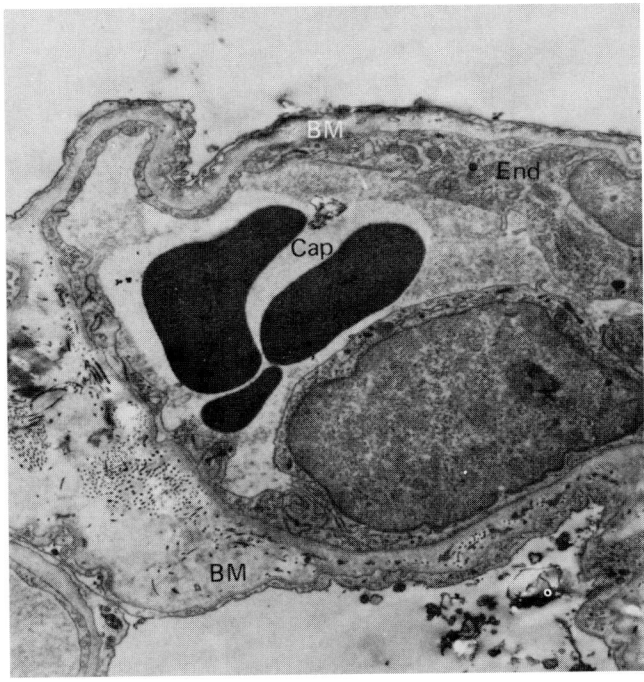

F

FIGURE 17-3

(Continued) E. Portion of an alveolar septum showing regenerating capillary (Cap), mast cell (M), and interstitial cell (I) in close proximity to collagen production (Col), 6 months postirradiation. (×742.)   *F.* Alveolar septum 12 months after 2000 cGy. The capillary (Cap) is patent, but the endothelium (End) and the basement membranes (BM) are markedly thickened. (×4950.)

Latent Period

This occurs soon after the radiotherapy and there is congestion of the alveolar walls and intraalveolar edema (Fig. 17-4) due to endothelial cell damage. An increased level of alveolar surfactant is one of the earliest detectable changes after experimental lung irradiation.[25] There is desquamation of epithelial cells from the alveolar walls and a consequent infiltration of chronic inflammatory cells.

Acute Pneumonitis

In the weeks to months after radiation the interstitial edema in the alveolar walls organizes into fibrosis. These changes may resolve in weeks to months but there may be hyperplasia of the type II pneumocytes with large bizarre hyperchromatic nuclei. These should not be confused with tumor cells. The interstitial infiltrate is lymphocytic and few neutrophil polymorphs are seen. There is also an increase in alveolar macrophages. Hyaline membranes may line a variable number of alveoli (Fig. 17-5) and are often difficult to demonstrate on biopsies. There is an influx of myoepithelial cells into arterioles. Some arteries show hyalinization with glassy, thick-walled vessels (Fig. 17-6). There may be foam cells in the intima but these are nonspecific.

Bronchiolitis obliterans has been documented in the early postirradiation period.[26]

Late Phase

This is characterized by progressive fibrosis of the alveolar walls (Fig. 17-7) and an increase in arteriolar changes, loss of hyaline membranes in some cases, and marked cellular atypia of the alveolar lining cells. There may be frayed elastic fibers in the thickened septae (Figs. 17-8 and 17-9). In the late stages alveoli collapse and are obliterated by connective tissue, leaving dense fibrosis. There is evidence of telangiectasia in the thickened alveolar walls (Fig. 17-10). In common with other radiation injury, radiation fibroblasts may be seen (Fig. 17-11). Bronchial walls are thickened by fibrosis and edema and also show destruction of the bronchial cartilage (Fig. 17-12).

Before ascribing all the changes to radiotherapy it is important to exclude any concomitant chemotherapy, which can also produce similar changes.

There is evidence that the pneumonitis and fibrosis are two separate phases and late events may not be preceded by acute effects.[27]

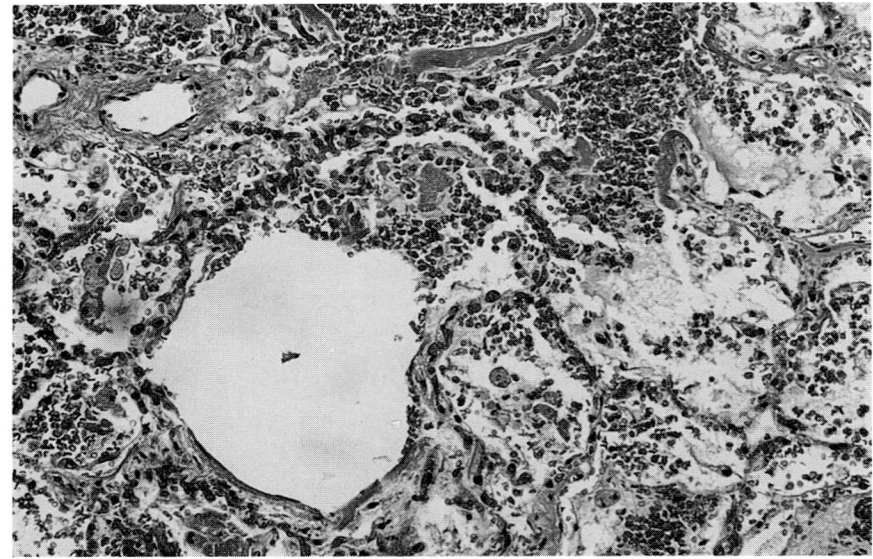

FIGURE 17-4
Acute radiation change after total body irradiation. There is intraalveolar edema, hemorrhage, and congestion of the alveolar wall. Scanty hyaline membranes are present. (H&E, ×125.)

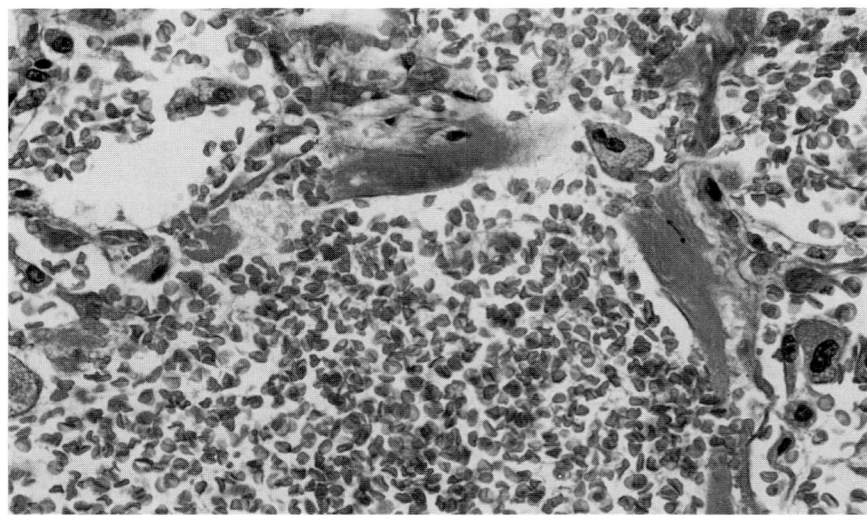

FIGURE 17-5
Prominent hyaline membranes lining an alveolar duct. A few cells with hyperchromatic nuclei are seen. Same patient as in Figure 17-4. (H&E, ×313.)

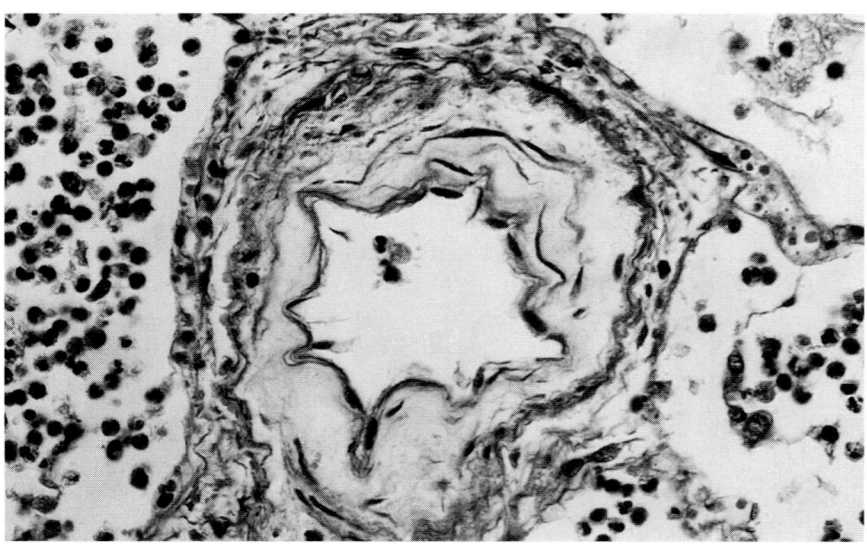

FIGURE 17-6
Chronic radiation damage. Pulmonary artery with loss of the normal media and replacement by a glassy, pale-staining material. (H&E, ×313.)

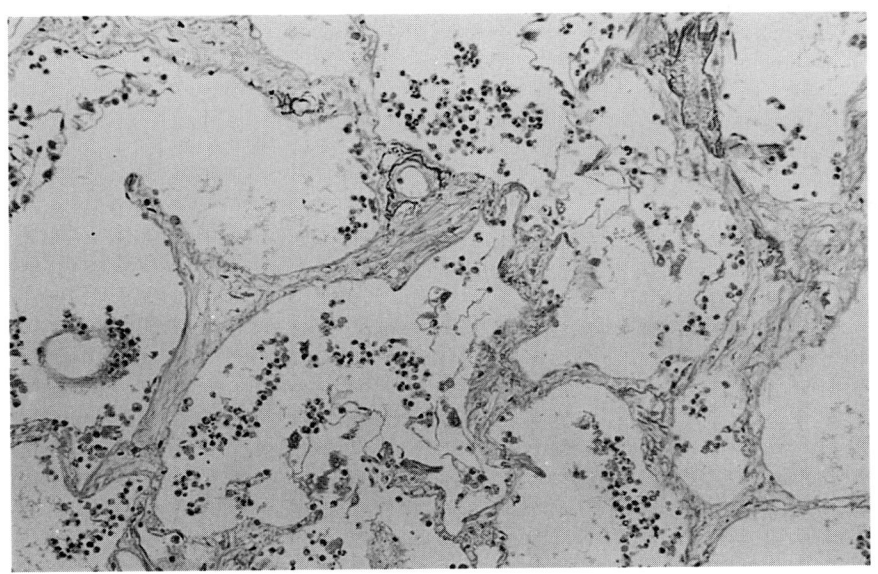

FIGURE 17-7
Chronic radiation damage. Interstitial pulmonary fibrosis with pale, loose, myxoid fibrosis. There is acute inflammation in the dilated alveolar spaces. (H&E, ×125.)

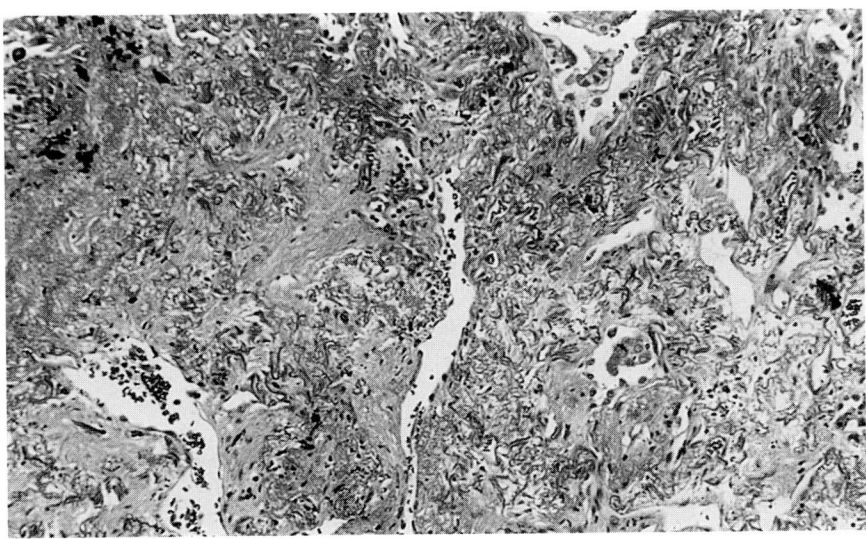

FIGURE 17-8
Chronic radiation damage. Loss of alveoli and replacement by confluent fibrosis and some elastic tissue. (H&E, ×125.)

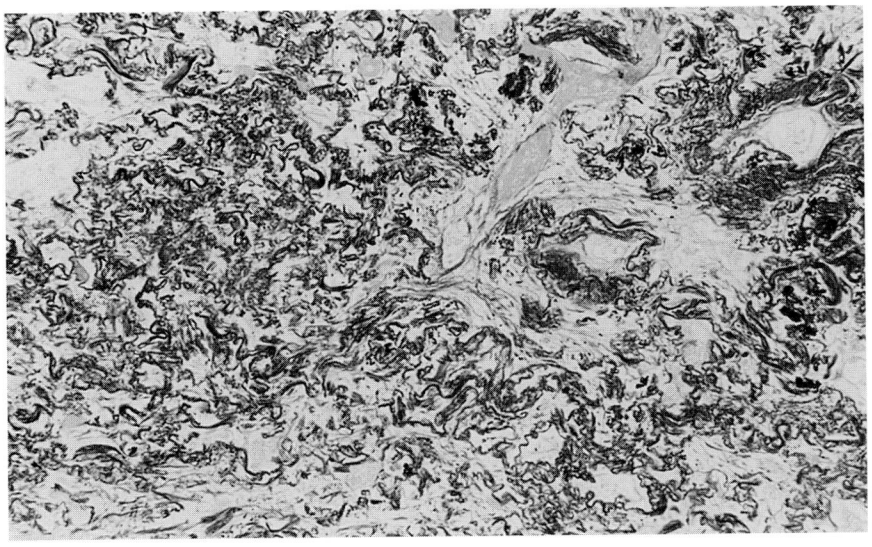

FIGURE 17-9
Chronic radiation damage. Obliteration of alveoli by fibrosis and prominent elastic tissue. (Elastic-van Gieson, ×125.)

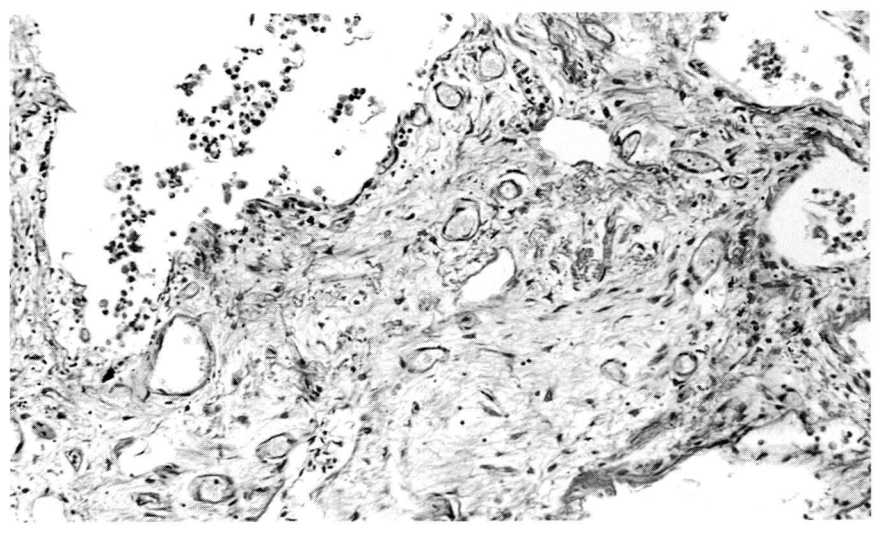

FIGURE 17-10
Chronic radiation damage. Telangiectasia in an area of fibrosis. (H&E, ×125.)

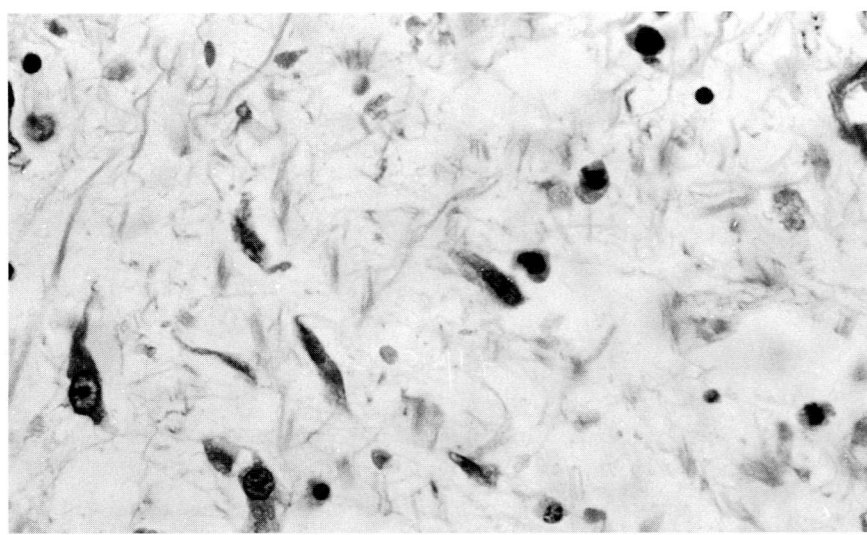

FIGURE 17-11
Chronic radiation damage. Radiation fibroblasts with prominent, hyperchromatic nuclei. (H&E, ×125.)

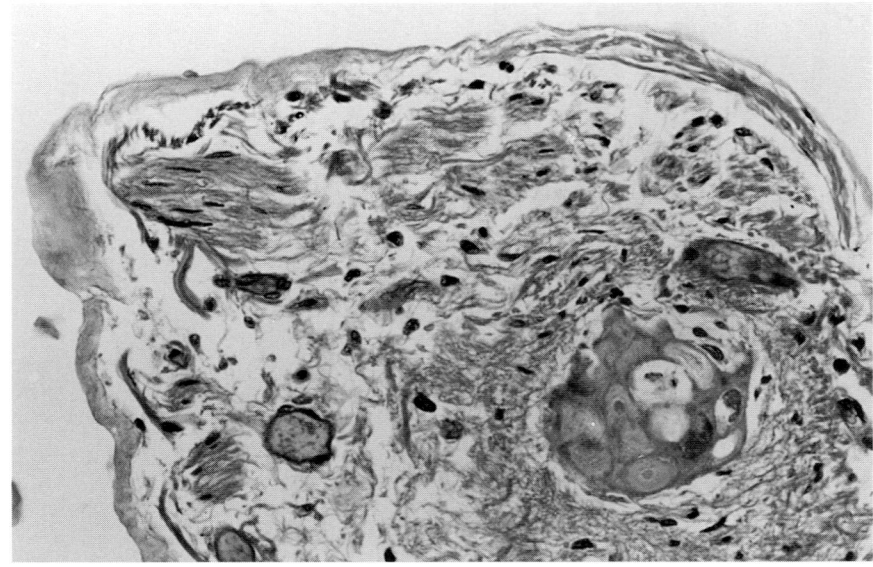

FIGURE 17-12
Chronic radiation damage. Bronchial wall with loss of epithelium, a thickened basement membrane, and fibrosis in the wall. There is destruction of the cartilage, which shows no viable nuclei. (H&E, ×125.)

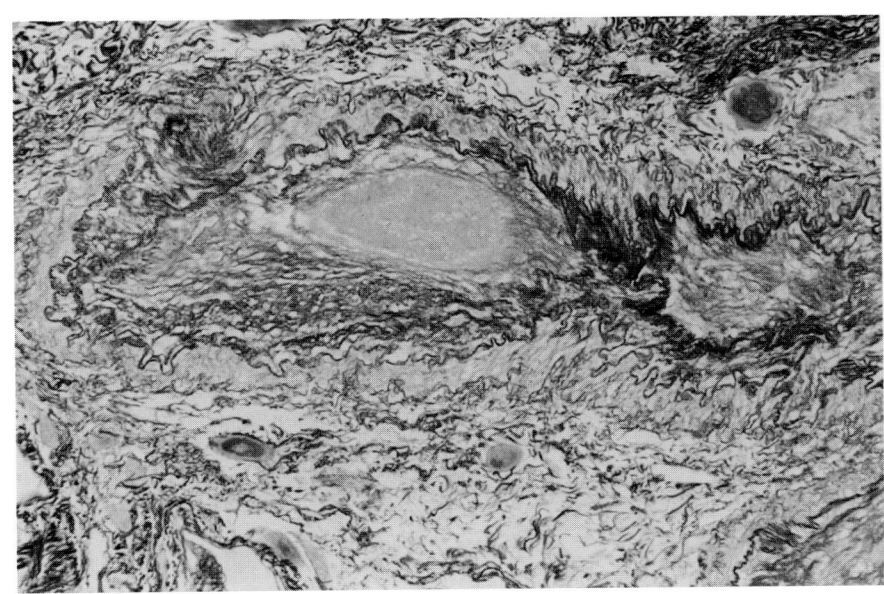

FIGURE 17-13
Chronic radiation damage. Muscular pulmonary artery with marked intimal fibrosis and recanalization. (Elastic-van Gieson, ×125.)

The vascular changes from irradiation[28] have been described in a series of patients receiving radiotherapy, the last dose being given between 3 months and 19 years. None of the cases showed any interstitial fibrosis or right ventricular hypertrophy.

The pulmonary arteries showed intimal fibrosis (Fig. 17-13) in both irradiated and nonirradiated areas of lung as well as intimal longitudinal muscle (Fig. 17-14). Some arteries in irradiated areas showed eccentric intimal fibrous plaques, as well as breaks in the internal elastic lamina (Fig. 17-15). Fibrin thrombi were seen in a few capillaries (Fig. 17-16).

Two forms of occlusive venous lesions were described, both of which were seen in irradiated and nonirradiated tissue. The first consisted of fibrous intimal tissue (Fig. 17-17). The second consisted of "collander" or recanalization lesions (Fig. 17-18) seen in old thrombi. There was a greater amount of concentric intimal thickening of the venous system in the irradiated areas as compared to the nonirradiated areas. The difference in the changes between arteries and veins was thought to be due to the fact that the pulmonary veins carry oxygen and one of the most important free radicals is the superoxide ion O^3.

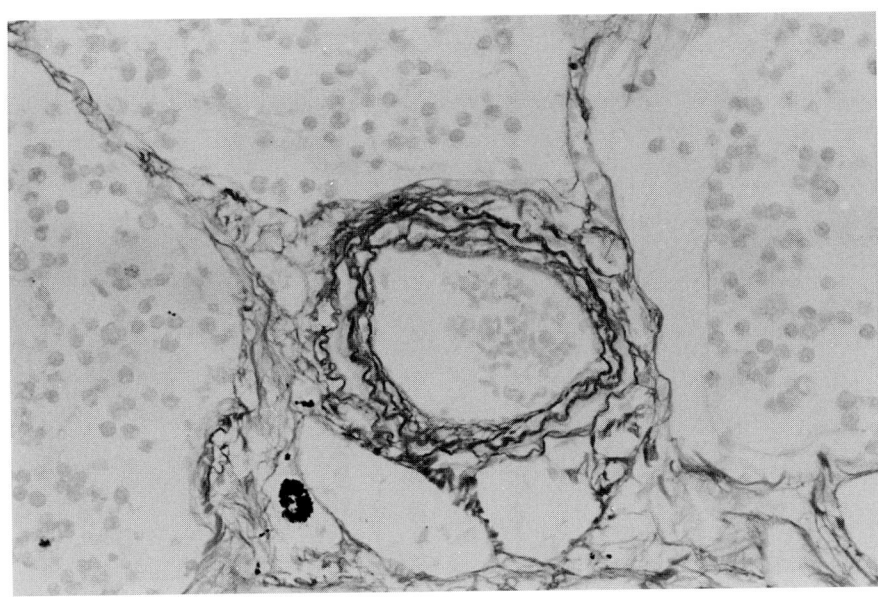

FIGURE 17-14
Chronic radiation damage. Intimal longitudinal muscle in a muscular pulmonary artery. (Elastic-van Gieson, ×313.)

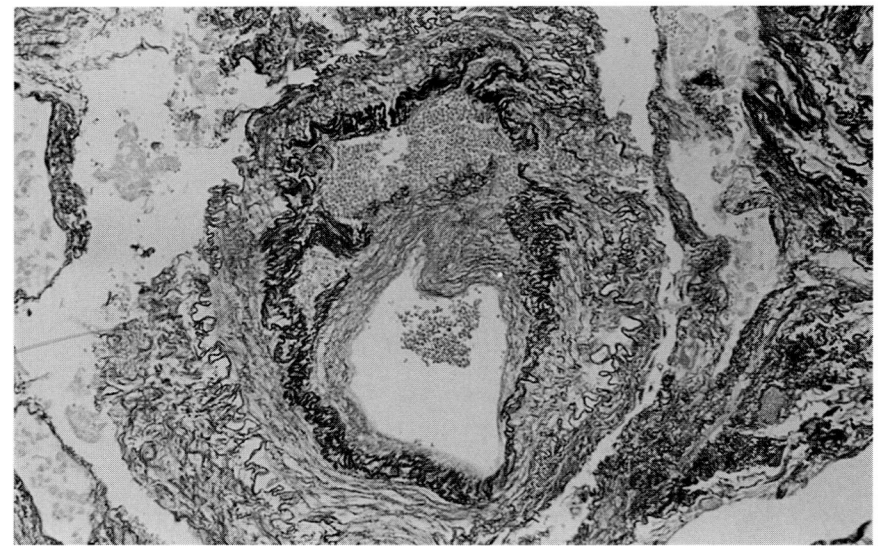

FIGURE 17-15
Chronic radiation damage. Muscular pulmonary artery with eccentric intimal fibrosis and a break in the internal elastic lamina. (Elastic-van Gieson, ×125.)

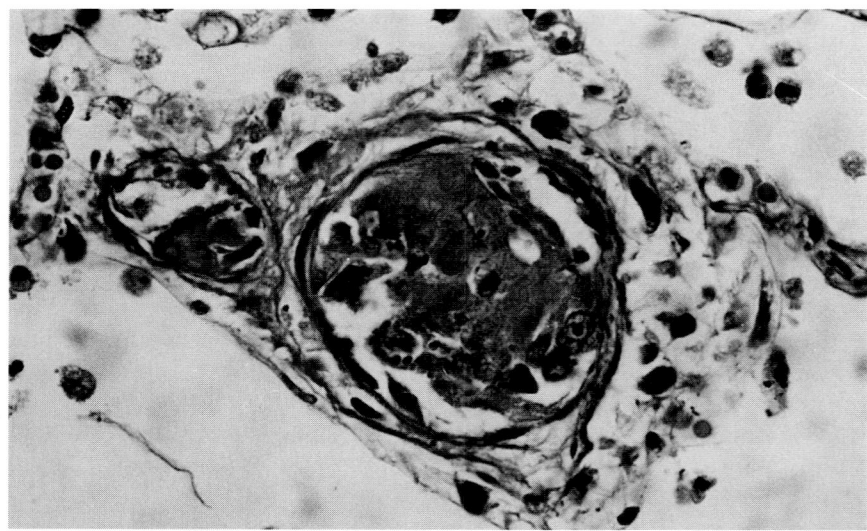

FIGURE 17-16
Chronic radiation damage. Fibrin thrombi in pulmonary capillaries. (H&E, ×313.)

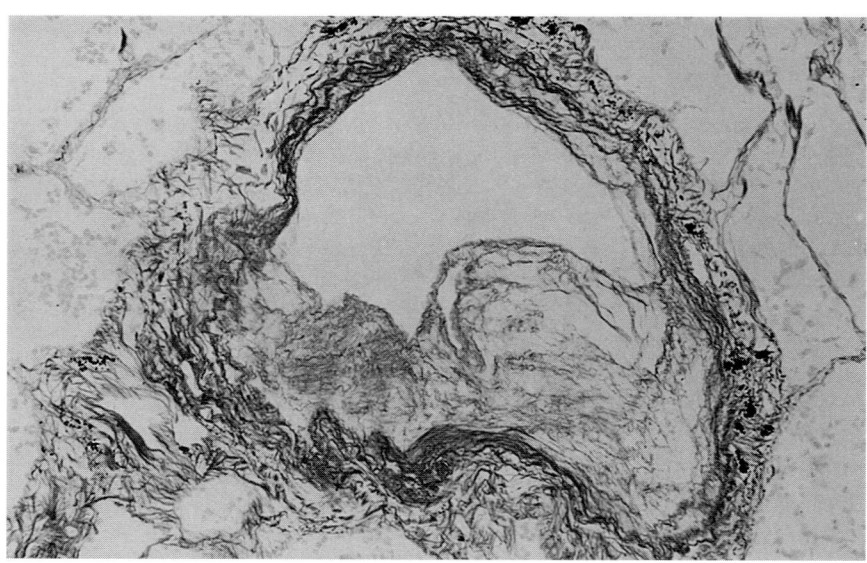

FIGURE 17-17
Chronic radiation damage. Eccentric intimal fibrosis in a pulmonary vein. (Elastic-van Gieson, ×125.)

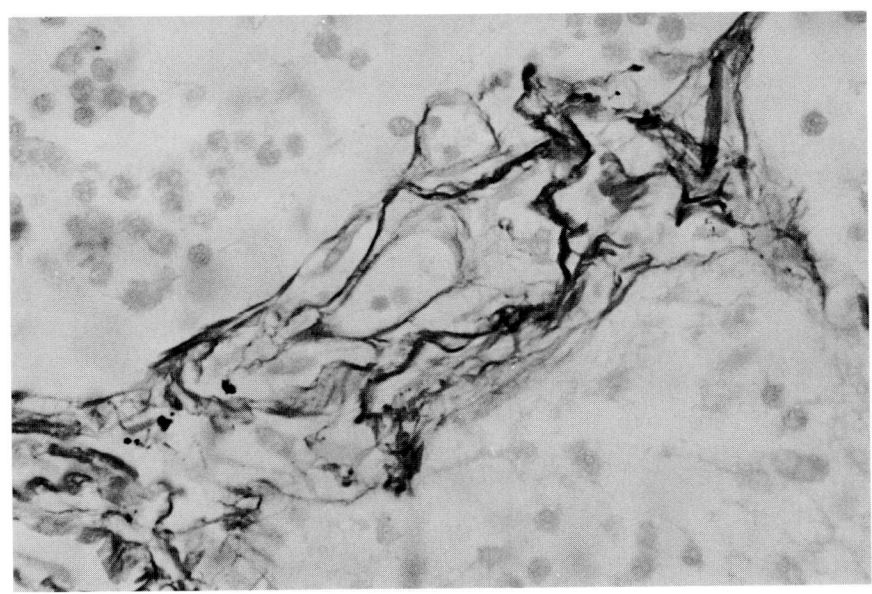

FIGURE 17-18
Pulmonary vein with some fibrosis and a small vascular channel in the middle of the vessel. (Elastic-van Gieson, ×125.)

A recent report[29] described pulmonary veno-occlusive disease 6 years after mantle irradiation for Hodgkin's disease.

PATHOPHYSIOLOGY

The etiology of the fibrosis is unknown. Blood vessel damage leads to exudation of plasma proteins into the extravascular spaces and this exudate may initiate fibrosis.[30] Cytokines may be involved. Alveolar macrophages and neutrophils could be important cells in the induction of radiation fibrosis through the mediation of fibronectin and alveolar macrophage-derived growth factor.[31] Mechanisms of pulmonary fibrosis have been described in previous sections on ARDS and interstitial pulmonary fibrosis.

Human data concerning the biology of radiation-induced lung injury are scarce. It has been suggested that it may be a hypersensitivity pneumonitis, because lavage performed during the phase of radiation pneumonitis in irradiated and nonirradiated lung fields both show the same pathology of an increased cell count as well as a lymphocytosis.[7] The lymphoctyes were of T-cell origin. This finding has been confirmed[32] in a study on women receiving postoperative radiotherapy for carcinoma of the breast. Premenopausal patients were given two cycles of chemotherapy before operation. Lavage showed an elevation of the lymphocyte count in both the irradiated and nonirradiated areas. Chemotherapy appeared to play no part, although the numbers studied were small.

Based on their lavage data, Gordier and colleagues[33] suggest that three pathogenetic mechanisms are responsible for radiation-induced lung injury. These are permeability edema, lymphocytic alveolitis, and collagenolytic activity in the alveolar structures exceeding antiprotease defenses in the lower respiratory tract. It is also possible that radiotherapy causes direct tissue damage, leading to antigen release[14] overriding the normal tolerance processes and producing sensitization of autoreactive lymphocyte clones, which migrate to the lung.

RADIATION CARCINOGENOSIS

Carcinoma of the lung may be a late complication of pulmonary fibrosis. The author has seen adenocarcinoma developing in cases treated by thoracic irradiation many years previously for Hodgkin's disease (Figs. 17-19 and 17–20). Cancer has been noted to be the only long-term cause of death increased by irradiation.[34,35] The evidence suggests that persons exposed at younger ages are at greater risk.[36]

As a general rule the *minimum* latent period for solid tumors to develop is about 10 years. Bilateral carcinoma of the lung has been described 25 months after starting therapy for Hodgkin's disease and it should be noted that the therapy included not only radiation but also nitrogen mustard, vinblastine, prednisone, and procarbazine.[37] A further tumor that may complicate radiation therapy is malignant fibrous histiocytoma.[38]

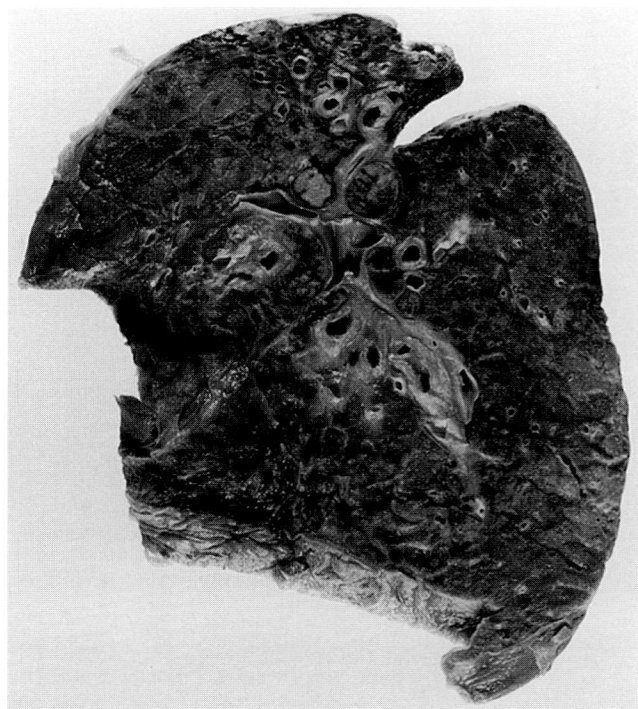

FIGURE 17-19
Patient who had Hodgkin's disease 20 years previously treated with radiotherapy. There was upper lobe fibrosis and histologic evidence of adenocarcinoma. Tumor is present in lymph nodes. In addition there is dilatation of some of the upper lobe bronchi due to radiation damage to the cartilage.

REFERENCES

1. Burt PA, O'Driscoll BR, Notley HM, Barber PV, Stout R: Intraluminal irradiation for the palliation of lung cancer with a high dose rate micro-Selectron. *Thorax* 45:765–768, 1990.
2. Trask CWL, Joannides T, Harper PG, et al: Radiation-induced lung fibrosis after treatment of small cell carcinoma of the lung with very high-dose cyclophosphamide. *Cancer* 55:57–60, 1985.
3. Withers HR: Biological basis of radiation therapy for cancer. *Lancet* 339:156–159, 1992.
4. Shapiro SJ, Shapiro SD, Mill WB, Campbell EJ: Prospective study of long-term pulmonary manifestations of mantle irradiation. *Int J Radiat Oncol Biol Phys* 19:707–714, 1990.
5. Rosiello RA, Merrill WW: Radiation-induced lung injury. *Clin Chest Med* 11:65–71, 1990.
6. Hall EJ: *Radiobiology for the Radiologist*, 2d ed. Philadelphia, Harper & Row 1978, p 87.
7. Gibson PG, Bryant DH, Morgan GW: Radiation-induced lung injury: A hypersensitivity pneumonitis? *Ann Intern Med* 109:288–291, 1988.
8. Rosiello RA, Merrill WW, Rockwell S, et al: Evolution of radiation injury in the lung assessed by bronchoalveolar lavage. *Am Rev Respir Dis* 135:A30, 1987.
9. Fajardo LF, Berthrong M: Radiation injury in surgical pathology. Part 1. *Am J Surg Pathol* 2:159–199, 1978.
10. Fajardo LF, Lee A: Rupture of major vessels after radiation. *Cancer* 36:904–913, 1975.
11. Tannock IF, Hayashi S: The proliferation of capillary endothelial cells. *Cancer Res* 32:77–82, 1972.
12. Phillips TL: Radiation fibrosis, in Fishman AP (ed), *Pulmonary Diseases and Disorders*, 2d ed. New York, McGraw-Hill, 1988, pp 773–792.
13. Maasilta P: Radiation-induced lung injury. From the chest physician's point of view. *Lung Cancer* 7:367–384, 1991.
14. Gross MJ: Pulmonary effects of radiation therapy. *Ann Intern Med* 86:81–92, 1977.

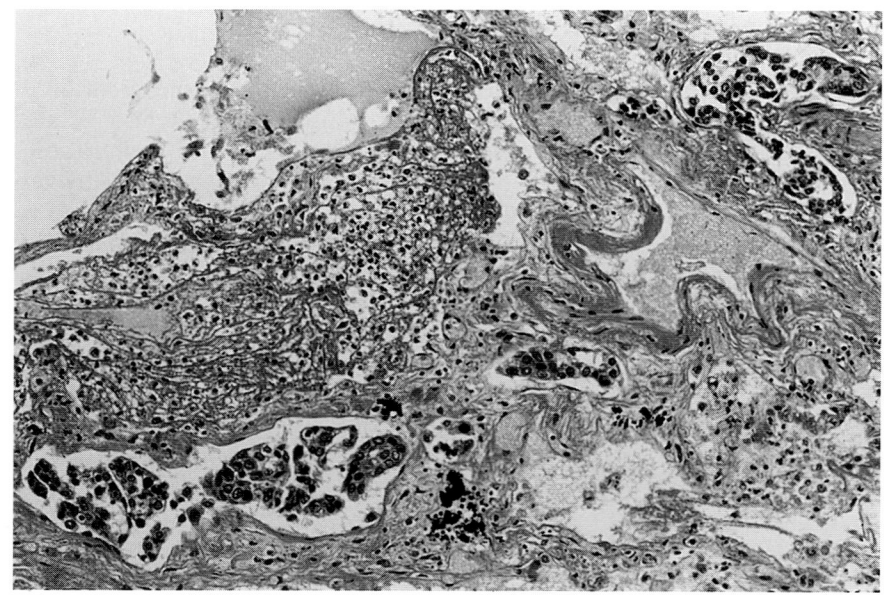

FIGURE 17-20
Chronic radiation lung damage. The lymphatics contain secondary carcinoma from a primary bronchial tumor. (H&E, ×125.)

15. Gross MJ: The pathogenesis of radiation-induced lung damage. *Lung* 159:115–125, 1981.

16. Roswit B, White DC: Severe radiation injuries of the lung. *Am J Roentgenol* 129:127–136, 1977.

17. Fulkerson WJ, McLendon RE, Prosnitz LR: Adult respiratory distress syndrome after limited thoracic radiotherapy. *Cancer* 57:1941–1946, 1986.

18. Yazdy AM, Tomashefski JR Jr, Yagan R, Kleinerman J: Regional alveolar damage (RAD): A localized counterpart of diffuse alveolar damage. *Am J Clin Pathol* 92: 10–15, 1989.

19. Byhardt RW, Abrams R, Almagro U: The association of adult respiratory distress syndrome (ARDS) with thoracic irradiation (RT). *Int J Radiat Oncol Biol Phys* 15:1441–1446, 1988.

20. Keane TJ, Van Dyk J, Rider WD: Idiopathic interstitial pneumonia following bone marrow transplantation: The relationship with total body irradiation. *Int J Radiat Oncol Biol Phys* 7:1365–1370, 1981.

21. Adamson IYR, Bowden DH, Wyatt JP: A pathway to pulmonary fibrosis: An ultrastructural study of mouse, rat following radiation to the whole body and hemithorax. *Am J Pathol* 58:481–487, 1970.

22. Phillips TL: An ultrastructual study of the development of radiation injury in the lung. *Radiology* 87:49–54, 1966.

23. Giri PG, Kimler BF, Giri UP, et al: Comparison of single, fractionated and hyperfractionated irradiation on the development of normal tissue damage in rat lung. *Int J Radiat Oncol Biol Phys* 11:527–534, 1985.

24. Slauson DO, Hahn FF, Benjamin SA, et al: Inflammatory sequences in acute pulmonary radiation injury. *Am J Pathol* 82:549–572, 1976.

25. Rubin P, Siemann DW, Shapiro DL, Finkelstein J, Penney DP: Surfactant release as an early measure of radiation pneumonitis. *Int J Radiat Oncol Biol Phys* 9:1669–1673, 1983.

26. Kaufman J, Komorowski R: Bronchiolitis obliterans: A new clinical-pathologic complication of irradiation pneumonitis. *Chest* 97:1243–1244, 1990.

27. Rubin P: Radiation toxicity: Quantitative radiation pathology for predicting effects. *Cancer* 39(suppl 2): 729–736, 1977.

28. Wilkinson MJ, MacLennan KA: Vascular changes in irradiated lungs: A morphometric study. *J Pathol* 158:229–232, 1989.

29. Kramer MR, Estenne M, Brekman N, et al: Radiation-induced pulmonary veno-occlusive disease. *Chest* 104:1282–1284, 1993.

30. Law MP: Vascular permeability and late radiation fibrosis in mouse lung. *Radiat Res* 103:60–76, 1985.

31. Rennard SI, Bitterman PB, Crystal RG: Current concepts of the pathogenesis of fibrosis: Lessons from pulmonary fibrosis, in *Myelofibrosis and the Biology of Connective Tissue.* New York, Liss, 1984, pp 359–377.

32. Roberts CM, Foulcher E, Zaunders JJ, et al: Radiation pneumonitis: A possible lymphocyte-mediated hypersensitivity reaction. *Ann Intern Med* 118:696–700, 1993.

33. Gordier JE, Mornex JF, Lasne Y, et al: Bronchoalveolar lavage in radiation pneumonitis. *Bull Eur Physiopathol Respir* 20:369–374, 1984.

34. Beebe GW, Kato H, Land CE: Mortality experience of atomic bomb survivors 1950–74. *Radiat Res* 75:138–201, 1978.

35. Smith PG, Doll R: Mortality from cancer and all causes among British radiologists. *Br J Radiol* 54:187–194, 1981.

36. Kato H, Schull W: Studies of the mortality of A-bomb survivors. Mortality, 1950–1978 Part 1. Cancer mortality. *Radiat Res* 90:395–432, 1982.

37. Kowlaski P, Rodziewicz B, Pejcz J: Bilateral bronchioloalveolar carcinoma of the lungs in a 7 year old girl treated for Hodgkins disease. *Tumori* 75:449–451, 1989.

38. Chowdhury LN, Swerdlow MA, Wellington J, Kathpalia S, Desser RK: Postirradiation malignant fibrous histiocytoma of the lung: Demonstration of alpha-1-antitrypsin-like material in neoplastic cells. *Am J Clin Pathol* 74:820–826, 1990.

Chapter *18*

Drug-Induced Lung Disease

J. Michael Kay

Drug-induced lung disease is a significant problem for clinicians and pathologists. It is increasing in frequency and the number of drugs implicated is growing. In 1972, one of the earliest reviews of this subject listed 19 therapeutic agents associated with parenchymal lung disease.[1] Twenty years later, an update published by the same author listed 60 agents known or suspected to cause pulmonary disease.[2] Descriptions of the pathology are available in only a small proportion of these drugs because tissue is examined in a minority of cases. A physician confronted with diagnosing a respiratory disorder should always consider the possibility of drug-induced disease.[3] The history should always elicit details of past and present administration of drugs. This may not be easy. Some patients do not consider substances such as aspirin, mineral oil, oral contraceptives, and L-tryptophan as drugs. This may be because taking the drug is a habit, or it was purchased "over the counter," or it was ingested intermittently.

Identification of drug-induced lung disease is also difficult because of the enormous variability in the signs and symptoms and the time that it takes for toxicity to develop. Some reactions occur within days after the commencement of treatment, whereas others may not be manifested until as late as 17 years after cessation of therapy.[4] Accordingly, if the patient's problem is consistent with an adverse drug reaction, the physician must make specific and wide-ranging inquiries about particular drugs. Another difficulty in the recognition of drug-induced lung disease is that the clinical, diagnostic imaging, and pathologic features are usually nonspecific. The patient may have a complex illness, in which a variety of unrelated pulmonary complications may occur. From the clinical viewpoint, the differential diagnosis may comprise recurrence or reactivation of the original disease, opportunistic infection, and iatrogenic conditions related to the administration of drugs or radiation.

The lungs respond to injury in a limited number of ways and none of the lesions associated with drugs are distinctive. The types of evidence that support a

TABLE 18-1

Evidence Supporting a Link between a Drug and an Adverse Clinical Event[5]

1. A plausible temporal sequence exists between administration of the drug and the event.
2. Withdrawal of the drug is followed by improvement.
3. Recurrence of the event follows reexposure to the drug.
4. Other causes of the clinical event are excluded.

link between a drug and an adverse clinical event are summarized in Table 18-1. It is difficult to prove that a drug caused a particular clinical event.[5] Irey has recommended the application of criteria that are useful not only in identifying drug toxicity in individual patients, but also in critically analyzing the medical literature.[6] In practice, such a rigorous approach is rarely feasible and thus many adverse drug reactions are unproven and most remain putative.

MECHANISMS OF TOXICITY

The lungs are frequently the site of adverse drug reactions.[7] This may be related to their role as the site of gas exchange. Lung cells are exposed to higher oxygen concentrations than any other cells in the body. Drugs that have the potential to cause damage by generating reactive oxygen species are likely to have this process facilitated in an oxygen-rich environment. It is now well recognized that the lungs are a site for the uptake, accumulation, and metabolism of numerous endogenous and exogenous chemicals.[8,9] Three distinctive properties of the pulmonary circulation are relevant to this role: it receives the entire cardiac output, it includes the largest capillary bed in the body, and it occupies a unique position between the systemic venous and arterial circulations.[10] The alveolar epithelium and capillary endothelium are very thin and located close together to facilitate gas exchange. This brings the alveolar epithelium in close proximity to blood, which allows rapid absorption, transfer, and metabolism of chemicals.[9]

The exact mechanisms of pathogenesis of most adverse pulmonary drug reactions are unknown. The possible adverse effects of cytotoxic[11] and noncytotoxic[12] drugs on the oxidant-antioxidant, immunologic, matrix repair, and proteolytic systems of the lung have been well reviewed. The following is a brief summary.

Oxygen free radicals have been implicated in the pathogenesis of a number of respiratory diseases.[13] There is some evidence that bleomycin, cyclophosphamide, carmustine (BCNU), and nitrofurantoin adversely affect the lung by interfering with the oxidant-antioxidant system. Bleomycin,[14] cyclophosphamide,[15] and nitrofurantoin[16] generate reactive oxygen species. Carmustine and cyclophosphamide may interfere with the body's antioxidant defense mechanisms by reducing glutathione stores.[17,18] A long list of drugs may produce pulmonary infiltrates accompanied by blood eosinophilia, suggesting an immune pathogenesis.[12,19] Changes in lung lymphocytes suggestive of an immune response have been reported in patients with adverse reactions to gold salts,[20] nitrofurantoin,[21] amiodarone,[22] and methotrexate.[23] The pulmonary-renal syndrome that may complicate penicillamine therapy appears to reflect an exacerbation of an autoimmune phenomenon.[24] Collagen deposition can be a useful process in the repair of tissues after injury. However, excessive collagen may cause irreversible pulmonary dysfunction. Modulation of fibroblast proliferation is an important method for controlling collagen production. Bleomycin alters the growth of fibroblasts.[25] Gold salts[26] and penicillamine[27] alter collagen structure, which may be relevant to the development of drug-induced lung disease.

Risk Factors

Factors that predispose to pulmonary toxicity are unknown for most agents associated with drug-induced lung disease. The development of drug-induced lung disease may be related to the dose, age of the patient, radiotherapy, oxygen therapy, and the use of other drugs.

Dose
Some drugs appear to be directly toxic to the lung and the adverse effects are related to the dose. Other adverse drug reactions appear to be idiosyncratic and are not dose dependent. A number of cytotoxic agents, including bleomycin, busulfan, and carmustine, appear to be directly toxic to the lung and commonly show enhancement of toxicity with increasing dose.[28] This suggests that deposition and accumula-

tion of the drug in the lung tissue may be one of the mechanisms of toxicity. The majority of adverse pulmonary reactions associated with noncytotoxic drugs appear to be idiosyncratic and are not related to the dose. An exception is amiodarone where pulmonary toxicity increases with higher maintenance doses although a cumulative dose does not appear to be an important factor.[29]

Age
A decrease in the effectiveness of the antioxidant defense mechanism may occur with increasing age.[30] Accordingly, it might be expected that older patients would be more likely to develop adverse reactions to bleomycin, cyclophosphamide, carmustine, and nitrofurantoin. In fact, bleomycin is the only drug for which such a relationship has been demonstrated.[28] In contrast, it has been suggested that younger patients may be more liable to develop pulmonary toxicity with carmustine.[31]

Radiotherapy
Lung damage caused by irradiation may be related to the production of reactive oxygen species so radiotherapy should enhance the toxicity of certain cytotoxic agents. This has been demonstrated with bleomycin,[32] busulfan,[33] and mitomycin.[34] Nitrofurantoin generates reactive oxygen species.[16] The reason a synergistic relationship with radiation has not been shown with it may be the lack of patients receiving chest radiotherapy while taking nitrofurantoin.

Oxygen Therapy
High concentrations of inspired oxygen result in the generation of reactive oxygen species. These might be expected to enhance the toxicity of bleomycin, cyclophosphamide, carmustine, and nitrofurantoin. However, bleomycin is the only drug in which such a relationship has been established in a clinical setting.[35] Synergism between oxygen therapy and pulmonary toxicity to cyclophosphamide[36] and nitrofurantoin[37] have been demonstrated in animal studies.

Other Drug Therapy
The cytotoxic drugs bleomycin,[38] cyclophosphamide,[38,39] methotrexate,[39] carmustine,[40] and mitomycin[41] appear to show increased pulmonary toxicity when administered as part of a combination chemotherapy regimen. Combination chemotherapy using these agents increases the incidence of pulmonary toxicity and lowers the dose at which lung damage occurs.

CLASSIFICATION OF DRUG-INDUCED LUNG DISEASE

A wide variety of drugs administered for therapeutic and illicit purposes may cause respiratory disease. Pulmonary complications of illicit drug use may be the most common form of drug-induced lung disease worldwide.[2] The adverse effects of drugs on the lungs may be primary or secondary. Secondary effects such as opportunistic infections complicating chemotherapy, aspiration, altered control of breathing, respiratory muscle paralysis, and embolic phenomena are outside the scope of this chapter. Drug-induced lung diseases can be classified according to the clinical syndromes produced (Table 18-2), the histologic patterns of tissue injury that have been identified (Table 18-3), and the type of drug implicated (Table 18-4). Some drugs may produce different types of adverse reaction in different patients. For example, nitrofurantoin may be associated with diffuse alveolar damage in some individuals and usual interstitial pneumonia (fibrosing alveolitis), desquamative interstitial pneumonia, eosinophilic pneumonia, or pulmonary hemorrhage in others (see Table 18-3).

The following sections review the pathology of the adverse reactions in the lung that have been reported for each drug. Morphologic features are discussed in detail only when they have not been addressed in other chapters. Several reviews of the clinical aspects[2,3,7,11,12,42] and pathologic aspects[19,43–47] of drug-induced lung disease have been published.

TABLE 18-2
Clinical Syndromes Associated with Drug-Induced Lung Disease[1,2,7]

Pulmonary edema
Pulmonary hemorrhage
Chronic interstitial lung disease
Acute pulmonary infiltrates with or without eosinophilia
Adult respiratory distress syndrome
Pulmonary hypertension
Obstructive bronchiolitis
Systemic lupus erythematosus
Bronchospasm
Neuromuscular blockade affecting respiratory muscles
Metastatic calcification
Pleural effusion

CYTOTOXIC DRUGS

Cytotoxic drugs comprise the largest group of agents associated with drug-induced lung disease.[11,28] Most of these drugs are used in the treatment of malignant neoplasms and leukemia, although some cytotoxic agents are also used in the treatment of nonneoplastic diseases. They can be classified as alkylating agents, antibiotics, nitrosoureas, antimetabolites, and miscellaneous agents (see Table 18-4).

Alkylating Agents

Alkylating agents are a chemically diverse group of substances whose therapeutic activity depends on the cross-linking of intracellular DNA molecules. This property is directed mainly against cells in an active state of division, whether malignant or not.[48] The lungs have a low activity of O^6-alkylguanine-DNA alkyltransferase compared with other tissues. This may make them relatively more sensitive to the cytotoxic effects of DNA alkylation.[49] Busulfan and cyclophosphamide are the two alkylating agents that most commonly cause drug-induced lung disease. Chlorambucil, ifosfamide, melphalan, and uracil mustard account for only a few cases of lung toxicity.

BUSULFAN

Busulfan is used in the treatment of myeloproliferative disorders, particularly chronic myelogenous leukemia, and in the preparation for bone marrow transplantation. It was the first cytotoxic drug reported to cause lung disease.[50] Pulmonary toxicity occurs in 4 to 6 percent of patients after latent periods ranging from 6 weeks to 12 years.[2,28] The disease commences an average of 41 months after initiation of therapy at an average cumulative dose of 2900 mg.[2] Although busulfan pulmonary toxicity does not appear to be directly dose dependent, there may be a threshold dose over which toxicity may occur.[28] No patient receiving less than 500 mg of the drug has developed lung disease unless exposed to radiation,[33] melphalan,[51] or uracil mustard.[52] These may potentiate the toxic effects of busulfan.[11] Although most cases have developed during busulfan therapy, there is a report of toxicity occurring 1 month after discontinuation of the agent.[53] Evidence of subclinical lung damage has been noted at autopsy in as many as 46 percent of patients treated with busulfan.[54] The prognosis for patients with busulfan-induced lung disease is poor, with an estimated mortality rate of 84 percent.[2]

Organizing diffuse alveolar damage[53–55] and

TABLE 18-3
Histologic Patterns of Drug-Induced Lung Disease

Diffuse Alveolar Damage

Amiodarone, azathioprine, bleomycin, busulfan, carmustine, colchicine, cyclophosphamide, cytosine arabinoside, dacarbazine, fludarabine, fotemustine, gold salts, hexamethonium, lomustine, mecamylamine, melphalan, methotrexate, mitomycin, nitrofurantoin, penicillamine, pentolinium, procarbazine, streptokinase, sulfasalazine, teniposide, vinblastine, vindesine, zinostatin

Usual Interstitial Pneumonia

Azathioprine, busulfan, chlorambucil, cyclophosphamide, ifosfamide, nitrofurantoin, pinodol, semustine, tocainide

Nonspecific Chronic Interstitial Pneumonia

5-Aminosalicylic acid, bepridil, busulfan, carmustine, cephalosporin, chlorambucil, chlorozotocin, cyclophosphamide, gold salts, melphalan, methotrexate, minocycline, nilutamide, phenytoin, propylthiouracil, sulphasalazine, tocainide, L-tryptophan, uracil mustard

   With granulomas

   Carbamazepine, methotrexate, nitrofurantoin, procarbazine, sulphasalazine

   With foam cells

   Amiodarone

Desquamative Interstitial Pneumonia

Nitrofurantoin, sulfasalazine

Giant Cell Interstitial Pneumonia

Nitrofurantoin

Lymphoid Interstitial Pneumonia

Nitrofurantoin, phenytoin

Bronchiolitis

Gold salts

Bronchiolitis Obliterans Organizing Pneumonia

Acebutolol, amiodarone, bleomycin, cromolyn sodium, cyclophosphamide, gold salts, methotrexate, mitomycin, nitrofurantoin, penicillamine, sulphasalazine, tocainide

Constrictive (Obliterative) Bronchiolitis

Lomustine, penicillamine

Eosinophilic Pneumonia

Ampicillin, bleomycin, carbamazepine, diclofenac, minocycline, naproxen, nitrofurantoin, phenylbutazone, piroxicam, procarbazine, pyrimethamine, sulfasalazine, sulfonamides, sulindac

Pulmonary Alveolar Proteinosis

Busulfan, cyclophosphamide

Pulmonary Hemorrhage

Amiodarone, amphotericin B, anticoagulants, cyclophosphamide, mitomycin, nitrofurantoin, penicillicamine

Pleuropulmonary Fibrosis

Bromocriptine, ergotamine, methysergide, practolol

Pulmonary Vascular Disease

   Pulmonary artery medial hypertrophy and intimal fibrosis

   Phendimetrazine

   Pulmonary arteriopathy with plexiform lesions

   Aminorex, dexfenfluramine, fenfluramine

   Occlusive pulmonary arteriopathy

   Cyclophosphamide

   Pulmonary vasculitis

   L-Tryptophan, phenytoin

   Pulmonary venoocclusive disease

   Bleomycin, carmustine, mitomycin, zinostatin

Pulmonary Edema

Cytosine arabinoside

TABLE 18-4
Pathology of Adverse Drug Reactions in the Lung

Drug	Lung Pathology	Drug	Lung Pathology
Cytotoxic Drugs		*Antimicrobials*	
Alkylating agents		Ampicillin	EP
Busulfan	DAD, CIP, UIP, PAP	Amphotericin B	PH
Cyclophosphamide	DAD, CIP, UIP, BOOP, PAP, PH, OPA	Cephalosporin	CIP
		Minocycline	CIP, EP
Chlorambucil	UIP, CIP	Nitrofurantoin	DAD, PH, CIPG, DIP, UIP, GIP, EP, LIP, BOOP
Ifosfamide	UIP		
Melphalan	DAD, CIP		
Uracil mustard	CIP	Pyrimethamine	EP
Antibiotics		Sulfasalazine	EP, CIP, CIPG, DAD, UIP, DIP, BOOP
Bleomycin	DAD, BOOP, EP, PVOD		
Mitomycin	DAD, BOOP, PH, PVOD	Sulfonamides	EP
Zinostatin	DAD, PVOD	*Anti-inflammatory agents*	
Nitrosoureas		5-Aminosalicylic acid	CIP
Carmustine (BCNU)	DAD, CIP, PVOD	Diclofenac	EP
Lomustine (CCNU)	DAD, CB	Gold salts	CIP, BOOP, B, DAD
Semustine (methyl-CCNU)	UIP	Naproxen	EP
Chlorozotocin	CIP	Penicillamine	CB, BOOP, PH, DAD
Fotemustine	DAD	Phenylbutazone	EP
Antimetabolites		Piroxicam	EP
Azathioprine	DAD, UIP	Sulindac	EP
Cytosine arabinoside	PE, DAD	*Anticonvulsants*	
Methotrexate	CIP, CIPG, BOOP, DAD	Carbamazepine	EP, CIPG
Miscellaneous		Phenytoin	CIP, LIP, PV
Dacarbazine	DAD	*Anorexigens*	
Fludarabine	DAD	Aminorex	PAPLX
Procarbazine	CIPG, EP, DAD	Dexfenfluramine	PAPLX
Vinblastine	DAD	Fenfluramine	PAPLX
Vindesine	DAD	Phendimetrazine	PAMHIF
Teniposide	DAD	*Miscellaneous Drugs*	
Noncytotoxic Drugs		Anticoagulants	PH
Antiarrythmic agents		Bromocriptine	PPF
Amiodarone	CIPF, BOOP, DAD	Cromolyn sodium	BOOP
Tocainide	CIP, UIP	Ergotamine	PPF
Antihypertensive agents		Methysergide	PPF
Acebutolol	BOOP, granulomas	Nilutamide	CIP
Hexamethonium	DAD	Propylthiourucil	CIP
Mecamylamine	DAD	Streptokinase	DAD
Pentolinium	DAD	L-Tryptophan	CIP, PV
Pindolol	UIP		
Practolol	PPF		
Calcium channel blocking agent			
Bepridil	CIP		

B = bronchiolitis; BOOP = bronchiolitis obliterans organizing pneumonia; CB = constrictive (obliterative) bronchiolitis; CIP = nonspecific chronic interstitial pneumonia; CIPF = chronic interstitial pneumonia with foam cells; CIPG = chronic interstitial pneumonia with granulomas; DAD = diffuse alveolar damage; DIP = desquamative interstitial pneumonia; EP = eosinophilic pneumonia; GIP = giant cell interstitial pneumonia; LIP = lymphoid interstitial pneumonia; OPA = occlusive pulmonary arteriopathy; PAMHIF = pulmonary artery medial hypertrophy and intimal fibrosis; PAP = pulmonary alveolar proteinosis; PAPLX = pulmonary arteriopathy with plexiform lesions; PE = pulmonary edema; PH = pulmonary hemorrhage; PPF = pleuropulmonary fibrosis; PV = pulmonary vasculitis; PVOD = pulmonary veno-occlusive disease; UIP = usual interstitial pneumonia

FIGURE 18-1
Busulfan lung in a 61-year-old man treated for chronic myeloid leukemia. Many areas of the lung are pale and firm due to fibrosis, which appears to be predominantly in the peripheral parts of the secondary lung lobules. The diaphragm is adherent to the inferior aspect of the lower lobe. Microscopically there was interstitial fibrosis with organizing intraalveolar exudate and atypical alveolar epithelial cells.

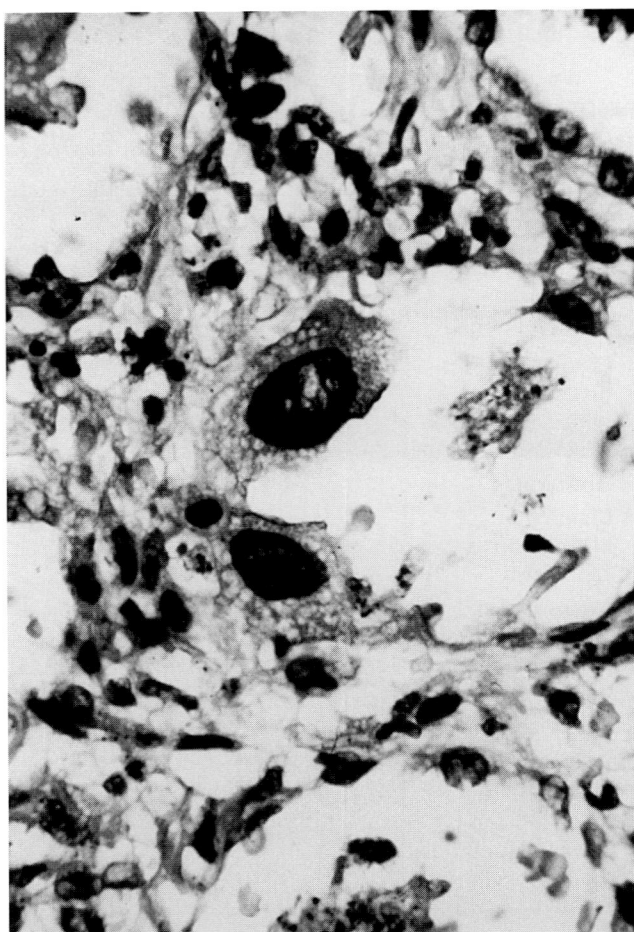

FIGURE 18-2
Busulfan lung. There is interstitial fibrosis and the alveoli are lined by atypical epithelial cells with large hyperchromatic nuclei and vacuolated cytoplasm. Electron microscopy confirmed that these cells were type II pneumocytes. Same case as Fig. 18-1. (H&E, ×800.)

usual interstitial pneumonia[50] are the most common manifestations of busulfan lung toxicity (Fig. 18-1). Nonspecific chronic interstitial pneumonia is a less common manifestation that has been reported only in individuals who have received large doses of the drug over relatively long periods of time.[56] Cytologic atypia affecting both the bronchiolar and alveolar epithelium (Fig. 18-2) is a characteristic feature of busulfan lung toxicity.[53,54] The epithelial changes include cytomegaly, nuclear pleomorphism, and prominent nucleoli. Cytologic atypia may occur without clinical, radiologic, or other morphologic evidence of lung disease.[54] Cytologic atypia alone is regarded as a marker of busulfan administration rather than evidence of significant pulmonary toxicity. The epithelial atypia has been noted in a variety of extrapulmonary sites including the urinary bladder, breast, pancreas, and uterine cervix.[57] Pulmonary alveolar proteinosis (Fig. 18-3) has been reported as a rare complication of busulfan therapy. This has a bad prognosis.[2,58,59] Osseous metaplasia[60] and adenocarcinoma[61] have been reported in the lungs of patients receiving busulfan therapy. These

changes probably represent incidental associations rather than complications of busulfan therapy.

CYCLOPHOSPHAMIDE

Lung disease has been reported in patients receiving cyclophosphamide for the treatment of both malignant[62] and nonmalignant disease.[63,64] Cyclophosphamide is a widely used drug and the small number of reported cases of lung disease suggests that pulmonary toxicity is uncommon, with an incidence below 1 percent.[11] No relationship has been established between the dose of the drug or duration of therapy and the development of lung disease. Cases have been reported after the administration of as little as 150 mg of the drug.[28] The latent period between the administration of cyclophosphamide to the onset of symptoms has varied from 2 weeks[65] to 13 years.[66] Lung disease may occur several months after discontinuation of cyclophosphamide therapy.[67] Symptoms may develop after reducing cortico-

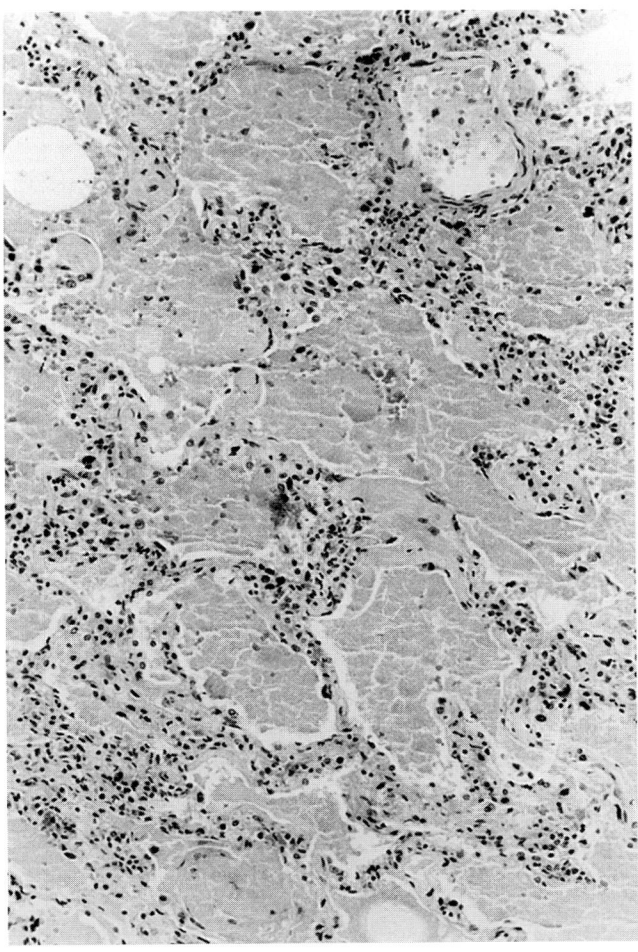

FIGURE 18-3
Pulmonary alveolar proteinosis in a patient with chronic myeloid leukemia treated with busulfan. The alveoli are filled with granular amorphous debris. There is slight thickening of the alveolar walls. (H&E, ×200.)

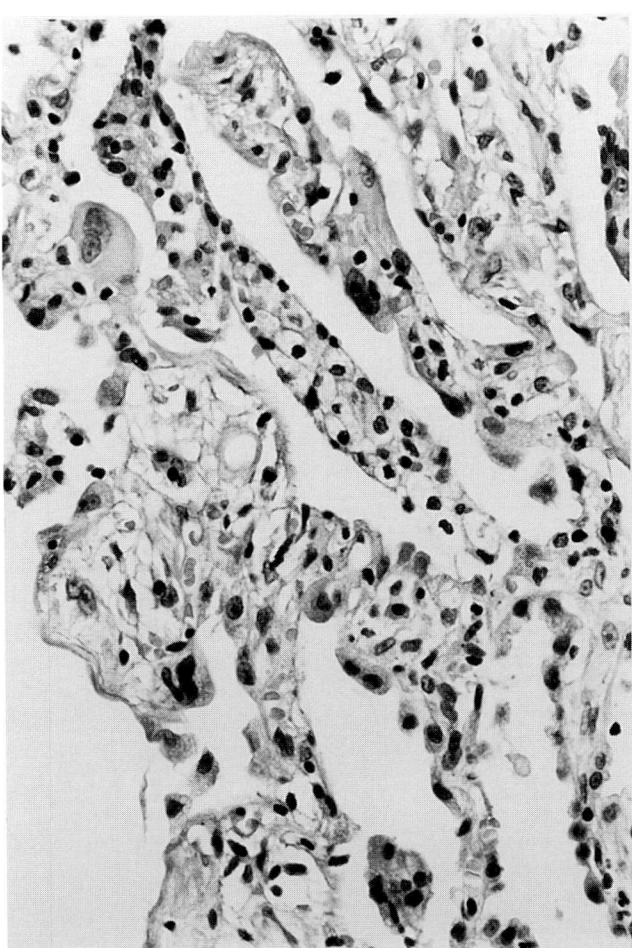

FIGURE 18-4
Cyclophosphamide-induced lung disease. Chronic interstitial pneumonia with atypical alveolar epithelial cells. (H&E, ×500.)

steroids, which are part of a treatment protocol. This suggests that the steroid therapy may mask cyclophosphamide-induced lung disease.[62,68] The prognosis for patients who develop cyclophosphamide-induced lung disease is variable. Approximately 60 percent of patients recover or remain stable with discontinuation of cyclophosphamide therapy.[11]

The most common manifestation of cyclophosphamide-induced lung disease is organizing diffuse alveolar damage.[62,64,69] Chronic interstitial pneumonia (Fig. 18-4),[63] usual interstitial pneumonia,[66] and pulmonary alveolar proteinosis (Fig. 18-5)[19,70] occur less frequently. Bronchiolitis obliterans and organizing pneumonia has been reported in a patient who developed lung disease after 5 months of combination chemotherapy involving bleomycin, cyclophosphamide, vincristine, and prednisone.[68] Areas of acute intraalveolar hemorrhage have been observed in autopsies on patients receiving high-dose intermittent cyclophosphamide therapy.[71] The hemorrhage was seen in more than half the patients dying

within the first month of treatment and may have been related to thrombocytopenia. There has been one report of fatal pulmonary hypertension developing in an infant during the 7-month period in which he received cyclophosphamide combined with alternating cycles of doxorubicin with either cisplatin of VM-26.[72] The chemotherapy was administered via a central venous catheter that was also used for intermittent parenteral feeding. The small pulmonary arteries showed medial hypertrophy with very extensive concentric and eccentric intimal fibrosis. The pulmonary veins were normal.

CHLORAMBUCIL

Chlorambucil, used primarily in the treatment of hematologic malignancies, is rarely associated with pulmonary toxicity.[11] In most of the reported cases, the patients were receiving chlorambucil alone. However, some were receiving combination chemotherapy that included methotrexate[39] and vincristine.[73] No relationship between the total drug dose and the development of pulmonary toxicity has

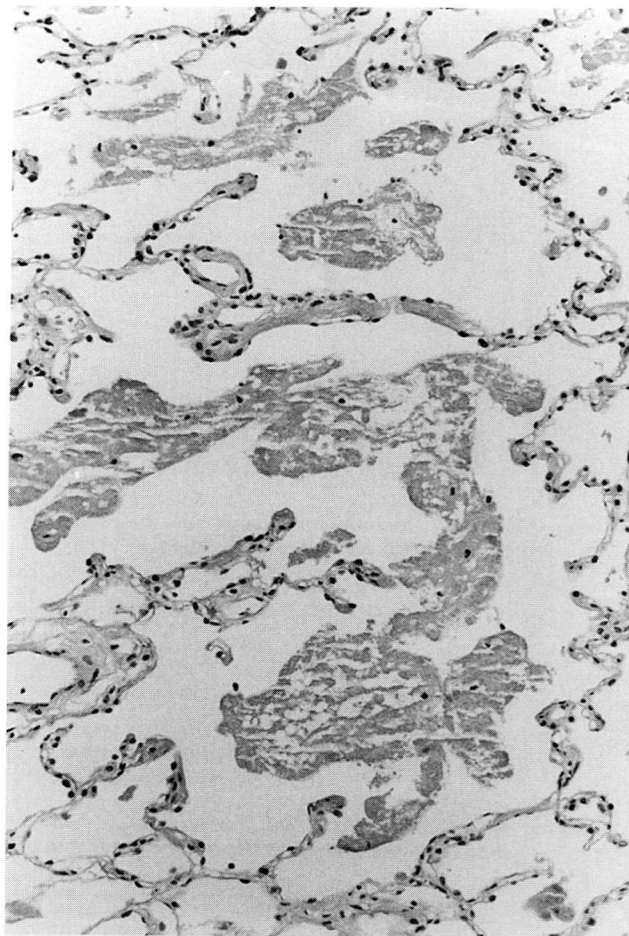

FIGURE 18-5
Pulmonary alveolar proteinosis in a patient receiving cyclophosphamide. The alveoli contain amorphous granular debris. The alveolar walls are not significantly thickened. (H&E, ×200.)

been established. Clinical manifestations of lung disease appear after 6 months to 3 years of therapy.[11] About half the patients with chlorambucil-induced lung disease have died. One patient improved dramatically with discontinuation of the drug and the commencement of steroid therapy.[74] The most common pathologic finding is chronic interstitial pneumonia.[74,75] Some cases resemble usual interstitial pneumonia, whereas in others there is a nonspecific chronic interstitial pneumonia that may be associated with organizing pneumonia.[74]

IFOSFAMIDE

Ifosfamide is a structural isomer of cyclophosphamide that is active against a variety of carcinomas, sarcomas, lymphomas, and testicular cancer. The toxicities of ifosfamide and cyclophosphamide are similar, but there has only been one report of lung disease associated with ifosfamide therapy.[76] The patient was a 58-year-old-woman who developed respiratory failure after receiving five cycles of a combination chemotherapy regimen including ifos-

famide, doxorubicin, and dacarbazine. She was treated with steroids but died of respiratory failure about 3 months later. At autopsy the lungs showed diffuse interstitial fibrosis with extensive honeycomb change. The alveolar epithelial cells were enlarged and atypical.

MELPHALAN

Melphalan is used primarily in the treatment of multiple myeloma. Lung disease complicating melphalan therapy has been reported in only a few patients,[77–81] some of whom also received cyclophosphamide[77] or busulfan.[79] Symptoms developed 1 to 4 months after the commencement of melphalan therapy.[11] Over half the reported patients died of respiratory failure despite discontinuation of the drug.[11] One patient recovered completely.[81] The pathologic changes comprise organizing diffuse alveolar damage and chronic interstitial pneumonia. The chronic interstitial pneumonia has been associated with organizing pneumonia and atypical alveolar epithelial cells.[77] Melphalan resembles busulfan in that it may cause cytologic atypia without associated lung disease.[51]

URACIL MUSTARD

Chronic interstitial pneumonia has been reported in a case of pulmonary toxicity recurring after a 6-week course of busulfan and after subsequent therapy with uracil mustard.[52]

Antibiotics

The cytotoxic antibiotics are a chemically diverse group of antineoplastic agents derived from bacterial cultures. Bleomycin, mitomycin, and zinostatin are cytotoxic antibiotics that have been associated with drug-induced lung disease.

BLEOMYCIN

Bleomycin is the generic name for a group of antibiotics isolated from *Streptomyces verticullis*.[82] It is used primarily in the treatment of lymphomas, squamous cell carcinomas, and testicular neoplasms.[82] Although the agent is highly effective in the treatment of certain neoplasms, skin and pulmonary toxicity limit the dose tolerated by many patients.[11] Bleomycin has been extensively studied because of its ability to produce interstitial pneumonia and pulmonary fibrosis in a wide range of animal species.[11]

Risk factors for the development of bleomycin-induced lung disease include: age of 70,[28,82] previous or concomitant thoracic radiotherapy,[32] high inspired concentrations of oxygen,[35] prior administration of bleomycin during the previous 6 months,[83] and combination chemotherapy in which bleomycin is combined with cyclophosphamide.[38,84] It has been suggested that continuous infusion of bleomycin results in a lower incidence of lung disease. Renal failure increases the sensitivity of the lung to bleomycin toxicity.[11]

The incidence of bleomycin-induced lung disease reported in the literature varies widely.[2,11] This is probably related to the variable presence of the risk factors listed above. In two large series involving 808 and 93 patients, the incidence of lung disease was 11 percent and 9 percent, respectively.[82,85] The risk of bleomycin-induced lung disease increases dramatically after a total cumulative dose of 450 units.[28,82] However, below this total dose there are well-documented cases of lung disease, particularly in association with other risk factors.[86] In fact, fatal pulmonary damage has been reported after the administration of 100 units bleomycin.[87]

Most patients with bleomycin-induced lung disease present with exertional dyspnea and cough associated with fine reticular and alveolar infiltrates in the lower zones of chest radiographs.[85] Such patients show the histologic features[19,88,89] and ultrastructural features[90] of the acute and organizing phases of diffuse alveolar damage (Fig. 18-6). Both the early and advanced changes are more prominent in the lower lobes with minimal or absent involvement of the upper lobes.[88] The changes tend to affect the subpleural lung parenchyma and some cases progress to honeycomb lung.[88] Alveolar hemorrhage and pleural fibrosis may be associated findings.[87] Some patients may develop pleural blebs, which may explain the rare occurrence of pneumothorax and pneumomediastinum in bleomycin-induced lung disease.[91] The mortality of acute lung injury due to bleomycin has been documented as 12 percent.[85] If a patient survives the acute phase, complete recovery of lung function may occur within 2 years.[85] A number of case reports describe nodules mimicking metastases in chest radiographs and computed tomography (CT) scans of patients taking bleomycin.[92,93] The patients were children and young adults being treated for osteogenic sarcoma and testicular neoplasms. Some of the nodules cavitated.[2] Open lung biopsies revealed a nodular form of bronchiolitis obliterans and organizing pneumonia (Fig. 18-7).[92,93] One of the cases showed a significant eosinophilic infiltrate.[92] Eosinophilic pneumonia has been reported as an uncommon manifestation of bleomycin toxicity in patients who were receiving combination chemotherapy including other cytotoxic agents.[19,94,95] An eosinophilic pleuritis was also present in one case.[95] The prognosis was good with some cases responding to steroid therapy.

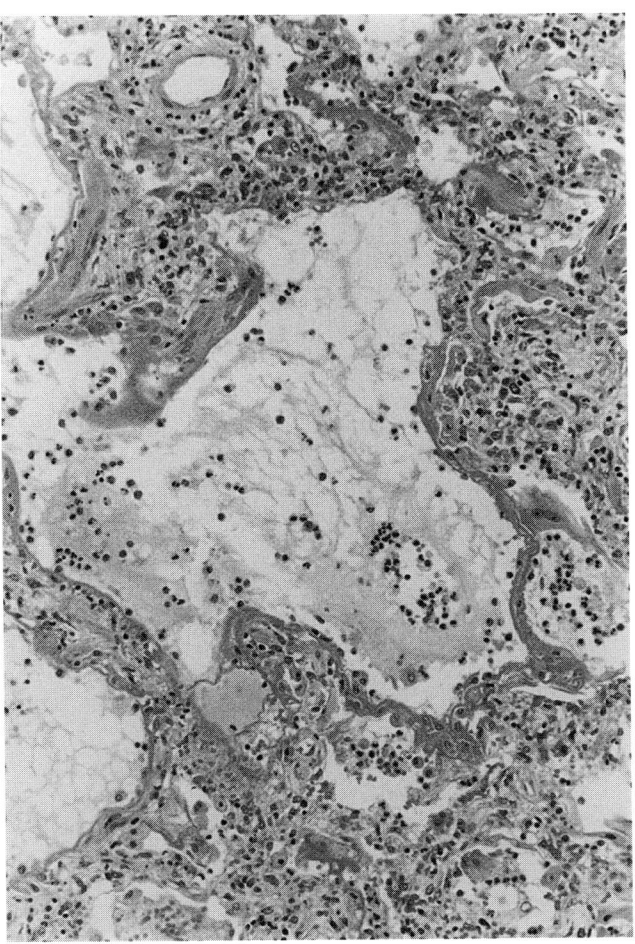

FIGURE 18-6
Diffuse alveolar damage in a patient who developed acute respiratory failure while receiving bleomycin. The air spaces are lined by hyaline membranes and they contain edema fluid intermingled with scanty neutrophil polymorphs. (H&E, ×200.)

Pulmonary venoocclusive disease has been attributed to bleomycin toxicity.[96–98] Some patients received combination chemotherapy including mitomycin, cisplatin, and vinblastine.[96,98] Pulmonary artery pressures were not measured but in two cases the pulmonary venoocclusive lesions were accompanied by right ventricular hypertrophy.[96,98]

MITOMYCIN

Mitomycin derived from cultures of *Streptomyces caespitous* is an alkylating agent used primarily in the treatment of carcinomas of the esophagus, pancreas, stomach, and colon. The drug is also used in the treatment of breast and lung malignancies.[11] The frequency of lung disease associated with mitomycin ranges from 3 to 12 percent.[99–101] Risk factors for mitomycin-induced lung disease are not well established although there is a suggestion that toxicity is enhanced by radiation[34] and combination chemotherapy.[41] Most patients with mitomycin-induced lung disease have received a total cumulative dose of at least 30 mg/m².[101]

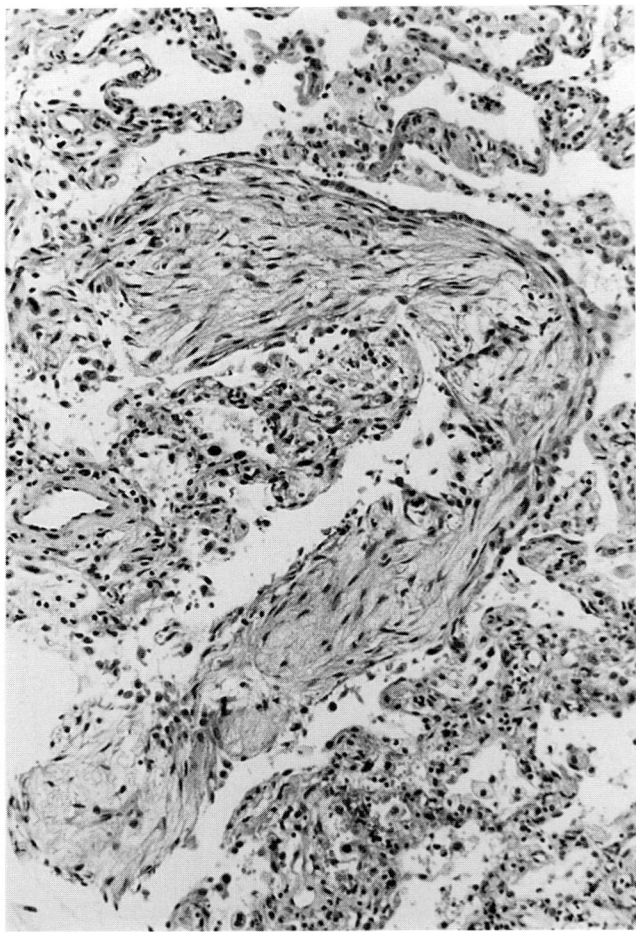

FIGURE 18-7
Organizing pneumonia in a patient receiving bleomycin. The alveolar spaces contain plugs of cellular fibrous tissue. Elsewhere, the open lung biopsy showed bronchiolitis obliterans. (H&E, ×200.)

Most patients with mitomycin-induced lung disease show varying stages of diffuse alveolar damage.[34,99,100] Early recognition of respiratory symptoms is important because complete recovery can occur after withdrawal of the drug and the administration of corticosteroids.[34,102] A less common manifestation of mitomycin-induced lung disease is bronchiolitis obliterans and organizing pneumonia.[99] Most of the reported patients had also received other potentially toxic drugs including vinca alkaloids, cyclophosphamide, and bleomycin. However, the course of their illnesses suggested that mitomycin was the most likely toxic agent.

A small percentage of patients treated with mitomycin develop a syndrome comprising microangiopathic hemolytic anemia, thrombocytopenia, and renal failure. Patients with this syndrome may develop pulmonary hypertension with fibrin thrombi in small pulmonary blood vessels.[103] Other patients may develop diffuse alveolar hemorrhage and angiomatoid changes of alveolar capillaries.[104] A

case of pulmonary venoocclusive disease has been attributed to mitomycin therapy.[105]

ZINOSTATIN

Fatal interstitial lung disease was reported in two patients receiving the polypeptide cytotoxic antibiotic zinostatin (neocarzinostatin).[106,107] No photomicrographs are included in the reports, but the parenchymal changes were said to be identical to bleomycin-induced pulmonary fibrosis. In addition, both cases showed evidence of pulmonary venoocclusive disease. There was thickening and occlusion of pulmonary veins with hypertrophy of muscular pulmonary arteries. In one case, there was thrombosis of small pulmonary veins.[106]

Nitrosoureas

The nitrosoureas are used in the treatment of malignant intracranial neoplasms, melanoma, gastrointestinal tract neoplasms, and hematologic malignancies.[108] Carmustine (BCNU), lomustine (CCNU), semustine (methyl-CCNU), chlorozotocin, and fotemustine have all been associated with drug-induced lung disease, but carmustine accounts for the vast majority of cases.

CARMUSTINE (BCNU)

Carmustine is commonly used alone in the treatment of primary brain neoplasms so its toxic effects can be assessed in a setting where the primary disease rarely affects the lungs. For patients with intracranial neoplasms treated with carmustine as a single agent, the incidence of drug-induced lung disease is approximately 20 to 30 percent.[108] In a series of 794 patients with hematologic malignancies treated with carmustine, the incidence of drug-induced lung disease was 1 percent.[40]

There is a direct relationship between total cumulative dose and lung toxicity.[31] Drug-induced lung disease may occur at lower doses in patients being treated with cyclophosphamide.[40] A history of cigarette smoking and preexisting lung disease have been associated with an increased risk of drug-induced lung disease.[31]

Symptoms may develop as early as 1 month after the administration of carmustine.[109] However, the majority of cases occur after a longer period of time commensurate with the cumulative dose received.[31] A unique feature of carmustine-induced lung disease is that pulmonary fibrosis may become apparent as many as 17 years after treatment.[4] The prognosis of carmustine-induced lung disease is variable. In a series of patients with intracranial neoplasms who developed pulmonary disease, the mortality rate was

16 percent.[31] In a series of patients with hematologic malignancies treated with carmustine and various other cytotoxic drugs, the mortality rate was 70 percent.[40] The majority of patients with carmustine-induced lung disease were already receiving corticosteroids, making the use of these agents in treatment questionable.[31,40]

The acute and organizing phases of diffuse alveolar damage are the most common manifestations of carmustine-induced lung disease.[110-113] Some cases show nonspecific chronic interstitial pneumonia with absent or minimal inflammatory cell infiltration.[4,114,115] This may be the result of reporting patients with end-stage disease. The interstitial fibrosis may be accompanied by cytologic atypia of the alveolar epithelium.[40,114,115] In some patients the parenchymal lung disease is associated with thickening of the interlobular fibrous septa[115] and pleura.[40,114] Involvement of the pleura may explain the occurrence of pneumothorax, which has been described in several patients with carmustine-induced lung disease.[40,114]

An unusual group of eight patients has been described who presented with upper zone pulmonary fibrosis 12 to 17 (mean 14) years after receiving carmustine for cerebral tumors.[4,116] The total dose of carmustine ranged from 770 to 1410 mg/m² (mean 1135 mg/m²).[11] Four cases also received vincristine and five had spinal radiotherapy, neither of which were thought to have contributed to the lung disease. Right lower lobe transbronchial biopsies from seven patients were studied by light and electron microscopy.[116] Light microscopy showed interstitial elastosis and intraalveolar fibrosis, which was often focal with an associated mild lymphoplasmacystic infiltrate, intraalveolar edema, macrophages and some neutrophills (Fig. 18-8). Ultrastructural studies showed electron lucency of type I pneumonocytes, with breaks in the cytoplasmic membranes leaving a bare basement membrane. The endothelial cells showed degenerative changes with lipofuscin deposition. One patient with pulmonary hypertension due to pulmonary-fibrosis died awaiting heart-lung transplantation. At autopsy the lungs were small. The cut surfaces showed diffuse fibrosis, which was especially prominent in the subpleural areas of the apices of both upper lobes (Fig. 18-9). There was no honeycomb change. Histologically there was interstitial fibrosis and elastosis accentuating the alveoli. There was also focal intraalveolar fibrosis (Fig. 18-10).

Pulmonary venoocclusive disease has been described in two young adults who received carmustine for the treatment of brain neoplasms.[117] Pulmonary hypertension and right ventricular

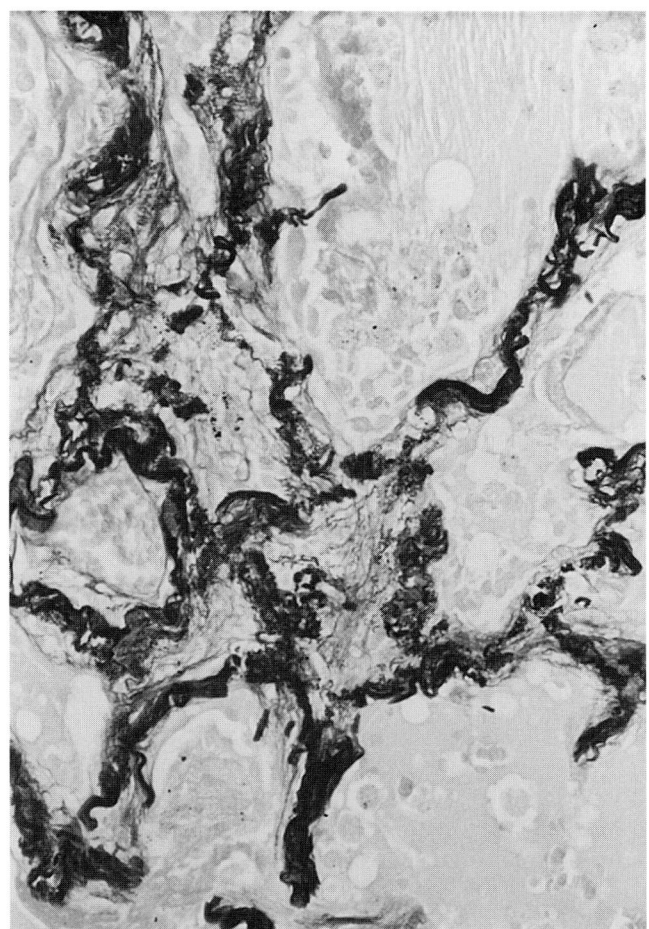

FIGURE 18-8
Late carmustine (BCNU)-induced lung disease. There is interstitial fibrosis and elastosis accentuating the alveoli. In the center of the picture there is focal intraalveolar fibrosis. (Elastic-van Gieson, ×313.) *(Reproduced with permission from Hasleton et al.[116]).*

enlargement were documented in one case but not the other.

OTHER NITROSOUREAS

Lomustine (CCNU) has been reported to produce changes consistent with diffuse alveolar damage in patients receiving high doses of the drug.[118-120] Some patients have received lomustine alone for the treatment of cerebral neoplasms,[118] whereas in others it has been combined with cyclophosphamide, vincristine, and busulfan for the treatment of chronic myeloid leukemia.[119] In one patient there was fibrous obliteration of bronchioles and bronchi.[118]

Usual interstitial pneumonia has been reported in a patient who received semustine (methyl-CCNU).[121]

Nonspecific chronic interstitial pneumonia has been described in patients who received chlorozo-

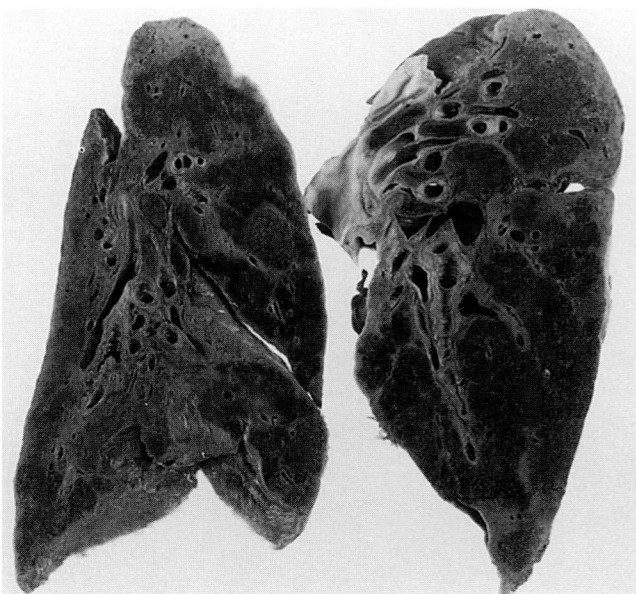

FIGURE 18-9
Late carmustine (BCNU)-included lung disease. Cut surface of lungs showing diffuse fibrosis, which is especially prominent in the subpleural areas of the apices of both upper lobes. There is no honeycomb change. *(Reproduced with permission from Hasleton et al.[116]).*

FIGURE 18-10
Late carmustine (BCNU)-induced lung disease. There is extensive intraalveolar fibrosis and interstitial elastosis that accentuates the preexisting alveolar walls. (Elastic-van Gieson, ×65.) *(Courtesy of Dr. P. S. Hasleton).*

tocin.[122] Discontinuation of the drug resulted in either improvement or stabilization of the pulmonary disease.

Fotemustine is a new drug containing a phosphonoalanine carrier grafted to the nitrosourea radical. It has shown promise in the treatment of disseminated malignant melanoma.[49] Fotemustine alone has not been associated with drug-induced lung disease. However, the adult respiratory distress syndrome developed in 2 of 60 patients treated with a combination of dacarbazine and fotemustine.[49] One patient died, whereas the other responded to high-dose corticosteroid therapy. Follow-up studies in another 10 patients showed a significant subclinical deterioration in lung function after chemotherapy.

Antimetabolites

Antimetabolites are analogues of metabolites normally required for cell function and replication. Pathologic changes in the lung have been described in patients receiving azathioprine, cytosine arabinoside, and methotrexate.

AZATHIOPRINE

Azathioprine is an immunosuppressive agent that has been used for antirejection therapy in patients receiving renal transplants and also in the management of several chronic diseases such as regional enteritis, ulcerative colitis, chronic active hepatitis, rheumatoid arthritis, systemic lupus erythematosus, and glomerulonephritis.

Azathioprine-induced lung disease is rare. Diffuse alveolar damage was diagnosed in an open lung biopsy from a 24-year-old woman who developed bilateral pulmonary infiltrates after being treated with azathioprine for membranoproliferative glomerulonephritis.[123] She recovered after discontinuation of azathioprine and treatment with corticosteroids. There are two reports of diffuse alveolar damage progressing to usual interstitial pneumonia in renal transplant recipients treated with azathioprine.[124,125] The patients with diffuse alveolar damage improved after replacement of azathioprine by cyclophosphamide. Those with usual interstitial pneumonia died of respiratory failure.[125] Diffuse alveolar damage was associated with doses from 2850 to 4355 mg, whereas usual interstitial pneumonia was associated with doses from 5600 to 28,625 mg azathioprine.[125]

CYTOSINE ARABINOSIDE

Noncardiogenic pulmonary edema occurs in about 13 percent of patients with acute leukemia treated with high-dose cytosine arabinoside.[126,127] The fatality rate is about 69 percent,[127] Tissue pathology shows an intense intra-alveolar proteinaceous edema with minimal inflammatory changes in the interstitium.[126,127] Diffuse alveolar damage has also been described.[127]

METHOTREXATE

Methotrexate is a folic acid analogue that inhibits cellular proliferation by causing an acute intracellular deficiency of folate coenzymes.[11] It can be administered orally, intravenously, intramuscularly, or intrathecally. The drug is used in a variety of malignant conditions including leukemia, osteogenic sarcoma, and choriocarcinoma. It is also finding greater use as an anti-inflammatory agent for many nonneoplastic conditions including rheumatoid arthritis,[128] psoriasis, and asthma.[2,11]

The frequency of lung toxicity in patients with neoplastic disease is about 7.6 percent.[129] The frequency of drug-induced lung disease may be increased when methotrexate is used in combination with other cytotoxic agents.[39,41,130] Peripheral blood eosinophilia occurs in approximately 40 percent of subjects. The prognosis is favorable with a mortality rate approaching 1 percent.[28,129] Some patients respond to corticosteroid therapy.[129] In some cases, the respiratory symptoms regress and the pulmonary infiltrates clear despite continuation of the drug.[129] Reinstitution of methotrexate does not always result in a recurrence of respiratory disease.[129]

Drug-induced lung disease has emerged as one of the most unpredictable and potentially serious adverse effects associated with the use of low-dose, pulse methotrexate in treating patients with rheumatoid arthritis.[128,131] The reported frequency of methotrexate-induced lung disease in this setting ranges from 0.3 to 11.6 percent.[131] A greater than expected proportion of patients have a smoking history and preexisting lung disease and are male. The prognosis is better in patients treated with corticosteroids.[128] Peripheral blood eosinophilia is not a feature of methotrexate-induced lung disease in patients with rheumatoid arthritis.[132]

The total dose of methotrexate received at the time of development of lung toxicity varies widely from 40 to 41,000 mg.[129,133] Onset of symptoms usually occurs from 12 to 196 days after the initiation of therapy.[129] However, fatal pulmonary disease occurred in one patient 10 h after receiving one dose of methotrexate,[134] whereas in another lung disease occurred after 5 years of weekly injections of the drug.[129]

Most lung biopsies from patients with methotrexate-induced lung disease show a chronic interstitial pneumonia characterized by nodular infiltrates of lymphocytes, plasma cells, mononuclear cells, and occasional giant cells.[128,129,133-138] Eosinophils are present in about half of the cases and there may be poorly formed nonnecrotizing granulomas. An uncommon manifestation of methotrexate-induced lung disease is diffuse alveolar damage (Fig. 18-11) progressing to interstitial fibrosis and honeycomb lung.[129,132,133] In some cases the alveoli are lined by hyperplastic, atypical epithelial cells (Fig. 18-12).[133] Bronchiolitis obliterans and organizing pneumonia (Fig. 18-13) sometimes associated with the presence of poorly formed nonnecrotizing granulomas has been reported in some patients receiving relatively low doses of methotrexate for rheumatoid arthritis or mycosis fungoides.[128,139,140]

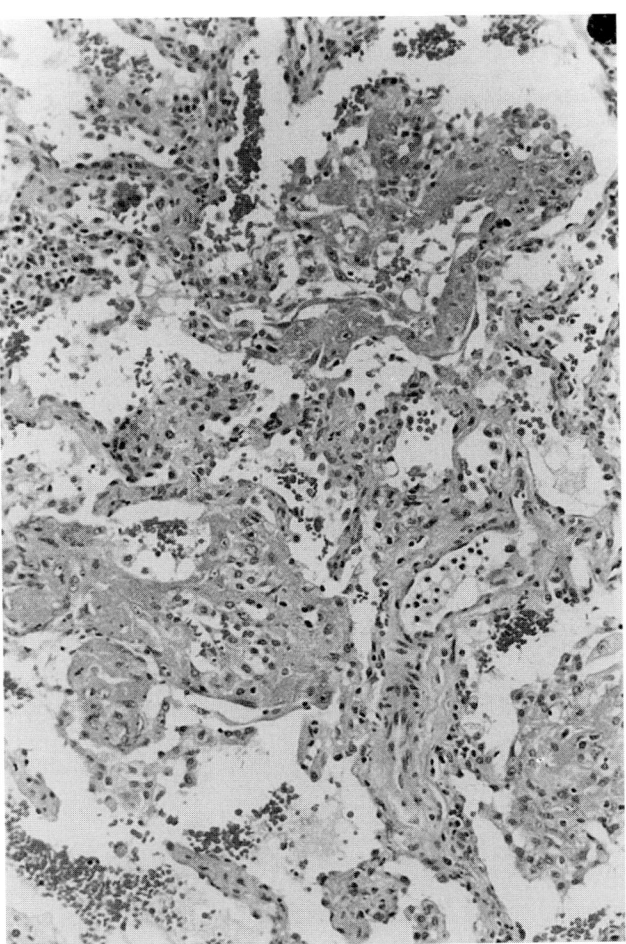

FIGURE 18-11
Organizing diffuse alveolar damage in an open lung biopsy from a patient with rheumatoid arthritis treated with methotrexate. (H&E, ×200.)

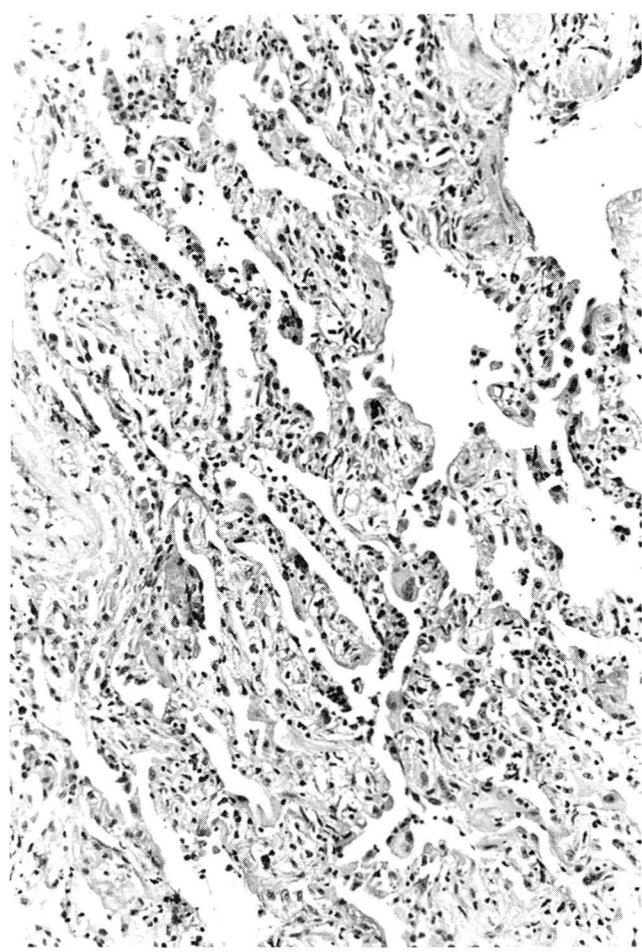

FIGURE 18-12
Methotrexate-induced lung disease. Interstitial fibrosis with atypical alveolar epithelial cells. (H&E, ×200.)

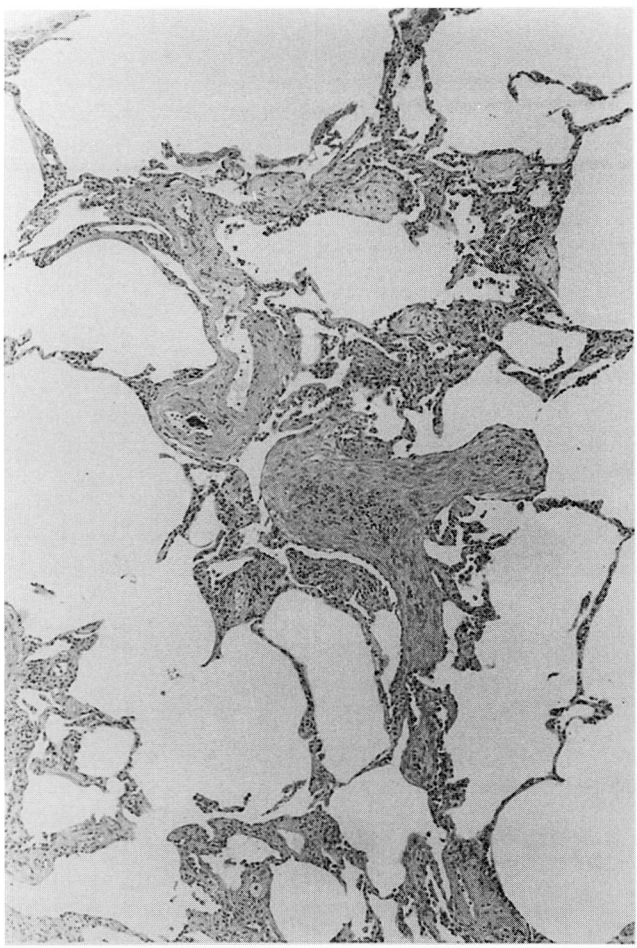

FIGURE 18-13
Bronchiolitis obliterans and organizing pneumonia in a patient with rheumatoid arthritis treated with methotrexate. (H&E, ×80.)

Miscellaneous Cytotoxic Drugs

Dacarbazine (DTIC) is used for treating disseminated malignant melanoma. When used alone, it is not toxic to the lungs. When it is combined with fotemustine, drug-induced lung disease may occur.[49] The adult respiratory distress syndrome developed in 2 of 60 patients with metastatic malignant melanoma treated with decarbazine and fotemustine. One patient died and the autopsy revealed diffuse alveolar damage. The second patient responded to high-dose corticosteroid therapy. Follow-up studies in another 10 patients showed a significant subclinical deterioration in lung function following chemotherapy.[49] Fotemustine alone is not associated with lung toxicity.[49]

Fludarabine is a novel nucleotide analogue used for the treatment of refractory chronic lymphocytic leukemia. Occasional cases of lung disease have been associated with its use. These have ranged in severity from mild interstitial lung disease to frank respira-tory failure.[141] At the time of development of lung disease, the patients were receiving fludarabine alone. However, up to 2 months before the use of fludarabine, they had received chlorambucil and cyclophosphamide.[141–143] An open lung biopsy in a patient with acute respiratory failure showed diffuse alveolar damage with atypical type II pneumonocytes.[141]

Procarbazine is a methylhydrazine derivative used primarily in the treatment of lymphoma. Procarbazine-induced lung disease is rare and has been reported in patients receiving combination chemotherapy for Hodgkin's disease.[144–147] The prognosis is excellent but in one case symptoms recurred after recommencement of procarbazine.[146] The lung pathology is characterized by a chronic interstitial pneumonia similar to that seen with methotrexate toxicity. There are nodular interstitial infiltrates of lymphocytes, plasma cells, mononu-

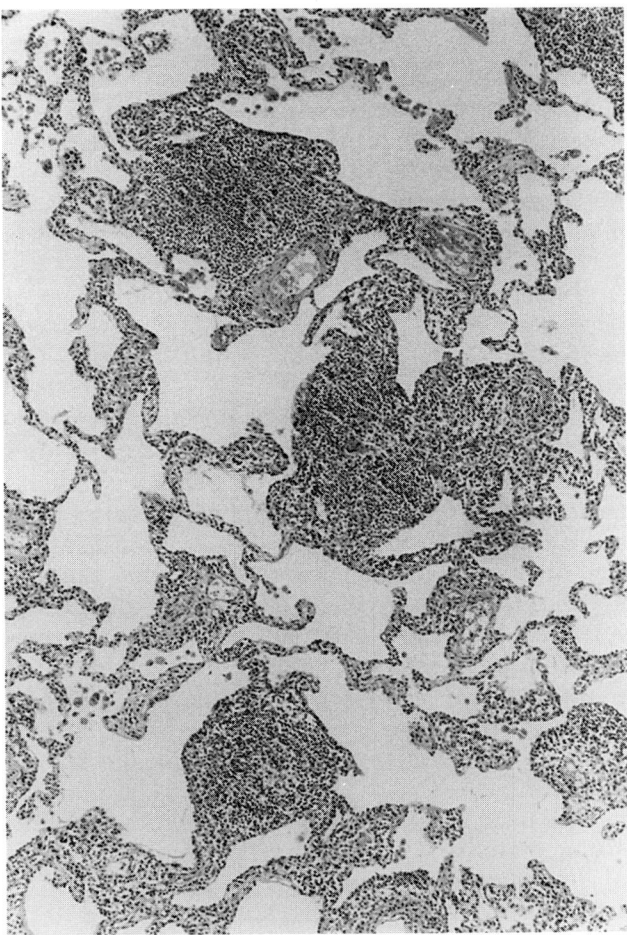

FIGURE 18-14
Procarbazine-induced lung disease. There is a chronic interstitial pneumonia characterized by nodular infiltrates of lymphocytes, mononuclear cells, plasma cells, and occasional giant cells. (H&E, ×80.)

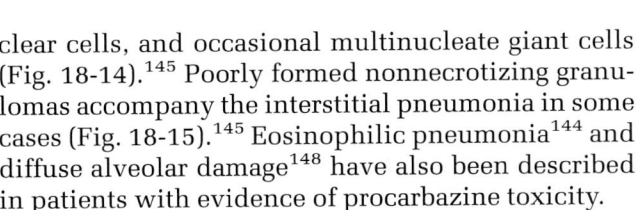

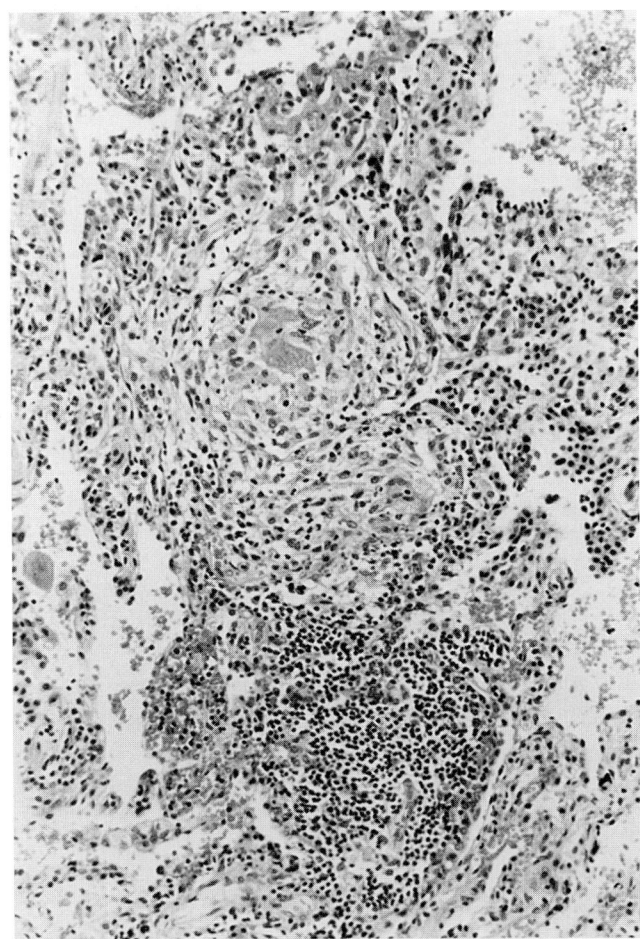

FIGURE 18-15
Procarbazine-induced lung disease. Chronic interstitial pneumonia including a poorly defined granuloma without necrosis. Same case as Figure 18-14. (H&E, ×200.)

clear cells, and occasional multinucleate giant cells (Fig. 18-14).[145] Poorly formed nonnecrotizing granulomas accompany the interstitial pneumonia in some cases (Fig. 18-15).[145] Eosinophilic pneumonia[144] and diffuse alveolar damage[148] have also been described in patients with evidence of procarbazine toxicity.

Vinblastine and vindesine are vinca alkaloids. They have been associated with acute diffuse pulmonary infiltrates and respiratory failure in patients treated with these agents combined with mitomycin for breast, ovarian, or lung cancer.[41,99,149–152] The incidence of pulmonary toxicity with this combination of drugs was as high as 39 percent in one study.[149] Unlike other cases of cytotoxic drug-induced lung disease, wheezing was noted in several patients and lung function tests showed an obstructive pattern.[41] An open lung biopsy in a patient who received vinblastine and mitomycin-C showed diffuse alveolar damage.[152] Organizing diffuse alveolar

damage with atypical epithelial cells has been seen in a patient with Kaposi's sarcoma treated with vinblastine (Fig. 18-16).

Teniposide (VM-26), a podophyllotoxin derivative, was reported to have produced diffuse alveolar damage in a patient who received 1800 mg of the drug over a 2-month period.[153] However, the patient had previously received 2102 mg carmustine and the contribution of this drug, known to be toxic to the lung, is unclear.

NONCYTOTOXIC DRUGS

Information is currently available on the pathology of drug-induced lung disease associated with more than 20 noncytotoxic agents. They can be classified as antiarrythmic agents, antihypertensive agents, calcium channel blocking agents, antimicrobials, anti-

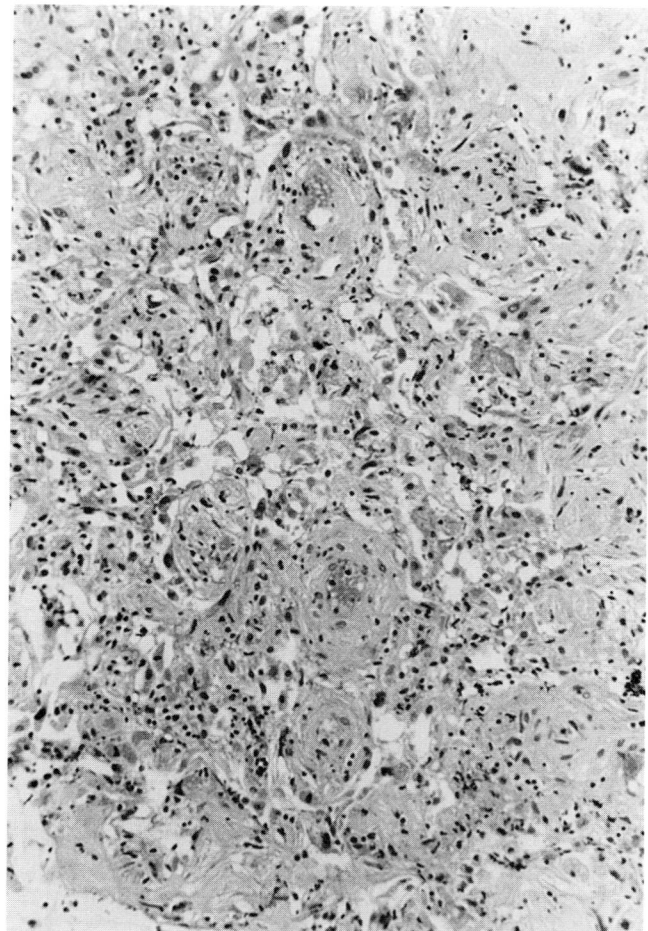

FIGURE 18-16
Vinblastine-induced lung disease. This man developed acute respiratory failure while being treated for Kaposi's sarcoma. The open lung biopsy showed organizing diffuse alveolar damage with atypical alveolar epithelial cells. (H&E, ×80.)

inflammatory agents, anticonvulsants, anorexigens, and miscellaneous drugs (see Table 18-4).

Antiarrhythmic Agents

AMIODARONE

Amiodarone is an iodinated benzofuran derivative, the chemical structure of which resembles thyroxine.[154] It is an effective antiarrhythmic drug with complex pharmokinetic properties and the potential for a wide range of adverse effects, the most serious of which is lung disease. Amiodarone lung toxicity occurs in about 6 percent of patients who take 400 mg or more of the drug a day.[155–157] In a large study of 573 patients treated with amiodarone, no patient receiving less than 305 mg a day developed pulmonary toxicity.[155] However, as with other drugs that show a dose relationship to toxicity, there have been scattered reports of toxicity with low doses of amiodarone.[158,159] Neither the duration of therapy nor the

total cumulative dose correlate with the risk of lung disease. Toxicity usually becomes evident between 6 days and 60 months of treatment with the highest incidence occurring during the first 12 months.[155] Older patients develop it more frequently and it is rare under the age of 40 years.[155]

Most patients present with the insidious onset of dyspnea and nonproductive cough.[156] However, about one third of patients present with the adult respiratory distress syndrome.[156,160–162] This syndrome has been described as a postoperative phenomenon in patients receiving amiodarone and undergoing surgical procedures requiring general anesthesia.[163–167] CT scanning has recently been shown to be useful in the diagnosis of amiodarone lung disease. This is because its iodine content increases lung density in areas where the drug accumulates.[2,168] Most patients with amiodarone-induced lung disease will recover after withdrawal of the drug.[160,161,169,170] However, the mortality rate of amiodarone-induced lung disease is 10 to 20 percent.[2,155] The value of corticosteroid therapy is uncertain.[12,156]

The most common lung tissue reaction to amiodarone toxicity is a nonspecific chronic interstitial pneumonia associated with the presence of foamy macrophages in the alveolar spaces (Fig. 18-17).[160,161,170,171] The macrophages are large with relatively small nuclei and finely vacuolated clear cytoplasm (Fig. 18-18).[172]

Ultrastructurally, the vacuolated appearance of the macrophages is due to the presence of intracytoplasmic lamellar inclusions (Fig. 18-19 and 18-20).[160,161,171,172] These cytoplasmic inclusions are also encountered in bronchiolar epithelial cells (Fig. 18-21), type II pneumonocytes, endothelial cells (Fig. 18-22), interstitial cells, and circulating leukocytes (Fig. 18-23).[172] These membrane-bound bodies measure approximately 1 μm in diameter. They are round or oval in shape and are composed of distinct, closely spaced, concentric parallel lamellae.[172] Focal granular densities are scattered among the lamellae, mostly at the periphery of the inclusion (see Fig. 18-20). Some large bodies are irregular in outline and are probably the result of the fusion of multiple inclusions (see Fig. 18-20).[172] Some inclusions show an eccentric irregular homogeneous dense core with a peripheral clear halo.

Distinguishing surfactant bodies from those induced by amiodarone can be difficult. Surfactant bodies are restricted to type II pneumonocytes; they are usually homogeneous and lack the peripheral granular densities and frequent fusions found in amiodarone inclusion bodies. Foamy macrophages

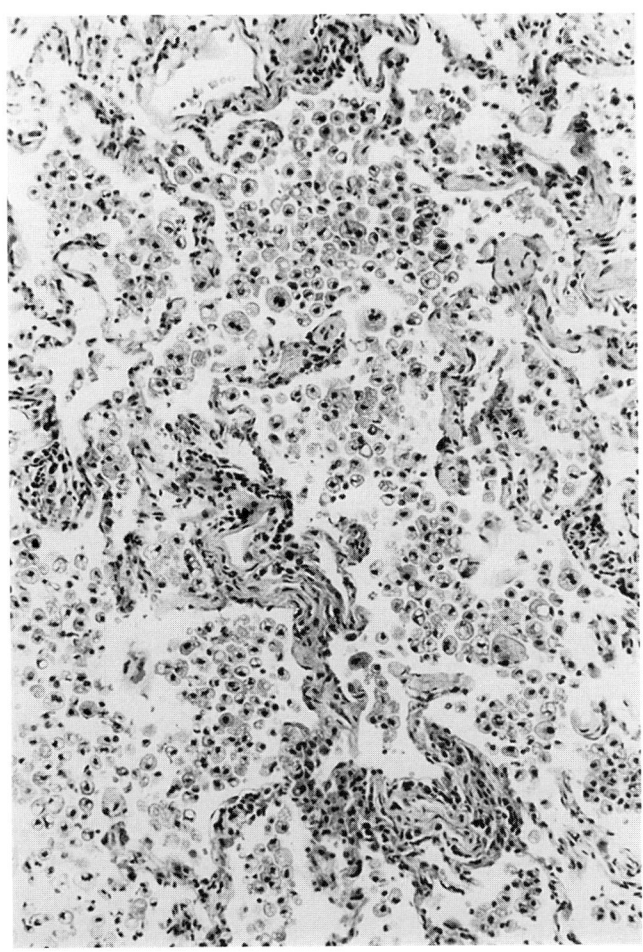

FIGURE 18-17
Amiodarone-induced lung disease. There is a mild chronic interstitial pneumonia. Abundant foamy macrophages are present in the alveoli. (H&E, ×200.)

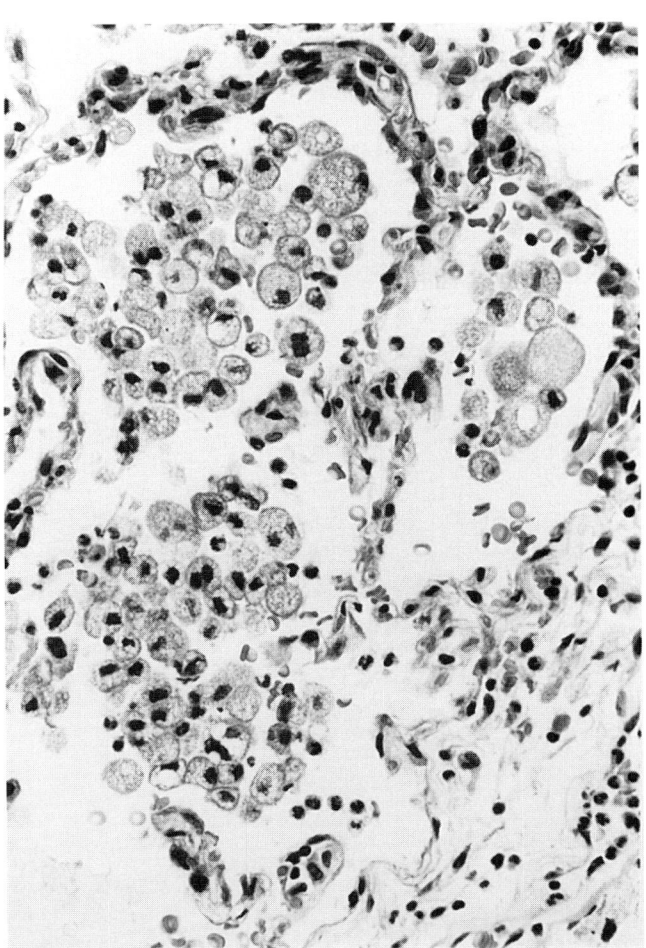

FIGURE 18-18
Amiodarone-induced lung disease. Mild chronic interstitial pneumonia with abundant intra-alveolar foam cells. (H&E, ×500.)

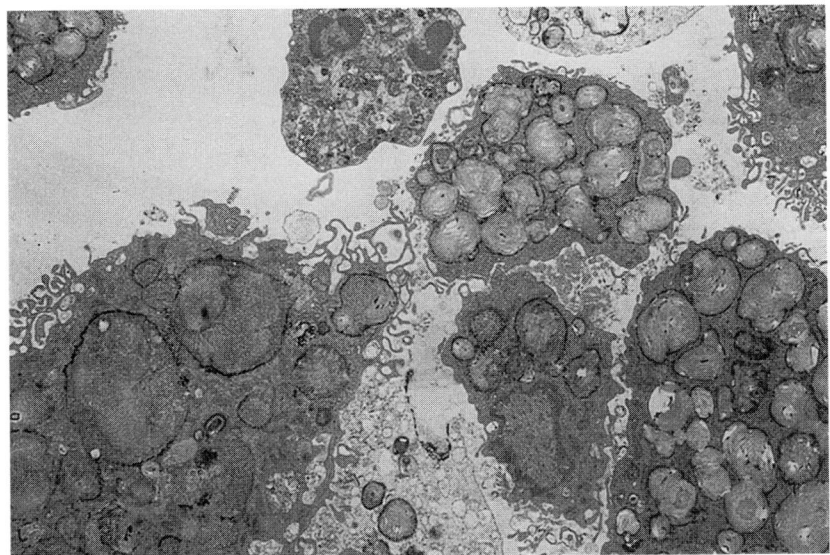

FIGURE 18-19
Amiodarone-induced lung disease. Numerous dense lamellar bodies are present in the cytoplasm of these alveolar macrophages. (Electron micrograph ×5100.) *(Reproduced with permission from Colgan et al.[172])*

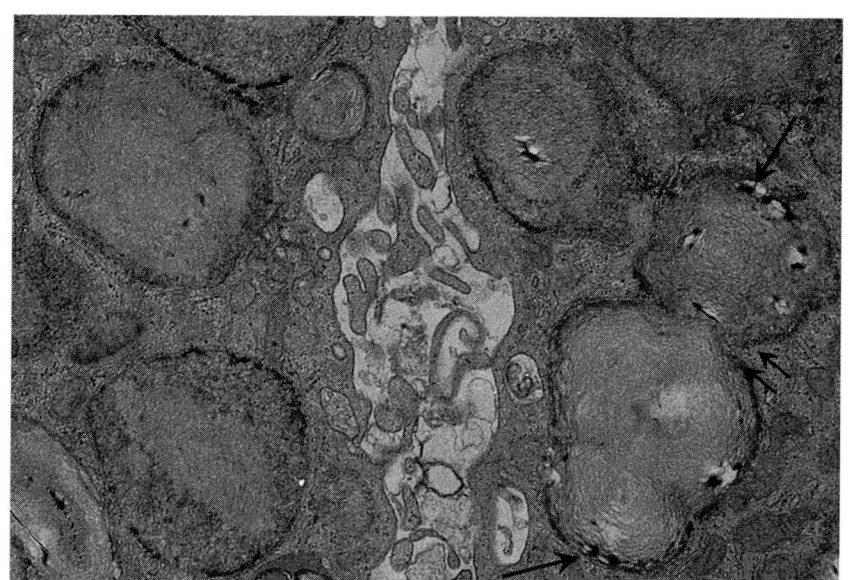

FIGURE 18-20
Amiodarone-induced lung disease. High-power view of alveolar macrophages. Lamellar bodies show distinct, concentric, parallel membranes, spaced about 5 to 10 nm apart. Peripheral granular densities are prominent (*single arrows*) in most of these bodies. Fusion of two bodies may occur (*double arrow*). (Electron micrograph ×28,500.) *(Reproduced with permission from Colgan et al.*[172]*)*

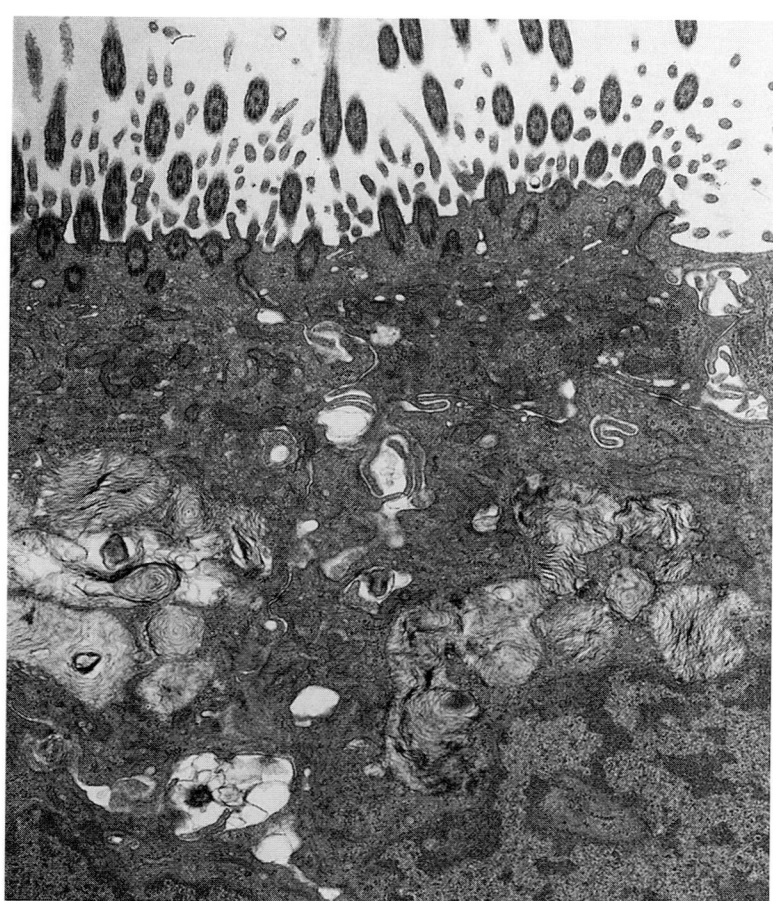

FIGURE 18-21
Amiodarone-induced lung disease. Lamellar bodies are present in the cytoplasm of these ciliated bronchiolar epithelial cells. Some of the large and irregular bodies suggest fusion of single lamellar bodies. (Electron micrograph ×17,000.) *(Reproduced with permission from Colgan et al.*[172]*)*

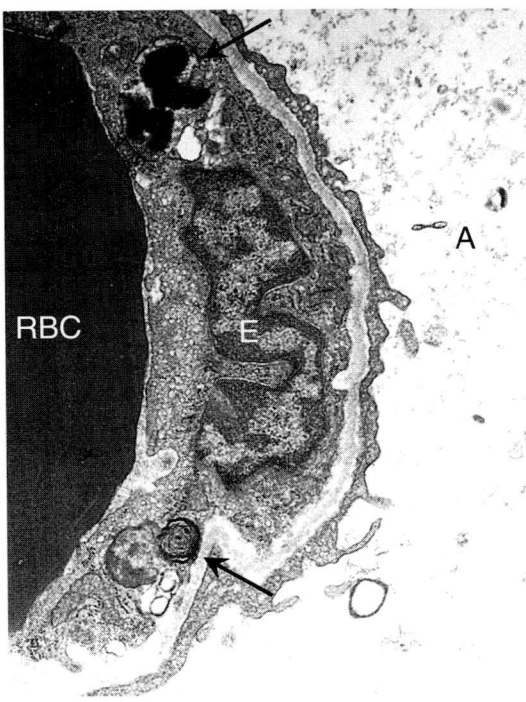

FIGURE 18-22
Amiodarone-induced lung disease. The alveolar space (A) is lined by a type I pneumonocyte (I). The endothelial cell (E) lining the capillary contains at least two lamellar bodies of heterogenous morphology (*arrows*). The capillary contains a red blood cell (RBC). (Electron micrograph ×14,400.) *(Reproduced with permission from Colgan et al.[172])*

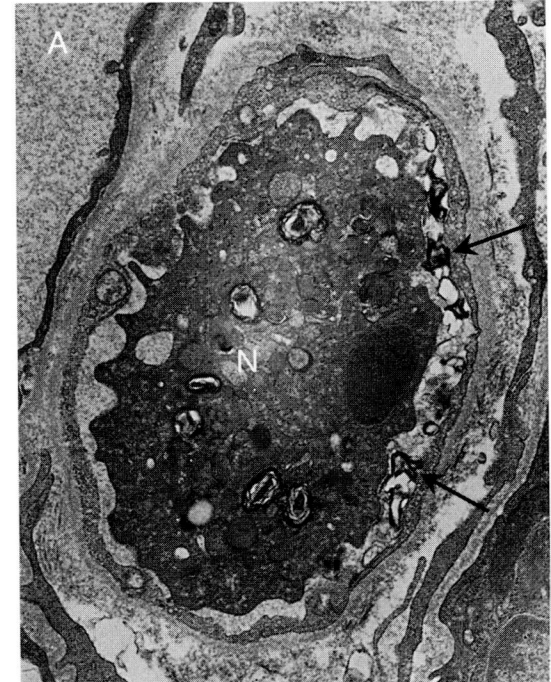

FIGURE 18-23
Amiodarone-induced lung disease. Alveolar septum. The neutrophil (N) in the capillary has several intracytoplasmic lamellar bodies. Note the lamellar bodies in capillary lumen between endothelial cell and neutrophil (*arrows*). The alveolar space (A) is at the top left corner. (Electron micrograph ×18,000.) *(Reproduced with permission from Colgan et al.[172])*

and cytoplasmic lamellar inclusions are frequently found in patients receiving amiodarone who do not have lung disease.[161,171,173] Accordingly, they are a cellular marker of amiodarone therapy but are not by themselves diagnostic of drug-induced lung disease.[156,157,171] The intracytoplasmic inclusions that occur in many of the cells of the body in amiodarone toxicity are identical to those described in the cells of patients and experimental animals receiving various neuroleptics, antidepressants, inhibitors of cholesterol synthesis, antihistamines, and anorectic agents.[174–177] Amiodarone and all these drugs have two structural features in common: a nonpolar ring system and a hydrophilic cationic side chain. These amphiphilic compounds may alter the substrates of lipid metabolism by complexing to polar lipids, resulting in the accumulation of various polar lipids within lysosomes.[174] There is evidence that amiodarone is directly toxic to the lung, but indirect inflammatory and immunologic processes may also be involved in the development of lung disease.[157,178]

Diffuse alveolar damage with hyaline membranes is encountered in a minority of patients in whom it is accompanied by the presence of intraalve-olar foamy macrophages (Fig. 18-24).[160,162,169,171] Bronchiolitis obliterans and organizing pneumonia is an uncommon manifestation of amiodarone toxicity (Fig. 18-25). It is usually associated with a chronic interstitial pneumonia and intraalveolar foamy macrophages.[169,171,172,179]

Apart from excluding infection, the contribution of bronchoalveolar lavage (BAL) to the diagnosis of amiodarone-induced lung disease is limited.[173,180] Foamy macrophages are invariably encountered in BAL fluid of patients who take amiodarone irrespective of the presence or absence of lung disease.[173] The differential cell count of BAL fluid in amiodarone-induced lung disease is variable, and no cellular pattern is diagnostic or predictive of a detrimental outcome.[180] The BAL differential cell count is normal in 20 to 30 percent of patients with amiodarone-induced lung disease.[156,180] In the remaining subjects there is an increase in the proportion of neutrophil polymorphonuclear leukocytes or lymphocytes or both.[156,173,180]

TOCAINIDE

Tocainide is a primary amine analogue of lidocaine administered orally for the treatment of ventricular

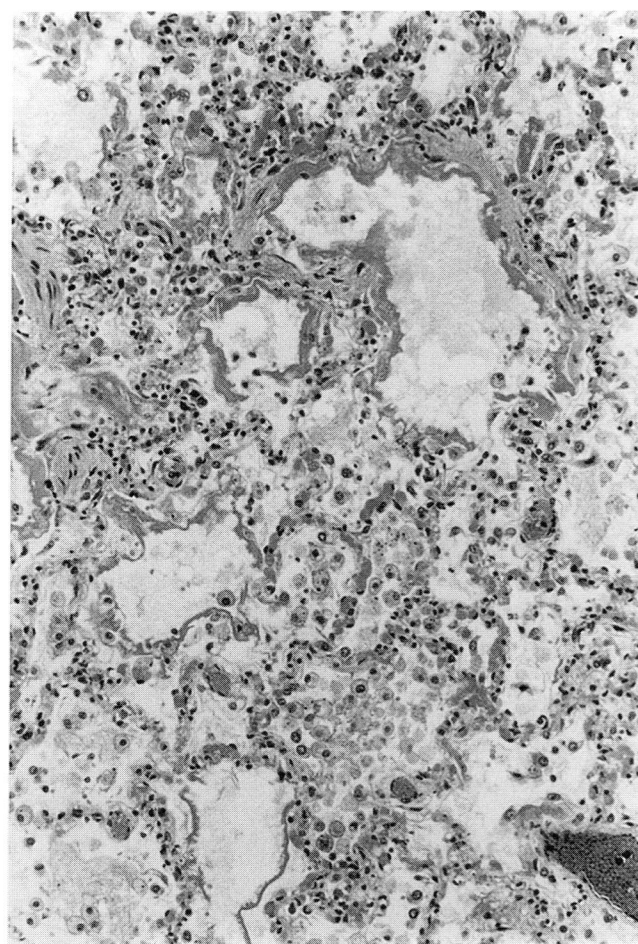

FIGURE 18-24
Diffuse alveolar damage in a patient who developed acute respiratory failure while receiving amiodarone. The alveoli are lined by hyaline membranes. Foamy macrophages are present in many of the alveolar spaces. (H&E, ×200.)

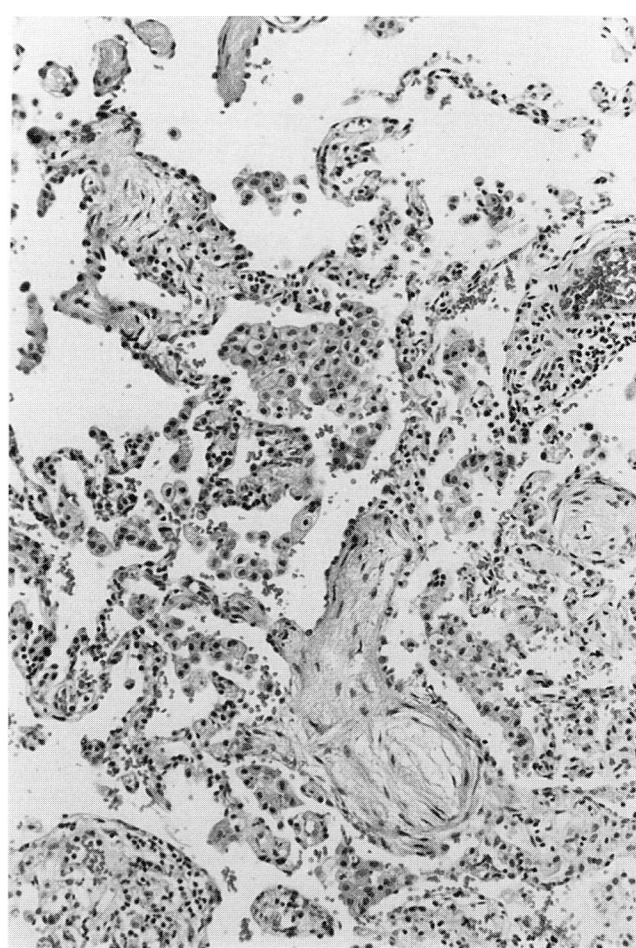

FIGURE 18-25
Amiodarone-induced lung disease. Bronchiolitis obliterans and organizing pneumonia. Foamy macrophages are present in many of the alveoli. (H&E, ×200.)

arrhythmias. Tocainide-induced lung disease is rare and its incidence has been estimated as 0.3 percent.[181] In the few cases reported in the literature, symptoms of chronic interstitial lung disease occurred from 1.5 to 14 months after initiation of therapy. These patients received from 600 to 1600 mg tocainide per day.[182–186] Most patients recovered after discontinuation of tocainide and the administration of corticosteroids. However, at least one fatality has occurred.[186]

There is little detailed information about the pathology of tocainide-induced lung disease. Transbronchial biopsies have shown organizing pneumonia[184] and interstitial pneumonia with focal macrophages and fibrin deposition.[185] An open lung biopsy in one case showed active alveolitis with chronic diffuse interstitial pneumonitis and early fibrosis.[183] The description and photomicrographs of an open lung biopsy from another case resemble usual interstitial pneumonia.[186] An autopsy on the latter case showed honeycomb lung.[186]

Antihypertensive Agents

β BLOCKERS

The literature on β blockers and drug-induced pleuropulmonary disease has recently been reviewed with respect to acebutolol, alprenolol, atenolol, celiprolol, labetalol, nadolol, pindolol, practolol, and propranolol.[187] Most patients recovered and only a few cases have been investigated by lung biopsy so details of the pathology are lacking with respect to most drugs. Practolol and pindolol are structurally different from propranolol and alprenolol in having a nitrogen atom attached to an aromatic ring, producing a stereochemical configuration similar to methysergide, which is known to produce fibrosing reactions.[188]

Acebutolol

Lung disease has been described in five men and five women aged 47 to 82 years who took acebutolol for periods ranging from 6 to 24 months.[187] In four patients the lung disease was accompanied by pleural effusion. All the patients recovered after withdrawal of the drug. An open lung biopsy in one case showed granulomatous inflammation with eosinophils.[189] A transbronchial biopsy in another case showed bronchiolitis obliterans organizing pneumonia.[179]

Pindolol

Usual interstitial pneumonia was diagnosed in an open lung biopsy from a 55-year-old man who took pindolol for 84 months.[188] Electron microscopy showed variable thickening of the alveolar septa with a few smooth muscle cells present between collagen fibers. Type II pneumocytes were prominent.

Practolol

This drug was introduced into clinical practice in 1970 but was withdrawn in 1976.[44] A relatively specific entity developed in some patients after exposure to practolol.[187] They developed pleural thickening or effusions and a restrictive lung function defect. At thoracotomy there was dense fibrosis of the pleura measuring up to 2 cm in thickness.[190] Radiologic evidence of honeycomb lung[191] was apparent and a lung biopsy in one patient showed interstitial fibrosis.[192]

GANGLION BLOCKING AGENTS

Hexamethonium,[193] mecamylamine,[194] and pentolinium[195] are ganglion blocking agents that used to be popular for treating systemic hypertension. They produced lung injury similar to that seen in uremia.[196,197] The changes consist of organizing intraalveolar fibrinous exudate and hyaline membranes.[193-195,198]

Calcium Channel Blocking Agent

BEPRIDIL

Bepridil is a calcium channel blocking agent that has been available for over 10 years. Its use has been associated with a few examples of parenchymal lung disease one of which was associated with eosinophilia.[199] A transbronchial biopsy was obtained from a 72-year-old man who received 200 to 400 mg bepridil a day for 9 weeks before developing cough, dyspnea, and pulmonary infiltrates. There was widening of the alveolar walls and an interstitial infiltrate of plasma cells, lymphocytes, and mononuclear cells.[199] He recovered after discontinuation of bepridil and the administration of prednisone.

Antimicrobials

With the exception of nitrofurantoin, drug-induced lung disease is a rare complication of antimicrobial therapy.

AMPICILLIN

One case of pulmonary infiltrates associated with peripheral blood eosinophilia has been associated with ampicillin therapy.[200] A transbronchial biopsy showed eosinophilic pneumonia. The patient showed an immediate skin test reaction to ampicillin but not to penicillin or cephalothin.

AMPHOTERICIN B

Amphotericin B, used in the treatment of systemic fungal infections, is thought to act through binding to sterol components of fungal cell walls.[201] There appears to be a synergistic toxicity between amphotericin B and leukocyte transfusions. Between 25 and 64 percent of patients who receive both agents develop pulmonary disease with dyspnea, hypoxemia, and hemoptysis.[202,203] Lung disease occurs within 4 days of starting amphotericin and it has a poor prognosis with a mortality rate of about 30 percent.[202] Lung biopsies show diffuse intraalveolar hemorrhage.[202]

CEPHALOSPORIN

Diffuse interstitial lung disease is a rare complication of cephalosporin therapy. Affected patients have improved dramatically with discontinuation of cephalosporin and the initiation of corticosteroid therapy.[204,205] The appearances in a transbronchial lung biopsy were described as interstitial pneumonia.[205] However, focal intra-alveolar myxomatous connective tissue was described, which raises the possibility of an organizing pneumonia.

MINOCYCLINE

Minocycline is a semisynthetic tetracycline derivative that is mainly used in the treatment of bronchopulmonary and urinary tract infections. It is also prescribed for the treatment of acne vulgaris in young persons. Twenty-nine cases of minocycline pneumonitis have been described. In the largest series of eight cases, the duration of treatment was 13 ± 5 days (mean $\pm$ SE).[206] The total dose was 2060 ± 540 mg. Patients presented with dyspnea, fever, dry cough, hemoptysis, chest pain, fatigue, and rash. The chest radiographs showed bilateral infiltrates in all cases. Hypoxemia was present in seven patients. Eosinophilia was demonstrated in the peripheral blood and BAL fluid in seven cases. Transbronchial lung biopsy specimens from two cases showed interstitial pneumonia. In one case the interstitial infiltrate included abundant eosinophils. No eosinophils were seen in the other case but some of the alveolar macrophages showed brown granular pigmentation with negative staining for hemosiderin.[206] The prognosis was good and all patients recovered when minocycline was terminated. Some patients with severe symptoms required corticosteroid therapy.

NITROFURANTOIN

Nitrofurantoin is an antiseptic agent used primarily in the treatment of urinary tract infections. The drug is administered orally and is thought to act by inhibiting bacterial enzyme systems. Nitrofurantoin is responsible for more reported cases of drug-induced lung disease than any other drug, and over 500 cases have been published since the first description in 1962.[12,207] The large number of cases reported is probably due to extensive use of the drug rather than to a high frequency of adverse reactions.[12] Nitrofurantoin may injure the lung through the production of reactive oxygen species.[16]

More than 90 percent of cases of nitrofurantoin-induced lung disease occur in women, probably because more women take the drug than men.[208,209] Age may be a risk factor because the majority of cases have been reported in patients 40 to 70 years of age.[209] This may reflect the more frequent use of the drug in this age group.[12]

Nitrofurantoin-induced lung disease may present in two distinct clinical patterns characterized by the length of treatment before development of the syndrome. The acute presentation occurs within a month of commencement of therapy, whereas the chronic presentation occurs after 2 months to 5 years of continuous therapy.[209]

The acute syndrome is more common and represents about 90 percent of reported cases.[208,209] Patients with the acute presentation usually develop symptoms within 2 weeks of continuous treatment with nitrofurantoin. However, symptoms may develop within 24 h of the first dose if patients have received previous nitrofurantoin therapy.[208,209] The length of treatment before the development of drug-induced lung disease appears to be inversely related to the number of prior exposures to nitrofurantoin.[208,209] Symptoms include dyspnea, nonproductive cough, and fever. Peripheral blood eosinophilia is present in 70 to 80 percent of cases.[208,209] The prognosis of acute nitrofurantoin-induced lung disease is good and most patients recover after withdrawal of the drug. Corticosteroid therapy does not appear to accelerate recovery. The mortality rate is about 1 percent. Lung biopsy is rarely necessary to establish the diagnosis in acute nitrofurantoin-induced lung disease. Eosinophilic pneumonia is generally considered to be the underlying pathology,[19] but no biopsy-proven case has been reported.

Uncommon pathologic manifestations of acute nitrofurantoin-lung disease include diffuse alveolar damage (Fig. 18-26),[210] desquamative interstitial pneumonia,[208] and diffuse alveolar hemorrhage.[211,212] There has also been a case of chronic interstitial pneumonia with loosely formed granulomas including scattered giant cells resembling extrinsic allergic alveolitis.[213]

The chronic form of nitrofurantoin-induced lung disease accounts for about 10 percent of cases. The patients present with insidious onset of exertional dyspnea and a nonproductive cough occurring after months to years of nitrofurantoin therapy.[208,209] The radiographic features mimic idiopathic chronic interstitial pulmonary fibrosis. Peripheral blood eosinophilia is less common than with the acute syn-

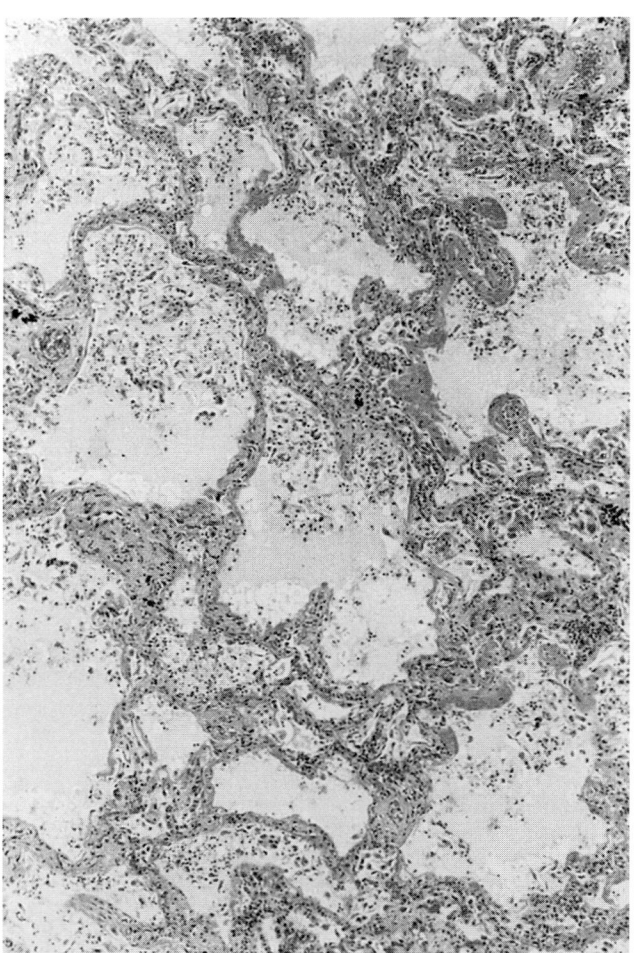

FIGURE 18-26
Nitrofurantoin-induced lung disease. Acute diffuse alveolar damage with abundant hyaline membranes and intraalveolar edema. (H&E, ×80.)

drome but has been reported in up to 40 percent of patients with chronic disease.[209] Many patients have low titers of antinuclear autoantibodies. The acute form of the disease does not appear to predispose to the chronic syndrome.[208] The mortality rate of chronic nitrofurantoin-induced lung disease is approximately 10 percent.[208] The lung tissue in the majority of patients with chronic nitrofurantoin-induced lung disease shows usual interstitial pneumonia.[208,214–218] Eosinophils may or may not be present but are rarely conspicuous. Unusual histologic manifestations of chronic nitrofurantoin-induced lung disease are desquamative interstitial pneumonia,[219] chronic eosinophilic pneumonia,[220] giant cell interstitial pneumonia,[213] lymphoid interstitial pneumonia,[19] and bronchiolitis obliterans organizing pneumonia.[221]

PYRIMETHAMINE

The increased use of malaria prophylaxis has led to more reports of drug-induced lung disease. In at least 14 reports pulmonary eosinophilia has been attributed to pyrimethamine.[222] Pyrimethamine has been taken in combination with sulfadoxine, dapsone, and chloroquin but it is believed that pyrimethamine is the cause of lung disease.[222] Patients have presented with progressive respiratory failure and radiology has shown bilateral patchy consolidation or widespread nodular infiltrates. Most patients have recovered after withdrawal of pyrimethamine. Some have required corticosteroid therapy. However, there has been at least one fatality.[223] Open lung biopsies in four patients have shown eosinophilic pneumonia.[222] In one patient the eosinophilic pneumonia was complicated by an eosinophilic vasculitis and pulmonary infarction.[222]

SULFONAMIDES

Sulfasalazine
Sulfasalazine has been widely used for the treatment of inflammatory bowel disease for almost 50 years and in the last decade has been increasingly used for the treatment of rheumatologic disorders. The drug is metabolized in the bowel to 5-aminosalicylic acid, which is excreted in the feces, and sulfapyridine, which is absorbed into the body.[12,224] Sulfapyridine is thought to be responsible for adverse reactions.[224] Sulfasalazine-induced lung disease is rare and 40 cases have been reported.[225] Two had Crohn's disease, six had rheumatoid arthritis, and the remainder had ulcerative colitis. The known association of interstitial lung disease with inflammatory bowel disease[225,226] and rheumatoid disease makes the

establishment of cause difficult; however, in the majority of reported cases, the symptoms and signs have regressed after discontinuation of the drug. Daily dosage of sulfasalazine ranged between 1.5 and 8 g, treatment duration between 2 weeks and 8 years, and total dose between 24 and 11,520 g.[225] Risk factors for the development of lung disease are unknown. There appears to be no relationship between the development of lung toxicity and duration of therapy, total cumulative dose, or daily maintenance dose. The prognosis of sulfasalazine-induced lung disease is good and most patients recover after discontinuation of the drug. Corticosteroids hasten improvement in some patients.[12] Three fatalities have been reported.[225,227]

Eosinophilic pneumonia is thought to be the underlying disease in most patients who present with cough, dyspnea, and peripheral blood eosinophilia. An open lung biopsy in one case showed eosinophilic pneumonia.[228] A chronic interstitial pneumonia without eosinophilia has been observed in transbronchial biopsy specimens from some other cases.[229,230] A transbronchial biopsy specimen from another patient showed a chronic interstitial pneumonia with granulomas including multinucleate giant cells resembling extrinsic allergic alveolitis.[231] Examples of desquamative interstitial pneumonia[232] and bronchiolitis obliterans organizing pneumonia[225,233] have been reported. Open lung biopsies on two patients who developed lung disease after receiving sulfasalazine for Crohn's disease and for rheumatoid arthritis were said to show usual interstitial pneumonia and acute interstitial pneumonia, respectively.[224] However, the patient said to have usual interstitial pneumonia recovered when the drug was discontinued. The patient said to have acute interstitial pneumonia made a rapid recovery when the drug was withdrawn and corticosteroids were given. This suggests that the patients had a chronic interstitial pneumonia rather than usual interstitial pneumonia or acute interstitial pneumonia. Another patient with Crohn's disease treated with salfasalazine, corticosteroids, and a short course of azathioprine developed an unusual perilobular pattern of interstitial fibrosis.[226] A fatal case of lung disease attributed to sulfasalazine occurred in a 47-year-old woman who was treated with the drug and prednisolone for ulcerative colitis. After 4 months treatment she developed shortness of breath and died 1 month later. The lungs showed organizing diffuse alveolar damage.[227]

Other Sulphonamides
There have been a few well-documented cases of pulmonary infiltrates and peripheral blood eosinophilia

complicating sulphonamide therapy administered by the oral and vaginal routes.[12] An open lung biopsy in one case revealed eosinophilic pneumonia.[234]

Anti-Inflammatory Agents

Several drugs such as gold, penicillamine, and non-steroidal anti-inflammatory agents used to treat rheumatoid arthritis and other inflammatory conditions have been associated with drug-induced lung disease.

GOLD

Current preparations of gold are aurous salts with sulfur attached to the metal. Chrysotherapy is most commonly used for rheumatoid arthritis, but it has also been administered for other diseases including asthma, osteoarthritis, and pemphigus. The therapeutic efficacy of gold may be related to inhibition of lysosomal enzymes and inhibition of macrophage phagocytosis after ingestion of the metal.[12] Suppression of cellular immunity, prevention of prostaglandin synthesis, and inhibition of histamine release may also play a role. The incidence of gold-induced lung disease appears to be low and is probably less than 1 percent.[12] There are no clearly defined risk factors for developing lung disease secondary to chrysotherapy. Respiratory disease has been reported after cumulative doses ranging from 120 to 1660 mg.[235] The average duration of chrysotherapy in patients who developed lung disease was 15 weeks.[235] Respiratory symptoms may develop from 6 hours to 1 month after receiving the last dose of gold.[236] The presentation may rarely be acute with fever, wheezing, and cough developing over several hours.[237] More commonly, the presentation is subacute with progressive dyspnea and a nonproductive cough developing over several weeks.[238] About 40 percent of the patients have fever and the frequency of eosinophilia ranges from 33 to 42 percent. A concurrent dermatitis is present in 44 percent of patients.[238] Although most patients have underlying rheumatoid arthritis, gold-induced lung disease has been reported in patients with asthma, osteoarthritis, and pemphigus.[235] Most patients with gold-induced lung disease recover following discontinuation of the drug. However, almost 50 percent of patients show some residual radiologic or physiologic abnormality.[12] The role of corticosteroid therapy is unclear.[12]

Most of the lesions associated with gold-induced lung disease are similar to the pulmonary manifestations of rheumatoid disease, and so it can be difficult to precisely identify a cause in a given patient.[237] The most commonly described reaction is non-

specific chronic interstitial pneumonia.[237,239,240] Bronchiolitis obliterans organizing pneumonia[221] has been described and is sometimes associated with chronic interstitial pneumonia.[241,242] A fatal case of bronchiolitis obliterans occurred in a 12-year-old girl with juvenile rheumatoid arthritis who received a 6-month course of intramuscular gold.[242] Cases of chronic bronchiolitis have been described. In one case the walls of the bronchioles showed a dense infiltrate of lymphocytes.[243] In another case, the walls of the bronchioles showed a chronic inflammatory reaction with the presence of poorly defined noncaseating granulomas.[244] Occasional fatal cases of diffuse alveolar damage have been described.[245]

Ultrastructural studies of lung biopsy tissue and BAL fluid have shown characteristic electron-dense deposits of gold within lysosomes of alveolar and interstitial macrophages.[236,240,246] These gold-laden lysosomes, which have been termed aurosomes, have been identified within alveolar macrophages in patients without lung disease. They may be present for as long as 2 years after cessation of chrysotherapy.[247]

PENICILLAMINE

Penicillamine is a dimethylcysteine derivative of penicillin, which is an effective chelator of copper, mercury, zinc, and lead.[12] In addition to being used to treat metal poisoning, penicillamine is also used in the management of Wilson's disease, cystinuria, rheumatoid arthritis, scleroderma, primary biliary cirrhosis, and other inflammatory disorders. The frequency of penicillamine-induced lung disease is low and less than 1 percent. A review of data on 3356 patients treated with penicillamine revealed only two cases of fatal bronchiolitis.[248] Many of the patients taking penicillamine who develop lung disease also have underlying rheumatoid arthritis, which may itself be associated with lung disease.

Risk factors for the development of penicillamine-induced lung disease are unclear.[12] No apparent relationship exists between lung toxicity and the duration of therapy, total cumulative dose, or daily maintenance dose. The majority of patients are middle-aged women with connective tissue disorders. However, this may relate to the fact that a high proportion of patients treated with penicillamine fall into these groups. The interval from commencement of therapy to onset of respiratory disease has ranged from 18 days to 7 years.[12,249,250] Total doses of the drug administered before the development of pulmonary disease range from 900 mg[251] to more than 1000 g.[251] Patients developing the pulmonary–renal syndrome have generally been treated the longest and have received the largest cumulative doses of penicillamine.[12]

The most common form of penicillamine-induced lung disease is constrictive bronchiolitis (Fig. 18-27).[252–254] This is a lesion characterized by alterations in the walls of membraneous and respiratory bronchioles. The histologic findings range from mild bronchiolar inflammation and scarring to progressive concentric submucosal fibrosis and occasionally, complete occlusion of the small airways.[255] An elastic-van Gieson stain may be required to detect this lesion (Fig. 18-28). Other findings include smooth muscle hyperplasia, bronchiolectasis with mucostasis and fibrosis of the small airway walls. In

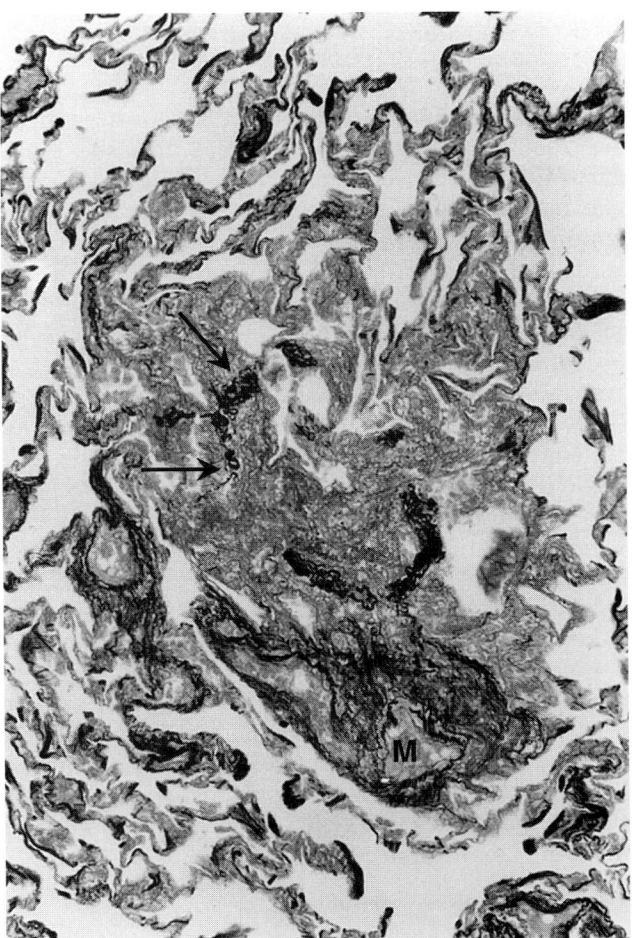

FIGURE 18-28
Constrictive bronchiolitis. An incomplete ring of black elastic tissue (arrows) outlines the wall of this bronchiole occluded by submucosal fibrosis. The accompanying muscular pulmonary artery (M) has been sectioned in various planes. (Elastic-van Gieson, ×200.)

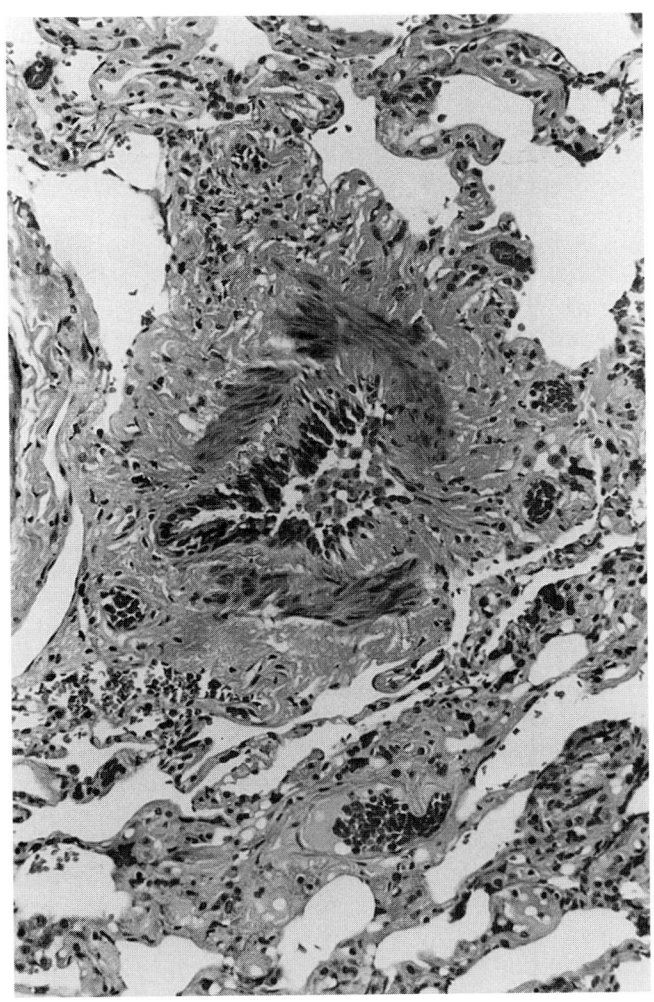

FIGURE 18-27
Constrictive bronchiolitis. The lumen of this membranous bronchiole is narrowed due to submucosal fibrosis, muscular hypertrophy, and adventitial fibrosis. There is no significant leukocytic infiltration of the wall. (Trichrome, ×200.)

some cases the submucosal infiltrate is associated with prominent lymphoid follicles and resembles follicular bronchiolitis.[254] Less commonly, the changes of bronchiolitis obliterans organizing pneumonia have been observed.[254] The majority of patients with presumed penicillamine-associated bronchiolitis are women who have never smoked. Breathlessness and cough begin within 3 to 24 months after the commencement of penicillamine therapy.[254] In one third of the cases in which the outcome is known, death was from progressive respiratory failure.[254] Corticosteroid therapy has not improved lung function or symptoms and 50 percent of patients so treated have died.[12] Bronchiolitis occurs in patients with rheumatoid disease who have not received penicillamine. Accordingly, whether bronchiolitis in patients receiving penicillamine represents drug-induced lung disease, or whether it is a manifestation of rheumatoid disease, is difficult to assess.

The pulmonary-renal syndrome has been described in patients with Wilson's disease,[250] primary biliary cirrhosis,[249] and rheumatoid arthritis[256] who received relatively high maintenance doses of penicillamine from 6 months to 7 years. The incidence of this complication has been estimated at 0.5 percent.[250,256] The mortality rate is high and autopsies have shown diffuse alveolar hemorrhage with no sign of vasculitis.[250] Immunofluorescent and ultrastructural studies on the lung in one case were negative.[257] Immunologic studies on the kidneys show no evidence of antibasement membrane antibody disease. The findings have been variable with some suggesting an immune complex disease.[256,257]

Some patients with rheumatoid arthritis being treated with penicillamine develop dyspnea and diffuse pulmonary infiltrates after 3 weeks to 12 months of therapy.[236,258] About half the patients have peripheral blood eosinophilia. All have improved or recovered after discontinuation of penicillamine. The disease is assumed to be pulmonary eosinophilia but biopsy specimens have not been examined during the acute phase.[258] Organizing diffuse alveolar damage has been described at autopsy in one patient receiving penicillamine.[259]

OTHER ANTI-INFLAMMATORY DRUGS

There are reports of cough, low-grade fever, dyspnea, and bilateral pulmonary infiltrates in patients receiving the nonsteroidal anti-inflammatory drugs oxyphenbutazone, phenylbutazone, sulindac, naproxen, tolfenamic acid, fenbufen, loxoprofen, nalfon, azopropazone, piroxicam, and diclofenac.[260,261] Peripheral blood eosinophilia was present in most cases. Transbronchial biopsies obtained from patients receiving diclofenac,[260] piroxicam,[261] naproxen,[262] phenylbutazone,[263] and sulindac[264] showed thickening of the alveolar walls and an interstitial infiltrate of eosinophils, lymphocytes, and mononuclear cells. All patients recovered after discontinuation of the offending agent.

5-Aminosalicylic acid (also known as mesalamine) is used for treating chronic ulcerative colitis and Crohn's disease. Drug-induced lung disease has been reported in seven patients who received a daily dose ranging between 0.75 and 3.6 g.[225] The duration of treatment ranged between 5 days and 44 months and the total cumulative dose was between 3.7 and 4746 g. Chest radiographs showed diffuse or basilar infiltrates. Two patients had eosinophilia. Lung biopsy was performed in two cases and showed interstitial pneumonia. Cessation of treatment with 5-aminosalicylic acid led to the resolution of symptoms.[225]

Anticonvulsants

Carbamazepine and phenytoin, which are thought to exert therapeutic effects through stabilization of neuronal membranes, are rarely associated with drug-induced lung disease.[12]

CARBAMAZEPINE

Carbamazepine is used in the treatment of epilepsy, neuralgic pain, and bipolar affective disorder. The drug has been widely used but only seven cases of carbamazepine-induced lung disease have been reported in the literature.[265] The patients presented with cough, dyspnea, and diffuse reticular nodular infiltrates a few weeks after commencement of carbamazepine therapy. Hilar prominence was noted in one patient[266] and bilateral pleural effusions in another.[265] Other symptoms and signs included fever, skin rash, generalized lymphadenopathy, and splenomegaly.[12,265] About half the cases showed peripheral blood eosinophilia.[265]

The prognosis of carbamazepine-induced lung disease is excellent. All patients showed pronounced symptomatic improvement within 1 to 2 weeks of discontinuation of the drug and ultimate full recovery.[12] A transbronchial biopsy from one patient showed eosinophilic pneumonia.[267] A transbronchial biopsy from another patient showed an interstitial pneumonia with a noncaseating granuloma.[265]

PHENYTOIN

Phenytoin is a rare cause of drug-induced lung disease. The reported patients have presented with fever, cough, and dyspnea 3 to 6 weeks after commencement of therapy. Peripheral blood eosinophilia and signs of liver dysfunction were present in all patients.[12] Some of the patients showed a maculopapular skin eruption and generalized lymphadenopathy. The chest radiographs showed a diffuse reticular nodular infiltrate that was sometimes complicated by hilar or mediastinal lymphadenopathy.[12] The prognosis of phenytoin-induced lung disease is excellent. All patients have shown marked symptomatic improvement within 1 to 2 weeks of discontinuation of the drug followed by ultimate full recovery.[12] A transbronchial lung biopsy in one case showed a chronic interstitial pneumonia.[268] Transbronchial lung biopsies from two other cases showed features resembling lymphoid interstitial pneumonia.[269,270] An autopsy on a 59-year-old man with cerebral vascular disease treated with phenytoin showed a necrotizing vasculitis with heavy eosinophilic infiltrates affecting the medium-sized renal, pulmonary, and splenic arteries.[271]

Anorexigens

Pulmonary hypertension has been reported as an adverse reaction to aminorex, fenfluramine, phendimetrazine, and dexflenfluramine.[272] Pulmonary hypertension was reported in two patients receiving propylhexedrine. It regressed after discontinuation of the drug, but no details of tissue pathology are available.[273]

AMINOREX

Aminorex is an appetite-suppressing drug that was available in Switzerland, Austria, and Germany from November 1965 to November 1968. Evidence suggests that it was responsible for an epidemic of unexplained pulmonary hypertension in these countries.[274,275] In a study of 17 fatal cases of pulmonary hypertension following aminorex ingestion, pulmonary arteriopathy with plexiform lesions was present in 5 cases (Fig. 18-29). Eleven cases of pulmonary arteriopathy with medial hypertrophy and intimal fibrosis (Fig. 18-30) and one of thrombotic pulmonary arteriopathy were reported.[276] Experience during the last 20 years has shown that the prognosis of animorex-related pulmonary hypertension is better than that of unexplained pulmonary hypertension. In aminorex pulmonary hypertension, prognosis depends on the age of the patient at the time of the anorectic treatment, the initial severity of pulmonary hypertension and right ventricular failure, and the amount of anorexigen ingested.[277] In some patients, withdrawal of aminorex led to partial remission of the pulmonary hypertension.[277]

FENFLURAMINE

Pulmonary hypertension has been described in patients receiving fenfluramine.[278–281] In a 5-year retrospective study of unexplained pulmonary hypertension, 20 percent of patients had used fenfluramine and there was a close temporal relation between use of the drug and the development of exertional dyspnea.[280] In some patients, the pulmonary hypertension has regressed after withdrawal of the drug. Lung tissue obtained from autopsies and after lung transplantation has shown pulmonary arteriopathy with plexiform lesions.[280,281]

PHENDIMETRAZINE

Pulmonary hypertension developed in a patient receiving phendimetrazine.[282] An open lung biopsy

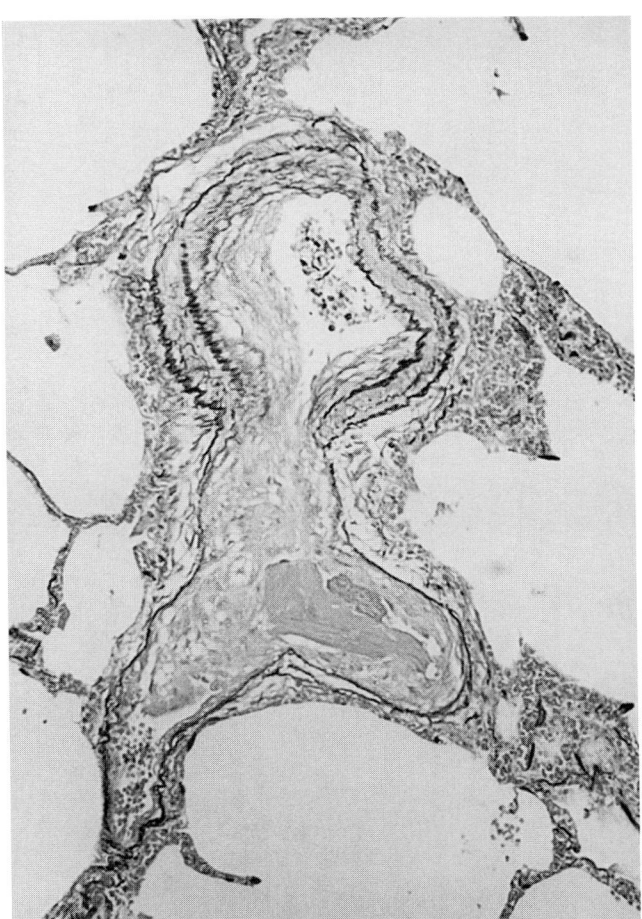

FIGURE 18-29
Plexiform lesion in a woman who presented with unexplained pulmonary hypertension after ingesting aminorex. The parent muscular pulmonary artery in the top half of the picture shows medial hypertrophy and intimal fibrosis. The lateral branch in the bottom half of the picture is occupied by a plexiform lesion. There is destruction of the internal elastic lamina and aneurysmal dilatation. The lumen contains a cellular proliferation and some thrombus. (Elastic-van Gieson, ×200.) *(Reproduced with permission from Kay et al.[275])*

showed pulmonary arteriopathy with medial hypertrophy and intimal fibrosis. The pulmonary hypertension regressed after discontinuation of the drug.

DEXFLENFLURAMINE

There have been two case reports of pulmonary hypertension occurring in patients receiving dexflenfluramine. In one of these, the pulmonary hypertension reversed when the drug was withdrawn.[283] The other case was fatal and the lungs showed pulmonary arteriopathy with plexiform lesions.[284]

Miscellaneous Noncytotoxic Drugs

Significant pulmonary hemorrhage is a rare complication of anticoagulant therapy using either heparin

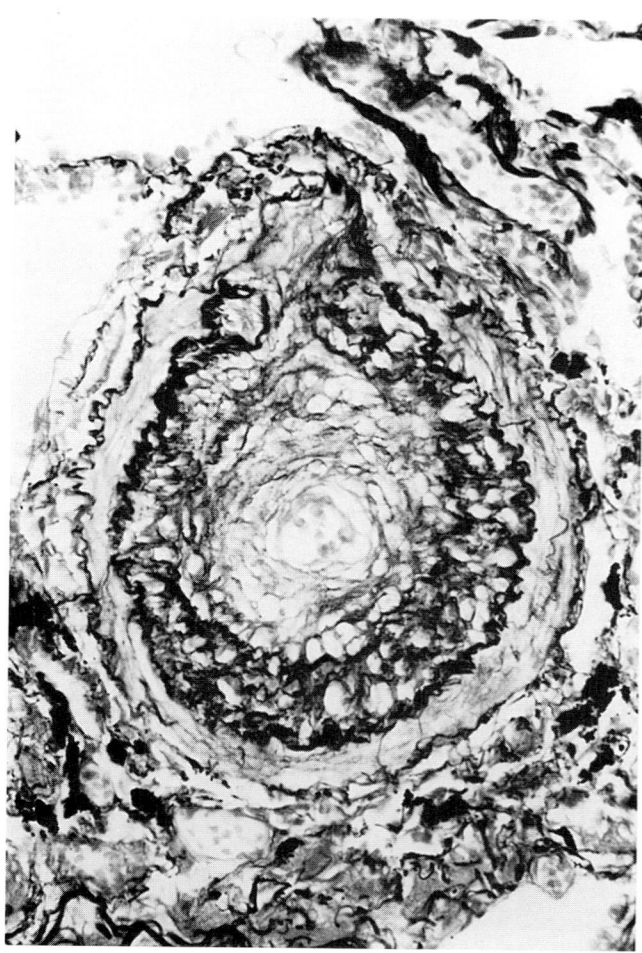

FIGURE 18-30
Muscular pulmonary artery from a woman who presented with unexplained pulmonary hypertension after ingesting aminorex. There is medial hypertrophy and severe narrowing of the lumen due to concentric laminar intimal fibrosis. (Elastic-van Gieson, ×200.)

or warfarin. Its occurrence has been estimated at under 1 percent. Most patients have recovered after complete or partial reversal of the anticoagulation. Diffuse alveolar hemorrhage was encountered at autopsy on one patient and BAL revealed hemosiderin-laden macrophages in another patient.[285,286]

Pleuropulmonary fibrosis has been described in patients receiving ergotamine[287] or methysergide[288] for the treatment of migraine. Open lung biopsies have shown fibrous thickening of the pleura associated with an infiltrate of lymphocytes and plasma cells. The underlying lung showed nonspecific interstitial fibrosis.[287] Regression of the lesions has been documented after the drugs were withdrawn. Bromocriptine is structurally related to ergotamine and methysergide. It is a long-acting dopamine agonist, which has been effective in the management of patients with hyperprolactinemia and Parkinson's disease. Pleuropulmonary fibrosis develops in 2 to 3 percent of patients with Parkinson's disease being

treated with relatively large doses of bromocriptine or its cogener, mesulergine.[289,290] Most of the patients have been men aged over 55 years who were receiving daily doses of bromocriptine ranging from 22.5 to 50.0 mg. The patients had a history of heavy cigarette smoking and the duration of treatment before the development of symptoms ranged from 9 months to 4 years.[289] Symptomatic improvement was noted in some patients after withdrawal of the drug. Pleuropulmonary fibrosis has not been described in patients with hyperprolactinemia who generally receive bromocriptine in doses of 2.5 to 7.5 mg per day, and rarely exceed 20 mg per day. Patients with Parkinson's disease receive much higher doses in excess of 20 mg per day. This suggests that pleuropulmonary fibrosis may be a dose-dependent side effect of bromocriptine therapy.[290] Pleural biopsies in two cases[289] and an open lung biopsy in one case[291] have shown fibrous thickening of the pleura with an infiltrate of lymphocytes and plasma cells. The underlying lung showed nonspecific interstitial fibrosis.

Granulomatous inflammation and bronchiolitis obliterans have been reported in a patient receiving cromolyn sodium.[292]

Interstitial lung disease has been reported in patients with prostate cancer being treated with nilutamide, a nonsteroidal antiandrogen drug.[293] Its reported frequency ranges from 2 to 12.8 percent of patients receiving 150 to 300 mg nilutamide per day.[293] The mean age of one series of 12 patients was 71.5 years and 35 percent had a history of lung disease. All patients developed dyspnea, cough, and fever within 4.7 months of commencing nilutamide. Symptoms resolved in 11 cases when the drug was withdrawn. Six patients received glucocorticoids.[294] Transbronchial lung biopsies have shown a chronic interstitial pneumonia characterized by an infiltrate of lymphocytes and plasma cells.[295,296]

Chronic interstitial pneumonia has been described in transbronchial lung biopsies derived from patients receiving propylthiouracil for Graves' disease.[297] Respiratory disease developed 3 weeks to 6 months after start of therapy. Symptoms and signs regressed after cessation of propylthiouracil and the administration of prednisolone.

Diffuse alveolar damage has been reported following streptokinase therapy.[298]

L-Tryptophan ingestion has been implicated as a causal agent in the eosinophilic myalgia syndrome. Most patients have presented with myalgia, fatigue, and muscle weakness and some have had skin lesions. There have now been several reports of patients developing interstitial lung disease after the ingestion of L-tryptophan.[272,299-301] Most patients have been women aged 34 to 65 years who presented with respiratory symptoms that began 1 to 9 months after ingestion of L-tryptophan. The chest radiographs showed bilateral interstitial infiltrates and most patients had peripheral blood eosinophilia. About half the patients have had pulmonary hypertension.[299,300] Some patients have presented with acute respiratory failure.[302] Most patients have recovered promptly after discontinuation of L-tryptophan and the administration of corticosteroid therapy.[299] However, pulmonary hypertension and dyspnea have persisted in some patients.[299,300] Open lung biopsies have shown chronic interstitial pneumonia and nonnecrotizing vasculitis.[272,299-302] The pneumonia was characterized by interstitial infiltrates of lymphocytes, occasional plasma cells, and rare giant cells. Eosinophils were present in all cases but prominent in only about half the patients. Some patients showed an intraalveolar exudate of fibrin intermingled with histiocytes and eosinophils. Small arteries and veins were surrounded by tight cuffs of lymphocytes mixed with small numbers of eosinophils (Fig. 18-31). The inflammatory infiltrate traversed the wall to involve the media and intima in some vessels. A granulomatous vasculitis was present in one case.[299] The etiology of the eosinophilia myalgia syndrome associated with L-tryptophan ingestion is still under investigation although a contaminant acting in conjunction with host factors is the favored hypothesis. T-cytotoxic/suppressor cells may be involved in the pathogenesis of the lung injury.[303]

Drug-Induced Systemic Lupus Erythematosus

Certain drugs may induce a lupus syndrome that is commonly accompanied by cardiopulmonary involvement. Some evidence indicates that the incidence of pleuropulmonary involvement in drug-induced lupus syndrome is greater than that seen in the spontaneously occurring disease.[12] The respiratory manifestations of drug-induced lupus syndrome include pleurisy and interstitial pneumonia.

Histologically, the lungs show a diffuse interstitial lymphocytic infiltrate without much fibrosis, which may account for the good response to corticosteroids.[19] The manifestations of drug-induced lupus syndrome are generally less severe than those of the naturally occurring disease.[19]

The following drugs have been identified as definite causes of drug-induced lupus syndrome: hydralazine, procainamide, practolol, D-penicillamine, isoniazid, diphenylhydantoin, ethosuximide,

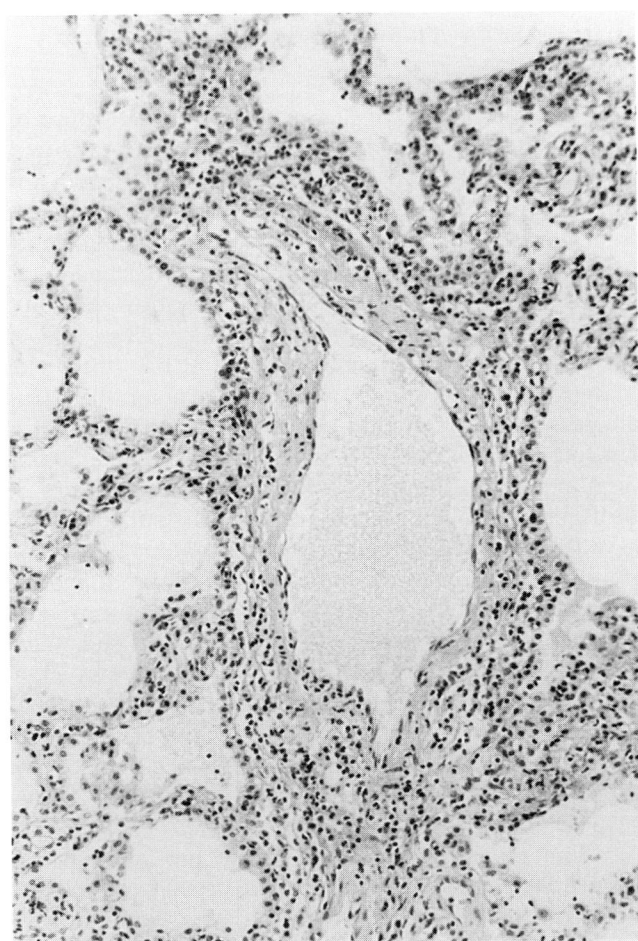

FIGURE 18-31

Open lung biopsy from a 34-year-old woman who presented with interstitial lung disease and pulmonary hypertension after ingesting L-tryptophan. There is a chronic interstitial pneumonia and nonnecrotizing vasculitis. The alveolar walls are infiltrated by lymphocytes, plasma cells, mononuclear cells, and eosinophils. A similar cellular infiltrate involves the adventitia, media and intima of a pulmonary vein. (H&E, ×200.) *(Courtesy of Dr. T. V. Colby).*

propylthiouracil, methylthiouracil, trimethadione, primidone, and chlorpromazine.[304] Of these, hydralazine, procainamide, practolol, and D-penicillamine should be considered drugs of high risk, such that 1 percent or more of the patients receiving them for more than 3 months might be expected to develop a lupus syndrome. Isoniazid, diphenylhydantoin, ethosuximide, the propylthiouracils, and trimethadione may be considered of moderate risk. Primidone and chlorpromazine are associated with a small risk.[304] The frequency of pleuropulmonary involvement is greater with procainamide than with other lupus-inducing drugs.[305]

DRUG ABUSE

The prevalence of drug abuse is thought to be increasing worldwide.[306] The pulmonary complications of illicit drug use may be the most common form of drug-induced lung disease.[2] Because the possession of controlled drugs is a criminal offense, and only a proportion of individuals are undergoing treatment, the number of drug abusers in a community can only be estimated indirectly. It has been estimated that in the United States there are 800,000 heroin users, 5 million regular cocaine users, and 10 million regular marijuana users.[307] It has been estimated that there are 500,000 intravenous drug abusers within the European community.[306] In the United Kingdom there are now an estimated 75,000 regular users of heroin, about one third of whom are thought to smoke rather than inject the substance.[306]

The patterns of drug abuse prevalent in a given population are determined by a variety of factors such as the cost or availability of particular substances, peer pressure, local customs, and legal pressures. The potential for respiratory system complications depends not only on the drug used but also on the route of administration, the origin of the drug, the presence of contaminants, whether or not there is sharing of paraphernalia, and the host response of the individual user.[307]

The pattern of cocaine abuse has changed.[308] Before 1980, the price of the drug provided a natural deterrent for its widespread use. The water-soluble form of the hydrochloride salt with a purity of 10 to 15 percent was injected intravenously or nasally insufflated. The drug was frequently injected in combination with other drugs such as heroin in a bolus termed a "speed ball." By the early 1980s, there was widespread manufacture of the alkaloidal form of cocaine hydrochloride known as "freebase" or "crack." It possessed a lower melting point than the hydrochloride compound and this property allowed it to be smoked. The alkaloidal form was 30 to 90 percent pure. When smoked it produced a rapid and intense euphoria that was severely addicting. The alkaloidal form was initially produced using highly volatile and explosive solvents such as ether. By 1985, it was manufactured more rapidly and less dangerously by mixing cocaine hydrochloride with baking powder and water. This process resulted in the formation of coarse "rocks" of freebase cocaine that produced an audible "crack" when smoked. The availability of an inexpensive, highly pure, and widely distributed crack cocaine has resulted in an epidemic of addiction, disease, and death.

The clinical aspects of the impact of drug abuse on the respiratory system have been reviewed recently.[306,307,309] The following paragraphs describe the pulmonary pathology associated with inhaled drug abuse and intravenous drug abuse.

Inhaled Drug Abuse

The two most commonly abused inhaled drugs are marijuana and cocaine. These substances may produce lesions in the conducting airways and lung parenchyma.

MARIJUANA

Marijuana smoking exposes the lungs to approximately 60 cannabinoid compounds and to many of the same respiratory irritants found in tobacco smoke.[310] Although most regular tobacco smokers consume more than 15 tobacco cigarettes a day, most current smokers of marijuana smoke less than one marijuana cigarette a day. Even among daily smokers of marijuana, the consumption of more than five marijuana cigarettes a day is unusual.[311] However, the habitual smoking of three or four marijuana cigarettes a day is associated with the same frequency of symptoms of acute and chronic bronchitis and the same type and extent of epithelial damage in the central airways as the regular smoking of more than 20 tobacco cigarettes a day.

Bronchial biopsy specimens from habitual marijuana smokers show squamous metaplasia, hyperplasia of goblet cells and basal cells, thickening of the epithelial basement membrane, and an infiltrate of lymphocytes and eosinophils in the lamina propria.[310] Compared with tobacco smoking, the smoking of marijuana is associated with a 5-fold greater increment in blood carboxyhemoglobin level, a 3-fold increase in the amount of tar inhaled, and retention in respiratory tract of more than one third more inhaled tar.

There are several reasons for the greater burden of particulate to the lungs from marijuana than from a similar quantity of tobacco. Tobacco cigarettes are more densely packed than marijuana cigarettes and most tobacco cigarettes have a filter tip; therefore, the filtration efficiency of the tobacco cigarettes is greater. Marijuana cigarettes are shorter than tobacco cigarettes, thereby further reducing the filtration efficiency of the marijuana cigarette.

Significant differences exist in the dynamics of smoking marijuana and tobacco. Marijuana smokers have an approximately two-thirds larger puff volume, a one-third greater depth of inhalation, and a 4-fold longer breath-holding time compared with tobacco smokers.[311] Smoking marijuana, regardless of tetrahydrocannabinol content, results in a greater respiratory burden of carbon monoxide and tar than smoking a similar quantity of tobacco. It is not known whether the pathologic changes in the bronchi will progress with continued marijuana smoking or regress with cessation of smoking. An increased risk of lung cancer has not yet been demonstrated for marijuana possibly because marijuana smoking by a significant proportion of the population is such a recent phenomenon that the effect has not yet become apparent.

COCAINE

Pulmonary granulomatosis due to celluose[312] and talc[313] have been reported following the nasal insufflation ("snorting") of cocaine hydrochloride. Cellulose particles are large and elongated (up to 120 μm in length). When the question of inhalation versus intravenous origin arises, the presence of particles larger than 5 μm strongly suggests an intravenous route of delivery because most inhaled particles more than 5 μm diameter are retained in the upper airways.[314] It was thought that cellulose particles were inhaled into the alveoli, like asbestos fibers, due to their aerodynamic properties.[312] Lung disease (Table 18-5) is much more common among crack smokers than other cocaine users.[315] The frequency of respiratory symptoms in cocaine freebase users ranges from 25 to 60 percent.[316,317] The most frequent problems are dyspnea (63 percent), cough (58 percent), cough with blood-stained or black sputum (34 percent), and ill-defined chest pain (25 percent).[317] In an autopsy study of 52 subjects with positive toxicologic tests for cocaine, the pulmonary histopathology was remarkably consistent irrespective of the method of use.[318] The lesions included acute hemorrhage (58 percent), alveolar hemosiderin-laden macrophages (40 percent), interstitial pneumonia with and without fibrosis (38 percent), vascular congestion (88 percent), and intraalveolar edema (77 percent).

Acute reversible pulmonary edema has been described after freebase cocaine smoking.[319,320] The

TABLE 18-5
Pulmonary Pathology Associated with Inhalation of Cocaine

Cellulose granulomatosis

Talc granulomatosis

Pulmonary edema

Pulmonary eosinophilia

Diffuse alveolar damage

Pulmonary hemorrhage

Chronic interstitial pneumonia/fibrosis

Bronchiolitis obliterans organizing pneumonia

Histiocytic interstitial pneumonia with crystalline silica

Black sputum and bronchoalveolar lavage fluid

pathogenesis of the pulmonary edema is not clear although an increase in the protein concentration of BAL fluid suggests damage to the alveolar epithelium or capillary endothelium.[319] A transbronchial lung biopsy in one case was normal on light microscopy.[319]

The term "crack lung" has been applied to two different types of acute disease following the smoking of freebase cocaine. Some cases of crack lung correspond to acute pulmonary eosinophilia.[321,322] Patients present with fever, bronchospasm, perihilar lung infiltrates, peripheral blood eosinophilia, and greatly elevated IgE levels. Treatment with prednisone results in rapid improvement of symptoms and dramatic clearing of the chest radiograph.[322] A transbronchial biopsy in one case showed focal interstitial collections of plasma cells, lymphocytes, and rare eosinophils.[321] The alveoli contained minute deposits of fibrin and showed focal hyperplasia of type II pneumonocytes. Eosinophils and Charcot-Leyden crystals were identified in bronchial washings. A transbronchial biopsy specimen from another case showed normal lung parenchyma but the bronchial wall was infiltrated by eosinophils. BAL showed a high eosinophil count.[322]

Crack lung has also been used to describe acute respiratory failure without peripheral blood eosinophilia, which developed in four patients after inhaling freebase cocaine.[320] Two patients had prolonged lung injury associated with fever, hypoxemia, hemoptysis, respiratory failure, and diffuse alveolar infiltrates. Both patients were treated with systemic corticosteroids and rapidly improved. An open lung biopsy in one case showed the organizing phase of diffuse alveolar damage with residual hyaline membranes and abundant hemosiderin-laden macrophages. The interstitial inflammatory infiltrate consisted of eosinophils, lymphocytes, and plasma cells. Immunofluorescence microscopy revealed deposition of IgE in lymphocytes and alveolar macrophages.[320] A transbronchial biopsy in the other case showed intraalveolar hemorrhage, hemosiderin-laden macrophages, and hyaline membranes. The inflammatory infiltrate consisted of eosinophils, lymphocytes, and neutrophil polymorphs. Immunofluorescence studies were not performed but the serum IgE level was normal. The other two cases appeared to represent acute pulmonary edema. The patients presented acutely with diffuse alveolar infiltrates associated with dyspnea and hypoxemia without fever and peripheral blood eosinophilia. Their symptoms and pulmonary infiltrates resolved spontaneously within 36 h.[320]

Hemoptysis is not unusual after inhalation of

freebase cocaine, even without evidence of acute lung injury. A retrospective study noted a 14 percent incidence of hemoptysis in young crack users.[323] The hemoptysis frequently occurs shortly after inhaling cocaine and generally resolves spontaneously. However, life-threatening alveolar hemorrhage may occur after repeated smoking of freebase cocaine.[324] An autopsy study of subjects who died acutely after inhaling cocaine revealed hemosiderin-laden macrophages in the alveoli in 33 percent of cases.[317] The mechanism of the alveolar hemorrhage is not clear. An open lung biopsy in one case showed acute alveolar hemorrhage with no evidence of vasculitis.[324] Electron microscopy showed intact type I pneumocytes and an increase in the number of type II pneumonocytes. The capillary and alveolar basement membranes were intact. Immunohistochemical staining for IgA, IgG, IgM, and complement was negative. The serum was negative for antiglomerular basement membrane antibodies.[324] It has been suggested that inhalation of cocaine causes vasoconstriction of the pulmonary vascular bed resulting in ischemic damage to the capillary endothelium.[324] This seems unlikely because acute hypoxic pulmonary vasoconstriction does not produce pulmonary hemorrhage in either human subjects or experimental animals.[325] It has also been suggested that cocaine is directly toxic to the capillary endothelium.[324] If this were the case, one would expect the open lung biopsy to show ultrastructural evidence of injury. At autopsy, evidence of intraalveolar hemorrhage is frequently accompanied by capillary congestion and edema suggesting that the hemorrhage might be the result of passive congestion secondary to cocaine-related cardiac disease.[318] However, autopsy studies have failed to show medial hypertrophy of muscular pulmonary arteries in subjects dying after inhalation of cocaine.[317,318] If the pulmonary hemorrhage were the result of chronic pulmonary venous congestion, it would be expected that the muscular pulmonary arteries would show medial hypertrophy.[325]

Interstitial pneumonia and interstitial fibrosis were encountered in 27 and 25 percent, respectively, of autopsies on 52 subjects from medical examiners' offices.[318] Interstitial fibrosis was usually seen in the context of interstitial pneumonia. When the interstitial pneumonia and fibrosis groups were combined, they accounted for 38 percent of subjects. Twenty-three percent of subjects primarily used cocaine intravenously and 12 percent primarily smoked it. In most cases, the mode of use was not known. The pulmonary pathology was consistent regardless of the method of use of cocaine. The findings raise the pos-

sibility that asymptomatic users of cocaine may be at increased risk for the eventual development of interstitial lung disease.[318]

There has been a report of chronic interstitial pneumonia with extensive accumulation of free silica within histiocytes in a 33-year-old woman who developed acute bilateral pulmonary infiltrates after the intense use of crack cocaine.[326] The origin of the silica was not clear but it was presumed to be a contaminant of the cocaine. The condition was unresponsive to prednisone and cyclophosphamide and the patient died 21 months after presentation.

A case of bronchiolitis obliterans organizing pneumonia has been attributed to the smoking of freebase cocaine.[327] The condition responded to methyl prednisolone.

People who smoke crack cocaine heavily may produce black sputum or BAL fluid.[328] Cytologic examination reveals excessive carbonaceous material in the cytoplasm of alveolar macrophages or in the extracellular compartment of the smears. The latter feature is not seen in people who smoke tobacco cigarettes. As crack is smoked, a dark, tarry residue forms on the inside of the bowl and barrel of the pipe. Many smokers consider this residue to be concentrated cocaine. They scrape it free, reheat it, and vigorously inhale it.[328] The carbonaceous residue in the lavage and sputum probably results from inhalation of nonvolatilized impurities that occur when crack and its tarry residue are smoked.[328]

Intravenous Drug Abuse

The pathogenesis of the pulmonary complications associated with intravenous drug abuse is related to the substances used and their method of preparation. Worldwide, heroin is the most common substance taken by intravenous drug abusers. In the United Kingdom, 84 percent of intravenous drug abusers use heroin and 6 percent use cocaine.[306] Many other drugs, including those intended for oral use (Table 18-6) may also be administered intravenously. They may be used alone or in combination with other drugs including heroin. Heroin is usually acetylated from the parent compound morphine and arrives in western countries in a pure form as a white powder. The pure heroin is progressively adulterated ("cut") by diluting it 20- to 100-fold with soluble "fillers" such as quinine, lactose, maltose, mannitol, baking soda, starch, barbiturates, and chloroquine.[329] The concentration of heroin in the product sold to the user by the pusher varies from 0 to 20 percent. The user mixes the dry white powder in unsterile water or in saliva. The mixture is heated in a spoon or bottle cap held over a lighted flame and removed from

TABLE 18-6
Fillers in Oral Medications Used by Intravenous Drug Abusers[338–344]

Drug	Fillers
Acetominophen/aspirin/ codeine	Microcrystalline cellulose
Amphetamines	Talc
Barbiturates	Cornstarch, talc
Dilaudid	Talc
Meperidine (Demerol)	Talc
Methadone	Cornstarch, talc
Methaqualone (Quaalude)	Microcrystalline cellulose
Methylphenidate (Ritalin)	Talc
Pentazocine (Talwin)	Cornstarch, microcrystalline cellulose
Percodan	Microcrystalline cellulose
Phenmetrazine (Preludin)	Talc
Propoxyphene napsylate (Darvon-N)	Microcrystalline cellulose
Propylhexedrine (Benzedrex)	Talc
Tripelennamine hydrochloride (Pyribenzamine)	Cornstarch, talc

the heat as soon as bubbles appear. Other forms of heroin such as "brown" heroin are poorly soluble in water and require acidification with substances such as lemon juice or vinegar before heating. The heroin mixture is aspirated into a syringe through a ball of cotton wool to filter out the larger impurities. The intravenous injection ("mainlining") is performed without sterilization of the skin, often in the presence of other users, who then share the syringe and needle without sterilization.[306]

Intravenous drug abusers are at risk from infection and a wide range of lung parenchymal and vascular lesions unrelated to infection (Table 18-7).

INFECTIONS

The risk of infections is increased due to general self-neglect, poor nutritional status, and the nonsterile techniques used to prepare the drug and administer the intravenous injection.

Human Immunodeficiency Virus (HIV) Infection

In the United States and western countries, the prevalence of HIV infection has increased disproportionately in intravenous drug abusers compared with other risk groups.[306,309] In the United States, the seroprevalence rate of HIV antibody is increasing by up to 14 percent a year among intravenous drug abusers.[330] The prevalence of HIV positivity varies widely among countries and cities and between ethnic

TABLE 18-7
Pulmonary Pathology Associated with Intravenous Drug Abuse

Infections
  Human immunodeficiency virus
  Pneumonia
    Community acquired
    Aspiration
    Septic pulmonary emboli
    Fungal
  Lung abscess
  Emphysema
  Tuberculosis

Vascular Lesions
  Foreign body embolism
    Talc
    Microcrystalline cellulose
    Cornstarch
    Cotton
    Mercury
    Needles
  Pulmonary artery medial hypertrophy
  Mycotic aneurysm

Parenchymal Lesions
  Pulmonary edema
  Progressive massive fibrosis
  Emphysema
  Interstitial pneumonia/fibrosis
  Hemosiderosis

groups within single cities.[331] Factors that increase the risk of HIV infection and may account for the variable seropositivity rates among different ethnic and social groups include sharing syringes and needles, withdrawal of blood into the syringe to mix with the drug, and injections of cocaine with or without heroin.[331]

Intravenous drug abuse is the second most common risk factor for the development of the acquired immunodeficiency syndrome (AIDS) in western countries. In the first 6 months of 1989, it accounted for 23 percent of cases of AIDS in the United States, compared with 15 percent in 1985 and 2 percent in 1983. Half the patients were female.[309] HIV-positive intravenous drug abusers have the same high incidence of opportunistic and nonopportunistic lung infections as other HIV-positive subjects. In contrast, pulmonary complications not directly related to infection, such as Kaposi's sarcoma and non-Hodgkin's lymphoma, are much less common in HIV-positive intravenous drug abusers than in other HIV-positive groups.[306,309]

Pneumonia

The incidence of lung infections is increased in intravenous drug abusers because of general self-neglect, poor nutrition, and the nonsterile techniques used to prepare and administer the drugs.

One of the most common pulmonary infectious complications of intravenous drug abuse is community-acquired pneumonia. In intravenous drug abusers, its incidence is increased 10-fold compared with the general population.[309] The infecting organisms and their antibiotic sensitivities are similar to those of community-acquired pneumonias in general.

Pneumonia may result from aspiration secondary to vomiting or narcotic-induced stupor.

Pneumonia may be acquired hematogenously as a result of septic pulmonary emboli. The organisms concerned are usually either *Staphylococcus aureus* or *Staphylococcus albus*. The source of infection is usually the skin at the injection site or tricuspid valve endocarditis.[309] The resulting pneumonia is often necrotizing and may be complicated by lung abscess, empyema, pneumothorax, and bronchopleural fistula. A pulmonary infarct, if present, may become infected and cavitate.[332]

Pulmonary candidiasis has been reported in intravenous drug abusers taking heroin.[333,334] The source of the infection was contamination by *Candida albicans* of the lemon used to acidify the drug before injection. Heroin may be contaminated by other fungi. Heroin addicts have a high prevalence

of serum precipitins to fungi such as *Aspergillus* species.[335] This suggests that heroin could cause other types of fungal pneumonia.

Tuberculosis

Pulmonary tuberculosis is more common in intravenous drug abusers than in the general population. This correlates with risk factors associated with their life-style such as self-neglect, poor nutrition, close contact, alcoholism, and depressed immunity.[336] HIV is an important risk factor for tuberculosis. The prevalence of *Mycobacterium tuberculosis* infection in intravenous drug abusers with AIDS is considerably higher than in homosexuals with AIDS who do not abuse drugs. In contrast, infection rates for atypical mycobacteria such as *Mycobacterium avium-intracellulare* are similar in the two groups.[309]

VASCULAR LESIONS

Foreign Body Embolism

Alterations in pulmonary function are frequently seen in intravenous drug abusers suggesting that foreign body emboli are common.[306,307] However, significant respiratory symptoms occur in less than 1 percent of subjects and foreign body embolism is rarely the cause of death in intravenous drug abusers.[306,337] Significant foreign body embolism is encountered in the small proportion (less than 5 percent) of intravenous drug abusers who inject aqueous suspensions of medications intended for oral use.[306,337] Significant foreign body embolism is rarely seen following the use of heroin alone because this drug is usually diluted with soluble "fillers" such as quinine, lactose, maltose, mannitol, and baking soda.[338,339]

Tablets and capsules prepared for oral administration frequently contain insoluble "fillers" such as talc (magnesium trisilicate) microcrystalline cellulose, and cornstarch (see Table 18-6). In some cases, talc accounts for three-fourths of the weight of the tablet. Drug addicts crush the tablets in a spoon or bottle top, add water, and heat the mixture. The fluid is drawn into a syringe through a filter of cotton fibers and injected intravenously. The suspended particles produce a variety of reactions in the lung. Talc[338-341] and microcrystalline cellulose[339,342,343] are the particles most frequently encountered in the lungs of intravenous drug abusers. Cornstarch particles are uncommon and usually associated with talc.[339] Talc, microcrystalline cellulose, and cornstarch are all birefringent when examined with polarized light. Cornstarch can be identified by its round shape and Maltese-cross biregringence.[339] Talc forms irregular, strongly birefringent plates varying in length from

5.7 μm to 70 μm. Talc is sometimes pale yellow in sections stained with hematoxylin and eosin. Microcrystalline cellulose forms long crystals (10 to 250 μm) that stain with periodic acid-schiff and diastase, methenamine silver, and Congo red.[339,343]

The pulmonary arterial lesions associated with intravascular talc and microcrystalline cellulose have been termed "angiothrombosis."[339] The lesions consist of intravascular foreign material mixed with thrombi in various stages of organization (Fig. 18-32). Weblike lesions and eccentric intimal fibrosis represent the late or end stages of organization of such thrombi. In cases with abundant foreign material there are clusters of multiple dilated vessels surrounded by fibrous tissue and foreign body granulomas.[339] These lesions consist of dilated tortuous small muscular pulmonary arteries in which the elastic laminae have been replaced by fibrous tissue. Human[339,343] and experimental[344] studies have shown that the foreign material migrates through the pulmonary arterial wall to produce perivascular and interstitial granulomatous lesions (Fig. 18-33). Subjects who have been taking dissolved tablets intravenously the longest tend to have granulomas located primarily in the pulmonary interstitium. Those with the shortest duration of addiction tend to have the granulomas in the lumens of pulmonary arteries.[341] At autopsy, lungs with massive amounts of foreign material are palpably gritty on gross examination and contain multiple hard white nodules (0.5 to 3 mm in diameter) scattered evenly throughout the parenchyma.[339,341,343,345] Pulmonary hypertension is a rare complication of intravenous drug abuse and develops in subjects who inject crushed

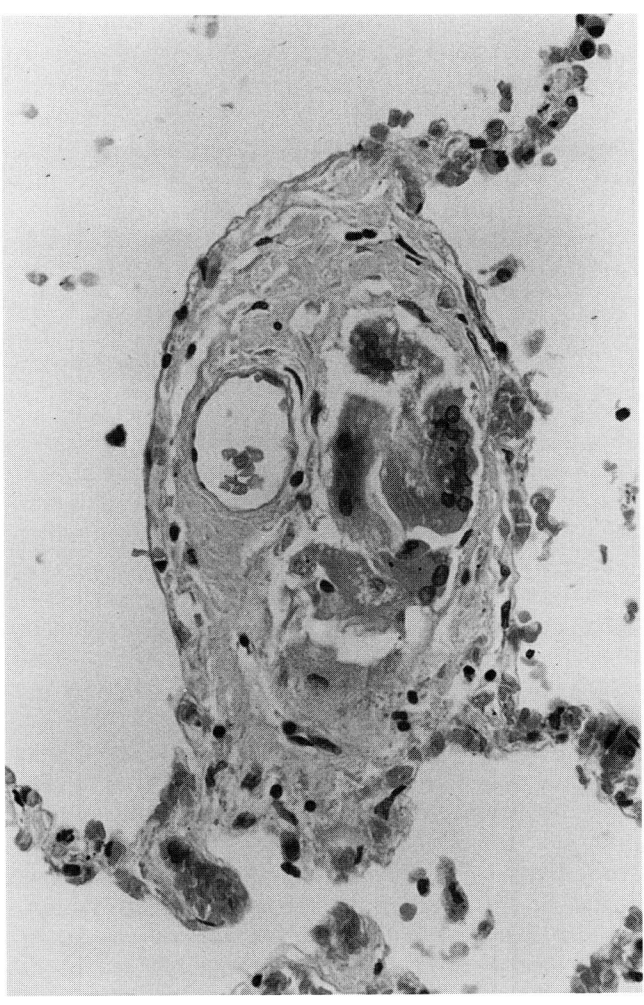

FIGURE 18-32
Intravenous drug abuse. Small pulmonary artery with foreign body giant cell granuloma in a recanalized thrombus. (H&E, ×500.) *(Courtesy of Dr. D. E. L. King).*

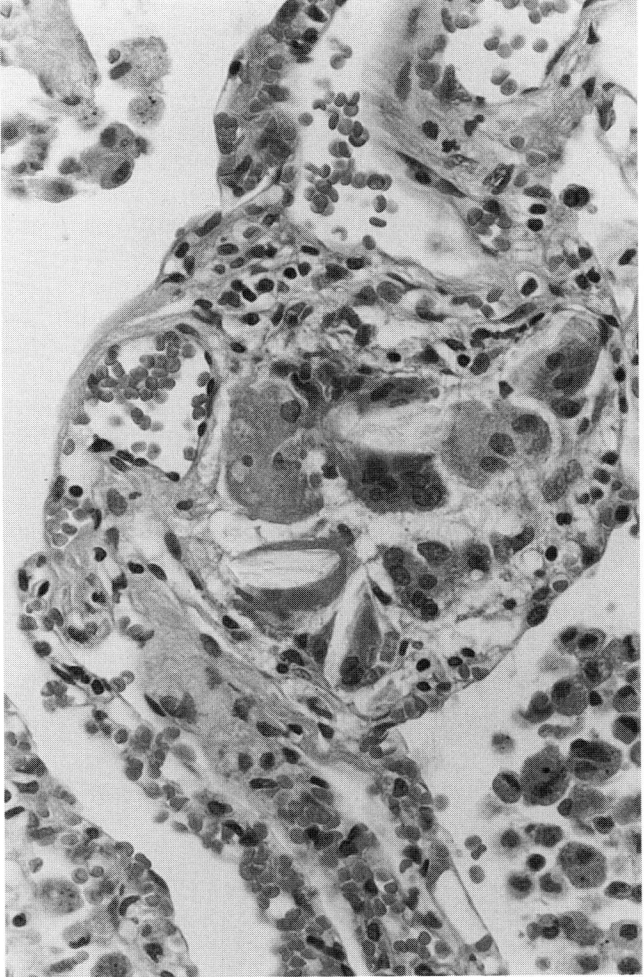

FIGURE 18-33
Intravenous drug abuse. An interstitial perivascular granuloma composed of foreign body giant cells and talc particles. (H&E, ×500.)

tablets containing either talc or microcrystalline cellulose for periods ranging from 1 to 20 years (average, 10 years).[338–343,345] In patients without evidence of pulmonary hypertension, reduction in diffusing capacity is common.[341,346] This change has been attributed to loss of pulmonary vascular bed secondary to widespread foreign body embolism.

Intravenous injection of pure cornstarch does not elicit intravascular granulomatosis but it may produce bland pulmonary thromboses.[339,340] If injected in sufficient quantity it may produce pulmonary hypertension.[340] Occasionally, a few microscopic cotton fibers from the filter contaminate the injection fluid and lead to cotton fiber embolism.[337,339] These cotton fiber emboli are asymptomatic. There are several reports of mercury emboli in the lungs of intravenous drug abusers.[347,348] Self-injection of mercury is mistakenly thought to improve athletic and sexual prowess. Needles have embolized to the lung of drug abusers using central venous sites for injections.[349]

Pulmonary Artery Medial Hypertrophy
Pulmonary artery medial hypertrophy without foreign particle microembolization has been described in association with intravenous cocaine abuse.[317] In half the cases, hemosiderin-laden macrophages were noted in the alveoli. The cause of the medial hypertrophy was not apparent and there was no mention as to whether it was associated with right ventricular hypertrophy or pulmonary hypertension. Another autopsy study of cocaine abusers failed to identify medial hypertrophy of muscular pulmonary arteries.[318] In another study of intravenous drug abusers, there were isolated instances of medial hypertrophy of muscular pulmonary arteries unassociated with large amounts of foreign material. However, morphometric analysis disclosed no significant difference in the medial thickness of small muscular pulmonary arteries in drug abusers with minimal foreign material compared with nonaddict controls.[339]

Mycotic Aneurysm
Mycotic aneurysms of the pulmonary arteries have been identified at autopsy in intravenous drug abusers.[350] The lesions were associated with pneumonia due to septic emboli.

PARENCHYMAL LESIONS

Pulmonary Edema
Acute noncardiogenic pulmonary edema is one of the serious complications of heroin abuse occurring in about 15 percent of addicts admitted to hospital after drug overdose.[306] The usual cause of drug over-dose is lack of awareness of the potency of the fix. Inconsistent adulteration of pure heroin leads to considerable variation in the potency of samples available on the street. Addicts returning to intravenous use after a period of abstinence, during which their tolerance has fallen, are specially prone to this complication. It also occurs in new addicts attempting to imitate their more experienced and tolerant colleagues. At autopsy, the lungs show hyaline membranes, an intraalveolar edema coagulum, and infiltration of the alveolar walls with neutrophil polymorphs.[337,351] The mechanism of the acute noncardiogenic pulmonary edema is obscure.[306,307] It has been suggested that nonnarcotic contaminants may cause an acute alveolitis.[306,307,351]

Progressive Massive Fibrosis
Intravenous drug abusers who inject themselves with large numbers of tablets containing talc for many years develop large mass lesions composed of huge numbers of particles with foreign body giant cells embedded in dense fibrous tissue (Fig. 18-34). These lesions, which correspond to the progressive massive fibrosis of pneumoconiosis, are typically upper zonal and may be bilateral.[340,352,353] At autopsy irregular but well demarcated, variegated, gray to gray-black areas of dense and gritty tissue measuring up to 6 cm in diameter are located in the upper and middle lobes and apical segments of the lower lobes.[352] The talc particles are larger than 5 μm in diameter indicating an intravenous route of delivery. Most inhaled particles more than 5 μm in diameter are retained in the upper airways.[314]

Emphysema
Emphysema, with or without bullae, is another complication of intravenous drug abuse using medications intended for oral use. In one study, six patients who injected methadone tablets containing talc were assessed over a period of 10 years or more from the time of initiation of the habit.[353] Despite discontinuation of the drug abuse, all developed severe respiratory disability and three died from their disease. The upper lobes of their lungs showed progressive massive fibrosis. The lower lobes showed emphysema with the formation of bullae and the development of pneumothorax. Severe panacinar emphysema that tended to be more severe in the lower zones was described in a group of seven young intravenous Ritalin abusers.[354] The emphysema was associated with microscopic talc granulomas (Fig. 18-35). Vascular involvement by talc granulomas was variable. There was no significant interstitial fibrosis. Five patients were tested for α1-antitrypsin deficiency and were found to be normal. These studies

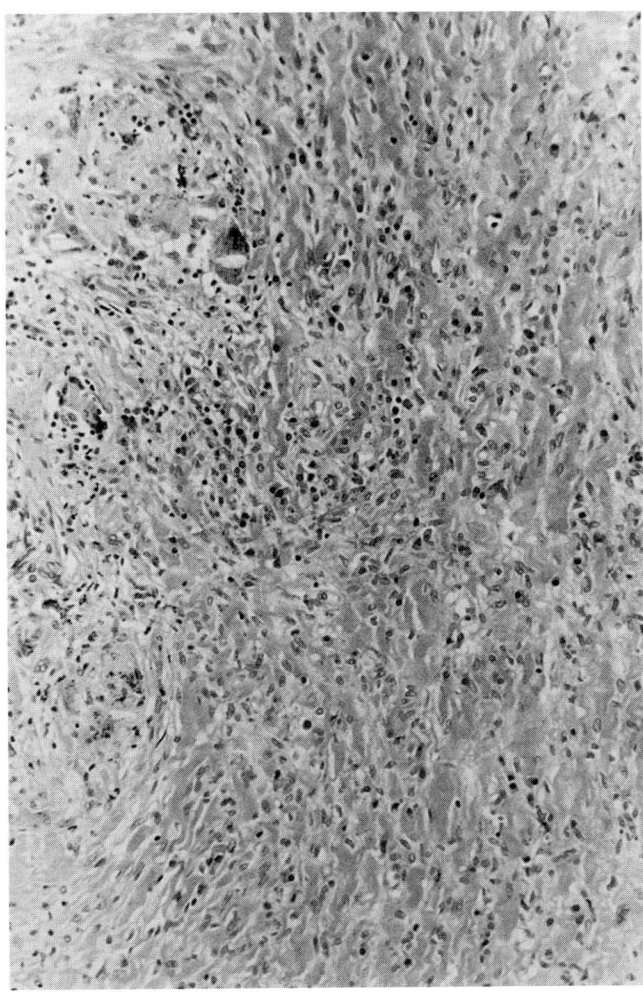

FIGURE 18-34
Progressive massive fibrosis in an intravenous drug abuser. The dense fibrous tissue includes foreign body giant cells containing talc particles. (H&E, ×200.)

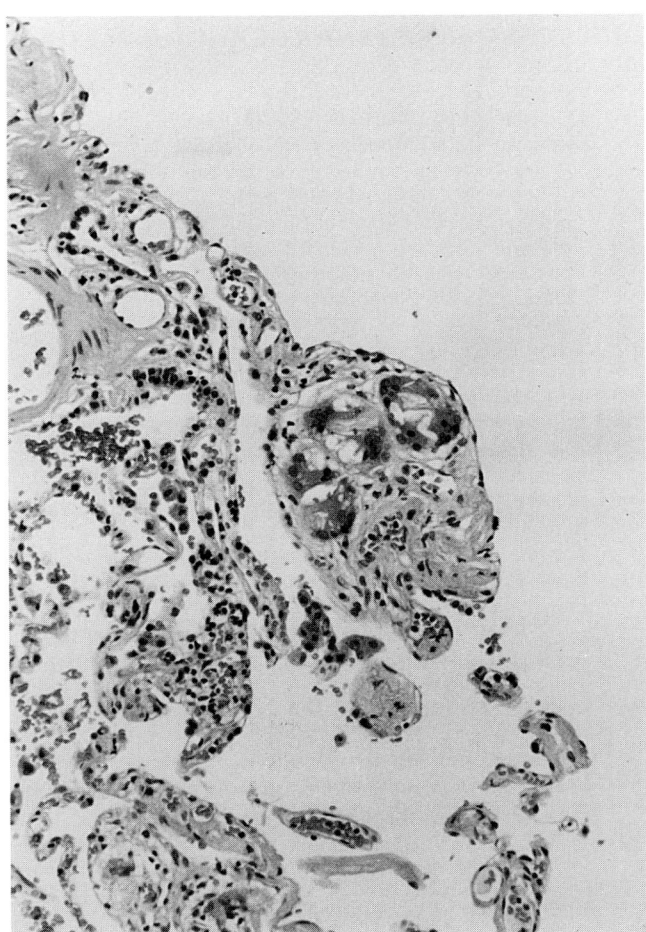

FIGURE 18-35
Emphysema in an intravenous drug abuser. An interstitial talc granuloma is in the center of the picture. (H&E, ×200.)

suggest that panacinar emphysema in intravenous drug abusers results from the filler in the tablets and not from the specific drug. The mechanism is uncertain but may be related to localized protease–antiprotease imbalances around the innumerable talc granulomas.[354]

Interstitial Pneumonia/Fibrosis
In an autopsy study of 52 cocaine abusers, interstitial pneumonia and interstitial fibrosis were encountered in 27 and 25 percent of cases, respectively.[318] Twelve subjects used the drug intravenously and six smoked it. In most cases, the method of use was unknown. Polarizable foreign material and microgranulomas were present in only six cases. These findings raise the possibility that cocaine abusers may be at increased risk for the development of interstitial lung disease. An alveolitis characterized by an interstitial infiltrate of lymphocytes, plasma cells, and eosinophils has been described in patients with talcosis resulting from intravenous injection of drugs intended for oral use.[341]

Hemosiderosis
Hemosiderin-laden macrophages have been identified in 35 to 40 percent of autopsies on cocaine abusers.[317,318] They also occur in patients with pulmonary talc granulomatosis resulting from the intravenous administration of crushed tablets.[341] The mechanism of the intraalveolar hemorrhage is not understood.

REFERENCES

1. Rosenow EC III: The spectrum of drug-induced pulmonary disease. *Ann Intern Med* 77:977–991, 1972.
2. Rosenow EC III, Myers JL, Swensen SJ, Pisani RJ: Drug-induced pulmonary disease. An update. *Chest* 102:239–250, 1992.
3. Davies PDB: Drug-induced lung disease. *Br J Dis Chest* 63:57–570, 1969.

4. O'Driscoll BR, Hasleton PS, Taylor PM, Poulter LW, Gattamaneni HR, Woodcock AA: Active lung fibrosis up to 17 years after chemotherapy with carmustine (BCNU) in childhood. *N Engl J Med* 323:378–382, 1990.

5. Achong MR: When is a clinical event an adverse drug reaction? *Can Med Assoc J* 119:1315–1316, 1978.

6. Irey NS: When is a disease drug induced? In Riddell RH (ed): *Pathology of Drug-Induced and Toxic Diseases.* New York, Churchill Livingstone, 1982, pp 1–18.

7. Kilburn KH: Pulmonary disease induced by drugs, in Fishman AP (ed): *Pulmonary Diseases and Disorders.* New York, McGraw-Hill, 1980, pp 707–724.

8. Boyd MR: Biochemical mechanisms in chemical-induced lung injury: Roles of metabolic activation. *CRC Crit Rev Toxicol* 7:103–176, 1980.

9. Bend JR, Serabjit-Singh CJ, Philpot RM: The pulmonary uptake, accumulation, and metabolism of xenobiotics. *Annu Rev Pharmacol Toxicol* 25:97–125, 1985.

10. Pang JA, Geddes DM: The biochemical properties of the pulmonary circulation. *Lung* 159:231–241, 1981.

11. Cooper JAD Jr, White DA, Matthay RA: Drug-induced pulmonary disease. Part 1: Cytotoxic drugs. *Am Rev Respir Dis* 133:321–340, 1986.

12. Cooper JAD Jr, White DA, Matthay RA: Drug-induced pulmonary disease. Part 2: Noncytotoxic drugs. *Am Rev Respir Dis* 133:488–505, 1986.

13. Freeman BA, Crapo JD: Biology of disease: Free radicals and tissue injury. *Lab Invest* 47:412–426, 1982.

14. Berend N: Protective effect of hypoxia on bleomycin lung toxicity in the rat. *Am Rev Respir Dis* 130:307–308, 1984.

15. Gurtoo HL, Hipkens JH, Sharma SD: Role of glutathione in the metabolism-dependent toxicity and chemotherapy of cyclophosphamide. *Cancer Res* 41:3584–3591, 1981.

16. Martin WJ II: Nitrofurantoin: Evidence for the oxidant injury of lung parenchymal cells. *Am Rev Respir Dis* 127:482–486, 1983.

17. Arrick BA, Nathan CF: Glutathione metabolism as a determinant of therapeutic efficacy: A review. *Cancer Res* 44:4224–4232, 1984.

18. Nathan CF, Arrick BA, Murray HW, DeSantis NM, Cohn ZA: Tumor cell anti-oxidant defenses: Inhibition of the glutathione redox cycle enhances macrophage-mediated cytolysis. *J Exp Med* 153:766–782, 1981.

19. Bedrossian CWM: Pathology of drug-induced lung diseases. *Semin Repir Med* 4:98–105, 1982.

20. Geddes D, Brostoff J: Pulmonary fibrosis associated with hypersensitivity to gold salts. *Br Med J* i:1444, 1976.

21. Back O, Liden S, Ahlstedt S: Adverse reactions to nitrofurantoin in relation to cellular and humoral mechanisms. *Clin Exp Immunol* 28:400–406, 1977.

22. Akoun GM, Gauthier-Rahman S, Milleron BJ, Perrot JY, Maynaud CM: Amiodarone-induced hypersensitivity pneumonitis. Evidence of an immunological cell-mediated mechanism. *Chest* 85:133–135, 1984.

23. White DA, Gellene R, Rankin JR, Gupta S, Cunningham-Rundles C, Stover DE: Methotrexate pneumonitis: Bronchoalveolar lavage findings suggest an immune mediated disorder. *Am Rev Respir Dis* 129:A64, 1984.

24. Sternlieb I, Bennett B, Scheinberg IH: D-Penicillamine induced Goodpasture's syndrome in Wilson's disease. *Ann Intern Med* 82:673–676, 1975.

25. Absher M, Hildebran J, Trombley L, Woodcock-Mitchell J, Marsh J: Characteristics of cultured lung fibroblasts from bleomycin-treated rats. Comparison with in vitro exposed normal fibroblasts. *Am Rev Respir Dis* 129:125–129, 1984.

26. Adam M, Kuhn K: Investigations on the reactions of metals with collagen in vivo. I. Comparison of the reaction of gold thiosulfate with collagen in vivo and in vitro. *Eur J Biochem* 3:407–410, 1968.

27. Ninmni BE, Bavetta LA: Collagen defect induced by penicillamine. *Science* 150:905–907, 1965.

28. Ginsberg SJ, Comis RL: The pulmonary toxicity of antineoplastic agents. *Semin Oncol* 9:34–51, 1982.

29. Marchlinski FE, Gansler TS, Waxman HL, Josephson ME: Amiodarone pulmonary toxicity. *Ann Intern Med* 97:839–845, 1982.

30. Scoggin CH: The cellular, biochemical and genetic basis of aging, in Schrier RW (ed): *Clinical Internal Medicine in the Aged.* Philadelphia, WB Saunders, 1982, pp 24–28.

31. Aronin PA, Mahaley MS, Rudnick SA, et al: Prediction of BCNU pulmonary toxicity in patients with malignant gliomas. An assessment of risk factors. *N Engl J Med* 303:183–188, 1980.

32. Samuels ML, Johnson DE, Holoye PY, Lanzotti VJ: Large-dose bleomycin therapy and pulmonary toxicity. A possible role of radiotherapy. *JAMA* 235:1117–1120, 1976.

33. Soble AR, Perry H: Fatal radiation pneumonitis following subclinical busulfan therapy. *AJR Am J Roentgenol* 128:15–18, 1977.

34. Buzdar AU, Legha SS, Luna MA, Tashima CK, Hortobagyi GN, Blumenschien GR: Pulmonary toxicity of mitomycin. *Cancer* 45:236–244, 1980.

35. Goldiner PL, Carlon GC, Cvitkovic E, Schweizer O, Howland WS: Factors influencing postoperative morbidity and mortality in patients treated with bleomycin. *Br Med J* i:1664–1667, 1978.

36. Hakkinen PJ, Whiteley JW, Witschi HR: Hyperoxia, but not thoracic irradiation, potentiates bleomycin and cyclophosphamide-induced lung damage in mice. *Am Rev Respir Dis* 126:281–285, 1982.

37. Boyd MR, Catignani GL, Sasasme HA, Mitchell JR, Stiko AW: Acute pulmonary injury in rats by nitrofurantoin and modification by vitamin E, dietary fat, and oxygen. *Am Rev Respir Dis* 120:93–99, 1979.

38. Skarin At, Rosenthal DS, Maloney WC: Combination chemotherapy of advanced non-Hodgkin's lymphoma (NHL) with bleomycin, adriamycin, cyclophosphamide, vincristine and prednisone. *Blood* 49:759–770, 1977.

39. White DA, Orenstein M, Godwin TA, Stover D: Chemotherapy-associated pulmonary toxic reactions during treatment for breast cancer. *Arch Intern Med* 144:953–956, 1984.

40. Durant JR, Norgard MJ, Murad TM, Bartolucci AA, Langford KH: Pulmonary toxicity associated with bischloroethylnitrosourea (BCNU). *Ann Intern Med* 90:191–194, 1979.

41. Luedke D, McLaughlin TT, Daughaday C, et al: Mitomycin C and vindesine associated pulmonary toxicity with variable clinical expression. *Cancer* 55:542–545, 1985.

42. Whitcomb ME: Drug-induced lung disease. *Chest* 63:418–422, 1973.

43. Gillett DG, Ford GT: Drug-induced lung disease, in Thurlbeck WM (ed): *The Lung.* Baltimore, Williams & Wilkins, 1978, pp 21–42.

44. Whimster WF, de Poitiers W: The lung, in Riddell RH (ed): *Pathology of Drug-Induced and Toxic Diseases.* New York, Churchill Livingstone, 1982, pp 167–200.

45. Hayes JA: Drug-induced lung disease. *Surv Synth Path Res* 2:115–119, 1983.

46. Myers JL: Pathology of drug-induced lung disease, in Katzenstein A-L A, Askin FB (eds): *Surgical Pathology of*

Non-Neoplastic Lung Disease, 2d ed. Philadelphia, WB Saunders, 1990, pp 97–127.

47. Colby TV, Carrington CB: Infiltrative lung disease, in Thurlbeck WM, Churg AM (eds): *Pathology of the Lung*, 2d ed. New York, Thieme, 1995, pp 589–737.

48. Haddow A. Discussion on the chemotherapy of the reticuloses. *Proc R Soc Med* 46:692–696, 1953.

49. Lee SM, Margison GP, Woodcock AA, Thatcher N: Sequential administration of varying doses of dacarbazine and fotemustine in advanced malignant melanoma. *Br J Cancer* 67:1356–1360, 1993.

50. Oliner H, Schwartz R, Rubio F, Dameshek W: Interstitial pulmonary fibrosis following busulfan therapy. *Am J Med* 31:134–139, 1961.

51. Schallier D, Impens N, Warson F, Van Belle S, De Wasch G: Additive pulmonary toxicity with melphalan and busulfan therapy. *Chest* 84:492–493, 1983.

52. Hankins DG, Sanders S, MacDonald FM, Drage CW: Pulmonary toxicity recurring after a six week course of busulfan therapy and after subsequent therapy with uracil mustard. *Chest* 73:415–416, 1978.

53. Littler WA, Kay JM, Hasleton PS, Heath D: Busulphan lung. *Thorax* 24:639–655, 1969.

54. Heard BE, Cooke RA: Busulphan lung. *Thorax* 23:187–193, 1968.

55. Kirschner RM, Esterly JR: Pulmonary lesions associated with busulphan therapy of chronic myelogenous leukemia. *Cancer* 27:1074–1080, 1971.

56. Koss L, Melamed M, Mayer K: The effect of busulfan on human epithelia. *Am J Clin Pathol* 44:385–397, 1965.

57. Feingold M, Koss L: Effects of long-term administration of busulphan. *Arch Intern Med* 124:66–71, 1969.

58. Aymard JP, Gyger M, Lavallee R, Legresley LP, Desy M: A case of pulmonary alveolar proteinosis complicating chronic myelogenous leukemia. A peculiar pathological aspect of busulfan lung? *Cancer* 53:954–956, 1984.

59. Bedrossian CWM, Luna MA, Conklin RH, Miller WC: Alveolar proteinosis as a consequence of immunosuppression: A hypothesis based on clinical and pathologic observations. *Hum Pathol* 11:527–535, 1980.

60. Kuplic JB, Higley CS, Niewoehner DE: Pulmonary ossification associated with long-term busulfan therapy in chronic myeloid leukemia. *Am Rev Respir Dis* 106:759–762, 1972.

61. Min KW, Gyorkey F: Interstitial fibrosis, atypical epithelial changes and bronchiolar cell carcinoma following busulfan therapy. *Cancer* 22:1027–1032, 1968.

62. Spector JI, Zimbler H, Ross JS: Early onset cyclophosphamide induced interstitial pneumonitis. *JAMA* 242:2852–2854, 1979.

63. Mark GJ, Lehimgar-Zadeh A, Ragsdale BD: Cyclophosphamide pneumonitis. *Thorax* 33:89, 1978.

64. Burke DA, Stoddart JC, Ward MK, Simpson CGB: Fatal pulmonary fibrosis occurring during treatment with cyclophosphamide. *Br Med J* 285:696, 1982.

65. Maxwell I: Reversible pulmonary edema following cyclophosphamide treatment. *JAMA* 229:137–138, 1974.

66. Abdel-Karim FW, Ayash RE, Allam C, Salem PA: Pulmonary fibrosis after prolonged treatment with low-dose cyclophosphamide. A case report. *Oncology* 40:174–176, 1983.

67. Alvarado CS, Boat TF, Newman AJ: Late-onset pulmonary fibrosis and chest deformity in two children treated with cyclophosphamide. *J Pediatr* 92:443–446, 1978.

68. Patel AR, Shah PC, Rhee HL, Sassoon H, Rao KP: Cyclophosphamide therapy and interstitial pulmonary fibrosis. *Cancer* 38:1542–1546, 1976.

69. Topilow A, Rothenberg S, Cottrell TS: Interstitial pneumonia after prolonged treatment with cyclophosphamide. *Am Rev Respir Dis* 108:114–117, 1973.

70. Koss LG: *Diagnostic Cytology and Its Histopathologic Basis*, 4th ed. Philadelphia, JB Lippincott, 1992, vol 1, p 759.

71. Slavin RE, Millan JC, Mullins GM: Pathology of high dose intermittent cyclophosphamide therapy. *Hum Pathol* 6:693–709, 1975.

72. Bentur L, Cullinane C, Wilson P, Greenberg M, O'Brodovich H, Silver MM: Fatal pulmonary arterial occlusive vascular disease following chemotherapy in a 9-month-old infant. *Hum Pathol* 22:1295–1298, 1991.

73. Rose MS: Busulphan toxicity syndrome caused by chlorambucil. *Br Med J* ii:123, 1975.

74. Cole SR, Myers TJ, Klatsky AU: Pulmonary disease with chlorambucil therapy. *Cancer* 41:455–459, 1978.

75. Godard P, Marty JP, Michel FB: Interstitial pneumonia and chlorambucil. *Chest* 76:471–473, 1979.

76. Baker WJ, Fistel SJ, Jones RV, Weiss RB: Interstitial pneumonitis associated with ifosfamide therapy. *Cancer* 65:2217–2221, 1990.

77. Codling BW, Chakera TMH: Pulmonary fibrosis following therapy with melphalan for multiple myeloma. *J Clin Pathol* 25:668–673, 1972.

78. Goucher G, Rowland V, Hawkins J: Melphalan-induced pulmonary interstitial fibrosis. *Chest* 77:805–806, 1980.

79. Schallier D, Impens N, Warson F, et al: Additive pulmonary toxicity with melphalan and busulfan. *Chest* 84:492–493, 1983.

80. Taetle R, Dickman PS, Feldman PS: Pulmonary histopathologic changes associated with melphalan therapy. *Cancer* 42:1239–1245, 1978.

81. Westerfield B, Michalski J, McCombs C, Light R: Reversible melphalan-induced lung damage. *Am J Med* 68:767–771, 1980.

82. Blum RH, Carter SK, Agre K: A clinical review of bleomycin—a new antineoplastic agent. *Cancer* 31:903–914, 1973.

83. Crooke ST, Einhorn LH, Comis RL, et al: The effects of prior exposure to bleomycin on the incidence of pulmonary toxicities in a group of patients with disseminated testicular carcinomas. *Med Pediatr Oncol* 5:93–98, 1978.

84. Bauer K, Skarin A, Balikian J, et al: Pulmonary complications associated with combination chemotherapy programs containing bleomycin. *Am J Med* 74:557–563, 1983.

85. Van Barneveld PWC, Sleijfer DT, Van Der Mark TW, et al: Natural course of bleomycin-induced pneumonitis. A follow-up study. *Am Rev Respir Dis* 135:48–51, 1987.

86. O'Neill TJ, Kardinal CG, Tierney LM: Reversible interstitial pneumonitis associated with low-dose bleomycin. *Chest* 68:265–267, 1975.

87. Iacovino JR, Leitner J, Abbas AK, Lokich JJ, Snider GL: Fatal pulmonary reaction from low doses of bleomycin. An idiosyncratic tissue response. *JAMA* 235:1253–1255, 1976.

88. Luna MA, Bedrossian CWM, Lichtiger B, Salem PA: Interstitial pneumonitis associated with bleomycin therapy. *Am J Clin Pathol* 58:501–510, 1972.

89. Krous HF, Hamlin WB: Pulmonary toxicity due to bleomycin: A case report. *Arch Pathol* 95:407–410, 1973.

90. Bedrossian CWM, Luna MA, MacKay B, Lichtiger B: Ultrastructure of pulmonary bleomycin toxicity. *Cancer* 32:44–51, 1973.

91. White D, Stover D: Severe bleomycin-induced pneumonitis. Clinical features and response to corticosteroids. *Chest* 86:723–728, 1984.

92. Santrach PJ, Askin FB, Wells RJ, Azizkan RG, Merten DF: Nodular form of bleomycin-related pulmonary injury in patients with osteogenic sarcoma. *Cancer* 64:806–811, 1989.

93. Cohen MB, Austin JHM, Smith-Vaniz A, Lutzky J, Grimes MM: Nodular bleomycin toxicity. *Am J Clin Pathol* 92:101–104, 1989.

94. Holoye PY, Luna MA, MacKay B, Bedrossian CWM: Bleomycin hypersensitivity pneumonitis. *Ann Intern Med* 88:47–49, 1978.

95. Yousem SA, Lifson JD, Colby TV: Chemotherapy-induced eosinophilic pneumonia. Relation to bleomycin. *Chest* 88:103–106, 1985.

96. Joselson R, Warnock M: Pulmonary veno-occlusive disease after chemotherapy. *Hum Pathol* 14:88–91, 1983.

97. Rose AG: Pulmonary veno-occlusive disease due to bleomycin therapy for lymphoma. Case reports. *S Afr Med J* 64:636–638, 1983.

98. Knight BK, Rose AG: Pulmonary veno-occlusive disease after chemotherapy. *Thorax* 40:874–875, 1985.

99. Gunstream SR, Seidenfeld JJ, Sobonya RE, McMahon LJ: Mitomycin-associated lung disease. *Cancer Treat Resp* 67:301–304, 1983.

100. Orwoll ES, Kiessling PJ, Patterson JR: Interstitial pneumonia from mitomycin. *Ann Intern Med* 89:352–355, 1978.

101. Verweij J, Van Zanten T, Souren T, Golding R, Pinedo HM: Prospective study on the dose relationship of mitomycin C-induced interstitial pneumonitis. *Cancer* 60:756–761, 1987.

102. Chang AY, Kuebler JP, Pandya KJ, Israel RH, Marshall BC, Tormey DC: Pulmonary toxicity induced by mitomycin-C is highly responsive to glucocorticoids. *Cancer* 57:2285–2290, 1986.

103. McCarthy JT, Staats BA: Pulmonary hypertension, hemolytic anemia, and renal failure. A mitomycin-associated syndrome. *Chest* 89:608–611, 1986.

104. Chang-Poon VY-H, Hwang WS, Wang A, Berry J, Klassen J, Poon M-C: Pulmonary angiomatoid vascular changes in mitomycin C-associated hemolytic-uremic syndrome. *Arch Pathol Lab Med* 109:877–888, 1985.

105. Waldhorn RE, Tsou E, Smith FP, Kerwin DM. Pulmonary veno-occlusive disease associated with microangiopathic hemolytic anemia and chemotherapy of gastric adenocarcinoma. *Med Pediatr Oncol* 12:394–396, 1984.

106. Seltzer SE, Griffin T, D'Orsi C, Tryka F, Herman PJ: Pulmonary reaction associated with neocarzinostatin therapy. *Cancer Treat Rep* 62:1271–1272, 1978.

107. Calvo DB, Legha SS, McKelvey EM, Bodey GP: Zinostatin-related pulmonary toxicity. *Cancer Treat Rep* 65:165, 1981.

108. Weiss RB, Poster DS, Penta JS: The nitrosoureas and pulmonary toxicity. *Cancer Treat Rev* 8:111–125, 1981.

109. Melato M, Tuveri G: Pulmonary fibrosis following low-dose 1,3-*bis*(2-chlorethyl)-1-nitrosourea (BCNU) therapy. *Cancer* 45:1311–1314, 1980.

110. Litam JP, Dail DH, Spitzer G, et al: Early pulmonary toxicity after administration of high-dose BCNU. *Cancer Treat Rep* 65:39–44, 1981.

111. Mitsudo SM, Greenwald ES, Banerji B, Koss LG: BCNU (2,3-*bis*-(2-chloroethyl)-1-nitrosourea) lung. Drug-induced pulmonary changes. *Cancer* 54:751–755, 1984.

112. Patten GA, Billi JE, Rotman HH: Rapidly progressive, fatal pulmonary fibrosis induced by carmustine. *JAMA* 244:687–688, 1980.

113. Selker RG, Jacobs SA, Moore PB, et al: 1,3-*Bis*(2-chlorethyl)-1-nitrosourea (BCNU)-induced pulmonary fibrosis. *Neurosurgery* 7:560–565, 1980.

114. Holoye PY, Jenkins DE, Greenberg SD: Pulmonary toxicity in long-term administration of BCNU. *Cancer Treat Rep* 60:1691–1694, 1976.

115. Case Records of the Massachusetts General Hospital: Case 41–1980. *N Engl J Med* 303:927–933, 1980.

116. Hasleton PS, O'Driscoll BR, Lynch P, et al: Late BCNU lung: a light and ultrastructural study on the delayed effect of BCNU on the lung parenchyma. *J Pathol* 164:31–36, 1991.

117. Lombard CM, Churg AM, Winokur S: Pulmonary veno-occlusive disease following therapy for malignant neoplasms. *Chest* 92:871–876, 1987.

118. Vats TS, Trueworthy RC, Langston CM: Pulmonary fibrosis associated with lomustine (CCNU): A case report. *Cancer Treat Rep* 66:1881–1882, 1982.

119. Cordonnier C, Vernant J-P, Mital P, Lange F, Bernaudin JF, Rochant H: Pulmonary fibrosis subsequent to high doses of CCNU for chronic myeloid leukemia. *Cancer* 51:1814–1818, 1983.

120. Stone MD, Richardson MG: Pulmonary toxicity of lomustine. *Cancer Treat Rep* 71:786–787, 1987.

121. Lee W, Moore RP, Wampler GL: Interstitial pulmonary fibrosis as a complication of prolonged methyl-CCNU therapy. *Cancer Treat Rep* 62:1355–1358, 1978.

122. Ahlgren JD, Smith FP, Kerwin DM, Sikic BI, Weiner JH, Schein PS: Pulmonary disease as a complication of chlorozotocin chemotherapy. *Cancer Treat Rep* 65:223–229, 1981.

123. Weisenburger DD: Interstitial pneumonitis associated with azathioprine therapy. *Am J Clin Pathol* 69:181–185, 1978.

124. Bedrossian CWM, Conklin RH, Kahan B, Sussman J: Azathioprine-induced interstitial pneumonitis in renal transplant patients. *Lab Invest* 46:9A, 1982.

125. Carmichael DJS, Hamilton DV, Evans DB, Stovin PGI, Calne RY: Interstitial pneumonitis secondary to azathioprine in a renal transplant patient. *Thorax* 38:951–952, 1983.

126. Jehn U, Göldel N, Reinmüller R, Wilmanns W: Non-cardiogenic pulmonary edema complicating intermediate and high-diose Ara C treatment for relapsed acute leukemia. *Med Oncol Tumor Pharmacother* 5:41–47, 1988.

127. Andersson BS, Luna MA, Yee C, Hui KK, Keating MJ, McCredie KB: Fatal pulmonary failure complicating high-dose cytosine arabinoside therapy in acute leukemia. *Cancer* 65:1079–1084, 1990.

128. Searles G, McKendry RJR: Methotrexate pneumonitis in rheumatoid arthritis: Potential risk factors. Four case reports and a review of the literature. *J Rheumatol* 14:1164–1171, 1987.

129. Sostman HD, Matthay RA, Putman CE, et al: Methotrexate-induced pneumonitis. *Medicine* 55:371–388, 1976.

130. Stutz FH, Tormey DC, Blom J: Nonbacterial pneumonitis with multidrug antineoplastic therapy in breast carcinoma. *Can Med Assoc J* 108:710–714, 1973.

131. Barrera P, Laan RFJM, van Riel PLCM, Dekhuijzen PNR, Boerbooms AMT, van de Putte LBA: Methotrexate-related pulmonary complications in rheumatoid arthritis. *Ann Rheum Dis* 53:434–439, 1994.

132. Carson CW, Cannon GW, Egger MJ, Ward JR, Clegg DO: Pulmonary disease during treatment of rheumatoid arthritis with low dose pulse methotrexate. *Semin Arthritis Rheum* 16:186–195, 1987.

133. Bedrossian CWM, Miller WC, Luna MA: Methotrexate-induced diffuse interstitial pulmonary fibrosis. *South Med J* 72:313–318, 1979.

134. Lascari AD, Strano AJ, Johnson WW, Collins JGP: Methotrexate-induced sudden fatal pulmonary reaction. *Cancer* 40:1393–1397, 1977.

135. Clarysse AM, Cathey WJ, Carteright GE, Wintrobe M: Pulmonary disease complicating intermittent therapy with methotrexate. *JAMA* 209:1861–1868, 1969.

136. Everts CE, Westcott JL, Bragg DG: Methotrexate therapy and pulmonary disease. *Radiology* 107:539–543, 1973.

137. Nesbit M, Krivit W, Heyn R, Sharp H: Acute and chronic effects of methotrexate on hepatic, pulmonary and skeletal systems. *Cancer* 37(suppl):1048–1057, 1976.

138. Case Records of the Massachusetts General Hospital: Case 6–1985. *N Engl J Med* 312:359–369, 1985.

139. Goldman GC, Moschella SL: Severe pneumonitis occurring during methotrexate therapy. Report of two cases. *Arch Dermatol* 103:194–197, 1971.

140. Cannon GW, Ward JR, Clegg DO, et al: Acute lung disease associated with low-dose pulse methotrexate therapy in patients with rheumatoid arthritis. *Arthritis Rheum* 26:1269–1274, 1983.

141. Kane GC, McMichael AJ, Patrick H, Ersler AJ: Pulmonary toxicity and acute respiratory failure associated with fludarabine monophosphate. *Respir Med* 86:261–263, 1992.

142. Hurst PG, Habib MP, Garewal H, Bluestein M, Paquin M, Greenberg BR: Pulmonary toxicity associated with fludarabine monophosphate. Invest New Drugs 5:207–210, 1987.

143. Cervantes F, Salgado C, Montserrat E, Rozman C: Fludarabine for prolymphocytic leukemia and risk of interstitial pneumonitis. *Lancet* 336:1130, 1990.

144. Dohner VA, Ward HP, Standord RE: Alveolitis during procarbazine, vincristine, and cyclophosphamide therapy. *Chest* 62:636–639, 1972.

145. Farney RJ, Morris AH, Armstrong JD, Hammar S: Diffuse pulmonary disease after therapy with nitrogen mustard, vincristine, procarbazine, and prednisone. *Am Rev Respir Dis* 115:135–145, 1977.

146. Jones SE, Moore M, Blank N, Castellino RA: Hyper-sensitivity to procarbazine (Matulane) manifested by fever and pleuropulmonary reaction. *Cancer* 29:498–500, 1972.

147. Lewis L: Procarbazine associated alveolitis. *Thorax* 39:206–207, 1984.

148. Horton LWL, Chappell AG, Powell DEB: Diffuse interstitial pulmonary fibrosis complicating Hodgkin's disease. *Br J Dis Chest* 71:44–48, 1977.

149. Ozols RF, Hogan WM, Ostchega Y, Young RC: MVP (mitomycin, vinblastine and progesterone): A second-line regimen in ovarian cancer with a high incidence of pulmonary toxicity. *Cancer Treat Rep* 67:721–722, 1983.

150. Israel RH, Olsen JP: Pulmonary edema associated with intravenous vinblastine. *JAMA* 240:1585, 1978.

151. Kris MG, Pablo D, Gralla RJ, Burke MT, Prestifilippo J, Lewin D: Dyspnea following vinblastine or vindesine administration in patients receiving mitomycin plus vinca alkaloid combination therapy. *Cancer Treat Rep* 68:1029–1031, 1984.

152. Konits PH, Aisner J, Sutherland JC, Wiernik PH: Possible pulmonary toxicity secondary to vinblastine. *Cancer* 50:2771–2774, 1982.

153. Commers JR, Foley JF: Pulmonary hyaline membrane disease occurring in the course of VM-26 therapy. *Cancer Treat Rep* 63:2093–2095, 1979.

154. Heger JJ, Prystowsky EN, Jackman WM, et al: Amiodarone. *N Engl J Med* 305:539–545, 1981.

155. Dusman RE, Stanton MS, Miles WM, et al: Clinical features of amiodarone-induced pulmonary toxicity. *Circulation* 82:51–59, 1990.

156. Martin WJ II, Rosenow EC III: Amiodarone pulmonary tox-icity. Recognition and pathogenesis (part I). *Chest* 93:1067–1075, 1988.

157. Martin WJ II, Rosenow EC III: Amiodarone pulmonary toxicity. Recognition and pathogenesis (part II). *Chest* 93:1242–1248, 1988.

158. Riley SA, Williams SE, Cooke NJ: Alveolitis after treatment with amiodarone. *Br Med J* 284:161–162, 1982.

159. Manicardi V, Bernini G, Bossini P, Bertorelli G, Pesci A, Bellodi G: Low-dose amiodarone-induced pneumonitis: Evidence of an immunologic pathogenetic mechanism. *Am J Med* 86:134–135, 1989.

160. Dean PJ, Groshart KD, Porterfield JG, Iansmith DH, Golden EB: Amiodarone-associated pulmonary toxicity. A clinical and pathological study of eleven cases. *Am J Clin Pathol* 87:7–13, 1987.

161. Kennedy JI, Myers JL, Plumb VJ, Fulmer JD: Amiodarone pulmonary toxicity. Clinical, radiologic and pathologic correlations. *Arch Intern Med* 147:50–55, 1987.

162. Jirik FR, Henning H, Huckell VF, Ostrow DVN: Diffuse alveolar damage associated with amiodarone therapy. *Can Med Assoc J* 128:1192–1195, 1983.

163. Kay GN, Epstein AE, Kirklin JK, Diethelm AG, Graybar G, Plumb VJ: Fatal postoperative amiodarone pulmonary toxicity. *Am J Cardiol* 62:490–492, 1988.

164. Nalos PC, Kass RM, Gang ES, Fishbein MC, Mandel WJ, Peter T: Life-threatening postoperative pulmonary complications in patients with previous amiodarone pulmonary toxicity undergoing cardiothoracic operations. *J Thorac Cardiovasc Surg* 93:904–912, 1987.

165. Tuzcu EM, Maloney JD, Sangani BH: et al: Cardiopulmonary effects of chronic amiodarone therapy in the early postoperative course of cardiac surgery patients. *Cleve Clin J Med* 54:491–497, 1987.

166. Greenspon AJ, Kidwell GA, Hurley W, Mannion J: Amiodarone-related postoperative adult respiratory distress syndrome. *Circulation* 84 (suppl 3):407–415, 1991.

167. Van Mieghaem W, Coolen L, Malysse I, Lacquet LM, Deneffe GJD, Demedts MGP: Amiodarone and the development of ARDS after lung surgery. *Chest* 105:1642–1645, 1994.

168. Kuhlman JE, Teigen C, Ren H, Hruban RH, Hutchins GM, Fishman EK: Amiodarone pulmonary toxicity: CT findings in symptomatic patients. *Radiology* 177:121–125, 1990.

169. Sobol SM, Rakita L: Pneumonitis and pulmonary fibrosis associated with amiodarone treatment: A possible complication of a new antiarrhythmic drug. *Circulation* 65:819–824, 1982.

170. Marchlinski FE, Gansler TS, Waxman HL, Josephson ME: Amiodarone pulmonary toxicity. *Ann Intern Med* 97:839–845, 1982.

171. Myers JL, Kennedy JI, Plumb VJ: Amiodarone lung: Pathologic findings in clinically toxic patients. *Hum Pathol* 18:349–854, 1987.

172. Colgan T, Simon GT, Kay JM, Pugsley SO, Eydt J: Amiodarone pulmonary toxicity. *Ultrastruc Pathol* 6:199–207, 1984.

173. Liu FL-K, Cohen RD, Downar E, Butany JW, Edelson JD, Rebuck AS: Amiodarone pulmonary toxicity: Functional and ultrastructural evaluation. *Thorax* 41:100–105, 1986.

174. Lullman H, Lullman-Rauch R, Wassermann O: Lipidosis induced by amphiphilic cationic drugs. *Biochem Pharmacol* 27:1103–1108, 1978.

175. Heath D, Smith P, Hasleton PS: Effects of chlorphentermine on the rat lung. *Thorax* 28:551–558, 1973.

176. Smith P, Heath D, Hasleton PS: Electron microscopy of chlorphentermine lung. *Thorax* 28:559–566, 1973.

177. Vijeyaratnam GS, Corrin B: Fine structural alterations in the lungs of iprindole-treated rats. *J Pathol* 114:233–240, 1974.

178. Martin WJ II, Howard DM: Amiodarone-induced lung toxicity. In vitro evidence for the direct toxicity of the drug. *Am J Pathol* 120:344–350, 1985.

179. Camus P, Lombard J-N, Perrichon M, et al: Bronchiolitis obliterans organizing pneumonia in patients taking acebutolol or amiodarone. *Thorax* 44:711–715, 1989.

180. Coudert B, Bailly F, Lombard JN, Andre F, Camus P: Amiodarone pneumonitis. Bronchoalveolar lavage findings in 15 patients and review of the literature. *Chest* 102:1005–1012, 1992.

181. Horn HR, Hadidian Z, Johnson JL, Vassallo HG, Williams JH, Young MD: Safety evaluation of tocainide in the American Emergency Use Program. *Am Heart J* 100:1037–1040, 1980.

182. Bastian BC, Macfarlane PW, McLauchlan JH, et al: Prospective randomized trial of tocainide in patients following myocardial infarction. *Am Heart J* 100:1017–1022, 1980.

183. Braude AC, Downar E, Chamberlain DW, Rebuck AS: Tocainide-associated interstitial pneumonitis. *Thorax* 37:309–310, 1982.

184. Perlow GM, Jain BP, Pauker SG, Zarren HS, Wistran DC, Epstein RL: Tocainide-associated interstitial pneumonitis. *Ann Intern Med* 94:489–490, 1981.

185. Stein MG, Demarco T, Gamsu G, Finkbeiner W, Golden JA: Computed tomography: pathologic correlation in lung disease due to tocainide. *Am Rev Respir Dis* 137:458–460, 1988.

186. Feinberg L, Travis WD, Ferrans V, Sato N, Bernton HF: Pulmonary fibrosis associated with tocainide: Report of a case with literature review. *Am Rev Respir Dis* 141:505–508, 1990.

187. Lombard JN, Bonnotte B, Maynadie M, et al: Celiprolol pneumonitis. *Eur Respir J* 6:588–591, 1993.

188. Musk AW, Pollard JA: Pindolol and pulmonary fibrosis. *Br Med J* 2:581–582, 1979.

189. Wood GM, Bolton RP, Muers MF, Losowsky MS: Pleurisy and pulmonary granulomas after treatment with acebutolol. *Br Med J* 2:936, 1982.

190. Hall DR, Morrison JB, Edwards FR: Pleural fibrosis after practolol therapy. *Thorax* 33:822–824, 1978.

191. Marshall AJ, Barritt DW, Griffiths DA, et al: Respiratory disease associated with practolol therapy. *Lancet* 2:1254–1257, 1977.

192. Erwteman TM, Braat MCP, Van Aken WG: Interstitial pulmonary fibrosis: A new side effect of practolol. *Br Med J* 2:297–298, 1977.

193. Perry HM, O'Neal RM, Thomas WA: Pulmonary disease following chronic chemical ganglionic blockade. *Am J Med* 22:37–50, 1957.

194. Rosketh P, Storstein O: Pulmonary complications during mecamylamine therapy. *Acta Med Scand* 167:23–27, 1960.

195. Hilden T, Krogsgaard AR, Vimtrump B: Fatal pulmonary changes during the medical treatment of malignant hypertension. *Lancet* 2:830–832, 1958.

196. Doniach I: Uraemic edema of the lungs. *AJR Am J Roentgenol* 58:620–628, 1947.

197. Hopps HC, Wissler RW: Uremic pneumonitis. *Am J Pathol* 31:261–267, 1955.

198. Doniach I, Morrison B, Steiner RE: Lung changes during hexamethonium therapy for hypertension. *Br Heart J* 16:101–108, 1954.

199. Vasilomanolakis EC, Goldberg NM: Bepridil-induced pulmonary fibrosis. *Am Heart J* 126:1016–1017, 1993.

200. Poe RH, Condemi JJ, Weinstein SS, Schuster RJ: Adult respiratory distress syndrome related to ampicillin sensitivity. *Chest* 77:449–451, 1980.

201. Hamilton-Miller JMT. Fungal sterols and the mode of action of the polyene antibiotics. *Adv Appl Microbiol* 17:109–134, 1974.

202. Wright DG, Robichaud KJ, Pizzo PA, Deisseroth AB: Lethal pulmonary reactions associated with the combined use of amphotericin B and leukocyte transfusions. *N Engl J Med* 304:1185–1189, 1981.

203. Dutcher JP, Kendall J, Norris D, Schiffer C, Aisner J, Wiernik PH: Granulocyte transfusion therapy and amphotericin B: Adverse reactions? *Am J Hematol* 31:102–108, 1989.

204. Grinblat J, Mechlis S, Lewitus Z: Organizing pneumonia-like process, an unusual observation in steroid responsive cases with features of chronic interstitial pneumonia. *Chest* 80:259–263, 1981.

205. Dreis DF, Winterbauer RH, Van Norman GA, Sullivan SL, Hammar SP: Cephalosporin-induced interstitial pneumonitis. *Chest* 86:138–140, 1984.

206. Sitbon O, Bidel N, Dussopt N, et al: Minocycline pneumonitis and eosinophilia. *Arch Intern Med* 154:1633–1640, 1994.

207. Israel HL, Diamond P: Recurrent pulmonary infiltration and pleural effusion due to nitrofurantoin sensitivity. *N Engl J Med* 266:1024–1026, 1962.

208. Holmberg L, Boman G: Pulmonary reactions to nitrofurantoin. 447 cases reported to the Swedish Adverse Drug Reaction Committee 1966–1976. *Eur J Respir Dis* 62:180–189, 1981.

209. Hailey FJ, Glascock HW, Hewitt WF: Pleuropneumonic reactions to nitrofurantoin. *N Engl J Med* 281:1087–1090, 1969.

210. Geller M, Dickie HA, Kass DA, et al: The histopathology of acute nitrofurantoin-associated pneumonitis. *Ann Allergy* 37:275–279, 1976.

211. Averbuch SD, Yungbluth P: Fatal pulmonary hemorrhage due to nitrofurantoin. *Arch Intern Med* 140:271–273, 1980.

212. Bucknall CE, Adamson MR, Banham SW: Non fatal pulmonary hemorrhage associated with nitrofurantoin. *Thorax* 42:475–476, 1987.

213. Magee F, Wright JL, Chan N, et al: Two unusual pathological reactions to nitrofurantoin: case reports. *Histopathology* 10:701–706, 1986.

214. Israel KS, Brashear RE, Sharma HM, Yum MN, Glover JL: Pulmonary fibrosis and nitrofurantoin. *Am Rev Respir Dis* 108:353–356, 1973.

215. Rosenow EC III, DeRemee RA, Dines DE: Chronic nitrofurantoin pulmonary reaction. *N Engl J Med* 279:1258–1262, 1968.

216. Strandberg I, Wengle B, Fagrell B: Chronic interstitial pneumonitis with fibrosis during long-term treatment with nitrofurantoin. *Acta Med Scand* 196:483, 1974.

217. Robinson GM, Bai TR, Steele RH: Nitrofurantoin induced chronic pulmonary reaction: Case report. *N Z Med J* 91:50–52, 1980.

218. Simonian SJ, Kroecker EJ, Boyd DP: Chronic interstitial pneumonitis with fibrosis after long-term therapy with nitrofurantoin. *Ann Thorac Surg* 24:284–288, 1977.

219. Bone RC, Wolfe J, Sobonya RE, et al: Desquamative interstitial pneumonia following long-term nitrofurantoin therapy. *Am J Med* 60:697–701, 1976.

220. Carrington CB, Addington WW, Goff AM, et al: Chronic eosinophilic pneumonia. *N Engl J Med* 280:787–798, 1969.

221. Cohen AJ, King TE Jr, Downey GP: Rapidly progressive bronchiolitis obliterans with organizing pneumonia. *Am J Respir Crit Care Med* 149:1670–1675, 1994.

222. Davidson AC, Bateman C, Shovlin M, Camerson IR, Burton GH: Pulmonary toxicity of malaria prophylaxis. *Br Med J* 297:1240–1241, 1988.

223. Whitfield D: Presumptive fatality due to pyrimethamine-sulfadoxine. *Lancet* 2:1272, 1982.

224. Hamadeh MA, Atkinson J, Smith LJ: Sulfasalazine-induced pulmonary disease. *Chest* 101:1033, 1992.

225. Camus P, Picard F, Ashcroft T, Gal AA, Colby TV: The lung in inflammatory bowel disease. *Medicine* 72:151–183, 1993.

226. Dawson A, Gibbs AR, Anderson G: An unusual perilobular pattern of pulmonary interstitial fibrosis associated with Crohn's disease. *Histopathology* 23:553–556, 1993.

227. Davies D, MacFarlane A: Fibrosing alveolitis and treatment with sulphasalazine. *Gut* 15:185–188, 1974.

228. Sullivan SN: Desensitization to sulfasalazine and treatment with acrylic coated 5-ASA and azodisalicylate. *J Clin Gastroenterol* 9:461–463, 1987.

229. Moseley RH, Barwick KW, Dobluler K, DeLuca VA: Sulfasalazine-induced pulmonary disease. *Dig Dis Sci* 30:901–904, 1985.

230. Wang KK, Bowyer BA, Fleming CR, Schroeder KW: Pulmonary infiltrates and eosinophilia associated with sulfasalazine. *Mayo Clin Proc* 59:343–346, 1984.

231. Kolbe J, Caughey D, Rainer S: Sulfasalazine-induced subacute hypersensitivity pneumonitis. *Respir Med* 88:149–152, 1994.

232. Teague WG, Sutphen JL, Fechner RE: Desquamative interstitial pneumonitis complicating inflammatory bowel disease of childhood. *J Pediatr Gastroenterol Nutr* 4:663–667, 1985.

233. Williams T, Eidus L, Thomas P: Fibrosing alveolitis, bronchiolitis obliterans and sulfasalazine therapy. *Chest* 81:766–768, 1982.

234. Fiegenberg DS, Weiss H, Kirshman H: Migratory pneumonia with eosinophilia associated with sulfonamide administration. *Arch Intern Med* 120:85–89, 1967.

235. Evans RB, Ettensohn DB, Fawaz-Estrup F, Lally EV, Kaplan SR: Gold lung: Recent developments in pathogenesis, diagnosis, and therapy. *Semin Arthritis Rheum* 16:196–205, 1987.

236. Smith W, Ball GV: Lung injury due to gold treatment. *Arthritis Rheum* 23:351–354, 1980.

237. Scott DL, Bradby GVH, Aitman TJ, Zaphiropoulos GC, Hawkins CF: Relationship of gold and penicillamine therapy to diffuse interstitial lung disease. *Ann Rheum Dis* 40:136–141, 1981.

238. Morley TF, Komansky HJ, Adelizzi RA, Giudice JC: Pulmonary gold toxicity. *Eur J Respir Dis* 65:627–632, 1984.

239. Winterbauer RH, Wilske KR, Wheelis RF: Diffuse pulmonary injury associated with gold treatment. *N Engl J Med* 294:919–921, 1976.

240. McCormick J, Cole S, Lahirir B, Knauft F, Cohen S, Yoshida T: Pneumonitis caused by gold salt therapy: Evidence for the role of cell-mediated immunity in its pathogenesis. *Am Rev Respir Dis* 122:145–152, 1980.

241. Fort JG, Scovern H, Abruzzo JL: Intravenous cyclophosphamide and methylprednisolone for the treatment of bronchiolitis obliterans associated with crysotherapy. *J Rheumatol* 15:850–854, 1988.

242. Pegg SJ, Lang BA, Mikhail EL, Hughes DM: Fatal bronchiolitis obliterans in a patient with juvenile rheumatoid arthritis receiving chrysotherapy. *J Rheumatol* 21:549–551, 1994.

243. O'Duffy JD, Luthra HS, Unni KK, Hyatt RE: Bronchiolitis in a rheumatoid arthritis patient receiving auranofin. *Arthritis Rheum* 29:556–559, 1986.

244. Lahdensos A, Mattila J, Vilppula A: Bronchiolitis in rheumatoid arthritis. *Chest* 85:705–708, 1984.

245. Gould PW, McCormack PL, Palmer DG: Pulmonary damage associated with sodium aurothiomalate therapy. *J Rheumatol* 2:252–260, 1977.

246. Pääkko P, Anttila S, Sutinen S, Hakala M: Lysosomal gold accumulations in pulmonary macrophages. *Ultrastruct Pathol* 7:289–294, 1984.

247. Garcia JGN, Munim A, Nugent KM, et al: Alveolar macrophage gold retention in rheumatoid arthritis. *J Rheumatol* 14:435–438, 1987.

248. Lyle WH: D-Penicillamine and fatal obliterative bronchiolitis. *Br Med J* i:105, 1977.

249. Matloff DS, Kaplan MM: D-Penicillamine-induced Goodpasture's-like syndrome in primary biliary cirrhosis. Successful treatment with plasmapheresis and immunosuppressives. *Gastroenterology* 78:1046–1049, 1980.

250. Sternlieb I, Bennett B, Scheinberg IH: D-Penicillamine induced Goodpasture's syndrome in Wilson's disease. *Ann Intern Med* 82:673–676, 1975.

251. Shettar SP, Chattopadhyay C, Wolstenholme RJ, Swinson R: Diffuse alveolitis on a small dose of penicillamine. *Br J Rheumatol* 23:220–224, 1984.

252. Epler GR, Snider GL, Gaensler EA, Cathcart ES, Fitzgerald MX, Carrington CB: Bronchiolitis and bronchitis in connective tissue disease. A possible relation to the use of penicillamine. *JAMA* 242:528–532, 1979.

253. Murphy KC, Atkins CJ, Offer RC, Hogg JC, Stein HB: Obliterative bronchiolitis in two rheumatoid arthritis patients treated with penicillamine. *Arthritis Rheum* 24:557–560, 1981.

254. King TE Jr: Bronchiolitis obliterans. *Lung* 167:69–93, 1989.

255. Kraft M, Mortensen RL, Colby TV, Newman L, Waldron JA Jr, King TE Jr: Cryptogenic constrictive bronchiolitis. A clinicopathologic study. *Am Rev Respir Dis* 148:1093–1101, 1993.

256. Turner-Warwick M: Adverse reactions affecting the lung: possible association with D-penicillamine. *J Rheumatol* 8(suppl 7):166–168, 1981.

257. Louie S, Gamble CN, Cross CE: Penicillamine associated pulmonary hemorrhage. *J Rheumatol* 13:963–966, 1986.

258. Davies D, Jones JKL: Pulmonary eosinophilia caused by penicillamine. *Thorax* 35:957–958, 1980.

259. Camus P, Degat OR, Justrabo E, Jeannin L: D-Penicillamine-induced severe pneumonitis. *Chest* 81:376–378, 1982.

260. Khalil H, Molinary E, Stoller JK: Diclofenac (Voltaren)-induced eosinophilic pneumonitis. *Arch Intern Med* 153:1649–1652, 1993.

261. Pfitzenmeyer P, Meier M, Zuch P, et al: Piroxicam induced pulmonary infiltrates and eosinophilia. *J Rheumatol* 21:1573–1577, 1994.

262. Nader DA, Schillaci RF: Pulmonary infiltrates with eosinophilia due to naproxen. *Chest* 82:280–282, 1983.

263. Thurston JGB, Marks P, Trapnell D: Lung changes associated with phenylbutazone treatment. *Br Med J* 2:1422–1423, 1976.

264. Park GD, Spector R, Headstream T, Goldberg M: Serious adverse reactions associated with sulindac. *Arch Intern Med* 142:1292–1294, 1982.

265. King GG, Barnes DJ, Hayes MJ: Carbamazepine-induced pneumonitis. *Med J Aust* 160:126–127, 1994.

266. Cullinan SA, Bower GC: Acute pulmonary hypersensitivity to carbamazepine. *Chest* 68:580–581, 1975.

267. Stephan WC, Parks RD, Tempest B: Acute hypersensitivity pneumonitis associated with carbamazepine therapy. *Chest* 74:463–464, 1978.

268. Michael JR, Rudin ML: Acute pulmonary disease caused by phenytoin. *Ann Intern Med* 95:452–454, 1981.

269. Chamberlain DW, Hyland RH, Ross JD: Diphenylhydantoin-induced lymphocytic pneumonia. *Chest* 90:458–460, 1986.

270. Munn NJ, Baughman RP, Ploysongsang Y: Bronchoalveolar lavage in acute drug-hypersensitivity pneumonitis probably caused by phenytoin. *South Med J* 77:1594–1596, 1984.

271. Yermakov VM, Hitti IF, Sutton AL: Necrotizing vasculitis associated with diphenylhydantoin: Two fatal cases. *Hum Pathol* 14:182–184, 1983.

272. Kay JM: Dietary pulmonary hypertension. *Thorax* 49(suppl):S33–S38, 1994.

273. Cameron J, Waugh L, Loadsman T, White P, Radford DJ: Possible association of pulmonary hypertension with anorectic drug. *Med J Aust* 140:595–597, 1984.

274. Gurtner HP, Gertsch M, Salzmann C, Scherrer M, Stucki P, Wyss F: Haufen sich die primar vascularen Formen des chronischen Cor pulmonale? *Schweiz Med Wochenschr* 98:1579–1587, 1695, 1968.

275. Kay JM, Smith P, Heath D: Aminorex and the pulmonary circulation. *Thorax* 26:262–270, 1971.

276. Heath D, Kay JM: Diet, drugs and pulmonary hypertension. In Yu PN, Goodwin JF (eds): *Progress in Cardiology*. Philadelphia, Lea & Febiger, 1978, vol 7, pp 125–140.

277. Gurtner HP: Aminorex pulmonary hypertension. In Fishman AP (ed): *The Pulmonary Circulation: Normal and Abnormal*. Philadelphia: University of Pennsylvania Press, 1990, pp 397–411.

278. Douglas JG, Munro JF, Kitchin AH, Muir AL, Proudfoot AT: Pulmonary hypertension and fenfluramine. *Br Med J* 283:881–883, 1981.

279. Pouwels HMM, Smeets JLRM, Cheriex EC, Wouters EFM: Pulmonary hypertension and fenfluramine. *Eur Respir J* 3:606–607, 1990.

280. Brenot F, Herve P, Petitpretz P, Parent F, Duroux P, Simonneau G: Primary pulmonary hypertension and fenfluramine use. *Br Heart J* 70:537–541, 1993.

281. McMurray J, Bloomfield P, Miller HC: Irreversible pulmonary hypertension after treatment with fenfluramine. *Br Med J* 293:51–52, 1986.

282. Nall KC, Rubin LJ, Lipskind S, Sennesh JD: Reversible pulmonary hypertension associated with anorexigen use. *Am J Med* 91:97–99, 1991.

283. Roche N, Labrune S, Braun J-M, Huchon GJ: Pulmonary hypertension and dexfenfluramine. *Lancet* 339:436–437, 1992.

284. Atanassoff PG, Weiss BM, Schmid ER, Tornic M: Pulmonary hypertension and dexfenfluramine. *Lancet* 339:436, 1992.

285. Finly TN, Aronow AB, Cosentino AM, Golde DW: Occult pulmonary hemorrhage in anticoagulated patients. *Am Rev Respir Dis* 112:23–29, 1975.

286. Santalo M, Domingo P, Fontcuberta J, et al: Diffuse pulmonary hemorrhage associated with anticoagulant therapy. *Eur J Respir Dis* 69:114–119, 1986.

287. Taal BG, Spierings ELH, Hilbering C: Pleuropulmonary fibrosis associated with chronic and excessive intake of ergotamine. *Thorax* 38:396–398, 1983.

288. Hindle W, Posner E, Sweetnam MT, Tan RSH: Pleural effusion and fibrosis during treatment with methysergide. *Br Med J* 1:605–606, 1970.

289. McElvaney NG, Wilcox PG, Churg AM: Pleuropulmonary disease during bromocriptine treatment of Parkinson's disease. *Arch Intern Med* 148:2231–2236, 1988.

290. Melmed S, Braunstein GD: Bromocriptine and pleuropulmonary disease. *Arch Intern Med* 149:258–259, 1989.

291. Wiggins J, Skinner C: Bromocriptine induced pleuropulmonary fibrosis. *Thorax* 41:328–330, 1986.

292. Burgher LW, Kass I, Schenken JR: Pulmonary allergic granulomatosis: A possible drug reaction in a patient receiving cromolyn sodium. *Chest* 66:84–86, 1974.

293. Harris MG, Coleman SG, Faulds D, Chrisp P: Nilutamide. A review of its pharmacodynamic and pharmacokinetic properties and therapeutic efficacy in prostate cancer. *Drugs Aging* 3:9–25, 1993.

294. Jonville AP, Diot E, Dutertre JP, Autret E: Toxicité pulmonaire du nilutamide (Anadron). Bilan coopératif des centres français du pharmacovigilance. *Therapie* 47:393–397, 1992.

295. Seigneur J, Trechot PF, Hubert J, Lamy P: Pulmonary complications of hormone treatment in prostate carcinoma. *Chest* 93:1106, 1988.

296. Gomez JL, Dupont A, Cusan L, et al: Simultaneous liver and lung toxicity related to the nonsteroidal antiandrogen nilutamide (Anandron): A case report. *Am J Med* 92:465–470, 1992.

297. Miyazono K, Okazaki T, Uchida S, et al: Propylthiouracil-induced diffuse interstitial pneumonitis. *Arch Intern Med* 144:1764–1765, 1984.

298. Martin TR, Sandblom RL, Johnson RJ: Adult respiratory distress syndrome following thrombolytic therapy for pulmonary embolism. *Chest* 83:151–153, 1983.

299. Tazelaar HD, Myers JL, Drage CW, King TE, Aguayo S, Colby TV: Pulmonary disease associated with L-tryptophan-induced eosinophilic myalgia syndrome. Clinical and pathologic features. *Chest* 97:1032–1036, 1990.

300. Bogaerts Y, Van Renterghem D, Vanvuchelen J, et al: Interstitial pneumonitis and pulmonary vasculitis in a patient taking an L-tryptophan preparation. *Eur Respir J* 4:1033–1036, 1991.

301. Strumpf IJ, Drucker RD, Anders KH, Cohen S, Fajolu O: Acute eosinophilic pulmonary disease associated with the ingestion of L-tryptophan-containing products. *Chest* 99:8–13, 1991.

302. Banner AS, Borochovitz D: Acute respiratory failure caused by pulmonary vasculitis after L-tryptophan ingestion. *Am Rev Respir Dis* 143:661–664, 1991.

303. Tazelaar HD, Myers JL, Strickler JG, Colby TV, Duffy J: Tryptophan-induced lung disease: An immunophenotypic, immunofluorescent, and electron microscopic study. *Mod Pathol* 6:56–60, 1993.

304. Lee SL, Chase PH: Drug-induced systemic lupus erythematosus: a critical review. *Semin Arthritis Rheum* 5:83–103, 1975.

305. Byrd RB, Schanzer B: Pulmonary sequelae in procainamide induced lupus-like syndrome. *Dis Chest* 55:170–172, 1969.

306. Hind CRK: Pulmonary complications of intravenous drug abuse. 1. Epidemiology and non-infective complications. *Thorax* 45:891–898, 1990.

307. Glassroth J, Adams GD, Schnoll S: The impact of substance abuse on the respiratory system. *Chest* 91:596–602, 1987.

308. Jentzen J: Medical complications of cocaine abuse. *Am J Clin Pathol* 100:475–476, 1993.

309. Hind CRK: Pulmonary complications of intravenous drug abuse. 2. Infective and HIV related complications. *Thorax* 45:957–961, 1990.

310. Gong H, Fligiel S, Tashkin DP, Barbers RG: Tracheobronchial changes in habitual, heavy smokers of marijuana with and without tobacco. *Am Rev Respir Dis* 136:142–149, 1987.

311. Wu T-C, Tashkin DP, Djahed B, Rose JE: Pulmonary hazards of smoking marijuana as compared with tobacco. *N Engl J Med* 318:347–351, 1988.

312. Cooper CB, Bai TR, Heyderman E, Corrin B: Cellulose gran-

ulomas in the lungs of a cocaine sniffer. *Br Med J* 286:2021–2022, 1983.

313. Oubeid M, Bickel JT, Ingram EA, Scott GC: Pulmonary talc granulomatosis in a cocaine sniffer. *Chest* 98:237–239, 1990.

314. Abraham JL, Brambilla C: Particle size for differentiation between inhalation and injection taclosis. *Environ Res* 21:94–96, 1980.

315. Brody SL, Slovis CM, Wrenn KD: Cocaine-related medical problems. Consecutive series of 233 patients. *Am J Med* 88:325–331, 1990.

316. Itkonen J, Schnoll S, Glassroth J: Pulmonary dysfunction in "freebase" cocaine users. *Arch Intern Med* 144:2195–2197, 1984.

317. Murray RJ, Smialek JE, Golle M, Albin RJ: Pulmonary artery medial hypertrophy in cocaine users without foreign particle microembolization. *Chest* 96:1050–1053, 1989.

318. Bailey ME, Fraire AE, Greenberg SD, Barnard J, Cagle PT: Pulmonary histopathology in cocaine abusers. *Hum Pathol* 25:203–207, 1994.

319. Cucco Ra, Yoo OH, Cregler L, Chang JC: Nonfatal pulmonary edema after "freebase" cocaine smoking. *Am Rev Respir Dis* 136:179–181, 1987.

320. Forrester JM, Steele AW, Waldron JA, Parsons PE: Crack lung: An acute pulmonary syndrome with a spectrum of clinical and histopathologic findings. *Am Rev Respir Dis* 142:462–467, 1990.

321. Kissner DG, Lawrence WD, Selis JE, Flint A: Crack lung: Pulmonary disease caused by cocaine abuse. *Am Rev Respir Dis* 136:1250–1252, 1987.

322. Oh PI, Balter MS: Cocaine induced eosinophilic lung disease. *Thorax* 47:478, 1992.

323. Tashkin DP, Simmons MS, Coulson AH, Clark VA, Gong H: Respiratory effects of cocaine "freebasing" among habitual users of marijuana with or without tobacco. *Chest* 92:638–644, 1987.

324. Murray RJ, Albin RJ, Mergner W, Criner GJ: Diffuse alveolar hemorrhage temporally related to cocaine smoking. *Chest* 93:427–429, 1988.

325. Kay JM: Vascular disease, in Thurlbeck WM, Churg AM (eds): *Pathology of the Lung*, 2d ed. New York, Thieme, 1995, pp 931–1066.

326. O'Donnell AE, Mappin FG, Sebo TJ, Tazelaar H: Interstitial pneumonitis associated with "crack" cocaine abuse. *Chest* 100:1155–1157, 1991.

327. Patel RC, Dutta D, Schonfeld SA: Free-base cocaine use associated with bronchiolitis obliterans organizing pneumonia. *Ann Intern Med* 107:186–187, 1987.

328. Greenebaum E, Copeland A, Grewal R: Blackened bronchoalveolar lavage fluid in crack smokers. A preliminary study. *Am J Clin Pathol* 100:481–487, 1993.

329. O'Gorman P, Patel S, Notcutt S, Wicking J: Adulteration of "street" heroin with chloroquine. *Lancet* 1:746, 1987.

330. Hahn RA, Onorato IM, Jones TS, Dougherty J: Prevalence of HIV infection among intravenous drug users in the United States. *JAMA* 261:2677–2684, 1989.

331. Schoenbaum EE, Hartel D, Selwyn PA, et al: Risk factors for human immunodeficiency virus infection in intravenous drug abusers. *N Engl J Med* 321:874–879, 1989.

332. Libby LS, King TE, LaForce FM, Schwartz MI: Pulmonary cavitation following pulmonary infarction. *Medicine (Baltimore)* 64:342–348, 1985.

333. Collignon JP, Sorrel TC: Disseminated candidiasis: Evidence of a distinctive syndrome in heroin abusers. *Br Med J* 287:861–862, 1983.

334. Hoy J, Speed B: Candidiasis in heroin abusers. *Br Med J* 287:1549, 1983.

335. Smith WR, Wells ID, Glauser FL, Novey HS: High incidence of precipitins in sera of heroin addicts. *JAMA* 232:1337–1338, 1975.

336. Brown SM, Stimmel B, Taub RN, et al: Immunologic dysfunction in heroin addicts. *Arch Intern Med* 134:1001–1006, 1974.

337. Siegel H: Human pulmonary pathology associated with narcotic and other addictive drugs. *Hum Pathol* 3:55–66, 1972.

338. Arnett EN, Battle WE, Russo JV, Roberts WC: Intravenous injection of talc-containing drugs intended for oral use. A cause of pulmonary granulomatosis and pulmonary hypertension. *Am J Med* 60:711–718, 1976.

339. Tomashefski JF Jr, Hirsch CS: The pulmonary vascular lesions of intravenous drug abuse. *Hum Pathol* 11:133–145, 1980.

340. Paré JAP, Fraser RG, Hogg JC, Howlett JG, Murphy SB: Pulmonary "mainline" granulomatosis: Talcosis of intravenous methadone abuse. *Medicine (Baltimore)* 58:229–239, 1979.

341. Waller BF, Brownlee WJ, Roberts WC: Self-induced pulmonary granulomatosis. A consequence of intravenous injection of drugs intended for oral use. *Chest* 78:90–94, 1980.

342. Lewman LV: Fatal pulmonary hypertension from intravenous injection of methylphenidate (Ritalin) tablets. *Hum Pathol* 3:67–70, 1972.

343. Tomashefski JF Jr, Hirsch CS: Microcrystalline cellulose pulmonary embolism and granulomatosis. A complication of illicit intravenous injection of pentazocine tablets. *Arch Pathol Lab Med* 105:89–93, 1981.

344. Puro HE, Wolf PL, Skirgaudas J, Vazquez J: Experimental production of human "blue velvet" and "red devil" lesions. *JAMA* 197:1100–1102, 1966.

345. Zientara M, Moore S: Fatal talc embolism in a drug addict. *Hum Pathol* 1:324–327, 1970.

346. Overland ES, Nolan AJ, Hopewell PC: Alteration of pulmonary function in intravenous drug misusers. *Am J Med* 68:231–237, 1980.

347. Clague JR, Gray HH, Kay PH: Self injection with mercury. *Br Med J* 299:1567, 1989.

348. Shaffer BA, Schmidt-Nowara WW: Multiple small opacities of metallic density in the lung. *Chest* 96:1179–1181, 1989.

349. Lewis TD, Henry DA: Needle embolus: A unique complication of intravenous drug abuse. *Ann Emerg Med* 14:906–908, 1985.

350. Navarro C, Dickinson PCT, Kondlapoodi P, Hagstrom JWC: Mycotic aneurysms of the pulmonary arteries in intravenous drug addicts: Report of three cases and review of the literature. *Am J Med* 76:1124–1131, 1984.

351. Byers JM, Soin JS, Fisher RS, Hutchins GM: Acute pulmonary alveolitis in narcotics abuse. *Arch Pathol* 99:273–277, 1975.

352. Crouch E, Churg A: Progressive massive fibrosis of the lung secondary to intravenous injection of talc. A pathologic and mineralogic analysis. *Am J Clin Pathol* 80:520–526, 1983.

353. Paré JP, Cote G, Fraser RS: Long-term follow-up of drug abusers with intravenous talcosis. *Am Rev Respir Dis* 139:233–241, 1989.

354. Schmidt RA, Glenny RW, Godwin JD, Hampson NB, Cantino ME, Reichenbach DD: Panlobular emphysema in young intravenous Ritalin abusers. *Am Rev Respir Dis* 143:649–656, 1991.

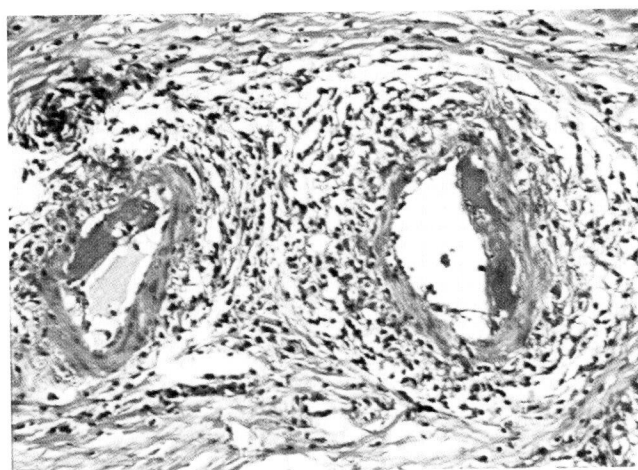

COLOR PLATE 1 Rocky Mountain spotted fever. Skin biopsy at day 10 of the illness, with two arterioles showing perivascular inflammation and mural thrombus.

COLOR PLATE 2 Spores being disturbed in a haystack. (*Reproduced with permission of Dr. A. Gibbs, Penarth Hospital, Wales.*)

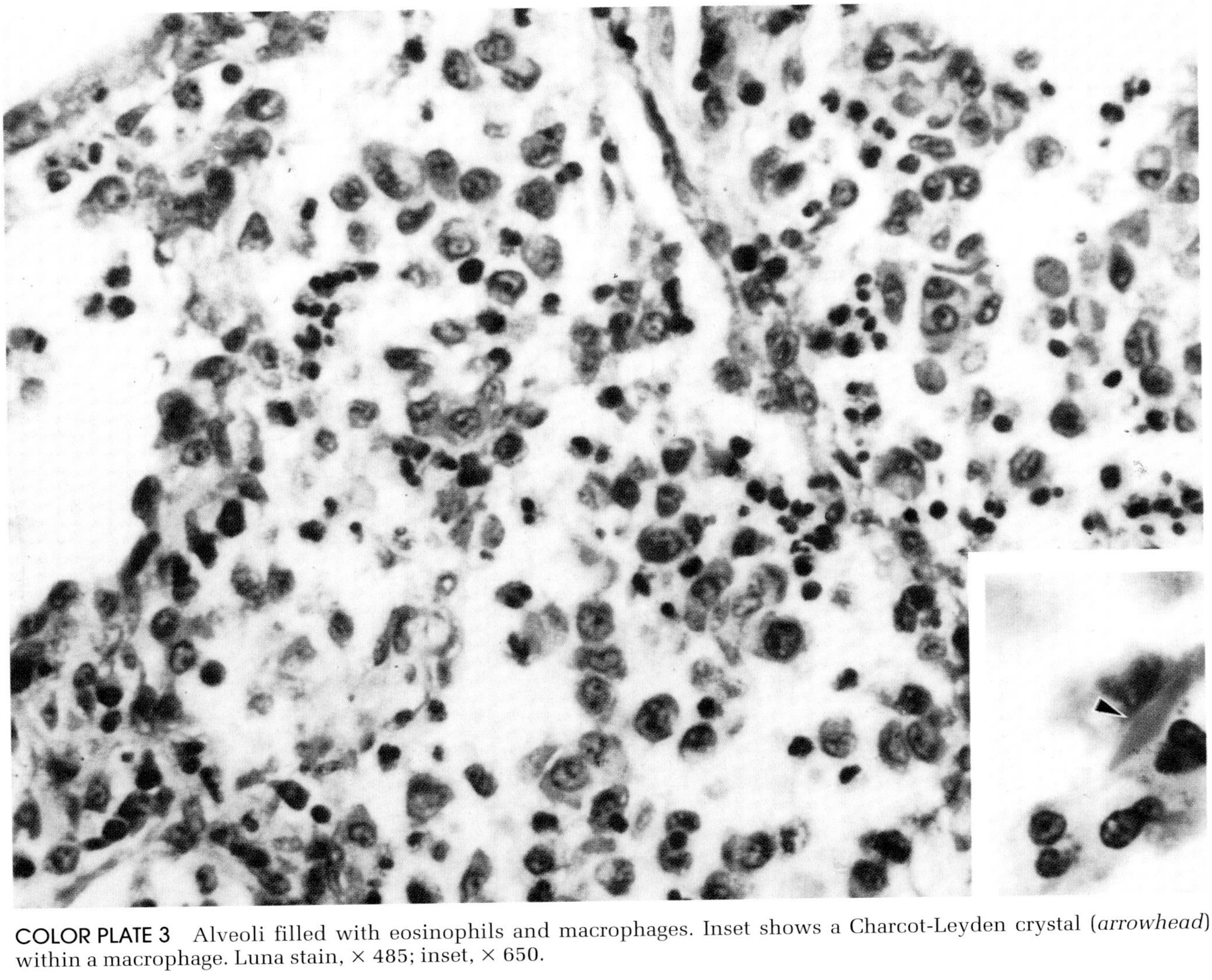

COLOR PLATE 3 Alveoli filled with eosinophils and macrophages. Inset shows a Charcot-Leyden crystal (*arrowhead*) within a macrophage. Luna stain, × 485; inset, × 650.

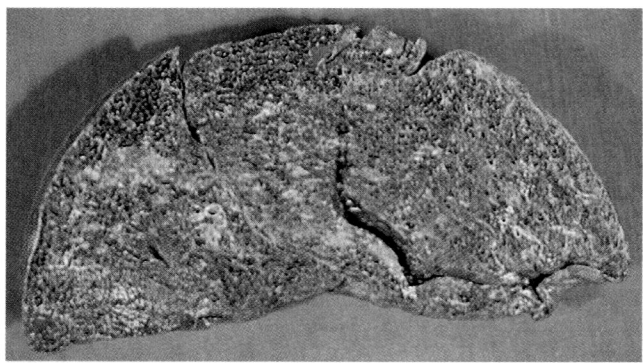

COLOR PLATE 4 The lung from a subject who had been exposed to ochers for many years showing a diffuse reddish brown appearance and multiple stellate gray-brown lesions.

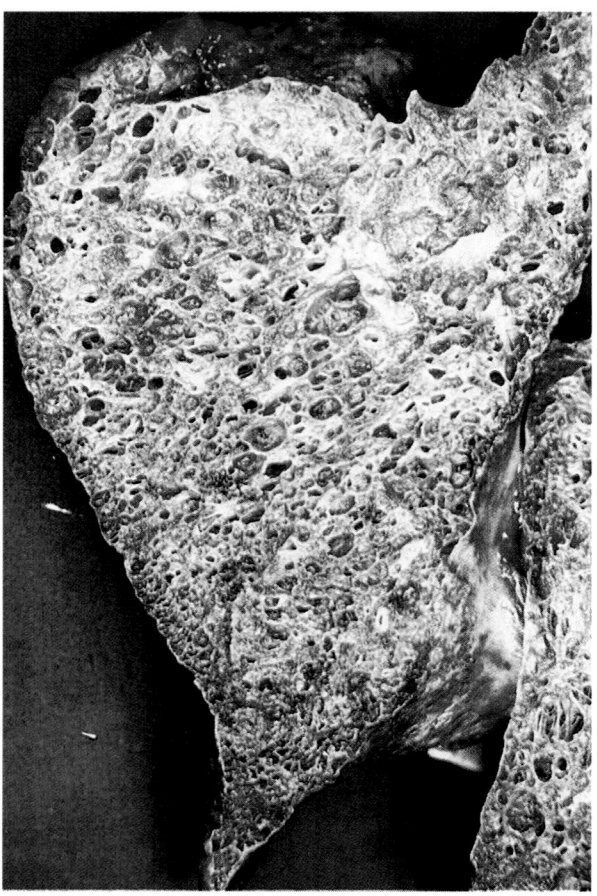

COLOR PLATE 5 LAM (lymphangioleiomyomatosis) macroscopic specimen with honeycomb appearance. (*Reproduced with permission of M. Burke, M.D., Harefield Hospital, London, England.*)

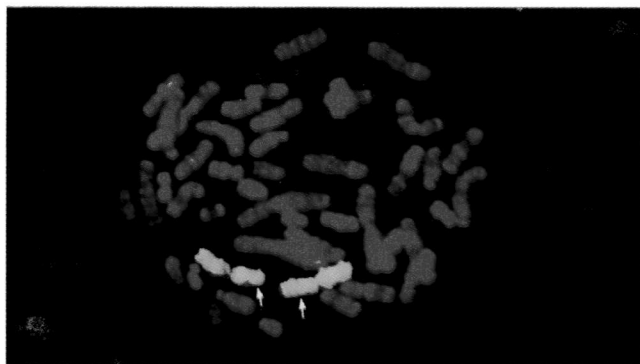

A

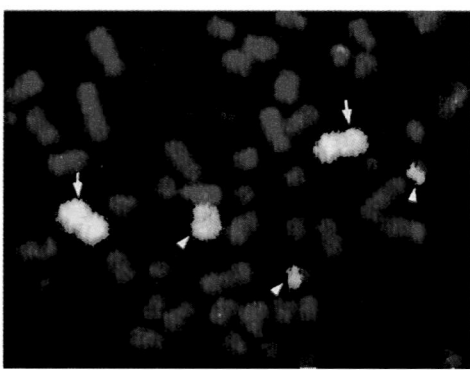

B

COLOR PLATE 6 Fluorescence in situ hybridization using chromosome 3 paints. Fluorescence in situ hybridization with chromosome 3 painting probe to a normal metaphase spread (*A*), showing two normal chromosome 3 homologs (*arrows*) and an abnormal metaphase spread prepared from a lung cancer cell line (COR L 361) (*B*). Two apparently normal chromosome 3 homologs (*arrows*) and three derivative chromosome 3s (*arrowheads*) are visible. The chromosomes are stained with propidium iodide (*red*) and the chromosome 3 paint signal is identified using fluorescein (*green/yellow*).

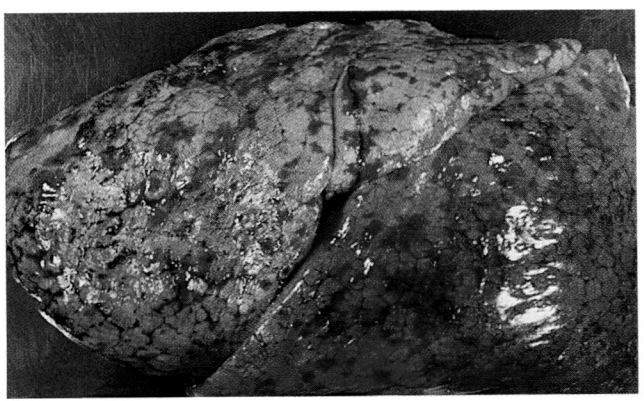

COLOR PLATE 7 Kaposi's sarcoma. The pleural surface shows patchy red, hemorrhagic streaks of tumor.

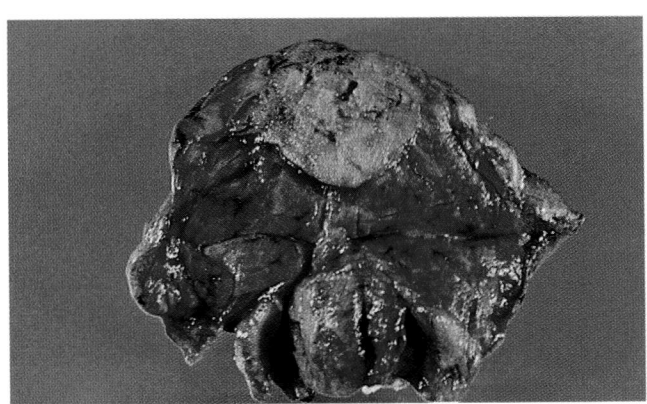

COLOR PLATE 8 Pulmonary blastoma. Cut surface of the lung shows a well-delimited, fleshy, minimally cavitated mass.

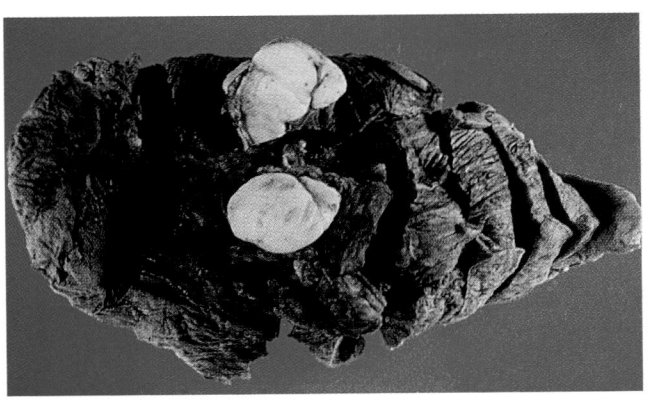

COLOR PLATE 9 Pulmonary carcinosarcoma. The gross specimen shows a fleshy mass protruding from the bronchial lumen.

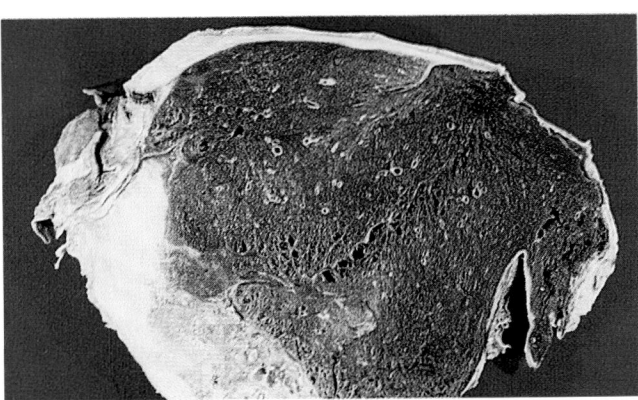

COLOR PLATE 10 Diffuse pleural fibrosis. *(Courtesy of Dr. A. Gibbs, Penarth Hospital, Wales, with permission.)*

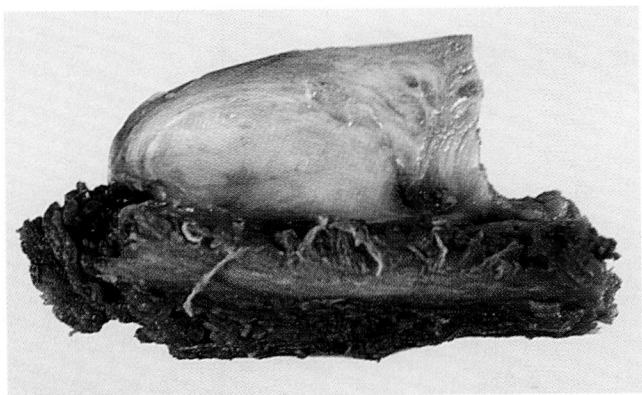

COLOR PLATE 11 Macroscopic view of benign mesothelioma (broad-based).

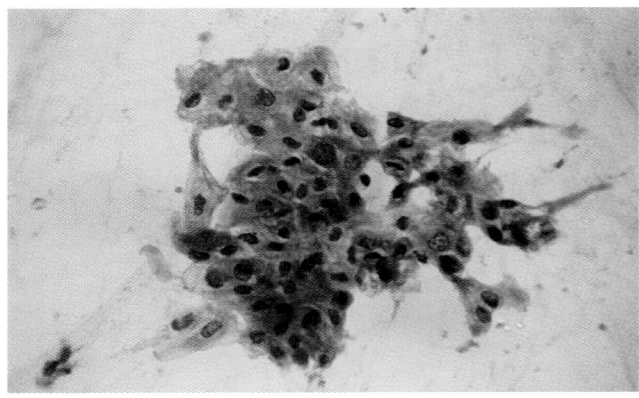

COLOR PLATE 12 Squamous metaplasia of bronchial epithelium. Sputum from a smoker. (Papanicolaou, moderate magnification)

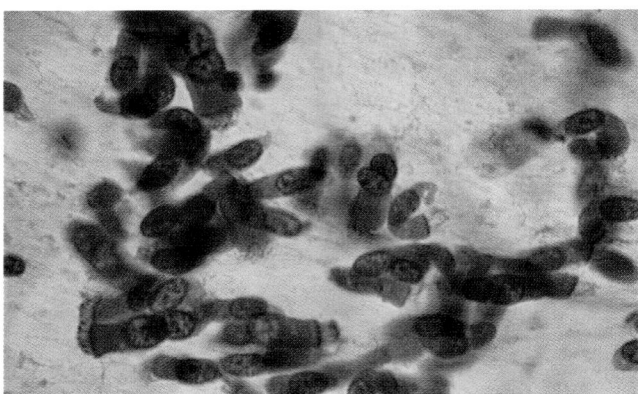

COLOR PLATE 13 Normal bronchial epithelial cells. BAL from a heart/lung transplant recipient. (Papanicolaou, high magnification)

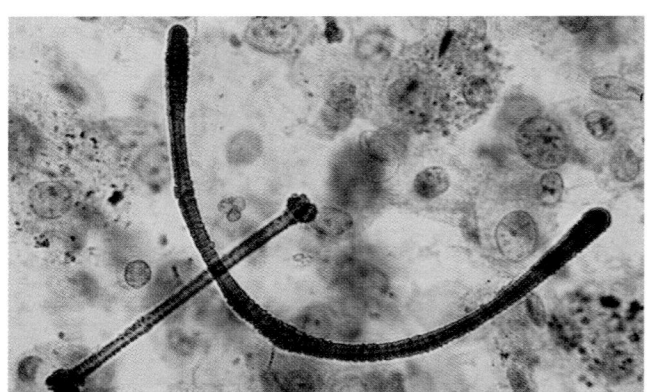

COLOR PLATE 14 Asbestos bodies against a background of alveolar macrophages and a pink-staining squamous cell in BAL fluid. (Papanicolaou, high magnification)

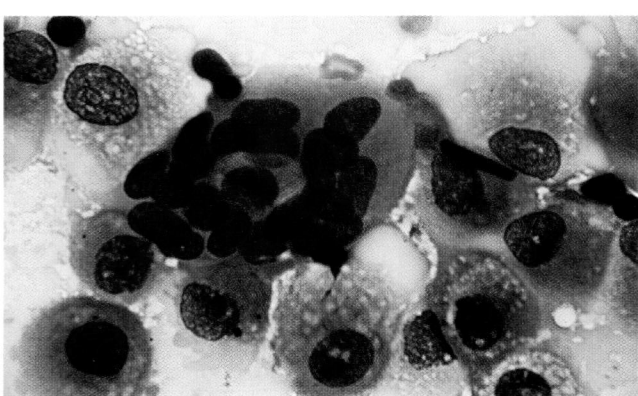

COLOR PLATE 15 Normal mononuclear alveolar macrophages and a large multinucleated cell. BAL from case of quiescent sarcoidosis. (May-Grunwald-Giemsa, high magnification)

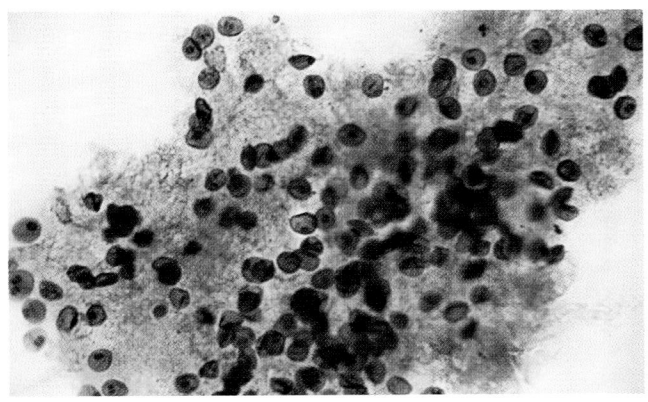

COLOR PLATE 16 Alveolar cast containing cystic forms of
P. carinii in BAL samples. (Grocott, high magnification)

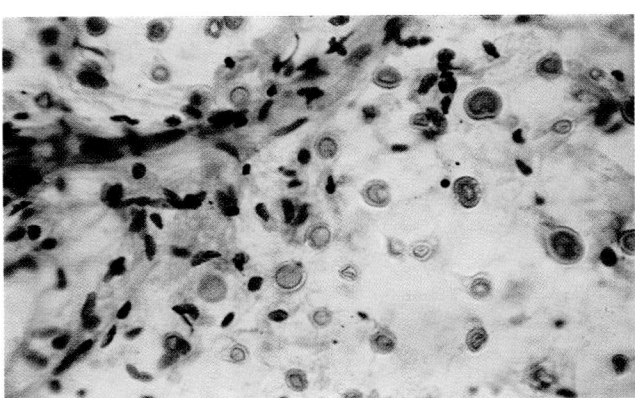

COLOR PLATE 17 Numerous encapsulated yeast forms of
C. neoformans, mixed with streaks of mucus and degenerate cells from FNA of lung. (Papanicolaou, moderate magnification)

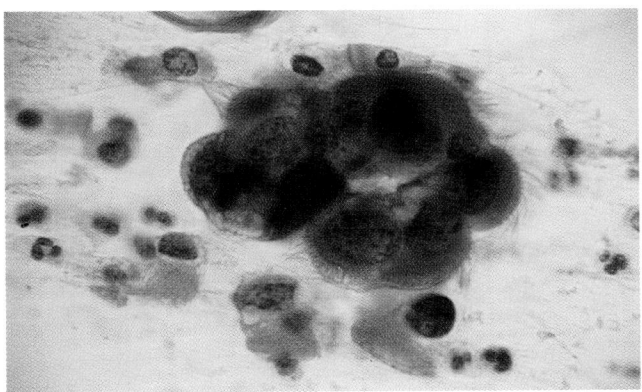

COLOR PLATE 18 Abnormal bronchial epithelial cells with
enlarged nuclei and well-preserved cilia. BAL from case of
respiratory distress following marrow transplantation.
(Papanicolaou, high magnification)

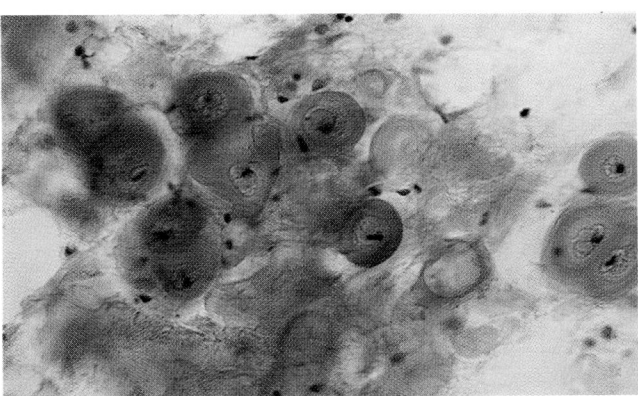

COLOR PLATE 19 Hamartoma. Cartilage cells mixed with
feathery mesenchymal material from FNA of lung.
(Papanicolaou, low magnification)

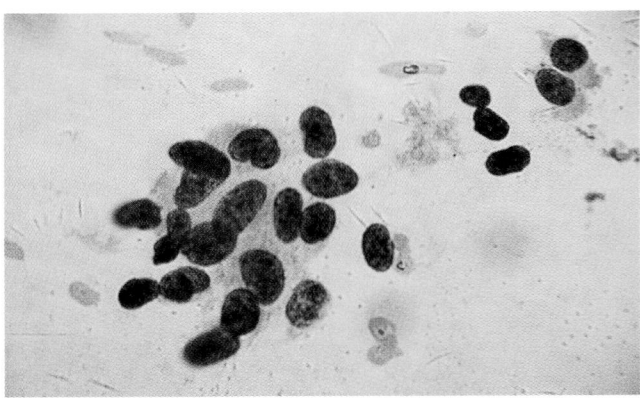

COLOR PLATE 20 Carcinoid tumor. Cluster of small neoplastic cells in bronchial brushing sample. (Papanicolaou, moderate magnification)

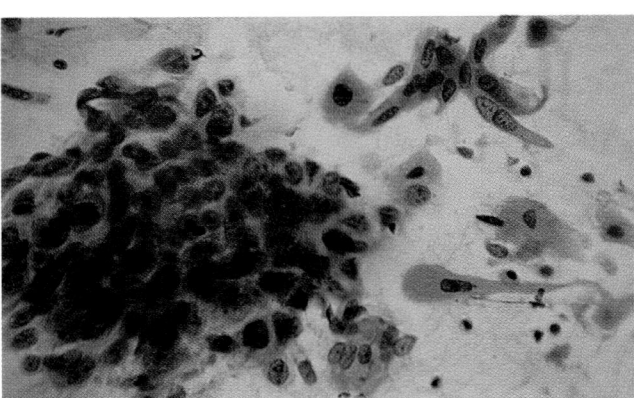

COLOR PLATE 21 Squamous cell carcinoma. Well-differentiated orangeophilic "tadpole" cell next to group of poorly differentiated malignant cells in bronchial brushing samples. (Papanicolaou, low magnification)

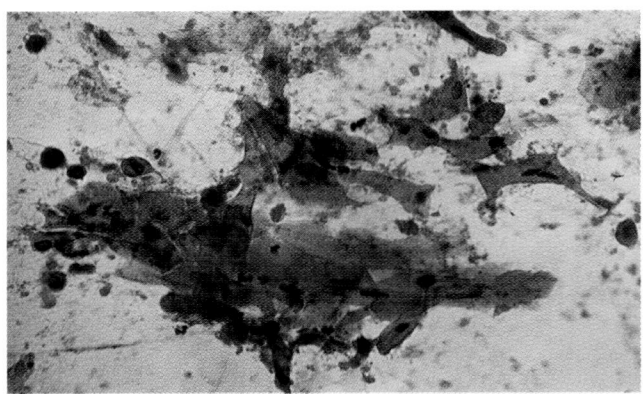

COLOR PLATE 22 Squamous cell carcinoma. Copious, mainly anucleate, orangeophilic debris in sputum specimen. (Papanicolaou, low magnification)

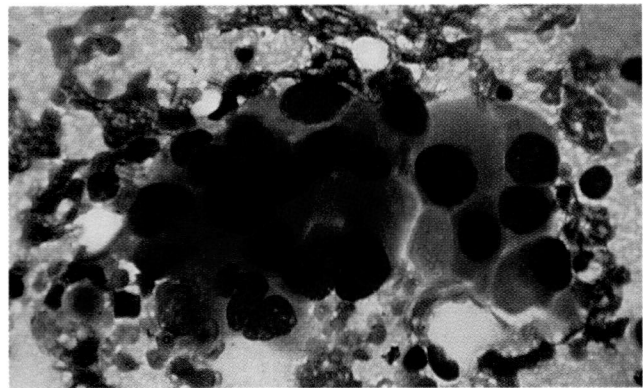

COLOR PLATE 23 Squamous cell carcinoma. Malignant cells with dense flat cytoplasm and sharply articular borders from FNA sample. (May-Grunwald-Giemsa, moderate magnification)

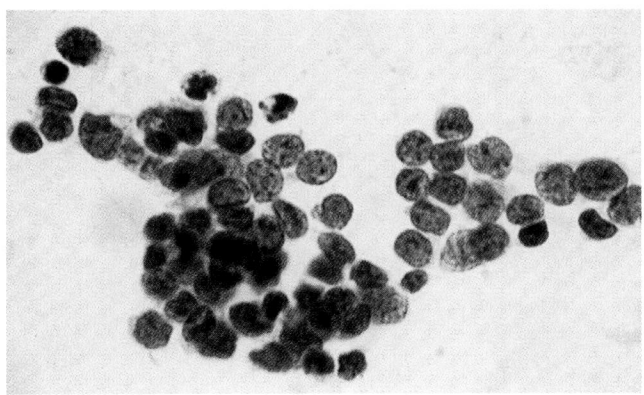

COLOR PLATE 24 Oat cell carcinoma. Cluster of small malignant cells with minimal cytoplasm and granular nuclei in sputum specimens. (Papanicolaou, high magnification)

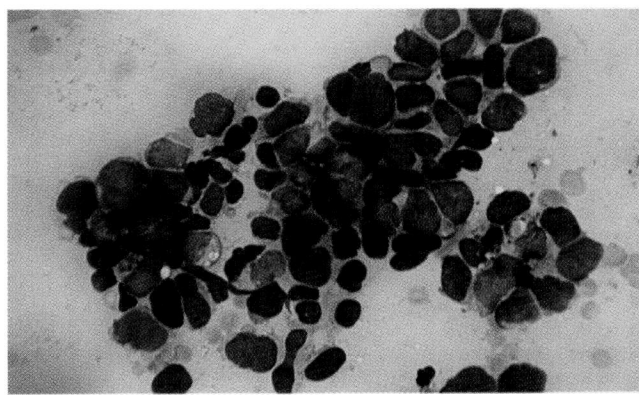

COLOR PLATE 25 Oat cell carcinoma. Deeply hyperchromatic nuclei show evidence of molding from FNA of lung. (May-Grunwald-Giemsa, moderate magnification)

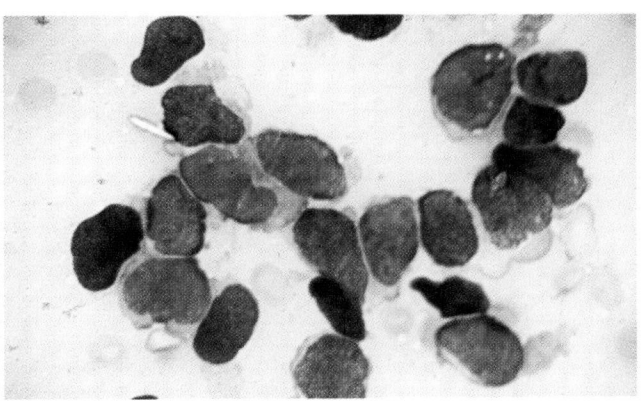

COLOR PLATE 26 Oat cell carcinoma. Small pleomorphic nuclei with finely granular chromatin and inconspicuous nucleoli from FNA of lung. (May-Grunwald-Giemsa, high magnification)

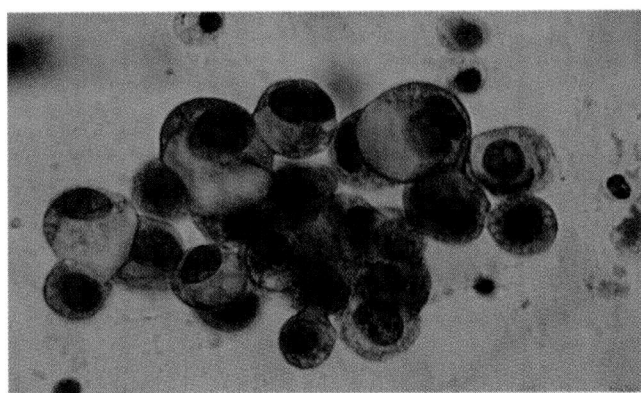

COLOR PLATE 27 Adenocarcinoma. Three-dimensional group of malignant cells with eccentric nuclei and copious vacuolated cytoplasm in sputum specimen. (Papanicolaou, high magnification)

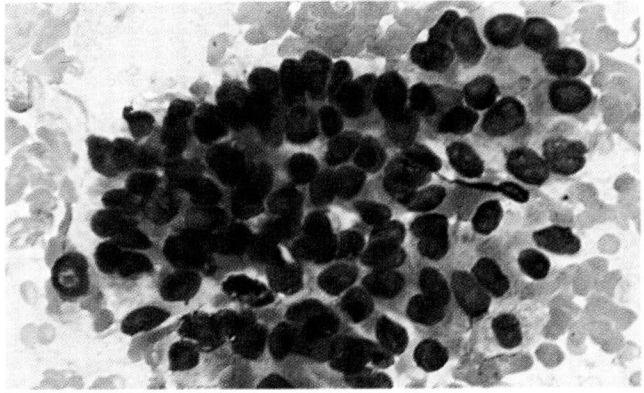

COLOR PLATE 28 Adenocarcinoma. Large cluster of malignant cells with discernible acinar pattern in FNA of lung. (May-Grunwald-Giemsa, low magnification)

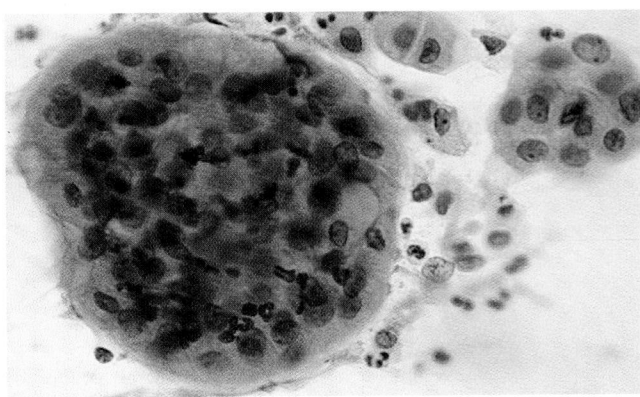

COLOR PLATE 29　Bronchoalveolar carcinoma. Two spherical groups of well-differentiated malignant cells in FNA of lung. (H & E, moderate magnification)

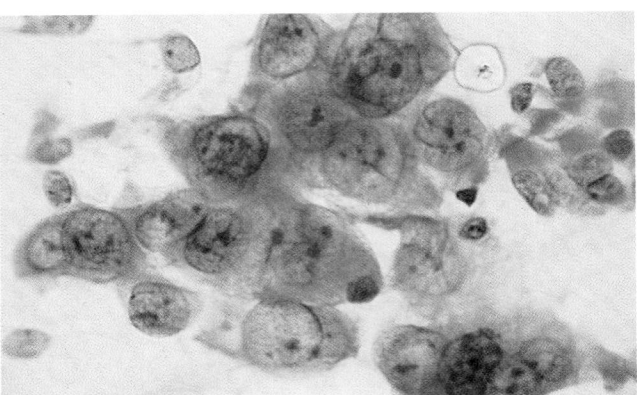

COLOR PLATE 30　Large cell carcinoma. Poorly cohesive large malignant cells show no identifiable pattern of differentiation; from a bronchial brushing. (Papanicolaou, high magnification)

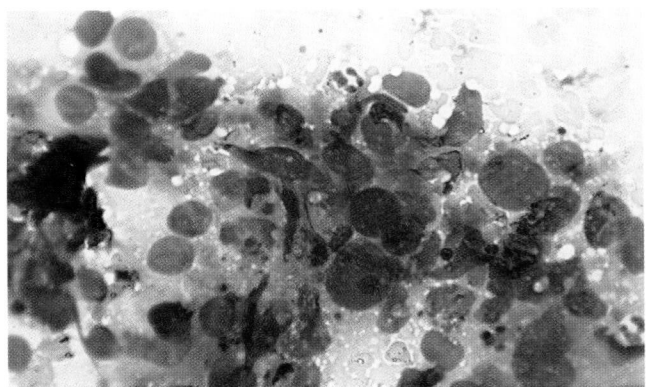

COLOR PLATE 31　Large cell carcinoma. Aggregate of poorly differentiated malignant cells with large nuclei and moderate quantities of cytoplasm from FNA of lung. (May-Grunwald-Giemsa, moderate magnification)

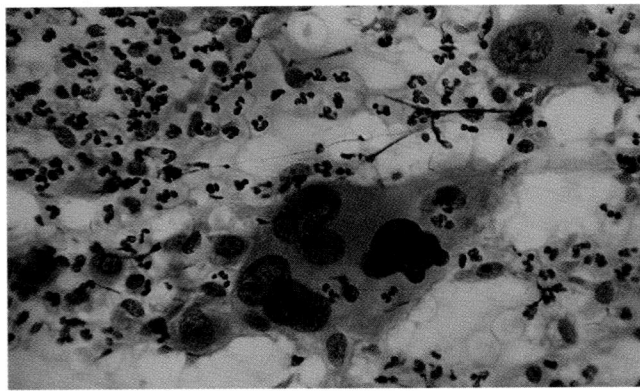

COLOR PLATE 32　Giant cell carcinoma. One of numerous very large multinucleated malignant cells plus smaller mononuclear malignant cells against a background of inflammatory debris from a bronchial brushing. (Papanicolaou, moderate magnification)

Chapter 19

Chronic Bronchitis, Emphysema, and the Pathological Basis of Chronic Obstructive Pulmonary Disease

D. Lamb

CLINICAL DEFINITION OF CHRONIC OBSTRUCTIVE PULMONARY DISEASE

There is no generally agreed clinical definition of chronic obstructive pulmonary disease (COPD). Burrows has suggested a useful description of COPD: "A chronic, slowly progressive airway obstructive disorder resulting from some combination of pulmonary emphysema and irreversible reduction in the calibre of small airways of the lung."[1] This describes the condition but does not help to diagnose COPD in life as neither small airway disease nor emphysema can be recognised physiologically. It does not help the pathologist to diagnose COPD after death as the basis for the reduction in caliber of small airways is still under discussion and the type of emphysema is not specified.

The term *COPD* is applied to patients who have an increased airflow resistance or airway obstruction without evidence of reversibility.[2,3] Asthma characterized by reversible airway obstruction is not included under the heading COPD. However, an asthmatic who develops a fixed airway obstruction associated with long-standing cigarette smoking would be considered as having COPD. Patients with COPD on the other hand may have a reversible component in addition to their fixed component, which may be of clinical relevance. Those patients with diseases such as cystic fibrosis or bronchiolitis obliterans who also develop airway obstruction are not included in the definition.

There have been three hypotheses about the nature of the airway obstruction in COPD:

- The "British Hypothesis" was based on the concept that chronic mucus hypersecretion was associated with recurrent respiratory infections which caused damage to the airways and/or alveoli to produce progressive airway obstruction.[4]
- The "Dutch Hypothesis" proposed that an "asthmatic constitution," consisting of allergy, airway hyper-responsiveness and eosinophilia was an essential prerequisite for the development of generalized airway disease. Smoking was only one of several extrinsic factors which could lead to airway obstruction.[5]

Neither hypothesis included emphysema as a prime abnormality, considering it a late complication of the underlying airway disease.

- The "American Hypothesis." Physicians in North America considered that patients with chronic airway obstruction were suffering from emphysema.

When similar criteria were used for assessing patients in Great Britain and the United States it was realized that there were similar clinical and physiological features in British patients with chronic bronchitis and American patients with emphysema.[6] The term *chronic obstructive lung disease* (COLD) was introduced to recognize this similarity[6] though in more recent years the term *chronic obstructive pulmonary disease* (COPD) has come to be used instead.

The two conditions, chronic bronchitis and emphysema, as they are currently defined, need not necessarily be accompanied by airway obstruction. They are therefore unsuitable terms to identify a population of patients whose fundamental abnormality is fixed airway obstruction.

Most British and North American physicians consider that asthma and asthmatic bronchitis progressing to fixed airway obstruction in the absence of smoking or another insult is uncommon. As it is a fundamentally different entity to smoking-related COPD, there will be no discussion of it in this chapter (see Chap. 22).

CLINICAL EPIDEMIOLOGY AND RISK FACTORS FOR COPD[7-11]

The clinical epidemiology of COPD has been confused by the varied approaches enshrined in the three hypotheses. This has been confounded by the varied terms under which the causes of death have been reported,[7,10] by the problems of studying epidemiologically a condition without either a clear definition or an accepted pathological basis.

In the United States, 3 percent of all male deaths are attributed to COPD and as many again have COPD as a contributing cause. In the United Kingdom some 6 percent of male deaths are attributed to COPD and in both countries about half as many deaths in women are attributed to this cause. These figures may be an underestimate of the size of the problem as there are a large number of people whose cause of death is lung cancer or coronary artery disease, both of which are smoking-related and may be accompanied by COPD. COPD is not necessarily recorded on the death certificate. Though the death rate for men appears to be static, that for women appears to be increasing.

There are marked variations in death rates in different countries, with the highest rates attributable to COPD being in Scotland, England, Wales, and Eastern European countries. The lowest death rates are in Southern European countries and Japan, with the United States lying in the middle. The difference between the highest and the lowest incidence is approximately four-fold.

There are geographic variants in death rates within the United Kingdom; the age-adjusted death rates of chronic respiratory disease varies by a factor of five-fold for men and ten-fold for women in different geographic areas. The general trend is for higher mortality rates in towns and major conurbations with the high rates being in the North West of England, Wales, and Scotland.[10]

Tobacco smoking is the most important etiologic factor in the production of COPD.[3,4,7,8,10,12] The relationship is dose-related, the decline in respiratory function is much reduced on stopping smoking.[4,14]

The prime abnormality in COPD is fixed airway obstruction, and a great deal of work has been carried out investigating physiologic measurements of the airway obstruction, in particular the FEV_1, both in cross-sectional and longitudinal studies. The most important work in this field is that of Fletcher and

Peto,[4] who showed clearly that whereas non-smokers had a steady decline in their FEV_1 values of approximately 30 mL per year, in patients who developed COPD this decline was accelerated to between 50 and 100 mL per year. The accelerated decline in those developing COPD was relatively constant and did not have the step-wise progression expected if the decline was a consequence of infectious episodes. They also showed that on stopping smoking the accelerated decline appeared to cease, reverting to a normal rate of decline. These findings must be explained by any investigation into the pathologic basis of the fixed airway obstruction. These authors also showed that the symptoms of cough and sputum, a defining clinical feature of chronic bronchitis, were not of statistical significance in relationship to the accelerated decline in FEV_1 or mortality from COPD.

Cigarette smokers show wide variation in the susceptibility to develop progressive airflow obstruction. Some smokers show no accelerated decline. Though there is a dose-related effect in that those who smoke more have a higher mortality from COPD, there is a two-fold increase between smokers of less than 15 and smokers of more than 25 cigarettes a day.[14] There is still a wide variation in the rate of decline of FEV_1 among smokers of similar amounts of tobacco. Though other risk factors are identified, these explain only a small part of the varying susceptibility to cigarette smoke.[11]

The other risk factors include:

- Occupation, particularly dusty occupations. The main quantitative data relate to coal mining, where quantitative dust exposures as well as smoking histories are available. The risk of a miner dying of chronic bronchitis and emphysema increases in proportion to his dust exposure.[16] The risk of a miner developing emphysema is proportional to his dust exposure during life. This holds true for both smokers and nonsmokers.[17] During life the FEV_1 declines in relation to a miner's dust exposure.[18,19]

- There is little evidence that chest infections initiate the progressive decline in airway function which is one basis of COPD.[4] Acute bronchopulmonary infections which cause a decline in short-term lung function and, may well precipitate acute-on-chronic respiratory failure. They are often responsible for death in advanced disease.

- Alpha-1-antitrypsin deficiency.[20] The relationship of alpha-1-antitrypsin deficiency was first reported in 1963 by Laurell and Ericksen.[21,22] Heterogeneity in the genes associated with alpha-

1-antitrypsin deficiency is associated with varying plasma concentrations of alpha-1-antitrypsin. When this is below about 1 g/liter there is a significantly increased risk factor for the development of emphysema and associated COPD. The risk for emphysema is greatly magnified by concomitant cigarette smoking (see below).

As not all cigarette smokers develop COPD, individual susceptibility factors, as yet unidentified, probably determine the risk from smoking.[23] The additional risk factors mentioned above do not solve this problem. The major unanswered research question in the epidemiology of COPD is the determination of the factors responsible for the individual's susceptibility to cigarette smoke.

CLINICAL FEATURES OF COPD

The clinical aspects of COPD have been extensively reviewed.[24–26] The symptoms experienced by patients developing COPD usually have an insidious onset. The development of a "smoker's cough" in the morning with gradually worsening exertional dyspnoea, particularly in damp, foggy winter weather may be the first features. There is often a history of an increasing frequency of winter chest infections with *H. influenzae* and *S. pneumoniae*. As the disease progresses there is increasing breathlessness, eventually on minimal exertion. This may become severe, with the patient being breathless at rest. The patient is severely disabled. In the latter stages of the disease peripheral edema may become apparent, secondary to the development of cor pulmonale.

The physical signs vary widely and to some extent depend on the severity of the disease. There is no good correlation between clinical signs and any pathologic or clinical measure of COPD severity. In general patients may show weight loss, cough, breathlessness, or hyperinflation of the chest. Central cyanosis may be present as well as peripheral edema. These signs tend to be more marked during infective exacerbations with associated hypoxemia.

The chest radiograph may show hyperinflation with an enlarged heart, if there is cor pulmonale. The hila may be prominent due to pulmonary hypertension and emphysematous bullae may be observed, particularly in the upper zones. Respiratory function tests reveal an obstructive pattern with a low FEV_1, FVC, and FEV_1/FVC ratio and a high TLC and increased RV/TLC ratio. The TLCO is also usually reduced and variable patterns of abnormal arterial blood gases may be observed (see below). Many

patients also have polycythaemia with significantly elevated hemoglobin concentrations and hematocrits. These are secondary to the increased levels of renal erythropoietin production due to hypoxemia. Patients with cor pulmonale will show evidence of right ventricular hypertrophy on ECG and echocardiography.

Patients with COPD may be divided based on their clinical features and arterial blood gas analyses into two groups.[6,27,28] The "pink puffing patient" is characteristically thin, breathless, with a markedly hy-perinflated chest but does not show peripheral cyanosis. Arterial blood gas analysis shows a low normal P_{O_2} and P_{CO_2} (type I respiratory failure). The second type of patient, or "blue bloater," may not be particularly breathless at rest but has severe central cyanosis and signs of cor pulmonale. The arterial P_{O_2} is low and the P_{CO_2} elevated (type II respiratory failure) although acidosis is only a feature during exacerbations. These patients also appear to be particularly susceptible to sleep apnea. Some patients have features of "pink puffers" and "blue bloaters."[6,27–29]

The physiologic basis of these two patterns of clinical disease is believed to relate to the patient's response to elevated P_{CO_2} levels. In the case of the "pink puffer" the normal P_{CO_2} dependent ventilatory drive is maintained. Therefore the patient maintains normal or low levels at the cost of persistent hyperventilation. "Blue bloaters" appear to lose the physiologic increased ventilatory response to a rising P_{CO_2} and depend on the weaker physiologic stimulus of hypoxia to drive respiration. These patients are often less breathless but remain in type II respiratory failure with a low P_{O_2} and elevated P_{CO_2}. This poor physiologic ventilatory drive may also be the cause of sleep apnea in these patients. The reason dictating which patients develop these two patterns of physiological response to COPD remains unclear.

The prognosis for patients with COPD remains poor, particularly those with type II respiratory failure and cor pulmonale, where the 5 year survival may be less than 30 percent. Some increased survival may be obtained by using continuous long-term, controlled oxygen therapy.

PATHOLOGIC FEATURES OF COPD

COPD is characterized by fixed and irreversible airflow limitation. It is characteristic that in COPD the decline in FEV_1 develops as an acceleration of the normal age-related decline in FEV_1.

Despite much research, the pathologic basis of the physiologic abnormalities in COPD is incompletely understood. There are three clinical or clinicopathologic conditions which are discussed in the context of pathologic basis of COPD. These are chronic bronchitis, emphysema, and small airways disease. In this chapter these three conditions will be dealt with separately.

The fixed nature of the airway obstruction in COPD cannot be overemphasized. If the airway obstruction is truly fixed then it is beyond pharmacologic therapy.

CHRONIC BRONCHITIS

Chronic bronchitis was defined by the Medical Research Council (MRC) in 1965 as "chronic or recurrent increase in the volume of bronchial secretions sufficient to cause expectoration (on most days for a minimum of three months of the year for not less than two successive years) which cannot be attributed to other cardiac or pulmonary disease."[31] This definition is functional and has been of major value in elucidating the epidemiology of mucus hypersecretion and its relationship to atmospheric pollution and smoking.[32]

The respiratory tract mucus is produced by the submucosal glands of the larger airways and the goblet cells of the surface epithelium. In nonsmokers the goblet cells do not extend peripherally as far as terminal bronchioles.[33]

In chronic bronchitis there is an increase in mass of the submucosal glands and this is closely related to the smoking history. The Reid Index is the ratio of the distance between the epithelial basement membrane to the cartilage and the thickness of the gland layer. It was an attempt to produce a simple method for assessing the gland mass, independent of variation in bronchial dimensions.[34] Though the Reid Index is increased in patients with COPD, the increase of submucosal glands is no greater than in smokers without COPD.

The increase in gland mass is associated with an increase in the components of individual glands. There is no increase in the number of glands.

Goblet cells are found in increased numbers in the peripheral airways of smokers and they may extend to the smallest peripheral bronchioles.[33] In the more proximal airways, the presence of metaplastic or dysplastic changes in the surface epithelium, induced by cigarette smoking, may obliterate or mask such changes.

The volume of the submucosal glands is much greater than the volume of goblet cells. It is believed that the glands contribute the bulk of the mucus secretions which form sputum. There is a relationship between the amount of submucosal gland tissue and the volume of sputum production.[34]

Experiments in which animals are exposed to irritant gases or tobacco smoke, in the absence of infection, have shown that these irritants can cause increase in the mass of submucosal glands and in the number and extent of surface goblet cells.[35–37]

Extensive epidemiologic studies of the symptomatology of chronic bronchitis have shown no relationship between rate of decline in the FEV_1 or mortality and the symptoms of chronic bronchitis.[4,15,30,38] For this reason the histologic features of the mucus hypersecretion must be considered noncontributors to the pathologic basis for the fixed airway obstruction in COPD.

EMPHYSEMA

Emphysema is not a single condition but can be subdivided into types or patterns depending on the distribution of the abnormal air spaces within the acinar unit. The different patterns of emphysema may have dissimilar distributions within the lung and different functional effects. The different patterns of emphysema may be caused by various pathogenetic mechanisms.

Definition

Emphysema was first defined at the CIBA Guest Symposium[39] as "a condition of the lung characterized by increase beyond the normal in size of the air spaces distal to the terminal bronchiole either from dilatation or from destruction of their walls."

This is an anatomical definition and defines emphysema in terms of the acinar unit. This definition did not distinguish between the enlargement of air spaces produced by overinflation, for example in an asthma attack or in the compensatory overinflation of a residual lung after pneumonectomy, and the destruction of the lung architecture which occurs in smoking-related emphysema. Since then the definition has been modified on two occasions to emphasize the destructive process which is part of our concept of smoking-related emphysema. In 1962 the American Thoracic Society[40] suggested that emphysema should be defined as: "A condition of the lung characterised by abnormal, permanent enlargement of air spaces distal to the terminal bronchiole accompanied by destruction of their walls." This was further modified by Snider et al. in 1985[41] as: "A condition of the lung characterised by abnormal, permanent enlargement of the air spaces distal to the terminal bronchiole, accompanied by destruction of their walls and without obvious fibrosis."

These later definitions exclude many conditions which were originally accepted as, and discussed under, the heading of emphysema.[42] In particular many pediatric conditions in which there has been no alveolar destruction but rather pre- or postnatal abnormalities of development, as in McLeod syndrome, are excluded by the definition. The definition includes the concept of "normal air space size" and "without obvious fibrosis." Both concepts assume a knowledge of normality of the lung which has not hitherto been defined. It is clear that there may be practical problems in the literal applications of the definition of emphysema in practice.

Diagnosis and Classification of Emphysema

At present the diagnosis of emphysema based on the current definitions relies on the interpretation of structural changes in lung tissue. It is therefore the responsibility of the pathologist. This limits the study of emphysema to those situations where sufficient lung tissue is available. This in practice has meant the availability of surgical resection specimens, postmortem material, or more recently lungs removed prior to lung transplantation. Though there are broad correlates between radiological studies, including CT functional correlates, and the severity of emphysema, none of these clinical investigations is at present capable of identifying and quantifying presence of emphysema or classifying it into its various patterns.

There are three components to the definition of emphysema, (1) the size of the air spaces, (2) evidence for a destructive process, and (3) an assessment of the amount of fibrosis. In practice pathologists rely entirely on assessment of air space size for the diagnosis of the presence of emphysema. The subclassification into types (see below) is based entirely on the recognition of patterns of distribution of abnormally large air spaces within the lung.

In the following sections on the diagnosis, classification, and quantitation of emphysema it is assumed that the lung has been properly fixed by distension with formol saline before being sliced. Examination of the fresh, unfixed lung by slicing in the postmortem room or of fixed but undistended lung tissue prohibits any sensible comment on the presence, absence, type, or severity of emphysema.

MACROSCOPIC DIAGNOSIS OF EMPHYSEMA

When a fixed distended lung is sliced one cannot recognize individual alveolar spaces with the naked eye or with a hand lens, though it may be possible to identify alveolar ducts (Fig. 19-2A). Alveolar spaces are about a quarter of a mm in diameter and when the lung is sliced surface tension folds the alveolar walls together leaving a finely granular surface. One can improve the ability to recognize air spaces by immersing the lung slice in clean water. Then using tangential illumination and a hand lens, normal air spaces are just visible. Visibility of the air spaces can be improved by increasing the contrast of the lung tissue by impregnation with barium sulphate.[43] Whichever technique is used, it is not possible to recognize minor variations in air space size.

By the time air spaces can clearly be seen by the naked eye they are usually about 1 mm in diameter. Such easily visualized air spaces indicate the presence of emphysema. This simple approach avoids the problem of defining accurately normal air space size and the necessity to measure the minor changes. The enlarged air spaces present a limited number of patterns of lung involvement.

MACROSCOPIC CLASSIFICATION OF EMPHYSEMA

The macroscopic classification of emphysema is based on the distribution of the abnormal air spaces within the acinar unit. The following types are recognized (Fig. 19-1): *centriacinar* (preacinar or centrilobular), *panacinar* (panlobular), *paraseptal* (distal acinar), and *scar emphysema*.[39,42,44]

The definition of emphysema refers to airspaces distal to the terminal bronchiole, i.e., in the acinar unit. It seems therefore preferable to describe the macroscopic patterns of emphysema in terms of the acinus. The terms *centriacinar* and *panacinar* are synonymous with the terms *centrilobular* and *panlobular* emphysema used by some authors.

This classification originally agreed on at the CIBA Symposium in 1958 has stood the test of time and is still in current use (Fig. 19-2A–F).[39]

Panacinar Emphysema (PAE)

The abnormally large airspaces are found evenly distributed across the acinar unit. Adjacent acinar units are usually involved to a similar degree giving a confluent appearance to the cut surface of the lung, with extensive areas being involved (Fig. 19-2D and 2E).

Centriacinar or *Proximal Acinar* Emphysema (CAE)

Where the abnormal airspaces are found initially to involve respiratory bronchioles. In more severe cases

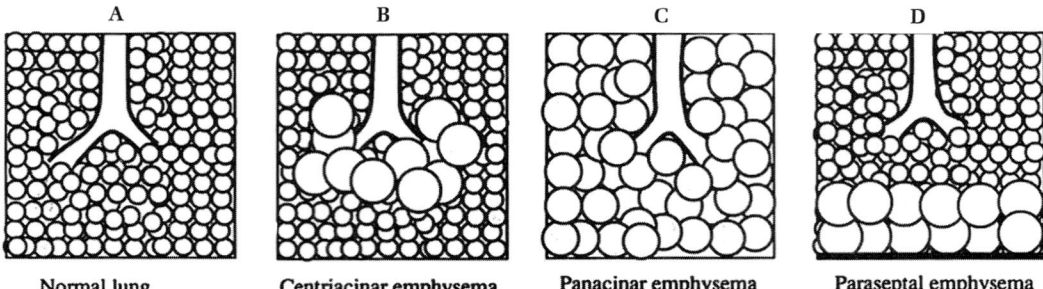

A · · · · · · B · · · · · · C · · · · · · D

Normal lung   Centriacinar emphysema   Panacinar emphysema   Paraseptal emphysema

FIGURE 19-1

A diagrammatic representation of the distribution of the abnormal airspaces within the acinar unit in the three major types of emphysema.   *A.* This represents the acinar unit from a normal lung, though illustrated as a clearly defined area for the purposes of this diagram it must be remembered that in the lung adjacent acinar units intercommunicate and are not necessarily demarcated by septa.   *B.* This shows the focal enlargement of the air-spaces around the respiratory bronchioles in centriacinar emphysema. *C.* This shows the confluent even involvement of the acinar unit in panacinar emphysema, and *D.* This shows the peripherally distributed enlarged airspaces where that portion of the acinar unit butts up against a fixed structure such as the pleura in paraseptal emphysema.

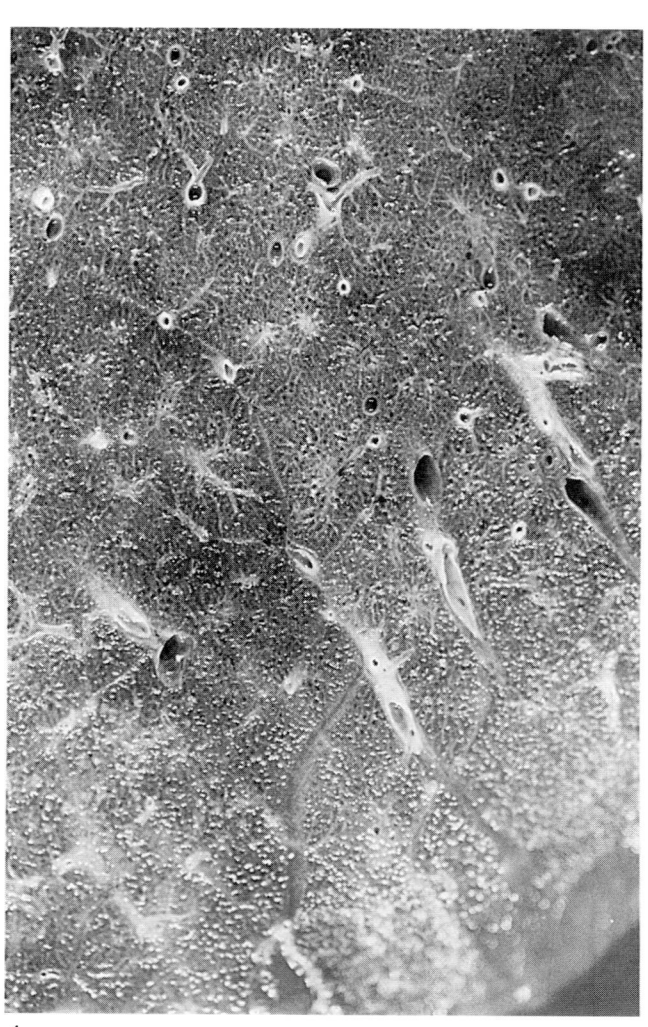

A

B

FIGURE 19-2

Examples of the different patterns of emphysema.   *A.* A normal non-smoking lung, showing the normal textured surface when a fixed lung slice is viewed in air (×1.32).   *B.* Widespread severe centriacinar emphysema showing the normal pattern of lung in between the focal lesions (×0.88).

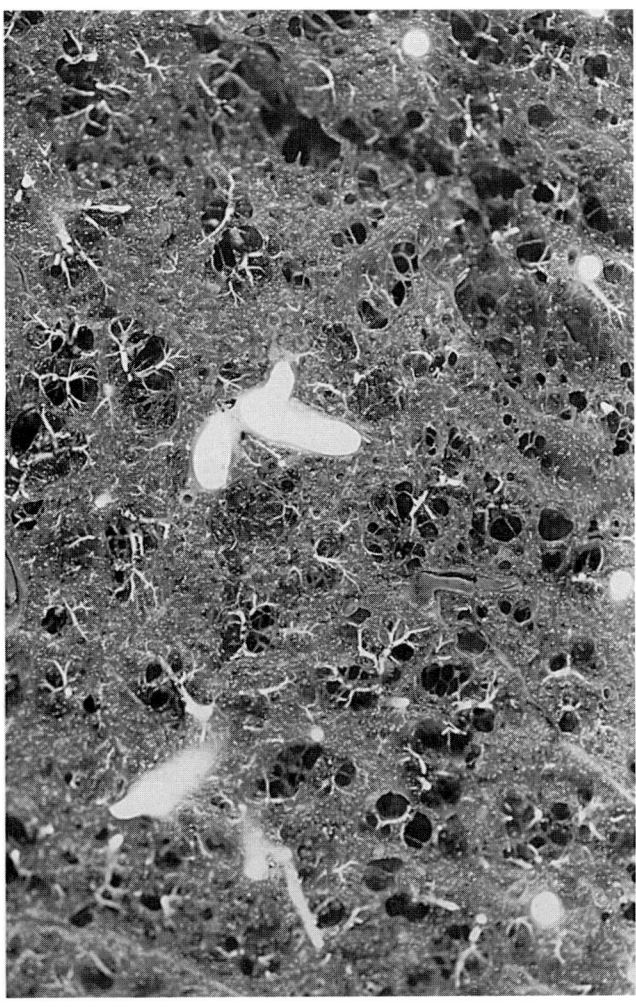

C

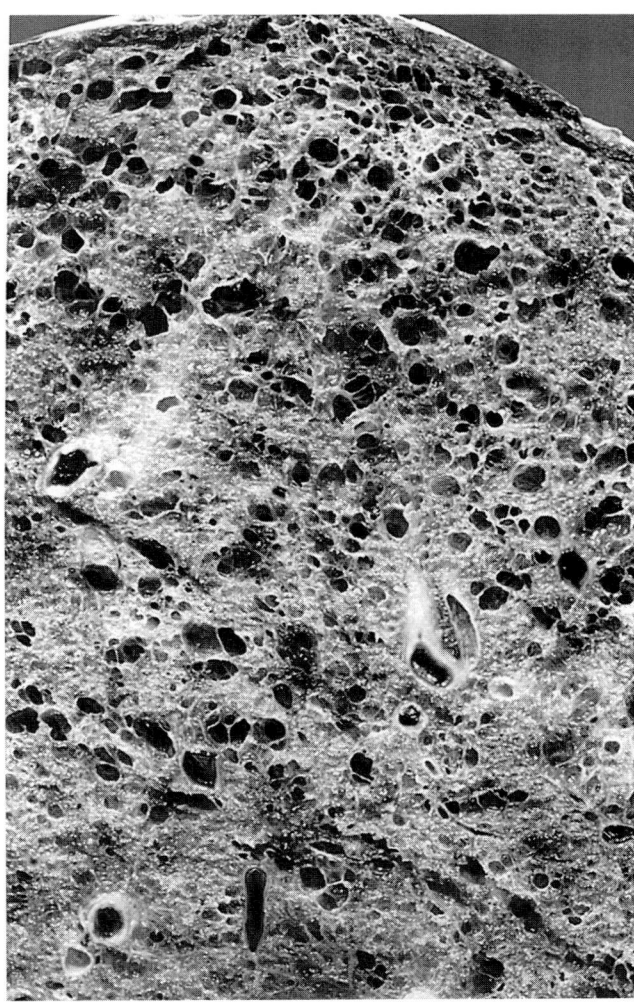

D

FIGURE 19-2 (*Continued*)
C. This specimen has had the arteries injected with a barium gelatine mixture to outline the vascular supply. The centriacinar lesions are seen to have a central vessel with major branches crossing the emphysematous spaces. The strands of tissue within the lesions are mainly vascular in nature (×1.32).   *D*. Mild

panacinar emphysema, the airspaces being 1 to 2 mm in size on average, though there are some larger airspaces there is no centriacinar accentuation. Note the atheromatous pulmonary vessels, more prominent than their adjacent airways. This patient has severe COPD and right ventricular hypertrophy (×0.88).

virtually the whole acinar unit may be affected. In centriacinar emphysema the focal nature of the lesions stands out against often apparently normal lung and quite small lesions can be identified (Fig. 19–2*B* and *C*). However, when centriacinar and panacinar emphysema co-exist and the panacinar emphysema is of some severity, it may be very difficult to identify the presence of associated centriacinar emphysema.

Periacinar or *Paraseptal* Emphysema
The abnormal air spaces run along the edge of the acinar unit. However, it is only diagnosed where it abuts against a fixed structure such as the pleura, a vessel or a septum (Fig. 19-2*F*).

Scar or *Irregular* Emphysema
The emphysematous spaces are found around the margins of a scar. As the scar itself may not be related to the anatomy of the acinar unit, this type of emphysema is not classified in relationship to the acinus. There is no clinical significance to scar emphysema.

Panacinar and centriacinar emphysema have differing distributions within the lung. Panacinar, involving confluent areas of lung tissue, may be found in the upper or lower lobe. Particularly in alpha-1 protease deficiency it is characteristically maximal at the base. Centriacinar emphysema has a clear preference for the upper zones of both the upper and lower lobes.[44] The only time in which centriacinar emphysema is seen to confluently involve acinar

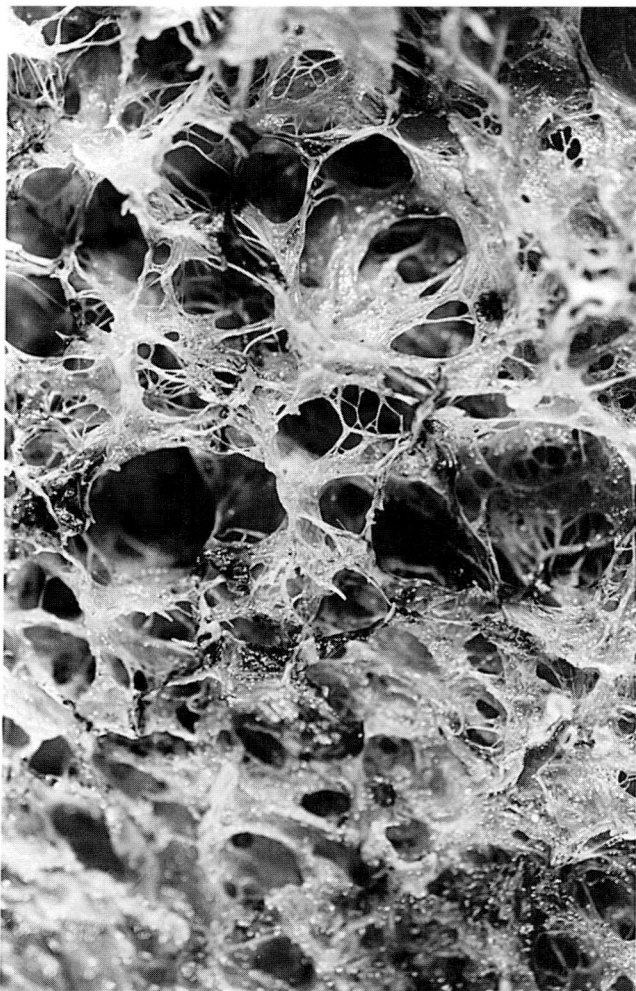

E

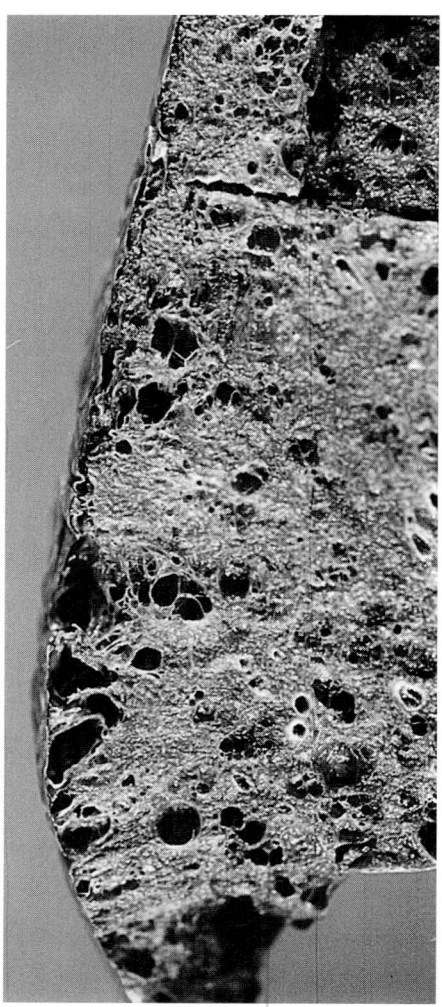

F

FIGURE 19-2 (*Continued*)
E. Severe panacinar emphysema with virtually complete destruction of the airspace walls. The residual structures are septae, vessels and airways associated with a degree of scarring around them. It is easy to see that in severe emphysema the airspaces continuously interconnect (×0.88). *F.* Paraseptal emphysema running along the posterior aspect of the lower lobe immediately beneath the pleura.

units from apex to base is in end-stage COPD and in coal-workers pneumoconiosis. Paraseptal emphysema has no particular distribution within the lung but is probably more common in a subpleural position than deeper in the lung. The differing types of emphysema are not exclusive and may occur alone in a lung or together. A major problem for the pathologist is the identification of the different types and grading their severity in a lung when more than one pattern of emphysema is present.

Although the classification of emphysema into types based on the macroscopic appearance is simple to understand and the individual types easy to recognize under ideal conditions of fixation there can be problems.

The lesions of centriacinar emphysema are easy to identify if they appear in the absence of panacinar emphysema. The focal areas of enlarged airspaces are clearly recognizable against the built-in control of the adjacent normal tissue. Due to the paucity of septae in the human lung it is not possible to delineate an acinar or lobular architecture just by looking at the cut surface of the lung. Some pathologists find it difficult to accept that the focal lesions of CAE are genuinely centriacinar in distribution. However, it is reasonable to assume that focal areas of emphysema, each of a few millimeters in diameter and separated by 5 to 10 mm of normal lung, are centriacinar in nature. Histological examination of small lesions can usually identify the relationship of the emphysematous spaces to the region of the respiratory bronchioles. Centriacinar lesions may be markedly pigmented, anthracotic pigment having a proximal acinar distribution. However it is possible to mistake pigmented macules having a centriacinar distribution for early centriacinar emphysema. Anthracotic pigment accumulates in patients who have been exposed to a dusty atmosphere, either through occupation or a life spent in an industrial and polluted geographical area. Cigarette smoking appears to increase dust retention but the anthracotic pigment retained in the lung is not derived from the cigarette smoke. Heavy smokers from rural areas may have no pigmentation and small centriacinar lesions in these patients may be particularly difficult to delineate against the adjacent normal lung tissue.

The major problem with the recognition of panacinar emphysema is identifying mild or early panacinar emphysema. When a fixed and inflated lung is sliced in the presence of mild panacinar emphysema the airspace walls may collapse, rendering the individual airspaces invisible. However there is a general impression that the surface of the lung slice is "falling in" and the margins of the slice may show rolling of the pleura. If such a lung slice is examined under water or by the barium impregnation technique the mild panacinar emphysema can usually be identified. Lungs from elderly patients have larger airspaces and the alveolar ducts may be clearly delineated. Whether this change is emphysema or a non-emphysematous aging change is discussed below.

The presence of centriacinar and panacinar emphysema in the same lung can give rise to problems. Widespread severe panacinar emphysema can obliterate any evidence of preexisting centriacinar emphysema. However in many cases the very severe panacinar emphysema does not involve the whole lung and it is often possible to identify the presence of small areas of lung with centriacinar lesions. This is particularly true in end-stage chronic obstructive airways disease, where centriacinar lesions, when present, usually extend from the apex to the base.

There is an entity often described as "confluent centriacinar emphysema" in which there is severe emphysema, often heavily pigmented, in the central and upper portion of the upper lobe and the apical portion of the lower lobe. This gives the impression by the association of adjacent centriacinar lesions that the centriacinar emphysema has become so severe that it has become confluent and hence, by definition, panacinar. It is the author's experience that this occurs when a lung is the seat of both centriacinar and mild panacinar emphysema. All the pathologist can do is to describe what is present in the lung being examined. If centriacinar lesions are so confluent that no normal or less involved lung can be recognized between them, this must be described as panacinar emphysema.

The linear pattern of paraseptal emphysema running along the fixed structures is usually easy to recognize. However, this too can be obliterated by widespread panacinar emphysema.

The term *scar emphysema* refers to the emphysematous spaces immediately adjacent to an area of scarring. However, an extensively scarred lung may be associated with widespread usually panacinar emphysema elsewhere. This may be due to an exaggeration of a mild panacinar emphysema by the shrinkage of scar tissue and over-distension of the non-scarred area. Alternatively the emphysema may be produced by the disease process which led to the scarring. It is likely that the former is the case.

MACROSCOPIC ASSESSMENT OF THE SEVERITY OF EMPHYSEMA

The macroscopic examination of distended lung tissue allows the presence or absence of emphysema to be identified but it is much more difficult to give a truly quantitative assessment of the severity of that emphysema. There are actually two components which should be taken into account; the extent of emphysema being the proportion of the lung which is involved by the emphysema and the severity of the emphysematous change being the size of the abnormal airspaces. As the severity of emphysema is often not homogeneous within the lung, it is difficult to produce an adequate quantitative expression taking both extent and severity into account.

Macroscopic assessment of emphysema is carried out by observation of the cut surface of a properly inflated and fixed lung or on Gough-Wentworth paper mounted thin sections.[45] These thin sections were much used in earlier research on emphysema. They have the advantages of forming a permanent, dry, easily handled specimen. Several sections from the same portion of lung can be prepared and they are easily stored. Their disadvantages are that there is a finite thickness of lung condensed down to a thin layer on the paper which obliterates minor changes. Non-pigmented lungs are very difficult to see and pigmented areas of lung tissue, especially pigmented scars, are given undue prominence. There is no three-dimensional view of the lesions and it is more difficult to separate centri- and panacinar emphysema, particularly when they occur together. At the CIBA Symposium in 1958 an attempt was made to standardize the assessment of emphysema.[39] Here the severity of the different patterns of emphysema was assessed by the area of involvement of lung for each lobe so the size of the individual emphysematous lesions was not taken into account. Mild emphysema was considered less than 25 percent involvement, moderate 25 to 50 percent, and severe more than 50 percent of involvement.

Heard and Ryder both described grading techniques whereby the lung surface was divided into zones and the amount of emphysema in each zone scored or graded.[46,47] These techniques were not truly quantitative and there were some difficulties in trying to record different patterns of emphysema.

Dunnill described the value of the point counting technique applied to the assessment of extent of macroscopic emphysema. This provided a quantitative assessment of the area involved by emphysema.[48,49] However, as Thurlbeck[44,50] noted, a major disadvantage of this technique was that it measured the extent of emphysema but not necessarily the severity of the disease, as the size of the emphysematous airspaces was ignored. A more fundamental problem was that there is no distinction between the types of macroscopic emphysema. It may be difficult to assess the pattern of emphysema relating to an abnormal area lying under a cross-hair or point on a point counting grid. In fact Dunnill did note that centriacinar lesions when occurring in a widespread manner throughout the lung, in the absence of panacinar emphysema, at a maximum gave a value of 10 to 15 percent of lung involvement.[48] Expression of centriacinar emphysema as a percentage involvement by a point counting technique may not therefore represent the severity of involvement of the lung. An alternative method for assessing the severity of centriacinar emphysema is to count the number of lesions on a lung slice and hence gain an idea of the number of acinar units involved by centriacinar emphysema. This can be combined with an assessment of severity by grouping the count according to the size of the emphysematous lesions.

Thurlbeck et al.[50–52] produced a simple technique for the grading of emphysema by using a series of paper-mounted Gough-Wentworth sections graded from 0 to 100, where 0 was normal lung, 50 an example of moderate emphysema, and 100 the worst case of emphysema encountered in a series of 500 cases reviewed. Standard pictures were produced for this grading system at intervals of 5 between 0 to 50 and at intervals of 10 from 60 to 100.[52]

Though this technique took into account both the area of involvement and the severity of the individual lesions, it did not separate the different patterns of emphysema. It should be stressed that the grades in this method represent "arbitrary intuitive milestones in the spectrum of severity of emphysema".[51] The values do not represent percentages.

They portray a nonlinear assessment of emphysema causing problems with statistical analysis.

Any comment about the presence of emphysema should include information on type, severity, the size of the abnormal air spaces, and the extent of lung involvement. For day to day autopsy reporting a simple system such as that recommended in the CIBA Symposium is adequate.

The author describes the presence or absence of macroscopic emphysema in the following style: "There is mild/moderate/severe CAE involving the upper half of the upper lobe, the apical segment of the lower lobe; an area of mild/moderate/severe PAE involving 20% of the lower lobe immediately above the diaphragm, and paraseptal emphysema involving 12 cm of the pleura anteriorly in the upper lobe." The statement that 20 percent of a lobe is involved by panacinar emphysema is clear and understandable but the statement that half a lobe is involved by centriacinar emphysema may be very misleading in that this may be a scattering of small lesions involving only some 2 to 3 percent by volume.

All macroscopic techniques disregard the fact that the criterion for recognition of the emphysema was visible airspaces with a size of approximately 1 mm. They ignore mild or early emphysema, which can only be identified microscopically.

MICROSCOPIC ASSESSMENT OF SEVERITY

Normal alveolar spaces are approximately 250 μm in diameter.[53,54] Therefore by the time the airspaces in panacinar emphysema have reached 1 mm and become visible to the naked-eye, approximately three-quarters of the alveolar surface area has been destroyed. If one is to study early emphysema and its relationship to pathogenesis or function this must be done at the microscopic stage (Fig. 19-3A–C).

Unfortunately the classification of emphysema based on macroscopic patterns depends on the variation in arrangement of the airspace abnormalities within the acinar unit. These changes can only be identified when several acinar units are viewed at the same time. Standard histologic sections are too small to allow the patterns of emphysema to be recognized. However, early centriacinar lesions can often be detected by comparison with the adjacent relatively normal lung. These become less easy to identify as they increase in size, and it is difficult to classify a combined CA and PA emphysema on microscope slides.

There have been recent attempts to separate centriacinar and panacinar emphysema at a microscopic

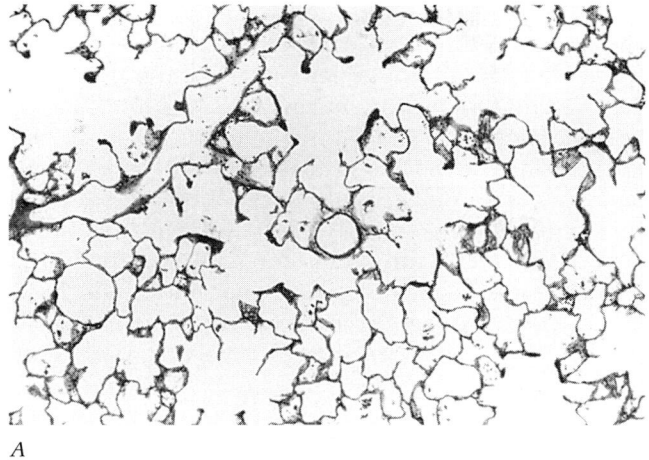

A

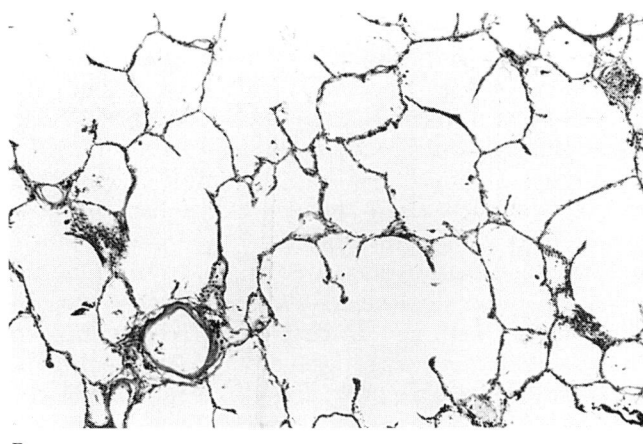

B

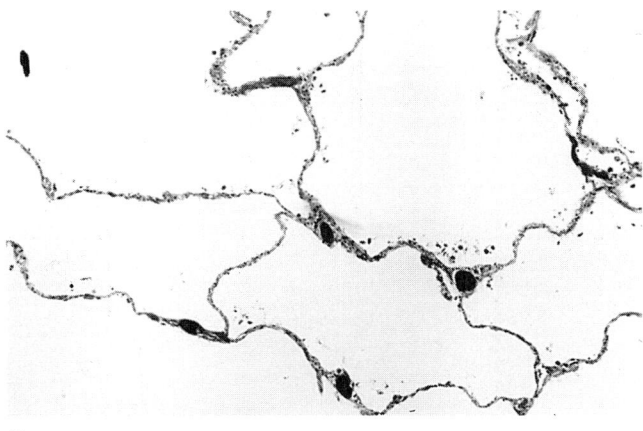

C

FIGURE 19-3

A. Normal airspace pattern from a male aged 30, life-long non-smoker.   *B.* Normal airspace pattern from a female aged 87, life-long nonsmoker.   *C.* A 70-year-old with smoking-related panacinar emphysema, macroscopically the airspace size was on average 2 mm in diameter. (H & E, ×36.5)

level.[55] The diagnosis of panacinar emphysema is made when the enlargement of airspaces involves the whole acinar unit and the distinction between alveoli and alveolar ducts is lost. Microscopic centriacinar emphysema is thought to be present when sharply demarcated emphysematous spaces are separated from the acinar periphery by intact normal alveolar ducts and alveolar sacs. Unfortunately these authors did not provide parallel data on the macroscopic patterns of emphysema in their cases. The method is non-quantitative in nature and no grading of severity of the two forms is involved.

Dunnill and Thurlbeck[49,56] both used microscopic techniques to quantify early abnormalities using the linear intercept technique to give an estimate of the diameter of airspaces. The mean linear intercept (Lm) is essentially the mean distance between alveolar walls. Lm is measured on histological sections using an eyepiece graticule. Intercepts between alveolar walls and the cross-hairs of the graticules (test lines) are counted. If the length of the test lines is known, then $Lm = N \times T/i$. Where N = the number of fields measured, T = the test line length, and i = the total number of intercepts. Lm can be used as an index of emphysema where Lm is increased with increasing airspace size.

Alternatively the surface area of alveolar walls can be calculated using a standard formula, $SA = 2v/Lm$, where v is the volume of lung tissue.[57] A word of caution is necessary in that some authors intending to identify the internal surface area of the lung have counted an intercept entering an alveolar wall and also when leaving it, such that a test line crossing an alveolar wall equals 2 and a test line entering a pleural surface equals 1 intercept. Other authors, however, have indicated only a single intercept for a test line transecting an alveolar wall. It is often not clear which criteria has been used when Lm data is quoted.

More recently with the introduction of automatic image analysis systems it has become easier to make accurate estimates of the surface area of the alveolated portion of the lung. Recently a fast, fully automatic method based on the linear intercept technique has been developed.[58,59] This allows a detailed analysis of large numbers of lung samples. These microscopic techniques produce data which can be expressed in terms of airspace size, e.g., linear intercept *(Lm),* or in terms of the surface area of alveolar or airspace wall which surround the enlarged airspaces. There are advantages in expressing the data in terms of surface area of airspace wall per unit volume (AWUV).[58–60] AWUV represents airspace wall per unit volume, that is, density of lung and is therefore directly comparable to the values produced by CT scan which gives a radiological tissue density. This has proved valuable in attempting to develop radiological techniques which can give true estimates of emphysema.[60,61] Furthermore tests of pulmonary function, in which the function may be related either to the surface for gas exchange or to the amount of tissue providing elasticity may best correlate with AWUV.

Having measured either airspace size or alveolar surface area there remains a problem: what is normal and what is abnormal? Alveoli multiply rapidly after birth[62–64]. The rapid multiplication of alveoli slows down at the age of 2 or 3 but new alveoli continue to be formed after this.[63,65] The exact age at which the adult alveolar number is achieved is unclear.[64,65] There has also been doubt as to whether in the adult, it is the number or size of alveoli which is constant in the population. If alveolar number stops increasing during childhood but the lung volume continues to increase in size, one would assume that the adult alveoli would vary in size.[62] However there is evidence that the size of the alveoli is not related to patient size[62,64,67] and that tall people have more alveoli than short people.[65] "Non-emphysematous lungs"[66] or lungs from life-long nonsmokers[68] have a narrow range of values for airspace size or alveolar surface area. This suggests that young adults have a broadly similar size of airspace which is independent of sex or body size. Gillooly et al.[68] studied a group of nonsmokers between the ages of 23 and 93 and defined a normal range (Fig. 19-4). These authors suggest defining normality as those values lying out-side the 95 percent confidence limits of their measurements. Alternatively one can normalize the data for age by calculating a percentage of the predicted value. This change in mean AWUV identifies an overall change in the whole lung and is probably appropriate for the identification of early panacinar emphysema as this appears to affect the whole lung. In a study of smokers they identified approximately 30 percent who had values for AWUV outside the normal range (Fig. 19-5) and have proposed that such cases of microscopically assessed emphysema (MAE) represented mild or early PAE involving the whole lung.[69]

Early workers quantifying emphysema microscopically noticed that there were clear cut differences between lungs from elderly and young patients (Fig. 19-3*A* and *B*). The changes in the senile lung were ascribed to emphysema, and the phrase "senile emphysema" was coined.[42,44] There appears to be a linear decline in alveolar surface area throughout adult life as a normal consequence of aging.[44] The CIBA Symposium recommended the term *senile emphysema* should not be used until the normal range of airspace size in the lung at different ages was established. This is now the case,[68] and in their study of life-long nonsmokers Gillooly et al. could identify no accelerated decline in alveolar surface area in nonsmokers with age or a subgroup which showed increased airspace size in the elderly. The decline in alveolar surface area and increase in the size of airspaces with age appear to progress in a linear and consistent manner between young adults and the most elderly patients. It appears to involve the whole population and there is probably no such entity as

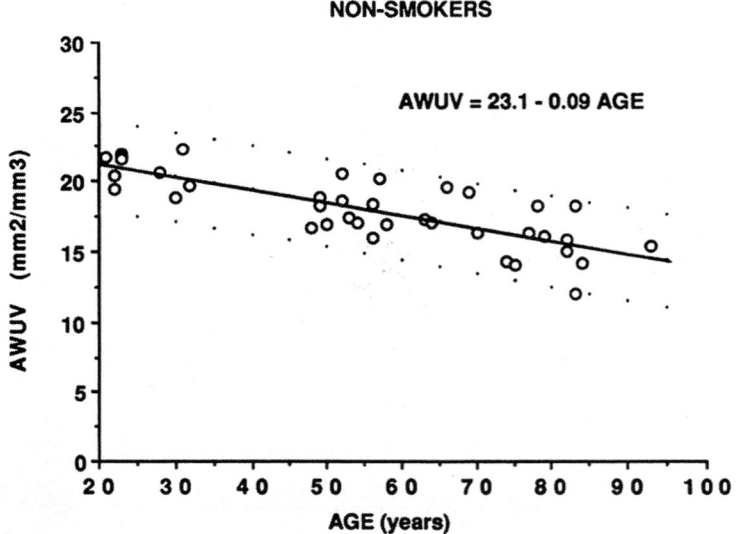

NON-SMOKERS

AWUV = 23.1 - 0.09 AGE

FIGURE 19-4
This figure shows the mean AWUV (expressed as mm^2/mm^3 lung tissue) of 38 life-long nonsmokers. The mean value varies with the age of the patient. The line indicates the regression line for a linear decline of AWUV with age (AWUV = 23.1 − 0.09 AGE). The dotted lines indicate the 95 percent prediction limits and indicate the limits of the normal mean AWUV values.

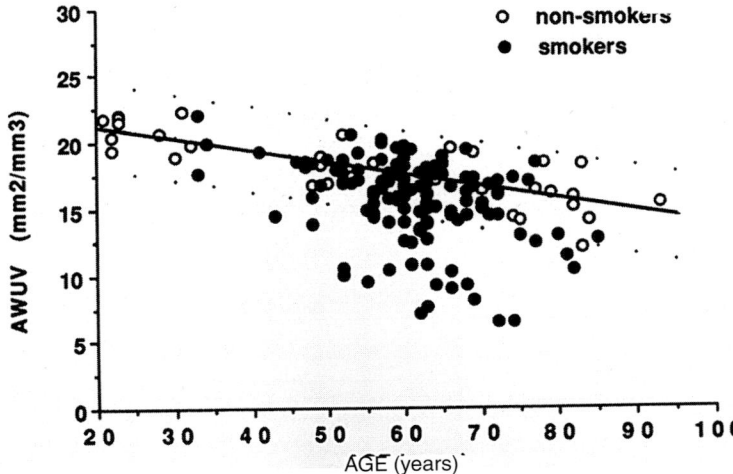

"senile emphysema." Under the current definition the changes in aged lungs are not "beyond the normal" for their age.

Centriacinar emphysema has a minimal effect on the overall mean AWUV or *Lm* value, as it involves a small proportion of the total lung volume. There are two ways of identifying early centriacinar emphysema. One method involves the examination of tissue sections and observing qualitatively the presence of focal abnormalities of centriacinar type. Alternatively one can observe the frequency distribution of the measurements (AWUV *Lm*) looking for evidence of a focal abnormality. This is revealed by increased coefficient of variation of individual measurements, and skewed distribution of values.[55,70,71]

The point of using a microscopic technique to assess the severity of emphysematous change is to ensure a quantitative linear measurement covering normality through to severe emphysema. It is not limited to the assessment of early emphysema. One must not confuse the "microscopic measurement" with emphysema that is only identifiable by a histologic technique.[69]

Alveolar Wall Support for Bronchioles

There is one aspect of the microscopic assessment of emphysema which may be of major importance in identifying the structural abnormality which gives rise to the fixed airway obstruction of COPD. Bronchioles and small bronchi are supported by, and owe their tubular integrity to, the attachment of adjacent alveolar walls (Figs. 19-6 and 19-7A–D).[72-76] The distance between the alveolar attachments to bronchioles is related to the size of the peripheral airspaces in the acinar unit.[76] The loss of such support may lead to distortion and irregularity of small airways and consequent airflow limitation, particularly on expiration (Fig. 19-7B and D).[72-83] Any interruption of the integrity of the alveolar walls adjacent to airways will fulfill the criteria of the definition of emphysema. An increase in the IAAD (interalveolar attachment distance) is recognized in emphysema.[72,80,82,83] Airways tend to run between acinar units and the alveolar walls that abut against the bronchiolar walls would therefore be those at the periphery of the acinus. These areas will be affected by panacinar or paraseptal emphysema but are

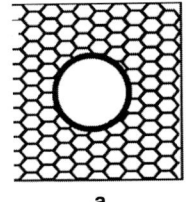
a

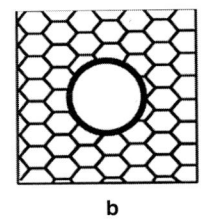
b

c

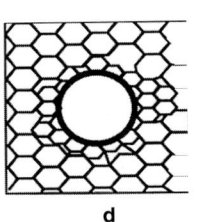
d

FIGURE 19-6

The diagrammatic representation of how variation in the interalveolar attachment distance (IAAD) may arise. *A.* Represents a normal lung with airway cut in cross-section. The alveoli are represented by the hexagons. In *B*, showing panacinar emphysema, the hexagons are larger and it is clear that the distance between alveolar walls attached to the outer aspect of the airway is increased. In *C*, there is selective loss of alveolar wall attachments adjacent to the airway despite the presence of normal alve-

oli a short distance away. This might occur in paraseptal emphysema. *D* This shows an appearance which is occasionally seen in what otherwise appears to be macroscopically panacinar emphysema. At the microscopic level the emphysema does not extend right up to the airway, leaving a narrow band of maintained alveoli giving a normal IAAD despite the loss of surface area elsewhere. Such a situation may conserve function.

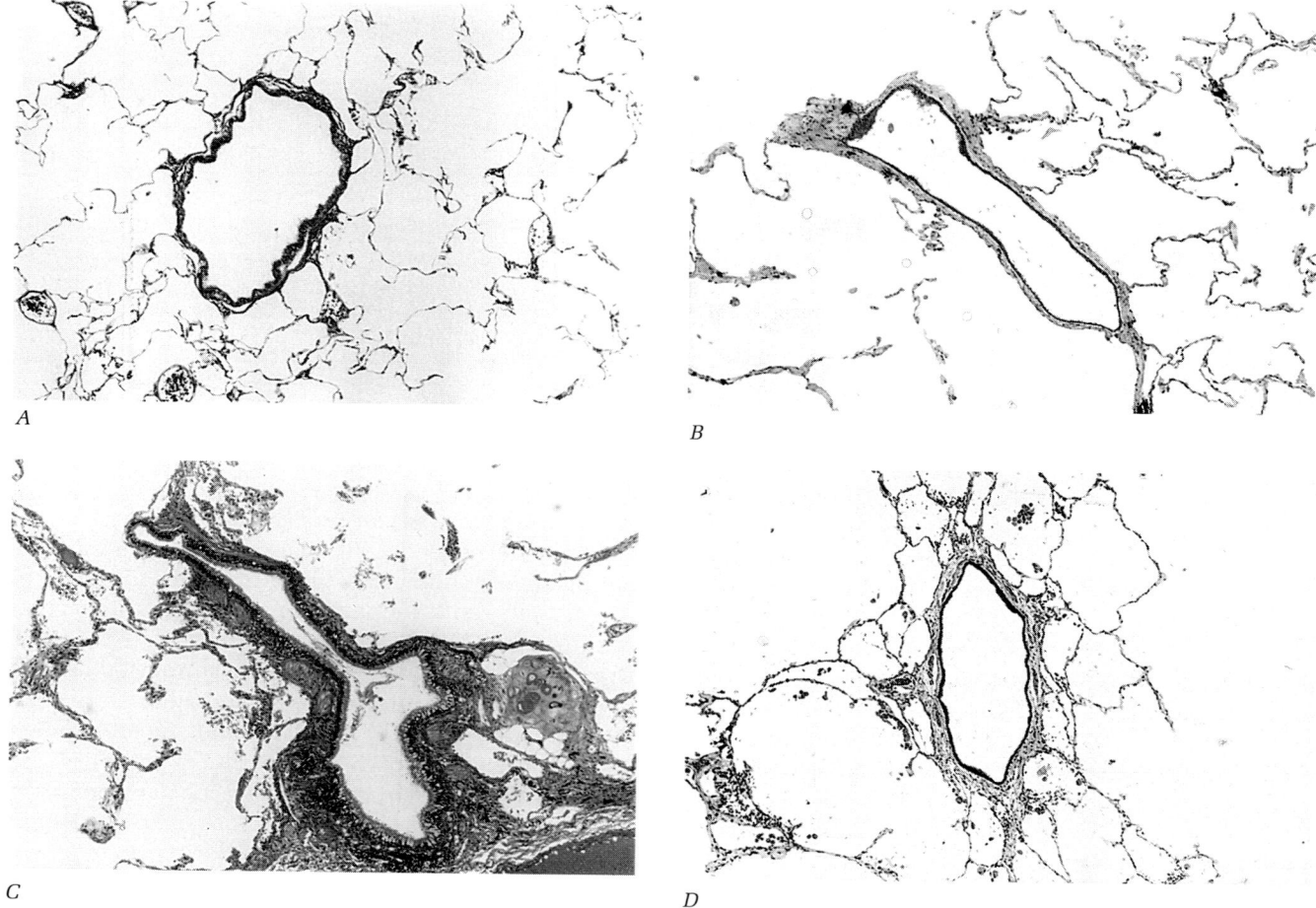

FIGURE 19-7

A. Normal nonsmoking lung showing a small bronchiole with the surrounding airspaces showing how the alveolar walls attached to its perimeter help to maintain its integrity as a tube. *B.* A similar airway in a patient with panacinar emphysema showing loss of alveolar wall attachments and the distortion of lumen outline. Such an airway is markedly affected by an increase in intrathoracic pressure in expiration and will collapse early. *C.* A small bronchus, note cartilage on the right-hand side, showing loss of alveolar support over much of the wall that is not supported by the pulmonary artery (to the bottom right) in a patient with mild panacinar emphysema. *D.* A small bronchiole in a patient with panacinar emphysema but with a rim of surviving alveolar walls giving some support to it (see Fig. 19-7(*D*). Magnifications: (*A, B* and *D* ×36.5 *C* ×18.25.

unlikely to be involved by early centriacinar emphysema. Our current classification of emphysema based on macroscopic appearances may not include all patterns of abnormality found at a microscopic level. It is not yet known whether loss of alveolar walls adjacent to airways (Fig. 19-6) is part of a recognized pattern of emphysema or can be a selective loss occurring as an isolated abnormality.[82]

Changes in alveolar wall support to small airways can be assessed by measuring the linear distance between the junction of alveolar walls with the outside of the airway walls. The distance is proportional to the diameter of the alveolar spaces which abut the airways.[76]

Various histologic techniques to quantitate the alveolar wall support to the peripheral airways have been described. The simplest is to count the number of alveolar walls attached to the outer aspect of airways and express the results as the number of attachments per airway.[72,80,83] It has the major disadvantage that the airways vary in size depending on whether they are proximal or distal and probably on the size of the patient. In addition, there are changes in apparent circumference based on tangential cutting of airways.

The simplest technique to give a truly quantitative estimate of the IAAD is to compare the number of alveolar walls attached to the outer aspect of an airway with its circumference. This is usually measured at the inner aspect of the epithelium of the basement membrane.[76,77,82–84] This gives a figure in microns for the mean IAAD for single airways or for a lung. However, in many small airways the adjacent pulmonary artery provides support to the wall directly with no intervening alveolar walls (Fig. 19-7C). A simple measure of circumference including this vascular portion can give erroneous values for the IAAD. Ideally the circumference should only include that portion of the wall not supported by the pulmonary artery.[85] In practice up to 30 airways have to be measured before a constant mean value is achieved.

It is possible to measure the individual IAAD between each pair of alveolar walls around the outside of an airway. This is very time consuming and there is no mathematical or statistical advantage using such detailed information over the simpler methods described above. In particular there is no evidence that individual large gaps in the alveolar support occurred independent of changes in the mean IAAD.

Gillooly et al. have shown that not only is there an increase in size of airspaces and a concomitant decrease in alveolar surface area with age but the IAAD shows a linear increase with age.[85] Such age-related changes make attempts to correlate structural changes with functional abnormalities in relationship to COPD extremely complex.

Differential Diagnosis of Emphysema

In the recent publication on the definition of emphysema, respiratory airspace enlargement was subdivided into three groups.[41] In addition to emphysema these groups were:

- Simple airspace enlargement, defined as enlargement of the airspaces without destruction of airspace walls. This may be congenital, as in infantile lobar emphysema or intralobar sequestration, or acquired, as in the overdistention of the remaining lung after pneumonectomy.
- Airspace enlargement with fibrosis. There is obvious fibrosis which is visible on a chest radiograph or macroscopically. This is usually consequent upon interstitial diseases such as sarcoid or infections such as tubercle.

Simple airspace enlargement after pneumonectomy is unlikely to cause diagnostic problems. There is more likely to be a problem in the assessment of the severity of coincidental emphysema, remembering that most patients having pneumonectomy for carcinoma are or have been smokers. In such a situation it is not possible to assess how much of the airspace enlargement is emphysema and how much compensatory dilatation.

Airspace enlargement with fibrosis visible to the naked eye should not be a problem provided the pathologist can recognize fibrosis (Fig. 19-8). There is no problem in the case of the cryptogenic fibrosing alveolitis with end-stage honeycomb lung. The disease has a characteristic subpleural position in the lower parts of the lung and the firm and almost rigid fibrosis is easily recognizable. More difficult to diagnosis are milder degrees of fibrosis of old sarcoid when the abnormal airspaces may have a gray or whitish appearance. This is because of the fibrous tissue in the walls. The fibrotic areas are palpable, compared to the soft, vanishing texture of emphysema.

There are two infiltrative conditions of the lung which are associated with loss of alveolar walls which can mimic emphysema. These are histiocytosis X and lymphangioleiomyomatosis.

Histiocytosis X is a nodular infiltrate of abnormal histiocytes associated with eosinophils and a lymphoid infiltrate. This undergoes maturation into less cellular and more fibrotic nodules, by which time the diagnosis may not, at that site, be evident. Different nodules mature at different rates. In later lesions there appears to be some loss of alveolar walls adjacent to the nodules with the production of irregular, large air spaces. The enlarged airspaces can resemble a rather bizarre centriacinar emphysema. In the later stages of the disease the presence of the widespread nodular infiltrates with an end-stage honeycomb pattern easily differentiate the condition from emphysema.

Lymphangioleiomyomatosis has true alveolar wall destruction to produce focal 'emphysematous' spaces which are centriacinar in early disease. These become confluent to produce a pattern of a widespread cystic lung at autopsy. At this stage the airspaces are 3 to 10 mm in size and may initially be diagnosed as widespread panacinar emphysema, particularly as the texture is soft and lacking in fibrosis. The appearances differ from emphysema in that the spaces are cyst-like (Fig. 19-8B) and one can see the back of the lesions. By contrast in severe panacinar emphysema the abnormal airspaces intercommunicate and one can look through the surface spaces deep into the tissues (Fig. 19-2E).

The pathogenesis of the airspace wall loss is unknown but is clearly related to the proliferation of smooth muscle. This can usually be identified at the edges of the enlarged airspaces microscopically and may be seen as small yellow pads or nodules in the lining of the cysts macroscopically (Fig. 19-8B). It is probably because of the development of these large airspaces that 50 percent of cases present with repeated pneumothoraces.[86,87] On CT the appearances resemble focal emphysema with an identifiably thick edge.

In early disease, when smooth muscle proliferation may be difficult to identify, the enlarged air-

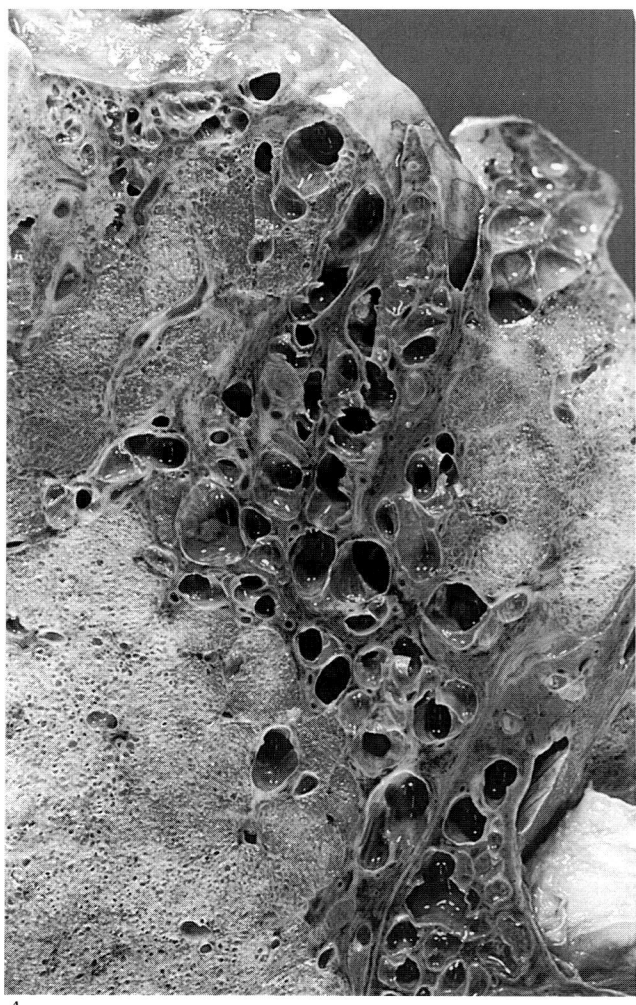

A

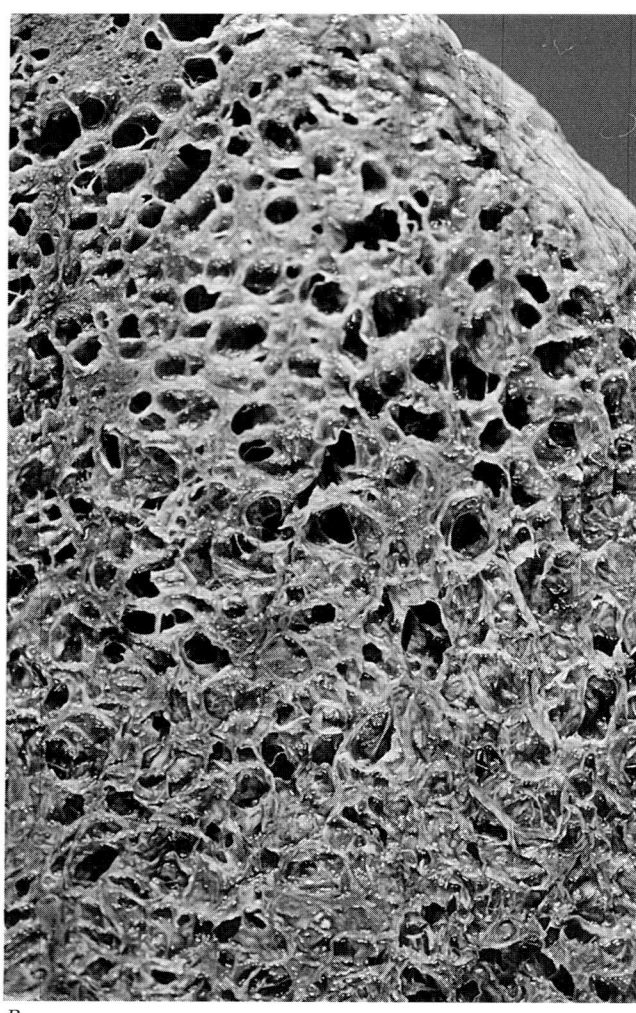

B

FIGURE 19-8
Differential diagnosis of airspaces in the lung. *A.* Shows airspaces surrounded by residual airway walls and fibrosis. This is an area of long-standing severe bronchiectasis with complete destruction of the adjacent parenchyma. Note the normal pattern of the adjacent lung. *B.* Lympangioleiomyomatosis, the rather thick-walled cysts have a characteristic appearance with pale bands of proliferating muscle in their walls.

spaces may be a prominent histological feature. One could argue that the lesions are really truly emphysema! There is certainly enlargement of airspaces due to destruction of preexisting alveolar walls and there is insignificant fibrosis. There are changes in the connective tissue with elastin destruction resembling those seen in emphysema.[88] Interestingly the pulmonary function in lymphangioleiomyomatosis shows obstructive as well as restrictive features. The association of the abnormal airspaces with a specific disease process probably excludes the term *emphysema*, but there are close similarities.

More of a problem is the situation where there is an apparent combination of emphysema and fibrosis. I am not referring to the mainly microscopic amounts of fibrous tissue seen in smoking-related emphysema, which is discussed in a later section, but macroscopic fibrosis. Emphysema around the margins of the lung, particularly the apices of the upper and lower lobes, often paraseptal in type, may be associated with marked fibrosis. This is primarily emphysema with secondarily acquired fibrosis, as a consequence of inflammatory episodes. Such changes are commonly seen in association with bullae. In a lung with widespread emphysema there may be areas which are clearly fibrotic though similar in pattern to the non-fibrotic areas. I interpret these in a similar way and accept that the underlying increase in airspace size is emphysema.

Evidence for Airspace Wall Destruction

Though the major component of the definition of emphysema is an enlargement of airspace size, this is defined as being accompanied by destruction of their walls. In the very earliest stage of emphysema evi-

612

dence of early destruction might be present at a time when increase in airspace size or actual loss of alveolar walls may be difficult to identify. Light and scanning electron microscopy have been used to identify an increase in size and number of the fenestrae in alveolar walls in smokers[89–92] Boren took an upper limit of 20 microns for a fenestra.[89] These enlarged fenestrae may be the earliest stage of smoking-related damage to airspace walls. Extension of this change may give rise to the grosser abnormalities of true emphysema.[91] Abnormal fenestrae or "holes" are associated with the development of centriacinar but not with panacinar emphysema.[91,92]

In an attempt to combine an estimate of airspace size with this element of destruction of alveolar walls, Saetta et al. described a destructive index (DI) which combined a qualitative assessment of airway walls with point counting to identify early emphysema. They used a combination of enlargement of airspaces and evidence of airspace wall damage.[93] Using this technique it has been possible to identify smoking-related alveolar wall damage by light microscopy in the absence of measurable emphysema.[93,94]

So far there is no proof that such damage to alveolar walls is a necessary stage in the development of either centriacinar or panacinar emphysema. However, if we are to elucidate pathogenetic mechanisms of the types of emphysema, then understanding of the earliest evidence of damage is important and more information is needed on the role of fenestrae or "holes" in early emphysema.

Fibrosis and Emphysema

The definition of emphysema describes emphysema as "being without obvious fibrosis." The emphysematous lung lacks substance and has a tendency to collapse. This characteristic appearance is fundamentally different from interstitial diseases, such as cryptogenic fibrosing alveolitis, where enlarged airspaces are associated with fibrosis to give a rigid honeycomb pattern. However, emphysema is in the broadest sense a condition associated with inflammatory damage in the lung and most forms of inflammation give rise to some fibrosis. Histologically, there is fibrosis in the region of terminal and respiratory bronchioles in smokers. The Respiratory Bronchiolitis described by Niewoehner may occasionally produce a true interstitial lung disease.[95,96] Nagai and Thurlbeck, using SEM, identified fibrosis associated with centriacinar emphysema.[97] On the other hand if the protease/anti-protease theory of the pathogenesis of emphysema is correct, one might expect a reduced collagen and elastin in the emphysematous pulmonary parenchyma, at least in the emphysema associated with alpha-1-antitrypsin deficiency. It is difficult to measure the amount of collagen within the lung parenchyma in relation to the alveolar walls. A large amount of the total collagen framework of the lung is associated with bronchovascular bundles, pleura, and septae and is not in the alveolar walls. Within the acinar unit the collagen and elastin are not evenly distributed but are concentrated in the alveolar ducts, around the mouths of alveolar spaces. When sensitive biochemical techniques are used, an increase in collagen can be identified in the parenchyma of smokers.[98] When emphysematous lungs are examined the amount of collagen is also increased when compared with less emphysematous areas of the same lung, or to nonemphysematous lungs.[99,100] One must be careful in the interpretation of such results because of the condensation of connective tissue framework around the mouths of the alveolar spaces which may remain after loss of alveolar walls. This provides a relative retention of the collagen and elastin framework of the lung. This could give rise to an increase in apparent collagen and elastin when these are compared with the amount of alveolar wall present. It is often assumed that the quality of the lung is constant but that the quantity is altered in emphysema. It is possible that both the quality and quantity of alveolar walls may change and these may both affect elastic recoil and support to small airway walls.

Bullae

The CIBA Guest Symposium (1959) defined a *bulla* as an emphysematous space with a diameter of more than 1 cm in the distended state. The term is usually limited to those subpleural emphysematous spaces which protrude above the surface of the adjacent pleura. Reid[42] classified bullae into three morphological types, based on the amount of lung involved and the size of the "neck" of the bulla (Fig. 19-9). The classification did not relate to the size of the bullae. Type 1 was a bulla arising from a small amount of lung showing marked overinflation and a narrow neck. The sack of the bulla was usually empty and without vascular or airway remnants. A type 2 bulla was from a less overinflated but still superficial portion of lung tissue with a broad neck. A type 3 bulla was overinflation of a large amount of lung tissue and was usually found in association with severe emphysema elsewhere in that lobe or lung. Types 2 and 3 had within them remnants of airways and blood vessels. This classification is helpful in understanding the production of bullae but bears little

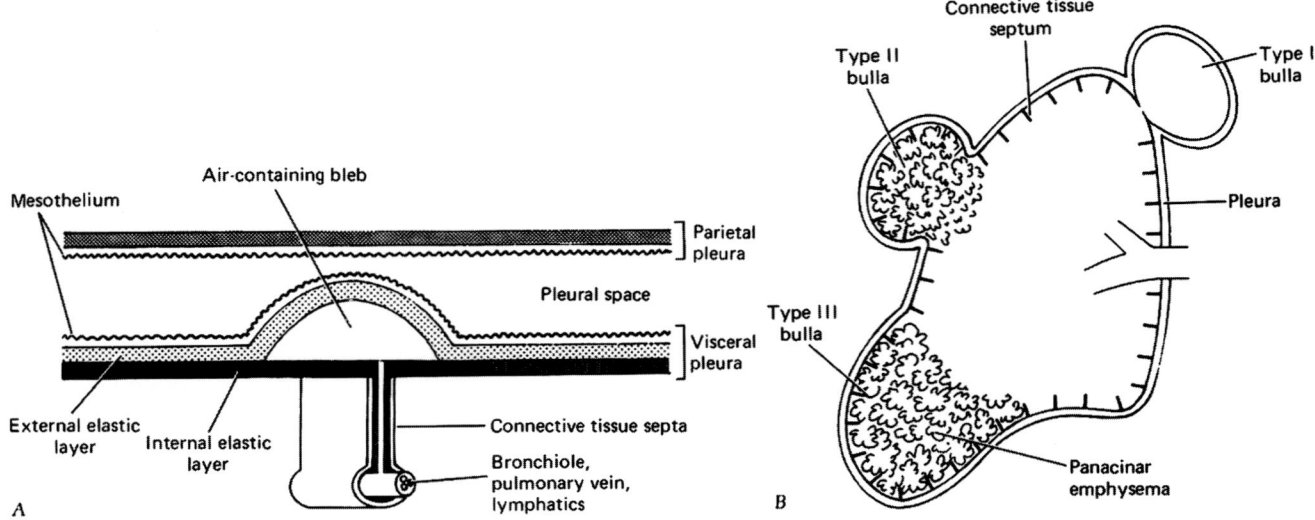

FIGURE 19-9

Blebs and bullae.   *A.* Development of a bleb. A bleb is an accumulation of air within the pleura that is not confined by connective tissue septa within the lung. Air that escapes within the substance of the lungs makes its way to the surface, separating the internal from the external elastic layers on the visceral pleura.   *B.* Different types of bullae. In contrast to a bleb, a bulla is confined by connective tissue septa of the lung and is deep to the internal elastic layer of the visceral pleura. Three different types of bullae are shown arising from a lung which has been removed from within the chest wall. A type I bulla is shown at the apex, a type II in the middle zone, and type III arising at the base. The short dark lines denote connective tissue septa. Panacinar emphysematous parenchyma is present within the types II and III bullae. (*After Reid,*[42] *with permission.*)

relationship to their clinical effects. When the pathologist sees a bullous lobe or lung the bullae protrude above the surface, and may be up to 10 or 15 cm in diameter. In life they cannot protrude above the adjacent lung in this manner but are pushed into the adjacent lung by the chest wall. The type 1 bullae is then almost surrounded by a double layer of pleura, its own covering and that of the adjacent lung, which is pushed inwards. Considering that the wall of the bulla itself is often fibrotic, this can often clearly be seen as a linear outline on the chest x-ray. This inpushing of the lung is also seen with a type 2 bulla but much less with the type 3. When the type 3 bulla occurs at the base of the lung, as in alpha-1-antitrypsin deficiency or end-stage COPD, the redundant lung does not invert as a single bulla. It does so as a series of irregular infoldings of the pleura which may give a variety of radiologic patterns.

Very occasionally there may be subpleural paraseptal emphysema of significant degree causing widespread subpleural bullous transformation. These may give rise to a pneumothorax.[168]

Radiological Assessment of Emphysema

In 1978 Thurlbeck and Simon described criteria for the diagnosis of emphysema using the chest x-ray based on indications of overinflation[101] and vascular pruning. They showed only moderate correlation

with macroscopic emphysema as assessed by the picture matching technique. When the CT scan became available it was possible to extract quantitative data on lung density. Emphysema is an increase in airspace size or conversely a decrease in the amount of alveolar tissue per unit volume, that is, a decrease in tissue density. One would expect a good correlation between CT measurements and the quantitative assessment of emphysema. When quantitative histologic assessment is compared with quantitative CT density there is a good correlation[60,61] even within the range of those showing normal age change or early emphysema. The AWUV measurements relate to volume values for 1 mm cubes of lung which are comparable to the sensitivity of modern CT scans.

Such CT assessment of emphysema has the advantage of being applied to a patient in life. However, all the problems that arise in the assessment of microscopic emphysema pathologically remain. Those measuring emphysema using CT scans must take into account whether a decrease in density is due to hyperinflation or true emphysema, what the age-related normal values are, and the fact that mean values are likely to relate to panacinar emphysema. The focal lesions of CAE need careful examination of frequency distributions of the density measurements or direct qualitative assessment in the identification of individual lesions.[102,103]

Unfortunately many attempts to use CT to assess the severity of emphysema have compared CT and lung pathology, using qualitative grading, scoring or picture matching techniques. However, both CT and pathological assessment are known to be inaccurate and non-linear. Those groups who have used true density measurements have shown better correlations of their lung density measurements ("emphysema") with pulmonary function tests, including gas transfer and FEV_1.[60,61,104,105]

Etiology and Pathogenesis of Emphysema[20,44,106,107]

ETIOLOGY AND RISK FACTORS

The clinical epidemiology of emphysema has been discussed earlier. In this section the discussion of etiology and risk factors is confined to emphysema as assessed pathologically.

All the evidence points to emphysema being a response by the lung to inhaled particulate or gaseous insults. The two situations where there is clear evidence based on epidemiologic pathology are tobacco smoking and occupational exposure, in particular, coal mining.

All large surveys of the incidence of emphysema are based on macroscopic techniques and many have not separately assessed the subtypes of emphysema. The following factors have been shown to be involved:

- *Smoking:* Emphysema is consistently found more frequently and with increased severity in smokers than in nonsmokers.[47,108,109–115] The large amount of work carried out before 1976 is well reviewed.[47] Most series show that centriacinar emphysema in smokers is commoner and more severe than panacinar emphysema. Emphysema does occur in nonsmokers and over half the elderly male nonsmokers have some centriacinar lesions, though very few have true panacinar emphysema.[44] There is a rise in air space size with age. It has been argued that as this is a normal age change, it should not be considered senile emphysema.[68] True panacinar emphysema with an air space size in excess of the age-related changes is very unusual in life-long nonsmokers.

- *Sex Difference:* Emphysema is commoner in men than in women, even when the amount smoked is taken into account.[44,114,116–119] Though this increased incidence in men has been ascribed to differences in smoking habits, type of tobacco, or inhalation patterns, these factors do not appear to

be complete explanations. One factor that may be relevant is the difference in body size and lung volume in the two sexes. It is possible that the size and extent of macroscopic emphysema lesions could be affected by the increase in stresses and strains within the lungs of taller patients.[120] These stresses, particularly in the upper part of the lung, have been used to explain the increase in number and size of centriacinar lesions in the upper zones of both sexes.

- *National Differences:* There are international differences in the incidence of emphysema.[109,112,114,121–123] Japan and New Guinea have the lowest incidence,[21,23] the United Kingdom and the United States highest.[112,114,122] The cause of these international variations is unclear, but they do not appear to be attributable solely to smoking. It is possible that rather like the clinical epidemiology of COPD, there are international differences associated with atmospheric pollution and regions of heavy industry.

- *Industrial Exposure:* Heppleston was the first to suggest a distinctive pattern of emphysema in coal workers with dilatation of respiratory bronchioles. This was associated with focal dust deposition involving all generations of respiratory bronchioles. This is in contrast to the centriacinar emphysema associated with smoking, where the earliest lesions develop around the second and third order respiratory bronchioles.[124] It can be argued that in coal workers the earliest "dilatation lesions" are not truly emphysematous under the current definition. By the time the lesions are more advanced it is not possible on morphological grounds to separate proximal acinar changes in coal miners from the centriacinar emphysema seen in non-mining smokers. The situation is complicated by the fact that many coal miners smoke. Thus they may have centriacinar lesions to which both dust and smoking have contributed. It is not possible by examining the lung in such cases to determine the relative contributions of smoking or industrial exposure to the emphysema.

  What is clear is that coal miners develop more emphysema than non-miners with similar smoking histories,[17,112,125] and there is a particular increase in centriacinar emphysema.[17,125] The risk of a miner having emphysema is proportional to the lifetime dust exposure, if he has fibrotic or nodular pulmonary lesions. This holds true even in nonsmokers.[13] Pure silicosis in coal workers appears to have a lower correlation with emphysema.

An increase in emphysema in gold miners in South Africa has also been reported but did not appear to be related to the grade of silicosis.[126]

• *Alpha-1-Antitrypsin Deficiency* (alpha-1-protease inhibitor, or α1-Pi). Congenital abnormality of the genes controlling α-Pi production, providing low levels of circulating α1-Pi, is associated with emphysema even in nonsmokers. In smokers the development of emphysema appears to be inevitable in patients with abnormal α1-Pi production and occurs at a much earlier age.[21,22]

Pathogenesis[20,106,123]

The current concepts about the pathogenesis of emphysema are based on two separate publications in 1963 and 1964.

• Laurell and Eriksson[21] reported an increase in the incidence of emphysema in patients with low levels of serum α1 globulins, later shown to be due to a decrease in α1-antitrypsin, and suggested that the decreased antiprotease activity could be part of the pathogenesis of emphysema.

• Gross et al. induced emphysema in experimental rats by instilling papain intratracheally.[128] This was later shown to be due to the elastolytic activity of the enzyme.

These independent findings led to a simple concept for the pathogenesis of emphysema. This was that an imbalance of the protease:antiprotease ratio could allow uncontrolled elastolytic activity with loss of the elastic framework and destruction of air space walls. This simple model fits the situation in patients with homozygote α1-Pi deficiency when very low levels of circulating serum α1-Pi are associated with the very early development of emphysema in smokers. Such patients show double the decline in FEV$_1$ as occurs in tobacco smokers with COPD but without α1-Pi deficiency.[22,129,130] Patients with homozygous α1-Pi who do not smoke *may* also develop emphysema but 10 to 20 years later than in smokers. It is still uncertain whether all patients with α1-Pi deficiency who do not smoke develop emphysema or if there is another stimulus as an alternative to cigarette smoke in these cases. Subsequent investigations have shown that intermediate levels of circulating α1-Pi are not associated with the development of emphysema, even in those who smoke.

A long series of animal models has confirmed that in an experimental situation elastolytic enzymes, including neutrophil elastase, introduced via the trachea, but not intravenously, can produce emphysema.[127,131,132] The severity of the emphysema produced is proportional to the dose of elastolytic enzyme. Exposure to tobacco smoke can increase the amount of emphysema.[133,134]

In smokers there is probably an increased elastolytic load due to a greater number of polymorphs attracted into alveoli by "smoke-affected" alveolar macrophages. These polymorphs release neutrophil elastase, which is not entirely neutralized by the normal levels of α1-Pi, causing progressive lung destruction. This simplistic hypothesis does not appear to fit the available evidence relating tobacco smoking to the development of emphysema in non-α1-Pi patients. There is no evidence that smokers have lower levels of α1-antitrypsin in their bronchoalveolar lavage (BAL) fluid than nonsmokers. Smokers do not appear to have excess of proteolytic or elastolytic enzymes in their BAL.[20,106,127,134] Though the numbers of polymorphs attracted into the lung is increased "many fold," the actual numbers in BAL fluid are still only some 5 percent of total cells, with 90 percent macrophages.[135] The role of macrophages in the pathogenesis of emphysema is still uncertain.

The theory has now been modified to include other effects of smoking. Tobacco smoke includes oxidants and induces oxidant production and their release by polymorphs and macrophages. These oxidants may produce tissue damage directly or by partially inactivating α1-Pi.[3,4,32,127,138] It has been suggested that other proteolytic enzymes or other proteinase inhibitors may be involved.[20]

Pathologically there is little to confirm the pivotal role of elastase in the production of smoking-induced emphysema. One study to Damiano et al.[136] identified increased quantities of elastase attached to lung elastin in emphysema using immunocytochemical electron microscopic techniques. The increase was proportional to the severity of the emphysema. This work has not been repeated and has been criticized on technical grounds.[137]

Much of the work attempting to elucidate the pathogenesis of emphysema carried out on human smokers in the last 20 years has been misdirected. Most series compare smokers with non-smokers and ignore the fact that only a proportion of smokers develop emphysema. The two populations of smokers and nonsmokers have usually been selected simply for their smoking history and not their ability to develop emphysema. The two populations may in fact be entirely identical in relation to risk factors. It is surprising that no one has utilized more suitable material. These include patients coming to surgical resection in which the presence or absence of emphysema can be confirmed; smokers who have demonstrated their ability to develop COPD and then stopped smoking. Ceasing this habit will arrest the decline in FEV_1 and presumably the development of emphysema.[20] None of the work has attempted to identify different pathogenetic mechanisms for the various patterns of emphysema produced by tobacco smoking.[44,58,72]

The pathogenesis of smoking-related emphysema appears to be multifactorial and is at present ill-understood. The possible involvement of oxidants producing tissue damage and perhaps having a significant role in the production of emphysema has opened a new area of research; the role of antioxidants being particularly important in that they are amenable to manipulation by dietary means.[127,139]

There has been little work on the pathogenesis of emphysema in coal workers, the general assumption being that there is in some way a potentiation of similar mechanisms to those involved in smoking-related emphysema. Nothing is known of the epidemiology or pathogenesis of paraseptal emphysema, probably due to its rarity and minimal clinical impact.

PATHOLOGIC BASIS FOR PULMONARY FUNCTIONAL ABNORMALITIES IN COPD

Patients with COPD show major abnormalities of pulmonary function, the most important being a fixed airway obstruction as shown by a fall in the FEV_1 or peak flow measurement. But associated with this are abnormalities of gas transfer and lung compliance, as shown by abnormal pressure/volume curves.[46] To identify the pathological changes within the lung and the pattern of pulmonary function abnormalities in COPD requires that appropriate tissue is available for study linked to recent pulmonary function tests. In practice this means study of either end-stage COPD with autopsy lungs being compared to function in life. Alternatives are the study of surgical resection specimens, resected usually for cancer or more recently lungs removed at the time of lung transplants, with preoperative functional studies. The earlier studies were predominantly on end-stage disease but in the last decade there have been a number of studies based on surgical resection specimens.

Earlier autopsy studies of COPD showed that emphysema was frequent and usually severe.[42,44,139,140] The increased frequency of emphysema was maintained when cases were compared with a random smoking population.[140–142] However, there was no clear-cut relationship between emphysema and COPD as there were patients with significant amounts of moderate or even severe emphysema without COPD in these series. These early studies were based on the macroscopic assessment of emphysema. As previously described the techniques for assessment of emphysema macroscopically fail to separately identify the different patterns of emphysema. In particular, though both centriacinar and panacinar emphysema are smoking-related and commonly found in COPD, there was no clear evidence as to which pattern of emphysema was the most important in producing the functional abnormalities. Most authors assumed that the frequency of centriacinar emphysema in COPD confirmed it had a significant role in explaining the pathophysiology.[44] Others felt that the panacinar emphysema was of preeminent importance introducing fixed airway obstruction.[42]

Pathologic Basis for the Decline in Measurement of Gas Transfer

Physiologic tests for gas transfer within the lung are abnormal in a variety of lung diseases, in particular the interstitial lung diseases as well as in COPD. In COPD there are two possible reasons why these tests should be abnormal. One is that there is heterogeneity in ventilation and perfusion in the lung. The other is that there is an increased distance for diffusion due to the destruction of alveolar walls. It should be remembered that in the lung the flow down the airways in inspiration/expiration ceases at the region of the respiratory bronchioles and first order alveolar ducts. In the more distal lung diffusion is of preeminent importance. Loss of alveolar walls and increase in size of air spaces in emphysema increases the diffusion distance.

When emphysema is assessed macroscopically, relationships with gas transfer results are identified.[42,104,143–145,149] However when emphysema was assessed microscopically, usually with more accurate and linear quantitative techniques, much improved correlations between gas transfer and these measurements of emphysema were identified.[61,63] These studies were carried out on surgical resection specimens and included smokers with and without emphysema. In such a group there was a linear relationship between tests of gas transfer and alveolar surface area which covered both normal and abnormal values of alveolar surface area. This very close correlation suggests that, at least in early smoking-related disease, nonventilated areas due to airway abnormalities are not essential for the abnormal gas transfer values and that panacinar emphysema is probably the most important.

Pathological Basis of Airway Obstruction[107,144–148]

The prime physiological abnormality in COPD is a fixed airflow obstruction. There are many causes of temporarily or reversible increase in airflow obstruction. These include bronchospasm, intraluminar mucus or mucopus and inflammatory thickening of the airways. These may be of clinical importance when superimposed upon the fixed component and may precipitate exacerbations of severity of the clinical condition. To account for a progressive fixed airway obstruction there appear to be only two main alternatives:

1. The airflow obstruction is due to narrowing and distortion of small airways due to scarring, itself a consequence of inflammatory changes induced by cigarette smoking. The is so-called small airway disease.
2. The airflow obstruction is a consequence of emphysema. The smaller bronchi and bronchioles are not self-supporting tubular structures but depend entirely on the support of the surrounding parenchyma and the alveolar walls. These are inserted into the peribronchiolar connective tissue and are responsible for their tubular integrity. The alveolar walls attach to the outer aspects of the airways and in inspiration exert a pull on the outer aspect, opening up and distending the airways. In expiration the elastic recoil of the lung maintains this lateral pull on airways and tends to keep them open despite the decreasing lung volume. As the intrapulmonary pressure rises the tendency is for small airways to narrow and eventually collapse. The closure of airways till the opposing surfaces touch is probably the limiting factor to normal expiration. In emphysema with decreased support and decreased lung elasticity there can be a functional airway obstruction. This causes airways to collapse and close at high lung volumes, even though the airway walls themselves may be structurally normal.

Small Airway Disease and Airway Obstruction

The concept of small airway disease, that the site of the fixed airway obstruction in COPD resided in the small parenchymal airways of around 2 mm or smaller, was introduced by Hogg and Macklem in 1968.[148] They envisaged an inflammatory process induced by smoking with the end result of structural narrowing of the airways.

A great deal of work has been done on identifying and quantifying the effects of smoking on the small peripheral airways, largely carried out on resection specimens. There is inflammation and scarring. Evidence for smoking-related inflammatory changes in small airways has steadily accumulated.[84,96,144,149–153] This involves the small airways and respiratory bronchioles and there can be recognizable scarring, inflammation, and increases in macrophages in young adult smokers. These changes may be so florid as to justify being called an "interstitial disease." The walls of small airways show an increase in inflammatory cells, which differs from the inflammation seen in asthma as there is no increase in the basement membrane, subepithelial connective tissue, or eosinophils. There is a problem in that it is possible for a patient with asthma to develop COPD if they smoke.

A whole range of other changes are seen in the small airways, including an increase in goblet cells, metaplastic changes in the epithelium, an increase in anthracotic pigment, and scarring. In an attempt to quantitate the smoking-related changes in small airways Cosio and colleagues scored the separate components depending upon their subjective severity and summed the total reached as a "small airway disease score" (SAD score).[153–155] The eight aspects were:

1. Occlusion of the lumen by mucus and cells
2. Presence or absence of mucosal ulceration
3. Goblet cell hyperplasia
4. Squamous cell metaplasia
5. Inflammatory infiltrate of the airway wall
6. Amount of fibrosis in the airway wall
7. Amount of muscle
8. Degree of pigmentation

The first two of these were directly assessed as a percentage of airways involved: the remaining six were scored 0 to 3 qualitatively and the score expressed as percentage of a maximum possible score of 18. The total of the three individual percentage scores was added to provide an overall small airway disease score (SAD) with a maximum of 300. The assessment of variables 3 through 8 is "semi-quantitative" by a subjective grading which became refined at a later date by use of a set of standard illustrations of the various grades of abnormality.[155]

Using such scoring systems, relationships between these and reduced FEV_1 have been shown.[151,154,156] The significance of these findings is not straightforward, because the scoring system is a summation of several individual semi-quantitative gradings. It is difficult to see how many of the individual components could be related in a cause and effect sense with airflow resistance unless they are associated with a decrease in airway lumen. Measurements of airway lumen dimensions have shown a decrease in diameter associated with changes in the FEV_1, but these changes have been minor.[157] Not all investigators have identified changes in small airway lumen.[82]

There is a major problem in identifying statistical links between the SAD score or other estimates of smoking-related airway damage and changes in FEV_1. It is accepted that the FEV_1 declines with age and smoking, and the airway abnormalities are recognized as smoking-related phenomena. It is not surprising that there is a broad relationship between them, though this may not be a cause and effect relationship. It is also possible that inflammatory changes may merely be indicators of later structural damage in the small airways or even to the development of emphysema. Cosio and coworkers have identified a relationship between the inflammatory changes and scarring in small airways and centriacinar but not panacinar emphysema or changes in FEV_1.[55] Epidemiologic studies of the decline in FEV_1 in patients with COPD[4,10,11] have shown that this accelerated decline appears to affect only some 20 percent of heavy smokers. Studies of the inflammatory changes in the small airways have not identified a susceptible group amongst smokers, which would account for these epidemiologic findings.

Emphysema and Airway Obstruction[44,107,159]

There are three sets of information linking emphysema in a causative way to the development of fixed airway obstruction:

- *Epidemiologic:* Emphysema is commonly widespread and severe at autopsy in patients with

COPD.[32,42,44,139,160] The amount of emphysema in patients with COPD is greater than that in populations of smokers without COPD.[27,32,139,140] It is the custom, particularly among pathologists, to ascribe a major functional role to emphysema in COPD. The data from such studies does not implicate any particular pattern of emphysema and there are smokers that have emphysema but not COPD.

- *Quantitative Comparisons of Emphysema and Pulmonary Function.* There appears to be a positive relationship between the severity of emphysema and the severity of airway obstruction.[32,139,140,159,160,161] Surprisingly none of these studies assessed centriacinar and panacinar emphysema separately. We therefore have no clear idea of the functional significance of these patterns of emphysema either alone or together. Recently a microscopic study of emphysema has suggested different functional roles for centri- and panacinar emphysema.[55]

- *Relationship of Alveolar Wall Attachments to Bronchioles and Airway Obstruction.* The work of Anderson, Foraker, and Linhartova has suggested a major role for the peribronchiolar alveolar attachments in maintaining airway functional integrity.[72,73,76] A relationship between a decrease in the number of alveolar attachments (or an increase in the interalveolar attachment distance) and a fall in FEV_1 has been seen in early disease.[55,82,162] However, in one survey the relationship of airway obstruction to measured emphysema in an autopsy population was stronger than the correlation to alveolar attachment loss.[163]

It is difficult to separate the role of emphysema and of the loss of peribronchiolar attachments in producing airway obstruction because of the interrelationship of these two structural changes. When both are measured in the same investigation, emphysema seems to be more important in later stages of the disease[163] and the attachments of prime importance in early disease.[82] It is probable we are looking at two aspects of one structural abnormality, emphysema. The two aspects are the overall decrease in lung elasticity and recoil produced by widespread emphysema and the local effects of the loss of alveolar walls adjacent to bronchioles. It seems possible that these two components are at least partly independent of each other, at least in early disease.[82] In lungs with widespread panacinar emphysema there are occasional cases where the peribronchiolar alveolar walls appear conserved rather than destroyed (Fig. 19-8*D*). In such cases the structure:function correlations suggest a lesser degree of FEV_1 loss than expected for the severity of the emphysema.[82] Loss of alveolar attachments is also associated with abnormalities of those physiologic measurements thought to identify abnormality of small airway function.[82,163]

Conclusion

There are two alternative theories about the structural basis for airway obstruction in COPD, small airway disease, and/or emphysema. It is clear that to decide which is the prime cause, both should be investigated on the same population. When this has been done in autopsy material, emphysema appears to be more important than airway changes. When investigated on surgical resection specimens, the loss of alveolar wall attachment appears to have a most important effect on the FEV_1.[82,139,140,159–161]

The most likely basis for the functional abnormalities in COPD is that panacinar emphysema and the loss of alveolar support to bronchioles, as estimated by an increase in the IAAD, relate to airway obstruction in both early and late disease. Though severe small airway damage is not seen in early disease, it is common in end-stage disease but its relative significance when compared to the usually severe emphysema present at that stage is still uncertain. Tests of gas transfer show that there is a close correlation with the severity of emphysema.

CLINICOPATHOLOGIC FEATURES OF COPD

Pneumothorax as a Complication of COPD

Pneumothorax can be classified as either spontaneous or traumatic. The spontaneous pneumothoraces can be divided into primary, with no clinically evident underlying pulmonary disease, and secondary, which occurs when the pneumothorax is associated with a clinically evident preexisting disease process. Eighty percent of primary spontaneous pneumothoraces occur in young adult males. Of the remaining 20 percent, half of those that occur in older patients occur in patients with COPD.[164] The commonest cause of spontaneous pneumothorax in over 40s is COPD.[165]

Though a recognized complication of COPD, pneumothorax is a rare complication, occurring in only 0.8 percent of patients admitted to hospital with an acute exacerbation of COPD.[166] This figure may be increased if the patient undergoes artificial ventilation when high inflation pressures have been used.

In young adults pneumothorax appears to be associated in many cases with the presence of blebs in the pleura. These blebs must be clearly differentiated from bullae (Fig. 19-9). The blebs are due to the presence of air within the pleura, i.e., due to interstitial emphysema. They are probably due to the rupture of alveoli along lobular septae with tracking of air into the pleura.[167] In patients with COPD the pneumothorax is associated with emphysema in the underlying lung. In those in whom surgical intervention occurs, usually to carry out a pleurodesis, the surgeon may remove areas of lung he thinks may be the source of the leak. These areas almost invariably consist of portions of scarred and emphysematous lung from the apex of the lower or upper lobe. The pattern of emphysema may be technically paraseptal but the severity of the associated scar tissue makes classification difficult, and they are best considered small bullae. The differentiation between blebs and bullae is important theoretically but probably requires histological examination of the underlying lung for absolute differentiation.

In the absence of COPD, paraseptal emphysema has been associated with serial pneumothoraces in cases of extensive and severe subpleural paraseptal emphysema. In the reported cases, the emphysema was so severe as to give a "soap bubble" frothy appearance to the chest x-ray.[168]

Polycythemia

Polycythemia is a physiologic adaptation due to chronic hypoxia and is mediated by an increased erythropoietin release from the kidney due to tissue hypoxia. Polycythemia is seen in patients with COPD and hypoxia.[169] The level of polycythemia can be affected by the patient's smoking. Smoking and raised carboxyhemoglobin levels alone can produce polycythemia and in COPD the red cell mass is affected by the mean carboxyhemaglobin levels.[170,171] Factors which potentiate the hypoxia are exercise, which adds to the overall hypoxic stimulus,[172] and the level and extent of nocturnal desaturation associated with sleep apnea.[175]

Long-term oxygen therapy is associated with a decrease in the polycythemia[174,175] though again the presence of high levels of carboxyhemaglobin limit this effect of oxygen therapy.

Polycythemia increases the risk of vascular thrombotic incidents in COPD. In COPD there is an

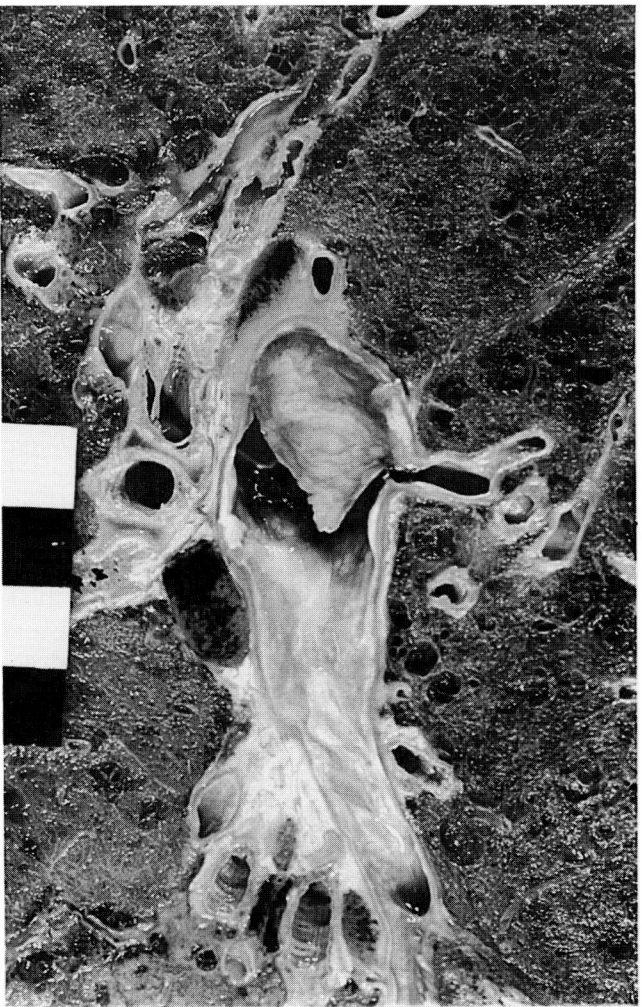

FIGURE 19-10
A sagittal section of lung adjacent to the hilum from a patient dying of COPD after five years long-term oxygen therapy. There was right ventricular hypertrophy. The pulmonary artery is markedly dilated and partially occluded by old laminated *in situ* thrombus. Compare its diameter to that of the adjacent lobar bronchus. The intima shows marked atheroma.

increase in deep vein thrombosis in patients admitted with exacerbations and subsequent pulmonary emboli.[175] We have observed extensive thrombosis of the dilated proximal pulmonary arteries in patients dying from COPD (Fig. 19-10).

Weight Loss in Patients with COPD

Weight loss occurs in many patients with COPD and has been associated with an increased morbidity and mortality.[176] The decreased body mass is not necessarily associated with the most severe respiratory disease.[177] The weight loss seen in patients with COPD has in the past been ascribed to increased muscular work of breathing. However, patients with similar increases in respiratory work load due to diseases other than COPD do not necessarily show a decrease in body mass.[178] It is not explained on simple malnutrition or negative energy levels as the weight loss occurs despite normal or increased dietary intake and attempts to reverse the weight loss by a variety of dietary supplements.

Several studies have shown that patients with COPD have an abnormally high resting energy expenditure, increased basic metabolic rates, and abnormal thermic responses to food.[179] There appears to be a close relationship between the increased resting energy expenditure and increased weight loss and more severe airway obstruction.[180] The weight loss and muscle wasting, including the respiratory muscles and diaphragm, appear similar to the weight loss seen in association with some tumors.

Several recent studies have implicated increased serum concentrations of tumor necrosis factor (TNFα) probably derived from "hypoxia-primed" monocytes.[180,181] The suggestion that weight loss and heart failure might be the result of altered cytokine activity mediated by hypoxia has opened the possibility of therapeutic interference.

Respiratory Muscles and Diaphragm

The respiratory muscles are more important for inspiration than expiration. Expiration is largely passive, dependent on the elastic recoil of the chest wall and lung. Inspiration is mainly due to contraction of the diaphragm. The muscle fibers shorten and the diaphragm moves down with an increase in intra-abdominal pressure and a decreased intrapleural pressure. The intercostal and neck muscles help to stabilize the chest but become increasingly involved as the work of respiration increases.

With increasing airway obstruction there is hyperinflation of the lung and chest wall and the diaphragm is particularly affected. The diaphragm shortens, loses its upwards curvature and becomes flattened. This loss of curvature was the characteristic feature identified by radiologists in their attempts to diagnose emphysema.[101] One might expect that the increased work required to work against the mechanical disadvantages of a hyperinflated chest, and the increased work load imposed by deteriorating respiratory function, would lead to hypertrophy of the diaphragm and the respiratory muscles. However, autopsies show that there is a decrease in the area, thickness, and volume of the diaphragm in COPD, and this is proportional to the amount of emphysema present.[182,183] Thurlbeck showed that though in normals the diaphragmatic weight and thickness were in proportion to body weight, in COPD diaphragmatic weight declined as a proportion of body weight.[184] This is even more surprising when one considers that body weight itself falls in later stages of COPD and the diaphragmatic loss appears to be greater than the generalized muscle loss. It is possible that the mechanical disadvantages of hyperinflation and the low flat position of the diaphragm means that it is less involved in respiration, with some of its functions being taken over by the abdominal muscles. Respiratory muscle fatigue may contribute to respiratory failure[185] particularly in exacerbations when respiratory stimulants have been given. One of the roles of artifical ventilation in such cases is to take over the role of the failing respiratory muscles.

Cor Pulmonale[186,187]

Clinicians often use the term *cor pulmonale* to indicate that there is right heart failure secondary to pulmonary heart disease or to the onset of peripheral edema or even to indicate the presence of pulmonary hypertension.

Because of the clinical problems inherent in the diagnosis of right heart failure in life, cor pulmonale was specifically defined in pathological terms by a WHO Committee as: "hypertophy of the right ventricle resulting from diseases affecting the function and/or structure of the lungs, except when these pulmonary alterations are the result of disease that primarily affects the left side of the heart as in congenital heart disease."[188] However, there is no attempt to define the term "normality" and the clinical use was severely limited by the inability to recognize right ventricular hypertrophy on radiologic or ECG terms in life.

At autopsy the assessment of right ventricular hypertrophy is best assessed by the technique of Fulton et al.[189,190] This involves dissection of the heart, removal of all fat, atria, and vessels and separate weighing of the free wall of the right ventricle, comparing it to the weight of the left ventricle and septum. There is a wide variation in ventricular weight due to the variation in body size and physical activity and this makes the assessment of minimal degree of right ventricular hypertrophy difficult. The most sensitive method is to compare the ratio between the weight of the left ventricle and septum to that of the free wall of the right ventricle. This ratio is normally greater than 2.2 but values as low as 1

associated with the weight of the free wall of the right ventricle as high as 160 g can be seen in hypoxic cor pulmonale due to COPD.

The original criteria for normality produced by Fulton et al. were based on hearts from patients under the age of 65, and in the elderly the normal weight of the right ventricle is significantly lower than in younger patients probably representing decline in physical activity.[190] Assessment of right ventricular hypertrophy in this population would require a new set of control data. The comparison of the ratio between left and right ventricular weight is based on the assumption of normality of the left ventricle and therefore is inappropriate in patients with valvular disease, hypertension, or previous myocardial infarcts. In such cases the weight of the free wall of the right ventricle taken alone is of value, the upper value for normal being about 75 g and values of 80 g and over are usually considered abnormal.

Measurement of the thickness of the right ventricular wall does not accurately reflect muscle mass, as the ventricular dilatation in cardiac failure gives a relatively thin wall. This underestimates the degree of hypertrophy. It is characteristic that the heart shadow on chest x-ray increases its diameter in exacerbations of COPD. This enlargement is largely due to right ventricular dilatation.

Whether true heart failure in terms of pump failure is a prerequisite of cor pulmonale is uncertain, as cardiac output is usually maintained.[191–193] The ankle edema so often seen in the later stages of COPD is of uncertain pathogenesis but may be due to problems in handling water and electrolytes rather than being a consequence of ventricular failure.

The pathogenesis of the right ventricular hypertrophy appears to be closely related to the levels of arterial hypoxemia[191–194] which can be reversed by long-term oxygen therapy, possibly explaining the improved prognosis of those treated by this method.[174,175] The relationship of right ventricular hypertrophy to hypoxemia is much closer than the relationship of right ventricular hypertrophy to either pulmonary artery pressure or pulmonary vascular resistance.[194] Left ventricular hypertrophy is common in emphysema and its cause is unknown.

The loss of vascular bed in emphysema which might be expected to increase pulmonary artery pressure does not appear to be related to cor pulmonale.[195–197] The sudden decrease in the vascular bed following resection for carcinoma of the bronchus may precipitate the onset of cor pulmonale in an otherwise compensated patient with COPD. The changes in the vessels in chronic hypoxia are given in Chap. 21.

Autopsy Findings in Patients with COPD

At autopsy the patient may show evidence of cachexia, edema of the legs and cyanosis.

- *Emphysema:* On opening the chest, the lungs fail to deflate due to air trapping and those with panacinar emphysema have a characteristic soft pillowy texture. Air can be easily moved around inside the lung due to the extensive large interconnecting airspaces. There may be bullae, most commonly associated with the apex, the anterior borders of the lobes, and the lung bases. In the last location, type III bullae are commonly seen in association with severe panacinar emphysema. If the lungs are fixed by distention it will be possible to assess the type, distribution, and severity of emphysema. Most commonly seen is severe widespread centriacinar emphysema either alone or associated with panacinar emphysema. In only a small proportion of cases, less than 20 percent, is panacinar emphysema found alone without evidence of centriacinar emphysema. It is characteristic that the lesions of centriacinar emphysema in end-stage disease involve the whole lung from apex to base.

- *Emphysema in α1-Pi deficiency:* The lungs of patients with severe COPD associated with α1-Pi deficiency usually show severe widespread panacinar emphysema, which is classically said to be worse in the lower lobes. However, in those lungs I have examined at autopsy or from transplant cases this difference has not been evident. My experience is that there is also often an associated or superimposed widely distributed centriacinar emphysema, sometimes only clearly seen in the areas least affected by panacinar emphysema. Macroscopically or microscopically there is no significant difference between the type of emphysema seen in α1-antitrypsin deficiency and in the more usual COPD, though it is true that in α1-antitrypsin deficiency the lung is more pillowy and less fibrotic than many end-stage COPD lungs. An active purulent bronchitis and bronchiolitis with patchy pneumonia has been common, even in transplant cases, and may be part of the disease process.

- *Infections:* It is common to find terminal chest infections with acute bronchitis, bronchiolitis, and bronchopneumonia. Pneumonia involving emphysematous lungs can produce unusual pat-

terns of lung involvement with failure of resolution. The lesions of centriacinar emphysema or larger individual spaces in panacinar emphysema may fill with pus to give a nodular appearance on the chest x-ray. Pus spilling into bullae can produce fluid levels on the chest x-ray and mimic lung abscesses.

- *Mucus plugging of large or small bronchi* can be seen though it is never as widespread or the mucus as solid and tenacious as in asthma. Occasionally mucus plugging of airways may produce mucus impaction in patients with severe mucus hypersecretion. The author has seen examples in which fungus, probably *Aspergillus* species, was growing within mucus plugs but without the eosinophilic infiltrate characteristic of an allergic bronchopulmonary aspergillosis.

- *Small airway disease with narrowed lumina* is difficult to identify in early COPD. Scarred small airways are particularly seen in areas of severe emphysema. The presence of a terminal chest infection makes any attempt to assess the significance of chronic airway inflammation and its role in the pathogenesis of COPD very difficult.

- *Capillary bed:* The capillary bed is reduced in emphysema due to the loss of alveolar walls but whether the capillary bed in the remaining lung is normal or reduced remains uncertain. It is certainly possible to see a normal pattern of capillary arcades in many severely emphysematous lungs when the alveolar walls are seen *en face* in histologic sections. However some lungs show focal loss or distortion of the residual capillaries (Fig. 19-11A–C).

- *Cor pulmonale and right ventricular hypertrophy* are usual and marked in those patients that have been hypoxic in life.

- *Hypoxic related changes in other organs:* Carotid body enlargement is associated with hypoxemia.[198,199] Enlargement of renal glomeruli and the juxtaglomerular body may also be seen.[200]

- *Smoking-related changes in airways* are usual. These consist of sub-mucosal gland hypertrophy, increase in goblet cell number, and increase in the distribution into the periphery of the lung. This increase in goblet cells may be masked by metaplastic changes in the proximal airways.

- *An atrophy in bronchial cartilage* is said to occur in association with emphysema and with mucus-gland enlargement.[201,202] The significance of such claims are uncertain and not all studies have confirmed them.[47] If atrophy of the cartilage was asso-

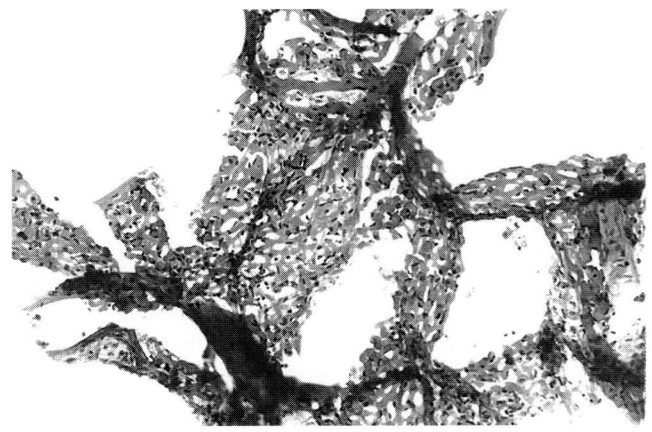

A

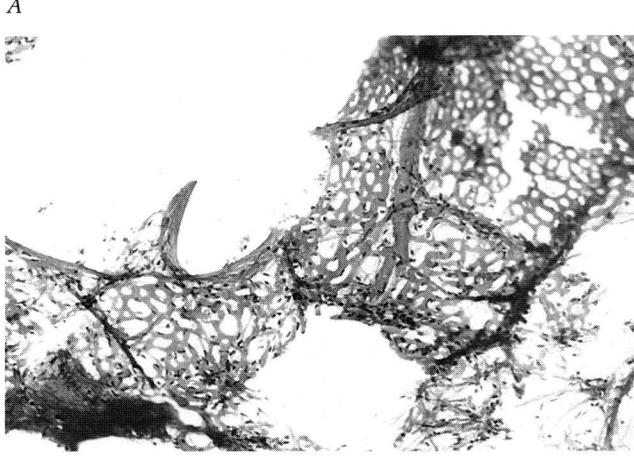

B

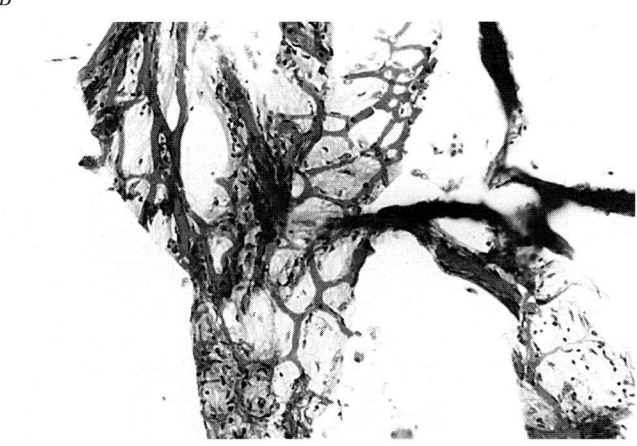

C

FIGURE 19-11

A. The capillary bed from the lung of a 70-year-old nonsmoker. Compare this with *B*, a capillary bed from a 60-year-old patient with severe COPD and marked emphysema. The vascular bed has a normal pattern, though there is some stretching of the capillary arcades. Overall the pattern is comparable to that in Fig. 19-11*A*. *C.* Another view of the same lung as Fig. 19-11*B* showing an area where the capillary arcades are grossly abnormal and many have been lost. Such "bare areas" may impair gas transfer. Such capillary loss is not specific to emphysema.

ciated with airway obstruction it would involve the proximal one-third of airway generations, where cartilage provides a major support to the airways. All evidence suggests that it is the smaller peripheral airways which are the seat of the fixed airways obstruction in COPD.

• *Cancer:* There is evidence of an increased incidence of lung cancer in COPD.[203,204]

Acknowledgment

I wish to thank Dr. W. A. H. Wallace for his help with the clinical aspects of this chapter.

REFERENCES

1. Burrows B: Course and prognosis in advanced disease, in Petty TL (ed), *Chronic Obstructive Pulmonary Disease.* New York, Marcel Dekker, 1985, pp 31–42.
2. American College of Chest Physicians/American Thoracic Society: Pulmonary terms and symbols. *Chest* 67:583–593, 1975.
3. Burrows B: Airway obstructive diseases: Pathogenetic mechanisms and natural history of the disorders. *Med Clin North Am* 74:547–560, 1990.
4. Fletcher CM, Peto R, Tinker CM, Speizer FE: *The Natural History of Chronic Bronchitis and Emphysema. An Eight Year Study of Working Men in London.* Oxford, Oxford University Press, 1976.
5. Orie NG, Sluiter HJ, de Vries K, et al: The host factor in bronchitis. in Orie NGM, Sluiter HJ (eds), *Bronchitis an International Symposium.* Assen, Royal Vangorcum, 1961, pp 43–69.
6. Burrows B, Nieden AH, Fletcher CM, et al: Clinical types of chronic obstructive lung disease in London and Chicago. *Am Rev Respir Dis* 90:14–27, 1964.
7. Higgins MD, Thom T: Incidence, prevalence and mortality: Intra- and intercountry differences, in Hensley MJ, Saunders NA (eds), *Clinical Epidemiology of Chronic Obstructive Pulmonary Diseases,* Vol. 43, *Lung Biology in Health and Disease.* New York, Marcel Dekker, 1989, pp. 23–43.
8. Coultas DB, Samet JM. Cigarette smoking, in Heasley NMJ, Saunders NA (eds), *Clinical Epidemiology of Chronic Obstructive Pulmonary Diseases,* Vol. 43, *Lung Biology in Health,* New York, Marcel Dekker, 1989, pp 109–138.
9. Garshick E, Schenker MB: Occupation and chronic airflow limitation, in Hensley MJ, Saunders NA (eds), *Clinical Epidemiology of Chronic Obstructive Pulmonary Diseases.* Vol. 43, *Lung Biology in Health and Disease.* New York, Marcel Dekker, 1989, pp 227–258.
10. Strachan DP: Epidemiology: A British perspective, in Calverley P, Pryde N (eds), *Chronic Obstructive Pulmonary Disease.* London, Chapman Hall, 1995, pp 50–67.
11. Pryde NB, Burrows B: Development of impaired lung function: Natural history and risk factors, in Calverley P, Pryde N (eds), *Chronic Obstructive Pulmonary Disease.* London, Chapman Hall, 1995, pp 69–91.
12. United States Department of Health and Human Services, Public Health Service: The health consequences of smoking:

Chronic obstructive lung disease. A report of the Surgeon General, US Government Printing Office, Washington DC, 1984.

13. United States Department of Health and Human Services, Public Health Service. The health benefits of smoking cessation. A report of the Surgeon General, US Government Printing Office, Washington DC, 1990.

14. Doll R, Peto R: Mortality in relation to smoking. Twenty years observation on male British doctors. *Br Med J* 4:1525–1536, 1976.

15. Peto R, Speizer FE, Cochrane AL, et al: The relevance in adults of airflow obstruction, but not of mucus hypersecretion, in mortality from chronic lung disease. *Am Rev Respir Dis* 128:491–500, 1983.

16. Miller BG, Jacobsen M: Dust exposure, pneumoconiosis and mortality of coal miners. *Br J Ind Med* 42:723–733, 1985.

17. Ruckley VA, Gauld SJ, Chapman JS, et al: Emphysema and dust exposure in a group of coal workers. *Am Rev Respir Dis* 129:528–532, 1984.

18. Rogan JM, Attfield MD, Jacobsen M, et al: Role of dust in the working environment in the development of chronic bronchitis in British coal miners. *Br J Ind Med,* 30:217–226, 1973.

19. Love R, Miller BD: Longitudinal study of lung function in coal miners, 1980. *Thorax* 37:193–197, 1980.

20. Stockley RA: Biochemical and cellular mechanisms, in Calverley P, Pryde N. (eds), *Chronic Obstructive Pulmonary Disease.* London, Chapman and Hall, 1995, pp 93–133.

21. Laurell CB, Eriksson S: The electrophoretical alpha-1 globulin pattern of serum in alpa-1-antitrypsin deficiency. *Scand J Clin Lab Invest,* 15:132–140, 1963.

22. Eriksson S: Studies in alpha-1-antitrypsin deficiency. *Acta Med Scand* 177:(Suppl 432), 1965.

23. Snider GL: Chronic obstructive pulmonary disease—A continuing challenge. *Am Rev Respir Dis* 133:942–944, 1986.

24. Benson MK: Chronic bronchitis, emphysema and chronic obstructive airways disease, in Weatherall DJ, Ledingham JGG, Warrell DA. (Eds), *Oxford Textbook of Medicine,* 2d ed. Oxford, Oxford Medical Publications, 1987, pp15.83–15.91.

25. Flenley DC: Chronic bronchitis and emphysema, in *Respiratory Medicine.* London, Bailliere Tindall, 1981, pp 91–109.

26. Rogers TK, Howard P: COPD: Clinical features in management, in Brewis RAL, Corrin B, Geddes DM, and Gibson GJ (eds): *Respiratory Medicine,* 2d ed. London, Saunders, 1995.

27. Burrows BE, Fletcher CM, Heard BE, et al: The emphysematous and bronchial types of chronic airways obstruction. A clinicopathological study of patients in London and Chicago, *Lancet* 1:830–835, 1966.

28. Dornhorst AC: Respiratory insufficiency. *Lancet,* 1:1185–1187, 1955.

29. Filley GF: Emphysema and chronic bronchitis: Clinical manifestations and their physiological significance. *Med Clin North Am* 51:283–292, 1967.

30. Fletcher CM, Peto R: The natural history of chronic airflow obstruction. *Br Med J* 1:1645–1648, 1977.

31. Medical Research Council: Definition and classification of chronic bronchitis for clinical and epidemiological purposes. A report to the Medical Research Council by their Committee on the Aetiology of Chronic Bronchitis. *Lancet,* 1:775–780, 1965.

32. Ryder RC, Dunnill MS, Anderson JA: A quantitative study of bronchial mucus gland volume, emphysema and smoking in a necropsy population. *J Pathol* 104:59–71, 1971.

33. Lumsden AB, McLean A, Lamb D: Goblet and Clara cells of human distal airways: Evidence for smoking-induced changes in their numbers. *Thorax* 39:844–849, 1984.

34. Reid L: Measurements of the bronchial mucous gland layer: A diagnostic yardstick in chronic bronchitis. *Thorax* 15:132–141, 1960.

35. Lamb D, Reid L: Goblet cell increase in rat bronchiolar epithelium after exposure to cigarette and cigar smoke. *Br Med J* 1:33–35, 1969.

36. Stephens RJ, Freeman G, Crane SC, Furiosi NJ: Ultrastructural changes in the terminal bronchiole of the rat during continuous low level exposure to nitrogen dioxide. *Exp Mol Pathol* 14:1–19, 1971.

37. Lamb D, Reid L: Mitotic rates, goblet cell increase and histochemical changes in mucus in rat bronchial epithelium during exposure to sulphur dioxide. *J Pathol Bacteriol* 96:97–111, 1968.

38. Peto R, Speizer FE, Moore CF, et al: The relevance in adults of airflow obstruction, but not of mucous hypersecretion to mortality from chronic lung disease. *Am Rev Respir Dis* 123:491–500, 1983.

39. Ciba Guest Symposium: Terminology, definitions and classification of chronic pulmonary emphysema. *Thorax* 14:286–299, 1959.

40. American Thoracic Society: Chronic bronchitis, asthma and pulmonary emphysema. A statement by the Committee on Diagnostic Standards for nontuberculous diseases. *Am Rev Resp Dis* 85:762–768, 1962.

41. Snider GL, Kleinerman J, Thurlbeck WM: The definition of emphysema. Report of the National Heart, Lung and Blood Institute, Division of Lung Diseases Workshop. *Am Rev Respir Dis* 132:182–185, 1985.

42. Reid L: *The Pathology of Emphysema.* London, Lloyd Luke, 1967, pp 1–21.

43. Heard BE: *The Pathology of Chronic Bronchitis and Emphysema.* London J and A Churchill, 1969.

44. Thurlbeck WM: Chronic airflow obstruction in lung disease, in Bennington JL (ed), *Major Problems in Pathology,* Series No 5, Vol. 98, Philadelphia, WB Saunders, 1976, p. 300.

45. Gough J, Wentwood JE: Use of thin sections of entire organs in morbid anatomical studies. *J R Microsc Soc* 69:231–235, 1949.

46. Heard BE: *The Pathology of Chronic Bronchitis and Emphysema.* London, J & A Churchill Ltd, 1969.

47. Ryder RC, Thurlbeck WM, Gough J: A study of interobserver variation in the assessment of the amount of pulmonary emphysema in paper mounted whole lung sections. *Am Rev Resp Dis* 99:354–364, 1969.

48. Dunnill MS: Quantitative methods in the study of pulmonary pathology. *Thorax* 17:320–329, 1962.

49. Dunnill MS: Evaluation of a simple method of sampling the lung for quantitative histological analysis. *Thorax* 19:443–448, 1964.

50. Thurlbeck WM. Measurement of pulmonary emphysema. *Am Rev Resp Dis* 95:752–764, 1967.

51. Thurlbeck WM, Horowitz I, Siemiatycki J et al: Intra- and interobserver variations in the assessment of emphysema. *Arch Environ Health* 18:646–659, 1969.

52. Thurlbeck WM, Dunnill MS, Hartung W, et al: A comparison of three methods of measuring emphysema. *Hum Pathol* 1:215–226, 1970.

53. Weibel ER: *Morphometry of the Human Lung.* Berlin, Springer Verlag, 1963.

54. Schreider JP, Raabe OG: Structure of the respiratory acinus. *Am J Anat* 162:221–232, 1957.

55. Kim WD, Eidelman DH, Izquiredo JL et al: Centrilobular and panlobular emphysema in smokers. Two distinctive morphological and functional entities. *Am Rev Resp Dis* 144:1385–1390, 1991.

56. Thurlbeck WM: The internal surface area of non-emphysema lungs. *Am Rev Respir Dis* 95:765–773, 1967.

57. Aherne WA, Dunnill MS: *Morphometry.* London, Edward Arnold, 1982.

58. McLean A, Warren PM, Gillooly M, et al: Microscopic and macroscopic measurements of emphysema: Relation to carbon monoxide gas transfer. *Thorax* 47:144–149, 1992.

59. Gillooly M, Lamb D. Farrow ASJ: A new automated technique for the assessment of emphysema on histological sections. *J Clin Pathol* 44:1007–1011, 1991.

60. Gould GA, MacNee WM, McLear A, et al: CT measurements of lung density in life can quantitate distal air space enlargement—An essential defining feature of human emphysema. *Am Rev Respir Dis* 137:380–392, 1988.

61. MacNee W, Gould G, Lamb D: Quantifying emphysema by CT scanning. Clinicopathological correlations. *Ann NY Acad Sci* 624:179–194, 1991.

62. Dunnill MS: Postnatal growth of the lung. *Thorax* 17:329–333, 1962.

63. Davies G, Reid L: Growth of the alveoli and pulmonary arteries in childhood. *Thorax* 26:669–681, 1970.

64. Angus GE, Thurlbeck WM: Number of alveoli in the human lung. *J Appl Physiol* 12:483–485, 1972.

65. Thurlbeck WM. Postnatal human lung growth. *Thorax* 37:564–571, 1982.

66. Thurlbeck WM: Internal surface area of non-emphysematous lungs. *Am Rev Respir Dis* 95:765–773, 1967.

67. Thurlbeck WM, Angus GE: Growth and ageing of the normal human lung. *Chest* 67(Suppl):3S–6S, 1975.

68. Gillooly M, Lamb D: Airspace size in lungs of lifelong non-smokers: Effect of age and sex. *Thorax* 48:39–43, 1993

69. Gillooly M, Lamb D: Microscopic emphysema in relation to age and smoking habit. *Thorax* 48:491–495, 1993.

70. Gillooly M, Lamb D: The relationship between centriacinar emphysema and microscopically assessed emphysema. *Am Rev Respir Dis* 147(Suppl):A864, 1993.

71. Lichros I, Kim WD, Saetta M, et al: Distribution of microscopic emphysema in whole lungs of smokers. *Am Rev Respir Dis* 143(Suppl):A536, 1991.

72. Linhartova A, Anderson AE, Foraker AG: Radical traction and bronchiolar obstruction in pulmonary emphysema. *Arch Pathol* 92:384–391, 1971.

73. Linhartova A, Anderson AE, Foraker AG: Nonrespiratory bronchiolar deformities. *Arch Pathol* 95:45–48, 1973.

74. Linhartova A, Anderson AE, Foraker AG: Topology of non-respiratory bronchioles of normal and emphysematous lungs. *Hum Pathol* 5:729–735, 1974.

75. Linhartova A, Anderson AE, Foraker AG: Further observations on luminal deformity and stenosis of nonrespiratory bronchioles in pulmonary emphysema. *Thorax* 32:53–59, 1977.

76. Linhartova A, Anderson AE, Foraker AG: Affixment arrangements of peribronchiolar alveoli in normal and emphysematous lungs. *Arch Pathol Lab Med* 106:499–502, 1982.

77. Pratt PC, Haque A, Klugh GA: Correlation of postmortem function and structure in normal and emphysematous lungs. *Am Rev Respir Dis* 83:856–865, 1961.

78. Anderson AE, Foraker AG: Relative dimensions of bronchioles and parenchymal spaces in lungs from normal subjects and emphysematous patients. *Am J Med* 32:218–226, 1962.

79. Saetta M, Ghezzo H, Kim WD, et al: Loss of alveolar attachments in smokers. A morphometric correlate of lung function impairment. *Am Rev Respir Dis* 132:894–900, 1985.

80. Petty TL, Silvers GW, Stanford RE: Radial traction and small airways disease in excised human lungs. *Am Rev Respir Dis* 133:132–135, 1986.

81. Wright JL, Hobson JE, Wiggs B, et al: Airway inflammation and peribronchiolar attachments in the lungs of nonsmokers, current and ex-smokers. *Lung* 166:277–286, 1988.

82. Lamb D, McLean A, Gillooly M, et al: Relation between distal airspace size, bronchiolar attachments and lung function. *Thorax* 48:1012–1017, 1993.

83. Nagai A, Yamawaki I, Takizawa T. Thurlbeck WM: Alveolar attachments in emphysema of human lungs. *Am Rev Respir Dis* 144:888–891, 1991.

84. Willems LNA, Kramps JA, Stijen T, et al: Relationship between small airways disease and parenchymal destruction in surgical lung specimens. *Thorax* 45:89–94, 1990.

85. Gillooly M, Lamb D: The effect of age on the alveolar support of peripheral bronchioles in non-smokers. *Am Rev Respir Dis* 147(Suppl):A864, 1993.

86. Corrin B, Leibow A, Friedman PJ: Pulmonary lymphangiomyomatosis: A review. *Am J Pathol* 79:347–382, 1975.

87. Taylor JR, Ryv J, Colby TV, Raffin TA: Lymphangioleiomyomatosis: Clinical course of 32 patients. *N Engl J Med* 323:1254–1260, 1990.

88. Fukuda Y, Kawamoto M, Yamamoto A, et al: Role of elastic fibre degeneration in emphysema-like lesions of pulmonary lymphangiomyomatosis. *Hum Pathol* 21:1252–1261, 1990.

89. Boren HG: Alveolar fenestrae: Relationship to pathology and pathogenesis of pulmonary emphysema. *Am Rev Respir Dis* 85:328–344, 1962.

90. Takaro T, Gaddy LR, Pirra S: Thin alveolar epithelial partitions across connective tissue gaps of the human lung: Ultrastructural observations. *Am Rev Respir Dis* 126:328–331, 1982.

91. Cosio MG, Shiner RJ, Saetta M, et al: Alveolar fenestrae in smokers. Relationship with light microscopic and functional abnormalities. *Am Rev Respir Dis* 133:126–131, 1986.

92. Kuhn C, Tavassoli F: The scanning electron microscopy of elastase-induced emphysema. *Lab Invest* 34:2–9, 1986.

93. Saetta M, Shiner RJ, Angus GE, et al: Destructive Index: A measurement of lung parenchymal destruction in smokers. *Am Rev Respir Dis* 131:764–769, 1985.

94. Eidelman DH, Ghezzo H, Kim WD, Cosio MG: A destructive index and early lung destruction in smokers. *Am Rev Resp Dis* 144:156–159, 1991.

95. Myers JL, Veal CF, Shin MS, et al: Respiratory bronchiolitis causing interstitial lung disease: A clinicopathologic study of six cases. *Am Rev Respir Dis* 135:880–884, 1987.

96. Niewoehner DE, Kleinamen J, Rice DB: Pathologic changes in the peripheral airways of young cigarette smokers. *N Eng J Med* 291:755–758, 1974.

97. Nagai A, Thurlbeck WM: Scanning electron microscopic observations of emphysema in humans: A descriptive study. *Am Rev Resp Dis* 144:901–908, 1991.

98. Lang MR, Fiaux GW, Hulmes DJS, et al: Quantitative studies

of human lung airspace wall in relation to collagen and elastin content. *Matrix* 13:471–480, 1993.

99. Lang MR, Fiaux GW, Gillooly M, et al: Collagen content of alveolar wall tissue in emphysematous and non-emphysematous lungs. *Thorax* 49:319–326, 1994.

100. Cardoso WV, Sekhon HS, Hyde DM, Thurlbeck WM: Collagen and elastin in human pulmonary emphysema. *Am Rev Respir Dis* 147:975–981, 1993.

101. Thurlbeck WM, Simon G: Radiographic appearance of the chest in emphysema. *Am J Roentgenol* 14:429, 1978.

102. Murata K, Itoh H, Todo T, et al: Centrilobular lesions of the lung. Demonstration by high-resolution CT and pathologic correlation. *Radiology* 161:641–646, 1986.

103. Bergin C, Müller N, Nichols DM, et al: The diagnosis of emphysema. A computed tomographic pathologic correlation. *Am Rev Resp Dis* 133:541–546, 1986.

104. Heremans A, Verschakaelen JA, Fraeyenhoven L, van Demedts M: Measurement of lung density by means of quantitative CT scanning. A study of correlations with pulmonary function test. *Chest* 102:805–811, 1992.

105. Hruban RH, Meziane MA, Zerhouni EA, et al: High resolution computer tomography of the inflation fixed lung. Pathologic-radiologic correlation of centrilobular emphysema. *Am Rev Respir Dis* 136:935–940, 1987.

106. Weissler JC. Pulmonary emphysema: Current concepts of pathogenesis. *Am J Med Sci* 293:125–138, 1987.

107. Snider GL. Emphysema: The first two centuries and beyond. A historical overview, with suggestions for future research. *Am Rev Respir Dis* 146:Part 1, 1334–1344; Part 2, 1613–1615, 1992.

108. Anderson AE, Hernandez JA, Beckert P, Foraker AG: Emphysema in lung macrosections correlated with smoking habits. *Science* 144:1025–1026, 1964.

109. Anderson AE, Hernandez JA, Holmes WL, Foraker AG: Pulmonary emphysema prevalence, severity and anatomical pattern in macrosections with respect to smoking habits. *Arch Environ Health* 12:569–577, 1966.

110. Petty TL, Ryan SF, Mitchell RS: Cigarette smoking and the lungs. Relation to postmortem evidence of emphysema, chronic bronchitis and black lung pigmentation. *Arch Environ Health* 14:172–177, 1967.

111. Ishikawa S, Bowden DH, Fisher V, Wyatt JP: The "emphysema profile" in two mid-western cities in North America. *Arch Environ Health* 18:660–666, 1969.

112. Ryder R, Dunnill MS, Anderson JA: A quantitative study of bronchial mucus gland volume, emphysema and smoking in a necropsy population. *J Pathol* 104:59–71, 1971.

113. Spain DM, Sieve LH, Bradess VA: Emphysema in apparently healthy adults. Smoking, age and sex. *JAMA* 224:322–325, 1973.

114. Thurlbeck WM, Ryder RC, Sternby N: A comparative study of the severity of emphysema in necropsy populations in three different countries. *Am Rev Respir Dis* 109:239–248, 1974.

115. Bignon J, Lenfant C, Scarpa GL: Emphysema: Past, present and future. *Bull Eur Physiol Pathol Respir* 16(Suppl 4):423–428, 1980.

116. Azcy A, Anderson AE, Foraker AG: The morphological spectrum of the ageing and emphysematous lungs. *Ann Intern Med* 57:1–17, 1962.

117. Anderson AE, Foraker AG: Centrilobular and panlobular emphysema: Two different diseases. *Thorax* 28:547–550, 1973.

118. Sutinen S, Vaajalahti P, Paakko P: Prevalence, severity and types of pulmonary emphysema in a population of deaths in a Finnish city. Correlation with age, sex and smoking. *Scand J Respir Dis* 59:101–115, 1978.

119. Dijkman JH: Morphological aspects, classification and epidemiology of emphysema. *Bull Eur Physiol Pathol Respir* 22:241S–243S, 1986.

120. West JB: Distribution of mechanical stress in the lung. A possible factor in localisation of pulmonary disease. *Lancet:* 1839–841, 1971.

121. Yamanaka A: Pulmonary emphysema in Japan. *Pathol Microbiol* 35:161–166 1970.

122. Hasleton PS: Incidence of emphysema at necropsy as assessed by point counting. *Thorax* 27:552–556, 1972.

123. Cooke R, Toogood I: Emphysema in Papua New Guinea, a pathological study. *Aust NZ J Med* 5:147–154, 1975.

124. Heppleston AG: The pathological anatomy of simple pneumoconiosis in coal workers. *J Pathol Bacteriol* 66:235–246, 1953.

125. Cockroft A, Wagner JC, Ryder R, et al: Postmortem study of emphysema in coal workers and non-coal workers. *Lancet* 2:600–603, 1983.

126. Becklake MR, Irwig L, Kielkowski D, et al: The predictors of emphysema in South African gold miners. *Am Rev Respir Dis* 135:1234–1241, 1987.

127. Evans MD, Pryor WA: Cigarette smoking emphysema and damage to alpha-1-proteinase inhibitor. *Am J Physiol* 266:L593–L611, 1994.

128. Gross P, Phitzer EA, Tolker E, et al: Experimental emphysema. Its production with papain in normal and silicotic rats. *Arch Environ Health* 11:50–58, 1964.

129. Burrows E, Earle RH: Course and prognosis of chronic obstructive lung disease. *N Engl J Med* 280:397–404, 1969.

130. Beuist AS, Sexton GJ, Azzan A-MH, Adams BE: Pulmonary function in heterozygotes for alpha-1-antitrypsin deficiency; a case controlled study. *Am Rev Respir Dis* 120:759–766, 1979.

131. Snider GL, Lucey EC, Stone PJ: Animal models of emphysema. *Am Rev Respir Dis* 133:149–169, 1986.

132. Dunnill MS: Aetiology of emphysema. *Bull Eur Physiol Pathol Respir* 15:1015–1029, 1979.

133. Hoidal JR, Niewoehner DE: Cigarette smoke inhalation potentiates elastase induced emphysema in hamsters. *Am Rev Respir Dis* 127:478–481, 1983.

134. Janoff A: Elastases and emphysema: Current assessment of the protease-antiprotease hypothesis. *Am Rev Respir Dis* 132:417–433, 1985.

135. Martin T, Raghu G, Maunder R, Springmeyer S: The effects of chronic bronchitis and chronic airflow obstruction on the lung cell populations recovered by bronchoalveolar lavage. *Am Rev Respir Dis* 132:254–266, 1986.

136. Damiano VV, Tsang A, Kucich U, et al: Immunolocalisation of elastase in human emphysematous lungs. *J Clin Invest* 78:482–93, 1986.

137. Fox B, Bull PB, Buz A, et al: Is neutrophil elastase associated with elastic tissue in emphysema? *J Clin Pathol* 41:435–40, 1988.

138. Hoidal JR, Fox RB, Le Marbe PA, et al: Altered oxidative metabolic responses in vitro of alveolar macrophages from asymptomatic cigarette smokers. *Am Rev Respir Dis* 123:85–89, 1981.

139. Nagai A, West WW, Paul JL, Thurlbeck WM: The National Institutes of Health Positive Pressure Breathing Trial:

Pathology Studies I: Interrelationship between morphologic lesions. *Am Rev Respir Dis* 132:937–945, 1985.

140. Nagai A, West WM, Thurlbeck WM: The National Institutes of Health Intermittent Positive Pressure Breathing Trial: Pathology Studies II. Correlations between morphologic findings, clinical findings and the evidence of respiratory airflow obstruction. *Am Rev Respir Dis* 132:946–953, 1985.

141. Mitchell RS, Standford R, Johnstone JM, et al: The morphologic features of the bronchi, bronchioles and alveoli in chronic airway obstruction. A clinico-pathological study. *Am Rev Respir Dis* 114:137–145, 1976.

142. Petty TL, Ryan SF, Mitchell RS: Cigarette smoking and the lungs. Relation to postmortem evidence of emphysema, chronic bronchitis and black lung pigmentation. *Arch Environ Health* 14:172–177, 1967.

143. Hale KA, Ewing SL, Gosnell BA, Niewoehner DE: Lung disease in long term cigarette smokers with and without chronic airflow obstruction. *Am Rev Respir Dis* 130:716–721, 1984.

144. Wright JL, Cagle P, Churg A, Colby TV, Myers J: State of the art: Diseases of the small airways. *Am Rev Respir Dis* 146:240–262, 1992.

145. Greaves IA, Colebatch HJH: Editorial: Observations on the pathogenesis of chronic airflow obstruction in smokers: Implications for the detection of "early" lung disease. *Thorax* 41:81–87, 1986.

146. Pryde NB: COPD: Pathophysiology, in Brewis RAL: Corrin B, Geddes DM, Gibson GJ (eds.): *Respiratory Medicine*, 2d ed. London, Saunders, 1995.

147. Groskin SA: Editorial: Emphysema: Fact, fiction, or just a lot of hot air? *Radiology* 183:319–320, 1992.

148. Hogg JC, Macklem PT, Thurlbeck WM: Site and nature of airway obstruction in chronic obstructive lung disease. *N Engl J Med* 278:1355–1360, 1968.

149. Cosio MG, Ghezzo H, Hogg JC, et al: The relations between structural changes and small airways and pulmonary function tests. *N Engl J Med* 298:1277–81, 1978.

150. Berend N, Woolcock AJ, Martin GE: Correlation between the function and structure of the lung in smokers. *Am Rev Respir Dis* 119:695–705, 1978.

151. Wright JL, Wiggs BR, Hogg JC: Airway disease in upper and lower lobes in lungs of patients with and without emphysema. *Thorax* 39:282–285, 1984.

152. Wright JL, Hobson J, Wiggs BR, et al: Effects of cigarette smoking on the structure of the small airways. *Lung* 165:91–100, 1987.

153. Cosio MG, Hale KA, Niewoehner DE: Morphologic and morphometric effects of prolonged cigarette smoking on the small airways. *Am Rev Respir Dis* 122:265–271, 1980.

154. Saetta M, Ghezzo H, Kim WD, et al: Loss of alveolar attachments in smokers. A morphometric correlate of lung function impairment. *Am Rev Respir Dis* 132:894–900, 1985.

155. Wright JL, Cosio M, Wiggs B, Hogg JC: A morphologic grading system for membranous and respiratory bronchioles. *Arch Pathol Lab Med* 109:163–165, 1985.

156. Matsuba K, Wright JL, Wiggs BR, et al: The changes in airway structure associated with reduced forced expiratory volume in one second. *Eur Respir J* 2:834–839, 1989.

157. Bosken CH, Wiggs BR, Paré PD, Hogg JC. Small airway dimensions in smokers with obstruction to airflow. *Am Rev Respir Dis* 142:563–570, 1990.

158. Lamb D, McLean A, Gillooly M, et al: Relation between distal airspace size, bronchiolar attachments and lung function. *Thorax* 48:1012–1017, 1994.

159. Greaves IA, Colebatch HJH: Observations of the pathogenesis of chronic airflow obstruction in smokers: Implications for the detection of "early" lung disease. *Thorax* 41:81–87, 1986.

160. Mitchell RS, Stanford RE, Johnstone JM, et al: The morphological features of the bronchi, bronchioles and alveoli in chronic airway obstruction: A clinicopathological study. *Am Rev Respir Dis* 114:137–145, 1976.

161. Hale KA, Ewing SL, Gosnell BA, Niewoehner DE: Lung disease in long term cigarette smokers with and without chronic airflow obstruction. *Am Rev Respir Dis* 130:716–721, 1984.

162. Petty TL, Silvers GW, Stanford RE: Radial traction and small airways disease in excised human lungs. *Am Rev Respir Dis* 133:132–135, 1986.

163. Nagai A, Yamawaki I, Takizawa T, Thurlbeck WM: Alveolar attachments in emphysema of human lungs. *Am Rev Respir Dis* 144:888–891, 1991.

164. Watt AG: Spontaneous pneumothorax—A review of 120 consecutive admissions to Royal Perth Hospital. *Med J Aust* 1:186–188, 1978.

165. George RB, Herbert SJ, Shames JM, et al: Pneumothorax complicating emphysema. *JAMA* 234:389–393, 1977.

166. Cabiran LR, Ziskind MM: Spontaneous pneumothorax in pulmonary emphysema. *Dis Chest* 46:571–577, 1964.

167. Coole JC, Gillespie JB: Mediastinal emphysema: Pathogenesis and management. Report of a case. *Dis Chest* 49:104–108, 1966.

168. Edge J, Simon D, Reid L: Peri-acinar (paraseptal) emphysema: Its clinical, radiological and physiological features. *Br J Dis Chest* 60:10–18, 1996.

169. Stradling JR, Laing DJ: Development of secondary polycythemia in chronic airways obstruction. *Thorax* 36:321–5, 1981.

170. Smith JR, Landaw SA: Smokers polycythemia. *N Engl J Med* 298:6–10, 1978.

171. Calverley PMA, Leggatt RGE, McElderry L, Flenley DC: Cigarette smoking and secondary polycythemia in hypoxic cor pulmonale. *Am Rev Respir Dis* 125:507–510, 1982.

172. Spence DPS, Hay JG, Carter J, et al: Oxygen desaturation and breathlessness during corridor walking in chronic obstructive pulmonary disease. *Thorax* 48:115–150, 1993.

173. Nocturnal Oxygen Therapy Trial Group: Continuous or nocturnal oxygen therapy in hypoxaemic chronic obstructive lung disease. *Ann Intern Med* 93:391–398, 1980.

174. Medical Research Council Working Party: Long-term domiciliary oxygen therapy in chronic hypoxic cor pulmonale complicating chronic bronchitis and emphysema. *Lancet* 1:681–686, 1981.

175. Winter JH, Buckler PW, Bautista A, et al: Frequency of venous thrombosis in patients with an exacerbation of chronic obstructive lung disease. *Thorax* 38:605–608, 1983.

176. Wilson DO, Rogers RM, Wright EC, Anthonisen NR: Body weight in chronic obstructive pulmonary disease: National Institutes of Health Intermittent to Positive Pressure Breathing Trial. *Am Rev Resp Dis* 139:1435–1438, 1989.

177. Green JH, Muers MF: Comparisons between basal metabolic rate and diet-induced thermogenesis in different types of chronic obstructive pulmonary disease. *Clin Sci* 83:109–116, 1992.

178. Danaghoe M, Rogers RM, Wilson DO, Pennock BE: Oxygen consumption of the respiratory muscles in normal and in malnourished patients with chronic obstructive pulmonary disease. *Am Rev Respir Dis* 140:385–391, 1989.

179. Schols AMWJ, Souters PB, Mostert R, et al: Energy balance in chronic obstructive pulmonary disease. *Am Rev Respir Dis* 143:1248–1252, 1991.
180. Sridhar MK: Editorial: Why do patients with emphysema lose weight? *Lancet* 345:1190–1191, 1995.
181. Di Francia M, Barbier D, Mege J, Orehek J: Tumour necrosis factor α and weight loss in chronic obstructive pulmonary disease. *Am J Resp Crit Care Med* 150:1453–1455, 1994.
182. Steele RH, Heard BE: Size of the diaphragm in chronic bronchitis. *Thorax* 28:55–60, 1973.
183. Butler C: Diaphragmatic changes in emphysema. *Am Rev Respir Dis* 114:155–159, 1976.
184. Thurlbeck W: Diaphragm and body weight in emphysema. *Thorax* 33:483–487, 1978.
185. Macklem PT, Roussos CS: Respiratory muscle fatigue a cause of respiratory failure? *Clin Sci* 53:419–422, 1977.
186. Wiedemann HP, Mattay RA: Cor pulmonale in chronic obstructive pulmonary disease. *Clin Chest Med,* 11:523–545, 1990.
187. MacNee W: Pulmonary circulation, cardiac function and fluid balance, in Calverley P, Pryde N (eds), *Chronic Obstructive Pulmonary Disease,* London, Chapman & Hall, 1995, pp 241–291.
188. World Health Organisation: Chronic cor pulmonale. A report of the Expert Committee. *Circulation* 27:594–598, 1963.
189. Fulton RM, Hutchinson EC, Jones AM: Ventricular weight in cardiac hypertrophy. *Br Heart J* 14:413–420, 1952.
190. Lamb D: Heart weight and assessment of ventricular hypertrophy, in Dyke SC (ed), *Recent Advances in Clinical Pathology,* Series 6. Edinburgh, Churchill Livingstone, 1973.
191. Flenley DC: Long term oxygen therapy and the pulmonary circulation, in Heath D (ed), *Aspects of Hypoxia.* Liverpool, Liverpool University Press, 1986.
192. Hodgkin JE: Prognosis in chronic obstructive pulmonary disease. *Clin Chest Med* 11:555–569, 1990.
193. Howard P: Drugs or oxygen for hypoxic cor pulmonale? *Br Med J Clin Res* 287:1159–1160, 1983.
194. Calverley PMA, Howatson R, Flenley DC, Lamb D: Clinicopathological correlations in hypoxic cor pulmonale: *Thorax* 47:494–498, 1992.
195. Hicken P, Brewer D, Heath D: The relation between the weight of the right ventricle of the heart and the internal surface area and the number of alveoli in the human lung in emphysema. *J Pathol Bacteriol* 92:529–546, 1966.
196. Hicken P, Heath D, Brewer D: The relation between the weight of the right ventricle and the percentage of abnormal airspace in the lung in emphysema. *J Pathol Bacteriol* 92:519–528, 1966.
197. Biernacki W, Gould GA, White KF, *et al:* Pulmonary haemodynamics, gas exchange and the severity of emphysema as assessed by quantitative CT scan in chronic bronchitis and emphysema. *Am Rev Respir Dis* 139:1509–1515, 1989.
198. Heath D, Smith P, Jago R: Hyperplasia of the carotid body. *J Pathol* 138:115–127, 1982.
199. Bee D, Howard P: The carotid body: A review of its anatomy, physiology and clinical importance. *Monaldi Arch Chest Dis* 48:48–53, 1993.
200. Campbell JL, Calverley PMA, Lamb D, Flenley DC: The renal glomerulus in hypoxic cor pulmonale. *Thorax* 37:607–611, 1982.
201. Thurlbeck WM, Pun R, Toth J, Fraser RG: Bronchial cartilage in chronic obstructive lung disease. *Am Rev Respir Dis* 109:73–80, 1974.
202. Tandon MK, Campbell AH: Bronchial cartilage in chronic bronchitis. *Thorax* 24:607–612, 1969.
203. Samet JM, Humble CG, Pathak DR: Personal and family history of respiratory disease and lung cancer risk. *Am Rev Respir Dis* 134:466–470, 1986.
204. Skillrid DM, Offord KP, Miller RD: Higher risks of lung cancer in chronic pulmonary disease: A prospective matched case control study. *Ann Intern Med* 105:503–507, 1986.

Chapter 20

Carotid Body and Pulmonary Glomera in Cardiorespiratory Disease

Paul Smith

FUNCTION AND DISSECTION OF THE CAROTID BODIES

The carotid bodies are small nodules of tissue situated in the angle of the bifurcation of the common carotid artery. They are chemoreceptors and monitor the partial pressure of oxygen in arterial blood. This chemoreceptor role was first deduced by de Castro[1] who demonstrated that the innervation of the carotid body by the glossopharyngeal nerve was afferent and hence sensory in nature. His interpretation of the histologic structure of the organ led him to infer that it should be sensitive to changes in blood gases. This hypothesis was confirmed by the physiologic studies of Heymans and colleagues.[2] These authors demonstrated that the carotid body responds to a reduction in the partial pressure of oxygen, an elevation of the partial pressure of carbon dioxide, and lowered pH in systemic arterial blood.

It has subsequently been shown that, in animals at least, the carotid body can also be stimulated by changes in temperature, osmolarity, and blood pressure.[3] The carotid body responds to these stimuli by initiating reflex effects on the lungs, heart, and blood vessels via the petrosal ganglion and respiratory centers in the brain. The strongest and most immediate of these reflexes is the response to hypoxia, which initiates a rapid increase in minute ventilation. That this effect is mediated by the carotid bodies has been demonstrated in animals by their removal, causing abolition of the ventilatory response to hypoxia. In human beings, surgical extirpation of the carotid bodies causes no change in ventilation when normal air is breathed. However, acute hypoxia fails to induce in such individuals the customary hyperventilation typical of subjects with intact glomera.[4-6] This is because the human carotid body contributes only some 10 to 15 percent of the ventilatory drive in normal air. As the oxygen tension in arterial blood falls the carotid bodies exert an exponentially more dominant influence on the respiratory centers. Hypercarbia also stimulates ventilation but a major part of this reflex is governed by the central nervous system. It has been estimated that the carotid bodies account for 30 percent of the response to hypercarbia, the remainder being centrally mediated.[5,6] Although the identity of these reflexes is now well documented, the precise mechanism by which the carotid body detects changes in the levels of oxygen and carbon dioxide and translates them into nervous impulses remains a mystery.[3,7-9]

The nonneoplastic carotid body has largely been ignored by pathologists. This reluctance to include examination of the carotid bodies at necropsy may be related to its small size and the mistaken belief that it is difficult to dissect out. Furthermore, the fact that its mechanism of action is uncertain may lead to a suspicion that little advantage can be gained from examining it.

The carotid body is not difficult to dissect and can be obtained by pinning the carotid bifurcation to a cork board and stripping off the adventitial fat and connective tissue with dissecting scissors, starting with the common carotid artery and working distally.[9] The carotid body rapidly becomes distinguishable as a reddish brown ovoid that remains attached to its artery of origin as the surrounding adventitial connective tissue is removed; the process takes no longer than 5 min. The short glomic artery can then be cut and flecks of adipose tissue removed from the carotid body before weighing and fixing it for histologic examination.

Although special staining techniques and immunohistochemistry may reveal some useful information, most of the pathologic changes described in this chapter are best seen on sections stained routinely with hematoxylin and eosin. These include the variations in structure that develop as part of the aging process and hyperplasia of the carotid body, which characterizes patients with prolonged hypoxemia secondary to diseases of the heart and lungs. This is followed by an account of the form and function of the "glomus pulmonale." The chapter commences with a description of the macroscopic and histologic appearances of the normal adult carotid body.

NORMAL STRUCTURE

Gross Anatomy

In most cases the carotid body is a single, spherical or ovoid pedunculated structure situated between the arms of the carotid bifurcation (Fig. 20-1). It is attached to the angle of the bifurcation by a short stalk, the ligament of Mayer, which contains the glomic artery. Dimensions of the organ in the human adult taken from 42 unselected cases[10] were 3.3 × 2.2 × 1.7 mm. When it undergoes hyperplasia its dimensions may almost double.

The above appearance is subject to certain anatomic variations (Fig. 20-2). In a significant minority the carotid bodies are bilobed with two distinct apical lobes fused at the base to share a common glomic artery. This form had incidences of between 4 and 9 percent.[11,12] Less commonly (3 to 7 percent) two distinct carotid bodies may be found on one or

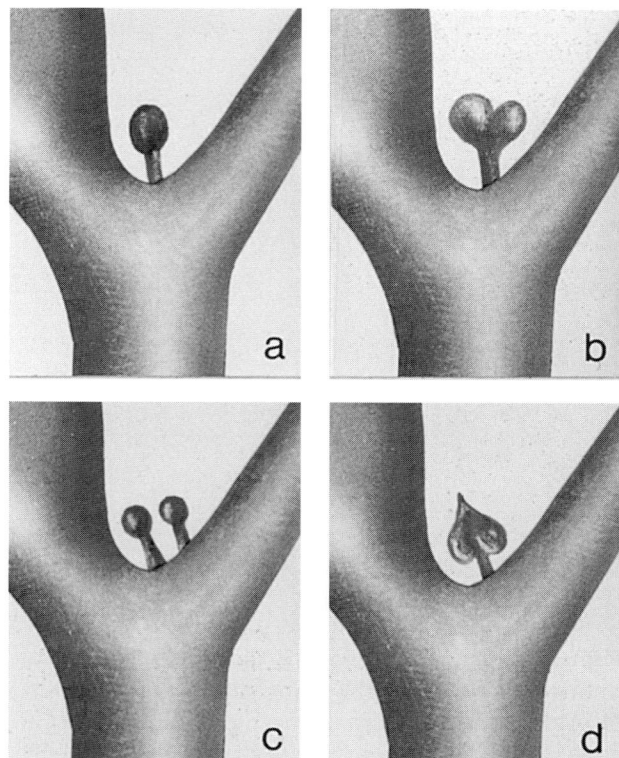

FIGURE 20-2
Diagrammatic representation of the four anatomic variants of the carotid body. *A.* The common ovoid form corresponding to Fig. 20-1. *B.* The bilobed variant. *C.* The double form. *D.* The rare leaf-shaped variant.

other side. Such double carotid bodies usually arise close together but may be located on different branches of the carotid bifurcation. A rare variant takes the form of a flattened leaflike shape.

There is also considerable variation in the location of the carotid bodies. Between 86 and 88 percent have the classic origin from the angle of the bifurcation, but sometimes the glomic artery originates from the external carotid artery or the carotid sinus. It may be rarely found divorced from the bifurcation on the ascending pharyngeal artery.[11,12]

Histology

The carotid body is one of a series of small paraganglia scattered throughout the body sharing similar histologic appearances. However, only the aortic and carotid bodies have been shown to serve a chemoreceptor role. These scattered collections of cells are collectively referred to as glomera and hence their parenchymal elements are called glomic tissue and their constituent cells are termed glomus or glomic cells. Glomic tissue in the carotid body is concentrated into several ovoid lobules that are embedded in a loose, fibrous stroma (Fig. 20-3). There is no capsule to delineate the boundary of the organ although the glomic tissue is sufficiently compact to permit recognition of the carotid body as a discrete nodule.

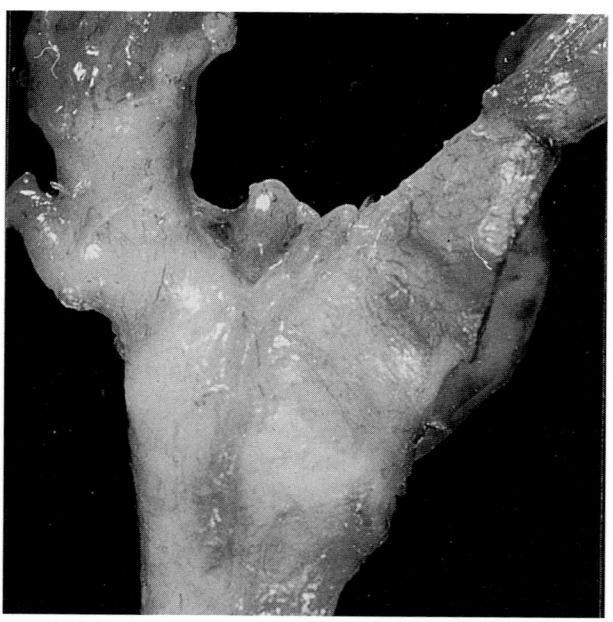

FIGURE 20-1
Right carotid bifurcation from a man aged 51 who died after a myocardial infarction. The bifurcation has been dissected to show the ovoid carotid body situated in its angle.

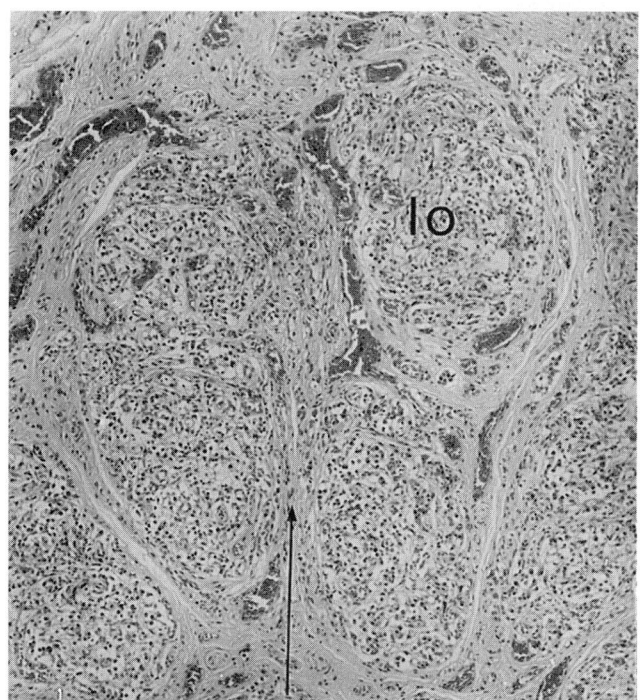

FIGURE 20-3
Part of the carotid body from an 11-year-old boy showing several lobules of glomic tissue (lo). The lobules are separated by narrow septa of fibrous tissue (*arrow*) in which there are numerous blood vessels. (H&E, ×99.)

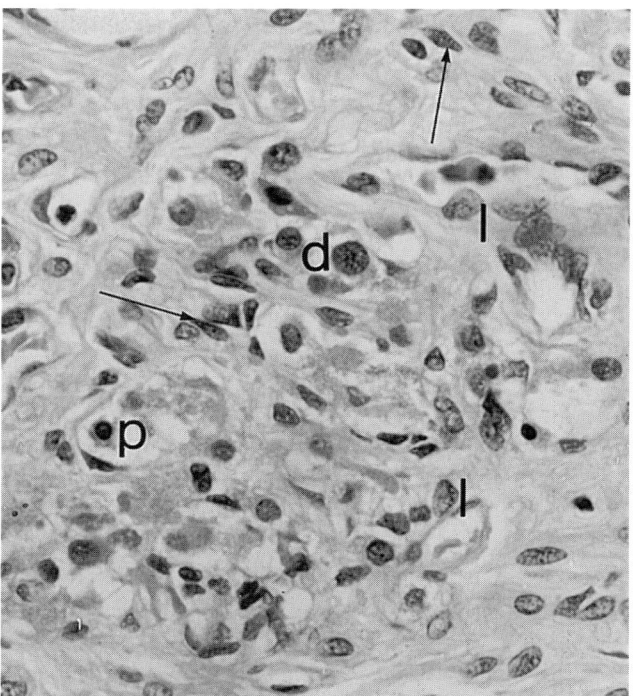

FIGURE 20-4
Detail of a cell cluster showing the constituent glomic cells. Chief cells consist mainly of the light variant (l) with fewer numbers of the dark (d) and progenitor variants (p). Sustentacular cells (*arrows*) are seen both within and surrounding the cell cluster. (H&E, ×736.)

The stroma contains the main glomic artery, which enters the proximal pole and then gives rise to 3 to 4 first order branches. These divide further to give smaller interlobular arteries. Also between the lobules are glomic veins (see Fig. 20-3) that drain into an extensive venous network, which is concentrated at the apical pole. Also prominent in the stroma are numerous bundles of myelinated nerves from which branches penetrate between and into the lobules.

Lobules of glomic tissue are usually round or oval in cross section (see Fig. 20-3) but may be elongated to form cords of cells. They have a clearly defined border and an average diameter between 250 and 565 μm.[11] They are composed of numerous small groups of glomic cells called cell clusters. These consist of predominantly round chief cells surrounded by a narrow, often discontinuous, rim of elongated sustentacular cells (Fig. 20-4). Cell clusters are separated by narrow, indistinct septa of connective tissue in which are situated small intralobular glomic arterioles and venules. Bundles of myelinated nerves enter the glomic lobules and branch among the cell clusters. In this location they are invested by Schwann cells. As they approach and surround the cell clusters the supportive role of Schwann cells is taken over by the sustentacular. As discussed below, the regular arrangement of glomus cells tends to become disorganized as the carotid body ages.

Features of Glomus Cells

CHIEF CELLS

Cell clusters comprise two types of glomic cell—chief and sustentacular cells. Three variants of the chief cells can be recognized on sections stained with hematoxylin and eosin (see Fig. 20-4). The most numerous is called the light cell because of the pale staining reaction of its nucleus. This is round or slightly oval in profile and larger than that of the other two variants (see Fig. 20-4). Its chromatin pattern usually consists of fine, branching strands enclosing pale vesicular areas. These strands often radiate from a single nucleolus, with a few small clumps of chromatin subjacent to the nuclear membrane. In some carotid bodies the nuclei of light cells may present a fine stippled pattern. The copious cytoplasm of light cells is eosinophilic and vacuolated and has ill-defined cytoplasmic borders producing the spurious impression of a syncytium. The vacuoles in light chief cells are larger than those in the other two variants and may approach the diameter of the nucleus. Such vacuoles are irregular in shape with disruption of internal cytoplasmic membranes suggesting autolytic change. Indeed this variant is particularly susceptible to autolysis, a fact that should be kept in mind when assessing its histology.

The nucleus of the dark variant of chief cell is generally slightly smaller than that of the light although an occasional cell may be encountered in which it is substantially larger (see Fig. 20-4). Its nuclear pattern consists of coarsely granular clumps of chromatin. The cytoplasm stains deep purple and has a clearly defined border. Vacuoles, where present, are small and inconspicuous. Cytoplasmic outline ranges from smooth and oval to elongated with long streamers. Dark cells account for only 5 percent of glomic cells in young adults. In children and certain hypoxemic states they are more numerous.

The third variant of chief cell is termed the progenitor cell. It has a small, intensely hematoxyphilic nucleus that is situated to one side of the purple, clearly defined cytoplasm (see Fig. 20-4). The cell is usually oval but, like the dark cell, may be thrown into long streamers. The term progenitor cell is of recent origin,[13] having been previously referred to as the "pyknotic cell."[10] However, this term is misleading because ultrastructural studies show that these cells are not effete and are especially numerous in the growing carotid bodies of infants.[13] Progenitor cells probably mature into the dark and light variants.[9] The use of the term progenitor is not intended to imply that such cells are undifferentiated stem cells that constitute the carotid body during early fetal development because fetal glomic cells have an altogether different appearance.[14]

At the electron microscopic level all three variants have typical dense-core vesicles. These organelles are circular in section and measure between 100 and 200 nm in diameter.[15] They consist of a central, electron-dense core surrounded by a narrow clear zone or "halo" and are bounded by a single limiting membrane. Dark and progenitor cells contain many more vesicles than the light variant. They show less disruption of their cytoplasmic membranes, emphasizing their viability. Chief cells contain catecholamines, such as dopamine, as well as biologically active peptides the most abundant of which are met- and leu-enkephalins.[16] These two peptides can be demonstrated by immunohistochemistry[17] and are confined almost exclusively to the dark and progenitor variants. Substance P, vasoactive intestinal polypeptide (VIP), and gastrin-releasing peptide are also found in the carotid body in smaller concentrations.[16] The first two peptides show a faint immunoreactivity in all three variants of chief cell. Gastrin-releasing peptide is localized to the periphery of smooth muscle cells in glomic arteries.[17]

SUSTENTACULAR CELLS

The second type of glomic cell is termed the sustentacular cell. It has also been referred to as the glomus type II, sheath, or supporting cell. Sustentacular cells are elongated and confined to the periphery of the clusters with a few within the cores of chief cells (see Fig. 20-4). Their elongated or triangular nuclei, although generally dark, show appearances ranging from dense basophilia to a stippled chromatin pattern. Their cytoplasm stains lightly with eosin but is usually difficult to define because it is finely attenuated and merges with adjacent fibrous tissue. At the ultrastructural level sustentacular cells contain nerve axons within simple mesaxons in their attenuated cytoplasm.[9] In this respect they resemble Schwann cells with the notable exception that they also encircle chief cells, conveying nerve axons close to their surfaces with which they make synaptic contact. At the light microscopic level it is often difficult to distinguish between sustentacular and Schwann cells. The latter usually have slightly broader, paler nuclei but there is a degree of overlap between their nuclear configurations. As the nerves branch within the lobules, their Schwann cells become closely apposed to sustentacular cells at the periphery of each cluster. The two cell types merge making precise identification impossible. In practice a precise distinction between the two types of cell is probably not important because both support nerves and both are probably neural in nature.[18,19]

Innervation of Glomic Tissue

The carotid body receives its nervous supply from two main sources. The carotid sinus nerve conveys afferent sensory impulses both from the carotid body and the carotid sinus to the glossopharyngeal nerve. The second source of innervation is via the ganglioglomerular nerve, some fibers of which enter the carotid body. Most of these sympathetic nerves terminate on glomic blood vessels and probably regulate blood flow through the glomic tissue.[20] However, about 5 percent of sympathetic axons are associated with chief cells and may exert an inhibitory influence on chemosensitivity.[21] Ganglion cells of the autonomic nervous system are frequently found within the connective tissue surrounding the carotid body.

Large bundles of myelinated nerves running between glomic lobules can be readily identified on light microscopy. They can be traced for a short distance into the lobules as parallel rows of Schwann cells. The fine axons that supply the cell clusters

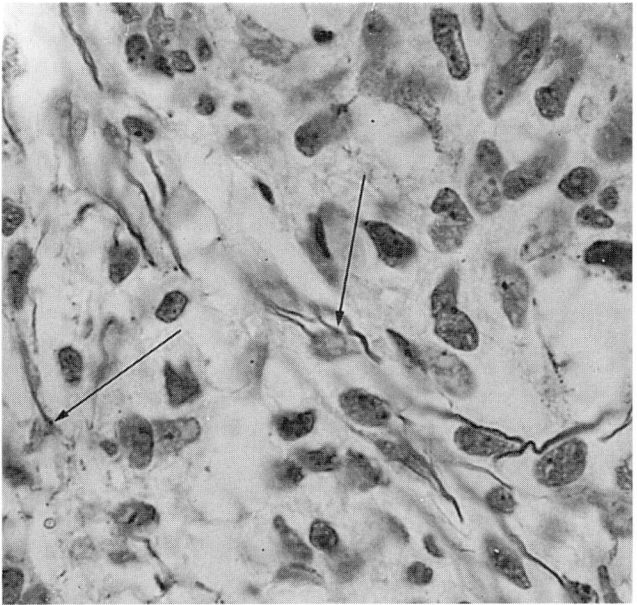

FIGURE 20-5
Developing cell cluster in the carotid body of a 30-week fetus. Numerous, dark axons ramify among the chief cells and some axons terminate on them in the form of small boutons (*arrows*). (Bodian's technique, ×1380.)

require special staining techniques to demonstrate them. The best method is that of immunofluorescence for neurofilament proteins applied to thick cryostat sections. They can also be identified by immunoperoxidase techniques using neuron-specific enolase or PGP 9.5. A useful alternative is the silver protargol method devised by Bodian,[22] which demonstrates that axons form a concentric plexus around each cell cluster associated with the sustentacular cells. From these peripheral nerves, several axons weave an irregular course into the center of the cluster, frequently from one or two poles. These nerves permeate the central cores of chief cells making contact with them, usually in the form of small boutons[23] (Fig. 20-5). Calyces, discs, and menisci are also encountered.[24]

Quantitative Histology

The normal histology of the carotid body described above can usually be distinguished from age changes or hyperplasia by qualitative assessment and the pathologist need not proceed any further. In some situations, such as research projects, it is useful to have a numerical yardstick against which abnormal carotid bodies can be compared. For example, the glomic lobules can be measured by one of two techniques. The first involves measuring the total area of glomic tissue on histologic sections either by point counting or by computerized image analysis. This produces an "area of functional parenchyma"[25] that

can be divided by the total area of the carotid body to give the percentage of glomic tissue.[25,26] The second and quicker method is simply to measure the maximum and minimum dimensions of each lobule and calculate the average diameter. The mean value of such measurements provides a reasonable estimate of the size of lobules. Similar measurements can be applied to cell clusters although this is less precise because the limits of cell clusters may be indistinct.

The proportions of glomic cells can be obtained by counting the individual types of cell within randomly selected high-power fields or within the squares of an eyepiece grid. If the count of each type of cell is expressed as a percentage of the total, a differential cell count is obtained. This value suffers from the disadvantage that if one type of cell increases in number the differential counts of the others are automatically depressed. An absolute count in which the number of a particular type of cell is divided by the area of section sampled overcomes this problem. Such data obtained from 57 normal adults provided a numerical definition of the carotid body derived from the 95 percent probability limits.[11] On this basis the normal carotid body has a combined weight of less than 30 mg, a lobule diameter less than 565 μm, a cell cluster diameter less than 110 μm, and a differential count of sustentacular cells less than 47 percent. Probability limits for chief cells were not calculated but their mean counts were 54 percent for light cells, 5 percent for dark cells, and 2 percent for progenitor cells.

THE EFFECTS OF AGING

The above description of the normal carotid body is based on a normal child or a young to middle-aged adult and has been termed the "basic" histologic pattern.[27] In subjects younger than 40 years there may be a superimposed "dark cell prominence." Such dark cells are often disposed in groups so that their differential count overall is only slightly greater than that of the basic pattern.[28] Between the ages of 50 and 60 years the carotid bodies undergo structural changes associated with aging. However, the occasional case may be encountered in which the basic pattern with dark cell prominence has persisted until the age of 90 years.[27]

Changes to the Area of Glomic Tissue

At low magnification the carotid bodies of the elderly have much more stroma and the individual lobules are more widely separated by fibrous tissue

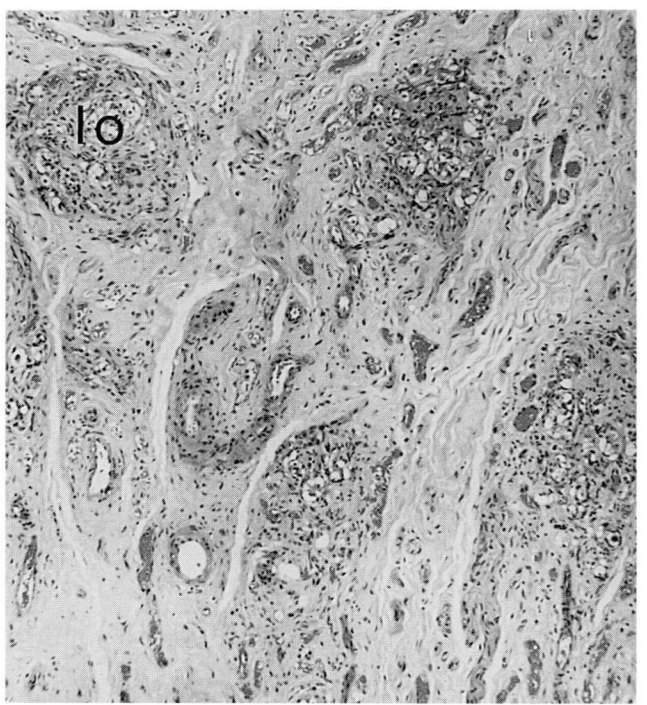

FIGURE 20-6
Part of the carotid body from an 80-year-old woman showing small, atrophic lobules (lo) widely separated by acellular collagen. This appearance contrasts sharply with that of a young carotid body shown in Fig. 20-3. (H&E, ×99.)

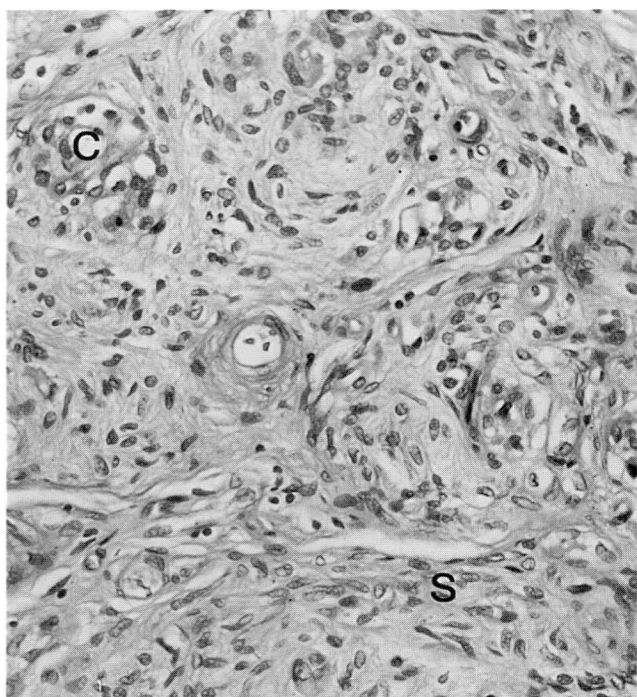

FIGURE 20-7
Sustentacular cell proliferation as an age change. Cell clusters (c) are small and are surrounded by numerous elongated cells (s) that also form irregular sheets between them. The elongated cells probably consist of a mixture of sustentacular and Schwann cells. (H&E, ×345.)

(Fig. 20-6). The lobules also show an irregular outline with considerable variation in size.

The functional area of glomic tissue has been measured in 47 adults ranging in age from 14 to 100 years.[27] It declined from 45 percent in the age range 14 to 40 years to 39 percent between 41 and 65 years, and was only 29 percent in subjects over the age of 66 years. This loss of glomic tissue was associated with a reduction in the diameter of the largest lobule and an increase in total area of cross section of the carotid body due mainly to an increase in connective tissue.

Sustentacular Cell Proliferation

A characteristic feature of the aging process is an increase in the number of sustentacular cells around the cell clusters. The thin rims of sustentacular cells become replaced by several concentric layers of plump elongated cells (Fig. 20-7). Broad bands of elongated cells running in various directions also separate the clusters. The overall appearance is that of a disorganized proliferation of sustentacular cells with a reduction in the size of the cores of chief cells (see Fig. 20-7). This histologic pattern is referred to as "sustentacular cell proliferation" although some of the elongated cells may in fact be Schwann cells.

Atrophy and Fibrosis

Sustentacular cell proliferation is superseded by "atrophy and fibrosis" in which the glomic lobules are partially replaced by fibroblasts and collagen. In the cell clusters, the cores of chief cells are atrophic and the broad rims of sustentacular cells become replaced by acellular collagen (Fig. 20-8). Ghost outlines of ablated cell clusters may be seen embedded within large sheets of collagen fibers that run from the lobules into the perilobular stroma (see Fig. 20-8). In some cases this collagen becomes brightly eosinophilic and amorphous or hyaline.

Although the stage of atrophy and fibrosis appears to develop from sustentacular cell proliferation there are often intermediate appearances between the two extremes. In most cases, however, one or other of the histologic patterns is predominant, making classification of the age-related changes unequivocal. By following these criteria, carotid bodies from 50 cases over the age of 60 years were allocated to one of the three histologic categories described above. Of these, 10 showed the basic pattern, 18 sustentacular cell proliferation, and 22 atrophy and fibrosis.[28] Quantitation revealed that the mean differential count of sustentacular cells increased from 37 percent in the basic pattern to 45 percent in sustentacular cell proliferation. The pop-

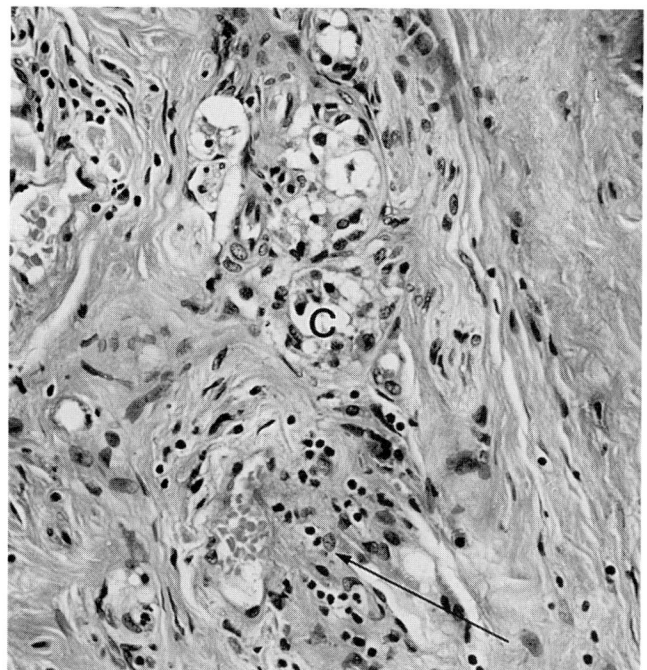

FIGURE 20-8
Atrophy and fibrosis as an age change. A glomic lobule is shown enclosed by dense, acellular collagen. In the upper part of the lobule a few cell clusters (c) persist but the lower part (*arrow*) has been largely replaced by fibrous tissue from the surrounding stroma. There is also a sparse infiltrate of lymphocytes. (H&E, ×345.)

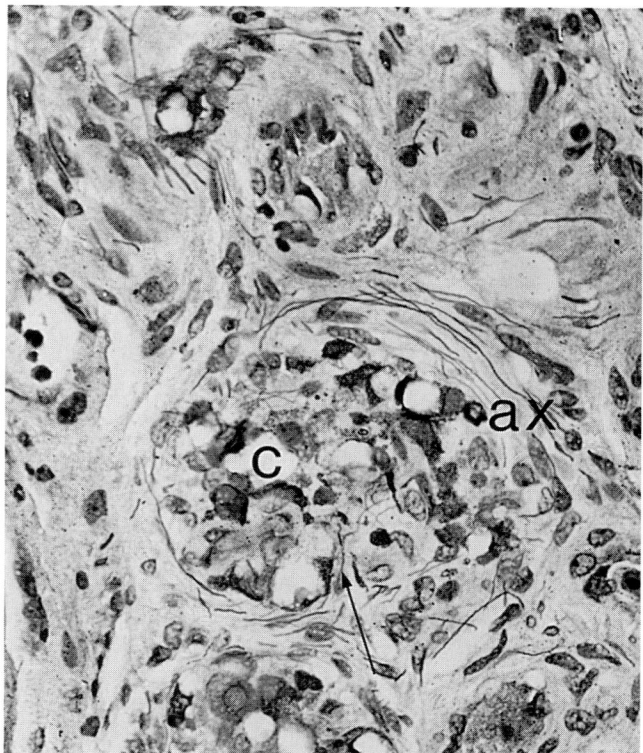

FIGURE 20-9
Sustentacular cell proliferation as an age change stained to demonstrate nerve axons. A cell cluster (c) is surrounded by numerous axons (ax). These are confined largely to the periphery of the cluster but a few enter it from one pole (*arrow*). (Bodian's technique, ×552.)

ulation of light chief cells changed little from the basic pattern but the numbers of dark and progenitor cells per unit area were roughly halved.[28]

Aging and Glomic Nerves

Because sustentacular cells are normally associated with nerve axons, it might be anticipated that in sustentacular cell proliferation the number of axons is increased. This was confirmed by staining the nerves.[28] The axons follow the same circular course as the sustentacular cells, forming a broad plexus of interweaving fibers surrounding each cell cluster (Fig. 20-9). From these peripheral nerves, fine axons penetrate the cores of the clusters and take an irregular path among the chief cells. Only a minority of axons appear to terminate as nerve endings on them. Large bundles of parallel axons also run through the stroma extending between and into the lobules. When sustentacular cell proliferation is superseded by atrophy and fibrosis the nerve supply is greatly reduced and focal in distribution. These qualitative impressions were quantified by measuring the total length of nerve axons per unit area of section in carotid bodies from 50 elderly subjects. The mean nerve density was 11.4 mm/mm^2 in the basic pattern,

18.6 mm/mm^2 in sustentacular cell proliferation and fell to 9.9 mm/mm^2 during atrophy and fibrosis.[28]

Against this background of changing structure should be viewed the reduced chemoreceptor function that develops with advancing age. Thus, the ventilatory response to hypoxia was reduced by 51 percent in eight elderly men aged 64 to 73 years when compared with eight men in their twenties.[29] The response to hypercapnia was also reduced by 41 percent in the same subjects. Furthermore, a slight reduction in hypoxic drive was detected as early as the age range of 31 to 40 years.[30] This became more pronounced over the age of 40. Reduced chemoreceptor function as a factor of age is readily explained in atrophy and fibrosis where there is loss of chief cells, fibrous replacement of glomic tissue, and reduced nerve density. However, sustentacular cell proliferation is associated with an increased nerve density, which might be expected to augment the traffic of sensory impulses up the carotid sinus nerve. This paradox may be reconciled on the grounds that there are relatively few nerve endings on chief cells and that there are fewer chief cells in the cores of the clusters. Thus, despite the increased network of nerves there are fewer than normal units capable of generating nervous impulses to travel along them.

Lymphocytic Infiltrates

At about the age of 50 years onward it is common to find a sparse infiltrate of lymphocytes in the carotid bodies.[27,31,32] This infiltrate involves mainly the stromal tissues but may also extend into the glomic lobules (see Fig. 20-8). The relationship between lymphocytic infiltrates and age is close. In a series of 47 cases only 2 of the 15 patients under age 50 years had lymphocytic infiltration, whereas of the 32 over this age they were found in all but 4 patients.[27]

Lymphocytes arranged in focal aggregates in the stroma may be encountered less commonly in the carotid bodies of the aged, involving 21 percent of cases over the age of 50 years.[32] Such aggregates, often cuff glomic arterioles, venules, and nerve bundles but they may also encroach on adjacent lobules. They vary in size from small perivascular collections to large fusiform sheets of lymphocytes measuring up to 400 μ in their long axis. The lymphocytes are predominantly of the small, nontransformed variety with a scattering of immunocytes. Immunohistochemistry reveals a mixture of T and B cells but with T lymphocytes always comprising at least one-half of the population. These appearances are regarded as forming a distinct pathologic entity that has been termed "chronic carotid glomitis."[32] Its similarity to autoimmune thyroiditis strongly suggests a basis in autoimmunity. This concept is supported by the finding in one case of numerous plasma cells.[33]

CAROTID BODY HYPERPLASIA

Next to chronic carotid glomitis, hyperplasia is the most common abnormality to be found in the carotid bodies. The enlargement is usually associated with chronic hypoxemia, which is believed to stimulate the carotid body leading to a proliferation of glomic cells. In systemic hypertension, the cause of the hyperplasia is less readily explained. It should be pointed out that division of glomic cells in human beings is inferred rather than proven because no studies have estimated the total number of cells present in normal and enlarged carotid bodies. However, those morphometric techniques that have been applied strongly suggest that the total number of glomic cells is increased. Furthermore, studies on hypoxic rats demonstrate that enlargement of the carotid bodies is due to both hypertrophy and hyperplasia of glomic cells[34] with the presence of mitotic figures in chief cells.[35]

In animals only one type of hyperplasia of glomic tissue has been described but in human beings two distinct histologic variants are recognized; one involves mainly chief cells and the other predominantly sustentacular cells.[9]

Chief Cell Hyperplasia

When enlargement of the carotid bodies involves predominantly chief cells it may be referred to as "chief cell hyperplasia."[9] It is characteristic of subjects who are permanently resident at high altitudes and in children and young adults with cardiopulmonary diseases that induce sustained hypoxemia.

HIGH ALTITUDE

At a meeting of the American Association of Pathologists and Bacteriologists in 1969, Arias-Stella reported that the carotid bodies of Quechua Indians born and living in the Andes of Peru were larger than those of mestizos living in Lima at sea level.[36] This important observation formed the basis for all subsequent studies of the carotid body unconnected with its neoplasm. In a more detailed report, Arias-Stella and Valcarcel[37] compared the carotid bodies of a group of patients from Lima with those of a group from Cerro de Pasco, at an altitude of 4330 m. They found that the dimensions and weight of carotid bodies from the highlanders were consistently greater than age-matched controls from sea level and that this difference became greater with increasing age. The histology of the carotid bodies revealed an increase in the size and number of lobules with a reduction in the quantity of connective tissue between them.[37,38] The increased size of the lobules was attributed to a proliferation of chief cells of such intensity as to form a homogeneous appearance. The hyperplastic chief cells also showed intense vacuolation.

More recently the carotid bodies of three adult Ladakhi highlanders, resident between 3300 and 4200 m, were studied and confirmed the findings of Arias-Stella.[39] A feature not mentioned by the Peruvian workers was that the lobules consisted of numerous cell clusters that were much smaller than usual (Fig. 20-10). They measured only 45 μm in average diameter as contrasted with 82 μm for normal carotid bodies. These miniclusters sometimes consisted of only three or four chief cells but were surrounded by thin rims of sustentacular cells in the normal fashion. The photomicrographs at high magnification in the paper of Arias-Stella and Valcarcel[38] show a similar arrangement of miniclusters within hypercellular lobules. Thus, chief cell hyperplasia is typified by an orderly architecture that is similar to normal in all but the size of the cell clusters. It is possible that in the elderly highlander the histologic

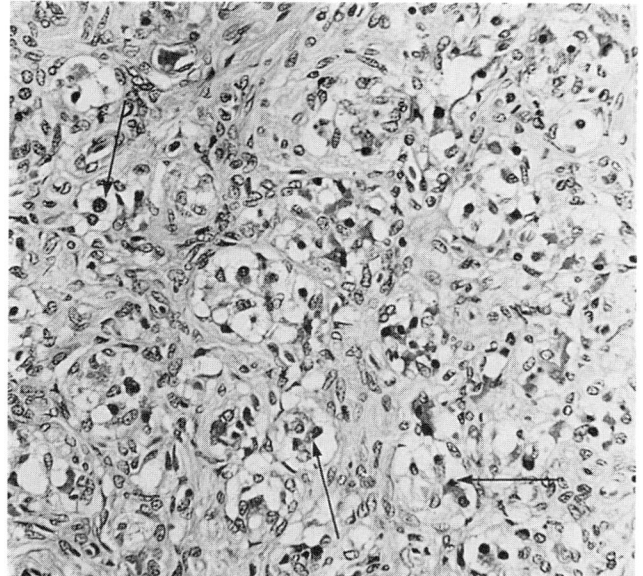

FIGURE 20-10
Part of a glomic lobule from a man of 35 years who lived at an altitude of 3500 m in Ladakh and was killed in a traffic accident. There are numerous miniclusters of chief cells (*arrows*) surrounded by sustentacular cells typical of chief cell hyperplasia. (H&E, ×345.)

picture may be modified by the same age-related changes that occur in subjects at sea level.[39]

CYANOTIC CONGENITAL HEART DISEASE

Patients with congenital cardiac defects are sometimes cyanotic and hypoxemic for the whole of their lives. This occurs in most cases of Fallot's tetralogy, or where there is reversal of an intracardiac shunt such as a large ventricular septal defect or patent ductus arteriosus. Such patients are young and their carotid bodies show chief cell hyperplasia.

The pathology of the carotid bodies in congenital heart disease was described in a series of papers by Lack and coworkers.[25,31,40] They studied a total of 213 patients of whom 17 had cardiac abnormalities associated with hypoxemia and were aged 30 years or less.[25] The carotid bodies from these patients were significantly heavier than those of age-matched controls[25,40] and their constituent lobules were larger.[31] The number of lobules present on sections taken through the greatest diameter was also increased from between 11 and 14 in control cases to between 18 and 25 in the cases with cardiac defects. The total area of glomic tissue was also increased, and there was a reduction in the intervening stroma with lobules packed closely together.[25] The lobules were composed entirely of cell clusters each with the normal architecture of a central core of chief cells surrounded by a narrow rim of sustentacular cells. The

proportions of the two types of glomic cell were not altered. Lack did not refer to the presence of miniclusters but his illustration of a glomic lobule from an 8-year-old patient with an interatrial septal defect shows numerous small clusters similar to those in highlanders from Ladakh.[40]

CYSTIC FIBROSIS

Thirty patients with this disease, all less than 30 years of age, were studied by Lack and colleagues.[25] The pathology of their carotid bodies was identical to that of patients with congenital heart disease described above. Thus chief cell hyperplasia in these young patients appears to present a consistent histologic picture irrespective of its cause.

At the ultrastructural level the carotid bodies of patients with cystic fibrosis show a reduction in the number of dense-core vesicles either uniformly or in focal areas.[25] This is associated with an increased quantity of rough endoplasmic reticulum with dilated cisternae containing a granular electron-dense material.

Sustentacular Cell Hyperplasia

A common form of hyperplasia involves mainly the sustentacular cells. This type of reaction is seen in patients with pulmonary emphysema associated with severe, prolonged alveolar hypoxia. It also occurs in cases of long-standing systemic hypertension.

Macroscopically, the enlarged carotid body in cases of chronic bronchitis and emphysema is congested and mauve in color. It may have a finely nodular appearance suggestive of nodular hyperplasia; however, a spurious impression of coarse nodularity may be produced when hyperplasia develops in a previously bilobed variant.[11,12]

HISTOLOGIC APPEARANCE

Even under low power, hyperplasia of the carotid body can usually be inferred because the glomic lobules are greatly enlarged with fusion and irregular profiles. There is an associated reduction in fibrous tissue. These hypercellular lobules show an increase in the number of elongated cells (Fig. 20-11). These form thick rings, several cells in thickness, around the edge of each cell cluster with a reduction in the size of the cores of chief cells[11,41] (Figs. 20-11 and 20-12). A proliferation of elongated cells is described earlier in this chapter as part of the aging process in the carotid body. However, this sustentacular cell proliferation is less regular, the cellularity is less intense, and the size of the lobules is either normal or

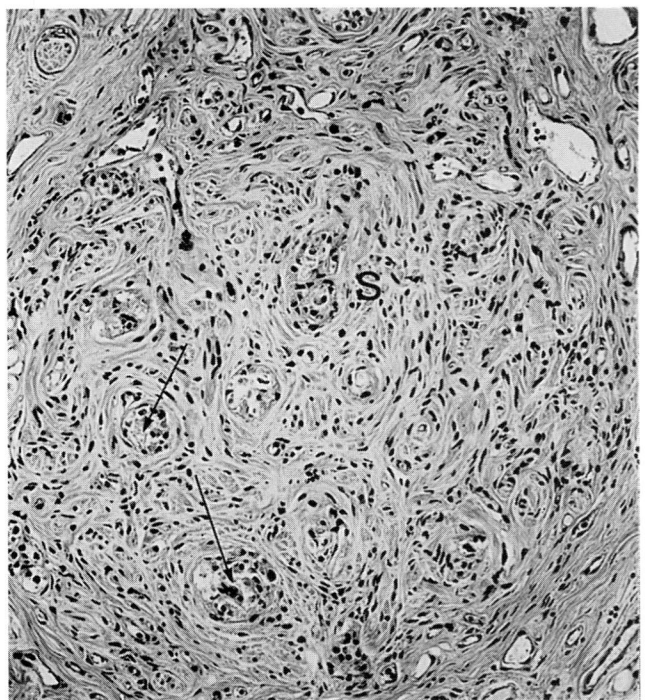

FIGURE 20-11
Part of an enlarged glomic lobule from a 72-year-old man with centrilobular emphysema, severe hypoxemia, and right ventricular hypertrophy. The atrophic cell clusters (*arrows*) are widely separated by concentric rings of sustentacular and Schwann cells (s) producing a whorled appearance characteristic of sustentacular cell hyperplasia. (H&E, ×138.)

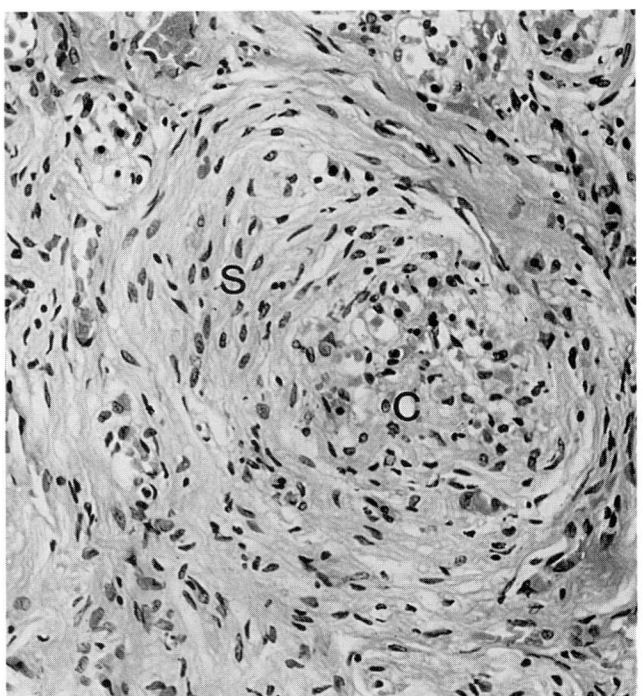

FIGURE 20-12
Detail of a cell cluster from the same case as in Figure 20-11. A central core of chief cells (c) is surrounded by a broad zone of sustentacular and Schwann cells (s) orientated concentrically around it. (H&E, ×345.)

reduced. In hyperplasia the elongated cells form complete circles around the clusters producing a characteristic whorled arrangement (see Fig. 20-11). The location of these whorls of elongated cells at the periphery of clusters suggests that they are sustentacular and the term "sustentacular cell hyperplasia" has been coined.[9] The nuclei of the elongated cells are of variable size and staining intensity and some are plump and pale like those of Schwann cells (see Fig. 20-12). Electron microscopy has confirmed that the outermost layer of elongated cells is associated with myelinated nerves.[42] Cells closest to the cores of chief cells have narrower nuclei and not only envelop axons but also chief cells and are thus unequivocal sustentacular cells. Between these extremes the precise nature of the cells is often uncertain. Hence, the elongated cells comprise a mixture of sustentacular and Schwann cells in indeterminate proportions. The term sustentacular cell hyperplasia is thus not wholly accurate but may be retained because it provides a visual picture of the histologic appearance.

The proliferation of sustentacular cells is associated with a greatly increased density of nerves affecting the fine axons around the cell clusters. The axons follow the same circular course as the sustentacular cells with which they form a simple mesaxonal relationship. Electron microscopy shows that each sustentacular cell may enfold as many as five axons

instead of the usual one or two.[42] Despite this luxuriant nerve supply, relatively few axons leave the peripheral plexus to enter the cores of chief cells. Those that do so establish nerve endings on chief cells in the form of boutons or calyces.[43] The increase in nerve density in sustentacular cell hyperplasia has been quantified by an intercept method of morphometry in which the total length of axon per millimeter^2 of section was calculated.[43] In five control cases this value was 12.6 but in eight examples of sustentacular cell hyperplasia the nerve density was almost four times as great at 44.3 mm/mm^2.

The histology of sustentacular cell hyperplasia is so distinctive that it can be recognized by qualitative examination. However, such impressions can be confirmed by measurement using the parameters defined earlier in this chapter. The weight of the carotid bodies is the least reliable criterion because, although many cases may have a combined carotid body weight in excess of the 30 mg limit, a significant proportion will be below this value. The diameters of glomic lobules can be measured rapidly and are increased although some overlap with control values occurs.[11] The most reliable measurement is the more time-consuming technique of counting the glomic cells to produce a differential cell count. Thus, of the 43 cases with sustentacular cell hyperplasia referred to above,[11] only 4 had a sustentacular cell count below the upper normal limit of 47 percent.

640

CHRONIC BRONCHITIS AND EMPHYSEMA

These diseases are frequently associated with hyperplasia of the carotid bodies induced by prolonged hypoxemia secondary to alveolar hypoxia. Hyperplasia is usually found in subjects with severe arterial oxygen desaturation, hypercarbia, and systemic edema, the so-called blue-and-bloated patients. These patients have right ventricular hypertrophy and muscularization of their pulmonary arterioles.[44] These pathologic markers of pulmonary arterial hypertension are closely related to the presence of severe alveolar hypoxia during life. Thus, in cases of pulmonary emphysema in which the right ventricle was not hypertrophied (weighing less than 80 g) the combined carotid body weight ranged from 27.2 to 37.6 mg with a mean of 32.4 mg. In cases with right ventricular hypertrophy the range of carotid body weight was 32.2 to 89.3 mg with a mean of 56.2 mg.[44] Glomic lobules may show gross enlargement in chronic bronchitis and emphysema with mean diameters of up to 1300 μm.[41]

Histologically the lobules show all the features of sustentacular cell hyperplasia described above. Differential cell counts of sustentacular cells may be as high as 55 or 62 percent in cases of emphysema with right ventricular hypertrophy.[11,41] An increase in weight of the carotid bodies with sustentacular cell hyperplasia has also been described in a 60-year-old man with the pickwickian syndrome.[41] In such patients alveolar hypoxia is secondary to hypoventilation brought about by gross obesity.

SYSTEMIC HYPERTENSION

In a study of 40 successive subjects coming to necropsy a statistical correlation was found between the left ventricular weight and the carotid body weight.[10] This observation was confirmed in a further study of 15 cases with left ventricular hypertrophy.[44] The combined carotid body weights in these cases ranged from 17.6 to 80.8 mg, of which 10 cases were above the upper limit of normal of 30 mg. In another series of 100 consecutive necropsies a highly significant statistical correlation was found between the weights of the left ventricles and the carotid bodies.[11] The increase in weight of the carotid bodies was related to a clinical history of prolonged systemic hypertension and not to left ventricular hypertrophy due to other causes, thus establishing a link between the carotid bodies and systemic hypertension.[11]

The histopathology of the enlarged carotid bodies in systemic hypertension is identical to that seen in emphysema and is based on sustentacular cell hyperplasia. Thus, the lobules are enlarged and cell clusters are surrounded by thick whorls of sustentacular and Schwann cells,[11] associated with a luxuriant supply of nerve axons.[43]

Habeck[45] found only insignificant differences between the carotid bodies of 27 hypertensive patients and controls. Unequivocal hyperplasia involving sustentacular cells occurred in only 2 cases with renal hypertension. These appearances contrasted sharply with those of nine patients with chronic lung diseases including emphysema, pulmonary fibrosis, and cystic fibrosis, eight of whom showed an extensive proliferation of sustentacular cells. The exception was a boy of 15 years with cystic fibrosis. The hypertensive subjects showed significant elevations of both diastolic and systolic blood pressures during life. The left ventricular weights were not quoted so that the duration of systemic hypertension cannot be assessed. A clinical history of systemic hypertension alone does not necessarily mean that the carotid bodies will be hyperplastic. Pathologic evidence in the form of a left ventricular weight in excess of 200 g, signifying prolonged, severe systemic hypertension, should be obtained because this change is associated with sustentacular cell hyperplasia in the vast majority of cases.[11]

At first sight hyperplasia of the carotid bodies in response to a raised intravascular pressure is contrary to what one might expect of an organ sensitive to oxygen tension. A possible explanation for this apparent exception might lie in the development of occlusive intimal fibrosis in glomic arteries and arterioles rendering the glomic tissues ischemic. This ischemia may stimulate the carotid bodies to undergo hyperplasia in the same way as they respond to high altitude or emphysema. Intimal fibroelastosis in interlobular glomic arteries is a common finding in systemic hypertension[9] and similar lesions have been studied in 13 cases.[46] In them there was thickening of the media of glomic arteries by eosinophilic hyaline with partial or total occlusion of the lumen by intimal fibrosis. However, not all examples of carotid body hyperplasia associated with systemic hypertension show such vascular lesions and obstruction to their blood supply may occur more proximally. The carotid bifurcation is a common site for atherosclerosis, which is exaggerated in hypertension. In some patients, it may partially block the orifice of the main glomic artery leading to ischemia.[9] Support for a vascular origin of hyperplasia in systemic hypertension comes from studies on spontaneously hypertensive rats. Their glomic arteries show striking intimal proliferations of myofibroblasts, which become progressively more occlusive with age.[47–49] These lesions appear to develop in response to a rapidly increasing systemic blood pressure and may explain the three-fold increase in volume of their carotid bodies when compared with control rats of the Wistar strain.[49]

The hypothesis that ischemia of glomic tissue in systemic hypertension leads to hyperplasia of the carotid bodies has the advantage that it links all causes of carotid body enlargement to the same mechanism of hypoxemia. However, it is not the sole possibility. There is strong experimental evidence that the carotid bodies are involved in salt and water regulation and subserve an endocrine role by secreting a natriuretic substance.[50,51] The tendency to retain sodium and water in emphysematous patients with the blue-and-bloated syndrome and in systemic hypertension may lead to an increased demand for such a natriuretic factor causing the carotid bodies to become hyperplastic.[9]

COARCTATION OF THE AORTA

In this congenital defect the upper parts of the body only are exposed to an elevated systemic blood pressure. It might be expected that the carotid bodies respond to this hemodynamic abnormality in the same way as they do to systemic hypertension. To date the carotid bodies from one patient with coarctation of the aorta have been described.[52] This 61-year-old man had severe left ventricular hypertrophy and his carotid bodies were greatly enlarged with a combined weight of 68.6 mg. Glomic lobules were enlarged by dense masses of elongated cells, which not only formed characteristic whorls around cell clusters but also formed large sheets of parallel cells (Fig. 20-13), a significant proportion of which were probably Schwann cells. They accounted for 73 percent of the total population of glomic cells. They were associated with numerous axons similar to sustentacular cell hyperplasia in systemic hypertension. Thus, although this case showed the features of sustentacular cell hyperplasia, it was more extreme with a greater degree of atrophy of chief cells. This is likely to be the result of a greatly elevated blood pressure in the carotid arteries, which is to be expected in coarctation of the aorta.

BRONCHIAL ASTHMA

Asthmatic patients may develop alveolar hypoxia as a result of airways obstruction and one might anticipate, therefore, that their carotid bodies are hyperplastic. Glomic tissue for confirming this hypothesis is available only from the few centers where unilateral glomectomy is practiced. This procedure is said to relieve dyspnea and bronchospasm.[53] Carotid bod-

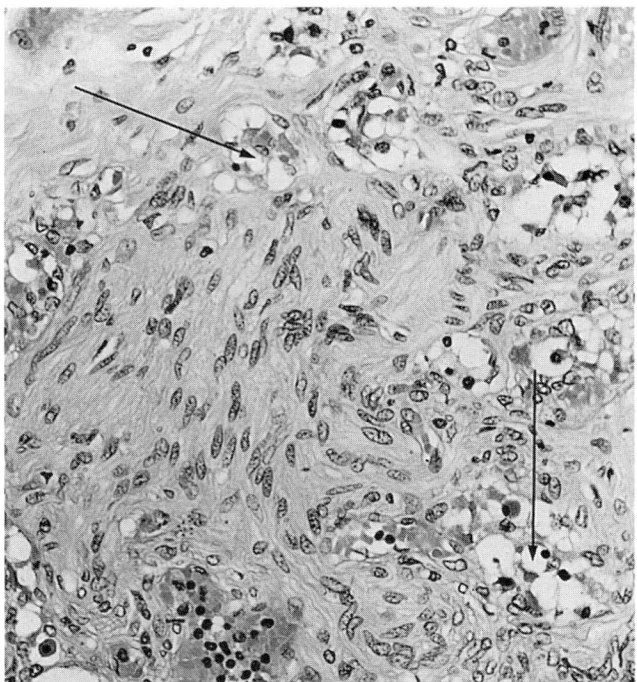

FIGURE 20-13
Carotid body from a 61-year-old man with coarctation of the aorta. Cell clusters (*arrows*) are atrophic and are surrounded by a profusion of elongated cells which form sheets. (H&E, ×345.)

ies thus resected have been studied in 50 patients all of whom suffered from episodes of dyspnea. These patients only received surgery if their daytime arterial oxygen tension was greater than 65 mmHg.[54] Eleven had a history of bronchial asthma for up to 5 years and their carotid bodies were compared with those from 10 patients who at necropsy had no cardiopulmonary disease. Morphometry showed that the carotid bodies of the asthmatics were not enlarged and the lobules were not increased in size. They were, however, irregular in shape, elongated, and crowded together.[54] The centers of the lobules were occupied by dense sheets of sustentacular cells, which also formed broad concentric rings around the cell clusters (Fig. 20-14). The differential count of sustentacular cells was 33 percent in the control cases and 68 percent in the asthmatics. An unusual feature of these cases was a prominence of the dark variant of chief cell (see Fig. 20-14). This variant accounted for 28 percent of all chief cells in the controls but averaged 43 percent in the cases of asthma. Thus, in asthmatic patients both the dark chief and sustentacular cells appear to undergo hyperplasia. The resulting histologic picture is similar but not identical to sustentacular cell hyperplasia. It appears to be produced by a prolonged stimulus of mild, intermittent hypoxemia.

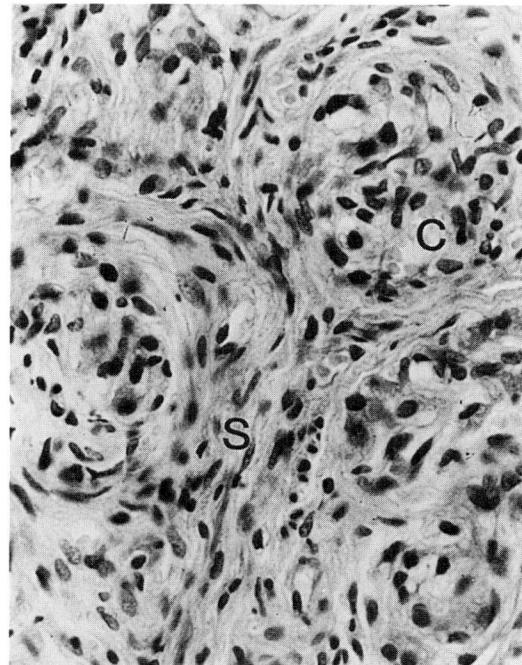

FIGURE 20-14
Part of a glomic lobule from a case of long-standing bronchial asthma. Cell clusters (c) are surrounded by dense masses of sustentacular cells (s). Unlike sustentacular cell hyperplasia associated with pulmonary emphysema the majority of chief cells are of the dark variety. (H&E, ×552.) *(By permission of Dr. C. Bencini).*

PATHOGENESIS OF SUSTENTACULAR CELL HYPERPLASIA

Animals that respond to chronic hypoxia by hyperplasia of their carotid bodies do so by a proliferation of all cellular elements of the glomus, a process analogous to chief cell hyperplasia. It is probable that in human beings sustentacular and Schwann cells proliferate in response to a need for increased axonal support. However, the reason for this increased branching of axons around cell clusters is obscure. A possible explanation may be simply that prolonged stimulation of the nerve endings in the carotid bodies provokes the axons to proliferate. There is strong experimental evidence that the nerve endings play a key role in transduction of the chemosensory signal.[3,9] If this is so, prolonged hyperstimulation of nerve endings might create a need for an increased nervous supply to the chief cells. This theory presupposes that the proliferating axons are sensory afferents but there is no evidence that this is the case. Furthermore, the dense plexus of axons surrounding each cell cluster does not appear to be associated with any increase in the number of nerve endings on chief cells. An alternative explanation might be that sustentacular cells proliferate not to support nerves but to phagocytose debris formed during wallerian

degeneration.[21] This explanation is unlikely in view of the fact that ultrastructural studies of sustentacular cell hyperplasia have not revealed the presence of residual bodies.[42]

The axonal proliferation in sustentacular cell hyperplasia is unusual in that it is disorganized, forming concentric rings from which only a few axons penetrate the cores of chief cells. The appearance is reminiscent of an amputation neuroma in which there is a profuse and uncoordinated proliferation of nerves in the stump of an amputated limb. Indeed, the histology of the carotid bodies in cases of emphysema has been described as "a neurinoma-like picture."[45] In an amputated limb, nerves proliferate because they have lost the trophic influence of their target organ. In the carotid body, too, axonal proliferation may be a response to loss of a target organ, in this case the chief cells. Atrophy of the central cores of chief cells is a usual feature of sustentacular cell hyperplasia. If it occurs as an early event, the axons that supply them might be expected to proliferate in a vain attempt to find their target organ, thereby simulating the appearance of a neuroma.

This hypothesis alone does not explain why natives to high altitude show chief cell hyperplasia, whereas patients at sea level with chronic obstructive lung disease develop a proliferation of nerves and sustentacular cells. It has been suggested that the factor that determines the nature of the histologic change in the carotid bodies is the age of the subject when the hypoxemic stimulus commences.[9] Thus, natives to high altitude are born into a hypoxic environment and their carotid bodies are hypoxemic while they are growing. These developing carotid bodies should respond to sustained stimulation by a heightened proliferation of chief cells, axons, and sustentacular cells in a coordinated fashion. The same situation applies to children and young adults with congenital heart disease or cystic fibrosis in whom there is also chief cell hyperplasia.[25] On the other hand, patients with chronic bronchitis and emphysema are usually middle aged or elderly. Hypoxemia with hypercarbia develops after the age-related changes of chief cell atrophy and sustentacular cell proliferation. It may be that the chief cells in their carotid bodies have lost the potential for division and are thus unable to respond to chronic stimulation by undergoing hyperplasia. The axons and sustentacular cells, on the other hand, are already active and respond by proliferating but, because the chief cells are atrophic, the axons are unable to establish adequate contact with their target organ. Consequently, they and the sustentacular cells continue to proliferate in an uncoordinated fashion to form a kind of neuroma. According to this hypothesis chief cell hyperplasia should be associated with young carotid bodies and sustentacular cell hyperplasia should be found in hypoxemic carotid bodies of the elderly.

VENTILATORY DRIVE IN CAROTID BODY HYPERPLASIA

The enlargement of the carotid bodies in permanent residents of high altitude is associated with a blunted ventilatory response to hypoxia.[55–57] Thus, although these subjects ventilate more than lowlanders they show a marked hypoventilation when compared with newcomers to high altitude.[58] Once acquired, this insensitivity to hypoxia is irreversible even after prolonged residence at sea level.[56,57]

The reason for this blunted response to hypoxia is obscure because the carotid bodies of high altitude natives show hyperplasia of chief cells.[38] Physiologic studies on animals have shown that prolonged hypoxia induces an increased synthesis of dopamine by chief cells[59] and that this amine accumulates in the carotid bodies to several times its normal concentration.[35,60] Because dopamine exerts a predominantly inhibitory influence on chemoreception, its increased concentration might explain the hypoxic insensitivity at high altitude. If this is the case, however, the altered biochemistry of the chief cells must become permanent to explain the fact that normal chemosensitivity is not restored on return to sea level. Furthermore, the permanent blunting of chemoreceptor drive at high altitude is acquired during the first few years of life.[56] It fails to develop in adults from sea level who move to high altitude for periods of up to 12 years.[61] A contributory factor may be that age changes in the carotid bodies of highlanders render them less sensitive to hypoxia. Thus, Arias-Stella and Valcarcel[38] noted that the increase in size of the carotid bodies with age was related to a progressive fall in pulmonary ventilation in high-altitude natives. Sustentacular cell proliferation consistent with age-related change has been described in a 54-year-old resident of Ladakh[39] and, as discussed earlier, aging, even at sea level, is associated with a reduced hypoxic ventilatory drive. The chronic hypoxemia in patients with cyanotic heart disease also blunts the response of the carotid bodies to hypoxia.[62] There is, however, conflicting evidence as to whether this effect persists after the defect has been corrected, with one group claiming that hypoxic insensitivity persists[63] and another that it is

abolished.[62] In neither study can the effects of age-related changes be invoked as a cause of the blunted response to hypoxia because the oldest patient was only 37 years and most were less than 25 years.

Sustentacular cell hyperplasia, which characterizes older patients with chronic bronchitis and emphysema, is also associated with a depressed sensitivity to hypoxia.[64,65] This effect is more readily explicable on the basis of atrophy of chief cells and a disorganized proliferation of axons. Thus, patients with a tendency to carbon dioxide retention show the greatest reduction in ventilatory response to hypoxia[64] and such patients commonly show sustentacular cell hyperplasia. Flenley[66] has measured the hypoxic drive in healthy subjects and recorded a sevenfold variation. He suggested that patients who develop the blue-and-bloated syndrome, and hence sustentacular cell hyperplasia, are those with a pre-existing reduction to their hypoxic drive. These patients would hypoventilate when they develop emphysema.

The influence of systemic hypertension on chemosensitivity is less clearly defined. The experiments of Trzebski and colleagues[67] on healthy young volunteers indicate that the chemoreceptor drive may be increased. They studied two groups of young male students, one with mild systemic hypertension and the other with normal systemic blood pressure. Progressive hypoxia was associated with an increase in minute ventilation in both groups but was exaggerated in the group with early hypertension. This augmentation of the hypoxic ventilatory response resulted in a greater tidal volume with a twofold increase in the force generated by the respiratory muscles. Similar studies on male medical students with mild, early systemic hypertension showed a significantly higher minute ventilation compared with controls even while breathing normal air.[68] It is not known whether the increased ventilatory drive in these young subjects is associated with structural modifications of their carotid bodies. It seems unlikely that such mild elevations of systemic blood pressure in young people could cause sufficient occlusion of glomic arteries to lead to carotid body hyperplasia. In view of the sevenfold variability in hypoxic drive detected by Flenley[66] it may be that individuals with a moderate elevation of systemic blood pressure have a drive in the upper range of normal.

Elderly patients with established, severe systemic hypertension show sustentacular cell hyperplasia of their carotid bodies. Little is known of their hypoxic ventilatory drive but on analogy with patients with chronic obstructive lung disease, whose carotid bodies are identical in appearance, one may predict that it is blunted. This prediction is supported by the studies of Tafil-Klawe and associates[30] who showed that the ventilatory response to progressive isocapnic hypoxia in patients of different ages with severe, essential hypertension reduces with increasing age. The above considerations led to the conclusion that carotid body hyperplasia, irrespective of its cause or nature, is associated with a blunted chemoreceptor response to hypoxia.

HYPERPLASIA OF DARK CHIEF CELLS

Numerous dark chief cells are a normal feature of youth. However, if a profusion of dark cells is encountered in a middle-aged or elderly subject the carotid bodies are likely to be abnormal. The first description of a diffuse hyperplasia of dark cells was in a woman of 62 years who had a ventricular septal defect with a recent reversal of the shunt inducing hypoxemia for only a few months.[69] In this case the carotid bodies were not enlarged but there was a proliferation of dark chief cells. At the ultrastructural level dark cells contained numerous dense-core vesicles and an abundance of mitochondria suggestive of active metabolism.[69] The active, proliferating dark cells also showed a strong immunoreactivity for leu- and met-enkephalin.[70] A proliferation of dark cells strongly immunoreactive for enkephalins has been described in high altitude natives to Ladakh[39] and, as described above, it is also a feature of bronchial asthma.[54] It appears to be a response of glomic tissue to mild or brief hypoxemia and may therefore represent an early stage in the development of chief cell hyperplasia.

This relationship between proliferating dark cells and early carotid body hyperplasia is illustrated by experiments on rabbits,[71] a species in which both light and dark variants of chief cell are normally present. Six Dutch rabbits were kept in a hypoxic chamber at a partial pressure of oxygen simulating the altitude of Cerro de Pasco in Peru (4330 m). Three were killed after 3 months and their carotid bodies were enlarged and showed a proliferation of dark cells that had a differential count of 31 percent instead of the normal value of 14 percent. The remaining three rabbits were hypoxic for 6 months, but their enlarged carotid bodies were composed mainly of light cells with a dark cell count of only 6 percent. These findings are consistent with the view that the early response of the carotid body to hypoxemia is a proliferation of dark cells, which later mature into light cells when the hypoxic stimulus is

more prolonged. In this respect it is worthy of note that rabbits born and living in Cerro de Pasco all of their lives show chief cell hyperplasia with very few dark cells.[71] Dark chief cells appear to be sensitive to hypoxia and respond by undergoing hyperplasia at oxygen tensions that are too high to affect other organs. Thus, when Dutch rabbits were kept at the Rifugio Torino at Monte Bianco, Italy (3370 m) for 3 months their carotid bodies also showed an increase in the number of dark chief cells (20 percent), which returned to normal (11 percent) after 6 months residence at altitude.[72]

Focal proliferations of dark and progenitor chief cells may be occasionally encountered on a background of sustentacular cell hyperplasia irrespective of whether this be secondary to chronic obstructive lung disease or systemic hypertension[73] (Fig. 20-15). These nodules range in diameter from 40 to 400 μm and some of the cells in them may be pleomorphic. It is assumed that such focal proliferations are caused by sudden exacerbations of a long-standing stimulus for hyperplasia such as a respiratory infection developing in a patient with chronic bronchitis and emphysema.[9]

Recently an interesting association between a proliferation of dark chief cells and cirrhosis of the liver has come to light. Thus, diseases of the gut per se are not associated with changes in the carotid bodies but only cirrhosis of the liver appears to bring

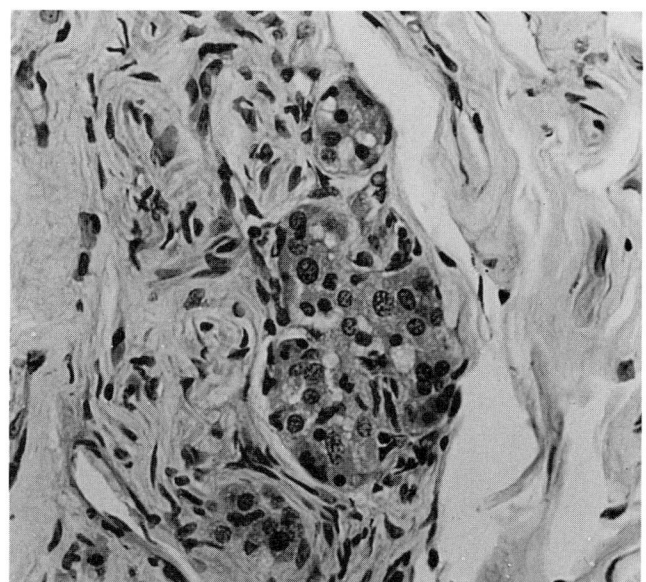

FIGURE 20-15

A focal proliferation of dark chief cells in an 80-year-old woman with systemic hypertension. The proliferation of dark cells is superimposed on a background of sustentacular cell hyperplasia. (H&E, ×552.)

about moderate glomic enlargement.[74] Furthermore, these carotid bodies show a highly significant increase in the number of dark cells distributed either diffusely or as focal groups.[75] Some cases of cirrhosis are associated with oxygen desaturation of arterial blood[76,77] probably due to portapulmonary anastomoses. In such cases dark cell proliferation may be due to systemic hypoxemia. However, an intriguing alternative explanation is possible. In some cases of cirrhosis there is elevation of plasma renin, angiotensin, and aldosterone leading to retention of sodium and water by the kidneys. There is also experimental evidence from animal studies that the carotid body releases a substance that inhibits renal tubular sodium absorption causing a natriuresis.[50,51] In cases of cirrhosis with sodium retention the demand for such a natriuretic factor should be increased, thereby stimulating a proliferation of glomus cells. The fact that it is the dark cell which is selectively increased in number allows us to tentatively suggest that it is this type of glomus cell that contains the hypothetical natriuretic factor.[75]

THE GLOMUS PULMONALE

There are many nodules of glomic tissue around the heart and great vessels, one of which is present on the dorsal aspect of the bifurcation of the pulmonary trunk. This structure is so constant in its location that Krahl[78] designated it the "glomus pulmonale" believing it to be derived from the sixth branchial arch from which the pulmonary arteries originate. Serial sections of the wall of the pulmonary trunk revealed the presence of glomic tissue (Fig. 20-16) in the precise position predicted by Krahl.[79] However, the presence of glomic tissue adjacent to a vessel does not necessarily imply that it is derived from it. For example, Comroe[80] was unable to demonstrate any chemoreceptor activity in the pulmonary arteries by injecting into them cyanide, a well-known stimulator of chemoreceptors. To prove that the glomus pulmonale is a true chemoreceptor of the pulmonary circulation, rather than one of the aortic bodies, it is essential that its blood supply should be demonstrated to originate from the pulmonary trunk.[80] It was subsequently found in rats and human beings that the intertruncal glomera have a solely systemic arterial supply from branches of the coronary arteries and the aorta.[79] There is, therefore, no true "glomus pulmonale" derived from the pulmonary trunk responsible for monitoring the oxygen tension in the pulmonary circulation.

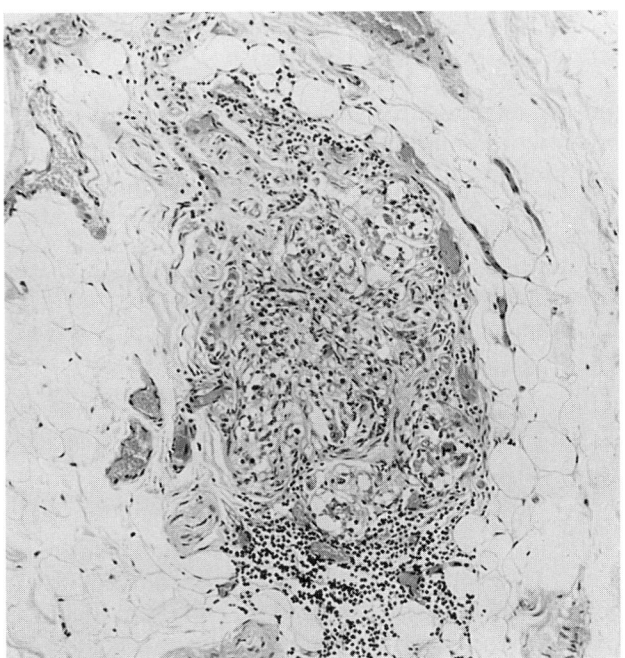

FIGURE 20-16
"Glomus pulmonale" from a woman aged 76. It consists of several cell clusters composed of chief cells and sustentacular cells similar to a lobule of the carotid body. At the lower border of the glomus is a focal collection of lymphocytes reminiscent of chronic cartoid glomitis. (H&E, ×144.)

REFERENCES

1. de Castro F: Sur la structure et l'innervation du sinus carotidien de l'homme et des mammifères. Nouveaux faits sur l'innervation et la fonction du glomus caroticum. Études anatomiques et physiologiques. *Trav Lab Rech Biol Univ Madr* 25:331–380, 1928.
2. Heymans C, Bouckhaert JJ, Dautrebande L: Sinus carotidien et réflexes respiratoires. II. Influences respiratoires réflexes de l'acidôse de l'alkalôse, de l'anhydride carbonique, de l'ion hydrogène et de l'anoxémie: Sinus carotidiens et échanges respiratoires dans les poumons et au delà des poumons. *Arch Int Pharmacodyn Ther* 39:400–448, 1930.
3. Eyzaguirre C, Zapata P: Perspectives in carotid body research. *J Appl Physiol* 57:931–957, 1984.
4. Holton P, Wood JB: The effects of bilateral removal of the carotid bodies and denervation of the carotid sinuses in two human subjects. *J Physiol (Lond)* 181:365–378, 1965.
5. Lugliani R, Whipp BJ, Seard C, Wasserman K: Effect of bilateral carotid body resection on ventilatory control at rest and during exercise in man. *N Engl J Med* 285:1105–1111, 1971.
6. Honda Y: Role of carotid chemoreceptors in control of breathing at rest and in exercise: Studies on human subjects with bilateral carotid body resection. *Jpn J Physiol* 35:535–544, 1985.
7. Eyzaguirre C, Fidone SJ: Transduction mechanisms in carotid body: Glomus cells, putative neurotransmitters, and nerve endings. *Am J Physiol* 239:C135–C152, 1980.
8. Pallot DJ: The mammalian carotid body. *Adv Anat Embryol Cell Biol* 102:1–91, 1987.
9. Heath D, Smith P: *Diseases of the Human Carotid Body.* London and Heidelberg, Springer-Verlag, 1992.

10. Heath D, Edwards C, Harris P: Post-mortem size and structure of the human carotid body. Its relation to pulmonary disease and cardiac hypertrophy. *Thorax* 25:129–140, 1970.
11. Smith P, Jago R, Heath D: Anatomical variation and quantitative histology of the normal and enlarged carotid body. *J Pathol* 137:287–304, 1982.
12. Khan Q, Heath D, Smith P: Anatomical variations in human carotid bodies. *J Clin Pathol* 41:1196–1199, 1988.
13. Heath D, Khan Q, Smith P: Histopathology of the carotid bodies in neonates and infants. *Histopathology* 17:511–520, 1990.
14. Scraggs M, Smith P, Heath D: Glomic cells and their peptides in the carotid body of the human fetus. *Pediatr Pathol* 12:823–834, 1992.
15. Grimley PM, Glenner GG: Ultrastructure of the human carotid body. A perspective on the mode of chemoreception. *Circulation* 37:648–665, 1968.
16. Heath D, Quinzanini M, Rodella A, Albertini A, Ferrari R, Harris P: Immunoreactivity to various peptides in the human carotid body. *Res Commun Chem Pathol Pharmacol* 62:289–293, 1988.
17. Smith P, Gosney J, Heath D, Burnett H: The occurrence and distribution of certain polypeptides within the human carotid body. *Cell Tissue Res* 261:565–571, 1990.
18. Ross LL: Electron microscopic observations of the carotid body of the cat. *J Biophys Biochem Cytol* 6:253–262, 1959.
19. Kondo H, Iwanaga T, Nakajima T: Immunocytochemical study on the localization of neuron-specific enolase and S-100 protein in the carotid body of rats. *Cell Tissue Res* 227:291–295, 1982.
20. O'Regan RG: Responses of carotid body chemosensory activity and blood flow to stimulation of sympathetic nerves in the cat. *J Physiol (Lond)* 315:81–98, 1981.
21. McDonald DM, Mitchell RA: The innervation of glomus cells, ganglion cells and blood vessels in the rat carotid body: a quantitative ultrastructural analysis. *J Neurocytol* 4:177–230, 1975.
22. Bodian D: A new method for staining nerve fibers and nerve endings in mounted paraffin sections. *Anat Rec* 65:89–97, 1936.
23. Smith P, Scraggs M, Heath D: The development of the nerve network in the fetal human carotid body and its subsequent function in cardiac disease. *Cardioscience* 4:143–149, 1993.
24. Eyzaguirre C, Gallego A: An examination of De Castro's original slides, in Purves MJ (ed): *The Peripheral Arterial Chemoreceptors.* London, Cambridge University Press, 1975, pp 1–23.
25. Lack EE, Perez-Atayde AP, Young JB: Carotid body hyperplasia in cystic fibrosis and cyanotic heart disease. A combined morphometric, ultrastructural and biochemical study. *Am J Pathol* 119:301–314, 1985.
26. Smith P, Greenberg S, Heath D, Gosney J: Effects of glucocorticoids on the rabbit carotid body. *Br J Exp Pathol* 68:251–258, 1987.
27. Hurst G, Heath D, Smith P: Histological changes associated with ageing of the human carotid body. *J Pathol* 147:181–187, 1985.
28. Scraggs M: *Studies on the Developing and Ageing Human Carotid Body.* Thesis for the degree of M. Phil., University of Liverpool, UK.
29. Kronenberg RS, Drage CW: Attenuation of the ventilatory and heart rate responses to hypoxia and hypercapnia with aging in normal men. *J Clin Invest* 52:1812–1819, 1973.
30. Tafil-Klawe M, Raschke F, Kublick A, Stoohs R, von Wichert P: Attenuation of augmented ventilatory response to hypoxia in essential hypertension in the course of aging. *Respiration* 56:154–160, 1989.
31. Lack EE: Carotid body hypertrophy in patients with cystic fibrosis and cyanotic congenital heart disease. *Hum Pathol* 8:39–51, 1977.
32. Khan Q, Heath D, Nash J, Smith P: Chronic carotid glomitis. *Histopathology* 14:471–481, 1989.
33. Heath D, Khan Q, Nash J, Smith P: Carotid body disease and the physician—chronic carotid glomitis. *Postgrad Med J* 65:353–357, 1989.
34. Dhillon DP, Barer GR, Walsh M: The enlarged carotid body of the chronically hypoxic and chronically hypoxic and hypercapnic rat: A morphometric analysis. *Q J Exp Physiol* 69:301–317, 1984.
35. Barer G, Wach R, Pallot D, Bee D: Almitrine, hypoxia, systemic hypertension and the carotid body, in Heath D (ed): *Aspects of Hypoxia.* Liverpool, England, Liverpool University Press, 1986, pp 113–129.
36. Arias-Stella J: Human carotid body at high altitudes. Item 150 in the sixty-ninth program and abstracts of the American Association of Pathologists and Bacteriologists, San Francisco, California, 1969.
37. Arias-Stella J, Valcarcel J: The human carotid body at high altitudes. *Pathol Microbiol* 39:292–297, 1973.
38. Arias-Stella J, Valcarcel J: Chief cell hyperplasia in the human carotid body at high altitudes. Physiologic and pathologic significance. *Hum Pathol* 7:361–373, 1976.
39. Khan Q, Heath D, Smith P, Norboo T: The histology of the carotid bodies in highlanders from Ladakh. *Int J Biometeorol* 32:254–259, 1988.
40. Lack EE: Hyperplasia of vagal and carotid body paraganglia in patients with chronic hypoxemia. *Am J Pathol* 91:497–516, 1978.
41. Heath D, Smith P, Jago R: Hyperplasia of the carotid body. *J Pathol* 138:115–127, 1982.
42. Jago R, Smith P, Heath D: Electron microscopy of carotid body hyperplasia. *Arch Pathol Lab Med* 108:717–722, 1984.
43. Fitch R, Smith P, Heath D: A quantitative study of nerve axons in carotid body hyperplasia. *Arch Pathol Lab Med* 109:234–237, 1985.
44. Edwards C, Heath D, Harris P: The carotid body in emphysema and left ventricular hypertrophy. *J Pathol* 104:1–13, 1971.
45. Habeck J-O: Morphological findings at the carotid bodies of humans suffering from different types of systemic hypertension or severe lung diseases. *Anat Anz* 162:17–27, 1986.
46. Habeck J-O, Waller H, Protze J: Pathological alterations of the arterial vessels of the carotid bodies in hypertensive humans. *Dtsch Gesundheitswes* 38:1970–1972, 1983.
47. Habeck J-O, Honig A, Pfeiffer C, Schmidt M: The carotid bodies in spontaneously hypertensive (SHR) and normotensive rats—a study concerning size, location and blood supply. *Anat Anz* 150:374–384, 1981.
48. Honig A, Habeck J-O, Pfeiffer C, et al: The carotid bodies of spontaneously hypertensive rats (SHR): A functional and morphologic study. *Acta Biol Med Germ* 40:1021–1030, 1981.
49. Smith P, Jago R, Heath D: Glomic cells and blood vessels in the hyperplastic carotid bodies of spontaneously hypertensive rats. *Cardiovasc Res* 18:471–482, 1984.

50. Honig A: Role of the arterial chemoreceptors in the reflex control of renal function and body fluid volumes in acute arterial hypoxia, in Acker H, O'Regan RG (eds): *Physiology of the Peripheral Arterial Chemoreceptors.* Amsterdam, Elsevier, 1983, pp 395–429.

51. Honig A: Peripheral arterial chemoreceptors and reflex control of sodium and water homeostasis. (A review). *Am J Physiol* 257:R1282–R1302, 1989.

52. Heath D, Smith P, Hurst G: The carotid bodies in coarctation of the aorta. *Br J Dis Chest* 80:122–130, 1986.

53. Bencini A: Reduction of reflex bronchotropic impulses as a result of carotid body surgery. *Int Surg* 54:415–423, 1970.

54. Bencini C, Pulera N: The carotid bodies in bronchial asthma. *Histopathology* 18:195–200, 1991.

55. Severinghaus JW, Bainton CR, Carcelan A: Respiratory insensitivity to hypoxia in chronically hypoxic man. *Respir Physiol* 1:308–334, 1966.

56. Sørensen SC, Severinghaus JW: Irreversible respiratory insensitivity to acute hypoxia in man born at high altitude. *J Appl Physiol* 25:217–220, 1968.

57. Lahiri S, Kao FF, Velasquez T, Martinez C, Pezzia W: Irreversible blunted respiratory sensitivity to hypoxia in high altitude natives. *Respir Physiol* 6:360–374, 1969.

58. Chiodi H: Respiratory adaptations to chronic high altitude hypoxia. *J Appl Physiol* 10:81–87, 1957.

59. Fidone S, Gonzalez C, Yoshizaki K: Effects of hypoxia on catecholamine synthesis in rabbit carotid body in vitro. *J Physiol (Lond)* 333:81–91, 1982.

60. Olson EB, Vidruk EH, McCrimmon DR, Dempsey JA: Monoamine neurotransmitter metabolism during acclimatization to hypoxia in rats. *Respir Physiol* 54:79–96, 1983.

61. Sørensen SC, Severinghaus JW: Respiratory sensitivity to acute hypoxia in man born at sea level living at high altitude. *J Appl Physiol* 25:211–216, 1968.

62. Edelman NH, Lahiri S, Braudo L, Cherniack NS, Fishman AP. The blunted ventilatory response to hypoxia in cyanotic congenital heart disease. *N Engl J Med* 282:405–411, 1970.

63. Sørensen SC, Severinghaus JW: Respiratory insensitivity to acute hypoxia persisting after correction of tetralogy of Fallot. *J Appl Physiol* 25:221–223, 1968.

64. Flenley D, Millar JS: Ventilatory response to oxygen and carbon dioxide in chronic respiratory failure. *Clin Sci* 33:319–334, 1967.

65. Bradley C, Fleetham J, Anthonisen N: Ventilatory control in patients with hypoxemia due to obstructive lung disease. *Am Rev Respir Dis* 120:21–30, 1979.

66. Flenley D: Long-term oxygen therapy and the pulmonary circulation, in Heath D (ed): *Aspects of Hypoxia.* Liverpool, England, Liverpool University Press, 1986, pp 45–59.

67. Trzebski A, Tafil M, Zoltowski M, Przybylski J: Increased sensitivity of the arterial chemoreceptor drive in young men with mild hypertension. *Cardiovasc Res* 16:163–172, 1982.

68. Quies von W, Claus T, Honig A: Die Hyperventilation der Hypertoniker, -eine Pilotstudie zum Verhalten der Reflexe der arteriellen Chemorezeptoren bei hypertensiven Erkrankungen. *Dtsch Gesundheitswes* 16:612–617, 1983.

69. Smith P, Hurst G, Heath D, Drewe R: The carotid bodies in a case of ventricular septal defect. *Histopathology* 10:831–840, 1986.

70. Khan Q, Smith P, Heath D: The distribution of enkephalins in human carotid bodies showing cellular proliferation and chronic glomitis. *Arch Pathol Lab Med* 114:1232–1235, 1990.

71. Smith P, Heath D, Fitch R, Hurst G, Moore D, Weitzenblum E: Effects on the rabbit carotid body of stimulation by almitrine, natural high altitude, and experimental normobaric hypoxia. *J Pathol* 149:143–153, 1986.

72. Smith P, Heath D, Williams D, Bencini C, Pulera N, Giuntini C: The earliest histopathological response to hypobaric hypoxia in rabbits in the Rifugio Torino (3370 m) on Monte Bianco. *J Pathol* 170:485–491, 1993.

73. Heath D, Smith P, Jago R: Dark cell proliferation in carotid body hyperplasia. *J Pathol* 142:39–49, 1984.

74. Heath D, Smith P: The carotid bodies enlarge in some cases of cirrhosis of the liver. *Cardioscience* 5:37–41, 1994.

75. Heath D, Smith P: Enlargement of the carotid bodies in cirrhosis of the liver. *Histopathology* 25:159–164, 1994.

76. Keys A, Snell AM: Respiratory properties of the arterial blood in normal men and in patients with disease of the liver: Position of the oxygen dissociation curve. *J Clin Invest* 17:59–67, 1938.

77. Blackburn CRB, Read J, McRae J, Colebatch AJ, Playoust MR, Holland RA. Veno-arterial shunting of blood in chronic liver disease. *Australas Ann Med* 9:204–208, 1960.

78. Krahl VE: The glomus pulmonale: its location and microscopic anatomy, in de Reuck AVS, O'Connor M (eds): *Ciba Foundation Symposium on Pulmonary Structure and Function,* 1962, pp 53–76.

79. Edwards C, Heath D: Site and blood supply of the intertruncal glomera. *Cardiovasc Res* 4:502–508, 1970.

80. Comroe JH. Discussion of a paper by Krahl. In de Reuck AVS, O'Connor M (eds): *Ciba Foundation Symposium on Pulmonary Structure and Function,* 1962, pp 70–73.

Chapter *21*

Pulmonary Vascular Disease

Donald Heath

The human pulmonary circulation is a low-resistance, low-pressure system composed of thin-walled vessels that are very susceptible to the effects of hypertension. There are many causes of raised pulmonary arterial pressure, and they are associated with different histologic changes in the walls of the blood vessels. Hence there are many types of pulmonary vascular disease, and these are listed in Table 21-1.

Three categories of pulmonary arterial vessels need to be distinguished. The elastic pulmonary arteries are characterized by multiple concentric elastic laminae down to a diameter of about 500 μm, which is half of the original lower diameter suggested by Brenner.[1] Their branches are the so-called muscular pulmonary arteries, which in fact have a *thin* media of circularly oriented smooth muscle sandwiched between inner and outer elastic laminae (Fig. 21-1). They range in diameter from 500 to 80 μm, a somewhat lower figure than the original 100 μm proposed by Brenner.[1] Pulmonary arterioles less than 80 μm in diameter have a wall consisting of a single elastic lamina (Fig. 21-1) except at their origin from their parent pulmonary arteries, where they have a thin cuff of circular smooth muscle (Fig. 21-2). A characteristic feature of the human pulmonary vasculature is its propensity to age-change intimal fibrosis (Fig. 21-3).

This account of pulmonary vascular disease starts with a description of plexogenic pulmonary arteriopathy, since this shows the complete gamut of pathologic changes to which the pulmonary vasculature is susceptible. A contrast is then made with the remodeling of the pulmonary circulation that occurs in response to chronic alveolar hypoxia. This may be due to a disorder such as chronic obstructive lung disease or may be the result of environmental circumstances, such as the hypobaric hypoxia to which healthy native highlanders are exposed. A progression through the various types of pulmonary vascular disease is then taken to include all the categories, extending finally to the rarer or more exotic ones,

such as congenital anomalies of the pulmonary vasculature and metabolic pulmonary vascular disease.

PLEXOGENIC PULMONARY ARTERIOPATHY

This type of pulmonary vascular disease has highly distinctive histologic features, and it is restricted to a comparatively small, well-defined group of conditions shown in Table 21-2. Most cases occur in instances of congenital cardiac shunt and primary pulmonary hypertension. These conditions provide most examples of the arteriopathy leading to combined heart-lung transplantation.[2] The arteriopathy may develop in posttricuspid shunts, such as ventricular septal defect, patent ductus arteriosus, persistent truncus arteriosus, and cardioaortic fistula. It also occurs in pretricuspid shunts, such as atrial septal defect with anomalous pulmonary venous drainage. The arteriopathy does not develop in congenital cardiac anomalies without a shunt to cause an initially high pulmonary blood flow. In fact, any cardiac disease leading to an elevation of pulmonary venous blood pressure, such as mitral stenosis, appears to prevent the development of plexogenic pulmonary arteriopathy. This vascular disease develops in only a small proportion of patients with congenital cardiac shunts. Thus, as early as 1971, Keith[3] found, from his survey of 1500 cases of isolated ventricular septal defect of all sizes, that only 5 percent of patients will ultimately die from the effects of advanced hypertensive pulmonary vascular disease. When it does occur, this progression runs a stereotyped course irrespective of the anatomic details of the underlying cardiac anomaly.

Histopathology

The development of the various histologic lesions found in plexogenic pulmonary arteriopathy is so stereotyped that it allowed Jesse E. Edwards and me to draw up a grading system so that complex collections of histologic changes could be expressed

TABLE 21-1
The Different Types of Pulmonary Vascular Disease

Disease	Etiologic Factors
Plexogenic pulmonary arteriopathy	Congenital cardiac shunts Primary pulmonary hypertension Aminorex fumarate Rare cases of cirrhosis of the liver and portal vein thrombosis
Hypoxic vascular remodeling	Chronic hypoxic lung disease, such as chronic bronchitis and pulmonary emphysema Native highlanders at high altitude
Dietary pulmonary hypertension	Aminorex fumarate (in humans) Pyrrolizidine alkaloids (in animals) Fenfluoramine and other appetite suppressants have been suspected.
Infantile subacute mountain sickness	Lack of initial acclimatization to high altitude in Han infants in Tibet
Pulmonary veno-occlusive disease	Viruses(?) Toxoplasmosis(?) Bleomycin
Pulmonary embolism	Thromboembolism Amniotic fluid Cotton-wool fragments Fat Bone marrow Air
Pulmonary thrombosis in cyanotic congenital heart disease	Fallot's tetralogy Tricuspid atresia
Pulmonary venous hypertension	Mitral stenosis Left atrial myxoma Chronic left ventricular failure
Pulmonary fibrosis	Interstitial fibrosis and honeycomb lung Massive fibrosis of pneumoconiosis Hematite lung
Granulomas	Tuberculosis Sarcoidosis Giant cell arteritis Takayasu's disease
Parasites	Bilharzia (in humans) *Aelurostrongylus abstrusus* (in cats) *Dirofilaria immitis* (in dogs)
Metabolic	Pulmonary calcinosis Toxic oil syndrome Fabry's disease
Tumors	Invasive hemangiomatosis of lung Sarcomas
Congenital anomalies	

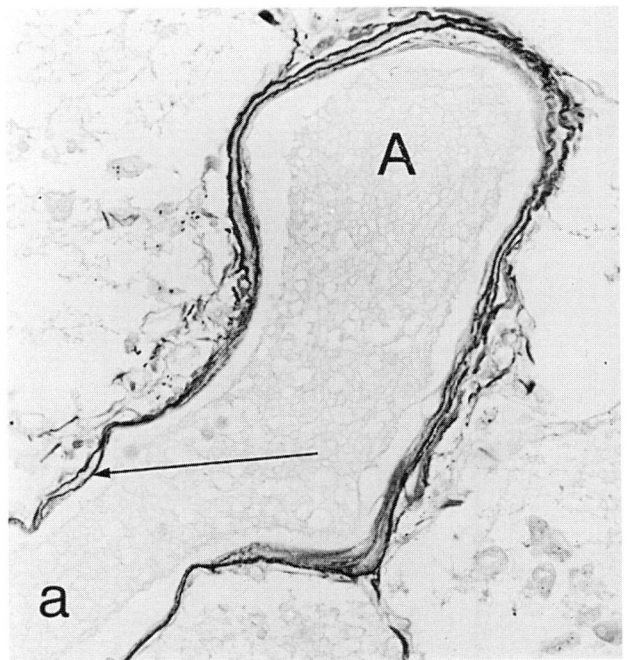

FIGURE 21-1
Oblique section through a muscular pulmonary artery (A) showing the characteristic thin media bounded by inner and outer elastic laminae. An arteriolar branch (a) arises from the parent vessel and passes downward and to the left. Its wall consists of a single elastic lamina except near its origin from its parent artery, where the remains of a thin media bounded by two elastic laminae can still be seen (*arrow*). (Elastic–van Gieson, ×284.)

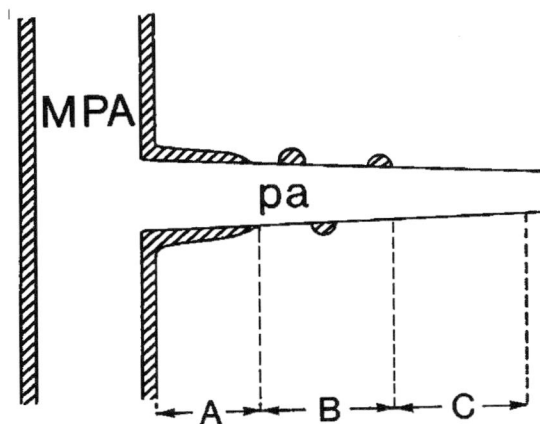

FIGURE 21-2
Diagram of structure of a pulmonary arteriole (pa) that arises at a side branch of a muscular pulmonary artery (MPA). *A.* In its proximal portion, the arteriole has a distinct media of circularly oriented smooth muscle, indicated by hatching. *B.* In the adjacent segment, this coat of smooth muscle becomes spiral and thus appears discontinuous in longitudinal sections of the vessel. *C.* In the distal part, the arteriole is devoid of a muscular media; its wall consists of a single elastic lamina.

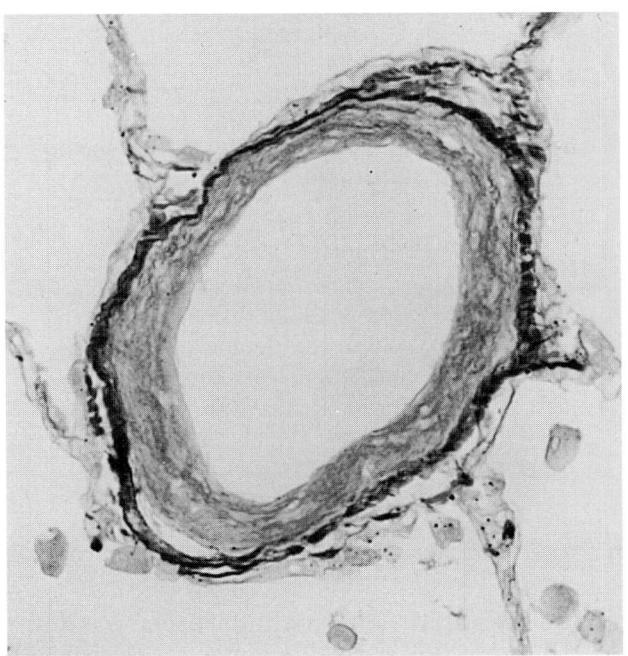

FIGURE 21-3
Transverse section of a normal pulmonary venule. The wall consists of a single elastic lamina. There is considerable intimal fibrosis due to age change. (Elastic–van Gieson, ×375.)

TABLE 21-2

Causes of Plexogenic Pulmonary Arteriopathy

Congenital cardiac shunts

Primary pulmonary hypertension

Aminorex fumarate

Rare cases of cirrhosis of the liver

Rare cases of portal vein thrombosis

succinctly as a single number.[4] In the original publication, the grading system (Table 21-3) was described as one of "hypertensive pulmonary vascular disease." The introduction of the term *plexogenic pulmonary arteriopathy* by Wagenvoort and me at a working party of the World Health Organization in Geneva had to wait another 15 years.[5] The grading system applies only to this arteriopathy and not to pulmonary vascular disease in general—a mistake that has been made many times in papers published on this subject. When Wagenvoort and I introduced this term, we were at pains to use the term *plexogenic* rather than *plexiform*, since this arteriopathy has the *potential* for forming plexiform lesions—which need not, however, be present to make the diagnosis in every case.

TABLE 21-3
Heath-Edwards Classification of Plexogenic Pulmonary Arteriopathy

Grade	Major Pathologic Feature
1	Muscularization of pulmonary arteries
2	Migration of dark muscle cells to intima to form cellular intimal reaction
3	Transformation of intimal myofibroblasts to form muscle, collagen, and elastin
4	Appearance of plexiform and other dilatation lesions
5	Rupture of dilated vessels to produce hemorrhages and exudative lesions
6	Necrotizing arteritis of pulmonary arteries

The Heath-Edwards grading system was introduced 35 years ago, and although the basic principles remain the same, they have been modified by advances in our knowledge over that period—particularly in electron microscopy and immunohistochemistry of the pulmonary endocrine system, unknown when the grading system was introduced. Moreover, advances in technology with regard to open heart surgery and heart-lung transplantation have lessened the application of the grading system for the clinical management of cases of congenital cardiac shunt.

GRADE 1: MUSCULARIZATION OF PULMONARY ARTERIES

In the solid, hypoxic lungs of the fetus, the elastic interlobular arteries, which will develop into the elastic pulmonary arteries of the adult lung, and the muscular intralobular arteries, which will form the "muscular pulmonary arteries" of the adult lung, are thick-walled.[6] In the healthy newborn at sea level, inflation of the lungs with air rich in oxygen causes the elastic and muscular pulmonary arteries to dilate and become thin-walled, while the pulmonary arterioles do not become muscularized. In contrast, in the newborn with a ventricular septal defect, the high pulmonary blood flow through the hole maintains the pulmonary hypertension and causes a thick-walled pulmonary arterial tree with hypertrophy of small pulmonary arteries (Fig. 21-4) and muscularization of pulmonary arterioles (Fig. 21-5).

Within these thick muscular arteries, two distinct types of vascular smooth muscle cell can be distinguished in electron micrographs. Most have copious electron-light cytoplasm, but a minority are more elongated with electron-dense cytoplasm, so-called dark muscle cells. There is intimate association of dark and light smooth muscle cells.

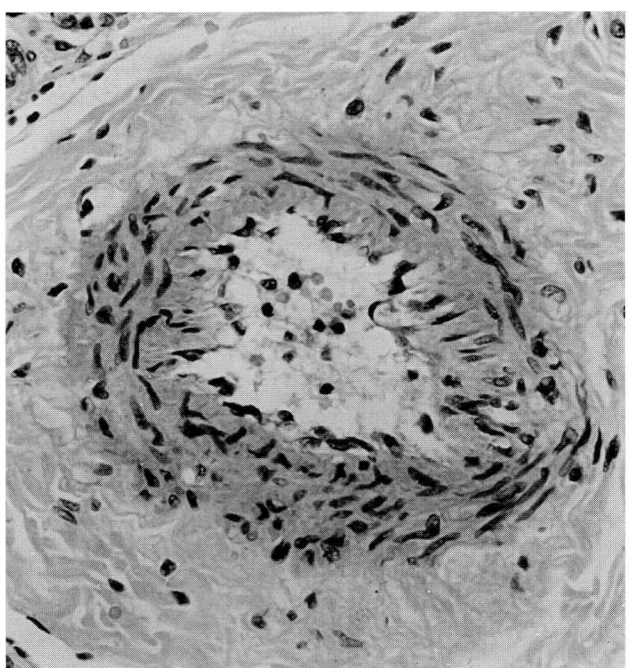

FIGURE 21-4
Transverse section of a small pulmonary artery from a 1 1/2-year-old boy with a ventricular septal defect complicated by pulmonary hypertension. There is hypertrophy of the media, whose thickness equals 22 percent of the external diameter of the vessel. The shapes of the nuclei reveal that most of the smooth muscle cells are oriented circularly. No form of intimal proliferation is present. Hypertensive pulmonary vascular disease (HPVD) grade 1. (H&E, ×375.)

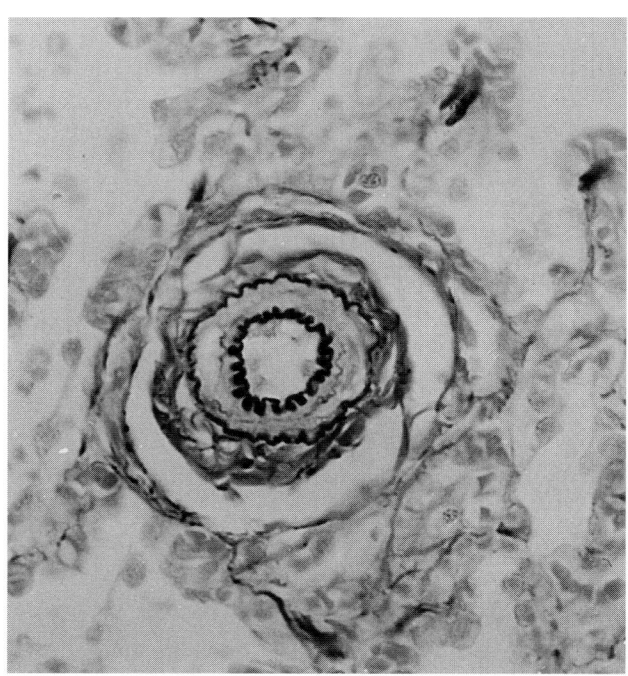

FIGURE 21-5
Transverse section of a muscularized pulmonary arteriole from the same case illustrated in Fig. 21-4. A distinct media of circular muscle is clearly delineated between inner and outer elastic laminae. The medial thickness is 25 percent of the external diameter of the vessel. There is no intimal proliferation. HPVD grade 1. (Elastic–van Gieson, ×600.)

Protuberances of dark cells often fit into depressions in the cytoplasmic borders of light muscle cells. This intimate association appears to precede the migration of dark cells toward the inner media and intima. These two types of dark cell can be identified in paraffin sections stained with phosphotungstic acid hematoxylin. Use of this staining method will demonstrate that two types of vascular smooth muscle cells are operating in the wall of the pulmonary artery in the early stages of plexogenic pulmonary arteriopathy. As part of the general muscularization in grade 1, fasciculi of longitudinal muscle form in the adventitia outside the outer elastic lamina (Fig. 21-6) and in the intima inside the inner elastic lamina.

GRADE 2: MIGRATION OF DARK MUSCLE CELLS TO THE INTIMA

Paraffin sections stained with phosphotungstic acid hematoxylin show dark muscle cells in the inner third of the media lying radially to the inner elastic lamina (Fig. 21-7). Ultrastructurally they appear elongated and spindle-shaped with electron-dense cytoplasm and have a cytoplasmic border free of pinocytotic vesicles.[7] Having migrated to the inner elastic lamina (Fig. 21-8), they pass through gaps in it. These gaps may be natural or may arise in pul-

monary hypertension as a result of elastase activity (Fig. 21-9).[8] Once the extremities of the cells have passed through these gaps and into the intima, they spread along this layer and into the vascular lumen. This blockage of small pulmonary arteries elevates pulmonary vascular resistance.

When Edwards and I introduced our grading system in 1958, before the routine use of electron microscopy in the pathology laboratory, we were uncertain of the nature of the intimal proliferation and referred to it rather vaguely as a "cellular intimal reaction." At the time, we did not appreciate its functional significance as a migration of vascular smooth muscle from the media into the intima.

The migration of the smooth muscle cells into the inner media and intima is associated with increased numbers of pulmonary endocrine cells, whose cytoplasm is rich in gastrin-releasing peptide, the human analogue of bombesin in amphibian skin, in the terminal bronchioles (Fig. 21-10).[9] The basis for this association is obscure, and it remains unclear whether the peptides in the pulmonary endocrine cells stimulate the migration of the pulmonary vascular smooth muscle cells or respond to it. While the largest numbers of bombesin-rich cells in the pulmonary endocrine cells are seen in this phase of the

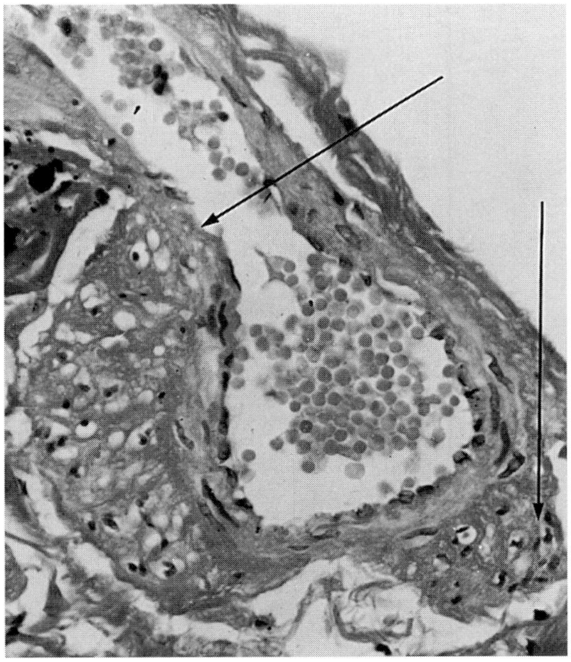

FIGURE 21-6
Transverse section of a small pulmonary artery from a 38-year-old woman with persistent truncus arteriosus and pulmonary hypertension. Large bundles of longitudinally oriented smooth muscle fibers (*arrows*) have developed outside the original media of circular muscle. HPVD grade 1. (H&E, ×450.)

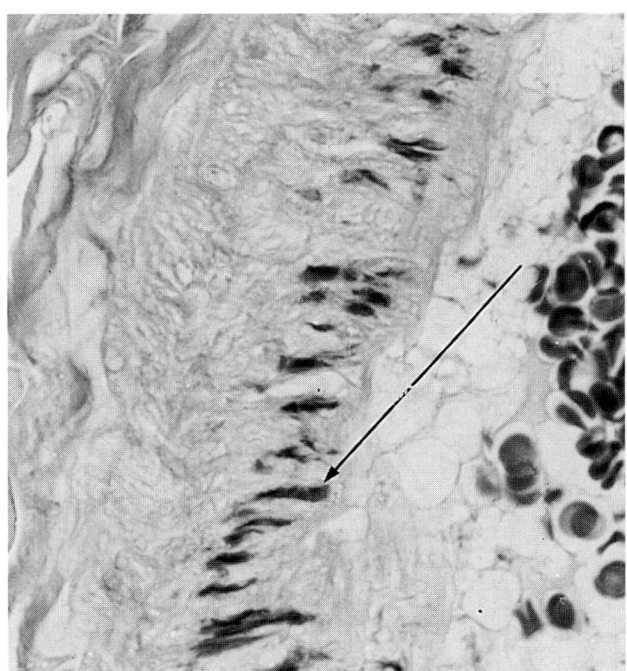

FIGURE 21-7
Section stained by phosphotungstic acid hematoxylin to differentiate between mature vascular smooth muscle cells (unstained) and dark variants, which have the capacity for migration toward the inner part of the media (stained black, *arrow*). This technique allows for rapid study of the distribution of the two variants of muscle cell. (×700.)

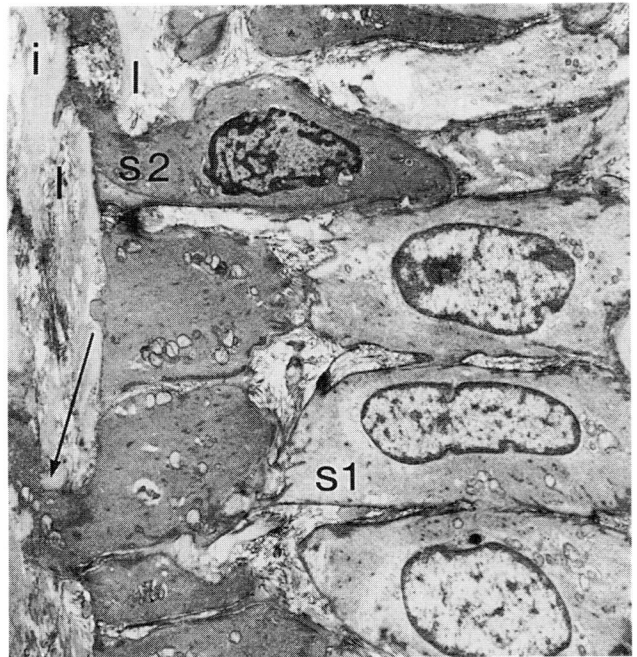

FIGURE 21-8
Electron micrograph of longitudinal section of a small pulmonary artery from a 23-year-old woman with primary pulmonary hypertension. The media is composed of smooth muscle cells sectioned transversely. Those in the outer media (S1) are pale with numerous peripheral attachment points between which the cytoplasm bulges outward. Smooth dark muscle cells in the inner media (S2) are denser and lack a ruffled border. Two of them can be seen in the process of sending out cytoplasmic extensions between gaps in the internal elastic lamina (l) into the intima (i). Here they are associated with elongated cytoplasmic processes from other smooth muscle cells (*arrow*). (×5600.)

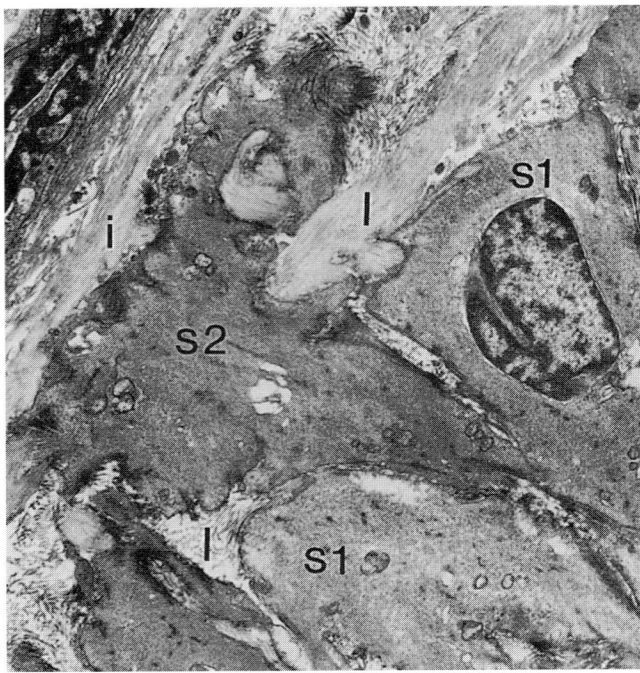

FIGURE 21-9
Part of the same vessel shown in the Fig. 21-8. It shows a dark smooth muscle cell (S2) in the process of inward migration between two light mature vascular smooth muscle cells (S1). The inner cytoplasmic extension of the dark cell has passed through a gap in the internal elastic lamina (l) to reach the intima (i). The cytoplasm of the dark smooth muscle cell is electron-dense and contrasts with the paler nonmigrating smooth muscle cells. (×7800.)

arteriopathy, considerable numbers are also seen in grade 3 and in the presence of cellular plexiform lesions in grade 4. The numbers fall off as the plexiform lesions age and as the activity of myofibroblasts in them diminishes.

GRADE 3: INTIMAL PROLIFERATION AND TRANSFORMATION OF INTIMAL MYOFIBROBLASTS

As dark vascular smooth muscle cells migrate into the intima, many become transformed into myofibroblasts; these swarm into the lumina of the pulmonary arteries, where they appear as collections of yellow cells in sections stained by the van Gieson method. Many myofibroblasts retain their cellular character and concentric pattern, referred to as "onionskin proliferation" (Fig. 21-11). The ultrastructure of this confirms that it is composed of myofibroblasts separated by copious amounts of proteoglycan (Fig. 21-12). Other myofibroblasts congregate in the intimal layer, where they manifest their multipotential capacity.[10] Many mature into smooth muscle cells, which form nodules adjacent to the inner elastic lamina. Others form collagen, so that the pulmonary arteries show progressive blockage by an

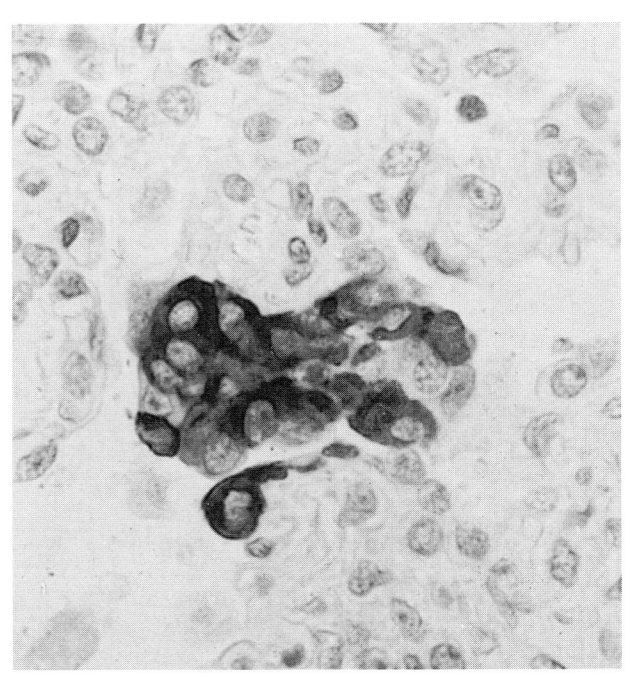

FIGURE 21-10
Endocrine cells immunoreactive for gastrin-releasing peptide in a terminal bronchiole from a 46-year-old woman with primary pulmonary hypertension: GRP cell count = 50/cm² of lung section. (Immunoperoxidase, × 1000.)

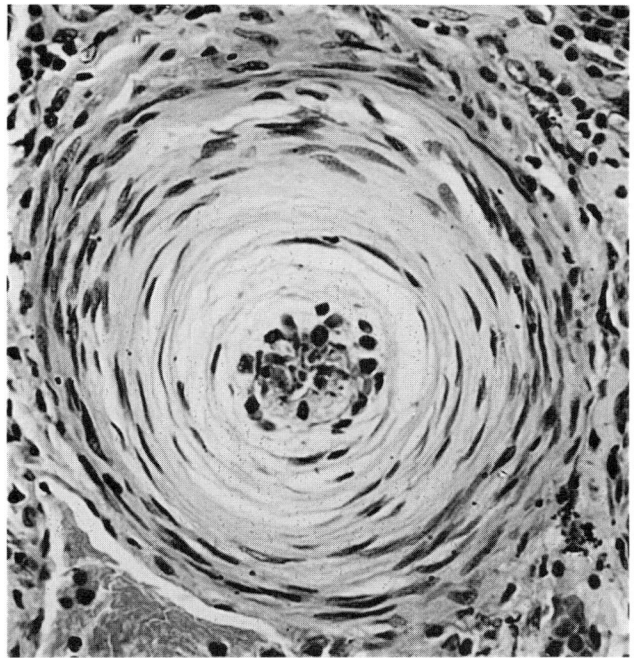

FIGURE 21-11
Transverse section of a small pulmonary artery from a 26-year-old woman with an aortopulmonary septal defect and pulmonary hypertension. There is pronounced concentric laminar intimal proliferation, the so-called onionskin proliferation. HPVD grade 3 (H&E, ×450.)

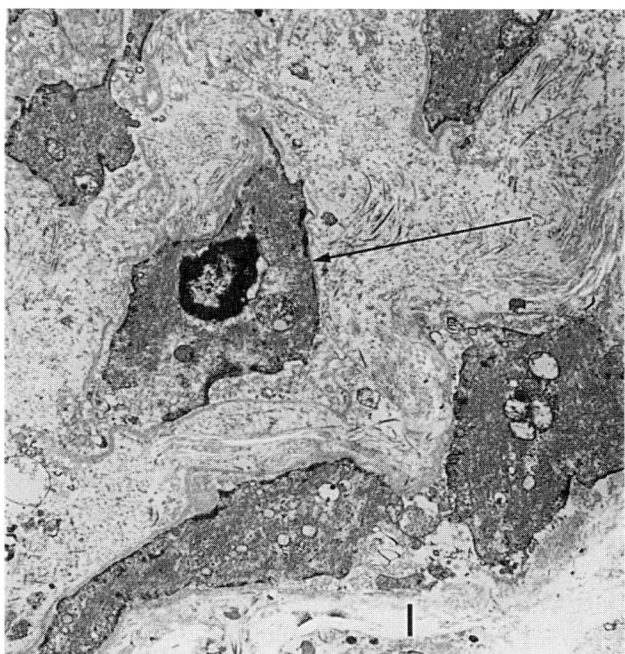

FIGURE 21-12
Electron micrograph of part of the wall of a small pulmonary artery from a 24-year-old man with severe pulmonary hypertension secondary to portal hypertension. On histologic examination the vessel showed the onionskin intimal proliferation characteristic of plexogenic pulmonary arteriopathy. The internal elastic lamina (l) can be seen, and beyond this is intimal proliferation consisting of myofibroblasts (*arrow*) that resemble smooth muscle cells. Between them is a mixture of elastin, collagen fibers, and an amorphous ground substance consisting of proteoglycans. (×6000.)

increasingly thick layer of intimal fibrosis that is at first cellular and later acelluar or even hyaline. Many myofibroblasts form elastin, so the pulmonary arteries become lined by a thick layer of fibroelastosis (Fig. 21-13). This progressive blockage of lumina of pulmonary arteries by these different forms of intimal proliferation is associated with a decline in the high levels of pulmonary blood flow that characterize congenital cardiac shunts with grades 1 and 2 plexogenic pulmonary arteriopathy. In grade 3 the levels of pulmonary blood flow approach the normal range.[11]

GRADE 4: DEVELOPMENT OF PLEXIFORM AND DILATATION LESIONS

The progressive blockage of the pulmonary arterial tree by these various forms of intimal proliferation and the accompanying rise of pulmonary arterial pressure lead to vascular dilatation. This may be generalized or localized. The arteriolar branches dilate and remain thin-walled, with a wall consisting of a single elastic lamina. These are so-called veinlike branches of pulmonary arteries. Less commonly these dilated arterial branches become complex, to form collections resembling small angiomas (Fig 21-14). These angiomatoid lesions and the other dilatation lesions occur proximal to obstructed segments of pulmonary arteries and bypass those to effect communications with the pulmonary capillary bed.

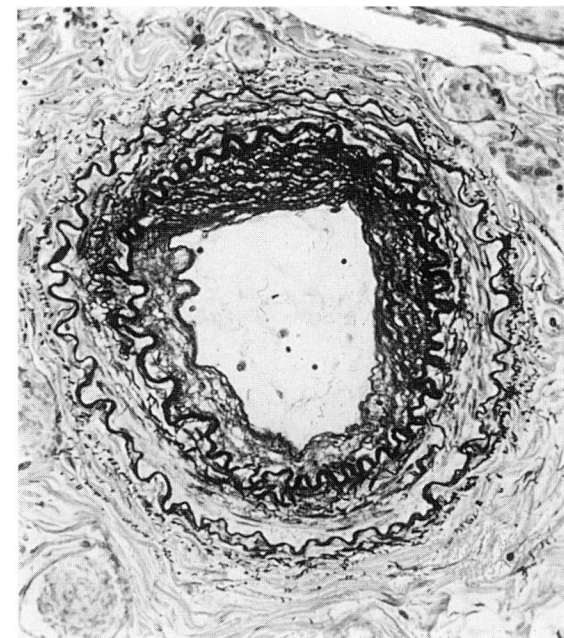

FIGURE 21-13
Transverse section of a small pulmonary artery from a 26-year-old woman with a widely patent ductus arteriosus and severe pulmonary hypertension. Severe concentric laminar fibroelastosis is present. There is also medial hypertrophy and crenation of elastic laminae suggestive of vasoconstriction. HPVD grade 3. (Elastic–van Gieson, ×185.)

Plexiform lesions now occur. The potential for forming these structures gives the arteriopathy its name, although as we have already seen, their presence is not necessary for the diagnosis to be made. It seems likely that these develop at the site of arteriolar branches of pulmonary arteries. These dilated sacs form at terminal and lateral branches of parent pulmonary arteries and form the microanatomic limits of the vascular structure, in which a complex series of cellular events take place (Fig. 21-15). The process appears to start with patchy fibrinoid necrosis in the media of the parent pulmonary artery, and fibrinoid substances can be detected in the sac with appropriate stains. These substances appear to stimulate a proliferation of minute vessels that assume a plexiform pattern—hence the name of the lesion (Fig. 21-16). These vessels are not true capillaries, for their lining cells do not contain Palade bodies.

Myofibroblasts migrate from the damaged media to infiltrate between the diminutive blood vessels. Other cells infiltrating the walls of these vessels are primitive vasoformative cells, which contain large numbers of cytoplasmic fibrils and hence are called fibrillary cells. Plexiform lesions age. As they mature, the walls and septa of the plexiform vessels become thicker as a result of fibrosis and muscular-

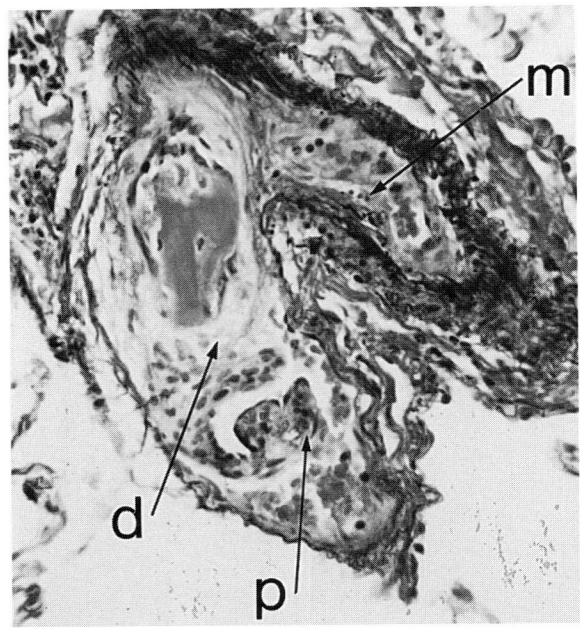

FIGURE 21-15
Plexiform lesion in a section of lung from a 21-year-old woman with a ventricular septal defect and severe pulmonary hypertension. This section was stained to demonstrate elastin and thus the origin of the dilated branch forming the plexiform lesion. A small muscular pulmonary artery (arrow m) has a dilated branch (*arrow d*) in which there has been a proliferation of cells arranged in a plexiform manner (*arrow p*). HPVD grade 4. (Elastic–van Gieson, ×160.)

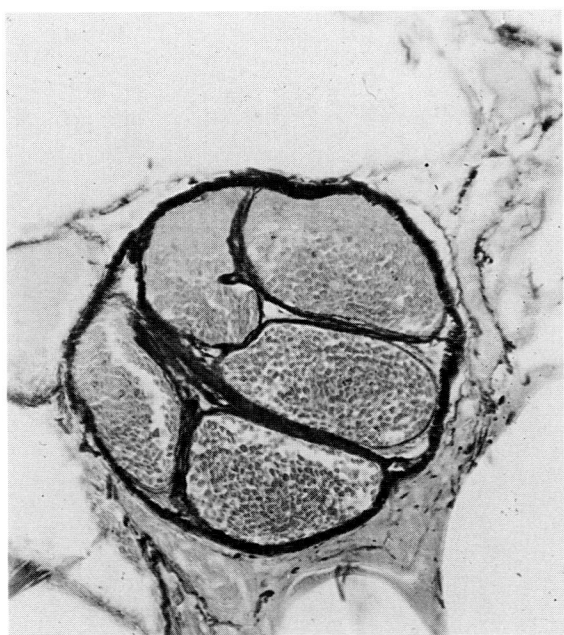

FIGURE 21-14
Small angiomatoid lesion from a 21-year-old woman with a large ventricular septal defect and severe pulmonary hypertension. Its wall consists of a thick single elastic lamina, while its constituent vessels have very thin walls composed of a delicate elastic lamina. HPVD grade 4. (Elastic–van Gieson, ×140.)

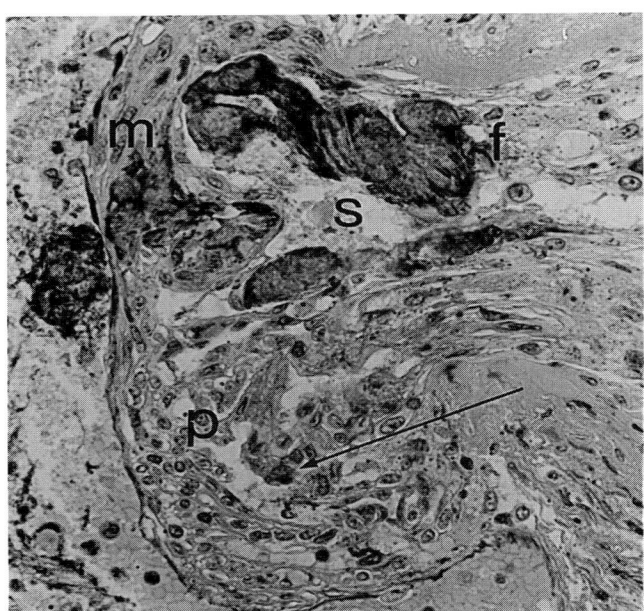

FIGURE 21-16
Cellular plexiform lesion from the same case, stained to show fibrin and its close association with cellular components in the formation of the lesion. Cells arranged in a plexiform pattern (p) have grown into a dilated sac (s) derived as a branch of a small pulmonary artery as shown in the preceding figure. The media of the parent artery (m) shows fibrinoid necrosis. Adjacent to it is a clump of dark-staining fibrin (f), which extends into one of the vascular channels of the plexiform lesion. Small fragments of fibrin (*arrow*) are also scattered among cells of the plexiform lesion. (Phosphotungstic acid hematoxylin, ×375.)

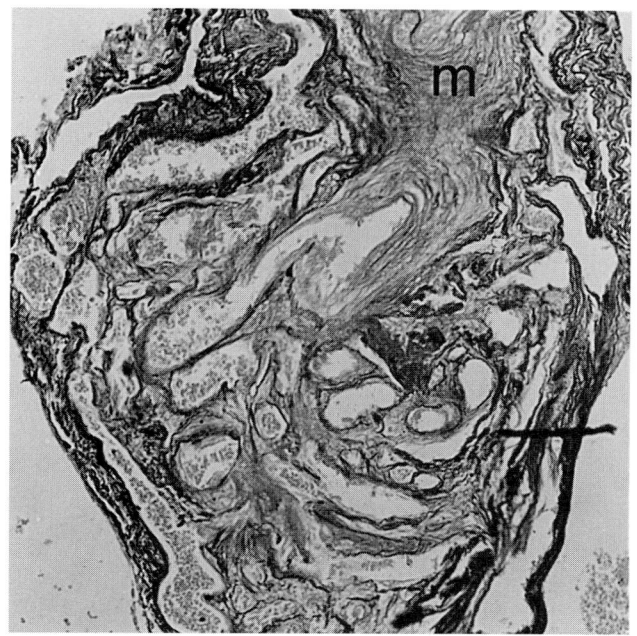

A

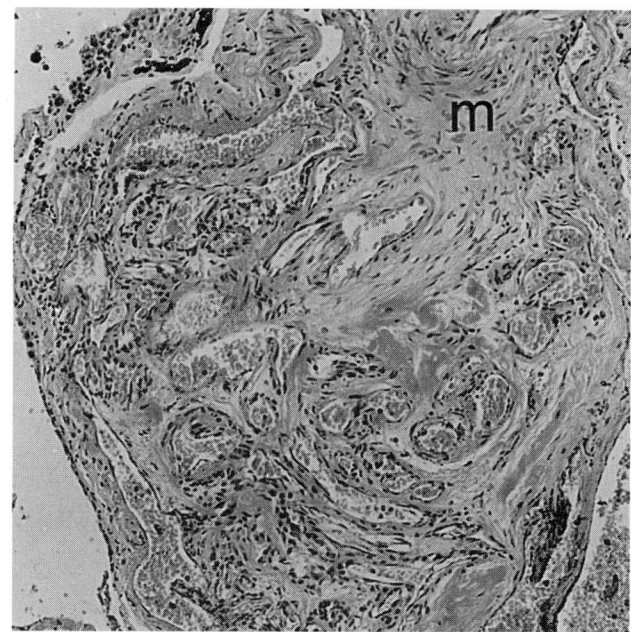

B

FIGURE 21-17
A large, mature plexiform lesion from the same case illustrated in Fig. 21-16. The parent muscular pulmonary artery (m) shows severe intimal fibroelastosis. It gives rise to a complex series of branches that pass downward and to the left. The vascular channels are lined by cells resembling, but not identical with,

endothelial cells and have thick walls. Such lesions may be mistaken by those not familiar with pulmonary vascular pathology for recanalization of thrombus.
(*A*. Elastic–van Gieson, ×125; *B*. H&E, ×125.)

ization. In this way, the plexiform lesions may mimic the colander lesions of recanalized thrombus described below (Fig. 21-17). This is an important point in differential diagnosis for the histopathologist not experienced in pulmonary vascular pathology. Later on, these mature plexiform lesions may show progressive elastosis.

Plexiform lesions have different distributions in primary and secondary pulmonary hypertension.[11a] Preacinar plexiform lesions were seen more commonly in patients with pulmonary hypertension secondary to congenital cardiac malformations (67 percent) than in primary plexogenic pulmonary arteriopathy (the distribution of plexiform lesions within the pulmonary arteriole tree therefore depends on the etiology). Plexiform lesions rarely, if ever, arise in bronchial arteries.

The appearance of plexiform lesions is of physiologic and clinical importance. They signify that the pulmonary blood flow falls into the normal range and then becomes abnormally low. This is associated with an inexorable rise of pulmonary arterial hypertension and pulmonary vascular resistance.[11] In the days before the introduction of combined heart-lung transplants for the treatment of congenital cardiac shunts and associated plexogenic pulmonary arteriopathy, lung biopsy specimens were commonly examined to assess the feasibility of corrective heart surgery. The appearance of plexiform lesions was

generally accepted as the histopathologic hallmark of the transition from a largely functional elevated pulmonary vascular resistance to one with a fixed organic basis. The transition was held to be a contraindication to surgical closure of ventricular septal defects.[12] If fragments of cotton wool are inadvertently introduced into the pulmonary circulation in the course of cardiac catheterization or another intravenous procedure, they may become trapped in the minute vessels of plexiform lesions, where they might evoke a foreign body giant cell reaction (Fig. 21-18).[13] Such a histologic appearance should not be misinterpreted as a form of naturally occurring granulomatous arteritis.

GRADE 5: RUPTURE OF DILATED VESSELS

The multiple thin-walled vessels throughout the lungs commonly rupture and give rise to small hemorrhages and subsequent foci of hemosiderosis. It is not generally appreciated that the small pulmonary veins and venules are commonly affected in plexogenic pulmonary arteriopathy. They frequently show significant intimal fibrosis, which may be cellular or hyaline (Fig. 21-19). The combination of this partial blockage of pulmonary veins and the rupture of the thin-walled vessels throughout the lung predisposes to pulmonary edema and a variety of exudative lesions.[2] The focal edema gives rise to an accumula-

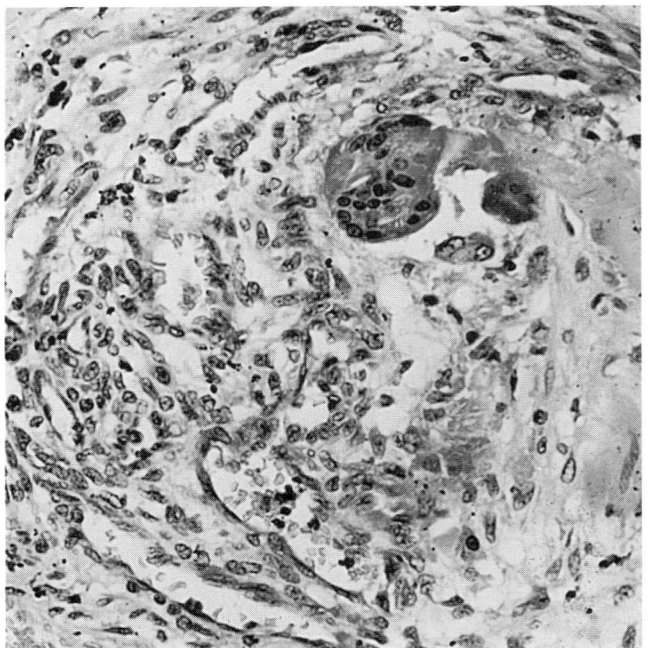

FIGURE 21-18
Plexiform lesion from a 26-year-old woman with an aortopulmonary septal defect who had been subjected to cardiac catheterization. It is composed of thin-walled vessels separated by septa of plump cells with rounded nuclei. Also present are multinucleated foreign body giant cells that have formed around fragments of cotton wool, which were brightly double refractile in polarized light. (H&E, ×190.)

tion of mast cells in the lung parenchyma.[14] Less commonly the lung edema may become associated with the appearance of small nodules around the pulmonary venules, which have a striking histologic resemblance to the arachnoid granulations of the dural venous sinuses (Fig. 21-20).[15] Perivenous arachnoid nodules in the lung may have a similar function of maintaining hydration of alveolar walls at a normal level and discharging excess edema fluid into the pulmonary venules (Fig. 21-21).[15] This is comparable to the function of discharging excess cerebrospinal fluid into the dural venous sinuses by the arachnoid granulations on the brain.

Other exudative pulmonary lesions are accumulations of alveolar macrophages and siderophages, cholesterol granulomas, focal proliferation of granular (type II) pneumonocytes, dystrophic calcification, and osseous nodules.[2]

GRADE 6: NECROTIZING ARTERITIS

Finally there may be development of extensive frank fibrinoid necrosis of the pulmonary arteries. Fibrinoid substances in the media may be demonstrated by the usual stains for fibrin, and there is commonly disruption of the inner and outer elastic

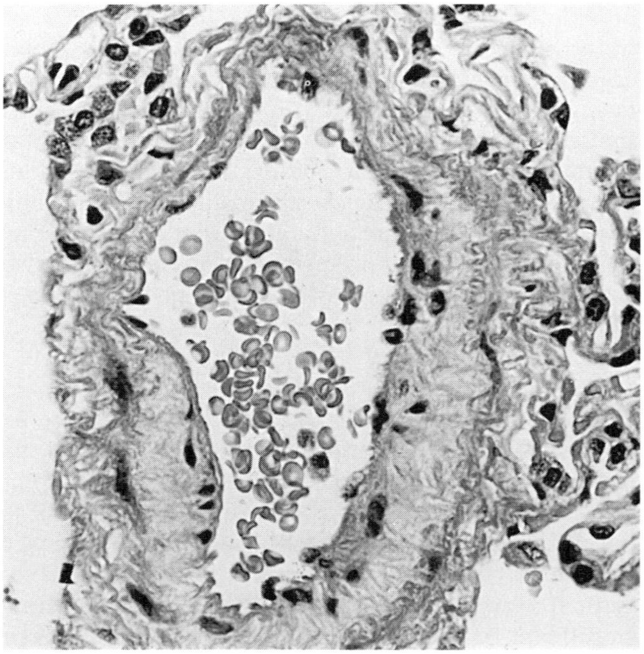

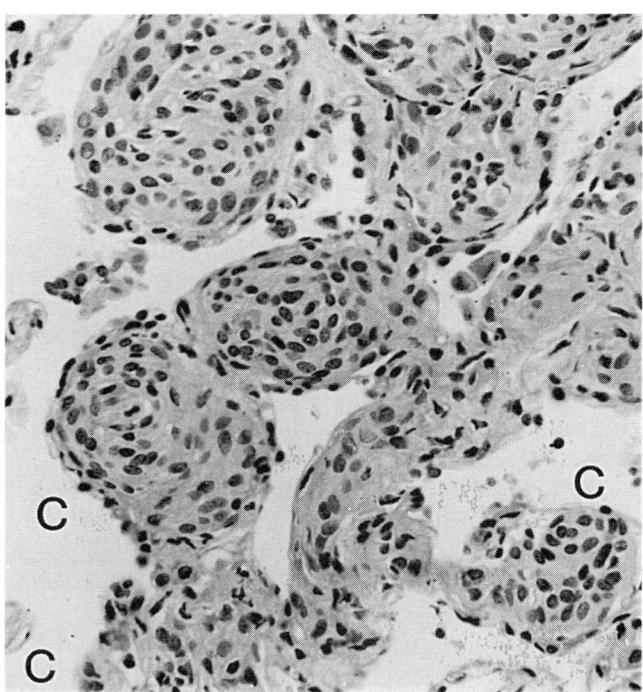

FIGURE 21-19
Small pulmonary vein from a 24-year-old man with a large ventricular septal defect and complete transposition of the great arteries, complicated by plexogenic pulmonary arteriopathy. There is a severe intimal fibrosis and proliferation of myofibroblasts. (H&E, ×570.)

FIGURE 21-20
Part of a nodule resembling an arachnoid villus surrounding a pulmonary venule in a case of primary plexogenic arteriopathy occurring in a 30-year-old woman. It shows cell clusters exhibiting whorling (C), which are strikingly similar to those of arachnoid granulations and meningiomas. (H&E, ×375.)

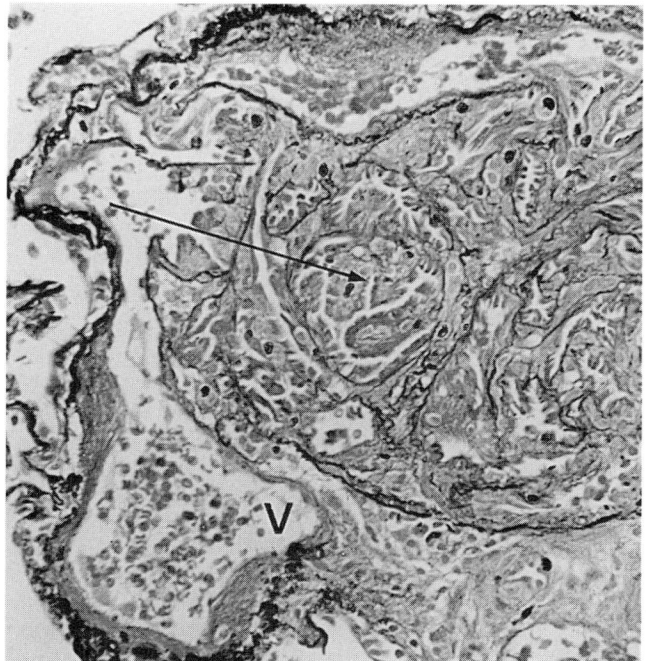

FIGURE 21-21
Part of one of the nodules from the case illustrated in Fig. 21-20 to show its vascular connections. It arises from the wall of a pulmonary venule (V). Much of the outer limit of the nodule is demarcated by a single elastic lamina. The mass has a complex structure, with groups of cells surrounding multiple channels (*arrow*). Small blood vessels of capillary size also permeate the nodule. (Elastic–van Gieson, ×375.)

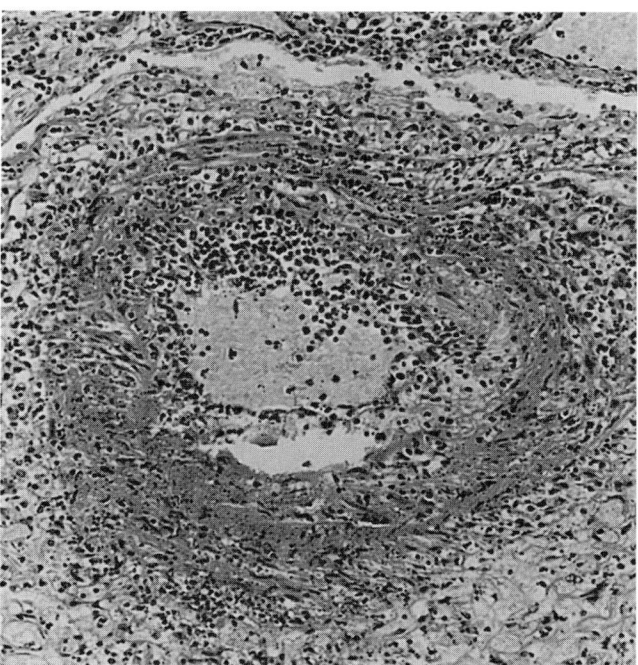

FIGURE 21-22
Transverse section of a small pulmonary artery from a 25-year-old woman with a wide patent ductus arteriosus and severe pulmonary hypertension. It shows necrotizing arteritis. The media is necrotic and is infiltrated by fibrinogen, which has then been precipitated as fibrin. The dead nuclei and fibrin have stimulated an intense acute inflammatory response. (H&E, ×150.)

laminae. Commonly the media is infiltrated by neutrophil polymorphs to present the classic picture of necrotizing arteritis (Fig. 21-22).

Sometimes this process may become subacute, with the formation of granulomas in the walls of pulmonary arteries.[4] Such extensive necrotizing arteritis has to be distinguished from the patchy fibrinoid necrosis thought to stimulate the development of plexiform lesions in grade 4. Some workers have been confused over this difference, and the early appearance of focal fibrinoid necrosis in this earlier phase has even been considered to invalidate the grading system. It should be appreciated that the extensive necrotizing arteritis of grade 6 is associated with the final fall of pulmonary blood flow to very low levels of pulmonary arterial resistance. Thus, the features of grade 6 have a totally different physiologic connotation from those of the patchy fibrinoid necrosis of grade 4.

REMODELING OF THE PULMONARY VASCULATURE IN STATES OF CHRONIC HYPOXIA

Chronic alveolar hypoxia brings about an entirely different set of structural modifications in the pulmonary vasculature from those found in plexogenic pulmonary arteriopathy. These changes develop irrespective of whether there is associated hypercarbia, as in patients with chronic bronchitis and pulmonary emphysema, or hypocarbia, as in native highlanders of the Andes. This suggests that the stimulus for their development is a diminished partial pressure of oxygen in the alveolar spaces. Since identical histologic changes occur in subjects with chronic hypoxic lung disease and in healthy native highlanders, it seems that the histologic changes in the pulmonary vasculature should be regarded not as pathologic but as a remodeling in response to chronic alveolar hypoxia.

The most characteristic feature of remodeling in states of chronic hypoxia is muscularization of the most peripheral portions of the pulmonary arterial tree. In the normal adult human lung the pulmonary arteriole has a cuff of circular smooth muscle at its origin that extends from the media of its parent artery. This muscular coat soon becomes discontinuous, so that it extends spirally around the arteriole. It is then lost altogether, and the wall of the pulmonary arteriole, less than 80 μm in diameter, eventually consists of a single elastic lamina.[16] In chronic hypoxic lung disease such as pulmonary emphysema, there is a peripheral extension of vascular smooth muscle. In transverse section the muscular-

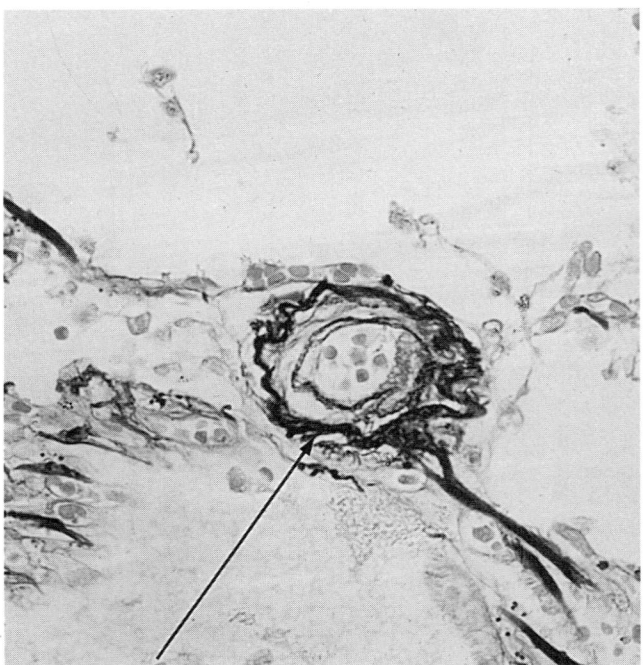

FIGURE 21-23
Transverse section of pulmonary precapillary vessel from a 64-year-old man with centrilobular emphysema. It shows the original elastic lamina (*arrow*). Inside this is a layer of circular muscle bounded on its inner aspect by a thin, freshly formed inner elastic lamina. (Elastic–van Gieson, ×600.)

ized pulmonary arteriole has a distinct media of circularly oriented smooth muscle, bounded on its outer and inner aspects by elastic laminae (Fig. 21-23). The outer lamina is usually thick and appears to be the original one of the arteriole, while the inner lamina is thin and pale-staining and appears to be newly formed by the inner new muscle coat (Fig. 21-23).[17] This muscularization extends to the far periphery of the pulmonary arterial tree to involve minute precapillary vessels in the alveolar walls (as small as 20 μm in diameter). The minute size of these muscularized vessels indicates that they are the result not of hypoxic vasoconstriction of parent arteries but of the laying down of a new muscle coat.

Muscularization of pulmonary arterioles should not be regarded as a manifestation of pulmonary vascular pathology, for it was reported in healthy Quechua Indians in the Peruvian Andes by Arias-Stella and Saldaña in a classic paper.[18] We were able to confirm their findings in later studies of Aymara Indians in the Bolivian Andes.[19,20] Ethnic factors seem to play a role, since arteriolar muscularization was found most commonly in Aymara Indians; less so in Mestizos, in whom there is a mixture of Spanish and Indian blood; and least of all in whites domiciled permanently at high altitude.[21] Muscularization of pulmonary arterioles is not to be anticipated in humans living below an altitude of 3500 m. It is a

marker of chronic alveolar hypoxia rather than a sign of pulmonary vascular disease. The elevation of pulmonary arteriolar resistance is both mild and benign in patients with chronic hypoxic lung disease and in native highlanders. In both groups the pulmonary vascular remodeling does not affect longevity or the ability to undertake normal daily activity.

Another characteristic feature of the pulmonary vascular remodeling found in states of chronic alveolar hypoxia is a laying down of longitudinally oriented smooth muscle in the intima of small pulmonary arteries and, to a lesser extent, in that of pulmonary arterioles (Fig. 21-24).[17] This emphasizes that the vascular remodeling is based on proliferation rather than constriction of vascular smooth muscle.[10] In patients with pulmonary emphysema and in native highlanders, the original media of circular smooth muscle of small pulmonary arteries remains thin, while there is a development of longitudinal muscle in the intima. At first this layer is cellular, with the muscle fibers being separated by thin elastic fibrils. The intimal layer of longitudinal muscle becomes progressively thicker and eventually may significantly compromise the vascular lumen. At the same time, the elastic fibers become thicker and more prominent, and eventually the whole layer may become predominantly elastotic. Severe intimal fibrosis may result, but a few longitudinal muscle fibers can still be seen. Their presence confirms that the histologic appearances are the end point of a long chain of events rather than indicative of a nonspecific intimal fibrosis.

The longitudinal muscle arises by the migration of vascular smooth muscle cells from media to intima.[22] They reach the intima by passing through naturally occurring gaps in the inner elastic lamina of the pulmonary artery. This is the same migration route as in plexogenic pulmonary arteriopathy. There the similarity ends, however, for in the arteriopathy the vascular smooth muscle cells are elongated and electron-dense and pursue an aggressive course that leads to severe pulmonary hypertension. In contrast, the migrating smooth muscle cells in chronic hypoxic lung disease closely resemble both histologically and ultrastructurally the mature vascular smooth muscle cells of the media. When these migrating cells reach the intima, they remain confined to it. They do not change into myofibroblasts and do not form the basis for a wide range of intimal proliferations that might lead to significant vascular occlusion. As a result, in chronic hypoxic lung disease the associated pulmonary hypertension remains mild and benign. The original suggestion by Dunnill[23] that the intimal muscle was due to longitu-

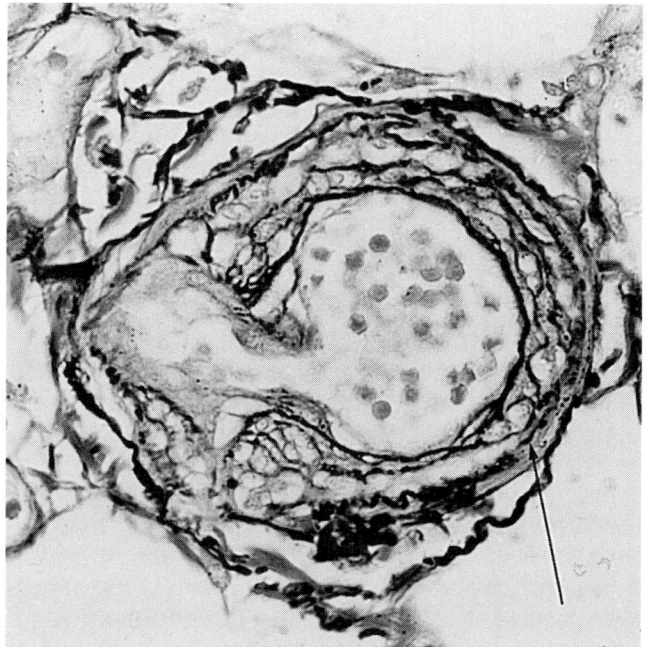

FIGURE 21-24
Transverse section of small pulmonary artery from a 64-year-old man with centrilobular emphysema. The original media of circularly oriented smooth muscle (*arrow*) is thin. In the intima there is a thick layer of longitudinal muscle; many of the fibers are separated by elastic fibrils. (Elastic–van Gieson, ×768.)

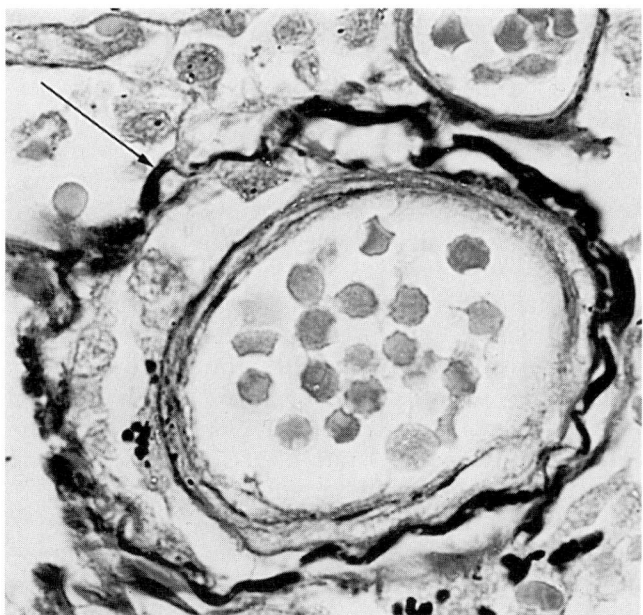

FIGURE 21-25
Transverse section of pulmonary arteriole from a 64-year-old man with centrilobular emphysema. The original thick elastic lamina is indicated by an arrow. Inside this lamina are groups of myofibroblasts. The arteriole is lined by a new layer of circular smooth muscle sandwiched between distinct internal and external elastic laminae. (Elastic–van Gieson, ×1500.)

dinal stretching of small pulmonary arteries around emphysematous spaces must now be regarded as invalid. Rather, the limited migration of smooth muscle cells into the intimal seems to be directly caused by chronic alveolar hypoxia.

Chronic alveolar hypoxia and hypoxic remodeling of the pulmonary arterial tree do not stimulate the appearance of increased numbers of bombesin-rich pulmonary endocrine cells.[9,24] As we have noted above, this is in striking contrast to what occurs in plexogenic pulmonary arteriopathy.

The third component of hypoxic pulmonary vascular remodeling in patients with chronic hypoxic lung disease and in native highlanders is the development of inner muscular tubes (Fig. 21-25) in small pulmonary arteries.[17,20] In affected pulmonary arterioles, a zone of myofibroblasts forms inside the original thick elastic lamina. Inside this is a coat of circular muscle bound by two elastic laminae. Rarely thick, like its limiting elastic fibrils it stains pale, suggesting that it is of recent formation. In some pulmonary arterioles, several muscular tubes are found; they run parallel to one another through the vascular lumen. The zone containing myofibroblasts is eventually replaced by fibroelastic tissue. Such inner muscular tubes range between 10 and 45 μm in diameter and are found in pulmonary arterioles up to 70 μm in diameter. In small pulmonary arteries, a thin layer of circularly orientated smooth muscle cells is usually situated adjacent to the endothelium. It probably accounts for the formation of the inner muscle tubes, when it becomes enclosed by newly formed thin elastic laminae.[22]

A proliferation of vascular smooth muscle occurs in all classes of small pulmonary blood vessels in the area adjacent to hypoxic alveolar spaces. Thus, muscularization is found in the pulmonary veins and venules in patients with chronic bronchitis and pulmonary emphysema[25] and in native Aymara highlanders from the Bolivian Andes.[20] In pulmonary veins the proliferation comprises individual smooth muscle cells separated by much collagen. This reflects the lack of arterial pulsation that leads to a more compact layer of muscle in pulmonary arteries in chronic hypoxia. Wagenvoort and Wagenvoort[26] also found that the pulmonary veins, especially smaller ones, of healthy residents at high altitude had a significantly thicker media than those of lowlanders. They noted a parallel increase in muscularity of pulmonary arteries and veins within the same lungs, suggesting a common etiologic factor such as hypoxia. It is clear from all these considerations that the remodeling of the pulmonary vasculature in states of chronic hypoxia is far more complex than the well-known muscularization of pulmonary arterioles.

Effects of Oxygen and Pulmonary Vasodilators

It is thought by some that treatment with pulmonary vasodilators such as calcium antagonists may be a valuable adjunct to ongoing long-term oxygen therapy for chronic obstructive lung disease. If one considered that chronic alveolar hypoxia brought about a simple constriction of small pulmonary arteries and muscularized pulmonary arterioles, it would be easy to see how pulmonary vasodilators and oxygen might act on such a theoretical model. As we have just seen, however, remodeling of the pulmonary vasculature is far more complex than this, entailing structural modifications in the microanatomy. It is not easy to see how vasodilators might affect such a complex system of tissue proliferation.[27]

To investigate this problem, a pathophysiologic study was carried out on 10 cases of hypoxic cor pulmonale, half of which had been treated by long-term oxygen therapy for 2 to 8 years.[17] In spite of this treatment, all five cases showed musculoelastic thickening of the intima, while four showed muscularization of the pulmonary arterioles or precapillaries. Muscular tubes were found in pulmonary arterioles in two of the five.[17] It was apparent that long-term oxygen therapy had no effect in reversing the pulmonary vascular changes in these patients.

In the same way, inhalation of oxygen will cause only a partial immediate lowering of pulmonary arterial pressure in native highlanders showing remodeling of the pulmonary arterial tree due to chronic hypoxia. This is presumably the effect of the gas in relaxing hypoxic pulmonary vasoconstriction. Complete reversal of elevated pulmonary arterial pressure and resistance takes much longer to achieve. This is likely to be the time required for the regression of the microanatomic remodeling described above. The ineffectiveness of oxygen inhalation alone in reversing the pulmonary hypertension in residents at high altitude was reported in two investigations at Leadville, Colorado (3100 m).[28,29] Total regression of pulmonary vascular remodeling in highlanders may take 2 years to achieve. Even longer periods may be required in patients with chronic hypoxic lung disease on long-term oxygen therapy, and there is no proof that regression takes place under these circumstances.

Recently considerable interest has been focused on the inhalation of nitric oxide (NO) for the alleviation of pulmonary hypertension[29a] NO is thought to act on the smooth muscle of pulmonary vessels. It stimulates production of cyclic guanosine monophosphate (cGMP), with subsequent alterations in intracellular calcium leading to relaxation of the vessel wall. Unlike other agents, NO inhalation reduces intrapulmonary shunting.

A Poor Animal Model for the Study of Human Pulmonary Hypertension

The pulmonary circulation of the rat is widely used as an animal model for studies of human pulmonary hypertension. This species is attractive for research, as it is a small laboratory animal that is readily accommodated in decompression chambers for studies of simulated high altitudes that mimic hypoxic lung disease. It is also susceptible to the effects of pyrrolizidine alkaloids, which produce a type of pulmonary vascular disease similar, but not identical, to plexogenic pulmonary arteriopathy.

However, the reaction of the pulmonary vasculature of the rat to hypoxia is very different from that of the human lung.[30] The pulmonary arteries of the rat respond by the formation of muscular evaginations. These are bulbous expansions of muscle cell cytoplasm that form clear structures devoid of myofilaments and organelles between attachment points on the plasmalemmal surface for actin and myosin filaments.[31,32] The evaginations pass through gaps in the elastic laminae that demarcate the media and then extend into both the intima of pulmonary veins and the adventitia of pulmonary arteries. Pulmonary arteries subsequently develop severe medial hypertrophy, with crenation of elastic laminae and muscularization of pulmonary arterioles.

Such evaginations are the marker of intense vasoconstriction, which leads to the rapid development of severe pulmonary hypertension and pronounced right ventricular hypertrophy. This may prove fatal if the rats are left exposed to the hypobaric hypoxia without relief. This process is very different from hypoxic remodeling of the human pulmonary arterial tree, which, as we have seen, is characterized by migration of vascular smooth muscle cells into the intima and their proliferation there, which may lead to benign pulmonary hypertension.

The reactions of the pulmonary arterial tree of the rat and that of humans represent two extremes—one intensely vasoconstrictive, leading to a malignant pulmonary hypertension, and the other proliferative, leading to a benign, mild elevation of pulmonary arterial pressure. For this reason, the rat is a poor animal model for the study of human pulmonary hypertension, and its use can lead to erroneous conclusions about the nature and behavior of the remodeling of the human pulmonary arterial tree in association with chronic alveolar hypoxia. It is also a poor model for primary pulmonary hypertension, for the administration of pyrrolizidine alkaloids such as monocrotaline to the rat induces medial hypertrophy and necrotizing arteritis in a substantial minority, but without the intimal proliferation or plexiform lesions that occur in plexogenic pulmonary arteriopathy in humans.

PRIMARY AND DIETARY PULMONARY HYPERTENSION

Plexogenic pulmonary arteriopathy may occur in the absence of congenital cardiac shunts or liver disease, and the resulting clinical disorder is known as primary pulmonary hypertension. Nearly a third of the cases reported as examples of this syndrome have been misdiagnosed pathologically. When the Wagenvoorts[33] histologically examined lung tissue from 156 cases reported as primary pulmonary hypertension in the world literature, they found that the diagnosis was correct in only 110 of them. Recurrent pulmonary thromboembolism should not be regarded as an etiologic factor in solitary plexogenic pulmonary arteriopathy, as it is an entirely different disease.[34]

Suggested Etiologic Factors for Primary Pulmonary Hypertension

There have been many reports of familial primary pulmonary hypertension,[34] and some of these have been explained on the basis of an autosomal dominant mode of inheritance.[35,36] Others have been interpreted as exhibiting a recessive mode of inheritance in which only one generation was affected.[37] Another view is that familial primary pulmonary hypertension is an expression of numerous genetic and environmental factors.

One of the central components in the histopathology of plexogenic pulmonary arteriopathy is fibrinoid necrosis of the pulmonary arteries. This disease is typically associated not only with severe pulmonary hypertension resulting from vasoconstriction but also with hypersensitivity and immunologic disturbance. It is of interest, therefore, that primary plexogenic pulmonary arteriopathy commonly develops in association with collagen diseases and autoimmune disorders. Thus the association has been reported in scleroderma, dermatomyositis, disseminated lupus erythematosus, rheumatoid disease, Hashimoto's disease, and primary biliary cirrhosis.[34]

All these diseases, like primary pulmonary hypertension, are more common in women.

Raynaud's disease, a vasoconstrictive disorder of the systemic circulation, is found in up to one-third of patients showing a combination of collagen disease and primary pulmonary hypertension. This close association of a vasoconstrictive disorder of the systemic circulation emphasizes the importance of vasoconstriction in primary pulmonary hypertension, whereas the close association of collagen and autoimmune diseases suggests an underlying immunologic mechanism. In several of these autoimmune diseases, fibrinoid necrosis and necrotizing arteritis may occur in both systemic and pulmonary circulations. If these changes are found only in the pulmonary arteries, however, it is likely that they are related to the severe pulmonary hypertension rather than to an autoimmune disease such as polyarteritis nodosa. The immunologic abnormalities found in primary pulmonary hypertension are elevated levels of autoimmunoglobulins G and M.

The early studies on primary pulmonary hypertension emphasized structural changes in the pulmonary blood vessels, and frequent reference was made to "primary pulmonary vascular sclerosis." I shall not refer here to those papers, which obviously relate to a heterogeneous group of conditions; the reader is referred to the studies of Brenner,[1] who included a comprehensive bibliography on the subject. Since that time some authors have reported other vascular changes in primary pulmonary hypertension and have regarded them as being of etiologic significance. This view is perhaps prompted by the thought that primary pulmonary hypertension in infancy is related to some congenital abnormality in the pulmonary arterial bed. It is more likely that the microanatomic changes reported are the result rather than the cause of the altered hemodynamics.

Hormonal influences may be of importance. Once the menarche is passed, women appear to become more susceptible to primary pulmonary hypertension than men. Its nature remains obscure, however. Plexogenic pulmonary arteriopathy becomes exaggerated in women taking oral contraceptives. Pregnancy is often associated with the onset of the disease, which is then poorly tolerated. Some cases could be explained by amniotic fluid embolism.

Dietary factors have assumed considerable theoretical significance in the proposed etiology of primary pulmonary hypertension since the epidemic of the disease erupted in Western Europe from 1967 to 1970. In particular, the ingestion of certain appetite suppressants appeared to be of importance. The anorexigen aminorex fumarate (Menocil) was held to be responsible for the epidemic.

Aminorex (2-amino-5-phenyl-2-oxazoline) resembles epinephrine and amphetamine in its chemical structure and suppresses the appetite by acting directly on the brain. The drug was introduced into the Swiss market in November 1965, in West Germany in April 1966, and in Austria in August 1966. A few months later the epidemic of primary pulmonary hypertension began in those three countries.

Gurtner[38] noted that of 62 female and eight male patients with primary pulmonary hypertension investigated between January 1967 and March 1970, 55 gave a history of having taken aminorex. Their symptoms first appeared 6 to 12 months after they had begun taking the drug. Although classic primary pulmonary hypertension is predominantly a disease of young women, the epidemic associated with aminorex was seen in women of all ages and in a significant number of men as well. I visited several medical centers in the area affected by the epidemic and found that the disease was plexogenic pulmonary arteriopathy whose histologic features were identical to those of the classic variety. The temporal and geographic links of aminorex with the epidemic suggested that the drug was the major causative factor. The effect of the drug on the pulmonary circulation appears to be specific to humans, since high doses of menocil did not induce the arteriopathy in rats or dogs.[39] Nor did it induce pulmonary vascular disease at high altitude in cattle whose pulmonary arterial trees were already sensitive to the vasoconstrictive effects of ambient hypoxia.[40] Nevertheless, the epidemiologic evidence for implicating aminorex as the cause of the epidemic was strong and remains so today. When the drug was withdrawn from the market, the epidemic disappeared within a few months. Other drugs, such as fenfluoramine, and other anorexigens have been suspected as causing primary pulmonary hypertension.

Dietary Pulmonary Hypertension in Animals

The concept of dietary pulmonary hypertension in humans is greatly strengthened by the fact that the same phenomenon is to be found in animals. In a range of animal species from rats to primates, severe pulmonary vascular disease is induced by the ingestion of the seeds of foliage of several species of the genera *Crotalaria* and *Senecio*. The lesions produced have certain features reminiscent of some of those of plexogenic pulmonary arteriopathy. Indeed, menocil is frequently used as an animal model for the human disease. As has been pointed out above, the rat is a poor animal model because it shows neither intimal proliferative lesions nor plexiform lesions in response to ingestion of the alkaloids, both of which are highly characteristic of human plexogenic pulmonary arteriopathy.

In rats the pulmonary vascular disease of pyrrolizidine alkaloids comprises medial hypertrophy of pulmonary arteries, muscularization of pulmonary arterioles, and necrotizing arteritis. Examples of plants capable of inducing this disease are *Crotalaria spectablis*, containing monocrotaline, in the United States[41]; *C. fulva*, containing fulvine, in Jamaica[42]; and *Senecio jacobaea*, containing six alkaloids, in Europe.[43] In humans fulvine produces venoocclusive disease of the liver, but in rats it produces pulmonary arterial disease. Ragwort (*S. jacobaea*) induces liver disease in cattle and pulmonary arterial disease in rats and is sold in some health stores for human consumption in the United Kingdom.[43] On a visit to Tanzania, I was shown sections of lung from a 19-year-old African who had died from primary pulmonary hypertension: they showed classic plexogenic pulmonary arteriopathy. Before being admitted to hospital, he had been treated by a witch doctor, who had administered brews of local plants. A search of the area discovered the species *C. laburnoides*. When the seeds of this plant were fed to rats in our laboratory, they induced pyrrolizidine pulmonary vascular disease.[44] The ingestion of these seeds in the witch doctor's brew may have led to primary pulmonary hypertension in the young African, but the association was never proved scientifically.

Plexogenic Pulmonary Arteriopathy in Liver Disease

The association of disease of the liver and the pulmonary circulation in animals poisoned with pyrrolizidine alkaloids is reflected in human pulmonary vascular disease, in which plexogenic pulmonary arteriopathy is sometimes found with cirrhosis of the liver or portal vein thrombosis. A moderate degree of pulmonary hypertension is common in patients with portal hypertension due to cirrhosis of the liver. The raised pulmonary arterial pressure is passive, probably secondary to a high cardiac output. In a small minority, however, plexogenic pulmonary arteriopathy develops with severe pulmonary hypertension, which is usually fatal.[45] The basis of this relation remains unclear, but because it is found in cases of portal vein occlusion as well as hepatic cirrhosis, it is possible that the removal of the liver as a detoxifying organ prevents removal of deleterious substances coming from the gut and portal system, allowing them to reach the pulmonary circulation. The histopathology in these cases is that of classic plexogenic pulmonary arteriopathy. The association between liver disease and an elevated pulmonary arterial pressure has been recognized for more than 40 years.[46] There is considerable literature on pulmonary vascular disease in cirrhosis of the liver.[47-50] Pyrrolizidine alkaloids affect both the liver and the pulmonary circulation.

PULMONARY VASCULAR DISEASE IN TIBET

In the account so far, I have presented hypoxic pulmonary vascular remodeling as seen in patients with chronic obstructive lung disease and native highlanders as a distinct entity, to be sharply contrasted with plexogenic pulmonary arteriopathy with infiltration of the intima by myofibroblasts and the development of severe pulmonary hypertension. There is one interesting exception to this distinction that came from an unexpected source. On a visit to Tibet in 1987, Harris and Anand came across a new form of mountain sickness that was familiar to their Chinese and Tibetan hosts but unknown to Western doctors.[51] This appeared to be the result of the policy of the Chinese government to introduce large numbers of unacclimatized lowlanders of Han origin to live among the native Tibetan highlanders of Lhasa (3600 m) and the Tibetan plateau. Infants of Han origin, taken by their parents up to Lhasa, may develop a pulmonary vascular disease that leads to right ventricular hypertrophy and failure. On average, affected infants survive only 2.1 months at high altitude.[51] The vascular disease comprised medial hypertrophy of small pulmonary arteries, muscularization of pulmonary arterioles, and migration of myofibroblasts into the intima and lumina of pulmonary arterioles (Fig. 21-26) and venules (Fig. 21-27). Thus it had features of both vascular remodeling in chronic alveolar hypoxia and migration of myofibroblasts with a fatal outcome.

In this disease, considerable numbers of bombesin-rich pulmonary endocrine cells appeared in the terminal bronchioles. As noted above, this finding is typical of the free migration of myofibroblasts in plexogenic pulmonary arteriopathy but not of the limited migration of vascular smooth muscle cells in states of chronic hypoxia.[52] This interesting disorder thus appears to be a hybrid of these two major groups of pulmonary vascular disease. It has been called "subacute infantile mountain sickness"[51]

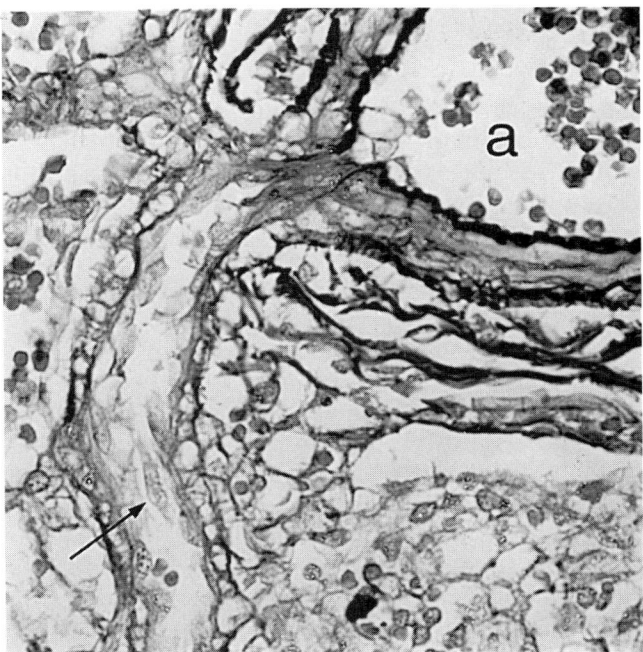

FIGURE 21-26
Section of lung from a male infant in Lhasa (3600 m) with subacute mountain sickness. An arc of a small pulmonary artery (a) is included. This vessel shows medial hypertrophy without any form of intimal proliferation. An arteriole arises from its parent artery and passes downward. Within its lumen is a proliferation of elongated cells (arrow) considered to be myofibroblasts. (Elastic–van Gieson.)

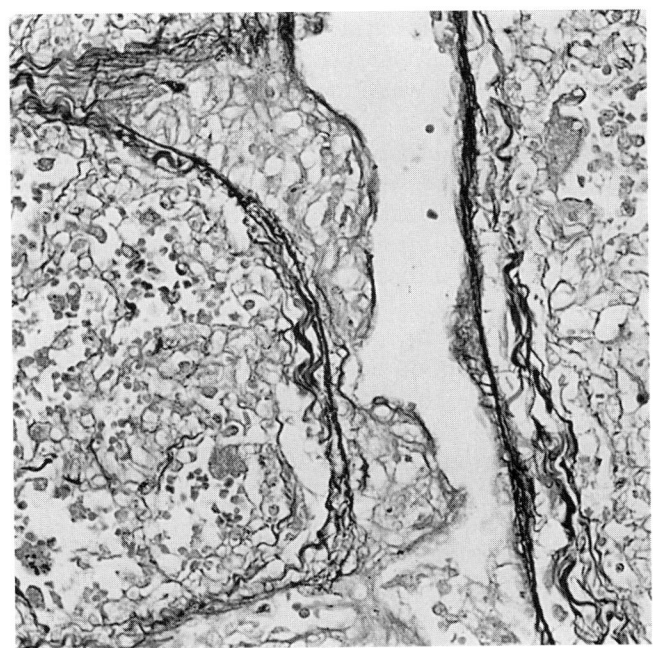

FIGURE 21-27
Longitudinal section of pulmonary vein from the same case as in Fig. 21-26. There is prominent intimal proliferation of myofibroblasts. (Elastic–van Gieson.)

and seems to represent a failure to achieve initial acclimatization to the hypobaric hypoxia of high altitude as in calves that develop brisket disease in the Wasatch mountains in Utah.

PULMONARY VENOOCCLUSIVE DISEASE

The development of intimal fibrosis as an age change in pulmonary veins and venules in the human lung was recognized by Brenner as long ago as 1935.[1] Much later, ultrastructural studies demonstrated that this highly characteristic feature was due to the movement of myofibroblasts into the intima. In the pulmonary arteries, where they were subjected to pulsation, they assumed the ultrastructural features of smooth muscle cells.[53] In pulmonary veins, in the absence of such hemodynamic forces, they developed into fibroblasts.[53]

It has already been pointed out that the pulmonary veins and venules may show a cellular intimal proliferation of fibrous tissue in plexogenic pulmonary arteriopathy, hypoxic remodeling, and subacute infantile mountain sickness. However, there is one disease of the pulmonary vasculature in which intimal fibrous proliferation in the pulmonary veins and venules is the central and dominant feature of the pathology. In the 1930s, case reports began to appear of a disease in which the pulmonary veins appeared to be primarily involved.[54] By 1960, such cases were being described as examples of "primary pulmonary hypertension with obstructive venous lesion."[55] In 1966, we reported such a case and stated that "the histological features in the lung are so distinctive as to constitute a distinct disease entity which should be separated from the main group of patients with classical primary pulmonary hypertension and called pulmonary venoocclusive disease."[56]

It is mainly a disease of young adults and children, although several cases have been found in infancy. The overall age range of reported cases has been 8 weeks[57] to 64 years.[58] There is no particular sex prevalence—in contrast to primary plexogenic pulmonary arteriopathy and thromboembolic pulmonary hypertension, both of which are more common in females. The onset is usually insidious, but there is often a history of a recent short febrile illness.[55] Once established, the disease is progressive; most patients die within 2 years of the onset of symptoms, although survival up to 7 years has been reported.[56]

The brunt of the pathology falls on the pulmonary veins and venules, which show a pronounced proliferation from the intima of a loose, often basophilic fibrous tissue (Fig. 21-28).[54–56,59,60]

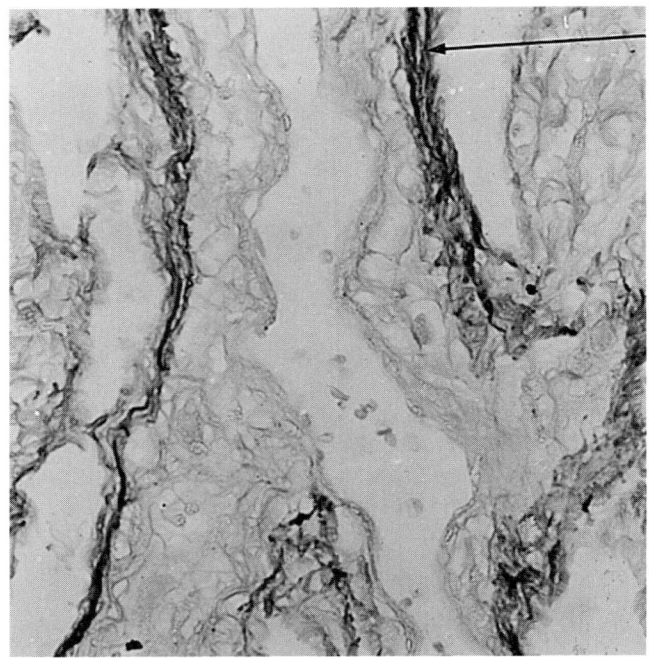

FIGURE 21-28
Longitudinal section of a small pulmonary vein in a 16-year-old girl with pulmonary venoocclusive disease. The vein receives two venules. Both the vein and its tributaries show extensive blockage by loose, basophilic fibrous tissue. The remaining lumen, much reduced in size, passes through the vein and one of its tributaries. There is some "arterialization" of the venous media (*arrow*). (Elastic–van Gieson, ×375.)

This severely compromises the lumen of the affected vessels and may occlude them, hence the name of the disease. The affected veins and venules may remain partly patent only through a tortuous narrow channel extending along their length. In oblique sections the central canal may be cut transversely more than once, and it is seen to have an irregular outline bordered by a rim of condensed collagenous tissue. Around this rim the intimal proliferation is often loose and edematous. As the lesions age, they tend to develop elastic fibrils and better-defined collagen fibers. The thickened zone around the remaining channels becomes better defined.

In some veins the intimal fibrosis takes the form of eccentrically situated nodules, and appearances of this type have been taken to indicate that the lesions may arise as a result of organization of thrombus. Sometimes fresh thrombus showing no or early organization can be found adherent to the nodules of fibrosis.[60] Histologic appearances of recanalization are common. Wide, thin-walled vascular channels are demarcated by very thin fibrous septa to present as "colander lesions" (Fig. 21-29).[56,60,61] The intimal lesions affect all sizes of vessel, ranging from large pulmonary veins down to the smallest postcapillary venules.[56,59] The occlusive process may be widespread, and in one case that we studied, no fewer than 95 percent of the pulmonary veins and venules

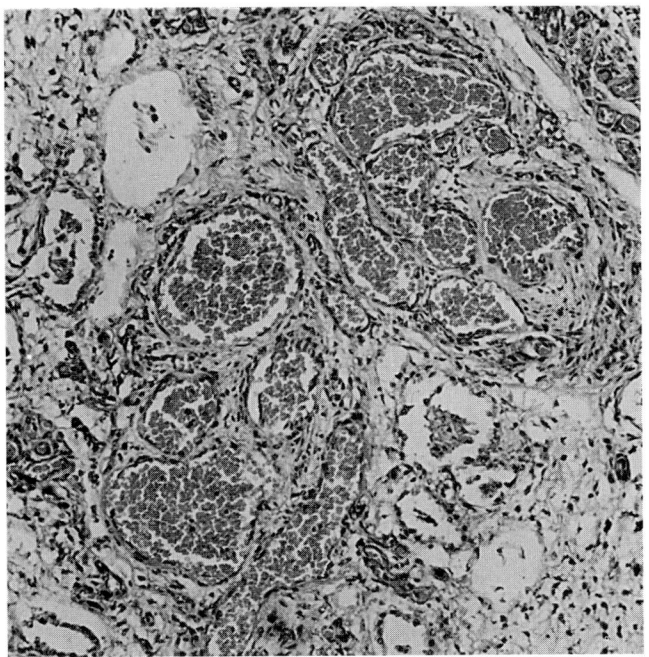

FIGURE 21-29
Transverse sections of two pulmonary veins from a 34-year-old woman with pulmonary venoocclusive disease showing multiple recanalization to give rise to colander lesions. The remaining channels are wide and separated by thin fibrous septa. (H&E, ×102.)

fibrous thickening of the alveolar walls but not honeycomb lung. There is distention of the lymphatic vessels in the connective tissue septa of the lung and around bronchi and pulmonary blood vessels. Small spicules of bone may be found in the alveolar spaces.

There is marked engorgement of the pulmonary capillary bed, but this is often focal and sharply demarcated (Fig. 21-30). Perhaps these areas of capillary engorgement are drained by a recently occluded pulmonary vein. Diapedesis from these dilated vessels leads to hemosiderin-laden macrophages in the alveolar spaces and sometimes to encrustation of elastic laminae by ferric iron. The hemosiderosis may be mild or prominent. In some cases the focally dilated vessels are much larger and form clusters of a significant size, resembling small angiomas. In another histologic variant the venous intimal proliferation shows a muscular change, so that the occluding tissues look more rigid and substantial than the customary loose, basophilic fibrous tissue. Inflammation is uncommon in pulmonary venoocclusive disease.[62] A case with granulomatous venulitis has been documented.[63]

were involved.[56] Even the bronchial veins may be affected.

There is often muscularization of the usually poorly defined media of the small pulmonary veins. They come to have a distinct muscular coat, bounded by internal and external elastic laminae. This "arterialization" of small pulmonary veins is similar to what occurs in mitral stenosis and other diseases, leading to pulmonary venous hypertension.

Pulmonary arterial hypertension follows the occlusion of the pulmonary veins, and hypertensive changes develop in the pulmonary arteries. These include medial hypertrophy of the small pulmonary arteries and muscularization of the pulmonary arterioles. Fibrinoid necrosis of the small pulmonary arteries has been described.[59,60] As in other forms of pulmonary arterial hypertension secondary to pulmonary venous obstruction, dilatation lesions, such as plexiform or angiomatoid lesions, do not occur. The elastic pulmonary arteries become atheromatous.

Striking changes may develop in the lung parenchyma.[56,59] The occlusive lesions in the small pulmonary veins and venules appear to induce a persistent edema of the alveolar walls. This leads to a proliferation of granular pneumocytes, so that the alveolar spaces develop a prominent cellular lining. The pneumocytes may accumulate in alveolar spaces to give an appearance resembling that of desquamative interstitial pneumonia. This is followed by

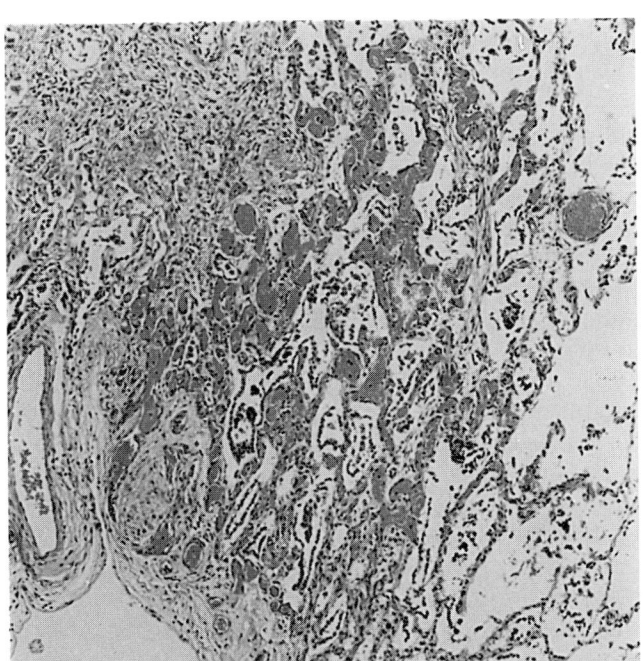

FIGURE 21-30
A focal area of pronounced dilatation and engorgement of the pulmonary capillary bed in a 34-year-old woman with pulmonary venoocclusive disease. This area is sharply demarcated from the surrounding lung parenchyma, and such an appearance is characteristic of this disease. It may be the result of distention of pulmonary venules and capillaries in the area drained by an occluded pulmonary vein. (H&E, ×600.)

Etiologic Factors

The occasional presence of fresh thrombus, the commonly nodular and eccentric configuration of the intimal fibrosis, and the development of colander lesions in the pulmonary veins in this disease are consistent with its being based on organization and recanalization of thrombi in these vessels. There are many causes of thrombosis and fibrosis in pulmonary veins, and this has led Wagenvoort[64] and Hasleton and coworkers[65] to suggest that pulmonary venoocclusive disease should be regarded as heterogeneous and more of a syndrome than a pathologic entity.

In several of the reported cases, there has been a history of a previous illness resembling influenza[55] or of a preceding chest infection.[59,66,67] It is commonly believed that the infectious agent is a virus. In one case, in a 39-day-old infant, the infection and the disease of the pulmonary veins appeared to have been acquired in utero, as the mother gave a history of an upper respiratory infection during the second trimester of pregnancy. The infant showed a simultaneous subacute myocarditis, suggesting that a viral infection might have been responsible for the lesions in the heart and pulmonary veins.

Such a viral infection could account for the thrombosis in the pulmonary veins. It might attack the pulmonary venous endothelium and deplete its plasminogen activator, already low in these vessels.[68] This would inhibit lysis and result in thrombus formation. Increased platelet adhesiveness has been reported in a case of pulmonary venoocclusive disease.[69] The presence of what were regarded as irregular deposits of IgG and complement was found in alveolar walls in a lung biopsy specimen from a patient with the disease.[70] It was suggested that immune complexes, possibly arising from a viral infection, may have initiated thrombosis by activation of contact clotting factors or platelets. The reservation has to be made that their illustrations of electron-dense deposits resemble those found in the pulmonary capillary wall in mitral stenosis, which were considered to be disintegrating extravasated erythrocytes.[71] As noted above, diapedesis of red cells through pulmonary capillaries and venules, with the formation of pulmonary hemosiderosis, is characteristic of pulmonary venoocclusive disease. In one case there were raised serologic titers for toxoplasmosis, but they were not high enough to be diagnostic.[60]

Pulmonary venoocclusive disease may follow bleomycin therapy for non-Hodgkin's lymphoma,[72] and it has been reported in a case of metastatic squamous cell carcinoma treated by bleomycin, mitomycin-C, and cis-platinum.[73] Pulmonary venoocclusive disease has been reported in association with hypertrophic cardiomyopathy in a 19-year-old man.[74] The disease may have a familial pattern,[61,65,75] but it does not appear to be sex-linked.[76] Pulmonary venoocclusive disease may be associated with a specific connective tissue disease, such as the CREST syndrome (calcinosis, Raynaud's phenomenon, esophageal hypomotility, sclerodactyly, and telangiectasia), systemic lupus erythematosus, or scleroderma.[77] It has been associated with chronic active hepatitis, celiac disease,[65] and membranoproliferative disease[78]—diseases in which immunologic mechanisms may be important.

Ultrastructural study reveals that the occluded pulmonary veins are lined by intact endothelial cells, beneath which is a haphazard proliferation of collagen fibrils and smooth muscle cells.[79] These cells of muscular pedigree are probably best regarded as myofibroblasts with a potential for translation into mature muscle cells and fibroblasts. There is an increase in the number of cytoplasmic processes of pericytes around alveolar capillaries, which showed thickening of the basement membrane of the endothelial cells. The thickened membrane showed electron-dense deposits, considered by the authors to represent disintegrating extravasated erythrocytes rather than immune complexes because immunofluorescence microscopy showed no immunoglobulin or complement deposition within the lung.[79]

The clinical features of pulmonary venoocclusive disease and their differential diagnosis from those of primary pulmonary hypertension are considered in detail by Thadani and colleagues.[80]

RECURRENT PULMONARY THROMBOEMBOLISM

This completes the triad of conditions that may form the basis for clinically unexplained pulmonary hypertension, the other two being primary plexogenic arteriopathy and pulmonary venoocclusive disease.

Many years ago, Goodwin and coworkers[81] divided patients with thromboembolic pulmonary hypertension into two clinicopathologic groups according to the presence or absence of clinical evidence of pulmonary embolism or infarction. Their first group comprised patients who presented acutely with dyspnea, pleuritic chest pain, hemoptysis, and clinical evidence of pulmonary hypertension. At necropsy these patients had right ventricular hypertrophy, thrombus in main pulmonary arteries, and identifiable sources of emboli in the calf veins, pelvic veins, or right ventricle. Patients in the second group presented insidiously with dyspnea, syncope, chest pain, and clinical evidence of pulmonary hypertension. At necropsy there was right ventricular hypertrophy, thrombotic lesions in small pulmonary arteries, but no obvious source of emboli in the systemic veins or heart. The average age of the 31 patients studied by the Wagenvoorts[33] was 44 years, and 71 percent were female.

Thromboembolic pulmonary hypertension is characterized by organic, obstructive lesions in the small pulmonary arteries and arterioles (Fig. 21-31). These lesions are composed of recent or organizing thrombus that eventually gives rise to cushionlike eccentric patches of intimal fibrosis. Sometimes the intimal fibrosis may be circular and so severe as to cause occlusion of the lumen.

There is no laminar or onionskin arrangement as occurs in plexogenic pulmonary arteriopathy. Recanalization may give rise to colander lesions, with fibrous septa running between the new channels. At first the septa are thick and cellular, but as the lesions age, the septa become progressively thinner and acellular and the channels enlarge. There is muscularization of pulmonary arterioles. The media of the small pulmonary arteries, although thicker than normal, rarely shows pronounced hypertrophy. Fibrinoid necrosis and subsequent arteritis are exceedingly rare in these cases and are almost certainly the result of an infected embolus. Dilatation and plexiform lesions do not occur, although sometimes intraarterial septa formed during the process of organization of thrombus may simulate mature plexiform lesions. The alveolar capillaries and pulmonary venules and veins are normal. Fibroelastic lattices may occur in major or large elastic pulmonary arteries as a result of recanalization of thrombus (Fig. 21-32).

There is no doubt that pulmonary thromboembolism is very common, and the lungs have an incredible capacity to dispose of these emboli. Morrell and Dunnill[82] found pulmonary thromboemboli in no fewer than 51.7 percent of 263 right lungs in a specially conducted postmortem investigation. This contrasted with an incidence of 11.8 percent found by the service pathologist in the left lungs of the same subjects. Small fresh thrombi are disposed of largely by fibrinolysis, whereas cellular processes with organization and recanalization are more important in the disposal of larger and older thromboemboli. It is likely, from analogy with observations

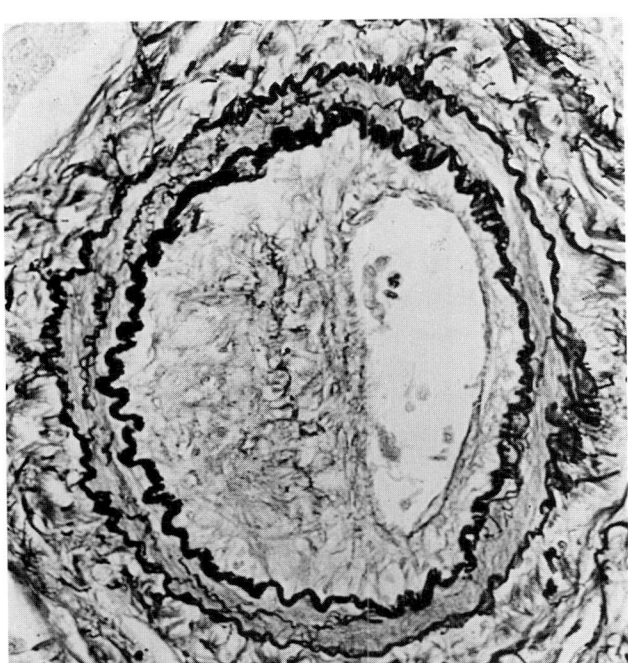

FIGURE 21-31
Pulmonary thromboembolism in a 27-year-old woman with congestive cardiac failure. This transverse section of a small pulmonary artery shows hypertrophy of the media with crenation of the elastic laminae, consistent with some degree of vasoconstriction. Apart from one eccentric recanalization channel, the lumen is occluded. Contrast this type of intimal proliferation with the concentric-laminar, "onionskin" intimal proliferation of plexogenic pulmonary arteriopathy found in primary pulmonary hypertension. (Elastic–van Gieson, ×375.)

FIGURE 21-32
Large elastic pulmonary artery in a woman aged 37 with recurrent pulmonary thromboembolism and pulmonary hypertension. A probe has been placed beneath a fibrous band representing an organized embolus.

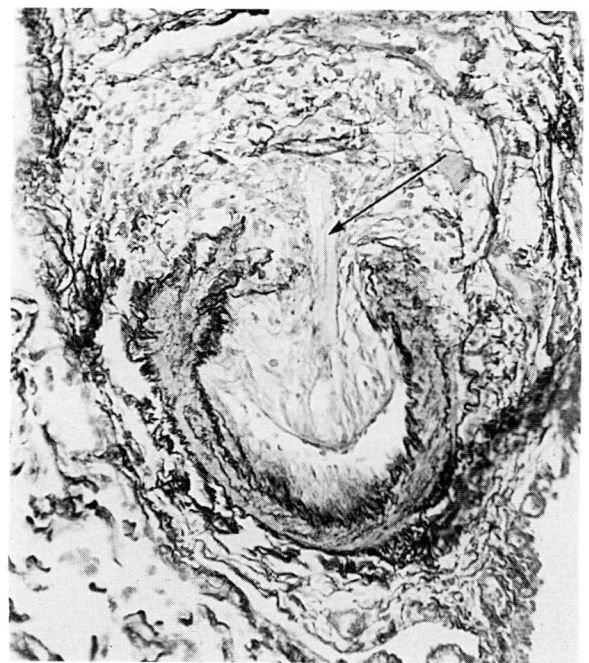

FIGURE 21-33
Cotton-wool granuloma of a small pulmonary artery in a 49-year-old woman who had ligation of a patent ductus arteriosus after investigation by cardiac catheterization. The catheter had been sterilized by autoclaving in thick cotton-wool gauze. The figure shows a transverse section of a hypertrophied small pulmonary artery. A cotton-wool fiber, indicated by an arrow, has impacted in the wall of the artery and has penetrated the media. A granulomatous reaction has taken place around the fiber, forming a dumbbell-shaped mass protruding into the lumen and adventitia of the artery. Such appearances may be simulated readily in experimental rats. (Elastic–van Gieson, ×150.)

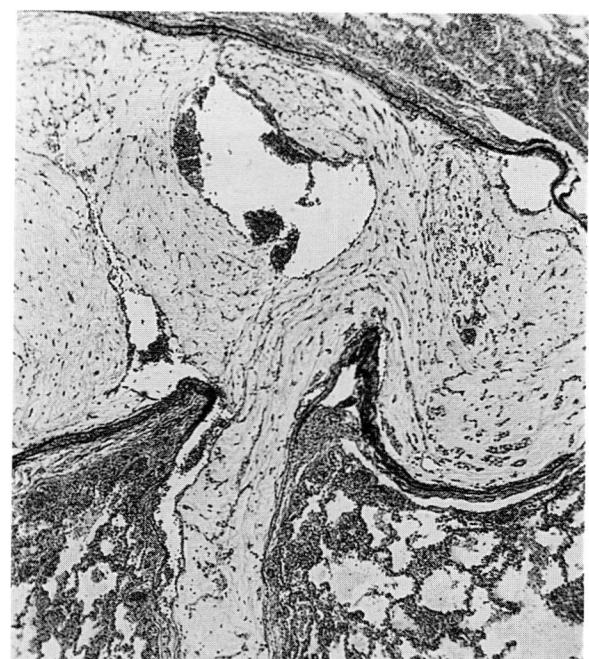

FIGURE 21-34
Section of lung from a 35-year-old woman with a myxoma of the right atrium that had given rise to pulmonary emboli and pulmonary hypertension. Included in the figure is part of a transverse section of a large pulmonary artery and a thin-walled arteriolar branch that passes downward. Both are largely occluded by an embolus of atrial myxoma that contains large thin-walled vessels. (H&E, ×40.)

on the lungs of animals exposed to experimental thromboembolism, that the pulmonary emboli become partly covered by adherent neutrophil polymorphs within a few hours of impaction. Early organization takes place during the following 3 or 4 days. It is conceivable that chronic pulmonary thromboembolism is in part caused not by the production of an excessive number of thromboemboli but by an intrinsic inability of the pulmonary circulation to deal with them. Such inadequacy of fibrinolytic activity would be likely to lie in the pulmonary arterial intima, which is normally a reservoir or plasminogen activator.

The lung is the great filter of the body, and all manner of particulate matter may become impacted in the small pulmonary arteries and arterioles, including amniotic fluid, cotton wool and other foreign bodies (Fig. 21-33), fat, fat cells, bone marrow, megakaryocytes, air, and tumor fragments (Fig. 21-34).[83] Each of these has its own characteristic histologic findings and clinical associations (Table 21-4). The histopathology of some of these conditions—especially pulmonary embolism due to cottonwool fragments—can be mimicked and studied experimentally.[84]

PULMONARY THROMBOSIS IN CYANOTIC CONGENITAL HEART DISEASE

There is thrombosis throughout the pulmonary arterial and venous trees in cyanotic congenital heart disease with pulmonary stenosis.[85] The process is continuous, so fresh thrombi and their organization into eccentric nodules are common. There is widespread recanalization with the formation of single or multiple channels separated by intraluminal fibrous septa, which are at first thick and cellular and later thin and acellular. The multiple channels are described as "colander lesions," after the culinary device (Fig. 21-35). When the underlying hematologic and hemodynamic factors are favorable for thrombus formation, the great majority of pulmonary arteries and veins become involved in the thrombotic process. Widespread thrombosis is common in Fallot's tetralogy, with certain reservations referred to below, and is even more common in tricuspid atresia. The etiologic factors appear to be the high hematocrit and the sluggish pulmonary blood flow.[86] Pulmonary thrombosis is seen most extensively in cases with the highest hematocrit levels and the most sluggish pulmonary arterial flow and pulsation. It is

TABLE 21-4
Causes of Pulmonary Embolism Other Than Thromboembolus

Nature of Embolus	Incidence	Clinical Significance	Histologic Features
Amniotic fluid	Rare, during or in days after childbirth	Serious; may prove fatal after a short period of dyspnea, cyanosis and shock	Amniotic fluid constituents such as vernix, lanugo hairs, meconium, and trophoblast cells in pulmonary capillaries and arterioles, inducing thrombosis
Fragments of trophoblast	Frequently found in lungs of pregnant women	Usually asymptomatic, may be clinically severe in hydatidiform mole	Large syncytial multinucleated masses; may induce thrombosis and lung infarction in hydatidiform mole
Cotton-wool fragments	Occasionally introduced into systemic veins during medical procedures	None	May mimic granulomatous pulmonary arteritis around refractile fragments of cotton wool; easily reproduced experimentally
Talcum or starch particles	Relatively rare; follows intravenous injection of material by drug addicts	Rare cause of pulmonary hypertension	Particles of talc and other introduced material
Fat	Common in crushing of bone marrow and fatty tissue in serious accidents	May be asymptomatic; May be fatal following dyspnea, cyanosis, fever, delirium, and coma due to cerebral fat embolism	Fat droplets in pulmonary capillaries and arteries demonstrated by oil red O
Adipose tissue or isolated fat cells	Common after trauma or operation	None	Lipid-laden macrophages in pulmonary capillaries
Air	Rare, usually iatrogenic	Small quantities asymptomatic; Large quantities may be fatal	Frothy mass in pulmonary artery, which should be opened under water for demonstration of air
Tumor	Common	Occasionally reveal their presence by inducing pulmonary hypertension before patient dies of underlying malignancy	Usually typical histology of underlying tumor; in rare cases, histology of the tumor may be modified in the pulmonary arteries
Ova of parasites	Common in areas, such as East Africa and Brazil, infested with bilharzia	Pulmonary hypertension	Typical features of the parasitic ovum with surrounding granulomatous and hypersensitivity reaction

much less pronounced in cases of Fallot's tetralogy, with a large ventricular septal defect hardly balanced by the degree of pulmonary valvular stenosis. In these cases the cyanosis is not marked, and there may even be a slight degree of pulmonary hypertension.

The low pulmonary arterial pressure leads to thinning of the media of small pulmonary arteries. The media of the pulmonary trunk becomes thin and its elastic fibrils clumped.[87] The widespread occlusion of the pulmonary arteries stimulates the development of a collateral flow through the bronchial arteries.

Bronchial Arteries

The capacity to develop a collateral blood supply to the lung in conditions in which there has been significant occlusion of the pulmonary arterial tree by organized thrombi or other processes depends on the fact that the lung has a dual blood supply, with pulmonary arteries originating from the right ventricle and bronchial arteries arising from the aorta and thus indirectly from the left ventricle. Since the bronchial arteries are part of the systemic circulation, they carry blood at a pressure some six times greater than

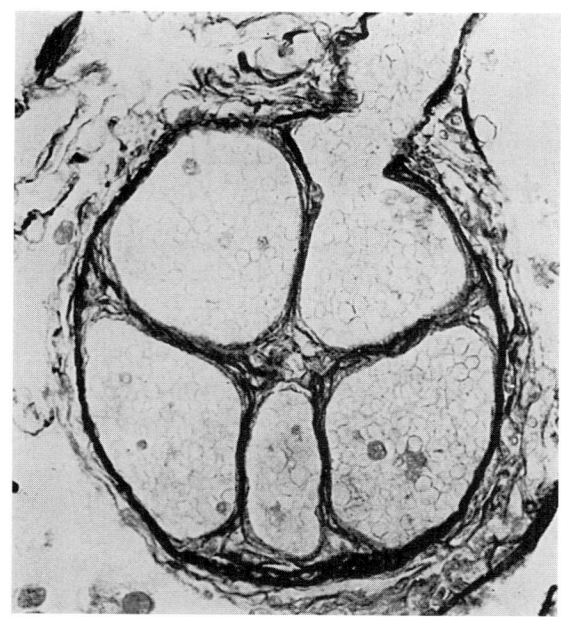

FIGURE 21-35
Small pulmonary artery from a 19-year-old woman with tetralogy of Fallot. Delicate septa divide the lumen into five compartments. The intima is not thickened except at the sites of attachment of the septa. The media is extremely thin. (Elastic–van Gieson, ×350.)

that in the pulmonary circulation. As a consequence, they have thick muscular walls, which may confuse the histopathologist unfamiliar with pulmonary vascular pathology. For this reason, we shall briefly consider the origin, course, and structure of the bronchial vasculature.

In most cases, the bronchial arteries arise from the inferior aspect of the aortic arch and descending portion of the aorta at about the level of the fifth dorsal vertebra. The left bronchial artery or arteries usually arise from the aorta, and so may the right bronchial vessels.

However, one or more right bronchial arteries may arise from some other vessel, such as the first right intercostal artery, the second right intercostal, or the right internal mammary arteries. Passing laterally from the aorta, the bronchial arteries cross the esophagus anteriorly and posteriorly. When they reach the hilum, an annulus is formed around the main bronchus. From this anastomosing network two distinct sets of branches arise: the "visceral pleural" and "true bronchial" arteries. Visceral pleural branches give rise to hilar vessels that supply the mediastinal pleura and to lateral vessels that supply the anterior lateral and interlobar visceral pleura. Visceral pleural bronchial arteries occupy a superficial position and independent course from the bronchial tree.

In contrast, "true bronchial arteries" follow the course of the bronchial tree. The main true bronchial artery arises from the hilar annulus and enters the

lung. The characteristic course of bronchial arteries is spiral, in contrast to the rectilinear one of the pulmonary arteries. They pass for a considerable distance into the lung before coming into contact with lung parenchyma. The true bronchial arteries follow the course of the bronchial tree, bifurcating with the bronchi.

There are two bronchial arteries with each bronchus, one along each side of its wall. While the pulmonary arteries are true end-arteries, the intrapulmonary bronchial arteries branch and anastomose repeatedly around the bronchi, forming peribronchial and intramural networks like arcades of the mesenteric arteries. Because of this anatomic arrangement, it is practically impossible to cut off the systemic arterial supply of a bronchus; the ligation of one bronchial artery is rapidly followed by the growth of collaterals.

Bronchial veins are also of the "true" bronchial or pleurohilar type. They communicate freely with the pulmonary veins throughout their course, and they drain into the left atrium either directly or indirectly via the pulmonary veins.

The bronchial arteries show considerable changes in histologic structure as they pass from the aorta to the periphery of the lung. This is a potential source of difficulty in interpretation of various pulmonary arteries by the uninitiated histopathologist. Immediately after arising from the aorta, the bronchial arteries have the characteristic structure of systemic arteries and consist of 10 to 20 layers of smooth muscle fibers between a thick internal and a very thin external elastic lamina. When the artery enters the bronchial wall, it often shows, inside the normal layer of circular muscle, a thick layer composed largely of elastic fibrils, probably derived from splitting of the internal elastic lamina. Between these elastic fibrils are a few longitudinal muscle fibers. In other bronchial arteries a very thick layer of longitudinal muscle is seen inside the circular muscle coat. This may almost block the vessel. In some bronchial arteries and arterioles, the layer of longitudinal muscle fibers may become well developed, with virtual loss of the circular muscle coat. Other bronchial arterioles have the structure of a typical systemic artery, with a comparatively thick media of circular smooth muscle situated between internal and external elastic laminae. In the normal lung, some bronchial arteries have been described as having a thick, well-formed musculoelastic zone adjacent to the intima. These vessels are said to have an enormously thickened media, which appears as massive muscular knots due to the spiral course of the dense muscle around the periphery of the media. This peculiar structure may be maintained over short or long segments of the vessel before returning to the normal structure of the bronchial artery. These "sperrarterien" have been considered to be abnormally sclerosed bronchial arteries.

Communications between bronchial and pulmonary arteries occur in the normal adult and in normal fetal and perinatal lungs. There are three main stimuli for the production of bronchopulmonary arterial anastomoses in disease: a diminished pulmonary blood flow, the formation of granulation tissue as part of a chronic inflammatory process in the lung, and the development of tumors such as bronchial carcinoma. Details of the histology of these various anastomoses are too specialized to be considered here. Congenital aortopulmonary anastomoses are a particular feature of pulmonary atresia.

PULMONARY VENOUS HYPERTENSION

While our understanding of much of the pulmonary vascular disease described so far in this chapter is recent and still expanding, the lesions in the pulmonary veins, lung parenchyma, and pulmonary arteries in disease associated with pulmonary venous hypertension are very familiar to histopathologists. Indeed, the lesions in the lung substance have passed into classic pathology as *brown induration of the lung*. The causes of chronic pulmonary venous hypertension are many.

In the past, rheumatic mitral stenosis and incompetence figured prominently, but this is no longer the case in Western Europe and North America with the decline in the incidence of rheumatic fever. Mitral stenosis still remains common, however, in many developing countries. Most patients with severe pulmonary venous hypertension now being referred for cardiac transplantation have ischemic heart disease. Causes of nonrheumatic mitral incompetence include ischemic heart disease, especially that causing rupture of the chordae tendineae or papillary muscle dysfunction, "floppy" mitral valve, senile calcification of the mitral valve annulus, and bacterial endocarditis. Many forms of congenital cardiac anomaly lead to pulmonary venous hypertension, including cor triatriatum, congenital stenosis of all the pulmonary veins at the left atrial ostia, constrictive endocardial sclerosis, and congenital mitral stenosis and incompetence. An interesting combination of pulmonary venous hypertension and a congenital cardiac shunt is mitral atresia with patent foramen ovale and a single ventricle. Another, much rarer cause of chronic pulmonary venous hyperten-

sion is mediastinal granuloma constricting pulmonary veins.[88]

Persistently raised left atrial pressure is also common when the left ventricle is overloaded or abnormal. Overload may be due to disease of the aortic valve or to systemic arterial hypertension. The most common cause of left ventricular damage is ischemia. Rare cardiomyopathies may prevent diastolic filling of the left ventricle by rendering it rigid or may so weaken the left ventricle that it becomes dilated.[88] Work hypertrophy of the left ventricle may also end with an ineffective, weakened myocardium. All these conditions lead to an increased end-diastolic pressure in the left ventricle and, thereby, to increased left atrial and pulmonary venous pressure.[88] A large left atrial myxoma may increase pressure in this cardiac chamber and in the pulmonary veins.

The basis of the pulmonary vascular disease in pulmonary venous hypertension lies in the ultrastructural changes of the alveolar-capillary wall. The thickness of the basement membrane, normally 8 to 10 nm, is commonly increased to 30 nm and in rare instances may reach 50 nm.[89] This thickening does not affect the convex portions of the blood-airway barrier bulging into the alveolar lumen. Sometimes the thickened basal lamina is split to enclose disintegrating fragments of extravasated erythrocytes.[71] Sometimes there is a subendothelial infiltration of a hyaline substance, almost certainly a ground substance of mucopolysaccharides. The endothelial cells of the pulmonary capillaries become edematous and swollen and may develop irregularly shaped vesicles that may be as large as 400 nm in diameter.[71] The endothelial swelling may be so severe that it gives rise to intraluminal projections of clear, edematous cytoplasm up to 1600 nm in diameter. Interstitial edema is an important ultrastructural feature of the alveolar-capillary wall in mitral stenosis. This occurs in the thicker part of the wall, which is concerned with water transport rather than gas exchange. Over wide areas, membranous pneumocytes are replaced by the granular variety. This is the basis for the complex pattern of change that develops in pulmonary venous hypertension.

The edema may become so pronounced as to present as basal horizontal lines (Kerley B lines) in chest radiographs. They are thought to be caused by edema of the connective tissue of the lung septa together with distention of the lymphatics.[90] Another radiologic feature of pulmonary venous hypertension is Fleischner's lines, which are permanent and largely due to the deposition of hemosiderin in the fibrous tissue septa between the secondary lung lobules.[91]

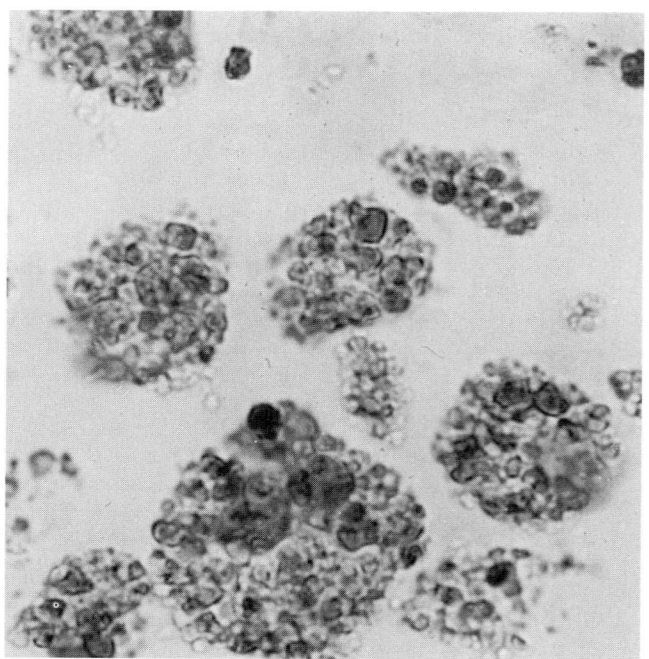

FIGURE 21-36
A group of hemosiderin-laden macrophages in an alveolar space in a woman 25 years of age with mitral stenosis. (H&E, ×1500.)

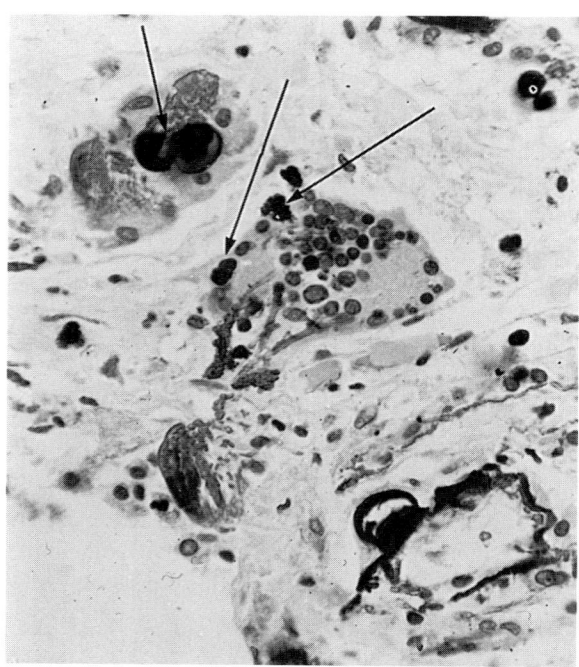

FIGURE 21-37
Section of lung from a 25-year-old woman with mitral stenosis showing deposition of ferric iron salts on elastic and reticulin fibers. Some of the deposits (*arrows*) have provoked around them a foreign-body giant cell reaction. (Perls' stain, phloxine and tartrazine, ×250.)

Transient basal horizontal lines appear when the mean left atrial pressure exceeds 25 mmHg.[92] Distended pulmonary lymphatics may be readily identified in lung biopsy specimens.

In mitral stenosis the pulmonary capillaries are distended, and there is diapedesis into the alveolar spaces. The fragments of disintegrating erythrocytes are ingested by macrophages, which aggregate in the lung to give rise to pulmonary hemosiderosis. If it is severe enough, this may be detected radiographically. The hemosiderin ingested by alveolar macrophages appears as dark-brown granules, which readily take up Perls' stain for ferric iron salts (Fig. 21-36). These siderophages are scattered throughout the lung, but they characteristically accumulate in small groups that often lie beneath the pleura, around bronchi and pulmonary arteries. Sometimes there is such a concentration of iron salts locally in the lung that there may be iron encrustation of reticulin and elastic fibrils in the lung parenchyma, and of the elastic laminae of pulmonary arteries, arterioles, and veins. In rare instances the encrusted fibrils provoke a foreign-body giant cell reaction (Fig. 21-37).

There is a gradual change from pulmonary hemosiderosis to siderofibrosis. The progressive fibrosis of the alveolar walls leads to obliteration of pulmonary capillaries. Superimposed on this developing fibrosis is hemosiderosis, which gives the firm,

fibrotic lung a brown color so that in classic morbid anatomy the disease has come to be called "brown induration of the lung."

When Ehrlich discovered mast cells in the last century, he soon realized that they were common in brown induration of the lung. We carried out lung mast cell counts on 27 necropsy specimens of lung from cases of mitral stenosis and chronic left ventricular failure.[93] The mean mast cell count was $3.67/mm^2$ of lung in cases with little or acute edema, $12.36/mm^2$ in subacute pulmonary edema, and $41.90/mm^2$ in brown induration.[93]

Osseous nodules are common in the alveolar spaces in mitral stenosis.[94] They probably arise as the result of persistence of fibrin clumps in the alveolar ducts and spaces. The nodules are composed of lamellar bone and may grow to a diameter of several millimeters. They have a smooth outline and provoke little or no tissue reaction in the alveolar spaces. A much rarer manifestation in the lungs of patients with mitral stenosis is microlithiasis (Fig. 21-38), in which extensive calcification occurs around foci of alveolar exudate to form numerous microliths or calcospherites. These are uniform in size and consist largely of stratified concretions of calcium phosphate that dissolve in hydrochloric acid to leave an onion-shaped organic envelope containing iron.[95] Each calcospherite is some 250 to 750 μm in diameter, is

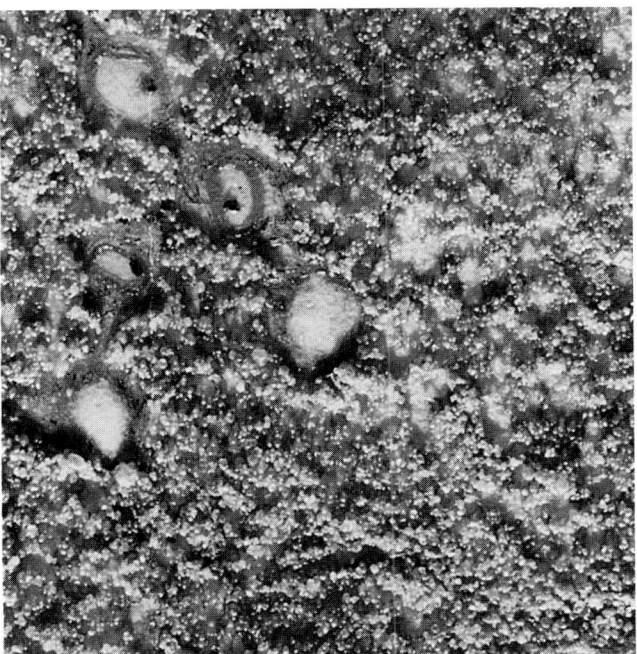

FIGURE 21-38
Part of a Gough-Wentworth section of lung from a case of microlithiasis in a 30-year-old woman with mitral stenosis. The alveolar spaces are filled with microliths, which give the lung the consistency of sandpaper. (×4.)

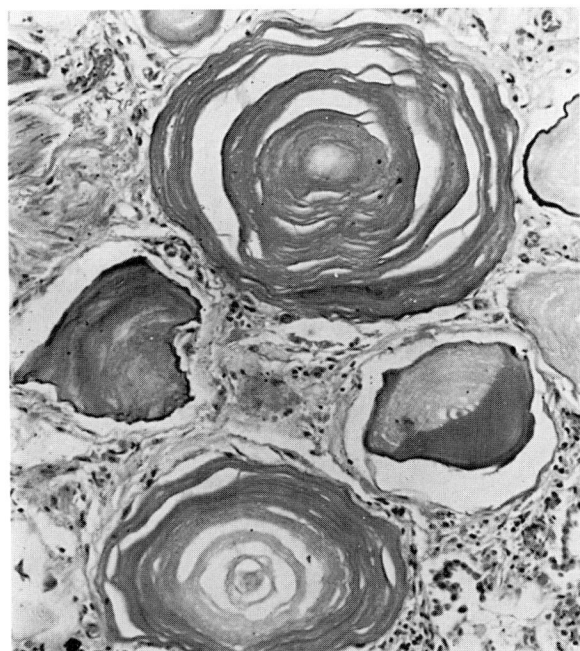

FIGURE 21-39
Section of lung from a woman aged 50 with rheumatic mitral stenosis and microlithiasis alveolaris pulmonum. The alveolar spaces contain lamellae of calcium phosphate. There is no reactive fibrous thickening of the adjacent alveolar walls. (H&E, ×125.)

roughly spherical, and contains about 15 rings of calcium phosphate. The microliths produce miliary opacities in radiographs (Fig. 21-39).

The sustained elevation of pressure in the pulmonary veins and venules leads to pulmonary arterial hypertension and vascular disease affecting the pulmonary arteries and arterioles. The pulmonary vascular disease of mitral stenosis differs from plexogenic pulmonary arteriopathy in that it does not include plexiform lesions or concentric-laminar ("onionskin") intimal proliferation. In fact, it appears that sustained pulmonary venous hypertension in some way protects against the development of plexiform lesions. In mitral stenosis the small pulmonary arteries show pronounced medial hypertrophy. This is often exaggerated by the development of longitudinal muscle in the adventitia outside the outer elastic lamina and in the intima inside the inner elastic lamina (Fig. 21-40). There is often crenation of the internal elastic lamina consistent with constriction. Fibrinous vasculosis occasionally occurs in the small pulmonary arteries in mitral stenosis.[96] In this disorder, fibrin traverses the media radially and is held up at the elastic laminae. It usually presents in the acute form as fibrinoid necrosis or necrotizing arteritis.[97–99] Less commonly it is seen in the healing, granulomatous state.[88] In the subacute stage, arcs of the media may be destroyed and replaced by granulation

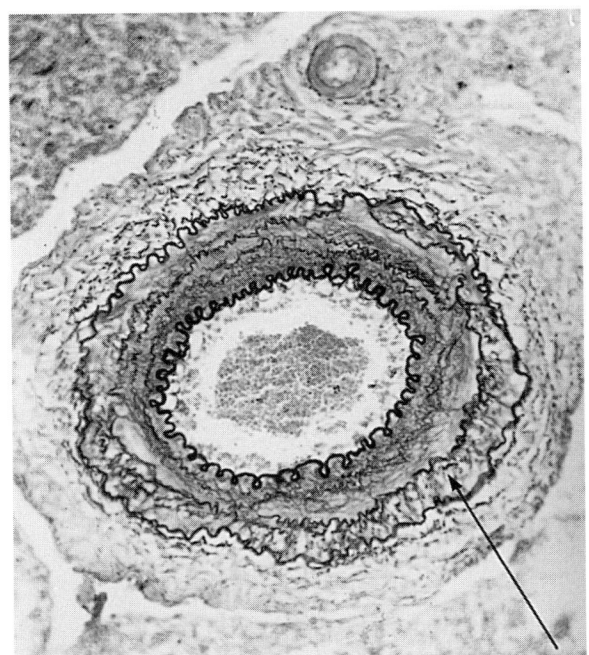

FIGURE 21-40
Transverse section of small pulmonary artery from a 6½-year-old boy with congenital mitral stenosis. There is pronounced hypertrophy of the original media of circularly oriented smooth muscle. There is crenation of the internal elastic lamina consistent with constriction or collapse. The medial hypertrophy has been exaggerated by the development of longitudinal muscle (*arrow*) in the adventitia, outside the original outer elastic lamina. (Elastic–van Gieson, ×140.)

675

and fibrous tissue, often with thrombotic occlusion of the vascular lumen. There is muscularization of the pulmonary arterioles in cases of pulmonary venous hypertension.

The small pulmonary arteries in mitral stenosis commonly show pronounced intimal proliferation due to myofibroblasts, but this is not of the "onion-skin type" found in plexogenic pulmonary arteriopathy. These multipotential cells may lead to the presence of collagen, muscle, elastin, and proteoglycans in the intima.

Pulmonary veins, like arteries, show medial hypertrophy as a reaction to the elevation in their intraluminal pressure. In a normal pulmonary vein, the muscle in the wall is arranged in an irregular fashion intermingled with collagen and elastic fibrils. In the presence of severe, chronic pulmonary venous hypertension, the muscle becomes condensed with a distinct muscular media, devoid of collagen and fragmented elastic fibrils, and sandwiched between internal and external elastic laminae. As a result, the media of the pulmonary vein may come to resemble that of an artery, the so-called arterialization. Intimal fibrosis is also common in the pulmonary veins in states of chronic pulmonary venous hypertension. It is usually acellular and looks like that due to age change.

In the presence of moderate pulmonary hypertension, ultrastructural changes in the endothelial cells in the small pulmonary arteries in mitral stenosis are slight. There is infiltration of the subendothelial zone, with a hyaline material similar to that found in pulmonary capillaries.[89] With more severe hypertension there is pronounced cytoplasmic vacuolation in the endothelial cells. Subendothelial cells with the ultrastructural features of myofibroblasts, which are responsible for the intimal proliferation, ingest hemosiderin and secrete ground substance. The fibrils of some of these myofibroblasts increase to mimic frank muscle cells.

In mitral stenosis the elastic pulmonary arteries tend to be more dilated in the upper than in the lower lobes.[100] The small pulmonary arteries are more severely hypertrophied in the lower than in the upper lobes. In contrast, pulmonary congestion, edema, and hemosiderosis are maximal in the posterior part of the upper lobe, where there is minimal arterial hypertrophy and narrowing.[100]

PULMONARY FIBROSIS

Fibrosis of the lung is a common and important complication of many pulmonary diseases of widely differing etiology. It involves the pulmonary vasculature to a greater or lesser extent and represents one of the main mechanisms by which chronic lung disease gives rise to pulmonary hypertension. Pulmonary fibrosis exerts its deleterious effects on the pulmonary circulation in a variety of ways. It may infiltrate and eventually obliterate elastic and small pulmonary arteries and pulmonary veins. Even large vessels may be engulfed in the massive fibrosis of the pneumoconioses, such as silicosis. On the other hand, the fibrotic process may block or destroy smaller vessels and capillaries in the alveolar walls. Such involvement of individual alveolar-capillary walls is known as *fibrosing alveolitis* or *interstitial pulmonary fibrosis.* It disturbs oxygen diffusion to the pulmonary capillaries, although it does this by interfering with ventilation:perfusion ratios rather than by merely thickening the alveolar capillary wall (see Chap. 12). Pulmonary fibrosis may follow the formation of granulomas, which themselves may first involve the pulmonary vasculature, or it may originate in a proliferation of profibroblasts in alveolar spaces as in poisoning by paraquat.

Pulmonary Vascular Disease in Honeycomb Lung

With chronicity, and as a result of obliteration of some respiratory bronchioles and dilatation of others, interstitial fibrosis may evolve into the cystic condition of "honeycomb lung," which is associated with pronounced changes in the pulmonary vasculature.[101] This is characteristic of the cystic condition but is independent of the primary lung disease giving rise to it. There is first hypertrophy of the media of circular muscle and the formation of longitudinal muscle in the intima, both in the small pulmonary arteries and in the pulmonary arterioles.[101] This early stage of muscularization is almost certainly associated with the development of pulmonary hypertension. It is followed by a fibrous stage, characterized by pronounced intimal fibroelastosis and recanalization of thrombus.[101] There is fibrous atrophy of the media and finally total ablation of affected vessels. When honeycomb change is restricted to a narrow subpleural band, most of the pulmonary vascular changes are confined to this zone. In contrast, muscularization of pulmonary arterioles is found throughout the lung.

Interstitial pulmonary fibrosis and honeycomb lung occur frequently but not invariably in scleroderma, and the pulmonary vascular lesions described above occur in this disease. Another striking histologic change in the small pulmonary arteries in scleroderma is recanalization of thrombi, which is reminiscent of the inner muscular tubes found in pulmonary emphysema. Some believe that the pulmonary vascular lesions in scleroderma are specific for that disease.

Progressive Massive Fibrosis

Widespread involvement of the pulmonary circulation may occur when there is gross fibrosis in the lung, as develops in pneumoconiosis (see Chap. 15). The capillaries are destroyed and replaced by fibrous tissue. The small pulmonary arteries are occluded and frequently destroyed by fibrous tissue. The large elastic pulmonary arteries become obstructed at the point where they enter massive areas of fibrosis but are patent and dilated next to these. Hence the right ventricle is forced to maintain its output against an increased pulmonary peripheral vascular resistance.

In coalminer's pneumoconiosis, the presence and state of the capillaries depend on the type of fibrosis.[102] In the fibroanthracotic type, characterized by coal-dust pigment distributed through loosely arranged collagen, there are many large capillary channels lined by only a thin endothelium abutting directly on the fibrous tissue. In the fibrohyaline type of fibrosis, in which less numerous dust cells are embedded in tightly packed hyaline collagen fibers, there are only occasional small vessels with walls thicker than those in the fibroanthracotic masses.[102] At the edges of massive fibrotic lesions, there is a relative vascular abundance of thin-walled vessels.

In progressive massive fibrosis, the small pulmonary arteries and veins show occlusive cellular intimal fibrosis. There is often stenosis or distortion of these vessels by medial fibrosis. Eventually the lumen may be replaced by capillary-like channels, and the whole vessel is overrun by dust-bearing tissue, so that only elastic stains reveal the remains of pulmonary blood vessels. Vascular damage is disproportionately great in the presence of superimposed infection, especially tuberculosis. In many of the occupational chest diseases, specific dust particles or fibers, asbestos bodies, and dust cells may be demonstrated in the outer part of the media and in the perivascular lymphatics.

The large pulmonary arteries frequently show dilatation, which extends into the intrapulmonary branches. Many of the elastic pulmonary arteries are engulfed in areas of massive fibrosis and are obliterated. Patent vessels skirting around the edges of such fibrotic lesions are distorted and frequently deficient in smaller branches.[102] Atheroma is common in these elastic arteries, and ulceration of the intima occa-

sionally occurs in vessels near fibrous masses or hard lymph nodes. The media of the larger pulmonary arteries shows fibrosis, which may end in cicatricial stenosis or aneurysmal dilatation and rupture.[103] There is often a great affinity of the media for basic dyes and intense metachromasia with toluidine blue.[102] Thrombosis is common in the large pulmonary arteries of patients with pneumoconiosis.

Hematite Lung

Occasionally, other industrial mineral substances are deposited in the lung and lead to pulmonary vascular pathology. One such is hematite, which is iron sesquioxide. Only the hard rock mined at Whitehaven, in the north of England, which contains 10 percent silica in addition to the iron ore, leads to fibrosis. Thus, *hematite lung* is in reality silicosis with associated gross deposition of hematite dust. The changes in the pulmonary vasculature are those of silicosis, already described above, together with an intense accumulation of iron-containing dust in and around the pulmonary blood vessels. Pulmonary arteries in the fibrotic areas of lung at first show intimal fibrosis or fibroelastosis. Later the fibrosis spreads to the media, causing its focal atrophy or ablation. Commonly this is associated with severe fibrosis of the adventitia.

In the small elastic pulmonary arteries there are often recanalized thrombi, and within them or in the thick surrounding layer of intimal fibrosis there is often a great deal of hematite pigment, some within macrophages.[104] The overwhelming fibrous reaction to the silicohematite dust may produce striking appearances, with part of the circumference of the artery being totally engulfed by fibrous tissue. The pulmonary veins show severe intimal fibrosis, aggregates of hematite dust in the adventitia, and thrombosis with recanalization. There is commonly intimal longitudinal muscle in the small pulmonary arteries.

Granulomas in the Pulmonary Circulation

There is a large and heterogeneous group of diseases of the lung in which the pulmonary vasculature is invaded by granulomas or necrotizing arteritis before it is overtaken by reactive or reparative fibrosis. Both the pulmonary and the bronchial circulations may become impaired in tuberculosis. In areas of lung parenchyma severely damaged by this disease, there is a progressive obliteration of pulmonary arteries and veins and an associated increase in size of the bronchial arteries and the development of bronchopulmonary anastomoses. Elastic pulmonary

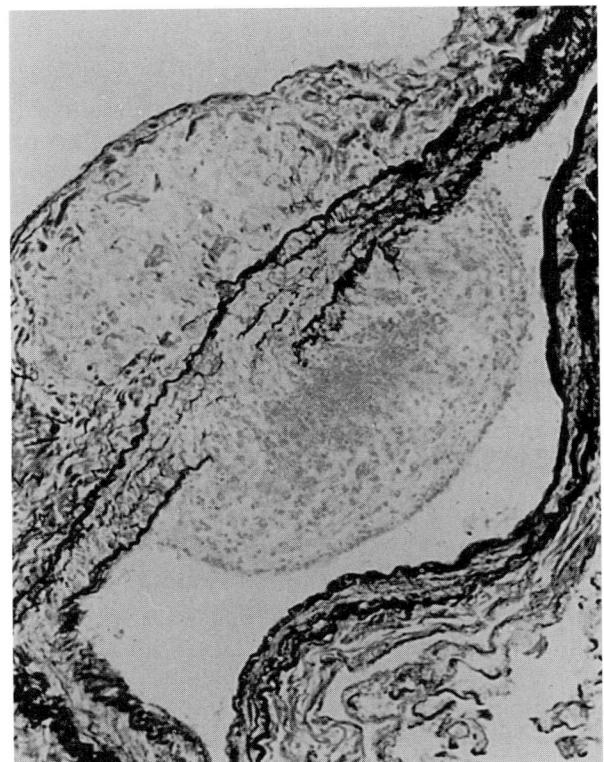

FIGURE 21-41
Oblique section of a small pulmonary artery from a 38-year-old man with pulmonary tuberculosis with involvement of the pericardium. There is a tuberculous granuloma in the adventitia, and another arises from the intima and bulges into the lumen. They are connected by an involved segment of the underlying media. The granulomas consisted of epithelioid cells, lymphocytes, and giant cells. (Elastic–van Gieson, ×140.)

arteries in or near areas of caseation and cavitation may show caseous necrosis of the media. In the walls of tuberculous cavities, such weakened pulmonary arteries may form aneurysms that may rupture and bleed, sometimes severely. Small pulmonary arteries may be totally engulfed in an area of caseation showing medial necrosis and occlusion of the lumen by caseous material and thrombus. Stains for elastic tissue may be needed to detect the remains of overwhelmed pulmonary vessels.

Sometimes there may be a specific tuberculous arteritis, so that the wall of the vessel is occupied by a granuloma consisting of epithelioid cells, lymphocytes, and Langhans'-type giant cells (Fig. 21-41). Such vascular granulomas often assume a dumbbell appearance, one cellular nodule in the intima bulging into the lumen and another in the adventitia with a connecting strand in the media. Sometimes a diffuse, nonspecific chronic inflammatory exudate is seen throughout the arterial wall. In areas of fibrocaseous tuberculosis, there is often much obliteration of the vascular lumen by a mixture of intimal fibrosis and thrombosis, rather vaguely termed *endarteritis obliterans.* Involvement and erosion of small pul-

monary veins by active tuberculous processes, especially caseation, may lead to miliary tuberculosis. Bronchial arteries near tuberculous areas become dilated and hypertrophy. Recanalization of thrombosed arteries in the vicinity of caseous areas together with proliferation of their vasa vasorum, which are derived from the bronchial circulation, leads to the formation of bronchopulmonary anastomoses.[105]

Pulmonary arteries and veins are frequently affected by granulomatous and fibrotic lesions in sarcoidosis just as in tuberculosis. Usually the granulomas lie in the vascular adventitia, distorting the vessel and pressing on the lumen. Less often they are found in the media or intima (see Chap. 17). The disseminated form of temporal arteritis and Takayasu's disease may affect the pulmonary arteries.[106,107] Temporal arteritis may be preceded by pulmonary vascular changes, with muscular pulmonary arteries showing focal medial destruction, acute or chronic inflammation, and some ill-defined adventitial granulomas.[108] These vascular changes cause pulmonary infarction.

PARASITES

Schistosomiasis

Physicians and pathologists practicing in Europe and North America encounter pulmonary hypertension mainly as a complication of heart and lung disease, and they may regard worms as a somewhat esoteric cause of elevated pulmonary arterial pressure. If we take a wider view, however, it is undeniable that the nematode *Schistosoma* is an important cause of pulmonary hypertension in tropical and subtropical parts of the African and South American continents.

The pulmonary circulation may become involved when adult flukes produce ova to maintain the life cycle (see Chap. 9). Ova from the small veins in the gut wall pass to the liver and hence into the lung to impact in the walls of the small pulmonary arteries and arterioles (Fig. 21-42). However, the pulmonary vascular pathology is far more complex than one of simple embolism. The obstruction is soon exaggerated by a granulomatous reaction, probably allergic, around the egg.[109–111] This consists of foreign-body giant cells, eosinophils, mononuclears, calcified remains of ova, and fibrosis (Fig. 21-42). The granuloma is probably induced by the ova's acting as antigenic stimuli. The granuloma may form in the lumen, where it frequently provokes surrounding thrombosis, or it may penetrate the arterial wall.[112,113] Foci of necrotizing arteritis in the walls of

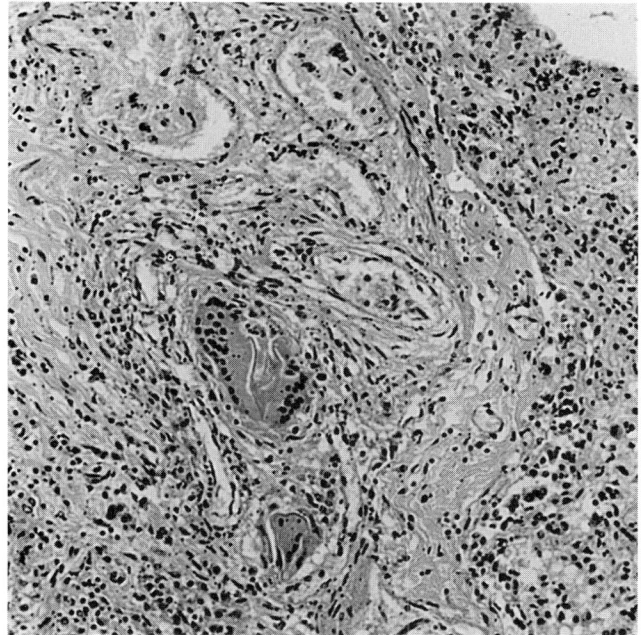

FIGURE 21-42
Section of lung from an East African who was dying from pulmonary schistosomiasis. There has been a vigorous giant-cell reaction around the remains of ova of *Schistosoma mansoni*, which had impacted in small pulmonary arterial vessels. Eosinophils and lymphocytes are also prominent in the surrounding inflammatory cellular exudate, as would be anticipated in a pulmonary vascular reaction in which hypersensitivity and immunologic mechanisms are involved. (H&E, ×177.)

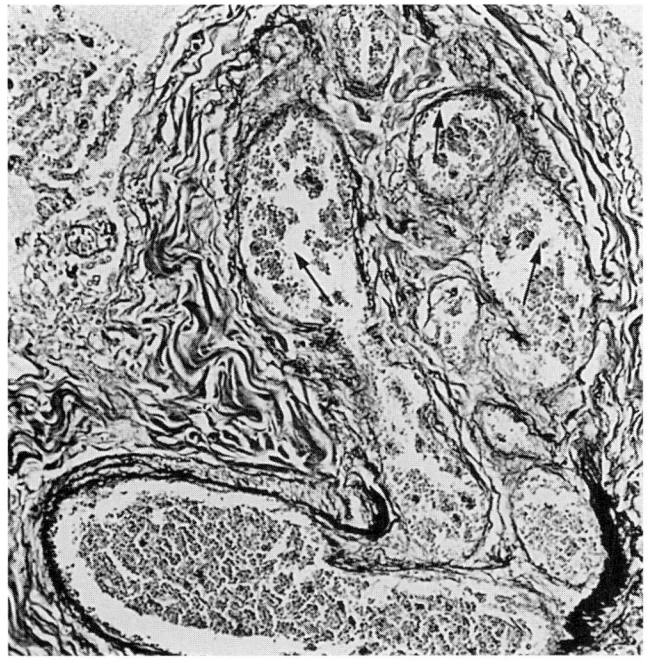

FIGURE 21-43
Bilharzial argiomatoid lesion from the same case as in Fig. 21-42. The parent muscular pulmonary artery gives rise to multiple thin-walled vessels (*arrows*), which form the argiomatoid lesion. (Elastic–van Gieson, ×150.)

the small pulmonary arteries represent another manifestation of the allergic response to the ova. The presence of a vasoconstrictive factor in bilharziasis has been reported,[114] and this may be an additional factor in bringing about the fibrinoid necrosis. Small pulmonary arteries proximal to the granulomas and associated thrombosis develop medial hypertrophy and intimal fibrosis as a response to irritant substances derived from the ova. This causes a rise in pulmonary arterial pressure brought about by the progressive pulmonary vascular obstruction.

A prominent feature of the histopathology of the pulmonary arteries in schistosomiasis of the lung is the angiomatoid lesion (Fig. 21-43).[109,110,115] It probably arises in the same way as the dilatation lesions of plexogenic pulmonary arteriopathy associated with congenital cardiac septal defects in that it forms secondary to widespread pulmonary arterial obstruction, although the histologic components of that obstruction are different. It is also of interest that both in congenital septal defects and in schistosomiasis the pulmonary arteries show fibrinoid necrosis, although in these diseases the necrosis is based on hypertension on the one hand and allergic hypersensitivity on the other.

Filariasis and Pulmonary Hypertension

During the period 1967 to 1972, there was an increased incidence of unexplained pulmonary hypertension in Sri Lanka.[116,117] In those five years no fewer than 65 patients were found to have primary pulmonary hypertension in a consecutive series of 2500 patients investigated for cardiovascular disease.[117] There were, however, unusual features in the cases occurring in Sri Lanka, suggesting that they were unlikely to be examples of classic primary pulmonary hypertension. There was, to begin with, a disproportionately large number of men in the series. There was a frequent association with eosinophilia, which diminished after treatment with diethyl carbamazine. Such features led to the suspicion that filariasis due to *Wuchereria bancrofti* was responsible for the elevated pulmonary arterial pressure. However, it has to be kept in mind that infestation with this parasite is common in Sri Lanka, and eosinophilia would be anticipated in the subjects studied no matter what other disease was present. This issue has remained controversial.

Parasites and Pulmonary Vascular Disease in Animals

Filarial worms are known to cause pulmonary hypertension in animals. *Brugia buckleyi* was found in the right ventricle and pulmonary arteries of the Ceylon wild hare (*Lepus nigricollis singhala*).[118] Such animals develop right ventricular hypertrophy indicative of pulmonary hypertension. Similar findings have been reported in the dog, due to infestation with the nematode *Dirofilaria immitis*.[119] Cats may become infested with the nematode lung worm *Aelurostrongylus abstrusus*,[120] which leads to massive hypertrophy of the pulmonary arteries.

METABOLIC PULMONARY VASCULAR DISEASE

The pulmonary blood vessels may show pathologic changes due to disturbances in body metabolism. The metabolic abnormalities may be induced by disease elsewhere in the body, by the ingestion of poisonous substances that affect body metabolic processes, or by inborn errors of metabolism. Human pulmonary vascular disease is found in each of these three categories. Metastatic calcification affects pulmonary vessels, veins more than arteries (see Chap. 25).

Toxic Oil Syndrome

An epidemic of pulmonary hypertension erupted in Spain in March 1981; no fewer than 18,000 cases were seen in the ensuing year. The epidemic appeared to be related to the ingestion of toxic cooking oil. Probably a much greater number of subjects were exposed to the toxic substance, since up to 35 percent of symptomless relatives of patients had a high eosinophil count, an abnormality characteristic of this so-called toxic oil syndrome occurring in 98 percent of patients who developed the disease. By September 1982, 327 affected subjects had died.[121]

When this epidemic erupted, the disease was initially regarded as a form of atypical pneumonia and the cause was claimed at first to be *Mycoplasma,* then poisonous strawberries, and finally infectious agents from caged birds. In June 1981, six weeks after the outbreak of the epidemic, the disease was shown to be related to the ingestion of unlabeled olive oil containing rape seed oil. At that time rape seed oil was imported from France, and could be marketed as a lubricant or edible oil. If it was to be used as a lubricant, it had to be labeled as an industrial rape seed oil. Commonly it was mixed with other oils. When aniline is added to industrial rape seed oil, it produces fatty acid anilides. Chemical analysis reveals such substances as ortho-, meta-, and paratoluidines and *N*-methyl anilide. These substances have been regarded by many as the cause of the bizarre clinical features of the toxic oil syndrome. It has been suggested that this syndrome is a "free radical disease."

The ingested oil may lead to the liberation of free radicals, such as superoxide, which are thought to damage cell membranes and enzymes.

Pulmonary involvement in toxic oil syndrome occurs in both the acute and the chronic phases of the disease.[122] In the acute phase, more than 90 percent of affected patients had a syndrome of noncardiogenic pulmonary edema with dyspnea, cough, or chest pain. There were radiographic infiltrates resembling pulmonary edema. In most patients these findings resolved within a few weeks[122]; 2 to 3 percent of the affected patients died of acute respiratory failure, and death usually occurred within 1 to 3 weeks of onset.[123]

The pathology of the toxic oil syndrome, regardless of the organ system affected, can be summarized as follows: chronic inflammation in connective tissue, nonnecrotizing microangiitis, endothelial injury followed by intimal proliferation, and fibrosis.[124]

Pulmonary changes in the acute phase include endothelial swelling and vacuolization in small pulmonary arteries, intimal and mural vascular infiltration by mixed inflammatory cells, including lymphocytes, histiocytes, and occasional neutrophils and eosinophils, perivascular infiltrates extending into the interstitium, and interstitial and intraalveolar edema.[123,124] There may be marked intimal proliferation.[125] Hyaline membranes are present in some cases. Type II cell proliferation may be prominent. In early cases, electron microscopy shows hydropic degeneration necrosis and sloughing of pneumocytes.[124]

In the chronic stage of the toxic oil syndrome, the small pulmonary arteries show medial hypertrophy. This may be considerable, but there is commonly much variation in medial thickness. Some areas of the media may be very thin because of destruction of the muscle coat, which is not due to fibrinoid necrosis. Instead the elastic and small pulmonary arteries the muscularized pulmonary arterioles, and the pulmonary veins show a peculiar form of intimal proliferation that is highly characteristic of the disease. This is termed *spumous (foamy or frothy) intimal proliferation,* as the tissue is composed of cells with a faintly basophilic cytoplasm, packed with vacuoles, which displaces the nucleus to one side (Fig. 21-44).[125] The prominent endothelial cell lining covers the intimal proliferation, which may form an eccentric nodule or a continuous rim around the vessel. Many small pulmonary arteries are completely occluded by accumulations of spumous cells. These vessels are often surrounded by a pale-staining basophilic zone infiltrated by a mixed cellular exudate composed of plasma cells, eosinophils, lymphocytes, and histiocytes. It has

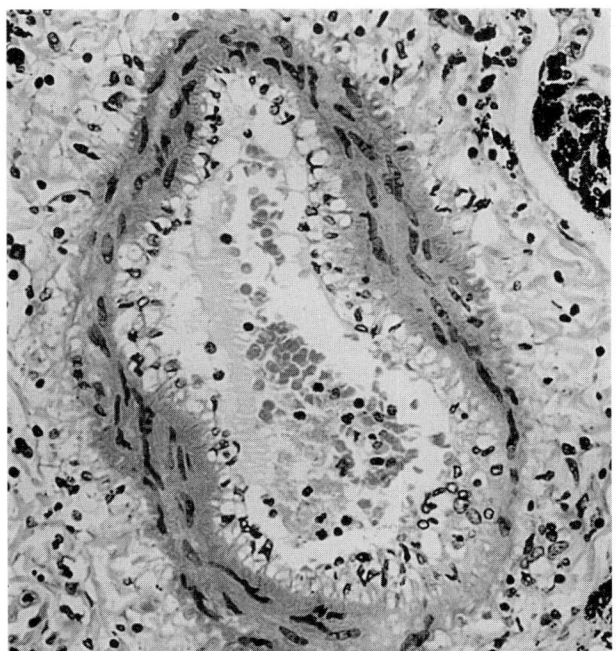

FIGURE 21-44
Toxic oil syndrome. Transverse section of a small pulmonary artery from a young Spanish woman who had ingested adulterated rape seed oil. There is medial hypertrophy with patchy areas of destruction (*lower right*). The vessel shows a spumous intimal proliferation, characteristic of the syndrome. (H&E, ×287.)

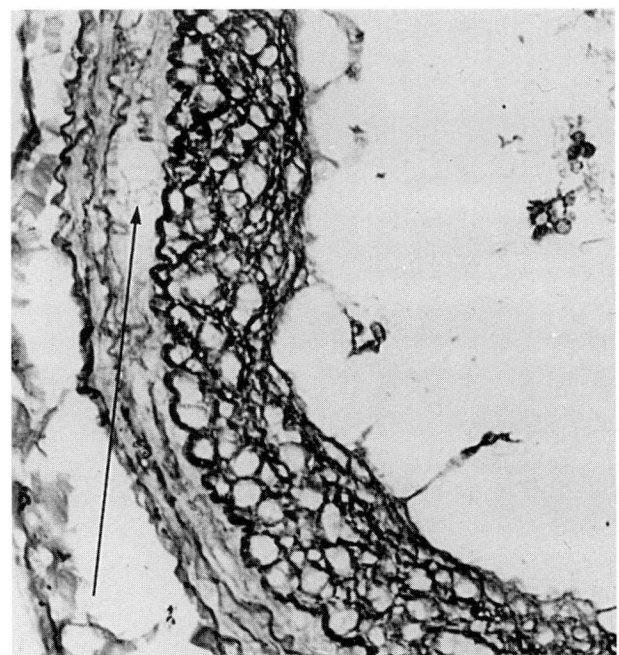

FIGURE 21-45
Fabry's disease in a 52-year-old man. Transverse section of a segment of a small pulmonary artery showing a bubbly appearance in smooth muscle cells in the media (*arrow*) and in myofibroblasts in a zone of intimal proliferation due to deposition of glycolipid. The myofibroblasts have formed elastic fibrils around themselves. (Elastic–van Gieson, ×432.)

been pointed out above that raised eosinophil counts are characteristic of patients with the disease and their symptomless relatives.

In one series of 15 patients studied by cardiac catheterization,[126] no fewer than 13 showed an elevated mean pulmonary arterial pressure and total pulmonary resistance. Both hemodynamic abnormalities decreased in most subjects after the administration of one of several pulmonary vasodilators, such as tolazoline or phentolamine.

Fabry's Disease

Anderson-Fabry's disease is a rare X-linked metabolic disorder due to alpha-galactosidase deficiency, leading to accumulation of the sphingoglycolipid ceramide trihexoside.[127] Although this can affect every organ, it frequently accumulates within the smooth muscle and endothelium of systemic blood vessels. Ceramide trihexoside may also collect in the pulmonary vasculature.[127] In such cases the media of the small pulmonary arteries is hypertrophied and presents a striking bubbly appearance due to collections of small clear vesicles in the cytoplasm of the distended vascular smooth muscle cells. They stain positively with Luxol fast blue and faintly with periodic acid–Schiff in paraffin sections and with Sudan black in frozen sections. The vesicles are separated from one another by elastic fibrils. The same histologic and staining reactions are seen in a thick zone of

proliferation of myofibroblasts in the intima (Fig. 21-45). In muscle cells sectioned transversely, the nucleous is displaced to one side by the cytoplasmic lipid. Vesicles are also seen in pulmonary arterioles, venules, and veins.

Electron microscopy shows numerous electron-dense inclusions within all classes of pulmonary blood vessel and alveolar walls. Most consist of dark, curved bands, loops, or rings, each of which is enclosed by a single membrane and composed of a series of closely adjacent concentric lamellae reminiscent of myelin. In some inclusions the dense structures are arranged as a series of straight parallel bands alternating with clear spaces, conferring on them a striped appearance and the designation *zebra bodies* (Fig. 21-46). The dark bands are composed of lamellae, as in the curved variants, but at their terminations the lamellae spread out in a fanlike series of fine strands, giving them a frayed appearance.[127] Zebra bodies and the other lamellar inclusions are found in profusion in the smooth muscle cells of the medias of small pulmonary arteries and veins. They are also identified in intimal myofibroblasts, where they form compact aggregates in the center of each cell, displacing the myofilaments to the periphery. Numerous lamellar inclusions, especially the striped zebra bodies, may be conspicuous in the distended endothelial cells of pulmonary arteries and veins.

There is controversy about whether the accumulations of glycolipid in the lung affect respiratory

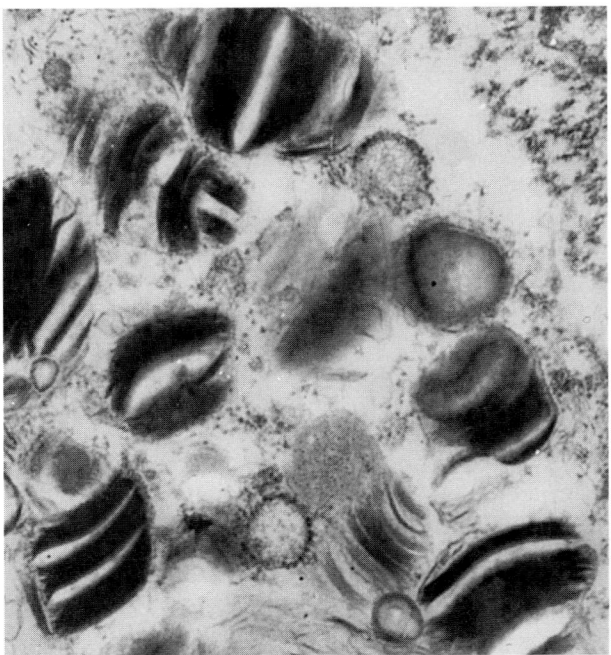

FIGURE 21-46
Fabry's disease. Electron micrograph of part of an endothelial cell from a pulmonary vein is distended by lamellar inclusions. Many of these are typical striped "zebra bodies." (×42,000.)

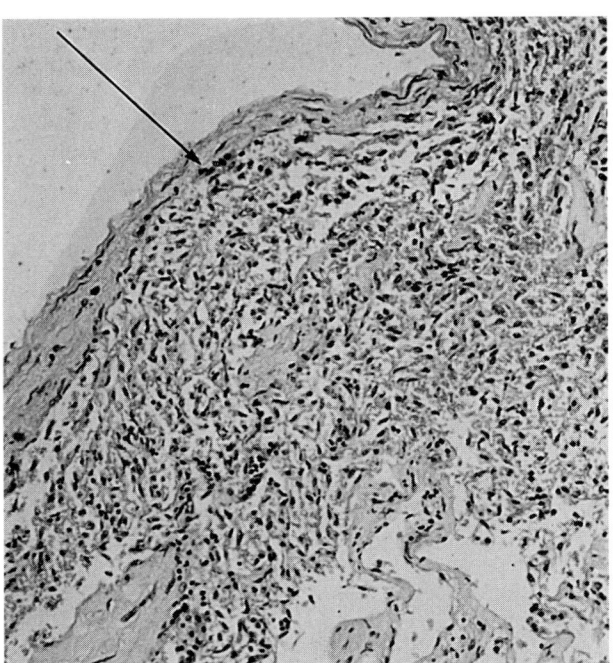

FIGURE 21-47
Section of lung from a 22-year-old Cypriot man with pulmonary capillary hemangiomatosis. Vessels of minute capillary dimension extend from the lung parenchyma and pleura to invade the wall of a large pulmonary vein in the upper left of the figure. (H&E, ×135.)

function. The most frequently described changes are those of chronic obstructive airways disease. Sometimes a brush biopsy will reveal zebra bodies in bronchial epithelial cells.

NEOPLASMS RELATED TO THE PULMONARY CIRCULATION

Tumors of the lung are dealt with extensively elsewhere in this book. In this chapter reference is made only to neoplasms specifically related to the pulmonary circulation that give rise to pulmonary vascular disease and pulmonary hypertension.

Pulmonary Capillary Hemangiomatosis

This is an obscure disorder in which great sheets of delicate blood vessels looking like capillaries spread throughout the lung. These thin-walled vessels come into intimate contact with, and appear to infiltrate, the walls of pulmonary blood vessels. This applies especially to pulmonary veins and venules. The disease is extremely rare. When Wagenvoort and his colleagues reported such a case in 1978, occurring in a woman 71 years of age, they could find no comparable report in the literature.[128]

There is variation in the size of the thin-walled vessels in individual cases. Thus, in a man 21 years old, the hemangiomatous tissue was of capillary dimension (Fig. 21-47).[129] In this case the small ves-

sels infiltrated the walls of the affected blood vessels so that, for example, the elastic fibrils had a ragged, "moth-eaten" appearance due to their disruption by the hemangiomatous tissue (Fig. 21-48). Their infiltration also led to wide separation of the vascular smooth muscle cells of the pulmonary vein.[129] Characteristically, the smaller pulmonary veins were surrounded by haloes formed of clusters of capillary vessels (Fig. 21-49). Similar angiomatous overgrowths were found in the intralobular fibrous septa and in the walls of pulmonary arteries and bronchi. Even in pulmonary venules, capillaries could be seen extending from the surrounding halo through interruptions in the elastic fibrils in the walls of the veins and venules to insinuate into the intima. There it provoked a severe intimal proliferation of fibrous tissue that closely resembled that of pulmonary venoocclusive disease (Fig. 21-50). In fact, so close was the histologic resemblance that we initially considered the diagnosis to be one of pulmonary venoocclusive disease.

In this disease the endothelial cells lining the capillary-like channels show hyperchromatism with pleomorphism of their nuclei.[128] The individual capillaries in the sheets of vascular tissue are closely arranged and separated by reticulin fibrils. Wagenvoort and colleagues[128] regarded this process as a neoplastic hemangiomatosis involving capillaries rather than a reactive vascular proliferation. The vasoformative tissue infiltrates the walls of pul-

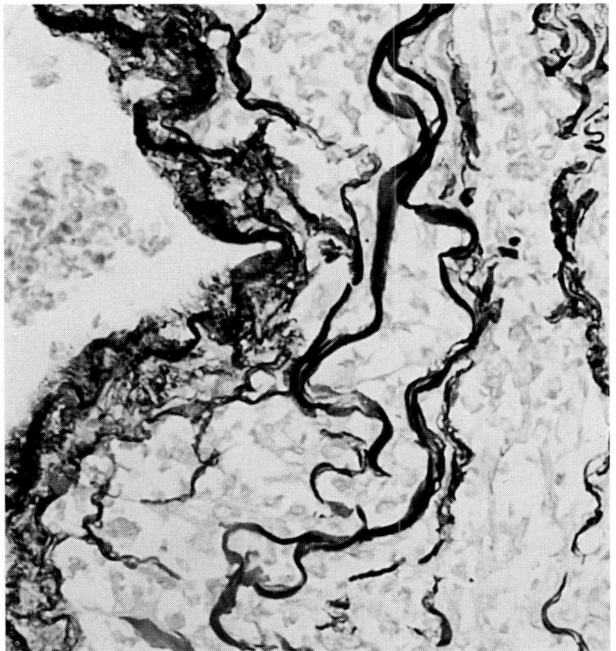

FIGURE 21-48
Section of lung from the same case illustrated in Fig. 21-47. There is disruption of the wall of an elastic pulmonary artery by the hemangiomatosis. The elastic laminae are ruptured and separated by capillary vessels showing expansion within the media. (Elastic–van Gieson, ×375.)

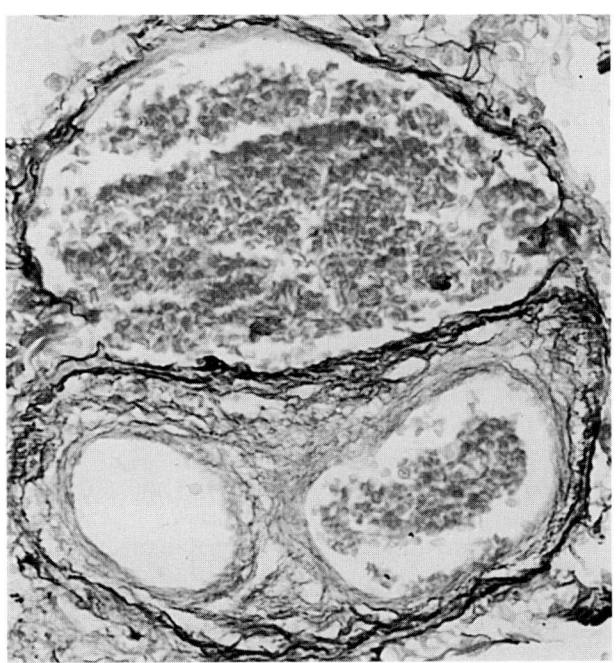

FIGURE 21-50
Oblique section of small pulmonary vein from the same case of pulmonary capillary hemangiomatosis as in previous three illustrations. The vein shows "arterialization" and severe intimal proliferation, reminiscent of pulmonary venoocclusive disease. (Elastic–van Gieson, ×337.)

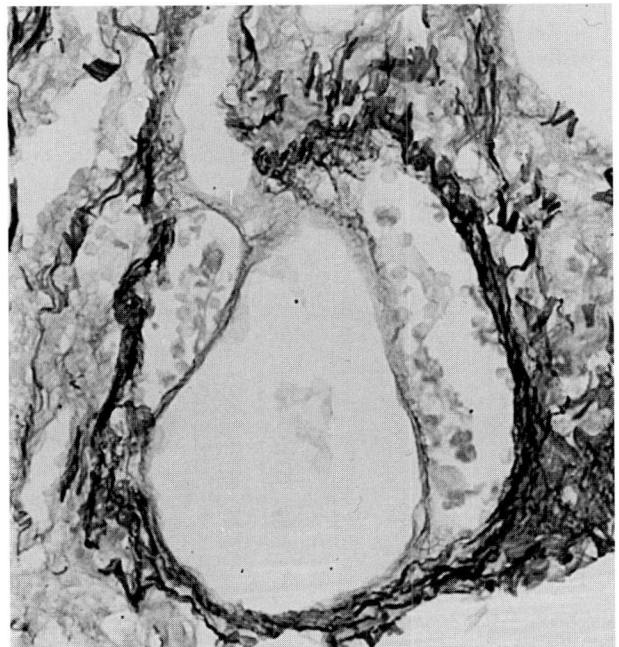

FIGURE 21-49
Section of lung from the same case as in Figs. 21-47 and 21-48. Transverse section of pulmonary vein showing dilated thin-walled vessels in the intima resulting from infiltration of the vessel by pulmonary capillary hemangiomatosis. (Elastic–van Gieson, ×345.)

monary vessels not only to affect the adventitia but regularly to destroy the media and the fibrotic intima of pulmonary veins. The obstructive lesions in the pulmonary veins and venules were thought by Wagenvoort and colleagues to be secondary.

The combination of rupture of vessels in the sheets of hemangiomatous tissue and the obstructive changes in the pulmonary veins and venules, with consequent distention of pulmonary capillaries, leads to alveolar hemorrhage. The red blood corpuscles are broken down to hemosiderin, which is engulfed by macrophages. These may become so prominent that there may be confusion with Ceelen's hemosiderosis in a lung biopsy specimen. The iron salts liberated impregnate elastic and reticulin fibrils in the walls of alveolar septa and of pulmonary blood vessels. A giant cell reaction to such areas of iron incrustation may occur. Such widespread hemosiderosis in the lung may again closely mimic pulmonary venoocclusive disease. There is ample evidence of medial hypertrophy of small pulmonary arteries and muscularization of pulmonary arterioles to indicate the development of pulmonary arterial hypertension secondary to occlusive changes in the pulmonary veins and venules.

The young man we studied[129] had suffered from repeated heavy hemoptyses, which could be attributed to the sheets of fragile tiny vessels throughout the lung substance and to the infiltration of the walls of bronchi by angiomatous tissue. This patient had also suffered bilateral hemothoraces due to widespread infiltration of the visceral pleura by hemangiomatous tissue.

In a second case, in a 51-year-old man, the thin-walled vessels were larger, measuring up to 180 μm in diameter.[130] These vessels had no elastic laminae in their walls. Apart from the size of the vessels, the other histologic features were similar to those encountered in the first, with infiltration of the walls of pulmonary blood vessels, bronchi, and pleura. In some small pulmonary arteries, large thin-walled vessels of venular size had infiltrated the media to reach the intima and occlude the lumen.

Recognition of invasive pulmonary hemangiomatosis is important, for it is another cause of clinically unexplained pulmonary hypertension. Clinical clues to the correct diagnosis are a history of repeated hemoptyses, sometimes heavy, and the development of a hemothorax.

Pulmonary Vascular Disease Due to Cardiac Myxoma

Cardiac myxomas can lead to pulmonary hypertension and pulmonary vascular disease. The mechanism of the raised pulmonary arterial pressure and the nature of the lesions in the pulmonary arteries differ according to the atrium in which the tumor originates. When the myxoma grows in the right atrium, the resulting pulmonary hypertension is due to the repeated impaction of emboli from the tumor into the pulmonary arteries.[131] The histologic appearances are different from those of recurrent pulmonary thromboembolism. Myxoma emboli are jellylike and stain vividly with alcian blue, demonstrating their high content of acid proteoglycans. The fragments of jelly are squeezed into the media by the force of the pulmonary hypertension (Fig. 21-34). The long axes of the cells and their nuclei are radially oriented as the myxoma streams through the media, disrupting the elastic laminae that lie in its path. The myxoma bulges in the adventitia to form dumbbell-shaped masses. It is important not to confuse these histologic appearances, which are the result of altered hemodynamics, with evidence of malignancy of the myxoma.[131]

In contrast, myxomas of the left atrium lead to pulmonary hypertension by elevating pressure in that cardiac chamber and in the pulmonary veins. It therefore acts in a manner similar to that of mitral stenosis. There is medial hypertrophy of small pulmonary arteries, muscularization of pulmonary arterioles, and hemosiderosis of the lung parenchyma.[132]

Sarcoma (see Chap. 32)

Sarcoma of the pulmonary artery is a very rare tumor. It forms a cream-colored or gray mass that may be flecked with small hemorrhages or areas of myxomatous change.[133] Often there is some adherence to the cusps of the pulmonary valve. The media and adventitia of the artery are infiltrated by the tumor, which also extends distally by continuity along the intima. Small pulmonary arteries may become blocked by emboli of the tumor and associated thrombus. Dissemination of the sarcoma to the regional lymph nodes, the lung parenchyma, mediastinum is common.[134]

Histologically the tumor of the pulmonary artery may be a spindle cell fibrosarcoma with much intercellular connective tissue and moderate pleomorphism.[134,135] It may include many multi-nucleated giant cells. Tumors reported as fibro-myosarcoma, malignant mesenchymoma, and leiomyosarcoma (Fig. 21-51) are considered by us in greater detail elsewhere.[133] Tumors may be osteo- or chondrosarcomas and the appropriate immunoperoxidase stain may be needed (Chap. 32). A leiomysarcoma of the pulmonary vein had its origin in the left upper lobe and formed a large mass in the left atrium.[136] Clinically, sarcomas of the pulmonary artery lead to dyspnea, syncope, and pericardial pain, terminating in progressive right ventricular failure.

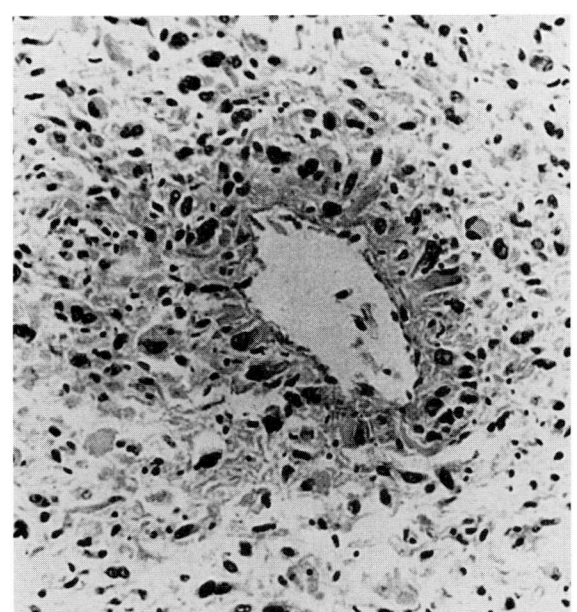

FIGURE 21-51
Section of a sarcoma of pulmonary trunk in a 40-year-old woman. The tumor is a leiomyosarcoma and is composed of irregular bundles of elongated and giant cells. In this figure, tumor cells can be seen to have formed a palisade arrangement around a small blood vessel. (H&E, ×140.)

Carcinoma

Invasion of pulmonary veins by bronchial carcinoma is very common, and even gross invasion of these vessels occurs in some 40 percent of cases of the tumor.[137] Pryce and Walter[137] also showed that pulmonary arteries are invaded with the same frequency as the veins. The incidence of gross infiltration of veins and arteries is the same irrespective of the site of origin or the histologic type of bronchial carcinoma.

In general, when elastic or small pulmonary arteries come into contact with carcinoma of any type, the adventitia becomes infiltrated first. Subsequently there is invasion of the media and intima, where reactive fibrosis takes place. This process is progressive, so the lumen may become occluded by tumor tissue and associated fibrosis.[137] Portions of tumor reaching the arterial intima commonly break away to plug pulmonary arterioles.

Having spread to the pulmonary arteries, metastases may grow along the intima, lining the arteries with neoplastic tissue. In a 39-year-old man with squamous cell carcinoma of the renal pelvis and carcinomatosis, the metastases were mainly within the branches of the pulmonary arteries, which became lined by mucoepidermoid carcinoma (Fig. 21-52).[138] In a similar case, metastases from a squamous cell carcinoma had grown along adrenal veins and pulmonary arteries, where they showed the histologic appearance of transitional epithelium.

ANEURYSMS OF THE PULMONARY ARTERIES

Aneurysms of the major pulmonary arteries are very rare. In one exhaustive study,[139] only eight such cases were found in more than 100,000 necropsies. They differ from aortic aneurysms in a variety of ways. Thus, whereas aortic aneurysm is predominantly a disease of the male, aneurysms affecting the pulmonary trunk occur with equal incidence in the two sexes. Pulmonary aneurysms occur in the young: 30 percent of the reported cases were in patients under 30 years of age. The average age at presentation is about 40 years. More than 80 percent of pulmonary aneurysms affect the pulmonary trunk.

Most aneurysms of the pulmonary trunk appear to be due to pulmonary hypertension, which causes accumulation of acid mucopolysaccharides in its media. Sections of the affected artery show metachromasia with toluidine blue and stain positively with alcian blue. Excessive amounts of proteoglycans are present in these large elastic arteries in old age and in the rare Marfan's disease. Areas of media packed with acid mucopolysaccharide tend to become cystic.[140,141] The term *cystic medial necrosis* is classically applied to this lesion by pathologists, but it is an unfortunate designation because necrosis forms no part of this process. The development of this disease is serious, since the affected parts of the pulmonary trunk have the consistency of wet blotting paper at the same time that the vessel is exposed to increased tension. Rupture of the pulmonary trunk may occur under these conditions.

Early reports of such rupture were in cases of patent ductus arteriosus,[140,142] mitral stenosis,[143] chronic pulmonary embolism,[144] and primary pulmonary hypertension.[145] In Marfan's disease, in which an inborn error of metabolism leads to disruption of the elastic fibrils of the media, producing a so-called moth-eaten appearance due to excessive accumulation of acid mucopolysaccharide, the pulmonary trunk may rupture in the presence of a normal pulmonary arterial pressure.[146]

Early reports of pulmonary aneurysms suggested that about half were congenital.[139,147] Subsequent studies, however, have made it clear that congenital aneurysms of the pulmonary trunk and its main branches are in fact extremely rare. There is no doubt that most aneurysms so reported are due to the effects of pulmonary hypertension on a media weakened by accumulations of proteoglycans.

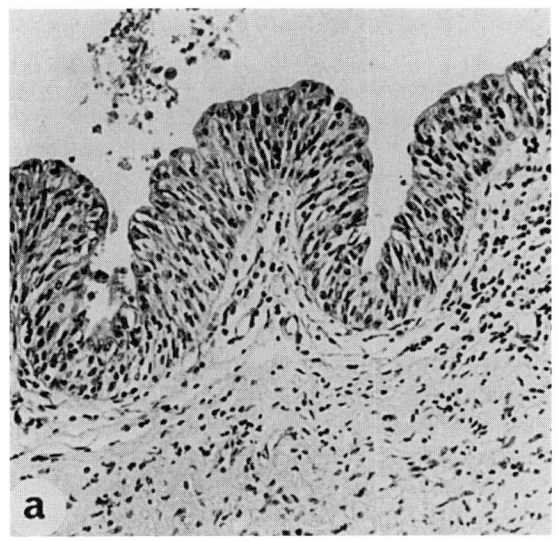

FIGURE 21-52
Metastatic growth of tumor along the lining of a large elastic pulmonary artery in a 58-year-old man who had a squamous cell carcinoma of the left renal pelvis. The tumor has metastasized to form a layer along the intima of the pulmonary artery. Unlike the primary growth, the metastasis shows the characteristic pattern of transitional epithelium. (H&E, ×140.)

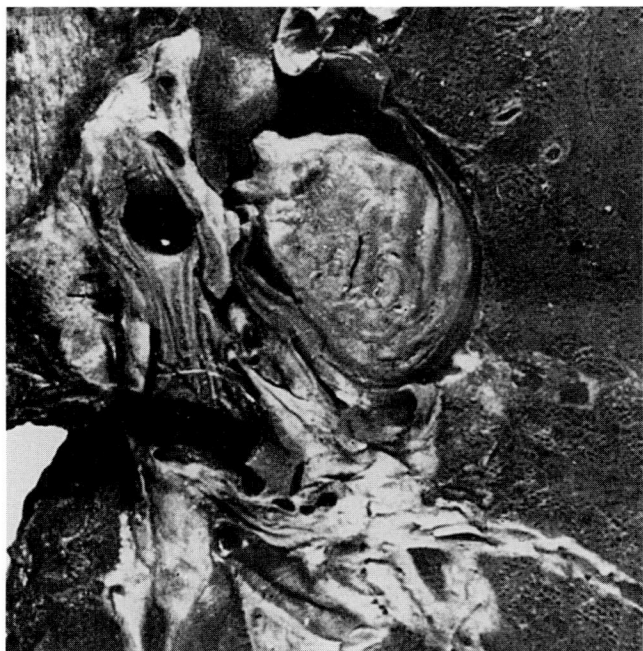

FIGURE 21-53
Mycotic aneurysm of a large intrapulmonary branch of the left main pulmonary artery in a 42-year-old man with subacute bacterial endocarditis of the tricuspid valve and a small ventricular septal defect. The aneurysm contains laminated thrombus.

Other rare causes of pulmonary aneurysm are syphilis,[147] Marfan's disease,[148] and trauma.[149] Mycotic aneurysms (Fig. 21-53) are more likely to affect large intrapulmonary arteries.[150,151] They are commonly secondary to congenital heart disease complicated by bacterial endocarditis.[151] Exceptionally, they may follow septicemia in the absence of any cardiovascular abnormality[152]; in these cases the wall of the pulmonary artery is damaged by such lesions as medial fibrosis and calcification.[152]

Dissecting aneurysm of the pulmonary arteries is rare. It, too, is usually due to pulmonary hypertension secondary to congenital heart disease, with the accumulation of acid mucopolysaccharides in the weakened media.[153]

CONGENITAL VASCULAR ANOMALIES

Developmental anomalies affect the larger vessels of the lung and may involve the pulmonary trunk or its main branches or the large pulmonary veins.

Pulmonary Trunk

The pulmonary trunk may show congenital idiopathic dilatation, congenital stenosis, or atresia. It may communicate with the aorta. Finally, it may give origin to the coronary arteries.[154] In some cases, dilatation of the pulmonary trunk may occur in the absence of pulmonary hypertension and cystic

medial necrosis of the wall of the vessel.[155] Congenital membranous stenosis may occur immediately above the pulmonary valve. Stenosis of the bifurcation has been reported and designated *coarctation of the pulmonary artery*.[156] Atresia may affect the root or whole length of the pulmonary trunk, which may thus be reduced to a cordlike structure, while the main pulmonary arteries remain recognizable as confluent vessels. In congenital atresia of the pulmonary valve, the pulmonary trunk is hypoplastic.

When the spiral (truncoconal) septum fails to develop within the truncus, this primitive vessel remains a single artery, with a single semilunar valve, which arises from the ventricular part of the heart and gives rise to the aorta, the coronary arteries, and, with one exception, the pulmonary arteries. There are various types, and the functional consequences of a *persistent truncus arteriosus* depend on the nature of the blood supply to the lung.[157] Readers are referred to the article by Collett and Edwards for more detailed information. An *aortopulmonary septal defect* may be regarded as a partial form of persistent truncus, the defect representing a focal absence of the truncoconal septum. Elsewhere the septum and its usual derivatives are formed, so that separate aortic and pulmonary valvular orifices occur, distinguishing this disorder from persistent truncus.

There may be *anomalous origin of the left coronary artery from the pulmonary trunk*, which leads to serious ischemia of the left ventricle in infancy. In one case[158] the left ventricle showed extensive myocardial fibrosis. In fetal life there is physiologic pulmonary hypertension, and blood flow may be from the pulmonary trunk into the left coronary artery. After birth, with a decrease in pressure in the pulmonary arteries, the pressure in the left coronary artery falls. The right coronary artery enlarges, and anastomoses develop between the two coronary circulations. The blood flow into the left coronary artery is largely from the right coronary artery into the left and then into the pulmonary trunk. This may lead to pulmonary hypertension and hypertrophy of the small pulmonary arteries.[158] In exceptional cases, there may be anomalous origins of the right coronary artery from the pulmonary trunk.

Pulmonary Arteries

The right pulmonary artery may arise below and to the left of the left pulmonary artery. As a result, the left artery has to cross the front of its right counterpart as it passes on its way to the left lung. For this reason, this congenital anomaly is usually referred to as *crossed pulmonary arteries*.[159]

Very rarely, the left pulmonary artery arises from the right main pulmonary artery proximal to its branches.[160,161] The left pulmonary artery passes back across the top of the right main bronchus, pressing down on it and perhaps leading to the retention of secretion in the right lung. The left pulmonary artery then curves to the left behind the trachea and in front of the esophagus. It is commonly referred to as a *vascular sling.* This developmental anomaly does not exert deleterious effects on the pulmonary circulation but is of clinical significance in obstructing the right bronchus.

Sometimes the pulmonary trunk runs as a single pulmonary artery to supply one of the lungs; the other lung receives an anomalous systemic supply that may originate from the ascending aorta or from the aortic arch by way of the ductus arteriosus or the innominate artery. In some of these cases, pulmonary hypertension may develop in the lung receiving the normal pulmonary arterial supply even when there is no associated congenital cardiac shunt. The explanation may be that subjects with unilateral absence of a pulmonary artery are exposed to the entire cardiac output from fetal life, when the pulmonary arteries are normally thick-walled.

Just as stenosis may affect the pulmonary trunk, so it may be found in its arterial branches. These congenital narrowings constitute a rare cause of pulmonary hypertension.[162] Four variants can be recognized: localized stenosis with poststenotic dilatation, segmental stenosis, diffuse tubular hypoplasia, and multiple peripheral stenoses. The stenosis is due to focal intimal proliferation and medial thickening of involved segments of vessel.[163] This developmental anomaly is particularly associated with congenital rubella. It is therefore not surprising that pulmonary arterial stenoses are commonly associated with other congenital cardiac anomalies, such as coarctation of the aorta and patent ductus arteriosus.

Pulmonary Veins

The lungs are derived by division of the foregut and share their blood supply, the splanchnic plexus, with it. This plexus drains into the systemic pre- and postcardial veins and the visceral veins of the abdomen, the umbilicovitelline system, with associated hepatic sinusoids. Thus, at this early stage of development, the veins of the primordia of the lungs do not drain directly into the heart but drain into the venous primordia that will form the superior vena cava, coronary sinus, innominate veins, proximal portion of the subclavian veins, portal vein, gastric veins, ductus venosus, and inferior vena cava. Later

on, out-pouching from the heart, termed the *common pulmonary vein*, extends into that part of the splanchnic plexus related to the lungs. It is necessary to understand this brief account of the normal development of the pulmonary veins to understand their common congenital anomalies.

Anomalous pulmonary venous drainage occurs when the pulmonary veins drain into the right atrium or the systemic or portal veins. It may be partial when some of the pulmonary venous drainage is into the left atrium or total when none of the drainage is normal. Partial anomalous venous drainage usually involves the veins of either upper pulmonary lobe; when it involves the right lung, the anomalous venous drainage tends to be into the right atrium or superior vena cava. That from the left lung is inclined to be into the left innominate vein and, much less commonly, into the coronary sinus. In total anomalous pulmonary venous drainage, an association with cardiac malformations is common.[164] Anatomic sites receiving anomalous pulmonary veins in order of descending frequency are the left innominate vein, right atrium, coronary sinus, right superior vena cava, portal vein, ductus venosus, inferior vena cava, and hepatic vein.[164]

When joining the right atrium, the anomalous veins may enter individually or first form a trunk. When the anomalous pulmonary veins enter a systemic vein, they converge superiorly to the left atrium in an extrapericardial location to form a vein or a chamber resembling the accessory chamber of cor triatriatum. In total anomalous pulmonary venous drainage, all the blood returning to the heart in systemic and pulmonary veins drains into the right atrium and thence through an atrial septal defect into the left atrium and systemic circulation. There is thus increased pulmonary blood flow. Pulmonary arterial hypertension complicates this anomaly.

In *cor triatriatum* an accessory atrial chamber is connected to the left atrium by an opening. The pulmonary veins drain into this chamber. It is possible that cor triatriatum is a malformation of the pulmonary venous system brought about by failure of the common pulmonary vein to become absorbed by differential growth into the left atrium. It remains instead as the accessory chamber. The hemodynamic and pathologic effects of cor triatriatum are like those of mitral stenosis.

In *the scimitar syndrome*, a hypoplastic bilobar right lung with a systemic arterial supply occurs in association with anomalous pulmonary venous drainage.[165] There is dextroposition of the heart. While the left pulmonary veins usually connect normally with the left atrium, the right pulmonary veins

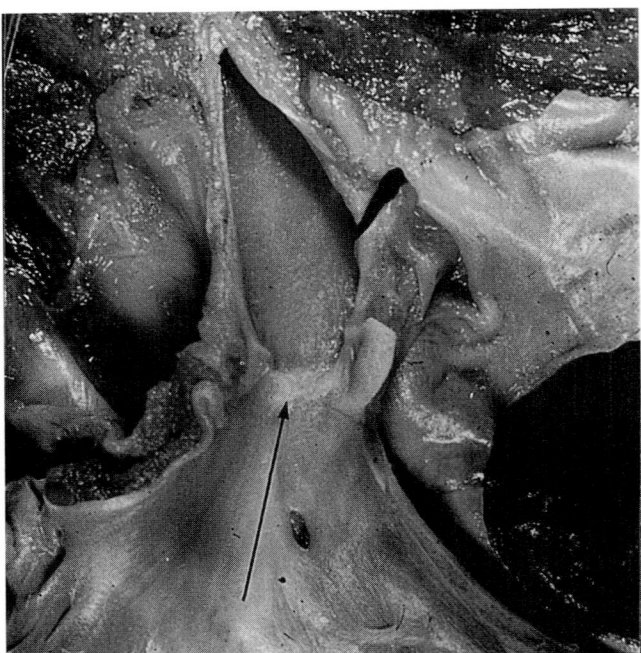

FIGURE 21-54
Congenital stenosis of the pulmonary vein, characteristically restricted to the venoatrial junction, indicated by an arrow.

FIGURE 21-55
Part of a bronchovascular corrosion cast showing multiple small arteriovenous fistulae of the lungs in a 17-year-old boy. In this single fistula from the specimen there is a peripheral plexus, which is flattened on its pleural surface. It has a stellate appearance, with many pulmonary arteries supplying it from the periphery and converging toward a central pulmonary vein. Several separate veins drain it from the opposite side. One component vessel of the plexus is greatly enlarged compared to the rest. (×10.)

drain into the inferior vena cava at the level of the diaphragm. This anomalous vein can be recognized on a chest radiograph as a crescent-shaped shadow resembling the scimitar, and hence the name of the syndrome. There is a pulmonary arterial supply to the upper lobe of the right lung, while one or more systemic arteries from the descending aorta supply the lower lobe. In the area of lung supplied by the aorta, the pulmonary arteries may show hypertensive changes, including plexiform lesions.[165]

Stenosis of pulmonary veins with normal connections may occur (Fig. 21-54). In some patients only a single vein or a few veins may be affected, but in others many or all of the pulmonary veins are stenotic.[166] The stenoses are the result of localized intimal fibrous proliferation, and they are characteristically restricted to the venoatrial junction. Sometimes, however, they are situated in the course of the pulmonary veins. Such obstructions in the pulmonary veins lead to pulmonary venous and arterial hypertension, with corresponding histologic changes in the lung parenchyma and pulmonary blood vessels.

Pulmonary Arteriovenous Fistula

Arteriovenous fistula of the lung, sometimes mistakenly referred to as a cavernous hemangioma, is probably the most common malformation of the pulmonary circulation. The fistulae vary in size from minute collections of dilated vessels, referred to as pulmonary telangiectases, to large vascular forma-

tions (Figs. 21-55 and 21-56). Most occur in the lower lobes. When the large fistulae are studied by means of a vinylite-plastic injection technique, they are found to consist of a distended thin-walled artery, distended efferent veins, and an intervening labyrinth of distended vessels.[167] A whole lobe may be replaced by a fistula; then the pulmonary artery supplying it is abnormally large, and both it and the affected lobe pulsate freely.[168]

The arteries forming the fistula are enlarged and may be tortuous. Vessels of more than one bronchopulmonary segment may be involved, and there is no regularity in the number and arrangement of these arteries and veins. Two separate aneurysmal loops may be present. The draining veins are larger than the arteries. There is no significant contribution from bronchial arteries. Histologically, the pulmonary arteriovenous fistula consists of cavernous, blood-filled spaces linked to one another. The vascular channels, which resemble pulmonary veins more than arteries, are up to 1.5 cm in diameter. They frequently show severe intimal fibrosis, and some contain organizing thrombi. They commonly rupture to give rise to fresh hemorrhage and hemosiderosis in the surrounding lung substance.

Patients with pulmonary arteriovenous fistulae often show vascular nevi elsewhere on the body—for

FIGURE 21-56
Part of the corrosion cast from the same case illustrated in Fig. 21-55. One large and two small plexuses are viewed from their venous side. Pulmonary arteries on the far side of the plexus are out of focus. (×20.)

example, on the lips, face, and soft palate—suggesting that many are a manifestation of Osler-Rendu-Weber's disease[168]; this is especially true when they are multiple. Large fistulae produce cyanosis as a result of deoxygenated blood's shunting directly from the pulmonary arteries into the pulmonary veins without first traversing the pulmonary capillary bed. Polycythemia and finger clubbing follow the systemic hypoxemia. High blood flow through the fistula leads to systolic murmurs and thrills. The fistulae may be detected on chest radiographs as collections of wormlike shadows extending from the hilum. The fistulae may be asymptomatic but may lead to fatal hemorrhage. If they become infected, they may give rise to a cerebral abscess.

REFERENCES

1. Brenner O: Pathology of the vessels of the pulmonary circulation. *Arch Intern Med* 56:211–237, 457–497, 724–752, 976–1014, 1189–1241, 1935.
2. Caslin AW, Heath D, Madden B, Yacoub M, Gosney JR, Smith P: The histopathology of 36 cases of plexogenic pulmonary arteriopathy. *Histopathology* 16:9–19, 1990.
3. Keith JD: Improved prognosis in ventricular septal defect, in Kidd BSL, Keith JD (eds), *The Natural History and Progress in Treatment of Congenital Heart Defects.* Springfield, Ill. Charles C Thomas, 1971, p 5, chap. 1.
4. Heath D, Edward JE: The pathology of hypertensive pulmonary vascular disease: A description of six grades of structural changes in the pulmonary arteries with special reference to congenital cardiac septal defect. *Circulation* 18:533–547, 1958.
5. Hatano S, Strasser T (eds): Classification and nomenclature. In *Primary Pulmonary Hypertension: Report on a WHO Meeting, October 15–17, 1973.* Geneva, World Health Organization, 1975, p 45.
6. Civin WH, Edwards JE: The postnatal structural changes in the intrapulmonary arteries and arterioles. *Arch Pathol* 51:192–200, 1951.
7. Heath D, Smith P, Gosney J: Ultrastructure of early plexogenic pulmonary arteriopathy. *Histopathology* 12:41–52, 1988.
8. Rabinovitch M: Elastase in the pathophysiology of pulmonary vascular disease. Paper presented to the International Conference on the Pulmonary Vasculature in Health and Disease, National Heart and Lung Institute, London, September 7–9, 1992.
9. Heath D, Yacoub M, Gosney JR, Madden B, Caslin AW, Smith P: Pulmonary endocrine cells in hypertensive pulmonary vascular disease. *Histopathology* 16:21–28, 1990.
10. Smith P, Heath D, Yacoub M, Madden B, Caslin A, Gosney J: The ultrastructure of plexogenic pulmonary arteriopathy. *J Pathol* 160:111–121, 1990.
11. Heath D, Helmholz FH Jr, Burchell HB, DuShane JW, Edwards JE: Graded pulmonary vascular changes and hemodynamic findings in cases of ventricular and atrial septal defect and patent ductus arteriosus. *Circulation* 18:1155–1166, 1958.
11a. Jamison BM, Michel RP: Different distribution of plexiform lesions in primary and secondary pulmonary hypertension. *Hum Pathol* 26:987–993, 1995.
12. Heath D, Helmholz FH Jr, Burchell HB, DuShane JW, Kirklin JW, Edwards JE: Relation between structural changes in the small pulmonary arteries and immediate reversibility of pulmonary hypertension following closure of ventricular and atrial septal defects. *Circulation* 18:1167–1174, 1958.
13. Heath D, Smith P: Plexiform lesions with giant cells. *Thorax* 37:394–395, 1982.
14. Heath D, Yacoub M: Lung mast cells in plexogenic pulmonary arteriopathy. *J Clin Pathol* 44:1003–1006, 1991.
15. Heath D, Smith P: Nodules resembling arachnoid villi in pulmonary venules in plexogenic pulmonary arteriopathy. *Cardioscience* 3:161–165, 1992.
16. Harris P, Heath D: The structure of the normal pulmonary blood vessels after infancy, in, *The Human Pulmonary Circulation,* 3d ed. Edinburgh, Churchill Livingstone, 1986, pp 30–47.
17. Wilkinson M, Langhorne CA, Heath D, Barer GR, Howard P: A pathophysiological study of 10 cases of hypoxic cor pulmonale. *Q J Med* 66:65–85, 1988.
18. Arias-Stella J, Saldaña M: The terminal portion of the pulmonary arterial tree in people native to high altitudes. *Circulation* 28:915–925, 1963.
19. Heath D, Williams D, Rios-Dalenz J, Calderon M, Gosney J: Small pulmonary arterial vessels of Aymara Indians from the Bolivian Andes. *Histopathology* 16:565–571, 1990.
20. Heath D, Williams D: Pulmonary vascular remodelling in a high-altitude Aymara Indian. *Int J Biometeorol* 35:203–207, 1991.
21. Heath D, Smith P, Rios-Dalenz J, Williams D, Harris P: Small pulmonary arteries in some natives of La Paz, Bolivia. *Thorax* 36:599–604, 1981.

22. Smith P, Rodgers B, Heath D, Yacoub M: The ultrastructure of pulmonary arteries and arterioles in emphysema. *J Pathol* 167:69–75, 1992.

23. Dunnill MS: An assessment of the anatomical factor in cor pulmonale in emphysema. *J Clin Pathol* 14:246–258, 1961.

24. Williams D, Heath D, Gosney J, Rios-Dalenz J: Pulmonary endocrine cells of Aymara Indians from the Bolivian Andes. *Thorax* 48:52–56, 1993.

25. Wagenvoort CA: The pulmonary veins in hypoxia, in Heath D (ed), *Aspects of Hypoxia*. Liverpool, Liverpool University Press, 1986, pp 19–27.

26. Wagenvoort CA, Wagenvoort N: Pulmonary veins in high altitude residents: A morphometric study. *Thorax* 37:931–935, 1982.

27. Heath D: Effects of pulmonary vasodilators on the remodelled pulmonary arterial tree in chronic alveolar hypoxia. *Eur Respir J* 3:1–2, 1990.

28. Grover RF, Vogel JHK, Voight GC, Blount SG Jr: Reversal of high altitude pulmonary hypertension. *Am J Cardiol* 18:928–932, 1966.

29. Hartley LH, Alexander JK, Modelski M, Grover RF: Subnormal cardiac output at rest and during exercise in residence at 3100 m altitude. *J Appl Physiol* 23:839–848, 1967.

29a. Barnes PJ, Belvisi MG: Nitric oxide and lung disease. *Thorax* 48:1034–1043, 1993.

30. Heath D: The rat is a poor animal model for the study of human pulmonary hypertension. *Cardioscience* 3:1–6, 1992.

31. Fay FS, Delise CM: Contraction of isolated smooth muscle cells: Structural changes. *Proc Natl Acad Sci USA* 70:641–645, 1973.

32. Heath D, Smith P: Electron microscopy of hypertensive pulmonary vascular disease. *Br J Dis Chest* 77:1–13, 1983.

33. Wagenvoort CA, Wagenvoort N: Primary pulmonary hypertension: A pathologic study of the lung vessels in 156 clinically diagnosed cases. *Circulation* 42:1163–1184, 1970.

34. Harris P, Heath D: Primary pulmonary hypertension, in, *The Human Pulmonary Circulation*, 3d ed. Edinburgh, Churchill Livingstone, 1986, pp 414–432.

35. Fleming H: Primary pulmonary hypertension in eight patients including a mother and her daughter. *Aust Ann Med* 9:18–28, 1960.

36. Thompson P, McRae C: Familial pulmonary hypertension: Evidence of autosomal dominant inheritance. *Br Heart J* 32:758–760, 1970.

37. Hood WB Jr, Spencer H, Lass RW, Daley R: Primary pulmonary hypertension: Familial occurrence. *Br Heart J* 30:336–343, 1968.

38. Gurtner HP: Hypertensive pulmonary vascular disease: Some remarks on its incidence and etiology. In *Proceedings of the 12th Annual Meeting of the European Society for the Study of Drug Toxicity, Uppsala, Sweden, June 1970*. Amsterdam, Excerpta Medica, 1971, pp 81–88.

39. Kay JM, Smith P, Heath D: Aminorex and the pulmonary circulation. *Thorax* 26:262–270, 1971.

40. Byrne-Quinn E, Grover RF: Aminorex (Menocil) and amphetamine: Acute and chronic effects on pulmonary and systemic haemodynamics in the calf. *Thorax* 27:127–131, 1972.

41. Kay JM, Heath D: Observations on the pulmonary arteries and heart weight of rats fed on Crotalaria spectabilis seeds. *J Pathol Bacteriol* 92:385–394, 1966.

42. Kay JM, Heath D, Smith P, Bras G, Summerell J: Fulvine and the pulmonary circulation. *Thorax* 26:249–261, 1971.

43. Burns J: The heart and pulmonary arteries in rats fed on *Senecio jacobaea*. *J Pathol* 106:187–194, 1972.

44. Heath D, Shaba J, Williams A, Smith P, Kombe A: A pulmonary hypertension–producing plant from Tanzania. *Thorax* 30:399–404, 1975.

45. Saunders JB, Constable DJ, Heath D, Smith P, Paton A: Pulmonary hypertension complicating portal vein thrombosis. *Thorax* 34:281–283, 1979.

46. Mantz FA Jr, Craige E: Portal axis thrombosis with spontaneous porta-caval shunt and resultant cor pulmonale. *Arch Pathol* 52:91–97, 1951.

47. Kerbel NC: Pulmonary hypertension and portal hypertension. *Can Med Assoc J* 87:1022–1026, 1962.

48. Segel N, Kay JM, Bayley TJ, Paton A: Pulmonary hypertension with hepatic cirrhosis. *Br Heart J* 30:575–578, 1968.

49. Chun PKC, San Antonio RP, Davia JE: Laennec's cirrhosis and primary pulmonary hypertension. *Am Heart J* 99:779–782, 1980.

50. Naeye RL: "Primary" pulmonary hypertension with coexisting portal hypertension: A retrospective study of six cases. *Circulation* 22:376–384, 1960.

51. Sui GJ, Liu YH, Cheng XS, Anand IS, Harris E, Harris P, Heath D: Subacute infantile mountain sickness. *J Pathol* 155:161–170, 1988.

52. Heath D, Harris P, Sui GJ, et al: Pulmonary blood vessels and endocrine cells in subacute infantile mountain sickness. *Respir Med* 83:77–81, 1989.

53. Smith P, Heath D: The ultrastructure of age-associated intimal fibrosis in pulmonary blood vessels. *J Pathol* 130:247–253, 1980.

54. Mallory TB: Case records of the Massachusetts General Hospital. Case 23511. *N Engl J Med* 217:1045–1049, 1937.

55. Brewer DB, Humphreys DR: Primary pulmonary hypertension with obstructive venous lesions. *Br Heart J* 22:445–448, 1960.

56. Heath D, Segel N, Bishop J: Pulmonary veno-occlusive disease. *Circulation* 34:242–248, 1966.

57. Wagenvoort CA, Losekoot G, Mulder E: Pulmonary veno-occlusive disease of presumably intrauterine origin. *Thorax* 26:429–434, 1971.

58. Dail DH, Liebow AA, Gmelich J, Carrington CB: A study of 43 cases of pulmonary veno-occlusive disease. *Lab Invest* 38:340–350, 1978.

59. Heath D, Scott O, Lynch J: Pulmonary veno-occlusive disease. *Thorax* 26:663–674, 1971.

60. Stovin PGI, Mitchinson MJ: Pulmonary hypertension due to obstruction of the intrapulmonary veins. *Thorax* 20:106–113, 1965.

61. Wagenvoort CA, Wagenvoort N: The pathology of pulmonary veno-occlusive disease. *Virchows Arch [A]* 364:69–79, 1974.

62. McDonnell PJ, Summer WR, Hutchins GM: Pulmonary veno-occlusive disease: Morphological changes suggesting a viral etiology. *JAMA* 246:667–671, 1981.

63. Crissman JD, Koss M, Carson RP: Pulmonary veno-occlusive disease secondary to granulomatous venulitis. *Am J Surg Pathol* 4:93–99, 1980.

64. Wagenvoort CA: Pulmonary veno-occlusive disease: Entity or syndrome? *Chest* 69:82–86, 1976.

65. Hasleton PS, Ironside JW, Whittaker JS, Kelly W, Ward C, Thompson GS: Pulmonary veno-occlusive disease: A report of four cases. *Histopathology* 10:933–944, 1986.

66. Crane JT, Grimes OF: Isolated pulmonary venous sclerosis: A cause of cor pulmonale. *J Thorac Cardiovasc Surg* 40:410–416, 1960.

67. Wagenvoort CA: Vasoconstrictive primary pulmonary hypertension and pulmonary veno-occlusive disease. *Cardiovasc Clin* 4:97–113, 1972.

68. BMJ Annotation: Pulmonary veno-occlusive disease. *Br Med J* 3:369, 1972.

69. Brown CH, Harrison CV: Pulmonary veno-occlusive disease. *Lancet* 2 61–65, 1966.

70. Corrin B, Spencer H, Turner-Warwick M, Beales SH, Hampslin JJ: Pulmonary veno-occlusion: An immune complex disease? *Virchows Arch [A]* 364:81–91, 1974.

71. Kay JM, Edwards FR: Ultrastructure of the alveolar-capillary wall in mitral stenosis. *J Pathol* 111:239–245, 1973.

72. Rose AG: Pulmonary veno-occlusive disease due to bleomycin therapy for lymphoma. *S Afr Med J* 64:636–638, 1983.

73. Joselson R, Warnock M: Pulmonary veno-occlusive disease after chemotherapy. *Hum Pathol* 14:88–91, 1983.

74. Rose AG, Learmonth GM, Benatar SR: Pulmonary veno-occlusive disease associated with hypertrophic cardiomyopathy. *Arch Pathol Lab Med* 108:267–268, 1984.

75. Voordes CG, Kuipers JRG, Elema JD: Familial pulmonary veno-occlusive disease: A case report. *Thorax* 32:763–766, 1977.

76. Davies P, Reid L: Pulmonary veno-occlusive disease in siblings: Case reports and morphometric study. *Hum Pathol* 13:911–915, 1982.

77. Case Records of the Massachusetts General Hospital: *N Engl J Med* 308:823–834, 1983.

78. Canny GJ, Arbus GS, Wilson GJ, Newth CJL: Fatal pulmonary hypertension following renal transplantation. *Br J Dis Chest* 79:191–195, 1985.

79. Kay JM, de Sa DJ, Mancer JFK: Ultrastructure of lung in pulmonary veno-occlusive disease. *Hum Pathol* 14:451–456, 1983.

80. Thadani U, Burrow C, Whitaker W, Heath D: Pulmonary veno-occlusive disease. *Q J Med* 44:133–159, 1975.

81. Goodwin JF, Harrison CV, Wilcken DEL: Obliterative pulmonary hypertension and thromboembolism. *Br Med J* 1:777–783, 1963.

82. Morrell MT, Dunnill MS: The postmortem incidence of pulmonary embolism in a hospital population. *Br J Surg* 55:347–352, 1968.

83. Harris P, Heath D: Pulmonary embolism, in, *The Human Pulmonary Circulation,* 3d ed. Edinburgh, Churchill Livingstone, 1986, pp 545–568.

84. Johnston B, Smith P, Heath D: Experimental cotton-fibre pulmonary embolism in the rat. *Thorax* 36:910–916, 1981.

85. Best PV, Heath D: Pulmonary thrombosis in cyanotic congenital heart disease without pulmonary hypertension. *J Pathol Bacteriol* 75:281–291, 1958.

86. Heath D, DuShane JW, Wood EH, Edward JE: The aetiology of pulmonary thrombosis in cyanotic congenital heart disease with pulmonary stenosis. *Thorax* 13:213–217, 1958.

87. Heath D, Wood EH, DuShane JW, Edwards JE: The structure of the pulmonary trunk at different ages and in cases of pulmonary hypertension and pulmonary stenosis. *J Pathol Bacteriol* 77:443–456, 1959.

88. Harris P, Heath D: Structural changes in the lung associated with pulmonary venous hypertension, in, *The Human Pulmonary Circulation*, 3d ed. Edinburgh, Churchill Livingstone, 1986, pp 329–344.

89. Hatt PY, Rouiller C: Les ultrastructures pulmonaires et le régime de la petite circulation: I. Au cours du rétrécissement mitral serré. *Semin Hôp Paris* 17:1371–1397, 1958.

90. Gough J: Correlation of radiological and pathological changes in some disease of the lung. *Lancet* 1:161–162, 1955.

91. Fleischner FG, Reiner L: Linear x-ray shadows in acquired pulmonary hemosiderosis and congestion. *N Engl J Med* 250:900–905, 1954.

92. Rossall RE, Gunning AL: Basal horizontal lines on chest radiograph: significance in heart disease. *Lancet* 1:604–606, 1956.

93. Heath D, Trueman T, Sukonthamarn P: Pulmonary mast cells in mitral stenosis. *Cardiovasc Res* 3:467–471, 1969.

94. Whitaker W, Black A, Warrack AJN: Pulmonary ossification in patients with mitral stenosis. *J Fac Radiol* 7:29–34, 1955.

95. Sharp ME, Danino EA: An unusual form of pulmonary calcification: "Microlithiasis alveolaris pulmonum." *J Pathol Bacteriol* 65:389–399, 1953.

96. Lendrum AC: Fibrinous vasculosis. *J Clin Pathol* 8:180–185, 1955.

97. Symmers WStC: Necrotizing pulmonary arteriopathy associated with pulmonary hypertension. *J Clin Pathol* 5:36–41, 1952.

98. Hicks JD: Acute arterial necrosis in the lungs. *J Pathol Bacteriol* 65:333–343, 1953.

99. Spain DM: Necrotizing and healing pulmonary arteritis with advanced mitral stenosis. *Arch Pathol* 62:489–493, 1956.

100. Harrison CV: The pathology of the pulmonary vessels in pulmonary hypertension. *Br J Radiol* 31:217–226, 1958.

101. Heath D, Gillund TD, Kay JM, Hawkins CF: Pulmonary vascular disease in honeycomb lung. *J Pathol Bacteriol* 95:423–430, 1968.

102. Wells AL: Pulmonary vascular changes in coal-worker's pneumoconiosis. *J Pathol Bacteriol* 68:573–587, 1954.

103. Schepers GWH: Comparative vascular pathology of occupational chest disease. *Arch Indust Health* 12:7–25, 1955.

104. Heath D, Mooi W, Smith P: The pulmonary vasculature in haematite lung. *Br J Dis Chest* 72:88–94, 1978.

105. Cudkowicz L: The blood supply of the lung in pulmonary tuberculosis. *Thorax* 7:270–276, 1952.

106. Wagenaar SSC, Westermann CJJ, Corrin B: Giant cell arteritis limited to large elastic pulmonary arteries. *Thorax* 36:876–877, 1981.

107. Lupi-Herrera E, Sanchez-Tores G, Marcushamer J, Mispireta J, Horwitz S, Vela JE: Takayasu's arteritis: Clinical study of 107 cases. *Am Heart J* 93:94–103, 1977.

108. Doyle L. McWilliam L, Hasleton PS: Giant cell arteritis with pulmonary involvement. *Br J Dis Chest* 82:88–92, 1988.

109. Shaw AFB, Ghareeb AA: The pathogensis of pulmonary schistosomiasis in Egypt with special reference to Ayerza's disease. *J Pathol Bacteriol* 46:401–424, 1938.

110. Marchand EJ, Marcial-Rojas RA, Rodríguez R, Polanco G, Díaz-Rivera RS: The pulmonary obstruction syndrome in *Schistosoma mansoni* pulmonary endarteritis. *Arch Intern Med* 100:965–980, 1957.

111. Garcia-Palmieri N: Cor pulmonale due to *Schistosoma mansoni*. *Am Heart J* 68:714–715, 1964.

112. De Faria JL: Pulmonary vascular changes in schistosomal cor pulmonale. *J Pathol Bacteriol* 68:589–602, 1954.

113. Chaves E: The pathology of the arterial pulmonary vasculature in Manson's schistosomiasis. *Dis Chest* 50:72–77, 1966.

114. Cavalcanti I de L, Tompson G, de Souza N, Barbosa FS: Pulmonary hypertension in schistosomiasis. *Br Heart J* 24:363–371, 1962.

115. Al-Naaman YD, Shamma AH, Damluji SF, El-Sayed HM: Angiologic manifestation of cardiopulmonary schistosomiasis, "bilharziasis." *Angiology* 17:40–45, 1967.

116. Obeyesekere I, De Soysa N: "Primary" pulmonary hypertension, eosinophilia, and filariasis in Ceylon. *Br Heart J* 32:524–536, 1970.

117. Obeyesekere I, Peiris D: Pulmonary hypertension and filariasis. *Br Heart J* 36:676–681, 1974.

118. Dissanaike AS, Paramananthan DC: On *Brugia* (*Brugiella* subgen. nov.) *buckleyi* n.sp., from the heart and blood vessels of the Ceylon hare. *J Helminthol* 35:209–220, 1961.

119. Patterson DF, Luginbuhl HR: Clinico-pathologic conference: *Dirofilaria immitis* infection in dog. *J Am Vet Med Assoc* 143:619–628, 1963.

120. Hamilton JM: Pulmonary arterial disease of the cat. *J Comp Pathol* 76:133–145, 1966.

121. Noriega AR, Toxic Epidemic Syndrome Study Group: Toxic epidemic syndrome, Spain, 1981. *Lancet* 2:697–702, 1982.

122. Alonso-Ruiz A, Calabozo M, Pérez-Ruiz F, Mancebo L: Toxic oil syndrome. *Medicine* 72:285–295, 1993.

123. Phelps RG, Fleischmajer R: Clinical, pathologic and immunopathologic manifestations of the toxic oil syndrome. *J Am Acad Dermatol* 18:313–324, 1988.

124. Martinez-Tello FJ, Navas-Palacios JJ, Ricoy JR, et al: Pathology of a new toxic syndrome caused by ingestion of adulterated oil in Spain. *Virchows Arch* 397:261–285, 1982.

125. Fernández-Segoviano P, Esteban A, Martínez-Cabruja R: Pulmonary vascular lesions in the toxic oil syndrome in Spain. *Thorax* 38:724–729, 1983.

126. Lopez-Sendon J, Coma-Canella I: Pulmonary hypertension following ingestion of toxic cooking oil. Paper presented to the Symposium on Primary Pulmonary Hypertension, European Society of Cardiology, Vienna, 1982.

127. Smith P, Heath D, Rodgers B, Helliwell T: Pulmonary vasculature in Fabry's disease. *Histopathology* 19:567–569, 1991.

128. Wagenvoort CA, Beetstra A, Spijker J: Capillary haemangiomatosis of the lungs. *Histopathology* 2:401–406, 1978.

129. Whittaker JS, Pickering CAC, Heath D, Smith P: Pulmonary capillary hemangiomatosis. *Diagn Histopathol* 6:77–84, 1983.

130. Heath D, Reid R: Invasive pulmonary haemangiomatosis. *Br J Dis Chest* 79:284–294, 1985.

131. Heath D, Mackinnon J: Pulmonary hypertension due to myxoma of the right atrium, with special reference to the behavior of emboli of myxoma in the lung. *Am Heart J* 68:227–235, 1964.

132. Fleming HA, Whitwell F, Heath D: Clinical-pathologic conference: Congestive cardiac failure complicating left atrial myxoma. *Am Heart J* 83:258–264, 1972.

133. Wagenvoort CA, Heath D, Edwards JE: Neoplasms and hamartomas of the pulmonary vasculature, in Wagenvoort CA, Heath D, Edwards JE (eds), *The Pathology of the Pulmonary Vasculature*. Springfield, Ill., Charles C Thomas, 1964, pp 440–477.

134. Elphinstone RH, Spector RG: Sarcoma of the pulmonary artery. *Thorax* 14:333–340, 1959.

135. Wolf PL, Dickenman RC, Langston JD: Fibrosarcoma of the pulmonary artery masquerading as a pheochromocytoma. *Am J Clin Pathol* 34:146–154, 1960.

136. Kidd BSL, Carson DJL, Lamont ES: Intra-atrial sarcoma. *Br Med J* 2:1476–1478, 1961.

137. Pryce DM, Walter JB: The frequency of gross vascular invasion in lung cancer with special reference to arterial invasion. *J Pathol Bacteriol* 79:141–146, 1960.

138. Wagenvoort CA, Morrow GE, Ten Cate HW: Squamous cell carcinoma of the renal pelvis with muco-epidermoid metastasis. *J Urol* 85:727–731, 1961.

139. Deterling RA, Clagett OT: Aneurysm of the pulmonary artery: Review of the literature and report of a case. *Am Heart J* 34:471–499, 1947.

140. Whitaker W, Heath D, Brown JW: Patent ductus arteriosus with pulmonary hypertension. *Br Heart J* 17:121–137, 1955.

141. Wade G, Ball J: Unexplained pulmonary hypertension. *Q J Med* 26:83–120, 1957.

142. Moench GL: Aneurysmal dilatation of pulmonary artery with patent ductus arteriosus. *JAMA* 82:1672–1673, 1924.

143. Thomas GC, Whitelaw DM, Taylor HE: Rupture of the pulmonary artery complicating rheumatic mitral stenosis. *Arch Pathol* 60:99–103, 1955.

144. Brettell HR, Hermann RE: Spontaneous rupture of the pulmonary artery in pulmonary hypertension. *Am Heart J* 59:263–276, 1960.

145. Rawson AJ: Incomplete rupture of the pulmonary artery based on cystic medionecrosis. *Am J Heart* 55:766–771, 1958.

146. Austin MG, Schaefer RF: Marfan's syndrome with unusual blood vessel manifestations: Primary medionecrosis, dissection of right innominate, right carotid and left carotid arteries. *Arch Pathol* 64:205–209, 1957.

147. Boyd LJ, McGavack TH: Aneurysm of the pulmonary artery: A review of the literature and report of two cases. *Am Heart J* 18:562–578, 1939.

148. Tung HL, Liebow AA: Marfan's syndrome: Observations at necropsy, with special reference to medionecrosis of the great vessels. *Lab Invest* 1:382–406, 1952.

149. Marble HC, White PD: Traumatic aneurysm of the pulmonary artery. *JAMA* 74:1778, 1920.

150. Charlton RW, du Plessis IA: Multiple pulmonary artery aneurysms. *Thorax* 16:364–371, 1961.

151. Davis BT, Davison PH, Heath D: Mycotic aneurysm of left main pulmonary artery. *Am Heart J* 65:261–266, 1963.

152. Deuvaert FE, Bouton JM, Goffin Y, Primo G: Mycotic aneurysm of the left main pulmonary artery in an infant. *J Cardiovasc Surg* 22:68–71, 1981.

153. Crumpton M: Congenital heart disease and dissecting aneurysm of pulmonary artery. *Br Med J* 1:1303, 1950.

154. Harris P, Heath D: Developmental anomalies of the large pulmonary blood vessels, in, *The Human Pulmonary Circulation*, 3d ed. Edinburgh, Churchill Livingstone, 1986, pp 12–27.

155. Greene DG, Baldwin E, Baldwin JS, Himmelstein A, Roh CE, Cournand A: Pure congenital pulmonary stenosis and idiopathic congenital dilatation of the pulmonary artery. *Am J Med* 6:24–40, 1949.

156. Søndergaard T: Coarctation of the pulmonary artery. *Dan Med Bull* 1:46–48, 1954.

157. Collett RW, Edwards JE: Persistent truncus arteriosus: Classification according to anatomical types. *Surg Clin North Am* 29:1245–1270, 1949.

158. Arnott M, Kearney MS, Heath D: Atrial septal defect with anomalous origin of the left coronary artery from the pulmonary trunk. *Am Heart J* 85:113–121, 1973.

159. Jue KL, Lockman LA, Edwards JE: Anomalous origins of pulmonary arteries from pulmonary trunk ("crossed pulmonary arteries"): Observations in a case with 18 trisomy syndrome. *Am Heart J* 71:807–812, 1966.

160. Niwayama G: Unusual vascular ring formed by the anomalous left pulmonary artery with tracheal compression. *Am Heart J* 59:454–461, 1960.

161. Jue KL, Raghib G, Amplatz K, Adams P Jr, Edwards JE:

Anomalous origin of the left pulmonary artery from the right pulmonary artery: A report of 2 cases and review of the literature. *Am J Roentgenol* 95:598–610, 1965.

162. D'Cruz IA, Agustsson MH, Bicoff JP, Weinberg M, Arcilla RA: Stenotic lesions of the pulmonary arteries: Clinical and hemodynamic findings in 84 cases. *Am J Cardiol* 13:441–450, 1964.

163. MacMahon HE, Lee HY, Stone PA: Congenital segmental coarctation of pulmonary arteries: An anatomic study. *Am J Pathol* 50:15–25, 1967.

164. Burroughs JT, Edwards JE: Total anomalous pulmonary venous connection. *Am Heart J* 59:913–931, 1960.

165. Neill CA, Ferencz C, Sabiston DC, Sheldon H: The familial occurrence of hypoplastic right lung with systemic arterial supply and venous drainage: "Scimitar syndrome." *Bull Johns Hopkins Hosp* 107:1–21, 1960.

166. Edwards JE: Congenital stenosis of pulmonary veins: Pathologic and developmental considerations. *Lab Invest* 9:46–66, 1960.

167. Lindskog GE, Liebow AA, Kausel H, Janzen A: Pulmonary arteriovenous aneurysm. *Ann Surg* 132:591–606, 1950.

168. Whitaker W: Cavernous haemangioma of the lung. *Thorax* 2:58–64, 1947.

Chapter 22

The Pathology of Bronchial Asthma

William R. Roche

The definition of bronchial asthma is a topic of continuing debate, which is of considerable importance both in clinical practice and as an epidemiologic tool. There is reasonable consensus that asthma is a disorder of reversible airways obstruction, associated with mucosal inflammation and increased responsiveness of the airways.[1] The reversibility may not be evident or persist in individual patients and bronchial hyperactivity is not a constant finding. The prevalence of bronchial asthma is increasing in most westernized countries.[2,3] Mortality continues to rise[4] despite, or even because of,[5] the development of new pharmacologic agents. There has been a major reduction in asthma morbidity as well as continued reduction in asthma mortality in New Zealand.[5a] A review of asthma outbreaks has been described in a recent editorial.[5b]

The pathologist may encounter asthma in a number of settings: presenting as unexplained sudden death or as fatal severe steroid-resistant asthma, in cytologic material where *Aspergillus* hyphae may also be sought, and in endoscopic biopsies taken as part of the diagnostic evaluation of a patient or for monitoring the effects of therapy. The increasing importance of such "medical" diseases in the workload of the "surgical" pathologist is a trend common to all organ systems. It reflects technical advances in endoscopic and other biopsy techniques and the demand for more specific diagnostic information to guide the choice of therapy and to monitor the progress of disease. Furthermore, increased understanding of the pathologic basis of asthma has directly influenced changes in treatment[6] and in the future may contribute to the prevention of this all too common disease.

The changes that occur in the airways in asthma and the mechanisms where they contribute to obstruction are reviewed. The complications of asthma and the necropsy appearances in asthma deaths are also discussed.

BRONCHIAL EPITHELIUM

The bronchial epithelium has a number of important roles to play in airway homeostasis: to act as a barrier to infective and noxious agents, to actively contribute to host defense mechanisms, to warm and humidify inspired air, and to play a role in inflammatory and neural networks. This major interface between the host and the environment is exposed to a wide variety of insults, including pollutants, viruses, and inhaled allergens. The shedding of the bronchial epithelium in asthmatics has been recognized by manifestations such as Creola bodies, which are epithelial clusters in sputum,[7] and the appearance of only residual basal cells remaining attached to the basement membrane in postmortem histology with the accumulation of sloughed epithelial cells in the bronchial lumen.[8]

The conducting airways are lined by a complex epithelium composed of superficial ciliated, goblet, and serous cells and underlying basal cells. This has been regarded as a "pseudostratified" epithelium. The results of comparative animal studies[9] and analysis of the effects of experimental epithelial damage[10] demonstrate that the epithelium is truly stratified with the superficial layer being dependent on the underlying basal cells for anchorage in the larger airways. The mechanism of epithelial cell damage in asthma can only really be understood in the light of this model of the bronchial epithelium as a truly stratified structure.

Bronchoalveolar lavage (BAL) fluid from asthmatic subjects contains epithelial clusters that are devoid of basal cells (Fig. 22-1), in keeping with autopsy findings of the shedding of superficial columnar epithelial cells with the presence of residual cells on the basement membrane. Furthermore, cells shed into bronchial lavage are viable, as demonstrated by persistent ciliary activity,[11] suggesting that epithelial loss may be related to a specific failure of intercellular adhesion mechanisms rather than generalized cell death.[12] The pattern of epithelial loss in asthma is consistent with experimental studies involving the exposure of tracheal epithelium to purified guinea pig eosinophil leukocyte-derived major basic protein. This induces intraepithelial separation at the basal-suprabasal cell junction at concentrations less than those that cause epithelial cell death.[13] The separation between basal and supra-

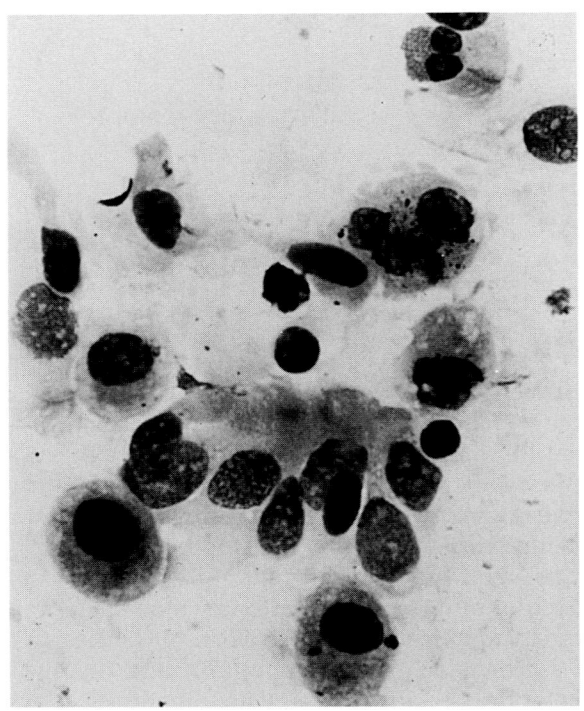

FIGURE 22-1
Ciliated bronchial epithelial cell cluster in bronchial washings.
(Cytocentrifuge, May-Grunwald-Giemsa, ×600.)

basal cells is also seen in virus-induced epithelial cell damage.[14]

In asthma, the eosinophil leukocyte is the most likely effector cell involved in this damage. The secretion of granule-derived cationic proteins into the epithelium has been detected by the immunohistochemical demonstration of their persistence in postmortem samples from patients dying of acute severe asthma.[15] The assessment of epithelial damage in mild asthma awaited the development of endoscopic biopsies and lavage. Although biopsies can be difficult to assess because of forceps-induced damage, the use of rigid bronchoscopy allowed for the demonstration of intraepithelial edema in the relatively large biopsies taken by this technique.[16] Similar findings were subsequently reported in fiberoptic bronchoscopy biopsies (Fig. 22-2) and were correlated with the number of eosinophil leukocytes that infiltrated the epithelium.[17] Other workers have produced evidence from BAL and bronchial washings of correlations between epithelial shed-

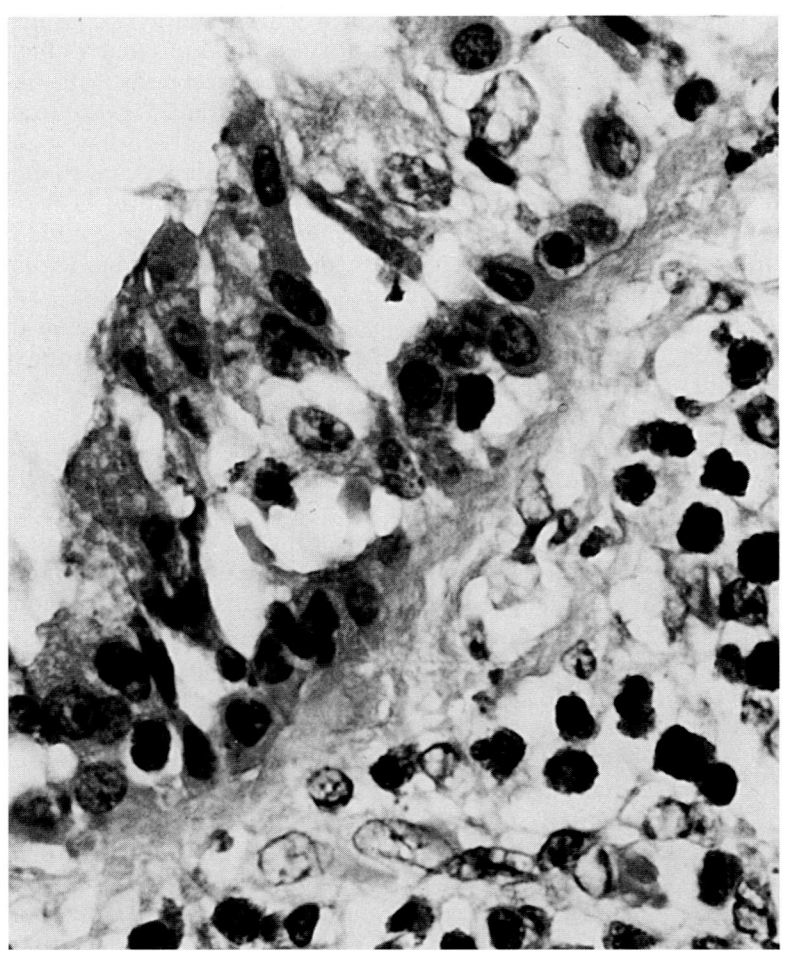

FIGURE 22-2
Separation of superficial columnar cells from basal cells in the bronchial epithelium and inflammatory cell infiltration of the underlying bronchial wall in asthma. (Paraffin embedded, H&E, ×600.)

ding and bronchial hyperreactivity.[18] However, the analyses included both normal and asthmatic subjects and there was a relatively restricted range of asthma severity.

The mucus-secreting goblet cells are confined to larger airways, except in asthma and chronic infective disorders, where they extend to the bronchioles. They contribute to the secretion of mucus, which aggravates airway obstruction.[19] Mucus is also secreted into the bronchial lumina from bronchial glands. These glands have ducts lined by ciliated epithelium, which connect with the collecting ducts that drain the secretory acini. The bronchial gland mass increases in asthma as a result of hyperplasia and the ducts may become so enlarged as to be described as bronchial diverticula or bronchial duct ectasia. Rupture of the ducts has been associated with the development of interstitial emphysema in asthma.[20]

CONNECTIVE TISSUE AND EXTRACELLULAR MATRIX

The basement membrane of the bronchial epithelium is composed of the nonfibrillar collagen IV, noncollagenous proteins, such as laminin and fibronectin, and proteoglycans.[21] In keeping with its stratified structure, the basement membrane is anchored to the underlying connective tissue by collagen VII fibers similar to the arrangement found in skin.[22] The basement membrane is supported by a layer of connective tissue collagen, the lamina reticularis. There is a population of specialized fibroblasts beneath this layer, which may have both structural and immunopathologic importance, because fibroblasts can secrete cytokines that contribute to the inflammatory networks in the mucosa. These cells have both synthetic and contractile cytoplasmic organelles and are termed "bronchial myofibroblasts."[23] Others have subsequently described similar cells in the rat trachea,[24] and this network of cells may be analogous to the network of contractile interstitial cells in the alveolar septae.

One of the most characteristic histologic features of bronchial asthma is an apparent thickening of the bronchial epithelial basement membrane (Fig. 22-3). This is an early event in asthma, occurring in patients with a short clinical history and mild disease and in children.[25,26] These appearances are due to the deposition of interstitial collagens III and V, fibronectin and, to a lesser extent, collagen I, beneath the bronchial epithelial basement membrane. This profile of extracellular matrix deposition would indicate

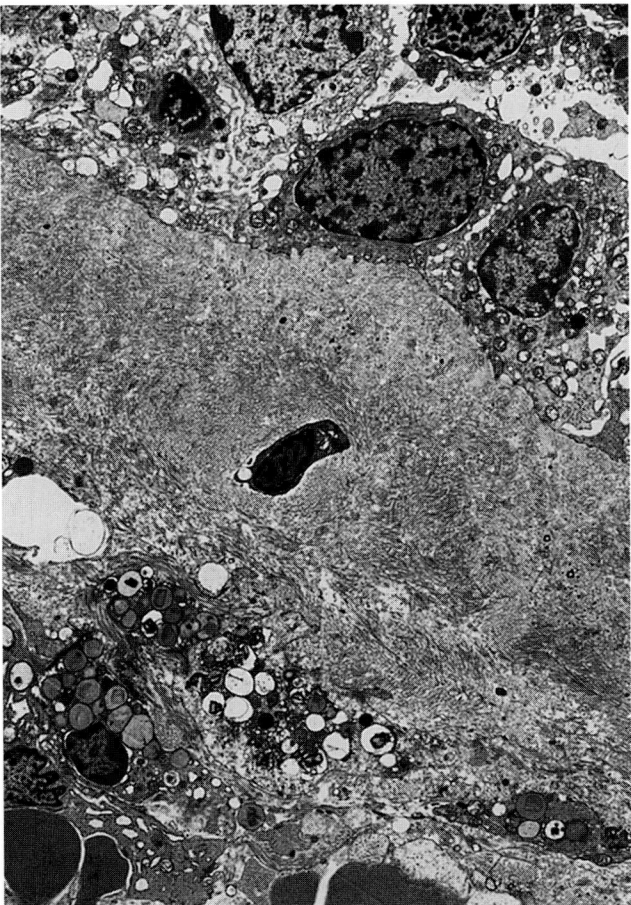

FIGURE 22-3
Bronchial basement membrane and superficial lamina propria showing residual adherent basal cells, dense collagen deposition in the lamina reticularis, and an underlying eosinophil-rich inflammatory cell infiltrate. (Transmission electron microscopy, uranyl acetate and lead citrate, ×4150.)

a mesenchymal origin of this matrix and increased numbers of subepithelial myofibroblasts have been associated with this collagen layer in asthmatic subjects.[23]

A superficial vascular plexus lies beneath the basement membrane. This is probably responsible for continuing the warming and humidification of inspired air, which mainly occurs in the nasal passages. It is difficult to measure the contribution of these vessels to the increased airway resistance in asthma but they frequently appear to be dilated in fatal acute severe asthma (Fig. 22-4). Although angiogenic mediators are expressed in this layer in asthma, it is uncertain whether there is an increase in the vascular network due to vascular proliferation.

BRONCHIAL SMOOTH MUSCLE

The bronchial muscle forms a spiraling network along the airways composed of bundles that form a

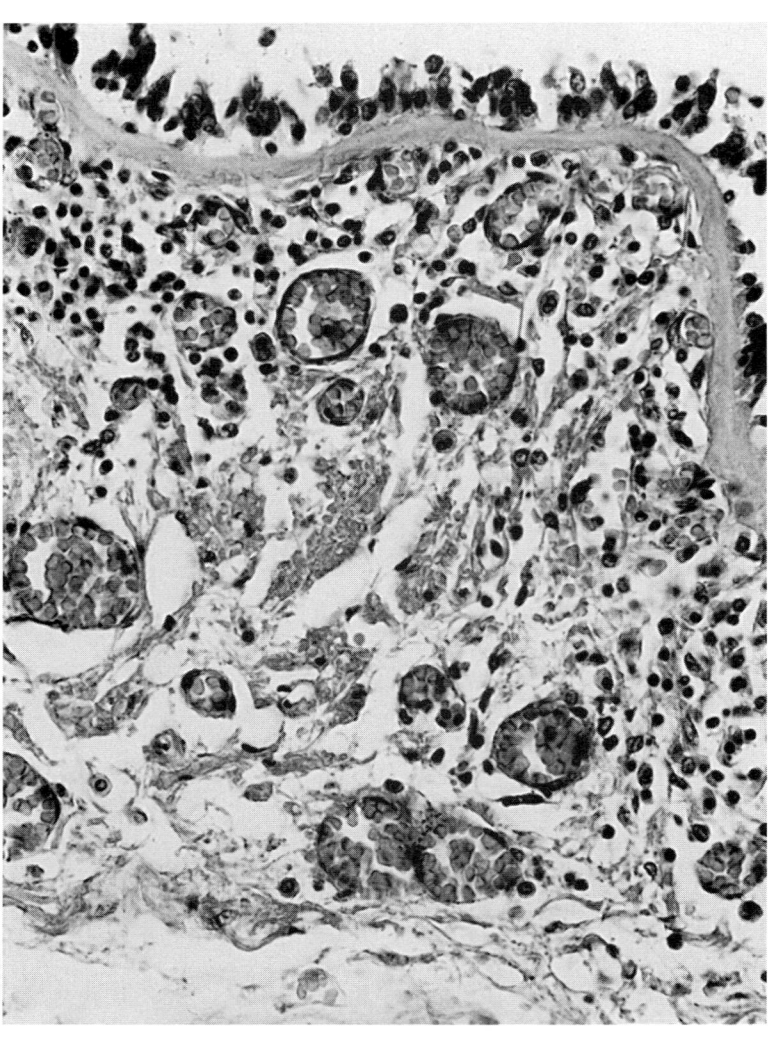

FIGURE 22-4
Fatal asthma showing congested vessels and inflammatory cells beneath the basement membrane. (H&E, ×240.)

perforated sheet in contrast to the continuous muscle layers that form the wall of the small intestine. Autopsy studies have shown an increase in the size of the muscle bundles, which is due to both cellular hyperplasia and hypertrophy,[28] but the function of individual muscle fibers is apparently normal. This increase is seen in both bronchi and bronchioles greater than 2 mm diameter[29] although it has been suggested that there are two patterns of increased muscle mass; the changes are confined to the more proximal large airways in some patients, whereas in others airway smooth muscle is thickened throughout the bronchial tree.[30] The reasons for this increase in smooth muscle are unknown. Possible mechanisms include the effect of continual cycles of contraction and relaxation due to IgE-mediated release of histamine and prostaglandins from mast cells, neural stimuli, and trophic stimuli, including cytokines, released in inflammatory responses.

INFLAMMATION OF THE BRONCHIAL MUCOSA

Two different approaches to the study of asthma in adults have focused on the importance of the inflammatory processes in the bronchial wall in the pathogenesis of asthma. First, the changes in airway resistance in response to allergens vary in their sensitivity to pharamacologic manipulation. Although the immediate asthmatic response is alleviated by β agonists, which inhibit mast cell degranulation, the subsequent responses to allergen challenge are refractory to these agents but respond to corticosteroids.[31] This would indicate that simple mast cell degranulation cannot explain all the components of allergen-induced airway obstruction. Second, the study of the cellular basis of asthma, which was previously confined to autopsy material, has been revolutionized by the access to the airway facilitated by fiberoptic bronchoscopy, BAL, and biopsy.[32]

The greatest expansion in the investigation of bronchial asthma in the last decade has been in the documentation of the inflammatory cells that infiltrate the bronchial wall, the mechanism of their recruitment, the mediators they express, and their role in clinical disease. Apart from focal collections of bronchus-associated lymphoid tissue, the bronchi have a constitutive population of mast cells and B lymphocytes and occasional granulocytes that clearly contribute to the normal host defense mechanisms at this site. The cellular infiltrate in asthma is composed of a slightly augmented T-cell population, markedly increased numbers of eosinophils,[33] and, in some circumstances, an influx of neutrophils.[34] Recent studies have yielded consistent evidence that a similar inflammatory population to that which has long been associated with fatal asthma is present even in the bronchial musosa, even in mild disease. Some data indicate that similar processes are active in childhood asthma.[35]

T Lymphocytes

The constitutive T lymphocyte population in the bronchus shows, at the most, a small increase in number in asthma.[36,37] These cells are CD3 positive, CD45RO positive, and are mainly CD-4 positive (Fig. 22-5), indicating that they are helper T cells that have had previous exposure to antigen. The bronchial T cells in asthma are activated to a greater extent than in nonasthmatic subjects, as evidenced by the expression of the interleukin-2 (IL-2) receptor marker CD25.[37] The T lymphocytes are not confined to the lamina propria but also infiltrate the epithelium, which may enhance their exposure to antigen and also be important in the generation of chemotactic gradients for other cells in the inflammatory infiltrate.

Most of the functions of T lymphocytes are mediated by cytokines, which are produced by gene transcription and translation in response to antigenic stimulation. There is increasing evidence for the division of the cytokine secretion profile of human helper T cells (T_H) into two categories, the T_{H1} and T_{H2} cell patterns. The T_{H1} pattern shows secretion of IL-2, tumor necrosis factor-α, (TNF-α), interferon-γ, and granulocyte-macrophage colony-stimulating factor (GM-CSF). The T_{H2} pattern is characterized by secretion of IL-4, IL-5, IL-6, TNF-α, and GM-CSF.[38]

The T_{H2} phenotype *broadly* corresponds with the immunopathologic features of bronchial asthma. The range of cytokines secreted favors the produc-

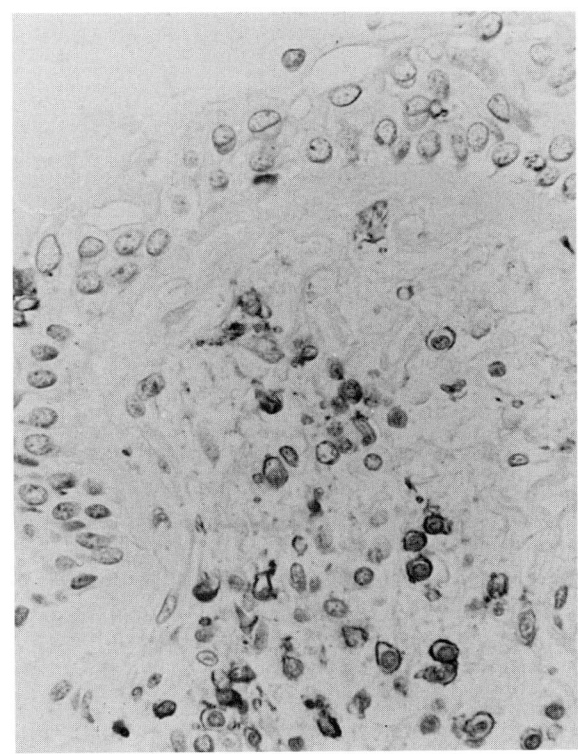

FIGURE 22-5
CD-4-positive T lymphocytes in the bronchial wall and epithelium in mild asthma. (Glycolmethacrylate embedded, immunoperoxidase, ×400.)

tion of IgE by the effect of IL-4 on β cells.[39] In addition, eosinophil leukocyte production, recruitment, survival, and activation are aided by the combined effects of GM-CSF and IL-5.[40–42] In situ hybridization for messenger RNA in BAL cells has shown an excess of cells positive for IL-2, IL-3, IL-4, IL-5, and GM-CSF in asthmatic subjects but no decrease in cells with interferon-γ message.[43] Thus, the observed range of T-cell cytokines associated with asthma does not entirely fit the proposed T_{H2} phenotype. However, the data are clearly in accordance with the view that the T lymphocyte plays an important role in the pathogenesis of the mucosal inflammation of the bronchi.

There is a correlation between T-cell numbers and eosinophil leukocyte infiltration of the airways.[44] This suggests either that eosinophil recruitment is T-lymphocyte dependent or that these cells share common mechanisms to enter the airways.[45] Atopic nonasthmatic individuals have eosinophil and T-lymphocyte numbers and activation at levels between those found in normal and asthmatic subjects, suggesting that there may be a critical threshold above which these allergen-mediated events find clinical expression.[36,37]

Eosinophil Polymorphonuclear Leukocytes

Eosinophil infiltration of the bronchial mucosa is characteristic of asthma in all ages. These cells are practically absent from the mucosa of normal subjects but can form an almost uniform sheet of cells in the mucosa of patients with severe asthma. These cells appear to have an important role in the clinical state because their number correlates with the severity of disease as measured by symptom scores and drug usage.[46] In contrast, there appear to be fewer neutrophils in the mucosa of stable asthmatics than in normal individuals. However, subgroups of patients with acute severe disease may have a neutrophil-predominant cellular infiltrate.[34] It is uncertain whether this pattern of neutrophil leukocyte infiltration is due to the effects of drug treatment modulating the inflammatory response or to acute allergen exposure that may induce a neutrophil response. This is probably due to the expression of nonspecific, vascular endothelial cell adhesion molecules.[47]

The presence of an excessive number of cells within a biopsy specimen is not necessarily evidence of increased cellular recruitment. The histologic appearances of an inflammatory cell infiltrate may be the result of increased numbers of circulating cells that can use the tissue recruitment mechanisms, enhanced or newly expressed recruitment, increased survival within the tissue compartment, or decreased traffic into another compartment, that is, the bronchial lumen.[48] Equally, increased recruitment with increased traffic through the tissue may not be evident from a biopsy specimen.

Nevertheless, there is considerable evidence for increased inflammatory cell recruitment into the bronchial mucosa in asthma. Ultrastructural studies have identified increased contact between intravascular leukocytes and the endothelium in asthma.[18] This probably reflects the rolling of leukocytes along the endothelium, which is mediated by the vascular selectins.[49] P selectin is stored in Weibel-Palade granules in endothelial cells and is expressed on the surface within minutes of stimulation, whereas the expression of E selectin is slower because of the need to induce synthesis before expression.[50] The selectin-mediated contact allows for further steps, including the interaction of leukocyte integrins with immunoglobulin supergene family adhesion molecules (ICAM-1, ICAM-2, and VCAM-1) on the endothelium and the action of chemoattractants that promote migration through the endothelium and within the extravascular compartment.

The presence of VCAM-1 on the endothelium allows for the selective binding of cells with expression of the β_2 integrin VLA-4. This enhances the recruitment of eosinophils, T cells, and basophils.[42] Although VCAM-1 can be selectively induced on the endothelium by IL-4, it is also expressed in response to the pleiotropic cytokines such as TNF-α, which also induce the nonspecific ligands E selectin and ICAM-1, which have been demonstrated in the bronchial mucosa in vivo.[47] Further selectivity may be exerted by specific chemoattractants such as capacity of platelet-activating factor to selectively induce the migration of eosinophils and not neutrophils through the vascular endothelium.[51] Similarly, the chemokine family of cytokines includes highly potent chemoattractants that are selective for cells bearing surface CD4 molecules such as T cells and eosinophils, but not neutrophils. Once present in the tissues, the eosinophils accumulate beneath the bronchial basement membrane in large numbers and from this site infiltrate the epithelium. This pattern of accumulation of eosinophils may be in part due to the potential for both myofibroblast- and epithelial-derived cytokines to enhance cell survival at this site. The subsequent migration into the epithelium may be the result of chemoattractants secreted by either epithelial cells or the intraepithelial T lymphocytes and mast cells.

The eosinophil contributes to the inflammatory process by the secretion of mediators that can be divided into three classes: the lipid-derived mediators, which induce acute inflammatory responses; the cytotoxic granule components; and the proinflammatory cytokines. The evidence for the release of eosinophil mediators in asthma includes the presence of Charcot-Leyden crystals, composed of lysophospholipase,[52] in bronchial lumina and secretions (Fig. 22-6) and the immunohistochemical demonstration of major basic protein[53] in association with epithelial damage.

Eosinophils secrete platelet-activating factor (PAF), which is in itself an eosinophil chemoattractant and activator. PAF also increases vascular permeability and induces smooth muscle spasm and may thus contribute to airway obstruction. Eosinophils are also major producers of leukotriene C_4 and also produce prostaglandin E_2 and 15-lipoxygenase products including 15-HETE, which contribute to further inflammation and mucus secretion.[54] The characteristic granules of eosinophils contain high concentrations of toxic proteins, including eosinophil cationic protein, major basic protein, eosinophil-derived neurotoxin, and eosinophil per-

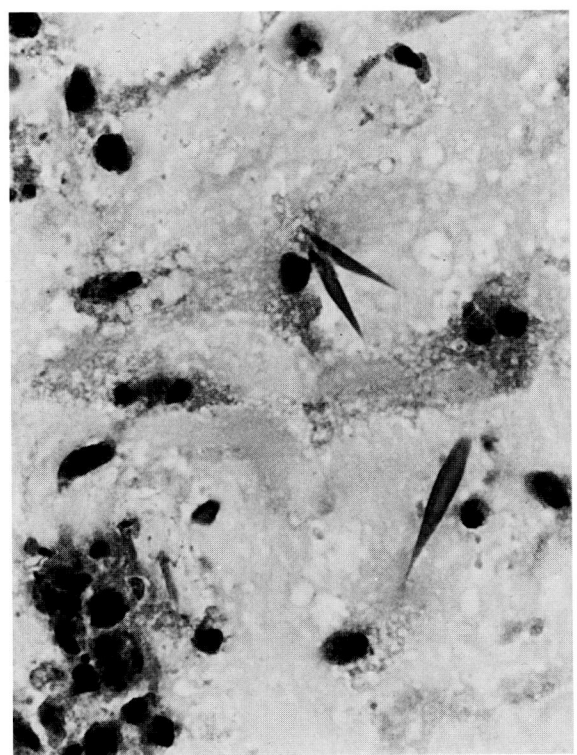

FIGURE 22-6
Eosinophils and Charcot-Leyden crystals in airway mucus. (H&E, ×600.)

oxidase. Major basic protein, which forms the characteristic cores of eosinophil granules, is toxic to a variety of cells, including bronchial epithelial cells, at concentrations that are achieved in vivo and has been detected in association with epithelial damage in asthmatic patients.[53] Eosinophil cationic protein is also toxic to epithelial cells and this protein also shares with major basic protein and eosinophil peroxidase the capacity to induce histamine release from mast cells.[55] Eosinophil peroxidase induces oxidative damage to cells in the presence of hydrogen peroxide and a halide and may also contribute to the cell damage seen in asthma.

Eosinophils express message for a range of cytokines, including TNF-α, transforming growth factors α and β, IL-1, and GM-CSF.[56–58] The secretion of these cytokines can directly, and by interactions with other cells in the mucosal cytokine network, enhance eosinophil recruitment, survival, and activation. Thus the eosinophil is not only an end-stage effector cell but can also contribute to the ongoing allergic inflammatory process.

Mast Cells

The greater understanding of the biology of the mast cell now requires that it must be regarded as both a rapid response cell and as a source of positive feed-back for IgE synthesis while also contributing to the mucosal inflammation and airway remodeling. Human mast cells may be divided into two types, based on the content of neutral proteases in their granules.[59] Mast cells that contain tryptase, a tetrameric trypsin-like protease, in their granules are termed MC_T whereas those that also possess chymase, carboxypeptidase-A, and cathepsin-G are called MC_{TC}. The biologic role of these proteases is uncertain but tryptase is present in large amounts in the mast cell (10 pg/cell in lung) and can enhance the effect of histamine on smooth muscle contraction,[60] activate the collagenase system, and enhance fibroblast proliferation.[61] Chymase can generate angiotensin II from angiotensin I, inactivate bradykinin, enhance mucus production, and degrade basement membranes.[62–64]

Both types of mast cell are present in the bronchus where they respond to allergen by surface IgE-mediated secretion of potent inflammatory mediators such as histamine, prostaglandin D_2, and leukotriene C_4. Release of the proteoglycan heparin from the granules stabilizes the tryptase tetramer and promotes the action of fibroblast growth factors. Mast cells can also contribute to the cytokine network in the mucosa by the expression of TNF-α and IL-3, IL-4, IL-5, and IL-6.[65]

The evidence for mast cell activation in asthma is based on the detection of histamine in BAL fluid[66] and ultrastructural and immunohistochemical studies (Fig. 22-7). Mast cells do not increase in number in the bronchial wall in asthma but in the disease state they infiltrate the bronchial epithelium.[33] This relocation of cells is accompanied by a characteristic ultrastructural appearance of the mast cell granules, which show dissolution of their electron-dense cores and partial degranulation. Little is known of the mechanisms of reconstitution of human mast cell granules and the biologic interpretation of the ultrastructural appearances in asthma is uncertain as are the relative proportion of mediators and the absolute composition of granules with this appearance.

The control of mast cell phenotype appears to be largely regulated by the cytokine stem cell factor, which regulates the production of mast cell precursors from stem cells and their subsequent differentiation into MC_{TC}.[67] This is a largely fibroblast-derived cytokine and there may be an interplay between mast cells and their surrounding mesenchymal cells with the production of trophic factors essential for mutual survival and function. In contrast, the MC_T cell is largely dependent on T-lymphocyte function[68] and again, the capacity for

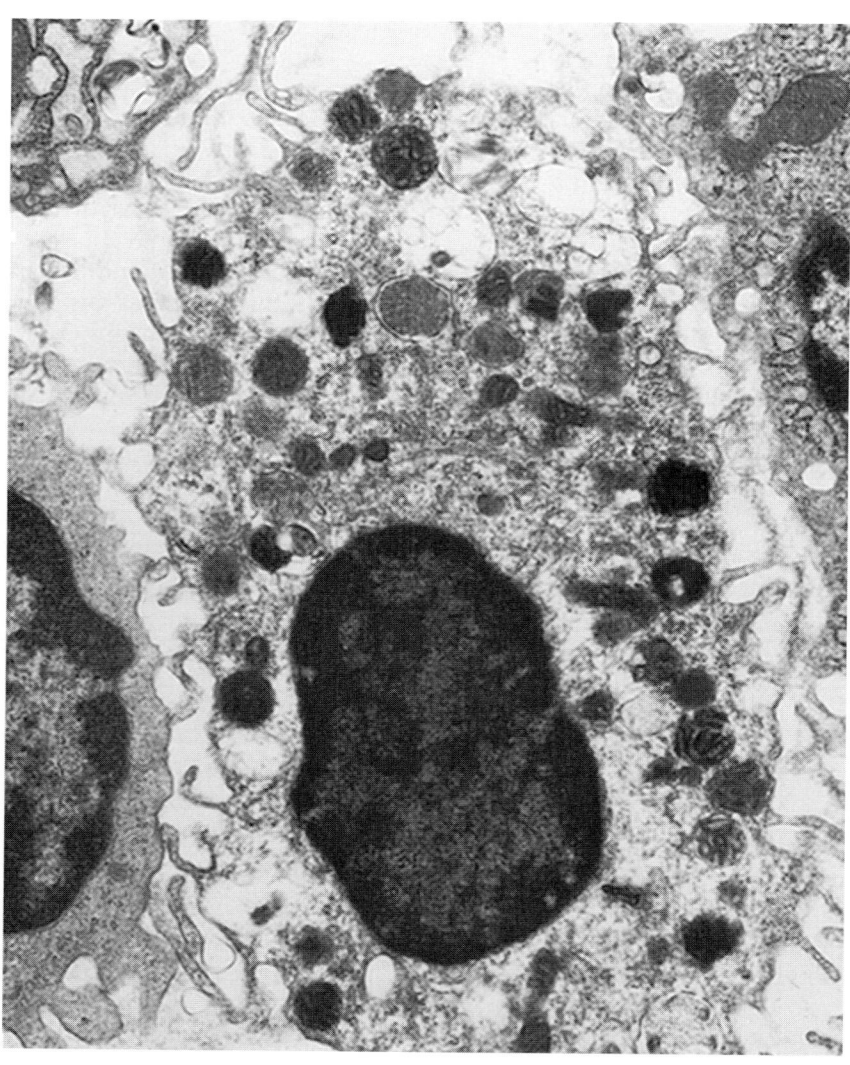

FIGURE 22-7
Mast cell showing piecemeal degranulation in the bronchial epithelium in asthma. (Transmission electron microscopy, uranyl acetate and lead citrate, ×11,700.)

cytokine secretion by the mast cell may mean that this interaction is also based on a two-way exchange of signals.

MECHANISMS OF AIRWAY OBSTRUCTION

The clinical manifestations of bronchial asthma are the result of immunologically mediated reactions occurring in a structurally and functionally abnormal airway wall. The airway responses to inhaled bronchoconstrictors, such as metacholine or histamine, are abnormal in asthma in both the extent of response to graded concentrations of provocant and the failure of the dose–response curve to reach a plateau before bronchoconstriction reaches dangerous levels. Computer modeling of the bronchial provocant response in asthma has shown that the loss of the plateau exhibited by normal subjects and the continual contraction of the bronchi with increasing doses of provocant can be predicted by a model that takes two factors into account, airway thickening and a loss of pulmonary recoil.[69] Although they show little effect in the basal state, these factors greatly aggravate the effect of a fixed amount of smooth muscle contraction on the increase in airway resistance. Thus, any process that contributes to bronchial wall thickening may be of considerable clinical importance determining not only the degree of bronchial hyperreactivity but also its potentially fatal nature.

The airway abnormalities that contribute to obstruction include epithelial cell damage, excess extracellular matrix accumulation, smooth muscle hyperplasia and hypertrophy, vascular congestion and edema, and accumulation of mucus, cellular debris, and fibrin in the airway lumina. Although the thickening of the lamina reticularis of the basement membrane is not sufficient to be of significance for airway flow in the large bronchi, it extends to the small bronchioles where it may contribute proportionately more to the increased resistance. Damage to the epithelium may have a number of physiologic consequences including the loss of modulation of smooth muscle function by epithelial-derived relaxants[70] and the induction of endothelin synthesis,[71] which contracts smooth muscle and is mitogenic for mesenchymal cells.

COMPLICATIONS OF ASTHMA

The complications of asthma, some of which may require pathologic evaluation, include acute severe asthma, fatal asthma, pneumothorax and pneumomediastinum, mucoid impaction and allergic bronchopulmonary aspergillosis, bronchiectasis, and irreversible loss of airway conductance. The complications of the treatment for asthma include the effects of inhaled and systemic corticosteroids, including candidiasis of the upper aerodigestive tract, and for systemic steroids: diabetes, osteoporosis, cataract, and hypertension. The cardiovascular effects of β_1-sympathomimetic stimulation may be aggravated by hypoxia and hypokalemia (for review of asthma therapeutics, see Barnes[6]). Theophyllines, which have a relatively low therapeutic index, may also contribute to cardiac events in some patients.

Asthma deaths may present as sudden unexpected deaths to the pathologist without knowledge of the clinical history. Patients may not have attended asthma clinics and often do not carry treatment cards. The discovery of a sympathomimetic medication inhaler, often in the absence of any other treatment, may alert the pathologist to possibility of asthma as a cause of death. Pneumothorax or pneumomediastinum is occasionally present and should be carefully sought. It should be remembered that the presence of air in the pleural cavity results in a macrophage- and eosinophil-rich inflammatory infiltrate, which is not evidence of underlying asthma.

In the absence of chronic irreversible airway obstruction there is usually no evidence of pulmonary hypertension or cor pulmonale. The lungs are often hyperinflated (Fig. 22-8) and the pleural surfaces show the markings of the ribs and intercostal spaces. The thoracic organs may show petechial hemorrhages on the serosal surfaces, indicative of forced inspiratory efforts in the presence of airway obstruction. There is usually mucoid plugging of large and small airways (Fig. 22-9), although this is not a constant feature. Bronchiectasis is occasionally present. The pulmonary parenchyma may show areas of atelectasis, which probably results from resorption collapse distal to mucoid impaction in segmental bronchi.

Mucoid plugging of the airways may be associ-

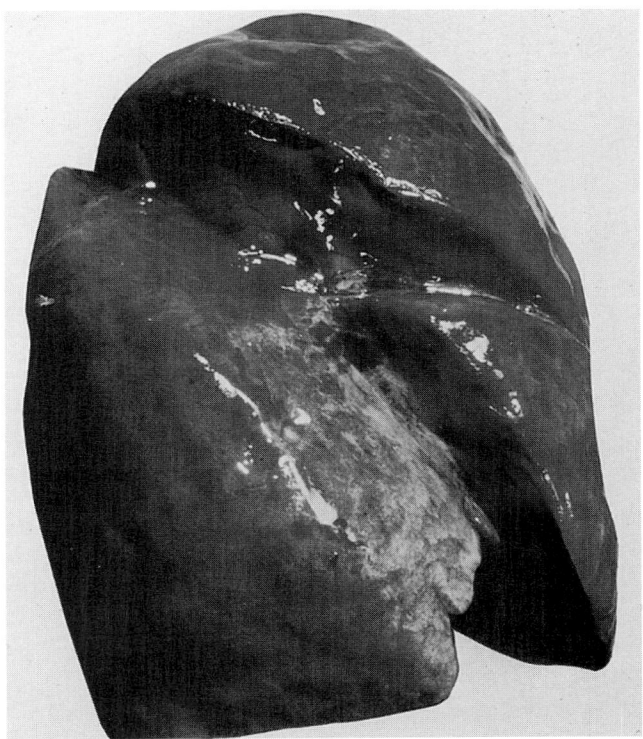

FIGURE 22-8
Pleural surface of the lung in fatal acute severe asthma, showing the absence of postmortem collapse.

FIGURE 22-9
Thickening of the bronchial walls and mucoid plugging (*) in fatal chronic severe asthma.

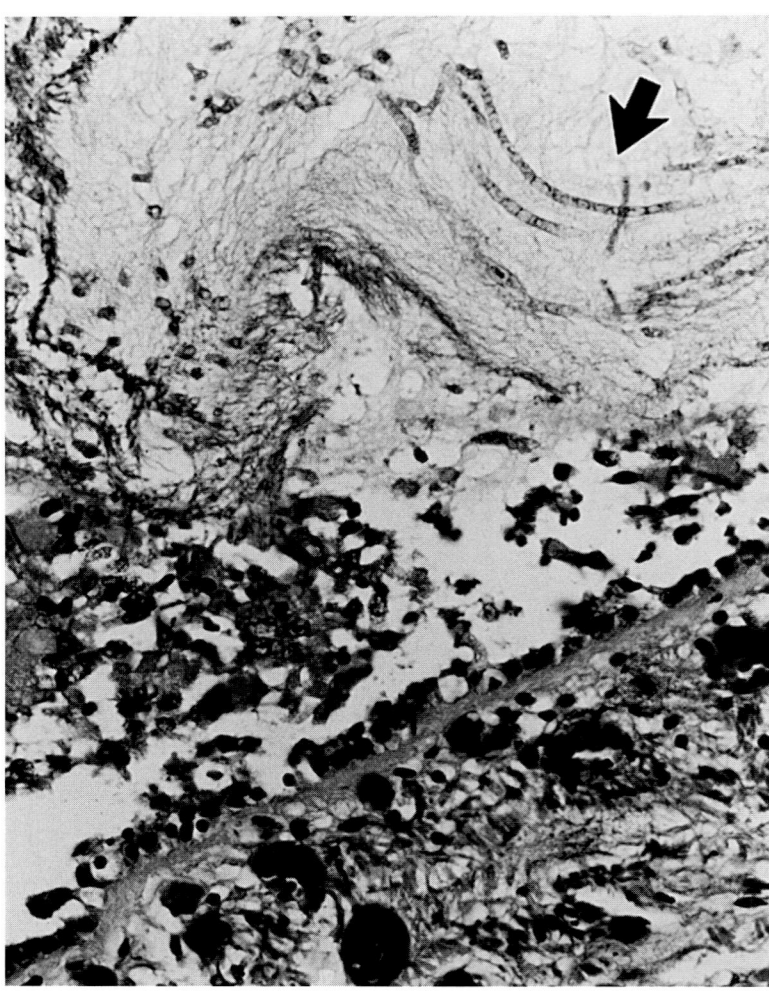

FIGURE 22-10
Fatal asthma showing epithelial cell shedding and hyphae of *Aspergillus* (*arrow*) in mucoid plugs. (Paraffin-embedded, H&E, ×240.)

ated with *Aspergillus* colonization of the bronchi, with increased eosinophil infiltration and increased *Aspergillus* antibody titers in the plasma. Fragments of *Aspergillus* hyphae may be found in expectorated mucus and in material suctioned from the airways at bronchoscopy or at autopsy (Fig. 22-10). Untreated, this may progress to bronchiectasis or fatal airway obstruction. The fungus is not invasive in this form of the disease and the clinical problems relate to the nature and extent of the host response, usually resolving with steroid treatment.

REFERENCES

1. Sheffer AL: Guidelines for the diagnosis and management of asthma. National Heart, Lung and Blood Institute National Asthma Education Program Expert Panel Report. *J Allergy Clin Immunol* 88: 425–534, 1991.
2. Haahtela T, Lindholm H, Bjorksten F, Koskenvou K, Laitinen LA: Prevalence of asthma in Finnish young men. *Br Med J* 301: 266–268, 1990.
3. Burney PG, Chinn S, Rona RJ: Has the prevalence of asthma increased in children? Evidence from the national study of health and growth 1973–86. *Br Med J* 300: 1302–1310, 1990.
4. Sears MR: Worldwide trends in asthma mortality. *Bull Int Union Tuber Lung Dis* 66: 79–83, 1991.
5. Beasley R, Crane J, Burgess C, Pearce N, Jackson RJ: Fenoterol and severe asthma mortality. *N Z Med J* 102: 294–295, 1989.
5a. Garrett J, Colby J, Richards G, et al: Major reduction in asthma morbidity and continued reduction in asthma mortality in New Zealand. What lessons have been learned? *Thorax* 50:303–311, 1995.
5b. Anto JM: Asthma outbreaks: An opportunity for research? *Thorax* 50:220–222, 1995.
6. Barnes PJ: A new approach to the treatment of asthma. *N Engl J Med* 321: 1517–1527, 1989.
7. Naylor B: The shedding of the mucosa of the bronchial tree in asthma. *Thorax* 17: 69–72, 1962.
8. Dunnill MS: The pathology of asthma with specific reference to changes in the bronchial mucosa. *J Clin Pathol* 13: 27–33, 1960.
9. Evans MJ, Cox RA, Shami SG, Wilson B, Plopper CG: The role of basal cells in attachment of columnar cells to the basal lamina of the trachea. *Am J Respir Cell Mol Biol* 1: 463–469, 1989.
10. Abdi S, Evans MJ, Cox RA, Lubbesmeyer H, Herndon DN, Traber DL: Inhalation injury to tracheal epithelium in an ovine model of cotton smoke exposure. Early phase (30 minutes). *Am Rev Respir Dis* 142: 1436–1439, 1990.
11. Montefort S, Roberts JA, Beasley R, Holgate ST, Roche WR: The site of disruption of the bronchial epithelium in asth-

matic and non-asthmatic subjects. *Thorax* 47: 499–503, 1992.

12. Stanley JR: Pemphigus and pemphigoid as paradigms of organ-specific, autoantibody-mediated diseases. *J Clin Invest* 83: 1443–1448, 1989.

13. Frigas SE, Loegering DA, Gleich GJ: Cytotoxic effects of the guinea pig major basic protein on tracheal epithelium. *Lab Invest* 42: 35–43, 1980.

14. Hers JFP: Disturbances of the ciliated epithelium due to influenza virus. *Am Rev Respir Dis* 93: 162–171, 1966.

15. Filley WV, Holley KE, Kephart GM, Gleich GJ: Identification by immunofluorescence of eosinophil major basic protein in lung tissues of patients with bronchial asthma. *Lancet* (i): 11–16, 1982.

16. Laitinen LA, Heino M, Laitinen A, Kava T, Haahtela T: Damage of the airway epithelium and bronchial reactivity in patients with asthma. *Am Rev Respir Dis* 131: 599–606, 1985.

17. Ohashi Y, Motjima S, Fukuda T, Makino S: Airway hyperresponsiveness, increased intracellular spaces of bronchial epithelium and increased infiltration of eosinophils and lymphocytes in bronchial mucosa in asthma. *Am Rev Respir Dis* 145: 1469–1476, 1992.

18. Beasley R, Roche WR, Roberts JA, Holgate ST: Cellular events in the bronchi in mild asthma and after bronchial provocation. *Am Rev Respir Dis* 139: 806–817, 1989.

19. Thurlbeck WM, Malaka D, Murphy K: Goblet cells in the peripheral airways in chronic bronchitis. *Am Rev Respir Dis* 112: 65, 1975.

20. Cluroe A, Holloway L, Thomson K, Purdie G, Beasley R: Bronchial gland duct ectasia in fatal bronchial asthma: Association with interstitial emphysema. *J Clin Pathol* 42: 1026–1031, 1989.

21. Yurchenco PD: Assembly of basement membrane. *Ann N Y Acad Sci* 580: 195–213, 1990.

22. Wetzels RHW, Robben HCM, Leigh AM, Schaofsma HE, Vooijs GP, Ramaeker FCS: Distribution patterns of type VII collagen in normal and malignant tissues. *Am J Pathol* 139: 451–459, 1991.

23. Brewster CEP, Howarth PH, Djukanovic R, Wilson JW, Holgate ST, Roche WR: Myofibroblasts and subepithelial fibrosis in bronchial asthma. *Am J Respir Cell Mol Biol* 3: 507–511, 1990.

24. Evans MJ, Guha SC, Cox RA, Moller PC: Attenuated fibroblast sheath around the basement membrane zone in asthma. *Am J Respir Cell Mol Biol* 8: 188–192, 1993.

25. Roche WR, Beasley R, Williams JH, Holgate ST: Subepithelial fibrosis in the bronchi of asthmatics. *Lancet* 1: 520–524, 1989.

26. Cutz E, Levison H, Cooper DM: Ultrastructure of airways in children with asthma. *Histopathology* 2: 407–421, 1978.

27. Laitinen LA, Laitinen A: The bronchial circulation: histology and electron microscopy, in Butler J (ed), *The Bronchial Circulation*. New York, Dekker, pp. 77–98, 1992.

28. Ebina M, Takakashi T, Chiba T, Motomiya M: Cellular hypertrophy and hyperplasia of airway smooth muscles underlying bronchial asthma. *Am Rev Respir Dis* 148: 720–726, 1993.

29. Carrol N, Elliot J, Morton A, James A: The structure of large and small airways in nonfatal and fatal asthma. *Am Rev Respir Dis* 147: 405–410, 1993.

30. Ebina M, Yaegashi H, Chiba R, Takahashi T, Motomiya M, Tanemura M: Hyperreactive site in the airway tree of asthmatic patients revealed by thickening of bronchial muscles. *Am Rev Respir Dis* 141: 1327–1332, 1990.

31. Cockcroft DW, Murdock KY: Comparative effects of inhaled salbutamol, sodium cromoglycate, and beclomethasone dipropionate on allergen-induced early asthmatic responses, late asthmatic responses, and increased bronchial responsiveness to histamine. *J Allergy Clin Immunol* 79: 734–740, 1987.

32. Djukanovic R, Roche WR, Wilson JW, et al: State of the art. Mucosal inflammation in asthma. *Am Rev Respir Dis* 142: 434–457, 1990.

33. Djukanovic R, Wilson JW, Britten KM, et al: Quantitation of mast cells and eosinophils in the bronchial mucosa of symptomatic atopic asthmatics and healthy control subjects using immunohistochemistry. *Am Rev Respir Dis* 142: 863–871, 1990.

34. Sur S, Crotty TB, Kephart GM, et al: Sudden-onset fatal asthma: A distinct entity with few eosinophils and relatively more neutrophils in the airway mucosa. *Am Rev Respir Dis* 148: 713–719, 1993.

35. Ferguson AC, Whitelaw M, Brown H: Correlation of bronchial eosinophil and mast cell activation with bronchial hyperresponsiveness in children with asthma. *J Allergy Clin Immunol* 90: 609–613, 1992.

36. Djukanovic R, Lai CWK, Wilson JW, et al: Bronchial mucosal manifestations of atopy: A comparison of markers of inflammation between atopic asthmatics, atopic non-asthmatics and healthy controls. *Eur Respir J* 5: 538–544, 1992.

37. Bradley BL, Azzawi M, Jacobson M, et al: Eosinophils, T-lymphocytes, mast cells, neutrophils and macrophages in bronchial biopsy specimens from atopic subjects with asthma: Comparison with biopsy specimens from atopic subjects without asthma and normal control subjects and relationship to bronchial hyperresponsiveness. *J Allergy Clin Immunol* 88: 661–674, 1991.

38. Romagnani S: Human T_{H1} and T_{H2} subsets; Doubt no more. *Immunol Today* 12: 256–257, 1991.

39. Del Prete G, Maggi E, Parronchi P, et al: IL-4 is an essential factor for the IgE synthesis induced in vitro by human T cell clones and their supernatants. *J Immunol* 140: 4193–4198, 1988.

40. Silberstein DS, Owen WF, Gasson JC, et al: Enhancement of human eosinophil cytotoxicity and leukotriene synthesis by biosynthetic (recombinant) granulocyte-macrophage colony-stimulating factor. *J Immunol* 137: 3290–3294, 1986.

41. Lopez AF, Sanderson CJ, Gamble JR, Campbell HD, Young IG, Vadus MA: Recombinant human interleukin 5 is a selective activator of human eosinophil function. *J Exp Med* 167: 219–224, 1988.

42. Dobrina A, Menegazzi R, Carlos TM, et al: Mechanisms of eosinophil adherence to cultured vascular endothelial cells. Eosinophils bind to the cytokine-induced endothelial ligand vascular cell adhesion molecule-1 via the very late activation antigen-4 integrin receptor. *J Clin Invest* 88: 20–26, 1991.

43. Robinson DS, Hamid Q, Sun Y, et al: Predominant Th2-like bronchoalveolar T lymphocyte population in atopic asthma. *N Engl J Med* 326: 298–304, 1992.

44. Bentley AM, Maestrelli P, Saetta M, et al: Activated T-lymphocytes and eosinophils in the bronchial mucosa in isocyanate-induced asthma. *J Allergy Clin Immunol* 89: 821–829, 1992.

45. Resnick MB, Weller PF: Mechanisms of eosinophil recruitment. *Am J Respir Cell Mol Biol* 8: 349–355, 1993.

46. Bousquet J, Chanez P, Lacoste JY, et al: Eosinophilic inflammation in asthma. *N Engl J Med* 323: 1033–1039, 1990.

47. Montefort S, Roche WR, Howarth PH, et al: Intercellular adhesion molecule-1 (ICAM-1) and endothelial leukocyte adhesion molecule-1 (ELAM-1) expression in the bronchial

mucosa of normal and asthmatic subjects. *Eur Respir J* 5: 815–823, 1992.

48. Wardlaw AJ, Dunnette S, Gleich GJ, Collins JV, Kay AB: Eosinophils and mast cells in bronchoalveolar lavage in subjects with mild asthma. *Am Rev Respir Dis* 137: 62–69, 1988.

49. Lawrence MB, Springer TA: Leukocytes roll on a selectin at physiologic flow rates: Distinction from and prerequisite for adhesion through integrins. *Cell* 65: 859–873, 1991.

50. Geng J, Bevilacqua MP, Moore, KL, et al: Rapid neutrophil adhesion to activated endothelium mediated by GMP-140. *Nature* 343: 757–760, 1990.

51. Morland CM, Wilson SJ, Holgate ST, Roche WR: Selective eosinophil recruitment by transendothelial migration and not by leukocyte-endothelial cell adhesion. *Am Rev Respir Dis* 6: 557–566, 1992.

52. Weller PF, Bach DS, Austen KF: Biochemical characterization of human eosinophil Charcot-Leyden crystal protein (lysophospholipase). *J Biol Chem* 259: 15100–15105, 1984.

53. Filley WV, Holley KE, Kephart GM, Gleich GJ: Identification by immunofluorescence of eosinophil granule major basic protein in lung tissues of patients with bronchial asthma. *Lancet* 2: 11–16, 1982.

54. Gleich GJ: The eosinophil and bronchial asthma: Current understanding. *J Allergy Clin Immunol* 85: 422–436, 1990.

55. Henderson WR, Chi EY, Klebanoff SJ: Eosinophil peroxidase-induced mast cell secretion. *J Exp Med* 152: 265–279, 1980.

56. Kita H, Onhishi T, Okubo Y, Weiler D, Abrams JS, Gleich GJ: Granulocyte/macrophage colony-stimulating factor and interleukin-3 release from human peripheral blood eosinophils and neutrophils. *J Exp Med* 174: 745–748, 1991.

57. Costa JJ, Matossian K, Resnik MB, et al: Human eosinophils can express the cytokine tumor necrosis factor-α and macrophage inflammatory protein-1α. *J Clin Invest* 91: 2673–2684, 1993.

58. Wong DTW, Elovic A, Matossian K, et al: Eosinophils from patients with blood eosinophilia express transforming growth factor β1. *Blood* 78: 2702–2707, 1991.

59. Irani AA, Schwartz LB: Neutral proteases as indicators of human mast cell heterogeneity. *Monogr Allergy* 27: 146–162, 1990.

60. Sekizawa K, Caughey GH, Lazarus SC, Gold WM, Nadel JA: Mast cell tryptase causes airway smooth muscle hyperresponsiveness in dogs. *J Clin Invest* 83: 175–179, 1989.

61. Ruoss SJ, Hartmann T, Caughey GH: Mast cell tryptase is a mitogen for cultured fibroblasts. *J Clin Invest* 88: 493–499, 1991.

62. Reilly CF, Tewksbury DA, Schechter NM, Travis J: Rapid conversion of angiotensin I to angiotensin II by neutrophil and mast cell proteinases. *J Biol Chem* 257: 8619–8622, 1982.

63. Briggaman RA, Schechter NM, Fraki J, Lazarus GS: Degradation of the epidermal-dermal junction by a proteolytic enzyme from human skin and human polymorphonuclear leukocytes. *J Exp Med* 160: 1027–1042, 1984.

64. Sommerhoff CP, Caughey GH, Finkbeiner WE, Lazarus SC, Basbaum CB, Nadel JA: Mast cell chymase. A potent secretagogue for airway gland serous cells. *J Immunol* 142: 2450–2456, 1989.

65. Bradding P, Roberts JA, Britten KM, et al: Interleukin-4, -5, and -6 and tumor necrosis factor-α in normal and asthmatic airways: Evidence for the human mast cell as a source of these cytokines. *Am J Respir Cell Mol Biol* 10: 471–480, 1994.

66. Agius RM, Godfrey RC, Holgate ST: Mast cell and histamine content of tumor bronchoalveolar lavage fluid. *Thorax* 40: 760–767, 1985.

67. Nocka K, Buck J, Levi E, Besmer P: Candidate ligand for the *c-kit* transmembrane kinase receptor: KL, a fibroblast-derived growth factor stimulates mast cells and erythroid progenitors. *EMBO J* 9: 3287–3294, 1990.

68. Irani AM, Craig SS, De Blois G, Elson CU, Schechter NM, Schwartz LB: Deficiency of the tryptase-positive, chymase-negative mast cell type in gastrointestinal mucosa of patients with defective T lymphocyte function. *J Immunol* 138: 4381–4386, 1987.

69. Wiggs BR, Boskou C, Pave PD, James A, Hogg JC: A model of airway narrowing in asthma and in chronic obstructive pulmonary disease. *Am Rev Respir Dis* 145: 1251–1258, 1992.

70. Morrison KJ, Vanhoutte PM: Airway epithelial cells in the pathophysiology of asthma. *Ann N Y Acad Sci* 629: 82–88, 1991.

71. Springhall DR, Howarth PH, Counihan H, Djukanovic R, Holgate ST, Polak JM: Endothelin reactivity of airway epithelium in asthmatic patients. *Lancet* 337: 697–701, 1991.

Chapter 23

Pulmonary Edema

H. Bachofen / E. R. Weibel

INTRODUCTION

Gas exchange is the main function of the lung, which is supported by the design of a delicate tissue barrier between air and blood.[1] In this gas exchange, a well-balanced transfer of fluid and water also takes place.[2–4] The control of this fluid balance is essential for maintaining the structural basis for gas exchange, namely a thin air-blood barrier and a large surface of air-blood contact. Both of these would be adversely affected by extravascular fluid accumulation, which is, by definition, pulmonary edema.

The problems of keeping the lung "dry" are appreciable. First, the blood perfusion rate of the lung is high, and the balance of forces is precarious despite a relatively low capillary pressure. This is because, unlike that of other organs, the pulmonary microvasculature is not encased in a solid mass of tissue but, rather, is loosely suspended in air over a large area with only a delicate tissue barrier for mechanical support. Furthermore, the cell layers separating the fluid spaces are vulnerable, so their permeability may become increased by various injuries. In reviewing the mechanisms and routes of abnormal fluid leakage, one has to consider the design of lung parenchyma and the different structural and functional safety measures that ensure maintenance of a well-balanced physiologic fluid exchange between intra- and extravascular fluid compartments.

THE NORMAL LUNG: FLUID SPACES AND BARRIERS

General Design

The gas-exchanging lung parenchyma contains three extracellular fluid spaces (Fig. 23-1): the blood space of the microvasculature, the interstitial space, and the alveolar fluid lining, which forms the hypophase for surfactant. None of these spaces are subdivided, but they form fluid continua that are separated by two thin but uninterrupted cell layers, the capillary endothelium and the alveolar epithelium, whose characteristics largely determine the fluid movement between the spaces. This three-compartment model is distinct from the two-compartment model with a single barrier that is commonly used to describe gas exchange (Fig. 23-2).

The fluid continuum of the microvasculature is obvious. Less well appreciated is the continuity of the alveolar fluid lining, as its volume is extremely small and unequally distributed over the surface. Only in alveolar corners can small fluid pools be recognized; on the plane parts of the alveolar surface there is only indirect evidence for an extremely thin coherent fluid film. Indeed, minute droplets of hydrophilic stains introduced into alveoli by micropipets reveal an immediate and homogeneous expansion over large distances.[5]

The continuity of the interstitial space is of particular importance for the homeostasis of extravascular lung fluid. This feature is ensured by the continuity of fine connective tissue fibers, both collagenous and elastic, which extend from the axial to the peripheral fiber system and are interlaced with the capillary network.[6,7] Thus, the alveolar septa, with their complex of fibers and capillaries, form a three-dimensional polyhedral network that is continuous over large regions of pulmonary parenchyma, essentially over the domain of an entire acinus or lobule. Within the septum the capillaries are arranged in an eccentric fashion so that the blood-gas barrier develops thin and thick parts on either side of the capillary. This conspicuous construction can be recognized on thin sections of alveolar septa (Fig. 23-1), and the functionality of the structure for both gas and fluid exchange becomes evident.

The *thin barrier parts* consist of only the endothelial and epithelial cell layers. Their basal laminae are fused so that this "restricted" (or, in fact, nonexistent) interstitial space cannot take up excess fluid. Hence, thin barrier parts are ideal for gas exchange.

In the *thick parts*, the basal laminae of the epithelial and capillary endothelial cell layers are

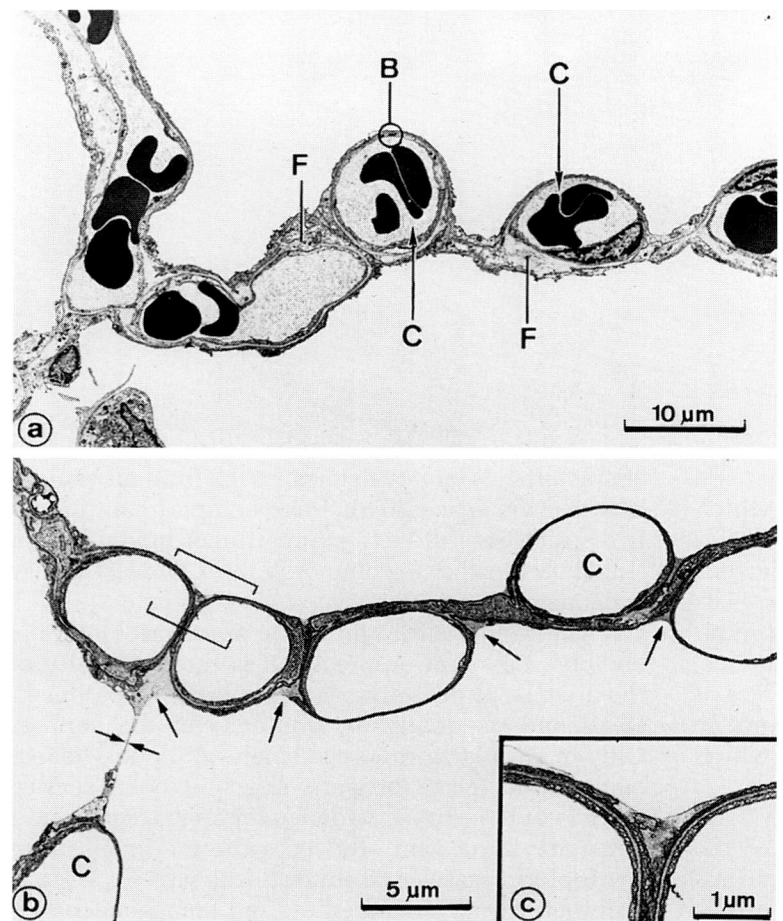

FIGURE 23-1
The fluid spaces in alveolar walls are shown in thin-section transmission electron micrographs. *A.* Septum from the human lung shows capillaries (C), with the blood separated from the air spaces by a very thin tissue barrier (B); fiber bundles (F) are interwoven with the capillary network and occur only on one side of each capillary where the barrier is thicker and contains an interstitial space. *B.* Similar septum from a rabbit lung fixed by vascular perfusion to retain the surface lining layer, which forms pools between bulging capillaries (*arrows*) and extends across an alveolar pore (*paired arrows*). *C.* High-power view of small pool of lining layer with surfactant film.

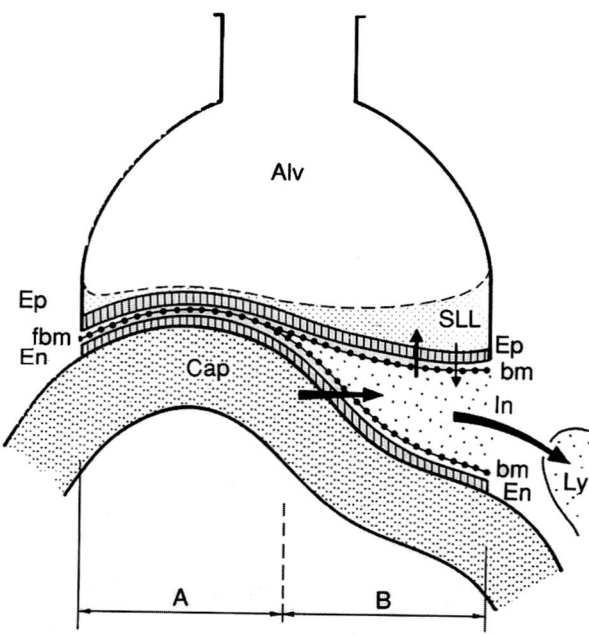

FIGURE 23-2
Model of fluid spaces and their barriers. The plasma space in the capillary (Cap) is separated by the endothelium (En) from the interstitial space (In), which is, in turn, separated by the epithelium (Ep) from the alveolar surface lining layer (SLL) bounded by surfactant film (broken line). The interstitial space is bounded by the endothelial and epithelial basement membranes (bm), which are fused in region A (fbm), excluding a true interstitial space from the primary gas exchange barrier. The interstitial space in region B is associated with fiber bundles and extends toward lymphatics in juxtaalveolar connective tissue.

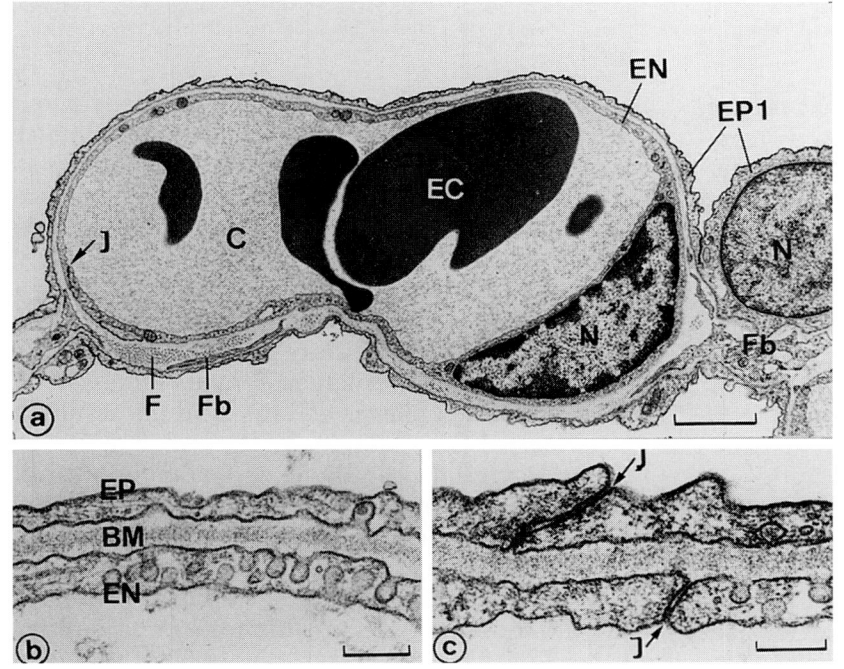

FIGURE 23-3
A. Alveolar capillary (C) with type I alveolar epithelial cell (EP1) and capillary endothelial cell (EN), whose nuclei (N) are surrounded by very little cytoplasm, which extends as thin leaflets to form the air-blood barrier. Note that the basement membranes (BM) of the epithelium and endothelium become fused in a minimal barrier. The interstitial space in the thick part of the barrier contains fibroblast processes (Fb) and bundles of collagen and elastic fibers (F). *B.* Minimal barrier with fused basement membranes and the four plasma membranes with some pinocytotic vesicles. *C.* Intercellular junctions (J) are long and tight in the epithelium and short and "leaky" in the endothelium.

separated by a true—albeit rather thin—interstitial space. It contains the fibrous "backbone" of the septa, together with associated cells (mainly fibroblasts), proteoglycans as ground substance, and in some locations small pools of free fluid. Because of its fibrous network, this nonrestricted interstitial space is continuous within the plane of the septum and extends toward the peripheral connective tissue; it can thus function as a pathway for water and solutes that have exuded from the capillary.[6] By this route excess fluid can flow to the interstitial fluid sumps associated with the connective tissue sheaths around conducting airways and blood vessels.[2,6,8,9] Lymphatic capillaries, which are found in these "juxta-alveolar" regions (but never in the alveolar septa), serve as second-order dripping channels.[10] Clearly, this structural arrangement allows for the liquid flow necessary for nutrition and defense of lung tissue, but it minimizes interference with gas exchange because the thick parts of the septum occupy merely half of the capillary gas exchange surface.

The three fluid space continua are separated by cellular barriers—the capillary endothelium and the alveolar epithelium (Fig. 23-3). The capillary endothelium is composed of seemingly uniform endothelial cells of simple construction.[11] The nucleus of the endothelial cell is surrounded by a slim rim of cytoplasm and a cytoplasmic leaflet 0.1 to 0.2 μm thick extends laterally to cover approximately 1200 μm^2 of capillary surface.[12,13] Endothelial cells contain only a modest set of cytoplasmic organelles. Their cytoplasmic leaflets are made up largely of the two plasma membranes, with some

cytoplasmic matrix interposed containing cytoskeletal fibrils. The cells contain a variable number of microvesicles, which are usually interpreted as pinocytotic vesicles related to macromolecular transport.[11,14] The outer membrane is associated with the capillary basement membrane. The junctions between neighboring endothelial cells are sealed by bandlike adhesions. However, these "tight junctions" are somewhat leaky, allowing the transfer of smaller macromolecules, such as serum albumin, between plasma and interstitial space, at least when the junctions become stretched by capillary distention. The endothelium therefore does not form a tight osmotically active barrier. The "pores" for the exchange of fluid between plasma and interstitial space are assumed to be located at the "leaky" endothelial junction lines.[15] The exchange of fluid between the plasma and the interstitial space is functionally important and does not interfere with gas exchange, provided this fluid can be efficiently removed from the septum into lymphatics, guided by the fiber system.

The alveolar epithelium is made of two cell types: the squamous type I cells form the actual lining, whereas the cuboidal type II cells are secretory cells and also serve as stem cells for the regeneration of the epithelium. Although these two cell types occur in about equal numbers, the type I cells cover more than 95 percent of the alveolar surface area. At first sight, the type I cells are similar to endothelial cells (Fig. 23-3): their small nucleus is likewise surrounded by a small amount of cytoplasm poor in organelles, and the cytoplasmic leaflets that extend from the perinuclear region are also about 0.1 μm thick. They consist essentially of two plasma membranes with pinocytotic vesicles and some interposed cytoskeletal filaments. The inner plasma membrane is tightly associated with the epithelial basement membrane. However, the epithelial cytoplasmic leaflets are much more extensive, those of a single cell covering an area approximately 5000 μm². Furthermore, the type I cell is not a simple squamous cell; its shape is complex, with the cytoplasmic leaflet showing many branchings.[12] A significant difference is also seen in the intercellular junctions (Fig. 23-3), which appear to be very tight in the epithelial lining, preventing a free exchange of solutes between the interstitial fluid and the alveolar lining layer.[12,16] In addition, the secretory type II cells are tightly integrated into the epithelial lining.

Because of the design of the cell barriers, the three fluid spaces are related to each other in different ways. Whereas the plasma and interstitial spaces can freely exchange solutes and even smaller macromolecules through the leaky endothelial junctions, the alveolar lining fluid is well separated from the interstitial space. The role of the pinocytotic vesicles in fluid exchange is still debated.[17,18] Between the fluid spaces is a dynamic but stable equilibrium. The regulation of the fluid exchange is not quite clear. However, several physiologic and structural safety factors prevent a deleterious flooding of alveoli.

SAFETY FACTORS OF PULMONARY FLUID BALANCE

Usually, a clear distinction is made between the physiologic and structural safety factors (Table 23-1); however, because of the complex structure-function relationships within the lung parenchyma, the boundaries are less easily recognizable. Among the

physiologic safety factors the low hemodynamic pressure in the pulmonary vessels, and in particular the pulmonary microvasculature, is the most important. Even at high perfusion rates, i.e., at a high cardiac output, the pressure increase in capillaries is relatively modest, presumably because some poorly perfused capillaries are "recruited" and because of a gentle distention of patent capillaries.[19,20] To some extent, the low transcapillary pressure difference is counterbalanced by differences in colloid-osmotic pressure between plasma and interstitium, resulting from a partial filter function of the endothelium; in contrast to systemic capillaries, the pulmonary capillaries are partly permeable to macromolecules, the reflection coefficient for albumin being less than 1.*[21,22] An increase in transendothelial fluid flow causes interstitial fluid to become diluted, increasing the colloid-osmotic pressure difference and limiting further fluid filtration.[22,23] Under otherwise normal conditions, however, the lungs can be perfused with isotonic salt solutions without the development of pulmonary edema.

The control and distribution of surface forces at the air-tissue interface are a further safety factor. Surface tension is kept near zero by the particular properties of the surfactant film; this results in a less negative interstitial pressure and, in turn, in a reduced transcapillary pressure difference. Small local variations in surface forces shift alveolar fluid away from the gas exchange barrier, and they tend to draw interstitial fluid along the fiber system toward alveolar corners[6] and eventually to the fluid sumps in the perivascular and peribronchial loose connective tissue. Abnormal increases in surface tension increase the capillary filtration rate by decreasing the

---

* The reflection coefficient (RC) is an index of the "leakiness" of a barrier to proteins; if RC = 1, there is no leakage; if RC = 0, proteins leak as rapidly as water.

TABLE 23-1
Safety Factors to Prevent Alveolar Flooding

Physiologic Factors	Structural Factors
Low hemodynamic pressures	Endothelial cell layer (leaky junctions)
High colloid-osmotic pressure of plasma	
Low colloid-osmotic pressure in the perimicrovascular space	Epithelial cell layer (tight junctions)
Low mechanical stress of alveolar septa owing to low alveolar surface tensions (surfactant)	Continuity of interstitial space for efficient drainage of excess fluid
	Efficient drainage by lymphatic system
Active resorption of excessive alveolar fluid	Fluid reservoirs in the peribronchial and perivascular connective tissue system

interstitial pressure; eventually they result in pulmonary edema.[24]

There is good evidence, both in vitro and in vivo,[25,26] that excessive alveolar fluid is actively reabsorbed. The mechanisms are not yet well understood. However, cell cultures of type II and Clara cells show a directed luminal to basal transport of sodium ions together with water. In patients with pulmonary edema, a continuous increase of the protein concentration of edema fluid could be shown during resorption of edema, which implies an active water transport even against a colloid-osmotic pressure difference.[26]

The single most important *structural safety factor* is the vitality of intact endothelial and epithelial cell layers as barriers. The epithelium is essentially impermeable to macromolecules such as plasma proteins; also, its permeability for electrolytes and water is much lower than that of the endothelium.[2,27] In fact, there is a continuous transendothelial fluid filtration that amounts to about 10 ml per hour in humans at rest.[22] Without an efficient drainage system, an increase in thickness of the blood-gas barrier of about 0.2 μm per hour would thus occur! Hence, other safety factors must be invoked that are related to the design of the alveolar septum. The interstitial spaces associated with fibers are capable of accepting excess fluid without affecting the thin part of the blood gas barrier, i.e., without impairment of gas exchange (Fig. 23-4A). Their compliance is substantial but not unlimited, because myofibroblast processes brace the space between the two basement membranes. This tends to force fluid to spread in the plane of the septum whereby the continuous fibrous meshwork establishes a pathway for rapidly draining fluid toward the fluid sumps in the loose peripheral connective tissue, where it reaches the lymphatics found in the surroundings of airways and vessels (Fig. 23-4B). If fluid filtration goes beyond the transport capacity of the extensive, highly organized lymphatic system of the lung, a transient safety factor comes into play: the considerable storage capacity of the loose connective tissue. The fluid "cuffs" around airways and vessels may amount to 0.5 L in the human lung.[28] However, this storage mechanism does not give full protection; alveolar edema may occur before the interstitial storage capacity is reached.[29]

PULMONARY EDEMA: BASIC MECHANISMS

Pulmonary edema is usually divided into two paradigmatic models according to the presumed particular pathogenetic mechanism. 1. *Hydrostatic pulmonary edema* is due to increases in pulmonary microvascular pressure. This results in transudation of fluid across the capillary walls into the interstitial spaces and eventually into the alveolar air spaces. It is commonly viewed as a passive filtration process with no or minimal injury to the cellular barriers, and the edema fluid usually shows relatively low concentrations of serum proteins. 2. *Permeability pulmonary edema* is characterized by direct injuries to the blood-gas barrier, i.e., to the endothelial and epithelial cell layers, with leakage of plasma into the interstitial and alveolar spaces. This basic concept has served well in the management of many clinical conditions in that hemodynamic disorders and the sequelae of direct lung tissue damage require different treatments. However, there are many situations in which the distinction between these two types of pulmonary edema is blurred, and recent observations challenge the simplistic pathogenetic models. In fact, even in "pure" hemodynamic pulmonary edema, there is evidence of structural alterations, such as disruption of the permeability barriers.[30–33] In "pure" permeability edema, a disequilibrium of forces regulating the fluid exchange may play an important role.[34–36]

Hydrostatic (Hemodynamic) Pulmonary Edema

The most common cause of interstitial and alveolar edema is an increase in capillary transmural pressure, due to congestive heart failure, overhydration, or both. Experimentally, the process of edema formation can best be observed in lungs fixed by vascular perfusion.[32] In mild alveolar edema, part of the excess fluid accumulates at alveolar sites with small radii of curvature, as in the "alveolar corners," or forms pools in the capillary meshes (Fig. 23-5). Together with the interstitial edema and the distention of the capillaries, the intercapillary fluid accumulation contributes to the widening of the alveolar septa without seriously interfering with gas exchange because, indeed, vast parts of the thin blood-gas barrier are still spared a fluid cover. It is interesting to note that fluid pools appear to be covered by a thin osmiophilic layer of surfactant that may still exert a stabilizing influence on the alveolar geometry.

With increasing transudation, a deeper fluid layer is formed that is still covered by an osmiophilic film. However, the capillaries are no longer molded by surface forces but, rather, bulge freely into the fluid layer, assuming the same configuration as in fluid-filled lungs.[37] This layer of edema now covers nearly the entire alveolar surface (Fig. 23-6); the alveoli lose their polyhedral shape and become spherical.[32] The only "dry" regions are the alveolar

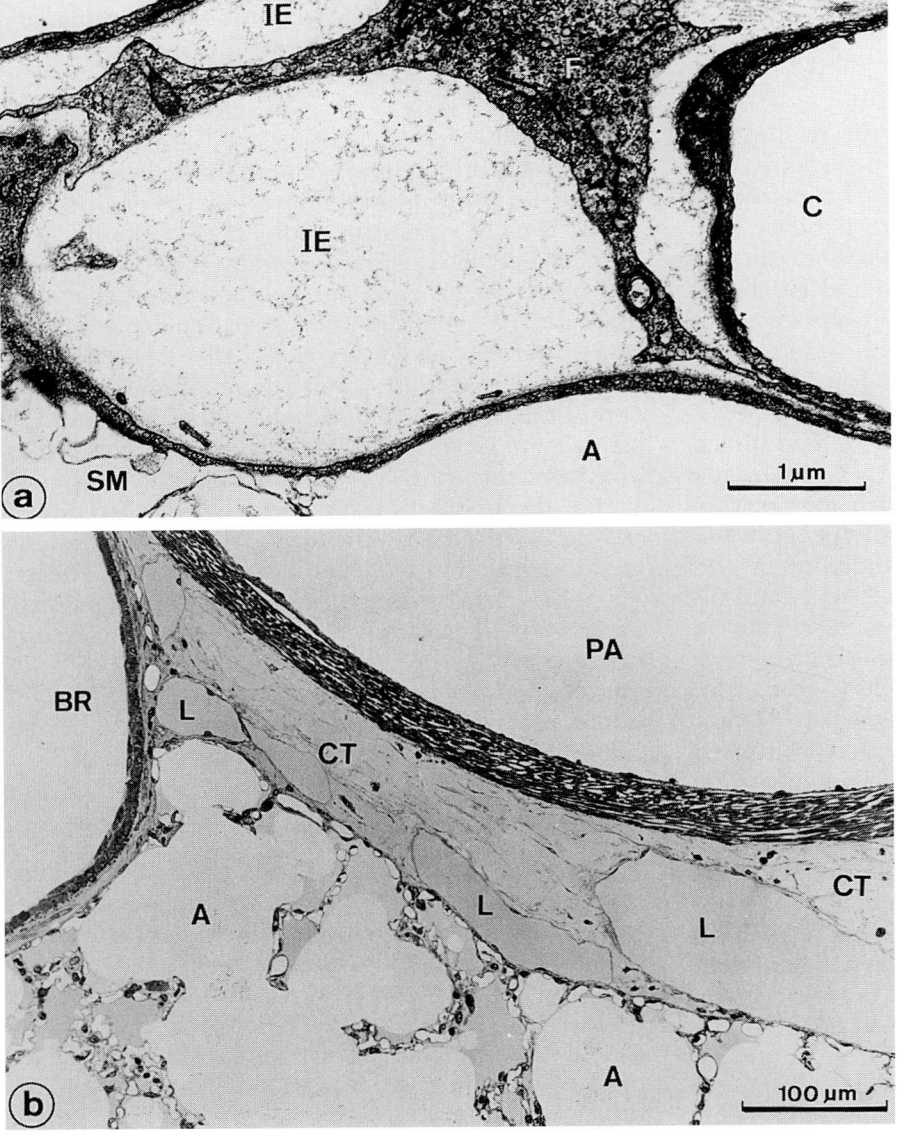

FIGURE 23-4
Interstitial lung edema. *A.* Electron micrograph of septum thickened by interstitial edema (IE), which is limited by myofibroblast processes (F) braced between the basement membranes (A, alveolar spaces; C, capillary; CF, collagen fibers; SM, surfactant material). *B.* Fluid sumps in the loose peripheral connective tissue (CT) surrounding vessels and airways (PA, pulmonary artery; BR, bronchiole; L, lymphatics; A, alveoli with small amounts of edema fluid).

entrance rings—the force-bearing elements of the alveolar ducts[38]—between which the fluid menisci appear to be expanded (Fig. 23-6). In accordance with the theoretical expectations,[39] both air-space stability and fluid equilibrium appear to be compromised at this stage of lung edema. Adjacent to relatively dry alveoli (Fig. 23-7), completely flooded air spaces can be observed, and the inhomogeneity is further enhanced by alterations of the air-space architecture, such as microatelectatic zones associated with large fluid accumulations on the one hand and widened alveolar ducts on the other. However, the unequal distribution of alveolar fluid might reflect not only problems in stability but also spatial differences in barrier permeability. This, in turn, could explain the apparent spatial differences in protein concentrations of interstitial and alveolar edema fluid (Fig. 23-8).

To sum up: the development of alveolar edema appears to be a gradual but inhomogeneous process. Alveolar edema is characterized by excessive alveolar fluid as well as changes in septal and alveolar structure. The structural adaptability of alveolar tissue can be viewed as an additional safety factor, in that appreciable amounts of fluid can be accommodated at first without seriously impairing gas exchange.

High-pressure edema has been thought to arise from the imbalance of forces affecting the normal process of fluid exchange, as described by the

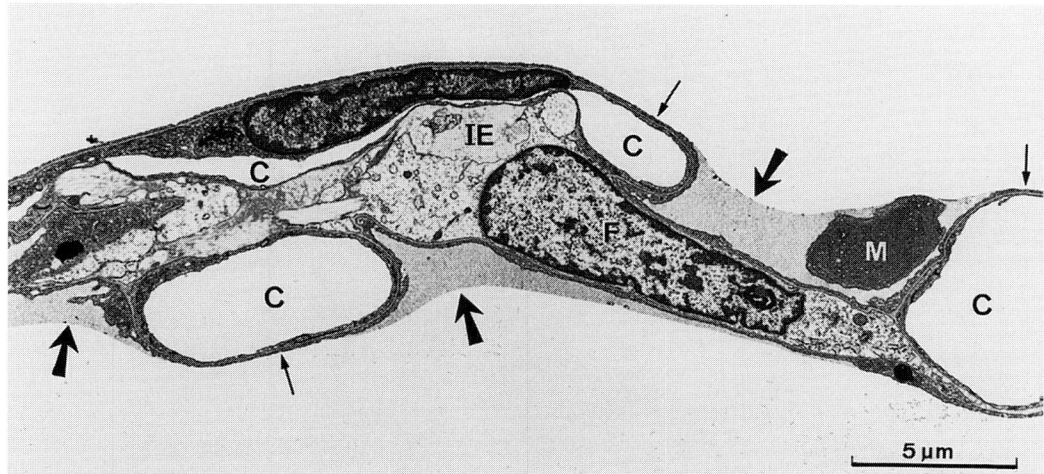

FIGURE 23-5
Mild lung edema in an excised rabbit lung fixed by vascular perfusion with glutaraldehyde and osmium tetroxide. Note small pools of edema fluid (*thick arrows*) in depressions between capillaries, which do not encroach on the thin gas-exchanging blood-gas barrier (*thin arrows*) (C, capillaries; F, fibroblast; IE, interstitial edema; M, macrophage).

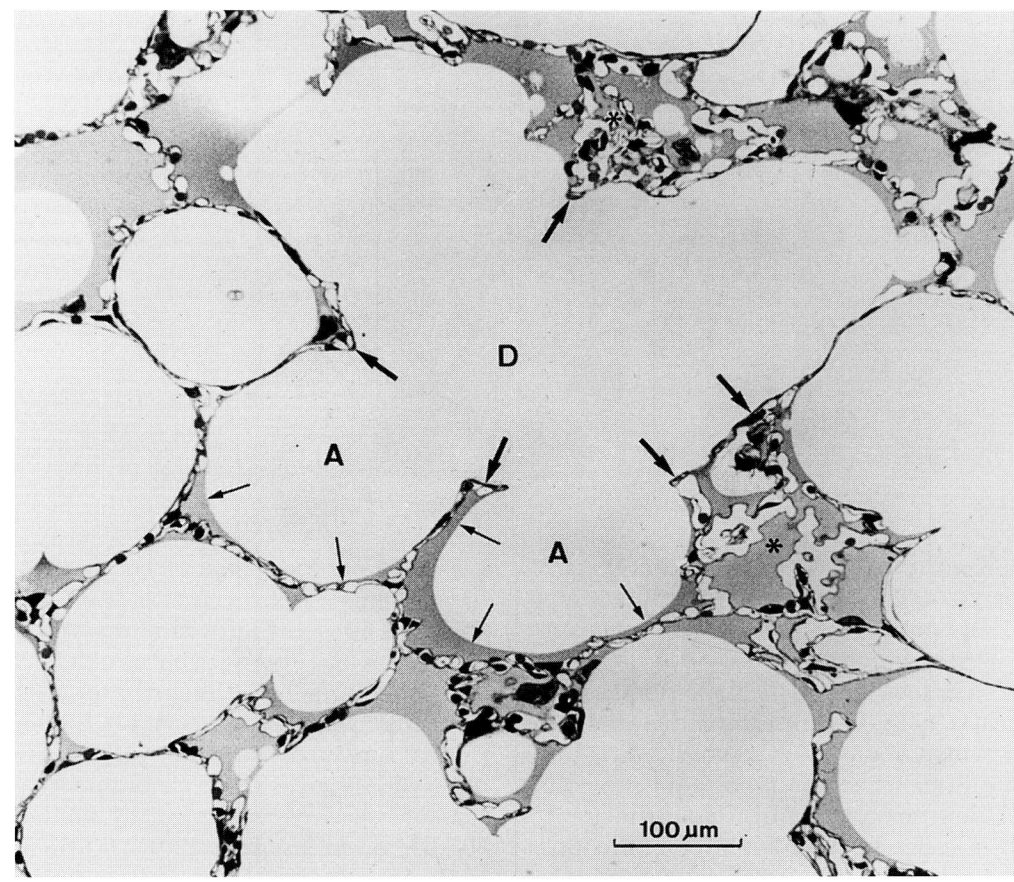

FIGURE 23-6
Low-power view of an excised rabbit lung with moderate degree of alveolar edema. Note smooth "spherical" air-fluid interfaces (*thin arrows*) of alveoli (A) and free alveolar entrance rings (*thick arrows*) around alveolar ducts (D) and the irregular arrangement of alveolar septa in larger fluid pools (*asterisks*).

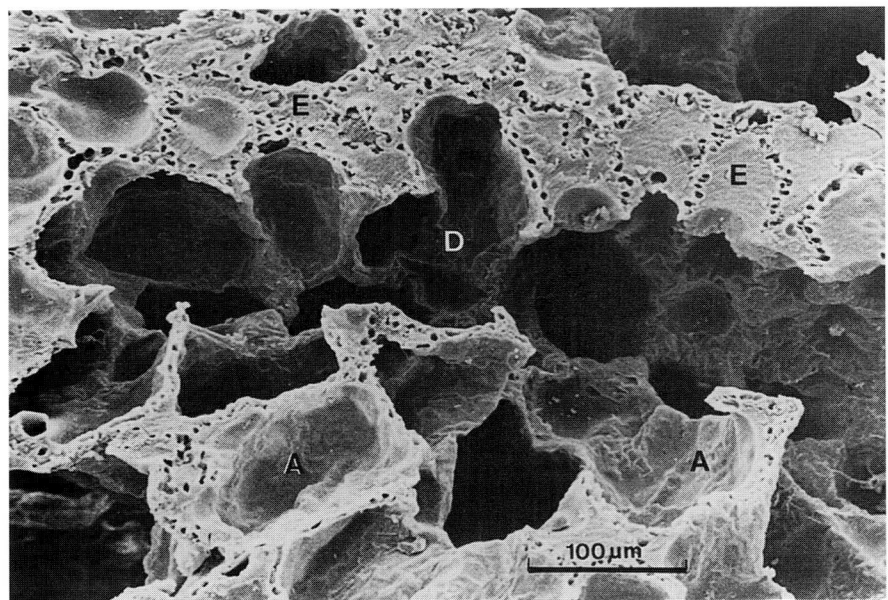

FIGURE 23-7
Scanning electron micrograph illustrating the inhomogeneous distribution of edema fluid. Within the same alveolar duct (D) are normally air-filled alveoli (A) and alveoli completely filled with edema fluid (E).

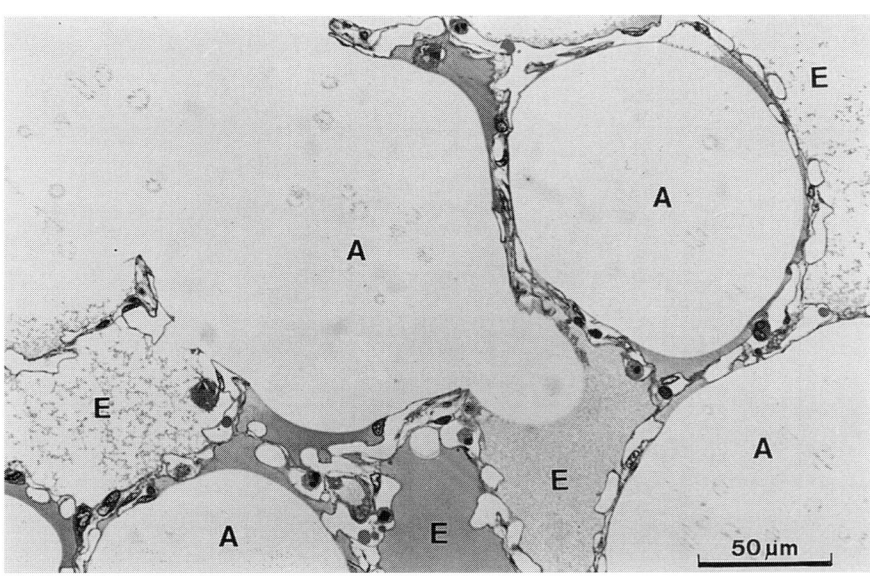

FIGURE 23-8
Differences in protein concentrations of alveolar edema fluid (E) between adjacent alveoli (A).

Starling hypothesis, rather than from structural alterations of the barriers. This concept was supported by two clinical observations: The first is the rapid termination of the outflow of edema fluid once the equilibrium of forces is reestablished. The second is the preservation of a selective permeability of the barrier, as indicated by the amount and type of the proteins in the edema fluid.[2–4] Most important, previous ultrastructural findings obtained in edematous animal lungs and in lungs from patients with left-sided heart failure revealed intact endothelial and epithelial cell layers with closed intercellular junctions.[40]

Several observations suggest that the distinction between permeability edema of acute lung injury and hydrostatic edema cannot be based exclusively on the presence or absence of barrier lesions. The conspicuous presence of numerous erythrocytes in the interstitial and alveolar spaces of hemodynamic lung edema in humans is inconsistent with the application of Starling's hypothesis (Fig. 23-9A,B). This finding is in line with the well-established clinical observation of "rusty" sputum and iron-laden macrophages expectorated by patients with left-sided heart failure (Fig. 23-9C). Furthermore, physiologic experiments showed an increased permeability of the barrier at high capillary pressure. Finally, in a recent experiment in rabbit lungs, which were made edematous by a moderate increase in capillary pres-

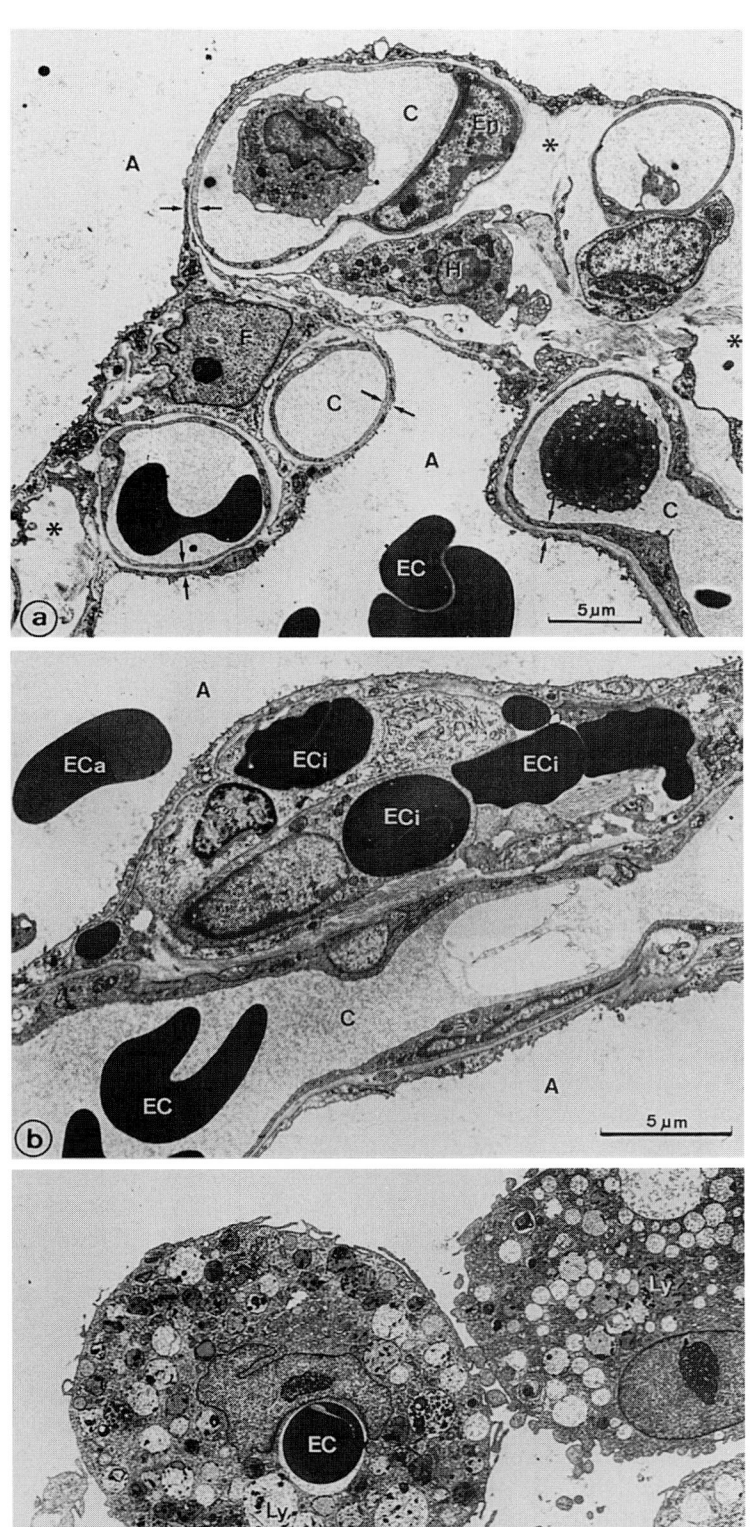

FIGURE 23-9
Acute hemodynamic edema in patient who died of left-heart failure.   *A.* Alveoli (A) are fluid-filled and show some sign of hemorrhage (EC, erythrocytes). Interstitial edema (*asterisk*) with histiocyte (H) is limited to thick barrier parts, while thin parts (*paired arrows*) are spared (C, capillaries; En, endothelial cell; F, fibroblast).   *B.* Erythrocytes are seen in capillary (EC), alveolar space (ECa), and interstitial space (ECi). *C.* Macrophages from alveolar fluid show signs of vigorous phagocytic activity, with numerous secondary lysosomes (Ly) and an ingested erythrocyte (EC).

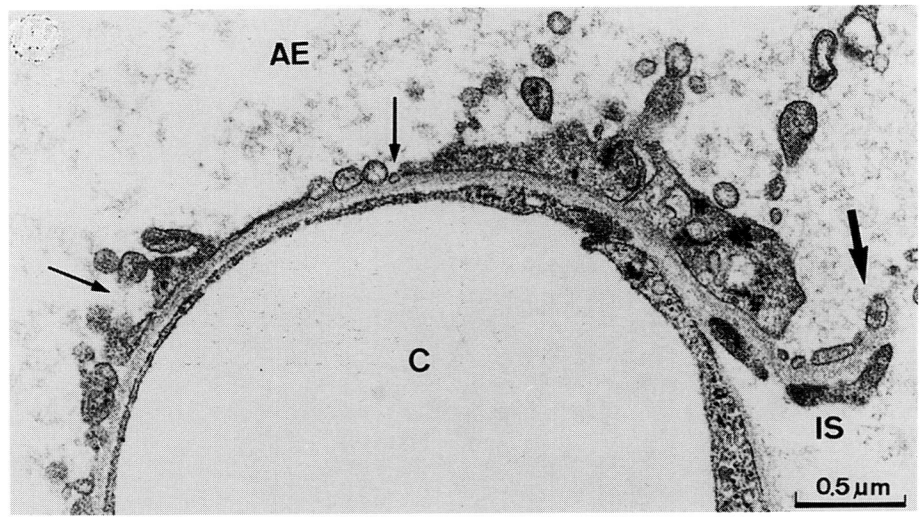

FIGURE 23-10
Epithelial breaks (*arrows*) on both thin and thick (*thick arrow*) parts of the barrier in rabbit lungs with experimental hydrostatic pulmonary edema. The interstitial space (IS) is distended by edema fluid (C, capillary; AE, alveolar edema).

sure (20 to 40 cm H$_2$O), frank discontinuities of the epithelial cell layer could be observed[33] on both the thin and thick sides of the barrier (Fig. 23-10). Thus, in both hydrostatic and permeability edema, structural alterations of the barrier play a role.

However, signs of inflammation are conspicuously rare in pure hemodynamic edema: neither intracapillary sequestrations and extravascular accumulations of leukocytes nor abnormal coagulation products, such as intravascular and extravascular fibrin deposits, can be seen. In contrast to the mechanical disruptions seen in lung tissue after extremely high pressures[31,41] or caused by pulmonary barotrauma[42] or to the toxic-inflammatory lesions of the barrier outlined in the following paragraph, the capillary basement membrane was found to be intact in all experiments studying hydrostatic pulmonary edema.[33] It was also intact in tissue samples obtained from patients with lung edema due to congestive heart failure. The escape of blood cells from the capillaries into interstitial and alveolar spaces suggests, however, that focal discontinuities in the basement membrane must occur, but they are rare and of transient nature. Such gaps can probably be quite small considering the enormous plasticity of leukocytes and erythrocytes in transit.

Permeability Pulmonary Edema—Acute Lung Injury Syndrome (Chap. 11)

With regard to its definition, causes, and pathogenesis, permeability edema is even more complicated than hemodynamic pulmonary edema.[43,44] Many causes of acute lung injury have been reported in humans and experimental animals (Chap. 11). The clinical hallmark is a high protein concentration in the alveolar edema fluid relative to plasma.[44–47] A second criterion is edema formation in the absence of

increased hemodynamic pressures and in particular at normal or low wedge pressures. In any case, the common initial lesions are direct lesions to the endothelial and epithelial cell layers with a corresponding increase in barrier permeability, which in turn results in profuse interstitial and alveolar edema.

In permeability edema induced by oleic acid, positron emission tomography could show that the increase in permeability as reflected by the transcapillary escape rate for transferrin closely correlates with the extent of damage to the endothelial and epithelial cell layers.[48] Almost instantaneously, the toxic agents and the resulting cellular lesions elicit inflammatory responses that amplify the destructive effect.[49] In particular, activated leukocytes accumulate and appear to assume a central role in destroying the barrier by releasing proteolytic enzymes, toxic oxygen species, and soluble mediators of inflammation. This vicious circle results in a pattern of structural alterations, which—at least in human lungs—does not reveal whether the overt leaks of the air-blood barrier were caused by direct toxic injury to the alveolar structures or by secondary injury elicited by the inflammatory response.

The complex interplay between the primary injurious agent and the damaging effect of secondary mechanisms is well illustrated in permeability edema elicited by pulmonary microemboli. Seemingly inert emboli, such as air and particles (e.g., glass beads and colloidal carbon particles), increase pulmonary microvascular permeability.[50] During air microembolization, the number of sequestered and marginated neutrophils increases severalfold in the pulmonary microcirculation. A layer of neutrophils is aggregated along the blood-air embolus interface. Superoxide dismutase with heparin prevents an increase in permeability.[47,51] This observation suggests that air emboli activate neutrophils, resulting in leakage of the permeability barriers due to released oxidants.

An even more perplexing model of microvascular injury induced by microemboli is the fat embolism syndrome.[52] This multisystem disorder, characterized by pulmonary and neurologic injuries, pyrexia, and a petechial rash, occurs most often after blunt trauma complicated by long-bone fractures. It can also be observed in association with acute pancreatitis, joint reconstructions, liposuction, and parenteral infusion of lipids. Pulmonary manifestations range from slight hypoxemia to acute respiratory failure caused by hemorrhagic pulmonary edema. The alveoli are filled with a proteinaceous fluid, red blood corpuscles, and also leukocytes, as markers of the ensuing inflammation. Adjacent to capillary fat emboli, there is not only damage and disruption of the endothelial and epithelial cell layers but also lysis of the fused basement membrane and hence a complete break of the blood-gas barrier (Fig. 23-11). The pathogenesis of this extreme case of injury to the permeability barrier is not clear. It is interesting to note that a small minority of patients with fat emboli have signs of lung injury. This suggests that in addition to local and direct injuries, inflammatory and humoral factors as well as individual susceptibility might play a role in some types of permeability edema.

In most conditions of permeability edema in human lungs, however, disruption of the epithelial cell layer is most conspicuous—i.e., that part of the barrier whose integrity is central to keeping the alveoli dry.[2] Usually, all stages of epithelial damage are visible, ranging from mere cytoplasmic swelling to total cellular destruction, leaving a bare but usually intact basement membrane that may become covered by hyaline membranes.[45,47,53,54] Overt endothelial defects are less frequently seen. Even in the immediate vicinity of completely destroyed epithelial cell extensions, separated by the basement membrane only, endothelial linings are usually continuous with morphologically intact cell junctions (Fig. 23-12). Occasional endothelial defects are covered with fibrin deposits. Endothelial gaps can also be observed at sites where blood cells, especially leukocytes, are about to escape into the extravascular space.

The discrepancy in the extent of damage between epithelium and endothelium may be explained by structural differences in the cytoplasmic extensions and the cell junctions on the one hand[27] and by the different repair potentials of the two cell layers on the other. Epithelial cell junctions are extremely tight, and pressure stresses result in

overdistention and disruption of the delicate cell extensions rather than in the opening of intercellular junctions. By contrast, the tight junctions between endothelial cells are to some extent leaky, leaving a reversible pathway for the passage of large molecules and blood cells without destruction of endothelial cells.

The hypothesis of a different repair capacity is supported by findings in lung biopsies of patients after severe hypovolemic-traumatic shock, before the development of full-blown permeability edema.[55] In these biopsies obtained shortly after injury, widespread lesions of endothelial cells have been observed, including cytoplasmic swelling, loss of organelles, and focal necrosis. The enormous plasticity and high repair capacity of endothelial cells have been demonstrated in numerous animal experiments[56–58]; these characteristics are supported by observations in human tissue samples. Focal fibrin deposits and small microthrombi are generally covered by tentacular endothelial cell extensions, and conspicuous fragmentation of the endothelial cell layer into multiple tightly joined segments may suggest a prompt repair of endothelial gaps (Fig. 23-13). On the other hand, the repair of the epithelial layer, and hence the reestablishment of a tight permeability barrier, requires more time, since the substitution of destroyed squamous epithelial cells occurs through proliferation and transdifferentiation of granular pneumocytes.[59,60]

The pathologic features of diffuse acute lung injury are the consequence of vast leakages in the air-blood barrier.[33] In the first exudative phase, the extravasation of protein-rich edema fluid, together with numerous inflammatory cells into the alveolar and interstitial spaces, dominates the pathologic pattern. The alveolar spaces are inhomogeneously filled with a proteinaceous and often hemorrhagic fluid containing neutrophils, macrophages, and cell fragments, beside amorphous material composed of plasma proteins, cell debris, fibrin strands, and remnants of surfactant. Occasionally, condensed sheet-like conglomerates of this exudate stick to alveolar surfaces wherever the latter are devoid of the epithelial lamina, forming hyaline membranes easily recognizable with light microscopy.

Typically, hyaline membranes are also located in alveolar ducts, where they cover the alveolar entrance rings. Most interstitial spaces are enlarged by accumulation of a cell-rich fluid, although a difference becomes apparent between the interstitial spaces of the peripheral and axial connective tissue system and those of the alveolar septa. In the former, particularly in the connective tissue cuffs around

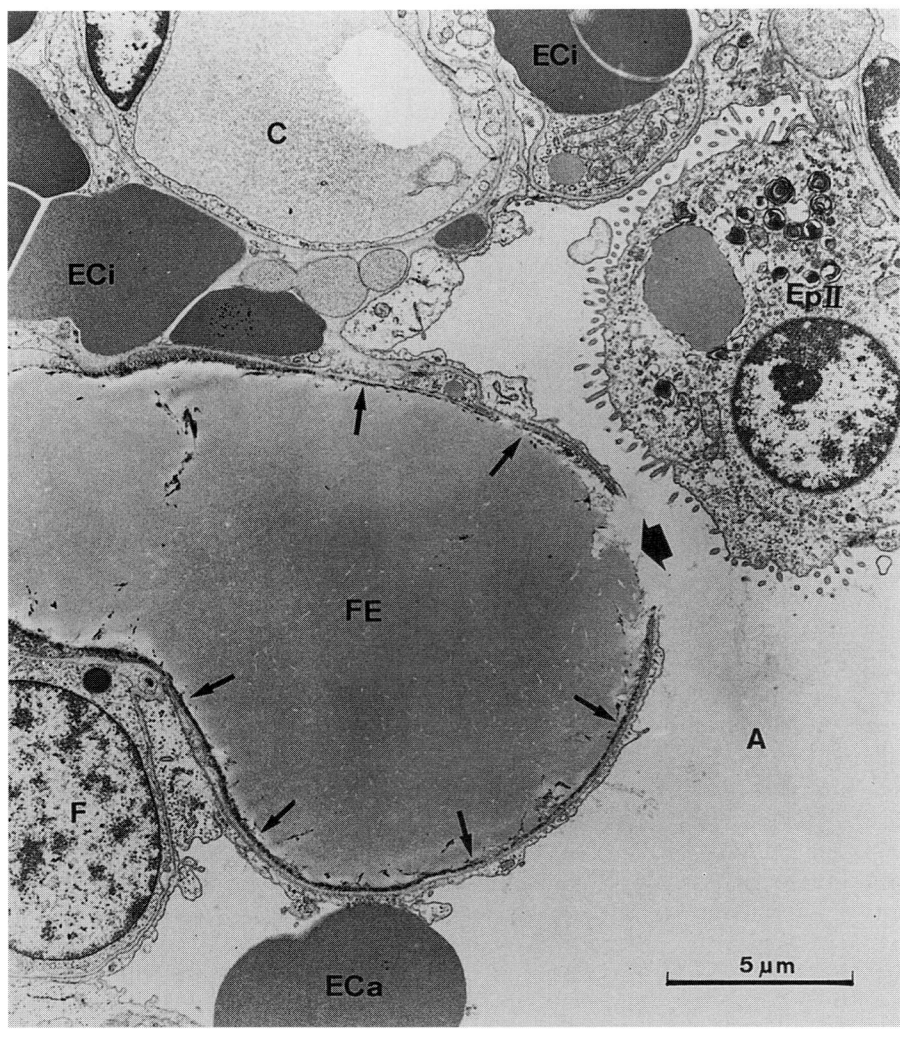

FIGURE 23-11
Fat embolism syndrome in human subject. Fat embolus (FE) in a lung capillary with a large break of the entire barrier (*thick arrow*). The endothelial cell layer adjacent to the fat embolus is destroyed (*thin arrows*). Erythrocytes in the interstitial space (ECi) and the alveolar space (ECa) are markers of diffuse hemorrhage (C, capillary; EpII, epithelial type II cell with ingested fat; F, fibroblast; A, alveolar space).

larger vessels and bronchioles, the abundant edema fluid rarely includes migrated blood cells or scraps of fibrin. The connective tissue of the alveolar septum, on the other hand, is engorged predominantly by the extravasation of cellular components, and local fibrin deposits give evidence of an activated coagulation process. At the exudative stage of the disease, the tissue compartment with the least spectacular structural alterations often is the capillary network.

Occasional obliterations resulting from microthrombi or capillary narrowing by voluminous interstitial edema cause, at best, a moderate capillary volume reduction as determined by morphometry.[53] Noteworthy, however, are increased numbers of intravascular granulocytes, which sometimes even exceed the number of red blood cells, especially in patients with septicemia.[53] At the worst, numerous capillaries are completely plugged by thrombi of sequestered neutrophils, seriously impairing capillary perfusion. As a further sign of disturbed microcirculation, erythrocytes can become highly con-

centrated in some parts of the capillary bed, whereas other parts contain plasma without cells, possibly as a consequence of capillary and precapillary microthrombi.

Detailed examination of the vascular bed of acutely injured lungs has shown that the vascular lesions are not restricted to the microvasculature, but that extensive alterations also occur in small extraalveolar vessels (both arteries and veins).[61] Vascular casts have revealed widespread stenoses and occlusions of small arteries, resulting in uneven perfusion of lung parenchyma.

PARTICULAR TYPES OF PULMONARY EDEMA

In part because of the association to particular disease states or situations and in part because of the apparently enigmatic pathogenesis, several types of pulmonary edema are often classified as distinct entities: high-altitude, neurogenic, and uremic pul-

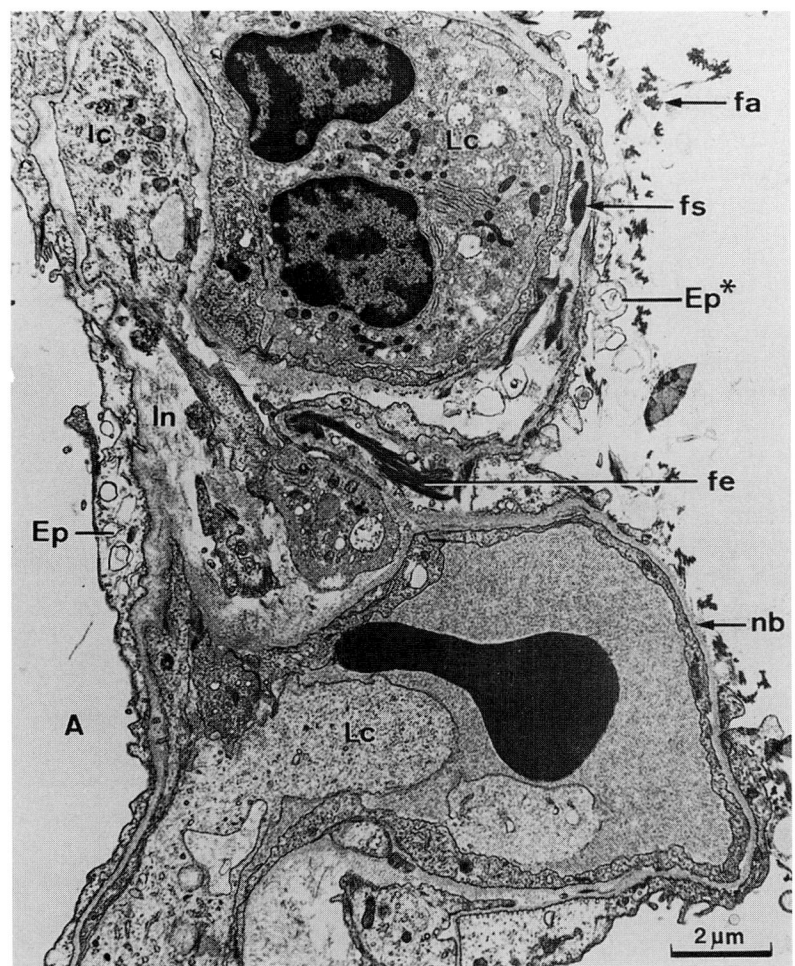

FIGURE 23-12
Alveolar septum from a patient with acute lung injury and permeability lung edema due to septicemia. Note the extensive damage to the epithelial cell layer (Ep*); in some places the basement membrane is denuded (nb). Although the endothelial cells appear intact, damage to the vascular endothelium is evidenced by the presence of fibrin beneath the endothelium (fs), beneath the epithelium (fe), and in the alveolar space (fa) (A, alveolar space; Ic, interstitial cell; In, interstitium; Lc, leukocytes in capillaries).

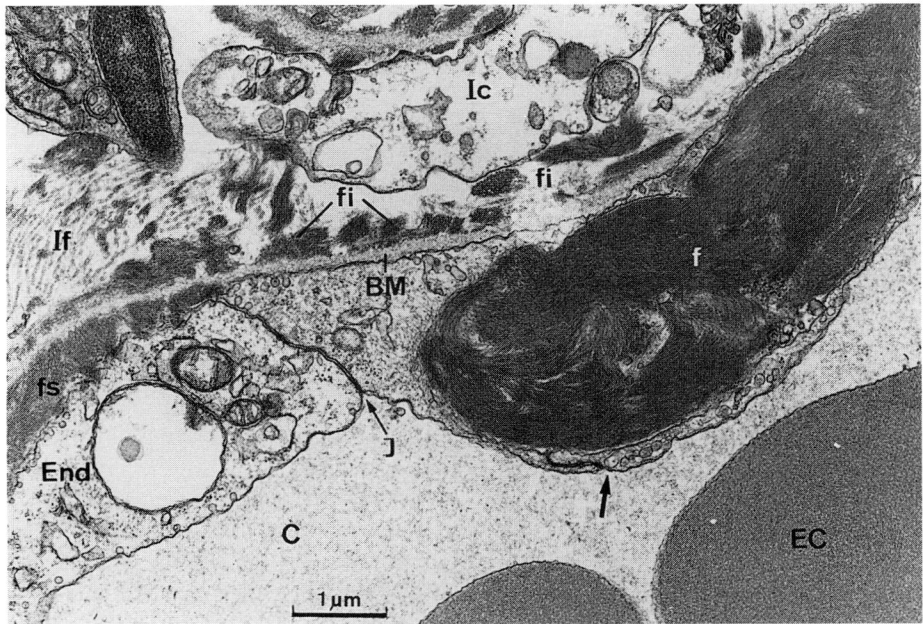

FIGURE 23-13
Capillary wall in a patient with acute lung injury due to septicemia. Fibrin specks are seen between the basement membrane (BM) and the endothelium (fs), as well as in the interstitial space (fi). A large fibrin mass (f) has been enwrapped by endothelial cell flaps, whose joined edges are seen at the heavy arrow (C, capillary; EC, erythrocyte; J, intercellular junction between endothelial cells [End]; If, interstitial fibers; Ic, interstitial cell).

monary edema. The categorization is of clinical importance for the management of individual patients, and these anomalies have served as useful models to elucidate the pathogenetic mechanisms of alveolar flooding.

High-Altitude Pulmonary Edema

High-altitude pulmonary edema (HAPE) falls into the category of noncardiogenic pulmonary edema that develops within the first days after rapid ascent to altitudes above about 3000 m in otherwise healthy subjects.[62,63] Under treatment with oxygen or after a rapid descent, the edema fluid swiftly reabsorbs without further sequelae. The exact mechanism of this disorder has not been fully clarified. However, it has been shown that the edema fluid has a high protein concentration and occasionally also contains numerous erythrocytes. These data suggest that HAPE is a permeability type of edema. On the other hand, there is a correlation between HAPE and the occurrence of pulmonary hypertension caused by hypoxic pulmonary vasoconstriction. On the assumption of an uneven hypoxic constriction of local arterioles, a high hydrostatic pressure could be transmitted to parts of the microvasculature—resulting in mechanical damage to parts of the permeability barrier, with an ensuing flooding of alveolar spaces with plasma and blood cells.[31] The question, then, is why the inflammatory reaction is minimal in such a mixed type of hydrostatic and permeability edema.

Neurogenic Pulmonary Edema

Neurogenic pulmonary edema is associated with acute brain injury and a rapid increase in intracranial pressure.[64–66] The often hemorrhagic edema fluid shows the characteristics of permeability edema. However, much experimental evidence suggests a cascade of pathogenetic events similar to that in HAPE. Both in experimental animals and in patients, it has been shown that a sudden increase in intracranial pressure may cause a massive adrenergic discharge, which in turn leads to pressure bursts in both systemic and pulmonary circulation. Together with the ensuing redistribution of blood from the systemic into the pulmonary circulation, an extreme mechanical (hydrostatic) stress on the microvasculature may cause disruptions of the permeability barrier. This hypothesis is well supported by electron-microscopic findings in lung tissue from experimental animals and from patients: the high-protein edema fluid is associated with conspicuous foci of interstitial and alveolar hemorrhage, and occasional disruptions of the endothelial and epithelial cell layers can be observed (Fig. 23-14).

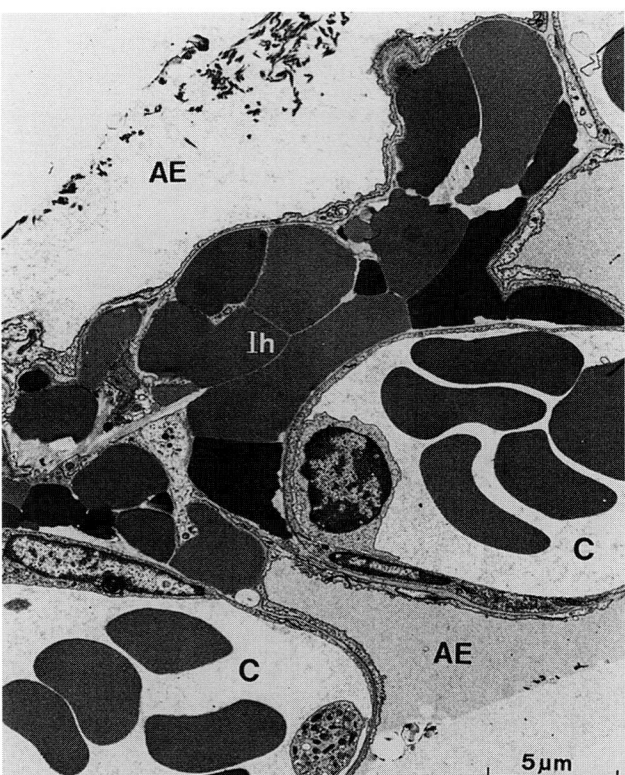

FIGURE 23-14
Neurogenic pulmonary edema in a patient after fatal head trauma. Endothelial and epithelial cell layers appear intact (AE, protein-rich alveolar edema with fibrin specks; Ih, interstitial hemorrhage; C, capillaries).

Uremic Pulmonary Edema

Quite frequently, pulmonary edema is a complication of uremia. On chest radiographs, it may occasionally show a "bat's wing" pattern: the perihilar alveolar regions are predominantly flooded, while the peripheries of the lung remain translucent. On the basis of analyses of edema fluid and histologic findings, this type of edema can be categorized as permeability edema. Indeed, the edema fluid contains high concentrations of plasma proteins.[67,68] Microscopically, the peripheral air spaces are filled with a proteinaceous fluid and a variable number of blood cells. Strands of fibrin deposition and hyaline membrane formation can be observed,[69,70] suggesting a leaky barrier that is possibly due to toxic byproducts of uremia.[71,72] This hypothesis is supported by the observation that uremic pulmonary edema is occasionally accompanied by an exudative pericarditis and pleuritis. However, in most cases of pulmonary edema associated with renal failure, numerous factors probably play a role. Because of the generally observed hypoproteinemia and hence the low colloid-osmotic pressure, one safety factor is impaired.[67] A hyperdynamic circulation, in part due to anemia, ventricular dysfunction, and, above all, substantial overhydration, imposes considerable hydrostatic stress on the pulmonary microvasculature. The importance of this factor is demonstrated by the fact that the reduction of the circulatory volume by dialysis is usually followed by a rapid resorption of pulmonary edema.

REFERENCES

1. Weibel ER: *The Pathway for Oxygen*. Cambridge, Mass., Harvard University Press, 1984.
2. Staub NC: Pulmonary edema. *Physiol Rev* 54:678–811, 1974.
3. Fishman AP, Renkin EM (eds): *Pulmonary Edema*. Bethesda, Md., American Physiological Society, 1979.
4. Matthay MA: Pathophysiology of pulmonary edema. *Clin Chest Med* 5:301–314, 1985.
5. Patrick G, Stirling C: A method for microinjection into subpleural alveoli of rat lung in situ. *J Appl Physiol* 60:307–310, 1986.
6. Weibel ER, Bachofen H: Structural design of the alveolar septum and fluid exchange, in Fishman AP, Renkin EM (eds), *Pulmonary Edema*. Bethesda, Md., American Physiological Society, 1979, pp 1–20.
7. Weibel ER: Functional morphology of lung parenchyma, in Macklem PT, Mead J (eds), *Mechanics of Breathing*, Part I: *Handbook of Physiology*, Section 3: *The Respiratory System*. Bethesda, Md., American Physiological Society, 1986, pp 89–111.
8. Fishman AP, Pietra GG: Hemodynamic pulmonary edema, in Fishman AP, Renkin EM (eds): *Pulmonary Edema*. Bethesda, Md., American Physiological Society, 1979, pp 79–96.
9. Staub NC, Nagano H, Pearce MC: Pulmonary edema in dogs, especially the sequence of fluid accumulation in lungs. *J Appl Physiol* 22:227–240, 1967.
10. Lauweryns JM, Baert JH: Alveolar clearance and the role of pulmonary lymphatics. *Am Rev Respir Dis* 115:625–683, 1977.
11. Simionescu M: Lung endothelium: Structure-function correlates, in Crystal RG, West JB, Barnes PJ, et al. (eds), *The Lung: Scientific Foundations*. New York, Raven Press, 1991, pp 301–312.
12. Weibel ER: Lung cell biology, in Fishman AP, Fisher AB (eds), *Circulation and Nonrespiratory Functions. Handbook of Physiology*, Section 3: *The Respiratory System*. Bethesda, Md., American Physiological Society, 1985, pp 47–91.
13. Stone KC, Mercer RR, Gehr P, Stockstil B, Crapo JD: Allometric relationships of cell numbers and size in the mammalian lung. *Am J Respir Cell Mol Biol* 6:235–243, 1992.
14. Nistor A, Simionescu M: Uptake of low density lipoproteins by the hamster lung: Interactions with capillary endothelium. *Am Rev Respir Dis* 134:1266–1272, 1986.
15. Chinard FP: Pulmonary edema, in McLellan H et al. (eds), *Advances in Physiological Research*. New York, Plenum Publishing, 1987, pp 353–376.
16. Schneeberger EE, Karnovsky MJ: Substructure of intercellular junctions in freeze-fractured alveolar-capillary of mouse lung. *Circulation Res* 38:404–411, 1976.
17. Palade GE: Role of plasmalemmal vesicles, in Crystal RG, West JB, et al. (eds), *The Lung: Scientific Foundations*. New York, Raven Press, 1991, pp 359–367.
18. Rippe B, Crone C: Pores and intercellular junctions, in Crystal RG, West JB, et al. (eds), *The Lung: Scientific Foundations*. New York, Raven Press, 1991, pp 349–357.
19. Gil J: Organization of microcirculation in the lung. *Ann Rev Physiol* 42:177–182, 1980.
20. Bachofen H, Wangensteen OD, Weibel ER: Surfaces and volumes of alveolar tissue under zone II and zone III conditions. *J Appl Physiol* 53:879–885, 1982.
21. Wangensteen OD, Lysaker E, Savaryn P: Pulmonary capillary filtration and reflection coefficients in the adult rabbit. *Microvasc Res* 14:81–97, 1977.
22. Taylor AE, Parker JC: Pulmonary interstitial spaces and lymphatics, in Fishman AP, Fisher AB (eds), *Handbook of Physiology*, Section 3: *The Respiratory System*. Bethesda, Md., American Physiological Society, 1985.
23. Gee MH, Spath JA: The dynamics of lung fluid filtration in dogs with edema. *Circ Res* 46:796–801, 1980.
24. Nieman GF, Bredenberg CE: High surface tension pulmonary edema induced by detergent aerosol. *J Appl Physiol* 58:129–136, 1985.
25. Check JM, Kim KJ, Crandall ED: Tight monolayers of rat alveolar cells: Bidirectional properties and active sodium transport. *Am J Physiol* 256:C688–694, 1989.
26. Matthay MA, Wiener-Kronish JP: Intact epithelial barrier function is critical for the resolution of alveolar edema in humans. *Am Rev Respir Dis* 142:1250–1257, 1990.
27. Schneeberger EE: Barrier function of intercellular junctions in adult and fetal lungs, in Fishman AP, Renkin EM (eds), *Pulmonary Edema*. Bethesda, Md., American Physiological Society, 1979, pp 21–38.
28. Staub NC: Alveolar flooding and clearance. *Am Rev Respir Dis* 127:544–551, 1983.
29. Zumsteg TA, Havill AM, Gee MH: Relationships among lung extravascular fluid compartments with alveolar flooding. *J Appl Physiol* 53:267–271, 1982.
30. Bachofen H, Bachofen M, Weibel ER: Ultrastructural aspects of pulmonary edema. *J Thorac Imag* 3:1–7, 1988.
31. West JB, Tsukimoto K, Mathieu-Costello O, Preditetto R: Stress failure in pulmonary capillaries. *J Appl Physiol* 70:1731–1742, 1991.
32. Bachofen H, Schürch S, Michel RP, Weibel ER: Experimental hydrostatic pulmonary edema in rabbit lungs: Morphology. *Am Rev Respir Dis* 147:989–996, 1993.
33. Bachofen H, Schürch S, Weibel ER: Experimental hydrostatic pulmonary edema in rabbit lungs: Barrier lesions. *Am Rev Respir Dis* 147:997–1004, 1993.
34. Meyrick B, Brigham KL: Acute effects of *Escherichia coli* endotoxin on the pulmonary microcirculation of anesthetized sheep: Structure-function relationships. *Lab Invest* 48:458–470, 1983.
35. Robin ED: Permeability lung edema, in Fishman AP, Renkin EM (eds), *Pulmonary Edema*. Bethesda, Md., American Physiological Society, 1979.
36. Newman JH: Sepsis and pulmonary edema. *Clin Chest Med* 6:371–391, 1985.
37. Gil J, Bachofen H, Gehr P, Weibel ER: Alveolar volume-surface area relation in air- and saline-filled lungs fixed by vascular perfusion. *J Appl Physiol* 47:990–1001, 1979.
38. Wilson TA, Bachofen H: A model for mechanical structure of the alveolar duct. *J Appl Physiol* 52:1064–1070, 1982.
39. Wilson TA: Effect of alveolar wall shape on alveolar water stability. *J Appl Physiol* 51:222–224, 1981.
40. Cotrell TS, Levine OR, Senior RM: Electron microscopic alterations at the alveolar level in pulmonary edema. *Circ Res* 21:783–798, 1967.

41. Tsukimoto K, Mathieu-Costello O, Prediletto R, Elliott AR, West JB: Ultrastructural appearances of pulmonary capillaries at high transmural pressures. *J Appl Physiol* 71:573–582, 1991.

42. Voegeli E, Bachofen M: Wichtige radiologische Befunde nach stumpfem Thorax-Trauma. *Roentgen Bl* 29:313–322, 1976.

43. Montaner JSG, Tsang J, Evans KG, et al: Alveolar epithelial damage: A critical difference between high pressure and oleic acid-induced low pressure pulmonary edema. *J Clin Invest* 77:1786–1796, 1986.

44. Robin ED: Permeability lung edema, in Fishman AP, Renkin EM (eds), *Pulmonary Edema.* Bethesda, Md., American Physiological Society, 1979, pp 217–228.

45. Bachofen M, Bachofen H, Weibel ER: Lung edema in the adult respiratory distress syndrome, in Fishman AP, Renkin EM (eds), *Pulmonary Edema.* Bethesda, Md., American Physiological Society, 1979, pp 241–252.

46. Gelb AP, Klein E: Hemodynamic and alveolar proteins studies in noncardiac edema. *Am Rev Respir Dis* 114:831–835, 1976.

47. Albertine KH: Ultrastructural abnormalities in increased permeability pulmonary edema. *Clin Chest Med* 6:345–369, 1985.

48. Velasquez M, Weibel ER, Kuhn C III, Schuster DP: PET evaluation of pulmonary vascular permeability: A structure-function correlation. *J Appl Physiol* 70:2206–2216, 1991.

49. Spragg RG, Smith RM: Biology of acute lung injury, in Crystal RG, West JB, et al. (eds), *The Lung: Scientific Foundations.* New York, Raven Press, 1991, pp 2003–2017.

50. Malik AB, Staub NC: Mechanism of lung microvascular injury. *Ann NY Acad Sci* 384:1–582, 1982.

51. Flick MR, Hoeffel JM, Staub NC: Superoxide dismutase with heparin prevents increased lung vascular permeability during air emboli in sleep. *J Appl Physiol* 55:1284–1291, 1983.

52. Levy D: The fat embolism syndrome: A review. *Clin Orthop* 261:281–286, 1990.

53. Bachofen M, Weibel ER: Alterations of the gas exchange apparatus in adult respiratory insufficiency associated with septicemia. *Am Rev Respir Dis* 116:589–615, 1977.

54. Bachofen M, Bachofen H: Acute lung injury: Parenchymal changes, in Crystal RG, West JB (eds), *Lung Injury.* New York, Raven Press, 1992, pp 259–269.

55. Schlag G, Redl HR: Morphology of the human lung after traumatic injury, in Zapol WM, Falke KJ (eds), *Acute Respiratory Failure.* New York, Marcel Dekker, 1985, pp 161–184.

56. Till GO, Johnson KJ, Kinkel R, Ward PA: Intravascular activation of complement and acute lung injury—dependency on neutrophils and toxic oxygen metabolites. *J Clin Invest* 69:1126–1135, 1982.

57. Reidy MA, Schwartz SM: Endothelial regeneration: III. Time course of intimal changes after small defined injury to rat aortic endothelium. *Lab Invest* 44:301–308, 1981.

58. Cunningham AL, Hurley JV: Alpha-naphthyl-thiourea-induced pulmonary oedema in the rat: A topographical and electron-microscope study. *J Pathol* 106:25–35, 1972.

59. Adamson IYR, Bowden DH: The type 2 cell as progenitor of alveolar epithelial regeneration: A cytodynamic study in mice after exposure to oxygen. *Lab Invest* 30:35–42, 1974.

60. Evans JM, Cabral LJ, Stephens RJ, Freeman G: Transformation of alveolar type 2 cells to type I cells following exposure to NO_2. *Exp Mol Pathol* 22:142–150, 1975.

61. Tomashefski JF: Pulmonary pathology of the adult respiratory distress syndrome. *Clin Chest Med* 11:593–619, 1990.

62. Hultgren HN: High altitude pulmonary edema, in Staub NC (ed), *Lung Water and Solute Exchange.* New York, Marcel Dekker, 1978, pp 437–470.

63. Schoene RB: Pulmonary edema at high altitude: Review, pathophysiology, and update. *Clin Chest Med* 6:491–508, 1985.

64. Theodore J, Robin ED: Speculation on neurogenic pulmonary edema (NPE). *Am Rev Respir Dis* 113:405–411, 1976.

65. Colice GL: Neurogenic pulmonary edema. *Clin Chest Med* 6:473–489, 1985.

66. Ell SR: Neurogenic pulmonary edema: A review of the literature and a perspective. *Invest Radiol* 26:499–506, 1991.

67. Rackow EC, Fein IA, Sprung C, Grodman RS: Uremic pulmonary edema. *Am J Med* 64:1084–1088, 1978.

68. Bush A, Gabriel R: The lungs in uremia: A review. *J R Soc Med* 78:849–855, 1985.

69. Hopps HC, Wissler RW: Uremic pneumonitis. *Am J Pathol* 31:261–273, 1953.

70. Henkin RI, Maxwell MH, Murray JF: Uremic pneumonitis: A clinical, physiological study. *Ann Intern Med* 57:1001–1008, 1962.

71. Crosbie WA, Snowden S, Parsens V: Changes in lung capillary permeability in renal failure. *Br Med J* 4:388–390, 1972.

72. Belcher NG, Rees PJ: Changes in pulmonary clearance of technetium labelled DTPA during hemodialysis. *Thorax* 41:381–385, 1986.

Chapter 24

Pulmonary Changes after Transplantation

P. S. Hasleton / H. M. Doran

Transplantation is now a well-established procedure and usually employs histopathologists in the assessment of rejection or infection in the recipient. All pathologists should have a knowledge of the pulmonary changes of transplantation, since such patients may present at their hospital and need investigation. Other patients may require postmortems, and a knowledge of the changes found after transplantation is important.

HEART-LUNG AND LUNG TRANSPLANTATION

More than 1708 heart-lung, 2465 single lung, 1344 double lung, and eight or more lobe transplants have been performed worldwide.[1,1a] The most common reasons for lung or heart-lung transplantation are given in Fig. 24-1. Most heart-lung transplants are done for congenital heart disease, cystic fibrosis, or primary pulmonary hypertension. Single lungs are usually transplanted for pulmonary fibrosis or emphysema and double lungs for emphysema, cystic fibrosis, idiopathic pulmonary fibrosis, bronchiectasis (including Kartagener's syndrome[2] and cystic fibrosis), or primary pulmonary hypertension. Other diseases requiring transplantation have included pulmonary thromboembolic disease, sarcoidosis, Langerhans' cell histiocytosis (histiocytosis X), giant cell interstitial pneumonitis, idiopathic pulmonary hemosiderosis, systemic lupus erythematosus, scleroderma, lymphangioleio-myomatosis, adult respiratory distress syndrome, and leiomyosarcoma of the pulmonary artery.

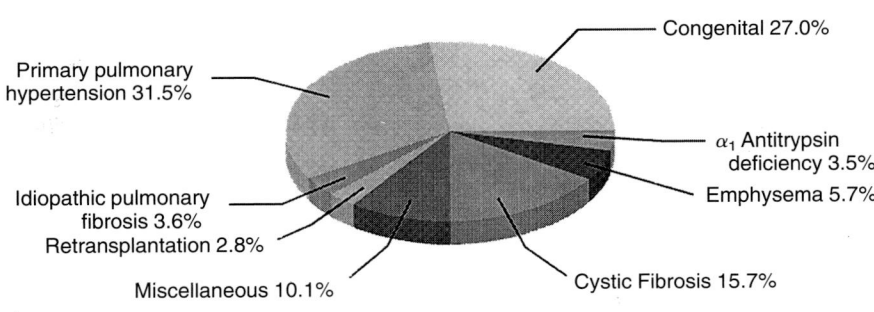

A

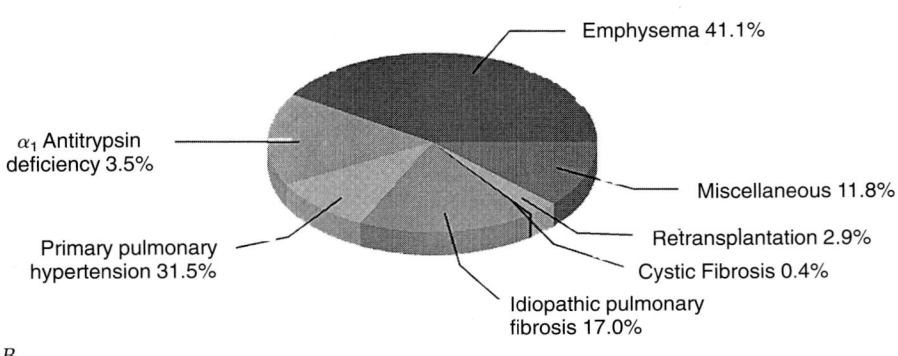

B

FIGURE 24-1
A. Indications for heart-lung transplantation.   *B.* Indications for single-lung transplantation. *(Reproduced with permission of Dr. J. O. Hosenpud et al., The Registry of the International Society, and the Journal for Heart and Lung Transplantation 13:561–570, 1994.)*

723

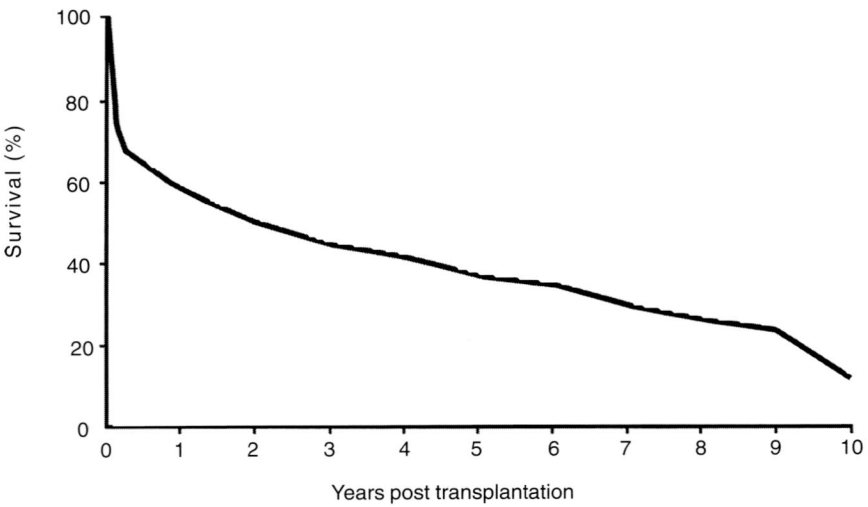

A

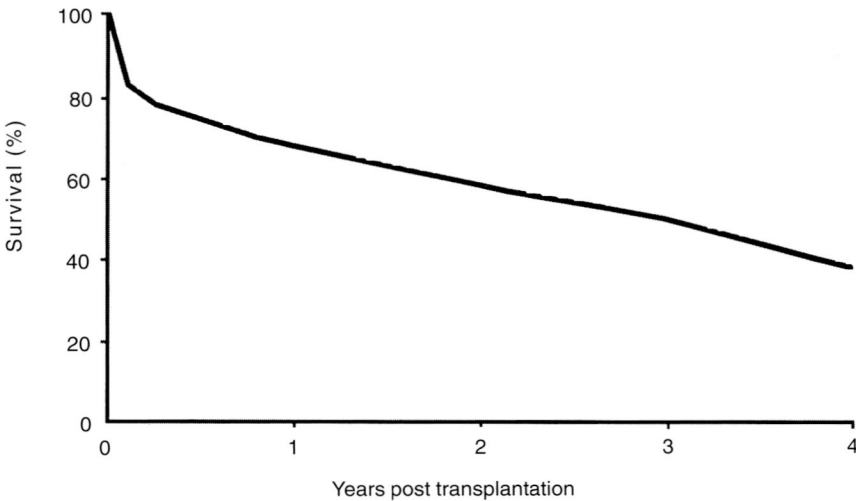

B

FIGURE 24-2
A. Actuarial analysis of heart-lung transplantation. *B.* Actuarial analysis of single-lung transplantation. *(Reproduced with permission of Dr. J. O. Hosenpud et al., The Registry of the International Society, and the Journal for Heart and Lung Transplantation 13:561–570, 1994.)*

The actuarial analysis for heart-lung transplantation is given in Fig. 24-2. Forty-four percent of adult patients with heart-lung transplants are alive at 5 years. The number of heart-lung and single lung transplants, as notified to the International Registry, is shown in Fig. 24-3. It can be readily seen that heart-lung transplantation is decreasing in popularity, but single lung transplants are increasing.

Although lung transplants were performed as early as 1963,[3] it was not until the Stanford group introduced combined heart-lung transplantation for primary pulmonary hypertension and were able to use cyclosporin A as a powerful immunosuppressant that the operation became established.[4] The long-term results of this group have recently been published.[5] Their series consisted of 109 patients operated on over an 11-year period. The complications this group encountered are the same as those documented below.

One of the main problems that face lung and heart-lung transplantation is that the organ block can be preserved for only 6 to 8 h. After organ retrieval, the lungs are flushed with donor blood or Euro-Collins solution and maintained in the latter solution. Recent work has suggested that bronchial arterial flush preservation as well as pulmonary artery perfusion will ensure better preservation of lungs for transplantation.[6]

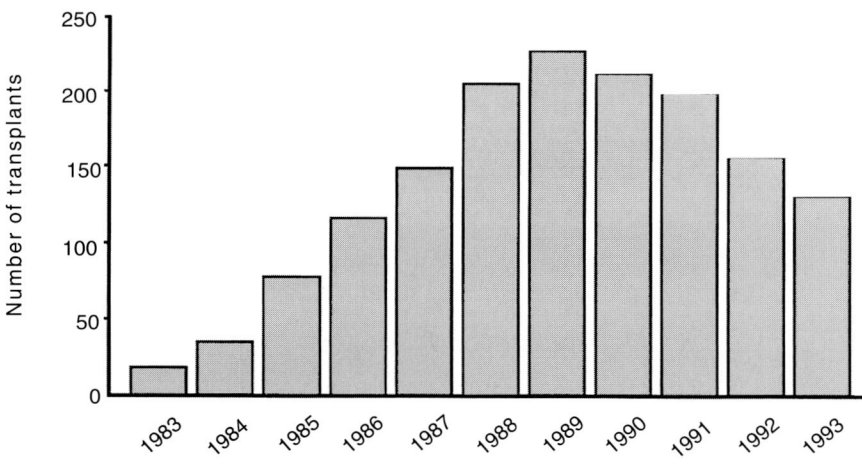

A

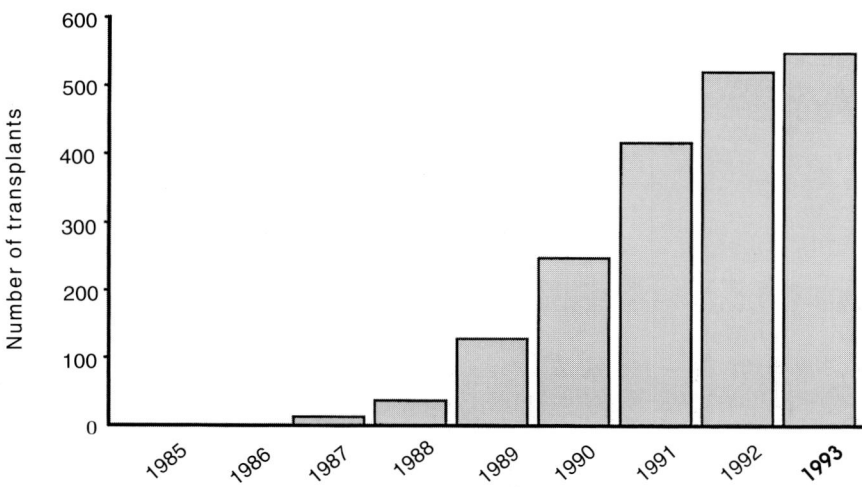

B

FIGURE 24-3
A. Numbers of heart-lung transplants performed, as received at the Registry of the International Society for Heart and Lung Transplantation. *B.* Numbers of lung transplants performed, as received at the Registry of the International Society for Heart and Lung Transplantation. *(Reproduced with permission of Dr. J. O. Hosenpud et al., The Registry of the International Society, and the Journal for Heart and Lung Transplantation 13:561–570, 1994.)*

It has been shown that serum interleukin-6 (IL-6) in the immediate postoperative period (4 h) is a useful marker of preservation injury in clinical lung transplantation. An IL-6 level of more than 1000 pg/ml is associated with poor outcome.[7]

Bronchial dehiscence and stenosis have been blamed on lack of vascularization. Direct revascularization by bronchial arteries may reduce the incidence of these complications.[8]

Because of an increasing shortage of cadaveric organ donors, a lung transplant using a living donor was performed at Stanford University in October 1990.[9] The use of a single lobe may reduce the risk to a donor, but if the transplanted lung does not grow, there may be insufficient pulmonary function in the growing pediatric patient. Long-term survival rates

with lobar transplants from living donors are unknown.

The Operation

Pathologists should have some knowledge of the operative procedure in heart-lung transplantation. It is performed through a median sternotomy; the phrenic nerves are visualized and kept intact. The patient is put on cardiopulmonary bypass, the aorta cross-clamped, and the heart removed by transection of the great vessels and atria. Then the lungs are removed. The left pulmonary vein and pulmonary artery are divided at the hilum, leaving the central portion of these vessels attached to the posterior mediastinum. The left bronchus is cut. The length of residual pulmonary vein is important, since torsion of the pulmonary vein may occur and cause venous infarction.

Similar dissection is carried out at the right hilum. The trachea is opened just superior to the carina.

The donor trachea is trimmed to one complete cartilaginous ring above the carina. Culture is taken from the trachea and the graft inserted into the chest beneath the respective phrenic nerves with the right lung posterior to the retained cuff of right atrium. The trachea is anastomosed to the recipient's main airway. The donor's right atrium is opened from the inferior vena cava up to the atrial appendage. The superior vena cava is ligated within the donor, preserving the sinoatrial node. The atrial anastomosis is sutured. It is next the turn of the aortic anastomosis; after this, warming is initiated and the patient is taken off bypass.

Double lung transplantation also requires cardiopulmonary bypass, but single lung transplantation does not.

One of the main posttransplant complications is hemorrhage due to anticoagulation, dense vascular adhesions secondary to previous pulmonary disease, or bronchopulmonary anastomoses. Cardiac tamponade may occur if the chest drains become obstructed by fibrin or clot.[10] Some transplanted lungs may be too large for the recipient, requiring further surgery to fit them inside the thorax.

Ischemic injury occurs if the transit time for the donor lungs is 6 h or more. Patients are also at increased risk for pulmonary edema, since at the time of operation they are usually in significant positive fluid balance. This lasts for at least 24 h, leading to pulmonary edema in both the native and transplanted lungs.

Early failures in lung and heart-lung transplantation were the result of bronchial dehiscence or strictures, and experimental studies[11] showed that omentopexy helped to revascularize the donor bronchus. More recently it has been shown that omentopexy or an internal mammary artery pedicle anastomosis does not improve bronchial healing in single lung transplantation.[12] As noted above, preservation of the bronchial arteries and their implantation may also keep the bronchi vascularized. The presence of gray slough or blackened mucosa at the level of the suture line presages dehiscence or stricture formation.

Late stricture formation may be a problem and can also occur after treatment of thoracic lymphomas when fibrosis narrows the bronchial lumen. A tracheal anastomosis of a heart-lung transplant heals more reliably than a bronchial anastomosis of a single or double lung because of collaterals between the coronary and bronchial circulations.[13]

The Pittsburgh experience with anastomotic problems has been reviewed.[14] The incidence of ischemic or stenotic complications has been reduced in 9 years from 32 to 12 percent.*

Reimplantation Response

The reimplantation response occurs in a single lung and in heart-lung transplants. It is usually seen within the first couple of postoperative days but may be detected up to 2 weeks after transplantation. There are radiologic opacities and temporary impairment of the ventilation:perfusion ratio.[15] The reimplantation response probably results from the combination of ischemia, reperfusion injury, and severance of the lymphatics at operation.

Histologically, there is diffuse peribronchiolar, periarterial, and perivenular dilated lymphatics, with interstitial pulmonary edema (Fig. 24-4) and a few neutrophil polymorphs in alveolar spaces and septa. There are increased neutrophils in the capillaries, as well as intraalveolar fibrin. In its most severe form, it may resemble the adult respiratory distress syndrome, and such cases may be due to long ischemic times. Severe ischemic lung injury may also manifest as acute airway inflammation.[16] In these patients the prognosis is poor.

Pulmonary Vein Thrombosis

Pulmonary vein thrombosis may occur as a mechanical fault in single lung transplants, due to marked angulation of the origin of pulmonary vein into the left atrium, causing venous infarction (Fig. 24-5).[17] Other vascular anastomotic complications have been documented.[14] On the arterial side they include short allograft artery, excessive length, and restrictive suture and clot. The last two factors may also operate on the venous side.

Adult Respiratory Distress Syndrome

The adult respiratory distress syndrome (ARDS) has been reported in other solid organ transplants, such as liver and renal transplants.[18,19] In these instances, cyclosporine was blamed for the ARDS. However, rejection may cause ARDS in 7 percent of cases,[19a] as has been shown by reversal with augmented steroid therapy. A long ischemic time (Figs. 24-6, 24-7, and

---

* Ischemia appears to be the most important factor influencing airway healing. Low pressure collateral bronchial blood flow from the pulmonary artery may be affected by low cardiac output, reperfusion edema, or rejection mucosal injury.[14a]

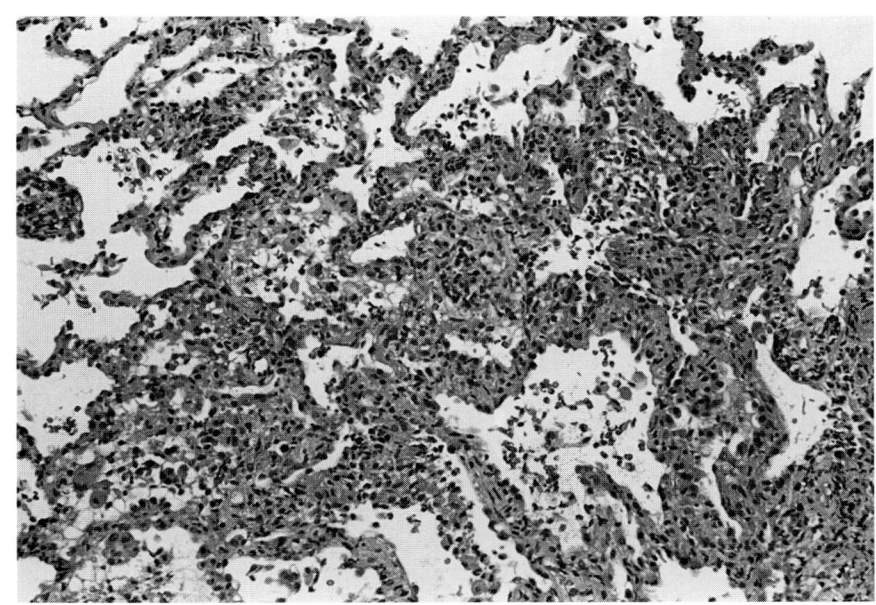

FIGURE 24-4
Reimplantation response with interstitial edema and polymorphs in capillaries. (H&E, ×80.)

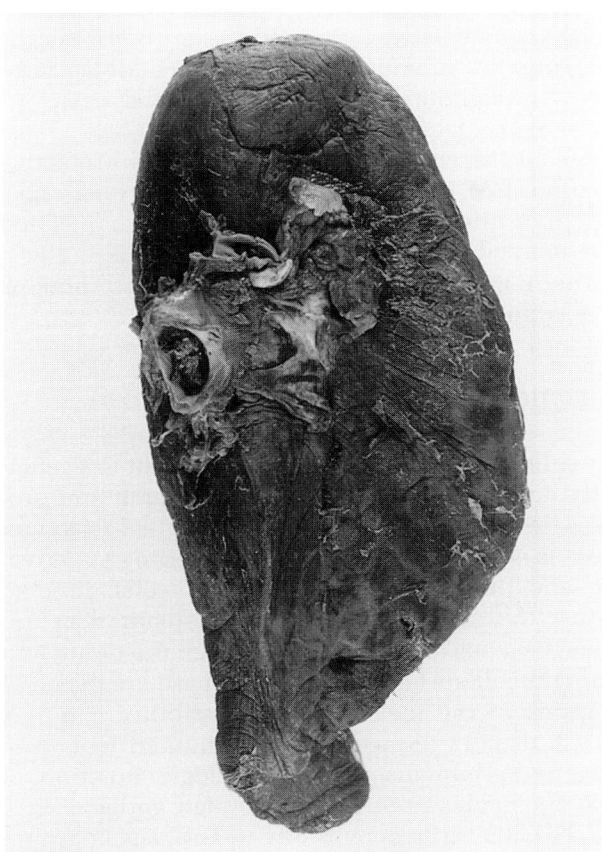

FIGURE 24-5

A. Venous infarction with thrombus in a pulmonary vein.

B. Venous infarction with marked hemorrhage.

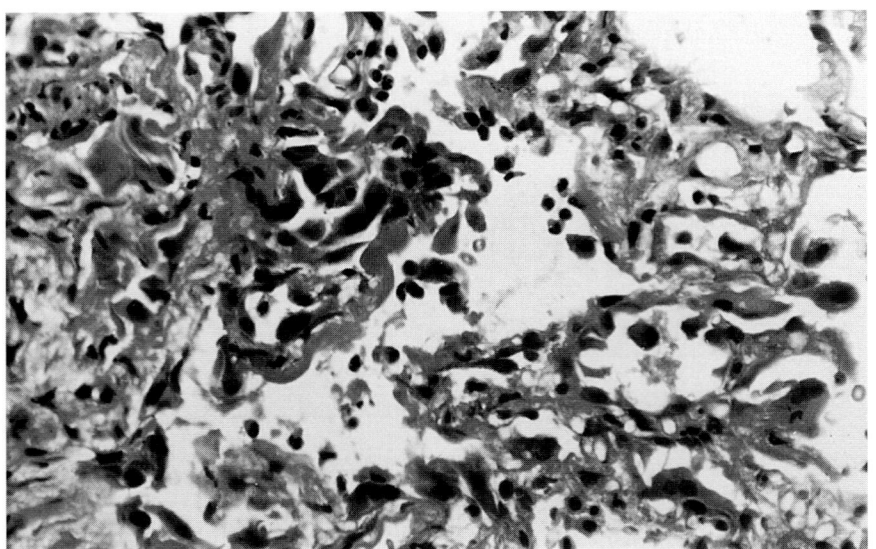

24-8), infection due to viruses or *Pneumocystis carinii*, bacteria in 52 percent of cases,[19a] and a reimplantation response may cause ARDS. In some cases there may be no obvious cause.[20] In one series survival was reduced in the presence of ARDS.[20a]

Depletion of Bronchus-Associated Lymphoid Tissue

Bronchus-associated lymphoid tissue (BALT) is the collection of submucosal lymphoid nodules scattered throughout the tracheobronchial tree.[21,22] These cells synthesize antibodies that are transported to the surface mucosa, aiding its protective function. BALT appears to be a target of lung rejection. Lung allografts from two patients with rejection have marked depletion of submucosal IgA- and IgG-bearing plasma cells compared with cases showing no rejection.[23]

REJECTION

Any solid organ transplanted from a donor to a genetically nonidentical recipient is an allograft and provokes the specific immune response called rejection. There is an immunologic cascade, leading to activation and proliferation of effector T cells directed against the donor cells. Most transplant rejection occurs because of differences between the donor and recipient cell surface molecules. These are encoded by genes in the major histocompatibility complex (MHC), usually termed the HLA (human leukocyte antigen) in humans. The physiologic function of MHC molecules is to present foreign antigens to T cells. T cells can respond only to antigens bound to the surface of antigen-presenting cells in association with class I or II MHC molecules. CD4+ (helper) lym-

FIGURE 24-7
Adult respiratory distress syndrome. The same case as in
Fig. 24-6 but 2 weeks after the biopsy illustrated. There is subpleural fibrosis and a fibrinous pleurisy.

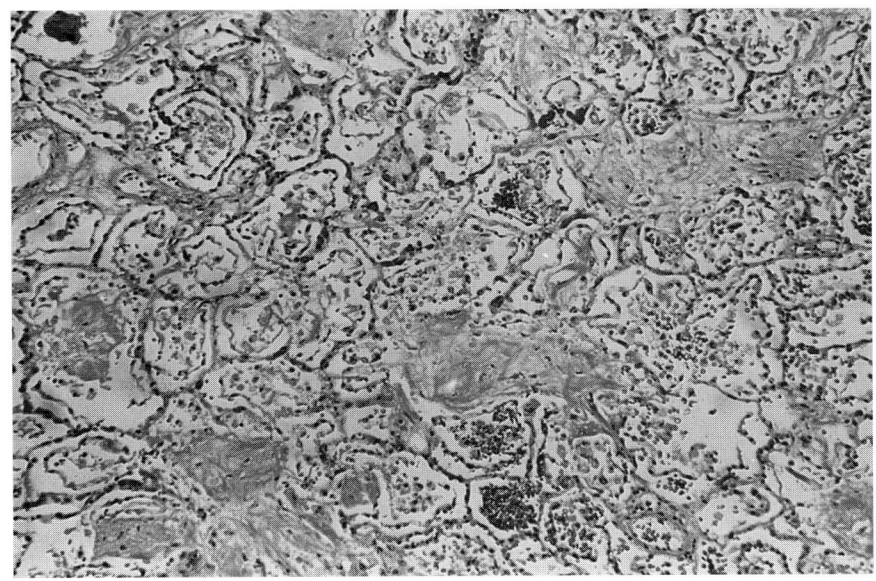

FIGURE 24-8
Histology of Fig. 24-7 with intraalveolar
fibrosis and type II cell proliferation.
These cells have artefactually shrunk from
the alveolar wall. (H&E, ×125.)

phocytes recognize antigens associated with class I MHC molecules, and cytotoxic (CD8+) T cells recognize those bound with class II molecules. A fuller description of the immunology of rejection can be obtained from immunology texts, such as that by Abbas and colleagues,[24] and from a review article.[25]

Graft rejection in most organs is classified histologically into hyperacute, acute, and chronic. Hyperacute rejection is caused by preexisting alloantibodies that bind to vascular endothelium in the donor organ, activate complement, and cause rapid thrombosis in the graft. This type of rejection has virtually been eliminated by screening recipients' serum for antibodies against a panel of cells representative of potential donors.

Hyperacute rejection is extremely rare after lung transplantation. It is probable that the only morphologic finding is a leukocytoclastic vasculitis. Such rejection is possible, as severe time restraints in lung transplantation preclude a direct, prospective crossmatch between the donor and recipient. Thus, antibodies against donor alloantigens not represented in the screening panel could be present and could cause hyperacute rejection.

Acute rejection is an integrated immune response stimulated by the recognition of alloantigens. Both specific cytotoxic cell lysis and delayed-type hypersensitivity mechanisms cause graft injury, but alloantibodies may also play a part.[26]

Acute rejection in lung transplantation is usually seen in the first 3 to 6 months.

Usually, acute pulmonary rejection precedes cardiac rejection in heart-lung transplantation. Experimental studies on baboons have confirmed that in inadequately immunosuppressed animals, acute pulmonary rejection precedes that in the heart. This observation has been confirmed clinically.[27] The Stanford group showed that in heart-lung transplants, rejection commonly occurred asynchronously.[5] The possibility of cardiac rejection, when there is a normal transbronchial biopsy but increasing breathlessness, should be considered.

The diagnosis of acute pulmonary rejection used to be made on increasing radiologic infiltrates, decreasing oxygenation, fever, and decreased perfusion. These are not specific for rejection and may be caused by infection, ARDS, or other factors.[28] The current method of distinguishing between infection and rejection is transbronchial biopsy. This has a low complication rate: only one patient in a series of 43 biopsies developed a small pneumothorax.[29]

The morbidity in 1200 transbronchial biopsies taken at Harefield Hospital, England,[30] over a 5-year period in 471 patients was 2 percent. The main complications were from bleeding and pneumothorax. On one occasion, we have seen pleural tissue in a transbronchial biopsy specimen.

Bronchoalveolar lavage (BAL) can be reliable in diagnosing infection if organisms are identified but cannot, on a simple differential count, distinguish it from rejection.[31] It has been suggested that lymphocytes in BAL fluid have increased donor-specific alloreactivity in a primed lymphocyte test, and in the future this may be useful in the diagnosis of pulmonary rejection.[32]

The optimum number of biopsies needed from each possible rejection episode for accurate diagnosis has varied between publications. The Papworth group[33] suggested that 18 specimens were required for a 95 percent confidence of diagnosing rejection.

In that study, 12.3 percent of patients had bleeding of more than 100 ml, but there were no long-term complications. The same group has revised this figure down to 10 specimens.[34] They[35] reported that a smaller number of transbronchial biopsies were needed to detect rejection in children. This was taken as reflecting a more patchy distribution of rejection in older patients.

Other studies have used smaller numbers of biopsies. The Lung Rejection Study Group[36,36a] recommended a minimum of five transbronchial biopsies containing pulmonary parenchyma. The histology of three adequate pieces of lung parenchyma correlated well with clinical rejection and infection (other than fungal), but was of limited value in bronchiolitis obliterans or fungal infection.[36b] In the Stanford series,[37] a range of one to eight biopsy specimens per case was taken. Since each case requires serial sections and a careful examination for other disease processes, such as *Pneumocystis carinii* infection, *Cytomegalovirus*, etc., the cumulative workload could become significantly increased with large numbers of biopsies. The grading of pulmonary rejection is given in Table 24-1.

Histological Features of Rejection

GRADE A0 (NO ACUTE REJECTION)

Normal parenchyma.

GRADE A1 (MINIMAL ACUTE REJECTION)

There are scattered, infrequent, perivascular lymphoplasmacytic infiltrates, not usually identified at

TABLE 24-1
Grading of Pulmonary Rejection

A. Acute Rejection
  Grade 0   No rejection
  Grade 1   Minimal rejection
  Grade 2   Mild rejection
  Grade 3   Moderate rejection
  Grade 4   Severe rejection
With/without
B. Airway Inflammation
C. Chronic Airways Rejection/Bronchiolitis Obliterans
  a. Active
  b. Inactive
D. Chronic Vascular Rejection

low magnification (×40 magnification) (Fig. 24-9). Infiltrates of small round plasmacytoid and transformed lymphocytes forming a ring of 2 to 3 cells in thickness especially surround venules but other blood vessels are affected.

GRADE A2 (MILD ACUTE REJECTION)

Perivascular mononuclear infiltrates frequently surround both arterioles and venules and are identified at low magnification (Fig. 24-10). The cells consist of activated and small round lymphocytes, plasma cells, plasmacytoid lymphocytes, macrophages, and eosinophils (Fig. 24-11).

There is frequent subendothelial infiltration by the mononuclear cells, with a hyperplastic or degenerative change in the endothelium termed *endothe-*

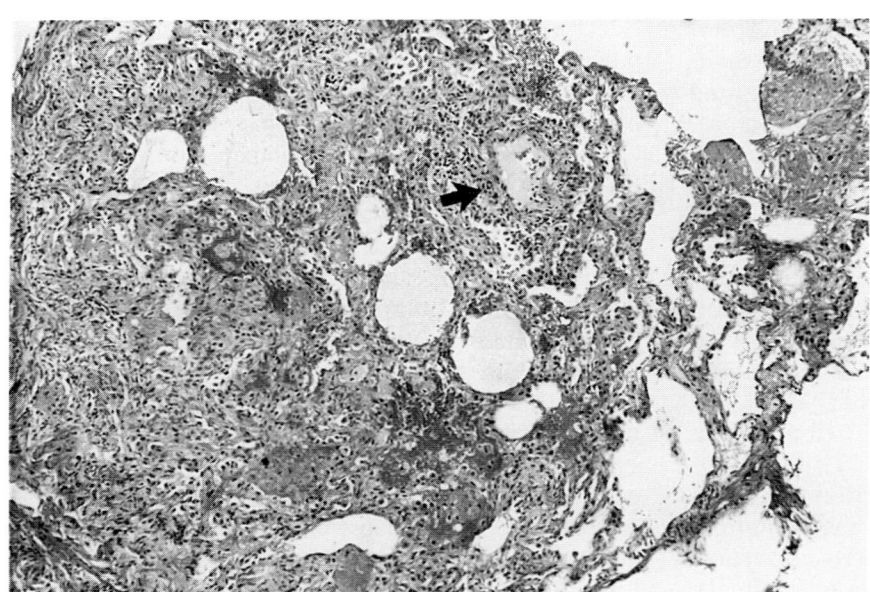

FIGURE 24-9
Minimal rejection (grade A1) with a focal perivascular infiltrate (*arrow*). The rest of the lung shows collapse, some intraalveolar edema, and focal hemorrhage. (H&E, ×80.)

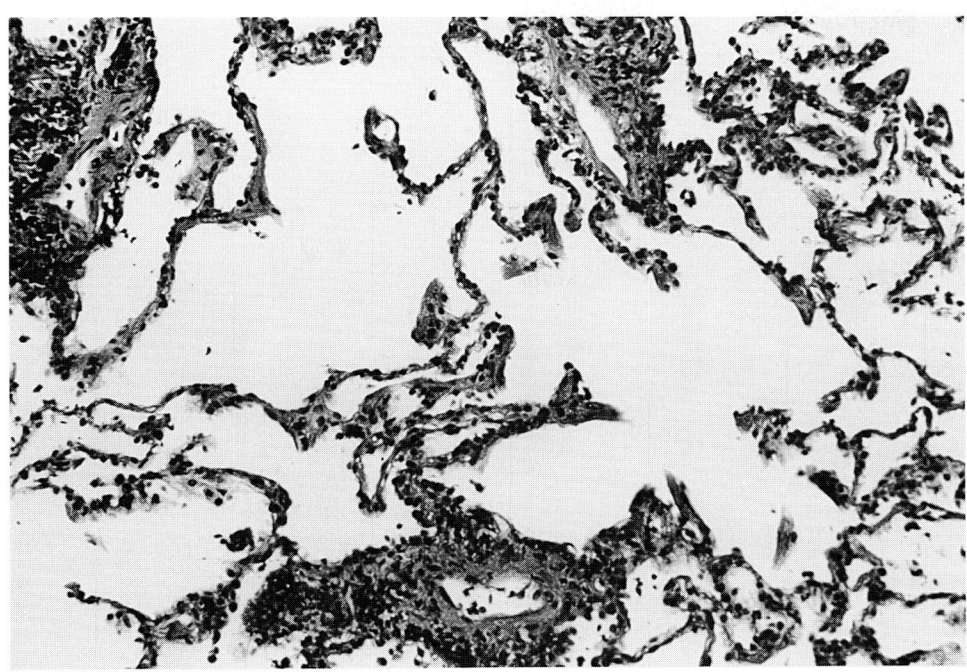

FIGURE 24-10
Mild acute rejection with cuffing of vessels (grade A2). (H&E, ×80.)

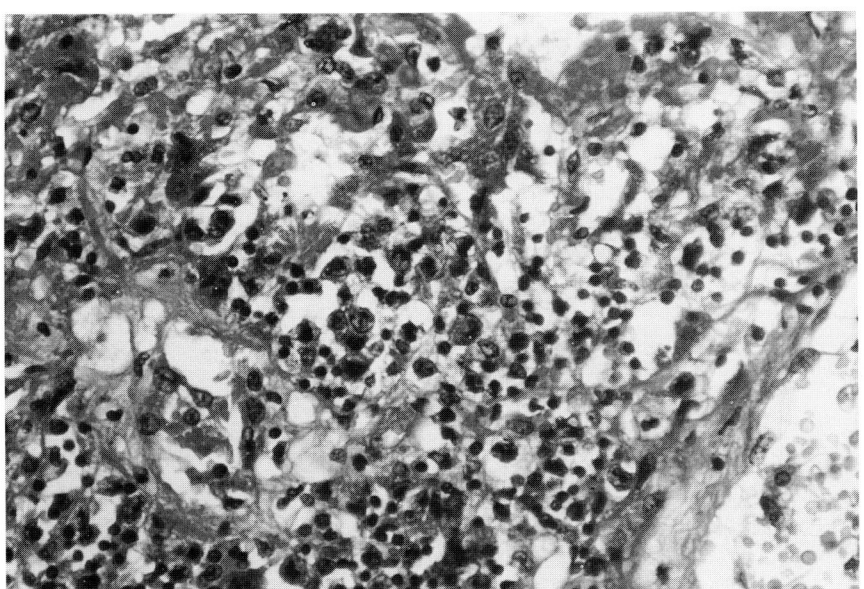

FIGURE 24-11
Higher magnification of Fig. 24-10 with vessel wall in bottom right-hand corner. There is a mixture of lymphocytes, plasma cells, and macrophages. (H&E, ×313.)

lialitis (Fig. 24-12). The inflammation extends into the interstitium.

GRADE A3 (MODERATE ACUTE REJECTION)

This grade shows readily recognizable cuffing of arterioles and venules by a dense perivascular infiltrate of mononuclear cells usually associated with endothelialitis (Fig. 24-13), eosinophils and occasionally neutrophils are common. The infiltrate extends into perivascular and peribronchiolar alveolar septa and into air spaces (Fig. 24-14).

GRADE A4 (SEVERE ACUTE REJECTION)

In this grade the infiltrate is diffuse, perivascular, intraalveolar, and interstitial and is also seen in the air spaces. It consists of mononuclear cells, hyaline membranes (Fig. 24-15), hemorrhage, and neutrophil polymorphs in small airways. There may be associated parenchymal necrosis, infarction, or a necrotizing arteritis. There is prominent alveolar pneumocyte damage usually associated with intraalveolar necrotic cells and its other features mentioned above.

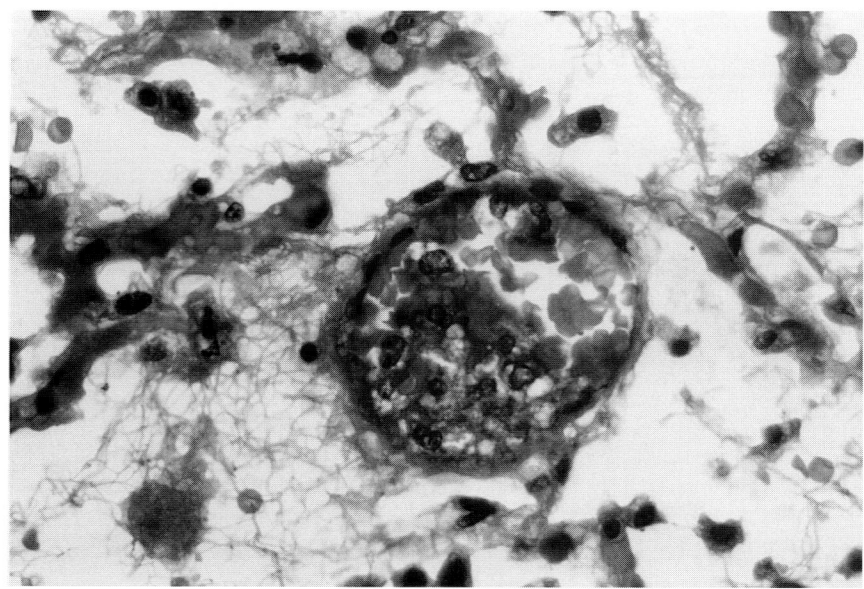

FIGURE 24-12
Endothelialitis with prominent endothelial cells lining this small vessel. (H&E, ×500.)

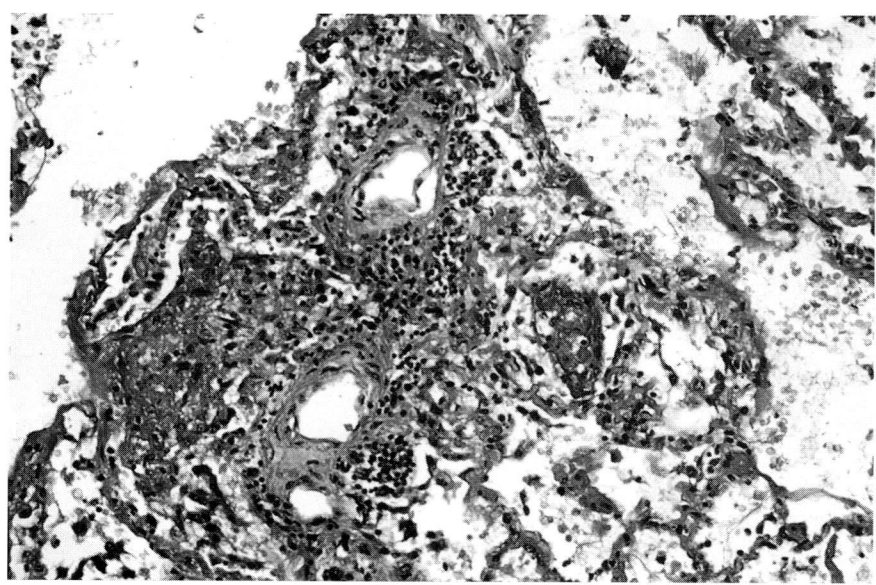

FIGURE 24-13
Moderate rejection (grade A3). Many of the vessels showed cuffing in this biopsy. (H&E, ×125.)

GRADE B (ACTIVE AIRWAY DAMAGE WITHOUT FIBROUS SCARRING)

In rejection, airway injury is usually noted first on bronchoscopy, and there is marked erythema of the main airways. There is either lymphocytic bronchitis (Fig. 24-16) or bronchiolitis (Fig. 24-17), with terminal and respiratory bronchiolar involvement. With lymphocytic bronchitis there is a mononuclear infiltrate in the mucosa and submucosa, with epithelial injury to submucosal glands (Fig. 24-18). Airway inflammation was divided into five grades. Many pathologists felt that airway inflammation on its own could not be used to grade rejection because of its frequent occurrence in airway infection.

GRADE C (CHRONIC AIRWAY REJECTION)

This is manifested as bronchiolitis obliterans.

GRADE D (CHRONIC VASCULAR REJECTION)

Grades C and D will be dealt with later in this chapter.

A recent study of airways disease in lung transplant recipients showed a neutrophilic infiltrate in both large and small airways.[16] In addition, there was mucus plugging, luminal dilatation, and granulation tissue formation. These changes were not specific for rejection and were seen in "harvest" injury, bronchiolitis obliterans, and infection. In the cases of rejec-

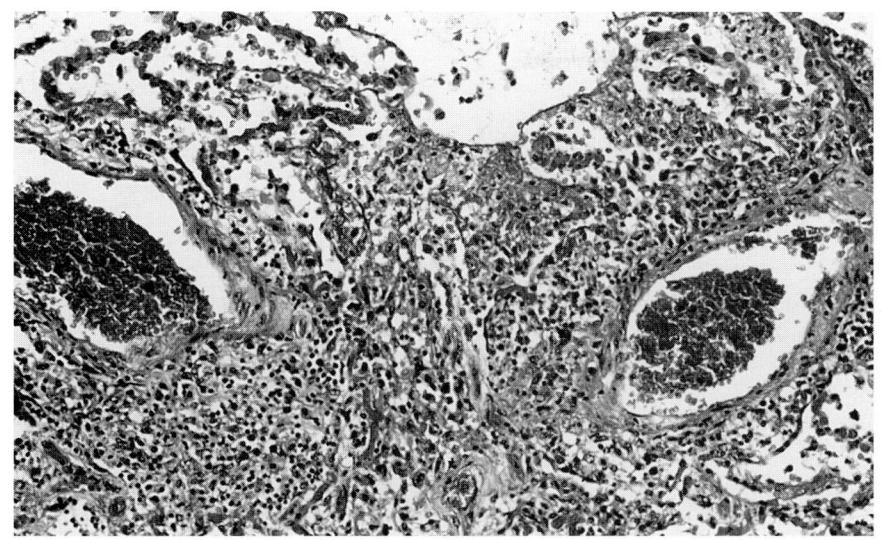

FIGURE 24-14
Grade A3 rejection with extension of lymphoplasmacytic infiltrate into the adjacent alveoli. (H&E, ×125.) *Reproduced with permission of Dr. S. Stewart, Papworth Hospital, Cambridge, England.*

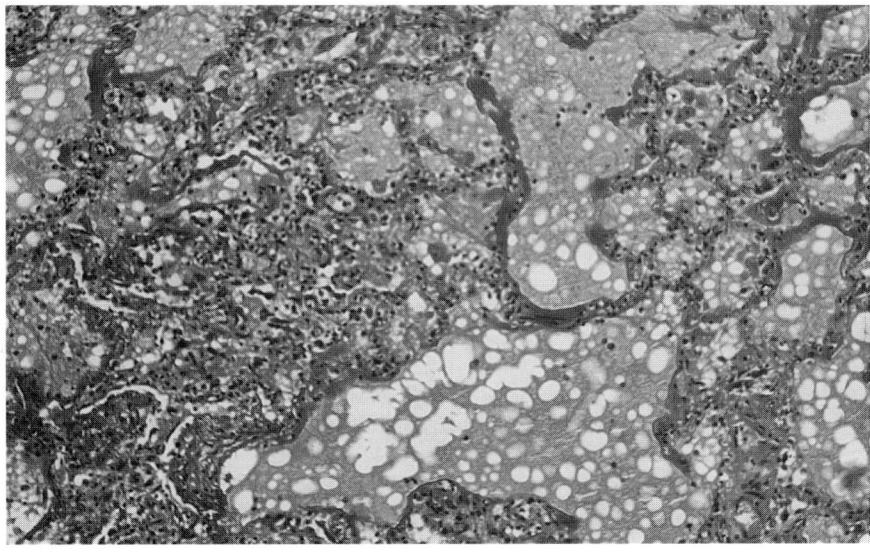

FIGURE 24-15
Severe acute rejection with prominent hyaline membranes. (H&E, ×125.) *Reproduced with permission of Dr. S. Stewart, Papworth Hospital, Cambridge, England.*

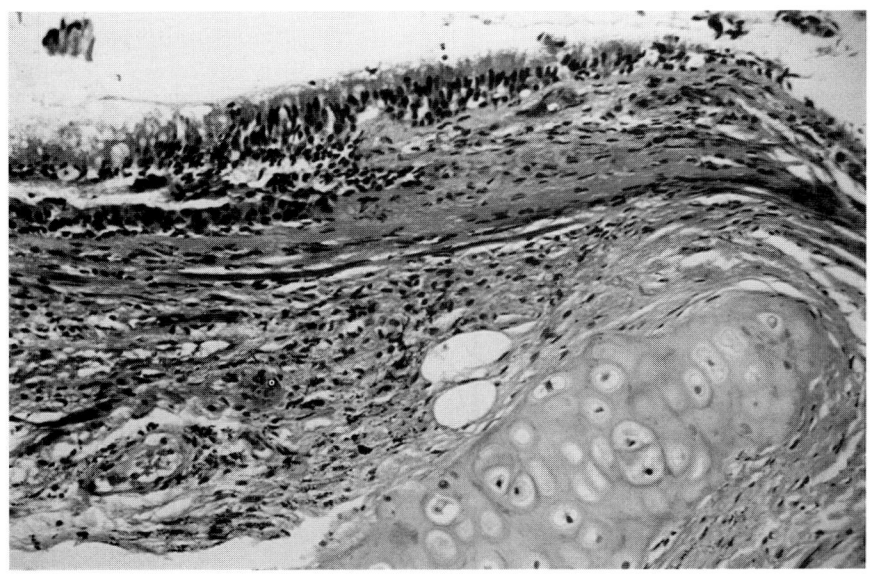

FIGURE 24-16
Bronchial inflammation (grade B1). No evidence of infection was found in this case. (H&E, ×125.)

733

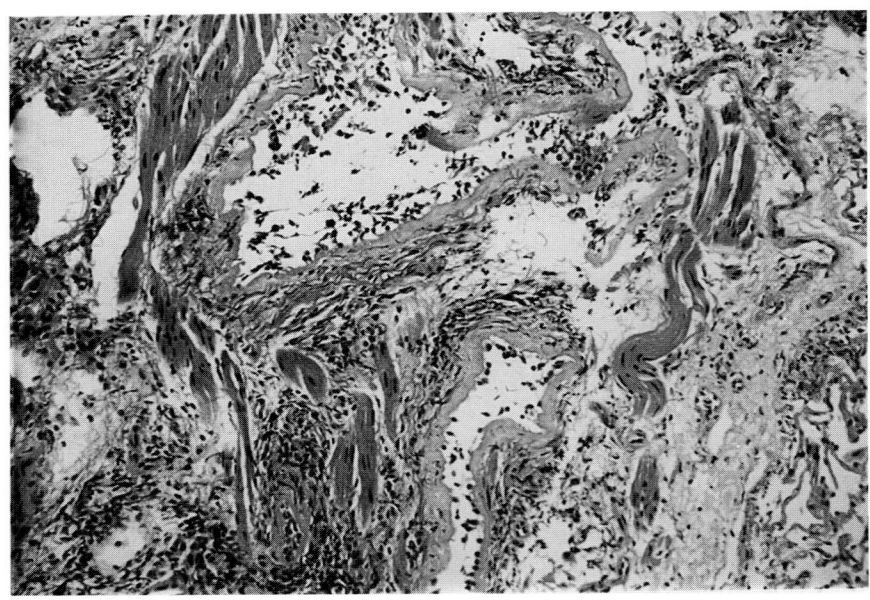

FIGURE 24-17
A disrupted bronchiole with a lymphocytic infiltrate. (H&E, ×125.)

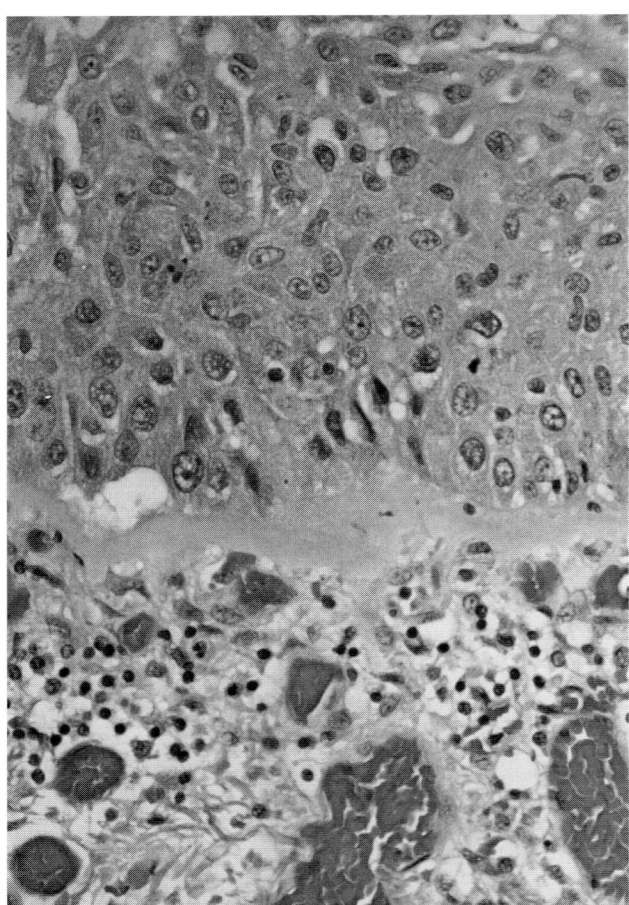

FIGURE 24-18
Squamous metaplasia of the bronchial epithelium and loss of
submucosal glands. (H&E, ×80.)

tion this airway disease represented a severe manifestation of immunological airway injury, with a predominantly lymphoplasmacytic response. It was
followed by bronchiolitis obliterans in five of seven
patients.

This is the revised (1996) version of the Working
Formulation for the classification and grading of lung
allograft rejection[36a] recommended by the Lung
Rejection Study Group as a simplification of their
previous system[36] which had been criticized as complicated and difficult to use.[38]

Recently, immunohistochemical analysis was
used to predict rejection in patients who responded
to steroids and those who did not.[39] The results in 21
patients suggested that large numbers of L26 positive
B cells in early acute rejection predict a lack of
response to conventional supplementary immunosuppressive therapy. There were greater numbers of
S100-positive cells in the resistant cases, suggesting
that the Langerhans' cells may increase the antigenicity of the graft in the early stages of rejection.

The Working Party suggested three levels should
be cut—for hematoxylin-eosin sections and a connective tissue stain, such as Masson's trichrome, and
an elastic–van Gieson—to evaluate submucosal
fibrosis and vascular damage. We also do a silver
stain for fungi and *P. carinii* and immunostains for
P. carinii and cytomegalovirus. However, the
immunoperoxidase for *P. carinii* rarely shows any
additional features beyond those seen on the routine
silver stains. Gram's stain yields little extra diagnostic information on hematoxylin-eosin sections and
has been abandoned in our laboratory.

In every transplant biopsy, it is important to compare the features with the previous biopsies. If there is rejection, this enables the pathologist to determine if it is ongoing or resolving. In the presence of rejection, it is important to state whether there is any change from the previous biopsy.

The sensitivity of transbronchial biopsy in the diagnosis of rejection is between 77 percent[36b] and 84 percent, and the specificity is 100 percent. This contrasts with a sensitivity of the chest radiograph of only 40 percent.[29]

Differential Diagnosis of Rejection

INFECTION

Infection with cytomegalovirus or *P. carinii* may present clinically as rejection, and any pulmonary infective process can cause perivascular cuffing. Many authors[36,40,41] are reluctant to diagnose rejection in the face of infection, especially with *P. carinii* or cytomegalovirus.

However, granulomatous inflammation and punctate zones of necrosis are unusual in rejection. CMV infection is suggested by perivascular edema being greater than the mononuclear infiltrate; a larger alveolar septal infiltrate than the perivascular cuffing and acute airway inflammation. The presence of the causative organism is obviously diagnostic. Follow-up biopsy should be obtained after appropriate antimicrobial therapy to adequately assess the presence of simultaneous rejection.[36a]

In a transplant setting, one must consider several disease processes, such as rejection, infection, and bronchiolitis obliterans, in any patient. The decision to treat for rejection as well as infection in the Stanford experience was based on the severity of infection, the intensity of the lymphoid infiltrate, and clinical judgment.[37] Occasional cases display chronic interstitial inflammation, and only repeated biopsies and electron microscopy demonstrate *P. carinii*. The significance of perivascular infiltrates in the submucosal bronchial vessels is not known. Lymphocytic bronchitis or bronchiolitis can be caused by rejection or infection and the histopathologist must know the bacteriological data.

EPSTEIN-BARR VIRUS–ASSOCIATED LYMPHOPROLIFERATIVE DISORDERS

The changes range from benign chronic interstitial inflammation to lymphoma and may be reversed by reduction of the immunosuppression and the appropriate antiviral agent. Epstein-Barr virus should be demonstrated, if possible, in the biopsy. A rising titer in the serum may help, but this investigation takes longer than immunohistochemistry. There is a high intrinsic risk for lymphoproliferative disorder in the Epstein-Barr virus–seronegative patient. This risk is highest in lung transplantation. Conversely the frequency of PTLD in EBV seronegative patients was extremely low.[41a]

BIOPSY SITE

A previous biopsy site shows chronic inflammation and granulation tissue (Fig. 24-19) and may give rise to misinterpretation as bronchiolitis obliterans–orga-

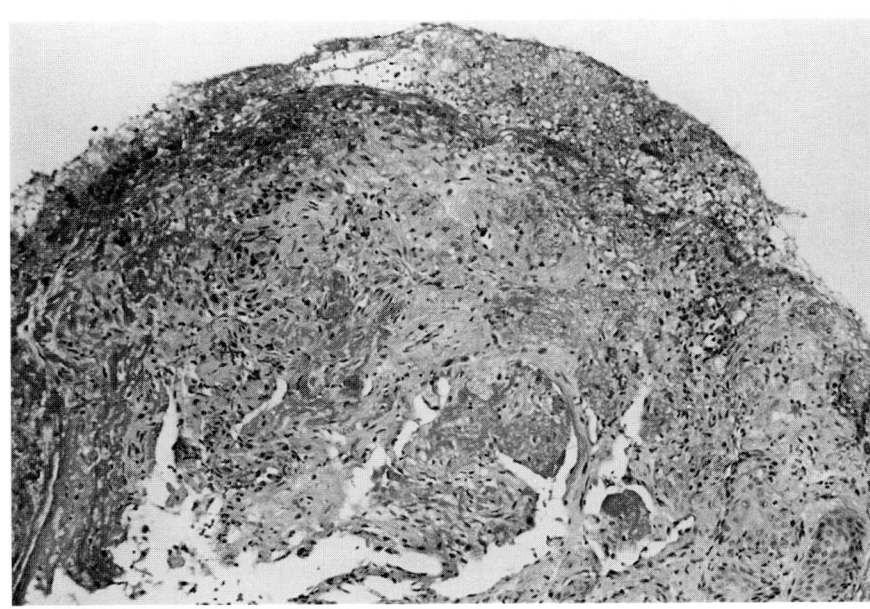

FIGURE 24-19
A biopsy site with fibrin and fibrosis. (H&E, ×80.)

nizing pneumonia, ARDS, or organization of acute rejection. BALT is normally seen in the subepithelial tissues and may be confused with lymphocytic bronchitis.

ARDS

ARDS may be a presenting feature of acute rejection but can occur in lungs that have had a long ischemic time. ARDS may have many other causes, including fungal, bacterial, and viral infection. In one series[41b] infection was the commonest cause of an acute alveolar injury picture as seen histologically. In such cases it is important to obtain a clinicopathologic correlation and to search for organisms before reaching a diagnosis of rejection.

RECURRENT DISEASE

Recurrent primary disease, such as sarcoidosis, Langerhans' cell histiocytosis (histiocytosis X), or interstitial pulmonary fibrosis, may be seen in transplanted lungs and cause a mis-diagnosis of rejection.

BRONCHIOLITIS OBLITERANS–ORGANIZING PNEUMONIA

Organizing pneumonia and bronchiolitis obliterans are a consequence of persistent infection, ischemic damage, or aspiration or, in some cases, an end result of rejection. In such instances, it is important to review the previous biopsies and obtain a full clinical history.

Graft eosinophilia has been found in patients receiving single lung and heart-lung transplants.[42]

Five cases were associated with moderate to severe cellular rejection. This group had eosinophilia early after transplantation that was associated with elevated white cell counts and occasional peripheral blood and BAL eosinophilia (Fig. 24-20). There was a mononuclear as well as an eosinophilic infiltrate around vessels and, in one case, intraalveolar eosinophils.

The second group had graft eosinophilia due to infection. Two of these four cases had underlying bronchiolitis obliterans and developed tissue eosinophilia late after transplantation. Lavage showed between 15 and 38 percent eosinophils. The infections were *Aspergillus* species, Coxsackie A2 virus, and *Pseudomonas maltophilia*. Histologically, the biopsies showed acute eosinophilic pneumonia.

In the third group, tissue eosinophilia was related to a penicillin allergy. The biopsy showed acute eosinophilic pneumonia. We have recently seen a case in which the transplanted lung developed allergic bronchopulmonary aspergillosis, with many eosinophils.

Natural History of Rejection in Transplant Recipients

Histologic grading of acute lung rejection has a predictive effect on the long-term outcome in heart-lung transplantation. Patients with higher grades of rejection have a higher risk of developing bronchiolitis obliterans.[43]

A detailed study has been carried out by the Stanford group.[37] Fifty-three lung transplant patients

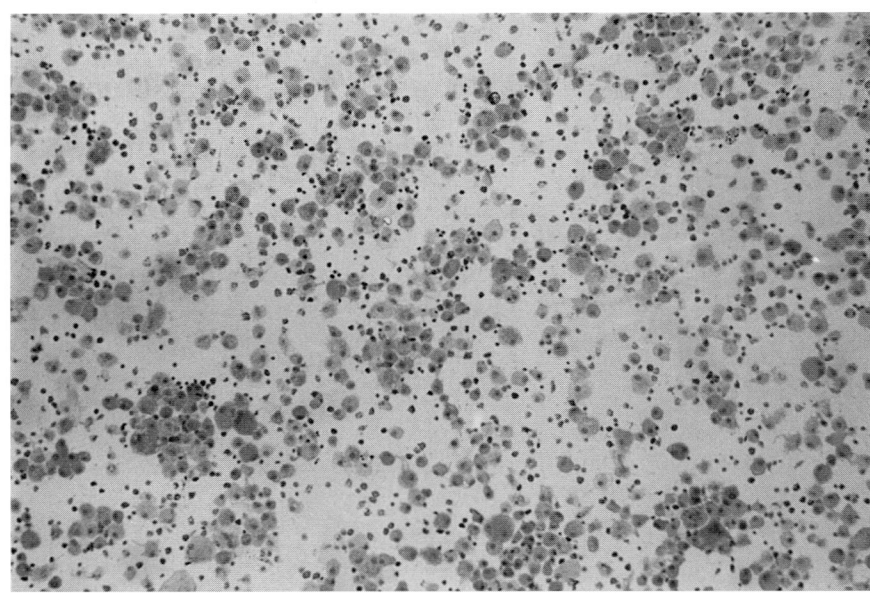

FIGURE 24-20
Bronchoalveolar lavage showing marked eosinophilia. (H&E, ×125.)

having 366 biopsies were analyzed. Twenty-five asymptomatic patients with grade 1 or 2 perivascular infiltrates did not receive antirejection therapy. Follow-up biopsy in these patients showed spontaneous resolution of the infiltrates in 19 and increased infiltrates in 6. Only two of these patients became ill, representing "progression" to clinical rejection in 8 percent of untreated patients. Of 17 patients with grade 1 infiltrates who were not initially treated with antirejection therapy, only five subsequently received treatment; the other 12 resolved without antirejection therapy.

The clinical significance of many grade 1 or minor infiltrates is still uncertain, and the following possibilities exist.

1. They may represent very low grade clinically symptomatic acute rejection.
2. These infiltrates may represent a very low grade, clinically asymptomatic acute rejection that, if persistent, would eventually result in chronic rejection. There is supportive evidence for this in that donor-specific alloreactivity has been detected by primed lymphocyte testing in lavage cells 3 months after development of abnormal pulmonary function tests.[44] Other solid organ transplants, such as kidney, pancreas, liver, and heart, may show chronic rejection without previous clinical acute rejection.

Many asymptomatic patients with grades 2 and 3 infiltrates were identified. Fifty-five percent of the grade 2 infiltrates and 75 percent of the grade 3 infiltrates occurred in patients who were clinically well or stable at the time of biopsy. Most infiltrates were detected in follow-up biopsies and were thought to represent resolving or asymptomatic ongoing rejection. The hypothesis was advanced that the cells were altered by antirejection therapy so that their products, which could be responsible for the tissue damage and clinical symptoms, were no longer active.

In the Stanford study, no action was planned if infiltrates were found on a protocol biopsy. It would be difficult to randomize patients with grades 2 and 3 changes onto a no-treatment arm, since there could well be an increase in incidence of bronchiolitis obliterans and chronic rejection. Seventy percent of grade 1 and 75 percent of grade 2 infiltrates not treated for rejection showed spontaneous resolution. Only six cases (24 percent) showed subsequent increased infiltrates.

Despite repeat antirejection therapy in some patients, these infiltrates persisted for an average of 30 days after diagnostic biopsy. Follow-up biopsies also showed asymptomatic infection, usually cytomegalovirus pneumonitis, which often persisted for weeks despite lack of symptoms. Perivascular infiltrates, similar to those seen in acute rejection, were also present in 38 percent of biopsies with evidence of infection. These resolved with antibiotics alone in nearly 50 percent of the patients with these features. Some patients in the Stanford study were, however, treated for both infection and rejection and responded to therapy. It is difficult to determine which of the two processes was more important in causing the histologic changes in these patients.

INFECTION

Infection is an important cause of morbidity and mortality in heart and heart-lung transplant patients. There are five main mechanisms predisposing to this complication.

Depletion of BALT Associated with Lung Allograft Rejection

BALT forms a local immune system that normally protects the lung from infection. Lung allografts from patients with rejection have a decrease in submucosal IgA- and IgG-bearing plasma cells, possibly contributing to the increased incidence of pulmonary infections in these patients.[23] Transplanted lungs present a larger surface area for direct interaction with the environment than do any other solid organ transplants.

Mucociliary Function after Lung Transplantation

Mucociliary clearance of radiolabeled microspheres is impaired in patients with lung transplants.[45] Shankar and colleagues[45] also examined ciliary beat frequency (CBF) proximal and distal to the anastomosis. The CBF was higher proximal to the anastomosis. On electron microscopy, ciliary ultrastructure was preserved. Similar results have been produced by the Newcastle group.[46] In the latter study, the reduction in CBF was seen in both the fastest- and the slowest-beating cilia.

The mechanisms of reduced CBF in transplanted lungs are still unclear. Factors that may also play a part include the effects of denervation, immunosuppressive drugs, and infection or rejection in the transplanted lung.

We have demonstrated a reduced number of neuroendocrine cells in the epithelium,[47] which may be related to denervation.

Aspiration

Aspiration predisposes to the development of gram-negative infection. Five of a series of 11 heart-lung transplant patients had chronic cough, probably associated with delayed gastric emptying or esophageal dyskinesia. All five had bronchiectasis, and three had bronchiolitis obliterans.[48] Three patients' condition improved after treatment to prevent reflux. Part of the reflux may stem from injury to the vagi during operation, due to their proximity to the posterior aspect of the hila of the lungs. Adult patients with cystic fibrosis may develop a meconium ileus, which will predispose to aspiration.

Influence of the Donor Lung

One study[49] compared the results from culture of the donor trachea with the type and prevalence of early intrathoracic infections in the recipients. Sixteen of 37 recipients (43 percent) had intrathoracic infections within 2 weeks of operation. Organisms isolated from the donor tracheal cultures were different from those associated with early infections—except in three out of four recipients with a heavy growth of *Candida* in donor tracheal cultures. They developed fatal invasive candidiasis. Another study of the donor lung showed the most common organisms were *Staphylococci* and *Enterobacter*.[50] Four donors had histologic evidence of pneumonia.

The only factor significantly associated with the onset of early infection was the presence of oral flora in the donor tracheal culture. This is further evidence that undetected aspiration may lead to early infection. The report referred only to early infections; infections occurring later in the course of the transplant are more likely to be related to immunosuppression.

Transfer of viral, and protozoal organisms from the donor may occur in cardiac transplantation, causing pulmonary infection in the recipient.[51] Therefore it can be seen that any infectious agents can be transferred from the donor in any transplant situation.

Immunosuppressive Drugs

Immunosuppressive drugs make the patient susceptible to infection. The larger the amount of immunosuppressive therapy, especially steroids, the more likely the patient is to develop lung infection.

INFECTIONS IN GENERAL

Many reports have described specific infections in recipients of heart-lung transplants. One of the largest series is that of the Stanford group.[37] Twenty-three percent of their biopsies showed infection. These authors found perivascular infiltrates, which could be diagnosed as acute rejection, in 38 percent of biopsy specimens with evidence of infection. These perivascular infiltrates resolved with antibiotic therapy in nearly 50 percent of the patients treated.

In a study by the Cambridge group,[52] eight of 23 patients had opportunistic infections. Five of these patients had cytomegalovirus pneumonitis, acquired from the donor. Two developed *P. carinii* pneumonia and *Aspergillus* pneumonitis, and one developed tuberculous empyema.

Patients receiving transplants for cystic fibrosis are at increased risk of developing infection, especially with *Pseudomonas*[53] (Fig. 24-21) and *Aspergillus*. It is impossible to sterilize the upper respiratory tracts of these patients before surgery. *P. cepacia* is associated with a poor prognosis, especially if acquired for the first time after transplantation.

Cytomegalovirus

Cytomegalovirus (CMV) is one of the most common organisms affecting lung transplants. Matching the CMV status of donor and recipients reduces the incidence of primary CMV infection from the donor organ. CMV reactivation is a more common problem, and it is important to distinguish viral inclusions in occasional cells, indicating "passenger virus" (Fig. 24-22), from active CMV pneumonitis (Fig. 24-23). Biopsy has a specificity of 91 percent and 83.5 percent sensivity.[36b] In situ hybridization may help distinguish the viral inclusions of CMV from herpes simplex virus.[54] In practice most laboratories use specific monoclonal antibodies to distinguish these two viruses.

Histologically, CMV pneumonitis shows a diffuse alveolitis with inclusions in type II pneumocytes, endothelial cells (Fig. 24-24), and pulmonary macrophages. There is associated perivascular and interstitial edema, neutrophil microabscesses, and chronic inflammation. CMV may coexist with rejection, although the virus alone can cause such a perivascular infiltrate.[41] Immunohistochemistry utilizing monoclonal antibodies to immediate-early and

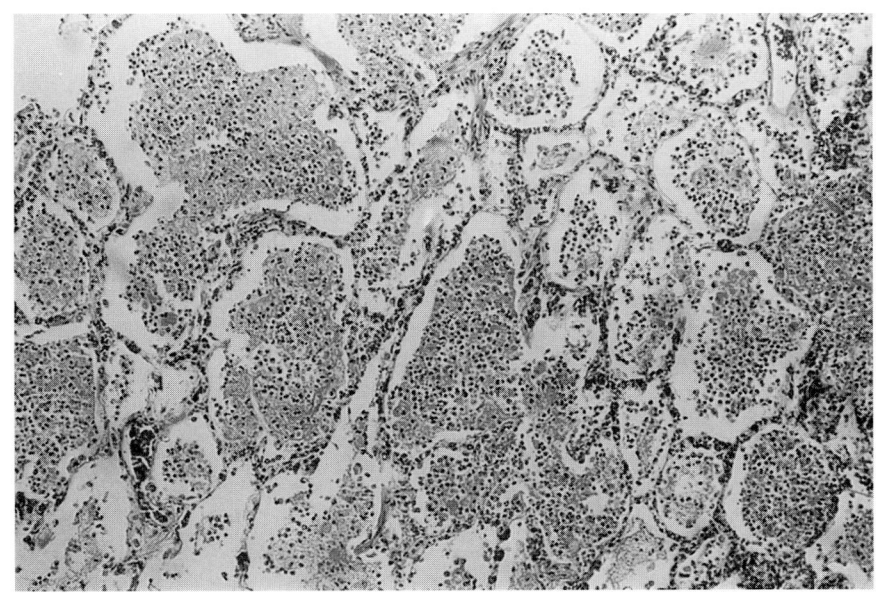

FIGURE 24-21
Pseudomonas pneumonia. (H&E, ×80.)

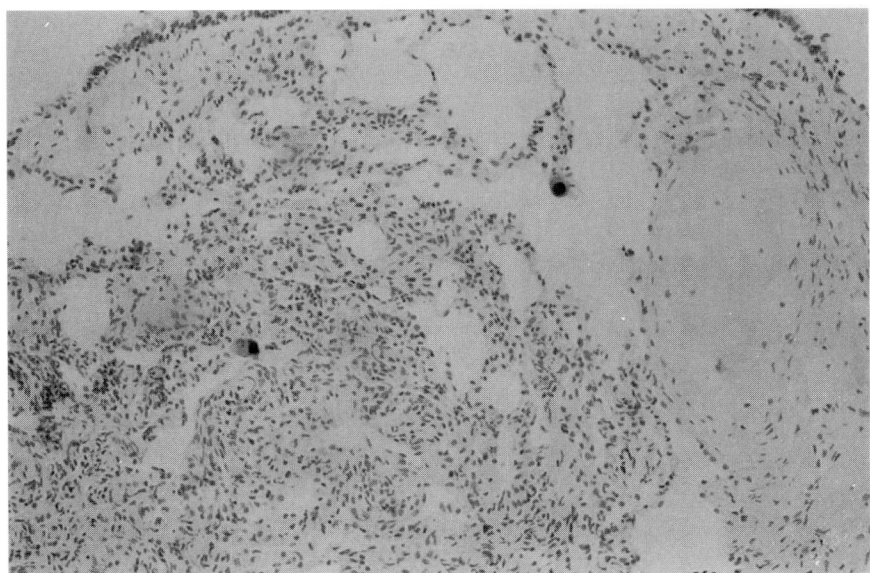

FIGURE 24-22
Passenger CMV infection with only a few viral inclusions. (Immunoperoxidase, ×125.)

early CMV nuclear antigens may indicate development of CMV pneumonitis before cytopathic changes are evident.[54a]

In transplant patients, CMV may present in other organs and be detected on the appropriate biopsy (Fig. 24-25).

Tests for detecting early antigen fluorescent foci (DEAFF test) for CMV have been used recently. Of 22 specimens positive on conventional culture and with a positive DEAFF test, only 45 percent (10 cases) showed histologic changes consistent with a herpesvirus infection.[55] Other studies trying to identify early CMV antigens suggest that this is a specific and rapid way of monitoring early CMV infection.[56] This method of diagnosis is especially useful in patients showing concomitant rejection.

The polymerase chain reaction has been used to semiquantitate CMV DNA in heart and heart-lung transplant patients.[57,58] Cases with and without acute rejection showed CMV DNA, which was also detected in two controls. There was a significant correlation between CMV culture and the polymerase

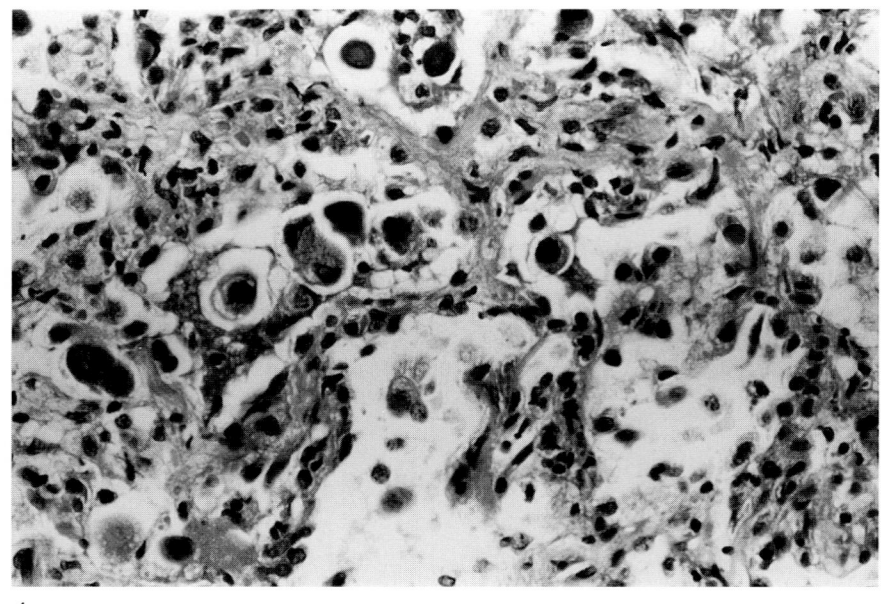

A

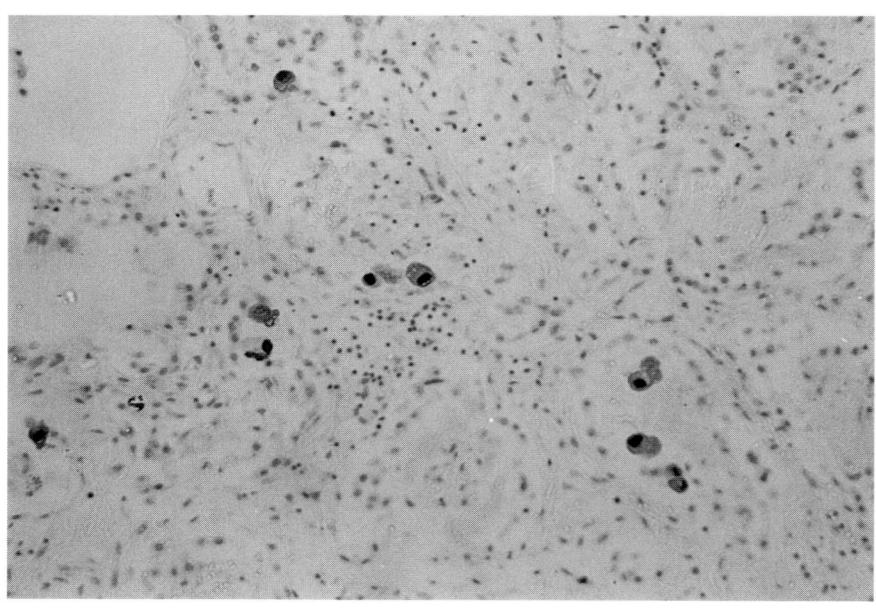

B

FIGURE 24-23
A. Active CMV infection. (H&E, ×313.)
B. Active CMV infection.
(Immunoperoxidase, ×125.)

chain reaction but it did not correlate with patient antibody status.

CMV pneumonitis may predispose lung or heart-lung transplant patients to subsequent bronchiolitis obliterans.[59,60] However, in the first series there was progressive bronchiolitis obliterans in four of the seven patients with no evidence of CMV infection. Most studies would now conclude that bronchiolitis obliterans is secondary to rejection rather than to CMV infection. Bronchiolitis obliterans is probably best considered as having many causes.

A larger series of patients with CMV infection after lung transplantation have been studied.[61] Fifty-nine patients surviving more than 30 days after lung transplantation were studied for mortality and morbidity in relation to CMV infection. This infection developed in 54 percent of cases and was more common in the preoperative CMV seropositive group than in the seronegative group (95 percent versus 38 percent). However, symptomatic infections, pneumonitis, and CMV-related mortality were higher in the seronegative (primary infection) group. Actuarial

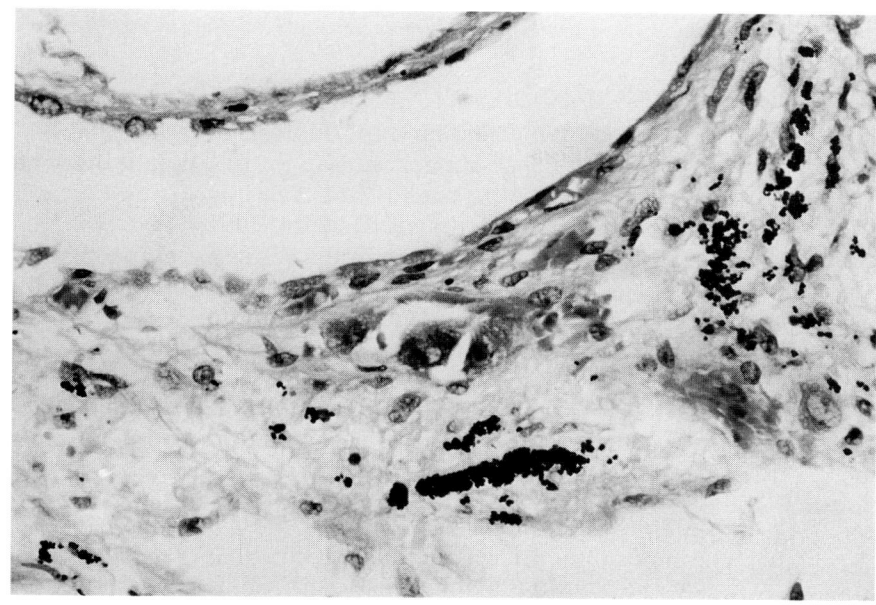

FIGURE 24-24
CMV in endothelial cells in transplanted lung. (H&E, ×313.)

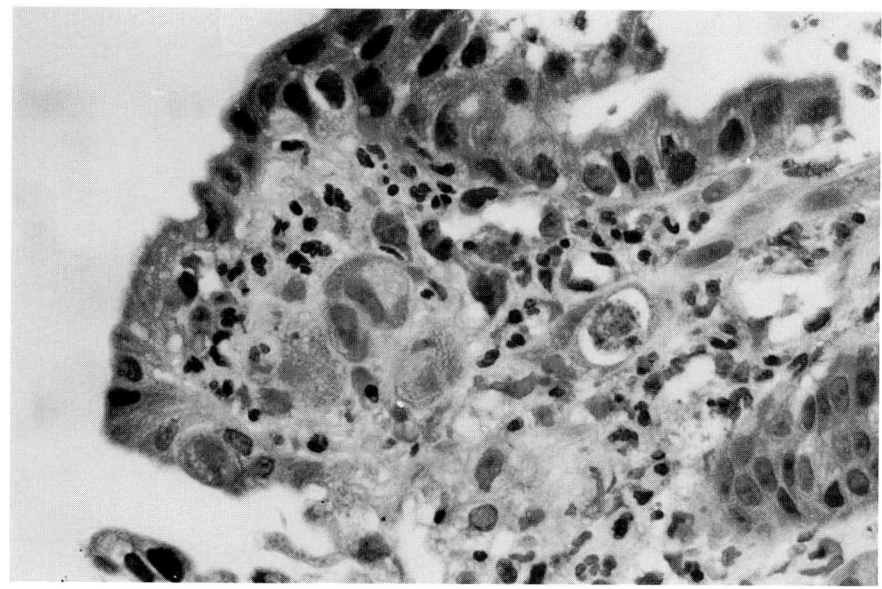

FIGURE 24-25
Biopsy of stomach with focal acute inflammation and CMV inclusions. (H&E, ×500.)

survival was lower in these patients—40 percent and 23 percent at 1 year and 5 years, respectively. The mortality of primary CMV infection was 54 percent and was associated with a significantly higher rate of pulmonary superinfections in the first year after transplantation.

CMV mismatching (donor positive, recipient negative) for the risk of primary infection, absence of CMV prophylaxis, and the development of CMV disease were the most significant risk factors for death, rejection, or infection, both early and later, after pulmonary transplantation in one study.[20a]

The incidence of late pulmonary infections was associated with the development of chronic rejection rather than CMV status. All primary infections were symptomatic. Three patients had a viral syndrome, and CMV pneumonitis developed in 11 of 13 (80 percent), with eight deaths. In three patients, CMV was disseminated to the stomach, pancreas, esophagus, and skin. The diagnosis of CMV pneumonitis was made by open lung biopsy in eight cases and BAL in three.

There were two distinct forms of pneumonitis. Seven patients had a mild to moderately severe illness with cough, fever, radiographic infiltrates, and impaired oxygenation but did not require ventilation. CMV was cultured from BAL or lung biopsy. None of these patients died of causes directly attributable to the virus. The remaining nine patients had severe pneumonitis necessitating ventilation, with multisystem failure. There were superimposed infections with gram-negative or -positive organisms, fungi, or protozoa. All but one patient in this group died of CMV infection. The higher incidence of CMV infection in lung transplant patients than in heart transplant patients probably reflects the high proportion of seronegative patients (66 percent) in a younger recipient population. Rejection and infection may also render the pulmonary allograft more susceptible to CMV infection and pneumonitis. The large amount of tissue transferred with the heart-lung block may represent a more significant inoculum of latent virus and therefore increase the chances that infection is transmitted from donor organs.

Primary infection with CMV causes not only graft disease but also systemic disease, with hepatitis, esophagitis, and encephalitis.

Herpes Simplex Virus

Herpes simplex virus (HSV) causes tracheobronchitis and necrotizing bronchopneumonia and may disseminate to other organs.[34] It may follow a period of augmented immunosuppression. Necrosis and lack of cytomegaly are useful in distinguishing it from CMV pneumonia. It is relatively uncommon.

Epstein-Barr Virus

Epstein-Barr virus (EBV) is more common in transplant recipients than in the general population because of immunosuppression. The Papworth group[62] studied 156 patients who had received heart or heart-lung transplants and had survived longer than 3 months after transplantation. They looked for an antibody titer rise of 16-fold or greater in the EBV capsid antigen (VCA IgG antibody indirect immunofluorescent test). 37.2 percent fulfilled this criterion; 8.3 percent had primary infections, and 24.4 percent had reactivated infection; 4.5 percent had antibody rises unconfirmed by other assays and were therefore regarded as suspicious of EBV infection.

A more recent study by the same group[62a] showed primary EBV infection in 6 percent of transplant patients whereas 17.4 percent had reactivation of a past infection. Of patients with serologically proven EBV, 52.9 percent had symptoms. These were malaise, fever, headache, and sore throat.

EBV VCA IgG or IgM antibody was also examined by indirect immunofluorescent testing, as were EBV nuclear antigen and an anti-EBV early antigen (EA) antibody. The EBV VCA IgG (IgM) and EA antibody titer rises were good indicators of reactivated infection. One patient in the series developed reactivated EBV infection and enlarged cervical lymph nodes due to non-Hodgkin's lymphoma.

The relationship between EBV infection and lymphomas will be considered below. EBV may be important in the development of posttransplant pulmonary fibrosis in a few cases.[63]

HIV Infection

Human immunodeficiency virus (HIV) may be transmitted to transplant patients from donated organs.[64] This is rare, however, because sera from all organ and blood donors are screened by enzyme-linked immunosorbent assay (ELISA). However, a negative HIV antibody test does not necessarily mean the donor is free from the infection, since it takes up to 6 weeks after exposure to produce an antibody response. HIV infection is associated with a poor prognosis—especially in organ transplant recipients, who are deliberately immunosuppressed.

Adenovirus

This infection is relatively rare in lung transplants and affects predominantly the pediatric-transplant population. There is a necrotizing bronchocentric pneumonia producing alveolar damage and organizing pneumonia. The typical smudgy intranuclear bodies are seen.[64a]

Toxoplasmosis

Toxoplasmosis is more of a problem in cardiac than in lung transplantation. This infection may occur in heart-lung transplant recipients, but it is unusual to see the organism in the lung (Fig. 24-26).[65]

P. carinii Pneumonia

P. carinii pneumonia (PCP) is a complication of both cardiac and lung transplants. In serial BAL, the prevalence of PCP was 88 percent (Fig. 24-27).[66] Ten episodes were associated with minimal or no symptoms. In three of six symptomatic episodes, a concurrent bacterial infection was also noted. We have seen one case after cardiac transplantation with persistent chronic alveolar wall inflammation and very scanty organisms. The organism was identified retrospec-

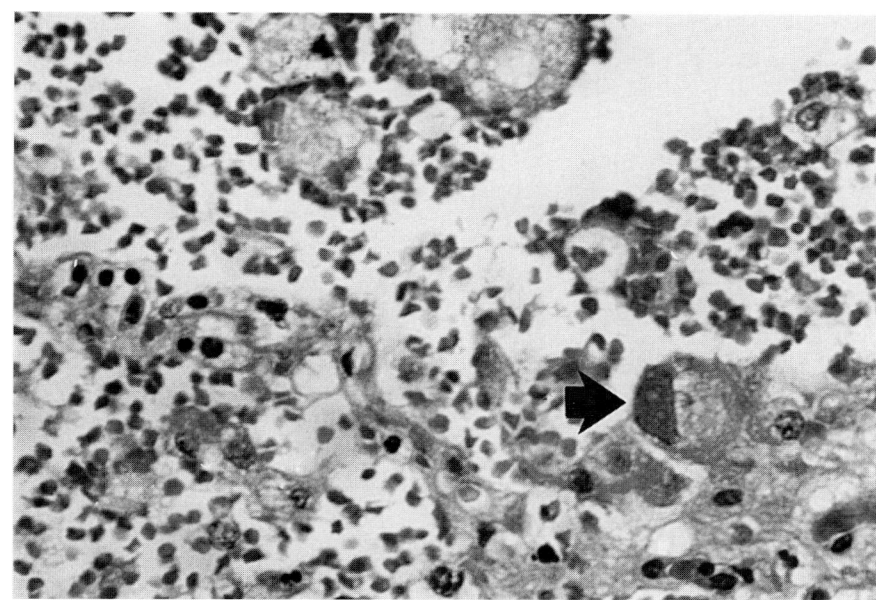

FIGURE 24-26
Toxoplasma (*arrow*) in lung with associated inflammation. (H&E, ×500.)

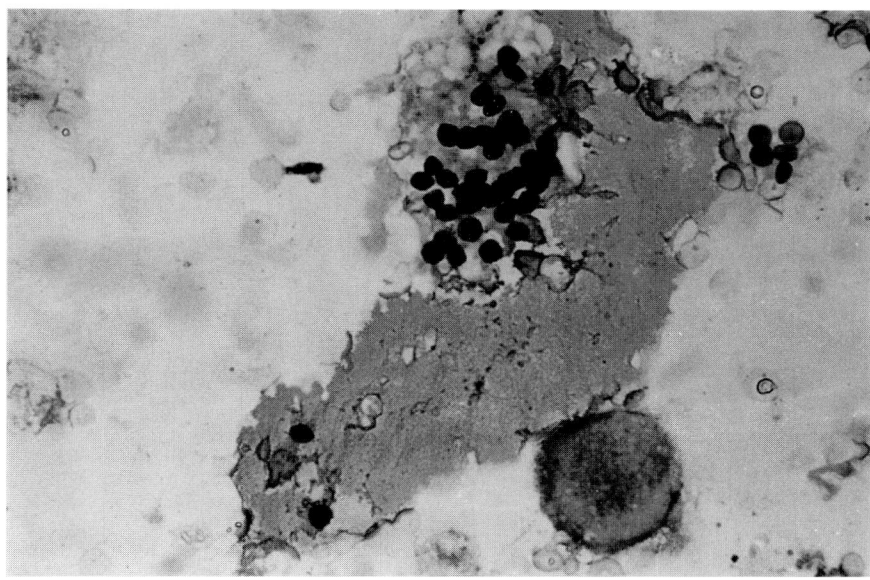

FIGURE 24-27
P. carinii as demonstrated on BAL. (Grocott, ×500.)

tively in one lavage fluid from an asymptomatic patient but not from subsequent lavages, suggesting that the infection may resolve without any therapy in some immunosuppressed patients. However, PCP may become chronic in some untreated patients. A granulomatous reaction to the organism may be seen in rare instances.[67,68]

In comparing four methods for the detection of *P. carinii* an indirect immunofluorescent assay was the most sensitive test and easiest to perform.[68a]

Aspergillosis

Aspergillosis is an uncommon infection in heart-lung transplant patients. It was identified in 4 percent of 75 heart-lung recipients at Stanford.[69] In this series there was usually tracheobronchitis with multiple ulcers. Pseudomembrane formation, disruption of the suture line, and necrosis of cartilage were identified (Fig. 24-28). In single lung transplants, *Aspergillus* was confined to the grafted side. In some

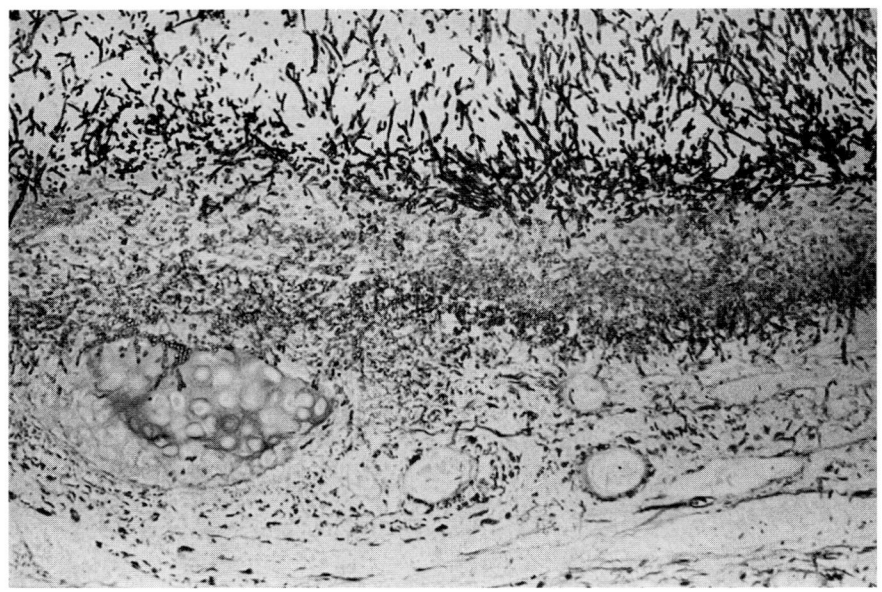

FIGURE 24-28
Aspergillus pseudomembrane formation. (Grocott, ×80.)

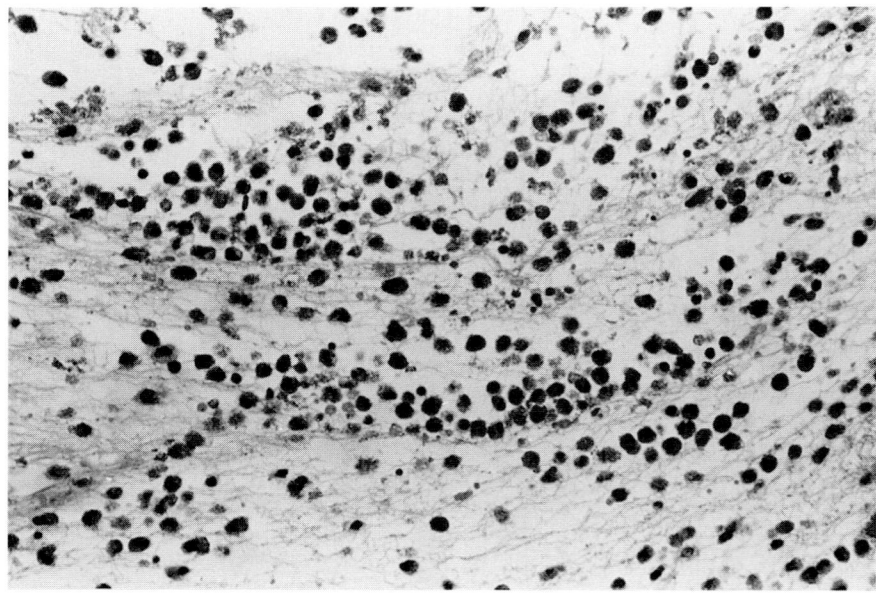

FIGURE 24-29
Eosinophils in a lavage in brochopulmonary aspergillosis developing in a transplanted lung. (H&E, ×125.)

cases, a granulomatous response may be seen in relation to the fungus.[70] Distal lung shows an eosinophilic pneumonia, and pulmonary artery invasion may occur. There may be allergic bronchopulmonary aspergillosis in the transplanted lung (Figs. 24-29 and 24-30).

Candida albicans

In the Pittsburgh series, *C. albicans* was the most common cause of fungal infection in lung recipients.[71] The authors thought that the frequency of

Candida was due to the absence of endemic infection with other fungal species in the Pittsburgh area. *Candida* often colonizes the donor lung. A mycotic aneurysm at the aortic anastomosis has been reported.[72]

Bacterial Infection

Bacterial pneumonia is one of the first major challenges that faces heart-lung transplant recipients in the early postoperative period. This complication

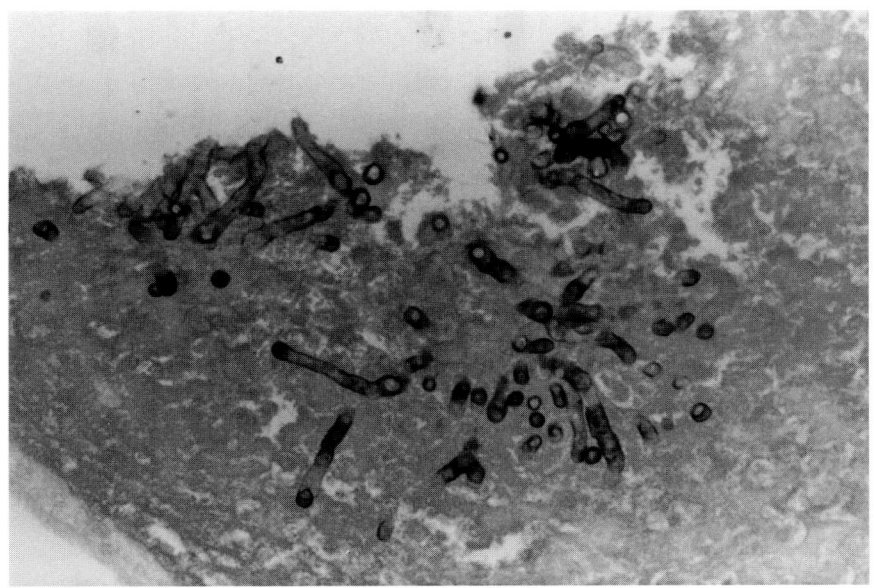

FIGURE 24-30
Aspergillus in a mucous plug. (Same case as illustrated in Fig. 24-29.) (Grocott, ×500.)

continues into the intermediate and late postoperative periods. Infection caused death in just over 25 percent of the Stanford series.[5] Diagnosis is made on either sputum or BAL and subsequent Gram stain and culture. Most pneumonias are caused by gram-negative rods, especially *Klebsiella pneumoniae*, *Escherichia coli*, enterobacteria, and *Pseudomonas*. Other organisms that cause pneumonia in transplant patients are *Staphylococcus aureus*, *Hemophilus pneumoniae*, and *Legionella pneumophila*. Mortality due to bacterial infection is 3 percent in the first 2 weeks after lung transplantation.[73] Tuberculosis may be seen in some cases, but unlike AIDS, a full granulomatous response occurs in the transplanted lung. The incidence is low. A review of all documented cases has been published.[73a]

Mediastinitis

This is rarely seen by the pathologist but can be a complication of transplantation. The increased risk in lung transplantation is probably related to complications of the airway anastomosis. Dauber and colleagues[71] divided mediastinitis into early and late. In rare instances, it may be due to *Mycoplasma*.[74]

Other Infections

Hall[75] described 13 intracranial infections in a total of 363 heart transplants and 54 heart-lung transplant recipients. These included *Nocardia* and *Aspergillus* abscesses; *Listeria, Cryptococcus, Toxoplasma*, and *Candida* infections; and one case with JC virus (a polyoma virus, not Creutzfeld-Jakob disease). One heart-lung recipient had a cerebral abscess, and another encephalitis.

Nonspecific Change

Twelve percent of the Stanford biopsies had nonspecific abnormalities.[37] These were defined as foci of alveolar or interstitial inflammation, fibrin deposits, and foamy or hemosiderin-laden macrophages (Fig. 24-31) without evidence of perivenular infiltrate, infection, diffuse alveolar damage, histologic bronchiolitis obliterans, or bronchiolitis obliterans–organizing pneumonia. Four other cases also had prominent peribronchial and endobronchial organizing hemorrhage, possibly representing injury from a previous biopsy.

BRONCHIOLITIS OBLITERANS

Bronchiolitis obliterans (BO) is one of the main long-term problems facing heart-lung and single lung transplant patients. Its development is now detected earlier, since patients have flow-volume loop measurements and biopsy performed weekly for the first three months after surgery. These investigations are continued at monthly intervals or, if respiratory symptoms develop, more frequently.

The early changes are cough, which may be productive or nonproductive, and dyspnea, which starts within months of the bronchitic manifestations. Severe dyspnea is a relatively late complication and is a reflection of extensive airways disease. Recurrent lower respiratory tract infections are common. Chest auscultation may be clear, or there may be early respiratory fine crepitations, occasional wheezes, or rhonchi.

The initial diagnosis is made physiologically. There is a disproportionate decrease in mean forced

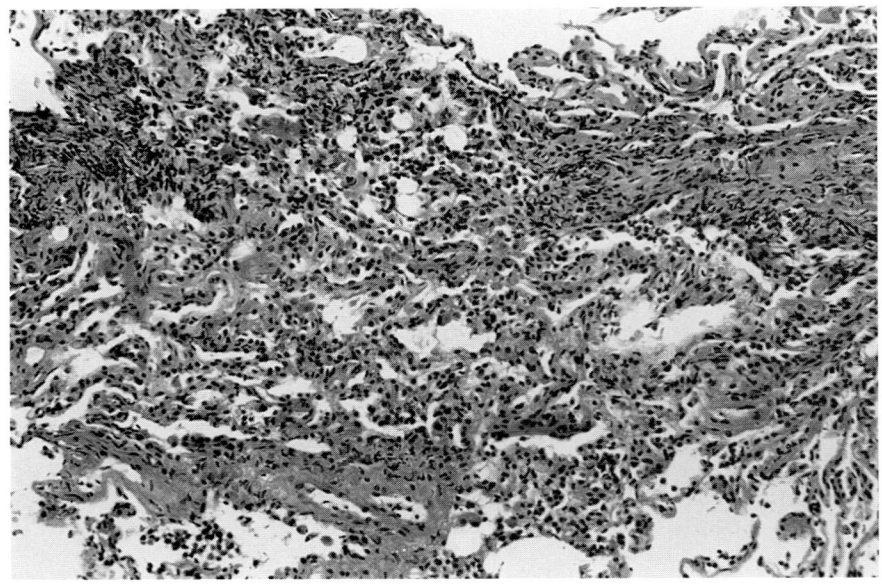

FIGURE 24-31
Nonspecific change with a mild increase in lymphoid cells in the alveolar wall. There is little perivascular cuffing. (H&E, ×80.)

expiratory volume (FEV) during the middle half of the forced vital capacity (FVC). A concavity of the expiratory flow pattern on the flow-volume loop (Fig. 24-32) is seen during the FVC. There are rapid decreases in FEV_1, FEV_1/FVC percent, and FVC. There is a variable restrictive ventilatory defect, and in time there may be hypoxemia. β-N-Acetylglucosaminidase, a lysosomal phagocytic enzyme, may

A   *B*

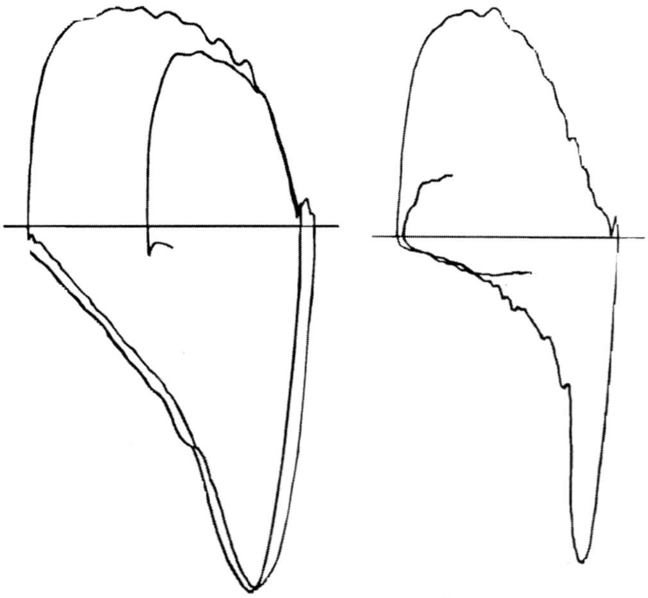

FIGURE 24-32
Flow-volume loops. *A.* Immediately after transplantation. *B.* With obliterative bronchiolitis (same patient, three years later). *(Courtesy of Dr. A. Woodcock, North-West Lung Centre, Wythenshawe Hospital, Manchester, England.)*

rise and plasma opsonic capacity (mean level of opsonin in the blood) may fall before the clinical onset of BO.[76]

The Working Party for the International Society for Heart and Lung Transplantation[77] use the term *bronchiolitis obliterans syndrome* to encompass graft deterioration secondary to progressive airways disease for which there is no other cause. "It is widely presumed, but unproved, that this is a manifestation of chronic rejection." The term *bronchiolitis obliterans* is reserved for histologically proven diagnoses only. Patients have to be 3 months or more posttransplant to be evaluable under the system the group proposes.

The incidence varies in different series.[78] In the initial Stanford series, 50 percent of long-term survivors with heart-lung transplants developed BO.[79] This figure has now been reduced to 38.6 percent.[80] The Papworth Group in Cambridge, England, report a 9.6 percent incidence of BO in long-term survivors, while the corresponding figure from Pittsburgh is 54 percent. In a recent series of single and double lung transplants reported from Toronto, the incidence of BO was 42.8 percent.[81]

Although BO occurs in single and double lung transplants, it is less common than with heart-lung transplants (Fig. 24-33). The Toronto Transplant Group, with a large experience with single and double lungs, had not initially observed this complication in their series.[82] However, 6 years later the incidence, as noted above, was 43 percent.[81]

BO is restricted histologically to membranous and respiratory bronchioles. The previous distinction between subtotal and total BO[36] was not now thought to be worthwhile by the working party.[36a]

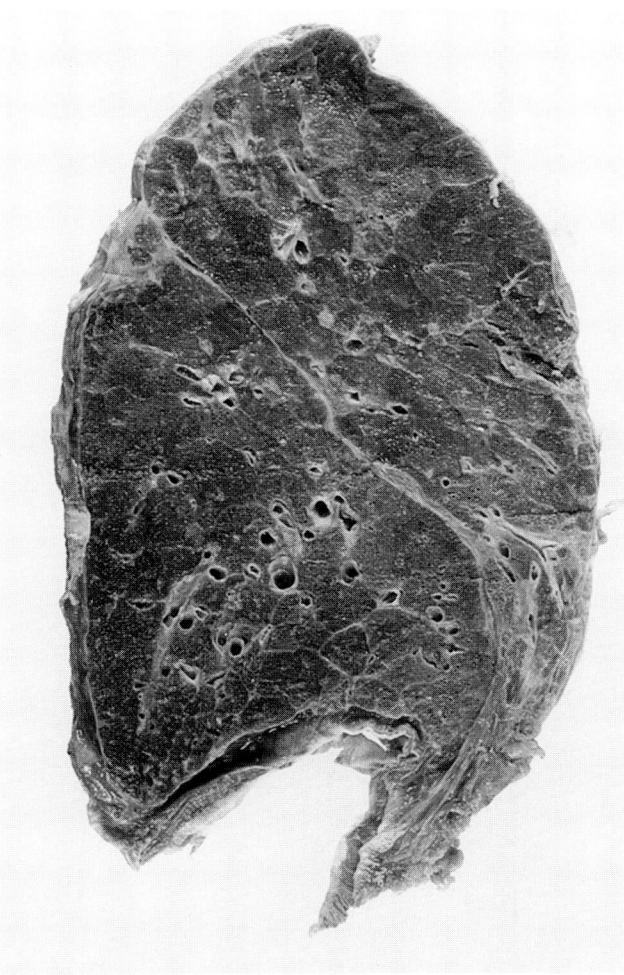

FIGURE 24-33
Obliterative bronchiolitis (autopsy specimen). It is difficult to convey in this photograph, but the lung was reduced in size. There is pleural and some focal pulmonary fibrosis.

They now felt an estimation of the relative activity of inflammatory infiltrate should be given. They graded BO as

a. *Active* In addition to the fibrosis there are intra- and/or peribronchiolar submucosal and peribronchiolar mononuclear infiltrates usually with ongoing epithelial damage.

b. *Inactive* There is dense fibrous scarring without any cellular infiltrate in the submucosa or peribronchiolar region (Figs. 24-34 to 24-37).

The Johns Hopkins Group[83] demonstrated two distinct histologic forms of BO. In patients with concurrent infections, aspiration, or large-airway obstruction, focal cellular BO extended into the distal alveolar spaces. In several cases there was associated aspirated material and foreign-body–type giant cells.

In the second group, in whom rejection was thought to be an important factor, the BO was characterized by a relatively acellular concentric fibrosing process limited to terminal bronchioles.

This distinction is not as useful as it may seem, since there is overlap in some cases; clearly, some transplant recipients may have concurrent rejection and infection or aspiration. In "pure" BO, which is probably rejection-related, the inflammatory process is predominantly lymphocytic, whereas in BO-organizing pneumonia the infiltrates consist of macrophages, lymphocytes, and fibroblasts.

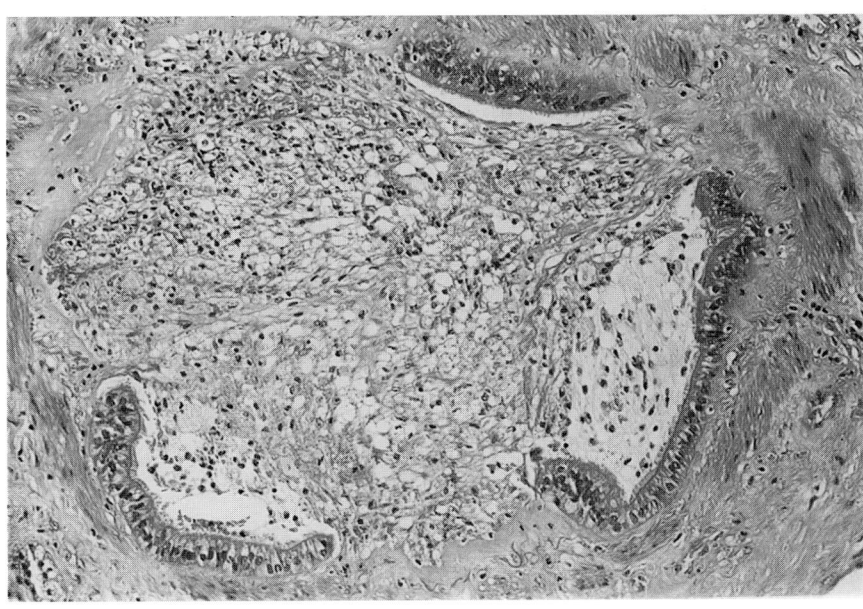

FIGURE 24-34
Partial obliterative bronchiolitis. (H&E, ×125.)

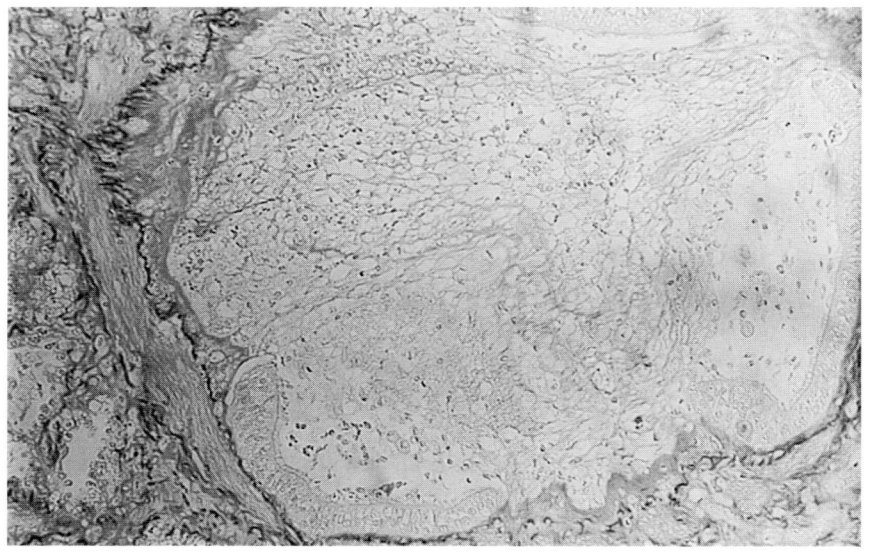

FIGURE 24-35
Partial obliterative bronchiolitis.
(Elastic–van Gieson, ×125.)

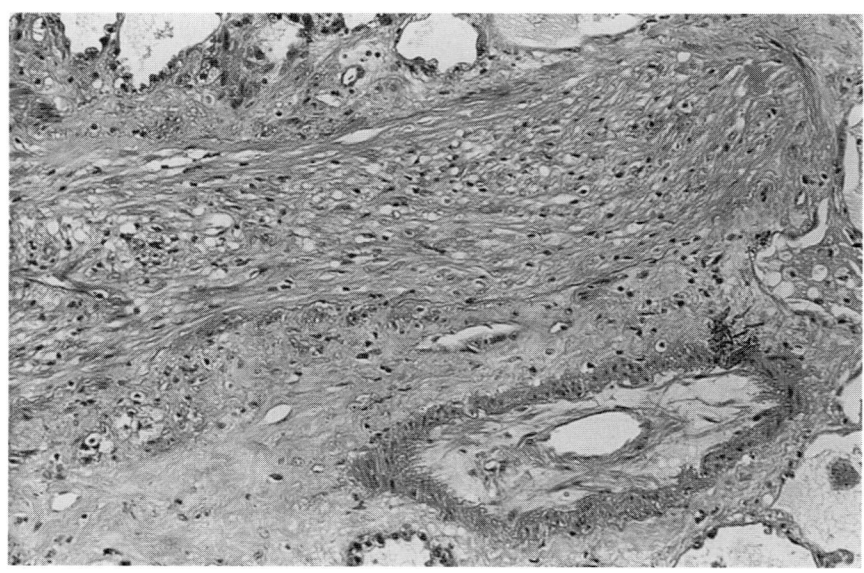

FIGURE 24-36
Total obliterative bronchiolitis. A pulmonary artery shows recanalized intimal fibrosis. (H&E, ×125.)

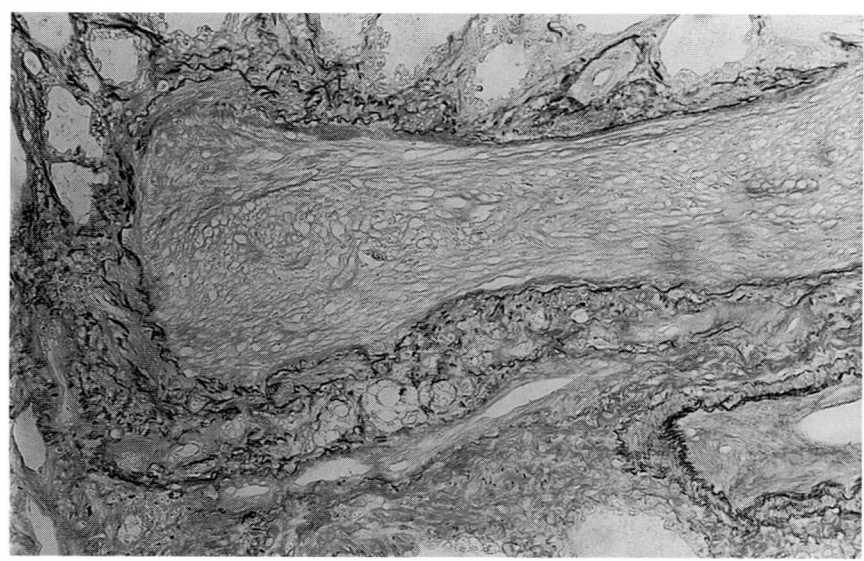

FIGURE 24-37
Total obliterative bronchiolitis. A different area of the case shown in Fig. 24-36. (Elastic–van Gieson, ×125.)

There are differing views on the value of transbronchial biopsy in the diagnosis of BO. Yousem and coworkers[84] reported a diagnostic transbronchial biopsy sensitivity rate of 66 percent in a small number of patients. In a more recent study of a larger number of patients, the rate increased to 87 percent.[85]

The Stanford study[37] found that only 3 percent of 366 lung biopsy specimens showed histologic evidence of BO. When larger volumes of tissue were available, including open biopsies, retransplantation specimens, and autopsies, BO was identified in nine of 13 patients. In a recent study, BO was detected on transbronchial biopsy in only 15.2 percent of cases.[86] Chamberlain and colleagues[81] obtained a sensitivity of 17.1 percent and a specificity of 94.5 percent. Similar figures were obtained by the Harefield Group (sensitivity, 27.7 percent; specificity, 75 percent).[36b]

Etiology and Pathogenesis of Bronchiolitis Obliterans

There are many causes of BO (see Chap. 7), and the most common in the context of lung transplantation is thought to be rejection. Posttransplant BO appears to be a specific disorder that follows a predictable clinical course. Theodore and coworkers[78] list the potential causes of BO in a transplant setting, including rejection, infection, lung denervation, bronchial ischemia, drug toxicity, altered mucociliary clearance, and impaired lower respiratory tract defense mechanisms. These factors may act individually or in concert.

BO is thought to be a manifestation of a chronic reaction to the allograft in that (1) lymphocytes accumulate in the subepithelial tissues of the airways in BO rather than around blood vessels, as in acute rejection; (2) it does not occur in all transplant recipients; (3) it usually begins 3 months or more after transplantation, when acute vascular rejection is unusual; and (4) it can be reversed by lower dosages of corticosteroids than are required for treating acute rejection.

HLA-A matching has shown a close relationship with the development of BO. In one series, patients having no or one mismatch on the HLA-A locus had less BO and a lower mortality than did patients with two mismatches on this locus.[87] Increased vascular HLA-DP expression is found in BO biopsies, as well as an increased frequency of bronchial HLA-DP.[87a]

There is also evidence to link BO with respiratory infection. The Pittsburgh group[88] showed that chronic rejection developed in 77 percent of patients after acute rejection, in 56 percent of recipients after

CMV infection, and in 50 percent of patients with PCP. There may be a relationship between the severity of rejection and the number of clinical infections. Burke and colleagues[79] noted that all three patients in their series with CMV infection developed BO. The distal respiratory epithelium expresses class II antigens.[89] BO can develop with acute airway rejection or persistent unresolved infection, and it may be exacerbated by infection, especially since class II antigens can be induced by CMV, EBV, or *P. carinii* infection. The lymphocyte alloreactivity may be nonspecifically up-regulated after infection. There are increased numbers of CD4+ cells.[87a] As noted in the section on CMV, mediators such as interferon γ are known to increase the expression of class II antigens,[90] provoking a response by activated T cells. It is thus possible that rejection is precipitated by infection, as a result of the effect on cell-surface antigen expression.

Changes in Bronchial Cartilage

These changes include necrosis with ischemic change (Fig. 24-38), fibrovascular ingrowth into normally acellular hyaline bronchial cartilage, ossification, and calcification (Fig. 24-39).[91] They are independent of small and large airways inflammation and may reflect a hypoperfusion state, since there is an interruption of bronchial artery supply to the main airways. The alterations occur as early as 53 days after transplantation, and in an uneventful transplant, collaterals from the coronary arteries and the omental vessels as well as the pulmonary arterial circulation supply blood to the bronchial wall.[92]

There must be some doubt whether the necrosis is all due to ischemia because omentopexy or direct implantation of the internal mammary artery has not improved the rate of stricture formation.[12] Accelerated pulmonary atherosclerosis (described below) in heart-lung transplants may limit cartilage perfusion.[93] Repeated infection in an area of diminished perfusion precipitates a cycle of injury that ends in bronchiectasis and bronchial cartilage destruction.

Bronchial cartilage destruction may cause diffuse acquired bronchomalacia.[94] The diagnosis of this disease is established with fiberoptic bronchoscopy. In most patients with this disease, the large-airway collapse is severe enough to cause almost complete tracheobronchial obstruction to enforced expiration. This airway collapse begins below the level of the tracheal anastomosis, extends well into the fourth- and fifth-generation bronchi, and is diffuse rather than segmental. Patients with

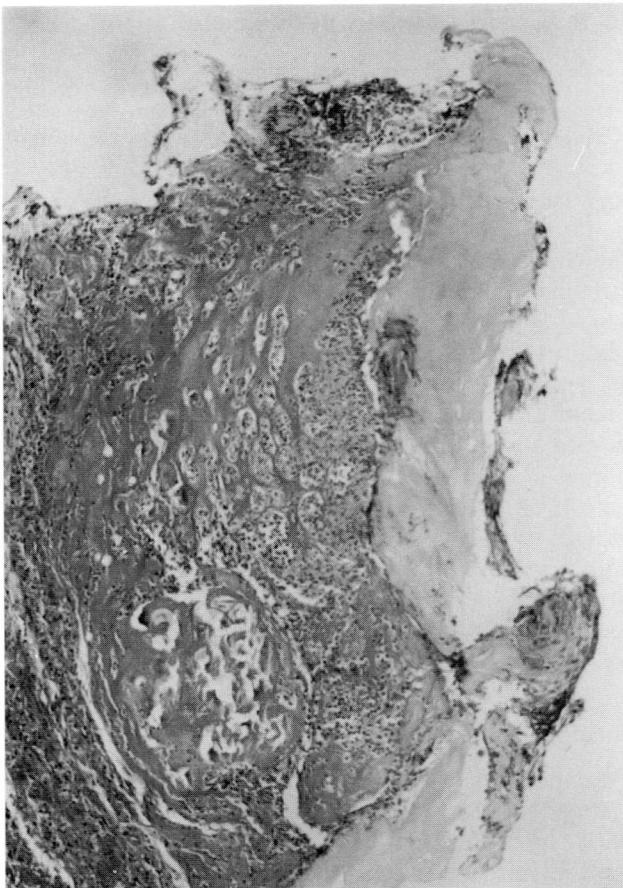

FIGURE 24-38
Ischemic change in bronchial cartilage with necrosis and active inflammation. (H&E, ×125.)

bronchomalacia have a characteristic "notch" in the expiratory spirogram and BO, but these changes could be due to a reduction in bronchial mucosal blood flow.

Chronic Vascular Rejection

This complication is seen both in the lungs and in the hearts of heart-lung transplant patients. Hearts in long-term survivors in one series were enlarged (average weight: 494 g).[20] All the hearts had evidence of graft coronary atherosclerosis, which was more extensive than the associated pulmonary vascular sclerosis. In some cases it has caused myocardial infarction. One case showed mild coronary artery endovasculitis, as well as large and small pulmonary vein endovasculitis. In these vessels, sparse numbers of lymphocytes, macrophages, and occasional plasma cells were present in the thickened intima.

In the pulmonary arteries there is mild to moderate medial hypertrophy (Fig. 24-40), but few other pulmonary vascular changes are seen apart from intimal fibrosis in small pulmonary veins (Fig. 24-41). The process may be active in the vessels with prominent cellularity and transmural infiltrates or "burned out," in that fibrosis is the only feature. This pulmonary vascular disease is often seen in patients with previous frequent and high-grade acute rejection. There is also an association with CMV infection.

A prospective study of 23 patients undergoing a total of 315 transbronchial biopsies[95] showed that intimal fibrosis and arteriolization of veins (Fig. 24-42) are significantly more common in heart-lung transplants with rejection and infection. The number

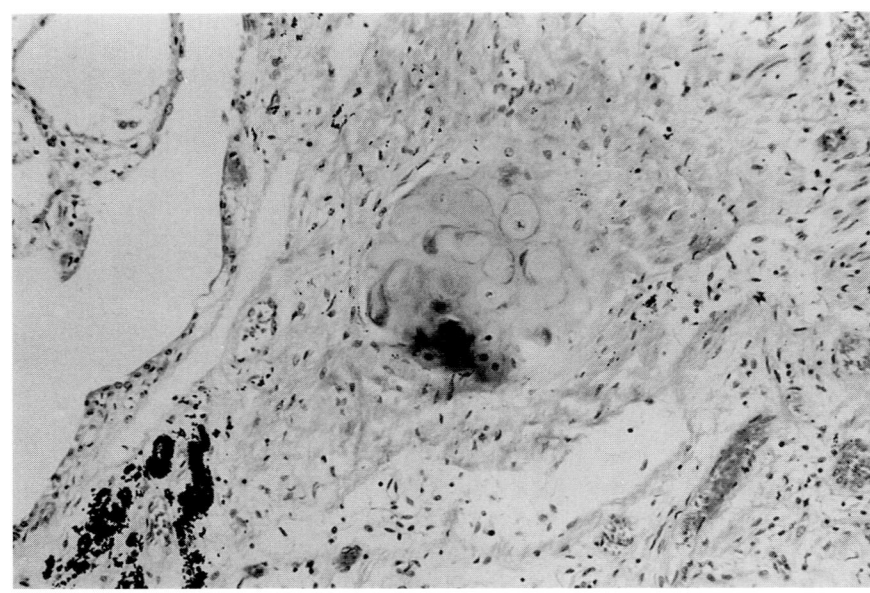

FIGURE 24-39
Focal calcification in bronchial cartilage. (H&E, ×125.)

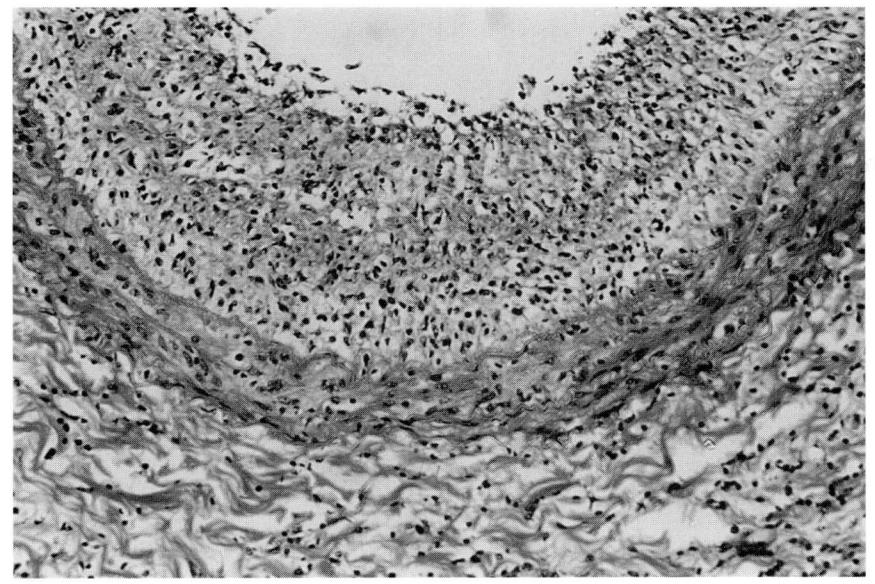

FIGURE 24-40
Medial hypertrophy of a muscular pulmonary artery. There is marked intimal fibrosis and chronic inflammation. This is from a case of chronic vascular rejection. (H&E, ×313.)

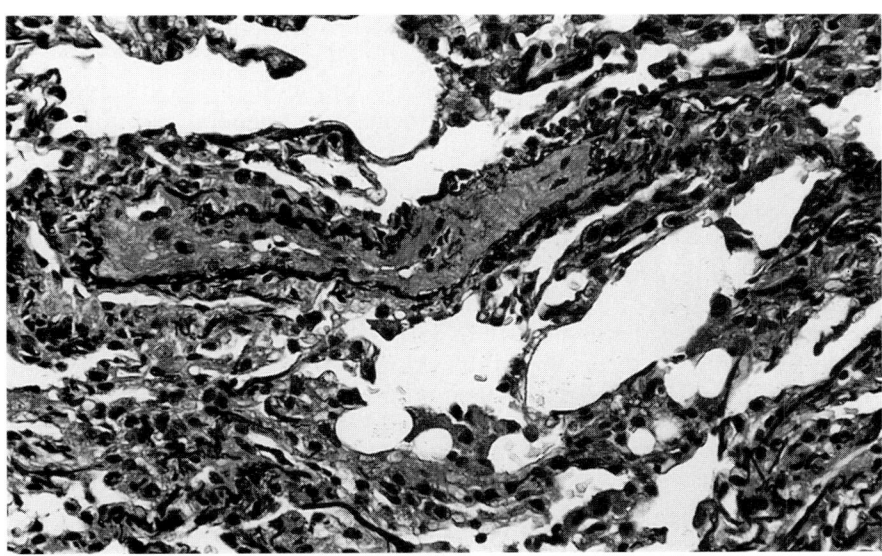

FIGURE 24-41
Chronic rejection. Intimal fibrosis occluding a small pulmonary vein. (Elastic–van Gieson, ×313.)

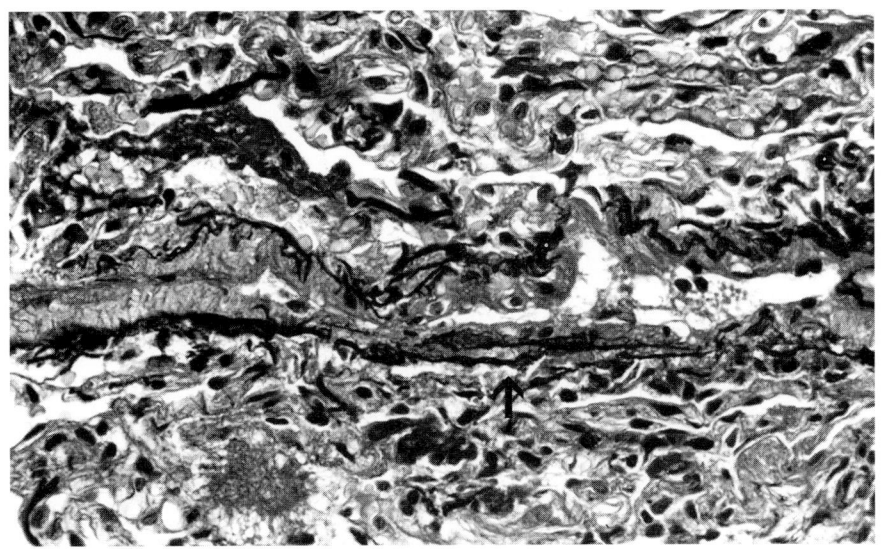

FIGURE 24-42
Arteriolization of a vein with marked intimal fibrosis and the focal development of an internal and external elastic lamina (*arrow*). (Elastic–van Gieson, ×313.)

of cases with BO and organizing pneumonia was too small for analysis. The venous changes were thought to be secondary to a combination of factors, including release of superoxide radicles in oxygenated blood, cytokine production in rejection and infection, immunosuppressive therapy, and lung preservation.

Posttransplant Lymphoproliferative Disease

Posttransplant lymphoproliferative disease (PTLD) is a leading cause of mortality outside the perioperative period. The overall incidence of thoracic PTLD over a 10-year period was 3.4 percent in heart transplant patients and 7.9 percent in lung transplant patients.[96] These figures appear to be higher than in renal (1 percent) and liver recipients (2 percent).[41a,97] Slightly lower figures have been given recently.[96a] Lymphoma is not the only cancer that develops in transplant patients: squamous carcinoma of the lip and skin, Kaposi's sarcoma, other sarcomas, carcinoma of the vulva, perineum, and kidney, and hepatobiliary tumors are more common in this group.

Of the lymphomas, most are non-Hodgkin's, accounting for 93 percent of lymphomas in transplant recipients compared with 65 percent in the general population.[98] They may present as a necrotic, ulcerative bronchitis after heart-lung transplants. (Figs. 24-43 and 24-44).[99] Some cases present with pulmonary infiltrates, others with masses in the lung or mediastinum or with disseminated pulmonary disease.[96]

Histologically, there is a wide spectrum of changes. The infiltrates may consist of mature plasma cells and may be mistaken for a rejection process. Typically the tumor is angiocentric, invades the full thickness of the vessel wall, and gives rise to necrosis, which may be the only histologic feature. Single-cell necrosis is common. At the other end of the spectrum are sheets of large blastic lymphoid cells with prominent nuclei and many mitoses. There may be a wide range of neoplastic cell types, including immunoblasts, transformed lymphoid cells, centrocytes and centroblasts, and histiocytes, which may show hematophagocytosis.[30] Vascular invasion is common.

The lesions range from polyclonal, benign B cell proliferations related to EB virus to frankly malignant, monoclonal B cell lymphomas.[97] Fourteen percent of the lymphomas are of T-cell origin, and fewer than 1 percent of null-cell origin.

After transplantation, non-Hodgkin's lymphomas differ from those in the general population in several respects. Extranodal and central nervous system involvement is more frequent in transplant recipients, and the lesions are often multicentric. The allograft itself is affected in 18 percent of cases.[100]

Lymphomas occur earlier in patients who have received cyclosporine than in those receiving azathioprine and cyclophosphamide. Lymphomas are similarly more common and occur earlier in patients treated with OKT3 and other monoclonal antibodies. Sixty-eight percent of the lymphomas occur within

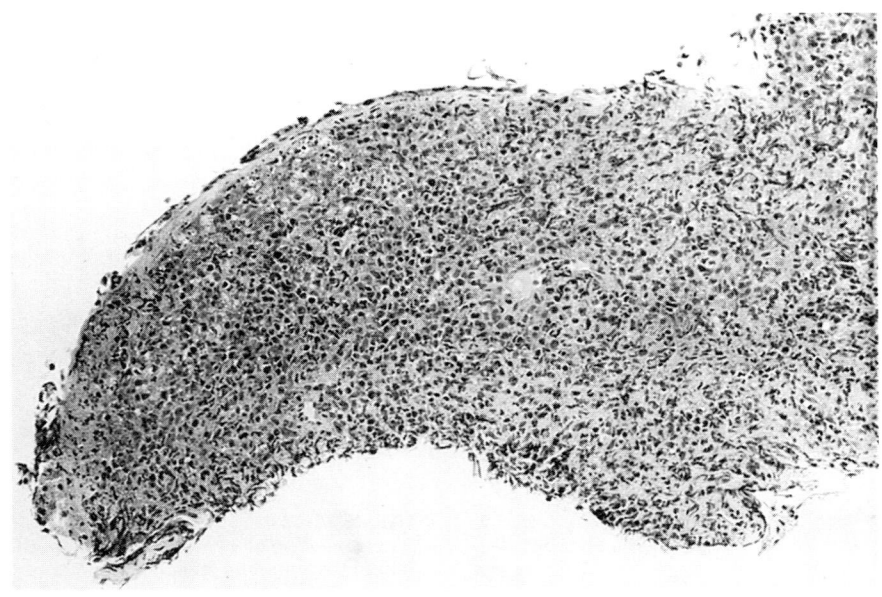

FIGURE 24-43
Non-Hodgkin's lymphoma in a bronchial biopsy. There is a diffuse infiltrate, which on low power may be diagnosed as chronic inflammation. (H&E, ×125.)

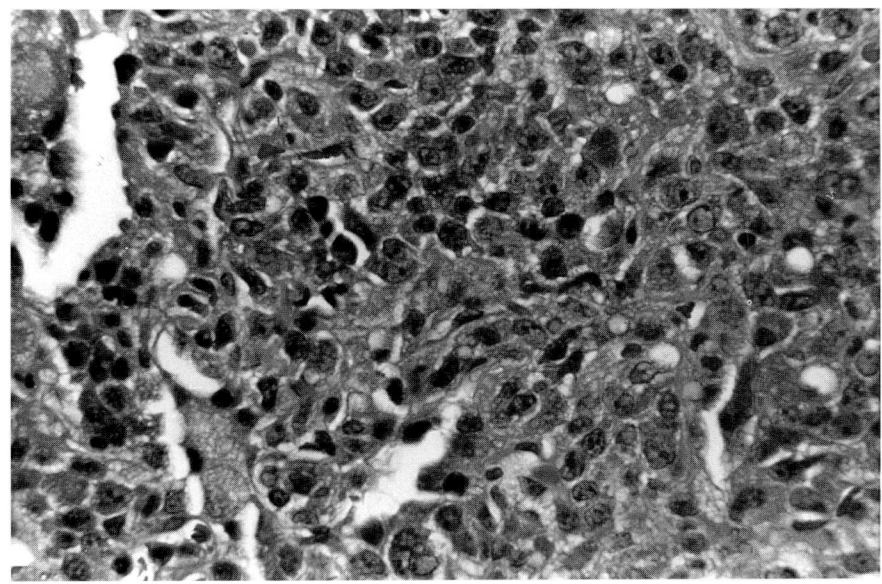

FIGURE 24-44
Non-Hodgkin's lymphoma with pleomorphic lymphoid cells. (H&E, ×500.)

the first 4 months, when immunosuppression is maximal. It has been suggested that immunosuppressive agents such as cyclosporine[101] may be particularly likely to induce lymphomas. A more plausible explanation is that no single agent is responsible but that in some patients who undergo a high degree of immunosuppression with many agents, non-Hodgkin's lymphomas develop as a consequence.[100]

PTLD may be difficult to distinguish from EBV pneumonitis.[102] Lymphomas may be related to EBV infection.[103] The risk is highest in seronegative recipients especially those who required antilymphocyte antibody therapy for rejection.[96a] Such proliferation may regress with antiviral agents and reduction of immunosuppression.

Recurrence of Primary Disease

This is rare. Sarcoid has recurred in the transplanted lung,[104–106] and patients with this disease have a higher incidence of rejection and lymphocytic bronchitis, as well as neutrophilia in their lavage. A recent case report described recurrent giant cell interstitial pneumonia in a single-lung transplant patient. Immunostains for CMV, herpes simplex viruses I and II, and measles virus were all negative, and there was no evidence of aspiration.[107] Pulmonary lymphangioleiomyomatosis[108] and diffuse panbronchiolitis[108a] have been shown to recur after transplantation. A fibrosing lung disease recurred after transplantation in another case.[109]

In a few cases, disease may develop that was present in the donor. Bronchial asthma and bronchopulmonary aspergillosis are two such diseases.

Bronchoalveolar lavage

This is useful in providing rapid diagnosis of infection, especially that from CMV, *P. carinii*, and *Aspergillus* species. The cell count in lavage fluid is no help in differentiating rejection from infection.[110] The lymphocyte count has no correlation with the grade of rejection seen on biopsy.

CARDIAC TRANSPLANTATION

Cardiac transplantation has become an established treatment of end-stage heart disease, but there may be pulmonary complications.

Complications

VASCULAR

Pulmonary artery torsion may in some cases cause right ventricular failure immediately after orthotopic heart transplantation.[111] This can be due to size mismatch between donor and recipient hearts and great vessels, resulting in malalignment of the main pulmonary artery stumps.

Some patients develop postoperative pulmonary hypertension secondary to pulmonary vascular changes due to preexisting heart failure. We have seen two cases with long-standing left ventricular failure causing muscularization of pulmonary arterioles, hypertrophy of muscular pulmonary arteries, arteriolization of veins with intimal longitudinal muscle (Fig. 24-45), and marked intimal fibrosis in pulmonary veins.[112] The picture is identical to that

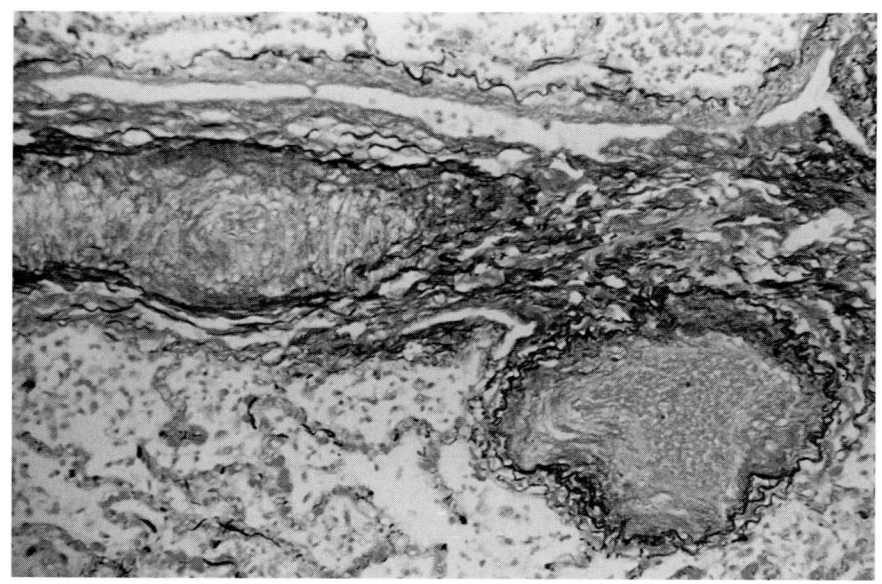

FIGURE 24-45
Arteriolization of veins secondary to chronic congestive cardiac failure. This patient died soon after a cardiac transplant and was noted to have a distended right ventricle before death. (Elastic–van Gieson, ×313.)

of pulmonary venoocclusive disease. In addition, there was interstitial fibrosis and many hemosiderin-laden macrophages.

We recently detected an increase in the number of hemosiderin-laden macrophages (Fig. 24-46) in BAL from cardiac transplant patients several months after surgery, when evidence of congestive cardiac failure had long disappeared. None of these patients had cardiac rejection or pulmonary infection. The number of macrophages was higher than in a control group with fibrosing alveolitis or in a group of heart-lung transplantation patients. The findings suggest that there is a microvascular leakage, in cardiac transplantation contributing to diffusion abnormalities.[113]

INFECTION

Cardiac transplant recipients are at increased risk of infection. Posttransplant reintubation and large doses of steroids are significant risk factors for pneumonia.[114]

The first period of increased risk for pneumonia is immediately after transplantation. This is when factors causing nosocomial pneumonia are seen in all intensive care patients. Gram-negative organisms are acquired by colonization of the repiratory tract during prolonged hospitalization. The factors causing this have been dealt with previously (Chap. 7).

The second period of increased risk for pneumonia has a bimodal distribution, peaking first at 2 to 3

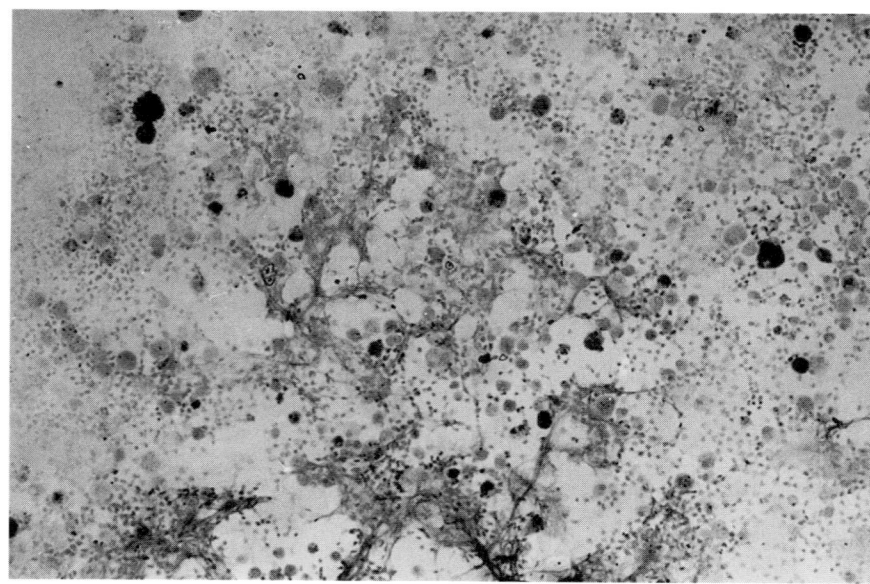

FIGURE 24-46
Increase in hemosiderin-laden macrophages in BAL after cardiac transplantation. (Perls', ×125.)

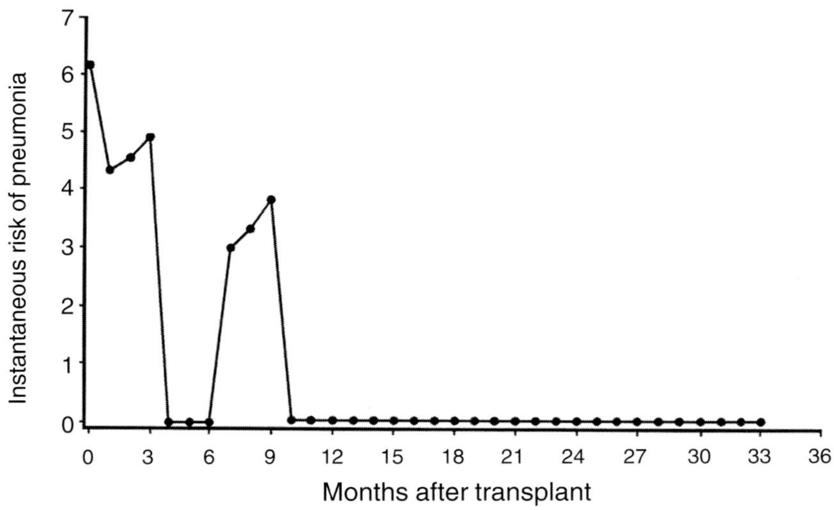

FIGURE 24-47
Peaks for infection after cardiac transplantation. *(Reproduced with permission of Dr. M. J. Gorensek et al. and Transplantation 46:860–865, 1988.)*

and then at 6 to 9 months after transplantation (Fig. 24-47). These are periods when the duration of immunosuppression has been sufficient to decrease cell-mediated immunity.

The incidence of infection after cardiac transplantation in a large series of cases was 31 percent, and the patients had one or more infective episodes.[115] Bacterial and viral infections were most common (47 and 41 percent, respectively), with fungal and protozoal infections accounting for 12 percent. The overall mortality in patients with infection was 13 percent, but the rate was higher with fungal infection (36 percent). The probability of infection was higher when OKT3 or antithymocyte globulin induction therapy was used.

In the immediate period after surgery, infections with usual nosocomial pathogens, including staphylococci and gram-negative bacili, are the most common. Opportunistic infections occur later. The specific infection varies according to the region, and cases of coccidioidomycosis and histoplasmosis are more common in America than in Europe. Although the lung is the most common site of infection after cardiac transplantation, the mediastinal wound, urinary tract, and then blood, spine, brain, and even gastrointestinal tract may be affected. The organisms that have been documented in cardiac transplantation are listed in Table 24-2.

Rhodococcus equi[117] and human parvovirus B19[118] have been described as causing pulmonary infection after cardiac transplantation, and it is likely that additional rare organisms will be described in the future. Most infection-related fatalities are caused by infection by bacteria, *Aspergillus,* or CMV.[119]

The risk factors for CMV infection after heart transplantation have been documented in a multi-institutional study.[120] The peak incidence of initial infection was at 2 months. In some patients the infection was recurrent. The primary site of infection was in the blood (43 percent), lung (30 percent), and gastrointestinal tract (23 percent). CMV pneumonia carried the highest mortality.

The primary risk factor by multivariate analysis was pretransplantation serology (positive donor, negative recipient [$p < .0001$], positive donor, positive recipient [$p + .0002$], and negative donor, positive recipient [$p = .02$]).

TABLE 24-2
Organisms Causing Infection in Cardiac Transplantation

Viral	
Cytomegalovirus	
Epstein-Barr virus	
Varicella zoster/Herpes simplex	
Bacterial	
S. aureus	*Pseudomonas* sp.
Pneumococcus	*Proteus* sp.
H. influenzae	Enterococcus
Streptococcus	*Legionella pneumophila*
Corynebacteria	*Listeria monocytogenes*
Enterobacter	*Nocardia asteroides*
E. coli	*M. tuberculosis*
Klebsiella sp.	
Fungal	
Aspergillus	Blastomycosis
Coccidiomycosis	Histoplasmosis
Phycomycosis (*Mucor* and *Rhizopus*)	*Pneumocystis carinii*
	Dactylaria constricta[116]
Cryptococcosis	
Candidiasis	
Protozoal	
Toxoplasma gondii	

CMV has been implicated by one paper[124] but not another[122] in the accelerated atheroma that occurs in the graft. No studies have looked specifically at the pulmonary vascular disease seen in lung transplantation and whether it has any relationship to CMV infection.

Regimens employing a combination of cyclosporine, steroids, and azathioprine cause less toxicity and lower infection rates than those that relied solely on cyclosporine and steroids or than the regimens employed in the 1970s. The pathology of the infections noted in Table 24-2 has been described in the appropriate section.

One organism, *Toxoplasma gondii*, should be discussed further. It appears to be acquired with the donor heart,[123] although occasionally it may be transmitted through a blood transfusion. The patients who seroconvert—in contrast to those who develop increasing titers of preexisting antibody—tend to have more severe disease. The organism may be detected in routine endomyocardial biopsy.[124] It may cause meningoencephalitis, cerebral abscess, pneumonitis, hepatitis, retinitis, or myocarditis.

Open lung biopsy has been advocated for cardiac transplant patients with bilateral interstitial infiltrates.[125] However, in most cases, the diagnosis can be made from BAL or transbronchial biopsy.

Malignancy

In a group of 275 patients who survived more than 1 month after heart transplantation, malignant tumors developed in 11.[126] The most common were squamous carcinomas affecting the skin; less commonly they affected the esophagus, anus, larynx, and lip. Also described were small-cell carcinoma of the bronchus and lymphoma, which was either a non-Hodgkin's lymphoma "thought to be B cell in origin" or a lymphoproliferative response to EB virus 4 months after transplantation. A recent report noted EB virus–associated high-grade B-cell pulmonary lymphoma less than 8 weeks after transplantation.[127] The prognosis in posttransplant lymphoma patients is generally poor, with a median survival of 14 months.[128]

A case of multiple myeloma 18 months after cardiac transplantation has been described.[129] There was initial improvement after withdrawal of cyclosporine. There was no evidence with either in situ hybridization or polymerase chain reaction (PCR) of any EBV genome.

PULMONARY DISEASES AFTER OTHER TRANSPLANTATION PROCEDURES

Marrow Transplantation

There have been several reviews of lung disease occurring after marrow transplantation.[130–135a] Figure 24-48 shows the time phase of the various complications that occur. The complications are divided into acute, occurring in the first 100 days after marrow transplantation (BMT), and chronic, seen more than 100 days after BMT.[136]

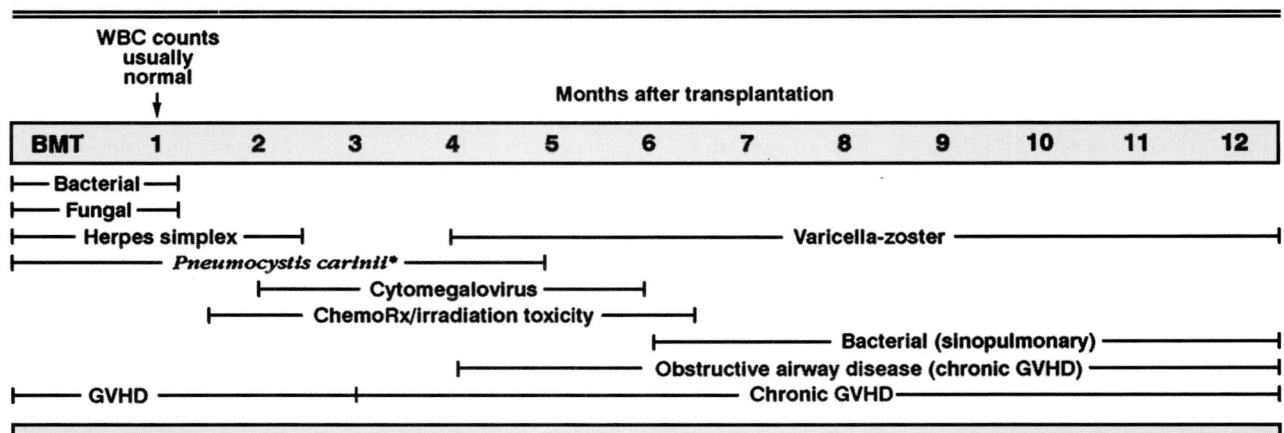

* *Very rare during trimethoprim-sulfamethoxazole prophylaxis*

FIGURE 24-48
Usual time of onset of the major complications after marrow transplantation (BMT x).
(Reproduced with permission of Dr. M. J. Krowka et al. and Chest 87:237–246.)

ACUTE

Infection

The infectious complications are similar to those occurring in any transplant patient—i.e., bacterial (especially gram-negative organisms as well as *S. pneumoniae* and *S. aureus*), fungal (especially *Aspergillus, Candida* species, and *P. carinii*), and viral (especially *H. simplex* and CMV).

One of the most common infections is CMV pneumonitis, which is seen in 15 percent of bone marrow recipients and has a fatality rate of 85 percent or greater.[137] The disease may progress to bronchiolitis obliterans–organizing pneumonia,[138] suggesting that CMV pneumonia is an immunopathologic process in marrow transplants.

Some of the causes predisposing to infection are similar to those with other transplants. Special factors operating in marrow recipients are reduced immunoglobulins, a defective response to specific antigenic challenge, and compromised cellular immunity caused by a reversed helper:suppressor ratio. This is due to reduction of helper cell number and function.

The use of laminar flow units and prophylactic antibiotics has reduced the incidence of *Aspergillus* and *P. carinii* infection, respectively.

Pulmonary edema

Pulmonary edema may occur in the first few weeks after marrow transplantation. The etiology of this edema is uncertain, but there may be hypoalbuminemia, sepsis, acute graft-versus-host disease, and occasional hepatic venoocclusive disease.[139]

Also implicated is elevated intravascular pressure, increased alveolar-capillary permeability, or both.[135] The increase in intravascular pressure may be due to fluid retention, caused by iatrogenic volume overload or sodium excess from antibiotic therapy. There may be cardiac dysfunction due to drugs (doxorubicin or cyclophosphamide), radiotherapy, or both. The increased capillary permeability may be caused by sepsis, total body irradiation, cyclosporin A, chemotherapy, and prophylactic granulocyte transfusions.

Alveolar hemorrhage

This may occur in the face of profound neutropenia. It has been noted in patients with normal coagulation measurements in the early phases of marrow recovery.[136] This complication has been reported in 20 percent of autologous marrow transplant recipients.[140] In some of these cases there is a polymorph infiltrate into the lung, giving rise to epithelial injury.[141] The mortality is high. Diagnosis is made

from BAL, followed by treatment with high-dose steroids. The cause of this condition is unknown.

Graft-versus-host disease

With lessening toxicity of drug regimes and better treatment of infections, graft-versus-host disease (GvHD) is emerging as one of the most serious complications of allogeneic bone grafting. Acute GvHD is seen in 30 to 60 percent of recipients of allogeneic marrow grafts.[142] The mortality is up to 50 percent.[143] Possible mechanisms of the disease are discussed in a recent review.[144]

In the lungs, GvHD causes lymphocytic bronchitis, where small lymphocytes infiltrate the proximal bronchial mucosa with ciliary loss and damage to submucosal glands, leading to lower respiratory tract infections. There is also evidence of mucosal necrosis, which may extend into submucosal glands. There is no correlation with interstitial lung disease, which may be a separate disorder in some of these patients.

Patients have a nonproductive cough and dyspnea, usually between days 25 and 75 after the BMT. On fiberoptic bronchoscopy there is a red tracheobronchial tree, and lavage shows a moderate increase in lymphocytes.

This lymphocytic bronchitis may not be a specific manifestation of acute or chronic GvHD but a complex interaction of marrow transplantation, chemotherapy, irradiation, and infection.

Interstitial pneumonitis

Thirty to 50 percent of marrow transplant recipients develop a pulmonary infiltrate in the first 6 months after the transplant.[132,137,145,146] Clinically the disease is indistinguishable from CMV pneumonia. Mortality may reach 80 percent. The pathologic findings are usually nonspecific.

A study of seven patients with late-onset respiratory symptoms and lymphocytic alveolitis has been reported.[147] All these patients had chronic GvHD.

Histology showed lymphoid interstitial pneumonia, which has been noted in this disease before.[148] There was moderate fibrosis of the alveolar walls (Fig. 24-49). In addition, one patient showed peribronchial, periarterial, and septal infiltration; another patient showed hyperplasia of type II pneumocytes. Interstitial fibrosis is an unusual pulmonary manifestation of chronic GvHD.

BAL demonstrated lymphocytosis ranging from 29 to 75 percent. These were all T cells, composed of CD8+-positive cells.

A role for drugs or viruses in these cases of late-onset pneumonitis could not be ruled out.

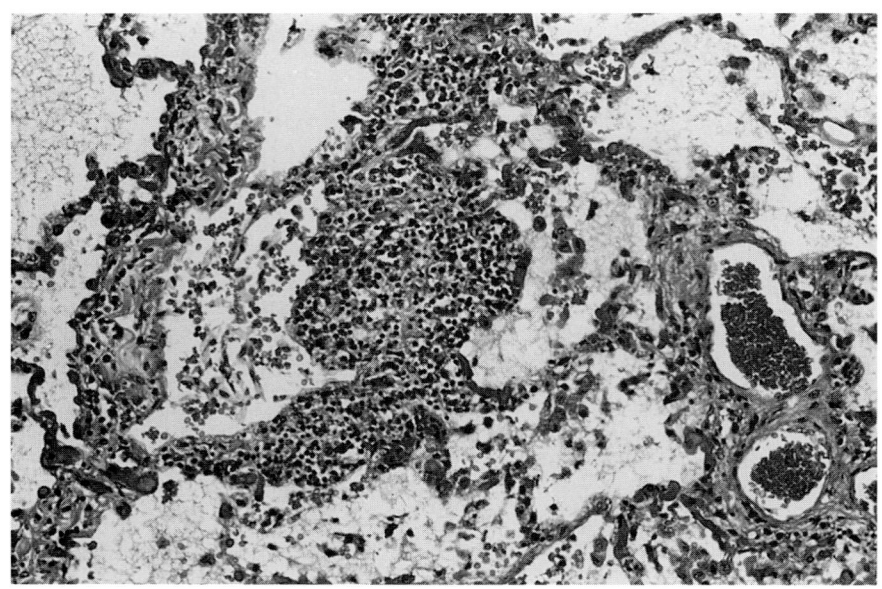

FIGURE 24-49
Lymphoid interstitial pneumonia after BMTx. (H&E, ×125.) *(Reproduced with permission of Dr. J. D. van der Walt, the Royal London Hospital, London, England.)*

CHRONIC

Infection

The main chronic infections in the chronic phase are *P. carinii Herpes simplex*, *Varicella zoster*, CMV, and bacteria such as pneumococcus and *S. aureus*. CMV is one of the main infectious complications in this phase.

Interstitial fibrosis

Interstitial fibrosis has been described above, as has BO, which usually occurs 6 to 12 months after marrow transplantation and in almost all cases is associated with GvHD. It is similar histologically to that seen in heart-lung transplantation (Fig. 24-50), and some cases may show an ARDS picture.

In the long term, interstitial fibrosis occurs in approximately 40 percent of patients 12 months or longer after BMT. As above in the short term, the changes are likely to be multifactoral and include chemotherapy, radiation, and effects of pulmonary infection and recurrent aspiration. As these patients present with an interstitial infiltrate, the possibility of recurrent disease or infection should be borne in mind. Recent PCR studies have shown higher levels of viral genes in the interstitial fibrosis after transplantation.[134]

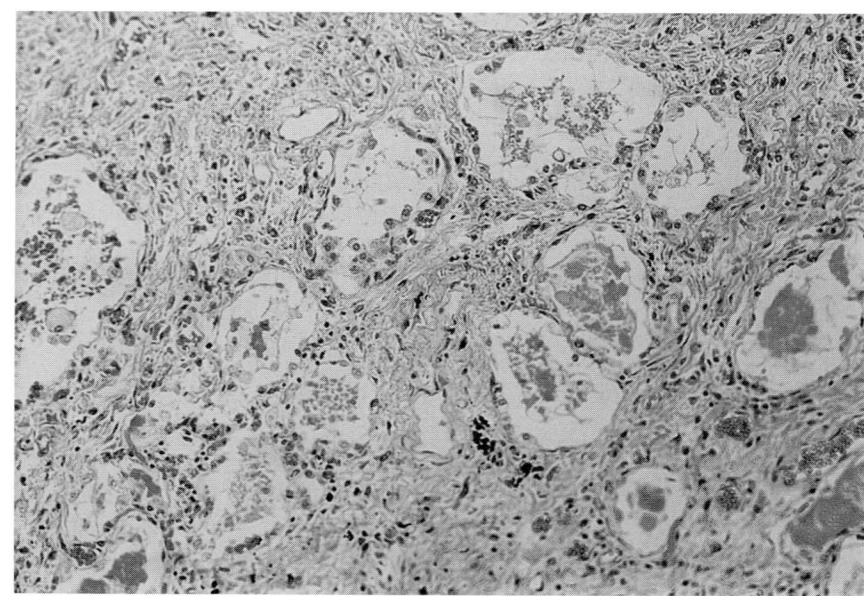

FIGURE 24-50
Severe interstitial fibrosis after BMTx. Only a few alveolar spaces remain, and these are lined by cuboidalized epithelium. (H&E, ×125.) *(Reproduced with permission of Dr. J. D. van der Walt, the Royal London Hospital, London, England.)*

Bronchiolitis obliterans

Several authors[149–152] have described progressive BO unrelated to lymphocytic bronchitis or interstitial lung disease occurring in approximately 10 percent of patients with chronic GvHD. This complication usually occurs 6 months after transplantation, with a range of 2 to 20 months.

It is similar to the disease seen in lung transplantation, and patients complain of nonproductive cough and dyspnea on exertion. Chest radiographs show hyperinflation with focal or diffuse infiltrates. The diagnosis is usually made on the findings of a reduced transfer factor and evidence of airflow obstruction.

The etiology of the disorder is not understood. Risk factors include immunosuppression with methotrexate, pre-BMT chemotherapy, irradiation, and low serum immunoglobulins after BMT. It may be caused by direct damage to small airways by GvHD, recurrent pulmonary infections, or repeated aspiration due to GvHD-associated esophagitis.

Graft-versus-host disease

GvHD may occur in the chronic phase, although it usually evolves from the acute form. There may be scleroderma-like changes in the skin, eye involvement resembling that in Sjögren's syndrome, and liver disease. In the lungs there is BO, lymphoid interstitial pneumonia, and interstitial pneumonitis, as described above. Lymphocytic bronchitis with a mononuclear infiltrate in the epithelium and a decrease in ciliated and goblet cells, as well as submucosal gland necrosis, have been reported[153] but is

regarded as a nonspecific abnormality. Marked intimal fibrosis may be seen in GvHD (Fig. 24-51).

Liver Transplantation

The noninfectious complications in liver transplantation reflect the prolonged general anesthesia, the extensive upper abdominal surgery, and the large quantities of blood transfused.[154] These complications include pulmonary atelectasis due to extensive right upper quadrant dissection and retraction of the right hemidiaphragm, causing diaphragmatic irritation or injury and resulting in abnormal diaphragmatic function. Pleural effusions, which are common after this procedure, further predispose to the development of atelectasis. They are usually right-sided and may be related to preoperative ascites or to reactive fluid formation as a result of the extensive subdiaphragmatic dissection.[155] Most effusions resolve spontaneously in the first few postoperative weeks.

ARDS

ARDS, usually due to sepsis, has been reported in up to 17.5 percent of liver transplant patients and is associated with a mortality approaching 80 percent.[156–158]

INFECTION

Infection is still a major complication after orthotopic liver transplantation.[159] The infections usually are fungal or bacterial and arise in the liver or biliary system. They are typically caused by aerobic gram-

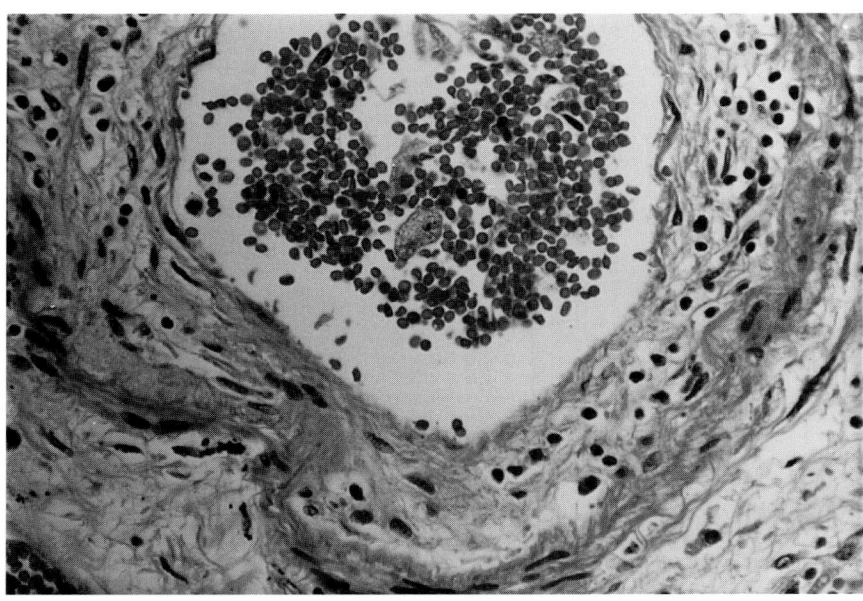

FIGURE 24-51
Intimal fibrosis in a pulmonary vein. There is some intimal fibrosis. BMT× patient. (H&E, ×313.) *(Reproduced with permission of Dr. J. D. van der Walt, the Royal London Hospital, London, England.)*

negative enteric bacteria or *Candida* species.[160] Liver injury by rejection, vascular damage, or defective biliary drainage increases the chances that organisms will invade the transplant recipient's bloodstream or peritoneum.

Postoperative atelectasis and aspiration predispose to gram-negative as well as gram-positive organisms colonizing the pulmonary tree and causing bronchopneumonia. CMV is also common, typically occurring in the fourth to sixth postoperative week. Fungal infection with organisms such as *Aspergillus, Candida,* and *P. carinii* used to be common, but improvements in surgery and postoperative care as well as prophylactic antibiotics have decreased their incidence.

CALCIFICATION

Lung calcification has been noted in liver transplant patients.[157,161,162] This complication may be asymptomatic, but patients with symptoms develop a nonproductive cough and dyspnea within several months of transplantation and may develop respiratory failure.

The diagnosis may be made on transbronchial biopsy, and histologically there are fine interstitial or perialveolar space microcrystals (Fig. 24-52) with occasional intraalveolar deposits. In severe cases, dense calcific deposits are accompanied by interstitial fibrosis. Risk factors such as ARDS and viral pneumonia have been implicated, but the exact mechanism causing the calcification is unknown.

A cause of the calcification maybe indicated in a recent paper demonstrating proliferating bile ducts producing parathyroid hormone–related peptide.[163] The patients also had high levels of circulating mid-molecule parathormone.[162]

Renal Transplantation

The most frequent noninfectious pulmonary complication is pulmonary edema.[164] This is caused by renal dysfunction associated with allograft rejection.

Pulmonary thromboembolism was found in 60 percent of patients before 1980.[164] This complication has decreased with earlier postoperative mobilization and improved surgical technique. However, manipulation of the pelvic veins during surgery and the use of high-dose steroids may predispose the patient to thrombosis.

MALIGNANCIES

One of the malignancies that occur after renal transplantation is B-cell non-Hodgkin's lymphoma.[165] The lung may be involved by transplant lymphomas, and in this instance there are focal nodular infiltrates. Pulmonary involvement by Kaposi's sarcoma may be seen.[166]

INFECTIOUS COMPLICATIONS

Infectious complications occur in renal transplant patients. As with other transplants, bacterial pathogens and CMV are the most important causes of

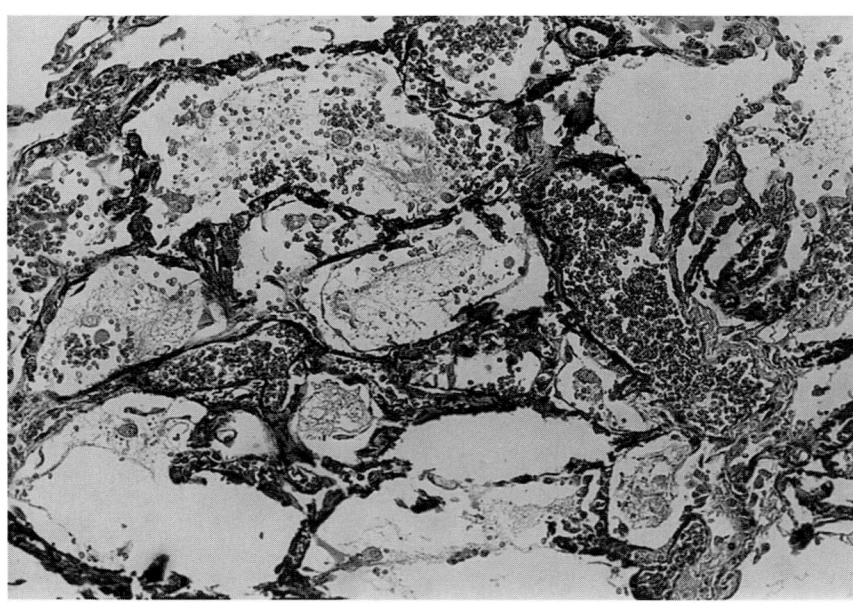

FIGURE 24-52
Calcification in the alveolar wall after liver transplantation. (H&E, ×125.) *(Reproduced with permission of Dr. S. Hubscher, University of Birmingham Medical School, Birmingham, England.)*

pneumonia, although *P. carinii, Aspergillus,* and mycobacteria have also been described. Some patients may have more than one infection. Lavage is a good diagnostic method for detecting the appropriate organisms.[167]

REFERENCES

1. Hosenpud JO, Novick RJ, Breen TJ, Daily OP: The Registry of the International Society for Heart and Lung Transplantation: Eleventh official report—1994. *J Heart Lung Transplant* 13:561–570, 1994.
1a. Hosenpud JD, Novick RJ, Breen TJ, Keck B, Daily P: The Registry of the International Society for Heart and Lung Transplantation: Twelfth Official Report—1995. *J Heart Lung Transplant* 14:805–815, 1995.
2. Graeter T, Schafers H-J, Wahlers T, Borst HG: Lung transplantation in Kartagener's syndrome. *J Heart Lung Transplant* 13:724–726, 1994.
3. Hardy JD, Webb WR, Dalton NL, Walker GR: Lung homotransplantation in man: Report of the initial case. *JAMA* 186:1064–1084, 1963.
4. Reitz BA, Wallwork J, Hunt SA, et al: Heart-lung transplantation: Successful therapy for patients with pulmonary vascular disease. *N Engl J Med* 306:557–563, 1982.
5. Sarris GE, Smith JA, Shumway NE, et al: Long-term results of combined heart-lung transplantation: The Stanford experience. *J Heart Lung Transplant* 13:940–949, 1994.
6. LoCicro J III, Massad M, Matano J, Greene R, Dunn M, Michaelis LL: Contribution of the bronchial circulation to lung preservation. *J Thorac Cardiovasc Surg* 101:807–815, 1991.
7. Pham SM, Yoshida Y, Aeba R, et al: Interleukin-6, a marker of preservation in clinical lung transplantation. *J Heart Lung Transplant* 11:1017–1024, 1992.
8. Schreinemakers HHJ, Weder W, Miyoshi S, et al: Direct revascularization of bronchial arteries for lung transplantation: An anatomical study. *Ann Thorac Surg* 49:44–54, 1990.
9. Goldsmith MF: Mother to child: First living donor lung transplant. *JAMA* 264:2724, 1990.
10. Todd TRJ: Early postoperative management following lung transplantation. *Clin Chest Med* 11:259–267, 1990.
11. Dubois P, Choiniere L, Cooper JD: Bronchial omentopexy in canine lung allotransplantation. *Ann Thorac Surg* 38:211–214, 1984.
12. Khaghani A, Tadjkarimi S, al-Kattan K, et al: Wrapping the anastomosis with omentum or an internal mammary artery pedicle does not improve bronchial healing after single lung transplantation: Results of a randomized clinical trial. *J Heart Lung Transplant* 13:767–773, 1994.
13. Sadeghi A, Guthaner DF, Wexler L, et al: Healing and revascularisation of the tracheal anastomosis following heart lung transplantation. *Surg Forum* 33:236–238, 1982.
14. Griffith BP, Magee MJ, Gonzalez IF, et al: Anastomotic pitfalls in lung transplantation. *J Thorac Cardiovasc Surg* 107:743–754, 1994.
14a. Shennib H, Massard G: Airway complications in lung transplantation. *Ann Thorac Surg* 57:506–511, 1994.
15. Siegelman SS, Sinha SBP, Veith FJ: Pulmonary re-implantation response. *Am Surg* 117:30, 1973.
16. Ohori NP, Iacono AT, Grgurich WF, Yousem SA: Significance of acute bronchitis/bronchiolitis in the lung transplant recipient. *Am J Surg Pathol* 18:1192–1194, 1994.
17. Sarsam MAI, Yonan NA, Beton D, McMaster D, Deiraniya AK: Early pulmonary vein thrombosis after single lung transplantation. *J Heart Lung Transplant* 12:17–19, 1993.
18. Powell-Jackson PR, Carmichael FJL, Calne RY, Williams R: Adult respiratory distress syndrome and convulsions associated with administration of cyclosporine in liver transplant recipients. *Transplantation* 38:341–343, 1984.
19. Carbone L, Appel GB, Benvenisty AI, Cohen DJ, Kunis CL, Hardy MA: Adult respiratory distress syndrome associated with oral cyclosporine. *Transplantation* 43:767–768, 1987.
19a. Chaparro C, Chamberlain D, Maurer J, De Hoyos A, Winton T, Kesten S: Acute lung injury in lung allografts. *J Heart Lung Transplant* 14:267–273, 1995.
20. Tazelaar HD, Yousem SA: The pathology of combined heart-lung transplantation: An autopsy study. *Hum Pathol* 19:1403–1416, 1988.
20a. Bando K, Paradis IL, Komatsu K, et al: Analysis of time-dependent risks for infection, rejection, and death after pulmonary transplantation. *J Thorac Cardiovasc Surg* 109:49–59, 1995.
21. Bienenstock J, Johnston N, Perey DYE: Bronchial lymphoid tissue: I. Morphologic characteristics. *Lab Invest* 28:686–692, 1987.
22. Beinenstock J, Johnston N, Perey DYE: Bronchial lymphoid tissue: II. Functional characteristics. *Lab Invest* 28:693–698, 1987.
23. Hruban RH, Beschorner WE, Baumgartner WA, et al: Depletion of bronchus associated lymphoid tissue associated with lung allograft rejection. *Am J Pathol* 132:6–11, 1988.
24. Abbas AK, Lichtman AH, Pober JS: Immunity to tissue transplants, in Abbas AK et al (eds), *Cellular and Molecular Immunology.* Philadelphia, W. B. Saunders, 1991, pp 317–334.
25. Krensky AM, Weiss A, Crabtree G, Davis MM, Parham P: T-lymphocyte–antigen interactions in transplant rejection. *N Engl J Med* 322:510–517, 1990.
26. Bradley JA, Bolton EM: The T-cell requirements for allograft rejection. *Transplant Rev* 6:115–129, 1992.
27. Cooper DKC, Novitzky D, Rose AG, Reichart BA: Acute pulmonary rejection precedes cardiac rejection following heart-lung transplantation in a primate model. *J Heart Transplant* 5:129–132, 1986.
28. Paradis IL, Duncan SR, Dauber JH, Yousem S, Hardesty R, Griffiths B: Distinguishing between infection, rejection and adult respiratory distress syndrome after human lung transplantation. *J Heart Lung Transplant* 11:5232–5238, 1992.
29. Higenbottam T, Stewart S, Penketh A, Wallwork J: Transbronchial lung biopsy for the diagnosis of rejection in heart-lung transplant patients. *Transplantation* 46:532–539, 1988.
30. Burke MM: Complications of heart and lung transplantation and cardiac surgery, in Anthony PP, MacSween RNM (eds): *Recent Advances in Histopathology*, Series 16. Edinburgh, Churchill Livingstone, 1994, pp 95–122.
31. Ettinger NA, Tsugi H, Muthuplackal J, Trulock EP: The use of bronchoalveolar lavage (BAL) to distinguish infection from rejection in single and double lung transplantation (abstract). *Am Rev Respir Dis* 143 (4, pt 2): A600, 1991.
32. Rabinowich H, Zeevi A, Paradis IL, et al: Proliferative

responses of bronchoalveolar lavage lymphocytes from heart-lung transplant patients. *Transplantation* 49:115–121, 1990.

33. Scott JP, Fradet G, Smyth R, et al: Prospective study of transbronchial biopsies in the management of heart-lung and single lung transplant patients. *J Heart Lung Transplant* 10:626–637, 1991.

34. Stewart S: Pathology of lung transplantation. *Semin Diagn Pathol* 9:210–219, 1992.

35. Scott JP, Higenbottam TW, Smyth RL, et al: Experience with transbronchial biopsies in children after heart/lung transplantation. *Pediatrics* 86:698–702, 1990.

36. Lung Rejection Study Group: A working formulation for the standardization of nomenclature in the diagnosis of heart and lung rejection. *J Heart Transplant* 9:593–601, 1990.

36a. Yousem SA, Berry GJ, Coyle PT, et al: A revision of the 1990 working formulation for the classification of lung allograft rejection. *J Heart Lung Transplant* 1996 (in press).

36b. Pomerance A, Madden B, Burke MM, Yacoub MH: Transbronchial biopsy in heart and lung transplantation: Clinico-pathologic correlations. *J Heart Lung Transplant* 14:761–773, 1995.

37. Sibley RK, Berry GJ, Tazelaar HD, et al: The role of transbronchial biopsies in the management of lung transplant recipients *J Heart Lung Transplant* 12:308–324, 1993.

38. Day JD, Hutchins GM, Hruban RH: Grading pulmonary rejection: A proposal for a simplified system. *J Heart Lung Transplant* 13:734–737, 1994.

39. Yousem SA, Martin T, Paradis IL, Keenan R, Griffith BP: Can immunohistological analysis of transbronchial biopsy specimens predict responder status in early acute rejection of lung allografts? *Hum Pathol* 25:525–529, 1994.

40. Nakleh RE, Bolman RM, Henke CA, et al: Lung transplant pathology: A comparative study of pulmonary acute rejection and cytomegalovirus infection. *Am J Surg Pathol* 15:1197–1201, 1991.

41. Tazelaar HD: Perivascular inflammation in pulmonary infections: Implications for the diagnosis of lung rejection. *J Heart Lung Transplant* 10:437–441, 1991.

41a. Walker RC, Paya CV, Marshall WF, et al: Pretransplantation seronegative Epstein-Barr virus status is the primary risk factor for posttransplantation lymphoproliferative disorder in adult heart, lung, and other solid organ transplantations. *J Heart Lung Transplant* 14:214–221, 1995.

41b. Chaparro C, Chamberlain D, Maurer J, De Hoyos A, Winton T, Kesten S: Acute lung injury in lung allografts. *J Heart Lung Transplant* 14:267–273, 1995.

42. Yousem SA: Graft eosinophilia in lung transplantation. *Hum Pathol* 23:1172–1177, 1992.

43. Clelland CA, Higenbottam TW, Otulana BA, et al: Histological grading of acute lung rejection on transbronchial biopsy helps predict long-term outcome of heart-lung transplants. *Transplant Proc* 22:1480, 1990.

44. Griffith BP, Paradis IL, Zealey A, et al: Immunologically mediated disease of the airways after pulmonary transplantation. *Ann Surg* 208:371–378, 1988.

45. Shankar S, Fulsham L, Read RC, et al: Mucociliary function after lung transplantation. *Transplant Proc* 23:1222–1233, 1991.

46. Veale D, Glasper PM, Gascoyne A, Dark JH, Gibson GJ,

Corris PA: Ciliary beat frequency in transplanted lung. *Thorax* 48:629–631, 1993.

47. Gosney J, Hasleton PS: Decrease in pulmonary neuroendocrine cells after lung transplantation (in preparation).

48. Reid KR, McKenzie FN, Menkis AH, et al: Importance of chronic aspiration in recipients of heart/lung transplants. *Lancet* 336:206–208, 1990.

49. Zenati M, Dowling RD, Dummer JS, et al: Influence of the donor lung on development of early infections in lung transplant recipients. *J Heart Transplant* 9:502–509, 1990.

50. Low DE, Kaiser LR, Haydock DA, Trulock E, Cooper JD: The donor lung: Infectious and pathologic factors affecting outcome in lung transplantation. *J Thorac Cardiovasc Surg* 106:614–621, 1993.

51. Wreghitt TG, Hakim M, Cory-Pearce R, English TAH, Wallwork J: The impact of donor-transmitted CMV and *Toxoplasma gondii* disease in cardiac transplantation. *Transplant Proc* 18:1375–1376, 1986.

52. Penketh ARL, Higenbottam TW, Hutter J, Coutts C, Stewart S, Wallwork J: Clinical experience in the management of pulmonary opportunist infection and rejection in recipients of heart-lung transplants. *Thorax* 43:762–769, 1988.

53. Snell GI, de Hoyos A, Krajden M, Winton T, Maurer JR: *Pseudomonas cepacia* in lung transplant recipients with cystic fibrosis. *Chest* 103:466–471, 1993.

54. Weiss LM, Movahed LA, Berry GJ, et al: In-situ hybridization studies for viral nucleic acids in heart and lung allograft biopsies. *Am J Clin Pathol* 93:675–679, 1990.

54a. Solans EP, Garrity ER Jr, McCabe M, Martinez R, Husain AM: Early diagnosis of cytomegalovirus pneumonitis in lung transplant patients. *Arch Pathol Lab Med* 119:33–35, 1995.

55. Wreghitt TG, Smyth RL, Scott JP, et al: Value of culture of biopsy material in diagnosis of viral infections in heart-lung transplant recipients. *Transplant Proc* 22:1809–1810, 1990.

56. Steinhoff G, Behrend M, Wagner TOF, Hoper MH, Haverich A: Early diagnosis and effective treatment of pulmonary CMV infection after lung transplantation. *J Heart Lung Transplant* 10:9–14, 1991.

57. Cagle PT, Buffone J, Holland VA, et al: Semiquantitative measurement of cytomegalovirus DNA in lung and heart-lung transplant patients by *in vitro* DNA amplification. *Chest* 101:93–96, 1992.

58. Flint A, Frank TS: Cytomegalovirus detection in lung transplant biopsy samples by polymerase chain reaction. *J Heart Lung Transplant* 13:38–42, 1994.

59. Burke CM, Glanville AR, Macoviak JA, et al: The spectrum of cytomegalovirus infection following human heart/lung transplantation. *J Heart Transplant* 5:267–272, 1986.

60. Keenan RJ, Lega ME, Dummer JS, et al: Cytomegalovirus serologic status and postoperative infection correlated with risk of developing chronic rejection after pulmonary transplantation. *Transplantation* 51:433–438, 1991.

61. Duncan AJ, Dummer JS, Paradis IL, et al: Cytomegalovirus infection and survival in lung transplant recipients. *J Heart Lung Transplant* 10:638–646, 1991.

62. Wreghitt TG, Sargaison M, Sutehall G, et al: A study of Epstein-Barr virus infections in heart and heart-lung transplant recipients. *Transplant Proc* 21:2502–2503, 1989.

62a. Gray J, Wreghitt TG, Pavel P, et al: Epstein-Barr virus infection in heart and heart-lung transplant recipients:

Incidence and clinical impact. *J Heart Lung Transplant* 14:640–646, 1995.

63. Egan J, Stewart J, Yonan N, et al: Epstein-Barr virus associated pulmonary fibrosis following heart/lung transplantation. *Eur Resp J.* (submitted).

64. Prompt CA, Reis MM, Grillo FM, et al: Transmission of AIDS in renal transplantation (letter). *Lancet* 2:672, 1985.

64a. Ohori NP, Michaels MG, Jaffe R, Williams P, Yousem SA: Adenovirus pneumonia in lung transplant recipients. *Hum Pathol* 26:1073–1079, 1995.

65. Wreghitt TG, Hakim M, Gray JJ, et al: Toxoplasmosis in heart and lung transplant recipients. *J Clin Pathol* 42:194–199, 1989.

66. Gryzan S, Paradis IL, Zeevi A, et al: Unexpectedly high incidence of *Pneumocystis carinii* infection after lung-heart transplantation: Implications for lung defense and allograft survival. *Am Rev Respir Dis* 137:1268–1274, 1988.

67. Stewart S, Cary N: Granulomatous pneumocystis pneumonia in immunosuppressed patients. *J Pathol* 163:164A, 1991.

68. Scoones DJ, Burke MM: The spectrum of histological appearances of *Pneumocystis carinii* pneumonia in transbronchial lung biopsies from heart and lung transplant recipients (abstr.). *J Pathol* 169:186, 1993.

68a. Tiley SM, Marriot DJ, Harkness JL: An evaluation of four methods for the detection of *Pneumocystis Carinii* in clinical specimens. *Pathology* 26:325–328, 1994.

69. Kramer MR, Denning DW, Marshall SE, et al: Ulcerative tracheobronchitis after lung transplantation: A new form of invasive aspergillosis. *Am Rev Respir Dis* 144:552–556, 1991.

70. Tazelaar H, Baird AM, Mill M, Grimes MM, Schulman LL, Smith CR: Bronchocentric mycosis occurring in transplant recipients. *Chest* 96:92–95, 1989.

71. Dauber JH, Paradis IL, Dummer JS: Infectious complications in pulmonary allograft recipients. *Clin Chest Med* 11:291–308, 1990.

72. Dowling RD, Baladi N, Zenati M, et al: Disruption of the aortic anastomosis following heart-lung transplantation. *Ann Thorac Surg* 49:118–122, 1990.

73. Deusch E, End A, Grimm M, Graninger W, Klepetko W, Wolner E: Early bacterial infections in lung transplant patients. *Chest* 104:1412–1416, 1993.

73a. Miller RA, Lanza LA, Kline JN, Geist LJ: *Mycobacterium tuberculosis* in lung transplant patients. *Am Rev Respir Crit Care Med.* 152:374–376, 1995.

74. Boyle EM Jr, Burdine J, Bolman RM III: Successful treatment of *Mycoplasma* mediastinitis after heart-lung transplantation. *J Heart Lung Transplant* 12:508–512, 1993.

75. Hall WA, Martinez J, Dummer JS, et al: Central nervous system infections in heart and heart-lung transplant recipients. *Arch Neurol* 46:173–177, 1989.

76. Hausen B, Dwenger A, Gohrbandt B, et al: Early biochemical indicators of the obliterative bronchiolitis syndrome in lung transplantation. *J Heart Lung Transplant* 13:980–989, 1994.

77. International Society for Heart and Lung Transplantation: A working formulation for the standardization of nomenclature and for clinical grading of chronic dysfunction in lung allografts. *J Heart Lung Transplant* 12:713–716, 1993.

78. Theodore J, Starnes VA, Lewiston MJ: Obliterative bronchiolitis. *Clin Chest Medicine* 11:309–321, 1990.

79. Burke EC, Theodore J, Baldwin J, et al: Twenty-eight cases of human heart/lung transplant. *Lancet* 1:517–519, 1986.

80. McCarthy P, Starnes VA, Theodore J, et al: Improved survival following heart-lung transplant. *J Thorac Cardiovasc Surg* 99:54–60, 1990.

81. Chamberlain D, Maurer J, Chaparro C, Idolor L: Evaluation of transbronchial lung biopsy specimens in the diagnosis of bronchiolitis obliterans after lung transplantation. *J Heart Lung Transplant* 13:963–971, 1994.

82. Baldwin J: Lung transplantation. *JAMA* 259:2286–2287, 1988.

83. Abernathy EC, Hruban RH, Baumgartner WA, Rietz BA, Hutchins GM: The two forms of bronchiolitis obliterans in heart-lung transplant recipients. *Hum Pathol* 22:1102–1110, 1991.

84. Yousem SA, Paradis IL, Dauber JH, Griffith BP: Efficacy of transbronchial lung biopsy in the diagnosis of bronchiolitis obliterans in heart lung recipients. *Transplantation* 47:893–895, 1989.

85. Yousem SA, Paradis I, Griffith BP: Can transbronchial biopsy aid in the diagnosis of bronchiolitis obliterans in lung transplant recipients? *Transplantation* 57:151–153, 1994.

86. Kramer MR, Stoehr C, Whang JL, et al: The diagnosis of obliterative bronchiolitis after heart-lung and lung transplantation: Low yield of transbronchial lung biopsy. *J Heart Lung Transplant* 12:675–681, 1993.

87. Harjula AL, Baldwin JC, Glanville AR, et al: Human leukocyte antigen compatibility in heart-lung transplantation. *J Heart Transplant* 6:162–166, 1987.

87a. Milne DS, Gasgoine AD, Wilkes J, et al: MHC class II and ICAM-1 expression and lymphocyte subjects in transbronchial biopsies from lung transplant recipients. *Transplantation* 57:1762–1766, 1994.

88. Paradis IL, Dummer S, Dauber J, et al: Risk factors for the development of chronic rejection in the human lung allograft. *Am Rev Respir Dis* 139:A529, 1989.

89. Yousem SA, Curley JM, Dauber JH, et al: HLA class II antigen expression in human heart-lung allografts. *Transplantation* 49:991–995, 1991.

90. Halloran P, Wadgyamr A, Autenried P: Regulation of expression of major histocompatibility complex products. *Transplantation* 41:413–420, 1986.

91. Yousem SA, Dauber JH, Griffith BP: Bronchial cartilage alterations in lung transplantation. *Chest* 98:1121–1122, 1990.

92. Griffith BP, Hardesty RL, Trento A, et al: Heart-lung transplantation: Lessons learned and future hopes. *Ann Thorac Surg* 43:6–16, 1987.

93. Yousem SA, Paradis IL, Dauber J, et al: Vascular abnormalities in long term heart-lung recipients. *Transplantation* 47:564–569, 1989.

94. Novick RJ, Ahmad D, Menkis AH, et al: The importance of acquired diffuse bronchomalacia in heart-lung transplant recipients with obliterative bronchiolitis. *J Thorac Cardiovasc Surg* 101:643–648, 1991.

95. Hasleton PS, Berry G, Billingham MD, Swindell R, Kramer MD: Pulmonary vascular changes in heart/lung transplantation. (In preparation)

96. Armitage JM, Kormos RL, Stuart RS, et al: Posttransplant

lymphoproliferative disease in thoracic organ transplant patients: Ten years of cyclosporine-based immunosuppression. *J Heart Lung Transplant* 10:877–887, 1991.

96a. Walker RC, Paya CV, Marshall WF, et al: Pretransplantation seronegative Epstein-Barr virus status is the primary risk factor for post-transplantation lymphoproliferative disorder in adult heart, lung and other solid organ transplants. *J Heart Lung Transpl* 14:214–221, 1995.

97. Nalesnik MA, Makowka L, Starzl TE: The diagnosis and treatment of posttransplant lymphoproliferative disorders. *Curr Probl Surg* 25:367–472, 1988.

98. Penn I: Why do immunosuppressed patients develop cancer? *Crit Rev Oncogene* 1:27–52, 1989.

99. Egan JJ, Hasleton PS, Yonan MY, et al: Necrotic ulcerative bronchitis the presenting feature of lymphoproliferative disease following heart-lung transplantation. *Thorax* 50:205–207, 1995.

100. Penn I: Cancers complicating organ transplantation. *N Engl J Med* 323:1767–1769, 1990.

101. Swinnen LJ, Costanzo-Nordin MR, Fisher SG, et al: Increased incidence of lymphoproliferative disorder after immunosuppression with the monoclonal antibody OKT3 in cardiac-transplant recipients. *N Engl J Med* 323:1723–1728, 1990.

102. Curtis F, Veal JR, Carr MB, et al: Diffuse pneumonia and acute respiratory failure due to infectious mononucleosis in a middle aged adult. *Am Rev Respir Dis* 41:502–504, 1990.

103. Nalesnik MA, Jaffe R, Starzl TE, et al: The pathology of post-transplant lymphoproliferative disorders occurring in a setting of cyclosporin A–prednisone immunosuppression. *Am J Pathol* 133:1273–1292, 1988.

104. Scott J, Higenbottam T: Transplantation of the lungs and heart for patients with severe complications from sarcoidosis. *Sarcoidosis* 7:9–11, 1990.

105. Johnson BA, Duncan SR, Ohori NP, et al: Recurrence of sarcoidosis in pulmonary allograft recipients. *Am Rev Respir Dis* 148:1373–1377, 1993.

106. Bjortuft O, Foerster A, Boe J, Geiran O: Single lung transplantation as treatment for end-stage pulmonary sarcoidosis: Recurrence of sarcoidosis in two different lung allografts in one patient. *J Heart Lung Transplant* 13:24–29, 1994.

107. Frost AE, Kellar CA, Brown RW, et al: Giant cell interstitial pneumonitis: Disease recurrence in the transplanted lung. *Am Rev Respir Dis* 148:1401–1404, 1993.

108. Nine JS, Yousem SA, Paradis IL, Keenan R, Griffith BP: Lymphangioleiomyomatosis: Recurrence after lung transplantation. *J Heart Lung Transplant* 13:714–719, 1994.

108a. Baz MA, Kassin PS, Van Trigt P, Davis RD, Roggli VL, Tapson VF: Recurrence of diffuse panbronchiolitis after lung transplantation. *Am J Respir Crit Care Med* 151:895–898, 1995.

109. Barberis M, Harari S, Tironi A, Lampertico P: Recurrence of primary disease in a single lung transplant recipient. *Transplant Proc* 24:2660–2662, 1992.

110. Clelland CA, Higenbottam TW, Monk JA, et al: Broncho-alveolar lavage lymphocytes in relation to transbronchial lung biopsy in heart-lung transplants. *Transplant Proc* 22:1479, 1990.

111. de Marchena D, Futterman L, Wozniak P, et al: Pulmonary artery torsion: A potentially lethal complication after orthotopic heart transplantation. *J Heart Transplant* 8:499–502, 1989.

112. Hasleton PS, Brookes NH: Severe pulmonary vascular change in patients dying from right ventricular failure after heart transplantation. *Thorax* 50:210–212, 1995.

113. Egan J, Martin N, Hasleton PS, Woodcock AA: Pulmonary capillary leakage/hemorrhage late after cardiac transplantation. *Thorax* 48:10:10703, 1993.

114. Gorensek MJ, Stewart RW, Keys TF, Mehta AC, McHenry MC, Goormastic M: Multivariate analysis of risk factors for pneumonia following cardiac transplantation. *Transplantation* 46:860–865, 1988.

115. Miller LW, Naftel DC, Bourge RC, et al: Infection after heart transplantation: A multi-institutional study. *J Heart Lung Transplant* 13:381–393, 1994.

116. Mancini MC, McGinnis MR: *Dactylaria* infection of a human being: Pulmonary disease in a heart transplant recipient. *J Heart Lung Transplant* 11:827–830, 1992.

117. Segovia J, Pulpon LA, Crespo MG, et al: *Rhodococcus equi:* First case in a heart transplant recipient. *J Heart Lung Transplant* 13:332–335, 1994.

118. Janner D, Bork J, Baum M, Chinnock R: Severe pneumonia after heart transplantation as a result of human parvovirus B19. *J Heart Lung Transplant* 13:336–338, 1994.

119. Linder J: Infection as a complication of heart transplantation. *J Heart Transplant* 7:390–394, 1988.

120. Kirklin JK, Naftel DC, Levine TB, et al: Cytomegalovirus after heart transplantation: Risk factors for infection and death: A multiinstitutional study. *J Heart Lung Transplant* 13:394–404, 1994.

121. Koskinen PK, Nieminen MS, Krogerus LA, et al: Cytomegalovirus infection and accelerated cardiac allograft vasculopathy in human cardiac transplants. *J Heart Lung Transplant* 12:724–729, 1993.

122. Skowronski EW, Mendoza A, Smith SC, Jaski BE: Detection of cytomegalovirus in paraffin-embedded postmortem coronary artery specimens of heart transplant recipients by the polymerase chain reaction: Implications of cytomegalovirus association with graft atherosclerosis. *J Heart Lung Trans-plant* 12:717–723, 1993.

123. Hakim M, Wreghitt TG, English TAH, et al: Significance of donor transmitted disease in cardiac transplantation. *J Heart Transplant* 4:302, 1985.

124. Wagner FM, Reichenspurner H, Uberfuhr P, Weiss M, Fingerle V, Reichart B: Toxoplasmosis after heart transplantation: Diagnosis by endomyocardial biopsy. *J Heart Lung Transplant* 13:916–918, 1994.

125. Miller R, Burton N, Karwande SV, Jones KW, Doty DB, Gay WA Jr: Early, aggressive open lung biopsy in heart transplant recipients. *J Heart Transplant* 6:96–99, 1987.

126. Couetil J-P, McGoldrick JP, Wallwork J, English TAH: Malignant tumors after heart transplantation. *J Heart Transplant* 9:622–626, 1990.

127. Scwend M, Tiemann M, Kreipe HH, et al: Rapidly growing Epstein-Barr virus–associated lymphoma after heart transplantation. *Eur Respir J* 7:612–616, 1994.

128. Morrison VA, Dunn DL, Manivel JC, Gajl-Peczalska KJ, Pterson BA: Clinical characteristics of post-transplant lymphoproliferative disorders. *Am J Med* 97:14–24, 1994.

129. Chucrallah AE, Crow MK, Rice LE, Rajagopalan S, Hudnall SD: Multiple myeloma after cardiac transplantation: An unusual form of posttransplant lymphoproliferative disorder. *Hum Pathol* 25:541–545, 1994.

130. Sloane JP, Depledge MH, Powles RL, Morgenstern GR, Trickey BS, Dady PJ: Histopathology of the lung after bone marrow transplantation. *J Clin Pathol* 36:546–554, 1983.

131. Hackman RC: Lower respiratory tract, in Sale GE, Shulman HM (eds), *The Pathology of Bone Marrow Transplantation*. Chicago, Year Book Medical Publishers, 1984, pp 156–170.

132. Krowka MJ, Rosenow EC III, Hoagland HC: Pulmonary complications of bone marrow transplantation. *Chest* 87:237–246, 1985.

133. Link H, Reinhard U, Walter E, et al: Lung diseases after bone marrow transplantation: Results of the clinical, radiological, histological, immunological and lung function study. *Klin Wochenschr* 64:595–614, 1986.

134. Clark JG, Hansen JA, Hertz MI, Parkman R, Jensen L, Peavey HH: Idiopathic pneumonia syndrome after bone marrow transplantation. *Am Rev Respir Dis* 147:1601–1606, 1993.

135. Breuer R, Lossos IS, Berkman N, Or R: Pulmonary complications of bone marrow transplantation. 87:571–579, 1993.

135a. Breuer R, Lossos IS, Berkman N, Or R: Pulmonary complications of bone marrow transplantation. *Respir Med* 87:571–579, 1993.

136. Chan CK, Hyland RH, Hutcheon MA: Pulmonary complications following bone marrow transplantation. *Clin Chest Med* 11:323–332, 1990.

137. Meyers JD, Flournoy N, Thomas ED: Nonbacterial pneumonia after allogenic marrow transplantation: A review of ten years' experience. *Rev Infect Dis* 4:1119–1132, 1982.

138. Chien J, Chan CK, Chamberlain D, et al: Cytomegalovirus pneumonia in allogenic bone marrow transplantation: An immunopathologic process. *Chest* 98:1034–1037, 1990.

139. Clark JG, Crawford SW: Diagnostic approaches to pulmonary complications of marrow transplantation. *Chest* 91:477–479, 1987.

140. Robbins RA, Linder J, Stahl MG, et al: Diffuse alveolar hemorrhage in autologous bone marrow transplant recipients. *Am J Med* 87:511–518, 1989.

141. Robbins RA, Thompson AB, Rennard S, et al: Association of neutrophils with diffuse alveolar hemorrhage in autologous bone marrow transplantation. *Clin Res* 36:373a, 1988.

142. Thomas ED: Marrow transplantation for malignant disease. *J Clin Oncol* 9:517–530, 1989.

143. Bortin MM, Rimm AA: Treatment of 144 patients with severe aplastic anemia using immunosuppression and allogeneic marrow transplantation: A report from the International Bone Marrow Transplantation Registry. *Transplant Proc* 13(pr 1):227–229, 1981.

144. Appleton AL, Sviland L: Current thoughts on the pathogenesis of graft versus host disease. *J Clin Pathol* 46:785–789, 1993.

145. Cordonnier C, Bernaudin FJ, Bierling P, et al: Pulmonary complications occurring after allogenic bone marrow transplantation: A study of 130 consecutive transplanted patients. *Cancer* 58:1047–1054, 1986.

146. Weiner RS, Hill R, Applebaum FR, et al: Interstitial pneumonitis following autologous bone marrow transplantation. *Transplantation* 42:515–517, 1986.

147. Leblond V, Zouabi H, Sutton L, et al: Late CD8+ lymphocytic alveolitis after allogenic bone marrow transplantation and chronic graft-versus-host disease. *Am J Respir Crit Care Med* 150:1056–1061, 1994.

148. Perreault C, Cousineau S, D'Angelo G, et al: Lymphoid interstitial pneumonia after allogenic bone marrow transplantation: A possible manifestation of graft-versus-host disease. *Cancer* 55:1–9, 1985.

149. Roca J, Granena A, Rodriguez-Roisin R, Alvarez P, Agusti-Vidal A, Rozman C: Fatal airway disease in an adult with chronic graft-versus-host disease. *Thorax* 37:77–78, 1982.

150. Kurzock R, Zaner A, Kanojia M, et al: Obstructive lung disease after allogenic bone marrow transplantation. *Transplantation* 37:156–160, 1984.

151. Chan CK, Hyland RH, Hutcheon MA et al: Small-airways disease in recipients of allogeneic bone marrow transplants: An analysis of 11 cases and a review of the literature. *Medicine* 66:327–340, 1987.

152. Clarke JG, Crawford SW, Madtes DK, et al: Obstructive lung disease after allogenic marrow transplantation. *Ann Intern Med* 111:368–376, 1989.

153. Beschorner WE, Saral R, Hutchins GM, et al: Lymphocytic bronchitis associated with graft-versus-host disease in recipients of bone-marrow transplants. *N Engl J Med* 299:1030–1036, 1978.

154. Krowka MJ, Cortese SE: Pulmonary aspects of liver disease and liver transplantation. *Clin Chest Med* 10:593–616, 1989.

155. Plevak DJ, Southorn PA, Narr BJ, Peters SG: Intensive-care unit experience in the Mayo liver transplantation program: The first 100 cases. *Mayo Clin Proc* 64:433–445, 1989.

156. Jensen WA, Rose RM, Hammer SM, et al: Pulmonary complications of orthotopic liver transplantation. *Transplantation* 42:484–490, 1986.

157. Thompson AB, Rickard KA, Shaw BW, et al: Pulmonary complications and disease severity in adult liver transplant recipients. *Transplant Proc* 20(suppl 1): 646S–649S, 1988.

158. Takaoka S, Brown MR, Paulsen W, Ramsay MAE, Lintmalm GB: Adult respiratory distress syndrome following orthotopic liver transplantation. *Clin Transplant* 3:294–299, 1989.

159. Afessa B, Gay PC, Plevak DJ, Swensen SJ, Patel HG, Krowka MJ: Pulmonary complications of orthotopic liver transplantation. *Mayo Clin Proc* 68:427–434, 1993.

160. Schroter GPJ, Hoelscher M, Putnam CW, Porter KA, Hansbrough JF, Starzl TE: Infections complicating orthotopic liver transplantation: A study emphasizing graft-related septicemia. *Arch Surg* 111:1337–1347, 1976.

161. Costello P, Williams CR, Jenkins RW, Jensen WA, Rose RM: The incidence and implications of chest radiographic abnormalities following orthotopic liver transplantation. *Can Assoc Radiol J* 38:90–95, 1987.

162. Munoz SJ, Nagelburg SB, Green PG, et al: Ectopic soft tissue calcium deposition following liver transplantation. *Hepa-tology* 8:476–483, 1988.

163. Roskams T, Campos RV, Drucker DJ, Desmet VJ: Reactive human bile ductules express parathyroid-related peptide. *Histopathology* 23:11–19, 1993.

164. Ramsey PG, Rubin RH, Tokoff-Rubin N, Cosimi AB, Russell PS, Greene R: The renal transplant patient with fever and pulmonary infiltrates: Etiology, clinical manifestations and management. *Medicine* 59:206–222, 1980.

165. Penn I: Risk of cancer in the transplant patient, in Flye M (ed), *Principles of Organ Transplantation*. Philadelphia, W. B. Saunders, 1989, pp 634–643.

166. Chanez P, Mourad G, Aubas P, et al: Kaposi's sarcoma of the bronchial tree in a renal transplant recipient. *Respiration* 53:259–261, 1988.

167. Sternberg RI, Baughman RP, Dohn MN, First MR: Utility of bronchoalveolar lavage in assessing pneumonia in immunosuppressed renal transplant recipients. *Am J Med* 95:358–364, 1993.

Chapter 25

Miscellaneous Conditions and Lung Diseases of Unknown Origin

Thomas V. Colby

PULMONARY HISTIOCYTOSIS X

Histiocytosis X (pulmonary eosinophilic granuloma, Langerhans' cell granulomatosis) refers to a group of disorders in which there is a proliferation of histiocytes resembling Langerhans' cells (Hx cells).[1-3] The clinical spectrum includes localized lesions at a variety of sites (e.g., eosinophilic granuloma of bone) and disseminated aggressive processes behaving in a clinically malignant fashion (e.g., Letterer-Siwe disease). The exact relationship among the various syndromes, and indeed whether or not histiocytosis X represents a reactive or a neoplastic proliferation, has not been fully clarified.[4]

Clinical Features

In close to 90 percent of adults with pulmonary involvement by histiocytosis X, the affliction is limited to the lungs as a bilateral interstitial lung disease[2,5-13]; only a small proportion of patients (15 percent or less) have extrapulmonary lesions or evidence of disseminated disease.[2,9] Among children, in whom histiocytosis X is much more often disseminated, the incidence of lung involvement is about 40 percent, but it does not by itself adversely affect outcome.[14] Isolated involvement of the lung is rare in children, but it has been reported.[15]

Most patients with pulmonary histiocytosis X are adults in the third and fourth decades, although there is a broad age range, from newborns into the seventh decade.[5-12,16] Women are affected about three times as often as men. As many as one-third of the patients are asymptomatic and found to have abnormal chest radiographs.[2,7,9,13] Asymptomatic radiographically occult cases have also been recognized when lung tissue is evaluated for some other problem, usually a lung tumor.[17] Some cases are identified during follow-up for a known disease, and the chest radiographs suggest the possibility of multiple metastatic nodules. Those patients with pulmonary symptoms complain of dyspnea, cough, chest pain, increased sputum production, and sometimes wheezing or hemoptysis.[2,5-7,9,10,18,19] In an early series, up to 20 percent of patients with pulmonary histiocytosis X presented with pneumothorax.[19] More recent series suggest the figure is closer to 5 percent.[9,13] Systemic symptoms, including fever, weight loss, and malaise, affect some patients. Clubbing may be present in patients with severe disease.[19] Because most adult patients (more than 90 percent) are smokers,[2,7,9] symptoms related to smoking and obstructive lung disease are commonly present.

Chest radiographs typically show reticulonodular infiltrates; in severe cases (up to 15 percent), honeycombing and cystic change are present.[9] These changes are better appreciated with conventional and computer-assisted tomography.[20-23] Nodules may be present and are rarely as large as 1.5 or 2 cm in diameter; most are less than 5 mm, and they may assume a stellate shape. Pulmonary histiocytosis X typically spares the costophrenic angles and is more severe in the upper lobes. In rare instances, chest radiographs are normal[24] or show a coin lesion.[25] Changes secondary to cigarette smoking and obstructive lung disease may also be present.

High-resolution CT scanning in some patients with pulmonary histiocytosis X is nearly pathognomonic, with the only differential diagnostic consideration being lymphangioleiomyomatosis.[2,20-23,26] High-resolution CT scanning shows cysts (up to 1 cm) and nodules (2 to 5 mm). Serial studies have shown that the nodules may cavitate and eventually form cysts.[23,26]

Pulmonary function testing in pulmonary histiocytosis X may be normal or may show restriction, obstruction, or combinations thereof.[2,5,8,10,27] Decreased diffusing capacity is the most common, and probably the most sensitive, abnormality.[13]

Bronchial reactivity to carbachol is increased in patients with pulmonary histiocytosis X as compared to controls.[13]

The laboratory findings in pulmonary histiocytosis X are nonspecific and rarely helpful in the diagnosis. Peripheral eosinophilia is not present. The findings in bronchoalveolar lavage (BAL) may be diagnostic, as described below.

Management and Prognosis

In terms of management, patients affected by pulmonary histiocytosis X should be encouraged to quit smoking, and this in and of itself may be associated with remission of the disease.[11] Since remission of the disease may occur spontaneously even in those who continue to smoke, the effect of therapy is often difficult to prove. Symptomatic patients are usually given steroid therapy; this also appears to show some effect, although no prospective, randomized studies have confirmed this. Schönfeld and colleagues showed that there was no disease progression among 36 patients who were put on steroids.[13] In the same study, 12 of 14 patients who had shown radiographically progressive disease subsequently showed radiographic improvement after steroids.[13]

The prognosis of pulmonary histiocytosis X is good: 90 percent of the patients show either stabilization or complete resolution of their disease.[2,7,9,13] Approximately 10 percent show progressive interstitial lung disease with fibrosis and eventually respiratory failure. Some patients may have slowly progressive disease manifesting primarily as an obstructive process. The prognosis in patients with multisystem histiocytosis X is significantly worse than in those with isolated lung involvement.[14,28,29]

Pulmonary histiocytosis X is analogous to sarcoidosis in many ways. A significant percentage of patients are asymptomatic, and the disease is an incidental finding on a chest radiograph. In addition, pulmonary histiocytosis X shows an active phase, followed by clearing of the process in most patients; only a minority of patients have some residue of the disease or develop progressive disease.

Pulmonary and extrapulmonary neoplasms have been reported in patients with pulmonary histiocytosis X.[30-33] In some cases, the patients have known tumors, and pulmonary histiocytosis X represents a fortuitous finding at routine follow-up.[30-32] In others, the tumor, usually a lung cancer, develops after a diagnosis of pulmonary histiocytosis X, with up to 20 years between the two.[31-33] The series are small, and there are confounding factors, especially cigarette smoking and background fibrosis (raising the possi-

bility of carcinomas associated with interstitial fibrosis), and the significance of the association of pulmonary histiocytosis X with various neoplasms remains to be clarified.

Macroscopic Pathology

The gross features of pulmonary histiocytosis X in biopsy material are distinctive but nonspecific (Fig. 25-1). Nodules up to 1 cm in diameter (occasionally 2 cm) may be associated with a central cavitation or surrounding emphysematous change in the lung tissue. The nodules are firm, and a stellate shape may be apparent. In more severe disease, particularly at autopsy (Fig. 25-2), honeycomb change may be prominent. The changes tend to be worse in the upper lobes, and there may be associated hyperinflation of the lungs.

Histology

Histologically, the appearance varies depending on the stage of the process (Figs. 25-3 through 25-6). In the early lesions there is a nodular proliferation of cells that centers on bronchioles and alveolar ducts: there is a mural or interstitial proliferation of Hx cells in and around the airway in association with increased alveolar macrophages, lymphocytes, eosinophils, and occasional neutrophils.[2,5,9] The relative numbers of the individual cell types in active lesions vary from individual to individual and from lesion to lesion in the same biopsy; some cases are dominated by eosinophils, whereas others are dominated by Hx cells. The nodules expand, and as they get larger the centrilobular distribution may be less apparent. The nodules are rounded or stellate and may show central cavitation[34] or emphysematous change in the surrounding lung tissue. These two changes account for

FIGURE 25-1
Lung biopsy in pulmonary histiocytosis X shows mild emphysematous change (*arrows*) and a small nodule (*curved arrow*) 4 mm in diameter.

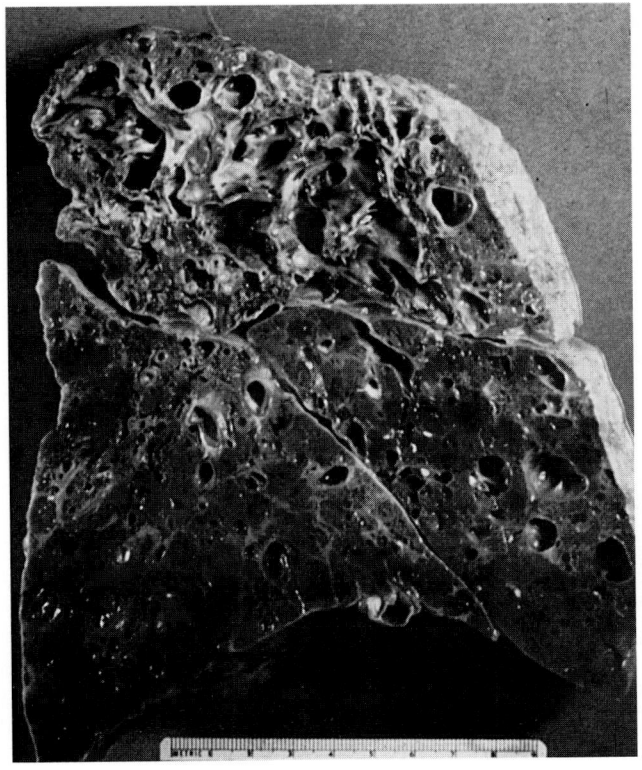

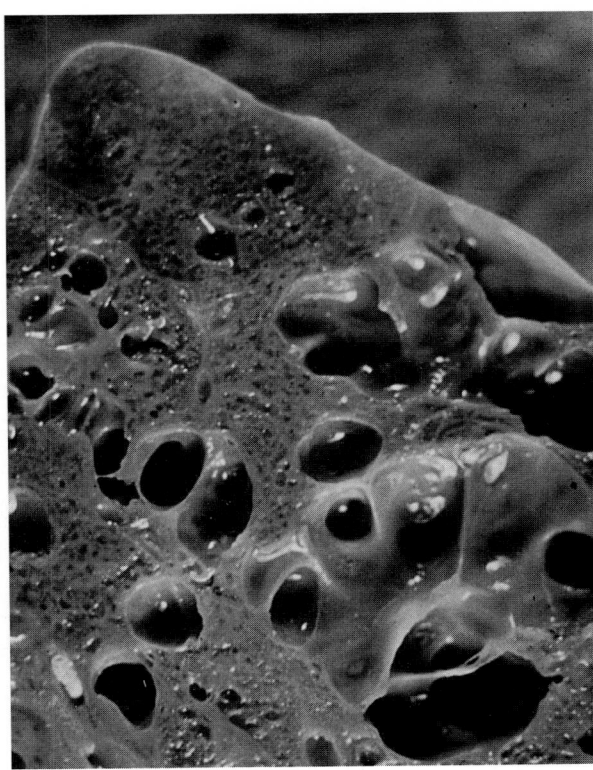

A   B

FIGURE 25-2
This case of fatal pulmonary histiocytosis X shows marked cystic change that is more severe in the upper lobes (*A, B*) with associated fibrosis, and the resultant contracted knobby appearance of the pleural surface is apparent (*A*).

the cystic change seen grossly and radiographically.

Some nodules are highly cellular, with numerous mitotic figures, and the appearance may even suggest the possibility of a lymphoma. Typically, the nodules have central fibrosis and are more cellular at the periphery; it is in the periphery that the Hx cells

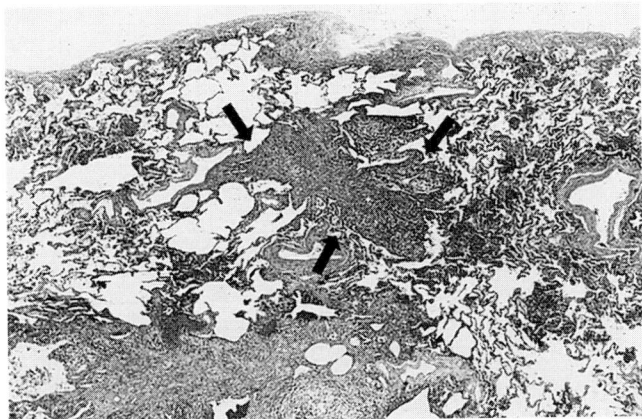

FIGURE 25-3
The classic lesions of pulmonary histiocytosis X are stellate nodules that may appear centrilobular in distribution with sparing of the pleura (*arrows*). The surrounding lung parenchyma shows an accumulation of alveolar macrophages (*right center*) and some air-space enlargement indicative of emphysematous change (*left center*).

should be sought. Within the fibrous tissue there are pigmented macrophages and small numbers of lymphocytes and occasional eosinophils. Necrosis is unusual and suggests another diagnosis. Eosinophilic abscesses are occasionally present. Modest vascular infiltration (angiitis) is also an occasional finding. In the lung tissue immediately surrounding the nodules, alveolar macrophages are often prominent. This feature, combined with smoker's (respiratory) bronchiolitis, which is also usually present, produces a pattern reminiscent of desquamative interstitial pneumonia.

A characteristic feature of pulmonary histiocytosis X in biopsy material is the finding of lesions of varying age (Fig. 25-4). Very early cellular lesions centering on respiratory bronchioles may be seen adjacent to stellate fibrotic lesions in which all the Hx cells have disappeared.

Pulmonary histiocytosis X has been traditionally considered a disorder associated with interstitial fibrosis. Intraluminal fibrosis (air-space organization) has been identified in more than 75 percent of cases histologically[2] and confirmed ultrastructurally[35] this finding, along with elastic fiber degeneration, has been implicated in the fibrotic remodeling of lung tissue.[35] These changes probably precede the interstitial fibrosis and honeycomb change.

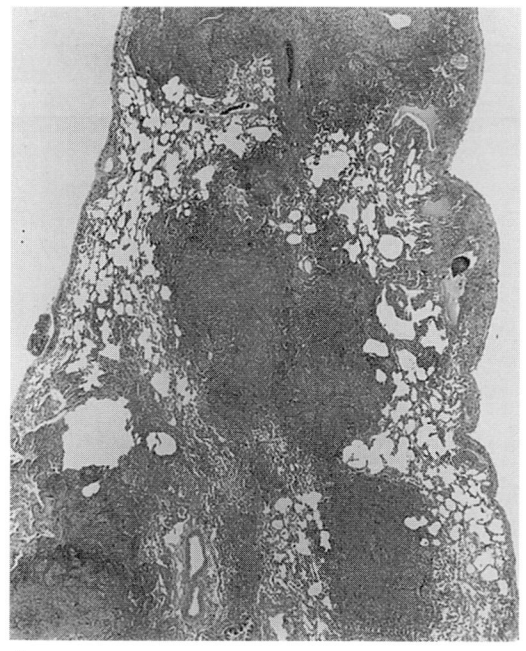
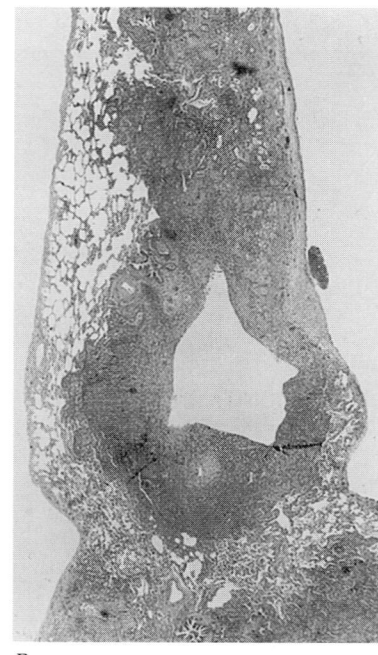

A

B

The late or healed phase of pulmonary histiocytosis X reveals stellate or rounded fibrotic nodules that may show a centrilobular distribution (Figs 25-3, 25-4, 25-6). The nodules contain scattered pigmented macrophages and metaplastic smooth muscle but are composed predominantly of fibrous tissue. When these are extensive and coalesce, honeycombing is produced (Fig. 25-2). The honeycomb or cystic spaces may be quite large and correlate with the cystic change seen radiographically. The effects of cigarette smoking, including smoker's bronchiolitis, chronic bronchitis, and emphysema, may be superimposed on any of the above findings. A hyperinflated yet fibrotic lung with honeycombing is often the end product in severe or progressive cases.

Pulmonary involvement in disseminated histiocytosis X is characterized by nodular interstitial infiltrates of Hx cells with relatively little fibrosis when compared to adult cases with histiocytosis X restricted to the lungs.

The Hx cell is a cytologically distinct cell of histiocytic origin (Fig. 25-5). The cytoplasm is quite

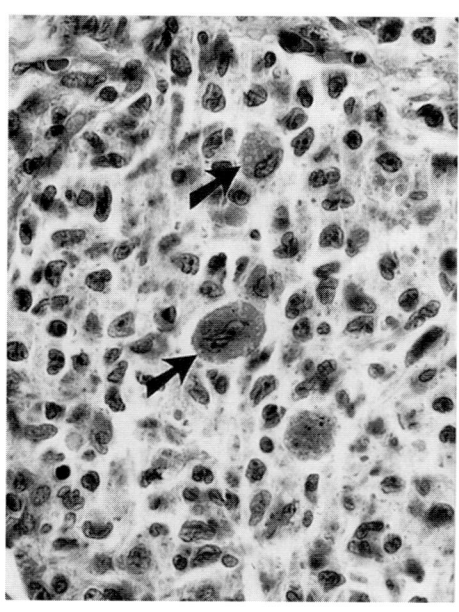
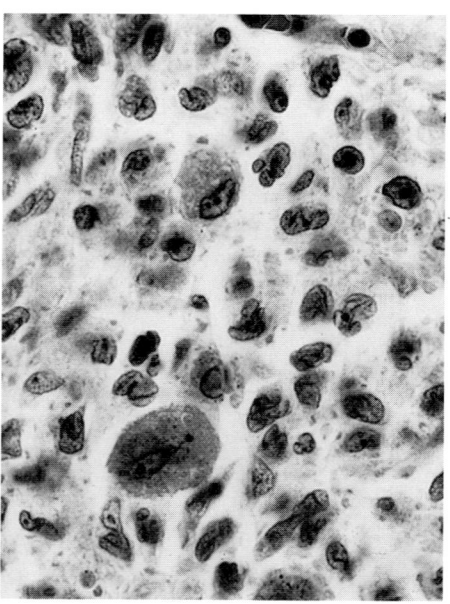

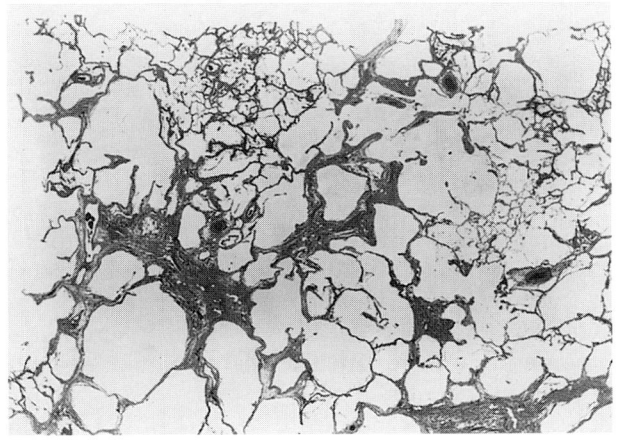

FIGURE 25-6
These sections are from a patient with biopsy-proven pulmonary histiocytosis X who developed slowly progressive and ultimately fatal obstructive lung disease over 10 years. At autopsy the lungs showed patchy fibrosis manifesting as the stellate scars (*center and left center*) representing the inactive remains of the Hx lesions, as well as marked air-space enlargement (emphysema).

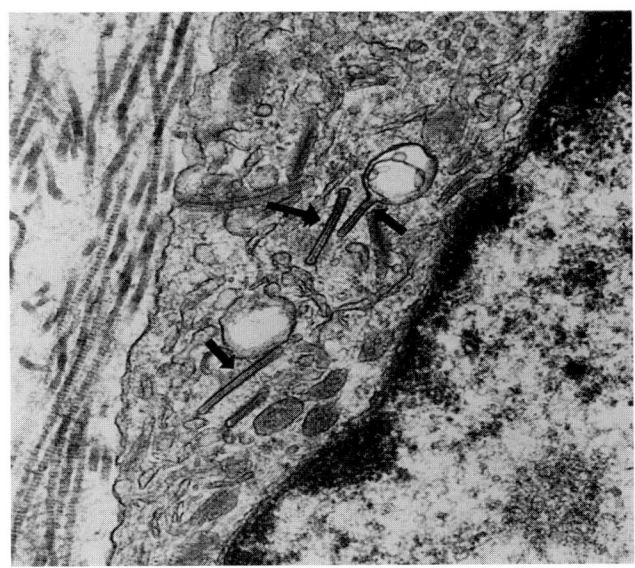

FIGURE 25-7
Ultrastructural detail of an Hx cell shows cytoplasmic trilaminar rod-shaped and racket-shaped bodies (*arrows*) known as Birbeck or Langerhans' cell granules.

abundant and is pale and eosinophilic with indistinct cell membranes. This contrasts with the more densely eosinophilic, sharply demarcated cytoplasm of pulmonary alveolar macrophages, which also usually contain flecks of brown or black debris. The nuclei of Hx cells tend to be irregular, elongated, and folded, with delicate nuclear membranes and small nucleoli. Hx cells stain positively with stains for S-100, CD-1, and HLA-DR.[2,36–39] CD-1 staining requires fresh cytospins or frozen sections.

Ultrastructural Pathology

The Hx cell is ultrastructurally unique (Fig. 25-7). Rod- or racket-shaped intracytoplasmic bodies called Langerhans' cells or Birbeck granules are seen.[2,5,6,8,11,40,41] These intracytoplasmic bodies are 420 to 450 Å thick and have a trilaminar membrane.[2] Their function is unknown, although their numbers are increased in active pulmonary hystiocytosis X.[42]

Etiology and Pathogenetic Factors

Hx cells are members of the monocyte-phagocyte system and share features with Langerhans' cells in the skin.[40-45] Langerhans' cells are accessory cells in antigen presentation; they can bind antigens and migrate to sites of lymphoid tissue.[46] Cells identical to Langerhans' cells are found normally in extracutaneous sites, including lymph nodes and lung. Hx cells in the lung can be identified in BAL fluid from normal subjects; they represent approximately 0.1 percent of the cells recovered.[47] In histologic sections stained with anti-CD1 monoclonal antibodies, Soler and coworkers[48] identified occasional Hx cells

in bronchiolar submucosa and alveolar walls of normal nonsmoking adults. The number of Hx cells recovered in BAL fluid and seen immunohistochemically in histologic sections was greatly increased in smokers in both of these studies.[47,48] Hx cells may be found in small numbers in examples of cryptogenic fibrosing alveolitis, extrinsic allergic alveolitis, desquamative interstitial pneumonia, sarcoidosis, and some tumors.[40,41,47,49,50] Hammar and colleagues showed that 93 percent of bronchioloalveolar carcinomas had ultrastructurally identifiable Langerhans' cells.[40]

In pulmonary histiocytosis X there is an abnormal proliferation of Hx cells, leading to the pathologic features described. Since most patients are smokers, the presumption is that some antigen, not yet identified, in the cigarette smoke somehow incites this proliferation. Some support for this is found in the fact that the Hx cells in pulmonary histiocytosis X are in a state of activation and interact with T cells.[42,44] Tazi's team concluded that Hx cells in pulmonary histiocytosis X are active in its pathogenesis as accessory cells in an immune response involving T cells,[42] although the Hx cell itself may be defective in antigen presentation in contrast to normal Langerhans' cells.[45] Aguayo and colleagues have demonstrated increased pulmonary neuroendocrine cells with bombesin-like activity in the lesions of pulmonary histiocytosis X and have suggested that the effects of bombesin-like activity may also be pathogenetically significant owing to its mitogenic effects on fibroblasts and chemotactic properties for inflammatory cells.[50]

Differential Diagnosis

The diagnosis of pulmonary histiocytosis X is made at two levels. It is suggested on the basis of the low-power appearance in which discrete nodules or stellate lesions show a centrilobular distribution. The nodules are fibrotic in the center and cellular at the periphery; it is in the cellular zones that one searches for Hx cells, which represent the second component necessary for the diagnosis. Even though the stellate centrilobular scarring is distinctive and may suggest the diagnosis, an absolute diagnosis should not be rendered without the identification of Hx cells. Immunohistochemical staining for S-100 protein may highlight these cells. In cases of healed pulmonary histiocytosis X, high-resolution CT scanning may be strongly supportive, and by combining radiographic findings, clinical findings, and histologic findings, one may make a clinicopathologic diagnosis of pulmonary histiocytosis X in such cases.

Most cases of pulmonary histiocytosis X have been diagnosed on the basis of open lung biopsy. Although as many as 40 percent of transbronchial biopsies done in patients with pulmonary histiocytosis X may be diagnostic,[2] a figure in the range of 10 to 20 percent is probably more reasonable in routine practice.[51] With high-resolution CT scanning there has been an increased awareness of this diagnosis, and it may often be suspected clinically or radiographically before any sort of biopsy. In such cases, diagnosis by bronchoalveolar lavage (with or without concomitant transbronchial biopsy) may be possible. While small numbers of Hx cells may be seen in BAL material from normal persons and smokers (up to about 3.5 percent), when more than 5 percent of the recovered cells are Hx cells, a diagnosis of pulmonary histiocytosis X is appropriate in the right clinical and radiographic setting.[39] While it is possible to identify Hx cells in sputum and BAL fluid by electron microscopy,[52] this is generally not a practical means of diagnosis.

The differential diagnosis of pulmonary histiocytosis X can be grouped as follows:

1. *Fibrosing interstitial pneumonias* (other than pulmonary histiocytosis X), especially cryptogenic fibrosing alveolitis. The nodular/stellate lesions of Hx show a centrilobular distribution, and the identification of appreciable numbers of Hx cells serve to distinguish it from most of these disorders. Small numbers of Hx cells are identifiable by special techniques in a variety of conditions (electron microscopy and immunohistochemistry), but they generally don't present a problem in differential diagnosis.

2. *Desquamative interstitial pneumonia* (DIP). DIP is a relatively uniform and diffuse process rather than a focal finding, and it lacks the nodular fibrotic zones typical of pulmonary histiocytosis X. The alveolar macrophages in DIP (as well as in other disorders with increased macrophages) sometimes have nuclear features that resemble those of Hx cells.

3. *Respiratory bronchiolitis.* Respiratory (smoker's) bronchiolitis (RB) is an inflammatory reaction centered on small airways associated with cigarette smoking and is a common finding in patients with pulmonary histiocytosis X. RB and the interstitial lung disease that is sometimes associated with it lack the nodular fibrosis and sheets of Hx cells seen in pulmonary histiocytosis X.

4. *Eosinophilic pneumonia.* Eosinophils are present in pulmonary histiocytosis X but are usually interstitial and only modest in number, whereas eosinophilic pneumonia is associated with airspace eosinophils in large numbers, peripheral eosinophilia, and distinctive radiographic features (alveolar filling), unlike those of pulmonary histiocytosis X.

5. *Eosinophilic pleuritis.* Eosinophilic pleuritis is a pleural reaction to the presence of air in the pleural space (pneumothorax); the proliferation of mesothelial cells and the eosinophils may suggest the possibility of pulmonary histiocytosis X, although the history of pneumothorax and the fact that no lesions are identified deep in the parenchyma should rule out histiocytosis. Obviously there is a possibility whereby pneumothorax occurs along with pulmonary histiocytosis X and both eosinophilic pleuritis and pulmonary histiocytosis X are present in the same case.

6. *Emphysema.* Emphysema may be prominent in pulmonary histiocytosis X, but the lack of discrete nodular and stellate scar zones should distinguish it from pulmonary histiocytosis X.

7. *Lymphangioleiomyomatosis* (LAM). LAM and pulmonary histiocytosis X are very similar on high-resolution CT scanning. Histologically, though, they are quite different. Some metaplastic, normal-appearing smooth muscle may be seen in the scars of pulmonary histiocytosis X, but it is readily separable from the distinctive immature-appearing (HMB-45–positive) smooth muscle seen surrounding the cysts in lymphangioleiomyomatosis.

PULMONARY ALVEOLAR PROTEINOSIS

Pulmonary alveolar proteinosis (pulmonary alveolar lipoproteinosis) is an idiopathic interstitial pneumonia associated with widespread intraalveolar accumulation of lipoproteinaceous material.[53–88] Clinical, radiographic, and functional manifestations of this disease are the result of air-space filling by this eosinophilic material. A reaction resembling pulmonary alveolar proteinosis may be seen in immunosuppressed patients associated with a variety of infections, in some patients with immunodeficiency syndromes, in patients exposed to certain dusts, and in association with hematolymphoid malignancies.

Clinical Features

Pulmonary alveolar proteinosis was originally described in 1958 by Rosen's team, based on a series of 27 patients.[53] This remains one of the best descriptions of the entity. Most patients are between 20 and 50 years of age, but the age range is from newborns to the seventh decade. Patients present with insidious dyspnea that has developed over months to years; some have cough with thick sputum. Late in the course, fatigue, chest pain, weight loss, fever, clubbing, and cyanosis may develop. There is considerable clinical variation among cases—from dramatic radiographic changes without symptoms to fulminant respiratory failure.[72,73] Most cases are sporadic, although some familial cases have been reported in children—in whom immunodeficiency syndromes with secondary infections and metabolic defects in surfactant metabolism should be carefully ruled out.[57,76,78,88]

Chest radiographs in pulmonary alveolar proteinosis typically show a butterfly pattern resembling pulmonary edema; reticular shadows may be present at the edge of the air-space infiltrates.[53,73] Radiographic asymmetry is unusual.[73] When the infiltrates clear, the clearing tends to begin at the periphery. High-resolution CT scanning shows ill-defined nodules and patchy to confluent consolidation.[86,89–91] When cavitation is present radiographically, a secondary infection is likely.[86]

Pulmonary function testing shows abnormal gas transfer, with decreased diffusion, decreased total lung capacity, normal flow rates, and often severe depression of the P_{O_2}.

The prognosis of pulmonary alveolar proteinosis is variable. Some cases clear spontaneously[66,89]; in early series, approximately one-third of the patients died of the disease.[53] Whole lung lavage appears to significantly affect the course and has improved the prognosis for patients with this disease,[75] although children tend to have less of a response to whole lung lavage.[57] Typically, abundant oily milky fluid is retrieved at the time of lavage, and the appearance may be diagnostic to experienced observers.

Macroscopic Pathology

Grossly, the lungs in pulmonary alveolar proteinosis show gray-yellow zones of consolidation that may appear purulent. In biopsies the zones of consolidation tend to be nodular, whereas at autopsy they become confluent and more extensive.

Histology

Pulmonary alveolar proteinosis is characterized by dense eosinophilic granular material filling alveoli (Figs. 25-8 and 25-9).[53,58,67] The material is PAS-positive and diastase-resistant and also stains positively with fat stains. Within this material are eosinophilic bodies, cholesterol clefts, a few foamy histiocytes,

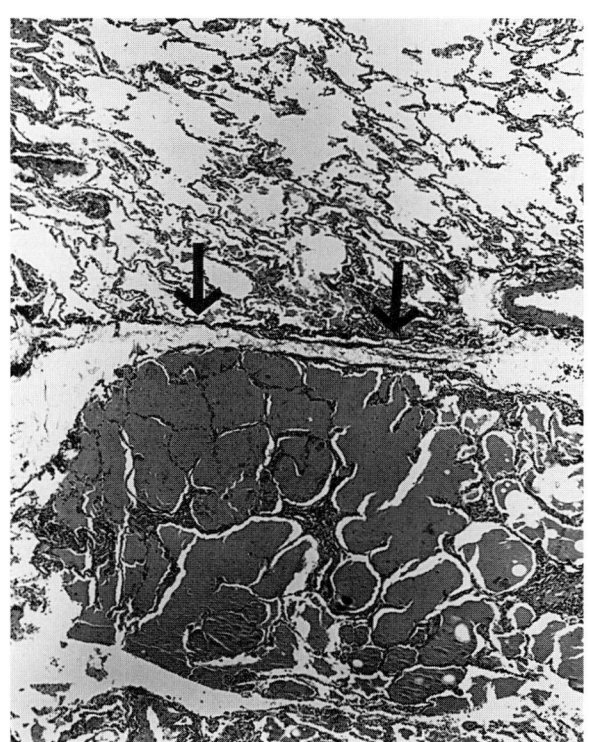

FIGURE 25-8
Low-power histologic appearance of pulmonary alveolar proteinosis in biopsy material shows nodular zones of airspace consolidation by dense granular eosinophilic material sharply demarcated by an interlobular septum (*arrows*).

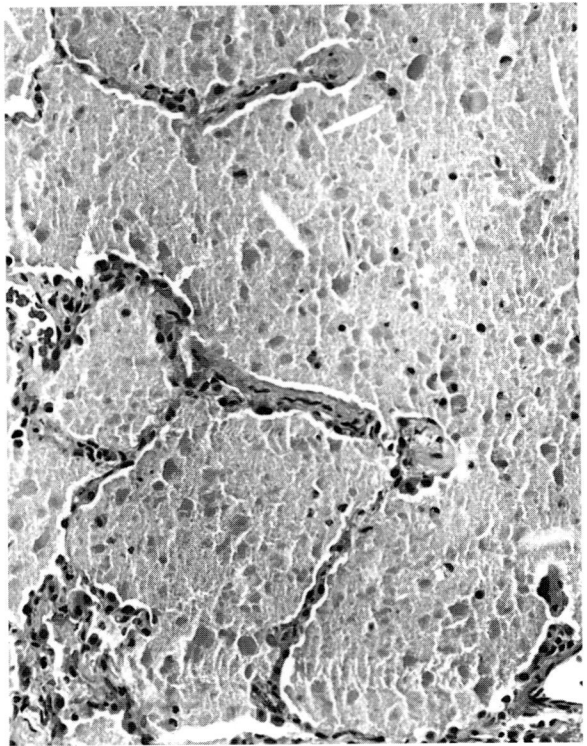

A

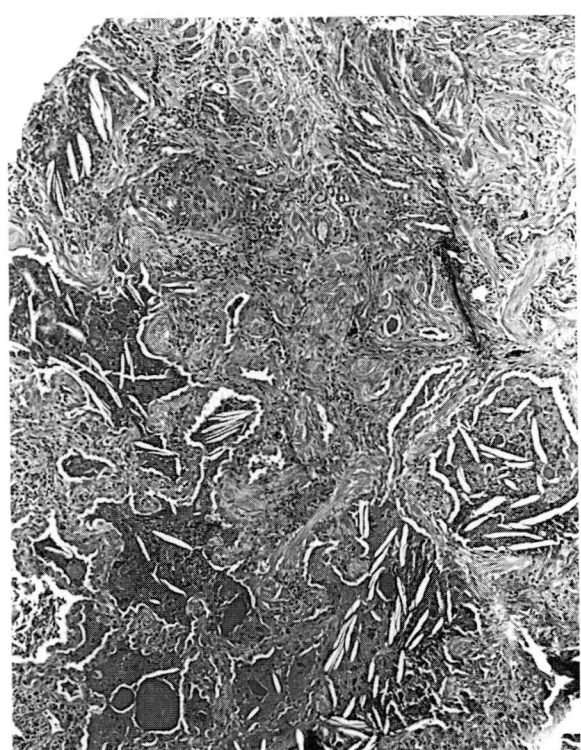

B

FIGURE 25-9

The classic appearance of pulmonary proteinosis (*A*) is of granular eosinophilic material filling air spaces with a minimal increase in cellularity in the accompanying alveolar walls. There is cell debris, nuclear debris, cholesterol clefts, and eosinophilic

blobs representing portions of cellular debris. Adjacent to such foci one may occasionally find focal scarring and numerous cholesterol clefts in air spaces (*B*).

and cell debris. Typically, there is a sharp demarction between affected and unaffected lung, often at interlobular septa. The intraalveolar eosinophilic material may extend into bronchioles. The alveolar walls show slight type II cell hyperplasia and a scant inflammatory infiltrate. Away from the zones of typical alveolar proteinosis, one sometimes sees clusters of giant cells with cholesterol clefts, foamy macrophages in the alveoli, and a mild interstitial infiltrate with prominent type II cells. Interstitial fibrosis is occasionally seen,[59,78] especially in chronic cases.

The intraalveolar material can be retrieved in sputum and BAL fluid, and examination of these specimens may be diagnostic in the appropriate clinical setting.[72,92] Cytologically the material is amorphous and granular, with a few alveolar macrophages and numerous anucleate eosinophilic bodies (Fig. 25-10).[92]

Ultrastructural Pathology

There have been a number of descriptions of the ultrastructural features of pulmonary alveolar proteinosis.[54,60,63,64,67,85,86,92,93] The intraalveolar material consists of degenerating cellular debris and numerous lamellar bodies up to 5 μm in diameter.

The lamellae may be parallel or concentric and may contain a central osmiophilic core or lipid material.

Etiology and Pathogenetic Factors

Biochemical analysis of the intraalveolar material in pulmonary alveolar proteinosis shows that it is a phospholipid: lecithin rich in palmitic acid and cholesterol, similar (and probably identical) to surfactant.[54,55,58,63–65,74,79,85,94] The material stains positively for surfactant apoprotein.[74,79]

The pathogenesis of pulmonary alveolar proteinosis is thought to entail an acquired defect in alveolar macrophages, which show impaired bactericidal activity, increased amounts of cytoplasmic fat, and morphologically abnormal lysosomes.[61,94,95,96] This is thought to result in impaired clearing of normal intraalveolar metabolic products (including surfactant) and cell debris, resulting in their intraalveolar accumulation.

Differential Diagnosis

The differential diagnosis of pulmonary alveolar proteinosis includes:

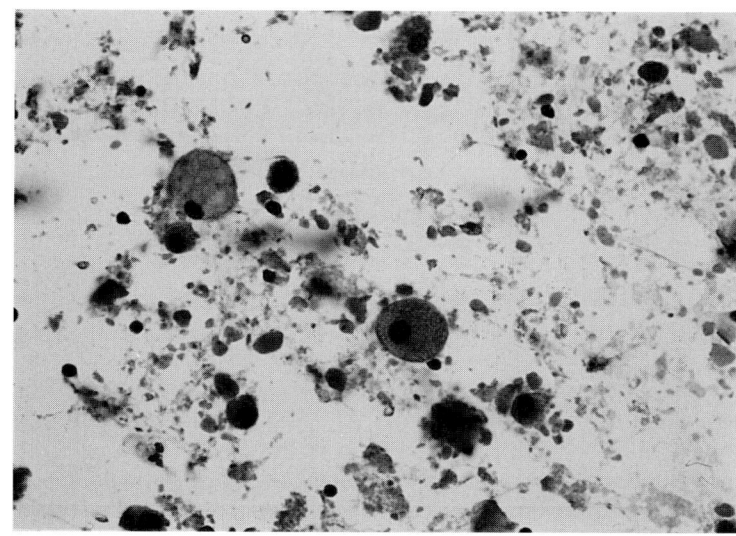

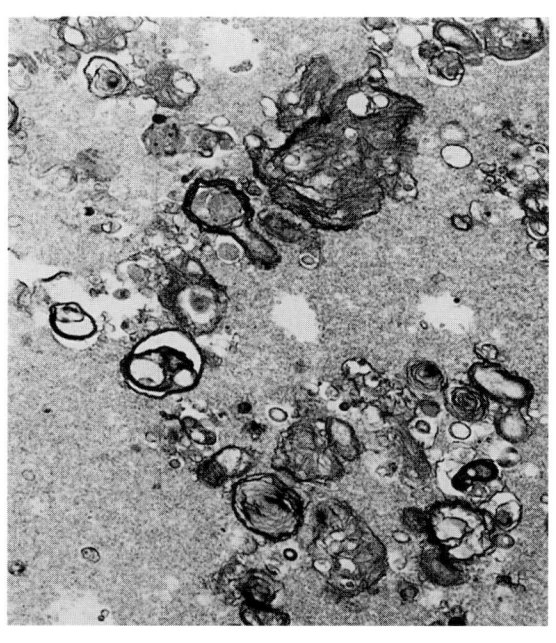

B

FIGURE 25-10
A. Pulmonary alveolar proteinosis. This hematoxylin and eosin stain of smear made from bronchoalveolar fluid shows intact macrophages, cell debris, and abundant granular material representing the alveolar contents in pulmonary alveolar proteinosis. *B.* Ultrastructurally numerous lavellar bodies are seen.

1. *Pulmonary edema fluid,* which is paler and less granular and lacks eosinophilic bodies and cholesterol clefts.
2. *Fibrin and hyaline membranes,* which are usually associated with more prominent interstitial changes; the typical mural location of hyaline membranes is also helpful, as is the lack of the granular appearance.
3. *Mucus* that often has pools of neutrophils and stains positively with mucin stains.

4. *Pneumocystis carinii pneumonia,* which is frothy as opposed to granular; positive silver stains for organisms.
5. *Other infections and clinical settings* with a pseudoproteinosis pattern as described below.

In any of the above conditions, there may be foci that are granular and reminiscent of pulmonary alveolar proteinosis; but in general these foci are not prominent, and careful evaluation of the entire slide reveals the nature of the process.

Pulmonary Alveolar Proteinosis–Like Reactions

Reactions resembling pulmonary alveolar proteinosis can be seen in a number of settings.[56,63,69,70,73,79,80,81,84,97–104] Practically speaking, these should be separated from idiopathic pulmonary alveolar proteinosis because of the different implications in pathogenesis and management. Reactions resembling pulmonary alveolar proteinosis can be seen in the following:

1. Acute silica exposure (silicotic alveolar proteinosis).[97]
2. Congenital immunodeficiency syndromes and defects of surfactant production.[88]
3. Hematologic malignancies.
4. Immunosuppressed patients and those on chemotherapy, particularly those with opportunistic infections.
5. Experimental animals exposed to a variety of agents, including silica, metallic aluminum, talc, bentonite, kaolinite, fine particulates of glass, asbestos in high concentration, antimony trioxide, 100 percent oxygen, chlorphentermine, and iprindole.[97]

In acute silicosis there is a strong resemblance to classic pulmonary alveolar proteinosis, although fine birefringent material may be identified within the intraalveolar material with polarization.[97]

Nearly 30 percent of children with pulmonary alveolar proteinosis have thymic alymphoplasia or some other immune deficiency.[62,76] In this situation, secondary infection should always be sought.

A reaction resembling pulmonary alveolar proteinosis, labeled secondary alveolar proteinosis, is seen in patients with hematologic malignancies—approximately 5 percent of patients overall with pulmonary symptoms, and 10 percent patients with myeloid disorders.[104] The changes may be reversible, especially with remission of the underlying disease. Cordonnier and colleagues suggest that defective macrophage function, as a result of the hematologic disease itself or its therapy, leads to secondary alveolar proteinosis.[104]

An alveolar proteinosis-like reaction is seen in immunosuppressed patients (including some on chemotherapeutic agents) and in acquired immunodeficiency states (including AIDS and immunoglobulin deficiencies). Usually there is an associated infection. The pattern may be very similar to that of idiopathic pulmonary alveolar proteinosis, but staining for surfactant apoprotein is said to be more focal,[79] and interstitial changes—including edema, inflammation, type II cell proliferation, and intraalveolar fibrin—are more prominent. The implicated infections are numerous and include cryptococcosis, tuberculosis, atypical mycobacteriosis, candidiasis, mucormycosis, nocardiosis, histoplasmosis, aspergillosis, pneumocystosis, streptococcal infections, cytomegalovirus infections, herpesvirus infections, and respiratory syncytial viral infections.[70,85,101] In assessing such a case in an immunosuppressed patient, a full battery of special stains should be performed in addition to all the routine cultures. In addition to the identification of the organism itself, clues to an infection-related pulmonary alveolar proteinosis-like reaction include inflammation (especially neutrophils), air-space organization, and necrosis.

Patients with idiopathic pulmonary alveolar proteinosis are themselves prone to infections, particularly nocardiosis,[105] since the intraalveolar material acts as a fertile culture medium. This is particularly true after steroid therapy.

In addition to the above conditions, a pulmonary alveolar proteinosis-like reaction may be seen as a focal incidental· finding adjacent to some other process. In this setting, the focal proteinosis reaction is clearly only an incidental histologic finding. An example is the case reported by Schiller and coworkers, in which there was a pulmonary alveolar proteinosis-like reaction adjacent to metastatic melanoma.[106]

PULMONARY AMYLOIDOSIS

Amyloidosis includes a spectrum of diseases associated with the extracellular deposition of amyloid, a family of glycoproteins.[107–112] Amyloidosis is classified according to the type of deposit: AL (Ig or light chain-derived) and AA (amyloid A protein-derived) are best known, but others are recognized (Table 25-1).

Amyloid protein deposition typically leads to firm, stiff, pale organs that may appear somewhat greasy on cut section.[113] Histologically, amyloid is pale eosinophilic ("waxy") extracellular material that may be deposited in a diffuse uniform interstitial fashion or in a more nodular fashion, particularly around vessels. Amyloid may be highlighted with many special stains; the Congo red stain is the best known and is associated with a characteristic apple-green color with birefringence (best appreciated in a dark room). AA amyloid loses its Congo red-associated birefringence after oxidation with permanganate solution, whereas AL amyloid retains this feature.[114] Structurally,[108] amyloid is a beta-pleated sheet of fibrils that ultrastructurally appear as a 75- to 100-Å woven fibrillar meshwork.[115,116] Hyalinized collagen may resemble amyloid on routine sections, but it lacks Congo red positivity, and often the wavy fibrils of collagen can be appreciated.

Amyloidosis in the lung is classified as primary pulmonary amyloidosis (when the amyloid is restricted to the lung); pulmonary amyloid deposition in systemic amyloid infiltration; or senile amyloidosis, which is usually an incidental finding. Systemic amyloidosis must be carefully excluded in suspected cases of primary pulmonary amyloidosis, since some cases of systemic amyloidosis can present as lung disease, and impairment of other organ systems may be relatively inconspicuous.

Primary pulmonary amyloidosis is traditionally divided into tracheobronchial amyloidosis, nodular parenchymal amyloidosis, and diffuse parenchymal (interstitial/diffuse alveolar septal) amyloidosis. The features of these three groups are compared in Table 25-2. There may be some overlap among them, and occasional cases are difficult to classify[117,118]; however, using a combination of the clinical and pathologic findings usually allows classification. Most cases of primary pulmonary amyloidosis have AL

TABLE 25-1
Classification of Amyloidoses

Clinical Designation	Amyloid Type
Primary or myeloma-associated (light-chain protein)	AL (immunoglobulin-derived)
Secondary	
Inflammation-associated	AA (amyloid A protein–derived)
Malignancy-associated	AA (amyloid A protein–derived)
	AE_t (associated with medullary thyroid carcinoma)
Familial (several subtypes)	AF (several subtypes)
Endocrine tissue–related (e.g., thyroid)	AE_t
Senile amyloid (various sites)	AS (several subtypes)
Cutaneous (dermal)	AD
Hemodialysis-associated	AH (β_2-microglobulin)

SOURCE: Modified from Colby TV, Carrington CB: Diffuse infiltrative lung diseases, in Thurlbeck WM, Churg A (eds), *Pathology of the Lung*, 2d ed. New York, Thieme Medical Publications, 1995.

amyloid deposition, although cases of AA amyloid presenting in the lung have been described.[119,120]

Tracheobronchial Amyloidosis

In trachebronchial amyloidosis, the amyloid is deposited in large proximal airways and sometimes in the larynx as nodules or plaques, with clinical and radiographic manifestations reflecting obstruction.[114,118,121–130] The deposits are usually extensive, involving multiple airways, but occasionally may be more localized. In some series, this is the most common form of primary pulmonary amyloidosis.

Symptoms of tracheobronchial amyloidosis are primarily the result of airway luminal compromise with either airflow obstruction or recurrent pneumonias. Some patients describe persistence of symptoms for several decades. Hoarseness or stridor may be present if the larynx is affected. Abnormal serum proteins are not found; if they are present, systemic amyloidosis should be considered.

The radiographic changes are nonspecific and reflect the effects of airways obstruction. Approximately one-fourth of patients have a normal chest radiograph. Bronchograms, though rarely done nowadays, may show dramatic multifocal involvement with nodules, plaques, and stenoses. Tracheal and bronchial narrowing may also be apparent on chest radiographs.[114]

The gross and bronchoscopic findings of tracheobronchial amyloidosis are characteristic (Fig. 25-11). There are nodules and plaques of varying size that may have overlying metaplastic or ulcerated epithelium. The largest lesions are often in the subglottic region.

Histologically, there is submucosal deposition of amyloid in and around vessels, bronchial glands, and connective tissue (Fig. 25-12). Secondary calcifica-

TABLE 25-2
Primary Amyloidosis of the Lungs

Type	Derivation	Cases (No.)	Male:Female	Mean Age: Yrs. (range)	Dyspnea (%)	Cough (%)	Hemoptysis (%)
					Patients Presenting with		
Tracheobronchial	AL	41	2:1	53 (16–76)	60	74	51
Nodular parenchymal	AL	31	1.3:1	65 (38–95)	12	30	4
Diffuse alveolar septal	AL	4	1:1	55 (39–67)	100	100	0

SOURCE: Modified from references 122 and 121.

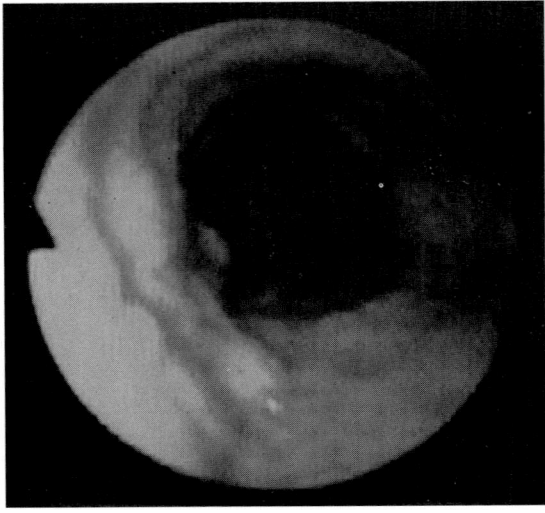

A

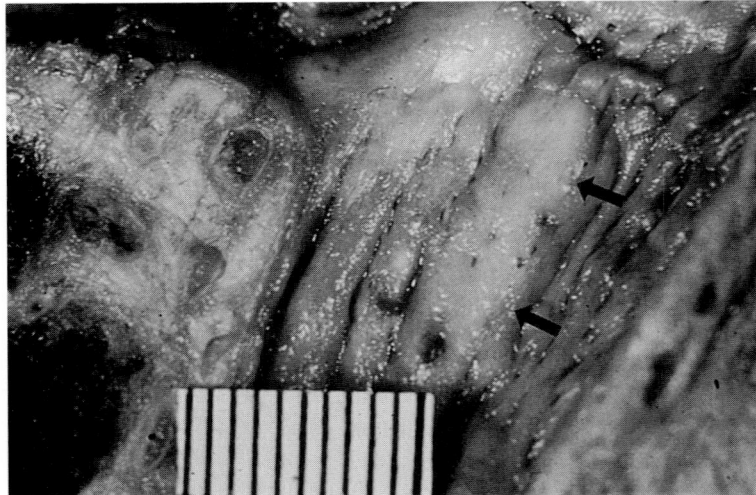

B

FIGURE 25-11
Tracheobronchial amyloidosis. *A.* The bronchoscopic appearance shows multiple nodules and plaques under the bronchial mucosa. *B.* The gross appearance at autopsy is of a whitish raised submucosal plaque (*arrows*).

tion and ossification may be present; when these changes are extensive, they may simulate tracheopathia osteochondroplastica. A giant cell reaction to the amyloid may be present. The overlying epithelium may be normal, ulcerated, or metaplastic. Lymphocytes and polyclonal plasma cells may be present; two of the four cases studied by Hui and coworkers had more lambda-positive than kappa-positive plasma cells,[114] but the significance of these findings is not clear.

The course of tracheobronchial amyloidosis is slow but relentless. Palliative surgery may offer some transient abatement of large individual lesions. One-fifth to one-third of the patients die of the effects of the disease. Rare spontaneous remissions have been described.[130] Affected patients do not develop systemic amyloidosis.[129]

The differential diagnosis of tracheobronchial amyloidosis is primarily with tracheopathia osteo-chondroplastica, which also manifests multiple submucosal nodules, although they are bony and lack the staining features of amyloid.

Nodular Parenchymal Amyloidosis

Nodular parenchymal amyloidosis produces grossly and radiographically apparent nodules in the lung parenchyma. They are generally single; less commonly, multiple nodules are identified. Nodular parenchymal amyloidosis is usually an incidental radiographic finding unless symptoms have resulted from local effects of the nodule.[114,118,121,124,125,132–143] Most cases come to biopsy or resection to exclude a neoplasm.

Only rarely are abnormal serum proteins found[114] in cases of nodular parenchymal amyloidosis. When they are present, a lymphoproliferative disorder should be ruled out.[144] In the few patients

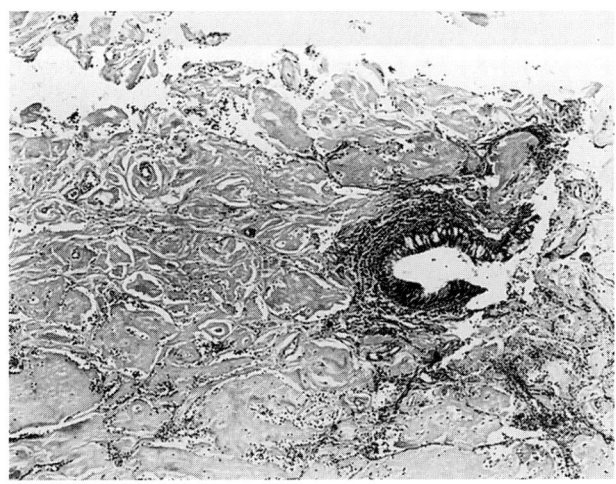

FIGURE 25-12
Bronchial biopsy in tracheobronchial amyloidosis. There is a portion of bronchiolar epithelium extending down the duct of a submucosal gland (*right center*). The surrounding submucosa is replaced by acellular amyloid deposits.

who have been examined, pulmonary function tests have been normal.[138]

Grossly and radiographically, the nodules are usually only a few centimeters in size, but they may become very large and even replace a hemithorax.[142] About one-fourth of the nodules are calcified. Grossly, the nodules are circumscribed and gray, yellowish, or translucent with a somewhat greasy cut section (Fig. 25-13). Calcification may be grossly apparent as chalky deposits.

Histologically, nodular parenchymal amyloidosis results in well-circumscribed masses of amyloid replacing the lung parenchyma (Figs. 25-13, 25-14, and 25-15). At the edge, there may be some interstitial and perivascular extension of the amyloid away from the main mass. A giant cell reaction may be present. Secondary calcification, ossification, and chondroid metaplasia are seen in some cases. A variable infiltrate of lymphocytes, polyclonal plasma cells, and histiocytes may be present.

The differential diagnosis of nodular parenchymal amyloidosis includes conditions associated with

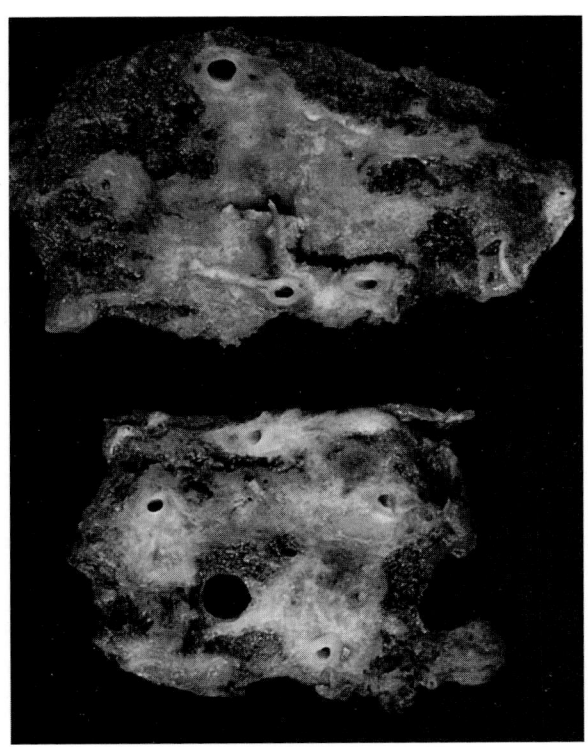

A

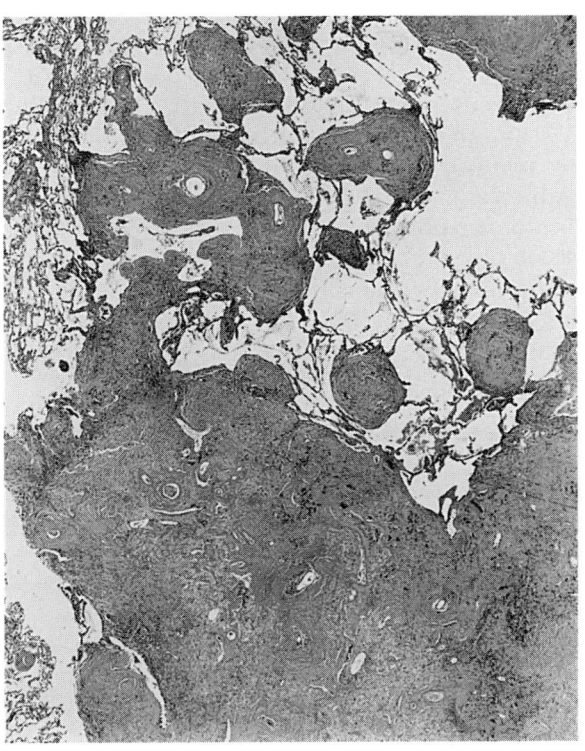

B

FIGURE 25-13
Nodular pulmonary amyloidosis. *A.* The gross appearance is that of a circumscribed grayish-white nodule that shows some extension around vessels at its periphery. *B.* This can also be appreciated in the low-power photomicrograph.

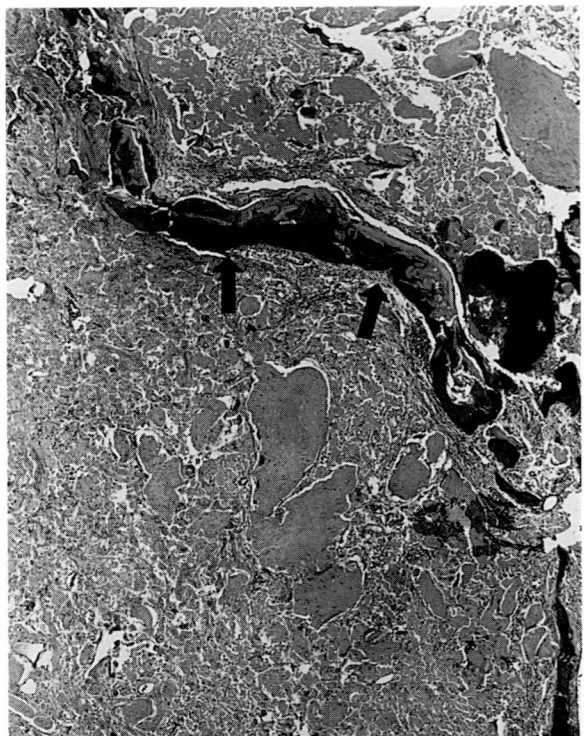

A

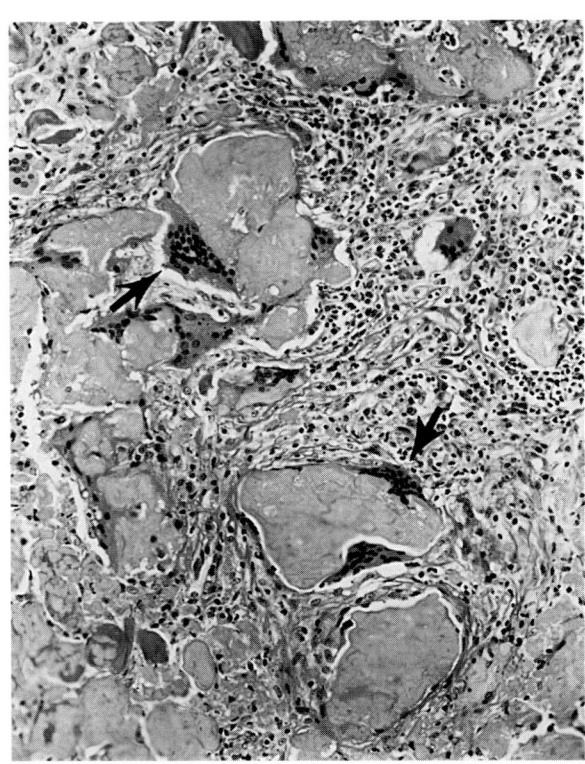

B

FIGURE 25-14
Nodular pulmonary amyloidosis. This case is composed of a mass of amyloid replacing lung parenchyma with some secondary ossification (*A, arrows*). In some regions a giant cell reaction to the amyloid is present (*B*), and the surrounding tissue shows a modest lymphoplasmacytic infiltrate.

acellular eosinophilic nodules in the lung, including old scars and infectious granulomas, silicotic nodules, hyalinized noninfectious granulomas (such as sarcoidosis and berylliosis), pulmonary hyalinizing granuloma, plasma cell granuloma, and nodular fibrosis representing the residue after chemotherapy or radiation treatment for neoplasms. These conditions can generally be ruled out on the basis of the character of the eosinophilic material deposited and the absence of apple-green birefringence with Congo red staining. Other pathologic conditions may be associated with nodular accumulations of amyloid, and these include lymphomas and plasma cell dyscrasias. In some cases the nodular accumulations of amyloid may be separate and discrete from the neoplastic lymphoid infiltrates, indicating the need for adequate sampling. Some cases of primary systemic amyloidosis may be associated with nodular deposits of amyloid.

The prognosis of nodular parenchymal amyloidosis is excellent. A few cases recur,[137] and for this reason a lobectomy is considered by some to be the treatment of choice. In patients in whom resection is impossible or who have multiple nodules, there may be slow progression as illustrated by the case of Eisenberg and Sharma,[142] but in the series reported by Hui and colleagues, none of the patients with nodular amyloid died, including several patients with multiple nodules.[114] Young described a patient with documented disease over 20 years but without adverse clinical effects.[143]

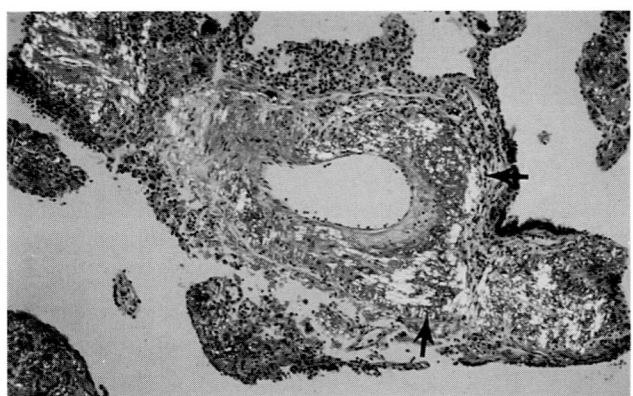

FIGURE 25-15
Nodular pulmonary amyloidosis. Positive birefringence of the amyloid (*arrows*) is seen on a Congo red stain of the perivascular amyloid deposits.

Diffuse Parenchymal (Interstitial/Alveolar Septal) Pulmonary Amyloidosis

Primary pulmonary amyloidosis with diffuse parenchymal deposition is the least common form of primary pulmonary amyloidosis.[114,118,121,122,145–150] The patients have clinical and radiographic features of interstitial lung disease, and all are symptomatic (Table 25-2).

Clinically, patients with diffuse parenchymal amyloidosis have shortness of breath and cough. They show evidence of restrictive lung disease with decreased diffusing capacity on pulmonary function testing. The chest radiograph shows features of interstitial lung disease with reticular or reticulonodular infiltrates. Pleural effusions may be present. In some patients, serum protein abnormalities are identified; if they are present, the possibility of systemic amyloidosis should be ruled out.

Grossly, the lungs of patients with diffuse parenchymal amyloidosis are pale and have a uniform firm and rubbery consistency.[149] Histologically, there is widespread interstitial amyloid deposition in alveolar walls, in and around vessels and airways, and in the pleura (Fig. 25-16). Although the changes are widespread, they may be somewhat patchy and vary in severity; micronodular formation may be present. Secondary calcification or ossification and a giant cell reaction to the amyloid may also be seen. A modest infiltrate of lymphocytes, polyclonal plasma cells, and histiocytes may be present. When lymphocytes or plasma cells are numerous, the possibility of a lymphoproliferative disorder should be considered.

The differential diagnosis of diffuse parenchymal amyloidosis is primarily systemic amyloidosis, either primary systemic amyloidosis or amyloidosis

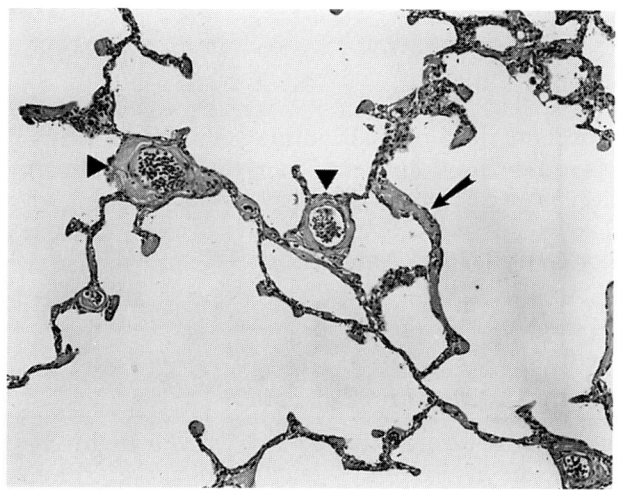

FIGURE 25-16
Diffuse alveolar septal amyloidosis. There is subtle thickening of the alveolar septa by acellular amyloid deposits (*arrow*). In addition to involvement of the alveolar septa, there is accumulation of the amyloid in vessel walls (*arrowheads*).

secondary to a lymphoproliferative disease. This distinction cannot be made on the basis of the histologic appearance of the lung. Fibrotic lung tissue may appear somewhat hyalinized and acellular and resemble amyloid, but special stains for amyloid are negative. Upper lobe fibrosis associated with bullous emphysema may be particularly eosinophilic and hyaline in appearance and resemble amyloid.

The course of most primary diffuse parenchymal amyloidosis of the lung is progressive, and the disease is fatal within a year. This in part reflects the fact that patients present late in the course of their disease, when they are symptomatic, and the full course of the disease is not well characterized, since insufficient numbers of cases have been identified in the presymptomatic phase.

Pulmonary Involvement in Multisystem or Systemic Amyloidosis

In systemic amyloidosis (primary systemic amyloidosis, amyloidosis associated with lymphoproliferative disorders, and amyloidosis secondary to chronic inflammatory processes) there is often involvement of multiple organs, and the clinical manifestations and organ systems affected most severely vary from patient to patient. Pulmonary disease is the dominant manifestation in some patients and an incidental, clinically insignificant finding in others.[107,109,110,115-117,122,151–154] Thirty to 90 percent of patients with systemic amyloidosis have lung involvement, but in only a minority of these does lung disease dominate the clinical picture. Clinically significant lung involvement tends to be more frequent with AL than AA amyloidosis.

The sex incidence is equal, and most patients are over 50 years of age, although patients as young as 16 have been reported.[122] Cough and dyspnea are the most frequent symptoms in patients who have symptoms. Pulmonary function tests show restriction and decreased diffusion, whose severity depends on the extent of lung involvement. A monoclonal gammopathy is usually present in primary systemic amyloidosis and amyloidosis associated with lymphoproliferative disorders; monoclonal gammopathy is usually not present in systemic AA amyloidosis.

Pathologically, most cases resemble diffuse parenchymal (interstitial/diffuse alveolar septal) amyloidosis, although single or multiple nodular lesions and even tracheobronchial involvement may be seen in some cases. The lungs may be grossly normal if impairment is minimal. Histologically, the amyloid invades the interstitium, vessels, airways, and pleura to varying degrees (Fig. 25-17). Giant

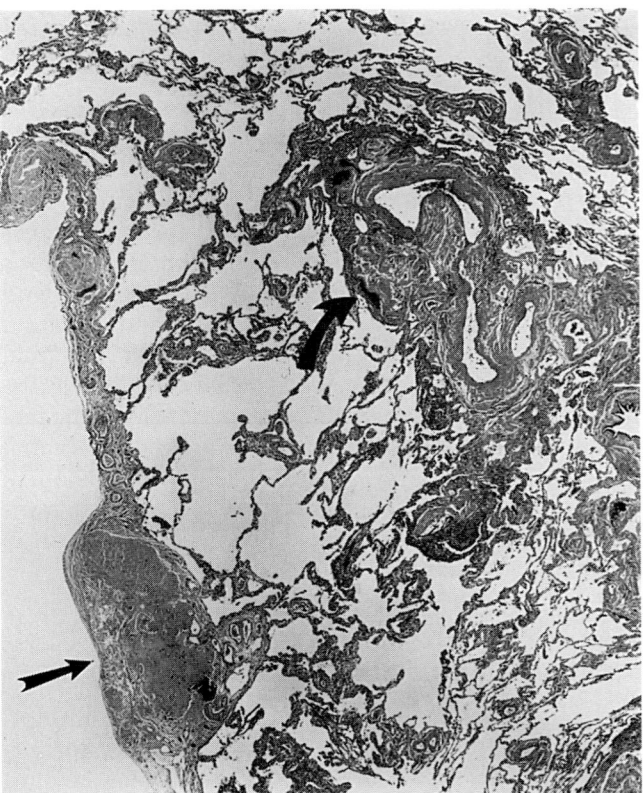

A

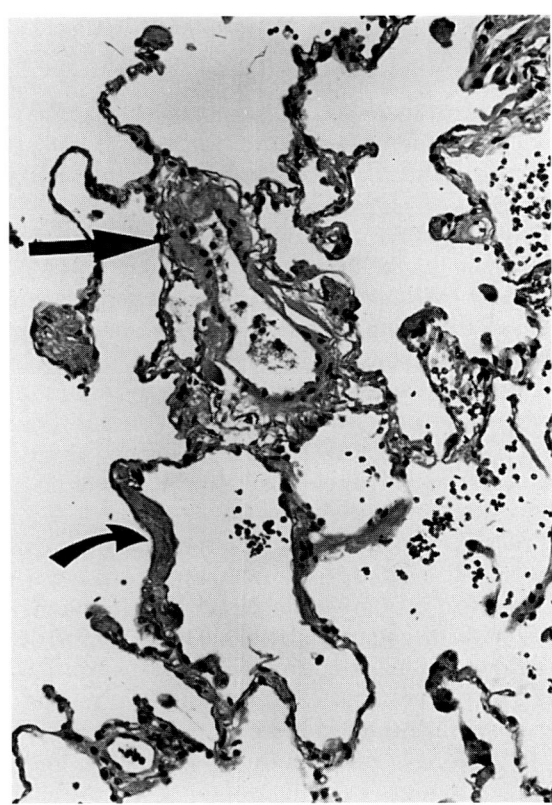

B

FIGURE 25-17
Pulmonary involvement in primary systemic amyloidosis.
A. There are pleural nodules of amyloid (*arrow*) as well as perivascular deposits (*curved arrow*). *B*. High-power microphoto-
graph also shows involvement of the alveolar septa (*arrow*) and

small vessels (*curved arrow*) by amyloid deposits, which are
indistinguishable from those in primary diffuse parenchymal
amyloidosis.

cells, secondary calcification, and ossification may
all be present. Amyloidosis due to a lymphoproliferative disorder may have an associated lymphomatous or plasmacytomatous infiltrate.

Occasional cases of primary systemic amyloidosis are associated with nodular infiltrates that are
apparent grossly and radiographically. In these cases
there is often more widespread diffuse parenchymal
infiltration.

Pulmonary amyloidosis may be associated with
malignancies other than malignant lymphomas and
plasma cell dyscrasias.[155,156] Renal cell carcinoma is
the tumor most often implicated, and AA amyloid is
the type usually found. If the tumor is successfully
resected, there may even be a decrease in the amyloid
deposits.

Orriols and coworkers described a case of extrinsic allergic alveolitis with associated amyloidosis in
the lung.[157] The amyloidosis resolved with treatment
of the extrinsic allergic alveolitis.

Senile Amyloidosis

Senile amyloid is an incidental finding at autopsy in
many organs.[107,109,158–160] Its incidence increases

with age: at least one site is involved in patients more
than 90 years of age. When it is widespread, the term
senile systemic amyloidosis has been applied.[160]
Lung involvement is usually associated with cardiac
involvement.[158]

Pleural Amyloidosis

Pleural amyloid deposition may occur with primary
pulmonary amyloidosis or in cases of systemic amyloidosis (see Chap. 34).[161–163]

Hilar Lymph Node Amyloidosis

The hilar lymph nodes are an occasional site for
localized amyloid deposition.[164]

Systemic Light-Chain Deposition

The lung is rarely affected in systemic light-chain
deposition disease.[165,166] In this disease, which is
associated with plasma cell dyscrasias, there is deposition of excess light chain in various organs, but the
deposits do not show all the features of amyloid. The
material is seen as eosinophilic, PAS-positive, interstitial material identical to amyloid on hematoxylin

and eosin sections (Fig. 25-18). It lacks apple-green birefringence with Congo red staining and ultrastructurally shows 11-to-14-nm fibrils without the typical random woven orientation of amyloid. Kijner and Yousem described a patient with bilateral reticulonodular infiltrates on chest radiograph due to systemic light-chain deposition disease.[166] Immunohistochemical staining showed that the material was kappa light chains.

PULMONARY LIPIDOSES AND RELATED DISORDERS

The lipidoses are associated with intracellular accumulation of lipids and related metabolic products. Central nervous system involvement is often the most prominent manifestation in many of these disorders, but lung involvement also occurs. In some patients, lung involvement is subclinical; in those with clinically manifest pulmonary disease, the findings are most commonly those of interstitial lung disease.

Gaucher's Disease

In Gaucher's disease, sphingolipid is deposited in cells in the reticuloendothelial system.[167–175] Gaucher's disease is a familial disorder with an autosomal recessive inheritance pattern. Pulmonary involvement is more often seen in infantile forms of Gaucher's disease than in adult forms. Patients with symptoms complain of shortness of breath, and the chest radiographs show reticular or reticulonodular infiltrates. Pulmonary function testing shows restriction. Manifestations of Gaucher's disease may or may not be present in other organs when lung involvement is diagnosed.

Grossly, the lungs in Gaucher's disease are pale and firm. Histologically, typical Gaucher's cells are found in the alveoli, interstitium, airways, pleura, and vessels (Fig. 25-19*A*). Some interstitial fibrosis in the lungs may also be present. Intravascular Gaucher's cells have been reported to be associated with pulmonary hypertension.[167,173] Because of the distinctive cytologic appearance of Gaucher's cells, diagnosis by fine needle aspiration is possible.[174]

Niemann-Pick Disease

In Niemann-Pick disease, sphingomyelin accumulates in cells in the reticuloendothelial and other organ systems.[168,176–179] Niemann-Pick disease is familial, with an autosomal recessive inheritance pattern. Infantile and adult forms of Niemann-Pick disease are recognized, and when pulmonary disease is evident, it is more often seen in the infantile form. Among 28 patients with Niemann-Pick disease described by Lachmann and colleagues, 18 had pulmonary findings described as nodular infiltrates on the chest radiograph; honeycombing was noted in some cases.[178] All 18 had presented in infancy, and all died before 8 years of age. None of the older children or adolescents in that series had pulmonary manifestations. Three adults with Niemann-Pick disease associated with pulmonary involvement were described by Long and coworkers: interstitial infil-

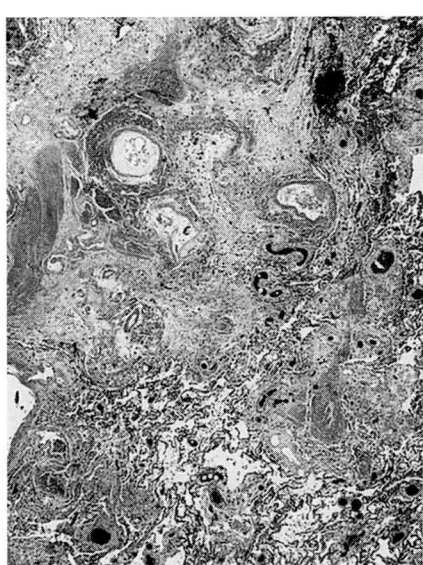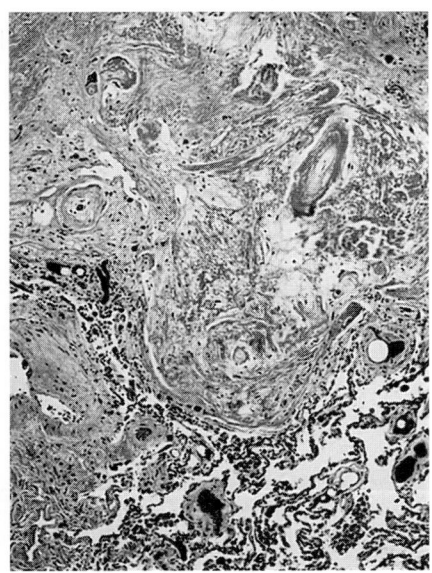

FIGURE 25-18
Pulmonary involvement in light-chain deposition disease. There is localized nodular interstitial and perivascular deposition of eosinophilic material similar to amyloid. (Courtesy of Dr. S. A. Yousem, M.D., Pittsburgh.)

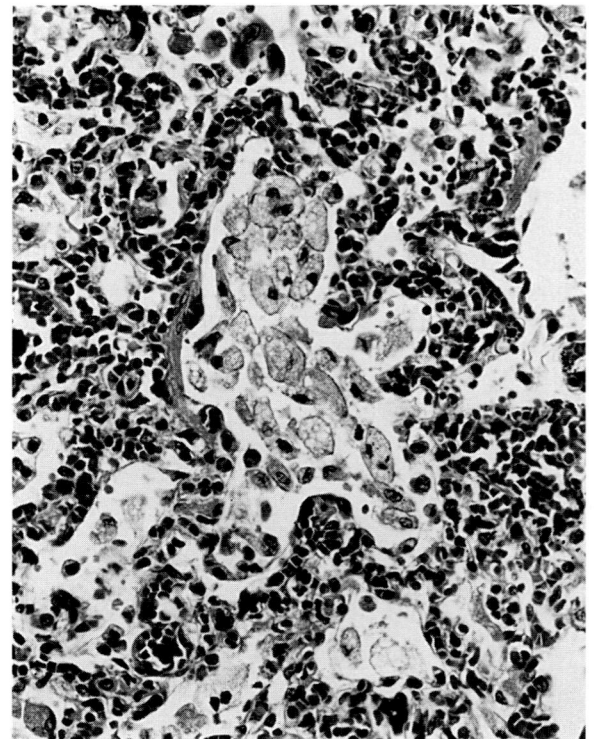 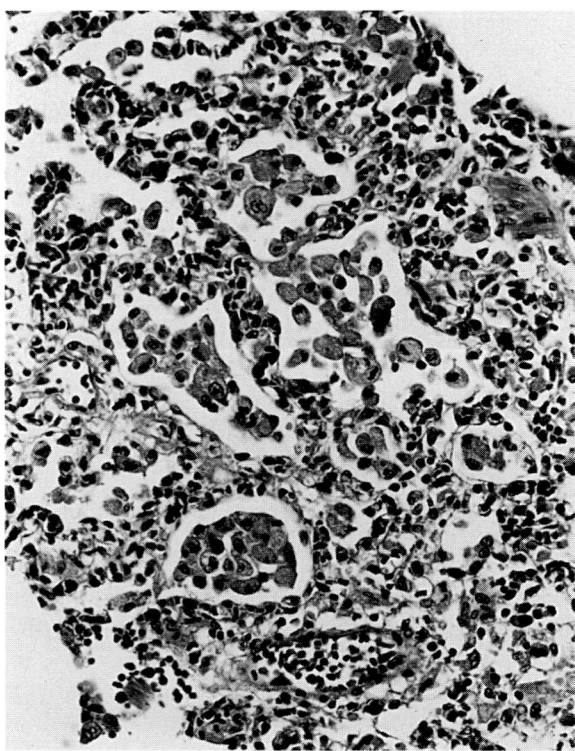

FIGURE 25-19
Pulmonary lipidoses. *A.* In Gaucher's disease there are alveolar accumulations of histiocytes showing typical cytoplasmic features of Gaucher's cells. *B.* Niemann-Pick disease is characterized by accumulation of lipid-filled histiocytes within alveoli.

trates with honeycombing on chest radiographs and decreased diffusion were the manifestations.[179]

Pathologically, the lungs of patients with Niemann-Pick disease are pale and heavy. They show an infiltrate of Niemann-Pick cells in the alveoli, interstitium, and pleura (Fig. 25-19*B*). Fibrosis is a late finding. Electron microscopy shows the characteristic features of Niemann-Pick cells.[177]

Hermansky-Pudlak Syndrome

Hermansky-Pudlak syndrome is an autosomal recessive disorder that includes oculocutaneous albinism, defects in platelet aggregation, and accumulation of ceroid-filled histiocytes in various tissues.[180–184] In the United States most patients are Puerto Rican, whereas in western Europe the majority are from the southern part of the Netherlands.[181] Some patients with Hermansky-Pudlak syndrome have a form of interstitial lung disease resembling cryptogenic fibrosing alveolitis. The pathologic findings in these cases are identical to those of cryptogenic fibrosing alveolitis except for the marked increase in ceroid-filled histiocytes in the alveoli and interstitium (Fig. 25-20).[181–183] Ceroid is a fine brown pigment that stains positively with digested PAS stains, weakly with Ziehl-Neelsen stains, and positively with Fontana's stain for myelin. Iron stains are negative. Orange fluorescence of the pigment is seen when it is

viewed with ultraviolet illumination. Ultrastructurally, ceroid-containing macrophages show amorphous particulate debris and lipid vacuoles. These cells may be identified in BAL material.[184] While in some cases the ceroid-filled histiocytes are prominent, in others they can be easily overlooked unless one is cognizant of the clinical history.

Marfan's Syndrome

Pulmonary lesions associated with Marfan's syndrome have included bullous emphysema, interstitial lung disease with honeycombing, cystic changes in the lungs, bronchial malformations, recurrent infections, bronchiectasis, and recurrent pneumothoraces.[185,186]

Ehlers-Danlos Syndrome

The pulmonary changes in Ehlers-Danlos syndrome are reported to include multiple spontaneous pneumothoraces, bronchial dilatation, arteriovenous anastomoses, cystic change in the lungs, and fibrous nodules thought to be the result of healing lung injury.[187]

Erdheim-Chester Disease

Erdheim-Chester disease is a peculiar disorder characterized by symmetric sclerotic lesions in the long

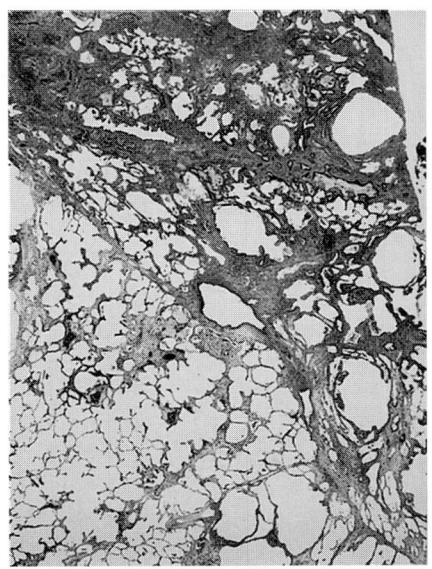 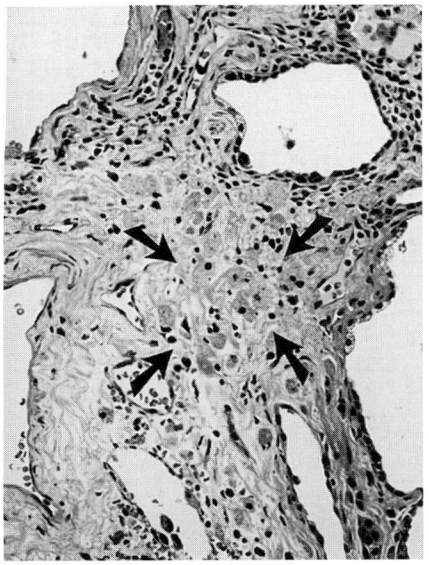

FIGURE 25-20
Fibrosing interstitial lung disease in Hermansky-Pudlak syndrome. There is a fibrosing interstitial pneumonia with honeycombing (left) identical to cryptogenic fibrosing alveolitis. A distinctive feature of Hermansky-Pudlak syndrome is the associated increase in ceroid-filled histiocytes, both interstitial (*arrows*) and intraalveolar.

bones and, less often, the flat bones with histologic infiltrates of lipid-filled histiocytes, giant cells, lymphocytes, plasma cells, and variable degrees of fibrosis.[188–193] Involvement of the lungs or pleura has been noted in a small proportion of the cases—four of the 19 cases reviewed by Miller's team.[191] In an autopsy case reported by Fink and colleagues, the lungs showed marked interstitial, peribronchial, and perivascular infiltrates that included epithelioid granulomas.[193] Pleural infiltration with nodular deposits has been described.[191] Other cases have shown exquisitely septal, pleural, and perivascular mixed inflammatory infiltrates composed of lymphocytes, plasma cells, histiocytes, and associated with variable degrees of fibrosis (Fig. 25-21) that may have been clinically significant and clinically progressive.

Idiopathic Cholesterol Pneumonitis

Accumulations of foamy alveolar macrophages are common and are seen in a variety of lung diseases, usually in those associated with airways obstruction or in chronic airway inflammatory disorders. Cholesterol can be identified in the foam cells. Cholesterol clefts within clusters of giant cells are also a common nonspecific finding in lung biopsy material. In some cases they may represent a residue of prior hemorrhage, although they may also be seen in conditions associated with lipid accumulation within air spaces, such as pulmonary alveolar proteinosis.

The term *idiopathic cholesterol pneumonitis* has been applied to localized disorders associated with the massive accumulation of foamy alveolar macrophages[194]; it is distinguished from secondary accumulations of foamy macrophages associated with bronchial obstruction and well-characterized chronic pulmonary inflammatory processes, such as bronchiectasis, chronic lung abscesses, and tuberculosis; it is also distinguished from exogenous lipoid pneumonia due to aspiration.[194]

Lawler described a case of idiopathic cholesterol pneumonitis and reviewed 49 cases previously reported in the literature.[194] According to this review, idiopathic cholesterol pneumonitis affects children and adults (age range: 12 to 67 years), with most cases presenting in the fifth decade. Males outnumbered females 9:1, and a smoking history was usually present. The patients presented with cough, hemoptysis, or chest pain; occasionally they were asymptomatic. The duration of symptoms varied from 2 to 12 months. Radiographically, there was a localized mass or infiltrate suspected to be a neoplasm; a single upper lobe was usually affected. Grossly the pleura showed thickening, and cut sections of the parenchyma often showed single or multiple abscess cavities. Histologically, there was a massive accumulation of foamy macrophages (which could be shown to contain cholesterol) filling air spaces.

According to Lawler, etiologic possibilities have included hypersensitivity states, incomplete infarction, association with other inflammatory processes (such as Wegener's granulomatosis), and incomplete resolution of a prior infection.[194]

Idiopathic cholesterol pneumonitis is probably not a specific entity but, rather, a common nonspecific change that can be associated with a number of processes. Personal experience suggests that cases that fit the definition of idiopathic cholesterol pneumonitis probably are the residue of a prior inflammatory process, usually an infection, that has not been recognized and for which no organism has been identified. Practically speaking, since the accumulation

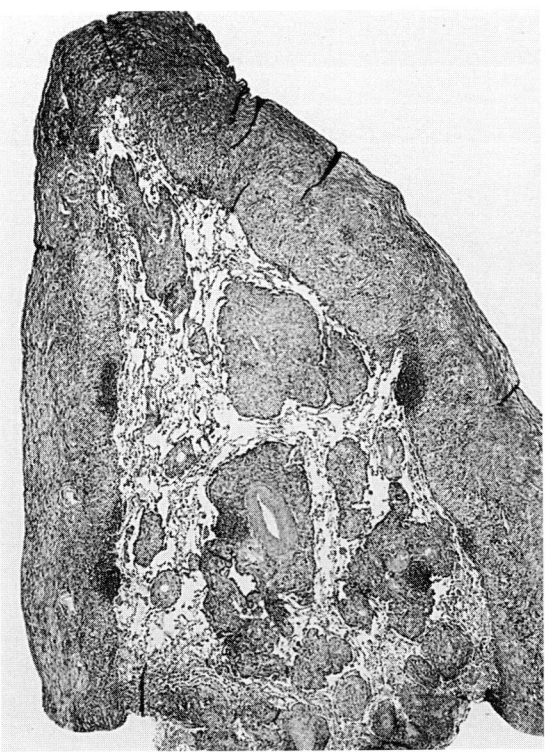

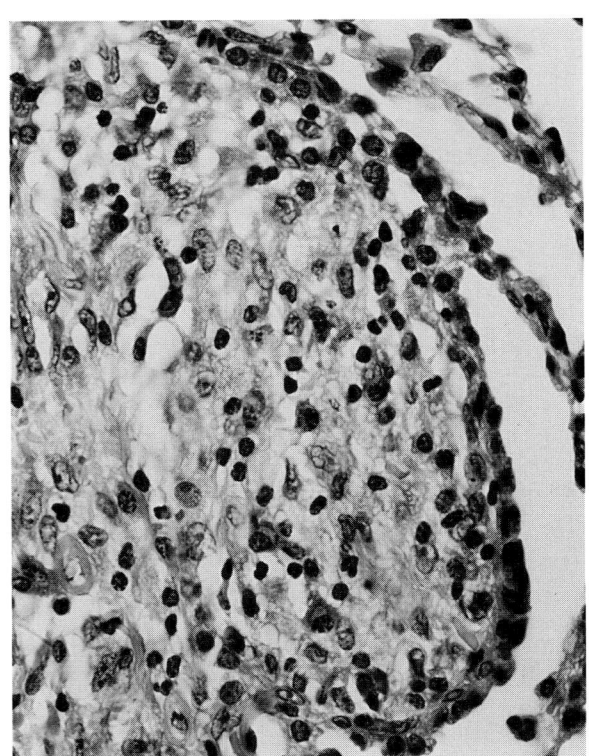

A

B

FIGURE 25-21
Erdheim-Chester disease with pulmonary involvement. There is marked thickening of pleural, perivascular, and septal regions (*A*) by a mixed infiltrate composed predominantly of histiocytes and small numbers of lymphocytes and plasma cells with associated fibrosis (*B*).

of foam cells may be seen in so many different disorders, it is imperative that in cases in which large numbers of foamy macrophages are seen the lesion be adequately sampled to identify any associated changes, such as small tumors occluding airways, abscesses, and bronchiectasis.

Other Metabolic Conditions

Rare metabolic diseases that have been implicated as causing lung disease also include Fabry's disease and airflow obstruction[195,196] (vascular lesions are discussed in Chap. 21); infantile GM₁ gangliosidosis with respiratory insufficiency from massive alveolar and interstitial foam cell accumulations[197]; renal tubular acidosis and interstitial lung disease resembling cryptogenic fibrosing alveolitis[198,199]; Krabbe's disease (globoid leukodystrophy), with progressive respiratory insufficiency from air-space infiltration by eosinophilic macrophages with amber brown (PAS-positive, Sudan black–positive) inclusions[200]; Pompe's disease (type II glycogenosis/acid maltase deficiency) with restrictive lung disease[201]; Salla disease (an autosomal recessive defect in sialic acid metabolism), leading to free sialic acid accumulation in many tissues and emphysema of the lungs[202]; cholesterol ester storage disease and pulmonary hypertension due to pulmonary arterial intimal cholesterol deposition[203]; and Farber's disease (disseminated lipogranulomatosis), with multiple nodular infiltrates of granulomatous tissue, including sheets of lymphocytes, histiocytes, and plasma cells associated with respiratory distress.[168,204]

PULMONARY ALVEOLAR MICROLITHIASIS

Pulmonary alveolar microlithiasis (PAM) is a histologically unique disorder in which there is diffuse bilateral deposition of intraalveolar laminated calcospherites (microliths).[205–228] PAM was first described in 1918 by Harbitz[229] and named in 1933 by Puhr.[230] Nearly 175 cases have been described since then, although the disease is still very rare.

Clinical Features

PAM affects people of all ages, from premature infants to persons in their 80s.[210,212,231] One large review reported the mean age at diagnosis to be 27 years.[228] The sex incidence is roughly equal; some series suggest a female predominance, particularly in familial cases,[210,215] while others show a male predominance.[228] The familial association has been confirmed in several series; familial cases tend to present at a younger age, and the familial association is usually restricted to siblings.[206,221,228] The number of familial cases varies from 38 to 58 percent.[228] A disproportionate number of cases of PAM have been described in Turks; one-quarter to one-third of the reported cases reviewed by Ucan and coworkers were in Turks.[228]

The cause of PAM is unknown, although there is an interesting report from Thailand of PAM's occurring in users of a particular snuff.[207] Serum electrolyte studies and calcium levels are normal in PAM.

Many patients with PAM are asymptomatic at the time of diagnosis—more than half in some series.[206–221, 228] Most patients come to medical attention because of abnormal chest radiographs. Patients with symptoms complain of cough, and as the disease progresses, shortness of breath and hemoptysis may develop; recurrent pneumonias and pneumothoraces also occur. In late stages there is cyanosis, clubbing, hypertrophic pulmonary osteoarthropathy, and cor pulmonale.[206,215,227]

The radiographic findings in PAM are considered diagnostic.[206,207,225] Chest radiographs show a "sandstorm" with basal predominance. The granular densities do not coalesce and show the same density as calcium. CT scans show similar uniform diffuse calcific densities, which are also characteristic.[222,227,232]

Pulmonary function studies in PAM parallel the symptoms.[214,215] They are nearly normal in patients who are asymptomatic, but as the disease progresses, affected patients develop restriction, diffusion impairment, and uneven ventilation.[209,211,212,214,215,228]

The course of PAM spans many years.[206,215,217] In some patients progression may appear to be arrested for a while or to be extremely slow—over 25 years in one case.[206,208] There is no proven treatment, although Göcmen and colleagues have suggested that diphosphonate treatment may be beneficial.[233]

Macroscopic Pathology

Grossly, the lungs are hard and maintain their shape; a saw may be required to cut them.[215,217] The cut surface, which is gritty because of the innumerable 0.2- to 0.3-mm sandlike grains, has been likened to sandpaper.[206]

Histology

Up to 80 percent of the alveoli in cases of PAM contain the distinctive laminated basophilic (or eosinophilic if they have been decalcified) calcospherites (Figs. 25-22 and 25-23).[215,217,224] They

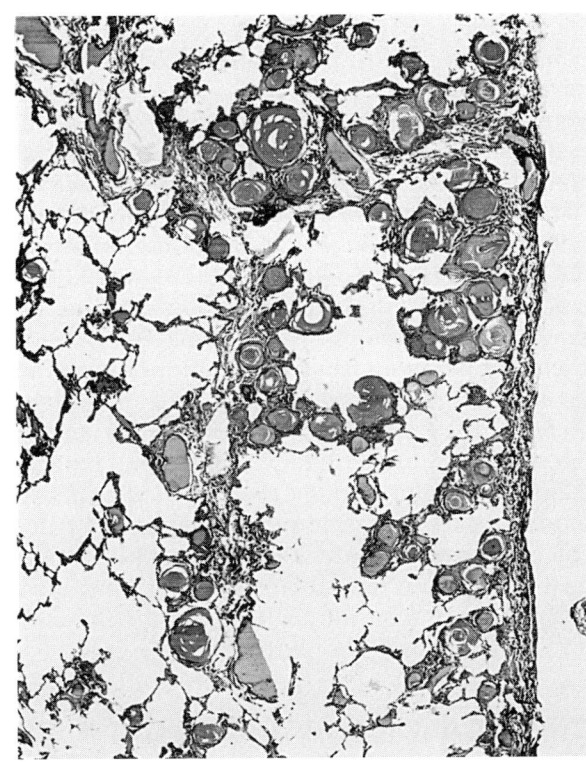
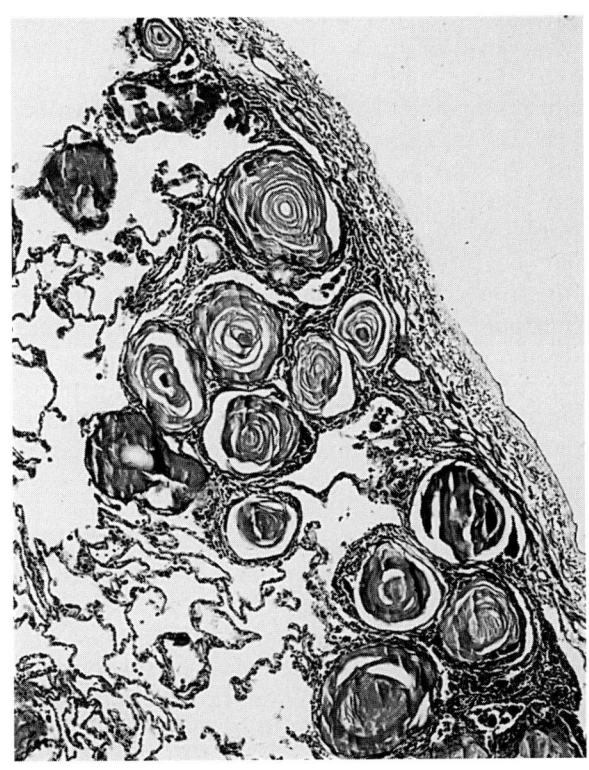

A *B*

FIGURE 25-22
Pulmonary alveolar microlithiasis. Many of the air spaces contain laminated eosinophilic to basophilic bodies representing the microliths (*A*, *B*). The surrounding lung parenchyma shows minimal inflammatory or fibrotic change.

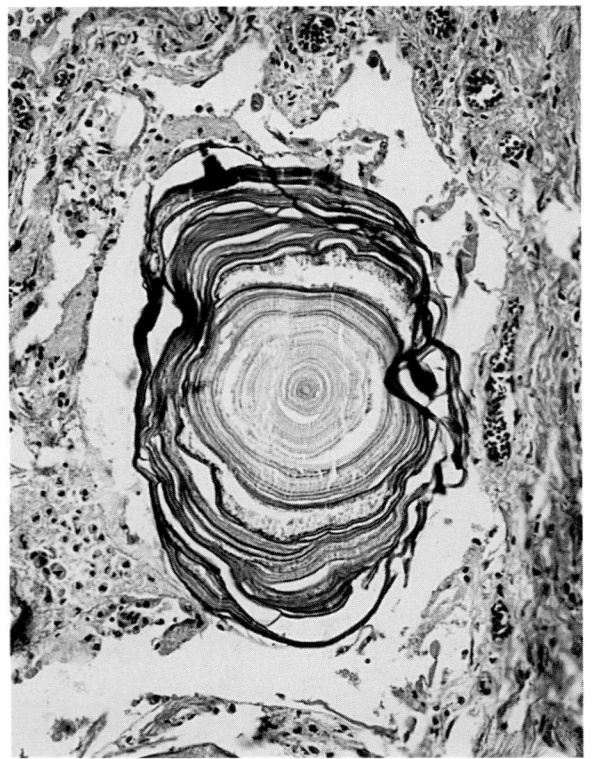
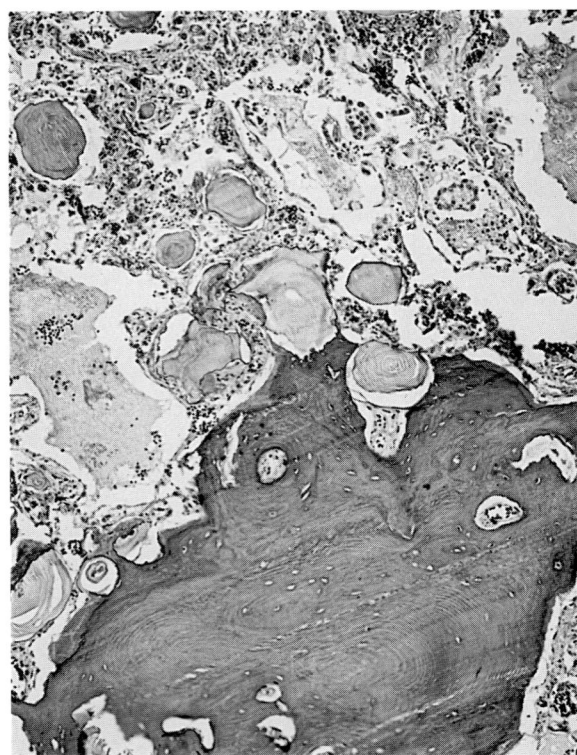

FIGURE 25-23
Pulmonary alveolar microlithiasis. *A*. Detail of a microlith demonstrates prominent multilayering by basophilic (calcific) laminations. *B*. A case with secondary ossification.

have a PAS-positive central polysaccharide core that is surrounded by basophilic calcific von Kossa–positive laminations. Their size usually varies from 50 to 200 μm, although some as large as 3 mm have been identified. The associated alveolar walls show mild inflammation and fibrosis. Occasional calcospherites may be found in the interstitium or in airway walls, and the latter feature might present problems in lung transplantation. Calcospherites may be birefringent with a Maltese-cross pattern; sometimes they are associated with giant cells. Confluent masses of calcospherites rarely occur, and in such cases one may find metaplastic bone.[205,222] Calcospherites are composed of calcium phosphate and calcium hydroxyapatite or carboxyapatite with traces of magnesium, silicon, iron, and aluminum.[215,219,220]

Pant and coworkers described a case of PAM with pleural calcification and nephrolithiasis.[226] Calcospherites have also been described in sympathetic ganglia and testes.[213]

Based on the radiographic findings and the positive family history in some of the cases, biopsy diagnosis is not always necessary. Nevertheless, the histologic findings are unique and may be identified in open lung biopsy material, in transbronchial biopsy material, and even in BAL fluid as in four of six cases reviewed by Ucan and colleagues.[228]

Differential Diagnosis

The differential diagnosis of PAM includes Schaumann bodies, blue bodies, corpora amylacea, psammoma bodies, and metaplastic bony nodules. Schaumann bodies are interstitial, usually seen as confluent clusters of basophilic laminated bodies in giant cells associated with granulomatous diseases. Calcium oxalates are often an accompanying finding with Schaumann bodies. Blue bodies are intraalveolar and usually associated with increased alveolar macrophages and giant cells; they are smaller than the calcospherites of PAM and have fewer laminations. Corpora amylacea are eosinophilic intraalveolar structures that are usually seen in only a few alveoli. They are not calcified and are only rarely lamellar; they are an incidental finding of no significance. Psammoma bodies associated with neoplasms tend to be smaller and much less numerous than the calcospherites of PAM. Pulmonary ossification may bear some gross resemblance to PAM, but histologically bone is seen rather than the distinctive laminated calcospherites of PAM.

CALCIFICATION (CALCINOSIS) OF THE LUNG

Regardless of the organ system affected, calcification is seen on hematoxylin-and-eosin slides as granular, lamellar, and platelike basophilic material that stains positively with von Kossa and alizarin red stains. Pulmonary calcification is generally divided into two forms: dystrophic calcification and metastatic calcification.

Dystrophic Calcification

Dystrophic calcification occurs in sites of prior injury or necrosis. It usually represents a focal incidental finding in the lung, and examples include old granulomas, healed pneumonias, old silicotic nodules, hilar nodes in cases of healed sarcoidosis, and many other conditions (Fig. 25-24). A peculiar form of dystrophic calcification is seen in some cases of *Pneumocystis carinii* pneumonia in AIDS: the cysts become calcified. This is usually only a histologic finding, but occasionally it is radiographically visible—as described by Srivatsa and colleagues in a patient who had been on pentamidine prophylaxis.[234] Dystrophic calcification may be associated with dystrophic ossification.

Metastatic Calcification of the Lung

Metastatic calcification refers to tissue deposition of calcium as the result of abnormal levels of serum calcium or phosphate (and possibly magnesium).[235-257] The deposition occurs in both normal and abnormal tissue, and sometimes the serum levels of calcium and phosphate are normal. Increased local tissue alkalinity may predispose to metastatic calcification; this explains why the stomach, kidney, and lung are relatively often affected.[241] Clinical settings in which metastatic calcification of the lung is seen are those that affect calcium and phosphate metabolism[235,236,241,244-256]—bone resorption or destruction (especially by tumors), chronic immobilization, chronic renal failure or renal dialysis, renal transplantation, hyperparathyroidism (especially primary and tertiary forms), hypervitaminosis D, steroid and phosphate therapy, milk-alkali syndrome, and hematolymphoid malignancies, including those associated with HTLV-1 infection. In a small percentage of cases, no obvious predisposing condition is found.[238,247,248]

Metastatic calcification usually occurs when calcium or phosphate levels are elevated.[241] Chemically, metastatic calcification may be hydroxyapatite or whitlockitt $[(CaMg)_3(PO_4)_2]$. Neff's team reported an unusual case of metastatic calcification due to calcium carbonate deposition that was

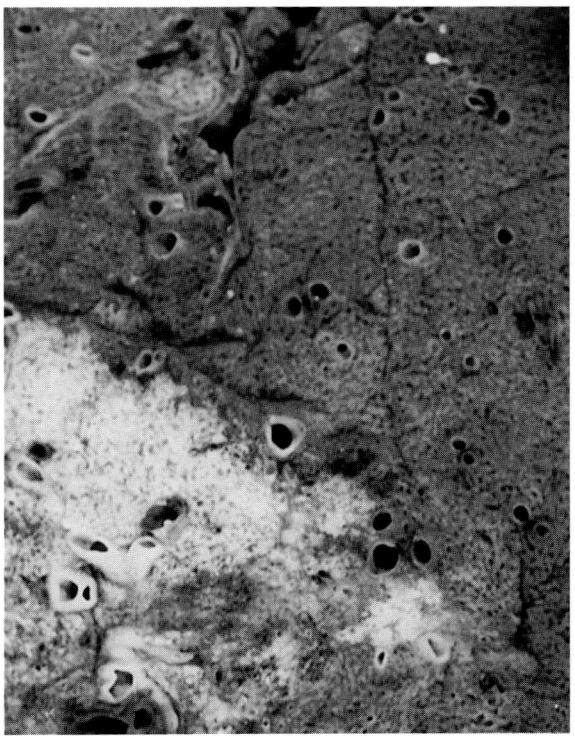

A

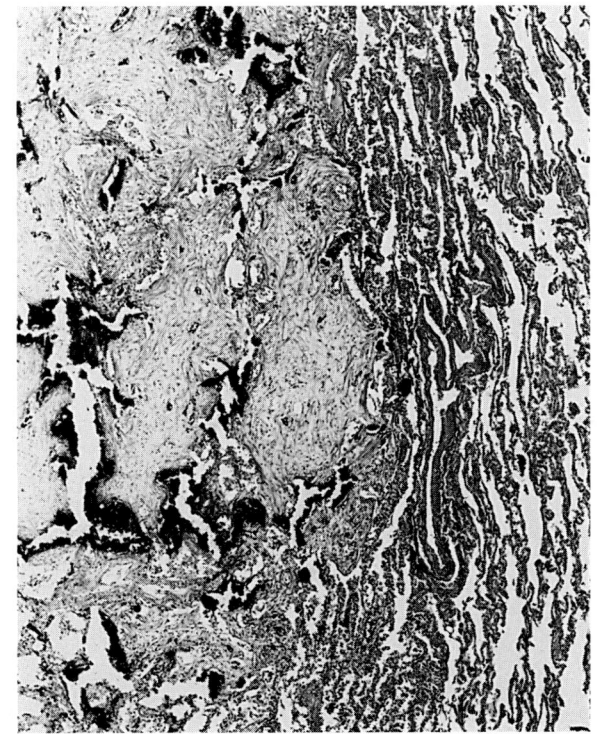

B

FIGURE 25-24
Dystrophic calcification. A focus of organizing pneumonia that became secondarily calcified presented grossly as a chalky zone of consolidation (*A, lower left*) and histologically as fibroblastic tissue with extensive dark basophilic calcification (*B*).

von Kossa–negative and alizarin red–positive; the patient had normal calcium and phosphate levels, and the authors offered no explanation for this peculiar case.[238]

Clinical Features

Metastatic calcification may be an incidental finding at autopsy or a life-threatening, clinically significant problem. Among 31 patients on chronic dialysis, Conger and colleagues found that of the 15 who died, nine had metastatic calcification histologically but only one of these had had radiographic abnormalities.[239] Nevertheless, patients on dialysis who develop clinical findings due to metastatic calcification have significant morbidity; seven of 13 patients in one series died of respiratory insufficiency.[241] Symptomatic patients have dyspnea and localized or diffuse radiographic infiltrates. Although radiographic changes may mimic pulmonary edema, they are distinctive for their unchanging nature.[236,241] The tissue calcification is rarely recognized as such on chest radiographs, but with CT scans the calcific density of the infiltrates can be appreciated,[254] usually as multiple nodular opacities.[257] Changes in pulmonary function, usually the development of restriction and abnormal diffusion, may precede manifestation of symptoms.[236] Clinical deterioration may be sudden,[236,240] and some patients have an abrupt fulminant course.[237,240,244]

Pathologic Findings

Grossly (Fig. 25-25), severely affected lungs are rigid and gritty on cut section.[236,242] Mild cases may appear grossly normal.[236] Histologically, there is cal-

cification of alveolar and vessel walls that may be focal or widespread.[236,238,241,243,245] The basophilic calcification is seen as plates or granules involving alveoli, vessels, and sometimes even airways; the vascular elastica are often affected early, and a giant cell reaction may be seen (Figs. 25-26 and 25-27). The affected tissue may be otherwise normal or show evidence of organization and alveolar injury. In some cases, it is impossible to determine whether the alveolar injury precedes or follows the calcification. Secondary ossification is only rarely seen. Electron microscopy shows that early deposits occur among elastic tissue fibers, and active growth is associated with rough edges in the small deposits that become smooth and polycyclic in larger (presumably more slowly growing) deposits.[242,245] Deposits larger than 100 μm are unusual.

BLUE BODIES AND CORPORA AMYLACEA

Blue Bodies

Blue bodies are intraalveolar pale-blue calcific lamellar bodies that occur singly and in clusters and are invariably associated with increased numbers of alveolar macrophages and giant cells (Fig. 25-28).[256,259] Blue bodies range up to 40 μm in diameter; they have a mucopolysaccharide matrix, are birefringent in unstained sections, are acid-soluble, and stain positively with PAS and Alcian blue stains. According to Koss and coworkers, they are a nonspecific finding related to the inflammatory process, possibly histiocyte metabolism.[259] Blue bodies are common in sputum and bronchial wash specimens (10.5 and 12 percent, respectively); Kung and col-

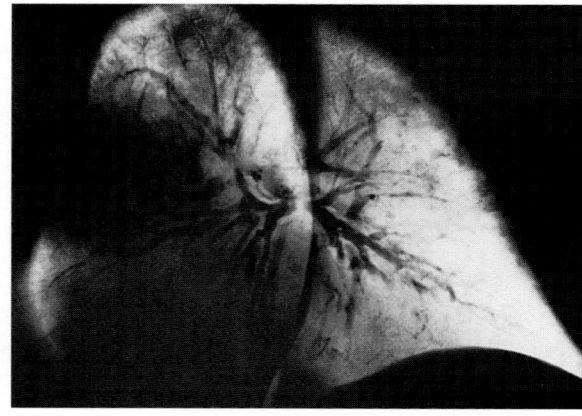

FIGURE 25-25
Metastatic calcification. This patient had widespread metastatic carcinoma of the breast with hypercalcemia. Secondary metastatic calcification developed in the lungs, which were grossly firm

and rigid without obvious consolidation (*A*). Radiographic specimen shows extensive calcium deposition, predominantly in the lower lobe (*B*).

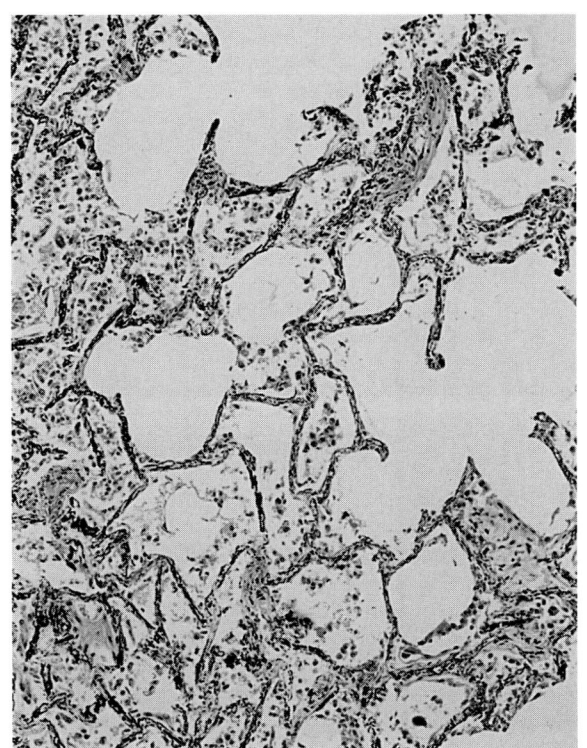

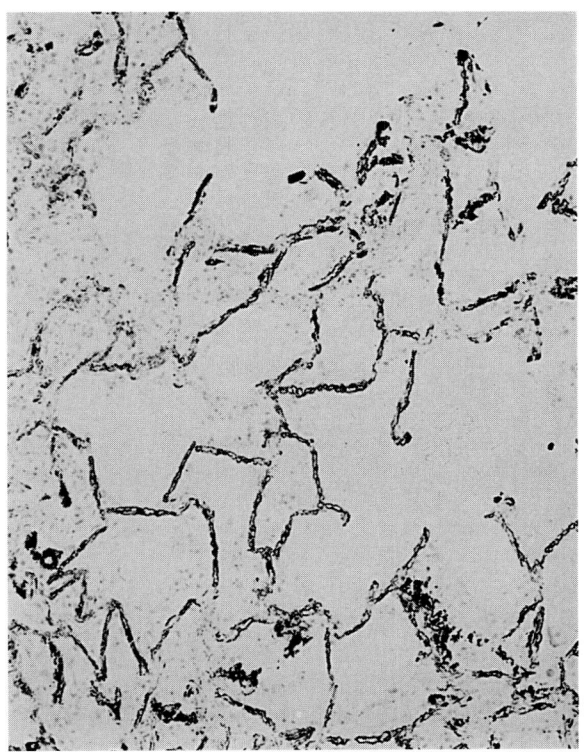

A   B

FIGURE 25-26
Metastatic calcification at autopsy. Fine diffuse calcification of the alveolar walls is seen as
basophilic deposits in the alveolar walls (*A*), which stain black with the von Kossa stain (*B*).

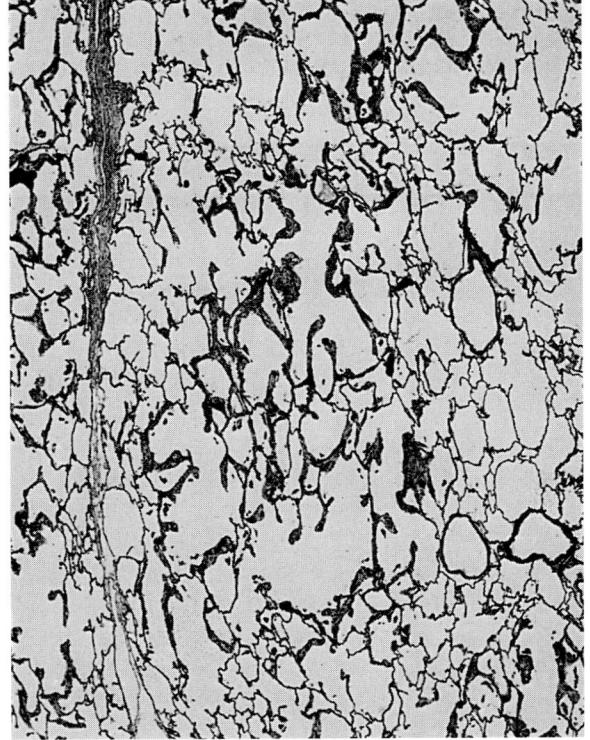

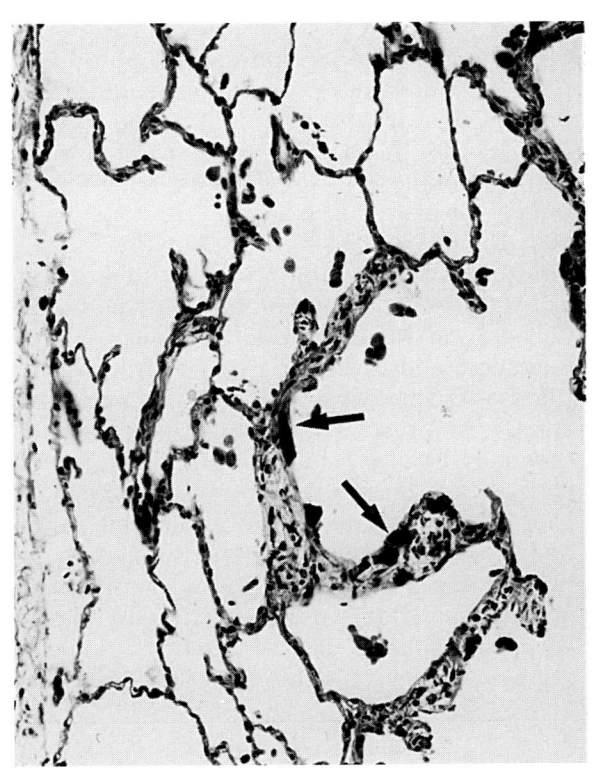

A   B

FIGURE 25-27
Metastatic calcification in a lung biopsy. This open lung biopsy
from a patient with chronic renal failure shows slight thickening
of the alveolar septa (*A*) by platelike deposits of calcium (*B*,
arrows), with no other significant changes in the lung
parenchyma.

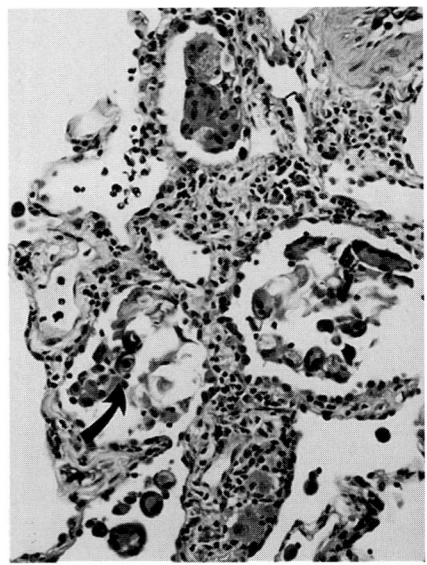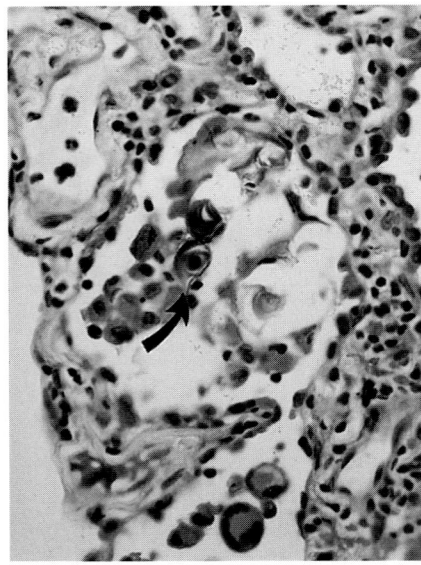

FIGURE 25-28
Pulmonary blue bodies. Blue bodies are clear to bluish, calcified, other laminar structures (*arrows*) seen within the alveolar spaces associated with macrophages and giant cells.

legues concluded that they were of no diagnostic significance in these specimens.[260] Blue bodies should be distinguished from pulmonary alveolar microlithiasis, Schaumann bodies, corpora amylacea, and psammoma bodies associated with primary and secondary papillary tumors in the lung.

Corpora Amylacea

Corpora amylacea are round eosinophilic acellular noncalcified intraalveolar bodies that represent clinically insignificant incidental findings in lung tissue (Fig. 25-29).[261–263] They may show lamination or radial cracking, and some are associated with a histiocytic or giant cell reaction. They vary from 30 to 200 μm in size. They are often PAS-positive and rarely stain for calcium. A small central nidus, 20 μm or less, is sometimes seen and may be birefringent. Corpora amylacea represent incidental findings in histologic and cytologic material from adult lungs. Their incidence was 0.6 and 3.8 percent in two autopsy studies.[261,263] Personal experience suggests they may be more common than that. The cause of corpora amylacea is unknown. Corpora amylacea should be distinguished from calcified intraalveolar bodies, especially those in pulmonary alveolar microlithiasis and blue bodies. Aspirated material (such as meat) does not show the smooth contour of corpora amylacea and is usually associated with more extensive acute inflammation and giant cells.

PULMONARY OSSIFICATION

Ossification in the lung takes a number of forms: dystrophic ossification and metaplastic bone, diffuse parenchymal ossification, ossification of tracheobronchial cartilage (which is an age-related phenomenon), and tracheopathia osteochondroplastica. Ossification may also occur in primary (both benign and malignant) and metastatic tumors of the lung and pleura, in some cases after radiation treatment or chemotherapy. In any of these types of ossification, the bone may be remarkably well formed and even contain marrow with hematopoietic elements.

Dystrophic Ossification

Like dystrophic calcification, dystrophic ossification (metaplastic bone) may occur in sites of prior injury

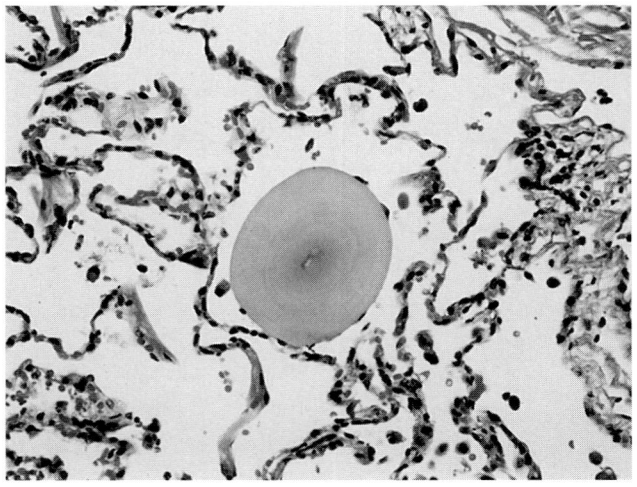

FIGURE 25-29
Pulmonary corpora amylacea. There is a large eosinophilic intraalveolar structure with concentric laminations, a few macrophages hugging the surface, and a small central nidus. The adjacent alveoli are normal.

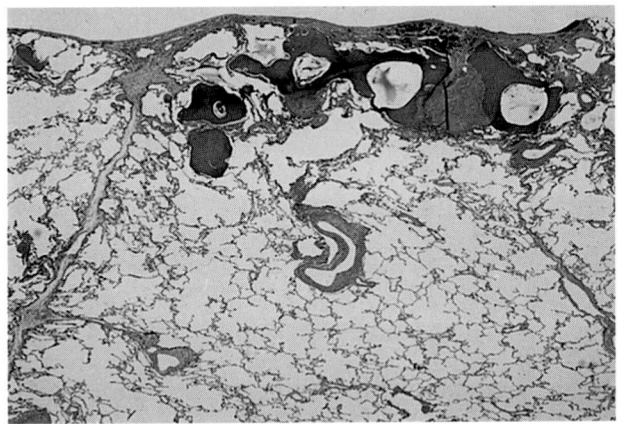

FIGURE 25-30
Pulmonary ossification. This case shows a focus of subpleural ossification for which no cause was apparent. Elsewhere in the lung, metastatic transitional cell carcinoma was present, but it was uncalcified and unossified and showed no relationship to the metaplastic bone, which was assumed to be an incidental finding.

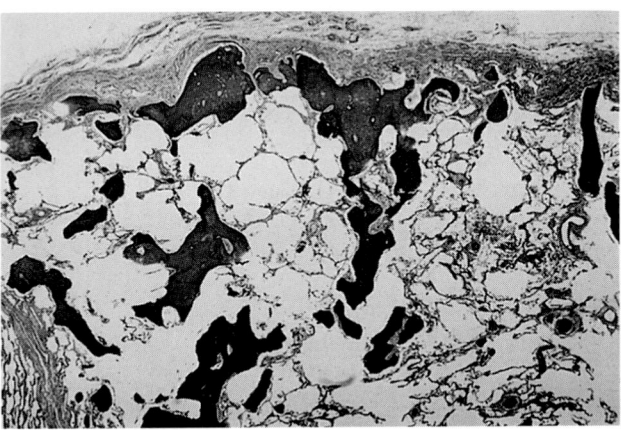

FIGURE 25-31
Dendriform pulmonary ossification. The lungs show multiple ramifying nodules of mature bone.

and scarring and may follow or be associated with dystrophic calcification (Fig. 25-30). Dystrophic ossification is usually a focal insignificant finding in biopsy and autopsy tissue, often associated with scarring. In occasional cases one finds metaplastic bone formation with no apparent cause whatsoever.

Diffuse Parenchymal Ossification

Diffuse parenchymal ossification shows two patterns—the dendriform and the diffuse nodular—although there is some overlap between them.[264]

As the name implies, *dendriform pulmonary ossification* shows extensive branching coral-like bony spicules that may be unilateral or, more often, bilateral. It is often associated with lung scarring.[264–274] The occurrence of dendriform pulmonary ossification in regions of scarring does not account for all cases, including several of those described by Felson and coworkers in which there was no apparent interstitial fibrosis.[270] Although men over 60 are most often affected, dendriform pulmonary ossification is seen in both men and women and affects persons of a wide age range.[270,273] Most patients have no symptoms directly attributed to the ossification, although they may have symptoms due to the underlying fibrotic process. Laboratory studies, including serum levels of calcium and phosphate, are normal. Dendriform ossification is relatively rare; Ndimbie and Lee found only 16 cases among 2800 autopsies.[272]

Radiographically, the lesions of dendriform pulmonary ossification are usually in the lower lobes and manifest as fine linear shadows 1 to 4 mm in thickness in branching and reticular arrays.[270,274]

Their calcific density is usually apparent, although some cases have been misinterpreted as bronchiectasis or interstitial fibrosis.[267,272]

Grossly, the lungs are gritty and firm on cut section with variable associated scarring. Histologically, branching arrays of mature bone are identified, and they may be associated with scarring and interstitial fibrosis (Fig. 25-31). The bone is usually in the interstitium and only occasionally in alveolar spaces. Marrow formation is relatively common.

The *diffuse nodular pattern* of pulmonary ossification is usually associated with chronic passive congestion, classically due to mitral stenosis.[264,275–280] Other conditions associated diffuse nodular pulmonary ossification are chronic pulmonary fibrosis, left atrial myxoma, constrictive pericarditis, coronary artery disease, systemic hypertension, aortic insufficiency, idiopathic hypertrophic subaortic stenosis, metastatic tumors, and healed adult respiratory distress syndrome.[277] In some cases, no predisposing condition is found.[278] Most of the reported cases have been in men (4:1) with an average age of 35 years; most have a history of chronic passive congestion of some type. Chest radiographs show disseminated nodular calcific densities, usually greater in the lower lobes. The densities are discrete and lack the branching and reticular features of dendriform ossification. Some cases may resemble healed granulomatous disease. In one study, half of the patients with mitral valve disease had some degree of diffuse nodular ossification at autopsy.[264]

Grossly, the lungs are gritty on cut section because of the 1-to-3-mm bony nodules.[279] Histologically, intraalveolar nodules of well-formed woven bone (with or without marrow) are identified (Fig. 25-32). Nodules as large as 1 cm in diameter are occa-

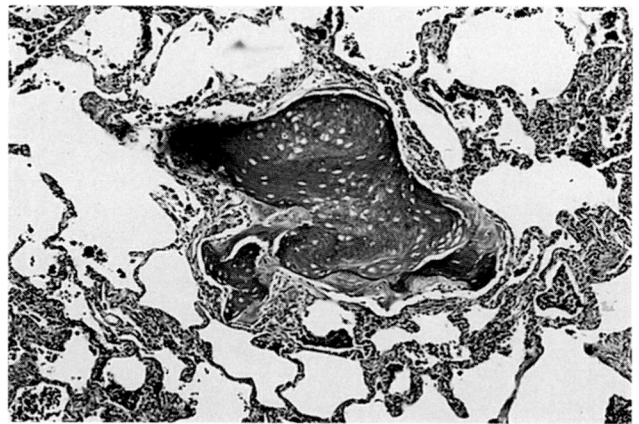

FIGURE 25-32
Diffuse nodular pulmonary ossification. This case of chronic vascular congestion due to mitral stenosis was associated with multiple nodules of woven bone within air spaces and the interstitium. The adjacent alveolar walls show mild thickening consistent with chronic passive congestion.

sionally seen. Concomitant changes of chronic possible congestion may be present.

Diffuse nodular pulmonary parenchymal ossification is thought to follow organization of intraalveolar exudates or to be related to intraalveolar hemosiderin accumulation.

Tracheopathia Osteochondroplastica

Tracheopathia osteochondroplastica is a peculiar disorder in which nodules of metaplastic bone and cartilage develop in the submucosa of the large cartilaginous airways and occasionally the larynx.[281-290] There is a predilection for men (3:1)[283]; the mean age at presentation is 42 years for men and 51 years for women.[282,290] Familial cases have been described.[289] Many patients are asymptomatic; those with symptoms have complaints related to airways obstruction with cough, dyspnea, chronic bronchitis, and hemoptysis. Radiographic studies show narrowing of the large airways, and on CT scans the submucosa bony nodules can actually be discerned; in such cases, CT scans may be diagnostic.[285,287]

Bronchoscopy reveals typical findings: submucosal nodules 1 to 3 mm in diameter that may coalesce to form plaques.[287] The overlying mucosa may be normal or show secondary ulceration and inflammatory changes. The nodules are primarily found in the trachea and proximal bronchi.

The gross findings parallel the bronchoscopic findings; rigid, hard white submucosal nodules and plaques are found in the trachea and proximal bronchi.[283,284] The nodules may be up to 1 cm in size.

Histologically,[289] one sees nodules of cartilage or bone in the submucosa (Fig. 25-33). They are associated with fibrous tissue, which in some cases appears continuous with the underlying perichondrium.[283]

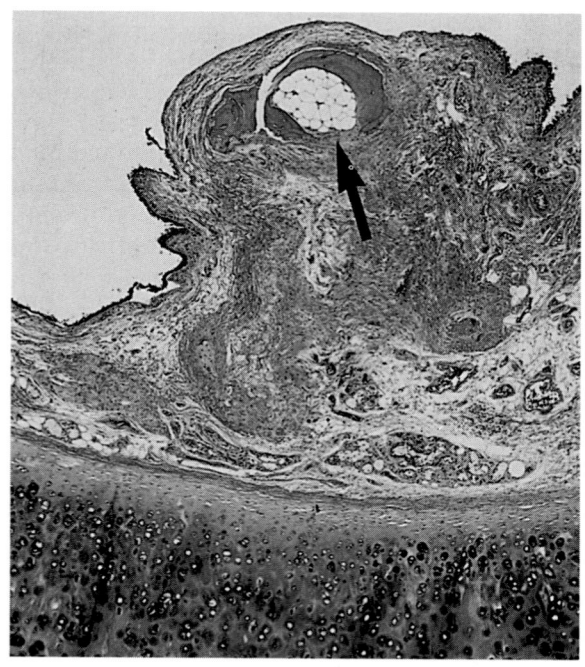

A

B

FIGURE 25-33
Tracheopathia osteochondroplastica. These two cases illustrate metaplastic bone in the tracheal submucosa seen as a tiny focus in *A* (*arrows*) and a much larger focus in *B*. In both cases, fatty marrow is present.

Secondary inflammatory and metaplastic changes may be seen in the overlying mucosa.

In most cases there is essentially no differential diagnosis. Tracheobronchial amyloidosis may show some resemblance to tracheopathia osteochondroplastica, but stains for amyloid are positive.

Since many cases are discovered at autopsy, the disease can be clinically silent. Cases recognized during life tend to have an indolent course with complications due to airways, obstruction, which sometimes requires palliation.

REFERENCES

1. Lichtenstein L: Histiocytosis X: Integration of eosinophilic granuloma of bone, "Letterer-Siwe disease," and "Schüller-Christian disease" as related manifestations of a single nosologic entity. *Arch Pathol* 56:84–102, 1953.
2. Travis WD, Borok Z, Roum JH, et al: Pulmonary Langerhans cell granulomatosis (histiocytosis X): A clinicopathologic study of 48 cases. *Am J Surg Pathol* 17:971–986, 1993.
3. Pritchard J, Broadbent V: Histiocytosis: An introduction. *Br J Cancer* 70:S1–S3, 1994.
4. Hage C, Willman CL, Favara BE, Isaacson PG: Langerhans' cell histiocytosis (histiocytosis X): Immunophenotype and growth fraction. *Hum Pathol* 24:840, 1993.
5. Basset F, Corrin B, Spencer H, et al: Pulmonary histiocytosis X. *Am Rev Respir Dis* 118:811–820, 1978.
6. Corrin B, Basset F: A review of histiocytosis X with particular reference to eosinophilic granuloma of the lung. *Invest Cell Pathol* 2:137–146, 1979.
7. Friedman PJ, Liebow AA, Sokoloff J: Eosinophilic granuloma of the lung: Clinical aspects of primary histiocytosis in the adult. *Medicine (Baltimore)* 60:385–396, 1981.
8. Lacronique J, Roth C, Battesti J-P, Basset F, Chretien J: Chest radiological features of pulmonary histiocytosis X: A report based on 50 adult cases. *Thorax* 37:104–109, 1982.
9. Colby TV, Lombard C: Histiocytosis X in the lung. *Hum Pathol* 14:847–856, 1983.
10. Marcy TW, Reynolds HY: Pulmonary histiocytosis X. *Lung* 163:129–150, 1985.
11. Hance AJ, Cadranel J, Soler P, Basset F: Pulmonary and extrapulmonary Langerhans' cell granulomatosis (histiocytosis X). *Semin Respir Med* 9:349–368, 1988.
12. Soler P, Kambouchner M, Valeyre D, Hance AJ: Pulmonary Langerhans' cell granulomatosis (histiocytosis X). *Annu Rev Med* 43:105–115, 1992.
13. Schönfeld N, Frank W, Wenig S, et al: Clinical and radiologic features, lung function and therapeutic results in pulmonary histiocytosis X. *Respiration* 60:38–44, 1993.
14. Ha, SY, Helms P, Fletcher M, Broadbent V, Pritchard J: Lung involvement in Langerhans' cell histiocytosis: Prevalence, clinical features, and outcome. *Pediatrics* 89:466–469, 1992.
15. Chatkin JM, Bastos JC, Stein RT, Jaiger AM: Sole pulmonary involvement by Langerhans' cell histiocytosis in a child. *Eur Respir J* 6:1226–1228, 1993.
16. Vade A, Hayani A, Pierce KL: Congenital histiocytosis X. *Pediatr Radiol* 23:181–182, 1993.
17. Pai U, Tomashefski JF: Occult pulmonary eosinophilic granuloma: An incidental finding in the lungs of cigarette smokers. *Am J Clin Pathol* 99:335, 1993.
18. Nezelof C; Histiocytosis X: A histological and histogenetic study. *Perspect Pediatr Pathol* 5:153–178, 1979.
19. Lewis JG: Eosinophilic granuloma and its variants with special reference to lung involvement: A report of 12 patients. *Q J Med* 33:337, 1964.
20. Müller NL, Miller RR: Computed tomography of chronic diffuse infiltrative lung disease (parts I and II). *Am Rev Respir Dis* 142:1206, 1990.
21. Hansell DM, Kerr IH: The role of high resolution computed tomography in the diagnosis of interstitial lung disease. *Thorax* 46:77, 1991.
22. Swensen SJ, Aughenbaugh GL, Douglas WW, Myers JL: High-resolution CT of the lungs: Findings in various pulmonary diseases. *AJR* 158:971, 1992.
23. Kulwiec EL, Lynch DA, Aguayo SM, Schwarz MI, King TE Jr: Imaging of pulmonary histiocytosis X. *Radiographics* 12:515–526, 1992.
24. Epler GR, McLoud TC, Gaensler EA, Mikus JP, Carrington CB: Normal chest roentgenograms in chronic diffuse infiltrative lung disease. *N Engl J Med* 298:934, 1978.
25. Fichtenbaum CJ, Kleinman GM, Haddad RG: Eosinophilic granuloma of the lung presenting as a solitary pulmonary nodule. *Thorax* 45:905–906, 1990.
26. Brauner MW, Grenier P, Mouelhi MM, Mompoint D, Lenoir S: Pulmonary histiocytosis X: Evaluation with high-resolution CT. *Radiology* 172:255, 1989.
27. Hoffman L, Cohn JE, Gaensler EA: Respiratory abnormalities in eosinophilic granuloma of the lung: Long-term study of five cases. *N Engl J Med* 267:577–589, 1962.
28. Lahey ME: Prognostic factors in histiocytosis X. *Am J Pediatr Hematol Oncol* 3:57, 1981.
29. Greenberger JS, Crocker AC, Vawter G, Jaffe N, Cassady JR: Results of treatment of 127 patients with systemic histiocytosis (Letterer-Siwe syndrome, Schuller-Christian syndrome and multifocal eosinophilic granuloma). *Medicine (Baltimore)* 60:311–338, 1981.
30. Sajjad SM, Luna MA: Primary pulmonary histiocytosis X in two patients with Hodgkin's disease. *Thorax* 37:110–113, 1982.
31. Lombard CM, Medeiros LJ, Colby TV: Pulmonary histiocytosis X and carcinoma. *Arch Pathol Lab Med* 111:339–341, 1987.
32. Tomashefski JF Jr, Khiyami A, Kleinerman J: Neoplasms associated with pulmonary eosinophilic granuloma. *Arch Pathol Lab Med* 115:499–506, 1991.
33. Sadoun D, Vaylet F, Valeyre D, et al: Bronchogenic carcinoma in patients with pulmonary histiocytosis X. *Chest* 101:1610–1613, 1992.
34. Brody AR, Kanich RE, Graham WG, Craighead JE: Cyst wall formation in pulmonary eosinophilic granuloma. *Chest* 66:576–578, 1974.
35. Fukuda Y, Basset F, Soler P, Ferrans VJ, Masugi Y, Crystal RG: Intraluminal fibrosis and elastic fiber degradation lead to lung remodeling in pulmonary Langerhans cell granulomatosis (histiocytosis X). *Am J Pathol* 137:415–424, 1990.
36. Flint A, Lloyd RV, Colby TV, Wilson BW: Pulmonary histiocytosis X: Immunoperoxidase staining for HLA-DR antigen and SIDO protein. *Arch Pathol Lab Med* 110:930, 1986.
37. Cagle PT, Mattioli CA, Truong LD, Greenberg SD: Immunohistochemical diagnosis of pulmonary eosinophilic granuloma on lung biopsy. *Chest* 94:1133, 1988.
38. Webber D, Tron V, Askin F, Churg A: S-100 staining in the diagnosis of eosinophilic granuloma of lung. *Am J Clin Pathol* 84:447, 1985.
39. Auerswald U, Barth J, Magnussen H: Value of CD-1–positive

cells in bronchoalveolar lavage fluid for the diagnosis of pulmonary histiocytosis X. *Lung* 169:305–309, 1991.

40. Hammar S, Bockus D, Remington F, Bartha M: The widespread distribution of Langerhans cells in pathologic tissues: An ultrastructural and immunohistochemical study. *Hum Pathol* 17:894–905, 1986.

41. Hammar S: Langerhans cells. *Pathol Annu* 2:293–328, 1988.

42. Tazi A, Bonay M, Grandsaigne M, Battesti J-P, Hance AJ, Soler P: Surface phenotype of Langerhans cells and lymphocytes in granulomatous lesions from patients with pulmonary histiocytosis X. *Am Rev Respir Dis* 147:1531–1536, 1993.

43. Soler P, Chollet S, Jacque C, Fukuda Y, Ferrans VJ, Basset F: Immunocytochemical characterization of pulmonary histiocytosis X cells in lung biopsies. *Am J Pathol* 118:439–451, 1985.

44. Colasante A, Poletti V, Rosini S, Ferracini R, Musiani P: Langerhans cells in Langerhans cell histiocytosis and peripheral adenocarcinomas of the lung. *Am Rev Respir Dis* 148:752–759, 1993.

45. Chu T, Jaffe R: The normal Langerhans' cells and the LCH cell. *Br J Cancer* 70(suppl):S4–S10, 1994.

46. Toews GB: Pulmonary dendritic cells: Sentinels of lung-associated lymphoid tissue. *Am J Respir Cell Mol Biol* 4:204–205, 1991.

47. Casolaro MA, Bernaudin J-F, Saltini C, Ferrans VJ, Crystal RG: Accumulation of Langerhans' cells on the epithelial surface of the lower respiratory tract in normal subjects in association with cigarette smoking. *Am Rev Respir Dis* 137:406–411, 1988.

48. Soler P, Moreau A, Basset F, Hance AJ: Cigarette smoking–induced changes in the number and differentiated state of pulmonary dendritic cells/Langerhans cells. *Am Rev Respir Dis* 139:1112–1117, 1989.

49. Kawanami O, Basset F, Ferrans VJ, Soler P, Crystal RG: Pulmonary Langerhans' cells in patients with fibrotic lung disorders. *Lab Invest* 44:227–233, 1981.

50. Aguayo SM, King TE, Waldron JA, Sherritt KM, Kane MA, Miller YE: Increased pulmonary neuroendocrine cells with bombesin-like immunoreactivity in adult patients with eosinophilic granuloma. *J Clin Invest* 86:838–844, 1990.

51. Housini I, Tomashefski JF, Cohen A, Crass J, Kleinerman J: Transbronchial biopsy in patients with pulmonary eosinophilic granuloma: Comparison with findings on open biopsy. *Arch Pathol Lab Med* 118:523–530, 1994.

52. Basset F, Soler P, Jaurand MC, Bignon J: Ultrastructural examination of broncho-alveolar lavage for diagnosis of pulmonary histiocytosis X: Preliminary report on 4 cases. *Thorax* 32:303–306, 1977.

53. Rosen SH, Castleman B, Liebow AA: Pulmonary alveolar proteinosis. *N Engl J Med* 258:1123–1142, 1958.

54. Kuhn C, Györkey F, Levine BE, Ramirez-Rivera J: Pulmonary alveolar proteinosis: A study using enzyme histochemistry, electron microscopy, and surface tension measurement. *Lab Invest* 15:492–509, 1966.

55. Ramirez RJ, Harlan WR Jr: Pulmonary alveolar proteinosis: Nature and origin of alveolar lipid. *Am J Med* 45:502–512, 1968.

56. Corrin B, King E: Pathogenesis of experimental pulmonary alveolar proteinosis. *Thorax* 25:230–236, 1970.

57. Colón AR Jr, Lawrence RD, Mills SD, O'Connell EJ: Childhood pulmonary alveolar proteinosis (PAP): Report of a case and review of the literature. *Am J Dis Child* 121:481–485, 1971.

58. Case Records of the Massachusetts General Hospital (Case 34-1974). *N Engl J Med* 291:464–469, 1974.

59. Hudson AR, Halprin GM, Miller JA, Kilburn KH: Pulmonary interstitial fibrosis following alveolar proteinosis. *Chest* 65:700–702, 1974.

60. Costello JF, Moriarty DC, Branthwaite MA, Turner-Warwick M, Corrin B: Diagnosis and management of alveolar proteinosis: The role of electron microscopy. *Thorax* 30:121–132, 1975.

61. Golde DW, Territo M, Finley TN, Cline MJ: Defective lung macrophage in pulmonary alveolar proteinosis. *Ann Intern Med* 85:304–309, 1976.

62. Carnovale R, Zornoza J, Goldman AM, Luna M: Pulmonary alveolar proteinosis: Its association with hematologic malignancy and lymphoma. *Radiology* 122:303–306, 1977.

63. Jacobovitz-Derks D, Corrin B: Degenerative processes in the pathogenesis of pulmonary alveolar lipoproteinosis. *Virchows Arch [A]* 376:165–174, 1977.

64. Hook GER, Bell DY, Gilmore LB, Nadeau D, Reasor MJ, Talley FA: Composition of bronchoalveolar lavage effluents from patients with pulmonary alveolar proteinosis. *Lab Invest* 39:342–357, 1978.

65. Ito M, Takeuchi N, Ogura T, et al: Pulmonary alveolar proteinosis: Analysis of pulmonary washings. *Br J Dis Chest* 72:313–320, 1978.

66. Rogers RM, Levin DC, Gray BA, Moseley LW Jr: Physiologic effects of bronchopulmonary lavage in alveolar proteinosis. *Am Rev Respir Dis* 118:255–269, 1978.

67. Hocking WG, Golde DW: The pulmonary-alveolar macrophage (second of two parts). *N Engl J Med* 301:639–645, 1979.

68. Bell DY, Hook GER: Pulmonary alveolar proteinosis: Analysis of airway and alveolar proteins. *Am Rev Respir Dis* 119:979–990, 1979.

69. Ranchod M, Bissell M: Pulmonary alveolar proteinosis and cytomegalovirus infection. *Arch Pathol Lab Med* 103:139–142, 1979.

70. Bedrossian CWM, Luna MA, Conklin RH, Miller WC: Alveolar proteinosis as a consequence of immunosuppression: A hypothesis based on clinical and pathologic observations. *Hum Pathol* 11:527–535, 1980.

71. Coleman M, Dehner LP, Sibley RK, Burke BA, L'Heureux PR, Thompson TR: Pulmonary alveolar proteinosis: An uncommon cause of chronic neonatal respiratory distress. *Am Rev Respir Dis* 121:583–586, 1980.

72. Martin RJ, Coalson JJ, Rogers RM, Horton FO, Manous LE: Pulmonary alveolar proteinosis: The diagnosis by segmental lavage. *Am Rev Respir Dis* 191:819–825, 1980.

73. Rubin E, Weisbrod GL, Sanders DE: Pulmonary alveolar proteinosis: Relationship to silicosis and pulmonary infection. *Radiology* 135:35–41, 1980.

74. Singh G, Katyal SL: Surfactant apoprotein in nonmalignant pulmonary disorders. *Am J Pathol* 101:51–61, 1980.

75. Smith LJ, Ankin MG, Katzenstein A-L, Shapiro BA: Management of pulmonary alveolar proteinosis: Clinical Conference in Pulmonary Disease from Northwestern University McGaw Medical Center, Chicago, *Chest* 78:765–770, 1980.

76. Webster JR Jr, Battifora H, Furey C, Harrison RA, Shapiro B: Pulmonary alveolar proteinosis in two siblings with decreased immunoglobulin A. *Am J Med* 69:786–789, 1980.

77. Miller PA, Ravin CE, Walker Smith GJ, Osborne DRS: Pulmonary alveolar proteinosis with interstitial involvement. *AJR* 137:1069–1071, 1981.

78. Teja K, Cooper PH, Squires JE, Schnatterly PT: Pulmonary alveolar proteinosis in four siblings. *N Engl J Med* 305:1390–1392, 1981.

79. Singh G, Katyal SL, Bedrossian CWM, Rogers RM: Pulmonary alveolar proteinosis: Staining for surfactant apoprotein in alveolar proteinosis and in conditions simulating it. *Chest* 83:82–86, 1983.

80. Harris JO: Pulmonary alveolar proteinosis: Specific or nonspecific response? (editorial) *Chest* 84:1–2, 1983.

81. Abraham JL, McEuen DD: Inorganic particulates associated with pulmonary alveolar proteinosis (abstract). *Am Rev Respir Dis* 119:196, 1979.

82. Clague HW, Wallace AC, Morgan WKC: Pulmonary fibrosis associated with alveolar proteinosis. *Thorax* 38:865–866, 1983.

83. Corsello BF, Choi H: Basophilic staining in pulmonary alveolar proteinosis: Report of three cases. *Arch Pathol Lab Med* 108:68–70, 1984.

84. Miller RR, Churg AM, Hutcheon M, Lam S: Pulmonary alveolar proteinosis and aluminum dust exposure. *Am Rev Respir Dis* 130:312–315, 1984.

85. Prakash UBS, Barham SS, Carpenter HA, Dines DE, March HM: Pulmonary alveolar phospholipoproteinosis: Experience with 34 cases and a review. *Mayo Clin Proc* 62:499–518, 1987.

86. Case Records of the Massachusetts General Hospital (Case 18-1988). *N Engl J Med* 318:1186–1194, 1988.

87. Schumacher RE, Marrogi AJ, Heidelberger KP: Pulmonary alveolar proteinosis in a newborn. *Pediatr Pulmonol* 7:178–182, 1989.

88. deMello DE, Hayman S, Phelps DS, et al: Ultrastructure of lung in surfactant protein B deficiency. *Am J Respir Cell Mol Biol* 11:230, 239, 1994.

89. Godwin JD, Müller NL, Takasugi JE: Pulmonary alveolar proteinosis: CT findings. *Radiology* 169:609–613, 1988.

90. Murch CR, Carr DH: Computed tomography appearances of pulmonary alveolar proteinosis. *Clin Radiol* 40:240–243, 1989.

91. Wilson DO, Rogers RM: Prolonged spontaneous remission in a patient with untreated pulmonary alveolar proteinosis. *Am J Med* 82:1014–1016, 1987.

92. Cardillo MR: Pulmonary alveolar proteinosis: A cytomorphological histochemical and ultrastructural study of one case. *Arch Anat Cytol Path* 37:259, 1989.

93. Schober R, Bensch KG, Kosek JC, Northway WH: On the origin of the membranous intra-alveolar material in pulmonary alveolar proteinosis. *Exp Mol Pathol* 21:246, 1974.

94. Masuda T, Shimura S, Sasaki H, Takishima T: Surfactant apoprotein-A concentration in sputum for diagnosis of pulmonary alveolar proteinosis. *Lancet* 337:580, 1991.

95. Gonzalez-Rothi FJ, Harris JO: Pulmonary alveolar proteinosis: Further evaluation of abnormal alveolar macrophages. *Chest* 90:656–661, 1986.

96. Hoffman RM, Dauber JH, Rogers RM: Improvement in alveolar macrophage migration after therapeutic whole lung lavage in pulmonary alveolar proteinosis. *Am Rev Respir Dis* 139:1030–1032, 1989.

97. Craighead JE, et al: Diseases associated with exposure to silica and nonfibrous silicate minerals. *Arch Pathol Lab Med* 112:673, 1988.

98. Ruben FL, Talamo TS: Secondary pulmonary alveolar proteinosis occurring in two patients with acquired immune deficiency syndrome. *Am J Med* 80:1187–1190, 1986.

99. Israel RH, Magnussen CR: Are AIDS patients at risk for pulmonary alveolar proteinosis? *Chest* 96:641–642, 1989.

100. Van Nhieu JT, Vojtek A-M, Bernaudin J-F, Escudier D, Fleury-Feith J: Pulmonary alveolar proteinosis associated with *Pneumocystis carinii*: Ultrastructural identification in bronchoalveolar lavage in AIDS and immunocomprised non-AIDS patients. *Chest* 98:801–805, 1990.

101. Tapson VF, Piantadosi CA: Pulmonary alveolar proteinosis and *Mycobacterium avium-intracellulare*: A frequent association (abstract). *Chest* 100:147S, 1991.

102. Case Records of the Massachusetts General Hospital (Case 14-1984). *N Engl J Med* 310:906–916, 1984.

103. Samuels MP, Warner JO: Pulmonary alveolar lipoproteinosis complicating juvenile dermatomyositis. *Thorax* 43:939–940, 1988.

104. Cordonnier C, Fluery-Feith J, Escudier E, Atassi K, Bernaudin JF: Secondary alveolar proteinosis is a reversible cause of respiratory failure in leukemic patients. *Am J Respir Crit Care Med* 149:788–794, 1994.

105. Case Records of the Massachusetts General Hospital. *N Engl J Med* 308:1147, 1983.

106. Schiller V, Aberle DR, Aberle AM: Pulmonary alveolar proteinosis: Occurrence with metastatic melanoma to lung. *Chest* 95:466, 1989.

107. Kyle RA, Bayrd ED: Amyloidosis: Review of 236 cases. *Medicine (Baltimore)* 54:271–299, 1975.

108. Glenner GG: Amyloid deposits and amyloidosis: The β-fibrilloses (part 1). *N Engl J Med* 302:1283–1292, 1980.

109. Glenner GG: Amyloid deposits and amyloidosis: The β-fibrilloses (part 2). *N Engl J Med* 302:1333–1343, 1980.

110. Cohen AS, Connors LH: The pathogenesis and biochemistry of amyloidosis. *J Pathol* 151:1–10, 1987.

111. Miura K, Shirasawa H: Lambda III subgroup immunoglobulin light chains are precursor proteins of nodular pulmonary amyloidosis. *Am J Clin Pathol* 100:561, 1993.

112. Colby TV, Carrington CB: Diffuse infiltrative lung diseases, in Thurlbeck WM, Churg A (eds), *Pathology of the Lung*, 2nd ed. New York, Thieme, 1995, pp 482–687.

113. Case Records of the Massachusetts General Hospital. *N Engl J Med* 327:1740, 1992.

114. Hui AN, Koss MN, Hochholzer L, Wehunt WD: Amyloidosis presenting in the lower respiratory tract. *Arch Pathol Lab Med* 110:212–218, 1986.

115. Rajan VT, Kikkawa Y: Alveolar septal amyloidosis in primary amyloidosis: An electron microscopic study. *Arch Pathol Lab Med* 89:521–529, 1970.

116. Toriumi J: The lung in generalized amyloidosis. *Acta Pathol Jpn* 22:141–153, 1972.

117. Gertz MA, Greipp PR: Clinical aspects of pulmonary amyloidosis (editorial). *Chest* 90:790–791, 1986.

118. Chen KTK: Amyloidosis presenting in the respiratory tract. *Pathol Annu* 24:253–273, 1989.

119. Planes C, Kleinknecht D, Brauner M, Battesti J-P, Kemeny J-L, Valeyre D: Diffuse interstitial lung disease due to AA amyloidosis. *Thorax* 47:323, 1992.

120. Beer TW, Edwards CW: Pulmonary nodules due to reactive systemic amyloidosis (AA) in Crohn's disease. *Thorax* 48:1287–1288, 1993.

121. Rubinow A, Celli BR, Cohen AS, Rigden BG, Brody JS: Localized amyloidosis of the lower respiratory tract. *Am Rev Respir Dis* 118:603–611, 1978.

122. Celli BR, Rubinow A, Cohen AS, Brody JS: Patterns of pulmonary involvement in systemic amyloidosis. *Chest* 74:543–547, 1978.

123. DaCosta P, Corrin B: Amyloidosis localized to the lower respiratory tract: Probable immunoamyloid nature of the tracheobronchial and nodular pulmonary forms. *Histopathol* 9:703–710, 1985.

124. Cordier JF, Loire R, Brune J: Amyloidosis of the lower respiratory tract: Clinical and pathologic features in a series of 21 patients. *Chest* 90:827–831, 1986.

125. Whitwell F: Localized amyloid infiltrations of the lower respiratory tract. *Thorax* 8:309–319, 1953.

126. Prowse CB: Amyloidosis of the lower respiratory tract. *Thorax* 13:308–320, 1958.

127. Antunes ML, Vieira da Luz JM: Primary diffuse tracheobronchial amyloidosis. *Thorax* 24:307–311, 1969.

128. Gottlieb LS, Gold WM: primary tracheobronchial amyloidosis. *Am Rev Respir Dis* 105:425–429, 1972.

129. Attwood HD, Price CG, Riddell RJ: Primary diffuse tracheobronchial amyloidosis. *Thorax* 27:620–624, 1972.

130. Hof DG, Rasp FL: Spontaneous regression of diffuse tracheobronchial amyloidosis. *Chest* 76:237–239, 1979.

131. Felix MA, Levy H, Feldman C, Abramowitz JA: Endobronchial appearance of tracheobronchial amyloidosis. *S Afr Med J* 75:241, 1989.

132. Holmes S, Desai JB, Sapsford RN: Nodular pulmonary amyloidosis: A case report and review of literature. *Br J Dis Chest* 82:414, 1988.

133. Weiss L: Isolated multiple nodular pulmonary amyloidosis. *Am J Clin Pathol* 33:318–329, 1960.

134. Cotton RE, Jackson JW: Localized amyloid "tumours" of the lung simulating malignant neoplasms. *Thorax* 19:97–103, 1964.

135. Hayes WT, Bernhardt H: Solitary amyloid mass of the lung: Report of a case with 6-year follow-up. *Cancer* 24:820–825, 1969.

136. Saab SB, Burke J, Hopeman A, Almond C: Primary pulmonary amyloidosis: Report of two cases. *J Thorac Cardiovasc Surg* 67:301–307, 1974.

137. Dyke PC, Demaray MJ, Delavan JW, Rasmussen RA: Pulmonary amyloidoma. *Am J Clin Pathol* 61:301–305, 1974.

138. Lee S-C, Johnson HA: Multiple nodular pulmonary amyloidosis: A case report and comparison with diffuse alveolar-septal pulmonary amyloidosis. *Thorax* 30:178–185, 1975.

139. Bignold LP, Martyn M, Basten A: Nodular pulmonary amyloidosis associated with benign hypergammaglobulinemic purpura. *Chest* 78:334–336, 1980.

140. Tamura K, Nakajima N, Makino S, Maruyama R, Kohno T, Koga Y: Primary pulmonary amyloidosis with multiple nodules. *Eur J Radiol* 8:128–130, 1988.

141. Schoen FJ, Alexander RW, Hood CI, Dunn LJ: Nodular pulmonary amyloidosis: Description of a case with ultrastructure. *Arch Pathol Lab Med* 104:66–69, 1980.

142. Eisenberg R, Sharma OP: Primary pulmonary amyloidosis. *Chest* 89:889–891, 1986.

143. Young WA: Bronchopulmonary amyloidosis—multiple tissue involvement and long follow-up. *Aust NZ J Med* 19:463, 1989.

144. Davis CJ, Butchart EG, Gibbs AR: Nodular pulmonary amyloidosis occurring in association with pulmonary lymphoma. *Thorax* 46:217, 1991.

145. Zundel WE, Prior AP: An amyloid lung. *Thorax* 26:357–363, 1971.

146. Eshun-Wilson K, Frandsen NE, Christensen HE: Pulmonary alveolar septal amyloidosis: A scanning and transmission electron microscopy study. *Virchows Arch [A]* 371:89–99, 1976.

147. Lewinsohn G, Bruderman I, Bohadana A: Primary diffuse pulmonary amyloidosis with monoclonal gammopathy. *Chest* 69:682–685, 1976.

148. Case Records of the Massachusetts General Hospital (Case 48-1977): *N Engl J Med* 297:1221–1228, 1977.

149. González-Cueto DM, Rigoli M, Gioseffi LM, Lancelle B, Martínez A: Diffuse pulmonary amyloidosis. *Am J Med* 48:668–670, 1970.

150. Poh SC, Tjia TS, Seah HC: Primary diffuse alveolar septal amyloidosis. *Thorax* 30:186–191, 1975.

151. Crosbie WA, Lewis ML, Ramsay ID, Doyle D: Pulmonary amyloidosis with impaired gas transfer. *Thorax* 27:625–630, 1972.

152. Kanada DJ, Sharma OP: Long-term survival with diffuse interstitial pulmonary amyloidosis. *Am J Med* 67:879–882, 1979.

153. Monreal FA: Pulmonary amyloidosis: Ultrastructural study of early alveolar septal deposits. *Hum Pathol* 15:338–339, 1984.

154. Beer TW, Edwards CW: Pulmonary nodules due to reactive systemic amyloidosis (AA) in Crohn's disease. *Thorax* 28:1287-1288, 1993.

155. Vanatta PR, Silva FG, Taylor WE, Costa JC: Renal cell carcinoma and systemic amyloidosis: Demonstration of AA protein and review of the literature. *Hum Pathol* 14:195–201, 1983.

156. Somer TP, Törnroth TS: Renal adenocarcinoma and systemic amyloidosis: Immunohistochemical and histochemical studies. *Arch Pathol Lab Med* 109:571–574, 1985.

157. Orriols R, Aliaga JL, Rodrigo MJ, Garcia F, Royo L, Morell F: Localized alveolar-septal amyloidosis with hypersensitivity pneumonitis. *Lancet* 339:1261–1262, 1991.

158. Smith RRL, Hutchins GM, Moore GW, Humphrey RL: Type and distribution of pulmonary parenchymal and vascular amyloid: Correlation with cardiac amyloidosis. *Am J Med* 66:96, 1979.

159. Kunze W-P: Senile pulmonary amyloidosis. *Pathol Res Pract* 164:413–422, 1979.

160. Pitkänen P, Westermark P, Cornwell GG III: Senile systemic amyloidosis. *Am J Pathol* 117:391–399, 1984.

161. Knapp MJ, Roggli VL, Kim J, Moore JO, Shelburne JD: Pleural amyloidosis. *Arch Pathol Lab Med* 112:57, 1988.

162. Kavuru MS, Adamo JP, Ahmad M, Mehta AC, Gephardt GN: Amyloidosis and pleural disease. *Chest* 98:20, 1990.

163. Kaw YT, Esparza AR: Solitary pleural amyloid nodules occurring as coin lesions diagnosed by fine-needle aspiration biopsy. *Diagn Cytopathol* 7:304, 1991.

164. Desai RA, Mahajan VK, Benjamin S, Van Ordstrand HS, Cordasco EM: Pulmonary amyloidoma and hilar adenopathy: Rare manifestations of primary amyloidosis. *Chest* 76:170, 1979.

165. Linder J, Vollmer RT, Croker BP, Shelburne J: Systemic kappa light-chain deposition: An ultrastructural and immunohistochemical study. *Am J Surg Pathol* 7:85–93, 1983.

166. Kijner CH, Yousem SA: Systemic light chain deposition disease presenting as multiple pulmonary nodules: A case report and review of the literature. *Am J Surg Pathol* 12:405–413, 1988.

167. Roberts WC, Fredrickson DS: Gaucher's disease of the lung causing severe pulmonary hypertension with associated acute recurrent pericarditis. *Circulation* 35:783–789, 1967.

168. Genereux GP: Lipids in the lungs: Radiologic-pathologic-correlation. *J Can Assoc Radiol* 21:1–15, 1970.

169. Kirkpatrick JA Jr, Capitanio MA: Pulmonary manifestations of systemic diseases in infants. *Semin Roentgenol* 7:149–172, 1972.

170. Wolson AH: Pulmonary findings in Gaucher's disease. *Am J Roentgenol* 123:712–715, 1975.

171. Lee RE, Peters SP, Glew RH: Gaucher's disease: Clinical, morphologic, and pathogenetic considerations. *Pathol Annu* 12:309–339, 1977.

172. Schneider EL, Epstein CJ, Kaback MJ, Brandes D: Severe pulmonary involvement in adult Gaucher's disease: Report of three cases and review of the literature. *Am J Med* 63:475–480, 1977.

173. Smith RRL, Hutchins GM, Sack GH Jr, Ridolfi RL: Unusual cardiac, renal and pulmonary involvement in Gaucher's disease. *Am J Med* 65:352–360, 1978.

174. Rahal F, McWilliams NB: Pulmonary Gaucher's disease diagnosed ante mortem. *Va Med* 108:186–187, 1981.

175. Links TP, Karrenbeld A, Steensma JT, Weits J, van der Jagt EJ, Postmus PE: Fatal respiratory failure caused by pulmonary infiltration by pseudo-Gaucher cells. *Chest* 101:265–266, 1992.

176. Crocker AC, Farber S: Niemann-Pick disease: A review of eighteen patients. *Medicine (Baltimore)* 37:1–95, 1958.

177. Skikne MI, Prinsloo I, Webster I: Electron microscopy of lung in Niemann-Pick disease. *J Pathol* 106:119–122, 1972.

178. Lachman R, Crocker A, Schulman J, Strand R: Radiological findings in Niemann-Pick disease. *Radiology* 108:659–664, 1973.

179. Long RG, Lake BD, Pettit JE, Scheuer PJ, Sherlock S: Adult Niemann-Pick disease: Its relationship to the syndrome of the sea-blue histiocyte. *Am J Med* 62:627–635, 1977.

180. Davies BH, Tuddenham EGD: Familial pulmonary fibrosis associated with oculocutaneous albinism and platelet function defect. *Q J Med* 45:219–232, 1976.

181. Garay SM, Gardella JE, Fazzini EP, Goldring RM: Hermansky-Pudlak syndrome: Pulmonary manifestations of a ceroid storage disease. *Am J Med* 66:737–747, 1979.

182. Hoste P, Williams J, Devriendt J, Lamont H, van der Straeten M: Familial diffuse interstitial pulmonary fibrosis associated with oculocutaneous albinism: Report of two cases with a family study. *Scand J Respir Dis* 60:128–134, 1979.

183. Schinella RA, Greco MA, Cobert BL, Denmark LW, Cox RP: Hermansky-Pudlak syndrome with granulomatous colitis. *Ann Intern Med* 92:20–23, 1980.

184. White DA, Walker Smith GJ, Cooper JAD Jr, Glickstein M, Rankin JA: Hermansky-Pudlak syndrome and interstitial lung disease: Report of a case with lavage findings. *Am Rev Respir Dis* 130:138–141, 1984.

185. Wood JR, Bellamy D, Child AH, Citron KM: Pulmonary disease in patients with Marfan syndrome. *Thorax* 39:780–784, 1984.

186. Sharma BK, Talukdar B, Kapoor R: Cystic lung in Marfan's syndrome. *Thorax* 44:978–979, 1989.

187. Corrin B, Simpson CGB, Fisher C: Fibrous pseudotumours and cyst formation in the lungs in Ehlers-Danlos syndrome. *Histopathology* 17:478, 1990.

188. Brower AC, Worsham GF, Dudley AH: Erdheim-Chester disease: A distinct lipoidosis or part of the spectrum of histiocytosis? *Radiology* 151:35, 1984.

189. Sherman JL, Citrin C, Johns T, Black J: Erdheim-Chester disease: Computed tomography in two cases. *AJNR* 6:444, 1985.

190. Evans S, Williams F: Case report: Erdheim-Chester disease: Polyostotic sclerosing histiocytosis. *Clin Radiol* 37:93, 1986.

191. Miller RL, Sheeler LR, Bauer TW, Bukowski RM: Erdheim-Chester disease. *Am J Med* 80:1230, 1986.

192. Lantz B, Lange TA, Heiner J, Herring GF: Erdheim-Chester disease. *J Bone Joint Surg* 71:456, 1989.

193. Fink MG, Levinson DJ, Brown NL, Sreekanth S, Sobel GW: Erdheim-Chester disease. *Arch Pathol Lab Med* 115:619, 1991.

194. Lawler W: Idiopathic cholesterol pneumonitis. *Histopathology* 1:385, 1977.

195. Bartimmo EE Jr, Guisan M, Moser KM: Pulmonary involvement in Fabry's disease: A reappraisal follow-up of a San Diego kindred and review of the literature. *Am J Med* 53:755–764, 1972.

196. Rosenberg DM, Ferrans VJ, Fulmer JD, et al: Chronic airflow obstruction in Fabry's disease. *Am J Med* 68:898–905, 1980.

197. Matsumoto T, Matsumori H, Taki T, Takagi T, Fukuda Y: Infantile GM_1-gangliosiodosis with marked manifestation of lungs. *Acta Pathol Jpn* 29:269–276, 1979.

198. Mason AMS, McIllmurray MB, Golding PL, Hughes DTD: Fibrosing alveolitis associated with renal tubular acidosis. *Br Med J* 4:596–599, 1970.

199. Zalin AM, Weeple J, Gumpel M: Fibrosing alveolitis and renal tubular acidosis (letter). *Br Med J* 4:804, 1970.

200. Clarke JTR, Ozere RL, Krause VW: Early infantile variant of Krabbe globoid cell leucodystrophy with lung involvement. *Arch Dis Child* 56:640–642, 1981.

201. Lightman NI, Schooley RT: Adult-onset acid maltase deficiency: Case report of an adult with severe respiratory difficulty. *Chest* 72:250–252, 1977.

202. Pääkkö P, Ryhänen L, Rantala H, Autio-Harmainen H: Pulmonary emphysema in a nonsmoking patient with Salla disease. *Am Rev Respir Dis* 135:979, 1987.

203. Cagle PT, Ferry GD, Beaudet AL, Hawkins EP: Clinicopathologic conference: Pulmonary hypertension in an 18-year-old girl with cholesteryl ester storage disease (CESD). *Am J Med Genet* 24:711, 1986.

204. Rutsaert J, Tondeur M, Vamos-Hurwitz E, Dustin P: The cellular lesions of Farber's disease and their experimental reproduction in tissue culture. *Lab Invest* 36:474–480, 1977.

205. Sharp ME, Danino EA: An unusual form of pulmonary calcification: "Microlithiasis alveolaris pulmonum." *J Pathol* 65:389–399, 1953.

206. Sosman MC, Dodd GD, Jones WD, Pillmore GU: The familial occurrence of pulmonary alveolar microlithiasis. *Am J Roentgenol* 77:947–1012, 1957.

207. Chinachoti N, Tangchai P: Case report section: Pulmonary alveolar microlithiasis associated with the inhalation of snuff in Thailand. *Dis Chest* 32:687–689, 1957.

208. Thomson WB: Pulmonary alveolar microlithiasis. *Thorax* 14:76–81, 1959.

209. Lebacq E, Lauweryns J, Billiet L: Pulmonary alveolar microlithiasis: Case report with lung function studies. *Br J Dis Chest* 58:31–35, 1964.

210. Caffrey PR, Altman RS: Pulmonary alveolar microlithiasis occurring in premature twins. *J Pediatr* 66:758–763, 1965.

211. O'Neill RP, Cohn JE, Pellegrino ED: Pulmonary alveolar microlithiasis—a family study. *Ann Intern Med* 67:957–967, 1967.

212. Fuleihan FJD, Abboud RT, Balikian JP, Nucho CKN: Pulmonary alveolar microlithiasis: Lung function in five cases. *Thorax* 24:84–90, 1969.

213. Coetzee T: Pulmonary alveolar microlithiasis with involvement of the sympathetic nervous system and gonads. *Thorax* 25:637–642, 1970.

214. Sears MR, Chang AR, Taylor AJ: Pulmonary alveolar microlithiasis. *Thorax* 26:704–711, 1971.

215. Ghavamian M: Pulmonary alveolar microlithiasis: A report of two cases. *Nebr Med J* 57:259–264, 1972.

216. Kino T, Kohara Y, Tsuji S: Pulmonary alveolar microlithiasis: A report of two young sisters. *Am Rev Respir Dis* 105:105–110, 1972.

217. Hossein E: Pulmonary alveolar microlithiasis. *Mich Med* 72:691–694, 1973.

218. Thind GS, Bhatia JL: Pulmonary alveolar microlithiasis. *Br J Dis Chest* 72:151–154, 1978.

219. Saputo W, Zocchi M, Mancosu M, Bonaldi U, Croce P: Pulmonary alveolar microlithiasis: A case report with a discussion of differential diagnosis. *Helv Pediatr Acta* 34:245–255, 1979.

220. Prakash UBS, Barham SS, Rosenow EC III, Brown ML, Payne WS: Pulmonary alveolar microlithiasis: A review including ultrastructural and pulmonary function studies. *Mayo Clin Proc* 58:290–300, 1983.

221. Kiatboonsri S, Charoenpan P, Vathesatogkit P, Boonpucknavig V: Pulmonary alveolar microlithiasis: Report of five cases and literature review. *J Med Assoc Thai* 68:672–677, 1985.

222. Chalmers AG, Wyatt J, Robinson PJ: Computed tomographic and pathological findings in pulmonary alveolar microlithiasis. *Br J Radiol* 59:408–411, 1986.

223. Hawass ND, Noah MS: Pulmonary alveolar microlithiasis. *Eur J Respir Dis* 69:199–203, 1986.

224. Barnard NJ, Crocker PR, Blainey AD, Davies RJ, Ell SR, Levison DA: Pulmonary alveolar microlithiasis: A new analytical approach. *Histopathology* 11:639–645, 1987.

225. Volle E, Kaufmann HJ: Pulmonary alveolar microlithiasis in pediatric patients: Review of the world literature and two new observations. *Pediatr Radiol* 17:439–442, 1987.

226. Pant K, Shah A, Mathur RK, Chhabra SK, Jain SK: Pulmonary alveolar microlithiasis with pleural calcification and nephrolithiasis. *Chest* 98:245–246, 1990.

227. Emri S, Cöplu L, Selçuk ZT, Sahin AA, Baris YI: Hypertrophic pulmonary osteoarthropathy in a patient with pulmonary alveolar microlithiasis. *Thorax* 46:145–146, 1991.

228. Ucan ES, Keyf AI, Aydilek R, et al. Pulmonary alveolar microlithiasis: Review of Turkish reports. *Thorax* 48:171, 1993.

229. Harbitz F: Extensive calcification of the lungs as a distinct disease. *Arch Intern Med* 21:139, 1918.

230. Puhr L: Mikrolithiasis alveolaris pulmonum. *Virch Arch Path Anat* 290:156, 1933.

231. Volle E, Kaufmann: Pulmonary alveolar microlithiasis in pediatric patients—review of the world literature and two new observations. *Pediatr Radiol* 17:439, 1987.

232. Winzelberg GG, Boller M, Sachs M, Weinberg J: CT evaluation of pulmonary alveolar microlithiasis. *J Comput Assist Tomogr* 8:1029, 1984.

233. Göcmen A, Toppare MF, Kiper N, Büyükpamukcu N: Treatment of pulmonary alveolar microlithiasis with a diphosphonate—preliminary results of a case. *Respiration* 59:250, 1992.

234. Srivatsa SS, Burger CD, Douglas WW: Upper lobe pulmonary parenchymal calcification in a patient with AIDS and *Pneumocystis carinii* pneumonia receiving aerosolized pentamidine. *Chest* 101:266, 1992.

235. Mulligan RM: Metastatic calcification. *Arch Pathol* 43:177–230, 1947.

236. Kaltreider HB, Baum GL, Bogaty G, McCoy MD, Tucker M: So-called "metastatic" calcification of the lung. *Am J Med* 46:188–196, 1969.

237. Mootz JR, Sagel SS, Roberts TH: Roentgenographic manifestations of pulmonary calcifications. *Radiology* 107:55–60, 1973.

238. Neff M, Yalcin S, Gupta S, Berger H: Extensive metastatic calcification of the lung in an azotemic patient. *Am J Med* 56:103–109, 1974.

239. Conger JD, Hammond WS, Alfrey AC, Contiguglia SR, Stanford RE, Huffer WE: Pulmonary calcification in chronic dialysis patients: Clinical and pathologic studies. *Ann Intern Med* 83:330–336, 1975.

240. Cohen AM, Maxon HR, Goldsmith RE, et al: Metastatic pulmonary calcification in primary hyperparathyroidism. *Arch Intern Med* 137:520–522, 1977.

241. Firooznia H, Pudlowski R, Golimbu C, Rafii M, McCauley D: Diffuse interstitial calcification of the lungs in chronic renal failure mimicking pulmonary edema. *Am J Roentgenol* 129:1103–1104, 1977.

242. Heath D, Robertson AJ: Pulmonary calcinosis. *Thorax* 32:606–611, 1977.

243. Justrabo E, Genin R, Rifle G: Pulmonary metastatic calcification with respiratory insufficiency in patients on maintenance haemodialysis. *Thorax* 34:384–388, 1979.

244. Margolin RJ, Addison TE: Hypercalcemia and rapidly progressive respiratory failure. *Chest* 86:767–769, 1984.

245. Bestetti-Bosisio M, Cotelli F, Schiaffino E, Sorgato G, Schmid C: Lung calcification in long-term dialysed patients: A light and electronmicroscopic study. *Histopathology* 8:69–79, 1984.

246. Sinniah D, Landing BH, Siegel SE, Laug WE, Gwinn JL: Pulmonary alveolar septal calcinosis causing progressive respiratory failure in acute lymphoblastic leukemia in childhood. *Pediatr Pathol* 6:439–448, 1986.

247. Breitz HB, Sirotta PS, Nelp WB, Ott S, Figley MM: Progressive pulmonary calcification complicating successful renal transplantation. *Am Rev Respir Dis* 136:1480–1482, 1987.

248. van der Bij W, Gouw ASH, Meinesz AF, van Ingen J, Postmus PE: Consolidation of both upper lobes. *Chest* 100:1685–1686, 1991.

249. Northcutt AD, Tio FO, Chamblin SA Jr, Britton HA: Massive metastatic pulmonary calcification in an infant with aleukemic monocytic leukemia. *Pediatr Pathol* 4:219, 1985.

250. Drachman R, Baillet G, Gagnadoux M-F, de Vernejoul P, Broyer M: Pulmonary calcifications in children on dialysis. *Nephron* 44:46, 1986.

251. Milliner DS, Lieberman E, Landing BH: Pulmonary calcinosis after renal transplantation in pediatric patients. *Am J Kidney Dis* 7:495, 1986.

252. Sinniah D, Landing BH, Siegel SE, Laug WE, Gwinn JL: Pulmonary alveolar septal calcinosis causing progressive respiratory failure in acute lymphoblastic leukemia in childhood. *Pediatr Pathol* 6:439, 1986.

253. Khafif RA, Delima C, Silverberg A, Frankel R, Groopman J: Acute hyperparathyroidism with systemic calcinosis. *Arch Intern Med* 149:681, 1989.

254. Kuhlman JE, Ren H, Hutchins GM, Fishman EK: Fulminant pulmonary calcification complicating renal transplantation: CT demonstration. *Radiology* 173:459, 1989.

255. Poe RH, Kamath C, Bauer MA, Qazi R, Kallay MC, Woll JE: Acute respiratory distress syndrome with pulmonary calcification in two patients with B cell malignancies. *Respiration* 56:127, 1989.

256. Senba M, Kawai K: Metastatic calcification due to hypercal-

cemia in adult T-cell leukemia-lymphoma (ATLL). *Zentralbl Pathol* 137:341, 1991.

257. Hartman TE, Muller NL, Primack SL, et al: Metastatic pulmonary calcification in patients with hyercalcemia: Findings on chest radiographs and CT scans. *AJR* 162:799-802, 1994.

258. Gardiner IT, Uff JS: "Blue bodies" in a case of cryptogenic fibrosing alveolitis (desquamative type)—an ultra-structural study. *Thorax* 33:806–813, 1978.

259. Koss MN, Johnson FB, Hochholzer L: Pulmonary blue bodies. *Hum Pathol* 12:258–266, 1981.

260. Kung ITM, Hsu C, Chan SCW, Leung BSY, Ng DWH: Frequency of "blue bodies" in pulmonary cytology specimens. *Diagn Cytopathol* 3:284–286, 1987.

261. Michaels L, Levene C: Pulmonary corpora amylacea. *J Pathol* 74:49, 1957.

262. Baar HS, Ferguson FF: Microlithiasis alveolaris pulmonum: Association with diffuse interstitial pulmonary fibrosis. *Arch Pathol* 76:659–666, 1963.

263. Hollander DH, Hutchins GM: Central spherules in pulmonary corpora amylacea. *Arch Pathol Lab Med* 102:629, 1978.

264. Popelka CG, Kleinerman J: Diffuse pulmonary ossification. *Arch Intern Med* 137:523–525, 1977.

265. Jacobs AN, Neitzschman HR, Nice CM: Metaplastic bone formation in the lung. *Am J Roentgenol* 118:344–346, 1973.

266. Müller K-M, Friemann J, Stichnoth E: Dendriform pulmonary ossification. *Pathol Res Pract* 168:163–172, 1980.

267. Ndimbie OK, Williams CR, Lee MW: Dendriform pulmonary ossification. *Arch Pathol Lab Med* 111:1062–1064, 1987.

268. Joines RW, Roggli VL: Dendriform pulmonary ossification: Report of two cases with unique findings. *Am J Clin Pathol* 91:398, 1989.

269. Chow LTC, Shum BSF, Chow WH, Tso CB: Diffuse pulmonary ossification—a rare complication of tuberculosis. *Histopathology* 20:435–437, 1992.

270. Felson B, Schwarz J, Lukin RR, Hawkins HH: Idiopathic pulmonary ossification. *Radiology* 153:303, 1984.

271. Gortenuti G, Portuese A: Disseminated pulmonary ossification. *Eur J Radiol* 5:14, 1989.

272. Ndimbie OK, Lee MW: Dendriform pulmonary ossification. *Am J Clin Pathol* 90:497, 1988.

273. Chow LTC, Shum BSF, Chow WH, Tso CB: Diffuse pulmonary ossification—a rare complication of tuberculosis. *Histopathology* 20:435–437, 1992.

274. Gevenois PA, Abehsera M, Knoop C, Jacobovitz D, Estenne M: Disseminated pulmonary ossification in end-stage pulmonary fibrosis: CT demonstration. *AJR* 162:1303, 1994.

275. Elkeles A, Glynn LE: Disseminated parenchymatous ossification in the lungs in association with mitral stenosis. *J Pathol* 58:517–522, 1946.

276. Daugavietis HE, Mautner LS: Disseminated nodular pulmonary ossification with mitral stenosis. *Arch Pathol* 63:7–12, 1957.

277. Buja LM, Roberts WC: Pulmonary parenchymal ossific nodules in idiopathic hypertrophic subaortic stenosis. *Am J Cardiol* 25:710–715, 1970.

278. Green JD, Harle TS, Greenberg SD, Weg JG, Nevin H, Jenkins DE: Disseminated pulmonary ossification: A case report with demonstration of electron-microscopic features. *Am Rev Respir Dis* 101:293–298, 1970.

279. Legge DA, Miller WE, Ludwig J: Pulmonary findings associated with mitral stenosis. *Chest* 58:403, 1970.

280. Kayser K, Stute H, Tuengerthal S: Diffuse pulmonary ossification associated with metastatic melanoma of the lung. *Respiration* 52:221–227, 1987.

281. Ashley DJB: Bony metaplasia in trachea and bronchi. *J Pathol* 102:186–188, 1970.

282. Härmä RA, Suurkari S: Tracheopathia chondro-osteoplastica: A clinical study of thirty cases. *Acta Otolaryngol* 84:118, 1977.

283. Young RH, Sandstrom RE, Mark GJ: Tracheopathia osteoplastica. *J Thorac Cardiovasc Surg* 79:537, 1980.

284. Pounder DJ, Pieterse AS: Tracheopathia osteoplastica: A report of four cases. *Pathology* 14:429, 1982.

285. Hirsch M, Goldstein J, Tovi F, Gerzof SG: Diagnosis of tracheopathia osteoplastica by computed tomography. *Ann Otol Rhinol Laryngol* 94:217, 1985.

286. Paaske PB, Taug E: Tracheopathia osteoplastica in the larynx. *J Laryngol Otol* 99:305, 1985.

287. Hodges MK, Israel E: Tracheobronchopathia osteochondroplastica presenting as right middle lobe collapse: Diagnosis by bronchoscopy and computerized tomography. *Chest* 94:842, 1988.

288. Tukiainen H, Torkko M, Terho EO: Lung function in patients with tracheobronchopathia osteochondroplastica. *Eur Respir J* 1:632–635, 1988.

289. Nienhuis DM, Prakash UBS, Edell ES: Tracheobronchopathia osteochondroplastica. *Ann Otol Rhinol Laryngol* 99:689–694, 1990.

290. Case Records of the Massachusetts General Hospital. *N Engl J Med* 327:1512, 1992.

Chapter **26**

The Lung in Connective Tissue Disorders

William D. Travis/Michael N. Koss/
Victor J. Ferrans

INTRODUCTION

Understanding the lung pathology of connective tissue disorders (CTDs) is difficult for several reasons. There are many CTDs (Table 26-1), each of which has its own spectrum of pulmonary and systemic manifestations.[1-6] However, the pulmonary and systemic manifestations of the various CTDs show considerable overlap. There is even a form of mixed CTD. In addition, the pulmonary pathologic manifestations in CTD are etiologically nonspecific. While characteristic pathologic patterns occur in some of the CTDs, one cannot diagnose a specific CTD on the basis of the pulmonary pathologic features.

The most important problems in analyzing the literature on the pulmonary manifestations of CTD are the lack of good pathologic documentation in many studies and the lack of application of current concepts in classification of the lung pathology. Another complicating feature of interpreting lung biopsies from patients is that often more than one lesion is present. In addition, many of the pulmonary lesions of CTD are clinicopathologic entities that cannot be diagnosed from the biopsy alone but require careful correlation with clinical and radiographic data. Furthermore, the immunosuppressive medications used to treat the patients can induce pulmonary toxic effects and predispose to opportunistic infections. Both drug toxicity and infections can cause some of the same histologic patterns of lung injury as those caused by the underlying CTD. For these reasons, it can be difficult and sometimes impossible to determine whether a given lung process is due to the underlying CTD, drug injury, or infection.

Awareness of the various clinical and pathologic pulmonary manifestations of CTD and their potential associated complications is important in assessing lung biopsies from patients. We will begin by focusing on the pathologic aspects of the major pulmonary lesions that occur in CTD and then review the features that are distinctive for each of the particular disorders.

CLINICOPATHOLOGIC LESIONS

Many of the pulmonary pathologic manifestations of the CTD, as listed in Table 26-2, especially the interstitial lesions, represent *clinicopathologic* entities. Therefore, the pathologic findings must be carefully correlated with the clinical, radiographic, and laboratory features before the diagnosis can be established. Thus, the pattern of lung injury seen in entities such as usual interstitial pneumonia and bronchiolitis obliterans organizing pneumonia occur not only in the idiopathic disease of the same names but they can also occur secondarily in other diseases such as CTD. For this reason, it is essential to interpret lung biopsies from patients with CTD not in a vacuum but in the context of all available clinical information.

TABLE 26-1

Connective Tissue Disorders: Clinical Syndromes

Rheumatoid arthritis
Systemic lupus erythematosus
Scleroderma
Polymyositis/dermatomyositis
Sjögren's syndrome
Mixed connective tissue disease
Ankylosing spondylitis
Relapsing polychrondritis
Primary biliary cirrhosis
Marfan's syndrome

Any of the components of the lower respiratory tract can be affected by CTD—the pleura, interstitium, alveoli (trachea, bronchi, bronchioles) and vessels (Table 26-2).[1]

Pleuritis

Pleuritis is the most common pleural lesion encountered in CTD. It can take the form of acute or chronic pleuritis.[1,2] Clinically, pleuritis can manifest as pleural effusion, pleuritic pain, or asymptomatic pleural thickening detected only by radiographic

TABLE 26-2
Pulmonary Pathologic Lesions in Connective Tissue Disorders

Pleura
  Pleuritis
  Fibrosis
Interstitium/parenchyma
  Diffuse alveolar damage
  Usual interstitial pneumonia
  Bronchiolitis obliterans–organizing pneumonia
  Lymphocytic interstitial pneumonia
  Granulomatous interstitial pneumonia
  Upper lobe fibrocystic changes
  Rheumatoid nodules
  Lymphoid hyperplasia
  Amyloid
  Aspiration pneumonia
Alveoli
  Diffuse pulmonary hemorrhage
  Eosinophilic pneumonia
  Alveolar proteinosis
Airways
  Chronic bronchiolitis
  Constrictive bronchiolitis
  Bronchiectasis
Vessels
  Vasculitis
  Pulmonary hypertension
Neoplasms
  Lung cancer
  Lymphoma
Indirect respiratory effects
  Respiratory muscle
  Renal
  Pericardial
  CNS

studies or pathologic examination. Both the visceral and the parietal pleura can be affected. Acute pleuritis may consist of prominent acute inflammation of the pleura, or it may be fibrinous. Fibrinous pleuritis consists of a layer of fibrin across the pleural surface

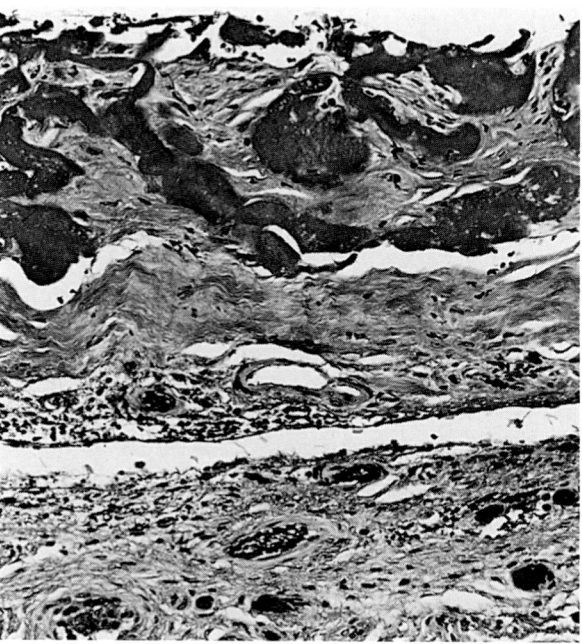

A

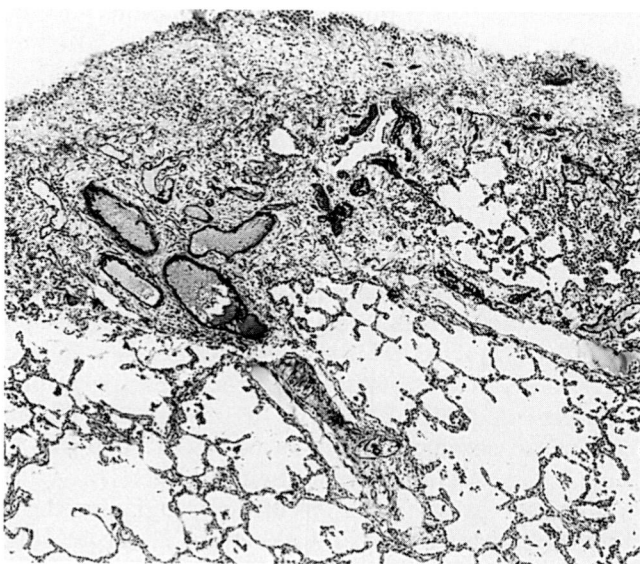

B

FIGURE 26-1

A. Acute fibrinous pleuritis. The pleural surface is covered with a layer of eosinophilic fibrin admixed with organizing fibrous connective tissue. (H&E, ×174.)   *B.* Chronic pleuritis. The pleura is thickened and densely infiltrated with chronic inflammatory cells. (H&E, ×43.)

(Fig. 26-1*A*). At the interface with the underlying pleura there is often a mixture of fibrin with proliferating mesothelial cells, fibroblasts, and granulation tissue. In chronic pleuritis, the pleura is infiltrated with varying numbers of lymphocytes and plasma cells (Fig. 26-1*B*). Pleural fibrosis may accompany either acute or chronic pleuritis. Infection must be included in the differential diagnosis of pleuritis, particularly acute pleuritis. In rare cases, an intense chronic pleuritis may raise the differential diagnosis of pleural infiltration by a low grade lymphoma (Fig. 26-2).

Interstitial Lesions

Virtually any interstitial lung disease can occur in CTD.[1–3,6] In many cases, more than one pulmonary lesion is present.[7] Therefore, pathologic analysis of lung biopsies can result in identification of primary and secondary lesions.[7]

Diffuse alveolar damage (DAD) is the pathologic counterpart to the adult respiratory distress syndrome (ARDS).[8] In patients with systemic lupus erythematosus (SLE), the term *acute lupus pneumonitis* has been used for this clinical syndrome,[9] but it can occur in most of the CTDs. DAD progresses through acute and chronic phases.[10] The acute phase is characterized by interstitial and alveolar edema, intra-alveolar hemorrhage, hyaline membranes, and acute inflammation (Fig. 26-3). As it progresses to the chronic phase, the hyaline membranes, edema, and hemorrhage resolve, and one begins to see type II

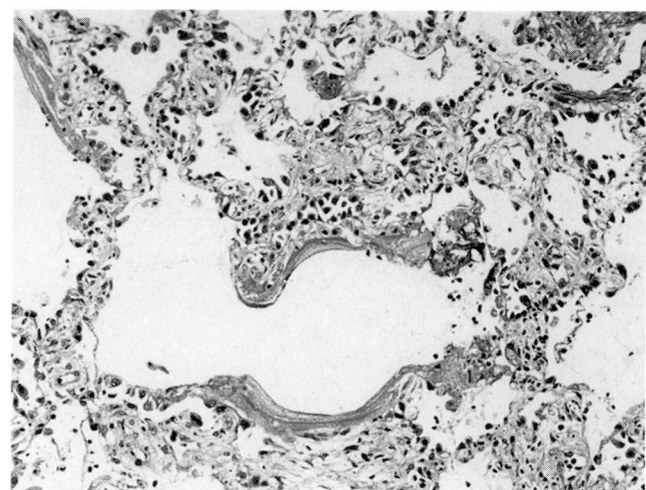

FIGURE 26-3
Diffuse alveolar damage in SLE. The alveolar walls are thickened by edema, loose connective tissue proliferation, extensive type II pneumocyte proliferation, and hyaline membranes. (H&E, ×60.)

pneumocyte proliferation and interstitial myofibroblastic proliferation. This can progress to a severely fibrotic process with honeycomb changes, or it may completely resolve.

Clinically, patients with *usual interstitial pneumonia* (UIP) have an insidious onset of shortness of breath, a radiographic picture of diffuse, bilateral pulmonary infiltrates, and restrictive pulmonary function.[11–14] Grossly, the lung in UIP shows patchy fibrosis that is often subpleural (Fig. 26-4). The typical histologic finding in UIP is patchy chronic interstitial fibrosis with normal lung adjacent to zones of fibrosis (Fig. 26-5). The fibrotic areas consist of dense

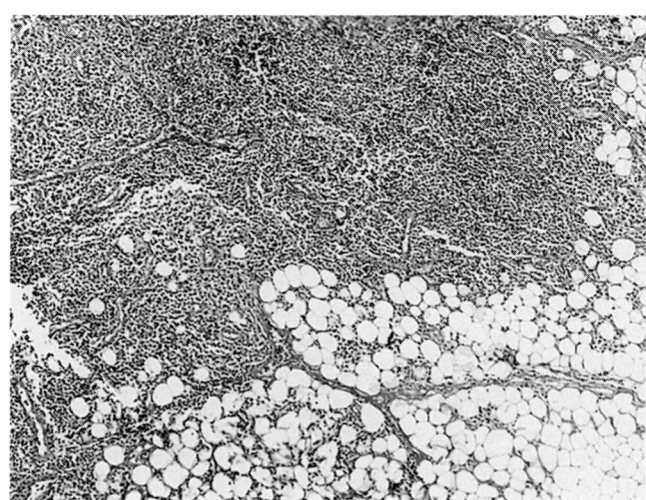

FIGURE 26-2
Marked chronic pleuritis in rheumatoid arthritis. The dense infiltration of the parietal pleural fat by chronic inflammation raises the question of a low grade lymphoma. (H&E, ×40.)

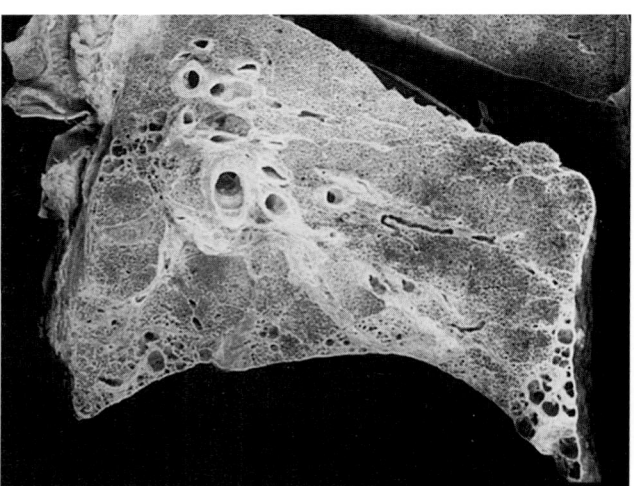

FIGURE 26-4
Gross features of usual interstitial pneumonia in scleroderma. The interstitial scarring in this lower lobe is mostly subpleural and shows prominent honeycomb cystic changes.

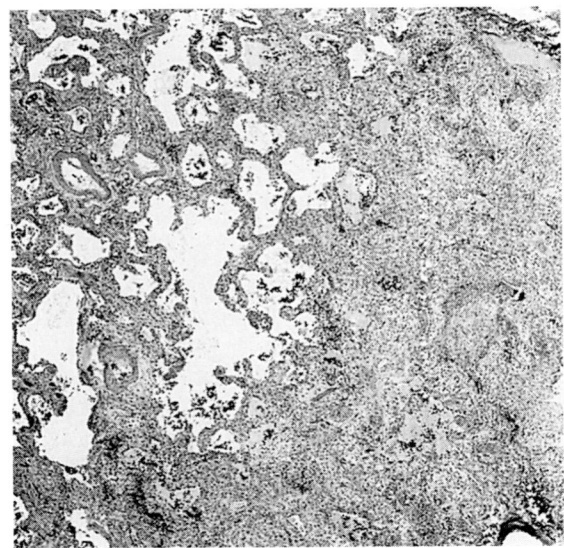

A

FIGURE 26-5

A. Usual interstitial pneumonia in scleroderma. The interstitial fibrosis is patchy with areas of dense scarring adjacent to areas

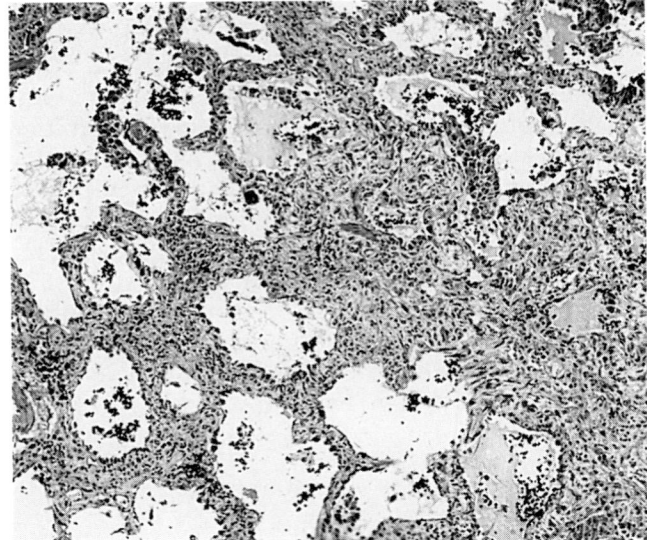

B

with relatively preserved alveolar architecture. (H&E, ×45.)
B. The interstitium is thickened by dense fibrosis accompanied by a mild chronic inflammatory infiltrate. (H&E, ×67.)

collagen causing remodeling of lung parenchyma. This remodeling often results in cystic or honeycomb change. Fibroblastic foci of young connective tissue may be seen at the edges of areas of dense fibrosis. There is a tendency for these fibroblastic foci to be prominent in the UIP of patients with CTD. These foci may be a clue to the presence of an underlying systemic disorder in cases in which pulmonary involvement is the initial manifestation.

Bronchiolitis obliterans–organizing pneumonia (BOOP) is a clinicopathologic entity character-

ized by a 4- to-6-week history of cough, shortness of breath, and bilateral nodular pulmonary infiltrates.[15–24] This disease is often preceded by a flu-like illness. BOOP is distinguished from UIP in that it is a steroid-responsive form of interstitial lung disease.[15–18,23,25–27] Histologically, lung biopsies show a patchy process with normal lung adjacent to areas of consolidation. It is characterized by evenly spaced plugs of loose, organizing intraluminal fibrosis protruding into distal airways, including bronchioles, alveolar ducts, and alveoli (Fig. 26-6). The

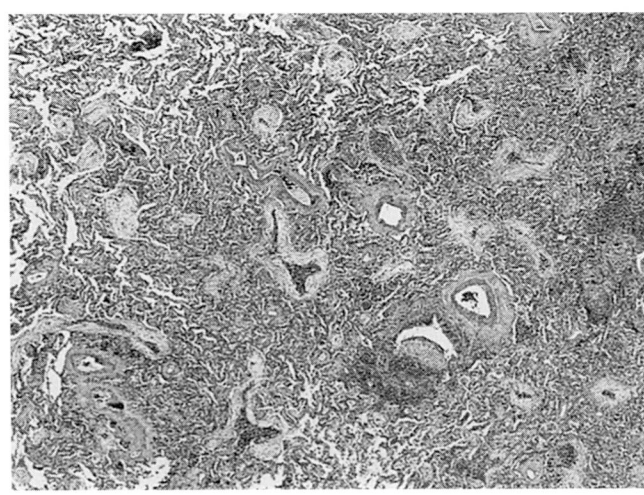

A

FIGURE 26-6

A. Bronchiolitis obliterans–organizing pneumonia in rheumatoid arthritis. This patchy nodular area was surrounded by relatively normal lung parenchyma. It is characterized by nodular intraluminal plugs of loose organizing fibrosis evenly spaced

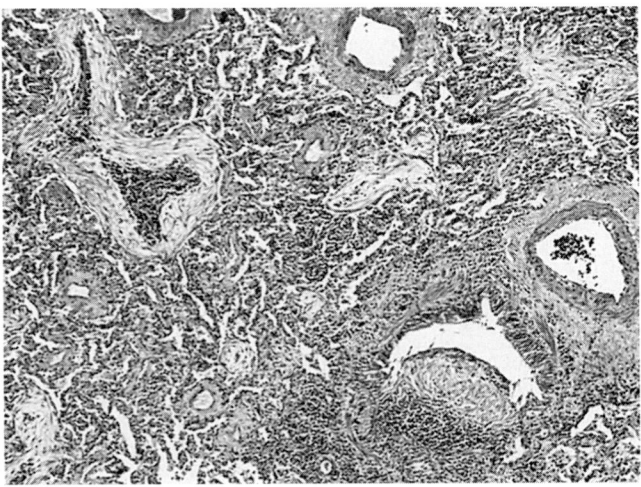

B

within distal airways. The architecture of the lung is preserved. (H&E, ×24.) *B.* The scattered plugs of intraluminal loose fibrosis are seen within alveoli, alveolar ducts, and bronchioles. There is mild interstitial chronic inflammation. (H&E, ×60.)

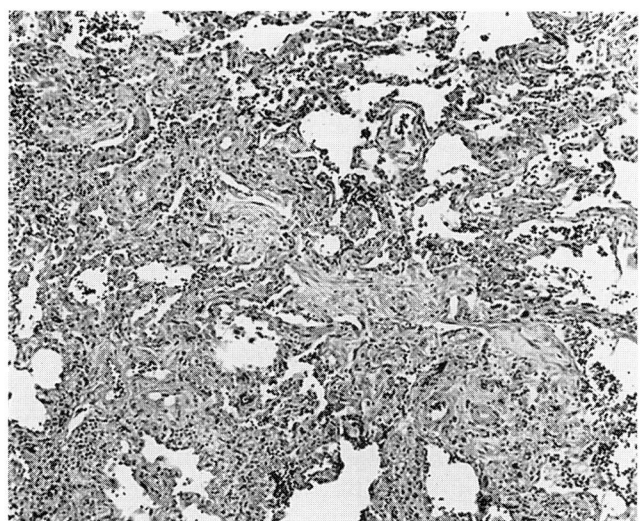

FIGURE 26-7
Focal intraluminal fibrosis. The few buds of intraluminal loose fibrosis were a clue that suggested an underlying collagen vascular disease in a biopsy that also showed fibrinous pleuritis, bronchiolitis, and focal granulomas (H&E, ×80.)

architecture of the lung is preserved, and there is no scarring or remodeling.

This pattern of lung injury is nonspecific and can be seen in other conditions, including infection and drug toxicity, and as a nonspecific reaction adjacent to such other lesions as abscesses, infarcts, granulomas, and neoplasms. It can also be easily confused with the organizing phase of DAD, in which the connective tissue may have the same morphologic features. In DAD, however, the connective tissue tends to affect the alveolar septa more than the air spaces, and one sees prominent type II pneumocyte

hyperplasia. The presence of features such as hyaline membranes can also be a helpful clue to the diagnosis. One should be cautious about suggesting the diagnosis of BOOP in the presence of abscesses, infarcts, granulomas, or neoplasms.[28]

Sometimes focal intraluminal fibrosis that falls short of BOOP is encountered in a lung biopsy that may have a constellation of other findings that suggest a collagen vascular disease (Fig. 26-7).

Lymphocytic interstitial pneumonia (LIP) usually presents with indolent onset of shortness of breath in a patient with diffuse bilateral pulmonary infiltrates.[29–31] Sjögren's syndrome is the setting in which LIP occurs most commonly,[29–32] but it can occur in SLE as well.[33,34] Histologically, LIP consists of extensive interstitial infiltrates of varying numbers of lymphocytes and plasma cells (Fig. 26-8). The differential diagnosis includes low-grade lymphocytic lymphoma,[35] infection (especially *Pneumocystis carinii*),[36,37] and human immunodeficiency virus (HIV) infection.[38]

Lymphoid hyperplasia is another potential manifestation of CTD. It is most often seen in rheumatoid arthritis.[7,39,40] Usually it consists of hyperplastic lymphoid follicles, which may be associated with bronchioles (Fig. 26-9). When the hyperplastic lymphoid follicles are situated adjacent to bronchioles, the term *follicular bronchiolitis* is sometimes used. It can occur in a pure form when it is the only cause of interstitial lung disease and the intervening lung parenchyma is normal. More commonly, however, it is seen as a secondary lesion, and other lesions, such

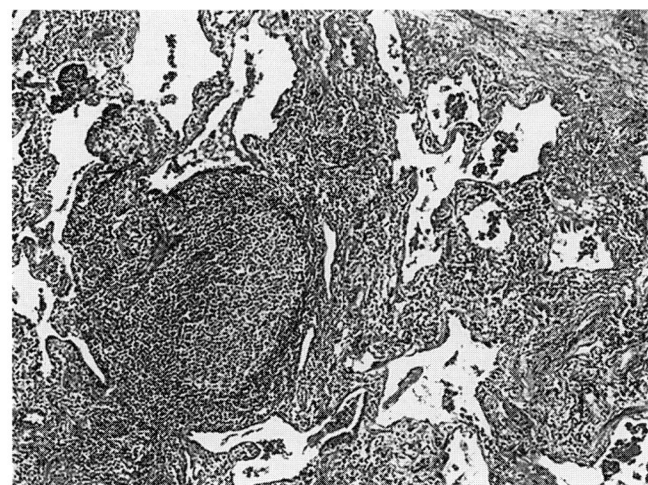

A

FIGURE 26-8
A. Lymphocytic interstitial pneumonia in SLE. This biopsy shows a hyperplastic lymphoid follicle in addition to a dense

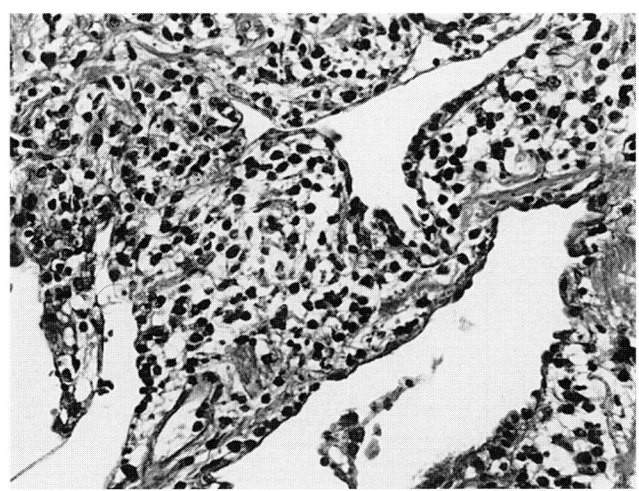

B

interstitial lymphocytic infiltrate in the adjacent alveolar septa. (H&E, ×60.) *B.* The inflammatory cells consist of lymphocytes and plasma cells. (H&E, ×240.)

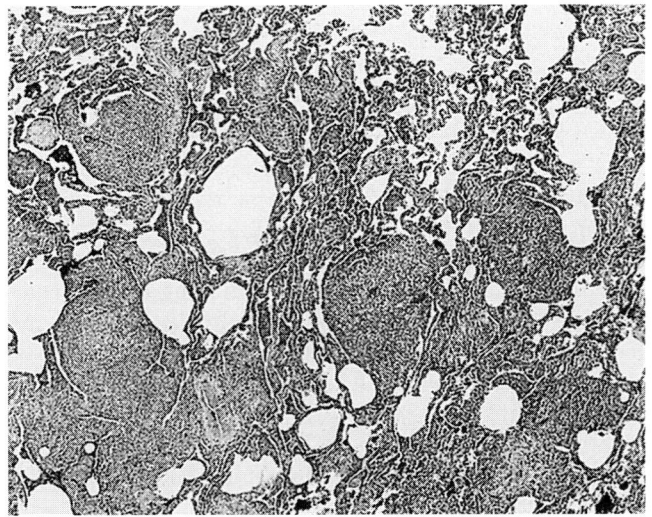

FIGURE 26-9
Lymphoid hyperplasia in SLE. In this specimen the primary finding was scattered hyperplastic lymphoid follicles, with relatively little lymphocytic infiltration of the intervening alveolar septa. (H&E, ×40.)

as BOOP and UIP, are the primary finding. Lymphoid hyperplasia can also present as a localized parenchymal nodule. Such cases are sometimes regarded as pseudolymphoma.

Cellular interstitial pneumonia (CIP) is a term used to describe lung specimens that show a nonspecific pattern of interstitial chronic inflammation with none of the classic patterns of interstitial lung disease.[7,41] In such cases, the differential diagnosis includes hypersensitivity pneumonitis (including drug toxicity), infection (especially *P. carinii*), and HIV infection.

Granulomatous interstitial pneumonia can take the form of a sarcoidlike process, a hypersensitivity pneumonitis picture, or randomly scattered giant cells or granulomas (Fig. 26-10A). Sarcoid-like granulomas consist of tightly packed collections of epithelioid cells or giant cells distributed along lymphatic routes. In pure sarcoid there is usually minimal associated interstitial chronic inflammation. In the setting of a CTD, however, inflammation may be prominent. Sarcoidal granulomas may also occasionally be found in enlarged hilar lymph nodes in CTD (Fig. 26-10B). In hypersensitivity pneumonitis there is a background of a mild to moderate chronic interstitial infiltrate associated with scattered poorly formed granulomas and buds of intraluminal fibrosis.[42] This picture can resemble that seen in extrinsic allergic alveolitis or drug-induced lung injury. The presence of granulomas should always raise the differential diagnosis of infection and prompt either cultures or careful examination of special stains for organisms.

Upper lobe fibrocystic changes are known to occur in rheumatoid arthritis[1,43–45] and ankylosing spondylitis.[46–49] These may become complicated by infection with tuberculosis or aspergillosis.

Rheumatoid nodules consist of necrotizing granulomas, which are frequently found in a subpleural location (Fig. 26-11A).[1,7] They may range from several millimeters to over 7 cm in diameter.[1,2] They may undergo cavitation, hemorrhage, infection, and rupture to form bronchopleural fistulas. In Caplan's

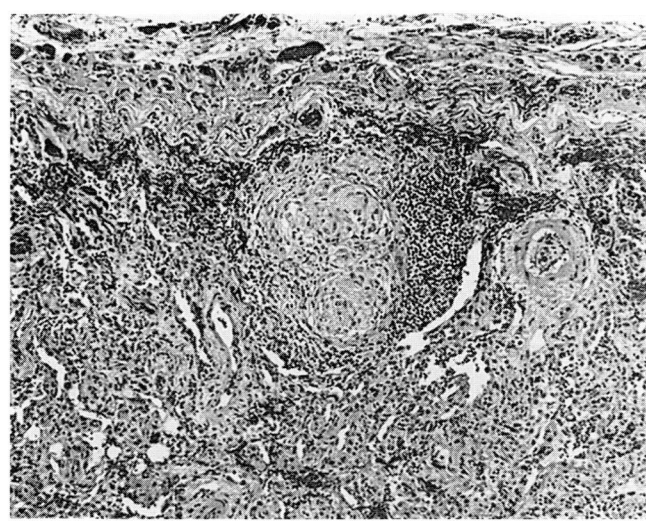

A

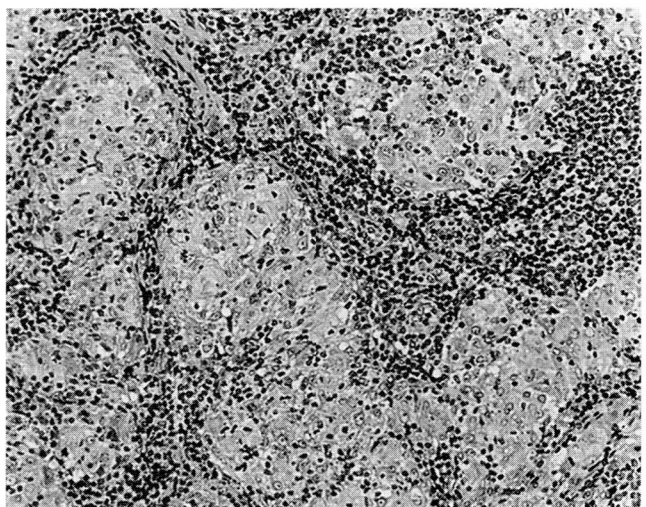

B

FIGURE 26-10
A. Granulomatous inflammation. This granuloma consists of a cluster of epithelioid and giant cells. It was a focal finding in a biopsy that showed other features suggestive of collagen vascular disease, including fibrinous pleuritis, bronchiolitis, and focal intraluminal, budding, fibrosis. (H&E, ×80.) *B.* In the same case a biopsy from an enlarged lymph node showed multiple epithelioid granulomas, giving a pattern reminiscent of sarcoid. (H&E, ×160.)

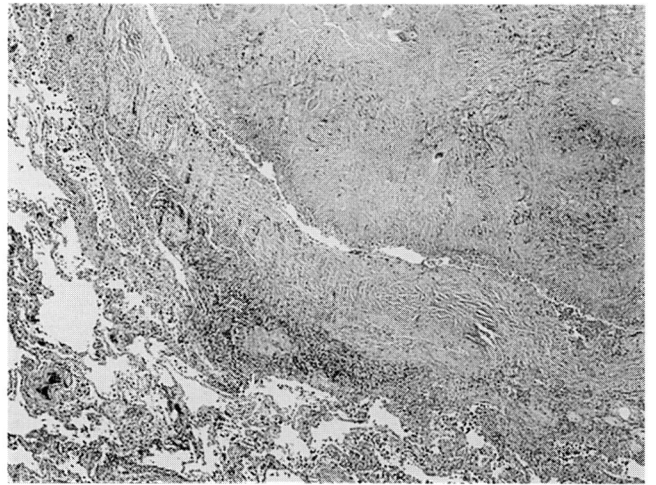

A

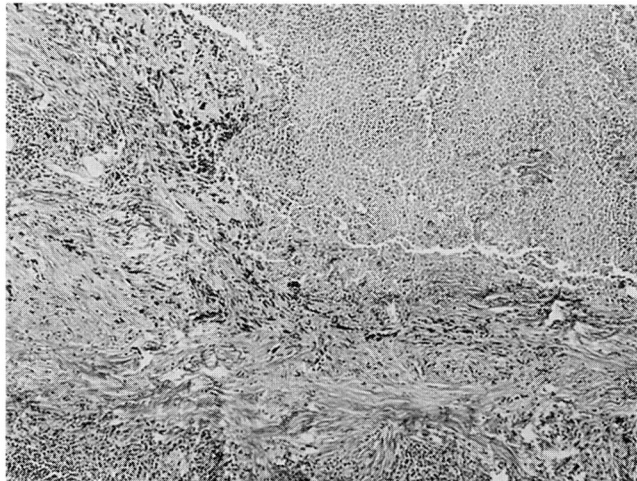

B

FIGURE 26-11

A. Rheumatoid nodule in Caplan's syndrome. The central necrotizing area of this rheumatoid nodule is surrounded by a thin fibrous wall showing chronic and granulomatous inflammation. (H&E, ×24.) *B*. The edge of the necrotic granuloma shows the anthracotic pigment of coal worker's pneumoconiosis signifying Caplan's syndrome. (H&E, ×60.)

syndrome, rheumatoid nodules are associated with a pneumoconiosis, such as coal worker's pneumoconiosis or silicosis.[1,50,51] In these patients, the anthracotic dust deposits may be found within the rheumatoid nodules (Fig. 26-11*B*).

Amyloidosis can occur in any of the CTDs. It has been reported in Sjögren's syndrome,[52,53] rheumatoid arthritis,[50,54,55] SLE,[56,57] scleroderma,[58] and ankylosing spondylitis.[59] It can take the form of nodular (Fig. 26-12*A*), alveolar septal,[55] or vascular deposits.[58] The deposits have an eosinophilic, amorphous, waxy appearance (Fig. 26-12*B*). Whether the amyloid is derived from SAA protein (AA type) or immunoglobulin light chains (AL type) is documented in few cases. In one patient with scleroderma, persistence of Congo red staining after potassium permanganate digestion favored AL amyloid, but positive immunohistochemical staining for AA protein favored AA amyloid.[58]

Aspiration pneumonia is a frequent complication of polymyositis due to respiratory muscle weakness and of scleroderma due to esophageal motility dysfunction.[1] The histologic findings in aspiration pneumonia may be acute or chronic. The acute pat-

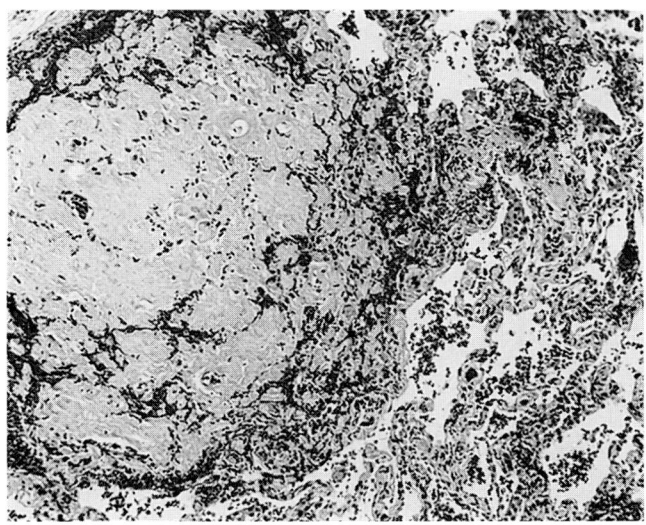

A

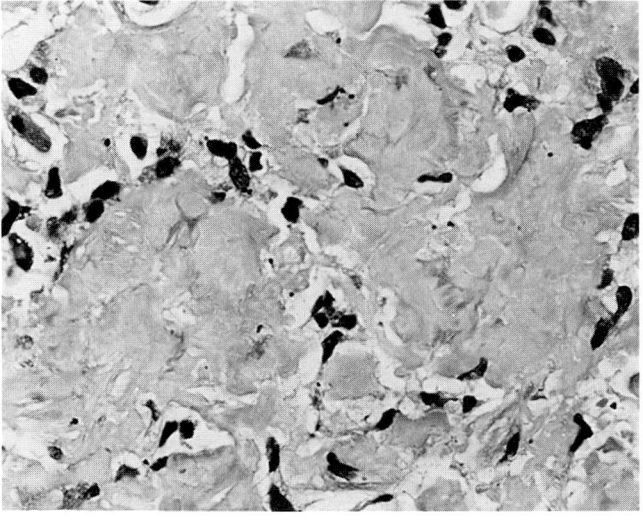

B

FIGURE 26-12

A. Amyloid in rheumatoid arthritis. This nodule of eosinophilic amorphous material is amyloid. (H&E, ×240.) *B*. At higher power the amorphous waxy appearance is characteristic of amyloid. (H&E, ×504.)

tern is of an acute bronchopneumonia with neutrophils within bronchi or bronchioles and the surrounding alveolar spaces. Food particles may be seen within the airways. In chronic aspiration, a significant component of chronic inflammation may be seen around airways and in the surrounding parenchyma. A foreign body giant cell reaction to the food particles may be seen.

Alveolar Lesions

Diffuse pulmonary hemorrhage is a potentially life-threatening complication of CTD. It occurs most commonly in SLE but can also occur in rheumatoid arthritis,[60–71] polymyositis/dermatomyositis, and scleroderma.[72] The hemorrhage may be acute or chronic. In acute hemorrhage, the biopsy typically shows intraalveolar accumulation of red blood corpuscles (Fig. 26-13). A neutrophilic capillaritis may be seen, especially with fulminant hemorrhage in SLE patients (Fig. 26-14).[73] In chronic hemorrhage, the primary finding is intraalveolar and interstitial hemosiderin-laden macrophages. Immune complexes are important in the pathogenesis of this type of hemorrhage in SLE. Electron microscopy may show interstitial electron dense deposits while immunofluorescence microscopy shows positive staining for immune complexes.[67] The main differential diagnosis is with Wegener's granulomatosis (WG), Goodpasture's syndrome, microscopic polyarteritis nodosa and idiopathic pulmonary hemorrhage.[61] To diagnose WG, one must identify areas of

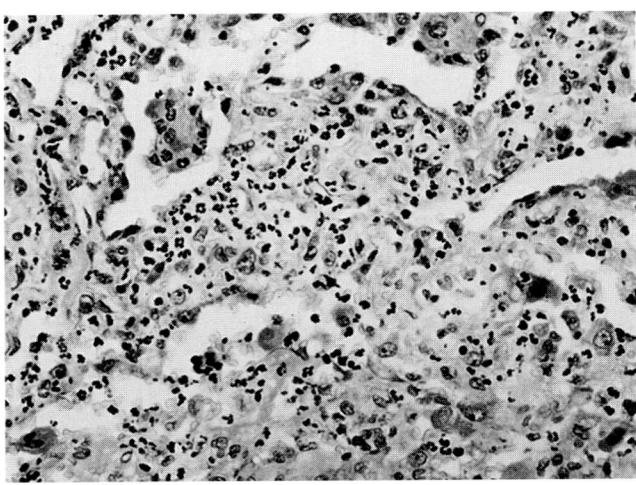

FIGURE 26-14
Neutrophilic capillaritis in SLE. These alveolar septa are thickened by dense interstitial infiltrates of neutrophils. (H&E, ×240.)

consolidation showing necrotizing granulomas, vasculitis affecting arterioles or venules, and neutrophilic microabscesses.[61] Other features that would support the diagnosis of WG are clinical evidence of sinus or renal disease and laboratory evidence of a serum cytoplasmic antineutrophil cytoplasmic antibody (C-ANCA) test.[74] In Goodpasture's syndrome, one must document the presence of anti–basement membrane antibody in serum or biopsies from the lung or kidney.[75] Idiopathic pulmonary hemorrhage is an exclusionary diagnosis based on the absence of evidence for WG, Goodpasture's syndrome, or underlying collagen vascular disease.[61] Both SLE and rheumatoid arthritis have presented as idiopathic pulmonary hemosiderosis.[76–78]

Eosinophilic pneumonia can occur in CTD. It may represent a drug reaction, or it may occur as an apparent manifestation of the underlying CTD.[79–82] It consists of intraalveolar accumulation of eosinophils, which may be accompanied by organizing intraluminal fibrosis (Fig. 26-15). This has best been described in rheumatoid arthritis.[80–82] A case of pneumonitis associated with polymyositis and eosinophilia has been reported.[83]

Alveolar proteinosis is rarely associated with CTD. It has been described in association with juvenile dermatomyositis.[84]

Airway Lesions

Constrictive bronchiolitis is a rare form of scarring fibrosis of bronchioles that causes marked narrowing of the airway lumina (Fig. 21-16)[85–92]; the intervening lung is frequently normal. It occurs most often in patients with rheumatoid arthritis, although it can occur in patients without known CTD. This lesion

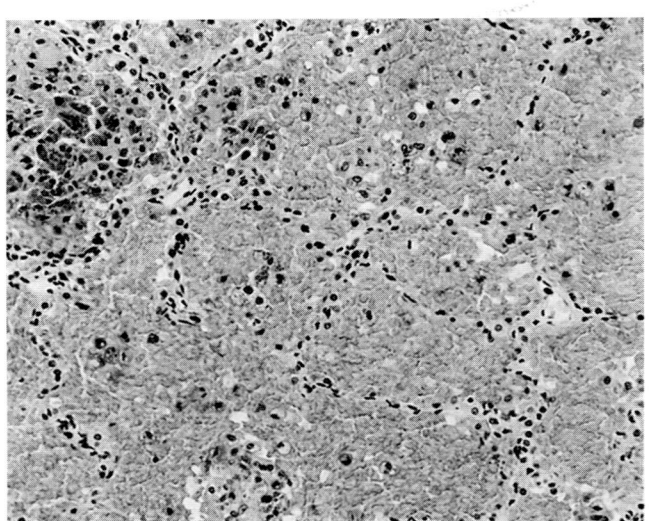

FIGURE 26-13
Diffuse alveolar hemorrhage in SLE. The alveoli contain numerous red blood corpuscles. Focal hemosiderin-laden macrophages are present (top left). (H&E, ×160.)

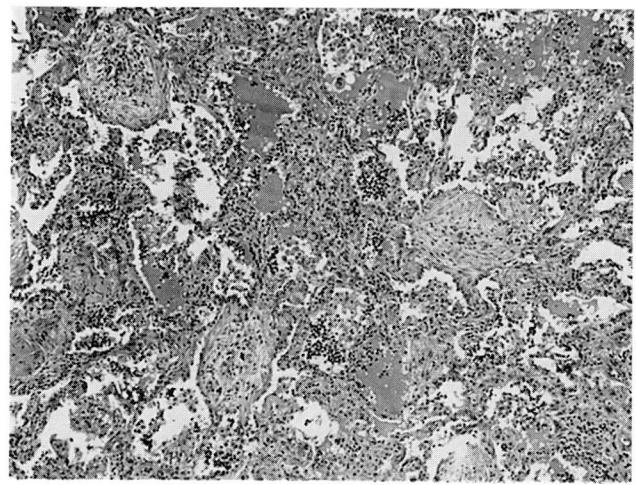

A

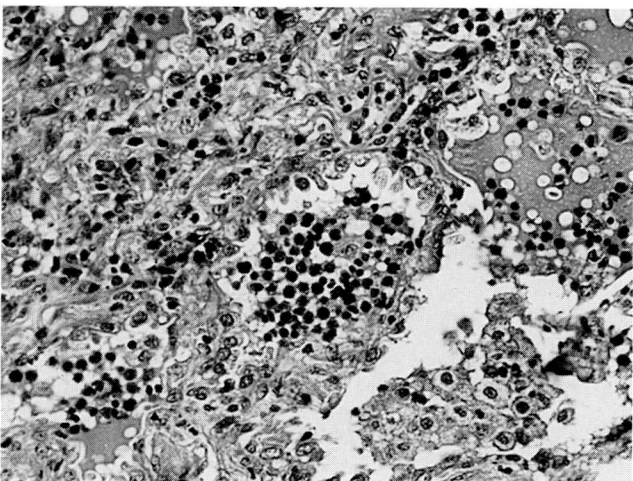

B

FIGURE 26-15

A. Eosinophilic pneumonia in rheumatoid arthritis. Numerous eosinophils are present within the alveolar spaces. In addition, there are scattered plugs of organizing intraluminal fibrosis, a common feature of eosinophilic pneumonia. (H&E, ×60.) *B.* At higher power the darkly staining cells within the alveolar space are eosinophils. (H&E, ×240.)

must be differentiated from BOOP, in which one sees polypoid plugs of intraluminal granulation tissue protruding into the lumens of bronchioles, alveoli, and alveolar ducts. It can be difficult to identify the airway lesions, and serial sections of a lung biopsy may be needed to identify them. Not every bronchiole is necessarily affected, and the intervening lung often looks normal, so one must search carefully for the affected airways. Patients usually have longstanding rheumatoid arthritis, severe obstruction by pulmonary function testing, and a rapidly progressive clinical course.

Bronchiectasis is a recognized complication of rheumatoid arthritis.[93,94] Pathologically, it is characterized by chronic inflammation and fibrosis of bronchi, resulting in marked ectasia of the lumina and thickening of the walls.

Bronchocentric granulomatosis (BCG) has been described in two cases of rheumatoid arthritis[95,96] and one case of ankylosing spondylitis.[97] BCG represents not a specific clinical entity but, rather, a pathologic pattern of pulmonary injury in which the bronchi or bronchioles are affected by necrotizing granulomatous inflammation. Most examples of BCG in nonasthmatics represent pulmonary infections such as tuberculosis or histoplasmosis; therefore, cultures and special stains are essential to exclude an organism. This would be especially true in patients with rheumatoid arthritis if they were being treated with immunosuppressive medications. Granulomatous lesions in the form of rheumatoid nodules and airway lesions in the form of bronchiolitis and bronchiectasis occur in rheumatoid arthritis. Thus it is possible that BCG could be a manifestation of pulmonary impairment in rheumatoid arthritis. In the absence of a specific cause, the term *idiopathic BCG* is appropriate, although in many of these cases an infectious cause may be suspected.[98]

Desiccation of the airways often occurs in Sjögren's syndrome and causes xerotrachea and chronic dry cough.[1,99,100]

Vascular Lesions

Vasculitis is a rare manifestation of CTD,[101] especially SLE,[102,103] rheumatoid arthritis,[104,105] and polymyositis.[106] Vasculitis in CTD may be associated

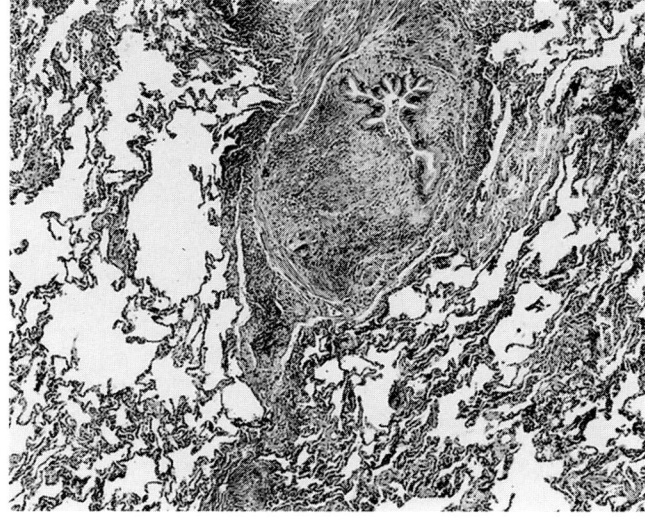

FIGURE 26-16

Constrictive bronchiolitis in rheumatoid arthritis. The lumen of this bronchiole is markedly narrowed by submucosal dense fibrosis and chronic inflammation. (H&E, ×80.)

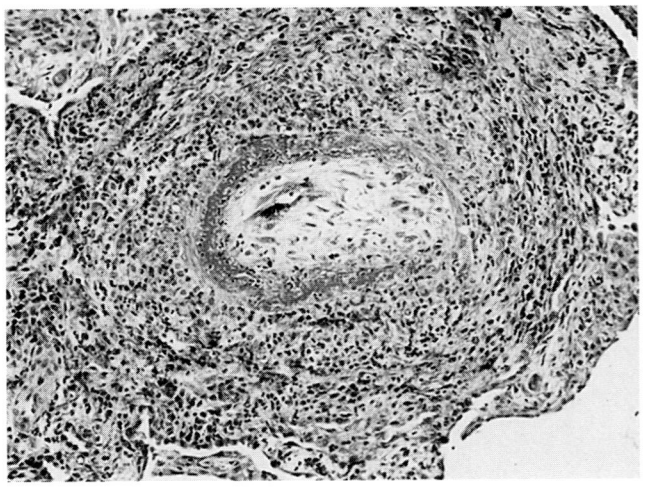

A

FIGURE 26-17

A. Necrotizing vasculitis and pulmonary hypertension in rheumatoid arthritis. This arteriole shows a fibrinoid necrotizing vasculitis with marked luminal narrowing and an intense perivascular chronic inflammatory infiltrate. (H&E, ×240.)

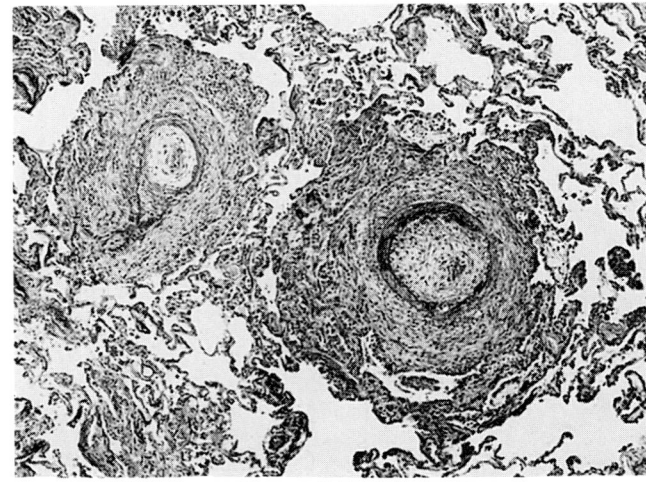

B

B. The marked hypertensive arterial changes are highlighted by this elastic stain, which demonstrates the concentric fibrotic narrowing of the lumen of these arterioles. (Elastic stain, ×60.)

with a rapidly progressive form of pulmonary hypertension.[102] In such cases, fibrinoid necrosis with severe chronic vasculitis may be seen in association with marked intimal fibrous proliferation and narrowing of the vascular lumens (Fig. 26-17). Some cases are diagnosed only at autopsy. Hypocomplementemic vasculitis may closely resemble SLE.[103]

Pulmonary hypertensive vascular changes are a potential pulmonary manifestation of CTD that can occur with or without clinical evidence of pulmonary hypertension. While secondary pulmonary hypertension is a potential complication of diffuse interstitial fibrosis, primary pulmonary hypertension

is relatively uncommon. Pulmonary hypertension occurs most commonly in patients with scleroderma (Fig. 26-18),[107,108] but it can occur in virtually any of the CTDs, including SLE,[109,110] rheumatoid arthritis,[111] mixed connective tissue disease,[112] Sjögren's syndrome,[113] dermatomyositis,[114] and polymyositis.[115–118] In pulmonary hypertension, arteries and arterioles show histologic evidence of intimal fibrosis, which sometimes causes obliteration of the vessel lumen (Fig. 26-18*A*), angiomatoid lesions, and thromboses. Secondary hypertensive arterial changes can be seen in UIP, but in primary pulmonary hypertension the lung parenchyma is rela-

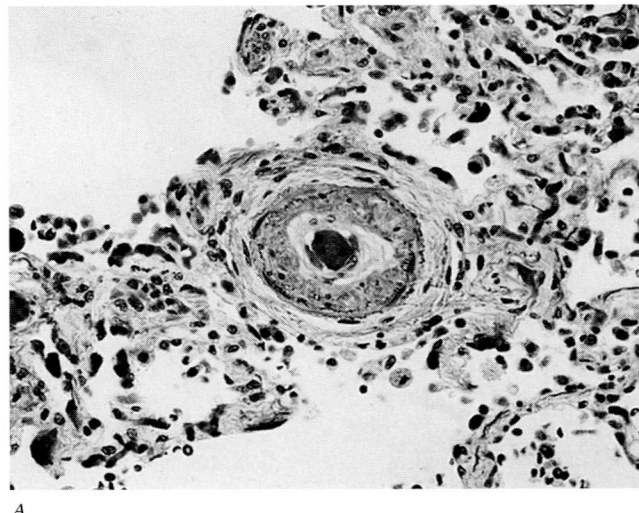

A

FIGURE 26-18

A. Hypertensive vascular changes in scleroderma. This arteriole shows concentric adventitial and intimal fibrosis causing narrowing of the vascular lumen. (H&E, ×240.) *B.* This arteriole

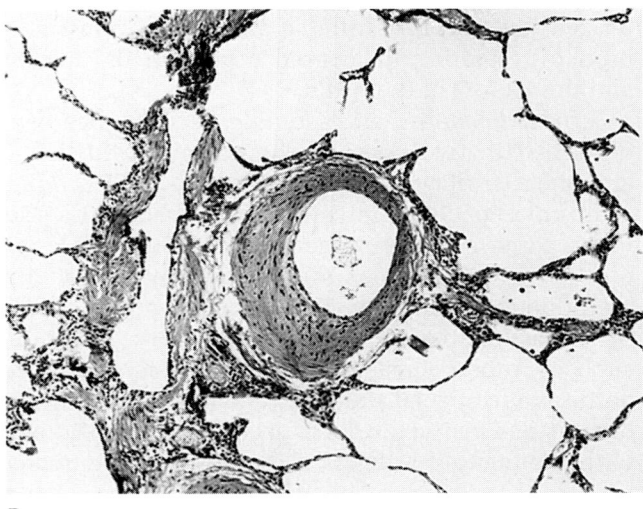

B

shows medial thickening. The surrounding lung parenchyma does not show fibrosis. (H&E, ×60.)

tively normal. In several cases of SLE, immunofluorescence has shown immune complexes in blood vessel walls[110]; other studies have failed to show immune complexes.[109] A pattern of venoocclusive disease has been reported in scleroderma.[107]

Indirect Respiratory Effects

A variety of indirect respiratory effects can occur, including respiratory muscle weakness and pulmonary edema due to renal, pericardial, or CNS disease. The shrinking lung syndrome or atelectasis has been linked in some studies to diaphragmatic dysfunction.[119,120]

Neoplastic Disorders

Neoplasms such as lung cancer (Fig. 26-19),[58,121,122] metastatic carcinomas to the lung, and lymphomas[123,124] are also known to be complications of certain CTDs.

CLINICAL AND CHARACTERISTIC PATHOLOGIC FEATURES

Rheumatoid Arthritis

Rheumatoid arthritis is characterized by a chronic, nonsuppurative arthritis that often affects the peripheral joints in a symmetric pattern. A broad spectrum of pleuropulmonary manifestations are recognized (Table 26-3). The most common pulmonary

TABLE 26-3
Pleuropulmonary Manifestations of Rheumatoid Arthritis

Pleural disease
  Pleuritis with or without effusion
  Sterile or septic empyema
Interstitial Lesions
  Usual interstitial pneumonia
  Diffuse alveolar damage
  Bronchiolitis obliterans–organizing pneumonia
  Apical fibrobullous disease
  Lymphocytic interstitial pneumonia
  Lymphoid hyperplasia
  Amyloid
Rheumatoid nodules
  Pneumoconiotic nodules (Caplan's syndrome)
Airway lesions
  Constrictive bronchiolitis (obliterative bronchiolitis)
  Bronchiectasis
Alveolar lesions
  Diffuse pulmonary hemorrhage
  Eosinophilic pneumonia
Vascular lesions
  Pulmonary vascular lesions
  Pulmonary hypertension
Indirect respiratory effects
  Thoracic cage immobility
  Upper airway dysfunction secondary to cricoarytenoid arthritis

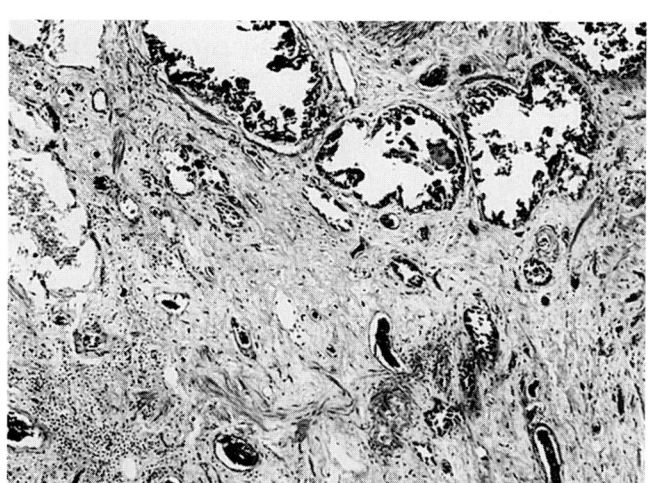

FIGURE 26-19
Adenocarcinoma associated with usual interstitial pneumonia in scleroderma. The dense fibrosis in this biopsy was associated with the picture of usual interstitial pneumonia. In addition, the atypical glands (top) with papillary luminal growth represent an adenocarcinoma. (H&E, ×60.)

manifestations of rheumatoid arthritis are pleural effusion or pleuritis. Pleuritis can be found in up to 40 percent of autopsy cases. Pleural effusions can be found in 3 to 5 percent of rheumatoid arthritis patients, although if one follows these patients through the course of their disease, clinical pleurisy can develop in up to 20 percent of them.[1,2] Pleuritis occurs predominantly in men (M:F = 5:1), although there is a strong female predominance in rheumatoid arthritis overall (M:F = 1:3). In 25 percent of cases, pleuritis may develop up to 6 months before the arthritis. Pleural disease tends to be associated with more severe arthritis, and approximately 80 percent of patients have subcutaneous nodules.[1,2] Compared to those in SLE, the pleural effusions of rheumatoid arthritis are less often symptomatic, less frequently bilateral, and usually larger.[1,2]

Diffuse interstitial fibrosis resembling UIP occurs in 1 to 4 percent of rheumatoid arthritis patients.[50,125–129] The frequency of subclinical fibrosis may be much higher, as suggested by the finding of a reduced diffusing capacity in up to 40 percent of patients. The fibrosis tends to be bibasilar. Similar to that in UIP, there is a poor prognosis, with a median survival of 3.5 years and a 39 percent 5-year survival.[4] UIP tends to be a relatively early manifestation of rheumatoid arthritis, since 70 percent of patients who develop UIP do so within 4 years of the onset of joint symptoms.[1] In the rare patients in whom the lung disease precedes arthritis, joint symptoms usually develop within 3 years.[1,130] HLA-B40 may be a marker for rheumatoid patients predisposed to developing lung disease.[131]

Rheumatoid nodules usually occur in men with advanced seropositive rheumatoid arthritis, who frequently have subcutaneous rheumatoid nodules.[1-3] The nodules can precede the onset of arthritis. Rheumatoid nodules do not cause symptoms unless they are complicated by infection, hemorrhage, or rupture and bronchopleural fistulae. Hemoptysis can occur when rheumatoid nodules cavitate and undergo hemorrhage. An accelerated form of rheumatoid nodulosis can sometimes occur while the patient is on methotrexate therapy.[132]

BOOP can occur in rheumatoid arthritis.[16,20,133,134] It has the same good response to steroids seen in patients without an underlying CTD.[16,20,134]

Constrictive bronchiolitis often presents in patients in the third to fourth decade of life with an abrupt onset and a rapid downhill course, culminating with death in 2 to 3 years.[44,87,89,91,135-141] The patients present with dyspnea and cough. On physical examination, they sometimes have a midinspiratory squeak. Pulmonary function tests show a severe obstructive defect. Patients often have a history of long-standing rheumatoid arthritis. It is strongly linked to penicillamine therapy, although it can occur in in patients who have not received this drug.[136]

Bronchiectasis is a rare complication of rheumatoid arthritis, occurring in fewer than 1 percent of patients.[93,94] In one recent study, it was suggested that bronchiectasis was a late complication of rheumatoid arthritis.[93]

Vasculitis is a rare complication of rheumatoid arthritis, and in some cases the lung can be affected.[105,111] Severe pulmonary arteritis may be associated with pulmonary hypertension.[105,111]

Pulmonary hemorrhage can occur in rheumatoid patients either as a manifestation of the underlying disease[111] or in association with penicillamine therapy. The hemorrhage can be acute or chronic.[61,78] The latter pattern may resemble idiopathic pulmonary hemosiderosis.[78] Antineutrophil cytoplasmic antibodies can be found in rheumatoid arthritis patients with associated pulmonary hemorrhage.[111] The differential diagnosis includes diffuse alveolar damage and infection.

Rare cases of chronic eosinophilic pneumonia have been reported in rheumatoid arthritis patients.[80-82] It can be associated with the histologic pattern of BOOP. Some cases are associated with drug therapy.[142]

Bronchocentric granulomatosis has been reported in a few patients with rheumatoid arthritis.[95-96] When this pattern of lung inflammation is encountered, the most important differential diagnosis is with infection such as tuberculosis or fungal disease.

Very few cases of primary pulmonary hypertension occur in rheumatoid arthritis.[111,116-118] The clinical picture is similar to that of idiopathic primary pulmonary hypertension.[111,116-118]

In rheumatoid arthritis, drug toxicity can be difficult to distinguish from underlying disease.[143] The drugs that are most commonly known to cause pulmonary toxicity in rheumatoid arthritis patients are methotrexate,[144] penicillamine,[138,139,142,145-151] and gold salts.[1,136,137,150,152] Lung toxicity in patients on gold therapy usually presents with dyspnea, dry cough, and diffuse bilateral infiltrates within 6 months of initiation of therapy.[143,148] In one-third of patients, systemic manifestations such as malaise, weight loss, proteinuria, and fever may occur. Peripheral eosinophilia can occur in 25 percent of cases, and an elevated serum IgE may be seen. Pulmonary function tests may show restriction, with reduced $D_{L_{CO}}$. Lung biopsies may show constrictive bronchiolitis, BOOP, chronic interstitial pneumonitis (nonspecific), or diffuse alveolar damage.[136] Bronchoalveolar lavage (BAL) may show lymphocytosis.

Penicillamine can be associated with constrictive bronchiolitis, BOOP, follicular bronchiolitis, DAD,[149] and diffuse pulmonary hemorrhage.[144,145,149,151,153,154] Constrictive bronchiolitis occurs in 1 to 3 percent of patients on penicillamine.[154,155] Respiratory symptoms usually occur within 10 months of the start of penicillamine therapy (range: 3 to 10 months).[155] Patients with constrictive bronchiolitis have a poor prognosis, and death has been reported in about 25 percent of cases.[154,155] Diffuse pulmonary hemorrhage also has a high mortality. Penicillamine may also cause peripheral eosinophilia[142] and drug-induced SLE.[151] Bronchiolitis obliterans is strongly linked to penicillamine, less so to gold.

Lung toxicity occurs in 5 to 10 percent of patients treated with methotrexate.[156,157] There is no consistent relationship to cumulative dose. The typical presenting features include a subacute febrile illness 3 to 4 months after initiation of therapy. Peripheral eosinophilia may be seen in up to 50 percent of cases. Most patients recover after the drug is discontinued, and steroids may be of help. Some patients may get better even if therapy is continued. It is thought that the lung injury is mediated by some immunologic reaction. Histologically, lung biopsies may show interstitial chronic inflammation, granulomas, or giant cells resembling a hypersensitivity pneumonitis. Eosinophils may be seen in up to 50 percent of specimens. Less commonly, one may encounter BOOP, DAD, or pulmonary edema.

Systemic Lupus Erythematosus

SLE is an immunologically mediated systemic disease in which positive antibodies to nuclear antigens and immune complex formation are key features. SLE can affect the lung in diverse ways (Table 26-4). Pulmonary involvement in SLE has been described in 20 to 90 percent of patients.[1–4,78,158–161]

TABLE 26-4
Pleuropulmonary Manifestations of Systemic Lupus Erythematosus

Pleural lesions
  Pleuritis with or without effusion
Interstitial lesions
  Diffuse alveolar damage (acute lupus pneumonitis)
  Bronchiolitis obliterans–organizing pneumonia
  Usual interstitial pneumonia
  Cellular interstitial pneumonia
  Granulomatous interstitial pneumonia
Alveolar lesions
  Diffuse pulmonary hemorrhage
Vascular disease
  Vasculitis
  Pulmonary hypertension
  Thromboembolism
Indirect respiratory effects
  Diaphragmatic dysfunction
  Shrinking atelectasis
Upper airway dysfunction
  Epiglottitis
  Laryngitis
  Cricoarytenoid arthritis

The most specific histologic finding for SLE is the LE cell. However, this is such an uncommon finding—encountered in only one of 120 autopsies of SLE patients[161]—that it is not a useful pathologic criterion for recognition of pulmonary SLE.

Pleuritis is one of the most common pleuropulmonary manifestations of SLE. Pleural inflammation or fibrosis can be found histologically at autopsy in 50 to 83 percent of SLE patients.[1,160,161]

Infection is the most common pulmonary manifestation of SLE. It is usually due to the immunosuppressive therapy. As a result, it is an important differential diagnosis in any SLE patient with acute respiratory failure. A variety of unusual infections can occur in SLE patients, including paragonimiasis,[162] *Nocardia asteroides*[163] and *Aspergillus terreus* infections,[164] and cavitary *P. carinii* pneumonia.[165]

Acute respiratory failure can occur in SLE from a variety of causes, including infection, congestive heart failure, uremia, drug reactions, DAD (ARDS), and diffuse pulmonary hemorrhage.[166] Lung biopsies may play an important role in excluding infection in order to avoid unneeded treatment with many antibiotics.[166] Acute lupus pneumonitis was described by Matthay and colleagues in a group of patients who pathologically appeared to have had DAD.[167]

Pulmonary hemorrhage is a potentially life-threatening complication of SLE.[60,66,68,73,168,169] Patients may have an acute or chronic presentation. Hemoptysis may or may not be present. A large percentage of patients who do not have systemic manifestations of SLE at presentation have acute alveolar hemorrhage. MR imaging may show a marked decrease in T2-weighted images in the lungs of SLE patients who have chronic or subacute pulmonary hemorrhage due to the paramagnetic effects of the ferric iron.[60] Clinical manifestations of SLE may develop as late as 4 years after presentation with pulmonary hemorrhage.[76,77] Immune complex deposition has been demonstrated by immunofluorescence or electron dense deposits by electron microscopy in many cases of pulmonary hemorrhage in SLE patients.[64,68]

Only a few cases of BOOP have been described in SLE, but it appears to be a distinct pulmonary manifestation.[18,170] One patient with SLE and BOOP also had Hunner's cystitis.[170] The pulmonary disease in these patients responded to corticosteroid therapy, as is characteristic of BOOP.[170]

Tubuloreticular inclusions can be demonstrated by electron microscopy in lung biopsies from SLE patients with pneumonitis.[171–173] They appear as clusters of branching and anastamosing microtubular structures measuring 20 to 22 nm in width and up to

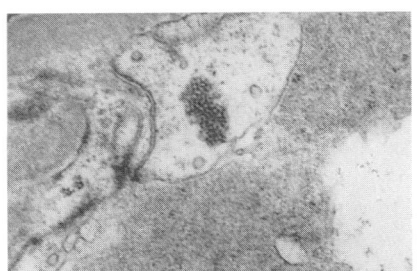

FIGURE 26-20
Tubuloreticular inclusion. The microtubular structure is reminiscent of a virus-like particle.

100 nm in length (Fig. 26-20). These inclusions can be seen in patients with CTD or other diseases, including infectious mononucleosis, malignant tumors, and HIV infection.[171] They are found more often in SLE than in Sjögren's syndrome or scleroderma. Tubuloretic-ular inclusions can be found in buffy-coat leukocytes or endothelial cells of lung, kidney, and skin.[171] Their significance is uncertain, but interferon can induce formation of these inclusions in some lymphoid cell lines. It has been proposed that tubuloreticular inclusions in SLE patients may reflect auto-immune activity.[171]

The shrinking lung syndrome is an uncommon but well-recognized manifestation of SLE.[119,120,174,175] Patients typically show elevated diaphragms on chest radiographs and small lung fields and may present with shortness of breath, cough, and restrictive pulmonary function.[119] It is not uncommon for patients to lack evidence of systemic involvement by SLE at the time of presentation.[119] Some studies have suggested that respiratory muscle dysfunction plays a major role in the pathogenesis of this disorder.[119,120] The disorder may respond well to treatment with corticosteroids, and patients have a good prognosis.

Sarcoidosis and SLE can occur together.[176,177] The diagnosis of sarcoid in a patient with SLE requires histologic confirmation of the presence of noncaseating granulomas in lung or lymph node tissue.[176,177] Additional clinical features, such as bilateral hilar adenopathy with or without pulmonary parenchymal infiltrates, are supportive of the diagnosis of sarcoidosis.[176,177] Special staining should be performed to exclude the presence of an infectious origin for the granulomas.

The frequency and spectrum of pulmonary manifestations of SLE in children are similar to those seen in adults.[178] The most common pulmonary manifestations in children include pleural effusions and pleuritis, acute and chronic pneumonitis, pulmonary hemorrhage, and shrinking lungs syndrome.[178] Diffuse pulmonary hemorrhage has a high mortality (50 percent) in children.[178] Maddison and coworkers

found a significantly increased incidence of pulmonary disease in elderly SLE patients.[179]

Pulmonary vascular changes as described by Fayemi in eight of 20 autopsies of SLE patients included acute and chronic lesions.[180] Acute lesions consist of fibrinoid necrosis and vasculitis. Chronic lesions include intimal fibrosis, hypertrophy of media, periadventitial fibrosis, and focal fragmentation and destruction of elastic laminae.[180] Muscular arteries were affected most often and severely. Arterioles, elastic arteries, and veins were also affected. However, none of the patients in the study by Fayemi had clinical evidence of pulmonary hypertension.[180] Necrotizing pulmonary vasculitis in SLE is rare and may be associated with pulmonary hypertension.[102]

Pulmonary hypertension in the absence of diffuse interstitial lung disease is uncommon in SLE. The precise incidence is difficult to determine; various studies report pulmonary hypertension in 0.5 to 14 percent of SLE patients.[109,110,181–184] Patients usually have a history of prior SLE, but in rare cases pulmonary hypertension may precede the onset of systemic manifestations.[184] Thus, pulmonary hypertension does not correlate with severity or activity of SLE. However, pulmonary hypertension in SLE patients has been associated with rheumatoid factor, lupus anticoagulant, and ribonucleic protein antibody.[102,183] Presenting pulmonary manifestations usually occur within several years of onset of SLE and include exertional dyspnea,[184] chest pain, nonproductive cough, and fatigue. On physical examination, a loud second pulmonic heart sound and a right ventricular lift may be found. The chest radiograph can show an enlarged heart with a prominent pulmonary outflow tract.[184] Raynaud's phenomenon is frequent, suggesting a systemic vascular defect.[109,183] The pathogenesis of pulmonary hypertension in SLE is not well understood. Vasoconstriction of pulmonary arteries similar to Raynaud's phenomenon, immune complex deposition, vasculitis, interstitial pneumonitis, or several of these features have been proposed.[102,109,183,184]

Pulmonary emphysema has been noted in SLE patients. Obstructive lung disease can occur in patients who develop an SLE-like syndrome manifested primarily by hypocomplementemic vasculitis.[103] Subpleural blebs and recurrent pneumothoraces are other rare pulmonary complications of SLE.[185]

Pulmonary neoplasms are uncommon in SLE patients. Kaposi's sarcoma in the lung occurred in one patient 3 years after presentation with SLE.[186] The patient received steroids for 3 years, but unfortunately no HIV testing was done in this case. Pulmonary lymphoma has been reported in a single SLE patient.[187]

Scleroderma

After the esophagus, the lung is the most common site of visceral impairment by scleroderma.[188,189] The pulmonary manifestations include a broad variety of lesions (Table 26-5). Most patients with scleroderma are adults, but about 1.5 percent of cases occur in the first decade of life and 7 percent in the second decade.[190] There is a female predominance of about 3–4:1, and most patients are between 30 and 50 years of age. The two major types of pulmonary involvement in scleroderma are a UIP pattern and pulmonary vascular disease. Much less common pulmonary lesions include lung carcinomas,[58,121,122,191–198] BOOP,[17] amyloid,[58,199,200] pulmonary hemorrhage,[72] upper lobe fibrocystic changes,[201] sarcoid,[202] aspiration pneumonia,[188] drug-induced pneumonitis,[188] pneumoconiosis,[188] and spontaneous pneumothorax.[188]

Diffuse interstitial fibrosis with the pattern of UIP is the most common type of pulmonary impairment in scleroderma.[1,188,203–209] It is present at autopsy in 74 percent of cases,[210] but pulmonary symptoms are found in 54 percent and radiographic abnormalities in 53 percent.[188] Pulmonary function studies may detect abnormalities in up to 92 percent of cases, suggesting that subclinical interstitial fibro-

TABLE 26-5
Pleuropulmonary Manifestations of Scleroderma

Pleura
  Pleuritis
  Fibrosis
  Spontaneous pneumothorax
Interstitium/parenchyma
  Usual interstitial pneumonia
  Bronchiolitis obliterans–organizing pneumonia
  Diffuse alveolar damage
  Upper lobe fibrocystic changes
  Aspiration pneumonia
Alveoli
  Diffuse pulmonary hemorrhage
Vessels
  Vasculitis
  Pulmonary hypertension
Neoplasms
  Lung cancer

sis is very common.[188] The clinical presentation is similar to that of UIP. Recognition of scleroderma precedes the onset of lung disease in the vast majority of cases[188]; rarely is lung disease the initial manifestation.[211] The most common pulmonary symptom is dyspnea on exertion, followed by nonproductive cough.[188,207]

Chest radiographic changes in patients with UIP in scleroderma resemble those in patients without an underlying CTD. Involvement of the lung bases progresses to include the lower two-thirds of the lung.[188] Eventually, cystic changes due to honeycomb fibrosis can be detected. The severity of the radiographic findings do not always correlate with the severity of pulmonary symptoms. High-resolution CT scans are more sensitive than chest radiographs in detecting early UIP in scleroderma patients and show an excellent correlation with lung biopsy changes.[204,206,212] The most common finding is a ground-glass appearance, followed by irregular pleural margins, septal subpleural lines, honeycomb changes, and subpleural cysts.[204]

Pulmonary function tests typically show a restrictive defect in 45 to 60 percent of scleroderma patients.[188,209,213,214] About 15 to 30 percent of patients have an obstructive dysfunction, but it is not clear whether this is due to the effects of cigarette smoking or to the underlying scleroderma.[188,214] About 15 percent of patients have a decreased diffusing capacity with normal spirometry. This is seen more often in patients with the CREST (calcinosis, Raynaud's phenomenon, esophageal dysmotility, sclerodactyly, and telangiectasia) syndrome than in those with diffuse cutaneous systemic sclerosis.[188] It is thought to correlate with pulmonary vascular disease.

BAL is a sensitive method for detecting pulmonary abnormalities in scleroderma patients.[206,209,213,215] BAL studies have identified factors such as fibronectin,[216] thrombin,[217] interleukin-1,[213] interleukin-6,[218] interleukin-8,[219] insulin growth factor–1,[220] and collagenase,[209] which may play a role in the pathogenesis of interstitial fibrosis in scleroderma. Several patterns of inflammatory cell alterations have been noted, including neutrophilia, lymphocytosis, and increased lymphocyte:neutrophil ratios.[209,221] Attempts have been made to correlate these patterns with clinical features.[188,208,209,213,221] Inconsistent observations may be due to differences in procedures (site of lavage, upper airway contamination), technical differences (processing of specimens), and spectrum of lung disease in the group of patients studied.[213,221]

Clinical subsets of scleroderma patients can be

defined by serologic profiles with differing frequency of lung involvement and varying prognosis.[222] About one-third of patients have anticentromere antibodies (ACA), a relatively slow progression of disease and infrequent lung fibrosis.[222] However, ACA may be found in patients with the CREST syndrome who are at risk for pulmonary hypertension without associated interstitial fibrosis.[188,223,224] Pulmonary involvement and anti–Scl-70 antibodies are found in about 25 percent of scleroderma patients and correlate with the fastest progression of disease.[222] Systemic sclerotic changes develop within 5 years from the presentation of Raynaud's syndrome, and almost 80 percent of patients have lung involvement.[222] A small subset of scleroderma patients have anti-SSA/Ro antibodies, and their cases are characterized by severe, rapidly progressive disease and consistent lung involvement.[222] Subjects with antinucleolar antibodies (ANoA) constitute about 10 percent of scleroderma patients, and their disease has a rapid course similar to that of patients with anti–Scl-70 antibodies, but 75 percent have pulmonary involvement.[222]

Pulmonary hypertension occurs in 15 to 30 percent of patients with diffuse scleroderma.[188,225] CREST syndrome patients are more commonly affected with pulmonary hypertension in up to 50 percent of cases.[225] Most cases of pulmonary hypertension are secondary to underlying interstitial fibrosis. However, pulmonary hypertension in the absence of interstitial fibrosis occurs more commonly in scleroderma than in any of the other CTDs. This is most likely to occur in the setting of the CREST syndrome, typically about 10 years after the onset of scleroderma. It can rapidly lead to respiratory failure, cor pulmonale, and death.[1,107,108,181,195,207,226,227] Pathologically, the muscular pulmonary arterioles are primarily affected, and the severity of the changes can be graded semiquantitatively.[226]

Medial hypertrophy of vascular smooth muscle and concentric intimal thickening with a myxoid connective tissue constitute grade 1 lesions. The vascular changes corresponding to grades 2 and 3 are when the concentric intimal fibrosis narrows the lumen of most arterioles by 50 and 75 percent, respectively. Pulmonary vascular changes were seen in all autopsies from CREST patients in one study, and they were the major pathologic finding in 58 percent of cases.[226] Histologic evidence of pulmonary vascular changes is more commonly seen at autopsy in patients with the CREST syndrome than in those with diffuse scleroderma.[108] Fibrinoid necrosis is notably absent in the pulmonary arteries of scleroderma patients, even in those with rapidly progressive pulmonary hypertension.[108,226] Pulmonary hypertension with the histologic pattern of veno-occlusive disease was reported in one patient with the CREST syndrome.[107] Another rare cause of pulmonary hypertension in scleroderma is chronic thromboembolization.[195]

Whether there is an increased risk of lung carcinoma in scleroderma is controversial.[58,121,122,191–198] While there are many reports of carcinomas, particularly of the lung, in these patients, some have argued that the association is fortuitous.[191] Nevertheless, it has been shown that lung carcinomas in scleroderma are more common (relative risk of 4.4 to 16.5) in patients with a previous history of interstitial fibrosis.[198,228] While bronchioloalveolar carcinomas are among the histologic subtypes of lung carcinoma seen in scleroderma,[58,191,196] other histologic subtypes can occur, including squamous cell carcinoma,[191,192] adenocarcinoma,[121] and small cell carcinoma.[122]

Pulmonary hemorrhage may be a complication of penicillamine therapy or, more rarely, a manifestation of scleroderma.[72,188,229–231] Hemoptysis has been reported in scleroderma due to endobronchial telan-giectasia.[232] It was thought that hemorrhage in this patient was due to the presence of bronchial telan-giectasia, with predisposition to bleeding from aspi-rin therapy and mild azotemia.[232]

BOOP is rare in scleroderma.[17] In the report of two cases by Bridges and colleagues, the patients presented with rapidly progressive pulmonary disease, restrictive pulmonary function, and reduced diffusing capacity.[17] One of the patients responded well to steroids; this patient had more patchy involvement of the lung on chest radiographs than the other patient, who died 12 weeks after presentation.[17]

Amyloid is very rare in scleroderma but has been observed in several patients, including one who also had a bronchioloalveolar carcinoma of the lung.[58,199,200] SAA protein, the precursor protein of secondary amyloid, has been shown to be elevated in scleroderma patients.[233]

Scleroderma has been described in association with pneumoconioses such as silicosis and coal worker's pneumoconiosis.[234,235]

Polymyositis/Dermatomyositis

Polymyositis and dermatomyositis (PM/DM) are forms of idiopathic inflammatory myopathy that are characterized by symmetric proximal muscle weakness, elevated serum muscle enzymes, electromyographic changes, and histologic evidence of myositis.[236,237] There is a female predominance, and two age peaks occur: one in children and the other in young to middle-aged adults.[236–238] Five to 10 new cases of PM/DM occur in the United States per million population per year.[237]

Lung involvement occurs in 5 to 20 percent of PM/DM patients and includes a wide variety of potential manifestations (Table 26-6).[238–240] If pulmonary function abnormalities are included, however, lung disease can be found in up to 45 percent of patients.[240] In at least one-third of patients, lung involvement precedes muscle or skin manifestations by months to years.[240]

Aspiration pneumonia is the most common pulmonary manifestation of PM/DM, occurring in 15 to 20 percent of patients.[238] Lakhanpal and colleagues found aspiration pneumonia in 9 percent of autopsied PM/DM patients.[106] Patients with PM/DM are predisposed to this complication because of weakness of the striated muscle of the upper aerodigestive tract. A variety of aspirated materials can cause pneumonia, including nontoxic material such as food, gastric acid, and bacteria-contaminated oropharyngeal secretions.[240] Improved therapy and patient management have reduced the morbidity and mortality of this complication; nevertheless, it remains one of the most common pulmonary complications of PM/DM. Early recognition and distinction from other types of interstitial lung disease are important so that proper therapy can be initiated.

Ventilatory insufficiency due to muscular weakness or hypostatic pneumonia occurs in 4 to 8 percent of patients. Histologic examination reveals inflammation and degenerative changes in the intercostal muscles, the diaphragm, and accessory muscles of respiration.[240] Most patients with this complication have severe generalized muscle weakness. Accompanying manifestations include dysphagia and rapid, shallow breathing, and respiratory failure may be precipitated by anxiety or dyspnea.[240] In rare instances, isolated diphragmatic paresis with interstitial lung disease may be the initial presenting finding of PM/DM.[241] Most of these patients have a good prognosis, although hospitalization is often necessary.

Interstitial lung disease can be encountered in 5 to 30 percent of PM/DM patients and can manifest a variety of patterns.[1,3,240,242] In the pathologic study of 65 autopsy cases by Lakhanpal and colleagues, interstitial lung disease was found in 42 percent of cases.[106] The pathologic features in these cases were primarily those of UIP.[106] The average age at presentation for PM/DM patients with pulmonary disease is 50 years of age (range: 5 to 77 years).[240] There is a 1.6:1 female-to-male predominance.[240] Three types of clinical presentation have been recognized: acute respiratory failure, chronic respiratory failure, and disease without pulmonary symptoms.[243] The most common presenting findings are dyspnea with or without nonproductive cough and bibasilar rales[1,3,240]; clubbing is uncommon. Raynaud's phenomenon can be seen in 15 percent of patients. Patients with interstitial lung disease often have arthritis.[240]

Chest radiographs of PM/DM patients with interstitial lung disease vary with the type of lung disease. UIP will show peripheral lower lobe interstitial fibrosis that progresses to honeycomb changes.[240] DAD typically will exhibit a granular or ground-glass shadow, which may be superimposed on an underlying reticulonodular pattern.[240] BOOP is characterized by patchy nodular densities, which may wax and wane or migrate.

Interstitial pulmonary fibrosis has been shown to have an adverse effect on prognosis in PM/DM patients.[244] In most studies, however, the lung pathologic findings are not classified according to current concepts.[242,244–248] These studies report a variable prognosis and response to therapy. Salmeron and colleagues made the general observation that therapeu-

TABLE 26-6
Pleuropulmonary Manifestations of Polymyositis and Dermatomyositis

Pleura

    Pleuritis with or without effusion

    Spontaneous pneumothorax

Interstitium/parenchyma

    Diffuse alveolar damage

    Usual interstitial pneumonia

    Bronchiolitis obliterans–organizing pneumonia

    Aspiration pneumonia (due to pharyngeal or esophageal dysfunction)

Alveoli

    Diffuse pulmonary hemorrhage

    Pulmonary alveolar proteinosis

    Alveolar proteinosis

Vessels

    Vasculitis

    Pulmonary hypertension

Neoplasms

    Lung cancer

Indirect respiratory effects

    Respiratory muscle dysfunction (hypostatic pneumonia, respiratory failure)

tic response seemed dependent on the degree of cellularity and interstitial fibrosis found on lung biopsy.[245] In the review by Tazelaar and coworkers, a series of lung biopsies from PM/DM patients were classified as DAD, BOOP, UIP, and CIP.[41] All the patients with DAD reported by Tazelaar's team died.[41] DAD was found by Lakhanpal and colleagues in 15 percent of autopsies of PM/DM patients.[106] Three of five of the patients with UIP reported by Tazelaar's team died.[41] A similarly poor prognosis has been observed in other studies that appear to have comprised mostly cases of UIP.[245,246,249] In contrast, studies that appear to have included lesions other than UIP reported a more favorable outcome.[243] One patient with CIP reported by Tazelaar and coworkers had resolution of the lung disease but died of leukemia 7 years later.[41]

A variety of antibodies against certain cytoplasmic aminoacyl-transfer RNA synthetases have been found to correlate with the presence of interstitial lung disease in PM/DM patients. The most common antisynthetase is anti-Jo1, found in 20 percent of PM/DM patients and directed against histidyl-tRNA synthetase.[250–254] Other antibodies associated with interstitial lung disease in myositis are anti-EJ,[252] anti-KJ,[236] and anti-OJ antibodies.[255]

Very few cases of BOOP are seen in PM/DM patients.[15,41] The cases described have had a relatively good prognosis. Five of the seven patients reported with BOOP are alive and well and responded to steroid therapy.[15,41] Unfortunately, other studies of interstitial lung disease in patients with PM/DM have not subclassified the lung biopsy findings to allow recognition of other cases of BOOP.

Pulmonary arterial hypertension is very uncommon in the absence of interstitial fibrosis.[256] It was reported in a 69-year-old woman, with a 20-month history of polymyositis and cardiac failure, who was found to have pulmonary hypertension shortly before death.[256]

Pulmonary vasculitis is not a recognized clinical manifestation of PM/DM. However, Lakhanpal's team found it in 8 percent of autopsies on patients who primarily had pulmonary symptoms and arthritis, associated interstitial lung disease, and highly elevated ESR.[106]

Infectious pneumonias are a significant cause of morbidity and mortality in PM/DM patients. This is reflected in the finding of bronchopneumonia in 54 percent of PM/DM patients coming to autopsy.[106] The causes of pneumonia include aspiration, opportunistic infection, and hypostatic pneumonia. Patients with PM/DM can be immunosuppressed as a result not only of treatment for their underlying CTD

but also of medication for their interstitial lung disease. Lakhanpal and coworkers found a variety of bacterial and fungal infections at autopsy in patients with PM/DM.[106] These included organisms such as *Pseudomonas aeruginosa*, *Staphylococcus aureus*, *Cryptococcus neoformans*, *Nocardia asteroides*, *Candida albicans*, Mucormycosis, and *P. carinii*.[106] In addition to the common bacterial organisms, a variety of infectious agents can cause pneumonia, including other fungal, protozoal, helminthic, and mycobacterial organisms.[240]

Pleuritis and pleural fibrosis in PM/DM occurs only in association with interstitial lung disease.[240] In this regard, PM/DM differs from other CTDs. Lakhanpal's team found fibrinous pleuritis in 12 percent of autopsies on patients with PM/DM.[106]

Malignancy occurs in 5 to 10 percent of patients with PM/DM,[1,3] although it has been reported in up to 34 percent.[257–261] It is debated whether the frequency of malignancy is increased over the general population.[106,262,263] The lung is one of the recognized sites of development of carcinoma.[258,259] Lung cancer may develop in the setting of preexisting interstitial fibrosis.[259]

Alveolar proteinosis is a rare complication of PM/DM and has been reported in a child with juvenile DM.[84] At the time of onset of alveolar proteinosis, the patient had active DM without evidence of interstitial lung disease, vasculitis, or underlying infection.[84]

Sjögren's Syndrome

Sjögren's syndrome (SS) is an autoimmune disorder characterized by keratoconjunctivitis, sicca, and xerostomia.[264] Primary SS occurs in the absence of associated CTD and primarily affects women in the fourth and sixth decades of life. Secondary SS is associated with a CTD, usually rheumatoid arthritis, and affects a more heterogeneous group of patients. The reported frequency of lung involvement in SS is quite variable, ranging from 9 to 60 percent.[32,99,265–269] One reason for the widely varied results is the differing definitions of lung involvement. The potential pulmonary manifestations of SS are summarized in Table 26-7. Respiratory involvement is more common in patients with secondary SS, perhaps because the lung disease represents a manifestation of the CTD.[267,270]

The most characteristic pulmonary manifestations of primary SS are xerotrachea (desiccation of the bronchial tree) and lymphocytic infiltration, which comprises a spectrum of lesions including lymphocytic alveolitis with mild interstitial chronic inflammation, LIP, pseudolymphoma, and malignant

TABLE 26-7
Pleuropulmonary Manifestations of Sjögren's Syndrome

Pleura
  Pleuritis with or without effusion
Interstitium/parenchyma
  Lymphocytic interstitial pneumonia
  Granulomatous interstitial pneumonia
  Lympyhoid hyperplasia
  Bronchiolitis obliterans–organizing pneumonia
  Usual interstitial pneumonia
  Lymphoid hyperplasia
  Amyloid
  Aspiration pneumonia
Vessels
  Vasculitis
  Pulmonary hypertension
Neoplasms
  Lymphoma
Desiccation of the tracheobronchial tree
  Xerotrachea (chronic dry cough)
  Chronic bronchitis with cough and production of tenacious sputum
  Atelectasis, middle lobe collapse
  Recurrent bronchopneumonia

lymphoma.[270] In the vast majority of patients, SS preceded pulmonary disease by a mean of 5.6 years (range: 1 month to 23 years). In about 5 percent of cases, however, patients may have lung symptoms before SS is diagnosed.[32] Pulmonary involvement by SS is manifest most often by cough, dyspnea, and recurrent pneumonitis.[32] Cough in SS is primarily due to xerotrachea.[270,271] Pulmonary involvement may be discovered on a chest radiograph in asymptomatic patients in 15 percent of cases.[32]

Kelly and coworkers, in studying 100 patients with primary SS, found abnormal pulmonary function in 24 patients.[272] In this study, repeat evaluation after 4 years in a subset of patients showed deterioration of pulmonary function.[272] Abnormal pulmonary function has been shown to correlate with an abnormal lip biopsy,[272] the presence of anti-Ro antibody,[272] and the sicca complex.[273] While there are reports of chronic obstructive airways disease in patients with primary SS, it is either infrequent or absent in most studies.[270] Despite several studies indicating small-airways disease in primary SS,[274] no differences were found in comparisons with control populations[275]; therefore, there are doubts whether this is a primary manifestation of SS.[270]

Radiographic examination in patients with pulmonary SS by chest radiography and high-resolution computerized tomography have documented interstitial infiltrates, interstitial fibrosis, pleural effusions, pleural thickening, and lymphadenopathy.[99,272] Pleural effusions and pleuritis occur primarily in patients with secondary SS associated with CTD, usually rheumatoid arthritis, rather than in those with primary SS.[270]

The lung can show a spectrum of pathologic lesions in SS, including bronchopneumonia,[32] bronchiectasis,[2] a variety of lymphoid lesions, BOOP,[19,21] and diffuse interstitial fibrosis with a UIP-like pattern.[1,2,32,99,264,270] Bronchial submucous glands show atrophy and chronic inflammation that results in tenacious secretions.[265,271,276,277] These can predispose patients to atelectasis, bronchiectasis, and recurrent pneumonias.[265] Bronchopneumonia was relatively common in early studies,[266] but perhaps because of improved therapy, it is encountered less often in recent studies.[270]

Amyloidosis is a well-recognized pulmonary complication of SS and may be associated with LIP or bullous lung disease.[32,52,53,278] There are several reports of multiple bullae in SS.[53,100] Sarcoid is rarely associated with SS.[279,280] Pulmonary hypertension is very rare in SS patients and may be associated with vasculitis.[265,281] One case of lymphangioleiomyomatosis, probably a coincidental finding, has been reported in SS.[282]

A T-cell lymphocytic alveolitis, detected in BAL, is often present in SS patients.[267,268,279] An increased percentage of BAL lymphocytes was demonstrated by Wallaert and colleagues in patients with primary SS, SS associated with primary biliary cirrhosis, and SS associated with other CTD.[267] Dalavanga and coworkers showed that about half of primary SS patients have a lymphocytosis greater than 15.2 percent in BAL.[268] In contrast to those without a lymphocytosis, these patients more frequently show cough, dyspnea, radiographic interstitial lung disease, lower total lung capacity, and lower diffusion capacity.[268] Wallaert's team also found an increased percentage of neutrophils in the minority of cases, but more often in secondary SS than in primary SS.[267] They also found a marked expansion of the T8 cell population in patients with a BAL neutrophilia.[267]

A spectrum of lymphoid lesions can be encountered, ranging from hyperplasia of the mucosa-associated lymphoid tissue, mild chronic interstitial pneumonitis,[279] and LIP[30,32,283] to pseudolymphoma[1,99,124,284,285] and malignant lymphoma.[99,123,124,286–290] Granulomas may accompany the lymphoid infiltrates, possibly suggesting a differential diagnosis of hypersensitivity pneumonitis, sarcoid, and infection. About 5 percent of SS patients develop a malignant lymphoma[264]; the risk of malignant lymphoma in SS patients is more than 40 times that in the general population.[291] Many cases previously thought to be pseudolymphomas are regarded currently as lymphomas, and some fall into the category of MALT lymphoma. (Chap. 33)

Mixed Connective Tissue Disease

Mixed connective tissue disease (MCTD) is characterized by features of SLE, scleroderma, and polymyositis and a serum antibody to the ribonuclear portion of the extractable nuclear antigen.[292] The reported frequency of pulmonary involvement varies from 20 to 85 percent.[293,294] In a retrospective study pulmonary involvement was found in 25 percent of cases, while in a prospective study it was found in 85 percent of patients.[294] The spectrum of pulmonary lesions in MCTD includes those seen in patients with SLE, scleroderma, and PM/DM (Table 26-8). The most common pulmonary presenting finding is dyspnea, followed by pleuritic pain and bibasilar rales.[294]

The most common form of lung disease in MCTD is interstitial lung disease, which usually takes the form of UIP. If the clinical features resemble scleroderma, the fibrosis tends to be more severe. About 70 to 90 percent of MCTD patients have abnormal pulmonary function with a reduced DLCO, while restrictive lung disease is seen in about half of patients.[293,295]

Pleuritis manifest by pleural effusion was encountered in 6 percent of patients in one series; in another, pleuritic chest pain was found in 40 percent of cases.[293,294,296] Pleural involvement is more common in MCTD patients who have SLE-like manifestations. When pleural effusions are encountered, they are usually small and can resolve spontaneously.[293,294,296] An unusual pulmonary presentation of MCTD consisted of bilateral exudative pleuritis.[297]

Among MCTD patients, pulmonary hypertension occurs mostly in those with sclerodermalike manifestations.[112,298–300] Potential causes include vasoconstriction, hypoxemia due to interstitial fibrosis, recurrent thromboemboli,[112] and plexogenic arteriopathy.[299] Pulmonary hypertension can be progressive, with a fatal outcome.[112,299,301] In one case it

TABLE 26-8
Pleuropulmonary Manifestations of Mixed Connective Tissue Disease

Pleura
  Pleuritis with or without effusion
  Fibrosis
Interstitium/parenchyma
  Diffuse alveolar damage
  Usual interstitial pneumonia
  Bronchiolitis obliterans–organizing pneumonia
  Lymphocytic interstitial pneumonia
  Granulomatous interstitial pneumonia
  Upper lobe fibrocystic changes
  Rheumatoid nodules
  Lymphoid hyperplasia
  Amyloid
  Aspiration pneumonia
Alveoli
  Diffuse pulmonary hemorrhage
  Eosinophilic pneumonia
  Alveolar proteinosis
Airways
  Constrictive bronchiolitis
  Bronchiectasis
Vessels
  Vasculitis
  Pulmonary hypertension and vasculitis
Neoplasms
  Lung cancer
  Lymphoma
Indirect respiratory effects
  Respiratory muscle with diaphragm dysfunction
  Renal
  Pericardial
  CNS

was associated with an immune-complex glomerulonephritis.[112]

Pulmonary vasculitis can occur in MCTD patients with or without associated pulmonary hypertension.[293,302] Fibrinoid necrosis can occur in patients with severe pulmonary arterial disease.[293]

Several cases of massive pulmonary hemorrhage have been reported in MCTD.[303,304] In one case the hemorrhage was rapidly fatal and was associated with necrotizing renal vasculitis and immune complex deposits in the glomeruli.[303] In another case it was associated with a tubulointerstitial nephritis, and immune complexes could be demonstrated.[304]

Aspiration pneumonia can be encountered in MCTD patients, since esophageal dysfunction is a frequent manifestation of scleroderma and PM/DM. Therefore, MCTD patients with prominent features of these disorders are predisposed to the development of this complication. Alveolar hemorrhage can occur in MCTD patients, especially those with SLE-like manifestations[303,304]; it is often associated with renal disease.[303,304] Myopathy in MCTD patients can lead to diaphragmatic dysfunction and respiratory failure, particularly in patients with features of PM/DM.[305,306]

Ankylosing Spondylitis

Ankylosing spondylitis is a seronegative spondyloarthritis disorder that causes inflammation primarily of the sacroiliac joints, but also can affect all joints of the axial skeleton. It has a prevalence of 0.15 percent in white men and there is a 10:1 male-to-female ratio.[264] HLA-B27 is identified in up to 95 percent of patients.

Pleuropulmonary abnormalities have been reported in up to 30 percent of patients, but they were found on chest radiographs by Rosenow and coworkers in only 1.3 percent of 2080 patients.[307] In this study, 25 patients had apical fibrobullous changes, two had pleural effusions, and one had both.[307] The various pulmonary manifestations described in ankylosing spondylitis are summarized in Table 26-9. Either the thoracic cage or pulmonary parenchyma can be affected.[264,308] Pulmonary involvement in ankylosing spondylitis usually occurs late in the course of disease and is typically asymptomatic.[307–309]

The spondylitis can cause thoracic cage fixation, which leads to reduction in pulmonary function, with vital capacity often reduced to 65 to 88 percent of predicted normal volumes.[264] Patients with severe kyphoscoliosis may have severe restrictive pulmonary function. Pulmonary function tests in one study indicated involvement of small airways.[308]

The most common pulmonary parenchymal finding in ankylosing spondylitis is apical fibrosis or fibrocavitary disease.[46–49,59,310,311] Such changes can be confused with tuberculosis or fungal disease.[309] About 20 percent of patients may have spontaneous pneumothoraces. The upper lobe fibrosis can be unilateral or bilateral.[309] The cause of this fibrosis is unknown, although an immune pathogenesis has been proposed.[309]

TABLE 26-9
Pleuropulmonary Complications of Ankylosing Spondylitis

Upper lobe fibrocystic disease

Pleural disease

    Pleuritis with or without pleural effusion

    Postoperative bronchopleural fistula or empyema

    Spontaneous pneumothorax

Miscellaneous processes

    Atelectasis

    Chronic aspiration

    Abnormal chest wall mobility

    Infection

        TB or atypical TB

        Aspergilloma

    Bronchocentric granulomatosis

    Scar carcinoma

    Mediastinal fibrosis

    Cricoarytenoid arthritis

    Amyloid

It is controversial whether ankylosing spondylitis predisposes to pulmonary infection. The apical changes resemble those in tuberculosis, which may have led to an exaggerated impression of the frequency of TB in these patients.[264] However, secondary infection of the fibrocavitary lesions—particularly that due to mycobacterial infection,[312] *Aspergillus* fungus balls, or invasive fungal infection—can lead to significant morbidity and mortality.[264,312] These fungal lesions can present with cough and hemoptysis, which may be massive.

A few lung carcinomas have been reported in patients with ankylosing spondylitis.[313,314] However, an increased frequency of lung cancer in ankylosing spondylitis has not been established. Increased lung cancer mortality was found in patients who received radiation therapy for ankylosing spondylitis.[315]

Cases of pulmonary hemorrhage[316] and amyloid have been reported in ankylosing spondylitis patients.[59]

Relapsing Polychrondritis

Relapsing polychondritis is characterized by inflammation of cartilaginous structures, such as the larynx, trachea, and large bronchi.[317–321] In patients with respiratory tract involvement, there is a 2.8:1 female-to-male predominance and an average age of 40 years (range: 2 to 73 years).[321] Patients present with hoarseness, dyspnea, cough, stridor, wheezing, and neck tenderness over the laryngotracheal cartilages.[321] The destructive inflammation can lead to airway narrowing in either the upper or lower respiratory tract. Pulmonary function tests may show evidence of airway obstruction.[322,323] Spirometry is important in defining the airway obstruction encountered in patients with this disease.[323] In patients who have fixed upper airway obstruction, computed tomography, bronchoscopy, cinetracheography, or laryngotracheography is helpful in locating the obstructive lesions.[321,322,324] Strictures can lead to fatal intramural tracheal or bronchial fibrosis.[264,325] This may occur even though patients are treated with tracheostomy or steroids.[321]

Histologically, in the acute phase of relapsing polychondritis the pericartilaginous tissue is surrounded by lymphocytes, plasma cells, and neutrophils. Subsequently, depletion of the matrix proteoglycans and chondrocyte degeneration and destruction occur.[264] After the cartilage is completely destroyed, it is replaced by granulation tissue and fibrosis.[264]

Marfan Syndrome

Marfan syndrome primarily affects the skeleton, cardiovascular system, and the eye.[326] The disorder is linked to abnormalities in synthesis of fibrillin, a component of connective tissue microfibrils. These fibrils become associated with elastin to form elastic fibers.[326a–d] Skeletal abnormalities, which can affect the thoracic cage and pulmonary function, include pectus excavatum, pectus carinatum, and scoliosis with or without kyphosis.[264,327,328] Mayfan's syndrome occurs in approximately four to six per 100,000 of the population.[264,326] Approximately 85 percent of cases represent an inherited form of CTD. The remaining 15 percent of cases result from a spontaneous mutation.[264,326] The abnormal gene appears to be located on chromosome 15.

Pulmonary parenchymal abnormalities include pneumothorax (which is often bilateral or recurrent),[329–333] emphysema,[334–338] apical bullae,[339,340] widespread bilateral thin-walled cysts (in children),[341] bronchiectasis,[342–343] congenital lobar malformations, recurrent infections,[344] and upper lobe fibrosis.[264] Emphysematous bullous cysts are uncommon, occurring in 5 percent of adult patients in one study.[328] It has been suggested that bullous emphysema and pneumothorax occur in Marfan syndrome as a result of weakness in the connective tissue of the lung.[340]

Pneumothorax may be the presenting manifestation of Marfan's syndrome.[330,331] Spontaneous pneumothorax in patients over 12 years of age was found in 4.4 percent of subjects in one study.[330] Three members of one family had striking multiple episodes of spontaneous bilateral pneumothorax in Marfan syndrome.[329] Pulmonary saccular aneurysms of the left and right pulmonary arteries can occur in Marfan syndrome.[345]

Bronchial hyperreactivity is a potential manifestation of Marfan syndrome in children.[346] Pulmonary hyperinflation and emphysema have been seen in infants with Marfan syndrome.[336,347]

Primary Biliary Cirrhosis

Primary biliary cirrhosis (PBC) is a chronic liver disease of unknown origin that primarily affects young to middle-aged women. It is characterized by granulomatous destruction of small intrahepatic bile ducts and is known to cause a variety of extrahepatic manifestations. Hallmark laboratory findings include an elevated serum alkaline phosphatase and antimitochondrial antibody. A number of pulmonary abnormalities have been reported.[348] One of the most notable findings is sarcoid-like granulomas in the lung.[349–355]

PBC can occur simultaneously with sarcoidosis, and in some cases an overlap syndrome has raised the question of a common pathogenesis.[349,352] BAL can show a cellular profile similar to that seen in sarcoidosis.[354] Other rare cases of UIP,[356,357] pulmonary hypertension,[358] lymphocytic interstitial pneumonia,[359] mild chronic interstitial pneumonitis,[360] and pulmonary hemorrhage in association with glomerulonephritis have been reported.[361] In one study pulmonary function abnormalities were found to be uncommon. Airflow obstruction is relatively rare, occurring in fewer than 10 percent of nonsmoking patients.[362] In addition, a correlation was found between the reduction in diffusing capacity with the severity of liver disease.[362] This differs from other data, which suggested that gas transfer abnormalities occur only in PBC patients with associated Sjögren's syndrome.[348] Another study noted that pulmonary function abnormalities were found in more than half of the patients, and they occurred primarily in those with symptomatic liver disease.[363] In addition, it was noted that the pulmonary function changes were similar to those found in sarcoidosis.[363] A subclinical alveolitis also has been detected in PBC patients with normal chest radiographs.[364]

Pulmonary complications may also occur in PBC patients after liver transplantation.[365] A patient with

PBC was also found to have the Churg-Strauss syndrome, temporal artery involvement, and polychondritis.[366]

REFERENCES

1. Wiedemann HP, Matthay RA: Pulmonary manifestations of the collagen vascular diseases. *Clin Chest Med* 10:677–722, 1989.
2. Hunninghake GW, Fauci AS: Pulmonary involvement in the collagen vascular diseases. *Am Rev Respir Dis* 119:471–503, 1979.
3. Harmon KR, Leatherman JW: Respiratory manifestations of connective tissue disease. *Semin Respir Infect* 3:258–273, 1988.
4. Turner-Warwick M: Connective tissue disorders and the lung. *Aust NZ J Med* 16:257–262, 1986.
5. Boulware DW, Weissman DN, Doll NJ: Pulmonary manifestations of the rheumatic diseases. *Clin Rev Allergy* 3:249–267, 1985.
6. Eisenberg H: The interstitial lung diseases associated with the collagen-vascular disorders. *Clin Chest Med* 3:565–578, 1982.
7. Yousem SA, Colby TV, Carrington CB: Lung biopsy in rheumatoid arthritis. *Am Rev Respir Dis* 131:770–777,1985.
8. Katzenstein A-LA: Idiopathic interstitial pneumonia: Classi-fication and diagnosis, in Churg A, Katzenstein A-LA (eds), *The Lung: Current Concepts.* Baltimore, Williams & Wilkins, 1993, pp 1–31.
9. Myers JL, Colby TV: Pathologic manifestations of bronchiolitis, constrictive bronchiolitis, cryptogenic organizing pneumonia, and diffuse panbronchiolitis. *Clin Chest Med* 14:611–622, 1993.
10. Katzenstein A-LA, Bloor CM, Leibow AA: Diffuse alveolar damage—the role of oxygen, shock, and related factors. A review. *Am J Pathol* 85:209–228, 1976.
11. Wells AU, Hansell DM, Rubens MB, et al: Fibrosing alveolitis in systemic sclerosis. Bronchoalveolar lavage findings in relation to computed tomographic appearance. *Am J Respir Crit Care Med* 150:462–468, 1994.
12. duBois RM: Idiopathic pulmonary fibrosis. *Annu Rev Med* 44:441–450, 1993.
13. Hyde DM, King TE, Jr., McDermott T, et al: Idiopathic pulmonary fibrosis. Quantitative assessment of lung pathology. Comparison of a semiquantitative and a morphometric histopathologic scoring system. *Am Rev Respir Dis* 146:1042–1047, 1992.
14. Crystal RG, Fulmer JD, Roberts WC, Moss ML, Line BR, Reynolds HY: Idiopathic pulmonary fibrosis. Clinical, histologic, radiographic, physiologic, scintigraphic, cytologic, and biochemical aspects. *Ann Intern Med* 85:769–788, 1976.
15. Hsue YT, Paulus HE, Coulson WF: Bronchiolitis obliterans organizing pneumonia in polymyositis. A case report with long term survival. *J Rheumatol* 20:877–879, 1993.
16. Ippolito JA, Palmer L, Spector S, Kane PB, Gorevic PD: Bronchiolitis obliterans organizing pneumonia and rheumatoid arthritis. *Semin Arthritis Rheum* 23:70–78, 1993.
17. Bridges AJ, Hsu KC, Dias-Arias AA, Chechani V: Bronchiolitis obliterans organizing pneumonia and scleroderma. *J Rheumatol* 19:1136–1140, 1992.
18. Gammon RB, Bridges TA, al-Nezir H, Alexander CB, Kennedy JI Jr.: Bronchiolitis obliterans organizing pneumonia associated with systemic lupus erythematosus. *Chest* 102:1171–1174, 1992.
19. Usui Y, Kimula Y, Miura H, et al: A case of bronchiolitis obliterans organizing pneumonia associated with primary Sjögren's syndrome who died of superimposed diffuse alveolar damage. *Respiration* 59:122–124, 1992.
20. van Thiel RJ, van der Burg S, Groote AD, Nossent GD, Wills SH: Bronchiolitis obliterans organizing pneumonia and rheumatoid arthritis. *Eur Respir J* 4:905–911, 1991.
21. Matteson EL, Ike RW: Bronchiolitis obliterans organizing pneumonia and Sjögren's syndrome. *J Rheumatol* 17:676–679, 1990.
22. Guerry-Force ML, Müller NL, Wright JL, et al: A comparison of bronchiolitis obliterans with organizing pneumonia, usual interstitial pneumonia, and small airways disease. *Am Rev Respir Dis* 135:705–712, 1987.
23. Epler GR, Colby TV, McLoud TC, Carrington CB, Gaensler EA: Bronchiolitis obliterans organizing pneumonia. *N Engl J Med* 312:152–158, 1985.
24. Davison AG, Heard BE, McAllister WAC, Turner-Warwick ME: Cryptogenic organizing pneumonitis. *Q J Med* 207:382–394, 1983.
25. Colby TV, Myers JL: Clinical and histologic spectrum of bronchiolitis obliterans, including bronchiolitis obliterans organizing pneumonia. *Semin Respir Med* 13:119–133, 1992.
26. Epler GR: Bronchiolitis obliterans organizing pneumonia: Definition and clinical features. *Chest* 102 (Suppl):2S–6S, 1992.
27. Cordier J-F, Loire R, Brune J: Idiopathic bronchiolitis obliterans organizing pneumonia. Definition of characteristic clinical profiles in a series of 16 patients. *Chest* 96:999–1004, 1989.
28. Colby TV: Pathologic aspects of bronchiolitis obliterans organizing pneumonia. *Chest* 102 (Suppl):38S–43S, 1992.
29. Koss MN, Hochholzer L, Langloss JM, Lazarus AA, Wehunt WD: Lymphoid interstitial pneumonitis: Clinicopathologic and immunopathologic findings in 18 patients. *Pathology* 19:178–185, 1987.
30. Liebow AA, Carrington CB: Diffuse pulmonary lymphoreticular infiltrations associated with dysproteinemia. *Med Clin North Am* 57:809–843, 1973.
31. Pitt J: Lymphocytic interstitial pneumonia. *Pediatr Clin North Am* 38:89–95, 1991.
32. Strimlan CV, Rosenow EC, Divertie MB, Harrison EG Jr: Pulmonary manifestations of Sjögren's syndrome. *Chest* 70:354–361, 1976.
33. Yum MN, Ziegler JR, Walker PD, Ridolfo AS, Brashear RE: Pseudolymphoma of the lung in a patient with systemic lupus erythematosus. *Am J Med* 66:172–176, 1979.
34. Benisch B, Peison B: The association of lymphocytic interstitial pneumonia and systemic lupus erythematosus. *Mt Sinai J Med* 46:398, 1979.
35. Colby TV, Carrington CB: Lymphoreticular tumors and infiltrates of the lung. *Pathol Annu* (18 pt 1):27–70, 1983.
36. Travis WD, Pittaluga S, Lipschik GY, et al: Atypical pathologic manifestations of *Pneumocystis carinii* pneumonia in the acquired immune deficiency syndrome. Review of 123 lung biopsies from 76 patients with emphasis on cysts, vascular invasion, vasculitis, and granulomas. *Am J Surg Pathol* 14:615–625, 1990.
37. Murphy DM, Fox C, Travis WD, Koenig S, Fauci AS: Acquired immunodeficiency syndrome may present as

severe restrictive lung disease. *Am J Med* 86:237–240, 1989.

38. Travis WD, Fox CH, Devaney KO, et al: Lymphoid pneumonitis in 50 adult patients infected with the human immunodeficiency virus: Lymphocytic interstitial pneumonitis versus nonspecific interstitial pneumonitis. *Hum Pathol* 23:529–541, 1992.

39. Yousem SA, Colby TV, Carrington CB: Follicular bronchitis/bronchiolitis. *Hum Pathol* 16:700–706, 1985.

40. Kinoshita M, Higashi T, Tanaka C, Tokunaga N, Ichikawa Y, Oizumi K: Follicular bronchiolitis associated with rheumatoid arthritis. *Intern Med* 31:674–677, 1992.

41. Tazelaar HD, Viggiano RW, Pickersgill J, Colby TV: Interstitial lung disease in polymyositis and dermatomyositis. Clinical features and prognosis as correlated with histologic findings. *Am Rev Respir Dis* 141:727–733, 1990.

42. Colby TV, Coleman A: The histologic diagnosis of extrinsic allergic alveolitis and its differential diagnosis. *Prog Surg Pathol* 10:11–25, 1989.

43. Macfarlane JD, Franken CK, van Leeuwen AW: Progressive cavitating pulmonary changes in rheumatoid arthritis: A case report. *Ann Rheum Dis* 43:98–101, 1984.

44. McCann BG, Hart GJ, Stokes TC, Harrison BD: Obliterative bronchiolitis and upper-zone pulmonary consolidation in rheumatoid arthritis. *Thorax* 38:73–74, 1983.

45. Roschmann RA, Rothenberg RJ: Pulmonary fibrosis in rheumatoid arthritis: A review of clinical features and therapy. *Semin Arthritis Rheum* 16:174–185, 1987.

46. Hakala M, Kontkanen E, Koivisto O: Simultaneous presentation of upper lobe fibrobullous disease and spinal pseudarthrosis in a patient with ankylosing spondylitis. *Ann Rheum Dis* 49:728–729, 1990.

47. Rumancik WM, Firooznia H, Davis MS Jr., et al: Fibrobullous disease of the upper lobes: An extraskeletal manifestation of ankylosing spondylitis. *J Comput Tomogr* 8:225–229, 1984.

48. Gupta SM, Johnston WH: Apical pulmonary disease in ankylosing spondylitis. *NZ Med J* 88:186–188, 1978.

49. Wolson AH, Rohwedder JJ: Upper lobe fibrosis in ankylosing spondylitis. *Am J Roentgenol Radium Ther Nucl Med* 124:466–471, 1975.

50. Bély M. Apáthy A: Changes of the lung in rheumatoid arthritis—rheumatoid pneumonia. A clinicopathological study. *Acta Morphol Hung* 39:117–156, 1991.

51. Helmers R, Galvin J, Hunninghake GW: Pulmonary manifestations associated with rheumatoid arthritis. *Chest* 100:235–238, 1991.

52. Batra P, Collins JD, Magidson JG: Pulmonary nodular amyloidosis presenting as Sjögren's syndrome. *J Natl Med Assoc* 75:903–905, 1983.

53. Kobayashi H, Matsuoka R, Kitamura S, Tsunoda N, Saito K: Sjögren's syndrome with multiple bullae and pulmonary nodular amyloidosis. *Chest* 94:438–440, 1988.

54. Kyle RA, Bayrd ED: Amyloidosis: Review of 236 cases. *Medicine (Baltimore)* 54:271–299, 1975.

55. Melato M, Bianchi C: Pulmonary amyloidosis with unusual pathological features. *Morphol Embryol (Bucur)* 24:133–135, 1978.

56. Nomura S, Kumagai N, Kanoh T, Uchino H, Kurihara J: Pulmonary amyloidosis associated with systemic lupus erythematosus. *Arthritis Rheum* 29:680–682, 1986.

57. Chan CN, Li E, Lai FM, Pang JA: An unusual case of systemic lupus erythematosus with isolated hypoglossal nerve palsy, fulminant acute pneumonitis, and pulmonary amyloidosis. *Ann Rheum Dis* 8:236–239, 1989.

58. Benharroch D, Sukenik S, Sacks M: Bronchioloalveolar carcinoma and generalized amyloidosis complicating progressive systemic sclerosis. *Hum Pathol* 23:839–841, 1992.

59. Blavia R, Toda MR, Vidal R, Benet A, Razquin S, Richart C: Pulmonary diffuse amyloidosis and ankylosing spondylitis. A rare association. *Chest* 102:1608–1610, 1992.

60. Hsu BY, Edwards DK, Trambert MA: Pulmonary hemorrhage complicating systemic lupus erythematosus: Role of MR imaging in diagnosis. *AJR Am J Roentgenol* 158:519–520, 1992.

61. Travis WD, Colby TV, Lombard CM, Carpenter HA: A clinicopathologic study of 34 cases of diffuse pulmonary hemorrhage with lung biopsy confirmation. *Am J Surg Pathol* 14:1112–1125, 1990.

62. Nadorra RL, Landing BH: Pulmonary lesions in childhood onset systemic lupus erythematosus: Analysis of 26 cases, and summary of literature. *Pediatr Pathol* 7:1–18, 1987.

63. Mark EJ, Ramirez JF: Pulmonary capillaritis and hemorrhage in patients with systemic vasculitis. *Arch Pathol Lab Med* 109:413–418, 1985.

64. Desnoyers MR, Bernstein S, Cooper AG, Kopelman RI: Pulmonary hemorrhage in lupus erythematosus without evidence of an immunologic cause. *Arch Intern Med* 144:1398–1400, 1984.

65. Leatherman JW, Davies SF, Hoidal JR: Alveolar hemorrhage syndromes: Diffuse microvascular lung hemorrhage in immune and idiopathic disorders. *Medicine (Baltimore)* 63:343–361, 1984.

66. Marino CT, Pertschuk LP: Pulmonary hemorrhage in systemic lupus erythematosus. *Arch Intern Med* 141:201–203, 1981.

67. Churg A, Franklin W, Chan KL, Kopp E, Carrington CB: Pulmonary hemorrhage and immune-complex deposition in the lung. Complications in a patient with systemic lupus erythematosus. *Arch Pathol Lab Med* 104:388–391, 1980.

68. Eagen JW, Memoli VA, Roberts JL, Matthew GR, Schwartz MM, Lewis EJ: Pulmonary hemorrhage in systemic lupus erythematosus. *Medicine (Baltimore)* 57:545–560, 1978.

69. Mintz G, Galindo LF, Fernández-Diez J, Jiménez FJ, Robles-Saavedra E, Enríquez-Casillas RD: Acute massive pulmonary hemorrhage in systemic lupus erythematosus. *J Rheumatol* 5:39–50, 1978.

70. Edwards RL, Chalk SM, McEvoy JD, Donald KJ: Pulmonary haemorrhage in disseminated cardiac haemangiosarcoma. *Br J Dis Chest* 71:127–131, 1977.

71. Torralbo A, Herrero JA, Portoles J, Barrientos A: Alveolar hemorrhage associated with antineutrophil cytoplasmic antibodies in rheumatoid arthritis. *Chest* 105:1590–1592, 1994.

72. Griffin MT, Robb JD, Martin JR: Diffuse alveolar haemorrhage associated with progressive systemic sclerosis. *Thorax* 45:903–904, 1990.

73. Myers JL, Katzenstein A-LA: Microangiitis in lupus-induced pulmonary hemorrhage. *Am J Clin Pathol* 85:552–556, 1986.

74. Hoffman GS, Kerr GS, Leavitt RY, et al: Wegener's granulomatosis: An analysis of 158 patients. *Ann Intern Med* 116:488–498, 1992.

75. Kelly PT, Haponik EF: Goodpasture syndrome: Molecular and clinical advances. *Medicine (Baltimore)* 73:171–185, 1994.

76. Elliot ML, Kuhn C: Idiopathic hemosiderosis: Ultra-structural abnormalities in the capillary walls. *Am Rev Respir Dis* 102:895–904, 1970.

77. Kuhn C: Systemic lupus erythematosus in a patient with ultrastructural lesions of the pulmonary capillaries previ-

ously reported in the review as due to idiopathic pulmonary hemosiderosis. *Am Rev Respir Dis* 106:931–932, 1974.

78. Lemley DE, Katz P: Rheumatoid-like arthritis presenting as idiopathic pulmonary hemosiderosis: A report and review of the literature. *J Rheumatol* 13:954–957, 1986.

79. Partanen J, van Assendelft AH, Koskimies S, Forsberg S, Hakala M, Ilonen J: Patients with rheumatoid arthritis and gold-induced pneumonitis express two high-risk major histocompatibility complex patterns. *Chest* 92:277–281, 1987.

80. Cooney TP: Interrelationship of chronic eosinophilic pneumonia, bronchiolitis obliterans, and rheumatoid disease: A hypothesis. *J Clin Pathol* 34:129–137, 1981.

81. Seed WA, Fox B: Chronic eosinophilic pneumonia and rheumatoid arthritis—coincidental? (letter) *J Clin Pathol* 34:813, 1981.

82. Payne CR, Connellan SJ: Chronic eosinophilic pneumonia complicating long-standing rheumatoid arthritis. *Postgrad Med J* 56:519–520, 1980.

83. Lakhanpal S, Duffy J, Engel AG: Eosinophilia associated with perimyositis and pneumonitis. *Mayo Clin Proc* 63:37–41, 1988.

84. Samuels MP, Warner JO: Pulmonary alveolar lipoproteinosis complicating juvenile dermatomyositis. *Thorax* 43:939–940, 1988.

85. Kraft M, Mortenson RL, Colby TV, Newman L, Waldron JA Jr., King TE Jr: Cryptogenic constrictive bronchilitis. A clinicopathologic study. *Am Rev Respir Dis* 148:1093–1101, 1993.

86. Aquino SL, Webb WR, Golden J: Bronchiolitis obliterans associated with rheumatoid arthritis: Findings on HRCT and dynamic expiratory CT. *J Comput Assist Tomogr* 18:555–558, 1994.

87. Schwarz MI, Lynch DA, Tuder R: Bronchiolitis obliterans: The lone manifestation of rheumatoid arthritis? *Eur Respir J* 7:817–820, 1994.

88. Yam LY, Wong R: Bronchiolitis obliterans and rheumatoid arthritis. Report of a case in a Chinese patient on D-penicillamine and review of the literature. *Ann Acad Med Singapore* 22:365–368, 1993.

89. Jacobs P, Bonnyns M, Depierreux M, Duchateau J, Sergysels R: Rapidly fatal bronchiolitis obliterans with circulating antinuclear and rheumatoid factors. *Eur J Respir Dis* 65:384–388, 1984.

90. Holness L, Tenenbaum J, Cooter NB, Grossman RF: Fatal bronchiolitis obliterans associated with chrysotherapy. *Ann Rheum Dis* 42:593–596, 1983.

91. Corrin B, Turner-Warwick M, Geddes DM, Brewerton DA: Bronchiolitis obliterans. A new form of rheumatoid lung? (letter) *Chest* 73:244, 1978.

92. du Bois RM, Geddes DM: Obliterative bronchiolitis, cryptogenic organising pneumonitis and bronchiolitis obliterans organizing pneumonia: Three names for two different conditions (editorial). *Eur Respir J* 4:774–775, 1991.

93. Shadick NA, Fanta CH, Weinblatt ME, O'Donnell W, Coblyn JS: Bronchiectasis. A late feature of severe rheumatoid arthritis. *Medicine (Baltimore)* 73:161–170, 1994.

94. Hillarby MC, McMahon MJ, Grennan DM, et al: HLA associations in subjects with rheumatoid arthritis and bronchiectasis but not with other pulmonary complications of rheumatoid disease. *Br J Rheumatol* 32:794–797, 1993.

95. Bonafede RP, Benatar SR: Bronchocentric granulomatosis and rheumatoid arthritis. *Br J Dis Chest* 81:197–201, 1987.

96. Hellems SO, Kanner RE, Renzetti AD Jr: Bronchocentric granulomatosis associated with rheumatoid arthritis. *Chest* 83:831–832, 1983.

97. Rohatgi PK, Turrisi BC: Bronchocentric granulomatosis and ankylosing spondylitis. *Thorax* 39:317–318, 1984.

98. Travis WD, Koss MN: Vasculitis, in Dail DH, Hammar SP (eds), *Pulmonary Pathology*, 2d ed. New York, Springer-Verlag, 1994, pp 1027–1095.

99. Gardiner P, Ward C, Allison A, et al: Pleuropulmonary abnormalities in primary Sjögren's syndrome. *J Rheumatol* 20:831–837, 1993.

100. Inase N, Usui Y, Tachi H, et al: Sjögren's syndrome with bronchial gland involvement and multiple bullae. *Respiration* 57:286–288, 1990.

101. Grisham E, Spiera H: Vasculitis in connective tissue diseases, including hypcomplementemic vasculitis, in Churg A, Churg J (eds), *Systemic Vasculitides*. New York, Igaku-Shoin, 1991, pp 273–292.

102. Roncoroni AJ, Alvarez C. Molinas F: Plexogenic arteriopathy associated with pulmonary vasculitis in systemic lupus erythematosus. *Respiration* 59:52–56, 1992.

103. Wisnieski JJ, Emancipator SN, Korman NJ, Lass JH, Zaim TM, McFadden ER: Hypocomplementemic urticarial vasculitis syndrome in identical twins. *Arthritis Rheum* 37:1105–1111, 1994.

104. Vollertsein RS, Conn DL, Ballard DJ, Ilstrup DM, Kazmar RE, Silverfield JC: Rheumatoid vasculitis: Survival and associated risk factors. *Medicine* 65:365–375, 1986.

105. Baydur A, Mongan ES, Slager UT: Acute respiratory failure and pulmonary arteritis without parenchymal involvement. Demonstration in a patient with rheumatoid arthritis. *Chest* 75:518–520, 1979.

106. Lakhanpal S, Lie JT, Conn DL, Martin WJ: Pulmonary disease in polymyositis/dermatomyositis: A clinicopathological analysis of 65 autopsy cases. *Ann Rheum Dis* 46:23–29, 1987.

107. Morassut PA, Walley VM, Smith CD: Pulmonary veno-occlusive disease and the CREST variant of scleroderma. *Can J Cardiol* 8:1055–1058, 1992.

108. Young RH, Mark EJ: Pulmonary vascular changes in scleroderma. *Am J Med* 64:998–1004, 1978.

109. Kanemoto N, Sato M, Moriuchi J, Ichikawa Y, Goto Y, Sasadaira H: An autopsied case of systemic lupus erythematosus with pulmonary hypertension—a case report. *Angiology* 39:187–192, 1988.

110. Wakaki K, Koizumi F, Fukase M: Vascular lesions in systemic lupus erythematosus (SLE) with pulmonary hypertension. *Acta Pathol Jpn* 34:593–604, 1984.

111. Kay JM, Banik S: Unexplained pulmonary hypertension with pulmonary arteritis in rheumatoid disease. *Br J Dis Chest* 71:53–59, 1977.

112. Jones MB, Osterholm RK, Wilson RB, Martin FH, Commers JR, Bachmayer JD: Fatal pulmonary hypertension and resolving immune-complex glomerulonephritis in mixed connective tissue disease. A case report and review of the literature. *Am J Med* 65:855–863, 1978.

113. Hedgpeth MT, Boulware DW: Pulmonary hypertension in primary Sjögren's syndrome. *Ann Rheum Dis* 47:251–253, 1988.

114. Caldwell IW, Aitchison JD: Pulmonary hypertension in dermatomyositis. *Br Heart J*: 273–276, 1965.

115. Padeh S, Laxer RM, Silver MM, Silverman ED: Primary pulmonary hypertension in a patient with systemic-onset juvenile arthritis. *Arthritis Rheum* 34:1575–1579, 1991.

116. Young ID, Ford SE, Ford DM: The association of pulmonary hypertension with rheumatoid arthritis. *J Rheumatol* 16:1266–1269, 1989.

117. Onodera S, Hill JR: Pulmonary hypertension. Report of a

case in association with rheumatoid arthritis. *Ohio State Med J* 61:141–144, 1965.

118. Gardner DL, Duthie JJR, Macleod J, Allan WSA: Pulmonary hypertension in rheumatoid arthritis: Report of a case with intimal sclerosis of the pulmonary and digital arteries. *Scot Med J* 2:183–188, 1957.

119. Walz-Leblanc BA, Urowitz MB, Gladman DD, Hanly PJ: The shrinking lung syndrome in systemic lupus erythematosus—improvement with corticosteroid therapy. *J Rheumatol* 19:1970–1972, 1992.

120. Thompson PJ, Dhillon DP, Ledingham J, Turner-Warwick M: Shrinking lungs, diaphragmatic dysfunction, and systemic lupus erythematosus. *Am Rev Respir Dis* 132:926–928, 1985.

121. Henry DW, Rosenthal A, McCarty DJ: Adenocarcinoma of the lung associated with eosinophilia and hidebound skin (letter). *J Rheumatol* 21:972–973, 1994.

122. Sarma DP, Weilbaecher TG: Systemic scleroderma and small cell carcinoma of the lung. *J Surg Oncol* 29:28–30, 1985.

123. Hansen LA, Prakash UB, Colby TV: Pulmonary lymphoma in Sjögren's syndrome. *Mayo Clin Proc* 64:920–931, 1989.

124. Faguet GB, Webb HH, Agee JF, Ricks WB, Sharbaugh AH: Immunologically diagnosed malignancy in Sjögren's pseudolymphoma. *Am J Med* 65:424–429, 1978.

125. McDonagh J, Greaves M, Wright AR, Heycock C, Owen JP, Kelly C: High resolution computed tomography of the lungs in patients with rheumatoid arthritis and interstitial lung disease. *Br J Rheumatol* 33:118–122, 1994.

126. Fujii M, Adachi S, Shimizu T, Hirota S, Sako M, Kono M: Interstitial lung disease in rheumatoid arthritis; Assessment with high-resolution computed tomography. *J Thorac Imaging* 8:54–62, 1993.

127. Van Hoeyweghen RJ, De Clerck LS, van Offel JF, Stevens WJ: Interstitial lung disease and adult-onset Still's disease. *Clin Rheumatol* 12:418–421, 1993.

128. Banks J, Banks C, Cheong B, et al: An epidemiological and clinical investigation of pulmonary function and respiratory symptoms in patients with rheumatoid arthritis. *Q J Med* 84:795–806, 1992.

129. Popp W, Rauscher H, Ritschka L, et al: Prediction of interstitial lung involvement in rheumatoid arthritis. The value of clinical data, chest roentgenogram, lung function, and serologic parameters. *Chest* 102:391–394, 1992.

130. Sumida S, Nagata A, Kaneko K, et al: A case of interstitial pneumonia antedating rheumatoid arthritis. *Kurume Med J* 36:87–89, 1989.

131. Charles PJ, Sweatman MC, Markwick JR, Maini RN: HLA-B40: A marker for susceptibility to lung disease in rheumatoid arthritis. *Dis Markers* 9:97–101, 1991.

132. Kerstens PJ, Boerbooms AM, Jeurissen ME, Fast JH, Assmann KJ, van de Putte LB: Accelerated nodulosis during low dose methotrexate therapy for rheumatoid arthritis. An analysis of ten cases. *J Rheumatol* 19:867–871, 1992.

133. Flowers JR, Clunie G, Burke M, Constant O: Bronchiolitis obliterans organizing pneumonia: The clinical and radiological features of seven cases and a review of the literature. *Clin Radiol* 45:371–377, 1992.

134. Rees JH, Woodhead MA, Sheppard MN, du Bois RM: Rheumatoid arthritis and cryptogenic organising pneumonitis. *Respir Med* 85:243–246, 1991.

135. Hakala M, Paakko P, Sutinen S, Huhti E, Koivisto O, Tarkka M: Association of bronchiolitis with connective tissue disorders. *Ann Rheum Dis* 45:656–662, 1986.

136. O'Duffy JD, Luthra HS, Unni KK, Hyatt RE: Bronchiolitis in a rheumatoid arthritis patient receiving auranofin. *Arthritis Rheum* 29:556–559, 1986.

137. Lahdensuo A, Mattila J, Vilppula A: Bronchiolitis in rheumatoid arthritis. *Chest* 85:705–708, 1984.

138. Penny WJ, Knight RK, Rees AM, Thomas AL, Smith AP: Obliterative bronchiolitis in rheumatoid arthritis. *Ann Rheum Dis* 41:469–472, 1982.

139. Murphy KC, Atkins CJ, Offer RC, Hogg JC, Stein HB: Obliterative bronchiolitis in two rheumatoid arthritis patients treated with penicillamine. *Arthritis Rheum* 24:557–560, 1981.

140. Geddes DM, Webley M, Emerson PA: Airways obstruction in rheumatoid arthritis. *Ann Rheum Dis* 38:222–225, 1979.

141. Geddes DM, Corrin B, Brewerton DA, Davies RJ, Turner-Warwick M: Progressive airway obliteration in adults and its association with rheumatoid disease. *Q J Med* 46:427–444, 1977.

142. Smith DH, Scott DL, Zaphiropoulos GC: Eosinophilia in D-penicillamine therapy. *Ann Rheum Dis* 42:408–410, 1983.

143. Myers JL: Diagnosis of drug reactions in the lung. *Monogr Pathol* (36):32–53, 1993.

144. Searles G, McKendry RJ: Methotrexate pneumonitis in rheumatoid arthritis: Potential risk factors. Four case reports and a review of the literature. *J Rheumatol* 14:1164–1171, 1987.

145. Stein HB, Ruedy J, Atkins CJ, Offer RC: Penicillamine and other remittive agents in rheumatoid arthritis: Comparisons and interaction. *Clin Invest Med* 7:59–63, 1984.

146. Stein HB, Chalmers A, Schroeder ML, Dillon A: Selected adverse reactions of D-penicillamine. *Clin Invest Med* 7:73–76, 1984.

147. Epler GR, Snider GL, Gaensler EA, Cathcart ES, FitzGerald MX, Carrington CB: Bronchiolitis and bronchitis in connective tissue disease. A possible relationship to the use of penicillamine. *JAMA* 242:528–532, 1979.

148. Scott DL, Bradby GV, Aitman TJ, Zaphiropoulos GC, Hawkins CF: Relationship of gold and penicillamine therapy to diffuse interstitial lung disease. *Ann Rheum Dis* 40:136–141, 1981.

149. Camus P, Degat OR, Justrabo E, Jeannin L: D-Penicillamine-induced severe pneumonitis. *Chest* 81:376–378, 1982.

150. Stein HB, Ruedy J, Atkins CJ, Offer RC: Penicillamine compared to previous chrysotherapy in rheumatoid arthritis. *J Rheumatol* 10:319–322, 1983.

151. Chalmers A, Thompson D, Stein HE, Reid G, Patterson AC: Systemic lupus erythematosus during penicillamine therapy for rheumatoid arthritis. *Ann Intern Med* 97:659–663, 1982.

152. Winterbauer RH, Wilske KR, Wheelis RF: Diffuse pulmonary injury associated with gold treatment. *N Engl J Med* 294:919–921, 1976.

153. Haerden J, Coolen L, Dequeker J: The effect of D-penicillamine on lung function parameters (diffusion capacity) in rheumatoid arthritis. *Clin Exp Rheumatol* 11:509–513, 1993.

154. Stein HB, Patterson AC, Offer RC, Atkins CJ, Teufel A, Robinson HS: Adverse effects of D-penicillamine in rheumatoid arthritis. *Ann Intern Med* 92:24–29, 1980.

155. Wolfe F, Schurle DR, Lin JJ, et al: Upper and lower airway disease in penicillamine treated patients with rheumatoid arthritis. *J Rheumatol* 10:406–410, 1983.

156. Goodman TA, Polisson RP: Methotrexate: Adverse reactions

and major toxicities. *Rheum Dis Clin North Am* 20:513–528, 1994.

157. Schnabel A, Gross WL: Low-dose methotrexate in rheumatic diseases—efficacy, side effects, and risk factors for side effects. *Semin Arthritis Rheum* 23:310–327, 1994.

158. Koh WH, Boey ML: Open lung biopsy in systemic lupus erythematosus patients with pulmonary disease. *Ann Acad Med Singapore* 22:323–325, 1993.

159. Meyers OL: Pulmonary involvement in systemic lupus erythematosus. *Scand J Rheumatol* (Suppl) 54:19–23, 1984.

160. Miller LR, Greenberg SD, McLarty JW: Lupus lung. *Chest* 88:265–269, 1985.

161. Haupt HM, Moore GW, Hutchins GM: The lung in systemic lupus erythematosus. Analysis of the pathologic changes in 120 patients. *Am J Med* 71:791–798, 1981.

162. Kraus A, Guerra-Bautista G, Chavarria P: Paragonimiasis: An infrequent but treatable cause of hemoptysis in systemic lupus erythematosus. *J Rheumatol* 17:244–246, 1990.

163. Yenrudi S, Shuangshoti S, Pupaibul K: Nocardiosis: Report of 2 cases with review of literature in Thailand. *J Med Assoc Thai* 74:47–54, 1991.

164. Kimura M, Udagawa S, Shoji A, et al: Pulmonary aspergillosis due to *Aspergillus terreus* combined with staphylococcal pneumonia and hepatic candidiasis. *Mycopathologia* 111:47–53, 1990.

165. Chayakul P, Thammakumpee G, Mitarnun W: Lung cavities from *Pneumocystis carinii* in a patient with systemic lupus erythematosus. *J Med Assoc Thai* 74:310–312, 1991.

166. Carette S, Macher AM, Nussbaum A, Plotz PH: Severe, acute pulmonary disease in patients with systemic lupus erythematosus: Ten years of experience at the National Institute of Health. *Semin Arthritis Rheum* 14:52–59, 1984.

167. Matthay RA, Schwarz MI, Petty TL, et al: Pulmonary manifestations of systemic lupus erythematosus: Review of twelve cases of acute lupus pneumonitis. *Medicine (Baltimore)* 54:397–409, 1975.

168. Onomura K, Nakata H, Tanaka Y, Tsuda T: Pulmonary hemorrhage in patients with systemic lupus erythematosus. *J Thorac Imaging* 6:57–61, 1991.

169. Gould DB, Soriano RZ: Acute alveolar hemorrhage in lupus erythematosus. *Ann Intern Med* 83:836–837, 1975.

170. Mana F, Mets T, Vincken W, Sennesael J, Vanwaeyenbergh J, Goossens A: The association of bronchiolitis obliterans organizing pneumonia, systemic lupus erythematosus, and Hunner's cystitis. *Chest* 104:642–644, 1993.

171. Lyon MG, Bewtra C, Kenik JG, Hurley JA: Tubuloreticular inclusions in systemic lupus pneumonitis. Report of a case and review of the literature. *Arch Pathol Lab Med* 108:599–600, 1984.

172. Hammar SP, Winterbauer RH, Bockus D, Remington F, Sale GE, Meyers JD: Endothelial cell damage and tubuloreticular structures in interstitial lung disease associated with collagen vascular disease and viral pneumonia. *Am Rev Respir Dis* 127:77–84, 1983.

173. Fraire AE, Smith MN, Greenberg SD, Weg JG, Sharp JT: Tubular structures in pulmonary endothelial cells in systemic lupus erythematosus. *Am J Clin Pathol* 56:244–248, 1971.

174. Hoffbrand BI, Vianna J, Khamashta M, Hughes GR, Walters DV: Antiphospholipid antibodies and shrinking lungs in SLE (letter). *Lupus* 1:408, 1992.

175. Rubin LA, Urowitz MB: Shrinking lung syndrome in SLE—a clinical pathologic study. *J Rheumatol* 10:973–976, 1983.

176. Hunter T, Arnott JE, McCarthy DS: Features of systemic

lupus erythematosus and sarcoidosis occurring together. *Arthritis Rheum* 23:364–366, 1980.

177. Askari A, Thompson P, Barnes C: Sarcoidosis: Atypical presentation associated with features of systemic lupus erythematosus. *J Rheumatol* 15:1578–1579, 1988.

178. Delgado EA, Malleson PN, Pirie GE, Petty RE: The pulmonary manifestations of childhood onset systemic lupus erythematosus. *Semin Arthritis Rheum* 19:285–293, 1990.

179. Maddison PJ: Systemic lupus erythematosus in the elderly. *J Rheumatol* 14(Suppl 13):182–187, 1987.

180. Fayemi AO: Pulmonary vascular disease in systemic lupus erythematosus. *Am J Clin Pathol* 65:284–290, 1976.

181. Pronk LC, Swaak AJ: Pulmonary hypertension in connective tissue disease. Report of three cases and review of the literature. *Rheumatol Int* 11:83–86, 1991.

182. Simonson JS, Schiller NB, Petri M, et al: Pulmonary hypertension in systemic lupus erythematosus. *J Rheumatol* 16:918–925, 1989.

183. Quismorio FP Jr, Sharma O, Koss M et al: Immunopathologic and clinical studies in pulmonary hypertension associated with systemic lupus erythematosus. *Semin Arthritis Rheum* 13:349–359, 1984.

184. Schwartzberg M, Lieberman DH, Getzoff B, Ehrlich GE: Systemic lupus erythematosus and pulmonary vascular hypertension. *Arch Intern Med* 144:605–607, 1984.

185. Masuda A, Tsushima T, Shizume K, et al: Recurrent pneumothoraces and mediastinal emphysema in systemic lupus erythematosus. *J Rheumatol* 17:544–548, 1990.

186. Greenfield DI, Trinh P, Fulenwider A, Barth WF: Kaposi's sarcoma in a patient with SLE. *J Rheumatol* 13:637–640, 1986.

187. Milligan DW, Chang JG: Systemic lupus erythematosus and lymphoma. *Acta Haematol* 64:109–110, 1980.

188. Silver RM, Miller KS: Lung involvement in systemic sclerosis. *Rheum Dis Clin North Am* 16:199–216, 1990.

189. Rocco VK, Hurd ER: Scleroderma and scleroderma-like disorders. *Semin Arthritis Rheum* 16:22–69, 1986.

190. Tai CC, Lee P, Wood RE: Progressive systemic sclerosis in a child: Case report. *Pediatr Dent* 15:275–279, 1993.

191. Talbott JH, Barrocas M: Progressive systemic sclerosis (PSS) and malignancy, pulmonary and non-pulmonary. *Medicine (Baltimore)* 58:182–207, 1979.

192. Enzenauer RJ, McKoy J, Riel M: Case report: Rapidly progressive systemic sclerosis associated with carcinoma of the lung. *Milit Med* 154:574–577, 1989.

193. Benson CH, Pinkston WC, Woodliff J, Harisdangkul V: Pulmonary malignancy in a 21-year-old male with progressive systemic sclerosis. *J Miss State Med Assoc* 24:147–149, 1983.

194. Talbott JH, Barrocas M: Carcinoma of the lung in progressive systemic sclerosis: A tabular review of the literature and a detailed report of the roentgenographic changes in two cases. *Semin Arthritis Rheum* 9:191–217, 1980.

195. Servi RJ, Albertini RE, Torretti D: Pulmonary hypertension, hypoxemia, and death in a patient with scleroderma. *South Med J* 78:739–741, 1985.

196. Tesluk H: Progressive systemic sclerosis and pulmonary cancer (letter). *Arch Pathol Lab Med* 108:7–8, 1984.

197. Jin TS, Park MS, Shin DY, Jang YB, Shun DJ: Bronchioloalveolar carcinoma in progressive systemic sclerosis. *Korean J Intern Med* 2:52–55, 1987.

198. Peters-Golden M, Wise RA, Hochberg M, Stevens MB, Wigley FM: Incidence of lung cancer in systemic sclerosis. *J Rheumatol* 12:1136–1139, 1985.

199. Focan C, Swale JL, Borlee-Hermans G, Claessens JJ: Systemic sclerosis, aplastic anemia and amyloidosis associated with lung carcinoma (letter). *Acta Clin Belg* 40:204–205, 1985.

200. Holzmann H, Korting GW: Pericollagenous amyloid deposits in the skin and internal organs in scleroderma. *Klin Wochenschr* 47:390–391, 1969.

201. Kosaka Y, Akizuki M, Yoshida S, Mimori T, Homma M: Large cystic lesions of the upper lobes of the lungs in two patients with CREST syndrome. *Arthritis Rheum* 27:935–938, 1984.

202. Groen H, Postma DS, Kallenberg CG: Interstitial lung disease and myositis in a patient with simultaneously occurring sarcoidosis and scleroderma. *Chest* 104:1298–1300, 1993.

203. Harrison NK, Myers AR, Corrin B, et al: Structural features of interstitial lung disease in systemic sclerosis. *Am Rev Respir Dir* 144:706–713, 1991.

204. Warrick JH, Bhalla M, Schabel SI, Silver RM: High resolution computed tomography in early scleroderma lung disease. *J Rheumatol* 18:1520–1528, 1991.

205. Guidry GG, Baethge BA, Payne DK, Grafton WD: Pulmonary fibrosis as the initial manifestation of scleroderma. *J La State Med Soc* 142:33–36, 1990.

206. Harrison NK, Glanville AR, Strickland B, et al: Pulmonary involvement in systemic sclerosis: The detection of early changes by thin section CT scan, bronchoalveolar lavage and ⁹⁹ᵐTc-DTPA clearance. *Respir Med* 83:403–414, 1989.

207. McCarthy DS, Baragar FD, Dhingra S, et al: The lungs in systemic sclerosis (scleroderma): A review and new information. *Semin Arthritis Rheum* 17:271–283, 1988.

208. Owens GR, Paradis IL, Gryzan S, et al: Role of inflammation in the lung disease of systemic sclerosis: Comparison with idiopathic pulmonary fibrosis. *J Lab Clin Med* 107:253–260, 1986.

209. Konig G, Luderschmidt C, Hammer C, Adelmann-Grill BC, Braun-Falco O, Fruhmann G: Lung involvement in scleroderma. *Chest* 85:318–324, 1984.

210. D'Angelo WA, Fries JF, Masi AT, Shulman LE: Pathologic observations in systemic sclerosis (scleroderma). A study of fifty-eight autopsy cases and fifty-eight matched controls. *Am J Med* 46:428–440, 1969.

211. Case Records of the Massachusetts General Hospital. *N Engl J Med* 320:1333–1340, 1989.

212. Wells AU, Hansell DM, Corrin B, et al: High resolution computed tomography as a predictor of lung histology in systemic sclerosis (corrected and republished article originally printed in *Thorax* 47:508–512, 1992). *Thorax* 47:738–742, 1992.

213. Edelson JD, Hyland RH, Ramsden M, et al: Lung inflammation in scleroderma: Clinical, radiographic, physiologic and cytopathological features. *J Rheumatol* 12:957–963, 1985.

214. Owens GR, Fino GJ, Herbert DL, et al: Pulmonary function in progressive systemic sclerosis. Comparison of CREST syndrome variant with diffuse scleroderma. *Chest* 84:546–550, 1983.

215. Rossi GA, Bitterman PB, Rennard SI, Ferrans VJ, Crystal RG: Evidence for chronic inflammation as a component of the interstitial lung disease associated with progressive systemic sclerosis. *Am Rev Respir Dis* 131:612–617, 1985.

216. Kinsella MB, Smith EA, Miller KS, Leroy EC, Silver RM: Spontaneous production of fibronectin by alveolar macrophages in patients with scleroderma. *Arthritis Rheum* 32:577–583, 1989.

217. Ohba T, McDonald JK, Silver RM, Strange C, Leroy EC, Ludwicka A: Scleroderma bronchoalveolar lavage fluid contains thrombin, a mediator of human lung fibroblast proliferation via induction of platelet-derived growth factor alpha-receptor. *Am J Respir Cell Mol Biol* 10:405–412, 1994.

218. Crestani B, Seta N, De Bandt M, et al: Interleukin 6 secretion by monocytes and alveolar macrophages in systemic sclerosis with lung involvement. *Am J Respir Crit Care Med* 149:1260–1265, 1994.

219. Crestani B, Seta N, Palazzo E, et al: Interleukin-8 and neutrophils in systemic sclerosis with lung involvement. *Am J Respir Crit Care Med* 150:1363–1367, 1994.

220. Harrison NK, Cambrey AD, Myers AR, et al: Insulin-like growth factor–I is partially responsible for fibroblast proliferation induced by bronchoalveolar lavage fluid from patients with systemic sclerosis. *Clin Sci* 86:141–148, 1994.

221. Frigieri L, Mormile F, Grilli N, et al: Bilateral bronchoalveolar lavage in progressive systemic sclerosis: Interlobar variability, lymphocyte subpopulations, and functional correlations. *Respiration* 58:132–140, 1991.

222. Parodi A, Puiatti P, Rebora A: Serological profiles as prognostic clues for progressive systemic scleroderma: The Italian experience. *Dermatologica* 183:15–20, 1991.

223. de Rooij DJ, Van de Putte LB, Habets WJ, Van Venrooij WJ: Marker antibodies in scleroderma and polymyositis: Clinical associations. *Clin Rheumatol* 8:231–237, 1989.

224. Fritzler MJ, Kinsella TD: The CREST syndrome: A distinct serologic entity with anticentromere antibodies. *Am J Med* 69:520–526, 1980.

225. Ungerer RG, Tashkin DP, Furst D, et al: Prevalence and clinical correlates of pulmonary arterial hypertension in progressive systemic sclerosis. *Am J Med* 75:65–74, 1983.

226. Yousem SA: The pulmonary pathologic manifestations of the CREST syndrome. *Hum Pathol* 21:467–474, 1990.

227. Naeye RL: Pulmonary vascular lesions in systemic scleroderma. *Dis Chest* 44:374–380, 1963.

228. Roumm AD, Medsger TA Jr: Cancer and systemic sclerosis. An epidemiologic study. *Arthritis Rheum* 28:1336–1340, 1985.

229. Kallenbach J, Prinsloo I, Zwi S: Progressive systemic sclerosis complicated by diffuse pulmonary haemorrhage. *Thorax* 32:767–770, 1977.

230. Dornetzhuber V, Vagac M, Kristúfek P: Progressive systemic sclerosis (diffuse scleroderma) and Goodpasture's syndrome in 47-year-old woman. *Pneumonologie* 152:267–272, 1975.

231. Kumagai Y, Yamagata J, Hamamoto T, et al: [Autopsy case of progressive systemic sclerosis (PSS) associated with rapidly progressive glomerulonephritis (RPGN) and diffuse intra-alveolar pulmonary hemorrhage—differentiation from Goodpasture's syndrome]. *Ryumachi* 23:212–220, 1983.

232. Kim JH, Follett JV, Rice JR, Hampson NB: Endobronchial telangiectasias and hemoptysis in scleroderma. *Am J Med* 84:173–174, 1988.

233. Brandwein SR, Medsger TA Jr, Skinner M, Sipe JD, Rodnan GP, Cohen AS: Serum amyloid A protein concentration in progressive systemic sclerosis (scleroderma). *Ann Rheum Dis* 43:586–589, 1984.

234. Rodnan GP, Benedek TG, Medsger TA Jr, Cammarata RJ: The association of progressive systemic sclerosis (scleroderma) with coal miners' pneumoconiosis and other forms of silicosis. *Ann Intern Med* 66:323–334, 1967.

235. Sluis-Cremer GK, Hessel PA, Nizdo EH, Churchill AR, Zeiss

EA: Silica, silicosis, and progressive systemic sclerosis. *Br J Ind Med* 42:838–843, 1985.

236. Targoff IN, Arnett FC, Berman L, O'Brien C, Reichlin M: Anti-KJ: A new antibody associated with the syndrome of polymyositis and interstitial lung disease. *J Clin Invest* 84:162–172, 1989.

237. Plotz PH, Dalakas M, Leff RL, Love LA, Miller FW, Cronin ME: Current concepts in the idiopathic inflammatory myopathies: Polymyositis, dermatomyositis, and related disorders. *Ann Intern Med* 111:143–157, 1989.

238. Schwarz MI: Pulmonary and cardiac manifestations of polymyositis-dermatomyositis. *J Thorac Imaging* 7:46–54, 1992.

239. Romans B, Cohen S: A rheumatologist's view of polymyositis/dermatomyositis: Extracutaneous and extramuscular involvement and overlap syndromes. *Clin Dermatol* 6: 15–22, 1988.

240. Dickey BF, Myers AR: Pulmonary disease in polymyositis/dermatomyositis. *Semin Arthritis Rheum* 14:60–76, 1984.

241. Schiavi EA, Roncoroni AJ, Puy RJ: Isolated bilateral diaphragmatic paresis with interstitial lung disease. An unusual presentation of dermatomyositis. *Am Rev Respir Dis* 129:337–339, 1984.

242. Schwarz MI, Matthay RA, Sahn SA, Stanford RE, Marmorstein BL, Scheinhorn DJ: Interstitial lung disease in polymyositis and dermatomyositis: Analysis of six cases and review of the literature. *Medicine (Baltimore)* 55:89–104, 1976.

243. Frazier AR, Miller RD: Interstitial pneumonitis in association with polymyositis and dermatomyositis. *Chest* 65:403–407, 1974.

244. Arsura EL, Greenberg AS: Adverse impact of interstitial pulmonary fibrosis on prognosis in polymyositis and dermatomyositis. *Semin Arthritis Rheum* 18:29–37, 1988.

245. Salmeron G, Greenberg SD, Lidsky MD: Polymyositis and diffuse interstitial lung disease. A review of the pulmonary histopathologic findings. *Arch Intern Med* 141:1005–1010, 1981.

246. Takizawa H, Shiga J, Moroi Y, Miyachi S, Nishiwaki M, Miyamoto T: Interstitial lung disease in dermatomyositis: Clinicopathological study. *J Rheumatol* 14:102–107, 1987.

247. Songcharoen S, Raju SF, Pennebaker JB: Interstitial lung disease in polymyositis and dermatomyositis. *J Rheumatol* 7:353–360, 1980.

248. Rowen AJ, Reichel J: Dermatomyositis with lung involvement, successfully treated with azathioprine. *Respiration* 44:143–146, 1983.

249. Duncan PE, Griffin JP, Garcia A, Kaplan SB: Fibrosing alveolitis in polymyositis. A review of histologically confirmed cases. *Am J Med* 57:621–626, 1974.

250. Lohr HF, Bocher WO, Hermann E, et al: Interstitial alveolitis as early manifestation of anti-Jo-1 positive polymyositis. *Z Rheumatol* 52:307–311, 1993.

251. Leu CC, Lan JL: Anti-Jo-1 antibody in patients with polymyositis/dermatomyositis. *Chung Hua Min Kuo Wei Sheng Wu Chi Mien I Hsueh Tsa Chih* 25:41–47, 1992.

252. Targoff IN, Trieu EP, Plotz PH, Miller FW: Antibodies to glycyl-transfer RNA synthetase in patients with myositis and interstitial lung disease. *Arthritis Rheum* 35:821–830, 1992.

253. Nash P, Schrieber L, Webb J: Interstitial lung disease as the presentation of anti-Jo-1 positive polymyositis. *Clin Rheumatol* 6:282–286, 1987.

254. Yoshida S, Akizuki M, Mimori T, Yamagata H, Inada S,

Homma M: The precipitating antibody to an acidic nuclear protein antigen, the Jo-1, in connective tissue diseases. A marker for a subset of polymyositis with interstitial pulmonary fibrosis. *Arthritis Rheum* 26:604–611, 1983.

255. Targoff IN, Trieu EP, Miller FW: Reaction of anti-OJ autoantibodies with components of the multi-enzyme complex of aminoacyl-tRNA synthetases in addition to isoleucyl-tRNA synthetase. *J Clin Invest* 91:2556–2564, 1993.

256. Bunch TW, Tancredi RG, Lie JT: Pulmonary hypertension in polymyositis. *Chest* 79:105–107, 1981.

257. Callen JP: The value of malignancy evaluation in patients with dermatomyositis. *J Am Acad Dermatol* 6:253–259, 1982.

258. Hidano A, Torikai S, Uemura T, Shimizu S: Malignancy and interstitial pneumonitis as fatal complications in dermatomyositis. *J Dermatol* 19:153–160, 1992.

259. Wang YT, Singh D, Poh SC: Diffuse interstitial pulmonary fibrosis, dermatomyositis and lung cancer—a case report. *Singapore Med J* 21:778–780, 1980.

260. Manchul LA, Jin A, Pritchard KI, et al: The frequency of malignant neoplasms in patients with polymyositis-dermatomyositis. A controlled study. *Arch Intern Med* 145:1835–1839, 1985.

261. Bohan A, Peter JB, Bowman RL, Pearson CM: Computer-assisted analysis of 153 patients with polymyositis and dermatomyositis. *Medicine (Baltimore)* 56:255–286, 1977.

262. Bohan A, Peter JB: Polymyositis and dermatomyositis (pt 1). *N Engl J Med* 292:344–347, 1975.

263. Bohan A, Peter JB: Polymyositis and dermatomyositis (pt 2). *N Eng J Med* 292:403–407, 1975.

264. Tanoue LT: Pulmonary involvement in collagen vascular disease: A review of the pulmonary manifestations of the Marfan syndrome, ankylosing spondylitis, Sjögren's syndrome, and relapsing polychondritis. *J Thorac Imaging* 7:62–77, 1992.

265. Fox RI, Howell FV, Bone RC, Michelson P: Primary Sjögren syndrome: Clinical and immunopathologic features. *Semin Arthritis Rheum* 14:77–105, 1984.

266. Bloch KJ, Buchanan WW, Wohl MJ, Bunim JJ: Sjögren's syndrome. A clinical, pathological, and serological study of sixty-two cases. *Medicine* 44:187–231, 1965.

267. Wallaert B, Prin L, Hatron PY, Ramon P, Tonnel AB, Voisin C: Lymphocyte subpopulations in bronchoalveolar lavage in Sjögren's syndrome. Evidence for an expansion of cytotoxic/suppressor subset in patients with alveolar neutrophila. *Chest* 92:1025–1031, 1987.

268. Dalavanga YA, Constantopoulos SH, Galanopoulou V, Zerva L, Moutsopoulos HM: Alveolitis correlates with clinical pulmonary involvement in primary Sjögren's syndrome. *Chest* 99:1394–1397, 1991.

269. Fairfax AJ, Haslam PL, Pavia D, et al: Pulmonary disorders associated with Sjögren's syndrome. *Q J Med* 50:279–295, 1981.

270. Constantopoulos SH, Tsianos EV, Moutsopoulos HM: Pulmonary and gastrointestinal manifestations of Sjögren's syndrome. *Rheum Dis Clin North Am* 18:617–635, 1992.

271. Constantopoulos SH, Drosos AA, Maddison PJ, Moutsopoulos HM: Xerotrachea and interstitial lung disease in primary Sjögren's syndrome. *Respiration* 46:310–314, 1984.

272. Kelly C, Gardiner P, Pal B, Griffiths I: Lung function in primary Sjögren's syndrome: A cross sectional and longitudinal study. *Thorax* 46:180–183, 1991.

273. Segal I, Fink G, Machtey I, Gura V, Spitzer SA: Pulmonary

function abnormalities in Sjögren's syndrome and the sicca complex. *Thorax* 36:286–289, 1981.

274. Constantopoulos SH, Papadimitriou CS, Moutsopoulos HM: Respiratory manifestations in primary Sjögren's syndrome. A clinical, functional, and histologic study. *Chest* 88:226–229, 1985.

275. Andonopoulos AP, Constantopoulos SH, Drosos AA, Moutsopoulos HM: Pulmonary function of nonsmoking patients with rheumatoid arthritis in the presence and absence of secondary Sjögren's syndrome, a controlled study. *Respiration* 53:251–258, 1988.

276. Bariffi F, Pesci A, Bertorelli G, Manganelli P, Ambanelli U: Pulmonary involvement in Sjögren's syndrome. *Respiration* 46:82–87, 1984.

277. Newball HH.Brahim SA: Chronic obstructive airway disease in patients with Sjögren's syndrome. *Am Rev Respir Dis* 115:295–304, 1977.

278. Eri Z, Stanic J, Duric D, Klem I: Nodular amyloidosis of the lungs associated with Sjögren's syndrome. *Plucne Bolesti* 42:153–155, 1990.

279. Taniguchi M, Sato A, Honda K, Ohgo S, Yoshimura K: Bronchopulmonary manifestations of Sjögren's syndrome: The findings of the BAL, TBLB, and bronchoscopy. *Panminerva Med* 28:249–251, 1986.

280. Deheinzelin D, de Carvalho CR, Tomazini ME, Barbas Filho JV, Saldiva PH: Association of Sjögren's syndrome and sarcoidosis. Report of a case. *Sarcoidosis* 5:68–70, 1988.

281. Bucher UG, Reid L: Sjögren's syndrome: Report of a fatal case with pulmonary and renal lesions. *Br J Dis Chest* 53:273–276, 1959.

282. Desche P, Couderc LJ, Epardeau B: Sjögren's syndrome and pulmonary lymphangiomyomatosis (letter). *Chest* 94:898, 1988.

283. Anderson LG, Talal N: The spectrum of benign to malignant lymphoproliferation in Sjögren's syndrome. *Clin Exp Immunol* 10:199–221, 1972.

284. Tsuzaka K, Akama H, Yamada H, Akizuki M, Tojo T, Homma M: Pulmonary pseudolymphoma presented with a mass lesion in a patient with primary Sjögren's syndrome: Beneficial effect of intermittent intravenous cyclophosphamide. *Scand J Rheumatol* 22:90–93, 1993.

285. Greenerg SD, Heisler JG, Gyorkey F, Jenkins DE: Pulmonary lymphoma versus pseudolymphoma: A perplexing problem. *South Med J* 65:775–784, 1972.

286. Schuurman HJ, Gooszen HC, Tan IW, Kluin DM, Wagenaar SS, van Unnik JA: Low-grade lymphoma of immature T-cell phenotype in a case of lymphocytic interstitial pneumonia and Sjögren's syndrome. *Histopathology* 11:1193–1204, 1987.

287. Capron F, Audouin J, Diebold J, Ameille J, Lebeau B, Rochemaure J: Pulmonary polymorphic centroblastic type malignant lymphoma in a patient with lymphomatoid granulomatosis, Sjögren syndrome and other manifestations of a dysimmune state. *Pathol Res Pract* 179:656–665, 1985.

288. Ebihara Y, Sagawa H, Kitazawa Y: Lymphomatoid granulomatosis (Liebow). *Acta Pathol Jpn* 32:641–648, 1982.

289. Weisbrot IM: Lymphomatoid granulomatosis of the lung, associated with a long history of benign lymphoepithelial lesions of the salivary glands and lymphoid interstitial pneumonitis. Report of a case. *Am J Clin Pathol* 66:792–801, 1976.

290. Batsakis JG, Bernacki EG, Rice DH, Stebler ME: Malignancy and the benign lymphoepithelial lesion. *Laryngoscope* 85:389–399, 1975.

291. Kassan SS, Thomas TL, Moutsopoulos HM, et al: Increased risk of lymphoma in sicca syndrome. *Ann Intern Med* 89:888–892, 1978.

292. Sharp GC, Irvin WS, Tan EM, Gould RG, Holman HR: Mixed connective tissue disease–an apparently distinct rheumatic disease syndrome associated with a specific antibody to an extractable nuclear antigen (ENA). *Am J Med* 52:148–159, 1972.

293. Prakash UB: Lungs in mixed connective tissue disease. *J Thorac Imaging* 7:55–61, 1992.

294. Sullivan WD, Hurst DJ, Harmon CE, et al: A prospective evaluation emphasizing pulmonary involvement in patients with mixed connective tissue disease. *Medicine (Baltimore)* 63:92–107, 1984.

295. Derderian SS, Tellis CJ, Abbrecht PH, Welton RC, Rajagopal KR: Pulmonary involvement in mixed connective tissue disease. *Chest* 88:45–48, 1985.

296. Prakash UBS, Luthra HS, Divertie MB: Intrathoracic manifestations in mixed connective tissue disease. *Mayo Clin Proc* 60:813–821, 1985.

297. Hoogsteden HC, van Dongen JJ, van der Kwast TH, Hooijkaas H, Hilvering C: Bilateral exudative pleuritis, an unusual pulmonary onset of mixed connective tissue disease. *Respiration* 48:164–167, 1985.

298. Hosoda Y, Suzuki Y, Takano M, Tojo T, Homma M: Mixed connective tissue disease with pulmonary hypertension: A clinical and pathological study. *J Rheumatol* 14:826–830, 1987.

299. Ueda N, Mimura K, Maeda H, et al: Mixed connective tissue disease with fatal pulmonary hypertension and a review of literature. *Virchows Arch [A]* 404:335–340, 1984.

300. Rosenberg AM, Petty RE, Cumming GR, Koehler BE: Pulmonary hypertension in a child with mixed connective tissue disease. *J Rheumatol* 6:700–704, 1979.

301. Kobayashi H, Sano T, Ii K, Hizawa K, Yamanoi A, Otsuka T: Mixed connective tissue disease with fatal pulmonary hypertension. *Acta Pathol Jpn* 32:1121–1129, 1982.

302. Wiener-Kronish JP, Solinger AM, Warnock ML, Churg A, Ordonez N, Golden JA: Severe pulmonary involvement in mixed connective tissue disease. *Am Rev Respir Dis* 124:499–503, 1981.

303. Sanchez-Guerrero J, Cesarman G, Alarcon-Segovia D: Massive pulmonary hemorrhage in mixed connective tissue diseases. *J Rheumatol* 16:1132–1134, 1989.

304. Germain MJ, Davidman M: Pulmonary hemorrhage and acute renal failure in a patient with mixed connective tissue disease. *Am J Kidney Dis* 3:420–424, 1984.

305. Martyn JB, Wong MJ, Huang SH: Pulmonary and neuromuscular complications of mixed connective tissue disease: A report and review of the literature. *J Rheumatol* 15:703–705, 1988.

306. Martens J, Demedts M: Diaphragm dysfunction in mixed connective tissue disease. A case report. *Scand J Rheumatol* 11:165–167, 1982.

307. Rosenow E, Strimlan CV, Muhm JR, Ferguson RH: Pleuropulmonary manifestations of ankylosing spondylitis. *Mayo Clin Proc* 52:641–649, 1977.

308. Feltelius N, Hedenstrom H, Hillerdal G, Hallgren R: Pulmonary involvement in ankylosing spondylitis. *Ann Rheum Dis* 45:736–740, 1986.

309. Boushea DK, Sundstrom WR: The pleuropulmonary manifestations of ankylosing spondylitis. *Semin Arthritis Rheum* 18:277–281, 1989.

310. Hurwitz SS, Conlan AA, Krige LP: Fibrocavitating pulmonary lesions in ankylosing spondylitis. *S Afr Med J* 61:168–170, 1982.

311. Libshitz HI, Atkinson GW: Pulmonary cystic disease in ankylosing spondylitis: two cases with unusual superinfection. *J Can Assoc Radiol* 29:266–268, 1978.

312. Levy H, Hurwitz MD, Strimling M, Zwi S: Ankylosing spondylitis lung disease and *Mycobacterium scrofulaceum*. *Br J Dis Chest* 82:84–87, 1988.

313. Shankar PS: Ankylosing spondylitis with fibrosis and carcinoma of the lung. *CA Cancer J Clin* 32:177–179, 1982.

314. Ahern MJ, Maddison P, Mann S, Scott CA: Ankylosing spondylitis and adenocarcinoma of the lung. *Ann Rheum Dis* 41:292–294, 1982.

315. Weiss HA, Darby SC, Doll R: Cancer mortality following X-ray treatment for ankylosing spondylitis. *Int J Cancer* 59:327–338, 1994.

316. Molina M, Ortega G, Cuesta F, Marras C: Pulmonary hemorrhage associated with ankylosing spondylitis (letter). *An Med Interna* 11:150, 1994.

317. Gibson GJ, Davis P: Respiratory complications of relapsing polychondritis. *Thorax* 29:726–731, 1974.

318. Herman JH: Polychondritis. *Curr Opin Rheumatol* 3:28–31, 1991.

319. McAdam LP, O'Hanlan MA, Bluestone R, Pearson CM: Relapsing polychondritis: Prospective study of 23 patients and a review of the literature. *Medicine (Baltimore)* 55:193–215, 1976.

320. Michet CJ Jr, McKenna CH, Luthra HS, O'Fallon WM: Relapsing polychondritis. Survival and predictive role of early disease manifestations. *Ann Intern Med* 104:74–78, 1986.

321. Eng J, Sabanathan S: Airway complications in relapsing polychondritis. *Ann Thorac Surg* 51:686–692, 1991.

322. Mohsenifar Z, Tashkin DP, Carson SA, Bellamy PE: Pulmonary function in patients with relapsing polychondritis. *Chest* 81:711–717, 1982.

323. Krell WS, Staats BA, Hyatt RE: Pulmonary function in relapsing polychondritis. *Am Rev Respir Dis* 133:1120–1123, 1986.

324. Port JL, Khan A, Barbu RR: Computed tomography of relapsing polychondritis. *Comput Med Imaging Graph* 17:119–123, 1993.

325. Higenbottam T, Dixon J: Chondritis associated with fatal intramural bronchial fibrosis. *Thorax* 34:563–564, 1979.

326. Pyeritz RE, McKusick VA: The Marfan syndrome: Diagnosis and management. *N Engl J Med* 300:772–777, 1979.

326a. Eldadah ZA, Brenn T, Furthmayr H, Dietz HC: Expression of a mutant human fibrillin allele upon a normal human or murine genetic background recapitulates a Marfan cellular phenotype. *J Clin Invest* 95:874–880, 1995.

326b. Christiano AM, Uitto J: Molecular pathology of the elastic fibers. *J Invest Dermatol* 103:53S–57S, 1994.

326c. Goldstein C, Liaw P, Jimenez SA, Buchberg AM, Siracusa LD: Of mice and Marfan: Genetic linkage analyses of the fibrillin genes, Fbn1 and Fbn2, in the mouse genome. *Mamm Genome* 5:696–700, 1994.

326d. Pereira L. D'Alessio M, Ramirez F, et al: Genomic organization of the sequence coding for fibrillin, the defective gene product in Marfan syndrome. *Hum Mol Genet* 2:1762, 1993.

327. Streeten EA, Murphy EA, Pyeritz RE: Pulmonary function in the Marfan syndrome. *Chest* 91:408–412, 1987.

328. Wood JR, Bellamy D, Child AH, Citron KM: Pulmonary disease in patients with Marfan syndrome. *Thorax* 39:780–784, 1984.

329. Yellin A, Shiner RJ, Lieberman Y: Familial multiple bilateral pneumothorax associated with Marfan syndrome. *Chest* 100:577–578, 1991.

330. Hall JR, Pyeritz RE, Dudgeon DL, Haller JA Jr: Pneumothorax in the Marfan syndrome: Prevalence and therapy. *Ann Thorac Surg* 37:500–504, 1984.

331. Gawkrodger DJ: Marfan's syndrome presenting as bilateral spontaneous pneumothorax. *Postgrad Med J* 57:240–241, 1981.

332. Sensenig DM, LaMarche P: Marfan's syndrome and spontaneous pneumothorax. *Am J Surg* 139:602–604, 1980.

333. Lipton RA, Greenwald RA, Seriff NS: Pneumothorax and bilateral honeycombed lung in Marfan syndrome. Report of a case and review of the pulmonary abnormalities in this disorder. *Am Rev Respir Dis* 104:924–928, 1971.

334. Bamforth S, Hayden MR: Pulmonary emphysema in neonate with the Marfan syndrome (letter). *Pediatr Radiol* 18:88, 1988.

335. Day DL: Pulmonary emphysema in a neonate with Marfan syndrome (letter). *Pediatr Radiol* 18:179, 1988.

336. Day DL, Burke BA: Pulmonary emphysema in a neonate with Marfan syndrome. *Pediatr Radiol* 16:518–521, 1986.

337. Sayers CP, Goltz RW, Mottiaz J: Pulmonary elastic tissue in generalized elastolysis (cutis laxa) and Marfan's syndrome: A light and electron microscopic study. *J Invest Dermatol* 65:451–457, 1975.

338. Reye RD, Bale DM: Elastic tissue in pulmonary emphysema in Marfan syndrome. *Arch Pathol* 96:427–431, 1973.

339. Tewari SC, Jayaswal R, Jetley RK, Rao PB: Pulmonary bullous disease in Marfan syndrome. *J Assoc Physicians India* 38:587–589, 1990.

340. Turner JA. Stanley NN: Fragile lung in the Marfan syndrome. *Thorax* 31:771–775, 1976.

341. Sharma BK, Talukdar B, Kapoor R: Cystic lung in Marfan's syndrome. *Thorax* 44:978–979, 1989.

342. Foster ME, Foster DR: Bronchiectasis and Marfan's syndrome. *Postgrad Med J* 56:718–719, 1980.

343. Teoh PC: Bronchiectasis and spontaneous pneumothorax in Marfan's syndrome. *Chest* 72:672–673, 1977.

344. Jain VK, Kumar P, Beniwal OP, Pareek RP: Pulmonary tuberculosis in Marfan's syndrome. *J Indian Med Assoc* 84:119–120, 1986.

345. Disler LJ, Manga P, Barlow JB: Pulmonary arterial aneurysms in Marfan's syndrome. *Int J Cardiol* 21:79–82, 1988.

346. Konig P, Boxer R, Morrison J, Pletcher B: Bronchial hyperreactivity in children with Marfan syndrome. *Pediatr Pulmonol* 11:29–36, 1991.

347. Dominguez R, Weisgrau RA, Santamaria M: Pulmonary hyperinflation and emphysema in infants with the Marfan syndrome. *Pediatr Radiol* 17:365–369, 1987.

348. Rodriguez-Roisin R, Pares A, Bruguera M, et al: Pulmonary involvement in primary biliary cirrhosis. *Thorax* 36:208–212, 1981.

349. Maddrey WC: Sarcoidosis and primary biliary cirrhosis.

Associated disorders? (editorial). *N Engl J Med* 308:588–590, 1983.

350. Kayser K, Bohrer M, Kayser C, et al: Alteration of human lung parenchyma associated with primary biliary cirrhosis. *Zentralbl Pathol* 139:377–380, 1993.

351. Wallace JG Jr, Tong MJ, Ueki BH, Quismorio FP: Pulmonary involvement in primary biliary cirrhosis. *J Clin Gastro-enterol* 9:431–435, 1987.

352. Keeffe EB: Sarcoidosis and primary biliary cirrhosis. Literature review and illustrative case. *Am J Med* 83:977–980, 1987.

353. Fagan EA, Moore-Gillon JC, Turner-Warwick M: Multiorgan granulomas and mitochondrial antibodies. *N Engl J Med* 308:572–575, 1983.

354. Spiteri MA, Johnson M, Epstein O, Sherlock S, Clarke SW, Poulter LW: Immunological features of lung lavage cells from patients with primary biliary cirrhosis may reflect those seen in pulmonary sarcoidosis. *Gut* 31:208–212, 1990.

355. Leff JA, Ready JB, Repetto C, Goff JS, Schwarz MI: Coexistence of primary biliary cirrhosis and sarcoidosis, *West J Med* 153:439–441, 1990.

356. Osaka M, Aramaki T, Okumura H, Kawanami O: Primary biliary cirrhosis with fibrosing alveolitis. *Gastroenterol Jpn* 23:457–460, 1988.

357. Yap SH, Schillings PH, Van Tongeren JH: Acute interstitial pneumonia (fibrosing alveolitis) in a patient with primary biliary cirrhosis. *Neth J Med* 23:111–116, 1980.

358. Yoshida EM, Erb SR, Ostrow DN, Ricci DR, Scudamore CH, Fradet G: Pulmonary hypertension associated with primary biliary cirrhosis in the absence of portal hypertension. A case report. *Gut* 35:280–282, 1994.

359. Weissman E, Becker NH: Interstitial lung disease in primary biliary cirrhosis. *Am J Med Sci* 285:21–27, 1983.

360. Ichikawa Y, Saisho M, Koga H, Tokisawa S, Tokunaga N, Oizumi K: Lymphocytic alveolitis associated with asymptomatic primary biliary cirrhosis. *Kurume Med J* 40:59–63, 1993.

361. Bissuel F, Bizollon T, Dijoud F, et al: Pulmonary hemorrhage and glomerulonephritis in primary biliary cirrhosis. *Hepatology* 16:1357–1361, 1992.

362. Krowka MJ, Grambsch DM, Edell ES, Cortese DA, Dickson ER: Primary biliary cirrhosis: Relation between hepatic function and pulmonary function in patients who never smoked. *Hepatology* 13:1095–1100, 1991.

363. Uddenfeldt P, Bjerle P, Danielsson A, Nystrom L, Stjernberg N: Lung function abnormalities in patients with primary biliary cirrhosis. *Acta Med Scand* 223:549–555, 1988.

364. Wallaert B, Bonniere P, Prin L, Cortot A, Tonnel AB, Voisin C: Primary biliary cirrhosis. Subclinical inflammatory alveolitis in patients with normal chest roentgenograms. *Chest* 90:842–848, 1986.

365. Krowka MJ, Cortese DA: Pulmonary aspects of chronic liver disease and liver transplantation. *Mayo Clin Proc* 60:407–418, 1985.

366. Conn DL, Dickson ER, Carpenter HA: The association of Churg-Strauss syndrome with temporal artery involvement, primary biliary cirrhosis, and polychondritis in a single patient. *J Rheumatol* 9:744–748, 1982.

Chapter 27

Noninfectious Necrotizing Granulomatous Disorders

Trudi Roberts / Thomas V. Colby

Wegener's granulomatosis, bronchocentric granulomatosis, and allergic granulomatosis and angiitis (Churg-Strauss syndrome) are distinct noninfectious granulomatous disorders that affect the lung. Although these disorders may share some clinical and pathologic features they are recognized as distinct and unrelated clinocopathologic entities on the basis of their natural history, symptomatology, distribution of associated vasculitis, laboratory investigations, response to treatment and prognosis. Necrotizing sarcoid granulomatosis is included with sarcoidosis in Chap. 16.

WEGENER'S GRANULOMATOSIS

Clinical Features

The most well known of these conditions is Wegener's granulomatosis, which was first formally defined by Wegener in 1939,[1] although an earlier case had been described by Klinger.[2] Wegener's granulomatosis is a rare condition which has a peak incidence in the fourth to the sixth decades of life and a slight male preponderance.[3–5] Genetic factors are likely to be important in the development of Wegener's granulomatosis and an association with the genes *HLADR2* and *HLA-DQw7* has been reported.[6] Siblings with the disease have been described and at least in one report prolonged separation prior to the development of the vasculitis seemed to make an environmental cause unlikely.[7] Wegener's granulomatosis is a multi-system disease and has been reported to affect almost every organ in the body (Tables 27-1 and 27-2). However, the best known form involves the upper and lower respiratory tracts and kidneys to produce a classic pathologic and clinical triad.[3–5] Virtually all patients present with some evidence of respiratory tract involvement, sinusitis being the most common symptom. The maxillary sinuses are the most frequently affected, followed by the sphenoid and to

a lesser extent the ethmoid sinuses. Nasal involvement presents as rhinitis, rhinorrhea and epistaxis. Erosion of the nasal cartilage results in septal perforation, which if severe can lead to a complete collapse of the nasal bridge, resulting in a characteristic deformity (Fig. 27-1).

The granulomatous process, which can occur anywhere in the upper respiratory tract, may produce ulceration or even pseudotumours. Thus, otitis media due to obstruction of the Eustachian tube is a common presentation. Cough, dyspnea, chest pain, and hemoptysis may all occur. Less commonly, an endobronchial granuloma can result in airway obstruction and atelectasis. Spontaneous pneumothoraces with broncho-pleural fistulae, pleurisy, and pleural effusions are also seen.

Many patients with Wegener's granulomatosis have clinical signs of renal involvement. These include hypertension and signs of fluid overload, with edema and congestive heart failure. Hematuria (sometimes microscopic) and proteinuria are also common.

Ocular involvement in Wegener's granulomatosis may present as a painful eye, disturbances of vision, or chemosis. Retroorbital granulomata may produce proptosis.

Skin involvement in Wegener's granulomatosis may manifest as vasculitic-type rashes, and, rarely, a pyoderma gangrenosum-like picture is encountered. Cardiac involvement can lead to a complete heart block due to involvement of the conducting tissue or nodal arteritis. Pancarditis and coronary arteritis may occur. Vasculitis of the vasa nervorum may result in polyneuritis or mononeuritis multiplex and cranial nerve abnormalities may be a presentation of central nervous system granulomatous lesions. Vasculitis of the cerebral vessels, together with thrombosis or hemorrhage have been described.

A limited form of Wegener's granulomatosis without renal involvement has been described. It generally follows a more indolent course. However,

TABLE 27-1

Characteristic Features of Organ System Involvement in Wegener's Granulomatosis

Organ System	Approximate Frequency (%)	Typical Features
Nasopharynx	75	Necrotizing granuloma with mucosal ulceration; saddle nose deformity
Paranasal sinuses	90	Pansinusitis; necrotizing granuloma; secondary bacterial infection
Eyes	60	Keratoconjunctivitis; granulomatous sclerouveitis
Ears	35	Serous otitis media; secondary bacterial infection
Lungs	95	Multiple nodular cavitary infiltrates; necrotizing granulomatous vasculitis
Kidneys	85	Focal and segmental glomerulitis; necrotizing glomerulonephritis later in course
Heart	15	Coronary vasculitis; pericarditis
Nervous system	20	Mononeuritis multiplex; cranial neuritis
Skin	40	Dermal vasculitis with secondary ulcerations
Joints	50	Polyarthragias

SOURCE: Fauci et al.,[3] with permission.

some patients, originally thought to have this form of the disease on the grounds that there were no biochemical abnormalities of renal function, subsequently have abnormal renal biopsies. Limited Wegener's granulomatosis may initially present with pulmonary disease (Fig. 27-2), only to develop the renal complications over a period of several years.

Elevated erythrocyte sedimentation rate or plasma viscosity, anemia with leucocytosis and thrombocytosis, and a positive rheumatoid factor are frequently found in Wegener's granulomatosis. The serum immunoglobulins are often elevated and although this is usually confined to the IgG and IgA, sometimes the IgM and, on occasion, the IgE are also elevated. The acute phase reactant C-reactive protein (CRP) is also elevated. Complement levels, however, are normal. Urine analysis frequently reveals red blood cells and protein.

The prognosis for patients with Wegener's granulomatosis has improved markedly over the years.[3–5] In 1958 the median survival for patients with untreated Wegener's granulomatosis was five months, and 80 percent of patients were dead within one year,[8] but survival rates at one year in excess of 90 percent were shown by Reza and coworkers in 1975,[9] and Fauci and colleagues in 1983.[4] Treatment is usually with oral cyclophosphamide and steroids, although some groups use cyclophosphamide alone.

TABLE 27-2
Presenting Signs and Symptoms in Wegener's Granulomatosis

	Patients	
Signs or Symptoms	No.	%
Constitutional		
Joint	37	44
Fever	29	34
Weight loss	14	16
Anorexia or malaise	7	8
Respiratory		
Pulmonary infiltrates	60	71
Sinusitis	57	67
Cough	29	34
Rhinitis or nasal symptoms	19	22
Hemoptysis	15	18
Epistaxis	9	11
Chest discomfort	7	8
Shortness of breath or dyspnea	6	7
Pleuritis or effusion	5	6
Urinary Tract		
Renal failure	9	1
Ocular		
Ocular inflammation (conjunctivitis,		
uveitis, episcleritis, and scleritis)	14	16
Proptosis	6	7
Cutaneous		
Skin Rash	11	13
Nervous system		
Headache	5	6
Aural		
Otitis	21	25
Hearing loss	5	6
Oral		
Oral ulcers	5	6

SOURCE: Fauci et al.,[4] with permission.

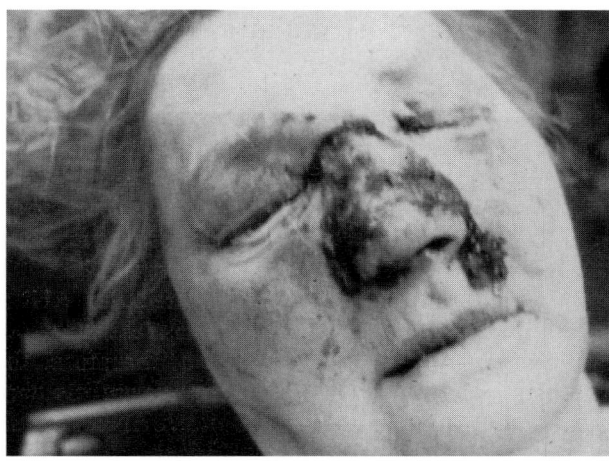

FIGURE 27-1
Wegener's granulomatosis showing nasal cartilaginous destruction. Appearance of the face postmortem.

marked granular staining and which is identical to the binding pattern originally described by Van der Woude. The second shows binding to ethanol-fixed neutrophils in a perinuclear distribution (p-ANCA). Both the c- and p-ANCA are specific for constituents of neutrophil azurophilic granules and monocyte lysozymes. The majority of patients with active Wegener's granulomatosis demonstrate a c-ANCA pattern of staining, whereas patients with the other forms of necrotizing vasculitis show a predominantly p-ANCA staining pattern.

Pathogenesis

Evidence that autoimmune mechanisms are involved in the systemic necrotizing granulomatous vasculitic disorders has emerged only within the last decade. In 1985, Van der Woude and coworkers showed that sera from patients with active Wegener's granulomatosis contained antibodies to neutrophil cytoplasmic components (ANCA).[10] Subsequently two forms of ANCA have been identified which are distinguishable by their indirect immunofluorescence pattern of binding to neutrophils.[11] One is the classic, or cytoplasmic, form of binding (c-ANCA) which has

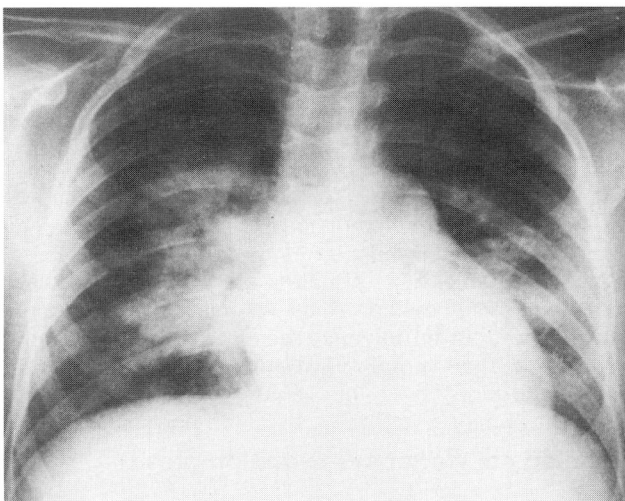

FIGURE 27-2
Wegener's granulomatosis. Chest x-ray showing multiple bilateral poorly defined peri-hilar densities with early cavitation.

Although initial reports suggested that the presence of c-ANCA had an 89 percent positive predictive value for Wegener's granulomatosis,[12] more recently it has been suggested that an overlap exists between Wegener's disease and microscopic polyarteritis in terms of the presence of c- or p-ANCA.[13] It may be that the specificity of a particular ANCA binding pattern in these conditions is less clear cut than was previously thought,[14] and, therefore, the diagnosis of a particular necrotizing vasculitis cannot be made solely on the basis of an ANCA result without corroborative clinical evidence.

Van der Woude et al.,[10] showed that ANCA levels correlated with disease activity and, more recently, Cohen Tervaert and coworkers[15] demonstrated that not only could relapses be predicted during follow-up by rising ANCA levels but that they could be aborted if cyclophosphamide was used when significant elevations occurred.

Thus the re-appearance of ANCA during long-term follow-up may be useful in identifying those patients who are at risk of relapse and who are most likely to benefit from long-term immunosuppressive therapy.

Although they may be of value in monitoring the disease activity, the pathogenic role of ANCA in systemic vasculitides remains unclear.[16] The antigenic target(s) of ANCAs have not been conclusively identified. However, there is increasing evidence that c-ANCA recognizes the neutrophil serine proteinase-3 and that p-ANCA recognizes myeloperoxidase. Purified myeloperoxidase has been shown to bind to endothelial cell membranes and, once bound, retains its antigenicity for p-ANCA. Purified proteinase-3 can also bind to endothelial cells and still be recognized by c-ANCA. The interactions between myeloperoxidase, proteinase-3, endothelial cell membranes and ANCA may therefore contribute to the vascular damage which is characteristic of these conditions.[17,18] However, some p-ANCA may also react with other neutrophil constituents such as elastase[19] or with the specific granule constituent lactoferin. Some c-ANCA do not react with purified proteinase-3, in which case the azurophil constituent CAP57 may be the target antigen.[19]

Pathology of Wegener's Granulomatosis

The many clinical patterns of Wegener's granulomatosis correlate with a diverse pathologic picture.[21] The pathologic findings also vary with time as lesions pursue their natural course or are modified by therapy.

MACROSCOPIC FINDINGS

The gross pathology of Wegener's granulomatosis parallels the histologic findings. Cases that manifest as acute alveolar hemorrhage show hemorrhagic consolidation of the lung tissue. Most cases show zones of consolidation which may include punctate yellow foci of necrosis or large coalescent zones of necrosis. Once extensive necrosis has developed, cavitation may occur when airways are eroded. Late in the course of the lesions and in some relatively indolent lesions, the necrosis may be surrounded by a relatively thin rim of viable tissue, and the appearance may be identical to an old infectious granuloma.

HISTOLOGY

The histologic findings of Wegener's granulomatosis vary with the clinical pattern.[21–27] Acute fulminant cases are dominated by alveolar hemorrhage and fibrinous exudates, whereas more indolent cases tend to show more extensive necrosis and scarring. Most histologic descriptions of Wegener's granulomatosis center on the distinctive features of the entity (i.e., the granulomatous inflammation and the vasculitis) and don't emphasize that Wegener's granulomatosis is an acute and chronic inflammatory process (the "inflammatory background" described below) that may have large zones that are quite nondescript histologically and lack granulomatous inflammation and vasculitis.

The distinctive and characteristic histologic features of Wegener's granulomatosis are reflected in several recent studies that have tried to identify histopathologic diagnostic criteria: those of Mark et al.,[23] Travis and coworkers,[25] and Galateau's group[27] are shown in Table 27-3.

The similarities of the studies in Table 27-3 are apparent, and all emphasize that Wegener's granulomatosis, while classified among the vasculitides, is more than a vasculitis and includes a component of necrosis (often granulomatous) and extensive tissue inflammation; it is the interplay of these features that leads to the varied and distinctive histologic picture of Wegener's granulomatosis that is never exactly the same from one case to the next.

Conceptually, the histologic manifestations of Wegener's granulomatosis can be viewed as having the following three components[21]:

Vasculitis
Necrosis/necrotic granuloma
Inflammatory background

TABLE 27-3
Histologic Diagnostic Features of Wegener's Granulomatosis

Mark et al.[23] (*N*=35)	Travis et al.[25] (*N*=87)	Galateau et al.[27] (*N*=40)
Major Discriminating Features:	*Major Manifestation:*	*Major Criteria:*
Palisading granulomas	Vasculitis of:	Polymorphonuclear microabscesses/geographic necrosis
In vessel walls	Arteries, veins, capillaries; acute, chronic, necrotizing granulomatous; nonnecrotizing granulomatous; fibrinoid necrosis, cicatricial	Angiitis of arteries, veins, and capillaries with focal crescent shaped mural microabscesses
In extravascular tissue		
Microabscess/fibrinoid necrosis in vessel walls		Polymorphous granulomas with giant cells
Leukocytoclastic capillaritis	Tissue necrosis:	*Minor Features of Lesser Significance:*
Granulomatous inflammation	Microabscesses/geographic	Acute/chronic inflammation with:
Diffuse granulomatous tissue	Granulomatous inflammation	Alveolar hemorrhage
Granulomatous bronchiolitis	Scattered giant cells	Endogenous lipid pneumonia
Minor Discriminating Features:	Poorly formed granulomas	Xanthomatous granulomas
Microabscesses	Microabscesses	Organizing pneumonia with alveolitis
Necrotizing bronchiolitis	Palisaded histiocytes	Bronchitis/necrotizing bronchiolitis
Lymphohistiocytic infiltrates in vascular walls	Sarcoid-like granulomas (rare)	
	Minor Manifestations:	Serofibrinous or neutrophilic pleural involvement with focal microabscesses
	Parenchymal	
	Nodular interstitial fibrosis	
	Endogenous lipoid pneumonia	
	Alveolar hemorrhage	
	Organizing intraluminal fibrosis	
	Lymphoid aggregates	
	Tissue eosinophils	
	Xanthogranulomatous lesions	
	Alveolar macrophage accumulation	
	Bronchial/Bronchiolar Lesions:	
	Chronic bronchiolitis	
	Acute bronchiolitis/bronchopneumonia	
	Bronchiolitis obliterans organizing pneumonia (BOOP)	
	Bronchocentric granulomatosis	
	Follicular bronchiolitis	
	Bronchial stenosis	
	Squamous metaplasia	

NOTE: None of the cases showed all these features.

The *vasculitis* of Wegener's granulomatosis typically affects medium-sized arteries and veins, although recently small vessel involvement has also been emphasized in the lung and given the label "capillaritis" (Figs. 27-3 through 27-6). Capillaritis in the lung is analogous to leukocytoclastic vasculitis in the skin (and other sites) and to the glomerular capillary lesions (glomerulitis) in the kidney.

Capillaritis is an inflammatory reaction involving the alveolar wall (capillaries or sometimes small venules) with a cuff of inflammation that is usually interstitial but may also spill into the adjacent airspaces. The inflammatory cells are typically neutrophils as well as cells resembling peripheral blood monocytes. Thrombosis and fibrinoid necrosis of the alveolar capillary may occur. But this is often hard to

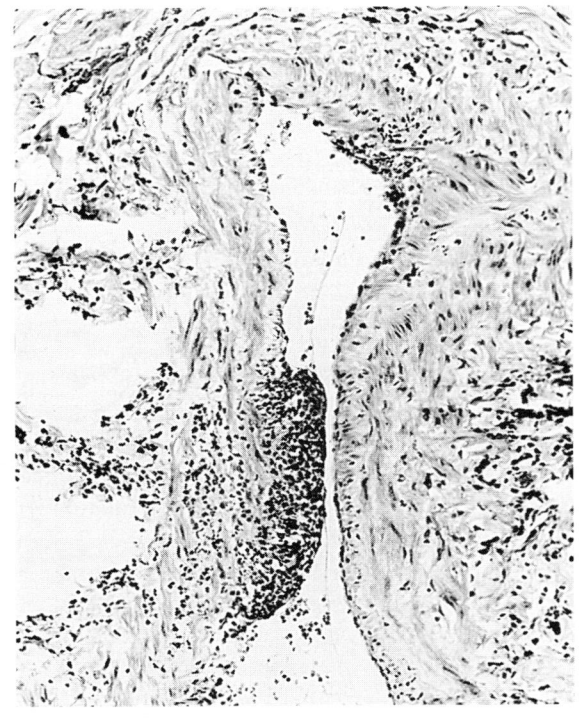

A

FIGURE 27-3

Vasculitis in Wegener's granulomatosis. Early vasculitic lesions are seen as intimal accumulations of mixed inflammatory cells including neutrophils, occasional eosinophils, and cells resembling peripheral blood monocytes (*A*). As the vessel is more

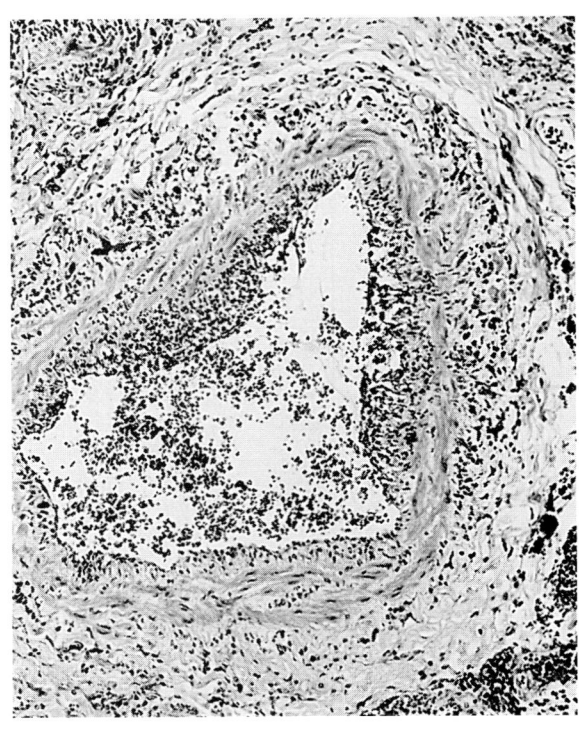

B

extensively involved, the infiltrate may become transmural (*B*). Panel *A*: (*Used with permission from: Churg A, Katzenstein ALA (eds): The Lung. Williams and Wilkins, Baltimore, 1993.*) (*A* and *B*, medium magnification.)

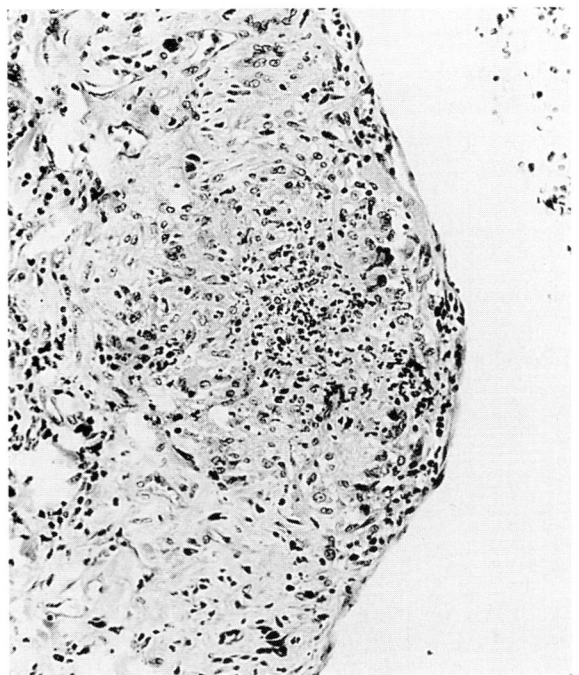

A

FIGURE 27-4

Vasculitis in Wegener's granulomatosis. In some cases, microabscesses develop within the vessel wall (*A*), and these may eventually become large intramural necrotic granulomas which typically are eccentric and crescentic in shape (*B*). The surrounding lung tissue shows characteristic geographic basophilic necro-

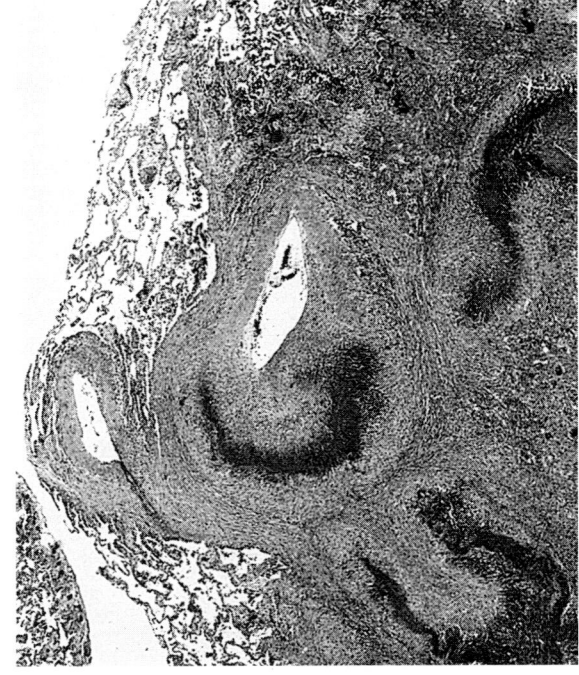

B

sis that is separate and distinct from that in the vessel wall (*B*). (*Used with permission from: Churg A, Katzenstein ALA (eds): The Lung. Williams and Wilkins, Baltimore, 1993.*) (*A*, high magnification; *B*, low magnification.)

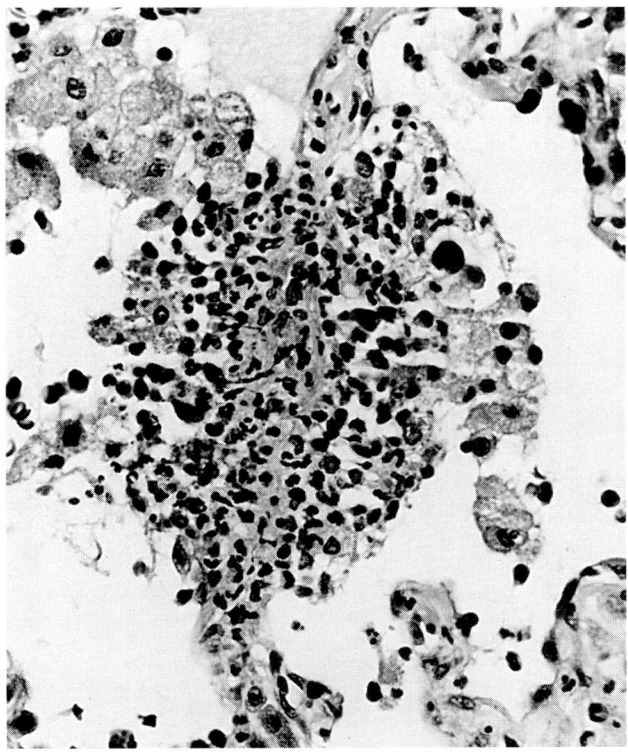

A

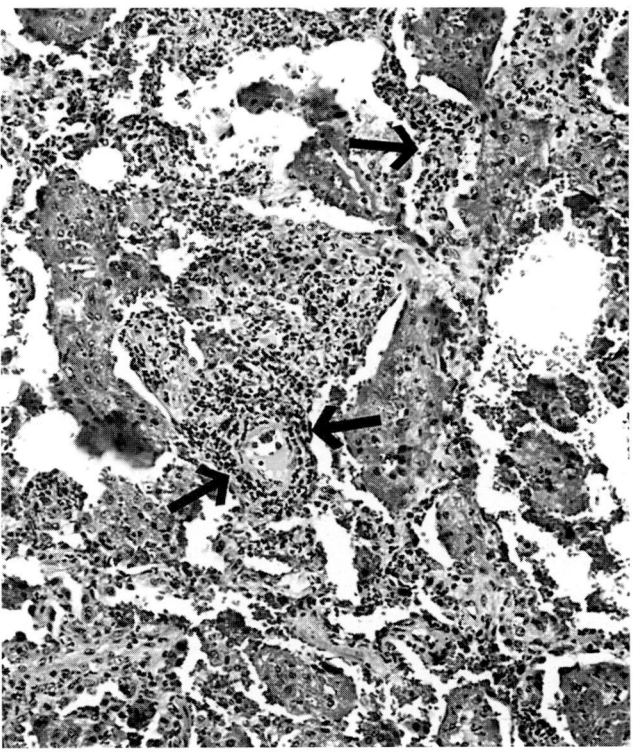

B

FIGURE 27-5

Capillaritis in Wegener's granulomatosis. Capillaritis is inflammation centered on the alveolar septum or around small veins and may be identifiable as a mixed infiltrate of neutrophils, occasional eosinophils, and cells resembling peripheral blood monocytes looking like a "swarm of bees" around the alveolar septum (*A*). Usually capillaritis is also associated with alveolar hemor-

rhage with alveolar spaces containing red blood cells, fibrin, and hemosiderin-filled macrophages, and the interstitium shows an increase in neutrophils either in alveolar septa or around small veins (*B, arrows*). (*Panel A used with permission from Churg A, Katzenstein ALA (eds): The Lung. Williams and Wilkins, Baltimore, 1993.*) (*A*, high magnification; *B*, medium magnification.)

appreciate in the lung, because it is difficult to distinguish from inflammation within the alveolar spaces. Capillaritis is most readily recognized histologically as a cuff of inflammatory cells resembling a "swarm of bees" around an alveolar septum or as an exquisitely interstitial accumulation of neutrophils with relative sparing of the adjacent alveolar spaces. It is this last feature which separates capillaritis from typical bacterial pneumonia with neutrophilic consolidation of airspaces. Occasionally, small venules and alveolar septa are the seat of a microgranulomatous reaction with palisaded histiocytes oriented toward the alveolar septum. Descriptively, this is a granulomatous capillaritis; it may also be observed at extrapulmonary sites.

Involvement of medium-sized arteries and veins in Wegener's granulomatosis typically begins as an intimal or transmural infiltrate of inflammatory cells, including neutrophils, occasional eosinophils, and cells resembling peripheral blood monocytes. The vessels are usually involved in a focal and segmental fashion. As the number of inflammatory cells increase, a small microabscess may develop resulting in an intramural necrotizing granuloma.

Uncommonly, the vasculitis of medium-sized arteries and veins manifests as a fibrinoid necrotizing vasculitis.

Necrosis in Wegener's granulomatosis is separate and distinct from the vasculitis and more than simply infarction of the tissue supplied by the affected vessels (Figs. 27-7*A* and *B*). The necrosis occurs in abnormal lung tissue, which shows inflammatory consolidation or fibrosis. The earliest foci of necrosis are seen as tiny clusters of neutrophils (early microabscesses) or as degenerative changes in the collagen. As these foci of necrosis expand, they become rimmed by a cuff of palisaded histiocytes and scattered giant cells and assume a granulomatous appearance. The microabcesses expand and coalesce to form the large zones of geographic basophilic necrosis that are so characteristic of Wegener's granulomatosis at low power magnification. The necrotic tissue is typically surrounded by large zones of consolidated (yet viable) lung tissue showing features of the inflammatory background described below. Although the necrosis is granulomatous, the presence of sarcoid-like non-necrotizing granulomas away from the zones of necrosis is distinctly unusual,

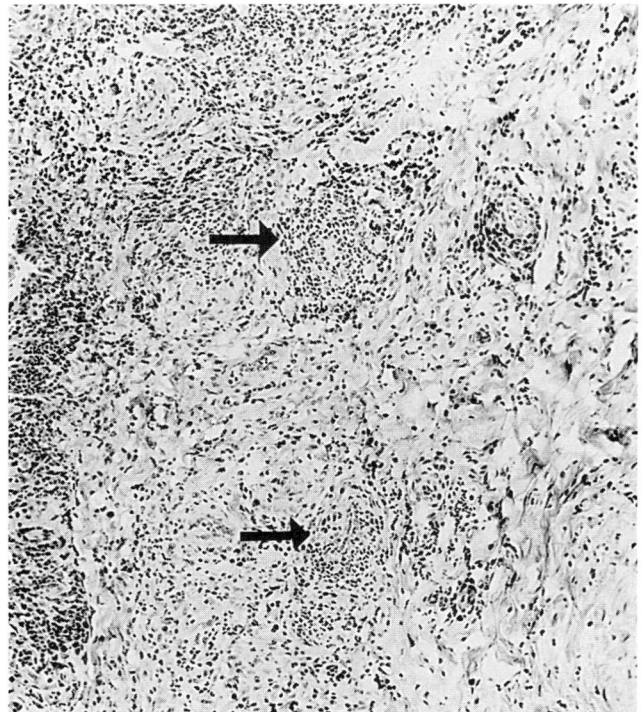

A

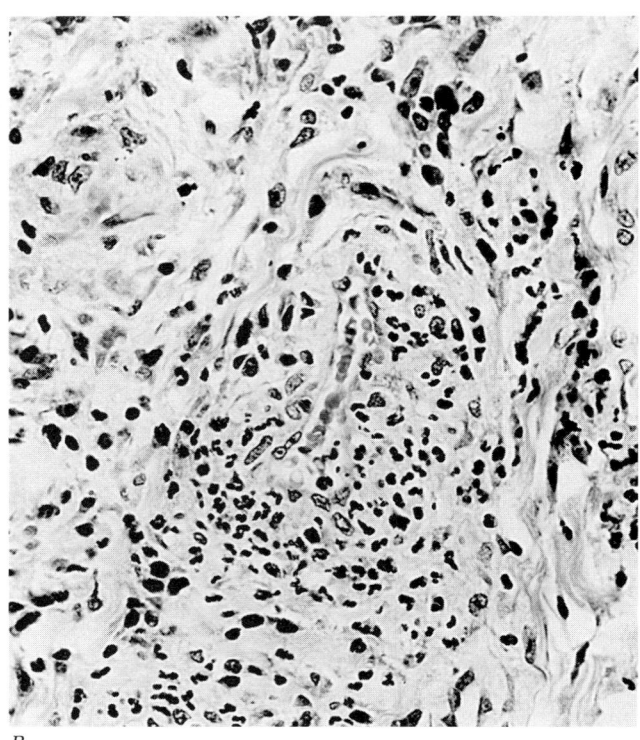

B

FIGURE 27-6

Capillaritis in Wegener's granulomatosis. The small vessels in the adventitia of a large pulmonary vein show capillaritis identical to leukocytoclastic vasculitis identified at other sites with clusters of neutrophils cuffing small vessels. (*A, arrows; B*) (*Used with*

permission from Churg A, Katzenstein ALA (eds.): The Lung. Williams and Wilkins, Baltimore, 1993.) (*A*, medium magnification; *B*, high magnification.)

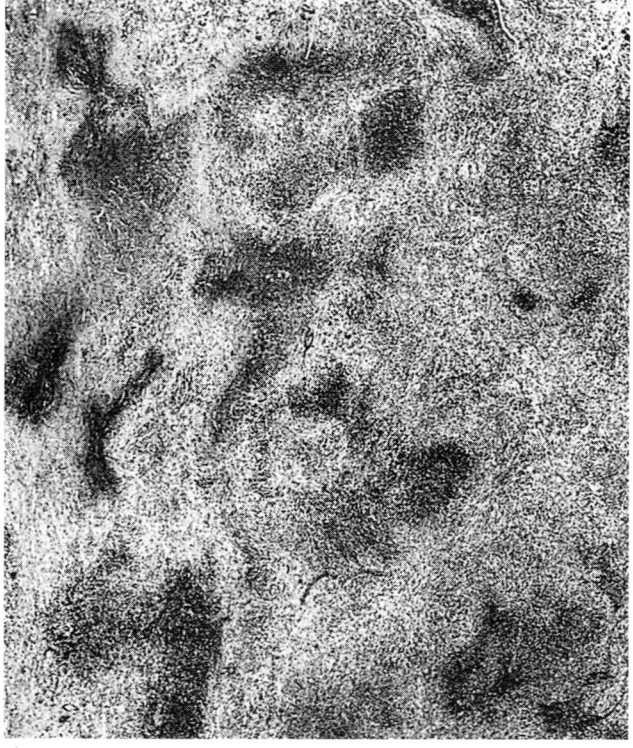

A

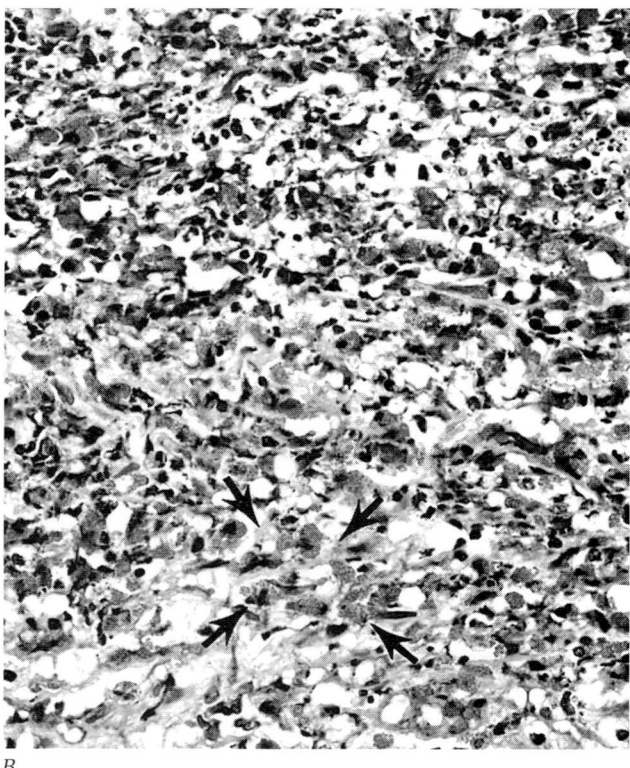

B

FIGURE 27-7

Necrosis in Wegener's granulomatosis. The necrosis in Wegener's granulomatosis is typically geographic and basophilic in appearance (*A*). In addition to numerous neutrophils, this case also had

large numbers of eosinophils (*B, arrows*), but clinical features and other histologic features of Churg-Strauss disease were lacking. (*A*, low magnification; *B*, high magnification.)

and this is a very helpful feature in separating Wegener's granulomatosis from most granulomatous infections.

The *inflammatory background* has been underemphasized in the pathologic descriptions of Wegener's granulomatosis (Figs. 27-8 through 27-11). Both the vasculitis and the necrosis develop in the setting of abnormal inflamed and consolidated tissue, which reflects the fundamental inflammatory nature of the disease. The reason that the inflammatory background has been underemphasized is that it tends to be relatively nonspecific in its appearance and resembles other inflammatory lesions in the lung. Nevertheless, the inflammatory background is the most prominent histologic finding in many cases of Wegener's granulomatosis, and it often manifests subtle histologic clues that separate it from other inflammatory reactions in the lung.

The inflammatory background of Wegener's granulomatosis has a varied histology, with some cases showing acute changes and others more chronic changes. Acute changes include alveolar hemorrhage, hyaline membranes, fibrinous exudates, neutrophilic capillaritis, and edema. Other cases have a subacute or organizing appearance with airspace organization resembling a nonspecific organizing pneumonia (also resembling bronchiolitis obliterans organizing pneumonia/cryptogenic organizing pneumonia). More chronic changes may be seen as collections of hemosiderin-filled macrophages, interstitial fibrosis (which may be hyalinized), and variable amounts of chronic inflammation with plasma cells. It is not unusual to see mixtures of the above features reflecting the ongoing activity of some cases.

The inflammatory cells comprising the inflammatory background of Wegener's granulomatosis generally include lymphocytes, plasma cells, occasional eosinophils, neutrophils (which may cluster to form early microabscesses), histiocytes, and, most distinctively, scattered darkly staining multinucleated giant cells.

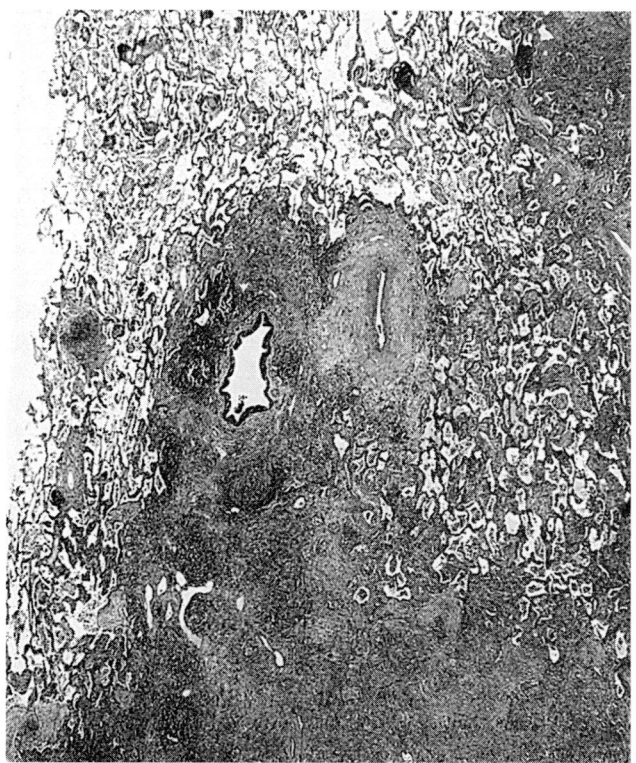

A

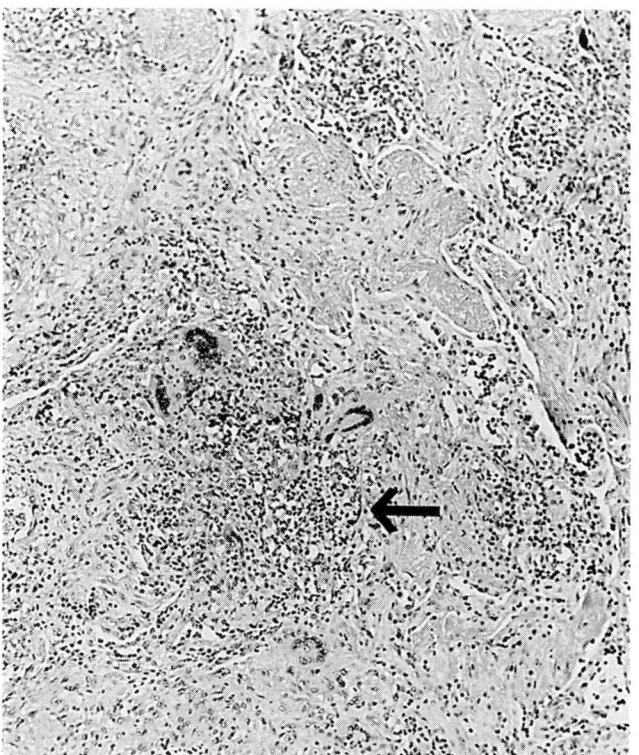

B

FIGURE 27-8

Wegener's granulomatosis. Wegener's granulomatosis typically produces consolidated and inflamed lung tissue which is separate and distinct from the vasculitis. In some cases, this involved tissue may spare vessels and even appear to center on airways (*A*). Within such tissue, one typically finds clusters of neutrophils (microabscesses, *arrow*) surrounded by palisaded histiocytes and scattered giant cells (*B*) without sarcoid-like non-necrotizing granulomas. (*Used with permission from Churg A, Katzenstein ALA (eds.): The Lung. Williams and Wilkins, Baltimore, 1993.*) (*A*, low magnification; *B*, medium magnification.)

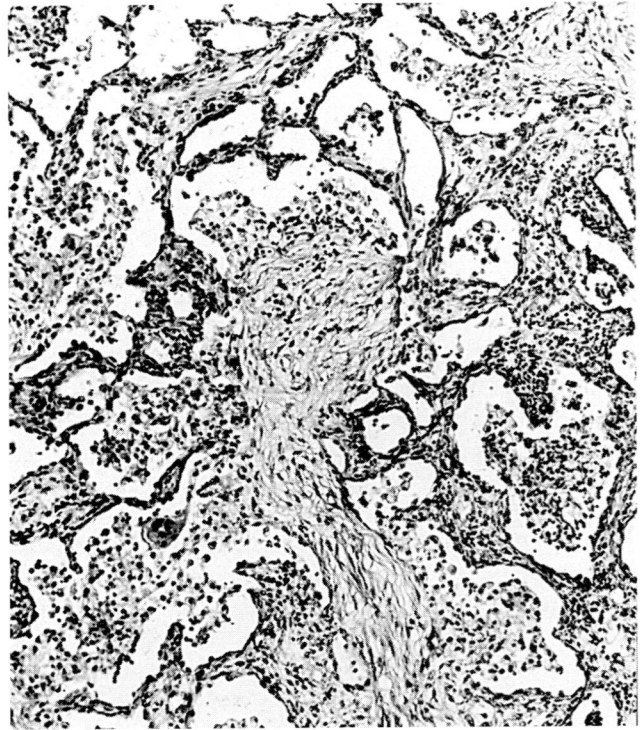

A

FIGURE 27-9
Wegener's granulomatosis. In some cases of Wegener's granulomatosis, large fields may resemble bronchiolitis obliterans organizing pneumonia/cryptogenic organizing pneumonia (*A*). In other cases, there is a dense sclerotic background of hyalinized fibrous tissue in which occasional clusters of neutrophils

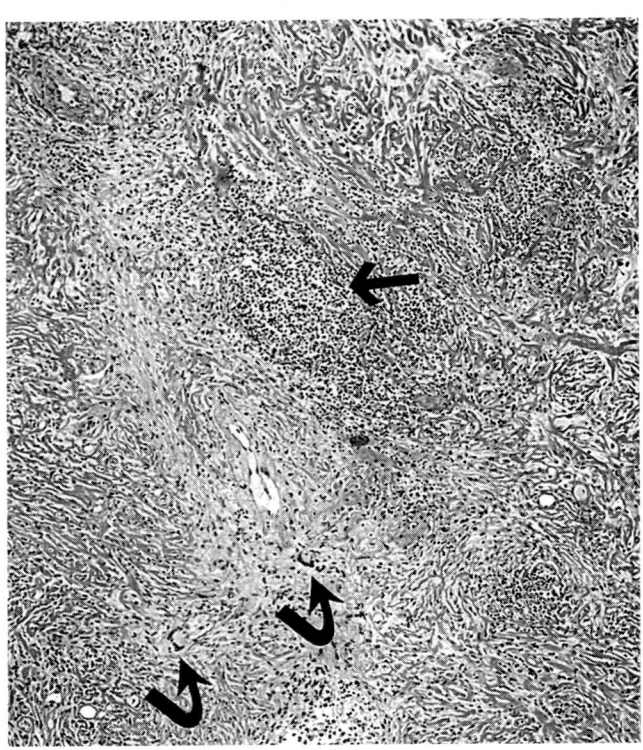

B

(microabscesses, *arrow)* and giant cells (*curved arrows*) may be found indicative of ongoing activity of the disease. (*Panel A used with permission from Churg A, Katzenstein ALA (eds.): The Lung. Williams and Wilkins, Baltimore, 1993.*) (*A* and *B*, medium magnification.)

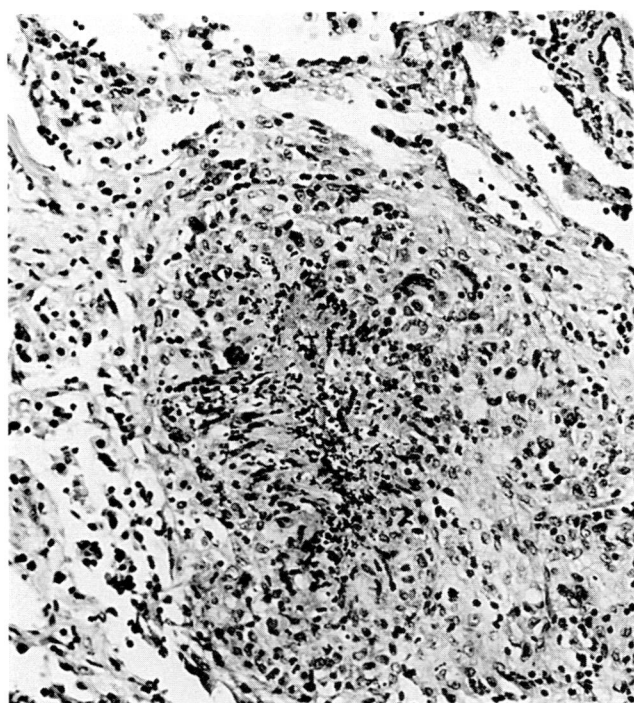

A

FIGURE 27-10
Early necrosis in Wegener's granulomatosis. Early necrosis is typically seen as microabscess formation (*A*). In occasional cases, the necrosis appears to start as a giant cell and palisaded histiocytic

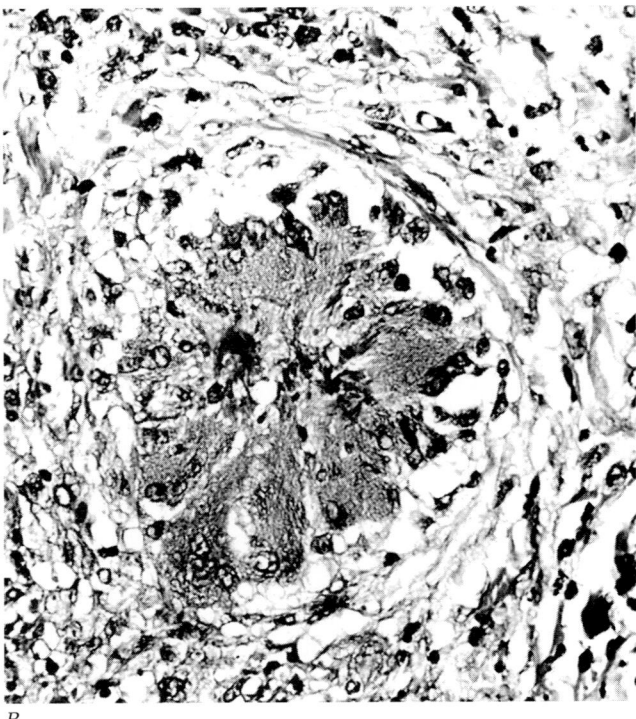

B

reaction around degenerated collagen (*B*). (*Used with permission from Churg A, Katzenstein ALA (eds.): The Lung. Williams and Wilkins, Baltimore, 1993.*) (*A* and *B*, high magnification.)

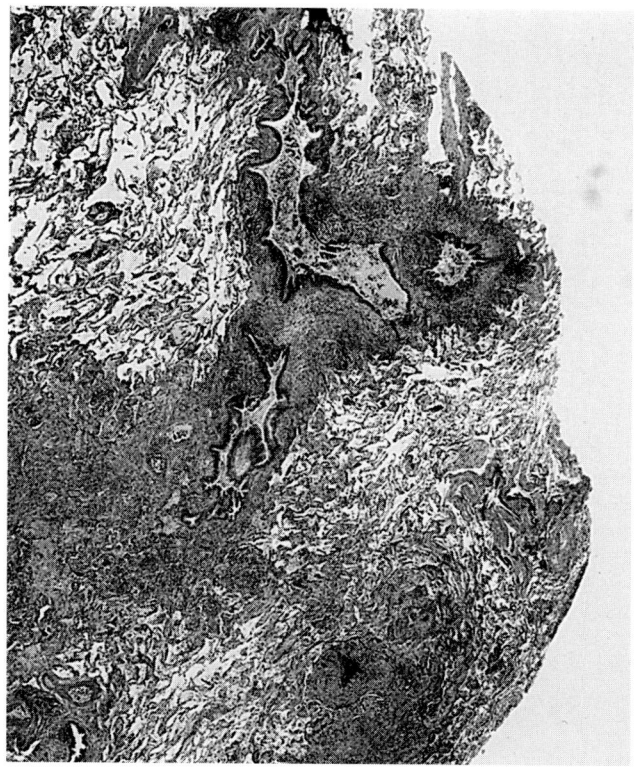

A

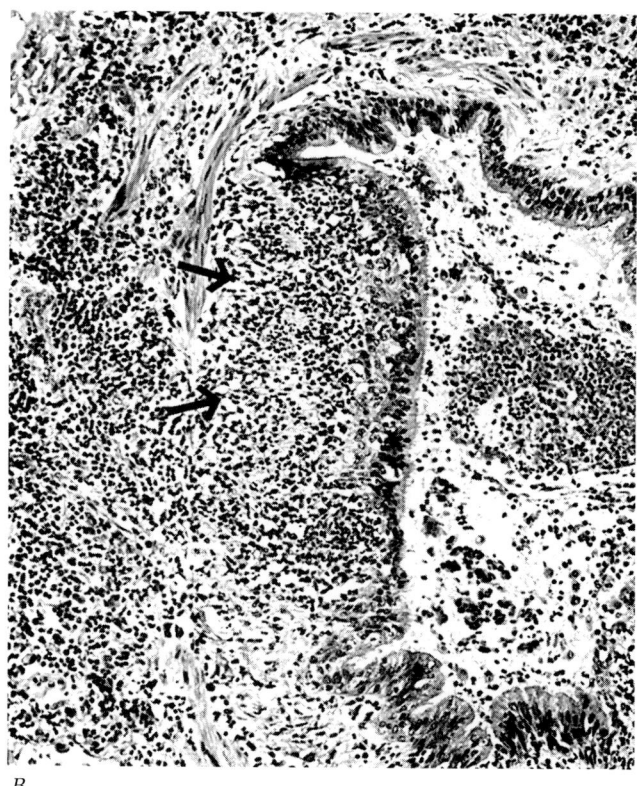

B

FIGURE 27-11

Bronchocentricity in Wegener's granulomatosis. Sometimes the inflammation in Wegener's granulomatosis appears to center on airways (A), and one may even see ulceration and microabscess formation (*arrows*) in the mucosa and submucosa of bronchi and bronchioles (B). (A, low magnification; B, medium magnification.)

The necrosis arises in the setting of this inflammatory background and is most typically seen to start as small clusters of neutrophils that eventually form microabscesses that then expand and coalesce to form the characteristic geographic necrosis. Less commonly necrosis is seen to develop as degeneration of collagen, sometimes within hyalinized fibrous tissue.

Viewing the histologic findings of Wegener's granulomatosis as an interplay of the vasculitis, necrosis, and inflammatory background, encompasses nearly all of the histologic findings that may be encountered. However, airway involvement does not fit well into this scheme unless one broadly calls it part of the inflammatory background. Although uncommon, airway involvement in Wegener's granulomatosis is occasionally seen and, when present, is quite distinctive. Early cases show scattered giant cells and pockets of inflammatory cells in the submucosa of bronchi and bronchioles. These may lead to ulceration and an appearance identical to bronchocentric granulomatosis (except elsewhere the lung tissue generally shows other typical features of Wegener's granulomatosis) (Figs. 27-11A and B). Involvement of airways is not surprising since this represents a mucosal site in continuity with mucosal sites in the head and neck, which are frequently the seat of ulcerative involvement by Wegener's granulomatosis.

The extent of vasculitis, necrosis, and inflammatory background (with or without airway lesions) varies from case to case, and, in any one case, one or more of the histologic changes may dominate over the others. Cases of Wegener's granulomatosis are recognized which lack necrosis or vasculitis, and, in such cases, knowledge of the clinical pattern and serologic findings (ANCA testing) are invaluable in helping to confirm the diagnosis.

The effects of therapy in Wegener's granulomatosis are rarely seen, although one must assume that the lesions almost entirely regress because chest radiographs may show complete clearing. In a few cases examined, the zones of active disease show fibrosis and only scattered giant cells with absent or only focal necrosis and marked diminution in the number of inflammatory cells (Figs. 27-12 and 27-13). Cases in complete remission may show only nonspecific old scarring.

Wegener's granulomatosis at extrapulmonary sites generally shows features similar to those seen in the lung, allowing for differences in the underlying anatomy of the organ (Figs. 27-14 through 27-18). As

845

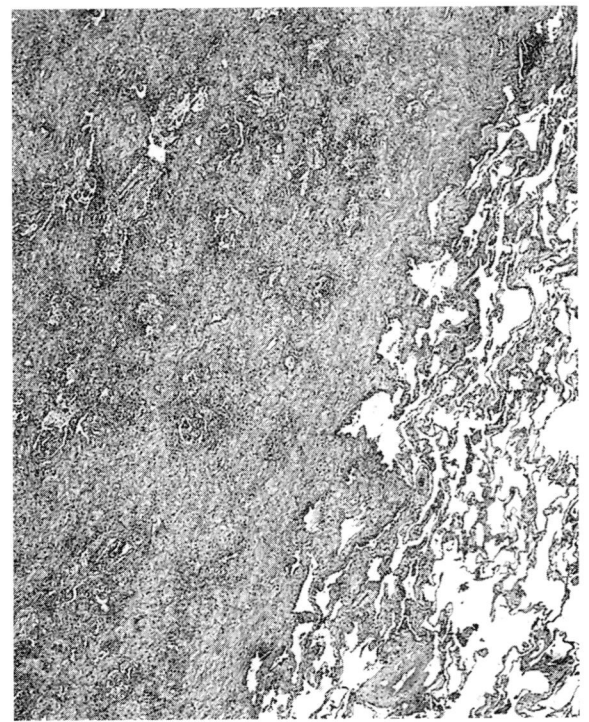

A

FIGURE 27-12
Treated Wegener's granulomatosis. This partially treated (biopsy a week or two after steroids were started) case of Wegener's granulomatosis had large zones of paucicellular fibrosis without

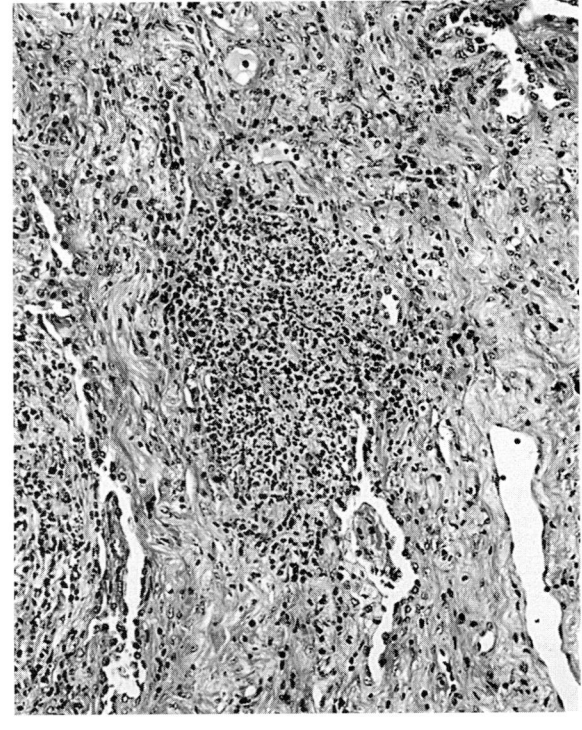

B

necrosis. Focal activity was found in some regions in the form of small microabscesses of neutrophils (*B*). (*A*, low magnification; *B*, medium magnification.)

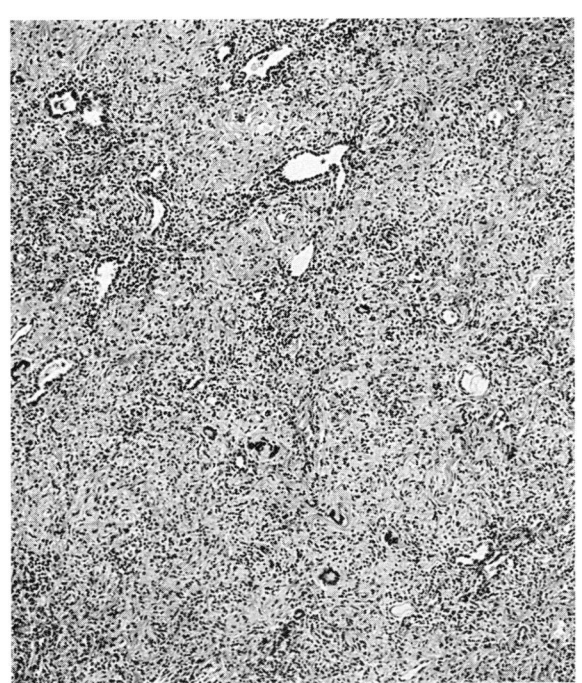

A

FIGURE 27-13
Treated Wegener's granulomatosis. Some cases of partially treated Wegener's granulomatosis may show only fibrous tissue with scattered giant cells without necrosis (*A*). Another case biopsied shortly after treatment was begun shows large zones resembling

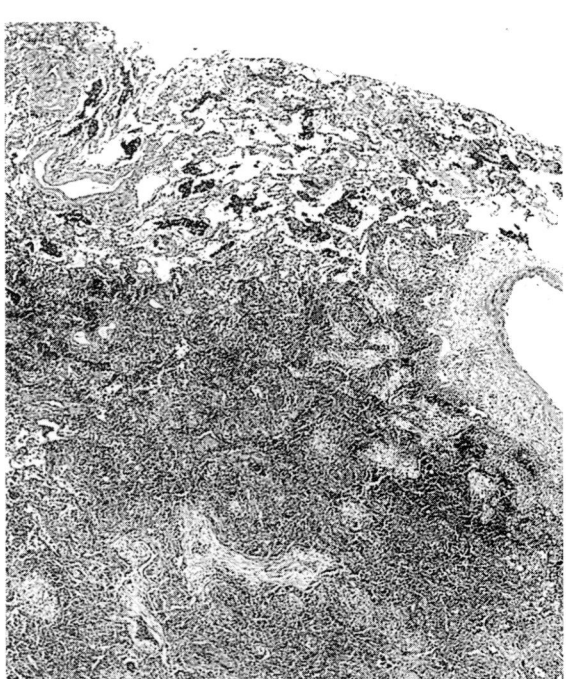

B

bronchiolitis obliterans organizing pneumonia (cryptogenic organizing pneumonitis) (*B*, bottom) as well as alveoli containing hemosiderin-filled macrophages indicative of prior hemorrhage (*B*, top). (*A* and *B*, low magnification.)

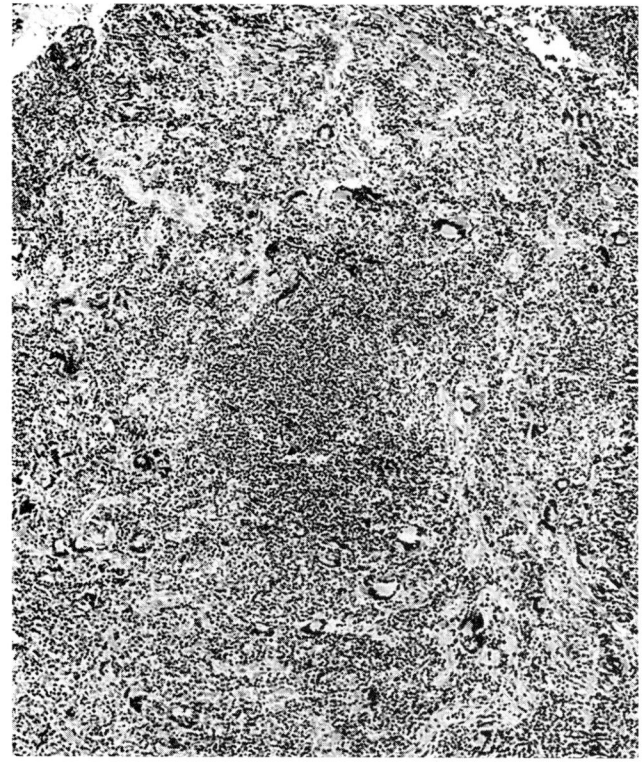

A

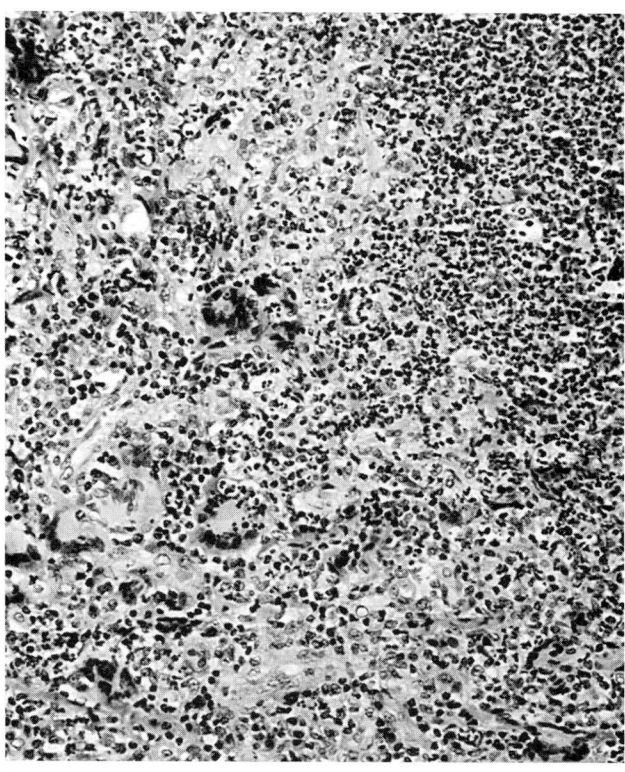

B

FIGURE 27-14
Nasal Wegener's granulomatosis. The most typical findings in nasal biopsies of Wegener's granulomatosis are inflammatory tissue with ulceration, necrosis, fibrosis, scattered giant cells, and the formation of neutrophilic microabscesses (*A*, center; *B*, upper right). (*A*, low magnification; *B*, medium magnification.)

in the lung, the histologic components can be divided into three groups: vasculitis, granulomatous necrosis, and inflammatory background. Biopsies from the nose, sinuses, and other regions in the head and neck as well as occasionally from other sites in the body, typically show foci of ulceration (if a mucosal site is involved), necrosis surrounded by palisaded histiocytes and scattered giant cells, acute and chronic inflammation, microabscess formation, and leukocytoclastic vasculitis. When medium-sized arteries are sampled, there may be difficulty in separating Wegener's granulomatosis from other forms of

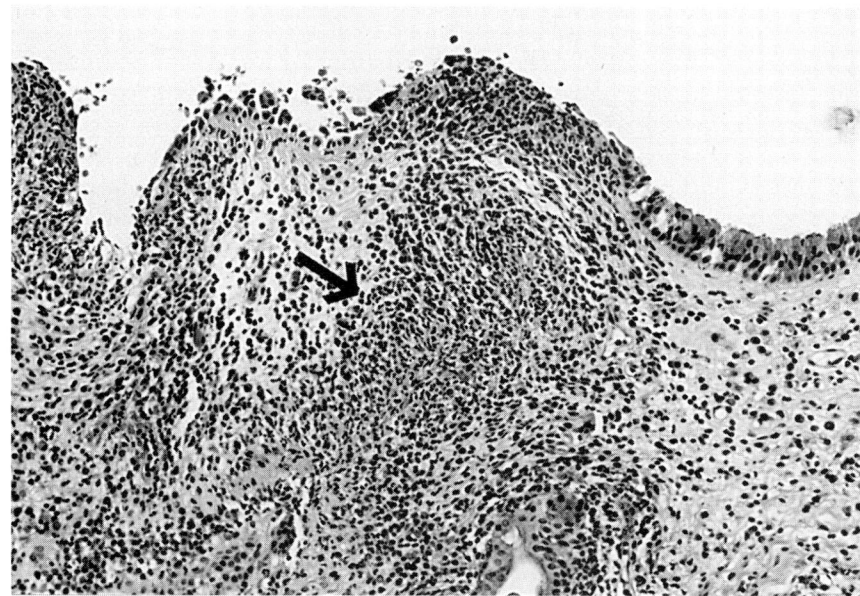

FIGURE 27-15
Nasal Wegener's granulomatosis. Early nasal Wegener's granulomatosis is seen as clusters of neutrophils in the submucosa (*arrow*) with focal overlying ulceration. (Medium magnification.)

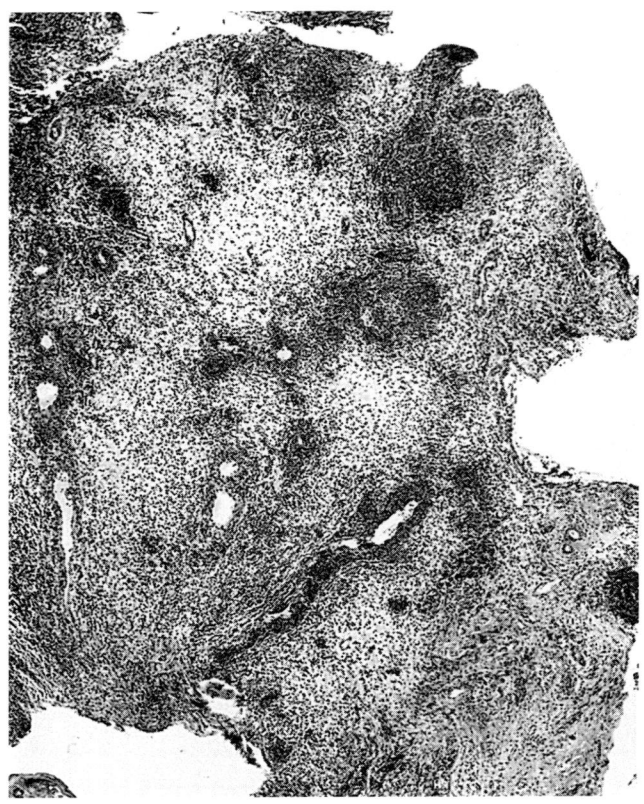

A

B

FIGURE 27-16

Sinus involvement by Wegener's granulomatosis. In the edematous and inflamed sinus tissue, one appreciates the predilection for the inflammatory cells to cuff small vessels (*A*). The cells typically include neutrophils, eosinophils, and cells resembling peripheral blood monocytes. Sometimes the mononuclear cells cluster to form microgranulomas (*B, arrows*). (*A*, low magnification; *B*, medium magnification.)

vasculitis on histologic grounds alone unless the distinctive segmental intramural necrotizing granuloma is identified.

Differential Diagnosis

From the histologic point of view, the differential diagnosis of Wegener's granulomatosis can be divided into three broad categories: alveolar hemorrhage syndromes, granulomatous infections, and other vasculitides.[21]

Some cases of Wegener's granulomatosis that present as alveolar hemorrhage are histologically indistinguishable from alveolar hemorrhage caused by other conditions, unless microabscess formation, scattered giant cells, and medium-sized vessel vasculitis are present.

The most common problem in the histologic diagnosis of Wegener's granulomatosis is separating it from a granulomatous infection. This distinction can sometimes be aided by the clinical pattern of disease, serologic testing, and, of course, positive cultures or special stains. While some cases of Wegener's

granulomatosis are sufficiently characteristic that a diagnosis can probably be rendered on the basis of routine hematoxylin-eosin sections, routine special stains to look for acid-fast and fungal organisms should be performed in all cases. The histologic clues that are helpful in distinguishing Wegener's granulomatosis from granulomatous infections are summarized in Table 27-4.

Wegener's granulomatosis is by far the most common vasculitis to involve the lung, and other classic forms of vasculitis only occasionally enter into the differential diagnosis. The American College of Rheumatology has developed a set of criteria that help distinguish Wegener's granulomatosis from other vasculitic syndromes,[28] and these are shown in Table 27-5. The presence of granulomatous necrosis, scattered giant cells among the inflammatory tissue, and the distinctive segmental necrotizing vasculitis with the formation of intramural necrotizing granulomas are all features that strongly favor Wegener's granulomatosis. The distinction of Wegener's granulomatosis from Churg-Strauss syndrome is discussed in the next section.

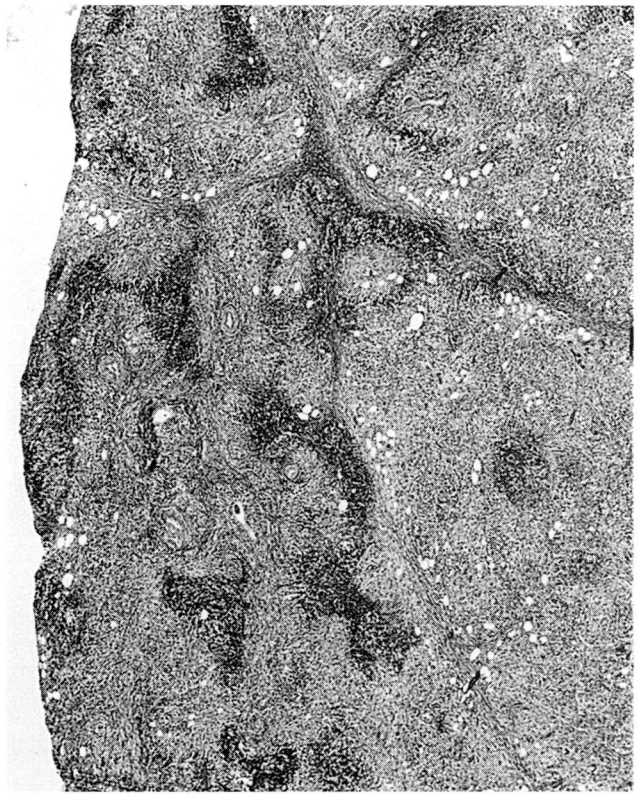

A

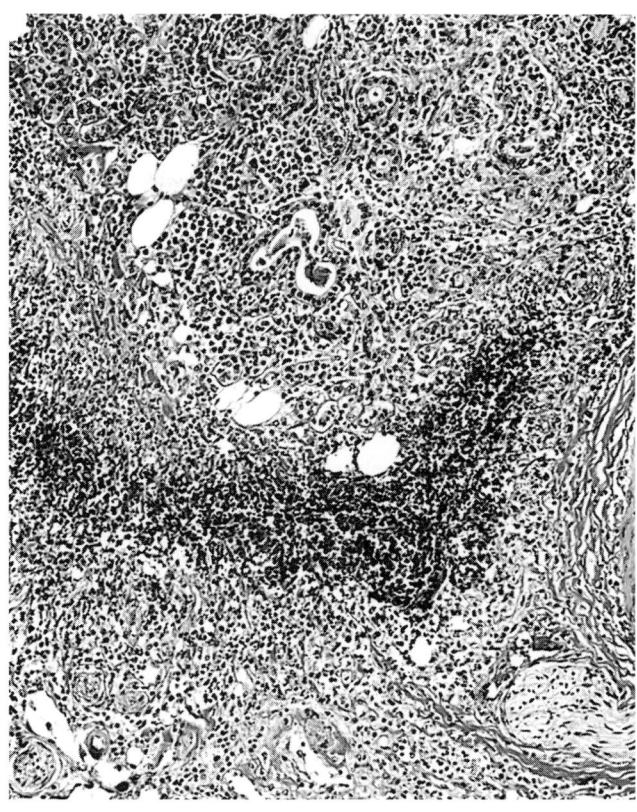

B

FIGURE 27-17
Salivary gland involvement in Wegener's granulomatosis. This case shows geographic necrosis at low power magnification similar to that seen in the lung, but with some preservation of the septal architecture (*A*). Higher power shows the basophilic necrosis involving atrophic salivary gland parenchyma with occasional residual ducts and acini identifiable (*B, center top*). (*A*, low magnification; *B*, medium magnification.)

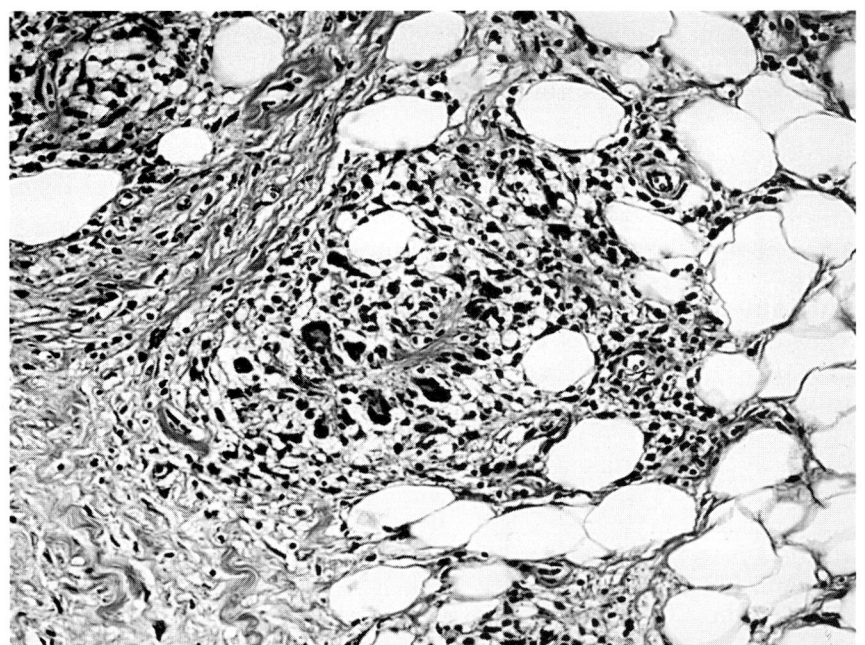

FIGURE 27-18
Soft tissue involvement in Wegener's granulomatosis. This patient had multiple lung nodules proven to be Wegener's granulomatous and a solitary soft tissue mass in the flank which was biopsied. The mass was composed of inflammatory tissue in which granulomatosis capillaritis could be identified. This was seen as giant cells and histiocytes palisaded around a small vessel (*center*). (Medium magnification.)

TABLE 27-4
Histologic Features of Wegener's Granulomatosis vs. Granulomatous Infection

Features Favoring Granulomatous Infection

Vasculitis tends to be secondary due to contiguity with necrotic foci; necrotic zones are rounded, often eosinophilic, with a relatively thin cuff of inflamed lung tissue; presence of sarcoid-like non-necrotizing granulomas away from the foci of granulomatous necrosis; scattered giant cells very unusual.

Features Favoring Wegener's Granulomatosis

Capillaritis/leukocytoclastic vasculitis (including granulomatous capillaritis); vascular lesions with eccentric and segmental involvement; origin of vascular lesions as an endothelialitis; geographic basophilic necrosis from merging and confluence of microabscesses; lack of sarcoid-like non-necrotizing granulomas; scattered giant cells typical; zones of necrosis are surrounded by extensive consolidated lung tissue (the "inflammatory background") that may show a variety of acute, subacute, or chronic inflammatory features.

TABLE 27-5
Sensitivity and Specificity of Potential Diagnostic Criteria in Wegener's Granulomatosis Compared to Other Vasculitides

Criterion	Sensitivity %	Specificity %
Hemoptysis	30	97
Paranasal sinus pain or tenderness	49	95
Abnormal sinus radiographs	61	89
Serum creatinine over 1.5 mg/dl	42	85
Urinary sediment	54	80
Eye inflammation	27	97
Otitis media or unilateral or bilateral deafness	40	94
Nasal or oral inflammation	73	88
Abnormal chest radiograph	66	89
Granulomatous inflammation on biopsy	75	89

SOURCE: Modified from Leavitt et al.[28]

Since Wegener's granulomatosis may include large zones of nondescript inflammatory tissue, biopsies in some cases may show nothing too specific. Nevertheless, Wegener's granulomatosis should not be excluded from consideration just because the more characteristic (and diagnostic) features are lacking.

Likewise, Wegener's granulomatosis has been included among the group of lesions that have been termed angiitis and granulomatosis.[29] While grossly these lesions bear some similarity to Wegener's granulomatosis (in that there is a nodule with central necrosis), the clinical and histologic differences among these lesions are readily recognizable and separable. Lymphomatoid granulomatosis is now recognized as a lymphoproliferative disorder. Low power evaluation shows sheets of lymphoid cells in contrast to the inflamed and fibrotic consolidated background tissue of Wegener's granulomatosis. The necrosis in lymphomatoid granulomatosis tends to be rounded instead of geographic, and it is usually eosinophilic instead of basophilic. Scattered giant cells, characteristic of Wegener's granulomatosis, are uncommon in lymphomatoid granulomatosis. Necrotizing sarcoid granulomatosis differs from Wegener's granulomatosis by the shear number of non-necrotizing (sarcoid-like) granulomas in the for-

mer. The granulomas are the major histologic finding and necrotizing sarcoid granulomatosis lacks the diffuse inflammatory consolidation so characteristic of Wegener's granulomatosis. Bronchocentric granulomatosis is discussed below.

The diagnosis of Wegener's granulomatosis has to be individualized in each case and based on evaluation of clinical data (e.g., pattern of disease, organ systems involved), serologic findings (ANCA tests and serologic tests for collagen vascular diseases and antiglomerular basement membrane antibody disease), and histologic findings. Special stains for fungi and acid fast organisms should be performed in all cases showing granulomatous necrosis. Cases can generally be placed into a differential diagnostic category based on the clinical and histologic findings; for example, an alveolar hemorrhage syndrome has one set of lesions in the differential diagnosis and a necrotizing granulomatous process has a different set of lesions to be considered. In any given case, one collects the clinical, serologic, and histologic data and makes a determination as to whether or not there is sufficient data present for a definitive diagnosis or at least a therapeutic decision. In some instances, overwhelming clinical features, characteristic serologic findings, and a small transbronchial biopsy showing some of the histologic changes described above may suffice.[30] In other cases, with involvement limited to the lungs and a negative c-ANCA test, a large open biopsy might be necessary to confirm a diagnosis of Wegener's granulomatosis.

BRONCHOCENTRIC GRANULOMATOSIS

The term *bronchocentric granulomatosis* was first used in 1972 by Liebow to describe a condition in which nectrotizing granulomata centered on bronchi and bronchioles.[31] It is probable that bronchocentric granulomatosis represents a non-specific response of the lung seen in a wide variety of conditions. Two forms have been recognized depending on the presence or absence of concomitant asthma.[32] Most cases in asthmatics are associated with allergic bronchopulmonary aspergillosis (ABPA) due to *Aspergillus fumigatus*, although other fungi are occasionally implicated.[32,33]

Clinical Features

Asthmatic patients with bronchocentric granulomatosis tend to be younger and male, whereas in non-asthmatics with this condition there is a greater predominance of women of older age. The typical symptoms of ABPA are cough, wheezing, dyspnea, chest pain and fever, which mimic exacerbations of asthma. In non-asthmatic individuals, although airway restriction is typically absent, patients frequently have chest symptoms, in particular a non-productive cough.

In patients with ABPA, raised IgE levels and a peripheral blood eosinophilia are found. Eosinophils are also found in the sputum and in the tissues. A leukocytosis is usually only mild, although in some it can be markedly elevated. The ESR or plasma viscosity may be mildly or moderately raised. Skin tests for aspergillus antigens may be positive in those patients with ABPA. Radiographically, nodules, solitary or multiple, and masses are seen in 20 to 60 percent of patients. The radiographic findings may be mistaken for malignant disease. Solitary lesions occur most commonly, but multiple small nodules or a fine reticular nodular pattern and alveolar or pneumonic infiltrates are also seen. Consolidation, emphysema, or cavitation have also been described.

The prognosis is generally excellent. Surgery, often carried out for supposed bronchial malignancy, can be curative if the disease is localized. Many other patients may benefit from oral steroids after the careful exclusion of an infectious etiology. Wegener's granulomatosis may be associated with bronchocentric granulomata,[26] and, thus, it is important to consider this diagnosis, as it requires more aggressive immunosuppression with cyclophosphamide. Most asthmatic patients with bronchocentric granulomatosis respond to treatment with steroids, although some continue to have episodic steroid-dependent asthma. In non-asthmatic patients, the finding of BCG should lend to a rigorous search for an infection.

Pathology of Bronchocentric Granulomatosis

Bronchocentric granulomatosis grossly resembles other necrotizing granulomas. The lesions may be large and coalescent or may show some branching indicative of involvement of airways. More proximal airways may show mucoid impaction and bronchiectasis. The lung tissue surrounding the necrosis is often consolidated.

The histopathology of bronchocentric granulomatosis is best understood as part of a continuum of pathologic changes that extend from the large airways into the distal lung parenchyma (Figs. 27-19 through 27-24).[32,34–37] Most cases of bronchocentric granulomatosis are associated with allergic bronchopulmonary fungal disease. There is mucoid impaction in proximal large airways. This may be found grossly or histologically or may have even been identified at the time of bronchoscopy in the form of mucus plugging. Airways with mucoid impaction are typically ectatic and filled with dense eosinophilic to basophilic mucus, which often shows a layered appearance with basophilic degenerated cells forming alternating layers with more eosinophilic mucus. The degenerating cells include eosinophils and neutrophils; Charcot-Leyden crystals are often numerous among the eosinophil debris. Within the mucin, one should try to identify fragments of fungal hypae, most easily found with methenamine silver. The involved airways often show features of asthma with goblet cell metaplasia, thickening of the basement membrane, smooth muscle hypertrophy, and a submucosal infiltrate rich in eosinophils.

The classic lesion of bronchocentric granulomatosis is a necrotizing granuloma centered on airways distal to mucoid impaction. The necrotizing lesions are thought to be allergic-mediated, caused by the fungal antigens present in the more proximal impacted mucus. The stage of development of the classic lesions can often be appreciated by scanning several small airways in the same case. Small and medium-sized bronchioles initially show a chronic

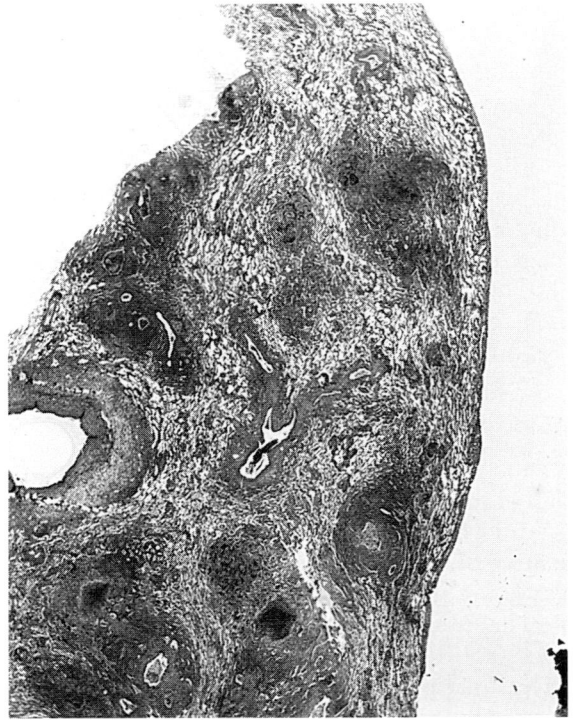

A

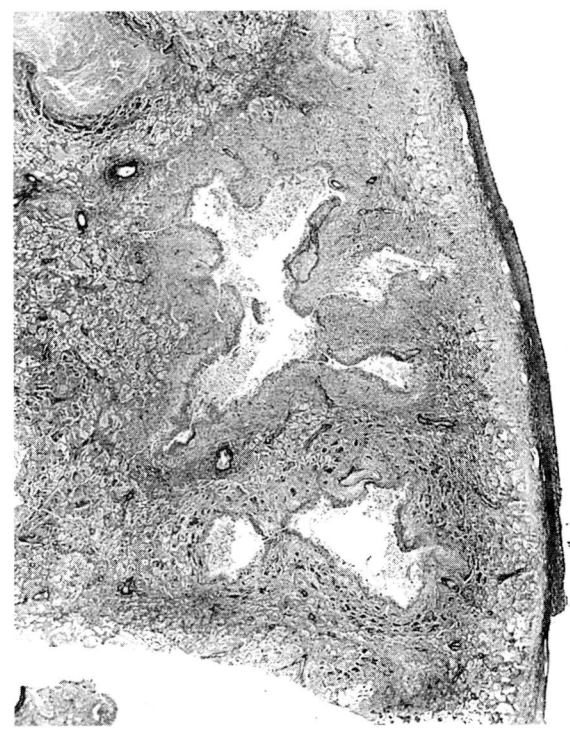

B

FIGURE 27-19

Bronchocentric granulomatosis. The early lesions of bronchocentric granulomatosis are seen an accute and chronic bronchiolitis (*A*). As the lesions develop, zones of necrosis develop which may be confluent or show an arborizing pattern that follows the anatomy of the small airways (*B*). (*A* and *B*, low magnification.)

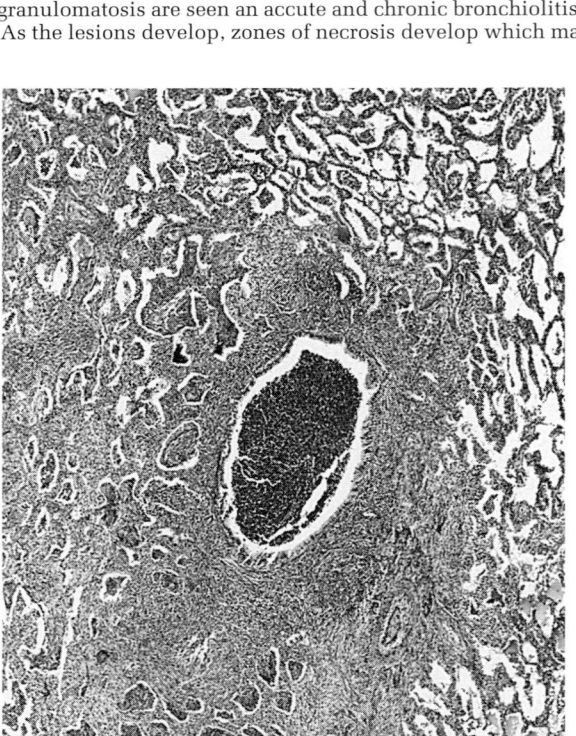

A

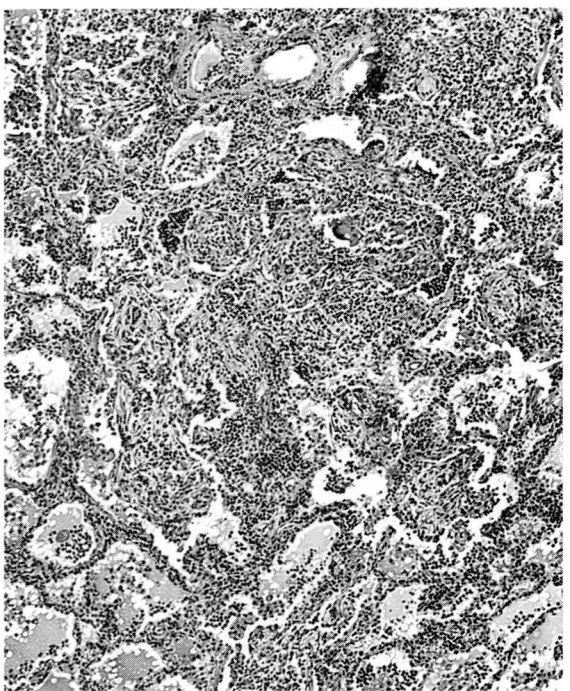

B

FIGURE 27-20

Bronchocentric granulomatosis. Early involvement of a bronchiole shows an acute and chronic bronchiolitis with inflammatory exudate (rich in eosinophils) in the lumen and surrounding inflammatory consolidation of the lung tissue (*A*). The consolidated lung tissue may show nonspecific effects of obstruction or features of eosinophilic pneumonia as in this case (*B*). (*A*, low magnification; *B*, medium magnification.)

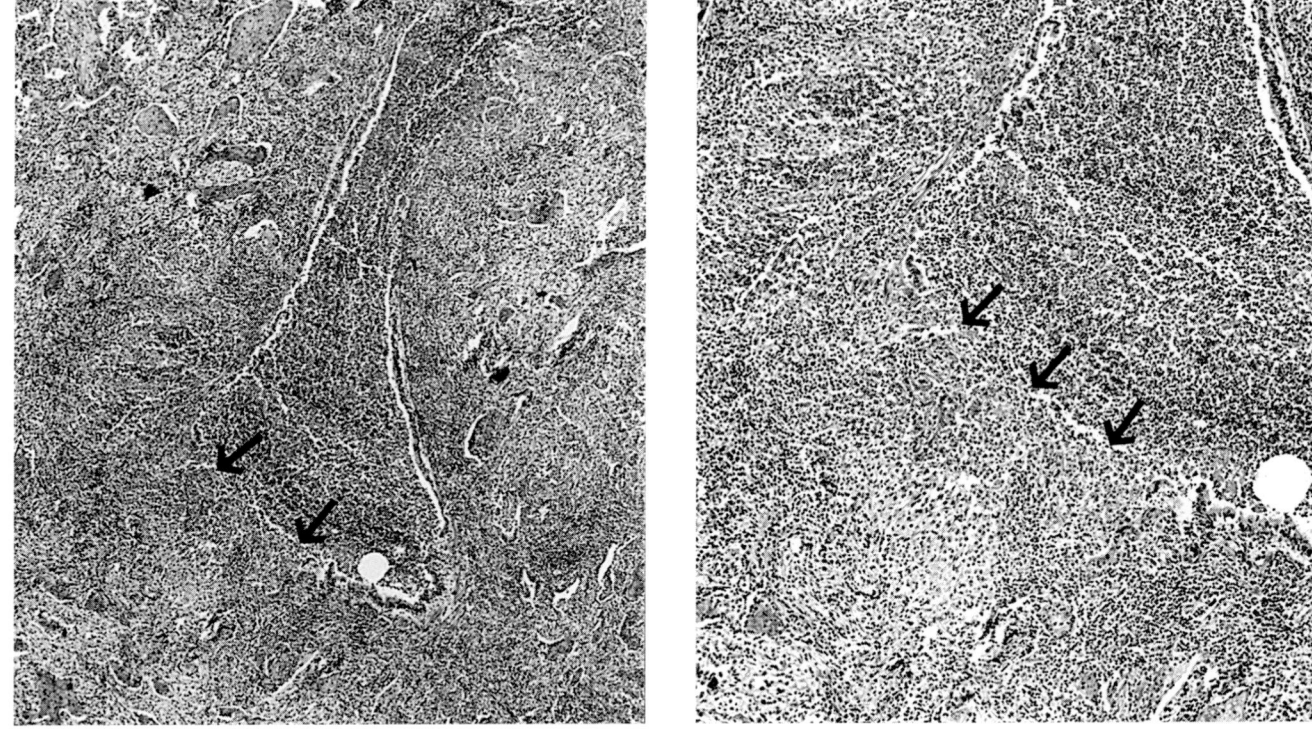

A

B

FIGURE 27-21
Bronchocentric granulomatosis. The earliest granulomatous change of bronchocentric granulo-
matosis is seen as focal erosion of the bronchial mucosa and replacement by palisaded histiocytes
(*A* & *B*, *arrows*). (*A* and *B*, low magnification.)

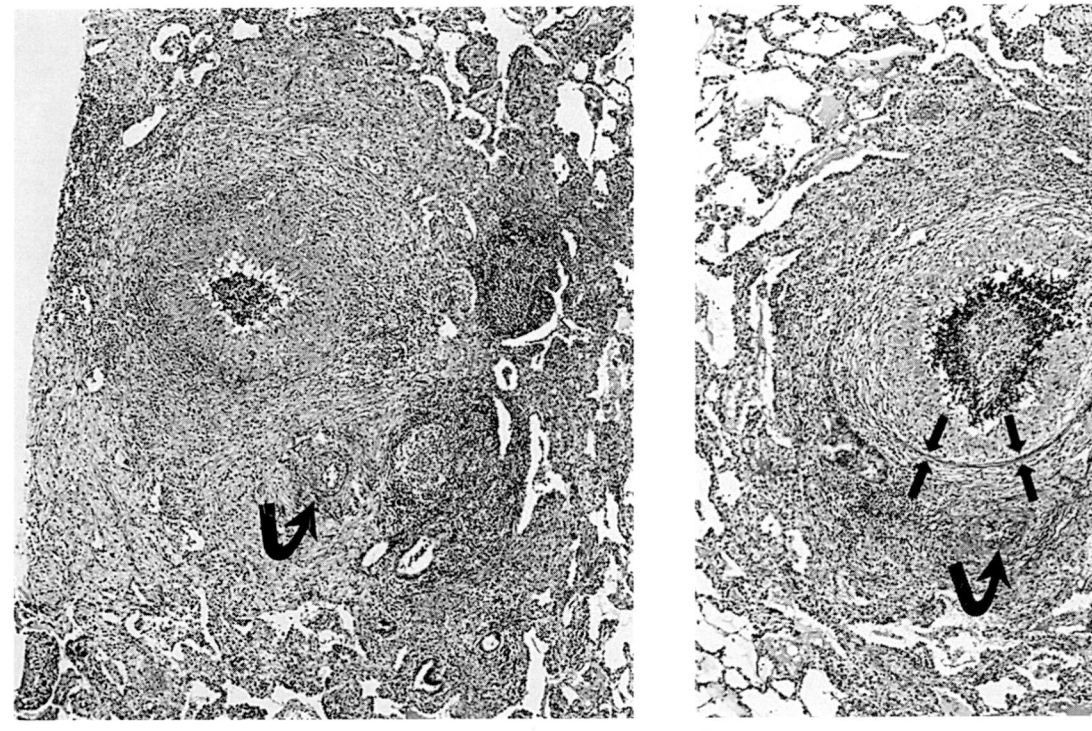

A

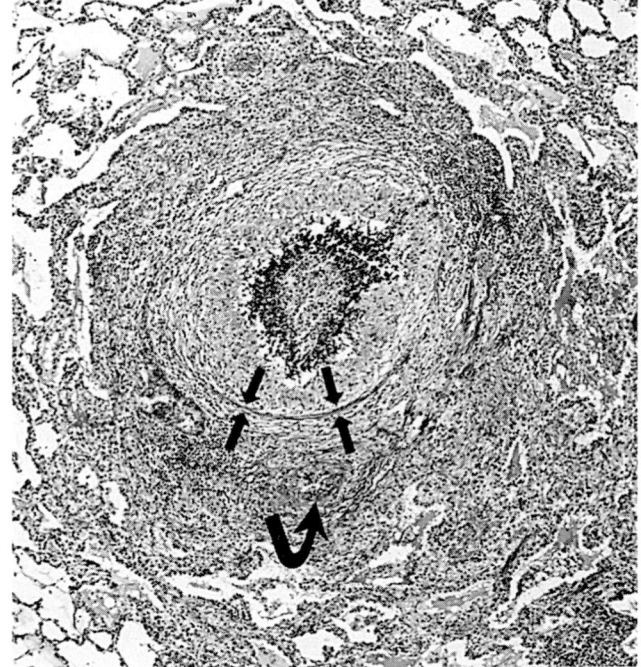

B

FIGURE 27-22
Bronchocentric granulomatosis. The classic lesions of bronchocentric granulomatosis (*A*, *B*) represent a cuff of palisaded histiocytes replacing the airway wall resembling a small necrotizing

granuloma adjacent an intact pulmonary artery (*curved arrows*).
Sometimes a small rim of residual smooth muscle of the airway
can be identified (*B*, *arrows*). (*A* and *B*, low magnification.)

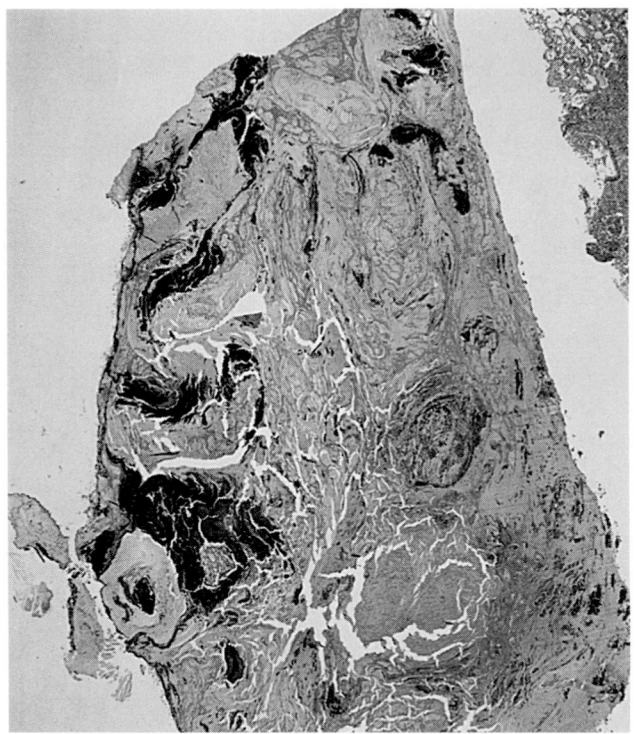

A

FIGURE 27-23

Mucoid impaction proximal to bronchocentric granulomatosis. Mucoid impaction typically occurs proximal to the necrotizing lesions of bronchocentric granulomatosis. The mucus is typically

eosinophilic in appearance with basophilic layering by clumps of inflammatory cells which typically include numerous eosinophils (*A, B*). Such is the appearance of "allergic mucin."

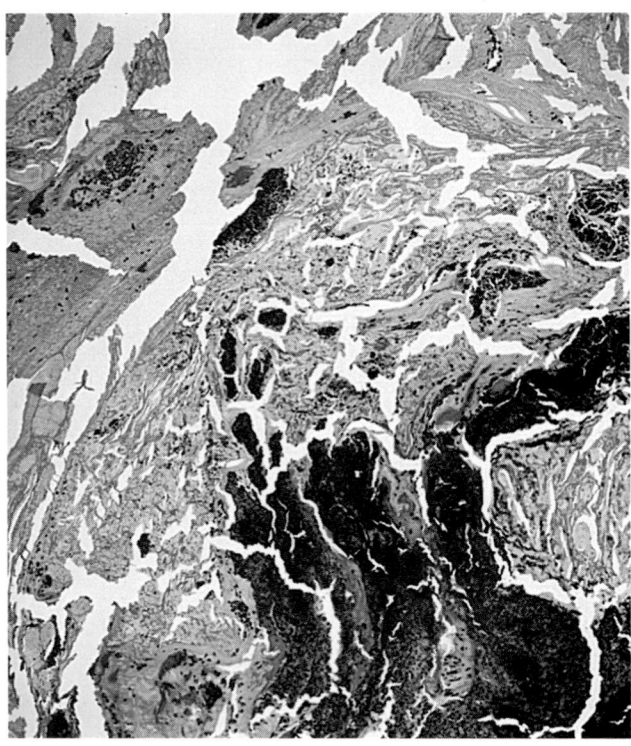

B

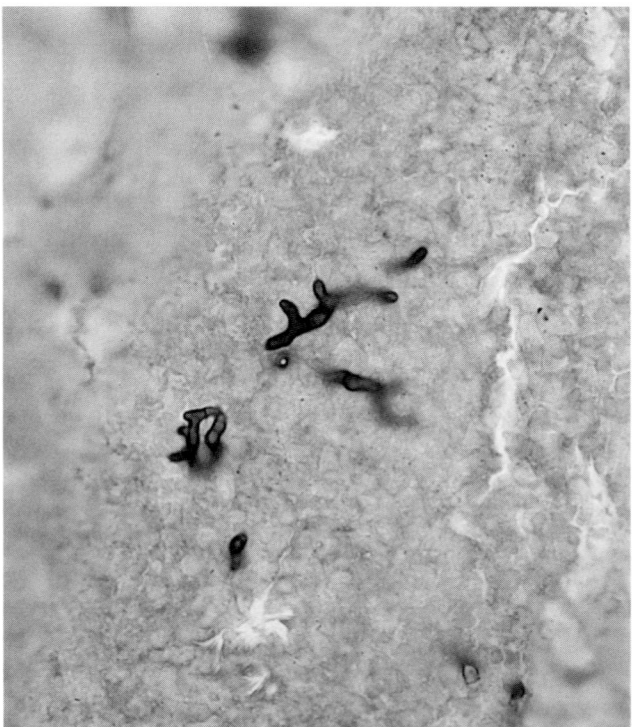

A

FIGURE 27-24

Mucoid impaction proximal to bronchocentric granulomatosis. Within the mucus, one may find fragments of fungal hypae, typically aspergillus (*A*). One finds large amounts of eosinophil

debris in addition to brightly eosinophilic elongated crystals, so-called Charcot-Leyden crystals (*B, arrows*). (*A*, high magnification; *B*, medium magnification.)

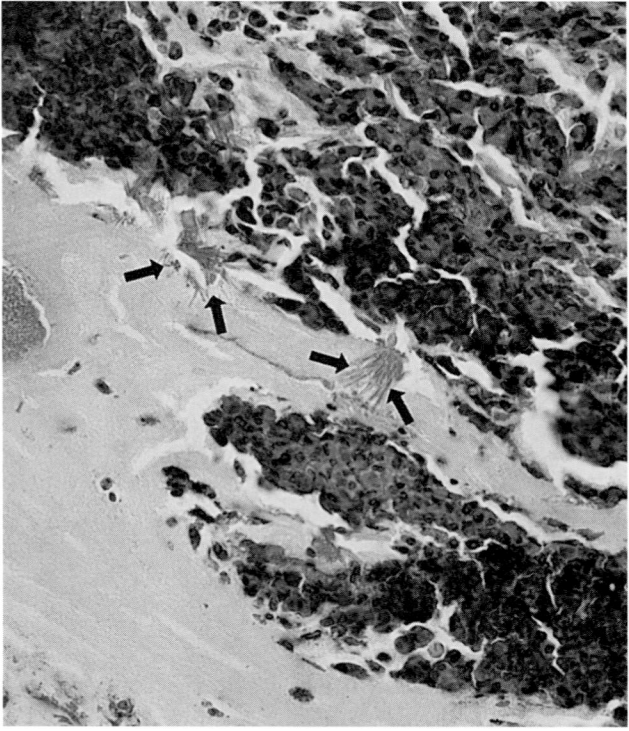

B

inflammation in the walls and an acute inflammatory exudate in the lumen. The mucosa is progressively destroyed and replaced by inflammatory tissue which ultimately becomes a rim of palisaded histiocytes replacing the bronchiolar wall. This palisading granuloma surrounds central necrotic debris that includes abundant degenerating eosinophils and neutrophils. The necrotic lesions may only be identifiable as bronchiolocentric because of the adjacent pulmonary artery. The bronchocentric origin of the necrosis may be difficult to appreciate in the large necrotic zones, although it may be recognizable at the periphery of necrotic regions. Elastic tissue stains are also helpful in identifying pulmonary arteries as well as remnants of the elastic tissue of their accompanying airways.

The pulmonary arteries commonly show nonspecific inflammatory changes, including some mural infiltration by inflammatory cells (secondary angiitis). The surrounding lung tissue is typically intensively inflamed, with acute and chronic inflammation, organization, and increase in airspace foam cells indicative of obstruction. The surrounding lung tissues may in addition show features of eosinophilic pneumonia with airspace pooling by large numbers of histiocytes and eosinophils and occasional non-necrotizing granulomas.

Differential Diagnosis

The differential diagnosis of bronchocentric granulomatosis includes both infectious and non-infectious processes, as shown in Table 27-6.

Studies by Myers and Katzenstein have shown that a variety of infections may produce necrotizing bronchocentric lesions similar or identical to those seen in classic bronchocentric granulomatosis associated with allergic bronchopulmonary fungal disease.[36,37] Such an appearance has been described with mycobacterial infections, histoplasmosis, and blastomycosis. Aspergillus and mucor have also been implicated, but in these, the necrosis was not thought to be allergic-mediated. Coccidiomycosis may also be associated with necrotizing bronchocentric lesions and, like classic BCG, may be associated with numerous tissue eosinophils. Bronchocentric fungal infections have also been described in transplant recipients.[38,39]

As noted above, bronchial and bronchiolar lesions are not uncommon in Wegener's granulomatosis and some cases of Wegener's granulomatosis are distinctively bronchocentric.[26] In Wegener's granulomatosis, the lesions are not restricted to the airways, and examination of the rest of the

TABLE 27-6
Conditions That May Resemble Bronchocentric Granulomatosis

Bronchocentric infections
  Fungal
  Mycobacterial
Bronchocentric Wegener's granulomatosis
Aspiration
Rare: rheumatoid nodules, necrotizing sarcoid
  granulomatosis, echinococcal cysts

SOURCE: In part from references 36 and 37.

parenchyma shows typical features of Wegener's granulomatosis, usually including vasculitis. Nevertheless, some foci of the bronchocentric lesions of Wegener's granulomatosis may be identical to those in bronchocentric granulomatosis, and the pulmonary arteries may be uninvolved.

Chronic aspiration pneumonia is associated with acute and chronic bronchiolitis and often a giant cell or granulomatous reaction. The aspirated materials are often apparent and the giant cells are generally found within the lumen of the airways; the distinctive destruction of the airway and its replacement by a palisaded cuff of histiocytes is usually lacking. Rarely, rheumatoid nodules, necrotizing sarcoid granulomatosis, and echinococcal infection may be associated with necrotizing lesions.[34] Necrotizing sarcoid granulomatosis is a controversial lesion that probably represents nodular sarcoid with necrosis. As such, the necrotic lesions do not show the bronchocentricity of bronchocentric granulomatosis, and the lesions also include confluent masses of non-necrotizing granulomas, a feature not seen in bronchocentric granulomatosis. Bronchocentric granulomatosis is exquisitely bronchocentric (in lesions in which there is sufficient maintenance of architecture to appreciate the distribution), whereas the lesions of nodular sarcoid tend to follow lymphatic routes and, when they undergo necrosis, may actually appear somewhat angiocentric in distribution, since the lymphatics follow bronchovascular structures.

It is important to maintain strict criteria for the diagnosis of bronchocentric granulomatosis. In the author's experience, the diagnosis may be suggested when an airway showing acute and chronic inflammation also has a few sarcoid-like granulomas in the wall. This is not bronchocentric granulomatosis. In the setting of bronchiectasis, the granulomas are not uncommonly due to secondary low grade infection by atypical mycobacteria.

ALLERGIC GRANULOMATOSIS AND ANGIITIS (AGA)

Clinical Features

AGA (also known as Churg-Strauss syndrome) is characterized by a marked eosinophilia in patients with a history of asthma who develop a systemic vasculitis, often with extra-vascular granulomata.[40-44] It was described in 1951 by Churg and Strauss who reported a group of young adults with a history of asthma and a peripheral eosinophilia who presented with a polyarteritis-like vasculitis.[40] As with Wegener's granulomatosis, the primary sites of involvement are the upper and lower respiratory tract (Fig. 27-25) but contrasts with Wegener's, which arises in previously healthy individuals. AGA patients invariably have a history of allergic conditions such as asthma or rhinitis. Occasionally the condition is preceded by sensitization to a drug.

There is a slight preponderance in males, and the age of diagnosis is usually late in the fourth decade. There appears to be no specific etiology. Eosinophilic infiltrates dominate the histologic picture. Eosinophilic cationic protein and eosinophil-protein-x are secreted by these activated cells. These proteins are toxic to cardiac muscle and other cells and thus may be responsible for the development of some of the clinical features.

The diagnostic triad is asthma (or allergic disease), eosinophilia, and evidence of a systemic vasculitis. The eosinophilia is marked (greater than $1.5 \times 10^9/l$), and vasculitic involvement of two or more extrapulmonary organs is required for a clinical diagnosis.[43] (The eosinophilia is greater than that normally seen in asthma but less than the level that can be found in the hyper-eosinophilic syndrome.) The mean duration of the preceding asthma is 3 to 8 years but has been as long as 30 years. The vascular lesions can be widespread and commonly affect the nervous system. A mononeuritis multiplex is seen in approximately two-thirds of all patients. Central nervous system involvement is rare but may include optic neuritis. Other head and neck manifestations include nasal obstruction, rhinorrhoea, nasal polyps, sinusitis, and septal perforation. The heart may also be affected, resulting in cardiac failure, pericarditis and hypertension. Abdominal pain is frequent. Vascular lesions in the bowel, which resemble eosinophilic enteritis, may cause obstruction and/or bloody diarrhea.

Involvement of the skin occurs in many cases either as nodules, often on the scalp or extensor surfaces of the limbs, or maculopapular, urticarial or purpuric eruptions. Livedo reticularis occurs infrequently. When present, skin nodules are useful for providing material for histologic diagnosis. Migratory arthralgia and myalgia are not infrequent, but true arthritis is rare. Muscle pain can be caused by a myositis or local ischemia due to the vasculitis. Renal involvement may result in hematuria and proteinuria, and a focal segmental glomerulonephritis may occur. This is not common, however, and renal failure is rare and less severe than that in Wegener's granulomatosis.

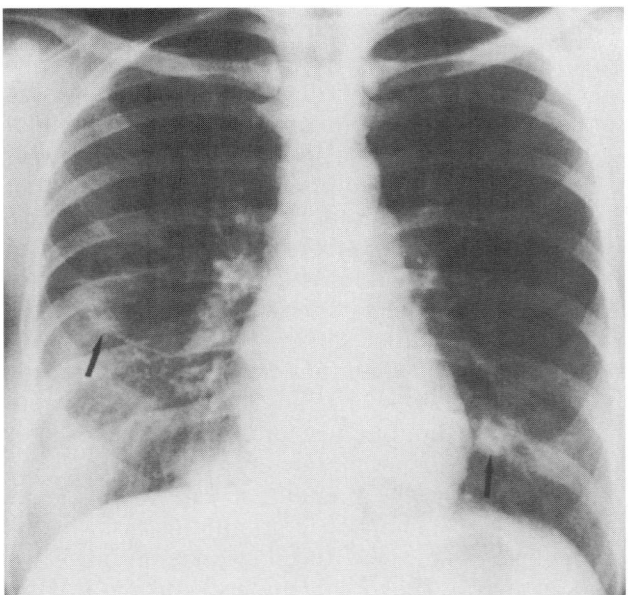

FIGURE 27-25
Allergic Granulomatosis and Angiitis. Chest radiograph of a 35-year-old woman with recent onset of variegated skin lesions, fever, and arthralgias. Arrows indicate areas of infiltration. Diagnosis established by lung biopsy. (*Courtesy of Dr. W. T. Miller.*)

Pathology of Allergic Granulomatosis and Angiitis

The classic histologic features of the AGA are vasculitis, necrotizing extravascular granulomas, and tissue infiltration by eosinophils.[40,45] Since all of these features may be seen in Wegener's granulomatosis, some have suggested that AGA is simply Wegener's granulomatosis occurring in an asthmatic patient. Nevertheless, (most) other investigators believe AGA is a distinct clinicopathologic entity with subtle histologic differences from Wegener's granulomatosis.

AGA is a rare disease. In addition, the diagnosis of AGA may sometimes be made on clinical grounds, and the patients successfully managed without biopsy. For these reasons, relatively little tissue comes to the pathologist for examination either in the form of biopsy material or autopsy material; very few pathologists see appreciable numbers of cases of AGA, and among the cases examined, very few show the full (classic) histologic spectrum described below.

Few gross descriptions of the pathologic findings of AGA are available. AGA in the lung tends to show ill-defined consolidation and nodules of varying size which undergo necrosis and cavitation. Vascular thrombi may be seen. The gross findings in other organ systems parallel the features seen in any systemic vasculitis: foci of hemorrhage and necrosis, infarction, and in the case of mucosal tissues, ulceration. Scarring is a late finding.

The classic histologic findings of AGA are: tissue infiltration by eosinophils, extravascular necrotizing granulomas, and vasculitis (Figs. 27-26 through 27-32).

The peripheral eosinophilia of AGA is reflected in the tissue by massive infiltration by eosinophils.

This is often the most obvious finding histologically. In the lung, this is seen as airspace, interstitial, and perivascular accumulations of eosinophils which may produce features identical to acute or chronic eosinophilic pneumonia. In some cases, especially those that have been on steroid therapy, the number of eosinophils may be reduced.

Granulomatous inflammation occurs as extravascular necrotizing granulomas and non-necrotizing sarcoid-like granulomas. Most characteristic (though not specific[46]) is a small extravascular necrotizing granuloma with central necrotic focus containing eosinophil debris. Less commonly large coalescing zones of granulomatous necrosis develop similar to those seen in Wegener's granulomatosis. Necrotizing granulomas may involve vessel walls. Non-necrotizing granulomas may also be scattered randomly or involve vessels; they tend to be single and do not coalesce as in sarcoidosis. Scattered giant cells, singly or in clusters, may also be seen.

Vasculitis in AGA varies from a cellular infiltration of the vessel wall rich in eosinophils to necro-

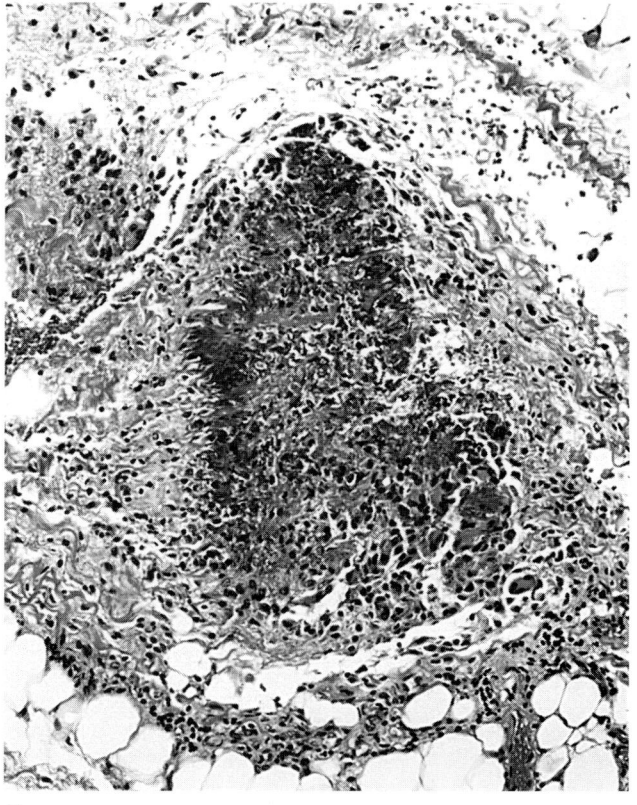

A

B

FIGURE 27-26

AGA. Low power magnification shows a nodule with geographic necrosis reminiscent of that seen in Wegener's granulomatosis (*A*). All the necrotic debris was composed of necrotic eosinophils. Another case shows an extravascular necrotizing granu-

loma involving the soft tissue around the trachea (*B*). All the necrotic material was eosinophil debris. (*A*, low magnification; *B*, medium magnification.)

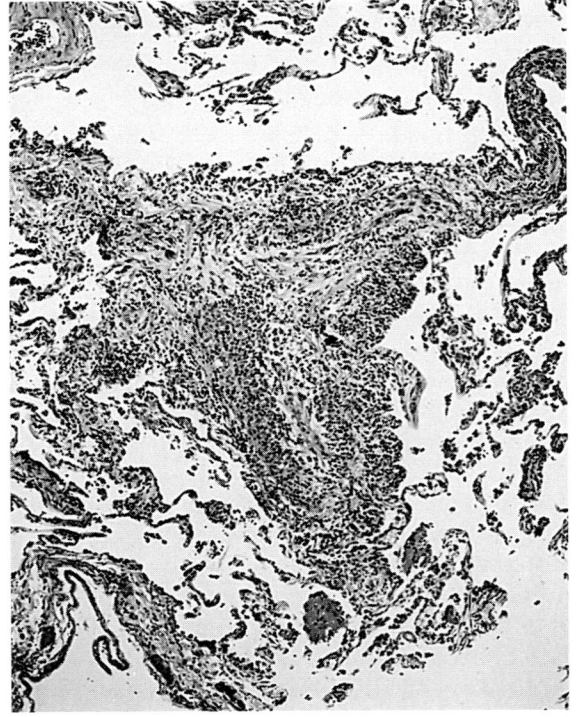

A

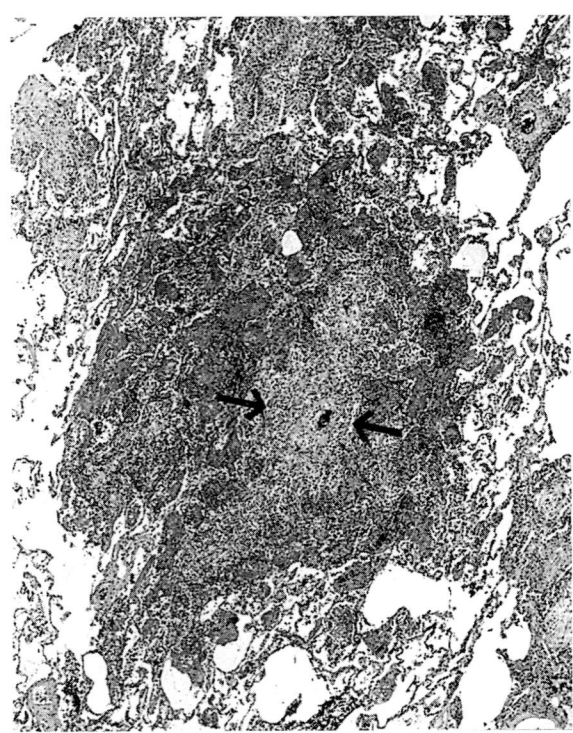

B

FIGURE 27-27
Pulmonary vasculitis in AGA may be seen as a inflammatory cuffing of small vessels with a mixed infiltrate of eosinophils and histiocytes (*A*). Another case shows a vascular and perivascular infiltrate (*arrows*) which is surrounded by a nodular zone of fibrinous exudate and eosinophils in airspaces (*B*). (*A* and *B*, medium magnification.)

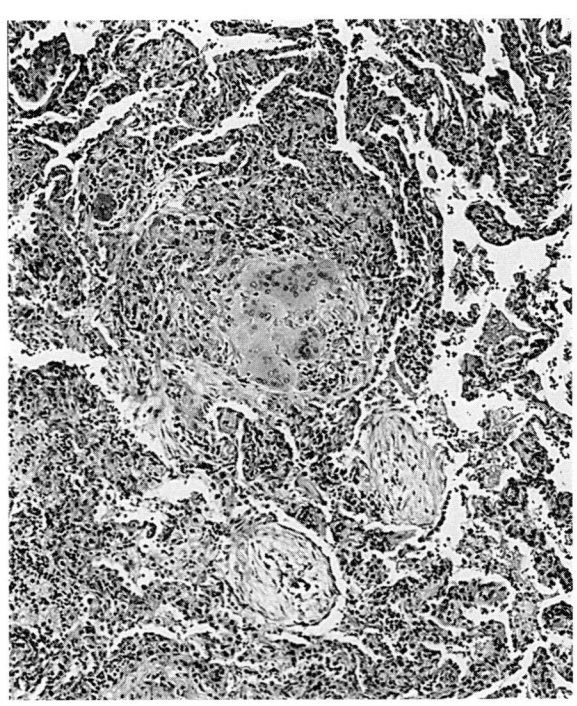

A

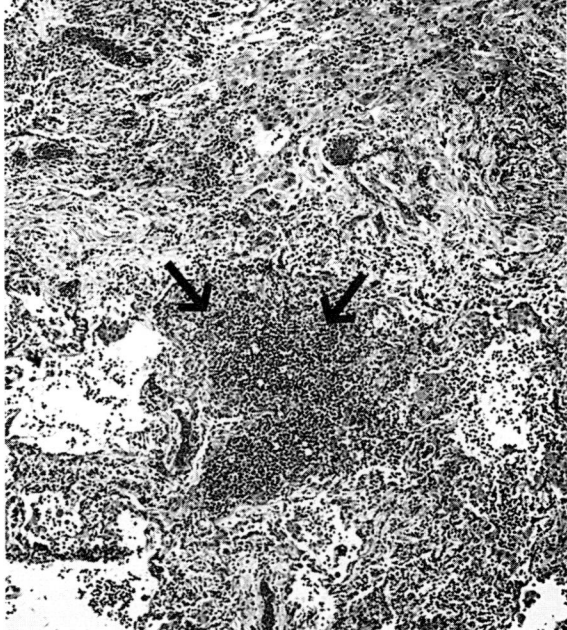

B

FIGURE 27-28
AGA. Some cases show scattered giant cells or small non-necrotizing epithelioid granulomas (*A, upper center*). This case also shows focal airspace organization (*A, lower center*). Other fields exhibited airspace filling by pools of eosinophils (*B, arrows*). (*A* and *B*, medium magnification.)

858

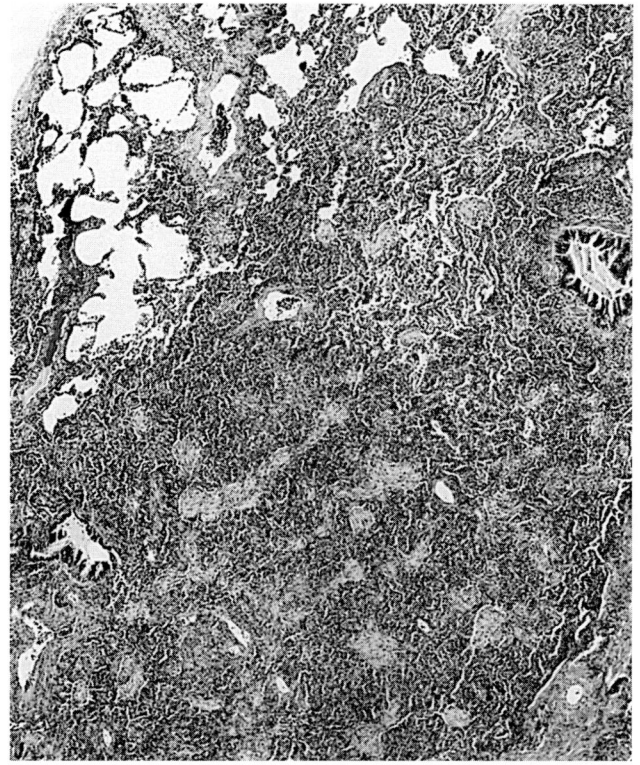

A

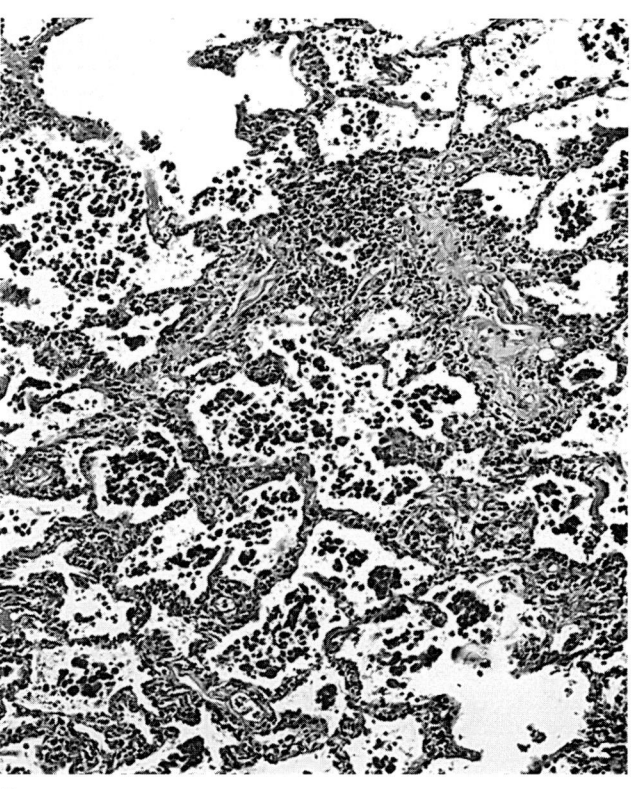

B

FIGURE 27-29

AGA. Some cases show large zones of airspace organization resembling bronchiolitis obliterans organizing pneumonia/cryptogenic organizing pneumonitis (*A*). Other cases show evidence of hemorrhage with hemosiderin-filled macrophages in airspaces (*B, bottom*); pools of eosinophils were also present (*B, upper center*). (*A*, low magnification; *B*, medium magnification.)

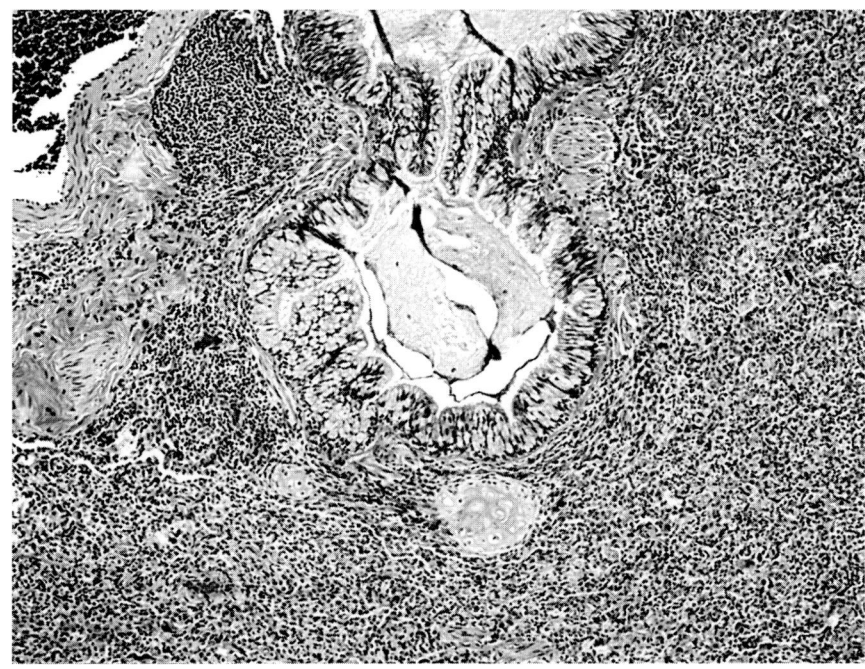

FIGURE 27-30

AGA. This small bronchus in a patient with AGA shows features of asthma with goblet cell metaplasia, mild smooth muscle hypertrophy, and mucus stasis in the bronchial lumen. (Medium magnification.)

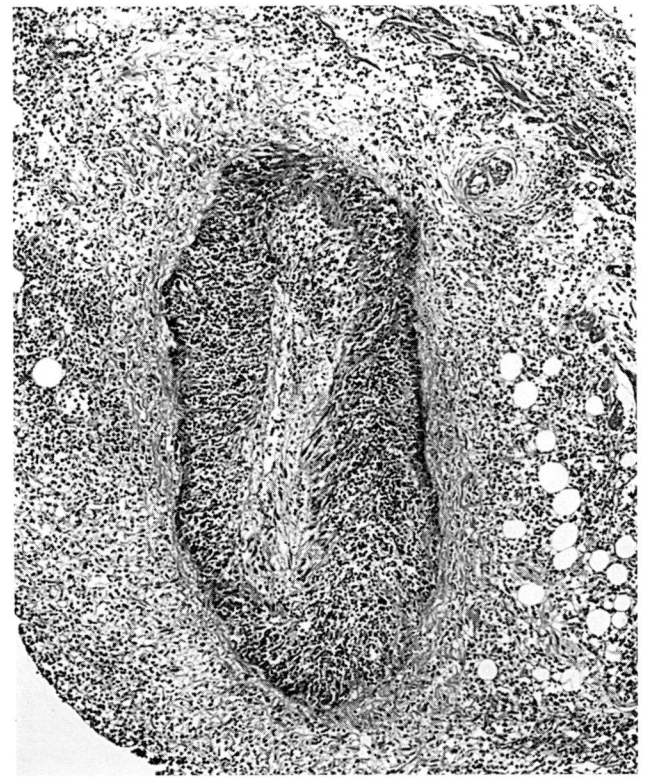

A

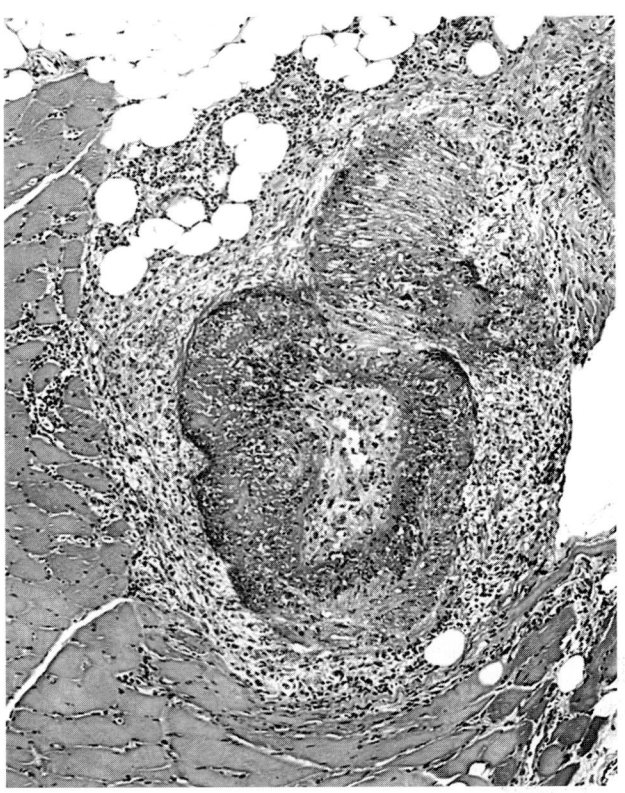

B

FIGURE 27-31
AGA. Involvement of a small epicardial coronary artery shows a necrotizing vasculitis with fibrinoid necrosis and transmural inflammatory infiltrate that was rich in eosinophils (*A*). Another case shows involvement of a muscular artery in a muscle biopsy; there is a fibrinoid necrotizing vasculitis that had only occasional eosinophils and was otherwise indistinguishable from other forms of arteritis (*B*). (*A* and *B*, low magnification.)

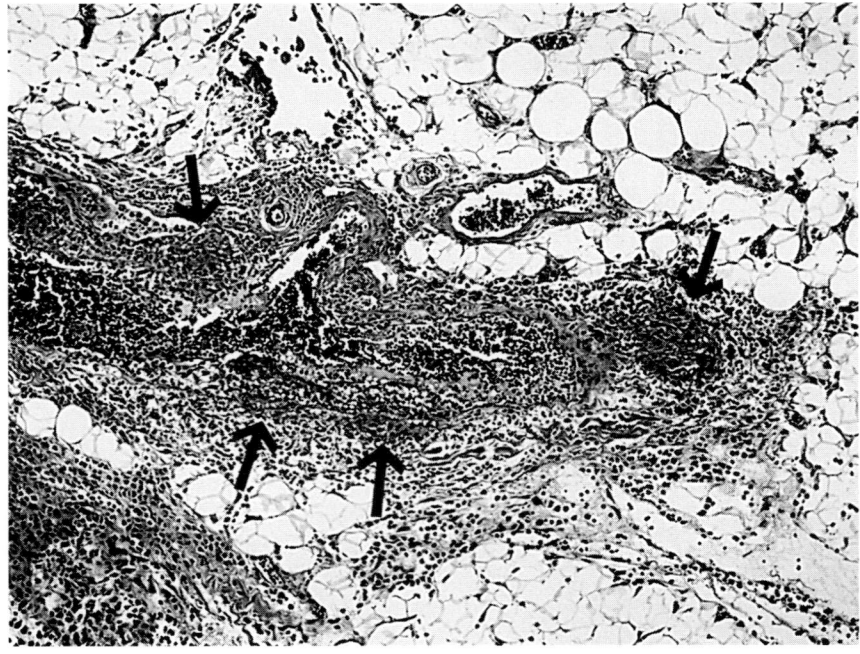

FIGURE 27-32
AGA. A subcutaneous vessel shows transmural inflammatory infiltrates rich in eosinophils as well as occasional small non-necrotizing granulomas (*arrows*). (Medium magnification.)

tizing fibrinoid vasculitis involving the entire circumference of the vessel. Occasionally intramural small necrotizing (or non-necrotizing) granulomas, similar to those seen in the extravascular tissue, are found. Depending on the clinical setting, some evidence of healing may be seen in the vascular lesions with a decrease in inflammatory cells, absence of necrosis, and scarring of the vessel wall. The vascular changes are better appreciated with elastic tissue stains.

A number of other histologic features may be found in the lung. Increased alveolar histiocytes commonly accompany the eosinophils just as in eosinophilic pneumonia. Interstitial inflammation with lymphocytes and plasma cells is also seen; neutrophils may also be present, especially in foci of necrosis. Since AGA characteristically involves patients with a history of asthma, the histologic changes of asthma are commonly present in the lung. These include mucostasis in small airways, goblet cell metaplasia in the bronchiolar mucosa, thickening of the subepithelial basement membrane region, submucosal edema, and inflammatory cell infiltrate including eosinophils, and smooth muscle hypertrophy. Occasional cases show alveolar hemorrhage that may be either acute or chronic.

The spectrum of lung pathology that may thus be encountered in AGA includes: asthma, eosinophilic pneumonia, parenchymal necrosis with eosinophils and neutrophils, granulomatous inflammation

TABLE 27-7
Criteria for the Diagnosis of Allergic Granulomatosis and Angiitis; Their Sensitivity and Specificity Versus Other Vasculitic Syndromes

Criterion	Sensitivity, %	Specificity, %
Asthma	100	96.3
Eosinophilia >10%	95	96.6
Neuropathy, mono or poly	75	79.8
Pulmonary infiltrates, nonfixed	40	92.4
Paranasal sinus abnormality	85.7	79.3
Extravascular eosinophils	81.3	84.4

SOURCE: Modified from Masi et al.[47]

(necrotizing, non-necrotizing, scattered giant cells), and vasculitis (with or without granulomas in the vessel wall), and combinations of these. Rarely, all of them are seen together in the same case.

Outside of the lung, biopsies in suspected cases of Churg-Strauss most commonly are from skin, nerve or muscle taken to confirm the presence of vasculitis. In addition to vasculitis, extravascular necrotizing granulomas may also be found. The vasculitis of medium-sized vessels may be identical to other forms of vasculitis but is typically rich in

TABLE 27-8
Differential Diagnosis of Allergic Granulomatosis and Angiitis (AGA)

Feature	AGA	Microscopic Polyarteritis	Classical Polyarteritis Nodosa	Wegener's Granulomatosis	Hypereosinophilic Syndrome
Male predominance	Slight	Slight	Marked	Slight	Marked
Upper respiratory tract involvement	Usual (allergic rhinitis/polyps)	Occasional	Uncommon	Common (destructive)	Uncommon
Asthma	Common	Uncommon	Uncommon	Uncommon	Uncommon
Pulmonary involvement	Common	Common	Uncommon	Common (destructive)	Uncommon
Cardiac	Common (pericardial and myocardial)	Uncommon	Occasional	Occasional	Common (endomyocardial)
Renal involvement glomerulonephritis	Common	Common	Occasional	Common	Rare
Renal failure	Rare	Common	Occasional	Common	Rare
Central nervous system	Occasional	Occasional	Occasional	Occasional	Common
Peripheral nervous system	Common (mononeuritis multiplex)	Occasional	Common	Occasional	Common (symmetrical neuropathy)
Skin disease	Common	Common	Occasional	Common	Rare

SOURCE: Modified from Churg and Churg.[48]

eosinophils. Small vessels, particularly in the skin, may show a leukocytoclastic vasculitis. Biopsies from the nose and sinus may show features of allergic polyposis. The heart may show eosinophilic myocarditis or evidence of coronary vasculitis.

Differential Diagnosis

The differential diagnosis of AGA is necessarily clinicopathologic; a number of patients diagnosed with this syndrome do not have any tissue biopsy taken to confirm it and those that do have tissue biopsies rarely show the full histologic spectrum described above. Criteria for the separation of AGA from vasculitis have been developed by the American College of Rheumatology (Table 27-7). The differential diagnosis of AGA with Wegener's granulomatosis, hypereosinophilic syndrome, polyarteritis nodosa, and microscopic polyarteritis are summarized in Table 27-8.

There are cases of AGA involving the lung in which a lung biopsy shows only nonspecific inflammatory changes with variable numbers of eosinophils. Some cases are histologically indistinguishable from chronic eosinophilic pneumonia. Wegener's granulomatosis is rarely a problem when all the clinical features are taken into account. Histologically, some cases of Wegener's granulomatosis have large numbers of tissue eosinophils and, field for field, may be indistinguishable from AGA. Nevertheless, Wegener's granulomatosis only rarely shows necrotizing extravascular granulomas surrounding eosinophilic debris that is so characteristic of AGA. Wegener's granulomatosis tends to be more destructive and have more extensive necrosis than AGA. The vasculitis of Wegener's granulomatosis also shows a characteristic eccentric mural lesion whereas fibrinoid necrosis and increased eosinophils in the vessel walls are more characteristic of AGA.

REFERENCES

1. Wegener F: Uber eine eigenartiger rhinogene Granulomatose mit besonderer Beteiligung des Arterien Systems und der Nieren. *Beitr Pathol Anat* 102:36–68, 1939.
2. Klinger H: Grenzformen der Periarteriitis nodosa. *Frankfurter Zeits Pathol* 42:455–480, 1931.
3. Fauci AS, Haynes BF, Katz P: Wegener's granulomatosis: Studies in eighteen patients and a review of the literature. *Medicine* 52:536–561, 1973.
4. Fauci AS, Haynes BF, Katz P, Wolff SM: Wegener's granulomatosis: prospective clinical and therapeutic experience with 85 patients over 21. *Ann Intern Med* 98:76–85, 1983.
5. Hoffman GS, Kerr GS, Leavitt RY, et al: Wegener granulomatosis: an analysis of 158 patients. *Ann Intern Med* 116:488–498, 1992.
6. Spencer SJW, Burns A, Gaskin G, et al: HLA class II specificities in the vasculitis with antibodies in neutrophil cytoplasmic antigens. *Kidney Int* 41:1059–1063, 1992.
7. Knudsen BB, Joergensen T, Munch-Jensen B: Wegener's granulomatosis in a family. *Scand J Rheumatol* 17:225–227, 1988.
8. Walton EW: Giant-cell granuloma of the respiratory tract (Wegener's granulomatosis). *Br Med J* 1:265–270, 1958.
9. Reza MJ, Dornfield L, Goldberg LS, et al: Wegener's granulomatosis: Long term follow-up of patients treated with cyclophosphamide. *Arthritis Rheum* 18:501–506, 1975.
10. Van der Woude FJ, Rasmussen N, Lobatto S, et al: Autoantibodies against neutrophils and monocytes: Tool for diagnosis and marker of disease activity in Wegener's granulomatosis. *Lancet* 1:425–429, 1985.
11. Specks U, Homburger HA: Antineutrophil cytoplasmic antibodies. *Mayo Clin Proc* 69:1197–1198, 1994.
12. Nolle B, Specks U, Ludemann J, et al: Anti-cytoplasmic autoantibodies: Their immunodiagnostic values in Wegener's granulomatosis. *Ann Intern Med* 111:28–24, 1989.
13. Venning MC, Quinn A, Broomhead V, Bird AG: Antibodies directed against neutrophils (cANCA and pANCA) are of distinct diagnostic valve in systemic vasculitis. *Q J Med* 284:1287–1296, 1990.
14. Davenport A, Lock RJ, Wallington TB, Feest TG: Clinical significance of anti-neutrophil cytoplasm antibodies detected by a standardized indirect immunofluorescence. *Q J Med* 87:291–299, 1994.
15. Cohen Tervaert JW, Huitema MG, Hene RJ: Prevention of relapses in Wegener's granulomatosis by treatment based on anti-neutrophil cytoplasm antibody titre. *Lancet* 2:709–711, 1990.
16. Fienberg R, Mark EJ, Goodman M, et al: Correlation of anti-neutrophil cytoplasmic antibodies with the extra-renal histopathology of Wegener's (pathergic) granulomatosis and related forms of vasculitis. *Hum Pathol* 24:160–168, 1993.
17. Savage CO, Gaskin G, Pusey CD, Pearson JD: Anti-neutrophil cytoplasm antibodies can recognize vascular endothelial cell-bound anti-neutrophil associated antibodies. *Exp Nephrol (Switzerland)* 1:190–195, 1993.
18. Ballieux BE, Zondervan KT, Kievit P, et al: Binding of proteinase 3 and myeloperoxidase to endothelial cells: ANCA-mediated endothelial damage through ADCC? *J Clin Exp Immunol* 97:52–60, 1994.
19. Falk RJ, Terrell RS, Charles LA, Jennette JC: Anti-neutrophil cytoplasmic autoantibodies induce neutrophils to degranulate and produce oxygen radicals in vitro. *Proc Nat Acad Sci (USA)* 87:4115–4119, 1990.
20. Godschmeding R, van der Schoot CE, ten Bokkel Huinink D: Wegener's granulomatosis autoantibodies identify a novel diisopropylfuorophosphate-binding protein in the lysosomes of normal human neutrophils. *J Clin Invest* 84:1577–1587, 1989.
21. Colby TV, Specks U: Wegener's granulomatosis in the 1990's —A pulmonary pathologists perspective, in Churg A, Katzenstein (eds), *The Lung, Current Concepts*. Baltimore, Williams & Wilkins, 1993 pp 195–218.
22. Mark EJ, Ramirz JF: Pulmonary capillaritis and hemorrhage in patients with systemic vasculitis. *Arch Pathol Lab Med* 109:413–418, 1985.
23. Mark EJ, Matsubara O, Tan-liu NS, Fienberg R: The pulmonary biopsy in the early diagnosis of Wegener's (pathergic) granulomatosis: A study based on 35 open lung biopsies. *Hum Pathol* 19:1065–1071, 1988.

24. Yousem SA, Lombard CM: The eosinophilic variant of Wegener's granulomatosis. *Hum Pathol* 19:682–688, 1988.

25. Travis WD, Hoffman SG, Leavitt RY, et al: Surgical pathology of the lung in Wegener's granulomatosis. *Am J Surg Pathol* 51:315–333, 1991.

26. Yousem SA: Bronchocentric injury in Wegener's granulomatosis: A report of five cases. *Hum Pathol* 22:535–540, 1991.

27. Galateau F, Loire R, Capron F, et at: Pulmonary lesions in Wegener's disease. Report of the French anatomo-clinical research group. Study of 40 pulmonary biopsies (English abstract). *Rev Mal Respir* 9:431–442, 1992.

28. Leavitt RY, Fauci AS, Bloch DA, et al: Classification of Wegener's granulomatosis. *Arthritis Rheum* 33:1101–1107, 1990.

29. Churg A: Pulmonary angiitis and granulomatosis revisited. *Hum Pathol* 14:868–883, 1983.

30. Lombard CM, Duncan SR, Rizk NW, Colby TV: The diagnosis of Wegener's granulomatosis from transbronchial biopsy specimens. *Hum Pathol* 21:838–842, 1990.

31. Liebow AA: Pulmonary angiitis and granulomatosis. *Am Rev Respir Dis* 108:1–18, 1973.

32. Katzenstein A-LA, Liebow AA, Friedman PJ: Bronchocentric granulomatosis, mucoid impaction and hypersensitivity reaction to fungus. *Am Rev Respir Dis* 111:497–537, 1975.

33. Greenberger PA, Patterson R: Allergic bronchopulmonary aspergillosis. Model of bronchopulmonary disease with defined serologic, radiologic, pathologic and clinical findings from asthma to fatal destructive lung disease. *Chest* 91:165S–167S, 1987.

34. Jelihovsky T: The structure of bronchial plugs in mucoid impaction, bronchocentric granulomatosis and asthma. *Histopathology* 7:153–167, 1983.

35. Bosken C, Myer J, Greenberger P, Katzenstein A-LA: Pathologic features of allergic bronchopulmonary aspergillosis. *Am J Surg Pathol* 12:216–222, 1988.

36. Myers J, Katzenstein AL: Granulomatous infection mimicking bronchocentric granulomatosis. *Am J Surg Pathol* 10:317–322, 1986.

37. Katzenstein ALA, Askin FB: *Surgical Pathology of Non-neoplastic Lung Disease*. Philadelphia, Saunders WB, 1990, pp 281–282.

38. Tazelaar HD, Baird AM, Mill M, et al: Bronchocentric mycosis occurring in transplant recipients. *Chest* 96:92–95, 1989.

39. Kramer MR, Denning DW, Marshall SE, et al: Ulcerative tracheobronchitis after lung transplantation. A new form of invasive aspergillosis. *Am Rev Respir Dis* 144:552–556, 1991.

40. Churg J, Strauss L: Allergic granulomatosis, allergic angiitis and periarteritis nodosa. *Am J Pathol* 27:277–302, 1951.

41. Chumbley LC, Harrison EG Jr, DeRemee RA: Allergic granulomatosis and angiitis (Churg-Strauss syndrome): Report and analysis of 30 cases. *Mayo Clin Proc* 52:477–484, 1977.

42. Koss MN, Antonovych T, Hochholzer L: Allergic granulomatosis (Churg-Strauss syndrome). Pulmonary and renal morphologic findings. *Am J Surg Pathol* 5:21–28, 1981.

43. Lanham JG, Elkon KB, Pusey CD, Hughes GR: Systemic vasculitis with asthma and eosinophilia: A clinical approach to the Churg-Strauss syndrome. *Medicine (Baltimore)* 63:65–81, 1984.

44. Sehgal M, Swanson JW, DeRemee RA, Colby TV: Neurologic manifestations of Churg-Strauss syndrome. *Mayo Clin Proc* 70:337–341, 1995.

45. Lie JT: Limited forms of Churg-Strauss syndrome, in: Rosen PP, Fechner RE (eds), *Pathology Annual*, Part 2, Vol. 38. Connecticut, Appleton & Lange, 1993, p.199.

46. Finan MC, Winkelmann RK: The cutaneous extravascular necrotizing granuloma (Churg-Strauss granuloma) and systemic disease: A review of 27 cases. *Medicine* 62:142–158, 1983.

47. Masi AT, Hunder GG, Lie JT, et al: The American College of Rheumatology 1990 criteria for the classification of Churg-Strauss syndrome (allergic granulomatosis and angiitis). *Arthritis Rheum* 33:1094, 1990.

48. Churg A, Churg J: *Systemic Vasculitides*. New York, Igaku-Shoin, 1991, pp 101–120.

Chapter 28

Pulmonary Hemorrhage Syndromes

F. Capron

Pulmonary hemorrhage syndromes represent a heterogenous group of diseases characterized clinically by hemoptysis, pulmonary infiltrates, and anemia. Pathologically there is hemorrhagic alveolitis. Many pathological conditions have some degree of alveolar hemorrhage, varying in severity. Thus the term *pulmonary hemorrhagic syndromes* includes a wide spectrum of diseases. Some of these diseases, such as collagen diseases and Wegener's granulomatosis are discussed elsewhere. A list of some of the causes of alveolar hemorrhage is given in Table 28-1.

Theoretically the definition of pulmonary hemorrhage could be limited to the presence of blood in the alveoli or as demonstrated by lavage. The phagocytosis of erythrocytes or hemoglobin by alveolar macrophages causes alveolar hemosiderosis, which is easily diagnosed histologically. In this section causes of such hemosiderosis, such as bleeding from bronchial tumors, will not be discussed.

Classification of pulmonary hemorrhage requires correlation with clinical and biologic data, such as antineutrophil or antimembrane antibodies and the presence of renal or other extrapulmonary vascular disease.

TABLE 28-1
Some Causes of Alveolar Hemorrhage

Mechanism	Example
Intact alveolar septa	
Aspiration of blood	Tracheobronchial neoplasm
Passive congestion	Pulmonary veno-occlusive disease
Bleeding diathesis	Disseminated intravascular coagulation
Injured alveolar septa	
Anti-basement membrane antibody	Goodpasture's syndrome
Immune complex deposition	Systemic vasculitis
Toxic	Trimellitic anhydride
Unknown	Idiopathic pulmonary hemosiderosis

CLINICAL PRESENTATION

Pulmonary hemorrhage is characterized by dyspnea, hemoptysis, anemia and diffuse radiological opacities singly or together. Dyspnea can be mild or present as acute respiratory distress. Hemoptysis is rarely massive and may be absent.[1–3]

The radiologic opacities can be focal[4] and of variable severity.[5] The CT scan is characterized by a ground glass appearance. The diffusion capacity of carbon monoxide is elevated in the acute phase.[5] There is little value in scintigraphic testing.[6]

PATHOPHYSIOLOGY AND MECHANISMS OF PULMONARY HEMORRHAGE

Pulmonary bleeding can be bronchial or parenchymal, localized or diffuse. Localized hemorrhage can be aspirated to the periphery of the lung, simulating localized parenchymal bleeding. Diffuse or localized parenchymal bleeding is due to either hemodynamic modifications of the capillary pulmonary blood flow or to pathological changes in the alveolar wall.

The erythrocytes can be forced through the alveolar wall by venous hypertension, as in mitral stenosis, left ventricular failure, and pulmonary venoocclusive disease.[8] The first two conditions are usually diagnosed on clinical and radiologic criteria. Pulmonary hemorrhage refers to nonhemodynamic diffuse bleeding secondary to lesions of the alveolar wall. Such causes must be excluded at the start of the investigation of these cases.

HEMOSIDERIN RESORPTION

Red blood corpuscles are phagocytosed by alveolar macrophages. The degradation of hemoglobin is followed by hemosiderin storage in alveolar macrophages. This can be easily visualized by Prussian blue or Perls' stain. Little information is available on the time course of hemosiderin formation and clearance from the lungs. In healthy volunteers, hemosiderin formation, as shown in alveolar macrophages recovered by lavage, appears in lavage fluid within 72 h.[9] The storage is at its peak 120 h later. Complete clearance takes place within 2 to 4 weeks. In some conditions, such as cardiac transplantation, there appears to be a permanent capillary leak, with hemosiderin-laden macrophages always identified in the alveolar lumina.[10]

DIAGNOSIS OF PULMONARY HEMORRHAGE

When pulmonary hemorrhage is severe, the clinical presentation is characterized by hemoptysis, dyspnea, a drop in the hemoglobin, and radiological infiltrates. Each of these signs can be very discrete and the diagnosis underestimated. A useful investigation is fiberoptic bronchoscopy with lavage. The recovery of a bloody fluid is diagnostic, if bronchial inflammation or trauma have been excluded. If the bleeding is chronic and scant, the resorption of red blood cells is occult, the diagnosis is only made if a proper evaluation of the iron content of alveolar macrophages is done.

Since Golde et al.[11] proposed a quantitative evaluation of hemosiderin load in pulmonary macrophages, it is clear the simple quantification of the fraction of positive macrophages seen on an iron stain may underestimate severe hemosiderin storage. Two hundred macrophages stained by Perls' stain are counted and the score given per 100. This gives a more precise evaluation of resorption.

The Cytological Diagnosis of Alveolar Hemorrhage and Alveolar Damage

Frank Hemorrhage (Fig. 28-1)

The fluid is pink or bloody and the cytologic smears are hemorrhagic. If the hemorrhage is very recent, the macrophage hemosiderin score can be low.[11,12]

Occult Alveolar Hemorrhage (Fig. 28-2)

The lavage fluid is clear, with a hemosiderin score of around 98. This evaluation has been validated by different authors.[12–14]

Cytology may yield important clues for the etiology of alveolar damage. There is hemorrhage, frank or occult, rare alveolar dystrophic epithelial cells, and a few polymorphonuclear cells and hyaline membranes. The presence of alveolar hemorrhage, with or without alveolar damage, requires a search for tumor cells or evidence of viral, parasitic (Fig. 28-3), fungal, or bacterial infection.[12,13] Special stains and a study of a paraffin block of the centrifuged lavage are very helpful.

Histological Diagnosis of Pulmonary Hemorrhage

Initially the alveolar lumina are filled with erythrocytes and macrophages containing red blood corpuscules. Iron laden macrophages fill the alveoli (Fig. 28-4). The intensity of the alveolitis is variable.

FIGURE 28-1
Bronchoalveolar lavage. Frank hemorrhage.

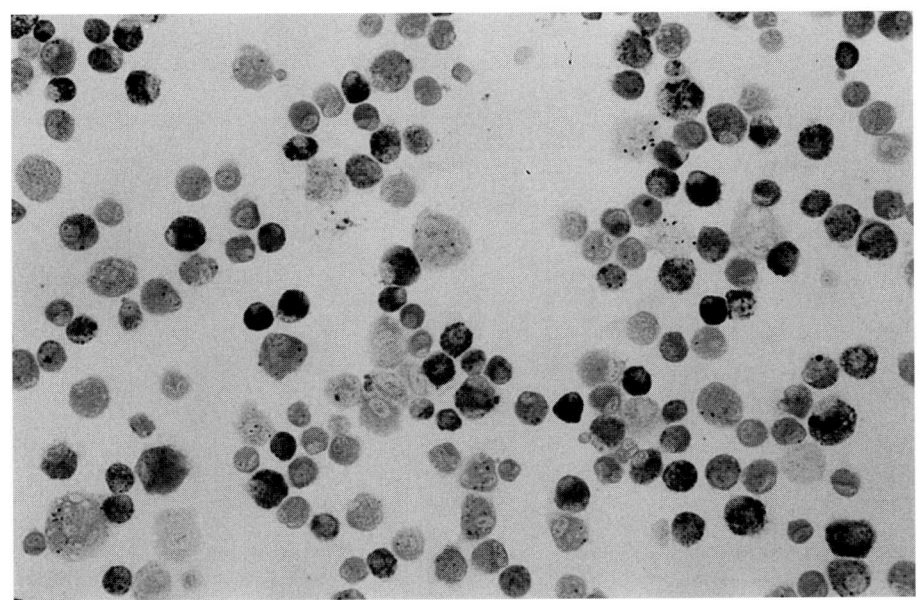

FIGURE 28-2
Bronchoalveolar lavage, occult hemorrhage, Perls' stain.

Hemosiderin deposition may be seen in the interstitium in chronic cases. There may be iron encrustation in the vascular elastic tissue. The elastic laminae are sometimes fragmented and surrounded by a foreign body giant cell reaction. Mild interstitial and intraalveolar fibrosis is seen in chronic hemorrhage. However, following acute hemorrhage, focal organizing alveolitis may supervene.

Hyaline membranes and epithelial desquamation are rare in alveolar hemorrhage. Their discovery helps to rule out traumatic hemorrhage due to the taking of the biopsy or bronchoscopy. A search for angiitis, granulomatous inflammation, necrosis, and pathogens such as *Diphyllobothrium latum* or, rarely, *Pneumocystis carinii* should be made. Recently there has been a case report of *Mycoplasma hominis* causing diffuse alveolar hemorrhage in a bone marrow biopsy recipient.[14]

In all cases of alveolar hemorrhage, a frozen section should be studied by immunofluorescence for IgG and the C3 fraction of complement.

Electron microscopy is useful to visualize pulmonary basement membrane deposits, though these are not always identified.

Capillaritis is a major histologic feature of pulmonary hemorrhage (Fig. 28-5), present in 88 percent

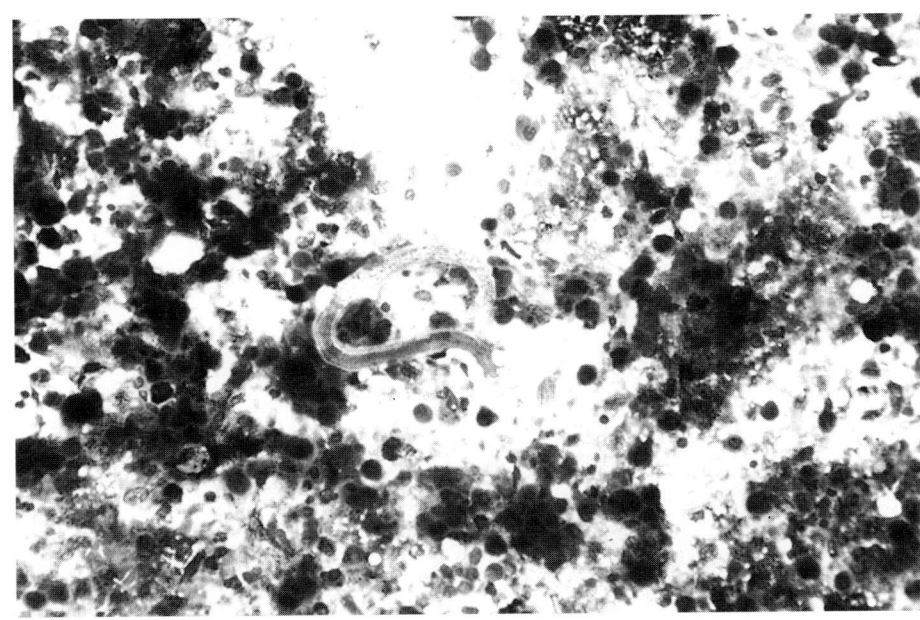

FIGURE 28-3
Bronchoalveolar lavage, ankylostomiasis.

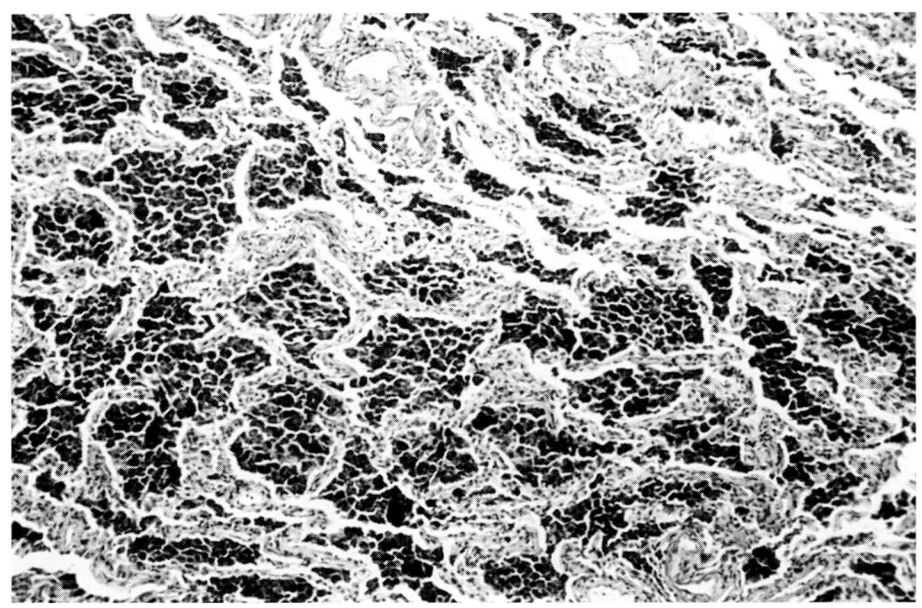

of the biopsies reviewed by Travis et al.[54] It has also been described in different diseases as hypersensitivity reaction, immune complex disease, and infection.

CLASSIFICATION AND ETIOLOGY OF PULMONARY HEMORRHAGIC SYNDROMES

Most cases of pulmonary hemorrhage are related to infective, toxic, hemodynamic, or local vascular causes. Others are considered as immune hemorrhagic syndromes. Vasculitides with antiglomerular basement membrane antibody, pulmonary lupus erythematosus, and idiopathic pulmonary hemosidero-

sis all belong to that group. Capillaritis is most frequently present in cases of Wegener's granulomatosis and microscopic polyangiitis that are characterized by the presence of anti-polymorphonuclear antibodies. Takayasu's disease may rarely cause pulmonary hemorrhage.[15]

Idiopathic Pulmonary Hemosiderosis

Idiopathic pulmonary hemosiderosis is a rare disease of young persons. It is characterized by hemoptysis, fleeting pulmonary infiltrates, and iron deficiency anemia. There is also fatigue and chest tightness, both probably related to anemia. It does not involve

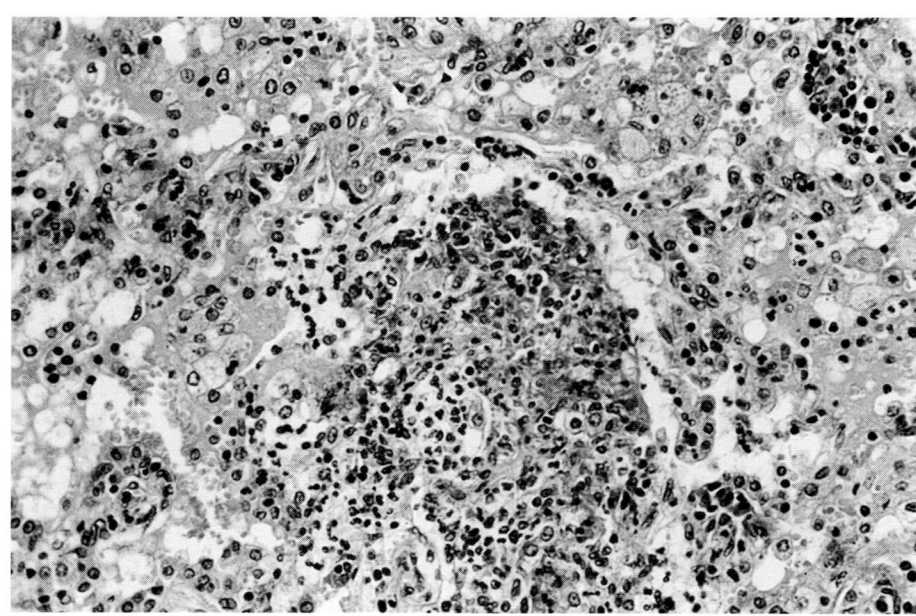

the kidney or any other extrapulmonary organs. Men are affected twice as often as women in patients less than 30 years of age. However, in one large series of 112 patients with the disease, only 19 percent were 16 years of age or older at its onset.[16]

On examination there is pallor and rales. Generalized lymphadenopathy has been described in 25 percent of patients and hepatosplenomegaly in 20 percent.[16]

Chest radiographs usually show bibasilar or perihilar infiltrates during episodes of hemorrhage. These resolve after several weeks. The pulmonary function tests vary from normal to showing a marked restrictive defect. In such cases there are diminished lung volumes, as well as a decreased diffusing capacity for carbon monoxide (DLCO). Some studies suggest improvement in the lung volumes and Pa_{O_2} during clinically evident remissions.

Histologically there is progressive filling of the alveoli by hemosiderin-laden macrophages, interstitial and intraalveolar fibrosis. There is minimal inflammation. Siderin impregnation and fractured elastic laminae in arterioles are common (Fig. 28-6). Pulmonary veins are normal.

Electron microscopic changes have been inconsistent, and nonspecific changes have been described.[20-24] These changes have been in elastic fibers, endothelial cells, capillary basement membranes, and loss of types I and II pneumocytes. There is also hyperplasia of type II pneumocytes.

The diagnosis is one of exclusion and requires negative antiglomerular basement membrane anti-body. The kidney must be normal and there should be no angiitis or collagen vascular disease.

The etiology of the disease is unknown. There is some resemblance to cow's milk hypersensitivity,[25] gluten-sensitive enteropathy,[26-28] and hemolytic anemia.[29]

Antibasement Membrane Antibody Disease

Goodpasture described a young man with pulmonary hemorrhage and glomerulonephritis in the influenza pandemic of 1919.[30] It is a matter of speculation as to how this case would be classified today, amidst the many diseases associated with pulmonary hemorrhage and renal diseases. This is important in view of the lack of immunologic tests available in Goodpasture's time. The disease is rare, with an estimated incidence of 0.32 per 100,000 per year.[31] For some time the association of alveolar hemorrhage and proliferative glomerulonephritis was considered diagnostic of Goodpasture's syndrome. Men are more commonly affected. There is a bimodal age distribution, peaking at between 30 and 60 years of age. The syndrome is thought to be an autoimmune disorder, consisting of glomerulonephritis, lung hemorrhage, and antiglomerular basement membrane antibody formation.

The clinical spectrum is wide, ranging from cases with massive pulmonary hemorrhage and little or no renal disease to cases with overt proliferative glomerulonephritis and very mild pulmonary hemorrhage. The nephrotic syndrome is unusual. A rela-

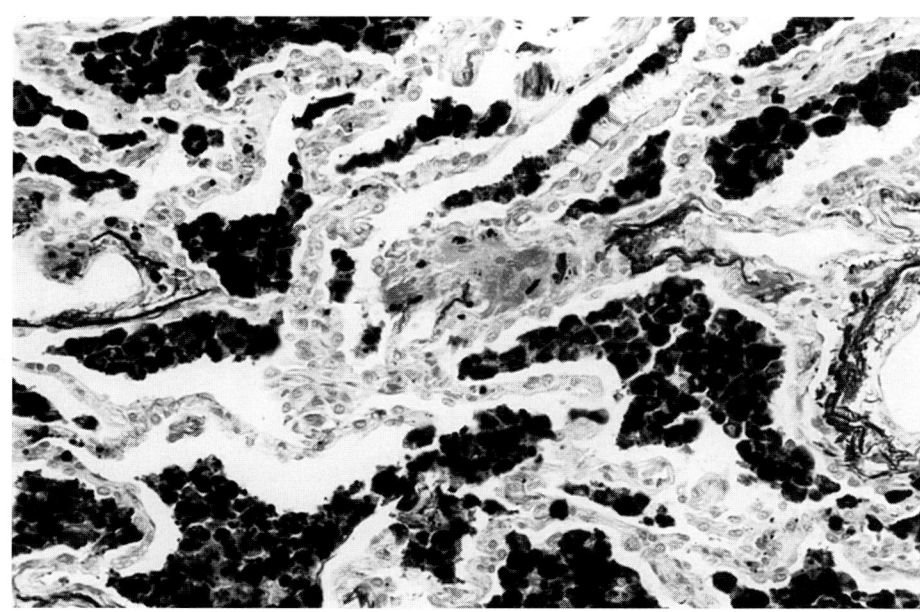

FIGURE 28-6
Idiopathic hemosiderosis, iron-laden macrophages in the alveoli, iron impregnation of the elastic membranes.

tionship between pulmonary hemorrhage and cigarette smoking has been described.[32] Symptoms are usually pulmonary, sometimes preceded by episodes of fever or weight loss. Many patients give a history of a flu-like illness. The main symptom related to the chest is hemoptysis but dyspnea and cough may also be seen. The disease may begin with either renal problems or hemoptysis, often occurring simultaneously. However, the interval between the two may be as long as one year. Arthralgia and arthritis may be prominent symptoms in 5 percent of patients. There is a hypochromic, microcytic anemia. Rarely, the disease seems purely renal or pulmonary in origin.[33,34]

The chest radiograph is normal between episodes of hemoptysis but with hemorrhage there are alveolar infiltrates.

Macroscopically, in acute hemorrhage there is hemorrhagic consolidation. The lung in chronic cases is brown due to hemosiderin deposition (Fig. 28-7).

A systematic evaluation of the surgical pathology of the lung in Goodpasture's syndrome has been reported.[35] In all cases the leading feature is massive alveolar hemorrhage with hemosiderin-laden macrophages. Alveolar capillaritis with fibrin thrombi were present in all cases. There may be hyaline membranes and therefore there is a resemblance to adult respiratory distress syndrome. As the disease resolves, the red blood corpuscles lyse and there may be Masson balls associated with interstitial fibrosis and proliferation of type II cells. Inflammation was mild, focal and limited to the interstitium. Some venulitis is described in the external part of the small venules. There is no thrombosis. A few eosinophils are detected in the interstitium and admixed with the neutrophils in the capillaries. There is no alveolar wall necrosis.[35]

The essential element in the diagnosis of Goodpasture's syndrome is antibody formation to an anti-collagen-alpha 3(IV).[36] The antibody is best isolated in the sera of patients with ELISA techniques, which have a high sensitivity.[37,38] The immunofluorescence seen in renal and lung biopsies,[21,39,40] shows linear deposits of mostly IgG, with C3 (Fig. 28-8) fragments. Some authors have reported positive results with transbronchial biopsies.[41–44] There are occasional false-negative studies, so a negative immunofluorescence does not exclude the diagnosis.[45,46]

Though the typical immunofluorescence shows IgG and C3, a case has been described with IgA deposition.[47] Occasional granular staining for IgA, IgM, or IgG may be found.

FIGURE 28-7
Goodpasture's disease. Whole section of the lung with hemorrhagic consolidation. (*Reproduced by kind permission of Dr. M. Schwartz, Department of Medicine, University of Colorado Health Sciences Center, Denver, Colorado.*)

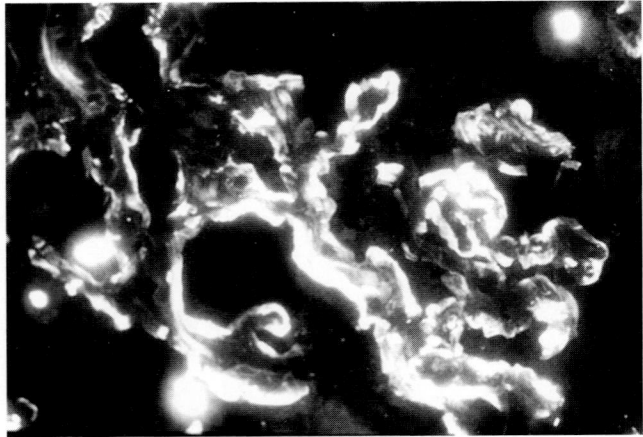

FIGURE 28-8
Goodpasture's disease. Immunofluorescence showing linear deposits of IgG. (*Reproduced by kind permission of Dr. M. Schwartz, Department of Medicine, University of Colorado Health Sciences Center, Denver, Colorado.*)

In all cases, as well as in serologic studies for anti-GBM antibodies, it is important to search for c- and p-ANCA (antineutrophil cytoplasmic antibodies) as well as tests for collagen vascular diseases.

The pathogenesis of the pulmonary lesions is obscure. In an experimental model of the disease, monkeys injected with sera from patients with Goodpasture's syndrome do not develop pulmonary lesions. Glomerular capillaries show fenestration, but pulmonary capillaries do not. This could explain the fact that after infections or toxic damage the lung is more receptive to the disease, as reported after inhalation of solvents or tobacco.[32,33,49,50] Cytokines play a role in the pathogenesis of human glomerular basement membrane disease.[51]

Pulmonary-Renal Diseases

Hemorrhagic syndromes are often related to severe glomerulonephritis causing progressive renal insufficiency. Pulmonary-renal syndromes have been classified on the basis of the type and topography of immune deposits in the renal biopsy.[52] Three groups are delineated.

1. Those with immune deposits either of the linear type (IgG, C3), such as are seen in Goodpasture's syndrome, or granular deposits, such as are seen in lupus erythematosus and related diseases.
2. Those without immune deposits, such as Wegener's disease, and microscopic polyangiitis.
3. Disease related to anti-polymorphonuclear antibodies.[53]

Pulmonary Hemorrhage and Angiitis

WEGENER'S DISEASE AND OTHER VASCULITIDES[54]

Among all vasculitides, Wegener's disease is the most likely to present as massive pulmonary hemorrhage.[14a] Cases presenting with massive hemorrhage and capillaritis with little in the way of angiitis or granulomas have been reported.[54a] Alveolar hemorrhage is a feature of acute onset of the disease and can be a major misleading clinical presentation with acute respiratory distress.[55-57] Alveolar hemorrhage is sometimes present in classical forms of the disease at the periphery of lesions or at distant sites, where it may represent more recent lesions. Such forms seem more prone to early renal lesions and present with little upper airway disease. Tests for anticytoplasmic neutrophil antibodies and pulmonary biopsy give the diagnosis.

Other vasculitides can be complicated with pulmonary hemorrhage. Included in this group are Churg-Strauss disease (allergic angiitis),[58] microscopic polyangiitis,[59,60] Henoch-Schönlein purpura,[61-63] and cryoglobulinemia.[64] Other diseases with immune complex mediation can present with pulmonary hemorrhage, such as endocarditis,[65] hepititis B,[66] and myasthenia gravis.[66]

Miscellaneous Diseases

INFECTIOUS DISEASE

Alveolar hemorrhage of variable intensity is a frequent feature of infective pneumonia. It has been reported and evaluated in immunocompromised patients by Kahn and coworkers.[12] The causative pathogens are, in order of decreasing frequency, invasive fungi, bacterial and viral pneumonia, *Pneumocystis carinii*, and mycobacteria. Other reports have incriminated gram-negative infection,[12] *Legionnella pneumophilia*, *Cytomegalovirus*, and *Candida albicans*,[68] leptospirosis,[69,70] and ankylostomiasis. Aspergillus infection can induce severe pulmonary bleeding.[12,71] Infection may provoke or increase alveolar hemorrhage in thrombocytopenic patients and in patients with a diseased lung parenchyma.

HEMATOLOGICAL DISORDERS

Defective coagulation is a predisposing factor in severe bleeding. Some reports have focused on the occurrence of hemorrhage in thrombocytopenic patients, those on anticoagulant therapy,[72] and those with disseminated intravascular coagulation.[73] Diffuse alveolar hemorrhage has been seen in patients with autologous marrow transplantation.[74,75] Some authors have discussed the possible role of dimethyl sulfoxide used for the cryopreservation of marrow as a possible inducer of endothelial injury.[75]

DRUG-INDUCED

Most drug-related pneumonias can be hemorrhagic. D-penicillamine[76,77] related toxic pneumonia, seen after doses of more than 750 mg per day for more than seven months, is always accompanied by renal disease. Many other drugs have been associated with alveolar hemorrhage. These include hydralazine,[72] aminoglutethimide,[78] amiodarone,[79] moxalactam,[80] and nitrofurantoin.[81]

CHEMICAL CAUSES

Chemically related diffuse alveolar damage can cause pulmonary hemorrhage. Trimellitic anhydride induces massive pulmonary hemorrhage.[82,83] Other substances, such as radiographic contrast media,[84] lipid embolism,[85] silicon injections,[86] and drug abuse[85,87] cause alveolar hemorrhage.

PULMONARY HEMORRHAGE RELATED TO NEOPLASIA

Alveolar hemorrhage is a constant feature of tumor invasion. Angiosarcomas[88] and choriocarcinomas produce severe bleeding.[89] Pulmonary hemorrhage is a major manifestation of Kaposi sarcoma[90–92] and of paraneoplastic microangiopathy.[65]

Tumor invasion in the lung is usely accompanied by some hemorrhage. It is common in the vicinity of a parenchymal nodule or can be aspirated to the periphery of a bronchiole. Hemorrhage can be a major problem in some cases of vascular invasion and lymphangitis carcinomatosis. Cytologic and histogic examination should always exclude malignant cause. A difficult situation concerns cases where alveolar hemorrhage is associated with alveolar damage. This diagnostic problem is specific to cytologic specimens, such as lavage fluid or transthoracic or bronchial aspirates. The desquamation of epithelial cells can cause the misdiagnosis of cancer. Some authors have proposed an immunohistochemical cytologic study with carcinoembryonic antigen. We have found this investigation unhelpful.

PULMONARY HYPERTENSION

Clinical and functional investigations should enable one to exclude hemorrhage due to pulmonary hypertension. The main cause used to be mitral stenosis.

REFERENCES

1. Bradley J: The pulmonary haemorrhage syndromes. *Clin Chest Med* 3:593–605, 1982
2. Eagen JW, Memoli VA, Roberts JL, Matthew GR, Schwartz MM, Lewis EJ: Pulmonary haemorrhage in systemic lupus erythematosus. *Medicine* 57:545–559, 1978.
3. Morgan PGM, Turner-Warwick M: Pulmonary haemosiderosis and pulmonary haemorrhage. *Br J Dis Chest* 75:225–242, 1981.
4. Palmer PES, Finley TN, Drew WL, Golde DW: Radiographic aspects of occult pulmonary haemorrhage. *Clin Radiol* 29:139–143, 1978.
5. Bowley NB, Hughes JMB, Steiner RE: The chest X ray in pulmonary capillary haemorrhage. Correlation with carbon monoxide uptake. *Clin Radiol* 30:413–417, 1979.
6. Degowin RL, Sorensen LB, Charleston DB, et at: Retention of radio-iron in the lungs of a woman with idiopathic pulmonary haemosiderosis. *Ann Intern Med* 69:1213–1220, 1968.
7. Samuels LD, Bass JC: Chromium-51 lung scan in idiopathic hemosiderosis. *J Nucl Med* 10:106–107, 1969.
8. Palevsky HI, Pietra GG, Fishman AP: Pulmonary veno-occlusive disease and its response to vasodilator agents. *Am Resv Respir Dis* 142:426–429, 1990.
9. Sherman JM, Winnie G, Thomassen MJ, Abdul-Karim FW, Boat FT: Time course of haemosiderin production and clearance by pulmonary macrophages. *Chest* 86:409–411, 1984.
10. De Lassence A, Fleury-Feith J, Escudier E, Beaune J, Bernaudin J-F, Cordonnier C: Alveolar hemorrhage. Diagnostic criteria and results in 194 immunocompromised hosts. *Am J Crit Care Med* 151:157–163, 1995.
11. Golde DW, Drew WL, Klein HZ, Finley TN, Cline MJ: Occult pulmonary haemorrhage in leukemia. *Br Med J* 2:166–168, 1975.
12. Kahn FW, Jones JM, England DM: Diagnosis of pulmonary haemorrhage in the immunocompromised host. *Am Rev Respir Dis* 136:155–160, 1987.
13. Grebski E, Hess T, Hold G, Speich R, Russi E: Diagnostic value of haemosiderin-containing macrophages in bronchoalveolar lavage. *Chest* 102:1794–1799, 1992.
14. Kane JR, Shenep JL, Krance RA, Hurwitz CA: Diffuse alveolar hemorrhage associated with *Mycoplasma hominis* respiratory tract infection in a bone marrow transplant recipient. *Chest* 105:1891–1892, 1994.
14a. Leatherman JW, Davies SF, Hoidal JR: Alveolar hemorrhage syndrome: Diffuse microvascular lung hemorrhage in immune and idiopathic disorders. *Medicine* (*Baltimore*) 63:343–361, 1984.
15. Kabayu S, Isaka N, Yada T, Konishi T, Nakano T: Severe respiratory failure caused by recurrent pulmonary hemorrhage in Takayasu's arteriris. *Chest* 104:1905–1906, 1993.
16. Soergel KH, Sommers SC: Idiopathic pulmonary hemosiderosis and related syndromes. *Am J Med* 32:499–511, 1962.
17. Fuleihan FJD, Abboud RT, Habaytar R: Idiopathic pulmonary hemosiderosis: case report with pulmonary function tests and review of the literature. *Am Rev Respir Dis* 98:93–97, 1968.
18. Allue X, Wise MB, Beaudry PH: Pulmonary function test studies in idiopathic hemosiderosis in children. *Am Rev Respir* 107:410–415, 1973.
19. Thaell JF, Griepp PR, Stubbs SE: Idiopathic pulmonary hemosiderosis: Two cases in a family. *Mayo Clin Proc* 53:113–118, 1978.
20. Elliot ML, Kuhn C: Idiopathic pulmonary hemosiderosis: Ultrastructural abnormalities in the capillary walls. *Am Rev Respir Dis* 102:895–904, 1970.
21. Donald KJ, Edwards RL, McEvoy JDS: Alveolar capillary basement lesions in Goodpasture's syndrome and idiopathic hemosiderosis. *Am J Med* 59:642–649, 1975.
22. Irwin RS, Cottrell TS, Hsu KC, et al: Idiopathic pulmonary hemosiderosis: An electron microscopic and immunofluorescent study. *Chest* 65:41–45, 1974.
23. Donlan CJ, Srodes CH, Duffy FD: Idiopathic pulmonary hemosiderosis: Electron microscopic, immunofluorescent, and iron kinetic studies. *Chest* 68:577–580, 1975.
24. Gonzalez-Crussi F, Hull MT, Grosfeld JL: Idiopathic pulmonary hemosiderosis: Evidence of capillary basement membrane abnormality. *Am Rev Respir Dis* 114:689–698, 1976.

25. Valassi-Adams H, Rouska A, Karpouyas J: Raised IgA in idiopathic pulmonary haemosiderosis. *Arch Dis Child* 50:320–322, 1975.

26. Wright PH, Menzies IS, Pounder RE, Keeling PWN: Adult idiopathic pulmonary haemosiderosis and coeliac disease. *Quart J Med* 197:95–102, 1981.

27. Reading R, Watson JG, Platt JW, Bird AG: Pulmonary haemosiderosis and gluten. *Arch Dis Child* 62:513–515, 1987.

28. Pachero A, Casanova C, Fogue L, Sweiro A: Long term follow-up of adult idiopathic pulmonary haemosiderosis and coeliac disease. *Chest* 99:1525–1526, 1991.

29. Rafferty JR, Cook MK: Idiopathic pulmonary haemosiderosis with autoimmune haemolytic anaemia. *Br J Dis Chest* 78:282–285, 1984.

30. Goodpasture EW: The significance of certain pulmonary lesions in relation to the etiology of influenza. *Am J Med Sci* 158:863, 1919.

31. Teague CA, Doak PB, Simpson IJ, Rainer SP, Herndson PB: Goodpasture's syndrome: An analysis of 29 cases. *Kid Int* 13:492–504, 1978.

32. Donaghy M, Rees AJ: Cigarette smoking and lung haemorrhage in glomerulonephritis caused by antibodies to glomerular basement membrane. *Lancet,* 2:1390–1393, 1983.

33. Robler A, Schurch E, Alterman HJ, Im Hof V: Anti-basement membrane antibody disease with severe pulmonary haemorrhage and normal renal function. *Thorax* 46:68–69, 1991.

34. Zimmerman SW, Varanasi VR, Hoff B: Goodpasture's syndrome with normal renal function. *Am J Med* 66:163–171, 1979.

35. Lombard DM, Colby TV, Elliot CG: Surgical pathology of the lung in anti-basement membrane antibody-associated Goodpasture's syndrome. *Hum Pathol* 20:445–451, 1989.

36. Hudson BG, Wieslander J, Wisdom BJ, Noelken ME: Goodpasture's syndrome: Molecular architecture and function of basement antigen. *Lab Invest* 61:256–269, 1989.

37. Brenner BM, Rector FC: Anti-membrane antibody-induced disease in man. *The Kidney,* in Brenner BM, Rector FC (eds): 2d ed. Saunders, Philadelphia, 81.

38. Wieslander J, Brygren P, Heinegard D: Antiglomerular basement membrane antibody: Antibody specificity in different forms of glomerulonephritis. *Kid Int* 23:855–861, 1983.

39. Scheer RL, Grossman MA: Immune aspects of the glomerulonephritis associated with pulmonary haemorrhage. *Ann Intern Med* 60:1009–1021, 1964.

40. Wilson CB, Dixon FJ: Antiglomerular basement membrane antibody-induced glomerulonephritis. *Kid Int* 3:74–89, 1973.

41. Beechler CR, Enquist RW, Hunt KK, Ward GW, Knieser MR: Immunofluorescence of transbronchial biopsies in Goodpasture's syndrome. *Am Rev Respir Dis* 121:869–872, 1980.

42. Abboud RT, Chase WH, Ballon HS, Grzbowski S, Magil A: Goodpasture's syndrome: Diagnosis by transbronchial biopsy. *Ann Intern Med* 89:635–638, 1978.

43. Johnson JP, Moore J, Austin HA, Balow JE, Antonovych TT, Wilson CB: Therapy of antiglomerular membrane antibody disease. Analysis of prognostic significance of clinical pathologic and treatment factors. *Medicine* 64:219–227, 1985.

44. Teichman S, Briggs WA, Knieser MR, et al: Goodpasture's syndrome: Two cases with contrasting early course and management. *Am Rev Respir Dis* 113:223–232, 1976.

45. Watters LC: Chronic alveolar filling disease, in Schwartz MI, King TE Jr (eds): *Interstitial Lung Disease.* St Louis, Mosby Year Book, pp310–323, 1993.

46. Scully RE, Mark EJ, McNeely WF, McNeely BU: Case records of the Massachusett's General Hospital (Case 16-1993). *N Engl J Med* 328:1183–1190, 1993.

47. Border WA, Baehler RW, Bhathena D, Glassock RJ: IgA, antibasement membrane nephritis with pulmonary hemorrhage. *Ann Int Med* 91:21–25, 1979.

48. Leaker B, Cambridge G, du Bois RM, Neild GH: Idiopathic pulmonary Haemosiderosis. A form of microscopic polyarteritis? *Thorax* 47:988–990, 1992.

49. Kleinchnecht D, Morel-Marogar L. Callard P, Adhemar J-P, Mathieu P: Antiglomerular basement nephritis after solvent exposure. *Arch Intern Med* 140:230–232, 1980.

50. Yamamoto T, Wilson CB: Binding of anti-basement membrane antibody to alveolar basement membrane after intratracheal gasoline instillation in rabbits. *Am J Pathol* 126:497–505, 1987.

51. Queluz TH, Pawlowski I, Brunda MJ, Brentjens JR, Viadutiu AO, Andres G: Pathogenesis of an experimental model of Goodpasture's hemorrhagic pneumonitis. *J Clin Invest* 85:1507–1515, 1990.

52. Couser WG: Rapidly progressive glomerulonephritis: classification, pathogenic mechanisms, and therapy. *Am J Kidney Dis* 11:449–464, 1988.

53. Jennette JC, Wilkman AS, Falk RJ: Anti-neutrophil cytoplasmic autoantibody-associated glomerulonephritis and vasculitis. *Am J Pathol* 135:921–930, 1989.

53a. Hoffman GS, Kerr GS, Leavitt RY, et al: Wegener's granulomatosis. An analysis of 158 patients. *Ann Intern Med* 116:488–498, 1992.

54. Travis WD, Hoffman GS, Leavitt RY, Pass HI, Fauci AS: Surgical pathology of the lung in Wegener's granulomatosis. Review of 87 open lung biopsies from 67 patients. *Am J Surg Pathol* 15:315–333, 1991.

54a. Myers JL: Diagnosis of drug reactions in the lung. *Monogr Pathol* 36:32–53, 1993.

55. Stokes C, McCann BG, Rees RT, Sims EH, Harriso S: Intrapulmonary haemorrhage in Wegener's granulomatosis. *Thorax* 37:315–316, 1982.

56. Misset B, Glotz D, Escudier B, et al: Wegener's granulomatosis presenting as diffuse pulmonary haemorrhage. *Intensive Care Med* 17:118–120, 1991.

57. Bax J, Gooszen HC, Hoorntje SJ: Acute fulminating alveolar hemorrhage as presenting symptom in Wegener's granulomatosis. *Eur J Respir Dis* 71:202–205, 1987.

58. Clutterbuck EJ, Pusey CD: Severe alveolar haemorrhage in Churg-Strauss syndrome. *Eur J Respir Dis* 71:158–163, 1987.

59. Even P, Dennewald G, Sors H, Reynaud P: Syndrome de Goodpasture et syndromes analogues. In: *Maladies ditessyst miques,* 2d ed, Kahn et Peltier Editeurs, Flammarion, Paris, 484–510, 1982.

60. Saltzman PW, West M, Chomet B: Pulmonary Hemosiderosis and glomerulonephritis. *Ann Intern Med* 56:409–421, 1962.

61. Weiss VF, Naidv S: Fatal pulmonary haemorrhage in Henoch-Schoenlein's purpura. *Cutis* 23:687–688, 1979.

62. Kathuria S, Cheifec G: Fatal pulmonary Henoch-Schoenlein syndrome. *Chest* 82:654, 1982.

63. Markus HS, Clark JV: Pulmonary haemorrhage in Henoch-Schoenlein purpura. *Thorax* 44:525–526, 1989.

64. Martinez JS, Kohler PF: Variant "Goodpasture's syndrome?" The need for immunologic criteria in rapidly progressive glomerulonephritis and hemorrhagic pneumonitis. *Ann Intern Med* 75:67–76, 1971.

65. Holdsworth S, Boyce N, Thomson NM, Atkins RC: the clini-

cal spectrum of acute glomerulonephritis and lung haemorrhage (Goodpasture's syndrome). *Quart J Med* 216:75–86, 1985.

66. Germouty J, Bonnaud F, Rouffaud J: L'hemosiderose pulmonaire idiopathique: . . . propos d'un cas. *Sem Hop Paris* 52:369, 1976.

66a. Mark EJ, Ramirez JF: Pulmonary capillaritis and hemorrhage in patients with systemic vasculitis. *Arch Pathol Lab Med* 109:413–418, 1985.

67. Kradin RL, Kiprov D, Dickersin R, Collins AB, Kradin L, Mark EJ: Immune complex disease with fatal pulmonary haemorrhage. Its occurence in a patient with myasthenia gravis. *Arch Pathol Lab Med* 105:582–585, 1981.

68. Drew WL, Finley TN, Golde DW: Diagnostic lavage and occult pulmonary hemorrhage in 434 thrombocytopenic immunocompromised patients. Am Rev Respir Dis 116:215–221, 1977.

69. Daoudal P, Mathieu P, Bloch B, Barale F: Leptospirose avec immunisation antimembrane basale glomrulaire. *Nouv Presse Med* 7:3535–3537, 1978.

70. Vuong TK, Laaban JP, Rabbat A, Capron F, Bouvet A, Rochemaure J: Leptospirose icterohemorragique avec syndrome de détresse et hémorragie pulmonaire. *Rev Mal Resp* 8:256–257, 1991.

71. Stover DE, Zaman MB, Hajdu SI, et al: Bronchoalveolar lavage in the diagnosis of diffuse pulmonary infiltrates in the immunocompromised host. *Ann Intern Med* 10:1–17, 1984.

72. Finley TN, Aronow A, Cosentino AM, et al: Occult pulmonary hemorrhage in anticoagulated patients. *Am Rev Respir Dis* 112:23–29, 1975.

73. Robboy SJ, Minna JD, Colman RW, et al: Pulmonary hemorrhage syndrome as a manifestation of disseminated intravascular coagulation: analysis of ten cases. Chest 63:718–721, 1973.

74. Robbins RA, Linder J, Stahl MG, et al: Diffuse pulmonary hemorrhage in autologous bone marrow recipients. *Am J Med* 87:511–518, 1989.

75. Chao NJ, Duncan Sr, Long GD, Horning SJ, Blumer KG: Corticosteroid therapy for diffuse alveolar hemorrhage in autologous bone marrow transplant recipients. *Ann Intern Med* 114:145–146, 1991.

76. Matloff DS, Kaplan MM: D-penicillamine-induced Goodpasture's-like syndrome in primary biliary cirrhosis: Successful treatment with plasmapheresis and immunosuppression. *Gastroenterology* 78:1046–1049, 1980.

77. Sternlieb I, Bennet B, Scheingerg IH: D-penicillamine induced Goodpasture's syndrome in Wilson's disease. *Ann Intern Med* 82:673–676, 1975.

78. Rodman DM, Hanley M, Parsons P: Aminoglutethimide,

alveolar damage, and hemorrhage. *Ann Intern Med* 105:633, 1986.

79. Dean PV, Groshart KD, Porterfield JG, Iansmith DH, Golden EB: Amiodarone-associated pulmonary toxicity. *Am J Clin Pathol* 87:7–13, 1987.

80. Branstetler RD, Tamarin FM, Rangraj MS, Ruiz M, Giampiero J: Moxalactam disodium-induced pulmonary hemorrhage. Chest 86:644–645, 1984.

81. Bucknall CE, Adamson MR, Banham SW: Non-fatal pulmonary haemorrhage associated with nitrofurantoin. *Thorax* 42:475–476, 1987.

82. Herbert FA, Oford R: Pulmonary hemorrhage and edema due to inhalation of resins containing trimelletic anhydride. *Chest* 76:546–551, 1979.

83. Abud-Mendoza C, Diaz-jouanen E, Alarcon-Segovia D: Fatal pulmonary haemorrhage in systemic lupus erythematosus. Occurence without haemoptysis. *J Rheumatol* 12:558–561, 1985.

84. Ahmad D, Morgan WKC, Patterson R, Williams T, Zeiss CR: Pulmonary haemorrhage and hemolytic anaemia due to trimellitic anhydride. *Lancet* 1:328–330, 1979.

85. Berrigan TJ, Carsky EW, Heitzman ER: Fat embolism. Roentgenographic/pathologic correlation in three cases. *Am Rev Roentgenol* 96:967–971, 1967.

86. Chastre J, Basset F, Viau F, Dournovo P, Bouchama A, Akesbi A, Gibert C: Acute pneumonitis after subcutaneous injections of silicone in transsexual men. *N Engl J Med* 308:764–767, 1983.

87. Murray RJ, Albin RJ, Mergner W, Criner GJ: Diffuse alveolar haemorrhage temporally related to cocaine smoking. Chest 93:427–429, 1988.

88. Carter ES, Bradburne RM, Shung JW, Ettensohn DB: Alveolar haemorrhage with epithelioid hemangioendothelioma. *Am Rev Respir Dis* 142:700-791, 1990.

89. Benditt JO, Farber HW, Wright J, Karnad AB: Pulmonary hemorrhage with diffuse alveolar infiltrates in men with high volume choriocarcinoma. *Ann Intern Med* 109:674–675, 1988.

90. Touboul JL, Mayaud CM, Fouret P, Akoun GM: Pulmonary lesions of Kaposi's sarcoma, intraalveolar haemorrhage and pleural effusion. *Ann Intern Med* 103:808–809, 1985.

91. Wash G, Fligiel S: Pathologic features of the lung in the acquired immune deficiency syndrome (AIDS): An autopsy study of seventeen homosexual males. *Am J Clin Pathol* 81:6–12, 1984.

92. Ognibene FP, Steis RG, Macher AB, et al: Kaposi's sarcoma causing pulmonary infiltrates and respiratory failure in the acquired immunodeficiency syndrome. *Ann Intern Med* 102:471–475, 1985.

Chapter 29

Benign Lung Tumors and Their Malignant Counterparts

P. S. Hasleton

Benign lung tumors account for only a small percentage of a histopathologist's workload. Despite this fact, they may be confused with malignant conditions. Because they are so varied, a large chapter is devoted to them. They may cause confusion radiologically in the differential diagnosis of a solitary pulmonary nodule, or in the case of bronchial carcinoids, they may be misdiagnosed as small-cell carcinoma on small biopsies.

BRONCHIAL "HAMARTOMA" (ADENOCHONDROMA, PULMONARY MESENCHYMOMA)

Bronchial hamartoma was defined originally as a tumor-like malformation that had a disorganized growth of tissues normally present in the affected organ. The malformations may consist of a change in the quantity, arrangement, or degree of differentiation or any combination of these.[1] The term is derived from the Greek *hamartia* ("defect") and *oma* ("tumor").

The chromosomal abnormalities, discussed below, as well as the age range of patients with these lesions favor a neoplasm rather than a true hamartoma.

The hamartomas may be either endobronchial or parenchymal. They were once regarded as separate entities,[2,3] but they are now considered identical and are thought to be mesenchymal in origin.[4,5]

Age, Sex Incidence, and Clinical Features

In six studies the average age of patients with pulmonary mesenchymoma was 54.6 years.[4-9] Endobronchial tumors are less common than intrapulmonary ones and more frequent in males. There is little difference between the sexes in incidence of intrapulmonary lesions. One of the largest series reported is that of McDonald and colleagues,[10] who in 8000 postmortems found 23 cases, giving an incidence in the adult population of 0.25 percent. This figure could be a low estimate, since small lesions may not be detected at necropsy. Van Den Bosch and coworkers,[7] in a Dutch surgical series, noted the annual incidence to be one per 100,000.

Clinical Features

Most parenchymal mesenchymomas are symptomless, although occasionally they may present with hemoptysis. Multiple pulmonary mesenchymomas have been described.[11] Endobronchial lesions cause pulmonary obstruction with cough, dyspnea, recurrent attacks of pneumonia, hemoptysis, and in some cases chest pain. The tumors are equally distributed between all the lobes.

Salminen[8] showed that Finnish male patients with this tumor were more often smokers (67.4 percent) than simply members of the general population (34 percent). Despite this, there was no significant difference in cancer risk between patients with pulmonary mesenchymoma and the general Finnish population.

Previously, Van Den Bosch[7] had suggested an association between mesenchymoma and bronchial carcinoma. The risk that bronchial carcinoma and other tumors will develop in patients with mesenchymoma has been found to be increased by a factor of 6.66. The tumors are either synchronous or metachronous. Thus, these patients should be regularly followed up.[12]

CARNEY'S TRIAD

This potentially lethal condition, usually seen in females below the age of 35 years, consists of gastric epithelioid leiomyosarcoma, pulmonary chondroma, and functioning extraadrenal paraganglioma.[19] The pulmonary tumors are usually asymptomatic.

Radiology

The tumors form peripheral opacities that are either rounded or lobulated and are usually homogeneous. Calcification is a variable feature and the so-called typical peppercorn calcification is no longer thought to be diagnostic of mesenchymoma. Previous chest radiographs had shown no visible tumor in 30 of 57 patients in periods ranging between 1 and 18 years before the mesenchymoma was detected.[7]

Genetics

Cytogenetic analysis of cultures from seven pulmonary mesenchymomas revealed an abnormal karyotype in six. The most characteristic aberration was an exchange of material between 6p21 and 14q24 in three tumors. Abnormalities of either 6p or 14q were seen in two other mesenchymomas.[13,14]

This abnormal karyotype in pulmonary mesenchymomas strongly suggests that they are genuine neoplasms. These clonal chromosomal abnormalities are not randomly distributed throughout the genome. There is a coexistence of a common 14q24 break point in uterine leiomyomas, indicating that a gene important to the genesis of both tumors exists in this band.

Macroscopic Features

Central endobronchial tumors are sessile or pedunculated and oval or fusiform and may be adherent to the bronchial wall. They range in size from less than 1 cm to nearly 4 cm in diameter. The boundary between endobronchial tumors (Fig. 29-1), and the bronchial wall is often poorly defined. Cut surface usually has a gray appearance, but it can be yellow if there is a large amount of fat. There is distal bronchiectasis and either lung abscess or bronchopneumonia.

Peripheral lesions (Fig. 29-2) also vary in color with the amount of fat. They are circular and firm and sometimes show cystic foci. They are usually 1.5 cm in diameter but may measure up to 8 cm (Fig. 29-3). They are frequently subpleural and may be shelled out from the surrounding parenchyma.

Histopathology

There is a marked similarity between the frequency of various tissues in the two types of tumor, endobronchial and intrapulmonary.[4] The only difference appears to be that there is more chondroid differentiation but less fat in the parenchymal tumors. Histologically there are foci of chondroid cartilage (Fig. 29-4), as well as osseous metaplasia (Fig. 29-5)

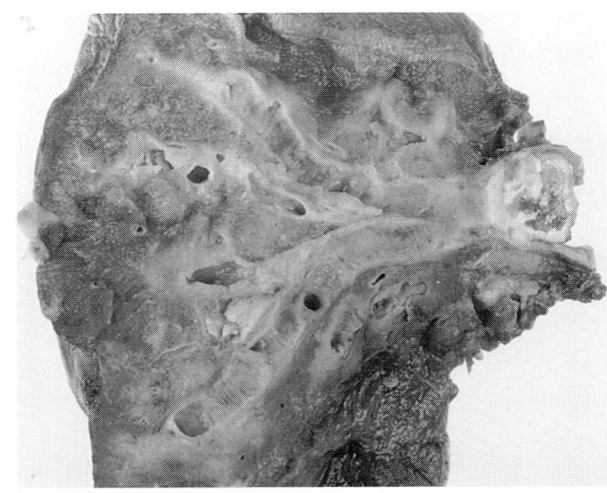

FIGURE 29-1
Calcified endobronchial mesenchymoma with distal bronchiectasis.

and even marrow precursors. Loose myxoid tissue merges with the cartilaginous lobules. The tumor may form dense papillary or lobular tufts. There are foci of fibrosis, fat (Fig. 29-6), smooth muscle, lymphoid, and plasma cells as well as epithelial-lined clefts. In endobronchial tumors the epithelium is

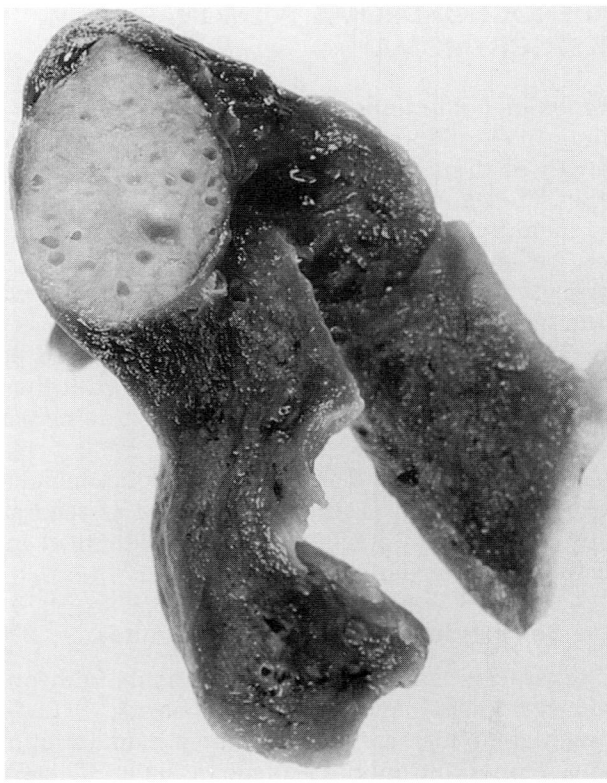

FIGURE 29-2
Peripheral mesenchymoma with some cystic change.

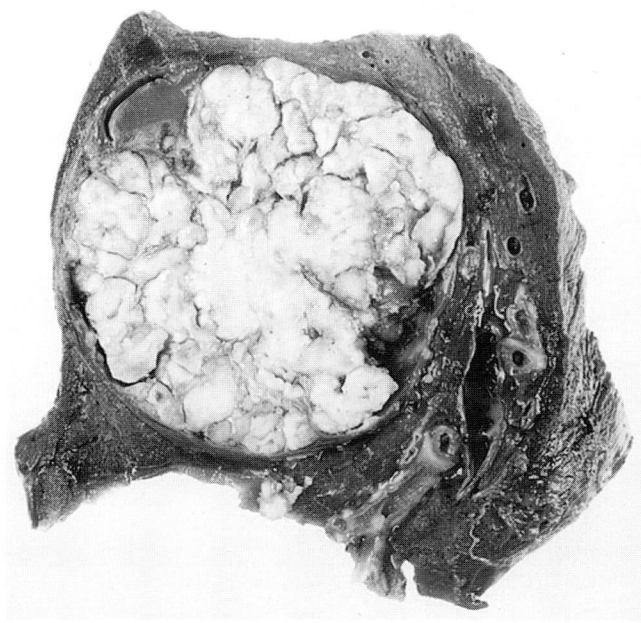

FIGURE 29-3
Large peripheral mesenchymoma with much calcification.

usually ciliated and columnar, whereas in parenchymal tumors it probably arises from type II pneumocytes, is usually nonciliated, and may even arise from mucus-producing bronchiolar cells.

In endobronchial tumors, seromucinous glands, as seen in the bronchial wall, are identified. The overlying epithelium may show squamous metaplasia or basal cell hyperplasia.

In rare cases, tumors may be composed entirely of adipose tissue, with some covering epithelium or epithelial clefts. These have been called *endo-*

bronchial lipomatous hamartomas or *endobronchial lipomas* (Fig. 29-7).[15] In such cases, cellular atypia should not lead to a misdiagnosis of liposarcoma. There is an overlap between this lesion and benign bronchial lipomas. Bateson[4] noted that endobronchial lesions do not usually contain epithelial cell–lined clefts or spaces of the type seen in intrapulmonary tumors. Some peripheral tumors may be associated with noncaseating granulomas, with no evidence of microorganisms or sarcoidosis either at thoracotomy or at follow-up.

Tumor doubling time was calculated using Schwartz's biomathematical method.[16] This was calculated in 24 cases[8] as 36.4 ± 27.7 months. The tumor was followed for more than a year in every case.

The tumors may recur 10 to 12 years after excision in a small number of patients.[7]

Ultrastructure

Several authors have described the ultrastructure of the tumor.[17,18] Whether the tumor is endobronchial or parenchymal, the epithelium appears the same in that there are cuboidal or columnar cells with rounded or flattened apices. These have variable numbers of small microvilli along the apical and lateral membranes. The cells are joined together along the lateral membranes by small desmosomes. In 10 percent of cases, the epithelial cells show unusual nuclear structures. These occupy up to half of the nuclear mass and are composed of round to oval

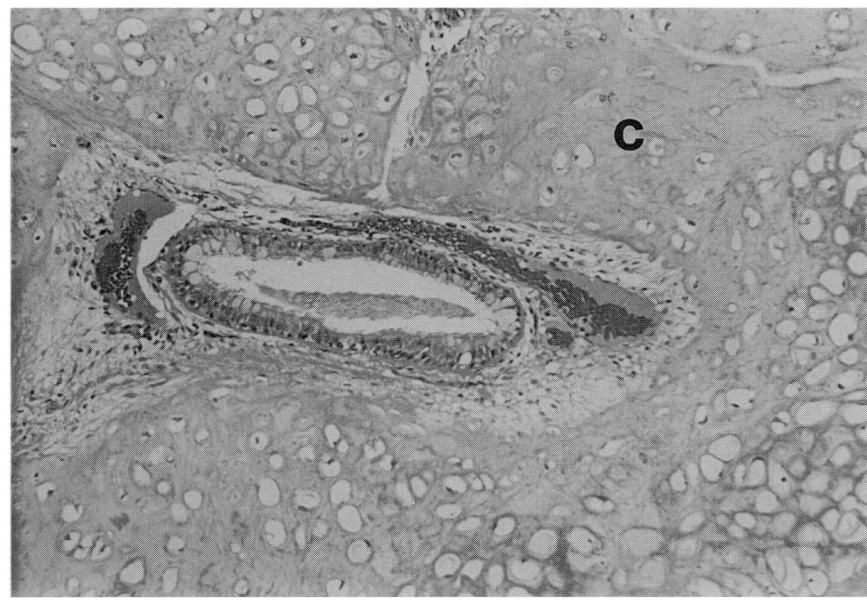

FIGURE 29-4
Mesenchymoma showing cartilage (C) and a central island of epithelium. (H&E, ×87.5.)

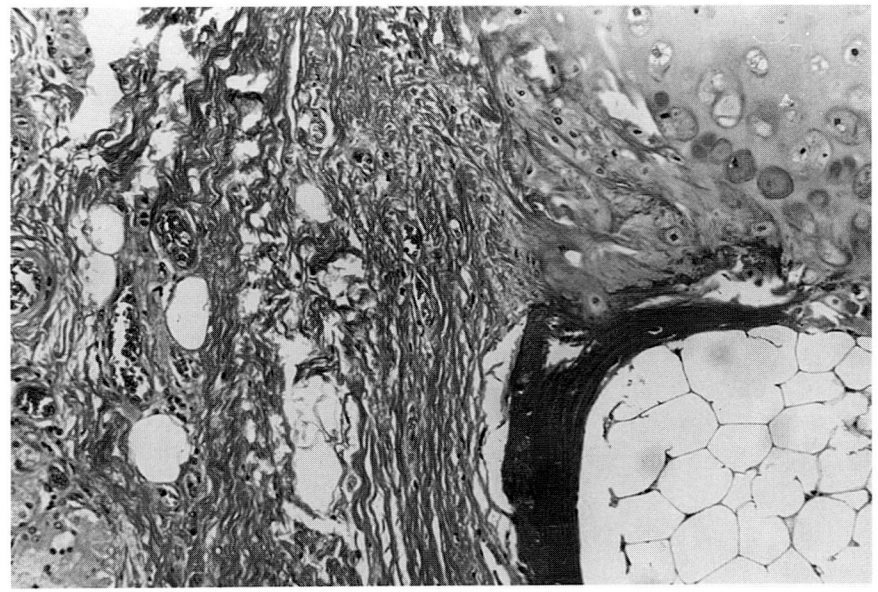

FIGURE 29-5
Mesenchymoma with cartilage, bone, and fat. (H&E, ×87.5.)

structures of varying size and shape. On section the body appears hollow or tubular and may contain finely dispersed material in the lumina. The individual hollow structures are limited by a simple membrane. Fibroblasts, chondrocytes, and fat cells show normal features.

Malignant Change in Cystic Hamartomas

One of the most convincing reports of this is rare complication is that of Basile and coworkers,[20] who described sarcomatous transformation in a long-standing, clinically silent hamartoma. Malignant change may have occurred shortly after resection, and the malignant tumor consisted of spindle cells with medium-sized or large round cells arranged in poorly oriented bundles. Cytokeratin, myoglobin, factor VIII, S-100, and carcinoembryonic antigen were all negative. The lesion was thought to be sarcoma of either cartilaginous or connective tissue origin. Adenocarcinomatous change has also been described.[21,22]

Differential Diagnosis

There is clearly overlap between mesenchymomas and benign chondromas, fibromas, and lipomas. The only other lesion that may cause confusion is metastasis from a germ cell tumor of either the ovary or testis after chemotherapy. Some of these secondary

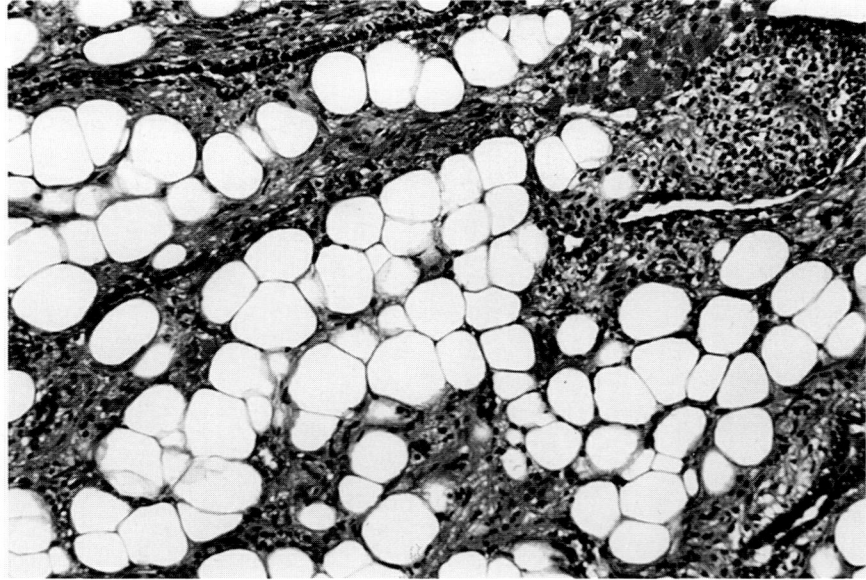

FIGURE 29-6
Mesenchymoma with fat, chronic inflammation, and some small compressed islands of epithelium. (H&E, ×87.5.)

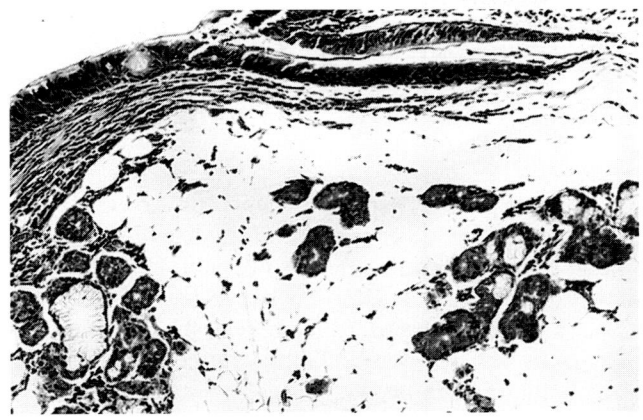

FIGURE 29-7
Subepithelial lipoma with admixed serous and mucous glands. (H&E, ×103.6.)

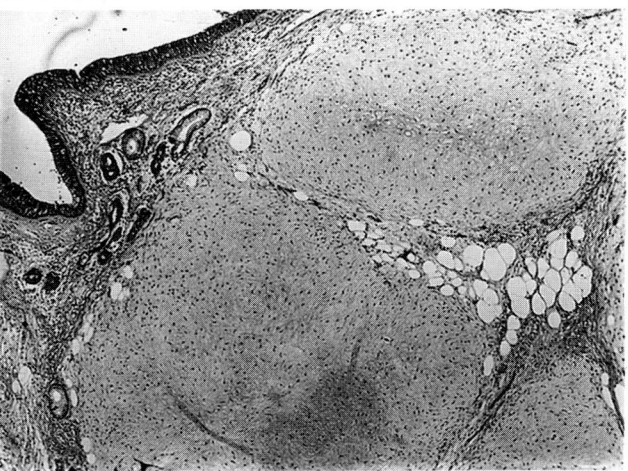

FIGURE 29-8
Chondroma of bronchus with large islands of cartilage. There is also some adipose tissue but no epithelial component is identified. (H&E, ×29.6.)

tumors consist entirely of neoplastic cartilage with no fibromyxoid material and entrapped alveolar epithelium, giving them a superficial resemblance to hamartomas. In such cases there is usually a history of previous orchidectomy or oophorectomy, and the cartilage is not mature as in hamartomas. Any lesion with smooth muscle proliferation, as in left ventricular failure—or, less likely, lymphangioleiomyomatosis—may be diagnosed as "hamartoma." A rare case of metastasis from a prostatic adenocarcinoma to a pulmonary mesenchyma has been described.[22a]

PULMONARY CHONDROMA

Pulmonary chondroma is a rare tumor. Most cases are "hamartomas" in which there are islands of cartilage, epithelium, fat, and sometimes bone marrow. Chondroma of the bronchus grows from fully formed bronchial cartilage, and no other tissue elements are seen in the tumor. Examples of true bronchial chondromas have been described by Davidson and other authors.[23–26] Carter and Eggleston[27] gave the incidence of this tumor as three in 40,000 autopsies.

Macroscopically, the tumors are usually about 1 to 2 cm and lobulated, with a smooth outer surface. They grow as sessile or polypoid lesions into the bronchial lumen and are often incidental findings. On cut surface they are firm and white and may have areas of calcification.

Histologically, the tumor consists of hyaline cartilage (Fig. 29-8) with a few elastic fibers. As in all chondromas, the chondrocytes are large, and they are arranged in a more disorderly fashion than in normal cartilage. Some chondromas may show myxoid changes. Others undergo ossification. With progressive enlargement, they obstruct the bronchus. As the

lesion is part of the cartilaginous plate, complete removal requires either resection of the cartilaginous ring or, if it is in the bronchial tree, a pneumonectomy. However, Davidson[23] successfully removed an obstructing endobronchial lesion by bronchoscopy.

Even rarer than chondromas are chondroblastomas and chondrosarcomas.[27,28]

LIPOMAS OF THE BRONCHUS AND LUNG

Adipose tissue is a normal constituent of the bronchial wall and may also be found beneath the pleura. Fat is identified in bronchi in airways as small as 1 mm in diameter.[29] It is located beneath the surface epithelial layer, lying between the cartilage and muscle and surrounding the mucous glands. Thus lipomas may arise from any adipose tissue within the lung and may be divided into bronchial and subpleural lipomas. Endobronchial lipomas account for 0.1 percent of all lung tumors.[30] The cases documented up to 1988 have been described.[31]

Clinically, most lipomas are seen in patients in their fifth and sixth decades. They usually arise in central bronchi, leading to obstruction and causing wheezing, pneumonitis, and bronchiectasis.

Macroscopically, lipomas are usually found in large bronchi. They are more frequent in the left main and lobar bronchi than in the corresponding bronchi on the right side.[32] The tumors often have a dumbbell shape and fill the bronchial lumen and subepithelial coat with a narrow neck lying in between. They usually project as smooth-walled polyps into the bronchus. As they grow, the normal consituents of the bronchial wall become attenuated. Muscle atrophies with time and is replaced by connective tissue. Most cases consist of mature adipose tissue, although some authors have described other cell types, including those of bone, cartilage and fibrosis.[33] This raises the possibility that these tumors are in fact part of a mesenchymoma.

Occasionally, pulmonary lipomas may have small giant cells with bizarre nuclei scattered amid the mature fat.[34] Despite this pleomorphic picture, follow-up for 18 months showed no recurrence.

Bronchial lipomas must be differentiated from hamartomas and also from intrathoracic and mediastinal lipomas, which can reach an enormous size and spread directly into parabronchial tissues, filling the lung hilum.

In a rare form of bronchial lipoma, the circumference of a medium-sized bronchus is replaced by the tumor.

The lipoma fills the subepithelial layer and extends over a considerable length of the bronchial wall. This may not be a true neoplasm and is regarded by some as a sequel to chronic bronchiectasis.

Subpleural Lipoma

Peripherally situated bronchial lipomas are rare.[35] Spencer described a case that spread over the surface of an upper lobe (Fig. 29-9).[36]

MESENCHYMAL CYSTIC HAMARTOMA OF THE LUNG

This entity was first described by Mark.[37] The patients' ages at thoracotomy ranged from 1.5 to 53 years. It has a characteristic clinical presentation, with lung cysts causing hemoptysis, pneumothoraces,[37a] hemothoraces, pleuritic chest pain, and dyspnea of mild to moderate degree. The disease may be multifocal and bilateral.

Histologically, the nodules consist of cysts that range from 5 to 100 mm in diameter and are lined by normal respiratory epithelium or a cuboidal epithelium (Fig. 29-10) showing squamous metaplasia. Small cysts have a connection with bronchioles. Beneath the epithelium is a bandlike layer of primitive mesenchymal cells, separated from the epithelium by a layer of collagen. These mesenchymal cells have dark oval nuclei, rare mitoses, and inconspicuous cytoplasm. Occasionally the mesenchyme forms a septum within the cyst. Hypertrophied systemic arteries are present in this mesenchymal layer, and

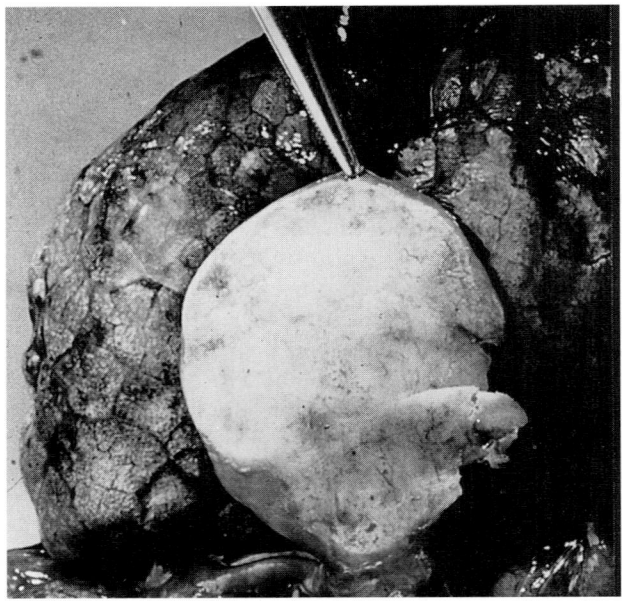

FIGURE 29-9
Subpleural lipoma.

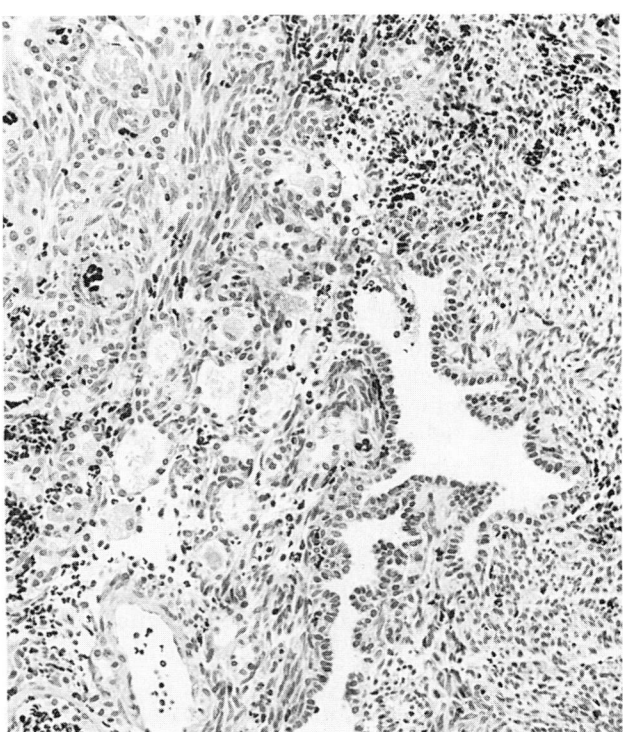

FIGURE 29-10
Mesenchymal cystic hamartoma with walls lined by cuboidal epithelium with underlying mesenchymal tissue. (H&E, ×180.) (*Reproduced with permission of Dr. W. D. Travis, Armed Forces Institute of Pathology, Washington, D.C.*)

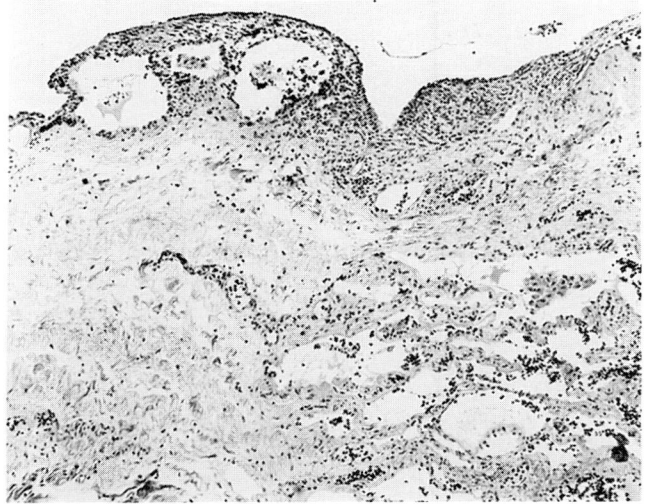

FIGURE 29-11
Mesenchymal cystic hamartoma with a focus of fibrosis. (H&E, ×77.) (*Reproduced with permission of Dr. W. D. Travis, Armed Forces Institute of Pathology, Washington, D.C.*)

there is evidence of hemorrhage inside the cysts and hemosiderin in adjacent alveoli.

The disease spreads and gives rise to multiple cysts and scarring (Fig. 29-11) in both lungs. One 1½-year-old child was disease-free 4 years after removal of the lesion. A further case has been associated with a vaginal myxoma.[38]

The differential diagnosis for this disorder is wide and includes sequestration of the lung, cystic adenomatoid malformation, cystic bronchiectasis, eosinophilic granuloma, thoracic endometriosis, cystic rhabdomyosarcoma, and metastatic uterine stromal sarcoma. The disease that theoretically could cause confusion is endometriosis, but this is rarely seen in the chest.

PAPILLOMA OF THE BRONCHUS AND BRONCHIOLES

Papillomas of the endobronchial tree have been classified into three main groups—solitary papillomas and multiple squamous papillomas, which are seen in juvenile laryngotracheobronchial papillomatosis, and inflammatory polyps associated with endobronchial foreign bodies, chronic bronchitis, or broncholithiasis. Solitary papillary squamous papillomas, all with in situ carcinoma, show no evidence of human papillomavirus[39]—unlike juvenile laryngotracheal papillomatosis (recurrent respiratory papillomatosis, described below).

Solitary bronchial papillomas are rare. (Fig. 29-12).[40–42] Fantone and colleagues[45] collected 59 endobronchial papillomas from the literature and added one columnar cell papilloma of their own. Solitary endobronchial papillomas tend to occur in middle age and are most common in men but have been described in children. The patients are often smokers.

Clinically they present with chronic airways obstruction, including productive cough, hemoptysis, wheezing, and dyspnea. Squamous papillomas are associated with dysplasia in almost a third of cases, as well as with invasive carcinoma. It is therefore wise to ensure that there is no recurrence by regular follow-up bronchoscopy.

Macroscopically, the tumors grow as wartlike lesions into the bronchial lumen. There are instances of tumors arising from terminal airways and presenting as cystic masses.

Histologically, solitary papillomas usually consist of squamous epithelium (Fig. 29-13), with approximately one-third showing carcinoma in situ (Fig. 29-14) or even invasive carcinoma. Ten of 33 cases in the study by Basheda and coworkers[42] showed dysplasia, but no mitoses or nuclear irregularity was identified in the columnar cell tumors.

Papillomas have a connective tissue stroma, often heavily infiltrated with lymphocytes and covered entirely with cuboidal (Fig. 29-15) or squamous epithelium. There is often a mixture of ciliated or nonciliated cuboidal and multilayered nonsquamous epithelium or even well-differentiated squamous epithelium. A papilloma of mixed epithelial type may sometimes be associated with an underlying monomorphic bronchial mucous gland cystadenoma.

Tumors arising in distal bronchi and bronchioles may spread to line adjacent alveolar spaces and either are filled with tumor cells or extend to form epithelial-lined spaces.

There are cases with an inverted type of bronchial epithelium resembling those seen in the nose (Fig. 29-16). They have been termed *papillary adenomas* or *transitional cell papillomas*.[43,44]

PAPILLARY ADENOMA OF THE LUNG

This is a rare lesion.[45-48] It is seen in men and women with an age range of 23 to 60 years. Three out of seven patients smoke. Radiology shows a well-defined density in the appropriate lung field. The tumor is seen either in a small bronchus[46] or in the lung

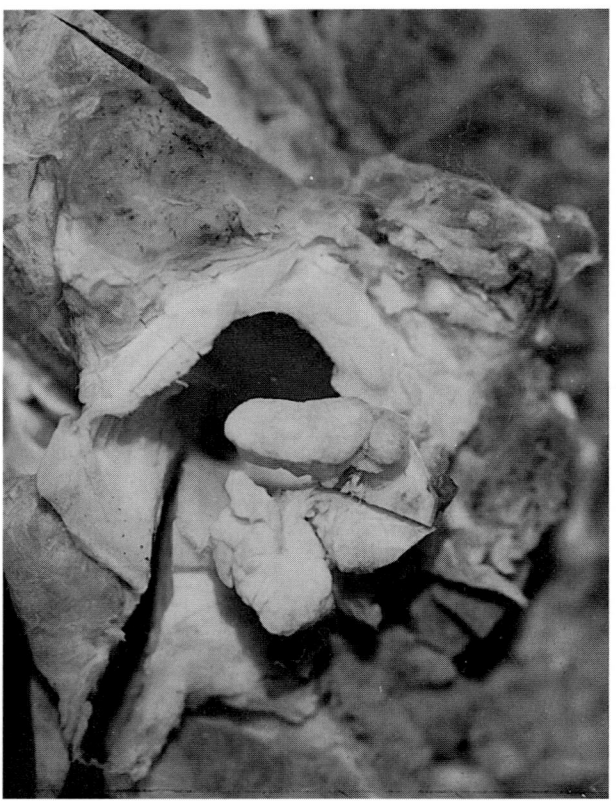

FIGURE 29-12
Squamous cell papilloma of bronchus projecting into the lumen of the bronchus near the cut margin.

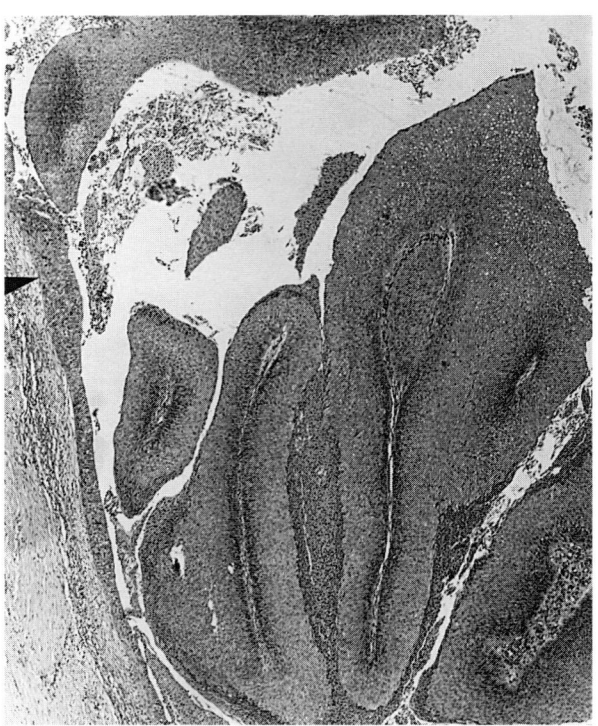

FIGURE 29-13
Squamous cell papilloma with dysplasia (arrow at one margin).

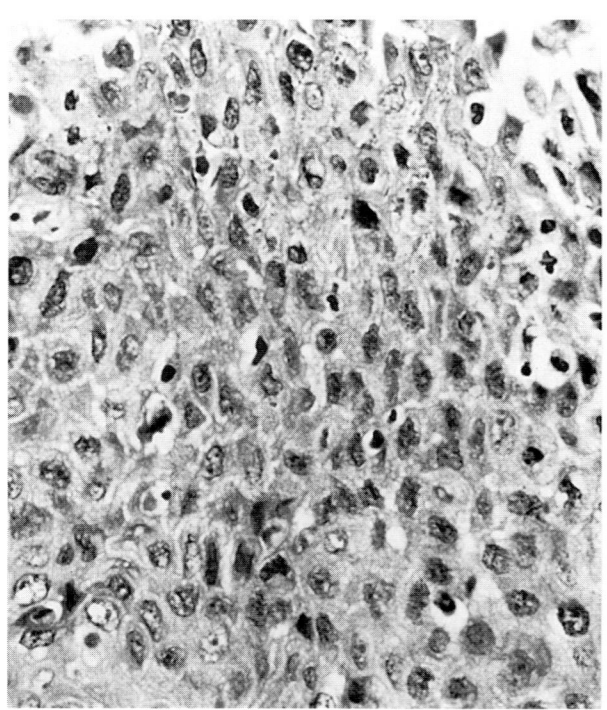

FIGURE 29-14
Dysplasia in squamous papilloma of bronchus. (H&E, ×266.)

parenchyma.[45,48] It is intrabronchial or parenchymal, well circumcribed, spherical, and soft. When parenchymal, the tumor is subpleural, brown, and well circumscribed and measures up to 2.5 cm in diameter.

Histologically, there is a papillary growth pattern (Figs. 29-17 and 29-18) composed of cuboidal and high columnar cells accompanied by edematous connective tissue. Some of the epithelial cells have microvilli on their free surfaces. The stroma is edematous and in some areas is infiltrated by lymphocytes, plasma cells (Fig. 29-19), and foamy histiocytes. Characteristically, there are no elastic fibers. Tumor compresses the surrounding normal parenchyma, but there is no distinct capsule. There is no nuclear pleomorphism, mitoses, or calcification. Immunocytochemistry for Clara cell antigen is positive.

Ultrastructurally, there are lamellar body–like intranuclear inclusions and intracytoplasmic granules containing myelin material. Microvilli are present on the surface, but cilia are absent. Surfactant apoprotein was not detected by immunostaining in the study by Hegg and colleagues.[47] In their study, the tumor was thought to have an origin in both Clara and alveolar type II cells. A second population of cells, containing prominent cytoplasmic mem-

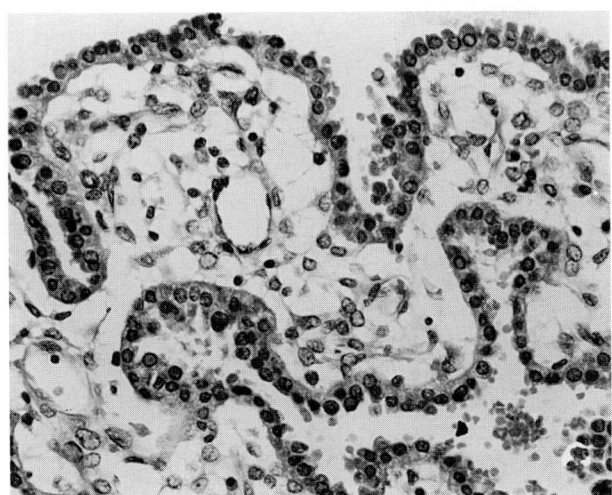

FIGURE 29-15
Peripheral bronchial adenoma with peg-shaped cells resembling
Clara cells. (H&E, ×87.5.)

brane–bound inclusions, was seen. In some cells
these inclusions contained concentric and electron-
dense membranous lamellae. Clara cell differentiation is not detected in all cases. Nuclear inclusions,
consisting of tubular structures 50 to 60 nm in diameter, were present in some tumor cells.

The differential diagnosis includes adenocarcinoma, which usually proliferates along the alveolar
walls and causes abnormal condensation of elastic
fibers or preserves the intract stromal network of the
lung. This tumor has no elastic fibers in its stroma. A

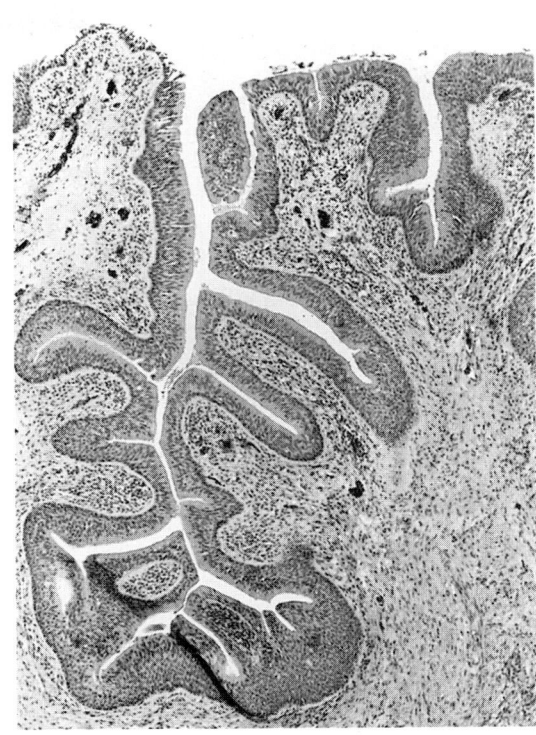

FIGURE 29-16
An inverted type of bronchial papilloma.

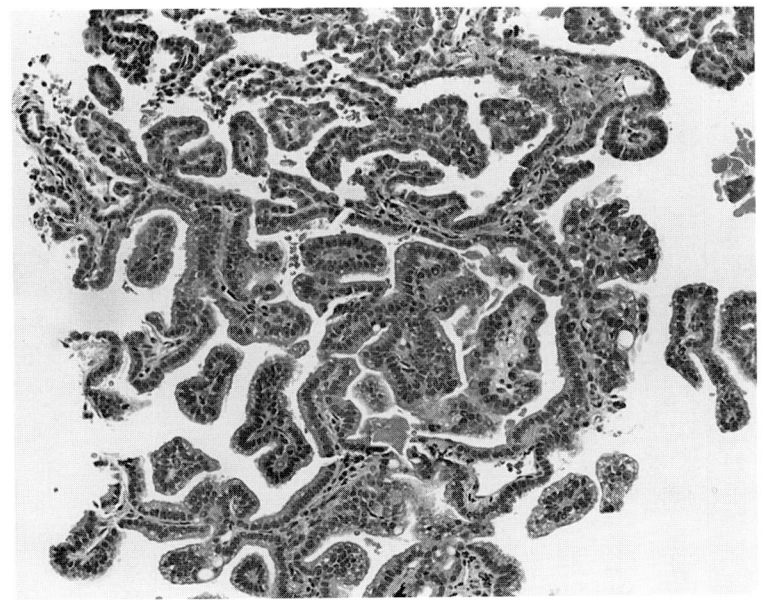

FIGURE 29-17
Papillary adenoma of lung with anastomosing bands
of epithelium and a little underlying stroma. (H&E,
×87.5.)

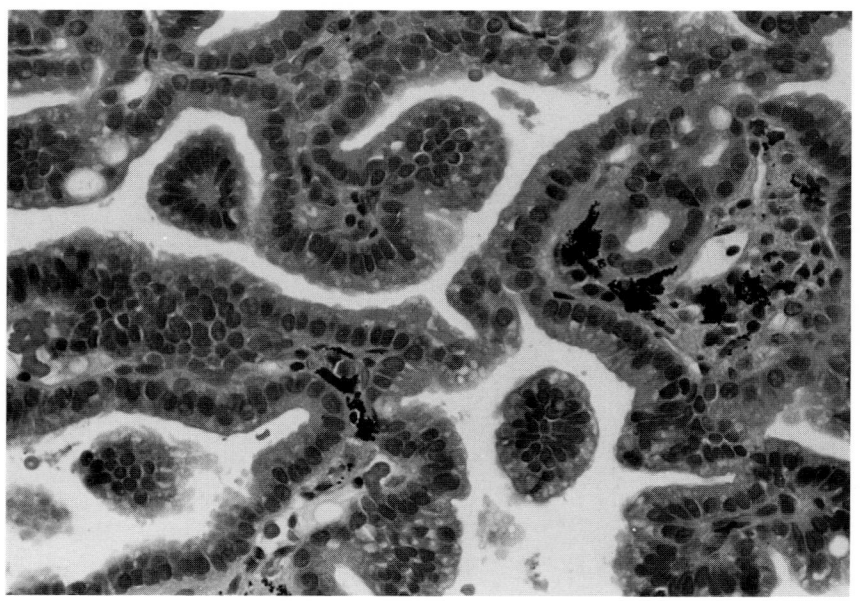

FIGURE 29-18
Same case as Fig. 29-17 showing columnar cells with basal nuclei. (H&E, ×219.)

papillary carcinoid may also cause diagnostic confusion, but the appropriate neuroendocrine stain should help in diagnosis.

Miller[49] described a bronchioloalveolar cell adenoma showing type II pneumocyte features. This is a small tumor, measuring up to 7 mm in diameter; it has an ill-defined margin and is multicentric (see Chap. 31). If the lung is fixed in Bouin's, the lesions can be seen with the naked eye. Most pathologists do not use this fixative, and so these tumors are an incidental histologic finding. It is now thought likely that such lesions are part of a field change seen in adenocarcinoma of lung.[50]

Papillary adenoma arises in close association with a bronchiole or small bronchus and is usually a single lesion. Without immunocytochemistry, it may be difficult to differentiate the lesion from the solitary papillomas mentioned above.

ALVEOLAR ADENOMA

This is a benign multicystic tumor that contains both alveolar epithelium and mesenchyme.[51] Few cases have been reported, and some of them have been termed *lymphangiomas*.[52,53]

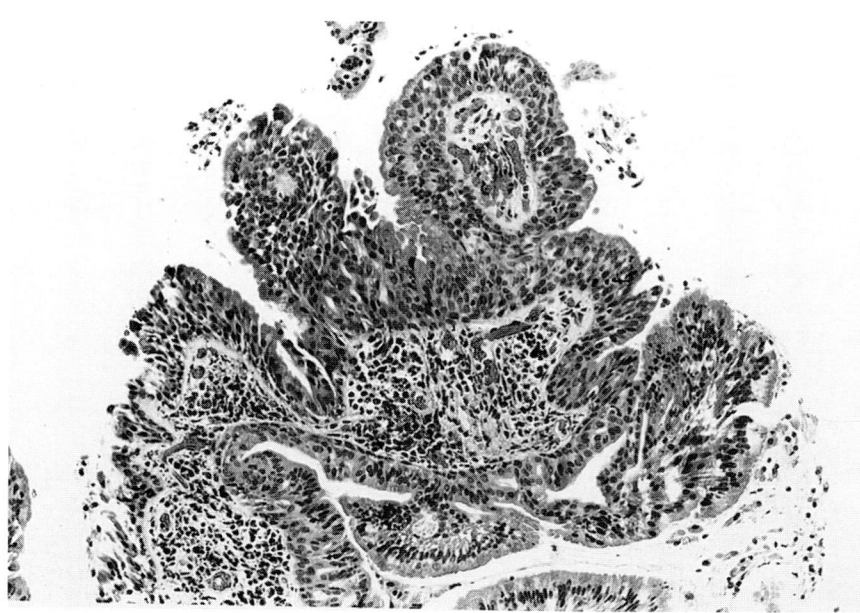

FIGURE 29-19
Papillary adenoma showing stroma with chronic inflammation. (H&E, ×87.5.)

Clinically, the lesion is usually asymptomatic and presents as a solitary nodule.

Macroscopically, the tumor is solitary, well circumscribed, and tan or grayish white. It is 1 to 2 cm in diameter.

Histologically, there are multiple cysts; these are of varying size and contain eosinophilic granular material, which is PAS-positive. The cysts are lined by an epithelium that varies from flattened to cuboidal (Fig. 29-20). When the cells are flat the lesion resembles a lymphangioma.

The cells stain with cytokeratin and focally for carcinoembryonic antigen.[54]

Ultrastructurally, the epithelial cells have the features of type II pneumocytes.[54]

The main differential diagnosis is lymphangioma, which is cytokeratin-negative, and bronchioloalveolar cell adenoma. The latter lesion is ill defined, may show nuclear atypia, and is often seen as part of an adenocarcinoma. Sclerosing hemangioma may form cystic spaces, but they contain blood, not eosinophilic material. This tumor also shows many different patterns in the same lesion.

JUVENILE LARYNGOTRACHEAL PAPILLOMATOSIS (RECURRENT RESPIRATORY PAPILLOMATOSIS)

This disease is caused by human papillomavirus types 6 and 11.[55] Human papilloma virus 11 has been detected, using the Southern blot and polymerase chain reaction, in both squamous cell papilloma and carcinoma in this condition.[56] The virus is acquired in the birth canal from a mother with genital warts. There is circumstantial evidence that adult disease may be acquired by oral-genital contact. The transmission rate is low, and remission, which either is spontaneous or occurs as a result of treatment, is unpredictable.

Papillomas involve the tracheobronchial tree (Fig. 29-21) and extend into the lung parenchyma to cause multiple cysts. Malignant change into invasive squamous cell carcinoma has been documented.[56]

The lung is affected in fewer than 1 percent of cases. Malignant transformation is rare in nonsmoking, nonirradiated patients and occurs about 15 years after the onset of papillomatosis. Diagnosis may be delayed because of long-standing respiratory dysfunction.

Clinically, recurrent respiratory papillomatosis usually presents with hoarseness of voice, and children have abnormal crying related to the laryngeal manifestations.[57] Stridor and upper airway obstruction are late manifestations.

The papillomas protrude into the bronchial lumen in a papillary pattern. They consist of hyperplastic squamous epithelium, with binucleate and koilocytotic cells suggestive of human papillomavirus infection (Fig. 29-22). In cases with carcinoma there is a continuum, whereby foci of moderate dysplasia merge with invasive squamous carcinoma. In some cases, it may be difficult to distinguish between the malignancy and florid papillomatosis. The well-differentiated squamous carcinomas spread

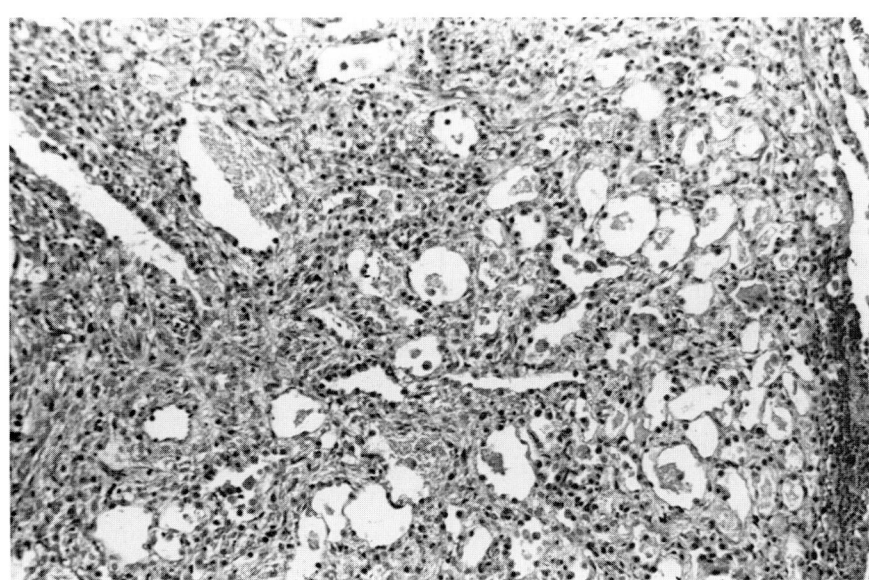

FIGURE 29-20
Alveolar adenoma with cystic spaces of varying size and lined by a low attenuated epithelium. (H&E, ×87.5.)

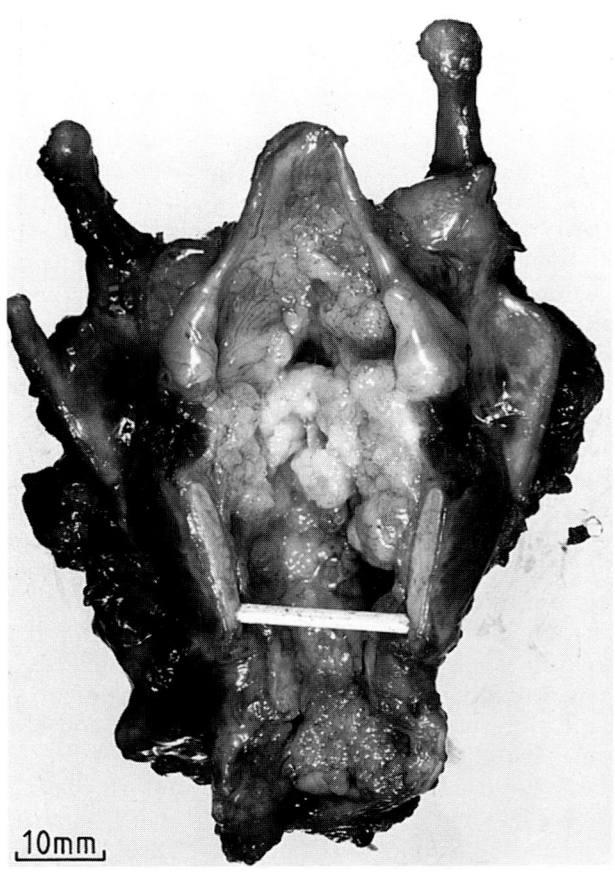

FIGURE 29-21
Juvenile laryngotracheal papillomatosis. This is a laryngectomy specimen from a woman aged 21. There is involvement of the epiglottis, glottis, subglottis, and trachea. (*Reproduced with permission of Professor L. Michaels from Pathology of the Larynx, Springer-Verlag, Berlin.*)

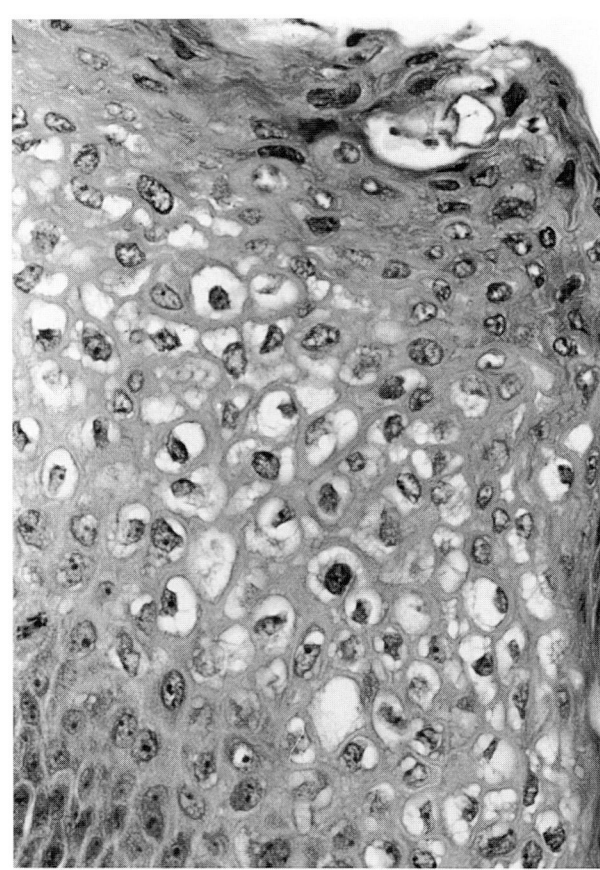

FIGURE 29-22
Juvenile laryngotracheal papillomatosis-koilocytotic cells. (H&E, ×219.)

into the peribronchial and mediastinal lymph nodes, chest wall, vertebrae, diaphragm, esophagus, and heart. In addition, there are distant metastases to the liver, bone, and adrenals. The carcinomatous change is often multicentric.

MUCOUS GLAND ADENOMA

Mucous gland adenomas are rare tumors that have been well documented by Heard and coworkers[58] and by Spencer.[59] These authors described a total of 12 cases. A recent review of the world literature and ten new cases have been described.[59a] The ages of the patients ranged from 7 to 67 years. Eight patients were male and four female. Few patients were smokers.

Clinically, the patients complained of chest pain, cough, and hemoptysis for up to 1 year. Some patients had increased sputum production and dyspnea. The adenomas may present as bronchogenic cysts.[60]

Tumors are always single, arise in lobar or segmental bronchi, and vary in size from 0.8–6.8 cm in maximum diameter. These tumors are rarely seen in the trachea, where they may be misdiagnosed as bronchial asthma.[61] On bronchoscopy they are polypoidal, glistening, pinkish-gray masses.

Pathology

Macroscopically, these tumors are round, grow in the bronchial lumen in a polypoid fashion, and compress adjacent lung tissue. In some cases they are internal to the bronchial cartilage. Cut surface is pale gray and slightly translucent.

Histologically, the tumors closely resemble normal bronchial mucous glands (Fig. 29-23), with a similar lobular arrangement of ducts and acini (more of the latter). Tubular and papillar foci may be seen.[59a] Acini are lined by mucus-secreting cells, and there is accumulation of mucus in the lumina, which sometimes forms microcysts. Occasionally, acicular

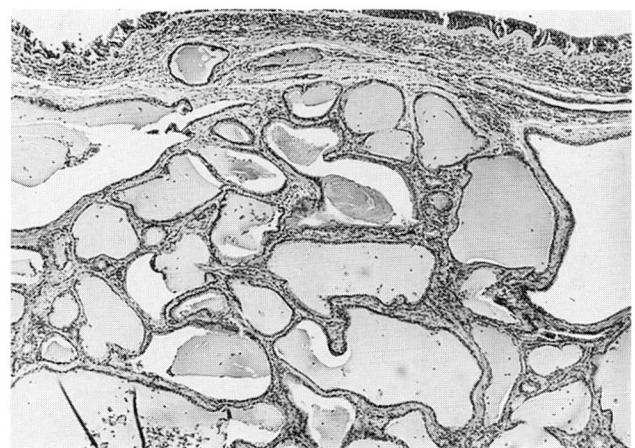

FIGURE 29-23
Mucous gland adenoma showing a similarity in structure to normal bronchial glands. (H&E, ×30.)

crystal clefts and fine calcium deposits are seen amid the mucin. All tumors stain positively with PAS with or without diastase and mucicarmine, but alcian blue is only focally positive. In two cases that have been reported, there was evidence that the acinar cells had abundant granular eosinophilic cytoplasm, suggesting oncocytic differentiation. Squamous metaplasia was seen in ducts in several cases, but most of the duct epithelium was ciliated and communicated with respiratory ciliated epithelium over the surface of the tumor. One case had several collections of medium-sized spindle-shaped cells that were larger and multinucleated. There was a variable amount of stromal connective tissue, and a little stromal amyloid was identified by Congo red staining. In all cases there was a mild chronic inflammatory infiltrate in the stroma.

Electron microscopy showed the main cell type to be the mucous cell, with membrane-bound mucous granules measuring 1 μm in diameter. The granules showed finely granular reticular or fibrillated material, and in some cases there were small electron-dense granules. In addition, many bundles of tonofilaments were identified in the cytoplasm. Neurosecretory granules were seen only in one case; this was situated below acinar lining cells, lying against the basement membrane.

Differential Diagnosis

MUCINOUS CYSTADENOMA OF THE LUNG

This tumor is discussed below but should not cause confusion, as it does not arise from the bronchial mucous glands.

MUCOEPIDERMOID TUMORS

In mucous gland adenomas, the squamous differentiation is limited. In mucoepidermoid carcinomas there is more squamous and intermediate cell differentiation. Mucoepi-dermoid tumors show a range of cytologic appearances that vary from relatively benign to more frankly malignant areas.

ONCOCYTIC CARCINOID

This lesion may cause little problem in differential diagnosis if neuroendocrine stains are positive. In cases that are negative, however, the diagnosis rests on electron microscopy and the demonstration of neurosecretory granules. One case diagnosed as a mucous gland adenoma contained these structures, but they were confined to the area adjacent to the basement membrane. There may also be confusion with an oncocytoma, and this very rare tumor could be a variant of mucous gland adenoma. Many pathologists believe it is an oncocytic carcinoid.

ONCOCYTOMA

This brochial tumor is probably a variant of mucous gland adenoma. In cases examined ultrastructurally there were no neurosecretory granules, and it has been regarded as distinct from an oncocytic carcinoid.[62–66]

There is a male predominance, with a mean age of 52 years (age range: 22 to 74 years). The tumor affects the right lung more commonly, and the patients present with cough or hemoptysis. Alternatively, a pulmonary shadow may appear in a patient hospitalized for another disease.

The tumors are bronchial-based and measure 1.0 to 3 cm in diameter. Their color varies from yellowish tan to reddish brown. One case reported was malignant and had multiple pulmonary nodules.[67] Focal hemorrhage and calcification have been described.

Histologically there is an alveolar arrangement of tumor cells with abundant finely granular eosinophilic cytoplasm and uniform nuclei (Fig. 29-24). Thin fibrous septa separate the cells, and in areas there is an organoid pattern. One case showed a microcystic and papillary pattern.[66] Focal positivity for Grimelius stain was seen in one case, casting doubt on the diagnosis. However, dense core neurosecretory granules were not detected ultrastructurally. There were numerous mitochondria and a few non–membrane-bound granules, smaller than neurosecretory granules. Some cases have serous

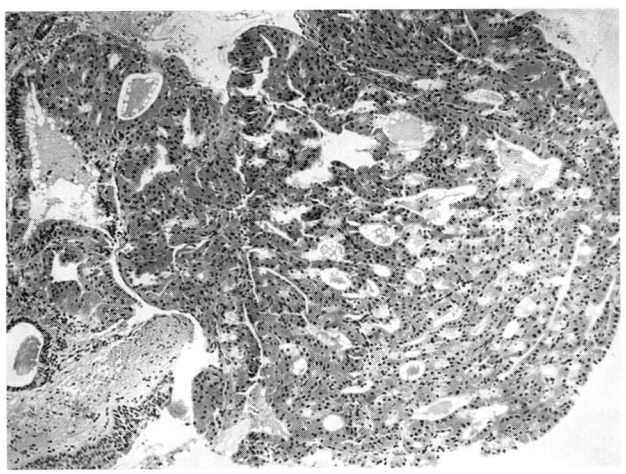

FIGURE 29-24
Oncocytic adenoma with strongly eosinophilic cells. (H&E, × approximately 126.)

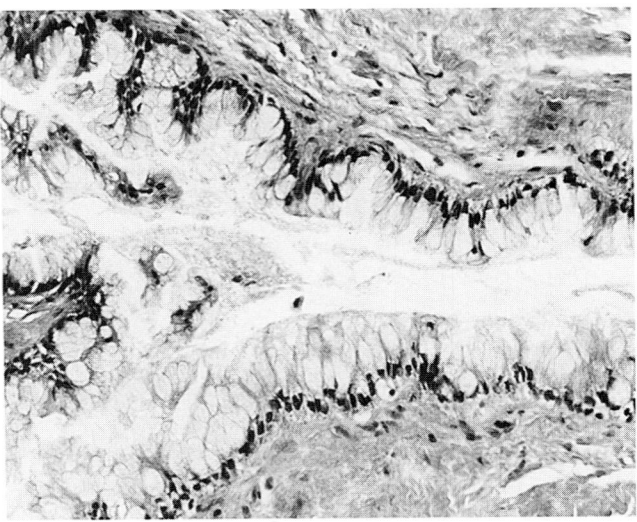

FIGURE 29-25
Mucinous cystic tumor of the lung with tall columnar mucinous epithelium. (H&E, ×154.) (*Reproduced with permission of Dr. W. D. Travis, Armed Forces Institute of Pathology, Washington, D.C.*)

granules. Rare goblet cells were identified by electron microscopy.

The differential diagnosis is bronchial carcinoid, which should have positive neuroendocrine immunocytochemistry and dense core granules, acinic cell tumor with dense round membrane-bound granules ranging from 225 to 950 nm, focal oncocytic change in a mucous gland adenoma, and granular cell myoblastoma that ultrastructurally shows osmiophilic granules, some lamellated and others membrane-limited.

MUCINOUS CYSTIC TUMORS OF THE LUNG

These have recently been well reviewed[68,69] in a total of 13 patients, eight women and five men. Ten patients smoked.

Clinically, some of the patients were asymptomatic, presenting with solitary pulmonary nodules. Others complained of cough, chest pain, weight loss, dyspnea, and, in one case, pneumothorax due to bullous disease.

Macroscopically, the tumors ranged in size from 1 to 15 cm, most being less than 10 cm in diameter. A fibrous-walled cyst was filled with clear gelatinous mucin. The cyst lining was often shaggy. Mucus dissection into the lung parenchyma was seen in half of the cases.

Histology

The tumor was lined by tall columnar mucinous epithelial cells with basal nuclei (Fig. 29-25). Some areas showed no cells and consisted of mucus within fibrous tissue (Fig. 29-26), while in others, cells were

heaped into tufts or papillae showing cytologic atypia and stratification. The well-differentiated areas showed goblet cells and surface cilia. Areas with nuclear pleomorphism, many mitoses, and frankly malignant epithelium were not identified. The surrounding lung showed variable inflammation with focal giant cells (Fig. 29-27). All mucin stains were strongly positive.

Carcinoembryonic antigen was focal in two cases, and one case was positive for chromogranin.

All cases were in living patients who were without evidence of disease after a mean follow-up of 4.7 years.

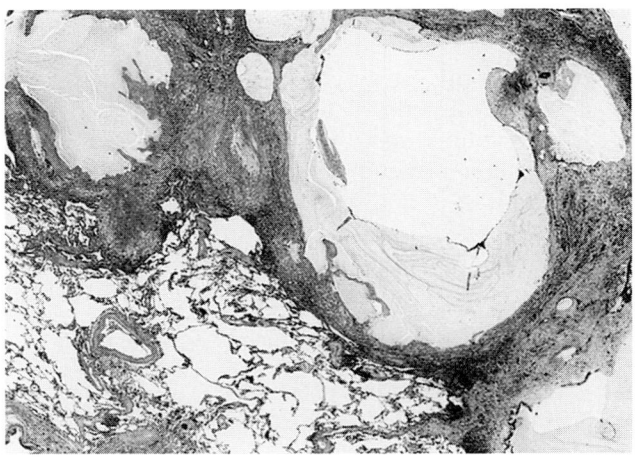

FIGURE 29-26
Mucinous cystic tumor of the lung. Mucus within fibrous tissue. (H&E, ×12.) (*Reproduced with permission of Dr. W. D. Travis, Armed Forces Institute of Pathology, Washington, D.C.*)

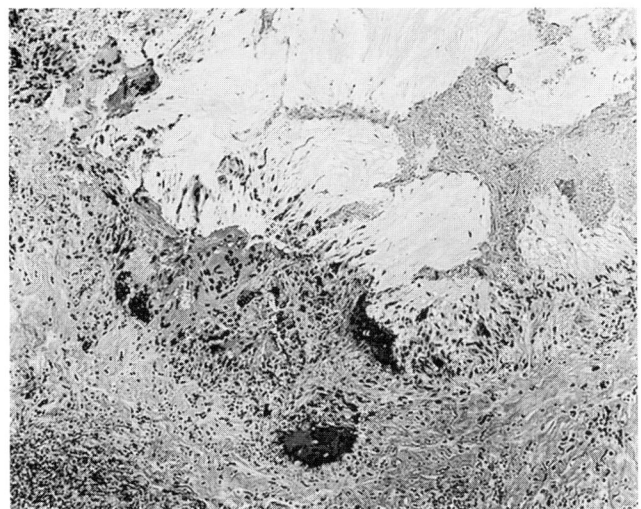

FIGURE 29-27
Fibrosis, chronic inflammation and focal giant cell formation in a mucinous cystic tumor. (H&E, ×12.3.) (By kind permission of Dr. W. D. Travis, Armed Forces Institute of Pathology, Washington, U.S.A.)

Differential Diagnosis

This is wide and should be considered in two categories.

NONNEOPLASTIC LESIONS

Nonneoplastic pulmonary cystic lesions that may cause confusion are bronchogenic cyst, adenomatoid malformation, and postinfectious bronchogenic cyst. All these cysts are lined by ciliated columnar epithelium and show only focal mucinous epithelium.

MALIGNANT TUMORS

Mucoepidermoid carcinoma shows squamous differentiation, a feature not seen in mucinous cystic tumors. Bronchial mucous adenoma is usually confined to the bronchial wall and does not display the heaping or nuclear irregularity seen in pulmonary mucinous tumors. Mucinous bronchioloalveolar carcinoma shows ultrastructural and immunohistochemical features identical to those of mucinous cystadenoma. Significant necrosis is not seen in mucinous cystic tumors of the lung, whereas cyst formation and necrosis may be seen in bronchioloalveolar carcinoma. Mucinous cystadenomas are well circumscribed nodules, whereas bronchioloalveolar carcinoma shows a lepidic spread along the alveolar wall. It is possible that some cases of bronchioloalveolar carcinoma have been included with these cystic pulmonary mucinous tumors, especially cases with "cavitary" lesions.[70]

A further differential diagnosis is secondary malignant teratoma, but most cases will have received a previous orchidectomy, oophorectomy, or thymectomy. Myxoma of the lung or pleura may cause confusion, especially if few nuclei are present. In most cases of myxoma there are thin stellate nuclei lying in a abundant matrix with no admixed epithelial elements.

ACINIC CELL CARCINOMA OF THE LUNG ("FECHNER TUMOR")

This is a rare tumor.[71–73] The lesion occurs predominantly in men, usually in the middle to later years (mean age: 56 years). In most cases, it is an incidental radiologic finding. None of the patients reported had had prior head and neck surgery or a previous salivary gland tumor. One patient presented with persistent cough and malaise, with an identifiable tumor on bronchoscopy.

Pathology

The tumor shows a predilection for the right lung and in some cases presented as well-circumscribed subpleural nodules. One case presented as a submucosal endobronchial mass.

The tumors ranged from 1.2 to 4 cm in diameter. They were fairly well circumscribed and tan-white to yellow in color and soft to rubbery in consistency.

Histologically, there was a solid pattern of sheets of large cohesive cells, traversed by bands of connective tissue forming islands of tumor (Figs. 29-28 and 29-29). In this connective tissue there was a prominent lymphoplasmacytic infiltrate, with occasional formation of small lymphoid follicles. In some cases the fibrous component was absent, and instead there was marked stromal vascularity, giving the tumor a neuroendocrine appearance (Fig. 29-30). In several cases, foci of acinar, microcystic (Fig. 29-31), or papillary appearances were present. The tumor cells were round with abundant granular cytoplasm (Fig. 29-32) and small, round, centrally placed nuclei. In some cases the nuclei were displaced to the periphery of the cell, giving a signet-ring appearance. Mitoses were rare.

Tumor infiltrated into adjacent airspaces, but there was no continuity with the lining of bronchial epithelium. The cells showed a weak granular cytoplasmic activity for PAS, which was mainly resistant with diastase.

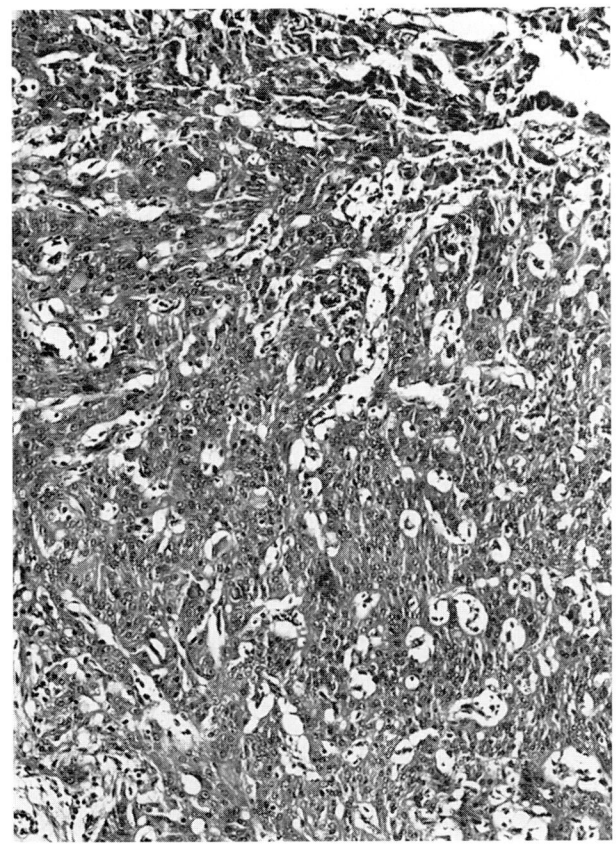

FIGURE 29-28
Acinic cell tumor with islands of cells separated by a little loose fibrous tissue with scattered inflammatory cells. (H&E, ×87.5.)

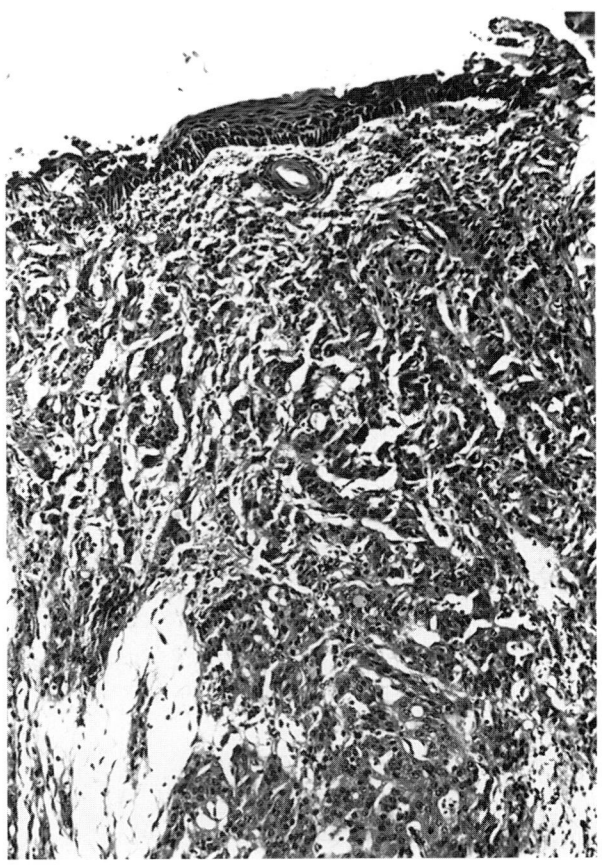

FIGURE 29-29
Acinic cell tumour with squamous metaplasia of the surface epithelium and a clump of tumor cells. Fibrosis and some chronic inflammation is seen at the margins of the tumor. (H&E, ×87.5.)

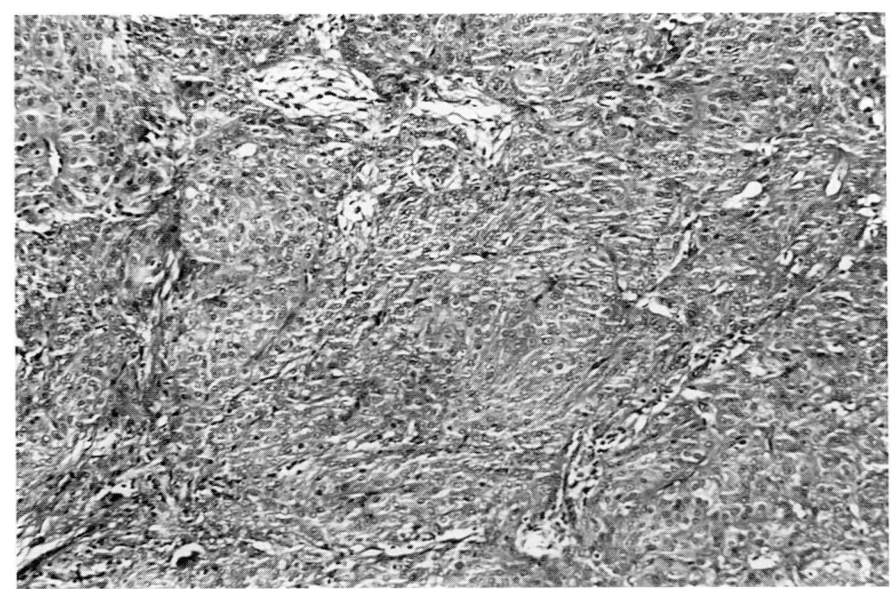

FIGURE 29-30
Carcinoid-like area in an acinic cell tumor. The insular areas are larger and less well defined than in a carcinoid. (H&E, ×87.5.)

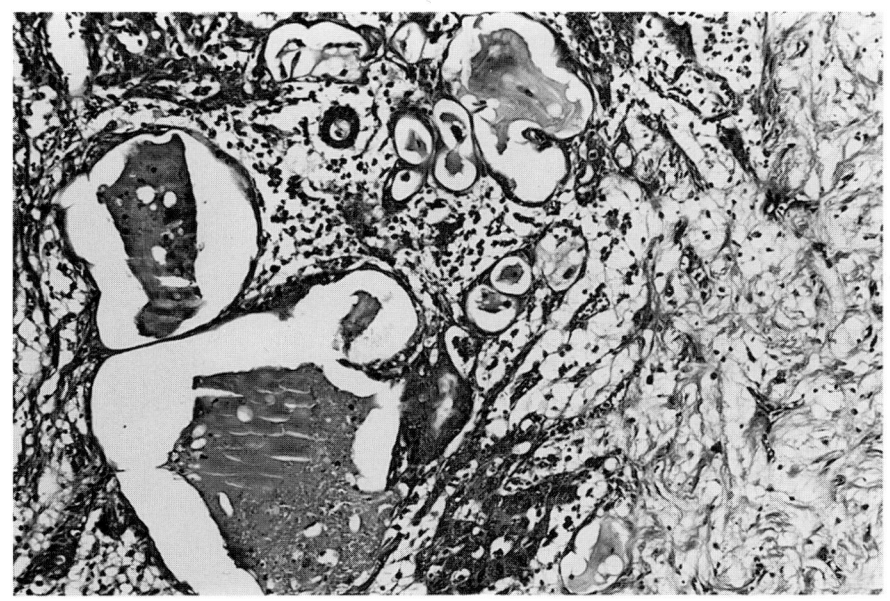

FIGURE 29-31
Microcystic foci in an acinic cell tumor.
(H&E, ×87.5.)

Immunohistochemistry was of diagnostic value in that the tumor cells were strongly positive with low-molecular-weight keratins (CAM 5.2). Only one case showed mild focal positivity with amylase.

Ultrastructural Features

No neurosecretory granules were identified, but dense round membrane-bound granules, ranging from 225 to 950 nm (Fig. 29-33), were identified.[73,74] The cytoplasm contained variable amounts of well-developed rough endoplasmic reticulum, occasional Golgi apparatuses, and small mitochondria. The cell surface had occasional short microvilli.

Differential Diagnosis

ONCOCYTIC CARCINOID

These tumors may occasionally form acini, and neuroendocrine stains are needed to differentiate them from acinic cell tumors. Ultrastructural studies of carcinoids show small dense-core neurosecretory granules ranging from 100 to 300 nm.

CLEAR CELL TUMOR OF THE LUNG ("SUGAR TUMOR")

This is rare, and the diagnosis should be made with caution until a primary renal tumor has been excluded. Clear cell tumors of the lung usually do not

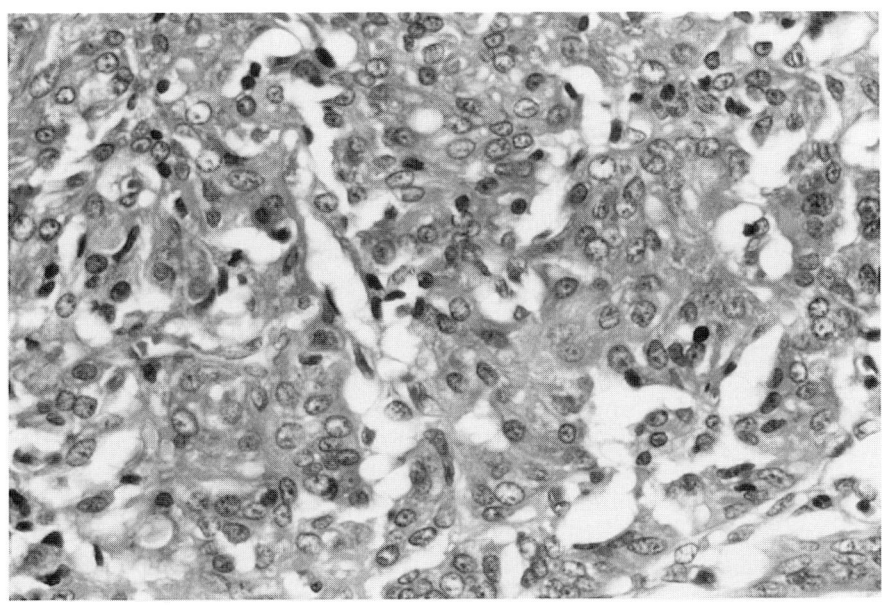

FIGURE 29-32
Cellular detail of an acinic cell tumor, loosely cohesive cells with an eosinophilic, finely granular cytoplasm and regular, central nuclei. (H&E, ×350.)

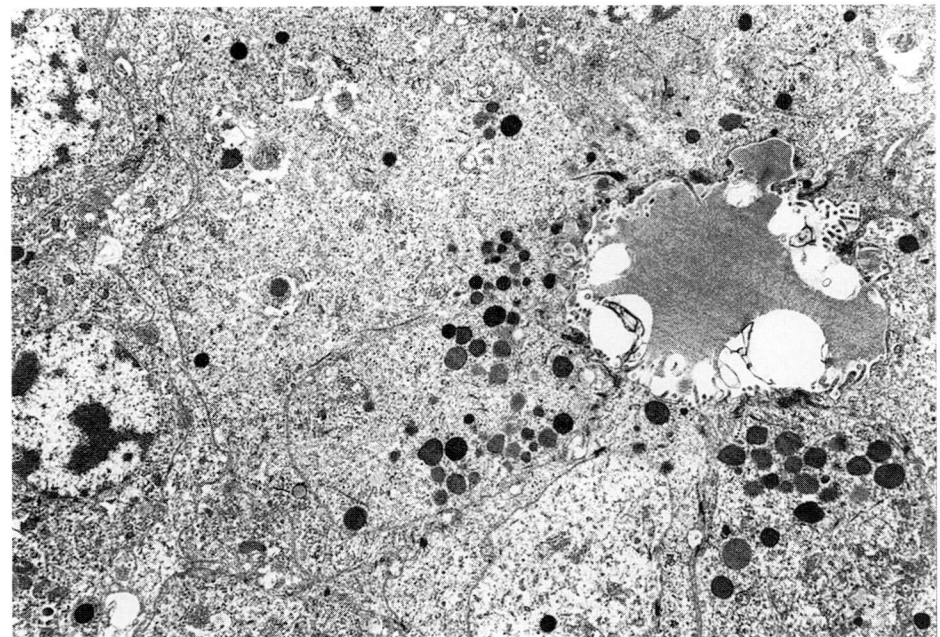

FIGURE 29-33
Electron micrograph of a tracheal acinic tumor. There are dense granules in the cytoplasm to the right of the picture. These are arranged around the lumen. (×4900.) (*Reproduced with permission of Professor B. Heard and Thorax.*)

show any evidence of epithelial differentiation.[75,76] Clear cell tumors have abundant intracytoplasmic glycogen and stain positively with HMB 45.

ONCOCYTOMA

Bronchial oncocytoma is a much blander tumor without a granular cytoplasm. Ultrastructurally there are numerous swollen mitochondria. Bronchial oncocytomas are usually bronchial-based lesions, whereas acinic cell tumors only rarely occur in main airways.

CLEAR CELL CARCINOMA

This usually shows evidence of squamous or glandular differentiation.

METASTATIC TUMOR

The important metastatic tumors are acinic cell tumor of the salivary gland and metastatic renal carcinoma. Rarer possibilities are secondaries from the thyroid, adrenal, or liver. It is likely that a hepatic primary will have declared itself. Alpha fetoprotein is usually positive in hepatomas.

BENIGN MIXED TUMOR

Benign mixed tumor or pleomorphic adenoma is usually seen in the trachea,[77–82] although a bronchial [83–85] or, more rarely, a parenchymal[86,87] origin

may be found. There are fewer than 40 cases in the literature.

The average age of patients up to 1994 was 48 years, with an age range of 26 to 71 years and a male predominance.

Tracheal tumors usually present with intermittent episodes of productive cough and shortness of breath, and the patient may be asthmatic. Other symptoms are dyspnea on exertion, hemoptysis, wheezing, stridor, hoarseness, dysphagia, and recurrent respiratory tract infection. Peripheral tumors may be identified only on routine radiology. Local recurrence has been seen up to 9 years later, and one case had a metastasis in the fourth lumbar vertebra.[84]

Most publications have not defined the actual size, which was noted as measuring between "cherries and plums." The largest tumor measured 4.5 by 2.5 by 2.5 cm. The tumors were generally sessile, whitish gray, firm, and polypoid. Cut surface was firm, gray, and slightly mucoid.

Tumors arose in the subepithelial tissues and were covered by a mixture of respiratory and squamous epithelium (Fig. 29-34). The glands had a tubular pattern, with varying amounts of eosinophilic material (Figs. 29-35 and 29-36). Foci of squamous metaplasia were identified in the tumor, and the epithelial elements were mixed with myxochondroid (Fig. 29-37). Foci of fibrosis were present (Fig. 29-38). The stroma was rich in alcian blue–staining mucopolysaccharide.

Tumor cells were positive for S-100 protein, cytokeratin, glial fibrillary acidic protein, actin, and vimentin.[86]

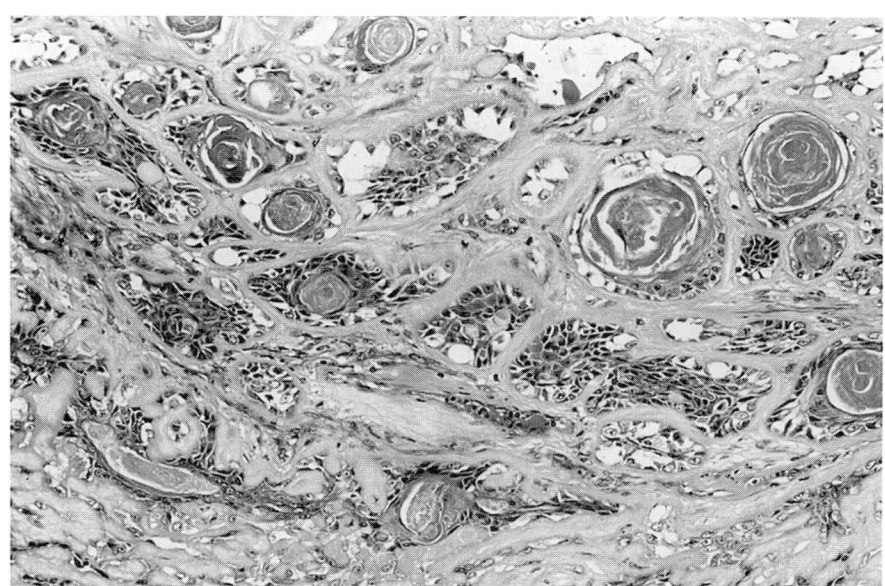

FIGURE 29-34
Benign mixed tumor with well-delineated squamous metaplasia. (H&E, ×87.5.)

Differential Diagnosis

The main differential diagnosis is adenoid cystic carcinoma. This tumor has cylinders of cells, and as in pleomorphic adenoma, the centers are filled with PAS-positive material. Cartilaginous foci are not seen in adnenoid cystic carcinoma, which tends to show perineural invasion and grow through and along the trachea and bronchus rather than as a discrete tumor, as in pleomorphic adenoma. The other differential diagnoses will be considered with the malignant variant described below.

Immunohistochemically, S-100 protein is identified in myoepithelial cells of plemorphic adenomas but not in adenoid cystic carcinoma. Gial fibrillary acidic protein is also usually demonstrated in pleomorphic adenomas, although in the malignant variety it may be absent.[88] Actin antibodies labeled many of the spindle cells, and there is strong cytoplasmic positivity with CAM 5.2, a cytokeratin marker.

A monomorphic tumor has been described[89] but shows cords and sheets of uniform epithelial cells with some focal S-100 positivity. There is a strong

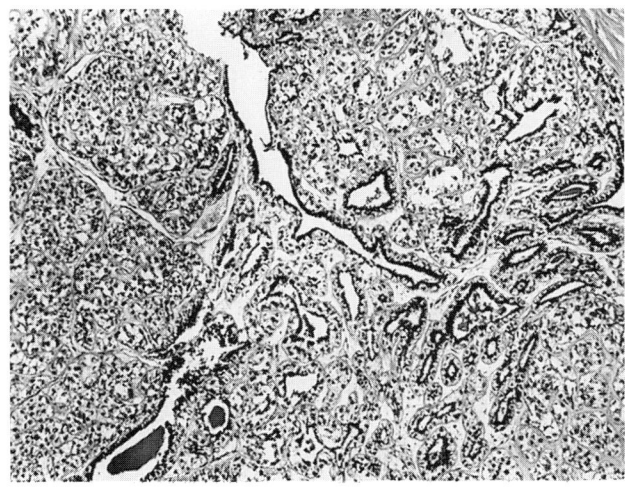

FIGURE 29-35
Pleomorphic adenoma with prominent tubular glands. (H&E, ×70.) (*Reproduced with permission of Virchows Archives of Pathological Anatomy.*)

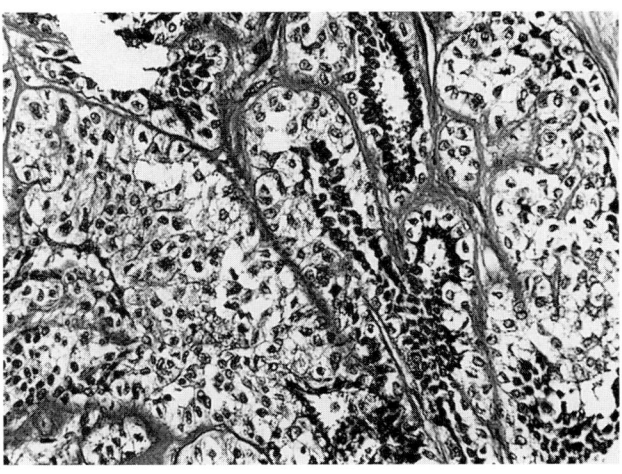

FIGURE 29-36
Higher power of Fig. 29-35 showing the glands lined by a well-defined basement membrane. (H&E, ×175.) (*Reproduced with permission of Virchows Archives of Pathological Anatomy.*)

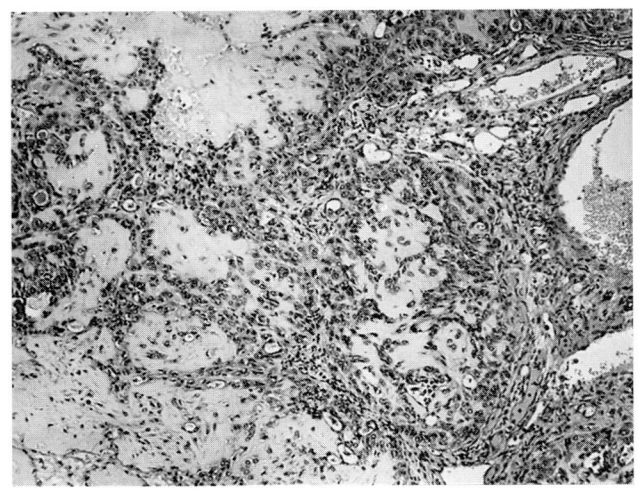

resemblance to a paraganglioma, which must cast some doubt on monomorphic adenoma as a distinct entity. A bronchial origin is unusual for a paraganglioma.

MALIGNANT PLEOMORPHIC ADENOMA (MALIGNANT MIXED TUMOR OF THE BRONCHUS)

Several malignant mixed pleomorphic adenomas have been described [88,90] as usually arising in the trachea or bronchus. A review of eight cases of benign and malignant mixed tumors of the lung has been published recently.[91]

The tumor is firm, pale yellow, and gelatinous, with elements of benign pleomorphic adenoma. In some cases, the lesions presented as peripheral parenchymatous nodules unrelated to a bronchus.

In addition, foci of squamous carcinoma with marked cellular atypia and complex tubulopapillary areas consistent with adenocarcinoma were seen (Fig. 29-39). Tumor cells were separated by an abundant myxoid martix. In some cases, angioinvasion was a prominent feature, as well as hemorrhage, necrosis, and raised mitotic counts—from five to 10 per 10 high-power fields.

Ultrastructurally, tumor cells showed myoepithelial characteristics, such as cytoplasmic filaments 4 to 6 nm in diameter and pinocytotic vesicles. In addition, cells with numerous tonofilaments indicating squamous differentiation were identified.

The differential diagnosis in the case of malignant mixed tumor may be difficult. The diagnoses to be considered are:

1. Secondary tumor from a primary salivary gland tumor. There is usually a history of previous surgery or a visible mass.

2. Pulmonary mesenchymoma. This has well-developed cartilaginous components sharply demarcated from the epithelial component. Malignant change in a benign mesenchymoma has only very rarely been recorded.

3. Pulmonary blastoma. In epithelial elements, a pattern of developing fetal lung is seen.

4. Carcinosarcoma is the most difficult differential diagnosis. Malignant salivary gland–type mixed

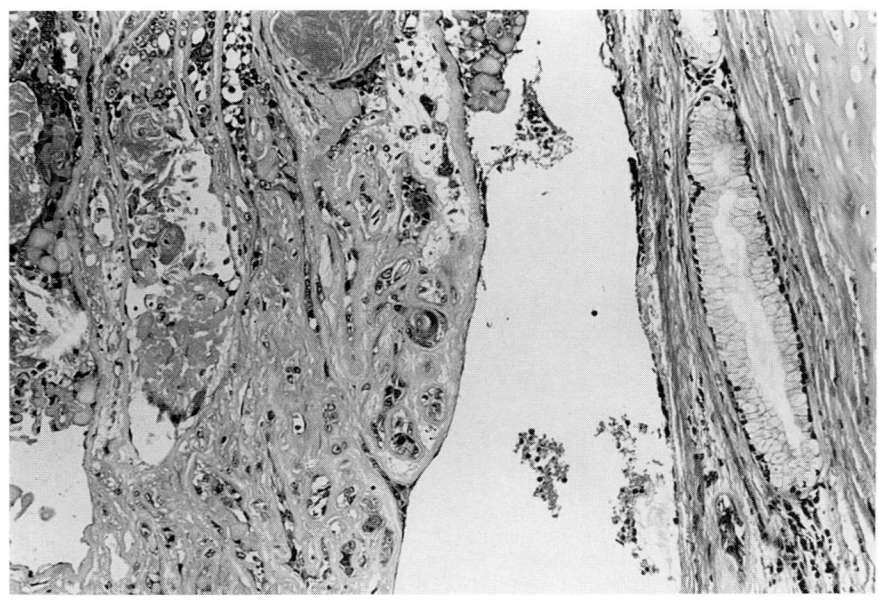

FIGURE 29-38
Glands trapped amidst fibrosis in a pleomorphic adenoma. (H&E, ×87.5.)

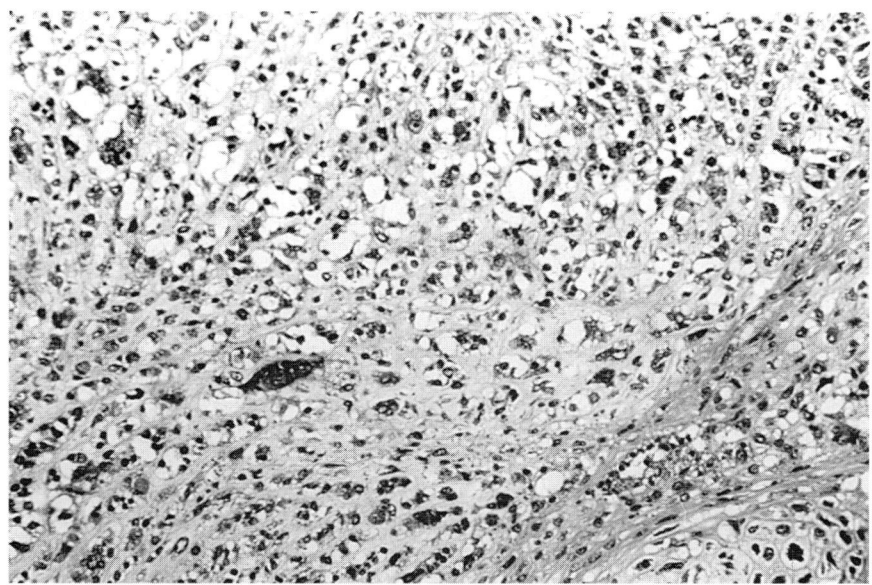

FIGURE 29-39
Malignant pleomorphic adenoma with cells showing nuclear pleomorphism and lying in a myxoid stroma. (H&E, ×87.5.)

tumors do not usually have a sarcomatous stromal component. The latter tumor shows myoepithelial cells with immunocytochemical stains.

5. Adenosquamous carcinoma and adenosquamous carcinoma with amyloid-like stroma[92] should also be considered. The latter entity does not have typical areas of a benign pleomorphic adenoma, and no myxoid stroma is seen.

The size of the tumor at presentation, extent of local infiltration, and the degree of mitotic activity are the most reliable prognostic features of these tumors.

NEUROENDOCRINE TUMORS OF THE LUNG

Bronchial carcinoids have previously been called "adenomas." There is no justification for continuing use of this term, as pure mucous gland adenomas occur in the lung and this entity will create confusion. In the past, the term encompassed three different bronchial tumors: carcinoids, mucoepidermoid tumors, and adenoid cystic carcinoma. These three lesions have different cells of origin, prognoses, and behavior.

Clinical Aspects

While carcinoid tumors account for approximately 85 percent of benign lung tumors, they account for only 1 to 2 percent of all lung tumors.[93] In several large series[94–97] there was a slight female predominance (F:M=1.2:1.0), with an age range of 12 to 82 years and a mean age of 50 years. Another large series (111 patients) had a slight male predominance.[96]

Carcinoids may be central or peripheral. If central, they classically present with hemoptysis or recurrent bronchial obstruction.[97] Chest pain, pleural effusion, cough, wheeze, and dyspnea may be presenting symptoms, but nearly 20 percent of cases may be asymptomatic.[98] Endocrine manifestations, including the carcinoid syndrome,[99] with left-sided valvular lesions are rare, with an incidence of between 1 and 7 percent.[100–102] This may be due to the small volume of tumor, which is insufficient to produce the mediators responsible for the syndrome. This low incidence is unexplained since tumors drain their products via the pulmonary veins into the left atrium. The syndrome is seen more commonly when there are metastases.[95,103] The peptides, prostaglandins, kinins, and other mediators produced by these tumors are rapidly inactivated by the lung and liver.[104]

Acromegaly,[105] Cushing's syndrome,[106] and insulin production with hypoglycemia have been documented.[107] The tumors may be part of the multiple endocrine adenomatosis syndrome.

Elevated levels of growth hormone–releasing hormone may be seen without acromegaly.[108] Hypercalcemia has been described in atypical bronchial carcinoid tumors. It is not known whether this was due to osteolytic metastases or some other tumor-associated cause.[109]

Carcinoid tumors may complicate cryptogenic pulmonary fibrosis as well as occurring synchronously with an adenocarcinoma of the lung.

Only 60 to 77 percent of tumors are seen through a bronchoscope. They are more common in the right lung and bleed easily; this feature has been given as a contraindication to biopsy.[110] A study based on 33 years of surgical experience found appreciable hemorrhage in only two cases, and in both it was easily controlled.[111]

Classification

Carcinoids are neuroendocrine tumors that originate in either neuroendocrine cells or neuroepithelial bodies (see Chap. 1). Warren and colleagues[112] classified these tumors as neuroendocrine neoplasms, i.e., tumors whose predominant pattern was neuroendocrine in contrast to squamous or glandular differentiation. These authors recognized four categories: bronchial carcinoids, well-differentiated neuroendocrine carcinomas, intermediate small-cell neuroendocrine carcinomas, and small-cell neuroendocrine carcinomas. The last two will be discussed with carcinoma of the lung.

Another and more easily workable classification of neuroendocrine tumors has been proposed.[113] This classification was reiterated at a recent international meeting.[114] The tumors have been placed in four categories: typical carcinoids, atypical carcinoids, large-cell neuroendocrine carcinoma, and small-cell carcinoma.

Typical and atypical carcinoids have similar histologic appearances, with organoid nests and other patterns as described below. The criteria for distinguishing atypical from typical carcinoids have been given by Arrigoni and coworkers[115] as follows:

1. Mitotic counts between 5 and 10 per 10 high-power fields. Any count above this lower number should raise the strong suspicion that the tumor is not an atypical carcinoid.
2. Focal areas of *punctate* necrosis. The presence of larger, more geographic islands of tumor necrosis militates against a diagnosis of atypical carcinoid.
3. Nuclear hyperchromatism and pleomorphism.
4. Disorganized architecture.

The criteria for large-cell neuroendocrine carcinoma are given on page 909.

This classification makes sense histopathologically. The problem is that relatively small numbers of neuroendocrine tumors have been analyzed in this fashion, and the reproducibility of the grading system has not been applied to a large group of neuroendocrine tumors. Table 29-1 compares the two classifications. Travis[113] and Gould[115a] et al.

The cause of pulmonary carcinoids is unknown, but a genetic susceptibility in some *duodenal* carcinoids has been described.[116]

Macroscopic Pathology

Central tumors are usually seen in main or lobar bronchi (Fig. 29-40); they are sessile and pedunculated and may completely occlude the bronchial lumen. Occasionally the tumor extends into adjacent tissues. On cut surface it is yellowish white with foci of hemorrhage and extends into dilated bronchi, with distal bronchiectasis. If bone is present, the tumor is focally hard and gritty. Peripheral tumors, which are usually well circumscribed (Fig. 29-41), have a similar coloration or may be hemorrhagic, possibly as a result of mobilization of the lobe. The rare oncocytic carcinoid has a soft consistency (Fig. 29-42).

Rare multiple peripheral bronchial carcinoid tumors, unassociated with gross lung disease (thereby excluding tumorlets), have been described.[117] These tumors are small and subpleural or subepithelial, and they may cause fibrosis. In some instances these lesions may occlude bronchioles, and there is associated pulmonary fibrosis (Figs. 29-43 and 29-44).

Histology

Typical carcinoids show variable patterns. The most common is mixed with insular (Fig. 29-45) and trabecular foci (Fig. 29-46). A pure acinar pattern is uncommon (Fig. 29-47). Tumors without these characteristics are termed *undifferentiated* (Fig. 29-48) and are classed with atypical carcinoids. The cells are small and uniform with regular central nuclei and a faintly eosinophilic cytoplasm.

TABLE 29-1
Classification of Neuroendocrine Tumors

New Classification	Old Classification
Typical carcinoid	Bronchial carcinoid
Atypical carcinoid	Well-differentiated neuroendocrine carcinoma
Large-cell neuroendocrine carcinoma	
Small-cell carcinoma of lung	Neuroendocrine carcinoma of intermediate cell type
	Neuroendocrine carcinoma of small cell type

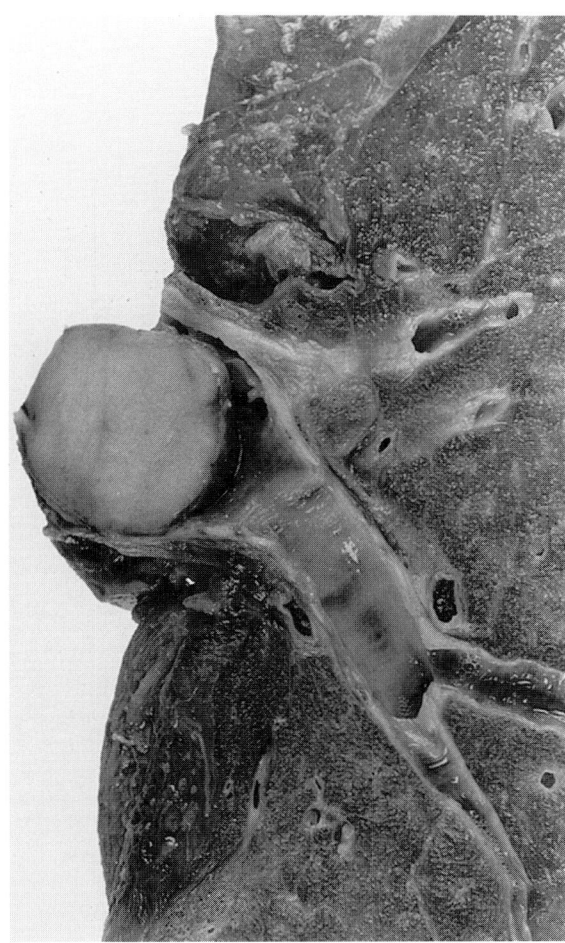

FIGURE 29-40
Endobronchial carcinoid.

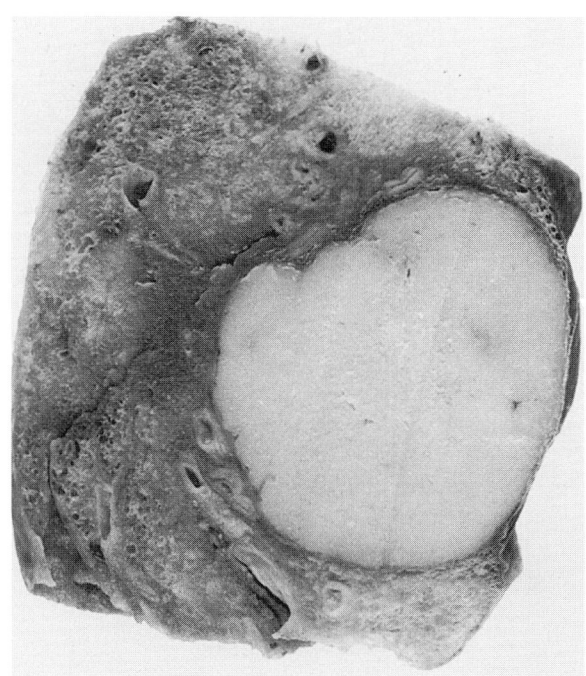

FIGURE 29-41
Peripheral carcinoid abutting the pleura.

FIGURE 29-42
Oncocytic carcinoid which had a soft consistency.

Oncocytic change may be focal or, in rare cases, diffuse; the tumor is thereby classified as an oncocytic carcinoid (Fig. 29-49). Oncocytic carcinoids show neurosecretory granules as well as mitochondrial hyperplasia. Ghadially and Block[118] described filamentous mitochondrial inclusions and calcified spherules that appear to develop in material secreted by tumor cells.

Both typical and atypical carcinoids contain sustentacular cells, as seen in paragangliomas.[119] These cells—which were stained by S-100 protein, glial fibrillary acidic protein, and nerve growth factor receptor—were most numerous in typical carcinoids with a well-developed insular pattern and less conspicuous in the atypical variant.

INDIVIDUAL CELL DEATH IN CARCINOID TUMORS

A second population of smaller cells with little cytoplasm and hyperchromatic nuclei is easily identified in all carcinoids (Fig. 29-50). These cells undergo necrosis or cell death ultrastructurally.[120]

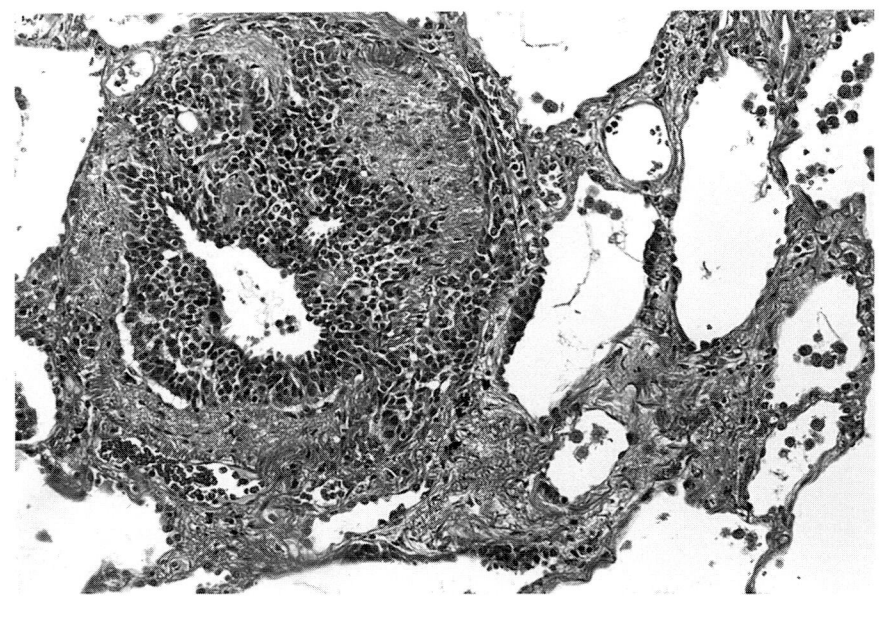

FIGURE 29-43
Intrabronchial proliferation of neuroendocrine cells. (H&E, ×87.5.)

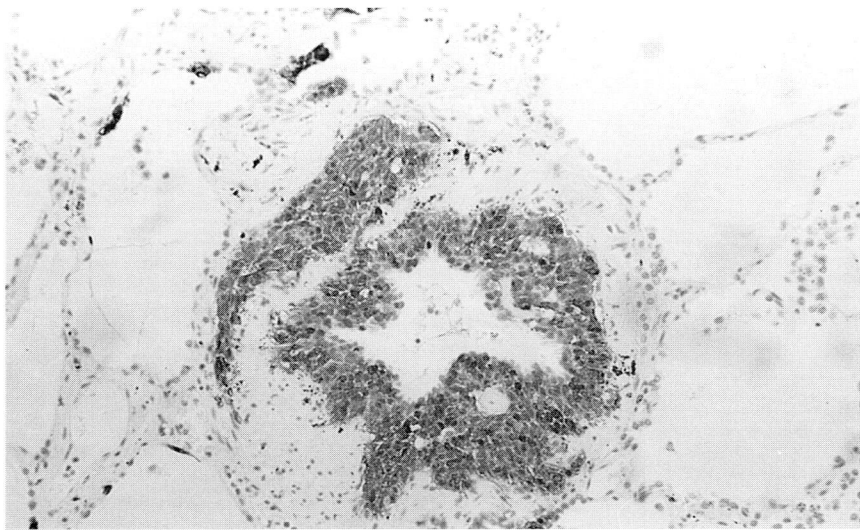

FIGURE 29-44
Same case as Fig. 29-43 but with positivity for PGP 9.5. (Immunoperoxidase, ×87.5.)

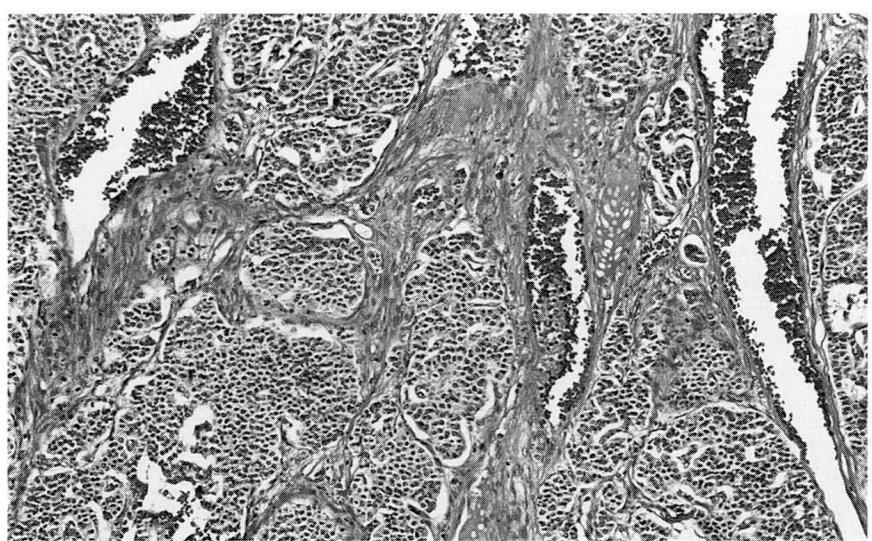

FIGURE 29-45
Typical carcinoid with an insular pattern. There are intervening bands of fibrous tissue and prominent blood vessels. (H&E, ×87.5.)

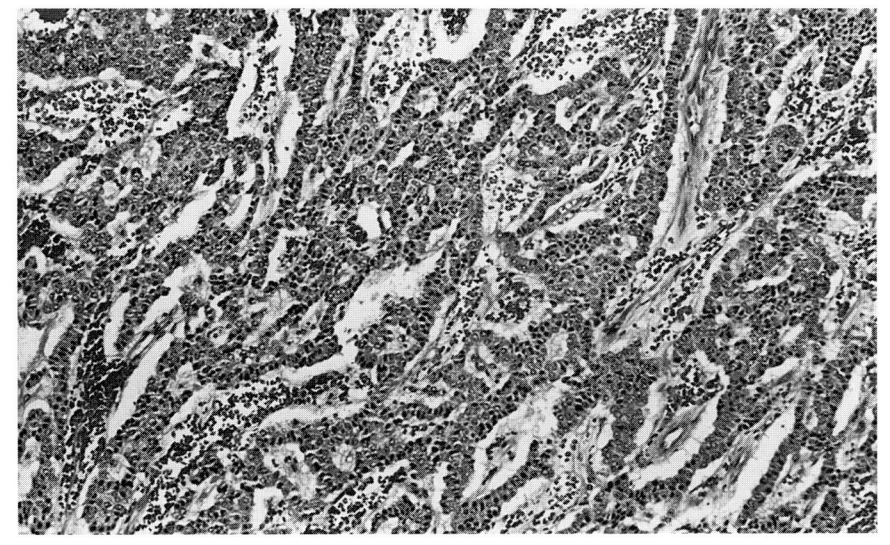

FIGURE 29-46
Typical carcinoid with a predominantly trabecular growth pattern. (H&E, ×87.5.) (*Reproduced with permission from Hasleton and Al-Saffar, Applied Pathology 7:205-218, 1989.*)

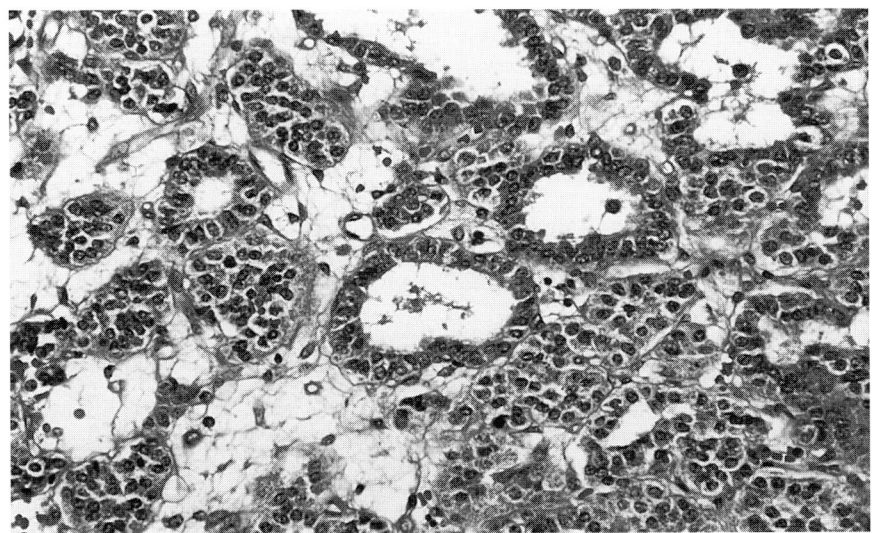

FIGURE 29-47
Typical carcinoid with foci showing an acinar pattern. (H&E, ×219.) (*Reproduced with permission from Hasleton and Al-Saffar, Applied Pathology 7:205–218, 1989.*)

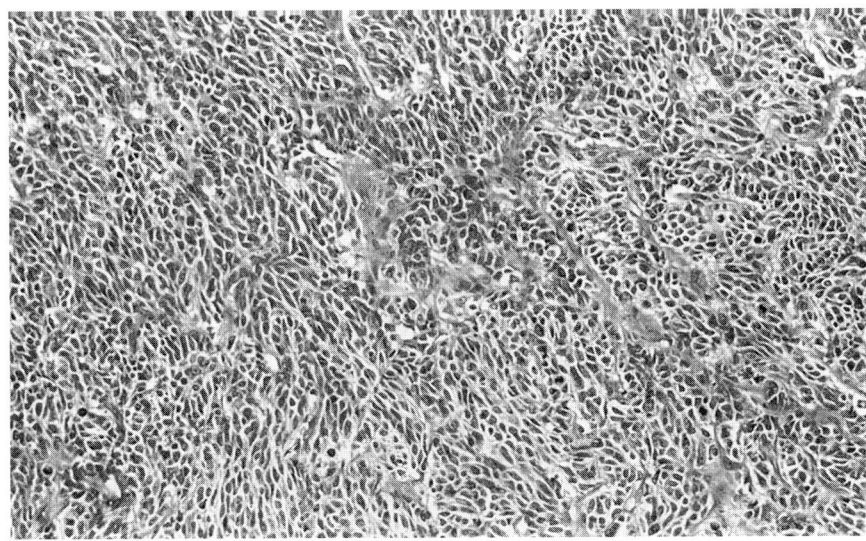

FIGURE 29-48
Typical carcinoid having a disorganized growth pattern. (H&E, ×87.5.)

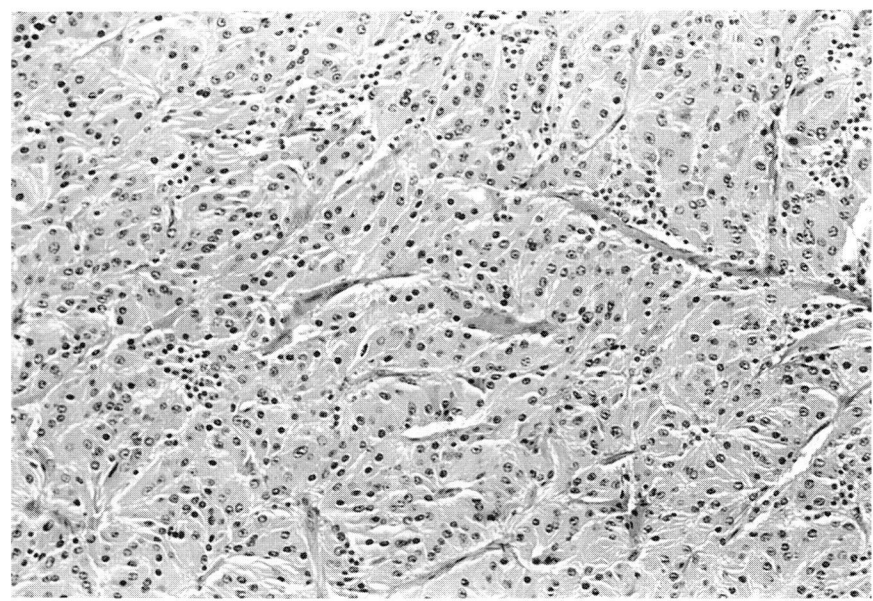

FIGURE 29-49
Typical oxyphil carcinoid tumor with large cells having a pale eosinophilic cytoplasm. (H&E, ×87.5.)

Morphometric analysis of individual cell death in bronchial carcinoids[120] showed the proportion of dead tumor cells was not significantly different for typical and atypical bronchial carcinoids (17 and 13 percent, respectively). In diploid tumors the proportion of dead cells was 18 percent; in aneuploid, 12 percent. Individual cell death did not appear to be positively associated with poor prognosis in the atypical carcinoids. The proportion of cells dead or dying was higher than the ranges quoted for "apoptotic" indices in rodent tumors (0.3 to 7 percent)[121] or in squamous carcinomas of the uterine cervix and bronchus (1.5 to 10 percent).[122]

OTHER FEATURES OF BRONCHIAL CARCINOID TUMORS

Papillary (Fig. 29-51), goblet cell, and melanin-containing or spindle cell carcinoids (Fig. 29-52) have been described.[123,124] Mark's team[125] described a papillary carcinoid, which must be distinguished from a papillary adenoma. Pure spindle cell carcinoids are usually subpleural and are uncommon.[126] They have monomorphic spindle cells, arranged in fascicles and loose groups. In some cases the organoid pattern of carcinoids is identified. There is no marked cytologic atypia or nuclear pleomorphism. Mitoses are few. Rare carcinoids include

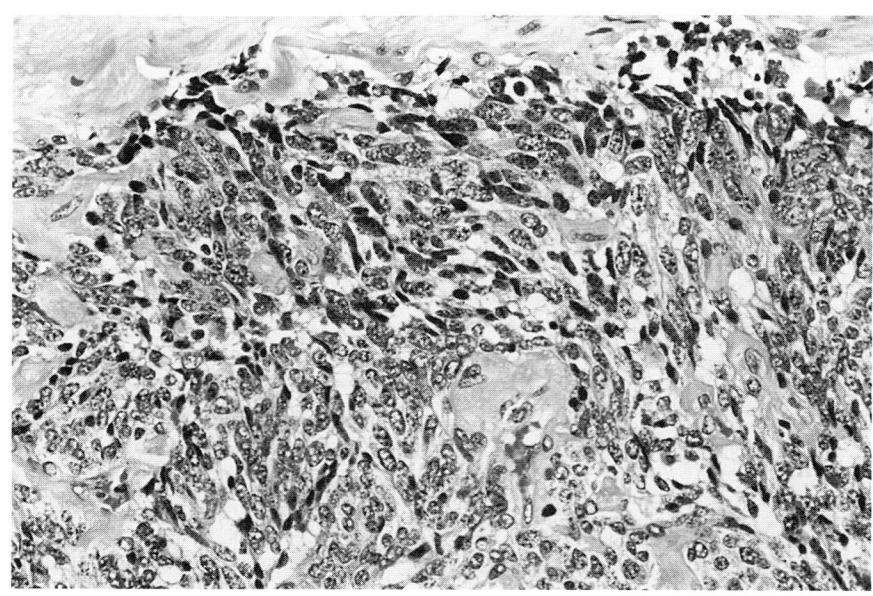

FIGURE 29-50
Bronchial carcinoid with a prominent component of dark, hyperchromatic cells. (H&E, ×350.)

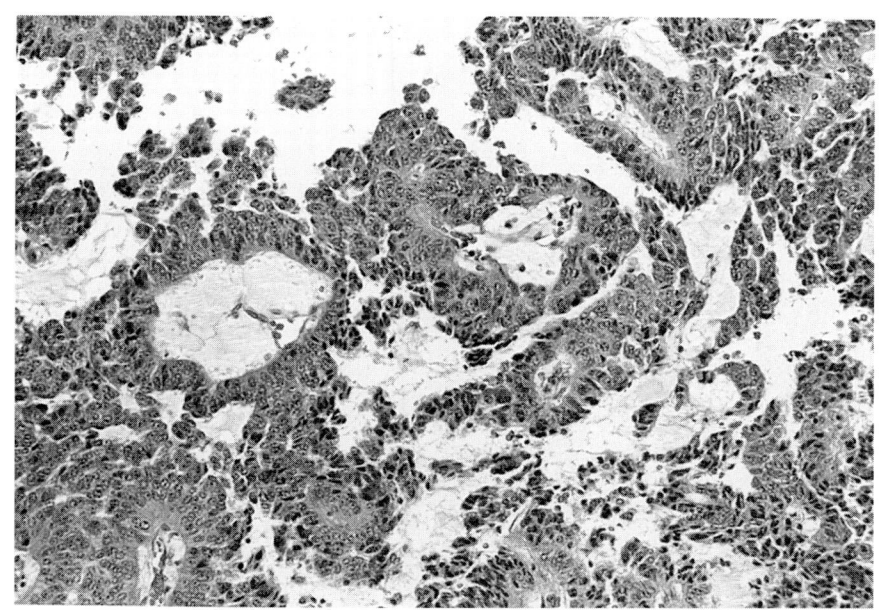

FIGURE 29-51
Bronchial carcinoid with a papillary growth pattern. (H&E, ×350.)

those with a signet-ring cell or bronchioloalveolar carcinoma pattern.[127]

Twenty-five percent of cases contain bone,[123] and if one excludes peripheral carcinoids, this figure rises to 30 percent. In some cases, bone is present in relation to bronchial cartilage. In many, however, there is no endochondral calcification, and bone is seen amid the tumor (Fig. 29-53). Cartilage may also be found in some tumors (Fig. 29-54) and may play a role in bone formation. We have seen a case with prominent cartilage proliferation in the bronchus (Fig. 29-55).

The mechanism in bone formation is unknown. Cooney and colleagues[128] believed it was due to locally produced calcitonin, having immunocytochemically demonstrated this hormone in four of seven bone-containing carcinoids. However, medullary carcinoma of the thyroid, a tumor rich in calcitonin, has calcification but no bone.[129] We identified carcinoids with no bone but calcitonin immunoreactivity.[123] Acid phosphatase and 5-nucleotidase have been suggested as initiators of the ectopic bone, but a recent case was negative for both these enzymes.[130]

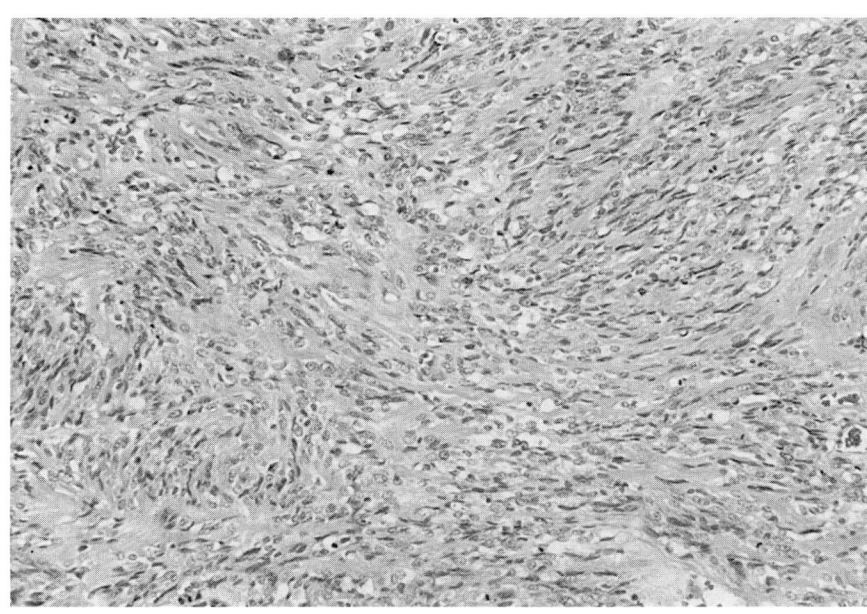

FIGURE 29-52
Bronchial carcinoid with a spindle cell pattern. (H&E, ×87.5.)

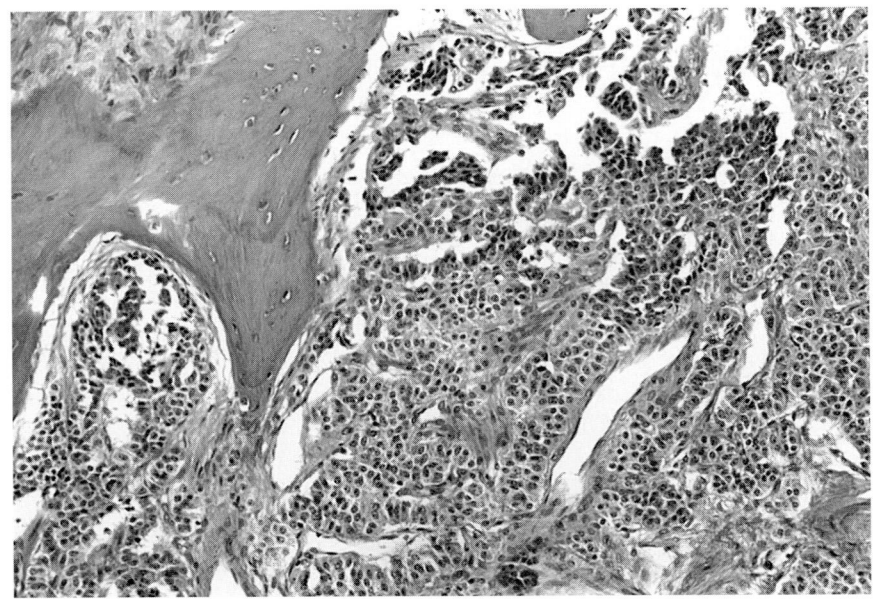

FIGURE 29-53
Typical carcinoid with an insular growth pattern and a well-defined focus of mature bone. (H&E, ×87.5.)

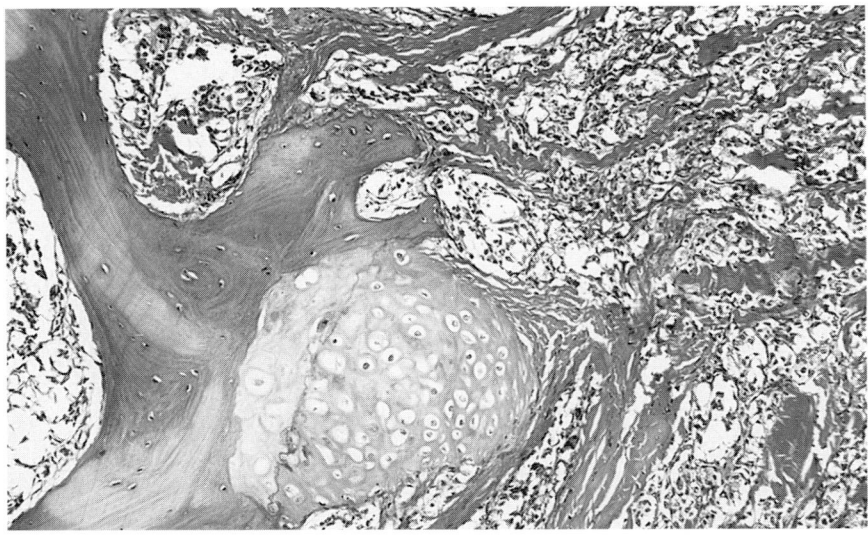

FIGURE 29-54
Bone and cartilage formation in a typical carcinoid. (H&E, ×87.5.) (*Reproduced with permission from P. S. Hasleton and Al-Saffar, Applied Pathology 7:205–218, 1989.*)

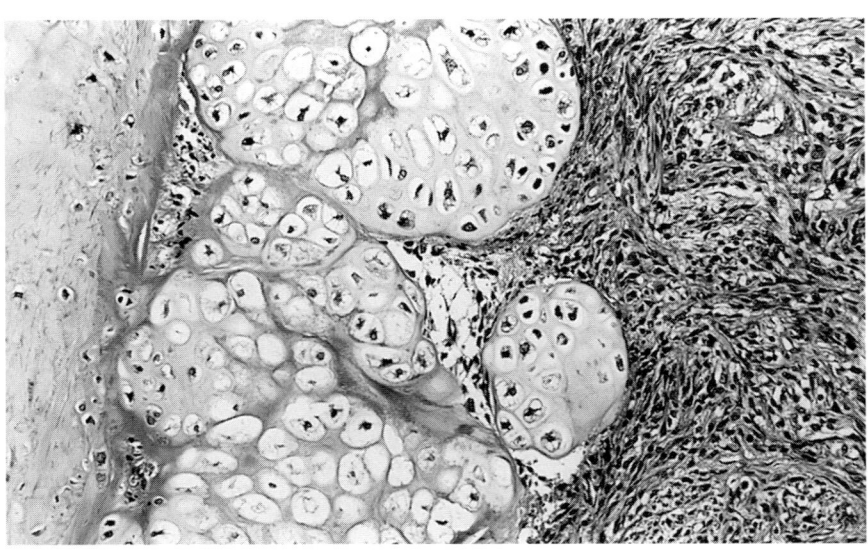

FIGURE 29-55
Spindle cell carcinoid with prominent proliferation of cartilage close to the cartilaginous ring. There is nuclear hyperchromatism in the cartilage. (H&E.)

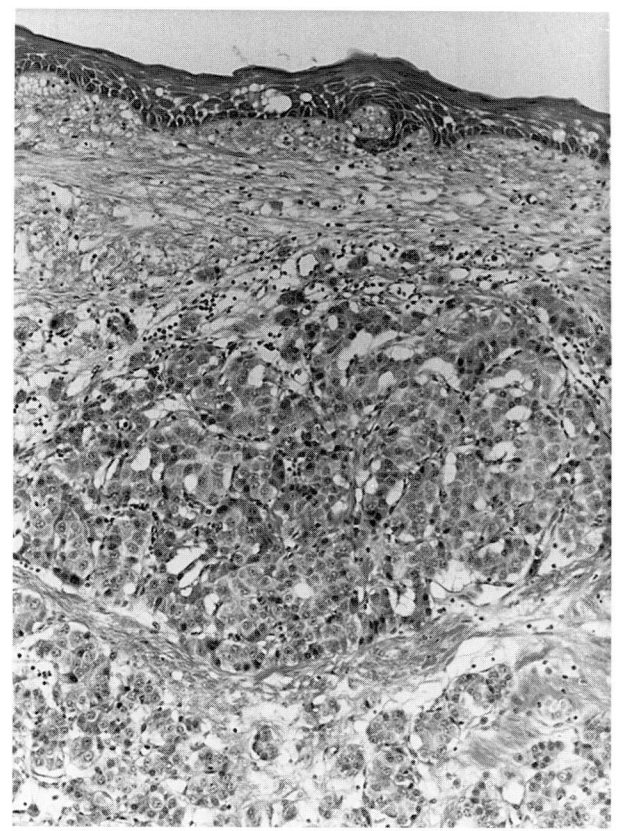

FIGURE 29-56
Bronchial carcinoid with prominent squamous metaplasia of the overlying epithelium. (H&E, ×56.)

The stroma is vascular, explaining the tendency to hemorrhage on bronchoscopy, and may undergo hyaline change and calcification and, rarely, may show amyloid.[131] Frequently there is squamous metaplasia of the overlying epithelium (Fig. 29-56), and in some cases it may appear mildly dysplastic. This is also a feature of small-cell carcinoma. In typical carcinoids, necrosis, mitoses, and nuclear pleomorphism are not seen. Any of these features should make one suspect an atypical carcinoid.

Mucin may be detected in carcinoids by both light and electron microscopy.

Atypical Carcinoids

There are clearly defined histologic features of atypical bronchial carcinoid: spotlike or punctate necrosis (Fig. 29-57), lymphatic (Fig. 29-58) or vascular permeation, fewer than 10 mitoses per 10 high-power fields, nuclear pleomorphism (Fig. 29-59), and an undifferentiated growth pattern.[123] An undifferentiated growth pattern may coexist with a trabecular, insular, or acinar pattern. It may be impossible to define one of these three main growth patterns. The presence of lymph node metastases, while suggesting an atypical carcinoid, should not be taken as evidence that the tumor is behaving in a malignant fashion in *all* cases. Occasionally patients with a typical carcinoid pattern and lymph node metastases may have a long postoperative survival.

A recent study of 25 bronchial carcinoids suggested that recurrence in bronchial carcinoids depends more on the histology of the tumor than on

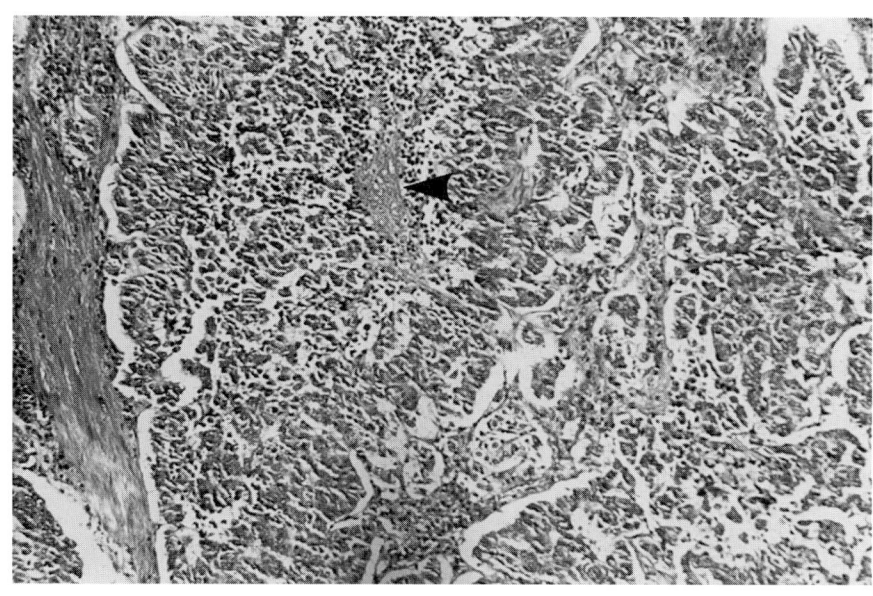

FIGURE 29-57
Atypical bronchial carcinoid with a focus of necrosis (*arrow head*) and a tendency to a disorganized growth pattern. (H&E, ×87.5.)

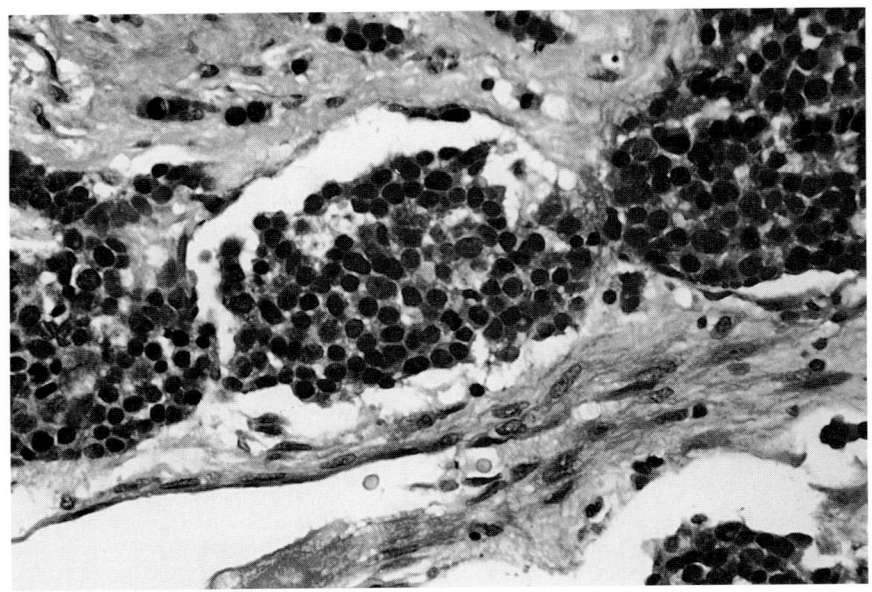

FIGURE 29-58
Atypical bronchial carcinoid with lymphatic permeation. (H&E, ×56.)

nodal status.[132] The incidence of lymph node metastasis in this series was unusual in that they were equally divided between the two types of carcinoids. Twelve typical carcinoids showed lymph node metastases. When larger series are analyzed, the presence of lymph node metastases is an adverse prognostic factor.

Endobronchial biopsies from atypical carcinoids may be misdiagnosed as small-cell carcinoma because of sampling error. As large a biopsy as possible should be obtained to avert this error. The prognosis for patients with atypical carcinoid is better than for small-cell carcinoma.[133,134]

SPECIAL STAINS FOR BRONCHIAL CARCINOID TUMORS

The most commonly used stain is Grimelius, which was positive in 46 of 63 cases in one series.[123] Diastase-PAS was positive in 13 cases. Carcinoembryonic antigen (CEA) was detected in approximately half of the cases, and there was a tendency for stronger staining in atypical tumors. CEA reactivity is a strongly significant factor for predicting treatment failure.[135]

A list of peptides present in bronchial carcinoids is given in Table 29-2.[136] However, if a combination of prealbumin and Grimelius[137] is used, approxi-

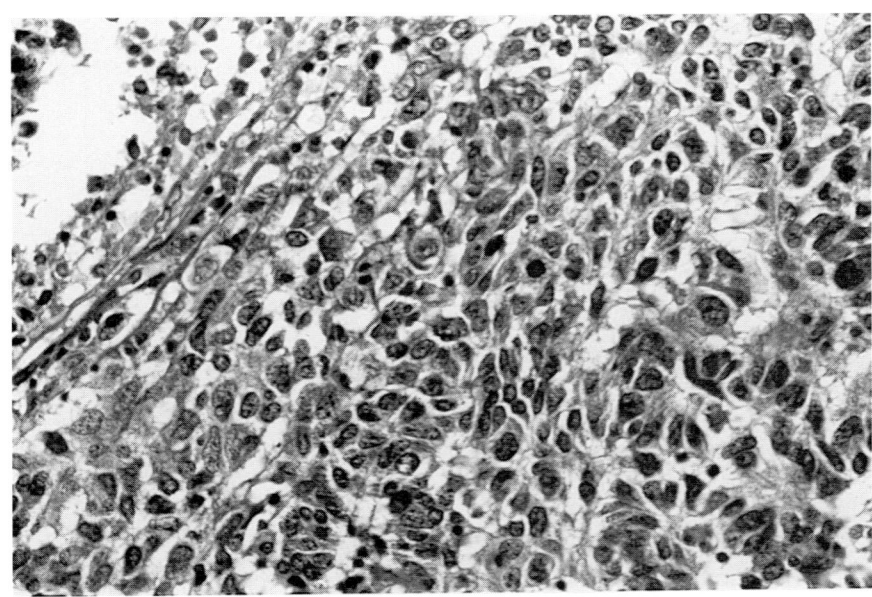

FIGURE 29-59
Atypical carcinoid with nuclear pleomorphism and an irregular growth pattern. (H&E, ×87.5.)

904

TABLE 29-2
Bronchial Carcinoid Neuroendocrine Protein Products

Peptides
  Amylase
  Bombesin
  Calcitonin
  Corticotropin
  Corticotropin-releasing hormone
  Gastrin
  Glucagon
  Growth hormone–releasing factor
  Human chorionic gonadotropin
  Insulin
  Melanocyte-stimulating hormone
  Motilin
  Neurotensin
  Pancreatic polypeptide
  Secretin
  Somatostatin
  Substance K
  Substance P
  Vasoactive intestinal polypeptide
  Vasopressin
Vasoactive amines
  Catecholamines
  Histamine
Kinins
  Bradykinins
  Tachykinins (neurokinin A)
Endorphins and enkephalins
  β-Endorphin
  Leu-enkephalin
  Met-enkephalin
  Pan-opioid
Other
  Kallikrein
  Prostaglandins

SOURCE: From Davila DG, Dunn WF, Tazelaar HD, Pairolero PC, Mayo Clin Proc 68:795–803, 1993.

mately 90 percent of all bronchial carcinoids will stain positively. Chromogranin A is useful in the diagnosis of bronchial carcinoids. In one study[138] chromogranin (type unspecified) was detected in all bronchial carcinoids but not in small-cell carcinomas. There is a tendency for immunoreactivity of bronchial carcinoids to various peptides to be less strongly expressed in atypical than in typical tumors (Figs. 29-60, 29-61, and 29-62).[139]

Neuron-specific enolase (NSE) is of little help in the diagnosis of carcinoids or small-cell lung cancer.[140] Bergh's team[140] found that 70 percent of small-cell carcinomas stained for NSE, and most (14 of 21) non–small-cell carcinomas also reacted with this antiserum.

Neurofilament proteins (NF) are expressed in carcinoid tumors, especially NF-70 (Fig. 29-63). However, this antibody does not detect more carcinoids than prealbumin and Grimelius.[141]

Carcinoids are cytokeratin-positive with AE1/3 and CAM 5.2. In a series of atypical carcinoids, most stained for synaptophysin and chromogranin A.[142] Desmoplakin is identified in carcinoid tumors.

The p53 protein is not seen in typical carcinoids and is strongly expressed in only a small number of atypical tumors.[143,144] This pattern of overexpression of p53 protein across this spectrum of tumors was thought to follow their increasing malignancy and suggests that these antioncogenes may be involved in the progression of atypical carcinoids. A nuclear proliferation antigen, KI-67, showed a higher proliferation index (greater than 4 percent).

A deletion in the retinoblastoma gene has been documented in atypical carcinoid tumors.[144]

Ultrastructure

The characteristic feature of carcinoid tumors is the presence of cytoplasmic dense core neurosecretory granules (Figs. 29-64 and 29-65).[145] There is considerable heterogeneity of granules with regard to size, configuration, and electron density. Mixed populations of granules within single tumor cells are also seen and range from 100 to 200 nm in diameter. It is possible to find evidence of squamous differentiation with distinct tonofilaments and well-developed desmosomes. In other cases there may be lumen formation, with typical microvilli and terminal bars. Rough endoplasmic reticulum, which may be arranged in parallel stacks and tumor cells, rests on well-defined focally thickened basal lamina.[146] As mentioned previously there is a prominent population of small pyknotic cells. (Fig. 29-66).

Ploidy in Bronchopulmonary Carcinoids

Carcinoids are recognized as tumors of low-grade malignant potential. Since cytogenetic changes are a recognized feature of many human tumors and they may be related to clinical behavior, DNA ploidy may be of prognostic value. DNA ploidy did not confer any independent prognostic information in bronchial carcinoids.[148]

There is a tendency for patients with DNA

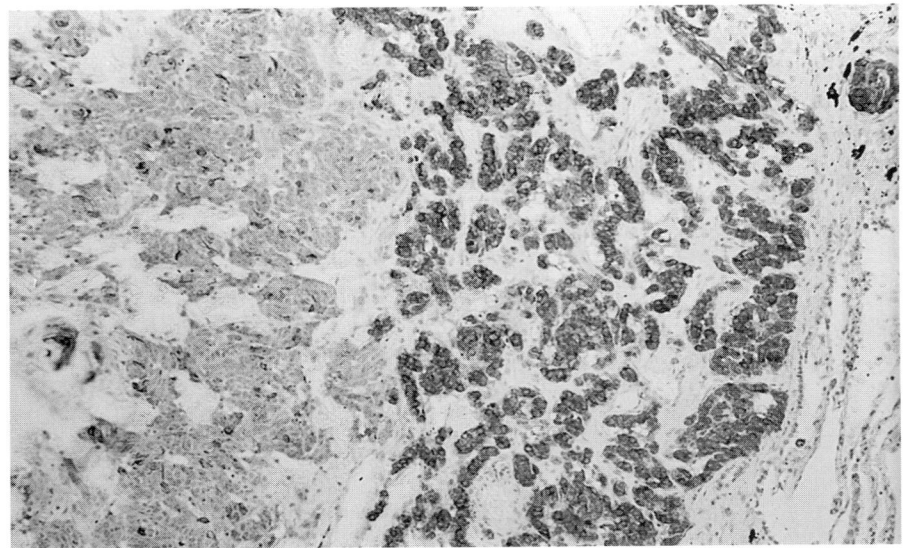

FIGURE 29-60
Bronchial carcinoid with strong positivity on the right hand side of the picture with an antibody to bombesin. (Immunoperoxidase, Bombesin, ×87.5.)

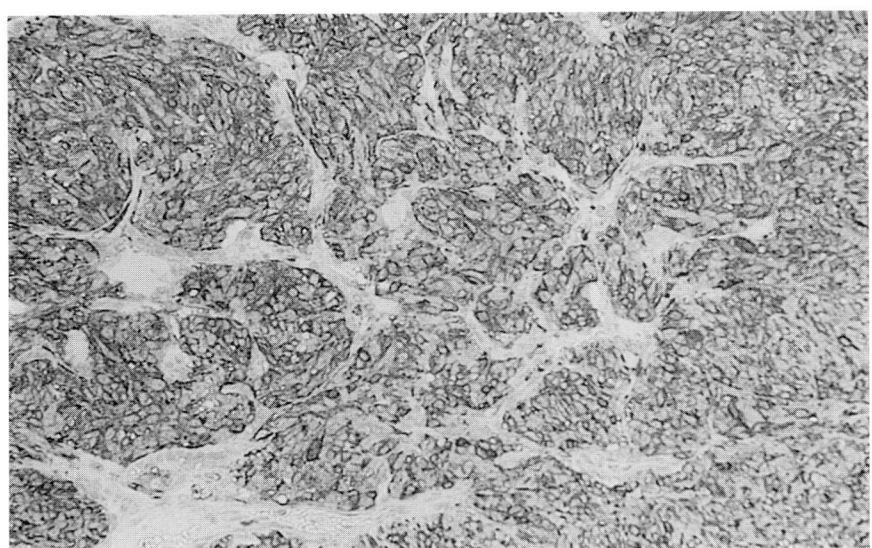

FIGURE 29-61
Bronchial carcinoid with strong positivity for ACTH. The patient had Cushing's syndrome, which responded to removal of the pulmonary tumor. (Immunoperoxidase ACTH, ×125.)

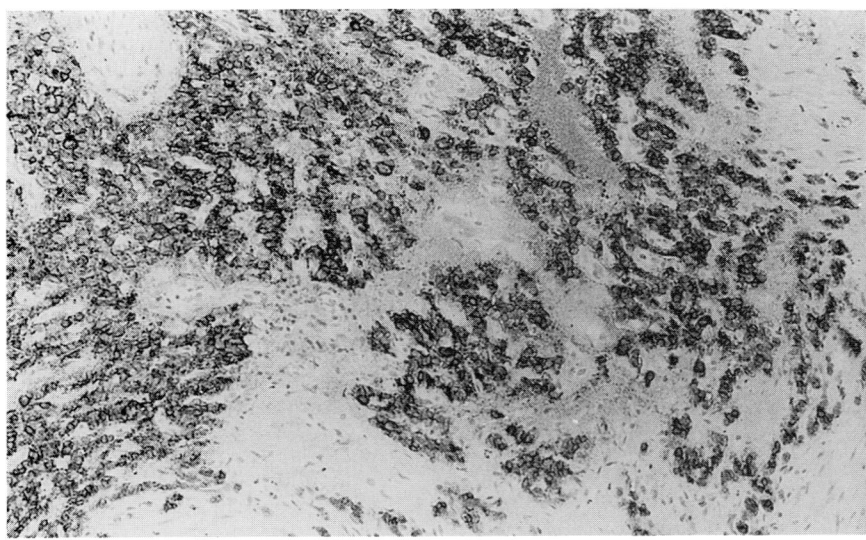

FIGURE 29-62
Dense staining with Leu 7 in an atypical carcinoid. (Immunoperoxidase, ×88.75.)

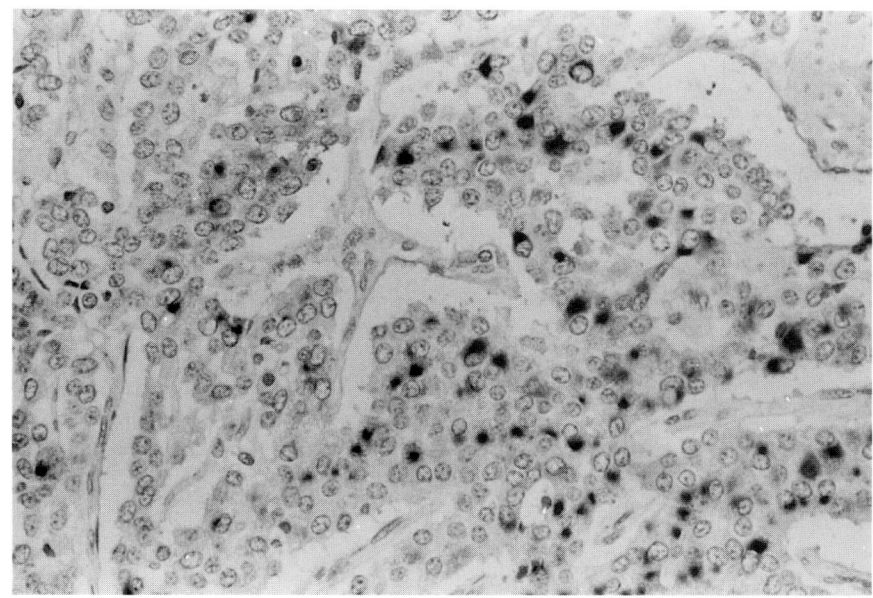

FIGURE 29-63
Bronchopulmonary carcinoid showing typical perinuclear staining with neurofilament kD70. The same pattern is seen with neurofilament kD150. (Immunoperoxidase, ×350.)

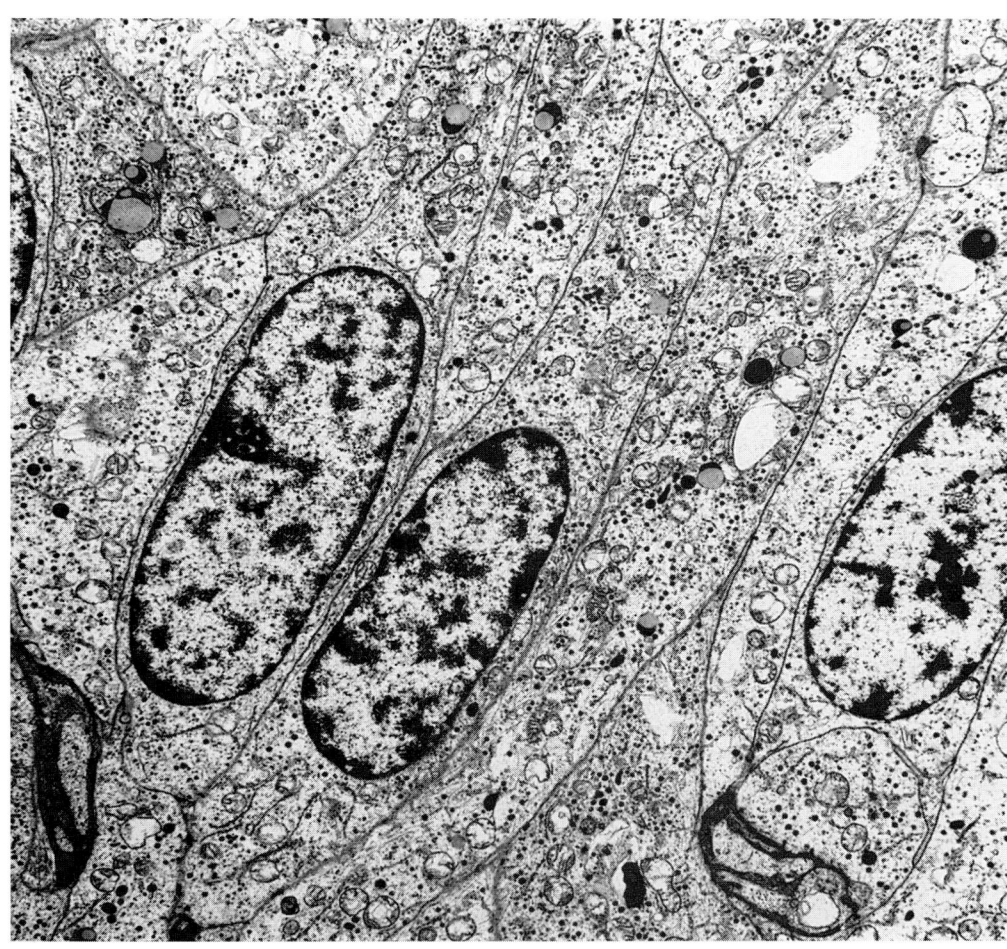

FIGURE 29-64
Electron micrograph of a carcinoid tumor with regular nuclei and the cytoplasm full of dense core granules. (Uranyl acetate and lead citrate, ×5250.)

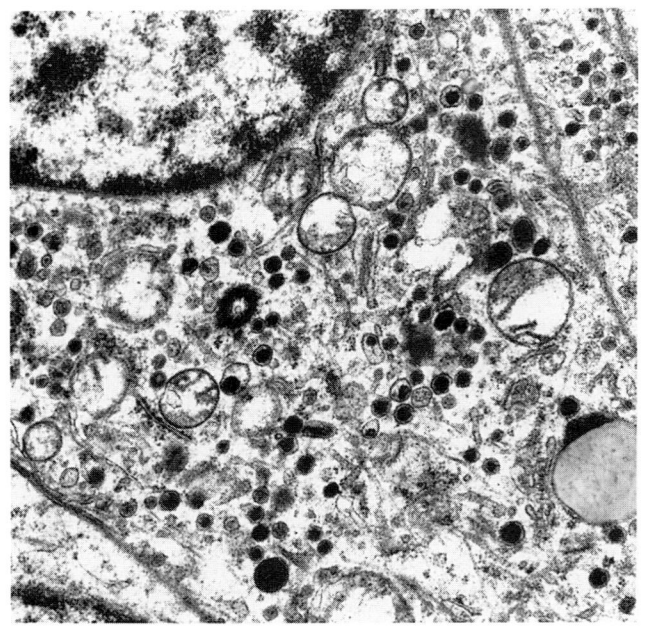

FIGURE 29-65
Higher magnification of the previous picture showing dense core granules, some with a "poached-egg" appearance. (Uranyl acetate and lead citrate, ×6750.)

diploid tumors to survive longer than those with DNA aneuploid tumors, but the difference is of borderline statistical significance (Fig. 29-67). It is possible that with an increased number of bronchopulmonary carcinoids (the above mentioned study was based on 53 patients) statistical significance may be obtained for survival of patients with tumors of aneuploid and diploid status. There is a tendency for the atypical tumors to be aneuploid in comparison with typical carcinoids (nine of 28 typical carcinoids were aneuploid as compared with 17 of 25 atypical carcinoids). The incidence of DNA aneuploidy in tumors with metastases was significantly higher than in those without nodal involvement. In a Cox multivariate analysis, the most powerful predictor of prognosis was the histologic growth pattern. Other variables of independent significance were the presence of affected lymph nodes and nuclear pleomorphism.

Another flow cytometric study generally confirmed the above-noted findings and showed that aneuploidy was related to tumors of 3 cm or greater in diameter.[149] This study had a lower percentage of diploid tumors than the previous, possibly because of underdetection of near-diploid aneuploidy.

Flow cytometry in large-cell neuroendocrine tumors[113] showed an aneuploid population in three of four cases and a single diploid tumor.

Prognostic Indices in Bronchial Carcinoids

The most significant clinical features affecting prognosis were increasing age, tumor size above 3 cm, T stage, N stage, lymph node involvement, number of lymph nodes involved, and number of cigarettes smoked per day. The last factor should be considered not a causal factor but an expression of the increased operative risk attendant on this habit.[150]

The histologic features related to survival were necrosis, increased mitotic count, nuclear pleomorphism, lymphatic and vascular invasion, and an undifferentiated growth pattern. With all these variables there are still cases that behave like true endocrine tumors and do not follow a set pattern.

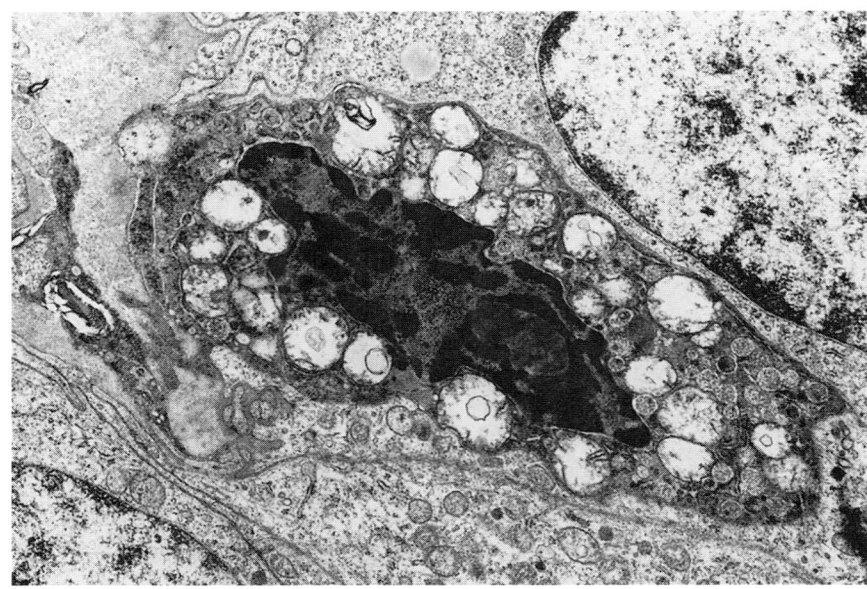

FIGURE 29-66
An apoptotic cell in a bronchopulmonary carcinoid. There is clumped nuclear chromatin, disintegrating mitochondria, and a few neurosecrtory granules. (Uranyl acetate and lead citrate, ×10,500.) (*Reproduced with permission from Hasleton and Al-Saffar, Applied Pathology 7:205–218, 1989.*)

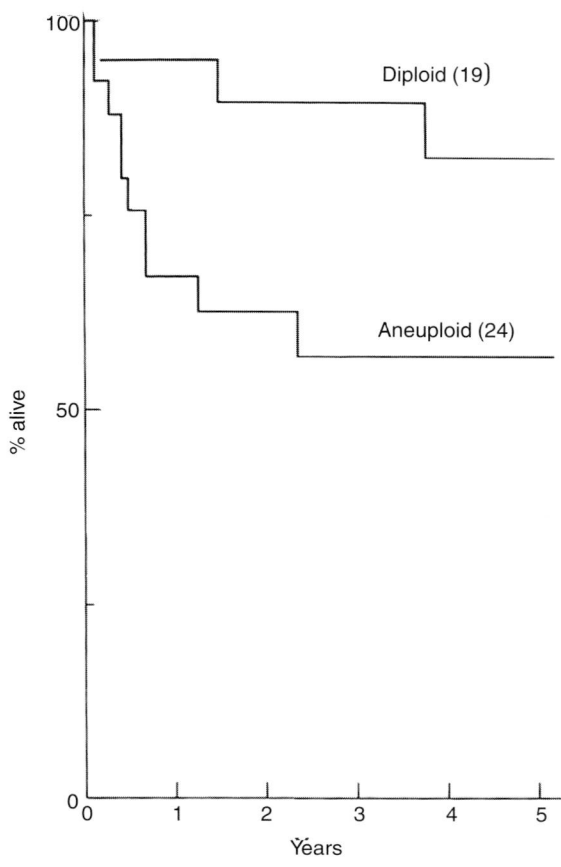

FIGURE 29-67
Survival status between diploid and aneuploid bronchopulmonary carcinoid tumors. (*Reproduced with permission of Thorax[148]*)

One such case of a typical carcinoid with regional metastases at operation was described by Bertelsen and coworkers.[94] Despite this adverse prognostic factor, the patient was alive and free from tumor 12 years later.

Large-Cell Neuroendocrine Tumors of the Lung

Large-cell neuroendocrine tumors, though a distinct entity, are often poorly recognized histologically. They are often classified as large-cell undifferentiated carcinomas.[113,147] The tumors average 3.2 cm in diameter and present with a firm circumscribed tan mass with foci of necrosis and hemorrhage; they are often peripheral. The tumor forms sheets of cells with organoid, trabecular, or palisading patterns. The cells are large and polygonal with hyperchromatic irregular pleomorphic nuclei, low nuclear: cytoplasmic ratios, coarsely granular chromatin, frequent nucleoli and mitoses, and abundant granular eosinophilic cytoplasm (Fig. 29-68). Necrosis is prominent. A mitotic count of more than 10 per 10 high-power field discriminated between large-cell neuroendocrine tumors and atypical carcinoids.

Usually the silver stain is negative (Fig. 29-69). There is variable reactivity with serotonin, bombesin, calcitonin, adrenocorticotropic hormone and leuenkephalin. Chromogranin is sometimes positive, and high- and low-weight cytokeratins can be demonstrated. Carcinoembryonic antigen was positive in all cases studied and is therefore of little help in discriminating a large-cell carcinoid from a carci-

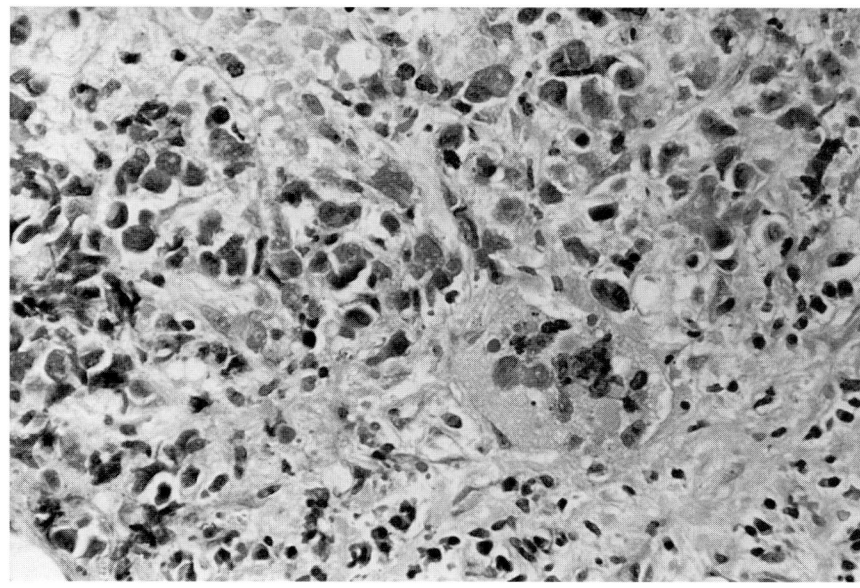

FIGURE 29-68
Large cell neuroendocrine carcinoma with a hint of an insular pattern and a tumor giant cell. (H&E, ×350.)

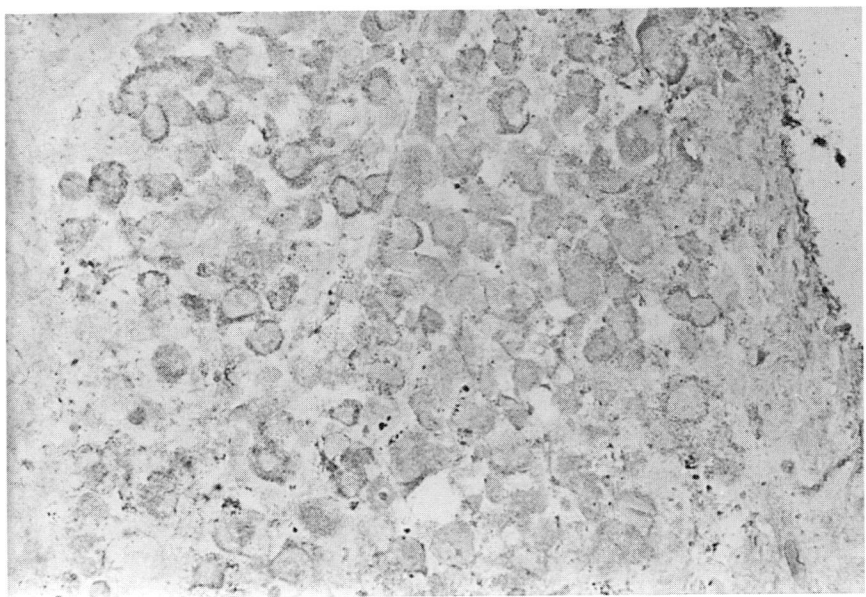

FIGURE 29-69
Another area of the case shown in Fig. 29-65. Grimelius stain is positive, an unusual finding in these tumors. (Grimelius, ×219.)

noma. The p53 mutations are identified in large-cell neuroendocrine carcinoma but not in typical or atypical carcinoid tumors.

Ultrastructurally there are usually fewer than five neurosecretory granules per cell, but more than 20 percent of the cells have granules, which vary in size from 100 to 270 nm. Glandular formation is rarely seen and tonofilaments are unusual, but desmosomes may be identified, suggesting squamous differentiation.

Prognosis is poor; all eight patients in one series[147] died of their disease. Three patients had no known metastases at the time of surgical resection but experienced tumor progression. In Travis's series,[113] two of the five patients were dead 15 and 25 months, respectively, after diagnosis. Autopsy in one case revealed metastases to the lung, kidney, and adrenals. The remaining three patients were alive at six months, one year, and eight years, respectively; all had metastases.

Much more research is needed on this subgroup of neuroendocrine tumors, since the number of cases reported is small and the response to surgery, radiotherapy, or chemotherapy is not known. The tumor must be differentiated from large-cell carcinoma of the lung without neuroendocrine features.

Differential Diagnosis

SMALL-CELL CARCINOMA

As mentioned above on a small biopsy, without differentiated carcinoid areas, accurate diagnosis of atypical carcinoid may be impossible.

OTHER SPINDLE CELL TUMORS

A spindle cell carcinoid is less likely to be confused with a spindle cell squamous carcinoma because of the lack of nuclear pleomorphism, mitoses, and keratine in the former tumor. Spindle cell carcinoids are usually positive with Grimelius or chromogranin. Another spindle cell tumor that may cause confusion is a leiomyoma, but smooth muscle actin and desmin are usually positive in this tumor. Inflammatory pseudotumor, with its storiform pattern, may cause confusion, as may a spindle cell sarcoma, which shows no endocrine features and does not have a uniform growth pattern.

PARAGANGLIOMA

This may be difficult to differentiate from a carcinoid tumor, but paragangliomas are rare in the bronchi. Carcinoid tumors are positive for cytokeratins and desmoplakins, whereas paragangliomas are negative for these two markers. Paraganglioma typically shows "Zellballen," with whorls and nests of cells separating a vascular stroma. In a typical carcinoid there is a trabecular or insular pattern. The presence of sustenacular cells in paraganglioma[151] does not help in diagnosis, since they are also identified in carcinoid tumors.[119]

HEMANGIOPERICYTOMA

This tumor should be considered when the reticulin surrounds individual tumor cells. In both tumors there is a rich vascular network. Neuroendocrine stains are negative in hemangiopericytoma, and no dense core granules are identified.

MULTIPLE MINUTE CHEMODECTOMAS

These form small grayish-white subpleural plaques measuring 1 to 3 mm in diameter. They are composed of interstitially located cell nests that expand the alveolar septa. Tumor cells are elongated or spindle-shaped, with eosinophilic cytoplasm and indistinct cell borders. They often form whorls of cells easily distinguishable from the surrounding stroma. S-100, NSE, and smooth muscle actin are negative.[152]

PAPILLARY ADENOMA OF THE LUNG

This tumor consists of cuboidal to columnar epithelial cells with an eosinophilic cytoplasm. It is thought to have a Clara cell origin. Immunocytochemistry with surfactant apoprotein or electron microscopy will distinguish this rare peripheral tumor from a papillary carcinoid.

GLOMUS TUMOR

Glomus tumors are often tracheal and resemble carcinoids histologically, but ultrastructurally there are intracytoplasmic microfilaments with dense bodies. No neurosecretory granules are seen. The reticulin pattern shows fine fibers surrounding individual tumor cells.

ACINIC CELL TUMOR

This tumor shows pale, clear cells with abundant granular cytoplasm growing as solid sheets, acinar, papillary, or microcystic foci, a few mitoses, necrosis, and foci of calcification. Electron microscopy shows dense zymogen-type granules up to 950 nm in diameter.

PRIMITIVE PERIPHERAL NEUROECTODERMAL TUMORS (pPNET)[153]

These tumors—also known as peripheral neuroepithelioma, Askin tumor, and peripheral or adult neuroblastoma—may be confused with atypical carcinoids in small biopsies. They show uniform, poorly differentiated round cells arranged in cords, nests, or clusters with occasional pseudorosettes without dendritic cytoplasmic projections. They are chest wall tumors, but they may spread to the lungs—although this is a rare or late event. Anti-human MIC2 gene product is positive in pPNETs and Ewing's sarcoma.

PARAGANGLIOMA

These tumors may be situated in the trachea[154,155] or lung. Only five cases have been documented in the

former site. They are thought to arise from tracheal paraganglial tissue. Four cases occurred in women. The usual presentation is either hemoptysis or dyspnea with a hoarse voice and difficulty in swallowing or stridor.

Eleven pulmonary paragangliomas have been described.[151,156–158] In the pulmonary tumors there is also a female predominance (nine females to two males). The average age of patients was 56 years.

In ten cases the tumor was in the right lung, with no preference for any lobe. Seven of the patients were asymptomatic, two complained of chest pain, one of mild exertional dyspnea, and one miner, with mild pneumoconiosis, had cough, dyspnea, and chest pain.

Five female patients had systemic hypertension, which is interesting in view of the catecholamine content of these tumors. In none of these cases was the epinephrine, norepinephrine, or renin content measured. In one normotensive patient with a tracheal tumor[154] there was no detectable epinephrine or dopamine on radioimmunoassay, but norepinephrine was detected at a concentration of 830 ng/mg wet weight of tumor. While extraadrenal tumors have the potential to secrete these hormones, catecholamine secretion appears uncommon.

Macroscopic Pathology

The tumor size ranged from 1 to 7 cm. Tumors were in most cases well circumscribed and lobulated, with a homogeneous grayish-pink, bulging cut surface. The malignant paraganglioma[151] completely surrounded and compressed the pulmonary artery and was separate from the right main bronchus. Metastatic tumor deposits were present in the subcarinal, mediastinal, and paratracheal nodes.

Histology

The histologic pattern most often seen is of a highly vascular, uniform proliferation of round or polygonal cells ("Zellballen") (Figs. 29-70 and 29-71). In view of the rarity of this tumor, some of the descriptions given below relate to extrapulmonary paragangliomas. Unlike the situation in normal paraganglia, the neoplastic cells are larger, irregular, and more atypical than benign chief cells.[159] The cytoplasm may be amphophilic, eosinophilic, or even clear. The cell nuclei are central, with chromatin varying from finely stippled to hyperchromatic (Fig. 29-72). Multinucleation and nuclear hyperchromatism may be seen in ischemic areas. Other growth patterns are trabecular and pseudoglandular, with pseudorosettes

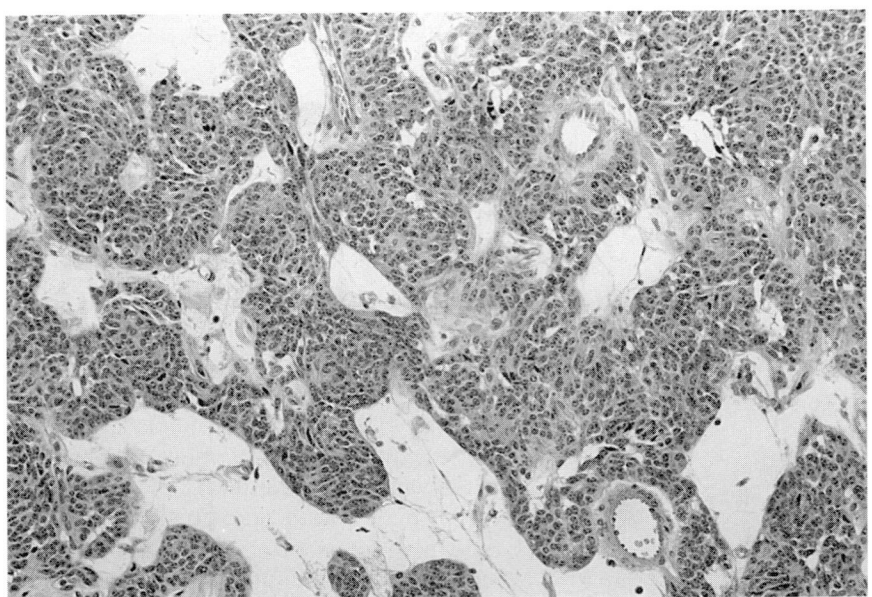

FIGURE 29-70
Paraganglioma of lung with well-formed "Zellballen" and loose connective tissue. (H&E, ×87.5.)

resembling carcinoids, neuroendocrine carcinomas, and even neuroblastic tumors. Some tumors may show spindle cell differentiation, a feature termed *sarcomatoid change.*

Most pulmonary tumors show a typical picture with "Zellballen" and sustentacular or type II cells. There is controversy about whether such cells are present in the neoplastic state.[160] However, most studies favor their existence, albeit in reduced numbers.[161] The presence of sustenacular cells would appear to be a good prognostic factor. Most pulmonary paragangliomas are closely associated with the pulmonary artery. Some authors, such as Singh and colleagues,[156] used this as a criterion for diagnosis.

Electron Microscopy

Since the advent of immunocytochemistry, this has been used less for diagnosis. The tumors lack tonofilaments, and desmosome-like junctions are common. The junctions are probably modified tight junctions, rather than true desmosomes, since recent immunohistochemical studies have revealed an absence of both cytokeratins and desmosome-specific proteins, such as the desmoplakins.[162] Intracytoplasmic glyco-

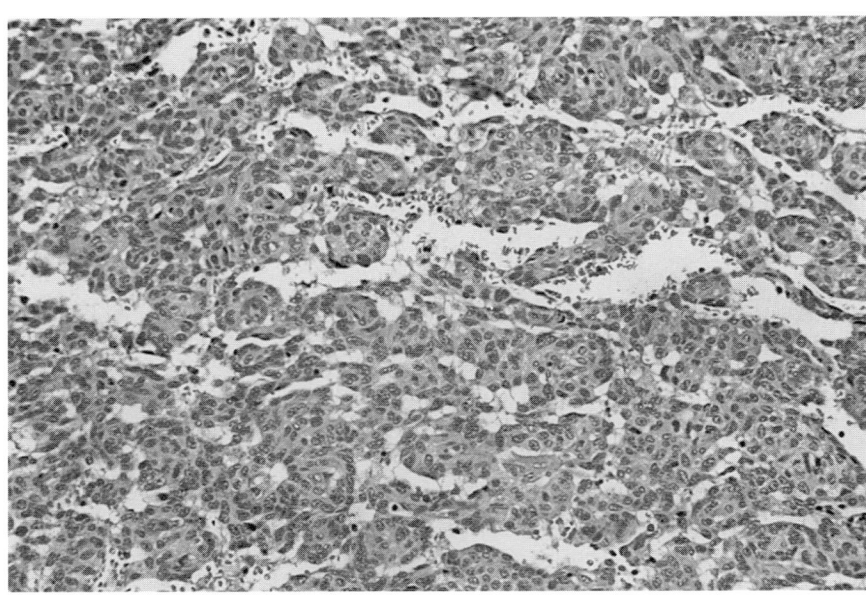

FIGURE 29-71
Paraganglioma of lung. A different area of the case shown in Fig. 29-70. There is a superficial resemblance to a carcinoid with a trabecular pattern. There is a tendency to a trabecular growth pattern. (H&E, ×87.5.)

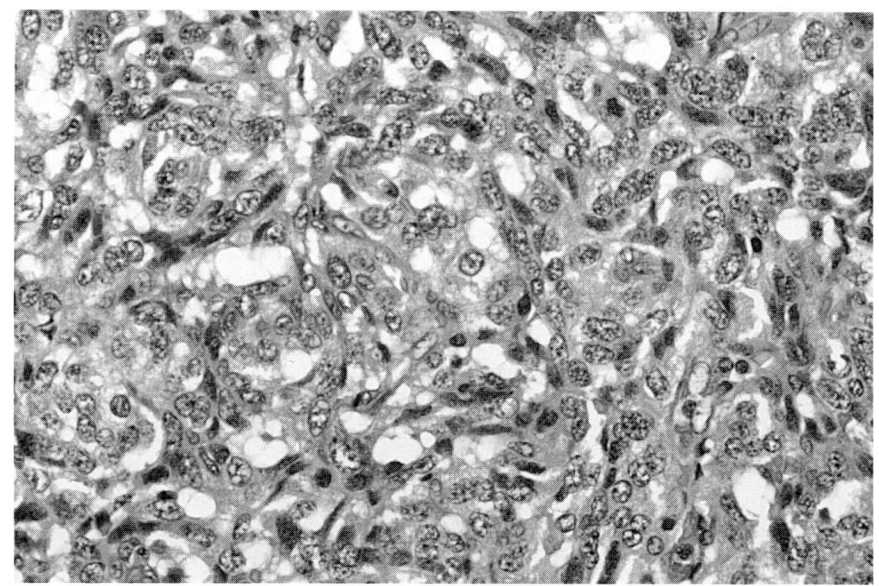

FIGURE 29-72
High magnification of a paraganglioma. The nuclear detail is shown. The chromatin varies from finely stippled to hyperchromatic. The whorled nature of the tumor is still apparent. (H&E, ×219.)

gen, lamellar bodies, and nuclear inclusions have been identified.[159] A constant feature is the presence of neurosecretory granules, which are more pleomorphic than in normal paraganglia, ranging from 40 to 400 nm.

Immunocytochemistry

The staining properties can be divided into markers staining chief cells and those that identify sustentacular cells. The markers for chief cells are predominantly those that usually identify carcinoid tumors, such as chromogranin, neurofilaments, and protein gene product 9.5 (PGP 9.5).

Sustentacular cells are identified by S-100 pro-

tein (Fig. 29-73), glial fibrillary acidic protein (GFAP), and nerve growth factor receptor. S-100 protein has also been identified in carcinoid tumors, ganglioneuromas, ganglioneuroblastomas, and neuroblastomas. Abundant S-100 protein–containing sustentacular cells are found in the more differentiated parts of the tumor and are associated with a good prognosis.[163] Benign paragangliomas usually contain sustentacular cells, and the malignant ones have rare or no type II cells.[164]

GFAP immunoreactivity parallels S-100 staining. Nerve growth factor receptor has yet to be established in the diagnosis of these tumors. Its immunoreactivity does not necessarily follow that of S-100 and GFAP.[159]

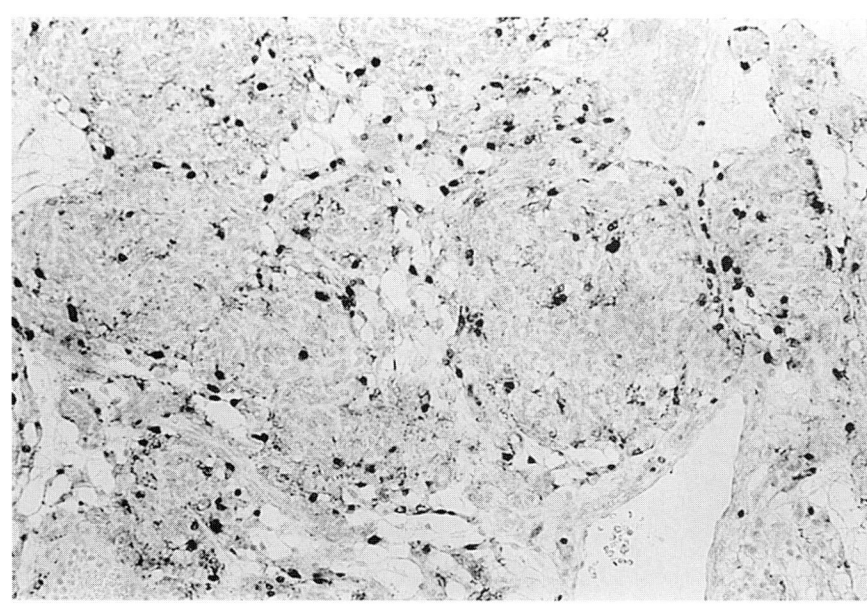

FIGURE 29-73
Paraganglioma with S100 positivity. (Immunoperoxidase, ×87.5.)

Multiple Paragangliomas Associated with Neurofibromatosis

This rare association has been described in a woman presenting with systemic hypertension at the age of 49 years. She had a pheochromocytoma, a glomus jugulare tumor, and multiple pulmonary paragangliomas. The pulmonary lesions were 1 cm subpleural nodules composed of endocrine-like tumor cells with occasional mitoses. There was no relationship to bronchial structures or pulmonary veins. These tumors were considered to be separate primary pulmonary paragangliomas rather than metastases from the pheochromocytoma or glomus jugulare tumor, since these primary tumors rarely spread to the lung.[165] The possibility that the pulmonary tumors were carcinoids could not be totally discounted, since desmoplakin, S-100 protein, or GFAP staining was not done. Figure 2 in the original paper of the lung nodule closely resembles a paraganglioma.

Differential Diagnosis

CARCINOID TUMOR

Carcinoid tumors are positive for cytokeratins and desmoplakins, whereas paragangliomas are negative for these two markers.

SMALL-CELL CARCINOMA

This tumor usually has a different growth pattern, with nuclear molding and a "salt and pepper" stippling in the nuclear chromatin. Difficulty may arise with the large-cell variant of neuroendocrine carcinoma. Small-cell carcinomas show cytokeratin immunoreactivity, a feature not seen in paragangliomas.[166]

MELANOMA

This is very rarely primary in the lung. A malignant melanoma would stain positively for S-100 protein. HMB-45 can be used as an additional marker (Figs. 29-74, 29-75, and 29-76). Malignant melanoma as a primary lung tumor has given rise to considerable debate about its existence. In any case labeled a *primary malignant melanoma of the lung*, it is very difficult to prove that the primary extrapulmonary tumor may not have undergone regression. The criteria suggested for diagnosis of a primary malignant melanoma[167–169] are junctional change with malignant melanoma beneath the bronchial epithelium, invasion of the epithelium by melanoma with no associated ulceration, a solitary pulmonary malignant melanoma, no demonstrable tumor elsewhere in the body, and a nevus-like lesion in the bronchial wall adjacent to the tumor. An accurate history or complete postmortem is essential in making this diagnosis. As stated above, however, this will not exclude a cutaneous melanoma that has undergone regression.

METASTATIC RENAL CELL CARCINOMA

This may cause diagnostic difficulty. If some tissue remains, a frozen section and a fat stain would be

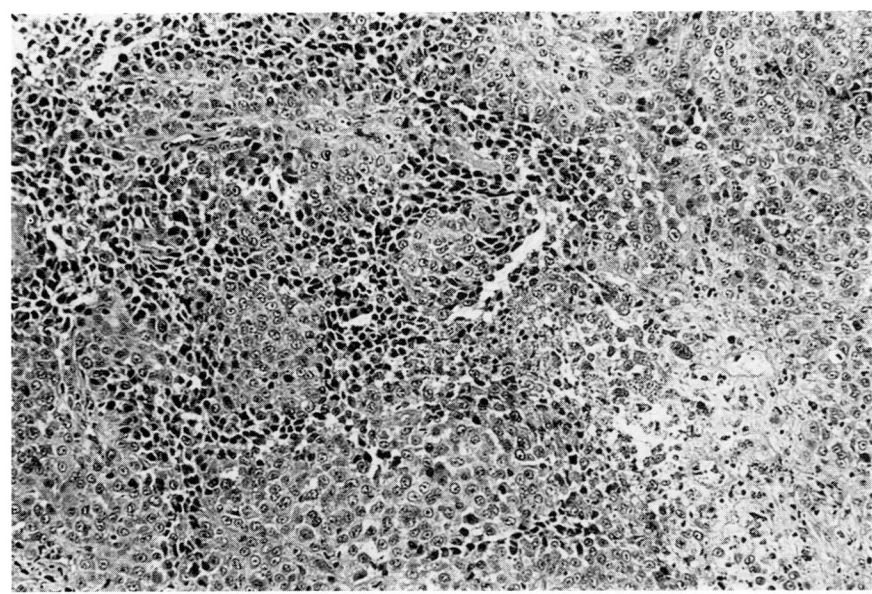

FIGURE 29-74
Malignant melanoma, found in a small bronchiole, after an episode of hemoptysis, in a 19-year-old boy. There was no other obvious primary site. The tumor had a very superficial resemblance to a poorly differentiated squamous cell carcinoma. Necrotic foci are present. (H&E, ×87.5.)

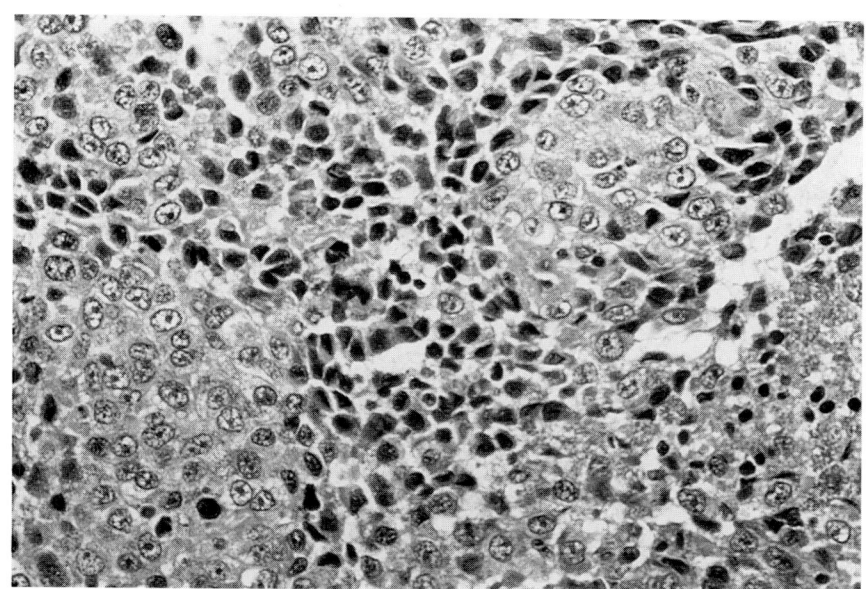

FIGURE 29-75
Malignant melanoma in the lung with large, open vesicular nuclei and prominent nucleoli. (H&E, ×219.)

positive in a renal tumor, whereas paragangliomas do not contain lipid. Hypernephromas contain cytokeratin, which is not present in paragangliomas.

ALVEOLAR SOFT-PART SARCOMA

This tumor may metastasize to the lung and is sometimes found in the chest wall. It may grow in an organoid pattern with alveolar areas and produce "Zellballen," causing difficulty in diagnosis. The pseudoalveolar pattern, if present, may help diagnosis. Alveolar soft-part sarcoma contains intracytoplasmic crystalline inclusions, sometimes seen in mitochondria, with a 20-nm repeating frequency.[167]

Usually patients with alveolar soft-part sarcoma are young, in the second or third decade.

Diastase or PAS-positive cytoplasmic crystalline material is the most helpful aid in the diagnosis of alveolar soft-part sarcoma. Immunohistochemistry may show variably positive desmin, S-100 and actin with a negative neuron-specific enolase (NSE). Neurosecretory granules have been identified in alveolar soft-part sarcoma.

MINUTE CHEMODECTOMA OF THE LUNG[171]

This entity was first reported by Korn and colleagues,[172] who demonstrated multiple peripheral

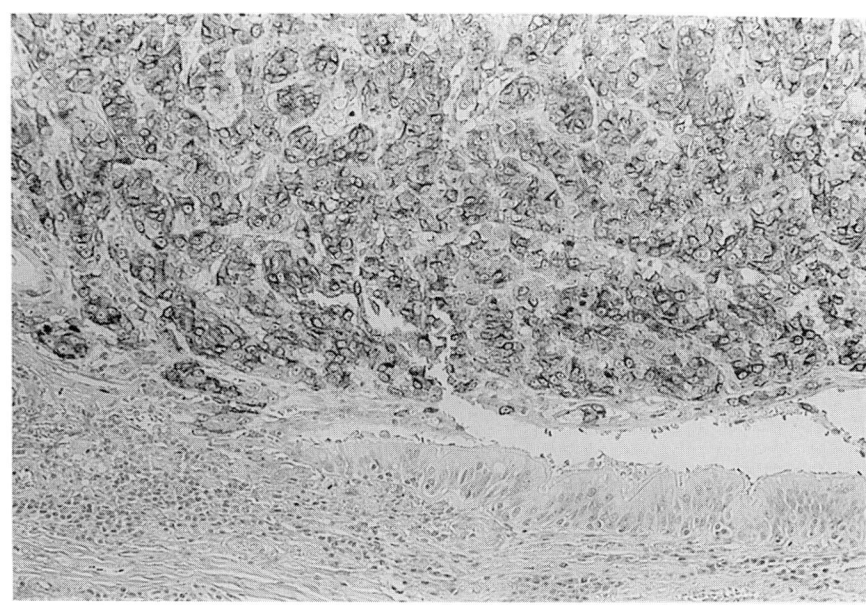

FIGURE 29-76
Same case as in Figs. 29-74 and 29-75. HMB 45 positive cells lie in the bronchial lumen and are partly replacing the bronchial wall. (Immunoperoxidase, ×219.)

grayish-pink nodules where the septa joined the pleura. These lesions have since been documented by other authors[173–175] and have been briefly described in this volume by Heath (Chap. 21). Torikata and Mukai[171] described seven cases, all incidental findings.

The incidence of such nodules is difficult to define. Some authors[176] have suggested that they may be found in one out of 40 lung specimens derived from necropsy or surgical histopathology. The postmortem incidence of multiple small chemodectomas has been estimated at 0.3 percent.[172] Similar frequencies of 0.3 percent[177] and 0.5 percent[178] have been recorded by other authors. When larger volumes of tissue were examined, the incidence rose to 3 percent.[178]

The lesions show a female predominance. Most cases are seen in patients' fifth and sixth decades, but this may be a reflection of the postmortem population. These nodules have recently been related to diseases such as those associated with chronic hypoxia, pulmonary edema, including mitral stenosis, plexogenic pulmonary arteriopathy, pulmonary thromboembolism, and chronic obstructive airways disease.

It has been postulated that the function of this tissue is similar to the arachnoid granulations in the brain, in that they transfer excess fluid from the alveolar spaces into the pulmonary veins. It is possible that they contribute to the avoidance of excessive hydration of the interstitial tissue of the alveolar wall, thus preventing pulmonary edema.[180]

Macroscopically, they are usually 2 mm in diameter and are often an incidental finding on light microscopy. Occasionally the tumor is seen in the interlobular connective tissue, especially where the lobules join the pleura. They may be confined to the adventitia of the venule.

The nodules consist of circular or oval interstitial nests (Fig. 29-77) of cells separated from airspaces by a rim of capillaries or fibrous connective tissue. The nodules do not extend into the alveolar capillaries.[172,173,179] The alveolar surface is covered with cuboidal alveolar lining cells.

The cell nests consist of large bland cells with ill-defined borders and an eosinophilic cytoplasm. The cells at the periphery have a characteristic whorled appearance. The nuclei may be slightly indented and round, oval, or elongated. The striking resemblance of the cells to arachnoid villi protruding into the dural sinuses overlying the brain has been noted.[180]

Immunohistochemically, they stain for vimentin and myosin. Cytokeratin, S-100 protein, NSE, and actin are negative.

Ultrastructurally, these lesions consist of a single type of cell with an irregular cytoplasmic membrane process showing a jigsaw puzzle–like arrangement

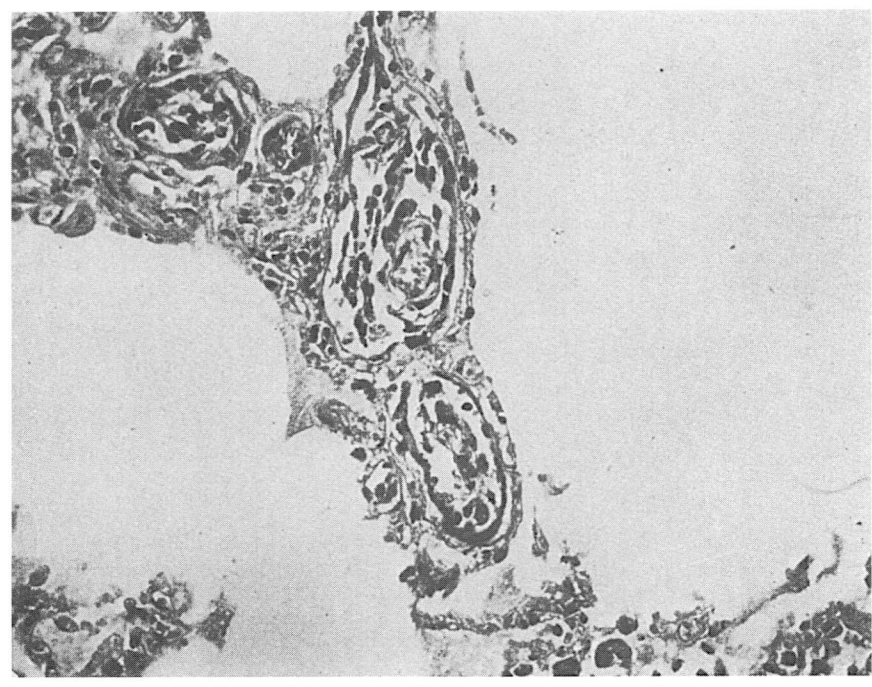

FIGURE 29-77
Multiple small chemodectomata lying in the interstitium.

lined by numerous well-formed desmosomes and desmosomal-like junctions. In some areas, floculant material is noted between opposing cell membranes, resembling basal lamina. Most of these cells contain abundant filaments 6 nm in diameter oriented in the long axis of the cell. These filaments give the cells an occasional dense, patchlike appearance, but typical actin-myosin complexes are not detected. The appearances are similar to those of meningeal arachnoid granulations.

Differential Diagnosis

MENINGIOMA OF THE LUNG

This is usually a larger tumor although confusion may occur with smaller single lesions.

HEMANGIOPERICYTOMA

The critical difference between the two cell types is the presence or absence of a basal lamina. Normal pericytes can be identified by electron microscopy. However, it is difficult to differentiate pericytes from meningeal cells by means of immunohistochemistry.

TUMORLETS

These lesions are usually found in or adjacent to scar tissue and will stain positively with neuroendocrine markers.

PULMONARY MENINGIOMA

Pulmonary meningiomas are extremely rare. Approximately 10 case reports can be found in the world literature.[181-188] The tumors are more common in women than men. The age range of patients is from 41 to 74 years, with an average of 53.1 years.

All but one of the patients presented with an asymptomatic radiographic nodule. The patient with symptoms complained of weakness, anorexia, and weight loss, but these did not appear to be related to the pulmonary lesion. All patients were treated surgically and followed for 1 to 7 years, with good results. One patient had a coexistent endometrial carcinoma and a melanoma of the left eye. Another case, followed for three years,[181] showed a slow increase in size. One patient died from pulmonary embolism, without operation.

Macroscopic Features

The tumors are well circumscribed with a granular white or yellowish-gray cut surface. They are typically located in the pulmonary parenchyma and may

be subpleural. They range in size from 1.8 by 1 to 4 cm in diameter. There is usually no bronchial association.

Histology

The tumor is composed of ill-defined bundles of spindle cells, often arranged in whorls (Figs. 29-78 and 29-79). The nuclei are enlarged, and there is often a small nucleolus. The cytoplasm is eosinophilic with poorly defined borders. Some thick strands of collagen, blood vessels, and a moderate number of psammoma bodies are present. The tumor is in contact with the adjacent lung tissue but has no capsule or bronchiolar origin. Mitoses are not usually found.

The center of the whorls is intensely stained with a van Gieson stain because of hyaline connective tissue. In a few areas the margin of the tumor is not well circumscribed. There may be looser areas, in which tumor cells appear to be spinning off around blood vessels.

Immunohistochemical stains show that the tumor is negative for high- and low-molecular-weight keratins, desmin, and chromogranin. Epithelial membrane antigen is weakly positive in the cytoplasm, and S-100 protein is focally positive. There is moderate staining with vimentin. Bombesin, chromogranin, NSE, and actin are all negative.

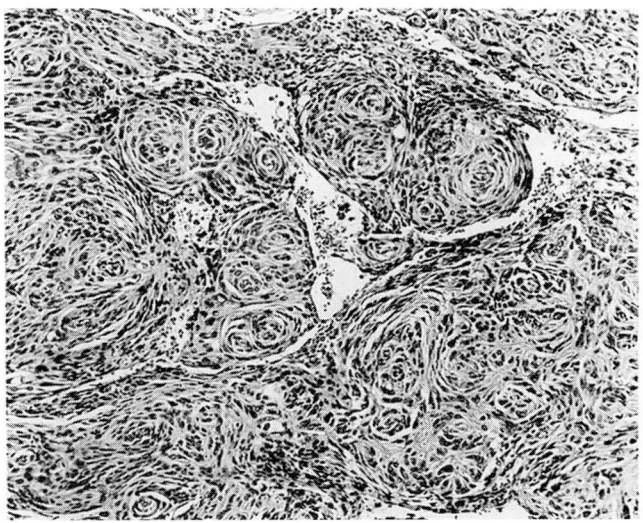

FIGURE 29-78
Meningioma of lung with characteristic whorls of cells. (H&E, ×77.) (*Reproduced with permission of W. D. Travis, M.D., Armed Forces Institute of Pathology, Washington, D.C.*)

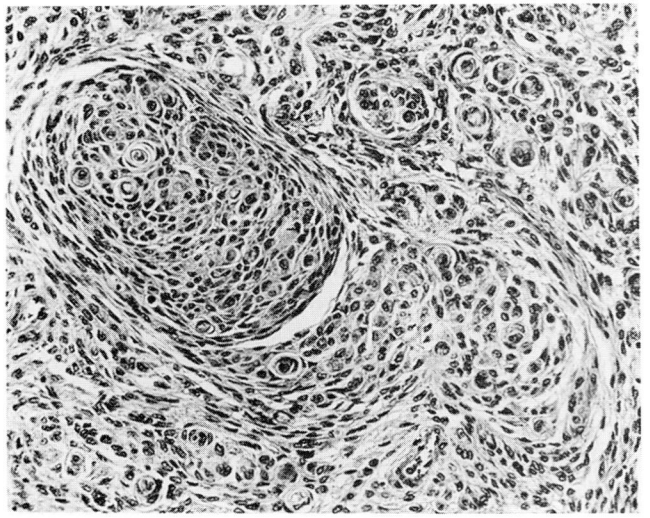

FIGURE 29-79
Meningioma, clumps of cells with ill-defined cell borders. No psammoma bodies are identified. (H&E, ×154.) (*Reproduced with permission of W. D. Travis, M.D. Armed Forces Institute of Pathology, Washington, D.C.*)

Electron Microscopy

Electron microscopy shows complex interdigitations of cell membranes and numerous desmosomes, although some cases have few or they are poorly formed. Basal lamina are not identified. Moderate numbers of mitochondria and some rough endoplasmic reticulum are seen. Rare cytoplasmic filaments without dense bodies have been identified.

Differential Diagnosis

SECONDARY MENINGIOMA

Before a pulmonary meningioma is diagnosed, extension from a primary intracranial meningioma or a metastasis must be ruled out. Secondary meningiomas may have a long latent period; one such case has been described by Kodama and coworkers.[183]

HEMANGIOPERICYTOMA

Pulmonary meningiomas may have a hemangiopericytomatous component. In meningioma, however, there are usually psammoma bodies, and the ultrastructure of the two tumors is different.

NEURILEMMOMA

Areas of a meningioma may show pallisading, but the overall pattern is not that of a neural tumor and S-100 protein is uniformly negative, unlike that in a nerve sheath tumor.

Paraganglioma

This tumor stains positively with neuroendocrine stains. If these are negative, electron microscopy may show dense core neurosecretory granules, not identified in meningiomas.

Histogenesis

The cell of origin is likely to be the ectopic arachnoid cell described above. The meningothelial-like nodules are multiple, whereas most meningiomas are solitary.

GLOMUS TUMORS

There have been few reports of tracheal glomus tumors.[189–193] They are rarer in the lung.[194,195] The tracheal lesion presents mainly in middle-aged people, often with late-onset "asthma." The pulmonary lesions may present as solitary nodules, although one case was in a 19-year-old who had had a resected infiltrating gluteal tumor mass,[196] and the lesion was considered a metastasis.

Typical glomus tumors arise from the glomus apparatus or Sucquet-Hoyer canal, which is an arteriovenous anastomosis associated with blood flow and temperature control.[197] The tumor is thought by some to be a hyperplasia or hamartomatous process rather than a neoplasm. The tumors are usually benign.

Macroscopic Features

Tracheal tumors are usually polypoid and grayish pink and occupy much of the lumen. There is a lobulated surface and a broad base, often in the posterior wall. The pulmonary lesions are well circumscribed peripheral lesions, either yellowish tan or gray, with a firm cut surface. Large blood vessels are identified in the tumor. The largest pulmonary tumor measured 6.5 by 5 by 5 cm and was ill defined. Tracheal tumors are usually 2 cm or less in diameter.

Histology

There are irregular masses of medium-sized, uniform polyhedral cells with central round nuclei surrounded by a small amount of moderately eosinophilic cytoplasm (Figs. 29-80 and 29-81). Oncocytic change may be seen.[198] In the trachea, tumor cells are present in the subepithelial connective tissue, and the tracheal muscle coat extends to the edge of the specimen. Some of these cells have a

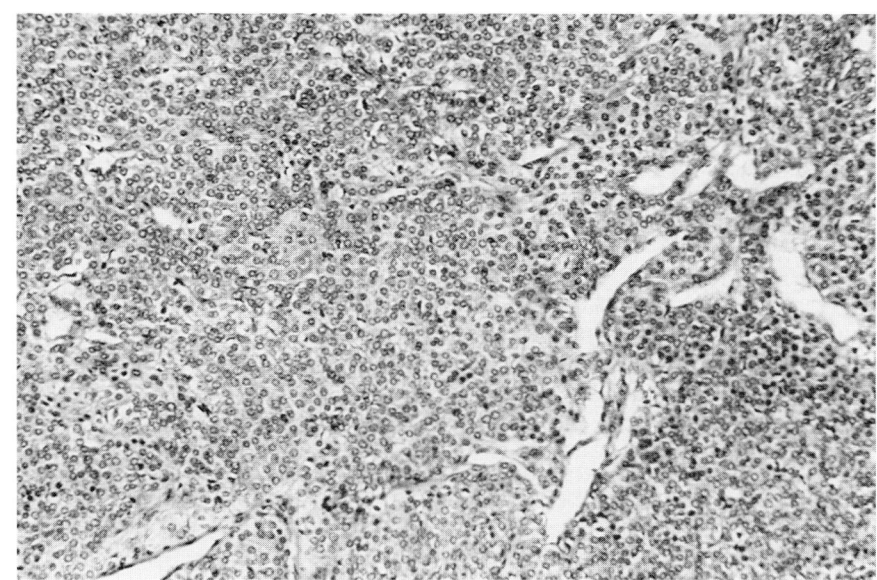

FIGURE 29-80
Glomus tumor composed of small cells and prominent blood vessels. (H&E, ×56.)

clear cytoplasm, and it is often difficult to appreciate the cell membrane. In some cases, the cytoplasm has a slightly granular tinge. The covering respiratory epithelium shows squamous metaplasia. The vascular structures within the tumor vary from small channels to huge, dilated, angulated, interlacing capillary spaces. There is no necrosis. Fine reticulin fibers encircle individual tumor cells or small clusters of cells.

Most blood vessels are surrounded by only a thin reticulin network (Fig. 29-82), but some have a small amount of collagen and a few smooth muscle fibers. Small vessels with somewhat thicker walls are occa-

sionally found. In central areas of the tumor there is a more hyalinized stroma with embedded tumor cells.

Immunocytochemistry is negative with neuroendocrine markers but positive for vimentin.

Ultrastructure

Ultrastructurally, there are no neurosecretory granules. Intracytoplasmic microfilaments with dense bodies have been identified (Fig. 29-83). Along the plasma membranes are rows of small regular pinocytotic vesicles and some small flat, dense plaques parallel to the membrane. A limited amount of basement

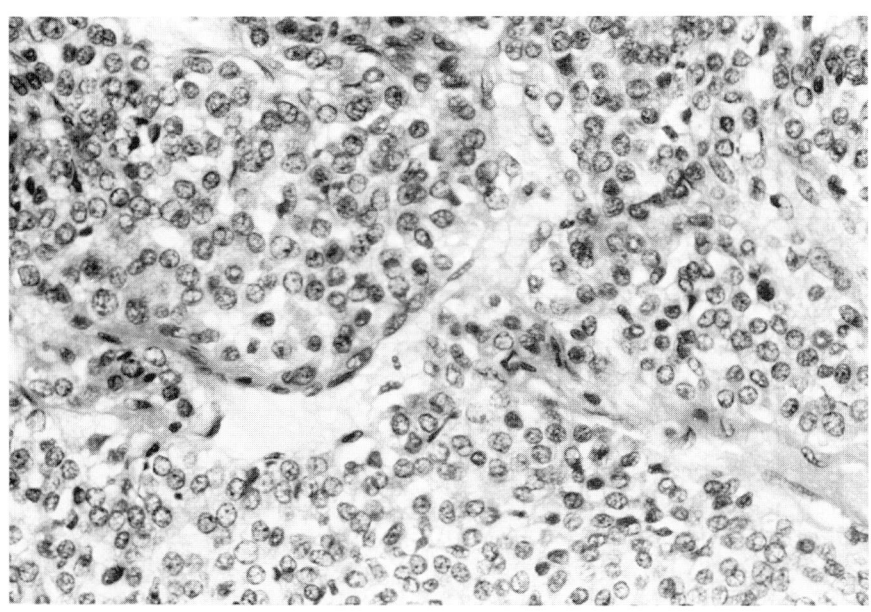

FIGURE 29-81
Glomus tumor with small regular cells and little nuclear pleomorphism. There is no evidence of an insular or trabecular pattern to suggest a carcinoid tumor. (H&E, ×219.)

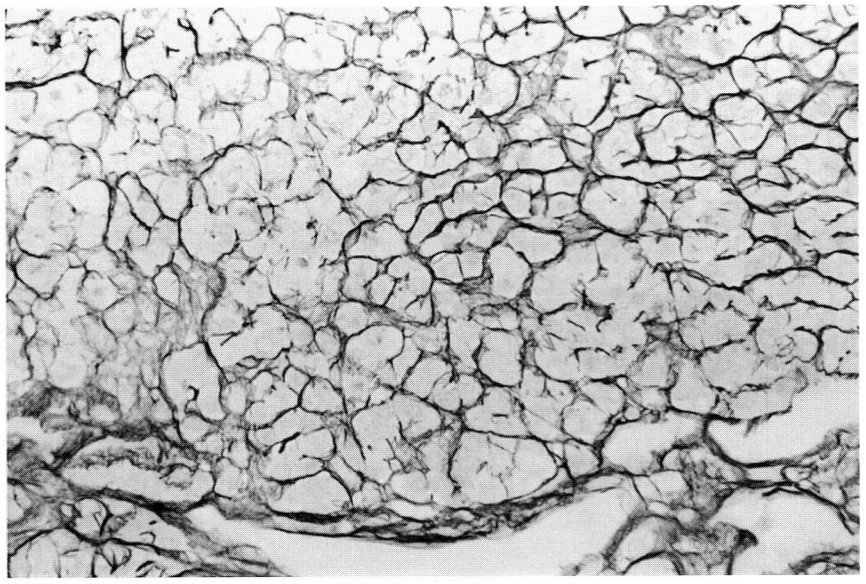

FIGURE 29-82
Glomus tumor with reticulin fibers surrounding individual cells. (Reticulin, ×219.)

membrane substance is present between the tumor cells in places. These features occur together only in smooth muscle and glomus cell tumors.

Differential Diagnosis

CARCINOID TUMOR

Glomus tumors may present as typical carcinoid tumors, and therefore one would expect to find positive staining with neurofilament 70, transthyretin (prealbumin), or any of the neuroendocrine markers. If material is available, electron microscopy shows intracytoplasmic microfilaments with dense bodies lying along the cell membrane in glomus tumors.

HEMANGIOPERICYTOMA

This tumor is similar histologically to a glomus tumor. There is, however, more variation in the cell size, and there may be spindle cell foci. Hemangiopericytoma classically has a "staghorn" pattern of the vascular channels on reticulin staining, whereas in a glomus tumor the vascular spaces vary considerably in size and have no characteristic pat-

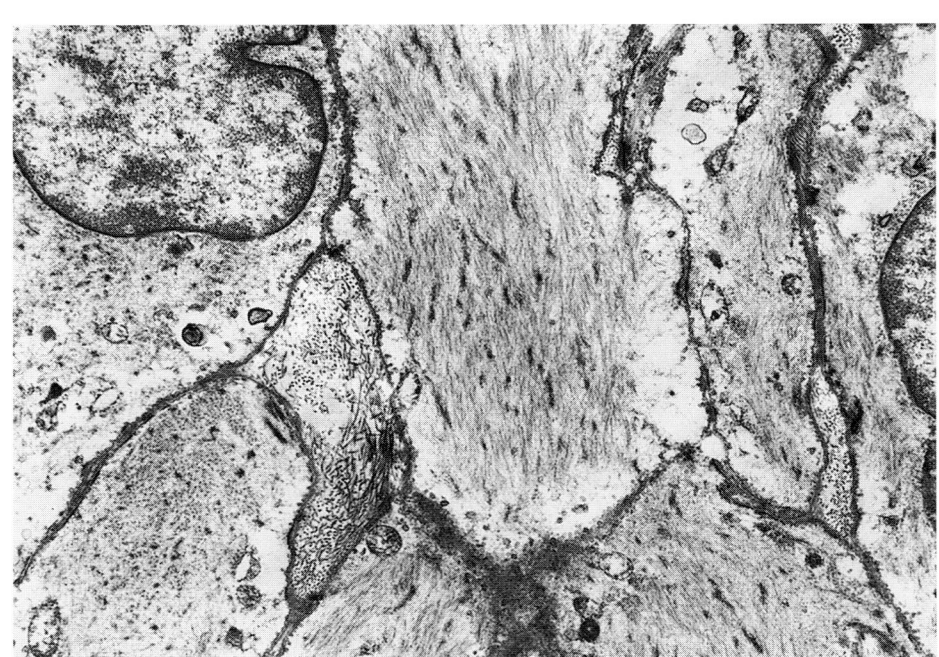

FIGURE 29-83
Glomus tumor of the trachea with intracytoplasmic fibrils, amidst which are dense bodies. Dense plaques are seen at the junctions of the cells. (Electron micrograph, original magnification, ×8540.) (*Reproduced with permission of Professor B. Heard and Thorax.*)

tern. In any individual case, relying solely on the reticulin pattern can be misleading. Ultrastructural studies may be useful in distinguishing the two tumors. A comparative ultrastructural study of hemangiopericytoma and a glomus tumor[199] showed that hemangiopericytoma had light osmiophilic material resembling basement membrane in the intercellular spaces. No intracytoplasmic microfilaments with dense plaques were seen in hemangiopericytoma.

PARAGANGLIOMA

It is unlikely that this tumor would be confused with a glomus tumor, since the cell size of a paraganglioma is larger and there are "Zellballen." In addition, the neuroendocrine stains are positive and the sustentacular cells would stain positively with S-100 and GFAP.

LEIOMYOBLASTOMA AND EPITHELIOID LEIOMYOSARCOMA

These tumors show more nuclear pleomorphism and mitoses. Ultrastructural study may help in diagnosis.

MUCOEPIDERMOID TUMOR

This is a rare tumor; only about 130 cases have been described in the world literature. It accounts for 0.2 percent of all lung tumors, and 1 to 5 percent of tumors arise from either the bronchi or trachea.[200] In the Mayo Clinic, a specialist referral center, 15 percent of a series of 80 patients with rare lung tumors had mucoepidermoid carcinoma.[201] In one series, four of 18 cases had tracheal tumors.

There appear to be two distinct groups. The first, "low-grade" tumors, occur in childhood and are usually benign. In one large series of pediatric cases, with a literature review, only one case had a peribronchial lymph node metastasis.[202] "High-grade" tumors are less common, are more often found in adults, and carry a poor prognosis. The grading is by histology, on the basis of necrosis, mitotic activity, and nuclear pleomorphism.

Clinical Features

The age range is wide, from 3 months to 78 years (the papillary variant has been excluded in these figures), but the tumor is uncommon in patients over the age of 60 years.[203] It has an equal sex incidence. A history of smoking, when taken, is present in less than half of the patients. The clinical symptoms may commonly be due to tracheal involvement, presenting as bronchial asthma or bronchial irritation. There is usually cough, hemoptysis, and wheezing or, in some cases, fever due to a postobstructive pneumonitis. Some patients may have constitutional symptoms, including weight loss and chest pain, indicative of a more aggressive tumor. There are asymptomatic patients whose disease is diagnosed only on a chest radiograph.

Radiology shows a tumor nodule, often associated with an obstructive pneumonitis. Occasionally patients have normal chest radiographs. In rare cases, these tumors may be peripheral.[204]

Macroscopic Findings

Appearances vary according to the grade of the tumor. "Low-grade" tumors measure between 0.8 and 6 cm in diameter, with a faint yellow, white, or tan-white appearance (Fig. 29-84). They are usually exophytic, with both solid and cystic foci, and occlude the bronchial lumen to a variable degree. There is mucus in the cystic areas. In most cases the tumors are confined to the tracheal or bronchial mucosa or to the internal perichondrium.

"High-grade" tumors usually have an intrabronchial origin, but there is extensive local invasion. The macroscopic picture is similar in the "low-grade" tumor, but there is infiltration into the lung parenchyma in some cases. The size ranges from 1.5 to 4 cm. The trachea is rarely affected by this grade of mucoepidermoid carcinoma.

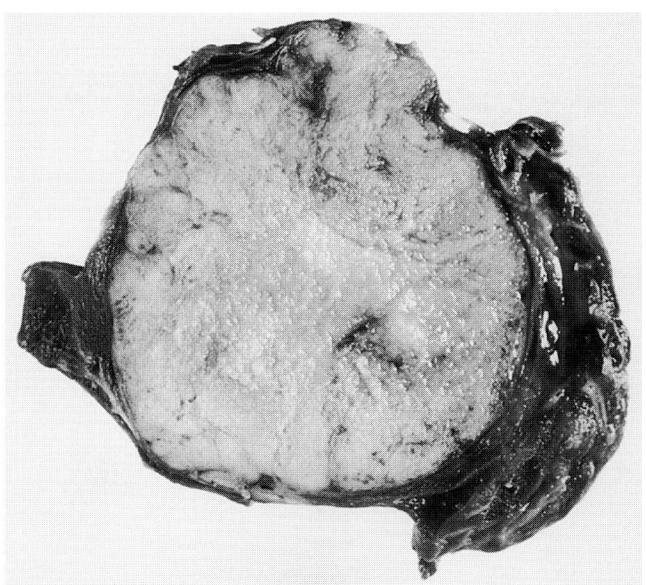

FIGURE 29-84
Mucoepidermoid carcinoma. This was a primary intrabronchial tumor.

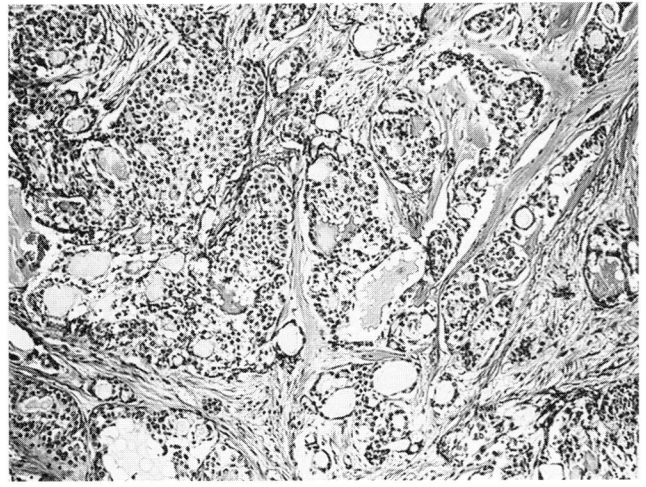

FIGURE 29-85
Low-grade mucoepidermoid carcinoma, showing islands of squamous and mucous cells. (H&E, ×71.)

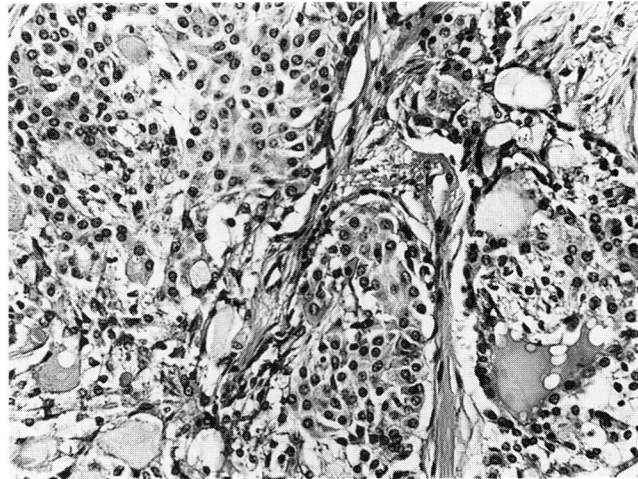

FIGURE 29-86
Higher power of Fig. 29-85 with both the squamous and glandular foci well delineated. (H&E, ×177.5.) (*Reproduced with permission of Virchows Archives of Pathological Anatomy.*)

Histology

The histology of a "low-grade" mucoepidermoid tumor usually poses little problem. There is ulceration and squamous metaplasia of the overlying respiratory epithelium. There is a mixture of cell types, with squamoid, intermediate (transitional), basiloid, clear, columnar (non–mucus-secreting), and well-formed mucous cells (Figs. 29-85, 29-86, and 29-87). Transitional cells are polygonal, with an eosinophilic cytoplasm and faint glycogen positivity. Solid foci are formed by squamoid and transitional cells. Cystic areas are lined by goblet or columnar cells. In a few cases, pseudopapillae may be formed when the overlying epithelium is thrown into folds by the tumor.

True papillae lined by goblet cells and with fibrovascular cores may be seen. Intercellular bridges may be present, but there is no keratinization in squamoid areas. The mucus-containing glands are of various sizes and shapes and are widely scattered in the tumor. Some tumors may contain oxyphil cells. Less than one mitotic figure per 20 high-power fields is seen.

Though the tumor is macroscopically well circumscribed, there is often invasion of the subepithelial tissues, but only rarely is there extension into the parenchyma. There is desmoplasia or, less commonly, elastosis in response to the tumor.

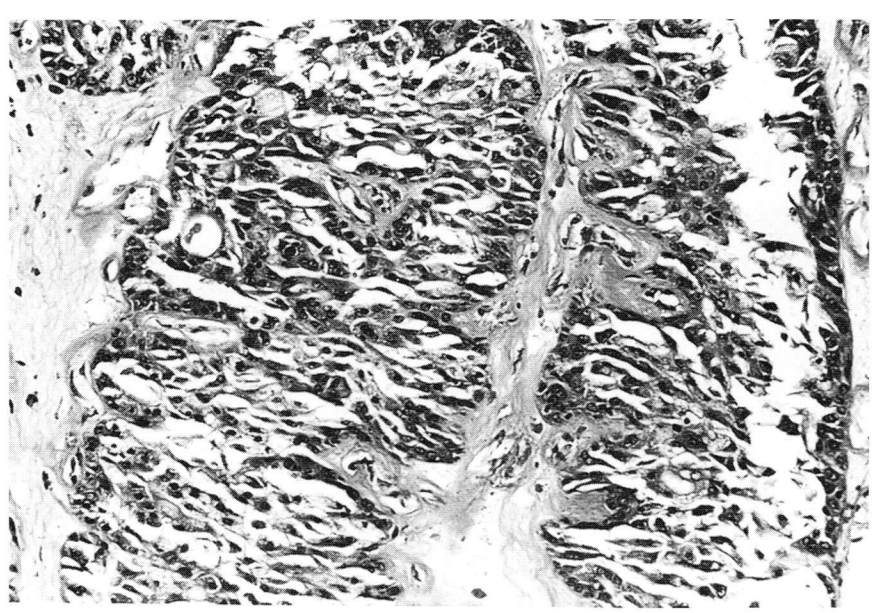

FIGURE 29-87
Mucoepidermoid carcinoma showing intermediate cells. (H&E, ×87.5.)

The mucin stains positively with mucicarmine, alcian blue, and PAS with and without diastase digestion (Fig. 29-88). No neutral mucin or glycogen is seen in the clear cells.

The features differentiating "low-grade" from "high-grade" mucoepidermoid carcinoma have been given by several authors[200,203,205,206] and are summarized in Table 29-3.

Lymph node metastases are rare in "low-grade" mucoepidermoid carcinoma.[202,203]

"High-grade" tumors show the same cellular constituents as noted above (Fig. 29-89). Abnormal giant cells may be identified. There is moderate to marked nuclear atypia (Fig. 29-90), with an average of four mitoses per 10 high-power fields. Necrosis may be seen. If there are cystic areas, they are lined by malignant columnar and goblet cells and contain mucin. The glands are distended by slightly eosinophilic material, giving some tumors a superficial resemblance to follicular carcinoma of the thyroid. All the mucin stains are positive in such foci. Most areas resemble a nonkeratinizing squamous cell carcinoma. Characteristically, there is abrupt transition from one cell type to another. In one series there were only two cases with hilar lymph node metastases.[203]

The cytology of a "low-grade" tumor based on fine needle aspiration has been documented.[207] The background of the smear was composed of large streaks of mucin, which some authors considered to be typical, intermingled with neutrophil polymorphs and a moderate number of alveolar macrophages. There were groups of cells with small papillary projections, some with a central fibrovascular core. The cells were polygonal with round, ovoid, or central nuclei and a finely dispersed nuclear chromatin. On the surface of these papillary groups was a single layer of mucinous cells with foamy or clear cytoplasm. Other, smaller groups were composed entirely of mucin-producing cells. The cytoplasm of the squamous cells was hyaline and eosinophilic, with no keratinization.

In the diagnosis of any mucoepidermoid carcinoma, a diastase with or without PAS is essential in

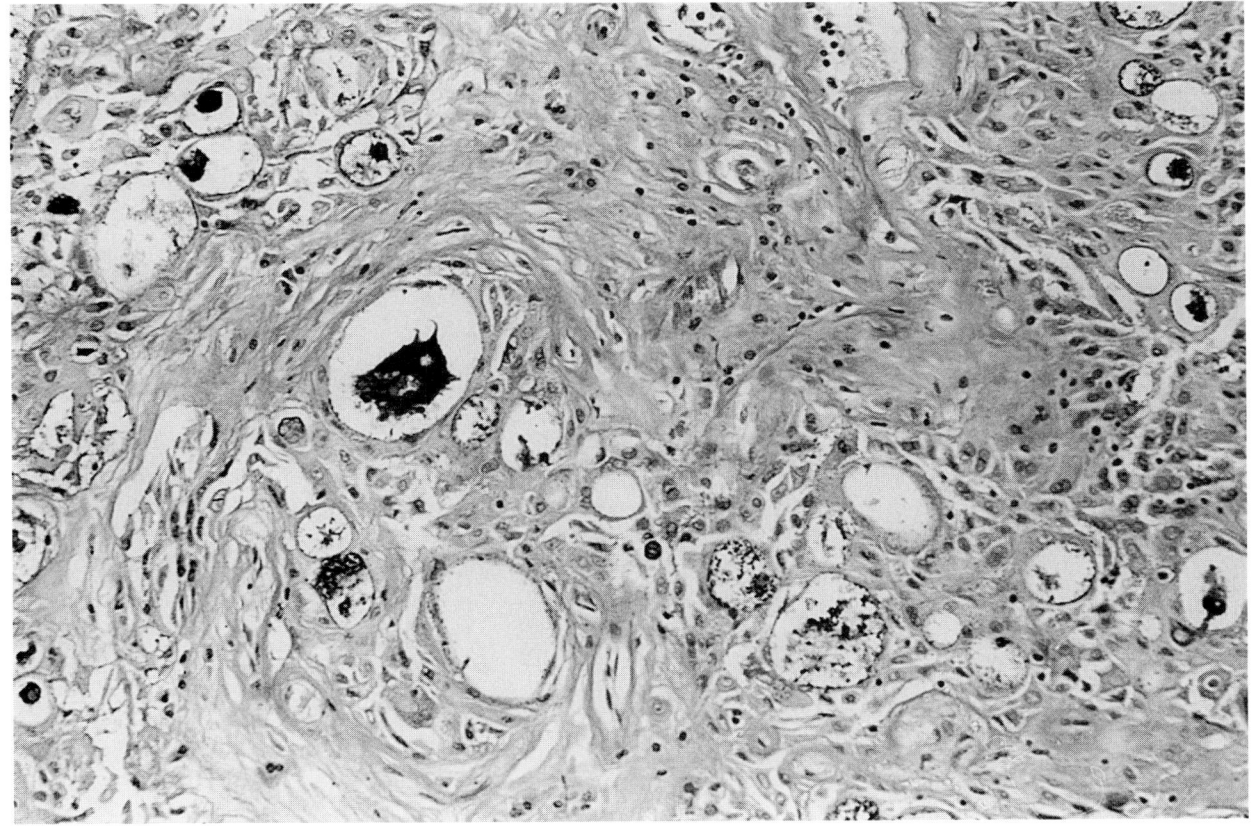

FIGURE 29-88
Low-grade mucoepidermoid carcinoma with mucin in some glandular spaces. (PAS/Diastase, ×87.5.)

TABLE 29-3
Features Differentiating Low- and High-Grade Mucoepidermoid Carcinomas

Low-Grade	High-Grade
Gross	
Exophytic bronchial mass with little extension to the underlying lung parenchyma	Bronchial tumor that is invading the lung parenchyma
Histologic Features	
There are sheets of squamous cells with no or few mitoses, many well-formed mucous glands; necrosis is not present	Cells are more pleomorphic with numerous mitoses, ranging from one to 20 per high-power field; nuclear pleomorphism is evaluated in terms of hyperchromatism, clumping of the chromatin, and enlargement of nucleoli; abrupt transition from one cell type to another
Ultrastructure	
Numerous well-formed goblet cells; abundant mitochondria and glyogen-rich cells with prominent microvilli; prominent gland formation; rare transitional cells	Rare goblet cells, infrequent mitochondria and glycogen-rich cells with poorly formed microvilli; infrequent glandular lumen formation; abundant transitional cells

differentiating the tumor from any other endobronchial lesion.

Electron Microscopy

Tumors have been divided at an ultrastructural level
into three groups: those with undifferentiated cells,
those with lumen-forming cells (of which there are
three different cell types), and those with squamous
cells (Figs. 29-91 and 29-92).

Undifferentiated cells are frequent in all types of
tumors. They have large ovoid nuclei with frequent

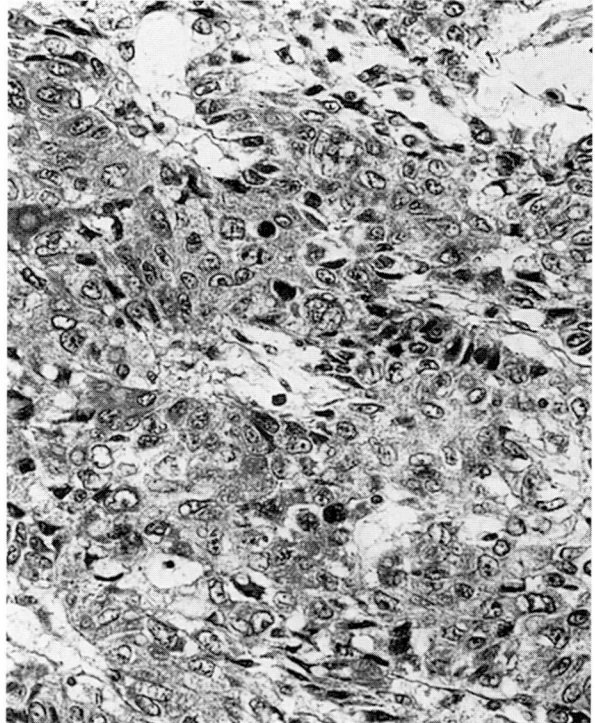

FIGURE 29-90
Higher magnification of the same case as in Fig. 29-89, showing
marked nuclear hyperchromatism and few glandular foci. (H&E,
×577.5.)

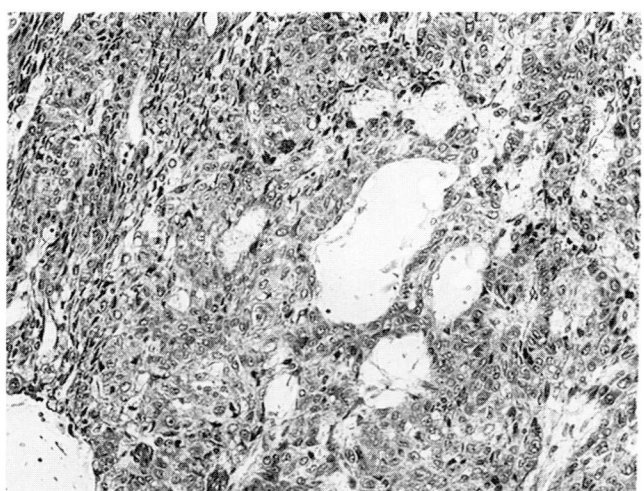

FIGURE 29-89
High-grade mucoepidermoid carcinoma. There are many squamous areas along with cystic spaces. (H&E, ×231.)

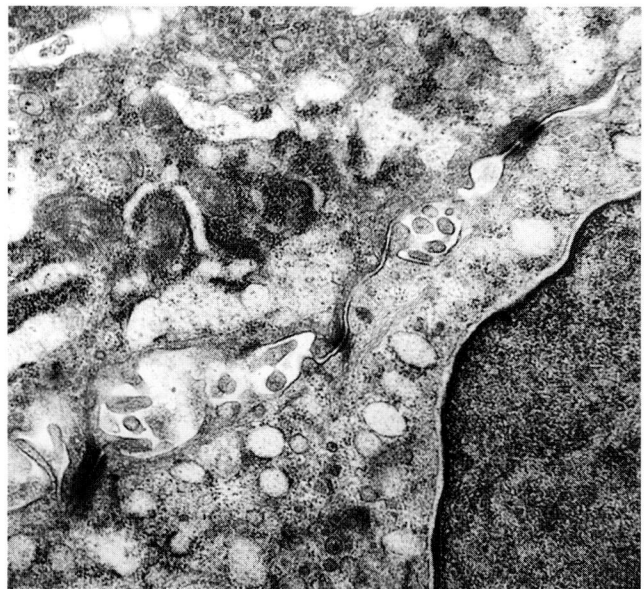

FIGURE 29-91
Mucoepidermoid carcinoma, low grade showing well-formed cell junctions. (Uranyl acetate and lead citrate, ×17,760.)

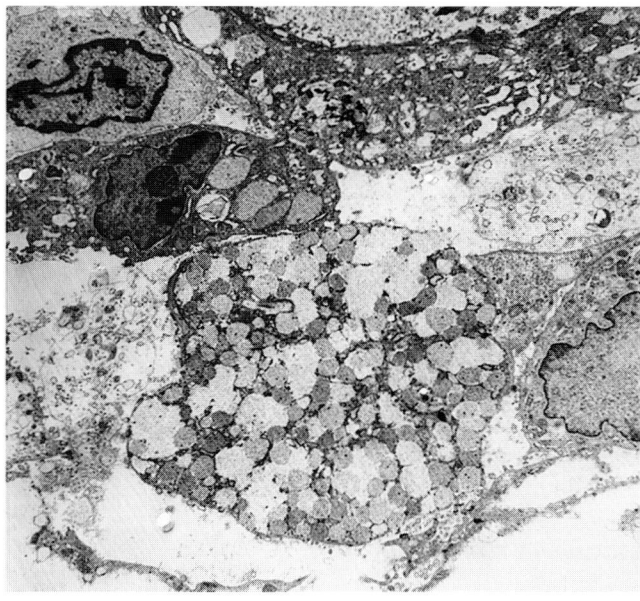

FIGURE 29-92
Low-grade mucoepidermoid carcinoma with a prominent mucous gland and an intermediate cell in the bottom right of the picture. (Uranyl acetate and lead citrate, ×2960.)

small cleftlike cytoplasmic indentations and a single prominent nucleolus. There are no distinguishing characteristic cytoplasmic features. All the usual intracytoplasmic structures—such as mitochondria, lysosomes, multivesicular bodies, and ribosomes—are present in moderate numbers.

Three types of cells form the lumina: goblet, transitional, and glycogen-rich cells. Mucin is present in all cell types, although the glycogen-rich cell usually contains scant amounts. Goblet cells are seen in smaller numbers in high-grade tumors, and their appearance is typical of goblet cells located anywhere else in the body. The transitional cell is rare in the low-grade but common in the high grade tumor. The cells are similar in appearance to undifferentiated cells, although they contain mucus, which varies in amount. The nuclei have indentations. In addition to mucus, the transitional cells are distinguished from undifferentiated cells by more granular cytoplasm and endoplasmic reticulum, which is frequently aligned in parallel arrays.

Natural History

Patients with low-grade tumors do uniformly well, and in children the tumors are considered benign. Unfortunately, in the one case of a low-grade tumor with hilar node metastases, the patient was lost to follow-up.[202] Some patients with low-grade histology may show high-grade biologic behavior, with dissemination of tumor to skin and bone.[206] One patient

with a low-grade tumor presented with a local recurrence at the bronchial resection margin 9 years after his original operation and went on to a complete resection.[200]

No patient with high-grade lesions younger than 30 years of age died or had recurrent disease. Three of 13 cases in one series[203] died of the disease. These patients had metastases in lung, bone, lymph nodes, and skin, as well as recurrence on the same side as the original tumor. Involvement of hilar and mediastinal lymph nodes at the time of initial resection indicates a poor prognosis.

Differential Diagnosis

ADENOSQUAMOUS BRONCHIAL CARCINOMA

This may be a difficult differential diagnosis in high-grade tumors. In adenosquamous carcinoma there is evidence of keratin production, epithelial pearls, and more in the way of intercellular bridges than is seen in mucoepidermoid carcinoma. Transitional areas from low- to high-grade tumors are seen in mucoepidermoid carcinoma but not in adenosquamous tumors. Squamous carcinoma in situ is not usually seen in adenosquamous carcinomas, and these tumors are usually much larger (and often peripheral) than mucoepidermoid tumors, which are usually central. However, adenosquamous carcinomas may occasionally be central.

It has been questioned whether stage 1 adenocarcinoma, squamous carcinoma, adenosquamous carcinoma, and high-grade mucoepidermoid should be separated, since all have a good prognosis. It is only through meticulous classification of tumors and clinical follow-up that biologic behavior will be correlated with histologic patterns of growth.

ADENOCARCINOMA

Adenocarcinoma may be diagnosed by the unwary as mucoepidermoid carcinoma in the absence of intercellular bridges and by relying solely on the presence of mucin. With low-grade tumors this is unlikely to be a problem, but with high-grade tumors some cases may be misdiagnosed as adenocarcinoma.

SECONDARY CARCINOMA

Secondary adenocarcinoma may be misdiagnosed as high-grade mucoepidermoid carcinoma. Perhaps one of the most common mistakes is made with a primary thyroid tumor. The use of thyroglobulin helps in differential diagnosis.

ADENOID CYSTIC CARCINOMA

This tumor has previously been called a *cylindroma*, but such a term should no longer be used. The tumor is thought to be a proliferation of both myoepithelial and secretory epithelial cells.[208] They are rare pulmonary tumors, accounting for only three of 1500 lung cancers reviewed.[209] Among 5500 patients with a diagnosis of primary lung cancer at the Sloan Kettering Memorial Hospital in New York, only five patients had a diagnosis of adenoid cystic carcinoma.[210] A North American series[211] from the Armed Forces Institute of Pathology, an international reference center for difficult pulmonary and other histologic problems, reported 16 cases. These figures are low when compared with a recent British survey of tracheal tumors, in which the incidence of adenoid cystic carcinoma was 10.6 percent.[212] However, that series concentrated on tracheal tumors. In the two other largest series of *tracheal* tumors, the incidence ranged from 25 to 40 percent.[213,214]

Clinical Features

The tumors occur in a wide age range (13 to 79 years), most cases being in patients around the age of 50 years. The carcinoma has rarely been seen in children.[215] There appears to be an equal distribution between the sexes, although some series show a male predominance.[211]

Proximal lesions cause wheezing, which may be confused with "late-onset asthma," cough, dyspnea, and hemoptysis, if there is ulceration of the underlying epithelium. Atelectasis may occur with bronchial obstruction. These tumors may present as asymptomatic peripheral nodules in up to 15 percent of recorded cases.[216–219] In rare instances they demonstrate ectopic release of adrenocorticotropic hormone and growth hormone–releasing hormone.[220] One patient had polyuria, lethargy, and acromegaly; high plasma cortisol levels were present, causing hyperglycemia. He developed cerebral metastases and a large sella tursica with subsequent pituitary infarction. No histology was demonstrated in this report.

Chromosomal Abnormalities

Two adenoid cystic carcinomas—one nasal, the other bronchial—showed rearrangement of 9p13 as a probable primary cytogenetic event[221]; 6q changes were thought to be secondary aberrations, as in salivary gland–type tumors.

Macroscopic Pathology

Tumors usually grow in a cylindric fashion in the main airways (Fig. 29-93). In one series,[222] tracheal tumors were located in the upper (two cases), middle (three cases), and lower third (four cases). There was a similar even spread between these sites in a national survey.[212] Three of those cases were noted to be extensive and involved more than one segment.

The tumors may be polypoid (Fig. 29-94)[223] and endobronchial, but more commonly they are seen as annular tumors with a firm white or pink appearance. They grow through the bronchial wall into the surrounding lung tissue. Tumors vary in size from 0.4 to 10 cm long.

Three growth patterns have been described.[222] Type I is entirely intraluminal, type II is predominantly intraluminal, and type III is predominantly extraluminal. To these tracheal- or bronchial-based tumors must be added 10 to 15 percent peripheral adenoid cystic carcinomas.[216]

In most cases the histologic grade (given below) correlates with the gross tumor type. Histologically, tumor is present at the surgical resection margin in over half of the cases examined.[222]

BIOLOGIC BEHAVIOR

The tumor extends through the bronchial walls, penetrates lymphatics, and is seen within perineural

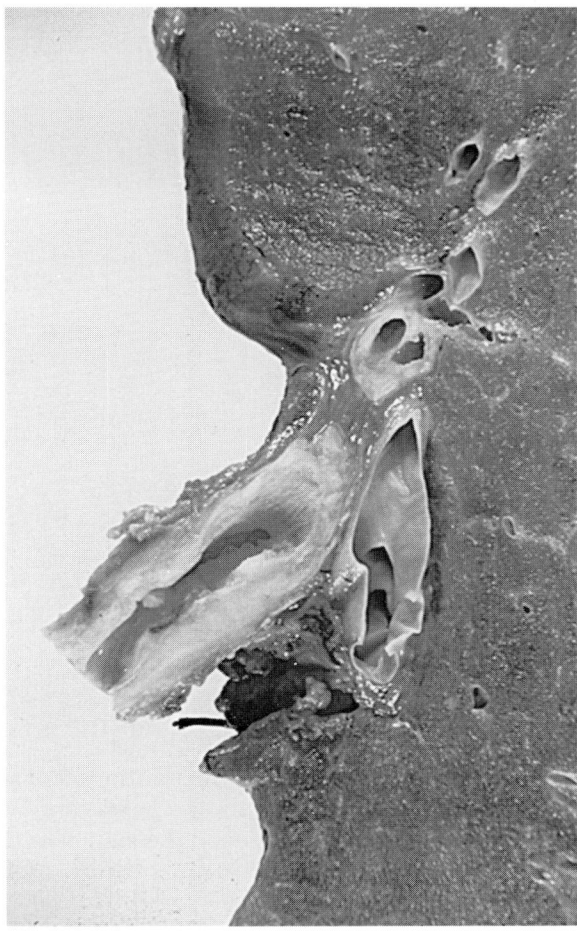

FIGURE 29-93
Adenoid cystic carcinoma showing an annular growth pattern.

FIGURE 29-94
Adenoid cystic carcinoma showing a polypoid growth pattern. The tumor is present at the carina and obstructs both bronchi. (*Reproduced with permission of E. Alvarez, M.D.*)

lymphatic channels. Metastases may be seen in lymph nodes, which may be extrathoracic. Kidney, liver, and, more rarely, other organs may be affected. The tumor often recurs locally, but one case presented as multiple pulmonary nodules 11 years after resection of the original tumor.[216]

In the AFIP series,[211] there was follow-up in 11 cases. Three patients were alive and well up to 12 years later, and three had recurrence or metastases at between 5 and 15 years later. Two patients had metastatic spread at the time of initial diagnosis and died 2 months and 1 year later, respectively, with metastases in lymph nodes, liver, spleen, kidney, and bone.

Histology

Three histologic subtypes have been identified.[211,222,224] These are tubular, cribriform (cylindromatous), and solid, although all types may be present in any single histologic section.

The tubular subtype is characterized by single-lumen tubular units with smaller nests than the cribriform subtype (Fig. 29-95). In some cases the tubular parts of the tumor may be partly lined by ciliated cells.

The cribriform subtype is characterized by small, stellate, dark or occasionally light tumor cells arranged in nests or sheets fenestrated by round or oval spaces—the so-called Swiss cheese pattern (Fig. 29-96). In these spaces there is PAS-positive and alcian blue–staining mucin (Fig. 29-97).

In the solid pattern the individual units are packed with small cells (Fig. 29-98) with oval uniform nuclei, scanty cytoplasm, and only a few lumina and thus no "Swiss cheese" pattern. Perineural invasion is prominent in this pattern, although it can be seen in the other histologic patterns as well (Fig. 29-99).

The stroma consists of two parts. The first corresponds to the layer of lamina propria that normally invests the mucous glands and frequently undergoes

927

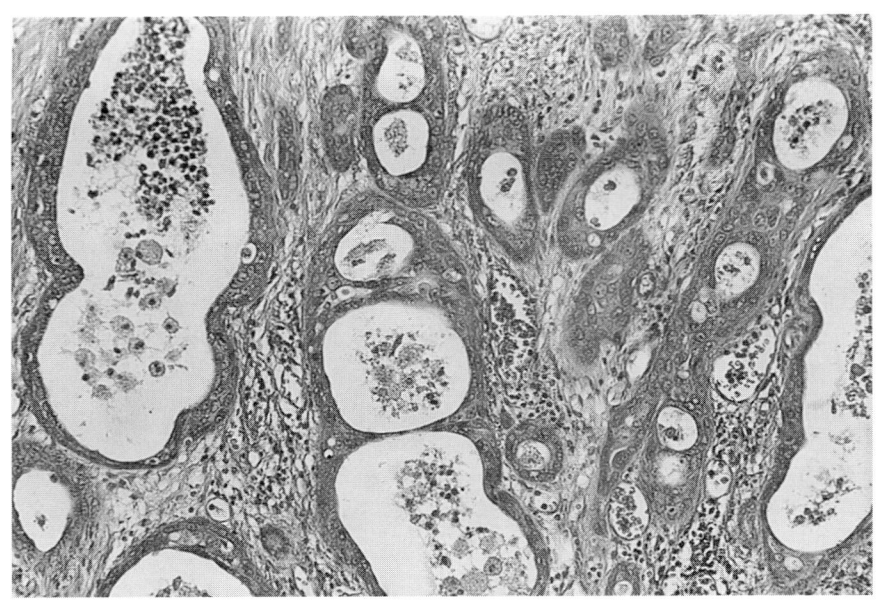

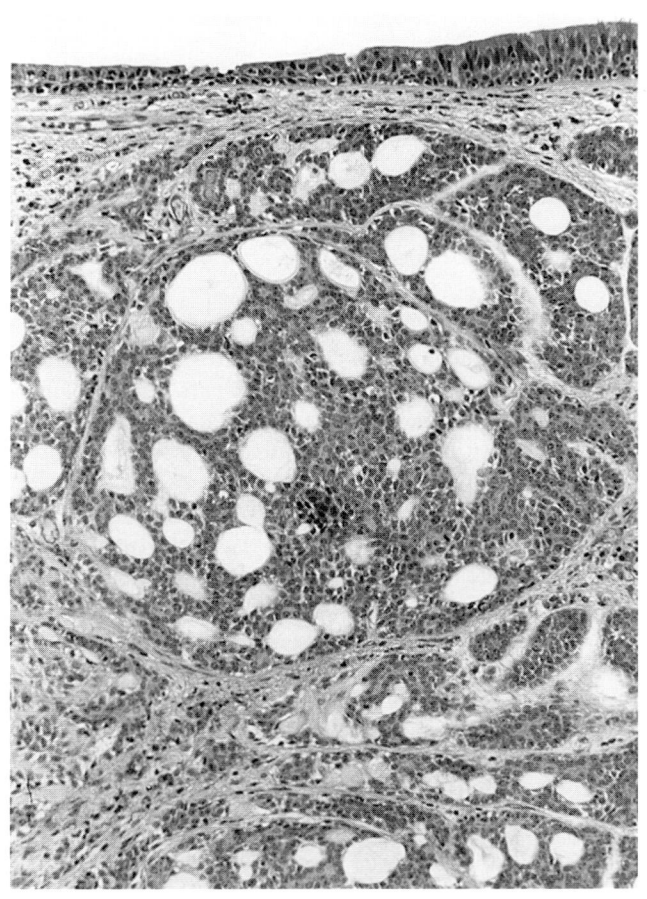

myxomatous change because of accumulation of acid mucopolysaccharides. This stains positively with alcian blue. The second, or true, collagenous stroma is formed by the connective tissue framework outside the lamina propria.

The three histologic subtypes are graded according to the area percentage of each pattern. Grade 1 is a tumor composed of tubular and cribriform subtypes, without any solid areas. Grade 2 consists of tubular and cribriform subtypes, with less than 20 percent solid subtype. Grade 3, the solid subtype, comprises more than 20 percent of the area of the section.

As mentioned above, the tumor extends beyond the cartilage plate (Fig. 29-100), infiltrating the submucosa and perineural lymphatics. Peripheral tumors show submucosal extension to proximal bronchi.[225]

In the tubular and cribriform subtypes, the epithelial cells stain positively for secretory component. The myoepithelial cells are demonstrated by smooth muscle actin (SMA) (Fig. 29-101), desmin, and S-100 protein.[224] There is also positivity for keratin. SMA is also positive in the rarer pleomorphic adenoma.

Laminin and fibronectin[226] are arranged in sheets between the tumor cells in adenoid cystic carcinomas in the breast and salivary gland. One tumor with disseminated metastases was negative for fibronectin and laminin at the time of initial biopsy, suggesting that lectins may be a prognostic guide to

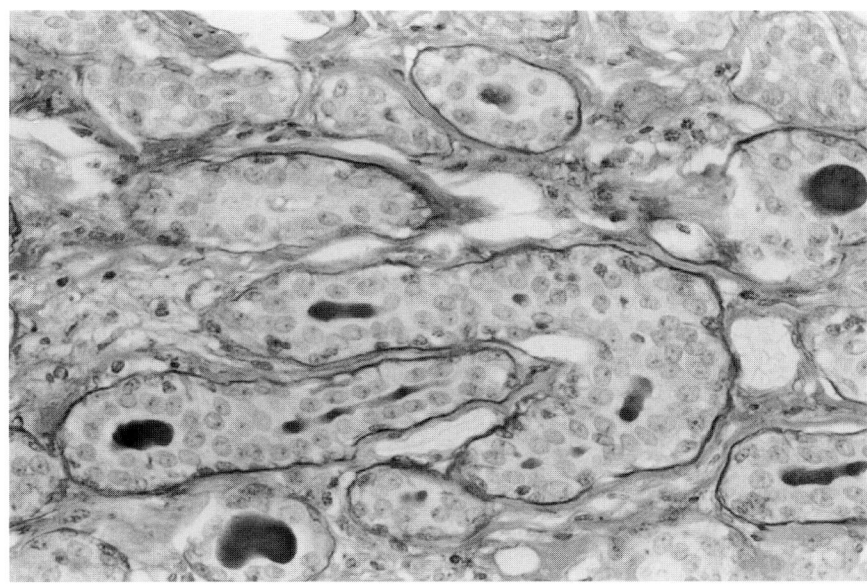

FIGURE 29-97
Adenoid cystic carcinoma. PAS-positive material in tubular spaces. (PAS, ×87.5.)

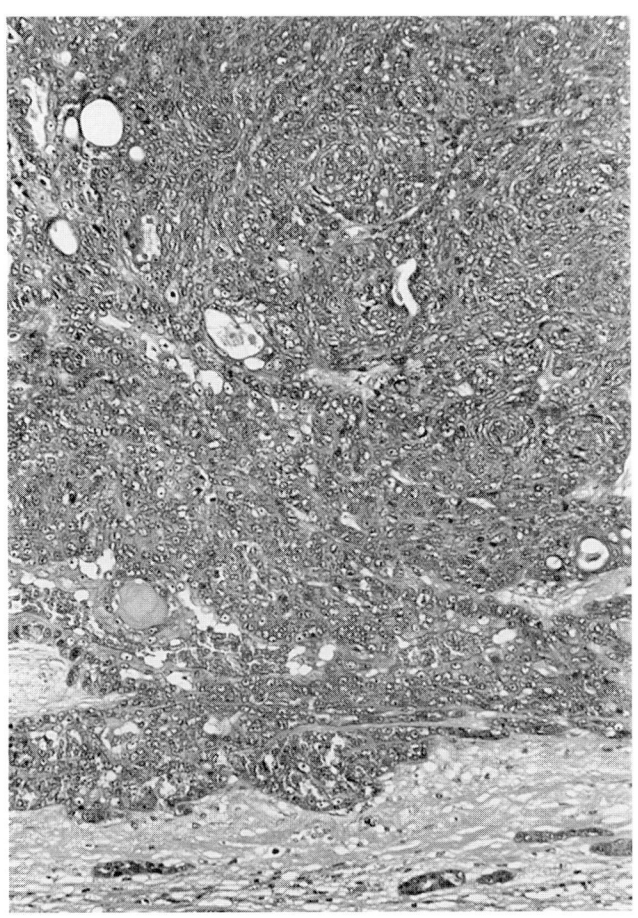

FIGURE 29-98
Adenoid cystic carcinoma with a predominantly solid pattern and a few microcystic foci. (H&E, ×92.5.)

the behavior of these tumors. Little is known of the lectin staining pattern in pulmonary tumors.

Electron Microscopy

An ultrastructural study[227] showed that the epithelial cells were encompassed by basal lamina. Both intercellular spaces and pseudocysts as identified by light microscopy were seen ultrastructurally. Pseudocysts (Fig. 29-102) are discrete rounded extracellular spaces, lined by tumor cells, that contain basal lamina (Fig. 29-103) and other extracellular material but do not have microvilli or features of true glandular lumina. The amount of basal lamina material varies from a single continuous strand to a highly replicated structure occurring around nests of cells and within some pseudocysts. In addition to basal lamina, the interstitial matrix and pseudocysts contain a mixture of amorphous flocculent material, periodic and aperiodic fibrils, stellate granules, and collagen. Amorphous flocculent material is the most prominent of these constituents. Fibrillar material is abundant in all cases.

True glandular lumina, lined by microvilli, were identified in all tumors studied.[227] The cells forming these lumina were occasionally joined by desmosomes and junctional complexes. Microvilli varied from thin and elongated to short and blunt. Myofilaments (Fig. 29-104) were seen in some of the cases studied but were localized to a small portion of the cytoplasm of the cells at the periphery and lining pseuodocysts. In these cases, small bands of thin,

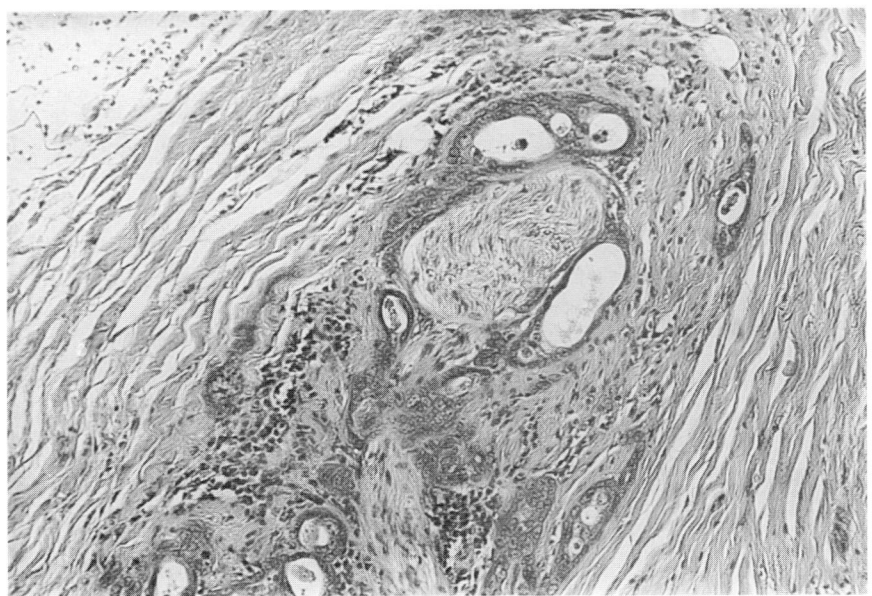

FIGURE 29-99
Adenoid cystic carcinoma with perineural invasion. (H&E, ×87.5.)

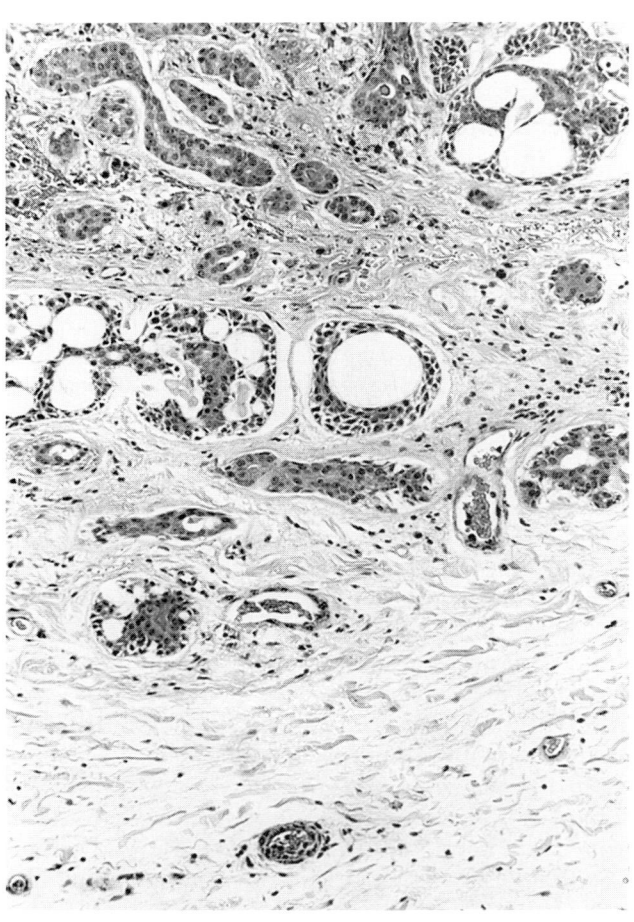

FIGURE 29-100
Adenoid cystic carcinoma, extension outside the cartilage and into the peribronchial tissue. (H&E, ×92.5.)

930

focally condensed myofilaments were distributed near the plasma membranes that adjoined basal lamina.

Differential Diagnosis

ADENOCARCINOMA

This may be a difficult differential diagnosis in the case of the solid variant. The monotonous nature of the tumor cells and the presence of myoepithelial cells are of help. The problem may arise with poorly differentiated carcinomas with some adenoid cystic foci. It is probably best to regard these as poorly differentiated adenocarcinomas.

SECONDARY ADENOID CYSTIC CARCINOMA

Secondary tumors may arise in the lung many years after resection of a primary tumor elsewhere. In one case, a pulmonary metastasis occurred 18 years after resection of a primary tumor from the scalp.[228]

SMALL-CELL CARCINOMA OF THE LUNG

If there are solid sheets of adenoid cystic carcinoma, it may superficially be confused with small-cell carcinoma. Attention to the nuclear detail will resolve the issue.

SCLEROSING HEMANGIOMA OF THE LUNG

This entity has been given several names. In one of Spencer's final publications the term *sclerosing*

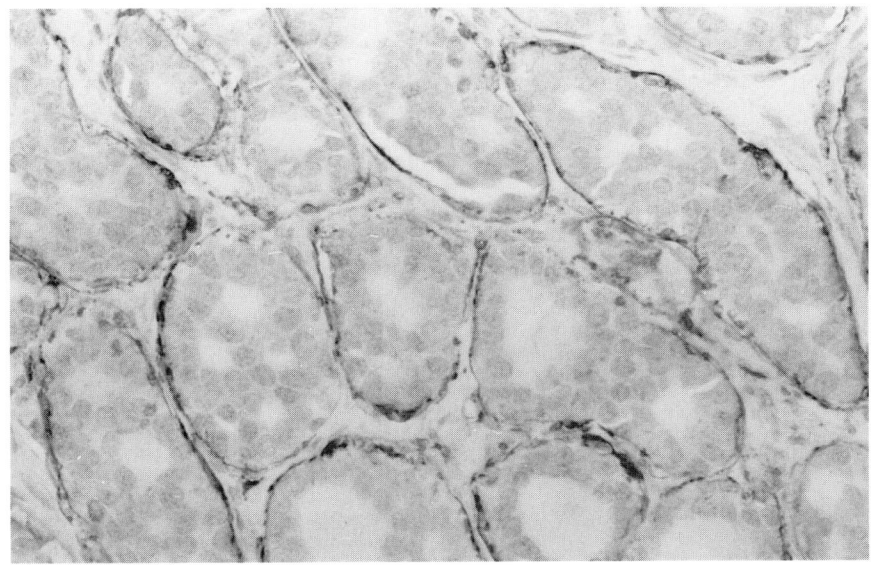

FIGURE 29-101
Adenoid cystic carcinoma. Smooth muscle actin outlining the glandular spaces. (Immunoperoxidase, ×219.)

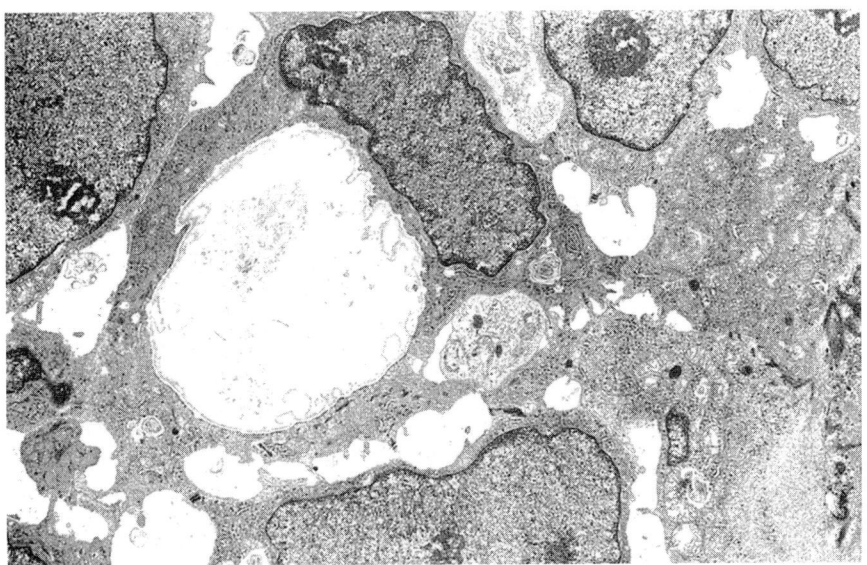

FIGURE 29-102
Adenoid cystic carcinoma showing a large vacuole with lucent contents. The periphery shows condensation of the contents. The nuclei are normal. Electron micrograph. (*Reproduced with permission of Professor Bruce McKay, MD Anderson Cancer Center, Houston, TX.*)

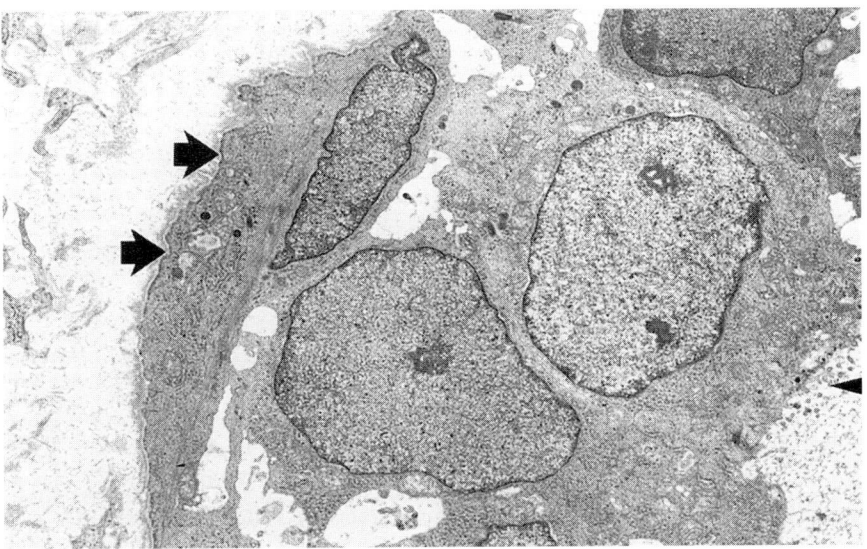

FIGURE 29-103
Collection of epithelial cells with microvilli (*single arrow*). Basal lamina is shown (*short arrows*). Focal densities are present near the basal lamina. The cell is consistent with a myoepithelial cell. (*Reproduced with permission of Professor Bruce McKay, MD Anderson Cancer Center, Houston, TX.*)

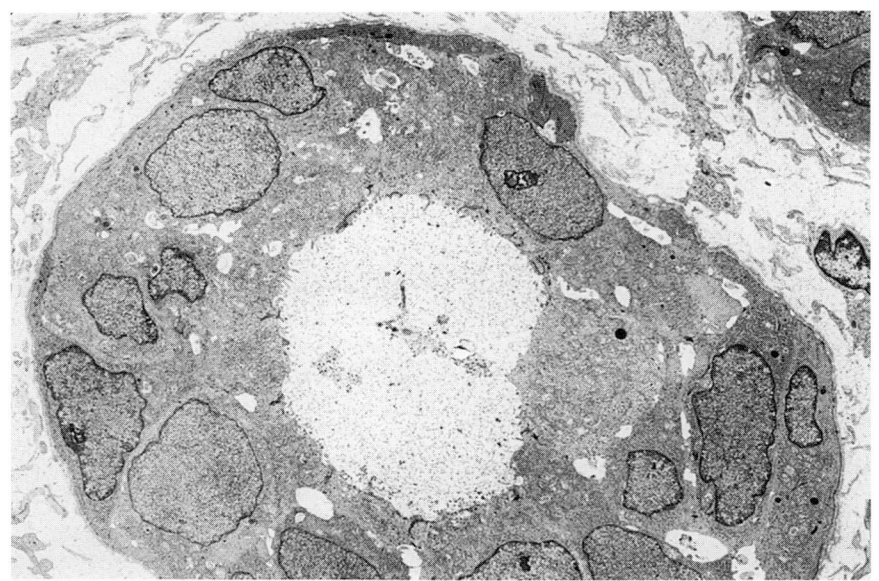

FIGURE 29-104
Adenoid cystic carcinoma. Electron micrograph showing a true glandular lumen lined by microvilli (*Reproduced with permission of Professor Bruce McKay, MD Anderson Cancer Center, Houston, TX.*)

hemangioma was retained.[229] A commentary on his paper suggested that the term *hemangioma* be retained, although the author realized that it would be changed in time.[230] An alternative term, *benign sclerosing pneumocytoma*, has been proposed by Chan's team.[231] As the cell of origin is most probably from a type II pneumocyte, a change in nomenclature is long overdue. The tumor is benign.

Clinical Findings

The entity appears more common in the Far East;[230,232,233] one series documented 196 cases.[234]

In a collected series,[229,231,234–244] a total of 299 cases were reported. Most patients were female. Their ages ranged from 15 to 83 years, with a mean, where individual ages were given, of 42.2 years.

The lesion was more common on the right side. One case was located entirely in the interlobar fissure, and in two it derived from either the right or left upper lobe and extended into the interlobar fissure. Some cases were multiple in both lung fields.

Most cases were asymptomatic, but some presented with hemoptysis, cough, chest pain, dyspnea, and pleurisy. One case[236] presented with back pain; recurrence was not recorded.

Radiologically the tumors are usually solitary masses, with no distinguishing features.

Macroscopic Features

The lesions range from 0.8 to 8.2 cm in diameter, but most are less than 3.5 cm in diameter.[234] They are usually well-defined, circumscribed, hemorrhagic nodules with a variegated tan-brown or white appearance (Fig. 29-105). There is a fleshy consistency. The tumor is often subpleural, and a thin,

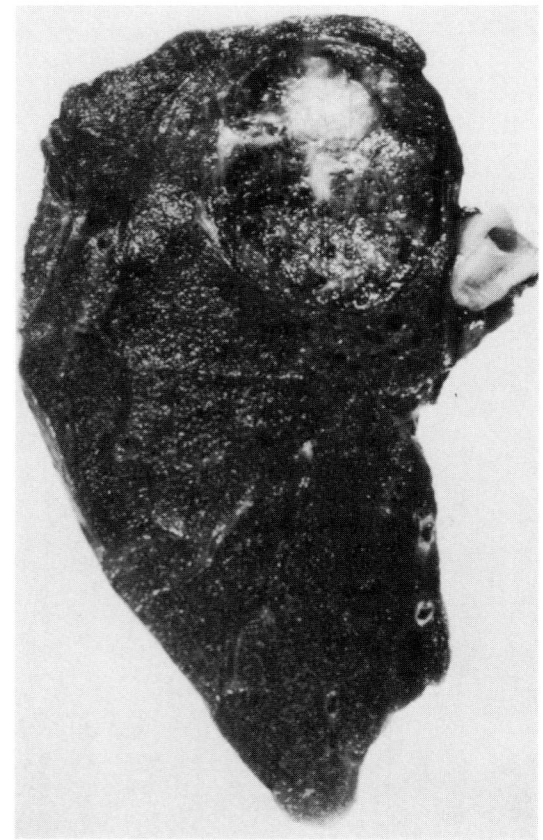

FIGURE 29-105
Sclerosing hemangioma situated in an upper lobe. The tumor has well-defined margin and cut surface is composed partly of fibrous and spongy hemorrhagic tissue.

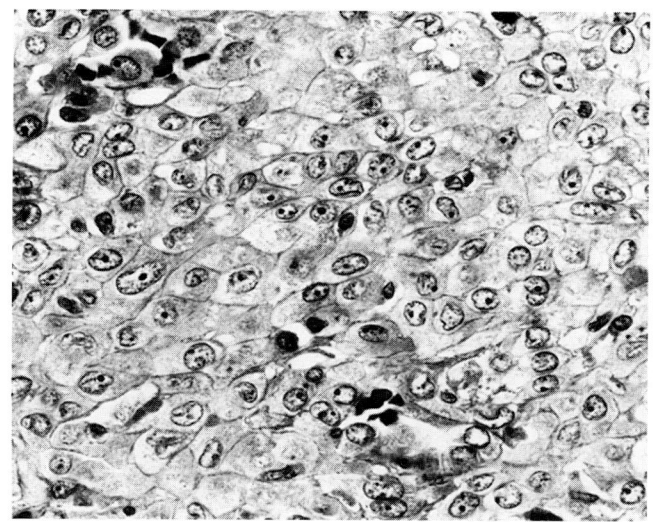

FIGURE 29-106
Sclerosing haemangioma with a solid growth pattern. The cells are regular, with open nuclei and prominent nucleoli. (H&E, ×485.) (*Reproduced with permission of W. D. Travis, M.D., Armed Forces Institute of Pathology, Washington, D.C.*)

fibrous pseudocapsule separates it from the adjacent compressed lung parenchyma. In a few cases, multiple tumors have been identified.[245,246] Some of these may be in the form of "satellite lesions."[228,237] Small, grossly inapparent, sclerosing hemangiomas may be identified near, but apparently distinct from, the main tumor in a few cases and are considered "incipient" tumors.

Histology

Four patterns are recognized: solid, hemorrhagic, papillary, and sclerotic (Figs. 29-106, 29-107, 29-108, 29-109, 29-110, 29-111, and 29-112).[237] In 47 of 51 tumors examined, there was a mixture of at least three of these patterns, and a solid area was present in all cases.

The cells are uniform and polygonal with a bland, oval or round nucleus. There is abundant eosinophilic cytoplasm, although some cases may have clear cytoplasm. Cell borders are indistinct. In a few cases, large intracytoplasmic vacuoles displace the nucleus and produce a close resemblance to signet ring cells. Mitoses may be identified but are rare.

In the solid pattern there are large nests and sheets of distinct round cells. Irregular spaces, lined by cuboidal entrapped alveolar cells, are easily distinguished from the tumor. Red blood corpuscles are also interspersed between the tumor cells and may form small lakes.

The hemorrhagic pattern is characterized by large dilated blood-filled spaces or irregular channels less densely packed with red blood corpuscles. Typical round cells are also found in the interstitium between the blood-filled spaces.

In the papillary pattern there are closely packed papillary projections lined by small cuboidal cells with dark nuclei, similar to the cells lining the entrapped spaces in the solid pattern. The central stalks of these papillae contain clusters of characteristic round cells, always sclerotic and acellular. There are distinct morphologic differences between the round cells in the stalk and those of the surface lining cells. Areas of dense sclerosis were found in all but one case reported.[237] These sclerotic foci were often adjacent to solid areas and between blood-filled spaces.

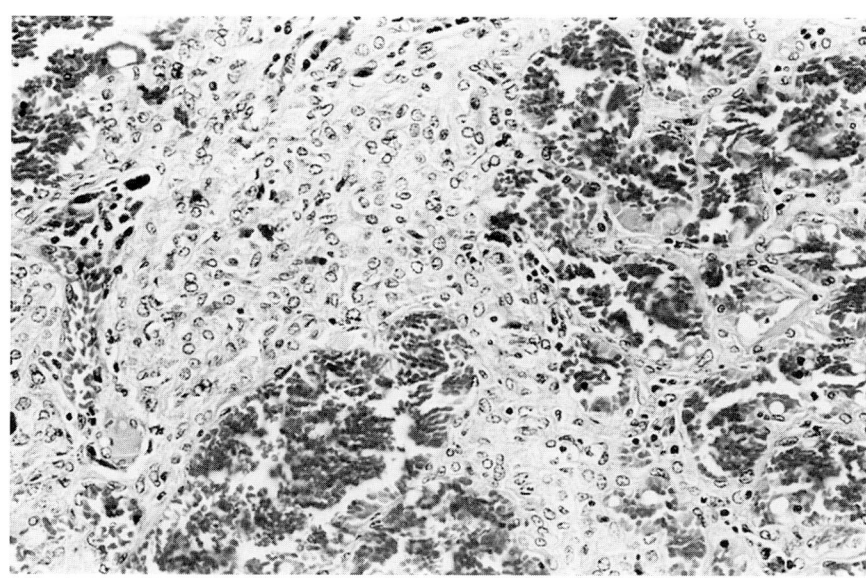

FIGURE 29-107
Sclerosing hemangioma with a hemorrhagic pattern. (*Reproduced with permission of Professor Bruce MacKay, MD Anderson Cancer Center, Houston, TX.*)

933

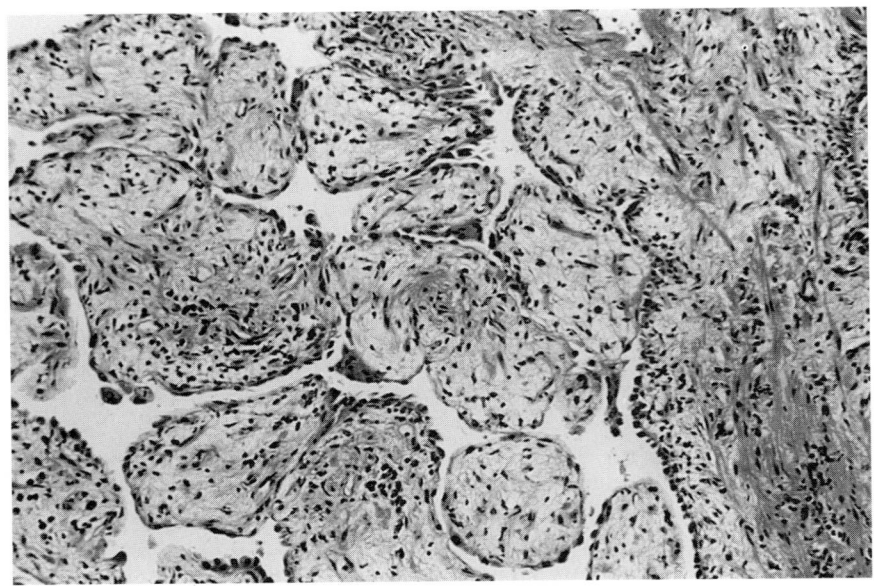

FIGURE 29-108
Sclerosing hemangioma, papillary pattern. The papillae are lined by a low, almost invisible layer of cells. The stroma consists of loose myxoid tissue. (H&E, ×87.5.)

In addition to chronic inflammation with lymphocytes and plasma cells, there are histiocytes with foamy cytoplasm and cholesterol clefts. Adjacent to these there may be foreign body granulomas. Hemosiderin deposition is seen in more than 50 percent of cases. Focal calcification is common, but psammoma bodies were identified in only one case. Small foci of necrosis, as well as some mature fat, may be seen in a few cases.

There have been few reports of fine-needle aspiration in the diagnosis of this disease.[247,248] In one report, the initial diagnosis was either metastatic or bronchioloalveolar carcinoma,[247] and differentiation from malignant tumors may be difficult.

Immunocytochemistry

Some studies have used immunocytochemistry to elucidate the nature of the tumor. One of the most comprehensive was that of Yousem and colleagues.[242] In all but one of their cases, cytokeratin, epithelial membrane antigen, vimentin, placental alkaline phosphatase, and *Ulex europaeus* agglutinin were positive. Surfactant apoprotein was

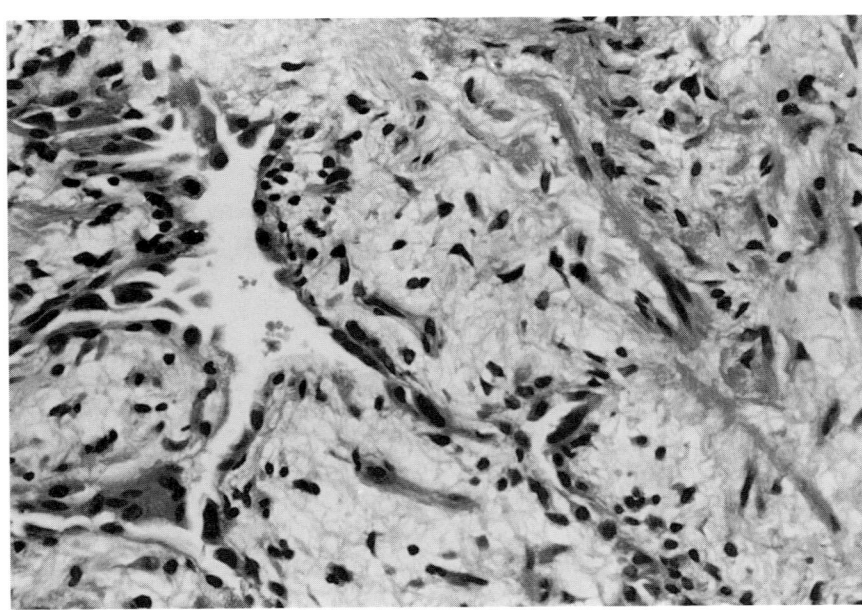

FIGURE 29-109
Sclerosing hemangioma showing a higher power of the papillae with the loose, edematous stroma. (H&E, ×219.)

FIGURE 29-110
Sclerosing hemangioma with a papillary process showing marked chronic inflammation. (H&E, ×219.)

present in all cases. Clara cell antigen was detected in all but three cases, as was carcinoembryonic antigen. Salivary amylase was identified in only three cases.

Factor VIII was not done. Other authors have identified surfactant apoprotein[239] in the solid tumor, the surface epithelium, and intranuclear areas. Secretory component may be demonstrated in some tumors.[241,243] Some authors[241] have identified this material in both the lining epithelium and solid areas. Using immunofluorescence rather than immunoperoxidase methods, Huszar and coworkers[240] demonstrated vimentin in all cells, but desmin was negative. Lysosyme has not been detected.[238,239] Factor VIII has been uniformly negative. Recently, in the largest study to date in addition to cytokeratin, EMA, S100 protein, Vimentin, alpha SMA positivity there was strong labeling for progesterone receptor and weak positivity in some cases for estrogen receptor protein.[248a]

These studies strongly suggest that the tumor is derived from epithelial cells, possibly from type II pneumocytes or Clara cells. However, the epithelial lineage does not correspond to any recognized respiratory epithelial cell.

Ultrastructure

There have been several ultrastructural studies.[229,236,237,243,249] In foci with papillary and cystic areas, the cells had dense inclusion bodies and many lamellar bodies, giving them a strong similarity to type II pneumocytes. There was a discontinuous basement membrane with interdigitating plasma membranes (Fig. 29-113). Long, slender microvilli were seen in some cases and showed a close similarity to mesotheliomas.

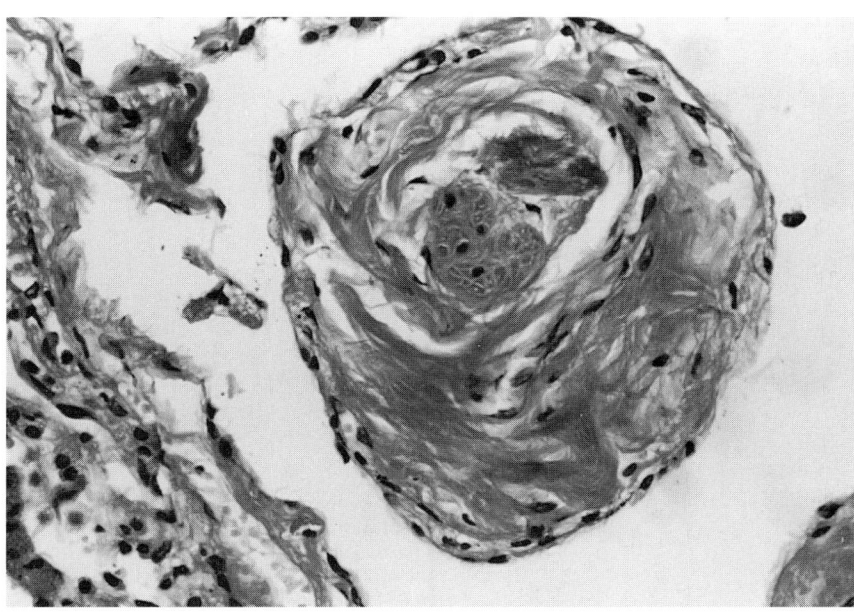

FIGURE 29-111
Sclerosing hemangioma with a papilla showing sclerosis. (H&E, ×219.)

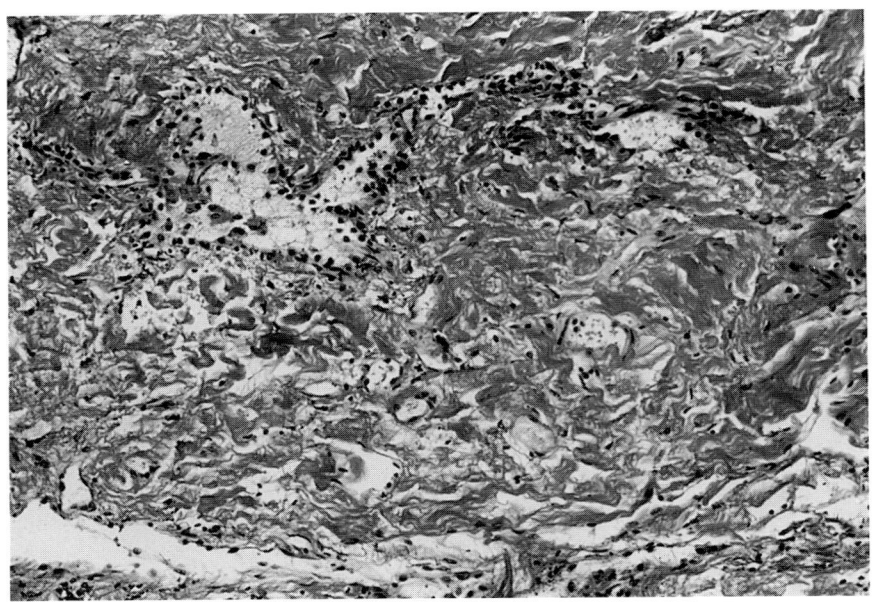

FIGURE 29-112
Sclerosing hemangioma with dense hyaline fibrous tissue and some entrapped alveolar tissue. (H&E, ×219.)

The solid areas had round to oval cells, and in some cases there were many mitochondria, although other authors describe few. There were scattered free ribosomes and small amounts of rough endoplasmic reticulum. Much glycogen was present. Scant intercellular spaces were seen connected to each other by primitive desmosome-like junctions (Fig. 29-114). Immunoelectron microscopy[243] demonstrated surfactant in these cells. No Weibel-Palade bodies (Fig. 29-115) or tonofilaments were identified.

In sclerotic areas there are ovoid to spindle-shaped cells amid dense collagen.

Differential Diagnosis

ADENOCARCINOMA AND OTHER MALIGNANCIES

This diagnosis may be suggested by papillary fronds, but sclerosing hemangioma has benign cytology, rare mitoses, and a lack of mucin production. However, some cases may show nuclear pleomorphism. Adenocarcinomas often show fibrosis/elastosis, which may be central. In sclerosing hemangioma, the fibrosis is irregular and the only pigment is iron. Sclerosing hemangioma may be confused with bronchioloalveolar cell carcinoma, secondary carcinoma

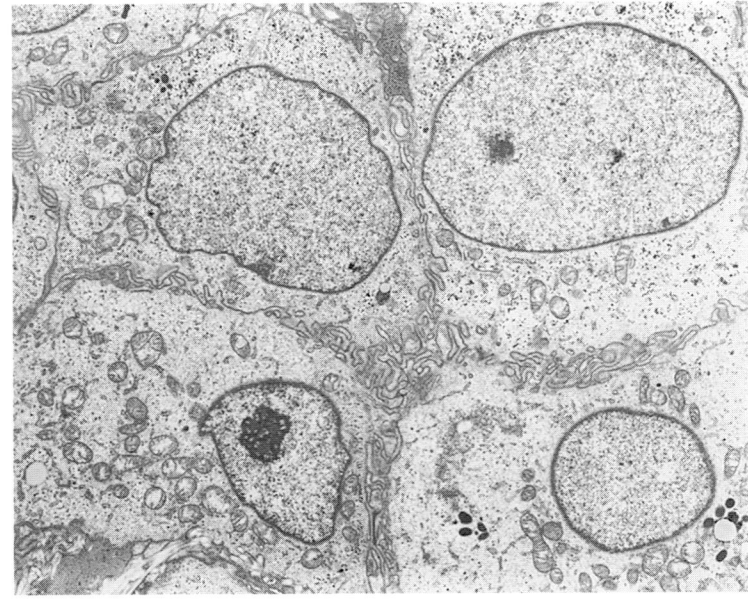

FIGURE 29-113
Sclerosing hemangioma. A group of four polygonal cells showing interdigitating cell membranes with some cells having abundant mitochondria. (Electron micrograph, ×2800.) (*Reproduced with permission of Y. K. Park and Journal of Korean Medical Science.*)

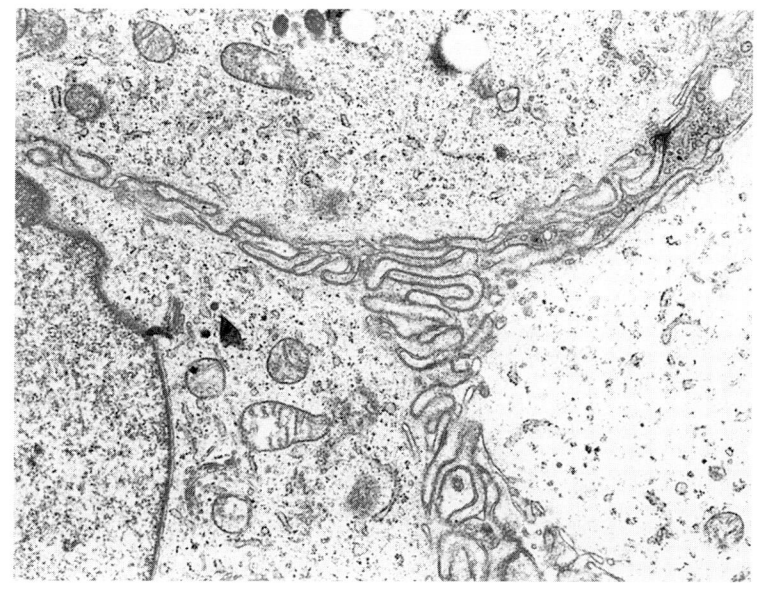

FIGURE 29-114
Sclerosing hemangioma with a higher magnification of the interdigitating cell membrane. (Electron micrograph, × approximately 8400.) (*Reproduced with permission of Y. K. Park and Journal of Korean Medical Science.*)

of the thyroid, and mesothelioma. Examination of the papillary stalks—with their varying diameter, regular lining cuboidal cells, content of characteristic small round cells, and variegated pattern—should help differentiate sclerosing hemangioma from the other tumors.

Sclerosing hemangioma may present as a tumor in the interlobar fissure and cause diagnostic problems.

ANGIOSARCOMA/EPITHELIOID HEMANGIOENDOTHELIOMA

Factor VIII and *Ulex* may be positive in some of these cases, but the vascular pattern is much more prominent than in sclerosing hemangioma. Epithelioid hemangioendothelioma utilizes the alveolar wall for its growth and ultrastructurally shows Weibel-Palade bodies.

It is unlikely to be confused with a benign angioma. These are hamartomas. They present in a different way clinically than sclerosing hemangiomia, in that if the anastomoses are large, there may be pulmonary or even pleural hemorrhage. The angiomas may be multiple in a third of cases. In most angiomas, the bronchial arteries do not communicate with the lesion. If there are multiple small angiomas or a single large one, blood will be shunted through the arteriovenous fistula, causing a varying degree of

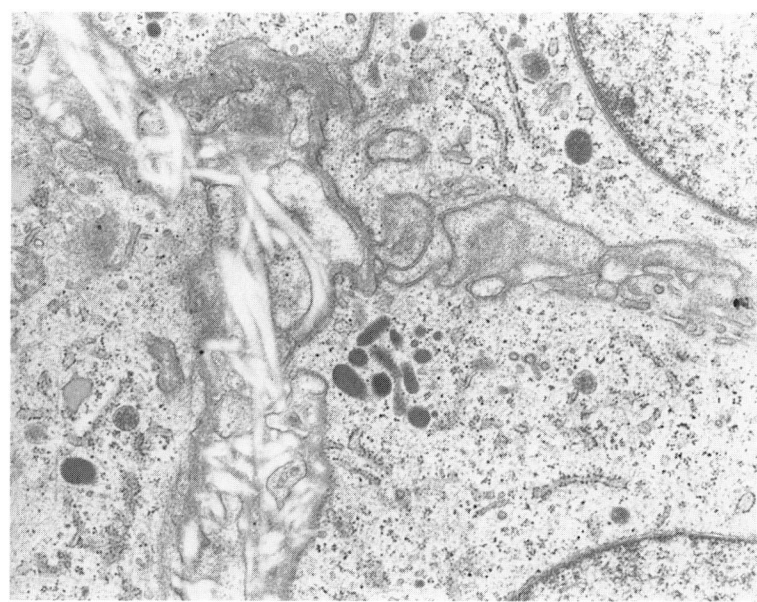

FIGURE 29-115
Sclerosing hemangioma with some ovoid and rod-shaped bodies, which simulate Weibel-Palade bodies. (Electron micrograph, × approximately 8400.) (*Reproduced with permission of Y. K. Park and Journal of Korean Medical Science.*)

oxygen desaturation. In addition, there may be secondary polycythemia. Right ventricular hypertrophy is rare because of the low vascular resistance. The macroscopic and histologic features are similar to those seen elsewhere in the body (Figs. 29-116 and 29-117).

GLOMUS TUMOR

Sclerosing hemangiomas show frequent infolding of nuclear membranes as well as multiple microvilli. Basal laminae surround individual cells in glomus tumors, whereas in sclerosing hemangiomas they outline individual groups of cells.

CARCINOID TUMOR

These may occasionally be difficult to distinguish from sclerosing hemangiomas if there is a solid or papillary pattern. Carcinoid tumors have a much more solid, cellular pattern, with little myxoid change. Neuroendocrine stains should help in differentiation.

INFLAMMATORY PSEUDOTUMOR

This lesion may cause problems in diagnosis. The "tumors" do not have the same strong female predilection as sclerosing hemangioma. Foam cells and chronic inflammation are more prominent, as is

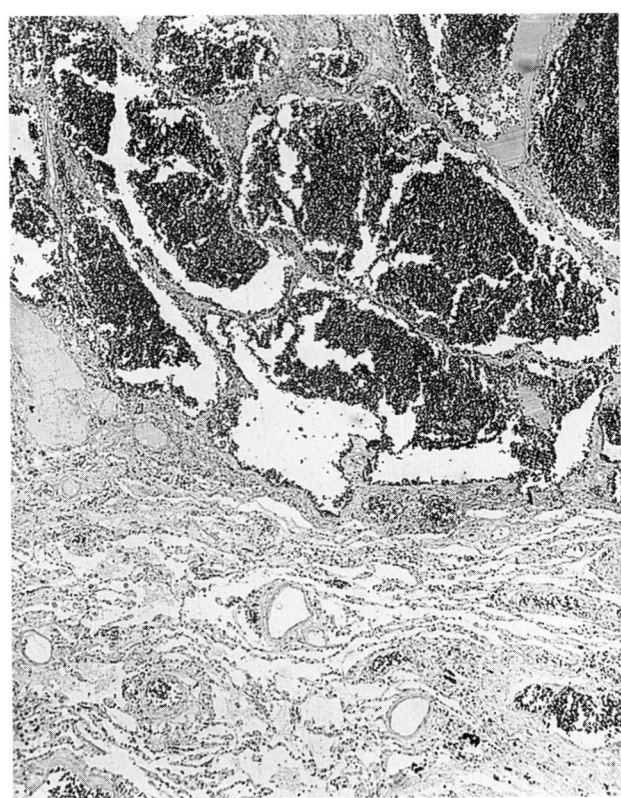

FIGURE 29-117
Well-defined angioma of lung. There is no associated inflammation, papillary, solid, or sclerotic areas. (H&E ,×38.)

sclerosis of small vessels. Inflammatory pseudotumor may have a well-marked storiform pattern not usually seen in sclerosing hemangioma, which may show a papillary pattern. Inflammatory pseudotumor is positive for smooth muscle actin and vimentin and focally positive for desmin. Cytokeratin is seen only in entrapped alveolar cells. In sclerosing hemangioma, cytokeratin, as well as surfactant apoprotein, is positive. Apart from vimentin, the other markers are negative.

BENIGN CLEAR CELL TUMORS OF THE LUNG

These tumors have no papillary areas, and the vascular spaces are sinusoidal rather than heterogeneous, as in sclerosing hemangioma. The uniform clear cells and the lack of variegation should differentiate this tumor from sclerosing hemangioma. HMB-45 is positive in benign clear cell tumors.

Histogenesis

The histogenesis of this tumor is probably epithelial, as demonstrated immunocytochemically. The following cells of origin have been proposed.

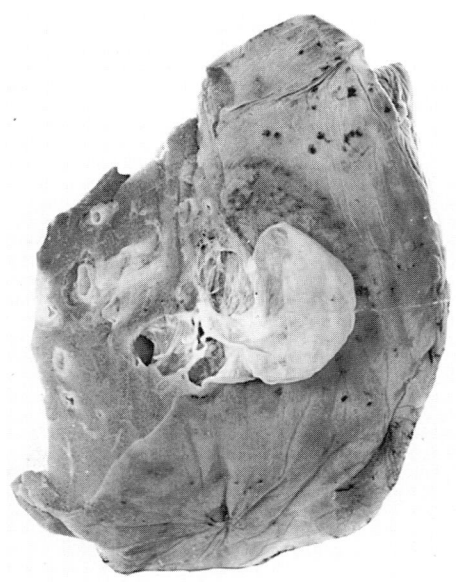

FIGURE 29-116
A subpleural angioma of the lung. Rupture of such a lesion could result in a large hemothorax. (*Reproduced with permission of Dr. K. F. W. Hinson.*)

ENDOTHELIAL CELLS[250-252]

The basis for calling this a vasoformative lesion was the presence of Weibel-Palade bodies. However, these could well be reactive endothelial cells mistaken for the neoplastic polygonal component, and such bodies have not been a constant feature in subsequent case reports of the lesion.

EPITHELIAL CELLS[229,231,236,241,242,248a,253]

The presence of carcinoembryonic antigen, surfactant apoprotein, Clara cell antigen, placental alkaline phosphatase, and salivary amylase suggests an epithelial origin. Epithelial membrane antigen and cytokeratin, while not specific for epithelial cells, provide additional support for an epithelial origin. There is coexpression of vimentin and cytokeratin in sclerosing hemangiomas. Vimentin may be seen in carcinomas of the lung.[254] Reactivity with *Ulex* is not specific for endothelial cells, and this lectin may also bind to epithelial cells. A cellular proliferation identified by *Ulex* must have no cytokeratin positivity to satisfy the supposition of an endothelial origin.

MESOTHELIAL CELLS

This has been proposed by Katzenstein[238] on the basis of glycosaminoglycan electrophoresis positivity for hyaluronic acid in one case and chondroitin sulfate in another. The electrophoretic pattern in these two cases was identical to that of mesotheliomas. The authors suggested that long, slender, uniform microvilli with branch points, as seen ultrastructurally, were suggestive of mesothelial origin. These microvilli resemble the "bushlike" microvilli described in epithelial mesothelioma. Moreover, intracellular lumina were similar to those seen in mesotheliomas. Studies of mesothelioma have shown that it is difficult to differentiate carcinoma from mesothelioma on the grounds of electron microscopy alone. The lamellar bodies found in these tumors, as well as the immunohistochemical findings, point to an epithelial rather than a mesothelial cell of origin. In addition, not all cases show microvilli.

MESENCHYMAL CELLS

A mesenchymal origin for sclerosing hemangioma has been suggested by Huszar and colleagues,[240] who based their evidence on the presence of antivimentin antibodies. However, these may be seen in epithelial cells.

Pathologists have concentrated on the cell of origin of sclerosing hemangioma, but we remain woefully ignorant of any factors causing this tumor.

PULMONARY LYMPHANGIOLEIOMYOMATOSIS

This desease was first described more than 50 years ago.[255,256] It mainly affects women, usually in their reproductive years.

It is manifestated by smooth muscle proliferation into peribronchial, perivascular, and other thoracic lymphatics. The term *lymphangiomyomatosis* was first used by Frack and colleagues.[257] Corrin's team[258] reviewed the subject and added 23 previously unpublished cases. Two recent reviews of a larger series has been compiled by workers at Stanford and the Mayo Clinic[259] and in Japan.[259a]

The disease is rare, with a possible worldwide occurrence of 100 new cases per year.[260]

Clinical findings

Classically, the patient is usually a woman of childbearing age, although cases have been reported in preadolescent girls[261] as well as in postmenopausal women.[262, 263]

In 40 percent of cases,[264] patients usually present with spontaneous pneumothoraces, chylothoraces, and hemoptysis, which was fatal in one case.[265] In addition, there is slowly progressive dyspnea and cough. Exertional dyspnea is one of the most common presenting symptoms. Chest pain, occasional hemoptysis, and a small pneumothorax may occur with the menses. The time from onset of symptoms to death ranges from 1.3 to 27.5 years, with a mean of 8.5 years.[259]

The disease may be found in other organ systems, giving rise to chylopericardium, chyloptysis, and chyluria. There may be abdominal distention, caused by enlarged intraabdominal nodes. The disease may present with a pneumoperitoneum[266] or inguinal lymphadenopathy.[267]

Silverstein and coworkers[268] divided patients into three groups—those with pulmonary parenchymal lymphangioleiomyomatosis (LAM) with mediastinal or retroperitoneal lymph node involvement or both, those with localized mediastinal or retroperitoneal LAM without pulmonary involvement, and those with pulmonary parenchymal and lymph node disease with tuberous sclerosis.

Wolff[269] listed the following conditions as associated with LAM: leiomyomas of the uterus and broad ligament, thyroid disease (including benign nodular colloid goiter, thyrotoxicosis, and papillary carcinoma of thyroid), pituitary hyperplasia, a history of hormone therapy with chorionic gonadotropin, meningioma, and (in two cases) recent pregnancy. More recently, a case report associated LAM with multiple soft-tissue tumors,[270] including a large solitary fibrous tumor of the lung, a large cavernous hemangioma of the liver, meningioma of the right cerebellopontine angle, and a focus of nodular stromal hyperplasia of the ovary. Papillary carcinoma of the thyroid and parathyroid adenoma were also found. The patient, who had received neck irradiation for tonsillitis at the age of 5 years, died from respiratory failure due to the LAM.

The best survival figures[248a] show 38 percent of patients are alive 8.5 years after the onset of disease. Patients with predominantly cystic lesions have a poor prognosis but the amount of smooth muscle proliferation does not correlate with survival.

ASSOCIATED CLINICAL DISORDERS

In the previous edition of this volume, LAM was given a separate section under tuberous sclerosis (epiloia, Bourneville's disease). There is a striking resemblance between the two pulmonary pathologies, and LAM is probably an incomplete expression of tuberous sclerosis.[271] The latter disease is rare, estimated to occur in one in 500,000 persons. It is inherited as an autosomal dominant trait with variable penetrance and a high rate of spontaneous mutation. This accounts for its phenotypic variability and frequently subtle clinical manifestations.

Recently, diagnostic criteria have been proposed for the disease.[272] Its main signs are mental retardation, epilepsy, renal angiomyolipomas, and sebaceous adenomas of the skin. Two cases with pulmonary involvement associated with tuberous sclerosis have been recorded in men.[273] More recently a case has been described with pulmonary nodules up to 0.8 mm in diameter.[274] These consisted of hyperplastic type II pneumocytes; most proliferated on the alveolar surface, but some cells formed clusters. Cytokeratin was strongly positive. There was associated emphysema but no smooth muscle proliferation. The pneumocyte hyperplasia was also present in alveolar septa. Peripheral lymphatics and alveoli were dilated, probably because of epithelial proliferation.[274] Angiomyolipomas of the kidney are seen in both tuberous sclerosis and LAM.[259a,275]

Only four of 355 patients with tuberous sclerosis

died of LAM.[276] The most common cause of death is renal failure due to angiomyolipomas, renal cysts, or both. Cardiac rhabdomyomas have been described. Status epilepticus and bronchopneumonia are also common causes of death. Cerebral lesions called tubers, which are hamartomas, have been found in 10 patients.[277] Recently, a pulmonary angiomyolipoma has been described in a 68-year-old woman without LAM or tuberous sclerosis.[277a]

An increased frequency of complex chromosomal rearrangements has been identified in peripheral lymphocytes of a patient with tuberous sclerosis, in comparison to a patient with isolated LAM.[278]

The chest radiograph in LAM usually shows a reticulonodular pattern, pleural effusions that are usually chylous, and pneumothoraces. The pulmonary densities may be reticular, reticulonodular, or miliary or may have a honeycomb pattern. There are cysts, bullae, and evidence of hyperinflation. The smooth muscle proliferation is evident only at a relatively late stage of the disease. Both thin-section CT (Fig. 29-118) and high-resolution CT add to the sensitivity of diagnosis.[279]

Pulmonary function tests show an elevated total lung capacity, differentiating this disease from other interstitial lung disorders. Other physiologic alterations are highly reminiscent of those seen in emphysema, suggesting that flow limitation in LAM is caused more by loss of alveolar support of the small airways than by excessive intramural smooth muscle proliferation.[280] Residual volume is also increased, owing to the multiple cysts, and bronchial obstruction is seen in most patients. FEV_1 is significantly reduced, as is $D_{L_{CO}}$. Alveolar volume is close to nor-

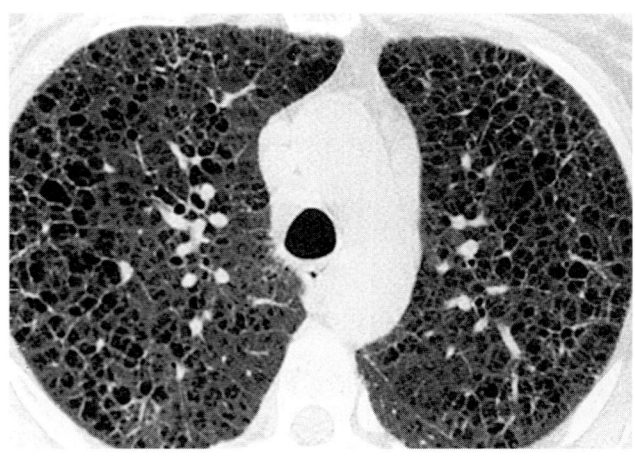

FIGURE 29-118
LAM-CT showing microcyst formation.

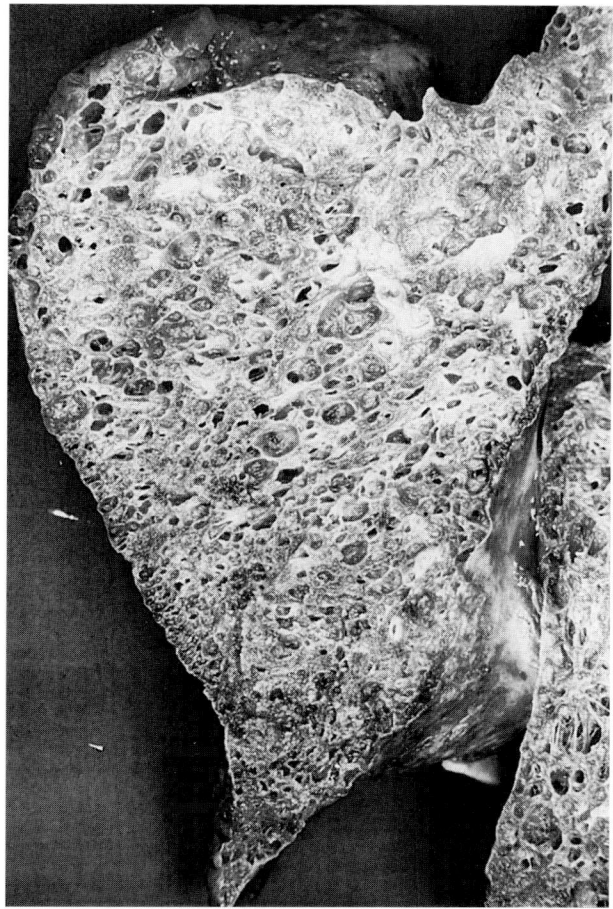

FIGURE 29-119
LAM macroscopic specimen with honeycomb appearance.
(*Reproduced with permission of M. Burke, M.D., Harefield Hospital, London, England.*) (*See also Color Plate 5.*)

mal, indicating that the alveolar capillary membrane is unaffected.[281]

Macroscopic Pathology

In a well-established case of lymphangioleiomyomatosis, the lung has a honeycomb appearance (Fig. 29-119), with a thickened pleura and septa surrounding distended air spaces. The cysts, measuring approximately 1 cm in diameter, are located predominantly in the upper lobes, though the lower lobes may in some cases be more affected. Normal parenchyma, as well as foci of emphysema, is seen between the cysts, which are filled with air and serosanguinous or occasionally chylous material. The outer surface of the lung often has blebs covering both lobes. The pleural cavities may be obliterated by dense adhesions. There may be associated right ventricular hypertrophy.

If hilar nodes are involved, they are enlarged and spongy, and the thoracic duct is dilated. This becomes spongelike and filled with chyle. Rupture of dilated mediastinal and pulmonary lymphatics causes an opalescent chylous pleural effusion. Dilated lymphatics extend below the diaphragm to involve the retroperitoneum as well as upward into the root of the neck.

Histology

There is a focal fibrous thickening in the pleura and subpleural intraalveolar septa. Aggregates of lymphocytes and hyperplastic smooth muscle with a spindle cell pattern are preserved in lung tissue adja-

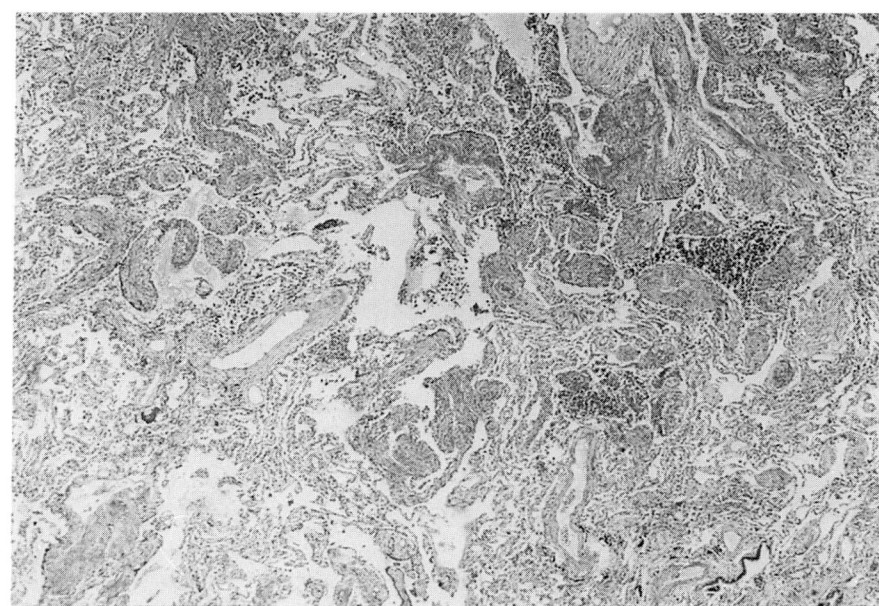

FIGURE 29-120
LAM showing a proliferation of spindle cells in the lumina of small airways. There is an associated chronic inflammatory reaction. (H&E, ×56.)(*Reproduced with permission of A. Gibbs, M.D., Penarth, Wales.*)

941

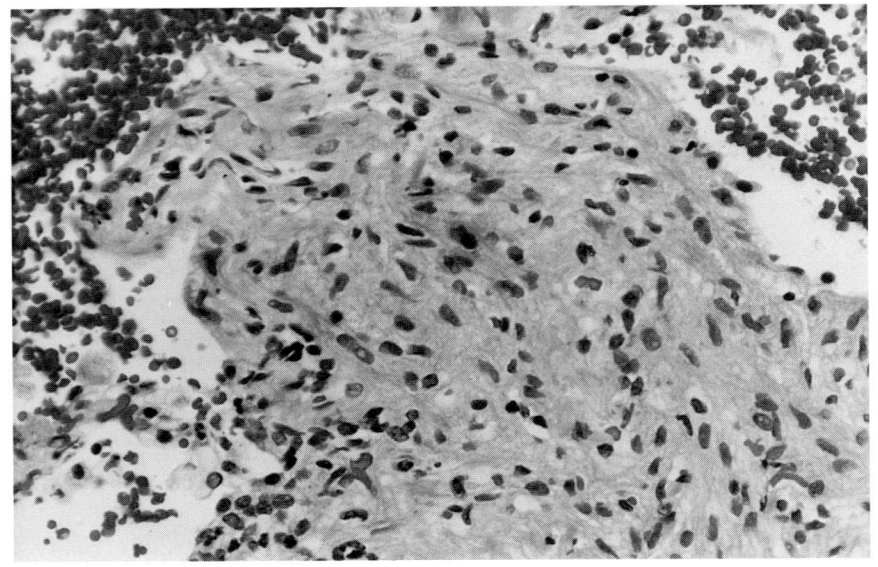

FIGURE 29-121
LAM. Same case as Fig. 29-120 but with spindle cells filling a larger airway, probably a dilated alveolar duct. (H&E, ×219.)

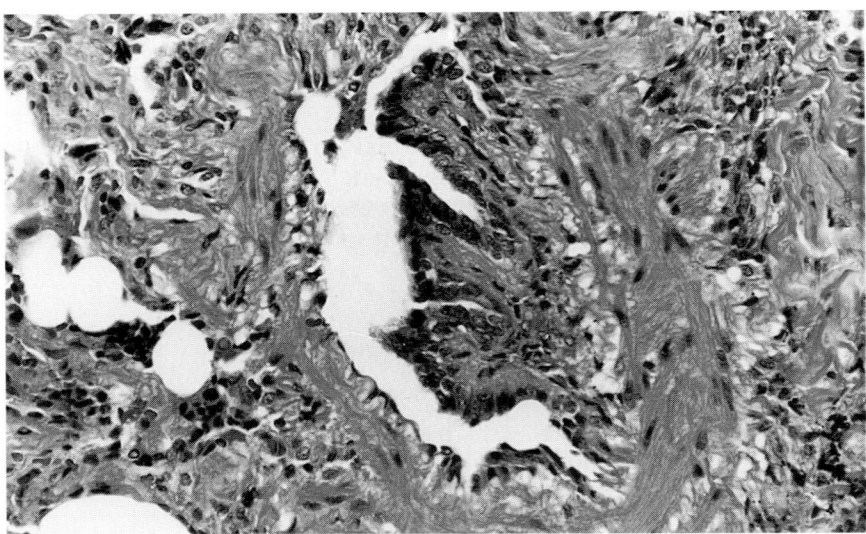

FIGURE 29-122
LAM. There is a marked increase in muscle in the wall of this bronchiole. Some muscle appears to be extending into the adjacent alveolar walls. (H&E, ×219.)

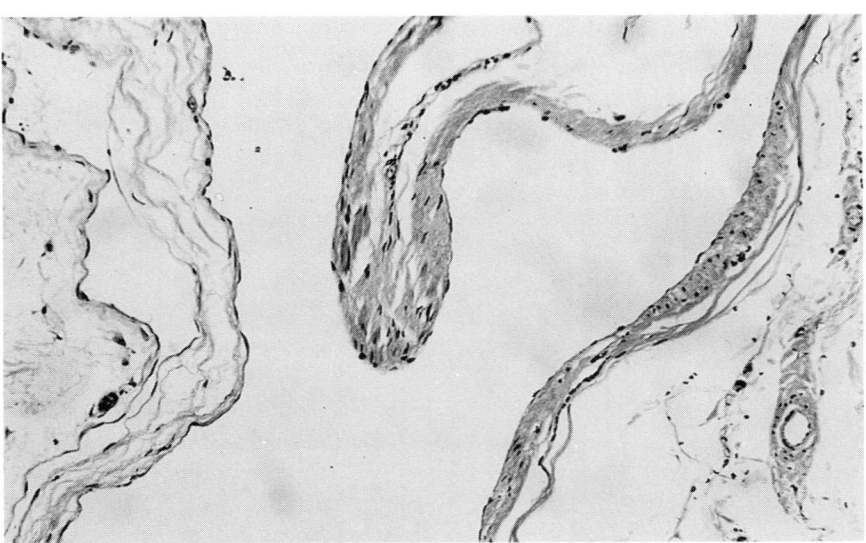

FIGURE 29-123
LAM. Smooth muscle proliferation in a lymphatic wall, with dilatation of the lymphatic. (H&E, ×219.)

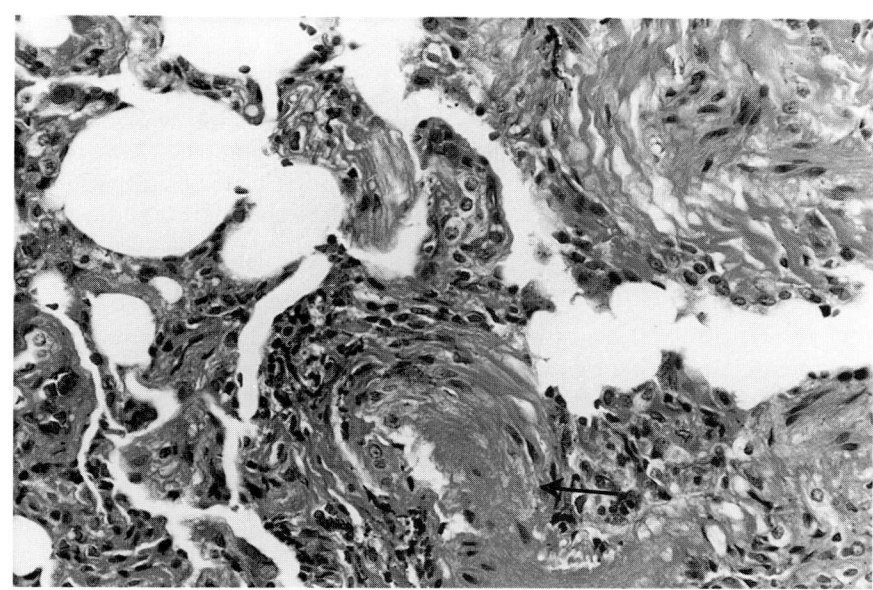

FIGURE 29-124
LAM. Smooth muscle proliferation in the alveolar wall as well as in the wall of blood vessel (*arrow*). (H&E, ×219.)

cent to areas of emphysema. Small airways of less than 1-mm diameter contain smooth muscle (Fig. 29-120, 29-121, and 29-122). These spindle cells proliferate in lymphatics and around small blood vessels (Figs. 29-123 and 29-124) as well as respiratory bronchioles. This causes their collapse, promoting emphysema. The spindle cells also proliferate inside alveoli and alveolar ducts (Fig. 29-125).

The spindle cells have elongated bland nuclei, which are closely packed and arranged in short pallisades. In early-stage disease, this proliferating smooth muscle, which can be detected by smooth muscle actin, is seen mainly around lymphatics and is probably derived from cells in their walls.[258] There are cystic spaces between the spindle cells, which contain lymph. In rare instances, proliferation of smooth muscle cells causes intrathoracic[282] or intraabdominal[283] tumors.

A quantitative study of the lesions of pulmonary lymphangioleiomyomatosis[280] showed centrilobular emphysema with decreased alveolar lung volume and internal surface area. The mean diameter of small airways was significantly decreased because of a large population of these airways measuring less than 0.35 mm in diameter. No bronchial gland enlargement was identified, and the amount of bronchial smooth muscle was within normal limits.

Centrilobular emphysema originates in the respiratory bronchiole, which is the site of termination of the deep pulmonary lymphatics as well as their

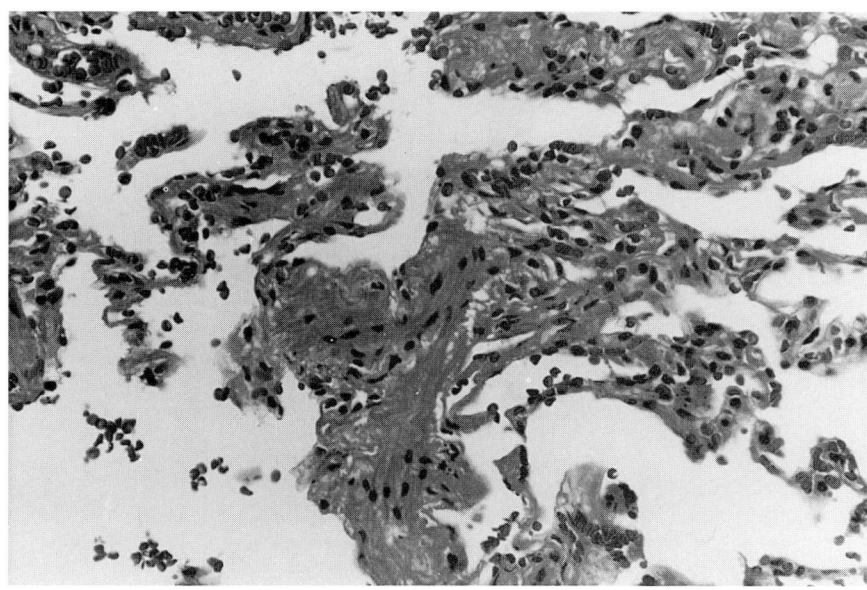

FIGURE 29-125
LAM. Smooth muscle proliferation in the wall of a probable respiratory bronchiole. (H&E, ×219.)

anastamoses with the pleural lymphatic plexus.[284] This is consistent with the origin of pulmonary LAM in pulmonary lymphatics.

Another reason has been proposed for the emphysema: Elastic fibers in areas of alveolar smooth muscle accumulation were decreased, disrupted, granular, or poorly delineated.[285] They accumulated in both early and late lesions, and α_1–antitrypsin antibody reacted with the elastic fibers located in areas of alveolar smooth muscle cell accumulation in both early and late lesions. Elastic fibers distant from the areas of smooth muscle cell accumulation in alveolar walls did not react with this antibody. The study's authors thought that the emphysema was related to an imbalance of the elastase–α_1–antitrypsin system, similar to that proposed in emphysema. This concept seems simplistic, however, because in emphysema there is evidence of macrophage dysfunction, usually due to cigarette smoking. The presence of α_1–antitrypsin in elastic fibers may be evidence of the degeneration of such tissue by elastase, but the presence or origin of elastase has not been elucidated in LAM. In addition to changes in elastin, there is partial degradation of collagen fibrils in LAM.

The lymphatics may be cystic, dilated, or reduced to small clefts.[286] Hemosiderin-laden macrophages are also present, due to venous obstruction by LAM.[258]

The diagnosis of lymphangioleiomyomatosis may be difficult on open lung or transbronchial biopsy because of sampling error. Stovin and colleagues[286] have suggested that a deep wedge biopsy is more likely to reveal the diagnosis. The initial diagnosis may be pulmonary fibrosis, although in one series,[259] three out of four transbronchial biopsies were considered diagnostic.

Immunohistochemistry

If pulmonary lymphangioleiomyomatosis is suspected, the best stains to use are alpha–smooth muscle actin, which is a major cytoplasmic component of the constituent cells, and HMB-45, a monoclonal antibody with a high specificity against junctional nevi and malignant melanomas. The latter immunostain was positive (Fig. 29-126) in three cases in the bundles of polygonal epithelioid cells as well as in spindle cells and intermediate cells.[287]

Other lung disorders, including cases with smooth muscle hyperplasia, granular cell myoblastoma, and leiomyosarcoma, were negative with HMB-45.

Some regard this immunoreactivity as indicating that the electron-dense granules share antigenic properties with premelanosomes,[288] similar to those seen in benign clear cell tumor. Others believe that these granules contain some kind of elastase.[285] Desmin and vimentin have also been identified.[289]

Angiomyolipoma and benign clear cell tumor, hamartomatous diseases related to LAM, also show smooth muscle positivity with HMB-45.[290] This marker may be useful in the future if these results can be repeated on a larger number of cases.

There is variability in endothelial cell markers.

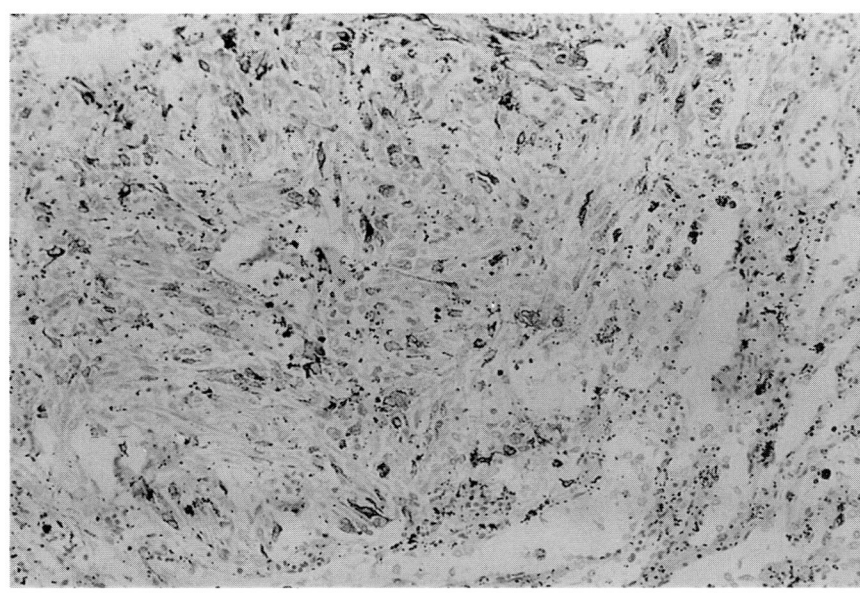

FIGURE 29-126
LAM. Histological section stained with HMB 45. (Immunoperoxidase, ×87.5.)

Factor VIII–related antigen and other endothelial markers (PALE and EN4) are reported to be strongly positive[267]; others have found factor VIII to be negative.[289]

Steroid receptors have been investigated in this condition, with varying results. Steroid receptors—specific for estrogen, progesterone, and glucocorticoids—have been identified in one study.[291] Even though the receptor levels were of a low absolute value, they were significant, since normal lung shows no receptors. Berger's team[292] demonstrated estrogen and progesterone receptors in two cases. In one case, the receptors were present in proliferative smooth muscle nodules as well as the interstitial tissue in close proximity to the muscle; in the other, only estrogen receptor was detected in some of the proliferative smooth muscle. The receptors were localized in the nuclei of the affected cells. Estrogen receptors were identified in a further case,[293] but this study did not examine for progesterone receptor.

Reports of treatment with antiestrogen therapy have been given but have not been fully evaluated.[294] Progesterone may produce clinical improvement.[295,296] The use of antiestrogen therapy with tamoxifen has so far been inconclusive in this disease.[259]

Electron Microscopy

Ultrastructural studies[269,285,289,293,297] have shown that smooth muscle cells have convoluted nuclei, prominent microfilament bundles with dense bodies, and rough endoplasmic reticulum. There are cytoplasmic inclusions, which are membrane-bound, electron-dense, and sometimes cross-striated; they are possibly lysosomal. Some cells with abundant rough endoplasmic reticulum, thought to be intermediate cells—possibly fibroblasts—are present.

The basement membranes are discontinuous around the smooth muscle cells. In addition, elastic fibers in such areas show irregularly outlined amorphous components with or without a few microfibrils. Some elastic fibers have many electron-dense granular deposits. Similar electron-dense deposits are seen in epithelial basement membranes, as well as in some collagen fibrils. The interstitial connective tissue is frequently abnormal. Collagen fibers vary in thickness and shape, and in longitudinal section there is separation or twisting of fibers. Alveolar epithelial and endothelial cell basement membranes are thickened, homogeneous, and electron-dense. Capillary basement membranes are sometimes reduplicated.

Energy-dispersive x-ray microanalysis showed calcium, iron, and phosphorus.

Etiology

The cause of this disease is unknown. The proliferating cells have been identified as smooth muscle cells, although they may be myofibroblastic. The clinical course of the disease may be influenced by estrogen and progesterone, since these hormones affect proliferation of smooth muscle cells in other parts of the body, such as the uterus, vascular smooth muscle, and skeletal muscle. The disease can become worse during pregnancy,[268] as well as after estrogen therapy.[298]

Differential Diagnosis

PULMONARY EMPHYSEMA

A superficial glance may not reveal the smooth muscle proliferation in the alveolar wall and may detect only centrilobular emphysema.

INTERSTITIAL FIBROSIS

In the late stages of the disease, because of misdiagnosed "honeycombing," a false diagnosis of interstitial pulmonary fibrosis may be made. Smooth muscle proliferation does occur in interstitial fibrosis, but it is patchy and does not form bundles inside the alveolar lumina and is not HMB-45 positive. Smooth muscle proliferation is seen secondary to left ventricular failure.

DIFFUSE PULMONARY LYMPHANGIOMATOSIS

This lesion does not show honeycombing or emphysema. There is no marked proliferation of smooth muscle extending into alveoli and around vessels and lymphatics as in LAM. Most of these cases are seen in children, whereas LAM shows a marked female predominance and occurs most often in women of childbearing age. HMB-45 is negative in diffuse pulmonary lymphangiomatosis.

PULMONARY METASTASES FROM A UTERINE LEIOMYOSARCOMA

This is a difficult differential diagnosis. In any woman with pulmonary infiltrates, a full pelvic examination should be carried out. HMB-45 positivity is a helpful feature.

HISTIOCYTOSIS X

This lesion is histologically dissimilar from pulmonary LAM in that there are small stellate areas that initially involve the lung and stain positively for

S100 protein. Only focal SMA and no HMB 45 staining is seen. There is no association between LAM and cigarette smoking.

CLEAR CELL TUMOR OF THE LUNG

The only reason for including this rare, usually solitary lesion in the differential diagnosis is that the cells are positive with HMB-45.[299]

LYMPHANGIOMA/HEMANGIOMA

These lesions are rare in the lung. They may be seen in children. It is important to distinguish hemangioma from an arteriovenous malformation or an alveolar adenoma, which stains positively for cytokeratin.

DIFFUSE PULMONARY LYMPHANGIOMATOSIS

This lesion, although sounding similar to LAM, is clinically distinct. It is rare, just one series having been recorded in North America.[300]

Clinical Features

There were seven males and one female, with a mean age at presentation of 10 years (age range: 1 month to 33 years). The presenting symptoms included wheezing, "asthma," recurrent pneumonia, cough, and mild exertional dyspnea. One case was identified because of abnormal spirometry.

The disease was progressive, especially in children. Two patients died, one from acute pulmonary hemorrhage and the other after heart-lung transplantation.

On examination, either there were no abnormal physical signs or breath sounds were decreased, with basal crepitations. There was clubbing in one case. Pulmonary function tests showed a restrictive defect.

Chest radiographs showed interstitial infiltrates, more marked in the lower lobes in all cases, and pleural effusions in five cases.

Pathology

There were no specific gross features, but histology was characterized by extensively anastomosing endothelial-lined spaces of variable size. These were in the line of the lymphatics, with a subpleural, paraseptal, perivascular, and peribronchial distribution (Fig. 29-127). The spaces contained lymph. Collagen, along with asymmetrically arranged bundles of spindle cells, with a clear to slightly eosinophilic cytoplasm and oval to cigar-shaped nuclei, were identified in all cases in relation to the lymphatics (Fig. 29-128). In the surrounding lung were hemosiderin-laden macrophages.

Vimentin was positive in both lining and spindle cells. Most of the cases were positive for actin and desmin. No HMB-45 or estrogen receptor was identified in the spindle cells. One case showed focal positivity for progesterone receptor. The lining cells stained for factor VIII–related antigen and in half the cases for *Ulex* as well.

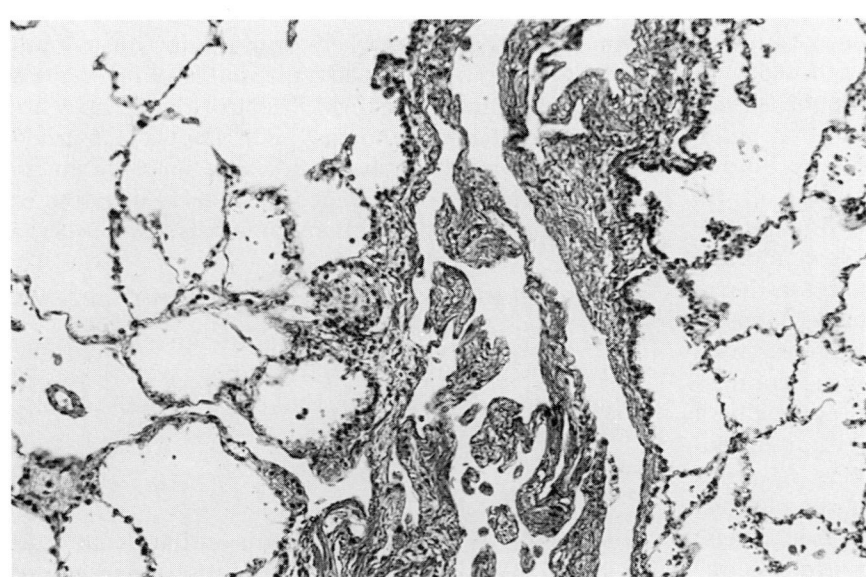

FIGURE 29-127
Lymphangiomatosis with a proliferation of spindle-shaped cells in a lymphatic. (H&E, ×87.5.) (*Reproduced with permission of T. Colby, M.D., Mayo Clinic, AZ.*)

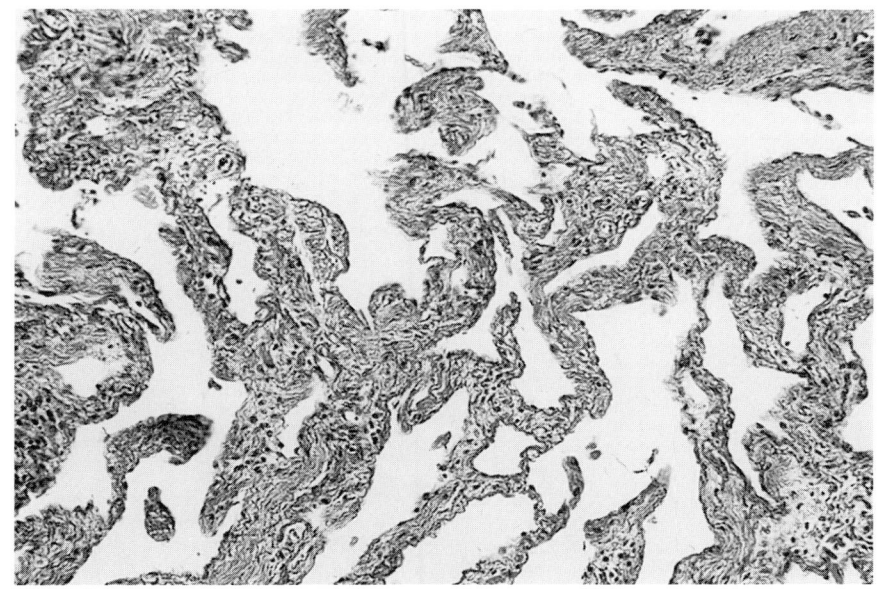

FIGURE 29-128
Lymphangiomatosis with a large area of spindle-shaped cells, forming vascular spaces. (H&E, ×87.5.) (*Reproduced with permission of T. Colby, M.D., Mayo Clinic, AZ.*)

Electron microscopy on one case showed that the spindle cells were invested with a basal lamina, consistent with smooth muscle cells. There were subplasmalemmal densities in the cytoplasm.

Differential Diagnosis

LAM

This presents usually in women of childbearing age and extends outside the existing lymphatics to invade alveoli, small vessels, and lymphatics. The spindle cells are HMB-45 positive. The lung shows emphysema or mass lesions.

PULMONARY CAPILLARY HEMANGIOMATOSIS

This disease shows a proliferation of small capillaries around blood vessels and bronchi. It is characterized by pulmonary hypertension, with the changes of pulmonary venoocclusive disease. No smooth muscle proliferation is seen in pulmonary capillary hemangiomatosis.

LYMPHANGIOMA

This produces a tumor, rather than being a diffuse pulmonary disease process.

KAPOSI'S SARCOMA

Differentiation between this and diffuse pulmonary lymphangiomatosis may be difficult, since occasional cases of the latter disease show red blood corpuscles in capillary spaces. The tumor cells in Kaposi's sarcoma are compact, are associated with smaller vasular spaces, show mitoses, and are more hyperchromatic than in diffuse pulmonary lymphan-

giomatosis. No PAS-positive globules are seen in the latter disease.

INTRAPULMONARY LYMPHANGIOMA

Mediastinal lymphangioma is a well-documented lesion, but only one case of a pulmonary lymphangioma has been described. This has been associated with mediastinal tumor of the same origin. It is possible that three further cases, all in adolescents,[301] represented this lesion, but the authors termed it *lymphangiectasis*. The case presenting in an adult[302] was in a 54-year-old male ex-smoker. He presented with a right hilar mass, which remained the same size for several years.

Thoracotomy showed a cystic mass intertwined and enveloping right hilar structures. Multiple lung cysts, measuring up to 5 cm in diameter, many filled with blood clot, were embedded in the hilum and perihilar parenchyma.

Histologically, the cysts were lined by flattened endothelium and had varying degrees of fibroblastic proliferation in their walls as well as focal aggregates of lymphocytes without germinal centers (Fig. 29-129). Collections of thin-walled, ectatic vascular spaces were seen. Similar structures following the lines of lymphatic sinuses were seen in a mediastinal node.

PULMON ARY FIBROMA

It is probably wrong to separate pulmonary fibroma from myxoma, since fibromatous and myxomatous areas coexist in some cases. These tumors are rare. Gilbert and coworkers[303] identified 39 fibromas

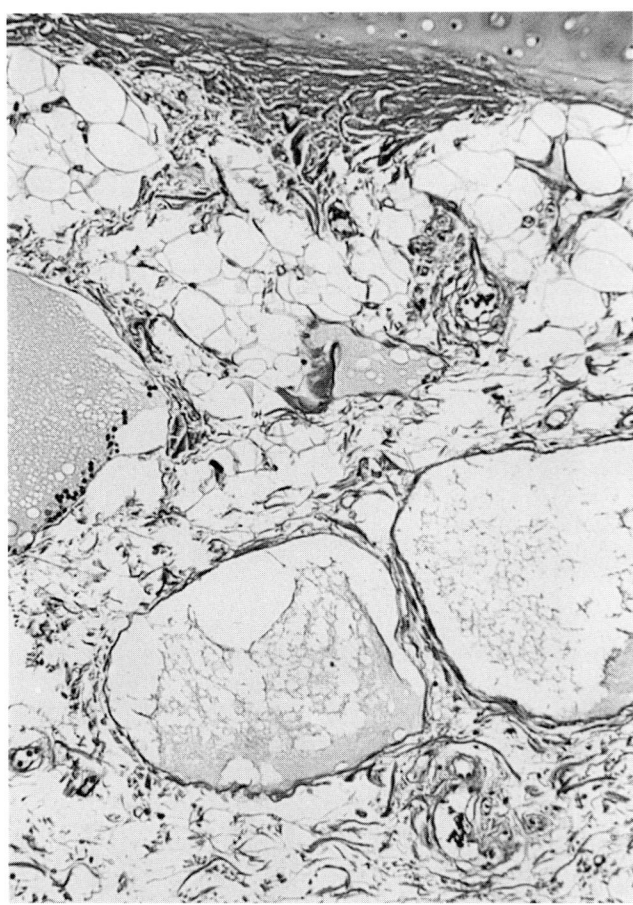

FIGURE 29-129
Lymphangioma with large, dilated lymphatics, lying outside the bronchial wall. (H&E, ×92.5.) (*Reproduced with permission of T. Colby, M.D., Mayo Clinic, AZ.*)

among 546 tracheal tumors reviewed. Recently, a series of 43 intrapulmonary cases were presented.[304]

Clinical Features

Clinically with tracheal tumors, there may be difficulty in breathing and the patient is suspected of having asthma. Hemoptysis, cough, and stridor are other symptoms. Tracheal tumors have a smooth, round surface, and they may create a ball-valve effect during respiration. They are asymptomatic until 50 to 70 percent of the trachea is occluded. The chest radiograph is usually normal.

Most patients with intrapulmonary localized fibrous tumors are asymptomatic and have solitary, circumscribed intrapulmonary masses. There is little difference between the sexes or the side affected. The age range of patients is 20 to 83 years, with a median of 58 years.

Pathology

Pulmonary fibromas are well-defined, circular, hard, and grayish white and often grow in the lung sub-

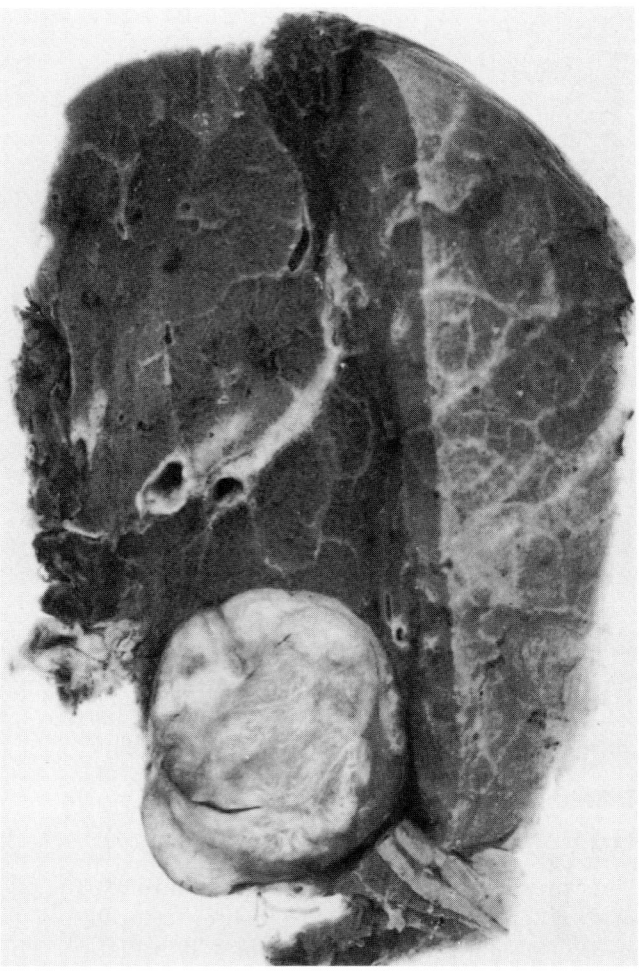

FIGURE 29-130
Peripheral fibroma, showing a well-defined tumor with a whorled appearance at the periphery of the lung. Differentiation from a benign pleural lesion may be difficult in a case such as this.

stance as tumors (Fig. 29-130). There is a range in the size of the tumors from 1.2 to 23 cm. Most are relatively small, however, with a median size of 3 cm.[304] They may be subpleural and are then difficult to distinguish from pleural fibromas.

Histologically, they are composed of spindle cells with a variable amount of collagen (Fig. 29-131) but may show evidence of myxomatous change.[305] Pure primary pulmonary myxomas have been described.[306] The matrix is finely particulate and basophilic, resembling mucin, and stains with alcian blue. Other patterns were seen in the large series mentioned above,[304] with storiform or hemangiopericytoma-like areas.

Four tumors were classified as malignant, based on the presence of at least two of the following criteria: necrosis (three of the four), more than four mitoses per 10 high-power fields (three), vascular

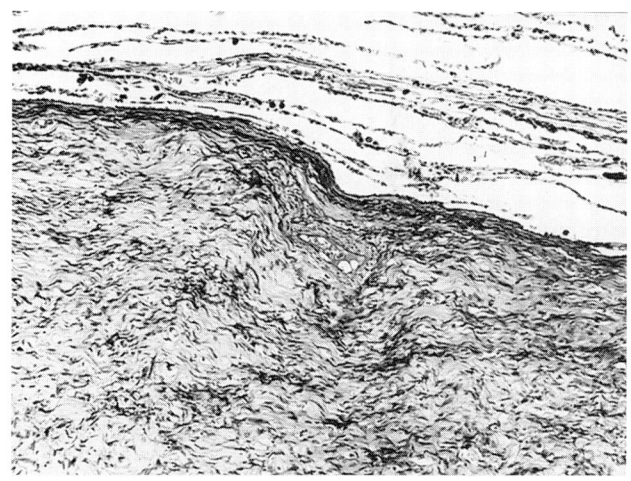

FIGURE 29-131
Peripheral fibroma with wavy bands of loose, myxoid fibrous tissue separated from the lung. (H&E, ×103.6.)

invasion (one), and increased cellularity (all four). These tumors had infiltrative borders.

Immunoctyochemically, the intrapulmonary localized fibrous tumors were positive for vimentin but not cytokeratin. However, only six cases were available for study. Five of the six cases were positive for CD34. This marker was negative in other spindle cell tumors and sclerosing hemangiomas.

Ultrastructurally, the spindle cells are without junctions or basal lamina; if there is myxoid change, there is an amorphous, focally collagenous background. The cells show irregular borders and contain numerous mitochondria, abundant, often dilated, rough endoplasmic reticulum, and frequent Golgi apparatuses. Numerous intermediate filaments without focal densities are present.

Differential Diagnosis

This includes nerve sheath tumors, which in the case of neurilemmoma should be positive for S-100 protein. Spindle cell carcinoma shows mitoses and nuclear pleomorphism and may be cytokeratin and CEA-positive. Leiomyomas have a characteristic nuclear pattern, but in problem cases SMA is helpful. Sclerosing hemangioma has spindle cell areas, but there are also other patterns, such as papillary and hemorrhagic. CD-34 is negative in sclerosing hemangioma.

INVASIVE FIBROUS TUMOR OF THE TRACHEOBRONCHIAL TREE

This rare entity has been described by the Boston Group.[307] It was seen in patients ranging from 8 to 68 years old, and there was a marked female predomi-

nance. The sites affected were the trachea, carina, left main bronchus, other main bronchi, and lobar bronchi. Five patients had symptoms of large-airway obstruction, often with wheeze or stridor and occasionally hemoptysis and chest pain. One patient presented with a radiologic nodule after mastectomy for infiltrating lobular carcinoma of the breast.

All patients were disease-free at periods ranging from 1 to 11 years after operation, although the lower tracheal tumor recurred after 1 year, necessitating further surgery.

Pathology

The tumor may grow circumferentially in the wall of the airway to form a nodular, eccentric expansion into the lumen. It may be sessile, measuring up to 3 cm in diameter, or polypoid, measuring up to 1.3 by 1 cm.

Histologically, the tumors are composed of proliferating fibroblasts. They grow beneath the surface epithelium, entrapping mucous glands and causing squamous metaplasia (Fig. 29-132). The fibroblasts grow in fascicles (Fig. 29-133), and in some cases a storiform pattern is present. Mild nuclear atypia was present in four cases reported and more moderate in three.[307] A few multinucleated tumor cells but no xanthoma cells were identified. There was negligible inflammation, and mitoses were less than two per 10 high-power fields. One case had a very myxoid stroma, resembling a myxoid malignant fibrous histiocytoma, but the nuclear changes were not those of a

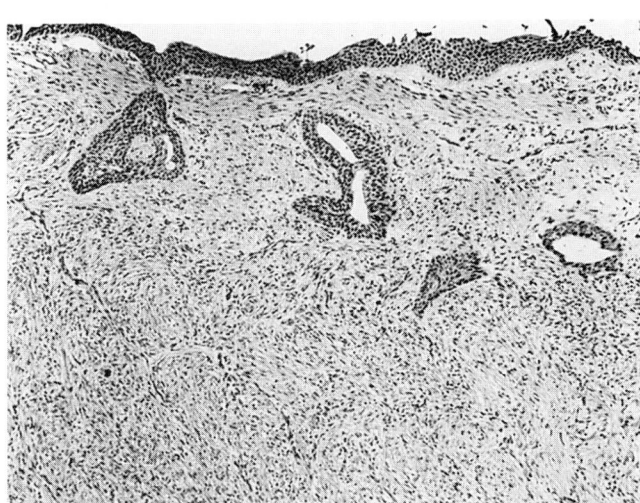

FIGURE 29-132
Invasive fibrous tumor of the trachea with overlying squamous metaplasia and regular fibroblastic tissue encircling the mucous glands. These also show squamous metaplasia. (H&E, ×77.)

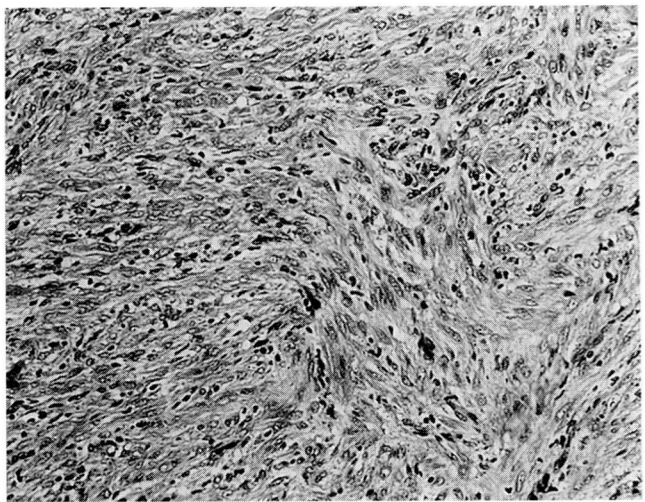

FIGURE 29-133
Invasive fibrous tumor of the trachea with fibroblastic tissue showing a fasicular growth pattern and a slight tendency to a storiform pattern. (H&E, ×154.)

high-grade malignancy. Another case had a circumferential growth pattern, with broad bands of densely collagenized stroma (Fig. 29-134) and few nuclei. It resembled a keloid in its deeper portion. There was peribronchial extension. All the tumors invaded at least to the inner perichondrium, although some extended into peribronchial adventitia.

Differential Diagnosis

INFLAMMATORY PSEUDOTUMOR

Invasive fibrous tumor of the tracheobronchial tree is endotracheal or endobronchial and usually causes stridor or wheezing. These features are unusual in inflammatory pseudotumor, which is usually a

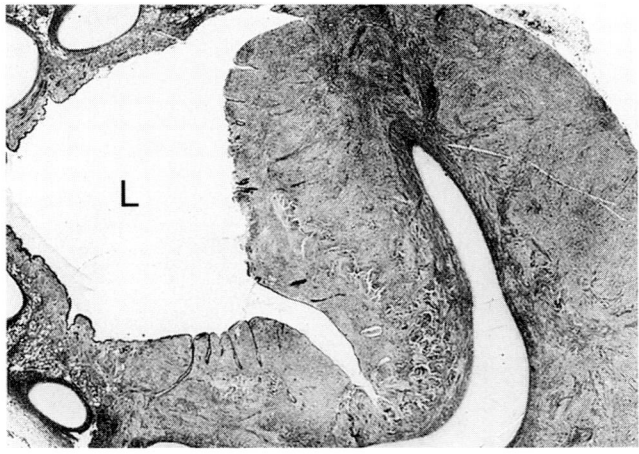

FIGURE 29-134
Invasive fibrous tumor of the trachea with the lumen (L) showing some collapse. There is replacement of the bronchial wall by fibroblastic tissue and loss of the glands as well as cartilage. (H&E, ×6.6.)

peripheral lesion. In addition, there is only a proliferation of spindle cells and no inflammation, as seen in a pseudotumor.

MALIGNANCY

Some cases were initially classified as malignant fibrous histiocytoma, fibrosarcoma, or malignant fibrous histiocytoma. Once again, the absence of xanthoma cells, multinucleated histiocytes, myxoid stroma, mitoses, and inflammation helps differentiate invasive fibrous tumor from these conditions. Invasive fibrous tumor is unlike a fibromatosis because the former invades by broad and blunt processes, the latter by thin, irregular strands.[308] Differentiation from a leiomyosarcoma is made by the absence of mitoses, and smooth muscle actin would be negative. In the presence of mitoses, it may be difficult to distinguish invasive fibrous tumor of the tracheobrochial tree from a low-grade fibrosarcoma. Tan-Liu and coworkers[307] could not exclude the possibility that the tumor was not a low-grade fibrosarcoma. This is unlikely, however, as no metastases were seen in their cases.

Endoscopic biopsy of an invasive fibrous tumor may be insufficient to distinguish it from inflammatory pseudotumor or malignant fibrous histiocytoma, because of the size of the biopsy.

CONGENITAL SUBGLOTTIC FIBROMATOSIS

The rare lesion has probably been described in the past as a fibroma. It is akin to the fibromatoses, well documented elsewhere.[309] Clinically, these lesions may present as tracheal "fibromas," with stridor at birth.[310] In one case a pedunculated, broad-based, pea-sized lesion arose beneath the left vocal cord and extended into the first and second tracheal rings. The lesion recurred 9 years later.

Histology showed dense fibrocollagenous tissue covered in part by respiratory and squamous epithelium. There were sparse lymphocytes and blood vessels. No granulation tissue was identified.

MYOEPITHELIOMA OF THE LUNG

Myoepitheliomas are found mainly in the salivary glands. A case report described one of these lesions in the lung;[311] it presented as an asymptomatic left upper lobe nodule in a 60-year-old man. A recent case report describing a similar tumor in a 66-year-old Japanese man, with no history of a salivary gland tumor, has been described.

Macroscopically, the tumor was tan-yellow and well circumscribed. It measured approximately 3 cm in diameter and had no bronchial origin.

Histologically there was a thin fibrous capsule, inside which were interdigitating bundles of spindle-shaped cells (Fig. 29-135) with no epithelial differentiation. The recent case[311a] showed solid and glandular areas. The tumor cells had nuclei and pale or clear cytoplasm. Small fascicular areas of spindle-shaped cells were also detected. The glandular areas showed ductal or papillary structures lined by inner columnar cuboidal epithelial cells and a basal layer of polygonal cells. Focally there were areas of pale acid mucopolysaccharide–rich matrix separating the cells. Nuclear pleomorphism was mild, and mitoses were rare. Reticulin fibers surrounded individual tumor cells. There was S-100 protein positivity in the cytoplasm and nuclei. Cytokeratins were negative in this case, but some myoepitheliomas may stain with cytokeratin.[312] Thus, the absence of intermediate filaments and lack of reactivity of the tumor to cytokeratin do not rule out a diagnosis of myoepithelioma.

This diagnosis is based on electron microscopy.

Ultrastructurally, the spindle-shaped cells showed numerous parallel and longitudinally oriented cytoplasmic myofilaments 6 nm in diameter. No well-defined focal densities or intermediate filaments were identified. Some of the cells were joined by desmosomes.

This tumor may represent a variant of pleomorphic adenoma or may be a distinct type of monomorphic adenoma.[313]

Differential Diagnosis

LEIOMYOMA

These have numerous myofilaments. Smooth muscle actin is positive.

LOCALIZED FIBROUS TUMOR

The myoepithelioma described above was an intrapulmonary tumor, not a pleural one. CD34 reactivity is very strong in a localized pulmonary fibroma. S-100 is positive in myoepithelioma.

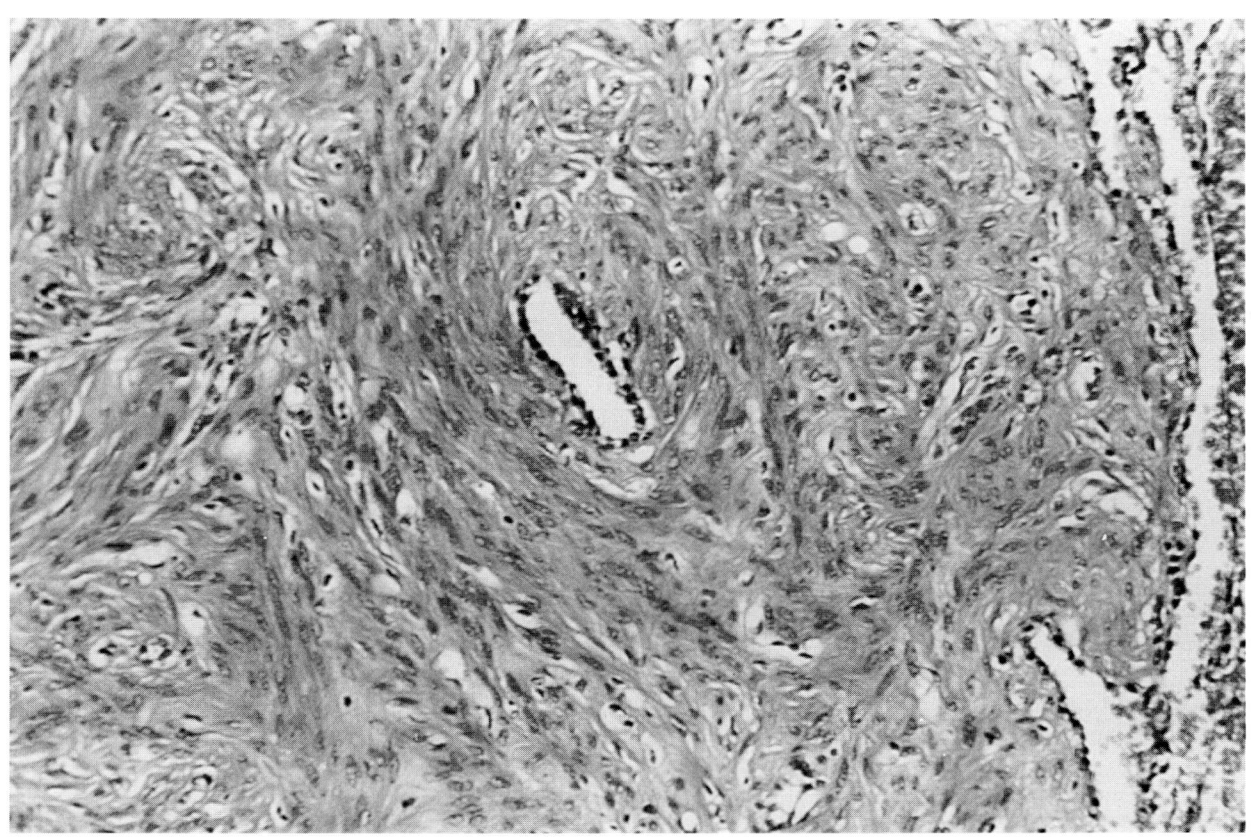

FIGURE 29-135
Myoepithelioma with bands of muscle with several well-defined glandular spaces. (H&E, ×87.5.)

NEURILEMMOMA AND NEUROFIBROMA

In neurilemmoma there are characteristic Antoni type A or B growth patterns. Myoepithelioma may be difficult to differentiate from neurofibroma, but the latter tumor does not demonstrate myofilaments.

MENINGIOMA OF THE LUNG

This tumor does not stain for S-100 protein and usually has psammoma bodies.

PRIMARY OR METASTATIC CHONDROSARCOMA

If one considered only immunocytochemistry, without recourse to clinical details or histology, this tumor would be positive for S100 but would show no desmosomes or evidence of myofilaments ultrastructurally.

LEIOMYOMA OF THE LOWER RESPIRATORY TRACT

Clinical Features

These are rare tumors, accounting for under 2 percent of benign pulmonary tumors.[314–317] Fewer than 75 cases have been reported.[318] In a review,[319] the ratio of females to males was 1.5:1. The preponderance of females[320] in earlier reports could have been due to underdiagnosis of metastatic low-grade uterine leiomyosarcomas.

The mean age was 35 years for bronchial and parenchymal tumors[321] and 40.6 for tracheal tumors.[314] One-third of patients present below the age of 20 years, and one case described was in a three-month-old infant.[322]

Leiomyomas are usually identified in the distal tracheobronchial tree. Only three of 51 cases reviewed by Orlowki and colleagues[321] arose from main-stem bronchi. In a literature review, White and coworkers[314] found 16 percent of these tumors to be in the trachea, 33 percent in the bronchus, and 51 percent in the lung parenchyma. Tracheal tumors usually involved the lower third of the posterior membranous portion.

These lesions may be asymptomatic and incidental radiologic findings.[320] Patients with bronchial tumors have obstructive symptoms, including cough, wheeze, chest pain, and recurrent pneumonia. Some cases may be initially diagnosed as asthma. There may be paroxysmal attacks of wheezing, precipitated or relieved by change in posture. Hemoptysis, orthopnea, and paroxysmal nocturnal dyspnea, without cardiac involvement, are additional clinical symptoms.

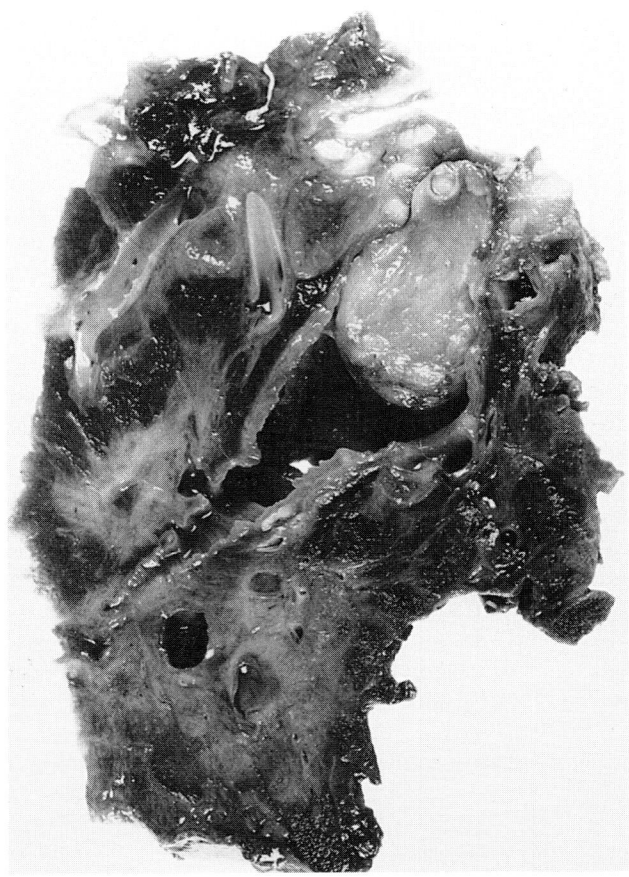

FIGURE 29-136
Leiomyoma as a central gray, tumor with distal dilatation of the bronchus and fibrosis in the lung.

There are no typical radiologic features of this tumor to distinguish it from other benign pulmonary lesions.

Macroscopic Pathology

Tumors in the trachea and main bronchi grow as broad-based polypoid lesions extending along the bronchial lumina as tongues of soft tissue. They may occlude the bronchial lumen (Fig. 29-136). Parenchymal tumors have a more fibrous consistency, with a whorled pattern resembling that of uterine leiomyomas.

Histology

Histology shows typical smooth muscle fibers with spindle cells, fusiform nuclei with typical cigar-shaped ends, and a bland chromatin pattern (Fig. 29-137). The bundles grow in whorls (Fig. 29-138) or bands, and there is no significant atypia or mitotic activity.

Smooth muscle actin and desmin are useful markers for diagnosis.

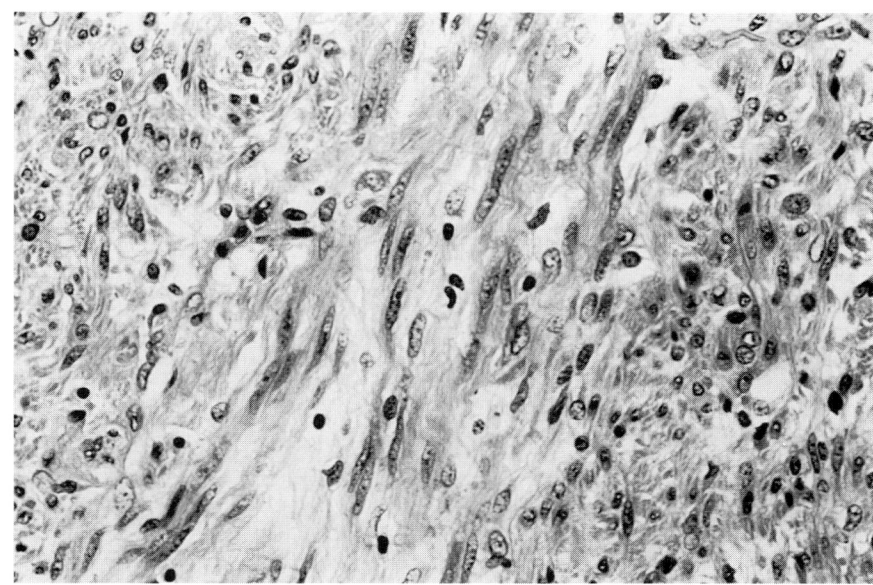

FIGURE 29-137
Leiomyoma showing cigar-shaped nuclei.
(H&E, ×219.)

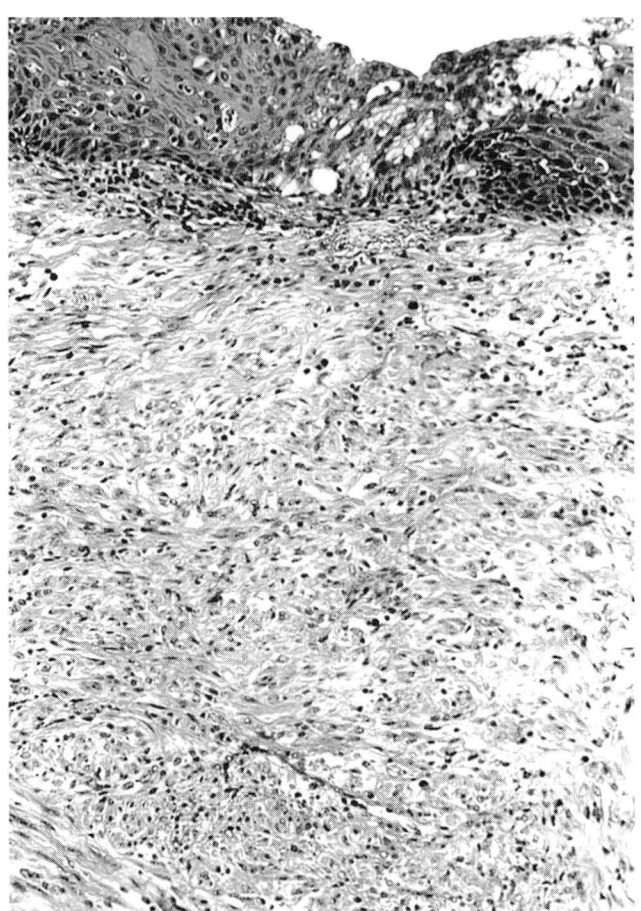

FIGURE 29-138
Leiomyoma with a whorled pattern and squamous metaplasia on the surface of the bronchus. (H&E, ×92.5.)

Electron Microscopy

Electron microscopy confirms the nuclear detail. There are cytoplasmic filaments with an average thickness of 70 Ångstroms, arranged both hapazardly and in parallel bundles. In some areas, especially below the cell membrane, the filaments are organized into dense bodies. The rest of the cytoplasm contains mitochondria, areas of dilated endoplasmic reticulum with ribosomes, and small amounts of glycogen. In the cell membrane are a large number of pinocytotic vesicles.

Histogenesis

These tumors are thought to arise from smooth muscle of the trachea, bronchi, bronchioles, or even blood vessels. However, in cryptogenic fibrosing alveolitis and chronic bronchitis and emphysema, in which there is smooth muscle proliferation, either in the lung parenchyma or in bronchi, there is no documented evidence of an increase in incidence of this tumor.

Differential Diagnosis

METASTASIZING FIBROLEIOMYOMA OF THE UTERUS[323]

These are probably deposits of well-differentiated leiomyosarcoma or, less likely, a multicentric benign leiomyomatous tumor. In the absence of a complete pelvic examination, it is extremely difficult in a female to rule out a secondary low-grade leiomyosarcoma when there is a solitary pulmonary nodule. In males with leiomyosarcoma elsewhere in the body, it may be difficult to distinguish this from a solitary pulmonary leiomyoma.

LEIOMYOMATOUS HAMARTOMAS[323a]

These contain bundles of smooth muscle, collagen, and glandular epithelial cells. These hamartomas may occasionally regress spontaneously during pregnancy.

NEURILEMMOMA

S-100 and neurofilament proteins should be positive in this tumor.

NEUROFIBROMA AND FIBROMA

Neurofibroma may be difficult to differentiate from a leiomyoma with areas of fibrosis. S-100 positivity is variable in this tumor, but smooth muscle actin is negative. A fibroma has larger areas of fibrosis and will not stain positively with smooth muscle actin. It is positive with CD34.

SPINDLE CELL CARCINOID

Spindle cell carcinoid should always be considered in the differential diagnosis of a spindle cell tumor.

SPINDLE CELL CARCINOMA

This tumor has areas with nuclear pleomorphism. Foci of necrosis are often present, and cytokeratin and CEA immunopositivity are seen.

LOCALIZED PLEURAL FIBROUS TUMOR

This tumor is pleural-based and would be vimentin positive. Ultrastructurally, the features are those of fibroblasts rather than mesothelial cells.

INFLAMMATORY PSEUDOTUMOR

Areas of this tumor (page 29-92) may have fascicles resembling smooth muscle fibers. However, the nuclear features are those of plasmacytoid cells.

MYOEPITHELIOMA

In this rare tumor there may be smooth muscle actin positivity but a lack of typical cigar-shaped nuclei. Electron microscopy shows very few myofilaments and dense bodies.

PULMONARY LEIOMYOSARCOMA

Clinical Features

These are rare tumors: Fewer than 100 cases of primary leiomyosarcoma have been reported.[314,315,324,325]

The tumor resembles primary bronchial carcinoma, with a male predominance and a hilar location, but the patients are younger, usually in their third or fourth decade. The tumor has been reported in children.[326] Patients have cough, hemoptysis, weight loss, and obstructive pneumonia.

Macroscopic Features

As with leiomyoma, the tumor may present as a polypoid endobronchial lesion or as a peripheral, apparently encapsulated yellow-white tumor. It may measure from 3 to 7 cm in diameter and spread to involve hilar structures of the lung, including nodes, pleura, and pericardium. Secondary tumor is seen in the liver, adrenals, and cerebrum, but the presence of metastases is not related to the size of the tumor.

Histology

The tumor has a fascicular growth pattern, with spindle cells containing oblong nuclei (Figs. 29-139 and 29-140). Some tumors may show cells with an epithelioid appearance, with oval to round nuclei (Fig. 29-141). Cellular borders are distinct, with an eosinophilic cytoplasm. Bizarre, multinucleated tumor giant cells may be found in some cases. Focal hemangiopericytomatous areas with "staghorn" vessels may be seen, as well as occasional nuclear palisading to suggest a neurilemmoma. Necrosis is present, as are mitoses, numbering up to 10 per 10 high-power fields.

Morgan and Ball[327] showed that the mitotic rate correlated well with survival, but as they studied relatively few cases, little import can be placed on this. This is shown by the fact that one of their patients with a low count survived for only one year. This may be explained by the fact that the tumor was large and had already invaded the mediastinum, as well showing scattered, bizarre, multinucleated cells. The latter feature is not a predictable one for leiomyosarcomas, if the analogy with uterine symplastic leiomyomas is confirmed in the lung.

PAS was positive in a few tumor cells but was removed by diastase. Reticulin was demonstrated around individual cells in some areas of the tumor.

The histologic growth pattern, nuclear shape, smooth muscle actin (Fig. 29-142), and desmin, as well as ultrastructural examination, confirm the diagnosis.

Differential Diagnosis

This is as given above for leiomyoma. It has been suggested that in women the diagnosis of leiomyosar-

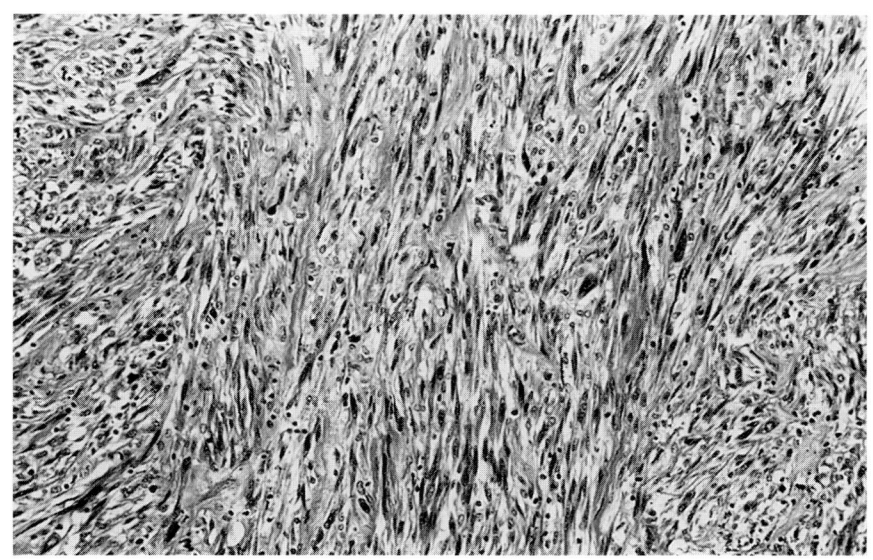

FIGURE 29-139
Leiomyosarcoma with a fascicular growth pattern and easily identifiable nuclear hyperchromatism. (H&E, ×87.5.)

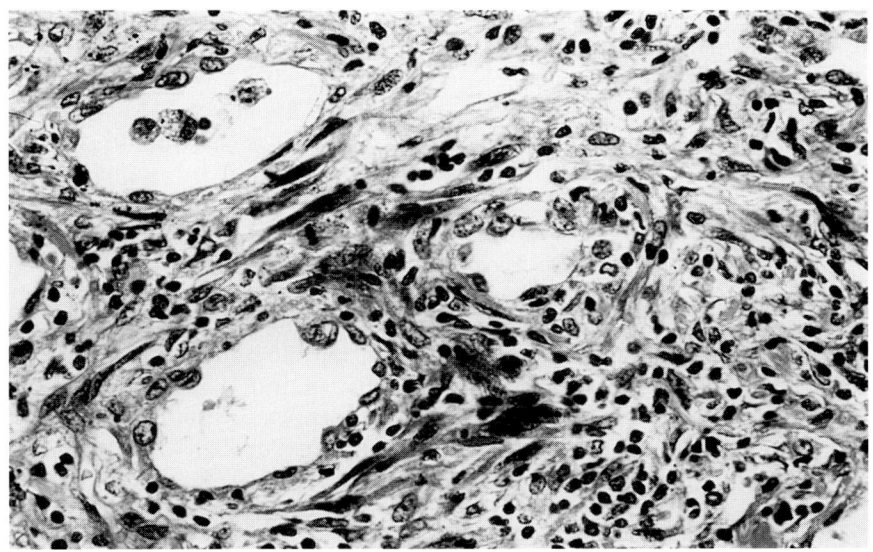

FIGURE 29-140
Leiomyosarcoma with marked nuclear hyperchromatism. (H&E, ×87.5.)

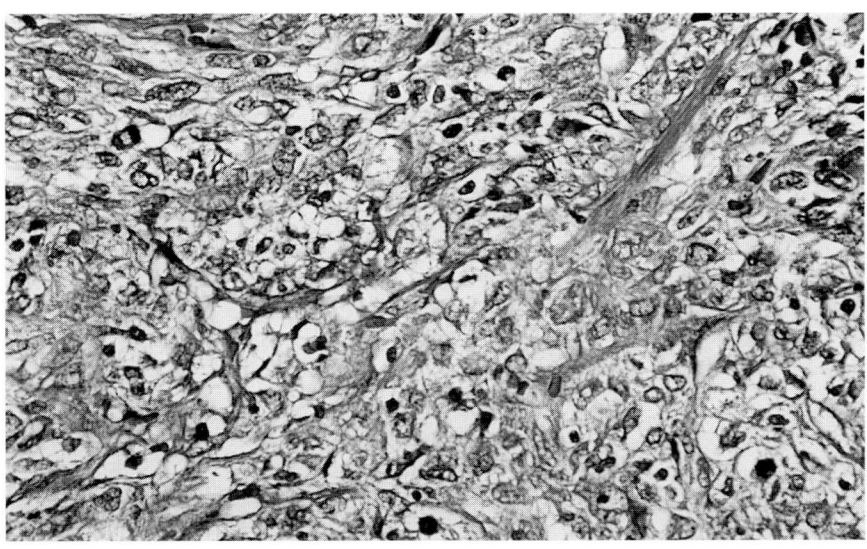

FIGURE 29-141
Leiomyosarcoma with an epithelioid growth pattern. (H&E, ×87.5.)

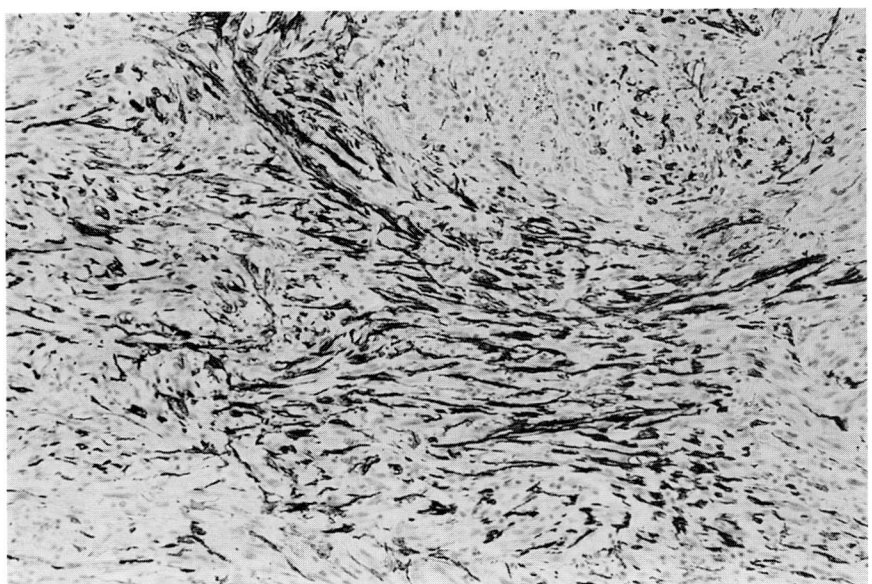

FIGURE 29-142
Leiomyosarcoma as shown in Figs. 29-139 and 29-140 stained for smooth muscle actin. (Immunoperoxidase, ×87.5.)

coma of the lung should never be considered as a primary tumor.

Leiomyosarcomas of the Pulmonary Artery (Chap. 32)[328]

These are rare tumors and present with malaise, weight loss, dyspnea, cough, hemoptysis, pleuritic pain, and congestive cardiac failure. The tumor is more common in females than in males. When it affects the main pulmonary artery, the hilar mass projects into the lung parenchyma in a pulmonary artery distribution. It is rare for the diagnosis to made preoperatively, and the initial diagnosis is usually pulmonary embolism. Prognosis is poor, and many patients die in the early postoperative period.

NEUROGENIC TUMORS

These include neurilemmoma, neurofibroma, and malignant nerve sheath tumor (malignant schwannoma).

NEURILEMMOMA

In rare instances, a neurilemmoma may be endotracheal.[329-332] Neurogenic pulmonary tumors, either bronchial or intrapulmonary, are also very uncommon.[333-335] Just over 25 cases have been described in the world literature, with just under half occurring in Japan.

Clinical Features

There is no sex predilection for endotracheal tumors, but bronchial or intrapulmonary neurilemmomas have a slight female predominance. The age range is wide, the youngest patient being 5 years old and the oldest 78 years old (mean: 39.5 years). Most of the reported intrapulmonary cases originated in the region of the terminal segmental bronchus. Some present as an intrabronchial polypoid mass.

Symptoms are usually mild, consisting of a dry or productive cough, fever, variable loss of voice, and hemoptysis. Chest pain, as in leiomyoma, which may be related to the position of the patient or aggravated by exercise, may be present. Sometimes there may be almost total tracheal obstruction.

Radiologically, the tumor appears as a round or ovoid lobulated homogeneous mass with a sharp outline.

Macroscopic Features[336]

These tumors are encapsulated, sometimes lobulated smooth-surfaced masses with a greatest diameter varying from 3 to 11 cm. Tracheal tumors may cross the intercartilaginous fibrous septa, giving the tumor a dumb-bell appearance. They may also use the weakest area, i.e., the posterior part of the bronchial wall, for their growth, which may also produce a dumbbell tumor (Fig. 29-143). On cut surface they are firm and yellow because of lipid. Some are multiloculated and cystic. If malignant change is present, the tumor is not encapsulated. Necrosis, hemorrhage, and calcification have been described, but these features are uncommon. Necrosis, hemorrhage, and lack of encapsulation are more suggestive of a malignant tumor.

Histology

A typical neurilemmoma has a thin fibrous capsule (Fig. 29-144), formed by compression of perineural

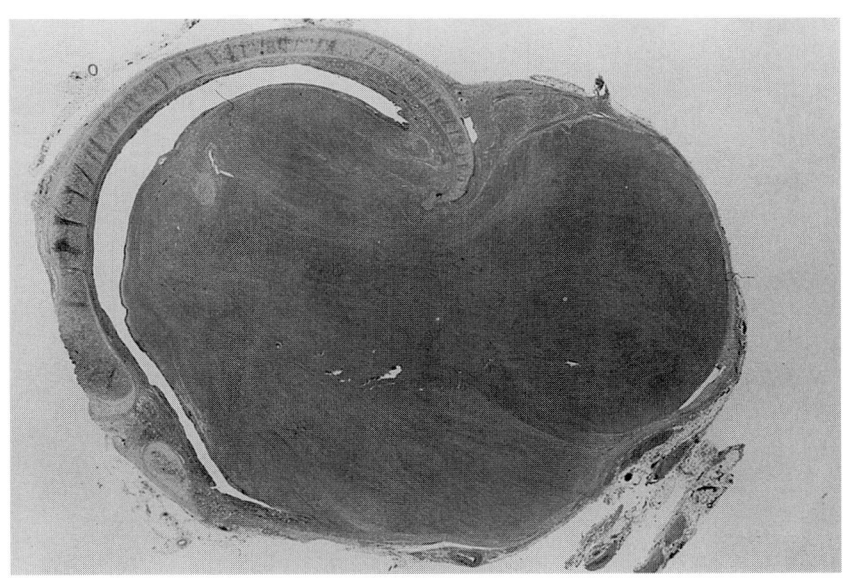

FIGURE 29-143
Neurilemmoma which has largely occluded the bronchial lumen and has also grown posteriorly through the posterior wall, the line of least resistance.

tissue. Antoni type A (cellular pattern) is formed of compact spindle cells with elongated nuclei arranged in parallel rows, creating a palisaded pattern (Fig. 29-145). Admixed with this or in separate areas, the tumor shows a less cellular or Antoni type B pattern, with elongated cells arranged in an irregular fashion and separated from one another by a matrix that stains poorly with hematoxylin and eosin. Hemorrhage, necrosis, and cystic change may be seen in benign tumors. Mitoses and nuclear pleomorphism are unusual.

Malignant transformation of a pulmonary neurilemmoma is extremely rare.[337] The following criteria have been proposed for malignant change: metastasis to a distant organ, nuclear pleomorphism, and many mitoses. Occasionally these tumors may present as pulmonary secondaries.[338]

Tumor cells are strongly positive for S-100 protein. This antibody is also positive in granular cell myoblastomas, malignant melanomas, and chondrosarcomas. A study of gastrointestinal schwannomas[339] also showed positivity for Leu 7, glial fibrillary acidic protein, and laminin.

Electron Microscopy

In contrast to the neurofibroma, in which there is a mixture of cell types, the neurilemmoma is derived from Schwann cells. These show numerous elongated cells with cytoplasmic processes, which tend to wrap around each other (Fig. 29-146). Well-defined basal lamina encompass tumor cells. Occasional desmosome-like junctions are seen. The

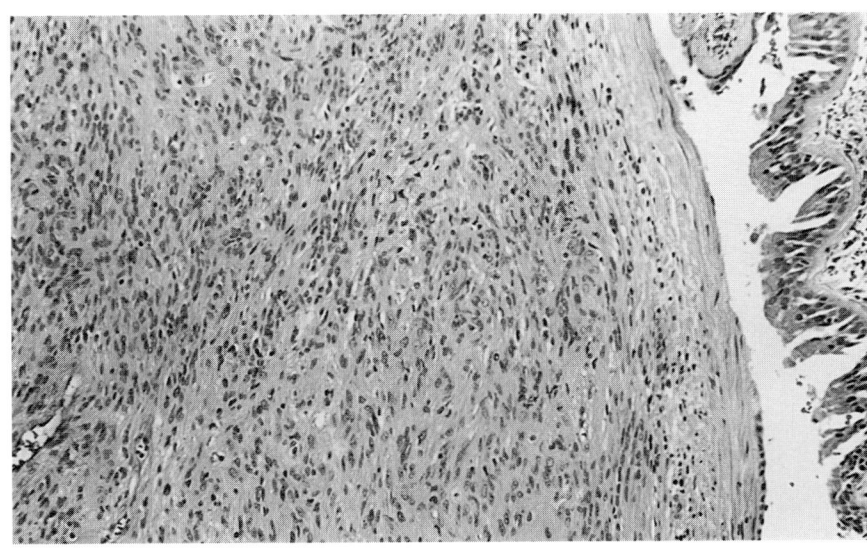

FIGURE 29-144
Neurilemmoma which has grown into the bronchial lumen and has a distinct fibrous capsule. (H&E, ×87.5.)

957

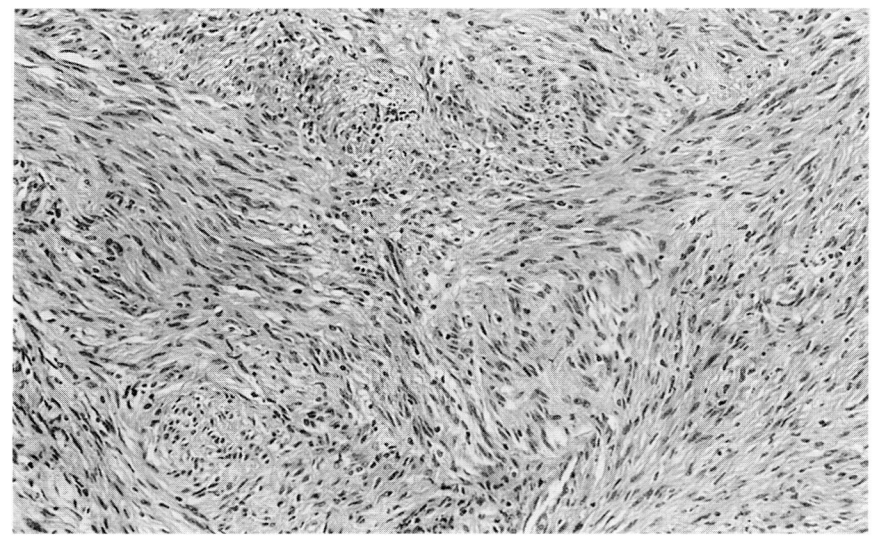

FIGURE 29-145
Neurilemmoma with well-marked pallisading. (H&E, ×87.5.)

cytoplasm contains scattered mitochondria, Golgi
apparatus, scarce mitochondria, and a small amount
of rough endoplasmic reticulum. Luse bodies may be
identified. (Fig. 29-147).

NEUROFIBROMA

The age range for these tumors is wide—6 to 63
years.[340,341] They may occasionally be found in the
trachea[330] and may present as pulmonary masses or
as part of neurofibromatosis. The latter disorder may
cause progressive tracheal and superior vena caval
obstruction[342] or present as discrete posterior mediastinal tumors. Pulmonary lesions have been seen in
a patient with neurofibromatosis.[343] In this case,
many subsegmental bronchi of the basal portion of
the left lower lobe had intramural, submucosal, circumferential small tan nodules. A small subpleural
tumor was found in the right lower lobe.

Histologically, neurofibromas show a mixture of
palisaded cells of neural origin and foci of fibrous tissue. There are areas composed of compact spindle
cells with many areas of nuclear palisading. They are
no different from such tumors seen elsewhere in the
body.

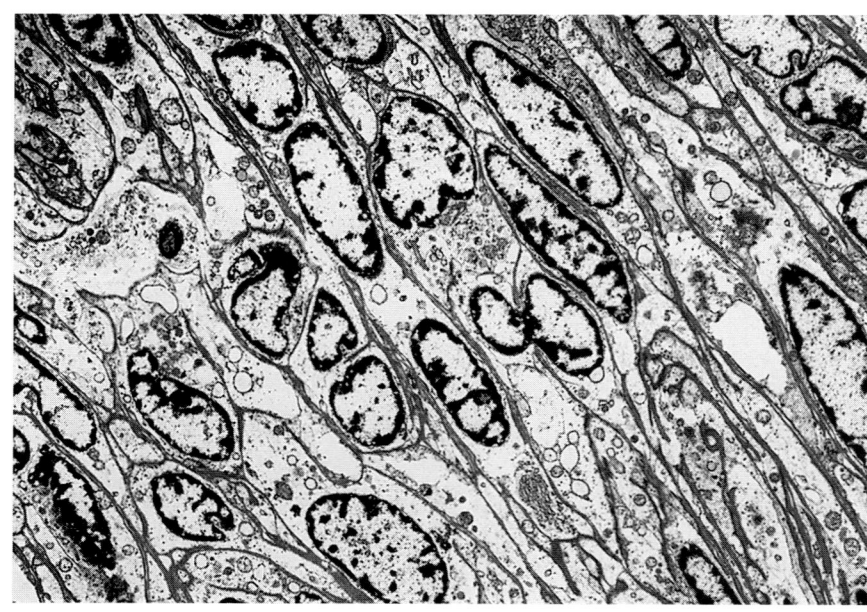

FIGURE 29-146
Neurilemmoma showing cells with well-
demarcated external lamina. No complex
folds are identified. (×2800.)

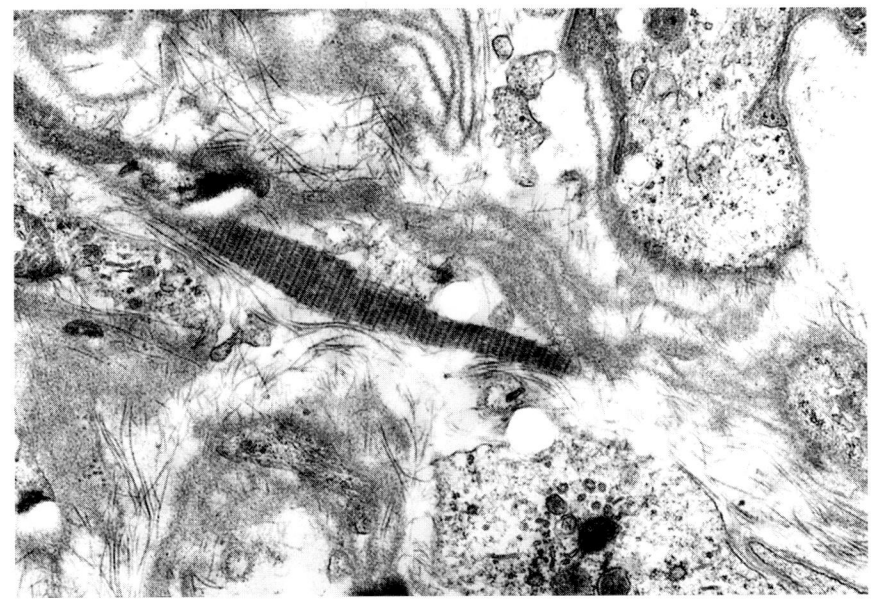

FIGURE 29-147
Neurilemmoma. A luse body, with the
typical banding is seen. (×15,750.)

S-100 protein positivity is present in neurofibroma but is not as prominent as in a neurilemmoma.
Electron microscopy shows both Schwann cells and
fibroblasts.

MALIGNANT NERVE SHEATH TUMOR (NEUROFIBROSARCOMA)

This tumor shows frequent mitotic figures, perineural invasion, indistinct cell borders, pleomorphic fusiform cells (Fig. 29-148), foci of necrosis, and
vascular invasion. The initial diagnosis may be a
spindle cell sarcoma NOS, and it is only with the
appropriate neural stain that the diagnosis becomes
clear.

Differential Diagnosis

SPINDLE CELL TUMORS (SEE "LEIOMYOMA," ABOVE)

In most cases of neurilemmoma, a palisaded pattern
makes diagnosis easy. S-100 protein is detected not
only in neurogenic tumors but also in granular cell
myoblastomas, malignant melanomas, and chondrosarcomas. The main tumor that could cause difficulty in this group is a malignant melanoma. HMB-45
should help differentiate malignant melanoma from
a primary neurogenic tumor.

SCLEROSING HEMANGIOMA

This can cause diagnostic problems. A positive
cytokeratin, seen in this lesion, will help to distinguish it from a neurilemmoma.

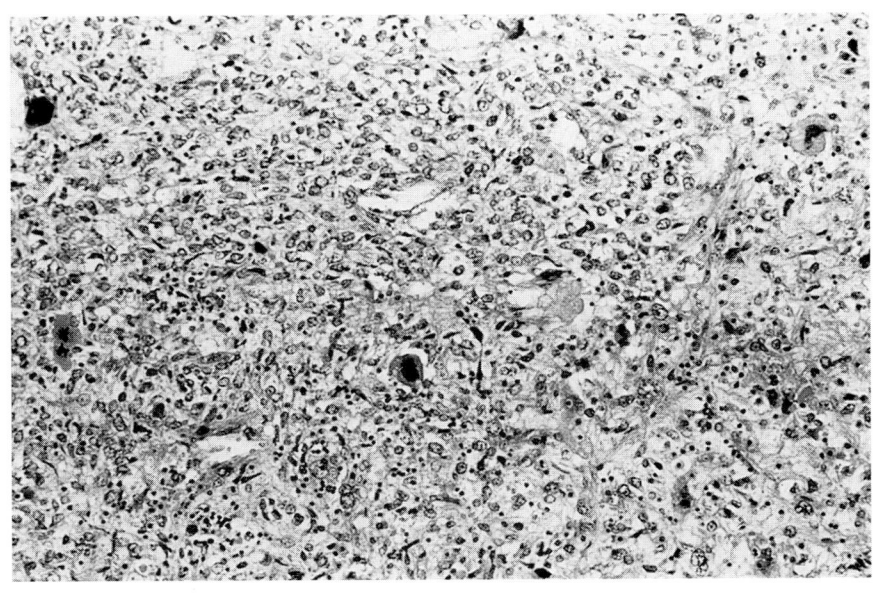

FIGURE 29-148
Malignant nerve sheath tumor with vacuolated cells and focal nuclear pleomorphism. S100 was strongly positive. (H&E,
×87.5.)

GRANULAR CELL MYOBLASTOMA

Clinical Features

This tumor has been found in various parts of the body, such as lung, skin, breast, and other soft tissues. It is rare in the respiratory tract, where it may be a solitary tumor in the trachea[344,345] or main bronchi.[346-348] It can be multicentric[349] or present as a coin lesion.[350,351] In one series of tracheal tumors,[345] all but one of the 10 cases described were in females. Robinson's team[352] collected 67 tracheobronchial granular cell myoblastomas and found an equal sex incidence. There has been a suggestion that the incidence is increasing, with the number of documented cases rising from 40 before 1975 to 90 within the next 10 years.[353] This increase coincided with the advent of fiberoptic bronchoscopy.

The age range was 6 to 59 years, with a mean of 36 years. There is a predilection for black races.

The right main bronchus is affected in 59 percent of cases; 15 percent of patients had multiple lesions. In one case reported, both the trachea and biliary tree were affected.[354] The tumor has been seen in an HIV-positive patient,[355] but this may have been a fortuitous association.

The most common symptoms are cough and chest pain with fever and night sweats. Some patients have pneumonia, a persisting pulmonary infiltrate, or unexplained dyspnea. In 15 percent of cases, the tumor may be an incidental bronchoscopic finding. The clinical picture reflects the degree of bronchial or tracheal obstruction. Some patients have no pulmonary symptoms and present with an asymptomatic nodule or mass.

The radiologic features are those of consolidation or aetelectasis or, in some cases, a pulmonary nodule. In 6 percent of cases, the chest radiograph is normal.[353]

Macroscopic Features

Few case reports document the tumor size, but where stated, it ranged from 0.4 to 6.5 cm. The tumors are sessile or polypoid lesions with smooth surfaces usually growing from the walls of the main airways (Fig. 29-149). In some cases, the tumor may present as a raised, plaquelike mucosal thickening or an infiltrating mass lesion. There have been reports of metachronous lesions at different sites in the tracheobronchial tree.[356,357] The tumor may in some cases be seen in the lung parenchyma as a coin lesion (Fig. 29-150). A distal obstructive pneumonitis is present in some cases.

Cut surface is usually yellowish gray or pink and often coarsely trabeculated.

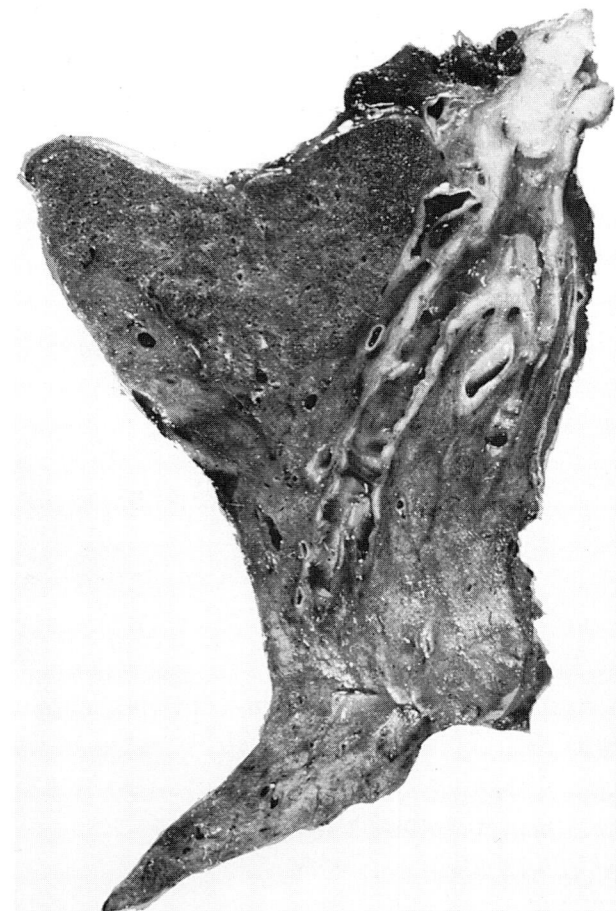

FIGURE 29-149
Granular cell myoblastoma with a central tumor, occluding the main bronchus.

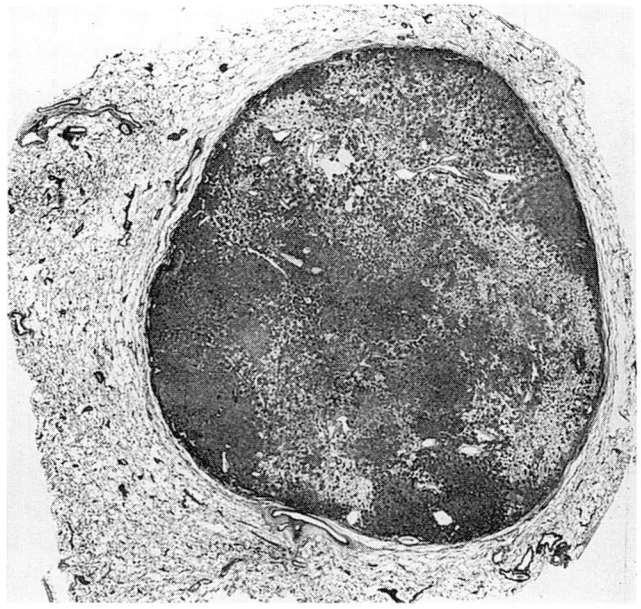

FIGURE 29-150
Granular cell myoblastoma with a peripheral, well-encapsulated tumor.

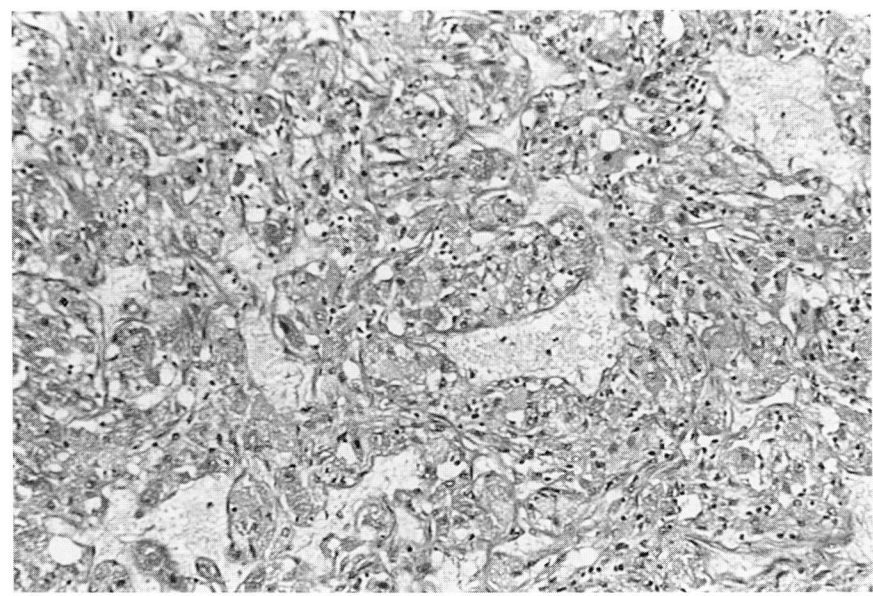

FIGURE 29-151
Granular cell myoblastoma showing a
syncytial pattern. (H&E, ×219.)

Histology

The tumor is composed of large granular foamy cells
with small dark nuclei, occasionally arranged in syncytial masses (Fig. 29-151) or, in some cases, cords
and nests. Some areas may show long fusiform cells
with elongated nuclei. There may be some variation
in size and shape, but the tumors behave in a benign
fashion. Usually, no mitoses are identified. There is a
varying degree of desmoplasia.

The cytoplasm has a coarse, granular appearance
(Fig. 29-152), and there are some small cells with
PAS-positive particles. Most of the tumor does not
stain with PAS.

The surface epithelium shows extensive metaplasia, which may be misinterpreted by the unwary
as squamous cell carcinoma. The tumor grows
around the glands in the wall and may occasionally
spread through the cartilaginous plates.

Immunocytochemistry shows variable positivity
with S-100 protein, neuron-specific enolase,
laminin, myelin protein,[358] and carcinoembryonic
antigen.

Electron Microscopy

Ultrastructurally,[359] the cytoplasm is filled with
osmiophilic granules of different sizes and shapes
(Fig. 29-153). The largest show a lamellated structure
(Fig. 29-154), and the smaller ones have a granular
content. All the granules are membrane-limited and
contain cell debris, including myelin figures (Fig. 29-
155), mitochondria, and axon-like structures. There
are some large vesicular structures with floculant
material.

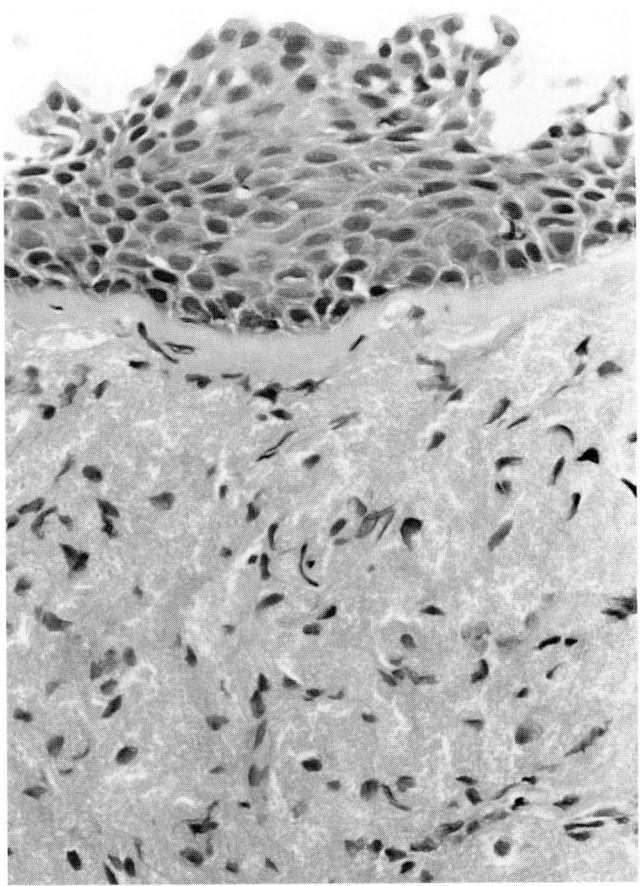

FIGURE 29-152
Granular cell myoblastoma with a well-marked granularity of the
cytoplasm and squamous metaplasia. (H&E, ×232.)

961

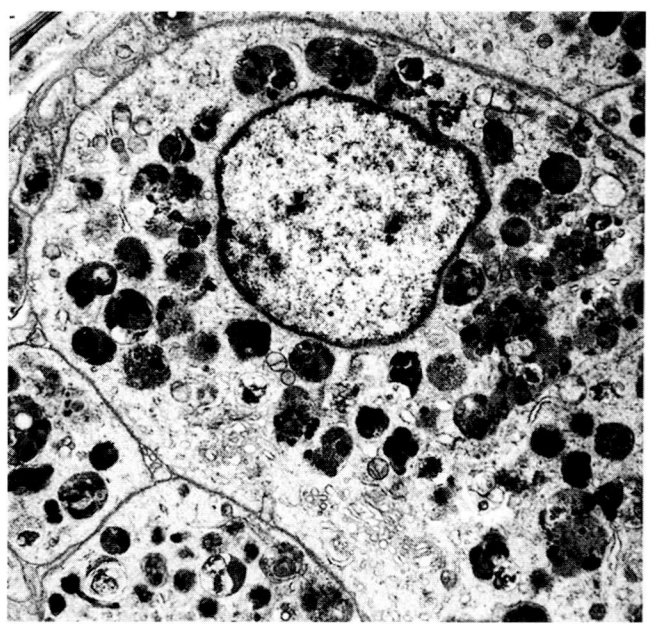

FIGURE 29-153
Granular cell myoblastoma with many osmiophilic inclusions of varying sizes. (Electron microscopy, ×10,000.)

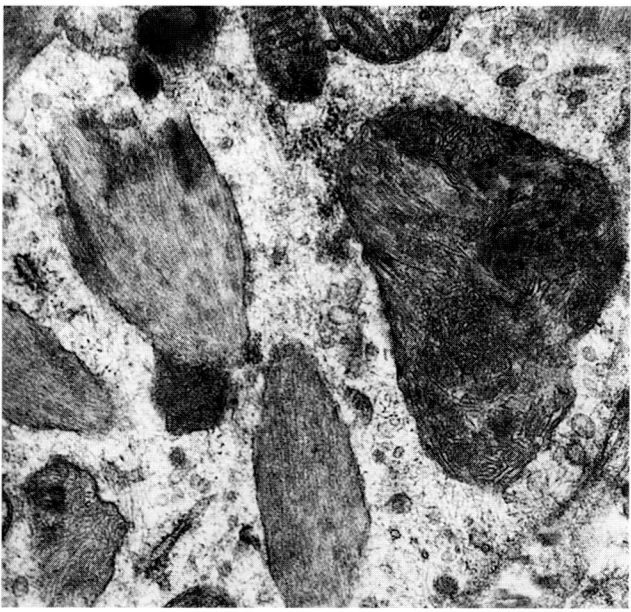

FIGURE 29-154
Granular cell myoblastoma with some of the osmiophilic inclusions, showing a lamellated pattern. (Electron microscopy, ×29,200.)

Clinical Behavior

These tumors grow slowly, often remaining radiographically unchanged for years. Although malignancy has been documented in extrathoracic tumors,[360] as far as the author is aware no such cases have been documented in the lung. However, malignant granular cell tumors may metastasize to the lung. Surgical excision of the tumor mass results in the highest cure rate, with only one patient in a series of 20 having a symptomatic recurrence.[353] The optimum extent of surgery necessary is not clear, but segmental or lobar resection is the most frequently used approach. Local resection may result in recurrence.

Factors affecting recurrence have not been fully identified, but only tumors greater than 1 to 2 cm are likely to recur.[361] Full-thickness invasion of the bronchial wall is found only when tumors exceed 8 mm in diameter. One case showed spontaneous resolution after biopsy, although the lesion may have been completely excised by this maneuver; the size of the lesion was not noted on bronchoscopy.[353]

Differential Diagnosis

SECONDARY GRANULAR CELL TUMOR

Multiple pulmonary granular cell tumors may behave in a benign fashion. This is not the situation outside the lung. Cases have been reported with pul-

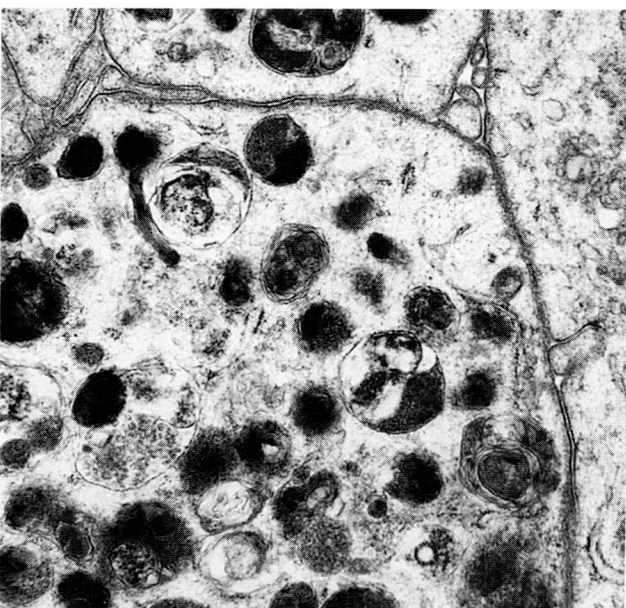

FIGURE 29-155
Granular cell myoblastoma with some of the osmiophilic inclusions containing myelin figures. (Electron microscopy, ×17,520.)

monary metastases[362] when the original lesion had been excised 5 years earlier.

SQUAMOUS CELL CARCINOMA

This erroneous diagnosis may be made because of the overlying pseudoepitheliomatous hyperplasia.

NEURAL TUMOR

If the tumor has a largely vesicular pattern with spindle cell areas, it may be misdiagnosed as a neurilemmoma.

MALAKOPLAKIA

Malakoplakia has some large granular cells, but with a Von Kossa stain, Michaelis-Gutmann bodies, up to 20 μm in diameter, should be identifiable. Their presence can be confirmed ultrastructurally. The granules in malakoplakia are smaller than in granular cell myoblastoma. There is associated inflammation in malakoplakia, but this is of little diagnostic help, since it may be seen around neoplasms.

ONCOCYTIC CARCINOID

This has a neuroendocrine pattern. It contains neurosecretory granules and should be positive with neuroendocrine stains.

ONCOCYTOMA

Oncocytoma does not have the coarse granular cytoplasm seen in granular cell myoblastoma. Ultrastructurally there are few non–membrane bound granules. Some cases have serous granules with mitochondria containing glycogen granules. No myelin figures are seen. In practice this is a very rare tumor, and most cases are oncocytic carcinoids.

Cell of Origin

Many authors would now regard this tumor as of Schwann cell lineage, possibly altered by a lysosomal defect.[363] This does not explain why the tumor develops at more than one site.

PULMONARY HYALINIZING GRANULOMA

This lesion was first described in 1977.[364] It is rare, and apart from Leibow's study documenting 20 cases, there is only one other large series, consisting of 24 cases.[365]

The lesion is more common in men. The age range is 24 to 77 years, with a mean of 43.4 years.

Clinical Features

The patients may present with cough, shortness of breath, hemoptysis, fatigue, fever, or pleuritic pain. Weight loss and, less frequently, dysphagia, fatigue, sinusitis, abdominal pain, and sore throat have been noted. Patients are often asymptomatic, and the lesion is then detected radiologically.

In Engleman's series[364] there was no evidence of rheumatoid or other collagen disease, but more sophisticated autoantibody screening tests have since become available. This is reflected in the later series of Yousem and Hochholzer,[365] in which positive antinuclear antibody titers were found in four of six patients tested, and one late respondent had a positive rheumatoid factor; three patients had a Coombs-positive hemolytic anemia. Thus, six of 10 patients tested had evidence of autoimmune phenomena. One additional patient had anti–smooth muscle and antimicrosomal antibodies. Seven of 14 patients were positive for tuberculin, histoplasmin, or coccidioidin skin testing.

Four patients had associated sclerosing mediastinitis. In three it was simultaneous, but not in continuity with the pulmonary hyalinizing granuloma. A case has been described associated with retroperitoneal fibrosis,[366] and a possible reaction to hydralazine was considered.

Chest radiographs show bilateral lesions in many cases and unilateral lesions in some. They were usually nodular and often described as a pneumonic infiltrate. The nodules were homogeneous, well circumscribed, but ill defined. Some showed evidence of speckled calcification. The nodules grew slowly, with a doubling time of 2.5 years for a lesion in one of the patients with extensive radiographic follow-up.[364]

Macroscopic Features

The lesions tend to be discrete, subpleural, or parenchymal. They are nodular, unencapusulated, and firm, and cut appearance ranged from pearly white to waxy translucent. The sizes range from 0.5 to 9 cm in diameter, with a small right-sided predominance.

Histology

There is a distinctive pattern, with a central area of branching, hyalinized collagen forming lamellae, and a coarse storiform patern. The collagen has a characteristic pattern, with thick, "ropey," whorled collagen bundles separated by clear spaces (Fig. 29-156). Capillaries and blood vessels may be obliter-

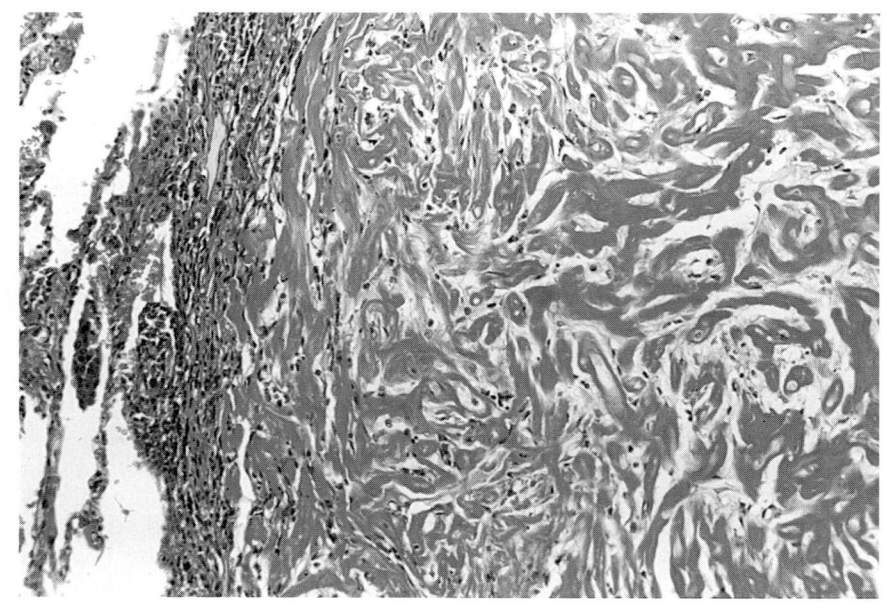

FIGURE 29-156
Pulmonary hyalinizing granuloma. This is the edge of the lesion. Beneath are many broad bands of thick, collagen-forming syncytial masses. (H&E, ×87.5.)

ated within the lesion. Bronchi and bronchioles were infrequent amid the nodules. Numerous large blood vessels were identified by an elastic–van Gieson stain, which shows marked intimal fibroelastosis and medial hyalinization. There is a mononuclear infiltrate through the vessel walls.

The lesion is surrounded by lymphocytes, some with germinal centers, plasma cells, histiocytes, occasional eosinophils, and spindle-shaped fibroblasts (Fig. 29-157). At the periphery are metaplastic alveolar pneumocytes forming slitlike or cystic spaces and foreign body giant cells. Bronchovascular structures adjacent to the advancing edge show evi-

dence of inflammation. Two cases in Yousem and Hochholzer's series[365] had discrete areas of organizing pneumonia, and another two showed diffuse lymphoid hyperplasia.

In approximately 50 percent of cases there is ischemic necrosis (Fig. 29-158), which tends to insinuate itself between collagen bundles, filling the "clear" spaces noted above. It does not usually form discrete foci. No definite fibrinoid necrosis is identified. Nuclear debris and microcalcifiation are frequent within the clear spaces. If the lesion abuts onto the pleura, there is associated chronic inflammation.

There is variation in the incidence of amyloid in

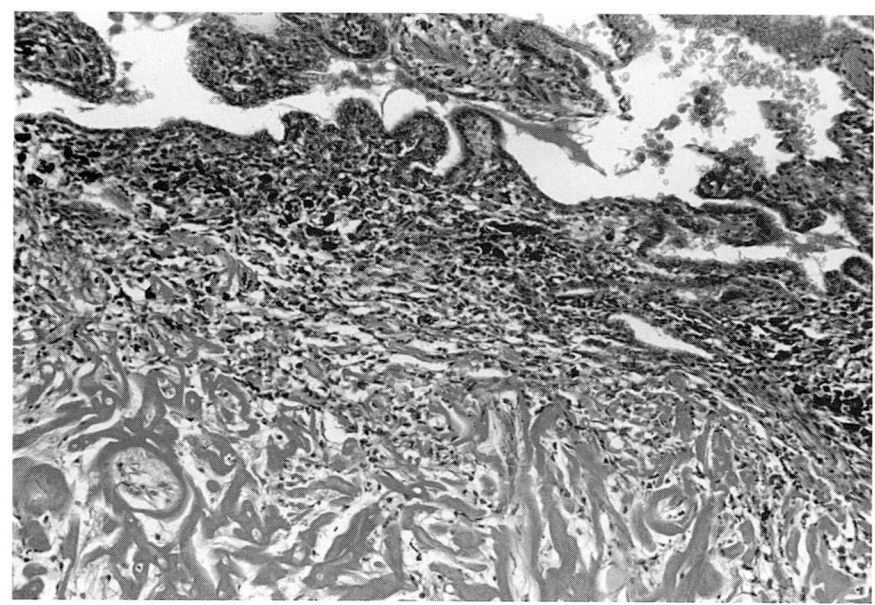

FIGURE 29-157
Pulmonary hyalinizing granuloma with some chronic inflammation at the edge of the lesion. (H&E, ×87.5.) (*Reproduced with permission of P. Stovin, M.D., Cambridge, England.*)

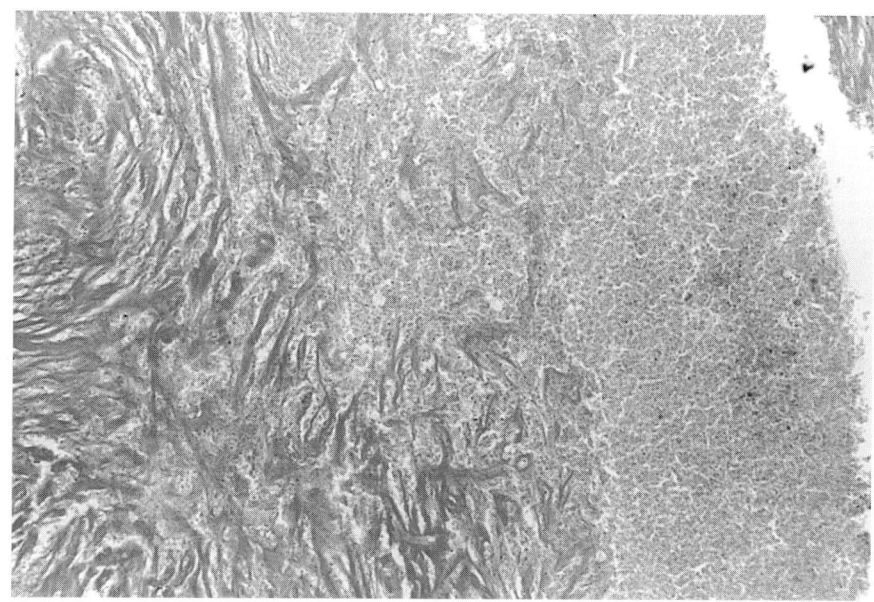

FIGURE 29-158
Pulmonary hyalinizing granuloma with a large focus of necrosis. (H&E, ×87.5.) (*Reproduced with permission of P. Stovin, M.D., Cambridge, England.*)

different series. In the early paper,[364] 11 of 18 cases examined showed "apple green" birefringence with Congo red under polarized light. Yousem and Hochholzer[365] failed to demonstrate amyloid in any of their cases.

Guccion and coworkers[367] failed to demonstrate any IgA, IgG, or IgM in the hyaline lamellae. Their patient had circulating immune complexes containing IgA.

Electron Microscopy

This showed[367] hyaline lamellae composed of electron-dense, compact, homogeneous amorphous material. Swollen collagen fibrils were scattered throughout them, and amyloid was not identified.

Differential Diagnosis

CHRONIC GRANULOMATOUS DISEASE

By definition, chronic granulomatous disease has been excluded in all cases studied—including tuberculosis, histoplasmosis, and coccidiomycosis. However, exposure to mycobacteria and fungal antigens was still identified by skin testing in some cases. It is possible that some of these cases represent old infectious granulomas, with resolution of the giant cells leaving only hyalinized tissue.

INFLAMMATORY PSEUDOTUMORS

This may be a difficult differential diagnosis. Inflammatory pseudotumors, especially the plasma cell granuloma variant, usually occur in younger people and are solitary, with a diffuse distribution of plasma cells and active fibroblasts. None of the coarse colla-

gen lamellae of pulmonary hyalinizing granuloma are seen.

HODGKIN'S DISEASE

This would require the presence of Reed-Sternberg or lacunar cells for a positive diagnosis.

NODULAR AMYLOIDOSIS

As mentioned above, there is a discrepancy in a reported series as to the presence of absence of amyloid. Unless amyloid is identified with a number of stains, this diagnosis cannot be considered.

RHEUMATOID NODULE

This does not show the collagen distribution seen in pulmonary hyalinizing granuloma. It has focal areas of necrosis with a palisaded, granulomatous rim. In pulmonary hyalinizing granuloma, the giant cells are scattered.

WEGENER'S GRANULOMATOSIS

This shows geographic necrosis, vasculitis, and well-formed giant cells. Even when it is healed, the collagen does not show the whorling or coarse bundles that are seen in pulmonary hyalinizing granuloma.

Etiology

The cause of this disease is unknown. It may represent an end stage of inflammatory pseudotumor. There is an association with immunologic abnormalities, and 60 percent of patients in one series[365] had positive autoantibodies. All these cases were associated with bilateral pulmonary involvement on the chest radiographs. Some cases had associated sclerosing mediastinitis.

Pulmonary hyalinizing granuloma may be an unusual response to a bacterial or fungal infection, such as tuberculosis, histoplasmosis, or coccidioidomycosis.

INFLAMMATORY PSEUDOTUMOR[368-370]

This entity has many synonyms, the most common being *plasma cell granuloma*. Other terms that have been used are *postinflammatory tumor of the lung*,[371] *histiocytoma of the bronchus*,[372] *benign histiocytic tumor*,[373] *vascular endothelioma*,[374] *plasmacytoma*,[375] *inflammatory fibrous histiocytoma* (? xanthogranuloma),[376] *fibrous histiocytoma*,[377,378] and *inflammatory myofibroblastic tumor* (plasma cell granuloma).[379]

The histologic findings vary according to the proportions of lymphocytes and histiocytic cells, and various authors probably describe the same lesion.

Inflammatory pseudotumor is not confined to the lung and has been described in other organs, such as the skin, soft tissues, larynx, liver, mesentery, stomach, small intestine, brain, spinal cord, kidney, renal pelvis, urinary bladder, major salivary glands, pancreas, thyroid, spleen, lymph nodes, and breast.[379,380]

Clinical Features

There have been several large series describing this disorder,[381–383] but the lesion is still relatively rare, with fewer than 150 recorded cases. A recent report from the Armed Forces Institute of Pathology described 50 cases.[384] The age range of patients is 1 to 77 years, with a mean of 28.3 years. The lesions constitute up to 56 percent of benign lung tumors in children.[385,386] There was no significant difference in incidence between sexes or between the right and left sides. However, the AFIP group found a slight male predominance (M=27, F=22).

The lesions may be asymptomatic in up to 50 percent of cases, and chest radiographs show a solitary coin lesion with occasional calcification or cavitation. In some patients there are multiple radiologic nodules. Symptoms consist of chest pain, fever, upper respiratory tract infection, a history of recurrent pneumonia, cough, hemoptysis, intermittent blood-streaked sputum, arthralgias, fatigue, empyema, pneumothorax, asthma, and difficulty in breathing. A history of a respiratory tract infection weeks to months before radiographic detection of the tumor may be noted. Previous pulmonary disease, including pulmonary infarcts and cryptococcosis, has also been documented. Rare "recurrences" have been noted.[379,387]

Weight loss may be a feature, and the lesion can invade the hilum of the lung, causing bronchial obstruction and associated atelectasis.[388] Intrabronchial growth has been documented.[389] There may be cortical destruction of a rib.[383]

Esophageal involvement causes dysphagia.[379,390,391] Extrapulmonary extension to the mediastinum or pericardium or into a thoracic vertebra[392] has been noted. The tumor may present as a posterior mediastinal mass.[379,393]

Macroscopic Features

The lesions range from 1.2 to 15 cm in diameter, although the average is closer to 5 cm. They are usually well circumscribed with no capsule. On cross section the tumors are grayish yellow, grayish white, or tan (Fig. 29-159). Hemorrhage, necrosis, and calci-

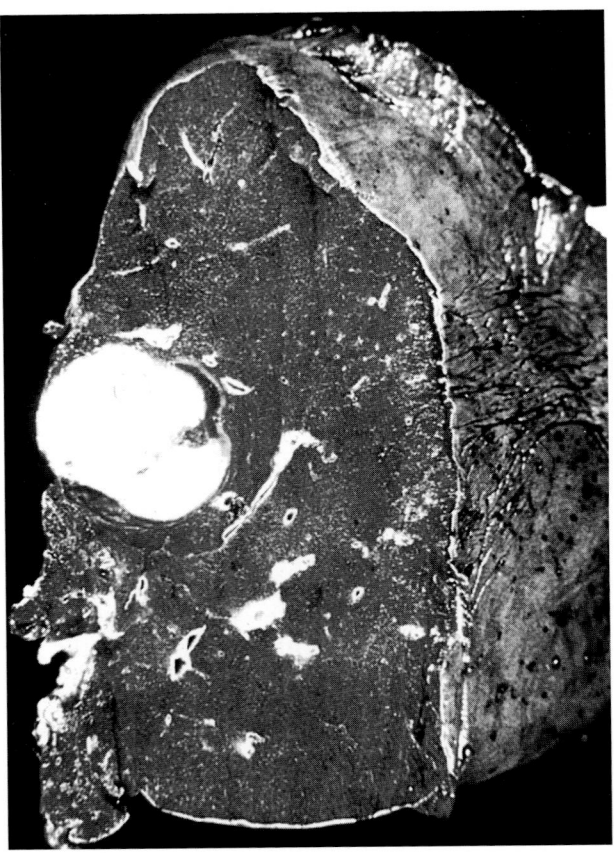

FIGURE 29-159
Inflammatory pseudotumor showing a well-defined margin.

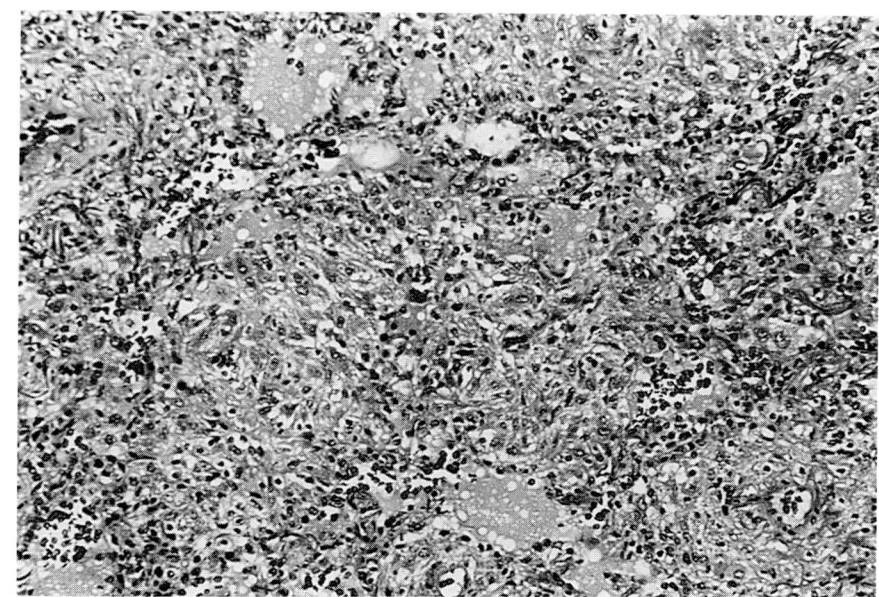

FIGURE 29-160
Inflammatory pseudotumor-organizing pneumonia with an intraalveolar lympho-histiocytic infiltrate. (H&E, ×87.5.)

fication are unusual. The tumors usually are intra-pulmonary, but some are endobronchial. At the hilum, several cases extended to the pleura.

Histology

The histologic features have been divided into three main types: organizing pneumonia, fibrous histiocytoma, and lymphoplasmacytic.[383] In practice there is a spectrum of changes encompassing all three types. Common to all is proliferation of fibroblasts, granulomatous and lymphocytic inflammation, lymphoid hyperplasia, and intraalveolar fibrosis at the periphery of the lesion. The border with the adjacent lung is ill defined.

ORGANIZING PNEUMONIA TYPE

In this variant there is intraalveolar lymphohistio-cytic inflammation and fibrosis at the edge of the tumor (Fig. 29-160). This becomes incorporated into interstitial tissue toward the center of the lesion. Fibroblasts and inflammatory cells are seen in small airways and bronchioles (Fig. 29-161). There is no purulent or granulomatous pneumonia suggesting active infection. In the center of the lesion, the collagen becomes hyaline. In rare cases there is calcium deposition in alveolar walls. Microabscess formation consisting of polymorphs, sometimes admixed with lymphocytes and plasma cells, may be identified both in the center and at the periphery of the lesions.

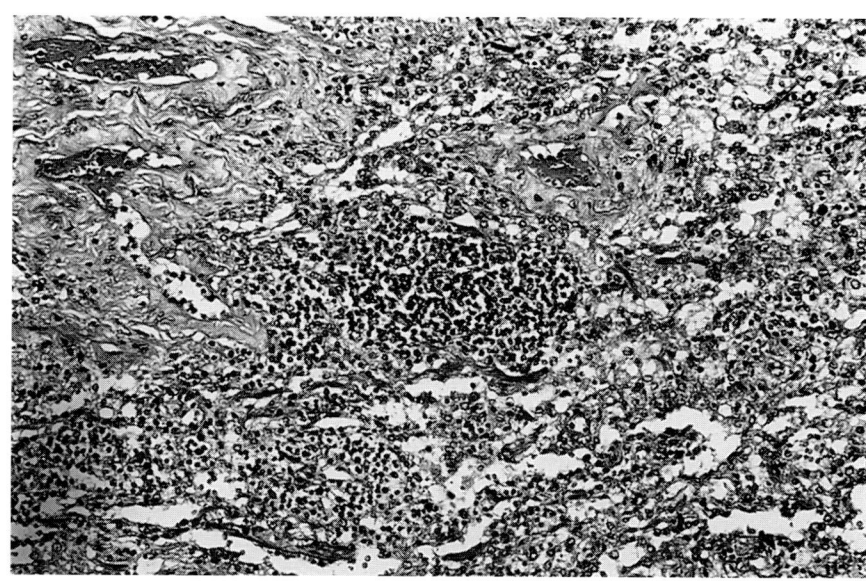

FIGURE 29-161
Inflammatory pseudotumor-inflammation in small airways. (H&E, ×87.5.)

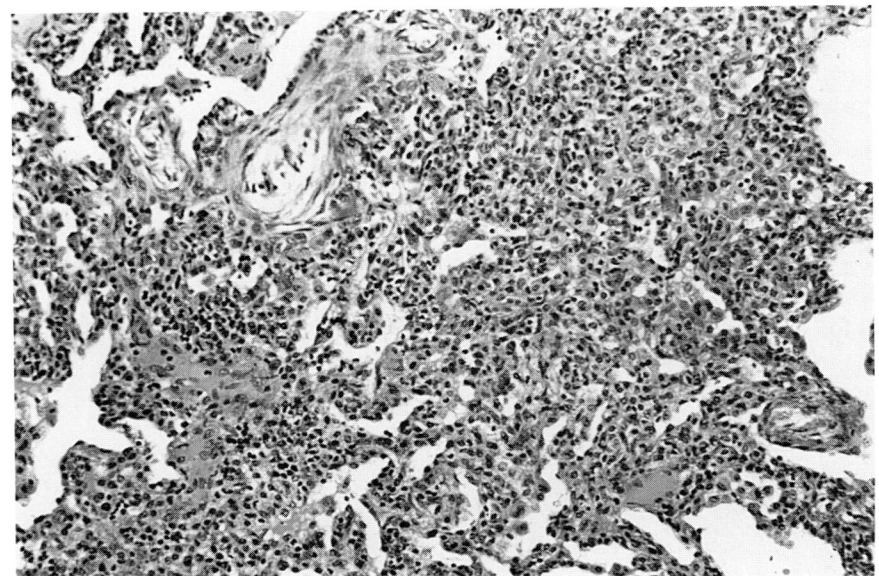

FIGURE 29-162
Inflammatory pseudotumor with some giant cells, chronic inflammation, and many macrophages. (H&E, ×87.5.)

Cavitation may occur in some instances. In the surrounding alveoli, there are vacuolated histiocytes and prominent type II pneumocytes (Fig. 29-162). Large lymphocytic aggregates with germinal centers and multinucleated Touton-type giant cells are seen in more than half of the cases.

FIBROUS HISTIOCYTOMA TYPE

This has the same pattern as an intradermal fibrous histiocytoma, with spindle cells arranged in a storiform pattern (Fig. 29-163). The nuclei are normochromatic or occasionally hyperchromatic. Mitoses are rare. There is variable collagen, and elastic fibers are preserved to form an alveolar architecture. There may be calcification, osseous metaplasia, and myxomatous change. At the periphery there are neutrophils, lymphocytes, plasma cells, and lipid-laden histiocytes.[373]

LYMPHOPLASMACYTIC TYPE

Plasma cells and lymphocytes may constitute over 50 percent of the tumor. The lymphoid aggregates become confluent with large germinal centers (Fig. 29-164). The spindle-celled fibroblastic components are conspicuous in areas. At the edge of the tumor there is also a lymphoplasmacytic/histiocytic and slightly neutrophilic response. Cavitation may occur.

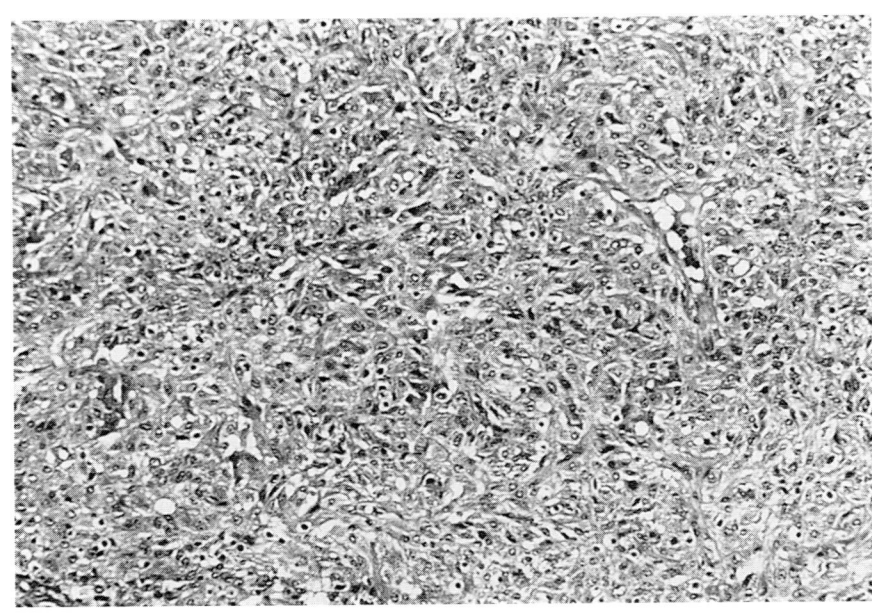

FIGURE 29-163
Inflammatory pseudotumor-fibrous histiocytoma pattern. (H&E, ×219.)

968

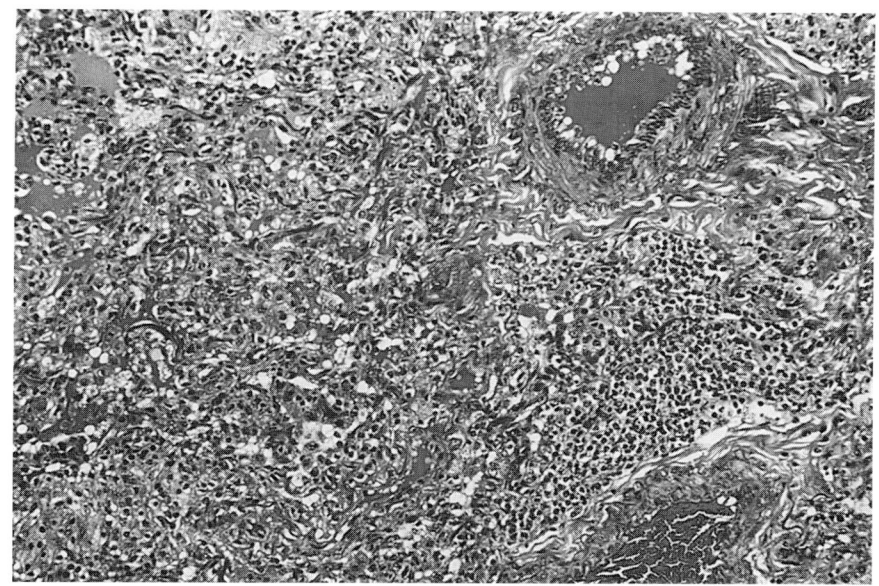

FIGURE 29-164
Inflammatory pseudotumor-lymphohistiocytic type with early germinal center formation. (H&E, ×87.5.)

Spencer[382] described malignant change in two cases. These showed pleomorphic cells and nuclear hyperchromatism. The cells contained large vacuolated nuclei. Most cases of inflammatory pseudotumor are considered benign, and it is possible that the two cases described by Spencer were initially malignant fibrous histiocytomas. There is a *histologic* overlap between inflammatory pseudotumor, fibrohistiocytic lesions of borderline malignancy, and unequivocal malignant fibrous histiocytoma.[394]

Immunohistochemical Findings[379]

The spindle cells stain strongly positively for vimentin and show reactivity for smooth muscle actin. Some cells are focally reactive for desmin. Cytokeratin and epithelial membrane antigen are positive only in entrapped small airways. S-100, Leu 7, and factor VIII–related antigen are negative. The plasma cells are polymorphic, staining for both kappa and lambda light chains. Most plasma cells are positive for IgG, although one study[395] showed only a few cells containing IgG, and many contained IgA and IgM.

Ultrastructural Studies[368,379,389]

Ultrastructural studies confirmed the presence of plasma cells with abdundant and dilated rough endoplasmic reticulum. Spindle cells are predominantly fibroblastic and arranged to form parallel streams or whorls. Most of their nuclei are characteristically indented. The cytoplasm is prominent, with conspicuous rough endoplasmic reticulum. Some of these cells occasionally contained bundles of cytoplasmic filaments with occasional dense bodies, mainly arranged parallel to the cell axis in a perinuclear distribution, a feature of myofibroblasts. These thin filaments were approximately 8 nm in width, parallel to the long axis of cells, and had intercalated dense bodies. No basal lamina was identified.

The histiocytic variety showed large cells with ultrastructural features of lymphocytes. There were irregular slender filopodia at random intervals around the entire cell circumference. The cytoplasm was abundant, with a variety of different organelles. Some cells were packed with lysosomes; others had large numbers of mitochondria. Many of these had unusual curved or elongated forms, and some contained round electron-dense inclusions. The cytoplasm of some cells was packed with lipid vacuoles.

Prognostic Features

Inflammatory pseudotumor has an excellent prognosis, and deaths are rare. Some fatalities have been recorded.[372,381] Vascular invasion has been identified.[396]

Gal and colleagues[394] studied a group of patients with inflammatory pseudotumor (IP), malignant fibrous histiocytoma (MFH), and borderline lesions. IP tended to occur in younger patients, with a median age of 30 years, than MFH, with a median age of 54 years. Age on its own was no guide, since there was an overlap in the ages of patients with these two diseases. All 15 patients with IP survived, as did the three with borderline malignant appearance. Local recurrence developed in two of 15 patients with IP and six of 13 cases with MFH. Distant metastases developed in seven MFH patients, and eight died of their disease.

969

Several histologic features were prognostically significant: necrosis, bizarre giant cells, mitoses equal to or greater than three per 50 high-power fields, and poor circumspection. Atypical mitoses, nuclear pleomorphism, the degree of fibrosis, and tumor size played no part in prognosis.

Differential Diagnosis

INFECTION

One case has been described[397] in which *Nocardia asteroides* caused the disease. In previous cases, *Corynebacterium equi*, mycobacterial infection, and *Coxiella burnetii* were causative agents. In most cases it is difficult to identify an organism, but this is still most probably a postinfective disorder.

BOOP

The causes of bronchiolitis obliterans–organizing pneumonia (BOOP) are given in Chap. 7. There are many causes of BOOP, and the infective processes overlap with inflammatory pseudotumor. Antecedent infection was documented in approximately one-third of the patients with inflammatory psedotumor. In cryptogenic organizing pneumonia and BOOP, the attempted resolution is incomplete and results in fibrosis.

The process of organization in inflammatory pseudotumors is similar to that in BOOP, but the diseases differ clinically in that 40 percent of patients with the former are asymptomatic. They do not have malaise, cough, dyspnea, crackles on auscultation, restrictive or obstructive pulmonary function, or a reduction in K_{CO}. Similarly, they do not have ground-glass densities on chest radiograph.

Pathologically, cryptogenic organizing pneumonia and BOOP produce many small scars, whereas inflammatory pseudotumor gives rise to a single lesion.

PLASMACYTOMA

Solitary plasmacytoma has been seen in the lung and can be differentiated from inflammatory pseudotumor by definition of the light-chain phenotype.

CARCINOMA OR ANOTHER SPINDLE CELL TUMOR

For a pure benign histiocytic lesion, the differential diagnosis may be that of a spindle cell tumor. Ideally, primary or secondary lung tumors should show epithelial differentiation, cytokeratins, and carcinoembryonic antigen, and in adenocarcinomas mucin positivity may be identified. Cytokeratin and epithelial membrane antigen are both negative in inflammatory pseudotumor. Entrapped alveoli may prompt the erroneous diagnosis of adenocarcinoma. Ultrastructurally, the benign histiocytic tumor shows lack of epithelial differentiation.

Bates and Hull[372] interpreted a benign histiocytic tumor as anaplastic squamous cell carcinoma on frozen section. Other secondary malignancies—such as hypernephroma, hepatoma, adrenal carcinoma, and malignant melanoma—must be differentiated from inflammatory pseudotumor.

A tumor that can cause difficulty in differential diagnosis is inflammatory sarcomatoid carcinoma of the lung.[398] This tumor has a bland appearance with reactive chronic inflammation. Macroscopically, the tumors are irregular masses measuring from 2.5 to 5.5 cm in diameter. Histologically, there is only modest nuclear pleomorphism and mitotic activity. The tumor consists of a spindle cell proliferation, arranged in fascicles, a storiform pattern, or haphazardly arranged bundles of cells. There are interspersed lymphocytes and plasma cells. Areas with a keloid pattern to the collagen have been identified. Vascular invasion was seen in two out of three cases. No mitoses or vascular invasion was identified in the inflammatory pseudotumors. Cytokeratin and epithelial membrane antigen were positive in inflammatory sarcomatoid carcinoma but not in inflammatory pseudotumor. Smooth muscle actin is present in the proliferating spindle cells of the inflammatory pseudotumor but not in the carcinomas. In the interpretation of the peroxidase, both with the cytokeratins and with smooth muscle actin, care must be taken to avoid overdiagnosis of entrapped alveoli or some resident smooth muscle elements. Clearly, great care should be taken in frozen section diagnosis of inflammatory pseudotumor. Ritter and Wick[398] recommended that frozen section not be used to separate these two entities.

Other tumors that should be considered are spindle cell carcinoids, benign and malignant muscle and nerve sheath tumors, and malignant fibrous histiocytoma. The last lesion shows cytologic atypia and nuclear pleomorphism, whereas in inflammatory pseudotumor the spindle cell component is homogeneous, with little nuclear pleomorphism or mitotic activity.

BENIGN CLEAR CELL TUMOR

This tumor is extremely rare, and metastatic tumor should be rigorously excluded. It contains abundant glycogen and little or no lipid. There is positivity with HMB-45. Benign histiocytic tumor contains little glycogen but much lipid.

ONCOCYTIC CARCINOID

This tumor should be positive with neuroendocrine markers and contain neurosecretory granules.

PULMONARY GRANULAR CELL MYOBLASTOMA

This has a characteristic appearance and will be positive with S-100 protein.

PULMONARY ONCOCYTOMA

This is positive for cytokeratin markers, and the epithelial cells are full of mitochondria. The tumor usually arises in the main airways.

Etiology

This is most probably a postinfective disease. Organisms have been demonstrated in a few cases, and in some series there was a history of previous infection. Recently Epstein-Barr virus has been demonstrated in 18 inflammatory pseudotumors outside the lung, including lymph nodes, spleen, and liver.[398a] One case was studied by lavage before antibiotic treatment.[399] This case had only a transbronchial biopsy, which showed granulomatous inflammation with plasma cells and xanthomatous histiocytes and lymphocytes. No specific pathogenic organism was identified. Before treatment the number of neutrophils and the neutrophil chemotactic activity were increased in the lavage fluid from the affected region. It was also moderately increased in the unaffected area of the opposite lung. The number of neutrophils and the neutrophil-chemotactic activity, as well as C5 and C5a des ARG (neutrophil chemotactic factors) in the lavage fluid from both these regions decreased during treatment. The authors considered that plasma cell granuloma was a chronic immune and inflammatory reaction. They suggested that neutrophils are involved in the development of the signs and symptoms of the disease.

The conclusion is that inflammatory pseudotumor is related to some infective cause in many cases. In some instances perhaps the causative organism has not been identified, it has disappeared because of previous antibiotic treatment, or there is a local deficiency in the immune response of the lung.

PRIMARY CHORIOCARCINOMA OF THE LUNG

Choriocarcinomas may arise outside the genital tract, although they most commonly affect the lung as a secondary tumor (Figs. 29-165 and 29-166). These tumors possibly arise from retained primordial germ cells that fail to complete migration along the urogenital tract. They have been found in both men and

FIGURE 29-165
Secondary choriocarcinoma with large hemorrhagic nodules in the lung and hemorrhage in the surrounding parenchyma.

FIGURE 29-166
Testis from case demonstrated in Fig. 29-165. There is a small nodule, which histologically consisted of necrotic tissue. This was the presumed site of the primary tumor.

women. Men may have gynecomastia and hemoptysis. The tumor has been recorded in a female infant[400] and in postmenopausal women.[401] Pushchak and Farhi[401] cited 15 examples in their review. Their reported case was one of the most complete, since a postmortem in the perioperative period showed no extrapulmonary primary site for the tumor. In some cases, as in the one illustrated above, the primary tumor may be very small or have regressed.

The tumor measures up to 11 by 10 by 8.5 cm and is a partly necrotic, hemorrhagic mass in the main-stem bronchus and pleura. In viable areas there are two cell types. The first are smaller cells with distinct cell borders and clear cytoplasm; the second are bizarre, giant multinucleated cells with ill-defined borders and pink cytoplasm (Fig. 29-167). These cells stained positively with human chorionic gonadotropin (HCG). The patient had a postoperative serum beta subunit HCG of 2000 U/L (normal: 0.10 U/L).

BENIGN CLEAR CELL (SUGAR) TUMOR OF THE LUNG

This entity was first described by Liebow and Castleman in 1963.[402] Since then, approximately 40

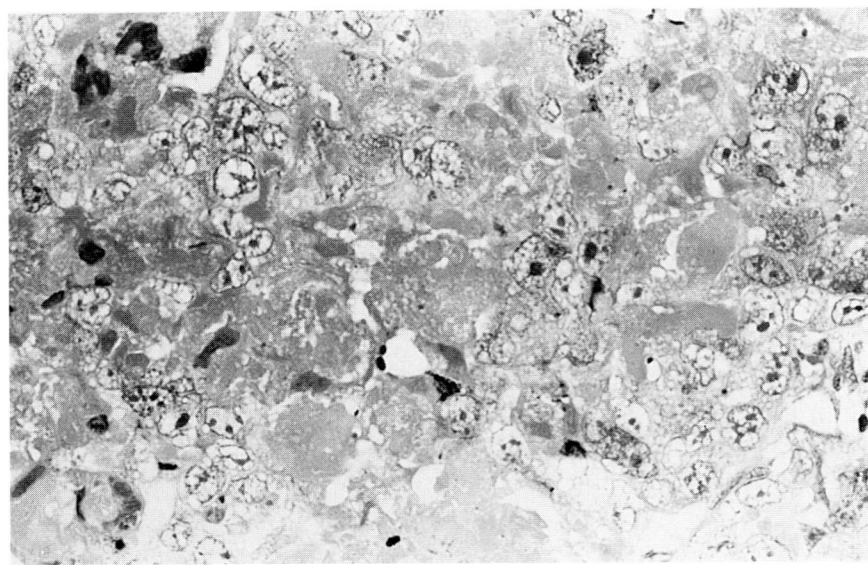

A

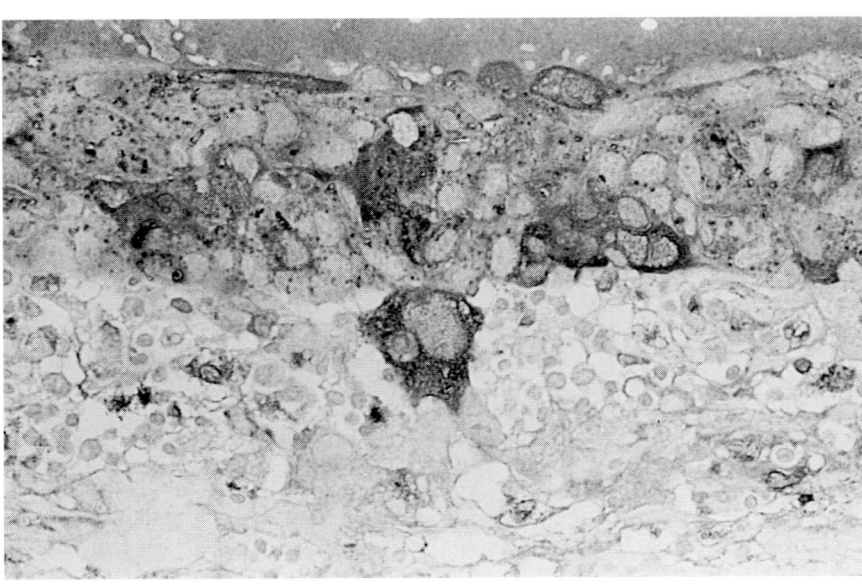

B

FIGURE 29-167
Histology of choriocarcinoma. *A.* Choricarcinoma primary in the lung with syncytiotrophoblast. (H&E, ×250.) *B.* Same case as in A but stained with antibody to βHCE. (Immunoperoxidase, ×250.)

cases have appeared in the world literature.[403] The other large series documenting this tumor is that of Gaffey and coworkers.[404]

The patients' ages ranged from 8 to 69 years, with a mean of 47.4 years. There were 19 males and 21 females. The vast majority of the patients presented with asymptomatic lesions, which were seen on incidental chest radiographs. When symptoms are present they include cough, chest pain, and dyspnea. Pneumonia may be present. There is no distinction between right and left side in the origin of the lesion.

The tumors were single rubbery well-circumscribed tan-white nodules that in some cases extended to the visceral pleura and measured from 0.7 to 6.5 cm in diameter.

Histopathology

The tumor is well demarcated from the adjacent lung parenchyma with only a thin fibrous pseudocapsule. Tumor cells are polygonal, arranged in a nested and insular pattern, and surrounded by a delicate capillary network (Fig. 29-168) well demarcated by reticulin. The abdundant clear to eosinophilic cytoplasm contains fine vacuoles with finely dispersed PAS positivity. This is in the form of granules and disappears after diastase predigestion. The nuclei are round to oval and have a finely stippled chromatin and indistinct nucleoli. In one case there was mild nuclear pleomorphism. Mitotic figures and necrosis are not seen.

There may be hemorrhage, which can be attributed to surgery. In some cases there may be inciden-tal noncaseating sarcoidlike granulomas of the periphery of the tumor and within the draining lymph nodes. One case showed a minute peribronchiolar carcinoid tumor.

Immunohistochemistry

This tumor is usually positive with HMB-45, cathepsin B, α_1-antitrypsin, and vimentin. However, the vimentin is variable according to the supplying agent and is positive with the Biogenex product but not with Dako. One case in the series of Gal and colleagues[403] did not stain positively for HMB-45. S-100 protein is identified in three out of five cases.

Cytokeratin markers are uniformly negative, as are chromogranin, synaptophysin, CEA, and GFAP. However, several cases have been reported positive with synaptophysin and Leu 7.[404]

Electron Microscopy

This tumor has been studied ultrastructurally by several authors.[403,404,406–408] These cells have abundant free and membrane-bound glycogen, which appears specific for this tumor. In addition, there were melanosomes in three of eight cases, suggesting a melanocytic tumor.[404] Becker and Soifer[406] detected membrane-bound dense core (neurosecretory) granules, suggesting an endocrine origin. These granules have not been seen in all cases of benign clear cell tumor. They have been identified, however, by other authors.[403,407,409,410] It is possible that some cases represent smooth muscle or pericyte prolifera-

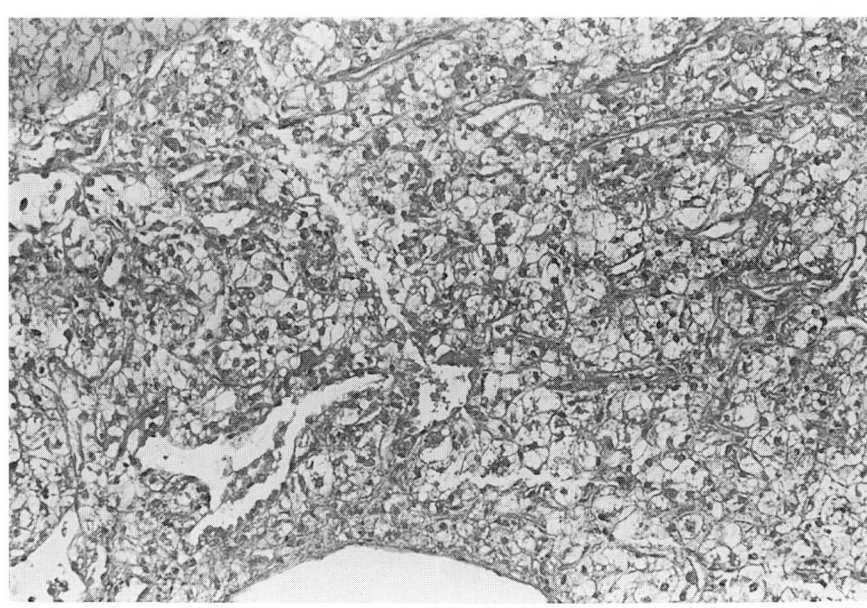

FIGURE 29-168
Benign sugar tumor with well-marked clear cells and a prominent vascular pattern. The tumor is well-defined. (H&E, ×87.5.)

tion.[411,412] Necrosis may be detected ultrastructurally in some areas of the tumor.[411,413]

Differential Diagnosis

METASTATIC CARCINOMA

It is important with any clear cell tumor to rule out metastatic hypernephroma. Metastatic hypernephromas may present as pulmonary lesions before the renal tumor has declared itself.[414] A case initially diagnosed as a benign clear cell tumor of the lung underwent diagnostic revision 18 months later when serial urograms detected renal carcinoma.[415] A metastatic renal carcinoma would contain fat as well as glycogen. HMB-45 would be negative.

CLEAR CELL CARCINOMA

This does not have the typical sinusoidal vascularity or the glycogen seen in benign clear cell tumors. Benign clear cell tumors contain only glycogen. Clear cell carcinomas of the lung[416] show no membrane-bound glycogen. In most cases, electron microscopy of these tumors shows either squamous or adenocarcinomatous differentiation.

Histogenesis

Melanocytic differentiation of benign clear cell tumor has been suggested on the basis of positivity with HMB-45 and evidence of melanosomes.[404,417] There is an immunophenotypic similarity between benign clear cell tumor of lung, clear cell sarcoma of tendon sheath, and malignant melanoma. However, HMB-45 is not specific to melanocytic tumors and may be detected in plasmacytoma, metastatic adenocarcinoma, hepatocellular carcinoma, and breast carcinoma.[418–420]

Other authors have postulated an epithelial origin for this tumor. This is based on the findings of short, irregularly branched microvilli, macula occludens‾type junctions, and a plasma membrane with complex interdigitations.[408] Failure to demonstrate keratins or carcinoembryonic antigen does not support this concept.

PRIMARY PULMONARY THYMOMA

This is a very rare group of tumors; fewer than 20 cases have been reported in the English literature.[51] Most cases presenting to a surgical pathologist will be secondary to a primary mediastinal thymoma.

Clinically, the tumors present as either a pulmonary or a pleural mass. There is no difference in the sex distribution for the pulmonary tumor. The age range is wide—19 to 74 years—with a mean of 55 years. Three of the patients in one series presented with myasthenia gravis. Most were asymptomatic. One patient had cough and fever and another hemoptysis.

There is a similar median age for pleural tumors: 51 years. The age range was 37 to 71 years.[421–423] As with pulmonary tumors, the patients may be asymptomatic but they may have chest pain, dyspnea, and weight loss. The tumors may occasionally mimic mesothelioma.[423]

Macroscopically, the pulmonary lesions can be divided into peripheral and hilar tumors.[424] Most of the peripheral tumors are located in the right lung and the hilar ones in the left. However, with such a small number of cases on record, it is not possible to be sure of the significance of this finding. The average size of these lesions is 3 cm, with a range of 1.7 to 12 cm. They are encapsulated nodules (Fig. 29-169). The cut surface is usually a tan color and is soft and fleshy. The lesions may be red, due to hemorrhage, or firm and white.

Histologically, most pulmonary tumors are encapsulated and well circumscribed, whereas the pleural ones are infiltrating in nature. As in a mediastinal thymoma, the tumor grows in a lobular pattern with a fibrous stroma between (Fig. 29-170). The same histologic types are seen as in the mediastinum.

Immunocytochemically, the epithelium stains with cytokeratin markers and the lymphocytes with UCHL-1.[51]

Differential Diagnosis

SECONDARY TUMOR FROM A MEDIASTINAL THYMOMA

This can be resolved only by recourse to radiologic, surgical, or, rarely, autopsy findings. In the pleura this may be a difficult problem. However, mesothelioma does not usually have a lobular growth pattern.

MALIGNANT LYMPHOMA

In practice this is not a difficult problem, since there should be no epithelial cells in a lymphoma.

SECONDARY CARCINOMA

If there is a lymphocyte-rich carcinoma, there could be difficulties in diagnosis. However, there is not usually a lobular pattern in carcinoma, and the latter tumor has more nuclear atypia than a thymoma.

SPINDLE CELL TUMORS

These can be separated from thymomas with this pattern by immunostains. As in the secondary carcinoma section above, the lobular pattern and the

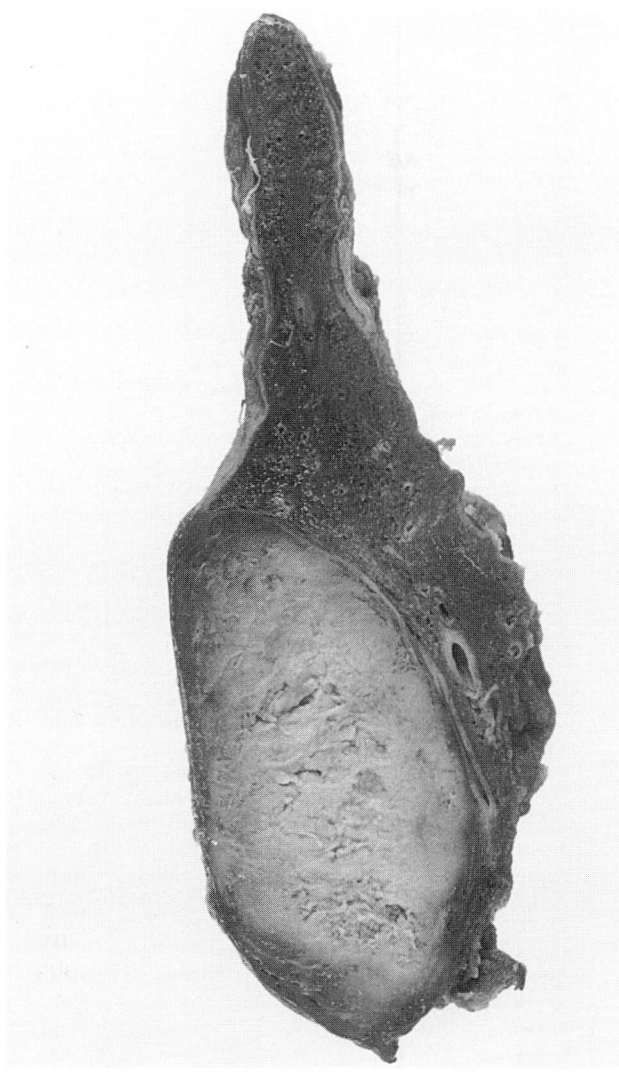

FIGURE 29-169
Thymoma in the lung. This was a secondary tumor, from a previously resected mediastinal primary.

nuclear atypia will help differentiate carcinoma from thymoma.

EPENDYMOMA

There has been only one case report of this tumor.[425] The tumor was small, measuring 2.0 by 1.0 by 0.8 cm. It was circumscribed but not encapsulated and had a gray lobulated cut surface.

Histologically, there were spindle to oval cells with an eosinophilic fibrillary cytoplasm. There was nuclear pleomorphism and many mitoses. Perivascular pseudorosettes were present, as was focal calcification.

There was strong positivity for glial fibrillary acidic protein and some positive staining for vimentin, Leu-7, and S-100 protein.

In the differential diagnosis the tumor is so rare that it would probably be called a neuroendocrine tumor, but the absence of strong positivity with the usual neuroendocrine stains would exclude this diagnosis. Secondary ependymoma should have declared itself in the history.

REFERENCES

1. Albrecht E: Ueber Hamartome. *Verh Deutsch ges Pathol* 7:153–157, 1904.
2. Liebow AA: Tumors of lower respiratory tract, in *Atlas of Tumor Pathology* (Fasc. 17) Washington D.C., Armed Forces Institute of Pathology, 1952, p 119.
3. Spencer H: *Pathology of the Lung*, Vol. 2, 3d ed. New York, Pergamon Press, 1977, p 892.
4. Bateson EM: Relationship between intrapulmonary and endobronchial cartilage-containing tumours (so called hamartomata). *Thorax* 20:447–461, 1965.

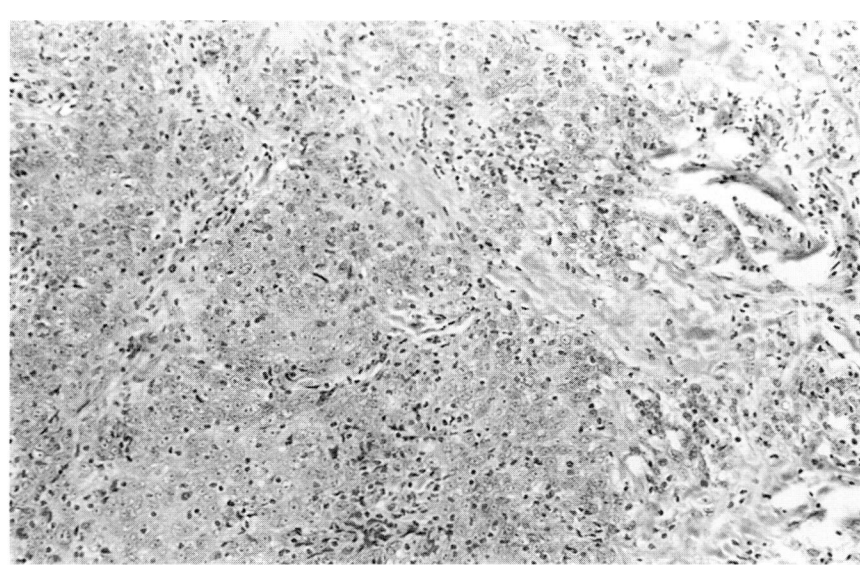

FIGURE 29-170
Thymoma secondary to the lung. There is a mixture of epithelial and lymphocytic cells. (H&E, ×87.5.)

5. Tomashefski JF Jr: Benign endobronchial mesenchymal tumors: Their relationship to parenchymal pulmonary hamartomas. *Am J Surg Pathol* 6:531–540, 1982.

6. Hernandez E, Reyes CV: A review of 64 cases: Hamartoma of the lungs. *IMJ* 158:331–334, 1980.

7. Van Den Bosch JM, Wagenaar Sj Sc, Corrin B, Elbers JRJ, Knaepen PJ, Westermann CJJ: Mesenchymoma of the lung (so called hamartoma): A review of 154 parenchymal and endobronchial cases. *Thorax* 42:790–793, 1987.

8. Salminen U-S: Pulmonary harmartoma: A clinical study of 77 cases in a 21-year period and review of literature. *Eur J Cardiothorac Surg* 4:15–18, 1990.

9. Hansen CP, Holtveg H, Francis D, Rasch L, Bertelsen S: Pulmonary hamartoma. *J Thorac Cardiovasc Surg* 104:674–678, 1992.

10. McDonald JR, Harrington SW, Clagett OT: Hamartoma (often called chondroma) of the lung. *J Thorac Surg* 14:128–143, 1945.

11. Talmadge EK Jr, Christopher KL, Schwarz MI: Multiple pulmonary chondromatous hamartomas. *Hum Pathol* 13:496–497, 1982.

12. Ribet M, Jaillard-Thery S, Nuttens MC: Pulmonary hamartoma and malignancy. *J Thorac Cardiovasc Surg* 107:611–614, 1994.

13. Johansson M, Dietrich C, Mandahl N, et al: Recombinations of chromosomal bands 6p21 and 14q24 characterise pulmonary hamartomas. *BR J Cancer* 67:1236–1241, 1993.

14. Fletcher JA, Pinkus GS, Donovan K, et al: Clonal rearrangement of chromosome band 6p21 in the mesenchymal component of pulmonary chondroid hamartoma. *Cancer Res* 52:6224–6228, 1992.

15. Case records of the Massachusetts General Hospital. *N Engl J Med* 315:1277–1285, 1986.

16. Schwartz M: A biomathematical approach to clinical tumor growth. *Cancer* 14:1272–1293, 1961.

17. Incze JS, Lui PS: Morphology of the epithelial component of the human lung hamartomas. *Hum Pathol* 8:411–419, 1977.

18. Stone FJ, Churg AM: The ultrastructure of pulmonary hamartoma. *Cancer* 39:1064–1070, 1977.

19. Carney JA: The triad of gastric epithelioid leiomyosarcoma, functioning extra-adrenal paraganglioma, and pulmonary chondroma. *Cancer* 43:374–382, 1979.

20. Basile A, Gregoris A, Antoci B, Romanelli M: Malignant change in a benign pulmonary hamartoma. *Thorax* 44:232–233, 1989.

21. Hayward RH, Carabasi RJ: Malignant hamartoma of the lung: Fact or fiction? *J Thorac Cardiovasc Surg* 53:457–466, 1967.

22. Poulson JT, Jacobson M, Francis D: Probable malignant transformation of a pulmonary hamartoma. *Thorax* 34:557–558, 1979.

22a. King TC, Myers J: Isolated metastasis to a pulmonary hamartoma. *Am J Surg Pathol* 19:472–475, 1995.

23. Davidson M: A case of primary chondroma of the bronchus. *Br J Surg* 28:571–574, 1941.

24. Liebow AA: *Atlas of Tumor Pathology.* Washington, D.C., Armed Forces Institute of Pathology, 1952.

25. Hochberg LA, Schacter B: Benign tumors of the bronchus and lung. *Am J Surg* 89:425–438, 1955.

26. Walsh TJ, Healy TM: Chondroma of the bronchus. *Thorax* 24:327–329, 1969.

27. Carter D, Eggleston JC: *Tumors of the Lower Respiratory Tract.* Washington, D.C., Armed Forces Institute of Pathology, 1980, p. 236.

28. Daniels AC, Conner GH, Straus FH: Primary chondrosarcoma of the tracheobronchial tree. Report of a unique case and brief review. *Arch Pathol* 84:615–624, 1967.

29. Watts CF, Clagett OT, McDonald JR: Lipoma of the bronchus: Discussion of benign neoplasms and report of a case of endobronchial lipoma. *J Thorac Surg* 15:132–144, 1946.

30. Schraufnagel DE, Morin JE, Wang NS: Endobronchial lipoma. *Chest* 75:97–99, 1979.

31. Remigo PA, Cruz MDL: Endobronchial lipoma. *NY State J Med* 88:550–501, 1988.

32. McCall RE, Harrison W: Intrabronchial lipoma: Case report. *J Thorac Surg* 29:317–322, 1955.

33. MacArthur CGC, Cheung DLC, Spiro SG: Endobronchial lipoma: A review of four cases. *Br J Dis Chest* 75:93–100, 1977.

34. Matsuba K, Saito T, Ando K, Shirakusa T: Atypical lipoma of the lung. *Thorax* 46:685, 1991.

35. Shapiro R, Carter MG: Peripheral lipoma of the lung. *Am Rev Tuberc* 69:1042–1044, 1954.

36. Spencer H: *Pathology of the Lung*, 4th ed. Oxford, Pergamon Press, 1985, p 976.

37. Mark EJ: Mesenchymal cystic hamartoma of the lung. *N Engl J Med* 315:1255–1259, 1986.

37a. van Klaveren RJ, Hassing HHM, Wiersma-van Tilberg JM, Lacquet LK, Cox AL: Mesenchymal cystic hamartoma of the lung: A rare cause of relapsing pneumothorax. *Thorax* 49:1175–1176, 1994.

38. Mushtaq M, Ward SP, Hutchinson JT, Man JS: Multiple cystic pulmonary hamartomas. *Thorax* 47:1076–1077, 1986.

39. Carey FA, Salter DM, Kerr KM, Lamb D: An investigation into the role of human papillomavirus in endobronchial papillary squamous tumors. *Respir Med* 84: 445–447, 1990.

40. Laubscher FA: Solitary squamous cell papilloma of bronchial origin. *Am J Clin Pathol* 52:599–603, 1969.

41. Trillo A, Guha A: Solitary condylomatous papilloma of the bronchus. *Arch Pathol Lab Med* 112:731–733, 1988.

42. Basheda S, Gephardt GN, Stoller JK: Columnar papilloma of the bronchus: Case report and literature review. *Am Rev Respir Dis* 144: 1400–1402, 1991.

43. Assor D: A papillary transitional cell tumor of the bronchus. *Am J Clin Pathol* 55:761–764, 1971.

44. Smith PS, McClure J: A papillary endobronchial tumor with a transitional cell pattern. *Arch Pathol Lab Med* 106:503–506, 1982.

45. Fantone JC, Geisinger KR, Appelman HD: Papillary adenoma of the lung with lamellar and electron dense granules. An ultrastructural study. *Cancer* 50:2839–2844, 1982.

46. Noguchi M, Kodama T, Shimosato Y, et al: Papillary adenoma of type II pneumocytes. *Am J Surg Pathol* 10:134–139, 1986.

47. Hegg CA, Flint A, Singh G: Papillary adenoma of the lung. *Am J Clin Pathol* 97: 393–397, 1992.

48. Fukuda T, Ohnishi Y, Kanai I, et al: Papillary adenoma of the lung: Histological and ultrastructural findings in two cases. *Acta Pathol Jpn* 42:56–61, 1992.

49. Miller RR: Bronchioloalveolar cell adenomas. *Am J Surg Pathol* 14:904–912, 1990.

50. Carey FA, Wallace WAH, Fregusson RJ, Kerr KM, Lamb D: Alveolar atypical hyperplasia in association with primary

pulmonary adenocarcinoma: A clinicopathological study of 10 cases. *Thorax* 47:1041–1043, 1992.

51. Colby TV, Koss MN, Travis WD: Thymoma, in Carter D, Eggleston JC (eds), *Atlas of Tumor Pathology: Tumors of the Lower Respiratory Tract.* Washington, D.C., Armed Forces Institute of Pathology, 1995, pp 482–483.

52. Al-Hilli F: Lymphangioma (or alveolar adenoma?) of the lung. *Histopathology* 11:979–980, 1987.

53. Wada A, Tateishi R, Terazawa T, Matsuda M, Hattori S: Case report. Lymphangioma of the lung. *Arch Pathol* 98:211–213, 1974.

54. Yousem SA, Hochholzer L: Alveolar adenoma. *Hum Pathol* 17:1066–1071, 1986.

55. Mounts P, Shah KV, Kashima H: Viral etiology of juvenile and adult onset squamous papilloma of the larynx. *Proc Natl Acad Sci USA* 79:5425–5429, 1982.

56. Guillon L, Sahli R, Chaubert P, Monnier P, Cuttat JF, Costa J: Squamous cell carcinoma of the lung in a nonsmoking, nonirradiated patient with juvenile laryngotracheal papillomatosis. Evidence of human papillomavirus–11 DNA in both carcinoma and papillomas. *Am J Surg Pathol* 15:891–898, 1991.

57. Kramer SS, Wehunt WD, Stocker JT, Kashima H: Pulmonary manifestations of juvenile laryngotracheal papillomatosis. *AJR* 144:687–689, 1985.

58. Heard BE, Corrin B, Dewar A: Pathology of seven mucous cell adenomas of the bronchial glands with particular reference to ultrastructure. *Histopatholology* 9:687–701, 1985.

59. Spencer H: Bronchial mucous gland tumours. *Virchows Arch [A]* 383:101–115, 1979.

59a. England DM, Hochholzer L: Truly benign "Bronchial adenoma." Report of 10 cases of mucous gland adenoma with immunohistochemical and ultrastructural findings. *Am J Surg Pathol* 19:887–899, 1995.

60. Dickerstein PJ, Amaral SM, Silva AM, Daltro PA, Ferreira AJ, Aiex A: Bronchial mucous gland adenoma presenting as bronchogenic cyst. *Pediatr Pulmonol* 16:370–374, 1993.

61. Ferguson CJ, Cleeland JA: Mucous gland adenoma of the trachea: Case report and literature review. *J Thorac Cardiovasc Surg* 95: 347–350, 1988.

62. De Jesus M, Poon TP, Chung K-Y: Pulmonary oncocytoma. *NY State J Med* 89: 477–480, 1989.

63. Santos-Briz A, Terron J, Sastre R, et al: Oncocytoma of the lung. *Cancer* 40:1330–1336, 1977.

64. Tesluk H, Dajee A: Pulmonary oncocytoma. *J Surg Oncol* 29:173–175, 1985.

65. Cwierzyk TA, Glasberg SS, Virshup MA, et al: Pulmonary oncocytoma: Report of a case with cytologic, histologic and electron microscopic study. *Acta Cytol* 29:620–623, 1985.

66. Warter A, Walter P, Sabountchi M, Jory A: Oncocytic bronchial adenoma. *Virchows Arch [A]* 392:231–239, 1981.

67. Nielsen AL: Malignant bronchial oncocytoma: Case report and review of the literature. *Hum Pathol* 16:852–854, 1985.

68. Kragel PJ, Devaney KO, Meth BM, Linnoila I, Frierson HF Jr, Travis WD: Mucinous cystadenoma of the lung. A report of two cases with immunohistochemical and ultrastructural analysis. *Arch Pathol Lab Med* 114:1053–1056, 1990.

69. Graeme-Cook F, Mark EJ: Pulmonary mucinous cystic tumors of borderline malignancy. *Hum Pathol* 22:185–190, 1991.

70. Donaldson JC, Kaminsky DB, Elliott RC: Bronchiolar carcinoma: Report of 11 cases and a review of the literature. *Cancer* 41:250–258, 1978.

71. Fechner RE, Bentinck BR, Askew JB Jr: Acinic cell tumor of the lung. A histologic and ultrastructural study. *Cancer* 29:501–508, 1972.

72. Katz DR, Bubis JJ: Acinic cell tumor of the bronchus. *Cancer* 38:830–832, 1976.

73. Moran CA, Suster S, Koss MN: Acinic cell carcinoma of the lung ("Fechner tumor"). A clinicopathologic, immunohistochemical, and ultrastructural study of five cases. *Am J Surg Pathol* 16:1039–1050, 1992.

74. Heard BE, Dewar A, Firmin RK, Lennox SC: One very rare and one new tracheal tumour found by electron microscopy: Glomus tumour and acinic cell tumour resembling carcinoid tumours by light microscopy. *Thorax* 37:97–103, 1982.

75. Gaffey MJ, Mills SE, Askin FB, et al: Clear cell tumor of the lung. A clinicopathologic, immunohistochemical, and ultrastructural study of eight cases. *Am J Surg Pathol* 14:248–259, 1990.

76. Gaffey MJ, Mills SE, Zarbo RJ, Weiss LM, Gown AM: Clear cell tumor of the lung. Immunohistochemical and ultrastructural evidence of melanogenesis. *Am J Surg Pathol* 15:644–653, 1991.

77. Chan K, Fine G, Lewis J, Lee MW: Benign mixed tumor of the trachea. *Cancer* 44:2260–2266, 1979.

78. Sano T, Hirose T, Hizawa K, et al: A case of pleomorphic adenoma of the trachea. *Jpn J Clin Oncol* 14:93–98, 1984.

79. Sakurai H, Tsuchida A, Takakura H, Amemiyar R, Oho K, Hayata Y: Benign mixed tumor in the trachea. *Kyobu Geka* 37:23–27, 1984.

80. Hasegawa M, Ebihara H, Shibusawa M, et al: Pleomorphic adenoma of the trachea. *J Laryngol Otol* 102:560–561, 1988.

81. Ma CK, Fine G, Lewis J, Lee MW: Benign mixed tumor of the trachea. *Cancer* 44: 2260–2266, 1979.

82. Yan H-C, Shen C-Y, Chiang C-H, et al: Pleomorphic adenoma of the trachea: Report of two cases. *J Formosan Med Assoc* 90:1124–1127, 1991.

83. Payne WS, Schier J, Woolner LB.: Mixed tumors of the bronchus (salivary gland type). *J Thorac Cardiovasc Surg* 49:663–668, 1965.

84. Spencer H: Bronchial mucous gland tumours. *Virchows Arch [A]* 383:101–115, 1979.

85. Davis PW, Briggs JC, Seal RM, Starring FK: Benign and malignant mixed tumours of the lung. *Thorax* 27:657–673, 1972.

86. Sakamoto H, Uda H, Tanaka T, Oda T, Morino H, Kikui M: Pleomorphic adenoma in the periphery of the lung: Report of a case and review of the literature. *Arch Pathol Lab Med* 115:393–396, 1991.

87. Wright ES, Pike E, Couves CM: Unusual tumors of the lung. *J Surg Oncol* 24:23–29, 1983.

88. Hayes MM, van der Westhuizen NG, Forgie R: Malignant mixed tumor of bronchus. A biphasic neoplasm of epithelial and myoepithelial cells. *Mod Pathol* 6:85–88, 1993.

89. Horinouchi H, Ishihara T, Kawamura M, et al: Epithelial myoepithelial tumour of the tracheal gland. *J Clin Pathol* 46:185–187, 1993.

90. Hemmi A, Hiraoka H, Mori Y, et al: Malignant pleomorphic adenoma (malignant mixed tumor) of the trachea: Report of a case. *Acta Pathol Jpn* 38:1215–1226, 1988.

91. Moran CA, Suster S, Askin FB, Koss MN: Benign and

malignant salivary gland-type mixed tumors of the lung. *Cancer* 73:2481–2490, 1994.

92. Yousem SA: Pulmonary adenosquamous carcinomas with amyloid-like stroma. *Mod Pathol* 2:420–426, 1989.

93. Carter D, Eggleston JC: Careinoid tumors, in *Atlas of Tumor Pathology: Tumors of the Lower Respiratory Tract.* Washington, D.C., Armed Forces Institute of Pathology, 1983, p 162.

94. Bertelsen S, Aasted A, Lund C, et al: Bronchial carcinoid tumors. A clinicopathologic study of 82 cases. *Scand J Thorac Cardiovasc Surg* 19:105–111, 1985.

95. McCaughan BC, Martini N, Bains MS: Bronchial carcinoids. Review of 124 cases. *J Thorac Cardiovasc Surg* 89:8–17, 1985.

96. Torre M, Barberis M, Barbieri B, Bonacina E, Belloni P: Typical and atypical bronchial carcinoids. *Respir Med* 83:305–308, 1989.

97. Martensson H, Bottcher G, Hambraeus G, et al: Bronchial carcinoids: An analysis of 91 cases. *World J Surg* 11:356–364, 1987.

98. Todd TR, Cooper JD, Weissberg D, Delarue NC, Pearson FG: Bronchial carcinoid tumors. Twenty years' experience. *J Thorac Cardiovasc Surg* 79:532–536, 1980.

99. Knott-Craig CJ, Schaff HV, Mullany CJ, et al: Carcinoid disease of the heart. Surgical management of ten patients. *J Thorac Cardiovasc Surg* 104:475–481, 1992.

100. Robbins SL, Cotran RS, Kumar V: *The Pathological Basis of Disease.* 3d ed. Philadelphia, W. B. Saunders, 1984.

101. Wilkins ER Jr, Darling RC, Soutter L, Sniffen RC: A continuing survey of adenomas of the trachea and bronchus in a general hospital. *J Thorac Cardiovasc Surg* 46:279–291, 1963.

102. Wynn SR, O'Connell EJ, Frigas E, Paynes WS, Sachs MI: Exercise-induced "asthma" as a presentation of bronchial carcinoid. *Ann Allergy* 57:139–141, 1986.

103. Ricci C, Patrassi N, Massa R, et al: Carcinoid syndrome in bronchial adenoma. *Am J Surg* 126:671–677, 1973.

104. Grahame-Smith DG: *The Carcinoid Syndrome.* London, Heinemann, 1972.

105. Shalet SM, Beardwell CG, MacFarlane IA, et al: Acromegaly due to production of a growth hormone releasing factor by a bronchial carcinoid tumour. *Clin Endocrinol* 10:61–67, 1979.

106. Yamashina M: An 18-year history of a corticotrophin-secreting spindle cell carcinoid in the lung. *Arch Pathol Lab Med* 109:673–675, 1985.

107. Shames JM, Dhurandar NR, Blackard WG: Insulin-secreting bronchial carcinoid tumor with widespread metastases. *Am J Med* 44:632–637, 1968.

108. Huber RM, Schopohl J, Losa M, et al: Growth-hormone releasing hormone in a bronchial carcinoid. *Cancer* 67:2538–2542, 1991.

109. Allen MB, Shamash J, Kerr KM, Leitch AG: Hypercalcemia in typical bronchial carcinoid tumors. *Chest* 96:1206–1208, 1989.

110. Okike N, Bernatz PF, Woolner LB: Carcinoid tumors of the lung. *Ann Thorac Surg* 22:270–275, 1976.

111. Hurt R, Bates M: Carcinoid tumours of the bronchus: A 33 year experience. *Thorax* 39:617–623, 1984.

112. Warren WH, Faber LP, Gould VE: Neuroendocrine neoplasms of the lung: A clinicopathologic update. *J Thorac Cardiovasc Surg* 98:321–332, 1989.

113. Travis WD, Linnoila RI, Tsokos MG, et al. Neuroendocrine tumors of the lung with proposed criteria for large-cell neuroendocrine carcinoma: An ultrastructural, immuno-histochemical and flow cytometric study of 35 cases. *Am J Surg Pathol* 15:529–533, 1991.

114. Travis WD: Classification of neuroendocrine tumors of the lung. *Lung Cancer* 11(suppl 2):197–198, 1994.

115. Arrigoni MG, Woolner LB, Bernatz PE: Atypical carcinoid tumors of the lung. *J Thorac Cardiovasc Surg* 64:413–421, 1972.

115a. Gould VE, Linnoila RI, Memoli WH: Neuroendocrine cells and neuroendocrine neoplasms of the lung. *Path Annu* 18:287–330, 1983.

116. Yeatman TJ, Sharp JV, Kimura AK: Can susceptibility to carcinoid tumors be inherited? *Cancer* 63:390–393, 1989.

117. Miller MA, Mark GJ, Kanarek D: Multiple peripheral pulmonary carcinoids and tumorlets of carcinoid type, with restrictive and obstructive lung disease. *Am J Med* 65:373–378, 1978.

118. Ghadially FN, Block HJ: Oncocytic carcinoid of the lung. *J Submicrosc Cytol* 17:435–442, 1985.

119. Gosney JR, Denley H, Resl M, Beesley C: Sustentacular cells in bronchial carcinoid tumours (abstract). *J Pathol* 173(suppl):122, 1994.

120. al-Saffar N, Moore JV, Hasleton PS: A morphometric analysis of individual cell death in bronchial carcinoids. *Cell Tissue Kinet* 23:325–330, 1990.

121. Moore JV: Death of cells and necrosis of tumours, in Potten CS, ed, *Perspectives on Mammalian Cell Death.* Oxford, Oxford University Press, 1987, pp 295–325.

122. Moore JV, Hasleton PS, Buckley CH: Tumour chords in 52 bronchial and cervical squamous cell carcinomas: Inferences for their cellular kinetics and radiobiology. *Br J Cancer* 51:407–413, 1985.

123. Hasleton PS, al-Saffar N: The histological spectrum of bronchial carcinoid tumours. *Appl Pathol* 7:205–218, 1989.

124. Grazer R, Cohen SM, Jacobs JB, Lucas P: Melanin-containing peripheral carcinoid of the lung. *Am J Surg Pathol* 6:73–78,1982.

125. Mark EJ, Quay SC, Dickersin GR: Papillary carcinoid tumor of the lung. *Cancer* 48:316–324, 1981.

126. Ranchod M, Levine GD: Spindle cell carcinoid tumors of the lung: A clinicopathologic study of 35 cases. *Am J Surg Pathol* 4:315–331, 1980.

127. Wise WS, Bonder D, Aikawa M, Hsieh CL: Carcinoid tumor of lung with varied histology. *Am J Surg Pathol* 6:261–267, 1982.

128. Cooney T, Sweeney EC, Luke D: Pulmonary carcinoid tumours: A comparative regional study. *J Clin Pathol* 32:1100–1109, 1979.

129. Williams ED, Brown CL, Doniach I: Pathological and clinical findings in a series of 67 cases of medullary carcinoma of the thyroid. *J Clin Pathol* 19:103–113, 1966.

130. Vanmaele L, Noppen M, Frecourt N, Impens N, Welch B, Schandevijl W: Atypical ossification in bronchial carcinoid. *Eur Respir J* 3:927–929, 1990.

131. Al-Kaisi N, Abdul-Karim FW, Mendelsohn G, Jacobs G: Bronchial carcinoid tumor with amyloid stroma. *Arch Pathol Lab Med* 112:211–214, 1988.

132. Martini N, Zaman MB, Bains MS, et al: Treatment and prognosis in bronchial carcinoids involving regional lymph nodes. *J Thorac Cardiovasc Surg* 107:1–17, 1994.

133. Arrigoni MG, Woolner LB, Bernatz PE: Atypical carcinoid tumors of the lung. *J Thorac Cardiovasc Surg* 64:413–421, 1972.

134. Mills SE, Walker AN, Cooper PH, Kron IL: Atypical carcinoid tumor of the lung: A clinicopathological study of 17 cases. *Am J Surg Pathol* 6:643–654, 1982.

135. Bishopric GA Jr, Ordonez NG: Carcinoembryonic antigen in primary carcinoid tumors of the lung. *Cancer* 58:1316–1320, 1986.

136. Davila DG, Dunn WF, Tazelaar HD, Pairolero PC: Bronchial carcinoid tumors. *Mayo Clin Proc* 68:795–803, 1993.

137. Suresh UR, Wilkes S, Hasleton PS, et al: Prealbumin in the diagnosis of bronchopulmonary carcinoid tumours. *J Clin Pathol* 44:573–575, 1991.

138. Walts AE, Said JW, Shintaku IP, Lloyd RV: Chromogranin as a marker of neuroendocrine cells in cytologic material—an immunocytochemical study. *Am J Clin Pathol* 84:273–277, 1985.

139. al-Saffar N, White A, Moore M, Hasleton PS: Immunoreactivity of various peptides in typical and atypical bronchopulmonary carcinoid tumours. *Br J Cancer* 58:762–766, 1988.

140. Bergh J, Esscher T, Steinholtz L, Nilsson K, Pahlman S: Immunocytochemical demonstration of neuron specific enolase (NSE) in human lung cancers. *Am J Clin Pathol* 84:1–7, 1985.

141. Bostanci G, Hasleton PS, Vasudev K: Neurofilaments in typical and atypical carcinoids (in preparation).

142. Valli M, Corrin B, Fabris GA, Sheppard MN: Atypical carcinoid of the lung. (abstract) *J Pathol*(Suppl)107:284, 1993.

143. Denley H, Gosney JR, Resl M, Beesley C: p53 protein in bronchial carcinoid tumours (abstract) *J Pathol* 173 (Suppl): 120, 1994.

144. Hiyama K, Hasegawa K, Ishioka S, Takahashi N, Yamakido M: An atypical carcinoid tumor of the lung with mutations in the p53 gene and the retinoblatoma gene. *Chest* 104:1606–1607, 1993.

144a. Costes V, Marty-Ane C, Picot MC, et al: Typical and atypical bronchopulmonary carcinoid tumors: A clinicopathologic and KI-67-labelling study. *Hum Pathol* 26:740–745, 1995.

145. Warren WH, Memoli VA, Gould VE: Immunohistochemical and ultrastructural analysis of bronchopulmonary neuroendocrine neoplasms. I. Carcinoids. *Ultrastruct Pathol* 61:15–27, 1984.

146. Neal MH, Kosinski R, Cohen P, Orenstein JM: Atypical endocrine tumors of the lung: A histological, ultrastructural and clinical study of 19 cases. *Hum Pathol* 17:1264–1277, 1986.

147. Hammond ME, Sause WT: Large cell neuroendocrine tumors of the lung. Clinical significance and histopathologic definition. *Cancer* 56:1624–1629, 1989.

148. Jones DJ, Hasleton PS, Moore M: DNA ploidy in bronchopulmonary carcinoid tumours. *Thorax* 43:195–199, 1988.

149. el-Naggar AK, Ballance W, Karim FW, et al: Typical and atypical bronchopulmonary carcinoids: A clinicopathologic and flow cytometric study. *Am J Clin Pathol* 95:828–834, 1991.

150. Hasleton PS, Gomm S, Blair V, Thatcher N: Pulmonary carcinoid tumours: A clinicopathological study of 35 cases. *Br J Cancer* 54:963–967, 1986.

151. Hangartner JR, Loosemore TM, Burke M, Pepper JR: Malignant primary pulmonary paraganglioma. *Thorax* 44:154–156, 1989.

152. Gaffey MJ, Mills SE, Askin FB: Minute pulmonary meningothelial-like nodules: A clinicopathologic study of so-called minute pulmonary chemodectoma. *Am J Surg Pathol* 12:167–175, 1988.

153. Kushner BH, Hajdu SI, Gulati SC, Evlandson RA, Exelby PR, Lieberman PH: Extracranial primitive neuroectodermal tumors. The Memorial Sloan-Kettering Cancer Center experience. *Cancer* 67:1825–1829, 1991.

154. Liew S-H, Leong AS-Y, Tang HMK: Tracheal paraganglioma: A case report and review of the literature. *Cancer* 47:1387–1393, 1981.

155. Gallimore AP, Goldstraw P: Tracheal paraganglioma. *Thorax* 48:866–867, 1993.

156. Singh G, Lee RE, Brooks DH: Primary pulmonary paraganglioma. Report of a case and review of the literature. *Cancer* 40:2286–2289, 1977.

157. Fawcett FJ, Husband EM: Chemodectoma of lung. *J Clin Pathol* 20:260–262, 1967.

158. Heppleston AG: A carotid-body like tumour in lung. *J Pathol Bacteriol* 75:461–464, 1958.

159. Kliewer KE, Cochrane AG: A review of the histology, ultrastructure, immunohistology and molecular biology of extra-adrenal paragangliomas. *Arch Pathol Lab Med* 113:1209–1218, 1989.

160. Gallivan MV, Chun B, Rowden G, Lack EE: Laryngeal paraganglioma. Case report with ultrastructural analysis and literature review. *Am J Surg Pathol* 3:85–92, 1979.

161. Reed RJ, Caroca PJ Jr, Harkin JC: Gangliocytic paraganglioma. *Am J Surg Pathol* 1:207–216, 1977.

162. Moll R, Cowin P, Kapprell H-P, Franke WW: Desmosomal proteins: New markers for identification and classification of tumors. *Lab Invest* 54:4–25, 1986.

163. Misugi K, Aoki I, Kikyo S, Sasaki Y, Nakajima T, Tsunoda A: Immunohistochemical study of neuroblastoma and related tumors with anti-S-100 protein antibody. *Pediatr Pathol* 3:217–226, 1985.

164. Wolfe K, Gosney JR: Cell populations in paragangliomas (abstract). *J Pathol* 172:132A, 1994.

165. DeAngelis LM, Kelleher MB, Post KD, Fetell MR: Multiple paragangliomas in neurofibromatosis: A new neuroendocrine neoplasia. *Neurology* 37:129–133, 1987.

166. Googe PB, Ferry JA, Bhan AK, Dickersin GR, Pilch BZ, Goodman M: A comparison of paraganglioma, carcinoid tumor and small cell carcinoma of the larynx. *Arch Pathol Lab Med* 112:809–815, 1988.

167. Allen MS Jr, Drash EC: Primary melanoma of lung. *Cancer* 21:154–159, 1968.

168. Jensen OA, Egedorf J: Primary malignant melanoma of the lung. *Scand J Respir Dis* 48:127–135, 1967.

169. Jennings TA, Axiotis CA, Kress Y, Carter D: Primary malignant melanoma of the lower respiratory tract. Report of a case and a literature review. *Am J Clin Pathol* 94:649–655, 1990.

170. Heller DS, Frydman CP, Gordon RE, Jagirdar J, Schwartz IS: An unusual organoid tumor. Alveolar soft part sarcoma or paraganglioma. *Cancer* 67:1894–1899, 1991.

171. Torikata C, Mukai M: So-called minute chemodectoma of the lung. An electron microscopic and immunohistochemical study. *Virchows Arch [A]* 417:113–118, 1990.

172. Korn D, Bensch K, Liebow AA, Castleman B: Multiple minute pulmonary tumors resembling chemodectomas. *Am J Pathol* 37:641–672, 1960.

173. Kuhn C III, Askin FB: The fine structure of so-called "minute pulmonary chemodectomas.: *Hum Pathol* 6:681–691, 1975.

174. Zak FG, Chabe S: Pulmonary chemodectomatosis. *JAMA* 183:887–889, 1963.

175. Gaffy MJ, Mills SF, Askin FB: Minute pulmonary

meningothelial-like nodules. A clinico-pathologic study of so-called minute pulmonary chemodectoma. *Am J Surg Pathol* 12:167–175, 1988.

176. Dail DH: Uncommon tumors, in Dail DH, Hammar SP, eds, *Pulmonary Pathology*. New York , Springer Verlag, 1988, pp 903–910.

177. Ichinose H, Hewitt RL, Drapanas T: Minute pulmonary chemodectomas. *Cancer* 28:692–700, 1971.

178. Churg AM, Warnock ML: So-called "minute pulmonary chemodectomas." A tumor not related to paragangliomas. *Cancer* 37:1759–1769, 1976.

179. Heath D, Smith P: Nodules resembling arahnoid villi in pulmonary venules in plexogenic pulmonary arteriopathy. *Cardioscience* 3:161–165, 1992.

180. Heath D, Williams D: Arachnoid nodules in the lung of high altitude Indians. *Thorax* 48:743–745, 1993.

181. Chumas JC, Lorelle CA: Pulmonary meningioma. A light and electron microscopic study. *Am J Surg Pathol* 6:792–801, 1982.

182. Kemnitz P, Spormann H, Heinrich P: Meningioma of lung: First report with light and electron microscopic findings. *Ultrastruct Pathol* 3:359–365, 1982.

183. Kodama K, Doi O, Higashimaya M, et al: Primary and metastatic pulmonary meningioma. *Cancer* 67:1412–1417, 1991.

184. Strimlan CV, Golembiewski RS, Celko DA, Fino GJ: Primary pulmonary meningioma. *Surg Neurol* 29:410–413, 1988.

185. Drlicek M, Grisold W, Lorber J, Hackl H, Wuketich S, Jellinger K: Pulmonary meningioma. Immunohistochemical and ultrastructural features. *Am J Surg Pathol* 15:445–459, 1991.

186. Flynn SD, Yousem SA: Pulmonary meningiomas. A report of two cases. *Hum Pathol* 22:469–474, 1991.

187. Robinson PG: Pulmonary meningiomas. Report of a case with electron microscopic and immunohistochemical findings. *Am J Clin Pathol* 97:814–817, 1992.

188. Zhang FL, Cheng XR, Xhang XS, Ding JA: Lung ectopic meningioma: A case report. *Chin Med J* 96:309–311, 1983.

189. Hussarek M, Rieder W: Glomustumor der Luftrohre. *Krebsartz* 5:208–212, 1950.

190. Fabich DR, Hafez GR: Glomangioma of the trachea. *Cancer* 45:2337–2341, 1980.

191. Kim YI, Kim JH, Suh J-S, Ham EK, Suh KP: Glomus tumor of the trachea. Report of a case with ultrastructural observation. *Cancer* 64:881–886, 1989.

192. Heard BE, Dewar A, Firmin RK, Lennox SC: One very rare and one new tracheal tumour found by electron microscopy: Glomus tumour and acinic cell tumour resembling carcinoid tumours by light microscopy. *Thorax* 37:97–103, 1982.

193. Ito H, Motohiro K, Nomura S, Tahara E: Glomus tumor of the trachea: Immunohistochemical and electron microscopic studies. *Path Res Pract* 183:778–784, 1988.

194. Tang C-K, Toker C, Foris NP, Trump BF: Glomangioma of the lung. *Am J Surg Pathol* 2:103–109, 1978.

195. Alt B, Huffer WE, Belchis DA: A vascular lesion with smooth muscle differentiation presenting as a coin lesion in the lung: Glomus tumor versus hemangiopericytoma. *Am J Clin Pathol* 80:765–771, 1983.

196. McKay B, Legha SS, Pickler GM: Coin lesion of the lung in a 19-year-old male. *Ultrastruct Pathol* 2:289–284, 1981.

197. Goodman TF: Fine structure of the cells of the Suquet-Hoyer canal. *J Invest Dermatol* 59:363–369, 1972.

198. Shin DH, Park SS, Lee JH, Park MH, Lee JD: Oncocytic glomus tumor of the trachea. *Chest* 98:1021–1023, 1990.

199. Murad TM, Haam E von, Murthy MS: Ultrastructure of a hemangiopericytoma and a glomus tumor. *Cancer* 22:1239–1249, 1968.

200. Heitmiller RF, Mathisen DJ, Ferry JA, Mark EJ, Grillo HC: Mucoepidermoid lung tumors. *Ann Thorac Surg* 47:394–399, 1989.

201. Miller DL, Allen MS: Rare pulmonary neoplasms. *Mayo Clin Proc* 68:492–498, 1993.

202. Torres AM, Ryckman FC: Childhood tracheobronchial muco-epidermoid carcinoma: A case report and review of the literature. *J Pediatr Surg* 23:367–370, 1988.

203. Yousem S, Hochholzer L: Mucoepidermoid tumors of the lung. *Cancer* 60:1346–1352, 1987.

204. Green LK, Gallion TL, Gyorkey F: Peripheral mucoepidermoid tumour of the lung. *Thorax* 46:65–66, 1991.

205. Klacsmann PG, Olson JL, Eggleston JC: Mucoepidermoid carcinoma of the bronchus: An electron microscopy study of the low grade and the high grade variants. *Cancer* 43:1720–1733, 1979.

206. Barsky SH, Martin SE, Matthews M, Gazdar A, Costa JC: "Low grade" mucoepidermoid carcinoma of the bronchus with "high grade" biological behavior. *Cancer* 51:1505–1509, 1983.

207. Brooks B, Baandrup U: Peripheral low-grade mucoepidermoid carcinoma of the lung—needle aspiration cytodiagnosis and histology. *Cytopathology* 3:259–265, 1992.

208. Chaudhry AP, Leifer C, Cutler LS, Satchidanand S, Labay GR, Yamane GM: Histogenesis of adenoidcystic carcinoma of the salivary glands, light and electron microscopic study. *Cancer* 58:72–82, 1986.

209. Sweeney WB, Thomas JM: Adenoid cystic carcinoma of the lung. *Comtemp Surg.* 28:97–100, 1986.

210. Turnbull AD, Huvos AG, Goodner JT, Beattie EF Jr: The malignant potential of bronchial adenoma. *Ann Thorac Surg* 14:453–464, 1972.

211. Moran A, Suster S, Koss MN: Primary adenoid cystic carcinoma of the lung. A clinicopathologic and immunohistochemical study. *Cancer* 73:1390–1397, 1994.

212. Gelder CM, Hetzel MR: Primary tracheal tumours: A national survey. *Thorax* 48:688–692, 1993.

213. Eschapasse H: Les tumeurs tracheales primitives traitment chirurgical. *Rev Fr Mal Respir* 2:425–426, 1974.

214. Grillo HC, Mathisen DJ: Primary tracheal tumors; treatment and results. *Ann Thorac Surg* 49:69–77, 1990.

215. Ahel V, Zubovic I, Rozmanic V: Bronchial adenoid cystic carcinoma with saccular bronchiectasis as a cause of recurrent pneumonia in children. *Pediatric Pulmonal* 12:260–262, 1992.

216. Gallagher CG, Stark R, Teskey J, Kryger M: Atypical manifestations of pulmonary adenoid cystic carcinoma. *Br J Dis Chest* 80:396–399, 1986.

217. Payne WS, Ellis SH Jr, Woolner LB, Moersch HE: The surgical treatment of cylindroma (adenoid cystic carcinoma) and mucoepidermoid tumours of the bronchus. *J Thorac Cardiovasc Surg* 38:709–726, 1959.

218. Singh HM, Thomas DME: A clinical sign of adenoid cystic tumour of the trachea. *Thorax* 28:442–443, 1973.

219. Dalton ML, Gatling RR: Peripheral adenoid cystic carcinoma of the lung. *South Med J* 83:577–579, 1990.

220. Southgate HJ, Archbold GP, el-Sayed ME, Wright J, Marks V: Ectopic release of GHRH and ACTH from air adenoid

cystic carcinoma resulting in acromegaly and complicated by pituitary infarction. *Postgrad Med J* 64:145–148, 1988.

221. Higashi K, Jin Y, Johansson M, et al: Rearrangement of 9p13 as the primary chromosomal aberration in adenoid cystic carcinoma of the respiratory tract. *Genes Chromosome Cancer* 3:21–23, 1991.

222. Nomori H, Kaseda S, Kobayashi K, Ishihara T, Yanai N, Torikata C: Adenoid cystic carcinoma of the trachea and main-stem bronchus: A clinical, histopathologic and immunohistochemical study. *J Thorac Cardiovasc Surg* 96:271–277, 1988.

223. Chin HW, DeMeester T, Chin RY, Boman B: Endobronchial adenoid cystic carcinoma. *Chest* 100:1464–1465, 1991.

224. Ishida T, Nishino T, Oka T, et al: Adenoid cystic carcinoma of the tracheobronchial tree: Clinicopatholgy and immunohistochemistry. *J Surg Oncol* 41:52–59, 1989.

225. Inoue H, Iwashita A, Kanegae H, Higuchi K, Fujinaga Y, Matsumoto I: Peripheral adenoid cystic carcinoma with substantial submucosal extension to the proximal bronchus. *Thorax* 46:147–148, 1991.

226. d'Ardenne AJ, Kirkpatrick P, Wells CA, Davies JD: Laminin and fibronectin in adenoid cystic carcinoma. *J Clin Pathol* 39:138–144, 1986.

227. Lawrence JB, Mazur MT: Adenoid cystic carcinoma: A comparative pathologic study of tumors in salivary gland, breast, lung and cervix. *Hum Pathol* 13:916–924, 1982.

228. Pappo O, Gez E, Cracium I, Zajick G, Okon E: Growth rate analysis of lung metastases appearing 18 years after resection of cutaneous adenoid cystic carcinoma. Case report and review of the literature. *Arch Pathol Lab Med* 116:76–79, 1992.

229. Spencer H, Nambu S: Sclerosing haemangiomas of the lung. *Histopathology* 10:477–487, 1986.

230. Heard BE: Commentary: "Benign sclerosing haemangioma of the lung." *Histopathology* 10:541–542, 1986.

231. Chan K-W, Gibbs AR, Lo WS, Newman GE: Benign sclerosing pneumocytoma of the lung (sclerosing haemangioma). *Thorax* 37:404–412, 1982.

232. Hsu NY, Chen CY, Kwan PC, et al: Sclerosing hemangioma of the lung: A clinicopathologic study. *Chin Med J* 52:149–154, 1993.

233. Sugio K, Yokoyama H, Kaneko S, Ishida T, Sugimachi K: Sclerosing hemangioma of the lung: Radiographic and pathological study. *Ann Thorac Surg* 53:295–300, 1992.

234. Kimura H, Kusajima Y, Konishi I, et al: A case of sclerosing hemangioma of the lung and review of 196 cases in the Japanese literature. *J Jpn Soc Clin Surg* 49:1403–1409, 1988.

235. Liebow AA, Hubbell DS: Sclerosing hemangioma (histiocytoma, xanthoma) of the lung. *Cancer* 9:53–75, 1956.

236. Palachios JJN, Escribano PM, Toledo J, Garzon A, Larru E, Palomera J: Sclerosing hemangioma of the lung: An ultrastructural study. *Cancer* 44:949–955, 1979.

237. Katzenstein A-LA, Gmelich JT, Carrington CB: Sclerosing hemangioma of the lung: A clinicopathologic study of 51 cases. *Am J Surg Pathol* 4:343–356, 1980.

238. Katzenstein A-LA, Weise DL, Fulling K, Battifora H: So-called sclerosing hemangioma of the lung: evidence for mesothelial origin. *Am J Surg Pathol* 7:3–14, 1983.

239. Nagata N, Dairaku M, Ishida T, Sueishi K, Tanaka K: Sclerosing hemangioma of the lung. Immunohistochemical characterization of its origin as related to surfactant apoprotein. *Cancer* 55:116–123, 1985.

240. Huszar M, Suster S, Herczeg E, Geiger B: Sclerosing hemangioma of the lung. Immunohistochemical demonstration of mesenchymal origin using antibodies to tissue-specific intermediate filaments. *Cancer* 58:2422–2427, 1986.

241. Nagata M, Dairaku M, Sueshi K, Tanaka K: Sclerosing hemangioma of the lung. An epithelial tumor composed of immunohistochemically heterogeneous cells. *Am J Clin Pathol* 88:552–559, 1987.

242. Yousem SA, Wick MR, Singh G, et al: So-called sclerosing hemangiomas of lung: An immunohistochemical study supporting a respiratory epithelial origin. *Am J Surg Pathol* 12:582–590, 1988.

243. Satoh Y, Tsuchiya E, Weng S-Y, et al: Pulmonary sclerosing hemangioma of the lung. A type II pneumocytoma by immunohistochemical and immunoelectron microscopic studies. *Cancer* 64:1310–1317, 1989.

244. Salminen U-S, Halttunen P, Miettinen M, Mattila S: Benign mesenchymal tumours of the lung including sclerosing haemangiomas. *Ann Chir Gynaecol* 79:85–91, 1990.

245. Lee ST, Lee YC, Hsu CY, Lin CC: Bilateral sclerosing hemangiomas of the lung. *Chest* 101:572–573, 1992.

246. Joshi K, Gopinath N, Shanker SK, Kumar P, Chopra P: Multiple sclerosing haemangiomas of the lung. *Postgrad Med J* 56:50–53, 1980.

247. Wang SE, Nieberg RK: Fine needle aspiration cytology of sclerosing hemangioma of the lung, a mimicker of bronchioloalveolar carcinoma. *Acta Cytol* 30:51–54, 1986.

248. Hirano H, Miyagawa Y, Nagata N, et al: Transbronchial needle aspiration in the diagnosis of pulmonary sclerosing haemangioma. *Respir Med* 87:475–477, 1993.

248a. Leong AS-Y, Chan KW, Senevirate HSK: A morphological and immunohistochemical study of 25 cases of so-called sclerosing haemangioma of the lung. *Histopathology* 27:121–128, 1995.

249. Park YK, Yang MH: So-called sclerosing hemangioma of the lung—two cases report with ultrastructural study. *J Korean Med Sci* 4:179–183, 1989.

250. Kay S, Still WJ, Borochovitz D: Sclerosing hemangioma of the lung: An endothelial or epithelial neoplasm? *Hum Pathol* 8:468–474, 1977.

251. Haas JE, Yunis E, Totten RS: Ultrastructure of a sclerosing hemangioma of the lung. *Cancer* 30:512–518, 1972.

252. Carstens PHB, Schrodt GR: Ultrastructure of sclerosing hemangioma. *Am J Pathol* 77:377–382, 1974.

253. Kennedy A: "Sclerosing haemangioma" of the lung: An alternative view of its development. *J Clin Pathol* 26:792–799, 1973.

254. Gatter KC, Dunnill MS, Van Muijen GN, Mason DY: Human lung tumours may coexpress different classes of intermediate filaments. *J Clin Pathol* 39:950–954, 1986.

255. Burrell LST, Ross JM: A case of chylous effusion due to leiomyosarcoma. *Br J Tuberc* 31:38–39, 1937.

256. Rosendal TH: A case of diffuse myomatosis and cyst formation in the lung. *Acta Radiol* 23:138–145, 1942.

257. Frack MD, Simon L, Dawson BH: The lymphangiomyomatosis syndrome. *Cancer* 22:428–437, 1968.

258. Corrin B, Liebow AA, Friedman PJ: Pulmonary lymphangiomyomatosis. A review *Am J Pathol* 79:348–382, 1975.

259. Taylor JR, Ryu J, Colby TV, Raffin TA: Lymphangioleiomyomatosis. Clinical course in 32 patients. *N Engl J Med* 323:1254–1260, 1990.

259a. Kitaichi M, Nishimura K, Hok H, Izumi T: Pulmonary lymphangioleiomyomatosis: A report of 46 patients including a clinico-pathologic study of prognostic factors. *Am J Respir Crit Care Med* 151:527–533, 1995.

260. Viskum K: Pulmonary lymphangioleiomyomatosis. *Monaldi Arch Chest Dis* 48:233–236, 1993.

261. Sakano T, Hamasaki T, Kawaguschi Y, Tanaka Y, Ueda K, Hiramoto T: Pulmonary lymphangiomyomatosis in childhood? Marked smooth muscle cell proliferation of the lung in a preadolescent girl with repeated pneumothorax and progressive dyspnea. *Hiroshima J Med Sci* 38:147–149, 1989.

262. Sinclair W, Wright JL, Churg A: Lymphangioleiomyomatosis presenting in a postmenopausal woman. *Thorax* 40:475–476, 1985.

263. Baldi S, Papotti M, Valente ML, Rapellino M, Scappaticci E, Corrin B: Pulmonary lymphangioleiomyomatosis in postmenopausal women: Report of two cases and review of the literature. *Eur Respir J* 7:1013–1016, 1994.

264. Carrington CB, Cugell DW, Gaensler EA, et al: Lymphangioleiomyomatosis: Physiologic-pathologic-radiologic correlations. *Am Rev Respir Dis* 116:977–995, 1977.

265. Fliegel E, Chitkara RK, Azueta V, Steinberg H: Fatal hemoptysis in lymphangiomyomatosis. *NY State J Med* 91:66–67, 1991.

266. Tamura K, Satoh Y, Ikoma A, Miyazaki T: Pneumoperitoneum in a patient with lymphangiomyomatosis. *Respiration* 58:211–213, 1991.

267. Osterborg A, Christensson B, Silfverswärd C, et al: Lymphangiomyomatosis—immunohistochemical analysis of a case presenting with enlarged inguinal lymph nodes and without pulmonary involvement. *Acta Oncol* 28:287–289, 1989.

268. Silverstein EF, Ellis K, Wolff M, Jaretzki A III: Pulmonary lymphangiomyomatosis *Am J Roentgenol Radium Ther Nucl Med* 120:832–850, 1974.

269. Wolff M: Lymphangiomyoma. Clinico-pathologic study and ultrastructural confirmation of its histogenesis. *Cancer* 31:988–1007, 1973.

270. Cagnano M, Benharroch D, Geffen DB: Pulmonary lymphangioleiomyomatosis. Report of a case with associated multiple soft-tissue tumors. *Arch Pathol Lab Med* 115:1257–1259, 1991.

271. Capron F, Ameille J, Leclerc P, et al: Pulmonary lymphangioleiomyomatosis and Bourneville's tuberous sclerosis with pulmonary involvement: The same disease? *Cancer* 52:851–855, 1983.

272. Roach ES, Smith M, Huttenlocher P, Bhat M, Alcorn D, Hawley L: Diagnostic criteria: Tuberous sclerosis complex: Report of the Diagnostic Criteria Committee of the National Tuberous Sclerosis Association. *J Child Neurol* 7:221–224, 1992.

273. Harris JO, Waltuck BA, Swenson EW: The pathophysiology of the lungs in tuberous sclerosis. A case report and literature review. *Am Rev Respir Dis* 100:379–387, 1969.

274. Popper HH, Juettner-Smolle FM, Pongratz MG: Micronodular hyperplasia of type II pneumocytes—a new lung lesion associated with tuberous sclerosis. *Histopathology* 18:347–354, 1991.

275. Lie JT, Miller RD, Williams DE: Cystic disease of the lungs in tuberous sclerosis: Clinicopathologic correlation, including body plethysmographic lung function tests. *Mayo Clin Proc* 55:547–553, 1980.

276. Shepherd CW, Gomez MR, Crowson CS: Causes of death in patients with tuberous sclerosis. *Mayo Clin Proc* 66:792–796, 1991.

277. Kingsley DPE, Kendall BE, Fitz CR: Tuberous sclerosis: A clinicoradiological evaluation of 110 cases with particular reference to atypical presentation. *Neuroradiology* 28:38–46, 1986.

277a. Guinee DG Jr, Thornberry DS, Azume N, Przygodzki RM, Koss NM, Travis WD: Unique pulmonary presentation of an angiomyolipoma: Analysis of clinical, radiographic and histologic features. *Am J Surg Pathol* 19:476–480, 1995.

278. Popper HH, Gamperl R, Pongratz MG, Kullnig P, Juttner-Smolle F-M, Pfragner R: Chromosome typing in lymphangioleiomyomatosis of the lung with and without tuberous sclerosis. *Eur Respir J* 6:753–759, 1993.

279. Aberle DR, Hansell DM, Brown K, Tashkin DP: Lymphangiomyomatosis: CT, chest radiographic and functional correlations. *Radiology* 176:381–387, 1990.

280. Sobonya RE, Quan SF, Fleishman JS: Pulmonary lymphangioleomyomatosis: Quantitative analysis of lesions producing airflow limitation. *Hum Pathol* 16:1122–1128, 1985.

281. Burger CD, Hyatt RE, Staats BA: Pulmonary mechanics in lymphangioleiomyomatis. *Am Rev Respir Dis* 143:1030–1033, 1991.

282. Carlson KC, Parnassus WN, Klatt EC: Thoracic lymphangiomatosis. *Arch Pathol Lab Med* 111:475–477, 1987.

283. Bhattacharyya AK, Balogh K: Retroperitoneal lymphangioleiomyomatosis. A 36-year benign course in a postmenopausal woman. *Cancer* 56:1144–1146, 1985.

284. Lauweryns JM, Baert JH: Alveolar clearance and the role of pulmonary lymphatics. *Am Rev Respir Dis* 115:625–683, 1977.

285. Fukuda Y, Kawamoto M, Yamamoto A, Ishizaki M, Basset F, Masugi Y: Role of elastic fiber degradation in emphysema-like lesions of pulmonary lymphangiomyomatosis. *Hum Pathol* 21:1252–1261, 1990.

286. Stovin PGI, Lum LC, Flower CDR, Darke CS, Beeley M: The lungs in lymphangiomyomatosis and tuberous sclerosis. *Thorax* 30:497–509, 1975.

287. Bonetti F, Chioder L, Pea M, et al: Transbronchial biopsy in lymphangiomyomatosis of the lung. HMB for diagnosis. *Am J Surg Pathol* 17:1092–1102, 1993.

288. Bonetti F, Pea M, Martignoni G, Zamboni G, Iuzzolino P: Cellular heterogeneity in lymphangiomyomatosis of the lung. *Hum Pathol* 22:727–728, 1991.

289. Peyrol S, Gindre D, Cordier JF, Loire R, Grimaud JA: Characterization of the smooth muscle cell infiltrate and associated connective matrix of lymphangiomyomatosis. Immunohistochemical and ultrastructural study of two cases. *J Pathol* 168:387–395, 1992.

290. Chan JKC, Tsang WYW, Pau MY, Tang MC, Pang SW, Fletcher CDM: Lymphangiomyomatosis and angiomyolipoma: Closely related entities characterised by hamartomatous proliferation of HMB-45-positive smooth muscle. *Histopathology* 22:445–455, 1993.

291. Brentani MM, Carvalho RR, Saldiva PH, Pacheco BS, Oshima CTF: Steroid receptors in pulmonary lymphangiomyomatosis. *Chest* 85:96–99, 1984.

292. Berger U, Khaghani A, Pomerence A, Yacoub M, Coombes RC. Pulmonary lymphangioleiomyomatosis and steroid receptors. An immunocytochemical study. *Am J Clin Pathol* 93:609–614, 1990.

293. Schiaffino E, Tavani E, Dellafiore L, Schmid C: Pulmonary lymphangiomyomatosis. Report of a case with immunohistochemical and ultrastructural findings. *Appl Pathol* 7:265–272, 1989.

294. Urban T, Kuttenn F, Gompel A, Marsac J, Lacronique J: Pulmonary lymphangiomyomatosis. Follow-up and long-term outcome with antiestrogen therapy: A report of eight cases. *Chest* 102:472–476, 1992.

295. Case Records of the Massachusetts General Hospital. *N Engl J Med* 330:1300–1306, 1994.

296. Dishner W, Cordasco EM, Blackburn J, Demeter S, Levin H, Carey WD: Pulmonary lymphangiomyomatosis. *Chest* 85:797–799, 1984.

297. Basset F, Soler P, Marsac J, Corrin B: Pulmonary lymphangiomyomatosis: Three new cases studied with electron microscopy. *Cancer* 38:2357–2366, 1976.

298. Shen A, Iseman MD, Waldron JA, King TE: Exacerbation of pulmonary lymphangioleiomyomatosis by exogenous estrogens. *Chest* 91:782–785, 1987.

299. Pea M, Bonetti F, Zamboni G, Martignoni G, Fiore-Donati L, Doglioni C: Clear cell tumor and angiomyolipoma. *Am J Surg Pathol* 15:199–200, 1991.

300. Tazelaar HD, Kerr D, Yousem SA, Saldana MJ, Langston C, Colby TV: Diffuse pulmonary lymphangiomatosis. *Hum Pathol* 24:1313–1322, 1993.

301. Wagenaar SS, Swierenga J, Wagenvoort CA: Late presentation of primary pulmonary lymphangiectasis. *Thorax* 33:791–795, 1978.

302. Holden WE, Morris JF, Antonovic R, Gill TH, Kessler S: Adult intrapulmonary and mediastinal lymphangioma causing haemoptysis. *Thorax* 42:635–636, 1987.

303. Gilbert JG, Mazerella LA, Feit LJ: Primary tracheal tumors in the infant and adult. *Arch Otolaryngol* 58:1–9, 1953.

304. Wilson R, Fishback N, Colby T, Fleming M, Koss M, Travis W: Intrapulmonary localized fibrous tumor (ILFT): A clincopathologic analysis of 43 cases. *United States and Canadian Academy of Pathology Abstracts* 909, 1995.

305. Pollak ER, Naunheim, KS, Little AG: Fibromyxoma of the trachea: A review of benign tracheal tumors. *Arch Pathol Lab Med* 109:926–929, 1985.

306. Littlefield JB, Drash EC: Myxoma of the lung. *J Thorac Surg* 37:745–749, 1959.

307. Tan-Liu NS, Matsubara O, Grillo HC, Mark EJ: Invasive fibrous tumor of the tracheobronchial tree. Clinical and pathologic study of seven cases. *Hum Pathol* 20:180–184, 1989.

308. Lattes R: Fibroblastic tumors (fibromatoses). Malignant mesenchymal tumors, in Lattes R (ed), *Atlas of Tumor Pathology: Tumors of the Soft Tissues*. Washington, D.C., Armed Forces Institute of Pathology, 1982, pp 1–30, 123–146.

309. Mackenzie DH: *The Differential Diagnosis of Fibroblastic Disorders*. Oxford, Blackwell Scientific Publications, 1970.

310. O'Connell JE, Raafat F, Proops D: Congenital subglottic fibromatosis. *J Laryngol Otol* 103:983–985, 1989.

311. Strickler JG, Hegstrom J, Thomas MJ, Yousem SA: Myoepithelioma of the lung. *Arch Pathol Lab Med* 111:1082–1085, 1987.

311a. Tsuji N, Tateishi R, Ishiguro S, Tevao T, Higashiyama M: Adenomyoepithelioma of the lung. *Am J Surg Pathol* 19:956–962, 1995.

312: Schurch W, Potvin C, Seemayer TA: Malignant myoepithelioma (myoepithelial carcinoma) of the breast. An ultrastructural and immunocytochemical study. *Ultrastruct Pathol* 8:1–11, 1985.

313. Chaudhry AP, Satchidanand S, Peer R, Cutler S: Myoepithelial cell adenoma of the parotid gland: A light and ultrastructural study. *Cancer* 49:288–293, 1982.

314. White SH, Ibrahim NBN, Forrester-Wood CP, Jeyasingham K: Leiomyomas of the lower respiratory tract. *Thorax* 40:306–311, 1985.

315. Gal AA, Brooks JSJ, Pietra GG: Leiomyomatous neoplasms of the lung: A clinical, histologic and immunohistochemical study. *Mod Pathol* 2:209–215, 1989.

316. Yellin AR, Rosenman Y, Lieberman Y: Review of smooth muscle tumours of the lower respiratory tract. *Br J Dis Chest* 78:337–351, 1984.

317. Shahian DM, McEnany MT: Complete endobronchial excision of leiomyoma of the bronchus. *J Thorac Cardiovasc Surg* 77:87–91, 1979.

318. Naresh KN, Pai SA, Vyas JJ, Soman CS: Leiomyoma of the bronchus: A case report. *Histopathology* 22:288–289, 1993.

319. Vera-Roman JM, Sobonya RE, Gomez-Garcia JL, Sanz-Bondia JR, Paris-Romeu F: Leiomyoma of the lung: Literature review and case report. *Cancer* 52:936–941, 1983.

320. Taylor TL, Miller DR: Leiomyoma of the bronchus. *J Thorax Cardiovasc Surg* 57:284–288, 1969.

321. Orlowki TM, Stasiak K, Kolodziej J: Leiomyoma of the lung. *J Thorac Cardiovasc Surg* 76:257–261, 1978.

322. Furoughi E: Leiomyoma of the trachea. *Dis Chest* 42:230–232, 1962.

323. Steiner P: Metastasizing fibroleiomyoma of uterus: Report of a case and review of literature. *Am J Pathol* 15:89–110, 1939.

323a. Horstmann JP, Pietra GG, Harman JA, Cole NG, Grinspan S: Spontaneous regression of pulmonary leiomyomas during pregnancy. *Cancer* 39:314–321, 1977.

324. Wick MR, Scheithauer BW, Piehler JM, Pairolero PC: Primary pulmonary leiomyosarcomas. A light and electron microscopic study. *Arch Pathol Lab Med* 106:510–514, 1982.

325. Agnos JW, Starkey GWB: Primary leiomyosarcoma and leiomyoma of the lung: Review of the literature and report of two cases of leiomyosarcoma. *N Engl J Med* 158:12–17, 1958.

326. Merritt JW, Parker KR: Intrathoracic leiomyosarcoma. *Can Med Assoc J* 77:1031–1033, 1957.

327. Morgan PGM, Ball J: Pulmonary leiomyosarcomas. *Br J Dis Chest* 74:245–252, 1980.

328. Pain JA, Sayer RE: Primary leiomyosarcoma of the pulmonary artery. *Eur J Respir Dis* 65:139–143, 1984.

329. Inoue H, Tsuneyoshi M, Enjoji M, Ishida T, Yasumoto K, Sugimachi K: Endotracheal neurilemmoma with a lymphoid cuff. An ultrastructural and immunohistochemical study. *Acta Pathol Jpn* 39:407–412, 1989.

330. Davies MJ, Hall DR, Ross BA: Rare tracheal tumours: Two case reports of primary neurogenic tumours occurring in the trachea. *Respir Med* 87:145–146, 1993.

331. Pang LC: Primary neurilemmoma of the trachea. *South Med J* 82:785–787, 1989.

332. Robin J, Wilson AC: Polypoid neurilemmoma of the trachea: An unusual cause of major airway obstruction. *Aust NZ J Surg* 58:912–914, 1988.

333. Bosch X, Ramirez J, Font J, et al: Primary intrapulmonary benign schwannoma. A case with ultrastructural and immunohistochemical confirmation. *Eur Respir J* 3:234–237, 1990.

334. Imaizumi M, Takahashi T, Niimi T, Uchida T, Abe T, Fukatsu T: A case of primary intrapulmonary neurilemmoma and review of the literature. *Jpn J Surg* 19:740–746, 1989.

335. Felhaus RJ, Anene C, Bogard P: A rare endobronchial neurilemmoma (schwannoma). *Chest* 95:461–462, 1989.

336. Roviaro G, Montorsi M, Varoli F, Binda A, Ceccheto A: Primary pulmonary tumours of neurogenic origin. *Thorax* 38:942–945, 1983.

337. Crofts NF, Forbes GB: Malignant neurilemmoma of the lung metastasizing to the heart. *Thorax* 19:334–337, 1964.

338. Saleh HA, Beydoun R, Masood S: Cytology of malignant schwannoma metastatic to the lung. Report of a case with diagnosis by fine needle aspiration biopsy. *Acta Cytol* 37:409–412, 1993.

339. Daimaru Y, Kido H, Hashimoto H, Enjoji M: Benign schwannoma of the gastrointestinal tract. A clinicopathologic and immunohistochemical study. *Hum Pathol* 19:257–264, 1988.

340. Bartley TJ, Arean VM: Intrapulmonary neurogenic tumors. *J Thorac Cardiovasc Surg* 50:114–123, 1965.

341. Miura H, Kato H, Hayata Y, Ehihara Y, Gonullu U: Solitary bronchial mucosal neuroma. *Chest* 95:245–247, 1989.

342. el Oakley R, Grotte GJ: Progressive tracheal and superior vena caval compression caused by benign neurofibromatosis. *Thorax* 49:380–381, 1994.

343. Unger PD, Geller SA, Anderson PJ: Pulmonary lesions in a patient with neurofibromatosis. *Arch Pathol Lab Med* 108:654–657, 1984.

344. Thaller S, Fried MP, Goodman ML: Symptomatic solitary granular cell tumor of the trachea. *Chest* 88:925–928, 1985.

345. Muthuswamy PP, Alrenga DP, Marks P, Barker WL: Granular cell myoblastoma: Rare localization in the trachea. Report of a case and review of the literature. *Am J Med* 80:714–718, 1986.

346. Alvarez-Fernandez E, Carretero-Albinana L: Bronchial granular cell tumor: Presentation of three cases with tissue culture and ultrastructural study. *Arch Pathol Lab Med* 111:1065–1069, 1987.

347. Ostermiller WE, Comer TP, Barker WL: Endobronchial granular cell myoblastoma. *Ann Thorac Surg* 9:143–148, 1970.

348. DeClerq D, Van der Straeten M, Roels H: Granular cell myoblastoma of the bronchus. *Eur J Respir Dis* 64:72–76, 1983.

349. Redjaee B, Rohatgi PK, Herman MA: Multicentric endobronchial granular cell myoblastoma. *Chest* 98:945–948, 1990.

350. Schulster PL, Khan FA, Azueta V: Asymptomatic pulmonary granular cell tumor presenting as a coin lesion. *Chest* 68:256–258, 1975.

351. Teplick JG, Teplick SK, Haskin ME: Granular cell myoblastoma of the lung. *Am J Roentgenol Radium Ther Nucl Med* 125:890–895, 1975.

352. Robinson JM, Knoll R, Henry DA: Intrathoracic granular cell myoblastoma. *South Med J* 81:1453–1457, 1988.

353. Hernandez OG, Haponik EF, Summer WR: Granular cell tumour of the bronchus: Bronchoscopic and clinical features. *Thorax* 41:927–931, 1986.

354. Yang KL, Ortiz L: Granular cell myoblastoma involving multiple organs. *South J Med* 86:478–479, 1993.

355. Ganti S, Marino W: Granular cell myoblastoma in an HIV positive patient. *NY State J Med* 91:265–266, 1991.

356. Schwartzberg DG, Al-Bazzaz FJ, Cassel J, et al: Multiple granular cell tumors of the bronchi. *Am Rev Respir Dis* 120:193–196, 1979.

357. O'Connell DJ, MacMahon H, DeMeester TR: Multicentric tracheobronchial and esophageal granular cell myoblastoma. *Thorax* 33:596–602, 1978.

358. Nathrath WBJ, Remberger K: Immunohistochemical study of granular cell tumours: Demonstration of neurone specific enolase, S 100 protein, laminin and alpha-1-antichymotrypsin. *Virchows Arch [A]* 408:421–434, 1986.

359. Sobel HJ, Schwartz R, Marquet E: Light and electron-microscope study of the origin of granular-cell myoblastoma. *J Pathol* 109:101–111, 1973.

360. Usui M, Ishii S, Yamawaki S, Sasaki T, Minami A, Hizawa K: Malignant granular cell tumor of the radial nerve. *Cancer* 39:1547–1555, 1977.

361. Daniel TM, Smith RH, Faunce HF, Sylvest VM: Transbronchoscopic versus surgical resection of tracheobronchial granular cell myoblastoma. *J Thorac Cardiovasc Surg* 80:898–903, 1980.

362. Klima M, Peters J: Malignant granular cell tumor. *Arch Pathol Lab Med* 111:1070–1072, 1987.

363. Garancis JC, Komorowski RA, Kuzma JF: Granular cell myoblastoma. *Cancer* 25:542–550, 1970.

364. Engleman P, Liebow AA, Gmelich J, Friedman PJ: Pulmonary hyalinizing granuloma. *Am Rev Respir Dis* 115:997–1008, 1977.

365. Yousem SA, Hochholzer L: Pulmonary hyalinizing granuloma. *Am J Clin Pathol* 87:1–6, 1987.

366. Dent RG, Godden DJ, Stovin PJI, Stark JE: Pulmonary hyalinising granuloma in association with a retroperitoneal fibrosis. *Thorax* 38:955–956, 1983.

367. Guccion JG, Rohatgi PK, Saini N: Pulmonary hyalinizing granuloma: Electron microscopic and immunologic studies. *Chest* 85:571–573, 1984.

368. Chen HP, Lee SS, Berardi RS: Inflammatory pseudotumor of the lung: Ultrastructural and light microscopic study of a myxomatous variant. *Cancer* 54:861–865, 1984.

369. Daudi FA, Lees GM, Higa TE: Inflammatory pseudotumors of the lung: Two cases and a review. *Can J Surg* 34:461–464, 1991.

370. Beradi RS, Lee SS, Chen HP, Stines GJ: Inflammatory pseudotumors of the lung. *Surg Gynecol Obstet* 156:89–96, 1983.

371. Umiker WO, Iverson L: Post-inflammatory "tumors" of the lung. Report of four cases simulating xanthoma, fibroma or plasma cell tumor. *J Thorac Surg* 28:55–63, 1954.

372. Bates T, Hull OH: Histiocytoma of the bronchus: Report of a case in a 6-year-old child. *Am J Dis Child* 95:53–56, 1958.

373. Katzenstein A-LA, Maurer JJ: Benign histiocytic tumor of lung. A light and electron-microscopic study. *Am J Surg Pathol* 3:61–68, 1979.

374. Edwards AT, Taylor AB: Vascular endothelioma of the lung. *Br J Surg* 25:487–495, 1937-8.

375. Gordon J, Walker G: Plasmacytoma of the lung. *Arch Pathol* 37:222–224, 1944.

376. Kay S: Inflammatory fibrous histiocytoma (? xanthogranuloma): Report of two cases with ultrastructural observation in one. *Am J Surg Pathol* 2:313–319, 1978.

377. Viguera JL, Pujol JL, Reboiras SD, Larrauri J, De Miguel LS: Fibrous histiocytoma of the lung. *Thorax* 31: 475–479, 1976.

378. Sajjad SM, Begin LR, Dail DH, Lukeman JM: Fibrous histiocytoma of lung: A clinicopathological study of two cases. *Histopathology* 5:325–334, 1981.

379. Pettinato G, Manivel JC, De Rosa N, Dehner LP: Inflammatory myofibroblastic tumor (plasma cell granuloma): Clinico-pathologic study of 20 cases with immunohistochemical and ultrastructural observations. *Am J Clin Pathol* 94:538–546, 1990.

380. Anthony PP: Inflammatory pseudotumour (plasma cell granuloma) of lung, liver and other organs. *Histopathology* 23:501–503, 1993.

381. Bahadori M, Liebow AA: Plasma cell granulomas of the lung. *Cancer* 31:191–208, 1973.

382. Spencer H: The pulmonary plasma cell/histiocytoma complex. *Histopathology* 8:903–916, 1984.

383. Matsubara O, Tan-Liu NS, Kenney RM, Mark EJ: Inflammatory pseudotumors of the lung: Progression from organising pneumonia to fibrous histiocytoma or to plasma cell granuloma in 32 cases. *Hum Pathol* 19:807–814, 1988.

384. Zeren H, Travis W, Fleming MV, Gal AA, Koss MN: Inflammatory pseudotumor of the lung: A clinicopathological study of 50 cases. *United States and Canadian Academy of Pathology Abstracts* 913, 1995.

385. Hartman G, Schochat S: Primary pulmonary neoplasms of childhood. *Ann Thorac Surg* 36:108–119, 1983.

386. Scott L, Blair G, Taylor G, Dimmick J, Fraser G: Inflammatory pseudotumors in children. *J Pediatr Surg* 23:755–758, 1988.

387. Weinberg PB, Bromberg PA, Askin FR: "Recurrence" of a plasma cell granuloma 11 years after resection. *South Med J* 80:519–521, 1987.

388. Iwami F, Hirotsu Y, Wakimoto J, Fukunaga H: Inflammatory pseudotumor arising from the hilum and invading the subepithelium (letter). *Chest* 101:1742, 1992.

389. Nonomura A, Mizukami Y, Matsubara F, et al: Seven patients with plasma cell granuloma (inflammatory pseudotumor) of the lung, including two with intrabronchial growth: An immunohistochemical and electron microscopic study. *Inter Med* 31:756–765, 1992.

390. Hutchins Gm, Eggleston JC: Unusual presentation of pulmonary inflammatory pseudotumor (plasma cell granuloma) as esophageal obstruction. *Am J Gastroenterol* 71:501–504, 1979.

391. Abdul-Karim FW, Slim MS, Melhem RE: Pulmonary inflammatory pseudotumor with esophageal obstruction: Report of a case and a review of the literature. *Pediatr Surg Int* 1:138–142, 1986.

392. Hong HY, Castelli MJ, Walloch JL: Pulmonary plasma cell granuloma (inflammatory pseudotumor) with invasion of thoracic vertebra. *Mt Sinai J Med* 57:117–121, 1990.

393. Kim I, Kim WS, Yeon KM, Chi JG: Inflammatory pseudotumor of the lung manifesting as a posterior mediastinal mass. *Pedartr Radiol* 22:467–468, 1992.

394. Gal AA, Koss MN, McCarthy WF, Hochholtzer L: Prognostic factors in pulmonary fibrohistiocytic lesions. *Cancer* 73:1817–1824, 1994.

395. Hammer J, Gradel E, Signer E, Ohnacker H, Rinderknecht BP, Rutishauser M: Plasma cell granuloma of the lung: Associated laboratory findings and ultrastructural evidence of inflammatory origin. *Pediatric Pulmonol* 10:299–303, 1991.

396. Warter A, Statge D, Roeslin N: Angioinvasive plasma cell granulomas of the lung. *Cancer* 59:435–443, 1987.

397. Case Records of the Massachusetts General Hospital. *N Engl J Med* 325:1155–1165, 1991.

398. Ritter MR, Wick MR: Inflammatory sarcomatoid carcinomas of the lung: Comparison with inflammatory pseudotumors. *United States and Canadian Academy of Pathology Abstracts* 891, 1995.

398a. Arber DA, Kamel OW, van de Riju M, et al: Frequent presence of the Epstein-Barr virus in inflammatory pseudotumor. *Hum Pathol* 26:1093–1098, 1995.

399. Ozaki T, Haku T, Kawano T, Yasuoka S, Ogura T: Neutrophil recruitment in the respiratory tract of a patient with plasma cell granuloma of the lung. *Chest* 98:770–772, 1990.

400. Kay S, Reed WG: Chorioepithelioma of the lung in a female infant. *Am J Pathol* 21:555–561, 1953.

401. Pushchak MJ, Farhi DC: Primary chorioncarcinoma of the lung. *Arch Pathol Lab Med* 111:477–479, 1987.

402. Liebow AA, Castleman B: Benign clear cell tumors of the lung (abstract). *Am J Pathol* 43:13a, 1963.

403. Gal AA, Koss MN, Hochholzer L, Chejfec G: An immunohistochemical study of benign clear cell ("sugar") tumor of the lung. *Arch Pathol Lab Med* 15:1034–1038, 1991.

404. Gaffey MJ, Mills SE, Askin FB, et al: Clear cell tumor of the lung: A clinico-pathologic immunohistochemical, and ultrastructural study of eight cases. *Am J Surg Pathol* 14:248–259, 1990.

405. Gaffey MJ, Mills SE, Zarbo RJ, Weiss LM, Gown AM: Clear cell tumor of the lung: A melanocytic neoplasm (abstract). *Lab Invest* 64:115a, 1991.

406. Becker NF, Soifer I: Benign clear cell tumor (sugar-tumor) of the lung. *Cancer* 27:712–719, 1971.

407. Liebow AA, Castleman B: Benign clear cell (and "sugar") tumors of the lung. *Yale J Biol Med* 43:213–222, 1971.

408. Adrion A, Mazzucco G, Gugliotta P, Monga G: Benign clear cell ("sugar") tumor of the lung. A light microscopic, histochemical and ultrastructural study with a review of the literature. *Cancer* 56:2657–2663, 1985.

409. Harbin WP, Mark GJ, Greene RE: Benign clear-cell tumor ("sugar" tumor) of the lung. A case report and review of the literature. *Radiology* 129:595–596, 1978.

410. Nakanishi K, Kawai T, Suzuki M: Benign clear cell tumor of the lung: the histopathologic study. *Acta Pathol Jpn* 38:515–522, 1988.

411. Hoch SS, Patchefsky AS, Takeda M, Gordon G: Benign clear cell tumor of the lung: An ultrastructural study. *Cancer* 33:1328–1336, 1974.

412. Fukuda T, Machinami R, Joshita T, Nagashima K: Benign clear cell tumor of the lung in an 8-year-old girl. *Arch Pathol Lab Med* 110:664–666, 1986.

413. Sale GE, Coolander BG: Benign clear cell tumour of the lung with necrosis. *Cancer* 37:2355–2358, 1976.

414. Katzenstein A-LA, Purvis RJR, Gmelich J, Askin F: Pulmonary resection for metastatic renal adenocarcinoma: Pathologic findings and therapeutic value. *Cancer* 41:712–723, 1978.

415. Wills JS, Hewes AC: Benign clear cell tumor of the lung: A cautionary tale. *Urol Radiol* 2:255–257, 1981.

416. Edwards C, Carlisle A: Clear cell carcinoma of the lung. *J Clin Pathol* 38:880–885, 1985.

417. Pear M, Bonetti F, Zamboni G, Martignoni G, Fiore-Donatio L, Doglioni C: Clear cell tumor and angiomyolipoma. *Am J Surg Pathol* 15:199–202, 1991.

418. Bonetti F, Colombari R, Manfrin E, et al: Breast carcinoma with positive results for melanomarker (HMB 45): HMB 45 reactivity in normal and neoplastic breast. *Am J Clin Pathol* 92:491–495, 1989.

419. Kornstein MJ, Franco AP: Specificity of HMB-45 (letter). *Arch Pathol Lab Med* 114:450, 1990

420. Wang J, Dhillon AP, Sankey EA, Wightman AK, Lewin JF, Scheur PJ: "Neuroendocrine" differentiation in primary neoplasms of the liver. *J Pathol* 163:61–67, 1991.

421. Fukuyama M, Maeda Y, Funata N, et al: Pulmonary and pleural thymoma: Diagnostic application of lymphocyte markers to the thymoma of unusual site. *Am J Clin Pathol* 89:617–621, 1988.

422. Moran CA, Travis WD, Rosado-de-Christenson M, Koss MN, Rosai J: Thymomas presenting as pleural tumors. Report of eight cases. *Am J Surg Pathol* 16:138–144, 1992.

423. Payne CE, Morningstar WA, Chester EH: Thymoma of the pleura masquerading as diffuse mesothelioma. *Am Rev Respir Dis* 94:441–446, 1960.

424. Yeoh CB, Ford JM, Lattes R, Wylie RH: Intrapulmonary thymoma. *J Thorac Cardiovasc Surg* 51:131–136, 1966.

425. Crotty TB, Hooker RP, Swensen SJ, Scheithauer BW, Myers JL: Primary malignant ependymoma of the lung. *Mayo Clin Proc* 67:373–378, 1992.

Chapter 30

Genetics of Lung Tumors

Vasi Sundaresan /
Pamela H. Rabbitts

This chapter describes the genetic changes associated with the development of lung tumors. Because the commonly occurring epithelial tumors have been investigated more fully than the rarer tumors, most of the chapter pertains to these histologic subtypes. Inherited predisposition is an important component in the genesis of some tumors. However, for lung cancer, although there is some indication that there may be an inherited component, the majority of genetic changes arise during the development of the tumor. These somatic genetic changes account for the majority of the information in this chapter.

INHERITED GENETIC CHANGES

A variety of genetic changes have been described in lung tumors, most of which have arisen somatically. In this part of the chapter the evidence for the genetically inherited basis of lung cancer is reviewed before the more common somatically acquired genetic changes are discussed. In most of the common solid tumors including breast cancer and colonic and renal tumors, a subset of patients is predisposed to the disease.[1] This enables genetic linkage studies to be undertaken. No such large, readily available lung cancer families exist, precluding such analysis. More than circumstantial evidence suggests that an inherited component may predispose to lung cancer development.

Genetic Predisposition to Lung Cancer

The causal role of smoking in the pathogenesis of lung cancer is no longer in doubt, although industrial exposure to other carcinogens modifies its incidence. The majority of smokers, however, do not develop lung cancer; only 10 to 15 percent of people smoking 20 or more cigarettes per day will develop the disease.[2] This implicates host factors in altering the risk/predisposition to the development of lung cancer. It is therefore conceivable that the variation in the effect of smoking and the incidence of lung cancer may relate to resistance to the disease in the 85 percent without lung cancer. Conversely factors may relate to susceptibility to lung cancer in the 15 percent who develop tumors. Such inherited predisposition is difficult to prove owing to the overriding effects of environmental exposure. Furthermore, the lack of penetrance of the probable putative gene(s) would further reduce the incidence of lung cancer in the group at risk, diluting the contribution of any inherited component. This would necessitate the analysis of large number of cases before the identification of those cases/tumors associated with inherited lung cancer.

EPIDEMIOLOGIC STUDIES

Mounting circumstantial evidence suggests that some of the genetic changes predisposing to lung cancer may be inherited in a mendelian manner.[3,4] First degree relatives of lung cancer patients have a 2.4-fold increased risk of lung cancer[5,6] or other non-smoking-related cancers.[7] Furthermore, this risk of lung cancer is greater for nonsmoking relatives of patients, highly suggestive of an inherited component.[5] These observations, however, do not allow for environmental passive exposure to tobacco smoke. This is important because blood relatives of smokers are more likely to smoke, and hence those *non*smokers in the family are more likely to be subjected to passive smoking.

Sellers and colleagues[4] investigated 337 families to identify cases not associated with smoking. A detailed study was carried out of each member of families with three generation pedigrees where one member had died of lung cancer during a 4-year period (1976–1979). These patients were divided into two groups (under or over 60 years) because 1915 signaled the dramatic increase in use of tobacco. Detailed clinical histories obtained from family members permitted subsequent statistical analysis. The results of this study, allowing for individual tobacco exposure, suggested the patterns of disease were best accounted for by mendelian codominant inheritance of an allele that was responsible for early onset of disease. Furthermore, these authors[4] reported that the cumulative probability of lung cancer at 80 for a noncarrier of the gene, assuming average tobacco consumption, was close to zero. These results suggest that a genetic predisposition to lung cancer is inherited and that the trait is only expressed in the presence of an environmental insult.

FAMILIAL CLUSTERING OF LUNG CANCER

Recent communications have reported familial clustering of lung cancer. These include the synchronous presentation of bronchoalveolar carcinoma in twins[8] and four squamous cell carcinomas and a small cell lung cancer (SCLC) in four of eight siblings.[9] Lung cancer of varying histologic subtypes was similarly described in five of ten[9] and four of eight siblings.[10] Although these groups of patients represent "lung cancer families," given the high incidence of the disease and the fact that all patients were heavy smokers, chance alone could theoretically account for this tumor incidence.

The suggestion of a genetic predisposition to lung cancer is further reinforced by a variety of reports including anecdotal case reports. For instance, patients successfully treated for lung cancer have an increased risk of second lung tumors[11] or leukemia.[12] Two patients with SCLC who developed erythroleukemia after successful chemotherapy and radiotherapy for SCLC demonstrated del(3)(p14-p23) in the leukemic cells.[13] Conversely, patients treated or cured of other malignancies outside the respiratory tract have an increased risk of developing lung cancer.[11,14,15] Some of these observations may be due to field change, namely, tobacco carcinogen induced or due to the genotoxic chemotherapy. These findings may also provide circumstantial evidence for a genetic component for the predisposition to lung cancer.

Definite evidence of the linkage of heredity and lung cancer has been observed by Saunders and coworkers.[16] In a retrospective study in relatives of patients with retinoblastoma, the ideal model for inherited cancer, the risk for nonocular cancer was 10 times that observed in the normal population, with a 15-fold risk of lung cancer developing in carriers. More recently, lung cancer has been observed in some families with the Li Fraumeni syndrome,[17–19]

an inherited cancer-prone condition due to a germ line mutation of the p53 tumor suppressor gene (17p13). Both the retinoblastoma gene and the p53 gene are mutated or inactivated in the majority of SCLC and non-SCLC patients.[20–22] Hence such inherited germ line mutations may predispose to lung cancer.

Indirect genetic evidence for the predisposition of lung cancer may be obtained by other methods. These include studying patients/families with large deletions/translocations involving regions known to be mutated/implicated in lung cancer; for instance, the short arm of chromosome 3 (3p), a region frequently lost in lung cancer. The rarity of reports of such anomalies may be speculated to be due to the lethal nature of such a constitutional karyotype. A recent report of a SCLC in a 22-year-old woman with Soto's syndrome, an overgrowth syndrome, reinforces the probability of the genetic predisposition to lung cancer.[23] This particular patient was karyotypically normal, but a previous report of karyotypic abnormality consisting of a balanced translocation at 3p21, t(3;6) (p21; p21) was reported in a patient with Soto's syndrome.[24]

Familial mesothelioma has been reported in four pairs of patients, six men and two women, aged between 44 and 84 years. However, all the patients in this study had exposure to asbestos.[25] This report yet again highlights the difficulty in divorcing the environmental exposure to carcinogens from the hereditary component.

METABOLISM OF CARCINOGENS IN LUNG CANCER

The reason only about 10 percent of heavy smokers develop lung cancer may relate to their ability to metabolize or inactivate and excrete carcinogens. This is important because it is usually the metabolites of the environmental carcinogens that initiate malignancies, rather than the parent compound.[26,27] Environmental carcinogens are usually metabolized to an active form and a constitutional inability to perform this step may be characteristic of those heavy smokers who do not develop lung cancer.

Increased activity of the phase I enzymes and decreased activity of phase II enzymes may impair the rapid elimination of potentially toxic metabolites. Oxygenated phase I intermediate substances are capable of binding covalently to nucleic acids and proteins to form carcinogen DNA adducts that result in DNA mutations. For instance, extensive metabolizers of debrisoquine have an increased risk of lung cancer.[28–32] The gene involved in debrisoquine metabolism has been cloned, but the observed polymorphism does not entirely correlate with the observed phenotype.[28,32]

Differences in the ability to repair damaged DNA may also influence genetic susceptibility. Some evidence exists for heritable differences among individuals in the activity of the repair enzymes, uracil DNA glycosylase, O^6-methylguanine-DNA methyltransferase, and O^6-alkylguanine-DNA transferase.[33,34]

GENETIC LINKAGE STUDIES AND LUNG CANCER

The hallmarks of genetic predisposition to lung cancers and other malignancies include early age of onset, one or more first degree relatives, and multiple primary neoplasms.[1,22,35] Hence large kindred with renal tumors, familial adenomatous polyposis, neurofibromatosis, malignant melanoma, and breast cancer have been critical in identifying the regions with the susceptibility locus, and the subsequent cloning of the putative tumor suppressor genes including the adenomatous polyposis coli (APC) gene, NF1 and VHL gene.[1,22,35-39] A similar approach to identifying the susceptibility locus in lung cancer is prevented by the absence of large kindreds with lung cancer, with no exposure to tobacco, occupational or environmental contaminants, and pollutants known to predispose to lung cancer.

Possibly, this difficulty in identifying susceptibility loci in human beings may be overcome by the use of murine models for lung cancer. The A/J strain of mouse is genetically susceptible to adenocarcinomas of lung.[40] There is currently evidence for three major susceptibility loci in mice.[41] A recent report has mapped one such gene, pas 1 (pulmonary adenoma susceptibility locus 1) to mouse chromosome 6, a region that shows synteny to human chromosomes 3p25 and 12(p12-13).[42]

SOMATIC GENETIC CHANGES IN MALIGNANT TUMORS

Methods of Detection

Human chromosomes are distinguishable from each other by size, position of the centromere, and the chromosome-specific banding pattern produced after staining. Solid tumors, including lung tumors, are characterized by a grossly abnormal karyotype involving extra copies of some chromosomes, loss or partial loss of others, translocations, and amplification of particular regions. Some of these abnormalities are illustrated in Fig. 30-2, which compares a

normal human karyotype with that obtained from a lung cancer cell line. Chromosome spreads can be produced from fresh biopsy specimens, but the vast majority of cytogenetic analyses have used cell lines established in tissue culture. SCLC specimens have the benefit of a specifically derived serum-free medium and hundreds of cell lines have been isolated, particularly at the National Cancer Institute in Bethesda. Fewer, but still large, numbers of non-SCLC cell lines are also available.

The aim of cytogenetic analysis is to identify the abnormalities characteristic of a particular tumor or histologic subtype. Despite the complexity of lung tumor karyotypes, it has been possible to identify chromosome-specific abnormalities, particularly for SCLC (discussed later). The identification of chromosome bands and the abnormal banding patterns due to translocations and deletions are difficult to corroborate by independent means. However, chromosome-specific probes have recently been developed that allow a particular chromosome or its various derivatives if they exist, to be identified on chromosome spreads. The probes are fluorescently labeled and the effect of hybridization of such probes to a chromosome spread results in the target chromosomes fluorescing brightly. This explains why these probes are known as chromosome paints (Fig. 30-1).

The development of recombinant DNA technology has made a huge impact on the characterization of the normal human genome and its aberrant forms found in tumor cells. Using gene-specific probes it has been possible to determine which genes are lost, amplified, or mutated in tumor cells. In addition to probes corresponding to known genes, many probes detect anonymous DNA sequences that show polymorphic variation. If a patient is constitutionally heterozygous at a particular locus, this can be detected using a number of molecular genetic techniques and compared to the patient's tumor genotype. If loss of heterozygosity (LOH) can be demonstrated in the tumor, then it is equivalent to a deletion (Fig. 30-3). Many hundreds of probes are now available for each human chromosome making it possible to study the whole genome to identify common regions of allele loss.[43] Furthermore, using a high probe density a particular chromosome region can be studied in great detail in many samples of the same tumor type. In this way, the smallest common region of loss can be determined. It is by this means that regions encompassing tumor suppressor genes have been identified.[22,35]

Although several regions of the genome have been implicated in lung cancer development (dis-

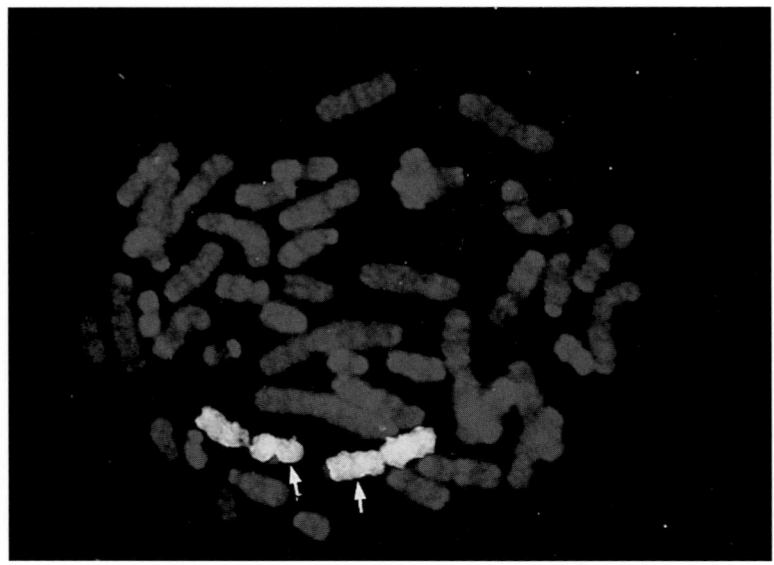

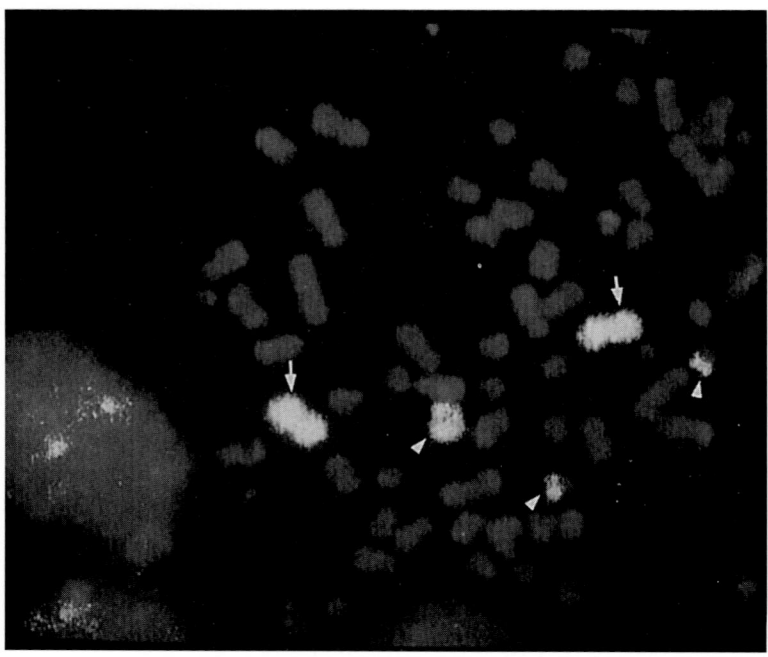

Fluorescence in situ hybridization using chromosome 3 paints. Fluorescence in situ hybridization with chromosome 3 painting probe to a normal metaphase spread (*A*), showing two normal chromosome 3 homologues (*arrows*) and an abnormal metaphase spread prepared from a lung cancer cell line (COR L 361) (*B*). Two apparently normal chromosome 3 homologues (*arrows*) and three derivative chromosome 3s (*arrowheads*) are visible. The chromosomes are stained with propidium iodide and the chromosome 3 paint signal is identified using fluorescein. (*See also Color Plate 6.*)

cussed later), it is not always known which gene within the region is important. However, involvement of 17p13 in lung cancer has been shown to be associated with mutations in the p53 gene, which maps within this locus.[44,45] A large number of point mutations have been detected in the coding region of the gene. The majority of these lead to an altered amino acid, which greatly enhances the stability of the protein. Normally, p53 has a very short half-life and is not detectable by immunohistochemistry. The mutated protein has a much longer half-life and can be detected by this method. This allows somatic genetic change to be "visualized" by a routine immunohistochemical procedure.[46,47]

Candidate Genes and Loci

SHORT ARM OF CHROMOSOME 3

Early work on somatic genetic changes in lung cancer mainly involved karyotype analysis of cell lines. Because SCLC was initially easier to establish in tissue culture, the majority of early cytogenetic analyses involved this histologic subtype.[48] One large study, mainly of cell lines but also including a few

Normal human karyotype

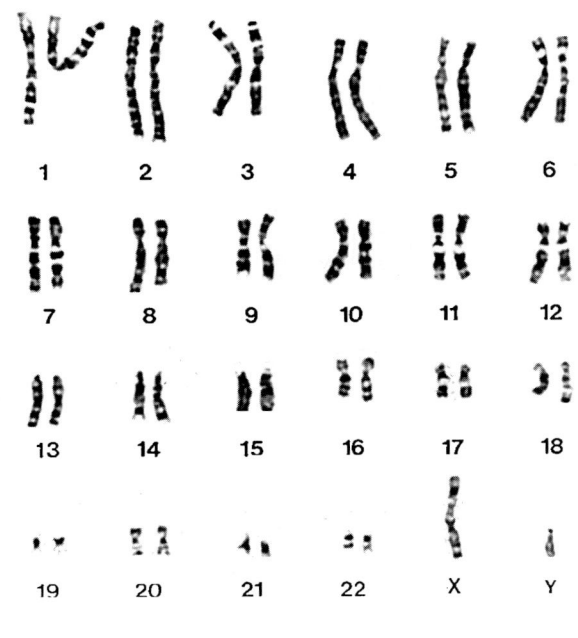

Lung tumour karyotype

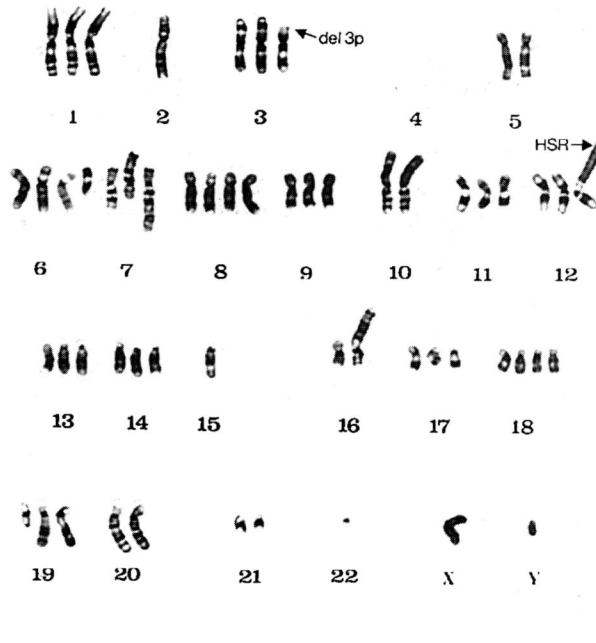

FIGURE 30-2

Comparison of a normal G banded human karyotype (*A*) with a G banded lung tumor karyotype (*B*). Several derivative and marker chromosomes are apparent in the tumor, for example, deletion of chromosome 3 material from the short arm (del 3p) and amplification of chromosomal material in the form of homogeneously staining regions (HSR).

direct preparations from biopsies, identified deletions of the short arm of chromosome 3 in all the samples.[49,50] These observations were not always reproducible.[51] Other cytogenetic analyses of smaller numbers of samples confirmed the observation of deletion on chromosome 3 but not in all samples.[52,53]

The advent of molecular genetic analysis using LOH determination, as described in the previous section, showed that genetic damage to chromosome 3 was a common event in SCLC (Table 30-1).[54–66] The disparity between the molecular genetic and cytogenetic evidence was explained when markers for the whole chromosome 3 were used to examine DNA from SCLC cell lines. This showed cell lines that appeared cytogenetically normal were genetically homozygous.[67] They had lost a whole chromosome 3 and duplicated the remaining one. Such a SCLC shows 3p allele loss but could have appeared cytogenetically normal.

Chromosome 3 loss has also been observed in non-SCLC by both cytogenetic and molecular genetic analysis (Tables 30-2 and 30-3).[54,55,57,62,63,65,68–79] As far as the different histologic subtypes are concerned, squamous cell carcinomas have been reported to show 100 percent loss in several studies (see Table 30-3). Non-SCLC samples, because of their heavy contamination with normal cells, are difficult to assess by LOH and for samples in which no LOH is recorded the possibility exists that these samples contain sufficient normal cells to obscure the LOH in the tumor cells. In our experience (unpublished results), when DNA for analysis is obtained from

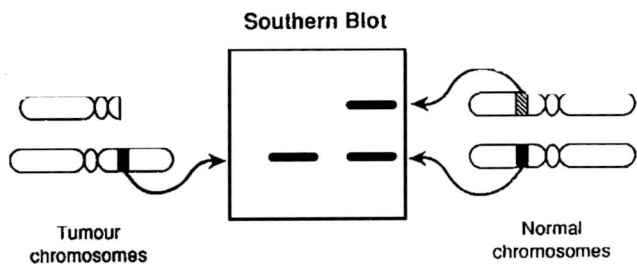

FIGURE 30-3

Demonstration of allelic loss by restriction fragment length polymorphism analysis. The normal chromosomes (*right*) have two distinguishable (i.e., polymorphic) alleles, identified by the black and a hatched box, which can be detected by Southern blot analysis of DNA, extracted from the chromosomes, as two bands of different sizes. The tumor chromosomes (*left*) have only one allele (black) due to deletion of part of the other chromosome. Southern blot analysis of DNA extracted from the tumor sample has only one band.

TABLE 30-1

Summary of Chromosome 3 Loss Studies in SCLC

Author	Reference	No. of 3p RFLP	No. of Tumors	3p loss (%)	No. Heterozygous for 3q	3q Loss
Brauch (1987)	70	3	13	13/13 (100)	4/13	0/4
Kok[a] (1987)	54	2	7	7/7 (100)	—	—
Naylor (1987)	55	3	9	9/9 (100)	7/9	0/7
Mooibroek (1987)	56	1	12	4/4 (100)	—	—
Yokota (1987)	57	1	47	7/7 (100)	—	—
Johnson (1988)	58	3	24	23/24 (93)	19/24	4/19
Sithanandam (1989)	59	4	11	10/11 (91)	—	—
Mori (1989)	60	4	15	15/15 (100)	6/15	0/6
Dobrovic (1988)	61	1	6	6/6 (100)	—	—
Leduc (1989)	62	2	17	16/17 (94)	—	—
Brauch (1990)	63	10	14	13/14 (92)	—	—
Daly (1991)	64	15	6	6/6 (100)	5/6	3/5
Hibi (1992)	65	13	9	9/9 (100)	—	—
Rabbitts (1993)	66	9	16	15/16 (94)	11/16	4/11

[a]Also analyzed unpaired cell lines of SCLC and tumors.

RFLP = restriction fragment length polymorphism; — = information not available, in publication

microdissected tissue, rather than a gross specimen, the proportion of squamous cell carcinoma with 3p allele loss is 100 percent.[79]

After the first demonstration of allele loss on 3p, several further studies using more tumors and probes sought to define the minimally deleted region on 3p. They appeared to produce conflicting results, which were resolved on the realization that there were probably at least three distinct regions of loss involved (Figs. 30-4 and 30-5). For two of these regions, viz 3p21 and 3p14 -> cen, regions of homozygous loss have been described.[67,80-82] Several independent studies have identified 3p21 homozygous deletions and it is likely that a gene with the properties of a tumor suppressor gene will probably be isolated from this region in the near future. These and other chromosomal loci demonstrated to have homozygous deletions are summarized in Table 30-4.[67,80-89] A tumor suppressor gene has been isolated from the 3p25 region—the gene for the von Hippel-Lindau syndrome.[39] As well as being involved in this inherited cancer syndrome, it appears to be mutated in renal cell carcinomas but very rarely in lung cancer.[90]

THE p53 GENE

The identification of abnormalities in the p53 gene did not follow the classic pattern of reverse genetics

TABLE 30-2
Summary of Chromosome 3 Loss Studies in non-SCLC

Author	Reference	No. of 3p RFLP	No. of Tumors	3p loss (%)	No. Heterozygous for 3q	3q Loss
Brauch (1987)	70	3	15	4/15 (26)	—	—
Kok[a] (1987)	54	2	17	17/17 (100)	—	—
Yokota (1987)	57	2	31	5/8 (62)	—	—
Weston (1989)	71	3	54	16/31 (52)	—	—
Becker (1989)	72	2	18	2/8 (25)	—	—
Leduc (1989)	62	2	25	9/25 (36)	—	—
Rabbitts (1989)	73	8	48	32/44 (75)	18	13/18
Brauch (1990)	63	10	24	13/24 (54)	—	—
Ludwig (1990)	74	3	37	5/8 (62)	—	—
Houle (1991)	75	3	27	12/27 (44)	—	—
Tsuchiya (1992)	76	1	53	20/37 (54)	40/53	6/40
Hibi (1992)	65	13	39	31/39 (79)	—	—
Yokoyama (1992)	77	19	67	51/67 (76)	—	—
Horio (1993)	78	5	70	34/70 (49)	—	—
Sundaresan (1994)[b]	79	5	25	25/25 (100)	—	—

[a]Analyzed unpaired cell lines of non-SCLC

[b]Includes results from references 47 and 175

RFLP = restriction fragment length polymorphism; — = information not available, in publication

as, when loss of 17p13 was detected in lung tumors, the p53 gene was already known to reside at that locus. Loss of one allele at 17p13 is common in lung tumors with mutation in the remaining allele. It is possible that even with retention of both alleles, if one is mutated, then this can have an oncogenic effect. This is often called a dominant negative effect (Tables 30-5 to 30-7),[91–105] because if both mutant and wild type proteins are produced, heterodimers of wild type and mutant polypeptide chains are formed, which substantially abrogate the function of wild type p53.

Although the p53 gene is relatively small, a wide spectrum of mutations has been observed. Missense mutations are most frequent but small deletions, splice junction, and nonsense mutations have also

TABLE 30-3
Summary of Published Results of LOH on 3p in non-SCLC by Histology

Author	Reference	All non-SCLC No. LOH/Inf[a]	%	Histologic subtypes Squamous carcinoma[a] No.	%	Adeno-carcinoma[a]	Large cell carcinoma[a]
Naylor (1987)	55	0/2	0	0/0	0	0/1	0/1
Brauch (1987)	70	4/15	26	1/2	50	3/10	0/2
Yokota (1987)	57	5/8	63	0/1	0	5/6	0/1
Kok[b] (1987)	54	14/14	**100**	6/6	**100**	7/7	1/1
Becker (1989)	72	2/8	25	1/2	50	1/6	0/0
Weston (1989)	71	16/29	55	9/14	64	7/13	0/2
Rabbitts (1989)	73	31/43	72	23/29	79	6/10	2/2
Tsuchiya[c] (1992)	76	20/37	54	7/7	**100**	12/29	1/1
Houle[c] (1991)	75	12/27	44	11/16	69	1/10	0/1
Hibi (1992)	65	31/39	79	14/20	70	13/15	4/4
Yokoyama (1992)	77	51/67	76	18/18	**100**	33/49	—
Horio (1993)	78	34/70	49	20/33	61	12/33	2/4
Sundaresan[d] (1994)	79	25/25	**100**	25/25	**100**	—	—

[a]Number of tumors showing loss of heterozygosity (LOH)/Number of informative cases

[b]Mainly unmatched cell lines

[c]Adenosquamous carcinomas included with adenocarcinoma group

[d]Includes results from references 47 and 175.

been detected (Table 30-8). These results contrast with the mutational spectrum observed in other tumor suppressor genes. For instance, in colorectal carcinoma, the APC gene is inactivated much more frequently by deletions, insertions, and nonsense mutations than by missense mutations (see Table 30-8). In lung tumors, the most common mutation is a transversion of guanine to thymidine[106] (Table 30-9).[45,98,107] This transversion can be elicited in vitro using carcinogens found in cigarette smoke suggesting that the mutations seen in p53 in lung cancer are a direct consequence of the mutational effect of these carcinogens. Interestingly lung tumors associated with survivors of the atom bomb show G:C to A:T transitions much more commonly than the G:C to T:A transversions. Such G to T transversions in lung tumors from smokers are more frequent in the nontranscribed DNA strand.[107]

Attempts have been made to determine if the presence of p53 mutation is of prognostic significance in lung cancer. Results have been equivocal, perhaps confounded by the possibility that p53 can be inactivated by agents other than mutation. For example, viral protein can bind to wild type p53

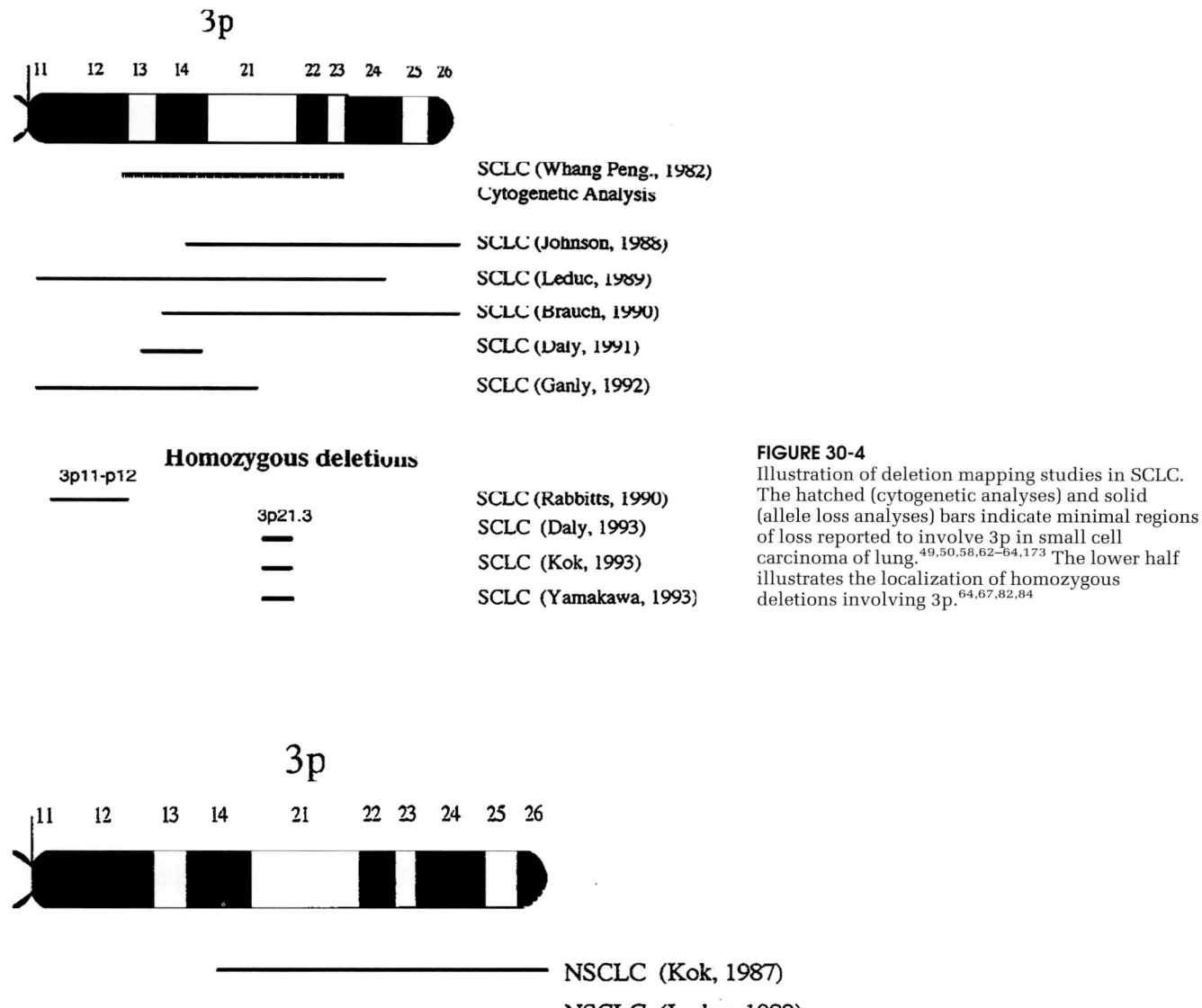

FIGURE 30-4
Illustration of deletion mapping studies in SCLC. The hatched (cytogenetic analyses) and solid (allele loss analyses) bars indicate minimal regions of loss reported to involve 3p in small cell carcinoma of lung.[49,50,58,62–64,173] The lower half illustrates the localization of homozygous deletions involving 3p.[64,67,82,84]

FIGURE 30-5
Illustration of deletion mapping studies in *non*-SCLC. The hatched (cytogenetic analyses) and solid (allele loss analyses) bars indicate minimal regions of loss reported to involve 3p in *non*-SCLC[54,62,65,67,68,77] and the location of the von Hippel Lindau (VHL) gene.[174]

TABLE 30-4
Summary of Homozygous Deletions in Pulmonary Cancer

Author	Reference	Histology	No. of Tumors	Location	Size of Deletion	Gene
Rabbitts (1990)	67	SCLC	1	3p11-12	8-10 Mb	—
Hensel (1990)	83	SCLC	2	13q14		Rb
Daly (1993)	80	SCLC	1	3p21	<0.5 Mb	—
Yamakawa (1993)	84	SCLC	5	3p21	<0.5	—
Kok (1993)	82, 85	SCLC	1	3p21		
Olapade (1993)	86	NSCLC Mesothelioma	4 1	9p21-22 9p21-22	2-3 Mb	—
Cheng (1993)	81	Mesothelioma	10	9p21-22	<1 Mb	—
Center (1993)	87	NSCLC Mesothelioma	1 1	9p21-22		—
Kohno (1994)	88	SCLC	1	2q33	>20Kb	—
Kamb (1994)	89	Lung Cancer	15	9p21		? MTS1-p16 gene

SCLC = small cell carcinoma of lung; NSCLC = non-small cell carcinoma of lung; — = gene(s) not identified

TABLE 30-5
Summary of p53 Gene Mutations in SCLC

Author	Reference	No. of Tumors Analyzed	p53 Mutations (%)	17p Loss (%)[b]
Hensel[a] (1991)	91	9	5 (55%)	—
Takahashi (1991)	92	15	11 (73%)	—
Sameshima[a] (1992)	93	27	23 (85%)	24 (96%)
Miller[a] (1992)	94	26	19 (73%)	24 (92%)

[a]More than one mutation present in some samples

[b]Results of allele loss studies of informative cases only

TABLE 30-6
Summary of p53 Gene Mutations in non-SCLC

Author	Reference	No. of Tumors Analyzed	p53 Mutations (%)	17p Loss (%)[b]
Iggo (1990)	95	3	3	—
Chiba[a,b] (1990)	96	51	24 (47%)	6 (30%)
Mitsudomi[a,b,c] (1992)	97	77	57 (74%)	50 (87%)
Vahakangas (1992)	98	18	6 (33%)	—
Miller[b] (1992)	94	13	3 (23%)	9 (70%)
Marchetti (1993)	99	45	20 (44%)	23 (51%)
Mitsudomi (1993)	100	120	51 (43%)	—
Horio (1993)	78	71	35 (49%)	—

[a]Results of allele loss studies in informative cases only

[b]More than one mutation present in some samples

[c]Includes results of NSCLC cell lines

TABLE 30-7

Summary of p53 Mutation by Histologic Subtype

Histology Reference	Iggo (1990) 95	Takahashi (1990) 101	Chiba (1990) 96	Suzuki (1992) 176	Hensel (1992) 91	Takahashi (1991) 92	Kishimoto (1992) 103	Mitsudomi (1992) 97	Miller (1992) 94	Sameshima (1992) 93	D'Amico (1992) 104	Mitsudomi (1993) 105	Horio (1993) 78
ADC	1/1	—	6/20	6/17	—	—	24/58	33/44	1/7	—	—	18/65	13/43
SQC	2/2	—	11/17	8/12	—	—	25/37	6/8	1/5	—	—	29/47	19/33
LCC	—	—	2/5	0/1	—	—	8/14	11/15	1/1	—	—	4/8	—
ADS	—	1/1	—	—	—	—	3/6	2/2	—	—	—	—	3/4
SCLC	—	1/1	—	—	6/10	11/15	—	—	20/27	23/27	32/36	—	—
BAC	—	—	3/7	—	—	—	—	—	—	—	—	—	—

ADC = adenocarcinoma; SQC = squamous cell carcinoma; LCC = large cell carcinoma; ADS = adenosquamous carcinoma; SCLC = small cell lung carcinoma; BAC = bronchioalveolar carcinoma

TABLE 30-8

Types of p53 Gene Mutation in Lung Cancer and Colorectal Carcinoma[45]

Gene	Tumor	Number of Tumors Analyzed	Silent Mutations (%)	Deletion and Insertions (%)	Nonsense Mutations (%)	Missense Mutations (%)
p53	Lung Cancer	1312	1	10	6	83
APC	Colon Cancer	37	0	54	37	9

APC = adenomatous polyposis coli

TABLE 30-9

Patterns of p53 gene mutation [45]

	No. of Tumors with p53 Mutation	Mutations at A:T → C:G (%)	→ G:C (%)	→ T:A (%)	Mutations at G:C → A:T (%) Non CpG	at CpG	→ T:A (%)	→ C:G (%)	Deletion (%)
Colon[45]	190	2	11	5	24	41	8	5	4
Lung cancer[45]	263	4	8	5	16	7	38	11	10
Lung cancer in uranium miners[98]	7	—	—	14	14[a]		28	14	28
Lung cancer in atom bomb survivors[107]	4	—	—	—	50[a]		25	25	—

[a]Information on mutational spectra at CpG islands not available

C = cytosine; G = guanine; A = adenine; T = thymidine

causing inactivation. This appears to be an important mechanism in cervical cancer where human papillomavirus infection is common. Some evidence suggests that such infection may be associated with lung tumors.[108–110]

RETINOBLASTOMA (RB) AND LUNG CANCER

A variety of somatic genetic changes have been identified in SCLC, including LOH at several loci, specifically 3p, 17p (p53 gene) and 13q. Hence to determine whether the Rb gene localized to 13q14 was involved in the molecular pathogenesis of lung cancer, in particular the presence of recessive mutations, Harbour and colleagues investigated the expression of Rb transcripts in lung tumors.[111] Structural abnormalities in size and expression of the 4.7kb mRNA were observed in 1 of 8 SCLC tumor biopsies and 4 of 22 SCLC cell lines. In addition gene rearrangement was identified in 1 of 4 pulmonary carcinoid tumor cell lines. No Rb gene rearrangement, however, was detected in any of the non-SCLC cases. Rb gene expression was absent in 60 percent of small cell lines contrasting with Rb transcripts being present in 90 percent of non-SCLC cell lines and in normal lung.

More recent molecular studies have identified 13q allele loss in virtually all SCLC tumors (Table 30-10).[57,58,60,71,72,76,83,112,113] This contrasts with 32 percent[76] to 55 percent of non-SCLC (see Table 30-10).[112] The observation of Becker and coworkers[72] of no allele loss at 13q in 16 non-SCLC tumors is unusual in the context of the above reports (see Table 30-10). This may be a reflection of the small number of samples analyzed.

Reissman and associates analyzed a large panel of tumors for Rb gene inactivation using Northern analysis and immunohistochemistry.[114] By Northern analysis, 22 of 219 non-SCLC tumors had altered or absent Rb transcript. However, using immunohistochemistry, absent or abnormal Rb protein staining was observed in 53 of 166 (32 percent) of evaluable cases, suggesting that Rb gene involvement in non-SCLC was greater than had previously been reported. The Rb gene is considerably larger than the p53 gene. Until recently, this has made the detailed analysis for evidence of mutational inactivation more difficult. Information on mutations that lead to inactivation of the remaining Rb allele has previously been limited by the lack of rapid and sensitive methods for their detection. Recently, Sachse and colleagues demonstrated that a much larger percentage of bronchial squamous cell carcinomas (75 percent) had LOH of 13q with accompanying Rb gene mutations. These

were identified by band shift by single-stranded conformational polymorphism analysis.[113,115] There was, however, no evidence of LOH in adenocarcinomas.

RAS MUTATION AND LUNG CANCER

Alterations to the *ras* family of oncogenes, H-*ras*, K-*ras* and N-*ras*, are infrequently detected in SCLC. However, K-*ras* is mutated in non-SCLC,[21] particularly in adenocarcinomas. K-*ras* mutation in non-SCLC correlates with poor clinical prognosis.[116–122] Harrada and coworkers evaluated the significance of *ras* oncogene expression in non-SCLC in 116 surgically managed patients.[120] Archival histologic sections were stained with an anti-*ras* p21 monoclonal antibody, rp-35. Patients undergoing definitive curative surgery who were p21 negative had a significantly higher survival rate than those with p21 positive tumors.

It is, however, apparent from Table 30-11[98,102,116,117,119,121–126] that the incidence of *ras* mutation in lung cancer varies markedly in different studies, from 57 percent[124] to 28 percent.[102,116] Similarly, the incidence of K-*ras* mutations in squamous cell carcinoma shows a wide range, from no mutations[98,117,125] to 10[124] and 20 percent of tumors.[121] Such wide discrepancies may in part be related to differences in the sensitivity of the methods used to identify mutations. They may also in part be due to the variations in the spectra of *ras* mutation associated with different population groups.

Mao and colleagues have very elegantly exploited the use of archival material and cytology specimens to demonstrate the feasibility of the early diagnosis of lung cancer using molecular markers.[126] Using archival material from patients with adenocarcinoma of the lung treated during the Johns Hopkins Lung Project, *ras* mutations were detected in 8 of 15 tumors. Mao and coworkers were able to demonstrate the detection of tumor-specific mutations in the sputum samples obtained from these patients, months before the definitive diagnosis.

The detection of K-*ras* mutations in approximately 30 percent of adenocarcinomas suggests K-*ras* mutation may represent a candidate for the early detection of lung cancer. At present, the cause of the reduced survival in patients with tumors carrying K-*ras* mutations is not clear.

MYC GENE

The *myc* family of oncogenes has been reported to be overexpressed in lung cancer, particularly in

TABLE 30-10
Rb Locus (13q) Involvement in Lung Cancer

Author	Reference	No of Markers Used	Histology	No. of Cases Showing LOH (%)[a]	Mutations Detected/ Immuno-Detection of Rb Protein[b]	Overall % of Lung Tumors Showing Involvement of Rb Gene
Yokota (1987)	57	4	11 SQC	4/11 (37)	—	48
			15 ADC	4/15 (27)		
			3 LCC	1/3 (33)		
			2 ADS	1/2 (50)		
			11 SCLC	10/11 (91)		
Yokota[c] (1988)	112		2 LCC	1/2	—	55
			4 ADC	0/4		
			3 SQC	0/3		
			9 SCLC	8/9	0/9	
Johnson (1988)	58	—	23 SCLC	18/23 (80)		80
Weston (1989)	71	1	23 SQC	4/9	—	43
			23 ADC	4/9		
			8 LCC	1/3		
Becker (1989)	72	4	4 SQC	0/4	—	8
			12 ADC	0/8		
			1 SCLC	1/1 (100)		
Mori (1989)	60	4	16 SCLC	15/16 (94)	—	94
Hensel (1990)	83	1	26 SCLC	6/6 (100)	2/14	100
Tsuchiya (1992)	76	1	11 SQC	5/9 (56)		32
			36 ADC	5/26 (19)	—	
			3 LCC	1/1 (100)		
			3 ADS	1/2 (50)		
Sachse[d] (1994)	113	2	13 SQC	6/8 (75)	2/8[e]	—
			25 ADC	0/16		
			3 ADS	NI	1/25[f]	

[a]Number of cases showing LOH expressed as fraction of informative cases

[b]No. of cases with immunodetectable protein/total no. examined

[c]cDNA probe used to assess homo- or hemizygosity. Paired tumor and normal samples assessed for 3 SCLC tumors reported in Yokota 1987. Allele loss results in others based on densitometry of Southern blots.

[d]SSCP band shift used to assess LOH

[e]Mutations observed in remaining allele of two SQCs showing LOH

[f]Point mutation identified in an uninformative (NI) case of ADC.

SQC = squamous cell carcinoma; ADC = adenocarcinoma; LCC = large cell carcinoma; ASC = adenosquamous carcinoma; SCLC = small cell lung carcinoma

SCLC, associated with increased aggressive behavior of these tumors.[127-132] Such observations, however, have not always been consistent.[133,134] These differences may in part be related to the different methods of analysis used and the nature of the tumor samples analyzed, that is, cultured versus frozen tissue. It is conceivable that tumors with amplified *myc* genes are more easily established in tissue culture. N-*myc* oncogene expression was increased in primary biopsies obtained from 15 untreated patients with lung

TABLE 30-11

ras Gene Mutations in Lung Cancer

Author and Reference	Methods of Detection	Histology		Mutations No.	%	Total no. of *ras* Mutations in LC (%)
Rodenhuis (1987) 117	Dot blot oligo-nucleotide hybridization	ADC	10	5/10	50	5/36 (14)
		SQC	15	—	—	All in K-*ras* condon 12,
		LCC	10	—	—	Other K-*ras*, N-*ras* and
		Car	1	—	—	Ha-*ras* hot spots
		Mets	2	—	—	were wild type
		SCLC	1	—	—	
Slebos (1990) 116	Dot blot oligo-nucleotide hybridization	ADC	69	19/69	28	19/69 (28) Only codon 12 of k-*ras* was analyzed
Suzuki (1990) 102	SSCP and direct sequence analysis	ADC	66	18/66	27	22/129 (17)
		SQC	36	2/36	6	16 in K-*ras*, 3 in Ha *ras*
		LCC	14	2/14	14	and 3 in N-*ras*
		ADS	5	—	—	
		SCLC	8	—	—	
Mitsudomi (1991) 119	RFLP	ADC	61	9/32	28	21/103 (20)
		SQC	32	2/6	33	19 in K-*ras*, 1 in Ha
		LCC	6	9/13	69	*ras* and 1 in N-*ras*
		Car	5	1/5	20	
		SCLC	42	0/42	0	
Vahakangas[a] (1992) 98	Sequence analysis	ADC	2	—	—	0/20 (0)
		SQC	6			K-*ras* codons 12 and
		LCC	4			13 sequenced
		SCLC	8			
Sugio (1992) 123		ADC	115	18/115	16	All mutations in codon 16 of K-*ras*
Husgafvel-Pursianen[b] (1993) 124	Dot blot oligo-nucleotide hybridization	ADC	21	12/21	57	14/48 (29)
		SQC	21	2/21	10	All mutations in K-*ras*
		LCC	2	0/2		None in Ha or N-*ras*
		SCLC	4	0/4		
Rossell (1993) 121	Dot blot oligo-nucleotide hybridization	ADC	22	3/22	13	13/66 (20)
		SQC	38	8/38	21	Only K-*ras* oncogene
		LCC	6	2/6	33	analyzed
Rossell (1994) 122	Dot blot oligo-nucleotide hybridization	46 non-SCLC		13/44	30	13/44 (30) All mutations in K-*ras*, 5 in codon 61, and 8 in codon 12
Kashii (1994) 125	SSCP and sequence analysis	ADC	10	3/10	30	4/32
		SQC	5	0/5	—	(12.5)
		LCC	3	1/3	—	
		SCLC	14	0/14	—	
Mao (1994) 126	PCR, cloning, and sequence analysis	ADC	15	8/15	53.3	53

[a]Study of patients with radon exposure

[b]Study of patients with asbestos exposure

ADC = adenocarcinoma; SQC = squamous cell carcinoma; LCC = large cell carcinoma; SCLC = small cell lung cancer; Car = carcinoid tumor; Mets = metastatic carcnoma; ADS = adenosquamous carcinoma; SSCP = single-stranded conformational polymorphism; PCR = polymerase chain reaction; RFLP = restriction fragment length polymorphism.

cancer. This correlated with a poor response to chemotherapy and short survival.[131] Clinical follow-up studies have correlated the different L-*myc* restriction fragment length polymorphism alleles with an adverse clinical outcome in patients with non-SCLC.[132] However, other studies contradict these observations.[135,136]

OTHER SOMATIC GENETIC CHANGES IMPLICATED IN LUNG CANCER

Cytogenetic studies of lung tumors suggest the presence of considerably more somatic genetic changes than those regions described in paragraphs above to be involved by molecular analysis.[68,69] It therefore comes as no surprise that many changes have been described in lung cancer involving chromosomes 1q, 2q, 5q, 8q, 9p, 11p, 19p, and the Y chromosome (Table 30-12).[71,72,74,76,86,88,137–142] The allelotyping of lung cancer carried out by Tsuchiya illustrates this phenomenon (see Table 30-12).[76] The candidate region of 5q, which includes the adenomatous polyposis coli/mutated in colon cancer (APC/MCC) gene cluster, frequently shows evidence of allele loss.[139,141] Such consistent allele loss may reflect the involvement of the APC or MCC genes. By analogy, the observation of LOH on 9p in 43 percent and 67.5 percent of lung tumors has coincided with the description of homozygous deletions in lung tumors.[81,86,142] Recently, a new tumor suppressor gene named MTS1 on 9p21[143] has been identified to be homozygously deleted in up to 35 percent of lung cancer cell lines (see Tables 30-4 and 30-12). Other regions showing similar consistent somatic genetic changes may also yield more tumor suppressor genes implicated in the molecular evolution of lung cancer.

The 11p loss described in cytogenetic studies[68,144] has been confirmed at the molecular level, being observed in 22 to 72 percent of lung tumors.[71,74,145–147] In the most recent study (see Table 30-12) of non-SCLC,[148] three distinct regions on 11p have been identified to show LOH, analogous to 3p. These observations are of interest as up to 80 percent of squamous cell carcinomas showed evidence of allele loss on 11p. This again implicates as yet undiscovered putative tumor suppressor genes. The WT1 gene has not been investigated in detail. However, allele loss studies tend to exclude the involvement of the WT1 locus in all non-small cell lung tumor histologies with the exception of adenocarcinoma.[149]

SOMATIC GENETIC CHANGES IN PREINVASIVE LESIONS

A variety of bronchial epithelial lesions showing increasing nuclear disorder are associated with squamous cell carcinoma. No longitudinal studies have demonstrated these abnormal noninvasive morphologic lesions precede invasive tumor development. However, circumstantial evidence strongly supports this suggestion. These bronchial epithelial lesions do not have a macroscopic appearance that distinguishes them from normal mucosa, so they are difficult to biopsy. Such lesions are only detectable by microdissection of formalin-fixed material obtained at surgical resection of malignant tumors. The size of the sample for genotype analysis is consequently very small. Thus the DNA isolated from it must be amplified at specific regions using polymerase chain reaction (as described previously) before genotypes can be assigned. This approach has identified allele loss on chromosome 3 and mutation in the p53 gene in areas of severe dysplasia. These foci are adjacent to but distinct from tumors carrying genetic damage of the same type as the adjacent tumor (Table 30-13).[47,98,126,150–154] In one study[47] three patients had severe dysplasia only, with no evidence of invasion. These specimens also showed somatic genetic changes in chromosome 3 and in the p53 gene. In another similar study using fresh material, it was possible to determine the genotype of two dysplastic samples using both molecular and cytogenetic techniques.[152] Evidence of p53 mutations was obtained; of great interest was the observation that although deletion of part of the short arm of chromosome 17 was observed, the karyotypes were much simpler than in fully malignant tumors.[150,152]

The studies discussed so far have used severe dysplasia as their starting point. However, it is important to identify the earliest morphologic change carrying an irreversible genetic change. So far this question has only been addressed using antibodies to p53 as an indicator of somatic genetic change (as described previously). Bennett and his coworkers[153] identified a series of morphologic changes and assessed each for positive staining with anti-p53 antibody. The results showed an increase in positivity from 6.7 percent of squamous metaplasia to 67.5 percent in microinvasion. It is becoming clear that mechanisms other than mutation can lead to p53 accumulation so studies that use only immunohistochemistry need to be confirmed by additional molecular genetic studies.[155] The detection of molecular

TABLE 30-12
Other Somatic Genetic Changes in Lung Cancer

Author and Reference	Histology No.	Gene/ Chromosomal Region	No. of Cases Showing LOH (%)[a]	Overall % of Lung Tumors Showing Somatic Genetic Changes in Study
Weston (1989) 71	22 SQC 22 ADC 7 LCC	11 pter-p15.5 and 11p13-q13	14/22 (63) 10/22 (45) 3/7 (43)	53
Becker (1989) 72	2 SQC 8 ADC 2 ADS 1 Car 1 SCLC	11p15	0/14	0
Ludwig (1990) 74		11	21/29 (72)	72
Emi (1992) 137	59 NSCLC	8p	14/35 (40)	40
Tsuchiya (1992) 76	11 SQC 36 ADC 3 LCC 3 ADS	1q 2q 5q 8q 19p	10/27 (37) 13/42 (31) 7/23 (30) 4/13 (31) 23/37 (62)	37 31 30 31 62
D'Amico (1992) 139	23 SCLC 7 NSCLC	MCC/APC 5q21	17/21 (80) 2/5 (40)	73
Centre (1993) 140	12 NSCLC	Yp	9/12 (75)	75
Olopade (1993) 86	56 Lung tumours	IFN or MTAP gene/9p	24/56 (43)	43
Hosoe (1994) 141	20 NSCLC 39 SCLC	5q	48/59	81
Merlo (1994) 142	16 SQC 18 ADC 6 LCC	9p	27/40	67.5
Kohno (1994) 88	17 SCLC 37 NSCLC	2q	5/17 (29) 12/32 (37.5)	35

[a]Number of tumors showing LOH/Number informative

SQC = squamous cell carcinoma; ADC = adenocarcinoma; LCC = large cell carcinoma; ADS = adenosquamous carcinoma; SCLC = small cell lung cancer; NSCLC = non-small cell lung cancer; MCC = mutated in colorectal carcinoma; APC = adenomatous polyposis coli; Yp = short arm of Y chromosome; IFN = interferon gene; MTAP = methylthioadenosine phosphorylase genes

changes that are correlated with lung tumor development, and may in fact precede it, has prompted the suggestion that detection of these changes could be relevant to patient management.

OTHER LUNG TUMORS

Mesotheliomas

Mesotheliomas have not been as intensively analyzed for somatic genetic changes as lung cancer. The cytogenetic abnormalities described in mesotheliomas are summarized in Table 30-14.[68,138,156–160] The cytogenetic and molecular studies of mesothelioma suggest the involvement of chromosomal regions that are common to both lung cancer and mesothelioma. These include chromosome 1, 3p, and 9p.[81,161] Numerous homozygous deletions from the 9p21 region have been identified in mesothelioma cell lines (see Table 30-14) and, in common with lung cancer, implicate the MTS1 gene in mesothelioma development.

TABLE 30-13
Summary of Early Somatic Genetic Changes in Lung Cancer Development

Author and Reference	Lesion	Genetic change
Sozzi (1991) 150	BE from patients with lung cancer	Clonal and nonclonal karyotypic abnormalities including 3p loss. Abnormalities in common with but simpler than the tumors. No morphology carried out in material analyzed.
Sundaresan (1991) 151	Severely dysplastic BE and tumors	3p loss in dysplastic BE and 17p loss at informative loci
Sundaresan (1992) 47	Severely dysplastic BE in patients without invasive tumors	3p and 17p loss in dysplastic BE
		p53 mutation and p53 IHC positivity in dysplastic BE from patients with invasive disease
Vahakangas (1992) 98	Severely dysplastic BE and tumor	Same p53 mutation in severely dysplastic BE and tumor
Sozzi (1992) 152	Severely dysplastic BE and tumor	Clonal karyotypic del(17)(p13) in BE, IHC staining and p53 mutation in BE in common with tumor
Bennett (1994) 153	Severely dysplastic BE and tumor	IHC evidence of mutant p53 in dysplastic BE and tumor
Mao (1994) 126	Cytologically negative for cancer sputum samples	p53 or *ras* mutations detected 1–3 months before diagnosis
Walker (1994) 154	Mild to severely dysplastic BE	Immunohistochemical evidence of mutant p53 in 14% of mild to 59% of severe BE dysplasia

BE = bronchial epithelium; IHC = immunohistochemical

TABLE 30-14
Cytogenetic Abnormalities in Mesothelioma

Author and Reference	No. of Cases	Most Common Clonal Abnormality
Gibas (1986) 156	12	Chromosome 3, del 3p14-p23, del 13p13
Popescu (1988) 157	9	Abnormalities of 3p, 1, and chromosome 7
Flejter (1989) 158	5	Losses involving chromosome 1 (especially 1p21-p22), and chromosome loss involving 3p21
Hagemeijer (1990) 159	40	Losses of chromosome 4, 22, 3p, 9p and gain of 7, 5, 20
Whang-Peng (1991) 68	3	Deletions and/or structural abnormality of chromosome 3
Taguchi (1993) 138	23	Losses of 1p (1p21-22), 9p (9p21-22), 3p21, 6q(q15-21)

Deletion mapping studies suggest 3p involvement in up to 65 percent of mesotheliomas,[161] with a common region of loss involving 3p21, a region that is clearly important in lung tumors. Involvement of the p53 locus has been suggested by the demonstration of 17p loss or p53 mutation in 2 of 4 cases[162] and in 3 of 20 mesothelioma cell lines.[163] However, larger numbers of cases need to be analyzed to evaluate the role of p53 in the development of mesothelioma. Involvement of the WT1 tumor suppressor gene[164] has been described in mesotheliomas in isolated cases. There was no evidence of K-*ras* oncogene involvement in a study of 20 mesothelioma cell lines.[163]

Pulmonary Hamartomas

The presence of clonal chromosomal abnormalities in tumors supports their neoplastic nature.[165] Occasionally, clonal chromosomal abnormalities have been described in nonneoplastic cells.[166] Therefore the observation of clonal chromosomal abnormalities described in seven of nine pulmonary hamartomas[166–168] suggests these lesions are benign tumors.

Carcinoid Tumors

The numbers of carcinoid tumors analyzed for somatic genetic changes are small. However these tumors have been shown to have changes that are in common with those observed in lung cancer. Some of these results are summarized in Tables 30-11 and 30-12. Cytogenetic analyses reveal relatively simple karyotypic abnormalities in carcinoid tumors, whereas atypical carcinoid tumors have a much more complex karyotype.[169,170] There is molecular evidence of 3p loss,[73] involvement of the Rb gene,[111] and p53 gene mutations in carcinoid tumors.[171,172] Unlike the non-SCLC, however, K-*ras* mutations are unusual in carcinoid tumors.[117,119,160]

SUMMARY

Although a variety of genetic changes have been described in the various histologic subtypes of lung cancer, there is no clear association between specific genetic changes and phenotype. This may be a reflection of an underlying common histogenesis of the major lung tumors or it may simply reflect the immaturity of this field. At present, for most genetic lesions involved in lung cancer the underlying mutations are not known. However, as more genes involved in lung cancer development are isolated, the utilization of automated DNA technology, and the use of archival material as a source of samples will allow large numbers of patient samples to be characterized and this may reveal patterns not yet evident from the small study numbers.

Furthermore, no clear pattern of accumulating genetic change has been identified in the development of the disease, although it is clear that genetic changes do occur early in the disease process. Lung tumors, like many of the tumors of epithelial origin, appear to be multifocal in origin. It is as yet unknown whether there is more than one route to the final malignant phenotype. This information will be particularly important if genetic changes are to serve as markers for the early detection of lung cancer.

REFERENCES

1. Ponder BAJ: Inherited predisposition to cancer. *Trends Genet* 6:213–218, 1990.
2. Law MR: Genetic predisposition to lung cancer. *Br J Cancer* 61:195–206, 1990.
3. Samet JM, Humble CG, Pathak DR: Personal and familial history of respiratory disease and lung cancer risk. *Am Rev Respir Dis* 134:466–470, 1986.
4. Sellers TA, Bailey-Wilson JE, Elston RC, et al: Evidence for mendelian inheritance in the pathogenesis of lung cancer. *J Natl Cancer Inst* 82:1272–1279, 1990.
5. Tokuhata GK, Lilienfeld AM: Familial aggregation of lung cancer in humans. *J Natl Cancer Inst* 30:289–312, 1963.
6. Ooi WL, Elston RC, Chen VW, Bailey-Wilson JE, Rothschild H: Increased familial risk for lung cancer. *J Natl Cancer Inst* 76:217–222, 1986.
7. Lynch HT, Kimberling WJ, Markvicka SE, et al: Genetics and smoking-associated cancers. *Cancer* 57:1640–1646, 1986.
8. Joishy SK, Cooper RA, Rowley PT: Alveolar cell carcinoma in identical twins. *Ann Intern Med* 87:447–450, 1977.
9. Brisman R, Baker RR, Elkins R, Hartmann WH: Carcinoma of lung in four siblings. *Cancer* 20:2048–2059, 1967.
10. Biran H, Goldstein J, Cohen Y: A cancer-prone kindred with four siblings afflicted by aggressive poorly differentiated bronchogenic carcinoma. *Lung Cancer* 7:345–353, 1991.
11. Johnson BE, Ihde DE, Matthews MJ, et al: Non-small-cell lung cancer. Major cause of late mortality in patients with small cell lung cancer. *Am J Med* 80:1103–1110, 1986.
12. Bradley EC, Schechterp G, Mathews MJ, et al: Erythroleukaemia and other hematological complications of intensive therapy in long term survivors of small cell lung cancer. *Cancer* 49:221–223, 1982.
13. Whang-Peng J, Lee EC, Minna JD, et al: Deletion of 3(p14-23) in secondary erythroleukemia arising in long-term survivors of small cell lung cancer. *J Natl Cancer Inst* 80:1253–1255, 1988.
14. Abernathy D, Beltran G, Stuckey WJ: Lung cancer following treatment for lymphoma. *Am J Med* 81:215–218, 1986.
15. Henry AM: Second cancer after the treatment for Hodgkin's disease: A report from the International Database on Hodgkin's disease. *Ann Oncol* 4:117–128, 1992.
16. Saunders BM, Jay M, Draper GJ, Roberts EM: Non-ocular cancer in relatives of retinoblastoma patients. *Br J Cancer* 60:358–365, 1989.
17. Li FP: Familial cancer syndromes and clusters. *Curr Probl Cancer* 14:73–114, 1990.
18. Malkin D, Li FP, Strong LC, et al: Germ line p53 mutations in a familial syndrome of breast cancer, sarcomas, and other neoplasms. *Science* 250:1233–1238, 1990.
19. Srivastava S, Zhiqiang Z, Pirollo K, Blattner W, Chang EH: Germ-line transmission of a mutated p53 gene in a cancer-prone family with Li-Fraumeni syndrome. *Nature* 348:747–749, 1990.
20. Gazdar AF: The molecular biology of lung cancer. *Tohoku J Exp Med* 168:239–245, 1992.
21. Gazdar AF: Molecular markers for the diagnosis and prognosis of lung cancer. *Cancer* 69:1592–1599, 1992.
22. Knudson AG: Antioncogenes and human cancer. *Proc Natl Acad Sci USA* 90:10914–10921, 1993.
23. Cole TR, Hughes HE, Jeffreys MJ, Williams GT, Arnold MM: Small cell lung carcinoma in a patient with Sotos syndrome: are genes at 3p21 involved in both conditions? *J Med Genet* 29:338–341, 1992.
24. Schrander SC, Fryns JP, Hamers GG: Sotos syndrome and de novo balanced autosomal translocation (t(3;6)(p21;p21)). *Clin Genet* 37:226–229, 1990.
25. Bianchi C, Brollo A, Zuch C: Asbestos-related familial mesothelioma. *Eur J Cancer Prev* 2:247–250, 1993.
26. Miller EC: Some current perspectives on chemical carcinogenesis in humans and experimental animals: Presidential address. *Cancer Res* 38:1479, 1978.

27. Farber E: Chemical carcinogenesis. *N Engl J Med* 305:1379, 1981.

28. Gonzalez FJ, Skoda RC, Kimura S, et al: Characterization of the common genetic defect in humans deficient in debrisoquine metabolism. *Nature* 331:442–446, 1988.

29. Law MR, Hetzel MR, Idle JR: Debrisoquine metabolism and genetic predisposition to lung cancer. *Br J Cancer* 59:686–687, 1989.

30. Caporaso N, Hayes RB, Dosemeci M, et al: Lung cancer risk, occupational exposure, and the debrisoquine metabolic phenotype. *Cancer Res* 49:3675–3679, 1989.

31. Caporaso NE, Tucker MA, Hoover RN, et al: Lung cancer and the debrisoquine metabolic phenotype. *J Natl Cancer Inst* 82:1264–1272, 1990.

32. Sugimura H, Capraso NE, Shaw GL, et al: Human debrisoquine hydroxylase gene polymorphisms in cancer patients and controls. *Carcinogenesis* 11:1527–1530, 1990.

33. Harris CC: Interindividual variation among humans in carcinogen metabolism, DNA adduct formation and DNA repair. *Carcinogenesis* 10:1563–1566, 1989.

34. Vahakangas K, Trivers GE, Plummer S: O^6-methylguanine-DNA methyl transferase and uracil DNA glycosylase in human broncho-alveolar lavage cells and peripheral blood mononuclear cells from tobacco smokers and non-smokers. *Carcinogenesis* 12:1389–1394, 1991.

35. Marshall CJ: Tumour suppressor genes. *Cell* 64:313–326, 1991.

36. Hall JM, Lee MK, Newman B, et al: Linkage of early-onset familial breast cancer to chromosome 17q21. *Science* 250:1684–1689, 1990.

37. Joslyn G, Carlson M, Thliveris A, et al: Identification of deletion mutations and three new genes at the familial polyposis locus. *Cell* 66:589–600, 1991.

38. Groden J, Thliveris A, Samowitz W, et al: Identification and characterization of the familial adenomatous polyposis gene. *Cell* 66:589–600, 1991.

39. Latif F, Tory K, Gnarra J, et al: Identification of the von Hippel-Lindau disease tumour suppressor gene. *Science* 260:1317–1320, 1993.

40. Malkinson AM: Primary lung tumors in mice: an experimentally manipulable model of human adenocarcinoma. *Cancer Res* 52:2670S–2676S, 1992.

41. Malkinson AM, Nesbitt MN, Skamene E: Susceptibility to urethan-induced pulmonary adenomas between A/J and C57L/6J mice: use of AXB and BXA recombinant inbred lines indicating a three-locus genetic model. *J Natl Cancer Inst* 75:971–974, 1985.

42. Gariboldi M, Manenti G, Canzian F, et al: A major susceptibility locus to murine lung carcinogenesis maps on chromosome 6. *Nat Genet* 3:132–136, 1993.

43. Gyapay G, Morissette J, Vignal A, et al: The 1993–94 Généthon human genetic linkage map. *Nat Genet* 7:246–249, 1994.

44. Hollstein M, Sidransky D, Vogelstein B, Harris CC: p53 mutations in human cancers. *Science* 253:49–53, 1991.

45. Harris CC, Hollstein M: Clinical implications of the p53 tumor-suppressor gene. *N Engl J Med* 329:1318–1327, 1993.

46. Nigro JM, Baker SJ, Preisinger AC, et al: Mutations in the p53 gene occur in diverse human tumour types. *Nature* 342:705–708, 1989.

47. Sundaresan V, Ganly P, Hasleton P, et al: p53 and chromosome 3 abnormalities, characteristic of malignant lung tumours, are detectable in preinvasive lesions of the bronchus. *Oncogene* 7:1989–1997, 1992.

48. Wurster-Hill DH, Maurer LH: Cytogenetic diagnosis of cancer: Abnormalities of chromosomes and ploidy levels in the bone marrow of patients with small cell anaplastic carcinoma of the lung. *J Natl Cancer Inst* 61:1065–1075, 1978.

49. Whang-Peng J, Kao-Shan CS, Lee EC, et al: Specific chromosome defect associated with human small-cell lung cancer: Deletion 3p(14-23). *Science* 215:181–182, 1982.

50. Whang-Peng J, Bunn PA, Kao-Shan CS, et al: A non-random chromosomal abnormality, del 3p(14-23), in human small cell lung cancer (SCLC). *Cancer Genet Cytogenet* 6:119–134, 1982.

51. Wurster-Hill DH, Cannizzaro LA, Pettengill OS, Sorenson GD, Cate CC, Maurer HL: Cytogenetics of small cell carcinoma of the lung. *Cancer Genet Cytogenet* 13:3303–3307, 1984.

52. Waters JJ, Ibson JM, Twentyman PR, Bleehan NM, Rabbitts PH: Cytogenetic abnormalities in human small cell lung carcinoma: Cell lines characterized for *myc* gene amplification. *Cancer Genet Cytogenet* 30:213–223, 1988.

53. de Fusco P, Frytak S, Dahl RJ, Weiland LH, Unni KK, Dewald GW: Cytogenetic studies in 11 patients with small cell carcinoma of the lung. *Mayo Clin Proc* 64:168–176, 1989.

54. Kok K, Osinga J, Carritt B, et al: Deletion of a DNA sequence at the chromosomal region 3p21 in all major types of lung cancer. *Nature* 330:578–581, 1987.

55. Naylor SL, Johnson BE, Minna JD, Sakaguchi AY: Loss of heterozygosity of chromosome 3p markers in small-cell lung cancer. *Nature* 329:451–454, 1987.

56. Mooibroek H, Osinga J, Postmus PE, Carritt B, Buys CH: Loss of heterozygosity for a chromosome 3 sequence presumably at 3p21 in small cell lung cancer. *Cancer Genet Cytogenet* 27:361–365, 1987.

57. Yokota J, Wada M, Shimosato Y, Terada M, Sugimura T: Loss of heterozygosity on chromosomes 3, 13, and 17 in small-cell carcinoma and on chromosome 3 in adenocarcinoma of the lung. *Proc Natl Acad Sci USA* 84:9252–9256, 1987.

58. Johnson BE, Sakaguchi AY, Gazdar AF, et al: Restriction fragment length polymorphism studies show consistent loss of chromosome 3p alleles in small cell lung cancer patients' tumors. *J Clin Invest* 82:502–507, 1988.

59. Sithanandam G, Dean M, Brennscheidt U, et al: Loss of heterozygosity at the c-*raf* locus in small cell carcinoma. *Oncogen* 4:451–455, 1989.

60. Mori N, Yokota J, Oshimura M, et al: Concordant deletions of chromosome 3p and loss of heterozygosity for chromosomes 13 and 17 in small cell lung carcinoma. *Cancer Res* 49:5130–5135, 1989.

61. Dobrovic A, Houle B, Belouchi A, Bradley WE: *erbA*-related sequence coding for DNA-binding hormone receptor localized to chromosome 3p21-3p25 and deleted in small cell lung carcinoma. *Cancer Res* 48:682–685, 1988.

62. Leduc F, Brauch H, Hajj C, et al: Loss of heterozygosity in a gene coding for a thyroid hormone receptor in lung cancers. *Am J Hum Genet* 44:282–287, 1989.

63. Brauch H, Tory K, Kotler F, et al: Molecular mapping of deletion sites in the short arm of chromosome 3 in human lung cancer. *Genes Chromosom Cancer* 1:240–246, 1990.

64. Daly MC, Douglas JB, Bleehen NM, et al: An unusually proximal deletion on the short arm of chromosome 3 in a patient with small cell lung cancer. *Genomics* 9:113–119, 1991.

65. Hibi K, Takahashi T, Yamakawa K, et al: Three distinct regions involved in 3p deletion in human lung cancer. *Oncogene* 7:445–449, 1992.

66. Rabbitts PH, Daly M, Douglas J, Ganly P, Heppell-Parton A, Sundaresan V: Deletion mapping of chromosome 3 in lung tumours. *Lung Cancer* 9:69–74, 1993.

67. Rabbitts P, Bergh J, Douglas J, Collins F, Waters J: A submicroscopic homozygous deletion at the D3S3 locus in a cell line isolated from a small cell lung carcinoma. *Genes Chromosom Cancer* 2:231–238, 1990.

68. Whang-Peng J, Knutsen T, Gazdar A, et al: Non random structural and numerical chromosome changes in non-small cell lung cancer. *Genes Chromsom Cancer* 3:168–188, 1991.

69. Lukeis R, Irving L, Garson M, Hasthorpe S: Cytogenetics of non-small cell lung cancer: Analysis of consistent non-random abnormalities. *Genes Chromosom Cancer* 2:116–124, 1990.

70. Brauch H, Johnson B, Hovis J, et al: Molecular analysis of the short arm of chromosome 3 in small-cell and non-small-cell carcinoma of the lung. *N Engl J Med* 317:1109–1113, 1987.

71. Weston A, Willey JC, Modali R, et al: Differential DNA sequence deletions from chromosomes 3, 11, 13, and 17 in squamous-cell carcinoma, large-cell carcinoma, and adenocarcinoma of the human lung. *Proc Natl Acad Sci USA* 86:5099–5103, 1989.

72. Becker D, Sahin AA: Loss of heterozygosity at chromosomal regions 3p and 13q in non-small-cell carcinoma of the lung represents low-frequency events. *Genomics* 4:97–100, 1989.

73. Rabbitts P, Douglas J, Daly M, et al: Frequency and extent of allelic loss in the short arm of chromosome 3 in non small-cell lung cancer. *Genes Chromosom Cancer* 1:95–105, 1989.

74. Ludwig CU, Dalquen P, Stulz P, Stahel R, Obrecht JP: DNA sequence deletions from chromosome 11, 3, and 17 in human non-small cell lung carcinoma (NSCLC). *Proc Am Assoc Cancer Res* 31:319, 1990.

75. Houle B, Leduc F, Bradley WEC: Implication of *RARB* in epidermoid (squamous) lung cancer. *Genes Chromosom Cancer* 3:358–366, 1991.

76. Tsuchiya E, Nakamura Y, Weng SY, et al: Allelotype of non-small cell lung carcinoma-comparison between loss of heterozygosity in squamous cell carcinoma and adenocarcinoma. *Cancer Res* 52:2478–2481, 1992.

77. Yokoyama S, Yamakawa K, Tsuchiya E, Murarta M, Sakiyama S, Nakamura Y: Deletion mapping on the short arm of chromosome 3 in squamous cell carcinoma and adenocarcinoma of the lung. *Cancer Res* 52:873–877, 1992.

78. Horio Y, Takahashi T, Kuroishi T, et al: Prognostic significance of p53 mutations and 3p deletions in primary resected non-small cell lung cancer. *Cancer Res* 53:1–4, 1993.

79. Sundaresan V, Heppell-Parton AC, Coleman N, et al: Somatic genetic changes in lung cancer and precancerous lesions. *Ann Oncol* 1995, in press.

80. Daly MC, Xiang RH, Buchhagen D, et al: A homozygous deletion on chromosome 3 in a small cell lung cancer cell line correlates with a region of tumor suppressor activity. *Oncogene* 8:1721–1729, 1993.

81. Cheng JQ, Jhanwar SC, Lu YY, Testa JR: Homozygous deletions within 9p21-p22 identify a small critical region of chromosomal loss in human malignant mesotheliomas. *Cancer Res* 53:4761–4763, 1993.

82. Kok K, van den Berg A, Veldhuis PMJF, et al: A homozygous deletion in a small cell lung cancer cell line involving a 3p21 region with a marked instability in YACs. *Cancer* Res 54:4183, 1994.

83. Hensel CH, Hsieh C, Gazdar AF, et al: Altered structure and expression of the human retinoblastoma susceptibility gene in small cell lung cancer. *Cancer Res* 50:3067–3072, 1990.

84. Yamakawa K, Takahashi T, Horio Y, et al: Frequent homozygous deletions in lung cancer cell lines detected by a DNA marker located at 3p21.3-p22. *Oncogene* 8:327–330, 1993.

85. Kok K, Veldhuis PMJF, van den Berg A, van der Veen AY, Buys CHCM: Long range mapping date of 3p21 reveals the positin and the size of a homozygous deletion in an SCLC derived cell line, in Naylor S; *Proceedings of the Fifth International Chromosome 3 Workshop*, Ann Arbor, MI, 1993.

86. Olopade OI, Buchhagen DL, Malik K, et al: Homozygous loss of the interferon genes defines the critical region on 9p that is deleted in lung cancers. *Cancer Res* 1993.

87. Center R, Lukeis R, Dietzsch E, Gillespie M, Garson OM: Molecular deletion of 9p sequences in non-small cell lung cancer and malignant mesothelioma. *Genes Chromosom Cancer* 7:47–53, 1993.

88. Kohno T, Morishita K, Takano H, Shapiro DN, Yokota J: Homozygous deletion at chromosome 2q33 in human small cell lung carcinoma identified by arbitrarily primed PCR genomic fingerprinting. *Oncogene* 9:103–108, 1994.

89. Kamb A, Gruis NA, Weaver-Feldhaus J, et al: A cell cycle regulator potentially involved in genesis of many tumor types. *Science* 264:436–440, 1994.

90. Sekido Y, Bader S, Latif F, et al: Molecular analysis of the von Hippel-Lindau disease tumour suppressor gene in human lung cancer. *Oncogene* 9:1599–1604, 1994.

91. Hensel CH, Xiang RH, Sakaguchi AY, Naylor SL: Use of the single strand conformation polymorphism technique and PCR to detect p53 gene mutations in small cell lung cancer. *Oncogene* 6:1067–1071, 1991.

92. Takahashi T, Takahashi T, Suzuki H, et al: The p53 gene is very frequently mutated in small-cell lung cancer with a distinct nucleotide substitution pattern. *Oncogene* 6:1775–1778, 1991.

93. Sameshima Y, Matsuno Y, Hirohashi S, et al: Alterations of the p53 gene are common and critical events for the maintenance of malignant phenotypes in small-cell lung carcinoma. *Oncogene* 7:451–457, 1992.

94. Miller CW, Simon K, Aslo A, et al: p53 mutations in human lung tumors. *Cancer Res* 52:1695–1698, 1992.

95. Iggo R, Gatter K, Bartek J, Lane D, Harris AL: Increased expression of mutant forms of p53 oncogene in primary lung cancer. *Lancet* 335:675–679, 1990.

96. Chiba I, Takahashi T, Nau MM, et al: Mutations in the p53 gene are frequent in primary, resected non-small cell lung cancer. *Oncogene* 5:1603–1610, 1990.

97. Mitsudomi T, Steinberg SM, Nau MM, et al: p53 gene mutations in non-small-cell lung cancer cell lines and their correlation with the presence of *ras* mutations and clinical features. *Oncogene* 7:171–180, 1992.

98. Vahakangas KH, Samet JM, Metcalf RA, et al: Mutations of p53 and *ras* genes in radon-associated lung cancer from uranium miners. *Lancet* 339:576–580, 1992.

99. Marchetti A, Buttitta F, Merlo G, et al: p53 alterations in non-small cell lung cancers correlate with metastatic involvement of hilar and mediastinal lymph nodes. *Cancer Res* 53:2846–2851, 1993.

100. Mitsudomi T, Oyama T, Kusano T, Osaki T, Nakanishi R, Shirakusa T: Mutations of the p53 gene as a predictor of poor prognosis in patients with non-small-cell lung cancer. *J Natl Cancer Inst* 85:2018–2033, 1993.

101. Takahashi T, D'Amico D, Chiba I, Buchhagen DL, Minna JD: Identification of intronic point mutations as an alternative mechanism for p53 inactivation in lung cancer. *J Clin Invest* 86:363–369, 1990.

102. Suzuki Y, Orita M, Shiraishi M, Hayashi K, Sekiya T:

Detection of *ras* gene mutations in human lung cancers by single-strand conformation polymorphism analysis of polymerase chain reaction products. *Oncogene* 5:1037–1043, 1990.

103. Kishimoto Y, Murakami Y, Shiraishi M, Hayashi K, Sekiya T: Aberrations of the p53 tumor suppressor gene in human non-small cell carcinomas of the lung. *Cancer Res* 52:4799–4804, 1992.

104. D'Amico D, Carbone D, Mitsudomi T, et al: High frequency of somatically acquired p53 mutations in small-cell lung cancer cell lines and tumors. *Oncogene* 7:339–346, 1992.

105. Mitsudomi T, Lam S, Shirakusa T, Gazdar AF: Detection and sequencing of p53 gene mutations in bronchial biopsy samples in patients with lung cancer. *Chest* 104:362–365, 1993.

106. Harris CC, Hollstein M: Clinical implications of the p53 tumor-suppressor gene. *N Engl J Med* 329:1318–1327, 1993.

107. Takeshima Y, Seyama T, Bennett WP, et al: p53 mutations in lung cancers from non-smoking atomic-bomb survivors. *Lancet* 342:1520–1521, 1993.

108. Bejui-Thivolet F, Chardonnet Y, Patricot LM: Human papillomavirus type 11 DNA in papillary squamous cell lung carcinoma. *Virchows Arch A* 417:457–461, 1990.

109. Bejui-Thivolet F, Liagre N, Chignol MC, Chardonnet Y, Patricot LM: Detection of papillomavirus DNA in squamous bronchial metaplasia and squamous cell carcinomas of the lung by in situ hybridization using biotinylated probes in paraffin embedded specimens. *Hum Pathol* 21:111–116, 1990.

110. Kulski JK, Demeter T, Mutavdzic S, Sterrett GF, Mitchell KM, Pixley EC: Survey of specimens of human cancer for human papillomavirus types 6/11/16/18 by filter in situ hybridization. *Am J Clin Pathol* 94:566–570, 1990.

111. Harbour JW, Lai SL, Whang PJ, Gazdar AF, Minna JD, Kaye FJ: Abnormalities in structure and expression of the human retinoblastoma gene in SCLC. *Science* 241:353–357, 1988.

112. Yokota J, Akiyama T, Fung Y, et al: Altered expression of the retinoblastoma (RB) gene in small-cell carcinoma of the lung. *Oncogene* 3:471–475, 1988.

113. Sachse R, Murakami Y, Shiraishi M, Hayashi K, Sekiya T: DNA aberrations at the retinoblastoma gene locus in human squamous cell carcinomas of the lung. *Oncogene* 9:39–47, 1994.

114. Reissman PT, Koga H, Takahashi R, et al., Lung Cancer Study Group: Inactivation of the retinoblastoma suscepti-.bility gene in non-small-cell lung cancer. *Oncogene* 8:1913–1919, 1993.

115. Orita M, Iwahana H, Kanazawa H, Hayashi K, Sekiya T: Detection of polymorphisms of human DNA by gel electrophoresis as single-strand conformation polymorphisms. *Proc Natl Acad Sci USA* 86:2766–2770, 1989.

116. Slebos RJC, Kibbelaar RE, Dalesio O, et al: K-*ras* oncogene activation as a prognostic marker in adenocarcinoma of the lung. *N Engl J Med* 323:561–565, 1990.

117. Rodenhuis S, Wetering ML, Mooi MJ, Evers SG, Zandwijk N, Bos JL: Mutational activation of the K-*ras* oncogene: A possible pathogenetic factor in adenocarcinoma of the lung. *N Engl J Med* 317:929–935, 1987.

118. Sugio K, Ishida T, Yokoyama H, Takashi I, Sugimachi K, Sasazuki T: *Ras* gene mutations as a prognostic marker in adenocarcinoma of the human lung without lymph node metastasis. *Cancer Res* 52:2903–2906, 1992.

119. Mitsudomi T, Steinberg SM, Oie HK, et al: *Ras* gene mutations in non-small cell lung cancers are associated with shortened survival irrespective of treatment intent. *Cancer Res* 51:4999–5002, 1991.

120. Harada M, Dosaka-Akita H, Miyamoto H, Kuzamaki N, Kawakami Y: Prognostic significance of the expression of *ras* oncogene produce in non-small cell lung cancer. *Cancer* 69:72–77, 1992.

121. Rosell R, Li S, Skacel Z, et al: Prognostic impact of mutated K-*ras* gene in surgically resected non-small cell lung cancer patients. *Oncogene* 8:2407–2412, 1993.

122. Rosell R, Gomez CJ, Camps C, et al: A randomized trial comparing preoperative chemotherapy plus surgery with surgery alone in patients with non-small-cell lung cancer (see comments). *N Engl J Med* 330:153–158, 1994.

123. Sugio K, Ishida T, Yokoyama H, Takashi I, Sugimachi K, Sasazuki T: *Ras* gene mutations as a prognostic marker in adenocarcinoma of the human lung without lymph node metastasis. *Cancer Res* 52:2903–2906, 1992.

124. Husgafvel PK, Hackman P, Ridanpaa M, et al: K-*ras* mutations in human adenocarcinoma of the lung: association with smoking and occupational exposure to asbestos. *Int J Cancer* 53:250–256, 1993.

125. Kashii T, Mizushima Y, Monno S, Nakagawa K, Kobayashi M: Gene analysis of K-, H-*ras*, p53, and retinoblastoma susceptibility genes in human lung cancer cell lines by the polymerase chain reaction/single-strand conformation polymorphism method. *J Cancer Res Clin Oncol* 120:143–148, 1994.

126. Mao L, Hruban RH, Boyle JO, Tockman M, Sidransky D: Detection of oncogene mutations in sputum precedes diagnosis of lung cancer. *Cancer Res* 54:1634–1637, 1994.

127. Buchhagen DL: Molecular mechanisms in lung cancer pathogenesis. *Biochem Biophys Acta* 1072:159–176, 1991.

128. Minna JD: Genetic events in the pathogenesis of lung cancer. *Chest* 96:17–23, 1989.

129. Johnson BE, Kelley MJ: Overview of genetic and molecular events in the pathogenesis of lung cancer. *Chest* 103:1–3, 1993.

130. Funa K, Steinholtz L, Nou E, Bergh J: Increased expression of N-*myc* in human small cell lung cancer biopsies predicts lack of response to chemotherapy and poor prognosis. *Am J Clin Pathol* 88:216–220, 1987.

131. Brooks B, Battey J, Nau M, Gazdar A, Minna J: Amplification and expression of the *myc* gene in small cell lung cancer. *Adv Viral Oncol* 7:155–172, 1987.

132. Kawashima K, Nomura S, Hirai H, et al: Correlation of L-*myc* RFLP with metastasis, prognosis and multiple cancer in lung cancer patients. *Int J Cancer* 50:557–561, 1992.

133. Gosney JR, Field JK, Gosney MA, Lye MDW, Spandidos DA, Butt SA: C-*myc* oncoprotein in bronchial carcinoma: expression in all major morphological types. *Anticancer Res* 10:623–628, 1990.

134. Volm M, Efferth T, Mattern J: Oncoprotein (c-*myc*, c-*erb*B1, c-*erb*B2, c-*fos*) and suppressor gene product (p53) expression in squamous cell carcinomas of the lung. Clinical and biological correlations. *Anticancer Res* 12:11–20, 1992.

135. Tamai S, Sugimura H, Caporaso NE, et al: Restriction-fragment-length polymorphism analysis of the L-*myc* gene locus in a case-control study of lung cancer. *Int J Cancer* 46:411–415, 1990.

136. Tefre T, Borresen AL, Aamdal S, Brogger A: Studies of the L-*myc* DNA polymorphism and relation to metastasis in Norwegian lung cancer patients. *Br J Cancer* 62:809–812, 1990.

137. Emi M, Fujiwara Y, Nakajima T, et al: Frequent loss of heterozygosity for loci on chromosome 8p in hepatocellular carcinoma, colorectal cancer, and lung cancer. *Cancer Res* 52:5368–5372, 1992.

138. Taguchi T, Jhanwar SC, Siegfried JM, Keller Sm, Testa JR: Recurrent deletions of specific chromosomal sites in 1p, 3p, 6q, and 9p in human malignant mesothelioma. *Cancer Res* 53:4349–4355, 1993.

139. D'Amico D, Carbone DP, Johnson BE, Meltzer SJ, Minna JD: Polymorphic sites within the *MCC* and *APC* loci reveal very frequent loss of heterozygosity in human small cell lung cancer. *Cancer Res* 52:1996–1999, 1992.

140. Center R, Lukeis R, Vrazas V, Garson OM: Y chromosome loss and rearrangement in non-small-cell lung cancer. *Int J Cancer* 55:390–393, 1993.

141. Hosoe S, Ueno K, Shigedo Y, et al: A frequent deletion of chromosome 5q21 in advanced small cell and non-small cell carcinoma of the lung. *Cancer Res* 54:1787–1790, 1994.

142. Merlo A, Gabrielson E, Askin F, Sidransky D: Frequent loss of chromosome 9 in human primary non-small cell lung cancer. *Cancer Res* 54:640–642, 1994.

143. Kamb A, Gruis NA, Weaver-Feldhaus J, et al: A cell cycle regulator potentially involved in genesis of many tumor types. *Science* 264:436–440, 1994.

144. Miura I, Siegfried JM, Resau J, Keller SM, Zhou JY, Testa JR: Chromosome alterations in 21 non-small cell lung carcinomas. *Genes Chromosom Cancer* 2:328–338, 1990.

145. Shiraishi M, Morinaga S, Noguchi M, Shimosato Y, Sekiya T: Loss of genes in the short arm of chromosome 11 in human lung carcinomas. *Jpn J Cancer Res (Gann)* 78:1302, 1987.

146. Skinner MA, Vollmer R, Huper G, Abbott P, Iglehart JD: Loss of heterozygosity for genes on 11p and the clinical course of patients with lung carcinoma. *Cancer Res* 50:2303–2306, 1990.

147. Ludwig CU, Raefle G, Dalquen P, Stulz P, Stahel R, Obrecht JP: Allelic loss on the short arm of chromosome 11 in non-small-cell lung cancer. *Int J Cancer* 49:661–665, 1991.

148. Bepler B, Garcia-Blanco MA: Three tumor-suppressor regions on chromosome 11p identified by high-resolution deletion mapping in human non-small cell lung cancer. *Proc Natl Acad Sci USA* 91:5513–5517, 1994.

149. Fong KM, Zimmerman PV, Smith PJ: Correlation of loss of heterozygosity at 11p with tumor progression and survival in non-small cell lung cancer. *Genes Chromosom Cancer* 10:183–189, 1994.

150. Sozzi G, Miozzo M, Tagliabue E, et al: Cytogenetic abnormalities and overexpression of receptors for growth factors in normal bronchial epithelium and tumour samples of lung cancer patients. *Cancer Res* 51:400–404, 1991.

151. Sundaresan V, Ganly P, Rudd R, Sinha G, Bleehen NM, Rabbitts PH: Genetic changes identified in pre-invasive lesions of the respiratory tract. *Lung Cancer* 7 (suppl):A55, 1991.

152. Sozzi G, Miozzo M, Donghi R, et al: Deletions of 17p and p53 mutations in preneoplastic lesions of the lung. *Cancer Res* 52:6079–6082, 1992.

153. Bennett WP, Colby TV, Travis WD, et al: p53 protein accumulates frequently in early bronchial neoplasia. *Cancer Res* 53:4817–4822, 1993.

154. Walker C, Robertson LJ, Myskow MW, Pendleton N, Dixon GR: p53 expression in normal and dysplastic bronchial epithelium and in lung carcinomas. *Br J Cancer* 70:297–303, 1994.

155. Barnes DM, Hanby AM, Gillett CE, et al: Abnormal expression of wild type p53 protein in normal cells of a cancer family patient. *Lancet* 340:259–263, 1992.

156. Gibas Z, Li FP, Antman K, Bernal S, Stahel R, Sandberg AA: Chromosome changes in malignant mesothelioma. *Cancer Genet Cytogenet* 20:191–201, 1986.

157. Popescu NC, Chahinian AP, DiPaolo JA: Nonrandom chromosome alterations in human malignant mesothelioma. *Cancer Res* 48:142–147, 1988.

158. Flejter WL, Li FP, Antman KH, Testa JR: Recurring loss of chromosomes 1, 3 and 22 in malignant mesothelioma: Possible sites of tumor suppressor genes. *Genes Chromosom Cancer* 1:148–154, 1989.

159. Hagemeijer A, Versnel MA, Van DE, et al: Cytogenetic analysis of malignant mesothelioma. *Cancer Genet Cytogenet* 47:1–28, 1990.

160. Wagner SN, Muller R, Boehm J, Putz B, Wunsch PH, Hofler H: Neuroendocrine neoplasms of the lung are not associated with point mutations at codon 12 of the Ki-*ras* gene. *Virchows Arch B Cell Pathol* 63:325–329, 1993.

161. Lu YY, Jhanwar SC, Cheng JQ, Testa JR: Deletion mapping of the short arm of chromosome 3 in human malignant mesothelioma. *Genes Chromosom Cancer* 9:76–80, 1994.

162. Cote RJ, Jhanwar SC, Novick S, Pellicer A: Genetic alterations of the p53 gene are a feature of malignant mesotheliomas. *Cancer Res* 51:5410–5416, 1991.

163. Metcalf RA, Welsh JA, Bennett WP, et al: p53 and Kirsten-*ras* mutation in human mesothelioma cell lines. *Cancer Res* 52:2610–2615, 1992.

164. Park S, Schalling M, Bernard A, et al: The Wilms' tumour gene WT1 is expressed in murine mesoderm-derived tissues and mutated in a human mesothelioma. *Nat Genet* 4:415–420, 1993.

165. Nowell PC: The clonal evolution of tumor cell populations. *Science* 194:23–28, 1976.

166. Johansson M, Heim S, Mandahl N, Johansson L, Hambraeus G, Mitelman F: t(3;6;14) (p21:p21:124) as the sole clonal chromosome abnormality in a hamartoma of the lung. *Cancer Genet Cytogenet* 60:219–220, 1992.

167. Fletcher JA, Pinkus GS, Weidner N, Morton CC: Lineage restricted clonality in biphasic solid tumors. *Am J Pathol* 138:1199–1207, 1991.

168. Johansson M, Dietrich C, Mandahl N, et al: Recombinations of chromosomal bands 6p21 and 14q24 characterise pulmonary hamartomas. *Br J Cancer* 67:1236–1241, 1993.

169. Teyssier JR, Ferré D: Frequent clonal chromosomal changes in human non-malignant tumors. *Int J Cancer* 44:828–832, 1989.

170. Johansson M, Heim S, Mandahl N, Hambraeus G, Johansson L, Mitelman F: Cytogenetic analysis of six bronchial carcinoids. *Cancer Genet Cytogenet* 66:33–38, 1993.

171. Mitsudomi T, Oyama T, Gazdar AF, Minna JD, Okabayashi K, Shirakusa T: Mutations of *ras* and p53 genes in human non-small cell lung cancer cell lines and their clinical significance. *Nippon Geka Gakkai Zasshi* 93:944–947, 1992.

172. Lohmann DR, Fesseler B, Putz B, et al: Infrequent mutations of the p53 gene in pulmonary carcinoid tumors. *Cancer Res* 53:5797–5801, 1993.

173. Ganly PS, Jarad N, Rudd R, Rabbitts PH: PCR based RFLP analysis allows genotyping of the short arm of chromosome 3 in small biopsies from patients with lung cancer. *Genomics* 12:221–228, 1992.

174. Sekido Y, Bader S, Latif F, et al: Molecular analysis of the von Hippel-Lindau disease tumour suppressor gene in human lung cancer cell lines. *Oncogene* 9:1599–1604, 1994.

175. Sundaresan V, Ganly P, Hasleton P, Bleehen NM, Rabbitts P: Paraffin wax-embedded material as a source of DNA for the detection of somatic genetic changes. *J Pathol* 169:43–52, 1993.

176. Suzuki H, Takahashi T, Kuroishi T, et al: p53 mutations in non-small cell lung cancer in Japan: association between mutations and smoking. *Cancer Res* 52:734–736, 1992.

Chapter 31

Common Lung Cancers

W. J. Mooi

INCIDENCE AND ETIOLOGY

In recent years, carcinoma of the lung has replaced gastric cancer as the most common malignant tumor worldwide. It is estimated that globally, one death due to lung cancer occurs every minute.[1] At present, the incidence rates are highest in Europe, North America, and Australia and lowest in Southeast Asia and Africa.[2] In England and Wales, yearly death rates are currently about 60 and 20/100,000 population for males and females, respectively.[3] In developed and affluent countries, a substantial increase in lung cancer incidence has occurred, starting in the 1930s; in other parts of the world, the increase in incidence is a more recent event. The incidence of lung cancer is closely related to tobacco consumption, especially cigarette smoking. In the past few years, the incidence in men has shown a slight decline in the United States and in several European countries, but overall, the incidence in women continues to rise (Fig. 31-1). This reflects the increase in smoking among women, postdating that among men by several decades. At present, the incidence of lung cancer is 3 to 10 times higher for men than for women, but because of the trends in time, the difference is gradually decreasing.

Lung cancer is largely a disease of late adult life. The incidence almost doubles every 5 years from age 35 to 60, and then rises more slowly until the age of 75 (Fig. 31-2). A subsequent apparent decrease in incidence is probably due to underdiagnosis and incomplete reporting in the very old.[4]

Lung cancer carries a poor prognosis. Over half the patients die within 1 year after diagnosis. Only radical surgery, feasible in about a third of cases offers a significant chance of cure, which lies in the order of 40 to 50 percent overall. Large scale lung cancer screening programs aimed at detection of tumors at an early, resectable stage, have had little effect in terms of prevention of cancer deaths, and appear to be unfavorable from a cost–benefit viewpoint. Because cigarette smoking is responsible for the large majority of lung carcinomas, the most important option in the battle against this disease is primary prevention. Special emphasis should be targeted to children and adolescents because the addiction to cigarette smoking is usually acquired before the age of 20 years. In addition, lung cancer risk is closely related to starting smoking at an early age.

It is estimated that about 80 to 90 percent of lung carcinomas are caused by tobacco.[5] Cigarette smoking is most important in this respect because inhala-

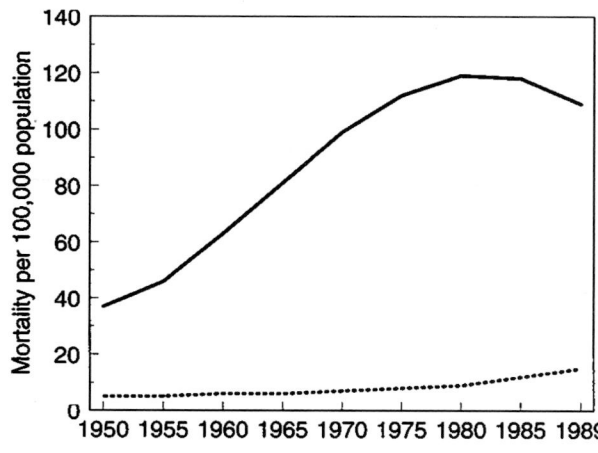

FIGURE 31-1
Lung cancer mortality per 100,000 person-years in the Netherlands, 1950–1989. As in many other countries, lung cancer incidence in Dutch males (——) has increased markedly over several decades, but has been slightly declining over the past 15 years. The incidence of lung cancer among females (------) is still rising. *(Source: Netherlands Cancer Registry).*

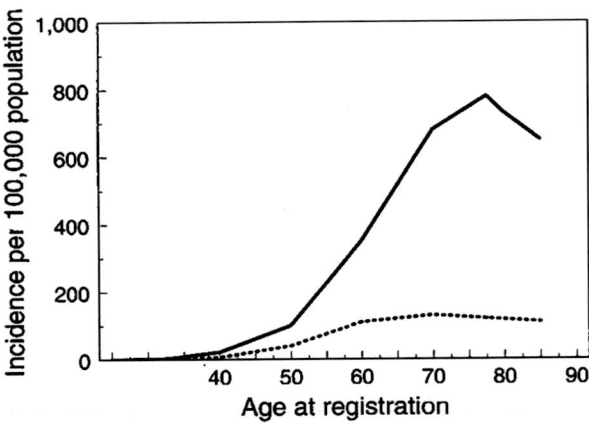

FIGURE 31-2
Lung cancer incidence rates by age and sex in England, 1992. Lung cancer incidence peaks after the age 70 in males (———) as well as females (------). The apparent decrease in incidence in the very old may well be due to underdiagnosis and incomplete reporting in this age group. [SOURCE: OPCS (ICD 612).]

tion of the cigarette smoke results in deep penetration of the carcinogenic agents. On the basis of data from eight prospective studies, it has been estimated that the relative risk of lung cancer is about 10 for male smokers compared with male nonsmokers.[6] The increase in lung cancer risk caused by tobacco smoke is related to the age at which the smoking habit was acquired, the number of years smoked, and the number of cigarettes smoked per day. Those who started smoking at 15 years of age or younger have, after 50 years of smoking, a 100-fold increased risk of lung cancer compared to life-long nonsmokers. For those who started later, the increased risk after the same period of smoking is a factor of 10.[7] Smoking 1 to 14 cigarettes per day results in a 5-fold risk of lung cancer; smoking 20 or more cigarettes per day is associated with a 10 to 25-fold risk. The total number of years smoked is also a central factor: smoking 10 to 25 cigarettes for 40 years results in a risk of 2.5 times that of smoking more than 25 cigarettes for 30 years.[8]

Evidence suggests that low-tar cigarettes are associated with a lower increase in lung cancer risk.[9] The effect is difficult to quantitate because most persons smoking these cigarettes have previously smoked nontipped, high-tar brands.[10] High serum levels of vitamins A and E, and β carotene are associated with a decreased lung cancer risk.[11] It has been postulated that a high intake of these substances may lower to some extent lung cancer risk. This is now the subject of a number of prospective trials.

Cessation of smoking results in a very gradual and prolonged decrease in risk of lung cancer, so that after 20 years the risk approaches that of nonsmokers.[5,7] Westra and colleagues[12] found that K-*ras* mutations, which are closely related to tobacco-induced lung cancers, occur equally often in tumors of former smokers as in those of current smokers. This pattern was independent of the duration of ces-

sation of smoking, indicating that tobacco-induced genetic damage with oncogenic potential may persist in lung epithelial cells for many years.

Passive smoking, that is, exposure to cigarette smoke exhaled by a smoker as well as the smoke escaping from the side of a burning cigarette ("sidestream smoke"), results in an increased lung cancer risk. This was demonstrated conclusively in a recent meta-analysis of 11 American studies addressing this issue,[13] where the overall increased risk was estimated to be about 20 percent. However, exact figures are difficult to obtain because nearly all persons have at some stage of their lives been passive smokers. This results in an underestimation of the increased risk, when it is calculated from the incidence of lung cancer in nonsmoking spouses of smokers. Similar to active smoking, there is probably a dose–risk relationship, and passive smoking during childhood and adolescence may be especially significant, in view of the high risk associated with active smoking at an early age. Indeed, in a recent study, which appeared too late to be included in the Environmental Protection Agency meta-analysis, Stockwell and colleagues[14] demonstrated a relative risk of 2.4 (95 percent confidence interval: 1.1–5.4) for nonsmoking women exposed to passive smoking for 22 or more years in childhood and adolescence.

Tobacco is associated with an increased risk of cancers of a variety of extrapulmonary sites, including the larynx, oral cavity, pharynx, pancreas, kidney, urinary bladder, and uterine cervix. In total, cigarette smoking is estimated to cause about one third of cancer deaths. The risk of tobacco-induced lung cancer is influenced by genetic factors. Familial clustering of cancer in lung cancer patients has been noted. In a recent study, this association was most prominent in female lung cancer patients under 57 years of age and in individuals who smoked for fewer than 20 years.[15] Lung cancer risk appears to be related to some HLA allelic haplotypes and a polymorphic locus of the tumor growth factor β gene.[16] These issues are considered in detail in Chap. 30.

Several other etiologic factors have been identified. Many of these are related to occupational exposure and often there is a synergistic effect with cigarette smoking.[17] Detailed lists of the relative risk of lung cancer in a large variety of professional occupations, based on case-control studies, have been published[18,19] (Table 31-1). Asbestos exposure greatly increases lung cancer risk in smokers. In one study, the relative risk of lung cancer of cigarette-smoking asbestos workers was 53 when compared to nonsmoking subjects without a history of asbestos exposure.[20] Polycyclic aromatic hydrocarbons, which are released during combustion of fossil fuels, are probably responsible for the doubling of lung cancer risk among coal gas industry workers.[21] Chlorated polycarbon compounds, chloromethyl ether,[22] and some other organic chemicals are similarly associated with an increase in lung cancer risk. Arsenic,[23] nickel,[24] and cadmium[25] probably account for the increased incidence of lung cancer associated with a variety of occupations.

Ionizing radiation constitutes an important risk factor among uranium, hard rock, fluospar, and hematite miners. Again, smoking increases the risk substantially. Inhalation of radon gas, a volatile radioactive uranium decay product emitted from the surface of uranium-containing ore and hard rock, results in exposure of the bronchopulmonary epithelium to the alpha radiation. The possible risk associated with exposure to radon gas within homes and public buildings in some areas of the world has caused some concern and has led to an ongoing debate with respect to the maximum acceptable indoor levels. There may be an increased risk associated with insulation and poor ventilation of private and public buildings.[26] A variety of other causes of radiation exposure of the lungs, ranging from the atomic bombs on Hiroshima and Nagasaki[27] to radiation therapy for breast cancer,[28] increase the risk of lung cancer, especially in cigarette smokers. Radioactive polonium and lead, present in cigarette smoke, result in a significant radiation exposure to the bronchial epithelium, contributing to the carcinogenic effect of cigarette smoking.[29]

Diffuse pulmonary fibrosis, regardless of its etiology, is associated with an increased incidence of lung cancer.[30–33] An increased incidence of lung cancer in the CREST (calcinosis, Raynaud's phenomenon, esophageal dysfunction, sclerodactyly, telangiectasia) syndrome was noted by Yousem.[34]

Human papillomavirus (HPV) has been incriminated as a causal factor in lung cancer, largely on the basis of the detection of genetic material of HPV, of high-risk and low-risk types, in some lung carcinomas.[35] Most evidence was based on in situ hybridization. However, in a recent study, Szabó and colleagues,[36] using polymerase chain reaction, failed to detect HPV in a series of lung carcinomas and have shed some doubt on the role of HPV in lung carcinogenesis. Further data are needed to resolve this issue. In the rare bronchial squamous carcinoma arising in patients with juvenile laryngotracheal papillomatosis,[37] HPV was convincingly demonstrated.[38,39] The Epstein-Barr virus (EBV) probably plays a causal role in the pathogenesis of the rare lymphoepithelioma-like large cell carcinoma of the lung (see p. 1047).

Lung Cancer Screening and Early Detection

Several studies have addressed the cost–benefit effects of large scale screening for occult lung cancer by chest radiographs or sputum cytology. A reduction in numbers of cancer deaths was suggested in one early British study.[40] In three subsequent studies,[41–43] together involving over 30,000 male smokers over 44 years of age, significant numbers of clinically occult, early lung carcinomas were detected. Disappointingly, this did not result in significant differences in mortality rates between the screened and nonscreened groups. Therefore, it appears that large scale screening for early lung cancer is not justified because it is not effective if measured by its impact on mortality.[44] In a recent Japanese study,[45] based on

TABLE 31-1

Occupations Associated with Elevated Lung Cancer Risk[18]

Profession	Odds Ratio
Excavating/mining	4.01
Furnace workers	3.11
Armed service personnel	3.10
Agricultural workers	2.05
Driver sales	2.21
Mechanics	1.72
Painters	1.96
Drivers	1.88

Industries	Odds Ratio
Farming	1.88
Mining	2.98
Primary ferrous metals manufacturing	2.43

yearly chest radiographs of more than 300,000 residents over 40 years of age, only half of the cancers detected were stage I, and there were significant numbers of interval cancers during the study. These further disappointing results contribute to the argument against mass screening by chest radiography. However, the results of all studies are obviously closely related to population factors, and the feasability of a 10 percent reduction in lung cancer mortality is presently being explored in a large scale National Cancer Institute trial, which includes chest radiography as one of the screening modalities.[46]

CLINICAL FEATURES

Many of the local signs and symptoms of lung cancer are directly related to the physical presence of a mass distorting and destroying the normal bronchopulmonary architecture. This is especially true for centrally located tumors obstructing large bronchi. Peripheral lung carcinomas tend to remain clinically silent for a longer period and more commonly manifest themselves through symptoms related to metastatic disease or general symptoms of malignant disease.

Signs and symptoms of centrally located tumors, which arise in main, lobar, or segmental bronchi, are often related to stenosis or occlusion of the bronchial lumen (Fig. 31-3). Persistent cough, wheezing due to partial occlusion of the bronchus, and recurrent bronchopneumonia are early signs. Lung collapse and endogenous lipid pneumonia due to bronchial occlusion may cause dyspnea and arterial hypoxemia due to shunting. Erosion of bronchial and pulmonary blood vessels leads to hemoptysis. This presents as blood-tinged, "rusty" sputum, or massive hemorrhage, which may cause death. Some tumors outgrow their blood supply, leading to central necrosis. The necrotic debris may be evacuated through the bronchus, resulting in central cavitation and a picture resembling seconday tuberculosis. In these instances, hemoptysis is common. Rarely, tuberculosis may coexist with the tumor. More commonly, fungal infection may supervene; in cavitating tumors, a central aspergilloma may form. In such instances, fine needle aspiration (FNA) material usually reveals the fungi but occasionally, recognizable tumor cells are absent from the specimen.

Invasion of mediastinal structures, by the primary tumor or regional metastases, may lead to hoarseness due to recurrent nerve involvement, unilateral paralysis of the diaphragm resulting from phrenic nerve involvement, or superior vena cava syndrome, dysphagia, cardiac arhythmias, and cardiac tamponade.

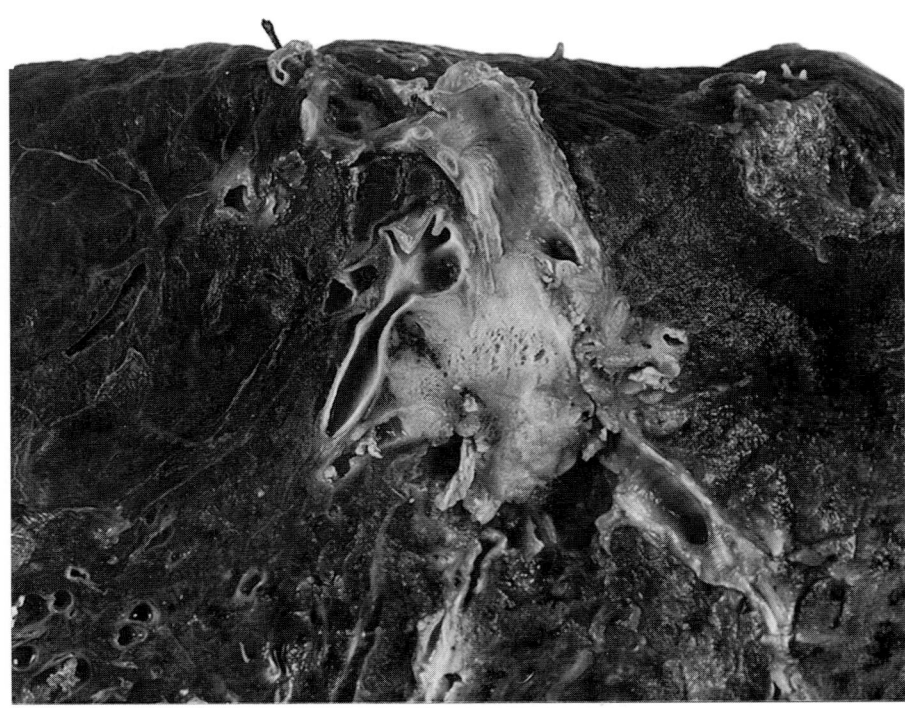

FIGURE 31-3
Central lung carcinoma occluding and destroying a large bronchus. (*Reproduced by permission of Lung Cancer Therapy.*)

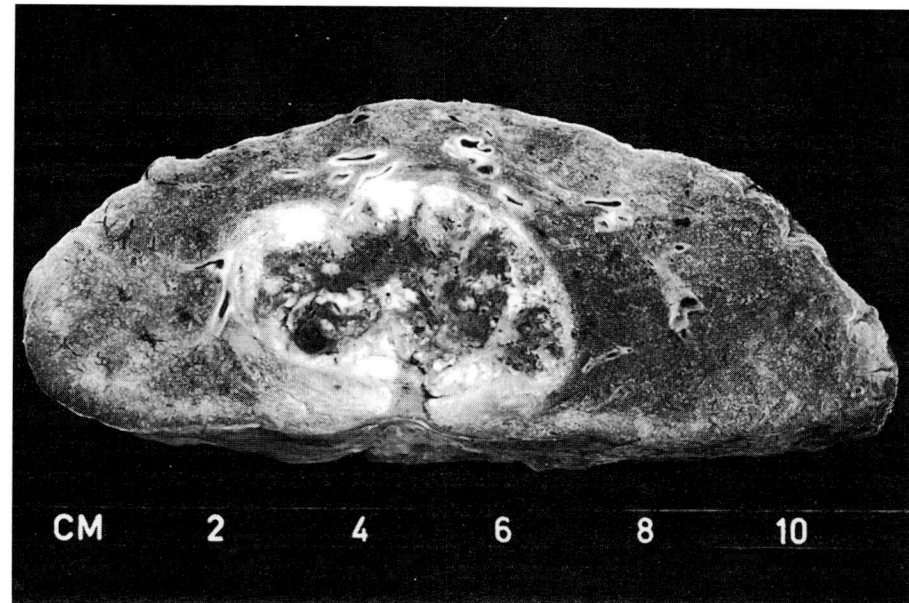

FIGURE 31-4
Peripheral, subpleural pulmonary adenocarcinoma. There is a subpleural area of scarring, with indentation of the overlying pleura. The lung parenchyma surrounding the tumor does not show any obvious abnormalities.

Peripheral tumors (Fig. 31-4) tend to remain clinically silent for longer periods. These tumors are often discovered incidentally on a routine chest radiograph. Local pain may result from pleural and thoracic wall involvement. Peripheral squamous cell carcinomas may grow by direct extension through the pleural blades into the thoracic wall. When the tumor is apical in location, invasive growth into the soft tissues of the thoracic inlet may compromise the eighth cervical and the first two thoracic nerves, resulting in shoulder pain radiating to the ulnar side of the arm (Pancoast's or superior sulcus syndrome). This is commonly seen in combination with ipsilateral Horner's syndrome (enophthalmos, ptosis, miosis, and ipsilateral decreased sweating) due to involvement of the sympathetic plexus. Adenocarcinoma, invading through the pleura, generally does not show this tendency to invade the thoracic wall. Instead, it often causes pleural effusions, leading to dyspnea.

Finally, the tumor may manifest itself through the general signs of malignant disease, such as loss of appetite, fatigue, and weight loss, paraneoplastic symptoms, or through symptoms related to distant metastatic disease. The skeleton and the central nervous system are favored sites of metastatic lung cancer.

Paraneoplastic Syndromes

Lung cancer is a common cause of paraneoplastic syndromes, a heterogeneous group of disorders associated with malignant disease. Paraneoplastic syndromes are subdivided into endocrine, neuromuscular, cutaneous, metabolic, hematologic, immunosuppressive, and collagen vascular disorders.[47] Many of these syndromes are most commonly associated with small cell carcinoma, and their presence is associated with a poor prognosis.[48,49]

In some instances, the pathophysiology is well understood, but in many, it still remains to be elucidated at the molecular level. A detailed treatment of the subject is outside of the scope of this chapter.[50] Only a brief discussion of some paraneoplastic syndromes commonly associated with lung cancer will be given.

Hypercalcemia is most commonly associated with squamous cell carcinoma. Although more than one mechanism is probably involved in its pathogenesis, the most important factor is parathyroid hormone-related protein (PTHrP), synthesized and excreted by the tumor cells.[51] PTHrP is homologous to parathyroid hormone (PTH) in its amino terminal part, and is able to activate PTH receptors in the bone, leading to increased bone reabsorption. It also activates renal PTH-sensitive adenylyl cyclase, resulting in increased calcium and decreased phosphate reabsorption in the kidney. Thus, the endocrine effects of increased circulating PTHrP are similar to those of hyperparathyroidism. Other endocrine paraneoplastic syndromes commonly encountered in lung cancer patients include inappropriate antidiuretic hormone secretion, and Cushing's syndrome, which is caused by corticotropin production by the tumor cells.[52]

Hypertrophic pulmonary osteoarthropathy and finger clubbing are common in lung cancer[53] and may occasionally constitute the presenting symptom.

A relationship with increased plasma levels of growth hormone, possibly due to the synthesis of growth hormone or growth hormone-related hormone-like substances by the tumor cells, has been demonstrated.[54] Similar abnormalities are encountered in association with a variety of nonneoplastic diseases such as congenital cardiac defects, cirrhosis of the liver, and Crohn's disease.

A wide variety of neuromuscular syndromes is associated with small cell lung cancer (SCLC). The Eaton-Lambert syndrome is characterized by proximal weakness and myalgias. It is caused by antibodies to SCLC cells, which cross-react with voltage-gated calcium channel proteins present in the presynaptic membrane. Other neurologic paraneoplastic syndromes include peripheral neuropathies, subacute cortical cerebellar degeneration, and polymyositis. Circulating antibodies reacting with epitopes on normal neurons appear to be causative in the pathogenesis of several of these syndromes.[55]

General symptoms of malignancy with an incompletely understood pathogenesis include fever, weight loss and decreased appetite, leading to wasting (cachexia) of the patient. Interleukin-1, tumor necrosis factor (TNF), and interferon-γ are probably important in the pathogenesis of tumor-associated fever and cachexia.[56,57] Bronchopneumonia or other infections should obviously be considered as possible causes of the fever.

DIAGNOSTIC TECHNIQUES

Noninvasive diagnostic techniques, such as chest radiography and computed tomography, establish the presence of an abnormal pulmonary mass. The subsequent management depends in part on the a priori likelihood that the mass is malignant. Overall, an asymptomatic solitary pulmonary nodule (a density completely surrounded by aerated lung, with circumscribed margins, of any shape, and usually 1 to 6 cm in diameter) is malignant in 35 percent of cases, but in a nonsmoker under age 35 years, this chance is less than 1 percent.[58] Repeat chest radiographs every 3 months for the first year and then yearly are recommended for the latter group of patients; lack of growth for 2 years appears to be a reliable indicator of benignity. In addition, certain types of radiographic calcification argue strongly in favor of a benign diagnosis. In smokers and persons older than 35 years, as well as in patients with expanding lesions, it is mandatory to obtain a tissue diagnosis.

Sputum cytology may provide definitive evidence of a malignancy in patients with a known or suspected lung tumor. It may be useful in asymptomatic patients and even occasionally in the absence of radiologic evidence of tumor. In screening programs of asymptomatic heavy smokers, about 20 percent of occult lung carcinomas were detected by sputum cytology, 72 percent by chest radiograph, and 6 percent by both methods.[58] Central squamous cell carcinomas are usually detectable by sputum cytology. The diagnostic yield is closely related to the quality of the sputum and the expertise of the cytologist. The highest diagnostic yield is obtained with early morning sputum, produced after thorough gargling and rinsing of the mouth. Pooling of sputa in an alcohol container over at least 3 days is advocated.[59] In the clinic, nebulizers may promote sputum production; at home, hot water or a humidifier may be helpful. The sputum can be smeared directly on a slide or embedded in paraffin. Alveolar macrophages indicate that the sputum derives from the deeper part of the respiratory tract and are therefore an important indicator of adequate specimen quality.

Bronchoscopy has a central place in the diagnosis of lung cancer. The fiberoptic bronchoscope allows inspection and biopsy of bronchi up to the level of the segmental and some subsegmental bronchi. The biopsy forceps can be advanced even further. In addition, brochial brushing, transbronchial FNA, and bronchial lavage provide specimens for cytology. Sputum cytology after bronchoscopy sometimes reveals the tumor cells, even when biopsies and brushings have been negative, and is therefore a worthwhile routine procedure.

Percutaneous transthoracic FNA biopsy under radiologic guidance is useful to obtain material from peripheral pulmonary tumors beyond bronchoscopic reach. Pneumothorax occurs in about 5 percent of cases; risk factors for this include advanced age and concurrent chronic lung disease.[60]

Mediastinoscopy allows biopsy of mediastinal lymph nodes. It is most effective on the right side; on the left, the mediastinoscope cannot reach all parts due to the presence of several vital structures. Left-sided mediastinotomy provides access to lymph nodes that are not adequately sampled by the mediastinoscope.

In some instances, no tumor tissue is obtainable before thoracotomy, so that the first material that becomes available is the surgical resection specimen. Frozen section diagnosis is indicated when the intraoperative histologic diagnosis would influence the extent of the surgical procedure. Frozen sections of bronchial resection margins are especially important in the case of a central squamous cell carcinoma and when margins of excision are small.

The approach to the macroscopic investigation and dissection of resection specimens should be systematic.[61,61a] Specimen type and size should be recorded, together with a statement with respect to the appearance of resection margin and pleura. The tumor should be palpated and the specimen should be dissected into parallel slices to disclose the tumor. Special attention should be paid to tumor size (in three directions), location, relation to bronchi, distance from the bronchial resection margin and pleural surface, or possible pleural involvement. At least two blocks should be taken from the tumor, but more blocks are necessary when the macrosopic features vary between different areas of the tumor. In case of possible pleural involvement, elastic stains should be requested, because these are indispensable for a proper evaluation of ingrowth into the visceral pleura.[62] A systematic search for hilar and parabronchial lymph nodes should be conducted and all nodes should be processed. The presence of any abnormality in the lung tissue, such as atelectasis or obstructive pneumonitis, should be noted and investigated histologically. A systematic search for intrapulmonary metastases or a second primary tumor should be undertaken because, also when preoperative investigations are negative, these are present in a few percent of cases.[63,64]

STAGING

Tumor stage is the single most important prognostic factor in lung cancer of all histologic types. The feasibility of curative surgery or radiotherapy depends primarily on tumor stage. For non-SCLC, the TNM staging system with four groupings is most widely used[65,67] (Table 31-2 and 31-3). A precise mapping of regional lymph nodes is facilitated by the Naruke

TABLE 31-2
International Staging System for Lung Cancer[65,66]

Primary tumor (T)		T4	Any size with any of following:
TX	Tumor that cannot be assessed, or tumor proven by cytology but not visualized by imaging or bronchoscopy		• Invasion of carina, mediastinum, heart, great vessels, trachea, esophagus, vertebral body • Malignant pleural effusion
T0	No evidence of primary tumor	*Note:*	• pT3 or less: No gross tumor at the resection margins (with or without microscopical involvement)
Tis	Carcinoma in situ		
T1	Tumor 3 cm or less in greatest diameter, surrounded by lung or visceral pleura, without evidence of invasion more proximal than lobar bronchus (exception: superficially spreading tumors that have an invasive component limited to the bronchial wall extending to the main bronchus are classified as T1)		• pT4: microscopic confirmation of invasion of carina, mediastinum, heart, great vessels, trachea, oesophagus, or vertebral body, or malignant cells in pleural fluid. However, even if malignant cells are not identified, pleural effusion often indicates unresectable disease
T2	Tumor with *any* of the following features: • Greatest diameter exceeding 3 cm • Invasion of visceral pleura • Atelectasis or obstructive pneumonia extending to the hilar region, but not involving the entire lung • Involvement of main bronchus, the proximal margin being at least 2 cm from the carina	*Nodal involvement (N)*		
		NX	Regional lymph nodes cannot be assessed
		N0	No regional node involvement
		N1	Metastasis to, or direct extention into, ipsilateral peribronchial or hilar nodes
		N2	Metastasis to ipsilateral mediastinal or subcarinal nodes
T3	Any size with any of following: • Invasion of the chest wall, diaphragm, mediastinal pleura, or parietal pericardium • Atelectasis or obstructive pneumonia extending to the hilar region, but not involving the entire lung • Proximal extent within 2 cm of the carina, but not involving the carina	N3	Metastasis to contralateral mediastinal or hilar nodes, or to scalene or supraclavicular nodes
		Note:	Mediastinal lymphadenectomy specimens will ordinarily include at least six lymph nodes.
		Distant metastasis (M)	
		MX	Presence of distant metastasis cannot be assessed
		M0	Distant metastasis absent
		M1	Distant metastasis present (specify site)

TABLE 31-3
International Staging System for Lung Cancer: Stage Groupings of TNM Subsets[67]

Occult carcinoma	TX	N0	M0
Stage 0	Tis	N0	M0
Stage I	T1	N0	M0
	T2	N0	M0
Stage II	T1	N1	M0
	T2	N1	M0
Stage IIIA	T3	N0	M0
	T3	N1	M0
	T1–3	N2	M0
Stage IIIB	T1–4	N3	M0
	T4	N0–3	M0
Stage IV	T1-4	N0–3	M1

nodal map[68] (Table 31-4). Paratracheal and subcarinal nodes can be biopsied at preoperative mediastinoscopy to select patients for surgery.

The staging of SCLC is based on a simple division into two groups. "Limited disease" tumors are confined to one hemithorax, including cases with ipsilateral malignant pleural effusion or involvement of ipsilateral mediastinal and supraclavicular lymph nodes. All tumors that have spread beyond these limits are designated "extensive disease."

TABLE 31-4
Naruke Lymph Node Map[68]

N2 nodes
Superior mediastinal nodes
1   Highest mediastinal
2   Upper paratracheal
3   Pre- and retrotracheal
4   Lower paratracheal
Aortic nodes
5   Subaortic (aortic window)
6   Paraaortic (ascending aorta or phrenic)
Inferior mediastinal nodes
7   Subcarinal
8   Paraesophageal
9   Pulmonary ligament
N1 nodes
10   Hilar
11   Interlobar
12   Lobar
13   Segmental

PREMALIGNANT LESIONS

An increased awareness of the early signs and symptoms of lung cancer and the use of fiberoptic bronchoscopy have led to the increased detection of subtle atypical bronchial epithelial changes, thought to precede the development of lung cancer. In addition, lung cancer screening programs have yielded significant data. Saccomanno and coworkers[69] performed a longitudinal study of sputum cytology of uranium miners in the United States. They demonstrated squamous metaplasia with increasing cellular atypia, several years before the development of invasive carcinoma, usually SCLC or squamous cell carcinoma.

Experimentally, proliferative changes can be induced by damaging the bronchial lining epithelium with a variety of physical or toxic agents. The earliest phase consists of an increased number of small, basally located cells. Whether these cells derive from basal cells or from mucous cells remains a moot point. It is of interest that in animal models, mucous cells are the important progenitors of invasive carcinoma.[70] The same may perhaps be true for human beings because the production of small traces of mucin and the expression of keratin subtypes, usually associated with a glandular phenotype, are common in carcinomas of various histologic types.[71,72] The proliferating cells of the damaged bronchial epithelial lining acquire an increased amount of eosinophilic cytoplasm, and flattening of cells is seen in the more superficial cell layers. The appearance of intercellular bridges indicates squamous metaplasia. In time, the original luminal layer of ciliated cells is lost.

Dysplasia is hallmarked by increasing grades of nuclear enlargement, hyperchromatism, and pleomorphism. When mitoses are found at all levels of the epithelium, resulting a highly atypical and disorganized epithelium, squamous carcinoma in situ is present[68] (Figs. 31-5 and 31-6). The in situ carcinoma may involve the bronchial glands. Keratinization is sometimes seen, often in the form of solitary dyskeratotic cells. Total surgical removal of in situ carcinoma, which often requires a lobectomy but may be achieved with segmentectomy or sleeve resection in selected cases, is curative.[73]

Small numbers of bronchial neuroendocrine cells are normally present in the basal layer. These cells increase in number as a response to irritants and toxic substances, including tobacco smoke. Small nests of neuroendocrine cells may then be formed. Occasionally, proliferating neuroendocrine cells form small nodules known as "tumorlets."[74] These

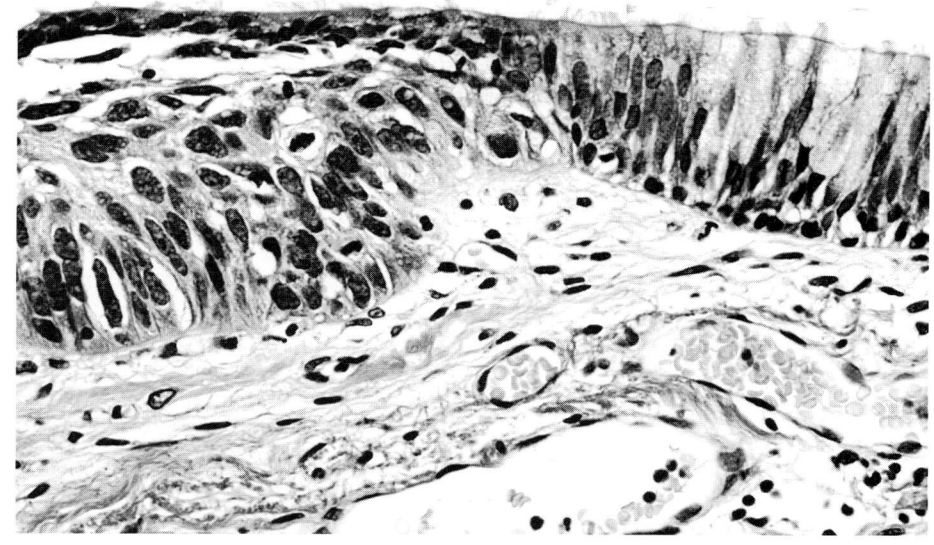

FIGURE 31-5
Squamous carcinoma in situ. The normal ciliated pseudostratified epithelium (*right*) is undermined by highly atypical stratified epithelium (*left*).

are common in bronchiectasis. The relation between such small nodules and malignant neuroendocrine tumors of the lung remains conjectural.

Proliferation of bronchiolar and alveolar epithelium occurs in a wide variety of inflammatory and fibrosing lung diseases. Bronchiolar-type epithelium shows a tendency to extend along alveolar septa, and the proliferating cells may show some nuclear irregularities and sometimes squamous metaplasia. In diffuse pulmonary fibrosing conditions such as cryptogenic fibrosing alveolitis, an increased incidence of lung carcinoma has been found.[30–33] It seems reasonable to assume that the increased proliferative activity of bronchiolar and alveolar epithelium mentioned above is causally related to this increased carcinoma risk.

Also in the absence of pulmonary inflammatory and fibrosing disorders, putative precursor lesions of peripheral pulmonary adenocarcinoma have been identified.[75–80] These take the form of discrete nodules, usually 1 to 7 mm in diameter, but sometimes larger, consisting of proliferations of atypical cuboidal or columnar cells, with the ultrastructural features of type II pneumocytes or Clara cells. Most are situated near the pleura. They are commonly multiple; rarely, the lung parenchyma is studded with such small nodules. An association with smoking was noted in one study.[80] Macroscopically, the small lesions are best identified after lung distention with Bouin's fixative. Systematic studies of lung cancer resection specimens yielded a prevalence of about 5 to 20 percent.[78–80] These lesions are most

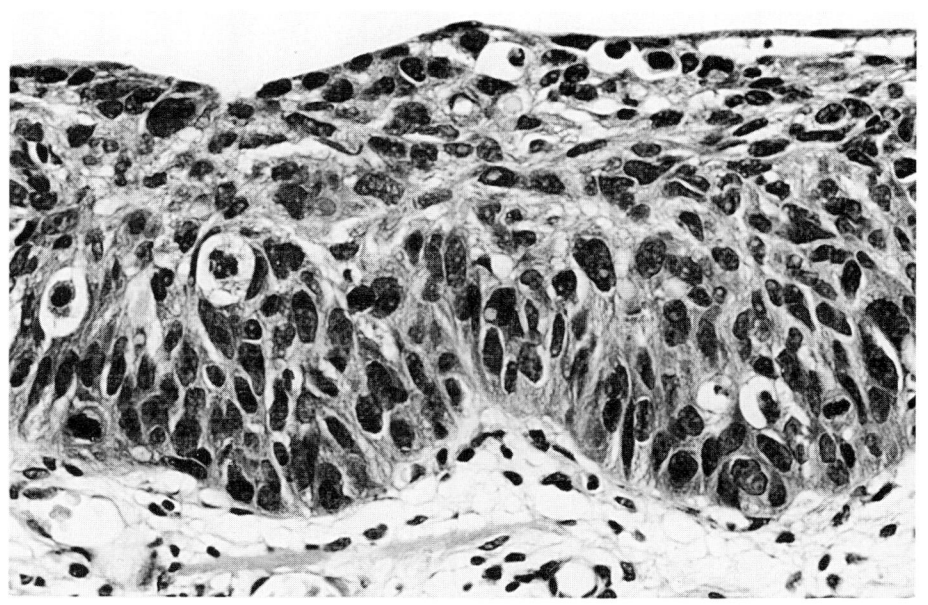

FIGURE 31-6
Squamous carcinoma in situ. Nuclei are enlarged, pleomorphic, and variably hyperchromatic. Mitoses are seen in suprabasal layers. Although there is some flattening of cells in the upper layers, nuclear orientation is irregular. The basal border of the atypical epithelium is sharp, and the underlying stroma has a loose quality (compare with the desmoplastic stroma seen in infiltrating squamous carcinoma in Fig. 31-8.)

common in association with pulmonary adenocarcinomas and may constitute precursor lesions of such tumors.[80] Because the dominant tumors of the resection specimens generally show different histologic features or more marked cellular atypia, intrapulmonary metastasis from these tumors can usually be excluded.

These lesions have been called papillary adenomas of type II pneumocytes, bronchioloalveolar adenomas, or atypical adenomatous hyperplasias. They may represent cancer precursor lesions, which would explain the high prevalence of multicentric pulmonary primaries, which may be as high as 19 percent,[78] though others reported lower figures.[81] The differences are probably related to thoroughness of the search for small additional tumor nodules as well as to the criteria used to diagnose very small carcinomas. Indeed, the various reports on this subject clearly illustrate the problem of distinguishing between premalignant lesions and small malignant tumors. For practical purposes, but somewhat arbitrarily, Miller[78] defined carcinoma as a lesion with marked nuclear atypia measuring more than 5 mm in diameter.

Histologically, these small hyperplastic lesions consist of a proliferation of epithelial cells with abundant cytoplasm, which spread in a lepidic fashion along alveolar septa. Central scarring and vascular invasion are absent. Some nuclear atypia may been seen, especially in larger lesions. In such instances, the distinction from well differentiated peripheral adenocarcinoma is problematic.[82] Electron microscopy often reveals lamellar intracytoplasmic surfactant bodies, characteristic of type II pneumocytes.[78] Clara cell granules were found in the cells of some lesions with atypia,[79] indicating that bronchiolar epithelium may participate. Morphometric analysis of nuclear area showed that the atypical hyperplastic epithelial lesions had smaller and less pleomorphic nuclei than adenocarcinomas with Clara cell features.[79,83] However, loss of expression of blood group antigens was seen in atypical epithelial hyperplastic lesions as well as in Clara cell adenocarcinomas, but was absent from hyperplastic lesions lacking nuclear atypia.[79] The picture emerging from these findings is not entirely clear. It seems fair to say that at present, the differential diagnosis between small adenocarcinomas and hyperplastic lesions with possible premalignant potential has not yet been elucidated completely.

These lesions should not be confused with the "alveolar adenoma" described by Yousem and Hochholzer.[84] The latter is a different tumor, generally detected clinically as an asymptomatic coin lesion with an average diameter of 1.8 cm, well circumscribed, with a characteristic central cystic area surrounded by a peripheral rim with multiple smaller cysts lined by type II pneumocytes. Some of these tumors show a central scar. The epithelial nature of the cyst-lining cells is one of the features distinguishing it from pulmonary lymphangioma, which it may resemble.

Models of Lung Cancer Histogenesis

Various models for the histogenesis of lung carcinoma have been put forward. It is tempting to speculate that squamous cell carcinoma arises from metaplastic squamous bronchial epithelium with atypia. These cells in turn may be derived from proliferating mucous or basal cells. Similarly, SCLC may be thought to arise from proliferating bronchial neuroendocrine cells, and pulmonary adenocarcinomas may be related to proliferating bronchial mucous cells, bronchiolar Clara cells, or alveolar type II pneumocytes. However, such a model of lung cancer histogenesis is to a large degree conjectural and certainly constitutes an oversimplification. The metaplastic potential inherent in pulmonary epithelium, the frequent occurrence of lung cancers combining more than one type of differentiation, and the changes in differentiation type during progression of some lung carcinomas preclude any scheme of lung cancer histogenesis based on similarities in cell type between tumors and preexistent bronchopulmonary epithelium.[85]

Sunday and colleagues,[86] pointing to the overlap in hormone production and differentiation-related surface protein expressed in various lung cancer types, postulated that all subtypes of lung cancer are derived from a common type of progenitor cell. Yesner[87] went further and proposed that the various types of lung cancer all originate from an immature precursor with an undifferentiated small cell phenotype. Apparent support for this idea was obtained in an autopsy study of 40 biopsy-proven SCLCs; at autopsy, the tumor histology had changed completely to squamous carcinoma (three cases), adenocarcinoma, or large cell carcinoma (one case each), whereas in a further six cases, a combination of small cell and non-small cell histologies had developed.[88] In in vitro and xenotransplant studies, loss of neuroendocrine differentiation parameters and a change to a large cell phenotype occurred in SCLC cell lines.[89] However, although SCLC can clearly change to non-small cell phenotypes, it seems unlikely that SCLC is the general precursor of other lung cancer types. The available data on putative precursor lesions in the bronchus and peripheral lung tissue do not support the concept of a single type of cancer precursor with the phenotype of SCLC.

HISTOLOGIC TYPES

Three main patterns of differentiation occur in lung carcinoma: squamous, glandular, and neuroendocrine. Each of these may be present alone or in combination with either or both of the other two types. When immunohistochemistry and electron microscopy are included in the investigation, combinations of differentiation types are found in the majority of cases.[90–93] Histologic patterns commonly vary between different areas of the tumor and may change during the course of disease. As a result, any histologic classification of lung tumors based on a limited number of subtypes, defined according to the predominant differentiation type, is an oversimplification. Previously, these considerations have been used as an argument that lung cancer is best viewed as a single disease, encompassing a continuous spectrum of histologic types.[94]

The World Health Organization (WHO) histologic classification of lung tumors,[95] which is most commonly used, is based on a light microscopic assessment of patterns of differentiation. There are a number of additional criteria, mainly with respect to the diagnosis of SCLC, the neuroendocrine differentiation phenotype of which is too poorly developed to allow reliable light microscopic detection (Table 31-5).

The main histologic subtypes are: squamous cell, adenocarcinoma, adenosquamous, large cell, and small cell carcinoma. As will be discussed below, SCLC is a poorly differentiated neuroendocrine lung carcinoma. Large cell carcinoma constitutes a heterogeneous group of tumors, which may express traces of squamous, glandular, and neuroendocrine differentiation, alone or in various combinations. This becomes apparent when these tumors are investigated ultrastructurally.[96–99]

In the WHO classification, each main type is subdivided into further categories (see Table 31-5). However, this presents problems. Spindle cell, giant cell, and clear cell change may occur in any of the non-SCLC types. The subdivision of adenocarcinoma is greatly hampered by the architectural and cytologic variability common in these tumors, so that elements of more than one subtype are often present in an individual tumor. The subclassification of SCLC has been changed to include small cell–large cell carcinoma as a distinct subtype. Furthermore, the distinction between oat cell and intermediate cell type

TABLE 31-5

Histologic Classification of Lung Carcinomas

1 Squamous cell carcinoma
   Variant: squamous cell carcinoma with spindle cell change
2 Small cell carcinoma
   a   Oat cell carcinoma
   b   Intermediate cell type
   c   Combined oat cell carcinoma
3 Adenocarcinoma
   a   Acinar adenocarcinoma
   b   Papillary adenocarcinoma
   c   Bronchioloalveolar carcinoma
   d   Solid Carcinoma with mucus formation
4 Large cell carcinoma Variants:
   Giant cell carcinoma
   Clear cell carcinoma
5 Adenosquamous carcinoma
6 Carcinoid tumors (see Chapter 29)
7 Bronchial gland carcinomas (see Chapter 29)
   a   Adenoid cystic carcinoma
   b   Mucoepidermoid carcinoma
   c   Others
8 Others

Based on the World Health Organization *Histological Classification of Malignant Epithelial Tumors of the Lung*, 1981[95]

SCLC is greatly influenced by the tissue preservation state. Intraobserver disagreement ranged from 2 to 20 percent in a study where a set of 50 lung carcinomas was classified twice by five pathologists.[100] Significant intra- and interobserver variation was also apparent from a recent study of 208 lung carcinomas typed independently by eight pathologists.[27] Consensus, defined as agreement by at least six pathologists, was reached in 76.4 percent of cases. To circumvent some of these problems and to incorporate the conclusions of some recent studies on the histology of lung cancer, discussed later in this chapter, we use a modification of the WHO classification, summarized in Table 31-6.

The classification of neuroendocrine carcinoma is under review. It is likely that small and intermediate cell types will probably be classed as small cell, and large cell neuroendocrine carcinoma will form a separate group (Chap. 29).

Relative Incidences of the Histologic Types of Lung Cancer

In Great Britain and many countries in Europe, squamous cell carcinomas account for about 40 percent of

TABLE 31-6

Histologic Classification of Lung Carcinomas as Used in This Chapter

1 Squamous cell carcinoma
2 Small cell carcinoma
    a Pure small cell carcinoma
    b Small cell–large cell carcinoma
    c Combined small cell carcinoma (with areas of squamous or glandular differentiation)
3 Adenocarcinoma
    Variant: Bronchioloalveolar carcinoma
4 Large cell carcinoma
5 Adenosquamous carcinoma

Note:

1 The presence of giant cell, spindle cell or clear cell change does not alter the diagnosis, but should be noted in the pathology report (see p. 1020).
2 Recent studies indicate that carcinomas with "basaloid" appearance, with or without a component showing one of the conventional non-small cell carcinoma types (see p. 1048) may be regarded as a distinct subtype of non-small cell lung carcinoma, with a poor prognosis.

all lung cancers. Adenocarcinomas and SCLC constitute roughly 15 to 25 percent each, large cell carcinomas and a small number of rarer tumor types accounting for the remaining 10 to 20 percent. The relative incidences of the various types of lung carcinoma vary significantly between countries. In addition, there have been changes in time. Thus, in the United States and Japan, adenocarcinoma is now the most common histologic type. In part, shifts in time may be related to alterations in criteria used for lung tumor typing, but much of the change is no doubt real.[101] In young patients, especially those who are females and nonsmokers, adenocarcinomas are relatively common. Obviously, the percentages depend in part on the material on which they are based; because resectability and cure rate of squamous cell carcinoma are higher than in other types, these tumors tend to be overrepresented in series of surgically treated patients and underrepresented in autopsy series.[102] Finally, some etiologic factors are specifically associated with certain histologic types of carcinoma. Small cell and squamous carcinoma are most common in uranium miners.[103,104] Exposure to chloromethyl ether and some other organic chemicals is also associated with SCLC.[22]

SQUAMOUS CELL CARCINOMA

In many countries, this is the most common type of lung cancer. Most arise in a large bronchus, the remainder are located in the lung periphery, often in the upper lobe. Of all lung cancer types, squamous cell carcinoma is most commonly confined to the lung and regional nodes at the time of diagnosis.

Pathology

Centrally located squamous cell carcinomas, which arise in main, lobar, or segmental bronchi, can usually be visualized bronchoscopically. Early bronchial squamous cell carcinomas have been subdivided according to their growth pattern.[105] One type grows longitudinally along the bronchial lumen, with irregular margins that are commonly difficult to recognize, so that there is a significant risk of incomplete removal at surgery. The other type has a propensity for transmural penetrating growth and is associated with a higher likelihood of metastasis. In a study by Nagamoto and colleagues,[106] metastases were found in 3 of 20 carcinomas of the superficial, "creeping" type and in 3 of 5 of the "penetrating" type.

In addition to the tumor, areas of squamous metaplasia with varying degrees of cytologic and architectural atypia are commonly present elsewhere in the bronchial tree. This reflects the multifocal occurrence of tobacco-induced preneoplastic changes throughout the bronchial mucosa. Frozen sections of surgical margins of the bronchus are important to ensure complete excision. Characteristically, the superficial tumor growth is situated within the epithelial compartment (in situ), replacing the original bronchial epithelial lining (Fig. 31-7). This contrasts with the preferred mode of intrabronchial growth of pulmonary adenocarcinoma, SCLC, and metastatic tumors, which more commonly grown in the subepithelial tissues, while the nonneoplastic bronchial surface epithelium remains intact.

Peripheral squamous cell carcinomas show a propensity to invade the pleura and grow per continuitatem through both pleural blades into the thoracic wall. Growth of an apical tumor cranially into the thoracic inlet commonly causes Pancoast's syndrome (see p. 1013).

Macroscopically, squamous cell carcinoma is usually solid and firm, with a pale gray, gritty cut surface. Occasionally, cavitation is present. Hemorrhage into the tumor, bronchial tree, and surrounding lung parenchyma occasionally results in a variegated appearance. In central tumors, fragments of cartilage and anthracotic lymph nodes are often identified. Poststenotic accumulation of bronchial mucus, bronchiectasis, and consolidation of the pulmonary parenchyma due to endogenous lipid pneumonia or

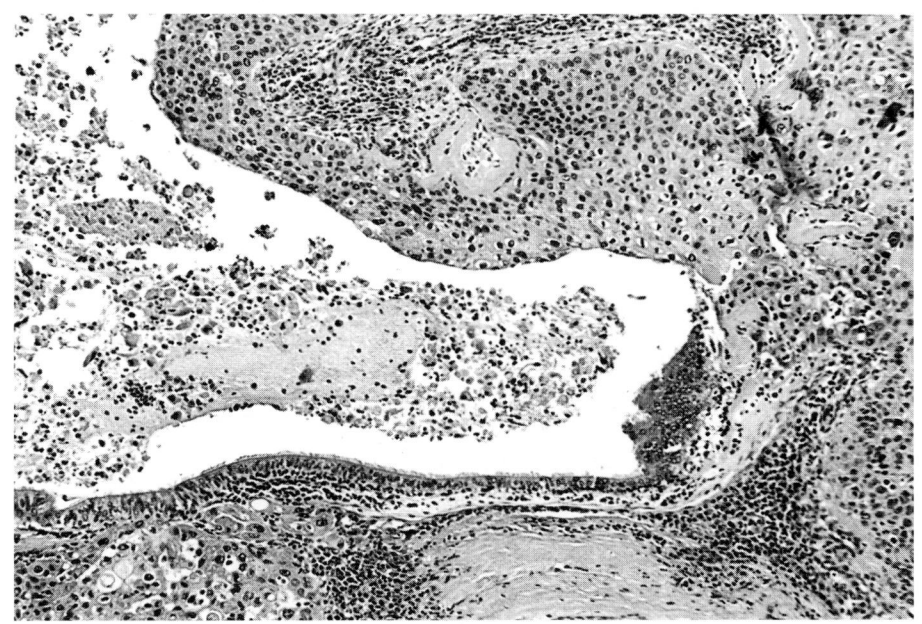

bronchopneumonia are common. Peripheral tumors
may show central or diffuse scarring, with indentation of the overlying pleura, and irregular anthracotic
pigmentation.

Histologically, the tumor consists of a proliferation of epithelial cells with squamous differentiation,
the light microscopic hallmarks of which are intercellular bridges and keratinization. Often, the tumor
cells form solid strands and nodules lying within cellular or sclerotic desmoplastic stroma, which often
contains inflammatory cells (Fig. 31-8). The tumor
may bulge into the bronchial lumen (Fig. 31-9) and
invades surrounding tissues. Rarely, the tumor shows

a papillary endobronchial architecture. Such neoplasms, which are nearly alway T1N0, tend to remain
confined to the bronchus, which may progressively
increase in diameter to accomodate the expanding
tumor (Fig. 31-10 and 31-11). Such tumors have no
better prognosis than other stage I tumors. Papillary
squamous cell carcinoma should be distinguished
from the solitary bronchial squamous papilloma of
the adult.[107] The presence of invasion establishes the
diagnosis of carcinoma.[108]

The tumor cells vary markedly in size and shape,
between tumors, and not uncommonly within an
individual tumor. The nuclei are often irregular, with

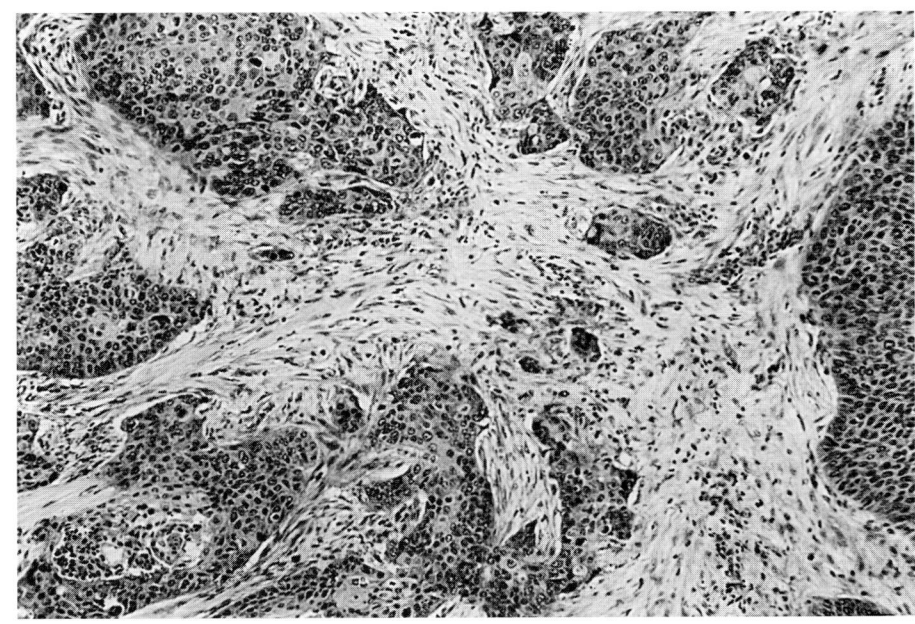

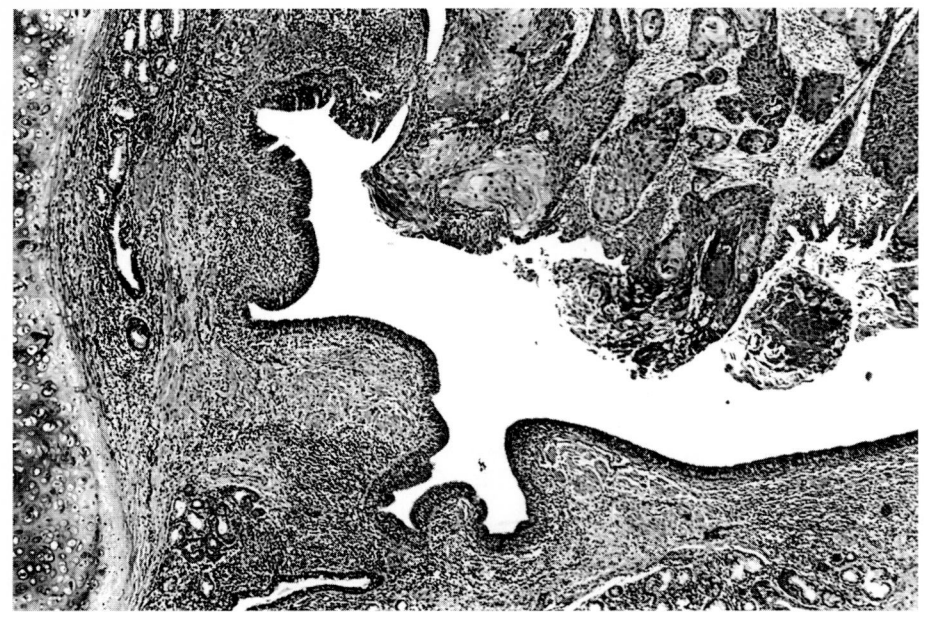

FIGURE 31-9
Squamous cell carcinoma. Tumor tissue bulges into the bronchial lumen; on the surface, tumor cells dissociate from the main tumor mass. Such tumor cells are usually demonstrable in the sputum.

coarsely clumped chromatin and an irregular envelope. One or several large nucleoli may be seen. Some tumors contain scattered giant tumor cells with greatly enlarged, bizarre nuclei (Fig. 31-12). The cytoplasm may be scanty, but more commonly it is well developed, usually eosinophilic, and with a slightly fibrillar texture in some cells. In some tumors the cytoplasm is amphophilic, pale, or water clear. The latter appearance is most commonly caused by glycogen accumulation. Rarely, large intracytoplasmic vacuoles, negative for mucin, are seen. These clear cell changes, which do not influence the prognosis, may also occur in other histologic types of lung cancer (see p. 1052).

Intercellular bridges ("prickles") appear as multiple and regularly spaced thin, threadlike connections between adjacent tumor cells (Fig. 31-13). They may be more easily identified by slightly racking down the condenser. They indicate the presence of large numbers of well developed desmosomes link-

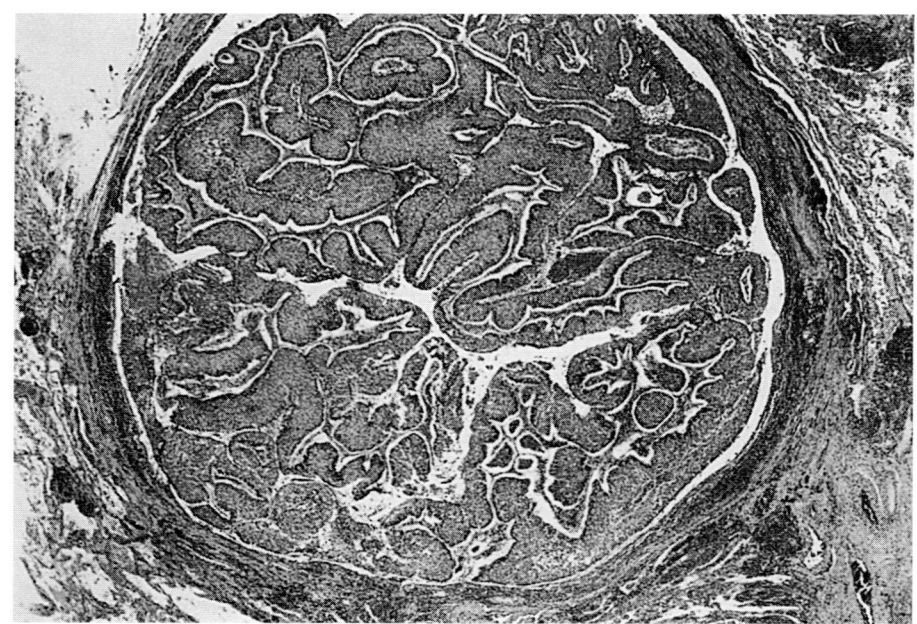

FIGURE 31-10
Squamous cell carcinoma with an endobronchial, papillary growth pattern. The bronchus is greatly expanded to accomodate this well-circumscribed, well-differentiated squamous cell carcinoma with a papillomatous architecture. The surrounding pulmonary parenchyma is compressed, but is not invaded by the tumor.

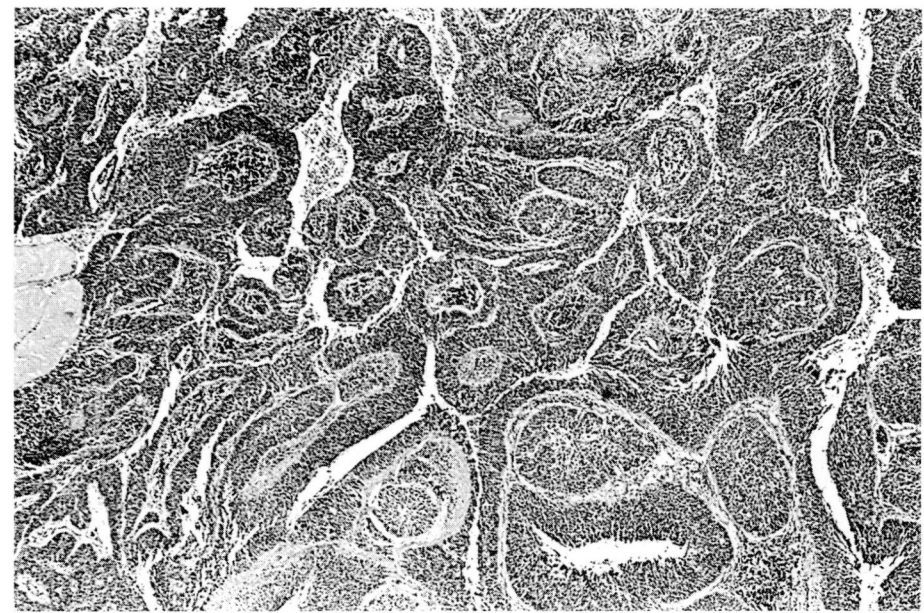

FIGURE 31-11
Squamous cell carcinoma with a papillary architecture. The mutilayering of cells distinguishes this growth pattern, seen in some squamous and large cell carcinomas, from papillary adenocarcinoma (see Fig. 31-32). At higher magnification intercellular bridges were seen, confirming the squamous differentiation.

ing tumor cells to each other (Fig. 31-14 and 31-15). When intercellular fluid accumulates or the cells shrink, as in tissue processing, the cells remain attached to each other via these desmosomes, resulting in the formation of the intercellular bridges. The desmosomes are too small for light microscopic visualization.

Ultrastructurally, the tumor cells show many well developed tonofilament bundles (Fig. 31-16). Light microscopically, these impart the slightly fibril-

lar, eosinophilic quality to the cytoplasm of squamous cells. However, this latter feature is not sufficiently distinctive to be of practical use in lung tumor typing.

Keratinization represents a complex process, leading to the formation of squames, consisting of dead cells with thickened cell membranes and filled with tonofilament bundles. When the process is disorganized, as in squamous cell carcinoma, irregular groups and single cells keratinize. This results in the

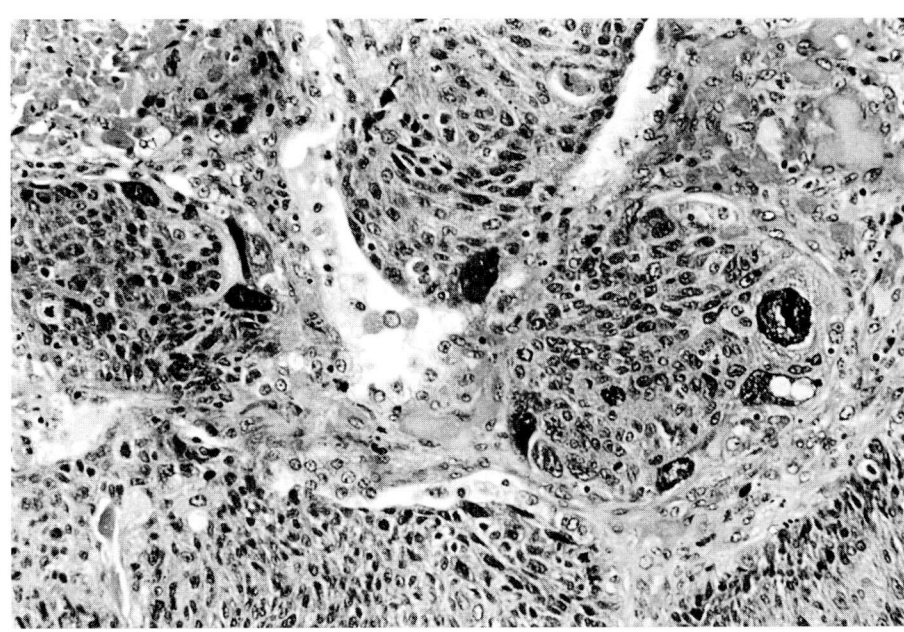

FIGURE 31-12
Squamous cell carcinoma with scattered giant cells exhibits greatly enlarged, bizarre nuclei.

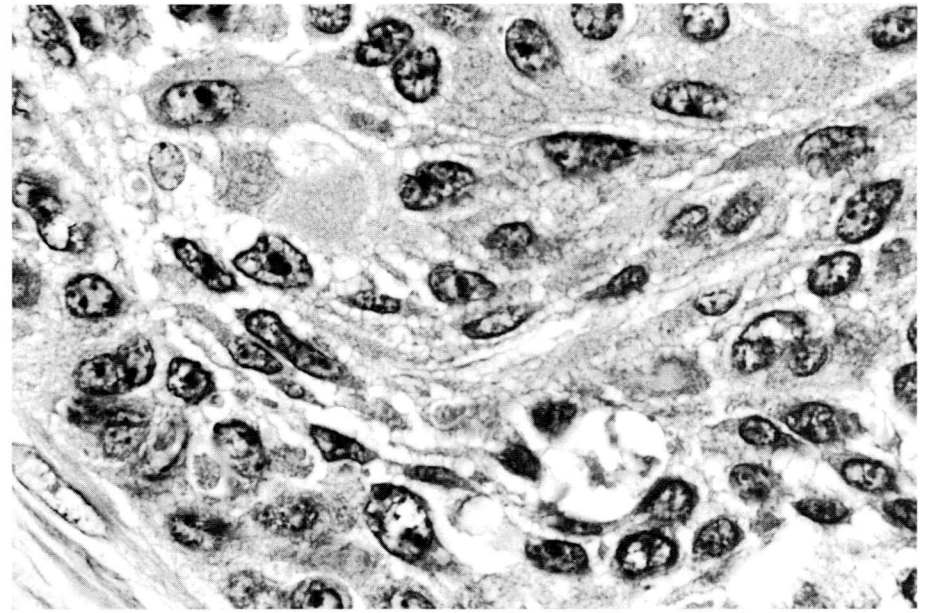

FIGURE 31-13
Squamous cell carcinoma. Intercellular bridges are evident between tumor cells. Their regular spacing, resulting in a resemblance to railway lines, distinguishes them from nonspecific cellular extensions present in carcinomas of all histologic types.

formation of concentrically arranged masses of keratinized cells, so-called epithelial pearls (Fig. 31-17) and single, individually keratinizated, "dyskeratotic" cells.

Keratinized squames, or groups of keratinized cells, are easily recognizable as such. Single dyskeratotic cells must be distinguished from individually degenerating or apoptotic cells. A keratinizing cell exhibits a slightly refractile and strongly eosinophilic cytoplasm. A central wrinkled, shrunken, strongly hyperchromatic nucleus is commonly sur-

rounded by a thin pale halo, which in turn is enveloped by the darkly eosinophilic cytoplasm. Usually, some adjacent tumor cells possess abundant pale eosinophilic cytoplasm with a waxy texture, indicating earlier stages of the keratinization process.

At the ultrastructural level, keratinization is a highly distinctive process.[109] Tonofibrils accumulate in the cytoplasm, while keratohyaline granules and membrane-coating granules (Odland bodies) appear (Fig. 31-18). In the final stage, the sulfhydryl-rich content of the latter is deposited along the inner sur-

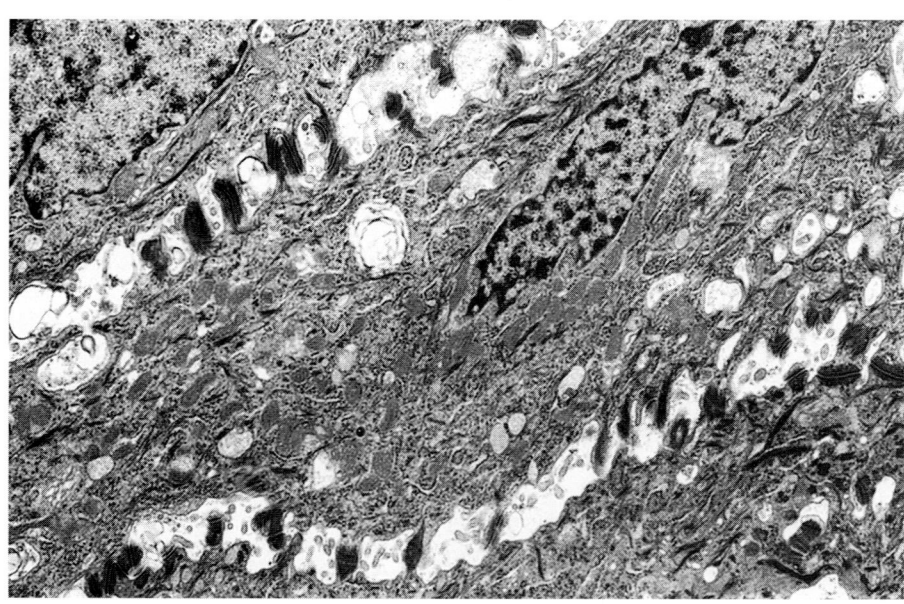

FIGURE 31-14
Squamous cell carcinoma. Electron micrograph demonstrates the presence of many large and well developed desmosomes between adjacent tumor cells and crossing the intercellular spaces.

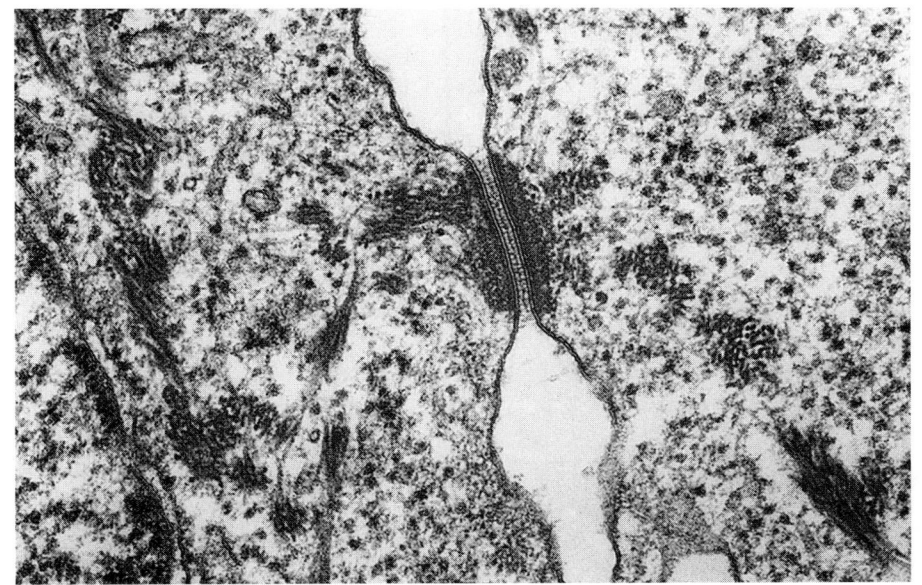

FIGURE 31-15
Electron micrograph of a large and well-developed desmosome indicating squamous differentiation. The distinctive features are the presence of a central electron-dense line between the cell membranes of the two adjacent tumor cells and the tonofilaments attached to the desmosome. Other types of intercellular junctions lack these features.

face of the cell membrane, resulting in marked thickening of this structure (Fig. 31-19) and death of the cell. This then becomes an electron-dense rounded or elongated mass in which remnants of organelles can only just be discerned.

Immunohistochemically, squamous cell carcinomas express involucrin[110] and, to varying degrees, keratins 4, 8, 13, 14, 15, 16, 17, and 18[111] according to Moll's catalogue.[112] Many of these keratins can also be found in other types of lung cancer. However, stable keratin 14 expression is found in all squamous carcinomas including the poorly differentiated ones, whereas pulmonary adenocarinomas and neuroendocrine tumors are consistently keratin 14 negative.[111,113] In addition, expression of the L1 antigen was strongly associated with a squamous phenotype in a study of 139 lung carcinomas of various types.[114] Low molecular weight keratins, carcinoembryonic antigen (CEA), and epithelial membrane antigen (EMA) are also commonly expressed, but they do not indicate squamous differentiation. Many keratin antigens are vulnerable to routine formalin fixation.

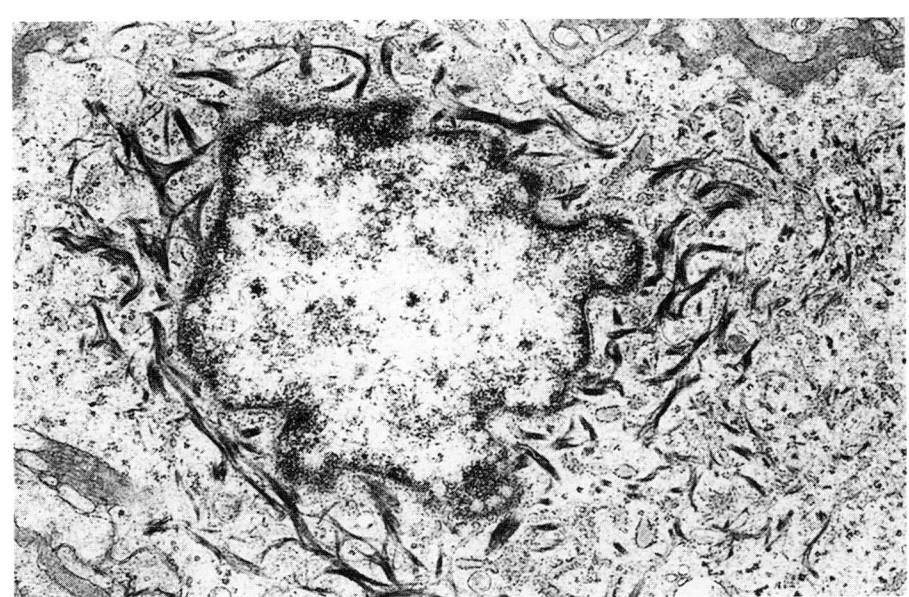

FIGURE 31-16
Squamous cell carcinoma. High power electron micrograph shows many well developed bundles of electron-dense tonofilaments arranged in concentric array around the nucleus. The periphery of the cell contains polyribosomes, a few small mitochondria, and a little endoplasmic reticulum. The paucity of organelles contrasts with the organelle-rich cytoplasm of adenocarcinoma cells.

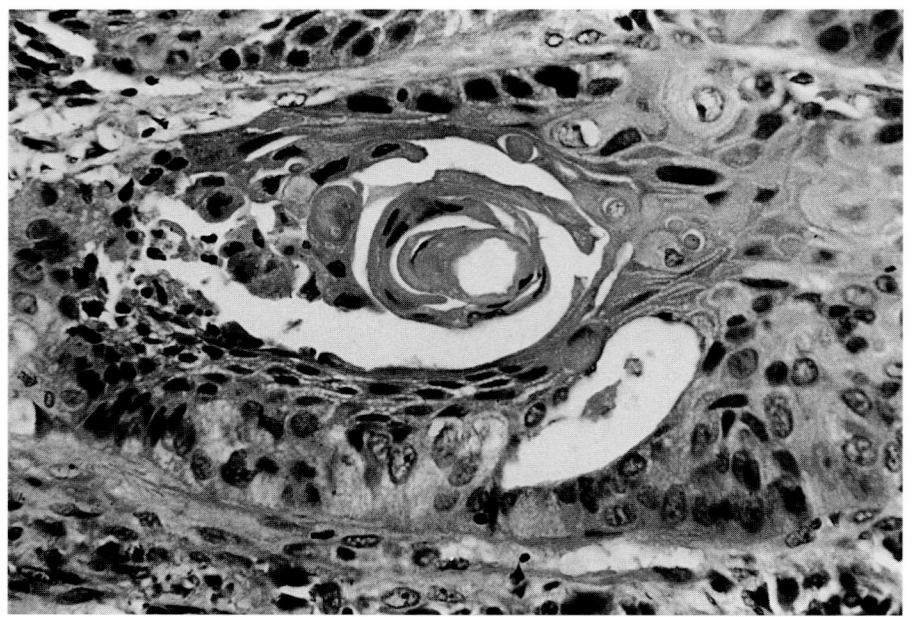

FIGURE 31-17
Epithelial pearl in a squamous cell
carcinoma. The tumor cells are
arranged around a fully keratinized
center consisting of cells with densely
eosinophilic, slightly refractile
cytoplasm and a shrunken, darkly
basophilic pyknotic nucleus.
Intercellular bridging is focally
discernible between the surrounding
tumor cells.

Tubules, papillary structures, and the production of significant amounts of epithelial mucin are by
definition absent. Their presence along with intercellular bridges and keratinization points to a diagnosis
of adenosquamous carcinoma (see p. 1048). A subtle
degree of glandular differentiation, usually
evidenced by the presence of small traces of
intracellular mucin, is commonly present in peripheral squamous cell carcinomas. This is usually considered insufficient to establish a diagnosis of
adenosquamous carcinoma.

True glandular differentiation by tumor cells
should be distinguished from entrapment of alveolar
epithelium (Fig. 31-20), a common phenomenon
resulting in the presence of epithelial-lined tubules
within the tumor.[115,116] Entrapped alveoli are most
common in tumors that induce relatively small
amounts of stroma, and they are preferentially
located along stromal septa. Some of the alveolar
remnants are very small, and single surviving type II
cells may even form intracellular lumina (Fig. 31-21).
Because the entrapped alveolar type II cells may

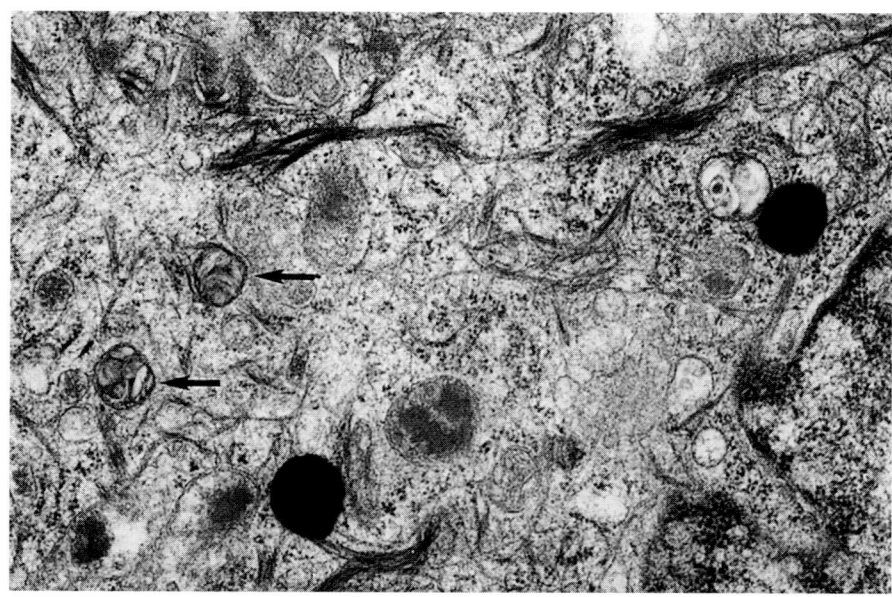

FIGURE 31-18
High power electron micrograph of the
cytoplasm of a squamous cell
carcinoma tumor cell. Amid the
tonofibrils, two very dark keratohyalin
granules and several membrane-
coating granules (*arrows*) with an
irregularly lamellar content are
present. The latter granules should not
be mistaken for surfactant granules,
which would indicate either glandular
(pneumocytic) differentiation or
entrapped alveolar cells.

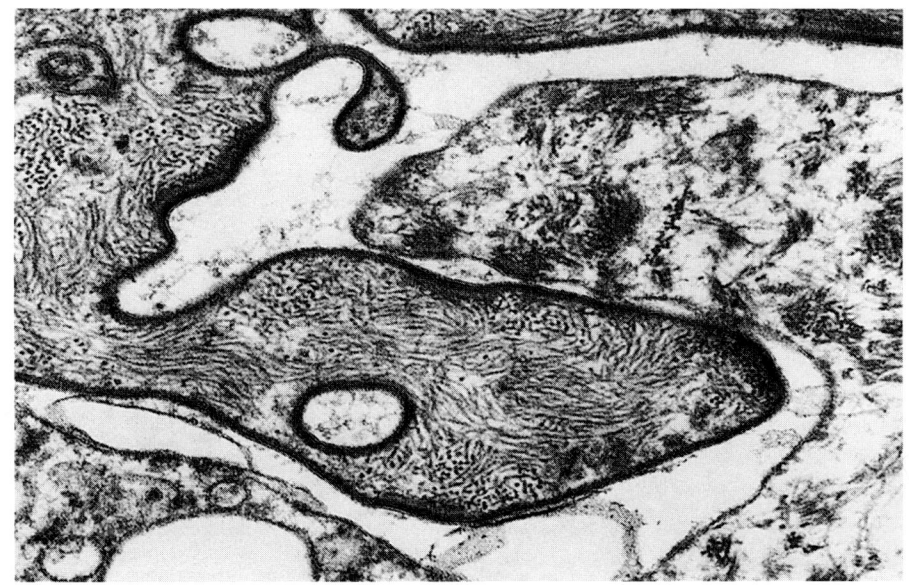

FIGURE 31-19
Keratinized squamous cell carcinoma tumor cell (*upper left and center*) exhibits a greatly thickened cell membrane due to apposition of electron-dense material (derived from membrane-coating granules) along its inner surface.

exhibit some reactive nuclear atypia, careful evaluation of the cytologic detail is necessary to recognize the two cell populations: neoplastic squamous cells with a solid growth pattern, on the one hand, and reactive epithelial cells lining the tubular structures, on the other. Ultrastructurally, the lumina of entrapped alveoli contain surfactant (Fig. 31-22) and the cytoplasm of the lining cells generally contains lamellated surfactant granules.

Necrosis in the center of solid tumor nodules, with subsequent dissolution and resorption of the necrotic debris, may result in structures resembling glandular differentiation. This may cause the erroneous diagnosis of adenosquamous carcinoma.

Apart from the characteristics of squamous differentiation, electron microscopy commonly shows a subtle degree of concomitant glandular differentiation (see p. 1034). However, such minor degrees of glandular differentiation do not change the diagnosis to adenosquamous carcinoma. The practical consequences of these subtle distinctions are limited. Squamous cell and adenosquamous carcinoma are treated identically. Although the prognosis of adenosquamous carcinoma, as a group, is slightly

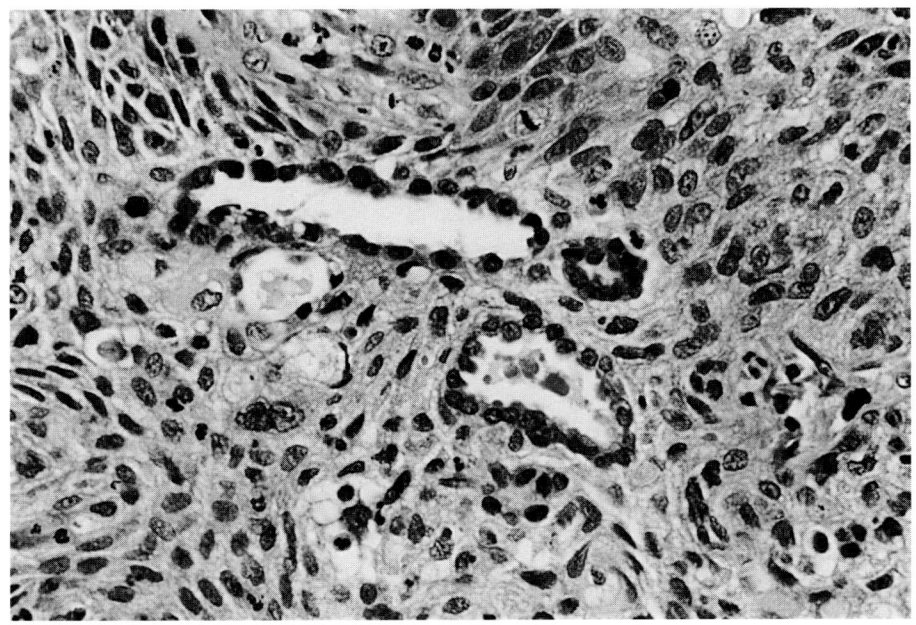

FIGURE 31-20
Remnants of alveoli lined by type II pneumocytes, buried deeply within a squamous cell carcinoma. The pneumocytes are usually smaller and more regular than the surrounding carcinoma cells. These tubular structures are commonly found next to stromal septa, most commonly in tumors lacking a marked desmoplastic stromal response. These features aid in their distintion and prevents an incorrect diagnosis of adenosquamous carcinoma.

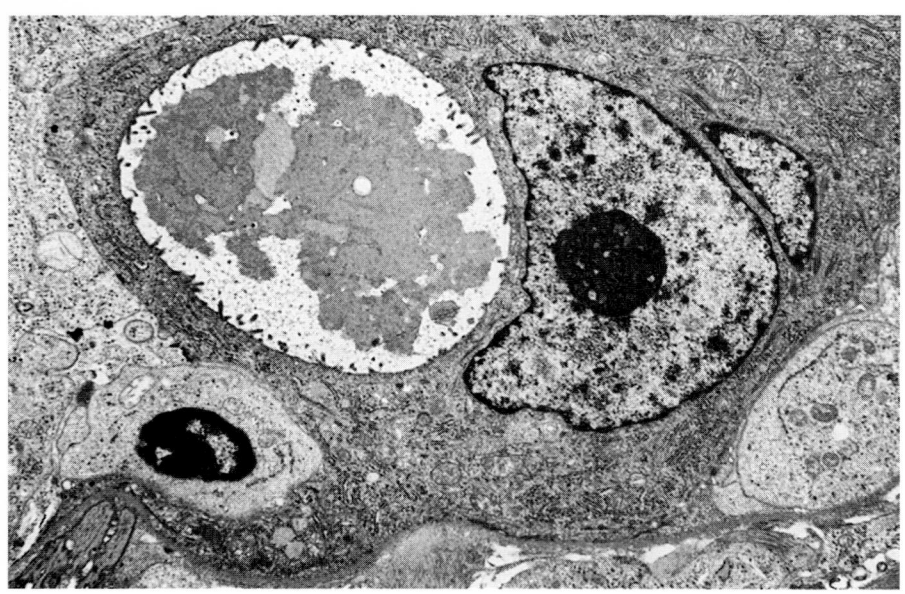

FIGURE 31-21
Alveolar pneumocyte entrapped within a squamous carcinoma. This cell forms an intracytoplasmic lumen containing secretory material.

worse than that of squamous cell carcinoma, there is much variation in tumor aggressiveness in both groups, and the differences are therefore irrelevant at the individual level.

Early necrosis is often confined to the centers of tumor nodules, farthest removed from the nutrient stromal vessels. Large geographic areas of necrosis, also affecting the stromal tissues, are present in many squamous cell carcinomas. They are presumably the result of tumor growth exceeding the rate of neovascularization and of fibrotic narrowing of vascular lumina due to obliterative intimal fibrosis.

In some tumors, the stroma contains multinucleated histiocytic giant cells (Fig. 31-23). This feature has no apparent clinical relevance. The histiocytic giant cells should be distinguished from tumor cells with multiple nuclei or bizarre nuclear shapes (Fig. 31-24).

Grading of squamous cell carcinoma is based on a subjective assessment of the degree of maturation (extent of keratinization) and the degree of cellular atypia, as evidenced by nuclear enlargement, hyperchromasia, anisonucleosis, and mitotic activity. The degree of maturation of most pulmonary squamous

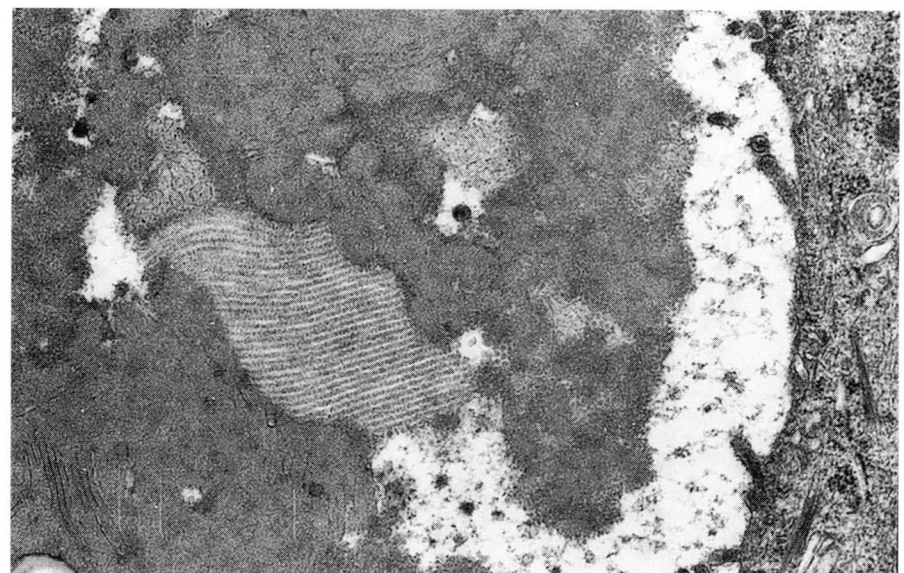

FIGURE 31-22
Secretory material in lumen lined by entrapped pneumocyte. Part of the secretory material shows a characteristic gridlike appearance indicating that it consists of surfactant rather than mucin.

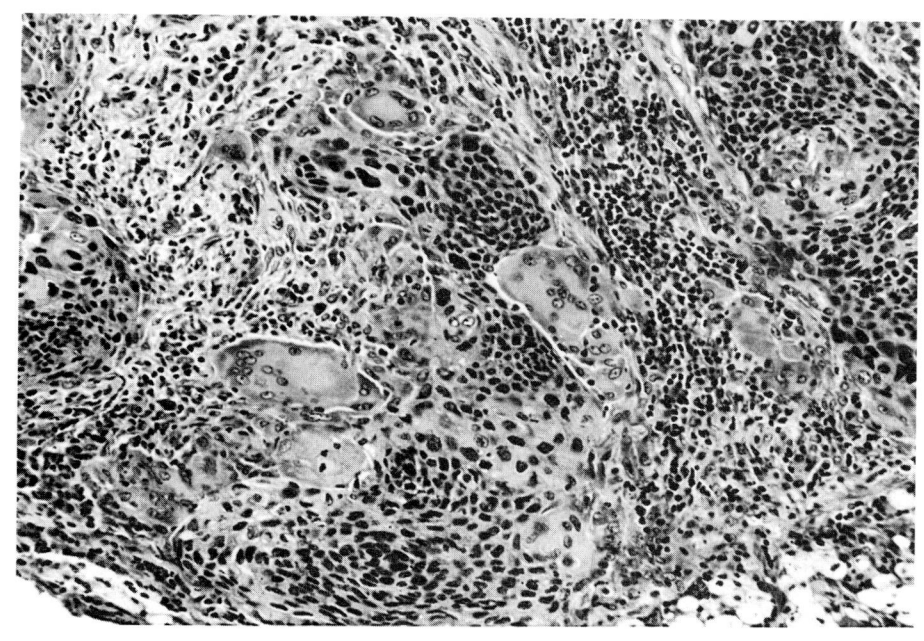

FIGURE 31-23
Squamous cell carcinoma with mutinucleated histiocytic stromal giant cells.

cell carcinomas is less than that seen in most similar tumors arising in squamous epithelia, such as the epidermis or the oral mucosa. As a consequence, most tumors are considered moderately or poorly differentiated.

Areas of dedifferentiation to a undifferentiated phenotype, with spindle or pleomorphic cells, similar to the tumor cells of pleomorphic malignant fibrous histiocytoma, is occasionally seen.[117] This phenomenon will be discussed later (see p. 1050). True metaplasia to a specific mesenchymal cell type (e.g., chondrocyte or rhabdomyoblast) is rare. Such carcinosarcomas are discussed in Chap. 32.

Occasionally, squamous cell carcinomas consist of small, relatively monomorphic cells with little cytoplasm and medium-sized, dark nuclei. Such tumors may resemble SCLC to such a degree that some authors have included them under the heading of SCLC.[118] There are subtle light microscopic differences between these tumors and typical neuroendocrine SCLCs. The squamous cell carcinomas with a small cell type show less nuclear molding, less crush

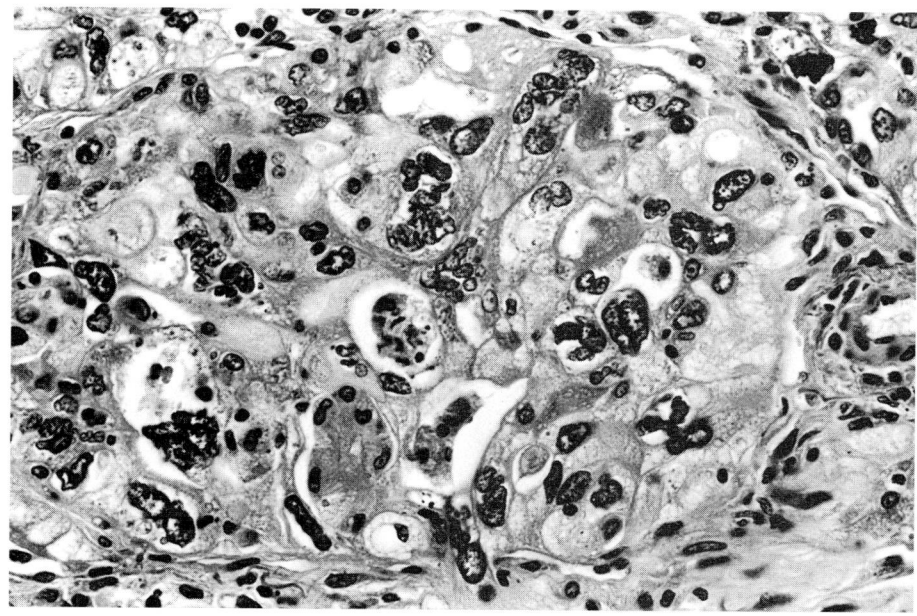

FIGURE 31-24
Squamous carcinoma with bizarre multinucleated tumor cells. These should be distinguished from the multinucleated histiocytic giant cells depicted in Fig. 31-23.

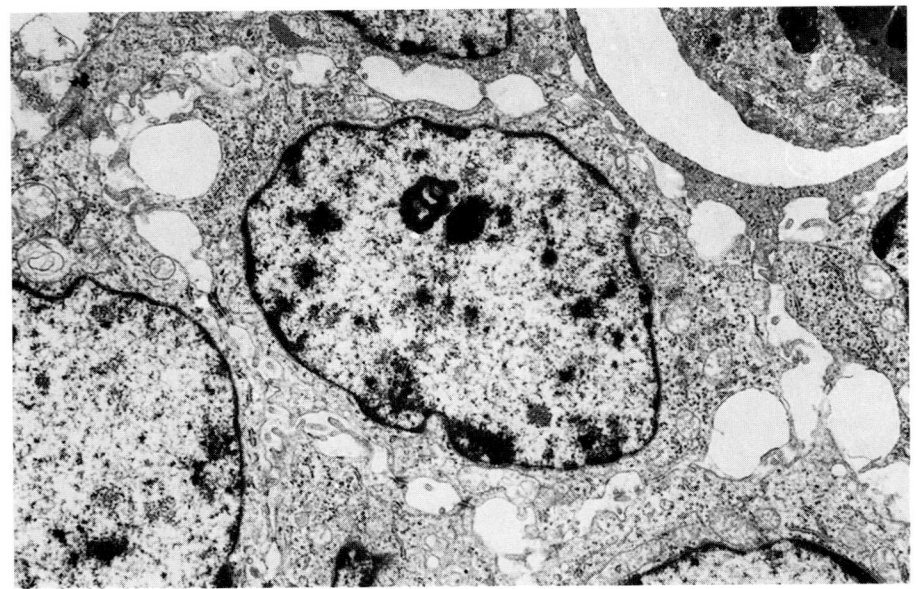

FIGURE 31-25
Poorly differentiated squamous cell carcinoma with small tumor cells. Neuroendocrine granules are absent. There is a subtle degree of intercellular bridging based on the presence of well developed desmosomes, indicating minor squamous differentiation.

artifact, and more crisp and darkly staining cytoplasm, sometimes with discrete cell borders. Focal peripheral palisading of tumor cells may be seen. If in doubt, electron microscopy will reveal the differentiation type (Fig. 31-25 and 31-26). This is not an academic exercise because the morphologic resemblance of these squamous cell carcinomas to small cell tumors is not paralleled by their clinical behavior. In our experience, such tumors exhibit a biologic behavior more indicative of non-SCLC than of SCLC.[119] Recently, Brambilla and colleagues[120] proposed a separate group of "basaloid carcinomas," which included these squamous carcinomas with adenocarcinomas and large cell carcinomas, which also resembled small cell carcinoma. This group of tumors is considered later in the chapter (see p. 1048).

Differential Diagnosis

Apart from the possible confusion with SCLC, the differential diagnosis usually poses no problems. Metastasis from a known or occult squamous cell carcinoma from the head and neck region or the uterine cervix should be excluded on clinical grounds. Rarely, malignant thymoma extending into the pleura[121] or metastatic to the lung, or the rare pri-

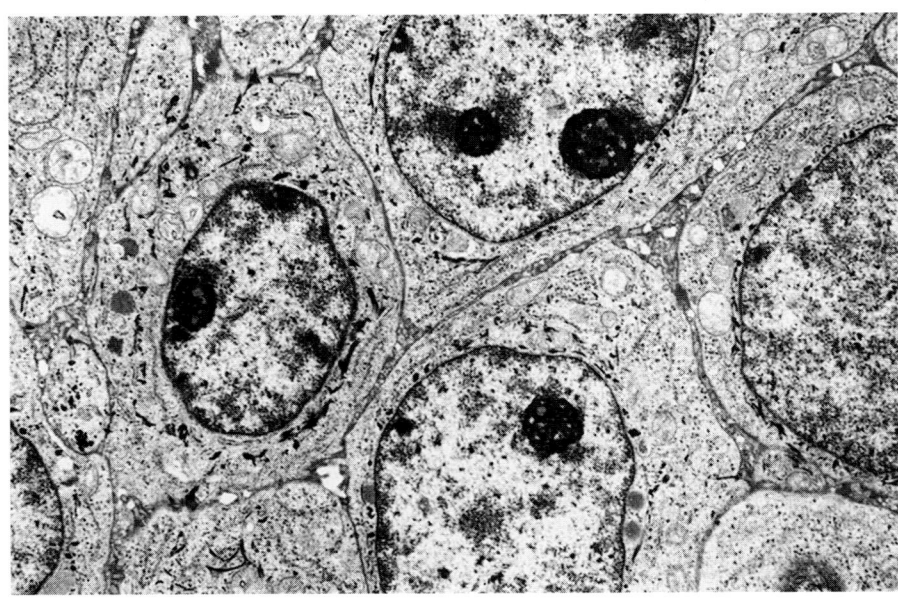

FIGURE 31-26
Squamous cell carcinoma with small tumor cells, resembling small cell carcinoma. Neuroendocrine granules were completely absent. Ultrastructurally, tonofilaments bundles are seen, indicating squamous differentiation.

mary pulmonary thymoma[122] should be considered. The diagnosis of pulmonary adenosquamous carcinoma rests on the detection of concomitant glandular differentiation in the tumor. In rare instances, pleural mesothelioma enters the differential diagnosis.

Some squamous cell carcinomas closely resemble angiosarcoma because of the presence of irregularly anastomosing spaces containing erythrocytes, lined by atypical epithelioid cells.[123,124] Focally, the typical features of squamous cell carcinoma are identified, and the tumor cells are negative for endothelial markers such as von Willebrand factor, CD 31, and CD 34, and do not bind Ulex europaeus I lectin.[124]

ADENOCARCINOMA

In many countries, there has been a recent rise in the relative incidence of adenocarcinoma, mainly at the expense of squamous cell carcinoma. Bronchioloalveolar carcinoma, a subtype of peripheral adenocarcinoma with characteristic histologic features (see p. 1037) appears to account for a large part of this increased incidence. In one study, the proportion of bronchioloalveolar carcinomas among resection specimens rose from 5 percent in 1955 to 24 percent in 1990.[125] In the United States and Japan, adenocarcinoma has become the most common type of lung cancer. In Great Britain and several other European countries, squamous cell carcinoma remains the most common lung cancer type. The causes of these shifts in time and differences between countries are unclear. A disproportionate number of adenocarcinomas are seen among females, patients under 50 years of age, and nonsmokers. Nonetheless, the large majority of adenocarcinomas arise in smokers or ex-smokers.

Most pulmonary adenocarcinomas are peripheral, so that local symptoms develop relatively late. Consequently, a relatively large number are discovered incidentally on routine chest radiographs or first manifest themselves through symptoms caused by metastases or the general symptoms of malignant disease. Some mucinous bronchioloalveolar carcinomas present with consolidation of large parts of one or both lungs, leading to increasing respiratory dysfunction and the production of copious amounts of mucoid sputum.

Pathology

Macroscopically, the tumor is moderately firm or soft. Mucin-producing tumors often have a glistening cut surface and a mucoid consistency. There may be a central scar, which has led previous investigators to assume that such tumors arose in an area of fibrosis (see p. 1040). Cavitation is rare. The overlying pleura is often indented. Adenocarcinomas of the bronchioloalveolar variety may consolidate large areas of one or both lungs or form multiple nodules. Rarely, primary pulmonary adenocarcinoma presents with a diffuse or multinodular pleural thickening similar to diffuse pleural mesothelioma. Conversely, pleural mesothelioma occasionally presents as a localized pleural mass.[126] Because of this, the possibility of pleural mesothelioma should routinely be considered in the differential diagnosis of adenocarcinoma involving the pleura (see p. 1036).

Multicentric adenocarcinoma of the lung is not uncommon and may become apparent only at dissection of the surgical specimen. McElvaney and colleagues[63] found two or more tumors in 12 of 62 lung adenocarcinoma resections specimens. In only 2 of these 12 instances was the presence of a second tumor suspected preoperatively.

Histologically, adenocarcinoma is characterized by glandular differentiation, which is hallmarked by the formation of tubular or papillary structures, or the synthesis of epithelial-type mucins. In tumors with a solid growth pattern, mucin production suffices for a classification as adenocarcinoma.

The histologic spectrum of adenocarcinoma is varied, probably more than any other lung cancer. Some tumors produce large amounts of mucin or consist largely of signet ring cells with a diffuse growth pattern.[127] Apart from metastatic adenocarcinoma, often from an abdominal site, the differential diagnosis of these rare tumors should include metastatic melanoma, which may present in the lung with conspicuous signet ring-shaped tumor cells and which may be S-100 negative.[128] Positivity with the monoclonal antibody HMB-45 and negativity for keratins and epithelial mucins indicate the correct diagnosis.

The tumor cells commonly show significant variability in size, shape, nuclear features, and cytoplasmic staining characteristics. Indeed, the histologic features of primary adenocarcinoma of the lung are almost as varied as that of metastatic adenocarcinoma in the lung. As a consequence, the histologic morphologic features generally do not allow a confident distinction between primary and metastatic adenocarcinoma (see p. 1056).

Attempts to subclassify adenocarcinoma of the lung into a number of distinct subtypes have suffered from these variations in histologic appearance within individual tumors. Because the growth pattern and

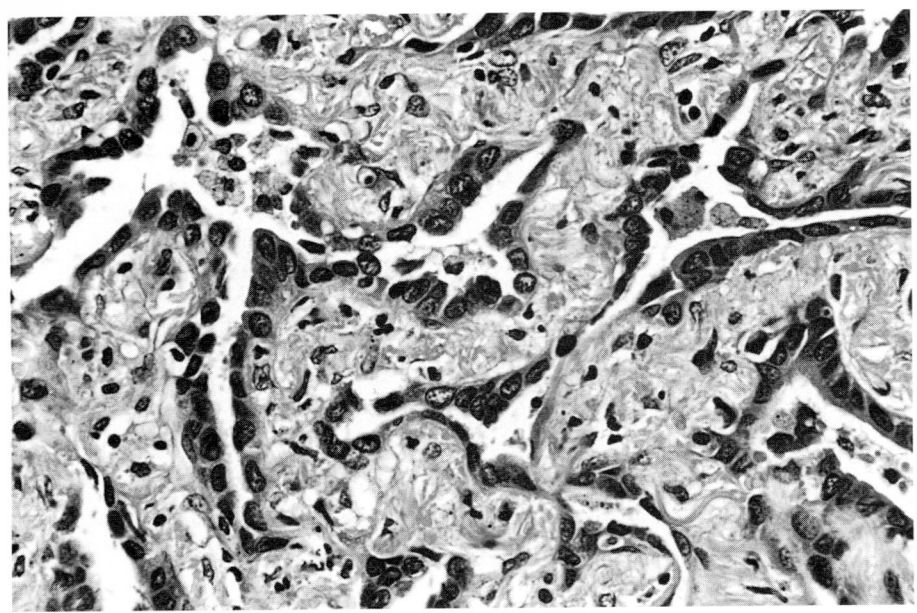

FIGURE 31-27
Adenocarcinoma exhibiting irregular, angular tubules lined by small atypical cuboidal to low columnar tumor cells. See also Fig. 31-28.

cell type may vary significantly between areas of the same tumor (Fig. 31-27 and 31-28), attempts at subtyping pulmonary adenocarcinoma have met with limited success. Acinar, papillary, and solid growth patterns, which are the bases of specific subtypes of adenocarcinoma in the WHO classification, occur frequently in different areas of the same tumor. Endoscopic biopsy specimens or similar small tissue fragments certainly do not allow subtyping or grading of adenocarcinoma. Edwards,[129] reviewing 106 pulmonary adenocarcinomas, provided a more simple subclassification into parenchymal adenocar-

cinoma (67 percent), bronchial adenocarcinoma (13 percent), and adenocarcinoma of uncertain origin (20 percent). Interestingly, this relatively straightforward subclassification may have prognostic implications, because in his series, the bronchial adenocarcinomas were associated with a shorter postoperative survival.

The tubules and acinar structure may be quite regular or very irregular and angulated in outline (Fig. 31-29). Occasionally, the features are similar to adenocarcinoma of the large intestine, including pseudostratification of the lining columnar cells, the

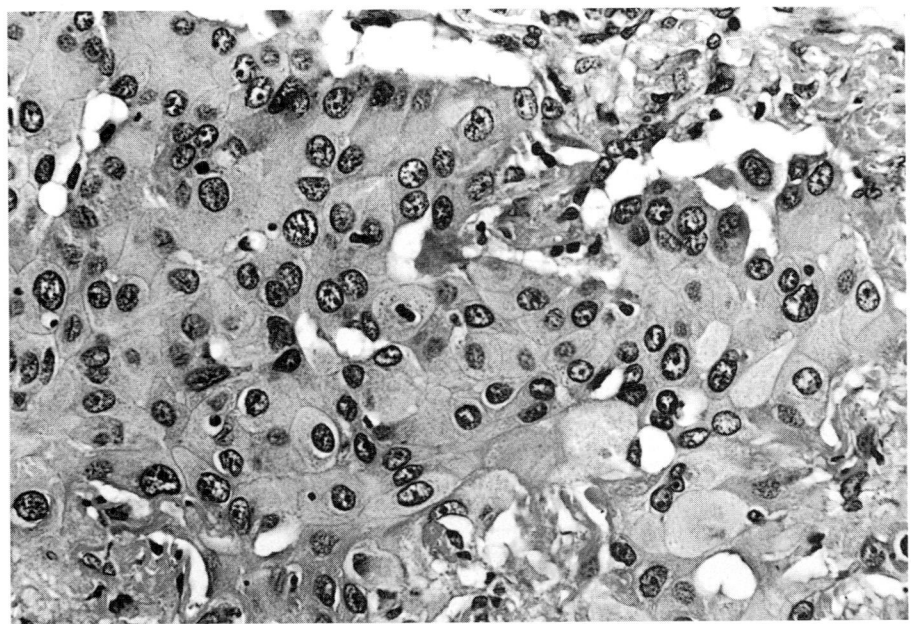

FIGURE 31-28
Same tumor as depicted in Fig. 31-27. In this field, the tumor cells are larger and more pleomorphic, and the growth pattern is solid. Such variations in cell type and growth pattern, which are relatively common in pulmonary adenocarcinoma, preclude a simple subdivision of adenocarcinoma into well defined subtypes.

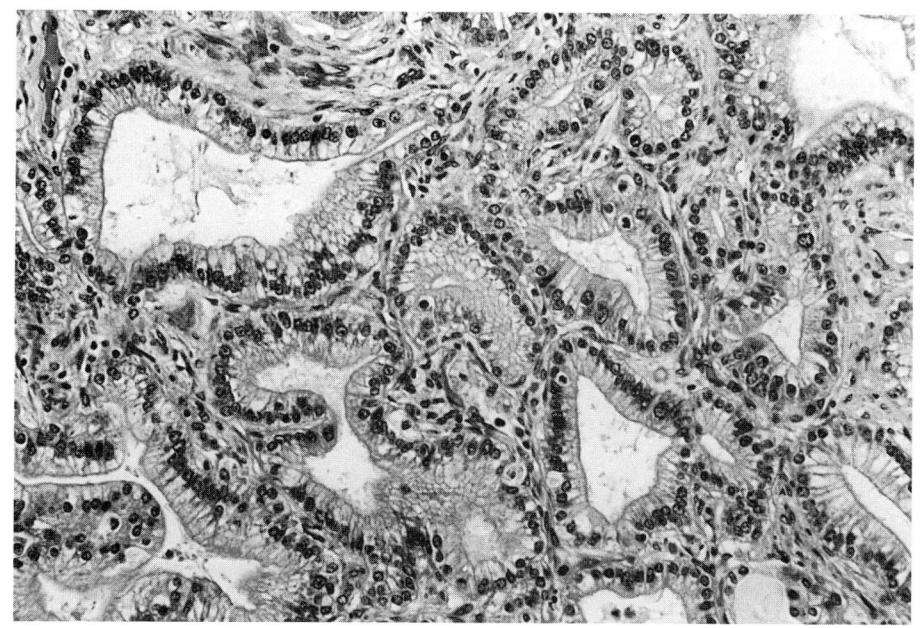

presence of "dirty" cellular debris with basophilic nuclear remnants within the lumina, and necrosis of tumor cells along part of the circumference of the tubule (Fig. 31-30). Although such findings should alert the pathologist to the possibility of an occult abdominal tumor, they are also compatible with a primary tumor of the lung. Cribriform (Fig. 31-31) and papillary structures (Fig. 31-32) are common in pulmonary adenocarcinomas.

Epithelial-type mucins are demonstrated with periodic acid-Schiff or mucicarmine stains. Alcian blue positivity is seen in many adenocarcinomas but also in mesothelioma (see p. 1036). Because mucin production is sufficient evidence to conclude that glandular differentiation is present, the appropriate stains are mandatory for typing all pulmonary large cell carcinomas and squamous cell carcinomas. Occasionally, brightly eosinophilic intracytoplasmic globules of uncertain nature are seen, especially near areas of necrosis. These globules appear to represent accum-ulation of glycoproteins within the endoplasmic reticulum.[130]

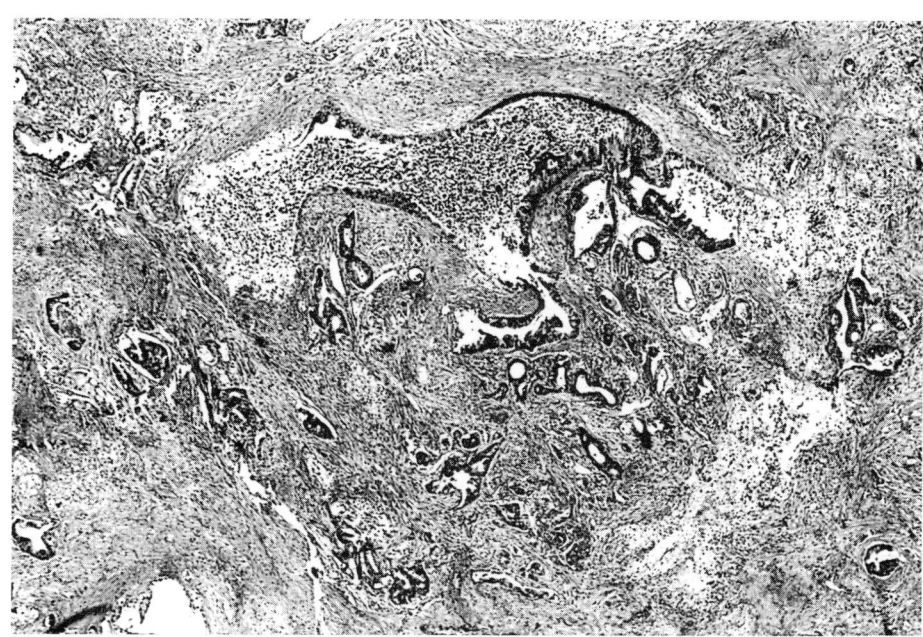

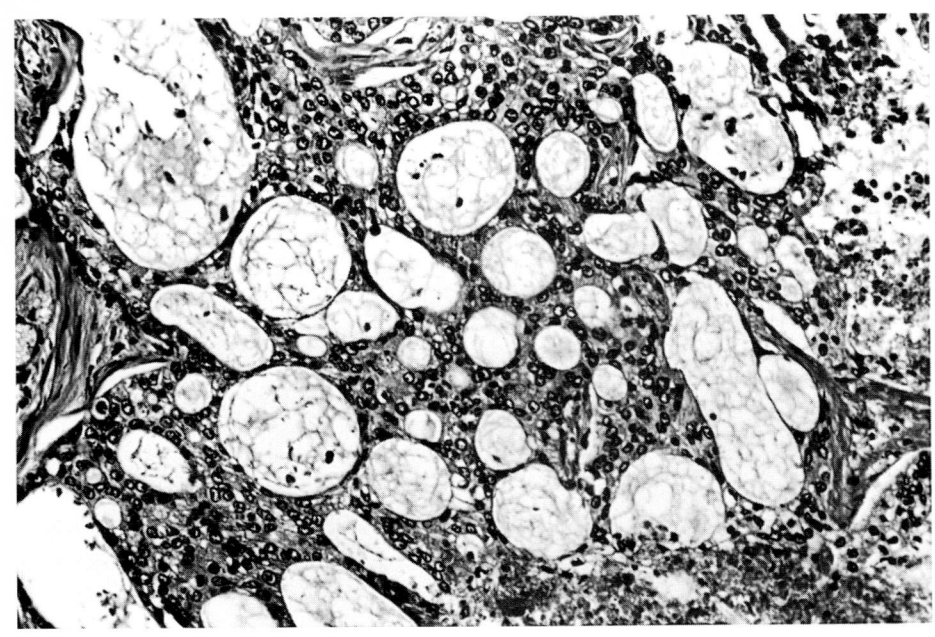

FIGURE 31-31
Cribriform growth pattern in
adenocarcinoma. An irregular nodule
of tumor cells contains multiple
rounded lumina, some of which
contain secretory material.

Ultrastructurally, the tumor cells exhibit intra-
or intercellular lumina, mucin production, or Clara
cell or type II pneumocyte differentiation.[99] Many
adenocarcinomas form tubules lined by multiple
tumor cells (Fig. 31-33) or much smaller lumina,
which lie entirely within one tumor cell (Fig. 31-34).
The luminal side of the cell membrane is characterized by the presence of numerous microvilli, which
are often slender and curved, although not to the
degree seen in mesothelioma. The microvilli may be
short and stubby, as in adenocarcinomas of the gut.

Interestingly, microvilli projecting into intracellular
lumina are generally very straight (see Fig. 31-34).
The cytoplasm is usually organelle rich and contains
many mitochondria, a well-developed rough endoplasmic reticulum and Golgi system, as well as secretory granules. Mucin granules contain flocculent
material and show a tendency to coalesce. Serous
granules, resembling the granules of bronchiolar
Clara cells, are more electron dense and remain intact
as single units. Tumors with type II pneumocyte differentiation show characteristic intracytoplasmic

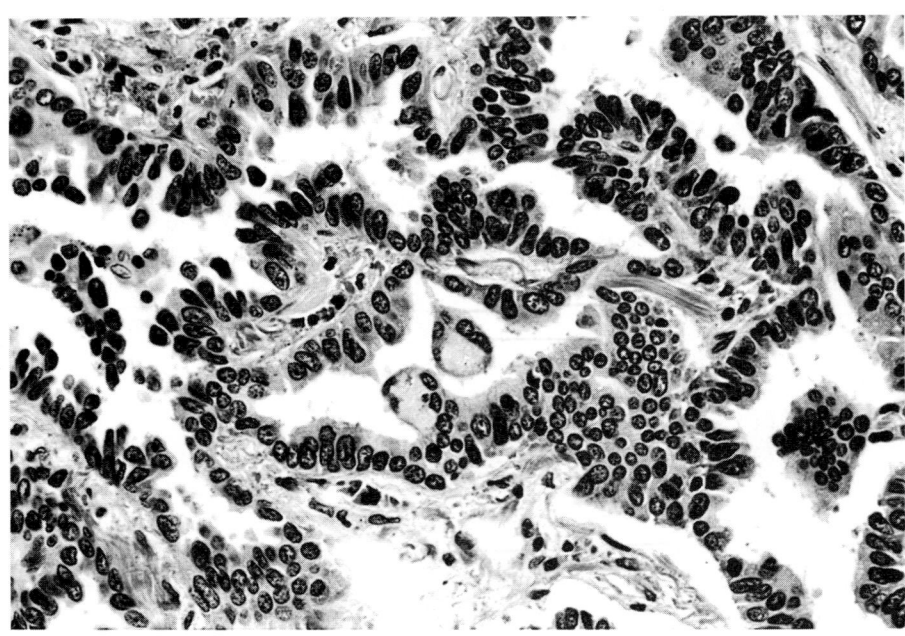

FIGURE 31-32
Adenocarcinoma with a papillary
growth pattern. Small columnar tumor
cells line slender stromal papillae.

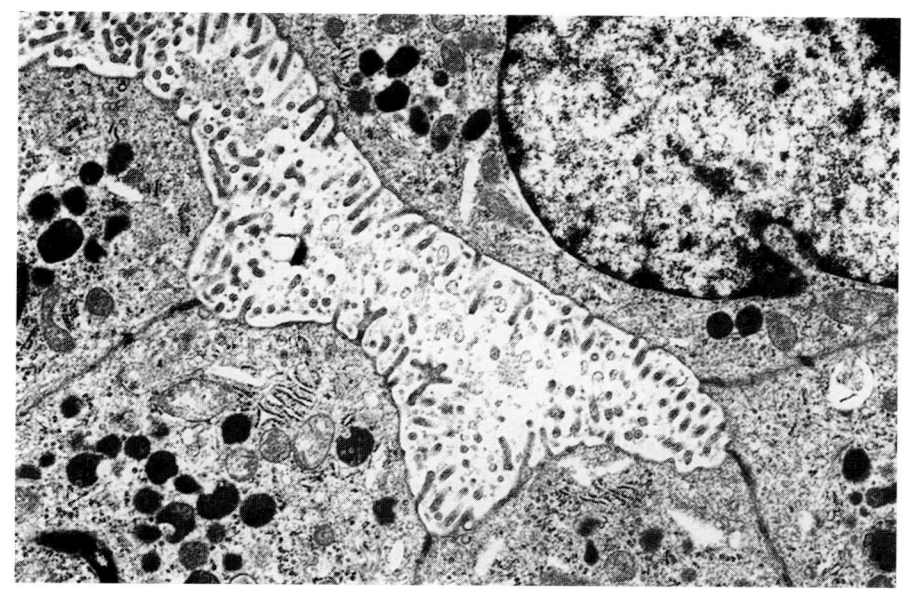

FIGURE 31-33
Adenocarcinoma. Electron micrograph shows an intercellular lumen (tubule) lined by tumor cells. Many microvilli project into the lumen, which contains some secretory material. In the organelle-rich cytoplasm of the tumor cells, multiple electron-dense exocrine secretory granules are seen.

lamellated surfactant granules. Occasionally, small exocrine secretory granules closely resemble neuroendocrine granules (see p. 1042). Their presence near the apical cell membrane with its characteristic microvilli, as well as their more irregular sizes, contours, and contents, allow their recognition as exocrine granules (Fig. 31-35).

Differential Diagnosis

Similar to some squamous tumors (see p. 1029), an occasional adenocarcinoma shows small and relatively monomorphic tumor cells, with scanty cytoplasm and a solid growth pattern, so it has to be distinguished from small cell carcinoma. Small amounts of mucin as well as the presence of small lumina, although not completely excluding small cell carcinoma of the "combined" subtype, argue in favor of adenocarcinoma. The nuclei are slightly more vesicular, with a more distinct nucleolus in paraffin sections, or they show a slightly more granular chromatin pattern. Nuclear molding is less apparent than in SCLC. In small biopsies, crush artifact is less conspicuous than in most SCLCs, lymphomas, and inflammatory infiltrates. In case of doubt, elec-

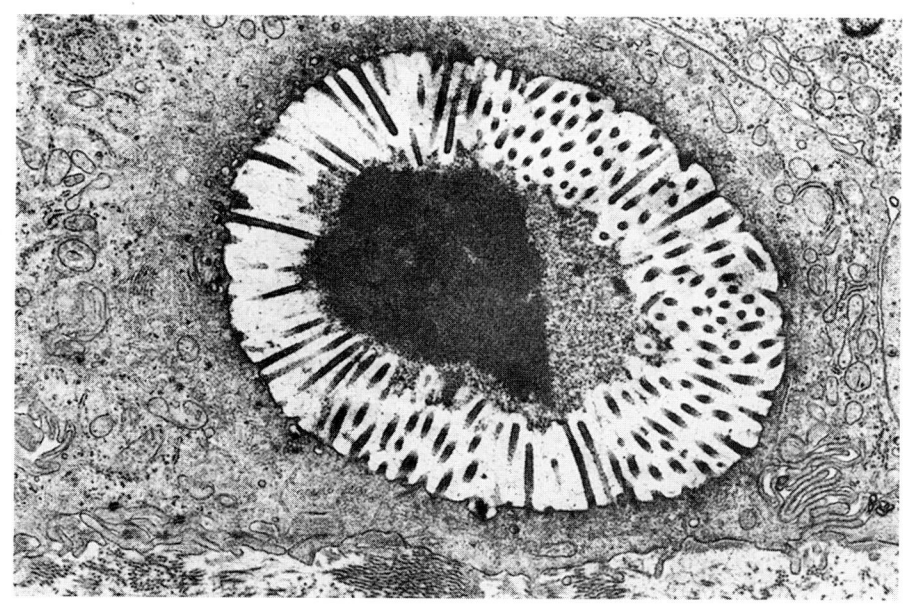

FIGURE 31-34
Adenocarcinoma, electron micrograph. Conspicuous straight microvilli project into an intracellular lumen, in which some secretory material has accumulated.

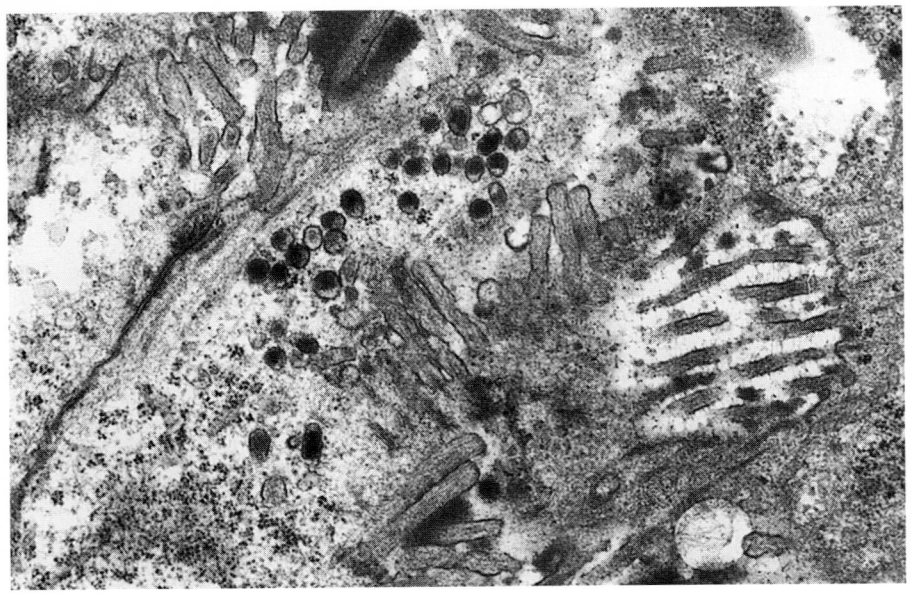

FIGURE 31-35
High power electron micrograph shows small electron-dense secretory granules in an adenocarcinoma cell. The presence of straight, intestinal-type microvilli (some with rootlets, *far right*) indicates the apical location of these granules. Their content is irregular, with an eccentric electron-dense core. These features contrast with the appearances of neuroendocrine granules (see Fig. 31-45).

tron microscopy will usually reveal the histologic type.

A biphasic architecture, with a glandular and spindle cell component, is occasionally present in peripheral adenocarcinoma of the lung and is especially significant with respect to the differential diagnosis of pleural mesothelioma.[131] This differential diagnosis, which generally requires the application of special techniques, is discussed in the next section.

Some rare salivary gland-type tumors, probably arising in bronchial glands, enter the differential diagnosis of pulmonary adenocarcinoma. Primary adenoid cystic carcinoma of the lung[132] mostly shows an endobronchial growth pattern, leading to bronchial obstruction. Histologically, cribriform, tubular, and solid growth patterns often occur in combination. There may be a prominent myxoid background. The tumor cells are relatively small and monomorphic; mitoses are absent or rare. A prominent myoepithelial component is hallmarked by expression of S-100, actin, vimentin, and high molecular weight keratin and aids in the distinction from adenocarcinoma. Benign and malignant salivary gland-type mixed tumors[133] are usually associated with a bronchus and may grow endobronchially. A biphasic architecture and a prominent myxoid and often chondroid matrix are important histologic features. Expression of keratins, but also actin and vimentin, and variable positivity for S-100 and glial fibrillary acidic protein aid in the distinction. In the series of Moran and colleagues,[133] two of eight cases showed hemorrhage, necrosis, and extensive invasion of surrounding lung tissue; these tumors later recurred.

Rare tumors entering the differential diagnosis include the pulmonary mucinous cystic tumor recently described by Dixon and coworkers[134] and the so-called pulmonary endodermal tumor, which is probably related to pulmonary blastoma and resembles the tubular stage of the fetal lung; the optically clear nuclei of these tumors were recently found to contain biotin, which may prove to be a diagnostically useful feature.[135] Finally, a pulmonary metastasis of a known or clinically occult adenocarcinoma elsewhere in the body can closely mimic primary pulmonary adenocarcinoma.

Immunohistochemical Features: Differential Diagnosis with Pleural Mesothelioma (Chap. 34)

Adenocarcinomas typically express keratins of simple epithelia: CK 7, 8, 18, and 19.[111] Other keratins are also not uncommonly found. They may indicate the sensitivity of immunohistochemistry in detecting low levels of concomitant squamous differentiation. Extensive expression of keratin 7[136] and absence of keratin 14[111] are probably the most reliable indicators of glandular differentiation and absence of squamous differentiation. Dipeptidyl aminopeptidase IV (DAP IV) was detected in 27 of 29 adenocarcinomas and in none of 16 lung carcinomas of other histological types in a recent study.[137] Vimentin is expressed to detectable levels in a minority of adenocarcinomas,[138] an important point with respect to the differential diagnosis of pleural mesothelioma, which is typically vimentin positive. Surfactant apoproteins or Clara cell protein can be demonstrated positive in a substantial number of adenocarcinomas.[139–141] With monoclonal antibody PE-10, surfactant apoprotein immunoreactivity was demonstrated in paraffin sections of 36 of 75 pulmonary adenocarcinomas but in none of 37 other lung tumor types or metastatic tumors from extrapulmonary primaries in one study.[140] PE-10 is positive in 10 percent of large cell undifferentiated adenocarcinomas, 20 percent of small cell carcinomas, and 40 percent of atypical carcinoids.[140a] Most of the positive adenocarcinomas showed features of Clara cell differentiation. Expression of secretory component was reported in 25 of 40 adenocarcinomas and none of 11 mesotheliomas by Kondi-Paphitis and Addis.[142]

Carcinoembryonic antigen is expressed in about three quarters of pulmonary adenocarcinomas, contrasting with pleural mesotheliomas, which are usually CEA negative (Chap. 34). Cross-reaction with nonspecific cross-reacting antigen, which may be present in malignant mesothelioma, may pose a problem when an antiserum lacking complete CEA specificity is used. In addition, monoclonal antibodies Leu-M1, Ber-EP4, and especially B 72.3, which are positive in most adenocarcinomas, but negative in the large majority of pleural mesotheliomas, are useful in this differential diagnosis.[143–148] Monoclonal antibody MOC-31, recognizing a membrane glycoprotein present on carcinoma cells of various organs and histologic types, was negative on reactive or neoplastic mesothelial cells. This may prove to have a similar diagnostic application.[149] Monoclonal antibody K1, applicable only on frozen sections, was found to stain all 19 epithelial or mixed mesotheliomas but none of 23 pulmonary adenocarcinomas investigated.[150] Expression of ABH blood group-related antigens and *Helix pomatia* agglutinin is common in well differentiated pulmonary adenocarcinoma but was absent in all 5 reactive and 29 malignant mesothelial lesions tested by Kawai and coworkers.[151]

Positivity with the periodic acid-Schiff stain after diastase pretreatment argues strongly in favor of adenocarcinoma, but may not, by itself, provide conclusive evidence to rule out mesothelioma.[152] Hyaluronic acid is commonly present in pleural mesothelioma, but may also be present in adenocarcinoma.

The ultrastructural features of mesothelial cells are sufficiently distinctive to be of help in the distinction between adenocarcinoma and mesothelioma. Mesothelial cells, including epithelial mesothelioma cells, possess large numbers of very long, wavy, and sometimes branching microvilli, which differ from those seen in adenocarcinoma.[146,153] A length–diameter ratio exceeding 15 has been taken to indicate mesothelioma rather than adenocarcinoma, but

shorter microvilli do not exclude that diagnosis.[154] Furthermore, numerous tonofibrils are commonly present in mesothelioma, but they may also be seen in pulmonary adenocarcinomas coexpressing some degree of squamous differentiation. Finally, El-Naggar and colleagues,[155] studying 23 pleural mesotheliomas and 41 pulmonary adenocarcinomas, noted that 78 percent of mesotheliomas were diploid, contrasting with only 12 percent of pulmonary adenocarcinomas.

Bronchioloalveolar Carcinoma

Pathology

At their periphery, many peripheral pulmonary adenocarcinomas tend to spread in a lepidic fashion along existing alveolar septa. The architecture of the lung parenchyma is thus retained at the periphery of the tumor. This becomes strikingly apparent when antibodies to basement membrane compounds, such as collagen IV and laminin, are used.[156] Toward the center, reactive desmoplastic stroma and inflammation usually result in more irregularly shaped tumor tissue, in which the original pulmonary architecture is no longer evident.

Some well-differentiated peripheral adenocarcinomas elicit little inflammation or desmoplastic response. They show a conspicuous tendency to spread along existing stuctures, so that throughout much of the tumor, the lung architecture is retained (Fig. 31-36 and 31-37). These tumors are designated bronchioloalveolar carcinoma.[157]

A diagnosis of bronchioloalveolar carcinoma is best based on an assessment of the architecture of the entire tumor. However, in some instances, a combination of the characteristic clinical, radiologic, and biopsy findings allows a presumptive diagnosis of bronchioloalveolar carcinoma based on endoscopic biopsy specimens. Pulmonary metastases from adenocarcinomas of other sites of the body, such as the pancreas, stomach, colon, ovary, and breast, can produce an identical histologic picture.[158,159] Because the term bronchioloalveolar carcinoma implies a primary pulmonary neoplasm, this latter point is of immediate clinical importance.

Bronchioloalveolar carcinoma has been subdivided into three groups. The first, accounting for about a quarter of cases, consists of high columnar, mucin-producing cells showing little nuclear atypia (Fig. 31-38). These tumors have a marked propensity to spread via the airways so that one or more lobes are diffusely affected. Accumulation of large amounts of mucus in the pulmonary parenchyma results in a

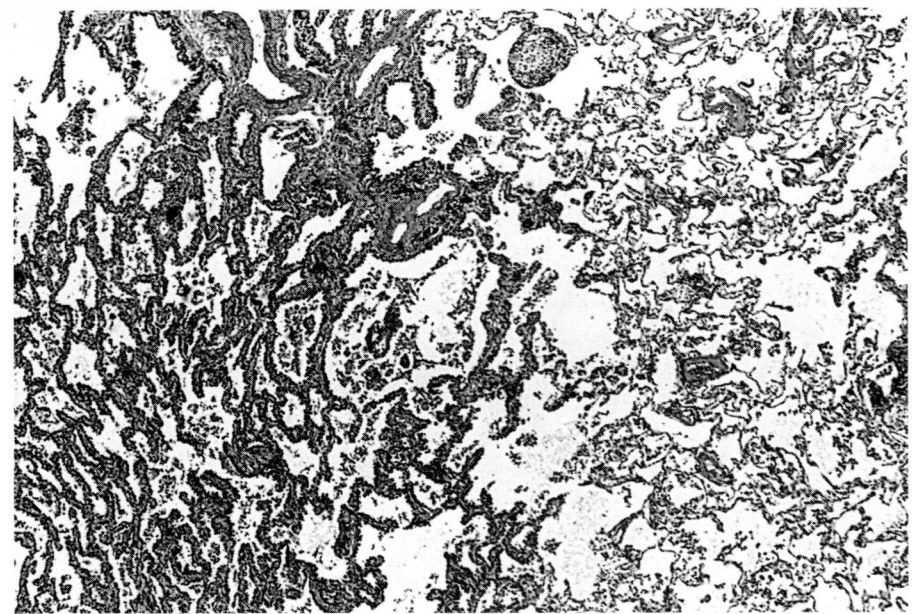

pneumonia-like clinical picture.[160] Macroscopically, consolidation of large parts of a lobe or an entire lung, with a glistening, slimy cut surface, is characteristic of this subtype. The second group of bronchioloalveolar carcinomas, which constitutes the majority of cases, exhibits cuboidal or low columnar tumor cells, with a higher degree of nuclear atypia, and producing little or no mucin[161] (Fig. 31-39). Rarely, numerous psammoma bodies are present.[162] This type of bronchioloalveolar carcinoma also disseminates via the airways,[163] although to a lesser degree than its mucin-producing counterpart. As with other lung carcinomas, prognosis of bronchioloalveolar carcinoma is primarily related to stage, so that the mucin-producing tumors are associated with a poor prognosis, despite their bland cytologic appearance. The third, more recently recognized subgroup is characterized by the presence of a central area of sclerosis.[164] Complete absence of collagen IV was noted in the sclerosing center, contrasting with nonsclerosing bronchioloalveolar carcinomas.[165] Multi-focality appears to be especially common in these

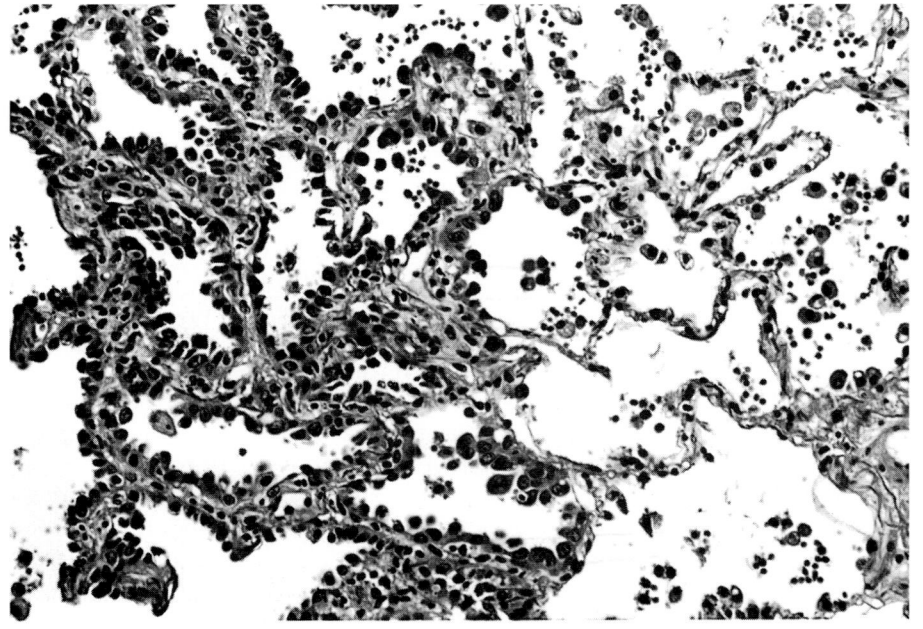

FIGURE 31-37
Bronchioloalveolar carcinoma. Tumor cells replace the alveolar epithelium. In contrast to reactive pneumocyte hyperplasias, as in, for instance, interstitial pneumonias, there is a very sharp border between the population of atypical cells and the normal alveolar pneumocytes. In this variant of bronchioloalveolar carcinoma, the tumor cells have scanty cytoplasm, contrasting with reactive proliferating alveolar type II cells which more commonly exhibit copious amounts of cytoplasm.

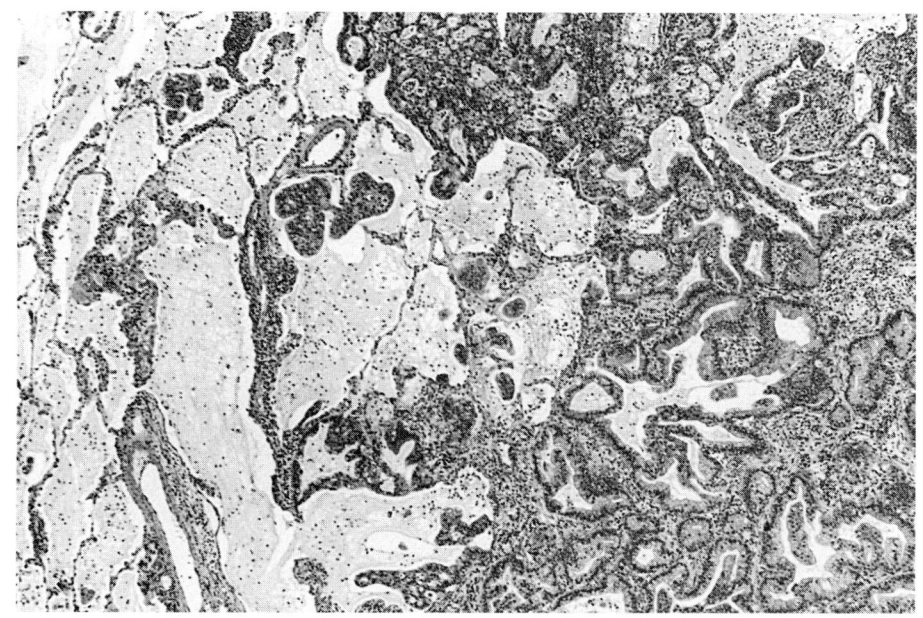

FIGURE. 31-38
Bronchioloalveolar carcinoma, mucin-producing subtype. Mucin-producing cylindrical tumor cells spread along preexistent alveolar septa. Mucin accumulates within the surrounding alveoli. Small groups of tumor cells are floating within these pools of mucin. Spread of such dissociating groups of cells via airways is thought to contribute to the diffuse, pneumonia-like pattern of spread, characteristic of this tumor.

tumors,[166] which is reflected in a poor prognosis.

Incidental reports on type II pneumocyte proliferations forming small tumor nodules[78,79,167] have blurred the border between malignancy and premalignant lesions. At present, there is some confusion about the correct diagnosis of such small lesions. This issue is discussed in more detail in a previous paragraph (see p. 1017). Multifocal lesions rather than aerogenous spread of bronchioloalveolar carcinoma may account for some cases with numerous nodules.[166]

Histologically, bronchioloalveolar carcinoma shows an intriguing resemblance to so-called jaagsiekte (pulmonary adenomatosis),[168] a disease of sheep, probably of viral etiology, common in some sheep-farming lands.[169] However, there is no evidence to suggest that bronchioloalveolar carcinoma in human beings has an infectious etiology. With respect to its association with smoking, including the presence of *ras*-mutations in some tumors, this group of neoplasms is very similar to pulmonary adenocarcinoma in general.

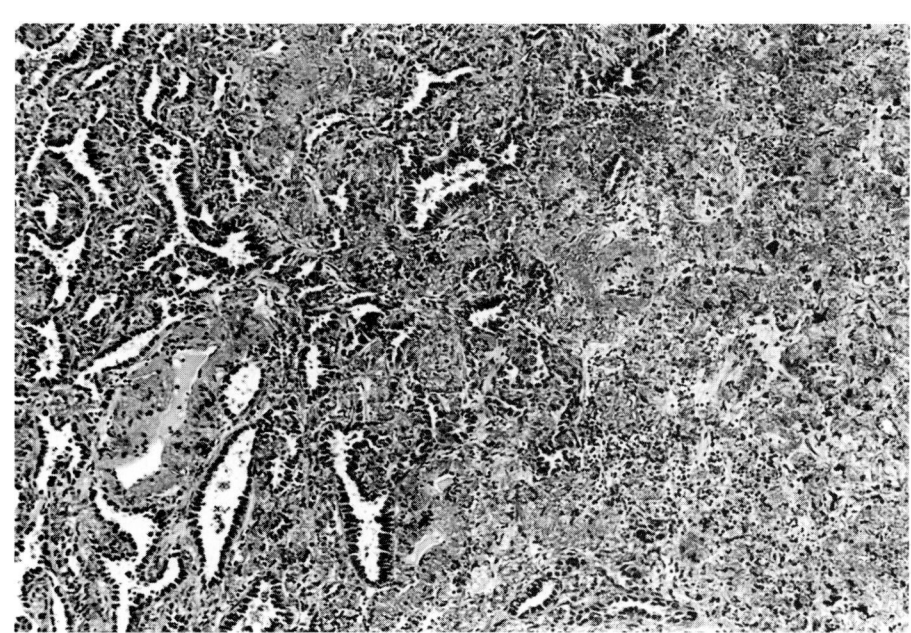

FIGURE 31-39
Adenocarcinoma. An area of central paucicellular scar tissue (*right*) with anthracotic pigmentation is surrounded by irregular tubules lined by tumor cells.

"Scar Cancer"

Proliferation of alveolar and bronchiolar epithelium is common in association with diffuse pulmonary inflammatory and fibrosing disorders. It is seen to a lesser extent in the vicinity of pulmonary scars, as caused by infarction, tuberculosis, silicoanthracosis, and others. Because a central area of scarring is noticed in the center of many pulmonary adenocarcinomas (Fig. 31-39), a causal relationship has previously been assumed to exist between pulmonary scarring and the development of lung cancer, especially adenocarcinoma.[170]

However, it has now become clear that in the large majority of cases, pulmonary adenocarcinoma arises in the absence of an existing scar. Areas of central fibrosis are commonly present in the center of adenocarcinoma and other types of lung cancer, but they are the result rather than the cause of the tumor. Shimosato and colleagues[171] studied 58 peripheral lung cancers with central scarring, and gave six arguments against the scar being the tumor precursor. These were:

1. The amount of fibrosis was generally related to tumor size.
2. Adenocarcinoma often developed in lungs that lacked any evidence of scarring or fibrosis.
3. Central fibrosis was also seen in the metastases of pulmonary adenocarcinomas.
4. Lung metastases of certain tumors, such as some breast carcinomas, also form central scars.
5. Psammoma bodies may be found in the scar and within adenocarcinoma of the lung, but are extremely rare in nonneoplastic conditions.
6. Retrospective studies of chest radiographs did not indicate preexistent scarring.

In addition, the increased amount of type III collagen found in the fibrotic tissue has been taken as an argument in favor of a host response to the tumor.[172] The pathogenesis of the central scarring is probably similar to central scarring seen in some other carcinomas, such as "scirrhous" infiltrating ductal carcinoma of the breast. In addition, infarction, due to obliterative changes in pulmonary arteries, commonly present in lung carcinomas, and collapse of the alveolar elastic framework have been incriminated as contributory causes.[173,174]

Scarring may be an adverse prognostic factor. In one study, the degree of scarring was significantly related to tumor associated death.[171] Cagle and colleagues[175] subdivided the central scars in an early form, with many fibroblasts and small or moderate amounts of collagen, and an advanced form, with scanty cellularity, abundant collagen, or areas of hyalinization. In a series of 21 resected tumors followed for at least 10 years, 1 of 6 tumors with an "early scar" led to the death of the patient, contrasting with 4 of 8 tumors with an "advanced scar."[175]

Although it has now been accepted that central scarring in a pulmonary adenocarcinoma represents a desmoplastic response to the tumor rather than its cause, there are rare instances where the evidence of a preexistent scar is compelling. Isolated cases of adenocarcinoma intimately associated with a known scar caused by tuberculosis or trauma are on record. The association with diffuse fibrosing conditions of the lung has been mentioned above. However, in general, it is wise to assume a skeptical attitude when considering such a causal relationship.

SMALL CELL CARCINOMA

Small cell lung carcinoma ("oat cell carcinoma") has a neuroendocrine phenotype. Other neuroendocrine lung tumors include typical and atypical pulmonary carcinoid (Chap. 29) and a subset of large cell carcinomas (see p. 1047). Together, these neuroendocrine tumors constitute a spectrum of neoplasms, ranging from the mature and clinically indolent typical bronchial carcinoid, via the unpredictable atypical carcinoid, to the immature SCLC, with its characteristic highly aggressive course of disease.

Clinical Features

Clinically, SCLC is characterized by rapid growth, early and widespread metastasis, and an often gratifying initial response to polychemotherapy and radiotherapy. There is, however, a high relapse rate, leading to death within 3 years in the large majority of cases. Survival beyond 5 years is achieved in about 10 to 20 percent of patients with limited disease, and less than 1 percent of patients with extensive disease.[176,177] The figures vary somewhat among series, especially in the limited disease subgroup, probably as a result of differences in patient selection criteria. Decreased chemo- and radiosensitivity is common in recurrent tumors, and the histology may change to another type of lung carcinoma, most commonly large cell carcinoma. A small percentage of SCLCs are amenable to surgery. In these, a 5-year-survival rate of about 20 percent is achieved.

Paraneoplastic syndromes are most commonly associated with this type of lung cancer. Various neurologic syndromes, such as the Eaton-Lambert myasthenic syndrome (see p. 1014) are probably caused by antibodies to SCLC cross-reacting with normal tissues. Inappropriate secretion of various peptide hormones such as corticotropin, antidiuretic hormone, calcitonin, and growth hormone account for a variety of paraneoplastic endocrine syndromes.

Pathology

In SCLC, neuroendocrine differentiation is too poorly developed to be reliably recognized by light microscopy. Thus, the histologic diagnosis is based on the assessment of a number of features unrelated to the characteristic differentiation pattern. As a consequence, the diagnostic criteria are not as distinctive as in squamous cell or adenocarcinoma. This is unfortunate because the distinction between SCLC and other types of lung cancer is most important in view of the differences in the therapeutic approach.

Small cell carcinoma usually arises near the lung hilum, rapidly invades surrounding tissues, and metastasizes to regional lymph nodes and beyond. Macroscopically, the tumor is soft or moderately firm, usually with a homogeneous pink or tan cut surface. Not uncommonly, the primary tumor is much smaller than its mediastinal and distant metastases. Indeed, the tumor was originally thought to represent a mediastinal rather than a pulmonary neoplasm.[178]

Histologically, SCLS consists of a proliferation of relatively monomorphic middle-sized cells, with hyperchromatic nuclei and scanty cytoplasm, lacking the hallmarks of squamous or glandular differentiation[179-181] (Fig. 31-40). Although they are smaller than the cells of most other types of lung carcinoma, the tumor cells are middle sized rather than small, having a diameter roughly three times that of a lymphocyte. The cells and their nuclei are round, oval, or elongated. In a minority of cases, there is an admixture of larger tumor cells, which may or may not show features of squamous or glandular differentiation (see p. 0000). Depending on the tissue processing conditions, nuclei have a smudged, somewhat vacuolated or very finely granular chromatin pattern ("salt and pepper" pattern, see Fig. 31-40). There is usually a single nucleolus, which is inconspicuous in paraffin sections or FNA smears. This is because it does not stand out against the surrounding dense chromatin. In plastic sections, however, the nucleolus can be striking, which may lead to an erroneous diagnosis of non-SCLC. Occasionally, scattered cells with bizarre, greatly enlarged nuclei are present in an otherwise typical SCLC.

There is only a small amount of cytoplasm, lacking distinguishing features, resulting in a high nuclear-cytoplasmic ratio. The poorly developed cytoskeleton confers little physical strength to the cell. Thus, it is easily traumatized during the biopsy procedure or the preparation of direct smears for cytologic investigation. In histologic sections, adjacent tumor cell nuclei often mold to each other. The cellular features of SCLC are especially evident in

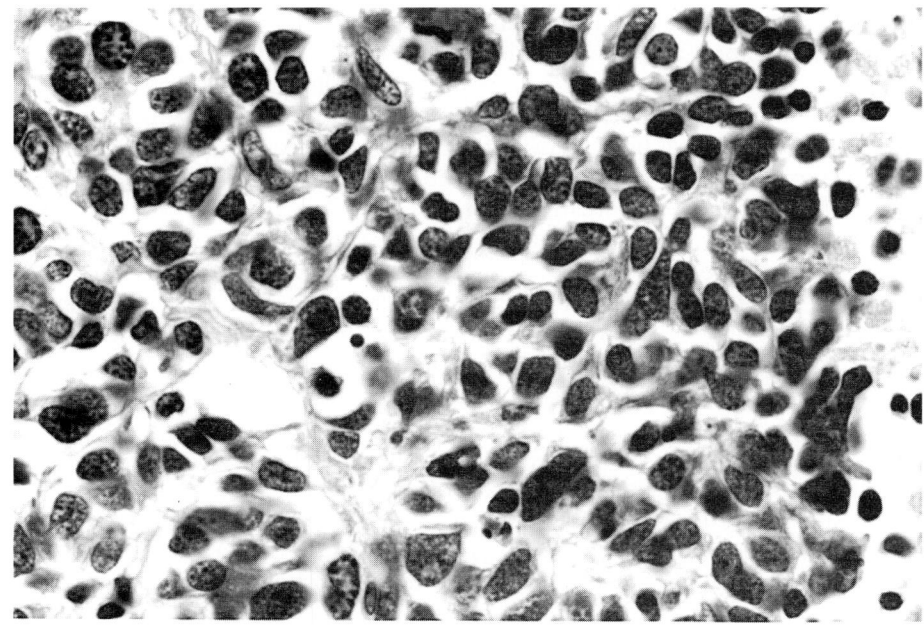

FIGURE 31-40
SCLC. The nuclei of tumor cells show finely dispersed chromatin. Nucleoli are inconspicuous. Because there is little cytoplasm, nuclei lie closely adjacent to one another; there is a hint of nuclear molding.

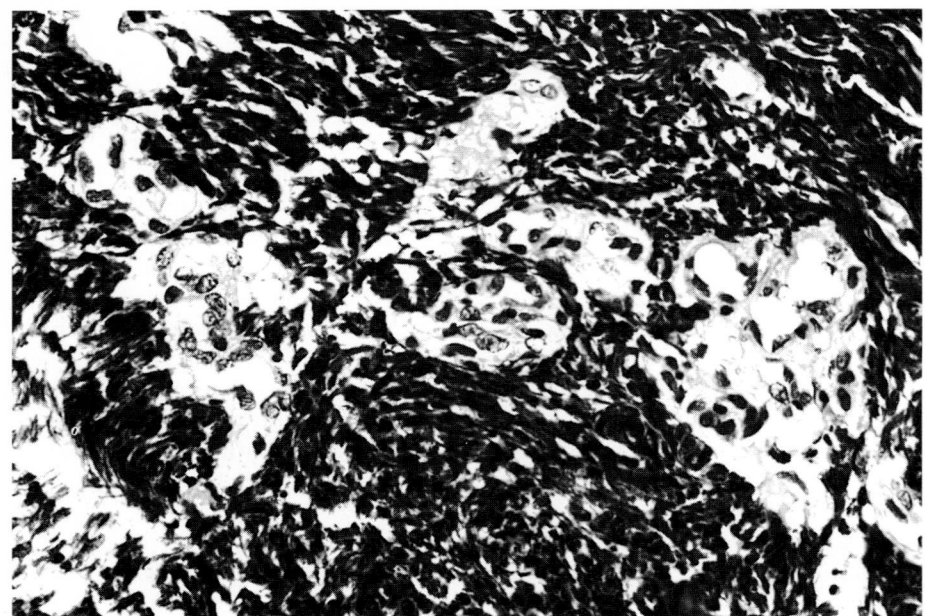

FIGURE 31-41
Bronchial biopsy specimen of SCLC shows severe crush artifact. Note endothelial proliferation and hypertrophy, which is common in SCLC biopsies. In contrast to the tumor cells, the endothelium shows very little crush artifact.

cytologic specimens, provided that they are not too heavily traumatized (see Chap. 36).

The tumor cells often diffusely infiltrate the surrounding lung tissue, forming large, ill-defined aggregates which, as indicated above, are easily traumatized by the biopsy forceps. As a result, all that remains of the tumor in a biopsy specimen may be a mass of basophilic nuclear material, where it is impossible to discern cellular details (Fig. 31-41). This is in stark contrast to the adjacent bronchial and lung tissue, which generally lacks any sign of mechanical trauma. When investigating a biopsy containing such a suspicious population of heavily traumatized cells, one should search for small groups of intact malignant cells lying apart from the main tissue mass. A note of caution is in order. In biopsies with crush artifact, the pathologist should be wary, in the absence of any viable cells, of making a diagnosis of SCLC. Tuberculosis, other causes of chronic inflammation, non-Hodgkins lymphoma, and Hodgkin's disease can all cause diagnostic problems. Immunohistochemistry may be helpful: inflammatory infiltrates and malignant lymphomas are usually strongly LCA (leukocyte common antigen, CD 45) positive, even in the presence of severe crush artifact. SCLCs are LCA negative but are usually positive with antibodies recognizing keratins and other epithelial antigens, although low levels of expression and decreased immunoreactivity in paraffin sections may lead to negative results.[182] SCLC, as well as other types of neuroendocrine lung cancer, most commonly express keratins 8 and 18. A monoclonal antibody to polysialylated neural cell adhesion molecule (N-CAM), MoAb 735 (see p. 1056), which is generally strongly positive in SCLCs, is helpful in the distinction between carcinoma cells and (neoplastic or reactive) lymphoid cells.

Within bronchi, SCLC tends to grow diffusely within the lamina propria, remaining covered by residual bronchial surface epithelium (Fig. 31-42 and 31-43). Generally, the tumor cells of SCLC do not line the bronchial lumen. In roughly half the cases, the covering nonneoplastic surface epithelium shows squamous metaplasia. In a systematic study of 128 biopsies of SCLC, Yoneda and Boucher[183] found squamous metaplasia in 47 percent and atypical squamous metaplasia and squamous carcinoma in situ in 9 percent each. Positivity of the SCLC tumor cells with antibodies against epidermal growth factor was significantly associated with squamous metaplsia and increasing cellular atypia, suggesting that these changes in the overlying epithelium may perhaps be directly attributable to epidermal growth factor secreted by the tumor. This hypothesis offers an interesting alternative to the concept of squamous carcinoma in situ as a precursor of SCLC, as previously put forward by Saccomanno and colleagues.[69]

Another distinctive feature of SCLC, less commonly appreciated, is the presence of endothelial hypertrophy and hyperplasia (see Fig. 31-41), which in some instances may be reminiscent of the endothelial changes encountered in astrocytomas.

Feulgen-positive, basophilic material derived from the nuclei of degenerating tumor cells may be deposited along vessels, resulting in basophilic streaks in and along these structures (Figure 31-44). This feature, known as Azzopardi's phenomenon, is common in SCLC, but it is not characteristic because

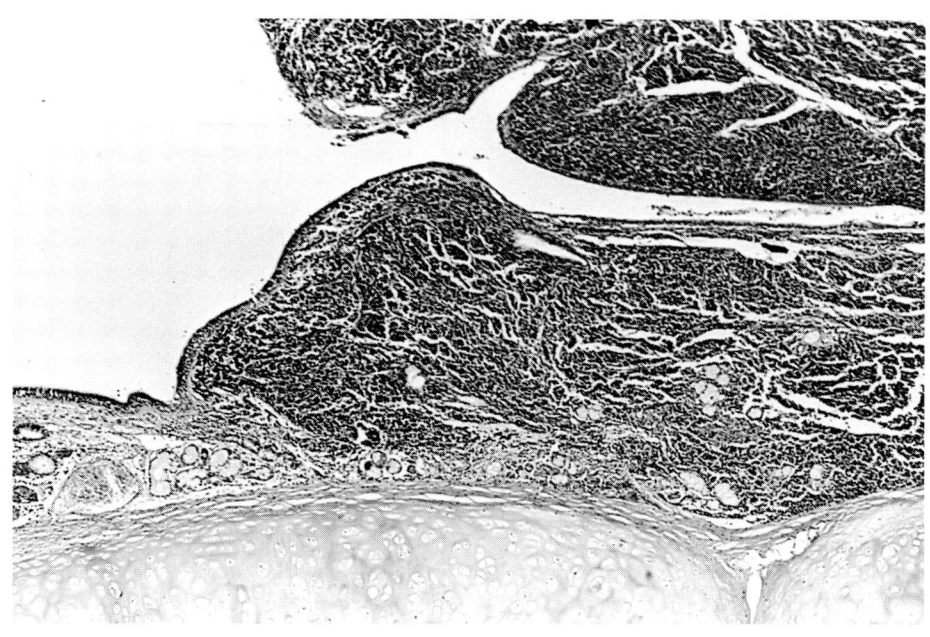

FIGURE 31-42
SCLC, diffusely infiltrating the lamina propria of a large bronchus.

it may also occur in some other fast-growing tumors with a high rate of cell death. It is seen more commonly at postmortem than in biopsy specimens.

Ultrastructurally, neuroendocrine differentiation is evident, although only to a limited degree. In well preserved biopsy specimens, intracytoplasmic dense-core neuroendocrine granules can practically always be found.[184] Neuroendocrine granules contain a homogeneously electron-dense core, surrounded by an electron-lucent halo, which in turn is surrounded by a rounded or somewhat hexagonal, smooth unit membrane (Fig. 31-45). The diameter varies between 50 and 200 nm; most are about 100 nm. Sometimes, the granules are concentrated in small cellular extensions, reminiscent of abortive neuritic processes. Occasionally, neurotubules are encountered (Fig. 31-46).

Apart from neuroendocrine differentiation, some SCLCs exhibit subtle ultrastructural signs of squamous or glandular differentiation (Fig. 31-47).

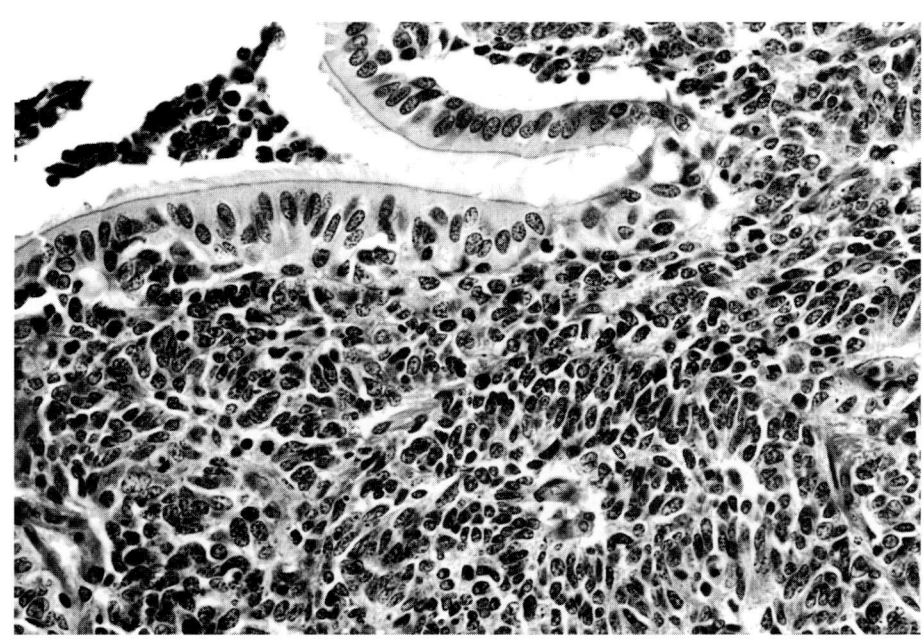

FIGURE 31-43
SCLC. A layer of residual ciliated bronchial surface epithelium covers diffuse sheets of tumor cells. Note the elongated shape of the tumor cell nuclei.

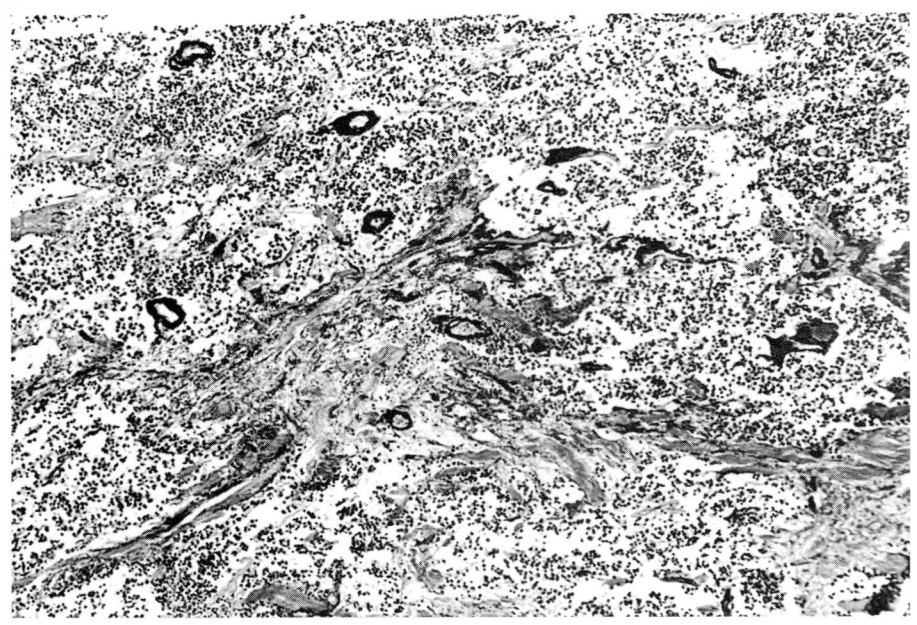

However, when only squamous or glandular differentiation is detected, the diagnosis of SCLC should be
critically reconsidered. We have encountered several
cases where, in retrospect, the light microscopic features were not entirely typical of SCLC, and the subsequent course of disease was more indicative of
non-SCLC.[119] In such problem cases, ultrastructural
assessment of the tumor has some practical value.
Because diagnostic electron microscopy is not generally available, it is hoped that immunohistochemical
assessment of neuroendocrine differentiation will
have a similar diagnostic application. Antibodies
against chromogranin A and B often yield negative
results.[185] This is undoubtedly related to the paucity
and small size of the neuroendocrine granules of
SCLC. Synaptophysin also lacks sufficient sensitivity. Neuron-specific enolase is usually positive in
SCLC,[186] but it lacks sufficient specificity.

Uncommonly, the tumor shows two distinct
components, namely, SCLC together with one of
the non-SCLC types[181,187] (Fig. 31-48). In such
instances, the small cell component may or may not
show ultrastructural evidence of neuroendocrine differentiation. If it does, the tumor is a subtype of
SCLC. If only squamous or glandular differentiation
is found, it is advisable to classify the tumor accord-

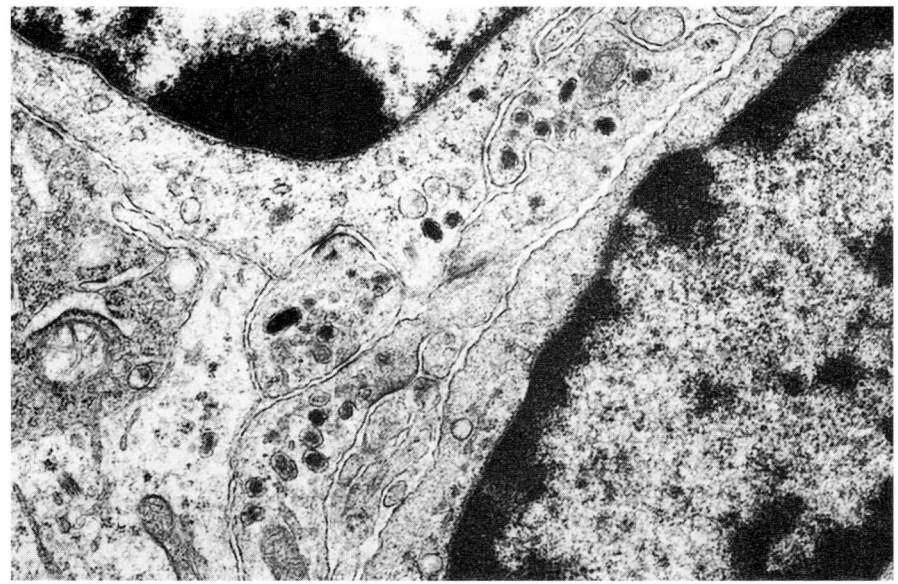

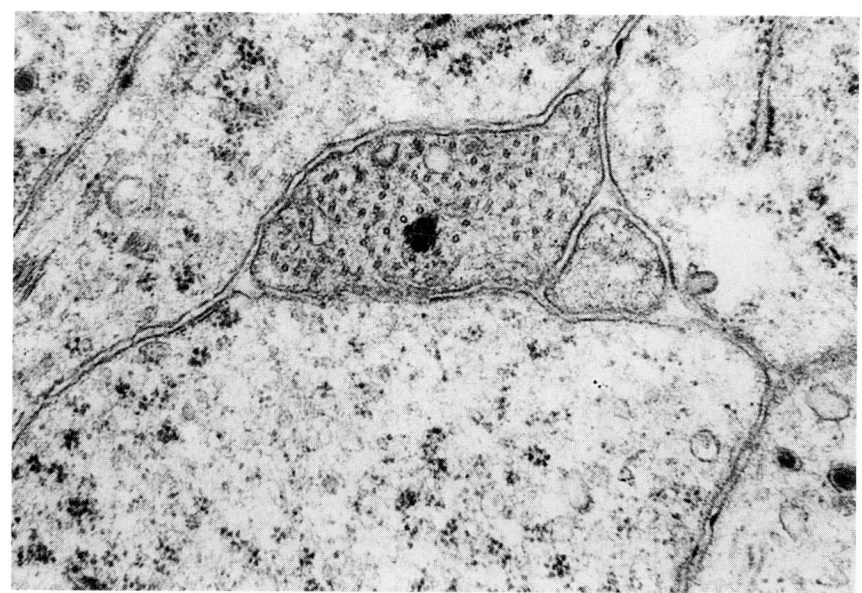

FIGURE 31-46
SCLC. Electron micrograph showing cytoplasmic extension with multiple neurotubules. Note the presence of a few neuroendocrine granules and the paucity of filaments and organelles in the surrounding cells.

ing to the component exhibiting squamous or glandular features.

Changes in cell type are not uncommon during the treatment of SCLC. This is illustrated in the autopsy study of 40 biopsy-proven SCLCs, mentioned previously (see p. 1018), where a complete or partial change to a non-small cell histology was noted in 11 instances.[88]

Attempts to subdivide SCLC into clinically relevant subgroups have so far met with limited success. The WHO subdivision of SCLC into oat cell, intermediate-type small cell (the cells of which are slightly larger than those of oat cell carcinoma), and combined oat cell carcinoma (which combine an oat cell carcinoma component with areas of adenocarcinoma or squamous cell carcinoma) is falling into disuse. This is mainly because of the difficult and problematic distinction between oat cell and intermediate-

type SCLC. The resemblance of oat cell carcinoma cells to oat grains is less than striking. Furthermore, the term "intermediate-type small cell carcinoma" may create misunderstandings with clinical colleagues, who may not be completely familiar with this somewhat confusing terminology. The term intermediate does not indicate "intermediate" between "oat cell type" and "combined oat cell carcinoma," the third subtype defined by the WHO. More importantly, the subtle morphologic differences between oat cell carcinoma and intermediate-type SCLC are, at least in part, the result of variations in tissue handling and processing rather than a reflection of phenotypic differences between distinct tumor types. Not surprisingly, there is much interobserver variation in the distinction between oat cell and intermediate-type SCLC.[188] There have been no convincing reports of differences

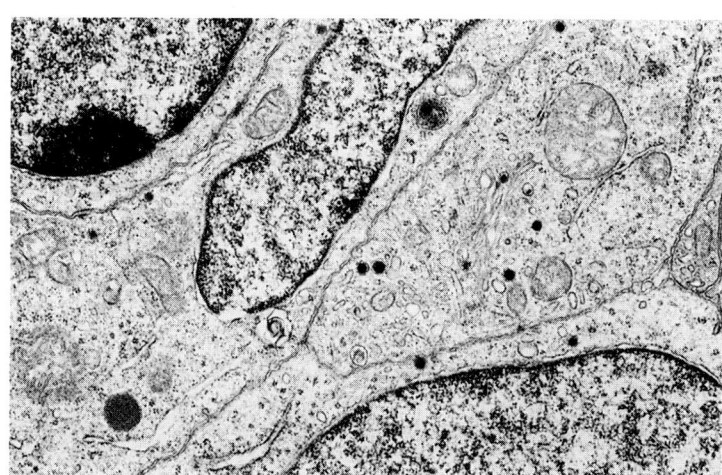

FIGURE 31-47
SCLC showing neuroendocrine granules and a subtle degree of glandular differentiation, hallmarked in this field by the presence of many intracytoplasmic organelles, including rough endoplasmic reticulum and a well developed Golgi apparatus.

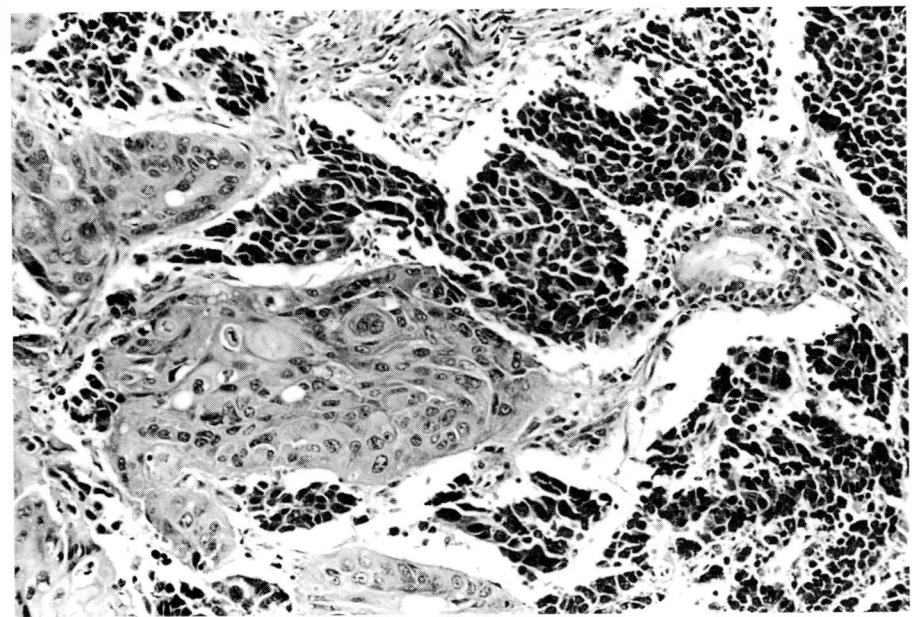

FIGURE 31-48
Combined SCLC, with areas of squamous carcinoma intermingled with SCLC. At the ultrastructural and immunohistochemical level, not all of such tumors show neuroendocrine differentiation in the small cell component. Further data are needed to clarify the relationships between light microscopic morphology, cell types, and clinical behavior in these rare tumors.

in presentation or clinical behavior between the two groups.[189]

Histologically, combined SCLC appears to be a distinct tumor type. The glandular or squamous component of this tumor type is diagnosed according to the guidelines given in the sections on squamous cell and adenocarcinoma. The SCLC component of this combined tumor type is identical to that of the more common "pure" SCLC.

A recently proposed subgroup of SCLC is "small cell–large cell carcinoma." This is characterized by solitary or grouped large tumor cells, with large, vesicular nuclei containing distinct nucleoli, and with varying amounts of eosinophilic cytoplasm and distinct cell borders, set against a background of typical small cell carcinoma.[190] When strict criteria are applied, such a large cell component is rare in untreated small cell carcinomas, to a degree that some authorities have questioned its very existence.[191] Others have reported percentages over 10 percent,[192] probably reflecting more liberal criteria in the diagnosis of this tumor, and possibly the effects of chemotherapy. Indeed, an admixture of large tumor cells not uncommonly appears after chemotherapy. The WHO classification of lung tumors is about to be reviewed at the time of writing. It is likely that the small cell–large cell will be split into small cell and large cell neuroendocrine carcinoma (Chap. 29).

Some investigators reported a poor response to chemotherapy for such small cell–large cell tumors,[192-194] but this was not found by others. In an early study, 14 percent of 200 small cell carcinomas were of the small cell–large cell variety and in these,

median survival was 168 days, whereas it was 280 days for the "pure" small cell group.[192] Fushimi and colleagues[194] studied 430 SCLCs and diagnosed the small cell–large cell variant in 25 of 299 brushings, 75 of 400 FNAs, and 8 of 232 sputa. Median survivals of the small cell–large cell versus the pure small cell groups were 144 days versus 285 days (biopsy material), 160 versus 275 days (cytology material), and 47 versus 259 days (sputa). Because a significant difference in survival was found when tumors were subdivided according to each of these sample types, the authors concluded that the diagnosis of small cell–large cell carcinoma can be made on the basis of each of these sample types. In contrast, Bepler and associates,[195] studying 550 extensive disease SCLCs, did not find a shorter survival in the small cell–large cell group (24 cases). In addition, these authors found a concordance between two pathologists in diagnosis of this subtype in only 11 instances. Further studies are clearly needed to establish the observer variability and clinicopathologic correlations in this type of SCLC.

At present, the author adopts the recommendations of the pathology committee of the International Association for the Study of Lung Cancer[196] to subdivide SCLC as follows:

1. SCLC (over 90 percent of cases)
2. Mixed small cell–large cell carcinoma (possibly associated with poor prognosis and poor response to therapy, estimated incidence 4 to 6 percent of untreated SCLCs)
3. Combined SCLC: admixed with areas of squamous cell or adenocarcinoma.

Differential Diagnosis

The histologic differential diagnosis of SCLC includes a variety of pulmonary and extrapulmonary tumors. The resemblance to SCLC of some squamous carcinomas (see p. 1029), adenocarcinomas (see p. 1035), and large cell carcinomas (see p. 1047), and the concept of basaloid carcinoma (see p. 1048), are discussed elsewhere in this chapter. Because the large majority of SCLCs occur in smokers, alternatives should be seriously considered in the non-smoking patient. The clinical presentation and the histology together with a characteristic coexpression of epithelial and neural-neuroendocrine markers exclude the large majority of entities entering the differential diagnosis. Special mention should be made of non-Hodgkin's lymphomas and Hodgkin's disease; when necessary, this possibility should be investigated immunohistochemically. Some pleural malignant mesotheliomas show areas resembling SCLC.[197] The characteristic site, clinical presentation, and more typical areas in other blocks point to the correct diagnosis. Small cell tumors of childhood are usually excluded on clinical grounds and, if the need arises, by the use of special techniques, especially immunohistochemistry and electron microscopy.

LARGE CELL CARCINOMA

The diagnosis of large cell lung carcinoma is based on a series of negative findings. There is no squamous or glandular differentiation and none of the distinctive histologic features of SCLC are seen (Fig. 31-49). Absence of glandular differentiation implies negative mucin stains, which are therefore mandatory for the diagnosis.

As is the case with several other tumor types defined by negative criteria, this entity constitutes something of a dustbin of poorly differentiated tumors, mostly poorly differentiated squamous, adenocarcinomas, and adenosquamous carcinomas. This becomes strikingly apparent when large cell carcinomas are studied by electron microscopy.[97–99,198,199] Indeed, ultrastructurally, any of the three main patterns of differentiation can occur in this tumor type, alone or in combination. Although a relationship between the ultrastructural cell type and prognosis was found in one study,[97] this would not influence treatment. Therefore, there appears to be no practical need for such additional studies. Immunohistochemically, there are no positive distinguishing features. Epithelial differentiation can be demonstrated with keratin antibodies. Vimentin also is commonly expressed.[138]

Neuroendocrine differentiation is not uncommon.[86,200–203] Travis and colleagues[202] described a subgroup of "large cell neuroendocrine carcinomas," characterized by immunohistochemical or ultrastructural features of neuroendocrine differentiation, while the histology showed large polygonal tumor cells, with a low nuclear-cytoplasmic ratio, arranged in trabecular, palisading, or rosette patterns. Coarse chromatin, frequent nucleoli, high mitotic rate (> 10/10 high power fields), and frequent necrosis fur-

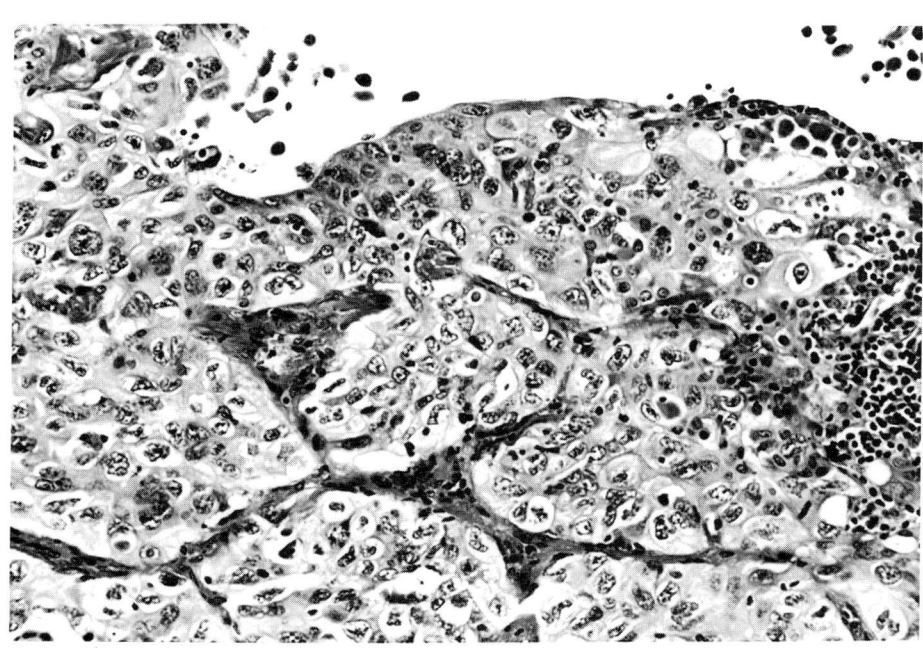

FIGURE 31-49
Large cell carcinoma. Pleomorphic tumor cells with well-developed cytoplasm and a solid pattern of growth. Intercellular bridging and keratinization are absent. Mucin stains were negative.

ther characterize these tumors, which in their material had a prognosis intermediate between atypical carcinoid and SCLC.

Occasional large cell carcinomas show a histologic picture identical to the nasopharyngeal lymphoepithelioma-like carcinoma associated with the EBV.[204–206,206a] In these tumors, islands and sheets of undifferentiated, large tumor cells are surrounded by stroma containing a very dense infiltrate of lymphocytes and plasma cells. With in situ hybridization, EBV was detected in the tumor cells.[205,206,206a] In two instances, clonal episomal EBV was demonstrated, indicating that the EBV infection preceded the clonal expansion.[206] Similar carcinomas have now been recognized in a variety of other body sites, including the salivary gland, thymus, uterine cervix, skin, and stomach.[207]

As has been noted earlier in this chapter, many squamous cell and adenocarcinomas of the lung only focally show distinguishing features. Therefore, a noncommittal diagnosis of non-SCLC rather than large cell carcinoma is better used when a biopsy does not show typical features of squamous cell carcinoma, adenocarcinoma, or SCLC.

Macroscopically, large cell carcinomas are commonly peripheral, solid, with a homogeneous grayish cut surface. Scarring is absent or minimal in most cases. Histologically, solid strands and nodules of moderately sized or large, variably pleomorphic tumor cells are seen within a varying amount of desmoplastic stroma (see Fig. 31-49) or ill-defined sheets of tumor cells are formed. Necrosis is common.

In the WHO classification of lung tumors, giant cell and clear cell carcinoma are subtypes of large cell carcinoma. However, as will be discussed below, giant and clear cell change may also occur in other types of non-SCLC and therefore do not identify distinct histologic types. Furthermore, spindle cell and giant cell change often occur together. Both probably represent manifestations of dedifferentiation,[208] which may occur in any of the main types of non-SCLC: squamous cell carcinoma, adenocarcinoma, and large cell carcinoma.

BASALOID CARCINOMA

Brambilla and colleagues[120] recently reported on 38 "basaloid" carcinomas selected from a total of 115 resected poorly differentiated lung carcinomas. The distinguishing features of this subgroup, which they designated "basal cell (basaloid) carcinoma," were lobular or trabecular growth pattern with peripheral palisading of cells, small cuboidal or fusiform tumor cells, moderately hyperchromatic nuclei lacking prominent nucleoli, a high mitotic rate, and scanty but distinctly visible cytoplasm, and absence of molding. In half the cases, there were areas of squamous, adenocarcinoma or large cell carcinoma. Thus, the distinctive features of basaloid carcinoma can be present in a number of main lung tumor types. Because of this, basaloid carcinoma cannot be considered a separate and distinct entity, but "basaloid features," as described, may carry significant prognostic implications, as indicated below. Neuro-endocrine markers yielded inconsistent results. Ultrastructurally, neuroendocrine granules were absent, whereas a subtle degree of squamous or glandular differentiation was present. However, expression of a number of neuroendocrine markers was detected immunohistochemically in some instances.[209,210]

Prognosis of this group of tumors was poor in the original study of Brambilla and associates, with a median survival of 22 months for stage I and II tumors. The 5-year actuarial survival rate was 15 percent, which was intermediate betweeen SCLC and resectable non-SCLC.[211] Two cases of primary basaloid squamous cell carcinoma of the trachea have been described.[211a]

ADENOSQUAMOUS CARCINOMA

Adenosquamous carcinoma exhibits both squamous and glandular differentiation. As discussed before, combinations of differentiation are exceedingly common at the ultrastructural level (Fig. 31-50), but usually, the least pronounced differentiation type is not evident light microscopically. In adenosquamous carcinoma, both should be present to a degree where both are evident on light microscopy in a substantial portion of the tumor (Figs. 31-51 and 31-52). Such tumors account for only a small percentage of lung tumors. The clinical presentation and behavior of adenosquamous carcinoma is similar to adenocarcinoma. The metastases of adenosquamous carcinoma generally show the same combination of squamous and glandular differentiation characteristic of the primary tumor.

Adenosquamous carcinoma most commonly presents as a peripheral tumor and may be associated with central scarring or indentation of the overlying pleura. No macroscopic features distinguish the tumor from the more common pulmonary adenocarcinoma.

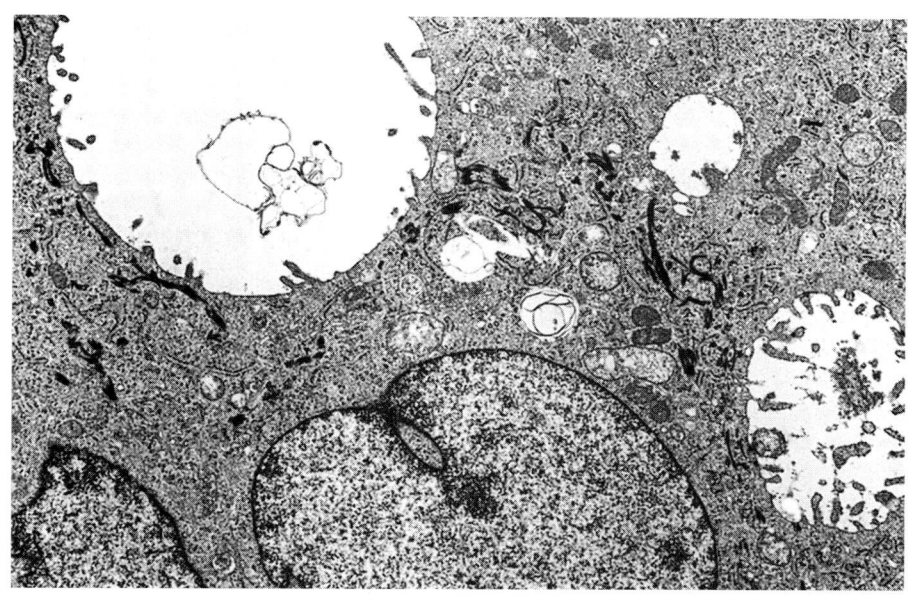

FIGURE 31-50
Combined squamous and glandular differentiation at the ultrastructural level. Multiple well developed tonofibrils and two intracellular lumina lined by microvilli are present.

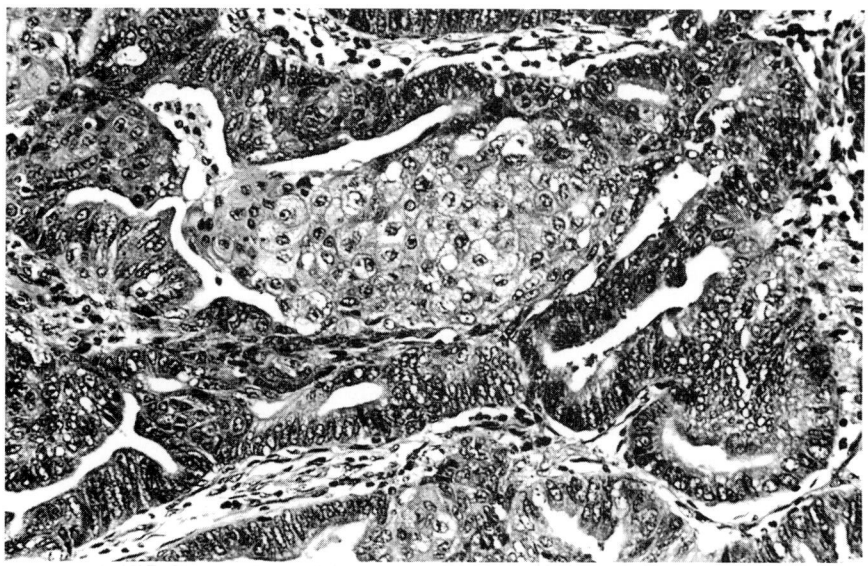

FIGURE 31-51
Adenosquamous carcinoma. Solid groups of tumor cells with squamous differentiation, as evidenced by areas of intercellular bridging (visible at higher power) and occasional keratinizing tumor cells, and a glandular component consisting of atypical cylindrical cells are present.

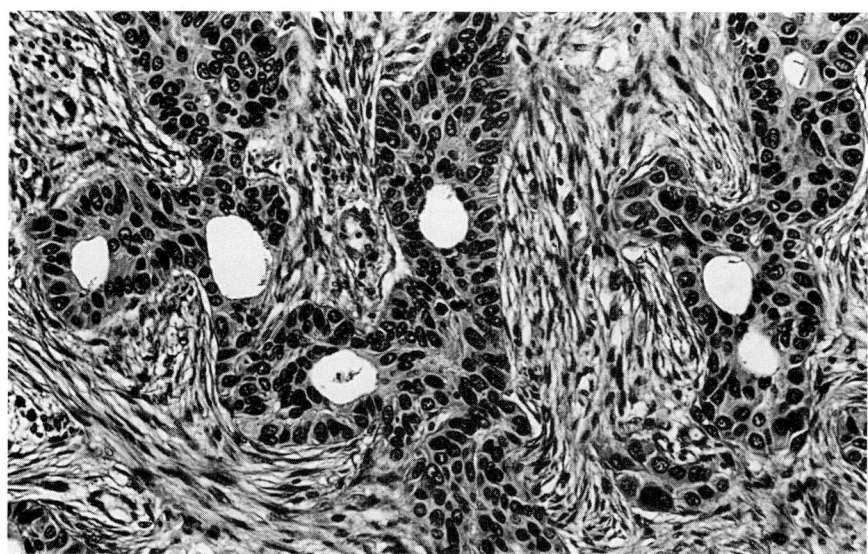

FIGURE 31-52
Adenosquamous carcinoma. The tumor consists of irregular strands of tumor cells with intercellular bridging (not evident at this magnification) and contains multiple lumina with the nuclei of surrounding tumor cells orientated toward the lumen.

Differential Diagnosis

As discussed above (see p. 1026), entrapment of alveolar epithelium and resorption of necrotic debris in squamous cell carcinoma should not be mistaken for evidence of glandular differentiation. Adenosquamous carcinoma should also be distinguished from mucoepidermoid carcinoma arising from bronchial glands, which likewise shows a combination of squamous and glandular elements. Low-grade mucoepidermoid carcinoma of bronchial glands shows histologic features identical to the more common salivary gland counterpart, with characteristic cysts lined by mucinous epithelium, a close intermingling of mucinous, squamous, and intermediate cells, and a mild degree of nuclear atypia. In the author's opinion, the histologic features of high-grade mucoepidermoid carcinoma cannot be distinguished reliably from adenosquamous carcinoma and this distinction is therefore better not attempted in lung tumors.

CARCINOMAS WITH GIANT CELL OR SPINDLE CELL CHANGE

The hallmarks of giant cell change are marked nuclear atypia, a high mitotic rate, and the presence of bizarre, sometimes multinucleated tumor cells. Tumors showing these features often contain a dense mixed inflammatory infiltrate, which includes neutrophils. This infiltrate generally penetrates the sheets of tumor cells extensively, and inflammatory cells come to lie entirely within giant tumor cells (emperipolesis) and subsequently degenerate. Another characteristic feature of these tumors is loss of intercellular cohesiveness (Fig. 31-53).

In a minority of cases, the entire tumor exhibits these features. More commonly, it is a focal change in a large squamous, or, rarely, adenocarcinoma. This means that this phenomenon, which is not related to cell type, does not by itself identify a homogeneous group of tumors. Several authors reported an association with poor prognosis,[208,212] but this has not been confirmed by all.[213] An increased incidence of gastrointestinal metastasis was noted in one study.[213]

Some lung carcinomas show transitions to a malignant spindle cell tumor[117,131,214,215] (Fig. 31-54 and 31-55). Squamous cell carcinomas exhibiting this feature often possess a polypoid intrabronchial component. Apart from vimentin, which is usually present in the tumor cells, focal expression of keratin is detectable.[215–217]

In rare instances, a true metaplasia to a different type of differentiation, such as cartilage, striated muscle, or bone occurs. Such tumors are designated carcinosarcomas.[216] In the case of bone formation within the tumor, a distinction has to be made between osteoid produced by malignant cells, as in carcinosarcoma, and metaplastic ossification of stroma, as can rarely be seen in other types of lung cancer, most commonly in bronchial carcinoids (see Chap. 29).

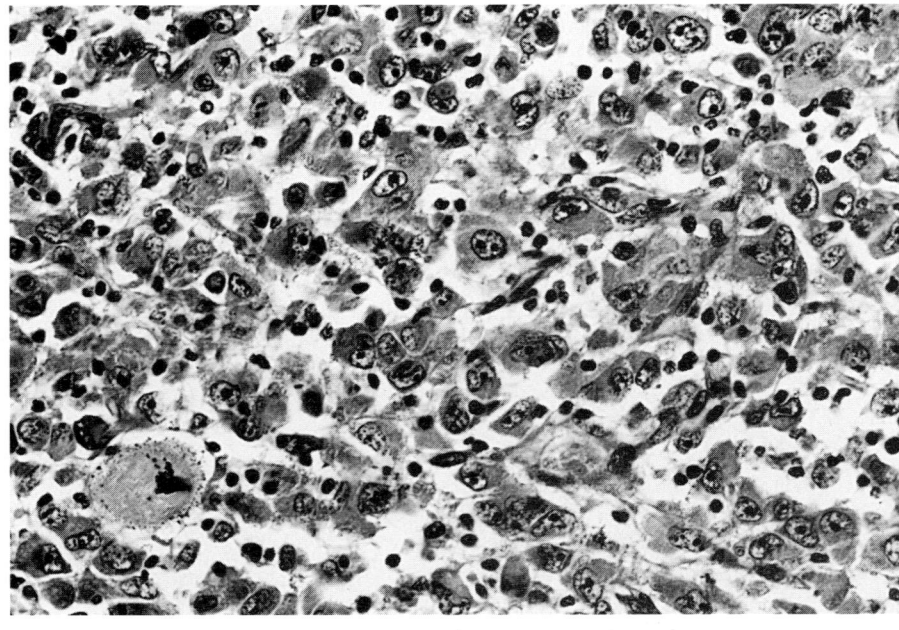

FIGURE 31-53
Giant cell carcinoma. Pleomorphic tumor cells show characteristic dissociation and a diffuse inflammatory infiltrate.

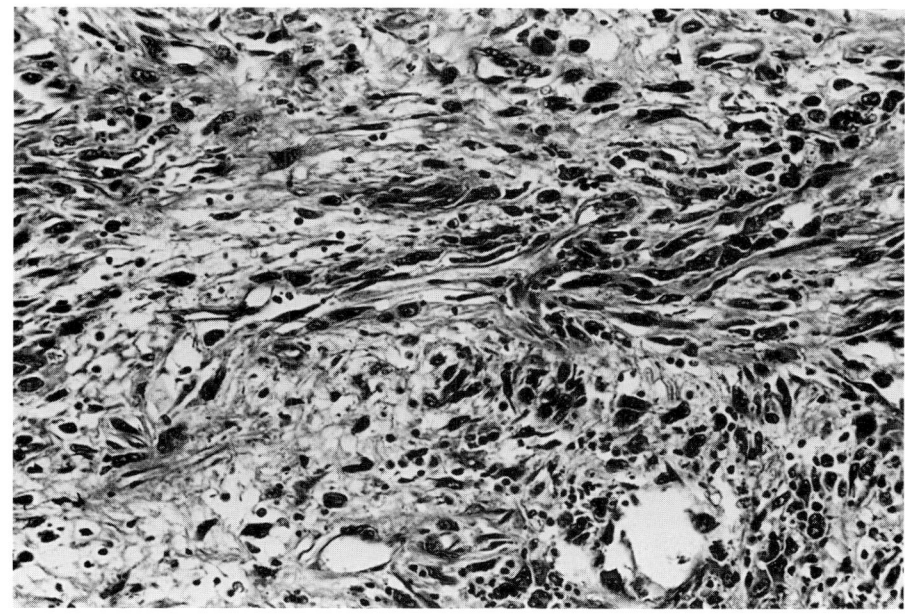

FIGURE 31-54
Poorly differentiated adenocarcinoma with spindle cell change. In this field, a pleomorphic spindle cell tumor lacking further distinguishing features is seen. Same tumor as in Fig. 31-55.

When only spindle cell change is present, the tumor is best called (squamous cell, adenocarcinoma, or large cell) carcinoma with spindle cell change, or sarcomatoid carcinoma. The term "metaplastic carcinoma" has often been used to designate these tumors, but it is not entirely correct because the spindle cell change by itself does not indicate a true change to another cell type. The term "carcinoma with sarcomatoid stroma" is conceptually incorrect: the tumor cells rather than the stromal, nonneoplastic cells undergo the "sarcomatoid" changes.

Giant cell and spindle cell change often occur together. Recently, Fishback and colleagues[208] proposed spindle cell and giant cell carcinoma as a single histologic type: "pleomorphic carcinoma." These authors studied 78 resection specimens with spindle cell or giant cell change. Areas of spindle cell and giant cell carcinoma occurred together in 38 percent. In 78 percent of cases, a squamous cell, adenocarcinoma, or large cell carcinoma was also present. The prognosis was poor, with a postoperative 5-year survival rate of 10 percent. This indicates that although

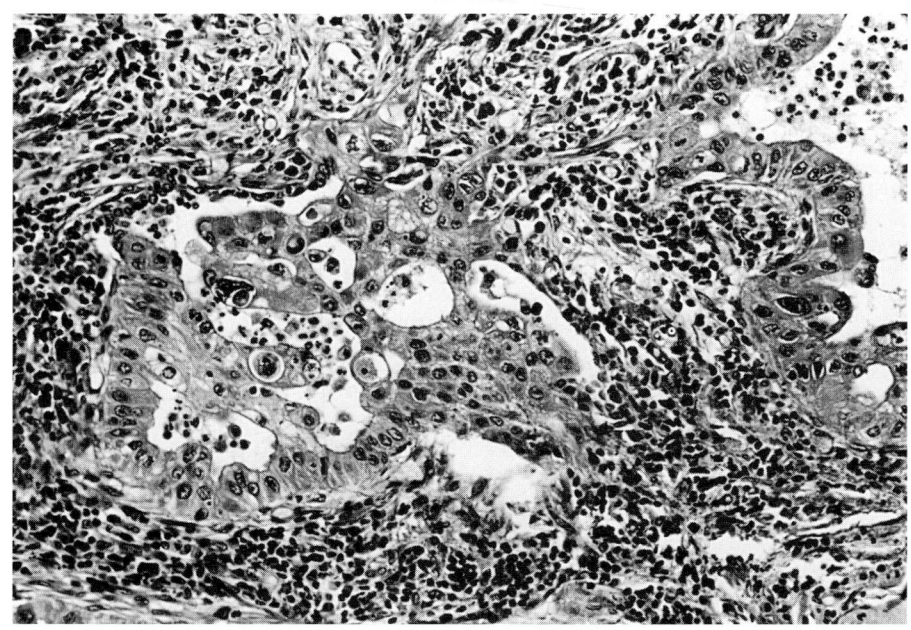

FIGURE 31-55
Poorly differentiated adenocarcinoma with spindle cell change. In this field of the same tumor as illustrated in the previous figure, the glandular differentiation is easily appreciated.

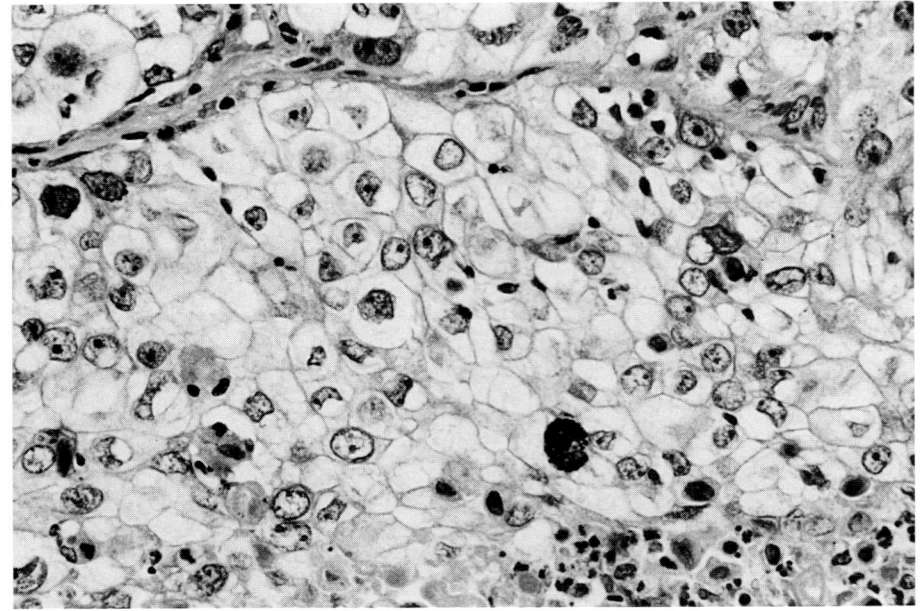

FIGURE 31-56
Clear cell carcinoma. Note pleomorphic nuclei, water-clear cytoplasm, and well defined cell borders. A periodic acid-Schiff stain was strongly positive but mucin and fat stains were negative.

"pleomorphic carcinoma" cannot be considered a distinct entity, because such "pleomorphic" changes occur in several distinct WHO-defined subgroups of lung tumor, this change has a significant prognostic connotation. The situation with respect to classification and prognosis shows a parallel to "basaloid carcinoma" (see p. 1048).

Differential Diagnosis

The differential diagnosis of these tumors includes a variety of (primary or, more commonly, metastatic) pleomorphic or spindle cell sarcomas and pleural mesothelioma. The latter diagnosis may be suggested in the case of a peripheral adenocarcinoma with a malignant spindle cell component.[131] This differential diagnosis is discussed elsewhere in this chapter (see p. 1036).

CARCINOMAS WITH CLEAR CELL CHANGE

The tumor cells of some non-SCLCs of any histologic type show water-clear cytoplasm (Fig. 31-56) due to glycogen accumulation (Figure 31-57). This feature may be diffuse or focal. Tumors lacking features of squamous or glandular differentiation are called clear cell carcinoma.[218] This is one of the subtypes of large cell carcinoma in the WHO classification. However, like giant cell change, clear cell change does not identify a distinct and homogeneous group of tumors because it may occur in any type of non-SCLC.[219] Contrary to early claims of a better prognosis, this cellular cell change does not appear to influence the prognosis.

From a practical clinical viewpoint, the clear cell change raises the possibility of an occult renal carcinoma metastatic to the lung. The tumor cells of renal

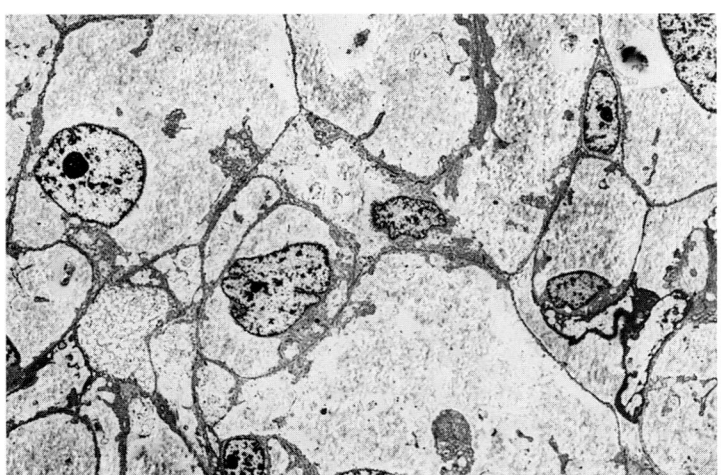

FIGURE 31-57
Clear cell carcinoma. Electron micrograph shows large quantities of intracellular glycogen, paucity of organelles, and absence of fat vacuoles.

carcinoma and its metastases often contain intracellular lipid, which is absent or minimal in primary clear cell carcinoma of the lung. Conversely, the demonstration of intracellular mucin rules out metastatic renal carcinoma, resulting in a diagnosis of pulmonary adenocarcinoma with clear cell change. If special stains are inconclusive, the appropriate clinical and radiologic studies to rule out a renal tumor are necessary.

Pulmonary clear cell carcinoma should be distinguished from the rare benign clear cell tumor of the lung.[220] The presence of more than a very few mitoses, necrosis, nuclear atypia, and the presence of thick-walled vessels rather than the sinusoidal vessels characteristic of benign clear cell tumor aid in this distinction. Interestingly, Gaffey and colleagues,[221] investigating a series of benign clear cell tumors, found positivity with melanocyte marker HMB-45 and ultrastructural evidence of melanogenesis in seven of eight and three of eight cases, respectively. These findings have obvious diagnostic potential.

TERMINOLOGY AND SAMPLING ERROR

The histologic spectrum of lung cancer is varied and complex, so the pathology report is necessarily an oversimplification. The diagnostic pathologist should guide the clinician rather than baffle and confuse him with complex reports containing a plethora of histologic findings that are of little relevance with respect to the classification of the tumor. Terms such as "large cell squamous carcinoma" and "small cell squamous carcinoma," should be discouraged because they are misleading.

About two thirds of all lung cancers are unsuitable for resection, so histologic diagnosis and classification are based entirely on small biopsies. The clinician should be aware of the limitations of this method. The difficulties associated with the typing of poorly differentiated tumors are compounded when only biopsy material is available. Furthermore, differences in histologic appearance between different areas of the tumor cannot be appreciated, resulting in sampling error.[222] Indeed, Nash[223] complained that "at a time when the expectations and demands of our clinical colleagues are increasing, the material on which we must base our findings is vanishing."

Because traces of intercellular bridging or mucin production are often discernible only focally in a lung tumor, such hallmarks of differentiation are far more likely to be detected in resection specimens than in small biopsies. Because of this, the diagnosis

of large cell carcinoma should be reserved for resection specimens, where the absence of differentiation can be assessed throughout the tumor. The term "non-small cell carcinoma" is useful in those instances where a biopsy does not include the diagnostic features of glandular or squamous differentiation.[222] A diagnosis of large cell carcinoma based on such biopsies would lead to many discrepancies between biopsy and resection specimen diagnoses (e.g., large cell carcinoma changed to poorly differentiated squamous cell carcinoma). Obviously, even when the diagnosis of large cell carcinoma is limited to resection specimens, this group of tumors continues to be heterogeneous at the ultrastructural level (see p. 1047). The common denominator is the fact that the tumor *in its entirety* lacks histologic features of differentiation. Similarly, we prefer to reserve the diagnosis of bronchioloalveolar carcinoma for resection specimens, where the characteristic growth pattern can be investigated throughout the tumor. When it is seen in a biopsy specimen, we diagnose adenocarcinoma, and add a comment stating that the features are compatible with, but not diagnostic of, bronchioloalveolar carcinoma.

Ultrastructurally, many lung carcinomas show several cell types, most commonly a combination of glandular and squamous differentiation.[93,224] Mooi and colleagues[93] studied neuroendocrine, squamous, and glandular differentiation in paired samples taken from 44 primary lung tumors, using a standardized criteria for the grading of the degree of differentiation. Twenty-five tumors showed ultrastructural evidence of more than one type of differentiation in one or both samples. In two instances, the predominant type of differentiation differed between the two samples, while in a further nine cases, two types of differentiation were present to the same degree in one sample, whereas one predominated in the other. Thus, intratumor heterogeneity of differentiation is obvious ultrastructurally as well as histologically. Thus, although ultrastructural investigation of bronchial biopsies yields additional information with respect to the tumor type,[225] there appears to be no need for this from a practical, diagnostic viewpoint.

Histologic grading of lung carcinomas can only be done with any degree of certainty in resection specimens. In every lung tumor, the degree of differentiation varies between areas to such a degree that grading based on biopsy samples becomes a futile exercise. Therefore, it is advisable not to include a statement on tumor grade in reports of biopsy specimens. The growth pattern and the degree of mucin production or keratinization are similarly subject to sampling error and do not affect treatment.

PROGNOSIS OF LUNG CANCER

The number of prognostic factors of lung cancer that have been put forward now exceeds 100.[226] This number precludes a detailed analysis and discussion of each one. Clinical features such as performance status, weight loss, the presence and type of paraneoplastic syndromes, and a variety of blood assays fall outside the scope of this chapter. Other parameters, however, are assayed in the tumor tissue itself. The prognostic significance of many of these is often based on postoperative survival data and therefore generally concern stage I and II non-SCLCs. Many prognostic factors need to be further substantiated and for practical diagnostic purposes, it is acceptable at present to limit the search for prognostic features to a few parameters, the significance of which has been well corroborated.

The most important feature remains the tumor stage (see p. 1015). Staging determines the feasibility of curative surgery, which at present offers the best chance of cure. Within the subgroup of resectable tumors, small carcinomas without lymph node metastases carry a favorable prognosis. In a recent series of 56 resected stage I peripheral adenocarcinomas smaller than 2 cm in diameter, 51 patients (91 percent) were free of recurrence after 5 years.[227] Indeed, in a study of 92 radiographically occult lung carcinomas, all 59 tumors 2 cm or less in diameter and 11 in situ carcinomas and 4 tumors "suspicious for invasion" were free of lymph node metastases, which led the authors to suggest that lymph node dissection may not be necessary for these patients.[106] Interestingly, intrapulmonary metastases, which escape preoperative detection and are encountered by the pathologist investigating the resection specimen of a primary lung carcinoma, may not affect the prognosis unfavorably. Shimizu and colleagues[64] found such metastases in 42 of 839 resection specimens, most commonly in association with adenocarcinoma; when the primary tumor was 3 cm or less in diameter, and when lymph node metastases were absent, the finding of an intrapulmonary metastasis did not influence postoperative survival.

The influence of histologic type on prognosis is less distinct. The prognosis of resected SCLC is worse than that of resected non-SCLC. However, within the group of resected non-SCLCs, tumor type appears not to carry a major prognostic significance, after correction for tumor stage. In a series of 1121 resected non-

SCLCs, 5-year survival figures for squamous cell carcinoma and adenocarcinoma were 65 and 55 percent, respectively.[228] However, in some studies, such as the one reported by Kayser and colleagues,[229] adenocarcinomas had a more favorable prognosis, or there was no correlation between tumor type and chance of recurrence.[230]

A standardized histologic grading system of lung tumors is not available. Most pathologists grade non-SCLCs loosely on the basis of the degree of morphologic evidence of the cell type, and on nuclear parameters such as size, staining characteristics, the degree of pleomorphism, and mitotic rate. Many of these parameters are subjective and the difficulties are compounded by the differences in grade between areas of the same tumor and the apparently undifferentiated areas occurring in many carcinomas.[231] Perhaps not surprisingly, no consistent picture emerges from studies comparing tumor grade and clinical course of disease, some authors reporting decreased survival in poorly differentiated carcinomas, and others failing to find a significant correlation. Indeed, the clinical relevance of any histologic grading system that cannot be rigidly standardized and is subject to substantial sampling error is questionable. In lung cancer, tumor grade does not influence treatment planning.

Other histologic features of potential prognostic significance include the degree of lymphoid infiltrate[232] and the number of microvessels in the tumor tissue.[233] Blood vessel invasion was associated with a markedly increased likelihood of tumor recurrence in a recent study of 128 resected non-SCLCs.[234] In a study of 41 resected lung carcinomas, the presence of more than two Langerhans cells per high power field, as detected by CD1 immunostaining, was correlated with a markedly worse survival.[235] A recent review has documented the molecular and biological factors affecting prognosis in non-small cell lung cancer.[234a] Further studies are needed to corroborate these findings.

Markers of Cellular Proliferation

In several studies,[234,236] expression of high levels of proliferating cell nuclear antigen (PCNA) was associated with a decreased disease-free survival after resection of non-small cell carcinoma. Marked intratumor variations in PCNA immunoreactivity were found by Carey and colleagues[237] and was associated with intratumor heterogeneity of DNA content. Systematic tumor sampling is therefore mandatory to obtain meaningful results.

In a study of 85 resected T1N0 tumors, the $S+G_2M$ phase fraction did not differ between recurrent and nonrecurrent tumors.[230] In contrast, in another study, all 22 patients with an S phase fraction of 6 percent or less survived 5 years, contrasting with only 9 of 22 patients whose tumors had an S phase fraction of more than 6 percent.[238] Differences in disease-free survival in the first years after resection may not be indicative of a difference in cure rate. They may reflect differences in tumor cell kinetics rather than the likelihood of subclinical metastases at the time of surgery, which will determine ultimate treatment failure. Radiologically, tumor doubling times can be estimated and correlate with tumor histology. Thus, the average radiologic doubling times for adenocarcinomas, squamous carcinomas/large cell carcinomas, and small cell carcinomas are 183 days, 100 days, and 55 days, respectively.[239]

DNA Ploidy

During malignant transformation and tumor progression, the total amount of DNA contained in the tumor cells often becomes abnormal. This abnormality can be assessed by image analysis of cytospin preparations or by flow cytofluorimetry. The latter technique has the advantage of measuring very large numbers of cells in a short time. However, it does not distinguish between tumor cells and diploid nonmalignant cells present in each sample of tumor tissue. When the latter greatly outnumber the tumor cells, an aneuploid population of cells may be obscured in the DNA histogram.

With a number of interesting exceptions, aneuploid tumors tend to carry a worse prognosis than their diploid counterparts. In lung carcinoma, a correlation with poor prognosis was noted by some[226,238] but not all.[240] Not uncommonly, more than one aneuploid peak is found,[241] indicating the presence of several aneuploid stem lines. A correlation with clinical data indicates that it may be important to sample multiple sites for flow cytometry to obtain optimal prognostic information.

Various Molecular Prognostic Markers

Overexpression of *bcl*-2 was recently demonstrated in 20 of 80 squamous cell carcinomas and 5 of 42 adenocarcinomas of the lung and was associated with a favorable postoperative prognosis, especially in the subgroup of patients over 60 years of age with squamous cell carcinomas.[242] However, this favorable prognostic influence was not confirmed by others.[243] The numbers of patients reported are small and further investigation is needed.

Abnormalities of p53, as evidenced by accumulation of p53 protein or mutational analysis, are found in about two thirds of non-SCLCs and the large majority of SCLCs. Several groups have reported an association with poor prognosis in non-SCLC.[244,245]

A minority of non-SCLCs overexpress the *neu* (c-*erbB*-2) gene,[246,247] which encodes a transmembrane tyrosine kinase with extensive homology to the epidermal growth factor receptor. *Neu* overexpression was found in 5 of 29 resected squamous cell carcinomas and in 10 of 29 pulmonary adenocarcinomas and was associated with poor prognosis in the adenocarcinoma group.[247] Amplification of the c-*erbB*-2 gene is very rare in lung cancer,[248] so that the cause of the overexpression remains to be elucidated.

Increased expression of cathepsin B, a lysosomal cystein proteinase, was associated with decreased postoperative survival in non-SCLC, also after stratification for stage, in a recent study by Sukoh and colleagues.[249]

An association between loss of tumor cell expression of blood group antigens and unfavorable prognosis has been found in a variety of human cancers,[250–252] including carcinomas of the lung. Loss of expression of blood group antigen A by tumor cells, of patients with blood group A or AB, was associated with an unfavorable postoperative course of disease in a series of 164 resected non-SCLCs.[253] Patients with stage I tumors of this latter group had a median postoperative survival of 15 months, similar to those with stage II or III disease. In contrast, patients with blood group A or AB with stage I tumors retaining expression of blood group A antigen had a more favorable median postoperative survival of 71 months. Expression of blood group antigens B or H (in patients with blood group B or O) did not influence survival (median survival: 39 months) in this study. However, in another study of 87 resected non-SCLCs, loss of blood group B antigen expression, present in 42 cases, was associated with decreased postoperative survival.[254] With respect to blood group antigens A and H, the differences did not reach statistical significance.

Expression of H/Le^y/Le^b antigens, as evidenced immunohistochemically by positive staining with monoclonal antibody MIA-15-5, was inversely correlated with 5-year survival in a series of 149 resected non-SCLCs.[255] The MIA antigen constitutes a precursor of A and B blood group antigens. Interestingly, in an in vitro model system, the MIA-15-5 antibody was found to inhibit the mobility and metastatic potential of tumor cells expressing the MIA antigen.[256] Five-year survival of patients with MIA-15-5 positive tumors (91 cases) was 20.9 percent, contrasting with 58.6 percent in the group of 58 MIA-15-5 negative tumors.[255] In this series, the MIA-15-5 staining result was an even stronger prognostic factor than lymph node status and tumor size. The most marked difference was found in the group of squamous cell carcinomas, where 5-year survival figures of cases positive and negative for MIA-15-5 were 10.5 and 62.1 percent, respectively. The difference in survival was significant in patients with blood groups A or AB but not in those with blood groups B or O. However, again, dissimilar findings were reported. Kawai and coworkers,[257] studying 102 stage I to IV pulmonary adenocarcinomas, did not find an association between H/Le^y/Le^b antigen positivity and decreased postoperative survival.

Ogawa and colleagues[234] studied expression of Sialyl Lewis^x, a carbohydrate ligand of the endothelial cellular adhesion molecule 1 (E selectin) in 128 stage I non-SCLCs; positive tumors had a significantly shorter disease-free survival.

Neuroendocrine Differentiation and N-CAM Expression in non-SCLC

As has been described in a previous paragraph (see p. 1019), some non-SCLCs of all major histologic types exhibit features of neuroendocrine differentiation.[258] More aggressive behavior has been found in some studies[200,259] but not in all.[260]

This matter is complicated for several reasons. Firstly, neuroendocrine differentiation is not a single and straightforward parameter. The presence of neuroendocrine granules is commonly taken as the most specific feature. However, their assessment requires electron microscopy because the granules may be too small and few in number to be identified immunohistochemically, as is usually the case in SCLC. Expression of a variety of other genes, whose products are not localized in granules are associated with neuroendocrine differentiation: neuron-specific enolase, synaptophysin, gastrin-releasing peptide, protein gene product 9.5 (PGP 9.5) and Leu-7.[261] Because none of these markers is ideal in terms of specificity and sensitivity,[203] it is important to use panels of antibodies. However, it is far from obvious what the minimal requirements should be to establish the presence of neuroendocrine differentiation. Second, and perhaps even more importantly, neuroendocrine differentiation can be associated with both aggressive (SCLC) and indolent behavior (typical bronchial carcinoid). Neuroendocrine differentiation per se does not indicate tumor aggressiveness.

It is possible that compound parameters such as neuroendocrine or indeed any type of differentiation, are, by their very nature, too complex. Studies of individual genes whose products are relevant with respect to growth regulation, cell motility, and other factors pertinent to tumor growth and metastasis may provide more insight into the determinants of clinical tumor behavior.

Practically all neuroendocrine lung carcinomas express the neural cellular adhesion molecule, N-CAM.[262] Interestingly, different types of N-CAM result from alternative mRNA splicing and posttranslational modifications. These isoforms are related to differences in cell motility and intercellular adhesiveness under physiologic circumstances. During early embryonic development, highly sialylated "embryonal-type" N-CAM is expressed by cells migrating from the neural crest to distant sites. In postnatal life, other, "adult-type" N-CAM isoforms with much shorter polysialic acid side chains, predominate. Interestingly, the predominant isoform of N-CAM varies between different types of neuroendocrine lung tumors. In SCLC, the highly sialylated, embryonal-type of N-CAM is expressed, whereas in typical carcinoids, this type of N-CAM is rare, but the adult-type N-CAM is abundant. It is tempting to speculate that such differences in N-CAM isoforms, which regulate cell migration and intercellular adhesion in the embryo and later in life, are related to the difference in biologic aggressiveness between carcinoids and SCLCs.

The N-CAM is also expressed by a minority of non-SCLCs that show no obvious signs of neuroendocrine differentiation.[263] Its expression was associated with an unfavorable postoperative clinical course of disease in two recent studies.[264,265]

LUNG METASTASES

The lungs receive the entire cardiac output and have a very extensive capillary bed, which is the first site to be reached by tumor cells entering the venous side of the systemic circulation from many sites of the body. These factors contribute to the common occurrence of lung metastases. Occasionally, a large number of lung vessels become blocked with tumor emboli, leading to respiratory dysfunction and pulmonary hypertension.[266] Rarely, large emboli may cause pulmonary infarction or even acute cardiac failure.

When the patient is known to have had a previous malignant tumor, a pulmonary shadow commonly represents a metastasis. However, because primary lung cancer is a common disease, the occurrence of a second primary carcinoma in the lung is by no means rare. Indeed, some tumors, such as squamous cell carcinomas of the head and neck, are associated with a substantially increased chance of second primary carcinoma in the lung.

In clinical practice, several findings and considerations aid in the distinction between primary and secondary lung tumors. Data regarding the first primary (site, size, grade, involvement of regional nodes) as well as number of recurrence-free years will allow an estimate of the likelihood of a metastasis from that tumor. Age and smoking history aid in estimating the likelihood of lung cancer. Finally, radiographic findings may provide significant information: metastases are more commonly multiple, of various sizes, and located preferentially in the middle and lower lung fields. Some show a characteristic polycyclic contour. The greatest diagnostic difficulties concern solitary pulmonary metastases, constituting a small minority of cases, and that often originate from carcinoma of the colon, breast, or kidney. Bronchoscopic findings are also important: primary squamous cell carcinoma tends to involve the surface epithelium, whereas metastases are more often extrabronchial, covered by intact residual bronchial epithelium. The previous histology should always be compared with that of the current lung tumor. However, due to the different microenvironment or tumor progression, the histology of a metastasis may be dissimilar from that of the primary tumor. In the coming years, molecular biologic techniques may become an important diagnostic adjunct in this respect.

If there is no evidence of extrapulmonary tumor and the patient's history is negative for malignancy, a solitary lung neoplasm may still represent a metastasis from an occult primary. The histology of the lung tumor is of importance in directing the search for such a possible occult primary. Squamous cell carcinoma may represent a metastasis of an occult tumor of the head and neck region, for instance in the tonsil or the nasopharynx. Adenocarcinoma may point to an intraabdominal occult primary tumor. In some instances, the production of specific compounds, such as prostatic acid phosphatase, prostate-specific antigen, or thyroglobulin, may point to the primary tumor elsewhere. Surfactant apoprotein or Clara cell protein within the tumor cells provides a strong argument in favor of primary lung cancer[141] (see p. 1036).

Some metastases closely simulate primary lung tumors, for example, by closely mimicking the archi-tecture of bronchioloalveolar carcinoma[158] or by growing exophytically within the confines of a bronchus.[267] Conversely, the histology of primary lung cancer may simulate metastatic tumor, for example, primary pulmonary adenocarcinoma of intestinal type, mimicking metastatic colorectal carcinoma (see p. 1033), and pulmonary clear cell carcinoma, simulating metastatic renal carcinoma (see p. 1052). The histology of a given lung tumor, when judged in isolation, rarely provides conclusive evidence that the tumor is either primary or metastatic.

Finally, it should be borne in mind that not all malignant tumors encountered in the lung are epithelial. Amelanotic metastases of (known or occult) melanoma may closely resemble carcinoma in biopsy specimens, and the same applies to some germ cell tumors, sarcomas, and lymphomas. Careful assessment of the cytologic and architectural features together with additional stains and immunohistochemistry will generally establish the correct diagnosis.

ACKNOWLEDGMENTS

Dr. K. P. Dingemans, University of Amsterdam, kindly provided the electron micrographs depicted in this chapter. Figure 31-4 was a gift of Professor J. Weening, University of Amsterdam. The author thanks Dr. F. van Leeuwen (NKI Amsterdam) and Dr. N. van Zandwijk (NKI Amsterdam) for critical comments and suggestions, and Mr. W. Meun, University of Amsterdam, for expert photographic advice and assistance.

REFERENCES

1. Hanson HH: Strategies against tobacco use. The physician's view. *Lung Cancer* 4S:20, 1988.
2. Muir C, Waterhouse J, Mack T, Powell J, Whelan S: *Cancer Incidence in Five Continents vol V.* IARC Sci Publ #88. Lyon, IARC, 1987.
3. Whelan SL, Parkin DM, Masuyer E, (eds): *Patterns of Cancer Incidence in Five Continents. A Picture Book.* IARC Sci Publ #102. Lyon, IARC, 1990.
4. Coebergh JWW, Van Leeuwen FE, Wagenaar SjSc: Lung cancer, in *Incidence of Cancer in the Netherlands.* Hoonte, Netherlands Cancer Registry, 1989.
5. Doll R, Peto R: Mortality in relation to smoking: 20 years' observation on male British doctors. *Br Med J* 2:1525–1536, 1976.
6. United States Public Health Service. *The Health Consequences of Smoking—Cancer: A Report of the*

Surgeon General. Rockville, MD, DHHS Office on Smoking and Health, 1982.

7. Pathak DR, Samet JM, Humble CG, Skipper BJ: Determinants of lung cancer risk in cigarette smokers in New Mexico. *J Natl Cancer Inst* 76:597–604, 1986.

8. Peto R: Influence of dose and duration of smoking on lung cancer rates, in Zaridze D, Peto R (eds), *Tobacco, a Major International Health Hazard.* IARC Sci Publ #74:23-33, Lyon, IARC, 1986.

9. Aitio A, Day NE, Heseltine E, et al: *Cancer: Causes, Occurrence and Control.* IARC Sci Publ #100, Lyon, IARC, 1990.

10. Tomatis L, Aito A, Day NE, et al (eds): *Cancer: Causes, Occurrence and Control.* IARC Sci Publ #100, Lyon, IARC, 1990.

11. Menkes MS, Comstock GW, Vuilleumier JP, Helsing KJ, Rider AA, Brookmeyer R: Serum beta carotene, vitamins A and E, selenium, and risk of lung cancer. *N Engl J Med* 315:1250–1254, 1986.

12. Westera WH, Slebos RJC, Offerhaus GJA, et al: K-*ras* oncogene activation in lung adenocarcinomas from former smokers. Evidence that K-*ras* mutations are an early and irreversible event in the development of adenocarcinoma of the lung. *Cancer* 72:432–438, 1993.

13. Environmental Protection Agency. *Respiratory Health Effects of Passive Smoking: Lung Cancer and Other Disorders.* Washington DC, December 1992.

14. Stockwell HG, Goldman AL, Lyman GH, et al: Environmental tobacco smoke and lung cancer risk in nonsmoking women. *J Natl Cancer Inst* 84:1417–1422, 1992.

15. Ambrosone CB, Roa U, Michalek AM, Cummings KM, Mattlin CJ: Lung cancer histologic types and family history of cancer. Analysis of histologic subtypes of 872 patients with primary lung cancer. *Cancer* 72:1192–1198, 1993.

16. Shimura T, Hagihara M, Takebe K, et al: The study of tumor necrosis factor beta gene polymorphism in lung cancer patients. *Cancer* 73:1184–1188, 1994.

17. Dubrow R, Wegman D: Setting priorities for occupational cancer research and control: Synthesis of the results of occupational disease surveillance studies. *J Natl Cancer Inst* 71:1123–1142, 1983.

18. Burns PB, Swanson GM: The occupational cancer incidence surveillance study (OCISS): Risk of lung cancer by usual occupation and industry in the Detroit metropolitan area. *Am J Ind Med* 19:655–671, 1991.

19. Morabia A, Markowitz S, Garibaldi K, Wynder EL: Lung cancer and occupation: Results of a multicentre case-control study. *Br J Ind Med* 49:721–727, 1992.

20. Hammond EC, Selikoff IJ, Seidman H: Asbestos exposure, cigarette smoking and death rates. *Ann NY Acad Sci* 330:473–490, 1979.

21. Doll R, Vessey MP, Beasley RWR: Mortality of gasworkers—final report of a prospective study. *Br J Ind Med* 29:394–406, 1972.

22. Weiss W, Moser RL, Auerbach O: Lung cancer in chloromethyl ether workers. *Am Rev Respir Dis* 120:1031–1037, 1979.

23. Enterline PE, Marsh GM: Cancer among workers exposed to arsenic and other substances in a copper smelter. *Am J Epidemiol* 116:895–911, 1982.

24. Pedersen E, Andersen A, Hogetveit A: Second study of the incidence and mortality of cancer of the respiratory organs among workers at a nickel refinery. *Ann Clin Lab Sci* 8:503–504, 1978.

25. Thun MJ, Schnorr TM, Smith AB, Halperin WE, Lemon RA: Mortality among a cohort of U. S. cadmium production workers: an update. *J Natl Cancer Inst* 74:325–333, 1985.

26. Samet JM, Nero AV: Sounding board: Indoor radon and lung cancer. *Engl J Med* 320:591–593, 1989.

27. Keehn R, Auerbach O, Nambu S, et al: Reproducibility of major diagnoses in a binational study of lung cancer in uranium miners and atomic bomb survivors. *Am J Clin Pathol* 101:478–482, 1994.

28. Neugut AI, Murray T, Santos J, et al: Increased risk of lung cancer after breast cancer radiation therapy in cigarette smokers. *Cancer* 73:1615–1620, 1994.

29. Winters TH, DiFranza JR: Radioactivity in cigarette smoke. *N Engl J Med* 306:364–365, 1982.

30. Fox B, Risdon RA: Carcinoma of the lung and diffuse interstitial pulmonary fibrosis. *J Clin Pathol* 21:486–491, 1968.

31. Fraire AE, Greenberg SD: Carcinoma and diffuse interstitial fibrosis of lung. *Cancer* 31:1078–1086, 1973.

32. Turner-Warwick M, Lebowitz M, Burrows B, Johnson A: Cryptogenic fibrosing alveolitis and lung cancer. *Thorax* 35:496–499, 1980.

33. Kawai T, Yakumaru K, Suzuki M, Kageyama K: Diffuse interstitial pulmonary fibrosis and lung cancer. *Acta Pathol Jpn* 37:11–19. 1987.

34. Yousem SA: The pulmonary pathologic manifestations of the CREST syndrome. *Hum Pathol* 21:467–474, 1990.

35. Béjui-Thivolet F, Liagre N, Chignol MC, Chardonnet Y, Patricot LM: Detection of human papillomavirus DNA in squamous bronchial metaplasia and squamous cell carcinomas of the lung by in situ hybridization using biotinylated probes in paraffin-embedded specimens. *Hum Pathol* 21:111–116, 1990.

36. Szabó J, Sepp R, Nakamoto K, Maeda M, Sakamoto H, Uda H: Human papillomavirus not found in squamous and large cell lung carcinomas by polymerase chain reaction. *Cancer* 73:2740–2744, 1994.

37. Helmuth RA, Strate RW: Squamous carcinoma of the lung in a nonirradiated nonsmoking patient with juvenile laryngotracheal papillomatosis. *Am J Surg Pathol* 11:643–650, 1987.

38. Byrne JC, Tsao MS, Fraser RS, Howley PM: Human papillomavirus-11 DNA in a patient with chronic laryngotracheobronchial papillomatosis and metastatic squamous-cell carcinoma of the lung. *N Engl J Med* 317:873–878, 1987.

39. Guillou L, Sahli R, Chaubert P, Monnier P, Cuttat J-F, Costa J: Squamous cell carcinoma of the lung in a nonsmoking, nonirradiated patient with juvenile laryngotracheal papillomatosis. Evidence of human papillomavirus-11 DNA in both carcinoma and papillomas. *Am J Surg Pathol* 15:891–898, 1991.

40. Nash FA, Morgan JM, Tomkins JG: South London cancer study. *Br J Med* 2:715–721, 1968.

41. Flehinger BJ, Melamed MR, Zaman MB, Heelan RT, Perchick WB, Martini N: Early lung cancer detection: results of the initial (prevalence) radiologic and cytologic screening in the Memorial Sloan-Kettering study. *Am Rev Respir Dis* 130:555–560, 1984.

42. Fontana RS, Sanderson DR, Taylor WF, et al: Early lung cancer detection: Results of the initial (prevalence) radiologic and cytologic screening in the Mayo Clinic study. *Am Rev Respir Dis* 130:561–565, 1984.

43. Frost JK, Ball WC, Levin ML, et al: Early lung cancer detection: Results of the initial (prevalence) radiologic and cytologic screening in the Johns Hopkins study. *Am Rev Respir Dis* 130:549–554, 1984.

44. Eddy D: Screening for lung cancer. *Ann Intern Med* 111:232–237, 1989.

45. Soda H, Tomita H, Kohno S, Oka M: Limitation of annual screening chest radiography for the diagnosis of lung cancer. *Cancer* 72:2341–2346, 1993.

46. Smart CR: Annual screening using chest X-ray examination for the diagnosis of lung cancer. *Cancer* 72:2295–2298, 1993.

47. Mendelsohn J: Principles of neoplasia, in Isselbacher KJ, Braunwald E, Wilson JD, Martin JB, Fauci AS, Kasper DL (eds): *Harrison's Principles of Internal Medicine,* 13th ed. New York, McGraw-Hill, 1994, pp 1814–1826.

48. De la Monte SM, Hutchins GM, Moore GW: Paraneoplastic syndromes and constitutional symptoms in prediction of metastatic behavior of small cell carcinoma of the lung. *Am J Med* 77:851–857, 1984.

49. Dimopoulos MA, Fernandez JF, Samaan NA, Holoye PY, Vassilopoulou-Sellin R: Paraneoplastic Cushing's syndrome as an adverse prognostic factor in patients who die early with small cell lung cancer. *Cancer* 69:66–71, 1992.

50. Carr DT, Holoye PY, Hong WK: Bronchogenic carcinoma, in Murray JF, Nadel JA (eds), *Textbook of Respiratory Medicine,* 2d ed. Philadelphia: WB Saunders, 1994, pp 1528–1596.

51. Ratcliffe WA: Parathyroid hormone-related protein and hypercalcaemia of malignancy (editorial). *J Pathol* 173:79–80, 1994.

52. Collichio FA, Woolf PD, Brower M: Management of patients with small-cell carcinoma and the syndrome of ectopic corticotropin secretion. *Cancer* 73:1361–1367, 1994.

53. Kochersberger GG, Lyles KW: Skeletal disorders in malignant disease. *Clin Geriatr Med* 3:561–574, 1987.

54. Gosney MA, Gosney JR, Lye M: Plasma growth hormone and digital clubbing in carcinoma of the bronchus. *Thorax* 45:545–547, 1990.

55. Drlicek M, Grisold W, Liszka U: Correlation of circulating antineuronal antibodies (CANA) with paraneoplastic syndromes in lung cancer. *Lung Cancer* 8: 245–258, 1993.

56. Dinarello CA, Cannon JG, Wolff SM, et al: Tumor necrosis factor (cachectin) is an endogenous pyrogen and induces production of interleukin 1. *J Exp Med* 163:1433–1450, 1986.

57. Sato K, Fujii Y, Kakiuchi T, et al: Paraneoplastic syndrome of hypercalcemia and leukocytosis caused by squamous carcinoma cells (T3M-1) producing parathyroid hormone-related protein, interleukin 1 alpha, and granulocyte colony-stimulating factor. *Cancer Res* 49:4740–4746, 1989.

58. Minna JD: Neoplasms of the lung, in Isselbacher KJ, Braunwald E, Wilson JD, Martin JB, Fauci AS, Kasper DL (eds), *Harrison's Principles of Internal Medicine,* 13th ed. New York, McGraw-Hill, 1994, pp 1221–1229.

59. Kato H, Konaka C, Ono J, Takahashi M, Hayata Y: *Cytology of the Lung. Techniques and Interpretation.* Tokyo, Igaku-Shoin, 1983, p 2.

60. Kato H, Konaka C, Ono J, Takahashi M, Hayata Y: *Cytology of the Lung. Techniques and Interpretation.* Tokyo, Igaku-Shoin, 1983, p 20.

61. Gibbs AR, Seal RME: Examination of lung specimens. Broadsheet 123. *J Clin Pathol* 43:68–72,1990.

61a. Association of Directors of Anatomic and Surgical Pathology: Recommendations for the reporting of resected primary lung carcinomas. *Hum Pathol* 26:937–939, 1995.

62. Gallagher B, Urbanski SJ: The significance of pleural elastica invasion by lung carcinomas. *Hum Pathol* 21:512–517, 1990.

63. McElvaney G, Miller RR, Mullere NL, Nelems B, Evans KG, Ostrow DN: Multicentricity of adenocarcinoma of the lung. *Chest* 95:151–154, 1989.

64. Shimizu N, Ando A, Date H, Teramoto S: Prognosis of undetected intrapulmonary metastases in resected lung cancer. *Cancer* 71:3868–3872, 1992.

65. Spiessl B, Beahrs OH, Hermanek P, et al (eds). *TNM Atlas, Illustrated Guide to the TNM/pTNM Classification of Malignant Tumours,* UICC, 3d ed. Berlin, Springer-Verlag, 1992, pp 142–151.

66. Hermanek P, Henson DE, Hutter RVP, Sobin LH: *TNM Supplement 1993. A Commentary on Uniform Use,* UICC. Berlin, Springer-Verlag, 1993, p 57.

67. Mountain CF: A new international staging system for lung cancer. *Chest* 89:225S–233S, 1986.

68. Naruke T, Suemasu K, Ishikawa S: Lymph node mapping and curability at various levels of metastasis in resected lung cancer. *J Thorac Cardiovasc Surg* 76:832–839, 1978.

69. Saccomanno G, Archer VE, Auerbach O, Saunders RP, Brennan LM: Development of carcinoma of the lungs as reflected in exfoliated cells. *Cancer* 33:256–270, 1974.

70. Trump BF, McDowell EM, Glavin F, et al: The respiratory epithelium. III. Histogenesis of epidermoid metaplasia and carcinoma in situ in the human. *J Natl Cancer Inst* 61:563–575, 1978.

71. Ramaekers F, Huysmans A, Moesker O, et al: Monoclonal antibody to keratin filaments, specific for glandular epithelial and their tumors. Use in surgical pathology. *Lab Invest* 49:353–361, 1983.

72. Blobel GA, Moll R, Franke WW, Vogt-Moykopf I: Cytokeratins in normal lung and lung carcinomas. *Virchows Arch (Cell Pathol)* 45:407–429, 1984.

73. Nagamoto N, Saito Y, Sato M, et al: Clinicopathological analysis of 19 cases of isolated carcinoma in situ of the bronchus. *Am J Surg Pathol* 17:1234–1243, 1993.

74. Ranchod M: The histogenesis and development of pulmonary tumorlets. *Cancer* 39:1135–1145, 1977.

75. Noguchi M, Kodama T, Shimosato Y, et al: Papillary adenoma of type 2 pneumocytes. *Am J Surg Pathol* 10: 134–139, 1986.

76. Weng SY, Tsuchiya E, Matsubara T, Nakagawa K, Kinoshita I, Kitagawa T: Histological classification of atypical adenomatous hyperplasia (AAH) and their immunohistochemical characterization in comparison with adenocarcinomas of the lung. *Lung Cancer* 4:A38, 1988.

77. Weng S, Tsuchiya E, Satoh Y, Kitagawa T, Nakagawa K, Sugano H: Multiple atypical adenomatous hyperplasia of type II pneumocytes and bronchiolo-alveolar carcinoma. *Histopathology* 16:101–103, 1990.

78. Miller RR: Bronchioloalveolar cell adenomas. *Am J Surg Pathol* 14:904–912, 1990.

79. Nakanishi K: Alveolar epithelial hyperplasia and adenocarcinoma of the lung. *Arch Pathol Lab Med* 114:363–368, 1990.

80. Carey FA, Wallace WAH, Gergusson RJ, Kerr KM, Lamb D: Alveolar atypical hyperplasia in association with primary pulmonary adenocarcinoma: A clinicopathological study of 10 cases. *Thorax* 47:1041–1043, 1992.

81. Wu SC, Zin ZQ, Xu CW, Koo KS, Huang OL, Xie DQ: Multiple primary lung cancers. *Chest* 92:892–896, 1987.

82. Mori M, Chiba R, Takahashi T: Atypical adenomatous hyperplasia of the lung and its differentiation from adenocarcinoma. *Cancer* 72:2331–2340, 1993.

83. Kodama T, Biyajima S, Watanabe S, Shimosato Y: Morphometric study of adenocarcinomas and hyperplastic epithelial lesions in the peripheral lung. *Am J Clin Pathol* 85:146–151, 1986.

84. Yousem SA, Hochholzer L: Alveolar adenoma. *Hum Pathol* 17:1066–1071, 1986.

85. McDowell EM, Beal TF: *Biopsy Pathology of the Bronchi.* London, Chapman & Hall, 1986, p 280.

86. Sunday ME, Choi N, Spindel ER, Chin WW, Mark EJ: Gastrin-releasing peptide gene expression in small cell and large cell undifferentiated lung carcinomas. *Hum Pathol* 22:1030–1039. 1991.

87. Yesner R: Spectrum of lung cancer and ectopic hormones. *Pathol Annu* 13(I): 217–240, 1978.

88. Abeloff MD, Eggleston JC, Mendelsohn G, Ettinger DS, Baylin SB: Changes in morphologic and biochemical characteristics of small cell carcinoma of the lung. A clinico-pathologic study. *Am J Med* 66:757–764, 1979.

89. Goodwin G, Shaper JH, Abeloff MD, Mendelsohn G, Baylin SB: Analysis of cell surface proteins delineates a differentiation pathway linking endocrine and nonendocrine human lung cancers. *Proc Natl Acad Sci USA* 80:3807–3811, 1983.

90. McDowell EM, McLaughlin JS, Merenyi DK, Kieffer RF, Harris CC, Trump BF: The respiratory epithelium. V. Histogenesis of lung carcinomas in the human. *J Natl Cancer Inst* 61:587–606, 1978.

91. Dunnill MS, Gatter KC: Cellular heterogeneity in lung cancer. *Histopathology* 10:461–475, 1986.

92. Broers JLV, Klein Rot M, Oostendorp T, et al: Immunocytochemical detection of human lung cancer heterogeneity using antibodies to epithelial, neuronal, and neuroendocrine antigens. *Cancer Res* 47:3225–3234, 1987.

93. Mooi WJ, Dingemans KP, Wagenaar SS, Hart AAM, Wagenvoort CA: Ultrastructural heterogeneity of lung carcinomas: Representativity of samples for electron microsopy in tumour classification. *Hum Pathol* 21:1227–1234, 1990.

94. Willis RA: *Pathology of Tumours,* 4th ed. London, Butterworths, 1967, p 362.

95. World Health Organization: *Histological Typing of Lung Tumours,* 2d ed. Geneva, World Health Organization, 1981.

96. Hammar SP, Bockus D, Wheelis RF, Hill L: Electron-microscopic studies of undifferentiated lung tumors. *Chest* 72:400, 1977.

97. Horie A, Ohta M: Ultrastructural features of large cell carcinoma of the lung with reference to the prognosis of patients. *Hum Pathol* 12:423–432, 1981.

98. Hammar SP, Bolen JW, Bockus D, Remington F, Friedman S: Ultrastructural and immunohistochemical features of common lung tumors. *Ultrastruct Pathol* 9:283–318, 1985.

99. Hammar S: Adenocarcinoma and large cell undifferentiated carcinoma of the lung. *Ultrastruct Pathol* 11:251–274, 1987.

100. Feinstein AR, Gelfman NA, Yesner R: Observer variability in the histopathologic diagnosis of lung cancer. *Am Rev Respir Dis* 101:671–684, 1970.

101. Johnston WW: Histologic and cytologic patterns of lung cancer in 2,580 men and women over a 15-year period. *Acta Cytol* 32:163, 1988.

102. Carter D: Squamous cell carcinoma of the lung: An update. *Semin Diagn Pathol* 2:226, 1985.

103. Saccomanno G, Archer VE, Auerbach O, Kuschner T, Saunders RP, Klein MG: Histologic types of lung cancer among uranium miners. *Cancer* 27:515–523, 1971.

104. Horacek J, Placek V, Sevc J: Histologic types of bronchogenic cancer in relation to different conditions of radiation exposure. *Cancer* 40:832–835, 1977.

105. Nagamoto N, Saito Y, Suda H, et al: Relationship between length of longitudinal extension and maximal depth of transmural invasion in roentgenographically occult squamous cell carcinoma of the bronchus (nonpolypoid type). *Am J Surg Pathol* 13:11–20, 1989.

106. Nagamoto N, Saito Y, Ohta S, et al: Relationship of lymph node metastasis to primary tumor size and microscopic appearance of roentgenographically occult lung cancer. *Am J Surg Pathol* 13:1009–1013, 1989.

107. Trillo A, Guha A: Solitary condylomatous papilloma of the bronchus. *Arch Pathol Lab Med* 112:731–733, 1988.

108. Dulmet-Brender E, Jaubert F, Huchon G: Exophytic endobronchial epidermoid carcinoma. *Cancer* 57:1358–1364, 1986.

109. Dingemans KP, Mooi WJ: Ultrastructure of squamous cell carcinoma of the lung. *Pathol Annu* 19(I): 249–273, 1984.

110. Said JW, Nash G, Sassoon AF, Shintaku IP, Banks-Schlegel S: Involucrin in lung tumors. A specific marker of squamous differentiation. *Lab Invest* 49:563–568, 1983.

111. Schaafsma HE, Ramaekers FCS: Cytokeratin subtyping in normal and neoplastic epithelium: Basic principles and diagnostic applications. *Pathol Annu* 29(I): 21–62, 1994.

112. Moll R, Franke WW, Schiller DL, Geiger B, Krepler R: The catalog of human cytokeratins: Patterns of expression in normal epithelia, tumors and cultured cells. *Cell* 31:11–24, 1982.

113. Wetzels RHW, Schaafsma HE, Leigh IM, et al: Laminin and type VII collagen distribution in different types of human lung carcinoma: Correlation with expression of keratins 14, 16, 17 and 18. *Histopathology* 20:295–303, 1992.

114. Dale I, Brandtzaeg P: Expression of the epithelial L1 antigen as an immunohistochemical marker of squamous cell carcinoma of the lung. *Histopathology* 14:493–502, 1989.

115. Alvarez-Fernandez E: Alveolar trapping in pulmonary carcinomas. *Diagn Histopathol* 5:59–64, 1982.

116. Singh G, Kartyal SL, Ordinez NG, et al: Type II pneumocytes in pulmonary tumors. Implications for histogenesis. *Arch Pathol Lab Med* 108: 44–48, 1984.

117. Suster S, Huszar M, Herczeg E: Spindle cell squamous carcinoma of the lung. Immunocytochemical and ultrastructural study of a case. *Histopathology* 11:871–878, 1987.

118. Churg A, Johnston WH, Stulbarg M: Small cell squamous and mixed small cell squamous-small cell anaplastic carcinomas of the lung. *Am J Surg Pathol* 4:255–263, 1980.

119. Mooi WJ, Van Zandwijk N, Dingemans KP, Koolen MGJ, Wagenvoort CA: The "grey area" between small cell and non-small cell lung carcinomas: Light and electron microscopy versus clinical data in 14 cases. *J Pathol* 149:49–54, 1986.

120. Brambilla E, Moro D, Veale D, et al: Basal cell (basaloid) carcinoma of the lung: A new morphologic and phenotypic entity with separate prognostic significance. *Hum Pathol* 23:993–1003, 1992.

121. Moran CA, Travis WD, Rosado-de-Christenson M, Koss MN, Rosai J: Thymomas presenting as pleural tumors. Report of eight cases. *Am J Surg Pathol* 16:138–144, 1992.

122. James CL, Iyer PV, Leong AS-Y: Intrapulmonary thymoma. *Histopathology* 21:175–177, 1992.

123. Banerjee SS, Eyden BP, Wells S, McWilliam LJ, Harris M: Pseudoangiosarcomatous carcinoma: A clinicopathological study of seven cases. *Histopathology* 21:13–23, 1992.

124. Nappi O, Swanson PE, Wick MR: Pseudovascular adenoid squamous cell carcinoma of the lung: Clinicopathologic study of three cases and comparison with true pleuropulmonary angiosarcoma. *Hum Pathol* 25:373–378, 1994.

125. Barsky S, Cameron R, Osann KE, Tomita D, Holmes EC: Rising incidence of bronchioloalveolar lung carcinoma and its unique clinicopathologic features. *Cancer* 73:1163–1170, 1994.

126. Crotty TB, Myers JL, Katzenstein A-LA, Tazelaar HD, Swensen SJ, Churg A: Localized malignant mesotheliomas. A clinicopathologic and flow cytometric study. *Am J Surg Pathol* 18: 357–363, 1994.

127. Kish JK, Ro JY, Ayala AG, McMurtrey MJ: Primary mucinous adenocarcinoma of the lung with signet-ring cells. A histochemical comparison with signet-ring cell carcinomas of other sites. *Hum Pathol* 20:1097–1102, 1989.

128. Bonetti F, Colombari R, Zamboni G, Chilosi M: Signet-ring melanoma, S-100 negative. *Am J Surg Pathol* 13:522–523, 1989.

129. Edwards CW: Pulmonary adenocarcinoma: Review of 106 cases and proposed new classification. *J Clin Pathol* 40:125–135, 1987.

130. Scroggs MW, Roggli VL, Fraire AE, Sanfilippo F: Eosinophilic intracytoplasmic globules in pulmonary adenocarcinomas: A histochemical, immunohistochemicial and ultrastructural study of six cases. *Hum Pathol* 20:845–849, 1989.

131. Cagle PT, Alpert LC, Carmona PA: Peripheral biphasic adenocarcinoma of the lung: Light microscopic and immunohistochemical findings. *Hum Pathol* 23:197–200, 1992.

132. Moran CA, Suster S, Koss MN: Primary adenoid cystic carcinoma of the lung. A clinicopathologic and immunohistochemical study of 16 cases. *Cancer* 73:1390–1397, 1994.

133. Moran CA, Suster S, Askin FB, Koss MN: Benign and malignant salivary gland-type mixed tumors of the lung. Clinicopathologic and immunohistochemical study of eight cases. *Cancer* 73:2481–2490, 1994.

134. Dixon AY, Moran JF, Wesselius LJ, McGregor DH: Pulmonary mucinous cystic tumor. Case report with review of the literature. *Am J Surg Pathol* 17:722–728, 1993.

135. Nakatani Y, Kitamura H, Inayama Y, Ogawa N: Pulmonary endodermal tumor resembling fetal lung. The optically clear nucleus is rich in biotin. *Am J Surg Pathol* 18: 637–642, 1994.

136. Van de Molengraft FJJM, Van Niekerk CC, Jap PHK, Poels LG: OV-TL 12/30 (keratin 7 antibody) is a marker of glandular differentiation in lung cancer. *Histopathology* 22:35–38, 1993.

137. Asada Y, Aratake Y, Kotani T, et al: Expression of dipeptidyl aminopeptidase IV activity in human lung carcinoma. *Histopathology* 23:265–270, 1993.

138. Upton MP, Hirohashi S, Tome Y, Miyazawa N, Suemasu K, Shimosato Y: Expression of vimentin in surgically resected adenocarcinomas and large cell carcinomas of lung. *Am J Surg Pathol* 10:560–567, 1986.

139. Singh G, Katyal SL, Torikata C: Carcinoma of type II pneumocytes. Immunodiagnosis of a subtype of "bronchioloalveolar carcinoma." *Am J Pathol* 102:195–208, 1981.

140. Mizutani Y, Nakajima T, Morinaga S, et al: Immunohisto-chemical localization of pulmonary surfactant apoproteins in various lung tumors, with special reference to lung adenocarcinoma subtypes. *Cancer* 61:532–537, 1988.

140a. Nicholson AG, McCormick CJ, Shimosata Y, Butcher DN, Sheppard NM: The value of PE-10, a monoclonal antibody against pulmonary surfactant, in distinguishing primary and metastatic lung tumours. *Histopathology* 27:57–60, 1995.

141. Linnoila R, Jensen SM, Steinberg SM, Mulshine JL, Eggleston JC, Gazdar AF: Peripheral airway cell marker expression in non-small cell lung carcinoma. *Am J Clin Pathol* 97–233–243, 1992.

142. Kondi-Paphitis A, Addis BJ: Secretory component in pulmonary adenocarcinoma and mesothelioma. *Histopathology* 10:1279–1287, 1986.

143. Sheibani K, Battifora H, Burke JS: Antigenic phenotype of malignant mesotheliomas and pulmonary adenocarcinomas. An immunohistologic analysis demonstrating the value of LeuM1 antigen. *Am J Pathol* 123:212–219, 1986.

144. Sheibani K, Shin SS, Weis LM: Ber-EP4 antibody as a discriminant in the differential diagnosis of malignant mesothelioma vs. adenocarcinoma. *Am J Surg Pathol* 15:779–784, 1991.

145. Otis CN, Carter D, Cole S, Battifora H: Immunohistochemical evaluation of pleural mesothelioma and pulmonary adenocarcinoma. A bi-institutional study of 47 cases. *Am J Surg Pathol* 11:445–456, 1987.

146. Wick MR, Loy T, Mills SE, Legier JF, Manivel JC: Malignant epithelioid pleural mesothelioma versus peripheral adenocarcinoma: A histochemical, ultrastructural, and immunohistologic study of 103 cases. *Hum Pathol* 21:759–766, 1990.

147. Brown RW, Clark GM, Tandon AK, Allred DC: Multiple-marker immunohistochemical phenotypes distinguishing malignant pleural mesothelioma from pulmonary adenocarcinoma. *Hum Pathol* 24:347–354, 1993.

148. Skov BG, Lauritzen AF, Hirsch F, Nielsen HW: The histopathological diagnosis of malignant mesothelioma vs. pulmonary adenocarcinoma: Reproducibility of the histopathological diagnosis. *Histopathology* 24:553–557, 1994.

149. Ruitenbeek T, Gouw ASH, Poppema S: Immunocytology of body cavity fluids. MOC-31, a monoclonal antibody discriminating between mesothelial and epithelial cells. *Arch Pathol Lab Med* 118: 265–269, 1994.

150. Chang K, Pai LH, Pass H, et al: Monoclonal antibody K1 reacts with epithelial mesothelioma but not with lung adenocarcinoma. *Am J Surg Pathol* 16:259–268, 1992.

151. Kawai T, Suzuki M, Torikata C, Suzuki Y: Expression of blood group-related antigens and *Helix pomatia* agglutinin in malignant pleural meothelioma and pulmonary adenocarcinoma. *Hum Pathol* 22:118–124, 1991.

152. Allred CD: Reply. *Hum Pathol* 24:1036, 1993.

153. Warhol MJ, Hickey WF, Corson JM: Malignant mesothelioma. Ultrastructural distinction from adenocarcinoma. *Am J Surg Pathol* 6:307–314, 1982.

154. Battifora H: The Pleura, in Sternberg S (ed), *Diagnostic Surgical Pathology*, 2d ed. New York, Raven Press, 1994, pp 1095–1123.

155. El-Naggar AK, Ordonez NG, Garnsey L, Batsakis JG: Epithelioid pleural mesotheliomas and pulmonary adenocarcinomas: A comparative DNA flow cytometric study. *Hum Pathol* 22:972–978, 1992.

156. Pääkö P, Risteli L, Autio-Harmainen H: Immunohistochemical evidence that lung carcinomas grow on alveolar basement membranes. *Am J Surg Pathol* 14:464–473, 1990.

157. Liebow AA: Bronchiolo-alveolar carcinoma. *Adv Intern Med* 10:329–358, 1960.

158. Rosenblatt MB, Lisa JR, Colliet F: Primary and metastatic broncho-alveolar carcinoma. *Dis Chest* 52:147–152, 1967.

159. Foster CS: Mucus-secreting 'alveolar-cell' tumour of the lung: A histochemical comparison of tumours arising within and outside the lung. *Histopathology* 4:567–577, 1980.

160. Beer DJ, Mark EJ: A 57-year old man with a chronic productive cough, dyspnea, and extensive bilateral air-space disease. Case records of the Massachusetts General Hospital. *N Engl J Med* 330:1599–1606, 1994.

161. Miller WT, Husted J, Freiman A, Atkinson B, Pietra GG: Bronchioloalveolar carcinoma: Two clinical entities with one pathologic diagnosis. *AJR Am J Roentgenol* 130:905–912, 1978.

162. Salisbury JR, Darby AJ, Whimster WF: Papillary adenocarcinoma of lung with psammoma bodies: Report of a case derived from type II pneumocytes. *Histopathology* 10:877–884, 1986.

163. Clayton F: Bronchioloalveolar carcinoma: Cell types, patterns of growth, and prognostic correlates. *Cancer* 57:1555–1564, 1986.

164. Clayton F: The spectrum and significance of bronchioloalveolar carcinomas. *Pathol Annu* 23:361–394, 1988.

165. Ohori NP, Yousem SA, Griffin J, et al: E: Comparison of extracellular matrix antigens in subtypes of bronchioloalveolar carcinoma and conventional pulmonary adenocarcinoma. An immunohistochemical study. *Am J Surg Pathol* 16:675–686, 1992.

166. Barsky SH, Grossman D, Ho J, Holmes EC: The multifocality of bronchioloalveolar lung carcinoma (BAC): Evidence and implications of a multiclonal origin (abstract). *Mod Pathol* 5:112A, 1992.

167. Miller RR, Nelems B, Evans KG, Muller NL, Ostrow DN: Glandular neoplasia of the lung. *Cancer* 61:1009–1014, 1988.

168. Bonne C: Morphological resemblance of pulmonary adenomatosis ("jaagsiekte") in sheep and certain cases of cancer of the lung in man. *Am J Cancer* 35:491–501, 1939.

169. Hod I, Herz A, Zimber A: Pulmonary carcinoma (jaagsiekte) of sheep. Ultrastructural study of early and advanced tumor lesions. *Am J Pathol* 86:545–558, 1977.

170. Auerbach O, Garfinkel L, Parks VR: Scar cancer of the lung: Increase over a 21-year period. *Cancer* 43:636–642, 1979.

171. Shimosato Y, Hashimoto T, Kodama T, et al: Prognostic implications of fibrotic focus (scar) in small peripheral lung cancers. *Am J Surg Pathol* 4:365–373, 1980.

172. Madri JA, Carter D: Scar cancers of the lung: Origin and significance. *Hum Pathol* 15:625–631, 1984.

173. Kung ITM, Lui IOL, Loke SL, et al: Pulmonary scar cancer. A pathologic reappraisal. *Am J Surg Pathol* 9:391, 1985.

174. Kolin A, Koutoulakis T: Role of arterial occlusion in pulmonary scar cancers. *Hum Pathol* 19:1161–1167, 1988.

175. Cagle PT, Cohle SD, Greenberg SD: Natural history of pulmonary scar cancer. Clinical and pathologic implications. *Cancer* 56:2031–2035, 1985.

176. Albain KS, Crowley JJ, Livingston RB: Long-term survival and toxicity in small cell lung cancer. Expanded Southwest Oncology Group Experience. *Chest* 99:1425–1432, 1991.

177. Hansen HH, Kristjansen PEG: Chemotherapy of small cell lung cancer. *Eur J Cancer* 27:342–349, 1991.

178. Bernard WG: The nature of the "oat-celled sarcoma" of the mediastinum. *J Pathol Bacteriol* 29:241–244, 1926.

179. Azzopardi JG: Oat-cell carcinoma of the bronchus. *J Pathol Bacteriol* 78: 513–519, 1959.

180. Yesner R: Small cell tumors of the lung. *Am J Surg Pathol* 7:775–785, 1983.

181. Carter D: Small-cell carcinoma of the lung. *Am J Surg Pathol* 7:787–795, 1983.

182. Moss F, Bobrow LG, Sheppard MN, et al: Expression of epithelial and neural antigens in small cell and non small cell lung carcinoma. *J Pathol* 149:103–111, 1986.

183. Yoneda K, Boucher LD: Bronchial epithelial changes associated with small cell carcinoma of the lung. *Hum Pathol* 24:1180–1183, 1993.

184. Mooi WJ, Dingemans KP, Van Zandwijk N: Prevalence of neuroendocrine granules in small cell lung carcinomas: Usefulness of electron microscopy in lung cancer classification. *J Pathol* 149:41–47, 1986.

185. Tötsch M, Müller LC, Hittmair A, Öfner D, Gibbs AR, Schmid KW: Immunohistochemical demonstration of chromogranins A and B in neuroendocrine tumors of the lung. *Hum Pathol* 23:312–316, 1992.

186. Sheppard MN, Corrin B, Bennett MH, Marangos PJ, Bloom SR, Polak JM: Immunocytochemical localization of neuron specific enolase in small cell carcinomas and carcinoid tumours of the lung. *Histopathology* 8: 171–181, 1984.

187. Griffiths AP, Mearns A, Horsfield GI: Combined small cell and bronchiolo-alveolar cell carcinoma. *Histopathology* 17:380–381, 1990.

188. Vollmer RT, Birch R, Ogden L, Crissman JD: Subclassification of small cell cancer of the lung: the Southeastern Cancer Study Group experience. *Hum Pathol* 16:247–252, 1985.

189. Davis S, Stanley KE, Yesner R, Kuang DT, Morris JF: Small-cell carcinoma of the lung: survival according to histologic subtype: A Veterans Administration Lung Group study. *Cancer* 47:1863–1866, 1981.

190. Yesner R: Classification of lung-cancer histology. *N Engl J Med* 312:652–653, 1985.

191. Mackay B, Lukeman JM, Ordóñez NG: *Tumors of the Lung.* Philadelphia, WB Saunders, 1991, p 230.

192. Hirsch FR, Osterlind K, Hansen HH: The prognostic significance of histopathologic subtyping of small cell carcinoma of the lung according to the World Health Organization classification. *Cancer* 52:2144–2160, 1983.

193. Radice PA, Matthews MJ, Ihde DC, et al: The clinical behavior of "mixed" small cell/large cell bronchogenic carcinoma compared to "pure" small cell subtypes. *Cancer* 50:2894–2902, 1982.

194. Fushimi H, Kihui M, Morino H, et al: Detection of large cell component in small cell lung carcinoma by combined cytologic and histologic examinations and its clinical implications. *Cancer* 70:599–605, 1992.

195. Bepler G, Neumann K, Holle R, Havemann K, Kalbfleisch H: Clinical relevance of histopathologic subtyping in small cell lung cancer. *Cancer* 64:74–79, 1989.

196. Hirsch FR, Matthews MJ, Aisner S, et al: Histopathological classification of small cell lung cancer. Changing concepts and terminology. *Cancer* 62:973–977, 1988.

197. Mayall FG, Gibbs AR: The histology and immunohistochemistry of small cell mesothelioma. *Histopathology* 20:47–51, 1992.

198. Churg A: The fine structure of large cell undifferentiated carcinoma of the lung. Evidence for its relation to squamous cell carcinomas and adenocarcinomas. *Hum Pathol* 9:143–156, 1978.

199. Albain KS, True LD, Golomb HM, Hoffman PC, Little AG: Large cell carcinoma of the lung. Ultrastructural differentiation and clinicopathologic correlations. *Cancer* 56:1618–1623, 1985.

200. Hammond ME, Sause WT: Large cell neuroendocrine tumors of the lung: Clinical significance and histopathologic definition. *Cancer* 56:1624–1629, 1985.

201. Berendse HH, De Leij L, Poppema S: Clinical characterization of non-small cell lung cancer tumors showing neuroendocrine differentiation features. *J Clin Oncol* 7:1614–1620, 1989.

202. Travis WD, Linnoila I, Tsokos MG, et al: Neuroendocrine tumors of the lung with proposed criteria for large-cell neuroendocrine carcinoma. An ultrastructural, immunohistochemical, and flow cytometric study of 35 cases. *Am J Surg Pathol* 15:529–553, 1991.

203. Loy TS, Darkow GVD, Quesenberry JT: Immunostaining in the diagnosis of pulmonary neuroendocrine carcinomas. An immunohistochemical study with ultrastructural correlations. *Am J Surg Pathol* 19:173–182, 1995.

204. Begin LR, Eskandari J, Joncas J, Panasci L: Epstein-Barr virus related lymphoepithelioma-like carcinoma of lung. *J Surg Oncol* 36:280–283, 1987.

205. Butler AE, Colby TV, Weiss L, Lombard C: Lymphoepithelioma-like carcinoma of the lung. *Am J Surg Pathol* 13:632–639, 1989.

206. Pittaluga S, Wong MP, Chung LP, Loke S-L: Clonal Epstein-Barr virus in lymphoepithelioma-like carcinoma of the lung. *Am J Surg Pathol* 17:678–682, 1993.

206a. Chan JKG, Hui P-K, Tsang WYW, et al: Primary lymphoepithelioma-like carcinoma of the lung: A clinicopathologic study of 11 cases. *Cancer* 76:413–422, 1995.

207. Gaffey MJ, Weiss LM: Association of Epstein-Barr virus with human neoplasia. *Pathol Annu* 271:55–74, 1992.

208. Fishback NF, Travis WD, Moran CA, Guinee DG, McCarthy WF, Koss MN: Pleomorphic (spindle/giant cell) carcinoma of the lung. A clinicopathologic correlation of 78 cases. *Cancer* 73:2936–2945, 1994.

209. Geddy PM, Gouldesbrough DR: Basal cell (basaloid) carcinoma of the lung. *Hum Pathol* 24:452, 1993.

210. Brambilla E, Moro D: Reply. *Hum Pathol* 24:452–453, 1993.

211. Moro D, Brichon PY, Brambilla E, Veale D, Labat F, Brambilla C: Basaloid bronchial carcinoma. A histologic group with a poor prognosis. *Cancer* 73:2734–2739, 1994.

211a. Saltarello MG, Fleming MV, Wenig BM, Gal AA, Mansour KA, Travis WD: Primary basaloid squamous cell carcinoma of the trachea. *Am J Clin Pathol* 104:594–598, 1995.

212. Addis BJ, Dewar A, Thurlow NP: Giant cell carcinoma of the lung—immunohistochemical and ultrastructural evidence of dedifferentiation. *J Pathol* 155:231–240, 1988.

213. Ginsberg SS, Buzaid AC, Stern H, Carter D: Giant cell carcinoma of the lung. *Cancer* 70:606–610, 1992.

214. Addis BJ, Corrin B: Pulmonary blastoma, carcinosarcoma and spindle-cell carcinoma: An immunohistochemical study of keratin intermediate filaments. *J Pathol* 147:291–301, 1986.

215. Matsui K, Kitagawa M, Miwa A: Lung carcinoma with spindle cell components: Sixteen cases examined by immunohistochemistry. *Hum Pathol* 23:1289–1297, 1992.

216. Humphrey PA, Scroggs MW, Roggli VL, Shelburne JD: Pulmonary carcinomas with a sarcomatoid element: An

217. Ro JY, Chen JL, Lee JS, Sahin AA, Ordóñez NG, Ayala AG: Sarcomatoid carcinoma of the lung. Immunohistochemical and ultrastructural studies of 14 cases. *Cancer* 69:376–386, 1992.

218. Edwards C, Carlisle A: Clear cell carcinoma of the lung. *J Clin Pathol* 38: 880–885, 1985.

219. Katzenstein A-LA, Prioleau PG, Askin FB: The histologic spectrum and signifance of clear-cell change in lung carcinoma. *Cancer* 45:943–947, 1980.

220. Gaffey MJ, Milles SE, Askin FB, et al: Clear cell tumor of the lung. A clinicopathologic, immunohistochemical, and utrastructural study of eight cases. *Am J Surg Pathol* 14:248–259, 1990.

221. Gaffey MJ, Mills SE, Zarbo RJ, Weiss LM, Gown AM: Clear cell tumor of the lung. Immunohistochemical evidence of melanogenesis. *Am J Surg Pathol* 15:644–653, 1991.

222. Chuang MT, Marchevsky A, Teirstein AS, Kirschner PA, Kleinerman J: Diagnosis of lung cancer by fibreoptic bronchoscopy: Problems in the histological classification of non-small cell carcinomas. *Thorax* 39:175–178, 1984.

223. Nash G: The diagnosis of lung cancer in the 80s. Will routine light microscopy suffice? *Hum Pathol* 14:1021–1023, 1983.

224. Mooi WJ, Dingemans KP Wagenaar SS, Van den Bergh Weerman MA: Heterogeneity of lung cancer: The problem of sample error in diagnostic electron microscopy. *Eur J Respir Dis* 70:45S–52S, 1987.

225. Elema JD, Keuning HM: The use of electron microscopy for the diagnosis of cancer in bronchial biopsies. *Hum Pathol* 19:304–308, 1988.

226. Buccheri G, Ferrigno D: Prognostic factors in lung cancer: Tables and comments. *Eur Respir J* 7:1350–1364, 1994.

227. Kurokawa T. Matsuno Y, Noguchi M, Mizuno S, Shimosato Y: Surgically curable "early" adenocarcinoma in the periphery of the lung. *Am J Surg Pathol* 18: 431–438, 1994.

228. Mountain CF, Lukeman JM, Hammar SP: Lung cancer classification: The influence of disease extent and cell type to survival in clinical trials population. *J Surg Oncol* 35:147, 1987.

229. Kayser K, Bulzebruck H, Probst G, Vogt-Moykopf I: Retrospective and prospective tumor staging evaluating prognostic factors in operated bronchus carcinoma patients. *Cancer* 59:355–361, 1987.

230. Schmidt RA, Rush VW, Piantadosi S: A flow cytometric study of non-small cell lung cancer classified as T1N0. *Cancer* 69:78–85, 1992.

231. Lamb D: Histological classification of lung cancer. *Thorax* 39:161–165, 1984.

232. Lee T, Horner RD, Silverman JF, Chen Y, Jenny C, Scarantino LW: Morphometric and morphologic evaluations in stage III non-small cell lung cancers. Prognostic significance of quantitative assessment of infiltrating lymphoid cells. *Cancer* 63:309–316, 1989.

233. Macchiarini P, Fontanini G, Hardin MJ, Squantini F, Angeletti CA: Relation of neovascularization to metastasis of non-small cell lung cancer. *Lancet* 340:145–146, 1992.

234. Ogawa J, Tsurumi T, Yamada S, Koide S, Shohtsu A: Blood vessel invasion and expression of sialyl Lewis and proliferating cell nuclear antigen in stage I non-small cell lung cancer. Relation to postoperative recurrence. *Cancer* 73:1177–1183, 1994.

234a. Kanters SDJM, Lammers J-WJ, Voest EE: Molecular and biological factors in the prognosis of non-small cell lung cancer. *Eur Respir J* 8:1389–1387, 1995.

235. Fox SB, Jones M, Dunnill MS, Gatter KC, Mason DY: Langerhans cells in human lung tumours: An immunohistological study. *Histopathology* 14:269–275, 1989.

236. Fontanini G, Macchiarini P, Pepe S, et al: The expression of proliferating cell nuclear antigen in paraffin sections of peripheral, node-negative non-small cell lung cancer. *Cancer* 70:1520–1527, 1992.

237. Carey FA, Fabbroni G, Lamb D: Expression of proliferating cell nuclear antigen in lung cancer: A systematic study and correlation with DNA ploidy. *Histopathology* 20:499–503, 1992.

238. Filderman AE, Silvestri GA, Gatsonis C, Luthringer DJ, Honig J, Flynn SD: Prognostic significance of tumor proliferative fraction and DNA content in stage I non-small cell lung cancer. *Am Rev Respir Dis* 146:707–710, 1992.

239. Straus MJ, Moran RE, Schackney SE: Growth characteristics of lung cancer, in Straus MJ (ed), *Lung Cancer: Clinical Diagnosis and Treatment,* 2d ed. New York, Grune & Stratton, 1982, p 24.

240. Van Bodegom PC, Baak JP, Stroet-van Galen C, et al: The percentage of aneuploid cells is significantly correlated with survival in accurately staged patients with stage I resected squamous cell lung cancer and long term follow up. *Cancer* 63:143, 1989.

241. Tirindelli-Danesi D, Teodori L, Mauro F, et al: Prognostic significance of flow cytometry in lung cancer: A 5-year study. *Cancer* 60:844–851, 1987.

242. Pezzella F, Turley H, Kuzu I, et al:*Bcl-2* protein in non-small-cell lung carcinoma. *N Engl J Med* 329:690–694, 1993.

243. Gaffney EF, O'Neill AJ, Staunton MJ: *Bcl-2* and prognosis in non-small-cell lung carcinoma. Letter to the editor. *N Engl J Med* 330:1757–1758, 1994.

244. Quinlan DC, Davidson AG, Summers CL, Warden HE, Doshi HM: Accumulation of p53 protein correlates with a poor prognosis in human lung cancer. *Cancer Res* 52:4828–4831, 1992.

245. Mitsudomi T, Oyama T, Kusano T, Osaki T, Nakanishi R, Shirakusa T: Mutations of the p53 gene as a predictor of poor prognosis in patients with non-small cell lung cancer. *J Natl Cancer Inst* 85:2018–2023, 1993.

246. Weiner DB, Nordberg J, Robinson R, et al: Expression of the *neu* gene-encoded protein (p185[neu]) in human non-small cell carcinomas of the lung. *Cancer Res* 50:421, 1990.

247. Kern JA, Schwartz DA, Nordberg JE, et al: p815[neu] expression in human lung adenocarcinoma predicts shortened survival. *Cancer Res* 50:5184–5191, 1990.

248. Shiraishi M, Noguchi M, Shimosato Y, Sekiya T: Amplification of protoocogenes in surgical specimens of human lung carcinomas. *Cancer Res* 49:6474–6479, 1989.

249. Sukoh N, Abe S, Ogura S, et al: Immunohistochemical study of cathepsin B. Prognostic significance in human lung cancer. *Cancer* 74:46–51, 1994.

250. Davidsohn I, Kovarik S, Lee CL: A, B, and O substances in gastrointestinal carcinoma. *Arch Pathol* 81:381–390, 1966.

251. Davidsohn I, Kovarik S, Ni LY: Isoantigens A, B, and H in benign and malignant lesions of the cervix. *Arch Pathol* 87:306–314, 1969.

252. Orntoft TF, Wolf H, Clausen H, Dabelsteen E, Hakomori S: Blood group ABH-related antigens in normal and malignant bladder urothelium: Possible structural basis for the deletion of type-2 chain ABH antigens in invasive carcinomas. *Int J Cancer* 43:774–780, 1989.

253. Lee JS, Ro JY, Sahin AA, et al: Expression of blood-group antigen A—a favourable prognostic factor in non-small cell lung cancer. *N Engl J Med* 324:1084–1090, 1991.

254. Matsumoto H, Muramatsu H, Shimotakahara T, et al: Correlation of expression of ABH blood group carbohydrate antigens with metastatic potential in human lung carcinomas. *Cancer* 72:75–81, 1993.

255. Miyake M, Taki T, Hitomi S, Hakomori SI: Correlation of expression of H/Le[y]/Le[b] antigens with survival in patients with carcinoma of the lung. *N Engl J Med* 327:14–18, 1992.

256. Miyake M, Hakomori SI: A specific cell surface glycoconjugate controlling cell motility: Evidence by functional monoclonal antibodies that inhibit cell motility and tumor cell metastasis. *Biochemistry* 30:3328–3334, 1991.

257. Kawai T, Suzuki M, Kase K, Ozeki Y: Expression of carbohydrate antigens in human pulmonary adenocarcinoma. *Cancer* 72:1581–1587, 1993.

258. Mooi WJ, Dewar A, Springall D, Polak JM, Addis BJ: Non-small cell lung carcinomas with neuroendocrine features: A light microscopic, immunohistochemical and ultrastructural study of 11 cases. *Histopathology* 13:329–337, 1988.

259. Wick MR, Berg LC, Hertz MI: Large cell carcinoma of the lung with neuroendocrine differentiation—a comparison with large cell "undifferentiated" pulmonary tumors. *Am J Clin Pathol* 97:796–805, 1992.

260. Graziano SL, Tatum AH, Newman NB, et al: The prognostic significance of neuroendocrine markers and carcinoembryonic antigen in patients with resected stage I and II non-small cell lung cancer. *Cancer Res* 54:2908–2913, 1994.

261. Sheppard MN: Neuroendocrine differentiation in lung tumours. *Thorax* 46:843–850, 1991.

262. Kibbelaar RE, Moolenaar CEC, Michalides RJAM, Bitter-Suermann D, Addis BJ, Mooi WJ: Expression of the embryonal neural cell adhesion molecule N-CAM in lung carcinoma. Diagnostic usefulness of monoclonal antibody 735 for the distinction between small cell lung cancer and non-small cell lung cancer. *J Pathol* 159:23–28, 1989.

263. Mooi WJ, Wagenaar SS, Schol D, Hilgers J: Monoclonal antibody 123C3:Its value in lung tumour classification. *Mol Cell Probes* 2:31–37, 1988.

264. Kibbelaar RE, Moolenaar KEC, Michalides RJAM, et al: Neural cell adhesion molecule expression, neuroendocrine differentiation and prognosis in lung carcinoma. *Eur J Cancer* 27:431–435, 1991.

265. Pujol JL, Simony J, Demoly P, et al: Neural cell adhesion molecule and prognosis of surgically resected lung cancer. *Am Rev Resp Dis* 148:1071–1075, 1993.

266. Kane RD, Hawkins HK, Miller JA, Noce PS: Microscopic pulmonary tumor emboli associated with dyspnea. *Cancer* 36:1473–1482, 1975.

267. Bourke SA, Henderson AF, Stevenson RD, Banham SW: Endobronchial metastases simulating primary carcinoma of the lung. *Respir Med* 83:151–152, 1989.

Chapter 32

Pulmonary Sarcomas, Blastomas, Carcinosarcomas, and Teratomas

Michael Koss / William Travis / Cesar Moran

Primary pulmonary sarcomas are rare in comparison with carcinomas of the lung and are even infrequent in comparison with metastatic sarcomas to the lung. In general, the frequency of histologic subtypes mirrors that of their counterparts in the soft tissues. Thus, the most common forms are malignant fibrous histiocytoma and leiomyosarcoma; the epidemic of acquired immunodeficiency syndrome (AIDS) has also produced a remarkable increase in pulmonary involvement by Kaposi's sarcoma. Other rarer pulmonary sarcomas include chondrosarcomas and osteosarcomas, rhabdomyosarcoma, fibrosarcoma, liposarcomas, and angiosarcoma. Many histologic subtypes of sarcoma can also be seen in pulmonary artery sarcomas, including leiomyosarcoma, rhabdomyosarcoma, and fibrosarcoma.

Malignant schwannomas and neurofibrosarcomas are vanishingly rare as primary lung tumors, outside the setting of von Recklinghausen's disease, and as discussed in Chap. 29. Leiomyosarcoma will be discussed in the same chapter.

FIBROUS AND FIBROHISTIOCYTIC TUMORS

Fibrosarcomas

Fibrosarcomas are malignant tumors composed of spindle cells that demonstrate a fibroblastic phenotype. In early studies, fibrosarcomas were considered to be one of the more common types of pulmonary sarcoma,[1,2] but with the systematic use of immunohistochemistry and electron microscopy, this tumor appears to be rare. For example, at least some pediatric "fibrosarcomas"[3–5] have been reclassified as congenital myofibroblastic tumors arising from the peribronchial mesenchyme ("congenital peribronchial myofibroblastic tumor") after immunohistochemical studies.[6] In addition, some fibrosarcomas that were thought to be primary in lung were actually pulmonary artery sarcomas.

The tumor appears in adults who range in age from 23 to 69 years, with a median age of 49 years.[1,7] Their symptoms vary depending on the location of the neoplasm. Those with endobronchial tumors have cough, dyspnea, and hemoptysis, whereas patients with intrapulmonary masses are most often asymptomatic. Chest radiographs usually show a well defined mass that may or may not be cavitated.

Grossly, endobronchial tumors are small (1 to 3 cm), polypoid or pedunculated lesions occurring in the lobar or main stem bronchi. Smaller tumors are more likely to be strictly endobronchial, whereas large ones tend to spread into the contiguous lung. Parenchymal tumors are usually larger than their endobronchial counterparts, measuring up to 2 to 16 cm in diameter. In general, they are solitary, well circumscribed, and unencapsulated masses with gray-white or yellow, firm cut surfaces, but there may be areas of hemorrhage or cyst formation.

Microscopically, as noted above, polypoid endobronchial tumors are usually confined to the bronchial wall (Fig. 32-1); however, they may extend beyond the cartilage plates into contiguous lung. The respiratory mucosa covering the polypoid growth may be intact or eroded. There may be foci of necrosis and hemorrhage within the neoplasm that are responsible for the clinical findings of hemoptysis.

Intrapulmonary tumors are usually well demarcated from the surrounding lung (Fig. 32-2). They surround and obliterate the underlying lung structures. Peripheral tumors may occasionally impinge on the visceral pleura.[1]

FIGURE 32-1
Whole mount view of polypoid, intrabronchial fibrosarcoma. (Figure 3 from Guccion and Rosen,[1] with permission.)

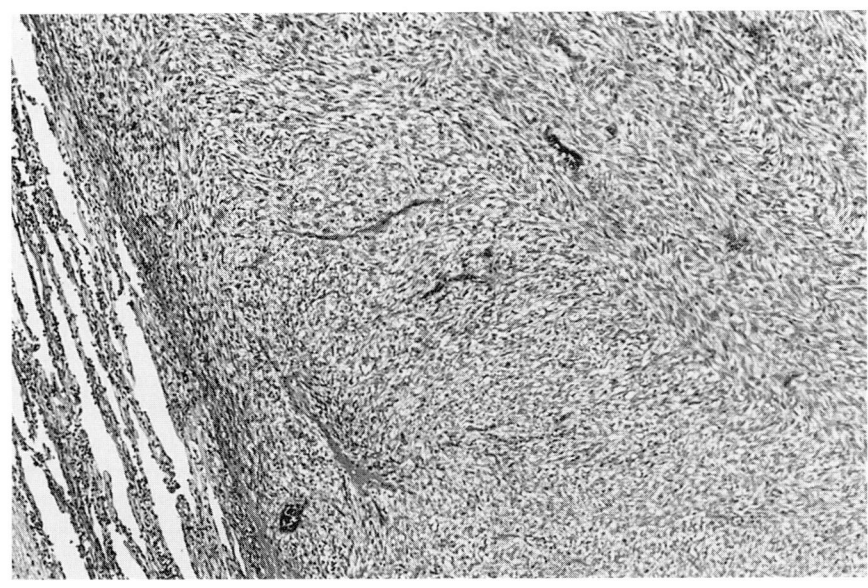

FIGURE 32-2
Intraparenchymal fibrosarcoma. The tumor consists of a solid spindle cell neoplasm well delineated from the surrounding lung parenchyma. (Low magnification.)

Fibrosarcomas are generally highly cellular. They consist of spindle cells with pointed cell bodies and nuclei growing in a herringbone or fascicular pattern. The amount of stromal collagen, albeit variable, is usually scant[8] (Figs. 32-3 and 32-4). There may be areas of necrosis. Mitoses vary from three or fewer (particularly in endobronchial tumors) to 8 to 40 or more per 10 high power fields. Reticulin stains show a fine network of fibers around individual cells. The malignant cells react with antibody to vimentin.

Surgical resection is the best opportunity for cure. Chemotherapy and radiotherapy are reserved for tumors that cannot be excised. The patients frequently develop distant metastases and have a high mortality rate.[1,7] In one series of patients, only two (22 percent) of nine achieved long-term survival and all patients with tumors greater than 5 cm died.[7] The size of the tumor at clinical presentation is linked to some extent with its location. Endobronchial tumors, because they provoke symptoms relatively early in the course, tend to be smaller on discovery and have a correspondingly better prognosis. The number of mitoses (>8/10 high power fields) may also be a prognostic factor.

Perhaps the most important point to remember in differential diagnosis is that metastatic sarcomas are far more frequent than primary pulmonary sarcomas. A diagnosis of sarcoma of lung should be con-

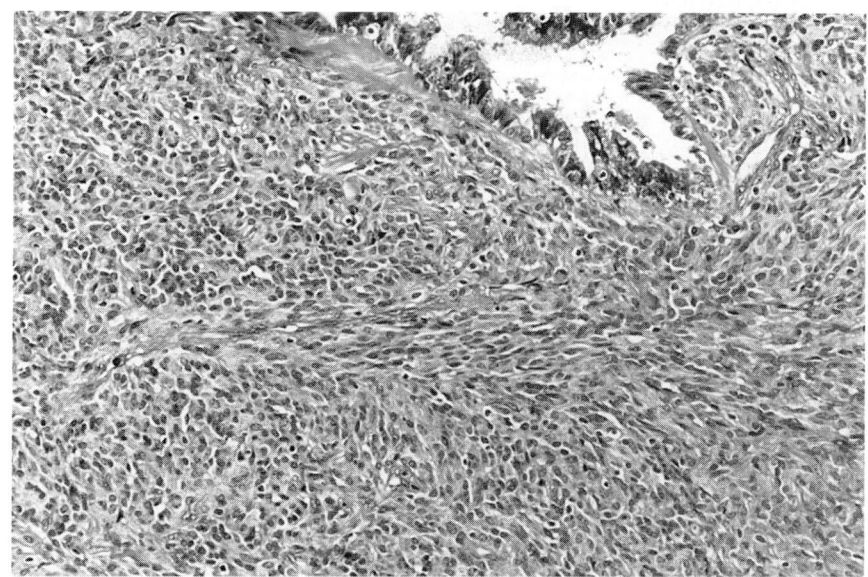

FIGURE 32-3
Pulmonary fibrosarcoma. This view shows the spindle cells arrayed in a typical fascicular pattern. The tumor invades around a small bronchus. (Low magnification.)

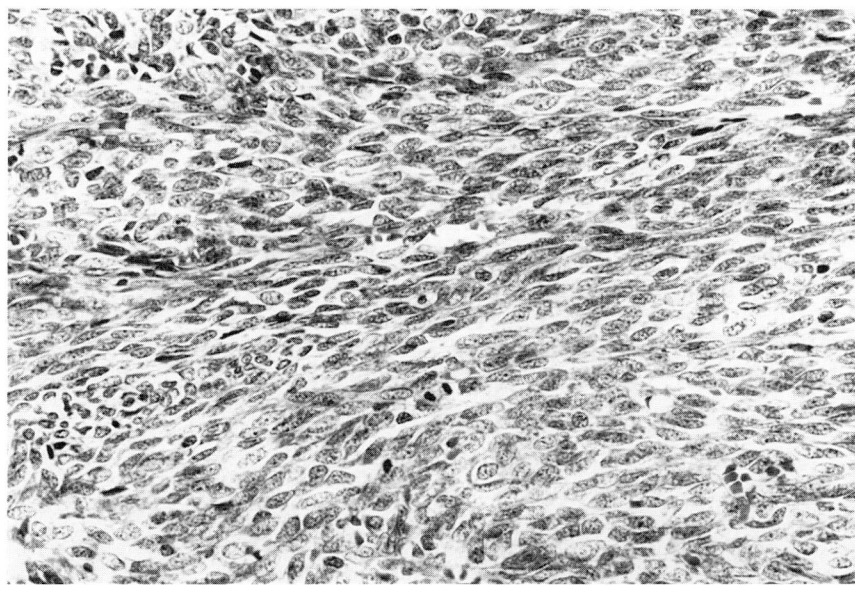

FIGURE 32-4
Pulmonary fibrosarcoma. The tumor shows marked crowding of the spindle cells with an easily identifiable mitotic figure. (Intermediate magnification.)

sidered only after rigorous exclusion of an origin in other primary sites. This is particularly true of metastatic monophasic synovial sarcomas, whose histologic appearance can mimic fibrosarcoma. The finding of diffuse and strong cytokeratin in a "fibrosarcoma" by immunohistochemical methods requires one to exclude synovial sarcoma. The differential diagnosis also includes other rare spindle cell sarcomas of lung, such as leiomyosarcoma, as well as localized fibrous tumor of the pleura, intrapulmonary fibrous tumor, and inflammatory pseudotumor. A careful application of immunohistochemical stains and electron microscopical observations helps exclude the first two possibilities, whereas a high

mitotic rate, foci of necrosis, or high cellularity is unusual in inflammatory pseudotumor.

Malignant Fibrous Histiocytoma

Malignant fibrous histiocytoma (MFH) is a sarcoma showing a variety of histologic patterns and composed of a mixture of pleomorphic tumor giant cells, fibroblasts, histiocyte-like cells, and undifferentiated mesenchymal cells. The tumor is most frequent in the soft tissues of the extremities, retroperitoneum, and trunk. It is rare as a primary lung neoplasm. By 1988, there were approximately 40 cases of primary pulmonary MFH supported by electron microscopic or immunohistochemical study.[9-13]

Patients are most often 50 to 70 years old, and the sexes are affected equally.[13] Sputum cytologies are usually negative for malignant cells. Bronchoscopy typically shows no gross endobronchial lesions although brushings may occasionally demonstrate malignant cells.[10] Chest radiographs demonstrate a solitary mass.

Grossly, the tumor is most often solitary, gray white, occasionally cavitated, and measures 2 to 10 cm. Microscopically, the neoplasm shows a polymorphous cell population arranged in storiform, pleomorphic, or fascicular patterns. Often more than one of these is present in a given case. The storiform pattern is characterized by interlacing bundles of spindle cells arranged in a cartwheel manner (Fig. 32-5). The pleomorphic malignant fibrous histiocytomas show a sheetlike growth of spindle, histiocyte-like, and pleomorphic giant cells. The fascicular pattern consists of parallel arrays of spindle (fibroblast-like) cells (Fig. 32-6). The tumor cells include spindled fibroblastic and myofibroblastic cells, numbers of cytologically atypical histiocyte-like cells showing moderate eosinophilic cytoplasm, with single vesicular nuclei, irregular shaped xanthoma cells, multinucleated giant cells, and variable numbers of lymphocytes and plasma cells (Fig. 32-7). The stroma is usually scant; myxoid areas may be present. Two important findings are numerous mitoses, including atypical mitoses, and extensive foci of necrosis. Neutrophils can be seen around areas of necrosis; occasionally, there are numerous, diffusely interspersed neutrophils. Histochemical stains for mucin and glycogen are negative, except near areas of necrosis, where intracellular glycogen may be found.

The ultrastructural appearance reflects the light microscopic diversity of this tumor.[9,10] There are spindle-shaped fibroblasts with numerous cisternae of rough endoplasmic reticulum and simplified or intermediate-type cellular junctions. Myofibroblasts, cells combining the abundant rough endoplasmic reticulum of fibroblasts with cytoplasmic aggregates of thin filaments, lateral dense plaques, and pinocytotic vesicles, also occur.[9,10] Histiocyte-like cells with numerous lysosomes, lipid droplets, occasional profiles of rough endoplasmic reticulum, and ruffled cell surfaces (pseudopodia) are present. Finally, there can be undifferentiated mesenchymal cells and cells showing features intermediate between fibroblasts and histiocytes.

The tumor cells stain diffusely with immunohistochemical stains for vimentin and more focally for α 1-antitrypsin[10,12,14] and α 1-antichymotrypsin.[10] Stains for keratin, epithelial membrane antigen (EMA), carcinoembryonic antigen (CEA), S-100 protein, desmin, and myoglobin are negative.

As far as differential diagnosis is concerned, metastatic MFH from the soft tissues is far more common than primary MFH of the lung. Still, only 0.5 percent of MFHs metastasize to the lung before the primaries are discovered.[15]

Pleomorphic (spindle and giant cell) carcinoma is far more frequent as a primary lung tumor than is MFH. Further, it may show spindle cells, multinucleated giant cells, a storiform pattern, numerous

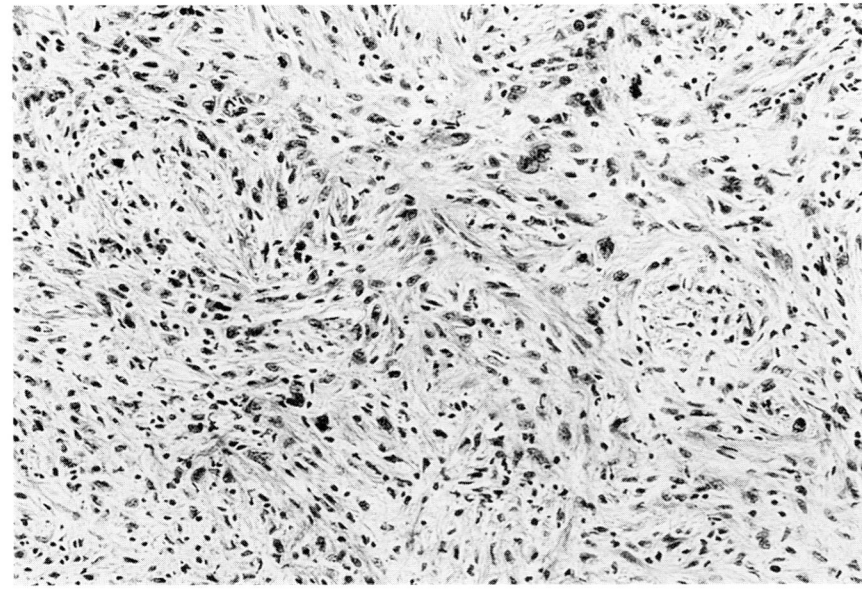

FIGURE 32-5
Malignant fibrous histiocytoma in lung. The tumor shows a storiform pattern of spindled malignant cells. (Intermediate magnification.)

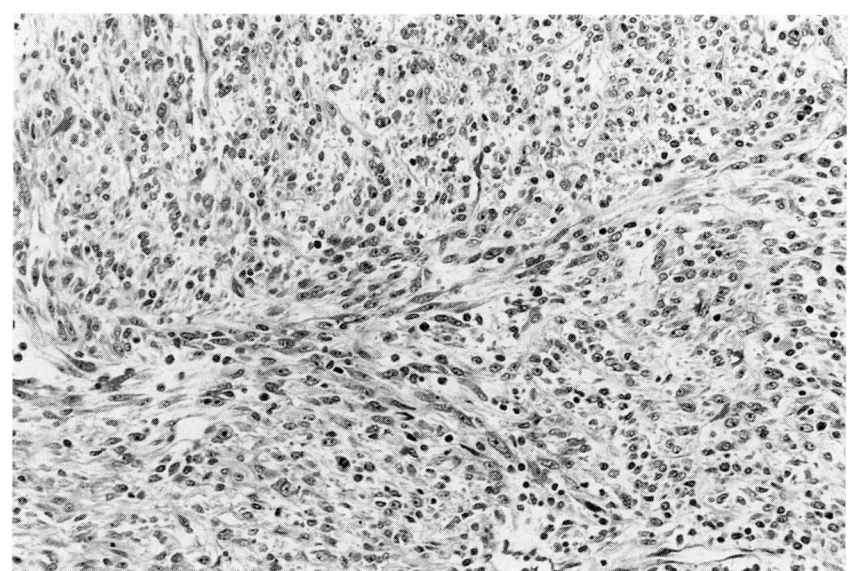

FIGURE 32-6
Fascicular pattern in pulmonary malignant fibrous histiocytoma shows parellel arrays of spindle (fibroblast-like) cells. (Intermediate magnification.)

mitoses, and areas of necrosis. Cohesive epithelial cell nests, squamous or glandular differentiation, or intracellular mucin suggest carcinoma, but it may be impossible to distinguish the tumors by light microscopy alone, a fact that makes it difficult to accept some published cases of pulmonary MFH.[16] The finding of intracellular keratin, CEA, or other epithelial epitopes by immunohistochemical methods or the presence of desmosomes, junctional complexes, microvilli within glands, or cytoplasmic tonofilaments by electron microscopy supports the diagnosis of carcinoma over MFH. Because of the difficulty in distinguishing spindle and giant cell carcinoma by light microscopy alone, we recommend that

all cases of putative MFH be evaluated by either immunohistochemistry or electron microscopy for epithelial features.

Malignant fibrous histiocytoma must also be distinguished from the fibrohistiocytic variant of inflammatory pseudotumor. The latter typically lacks anaplasia, necrosis, and the numerous mitoses of MFH. However, there appears to be a gamut of fibrohistiocytic proliferations that lie between benign fibrohistiocytic lesions and MFH. These lesions may show increased cellularity, mitotic rates up to 1/50 high power fields, small foci of necrosis, or bizarre giant cells. Still, a definite determination of malignancy may not be possible on histologic

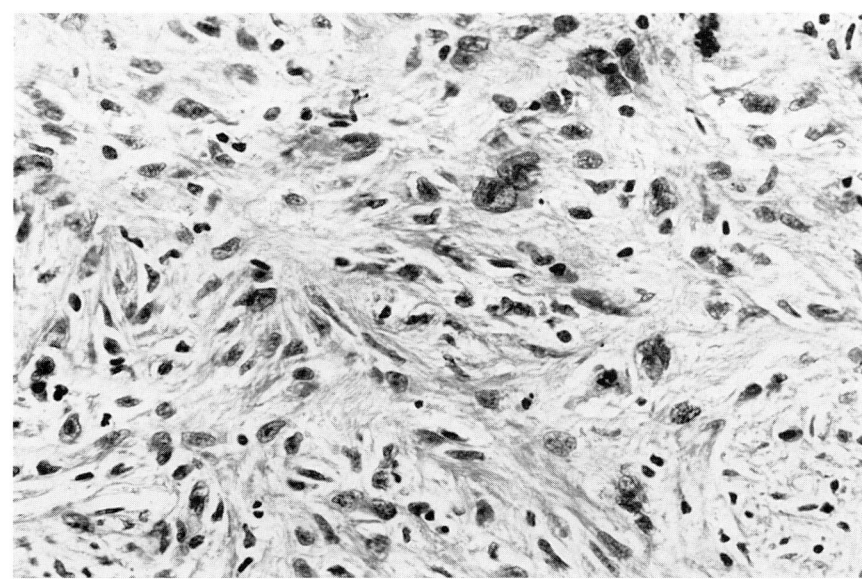

FIGURE 32-7
Malignant fibrous histiocytoma. The tumor consists of a mixture of spindle-shaped cells with a fibroblastic appearance, cytologically atypical histiocyte-like cells, and scattered inflammatory cells, principally lymphocytes and plasma cells. (High magnification.)

grounds. In our experience, none of these patients has died of disease.

Finally, MFHs located in the lung periphery invite a differential diagnosis with localized fibrous tumors of pleura, either benign or malignant. The latter tumors, even when invasive, retain a predominantly pleural location. Histologically, benign tumors show spindle cells set among ropy bundles of collagen, and while a storiform pattern can be seen, it is usually focal, and the typical growth appearance is the so-called patternless pattern. These tumors also stain routinely for CD34, which is not a typical feature of MFH. Malignant fibrous tumors of pleura consist of closely packed oval or spindle cells arrayed in fascicles or in a random fashion, a histologic pattern more closely resembling fibrosarcoma than MFH. Multinucleated giant cells and a dominant storiform pattern support the diagnosis of MFH.

Most patients with pulmonary MFH have localized disease when diagnosed: Only 20 to 25 percent present with metastatic disease. As far as treatment is concerned, the first line of therapy is surgical, most frequently lobectomy; adjuvant chemotherapy or radiation therapy has also been used in a few cases. Approximately 60 to 70 percent of patients develop recurrent or metastatic disease, and between 65 and 75 percent of the patients die, most often within 24 months.[10,13] Only 13 percent of patients survive more than 5 years. Perhaps the most significant prognostic indicator is advanced stage of disease at diagnosis.[13]

MYOGENIC TUMORS

Rhabdomyosarcoma

Primary neoplasms of muscle origin occurring in the lung are rare; in particular, rhabdomyosarcomas in adults are far more commonly found in the soft tissues.[17] Nonetheless, these tumors have been documented in unusual locations such as mediastinum,[18] kidney,[19] urinary bladder,[20] and also in the lung.[21-31] In 1955, Gordon and Boss,[24] in a presentation of a case in a 20-year-old woman, reviewed the existing literature and concluded that of four previously published cases, two could have been teratomas with a striated muscle component. Subsequently, however, there were additional reports of rhabdomyosarcomas in adults presenting in different anatomic regions within the lung. In 1955, Forbes described data on a 68-year-old man with a bronchial rhabdomyosarcoma.[25] In 1965, Conquest and collaborators reported data on three adults with intrapulmonary rhabdomyosarcomas.[23] Similarly, Lee and collaborators described data on a patient with a primary rhab-

domyosarcoma arising from the pulmonary trunk (probably a pulmonary artery sarcoma).[27] More recently, we have found three cases of intraparenchymal pulmonary rhabdomyosarcomas in adults between the ages of 57 and 78 years.[32]

The origin of pulmonary rhabdomyosarcomas is uncertain, but it has been suggested that rhabdomyosarcomas may arise from striated muscle within the lung parenchyma. In keeping with this possibility, Alderman and Patel[33] were able to document striated muscle within an extralobar pulmonary sequestration.

Anatomically, rhabdomyosarcomas may present in four distinct locations: (1) the bronchus; (2) the pulmonary trunk (i.e., pulmonary artery sarcoma); (3) associated with congenital cystic malformation[29,34]; and (4) intraparenchymatous tumor. It appears that rhabdomyosarcomas associated with congenital cystic malformations are more often seen in children and are better termed cystic blastomas (see pulmonary blastomas, below). Rhabdomyosarcomas arising in the bronchus and within the pulmonary parenchyma are most often seen in young individuals and adults. Clinically, rhabdomyosarcomas elicit symptoms related to the location of tumor. Tumors presenting as endobronchial lesions or affecting the bronchial tree tend to show symptoms related to pulmonary obstruction, whereas tumors located in the periphery of the lung present with chest pain.

Grossly, intrapulmonary rhabdomyosarcomas present as solid masses with areas of hemorrhage or necrosis. The size of the tumor varies depending largely on the location of the tumor. Endobronchial lesions tend to be smaller than intraparenchymatous neoplasms. It is likely that endobronchial tumors are smaller at resection because they tend to produce symptoms earlier in the clinical course than do their intraparenchymal counterparts.

Histologically, rhabdomyosarcomas show three well known growth patterns: alveolar, embryonal, and pleomorphic. In recent years the existence of the pleomorphic variant has been debated because of its similarity to MFH. Of the histologic subtypes, the pleomorphic variant is most often found in adults. In cases we have reviewed, at low magnification the tumor shows spindle cells arranged in tight fascicles with areas of necrosis and hemorrhage[32] (Fig. 32-8). At higher magnification, the tumor shows a storiform pattern composed of fusiform cells with moderate amounts of eosinophilic cytoplasm and oval-shaped nuclei. There may also be prominent nucleoli. Cellular pleomorphism in the form of large bizarre cells is marked (Figs. 32-8, 32-9, and 32-10). Rhabdomyoblasts are often identified (see Figs. 32-8

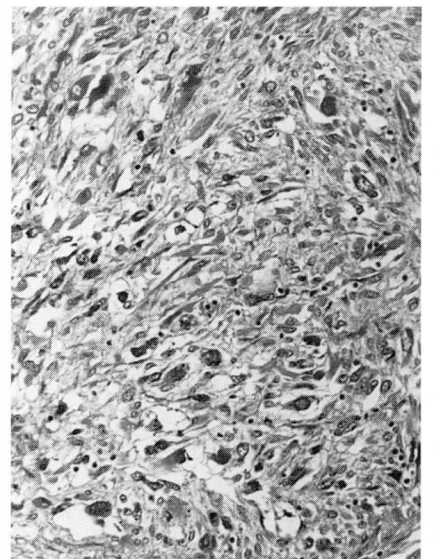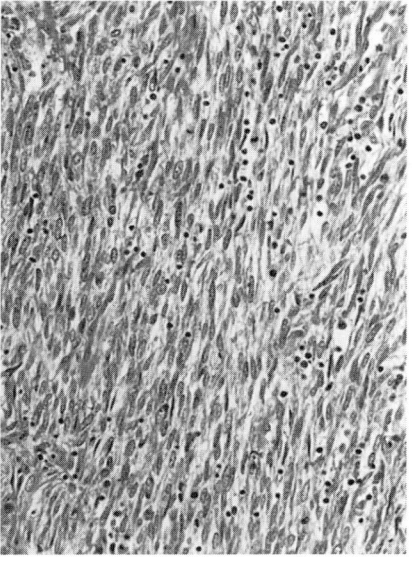

FIGURE 32-8
Pleomorphic rhabdomyosarcoma. Varying
histologic features may be encountered.
Left. Note scattered rhabdomyoblasts and
large atypical cells. *Right.* Spindle cell
proliferation. (Intermediate
magnification/intermediate magnification.)

and 32-10) and cross-striations may be relatively easily found. Mitoses are frequent. In cases of alveolar
and embryonal rhabdomyosarcoma, the histology is
the same as that observed when the tumors occur in
soft tissues. Immunohistochemically, the tumor cells
demonstrate reactivity for desmin and myoglobin.
They also may show Z-band material by electron
microscopy.

Because rhabdomyosarcomatous components
can be seen in carcinosarcomas and blastomas of the
lung, it is necessary to carefully sample a putative
rhabdomyosarcoma of lung before accepting it as
such. On the other hand, rhabdomyosarcoma, particularly the pleomorphic variant, also needs to be separated from other pleomorphic sarcomatous lesions

such as malignant fibrous histiocytoma and high-
grade leiomyosarcomas.

The prognosis of these tumors is poor. In the
cases described in the literature and the ones
reviewed by us, the majority of the patients have followed a fatal course in a period of no more than 2
years.

VASCULAR TUMORS

Angiosarcoma

Pulmonary angiosarcomas are rare; the vast majority
of angiosarcomas in the lung are metastatic.[35] For

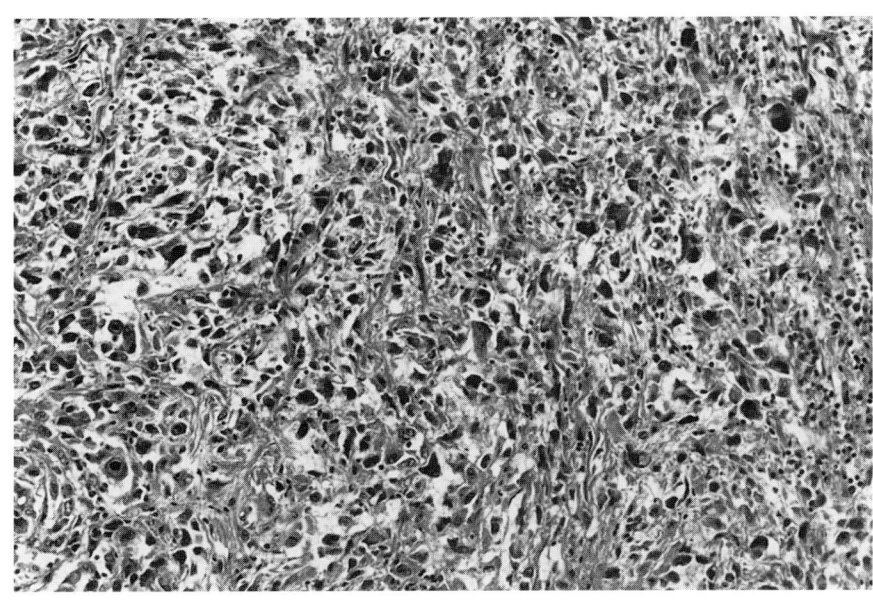

FIGURE 32-9
Rhabdomyosarcoma showing marked
cellular pleomorphism. (Intermediate
magnification.)

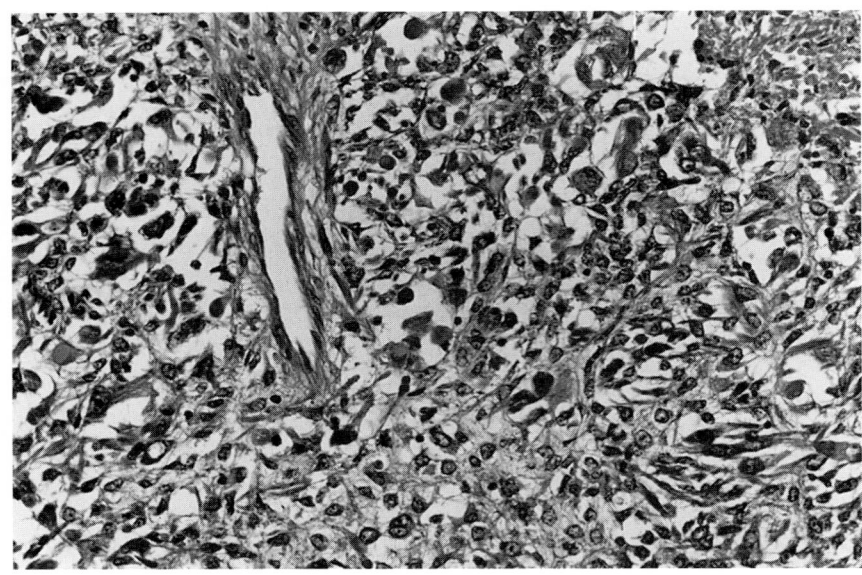

FIGURE 32-10
High-magnification view of a pleomorphic rhabdomyosarcoma showing numerous rhabdomyoblasts. (High magnification.)

example, all 15 cases of angiosarcoma of the lung recently reviewed by Patel and Rye were metastatic.[35] In fact, the lungs are one of the most common sites of metastases for angiosarcomas arising in the heart or pericardium,[36,37] breast, liver, skin, bone, and soft tissues.[38]

The diagnosis of primary angiosarcoma of the lung therefore requires exclusion of other possible sites of origin. Altogether, fewer than 10 cases of putative primary angiosarcomas of the lung have been reported.[39–43] In a review of 10,134 neoplasms at the Mayo Clinic over a 10-year period, primary angiosarcomas of the lung accounted for only 0.02 percent of cases,[44] and Dail and colleagues questioned whether some of these cases could have been metastases rather than primary lung tumors.[45] Dail agreed with Spencer's suggestion that if primary angiosarcoma of the lung exists, it is extraordinarily rare.[45]

The median age of patients with pulmonary angiosarcoma is 45 years (range, 5 to 71 years). There is no sex predilection.[35] The most common symptom is hemoptysis, followed by weight loss, cough, chest pain, dyspnea, and fever. Up to 20 percent of patients may be asymptomatic and another 20 percent may not be diagnosed until autopsy.[35] Angiosarcoma in lung can present with pneumothorax,[46,47] hemopneumothorax,[48] pleural effusion,[49] and diffuse pulmonary hemorrhage,[36,40] which may be rapidly fatal.[42] Chronic hemorrhage may lead to iron-deficiency anemia and, rarely, to persistent reticulocytosis.[50] The clinical and pathologic features of cases of putative primary and of metastatic pulmonary angiosarcoma are not significantly different.

Chest radiographs in patients with pulmonary involvement by angiosarcoma show bilateral pulmonary nodules in 70 percent of cases, pulmonary hemorrhage in 20 percent, linear and nodular infiltrates in 20 percent, pneumothorax in 7 percent and normal results in 7 percent.[35] Several unusual radiographic findings have been reported, including cavitation,[51,52] thin-walled cavities,[53,54] calcification,[55] and the computed tomography (CT) halo sign, which consists of a halo of ground glass attenuation surrounding a hemorrhagic nodule.[56]

There are a variety of predisposing factors for angiosarcomas, including chronic lymphedema (e.g., as in the postmastectomy, Stewart-Treves syndrome,[49] exposure to polyvinyl chloride or thorotrast, and prior mastectomy or radiation.[38]

By gross examination, most angiosarcomas in the lung show multiple, bilateral lesions. Rarely, angiosarcomas presenting in the lung consist of a solitary mass[57]; more frequently, they occur as multiple nodules; and sometimes they appear as red to purple streaks along bronchovascular bundles. The lesions are often hemorrhagic. Angiosarcomas can appear grossly as diffuse pulmonary hemorrhage.

Histologically, angiosarcomas infiltrate along lymphatic routes such as the bronchovascular bundles, pleura, and interlobular septa (Fig. 32-11). The tumor may form discrete nodules or subtle interstitial infiltrates (see Fig. 32-11). The neoplastic cells may grow in solid sheets, but more characteristically they form anastomosing vascular spaces lined with cytologically atypical endothelial cells, which often have a hobnail appearance (Fig. 32-12). The cells may have varying amounts of eosinophilic cytoplasm; when the cytoplasm is abundant, the tumor may be readily confused with a carcinoma.[57] Acute and organizing thromboemboli may be prominent and intravascular tumor growth is often present. Within the lumen of

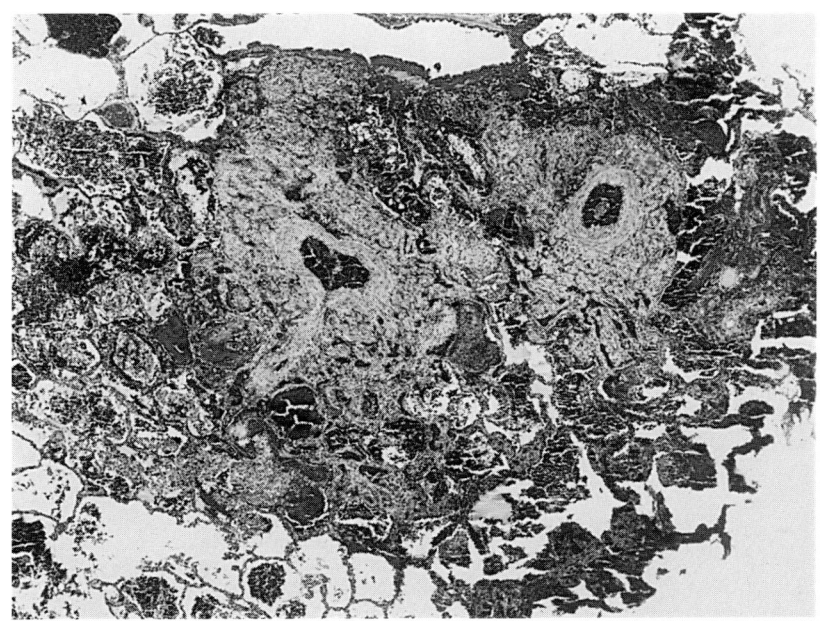

FIGURE 32-11
Angiosarcoma. This angiosarcoma is characterized by perivascular cellular infiltrates surrounded by intraalveolar hemorrhage. (Panoramic view.)

the blood vessels, the tumor cells may grow in papillary tufts (Fig. 32-13). When the pleura, interlobular septa, and peribronchial or perivascular connective tissue is infiltrated, the tumor may appear to elicit a desmoplastic response. Necrosis is often seen and hemorrhage is frequently present in the form of red blood corpuscle accumulation or hemosiderin deposition. The tumor cells may be hyperchromatic and have prominent nucleoli (see Fig. 32-13). Mitotic figures may be conspicuous. The chronic hemorrhage may result in encrustation of blood vessels leading to fragmentation of the elastica and a foreign body giant cell reaction (so-called endogenous pneumoconiosis).

The diagnosis of angiosarcoma may be difficult to establish on transbronchial biopsy. Nonspecific findings such as hemosiderosis may be seen.[50] A presumptive diagnosis of metastatic pulmonary angiosarcoma can be made in some cases where chest radiographs show multiple bilateral pulmonary lesions and a pathologic diagnosis can be established based on biopsies from other more readily accessible sites, such as the skin.[35]

Pleural fluid cytology of angiosarcomas may show red blood cells or hemosiderin. The tumor cells may be single or loosely clustered, and they may show finely vacuolated cytoplasm and distinct bor-

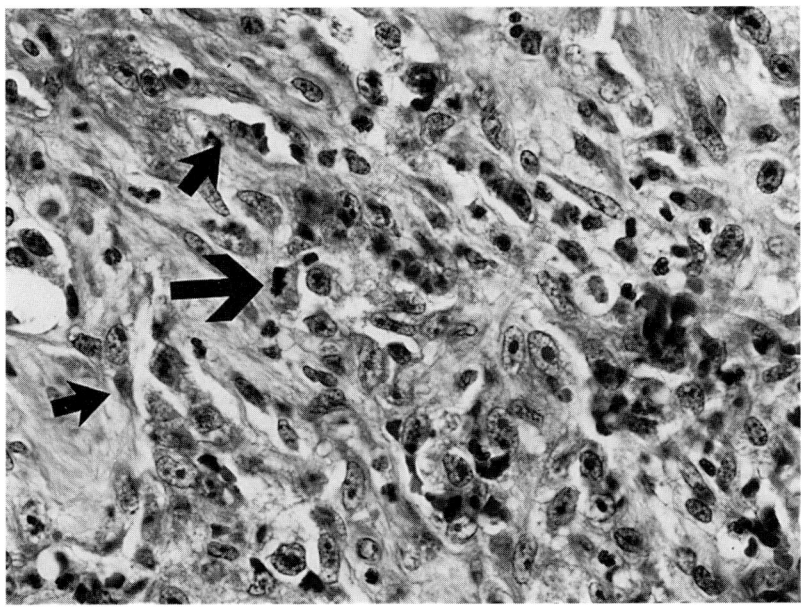

FIGURE 32-12
Angiosarcoma. The tumor is very cellular with large hyperchromatic nuclei and prominent nucleoli. There are focal slit-like spaces (*arrowheads*). A mitosis is present (*arrow*). (Intermediate magnification.)

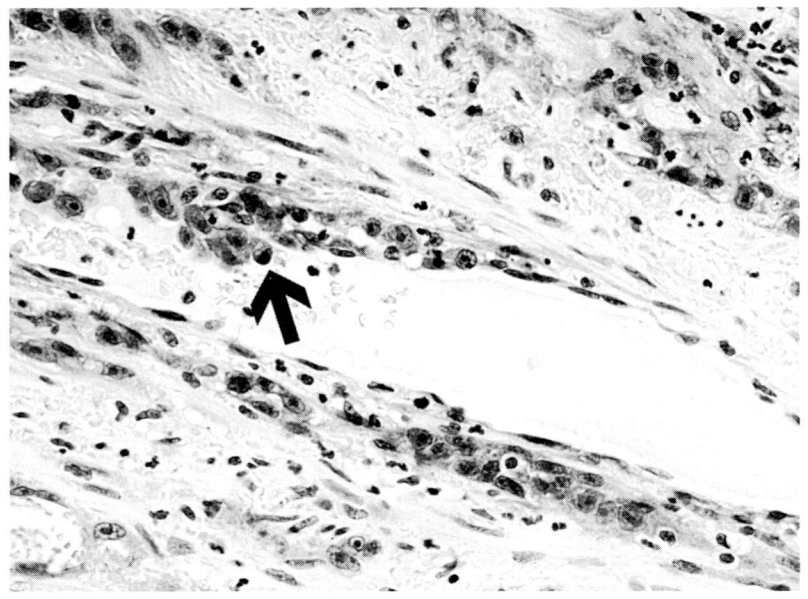

FIGURE 32-13
Angiosarcoma. The tumor is infiltrating around this venule. The atypical endothelial cells are growing along the luminal surface and form papillary tufts (*arrow*). (Intermediate magnification.)

ders. Nuclei may have irregular nuclear borders, coarse chromatin, and prominent nucleoli.[58] The diagnosis of angiosarcoma in the lung can also be established by transthoracic needle aspiration if diagnostic tissue is obtained in the biopsy specimen.[52]

Angiosarcomas typically stain for one or more of the following antigens: factor VIII-related antigen, Ulex europaeus, or CD34. In poorly differentiated tumors, however, it may be difficult to demonstrate staining for these vascular markers. Epithelial markers such as keratin or epithelial membrane antigen should be absent. Still, some epithelioid angiosarcomas may stain for epithelial markers such as keratin[59–61] and B72.3.[62] In such cases, the diagnosis must be carefully considered using other parameters such as the vascular immunohistochemical markers, histologic features, and electron microscopy.

Characteristic ultrastructural features of angiosarcomas include the finding of endothelial cells with pinocytotic vesicles, lateral desmosome-like attachments, and paranuclear filaments. A basal lamina can be present around the endothelial cells and Weibel-Palade bodies are a distinctive feature of endothelial differentiation.[38]

The differential diagnosis of angiosarcoma in the lung includes organizing thrombi, plexogenic hypertensive pulmonary arteriopathy, carcinoma, Kaposi's sarcoma, and epithelioid hemangioendothelioma.[41] Organizing thrombi may overshadow the neoplastic cells of an angiosarcoma. Whenever organizing thrombi are seen, a careful search should be made for intravascular malignancy. Plexogenic hypertensive arteriopathy can show intraarterial tufting and periarterial proliferation of capillary vas-

cular channels; however, it lacks significant cytologic atypia, mitotic activity, lymphatic distribution of cellular infiltrates and necrosis. The lymphatic distribution of angiosarcomas makes it important to exclude the possibility of lymphangitic carcinoma. Slitlike vascular channels, papillary tufts, the absence of keratin, and presence of factor VIII staining favor angiosarcoma over carcinoma. Carcinomas rarely form vascular spaces; however, a pseudovascular variant of squamous cell carcinomas has recently been described.[57,63] The latter is favored by the presence of a solitary mass, histologic demonstration of foci of typical carcinoma, and staining with antibodies to keratin rather than factor VIII-related antigen.[57,63] Epithelioid angiosarcoma may express immunoreactive keratin; in such a case, one must interpret the results in the context of other histologic, immunohistochemical, and ultrastructural data. Epithelioid hemangioendotheliomas generally have a distinctive appearance, as described below, and are usually not difficult to separate from angiosarcomas, which characteristically have a lymphatic distribution, greater cellularity, minimal intercellular stroma, and slitlike vascular channels.

The prognosis for angiosarcomas presenting in the lung is very poor, with an average survival of 9 months (range 1–3 years).[35] There is no known effective therapy for angiosarcomas involving the lung. Radiation therapy, chemotherapy, and surgical resection have been attempted with little success.[35]

Epithelioid Hemangioendothelioma

Pulmonary epithelioid hemangioendotheliomas are rare lung tumors that were previously known as intravascular bronchioloalveolar tumors.[64–68] Pulmonary epithelioid hemangioendothelioma is a low-grade vascular sarcoma that has distinctive clinical and pathologic features.

Most pulmonary epithelioid hemangioendotheliomas occur in young women. The female-to-male ratio is 4:1 and the mean age is 36 years (range 12–61 years).[66–68] Presenting symptoms include pleuritic chest pain, dyspnea, mild nonproductive cough, hemoptysis, and clubbing. Up to half the patients may be asymptomatic. Alveolar hemorrhage is a rare presentation of pulmonary epithelioid hemangioendothelioma.[69,70] Chest radiographs in the majority of patients show multiple, bilateral, sometimes calcified nodules measuring 1 to 2 cm in diameter. However, there may only be a solitary lung mass. Rarely, the predominant radiologic findings are pleural thickening and pleural effusions.

Epithelioid hemangioendothelioma may present with only lung involvement. The tumor can also occur in liver, bones, and soft tissues.[71–75] In these cases, lung involvement can appear simultaneously or subsequently. In some instances, the lung appears to be the site of metastases; in others, the primary site is not clear.

The majority of pulmonary epithelioid hemangioendotheliomas appear grossly as 0.3 to 2.0 cm, white-tan, lobulated nodules with a cartilaginous texture.[66,73] Although most cases present as bilateral multiple nodules, up to one third of them may appear as a solitary lung mass. Rarely, the tumor infiltrates the pleura diffusely in a pattern resembling malig-

nant mesothelioma.[76] Histologically, the tumor nodules are round or oval with a stroma-rich, hypocellular center and a cellular periphery (Fig. 32-14). Calcification and ossification may occur within the nodules. At the edge of the nodules, the tumor typically spreads into adjacent alveolar spaces in a micropolypoid fashion (Fig. 32-15). Intravascular and intrabronchiolar spread is characteristic. There is characteristically abundant myxoid stroma between the tumor cells (Fig. 32-16). Intracytoplasmic vacuoles forming vascular lumina can produce a signet ring appearance (Fig. 32-17). The nuclei are round to oval with delicate chromatin and occasional inconspicuous nucleoli. Prominent necrosis and cytologic atypia may occur in pulmonary epithelioid hemangioendothelioma, but they are uncommon (Fig. 32-18).

Pulmonary epithelioid hemangioendotheliomas demonstrate factor VIII and CD34, consistent with a vascular tumor.[77–82] Immunohistochemical staining for estrogen and progesterone receptors was absent in one study although a single case expressed 17-β-estradiol. By electron microscopy, the tumor cells are surrounded by an external basal lamina.[78,81,83,84] Occasional tight junctions may be seen. Pinocytotic vesicles and cytofilaments are characteristic. Weibel-Palade bodies are a distinctive marker of endothelial differentiation but can be hard to find.

The indolent nature of pulmonary epithelioid hemangioendothelioma is indicated by a mean survival of 4.6 years (range, < 1 year to 24 years). Poor

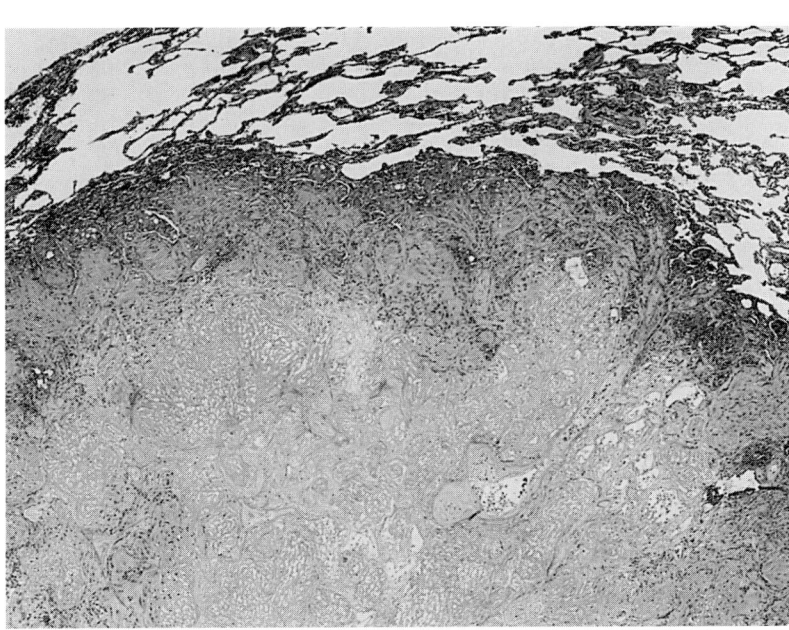

FIGURE 32-14
Pulmonary epithelioid hemangioendothelioma. At low power, this nodule shows a cellular periphery and a hypocellular and necrotic center. (Panoramic view.)

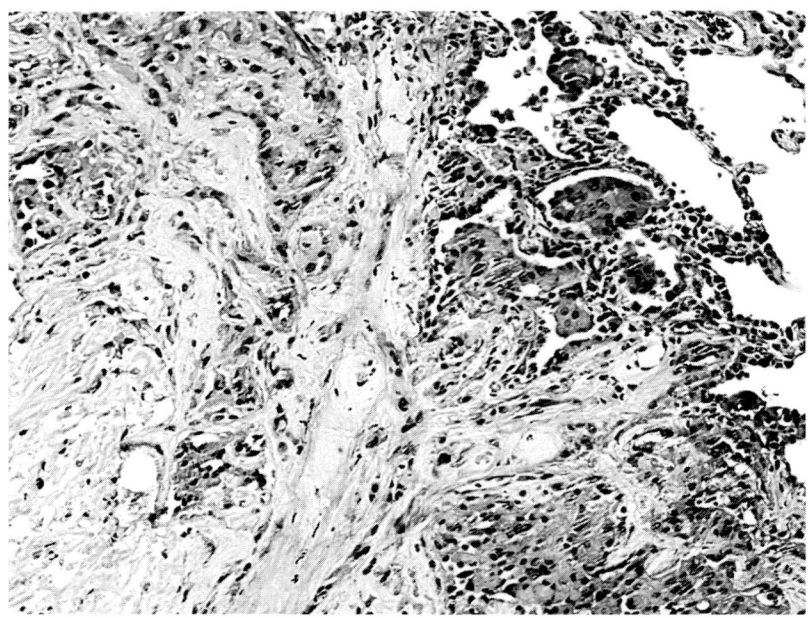

FIGURE 32-15
Pulmonary epithelioid
hemangioendothelioma. At the periphery of
this tumor nodule there is micropolypoid
extension into alveoli. The tumor cells are
embedded in an eosinophilic stroma. (Low
magnification.)

prognostic factors include presentation with symptoms, prominence of liver nodules, peripheral adenopathy, as well as extensive intravascular, endobronchial, interstitial, and pleural spread of tumor.[66] There is little evidence that chemotherapy or radiation therapy is of benefit in these patients. Asymptomatic patients can be managed without treatment unless their tumor is resectable.

Pulmonary epithelioid hemangioendothelioma must be distinguished from several benign nonneoplastic conditions such as granulomas, organizing infarcts, and amyloid nodules, as well as malignant tumors including mesothelioma, adenocarcinoma, sclerosing hemangioma, and angiosarcoma. When considering each of these diagnoses, it is important to think of the possibility of a pulmonary epithelioid hemangioendothelioma and look for characteristic histologic features, including nodules with a cellular periphery, a central hypocellular zone with a prominent hyaline matrix, and tumor cells with abundant eosinophilic cytoplasm and intracytoplasmic lumina. Metastatic epithelioid hemangioendothelioma to the lungs should be also considered and a search performed for other lesions in the liver, soft

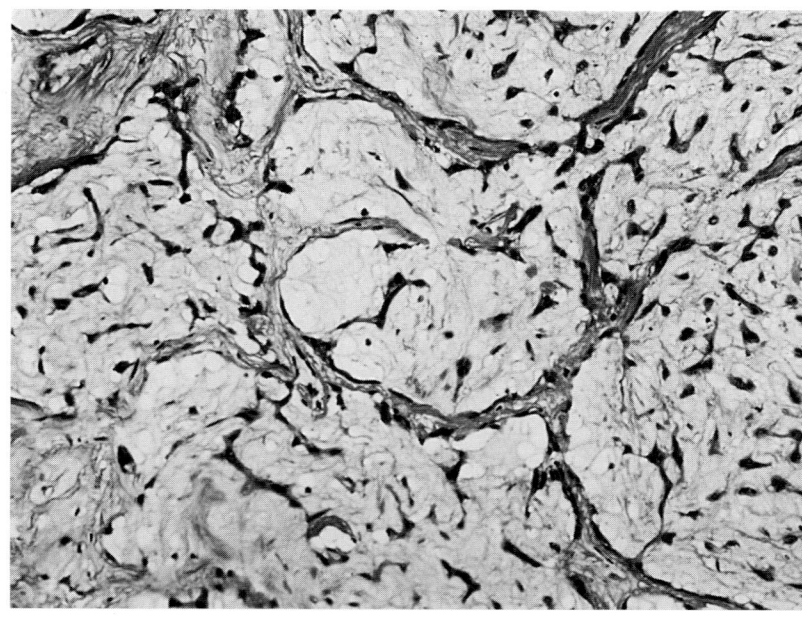

FIGURE 32-16
Pulmonary epithelioid hemangioendothelioma.
The tumor fills the alveolar spaces and shows a
myxoid stroma. (Intermediate magnification.)

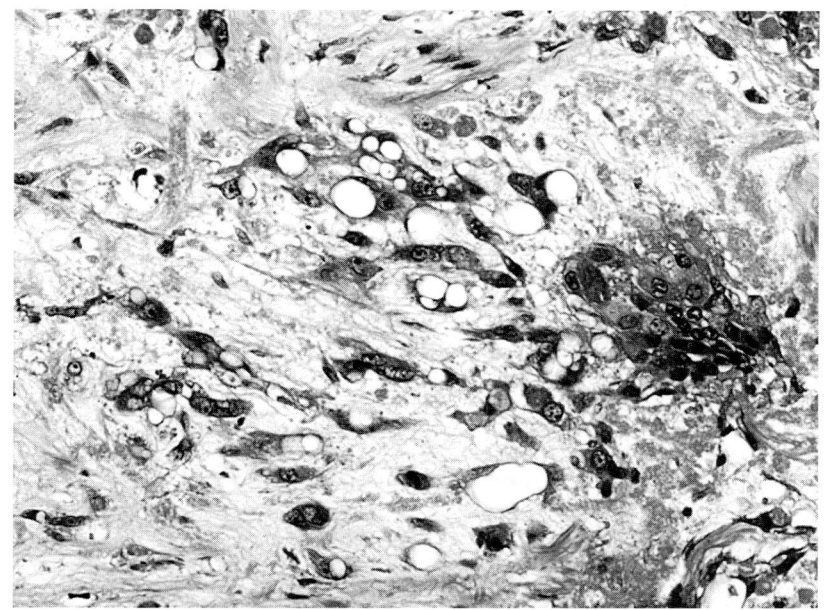

FIGURE 32-17
Pulmonary epithelioid hemangioendothelioma. Several tumor cells have prominent intracytoplasmic vacuoles resembling signet ring cells. There is a prominent eosinophilic stroma. (High magnification.)

tissues, or bones. In the presence of a dominant mass in an extrapulmonary site, the lung involvement may represent metastatic disease.

Kaposi's Sarcoma

Kaposi's sarcoma is an unusual type of vascular sarcoma that was described by Moritz Kaposi in 1872. It occurs in four forms: (1) classical variant; (2) African endemic variant; (3) Kaposi's sarcoma in iatrogenically immunosuppressed patients; and (4) epidemic Kaposi's sarcoma.[85] Classical Kaposi's sarcoma is a benign, indolent disease that occurs in elderly men of Jewish and Italian descent and is often limited to the lower extremities in the form of one or more skin lesions.[85] African endemic Kaposi's sarcoma accounts for up to 9 percent of all cancers in some regions and can be manifest in four different forms: (1) a benign nodular type; (2) an aggressive cutaneous type; (3) a rapidly widely disseminated form; (4) and a virulent lymphadenopathic form.[85] Because of the AIDS epidemic, Kaposi's sarcoma has become a relatively common disorder; recently however, the incidence of the disease in AIDS patients has been declining. In the early 1980s, at the beginning of the

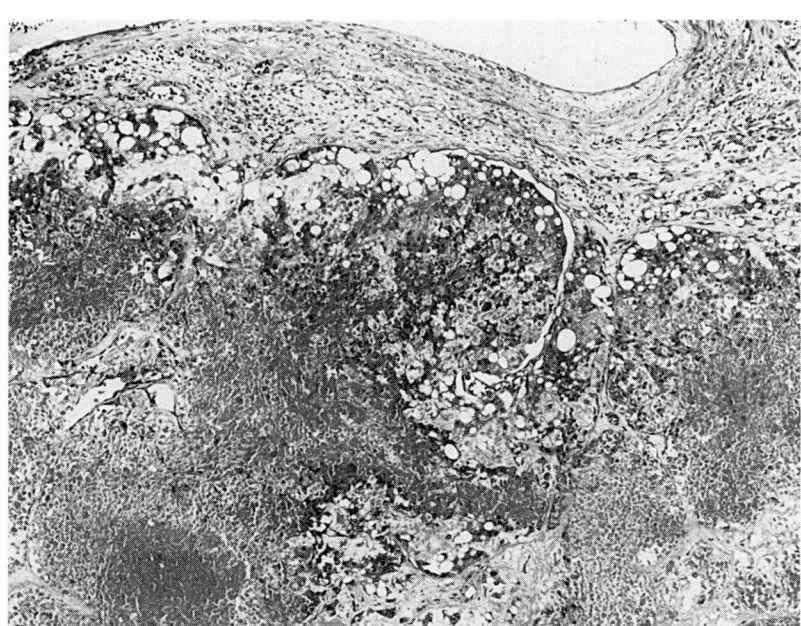

FIGURE 32-18
Pulmonary epithelioid hemangioendothelioma. This tumor shows prominent necrosis, an unusual feature in pulmonary epithelioid hemangioendotheliomas. (Low magnification.)

AIDS epidemic, Kaposi's sarcoma occurred in 20 to 25 percent of AIDS cases; however, in 1989 it was reported in only 10 to 20 percent of cases.[86] In addition to AIDS, Kaposi's sarcoma also occurs in other immunosuppressed states such as those following organ transplantation, especially renal transplantation,[87–89] or in systemic lupus erythematosus.[90]

Pulmonary Kaposi's sarcoma was found in 3.4 percent of cases in a large study of patients with known or suspected AIDS and pulmonary symptoms.[91] Homosexual behavior is the most common risk factor, accounting for 95 percent of cases.[85] Pleuropulmonary involvement can be detected in 20 percent of AIDS patients with systemic Kaposi's sarcoma.[92,93] At autopsy, Kaposi's sarcoma may be found in 7 percent of AIDS patients[94]; when systemic Kaposi's sarcoma is present, the lung is involved in 37 to 47 percent of autopsied cases.[95,96] Pulmonary involvement by Kaposi's sarcoma tends to be a late complication of AIDS.

Kaposi's sarcoma occurs in 3 to 6 percent of renal transplant patients—400 to 500 times the frequency encountered in the general population.[88,89] Pulmonary Kaposi's sarcoma occurred in 0.6 percent of renal transplant patients in one study.[88] It has been suggested that renal transplant patients treated with cyclosporin A may have a greater frequency of Kaposi's sarcoma than those treated with conventional immunosuppressive therapy and that the disease may be aggressive, resembling that seen in AIDS patients.[89]

Most patients with pulmonary Kaposi's sarcoma present with cough or dyspnea.[93] Fever,[93,97] chest pain,[98] and hemoptysis[99] can occur. Pulmonary infections may be found in up to 50 percent of AIDS patients with pulmonary Kaposi's sarcoma.[93]

The most common chest radiographic findings of pulmonary Kaposi's sarcoma are nonspecific bilateral hilar infiltrates, which may be linear or nodular.[100–102] Pleural effusions can be seen in 35 to 60 percent of cases.[92,102,103] Rarely in AIDS patients, cavitary lesions[104] or a solitary coin lesion[105] may be seen. CT scans of the lungs are more distinctive, showing hilar densities with multiple, bilateral, flame-shaped, or nodular lesions extending into the lung along bronchovascular bundles.[101,103] Discrete marginated nodules can be seen in 40 percent of cases.[100] Hilar adenopathy is uncommon and was seen in only 20 percent of cases in one study.[102]

Pleural effusions are usually serosanguinous or hemorrhagic[92]; rarely they are chylous.[92,106] Bronchoalveolar lavage shows hemosiderin-laden macrophages in 70 percent of patients, but this is not a specific finding because lavages in 45 percent of patients with other pulmonary diagnoses (including infections) yield similar results.[107]

Endobronchial Kaposi's sarcoma has a distinctive appearance at endoscopy. There are either diffuse, red, friable mucosal streaks or violaceous slightly red smooth lesions.[95,108,109] These features are so distinctive that some regard them as diagnostic if the patient already has a diagnosis of Kaposi's sarcome based on a biopsy from a more available site, such as the skin.

On gross examination, the visceral pleura shows violaceous streaks or plaques. The underlying parenchyma may have red, hemorrhagic infiltrates or nodules surrounding the bronchovascular bundles. The nodules tend to be small, but they may become confluent (Fig. 32-19). The bronchial mucosa can show cherry red mucosal streaks or plaques.

Histologically, Kaposi's sarcoma infiltrates along lymphatic routes, namely, the bronchovascular bundles, the pleura, and interlobular septae[110,111] (Figs. 32-20 and 32-21). Kaposi's sarcoma can have several patterns: (1) classic,[110] (2) cavernous-angiomatous,[45] (3) pleomorphic, and (4) inflammatory.[110] Classic disease is characterized by infiltrates consisting of spindle cells with slitlike spaces, which

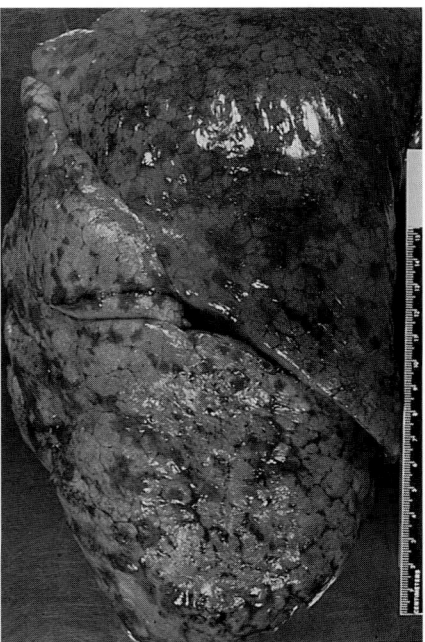

FIGURE 32-19
Kaposi's sarcoma. The pleural surface shows patchy red, hemorrhagic streaks of tumor. (*See also Color Plate 7.*)

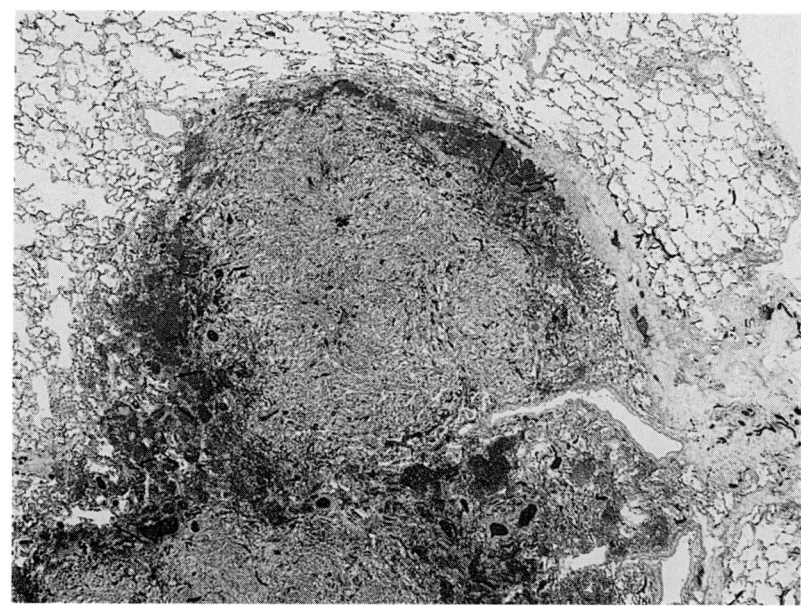

FIGURE 32-20
Kaposi's sarcoma. This nodule is situated adjacent to a vein and consists of spindle cells surrounded by hemorrhage. (Scanning magnification.)

contain extravasated blood cells (Fig. 32-22). The spindle cells are usually relatively bland, but they may show mild to moderate atypia. Mitoses are uncommon. The cavernous-angiomatous pattern consists of dilated, thin-walled vascular channels filled with red blood cells. Rarely, Kaposi's sarcoma may have a pleomorphic pattern with extensive atypia and a densely cellular appearance resembling MFH. In such cases, one may have to sample the tumor generously to identify recognizable Kaposi's sarcoma. The inflammatory pattern of Kaposi's sarcoma was described by Moskowitz and colleagues.[112]

These authors recognized polymorphous inflammatory infiltrates as early Kaposi's sarcoma. Using these criteria, they found Kaposi's sacoma in 94 percent of autopsies from AIDS patients, far higher than the 30 to 70 percent reported in other studies.[112] This discrepancy is probably due to the difference in diagnostic criteria used among these reports. Currently, it is more widely accepted to require the presence of some foci of classic Kaposi's sarcoma to establish the diagnosis.

There has been controversy about the use of transbronchial biopsies for the diagnosis of Kaposi's

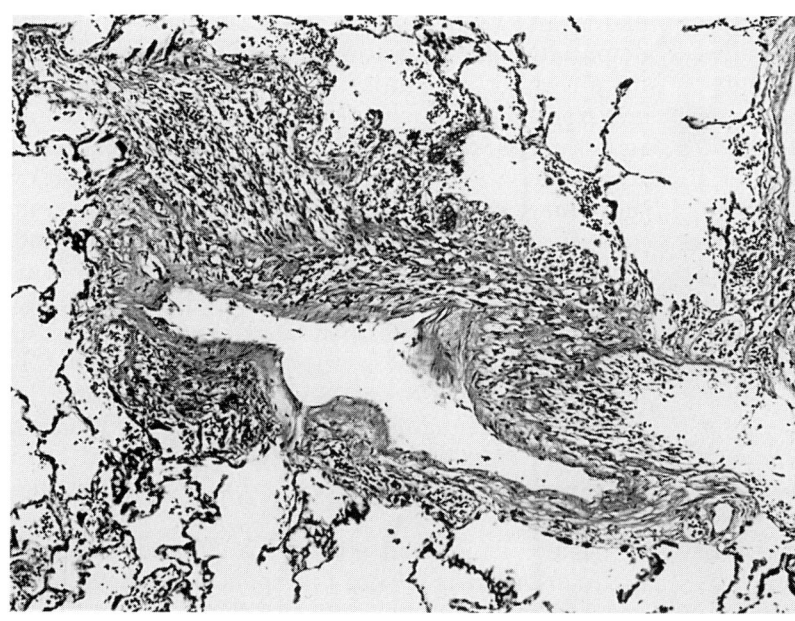

FIGURE 32-21
Kaposi's sarcoma. There is prominent spindle cell morphology to this tumor which infiltrates around this vein. (Low magnification.)

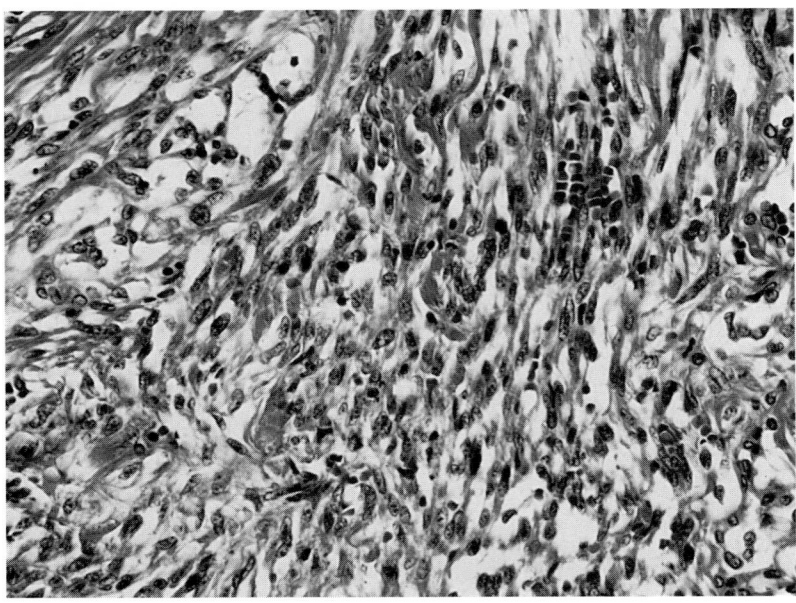

FIGURE 32-22
Kaposi's sarcoma. This tumor consists of prominent spindle cells with slitlike spaces filled with extravasated red blood cells. (Intermediate magnification.)

sarcoma due to the frequently low diagnostic yield, the potential complication of hemorrhage, and the distinctive clinical appearance of endobronchial lesions.[113] However, others have advocated transbronchial biopsies for the diagnosis of Kaposi's sarcoma and have not encountered the complication of hemorrhage.[111] Pleural biopsies usually show nonspecific histologic changes and have generally not been helpful for the diagnosis of Kaposi's sarcoma.[92] The premortem pathologic diagnosis of Kaposi's sarcoma is best established by open lung biopsy. The thoracoscopic approach is a less invasive way to obtain these open lung biopsies than open thoracotomy.[114]

The spindle cells of Kaposi's sarcoma stain positively for vascular markers such as factor VIII-related antigen[115] and CD34.[116] Stains for vimentin may also be positive. By electron microscopy, Kaposi's sarcoma shows variable mixtures of endothelial cells, pericytes, fibroblasts, and myofibroblasts.[117,118] Weibel-Palade bodies may be absent.[118]

Kaposi's sarcoma must be separated from spindle cell carcinoma, granulation tissue, organizing pneumonia, organizing diffuse alveolar damage, bacillary angiomatosis, and other sarcomas, particularly angiosarcomas. Carcinomas may have a spindle cell morphology, but they do not show slitlike spaces with extravasation of red blood cells. Immunohistochemical staining for keratin is present in spindle cell carcinomas and absent in Kaposi's sarcoma, whereas staining for factor VIII-related antigen is generally present in Kaposi's sarcoma and absent in carcinomas.

Bacillary angiomatosis occurs in AIDS patients and can cause endobronchial polyps, which histologically have an angiomatous appearance mixed with inflammatory cells resembling granulation tissue. However, it represents an unusual histologic response to infection by a bacillary organism called *Rochalimaea henselae*.[118–122] It can be demonstrated by Warthin-Starry stains, electron microscopy, and immunohistochemistry.[118,119]

Granulation tissue may occasionally be difficult to separate from Kaposi's sarcoma in a small biopsy. The former is usually characterized by a proliferation of well formed capillaries mixed with an inflammatory background; well formed capillaries are usually not seen in Kaposi's sarcoma. The proliferating connective tissue in organizing pneumonia and the organizing phase of diffuse alveolar damage might potentially be confused with Kaposi's sarcoma, but the airway or interstitial distribution is quite different from the bronchovascular, pleural, and paraseptal infiltration pattern of Kaposi's sarcoma. Angiosarcoma may show a lymphangitic pattern of pulmonary involvement, as in Kaposi's sarcoma, but most angiosarcomas have more cytologic atypia and do not show the pattern of spindle cells with slitlike spaces and red blood corpuscle extravasation seen in Kaposi's sarcoma. Further, angiosarcomas tend to invade medium-sized arteries or veins causing papillary tufting, a feature not seen in Kaposi's sarcoma. In addition, most angiosarcomas involving the lung are metastatic from extrapulmonary sites, especially the heart, skin, or soft tissues, so clinical correlation may be helpful.

There have been attempts at treating symptomatic pulmonary Kaposi's sarcoma with chemotherapy and radiation, but therapy represents a difficult challenge. A variety of conventional chemotherapy agents have been used such as doxorubicin, interferon-α, bleomycin, vincristine, and vinblastine, either individually or in combination. Preliminary data suggests that a liposomal formulation of doxorubicin[123,124] may be a safe and effective therapy for advanced Kaposi's sarcoma in AIDS patients. Radiation therapy is mainly used in patients with pulmonary Kaposi's sarcoma that is refractory to chemotherapy or in patients with severe disease where symptomatic relief is desired.[86] Early intervention with radiation therapy can provide symptomatic improvement for patients with pulmonary Kaposi's sarcoma and may play a palliative role.[86,125,126] Nevertheless, the outcome is poor for patients given radiation, with a median survival between 2 and 6 months.[86]

Hemangiopericytoma

Hemangiopericytoma is a rare form of sarcoma thought to be derived from the pericyte, a cell described by Zimmerman in 1923.[20,127] Approximately 10 percent of hemangiopericytomas are thought to arise in the lung.[128] Hemangiopericytomas accounted for 0.02 percent of 10,134 primary pulmonary neoplasms seen at the Mayo Clinic over a 10-year period.[26] Pulmonary hemangiopericytomas encompass a spectrum of benign and malignant tumors, as well as an intermediate group in which it is difficult to predict the clinical behavior.[20,128]

The mean age of patients with pulmonary hemangiopericytomas is 46 years (range, 18 to 71 years).[128,129] There is no sex predilection.[128,129] Patients most commonly present with chest pain (38 percent), hemoptysis (19 percent), cough (13 percent), dyspnea (13 percent), fever (6 percent), and clubbing (6 percent).[129] Patients are asymptomatic in 38 percent of cases.[129] Hypoglycemia and hypertension can occur.[128]

Chest radiographs typically show a well-circumscribed mass or homogeneous opacity with a round or oval contour.[130] The tumors tend to be large, with two thirds measuring more than 5 cm at presentation and one third larger than 10 cm.[130] However, there is a range of tumor size from small lesions to ones that replace the entire hemithorax. The tumors often compress rather than infiltrate surrounding structures.[131] Pleural effusions can occur,[130] but calcification is uncommon.[131] CT scans show a heterogeneous pattern in most cases.[131] Large necrotic tumors may show central areas of low attenuation[131]; this feature may be a better indicator of malignancy than infiltra-

tion by the tumor.[132] Because of the vascularity of the tumor, magnetic resonance imaging can be used to demonstrate the extent of hemangiopericytomas.[133]

At surgery, a minority of pulmonary hemangiopericytomas invade the chest wall or mediastinal structures.[129]

Gross examination of hemangiopericytomas typically reveals a circumscribed, noninfiltrative mass with a lobulated yellow, tan-brown cut surface. The median size is 5.4 cm (range, 2 to 14 cm).[128,129] The tumors may be encapsulated.[130,134] Areas of hemorrhage and necrosis are often seen, especially in larger lesions.

Microscopically, at low magnification hemangiopericytomas are circumscribed, tending to compress rather than infiltrate adjacent structures (Fig. 32-23). The tumors are characterized by tightly packed cells surrounding vascular spaces with a caliber ranging from small to large sinusoidal spaces (Figs. 32-24, 32-25, and 32-26).[20] These vessels are thin-walled and often have a staghorn, slitlike appearance (see Fig. 32-26).[20] The tumor cells are round to spindle shaped with round or oval nuclei and moderate amounts of eosinophilic cytoplasm (see Fig. 32-26). Nucleoli tend to be absent or inconspicuous. Myxoid changes and stromal hyalinization may be present in a minority of cases.[129]

Yousem and Hochholzer proposed the following criteria for malignancy: (1) infiltration of the chest wall or mediastinum, (2) vascular invasion, and (3) clinical evidence of recurrence or metastases.[129] Vascular invasion was defined as tumor penetration of a vessel wall, adherence to the endothelium, and incorporation of tumor into a thrombus. According to these criteria, 61 percent of the cases in their series were classified as malignant. All tumors with pleural invasion, bronchial destruction and a size 8 cm or greater were malignant. It was also suggested that tumors with greater than 3 mitoses per 10 high power fields were more likely to be malignant.[129] Necrosis and infiltrative margins were inconsistent criteria for malignancy.

Although the endothelial cells of hemangiopericytomas stain for factor VIII-related antigen and CD34, the tumor cells themselves either fail to stain[20] or stain only focally.[135] Soft tissue hemangiopericytomas as well as normal pericytes have recently been shown to stain for HLA-DR.[135] Vimentin and actin may also be demonstrable, but desmin and myoglobin are absent.

By electron microscopy, the tumor cells are situated around endothelial-lined vascular spaces. The cytoplasm contains a few organelles, such as rough endoplasmic reticulum, mitochondria, and some

FIGURE 32-23
Hemangiopericytoma. A circumscribed nodule of tumor compresses the adjacent lung. (Panoramic view.)

microfilaments. Elongated cellular processes are prominent. Pinocytotic vesicles are conspicuous, but desmosomes are uncommon. A multilayered basal lamina is found between the endothelial cells and the tumor cells.[20]

The differential diagnosis of pulmonary hemangiopericytoma is broad and includes spindle cell carcinoma, carcinoid, intrapulmonary fibrous tumor, synovial sarcoma, and mesenchymal chondrosarcoma.[132] Keratin is usually expressed in spindle cell carcinomas but not in hemangiopericytomas. Carcinoid tumors may show spindle cells and they may have a prominent vascular pattern; however, they also show an organoid nesting pattern and finely granular nuclear chromatin. In addition, carcinoid tumors show evidence of neuroendocrine differentiation by immunohistochemistry (chromogranin, synaptophysin and Leu-7 staining) or electron microscopy (dense core granules). Biphasic synovial sarcomas show glandlike epithelial differentiation, which is not seen in hemangiopericytomas, but separation from a monophasic variant of synovial sarcomas may be more difficult. In contrast to hemangiopericytomas, monophasic synovial sarcomas may express immunoreactive keratin and may show alternating hemangiopericytomatous and

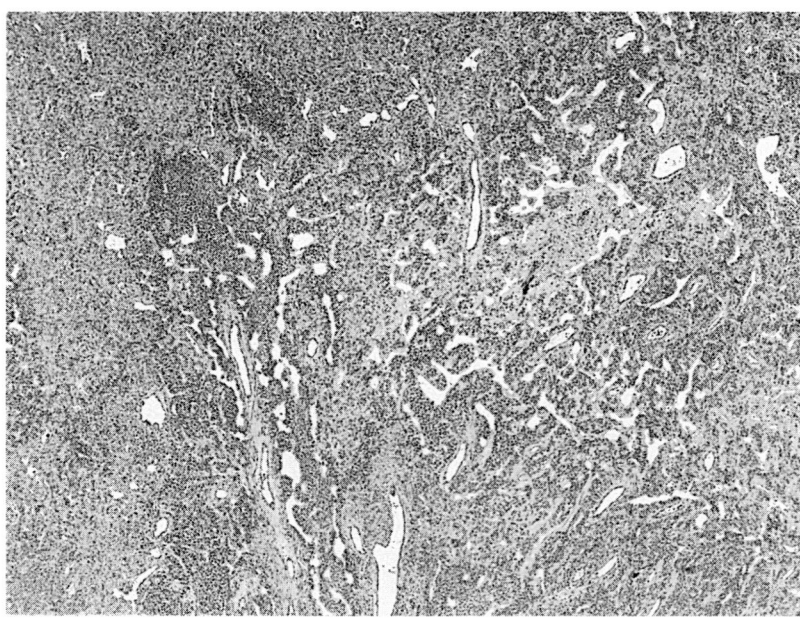

FIGURE 32-24
Hemangiopericytoma. The tumor consists of spindle cells growing around slitlike vascular spaces. (Low magnification.)

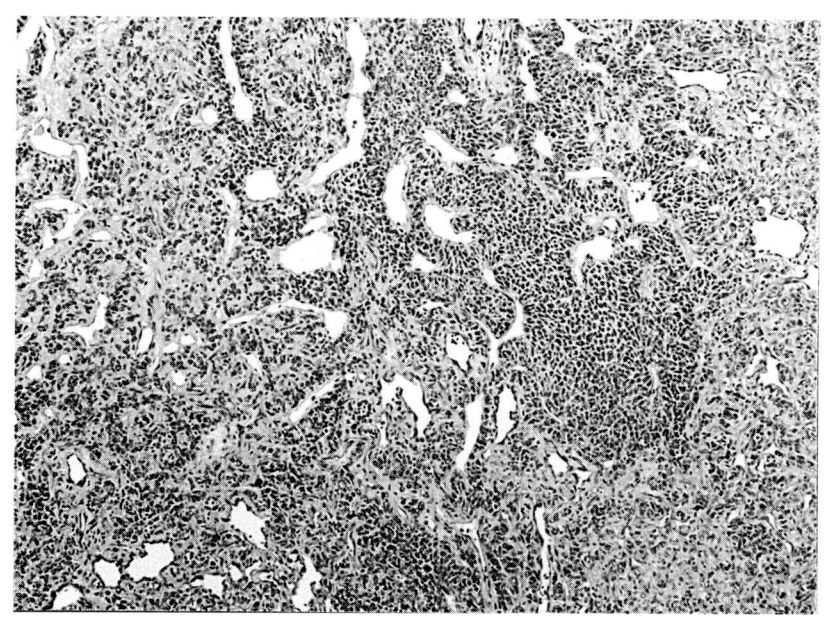

FIGURE 32-25
Hemangiopericytoma. The tumor is composed of round to oval cells and the vascular spaces are thin-walled with some branched, anastomosing vascular spaces (Low magnification.)

fibrosarcoma-like spindle cell areas and small areas of calcification.[132] Mesenchymal chondrosarcomas will have focal areas of malignant cartilage admixed with hemangiopericytomatous zones.[132]

Because a pericytomatous vascular pattern can be seen focally in a variety of tumors such as intrapulmonary fibrous tumor, synovial sarcoma, and mesenchymal chondrosarcoma, one must carefully search for characteristics of each of these tumors before making a diagnosis of hemangiopericytoma.

Complete surgical resection is the primary therapeutic approach. Benign pulmonary hemangiopericytomas may be cured by complete surgical resection, but the majority of malignant tumors recur locally. In one series, recurrence appeared in 67 percent of the malignant tumors after a mean of 55 months (range, 16 to 144 months).[129] Most recurrences were local ones within the thorax, but extrathoracic metastases occurred in one patient.[129] Perioperative rupture puts the patient at risk for early recurrence.[136]

Chemotherapy has not proven to be effective in pulmonary hemangiopericytomas. Results in a few cases suggest that complete surgical resection with

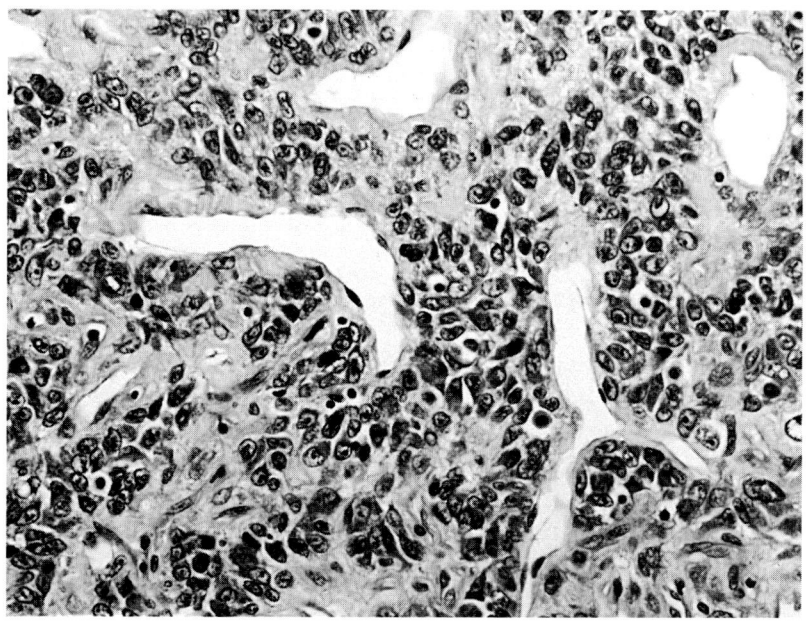

FIGURE 32-26
Hemangiopericytoma. The tumor cells are round to oval with mild to moderate cytoplasm and nuclei with vesicular chromatin and occasional faint nucleoli. The tumor cells grow subjacent to the vascular endothelium compressing the vascular spaces into a staghorn shape. (Intermediate magnification.)

intraoperative and postoperative radiation therapy may be useful in patients with very large tumors that are difficult to resect.[133,137]

Pulmonary Artery Sarcoma

Sarcomas of the major pulmonary vessels arise mostly in the pulmonary artery,[138–140] but they can occur in the pulmonary vein.[141] They are highly malignant and survival is very poor. As a result, in the past most cases were diagnosed at autopsy; more recently, cases have been diagnosed based on surgical or bronchoscopic biopsy.[138,139]

Patients with pulmonary artery sarcomas are older and mostly women. The median age is 52 years (range, 22 to 81 years).[139] Women predominate 2:1. The most common symptom is dyspnea (70 percent of cases), followed by systolic murmur (61 percent), chest pain (48 percent), cyanosis (36 percent), cough (34 percent), edema (32 percent), hemoptysis (30 percent), syncope (25 percent), and diastolic murmur (7 percent).[139]

Chest radiographs show pulmonary lesions in 46 percent of cases, cardiomegaly in 39 percent, a hilar mass in 39 percent, an enlarged pulmonary artery in 25 percent, and decreased pulmonary vascularity in 23 percent. An angiographic lesion can be demonstrated in all patients.[139]

Pulmonary artery sarcomas all involve the pulmonary trunk.[138,139] The pulmonary valve is affected in 57 percent of cases, and the right and left pulmonary arteries are involved in 67 percent and 60 percent of cases, respectively.[139] The heart is involved in 37 percent of cases and the right ventricular outflow tract in 25 percent. A polypoid appearance within the vascular lumen is present in virtually all cases. A myxoid appearance can be seen in 63 percent of cases. Metastases are present in 78 percent of cases, with involvement of the lung (67 percent), lymph nodes (20 percent), mesentery (5 percent), pancreas (5 percent), adrenals (3 percent), brain (3 percent), jejunum, kidney, or thyroid (2 percent), and liver by direct extension (2 percent).[138,139]

Histologically, these tumors grow within the lumina of large pulmonary arteries (Fig. 32-27). The tumor can show a remarkable variety of histologic subtypes, including myogenic, fibrocytic, cartilaginous, and osteogenic (Table 32-1)(Fig. 32-28).

Depending on the differentiation of the tumor, the results of immunohistochemistry or electron microscopy may vary. Tumors with definite skeletal muscle differentiation may demonstrate myoglobin, actin (Fig. 32-29), desmin, or HHF-35.[142] Chondrosarcomas may stain for S-100 protein, Leu-7,

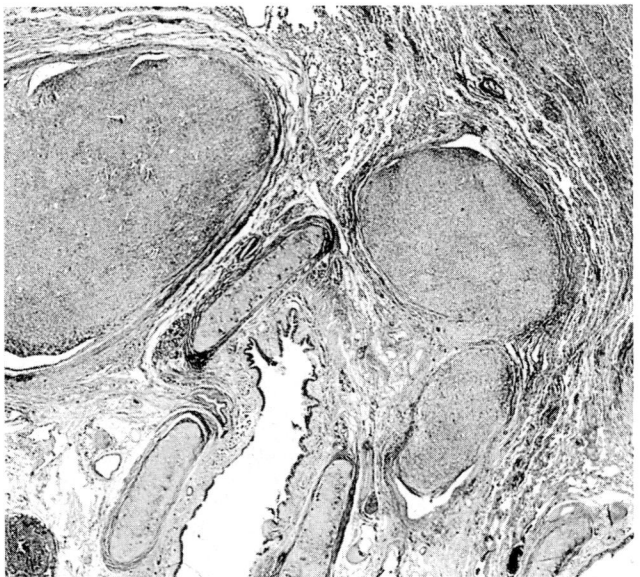

FIGURE 32-27
Pulmonary artery sarcoma. This tumor shows polypoid intraarterial growth within the pulmonary arteries surrounding this bronchus. (Panoramic view.)

and lysozyme. Angiosarcomas may stain with antibodies to factor VIII-related antigen or CD34. MFHs may demonstrate immunoreactive α-antitrypsin. Leiomyosarcomas may stain for smooth muscle actin. By electron microscopy, there can be a wide variety of features depending on the type of differentiation of the tumor.

In most cases of pulmonary artery sarcoma, the diagnosis is not difficult, particularly in those diagnosed at autopsy. However, the pulmonary artery origin of these tumors may be overlooked. Realization

TABLE 32-1
Histologic Differentiation in Pulmonary Artery Sarcomas[138,139,142]

Histologic Differentiation	% of Cases
Undifferentiated sarcoma	34
Leiomyosarcoma	20
Rhabdomyosarcoma	6
Fibrosarcoma	11
Fibromyxosarcoma	10
Chondrosarcoma	4
Osteosarcoma	3
Angiosarcoma	2
Hemangioendothelioma	1
Hemangiosarcoma	1
Malignant fibrous histiocytoma	2
Malignant mesenchymoma	3
Mesenchymoma	2
Mixed mesenchymal tumor	1

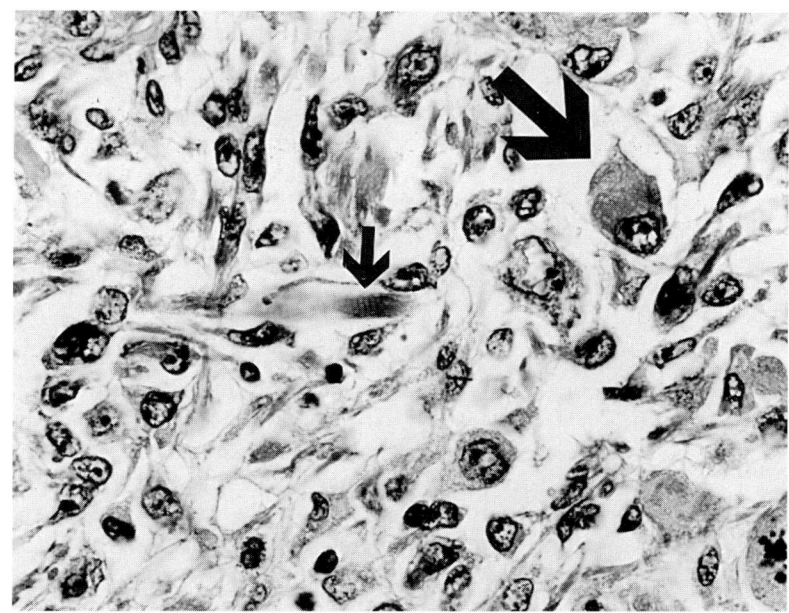

FIGURE 32-28
Pulmonary artery sarcoma. This tumor shows large rhabdomyoblasts with abundant eosinophilic cytoplasm (*large arrow*) and other cells with cross-striations. (High magnification.)

that pulmonary artery sarcomas can manifest a wide variety of types of differentiation means that this possibility should be excluded before the diagnosis of virtually any of the primary sarcomas of the lung. In summary, when examining a lung sarcoma, careful consideration should be given to the possibility of involvement of major pulmonary vessels.[142]

Because pulmonary artery sarcomas are rare, and in the past most cases have been diagnosed at autopsy, there is little information about therapy. Information about prognosis was previously inferred by analyzing the duration of symptoms, which was

10 months (range, 1 to 262 months).[139] Only 12 percent of patients survive more than 1 year after diagnosis.[139] A variety of chemotherapy agents have been given to these patients including adriamycin, methotrexate, and cyclophosphamide; the few patients given radiation have not responded well.[139]

With the advent of techniques such as CT scanning and magnetic resonance imaging,[143] earlier diagnosis may be possible allowing for more cases to be resected. In patients with unilateral tumors, endarterectomy is a potential method of resecting these tumors.[144]

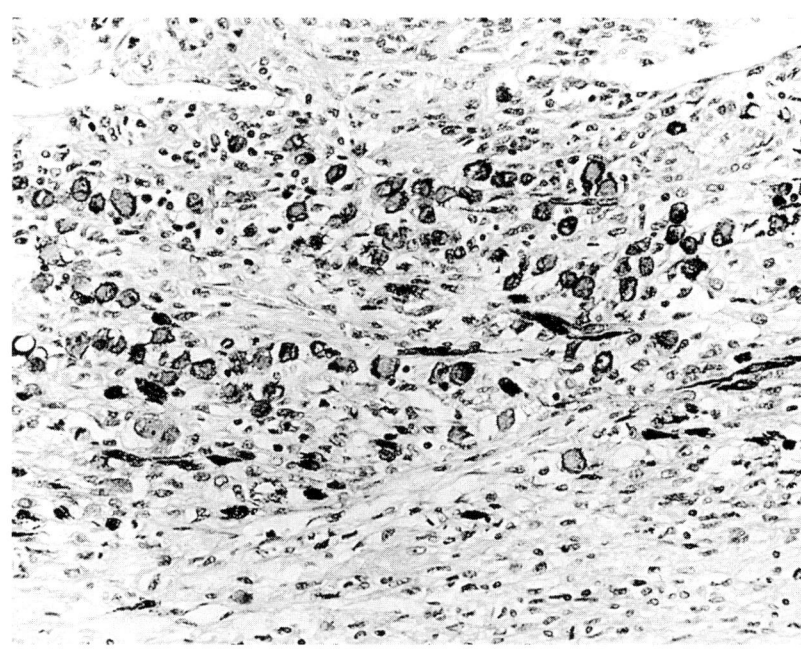

FIGURE 32-29
Pulmonary artery sarcoma. The large rhabdomyoblasts in this tumor stain strongly with antibody to muscle specific actin. (Intermediate magnification.)

TUMORS OF BONE, CARTILAGE, AND FAT

Chondrosarcoma

Chondrosarcomas of the lung are very rare, with less than 25 cases reported in the literature.[145] The lungs are one of the most common sites of metastases from bone or soft tissue sarcomas, so it is important to exclude a secondary before establishing the diagnosis of pulmonary chondrosarcoma. Pulmonary chondrosarcomas may primarily affect the lung parenchyma or the pleura.

The mean age of patients with pulmonary chondrosarcoma is 42 years (range, 15 to 82 years).[145] There is no sex predilection. Presenting symptoms include hemoptysis, dyspnea, anemia, and weight loss. Some cases may be discovered during routine chest radiograph examination.

Grossly, pulmonary chondrosarcomas are circumscribed, lobulated masses with gelatinous, necrotic, and hemorrhagic cut surfaces. They are large tumors with an average size of 10 cm (range, 2.5 to 30 cm) that may show endobronchial growth or cystic cavitation. They may be central or peripheral in location; in some cases, the size may be so large that it is impossible to localize the tumor to a central or peripheral origin. The tumors should be within lung, without involvement of the chest wall, mediastinum, heart, or great vessels.

Histologically, the tumors consist of lobulated masses of malignant cartilaginous tissue. The tumor cells tend to be large with abundant eosinophilic or clear cytoplasm and hyperchromatic, irregular nuclei (Fig. 32-30). Some areas of these tumors may have small cells with scant cytoplasm and small nuclei, or the cells may be spindled. The tumors may show chondroblastic or myxoid features. Rarely, pulmonary chondrosarcomas may show the pattern of mesenchymal chondrosarcoma with chondrosarcomatous foci scattered amidst a primitive round cell stroma with focal pericytomatous features.[146] Infiltrative growth, tumor necrosis, haphazard arrangement of atypical chondrocytes, and mitotic activity separate pulmonary chondrosarcomas from normal or benign cartilaginous lesions. None of the reported chondrosarcomas have been shown histologically to arise from bronchial cartilage or a preexisting hamartoma. Immunohistochemistry may be positive for S-100 protein. If diagnostic tissue is obtained in the biopsy specimen, the diagnosis of pulmonary chondrosarcoma can be established cytologically by fine needle aspiration.[147]

The diagnosis of primary pulmonary chondrosarcoma requires exclusion of a number of important entities that can cause confusion:

1. A late metastasis from a distant extrapulmonary site
2. Extension from an adjacent primary site, such as the mediastinum, rib, or vertebral bone
3. Pulmonary artery sarcoma

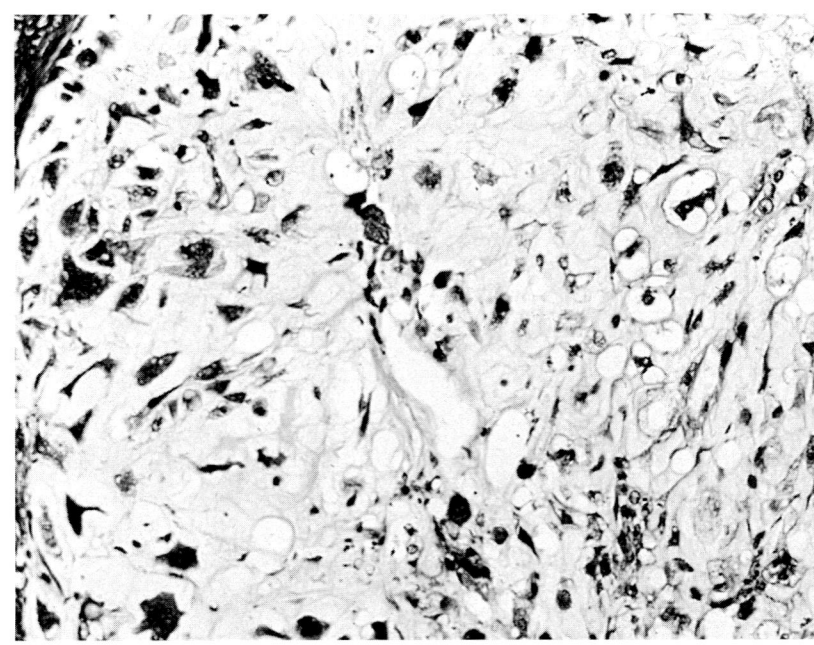

FIGURE 32-30
Chondrosarcoma. This chondrosarcoma consists of lobular nests of malignant chondrocytes showing hyperchromatic cells. The tumor cells have abundant clear cytoplasm and lacunar spaces filled with large atypical, hyperchromatic nuclei. (Intermediate magnification.)

4. Carcinosarcoma
5. Malignant mesothelioma with chondrosarcomatous features
6. Chondroid hamartoma
7. Osteosarcoma
8. Chondroid metaplasia within a primary or metastatic carcinoma
9. Epithelioid hemangioendothelioma

Chondrosarcomas may metastasize to the lungs up to 10 years after diagnosis of a soft tissue primary.[148] Pulmonary artery sarcomas can have chondrosarcomatous differentiation, but the prognosis is much worse than for primary chondrosarcomas of the lung.[139] In retrospect, two initial reports of primary pulmonary chondrosarcomas in patients who did very poorly were most likely pulmonary artery sarcomas.[139,149,150] It is important to extensively sample a tumor with chondrosarcomatous differentiation to exclude the presence of a carcinomatous component. The finding of a carcinomatous component in such a tumor would meet the criteria of carcinosarcoma.[151]

Surgical resection is the primary therapeutic approach. Recurrences may also be amenable to surgical excision. If the tumor cannot be controlled surgically, pulmonary chondrosarcomas may respond to interferon-α therapy.

Osteosarcoma

Primary osteosarcomas of the lung are very rare.[7] They can occur either as an intrapulmonary[152–155] or pleural[152,156–158] mass. Osteosarcoma accounted for 0.01 percent of 10,134 primary pulmonary neoplasms seen at the Mayo Clinic over a 10-year period.[44]

Most pulmonary or pleural osteosarcomas affect patients in the fourth to seventh decades of life,[153–156] but at least one occurred in a teenager.[152] There is no sex predilection. The patients may be asymptomatic or have manifestations of cough, fever, asthma, chest pain, or hemoptysis.[154,155] Chest radiographs reveal calcified masses. Effusions may be seen in cases with pleural involvement.

Pulmonary osteosarcomas may be found within the pleura or lung. Intraparenchymal tumors may also extend to involve the pleura. The tumors tend to be large, ranging between 4 and 16 cm and they may extend to replace an entire lobe of the lung. Bronchial obstruction can occur. Pleural involvement consists of thickening or studding by multiple firm, white, nodules. The cut surface shows lobular, white-tan,

firm masses that may be cystic, hemorrhagic, or necrotic. Minute spicules of bone may be visible macroscopically.

Histologically, the tumor consists of malignant spindle cells with areas of malignant hyaline, chondroid, and osseous tissue (Fig. 32-31). The malignant bone appears as irregular trabeculae surrounded by atypical osteoblasts and osteoclasts. By definition, no component of a carcinoma should be present.

Immunohistochemistry plays an important role in the diagnosis of pulmonary osteosarcoma by exclusion of a carcinosarcoma or malignant mesothelioma with osteosarcomatous differentiation. Thus, staining for epithelial markers such as keratin or EMA should be performed. If a malignant osseous tumor does not show epithelial differentiation, then it is consistent with pulmonary osteosarcoma.[154] Osteosaromas may express vimentin, but so may poorly differentiated carcinomas.

Intrapulmonary osteosarcomas must be separated from carcinosarcomas,[154] pulmonary artery sarcomas,[159] and carcinomas with osseous metaplasia.[160,161] Carcinosarcomas by definition should have a malignant epithelial component, in addition to osteosarcoma. The epithelial component may be squamous, adenocarcinoma, small, or large cell carcinoma, or it may show the pattern of spindle cell carcinoma. In the last of these, immunohistochemical staining for keratin or EMA may be particularly important to correctly identify the tumor. Extensive sampling of such tumors should be performed to search for a carcinomatous component before establishing the diagnosis of pulmonary osteosarcoma.

Pleural osteosarcomas must be separated from malignant mesotheliomas with osteosarcomatous differentiation.[162] Malignant mesotheliomas with osteosarcomatous differentiation show diffuse pleural thickening without significant intrapulmonary involvement. Further, most of these tumors will show focal immunohistochemical staining for keratin.

The prognosis for osteosarcomas of the lung is variable. Some patients die shortly after diagnosis or are diagnosed at autopsy. Others survive surgical resection and remain free of disease, but the follow-up in these patients has been relatively short, less than 14 months.[153]

Liposarcoma

Liposarcomas can present as pleural[163–167] or intrapulmonary[168–173] tumors. For a true pleural liposarcoma, ideally there should be neither mediastinal

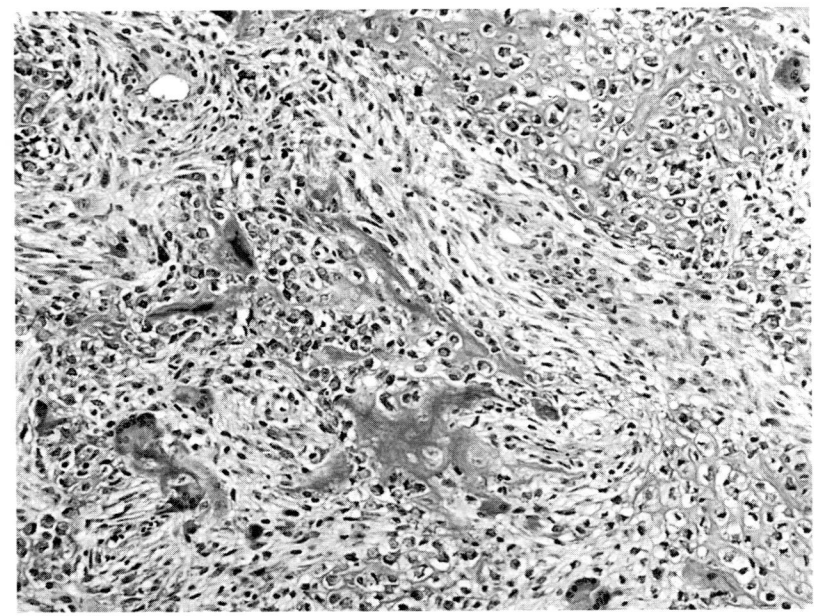

FIGURE 32-31
Osteosarcoma. This tumor has a spindle cell stroma with mitotic activity and irregular spicules of bone (*center*). Chondroblastic differentiation is also present (*right top* and *bottom*). (Intermediate magnification.)

nor pulmonary parenchymal involvement. However, extrapleural tumor has been present in some of the reported cases, with the tumors presenting primarily as a pleural mass. Primary tumors in the pleura or lung must be distinguished from metastatic liposarcomas from extrathoracic sites and tumors extending from the mediastinum.

Pleural liposarcomas are rare, with only nine cases reported in the English literature.[163] In this small group of cases, six were men and three women. One case occurred in a 23-year-old pregnant woman.[165] The patients had a median age of 51 years (range, 19 to 61 years). Presenting symptoms were nonspecific and included chest pain, dyspnea, and cough, but two patients were asymptomatic.[163]

Chest radiographs may show large pleural effusions and a homogeneous intrathoracic density with shift of mediastinal structures. By CT scanning, pleural or pulmonary liposarcomas appear as soft tissue densities, which may be homogeneous or inhomogeneous, depending on the mixture of histologic components. The CT density is usually greater than that of normal fat.[164] Calcification may also occur.[163]

Intrapulmonary liposarcomas are also very rare, with fewer than five reported in the English literature.[169,170,172] Patient ages range between 9 and 44 years and most patients are women.[172] Presenting manifestations include pleural effusion, dyspnea, cough, chest pain, and hemoptysis. Chest radiographs show a lung mass or pleural effusion or both.[169,172]

Pleural liposarcomas are usually discovered after they have become large. The weight of the tumors ranges from 900 to 4460 g. Pleural liposarcomas encase the lung in a fashion similar to malignant mesothelioma, without infiltration of the underlying lung parenchyma. The tumor can appear yellow white and lobulated with a fatty, myxoid, or firm consistency depending on whether it consists histologically of a well differentiated, myxoid, round cell, or pleomorphic pattern (Fig. 32-32).[163] The histologic subtype is not specified in several of the reported cases.[164,167,174,175] Common to each of these subtypes of liposarcoma is the presence of lipoblasts that consist of vacuolated cells with indentations of the nucleus. In one case, the tumor recurred with the pattern of a high-grade MFH, which probably represented dedifferentiation to pleomorphic liposarcoma.[166]

The differential diagnosis includes metastatic liposarcoma,[176] liposarcoma extending from the mediastinum, malignant mesothelioma,[177] pulmonary artery sarcoma,[170] lipoma,[178] atypical lipomatous hamartoma,[179] and alveolar adenoma.[180] It is important to exclude a metastasis by carefully examining the patient for other potential primary sites. This is particularly true because metastatic liposarcoma can present as a solitary lung mass more than 15 years after the primary diagnosis.[176] Exclusion of the possibility of a liposarcoma extending from the mediastinum to involve the lung requires good operative and radiologic information.

The difficulty in separating pulmonary artery sarcoma from a pulmonary liposarcoma is illustrated in the patient reported by Sawamura and coworkers.[170] This case may have represented a pulmonary artery sarcoma because the tumor appears to have involved the left and right main pulmonary arteries

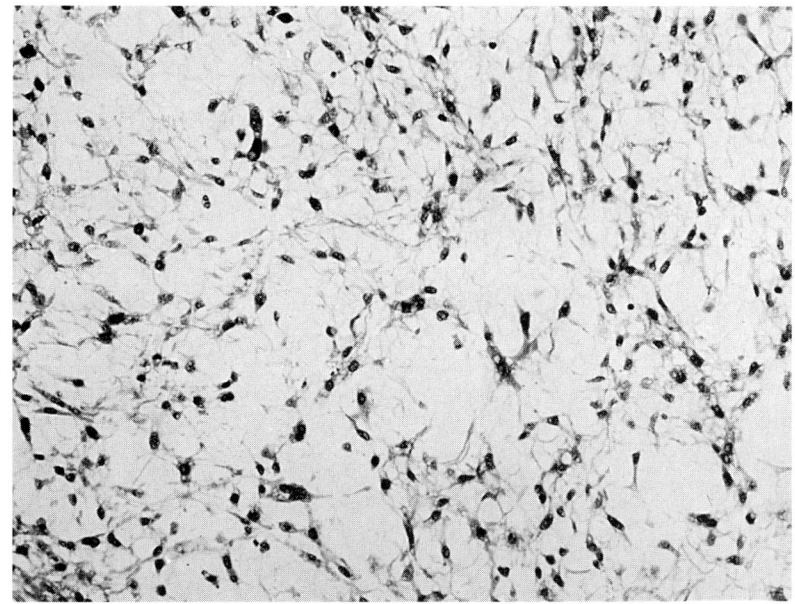

FIGURE 32-32
Pleural liposarcoma. This tumor has a prominent myxoid appearance with small oval to spindle-shaped cells and wispy fibrillar cytoplasm. (Intermediate magnification.)

with extension along pulmonary arteries into the periphery of the left lung. Also, the case reported by Krishna and Haqqani as liposarcomatous differentiation in diffuse pleural mesothelioma is unusual in that the liposarcoma and mesothelioma were not intermixed or contiguous.[177] It is also difficult to regard this case as a pleural liposarcoma because the liposarcoma formed a large left atrial mass with extension into pulmonary veins reminiscent of pulmonary artery sarcoma.

Lipomas of the lung may have atypical features such as bizarre, multiple nuclei similar to those seen in pleomorphic lipomas, but true lipoblasts are not seen.[178] Similarly, pulmonary hamartomas with a prominent lipomatous component may show cellular atypia, leading to consideration of the diagnosis of liposarcoma.[179] The presence of other components such as islands of bone or cartilage or epithelial-lined clefts points to the correct diagnosis of hamartoma. The unusual lipomatous tumor reported by Palvio and colleagues probably falls into this category.[179]

The reported liposarcomas of the pleura have had a variable prognosis. Surgical excision alone may be insufficient in some cases and adjunctive irradiation may be necessary. Of the nine patients reported in the literature, three were disease free at 5 months, 16 months, and 5.5 years.[163] Two individuals died of disease 1 year and 19 months after initial examination. Two patients had local relapses 2 and 4 years after surgery, respectively. The first died 7 years after failed surgery and the second is free of disease 4 years after salvage surgery and radiation.[163]

All of the patients with intrapulmonary primary liposarcomas have died of tumor. Most cases were

diagnosed at autopsy, but several underwent surgical resection.[170] Death occurred 3 to 6 months following diagnosis in those who survived the surgery.[170,172]

SARCOMAS OF UNCERTAIN HISTOGENESIS

Monophasic Synovial Sarcoma

Synovial sarcomas are well defined sarcomas that are most commonly found in the soft tissues. Still, they have also been described in numerous locations, such as mediastinum, abdominal wall, esophagus, parotid gland, oral cavity, and neck.[181–187] Histologically, these tumors have been divided into three subtypes: (1) biphasic, (2) monophasic spindle cell, and (3) monophasic epithelial type. The last of these is rare.

Recently, 25 cases of a primary intrapulmonary sarcoma have been reported, which show histologic and immunohistochemical features similar to those of monophasic spindle cell synovial sarcomas of soft tissues.[188] All of the tumors were intrapulmonary; none of the patients had a history of tumor elsewhere in the body. The patients ranged from 16 to 77 years of age with a fairly equal distribution among men and women. Grossly, the tumors varied in size from 0.6 to 20 cm in greatest dimension and were soft masses with areas of hemorrhage and necrosis; some of them had cystic changes.

Histologically and immunohistochemically, these neoplasms very much resemble synovial sarcomas of the soft tissues. They show a tight spindle cell proliferation with a solid growth pattern (Figs. 32-33

1089

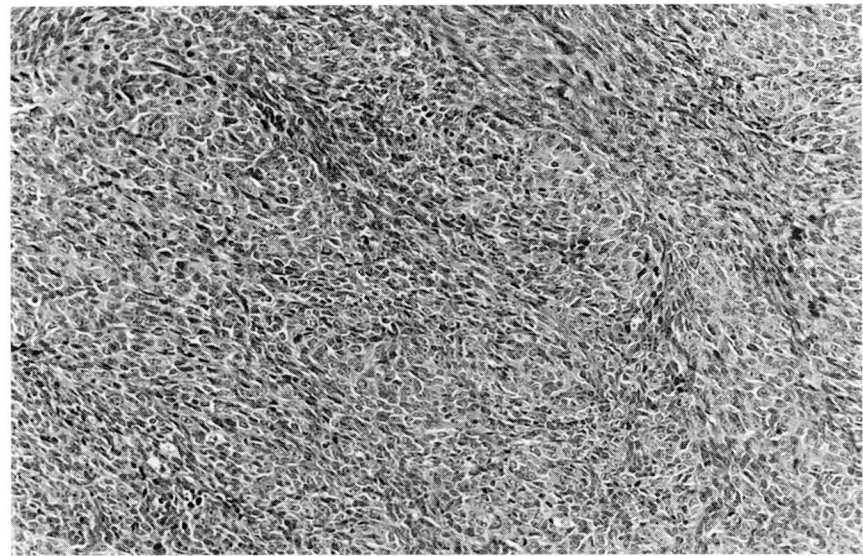

FIGURE 32-33
Pulmonary monophasic synovial sarcoma. Low-power view shows a highly cellular spindle cell proliferation. (Intermediate magnification.)

through 32-36). Myxoid, hemangiopericytic, and epithelioid areas are often seen in these tumors (Fig. 32-37). Necrosis, hemorrhage, and mitoses were frequently seen although they varied in proportion from case to case. Immunohistochemical studies demonstrate diffuse staining for EMA and keratin in nearly all the cases. Ultrastructural studies in a few cases show abundant rough endoplasmic reticulum and well developed desmosome-type intercellular junctions (Fig. 32-38).

The behavior of these pulmonary tumors appears to be similar to their counterparts in the soft tissues. One third of patients die from their tumors 2 to 20 years after diagnosis, whereas others were alive but have had recurrence.

The differential diagnosis includes numerous sarcomas that may present as primary or metastatic lesions in lung. The most important distinction is to determine whether the tumor is primary or metastatic. It is particularly important to exclude metastatic synovial sarcoma from the soft tissues because this tumor may have an indolent course with metastases years after initial discovery.

The difference between pulmonary monophasic synovial sarcoma and spindle cell carcinoma of the lung can be difficult. Spindle cell carcinoma is generally more infiltrative microscopically with greater cellular pleomorphism and higher mitotic activity. In addition, it may show areas of transition to more conventional foci of squamous cell carcinoma. Finally,

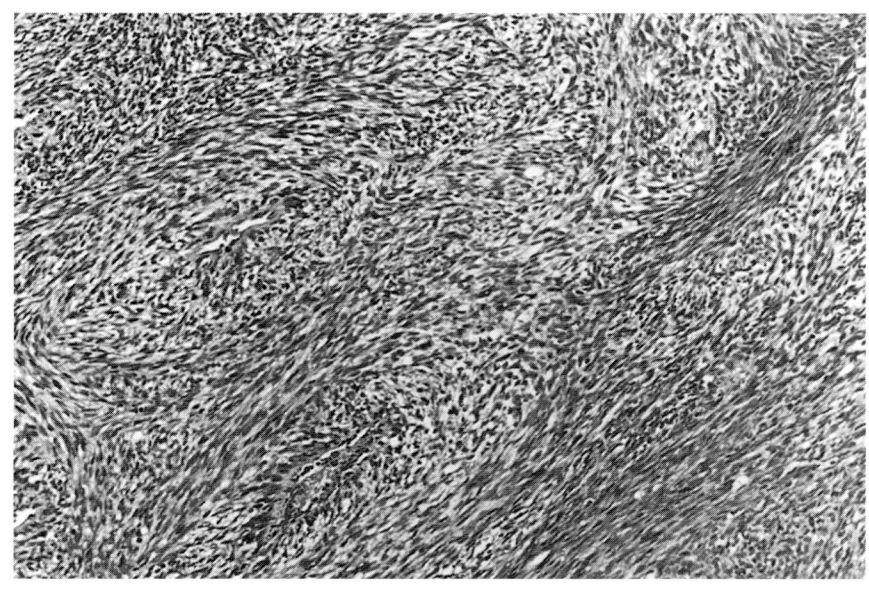

FIGURE 32-34
Pulmonary monophasic synovial sarcoma. Monophasic synovial sarcoma of lung shows interlacing fascicles of spindle cells. (Intermediate magnification.)

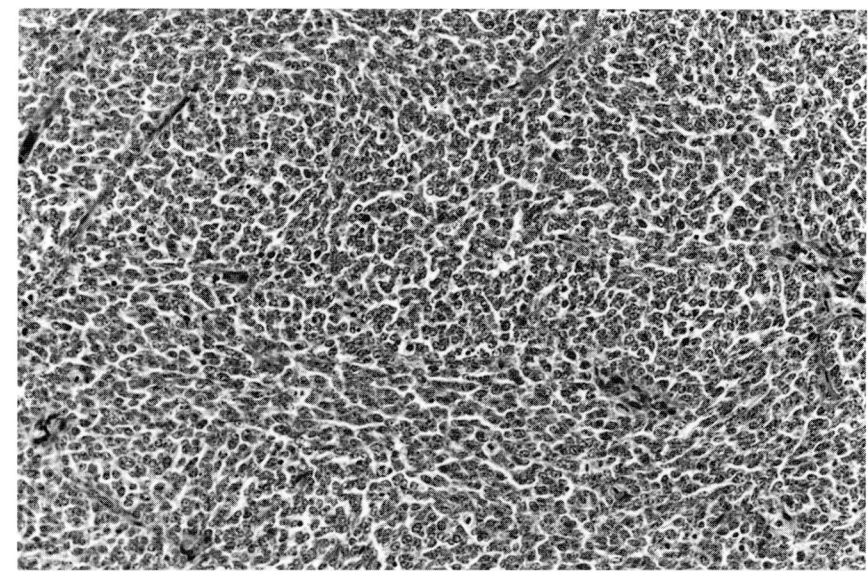

FIGURE 32-35
Monophasic synovial sarcoma of lung shows a proliferation of small oval and slightly spindled tumor cells. (Intermediate magnification.)

its natural history is similar to that of other carcinomas in the lung, that is, the tumor shows metastasis to regional lymph nodes and follows a more aggressive clinical course than does pulmonary synovial sarcoma. Immunohistochemical stains should be used with caution in the differential diagnosis because both tumors may show positive staining with epithelial markers (keratin or epithelial membrane antigen). Another epithelial marker, CEA, may be negative in some cases of spindle cell carcinoma.

Primitive Neuroectodermal Tumor (PNET)

Primary pulmonary peripheral neuroectodermal tumors are unusual tumors. PNETs are more commonly encountered in the soft tissues or in an extrapulmonary location within the thorax. The term PNET appears to encompass several other diagnoses, including extraskeletal Ewing's sarcoma, malignant small cell tumor of the thoracopulmonary region, paravertebral round cell tumor, and neuroepithelioma.[189–197]

We have recently studied several cases (unpublished data) in which the tumors appeared to arise within the lung parenchyma without evidence of bone and thoracic involvement. The patients were young and presented with symptoms of cough, hemoptysis, fever, shortness of breath, and chest pain.

Grossly, PNET presents as a mass measuring up to 8 cm in greatest dimension with areas of necrosis

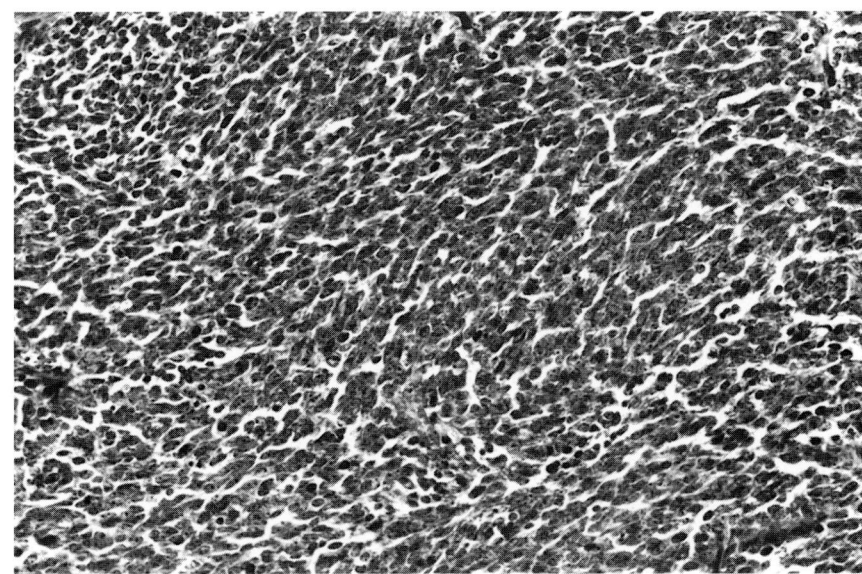

FIGURE 32-36
Higher magnification view of synovial sarcoma. The marked cellularity of this spindle cell tumor is evident. (Intermediate magnification.)

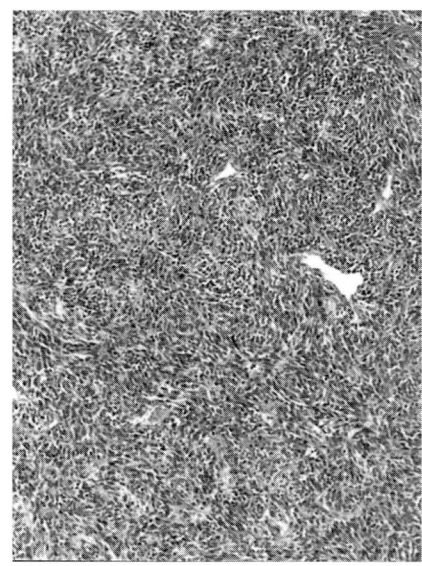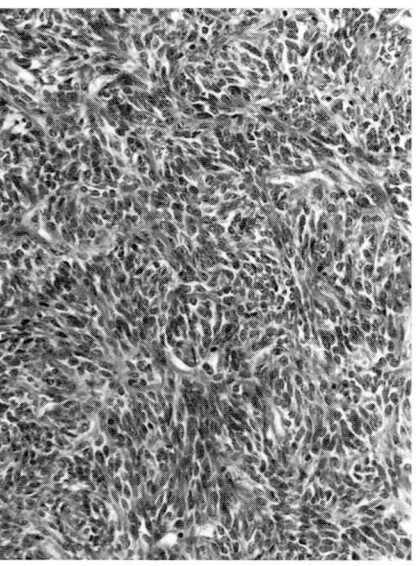

and hemorrhage. Histologically, at low magnification the tumor shows sheets of darkly stained, small cells with interspersed areas of hemorrhage or necrosis (Figs. 32-39 and 32-40). At higher magnification, the neoplastic cells are small and round with round or oval nuclei, scant cytoplasm, dispersed nuclear chromatin, and inconspicuous nucleoli (Fig. 32-41). In some areas, they may show clear cytoplasm whereas in others there are nests of smaller cells with pyknotic nuclei (Fig. 32-42). Mitoses are easily found and vary in number from case to case. Rosettes are usually present.

Histochemically, periodic acid-Schiff-stained sections show varying amounts of glycogen in tumor cells. The malignant cells fail to show immunohistochemical staining for epithelial and muscle markers, but may show vimentin or neuron specific enolase. More recently a new marker, MIC2, believed to be specific for Ewing's sarcoma, has been used in the evaluation of these neoplasms.[198,199]

The differential diagnosis of PNET includes other round cell sarcomas as well as epithelial and lymphoid tumors that may present as intrapulmonary neoplasms. These tumors include rhabdomyosarcoma, small cell carcinoma, and lymphoma. Careful histologic evaluation and use of immunohistochemical markers will lead to a correct interpretation.

MIXED EPITHELIAL AND MESENCHYMAL TUMORS

Pulmonary blastomas are tumors in which the histologic components of the neoplasm are primitive or embryonal in appearance.[200] Specifically, the glycogen-rich, nonciliated tubules or the stroma of the tumor resembles fetal lung between 10 and 16 weeks' gestation (the pseudoglandular stage of lung development).

Barnard[201,202] was the first to offer a pathologic description of blastomas, which he termed embryoma because of its resemblance to embryonic lung. Spencer[203] reported the microscopic features of additional cases in detail and used the name blastoma, assuming the tumor was the pulmonary analogue of nephroblastoma. In their descriptions, Spencer and others emphasized the presence of both malignant

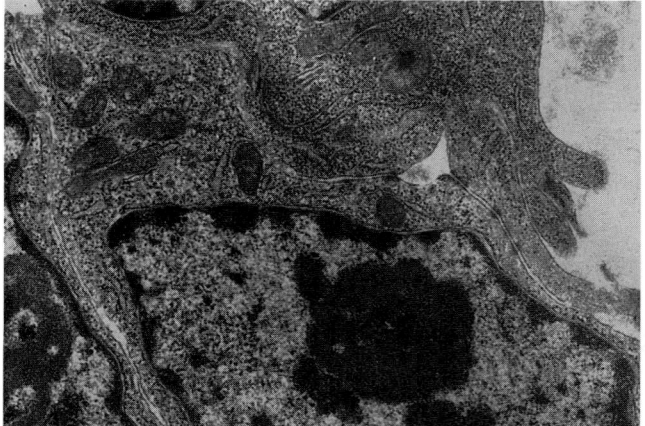

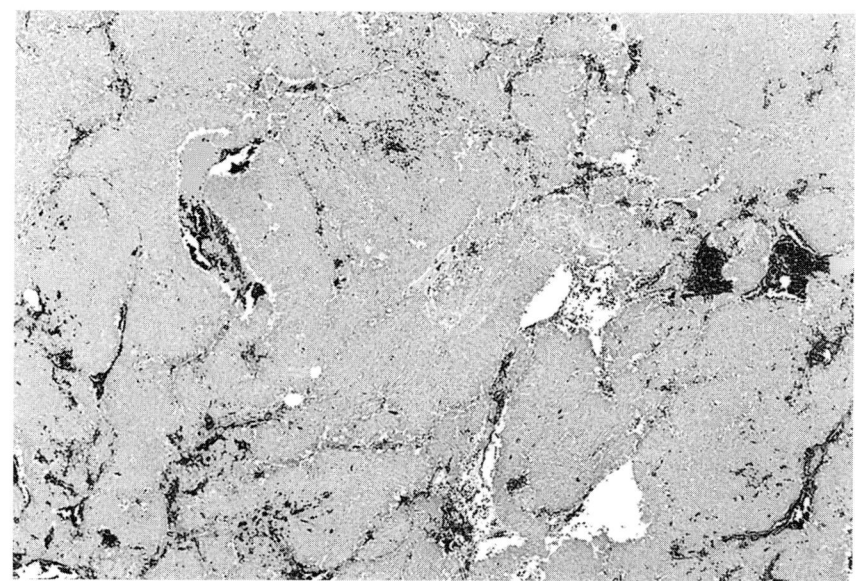

FIGURE 32-39
Primitive neuroectodermal tumor. This low-magnification view shows a highly cellular tumor that has a nodular and sheet-like pattern of growth with central areas of hemorrhage. (Low magnification.)

embryonic epithelium and fetal stroma in these unusual tumors.[203,204] Subsequently, a distinctive variant of pulmonary blastoma was discovered consisting of malignant but well differentiated fetal glands coexisting with a benign spindle cell stroma.[205] This tumor was given a variety of names: pulmonary endodermal tumor resembling fetal lung,[205–208] pulmonary adenocarcinoma of fetal type,[209] well differentiated adenocarcinoma simulating fetal lung tubules,[210] and well differentiated fetal adenocarcinoma.[211] Even though this tumor is a monophasic malignant epithelial neoplasm, because of its microscopic similarity to biphasic pulmonary blastoma, it is discussed in this section of the chapter.

Although the name blastoma implies a pediatric tumor, classically it was found that only a subset of patients had tumors in the first decades of life; most patients were adults.[212] On reconsideration of these cases, it appears that almost none of these childhood neoplasms are biphasic (as originally thought); rather, they are composed of a malignant embryonic stroma with entrapped, benign glandular epithelium of lung or pleural origin.[213,214] These neoplasms range from solid or focally cystic tumors involving the lung, pleura, or mediastinum—so-called pleuropulmonary blastomas—to largely cystic intrapulmonary malignancies.[213–215] The latter were recognized previously as sarcomas occurring in asso-

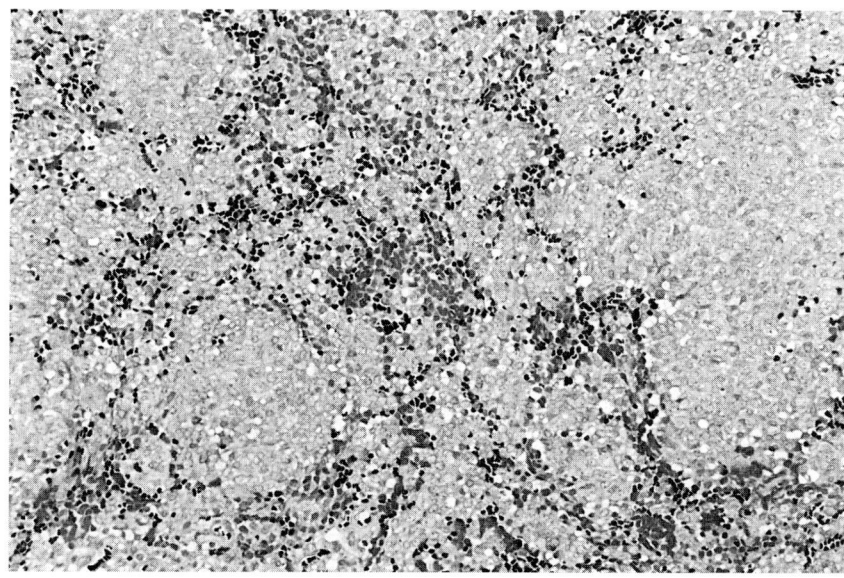

FIGURE 32-40
Primitive neuroectodermal tumor of lung. The tumor is composed of oval and round tumor cells with scant cytoplasm. (Intermediate magnification.)

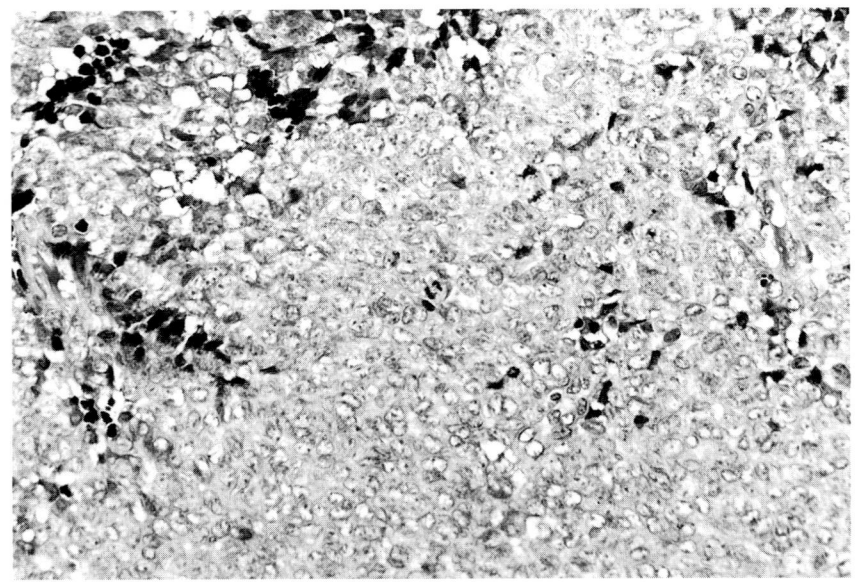

ciation with cysts of lung.[215] These childhood tumors appear to be clinically and pathologically distinct from the other forms of blastoma and will be discussed separately below.

Well Differentiated Fetal Adenocarcinomas (Pulmonary Endodermal Tumors)

This neoplasm is composed of neoplastic, glycogen-rich, nonciliated tubules that resemble the developing bronchial tubules of fetal lung between 10 and 16 weeks' gestation. The accompanying mesenchyme is histologically benign.[200] We are of the opinion that

the tumor is probably a variant of the earlier described biphasic blastomas.

Well differentiated fetal adenocarcinomas occur about as frequently as biphasic blastomas. Although there has been a suggestion that the tumor predominates in women, it most likely affects patients of both sexes about equally.[211] Despite the name blastoma, these tumors do not appear to occur in children, but rather affect adults, with an age peak in the fourth decade.[211] Most patients are smokers, suggesting a pathogenesis similar to that of the usual cacinomas of lung.[207,211]

Patients are usually asymptomatic, but fe-

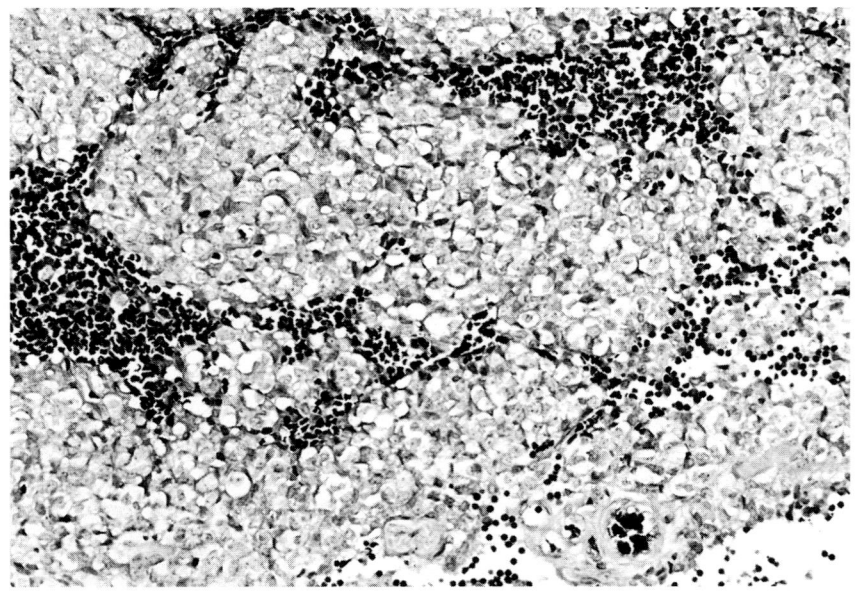

FIGURE 32-42
Primitive neuroectodermal tumor of lung.
The tumor shows focal clear cell changes as well as abundant hemorrhage. (Intermediate magnification.)

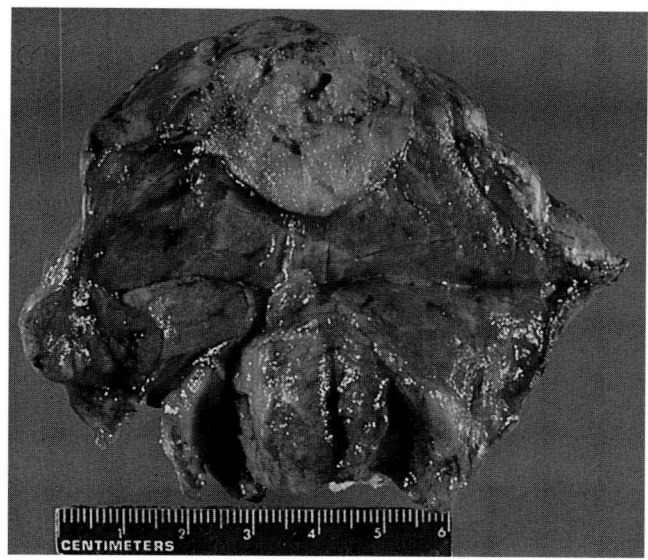

FIGURE 32-43
Pulmonary blastoma. Cut surface of the lung shows a well delimited, fleshy, minimally cavitated mass. (*See also Color Plate 8.*)

ver, cough, chest pain, and hemoptysis can occur.[203,207,211] The tumor is less likely to produce symptoms than biphasic blastomas because it is relatively smaller at presentation.[211]

The chest radiograph typically shows a solitary peripheral or mid-lung mass. The tumor occurs slightly more often in the upper lobes. Adenopathy and pleural effusion are rare or absent. The diagnosis may rarely be made by fine needle aspiration.[216]

Grossly, well differentiated fetal adenocarcinomas are solitary, well demarcated masses that range in size from 1 to 10 cm (mean, 4.5 cm).[211]

Infrequently, a dominant mass occurs with satellite lesions. The cut surface shows a bulging tan, white, or brown tumor with areas of cystic change and hemorrhage (Fig. 32-43); a polyploid intrabronchial component is seen infrequently.

Microscopically, well differentiated fetal adenocarcinoma shows branching tubules, sometimes arrayed in a cribriform pattern, in a scant spindle cell stroma (Fig. 32-44). The tubules are lined by pseudostratified, nonciliated columnar cells with clear or lightly eosinophilic cytoplasm and oval or round nuclei with little hyperchromasia or pleomorphism. These cells often have subnuclear and supranuclear cytoplasmic vacuoles due to abundant glycogen, producing a distinctly endometrioid appearance (Fig. 32-45). Rarely, there is nuclear pleomorphism in the form of large multinucleated cells.[217] In addition, there may be cords, ribbons, or solid epithelial nests containing minute rosette-like glands.[211] Small amounts of mucin can be present within glandular lumina. Ultrastructurally, the neoplastic glands show a basal lamina, apical junctional complexes, glycogen-free spaces, and apical microvillous surfaces.

Another distinctive endometrioid feature is the presence of morules—solid balls of squamoid cells with ample eosinophilic cytoplasm—at the bases of the glands[207,211,218] (Fig. 32-46). In about 50 percent of cases, the cells within morules contain optically clear nuclei. Optically clear nuclei contain biotin, a feature that they share with the optically clear nuclei of endometrial adenocarcinomas.[208] Ultrastructurally, this finding appears to correlate with the

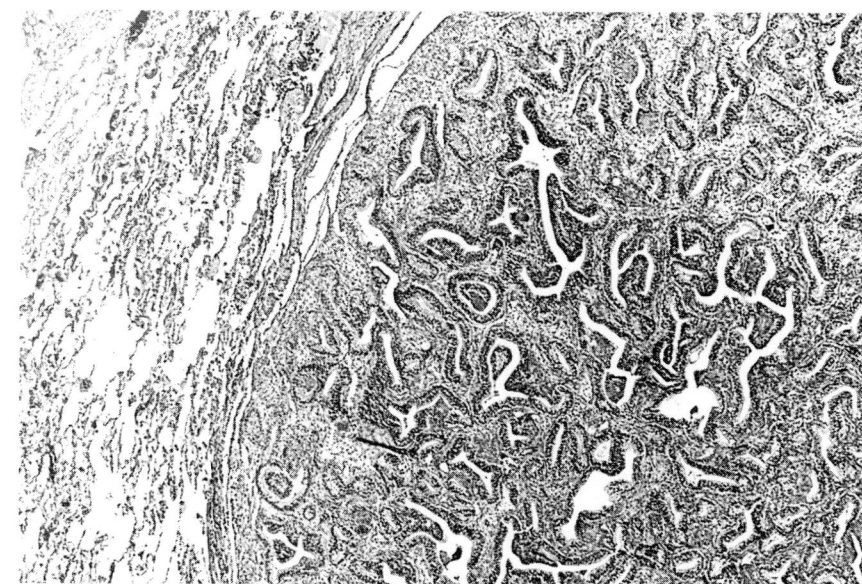

FIGURE 32-44
Well differentiated fetal adenocarcinoma (pulmonary endodermal tumor). At low magnification, the tumor is fairly well circumscribed from the surrounding lung parenchyma. It shows the typical appearance: Branching tubules in a scant stroma. (Low magnification.)

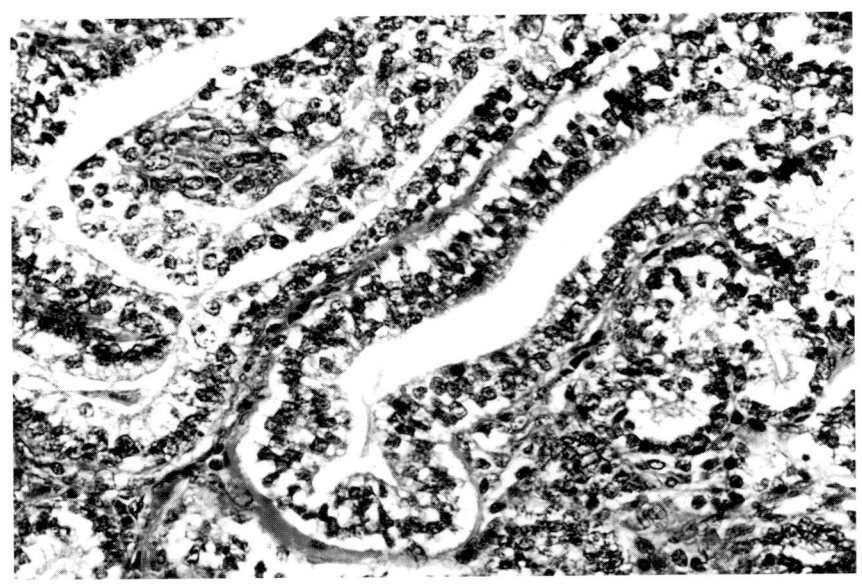

FIGURE 32-45
Well differentiated fetal adenocarcinoma.
The glands are lined by columnar cells,
which often contain subnuclear or
supranuclear cytoplasmic vacuoles.
(Intermediate magnification.)

intranuclear accumulation of 10 to 12 nm in diameter filaments.[208]

Neuroendocrine differentiation is a distinctive feature of well differentiated fetal adenocarcinoma. Argyrophil granules and chromogranin are present in a few glandular epithelial cells and within morular cells in 64 to 72 percent of cases[211] (Fig. 32-47). These cells show both dense-core granules and a number of specific amines and polypeptide hormones, including calcitonin, gastrin-releasing peptide, bombesin, leucine enkephalin and methionine enkephalin, synaptophysin, somatostatin, and serotonin.[206,207,210,211]

The malignant glands also stain with anti-

bodies to cytokeratin, CEA, EMA, and α-fetoprotein.[207,209, 217] Both Clara cell antigen and surfactant apoprotein are expressed in epithelial cells and particularly in morules.[207] This antigenic profile may also be found in developing fetal lung tubules, which show differentiation toward Clara cells beginning at 13 weeks of gestation and toward type II pneumocytes at 22 weeks.[217,219]

The stroma of well differentiated fetal adenocarcinoma is typically benign and scant and consists of spindled myofibroblastic cells[205–207,210,211] (see Figs. 32-44 and 32-46). The stromal cells show vimentin and muscle-specific actin positivity and they have typical myofibroblastic features, including well

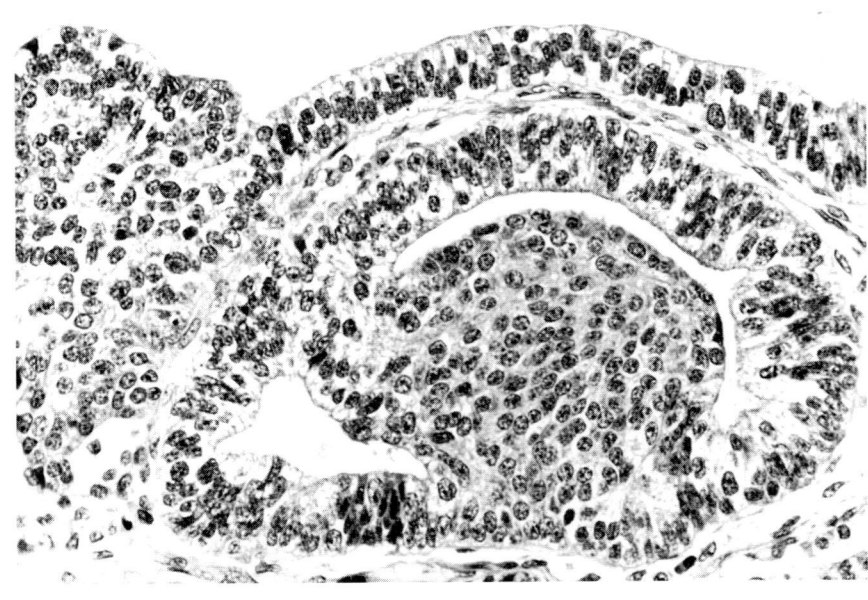

FIGURE 32-46
Morule in well-differentiated fetal
adenocarcinoma. The morule consists of a
solid nest of cells with oval nuclei.
(Intermediate magnification.)

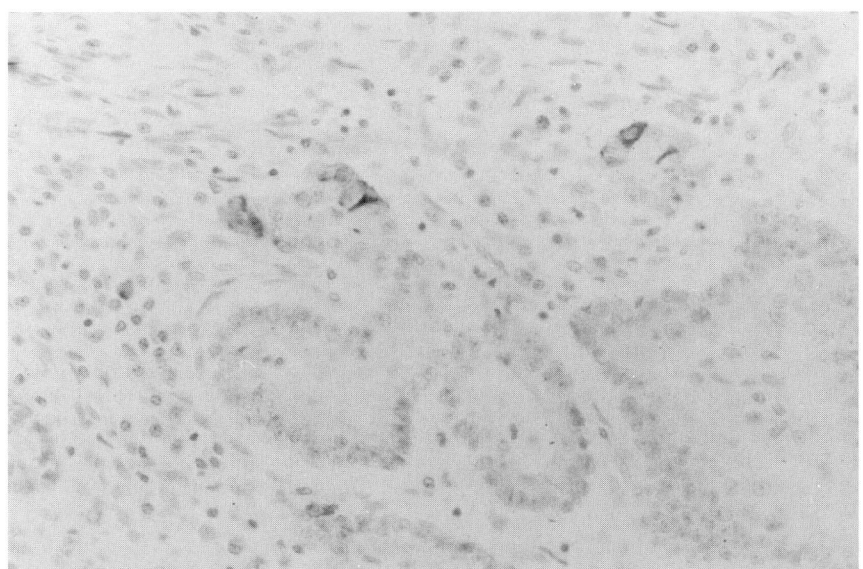

FIGURE 32-47
Well-differentiated fetal adenocarcinoma. Immunohistochemical staining for chromogranin. Scattered immunopositive cells are present in glands in this view. (Intermediate magnification.)

developed rough endoplasmic reticulum, peripheral cytoplasmic filaments forming dense bodies, pinocytotic vesicles, and an investing basal lamina.

Surgery is the primary treatment of this tumor. Combination chemotherapy is used when there is local or distant metastasis, where it may be of palliative benefit.[211] The prognosis of well-differentiated fetal adenocarcinoma initially appeared poor with up to 56 percent mortality[210]; however, these studies may have included adult-type clear cell adenocarcinomas. With stringent diagnostic criteria and greater numbers of cases, the prognosis appears very good, with tumor-associated mortality of only 10 to 14 percent (median follow-up, 95 months).[207,211] This low mortality rate may reflect the tendency of this neoplasm to recur in lung where it can be easily resected.[207,211]

Pulmonary Blastomas (Biphasic)

Biphasic pulmonary blastomas are composed of both malignant glands and malignant mesenchyme that are primitive or embryonal in appearance.[200] Pulmonary blastomas are rare, with a frequency of only 0.25 to 0.5 percent of all primary malignant lung tumors.[204,220] As in the case of well-differentiated fetal adenocarcinomas, most patients are smokers and nearly all are adults. In fact, there is a unimodal age of presentation in the fifth decade of life.[211] Over 80 percent of patients have symptoms such as cough, chest pain, dyspnea, or hemoptysis.[211]

The tumors are usually solitary peripheral masses that favor the upper lobes. When bronchoscopic and needle biopsies are done, they yield a cor-

rect or suggestive diagnosis in only one third of the cases. This low yield probably results from the histologic resemblance to other neoplasms and complex histology of the tumor.

Grossly, biphasic blastomas are solitary peripheral pulmonary tumors with a pale, fleshy appearance, areas of cystic change, and foci of hemorrhage. They are usually large, averaging 10 cm and sometimes measuring up to 27 cm in diameter.

Microscopically, the epithelial component shows the characteristic endometrioid glands in virtually all cases, but they are not as extensive as in well differentiated fetal adenocarcinomas (Fig. 32-48). Most tumors also show solid cords, ribbons, sheets, or nests of epithelial cells, sometimes with clear cytoplasm (Fig. 32-49). Morules are less common than in well differentiated fetal adenocarcinomas, occurring in less than 50 percent of cases.[211] Foci of squamous pearl formation may be present.[204,217,220]

The malignant stroma may be embryonic or "blastematous" with small, oval and spindled stromal cells lying in a myxoid stroma (see Fig. 32-48). Small foci of adult-type spindle cell sarcoma (most commonly showing a fascicular or storiform pattern) are also present in up to 83 percent of cases (see Fig. 32-49). About 25 percent of cases show areas of immature striated muscle and cartilage; about 5 percent of tumors have osseous differentiation[204,211] (Fig. 32-50). Some cases show short fascicles of "fetal-type" smooth muscle.[217] There are also rare cases that show a combined yolk sac component.[221]

The epithelial component of biphasic blastomas shares the antigenic profile of well differentiated

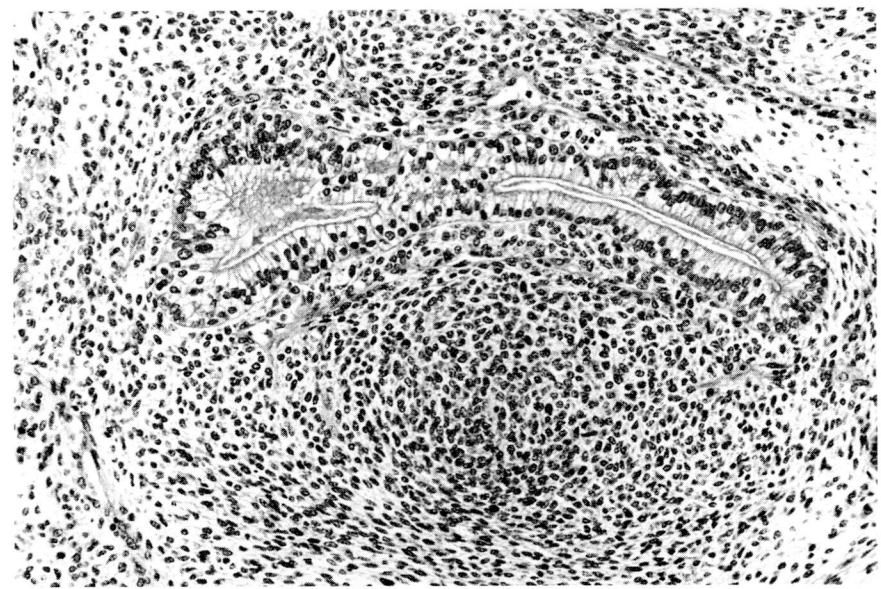

FIGURE 32-48
Biphasic pulmonary blastoma. The view shows both components—a typical gland with optically clear cytoplasm (*above*) and malignant embryonic or "blastematous" stroma with small, oval and spindled cells. (Intermediate magnification.)

fetal adenocarcinomas. The stromal cells contain vimentin and muscle-specific actin. Focally there may be desmin and myoglobin or S-100 when there is, respectively, striated muscle or cartilage. Although vimentin is found principally in the stromal component and keratin is present in the glandular epithelium,[211] one study reported vimentin in glands and focal staining for keratin in stromal cells.[217]

The tumors recur in 43 percent of cases, most commonly in regional thoracic lymph nodes, the lung or pleura, chest wall, diaphragm, and brain. The prognosis is as poor as that of common lung carcinomas.[204,211] Although occasional patients survive long term,[222] two thirds of them die within 2 years of diagnosis and only 16 percent survive 5 years.[204] The stage of the tumor may be prognostically important, with stage 1 "blastomas" having a 5-year survival of about 25 percent.[211] Other ominous prognostic factors include an initial diameter greater than 5 cm and tumor recurrence during the clinical course.

Cystic and Pleuropulmonary Blastomas of Childhood

Virtually all childhood blastomas differ microsopically from their adult counterparts. Specifically, these rare intrathoracic tumors are sarcomas, com-

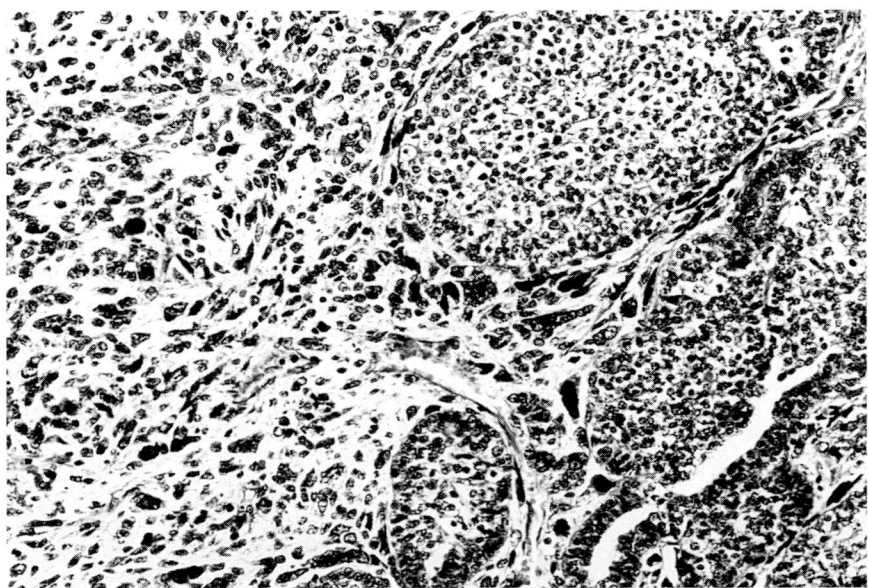

FIGURE 32-49
Biphasic pulmonary blastoma. In addition to solid nests of epithelium with focally clear cytoplasm, there is a pleomorphic, spindle cell sarcoma. (Intermediate magnification.)

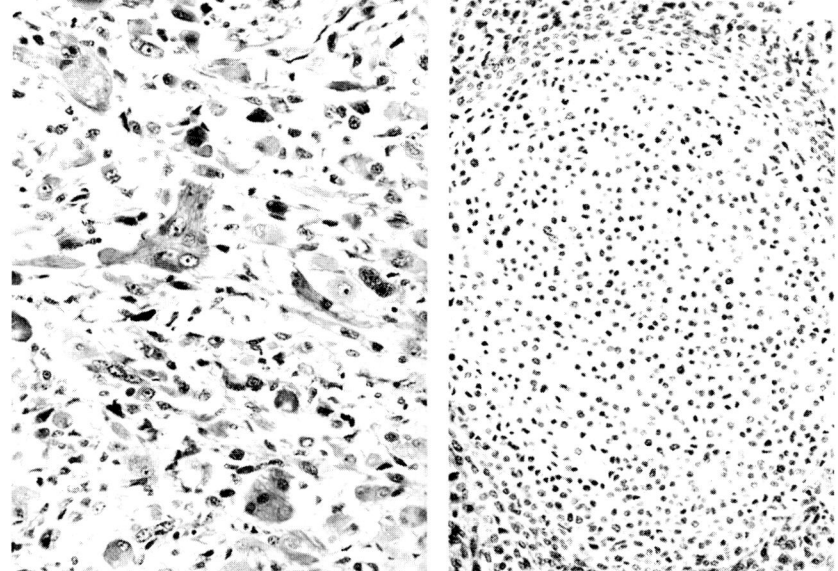

FIGURE 32-50
Biphasic pulmonary blastoma. Biphasic blastomas may show foci of immature striated muscle (*left*) and cartilage (*right*). (Low/intermediate magnification.) (Fig. 8 from Koss et al[211] with permission.)

posed of malignant, embryonic, or blastemal mesenchyme with either no epithelial component or with a nonneoplastic, presumably entrapped, epithelium. They may present as cystic lesions—thin-walled intrapulmonary cysts with an underlying cambium layer of embryonic and rhabdomyosarcomatous mesenchyme.[223] In the past, these cystic blastomas have been termed pulmonary sarcomas arising in mesenchymal cystic hamartoma, embryonal sarcoma, or pulmonary rhabdomyosarcoma arising in congenital cystic adenomatoid malformation or bronchogenic cyst.[213,224–226] By contrast, some childhood blastomas are predominantly or completely solid, involving not only lung but also mediastinum and pleura.[213] These pleuropulmonary blastomas probably encompass many of the "biphasic" blastomas of lung previously reported in children.[228]

Predominantly cystic lesions occur in children who are usually 1 to 9 years of age.[223] The patients may be asymptomatic or may have tachypnea, spontaneous pneumothorax, or respiratory distress.[223] Radiographically and by CT scanning, a single or multiloculated, peripheral cystic lesion, sometimes associated with a mass, is present. The mediastinum may be displaced to the side opposite the cyst. Younger patients with intrapulmonary, thin-walled cystic lesions are likely to have surgically resectable tumors and may have a good (>50 percent) long-term prognosis.[223]

Grossly, cystic tumors are single or multiloculated and may show nodular, thickened walls or pedunculated nodules. Histologically, they consist of one or more spaces lined by benign alveolar or ciliated columnar epithelial cells beneath which is a layer of primitive oval and spindled rhabdomyoblasts in a loose or dense fibrovascular stroma (Fig.

32-51). The nodular areas of thickening in the gross specimen are composed of embryonic, oval to stellate cells.[228]

Pleuropulmonary blastomas are largely solid, multilobulated masses that occur not only in lung (as in adult blastomas), but also in the mediastinum or pleura. The patients range from 30 months to 12 years of age; many are under 3 years of age.[214] Symptoms are universal and are due to the presence of a large intrathoracic mass that commonly opacifies an entire hemithorax, causing a contralateral shift of the medastinum, and may be associated with pleural effusion.[224] The symptoms include dry cough, fever, dyspnea, or chest pain.[213,214,215,219] The tumor-associated mortality rate is 50 to 75 percent.

Grossly, the tumors are multilobulated, white gray, and focally hemorrhagic masses that measure 8 to 23 cm in diameter and weigh 160 to 1100 g. They are predominantly solid but they may show a focal cystic component.[214]

Microscopically, pleuropulmonary blastomas consist of blastemal stromal cells, arranged in alternating bands of compact and loose cells in a myxoid matrix (Fig. 32-52). There may be anaplastic and pleomorphic mesenchymal cells with numerous mitoses. Areas of chondrosarcoma, rhabdomyosarcoma, and smooth muscle-like spindle cells, storiform areas, and, rarely, foci of lipoblastic differentiation may be found (Fig. 32-53). Cysts or smaller glandular spaces can also be present (Fig. 32-54). As in cystic blastomas, the cysts usually have a lining of histologically benign epithelial cells that are probably entrapped bronchiolar, alveolar, or mesothelial cells. Immunohistochemical staining mirrors the range of differentiation, with vimentin, histiocytic markers, or myoid antigens being com-

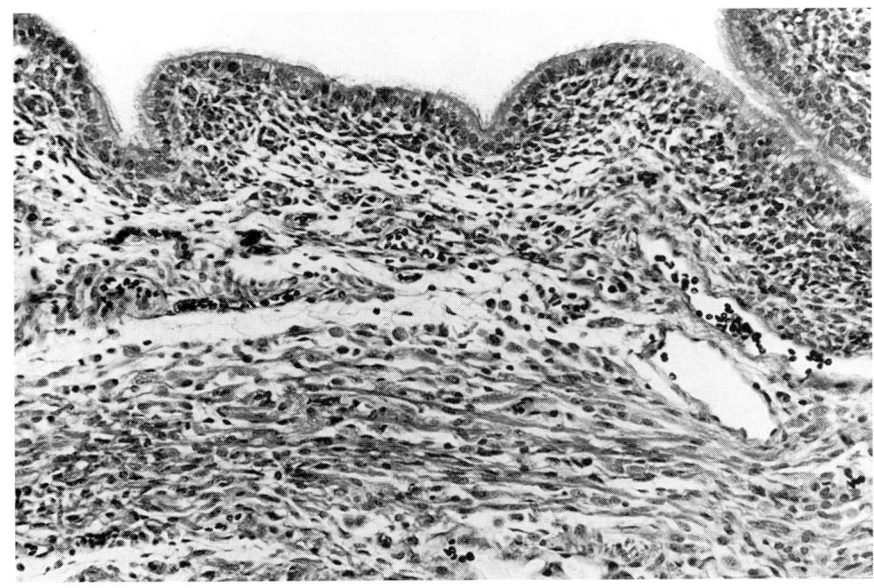

mon. Electron microscopic observation shows the spindle stromal cells to be a mixture of undifferentiated, myofibroblastic, fibroblastic, and histiocytic-like cells, analagous to the proliferation seen in malignant fibrous histiocytoma.[214]

Carcinosarcomas

The histologic definition of pulmonary carcinosarcoma, as adopted by the World Health Organization (WHO), is that of a tumor containing both malignant epithelial (carcinomatous) and mesenchymal (sarcomatous) elements of the type ordinarily seen in malignancies of adults.[231] Because these rare neoplasms are difficult to distinguish from the more common spindle cell carcinomas of lung,[229,230] the WHO definition[200] added the corollary that pulmonary carcinosarcomas should show differentiation of the mesenchymal component into specific heterologous tissues, such as neoplastic bone, cartilage, and striated muscle. Recently, immunohistochemistry and electron microscopy have been used as additional methods of demonstrating these differentiated mesenchymal elements.[230,231] These techniques have led to the view that carcinosarcomas are one end of a continuum of differentiation of spindle cell carcinomas. This, in turn, has led to the

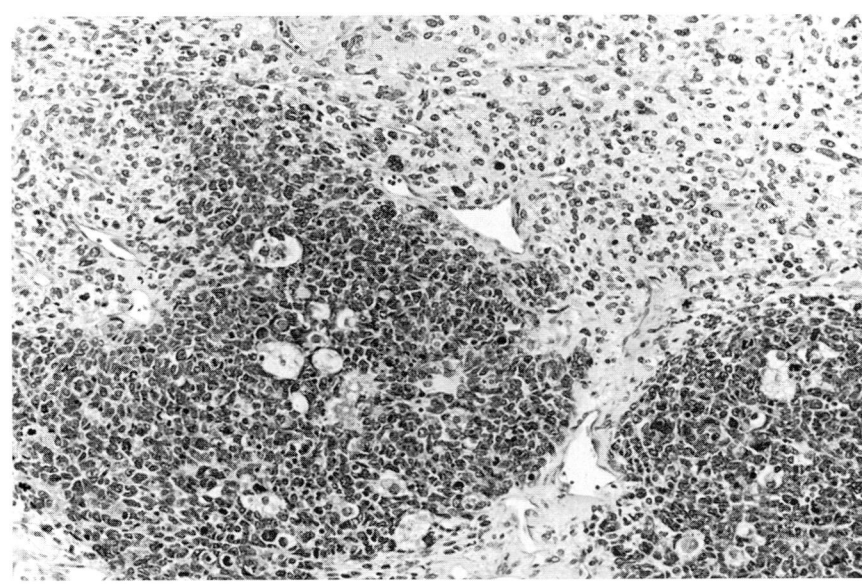

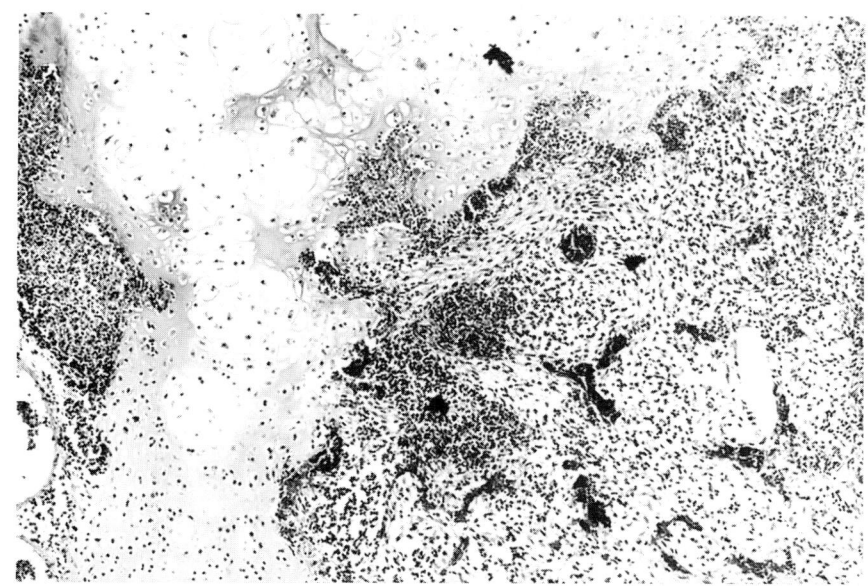

FIGURE 32-53
Pleuropulmonary blastoma. The tumor contains large, irregular islands of cartilage in a primitive stroma. (Intermediate magnification.)

use of unifying terms such as "carcinoma with a sarcomatoid element" or "pulmonary carcinoma with sarcoma-like lesion" to describe all of these tumors,[230,231] More recently, Nappi and Wick[232] have proposed monophasic sarcomatoid carcinoma as a synonym for spindle cell carcinoma and biphasic sarcomatoid carcinoma as a synonym for carcinosarcoma.

Many of the published cases diagnosed as pulmonary carcinosarcomas do not show heterologous sarcomatous elements and thus do not meet the WHO definition. There is some ambiguity, therefore, concerning the clinical and biologic behavior and patho-

logic features of these tumors. The following descriptions should therefore be read with this caveat in mind.

Men are affected far more frequently than women and most patients are middle-aged or older adults: Over 90 percent of them are between 50 and 80 years of age.[229–231,233-237] There is a strong association with smoking.[238]

The tumor typically presents as a well demarcated, lobulated mass in an upper lobe. One tumor with extensive pleural involvement has also been reported.[235] It appears that carcinosarcomas can be divided into parenchymal (peripheral) or central

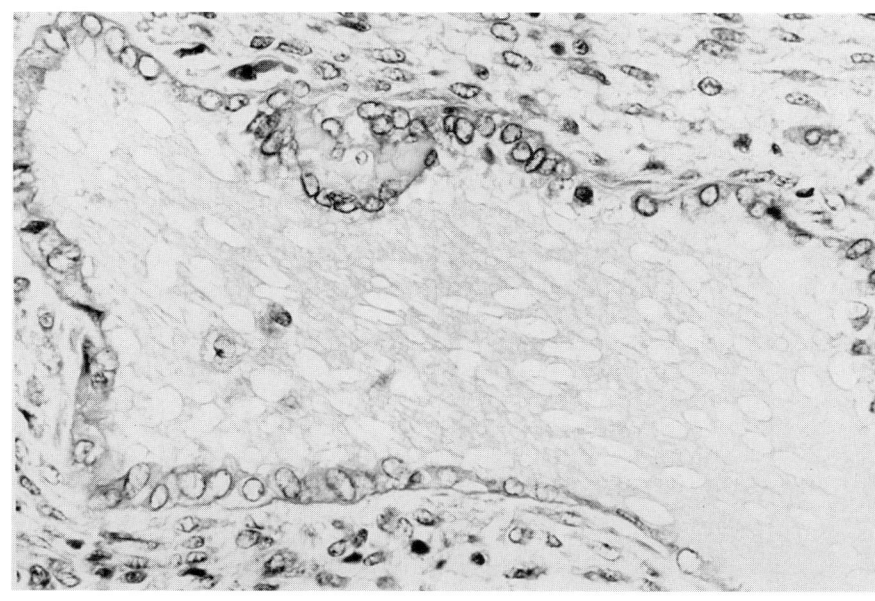

FIGURE 32-54
Pleuropulmonary blastoma. A small glandular space is lined by benign cuboidal cells that are presumably entrapped in the neoplasm. (High magnification.)

(endobronchial) tumors.[234] The solid parenchymal neoplasm typically presents as a large mid-lung or peripheral pulmonary mass. Because of its more distal location in lung, it may be asymptomatic in early stages, but later in its course it tends to invade mediastinum, pleura, and chest wall, producing thoracic pain.[231,236] About one third of carcinosarcomas fit into this category. By contrast, central tumors are usually pedunculated, endobronchial neoplasms involving lobar and segmental bronchi. They consequently are prone to produce early symptoms of an obstructing airway mass, such as cough, fever, dyspnea, and hemoptysis.[233,234] Two thirds of tumors present in this manner.

Sputum samples are rarely diagnostic.[238–240] When attempted, bronchoscopic procedures yield a diagnosis of malignancy in 40 to 66 percent of cases.[233,236–239] At least four peripheral tumors have been identified by transthoracic fine needle aspiration.[238,240] However, forceps biopsy or fine needle aspiration may be inaccurate because they produce tissue fragments that are too small or too necrotic to show both components of the neoplasm.[233] Fatal hemorrhage may also occur after bronchoscopic biopsy.[233]

Pneumonectomy or lobectomy is both diagnostic and the initial treatment in many cases. There is argument about whether endobronchial tumors have a more favorable outcome than peripheral tumors.[233,238,241] Ishida and colleagues,[230] concluded that definitive sarcomatous differentiation indicated a poorer prognosis, but this is by no means certain. Clinical or pathologic staging probably provides the best means for gauging prognosis.[238] Whatever the prognostic factors, there is agreement that the outcome of pulmonary carcinosarcomas is poor.[238] The average postoperative survival of patients is 9 months; only 27 percent of patients survive more than 6 months, and fewer than 10 percent of them survive 2 years.[222,240]

Grossly, carcinosarcomas are gray or white with areas of hemorrhage and necrosis. Peripheral tumors are solitary and well circumscribed, whereas central tumors are largely endobronchial and polypoid or may be parenchymal with an intrabronchial component[235,237] (Fig. 32-55). Tumors that meet the WHO criteria measure between 2 and 9 cm, with a mean diameter of 5 cm.[229–231,233–237]

Microscopically, these tumors are composed of an intimate admixture of carcinomatous and sarcomatous elements. The demarcation between the microscopic phases may be sharp or focally ill defined. The most frequent epithelial component is squamous cell carcinoma (Fig. 32-56). Most commonly, spindle cells and storiform areas are associated with foci of malignant bone, cartilage, or striated muscle. More than one differentiated mesenchymal cell type can be present in a given case.[234,237] Sometimes, the malignant stroma forms the bulk of the tumor, and only small foci of carcinoma are seen.[233]

The current WHO definition of carcinosarcoma is a light microscopic one that was proposed before there was any attempt at systematic application of immunohistochemical stains in these tumors. Stains for S-100 protein (chondroid) or desmin (myogenic)

FIGURE 32-55
Pulmonary carcinosarcoma. The gross specimen shows a fleshy mass protruding from the bronchial lumen. (*See also Color Plate 9.*)

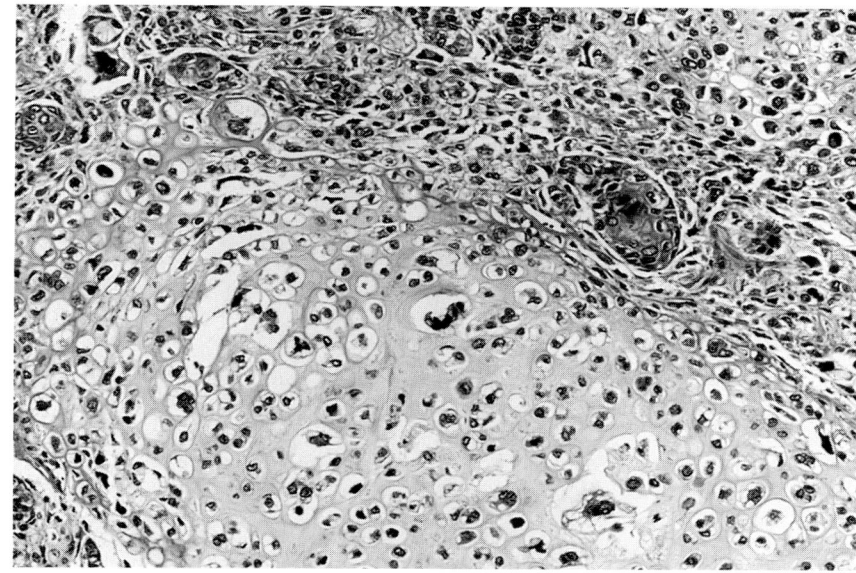

FIGURE 32-56
Pulmonary carcinosarcoma. The tumor consists of large islands of malignant cartilage intimately intermixed with foci of squamous cell carcinoma. (Intermediate magnification.)

may be helpful in marginal cases in which there is doubt about the presence of heterologous stromal elements.[232] More important, antibody studies may play a role in distinguishing carcinomas with a focal or extensive spindle cell component, a far more frequent tumor in lung, from carcinosarcoma.[232] Keratin, EMA, CEA, Ber EP4, and surfactant apoprotein can be used as markers to suggest that the spindle cell component of an apparently biphasic tumor is epithelial.[230,231,242] Electron microscopy to search for desmosomes, keratohyaline granules, or glandular differentiation in the spindle cells can also be used and shows good correlation with the immunohistochemical results.[230,231] However, the spindle cells of a carcinoma may fail to stain for keratin, and there are now reports that the spindle cell component of carcinosarcomas, showing heterologous elements, can express keratin.[232,239] As noted earlier, a growing consensus is that carcinosarcoma and spindle cell carcinoma are merely "windows" in a spectrum of biphasic lesions, that is, "carcinomas with a sarcomatoid element" or "pulmonary carcinomas with sarcoma-like lesion," and that the presence of epithelial and mesenchymal intermediate filaments reflects the direction of cellular differentiation rather than the cell of origin.[230–232,239]

Teratomas

Primary teratomas of the lung are exceedingly rare neoplasms.[243-254] Because these neoplasms are unusual as primary lung tumors, it is essential that metastases from sites such as mediastinum and testes, where these tumors occur more frequently, are excluded. On some occasions, the metastases from malignant germ cell tumors may present as mature teratomas in the lung; we have documented several instances of this phenomenon ourselves.[255] Most of these patients have received chemotherapy for primary germ cell neoplasms; nonetheless, there are rare instances of mature teratomas appearing in patients who have not received treatment.[256] Because of this, it is important to carefully eliminate the possibility of an extrapulmonary germ cell tumor in other locations before accepting origin in the lung.

The majority of pulmonary teratomas occur in adults, but cases in the pediatric age group have also been documented. Clinically, chest pain, cough, fever, hematemesis, weight loss, and hemoptysis are common symptoms. Grossly, most tumors occur within the lung parenchyma, but they may also be endobronchial in location. There appears to be a predilection for the left upper lobe. The tumors are typically single masses that may display features such as hair, teeth, and sebaceous material. The presence of more than one pulmonary nodule should alert one to the possibility of metastases.

Histologically, benign and malignant teratomas have been reported. As in other locations, these neoplasms show a wide variety of tissues belonging to the three germinal layers. The presence of a malignant epithelial or sarcomatous component as well as neural elements will determine the maturity or immaturity of these tumors.

Surgical resection appears to be the treatment of choice for the mature teratomas. Chemotherapy may be used for those tumors with immature components.

REFERENCES

1. Guccion J, Rosen S: Bronchopulmonary leiomyosarcoma and fibrosarcoma. *Cancer* 30:836–847, 1972.

2. Iverson L. Bronchopulmonary sarcoma. *J Thorac Surg* 27:130–148, 1954.

3. Haller JO, Kauffman SL, Kassner EG: Congenital mesenchymal tumour of the lung. *Br J Radiol* 50:217–219, 1977.

4. Pettinato G, Manivel JC, Saldana MJ, Peyser J, Dehner LP: Primary bronchopulmonary fibrosarcoma of childhood and adolescence: Reassessment of a low-grade malignancy. Clinicopathologic study of five cases and review of the literature. *Hum Pathol* 20:463–471, 1989.

5. Warren JS, Seo IS, Mirkin LD: Massive congenital mesenchymal malformation of the lung: a case report with ultrastructural study. *Pediatr Pathol* 3:321–328, 1985.

6. McGinnis M, Jacobs G, El-Naggar A, Redline RW: Congenital peribronchial myofibroblastic tumor (so-called "congenital leiomyosarcoma"). A distinct neonatal lung lesion associated with nonimmune hydrops fetalis. *Mod Pathol* 6:487–492, 1993.

7. Nascimento A, Unni K, Bernatz P: Sarcomas of the lung. *Mayo Clin Proc* 57:355–359, 1982.

8. Stout AP: Fibrosarcoma in infants and children. *Cancer* 15:1028–1040, 1962.

9. Lee JT, Shelburne JD, Linder J: Primary malignant fibrous histiocytoma of the lung. A clinicopathologic and ultrastructural study of five cases. *Cancer* 53:1124–1130, 1984.

10. McDonnell T, Kyriakos M, Roper C, Mazoujian G: Malignant fibrous histiocytoma of the lung. *Cancer* 61:137–145, 1988.

11. Mills SA, Breyer RH, Johnston FR, et al: Malignant fibrous histiocytoma of the mediastinum and lung: a report of three cases. *J Thorac Cardiovasc Surg* 84:367–372, 1982.

12. Tanino M, Odashima S, Sugiura H, Matsue T, Kajikawa M, Maeda S: Malignant fibrous histiocytoma of the lung. *Acta Pathol Jpn* 35:945–950, 1985.

13. Yousem S, Hochholzer L: Malignant fibrous histiocytoma of the lung. *Cancer* 60:2532–2541, 1987.

14. Chowdhury LN, Swerdlow MA, Jao W, Kathpalia S, Desser RK: Postirradiation malignant fibrous histiocytoma of the lung. Demonstration of alpha 1-antitrypsin-like material in neoplastic cells. *Am J Clin Pathol* 74:820–826, 1980.

15. Weiss SW, Enzinger FM: Malignant fibrous histiocytoma: An analysis of 200 cases. *Cancer* 41:2250–2266, 1978.

16. Misra DP, Sunderrajan EV, Rosenholtz MJ, Hurst DJ: Malignant fibrous histiocytoma in the lung masquerading as recurrent pulmonary thromboembolism. *Cancer* 51:538–541, 1983.

17. Hollowood K, Fletcher CDM: Rhabdomyosarcoma in adults. *Semin Diag Pathol* 11:47–57, 1994.

18. Suster S, Moran CA, Koss MN: Rhabdomyosarcomas of the anterior mediastinum. *Hum Pathol* 25:349–356, 1994.

19. Grignon DJ, McIsacc GP, Armstrong RF, et al: Primary rhabdomyosarcoma of the kidney. *Cancer.*62:2027–2032, 1988.

20. Goble M, Clarke T, Durrani A, et al: Pleomorphic rhabdomyosarcoma of the urinary bladder in association with recurrent urinary-tract infection. *Eur J Surg Oncol* 15:155–157, 1989.

21. Avagnina A, Elsner B, De Marco L, et al: Pulmonary rhabdomyosarcoma with isolated small bowel metastasis. *Cancer* 53:1948–1951, 1984.

22. Bleisch VR, Kraus FT: Polypoid sarcoma of the pulmonary trunk. *Cancer* 46:314–324, 1980.

23. Conquest HF, Thornton JL, Massie JR, et al: Primary pulmonary rhabdomyosarcoma. *Ann Surg* 161:688–692, 1965.

24. Gordon LZ, Boss H: Primary rhabdomyosarcoma of lung. *Cancer* 8:588–591, 1955.

25. Forbes GB: Rhabdomyosarcoma of bronchus. *J Pathol Bact* 70:427–431, 1955.

26. Drennan JM, McCormack RJM: Primary rhabdomyosarcoma of the lung. *J Pathol Bact* 79:147–149,1960.

27. Lee SH, Rengachary SS, Paramesh J: Primary pulmonary rhabdomyosarcoma. *Hum Pathol* 92–96, 1981.

28. Micallef-Eynaud PD, Goulden NT, Langdale-Brown B, et al: Intracerebral recurrence of primary intrathoracic rhabdomyosarcoma. *Med Pediatr Oncol* 21:132–136, 1993.

29. Murphy JJ, Blair GK, Fraser GC, et al: Rhabdomyosarcoma arising within congenital pulmonary cysts. *J Pediatr Surg* 27:1364–1367, 1992

30. Shariff S, Thomas JA, Shetty N, et al: Primary pulmonary rhabdomyosarcoma in a child with a review of the literature. *J Surg Oncol* 38:261–264, 1988.

31. Fallon G, Schiller M, Kilman JW: Primary rhabdomyosarcoma of the bronchus. *Ann thorac Surg* 12:650–654, 1971.

32. Przygodski RM, Moran CA, Suster S, Koss MN: Primary pulmonary rhabdomyosarcomas: A clinicopathologic and immunohistochemical study of three cases. *Mod Pathol* 8:658–661, 1995.

33. Aterman K, Patel S: Striated muscle in the lung. *Am J Anat* 128:341–358, 1970.

34. Ueda K, Gruppo R, Unger F, et al: Rhabdomyosarcoma of lung arising in congenital cystic adenomatoid malformation. *Cancer* 40:383–388, 1977.

35. Patel AM, Ryu JH: Angiosarcoma in the lung. *Chest* 103:1531–1535, 1993.

36. Edwards RL, Chalk SM, McEvoy JD, Donald KJ: Pulmonary haemorrhage in disseminated cardiac haemangiosarcoma. *Br J Dis Chest* 71:127–131, 1977.

37. Zwaveling JH, van Berkhout FT, Haneveld GT: Angiosarcoma of the heart presenting as pulmonary disease. *Chest* 94:216–218, 1988.

38. Enzinger FM, Weiss SW (eds). *Soft Tissue Tumors*, St. Louis, CV Mosby, 1995; pp 641–677.

39. Ott RA, Eugene J, Kollin J, Kanas RJ, Conston DE, Chi JC: Primary pulmonary angiosarcoma associated with multiple synchronous neoplasms. *J Surg Oncol* 35:269–276, 1987.

40. Segal SL, Lenchner GS, Cichelli AV, Promisloff RA, Hofman WI, Baiocchi GA: Angiosarcoma presenting as diffuse alveolar hemorrhage. *Chest* 94:214–216, 1988.

41. Yousem SA: Angiosarcoma presenting in the lung. *Arch Pathol Lab Med* 110:112–115, 1986.

42. Spragg RG, Wolf PL, Haghighi P, Abraham JL, Astarita RW: Angiosarcoma of the lung with fatal pulmonary hemorrhage. *Am J Med* 74:1072–1076, 1983.

43. Tralka GA, Katz S: Hemangioendothelioma of the lung. *Am Rev Respir Dis* 87:107–115, 1963.

44. Miller DL, Allen MS: Rare pulmonary neoplasms. *Mayo Clin Proc* 68:492–498, 1993.

45. Dail DH: Uncommon tumors, in Dail DH, Hammar SP (eds); *Pulmonary Pathology*, 2d ed, New York, Springer-Verlag, 1994; pp 1279-1461.

46. Sizer B: Bilateral pneumothorax as a presenting feature of metastatic angiosarcoma of the scalp (letter). *Br J Radiol* 64:72, 1991.

47. Lawton PA, Knowles S, Karp SJ, Suvana SK, Spittle MF: Bilateral pneumothorax as a presenting feature of metastatic angiosarcoma of the scalp (see comments). *Br J Radiol* 63:132–134, 1990.

48. Nomura M, Nakaya Y, Saito K, et al: Hemopneumothorax secondary to multiple cavitary metastasis in angiosarcoma of the scalp. *Respiration* 61:109–112, 1994.

49. Furue M, Yamada N, Takahashi T, et al: Immunotherapy for Stewart-Treves syndrome. Usefulness of intrapleural administration of tumor-infiltrating lymphocytes against massive

pleural effusion caused by metastatic angiosarcoma. *J Am Acad Dermatol* 30:899–903, 1994.

50. Rajdev N, Green R, Crosby WH: Angiosarcoma with pulmonary siderosis and persistent reticulocytosis. Steroid responsiveness suggests an immune basis. *Arch Intern Med* 138:1549–1551, 1978.

51. Barker AF, Smith JD: Rapidly developing cavitating pulmonary metastases. *Chest* 76:103–105, 1979.

52. Aronchick JM, Palevsky HI, Miller WT: Cavitary pulmonary metastases in angiosarcoma. Diagnosis by transthoracic needle aspiration. *Am Rev Respir Dis* 139:252–253, 1989.

53. Shioya H, Akiba T, Ohki T, et al: Case report of metastatic cutaneous angiosarcoma causing bilateral pneumothorax. *Nippon Kyobu Shikkan Gakkai Zasshi* 31:498–500, 1993.

54. Hashimoto K, Nagasaki F, Kim YC: A case of angiosarcoma of the scalp with numerous thin-walled cavitary pulmonary lesions. *Nippon Kyobu Shikkan Gakkai Zasshi* 31:534–537, 1993.

55. Garcia Rio F, Alvarez-Sala R, Caballero P: Calcified pulmonary metastases from a cardiac angiosarcoma (letter). *AJR Am J Roentgenol* 160:1147, 1993.

56. Primack SL, Hartman TE, Lee KS, Muller NL: Pulmonary nodules and the CT halo sign. *Radiology* 190:513–515, 1994.

57. Nappi O, Swanson PE, Wick MR: Pseudovascular adenoid squamous cell carcinoma of the lung: clinicopathologic study of three cases and comparison with true pleuropulmonary angiosarcoma. *Hum Pathol* 25:373–378, 1994.

58. Berry GJ, Anderson CJ, Pitts WC, Neitzel GF, Weiss LM: Cytology of angiosarcoma in effusions. *Acta Cytol* 35:538–542, 1991.

59. Wenig BM, Abbondanzo SL, Heffess CS: Epithelioid angiosarcoma of the adrenal glands. A clinicopathologic study of nine cases with a discussion of the implications of finding epithelial-specific markers. *Am J Surg Pathol* 18:62–73, 1994.

60. Alles JU, Bosslet K: Immunocytochemistry of angiosarcomas. A study of 19 cases with special emphasis on the applicability of endothelial cell specific markers to routinely prepared tissues. *Am J Clin Pathol* 89:463–471, 1988.

61. Gray MH, Rosenberg AE, Dickersin GR, Bhan AK: Cytokeratin expression in epithelioid vascular neoplasms. *Hum Pathol* 21:212–217, 1990.

62. Sirgi KE, Wick MR, Swanson PE: B72.3 and CD34 immunoreactivity in malignant epithelioid soft tissue tumors. Adjuncts in the recognition of endothelial neoplasms. *Am J Surg Pathol* 17:179–185, 1993.

63. Banerjee SS, Eyden BP, Wells S, McWilliam LJ, Harris M: Pseudoangiosarcomatous carcinoma: a clinicopathological study of seven cases. *Histopathology* 21:13–23, 1992.

64. Smith EAC, Cohen RV, Peale AR: Primary chondrosarcoma of the lung. *Ann Intern Med* 53:838–846, 1960.

65. Fainácci CJ, Blauw AS, Jennings EM: Multifocal pulmonary lesions of possible decidual origin (so-called pulmonary deciduosis): Report of a case. *Am J Clin Pathol* 59:508–514, 1973.

66. Dail DH, Liebow AA, Gmelich JT, et al: Intravascular, bronchiolar, and alveolar tumor of the lung (IVBAT). An analysis of twenty cases of a peculiar sclerosing endothelial tumor. *Cancer* 51:452–464, 1983.

67. Eggleston JC: The intravascular bronchioloalveolar tumor and the sclerosing hemangioma of the lung: Misnomers of pulmonary neoplasia. *Semin Diagn Pathol* 2:270–280, 1985.

68. Weiss SW, Ishak KG, Dail DH, Sweet DE, Enzinger FM: Epithelioid hemangioendothelioma and related lesions. *Semin Diagn Pathol* 3:259–287, 1986.

69. Carter EJ, Bradburne RM, Jhung JW, Ettensohn DB: Alveolar hemorrhage with epithelioid hemangioendothelioma. A previously unreported manifestation of a rare tumor. *Am Rev Respir Dis* 142:700–701 1990.

70. Struhar D, Sorkin P, Greif J, Marmor S, Geller E: Alveolar haemorrhage with pleural effusion as a manifestation of epithelioid haemangioendothelioma. *Eur Respir J* 5:592–593, 1992.

71. Dorfler H, Permanetter W, Küffer G, Häussinger K, Zöllner N: Sclerosing epitheloid angiosarcoma of bone and lung—intravascular sclerosing bronchioloalveolar tumor. *Klin Wochenschr* 68:388–392, 1990.

72. Ishak KG, Sesterhenn IA, Goodman ZD, Rabin L, Stromeyer FW: Epithelioid hemangioendothelioma of the liver: A clinicopathologic and follow-up study of 32 cases. *Hum Pathol* 15:839–852, 1984.

73. Verbeken E, Beyls J, Moerman P, Knockaert D, Goddeeris P, Lauweryns JM: Lung metastasis of malignant epithelioid hemangioendothelioma mimicking a primary intravascular bronchioalveolar tumor. A histologic, ultrastructural, and immunohistochemical study. *Cancer* 55:1741–1746, 1985.

74. Ekfors TO, Joensuu K, Toivio I, Laurinen P, Pelttari L: Fatal epithelioid haemangioendothelioma presenting in the lung and liver. *Virchows Arch A Pathol Anat Histopathol* 410:9–16, 1986.

75. Nerlich A, Berndt R, Schleicher E: Differential basement membrane composition in multiple epithelioid haemangioendotheliomas of liver and lung. *Histopathology* 18:303–307, 1991.

76. Yousem SA, Hochholzer L: Unusual thoracic manifestations of epithelioid hemangioendothelioma. *Arch Pathol Lab Med* 111:459–463, 1987.

77. Bhagavan BS, Dorfman HD, Murthy MS, Eggleston JC: Intravascular bronchiolo-alveolar tumor (IVBAT): A low-grade sclerosing epithelioid angiosarcoma of lung. *Am J Surg Pathol* 6:41–52, 1982.

78. Corrin B, Harrison WJ, Wright DH: The so-called intravascular bronchioloalveolar tumour of lung (low grade sclerosing angiosarcoma): Presentation with extrapulmonary deposits. *Diagn Histopathol* 6:229–237, 1983.

79. Miettinen M, Collan Y, Halttunen P, Maamies T, Vilkko P: Intravascular bronchioloalveolar tumor. *Cancer* 60:2471–2475, 1987.

80. Weldon-Linne CM, Victor TA, Christ ML: Immunohistochemical identification of factor VIII-related antigen in the intravascular bronchioloalveolar tumor of the lung (letter). *Arch Pathol Lab Med* 105:628–629, 1981.

81. Weldon-Linne CM, Victor TA, Christ ML, Fry WA: Angiogenic nature of the "intravascular bronchioloalveolar tumor" of the lung: An electron microscopic study. *Arch Pathol Lab Med* 105:174–179, 1981.

82. Sirgi KE, Wick MR, Swanson PE: B72.3 and CD34 Immunoreactivity in malignant epithelioid soft tissue tumors. Adjuncts in the recognition of endothelial neoplasms. *Am J Surg Pathol* 17:179–185, 1993.

83. Corrin B, Manners B, Millard M, Weaver L: Histogenesis of the so-called "intravascular bronchioloalveolar tumor." *J Pathol* 128:163–167, 1979.

84. Shirakusa T, Yoshida M, Tsutsui M, et al: Advanced intravascular bronchioloalveolar tumour and review of reports in Japan. *Respir Med* 83:127–132, 1989.

85. Buchbinder A, Friedman-Kien AE: Clinical aspects of epidemic Kaposi's sarcoma. *Cancer Surv* 10:39–52, 1991.

86. Doyle M, Johnstone PA, Watkins EB: Role of radiation therapy in management of pulmonary Kaposi's sarcoma. *South Med J* 86:285–288, 1993.

87. Siegel JH, Janis R, Alper JC, Schutte H, Robbins L, Blaufox MD: Disseminated visceral Kaposi's sarcoma. Appearance after human renal homograft operation. *JAMA* 207: 1493–1496, 1969.

88. Gunawardena KA, al-Hasani MK, Haleem A, al-Suleiman M, Al-Khader AA: Pulmonary Kaposi's sarcoma in two recipients of renal transplants. *Thorax* 43:653–656, 1988.

89. Margolius L, Stein M, Spencer D, Bezwoda WR: Kaposi's sarcoma in renal transplant recipients. Experience at Johannesburg Hospital, 1966–1989. *S Afr Med J* 84:16–17, 1994.

90. Greenfield DI, Trinh P, Fulenwider A, Barth WF: Kaposi's sarcoma in a patient with SLE. *J Rheumatol* 13:637–640, 1986.

91. Murray JF, Felton CP, Garay SM, et al: Pulmonary complications of the acquired immunodeficiency syndrome. Report of a National Heart, Lung, and Blood Institute workshop. *N Engl J Med* 310:1682–1688, 1984.

92. O'Brien RF, Cohn DL: Serosanguineous pleural effusions in AIDS-associated Kaposi's sarcoma. *Chest* 96:460–466, 1989.

93. Ognibene FP, Steis RG, Macher AM, et al: Kaposi's sarcoma causing pulmonary infiltrates and respiratory failure in the acquired immunodeficiency syndrome. *Ann Intern Med* 102:471–475, 1985.

94. Wallace JM, Hannah JB: Pulmonary disease at autopsy in patients with the acquired immunodeficiency syndrome. *West J Med* 149:167–171, 1988.

95. Meduri GU, Stover DE, Lee M, Myskowski PL, Caravelli JF, Zaman MB: Pulmonary Kaposi's sarcoma in the acquired immune deficiency syndrome. Clinical, radiographic, and pathologic manifestations. *Am J Med* 81:11–18, 1986.

96. Lemlich G, Schwam L, Lebwohl M: Kaposi's sarcoma and acquired immunodeficiency syndrome. Postmortem findings in twenty-four cases. *J Am Acad Dermatol* 16:319–325, 1987.

97. Bach MC, Bagwell SP, Fanning JP: Primary pulmonary Kaposi's sarcoma in the acquired immunodeficiency syndrome: A cause of persistent pyrexia. *Am J Med* 85:274–275, 1988.

98. Presant CA, Scolaro M, Kennedy P, et al: Liposomal daunorubicin treatment of HIV-associated Kaposi's sarcoma. *Lancet* 341:1242–1243, 1993.

99. Misra DP, Sunderrajan EV, Hurst DJ, Maltby JD: Kaposi's sarcoma of the lung: Radiography and pathology. *Thorax* 37:155–156, 1982.

100. Naidich DP, McGuinness G: Pulmonary manifestations of AIDS. CT and radiographic correlations. *Radiol Clin North Am* 29:999–1017, 1991.

101. Naidich DP, Tarras M, Garay SM, Birnbaum B, Rybak BJ, Schinella R: Kaposi's sarcoma. CT-radiographic correlation. *Chest* 96:723–728, 1989.

102. Sivit CJ, Schwartz AM, Rockoff SD: Kaposi's sarcoma of the lung in AIDS: Radiologic-pathologic analysis. *AJR Am J Roentgenol* 148:25–28, 1987.

103. Wolff SD, Kuhlman JE, Fishman EK: Thoracic Kaposi sarcoma in AIDS: CT findings. *J Comput Assist Tomogr* 17: 60–62, 1993.

104. Lai KK: Pulmonary Kaposi's sarcoma presenting as diffuse reticular nodular infiltrates with cavitary lesions. *South Med J* 83:1096–1098, 1990.

104. Roux FJ, Bancal C, Dombret MC, et al: Pulmonary Kaposi's sarcoma revealed by a solitary nodule in a patient with acquired immunodeficiency syndrome. *Am J Respir Crit Care Med* 149:1041–1043, 1994.

106. Pandya K, Lal C, Tuchschmidt J, Boylen CT, Sharma OP: Bilateral chylothorax with pulmonary Kaposi's sarcoma (letter). *Chest* 94:1316–1317, 1988.

107. Hughes-Davies L, Kocjan G, Spittle MF, Miller RF: Occult alveolar haemorrhage in bronchopulmonary Kaposi's sarcoma. *J Clin Pathol* 45:536–537, 1992.

108. Hanson PJ, Harcourt-Webster JN, Gazzard BG, Collins JV: Fibreoptic bronchoscopy in diagnosis of bronchopulmonary Kaposi's sarcoma. *Thorax* 42:269–271, 1987.

109. Pitchenik AE, Fischl MA, Saldana MJ: Kaposi's sarcoma of the tracheobronchial tree. Clinical, bronchoscopic, and pathologic features. *Chest* 87:122–124, 1985.

110. Purdy LJ, Colby TV, Yousem SA, Battifora H: Pulmonary Kaposi's sarcoma. Premortem histologic diagnosis. *Am J Surg Pathol* 10:301–311, 1986.

111. Hamm PG, Judson MA, Aranda CP: Diagnosis of pulmonary Kaposi's sarcoma with fiberoptic bronchoscopy and endobronchial biopsy. A report of five cases. *Cancer* 59:807–810, 1987.

112. Moskowitz LB, Hensley GT, Gould EW, Weiss SD: Frequency and anatomic distribution of lymphadenopathic Kaposi's sarcoma in the acquired immunodeficiency syndrome: An autopsy series. *Hum Pathol* 16:447–456, 1985.

113. Judson MA, Sahn SA: Endobronchial lesions in HIV-infected individuals. *Chest* 105:1314–1323, 1994.

114. Hill AD, Darzi A, Menzies-Gow N, Riordan JF: Thoracoscopic biopsy in the diagnosis of pulmonary Kaposi's sarcoma. *J Laparoendosc Surg* 3:571–572, 1993.

115. Millard PR, Heryet AR: An immunohistological study of factor VIII related antigen and Kaposi's sarcoma using polyclonal and monoclonal antibodies. *J Pathol* 146:31–38, 1985.

116. Yang GC, Brooks JJ, Roberts S, Gupta PK: The detection of acquired immunodeficiency syndrome-associated Kaposi sarcoma cells in pleural effusion by CD34 immunostain. *Cancer* 72:2260–2265, 1993.

117. Akhtar M, Bunuan H, Ali MA, Godwin JT: Kaposi's sarcoma in renal transplant recipients. Ultrastructural and immunoperoxidase study of four cases. *Cancer* 53:258–266, 1984.

118. Kostianovsky M, Lamy Y, Greco MA: Immunohistochemical and electron microscopic profiles of cutaneous Kaposi's sarcoma and bacillary angiomatosis. *Ultrastruct Pathol* 16: 629–640, 1992.

119. Reed JA, Brigati DJ, Flynn SD, et al Immunocytochemical identification of *Rochalimaea henselae* in bacillary (epithelioid) angiomatosis, parenchymal bacillary peliosis, and persistent fever with bacteremia. *Am J Surg Pathol* 16: 650–657, 1992.

120. Slater LN, Min KW: Polypoid endobronchial lesions. A manifestation of bacillary angiomatosis. *Chest* 102: 972–974, 1992.

121. Slater LN, Welch DF, Min KW: *Rochalimaea henselae* causes bacillary angiomatosis and peliosis hepatis. *Arch Intern Med* 152:602–657, 1992.

122. Foltzer MA, Guiney WB Jr, Wager GC, Alpern HD: Bronchopulmonary bacillary angiomatosis. *Chest* 104: 973–975, 1993.

123. Schürmann D, Dormann A, Grünewald T, Ruf B: Successful treatment of AIDS-related pulmonary Kaposi's sarcoma with liposomal daunorubicin. *Eur Respir J* 7:824–825, 1994.

124. Bogner JR, Kronawitter U, Rolinski B, Truebenbach K, Goebel FD: Liposomal doxorubicin in the treatment of advanced AIDS-related Kaposi sarcoma. *J Acquir Immune Defic Syndr* 7:463–468, 1994.

125. Nobler MP: Pulmonary irradiation for Kaposi's sarcoma in AIDS. *Am J Clin Oncol* 8:441–444, 1985.

126. Meyer JL: Whole-lung irradiation for Kaposi's sarcoma. *Am J Clin Oncol* 16:372–376, 1993.

127. Stout AP: Hemangiopericytoma. *Cancer* 2:1027–1042, 1949.
128. Meade JB, Whitwell F, Bickford BJ, Waddington JK: Primary haemangiopericytoma of lung. *Thorax* 29:1–15, 1974.
129. Yousem SA, Hochholzer L: Primary pulmonary hemangiopericytoma. *Cancer* 59:549–555, 1987.
130. Shin MS, Ho KJ: Primary hemangiopericytoma of lung: Radiography and pathology. *AJR Am J Roentgenol* 133:1077–1083, 1979.
131. Halle M, Blum U, Dinkel E, Brugger W: CT and MR features of primary pulmonary hemangiopericytomas. *J Comput Assist Tomogr* 17:51–55, 1993.
132. Enzinger FM, Smith BH: Hemangiopericytoma. An analysis of 106 cases. *Hum Pathol* 7:61–82, 1976.
133. Rusch VW, Shuman WP, Schmidt R, Laramore GE: Massive pulmonary hemangiopericytoma. An innovative approach toevaluation and treatment. *Cancer* 64:1928–1936, 1989.
134. Razzuk MA, Nassur A, Gradner MA, Martin J, Gohara SF, Urschel HC Jr: Primary pulmonary hemangiopericytoma. *J Thorac Cardiovasc Surg* 74:227–229, 1977.
135. Nemes Z: Differentiation markers in hemangiopericytoma. *Cancer* 69:133–140, 1992.
136. Van Damme H, Dekoster G, Creemers E, Hermans G, Limet R: Primary pulmonary hemangiopericytoma: early local recurrence after perioperative rupture of the giant tumor mass (two cases). *Surgery* 108:105–108, 1990.
137. Shimizu J, Murakami S, Hayashi Y, et al: Primary pulmonary hemangiopericytoma: A case report. *Jpn J Clin Oncol* 23:313–316, 1993.
138. McGlennen RC, Manivel JC, Stanley SJ, Slater DL, Wick MR, Dehner LP: Pulmonary artery trunk sarcoma: a clinicopathologic, ultrastructural, and immunohistochemical study of four cases. *Mod Pathol* 2:486–494, 1989.
139. Bleisch VR, Kraus FT: Polypoid sarcoma of the pulmonary trunk: Analysis of the literature and report of a case with leptomeric organelles and ultrastructural features of rhabdomyosarcoma. *Cancer* 46:314–324, 1980.
140. Shmookler BM, Marsh HB, Roberts WC: Primary sarcoma of the pulmonary trunk and/or right or left main pulmonary artery—a rare cause of obstruction to right ventricular outflow. Report on two patients and analysis of 35 previously described patients. *Am J Med* 63:263–272, 1977.
141. Kaiser LR, Urmacher C: Primary sarcoma of the superior pulmonary vein. *Cancer* 66:789–795, 1990.
142. Emmert-Buck M, Stay EJ, Tsokos M, Travis WD: Pleomorphic rhabdomyosarcoma arising in association with the right pulmonary artery. *Arch Pathol Lab Med* 118:1220-1222, 1994.
143. Smith WS, Lesar MS, Travis WD, et al: MR and CT findings in pulmonary artery sarcoma. *J Comput Assist Tomogr* 13:906–909, 1989.
144. Redmond ML, Shepard JW Jr, Gaffey TA, Payne WS: Primary pulmonary artery sarcoma. A method of resection. *Chest* 98:752–753, 1990.
145. Hayashi T, Tsuda N, Iseki M, Kishikawa M, Shinozaki T, Hasumoto M: Primary chondrosarcoma of the lung. A clinicopathologic study. *Cancer* 72:69–74, 1993.
146. Kurotaki H, Tateoka H, Takeuchi M, Yagihashi S, Kamata Y, Nagai K: Primary mesenchymal chondrosarcoma of the lung. A case report with immunohistochemical and ultrastructural studies. *Acta Pathol Jpn* 42:364–371, 1992.
147. Stanfield BL, Powers CN, Desch CE, Brooks JW, Frable WJ: Fine-needle aspiration cytology of an unusual primary lung tumor, chondrosarcoma: Case report. *Diagn Cytopathol* 7:423–426, 1991.
148. D'Ambrosio FG, Shiu MH, Brennan MF: Intrapulmonary presentation of extraskeletal myxoid chondrosarcoma of the

extremity. Report of two cases. *Cancer* 58:1144–1148, 1986.
149. Greenspan EB: Primary osteoid chondrosarcoma of the lung: Report of a case. *Am J Cancer* 18:603–609, 1933.
150. Lowell LM, Tuhy JE: Primary chondrosarcoma of the lung. *J Thorac Surg* 18:476–483, 1949.
151. Anonymous (ed): *Histological Typing of Lung Tumors* 2d ed, Geneva, World Health Organization, 1981.
152. Stark P, Smith DC, Watkins GE, Chun KE: Primary intrathoracic extraosseous osteogenic sarcoma: Report of three cases. *Radiology* 174:725–726, 1990.
153. Loose JH, el-Naggar AK, Ro JY, Huang WL, McMurtrey MJ, Ayala AG: Primary osteosarcoma of the lung. Report of two cases and review of the literature. *J Thorac Cardiovasc Surg* 100:867–873, 1990.
154. Colby TV, Bilbao JE, Battifora H, Unni KK: Primary osteosarcoma of the lung. A reappraisal following immunohistologic study. *Arch Pathol Lab Med* 113:1147–1150, 1989.
155. Reingold IM, Amromin GD: Extraosseous osteosarcoma of the lung. *Cancer* 28:491–498, 1971.
156. Connolly JP, McGuyer CA, Sageman WS, Bailey H: Intrathoracic osteosarcoma diagnosed by CT scan and pleural biopsy. *Chest* 100:265–267, 1991.
157. Stauss HK: Osteogenic sarcoma arising in traumatic hemothorax and hematoma of the thoracic wall. *Surgery* 29:917–918, 1951.
158. Cohn L, Hall AD: Extraosseous osteogenic sarcoma of the pleura. *Ann Thorac Surg* 5:545–549, 1968.
159. Kimura I, Kiuchi H, Sakamoto Y, Yamamoto K, Dohi Y, Takahama M: Primary osteogenic sarcoma of pulmonary artery. *Jpn J Med* 29:32–37, 1990.
160. Hall FM, Frank HA, Cohen RB, Ezpeleta ML: Ossified pulmonary metastases from giant cell tumor of bone. *AJR Am J Roentgenol* 127:1046–1047, 1976.
161. Rhone DP, Horowitz RN: Heterotopic ossification in the pulmonary metastases of gastric adenocarcinoma. Report of a case and review of the literature. *Cancer* 38:1773–1780, 1976.
162. Yousem SA, Hochholzer L: Malignant mesotheliomas with osseous and cartilaginous differentiation. *Arch Pathol Lab Med* 111:62–66, 1987.
163. Wong WW, Pluth JR, Grado GL, Schild SE, Sanderson DR: Liposarcoma of the pleura. *Mayo Clin Proc* 69:882–885, 1994.
164. Munk PL, Muller NL: Pleural liposarcoma: CT diagnosis. *J Comput Assist Tomogr* 12:709–710, 1988.
165. Carroll F, Kramer MD, Acinapura AJ, et al: Pleural liposarcoma presenting with respiratory distress and suspected diaphragmatic hernia. *Ann Thorac Surg* 54:1212–1213, 1992.
166. McGregor DH, Dixon AY, Moral L, Kanabe S: Liposarcoma of pleural cavity with recurrence as malignant fibrous histiocytoma. *Ann Clin Lab Sci* 17:83–92, 1987.
167. Evans AR, Wolstenholme RJ, Shettar SP, Yogish H: Primary pleural liposarcoma. *Thorax* 40:554–555, 1985.
168. Nanjo S, Nakamura K, Iioka S, et al: A case report of a primary liposarcoma of the lung. *Kyobu Geka* 33:543–546, 1980.
169. Ruiz-Palomo F, Calleja JL, Fogue L: Primary liposarcoma of the lung in a young woman. *Thorax* 45:298–299, 1990.
170. Sawamura K, Hashimoto T, Nanjo S, et al: Primary liposarcoma of the lung: report of a case. *J Surg Oncol* 19:243–246, 1982.
171. Sheppard MN: Primary liposarcoma of the lung in a young woman (letter). *Thorax* 45:908, 1990.
172. Wu JP, Gilbert EF, Pellett JR: Pulmonary liposarcoma in a

child with adrenogenital syndrome. _Am J Clin Pathol_ 62:791–796, 1974.

173. Yamagiwa H, Inamori S, Takeuchi T, Onishi T: Autopsy case of primary liposarcoma of the lung. _Naika_ 28:190–194, 1971.

174. D'Ambrosio V: First case of liposarcoma from the parietal pleura. _J Med Soc N J_ 71:17–19, 1974.

175. Wouters EF, Greve LH, Visser R, Swaen GJ: Liposarcoma of the pleura (letter). _Neth J Surg_ 35:192–193, 1983.

176. Going JJ, Brewin TB, Crompton GK, McLelland J: Soft tissue sarcoma: Two cases of solitary lung metastasis more than 15 years after diagnosis. _Clin Radiol_ 37:579–581, 1986.

177. Krishna J, Haqqani MT: Liposarcomatous differentiation in diffuse pleural mesothelioma. _Thorax_ 48:409–410, 1993.

178. Matsuba K, Saito T, Ando K, Shirakusa T: Atypical lipoma of the lung. _Thorax_ 46:685, 1991.

179. Palvio D, Egeblad K, Paulsen SM: Atypical lipomatous hamartoma of the lung. _Virchows Arch A Pathol Anat Histopathol_ 405:253–261, 1985.

180. Yousem SA, Hochholzer L: Alveolar adenoma. _Hum Pathol_ 17:1066–1071, 1986.

181. Ambla FR, Olsen KD, Nascimiento AG, Foote RL: Head and neck synovial sarcoma. _Otolaryngol Neck Surg_ 107:631–637, 1992.

182. Fetsch JF, Meis JM: Synovial sarcoma of the abdominal wall. _Cancer_ 72:469–477, 1993.

183. Massarelli G, Tanda F, Salis B: Synovial sarcoma of the soft palate. _Hum Pathol_ 9:341–345, 1978.

184. Novotny GM, Fort TC: Synovial sarcoma of the tongue. _Arch Otolaryngol_ 94:77–79, 1971.

185. Shaw GR, Lais CJ: Fatal intravascular synovial sarcoma in a 31-year-old woman. _Hum Pathol_ 24:809–810, 1993.

186. Sieberman R, Jenni R, Makel M, Oelz O, Turina M: Primary synovial sarcoma of the heart treated by heart transplantation. _J Thorac Cardiovasc Surg_ 99:567–568, 1990.

187. Witkin GB, Miettinen M, Rosai J: A biphasic tumor of the mediastinum with features of synovial sarcoma. A report of four cases. _Am J Surg Pathol_ 13:490–499, 1989.

188. Zeren H, Moran CA, Suster S, Fishback NF, Koss MN: Primary pulmonary sarcomas with features of monophasic synovial sarcoma: A clinicopatholigic, immunohistochemical and ultrastructural study of 25 cases. _Hum Pathol_ 26:474–480, 1995.

189. Angervall L, Enzinger FM: Extraskeletal neoplasm resembling Ewing's sarcoma. _Cancer_ 101:446–449, 1977.

190. Askin FB, Rosai J, Sibley R, Dehner LP, McAlister W: Malignant small cell tumor of the thoracopulmonary region in children. _Cancer_ 43:2438–2451, 1979.

191. Gould V, Jannson D, Warren W: Primitive neuroectodermal tumors (PNET) of the chest wall in adults. _Mod Pathol_ 4:115A, (abstract 681) 1991.

192. Hashimoto H, Enjoji M, Nakajima T, Kiryu H, Daimoru Y: Malignant neuroepithelioma (peripheral neuroblastoma). _Am J Surg Pathol_ 7:309–318, 1983.

193. Lane S, Ironside JW: Extraskeletal Ewin's sarcoma of the nasal fossa. _J Laryngol Otol_ 104:570–573, 1990.

194. Soule EH, Newton W, Moon TE, Tefft M: Extraskeletal Ewing's sarcoma with neuroblastoma-like features. _Cancer_ 3:99–102, 1978.

195. Suster S, Ronnen M, Husdar M: Extraskeletal Ewing's sarcoma of the scalp. _Pediatr Dermatol_ 5:126–128, 1988.

196. Tefft M, Vawter GF, Metus A: Paravertebral "round cell" tumors in children. _Radiology_ 92:1501–1509, 1969.

197. Wigger JH, Salazar G, Blanc WA: Extraskeletal Ewing's sarcoma. _Arch Pathol Lab Med_ 101:446–449, 1977.

198. Fellinger EJ, Garin-Chesa P, Triche TJ, Huvos AG, Rettig WJ: Immunohistochemical analysis of Ewing's sarcoma cell surface antigen p30/32MIC2. _Am J Pathol_ 139:317–325, 1991.

199. Ambros IM, Ambros PF, Strehl S, Kovar H, Gadner H, Salzer-Kuntschik M: MIC2 is a specific marker for Ewing's sarcoma and peripheral primitive neuroectodermal tumors. _Cancer_ 67:1886–1893, 1991.

200. Sobin L, Yesner R: _Histological Typing of Lung Tumors. International Histological Classification of Tumors No. 1._ Geneva, World Health Organization, 1981, pp 19-20.

201. Barnard W: Embryoma of the lung. _Thorax_ 7:229–234, 1952.

202. Barnett N, Barnard W: Some unusual thoracic tumors. _Br J Surg_ 32:447–457, 1945.

203. Spencer H: Pulmonary blastomas. _J Pathol Bacteriol_ 82:161–165, 1961.

204. Francis D, Jacobsen M: Pulmonary blastoma. _Curr Top Pathol_ 73:265–294, 1983.

205. Kradin R, Young R, Dickersin G, et al: Pulmonary blastoma with argyrophil cells lacking sarcomatous features (pulmonary endodermal tumor resembling fetal lung). _Am J Surg Pathol_ 6:165–172, 1982.

206. Manning J Jr, Ordonez N, Rosenberg H, Walker W: Pulmonary endodermal tumor resembling fetal lung. Report of a case with immunohistochemical studies. _Arch Pathol Lab Med_ 109:48–50, 1985.

207. Nakatani Y, Dickersin G, Mark E: Pulmonary endodermal tumor resembling fetal lung: A clinicopathologic study of five cases with immunohistochemical and ultrastructural characterization. _Hum Pathol_ 21:1095–1104, 1990.

208. Nakatani Y, Kitamura H, Inayama Y, Ogawa N: Pulmonary endodermal tumor resembling fetal lung. The optically clear nucleus is rich in biotin. _Am J Surg Pathol_ 18:637–642, 1994.

209. Muller-Hermelink HK, Kaiserling E: Pulmonary adenocarcinoma of fetal type: Alternating differentiation argues in favor of a common endocermal stem cell. _Virchows Arch Pathol Anat_ 409:195–210, 1986.

210. Kodama T, Shimosato Y, Watanabe S, et al: Six cases of well differentiated adenocarcinoma simulating fetal lung tissues in pseudoglandular stage: Comparison with pulmonary blastoma. _Am J Surg Pathol_ 8:725–744, 1984.

211. Koss M, Hochholzer L, O'Leary T: Pulmonary blastomas. _Cancer_ 67:2368–2381, 1991.

212. Dail DH: Uncommon tumors, in Dail DH, Hammar S (eds), _Pulmonary Pathology_, 2d ed; New York, Springer-Verlag, 1994, p 1313.

213. Manivel J, Priest J, Watterson J, et al: Pleuropulmonary blastoma. The so-called pulmonary blastoma of childhood. _Cancer_ 62:1516–1526, 1988.

214. Hachitanda Y, Aoyama C, Sato JK, Shimada H: Pleuropulmonary blastoma in childhood. A tumor of divergent differentiation. _Am J Surg Pathol_ 17:382–391, 1993.

215. Cohen M, Emms M, Kaschula ROC: Chilhood pulmonary blastoma; A pleuropulmonary variant of the adult-type pulmonary blastoma. _Pediatr Pathol_ 11:737–749, 1991.

216. Lee KG, Cho NH: Fine-needle aspiration cytology of pulmonary adenocarcinoma of fetal type: Report of a case with immunohistochemical and ultrastructural studies. _Diagn Cytopathol_ 7:408–414, 1991.

217. Yousem S, Wick M, Randhawa P, Manivel J: Pulmonary blastoma: An immunohistochemical analysis with compar-

ison with fetal lung in its pseudoglandular stage. *Am J Clin Pathol* 93:167–175, 1990.

218. Chejfec G, Cosnow I, Gould, NS, Husain AN, Gould VE: Pulmonary blastoma with neuroendocrine differentiation in cell morules resembling neuroepithelial bodies. *Histopathology* 17:353–358, 1990.

219. Dehner L: Tumors and tumor-like lesions of the lung and chest wall in childhood: Clinical and pathologic review, in Stocker J, Dehner L (eds); *Pediatric Pathology*. Philadelphia, JB Lippincott, 1992; p 232.

220. Jacobsen M, Francis D: Pulmonary blastoma. A clinico-pathologic study of eleven cases. *Acta Path Microbiol Scand* (A) 88:151–160, 1980.

221. Siegel R, Bueso-Ramos C, Cohen C, Koss M: Pulmonary blastoma with germ cell (yolk sac) differentiation: Report of two cases. *Mod Pathol* 4:566–570, 1991.

222. Gebauer C: The postoperative prognosis of primary pulmonary sarcomas. *J Thorac Cardiovasc Surg* 16:91–97, 1982.

223. Senac MO, Wood BP, Isaacs H, Weller M: Pulmonary blastoma: A rare childhood malignancy. *Radiology* 179: 743–746, 1991.

224. Becroft D, Jagusch M: Pulmonary sarcoma arising in mesenchymal cystic hamartomas (abstract). *Pediatr Pathol* 7:478, 1987.

225. Hedlund G, Bisset G, Bove K: Malignant neoplasms arising in cystic hamartomas of the lung in childhood. *Radiology* 173:77–79, 1989.

226. Krous H, Sexauer C: Embryonal rhabdomyosarcoma arising within a congenital bronchogenic cyst in a child. *J Pediatr Surg.* 16:506–508, 1981.

227. Ashworth T: Pulmonary blastomas: a true congenital neoplasm. *Histopathology* 7:585–594, 1983.

228. Stocker J: The respiratory tract, in Stocker J, Dehner L (eds); *Pediatric Pathology*; Philadelphia, JB Lippincott, 1992, p 559.

229. Addis B, Corrin B: Pulmonary blastoma, carcinosarcoma, and spindle-cell carcinoma: An immunohistochemical study of keratin intermediate filaments. *J Pathol* 147: 291–301, 1985.

230. Ishida T, Tateishi M, Kaneko S, et al: Carcinosarcoma and spindle cell sarcoma of the lung. Clinicopathologic and immunohistochemical studies. *J Thorac Cardiovasc Surg* 100:844–852, 1990.

231. Humphrey P, Scroggs M, Roggli V, Shelburn J: Pulmonary carcinomas with a sarcomatoid element: An immunocyto-chemical and ultrastructural analysis. *Hum Pathol* 19:155–165, 1988.

232. Nappi O, Wick MR: Sarcomatoid neoplasms of the respiratory tract. *Semin Diagn Pathol* 10:137–147, 1993.

233. Ludwigsen E: Endobronchial carcinosarcoma. A case with osteosarcoma of pulmonary invasive part, and a review with respect to prognosis. *Virchows Arch A Pathol Anat* 373: 293–302, 1977.

234. Moore T: Carcinosarcoma of the lung. *Surgery* 50:886–893, 1961.

235. Prive L, Tellem M, Meranze D, Chodoff R: Carcinosarcoma

of the lung. *Arch Pathol* 72:351–357, 1961.

236. Stackhouse E, Harrison E, Ellis F: Primary mixed malignancies of lung: Carcinosarcoma and blastoma. *J Thorac Cardiovasc Surg* 57:385–399, 1969.

237. Zimmerman K, Sobonya R, Payne C: Histochemical and ultrastructural features of an unusual pulmonary carcinosarcoma. *Hum Pathol* 12:1046–1051, 1981.

238. Davis M, Eagan R, Weiland L, Pairolero P: Carcinosarcoma of the lung: Mayo Clinic experience and response to chemotherapy. *Mayo Clin Proc* 59:598–603, 1984.

239. Cupples J, Wright J: An immunohistochemical comparison of primary lung carcinosarcoma and sarcoma. *Pathol Res Pract* 186:326–329, 1990

240. Cabarcos A, Gomez D, Lobo B: Pulmonary carcinosarcoma: A case study and review of the literature. *Br J Dis Chest* 79:83–94, 1985.

241. Bull J, Grimes O: Pulmonary carcinosarcoma. *Chest* 65: 9–12, 1974.

242. Ro J, Chen J, Lee J, et al: Sarcomatoid carcinoma of the lung: Immunohistochemical and ultrastructural studies of 14 cases. *Cancer* 69:376–386, 1992.

243. Zachar CK: Intrapulmonary dermoid cyst. *Semin Roentgenol* 22:231–232, 1987.

244. Walrond ER, Prussia PR: Pulmonary teratoma. *West Indian Med J* 36:39–42, 1987.

245. Trivedi SA, Mehta KN, Nanavaty JM: Teratoma of the lung. *Br J Dis Chest* 60:156–158, 1966.

246. Stair JM, Stevenson RD, Schaefer RF, Fullenwider JP, Campbell GS: Primary teratocarcinoma of the lung. *J Surg Oncol* 33:262–267, 1986.

247. Steler KJ: Benign cystic teratoma of the lung. *Postgrad Med* 83:85–91, 1988.

248. Saruk M, Stern H, Tronic B, Neuman RD, LiVolsi V: Intrapulmonary benign cystic teratoma. *Conn Med* 44: 687–689, 1980.

249. Pound AW, Willis RA: A malignant teratoma of the lung in an infant. *J Pathol* 98:111–113, 1968.

250. Jamieson MPG, McGowan AR: Endobronchial teratoma. *Thorax* 37:157–159, 1982.

251. Holt S, Deverall PB Boddy JE: A teratoma of the lung containing thymic tissue. *J Pathol* 126:85–88, 1978.

252. Gautman H: Intrapulmonary malignant teratoma. *Am Rev Resp Dis* 100:863–867, 1969.

253. Day DW, Taylor SA: An intrapulmonary teratoma associated with thymic tissue. *Thorax* 30:582–584, 1975.

254. Bateson EM, Hayes JA, Woo-Ming M: Endobronchial teratoma with bronchiectasis and bronchiolectasis. *Thorax* 23:69–71, 1968.

255. Moran CA, Travis WD, Carter D, Koss MN: Metastatic mature teratoma in lung following testicular embryonal carcinoma and teratocarcinoma. *Arch Pathol Lab Med* 117: 641–644, 1993.

256. Skinner DG, Leadbetter WF, Wilkins EW: The surgical management of testis tumors metastatic to the lung: A report of 10 cases with subsequent resection of from one to seven pulmonary metastases. *J Urol* 105:275–278, 1971.

Chapter 33

Pulmonary Lymphoproliferative Disorders and Related Conditions

M. Harris

NORMAL BRONCHOPULMONARY LYMPHOID TISSUE

Lymph nodes are found at the pulmonary hila and in relationship to the branches of the bronchial tree as far as the fourth order bronchi,[1] where they are situated in the peribronchial tissues. Occasionally they are encountered in the peripheral parts of the lung and have been recorded in the visceral pleura.[2] In all of these sites they have a conventional structure but commonly show minor nonspecific reactive changes and almost invariably in adults show some degree of anthracosis.

Small collections of lymphocytes and plasma cells lacking the organized structure of lymph nodes can be found in the pulmonary interstitium and the pleura. Bronchus-associated lymphoid tissue [BALT] is part of the mucosa-associated lymphoid tissue complex [MALT] whose best known component is gut-associated. BALT, in common with MALT in general, consists of B- and T-lymphoid cells, the B cells forming follicles and the whole situated beneath specialized cuboidal epithelium. The B cells of the marginal zones of the follicles infiltrate this epithelium.[3]

The occurrence of BALT varies in frequency between species, occurring, for example, in 100 percent of rabbits but being virtually absent in normal human lungs.[4] However, evidence suggests that BALT can develop in response to antigenic stimulation in the fetal and infant lung[5] and, to a lesser extent, in the adult lung.[6,7]

LYMPHOID HYPERPLASIA

Reactive proliferation of the lymphoid tissue of the lung (nodal, interstitial, and BALT) occurs in a variety of infective, autoimmune, and idiopathic inflammatory conditions and in acquired immunodeficiency syndrome (AIDS). They are described in other sections of this book but especially noteworthy are the lymphoid proliferations seen in follicular bronchiectasis and in follicular bronchiolitis, which occurs in association with Sjögren's syndrome.[8]

Familial pulmonary nodular lymphoid hyperplasia[9] is a newly described syndrome recorded in three members of a single family. They had chronic hypoxia, finger clubbing, hypergamma-globulinemia, and antinuclear antibodies; lung biopsies showed multiple lymphoid nodules in the walls of small airways. An autoimmune pathogenesis has been postulated.

Angiofollicular lymphoid hyperplasia (Castleman's disease) has occasionally been reported in the lung,[10–12] which is not entirely surprising because the mediastinum and pulmonary hila are common sites for this lesion. Its histology and clinical behavior appear to be no different from Castleman's disease at other sites. It is usually of the hyaline-vascular type and consists of a lymphoid mass in which the follicles have hyalinized, vascularized centers surrounded by a broad mantle of concentrically arranged small lymphocytes; the interfollicular tissue is rich in small blood vessels (Fig. 33-1). Sinuses are not present and the lesion is not thought to arise in a lymph node but is generally regarded as hamartomatous.

PSEUDOLYMPHOMA AND LYMPHOID INTERSTITIAL PNEUMONIA (LIP)

These confusing and misleading terms are hallowed by usage in the literature of pulmonary disease. Originally they were coined to describe proliferations of lymphoreticular cells that were deemed to be reactive.[13,14] Pseudolymphoma is a localized or

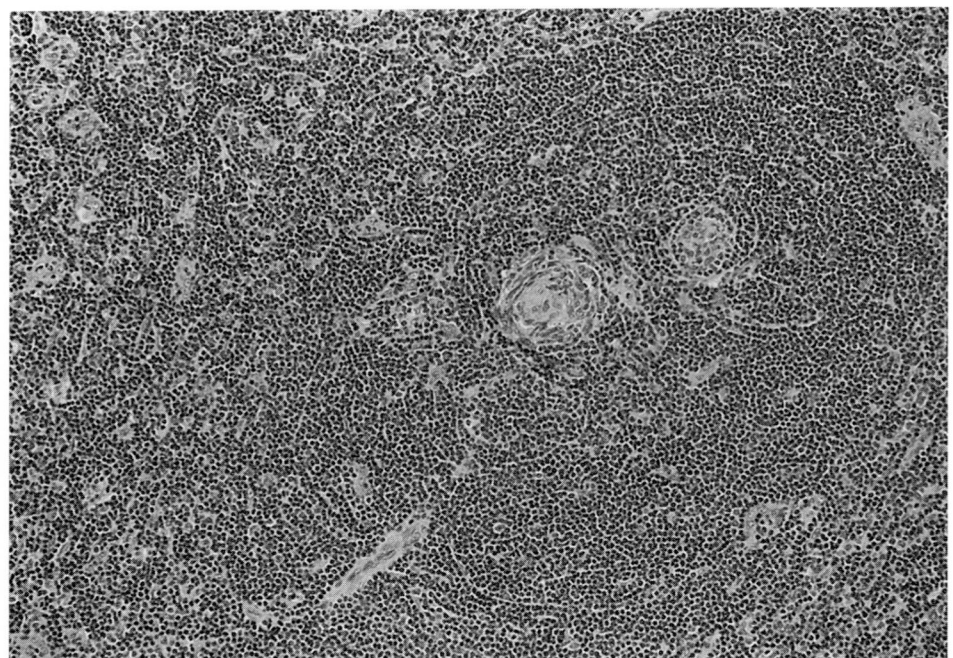

FIGURE 33-1
Hyaline-vascular type of Castleman's disease. The right half of the field is occupied by a follicle containing two hyalinized, vascularized centers. In places the surrounding mantle zone lymphocytes have a concentric pattern, but it is poorly developed in this case. The interfollicular tissue (*top left*) is very vascular. Note the radial vessel penetrating the follicle (*bottom center*). (H&E, low power.)

nodular form of the process, whereas in LIP the pattern is diffuse; an alternative terminology of nodular and diffuse lymphoid hyperplasia has been proposed[15] but does little to clarify the understanding of these conditions.

Over the years evidence has accumulated that both diseases may progress to frank non-Hodgkin's lymphoma and the view is now gaining ground that they both may be lymphomatous from the outset. Considerable confusion exists over the clinical implications of a diagnosis of pseudolymphoma or LIP. In diagnostic work the terms should be used with discretion and with an indication of their malignant propensities. However, it has to be admitted that in practice these terminological distinctions may not always be clinically important because, if they are lymphomas, they are of the mucosa-associated type and tend to remain localized to their site of origin for long periods. Surgical excision may then be effective treatment for pseudolymphoma, LIP, and low-grade mucosa-associated lymphomas alike.

Pseudolymphoma

The term "pseudolymphoma" was introduced in 1963 by Saltzstein[13] and about 14 percent of pulmonary lymphoproliferative lesions appear to fall into this histologic category.[16] Pseudolymphoma may occur at any age but is most frequent in the middle aged; most patients are asymptomatic but a minority have cough, fever, and weight loss.[16] Most lesions are solitary and peripheral but occasionally multinodular examples occur; macroscopically they

are usually rounded and well defined. Histologically the infiltrate is interstitial although this may only be obvious at the margins of the nodule; it consists of small lymphoid cells, plasma cells, and histiocytes, and B-cell follicles of reactive appearance are often present; bronchial walls, including cartilage, may be infiltrated. According to traditional criteria, the findings of polymorphism and prominent reactive follicles are useful in distinguishing pseudolymphoma from true non-Hodgkin's lymphomas, which were deemed to be more monomorphic and less likely to contain reactive follicles.[16] Newer information shows such criteria to be unreliable because polymorphism and reactive follicles are now known to be characteristic features of MALT lymphomas.[17] Therefore, in assessing the true nature of pseudolymphoma reliance must be placed on assessing clonality either by immunohistochemical demonstration of immunoglobulin light chain restriction or by molecular techniques demonstrating immunoglobulin gene rearrangements. Only a small number of reported cases have been subjected to adequate immunophenotypic or genotypic analysis but, of those that have, many appear to be monotypic and are thus to be regarded as true lymphomas, with other evidence indicating that they are of MALT type.[18–22]

It appears, therefore, that pseudolymphoma may well be, in most cases, a misnomer and a term ripe for abandonment. However, it is difficult to disprove the contention that even monotypic pseudolymphomas may be preceded by a polytypic, reactive phase, a situation analogous with myoepithelial sialadenitis (formerly known as "benign lymphoepithelial lesion") and MALT lymphoma of the parotid.

Lymphoid Interstitial Pneumonia

Carrington and Liebow[14,23] coined the term lymphoid interstitial pneumonia (LIP) to identify a histologic picture characterized by a diffuse interstitial infiltrate of small lymphocytes and plasma cells affecting large areas of lung (Figs. 33-2 and 33-3); lymphoid follicles of reactive type may be present and there may also be granulomas, interstitial fibrosis and amyloid deposition. Honeycomb lung may develop. It occurs mainly in middle-aged persons but may also occur in childhood and is usually associated with respiratory and systemic symptoms such as cough, dyspnea, and pyrexia. Associations with a variety of autoimmune diseases, most commonly Sjögren's syndrome, are well documented.[23-26] Polyclonal and monoclonal gammopathy can occur in association with LIP[26-28] and LIP can also occur in AIDS.[29-31]

The exact nature of LIP is currently under debate. As with pseudolymphoma, some cases proceed to frank lymphoma[32] and some have been shown to be monoclonal by immunohistochemical demonstration of light chain restriction and have

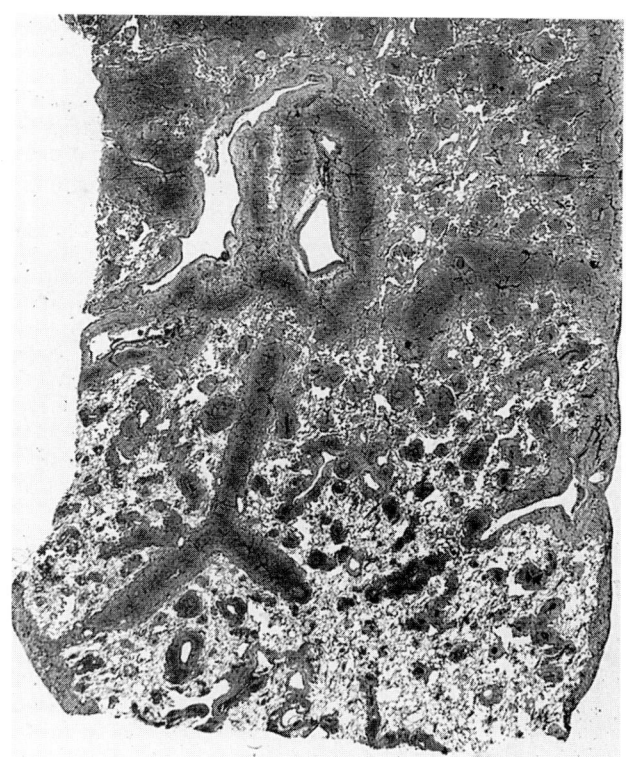

FIGURE 33-2
A large section of lung from a case of lymphocytic interstitial pneumonia shows the widespread infiltrative nature of the changes. (Reproduced courtesy of the Armed Forces Institute of Pathology, Washington, DC. Negative No. 59-2104) (H&E, low power.)

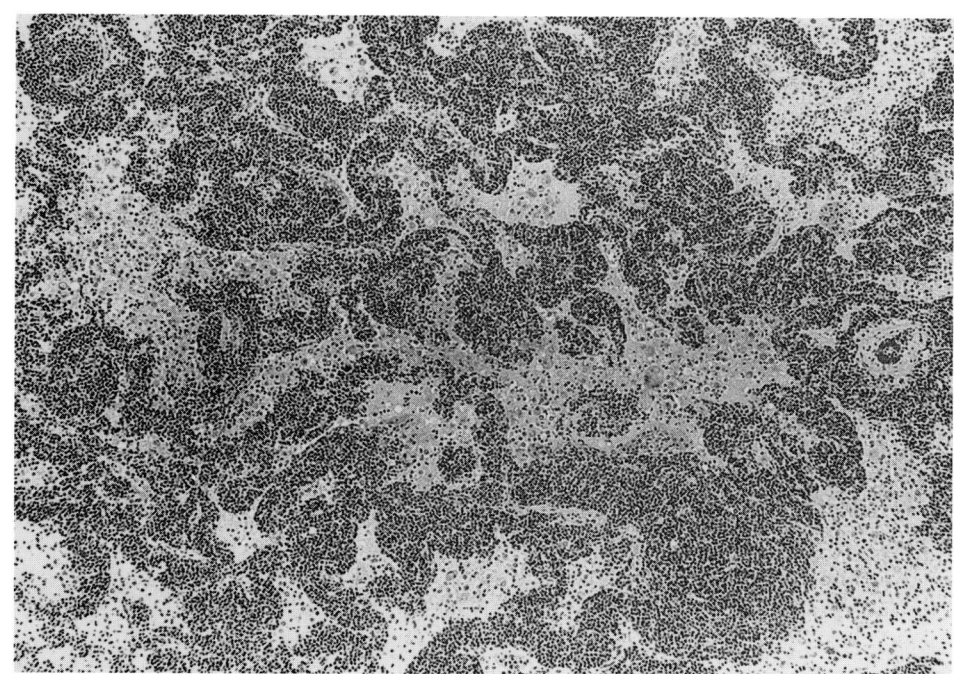

FIGURE 33-3
Lymphoid interstitial pneumonia. A heavy infiltrate of small lymphoid cells and plasma cells expands the alveolar walls. Follicles are not seen in this field. The alveolar spaces contain a histiocyte-rich exudate. (H&E, low power.)

1113

been deemed to be MALT lymphomas ab initio.[33] However, some cases have been found to have polytypic light chain expression, at least in the plasma cell population, and have therefore been regarded as nonneoplastic.[26] Recent evidence suggests that LIP and follicular bronchitis/bronchiolitis are related conditions with overlapping histologic features.[33a] In a recent study the interstitial component of the LIP was mainly of T lineage with B-cell aggregates centered on the airways,[33a] a finding in line with a previously reported single case with a predominantly T-cell infiltrate.[34] The B-cell component was judged to be polyclonal as a result of both staining for light chains and polymerase chain reaction amplification of immunoglobulin heavy chains.[33a] It seems probable, therefore, that most cases of LIP are reactive in nature but that a minority may be, or may become, neoplastic.

NON-HODGKIN'S LYMPHOMA

Secondary Pulmonary Non-Hodgkin's Lymphoma

The main topic for discussion in this section is primary pulmonary non-Hodgkin's lymphoma, but a brief note about secondary involvement is appropriate. This occurs either by direct spread from the mediastinum or as part of the picture of disseminated non-Hodgkin's lymphoma, which usually has presented in lymph nodes. As such, it is classifiable by any of the systems in current favor although the author's preference is for the revised Kiel classification[35] or the Revised European American Lymphoma (REAL) classification.[35a] The management and prognosis will then be related to the grade of the lymphoma as reflected by its histologic subtype and to the stage of the disease. The reader is referred to texts on lymphoma pathology for details. However, it is worth noting that certain uncommon forms of lymphoma have, on occasion, shown the ability to involve the lung; these include Waldenstrom's macroglobulinemia/malignant lymphoma, lymphoplasmacytic,[36] mycosis fungoides and Sezary syndrome,[37,38] and angiotropic lymphoma.[39] Pulmonary involvement occurs in approximately 8 percent of AIDS-related non-Hodgkin's lymphoma.[40]

Primary Pulmonary Non-Hodgkin's Lymphoma

This may be defined as non-Hodgkin's lymphoma presenting in the lung either unilaterally or bilaterally and showing no evidence of involvement of other sites (other than hilar lymph nodes) at the time of presentation or in the following 3-month period.[41]

Such lymphomas represent less than 0.5 percent of all primary lung tumors.[42]

As with non-Hodgkin's lymphomas at all sites, nodal and extra-nodal, terminology and conceptual changes make assessment of the older literature difficult. However, it seems clear from a recent detailed morphologic and immunohistochemical study of 62 cases by the Kiel group that low-grade B-cell lymphomas are overwhelmingly the most common type of primary pulmonary non-Hodgkin's lymphoma. These workers found that 58 of their 62 cases were of B lineage and only 2 were of T lineage, with 2 being high grade, unclassified. Of the 46 low-grade B-cell lymphomas, 43 were of BALT origin and 3 were of centroblastic/centrocytic type. Twelve cases (19 percent of the total) were of high-grade B-cell type and 5 of these were thought to have evolved from low-grade BALT lymphomas.[20] A strikingly similar result has also been reported in a retrospective study of 69 cases from France.[20a] This dominance of lymphomas of MALT is also supported by the smaller series of Addis and his colleagues who found that all 15 of their cases of pulmonary lymphoproliferative disease, which included examples formerly diagnosed as pseudolymphoma and LIP, could be classified as BALT lymphoma.[19] These findings are in line with the results of other groups using different classification systems who have also reported that most primary pulmonary lymphomas fall into small B-cell categories including small lymphocytic, lymphoplasmacytic, lymphoplasmacytoid, or small cleaved cell types.[16,41,43–45]

Against this background the clinical features of most cases of primary pulmonary non-Hodgkin's lymphoma are readily understood because they conform with the characteristics of MALT lymphoma in general. Most patients are middle aged and present with solitary or multiple nodules in the lung parenchyma, which are usually well defined radiologically. About 50 percent of patients in the Kiel series were asymptomatic, whereas the others presented a variety of respiratory and constitutional symptoms, including cough, dyspnea, pain, fever, weight loss, and night sweats.[20] In general, prognosis is excellent and survival for low-grade pulmonary B-cell lymphomas of all types and for BALT lymphomas in particular is not significantly different from that of a standard population; however, in the case of the BALT lymphomas the presence of B symptoms (fever, weight loss, night sweats) is associated with a worse prognosis.[20] This good prognosis is related to the property of MALT lymphomas, in general, of remaining localized to their site of origin for prolonged periods, a phenomenon probably related to the specific homing properties of the cells involved.[17,46]

PATHOLOGIC FEATURES

B-Cell Lymphomas

The non-BALT pulmonary lymphomas have morphologic appearances corresponding to types found at other sites and the reader is referred to texts of lymphoma pathology for details.

Low-grade BALT lymphomas usually present as solitary gray nodules in which cavitation is unusual. Histologically they have the features common to MALT lymphomas in general but adapted to their bronchopulmonary location. The infiltrate is basically interstitial (Fig. 33-4) although this may be obscured in the central parts of the mass. The neoplastic cells have a variable morphology that corresponds to the variants of the centrocyte-like cells, which characterize MALT lymphomas.[47] They are small with irregular indented nuclei although the irregularity may vary from slight (when they resemble small lymphocytes) to marked (Figs. 33-5 and 33-6); cytoplasm is variable in amount and may be scanty and basophilic or plentiful and pale staining when the cells resemble so-called monocytoid B cells. Variable numbers of blastic cells may be pre-

sent and plasma cell differentiation is often a feature (Fig. 33-7). A further finding of great diagnostic importance is the presence of lymphoepithelial lesions in which the neoplastic centrocyte-like cells infiltrate preexisting epithelial structures, in this case the bronchial epithelium (Figs. 33-8 and 33-9). A further characteristic is the occurrence within the lesion of follicles with germinal centers of reactive type (see Fig. 33-4); sometimes such follicles become colonized by the neoplastic cells and this can lead to an erroneous diagnosis of follicular lymphoma.[20,48] At the edge of the tumor mass there is often a picture resembling LIP, reflecting the tendency of the tumor cells to spread interstitially along the pathway of lymphatics (see Fig. 33-4).

Immunohistochemistry is important in establishing the B lineage of the lymphoma cells. This can readily be done using wax sections when they can be shown to be CD20 positive and CD3 negative although the latter will recognize the "reactive" T-cell infiltrate that is commonly present. If cryostat sections of fresh tissue are available a CD5-negative, CD10-negative phenotype can be demonstrated, distinguishing MALT lymphoma from other low-grade B-cell lymphomas. Light chain restriction can be established using either wax or frozen sections and is

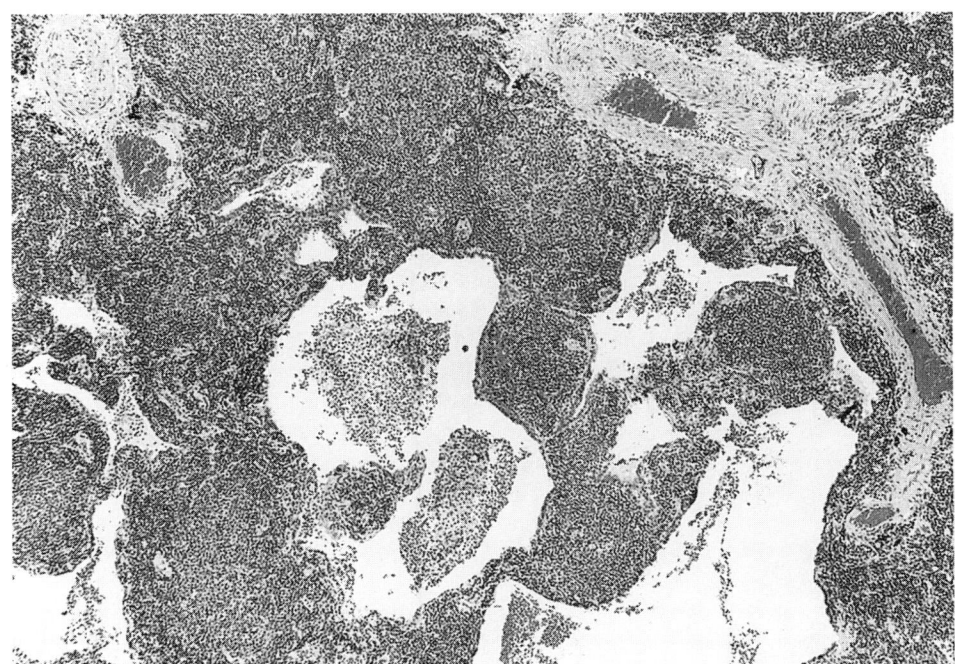

FIGURE 33-4
Low-grade non-Hodgkin's lymphoma of BALT. This is the periphery of the lesion where the infiltrate has an obvious interstitial pattern of infiltration. Note the presence of several germinal centers. (H&E, medium power.)

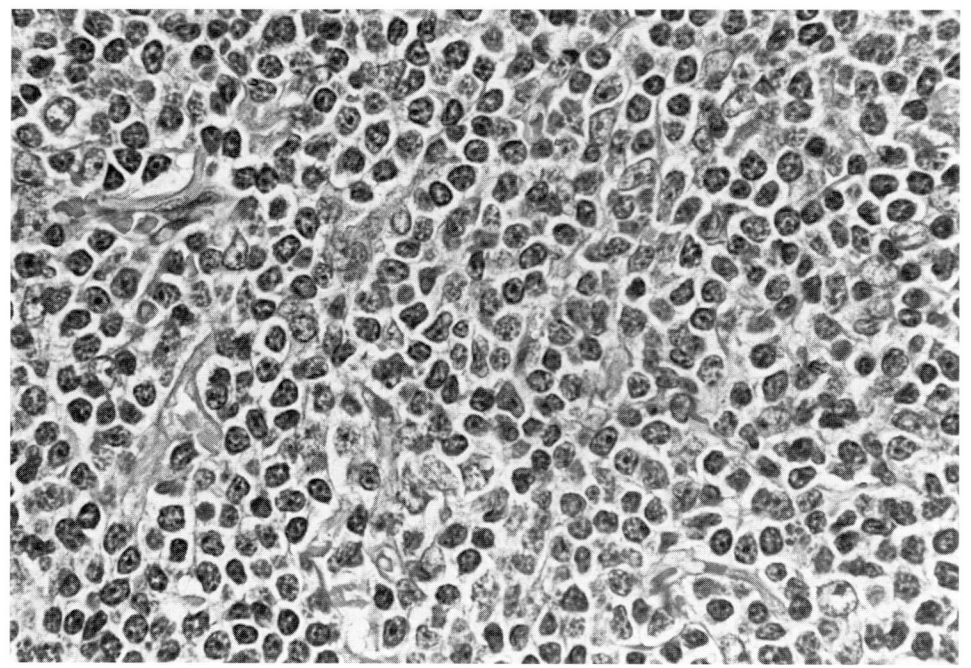

FIGURE 33-5
Low-grade BALT lymphoma. In this example the neoplastic lymphoid cells have relatively round nuclei with inconspicuous irregularities—the lymphocyte-like variant of the centrocyte-like cells that characterize low-grade BALT lymphoma. (H&E, high power.)

essential in cases where the diagnosis of malignancy is in doubt (Figs. 33-10 and 33-11).

High-grade BALT lymphomas appear to be rare and can only be recognized as such if coexistent low-grade BALT lymphoma is also present.

T-Cell Lymphomas (Including Lymphomatoid Granulomatosis)

Although precise figures are difficult to find it is clear that T-cell lymphomas form a small minority of primary pulmonary lymphomas. As mentioned above only 2 of the 62 cases reported by Li and his colleagues were of T lineage.[20]

However, a special form of T-cell lymphoma exists and deserves separate mention. This was first described in 1972 as *lymphomatoid granulomatosis* by Liebow and his colleagues who defined it as basically vasculitic and granulomatous but recognized that there was a relationship with lymphoma at least in some cases[49]; subsequent reports confirmed this

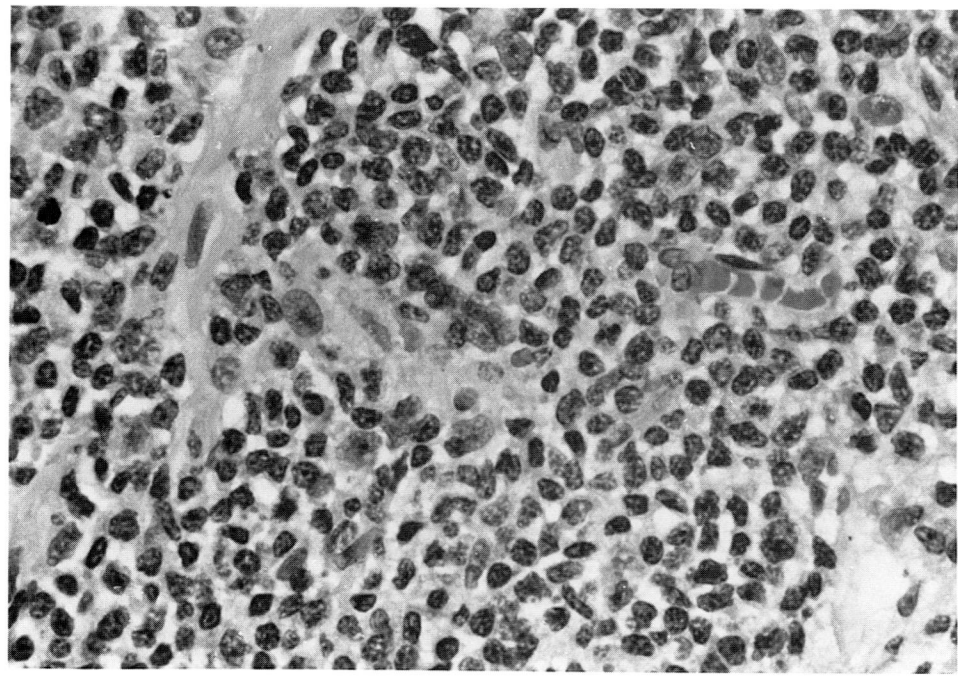

FIGURE 33-6
Low-grade BALT lymphoma. In this case the lymphoid cells show marked nuclear irregularity and are more obviously centrocyte-like than those in Fig. 33-5. (H&E, high power.)

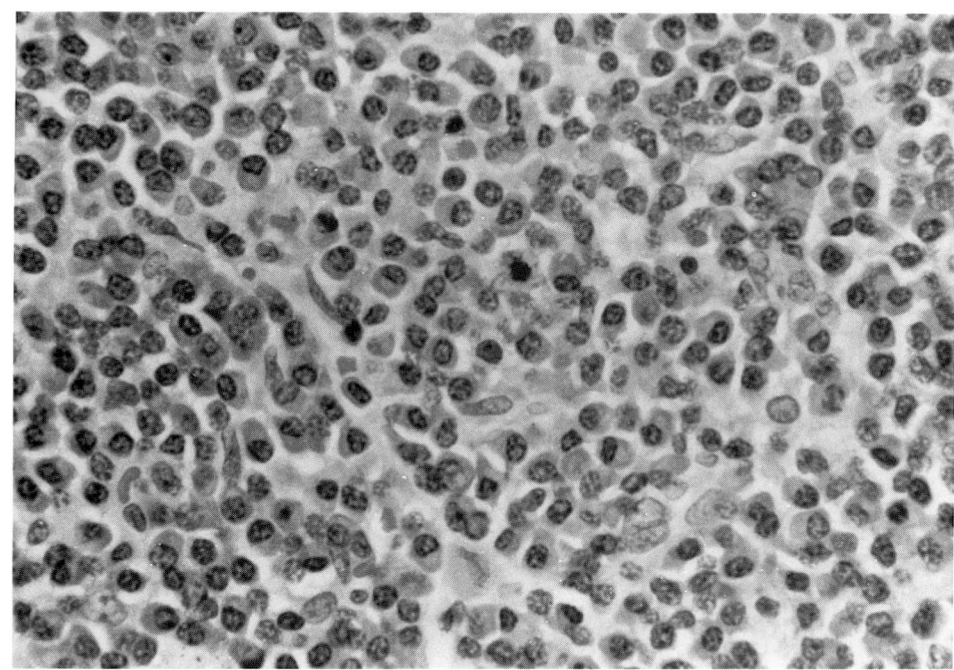

FIGURE 33-7
Low-grade BALT lymphoma. High power shows well developed plasma cell differentiation. (H&E, high power.)

association.[50–56] The application of modern insights to the morphologic features and the advent of immunophenotyping has clarified the confusion that persisted for many years about the nature of lymphomatoid granulomatosis and it is now generally accepted that it is a peripheral T-cell lymphoma of angiocentric type.[57,58] Similar lymphomas occur in the upper respiratory tract and these, too, were formerly regarded as granulomatous conditions and known as "lethal midline granuloma." However, a recent study has suggested an alternative view that lymphomatoid granulomatosis is an Epstein-Barr virus (EBV) driven proliferation of B cells with a prominent T-cell reaction.[59]

Most patients with lymphomatoid granulomatosis are middle aged and both respiratory and systemic symptoms occur. Typically the lesions are bilateral, multiple masses of variable size with cavitation occurring in up to 30 percent; lobar consolidation may be present. Commonly lesions are also

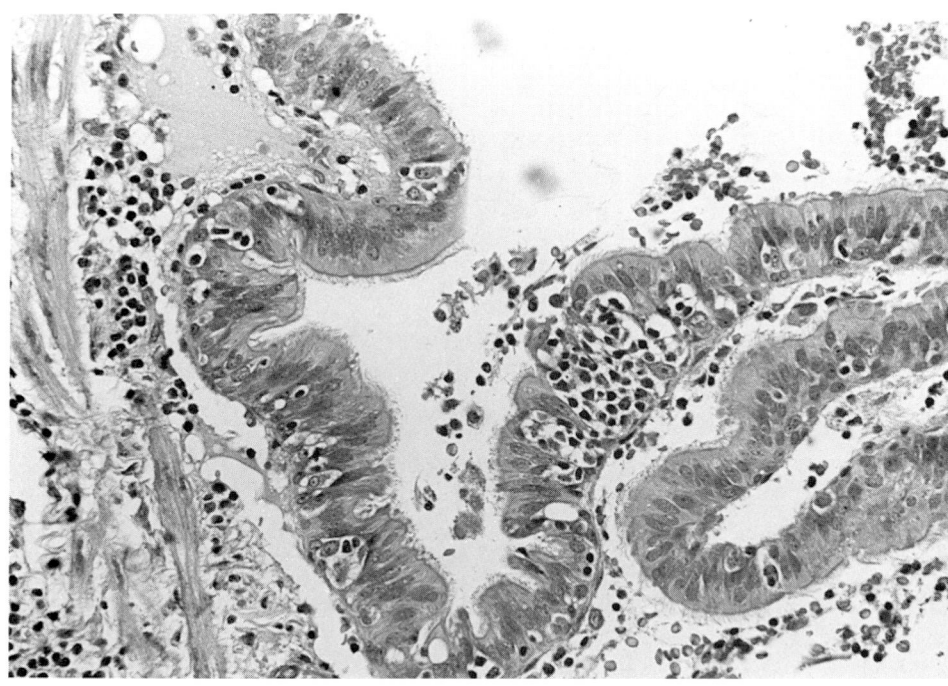

FIGURE 33-8
Low-grade BALT lymphoma illustrates typical lymphoepithelial lesions. One large and several small clusters of centrocyte-like cells are located within the bronchial epithelium. (H&E, high power.)

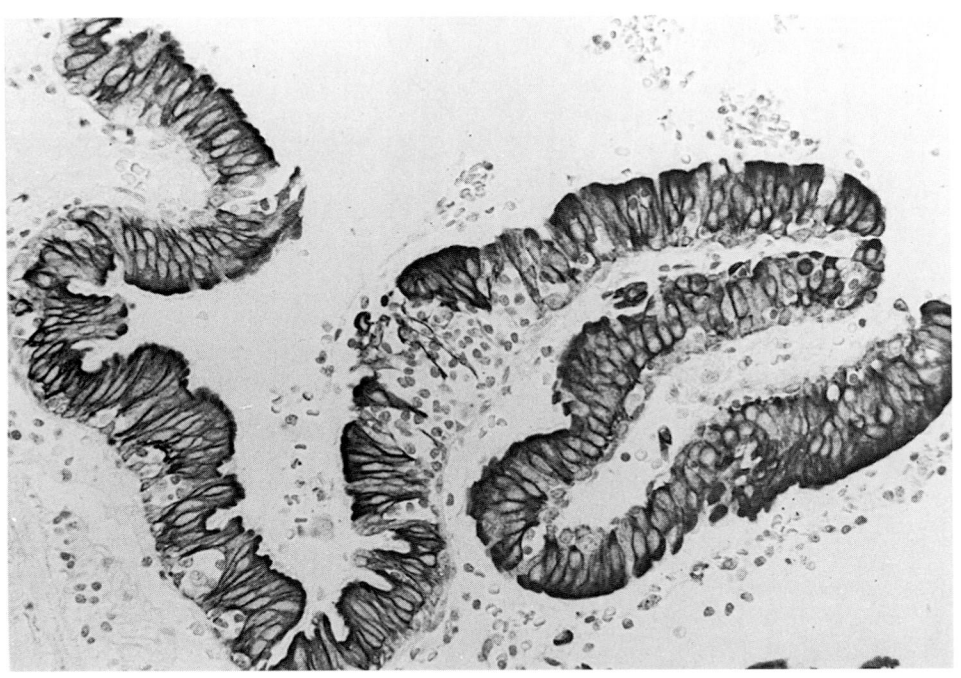

FIGURE 33-9
Low-grade BALT lymphoma. The same lymphoepithelial lesions as shown in Fig. 33-8 are shown in a consecutive section stained for cytokeratin that highlights the negative lymphoid clusters amongst the positive epithelial cells. (Immunoperoxidase stain, high power.)

present in the skin, central nervous system, and kidneys.

Histologically, the masses consist of a polymorphic population of lymphocytes, plasma cells, and histiocytes but eosinophils, giant cells, and granulomas are not usually present. The lymphocytes exhibit a variable degree of atypia ranging from small cells with irregular, dense nuclei to frankly blastic cells (Fig. 33-12). Lipford and colleagues have proposed three grades based on the degree of atypia.[60] Two additional cardinal features are also described: necrosis and angiocentricity. In the latter both arteries and veins may be involved, the vessel walls being infiltrated by the atypical lymphoid cells (Figs. 33-13 and 33-14). It was this feature that led to the view that angiitis was a key feature in the pathogenesis of the

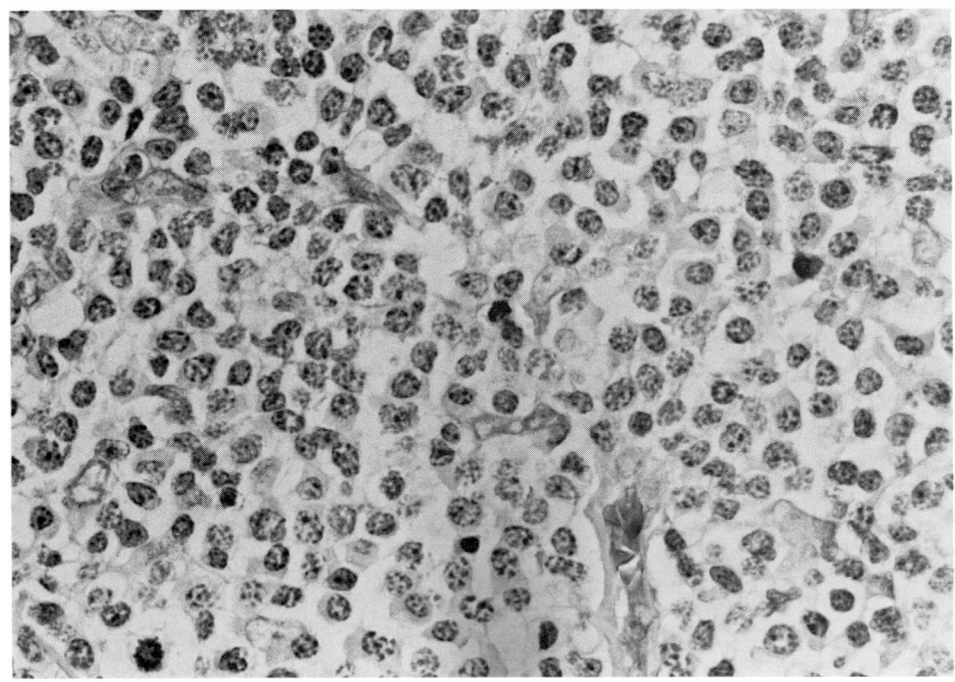

FIGURE 33-10
Low-grade BALT lymphoma with plasmacytic differentiation. Immunoperoxidase stain for lambda light chain is negative in the plasma cells [compare with Fig. 33-11]. (High power.)

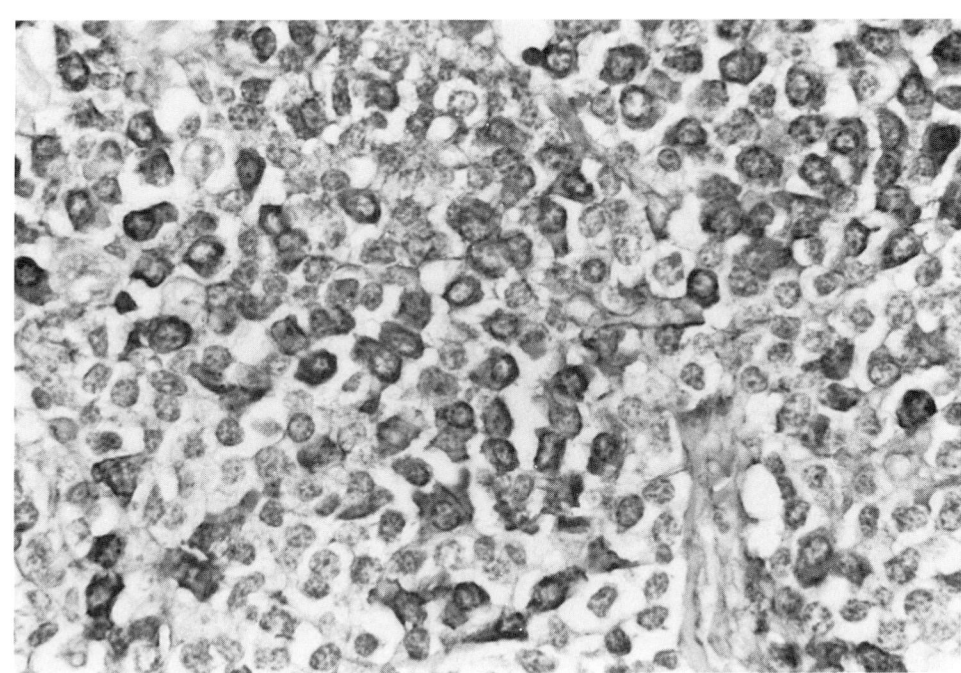

FIGURE 33-11
Low-grade BALT lymphoma with plasmacytic differentiation. Consecutive section to that in Fig. 30-10 stained for kappa light chain. The plasma cells are positive, that is, there is monotypic light chain expression. (High power.)

disease although fibrinoid necrosis is not usually present. It is now recognized that angiocentricity is a common feature of T-cell lymphomas but neither it nor necrosis can in any way be regarded as specific for the diagnosis of T-cell lymphoma of lymphomatoid granulomatosis type because both occur not uncommonly in B-cell lymphomas.[20] Immunophenotyping to establish cell lineage is therefore an essential step in establishing the diagnosis. The small lymphocytes in lymphomatoid granulomatosis are CD3 positive, establishing their T lineage, although the large cells may be either CD3 positive or negative (Figs. 33-15 and 33-16). However, it should be noted that molecular studies often fail to demonstrate T-cell receptor gene rearrangements in angiocentric immunoproliferative disorders.[61]

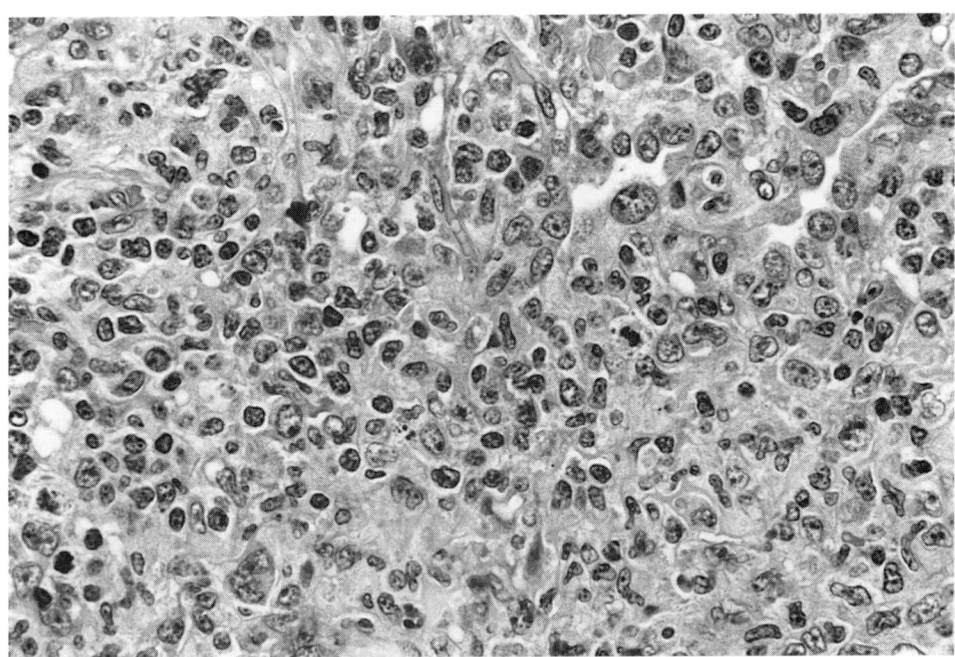

FIGURE 33-12
Angiocentric T-cell lymphoma (lymphomatoid granulomatosis). Diffuse infiltration by a mixed population of small lymphocytes, a smaller number of blastic lymphoid cells, and histiocytes. Some of the lymphoid cell nuclei have somewhat irregular nuclei. (H&E, high power.)

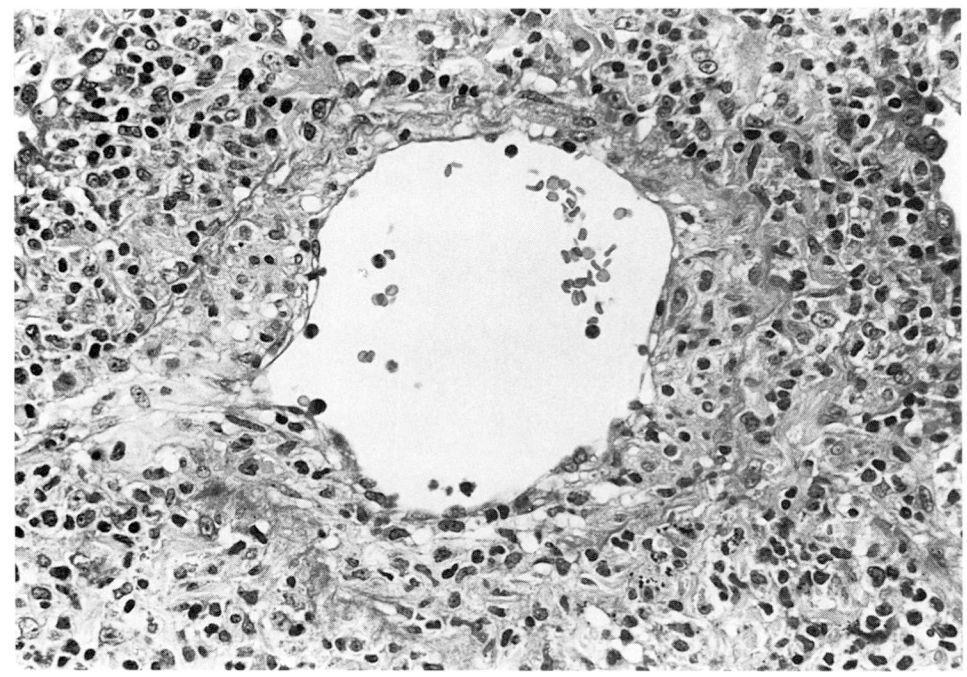

Frequently EBV genome can be demonstrated suggesting that it may have an etiologic role analagous to that in the B-cell lymphoproliferative disorders occurring in immunosuppressed patients.[61] Guinee and colleagues have recently suggested that the EBVirus is located in a small population of medium-sized to large atypical B cells and that the large T-cell population is reactive in nature, implying that lymphomatoid granulomatosis may not be an angiocentric T-cell lymphoma, contrary to the generally held view.[59]

The differential diagnosis of lymphomatoid granulomatosis includes other forms of non-Hodgkin's lymphoma (discussed in the preceding sections), mixed cellularity Hodgkin's disease, and Wegener's granulomatosis. Mixed cellularity

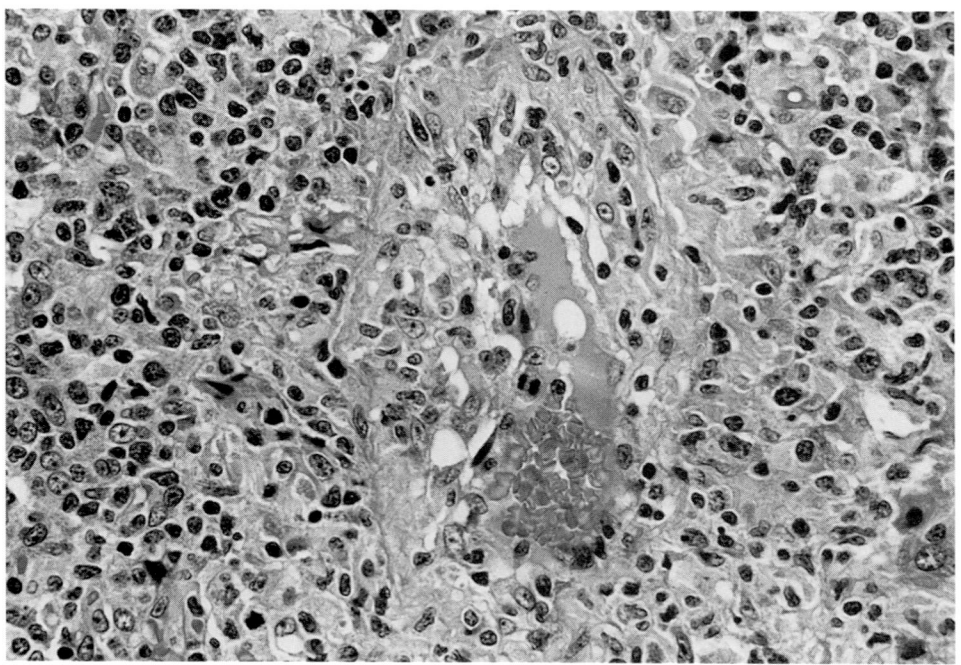

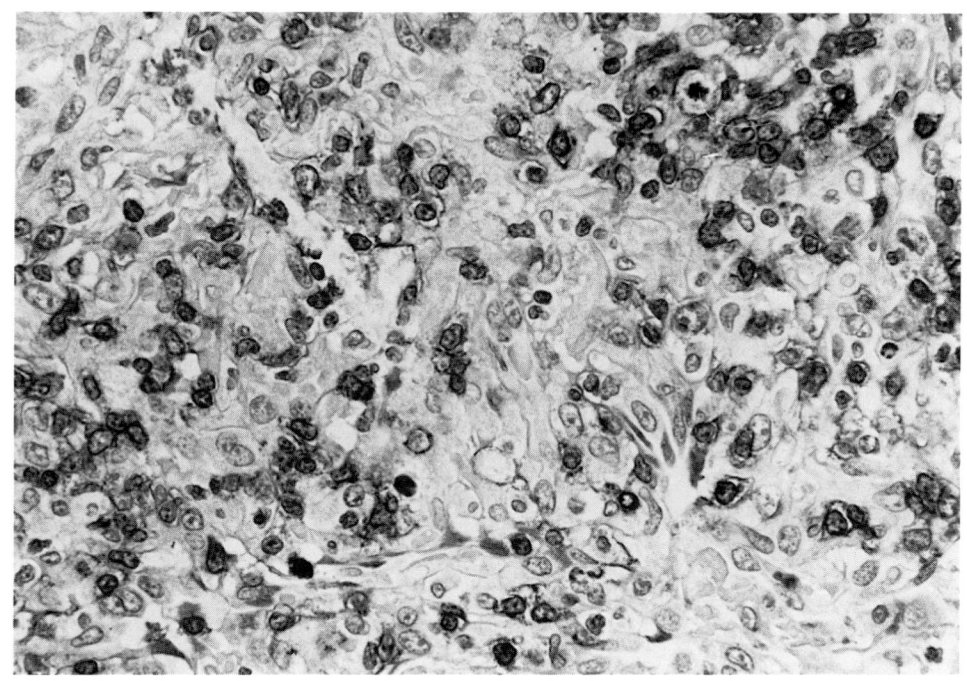

Hodgkin's disease should only be diagnosed when true Reed-Sternberg or Hodgkin's cells (or both) of acceptable morphology and correct immunophenotype (CD15 and CD30 positive) occur in the presence of an appropriate background cell population; if this rule is adhered to mistakes are generally avoidable. Wegener's granulomatosis is characterized by true vasculitis with fibrinoid necrosis (as opposed to the mere angiocentricity of lymphomatoid granulomatosis) and granulomas with giant cells (which are absent in lymphomatoid granulomatosis); furthermore, the cellular atypia of lymphomatoid granulomatosis is not present in Wegener's granulomatosis.

A special variant of angiocentric pulmonary T-cell lymphoma in which cellular atypia is minimal and necrosis is absent has been described under the

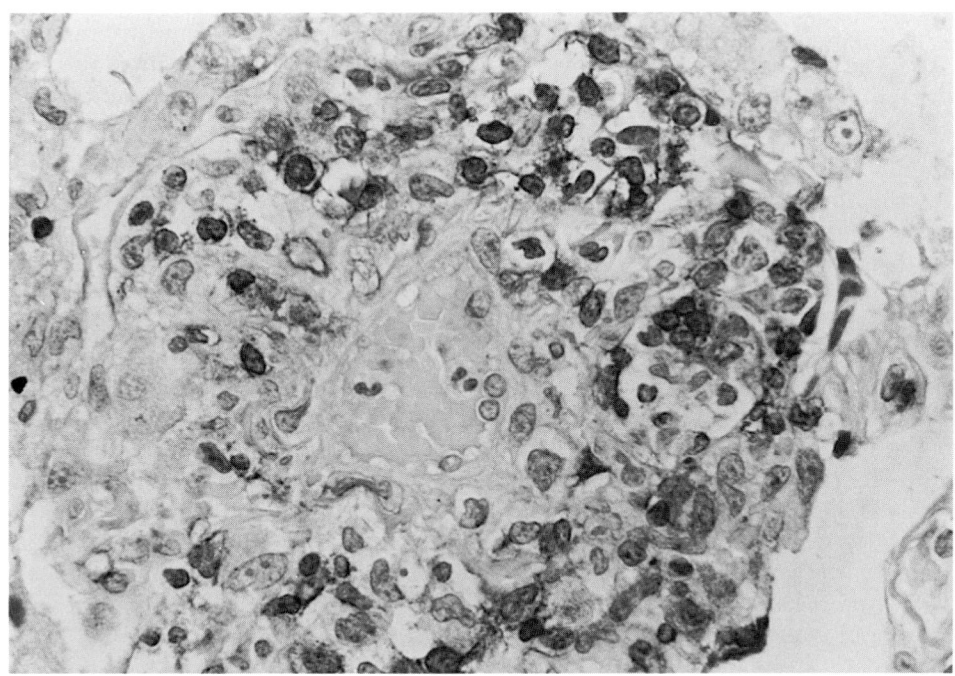

name "benign lymphocytic angiitis and granulomatosis"[51,62,63] and probably corresponds to the grade I angiocentric immunoproliferative lesions of Lipford and coworkers.[60] Whereas this variant is associated with an indolent course and a good prognosis, lymphomatoid granulomatosis tends to behave more aggressively.[53,56]

General Points Concerning the Differential Diagnosis of Lymphoid Infiltrates

Problems arising from needle biopsies The samples derived from closed biopsies create special problems in the diagnosis of lymphoproliferative disorders at any site and these are especially acute in the case of the lung where the majority of lymphomas are of low-grade B-cell type and have to be distinguished from the hyperplastic and reactive proliferations already discussed. The morphologic assessment of such problems is difficult due to the small sample size and frequent surgical trauma and should be supplemented by immunohistochemical or molecular studies (see below). However, even the latter may be problematic on small samples and wherever possible larger open biopsies should be obtained.

Lymphoma Versus Carcinoma Versus Metastatic Melanoma On occasions, especially when dealing with a small biopsy that may have been traumatized, the differentiation of lymphoma from small cell carcinoma may be difficult on morphologic grounds alone. Equally, high-grade lymphoma may be difficult to distinguish from undifferentiated carcinoma or metastatic melanoma. In these circumstances im-munohistochemistry is invaluable and a simple antibody panel of antileukocyte common antigen [CD45], anticytokeratin [e.g., CAM 5.2, AE1/3] and anti–S-100 protein will reliably differentiate the three tumor types (Table 33-1); this can be supplemented if necessary with additional antimelanoma antibodies such as HMB45 and NKIC3.

Reactive Versus Neoplastic Lymphoid Proliferation In the case of B-cell proliferations the conceptual and morphologic difficulties surrounding so-called pseudolymphoma and LIP and their relationship to low-grade BALT lymphoma have already been discussed. The key point in morphologically ambiguous cases is to establish the mono- or polyclonality of the proliferating B cells. This can be done either by immunohistochemical assessment of immunoglobulin light chain expression or by genotypic studies of immunoglobin gene rearrangements.

Dominance of expression of either kappa or lambda light chain signifies monoclonality and

TABLE 33-1

Immunohistochemical Differentiation of Lymphoma, Carcinoma, and Melanoma

Antigen	Lymphoma	Carcinoma	Melanoma
Leukocyte common antigen	+	−	−
Cytokeratin	−	+	[+/−]
S-100 protein	−	[+]	+

+ = usually positive; [+] = sometimes positive;
[+/−] = occasionally positive

therefore neoplasia, whereas a polytypic cell population indicates a reactive process. In the past, immunostaining for immunoglobulins was of limited value in wax-embedded tissue and only cytoplasmic immunoglobulins (e.g., in plasma cells) could reliably be assessed; frozen sections of unfixed tissue were needed to demonstrate surface immunoglobulins and this restricted the practical utility of the technique. However, recent advances in immunostaining technology now allow reliable demonstration of surface immunoglobulins in wax sections,[64] rendering the technique applicable in routine practice.

Molecular techniques have the advantage that they can be used to determine clonality in both B- and T-cell proliferations by examining immunoglobulin and T-cell receptor gene rearrangements, respectively.[65] This approach has also been limited in its practical application by the need for fresh tissue; however, this situation is also changing and the application of the polymerase chain reaction to DNA extracted from wax-embedded tissue holds great promise for the future.[66] Immunophenotyping and genotyping are applicable to cells recovered by bronchoalveolar lavage.[67,68]

HODGKIN'S DISEASE

As with non-Hodgkin's lymphoma, most cases of Hodgkin's disease involving the lung are secondary, representing either direct spread from the mediastinum or part of a disseminated disease process. In the past, secondary pulmonary involvement occurred in up to 43 percent of patients with Hodgkin's disease,[69] but in a more recent series it was only 11.6 percent.[70]

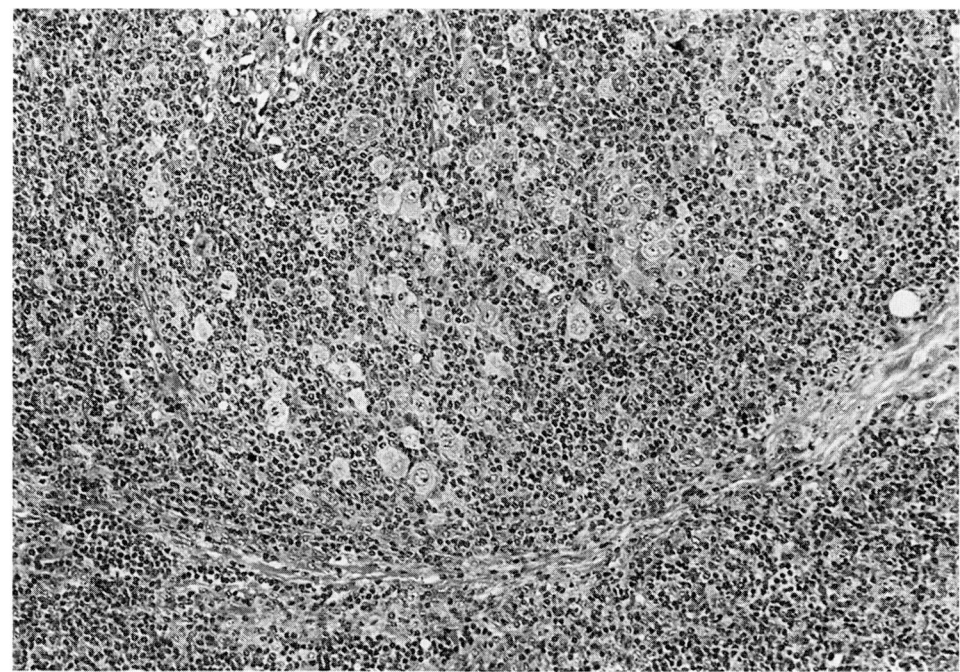

FIGURE 33-17
Nodular sclerosing Hodgkin's disease. Part of a nodule bounded by a thin collagenous septum. Within the nodule there is a mixed population of lymphocytes, eosinophils, and Reed-Sternberg or Hodgkin cells of lacunar type. (H&E, medium power.)

Primary Pulmonary Hodgkin's Disease

This is a very rare manifestation of Hodgkin's disease and in 1990 Radin was able to find only 60 recorded cases, adding one of his own.[70] More women than men are affected, their ages ranging from 12 to 82 years. The upper lobes are most often involved; the lesions may be solitary or multiple and cavitation is common. The most frequent histologic subtypes are nodular sclerosing (Fig. 33-17) and mixed cellularity, which is in line with the prevalence of subtypes in nodal Hodgkin's disease; it is notable that lymphocyte-predominant Hodgkin's disease rarely seems to present in the lung.

The diagnostic criteria do not differ from those of Hodgkin's disease in general and have already been briefly mentioned in the section of this chapter on the differential diagnosis of lymphomatoid granulomatosis; the reader is referred to standard works on lymphoma pathology for details. The differential diagnosis includes non-Hodgkin's lymphomas especially of T-cell type including lymphomatoid granulomatosis, and Wegener's granulomatosis. A further differential is Langerhans' cell histiocytosis. The typical morphology of Langerhans' cells with their folded or grooved nuclei and the absence of Reed-Sternberg cells should indicate the diagnosis of Langerhans' cell histiocytosis. This can be confirmed immunohistochemically because Langerhans' cells are positive for S-100 protein, whereas Hodgkin and Reed-Sternberg cells are negative; conversely Hodgkin/Reed-Sternberg cells in the mixed cellular-

ity and nodular sclerosing types of Hodgkin's disease are CD15 and CD30 positive and Langerhans' cells are negative. Pathognomonic Birbeck granules can be found by electron microscopy in Langerhans' cells but not in Reed-Sternberg cells.

PLASMA CELL TUMORS

Pulmonary parenchymal involvement in *multiple myeloma* is extremely rare occurring in only 4 of the 958 patients reviewed by Kintzer and his colleagues.[71]

Solitary plasmacytomas (Fig. 33-18) are well known in the upper respiratory tract although they are rare. They are even more uncommon in the lungs; Roikjaer and Thomsen found 20 recorded cases,[72] but Nonomura and his colleagues felt that there were only 9 convincing cases and thought that many of the reported examples were misdiagnosed plasma cell granulomas.[73]

Histologically, plasmacytomas consist of sheets of plasma cells with a slight but variable degree of cytologic atypia (Fig. 33-19); the cell population is monomorphic and there is little or no stroma except in those cases where amyloid is present. Immunoglobulin light chain restriction can readily be demonstrated immunohistochemically.

Plasma cell granuloma (inflammatory pseudotumor) is distinguished histologically from plasmacytoma by its more polymorphic cell population, which includes histiocytes and lymphocytes with

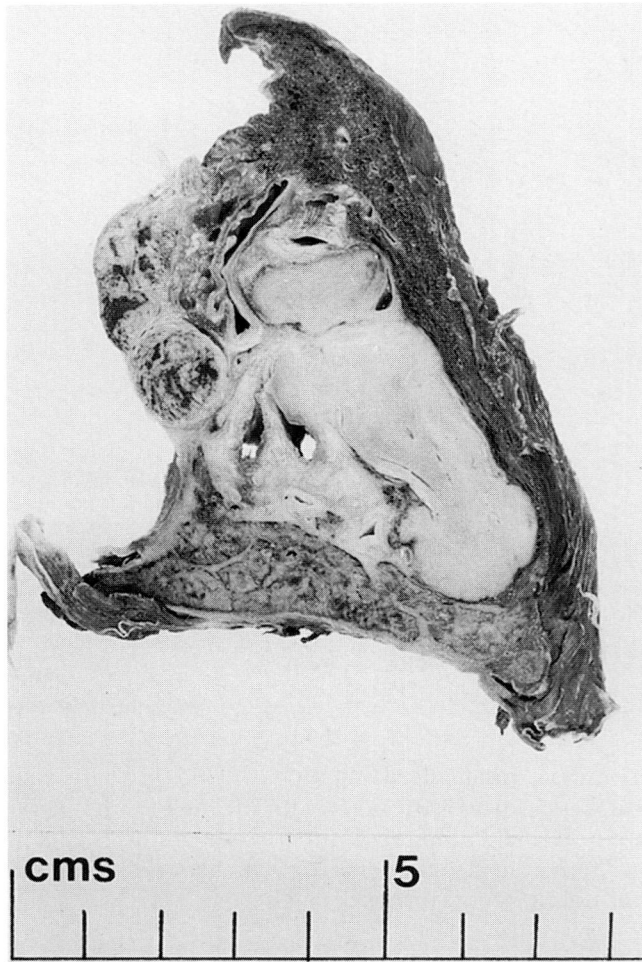

FIGURE 33-18
A pulmonary plasmacytoma. (Reproduced courtesy of Dr. J. S. Whittaker.)

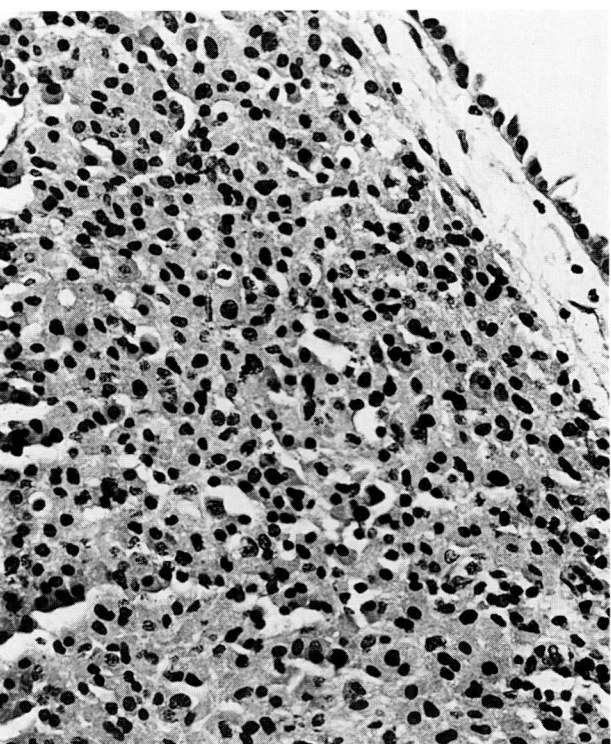

FIGURE 33-19
The same pulmonary plasmacytoma as shown in Fig. 33-18. The tumor consists of sheets of slightly pleomorphic plasma cells. (H&E, medium power.)(Reproduced courtesy of Dr. J. S. Whittaker.)

follicle formation, lack of cytologic atypia, and the presence of a fibroblastic/collagenous stroma.[74] Immunohistochemistry for light chains demonstrates the plasma cells to be polytypic and is, therefore, crucial in cases of difficulty.

LEUKEMIC INFILTRATION AND GRANULOCYTIC SARCOMA

Leukemic infiltration of the lungs can occur in any type of leukemia[75,76] and is said to be common.[77] In chronic lymphatic leukemia the infiltrate is perivascular and interstitial and may also involve bronchioles to produce small airways obstruction.[78,79] In myeloid leukemia respiratory failure due to pulmonary leukostasis has been recorded[80] (Fig. 33-20).

Granulocytic Sarcoma

This rare lesion occurs in a variety of soft tissue and visceral sites including, occasionally, the lung.[81,82] It

is a tumorous mass of leukemic tissue of myeloid or myelomonocytic type, which may precede the development of frank acute leukemia by a considerable period, even years. In the absence of overt leukemia, histologic diagnosis can be particularly difficult. The mass consists of sheets of blastic cells and can closely resemble high-grade non-Hodgkin's lymphoma (Fig. 33-21). Eosinophil myelocytes may be present and, when recognized, point to the correct diagnosis, which can then be readily confirmed by positive histochemical staining in wax sections for naphthol-ASD-chloroacetate esterase (Fig. 33-22) or muramidase.[83] The cells also stain positively for leukocyte common antigen [CD45] and for CD43 [antibody MT1], which may lead the unwary to a misdiagnosis of T-cell lymphoma[84]; however, other more specific T-cell markers are negative.

Mast Cell Tumors

Solitary mast cell tumor may occur in the lung as an extreme rarity (Fig. 33-23). In one case a mixed population of granular mast cells and clear cells of uncertain origin was described.[85] The nature of the mast cells can be confirmed by acid toluidine blue staining and electron microscopy.

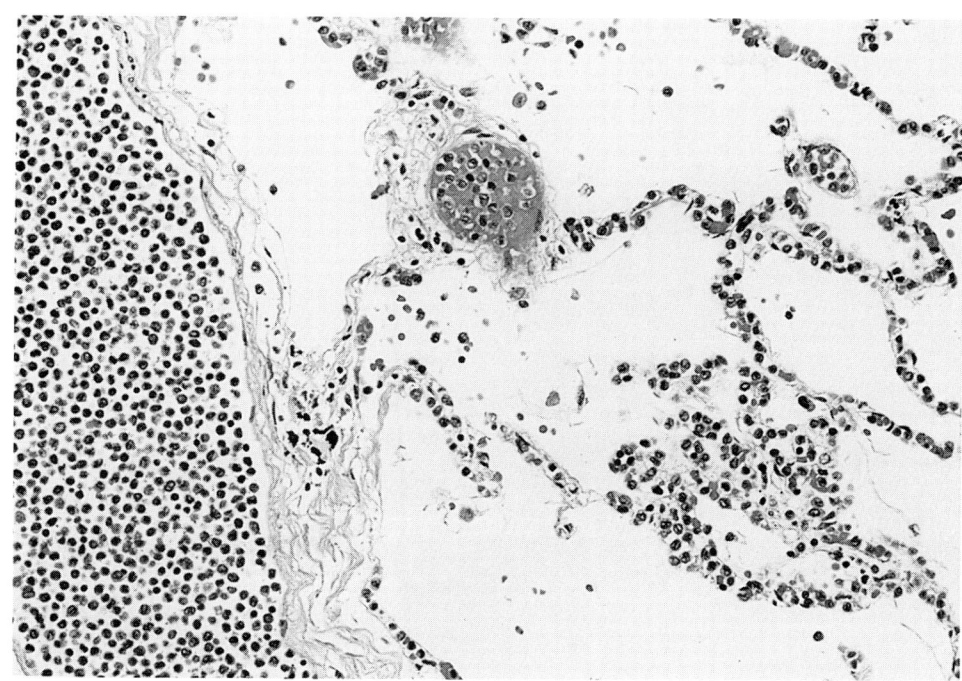

FIGURE 33-20
Acute myelomonocytic leukemia. Large vessels and alveolar capillaries are packed with leukemic cells. Autopsy specimen. (H&E, high power.)

Malignant Histiocytosis

Malignant histiocytosis (histiocytic medullary reticulosis) is a rare condition in which many organs are involved including lymph nodes, liver, bone marrow, central nervous system, kidney, and lungs. Most cases were described before the availability of immunostaining techniques and specific histiocytic markers and the validity of the diagnoses is questionable, many cases probably being non-Hodgkin's lymphomas of T lineage and large cell anaplastic type.

Involvement of the lungs has been emphasized by some workers.[88,89] It is characterized radiologically by bilateral reticulonodular or fluffy infiltrates. Histologically the infiltrates tend to follow the course

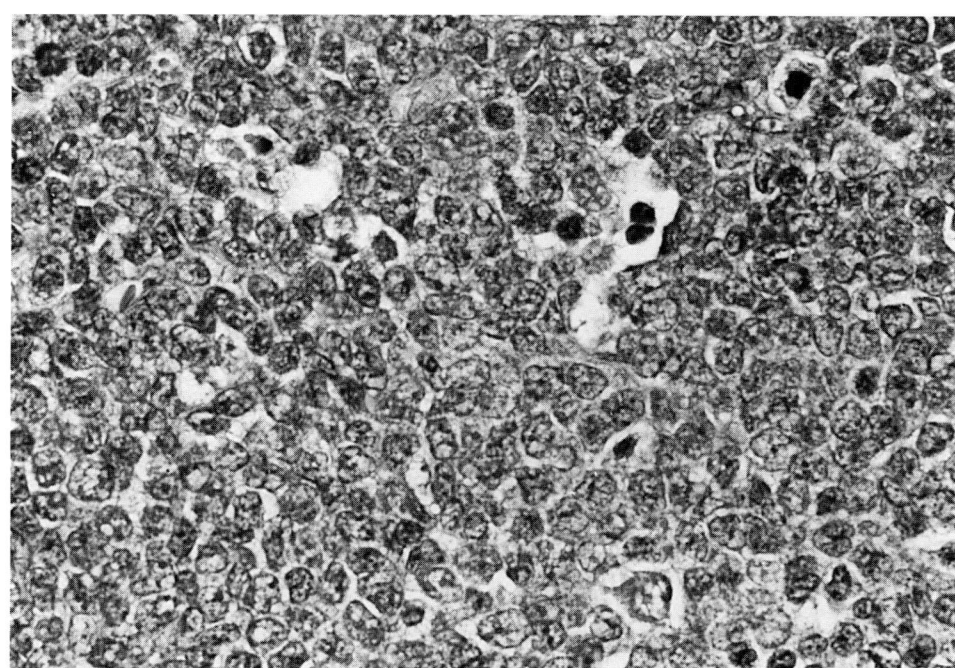

FIGURE 33-21
Granulocytic sarcoma. Sheets of large blastic cells that are difficult to recognize as myeloid as opposed to lymphoid in conventional stains. See also Fig. 33-22. (H&E, high power.)

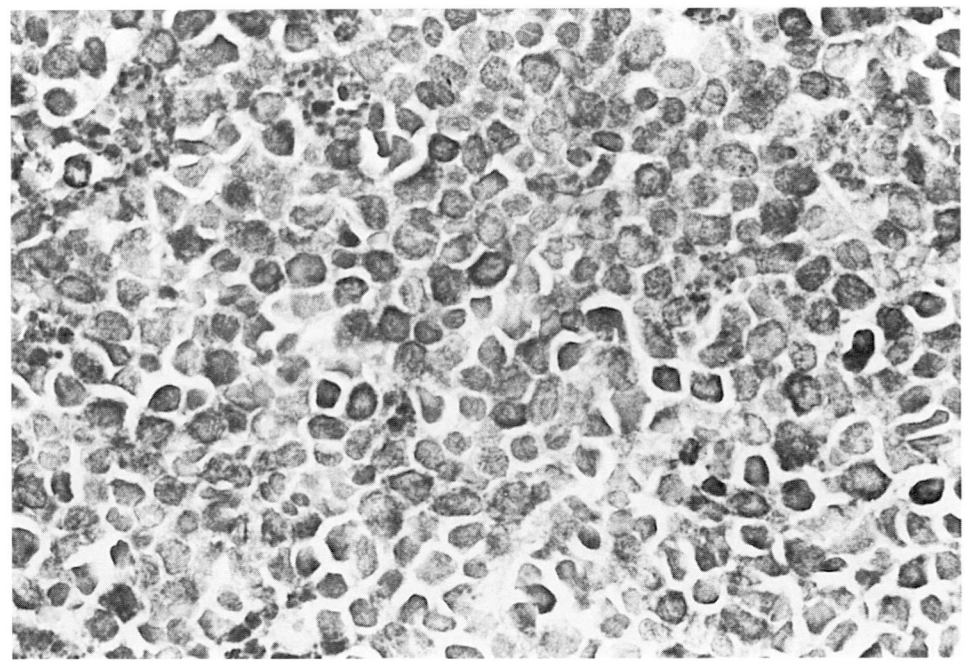

FIGURE 33-22
Granulocytic sarcoma. Same case as in Fig. 33-21 stained for naphthol-ASD-chloroacetate esterase shows granular cytoplasmic positivity. (High power.)

of lymphatics (as in some conventional non-Hodgkin's lymphomas) and to be accompanied by edema and fibrosis; in some cases small bronchi are occluded by the infiltrate. The infiltrates are polymorphic including lymphocytes, plasma cells, and macrophages as well as neoplastic cells. The latter vary in number and are sometimes inconspicuous; in addition, the degree of cytologic atypia varies from slight to marked. As suggested above, in modern practice a diagnosis of malignant histiocytosis should only be accepted if the malignant cells can be proven immunohistochemically not to be of B- or T-cell lineage and to express specific histiocyte-associated antigens using antibodies such as MAC387 or KP1 (CD68).

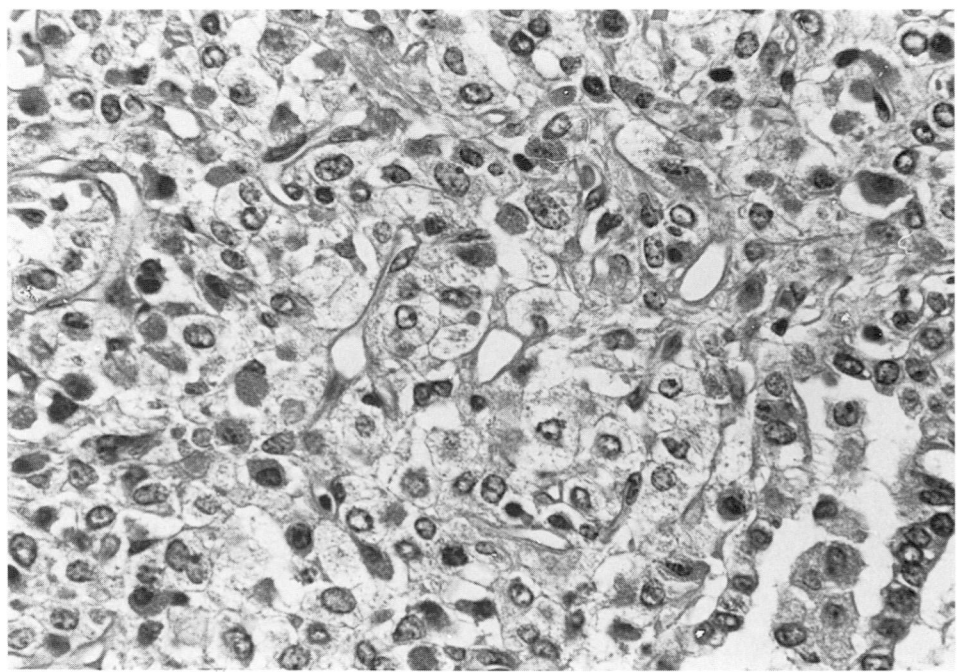

FIGURE 33-23
Mast cell tumor. Sheets of polygonal cells with abundant granular, partly cleared cytoplasm. (H&E, high power.) (Section supplied by Dr. Y. Shimosata, Tokyo.)

REFERENCES

1. Trapnell DH: Recognition and incidence of intrapulmonary lymph nodes. *Thorax* 19:44–50, 1964.
2. Kradin RL, Spirn PW, Mark EJ: Intrapulmonary lymph nodes. Clinical, radiologic, and pathologic features. *Chest* 87:662–667, 1985.
3. Spencer J, Finn T, Isaacson PG: Gut association lymphoid tissue: a morphological and immunocytochemical study of the human appendix. *Gut* 26:672–679, 1985.
4. Pabst R, Gehrke I: Is the bronchus-associated lymphoid tissue [BALT] an integral structure in the lung in normal mammals, including humans? *Am J Respir Cell Mol Biol* 3:131–135, 1990.
5. Gould SJ, Isaacson PG: Bronchus-associated lymphoid tissue [BALT] in human foetal and infant lung. *J Pathol* 169:229–234, 1993.
6. Richmond I, Pritchard GE, Ashcroft T, Avery A, Corris PA, Walters EH: Bronchus associated lymphoid tissue [BALT] in human lung: Its distribution in smokers and non-smokers. *Thorax* 48:1130–1134, 1993.
7. Delventhal S, Brandis A, Ostertag H, Pabst R: Low incidence of bronchus-associated lymphoid tissue [BALT] in chronically inflamed human lungs. *Virchows Arch B Cell Pathol* 62:271–274, 1992.
8. Yousem SA, Colby TV, Carrington CB: Follicular bronchitis/bronchiolitis. *Hum Pathol* 16:700–706, 1985.
9. Franchi LM, Chin TW, Nussbaum E, Riker J, Robert M, Talbert WM: Familial pulmonary nodular lymphoid hyperplasia. *J Pediatr* 121:89–92, 1992.
10. Keller AR, Hochholzer L, Castleman B: Hyaline vascular and plasma cell types of giant lymph node hyperplasia of the mediastinum and other locations. *Cancer* 29:670–683, 1972.
11. Mohamedani AA, Bennett MK: Angiofollicular lymphoid hyperplasia in a pulmonary fissure. *Thorax* 40:686–687, 1985.
12. Awoteda AA, Otulana BA, Ukoli CO: Giant lymph node hyperplasia of the lung (Castleman's disease) associated with recurrent pleural effusion. *Thorax* 45:775–776, 1990.
13. Saltzstein SL: Pulmonary malignant lymphomas and pseudolymphomas: Classification, therapy and prognosis. *Cancer* 16:928–955, 1963.
14. Carrington CB, Liebow AA: Lymphocytic interstitial pneumonia. *Am J Pathol* 48:36a, 1966.
15. Kradin RL, Mark EJ: Benign lymphoid disorders of the lung with a theory regarding their development. *Hum Pathol* 14:857–867, 1983.
16. Koss MN, Hochholzer L, Nichols PW, Wehmunt WD, Lazarus AA: Primary non-Hodgkin's lymphoma and pseudo-lymphoma of the lung: A study of 161 patients. *Hum Pathol* 14:1024–1038, 1983.
17. Isaacson PG: Lymphomas of mucosa-associated lymphoid tissue [MALT]. *Histopathology* 16:617–619, 1990.
18. Herbert A, Wright DH, Isaacson PG, Smith JL: Primary malignant lymphoma of the lung: Histopathologic and immunologic evaluation of nine cases. *Hum Pathol* 15:415–422, 1984.
19. Addis BJ, Hyjek E, Isaacson PG: Primary pulmonary lymphoma: a re-appraisal of its histogenesis and its relationship to pseudo-lymphoma and lymphoid interstitial pneumonia. *Histopathology* 13:1–17, 1988.
20. Li G, Hansmann M-L, Zwingers T, Lennert K: Primary lymphomas of the lung: Morphological, immunohistochemical and clinical features. *Histopathology* 16:519–531, 1990.
20a. Fiche M, Capran F, Berger F, et al: Primary pulmonary non-Hodgkin's lymphomas. *Histopathology* 26:529–538, 1995.
21. Wotherspoon AC, Soosay GN, Diss TC, Issacson PG: Low-grade primary B-cell lymphoma of the lung. An immunohistochemical, molecular, and cytogenetic study of a single case. *Am J Clin Pathol* 94:655–660, 1990.
22. Itoyama T, Sadamori N, Ichimaru M, et al: Evidence for neoplasia in a "pseudolymphoma" of lung. *Lancet* 335:668–669, 1990.
23. Liebow AA, Carrington CB: Diffuse pulmonary lymphoreticular infitrations associated with dysproteinemia. *Med Clin North Am* 57:809–843, 1973.
24. Strimlan CV, Rosenow EC, Divertie MB, Harrison EG: Pulmonary manifestations of Sjögren's syndrome. *Chest* 70:354–361, 1976.
25. Julsrud PR, Brown LR, Li C-Y, Rosenow EC, Crowe JK: Pulmonary processes of mature-appearing lymphocytes: Pseudo-lymphoma, well-differentiated lymphocytic lymphoma and lymphocytic interstitial pneumonia. *Radiol* 127:289–296, 1978.
26. Koss MN, Hochholzer L, Langloss JM, Wehunt WD, Lazarus AA: Lymphoid interstitial pneumonia: Clinicopathological and immunopathological findings in 18 cases. *Pathology* 19:178–185, 1987.
27. Montes M, Tomasi TB, Noebren TH, Culver GJ: Lymphoid interstitial pneumonia with monoclonal gammopathy. *Am Rev Respir Dis* 98:277–280, 1968.
28. Church JA, Isaacs H, Saxon A, Keens TG, Richards W: Lymphoid interstitial pneumonitis and hypogammaglobulinemia in children. *Am Rev Respir Dis* 124:491–496, 1981.
29. Grieco MH, Chinoy-Acharya P: Lymphocytic interstitial pneumonia associated with acquired immune deficiency syndrome. *Am Rev Respir Dis* 131:952–955, 1985.
30. Solal-Celigny P, Couderc LJ, Herman D, et al: Lymphoid interstitial pneumonitis in acquired immunodeficiency syndrome-related complex. *Am Rev Respir Dis* 131:956–960, 1985.
31. Saldana MJ, Manes JM: Lymphoid interstitial pneumonia in HIV infected individuals. *Prog Surg Pathol* 12:430, 1991.
32. Kradin RL, Young RH, Kradin LA, Mark EJ: Immunoblastic lymphoma arising in chronic lymphoid hyperplasia of the pulmonary interstitium. *Cancer* 50:1339–1343, 1982.
33. Herbert A, Walters MT, Cawley MID, Godfrey RC: Lymphocytic interstitial pneumonia identified as lymphoma of mucosa-associated lymphoid tissue. *J Pathol* 146:129–138, 1985.
33a. Nicholson AG, Wotherspoon AC, Diss TC, et al: Reactive pulmonary lymphoid disorders. *Histopathology* 26:405–412, 1995.
34. Kohler PF, Cook RD, Brown WR, Manguso RL: Common variable hypogammaglobulinaemia with T-cell nodular lymphoid interstitial pneumonitis and B-cell nodular lymphoid hyperplasia: Different lymphocyte populations with a similar response to prednisolone therapy. *J Allergy Clin Immunol* 70:299–305, 1982.
35. Stansfeld A, Diebold J, Noel H, et al: Updated Kiel classification for lymphomas. *Lancet* 1:292–293, 1988 (corrected table on p. 603).
35a. Harris NL, Jaffe ES, Stein H, et al: A revised European-American classification of lymphoid neoplasms: A proposal from the international lymphoma study group. *Blood* 84:1361–1392, 1994.
36. Rausch PG, Herion JC: Pulmonary manifestation of Waldenstrom macroglobulinaemia. *Am J Hematol* 9:201–209, 1980.

37. Rubin DL, Blank N: Rapid pulmonary dissemination in mycosis fungoides simulating pneumonia. *Cancer* 56:649–651, 1985.

38. Foster GH, Eichenhorn NS, van Slyck EJ: The Sezary syndrome with rapid pulmonary dissemination. *Cancer* 56:1197–1198, 1985.

39. Yousem SA, Colby TV: Intravascular lymphomatosis presenting in the lung. *Cancer* 65:349–353, 1990.

40. Irwin DH, Kaplan LD: Pulmonary manifestations of acquired immuno-deficiency syndrome-associated malignancies. *Semin Respir Infect* 8:139–148, 1993.

41. L'Hoste RJ, Filippa DA, Lieberman PH, Bretsky S: Primary pulmonary lymphomas. A clinicopathological analysis of 36 cases. *Cancer* 54:1397–1406, 1984.

42. Spencer H: *Pathology of the Lung*, 4th ed. Oxford, Pergamon Press, 1985, p 1033.

43. Le Tourneau A, Audouin J, Garbe L, et al: Primary pulmonary malignant lymphoma, clinical and pathological findings, immunocytochemical and ultrastructural studies in 15 cases. *Hematol Oncol* 1:49–60, 1983.

44. Kennedy JL, Nathwani BN, Burke JS, Hill LR, Rappaport H: Pulmonary lymphomas and other pulmonary lymphoid lesions. A clinicopathologic and immunologic study of 64 patients. *Cancer* 56:539–552, 1985.

45. Weiss LM, Youssem SA, Warnke RA: Non-Hodgkin's lymphomas of the lung. *Am J Surg Pathol* 9:480–490, 1985.

46. Pals ST, Horst E, Scheper RJ, Meijer CJLM: Mechanisms of human lymphocyte migration and their role in the pathogenesis of disease. *Immunol Rev* 108:111–133, 1989.

47. Isaacson PG, Spencer J: Malignant lymphoma of mucosa-associated lymphoid tissue. *Histopathology* 11:445–462, 1987.

48. Isaacson PG, Dogan A, Price SK, Spencer J: Immuno-proliferative small intestine disease: An immunohistochemical study. *Am J Surg Pathol* 13:1023–1033, 1989.

49. Liebow AA, Carrington CRB, Friedman PJ: Lymphomatoid granulomatosis. *Hum Pathol* 3:457, 1972.

50. Lee SC, Roth LM, Braschear RE: Lymphomatoid granulomatosis. A clinico-pathologic study of four cases. *Cancer* 38:846–853, 1976.

51. Saldana MJ, Patchefsky AS, Israel HI, Atkinson GW: Pulmonary angiitis and granulomatosis: The relationship between histological features, organ involvement and response to treatment. *Hum Pathol* 8:391–409, 1977.

52. Gibbs AR: Lymphomatoid granulomatosis—a condition with affinities to Wegener's granulomatosis and lymphoma. *Thorax* 32:71–79, 1977.

53. Katzenstein ALA, Carrington CB, Liebow AA: Lymphomatoid granulomatosis: a clinicopathological study of 152 cases. *Cancer* 43:360–373, 1979.

54. Fauci AS, Haynes BF, Costa J, Katz P, Wolff SM: Lymphomatoid granulomatosis. Prospective clinical and therapeutic experience over 10 years. *N Engl J Med* 306:68–78, 1982.

55. Sordillo PP, Epremian B, Koziner B, Lacher M, Lieberman P: Lymphomatoid granulomatosis. An analysis of clinical and immunologic characteristics. *Cancer* 49:2070–2076, 1982.

56. Koss MN, Hochholzer L, Langloss JM, Wehnunt WD, Lazarus AA, Nichols PW: Lymphomatoid granulomatosis: A clinicopathologic and immunopathologic study of 42 patients. *Pathology* 18:283–288, 1986.

57. Colby TV, Carrington CB: Pulmonary lymphomas: Current concepts. *Hum Pathol* 14:884–887, 1983.

58. Jaffe ES: Pathologic and clinical spectrum of post-thymic T-cell malignancies. *Cancer Invest* 2:413–426, 1984.

59. Guinee D, Jaffe E, Kingma D, et al: Pulmonary lymphomatoid granulomatosis. Evidence for a proliferation of Epstein-Barr virus infected B-lymphocytes with a prominent T-cell component and vasculitis. *Am J Surg Pathol* 18:753–764, 1994.

60. Lipford EH, Margolick JB, Longo DL, Fanci AS, Jaffe ES: Angiocentric immunoproliferative lesions: A clinico-pathologic spectrum of post-thymic T-cell proliferations. *Blood* 72:1674–1681, 1988.

61. Medeiros LJ, Peiper SC, Elwood L, Yano T, Raffeld M, Jaffee ES: Angiocentric immuno-proliferative lesions: A molecular analysis of eight cases. *Hum Pathol* 22:1150–1157, 1991.

62. Gracey DR, De Remee RA, Colby TV, Unni KK, Weiland LH: Benign lymphocytic angiitis and granulomatosis: Experience with 3 cases. *Mayo Clin Proc* 63:323–331, 1988.

63. Saldana MJ, Mones JM, Kahn E: Benign lymphocytic angiitis and granulomatosis of the lungs [BLAG]: Clinico-pathologic and immunohistologic observations in 6 cases. *Lab Invest* 62:88A, 1990.

64. Norton AJ, Isaacson PG: Detailed phenotypic analysis of B-cell lymphoma using a panel of antibodies reactive to routinely fixed wax-embedded tissue. *Am J Pathol* 128:225–240, 1987.

65. O'Connor NTJ, Gatter KC, Wainscoat JS, et al: Practical value of genotypic analysis for diagnosing lymphoproliferative disorders. *J Clin Pathol* 40:147–150, 1987.

66. Pan LX, Diss TC, Peng HZ, Isaacson PG: Clonality analysis of defined B-cell populations in archival tissue sections using microdissection and the polymerase chain reaction. *Histopathology* 24:323–327, 1994.

67. Betsuyaka T, Munakata M, Yamaguchi E, et al: Establishing diagnosis of pulmonary malignant lymphoma by gene rearrangement analysis of lymphocytes in bronchoalveolar lavage fluid. *Am J Respir Crit Care Med* 149:526–529, 1994.

68. Keicho N, Oka T, Takeuchi K, Yamane A, Yazaki Y, Yotsumoto H: Detection of lymphomatous involvement of lung by bronchoalveolar lavage. Application of immuno-phenotypic and gene re-arrangement analysis. *Chest* 105:458–462, 1994.

69. MacDonald JB: Lung involvement in Hodgkin's disease. *Thorax* 32:664–667, 1977.

70. Radin AI: Primary pulmonary Hodgkin's disease. *Cancer* 65:550–563, 1990.

71. Kintzer JS, Rosenow EC, Kyle RA: Thoracic and pulmonary abnormalities in multiple myeloma. A review of 958 cases. *Arch Intern Med* 138:727–730, 1978.

72. Roikjaer O, Thomsen JK: Plasmacytoma of the lung. A case report describing two tumors of different immunologic type in a single patient. *Cancer* 58:2671–2674, 1986.

73. Nonomura A, Mizukami Y, Shimuzu J, et al: Primary extramedullary plasmacytoma of the lung. *Intern Med* 31:1396–1400, 1992.

74. Bahadori M, Liebow AA: Plasma cell granulomas of the lung. *Cancer* 31:191–208, 1973.

75. Yoshioka R, Yamaguchi K, Yoshinaga T, Takatsuki K: Pulmonary complications in patients with adult T-cell leukemia. *Cancer* 55:2491–2494, 1985.

76. Doran HM, Sheppard MN, Collins PW, Jones L, Newland AC, Van der Walt JD: Pathology of the lung in leukaemia and lymphoma: study of 87 autopsies. *Histopathology* 18:211–219, 1991.

77. Gephardt GN, Tubbs, RR: Pulmonary lymphomas and other lymphoproliferative lesions, in Saldana MJ (ed); *Pathology of Pulmonary disease*. Philadelphia, JB Lippincott, 1994, p 637.

78. Palosaari DE, Colby TV: Bronchiolocentric chronic lymphocytic leukemia. *Cancer* 58:1695–1698, 1986.

79. Rollins SD, Colby TV: Lung biopsy in chronic lymphocytic leukemia. *Arch Pathol Lab Med* 112:607–611, 1988.
80. Myers TJ, Cole SR, Klatsky AU, Hild DH: Respiratory failure due to pulmonary leukostasis following chemotherapy of acute nonlymphocytic leukemia. *Cancer* 51, 1808–1813, 1983.
81. Callahan M, Wall S, Askin F, Delaney D, Koller C, Orringer EP: Granulocytic sarcoma presenting as pulmonary nodules and lymphadenopathy. *Cancer* 60:1902–1904, 1987.
82. Hicklin GA, Drevyanko TF: Primary granulocytic sarcoma presenting with pleural and pulmonary involvement. *Chest* 94:655–656, 1988.
83. Furebring-Freden M, Martinsson U, Sundstrom C: Myelosarcoma without acute leukaemia: Immunohistochemical and clinico-pathological characterization of 8 cases. *Histopathology* 16:243–250, 1990.
84. Elliott CJ, McCarthy KP, Carter RL, Davies P: Granulocytic sarcoma: Misleading immunohistological staining with MT1 and S100 protein antibodies. *J Clin Pathol* 42:188–190, 1989.
85. Kudo H, Morinaga S, Shimosato Y, et al: Solitary mast cell tumor of the lung. *Cancer* 61:2089–2094, 1988.
86. Kadin ME: T gamma cells: a missing link between malignant histiocytosis and T cell leukaemia-lymphoma? *Human Pathol* 12:771–772, 1981.
87. Delsol G, Al-Saati GA, Gatter KC: Co-expression of epithelial membrane antigen [EMA], Ki-1, and interleukin-2 receptors by anaplastic large cell lymphomas: Diagnostic value in so-called malignant histiocytosis. *Am J Pathol* 130:59–70, 1988.
88. Colby TV, Carrington CB, Mark GJ: Pulmonary involvement in malignant histiocytosis. A clinicopathologic spectrum. *Am J Surg Pathol* 5:61–73, 1981.
89. Aozasa K, Tsujimoto M, Inoue A: Malignant histiocytosis. Report of twenty five cases with pulmonary, renal and/or gastro-intestinal involvement. *Histopathology* 9:39–49, 1985.

Chapter 34

Pleural Disease

P. S. Hasleton

To most histopathologists, pleural disease is often synonymous with mesothelioma or secondary carcinoma. However, the main disease process in this cavity is a pleural effusion. In recent years, with the increase in disease related to acquired immunodeficiency syndrome (AIDS), pleural effusions due to tuberculosis are increasing. A full description of pleural disease is given by Sahn.[1] For texts on malignant mesothelioma and asbestosis, the reader is referred to several monographs.[2–4]

THE FORMATION OF PLEURAL FLUID

A description of the normal pleura is included at the end of the Chap. 1, on normal anatomy. The parietal pleura is supplied by systemic vessels, and the pressure in the pleural space is subatmospheric. Thus there is a pressure gradient from the interstitium of the parietal pleura into the pleural space. Probably the only important barrier to fluid exchange in this space is the microvascular endothelium.[5] The contribution of the visceral pleura to fluid and protein formation in the normal individual is likely to be minimal. This is because there is a lower filtration pressure between the visceral pleural microcirculation, as bronchial venules empty into the lower-pressure pulmonary veins.[6]

The initial filtrate seen in pleural fluid has a protein concentration of about 0.3 to 0.4 g/dL. Much of the filtered water is absorbed rapidly through the lower-pressure venules, concentrating the interstitial protein to a higher level of 1 to 1.5 g/dL. Pleural liquid is removed slowly. It enters the parietal pleural interstitial space and leaks through the mesothelium. Both pleural liquid and protein then exit via the parietal pleural stomata. There is a large reserve in the pleural lymphatics, so that an abnormal amount of pleural fluid can be removed quickly.

Multiple mechanisms are responsible for an increase in pleural fluid.[1] These can be summarized under the following headings:

1. *Increased hydrostatic pressure in the microvascular circulation.* Elevation of pulmonary capillary wedge pressure is probably the most important cause of pleural effusions in congestive heart failure. The fluid leaks into the pulmonary interstitium and moves across the mesothelial barrier along an interstitial-pleural pressure gradient.[7]

2. *Decrease in oncotic pressure in the microvascular circulation.* This occurs in patients with a low serum albumin, increasing the tendency to form pleural interstitial fluid. It would cause increased entry of this low-protein fluid into the pleural space. This is an unusual cause of large pleural effusions, because of a large functional reserve of the parietal pleural lymphatics.

3. *Decrease in pressure in the pleural space.* Clinically, large pleural effusions usually occur only with complete lung collapse. The increased negative pressure, in conjunction with the hydrostatic gradient across the parietal pleura, increase the tendency for pleural fluid formation. The separation of the lung from the chest wall may decrease pleural-space fluid movement and inhibit optimum pleural lymphatic drainage.

4. *Increased permeability of the microvascular circulation.* This is due to leaky vessels. It occurs in diseases such as pneumonia, where there is occlusion of parietal pleural stomata with fibrin, necrotic cells, and some swelling of the mesothelial cells.

5. *Impaired lymphatic drainage of the pleural space.* This is due to tumor or, less commonly, fibrosis. Increased fluid formation in malignancy may also be due to altered microvascular permeability.

6. *Movement of fluid from the peritoneal cavity.* Any condition causing ascites can lead to pleural effusion by passage of fluid from either diaphragmatic lymphatics or defects in this muscle. Such defects are usually small, being less than 1 cm in diameter. The movement of fluid from the peritoneal to the pleural cavities is due to a pressure gradient across the diaphragm.

PLEURAL EFFUSIONS AND THE IMMUNE RESPONSE

As mentioned above, pleural effusions may stem in part from local immunologic responses to infectious agents or tumors. In addition, they may be seen in various collagen diseases, such as systemic lupus erythematosus. There is variability in the non-selective transfer of immunoglobulins in the pleura and only a restricted role for local synthesis.[8] The concentration of immunoglobulins is slightly below that in serum. One of the exceptions is the elevated IgE concentration, associated with pleural eosinophilia or following trauma.[9]

In immunologically related diseases, the excessive filtration is due to increased vascular permeability of pleural capillaries. This may be due to circulating immune complexes. Antigen-antibody complexes generate a cleavage of complement products (C3A and C5A), which are chemotactic for neutrophils.

Most inflammatory pleural effusions have a high proportion of T cells and the ratio between T and B lymphocytes is greater than that in peripheral blood.[10] This feature is also seen in neoplastic pleural effusions. The accumulation of T cells in pleural tuberculosis results from a local cellular immune response to *Mycobacterium tuberculosis*, producing local activated T cells and macrophages. In malignant pleural effusions, the accumulation of T lymphocytes results from immune responses to tumor antigens. A predominance of lymphocytes in malignant effusions may indicate a better prognosis.[11]

Almost all patients with congestive heart failure have a total protein concentration in pleural fluid of 3 g/dL or less. If these patients have had chronic diuretic therapy, the protein concentration may be higher, with a range of 3 to 4 g/dL.[12] Tuberculous pleural effusions have protein concentrations above 4.0 g/dL and the range with effusions due to other pneumonias is variable, being from 2.5 to 6 g/dL.[13] Patients with malignant pleural effusions have protein concentrations varying from 1.5 to 8 g/dL.

The leukocyte count in pleural fluid is not diagnostic. Transudates usually have counts of less than 1000 neutrophil polymorphs per microliter. Counts above 10,000/uL indicate marked pleural inflammation, but this may be seen with a host of injuries, including pulmonary infarction and acute asbestos induced pleurisy.

A lymphocytosis in the pleural fluid, especially counts of 80 to 90 percent of the total, suggests tuber-culosis.[14] Other causes of a lymphocytosis in pleural fluid are lymphoma,[15] sarcoid,[16] and rheumatoid disease. In non-Hodgkin's lymphoma there may be atypical lymphocytes. Nearly one-third of patients with pleural transudates have a predominance of lymphocytes. However, histologically there are usually far fewer cells than are seen in the other conditions mentioned above, causing a pleural lymphocytosis. Carcinoma may also produce a lymphocytosis in two thirds of cases.[14]

Eosinophils are relatively uncommon in pleural fluid. If they are identified, the differential diagnosis includes pneumothorax, pulmonary infarction, hemothorax, fungal infection, parasitic disease, drugs, and asbestos induced pleurisy.[17] Rarely, eosinophils may be a manifestation of tuberculous pleurisy and malignancy. One-third of pleural effusions with eosinophilia may be "idiopathic." These cases may be due to occult pulmonary embolism or asbestos disease.[18]

A small number of plasma cells in a pleural fluid is non-diagnostic and may be seen in many conditions. A large number of plasma cells, especially if some are atypical and monoclonal, suggests multiple myeloma.

Red blood corpuscles are often seen in pleural effusions. These may be the result of a "bloody tap." If the blood has a traumatic origin, the fluid should clot within minutes and there will be no hemosiderin-laden macrophages on microscopy. The commonest cause of red blood corpuscles in pleural fluid is malignancy. Pulmonary embolism, post-cardiac-injury syndrome, and asbestos induced disease also cause serosanguinous effusions.

The biochemical nature of the fluid is beyond the scope of this book; for specific details of any single disease, the reader is referred to Sahn.[1] Pleural cholesterol was lower in transudates than in neoplastic or inflammatory exudates.[19] More routine tests that are used on pleural fluid are glucose, which is lowest in rheumatoid disease and empyema but is also low in malignancy. The pH is low in the same conditions as a low pleural glucose. Immunologic markers, such as rheumatoid factor and antinuclear antibody, are useful in the diagnosis of collagen diseases.

TECHNIQUES FOR OBTAINING PLEURAL BIOPSIES

Not infrequently a percutaneous pleural biopsy may reveal only muscle and some fibrous tissue. There may also be some crush artifact; a diagnosis of malignancy should not be made on such a biopsy.

After obtaining fluid for cytologic examination, it is usual to obtain a percutaneous pleural biopsy, often using a Abrams needle. The commonest cause for such a biopsy is an undiagnosed pleural effusion. A review of 14 papers from 1958 to 1985, where 2893 pleural biopsies had been studied histologically, gave a diagnostic yield of 57 percent with carcinoma involving the pleura and 75 percent with tuberculous pleurisy.[20]

Four or more biopsies at a single site may improve the diagnostic yield.[21] Canto and coworkers,[22] using thoracoscopy, showed a non-uniform location of pleural metastases. They suggested that their highest yield could be obtained from biopsies taken close to the diaphragm and midline. An increase in the number of biopsy sites at one sitting did not increase the diagnostic yield for either tuberculous disease or malignancy.[23]

Pleural fluid cytology may be more sensitive than biopsy for the diagnosis of malignancy; in one series, the yield with cytology approached 90 percent.[24] (See Chap. 36.)

It is rare to obtain a false-positive pleural biopsy. Confusion occurs when malignancy is diagnosed rather than a mesothelial proliferation occurring over a hemorrhagic pleural effusion or pulmonary infarction.

Thoracoscopy

This technique has much to offer in the diagnosis of disease based in the pleura. A rigid scope with a light source is placed into the thorax and allows a thorough exploration of the entire hemithorax as well as the mediastinum. The diagnostic yield is higher than with the Abrams needle; in a recent study of 412 thoracoscopies, 377 (92 percent) were positive.[20] This procedure can be done under local anesthetic, but it may require hospitalization.

Open Pleural Biopsy

This technique obviously gives the best yield but leaves the patient with a thoracotomy wound. By the time patients come to open pleural biopsy, they will have had a negative percutaneous biopsy and, less likely, a negative thoracoscopy. One of the main indications for open pleural biopsy is pursuit of a diagnosis of mesothelioma. This technique will obviously yield the largest amount of tissue. In our experience, thoracoscopy gives good tissue samples in this disease.

PLEURAL TRANSUDATES

1. *Congestive heart failure.* This is the commonest cause of a pleural transudate. Pleural effusions in heart failure may be unilateral, but at postmortem they are most commonly bilateral. If the effusions are unilateral, they are more prominently right-sided.[25] Pleural effusions in congestive heart failure are probably due to pulmonary venous hypertension. Patients with pulmonary emphysema and primary pulmonary hypertension, both having raised pulmonary arterial pressure but not usually pulmonary venous hypertension, do not have pleural effusions.[26]

In congestive heart failure, the pleural effusion is a transudate with a total protein of usually 3 g/dL. The majority of cells are mesothelial, with some lymphocytes but relatively few neutrophil polymorphs.

2. *Cirrhosis of the liver.* This occurs in approximately 6 percent of patients with cirrhosis and ascites.[27,28] The effusion is caused by ascitic fluid tracking through diaphragmatic defects or from the lymphatics draining from the abdomen through the diaphragm and into the pleural space. These small diaphragmatic defects are seen only when dyes or air-bubble techniques are used. Effusions may be located in the left pleural cavity or they may be bilateral.

The fluid is a transudate with a low leukocyte count, but occasionally it may be hemorrhagic if there is an underlying clotting defect.

3. *Peritoneal dialysis.* This may be associated with small bilateral pleural effusions but occasionally may cause large right-sided collections.[29] The effusion usually occurs within 30 h of the initiation of peritoneal dialysis, but it may be seen up to 2 years after beginning this procedure.[30]

The protein is usually less than 1 g/dL with a leukocyte count of less than 100/μL. In addition, the glucose concentration is 3 to 400 mL/dL, giving a clue to diagnosis.

4. *Urinothorax.* This is a pleural effusion secondary to an obstructive uropathy. It may follow or be associated with renal transplantation, carcinoma of the genitourinary system, renal stones, or trauma.[31,32] The pathway for the urine is via the retroperitoneal route, through diaphragmatic lymphatics. In addition, movement may take place through the diaphragmatic defects mentioned above.

The patients present with acute dyspnea and an effusion. If there is an obstructed kidney on one side, pleural effusion occurs on the same side.

The fluid is a transudate, which smells of urine. There is a low leukocyte count and low glucose, less than 65 mmol/dL in most cases.

5. *Nephrotic syndrome.* Pleural effusions are common in this syndrome, being found in just over one-fifth of cases.[33] The mechanism is a decreased oncotic pressure in the pleural microvasculature, caused by a low serum albumin as well as an increased hydrostatic pressure from salt and water retention. It is important to exclude renal vein thrombosis and pulmonary embolism.[1]

The fluid is a transudate with a small number of mononuclear cells, normal glucose concentration, and a pH greater than 7.40. The presence of blood and an increased protein concentration or polymorphonuclear cells suggests thromboembolism.

6. *Atelectasis.* This is a common cause of pleural effusion.[34] It follows upper abdominal surgery or obstruction of a large bronchus from any cause. The cause of the effusion is a decreased pleural pressure. When the lung collapses, there is a larger area separating the lung and the chest wall, creating an increase in negative pressure. The decreased pleural pressure helps to move fluid into the pleural space.

The fluid is a transudate with a small number of mononuclear cells and a normal glucose.

INFECTIVE PLEURAL EFFUSIONS

In the early stages of a bacterial pneumonia, there may be many neutrophil polymorphs, but the effu-sion can be sterile.[1] In a surgical specimen, after decortication of the lung, there is marked fibrosis, neovascularization, mesothelial cell proliferation, and fibrin on the surface. There is, depending on the stage of the disease, an acute or chronic inflammatory reaction, which goes through the pleural biopsy specimen (Fig. 34-1).

A misdiagnosis of mesothelioma may be made. The mesothelial cell proliferation, while not always on the surface, is never as florid as in an epithelial mesothelioma. There is never the hyperchromasia that may be seen in a fibroblastic or sarcomatous mesothelioma. The main differential diagnosis is a desmoplastic mesothelioma. This condition is described on p 1179, but in desmoplastic mesothelioma there is a storiform pattern, collagen necrosis, and extension of tumor around adipose tissue. These features are not identified in an inflammatory mesothelial process.

Empyema may be seen at postmortem. In the early stages, over consolidated lung, there are small flecks of pleural fibrin. With time, the fibrin increases and part or the entire pleural cavity may be obliterated by loose fibrin and fibrous tissue. This eventually forms dense pleural adhesions.

The organisms causing the pleurisy are as varied as those noted in the chapter on bacterial pneumonias (Chap. 7). They may be gram-positive or negative. Tuberculosis causes caseating granulomas. In tuberculous pleurisy, the fluid is an exudate with a total protein concentration usually greater than 5 g/dL. The leukocyte count is predominantly lymphocytic and above 5000/μL.

FIGURE 34-1
Resolving pleurisy with fibrosis, prominent capillaries, scattered chronic inflammation, and fibrin on the surface. (H&E stain, ×49.)

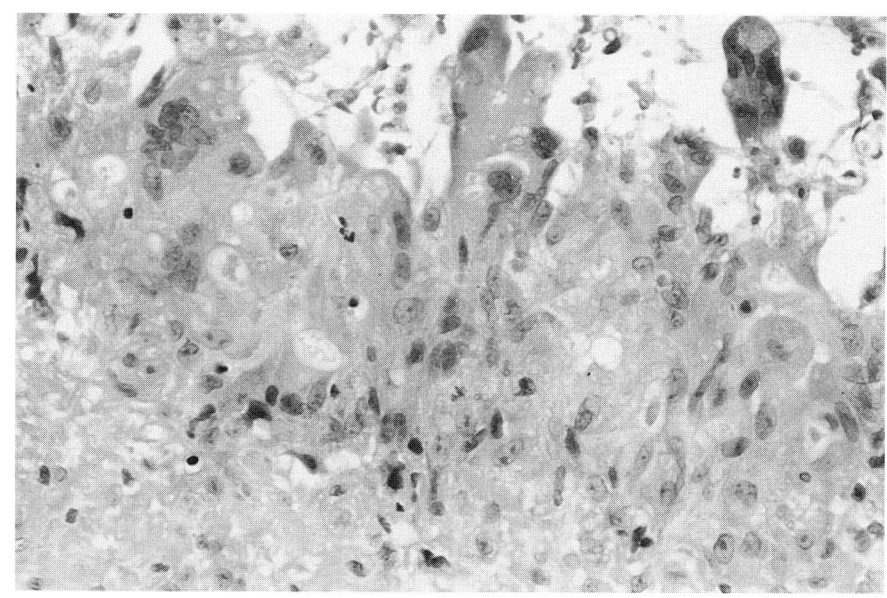

FIGURE 34-2
Giant-cell reaction in the pleura due to fungal infection. (H&E stain, ×219.)

A cause of confusion in tuberculous pleurisy may be a reactive eosinophilic pleuritis. This reaction is caused by air escaping into the pleura, causing the formation of *small* giant cells. These are not true caseous granulomata and there is no need, in the absence of caseation, to look for tubercle bacilli. If there are large confluent giant cells but no caseation, a diagnosis of sarcoidosis should be considered. A fungal pleuritis may also produce granulomata (Figs. 34-2 and 34-3). It is wise, when staining such pleural biopsies, to ask for a methenamine silver as well as a Ziehl-Neelsen stain. This is especially the case in an immunosuppressed patient.

Rarely, other organisms such as *Actinomyces, Pneumocystis carinii, Entamoeba histolytica,* and *Echinococcus granulosus* may involve the pleura. The pathology of these conditions is dealt with in the appropriate chapters.

Aspergillus empyema is not confined to the immunosuppressed. It may occur as a complication of tuberculosis, especially after pneumonectomy for this condition,[35] or after a spontaneous or in-

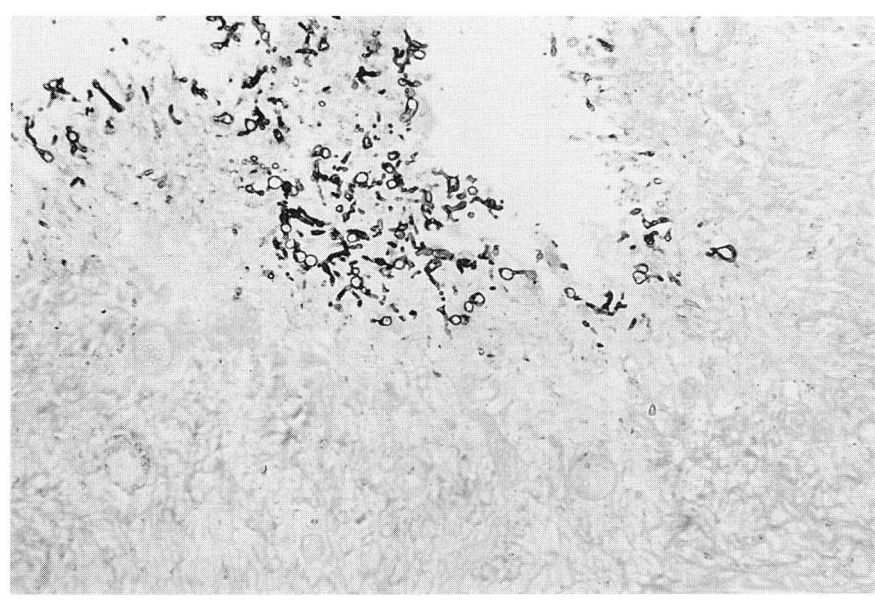

FIGURE 34-3
Same case as Fig. 34-2, with fungi staining prominently. (Methenamine silver stain, ×87.5.)

duced[36] pneumothorax. Rarely, cavitary pulmonary aspergillosis ruptures into the pleura to produce an *Aspergillus* empyema,[37] which has also been reported after pneumonectomy for bronchogenic carcinoma.[38]

SARCOIDOSIS

Pleural involvement is relatively uncommon in this condition and is seen in about 5 percent of patients.[16,39-43] Granulomata are more common than the presence of pleural effusions. In one study,[43] 200 patients with biopsy-proven sarcoidosis were studied. Ten percent had radiologic evidence of pleural thickening or effusion. In a large study and literature review[39] of the 73 cases reported since 1947, only 37 had been confirmed by pleural histology; 20 had an associated pleural effusion. It is likely that once a diagnosis of sarcoidosis has been made, the associated pleural lesions are accepted as part of the disease spectrum.

The pleural effusion is straw-colored or turbid. In 30 percent of effusions, the fluid may be a transudate, but the most characteristic finding is a lymphocytosis.

POST-CARDIAC-INJURY SYNDROME

This syndrome is characterized by fever, pleuropericarditis, and parenchymal infiltrates, which occur between 2 and 86 days following myocardial or pericardial inury.[44-47] The incidence is between 1 and 4 percent following myocardial infarction. It is seen with greater frequency following cardiac surgery, occurring in approximately 30 percent of patients.[48]

The etiology of this condition is obscure, but myocardial antibodies have been demonstrated in some patients.[49] Engle and coworkers[50] showed an association between high levels of antimyocardial antibodies and a rise in antiviral antibodies. These investigators thought that the syndrome was an autoimmune phenomenon, triggered by a viral illness. Histopathologically, there is usually a mild fibrinous pleurisy. A rare form that becomes chronic has been documented.[51]

YELLOW NAIL SYNDROME

This is seen in patients of around 40 years of age who present with yellow nails, lymphedema, and respiratory tract involvement. The latter consists of chronic pulmonary and sinus infection and pleural effusions.[52] The pleural effusion may be either bilateral or unilateral and can vary in size. The fluid is an exudate with a protein content often greater than 4 g/dL. The pleural fluid glucose is similar to the blood glucose. The pleural fluid pH is approximately 7.40.[1]

The syndrome may be due to impaired lymphatic drainage, based on lymphangiographic findings of hypoplastic or dilated lymphatics.[53] Electron microscopy confirms the presence of dilated but otherwise normal lymphatics. It has been suggested that there is obstruction of lymph flow, either in the major vessels or at lymph node level.[54]

COLLAGEN DISEASES

Systemic Lupus Erythematosus

Pleural involvement is common in systemic lupus erythematosus (SLE), being found in between 50 and 75 percent of patients.[55-57] There is pleuritic chest pain, which may be associated with pyrexia or pleural effusion. If there is an effusion, it is important to look for antinuclear antibodies and LE cells in the pleural fluid. The titer is usually 1 to 160 or greater.[58] There are an increased number of cells, which are either neutrophil polymorphs or mononuclear cells.

In autopsy studies, which will not necessarily reflect acute-stage disease, there is pleural fibrosis in one-third of cases. The more common presentation to the surgical pathologist is that of an acute fibrinous pleurisy with no evidence of underlying inflammation or definite vascular disease in 40 percent of cases.[59]

Rheumatoid Disease

Rheumatoid involvement of the pleura affects mainly males.[60] Patients, frequently in their 60s, develop pleural effusions often 5 years after the disease has begun. They may have pleuritic chest pain or dyspnea, but unlike the case in SLE, there is no fever. The effusions may be small or moderate in size and are often unilateral. They are exudative in nature with a protein up to 7.3 g/dL.[1] The cells vary and depend on the activity of the pleural disease. When it is acute, there are neutrophil polymorphs, but in the more chronic stage lymphocytes predominate.[61] The glucose is consistently low, as is the pH of the pleural fluid.

Though clinically rheumatoid pleurisy is found in approximately 5 percent of patients, its incidence at postmortem is much higher, being in the region of

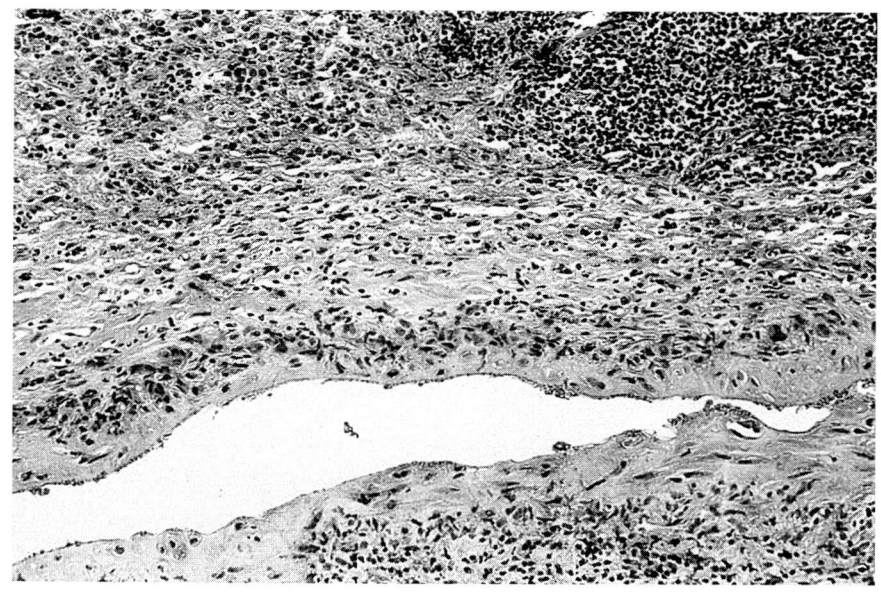

FIGURE 34-4
Rheumatoid pleurisy with prominent histiocytes along a free pleural border and marked chronic inflammation. (H&E stain, ×87.5.)

50 percent.[62–64] The reason for this number at necropsy is that pulmonary rheumatoid nodules may frequently be asymptomatic.

Histology can be extremely variable. There may be obliteration of the pleural space or patchy areas of visceral pleural fibrosis. Palisaded histiocytes and chronic inflammation are often seen (Fig. 34-4). Pleural nodules may be identified, with their characteristic features. A few cases may show prominent mesothelial proliferation, which may mimic mesothelioma. It is always prudent, before making this diagnosis, to enquire if the patient has a collagen disease.

A characteristic cytologic picture has been described.[65] There are large elongated cells as well as large multinucleated cells. There is a danger of diagnosing such cells as malignant.

WEGENER'S GRANULOMATOSIS

Chest pain may be a presenting feature in this disease.[66] In addition, there may be pleural effusion. Godman and Churg[67] showed that when the intrapulmonary lesions were close to the pleura, there was deposition of fibrin on the surface. Infarction, due to intravascular thrombi, has been suggested as a cause for these effusions.

AMYLOID

This is a rare cause of pleural disease. Most cases[68,69] present with dyspnea on exertion and pleural effusions, which are often right-sided. Macroscopically there is diffuse pleura inflammation of the panetal

pleura with light brown nodules measuring up to 5 mm in diameter. The visceral pleura is normal.[69a] The amyloid is often part of a systemic disease and pleural biopsy shows discrete deposits of amyloid; Congo red stain reveals apple-green birefringence. The congophilic deposits are also identified in blood vessel walls. The diagnosis is important, since pleural effusions and amyloid are usually attributed to cardiomyopathy secondary to amyloid infiltration.

Most reported cases of diffuse alveolar-septal amyloid do not describe pleural involvement.[70,71] It may involve the lung as multiple nodular pulmonary amyloidosis, amyloid affecting the tracheobronchial tree, amyloid involving the pulmonary vasculature, diffuse alveolar septal amyloid, or pleural amyloid. Heart failure due to amyloid is the commonest cause of pleural effusion in this condition.

PLEURA IN PNEUMOTHORAX (EOSINOPHILIC PLEURITIS)

Macroscopically, the pleura is thin. There may be focal thickening, but there is never the marked diffuse thickening associated with malignant mesothelioma or the fibrosis seen in a resolving empyema unless there has been associated infection. If underlying lung tissue is present, there is emphysema as well as small bullae.

Pneumothoraces may be associated with pneumatoceles in patients with AIDS, who usually have *Pneumocystis carinii* pneumonia.[71a,71b]

Histology shows varying degrees of mesothelial proliferation (Fig. 34-5), as well as variable numbers of eosinophils and chronic inflammatory cells. There is fibrin, and if the lesion is of long standing, there is

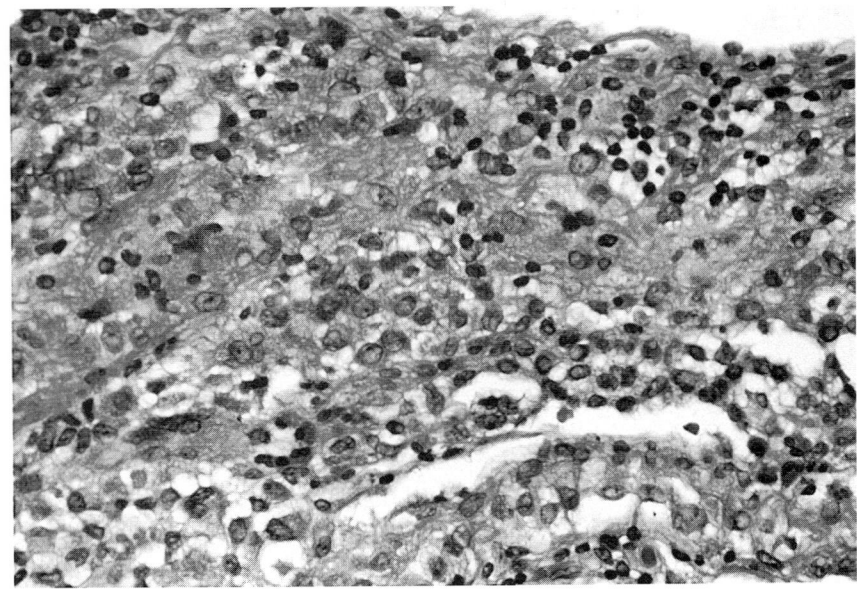

FIGURE 34-5
Pneumothorax. Proliferation of mesothelial cells along with chronic inflammation
in the pleura. (H&E stain, ×219.)

some pleural fibrosis. Multinucleated giant cells (Fig. 34-6) are present but these are usually small, relatively infrequent, and should not be confused with tuberculosis or rheumatoid disease, where the giant cells are larger. No caseation is identified. Calcification may be seen in AIDS cases.[71b] In the acute stages there are some neutrophil polymorphs. The mesothelial proliferation causes problems and, for the unwary, there is the potential misdiagnosis of malignant mesothelioma. The mesothelial cells tend to remain confined to the pleural surface, though occasional glandular spaces may be seen dipping down into the underlying tissue.[72] If materials such as talc have been previously instilled, it will be seen with a giant cell reaction around typical "Maltese

crosses" or birefringent material on microscopy (Fig. 34-7). In rare cases, cellophane may be identified in the pleural cavity if it was used to collapse lungs with tuberculous cavities (Fig. 34-8).

Differential Diagnosis

1. *Pulmonary eosinophilic granuloma.* Pneumothoraces may occur in this condition, but in practice there is established interstitial lung disease at the time of the pneumothorax. If pleural and lung biopsies are available there are, in active disease, large numbers of Langerhans' cells, which stain positively with S100 protein. On a pleural biopsy alone with mesothelial proliferation, some eosinophilia, and no

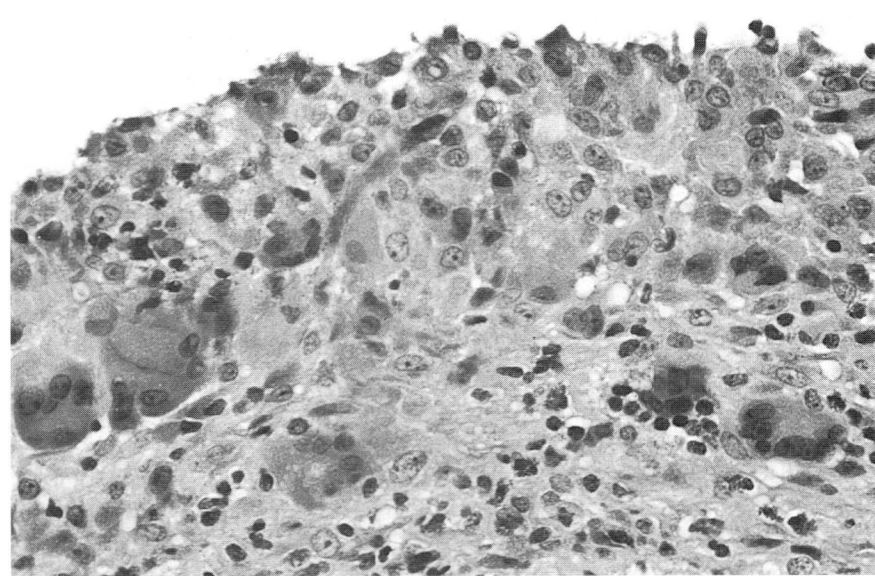

FIGURE 34-6
Reactive (eosinophilic) pleurisy with
many giant cells, mesothelial cells,
eosinophils, and lymphocytes. (H&E
stain, ×216.)

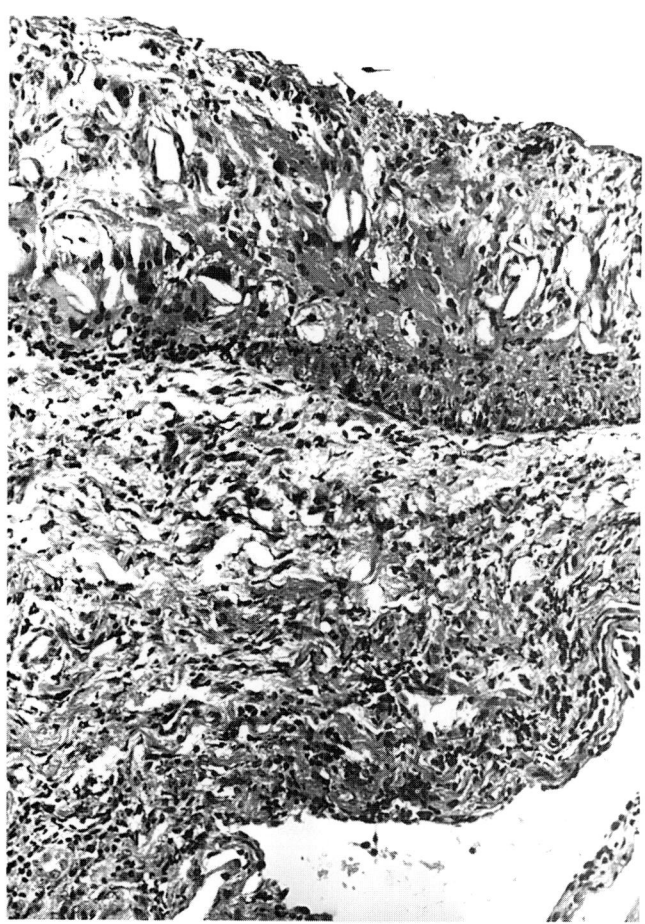

FIGURE 34-7
Foreign-body reaction due to talclike material inserted into the pleura to treat a pneumothorax. (H&E stain, ×92.5.)

FIGURE 34-8
Cellophane found in a pleural cavity, having been used to collapse an apical tuberculous cavity.

clinical history, it may be difficult to exclude a diagnosis of pulmonary eosinophilic granuloma.

A case report[73] describes a pneumothorax in a patient who had no initial pulmonary lesions but progressed to develop cystic lung disease. Biopsy during this first admission showed pleural blebs and subpleural bullae with reactive mesothelial cells and numerous eosinophils, mononuclear cells, and occasional multinucleated giant cells. There were, however, focal stellate and nodular interstitial aggregates as well as intraalveolar aggregates admixed with eosinophils and lymphocytes. Histiocytic cells in areas of eosinophilic granuloma were S100 protein–positive and lysosyme-negative. Mononuclear cells in areas of reactive eosinophilic pleuritis were S100–negative and lysosyme-positive.

Pulmonary disease predisposes to pneumothorax. A list of the causes is given in Table 34-1.

2. *The pleura in cystic fibrosis.* Patients with cystic fibrosis have small blebs or bullae. These may rupture and give rise to pneumothoraces. Most of the visceral pleura of a cystic fibrotic lung is structurally normal.[74] In a comparative study of cystic fibrosis cases and controls, the resected pleura from the former group showed vascular proliferation and myxoid change more commonly. Dunnill[75] described metaplastic parietal pleural columnar cells containing large cytoplasmic vacuoles. These showed dense peripheral staining with Alcian blue. A lightly stained brush border was also present. Subpleural air "cysts" in the lung in cystic fibrosis may be of three anatomic forms, bronchiectatic cysts, interstitial cysts, and emphysematous bullae.[76]

FOLDED-LUNG SYNDROME (SHRINKING PLEURITIS WITH ROUNDED ATELECTASIS)

The condition was first described by Blesovsky.[77] It presents in asymptomatic patients with extensive pleural disease. The clinical problem is that of radiologic, subpleural, round-lung densities. These are frequently accompanied by a cone-shaped density, described as a "comet tail." This latter component is thought to represent compressed vessels and bronchi entering the atelectatic zone. Such densities are seen more commonly in the posterior part of the lower lobes and most cases have been associated

TABLE 34-1
Underlying Lung Diseases Associated with Secondary Pneumothorax

Pneumonia

Staphylococcal septicemia

Pneumocystis carinii (AIDS)

Lung abscess

Tuberculosis

Coccidiomycosis

Hydatid disease

Chronic obstructive pulmonary disease

Asthma

Sarcoidosis

Berylliosis

Cancer of the lung

    Primary

    Secondary

Idiopathic pulmonary fibrosis

Cystic fibrosis

Scleroderma

Histiocytosis X

Lymphangioleiomyomatosis

Biliary cirrhosis

Marfan's syndrome

Ehlers-Danlos syndrome

Idiopathic pulmonary hemosiderosis

Pulmonary infarction

Rheumatoid disease

SOURCE: Anthonisen NR, Filuk RB: Pneumothorax, in Fishman AP (ed): *Pulmonary Diseases and Disorders*, 2d ed. New York: McGraw-Hill, 1988, p 217.

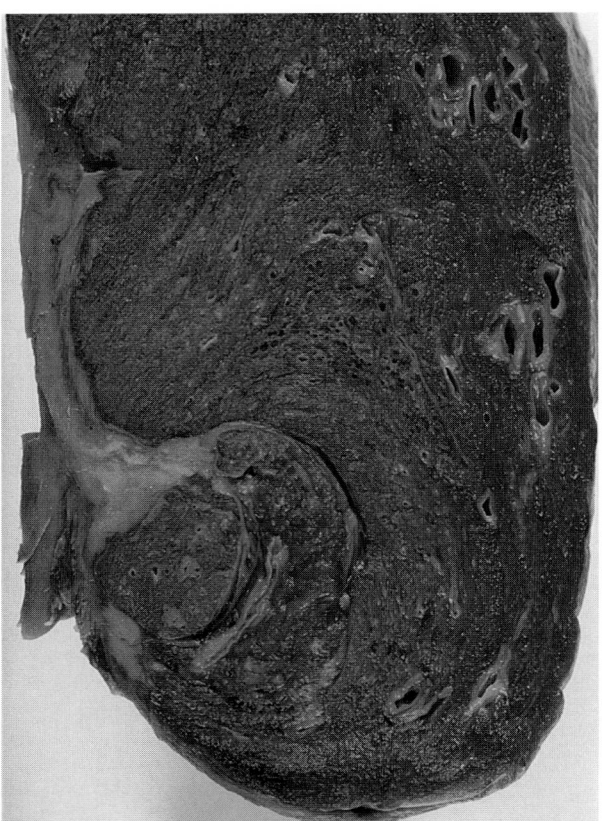

FIGURE 34-9
Rounded atelecasis with a pleural plaque and pleura which has become folded into the lung. The patient had been exposed to asbestos but there is no asbestosis.

with asbestos-induced pleural disease and asbestosis.[78,78a]

The pathogenesis has been suggested[79] as a pleural effusion causing infolding of part of the lung. This creates a cleft around the atelectatic zone. This is not the entire explanation, since pleural effusion is very common and rounded areas of atelectasis are unusual. More recently it has been suggested that the primary event is a pleuritis—the most superficial layer of the pleura thickens with progressive wrinkling of the underlying lung.[78a] In reality, the diagnosis can only be made by the exclusion of a malignant tumor. This may in some cases be difficult since the lesions have been known to increase in size.[80]

Pathology

It is unusual for the pathologist to have access to tissue from such a case. Mark[80] described the macro-

scopic pathology of a case. There were membranous adhesions and a firm, yellow, polygonal plaque, 3 cm in diameter, on the pleural surface of the lung. The plaque was retracted below the level of adjacent lung. Beneath the plaque the pleura was folded into the lung (Fig. 34-9). Cuts showed one fold perpendicular to the surface, then two folds and multiple more complex colinear folds. The lung adjacent to the pleural plaque and infolding was firm. Asbestos fibers are increased in both lavage fluid and lung tissue from these patients.[78a]

The visceral pleura dipped redundantly into the lung. The folded pleura had a normal complement of lymphatics, blood vessels and elastic fibers. There was only a minimal amount of fibrosis. The pleural plaque bridging the pleural folds was, for the most part, thick and hyalinized with no nuclei, vessels, or elastic fibers. The area of atelectasis showed parenchymal and visceral pleura twisted into a curl. There were large conducting airways and blood vessels lying abnormally close together in the artificial lobules of atelectatic lung. The presence of fat beneath the visceral pleura in the most severely atelectatic tissue indicated, to the author, atrophy of

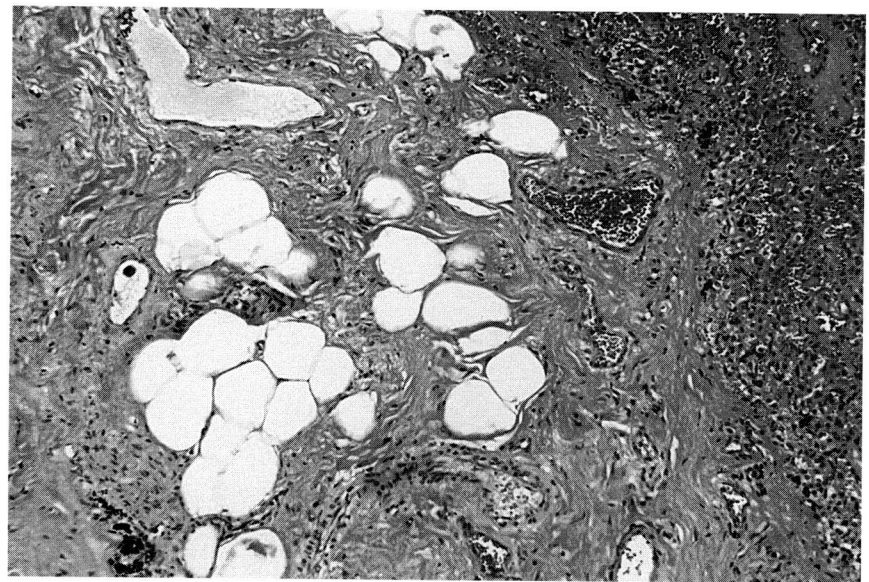

A

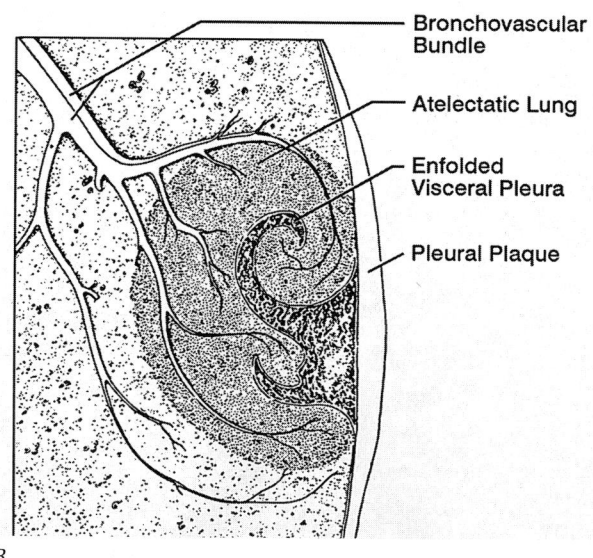

B

Bronchovascular
Bundle

Atelectatic Lung

Enfolded
Visceral Pleura

Pleural Plaque

FIGURE 34-10
A. Histology of folded lung with fat,
fibrous tissue, and collapsed lung on the
right of the picture. (H&E stain, ×56.)
B. Diagram of folded lung. (*From Mark,
with permission.*)

lung tissue (Fig. 34-10*A* and *B*). It is possible that this fat had originally been carried in by the infolding of the pleura.

MESOTHELIAL CELL INCLUSIONS IN MEDIASTINAL LYMPH NODES

This is not strictly a pleural disease. It may give rise to a misdiagnosis of metastatic carcinoma. Two cases have been reported.[81] The cells occurred in the sinuses and were predominantly single or in small clusters. The nuclei were those typically seen in benign mesothelial cells and were bland, with a nor-

mal nuclear/cytoplasmic ratio and well-defined cell borders. No mitoses were identified. Immunocyto-chemistry showed positive cytokeratin (AE1/3 and CAM 5.2) but no positivity for epithelial membrane antigen (EMA), Leu M1, or carcinoembryonic antigen.

The authors also concluded that the stronger immunoreactivity in a perinuclear location, with some fading at the cell periphery, was consistent with mesothelial cells. Ultrastructural analysis in both cases confirmed the presence of long microvilli consistent with mesothelial cell origin.

Both patients had a pleuritis, pleural effusion, and some mediastinal widening. In one case the exact cause of the benign pleural process was unknown. The second patient had a follicular lymphoma with pleural involvement. Extensive clinical workup of these patients did not reveal any primary carcinoma.

Mesothelial inclusions may also be seen in pelvic lymph nodes (Fig. 34-11).

PLEURAL PLAQUES

Clinically there are usually few features to denote patients with pleural plaques. There is however, an association with increasing duration of exposure to asbestos, increased stiffness of the lung, and decreased lung volumes.[82] While plaques are an indicator of asbestos exposure, they do not signify asbestosis. Some 67 percent of these men with confirmed exposure to asbestos were smokers or ex-smokers. The predominant finding in this study was a restrictive ventilatory defect. There was 15 to

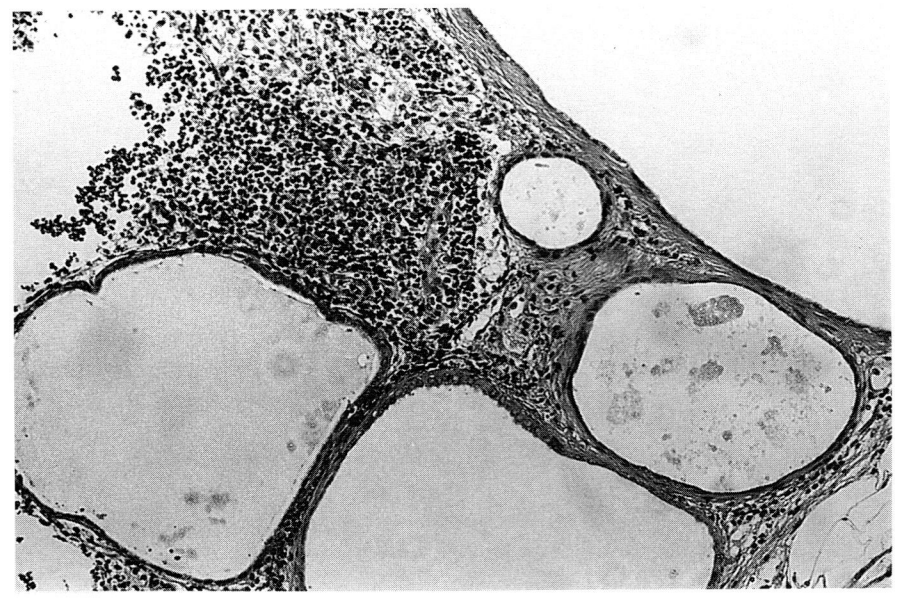

23 percent reduction in lung volumes and ventilatory capacity. The findings could not be explained by the presence of pleural plaques but rested upon the lung damage itself. The authors interpreted this association between extent of the pleural and pulmonary changes as evidence of slowly progressive disease in both the lungs and the pleura. The effect of smoking was not additive when statistics were used.

Plaques are seen on the parietal and diaphragmatic pleura. Occasionally they may be identified on the visceral pleura. Typically parietal pleural plaques are located on the posterolateral aspects of the lower part of the thorax and the upper part of the dome of the diaphragm. They lie parallel to the ribs (Fig. 34-12). Usually they are found in the apices of the lung or in the costophrenic angles. Plaques may be calcified and present as small, nodular, irregular, raised masses (Fig. 34-13). They may fuse and cover large areas of lung, in some cases resembling a mesothelioma (Fig. 34-14). Peritoneal plaques are rare.

Histologically, the plaques have a typical pattern with a "basketweave" of collagen (Fig. 34-15). This is accellular, but there may be occasional small foci of chronic inflammation. Any cells that are present have slender nuclei admixed with fibrosis. Fibrin may be present on the surface of the plaque. This is related to surgical removal. In other cases there may be a single lining of mesothelial cells. It is extremely rare to see asbestos bodies inside a pleural plaque. With time, they become heavily calcified. Noncalcified pleural plaques are more frequently detected at autopsy than radiographically.[83]

An autopsy study matched for age, race, and gender[84] showed significant correlations between the presence of pleural plaques and peribronchiolar

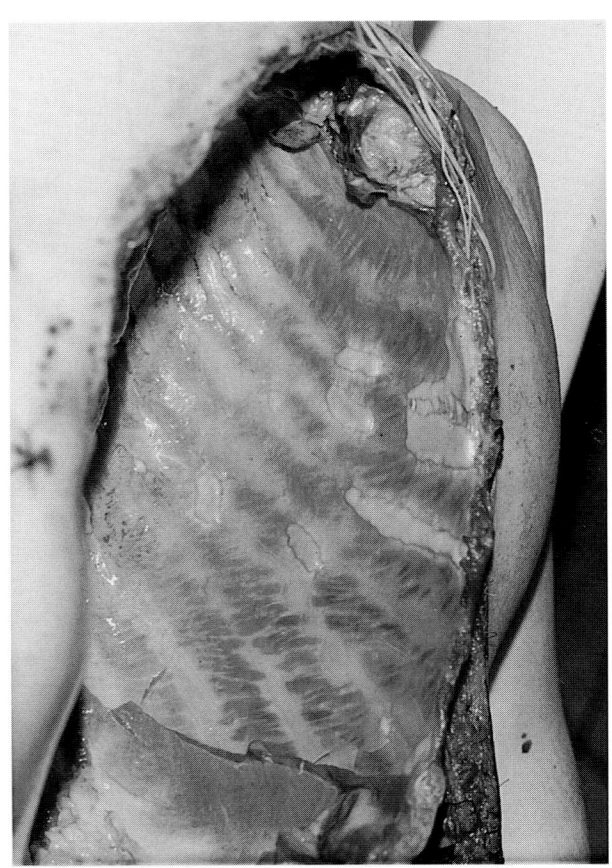

FIGURE 34-12
Parietal pleural plaques lying parallel to the ribs.

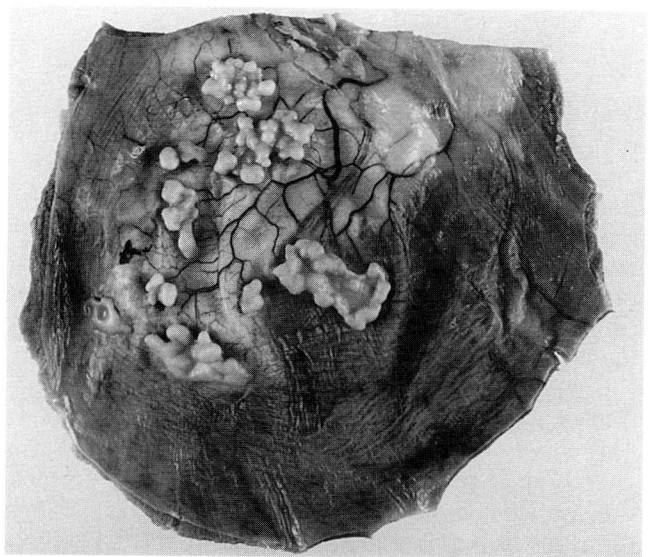

FIGURE 34-13
Raised nodular calcified diaphragmatic plaques.

FIGURE 34-14
Part of a large pleural plaque. The impression of the ribs is well seen.

fibrosis, alveolar fibrosis, large scars, scar-related emphysema, and pleural thickening. A history of smoking was also associated with pleural plaques. Interestingly interstitial fibrosis was not significantly different between the group with the pleural plaques and the control group. Peribronchiolar fibrosis, however, was seen in 49 (53 percent) of patients with and 36 (39 percent) without plaques. These data suggest that caution must be taken in extrapolating findings from the identification of pleural plaques to the diagnosis of asbestos-related pulmonary disease. Peribronchiolar fibrosis with associated asbestos

bodies was found in only 6 percent of patients with pleural plaques.

The levels of asbestos fibers in the lung have been documented using electron microscopy.[85-87] These authors showed a significant increase in the concentrations of commercial amphiboles (amosite and crocidolite) in the lungs of patients with plaques as compared to a control population. There were no significant differences for chrysolite or noncommercial amphiboles. Some 90 percent of commercial asbestos is chrysotile, but this is not a major fiber in lung tissue digests of patients with pleural plaques.

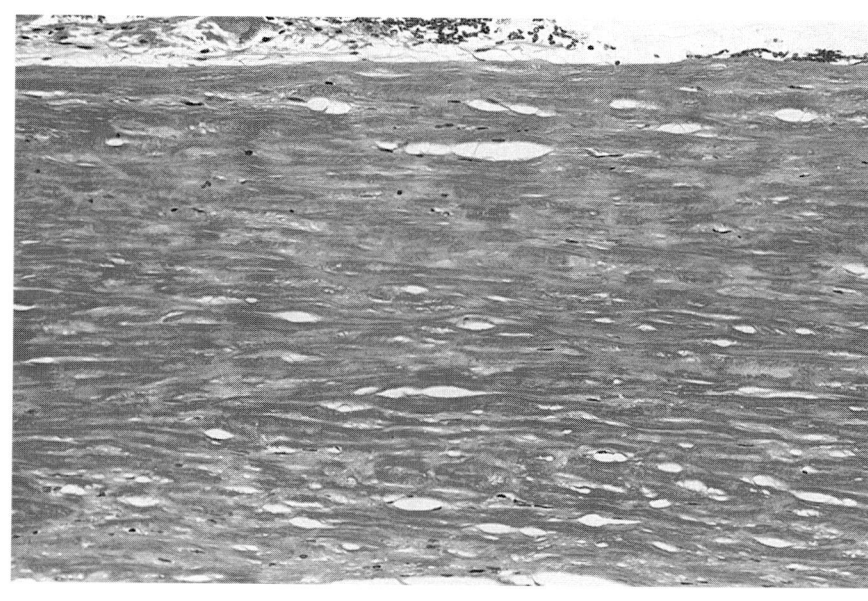

FIGURE 34-15
Fibrous pleural plaque with a basket-weave appearance. (H&E stain, ×87.5.)

Analysis of the plaques shows very occasional chrysotile fibers.

Other fibers have been implicated in plaque formation. These include mica; industrial talc, which contains asbestos fibers; fibrous diatomaceous earth; fibrous zeolite; other silicates; tremolite, anthophyllite; erionite; attapulgite; and wollastonite.[88,89] Other causes of plaque formation include old infection, especially tuberculosis; previous surgery; and hemopneumothorax. These plaques are usually unilateral, as opposed to the distribution of asbestos plaques.[90]

The interval between the initial exposure and the development of plaques is variable. A period of 10 to 20 years between exposure is commonly stated. The incidence in asbestos workers increases with age, so that more than 50 percent of a study group showed plaque formation radiologically 30 to 40 years after exposure.[91–93]

Plaques may also be found in families of asbestos workers and in people residing near asbestos mines and mills.[94] Four residents near a chrysotile mine had pleural plaques at postmortem, and though their lung chrysotile component was similar to that of a control group without plaques, there was a fourfold elevation in tremolite. One case had an elevated level of long titanium oxide fibers.[95] The source of the tremolite exposure was uncertain, but all four patients were engaged in occupations likely to generate dust from working with soil or rock (three farmers and one construction worker). None of the control cases had similar occupations.

Environmental exposure due to anthophyllite has been proposed as the cause of endemic pleural plaques in Finland,[96] though the more recent study from that country[87] showed that crocidolite/amosite fibers were more important. Anthophyllite has also been implicated in endemic pleural plaques in Bulgaria.[97] Erionite, a nonasbestos fiber, has been implicated as a cause of plaques in a small region of Turkey.[98] Environmental tremolite exposure has been suggested as the cause of plaques in Bulgaria,[97] Greece,[99] Austria,[100] and Turkey.[101]

Most of the environmental exposures where mineral sizing data are given show that the implicated fibers are relatively long, often exceeding 5 to 10 μm in length.[98,99,101]

The pathogenesis of parietal pleural plaques is at the moment poorly understood. As noted above, only less than half the patients with plaques have elevated asbestos body counts.[83] Plaques are a much more frequent radiologic and pathologic finding in patients exposed to asbestos dust than is interstitial fibrosis due to asbestos. The amount of asbestos exposure needed to cause plaques is significantly less than that required to produce pulmonary fibrosis.

Papers claiming an association between pleural plaques and carcinoma of the lung are in conflict.[102] Many papers have been published, of which only five have claimed a link between the risk of cancer and pleural plaques.[93,103–105] The remaining majority of reports have failed to show an association between the two conditions.[106–116]

The problem arises in some of the above studies that radiologic changes of pleural plaques are seen in only up to fifteen percent of patients, where they were detected at autopsy. Very few studies exist where there are adequately matched control subjects for asbestos exposure and smoking history.[114] There have been studies, quoted above, where plaque formation has been associated with smoking. There may be bias in smoking histories. The important link between the two conditions is asbestos. It is probably, therefore, best to regard pleural plaques as evidence of asbestos exposure.

Relatively few studies have looked at the risk of development of mesothelioma in patients with pleural plaques. A recent study by Hillerdal calculated the risk as 1 per 1700 per year.[103] Edge,[117] 18 years earlier, found a higher risk, 1 per 377 per year in patients with plaques. The risk is higher in this latter study, since these patients are likely to have been exposed to more amphiboles.

Plaques may also be identified on the pericardium.[118] Such patients may rarely have constrictive pericarditis.

Incidence of Pleural Plaques

Schwartz[119] collected a series of autopsy studies that had documented the presence of pleural plaques. In these 16 separate studies, which looked at 7085 routine autopsies, the prevalence of pleural plaques in the general population was 12.2 percent, with a range of 0.5 to 39.3 percent.

BENIGN ASBESTOS PLEURAL EFFUSIONS

This is a diagnosis by exclusion, since there should be a history of asbestos exposure and no other cause. Over 100 cases have been recorded in literature.[120] Epler and coworkers[121] studied 1135 asbestos-exposed workers and 707 controls radiologically. They found 35 idiopathic effusions among the asbestos workers but none in the control group. The increased frequency of asbestos exposure has recently been confirmed in men with idiopathic effusions.[122] These latter authors identified two additional features characteristic of asbestos pleural effusion: an initial chest radiograph showing converging pleural linear structures or rounded atelectasis and/or a history of recurrent pleural effusion. Patients may also have episodes of pleural pain.

The effusion may be unilateral, bilateral, or— more commonly— one side may follow the other.[123] Usually there are no symptoms, but if there is a large effusion there may be accompanying dyspnea.[124]

In rare cases, there may be an acute presentation, with fever, leukocytosis, and an elevated erythrocyte sedimentation rate (ESR).[125] In most cases, the clinical course is that of a benign self-limiting disease lasting 3 to 6 months. There is either complete resolution or—more commonly— the development of chronic pleural thickening (see below).

The fluid is an exudate, often blood-stained, with variable numbers of red blood corpuscles, macrophages, lymphocytes, and mesothelial cells.[126] Eosinophilia may be seen in up to one-quarter of patients.[127] There are, however, no diagnostic features in the pleural fluid.

The pathogenesis of benign asbestos pleural effusions is unknown. Asbestos has a direct cytotoxic effect on mesothelial cells, and these fibers reach the pleural space by means of macrophages and lymphatics. Once present in the pleural cavity in the rabbit, they stimulate chemotactic activity to form an exudative effusion. This causes an influx of neutrophil polymorphs and other inflammatory cells.[128]

DIFFUSE PLEURAL FIBROSIS

This condition has only recently become established as a specific entity related to this asbestos.[129–131]

Clinical features

Diffuse pleural fibrosis is usually found in men, often those who have had a substantial exposure to asbestos. In one series,[131] the seven patients had been exposed to asbestos for periods between 2 and 25 years. The age range was 58 to 80 years, with a mean age of 68.7 years. All the patients had chronic chest problems with dyspnea, wheeze, recurrent chest infections, and productive cough. Some patients had chronic pleuritic pain as well as intermittent pleural friction rubs.[132] Such pain is often seen in mesothelioma, but this diagnosis should be excluded by biopsy and long periods of follow-up. Of 7 patients where smoking history was documented, 6 were smokers.

Diffuse pleural fibrosis causes a greater reduction in FVC than circumscribed plaques.[133] It was significantly related to increasing age, the length of time after exposure to asbestos, and smoking. It was greater with larger amounts of radiological fibrosis, as judged by the International Labour Office (ILO) grade.

Macroscopic Findings

There was extensive visceral pleural fibrosis, most prominent at the bases but extending over the upper lobe (Fig. 34-16). The disease was bilateral and in many areas was greater than 5 mm in thickness. The fibrosis extended over half each lung and in two cases covered the entire lung. There were extensive adhesions between the visceral pleura and the chest wall. In 4 of 7 cases, discrete typical parietal pleural plaques were also identified. No effusions, empyema, or tumor was identified.

McLoud's study[130] was essentially radiologic. Of the 185 patients whose radiographs suggested diffuse pleural thickening, 29.7 percent had histologic confirmation. There were other associated diseases. These included 13 mesotheliomas and 8 bronchogenic carcinomas.

Histology

The features of diffuse pleural fibrosis are the same as those of pleural plaques, with a "basketweave" pat-

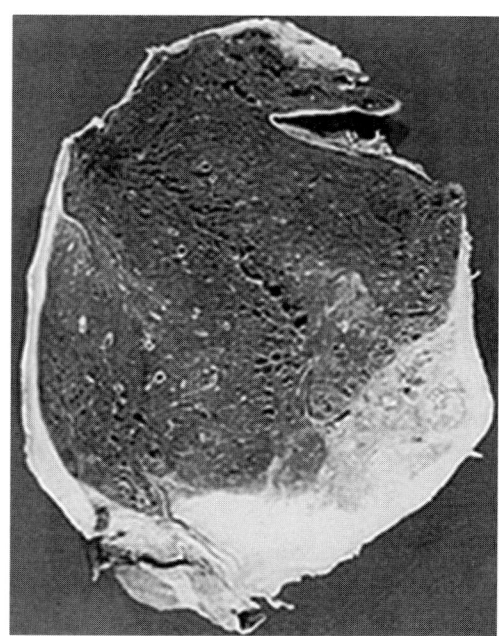

FIGURE 34-16
Diffuse pleural fibrosis. (*Courtesy of Dr. A. Gibbs, Penarth Hospital, Wales with permission.*) (*See also Color Plate 10.*)

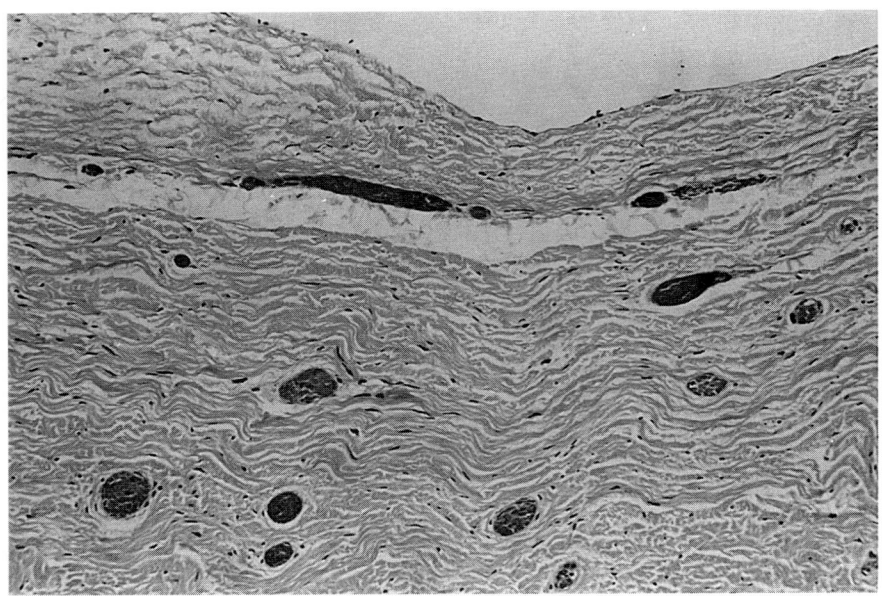

FIGURE 34-17
Histology of diffuse pleural fibrosis. (H&E stain, ×87.5.)

tern of fibrosis (Fig. 34-17). The external elastic lamina is preserved, along with submesothelial fat. The pleural space may be obliterated, as well as elastic laminae in some cases. No cellular fibroblastic proliferation or mesothelial hyperplasia was found. In all cases there was dense subpleural parenchymal interstitial fibrosis and fine honeycombing, which did not extend more than 1 cm into the underlying lung. The degree of interstitial fibrosis ranged from 1 to 3, using previously established criteria.[134]

Diffuse pleural fibrosis is dissimilar from pleural plaques in that the fibrosis is continuous and affects mainly the visceral as opposed to the discrete localized lesions confined to the parietal pleura in plaques. In addition the adhesions between the visceral and parietal pleura are common in diffuse fibrosis but relatively rare in plaque formation.

A small proportion of plaques may have visceral extension and coexist with diffuse pleural fibrosis, suggesting that borderline lesions may occur. Histology shows similarities between the two conditions. In diffuse pleural fibrosis, however, as mentioned above, there is fusion of the two pleural layers, with obliteration of the submesothelial elastic tissue and much of the fatty connective tissue. The findings suggest a previous severe inflammatory reation.

The total fiber count (electron microscopic) ranged from 9.2 to 83.5 × 10^6. There were high levels of amphiboles (crocidolite and amosite) in six cases, ranging from 2.5 to 28.08 × 10^6/gl dried lung.[131] In one case the amphibole count was within the normal range (less than 10^6/g of dried lung tissue). In non-occupationally exposed individuals, amphibole values were less than 2 × 10^6.[135] The significance of the control figures is discussed in the mesothelioma sec-

tion. Two more cases had high mullite counts, the significance of which was not clear.

The distribution of the asbestos fibers was documented in another paper from the Cardiff group.[136] There was wide case-to-case variation. There was no significant difference between samples taken from subpleural and central zones. Low asbestos counts were found in the pleura. More of the longer (>4 um) and thinner (<0.25 um) amphibole fibers were retained in the lung. Fibers greater than 8 um in length are thought to have an increased fibrogenic and oncogenic potential. Other authors have considered fibers greater than 5 um in length to have major pathogenic potential.[137]

Pathogenesis

Diffuse pleural fibrosis may be a sequel to benign recurrent pleural effusions associated with asbestos.[138] One-third of McLoud's cases[130] were associated with previous effusions. As in the study by Stephens and colleagues,[131] several cases showed subpleural honeycombing.

Other authors considered the fibrosis to be due to direct mesothelial damage, with subsequent effusion and fibrosis.[138] The presence of adhesions and obliteration in the external elastic laminae supports this contention but does not show whether these processes are due to recurrent acute episodes or chronic progressive change.

Immunologic abnormalities have been described in patients exposed to asbestos.[139,140] Chrysotile microfibrils migrate from alveoli to the interstitium and then, in some cases, to the pleural cavity.[141]

Asbestos fibers have occasionally been demonstrated in pleural effusions of asbestos workers.[142] Some authors have suggested that diffuse pleural thickening may be a direct extension of parenchymal fibrosis to the visceral pleura.[133] While this concept would explain the peripheral localization of asbestos fibers in the lung, it does not explain why diffuse pleural fibrosis is rare but asbestosis, in appropriately exposed individuals, is commoner. The role of long asbestos fibers in the causation of the disease has been noted above. The neutrophils and macrophages found in the pleural space may help to limit the amount of pleural fibrosis following asbestos exposure.[143]

Diffuse pleural fibrosis may be seen secondary to tuberculosis, pleurisy, empyema, rheumatoid disease,[144] or a traumatic hemothorax.[145]

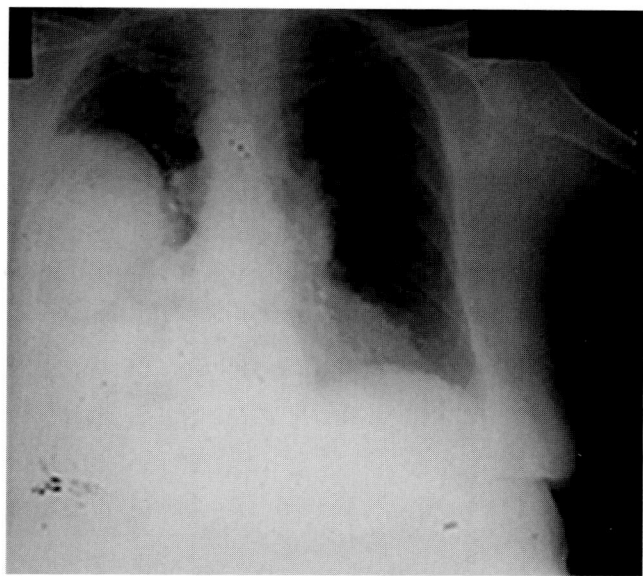

FIGURE 34-18
Chest radiograph showing a localized fibrous tumor.

LOCALIZED FIBROUS TUMOR

Over six hundred cases of localized fibrous tumors of the pleura have been described. These are still rare lesions, and most of the documented cases are described in four important papers.[146–149] The largest series came from the Armed Forces Institute of Pathology in Washington, D.C. This department described 223 cases of locaized fibrous tumor. Much of this material was submitted for review. Of those 223 cases, 42 percent came from five states: California, New York, Florida, Pennsylvania, and Illinois.

Clinical Features

The tumors, whether benign or malignant, were evenly distributed between the sexes, (73 women and 68 men). There was a wide age range, 9 to 86 years, with a median of 57 years. When seen, over half the patients were in their sixth or seventh decades. Over half the tumors had probably been present less than 2 years and three-quarters had been present less than 5.

Most cases are asymptomatic at the time of presentation, though some patients may complain of chest pain, cough, or dyspnea. Finger clubbing has been described.[150] Hypoglycemia was three times more frequent in females. In 10 of 12 cases with this biochemical abnormality, tumors were located in the right side of the chest. One patient in Briselli's series[146] had galactorrhea. Only one case had hyaline pleural plaques on chest radiographs. There is usually no history of occupational exposure to asbestos or of trauma.

Pleural effusion was commoner in the malignant group. Hemoptysis was limited to patients with malignant change. In this group, two of the three patients had tumors either in peripheral lung (an "inverted" tumor) or within a fissure.

Radiologic Features

Most of the tumors had sharply circumscribed borders; they were round or occasionally lobulated (Fig. 34-18). Most measured up to 10 cm in diameter, though in some cases they were greater than 15 cm, and the largest tumor weighed 3260 g. There may be pedicles attaching the tumor to the parietal pleura.

Macroscopic Pathology

Tumors were found throughout the thorax, including the mediastinum. Adhesions were present between visceral and parietal pleura on the surface of the tumor. In some cases the pericardium was also adherent. Most tumors were round, but some have been described as oblong, discoid, or trapezoid. Nearly half the tumors were attached to the pleura by a single pedicle, which contained prominent vessels. In the few cases where angiography has been done, the blood supply was from a branch of a phrenic or bronchial artery. Some tumors were broad-based (Fig. 34-19) with a "sessile" attachment to pleura or were found within fissures. Occasional tumors involved intercostal muscle, nerve, or soft tissue, and they eroded the outer cortex of the posterior ribs. In some cases the tumor was attached to the diaphragm.

On cut surface, the tumors were nodular, whorled, or lobulated (Fig. 34-20). They were firm, grayish/white, tan/yellow, or (uncommonly) brown. Occasionally small cysts were identified. These were

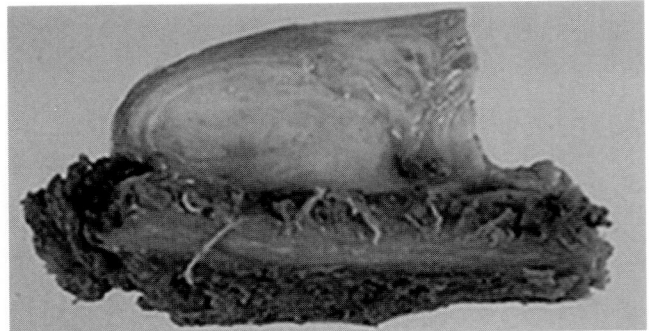

FIGURE 34-19
Broad–based benign pleural tumor attached to a rib. (*See also Color Plate 11.*)

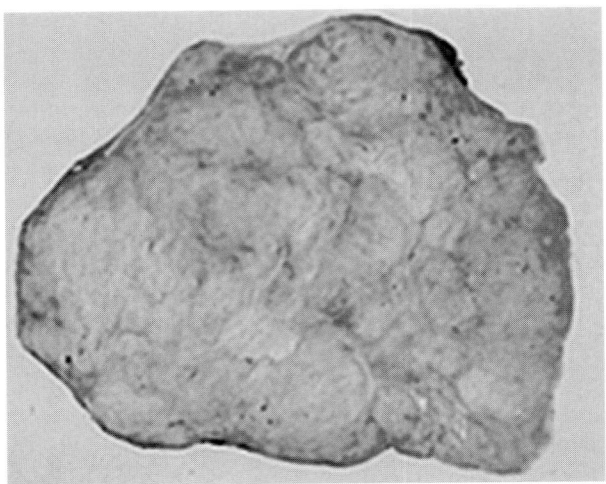

FIGURE 34-20
Cut surface of benign pleural tumor.

most common at the base of the tumor, near the pleural attachment.

Features distinguishing benign and malignant localized tumor are given in Table 34-2. Tumors greater than 10 cm in diameter were more likely to be malignant. The largest tumor in the series of England and colleagues,[148] was 39 cm in diameter. However, small size did not preclude malignancy, since over half the malignant tumors were less than 10 cm in diameter.

Histologic Features

England and coworkers[148] described three groups:

1. *"The patternless pattern."* This had a combination of fibroblast-like cells and connective tissue arranged in a disorderly or random pattern (Fig. 34-21). The fibroblasts showed no special features,

though at times the cytoplasm was vacuolated. Cellularity of the tumor varied from area to area. In some cases the collagen blended imperceptibly between tumor cells, creating a lacelike network. Calcification was seen in some cases, with dystrophic bone, cartilage, or rare psammoma bodies.

2. *Hemangiopericytoma-like tumor.* This was noted in a quarter of the cases and was usually combined with the patternless pattern. Closely packed tumor cells had amphophilic cytoplasm arranged around open or collapsed irregular branching capillaries and larger vessels (Fig. 34-22). Reticulin-stained fibers surrounded individual tumor cells and highlighted the blood vessels.

TABLE 34-2
Pathologic Features That Distinguish Benign and Malignant Localized Fibrous Tumor of the Pleura[a]

	Benign N=141		Malignant N=82	
	No.	%	No.	%
Gross				
Pedunculated	73	52	21	26
Atypical location[b]	67	48	55	67
Size(>10 cm)	34	24	45	55
Necrosis and hemorrhage	21	15	53	65
Microscopic				
Increased cellularity	18	13	62	76
Pleomorphism	14	10	69	84
Mitosis(> 4 mf/10 hpf)	2	1	63	77

[a]For all these values, the differences between the benign and malignant tumors are statistically significant by the chi-square test, $p<.05$).

[b]Tumor attached to parietal pleura, fissure, mediastinum, or inverted into peripheral lung.

SOURCE: From England et at.,[148] with permission.

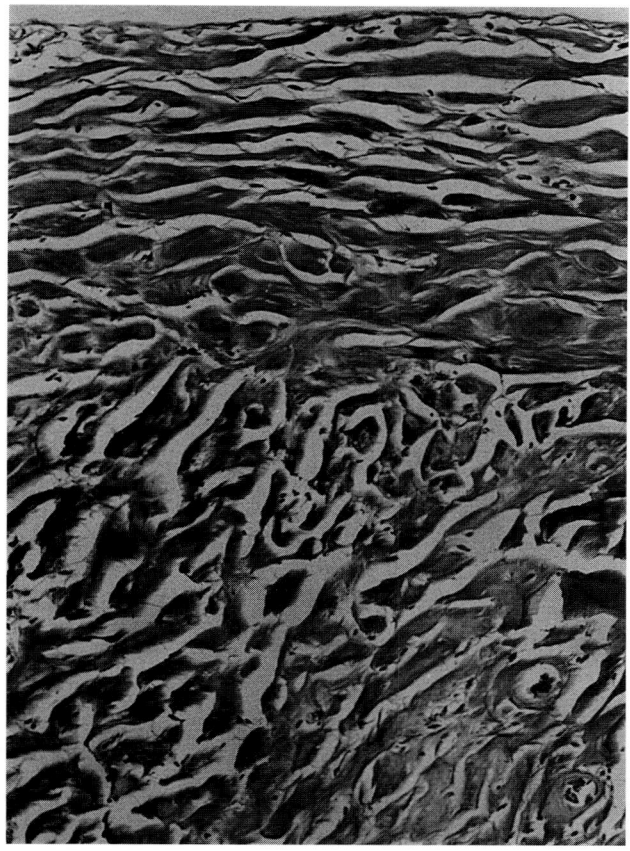

FIGURE 34-21
Localized fibrous tumor of pleura with dense, wirelike bands of collagen. (H&E stain, ×90.)

3. Less frequent patterns included a storiform (Fig. 34-23), herringbone, leiomyoma, or neurofibroma-like picture (Fig. 34-24). These histologic types were always seen in combination with the two most common types described above. Myxoid change may be seen (Fig. 34-25). Thick-walled blood vessels, usually seen in a neurofibroma, were not identified in solitary fibrous tumor. Chronic inflammation was identified in about one-fifth of the cases.

In the malignant variant, there were prominent nucleoli; giant cells were commoner, but these were occasionally identified in benign neoplasms. Two of the malignant tumors contained osteoclast-like giant cells.

Criteria for malignancy were a high cellularity with crowding and overlapping of nuclei, high mitotic activity with more than four mitotic figures per 10 high-power fields, and pleomorphism as based on nuclear size, irregularity, and nuclear prominence. Each tumor was graded by the most malignant area shown. Pleomorphic giant cells and abnormal mitotic activity were also indicative of malignancy.

Mitoses were few in the benign tumors, usually less than 3 per 10 high-power fields. There was a tendency for tumors with a high mitotic rate to show necrosis, hemorrhage, and myxomatous change.

Immunohistochemistry

Of 64 tumors studied by immunohistochemistry, 25 of 31 were positive for vimentin, 4 of 29 for desmin,

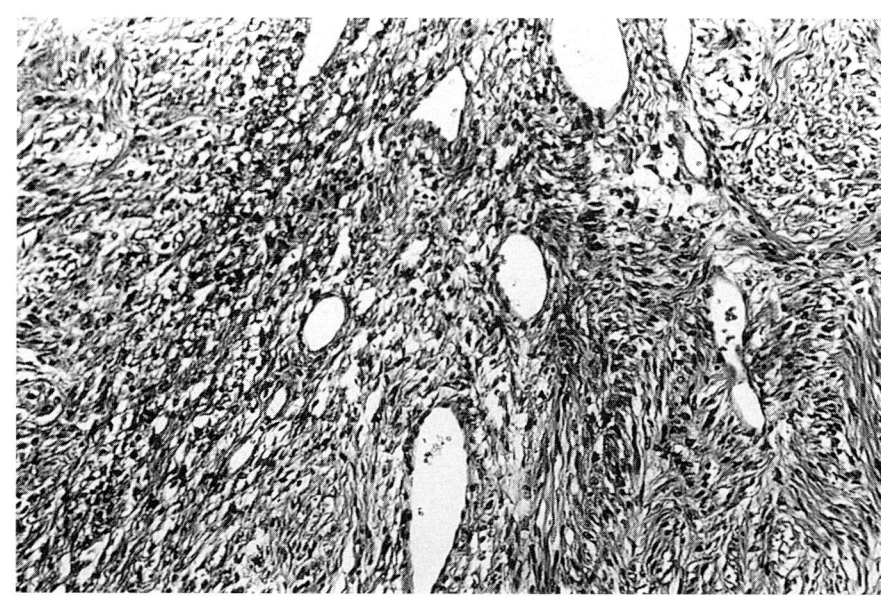

FIGURE 34-22
Localized fibrous tumor. Haemangiopericytomatous area. (H&E stain, ×87.5.)

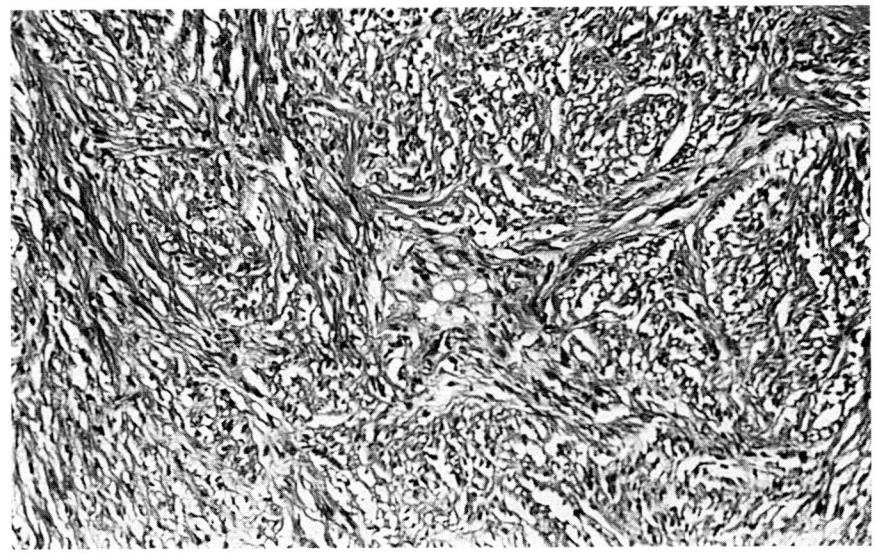

FIGURE 34-23
Localized fibrous tumor. Storiform area.
(H&E stain, ×87.5.)

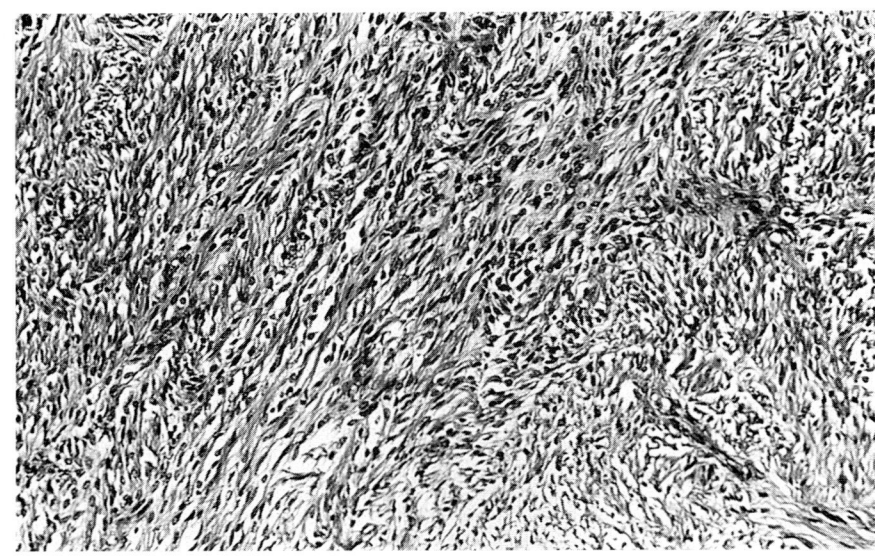

FIGURE 34-24
Localized fibrous tumor. Spindle-cell area.
(H&E stain, ×87.5.)

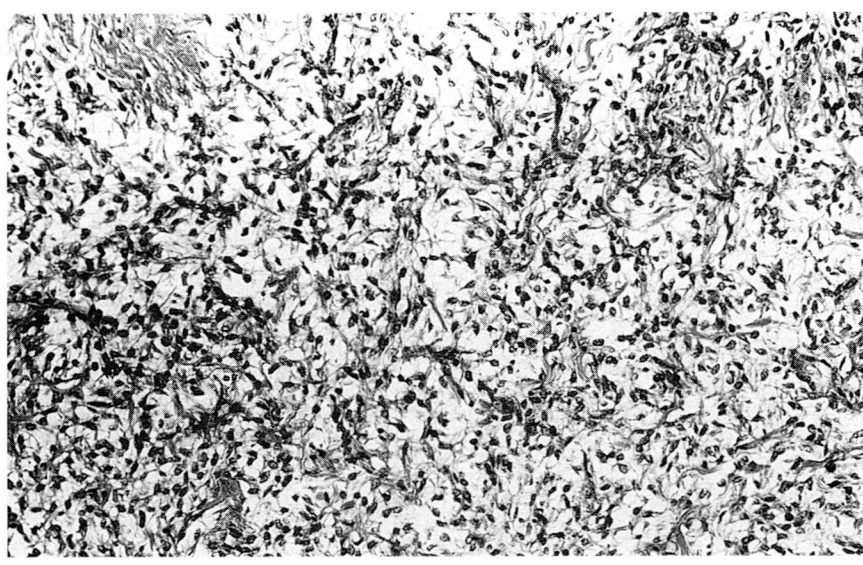

FIGURE 34-25
Localized fibrous tumor. Myxoid change.
(H&E stain, ×87.5.)

and 2 for alpha-1-antitrypsin. A further 2 tumors were positive for actin. Cytokeratins, epithelial membrane antigen (EMA), carcinoembryonic antigen (CEA), and neurofilaments were all negative. Malignant mesotheliomas are cytokeratin-positive.

Ultrastructural Changes

Electron microscopy this confirmed the presence of single fusiform to round tumor cells, interspersed between collagen bundles. The cells had scanty cytoplasm and indistinct cell borders. Although organelles were sparse, prominent rough endoplasmic reticulum was pres-ent. No brushlike microvilli were identified. In some cases, the cells were joined by primitive junctions of either the tight (zonula occludens) or intermediate (zonula adherens) type.

Follow-up of the group with benign tumors ranged from 1 to 317 months, with a mean of 57 months. All but two patients were free of disease. These two had a single recurrence, which required reexcision for cure.

Patients with malignant tumors had variable outcomes. Follow-up was obtained in 87 percent and ranged from 2 to 372 months (mean, 31 months). Of the total, 45 percent were cured. The other patients tended to have recurrence, which was multiple in some cases. Of the tumors that were refractory to treatment, 67 percent originated in the parietal pleura. Of 26 cases, 14 invaded the chest wall. Patients with no chest wall invasion but whose tumor had spread to intrathoracic organs or compressed vital structures in the mediastinum also had poor survival.

Some patients with localized malignant tumors had metastases. Tumor confined to the thorax spread to the ipsilateral or contralateral pleura and lung. Metastases to mediastinal nodes was seen in six patients. They extended like a malignant mesothelioma to the skin of the thoracotomy site in a small number of cases. Blood-borne spread was seen in the liver, central nervous system, spleen, peritoneum, adrenal gland, gastrointestinal tract, kidney, intraabdominal lymph nodes, and bone.

Six patients survived beyond the 5 year cure period, only to die of a later recurrence between 5.5 and 31 years. In general, this group had less aggressive disease initially and the tumors has fewer mitoses.

Differential Diagnosis

1. *Diffuse malignant mesothelioma.* If epithelial or biphasic, this tumor will cause little problem in diagnosis. However, a small biopsy from a fibrous malignant mesothelioma may be impossible to distinguish from a localized fibrous mesothelioma. In malignant mesothelioma, cytokeratin markers are usually positive, whereas in solitary fibrous tumors, they are negative.

2. *Fibrosarcoma.* This is a diagnosis of exclusion. With a pedunculated lesion, it would be an impossible differential diagnosis to make. In practice, fibrosarcomas are extremely rare in the pleural cavity.

3. *Leiomyoma or leiomyosarcoma.* These tumors have the typical cigar-shaped nuclei and a positive smooth muscle actin and desmin.[150a]

4. *Neurofibroma.* This tumor does not show either the typical pattern of a localized fibrous tumor or hemangiopericytoma. Immunocytochemistry may help in diagnosis, with a positive S100 protein.

5. *Hemangiopericytoma.* This can cause a problem in diagnosis. The reticulin pattern is the same in localized fibrous tumor of pleura as in hemangiopericytoma. However hemangipericytoma is a rare tumor, both in the lung and the pleura.

6. *Thymoma.* This tumor should have lymphoid and prominent epithelial components as well as a lobular growth pattern.

Cell of Origin

Localized fibrous tumors probably originate from submesothelial connective tissue cells. These may either be primitive fibroblasts or multipotential cells. This mesenchymal differentiation is supported by the finding of vimentin positivity in approximately 85 percent of tumors.

"LOCALIZED EPITHELIAL MESOTHELIOMA"

This presents as a papillary tumor, which is most probably a sclerosing hemangioma. Two cases have been described and termed mesothelioma.[151,152]

FIBROUS HISTIOCYTOMA

These lesions are described with benign lung tumors, but several cases arising in the pleural cavity, presenting as localized lesions, have been documented.[153–155] They are usually benign lesions, but the case described by Fuy and coworkers[153] was malignant.

TERATOMA OF THE PLEURA

One case is described.[156] This tumor filled the entire pleural cavity of a 17-year-old male. The tumor showed ectodermal components, including stratified squamous epithelium and cerebral tissue. In addition, mesodermal components—including blood vessels, cartilage, bone, and endodermal components such as bronchial and intestinal mucosa—were identified. Clearly these elements would not be confused with a mesothelioma. However, malignant mesothelioma is positive for alpha-fetoprotein, and this author has mistakenly called a mesothelioma a germ cell tumor on the basis of histology and positive beta-HCG and alpha-fetoprotein.

PLEURAL THYMOMAS

Rarely, thymomas may present as lesions based in the pleura.[157] One series contained five men and three women, aged between 38 and 71 years. Some patients were asymptomatic, others had respiratory difficulty, and further cases had nonspecific symptoms, such as fever and weight loss. There were either massive pleural effusions, pleural masses, plaquelike lesions of the pleura, or pleural thickening, with total encasement of the lung. In five cases the left pleural cavity was involved.

Macroscopic Features

In patients where surgical resection was performed, the tissue was described as multiple fragments of tan, soft tissue with a "fish-flesh" or brainlike appearance.

Histology

Histology showed, in most cases, the typical appearance of thymomas with a prominent lymphoid component and lobulation (Fig. 34-26). The lobules were separated by fibrosis. There was evidence of a biphasic cell population, with small round lymphocytes mixed with some epithelial cells. Two cases showed a predominance of the epithelial component, which exhibited nesting, spindling, and trabecular patterns. In some cases, pseudorosettes and irregular cystlike spaces were noted. In one case there was a hemangiopericytoma-like growth pattern.

Follow-up

One patient died in the postoperative period, one had an apparent inguinal secondary tumor, and two remained free of disease between 2 and 10 years after the initial diagnosis. There was no follow-up of the other 4 patients.

The tumors are probably ectopic thymomas[158] rather than arising from the mediastinal component, as has been the case in previous pleural based thymomas.[159–161]

Differential Diagnosis

1. *Pleural extension from a mediastinal thymoma.* This is the commonest presentation of thymoma, and the diagnosis may be given from either postmortem findings or computed tomography.

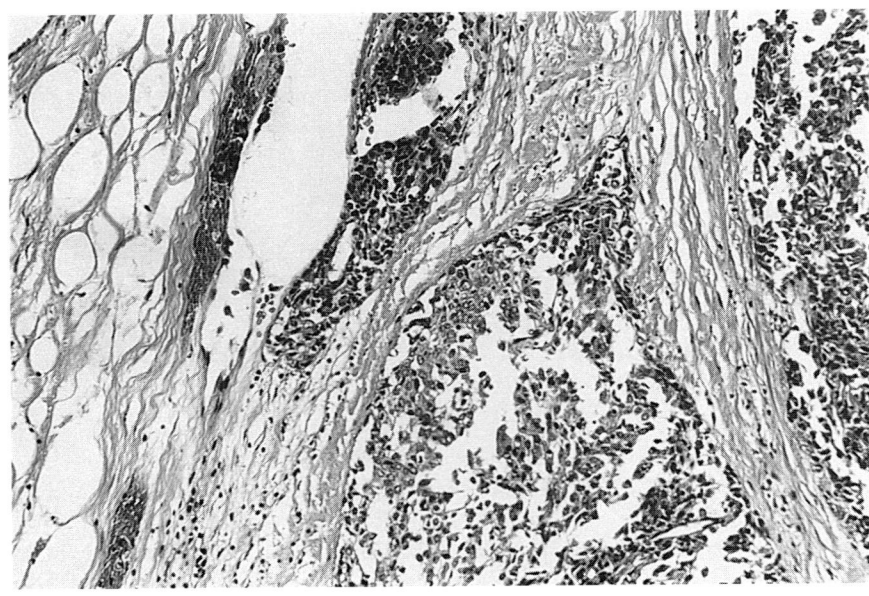

FIGURE 34-26
Pleural thymoma with a lobular growth pattern and dilated lymphatics. (H&E stain, ×87.5.)

2. *Malignant lymphoma.* It is possible to mistake a lymphocyte-predominant thymoma for a lymphoma. A cytokeratin immunostain, by demonstration of the epithelial components, helps in diagnosis. Mesothelial inclusions in mediastinal lymph nodes should also be considered.

3. *Epithelial mesothelioma.* This tumor very often has a tubular or papillary growth pattern. This picture is not identified in pleural lymphomas.

"AMYLOID TUMOR"

A rare case of an amyloid tumor in the left chest wall, associated with a plasmacytoma has been described.[156]

ENDOMETRIOSIS

This is associated with pelvic endometriosis, and the diagnosis may be difficult. Surgical resection may be necessary.[162,163]

LIPOMA

These tumors are rare in the pleura and mediastinum. They are usually discovered as asymptomatic masses on routine chest radiographs. They may be classified into tumors entirely within the pleural cavity and dumbell tumors, which show both intra- and extrathoracic growth.[164,165] Liposarcoma has been described as a primary pleural tumor.[166] These are extremely rare lesions; a literature review collected only eight cases.[166]

SMOOTH MUSCLE TUMORS[150a]

Moran and coworkers[150a] recently described five cases of smooth muscle tumors presenting as pleural neoplasms. There were three women and two men between the ages 21 and 69 years (mean = 45 years). Clinically one patient presented with chest pain, one with emphysema, and the other three were symptomatic.

The tumors varied from 10 to 18 cm in greatest dimension. In one case the tumor completely encased the lung resembling a mesothelioma. Three cases showed an atypical spindle cell proliferation with marked cellular pleomorphism, mitoses, and areas of hemorrhage and necrosis. The other two cases were characterized by a bland smooth muscle

proliferation with typical leiomyomatous features. Tests for smooth muscle actin and desmin were positive. One leiomyosarcoma was focally positive with a broad-spectrum cytokeratin marker.

Three of the patients had complete surgical excision and in the others it was incomplete. There was no recurrence in the first group or metastases in any of the patients 2 to 12 months after surgery.

The authors hypothesized that these tumors arose either from vascular smooth muscle from arteries supplying the pleura or from submesothelial stromal cells.

Sarcomatous mesotheliomas are also strongly positive for smooth muscle actin and muscle-specific action.[150b] The latter antibody was not studied by Moran et al.[150a] However, sarcomatous mesotheliomas also show cytokeratin positivity.

MYXOMA

This rare tumor has been described as a primary pleural lesion.[167] It consisted of spindle and stellate-shaped cells in a myxomatous stroma (Fig. 34-27). There were scattered blood vessels and scanty chronic inflammatory cells. No mitoses, pleomorphic cells or vascular invasion was present. Alcian blue was strongly positive. The tumor recurred 4 years later. On the contralateral side, there was a bronchial squamous cell carcinoma.

The differential diagnosis of myxoma includes a myxoid liposarcoma, myxoid chondrosarcoma, and myxoid malignant fibrous histiocytoma. The first and last diagnoses can be excluded on the basis of special stains. An extraskeletal myxoid chondrosarcoma (see below) usually occurs in the deep tissues of the extremities and has hyperchromatic nuclei and chondroblastic features.

EXTRASKELETAL MYXOID CHONDROSARCOMA

This tumor may rarely simulate a mesothelioma. A case presenting as a pleural-based tumor, encasing the right lung and extending into the major fissure, has been described.[168] The tumor was pale, gray, soft, and nodular, with a gelatinous texture.

Histology

The tumor showed the typical features of myxoid chondrosarcoma, with a nodular architecture. There were separate discrete nests of small oval to spindle-shaped cells arranged in anastomosing cords, sur-

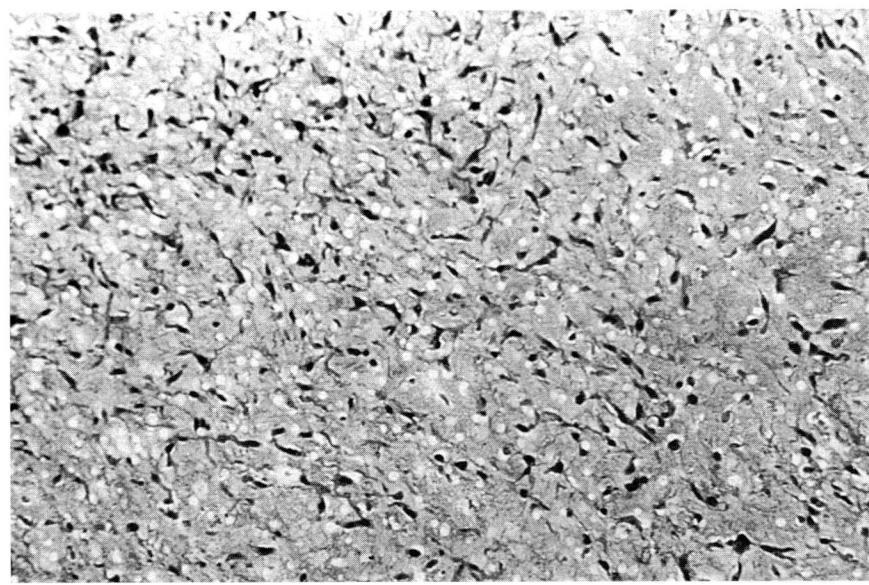

FIGURE 34-27
Pleural myxoma with thin stellate cells and much myxoid stroma. (H&E stain, ×87.5.)

rounded by an abundant myxoid matrix and a thin fibrous capsule. Many asbestos bodies were identified in lung adjacent to the tumor due to the previous asbestos exposure. This was thought to be a coincidental finding.

Immunohistochemistry

Glycogen was identified in tumor cells. The tumor was positive for vimentin. Tests for S100 and Leu 7 were negative. Chondrosarcomatous foci may be seen in malignant mesothelioma, but the cartilage forms only small areas amid the tumor cells.

OTHER MALIGNANT SARCOMAS

I have seen embryonal rhabdomyosarcoma and myxoid malignant fibrous histiocytoma presenting in the pleural cavity. Such cases are especially rare but may be misdiagnosed as malignant mesothelioma.

SQUAMOUS CARCINOMA OF THE PLEURA

This has been described in a 58-year-old patient with lingular bronchiectasis and a bronchopleural fistula.[169] The pleura was thickened and densely fibrotic, with foci of chronic inflammation. It was lined by hyperplastic and keratotic squamous epithelium with a prominent granular layer. In large areas, the squamous epithelium had undergone malignant change.

Squamous metaplasia of the pleura has been described.[170]

MALIGNANT MESOTHELIOMA

The key paper documenting the association between diffuse pleural mesothelioma and asbestos exposure was that of Wagner and coworkers.[171] The history of the discovery of the association between blue asbestos and mesotheliomas makes fascinating reading.[172] The association may now appear straightforward, but in the late 1950's, even the diagnosis of mesothelioma was queried. Willis,[173] one of the authorities on human tumors, gave such strict criteria for the identification of this tumor that it was virtually impossible to diagnose.

Asbestos transport in the airways is considered in Chap. 15.

Pleural mesothelioma is probably due to accumulation of fibers in the pleura. Morgan and colleagues[174] showed experimentally that the fibers travel though lung tissue to the pleura. Amphiboles migrated into the subserosa of rodent lungs over a period of months. Similarly, short fibers of chrysotile may be identified in visceral pleural tissue of occupationally exposed individuals.[175] The exact relevance of these observations to the pathogenesis of mesothelioma is questionable, since asbestos bodies also appear to be transported widely throughout the body via the lymphatics and bloodstream.[176]

Experimental Models of Mesothelioma Induction in Animals

Asbestos fibers migrate to the pleura after intratracheal injection in rats.[177] The number of chrysotile fibers able to reach the pleural cavity was greater than the number of crocidolite fibers. Most fibers recovered from the pleura were about 1 μm long.

Another experimental approach has been direct installation of the mineral into the pleural or peritoneal cavity.[178–180] In rats, the long, thin fibers, as compared with short fibers, cause a higher incidence of tumors.[181] Anthophyllite, fiberglass, and metal fibers are carcinogenic when introduced into the pleural cavities of animals. There is no evidence that the latter two materials cause human tumors. This suggests that in humans, transport of asbestos to the pleura is a critical factor in its pathogenicity.

Initially there is a foreign-body, giant-cell reaction; but, over a period of months, these lesions become increasingly cellular and ultimately malignant change occurs. The mesotheliomas produced show the same cytokeratin reactions as in human variants.

Asbestos-Induced Oncogenesis

A recent review[182] on the mechanism of asbestos carcinogenesis is summarized here. The amphibole hypothesis suggested that an important but not exclusive factor in asbestos carcinogenesis was the durability of fibers that lodged in the pleura or lung parenchyma over the long latency period of tumor development. Chrysotile fibers appeared to dissolve or fragment over time, whereas amphibole asbestos persisted at sites of tumor development. These fibers served as a chronic and necessary stimulus for neoplastic growth. The work of Brand and coworkers[183] looked at the mechanisms of "foreign body carcinogenesis," a phenomenon where plastic films or other foreign bodies caused sarcomas when implanted subcutaneously into rodents. In this model the tumors failed to develop if the implant was removed before a critical period of time.

It is now apparent that the cytotoxic, genotoxic, and proliferative effects of asbestos are mediated partly by active oxygen species. These are reactive metabolites of oxygen produced from phagocytic cells, particularly in response to long (greater than 5 μm) carcinogenic fibers or catalyzed by iron on the fibrous surface. Therefore the needle-like configuration and durability, as well as the increased iron content of crocidolite in comparison with amosite or chrysotile could govern its ability to produce chronic inflammation and proliferation.

The development of a sensitive human-hamster hybrid cell assay that detects both point mutations and large deletions that are often lethal to normal cells has facilitated the detection of large multilocus-type lesions caused by chrysotile and crocidolite asbestos.[184,185] These results are important, since they may explain the mechanism of asbestos-induced genotoxicity leading to cell death of human mesothelial cells—cells extremely sensitive to the cytotoxicity of asbestos by comparison with other cells of the respiratory tract.[186]

Recent studies[187–189] show the importance of active oxygen species in the carcinogenicity of asbestos. This hypothesis first stemmed from the finding that the cytotoxic activity of asbestos preparations containing long (greater than 10 μm) fibers could be blocked by antioxidant enzymes and other scavengers of active oxygen species. The cytotoxicity of shorter fibers and glass was unaffected.[190] Subsequent work has confirmed the importance of fiber length and geometry in the generation of active oxygen radicals by alveolar macrophages—longer, carcinogenic fibers (crocidolite, erionite) generate larger amounts of active oxygen species. Short fibers and particles are relatively inactive.[191] All types of asbestos may cause generation of some active oxygen species in cell-free systems by iron-catalyzed reactions on the fiber surface. This reveals a second mechanism of generation of oxygen by asbestos.[192,193]

Crocidolite is more catalytic than chrysotile on a surface-area basis.[186] The presence of iron seems to be a critical and necessary factor in asbestos-induced cytotoxicity,[194] lipid peroxidation,[195–197] and DNA breakage.[188,195,198] Chelation and subsequent mobilization of iron from asbestos enhance its biologic defects in vitro[187] and possibly, in the case of amphibole asbestos, occur over a protracted period of tumor development in vivo. The iron content of crocidolite (approximately 36 percent of weight) is substantially increased by comparison with that of chrysotile (only 2 to 3 percent in some chrysotile samples).[199] Therefore the decreased amounts and bioavailability of iron in chrysotile fibers may render them less biologically active over time.

Both chrysotile and amphibole asbestos cause chromosomal alterations in rodent and human mesothelial cells in culture.[186,200] Bronchial epithelial cells appear to be more resistant to the genotoxic and cytotoxic effects of asbestos.[186,201,202] Therefore asbestos appears to have more of a cocarcinogenic and promotor-like role in the development of lung cancer.[203]

Mossman's group[203] have concentrated of induction of oncogenes (c-*fos*, c-*jun*) implicated in the control of cell division in rodent mesothelial cell cultures, exposed to crocidolite or chrysotile. These genes encode protein complexes that interact with DNA sequences. These regulate gene transcription and are thought to participate in pathways that control entry of cells into the S phase of the cell cycle.[204]

The increases in c-*fos* and c-*jun* expression by asbestos in mesothelial cells are dose-dependent. This indicates a threshhold dose in that increases in protooncogenes are not seen at the lowest concentration of fibers tested. Changes are most pronounced with crocidolite asbestos and are not seen with non-asbestos fibers or chemically similar particles. More importantly induction of c-*fos* and c-*jun* by asbestos persists for many hours, unlike the rapid and transient increases with many classic tumor promotors.

The persistent protooncogene activation by asbestos illustrates one pathway of asbestos-induced cell proliferation. Other mechanisms of cell proliferation may be important. One example is that crocidolite and chrysotile stimulate the biosynthesis of polyamines, growth-regulatory nuclear molecules, that must accumulate in cells for cell division to occur.[205] This appears to be an oxidant-dependent phenomenon, as antioxidants block asbestos-mediated responses.[206] In addition, it is well documented that cytotoxic concentrations of amphibole asbestos cause cell death, which then stimulates proliferation of surviving cells (i.e., compensatory hyperplasia).[207] These cell responses are consistent with a chronically increased cell-division model that facilitates the accumulation of genetic errors during the process of tumorigenesis.[208,209]

The genetics of mesotheliomas is considered in Chapter 30. These changes are not consistent from case to case.[210] The commonest alterations affect chromosomes 1, 3p, and 9p. Homozygous deletions of the p16 gene have recently been described.[210a]

No mention has been made of the synergism between asbestos and other carcinogenic agents, such as cigarette smoke or fibrosis, as occurring in asbestosis. This is dealt with in Chap. 15. While pulmonary fibrosis and cigarette smoke are important causes of lung cancer, they play no part in the pathogenesis of mesothelioma.

Types of Asbestos and Other Fibers in the Development of Diffuse Mesothelioma.

This subject has been reviewed by Gibbs.[211]

CHRYSOTILE

Gibbs[211] takes the view that chrysotile on its own is an unusual cause of mesothelioma. Subsequent studies, showing an increased incidence of mesothelioma in patients exposed to this mineral, have identified that some of the chrysotile ore was contaminated with tremolite.[212–214] However, as mentioned below, the role of chrysotile in the development of mesothelioma is still open to question.

Peto and colleagues[215] identified 17 malignant mesotheliomas in 850 men who died more than 20 years after they were exposed to asbestos at a Rochdale (England) textile plant. These cases were initially attributed to chrysotile exposure, but it has become apparent that appreciable quantities of crocidolite were used at the factory.[216] Study of the lungs of these workers identified raised levels of crocidolite.[217] Women producing gas masks made of chrysotile did not develop mesotheliomas, whereas those working with crocidolite did.[218,219] Mesotheliomas seen in chrysotile miners may be produced by the tremolite contained in the chrysotile ore.

Churg[219] suggests that while chrysotile can induce mesothelioma in humans, the total number of cases is small and the doses required may be quite large. Berry[220] has taken a more extreme view and concluded there are only 10 acceptable examples of chrysotile-induced mesotheliomas in the world literature. He estimated that chrysotile alone produced a 0.3 percent incidence, amosite, 1.9 percent, and crocidolite 5.1 percent, a relative risk, at the most of about 15.

These data are indirect, and a role for the chrysotile fiber in causing mesothelioma still remains possible. This has recently been substantiated by an Australian group[221] who demonstrated that some cases of mesothelioma occurred in individuals who had been exposed only to chrysotile asbestos. They found a subgroup of cases where only chrysotile was identified in the lungs. A significant trend in the odds ratio was found with increasing fiber content. These authors noticed a significant dose-response effect for the shorter chrysotile fiber, independent of the long amosite and crocidolite fibers. It is difficult to compare the carcinogenicity of inhaled chrysotile with that of crocidolite, since white asbestos is cleared more quickly from the lungs.

In summary, it still remains difficult to identify the exact risk associated with chrysotile on its own. The prevailing, current view is that pure chrysotile does cause mesothelioma, but in relatively few cases.

CROCIDOLITE

The original association between crocidolite and malignant mesothelioma was noted by Wagner and coworkers.[171] The best data on this type of asbestos come from the mining town of Witenoom, Western Australia. Men who worked in the town between 1943 and 1966 were followed up to 1980. Of the 6506 men employed, 31 had developed mesothelioma.[222] Though this report shows that no deaths occurred in men who were employed for 90 days or less, Henderson and colleagues[3] reported a case of a man who was in the town for a period of only 1 week and subsequently developed a malignant mesothelioma.

Crocidolite is regarded as the most carcinogenic of the asbestos fibers.

AMOSITE

Less is known about amosite's potential to cause malignant mesothelioma than about that of chrysotile or crocidolite. This is because populations are rarely exclusively exposed to this mineral. In one series[223] only 2 of 232 cases of mesothelioma had exposure solely to amosite. One group of factory workers in America had exposure only to amosite.[224,225] Unfortunately these studies did not include lung fiber analyses; therefore it is not possible to ascertain if they were exposed to any other asbestos fibers.

Amosite is less carcinogenic than crocidolite, but exactly how carcinogenic is uncertain. Berry[220] has estimated that amosite produces 1.9 percent of mesotheliomas, as compared to 5.1 percent for crocidolite.

TREMOLITE/ACTINOLITE

Tremolite has assumed some importance since Baris and colleagues[226] described four patients with malignant pleural mesothelioma. These cases were from a small village in central Turkey. Tremolite was identified in the lung tissue of one patient, and it was thought that it had been inhaled from the stucco plaster on the walls of the houses. In this area there is also widespread use of erionite (p. 1158). Tremolite asbestos has little commercial value but often contaminates other minerals, some of which—such as chrysotile, vermiculite, and talc—are widely used industrially. Tremolite appears, dose for dose, more carcinogenic than chrysotile, but it is regarded by Churg[219] as a low-grade carcinogen. However, the long-fiber high-aspect-ratio (length to width) tremolite, such as that found in some vermiculite deposits or in areas of Greece, is a more potent inducer of the tumor.

The importance of tremolite has been underlined by McDonald and coworkers.[228] These authors felt that the tremolite contaminating many industrial minerals, including chrysotile, probably explained most of the cases in the Quebec mining region and perhaps up to 20 percent of cases elsewhere. Churg[219] has also indicated that tremolite alone has probably produced mesothelioma in humans. Tremolite has been implicated in malignant mesothelioma formation in other parts of the world, such as Corsica,[229] Cyprus,[230] and Greece.[99]

Tremolite may also be found with actinolite, but the exact role of this mineral in producing mesothelioma is at the moment unknown.

ANTHOPHYLLITE

This mineral has been used in Finland as a cheap filler and for insulation. A recent study of 19 cases of mesothelioma from Finland showed a predominance of anthophyllite in the lungs in 6 cases.[231] The drawback of this investigation was the use of scanning electron microscopy, which fails to detect crocidolite.

Fiber Concentration

ASBESTOS IN LUNG TISSUE IN PATIENTS WITH MALIGNANT MESOTHELIOMA

This is a difficult area, and the reader is referred to the monographs mentioned at the begining of this chapter. Results cannot easily be extrapolated from one laboratory to another. Many factors may affect fiber concentration (Table 34-3). Chrysotile fibers are physically cleared and dissolved more quickly from the lung than amphiboles. Chrysotile fibers are more difficult to visualize by light microscopy because of

TABLE 34-3
Factors Affecting Lung Asbestos Fiber Concentrations

Digestion procedures
    Chemical digestion
    Low-temperature plasma ashing
    Number of sites sampled
Recovery procedure
    Use of a centrifugation step
    Use of a sonification step
    Filtration (i.e., type of filter, pore size)
Analytical procedure
    Light (including phase contrast) or electron scanning, or transmission microscopy.
    Magnification used
    Sizes of fibers counted and other "counting rules"
    Numbers of fibers and/or fields counted
Reporting of results
    Asbestos bodies and/or fibers
    Sizes of fibers counted
    Concentration of fibers (per gram of dried lung tissue or per cubic centimeter)

SOURCE: Adapted from Roggli,[435] with permission.

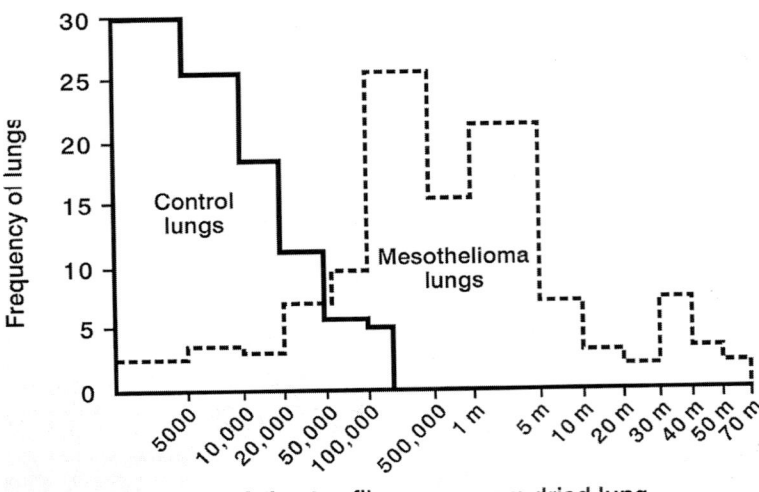

FIGURE 34-28
Histogram of fiber counts in mesothelioma and control cases. (*From Whitwell,*[241] *with permission.*)

their smaller diameters and lower relief.[232-234] Therefore a chrysotile fiber burden does not truly reflect past exposure to this mineral. In addition, the relative proportions of chrysotile and amphibole asbestos in lung tissue may not reflect their true proportions at the time of exposure. There may be variation in fiber counts between upper and lower lobes, between central and peripheral areas, and between samples separated by several centimeters.[235-238]

Values obtained by light and electron microscopy are difficult to compare. A comparison of the asbestos fiber count in mesothelioma and in normal controls is given in Fig. 34-28. Phase contrast microscopy cannot resolve fibers less than 0.2 μm in diameter. At least an order of magnitude more fibers can be identified by electron microscopy than by light.[238] Pooley and Ranson[239] showed that the light microscope was able to visualise only 5 percent of crocidolite, 26.5 percent of amosite, and 0.14 percent of chrysotile in lung tissue. Correlations of fibers visible by both light and electron microscopy was 0.79 for crocidolite, 0.74 for amosite, and 0.15 for chrysotile.

The asbestos fiber counts in reported cases of malignant mesothelioma are given in Table 34-4.[240-247]

There is difficulty in obtaining proper control samples. In some papers there are no full occupational histories. Many of these controls are based on hospital necropsies and therefore are a selected group. One study used coroners' cases, located in the region of a local shipyard. The coroner had jurisdiction over asbestos-related deaths, most of which were reported because of a history of lung cancer.[248]

Other Causes of Malignant Mesothelioma

1. *Mullite.* Mullite is found in on the island of Mull, Scotland. Otherwise it is seen only as a syn-thetic material and is commonly found in human lung.[249] Mullite may be formed when coals with a high clay content are burnt. Thus counts are found in lung tissue from workers in coal-fired electricity generating plants and occasionally workers from these plants have mesotheliomas. These cases had mullite as a major component of the dust in their lungs. Many of the fibers seen in these cases are short, but the possibility of selective retention of longer fibers is under study.

2. *Erionite.* This is part of the zeolite mineral group. It has been found in the Cappadocia region of Turkey, an area southeast of Ankara. Most of the cases of mesothelioma came from the village of Karain. As mentioned above,[226] the material was used by the villagers to stucco their houses and was also present in the local stones. Baris and coworkers[250] subsequently collected 126 cases of mesothelioma from this region in a 20 year period. They showed that the cases came from houses that contained loose volcanic tuff rocks containing erionite. This material was also used as a baby powder, as roof insulation, and to prevent water leaks.

3. *Synthetic mineral fibers.* It is unlikely these are more carcinogenic than asbestos. While there may be a possible increase in lung cancer due to these fibers, a recent editorial made no mention of any increase in malignant mesothelioma.[251]

4. *Familial mesothelioma.* Hammar and colleagues collected five reports of familial mesothelioma where the tumors occurred in two or more family members.[252] In the second family, the father had been occupationally exposed to asbestos and died 11 years before his son. All the cases, as well as a recent report,[253] had evidence of asbestos exposure in one member of the family group.

One interesting series is that of Risberg and colleagues[254] where the father, aged 61 at diagnosis, had

TABLE 34-4

Asbestos Content of Lung Tissue in Reported Series of Patients with Mesothelioma and Control Cases

Source	Number of Cases	Microscopic Method Used[a]	Asbestos Bodies per Gram of Dried Lung[b]	Uncoated Fibers per Gram of Dried Lung[b]
Mesothelioma cases				
Whitwell et al.[241]	100	PCLM	—	0.75 (0–70)
Gylseth et al.[242]	15	SEM	—	11.0 (2–490)
Mowe et al.[243]	14	SEM	—	2.4 (0.4–37)
Roggli et al.[4]	67	SEM[c]	13.9 (0.01–16,000)[d]	0.321[d] (0.012–93.1)
Roggli[244]	10	TEM	—	3.5 (0.1–85.2)
Churg and Wiggs[245]	10	TEM	—	3.5 (0.1–85.2)
Churg et al.[246]	6	TEM	—	238 (52–2190)
Guadichet et al.[247]	20	TEM[c]	3.2 (0.04–450)	18
Control Cases				
Whitwell et al.[241]	100	PCLM	—	0.007 (0–0.521)
Roggli[244]	19	TEM	0.020	—
Churg and Warnock[240]	20	TEM	0.28	—
Gaudichet et al.[247]	20	TEM[c]	0.18 (0–3.2)	—
Gibbs et al.[436]	55	TEM	0.3–83.1 × 10^6	—

[a]PCLM-phase-contrast light microscopy; SEM-scanning electron microscopy; TEM-transmission electron microscopy.

[b]Values reported are the median counts for thousands (10^3) of asbestos bodies or millions (10^6) of uncoated fibers per gram dried lung tissue. The study of Gaudichet et al. shows only the mean value for total fibers.

[c]In these two studies, fibers were counted by light microscopy.

[d]Values were multiplied by a factor of 10 (approx. ratio of wet to dry weight) for comparison.

SOURCE: Adapted from Roggli et al.,[4] with permission.

had no exposure, but two sons and one daughter developed the disease. Unfortunately no autopsy confirmation of the diagnosis is present in this series.

Despite the asbestos exposure in some of these cases, there is the suggestion that there are genetic factors involved in the etiology of some mesotheliomas. Blood typing, using to the ABO system, and HLA typing showed no definite linkage to mesothelioma.[255]

5. *Chronic inflammation and old scars.* Pleural scars, secondary to chronic empyema or a therapeutic pneumothorax, may cause malignant mesothelioma some decades later. Two cases of primary pleural tumors with such etiology have been described, and the literature review revealed 20 possible further cases.[256]

6. *Radiotherapy.* This factor has been incriminated as causing mesothelioma. One case[257] was described in a child who had received radiotherapy for a Wilm's tumor. This tumor was classified as an epithelial mesothelioma and therefore is unlikely to be confused a malignant fibrous histiocytoma, another tumor associated with radiotherapy. Two further mesotheliomas thought to be secondary to radiotherapy have been documented.[258,259]

7. *Herbicides.* These have been incriminated as initiating mesotheliomas. Donna[260] used atrazine, a triazinic compound, as a herbicide. Human mesothelial cells were then converted into a myogenic phenotype.[261] Donna suggested that the triazinic ring could play a role in the metabolism of DNA.

8. *Non-work-related mesotheliomas.* These account for between 10 and 20 percent of cases in the United Kingdom.[262] Cases have been documented in housewives, who have either dusted down or washed their husbands' clothing after a day's exposure to asbestos.[263,264] Clearly some of these cases may involve housewives with relatives who have worked with asbestos.

Some mesotheliomas appear to have no relation to asbestos.[265] Experimental evidence examined minerals, synthetic fibers, organic chemicals, and viruses. A case was quoted of a mesothelioma in a patient with a history of beryllium exposure who had worked as a metallurgist.[266] Two cases of mineral oil ingestion have been associated with mesotheliomas.[267,268]

Workers in the shoe industry,[269] petrochemical workers, engineering occupations without definite asbestos exposure,[270] stonecutters, and textile workers[271] have also developed mesotheliomas. Similarly, occupations involving contact with copper, nickel, fiberglass, glass dust[268] or rubber have been incriminated in the causation of the disease.[272] Cigarette smoking plays no part in the etiology or induction of malignant mesothelioma.[272–274]

Some of these studies were conducted in the "pre-immunoperoxidase" era, which may cast doubt on the diagnosis. In addition, for accurate diagnosis of occupationally induced mesothelioma, an asbestos fiber count is useful. However, the author has seen many cases of mesothelioma in the northwest of England. These had reliable occupational histories but low fiber counts. Despite these drawbacks, some mesotheliomas have proven to be non-asbestos-related.

OCCUPATION AND MESOTHELIOMA

A person's previous occupation may be of import in deciding asbestos exposure. Proportional mortality ratios are given in Table 34-5.[275] This list is not meant to be exhaustive, but recently there have been suggestions of some new risk groups.[276] These include brake mechanics, railroad workers, and personnel in the construction trades who work with asbestos-containing materials.

A study[277] looked at a group of patients who were in non-asbestos related occupations when they presented with their pleural tumors. All the other patients had had between 1 and 40 years of asbestos exposure. Also in this study were a group of 40 patients with mesothelioma but no asbestos-related occupation. There was no obvious reason for asbestos exposure in these patients.

EXPOSURE TO ASBESTOS IN BUILDINGS

This topic has been reviewed by Gaensler.[278] No cases of asbestos-related disease from simple school or building occupancy have been reported. A comparison between long-term residents of asbestos-containing buildings and those free from this mineral at a French University did not show any radiographic differences.[279] The workmen who had occupational exposure had the same risks as persons in similar trades elsewhere. People with only occasional exposure to asbestos, such as those engaged in minor repairs and cleaning in the boiler room, probably have no increased risk. Any pleural disease is related to their previous occupation.

The main concern has been regarding mesothelioma, and that it may arise following lower exposure from the buildings. This worry has been heightened by finding friable asbestos in 20 percent of public buildings. Gaensler[278] concluded that there was no epidemiologic or clinical support for asbestos-related disease from building occupancy. He noted, however, that only a few epidemiologic studies have contained quantitative estimates of exposure.

Epidemiology

INCIDENCE

World-wide statistics for the mortality of mesothelioma probably underestimate its incidence. This is because cases may well be described as adenocarcinoma or other malignant tumors involving the pleura and peritoneum. Unless immunohistochemistry and, in some cases, electron microscopy are available, the diagnosis will not be made. Assessment via cancer

TABLE 34-5

Proportional Mortality Ratios (PMR) of Men Aged 16 to 74 from
Mesothelioma in England and Wales, 1979–1980, 1982–1990[a]

Job	PMR, All Men= 100	No.	%	Cumulative, %
Metal-plate workers	700.4[b]	110	2.5	2.5
Vehicle-body builders	618.7[b]	35	0.8	3.2
Plumbers and gas fitters	442.8[b]	201	4.5	7.7
Carpenters	365.7[b]	258	5.7	13.5
Electricians	290.5[b]	161	3.6	17.0
Upholsterers	283.3[c]	19	0.4	17.5
Construction workers nec	255.6[b]	187	4.2	21.6
Boiler operators	253.9[b]	39	0.9	22.5
Electrical plant operators	253.5[b]	18	0.4	22.9
Chemical engineers and scientists	248.4[b]	18	0.4	22.3
Sheet-metal workers	233.2[b]	48	1.1	24.4
Scaffolders	225.6[d]	11	0.2	24.6
Production fitters	216.3[b]	304	6.8	31.4
Professional engineers nec	210.6[b]	105	2.3	33.7
Plasterers	202.8[b]	27	0.6	34.3
Welders	202.6[b]	70	1.6	35.9
Managers in construction	196.8[b]	40	0.9	36.8
Dockers and goods porters	195.1[b]	69	1.5	38.3
Electrical engineers	187.0[b]	39	0.9	39.2
Technicians nec	171.9[d]	24	0.5	39.7
Buildings and handymen	164.4[b]	98	2.2	41.9
Laboratory technicians	164.2[d]	27	0.6	42.5
Draughtsmen	160.6[d]	28	0.6	43.1
Machine tool operators	133.0[b]	179	4.0	47.1
Painters and decorators	131.0[d]	100	2.2	49.4

[a]The highest 25 occupational PMRs are based on 10 or more deaths.
[b]$p < .001$
[c]$p < .01$
[d]$p < .05$
SOURCE: Reproduced by kind permission of Professor J Peto and the *Lancet*.[275]

registry or histopathologic surveys may overestimate the true rate.[280]

The incidence is rising in the West[275,281] (Fig. 34-29) as well as in Australia. In the latter country, which has the highest incidence of mesothelioma in the world, annual notifications have almost tripled in the past decade, to 314 in 1990. It is calculated that a further 6000 cases will develop in the next 20 years. Thus the incidence will rise until the year 2010. The peak in Great Britain is predicted at 2700 deaths per

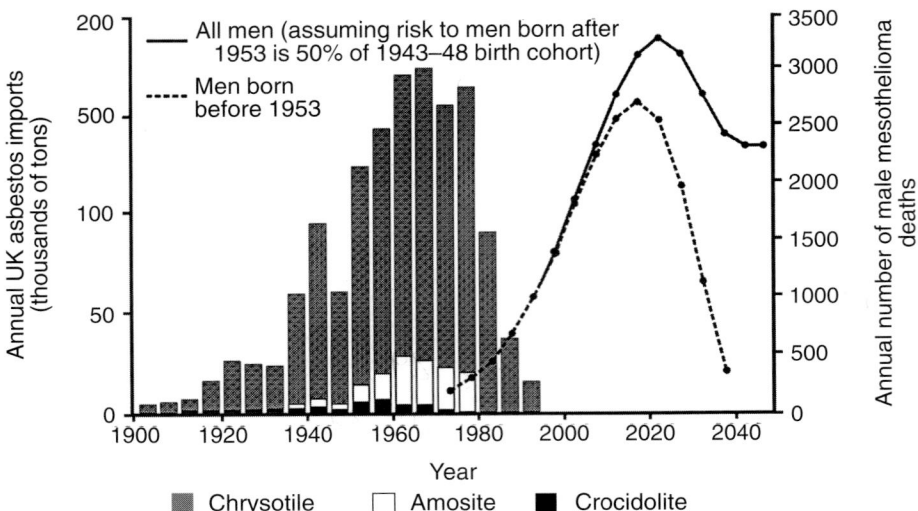

FIGURE 34-30
Mortality rates from mesothelioma as shown by counties in England, 1976–83. Note the high incidence close to dockyards. Deaths per million: □, 0–9.9; ▨, 10–19.9; ■, 20+. (*Courtesy of the Lung and Asthma Information Agency, St. George's Hospital Medcial School, London with permission.*)

annum and will tail off rapidly after 2020. There are, however, no data available for men born after 1958, so their risk is unknown. The mortality rates vary in different parts of the country (Fig. 34-30). "Excess cases" are seen in a few parts of the country, mainly in areas close to shipyards or areas where asbestos was used in manufacturing. Mesothelioma deaths by age and sex are shown in Fig. 34-31. There is a higher incidence in men than in women. The majority of deaths were seen in patients over the age of 45.

In the Netherlands the highest incidence was in men aged over 65, where the death rate was 10 to 15 per million.[281] These authors did not expect a decline in incidence until 2010. There has been a significant increase in mesothelioma in the United States,[282] where annual increases were around 12 percent. These rates are lower than those found in areas where there is naturally occurring erionite, such as Turkey. The incidence there rises to 216/100,000.

The highest rates are in countries where crocidolite is or was produced (i.e. Western Australia and South Africa). The highest rate among women is in Denmark; the cause for this is unknown. It is possible that the production at home of asbestos cement filler, which was a widespread practice, may play a causative role.[253]

The incidence in populations unexposed to asbestos is extremely low, being 1 to 2 per million.[283–285]

Clinical Features

PLEURAL MESOTHELIOMA

Cases have been reported in childhood. A review revealed a total of such 80 cases in the world literature. There were, however, only 22 cases with adequate material for histologic review—14 pleural, 7

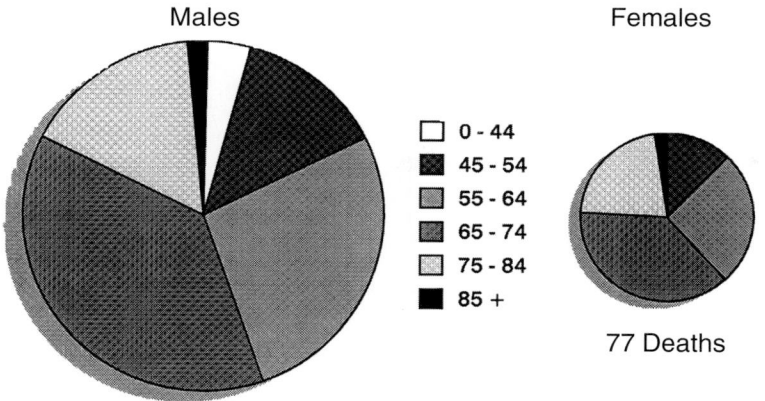

Males   Females

☐ 0 - 44
◼ 45 - 54
▨ 55 - 64
▨ 65 - 74
▨ 75 - 84
◼ 85 +

77 Deaths

428 Deaths   Source OPCS

FIGURE 34-31
Mesothelioma deaths by age and sex. England and Wales 1990. (*Courtesy of the Lung and Asthma Information Agency, St. George's Hospital Medical School, London, with permission.*)

peritoneal, and 1 pericardial. A panel of pathologists accepted only 10 cases. The ages of the patients with pleural mesotheliomas ranged from 2 to 12 years. A case has been described in a 19-month-old female.[286] Unusually, this tumor showed moderate intensity in some areas for CEA; ultrastructurally, there were a few microvilli on parts of the surface. There was no staining for keratin and no other cytochemical markers were used. Despite this, the histologic appearance of the light microscopic hematoxylin and eosin (H & E) section appears very convincing for a mesothelioma.

The disease does not usually present before the age of 40. There is a male predominance, and the presenting complaint is usually dyspnea, due to a pleural effusion. There may be chest pain, weight loss, cough, and fever. There are no characteristic clinical features, and any patient with pleural signs should be considered potentially at risk. Some patients may present with chronic "pleurisy," a purulent pleurisy, or a spontaneous pneumothorax.[287] Localized pain, which is often moderate, may well be the first sign. Review of 92 cases from the Mayo Clinic[288] noted, in addition to the above clinical features, chills or sweats in 33 percent of cases; weakness, fatigue, or malaise also in 33 percent; anorexia in 11 percent; and a sensation of heaviness or fullness in the chest in 7 percent. Other less common features were hoarseness, myalgias, and other symptoms including aphonia, dysphagia, abdominal distension, a sensation of pressure in the right upper quadrant, nausea, bad taste in the mouth, a perceived tachycardia, and headache. Increased sputum production was noted in almost 18 percent of the cases of Selikoff and coworkers[289]

The physical signs are those of a pleural effusion or a pleural mass. If the effusion is large, there may be displacement of the mediastinum towards the contralateral side. In advanced cases, tumor may erode through the chest wall and cause a localized tender

mass. Superior vena caval obstruction may occur, and there may be involvement of the phrenic or cervical nerves.[290] Some cases may be detected during routine radiologic examination of the chest.[289] A rare presentation is with miliary pulmonary parenchymal involvement, resembling miliary tuberculosis.[291a] Rarely, the disease may masquerade as sclerosing mediastinitis; it apparently develops after a flu-like illness.[292] Diffuse spinal meningeal thickening has been described, causing severe localized low back pain in the lumbosacral region.[293]

The mode of onset was insidious in all but a few cases and the mean interval before reaching hospital was 3.39 months for pleural tumors and slightly less for peritoneal tumors.[290]

The staging of mesothelioma has been given by Butchart and colleagues (Table 34-6).[294] A proposed new staging system has been recently described.[249a]

PERITONEAL MESOTHELIOMA

This disease is less common than pleural mesothelioma. In one series of 144 mesotheliomas, the primary site was predominantly peritoneal with 74

TABLE 34-6
Staging of Mesothelioma[437]

Stage I	Tumor confined in the "capsule" of the parietal pleura (i.e., involving only ipsilateral pleura, lung, pericardium, and diaphragm).
Stage II	Tumor invading chest wall or involving mediastinal structures (e.g., esophagus or heart. This stage includes intrathoracic lymph node involvement).
Stage III	Tumor that penetrates the diaphragm to involve the peritoneum. Involvement of the contralateral pleura. Extrathoracic lymph node involvement.
Stage IV	Distant hematogenous metastases.[437]

cases. There were 66 pleural mesotheliomas and the site was undetermined in 4.[295] The ratio of pleural to peritoneal sites showed continuous change when related to the year of first exposure. It varied from 5:1 pleural:peritoneal before 1921 to 1:3 after 1950. This strong temporal relationship probably reflects progressive dust suppression, including the nonfibrous dust present in insulation materials. A further factor was also the degree to which the fibers had been opened. Individual exposure was also another important factor, and the peritoneal site was preferentially associated with longer and heavier exposures. This was shown by the fact that the mean exposure of the pleural cases was less (113 months) than that of peritoneal cases (138 months). This difference did not reach a statistically significant level. A higher proportion of cases with peritoneal than pleural mesothelioma were associated with antemortem asbestosis. Elmes and Simpson[290] had similar results. Peritoneal mesothelioma is nearly always associated with pulmonary asbestosis.

The onset of symptoms is ill defined. There is usually vague discomfort, followed by a decrease in appetite and constipation. Weight loss is present and the patient complains of a full but not necessarily swollen, abdomen.

Physical signs may initially be few. There may be a full but not distended abdomen, with equivocal signs of free fluid. Masses may be palpable or visualized. At this stage there may be intestinal obstruction.

Pathology

PLEURAL MESOTHELIOMA

The pleura is the primary site in most recorded series. Peritoneal tumors are less common, and the pericardium and tunica vaginalis are extremely rare sites.[296]

Pleural mesotheliomas are more commonly seen on the right (right:left=3:2).[297] This may be explained by the greater size of the right pleural cavity. Though the lesion is often unilateral at presentation, it is not infrequent to find histologic evidence of mesothelioma in the contralateral pleural space. Macroscopic evidence of synchronous bilateral pleural tumors is rare. By the time the histopathologist sees the tumor, there is a dense thickening of the pleura by firm grayish-white or even yellow tissue, which may show cystic areas containing glairy, "mucoid-like" material (Figs. 34-32, 34-33, and 34-34). The tumor tissue extends into and often through the diaphragm, along the fissures, and into the underlying atelectatic lung. There may an associated pleural effusion.

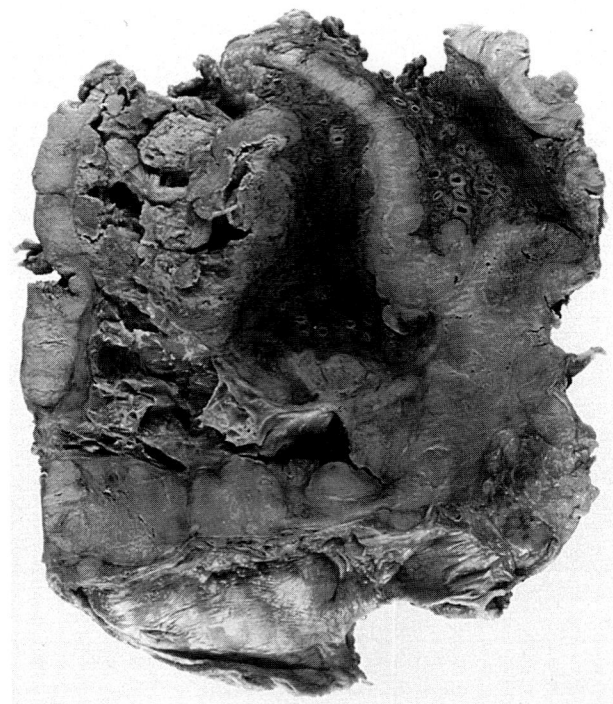

FIGURE 34-32
Pleural mesothelioma growing through the diaphragm and along the interlobar fissure. Much of the lung is collapsed.

FIGURE 34-33
Pleural aspect of a lung involved by mesothelioma.

FIGURE 34-34
Lower lobe involved by mesothelioma with an apical pleural plaque.

A

Since the pleural cavity is usually obliterated, it is not possible to define the origin of the tumor. In a few cases it may be confined as nodules to the parietal or less commonly the visceral pleura (Fig. 34-35A and B). Pleural plaques may be found either alongside the tumor or incorporated within it.

Chest wall invasion as well as direct invasion of the pericardium, myocardium, and mediastinum is identified (Fig. 34-36). Tumor encircles the aorta, esophagus, and other great vessels (Fig. 34-37). The tumor may have tracked along the site of the initial thoracotomy incision. If there is no effusion, the pleural space is totally obliterated by very dense tumor tissue. In some cases there is replacement of a lobe or an entire lung by tumor, making it difficult if not impossible to detect a possible primary bronchial carcinoma.

Tumors that are soft are usually epithelial, and similarly those that are cystic contain hyaluronic acid. Multicystic mesothelioma may affect the pleura in a similar way to the peritoneal cavity, though this is uncommon.[298] The very firm tumors are sarcomatous. When the tumor is advanced, it may extend into the mediastinum and push it into the contralateral side.

There is usually involvement of the hilar (Fig. 34-38) and mediastinal nodes with epithelial tumors.

In some cases subpleural fibrosis is associated with the tumor (Fig. 34-39).

SPREAD OF MESOTHELIOMA

The spread of the tumor depends on its histologic type. The epithelial tumors behave as carcinomas, whereas those with sarcomatous patterns grow like

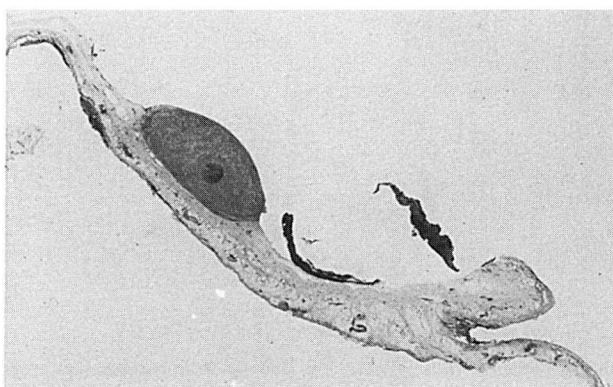

B

FIGURE 34-35
A. Mesothelioma confined to the visceral pleura and showing multiple nodules.   B. Nodule of mesothelioma on the pleura.

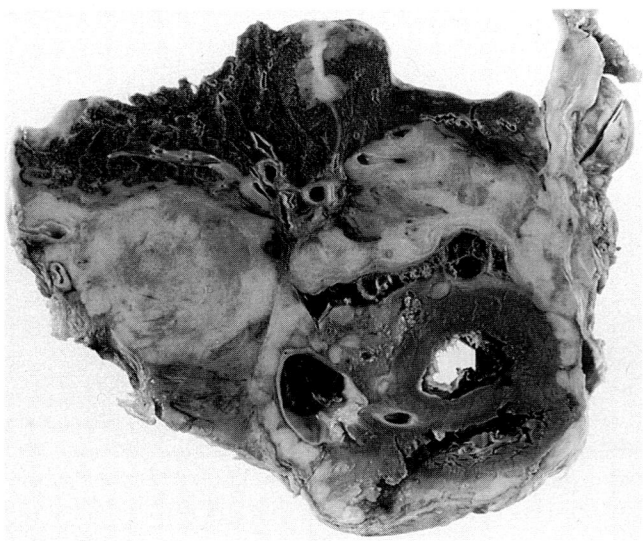

FIGURE 34-36
Mesothelioma surrounding the heart and aorta.

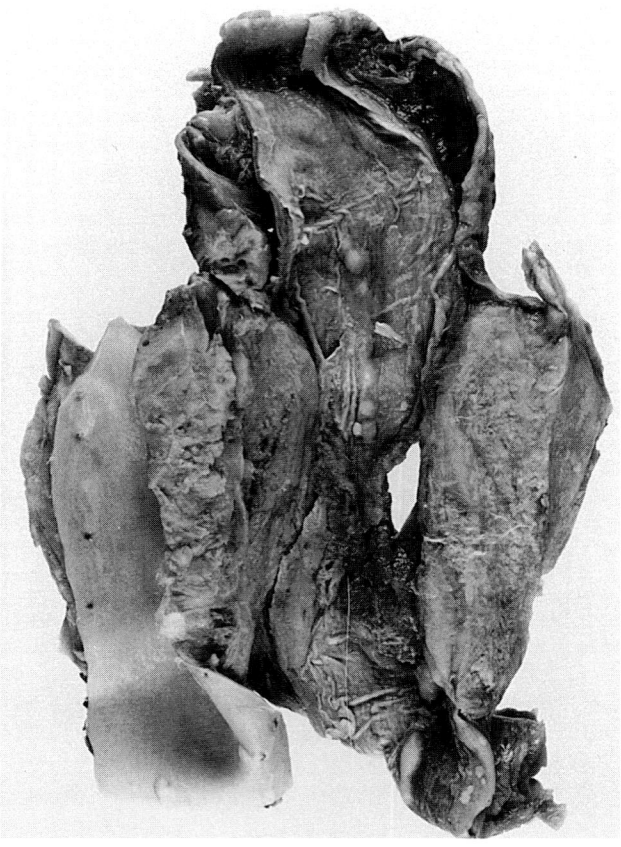

FIGURE 34-37
Tumor encircling the esophagus and the descending thoracic aorta.

sarcomas. Thus epithelial mesotheliomas spread to the adjacent lymph nodes and often have associated pleural effusions. In epithelial tumors, secondary deposits are found in the supraclavicular or axillary nodes and tend to extend directly into the pericardium and the contralateral pleura. Direct extension into the peritoneum is also commoner with epithelial mesotheliomas. Distant metastases are seen in the brain, vertebral column (Fig. 34-40), liver, pancreas, adrenals, kidney, and contralateral lung (Fig. 34-41). The liver may be affected by direct extension of tumor through the diaphragm. There may be a "lymphangitic picture" in some cases.

Sarcomatous tumors tend to involve the pericardium, the contralateral pleura, and the peritoneum less frequently than epithelial tumors. Mixed tumors are in between epithelial and sarcomatous variants in their biologic behavior. Distant metastases, however, are more common with sarcomatous mesotheliomas than epithelial ones.[299] Such metastases can cause diagnostic problems in the brain and osseous system.

EXAMINATION OF THE LUNGS FROM A MESOTHELIOMA PATIENT

Postmortems on cases of mesothelioma are difficult and time-consuming. A search of all the abdominal

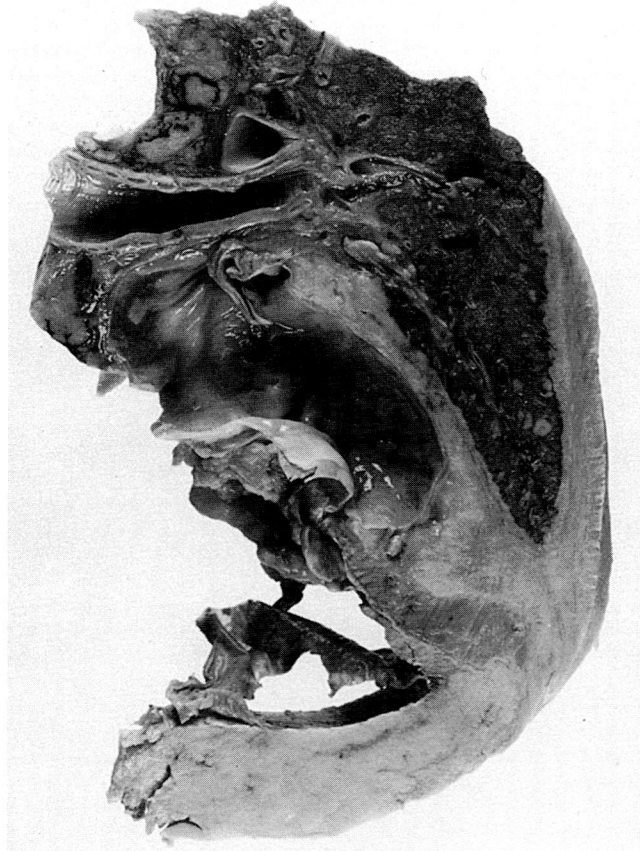

FIGURE 34-38
Mesothelioma surrounding the lung and diaphragm as well as the hilar nodes.

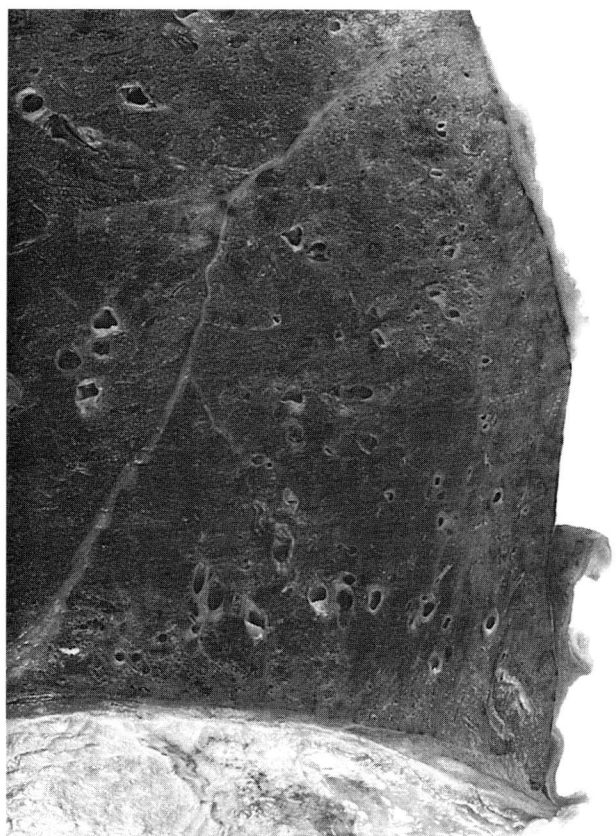

FIGURE 34-39
There is a thin layer of mesothelioma over the lower lobe and subpleural fibrosis.

viscera, including the gallbladder, liver, pancreas, stomach, and small and large intestines should be made to exclude a primary tumor. Removal of the affected lung is difficult. Probably the easiest way to remove such a lung is with a long, thin, sharp knife. This is placed between the rib cage and the affected pleura and the pleura is cut away from the chest wall.

After examination of the larynx and trachea, both lungs are distended with a perfusion pressure of 25 cm of water. The lung involved with the mesothelioma often has to be removed with some "pleural" deficits, and proper distension is often not possible. The lungs should be left for at least 48 h for proper fixation. If possible, the formalin should be changed after 24 h and excess blood at the bottom of the container removed.

The lungs are placed on a board that has three sides raised to a height of 1 cm. Search is made for any possible primary tumor. In practice, this is difficult, since nodules of tumor may replace a lobe, making it difficult to recognize any residual bronchi. A long, sharp knife is then used for slicing and blocks are taken (Fig. 34-42).

The origin of the blocks should be labeled for future reference. The blocks are taken from set sites as well as any areas of interest. This method was devised by the Pneumoconiosis Committee of the College of American Pathologists and the National Institute of Occupational Safety and Health.[94] In

FIGURE 34-40
Secondary tumor in vertebral column from a case with a pleural mesothelioma. Tumor is seen on the outside of the spine.

FIGURE 34-41
Discrete small metastases in contralateral lung with a case of mesothelioma.

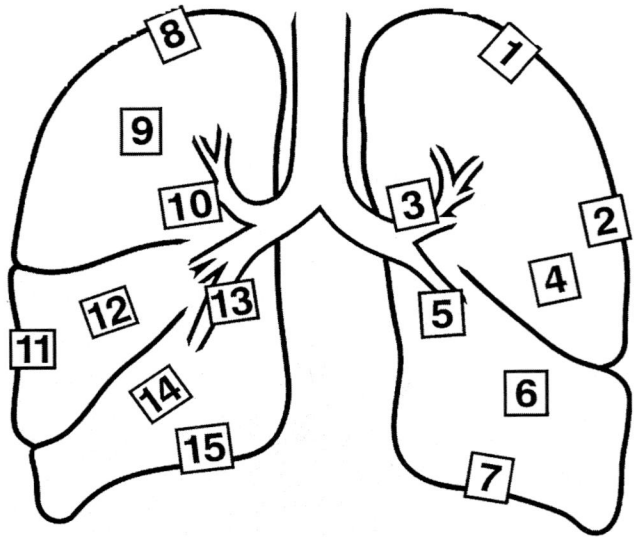

FIGURE 34-42
Routine blocks that should be taken from the lung in a case of possible asbestos exposure. (*Courtesy of Abraham, with permission.*)

addition to blocks from the apices of both upper lobes and diaphragmatic portions of the lower lobes, samples should include the visceral and subpleural space, deep parenchyma, and portions of the major bronchi with adjacent parenchyma from each lobe. Examination of the lungs in this manner is important, since asbestosis is irregularly distributed and tends to be more severe beneath the pleura and in the lower lobes.

At least four blocks of tumor tissue should be randomly sampled from the periphery of the lung. This enables a mixed-pattern mesothelioma to be detected. Some tumors show large areas of necrosis, and viable tissue will be seen if sections are taken from the growing edge. It is important to document lymph node metastases.

Blocks of tissue are taken for fiber counts. These should be sampled from the unaffected lung. It is my practice to take 5 g and place it in a universal container half full of formalin. The excess tissue is kept in case the fiber count needs to be repeated. Blocks should be taken from the apex of the upper lobe, the apex of the lower lobe, and the posterolateral aspect of the lower lobe.

Smaller samples can be used, including paraffin blocks, if electron microscopy is to be used for the fiber count. Some studies have reported on analysis of transbronchial biopsies,[234,300] but the size of such samples will cause sampling error.[4]

The techniques for demonstrating asbestos bodies from blocks of tissue from the unaffected lung have already been documented in the texts mentioned at the beginning of the chapter. In addition, key papers are those of Ashcroft and Heppleston and

others.[232,301–303] A control sample should be examined at the same time as the case in question. These techniques give reproducible results.[304] Some pathologists prefer to take 30 μm sections to demonstrate asbestos bodies. This method introduces artifacts and may complicate the evaluation of lung tissue. Screening with a ×10 eyepiece and ×10 objective enables the pathologist to detect all coated asbestos bodies present in routine 5 μm sections.

A little-used method is counting the asbestos fibers identified on Prussian blue–stained sections. This will not detect uncoated fibers. It is possible to calculate the number of asbestos bodies per gram using a conversion factor from the volume of paraffin-embedded tissue to wet weight of lung.[305,306] The formula applies only to sections 5 μm thick and fibers 35 μm in length. The more sections examined, the greater the accuracy of the estimated asbestos-body concentration.

PERITONEAL MESOTHELIOMA

Peritoneal mesotheliomas occur earlier than pleural ones and the patients die sooner. There is greater exposure to asbestos in these peritoneal tumors and the latent interval from exposure to development of the tumor is longer.

The peritoneum is the second most common site for involvement by malignant mesothelioma. These tumors are probably underdiagnosed and are labeled carcinomatosis. Terminally, there is encasement of the intraabdominal organs. The tumor usually grows in a multinodular pattern, producing large masses confined to the omentum. It then spreads to encase the abdominal viscera, causing great difficulty in dissection.

The tumor is usually firm and white with individual nodules studding the peritoneal surface. Matted masses of tumor are also identified. Rarely, peritoneal mesothelioma may cause multiple intestinal polyps, which cause intestinal obstruction. These polyps were seen in the lumen of the small intestine.[307]

In some cases it is impossible to be certain whether the tumor is pleural, peritoneal, or synchronous, since it has grown through the diaphragm.

An increasing number of reports of multicystic mesothelioma of the peritoneum are described.[308,309] Multicystic mesothelioma is a rare lesion that occurs most frequently in young to middle-aged women. It affects mainly the pelvic peritoneum, though Weiss and Tavassoli[308] described six cases in men. The mean age of the men who developed these tumors was 47 years, as compared to 38 years for women.

Previous surgery had been documented in 9 of 17 cases.

The patients complained of abdominal pain, tenderness, and distension, usually in association with a pelvic or abdominal mass. In two patients, there was a presentation with an acute abdomen. In three others, the masses were discovered incidentally at the time of cesarean section.

Macroscopically there were multiple translucent and membranous cysts that were grouped together to form a confluent mass or, more commonly, that studded the surface of the peritoneum in a discontinuous fashion.

PERICARDIAL MESOTHELIOMA

This is the least frequent primary site of this condition. It is probably most commonly seen after involvement from a tumor in the pleural cavity. Where there are advanced pericardial tumors, it may be difficult to distinguish them from a pleural primary. There may be a history of asbestos exposure.[310,311] The patients have dyspnea, cardiac tamponade,[312] pericarditis, cardiac failure, and/or arrythmias.[313]

Pericardial mesotheliomas involve both the parietal and visceral pericardium, and eventually the heart is encased in a mass of tumor tissue. There are few immunohistochemical and ultrastructural studies.[314]

Mesothelioma of the atrioventricular node is a misnomer. This tumor is probably derived from endoderm and is outside the scope of this text.[315,316]

MALIGNANT MESOTHELIOMA OF THE TUNICA VAGINALIS

This tumor has been associated with prior asbestos exposure, with reported lag times of 8 to 40 years between exposure and diagnosis. Antman and associates[317] described 6 patients and reviewed another 18. In 4 of their 6 cases, there were long histories of asbestos contact. This compared with only 2 of the first 17 reported cases. A recent series of 11 cases with a review of the literature has been documented.[317a]

The patients have an age range of 12 to 86 years. There is an enlarging intrascrotal mass and there may be an associated hydrocele. The tumor tends to occur locally; it spreads to regional lymph nodes and in some cases the lung.

Mesothelioma of the tunica vaginalis testis is usually epithelial, tubulopapillary in type though biphasic cases have been described. This tumor has recently been divided into two groups.[318] The first comprised low-grade or "benign" mesotheliomas.

These were predominantly papillary, with solid epithelial nests in several cases. There was only mild nuclear atypia and few mitoses. However, these were considered to be of borderline malignancy. The second group was high-grade mesothelioma, with a variegated pattern, including tubulopapillary and solid sheets of cells. There was much nuclear pleomorphism, frequent and abnormal mitoses, and tumor necrosis.

The condition must be distinguished from nodular mesothelial hyperplasia of hernial sacs. Nodular mesothelial hyperplasia of hernial sacs is also characterized by mesothelial proliferation with papillary projections, solid nests, or tubular structures (Fig. 34-43 and 34-44). These may form complex structures and simulate a mesothelioma. Psammoma bodies may be present.[319] It must also be distinguished from mullerian-type tumors, carcinoma of the rete testis, adenomatoid tumor and metastatic carcinoma.[317a]

Histology

Mesotheliomas were classified by the WHO[320] into epithelial, fibrous (spindle cell), and biphasic. The fibrous are also known as sarcomatous. There appear to be histologic differences between pleural and peritoneal mesothelioma, since the epithelial type is commoner in the peritoneum than in the pleura. This achieved statistical significance.[290,295,321]

The differential diagnosis differs in abdominal cavity tumors of females because of the female generative tract.

The epithelial pattern comprises approximately 50 percent of cases; sarcomatous, 20 percent; and mixed or biphasic, 30 percent.[270,297] These figures are probably not accurate, since the greater the number of blocks taken from any one tumor, the more likely it is that both epithelial and fibrous areas will be found.

EPITHELIAL VARIANT

Classically, epithelial mesotheliomas consist of tubulopapillary foci made up of cuboidal or polyhedral tumor cells. There are fibrous tissue cores to the papillae. The cells classically are arranged in a tubulopapillary pattern, but there may be simple papillary (Figs. 34-45 and 34-46), tubular, or in some cases solid sheets of polygonal cells (Fig. 34-47). Some of tubules form complex, branching patterns. In some cases the cells may have a squamoid pattern (Figs. 34-48 and 34-49). Between 5 and 10 percent of cases have psammoma bodies within the cores of proliferating cells (Fig. 34-50).[322] Microcystic foci are also identified, producing a lacelike network of flattened

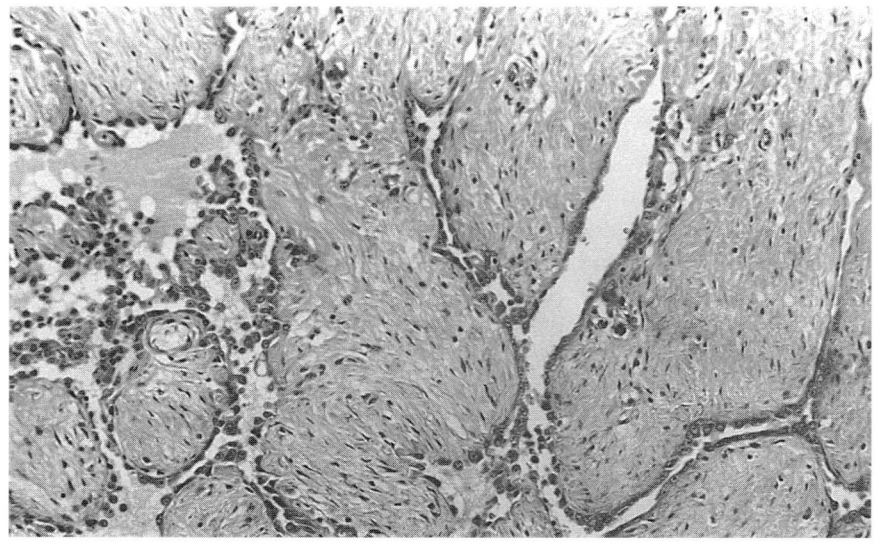

FIGURE 34-43
Nodular hyperplasia in a hernial sac. There are fibroblastic papillary processes covered with reactive mesothelial cells. (H&E stain, ×87.5.)

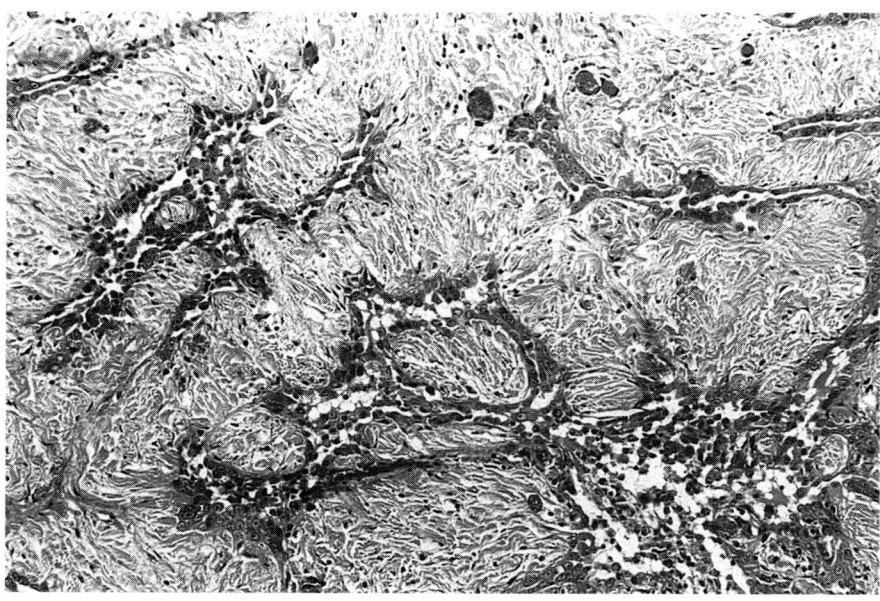

FIGURE 34-44
Nodular hyperplasia in a hernial sac. Infolded, proliferating mesothelial hyperplasia. (H&E stain, ×87.5.) (*Courtesy of Dr. M. Harris, Christie Hospital, Manchester, with permission.*)

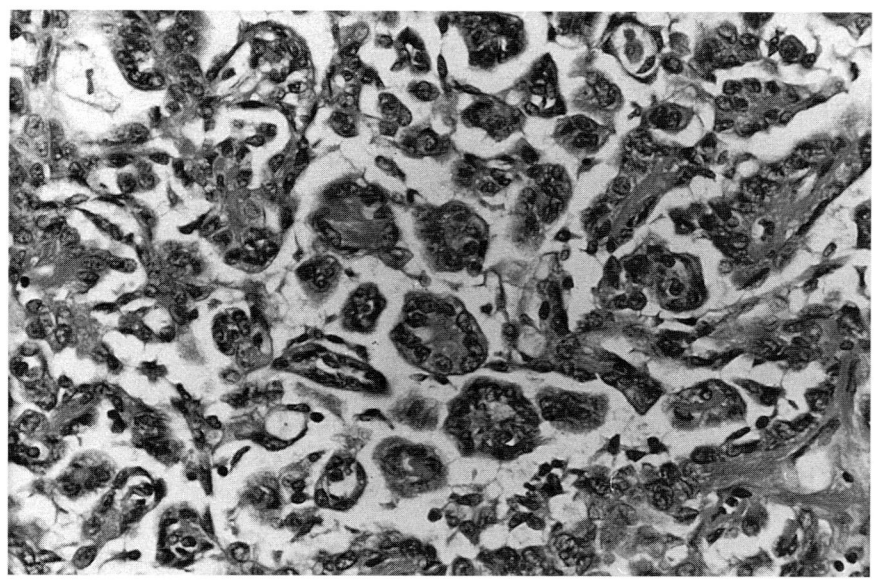

FIGURE 34-45
Tubulopapillary mesothelioma with well-marked papillae and fibrous cores. (H&E stain, ×219.)

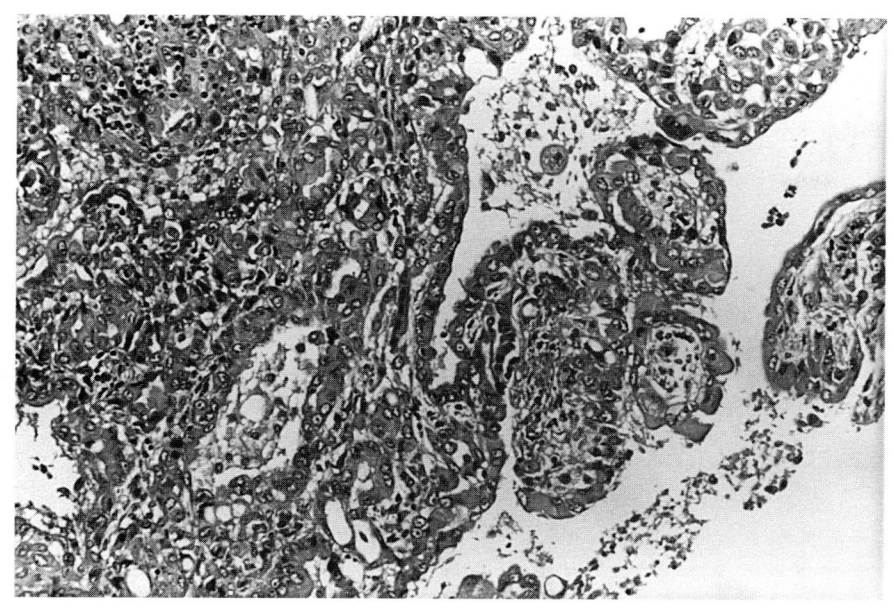

FIGURE 34-46
Mesothelioma with prominent papillae growing into lung. (H&E stain, ×87.5.)

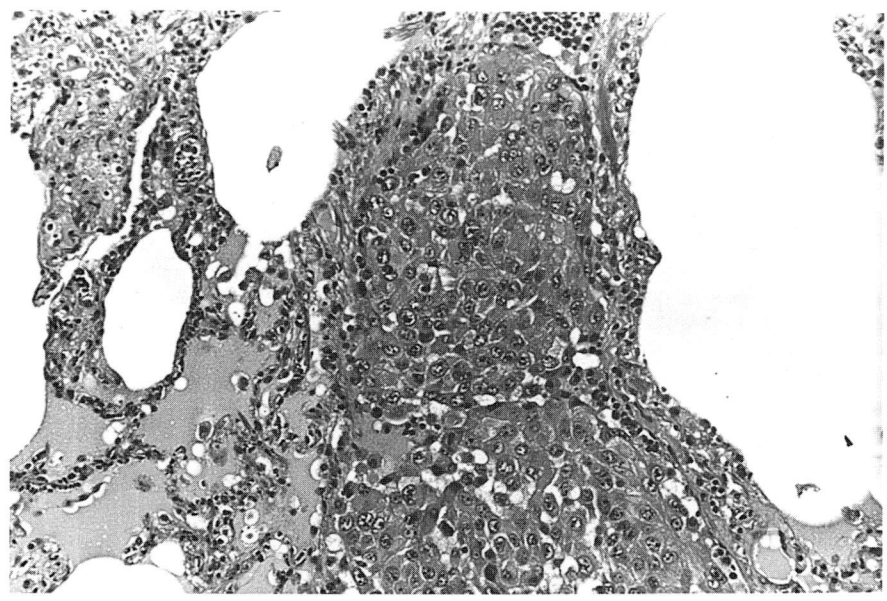

FIGURE 34-47
Mesothelioma with polygonal cells. (H&E stain, ×87.5.)

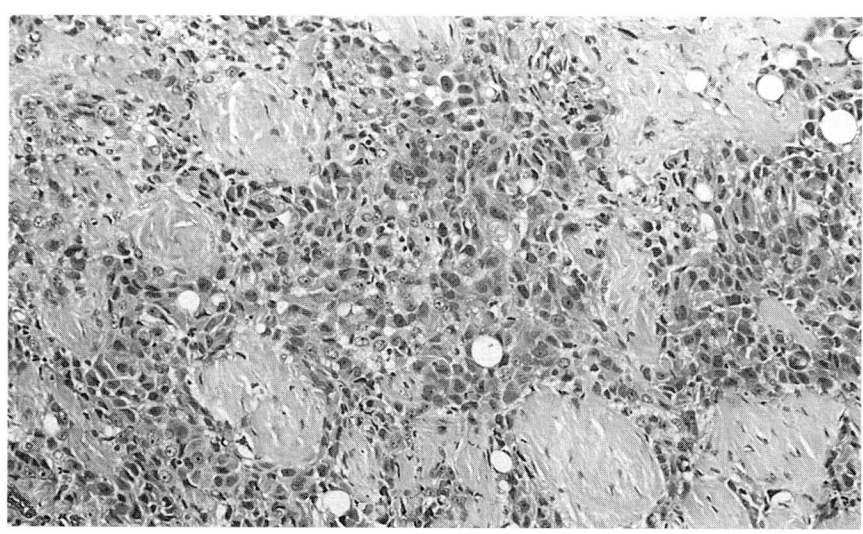

FIGURE 34-48
Area of squamous metaplasia in a mesothelioma. (H&E stain, ×87.5.)

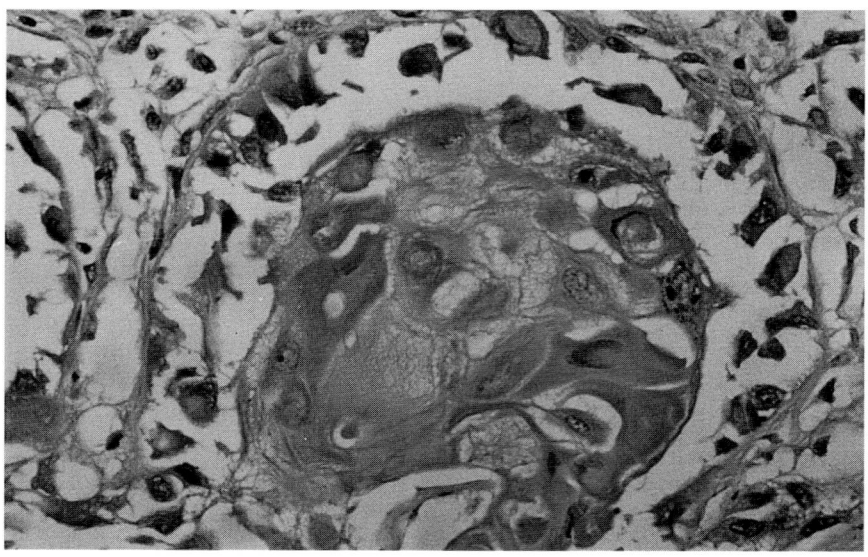

mesothelial cells (Fig. 34-51). The initial appearance resembles very strongly an adenomatoid tumor. Rare cases may show an adenoid cystic architecture.

A prominent feature is the presence of nuclei with open vesicular cytoplasm Fig. 34-52). There are one or two nucleoli. In keeping with the nuclear features, there is a constant nuclear/cytoplasmic ratio.[323] Mitoses, nuclear pleomorphism (Fig. 34-53), and atypical mitoses are unusual. The nuclear features, while not specific, are useful in differentiating mesothelioma from adenocarcinoma. The latter tumor has hyperchromatic and irregular neclei (Figs. 34-54 and 34-55).

The cytoplasm of mesotheliomas is often eosinophilic, and there are well-defined cell borders. In some cases the cytoplasm shows vacuolation, giving a signet-ring appearance (Fig. 34-56). These vacuoles do not contain any neutral mucosubstances and therefore should not be confused with adenocarcinoma. However, mesotheliomas with focal mucin positivity have been described.[313]

Tumor may grow in the underlying lung, simulating bronchioalveolar carcinoma. In addition, the tumor permeates lymphatics (Fig. 34-57A and B) as well as thin-walled pulmonary veins and arteries (Fig. 34-58). This leads to hematogenous spread, and rare cases have been reported as *presenting* as a miliary pattern.[324] The tumor may grow in the interstitium, between alveoli.

Nuclear pleomorphism, hyperchromasia, as well as mitoses and tumor giant cells may be found in some cases, but the presence of these features should raise the possibility of chemotherapy before the biopsy was taken.

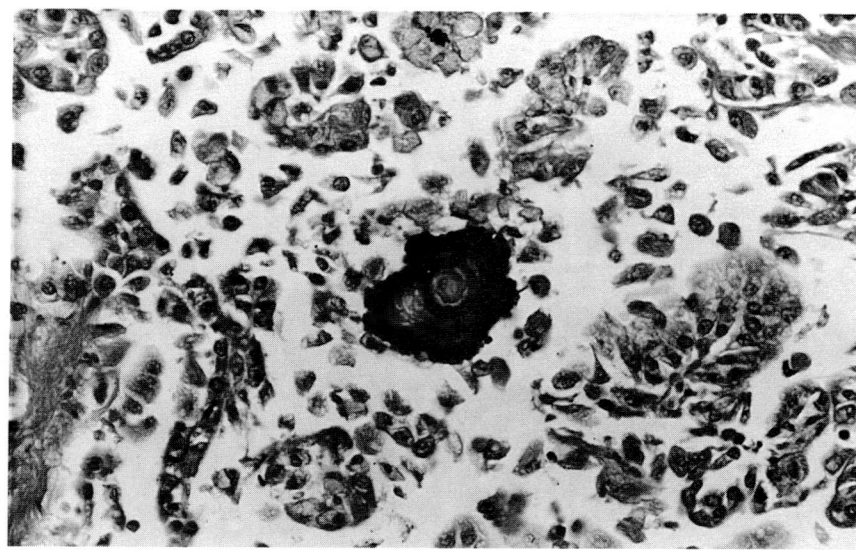

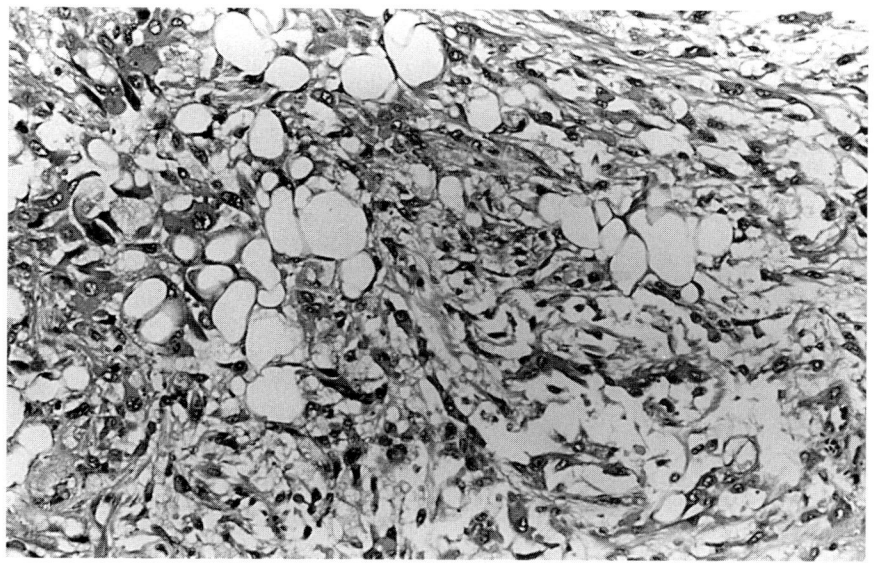

FIGURE 34-51
Mesothelioma with a microcystic area.
(H&E stain, ×87.5.)

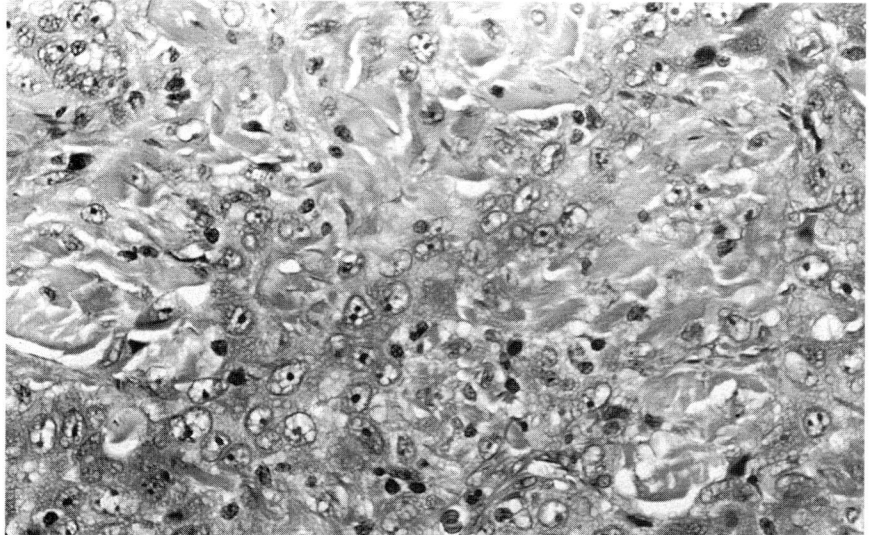

FIGURE 34-52
Nuclear detail in epithelial mesothelioma
with prominent nuclei with an open vesicular pattern and a small nucleolus. Some
nuclear hyperchromatism is seen. (H&E
stain, ×219.)

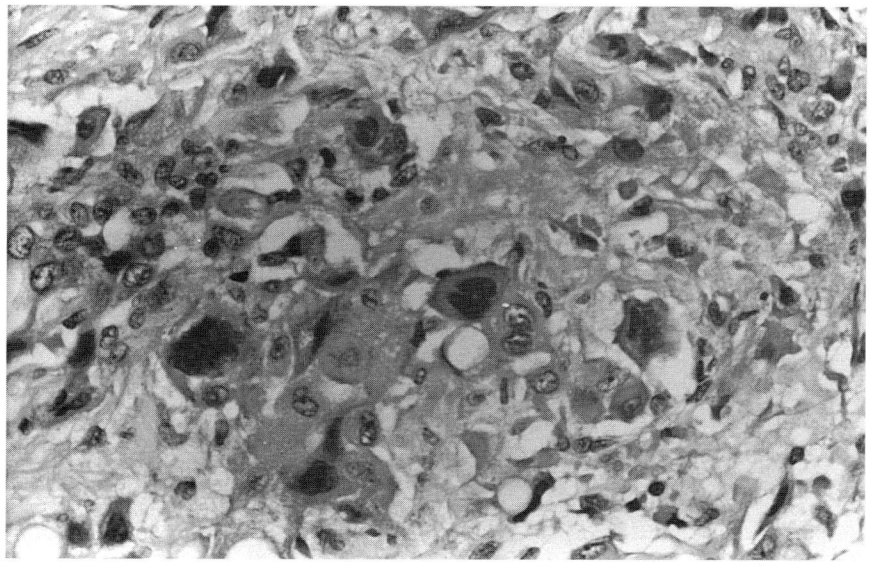

FIGURE 34-53
Mesothelioma with nuclear hyperchromatism. (H&E stain, ×219.)

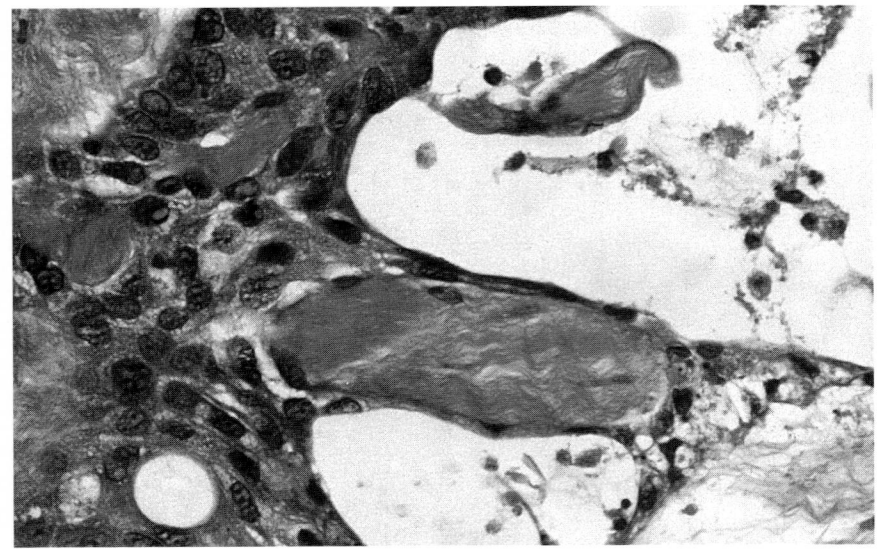

FIGURE 34-54
Adenocarcinoma involving the pleura. There is nuclear hyperchromatism. (×219.)

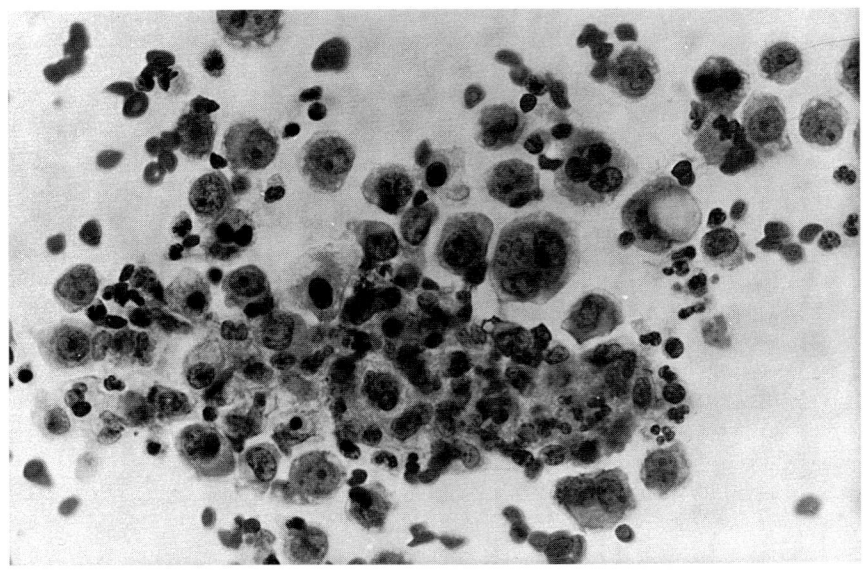

FIGURE 34-55
Pleural fluid with adenocarcinoma showing vacuolated cytoplasm and irregular hyperchromatic nuclei. (H&E stain, ×350.)

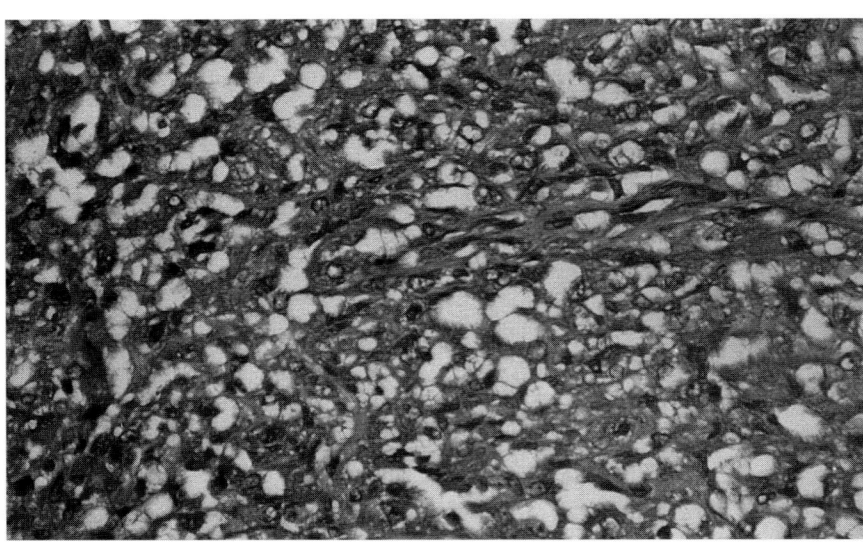

FIGURE 34-56
Mesothelioma with a signet-ring pattern, showing compressed nuclei. (H&E stain, ×219.)

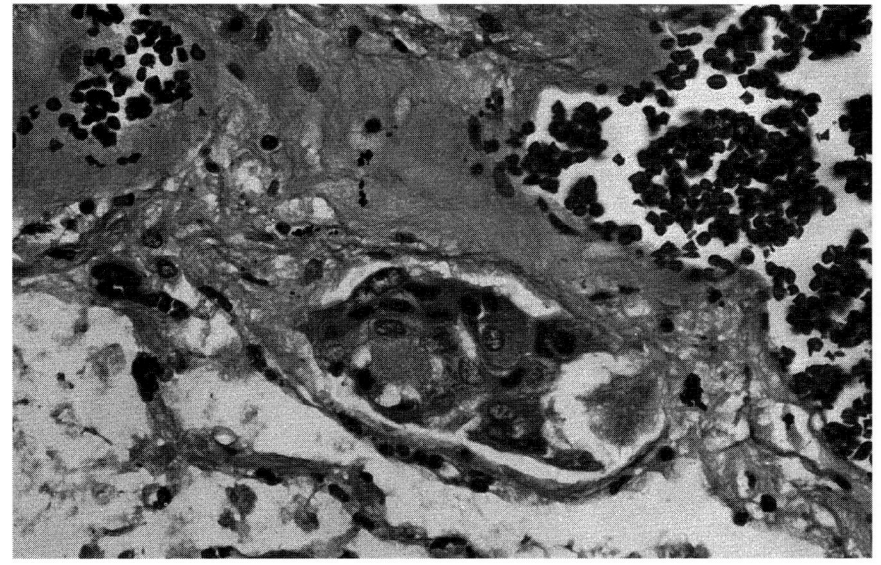

A

B

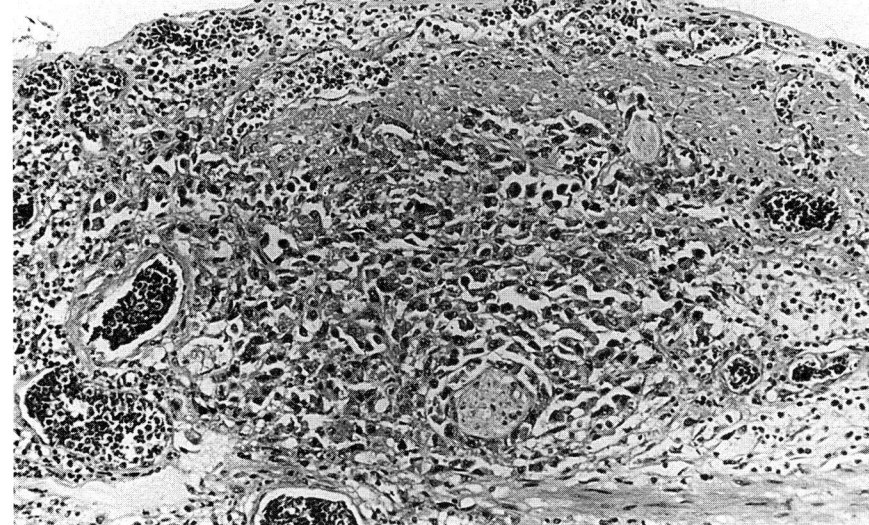

FIGURE 34-57
A. A clump of mesothelioma cells in a lymphatic. (H&E stain, ×219.)   *B.* Mesothelioma infiltrating the bronchial wall. The surface epithelium has been lost. (H&E stain, ×87.5.)

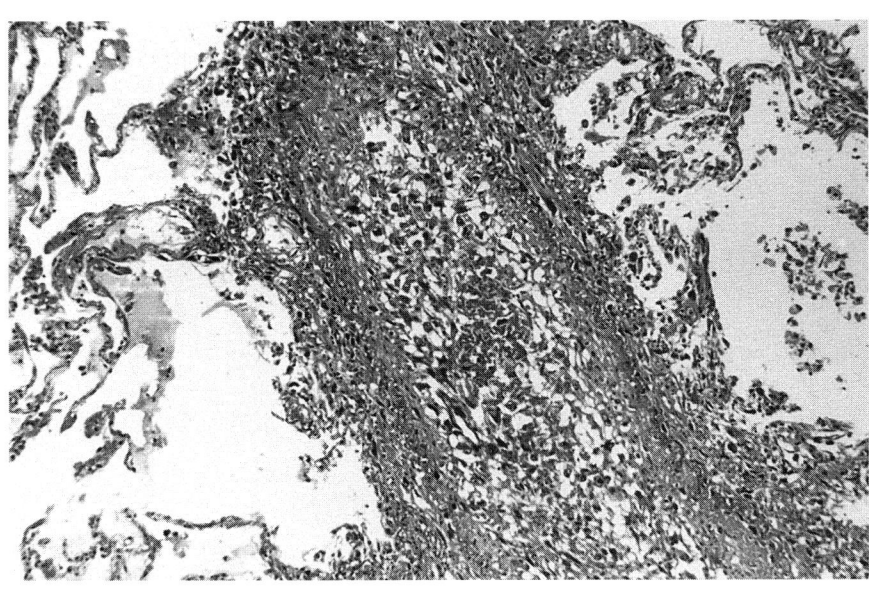

FIGURE 34-58
Intravascular invasion by a mesothelioma. (H&E stain, ×56.)

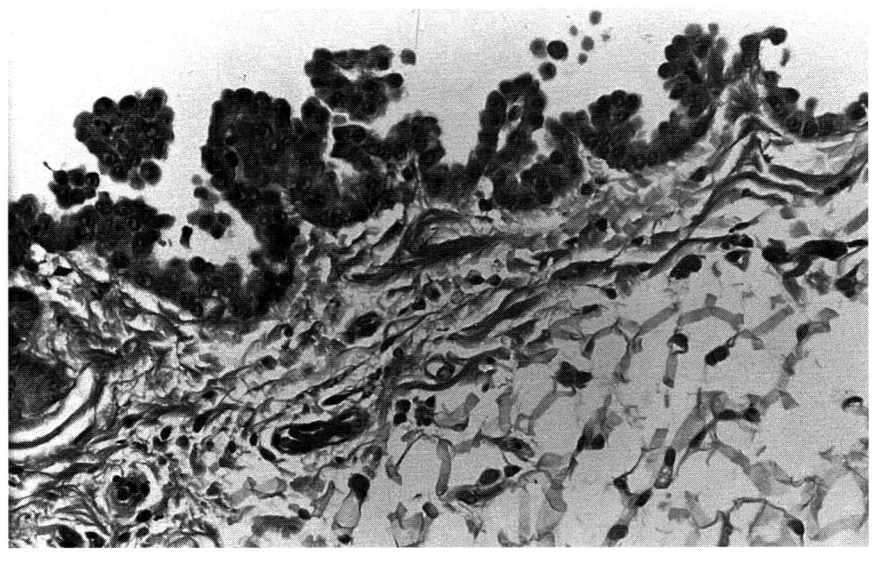

FIGURE 34-59
In situ mesothelioma. This was in the lung
contralateral to the tumor. (H&E stain,
×219.)

In situ mesothelial atypia has been described (Fig. 34-59).[3] This is, of necessity, a postmortem diagnosis. This condition should not be confused with the papillary processes seen in occasional cases of pleura taken from a pneumothorax (Fig. 34-60).

In epithelial mesotheliomas, the stroma is variable. It may consist of moderately cellular fibrous tissue (Fig. 34-61), with some myxoid change. This cellularity should not be confused with the sarcomatous variant. In some cases it may be difficult to differentiate the biphasic pattern from the epithelial because of stromal cellularity. Desmoplasia may be seen in some cases.

FIBROUS MESOTHELIOMA

This variant consists of spindle-shaped cells with varying nuclear pleomorphism and mitotic activity

(Figs. 34-62 to 34-64). The amount of collagen associated with the cells varies. It may be impossible to distinguish this tumor from the malignant variant of the localized pleural fibrous tumor. A storiform pattern may be present, causing confusion with the very rare pleural or pulmonary malignant fibrous histiocytoma. Osseous (Fig. 34-65A and B) and cartilaginous differentiation may be identified (Fig. 34-66).[325,326]

The largest series showing this change was that of Yousem and Hochholzer,[326] describing 10 cases. Among these, 3 were biphasic mesotheliomas, 7 showed cartilage, and all 10 had osseous differentiation. One case showed osseous metaplasia. These cases had the typical storiform and interweaving vesicular patterns; a transition into osseous differentiation was common, with neoplastic spindle cells enveloped by a sclerotic eosinophilic calcified osteoid matrix. In three cases, benign osteoclasts

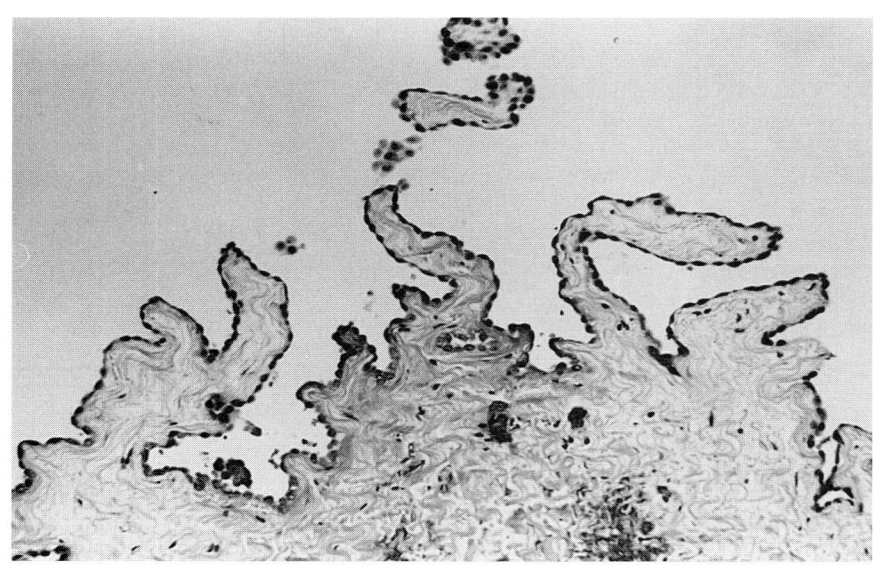

FIGURE 34-60
Papillary pattern seen in a reactive
mesothelioma due to a pneumothorax.
This is due to edema and does not show
the well-defined mesothelial cells seen in
Figure 34-59. (H&E stain, ×87.5.)

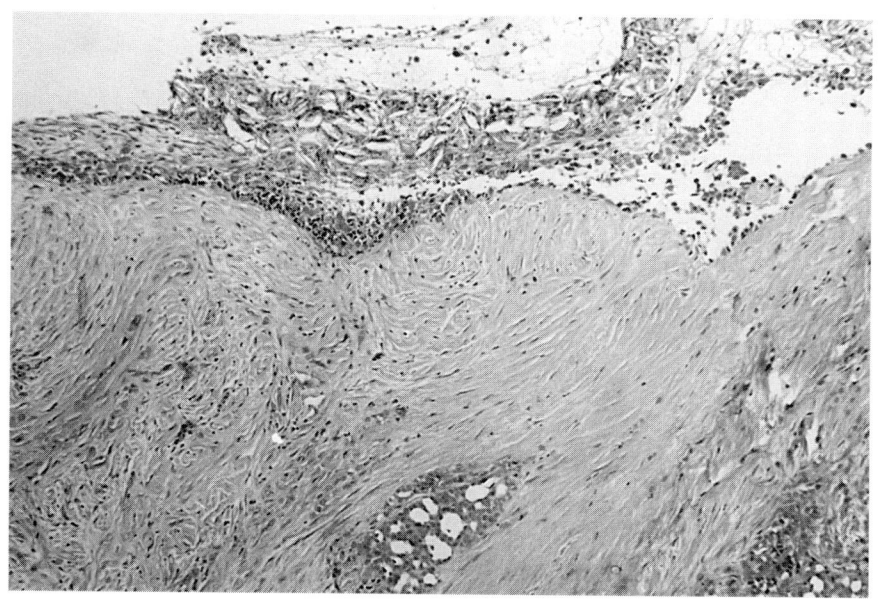

FIGURE 34-61
Mesothelioma with a fibroblastic stroma.
Part of this may be due to the foreign
material instilled to control the pleural
effusion. (H&E stain, ×87.5.)

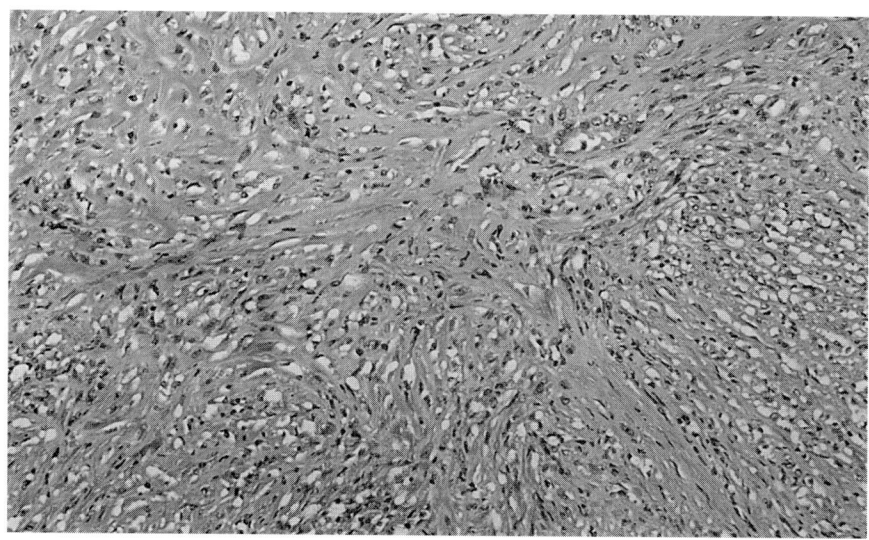

FIGURE 34-62
Fibrous mesothelioma with tendency to a
storiform pattern. (H&E stain, ×87.5.)

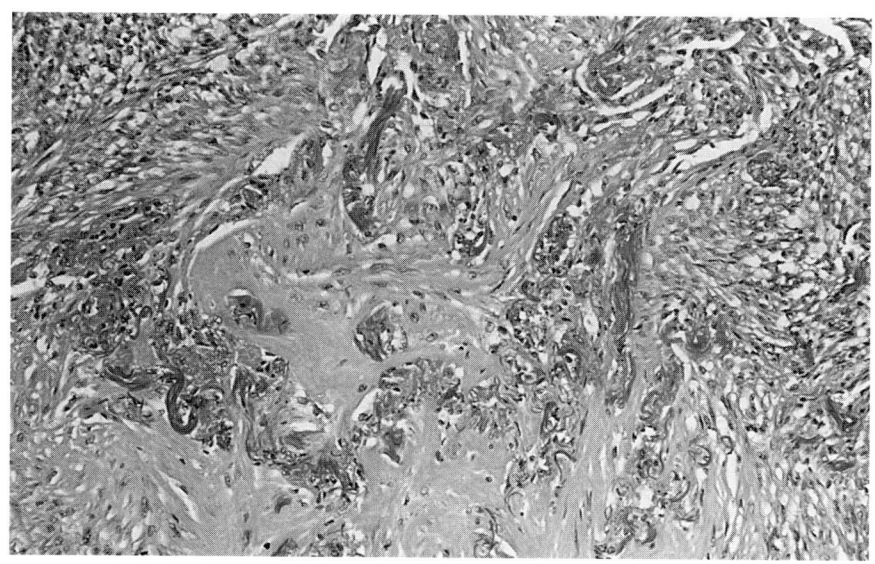

FIGURE 34-63
Fibrous mesothelioma with both hypercellular and accellular areas. (H&E stain,
×87.5.)

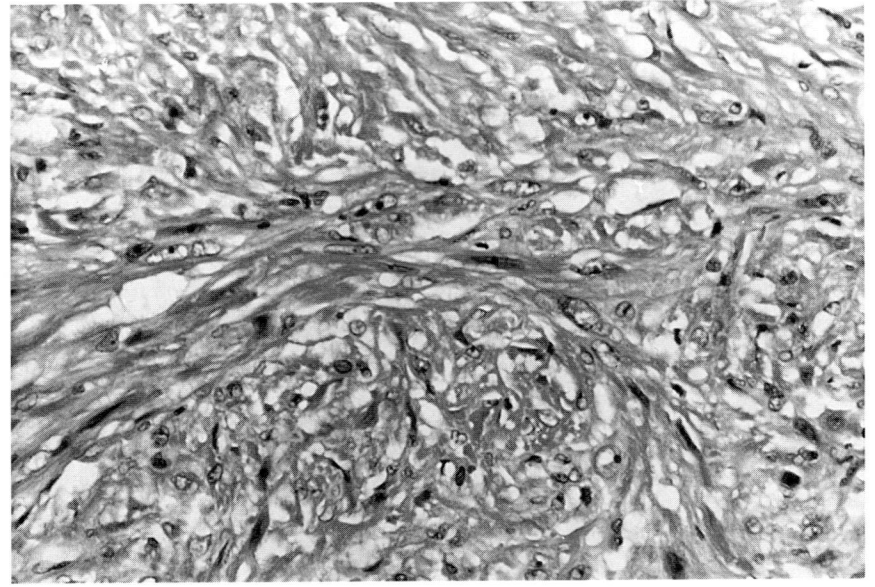

FIGURE 34-64
Fibrous mesothelioma with nuclear hyperchromatism and a storiform pattern. (H&E stain, ×219.)

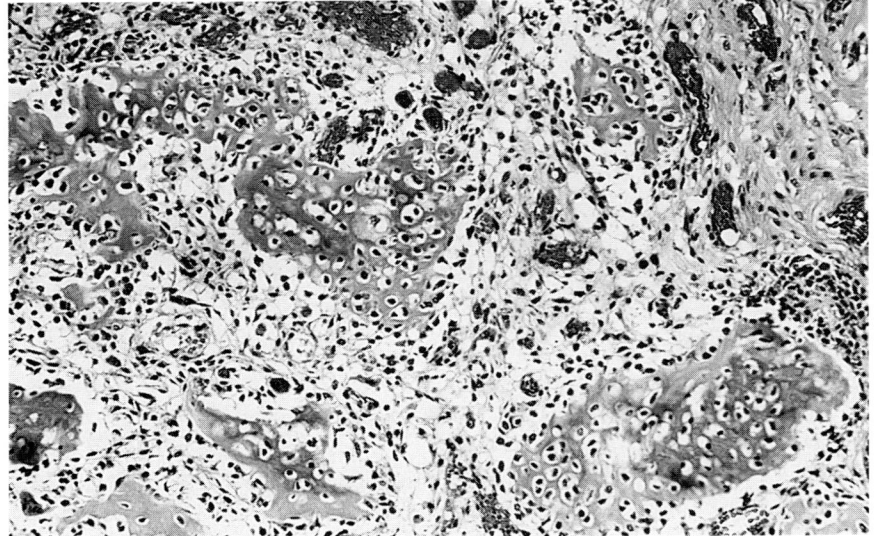

A

B

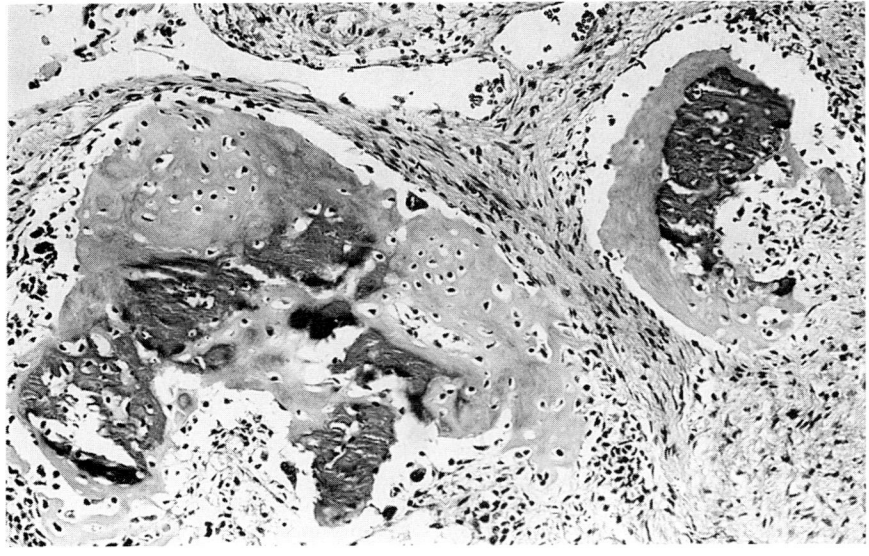

FIGURE 34-65
A. Osteoid in the middle of a mesothelioma. (H&E stain, ×87.5.) *B.* Foci of mature bone in relation to a mesothelioma. (H&E stain, ×87.5.)

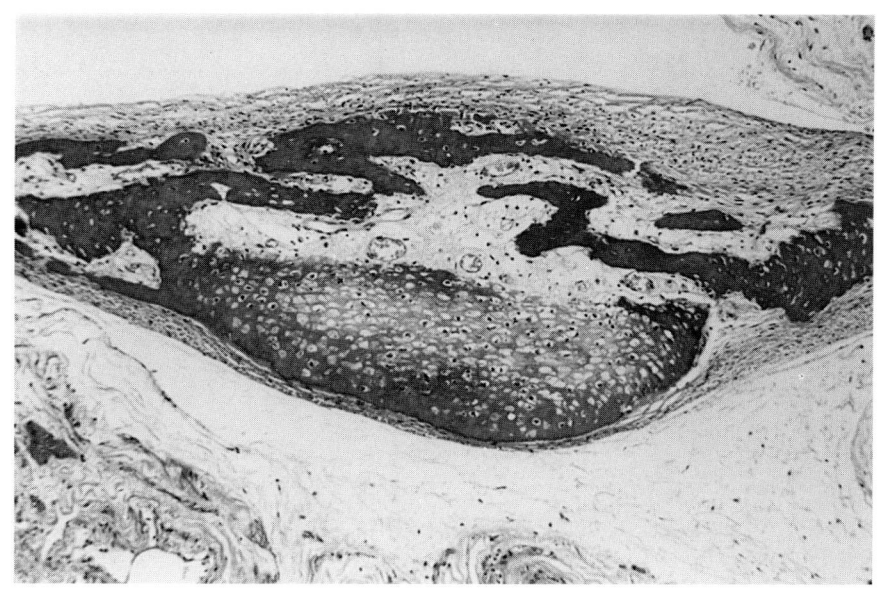

FIGURE 34-66
Cartilage and bone close to a mesothelioma (H&E stain, ×56.)

(Fig. 34-67) were admixed with malignant spindle cells. Endochondral ossification was observed in all cases where cartilage was present. In rare cases there is liposarcomatous differentiation in a fibrous mesothelioma (Fig. 34-68 and 34-69). This must be differentiated from a mesothelioma growing around existing fat spaces.

A variant of sarcomatous mesothelioma is desmoplastic malignant mesothelioma.[327–329] This subtype accounts for 6.6 percent of all mesotheliomas. In most cases the tumor cell type is sarcomatous. Histologically the arrangement of the collagenous areas varies. The most common pattern is a complex network consisting of anastomosing bands of often hyalinized collagen, with a prominent storiform pattern (Fig. 34-70). This is present in combination with more spindle-cell areas. The latter has features similar to the basketweave seen in benign pleural plaques. Cellular and desmoplastic areas often merge inperceptibly. A useful feature in diagnosis is collagen necrosis, seen in 70 percent of cases (Fig. 34-71). This is always bland without any associated inflammation and resembles fibrinoid necrosis on H&E sections. Mitoses are uncommon in desmoplastic malignant mesothelioma. This variant of mesothelioma is often difficult to diagnose. A further feature

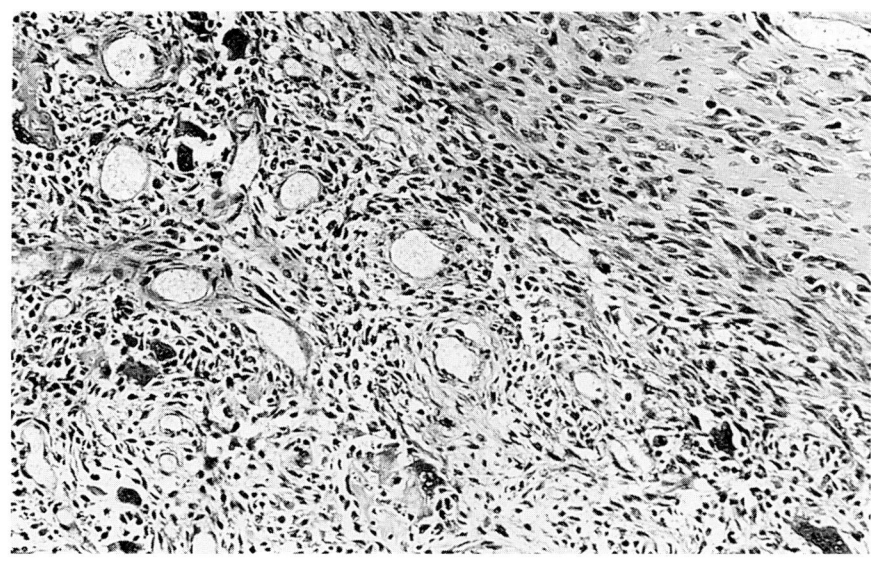

FIGURE 34-67
Mesothelioma with a fibrous pattern and
prominent osteoclastic cells. (H&E stain,
×87.5.)

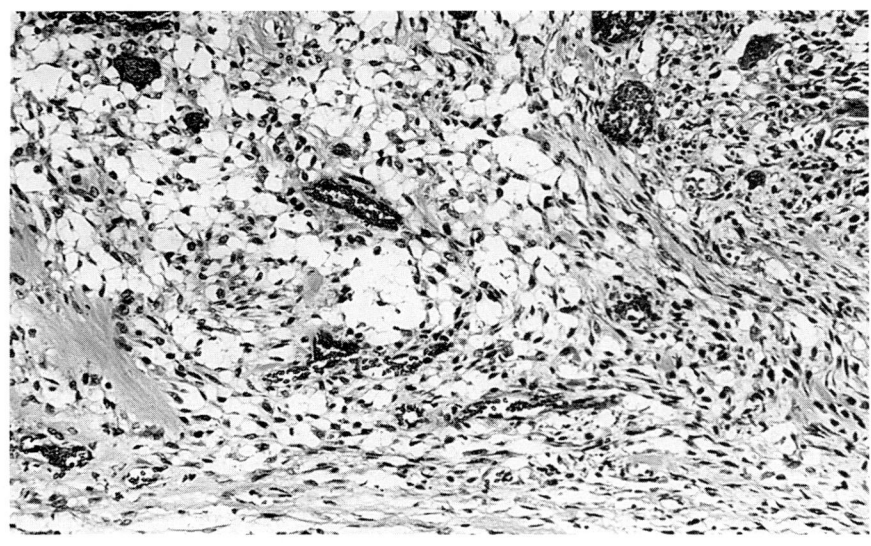

FIGURE 34-68
Liposarcomatous differentiation in the middle of a mesothelioma. (H&E stain, ×87.5.)

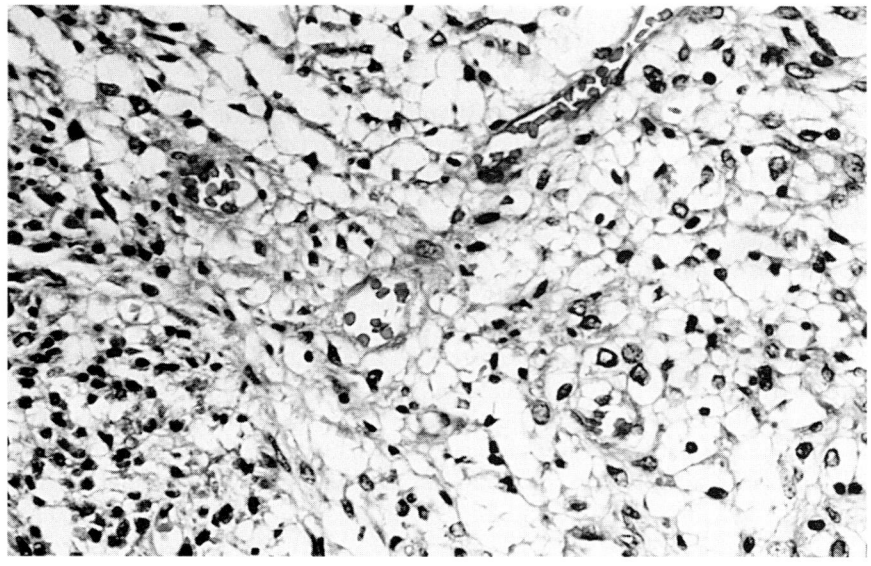

FIGURE 34-69
Higher magnification of Figure 34-68, showing nuclear hyperchromatism in the fat. Compare this with the bland fat spaces seen in Fig. 34-72. (H&E stain, ×219.)

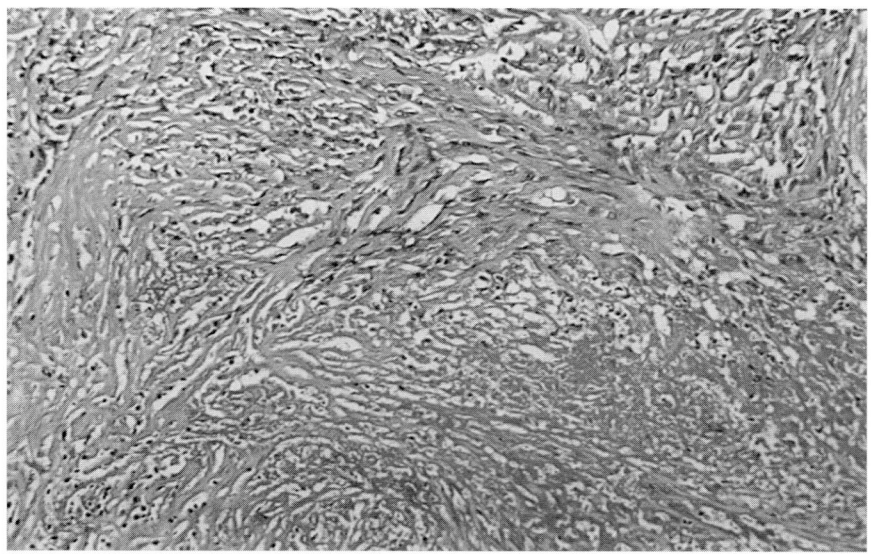

FIGURE 34-70
Desmoplastic mesothelioma with a storiform pattern as well as areas of necrosis. Some cellular foci are present. (H&E stain, ×87.5.)

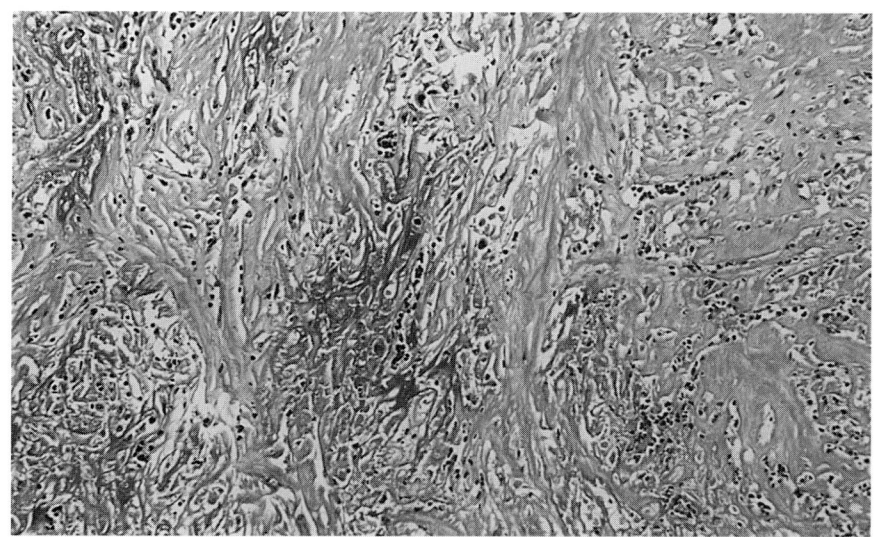

FIGURE 34-71
Desmoplastic mesothelioma with a prominent area of necrosis. (H&E stain, ×87.5.)

helpful in the diagnosis is its growth around the fat spaces in the parietal pleura (Fig. 34-72). The nuclei of the adipocytes are lost.

BIPHASIC OR MIXED MESOTHELIOMA

The mixed tumor is the easiest to diagnose since there are well-defined epithelial foci, with tubulopapillary patterns as well as sarcomatous areas (Fig. 34-73). As mentioned above, it may be difficult in some cases of epithelial mesothelioma to decide if the stroma is sarcomatous or part of the stromal reaction to the tumor. In the mixed pattern, there are usually large areas where only one histologic pattern predominates. After extensive sampling, the biphasic nature of the tumor is evident.

Other Cell Variants of Malignant Mesothelioma

1. *Lymphohistiocytoid mesothelioma.*[330] This is a rare variant, occurring in only 3 of 394 mesotheliomas studied by the Australian Mesothelioma Surveillance Program.[3] They may be misdiagnosed as lymphoma. The tumor tissue shows populations of large histiocytic cells, which vary from a round to a spindle shape (Figs. 34-74 and 34-75). The nuclei also vary from round to ovoid and contain finely divided chromatin with small prominent nucleoli. A diffuse but variable lymphoid infiltrate is also present in these cases. This consists of lymphocytes and plasma cells. Eosinophils may occasionaly be prominent. This gives an initial resemblance to Hodgkin's disease, but no Reed-Sternberg cells are identified.

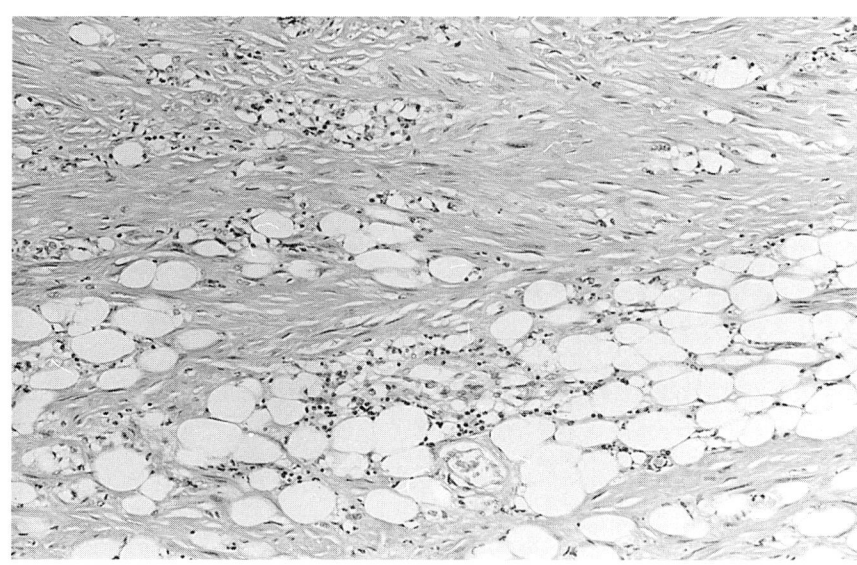

FIGURE 34-72
Desmoplastic mesothelioma growing round fat. (H&E stain, ×87.5.)

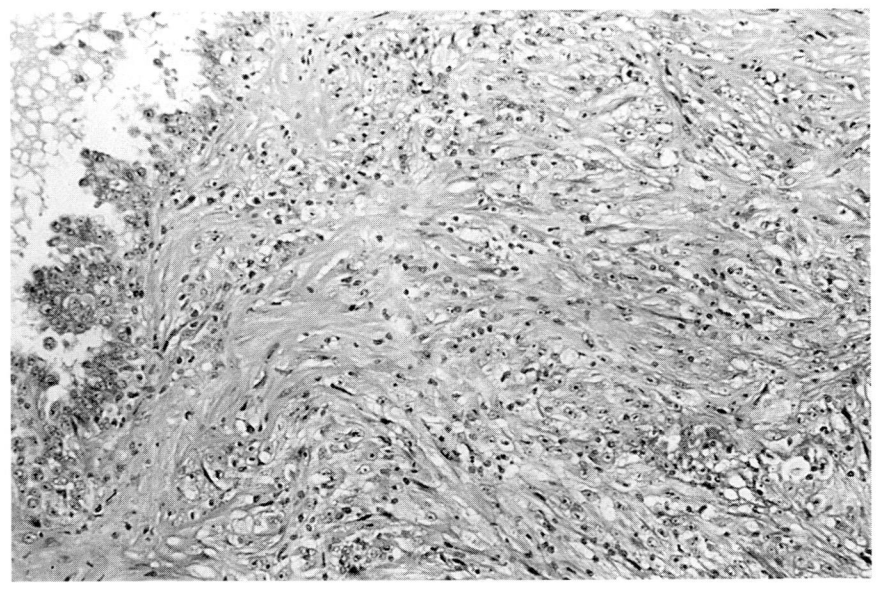

FIGURE 34-73
Biphasic mesothelioma. (H&E stain, ×87.5.)

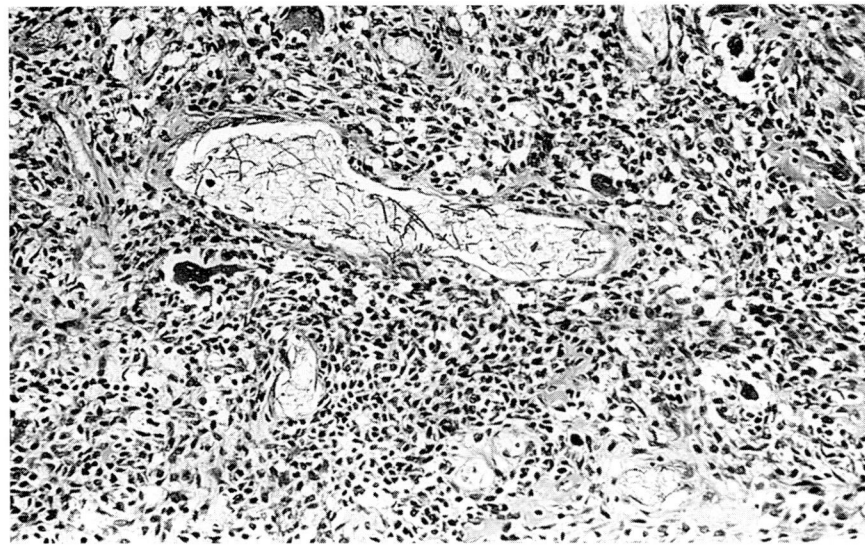

FIGURE 34-74
Lymphohistiocytoid mesothelioma, resembling a small-cell carcinoma or lymphoma. (H&E stain, ×87.5.)

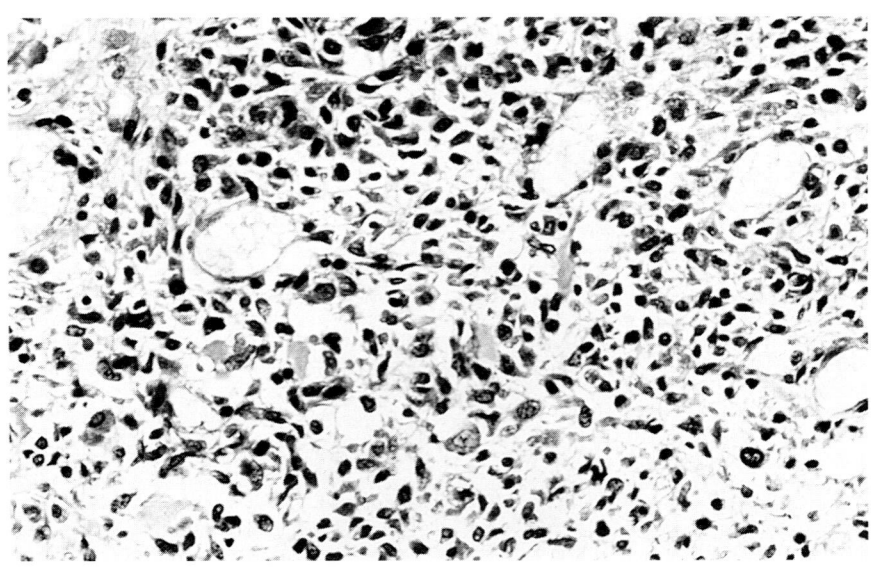

FIGURE 34-75
Higher magnification of Fig. 34-74 showing some cells with histiocytoid features. (H&E stain, ×219.)

The tumors are strongly positive for vimentin, cytokeratin, and low-molecular-weight cytokeratins (CAM 5.2). Tests for leukocyte common antigen are negative in all cases. In two cases, EMA was positive, and one case showed strong positivity for alpha-1 antichymotrypsin. Two cases showed weak positivity with S100 protein.

Electron microscopy of the lymphohistiocytoid variant showed predominantly mesenchymal cells with a mixed population of fibroblasts, fibrohistiocytic cells, histiocytes, myofibroblasts, rare xanthoma cells, and undifferentiated mesenchymal cells resembling those seen in malignant fibrous histiocytoma. The fibroblastic cells and fibrohistiocytes were large, bizarre cells with copious cytoplasm.

2. *Rhabdomyoblastic differentiation.* Rarely, there may be rhabdomyoblastic differentiation.[313]

3. *Giant cell.* A rare variant of malignant mesothelioma is the pleomorphic or giant-cell type. This may be confused with the giant-cell carcinoma of lung, and in such cases mucin and neuroendocrine stains should be done to exclude an adeno- or large-cell neuroendocrine tumor.

4. *Multicystic mesothelioma.* This tumor may affect the pleura in a similar way to the peritoneal cavity, though it is uncommon.[298] The lesion is considered in its entirety here for completeness. The lesions consist of thin-walled cysts separated by connective tissue and lined by a single layer of flattened and cuboidal epithelium (Fig. 34-76). These lesions are continuous with the parietal pleura and also adherent to the visceral pleura.

Histologically, the tumor is composed of multiple mesothelium-lined cystic spaces with an intervening fibrovascular stroma. The mesothelium ranges from flattened to plump cuboidal cells. Small buds or clumps of mesothelial cells are sometimes seen within the cystic spaces. In occasional cases, an adenomatoid pattern is identified. In some cases the tumor is indistinguishable from a conventional epithelial mesothelioma.

Villaschi and coworkers[309] showed positivity for AE1/3 and CK1 in stromal cells. Vimentin was weakly positive. The strongest staining was with AE1. Carcinoembryonic antigen, CA125 (a glycoprotein found in ovarian and intestinal tumors), and CA19-9 (an antigen found among glandular tumors) were all negative.

Ultrastructurally, the cells lining the cysts had the features of mesothelial cells, with a continuous basal lamina, numerous desmosomes, bundles of filaments, and a developed system of microvilli. The stroma between the cysts consisted of collagen with isolated elongated cells. These showed numerous microfilaments and dense bodies. Pinocytotic vesicles were visible, as well as an incomplete basal lamina. The features suggested that some of these cells originated from smooth muscle.

Follow-up. Where the lesion was solitary, 1 of 6 patients died of the disease. If the tumor was localized, 7 of 15 available for follow-up were alive. If the disease was diffuse, 1 of 16 patients who refused therapy died of the disease. There was therefore little correlation between the extent of the tumor and the patient's survival.

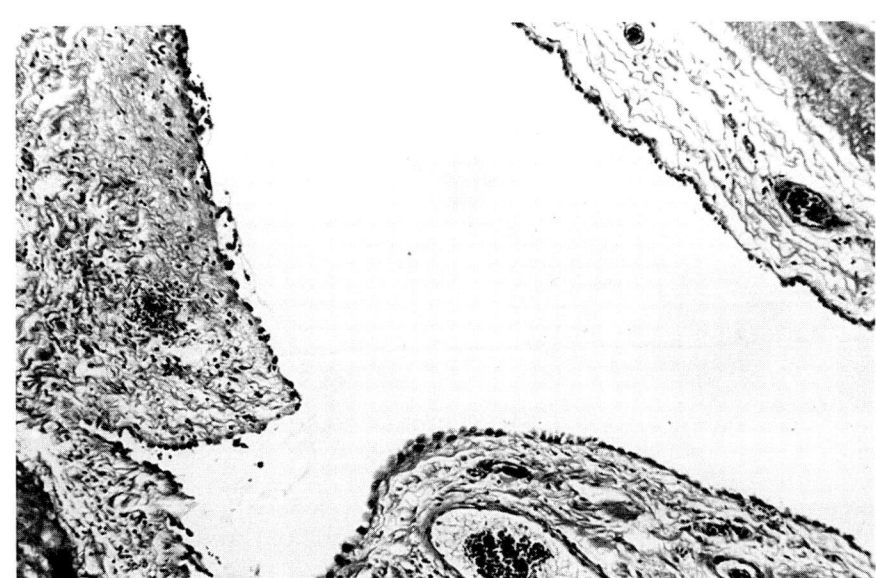

FIGURE 34-76
Multicystic mesothelioma with a low cuboidal mesothelial cell lining. (H&E stain, ×56.) (*Courtesy of Dr. H. Buckley, St. Mary's Hospital, Manchester, with permission.*)

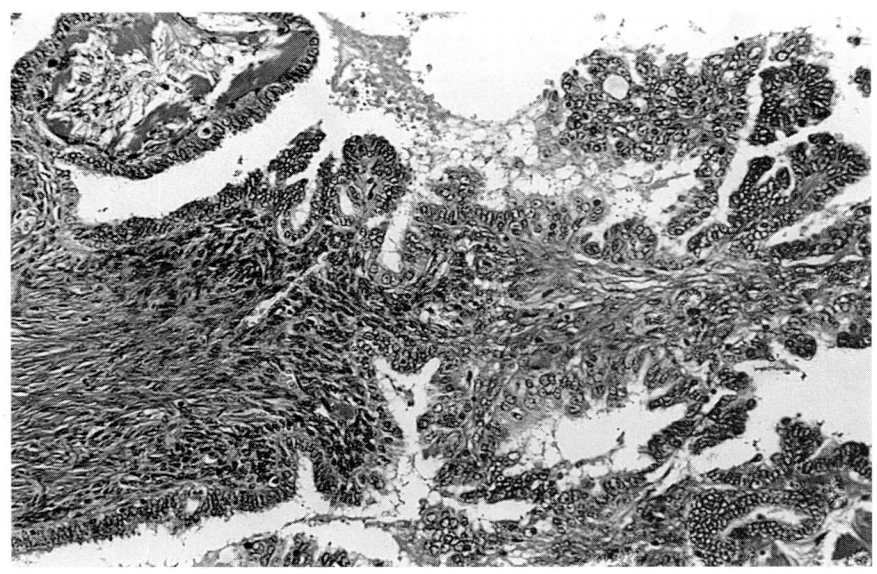

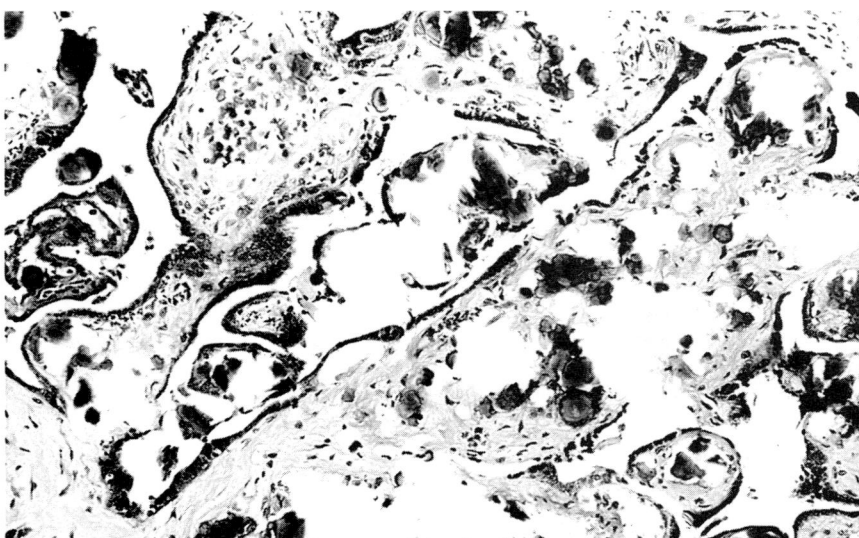

FIGURE 34-77
A. Endosalpingiosis with a well defined stroma and surface papillary projections. (H&E stain, ×87.5.) *B.* Endosalpingiosis with very prominent psammoma bodies. (H&E stain, ×87.5.) (*Courtesy of Dr. H. Buckley, St. Mary's Hospital, Manchester, with permission.*)

None of these patients with a cystic mesothelioma had been exposed to asbestos, though one such patient has been reported.[331]

The problem with this particular subgroup of tumors is whether to regard them as truly neoplastic and behaving in a similar way to mesothelioma in the other serous cavities. In view of the good survival and the association with chronic infection or previous surgery, it is likely that these are neoplasms but not mesotheliomas as understood in the rest of this chapter.

The differential diagnosis of cystic mesothelioma encompasses the following:

1. *Lymphangioma.* This tumor is lined by endothelial cells, which are sometimes positive for factor VIII. Cystic mesotheliomas are positive for cytokeratin.

2. *Endosalpingiosis.* These cysts are identical to the fallopian tube lining and may have areas of microcalcification (Fig. 34-77*A* and *B*). There is usually chronic salpingitis.

3. *Mesonephric remnants.* These are usually solitary lesions confined to the adnexae. They are lined by a transitional, squamoid, or clear epithelium.

4. *Epithelial inclusion cysts.* These are lined by serous or nonspecific cuboidal epithelium surrounding psammoma bodies (Fig. 34-78).

5. *Peritoneal serous micropapillomatosis.* Of low malignant potential, these are serous borderline tumors of peritoneum.[332] They have been described in women aged 16 to 67 years, with a mean age of 33 years. The patients may be asymptomatic or have chronic pelvic or abdominal pain and an adnexal

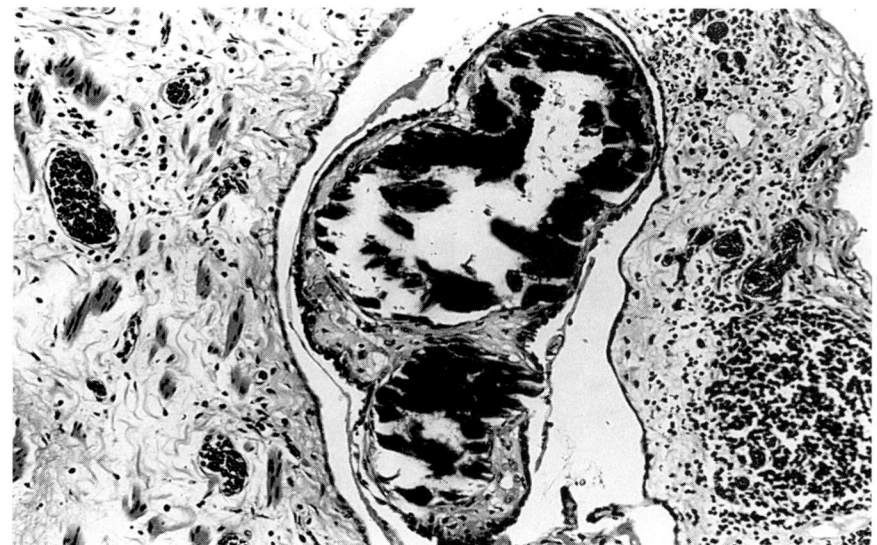

FIGURE 34-78
Epithelial inclusion cyst with psammoma body. (*Courtesy of Dr. H. Buckley, St. Mary's Hospital, Manchester, with permission.*)

mass causing obstruction of the small bowel. In just over half the cases, this is an incidental finding.

Macroscopically, the peritoneal lesions are either focal or diffuse, appearing as miliary granules. They may initially be diagnosed as peritoneal carcinomatosis. Histologically, there are desmoplastic and epithelial lesions or one of these components. Regardless of subtype, the peritoneal lesions consist of micropapillary clusters of serous epithelial cells. In all cases there are abundant psammoma bodies. In the desmoplastic lesions, there is a marked fibrous stromal reaction to the serous epithelial cells. The epithelium is relatively uniform, though all cases have either mild to moderate nuclear atypia. Most epithelial cells have cilia, but mitoses are infrequent.

Rare examples of well-differentiated papillary mesothelioma have been recorded in the pleura,[152] the epicardium,[333] and the tunica vaginalis of the testis.[334]

Electron microscopy shows cells with long, slender microvilli arranged in tufts and perinuclear bundles of cytoplasmic intermediate filaments. There are well-developed intact basement membranes and frequent desmosomal junctions. No asbestos bodies were identified, but the amount for sampling was small.

6. *Deciduoid peritoneal mesothelioma.* This rare, recently described entity affects young women and simulates exuberant, ectopic decidual tissue. One patient presented with vague pelvic pain, constipation, and early satiety. The other had a sudden onset of abdominal swelling, due to ascites, during pregnancy. These tumors run a rapidly fatal course.

Macroscopically, the omentum is replaced by a lobulated, granular, pink to gray tumor.

Histologically there is a proliferation of large round, ovoid, and polygonal cells. These have sharp cellular outlines, abundant glassy eosinophilic cytoplasm, and round vesicular nuclei with prominent nucleoli (Fig. 34-79). Binucleated cells are also seen, but mitoses are rare.

There is strong positivity for the cytokeratin markers, but CEA, Leu-M1, and germ-cell markers are negative.

The condition should be differentiated from any neoplastic or non-neoplastic lesion of intermediate trophoblast. Omental involvement by such trophoblastic cells would be unusual. There is usually a raised beta hCG; histologically, there are multiclefted nuclei and human placental lactogen in the trophoblastic cells.

Electron Microscopic Features of Mesotheliomas

With the advent of immunohistochemistry, ultrastructural studies of mesothelioma are now rarely done in most laboratories.

EPITHELIAL MESOTHELIOMAS

Epithelial mesotheliomas have similarities to normal mesothelium. Some have regarded the electron microscopy as of great value in exclusion of adenocarcinoma.[337] Six features have been given as being useful in the distinction between these two conditions.[338] These are (1) length and complexity of microvilli (Fig. 34-80 to 34-84); (2) microvillous/stromal collagen interdigitation (Fig. 34-85); (3) number and distribution of intermediate filament bundles (Fig. 34-86); (4) presence/absence of filamentous glycocalyx and glycocalyeal bodies; (5) presence/absence of microvillous core rootlets; (6) presence/

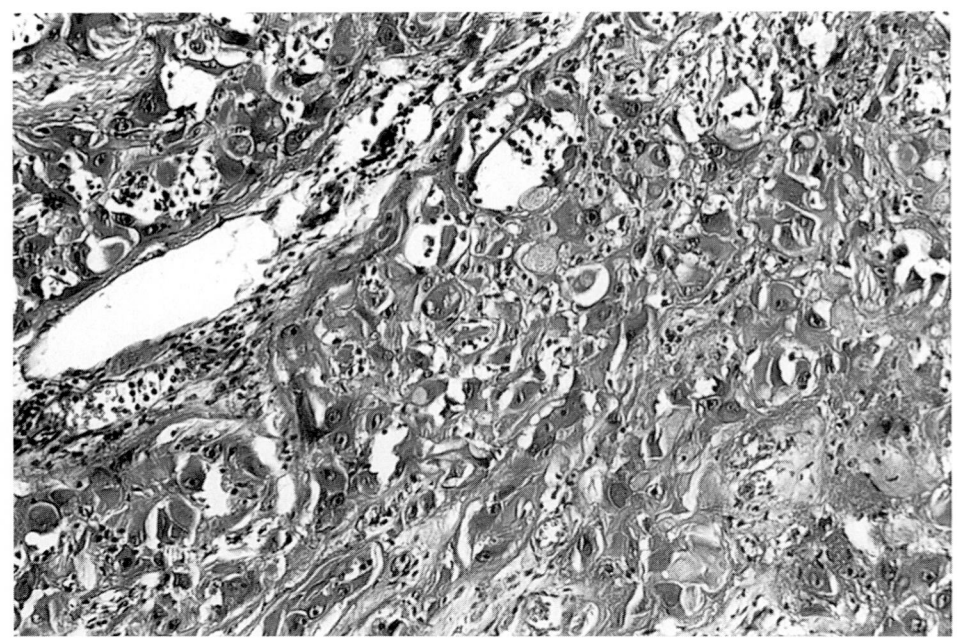

FIGURE 34-79
Deciduoid peritoneal mesothelioma with large polygonal cells and ample eosinophilic cytoplasm. Some binucleate cells are seen. (By courtesy of Professor C. D. Fletcher, Boston.) (H&E, ×190.)

absence of mucin granules and/or other secretory granules.

Tumor cells tend to show polarity related to the basal lamina. Intercellular junctions are common, desmosomes being the most frequent type. These are usually long.[339,340] However, less well-constructed desmosome-like junctions, intermediate cell junctions, and tight junctions can be identified.[341–345]

The cytoplasm of epithelial mesotheliomas is abundant, and intracytoplasmic vacuoles may occur. However, zymogen-like secretory granules are not identified. There are abundant intermediate filaments. These may be dispersed throughout the cell or arranged as perinuclear bundles.[341] These prominent intermediate filaments are responsible for the cytokeratin positivity of mesotheliomas. In semiquanti-

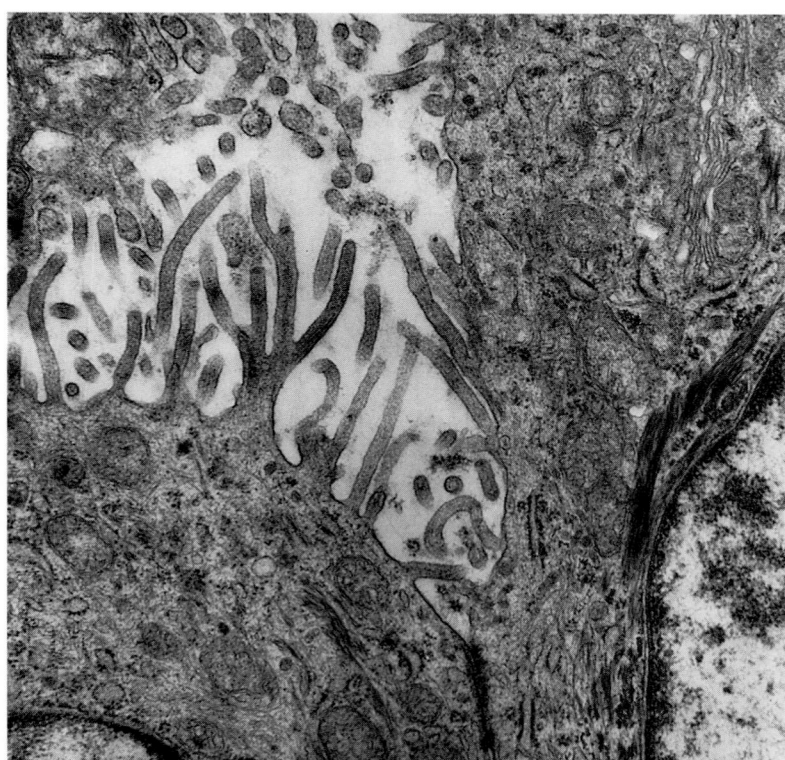

FIGURE 34-80
Epithelial mesothelioma with branching, cactus-like villi, prominent mitochondria, tonofilaments, and a prominent desmosome. (Uranyl acetate and lead citrate stain, ×18,000.)

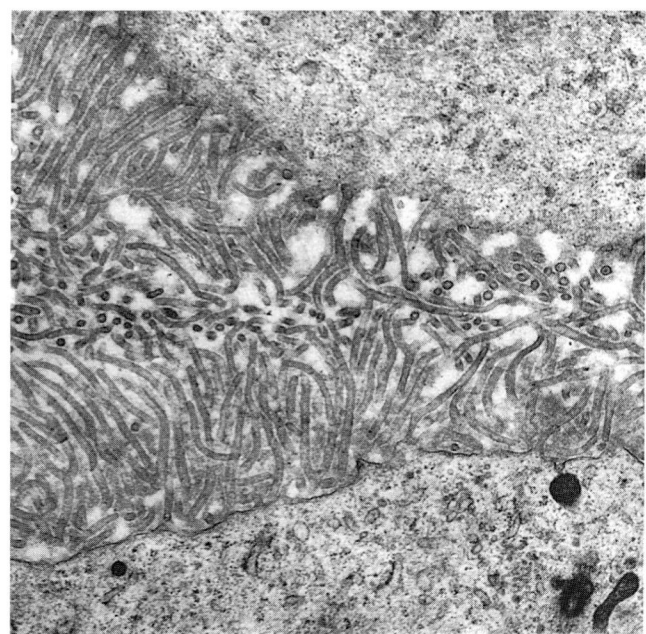

FIGURE 34-81
Complex intertwining long microvilli from two adjacent mesothelioma cells. (×11,840.)

tative comparative studies, the intermediate fiber content of mesotheliomas is significantly greater than that seen in adenocarcinomas. Few such bundles are seen in sarcomatous mesotheliomas.

Glycogen may be identified in mesotheliomas, being found in 78 percent of epithelial mesotheliomas in one series.[313]

Lysosomal granules, osmiophilic lamellar structures, and multivesicular bodies may occasionally be identified.[346] Membrane-bound, coarse secretory granules simulating neurosecretory granules have been identified in one study.[347]

The microvilli are a very useful identifying feature of mesothelial differentiation. In epithelial mesotheliomas, the microvilli are long and have a typical cactus-like appearance. Previous studies have separated adenocarcinomas from mesotheliomas by the length of the microvilli. Microvillus length is longer in mesotheliomas (mean ratio of length:diameter=15:7) as opposed to adenocarcinoma (8:7).[348] A microvillus length/diameter above 15 is highly suggestive of mesothelioma.[339] Adenocarcinomas usually have a value below 10.

In practice, this measurement is not commonly used and has been criticized.[349] This is because most studies have concentrated on well-differentiated mesotheliomas. In poorly differentiated tumors, the ratio may be markedly reduced. In addition, there is

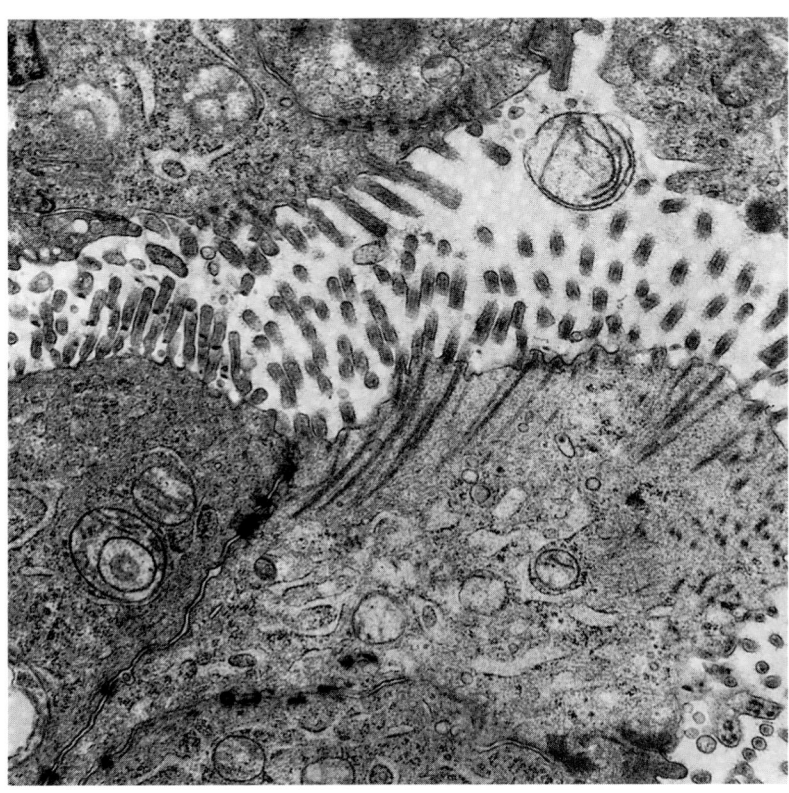

FIGURE 34-82
Small microvilli with well developed rootlets as seen in adenocarcinoma. (×18,000.)

1187

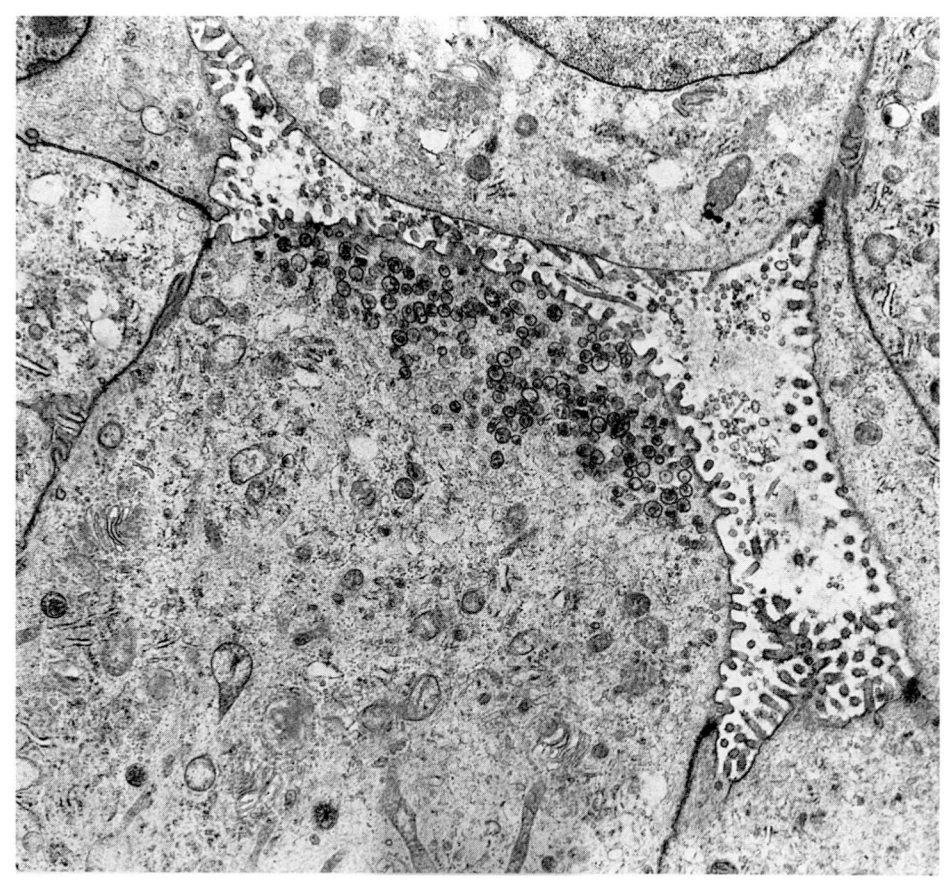

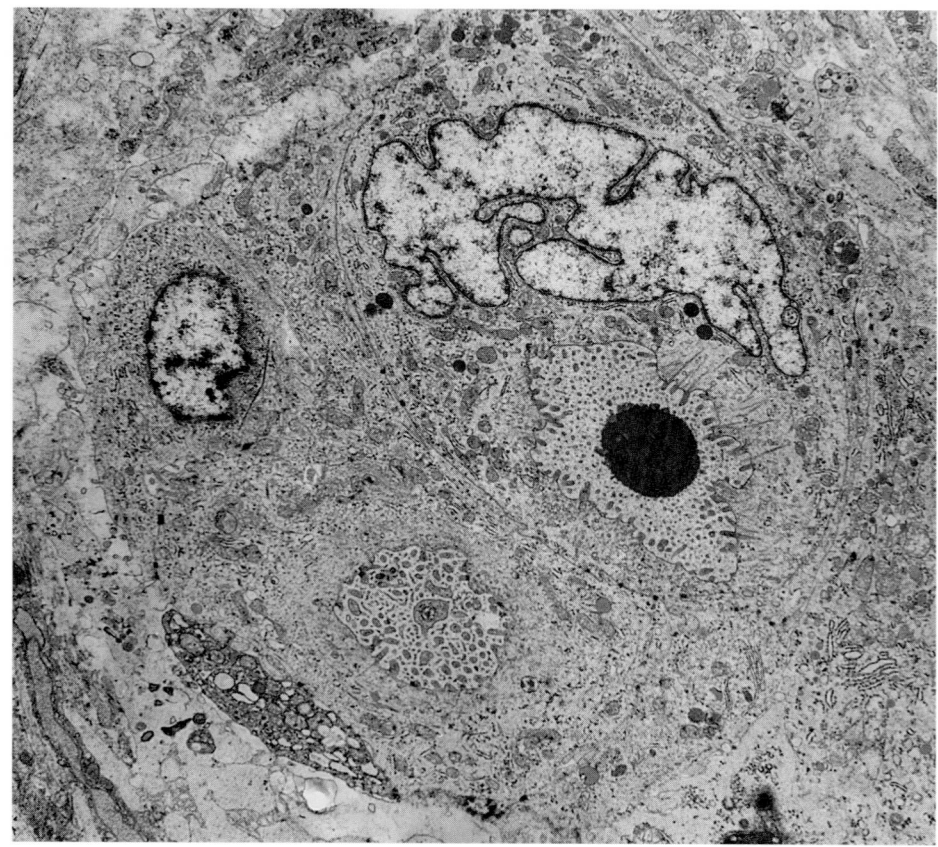

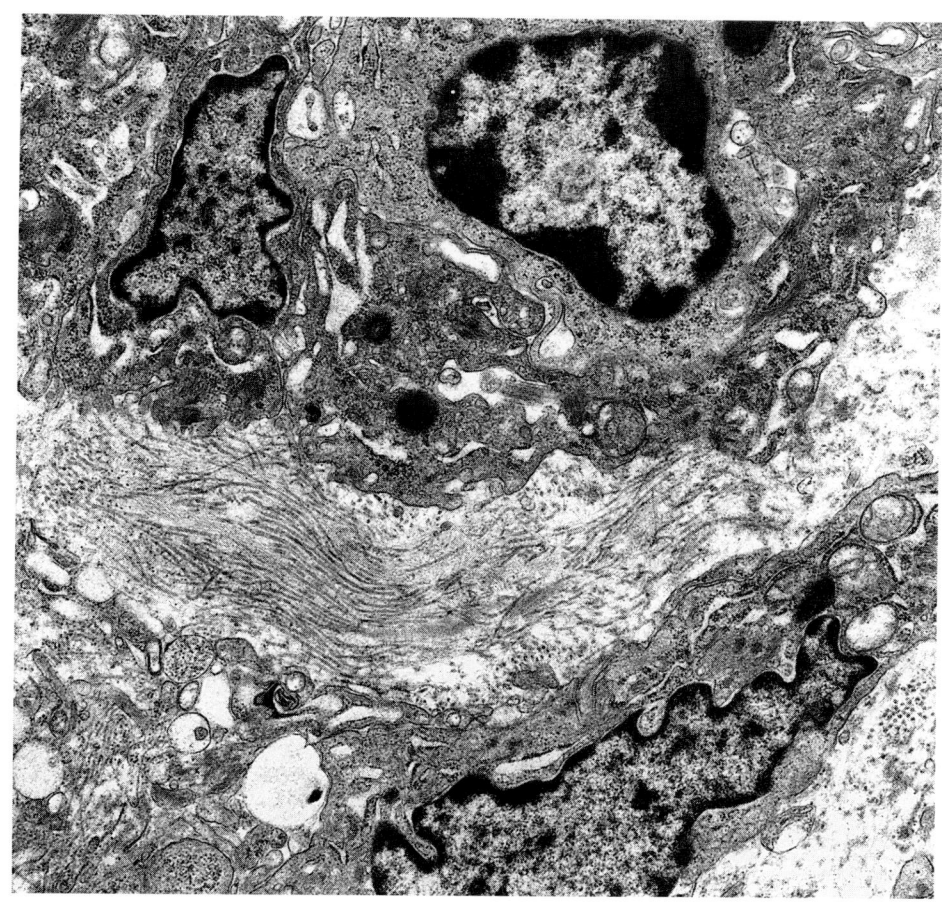

FIGURE 34-85
Focal stromal collagen-villous interdigitation. (×2920.)

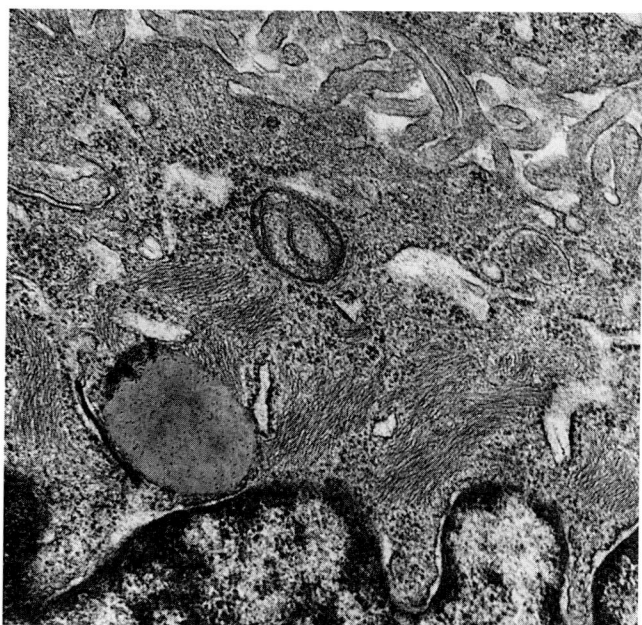

FIGURE 34-86
Mesothelioma with prominent intermediate filaments as well as a lipid droplet. (×29,600.)

a marked overlap in the range of microvillus length between mesotheliomas and adenocarcinomas. These authors concluded that "measurement of microvillus length is not a very sensitive parameter or a practical diagnostic discriminator, even in well-differentiated mesotheliomas."

Some adenocarcinomas may have long microvilli, but the complex elongated microvilli seen with secondary and tertiary branching in mesotheliomas are taken as characteristic. They project into the luminal spaces, into intracytoplasmic neolumina, or along intercellular spaces.[350] Some authors[340] have described microvilli making direct contact, through basement membrane deficiencies, with collagen fibers on the abluminal side of the tumor cells in most mesotheliomas.

Microvilli in adenocarcinomas are typically smaller in number and size and may have core rootlets with an underlying terminal filament. Such features are not identified in mesotheliomas. Some lymphomas, designated as anemone-cell tumors, may have elaborate microvillous processes.[337,351]

There are no intercellular junctions or intermediate filaments and their immunohistochemistry is different from that of mesothelioma.

Three distinct types of crystalloid formations in mesotheliomas have been described.[3] The first was extracellular, fernlike material thought to be due to the crystallization of hyaluronic acid.[352] Second, there was formation of elaborate sheets with a honeycomb structure arranged as cylinders, scrolls and sometimes enwrapping each other or microvilli.[353] The third type of crystalloid formation comprised crystallized bodies, which formed complex, multilayered sheaths, cylinders, or partially opened scrolls.

Hyaluronic acid may be seen as flocculent or granular material associated with microvilli or lying in intracytoplasmic lumina.[354]

SARCOMATOUS MESOTHELIOMAS

These show only rudimentary cell junctions and rare desmosomes. Microvilli, if present, are short and sparse. Formation of basal lamina is focal. The cytoplasm has abundant, randomly oriented intermediate filaments and microfilamentous bundles with focal densities (Figs. 34-87 and 34-88).[337] In some cases, tonofilaments may be sporadic. The cells have prominent rough endoplasmic reticulum, which may be dilated and contain osmiophilic material. The

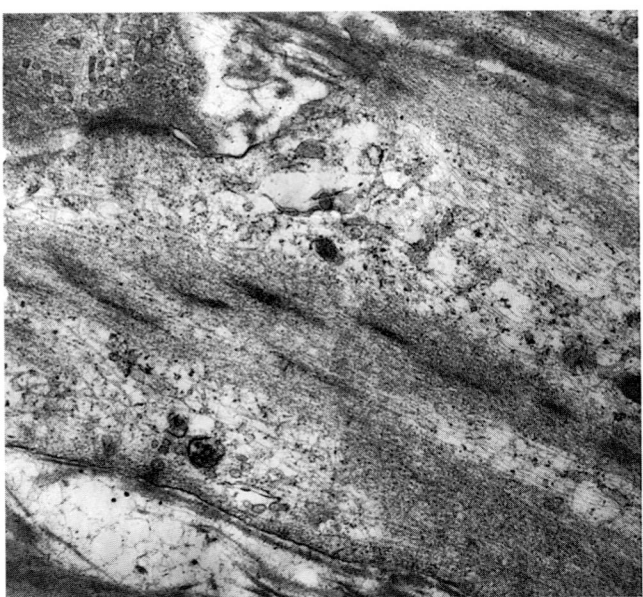

FIGURE 34-88
Desmoplastic mesothelioma with prominent intermediate filaments and focal densities as well as a well-defined cell junction. (×17,760.)

nuclei are often elongated and folded. These cells resemble fibroblasts or myofibroblasts with some epithelial characteristics.

Desmoplastic mesotheliomas are ultrastructurally similar to the fibrous ones, but there is a larger number of myofibroblasts.[341] Other authors have described the electron microscopy of the desmoplastic variants.[321]

BIPHASIC MESOTHELIOMAS

These show the two differentiation patterns described above, with both epithelial and sarcomatous foci. There are intermediate or transitional-cell forms as well as marked variation from area to area. It may be difficult to determine true mesothelial characteristics. Transitional-cell characteristics include epithelial features, such as basal lamina and desmosomal attachments to adjacent cells, in one region and cellular extension into the stroma in another.

The ultrastructure of poorly differentiated epithelial mesotheliomas has been described.[354] There is a mosaic pattern of closely associated tumor cells with long, narrow cytoplasmic processes. These lie parallel to adjacent tumor cells with long narrow cytoplasmic processes, lying parallel to adjacent plasma membranes. There was an absence of core filaments in the microvilli. The cells had abundant cytoplasm but limited numbers of organelles. Secretory granules were absent but prominent tonofilament bundles were identified in perinuclear

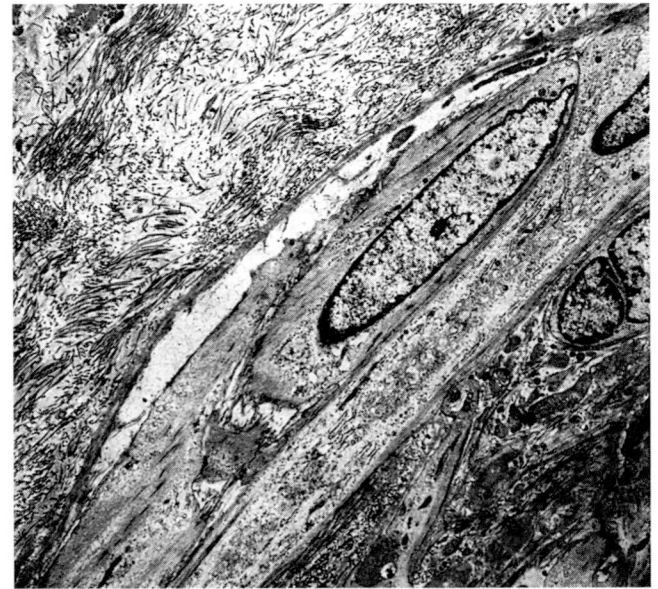

FIGURE 34-87
Mesothelioma with abundant collagen, fibroblastic cells, and focal densities in the cytoplasm. (×30,340.)

regions. This feature may be seen in only a small number of cells.

If the above ultrastructural features are reviewed, there is no single ultrastructural characteristic of malignant mesothelioma.[313] Thus one or more of the typical electron microscopic features of mesothelial differentiation may be absent in poorly differentiated tumors. As in most tumors, it is necessary to take into account the light microscopic, electron microscopic, and especially as will be seen in the next section the immunocytochemical findings.

Special Stains in Malignant Mesothelioma

The stains requested will depend on the morphology of the tumor. In the case of the epithelial tumor, it is important to exclude an adenocarcinoma. In practice, a simplified flow diagram (Fig. 34-89) is used. In epithelial mesotheliomas, it is mandatory to do a diastase/PAS to determine the presence or absence of neutral mucin. Diastase pretreatment is important, as both adenocarcinomas and some epithelial mesotheliomas contain glycogen. As has been noted above, some mesotheliomas contain *small foci* of neutral mucins.[313,337] The presence of D/PAS–positive material in intracytoplasmic lumina is convincing evidence of an adenocarcinoma. Only 39 percent of adenocarcinomas have a positive PAS reaction after diastase.[313]

We have not found any distinct advantage in using alcian blue with and without hyaluronidase digestion. This is because it is difficult to identify any cells on the slides used for the digestion.

Mucicarmine may occasionally be positive in mesotheliomas.[355,356]

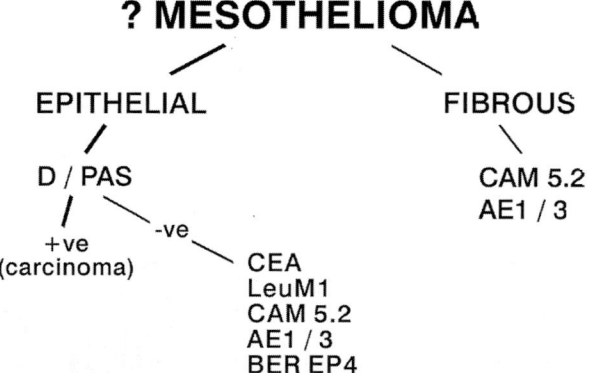

FIGURE 34-89
Flow diagram for stains used in diagnosis of mesothelioma.

Immunocytochemistry

Despite the previous description of the ultrastructural features of malignant mesothelioma, electron microscopy does not play a major role in the diagnosis of most cases of mesothelioma. Most histopathologists use a panel of markers for diagnosis. Immunocytochemistry is not easy. One of the main problems is that there is no one marker indicating mesothelial cell differentiation. The diagnosis is therefore one of exclusion. The pathologist must exclude carcinoma, one of the commonest differential diagnoses, but must also consider mesenchymal tumors. An excellent review of the role of immunocytochemistry in the diagnosis of malignant mesothelioma has been published.[357]

In practice, the histopathologist will approach the stains to be used after looking at routine H&E sections. The stains employed for sarcomatous or fibroblastic mesotheliomas can be modified from those used for epithelial tumors. Discussion of immunostains will center around stains used in the two main types of tumors. With regard to biphasic tumors, these will be described with the epithelial tumors.

SARCOMATOUS OR FIBROUS MESOTHELIOMAS

We use cytokeratin markers to detect mesothelial cells in the fibrous type of mesothelioma. Two markers, CAM 5.2, which detects low-molecular-weight cytokeratins, and AE1/3 (Fig. 34-90), an antibody to high- and low-molecular-weight keratins, are commonly used. These markers are also positive in epithelial mesothelioma (Fig. 34-91 and 34-92). It is important to use both high- and low-weight cytokeratins, since both are expressed by mesotheliomas. Some believe that there are different patterns of cytokeratin staining in carcinomas and mesotheliomas, but this is not generally accepted.[357] Others have shown that high- and low-molecular-weight cytokeratins are positive in mesotheliomas.[341,358–363]

Localized fibrous mesotheliomas are usually cytokeratin-negative,[364,365] though rare cases may be positive.[366] In some studies,[358,359] cytokeratin was absent from a large number of tumors that may mimic diffuse desmoplastic mesothelioma. The spindle cells tumors studied were malignant fibrous histiocytoma (MFH), malignant schwannoma, leiomyosarcoma, fibrosarcoma, fibromatosis, malignant hemangiopericytoma, angisarcoma, Kaposi's sarcoma, dedifferentated spindle-cell liposarcoma, osteosarcoma, and amelanotic spindle-cell malignant melanoma. Some soft tissue tumors, such as leiomyosarcoma and biphasic synovial sarcoma, may contain keratin.[367] Therefore it would be foolish to

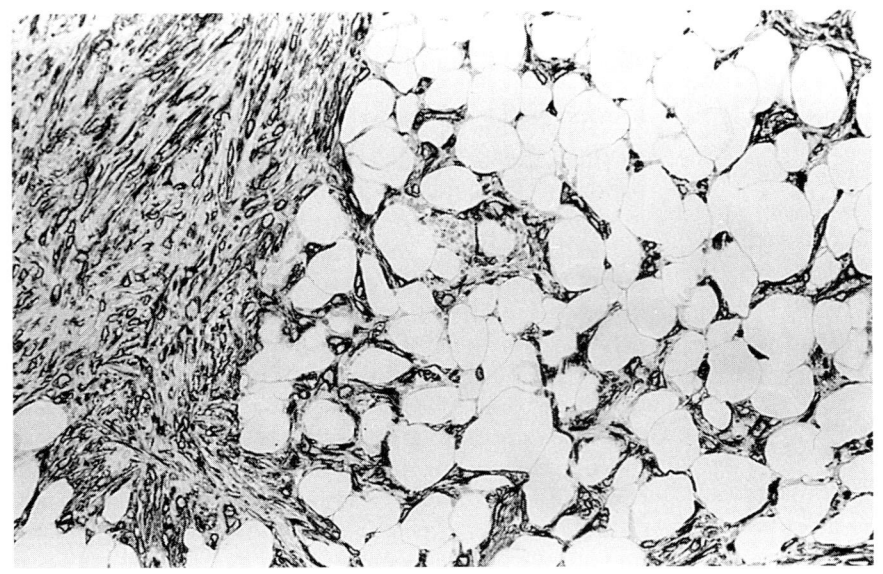

FIGURE 34-90
Fibrous mesothelioma extending into adipose tissue with cytokeratin positivity, as shown by AE1/3. (Immunoperoxidase stain, ×87.5.)

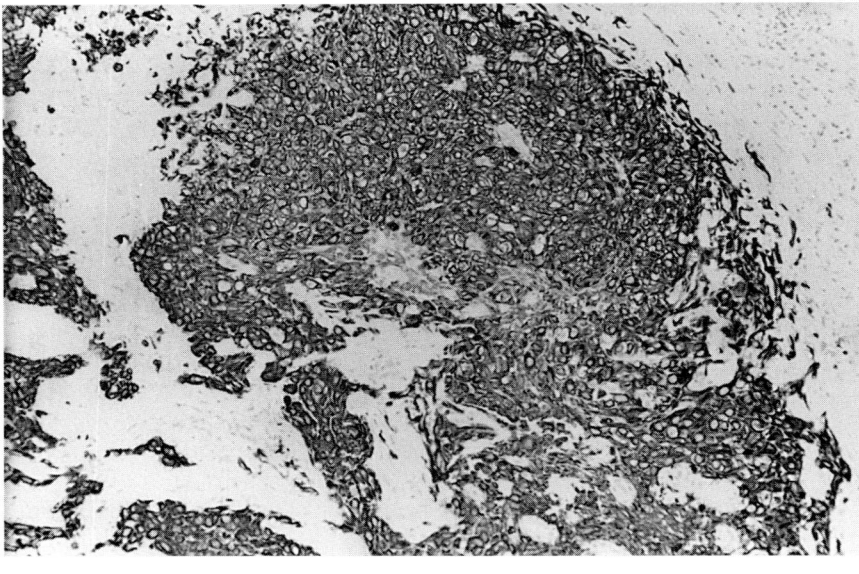

FIGURE 34-91
Epithelial mesothelioma with strong positivity to CAM 5.2. (Immunoperoxidase stain, ×87.5.)

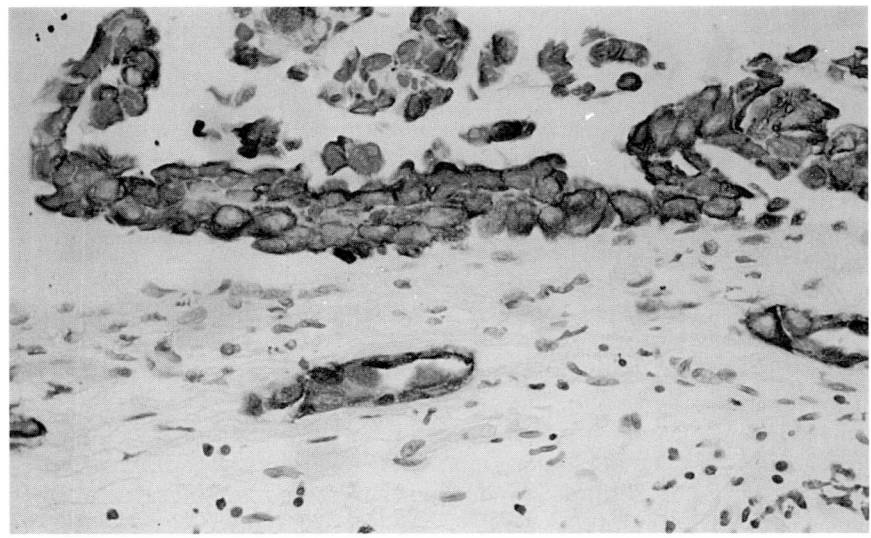

FIGURE 34-92
Epithelial mesothelioma with positivity to CAM 5.2. (Immunoperoxidase stain, ×219.)

rely on a single stain for confirmation of the diagnosis of diffuse sarcomatoid or fibrous malignant mesothelioma. Desmoplastic mesothelioma also stains positively with cytokeratin. Cytokeratins are identified in the desmoplasia induced by metastatic carcinoma to the pleura. This immunoreactivity should not be misinterpreted as a biphasic mesothelioma.

The use of vimentin (Fig. 34-93) in the diagnosis of malignant mesothelioma has given varying results and is not widely used diagnostically. Some authors[360,362,368] found that this antibody was good in differentating mesothelioma from adenocarcinoma. However, others identified vimentin in 46 percent[369] and 17 percent[370] of pulmonary adenocarcinomas. In the latter study, it was identified in only 41 percent of mesotheliomas. Double-staining immunohistologic methods, using vimentin and cytokeratins, showed that they were coexpressed in 11 of 12 mesotheliomas but in only 1 of 8 adenocarcinomas.[371] Though the level of vimentin staining is significantly greater in malignant mesothelioma than in pulmonary adenocarcinomas, it does not provide a clear-cut separation between these two groups. No study exists of the use of different vimentin antisera in the diagnosis of malignant mesothelioma.

Vimentin has been identified in normal reactive nonneoplastic subserosal cells, where it coexists with low-molecular-weight cytokeratin.[341]

Carcinomas of the kidney and thyroid and pleomorphic adenomas of salivary gland may coexpress both vimentin and cytokeratin.[372]

EPITHELIAL MESOTHELIOMAS

Carcinoembryonic Antigen

The absence of carcinoembryonic antigen (CEA) in mesotheliomas was first documented in 1979.[373] Since then the value of CEA-related antigens in malignant mesotheliomas has been a subject of continuing controversy in published reports. A recent study[374] summarized 40 studies using CEA in the diagnosis of mesotheliomas, pulmonary, and other carcinomas involving pleura or peritoneum; CEA was identified in 11 percent of mesotheliomas and 84 percent of carcinomas. The figure was lower in effusions, being 4 percent and 58 percent respectively. In serum as well as in pleural and ascitic fluid, significantly elevated levels of CEA were commonly associated with carcinomas but rarely with mesotheliomas.

Most of the published series have used polyclonal antibodies, but there is a family of glycoproteins with common antigenic epitopes and sometimes similar structural characteristics to CEA.[375] One of the commonest is nonspecific cross-reacting antigen (NCA) as well as a number of variants including NCA-95, NCA-55, NCA-160. Therefore the possibility of cross reactions with these CEA-related glycoproteins has to be considered when different studies are compared. In addition, neutrophil polymorphs react positively with NCA. MAb T84.66 binds weakly to neutrophil polymorphs and is a useful antibody.[376]

Churg,[377] in a review of the literature, found that 14 percent of mesotheliomas stained positively with

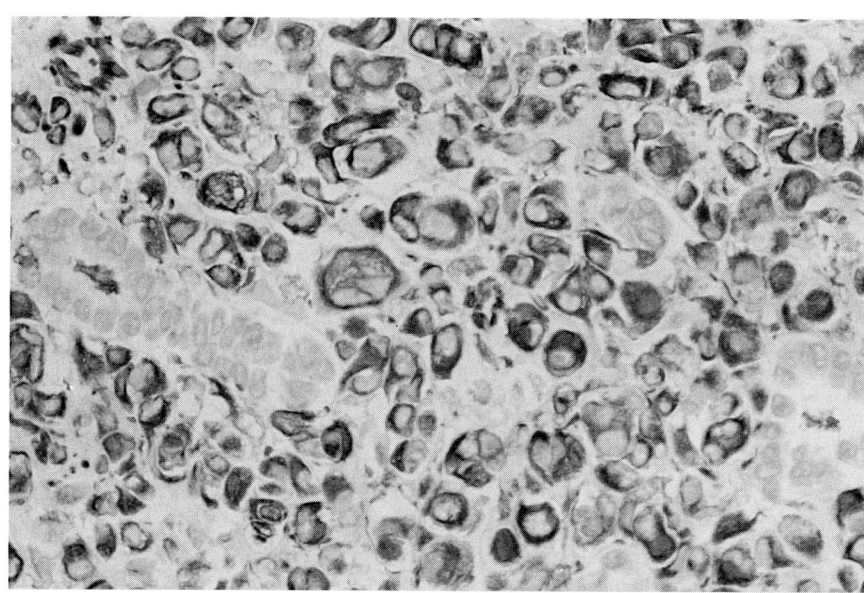

FIGURE 34-93
Epithelial mesothelioma with positivity to vimentin. (Immunoperoxidase stain, ×219.)

CEA. Henderson and colleagues,[3] in a similar review, gave a figure of 10.6 percent for mesotheliomas reacting to CEA, whereas 88 percent of pulmonary adenocarcinomas were CEA-positive.

In adenocarcinomas, the CEA positivity is usually strong, whereas in mesotheliomas the immunostaining is weak or equivocal. Henderson and coworkers[3] have demonstrated a few cases of mesothelioma with foci of coarse, granular staining often concentrated in the perinuclear areas. In addition, some cases had sporadic apical plasma membrane–related linear staining. This staining pattern coincides with the introduction of a monoclonal antibody to CEA (Biogenex).

In addition to staining neutrophil polymorphs, macrophages, foci of tumor necrosis, and alveolar cells entrapped amid the mesothelioma may also stain with CEA. There also appears to be a variability of the reliability of CEA between different laboratories. In our laboratory we find it of great use, whereas colleagues with similar experience of this tumor find it of little diagnostic value. It has recently been suggested[357] that any positive results obtained when a polyclonal anti-CEA antiserum is used without preabsorption with NCA (non-cross-reacting antibody) should be interpreted as nonspecific. These authors recommend that each antibody needs careful evaluation if nonspecific results are to be avoided. They use a monoclonal anti-CEA antibody (T84.86) developed in the Beckman Research Institute.[378] With this antibody, there was no evidence of expression of CEA by neoplastic cells in several hundred cases of malignant mesothelioma.[379] Seventy percent of adenocarcinomas were positive.[379] Ordonez,[380] using a mouse monoclonal antibody (Hybritech, San Diego, CA), found no mesotheliomas stained with CEA. Others have obtained similar results.[360] The anti-CEA used was specific for human carcinoembryonic antigen and did not bind to normal colon antigen (Zymed Lab., San Francisco).

Leu-M1 (CD-15 Antigen)

This monoclonal antibody reacts against the X-hapten (Lacto-N-fucose-pentosyl III) on a myelo-monocytic cell line.[381] This antibody also labels the neoplastic cells in Hodgkin's disease.[382] Meyloid cells, especially neutrophil polymorphs, are also positive. Leu-M1 is found in the majority of adenocarcinomas, including those of pulmonary origin.

Mesotheliomas are usually negative.[379,380,383,384] In the study of Sheibani and coworkers, 47 of 50 adenocarcinomas stained either focally or diffusely with Leu-M1. None of the 28 malignant mesotheliomas expressed this antigen.

This study was extended.[385] They studied 127 mesotheliomas and 268 adenocarcinomas, including 55 of pulmonary origin. All the mesotheliomas contained cytokeratins and lacked CEA and mucin. Leu-M1 was identified in the 199 of 268 (74 percent) adenocarcinomas. None of the 127 mesotheliomas showed Leu-M1 positivity. Other authors have confirmed these findings.[370,386]

We find Leu-M1, along with CEA, one of the most useful markers in the panel of antibodies used. It has to be taken in context of the clinical picture, including the operative or postmortem findings as well as the routinely stained H&E sections. In some adenocarcinomas, the immunoreactivity of Leu-M1, though intense, is focal. This causes a problem with small biopsies, where a false negative could be reported.

Monoclonal Antibody B72.3

This monoclonal antibody was produced by immunization of mice with cells from a human breast cancer.[387] The antibody may react with some normal adult tissues. It was positive in 86 percent of adenocarcinomas but in no mesotheliomas. The staining of adenocarcinomas may be strong, focal, and predominantly cytoplasmic. Unfortunately the monoclonal antibody has a low sensitivity, this being approximately 30 to 40 percent.[357]

Monoclonal Antibody Ber-EP4

This antibody was produced in 1990 by the immunization of mice with cells from a breast carcinoma line.[388] It is as yet unclear which antigen the antibody is recognizing. Eighty-seven percent of adenocarcinomas were positive but only 1 (1 percent) of the mesotheliomas showed any positivity.[389] Breast and renal carcinomas did not react with this antibody. The staining pattern in adenocarcinomas was intense and membranous, but occasional weak cytoplasmic staining was also identified.

There has been some confirmation of these findings in a recent series.[390] Of 120 adenocarcinomas, 103 (86 percent) showed positivity with this antibody. Reactivity was also seen in 2 of 9 adenomatoid tumors (22 percent) but 10 of 49 mesotheliomas (20 percent) were also positive. Staining in mesotheliomas was restricted to epithelial areas and was focal. However, in one case Ber-EP4 stained the majority of the neoplastic cells. This antibody is therefore restricted in its diagnostic ability and should be one of a panel of antisera.

Epithelial Membrane Antigen

Epithelial membrance antigen (EMA) is positive in both mesothelioma and adenocarcinoma.[391] It may be positive in desmoplastic mesothelioma though negative in localized fibrous mesothelioma.[392] The pleura is usually weakly positive, whereas epithelial cells of breast, sweat glands, salivary glands, and respiratory epithelium react strongly.[393] One series identified 78 percent of mesotheliomas as EMA-positive, with a smaller number (62 percent) of adenocarcinomas showing positivity.[394]

The main use of EMA has been in the possible differentiation of reactive pleural lesions from malignant mesothelioma. The staining in mesothelioma is usually strong whereas, it is weak in reactive lesions. In most cases this is not a major problem and the antibody is not part of our repertoire of special stains for the diagnosis of mesothelioma.

Human Milk Fat Globules

Antibodies reacting with human milk fat globules (HMFG-2) were identified in 1983.[395] These antibodies are similar in distribution to EMA and recognize oligosaccharide antigenic determinants, both in cytoplasm and on the cell membrane. Some have shown positivity in nearly all pulmonary adenocarcinomas and no staining with mesothelioma.[379] Absence of staining with HMFG-2 strongly favors a diagnosis of mesothelioma.

Other studies have cast doubt on the usefulness of this antibody. Wirth and coworkers[360] demonstrated both membrane and cytoplasmic staining with HMFG-2 in most adenocarcinomas, but in 26 percent of mesotheliomas there was only membrane staining. Reactive mesothelium in one study was negative with HMFG-1 and 2, but 12 of 16 mesotheliomas were positive. Staining was identified in the cytoplasm in some cases but was more common at the cell surface. Solid clumps of epithelial cells did not stain, but in clefts within the tumor the luminal surfaces of the cells became strongly positive. Three out of seven biphasic tumors showed strong granular cytoplasmic staining of the spindle-cell element.[396] Joglekar and coworkers[397] found HMFG-2 the least useful of the antisera they used in the diagnosis of mesothelioma.

Monoclonal Antibody 44-3A6

This has been generated from human pulmonary adenocarcinoma cells and recognizes a membrane-associated glandular differentiation antigen.[398] In one study, 10 adenocarcinomas stained strongly, whereas 10 of 43 mesotheliomas were only focally and weakly immunoreactive.[399] In a more recent study on cell blocks from cytologic specimens, 24 primary lung and 25 of 30 metastatic adenocarcinomas were positive, but only 3 of 36 cases of mesotheliomas showed reactivity.[400] In two of the mesotheliomas, a few

tumor cells were positive. In a third case, more than 50 percent of the tumor cells were strongly stained. This patient had had a contralateral right pleural malignant mesothelioma 1 year previously, which had stained diffusely with 44-3A6. The presence of one positive malignant mesothelioma again indicates caution before a diagnosis based on the use of this antibody is made.

Biotinylated Probe Specific for Hyaluronate

Azumi and coworkers developed a probe from the hyaluronate binding region of bovine cartilaginous proteoglycan.[401] Only 3 of 37 adenocarcinomas (8 percent) showed significant staining, but all the mesotheliomas were positive for hyaluronate. The staining reaction was classed as moderate or greater in 79 percent of mesotheliomas. Positivity was identified in sarcomatous, desmoplastic, and epithelial mesotheliomas. The procedure had a sensitivity of 79 percent and specificity of 92 percent. These initial findings, which deserve confirmation, indicate that the hyaluronate probe, especially when used in conjunction with immunohistochemistry and light microscopy, could become a powerful diagnostic tool for differentiating between mesothelioma and adenocarcinoma. It has an added advantage that it can be used on formalin-fixed, paraffin-embedded tissue.

Lectins

Lectins have received little research interest for the differentiation of mesothelioma and adenocarcinoma. Kawai[402] studied 6 reactive mesothelial lesions, 23 mesotheliomas of all cell types, and 28 well-differentiated pulmonary adenocarcinomas. Eight different lectins were utilized. Wheat germ and peanut agglutinin lectins were positive in 5 of 6 (83 percent) and 3 of 6 (53 percent) respectively in reactive mesothelial lesions. Of the malignant mesotheliomas, 57 percent showed a positive reaction with these two lectins.

In contrast, pulmonary adenocarcinomas showed positive reactions with peanut agglutinin (96 percent), *Ricinus communis* (93 percent), wheat germ (89 percent), and succinylated wheat germ (79 percent). Less than 9 percent of mesotheliomas expressed these last two lectins (*R. communis* and succinylated wheat germ agglutinin).

The problem is the same as identified above with regard to CEA and Leu-M1. This is that adenocarcinomas as well as some mesotheliomas stain positively, making application in an individual case difficult.

Basement Membrane Components

A small number of mesotheliomas examined for the basement membrane proteins laminin, two epitopes of the laminin receptor, and type IV collagen showed intracytoplasmic reactivity.[403] Normal mesothelium showed basement membrane staining with type IV collagen and laminin but little or no cytoplasmic activity. Extracellular staining was minimal in mesotheliomas but was seen in adenocarcinomas of breast and lung. Laminin receptors were identified on malignant mesothelioma cells by intense positive staining. Immunoreactivity was also seen in nonneoplastic mesothelial proliferations and adenocarcinomas, but there was greater heterogeneity and less intensity.

The p53 Protein

The p53 protein is encoded by a tumor suppressor gene discovered by Lane and Crawford.[404] Twenty reactive mesothelial proliferations showed no p53 reactivity, whereas 14 of 20 (70 percent) of malignant mesotheliomas were positive. Abnormalities of p53 are a common event in malignant mesothelioma and therefore may be of value in the distinction of malignant mesotheliomas from reactive hyperplasia.[405–407] p53 is of no value in differentiating mesothelioma from adenocarcinoma.[408]

Antimesothelioma Monoclonal Antibodies

To date these have not given consistent results. One monoclonal antibody, designated MAB-45, reacted with 4 of 7 mesothelioma cell lines but also with cells of some carcinomas, sarcomas, and melanomas. The antibody did not react with normal pleural or lung tissue.[409] Donna and colleagues[410] used a polyclonal antibody to mesothelial cells on conventionally processed tissues. All 16 malignant mesotheliomas were positive, and the antibody did not react with any of the 31 lung carcinomas studied. Unfortunately, this antibody has not become widely available.

Other Antibodies

Recently muscle actins have been demonstrated in mesotheliomas.[150b] All fibrous mesotheliomas were positive for both muscle specific actin (MSA) and smooth muscle actin (SMA). The epithelial cells were positive in 9/12 epithelial mesotheliomas and 2/5 biphasic mesotheliomas. All lung cancers were essentially negative for the muscle actins. Further cases need to be studied to assess the usefulness of these markers.

Parathyroid hormone–related protein (PTHrP) was demonstrated in the 84 percent of malignant mesotheliomas (type not specified) and 11 percent of adenocarcinomas. No PTHrP was identified in normal or reactive mesothelium.[410a]

In mesothelioma both the epithelial and fibrous components stain positively as shown in Fig. 1 of Clarke et al.[410a] However, when applied to a single case this antibody suffers the same drawbacks as many of those described above.

In summary we find that the most useful stains in differentiating mesothelioma from adenocarcinoma are D/PAS, CEA, Leu-M1, BER-EP4, CAM 5.2, and AE1/3.

Flow Cytometry Studies

Cytogenetic analysis of human tumors has provided information of clinical, prognostic, and epidemiologic importance. These studies have given varying results in mesotheliomas. Up to 78 percent of pleural epithelial mesotheliomas were diploid.[411,412] A larger series[413] found that only 26 percent were diploid and 43 percent aneuploid. In a recent paper[414] reporting a study of 70 mesotheliomas, the median S-phase fraction was 5.6 percent (range, 1.2 to 19.9 percent). Of the total number of tumors, 86 percent were diploid. Dazzi and associates[412] found the S phase percentage to be of greater significance than ploidy status.

The general findings are that, unlike other solid tumors, mesotheliomas tend to be diploid or near diploid. The presence of diploid content of this tumor cannot be taken to imply that it is a genetically stable tumor. As noted in Chap. 30, many chromosomal abnormalities have been identified in mesothelioma. The S phase is of prognostic value.

Prognostic Factors

A critique on the state of the art regarding prognosis in mesothelioma has been given by Van Meerbeeck.[415] Pleural mesothelioma is a terrible disease, with most treatments proving disappointing. There have been no randomized clinical trails. In most of the published studies, there is little use of a staging system.

Many studies show that the histologic type of tumor is important in prognosis. Patients with the epithelial subtype have a longer survival than those with fibrous tumors. Biphasic tumors lie in between.[288,299,416–422] Not all studies have shown a convincing survival advantage with the epithelial subtype.[423,424] However, there is generally a failure to state how many blocks of tissue were sampled. This is important, since the diagnosis may be made on small samples and tumors with a biphasic pattern may be underdiagnosed. In comparing results from various centers, tumor histology can be used as a stratification factor only if the same diagnostic modalities are used.

Other factors important in some studies in relation to improved survival are the duration of symptoms,[418,423] age at diagnosis,[416,418,423,420] female gender,[288,418,421] stage,[418–421,424] and performance status,[288,417,418,420,422,423] surgery,[417,423] chemotherapy,[423,424] and surgery and chemotherapy.[417,418]

Differential Diagnosis

The differential diagnosis of mesothelioma is large. A diagnosis of mesothelioma should not be made on a small needle biopsy, without a full clinical history and the radiologic findings. If it is not possible to give an unequivocal answer, it is wisest to ask for an open pleural biopsy. This can be done as a video-assisted thoracoscopic technique, causing little discomfort to the patient and providing the pathologist with an adequate piece of tissue for diagnosis. This facilitates rational treatment and enables the clinician to give some prognostic information to the patient.

The differential diagnosis falls into the following categories:

REACTIVE MESOTHELIAL LESIONS

These are not commonly misdiagnosed as mesothelioma. The commonest lesion is a pneumothorax, where there may be small inclusions of regular mesothelial cells but the pleura is thin, and there is inflammation consisting of lymphocytes, plasma cells, and varying numbers of eosinophils. Some small multinucleated giant cells are also present. Collagen diseases may also cause reactive pleural lesions, especially rheumatoid disease. Rare cases with marked mesothelial proliferation may lead to a misdiagnosis of mesothelioma.[425]

Rounded atelectasis may present as a pleura-based mass, causing some mesothelial proliferation. However, the pleura is infolded. Rarely, a resolving pleurisy may be confused with a mesothelioma. In this condition there is abundant fibrin on the surface, marked inflammation and fibrosis extending through the pleura, as well as new blood vessel formation. Nuclear pleomorphism is not a feature and there is no storiform pattern, as seen in fibrous mesotheliomas.

Recently atypical mesothelial hyperplasia has been associated with bronchogenic carcinoma[426] and may be confused with mesothelioma. Eight cases with excision of lung and pleura were described. During thoracotomy for removal of an intrapulmonary mass in one of these patients, the surgeon encountered a parietal pleural nodule, which was interpreted on frozen section as probable malignant mesothelioma. Each case showed mesothelial hyperplasia with varying degrees of atypicality. The mesothelial cells had hyperchromatic nuclei and enlarged nucleoli. The atypical mesothelial cells for the most part formed a single layer on the serosa, but the cells were occasionally piled up and large, with pleomorphic nuclei and abundant cytoplasm. In one case reactive mesothelial cells produced micropapillary and short tubular structures resembling an adenomatoid tumor.

Pleural papillary foci, even if small, are not benign but indicative of malignant mesothelioma. Such proliferations usually do not have atypical nuclei but are of the bland, open, vesicular type seen with malignant mesothelioma. Pulmonary infarction may cause a pleurisy with mesothelial hyperplasia and atypical mesothelial cells. Mesothelial hyperplasia may also accompany infection, especially with *M. tuberculosis*, where there are usually foci of caseation and giant cells.

In difficult cases, p53 antibody may differentiate malignant pleural lesions from reactive processes. AgNORs (Argyrophil nucleolar organizer regions) have been used, but in many laboratories have not given reproducible results.

SARCOMATOUS MESOTHELIOMA

The differential diagnosis in sarcomatous mesotheliomas is that of a spindle-cell tumor. This includes renal carcinoma, spindle-cell squamous carcinoma, secondary spindle-cell carcinoid, amelanotic melanoma, angiosarcoma, malignant hemangiopericytoma, extraosseous chondrosarcoma, osteogenic sarcoma, malignant nerve sheath tumor, malignant epithelioid hemangioendothelioma, leiomyosarcoma, fibrosarcoma, synovial sarcoma, malignant fibrous histiocytoma, and carcinosarcoma.

Pleural metastases are rare. It is unusual for most sarcomas to metastasize to the pleura without first declaring themselves at their primary site.

Specific antibodies and stains can be used for carcinoids, melanomas (HMB 45 and S100), malignant hemangiopericytoma (reticulin), angiosarcoma (some cases are positive with *Ulex europaeus* and factor VIII–related antigen), and malignant nerve sheath tumor (often, but not always S100-positive). Leiomyosarcoma may be desmin-, myoglobin,- and SMA-positive. In malignant epithelioid hemangioendothelioma, factor VIII, *Ulex*, CD 34, and thrombomodulin are positive. Myxoid chondrosarcoma has some vacuolated cells and is S100 protein–positive.

Synovial sarcoma may cause difficutly, since the epithelial component is keratin-positive. Ideally, a fibrosarcoma should be vimentin-positive and keratin-negative, but rare reports of spindle-cell sarcomas, especially leiomyosarcomas, being keratin-positive are described. Extraskeletal myxoid chondrosarcoma has glycogen-positive cells, alcian blue staining with or without hyaluronidase, and is negative with cytokeratin and muscle-associated antigens. Not all authors have identified S100-protein positivity. Leu-7 has been reported as positive in some cases.[427,428] Epithelioid sarcoma is rarely pleural-based, but these tumors are keratin-positive.[429]

Osteogenic sacoma may be confused with malignant mesothelioma when the latter tumor has osseous and cartilaginous differentiation. The histologic differentiation of mesothelioma with osseous differentiation and pleural osteogenic sarcoma may be difficult. The presence of an epithelial component and keratin-positive malignant cells should help distinguish a mesothelioma from an osteogenic sarcoma. Primary pleural osteogenic sarcoma would present as a large lesion and usually lacks a history of asbestos exposure.[430–432] These two features may not help in an individual case. In mesotheliomas, the foci of new bone or cartilage are relatively small.

Malignant fibrous histiocytoma (MFH) may cause confusion with fibrous mesothelioma. The former has a spindle-cell pattern; it may also show a storiform pattern and is keratin-negative. Since there are no specific stains for this tumor, differential diagnosis may be difficult. It is very rare as a primary pleural tumor. Rhabdomyoblastic differentiation has been described in mesotheliomas.[3]

Localized benign and malignant fibrous tumors of the pleura may cause confusion with malignant mesothelioma. The radiologic picture is that of a discrete tumor that is cytokeratin-negative but vimentin positive.

DESMOPLASTIC MESOTHELIOMA

The differential diagnosis of this condition includes nonspecific pleural fibrosis, seen in some cases of tuberculosis, pleurisy, and rheumatoid disease. In these conditions the clinical history may be helpful. Diffuse pleural fibrosis, associated with asbestos exposure, has a typical basketweave pattern, little inflammation, and no cellular fibroblastic proliferation or collagen necrosis. In addition, there is dense subpleural parenchymal interstitial fibrosis and the asbestos fiber counts are elevated.

Localized pleural plaques should not cause diagnostic difficulty as there is a regular basketweave arrangment and storiform or whorled areas are absent, as is collagen necrosis. Rarely, desmoplastic malignant mesothelioma masquerades as a sclerosing mediastinitis.[297] This latter diagnosis was initially made on sclerotic areas showing densely hyalinized fibrous tissue and necrotizing granulomas contain-ing organisms resembling *Histoplasma capsulatum* within mediastinal lymph nodes. At autopsy, the mediastinum contained a mass of sclerotic material with dense fibrous and sarcomatous tissue. The final diagnosis was a localized mediastinal desmoplastic mesothelioma.

LYMPHOHISTIOCYTOID MESOTHELIOMA

This may be misinterpreted as a lymphoma. The lymphocytes are often characterized by varying degrees of nuclear irregularity, but immunohistochemistry shows strong coexpression of cytokeratins and vimentins by the histiocytoid cells. There is no positivity for LCA.

A misdiagnosis of secondary neuroendocrine tumor may be made when examining small-cell tumors in the pleura (Fig. 34-94). This is because some mesotheliomas have dense core granules that may be detected by special stains.

Another tumor with many lymphocytes is the pleural thymoma. Macroscropically, this may mimic a diffuse epithelial mesothelioma. Pleural thymomas show a dense, prominent lymphoid component, some areas showing striking lobulation. The lobules are sharply separated by fibrous bands. There is a biphasic cell population, with small round lymphocytes admixed with scattered larger cells containing large vesicular nuclei. These latter cells are epithelial. Some pleural thymomas have a prominent epithelial component with nesting, spindling, and trabecular patterns. Immunocytochemistry shows a strong cytokeratin reaction in the epithelial component, giving a false diagnosis of mesothelioma. The degree of lobulation and the dense lymphocytic response is unusual in a malignant mesothelioma. This lymphocytic response may be seen in lymphohistocytoid mesothelioma.

EPITHELIAL MESOTHELIOMAS

As mentioned repeatedly above, the commonest difficulty in diagnosis is between mesothelioma and adenocarcinoma. Typically, an adenocarcinoma will show mucin, CEA, Leu M1, and Ber EP4 positivity.

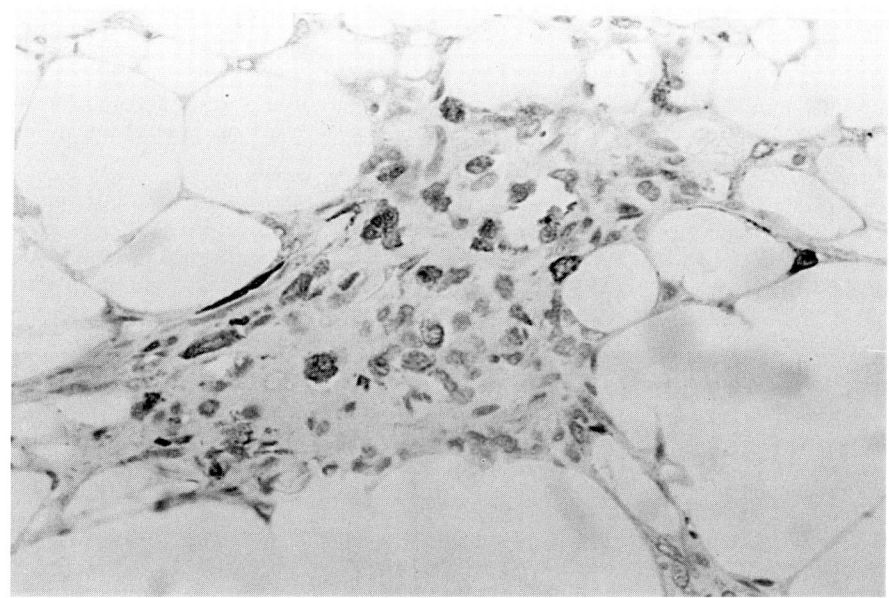

FIGURE 34-94
Mesothelioma with some cells showing positivity to PGP 9.5. (Immunoperoxidase stain, ×87.5.)

Occasionally, multicystic mesotheliomas may have an appearance almost like that of a germ cell tumor. This may cause problems, since beta-HCG and alpha-fetoprotein are positive in both germ cell tumors and mesotheliomas.

Multicystic mesotheliomas may be misdiagnosed as lymphangiomas. Positive staining with the endothelial marker UEA 1 does not exclude a mesothelial origin, since many epithelial surfaces express the H type II blood antigen, bound to this lectin.[433] Lymphangioma usually shows smooth muscle between the cysts. Electron microscopy of multicytic mesothelioma shows long, slender microvilli arranged in tufts, frequent desmosomal junctions, and perinuclear bundles of cytoplasmic intermediate filaments. The distinction between malignant mesothelioma of the cystic variety and lymphangioma is of practical importance because the mesotheliomas recur in nearly half the cases, whereas lymphangiomas rarely recur.

Another diagnostic problem is posed by mesothelial cell inclusions in mediastinal lymph nodes. These may mimic metastatic carcinoma.[81] The cells appear only in the nodal sinuses and occur predominantly as single individual cells arranged in small clusters. The nuclei are bland and the nuclear/cytoplasmic ratio is low. Cell borders are well defined and there are no mitoses. Immunohistochemically, the inclusions are positive for cytokeratins but negative for CEA and Leu-M1.

CARCINOSARCOMAS

This may be a very difficult diagnosis in a biphasic tumor. Mayall and Gibbs[434] described two carci-

nosarcomas presenting as pleural tumors. The biphasic histology closely mimicked mesothelioma, but the tumors showed neutral mucin, squamous, or neuroendocrine differentiation. One of the tumors encased the lung in a manner typical of mesothelioma. The second case showed tumor in the pleura of the left upper lobe. Osteoid was present in one case, surrounding an acinar, epithelial component.

Immunohistochemistry is of only limited use in these tumors, since cytokeratin was positive but CEA was only weakly positive in one case. Vimentin was negative in both pleural carcinosarcomas.

REFERENCES

1. Sahn SA: State of the art: The pleura. *Am Rev Respir Dis* 138:184–234, 1988.
2. Jones JSP: *Pathology of the Mesothelium.* London, Springer-Verlag, 1988.
3. Henderson DW, Shilkin KB, Langlois S Le P, Whitaker D: *Malignant Mesothelioma.* New York, Hemisphere, 1992.
4. Roggli VL, Greenberg SD, Pratt PC: *Pathology of Asbestos-Associated Diseases.* Boston, Little, Brown, 1992.
5. Nagel W, Kuschinsky W: Study of the permeability of the isolated dog mesentery. *Eur J Clin Invest* 1:149–154, 1970.
6. Staub MC, Wiener-Kronish JP, Albertine KH: Transport through the pleura: Physiology of normal liquids and solute exchange in the pleural space, in Chretien J, Bignon J, Hirsch A, (eds): *Pleura in Health and Disease.* New York, Marcel Dekker, 1986, pp 169–193.
7. Wiener-Kronish JP, Matthay MA, Callen PW, Filly RA, Gamsu G, Staub NC: Relationship of pleural effusions to pulmonary hemodynamics in patients with congestive cardiac failure. *Am Rev Respir Dis* 132:1253–1256, 1985.
8. Telver L, Jaubert F, Eyquem A, Andreux JP, Labrousse F, Chretien J: Study of immunoglobulins in pleura and pleural effusions. *Thorax* 34:389–392, 1979.

9. Beekman JF, Bosniak S, Canter HG: Eosinophilia and elevated IgE concentration in a serous pleural effusion following trauma. *Am Rev Respir Dis* 110:484–489, 1974.

10. Pettersson T, Klockars M, Hellstrom PE, Riska H, Wangel A: T and B lymphocytes in pleural effusions. *Chest* 73:49–51, 1978.

11. Yamagishi K, Tajima M, Suzuki A, Kimura K: Relation between cell composition of pleural effusions in patients with pulmonary carcinomas and their clinical courses. *Acta Cytol* 20:537–541, 1976.

12. Yamagishi K, Tajima M, Suzuki A, Kimura K: Relation between cell composition of pleural effusions in patients with pulmonary carcinomas and their clinical courses. *Acta Cytol* 20:537–541, 1976.

13. Light RW, MacGregor MI, Luchsinger PC, Ball WC: Pleural effusions: The diagnostic separation of transudates and exudates. *Ann Intern Med* 77:507–513, 1972.

14. Light RW, Erozan YS, Ball WC Jr: Cells in pleural fluid: Their value and differential diagnosis. *Arch Intern Med* 132:854–860, 1973.

15. Yam LT: Diagnostic significance of lymphocytes in pleural effusions. *Arch Intern Med* 66:972–982, 1967.

16. Chusid EL, Siltzbach LE: Sarcoidosis of the pleura. *Ann Intern Med* 81: 190–194, 1974.

17. Adelman M, Albelda SM, Gottlieb J, Haponik EF: Diagnostic utility of pleural fluid eosinophilia. *Am J Med* 77:915–920, 1984.

18. Mattson SB: Monosymptomatic exudative pleurisy in persons exposed to asbestos dust. *Scand J Respir Dis* 56:263–272, 1975.

19. Valdes L, Pose A, Suarez J, et al: Cholesterol: A useful parameter for distinguishing between pleural exudates and transudates. *Chest* 99:1097–1102, 1991.

20. Tomlinson JR, Sahn SA: Invasive procedures in the diagnosis of pleural disease. *Semin Respir Med* 9:30–36, 1987.

21. Mungall IPF, Cowen PN, Cooke NT, Roach TC, Cooke NJ: Multiple pleural biopsy with the Abrams needle. *Thorax* 35:600–602, 1982.

22. Canto A, Rivas J, Saumench J, Morena R, Moya J: Points to consider when choosing a biopsy method in cases of pleurisy of unknown origin. *Chest* 84:176–179, 1983.

23. Tomlinson JR, Miller KS, Lorch DG, Strange C, Wooton S, Sahn SA: Closed pleural biopsy: A prospective study of dual biopsy sites. *Am Rev Respir Dis* 133:A56, 1986.

24. Grunze H: The comparative diagnostic accuracy, efficiency and specificity of cytologic techniques used in the diagnosis of malignant neoplasm serous effusion of the pleural and pericardial cavities. *Acta Cytol* 8:150–164, 1964.

25. Race GA, Scheifler CH, Edwards JE: Hydrothorax in congestive cardiac failure. *Am J Med* 22:83–89, 1957.

26. Wiener-Kronish JP, Goldstein R, Matthay RA, et al: Lack of association of pleural effusion with chronic pulmonary arterial and right atrial hypertension. *Chest* 92:967–970, 1987.

27. Lieberman FL, Hidemura R, Peters RL, Reynolds TB: Pathogenesis and treatment of hydrothorax complicating cirrhosis with ascites. *Ann Intern Med* 64:341–351, 1966.

28. Johnson RF, Loo PV: Hepatic hydrothorax: Studies to determine the source of the fluid and report of 13 cases. *Ann Intern Med* 61:385–401, 1964.

29. Rudnick MR, Coyle JF, Beck LH, McCurdy DK: Acute massive hydrothorax, complicating peritoneal dialysis: Report of 2 cases and a review of the literature. *Clin Nephrol* 12:38–44, 1979.

30. Townsend R, Fragola JA: Hydrothorax in a patient receiving continuous ambulatory peritoneal dialysis: Successful treatment with intermittent peritoneal dialysis. *Arch Intern Med* 142:1571–1572, 1982.

31. Leung RW, Williams AJ, Oill PA: Pleural effusion associated with urinary tract obstruction: Support for a hypothesis. *Thorax* 36:632–633, 1981.

32. Carcillo J, Salcedoj JR: Urinothorax as a manifestation of non dilated obstructive uropathy following renal transplantation. *Am J Kidney Dis* 5:211–213, 1985.

33. Cavina C, Vitchi G: Radiological aspects of pleural effusions in medical nephropathy in children. *Am Radiol Diagn* 31:163–202, 1985.

34. Light RW, George RB: Incidence and significance of pleural effusion after abdominal surgery. *Chest* 69:621–625, 1976.

35. Krakowka P, Rowinska E, Halweg H: Infection of the pleura by *Aspergillus fumigatus*. *Thorax* 25:245–253, 1970.

36. Colp CR, Cook WA: Successful treatment of pleural aspergillosis and bronchopleural fistula. *Chest* 68:96–98, 1975.

37. Irani FA, Dolovich J, Newhouse MT: Bronchopulmonary and pleural aspergillosis. *Am Rev Respir Dis* 103:552–556, 1971.

38. Parry MF, Coughlin R, Zambetti FX: *Aspergillus* empyema. *Chest* 81:768–770, 1982.

39. Beekman JF, Zimmet SM, Chun BK, Miranda AA, Katz S: Spectrum of pleural involvement in sarcoidosis. *Arch Intern Med* 136:323–330, 1976.

40. Nicholls AJ, Friend JAR, Legge JS: Sarcoid pleural effusion: Three cases and review of the literature. *Thorax* 35:277–281, 1980.

41. Selroos O: Exudative pleurisy and sarcoidosis. *Br J Dis Chest* 60:191–196, 1966.

42. Sharma OP, Gordonson J: Pleural effusion in sarcoidosis: Report of six cases. *Thorax* 30:95–101, 1975.

43. Wilen SB, Rabinowitz JG, Ulreich S, Lyons HA: Pleural involvement in sarcoidosis. *Am J Med* 57:200–209, 1974.

44. Dressler W: The post myocardial infarction syndrome: Report of 44 cases. *Arch Intern Med* 103:28–42, 1959.

45. Engle MA, Ito T: The post pericardiotomy syndrome. *Am J Cardiol* 7: 73–82, 1961.

46. Lessof MH: Post cardiotomy syndrome: Pathogenesis and management. *Hosp Pract* 11:81–86, 1976.

47. Liem KL, ten Veen JH, Lie KI, Feltkamp TW, Durrer R: Incidence and significance of heart muscle antibodies in patients with acute myocardial infarction and unstable angina. *Acta Med Scand* 206:473–475, 1979.

48. Kaminsky ME, Rodan BA, Osborne DR, Chen J TT, Sealy WC, Putman CE: Post-pericardiotomy syndrome. *AJR* 138:503–508, 1982.

49. Van Der Geld H: Anti-heart antibodies in a post pericardiotomy and a post myocardial infarction syndromes. *Lancet* 2:617–621, 1964.

50. Engle MA, Gay WA Jr, McCabe J, et al: Post pericardiotomy syndrome in adults: Incidence, autoimmunity and virology. *Circulation* 64:(suppl 2): 58–60, 1981.

51. Stelzner TJ, King TE Jr, Anthony VB, Sahn SA: The pleuropulmonary manifestations of the post cardiac injury syndrome. *Chest* 84:383–387, 1983.

52. Pavlidakey GP, Hashimoto K, Blum D: Yellow nail syndrome. *J Am Acad Dermatol* 11:509–512, 1984.

53. Samman PD, White WF: The yellow nail syndrome. *Br J Dermatol* 76:153–157, 1964.

54. Solal-Celigny P, Cormier Y, Fournier M: The yellow nail syndrome: Light and electron microscopic aspects of the pleura. *Arch Pathol Lab Med* 107:183–185, 1983.

55. Alarcon-Segovia D, Alarcon DJ: Pleuro-pulmonary manifestations of systemic lupus erythematosus. *Dis Chest* 39:7–17, 1961.

56. Harvey AM, Shulman LE, Tumulty PA, Conley CL, Schoenrich EH: Systemic lupus erythematosus: Review of the literature and clinical analysis of 138 cases. *Medicine* 33:291–437, 1954.

57. Winslow WA, Ploss LN, Lotman B: Pleuritis in systemic lupus erythematosus: Its importance as an early manifestation in diagnosis. *Ann Intern Med* 49:70–88, 1958.

58. Good JT Jr, King TE, Anthony VB, Sahn SA: Lupus pleuritis: Clinical features and pleural fluid characteristics with special reference to pleural fluid antinuclear antibodies. *Chest* 84:714–718, 1983.

59. Purnell DC, Baggenstoss AH, Olsen AM: Pulmonary lesions in disseminated lupus erythematosus. *Ann Intern Med* 42: 619–620, 1955.

60. Walker WC, Wright V: Pulmonary lesions and rheumatoid arthritis. *Medicine* 47:501–519, 1968.

61. Sahn SA, Kaplan RL, Maulitz RM, Good JT Jr: Rheumatoid pleurisy: Observations on the development of low pleural fluid pH and glucose level. *Arch Intern Med* 140:1237–1238, 1980.

62. Baggenstoss AH, Rosenberg EF: Visceral lesions associated with chronic infectious (rheumatoid) arthritis. *Arch Pathol* 35:503–516, 1943.

63. Cruikshank B: Rheumatoid arthritis and rheumatoid disease. *Proc Assoc Med* 50:462–465, 1957.

64. Talbot JA, Calkins E: Pulmonary involvement in rheumatoid arthritis. *JAMA* 189:911–913, 1964.

65. Nosanchuk JS, Naylor B: A unique cytologic picture in pleural fluid from patients with a rheumatoid arthritis. *Am J Clin Pathol* 50:330–335, 1968.

66. Walton EW: Giant cell granuloma of the respiratory tract (Wegener's granulomatosis). *Lancet* 2:265–270, 1958.

67. Godman GC, Churg J: Wegner's granulomatosis: Pathology and review of the literature. *Am J Pathol* 58:533–553, 1954.

68. Kavuru MS, Adamo JP, Ahmad M, Mehta AC, Gephardt GN: Amyloidosis and pleural disease. *Chest* 98:20–23, 1990.

69. Knapp MJ, Roggli VL, Kim J, Moore JO, Shelburne JD: Pleural amyloidosis. *Arch Pathol Lab Med* 112:57–60, 1988.

69a. Bontemps F, Tillie-Leblond I, Coppin MC, et al: Pleural amyloidosis: Thoracoscopic aspects. *Eur Respir J* 8:1025–1027, 1995.

70. Celli BR, Rubinow A, Cohen S, Brody JS: Patterns of pulmonary involvement in systemic amyloidosis. *Chest* 74:543–547, 1978.

71. Smith RR, Hutchins GM, Moore GW, Humphrey RL: Type and distribution of pulmonary parenchymal and vascular amyloid: Correlation with cardiac amyloid. *Am J Med* 66:96–104, 1979.

71a. Metersky ML, Colt HG, Olson LK, Shanks TG: AIDS-related spontaneous pneumothorax: Risk factors and treatment. *Chest* 108:946–951, 1995.

71b. Shroeder SA, Beneck D, Dozor AJ: Spontaneous pneumothorax in children with AIDS. *Chest* 108:1173–1176, 1995.

72. Askin FB, McCann BG, Kuhn C: Reactive eosinophilic pleuritis: A lesion to be distinguished from pulmonary eosinophilic granuloma. *Arch Pathol Lab Med* 101:187–191, 1978.

73. McDonell TJ, Crouch E, Gonzalez JG: Reactive eosinophilic pleuritis: A sequela of haemothorax in pulmonary eosinophilic granuloma. *Am J Clin Pathol* 91:107–111, 1989.

74. Tomashefski JF Jr, Dahms B, Bruce M: Pleura in pneumothorax: Comparison of patients with cystic fibrosis and "idiopathic" spontaneous pneumothorax. *Arch Pathol Lab Med* 109:910–916, 1985.

75. Dunnill MS: Metaplastic changes in the visceral pleura in a case of fibrocystic disease of the pancreas. *J Pathol Bacteriol* 77:299–302, 1959.

76. Tomashefski JF Jr, Bruce M, Stern RC, et al: Pulmonary air cysts in cystic fibrosis: Relation of pathologic features to radiologic findings and history of pneumothorax. *Hum Pathol* 16:253–261, 1958.

77. Blesovsky A: The folded lung. *Br J Dis Chest* 60:19–22, 1966.

78. Mintzer RA, Cugell DW: The association of asbestos-induced pleural diseases and rounded atelectasis. *Chest* 81:457–460, 1982.

78a. Voisin C, Fisekci F, Voisin-Saltiel S, Ameille J, Brochard P, Pairon J-C: Asbestos-related rounded atelectasis: Radiologic and mineralogic data in 23 cases. *Chest* 107:477–481, 1995.

79. Hanke R, Kretzschmar R: Round atelectasis. *Semin Roentgenol* 15:174–182, 1983.

80. Case records of the Massachussetts General Hospital. *N Engl J Med* 308:1466–1472, 1983.

81. Brooks JSJ, Livolsi Va, Pietra GG: Mesothelial cell inclusions in mediastinal lymph nodes mimicking metastatic carcinoma. *Am J Clin Pathol* 93:741–748, 1990.

82. Fridriksson HV, Hedenström H, Hillerdal G, Malmeberg P: Increased lung stiffness in persons with pleural plaques. *Eur J Respir Dis* 62:412–424, 1981.

83. Wain SL, Roggli VL, Foster WL Jr: Parietal pleural plaques, asbestos bodies and neoplasia: A clinical, pathologic, and roentgenographic correlation of 25 consecutive cases. *Chest* 86:707–713, 1984.

84. Sisson RF, Hruban RH, Moore GW, Kuhlman JE, Wheeler PS, Hutchins GM: Pulmonary disease associated with pleural "asbestos" plaques. *Chest* 95:831–835, 1989.

85. Churg A: Asbestos fibers and pleural plaques in a general autopsy population. *Am J Pathol* 109:88–96, 1982.

86. Warnock M, Prescott B, Kuwarhara T: Numbers and types of asbestos fibers in subjects with pleural plaques. *Am J Pathol* 109:37–46, 1982.

87. Karjalainen A, Karhunen PJ, Lalu K, et al: Pleural plaques and exposure to mineral fibers in a male urban necropsy population. *Occup Environ Med* 51:456–460, 1994.

88. International Agency for Research on Cancer: *IARC Monographs on the Evaluation of the Carcinogenic Risk of Chemicals to Humans*: Vol 42. *Silica and Some Silicates.* Lyon; IARC, 1987, p 290.

89. Baris YI, Saracci R, Simonato L, Skidmore JW, Artvinli M: Malignant mesothelioma and radiological chest abnormalities in two villages in central Turkey. *Lancet* 1:984–987, 1981.

90. Parkes WR: *Occupational Lung Disorders* 2d ed. London, Butterworths, 1982.

91. Weiss W, Levin R, Goodman L: Pleural plaques and cigarette smoking in asbestos workers. *J Occup Med* 23:427–430, 1981.

92. Harries PG, MacKenzie FAS, Sheers G, et al: Radiological survey of men exposed to asbestos in naval dockyards. *Br J Ind Med* 29:274–279, 1972.

93. Meurman L: Asbestos bodies and pleural plaques in a finished series of autopsy cases. *Acta Pathol Microbiol Scand Suppl* 181:1–107, 1966.

94. Craighead JE, Abraham JL, Churg A, et al: The pathology of asbestos-associated diseases of the lungs and pleural cavi-

ties: Report of the Pneumoconiosis Committee of the American College of Pathologists and National Institute of Occupational Safety and Health. *Arch Pathol Lab Med* 106:544–596, 1982.

95. Churg A, DePaoli L: Environmental pleural plaques in residents of a Quebec chrysotile mining town. *Chest* 94:58–60, 1988.

96. Kiviluoto R, Meurman LO, Hakama M: Pleural plaques and neoplasia in Finland. *Ann NY Acad Sci* 330:31–33, 1979.

97. Burilkov T, Michailova L: Asbestos content of the soil and endemic pleural asbestosis. *Environ Res* 3:443–451, 1979.

98. Artvinli M, Baris YI: Malignant mesothelioma in a small village in the Anatolian region of Turkey: An epidemiologic study. *JCNI* 63:17–22, 1979.

99. Langer AM, Nolan RP, Costantopoulos SH, Moutsopoulos HM: Association of Metsovo lung and pleural mesothelioma with exposure to tremolite-containing whitewash. *Lancet* 1:965–967, 1987.

100. Neuberger M, Kundi M, Fiedl HP: Environmental asbestos exposure and cancer mortality. *Arch Environ Health* 39:261–265, 1984.

101. Yazicioglu S, Ilcayto R, Balci K, Sayli GS, Yorulmaz B: Pleural calcification, pleural mesotheliomas and bronchial cancers caused by tremolite dust. *Thorax* 35:564–569, 1980.

102. Smith DD: Plaques, cancer and confusion. *Chest* 105:8–9, 1994.

103. Hillerdal G: Pleural plaques and risk for bronchial carcinoma and mesothelioma: A prospective study. *Chest* 105:144–150, 1994.

104. Sheers G: Asbestos associated disease in employees of Devonport dockyard. *Ann NY Acad Sci* 330:281–287, 1979.

105. Liddell FDK, McDonald JC: Radiological findings as predictors of mortality in Quebec asbestos workers. *Br J Ind Med* 37:257–267, 1980.

106. Fletcher D: A mortality study of shipyard workers with pleural plaques. *Br J Ind Med* 29:142–145, 1972.

107. McDonald JC, Becklake MR, Gibbs GW, McDonald AD, Rossiter CE: The health of chrysotile asbestos mine and mill workers of Quebec. *Arch Environ Health* 28:61–68, 1974.

108. Edge J: The incidence of bronchial carcinoma in shipyard workers with pleural plaques. *Ann NY Acad Sci* 330:289–294, 1979.

109. Thiringer G, Blomquist N, Brolin I, Mattson S: Pleural plaques in chest x-rays of lung cancer patients and matched controls. *Eur J Respir Dis* 51(suppl 107):119–122, 1980.

110. Hillerdal G: Pleural plaques and risk for cancer in the county of Uppsala. *Eur J Respir Dis* 61(suppl 107):111–117, 1980.

111. Thiringer G, Blomquist N, Brolin I, Mattson SB: Pleural plaques in chest x-rays of lung cancer patients and matching controls. *Eur J Respir Dis* 61(suppl 107):119–122, 1980.

112. Mollo F, Andrion A, Colombo A, Segnan N, Pira E: Pleural plaques and risk of cancer in Turin, northwestern Italy. *Cancer* 54:1418–1422, 1984.

113. McMillan GHG, Rossiter CE: Development of radiological and clinical evidence of parenchymal fibrosis in men with non-malignant asbestos-related pleural lesions. *Br J Ind Med* 39:54–59, 1982.

114. Harber P, Mohsenifar Z, Oren A, Lew M: Pleural plaques and asbestos-associated malignancy. *J Occup Med* 29:641–644, 1987.

115. Sanden A, Jarvholm B: Cancer morbidity in Swedish shipyard workers, 1978–1983. *Int Arch Occup Enviorn Health* 59:455–462, 1987.

116. Partanen T, Nurminen M, Zitting A, et al: Localized pleural plaques and lung cancer. *Am J Ind Med* 22:185–192, 1992.

117. Edge JR: Asbestos-related disease in Barrow-in-Furness. *Environ Res* 11:244–247, 1976.

118. Davies D, Andrews MI, Jones JS: Asbestos induced pericardial effusion and constrictive pericarditis. *Thorax* 46:429–432, 1991.

119. Schwartz DA: New developments in asbesto-induced pleural disease. *Chest* 99:191–198, 1991.

120. Hillerdal G: Non-malignant asbestos pleural disease. *Thorax* 36:669–675, 1981.

121. Epler GR, McCloud TC, Gaensler EA: Prevalence and incidence of benign asbestos pleural effusion in a working population. *JAMA* 247:617–622, 1982.

122. Martensson G, Pettersson K, Thiringer G: Differentiation between malignant and non-malignant pleural effusion. *Eur J Respir Dis* 67:326–334, 1985.

123. Britton MG: Asbestos pleural disease. *Br J Dis Chest* 76:1–10, 1982.

124. Sheers G, Templeton AR: Effects of asbestos in dockyard workers. *Br Med J* 3:574, 1968.

125. Becklake MR: Asbestos-related disease of the lung and other organs: Their epidemiology and implications for clinical practice: State of the art. *Am Rev Respir Dis* 114:187–227, 1976.

126. American Thoracic Society: The diagnosis of non-malignant diseases related to asbestos. *Am Rev Respir Dis* 134:363–368, 1986.

127. Hillerdal G, Ozesmi M: Benign asbestos pleural effusion: 73 exudates in 60 patients. *Eur J Respir Dis* 71:113–127, 1987.

128. Shore BL, Daughaday CC, Spilburg I: Benign asbestos pleurisy in the rabbit: A model for the study of pathogenesis. *Am Rev Respir Dis* 128:481–485, 1983.

129. Davies D: Asbestos related diseases without asbestosis. *Br Med J* 287:164–165, 1983.

130. McLoud TC, Woods BO, Carrington CB, Epler GR, Gaensler EA: Diffuse pleural thickening in an asbestos-exposed population: Prevalence and causes. *AJR* 144:9–18, 1985.

131. Stephens M, Gibbs AR, Pooley FD, Wagner JC: Asbestos induced diffuse pleural fibrosis: Pathology and mineralogy. *Thorax* 42:583–588, 1987.

132. Miller A: Chronic pleuritic pain in four patients with asbestos induced pleural fibrosis. *Br J Ind Med* 47:147–153, 1990.

133. Schwartz DA, Fuortes LJ, Galvin JR, et al: Asbestos-induced pleural fibrosis and impaired lung function. *Am Rev Respir Dis* 141:321–326, 1990.

134. Hinson KFW, Otto H, Webster I, Rossiter CE: Criteria for the diagnosis and grading of asbestos, in Bogovski P (ed): *Biological Effects of Asbestos*. Lyons, World Health Organization, 1973, pp 54–57.

135. Jones JSP, Pooley FD, Clarke NJ, et al: The pathology and mineral content of lungs in cases of mesothelioma in the United Kingdom in 1976, in Wagner JC (ed): *Biological Effects of Mineral Fibres*. Lyons, IARC, 1980, pp 79–87.

136. Gibbs AR, Stephens M, Griffiths DM, Blight BJN, Pooley FD: Fiber distribution in the lungs and pleura of subjects

with asbestos related diffuse pleural fibrosis. *Br J Ind Med* 48:762–767, 1991.

137. Stanton MF, Layard M, Tegeris A, et al: Relation of particle dimension to carcinogenicity in amphibole asbestoses and other fibrous minerals. *J Natl Cancer Inst* 67:965–975, 1981.

138. Herbert A: Pathogenesis of pleurisy, pleural fibrosis and mesothelial proliferation. *Thorax* 41:176–189, 1986.

139. Lange A: An epidemiological survey of immunological abnormalities in asbestos workers: 1. Non-organ and organ specific autoantibodies. *Environ Res* 20:162–175, 1980.

140. Lange A: An epidemiological survey of immunological abnormalities in asbestos workers: 2. Serum and immunoglobulin levels. *Environ Res* 22:176–183, 1980.

141. Viallat JR, Raybuad F, Passarel M, Boutin C: Pleural migration of chrysotile fibers after intratracheal injection in rats. *Arch Environ Health* 41:282–286, 1986.

142. Bignon J, Jaurand MC, Sebastien P, Dufour G: Interaction of pleural tissue and cells with mineral fibers, in Chretien J, Hirsch A (eds): *Diseases of the Pleura.* New York, Masson, 1983, pp 198–207.

143. Sahn SA, Anthony VP: Pathogenesis of pleural plaques: Relationship of early cellular response and pathology. *Am Rev Respir Dis* 139:884–887, 1989.

144. Feagler JR, Sorensen GD, Rosenfeld MG, Osterland CK: Rheumatoid pleural effusion. *Arch Pathol* 92:257–266, 1971.

145. Barrett NR: The pleura—With special reference to fibrothorax. *Thorax* 25:515–524, 1970.

146. Briselli M, Mark EJ, Dickersin GR: Solitary fibrous tumors of the pleura: Eight new cases and review of 360 cases in the literature. *Cancer* 47:2678–2689, 1981.

147. Al-Izzi M, Thurlo NP, Corrin B: Pleural mesothelioma of the connective tissue type, localised fibrous tumor of the pleura and reactive submesothelial hyperplasia: An immunohistochemical comparison. *J Pathol* 158:41–44, 1989.

148. England DM, Hochholzer L, McCarthy MJ: Localised benign malignancy fibrous tumors of the pleura: A clinico-pathological review of 223 cases. *Am J Surg Pathol* 13:640–658, 1989.

149. Carter D, Otis CN: Three types of spindle cell tumors of the pleura: Fibroma, sarcoma and sarcomatoid mesothelioma. *Am J Surg Pathol* 12:747–753, 1988.

150. Clagett OT, McDonald JR, Schmidt HW: Localised fibrous mesothelioma of the pleura. *J Thorac Surg* 24:213–230, 1952.

150a. Moran CA, Suster S, Koss MN: Smooth muscle tumours presenting as pleural neoplasms. *Histopathology* 27:227–234, 1995.

150b. Kung ITM, Thallas V, Spencer EJ, Wilson SM: Expression of muscle actins in diffuse mesotheliomas. *Hum Pathol* 26:565–570, 1995.

151. Foster EA, Ackerman LV: Localised mesothelioma of the pleura: The pathologic evaluation of 18 cases. *Am J Clin Pathol* 34:349–364, 1960.

152. Yesner R, Hurwitz A: Localised pleural mesothelioma of epithelial type. *J Thorac Surg* 26:325–329, 1953.

153. Fuy YS, Gabbiani G, Kaye GI, Lattes R: Malignant soft tissue tumors of probable histiocytic origin (malignant fibrous histiocytomas): General considerations and electron microscopic and tissue culture studies). *Cancer* 35:176–198, 1975.

154. Kaye S: Inflammatory histiocytoma (xanthogranuloma). *Am J Surg Pathol* 2:313–319, 1978.

155. Kyriakos M, Kempson RL: Inflammatory fibrous histiocytoma. An aggressive and lethal lesion. *Cancer* 37:1584–1606, 1976.

156. Kawai T, Suzuki M: Non-malignant tumors of the pleura, in Chretien J, Bignon J, Hirsch A (eds): *The Pleura in Health and Disease.* New York, Marcel Dekker, 1985, p 540.

157. Moran CA, Travis WD, Rosado-de-Christenson M, Koss MN, Rosai J: Thymomas presenting as pleural tumors: Report of 8 cases. *Am J Surg Pathol* 16:138–144, 1992.

158. Asa SL, Dardick I, Van Nostarand AW, Bailey DG, Gullane PJ: Primary thyroid thymoma: A distinct clinico-pathological entity. *Hum Pathol* 19:1463–1467, 1988.

159. Hofman W, Moller P, Manke HG, Otto HF: Thymoma: A clinico-pathologic study of 98 cases with special reference to three unusual cases. *Pathol Res Pract* 179:337–353, 1985.

160. Thorburn JD, Stephen B, Grimes OF: Benign thymoma in the hilus of the lung. *J Thorac Surg* 24:240–243, 1952.

161. Honma K, Shimada K: Metastasising ectopic thymoma arising in the right thoracic cavity and mimicking diffuse pleural mesothelioma: An autopsy study of a case with review of the literature. *Wiener Klin Wochensch* 98:14–20, 1986.

162. Granberg I, Willems JS: Endometriosis of lung and pleura diagnosed by aspiration biopsy. *Acta Cytol (Baltimore)* 21:295–297, 1977.

163. Zaatari GS, Gupta PK, Bahagavan BS, Jarboe BR: Cytopathology of pleural endometriosis. *Acta Cytol (Baltimore)* 26:227–232, 1982.

164. Peacock MJ, Craddock DR, Allen PW: Pleural fibrolipoma: Report of a case. *Aust NZ J Surg* 44:117–119, 1974.

165. Rosenberg RF, Rubenstein BM, Messinger NH: Intra-thoracic lipomas. *Chest* 60:507–509, 1971.

166. Wong WW, Pluth JR, Grado GL, Schild SE, Sanderson DR: Liposarcoma of the pleura. *Mayo Clin Proc* 69:882–885, 1994.

167. Hasleton PS, Langdale-Brown B, Rahman A, Smyrniou N, Barber PV: Pleural myxoma associated with a pulmonary squamous cell carcinoma. *Respir Med* 83:443–444, 1989.

168. Goetz SP, Robinson RA, Landas SK: Extraskeletal myxoid chondrosarcoma of the pleura: Report of a case clinically simulating mesothelioma. *Am J Clin Pathol* 97:498–502, 1992.

169. Prabhakar G, Mitchell IM, Guha T, Norton R: Squamous cell carcinoma of the pleura following bronchopleural fistula. *Thorax* 44:1053–1054, 1989.

170. Crome L: Squamous metaplasia of the peritoneum. *J Pathol Bacteriol* 62:61–68, 1950.

171. Wagner JC, Sleggs CA, Marchand P: Diffuse pleural mesotheliomas and asbestos exposure in a north western Cape Province. *Br J Ind Med* 17:260–271, 1960.

172. Wagner JC: The discovery of the association between blue asbestos and mesotheliomas and the aftermath. *Br J Ind Med* 48:399–403, 1991.

173. Willis RA: *Pathology of Tumours.* London, Butterworths, 1967, p 182.

174. Morgan A, Evans JC, Holmes A: Deposition clearance of inhaled fibrous minerals in the rat: Studies using radioactive tracer techniques, in Walton WH, McGovern B (eds): *Inhaled Particles:* IV. Part I. Oxford, Pergamon Press, 1977, pp 259–274.

175. Sebastien P, Janson S, Bonnaud G, et al: Translocation of asbestos fibers through the respiratory tract and gastrointestinal tract according to fiber, type and size, in Lemen R, Dement JM (eds): *Dusts and Disease.* Park Forest South, IL, Pathotox, 1979; p 265.

176. Auerbach O, Conston AS, Garfinkel L, et al: Presence of asbestos bodies in organs other than the lung. *Chest* 77:133, 1980.

177. Rey F, Viallat JR, Farisse P, Boutin C: Pleural migration of asbestos fibers after intratracheal injection in rats. *Eur Respir Rev* 3. 11:145–147, 1993.

178. Davis JMG: The histopathology and ultrastructure of pleural mesotheliomas produced in the rat by injection of crocidolite asbestos. *Br J Exp Pathol* 60:642, 1979.

179. Wagner JC, Berry G, Timbrel V: Mesotheliomata in rats after inoculation with asbestos and other minerals. *Br J Cancer* 28:173, 1973.

180. Davis JMG: Histogenesis and fine structure of peritoneal tumors produced in animals by injections of asbestos. *J Natl Cancer Inst* 52:1823–1837, 1974.

181. Stanton MF, Laynard M, Tegaris A, et al: Carcinogenicity of fibrous glass: Pleural response in the rat in relation to fiber dimension. *J Natl Cancer Inst* 58:587–603, 1977.

182. Mossman BT: Mechanisms of asbestos carcinogenesis and toxicity: The amphibole hypothesis revisited. *Br J Ind Med* 50:673–676, 1993.

183. Brand KG: Foreign body induced sarcomas, in Becker FF (ed): *Cancer: a Comprehensive Treatise. Etiology: Chemical and Physical Carcinogenesis.* New York, Plenum Press, 1975, pp 485–511.

184. Hei TK, Piao CQ, He Zy, Vannals D, Waldren CA: Chrysotile fiber is a strong mutagen in mammalian cells. *Cancer Res* 52:6305–6309, 1992.

185. Hei TK, He Zy, Piao CQ, Waldren CA: The mutagenicity of mineral fibers, in Brown RC, Hoskins JA, Johnson NF (eds): *Mechanisms in Fiber Carcinogenesis.* New York, Plenum Press, 1990, pp 319–325.

186. Lechner JF, Tokiwa T, LaVeck M, et al: Asbestos-associated chromosomal changes in human mesothelial cells. *Proc Natl Acad Sci USA* 82:3884–3888, 1985.

187. Lund LG, Aust AE: Iron-catalysed reactions may be responsible for the biochemical and biological effects of asbestos. *BioFactors* 3:83–89, 1991.

188. Lund LG, Aust AE: Iron mobilization from crocidolite asbestos greatly enhances crocidolite-dependent formation of DNA single-strand breaks in OX174 FRI DNA. *Carcinogenesis* 13:637–642, 1992.

189. Kamp DW, Graceffa P, Pryor W, Weitzman S: The role of free radicals in asbestos-induced diseases. *Free Rad Biol Med* 112:673–720, 1992.

190. Mossman BT, Marsh JP, Shatos MA: Alteration of superoxide dismutase activity in tracheal epithelial cells by asbestos and inhibition of cytoxicity by antioxidants. *Lab Invest* 54:204–212, 1986.

191. Hansen K, Mossman BT: Generation of superoxide (0_2) from alveolar macrophages exposed to asbestiform and nonfibrous particles. *Cancer Res* 47:1681–1686, 1987.

192. Gulumian M, Van Wyk HA: Hydroxyl radical production in the presence of fibers by a fenton type reaction. *Chem Biol Interact* 62:89–97, 1987.

193. Weitzman SA, Graceffa P: Asbestos catalyzes hydroxyl and superoxide radical generation from hydrogen peroxide. *Arch Biochem Biophys* 228:373–376, 1984.

194. Waldren CA: Mutational analysis in cultured human hamster hybrid cells, in deSorres FJ (ed): *Chemical Mutagens: Principles and Methods for Their Detection.* New York, Plenum Press, 1983, pp 235–237.

195. Turver CJ, Brown RC: The role of catalytic iron in asbestos induced lipid peroxidation and DNA-strand breakage in C3H10T1/2 cells. *Br J Cancer* 56:133–136, 1987.

196. Weitzman SA, Weitberg AB: Asbestos-catalysed lipid peroxidation and its inhibition by desferrioxamine. *Biochem J* 225:259–262, 1985.

197. Goodlick LA, Pietras LA, Kane AB: Evaluation of the causal relationship between crocidolite asbestos induced lipid peroxidation and toxicity to macrophages. *Am Rev Respir Dis* 139:1265–1273, 1989.

198. Kasai H, Nishimura S: DNA damage by asbestos in the presence of hydrogen peroxide. *Gan To Kagaku Ryoho* 75:841–844, 1984.

199. Timbrell V: in Shapiro HA (ed): *Pneumoconiosis: Proceedings of the International Conference.* Cape Town, Johannesburg, University Press, 1970, pp 28–36.

200. Jaurand MC, Kheuang L, Magne L, Bignon J: Chromosome changes induced by chrysotile fibers and benzo-3,4-pyrene in rat pleural mesothelial cells. *Mutat Res* 169:141–148, 1986.

201. Fornace AJ Jr, Lechner JF, Graftsman RC, Harris CC: DNA repair in human bronchial epithelial cells. *Carcinogenesis* 3:1373–1377, 1982.

202. Mossman BT, Eastman A, Landesman JM, Bresnick E: Effects of crocidolite and chrysotile asbestos on cellular uptake, metabolism and DNA after exposure of hamster tracheal epithelial cells to benzopyrene (a). *Environ Health Perspect* 51:331–338, 1983.

203. Heinz NH, Janssen YM, Mossman BT: Persistent induction of c-*fos* and c-*jun* protooncogene expression by asbestos. *Proc Natl Acad Sci USA* 90:3299–3303, 1993.

204. Angel P, Karin M: The role of *jun*, *fos* and the AP-1 complex in cell-proliferation and transformation. *Biochimi Biophys Acta* 1072:129–157, 1991.

205. Marsh JP, Mossman BT: Mechanisms of induction of ornithine decarboxylase activity in tracheal epithelial cells by asbestiform minerals. *Cancer Res* 48:709–714, 1988.

206. Marsh JP, Mossman BT: Role of asbestos and active oxygen species in activation and expression of ornithine decarboxylase in hamster tracheal epithelial cells. *Cancer Res* 51:167–173, 1991.

207. Mossman BT, Craighead JE, MacPherson BE: Asbestos-induced epithelial changes in organ cultures of hamster trachea: Inhibition by retinyl methyl ether. *Science* 207:311–313, 1980.

208. Ames BN, Gold LS: Mitogenesis increases mutagenesis. *Science* 249:970–971, 1990.

209. Preston-Martin S, Pike MC, Ross RK, Jones PA, Henderson BE: Increased cell division as a cause of human cancer. *Cancer Res* 50:7415–7421, 1990.

210. Mossman BT, Bignon J, Corn MM, et al: Asbestos: Scientific developments and implications for public policy. *Science* 247:294–301, 1990.

210a. Lee WC, Altomare DA, Nobori T, Olopade OI, Buckler AJ, Testa JR: p16 alterations and deletion mapping of 9 p21–p22 in malignant mesothelioma. *Cancer Res* 54:5547–5551, 1994.

211. Gibbs AR: Role of asbestos and other fibers in the development of diffuse malignant mesothelioma. *Thorax* 45:649–654, 1990.

212. Churg A, Wiggs B: Fiber size and number in workers exposed to processed chrysotile asbestos, chrysotile miners and the general population. *Am J Ind Med* 9:143–152, 1986.

213. Pooley FD: Examination of the fibrous mineral content of asbestos lung tissue from the Canadian chrysotile mining industry. *Environ Rev* 12:281–298, 1976.

214. Rowlands N, Gibbs GW, McDonald JC: Asbestos fibers in the lungs of chrysotile miners and millers. *Ann Occup Hyg* 26:411–415, 1982.

215. Peto J, Doll R, Hermon C, Binns W, Clayton R, Goffe T: Relationship of mortality to measures of environmental asbestos pollution in an asbestos textile factory. *Ann Occup Hyg* 29:305–355, 1985.

216. Berry G, Gilson JC, Holmes S, Lewinsohn HC, Roach SA: Asbestosis: A study of dose-response relationships in an asbestos textile factory. *Br J Ind Med* 36:98–112, 1979.

217. Wagner JC, Berry G, Pooley FD: Mesotheliomas and asbestos type in asbestos textile workers: A study of lung contents. *Br Med J* 285:603–606, 1982.

218. Acheson ED, Gardner MJ, Pippard EC, Grime LP: Mortality of two groups of women who manufactured gas masks from chrysotile and crocidolite asbestos: A 40 year follow-up. *Br J Ind Med* 39:344–348, 1982.

219. Churg A: Chrysotile, tremolite and malignant mesothelioma in man *Chest* 93:621–628, 1988.

220. Berry G: Chrysotile and mesothelioma. *Accomp Oncol* 1:123–131, 1986.

221. Rogers AJ, Leigh J, Berry G, Ferguson DA, Moulder HB, Ackad M: Relationship between asbestos fiber type and concentration and relative risk of meosthelioma: A case-control study. *Cancer* 67:1912–1920, 1991.

222. DeKlerk NH, Armstrong BK, Musk AW, Hobbs MS: Cancer mortality in relation to measures of occupational exposure to crocidolite at a Witenoom Gorge in Western Australia. *Br J Ind Med* 46:529–536, 1989.

223. Webster I: Asbestos and malignancy. *S Afr Med J* 47:165–171, 1973.

224. Seidman H, Selikoff IJ, Gelb SK: Mortality experience of amosite factory workers: Dose-response, relationships 5 to 40 years after onset of short term exposure. *Am J Ind Med* 10:479–514, 1986.

225. Ribak J, Seidman H, Selikoff IJ: Amosite mesothelioma in a cohort of asbestos workers. *Scand J Work Environ Health* 15:106–110, 1989.

226. Baris YI, Artvinli M, Sahin AA, Bilir N, Kalyoncu F, Sebastien P: Non-occupational asbestos-related chest diseases in a small Anatolian village. *Br J Ind Med* 45:841–842, 1988.

227. McDonald JC, McDonald AD, Armstrong B, Sebastien P: Cohort study of mortality of vermiculite miners exposed to tremolite. *Br J Ind Med* 43:436–444, 1986.

228. McDonald JC, Armstrong B, Case B, et al: Mesothelioma and asbestos fiber type: Evidence from lung tissue analyses. *Cancer* 63:1544–1547, 1989.

229. Boutin G, Viallat JR, Steinbauer J, Dufour G, Gaudichet A: *Bilateral Pleural Plaques in Corsica: A Marker of Non-occupational Asbestos Exposure.* International Agency for Researching Cancer. pp 406–410, 1989.

230. McConnochie K, Simonato L, Mavrides P, Christofides P, Pooley FD, Wagner JC: Mesotheliomas in Cyprus: The role of tremolite. *Thorax* 42:342–347, 1987.

231. Tuomi T, Segerberg-Konttinen M, Tammilehto L, Tossavainen A, Vanhala E: Mineral fiber concentration in lung tissue of mesothelioma patients in Finland. *Am J Ind Med* 16:247–254, 1989.

232. Ashcroft T, Heppleston AG: The optical and electron microscopic determination of pulmonary asbestos fibre concentration and its relation to the human pathological reaction. *J Clin Pathol* 26:224–234, 1973.

233. Davis JMG: The pathology of asbestos-related disease. *Thorax* 39:801–808, 1984.

234. Dodson RF, Hurst MG, Williams MG Jr, Corn CJ, Greenberg SD: Comparison of light and electron microscopy for defining occupational exposure in transbronchial biopsy. *Chest* 94:366–370, 1988.

235. Churg A: Analysis of lung fibers from lung tissue: Research and diagnostic uses. *Semin Respir Med* 7:281–283, 1986.

236. Churg A, Wood P: Observations on the distribution of asbestos fibers in human lung. *Environ Res* 1:374–380, 1983.

237. Morgan A, Holmes A: Distribution and characteristics of amphibole asbestos fibres, measured with the light microscope, in the left lung of an insulation worker. *Br J Ind Med* 40:45–50, 1983.

238. Rogers AJ: Determination of mineral fiber in human lung tissue by light microscopy and transmission electron microscopy. *Ann Occup Hyg* 28:1–12, 1984.

239. Pooley FD, Ranson DL: Comparison of the results of asbestos fibre dust counts in lung tissue obtained by analytical electron microscopy and light microscopy. *J Clin Pathol* 39:313–317, 1976.

240. Churg A, Warnock ML: Asbestos fibers in the general population. *Am Rev Respir Dis* 122:669–678, 1980.

241. Whitwell F, Scott J, Grimshaw M: Relationship between occupations and asbestos fibre content of the lungs in patients with pleural mesothelioma, lung cancer and other diseases. *Thorax* 32:377–386, 1977.

242. Gylseth B, Mowe G, Skaug V, Wannag A: Inorganic fibers in lung tissue from patients with pleural plaques or malignant mesothelioma. *Scand J Work Environ Health* 7:109–113, 1981.

243. Mowe G, Gylseth B, Hartveit F, Skaug V: Fiber concentration in lung tissue of patients with malignant mesothelioma: A case control study. *Cancer* 56:1089–1093, 1985.

244. Roggli VL, Pratt PC, Brody AR: Asbestos content of lung tissue in asbestos associated diseases: A study of 110 cases. *Br J Ind Med* 43:18–28, 1986.

245. Churg A, Wiggs B: Fiber size and number in amphibole asbestos-induced mesothelioma. *Am J Pathol* 115:437–442, 1984.

246. Churg A, Wiggs B, DePaoli L, Kampe B, Stevens B: Lung asbestos content in chrysotile workers with mesothelioma. *Am Rev Respir Dis* 130:1042–1045, 1984.

247. Gaudichet A, Janson X, Monchaux G, et al: Assessment by analytical microscopy of total lung fiber burden in mesothelioma patients matched with four other pathological series. *Ann Occup Hyg* 32(suppl 1): 213–223, 1988.

248. Warnock ML, Kuwarhara TJ, Wolery G: The relation of asbestos burden to asbestos and lung cancer. *Pathol Annu* 18:109–145, 1983.

249. Wagner JC, Pooley FD: Mineral fibres and mesothelioma. *Thorax* 41:161–166, 1986.

250. Selcuk ZT, Coplu L, Kalyoncu AF, Sahin AA, Baris YI: Malignant pleural mesothelioma due to environmental mineral fiber exposure in Turkey: Analysis of 135 cases. *Chest* 102:790–796, 1992.

251. Enterline PE: Role of man-made mineral fibres in the causation of cancer. *Br J ind Med 47:145–146, 1990.*

252. Hammar SP, Bockus D, Remington F, Friedman S, Lazerte G: Familial mesothelioma.: A report of two families. *Hum Pathol* 20:107–112, 1989.

253. Otte KE, Sigstaard TI, Kjaerulff J: Malignant mesothelioma: Clustering in the family producing asbestos cement in their home. *Br J Ind Med* 47:10–13, 1990.

254. Risberg G, Nickels J, Wagermark J: Familial clustering of malignant mesotheliomas. *Cancer* 45:2422–2427, 1980.

255. Martensson G, Larsson S, Zettergren L: Malignant mesothelioma in two pairs of siblings: Is there a hereditary predisposing factor? *Eur J Respir Dis* 65:179–184, 1984.

256. Hillerdal G, Berg J: Malignant mesothelioma secondary to chronic inflammation and old scars: Two new cases and review of the literature. *Cancer* 55:1968–1972, 1984.

257. Fraire AE, Cooper S, Greenberg SD, Buffler P, Langston C: Mesothelioma of childhood. *Cancer* 62:838–847, 1988.

258. Anderson KA, Hurley WC, Hurley BT, Ohrt DW: Malignant pleural mesothelioma following radiotherapy in a 16 year old boy. *Cancer* 56:273–276, 1985.

259. Gilks B, Hegedus C, Freeman H, Fratkin L, Churg A: Malignant peritoneal mesothelioma after remote abdominal radiation. *Cancer* 61:2019–2021, 1988.

260. Donna A: Mesothelioma without exposure to mineral fibers. *Eur Respir Rev* 311:79–81, 1993.

261. Donna A, Betta PG, Bianchi V, Ribotta M, et al: A new insight into the histogenesis of "mesodermomas" malignant-mesotheliomas. *Histopathology* 19:239–243, 1991.

262. Gibbs AR, Jones JS, Pooley FD, Griffiths DM, Wagner JC: Non-occupational malignant mesotheliomas. IARC Sci Publ 90:219–228, 1989.

263. Dawson A, Gibbs AR, Pooley FD, Griffiths DM, Hoy J: Malignant mesothelioma in women. *Thorax* 48:269–274, 1993.

264. Huncharek M, Capotorto JV, Muscat J: Domestic asbestos exposure, lung fiber burden and pleural mesothelioma in a housewife. *Br J Ind Med* 46:354–355, 1989.

265. Peterson JT Jr, Greenberg SD, Buffler PA: Non-asbestos related malignant mesothelioma: A review. *Cancer* 54:951–959, 1984.

266. Oels HC, Harrison EG, Carr DT, Bernatz PE: Diffuse malignant mesothelioma of the pleura: A review of 37 cases. *Chest* 60:564–570, 1971.

267. Meynaird O, Boissonn AS, Laisne JE, Laroche C, Abelanet R: Chronic pneumonia due to paraffin oil and pleural modifications: Mesothelial hyperplasia, mesothelioma. *Rev Fr Mal Respir* 8:259–264, 1980.

268. Hirsch A, Brochard P, De Cremoux H, et al: Features of asbestos exposed and unexposed mesothelioma. *Am J Indust Med* 3:413–422, 1982.

269. Decoufle P: Mesothelioma amongst shoe workers. *Lancet* 1:259, 1980.

270. Roggli VL, McGavran MH, Subach J, Sybers HD, Greenberg SD: Pulmonary asbestos body counts and electron probe analysis of asbestos body cores in patients with mesothelioma: A study of 25 cases. *Cancer* 50:2423–2432, 1982.

271. Brenner J, Sordillo PP, Magill GB, Golbey RB: Malignant mesothelioma of the pleura: Review of 123 patients. *Cancer* 49:2431–2435, 1982.

272. McDonald AD, Harper A, McDonald JC: Epidemiology of primary malignant mesothelial tumors in Canada. *Cancer* 26:914–919, 1970.

273. McDonald AD, McDonald JC: Malignant mesothelioma in North America. *Cancer* 46:1650–1656, 1980.

274. Hammond EC, Selikoff IJ, Seidman H: Asbestos exposure, cigarette smoking and death rates. *Ann NY Acad Sci* 330:473–490, 1979.

275. Peto J, Hodgson JT, Matthews FE, Jones JR: Continuing increase in mesothelioma mortality in Britain. *Lancet* 345:535–535, 1995.

276. Huncharek M: Changing risk groups for malignant mesothelioma. *Cancer* 69:2704–2711, 1992.

277. Muscat JE, Wynder EL: Cigarette smoking, asbestos exposure and malignant mesothelioma. *Cancer Res* 51: 2263–2267, 1991.

278. Gaensler EA: Asbestos exposure in buildings. *Clin Chest Med* 13:231–242, 1992.

279. Cordier S, Ameille J, Brochard P, et al: Epidemiologic investigation of respiratory effects related to environmental exposure inside insulated buildings. *Arch Environ Health* 42:303–309, 1987.

280. Wright WE, Sherwin RP, Dicksen EA: Malignant mesothelioma: Incidence, asbestos exposure and reclassification of histopathology. *Br J Ind Med* 41:39–45, 1984.

281. Meijers MM, Planteydty HT, Slangen JJM, Swaen GMH, van Vliet C, Sturmans F: Trends and geographical patterns of pleural mesotheliomas in the Netherlands, 1970–1987. *Br J Ind Med* 47:775–781, 1990.

282. Spirtas R, Beebe GW, Connelly RR, et al: Recent trends in mesothelioma incidence in the United States. *Am J Ind Med* 9:397–407, 1986.

283. Berry G: *Epidemiology of mesothelioma,* in Preventing cancer, in Australian Professional Publications, Tattersall M (ed): Sydney, 1988, pp 35–44.

284. Peto J, Henderson BE, Pike MC: Trends in mesothelioma incidence in the United States and the forecast epidemic due to asbestos exposure during World War II, in Peto J, Schneiderman M (eds): Banbury Report: 9. Quantification of Occupational Cancer. Cold Spring Harbor Laboratory, New York, 1981, pp 51–73.

285. McDonald JC: Health implications of environmental exposure to asbestos. *Environ Health Perspect* 62:319–328, 1985.

286. Lin-Chu M, Lee YJ, Ho MY: Malignant mesothelioma in infancy. *Arch Pathol Lab Med* 113:409–411, 1989.

287. Buotin C, Vialat JR, Rey F, Astoul PH: Clinical diagnosis of pleural mesothelioma. *Eur Respir Rev* 311:18–21, 1993.

288. Adams VI, Unni KK, Muhm JR, et al: Diffuse malignant mesothelioma of the pleura: Diagnosis and survival in 92 cases. *Cancer* 58:1540–1551, 1986.

289. Ribak J, Lilis R, Suzuki Y, Penner L, Selikoff IJ: Malignant mesothelioma in a cohort of asbestos insulation workers: Clinical presentation, diagnosis and causes of death. *Br J Ind Med* 45:182–187, 1988.

290. Elmes PC, Simpson MJC: The clinical aspects of mesothelioma. *Q J Med* 45:427–449, 1976.

291. Hunchareck M: Miliary mesothelioma. *Chest* 106:605–606, 1994.

291a. Musk AW: More cases of miliary mesothelioma. *Chest* 108:587, 1995.

292. Crotty TB, Colby TV, Gay PC, Pisani JR: Desmoplastic malignant mesothelioma masquerading as sclerosing mediastinitis: A diagnostic dilemma. *Hum Pathol* 23:79–82, 1992.

293. Murray JB, Neilly JB, Hadley D, Moran F, McKean M: Diffuse meningeal thickening associated with pleural mesothelioma. *Thorax* 45:70–71, 1990.

294. Butchart EG, Ashcroft T, Barnsley WC, Holden MP: Pleuropneumonectomy in the management of diffuse

malignant mesothelioma of the pleura: Experience with 29 patients. *Thorax* 31:15–24, 1976.

294a. Rusch VW: A proposed new International TMN staging system for malignant pleural mesothelioma. *Chest* 108:1122–1129, 1995.

295. Browne K, Smither WJ: Asbestos-related mesothelioma: Factors discriminating pleural and peritoneal sites. *Br J Ind Med* 40:145–152, 1983.

296. Enzinger FM, Weiss SW: Soft tissue tumors, 2d ed, St Louis, Mosby, 1988, p 689.

297. Hillerdal G: Malignant mesothelioma, 1982: Review of 4710 published cases. *Br J Dis Chest* 77:321–343, 1983.

298. Ball NJ, Urbanski SJ, Green FHY, Kieser TH: Pleural multicystic mesothelial proliferation: The so-called multicystic mesothelioma. *Am J Surg Pathol* 14:375–378, 1990.

299. Law MR, Hodson ME, Herd BE: Malignant mesothelioma of the pleura: Relation between histological type and clinical behaviour. *Thorax* 37:810–815, 1982.

300. Kane PB, Goldman SL, Pillai BH, Bergofsky EH: Diagnosis of asbestosis by transbronchial biopsy: A method to facilitate demonstration of ferruginous bodies. *Am Rev Respir Dis* 115:689–694, 1977.

301. Williams MG, Dodson RF, Corn C, Hurst GA: A procedure for the isolation of amosite asbestos and ferruginous bodies from lung tissue and sputum. *J Toxicol Environ Health* 10:627–638, 1982.

302. Gylseth B, Baunan RH, Overaae L: Analysis of fibres in human lung tissue. *Br J Ind Med* 39:191–195, 1982.

303. Churg A, Sakoda N, Warnock ML: A simple method for preparing ferruginous bodies for electron microscopic examination. *Am J Clin Pathol* 68:513–517, 1977.

304. Churg A: Analysis of lung asbestos count. *Br J Ind Med* 48:649–652, 1991.

305. Roggli VL, Pratt PC: Numbers of asbestos bodies on iron-stained tissue sections in relation to asbestos body counts in lung tissue digests. *Hum Pathol* 14:355–361, 1983.

306. Vollmer RT, Roggli VL: Asbestos body concentrations in human lung: Predictions from asbestos body counts in tissue sections with a mathematical model. *Hum Pathol* 16:713–718, 1985.

307. Mayall FG, Gibbs AR: Malignant peritoneal mesothelioma giving rise to multiple intestinal polyps. *Histopathology* 18:480–482, 1991.

308. Weiss SW, Tavassoli FA: Multicystic mesothelioma: An analysis of pathologic findings and biologic behaviour in 37 cases. *Am J Surg Pathol* 12:737–746, 1988.

309. Villaschi S, Autelitano F, Santeusanio G, Balistreri P: Cystic mesothelioma of the peritoneum: A report of 3 cases. *Am J Clin Pathol* 94:758–761, 1990.

310. Beck B, Konetke G, Ludwig V, et al: Malignant pericardial mesotheliomas and asbestos exposure: A case report. *Am J Ind Med* 3:149–159, 1982.

311. Roggli VL: Pericardial mesothelioma after exposure to asbestos. *N Engl J Med* 304:1045, 1981.

312. Turk J, Kenda M, Kranjeck I: Cardiac tamponade caused by primary pericardial mesothelioma. *N Engl J Med* 325:814, 1991.

313. Roggli VL, Kolbeck J, Sanfilippo F, Shelburne JD: Pathology of human mesothelioma: Etiologic and diagnostic considerations. *Pathol Annu* 22: (part 2):91–131, 1987.

314. Nomori H, Shimosato Y, Tsuchiya R: Diffuse malignant pericardial mesothelioma. *Acta Pathol Jpn* 35:1475–1481, 1985.

315. Linder J, Shelburne JD, Sorge JP, et al: Congenital endodermal heterotopia of the atrio-ventricular node: Evidence for the endodermal origin of so-called mesotheliomas of the atrio-ventricular node. *Hum Pathol* 15:1093–1098, 1984.

316. Sopher IM, Spitz WE: Endodermal inclusions of the heart: So-called mesotheliomas of the atrio-ventricular node. *Arch Pathol* 92:180–186, 1971.

317. Antman K, Cohen S, Dimitrov NV, et al: Malignant mesothelioma of the tunica vaginalis testis. *J Clin Oncol* 2:447–451, 1984.

317a. Jones MA, Young RH, Scully RE: Malignant mesothelioma of the tunica vaginalis: A clinico-pathologic analysis of 11 cases with review of the literature. *Am J Surg Pathol* 19:815–825, 1995.

318. Grove A, Jensen ML, Donna A: Mesotheliomas of the tunica vaginalis testis and hernial sacs. *Virchows Arch A Pathol Anat Histopathol* 415:283–292, 1989.

319. Peritoneum, retroperitoneum and related structures, in Rosai J (ed): *Ackerman's Surgical Pathology,* 2d ed. St Louis, Mosby, 1989, p 1638.

320. Histological Typing of Lung Tumours. Geneva, World Health Organization, 1981.

321. Kannerstein MD, Churg J: Peritoneal mesothelioma. *Hum Pathol* 8:83–94, 1977.

322. Whitaker D, Papadimitriou JM, Walters MN-I: The mesothelium and its reactions: A review. *CRC Crit Rev Toxicol* 10:81–144, 1982.

323. Adams VI, Unni KK: Diffuse malignant mesothelioma of the pleura: Diagnostic criteria based on an autopsy study. *Am J Clin Pathol* 82:15–23, 1984.

324. Case records of Massachusetts General Hospital. *N Engl J Med* 316:1462–1470, 1987.

325. Goldstein B: Two malignant pleural mesotheliomas with unusual histological features. *Thorax* 34:375–379, 1979.

326. Yousem SA, Hochholzer L: Malignant mesotheliomas with osseous and cartilaginous differentiation. *Arch Pathol Lab Med* 111:62–66, 1987.

327. Kannerstein M, Churg J: Desmoplastic diffuse malignant mesothelioma in Fenoglio CM, Wolff M (eds): *Progress in Surgical Pathology,* vol 2. New York. Masson, 1980, pp 19–29.

328. Cantin R, Al-Jabi M, McCaughey WTE: Desmoplastic diffuse mesothelioma. *Am J Surg Pathol* 6:215–222, 1982.

329. Wilson GE, Hasleton PS, Chatterjee AK: Desmoplastic malignant mesothelioma: A review of 17 cases. *J Clin Pathol* 45:295–298, 1992.

330. Henderson DW, Atwood HD, Constance TJ, Shilkin KB, Steele RH: Lymphohistiocytoid mesothelioma: A rare lymphomatoid variant of predominantly sarcomatoid mesothelioma. *Ultrastruct Pathol* 12:367–384, 1988.

331. Blumberg NA, Murray JF: Multicystic peritoneal mesotheliomas: A case report. *S Afr Med J* 59:85–86, 1981.

332. Biscotti CV, Hart WR: Peritoneal serous micropapillomatosis of low malignant potential (serous borderline tumors of the peritoneum): A clinicopathological study of 17 cases. *Am J Surg Pathol* 16:467–475, 1992.

333. Larsen TE: Serosal papilloma of the epicardium. *Arch Pathol* 97:271–272, 1974.

334. Barbera V, Rubino M: Papillary mesothelioma of the tunica vaginalis. *Cancer* 10:183–189, 1957.

335. Talerman A, Montero JR, Chilcote RR, Okagaki T: Diffuse malignant peritoneal mesothelioma in a 13-year-old girl: Report of a case and a review of the literature. *Am J Surg Pathol* 9:73–80, 1985.

336. Nascimento AG, Keeney GL, Fletcher CDM: Deciduoid peritoneal mesothelioma: An unusual phenotype affecting young females. *Am J Surg Pathol* 18:439–445, 1994.

337. Coleman M, Henderson DW, Muhkerjee TM: The ultrastructural pathology of malignant pleural mesothelioma. *Pathol Annu* 24:303–353, 1989.

338. Warhol MJ, Hickey WF, Corson JM: Malignant mesothelioma: Ultrastructural distinction from adenocarcinoma. *Am J Surg Pathol* 6:307–314, 1982.

339. Burns TR, Greenberg SD, Mace ML, Johnson EH: Ultrastructural diagnosis of epithelial malignant mesothelioma. *Cancer* 56:2036–2040, 1985.

340. Dewar A, Valente M, Ring NP, Corrin B: Pleural mesothelioma of epithelial type and pulmonary carcinoma: An ultrastructural and cytochemical comparison. *J Pathol* 152:309–316, 1987.

341. Bolen JW, Hammar SP, McNutt MA: Reactive and neoplastic serosal tissues: A light microscopic, ultrastructural and immunocytochemical study. *Am J Surg Pathol* 10:34–47, 1986.

342. Bolen JW, Thorning D: Mesotheliomas: A light and electron-microscopical study concerning histogenetic relationships between the epithelial and mesenchymal variants. *Am J Surg Pathol* 4:451–464, 1980.

343. Henderson DW, Papadimitriou JM, Coleman M: *Ultrastructural Appearances of Tumours: Diagnosis and Classification of Human Neoplasia by Electron Microscopy*, 2d ed. New York, Churchill Livingstone, 1986.

344. Stoebner P, Brambilla E: Ultrastructural diagnosis of pleural tumors. *Pathol Res Pract* 173:402–416, 1982.

345. Suzuki Y, Churg J, Kannerstein M: Ultrastructure of human malignant diffuse mesothelioma. *Am J Pathol* 85:241–262, 1976.

346. Suzuki Y: Pathology of human malignant mesothelioma. *Semin Oncol* 8:268–282, 1981.

347. Echevarria RA, Arean VM: Ultrastructural evidence of secretory differentiation in a malignant pleural mesothelioma. *Cancer* 22:323–332, 1968.

348. Warhol MJ, Corson JM: An ultrastructural comparison of mesotheliomas with adenocarcinomas of the lung and breast. *Hum Pathol* 16:50–55, 1985.

349. Dardick I, Jabi M, McCaughey WT, et al: Diffuse epithelial mesothelioma: A review of the ultrastructural spectrum. *Ultrastruct Pathol* 11:503–533, 1987.

350. Warhol MJ: Electron microscopy in the diagnosis of mesothelioma with routine biopsy, needle biopsy and fluid cytology, in Antman K, Aisner J (eds): *Asbestos-related Malignancy*. Orlando, Grune & Stratton, 1987, pp 201–221.

351. Erlandson RA, Filippa DA: Unusual non-Hodgkin's lymphomas and true histiocytic lymphomas. *Ultrastruct Pathol* 13:249–273, 1989.

352. Hammar SP, Bolen JW: Pleural neoplasms, in Dail DH, Hammer SP (eds): *Pulmonary Pathology*. New York, Springer Verlag, 1988, pp 973–1028.

353. Bockus D, Remington F, Friedman S, Hammar S: Electron microscopy—What izzits. *Ultrastruct Pathol* 9:1–30, 1985.

354. Dardick I, Jabi M, McCaughey WTE, et al: Ultrastructure of poorly differentiated diffuse epithelial mesotheliomas. *Ultrastruct Pathol* 7:151–160, 1984.

355. Churg A: Malignant mesothelioma. *Chest* 89:367S, 1986.

356. Churg J, Rosen SH, Moolten S: Histological characteristics of mesothelioma associated with asbestos. *Ann NY Acad Sci* 132:614–622, 1965.

357. Sheibani K, Esteban JM, Bailey A, Battifora H, Weiss LM: Immunopathologic and molecular studies as an aid to the diagnosis of malignant mesotheliomas. *Hum Pathol* 23:107–116, 1992.

358. Montag AG, Pinkus GS, Corson JM: Keratin protein immunoreactivity of sarcomatoid and mixed types of diffuse malignant mesothelioma: An immunoperoxidase study of 30 cases. *Hum Pathol* 19:336–342, 1988.

359. Cagle PT, Truong LD, Roggli VL, Greenberg SD: Immunohistochemical differentiation of sarcomatoid mesotheliomas from other spindle cell neoplasms. *Am J Clin Pathol* 92:566–571, 1989.

360. Wirth PR, Legier J, Wright GL Jr: Immunohistochemical evaluation of seven monoclonal antibodies for the differentiation of pleural mesothelioma from lung adenocarcinoma. *Cancer* 67:655–662, 1991.

361. Ghosh AK, Gatter KC, Dunnill MS, Mason DY: Immunohistological staining of reactive mesothelium, mesothelioma, and lung carcinoma with a panel of monoclonal antibodies. *J Clin Pathol* 40:19–25, 1987.

362. Zeng L, Fleury-Feith J, Monnet I, Boutin C, Bignon J, Jaurand MC: Immunocytochemical characterisation of cell lines from human malignant mesothelioma: Characterisation of human mesothelioma cell lines by immunocytochemistry with a panel of monoclonal antibodies. *Hum Pathol* 25: 227–234, 1994.

363. Churg A: Immunohistochemical staining for vimentin and keratin in malignant mesothelioma. *Am J Surg Pathol* 9:360–365, 1985.

364. Said JW, Nash G, Banks-Schlegel S, et al: Localised fibrous mesothelioma: An immunohistochemical and electron microscopic study. *Hum Pathol* 15:440–443, 1984.

365. Dervan PA, Tobin B, O'Connor M: Solitary (localised) fibrous mesothelioma: Evidence against mesothelial proliferation. *Histopathology* 10:867–875, 1986.

366. Doucet J, Dardick I, Srigley JR, et al: Localised fibrous tumour of serosal surfaces: Immunohistochemical and ultrastructural analysis for a type of mesothelioma. *Virchows Arch Pathol Anat Histopathol* 409:349–363, 1986.

367. Brown DC, Theaker JM, Banks PM, Gatter KC, Mason DY: Cytokeratin expression in smooth muscle and smooth muscle tumours. *Histopathology* 11:477–486, 1987.

368. Al-Saffar N, Hasleton PS: Vimentin, carcinoembryonic antigen, keratin in the diagnosis of mesothelioma, adenocarcinoma and reactive pleural lesions of the lung. *Eur Resp J* 3:997–1001, 1990.

369. Jasani B, Edwards RE, Thomas MD, Gibbs AR: The use of vimentin antibodies in the diagnosis of malignant mesothelioma. *Virchow Arch Pathol Anat Histopathol* 406:441–448, 1985.

370. Wick MR, Loy T, Mills SE, et al: Malignant epithelial pleural mesothelioma versus peripheral pulmonary adenocarcinoma: A histochemical, ultrastructural and immunohistologic study of 103 cases. *Hum Pathol* 21:759–766, 1990.

371. Mullink H, Henzen-Longmans SC, Alons-van-Kordelaar JJM, et al: Simultaneous immunoenzyme staining of vimentin and cytokeratin with monoclonal antibodies as an aid in the differential diagnosis of malignant mesothelioma from pulmonary adenocarcinoma. *Virchows Arch B* 52:55–65, 1986.

372. Blobel GA, Moll R, Franke WW, et al: The intermediate filament cytoskeleton of malignant mesotheliomas and its diagnostic significance. *Am J Pathol* 121:235–257, 1985.

373. Wang N-S, Huang S-N, Gold P: Absence of carcinoembryonic antigen–like material in mesothelioma: An immunohistochemical differentiation from other lung cancers. *Cancer* 44:937–943, 1979.

374. Mezger J, Lamerz R, Permanetter W: Diagnostic signifi-

cance of carcinoembryonic antigen in the differential diagnosis of malignant mesothelioma. *J Thorac Cardiovasc Surg* 100:860–866, 1990.

375. Lamerz R: CEA and kreuzreagierende Antigene—Abgrenzung and Klinische Bedeutung. *Lab Med* 11:424–432, 1987.

376. Wagener C, Yang YHG, Crawford FG, Shively JE: Monoclonal antibodies for carcinoembryonic antigen and related antigens as a model system: A systematic approach for the determination of epitope specificities of monoclonal antibodies. *J Immunol* 130:2308–2315, 1983.

377. Churg A, Green SHY (eds): *Pathology of Occupational Lung Disease.* New York, Igaku-Shoin. 1988, p 315.

378. Battifora H, Kopinski M: Distinction of mesothelioma from adenocarcinoma: An immunohistochemical approach. *Cancer* 55:1679–1685, 1985.

379. Sheibani K, Battifora H, Burke J: Antigenic phenotype of malignant mesotheliomas and pulmonary adenocarcinomas: An immunohistologic analysis determining the value of Leu M1 antigen. *Am J Pathol* 123:212–219, 1986.

380. Ordonez NG: The immunohistochemical diagnosis of mesothelioma: Differentiation of mesothelioma and lung adenocarcinoma. *Am J Surg Pathol* 13:276–291, 1989.

381. Hanjan SNS, Kearney JF, Cooper MD: A monoclonal antibody (MMA) has identified a differentiation antigen on human myelo-monocytic cells. *Clin Immunopathol* 23:172–188, 1982.

382. Hsu SM, Jaffe ES: Leu-M1 peanut agglutinin stain the neoplastic cells of Hodgkin's disease. *Am J Clin Pathol* 82:29–32, 1984.

383. Otis CN, Carter D, Cole S, Battifora H: Immunohistochemical evaluation of pleural mesothelioma and pulmonary adenocarcinoma: A bi-institutional study of 47 cases. *Am J Surg Pathol* 11:445–456, 1987.

384. Sheibani K, Battifora H, Burke JS, et al: Leu M1 human neoplasms: An immunohistologic study of 400 cases. *Am J Surg Pathol* 10:227–236, 1986.

385. Sheibani K, Azumi M, Battifora H: Further evidence demonstrating the value of Leu M1 antigen in differential diagnosis of malignant mesothelioma and adenocarcinoma: An immunohistologic evaluation of 395 cases (abstr). *Lab Invest* 58:84a, 1988.

386. Wick MR, Mills SE, Swanson PE: Expression of "myelomonocytic" antigens in mesotheliomas and adenocarcinomas involving serosal surfaces. *Am J Clin Pathol* 94:18–26, 1990.

387. Szpak CA, Johnston WW, Roggli V, et al: The diagnostic distinction between malignant mesothelioma of the pleura and adenocarcinoma of the lung as defined by a monoclonal antibody(B72.3). *Am J Pathol* 122:252–260, 1986.

388. Latza U, Niedobitek G, Schwarting R, et al: Ber-EP4: New monoclonal antibody which distinguishes epithelia from mesothelia. *J Clin Pathol* 43:213–219, 1990.

389. Sheibani K, Shin SS, Kezirian J, Weiss LM: Ber-EP4 antibody as a discriminant in the differential diagnosis of malignant mesothelioma versus adenocarcinoma. *Am J Surg Pathol* 15:779–784, 1991.

390. Gaffey MJ, Mills SE, Swanson PE, Zarbo RJ, Shah AR, Wick MR: Immunoreactivity for Ber-EP4 in adenocarcinomas, adenomatoid tumors and malignant mesotheliomas. *Am J Surg Pathol* 16:593–599, 1992.

391. Walts AE, Said JW, Shintaku IP: Epithelial membrane antigen in the cytodiagnosis of effusions and aspirates: Immunocytochemical and ultrastructural localisation in

benign and malignant cells. *Diagn Cytolpathol* 3:41–49, 1987.

392. Epstein JI, Budin RE: Keratin and epithelial membrane antigen, immunoreactivity in nonneoplastic fibrous pleural lesions: Implications for the diagnosis of desmoplastic mesothelioma. *Hum Pathol* 17:514–519, 1986.

393. Sloane JP, Ormerod MG: Distribution of epithelial membrane antigen in normal and neoplastic tissues: Its value in diagnostic tumor pathology. *Cancer* 47:1786–1795, 1981.

394. Hammar SP, Bolen JW, Bockus D, et al: Ultrastructural and immunohistochemical features of common lung tumors: An overview. *Ultrastruct Pathol* 9:283–318, 1985.

395. Burchell J, Durbin H, Taylor-Papadimitriou A: Complexity of expression of antigenic determinants recognised by monoclonal antibodies HMFG-1 and HMFG-2, in normal and malignant human mammary epithelial cells. *J Immunol* 131:508–513, 1983.

396. Marshall RJ, Herbert A, Braye SG, Jones DB: Use of antibodies to carcino-embryonic antigen and human milk fat globules to distinguish carcinoma, mesothelioma and reactive mesothelium. *J Clin Pathol* 37:1215–1221, 1984.

397. Joglekar VM, Oliver D, Harris M: The value of anticarcinoembryonic antigen, human milk factor globulin, and antikeratin antibodies in differentiating mesothelioma from lung carcinoma. *Br J Ind Med* 48:34–37, 1991.

398. Radosevich JA, Ma Y, Lee I, et al: Monoclonal antibody 44-3A6 as a probe for a novel antigen found on human lung carcinomas with glandular differentiation. *Cancer Res* 45:5808–5812, 1985.

399. Lee I, Radosevich JA, Chejfec G, et al: Malignant mesotheliomas: Improved differential diagnosis from lung carcinomas using monoclonal antibodies 44-3A6 and 624A12. *Am J Pathol* 123:497–507, 1986.

400. Spagnolo DV, Whitaker D, Carrello S, Radosevich JA, Rosen ST, Gould VE: The use of monoclonal antibody 44-3A6 in cell blocks in the diagnosis of lung carcinoma, carcinomas metastatic to lung and pleura and pleural malignant mesothelioma. *Am J Clin Pathol* 95:322–329, 1991.

401. Azumi N, Underhill CB, Kagan E, Sheibani K: A novel biotinylated probe specific for hyaluronate: Its diagnostic value in diffuse malignant mesothelioma. *Am J Surg Pathol* 16:116–121, 1992.

402. Kawai T, Greenberg SD, Truong LD, Mattioli CA, Klima M, Titus JL: Differences in lectin binding of malignant pleural mesothelioma and adenocarcinoma of the lung. *Am J Pathol* 130:401–410, 1988.

403. Kallianpur AR, Carstens PHB, Liotta LA, Frey KP, Siegal GP: Immunoreactivity in malignant mesotheliomas with antibodies to basement membrane components and their receptors. *Mod Pathol* 3:11–18, 1990.

404. Lane DP, Crawford LV: T antigen is a host protein in SV40-transformed cells. *Nature* 278:261–263, 1979.

405. Kafiri G, Thomas DM, Shepherd NA, Krausz T, Lane DP, Hall PA: p53 expression is common in malignant mesothelioma. *Histopathology* 21:331–334, 1992.

406. Ramael M, Lemmens G, Eerdekens C, et al: Immunorectivity for p53 protein in malignant mesothelioma and nonneoplastic mesothelium. *J Pathol* 168:371–375, 1992.

407. Mayall FG, Goddard H, Gibbs AR: p53 immunostaining in the distinction between benign and malignant mesothelial proliferations using formalin-fixed paraffin sections. *J Pathol* 168:377–381, 1992.

408. Cagle PT, Brown RW, Lebowitz RM: p53 immunostaining in the differentiation of reactive processes from malig-

nancy in pleural biopsy specimens. *Hum Pathol* 25:443–448, 1994.

409. Anderson TM, Holmes EC, Kosaka CJ, et al: Monoclonal antibodies to human malignant mesothelioma. *J Clin Immunol* 7:254–261, 1987.

410. Donna A, Betta P, Jones JSP: Verification of the histological diagnosis of malignant mesothelioma in relation to the binding of an antimesothelial cell antibody. *Cancer* 63:1331–1336, 1989.

410a. Clarke SP, Chon ST, Martin TJ, Danks JA: Parathyroid hormone-related protein antigen localisation distinguishes between mesothelioma and adenocarcinoma of the lung, *J Pathol* 176:161–165, 1995.

411. el-Nagger AK, Ordonez NG, Garnsey L, Batsakis JG: Epithelial pleural mesothelioma and pulmonary adenocarcinomas: A comparative flow cytometric study. *Hum Pathol* 22:972–978, 1991.

412. Dazzi H, Hasleton PS, Thatcher N: DNA analysis by flow cytometry in malignant pleural mesothelioma: Relationship to histology and survival. *J Pathol* 162:51–55, 1990.

413. Burmer GC, Rabinovitch PS, Kulander BG, Rusch V, McNutt MA: Flow cytometric analysis of malignant pleural mesotheliomas. *Hum Pathol* 20:777–783, 1989.

414. Pyrhonen S, Laasonen A, Tammilehto L, et al: Diploid predominance and prognostic significance of S-phase cells in malignant mesothelioma. *Eur J Cancer* 27:197–200, 1991.

415. Van Meerbeeck JP: Prognostic factors in mesothelioma: Where do we go from here? *Eur Respir J* 7:1029–1031, 1994.

416. Van Gelder T, Damhuis RAM, Hoogsteden HC: Prognostic factors in malignant pleural mesothelioma. *Eur Respir J* 7:1035–1038, 1994.

417. Chahinian AP, Pajak TF, Holland JF, Norton L, Ambinder RM, Mandel EM: Diffuse malignant mesothelioma: Prospective evaluation of 69 patients. *Ann Intern Med* 96:746–755, 1982.

418. Antman K, Shemin R, Ryan L, et al: Malignant mesothelioma: Prognostic variables in a registry of 180 patients, the Dana-Farber Cancer Institute and Brigham and Women's Hospital experience over two decades, 1965–1985. *J Clin Oncol* 6:147–153, 1988.

419. Brenner J, Sordillo PP, Magill GB, Golbey RB: Malignant mesothelioma of the pleura: Review of 123 patients. *Cancer* 49:2431–2435, 1982.

420. Manzini VDP, Brollo A, Franceschi S, De Matthaeis M, Talamini R, Bianchi C: Prognostic factors of malignant mesothelioma of the pleura. *Cancer* 72:410–417, 1993.

421. Boutin C, Rey F, Gouvernet J, Viallat JR, Astoul PH, Ledoray V: Thoracoscopy in pleural malignant mesothelioma: A prospective study of 188 consecutive patients: Part 2: Prognosis and staging. *Cancer* 72:394–404, 1993.

422. Tammilehto L: Malignant mesothelioma: Prognostic factors in a prospective study of 98 patients. *Lung Cancer* 8:175–184, 1992.

423. Chailleux E, Dabuois G, Pioche D: Prognostic factors in diffuse malignant mesothelioma: A study of 167 patients. *Chest* 93:159–162, 1988.

424. Sridhar KS, Doria R, Raub WA Jr, Thurer RJ, Saldana M: New strategies are needed in diffuse malignant mesothelioma. *Cancer* 70:2969–2979, 1992.

425. Gibbs A, Seal R: *Atlas of Pulmonary Pathology* Lancaster, England, MTP, 1982.

426. Yokoi T, Mark EJ: Atypical mesothelial hyperplasia associated with bronchogenic carcinoma. *Hum Pathol* 22:695–699, 1991.

427. Wick MR, Burgess JH, Manivel JC: A re-assessment of "chordoid sarcoma": Ultrastructural and immunohistochemical comparison with chordoma and skeletal myxoid chondro-sarcoma. *Mod Pathol* 6:433–443, 1988.

428. Wick MR, Swanson PE, Manivel JC: Immunohistochemical analysis of soft tissue sarcomas: Comparisons with electron microscopy. *Appl Pathol* 6:169–196, 1988.

429. Chase DR, Enzinger FM, Weiss SW, Langloss JM: Keratin in epithelioid sarcoma: An immunohistochemical study. *Am J Surg Pathol* 8:435–541, 1984.

430. Greenspan EB: Primary osteoid chondrosarcoma of the lung: Report of a case. *Am J Cancer* 18:603–609, 1933.

431. Nosanchuk JS, Weatherbee L: Primary osteogenic sarcoma in lung. *J Thorac Cardiovasc Surg* 58:242–247, 1969.

432. Reingold IM, Amromin GD: Extraosseous osteosarcoma of the lung. *Cancer* 28:491–498, 1971.

433. Torrado J, Blasco E, Gutierrez-Hoyos A: Ulex Europaeus (UEA 1). *Am J Clin Pathol* 91:503, 1989.

434. Mayall FG, Gibbs AR: "Pleural and pulmonary carcinosarcomas." *J Pathol* 167:305–311, 1992.

435. Roggli VL: Human disease consequences of fiber exposures: A review of human lung pathology and fiber burden data. *Environ Health Perspect* 88:295–303, 1990.

436. Gibbs AR, Stephens M, Griffiths DM, Blight BJN, Pooley FD: Fibre distribution in the lungs and pleura of subjects with asbestos related diffuse pleural fibrosis. *Br J Ind Med* 48:762–770, 1991.

437. Butchart EG, Ashcroft T, Barnsley WC, Holden M: The role of surgery in diffuse malignant mesothelioma of the pleura. *Semin Oncol* 8:321–328, 1981.

Chapter 35

Techniques in Pulmonary Pathology

W. F. Whimster

The aim of pulmonary pathology is to provide structural explanations for abnormalities observed by clinical, imaging, and physiologic techniques.

For success, relevant cells or tissues from the trachea, conducting airways (bronchi and bronchioles), pulmonary parenchyma, vascular system, and supporting structures must be obtained. This can only be achieved by cytology, biopsy, or resection in life or removal at autopsy. The specimens must be appropriately prepared for examination, generally by the routine cytologic and histologic methods, sometimes with a special component for pulmonary tissue, such as inflation with air or distention with fixative.

Some methods, particularly in lung tumors, are used to assess prognosis. Others are used to satisfy curiosity and extend our knowledge of normal structure and its correlation with function, as well as etiology, pathogenesis, and morphology of pulmonary disease. Such techniques have recently been thoroughly reviewed by Gil.[1]

New techniques are being used to fill the gaps, such as quantitation, confocal microscopy, immunochemistry, and molecular biologic techniques. Nevertheless existing techniques have to be properly standardized and more are needed. New efforts are needed firstly to detect the three-dimensional relationships of pulmonary lesions such as those of emphysema and secondly to establish sampling rules for diffuse pulmonary lesions.

The pathologist depends, of course, on close collaboration with the physican or surgeon, not only for the appropriate clinical information but also for appropriate handling of the specimen(s).

HEALTH AND SAFETY AT WORK

Apart from the normal hazards of working in cytopathology and histopathology laboratories and in mortuaries, those dealing with biopsy, resection, or postmortem pulmonary tissues are particularly at risk of infection with tuberculosis. Furthermore, large quantities of fixative are often used, particularly to distend lungs, which may give concentrations of fixative in the atmosphere in excess of those permitted by the United Kingdom government regulations.[2,3] It is essential that the staff protect themselves from these risks by the use of masks and protective clothing, handling potentially infective tissues within appropriate protective cabinets, and following standard operating procedures. Fixative concentrations in the atmosphere should be checked regularly.

CYTOLOGY (Chap. 36)

Cytologic techniques are applied to sputum, brushings, washings, and lavage fluid obtained via the bronchoscope. In addition they are applied to pleural fluid and to fine needle aspiration (FNA) biopsies taken percutaneously under imaging control, usually to diagnose neoplastic disease, but also to diagnose infections.

These specimens are usually obtained and submitted to the pathology laboratory by clinical staff who are familiar with the techniques.[4] They are processed by the laboratory staff for microscopy by standard techniques. Certain points should be mentioned here.

Sputum needs *multiple deep* early morning specimens, which should be sent quickly to the laboratory where smears are made, fixed in 95% ethyl alcohol, and stained by the Papanicolaou technique. Prefixing techniques, for example, in 70% ethyl alcohol, produce less satisfactory results. Brushings are useful, but it is important to avoid air drying. Washings give a low yield of positives but are useful in difficult cases to exclude infection. Bronchial lavage, involving infusion and reaspiration of sterile saline, may be used as a diagnostic technique. The various cell counts (T4, T8, eosinophils) may also be useful.

Fine needle aspiration cytology, controlled by fluoroscopy or computed tomography (CT), is reliable if six aspirations are made. It also gives a good yield when used transbronchially.[5,6] The needle contents are expelled onto a slide, smeared, and either wet fixed and stained with Papanicolaou or air dried and stained with Giemsa. The reliability and limitations of transbronchial FNA have been reviewed by Wagner and colleagues.[7] Lung cancer metastases may also be diagnosed by FNA cytology.[8]

Cells obtained by any of these techniques may be suitable for the application of molecular biologic techniques, such as polymerase chain reaction[9] and in situ hybridization.

BIOPSY HANDLING

The biopsy procedure should be planned between the clinician and the pathologist to ensure that tissue is submitted fresh for microbiologic examination; for frozen section; certain histochemical, immunochemical, or molecular biologic procedures; culture or −70°C storage; or fixed appropriately.

Endobronchial and transbronchial biopsies are generally obtained via a fiberoptic bronchoscope. To obtain adequate uncrushed tissue requires multiple biopsies taken with wide cupped forceps. These should be carefully shaken with no handling or prodding to extract them into fixative (see below).

Transbronchial biopsies of parenchymal tissue may usefully be "distended" by shaking them first into 5 mL saline in a 10-mL syringe. Then negative pressure is applied with the plunger before transferring them to fixative. Transbronchial needle biopsies are reported to improve the yield of FNA cytology.[10] Bronchoscopists become frustrated when biopsies of clinically obvious tumors prove to be histologically negative. This is usually because the forceps have not penetrated the edematous mucosa proximal to the tumor or because the neoplasm is subepithelial.

Further biopsies, possibly with a rigid bronchoscope giving larger biopsies, are needed. The rigid bronchoscope may also occasionally be used to remove polypoid lesions, such as some carcinoid tumors, whole.

Percutaneous needle biopsy under visual control is used to obtain tissue from peripheral lesions. Fluoroscopy is used for large peripheral lesions, CT for smaller and less peripheral lesions, and ultrasound when there is bronchial obstruction and parenchymal collapse.

Open lung biopsy, once popular, is now used occasionally to elucidate the nature of diffuse lung disease, to diagnose or remove solitary peripheral nodules, for staging before lung transplantation, and rarely after lung transplantation. A video-assisted keyhole technique is increasingly used. The specimens obtained go straight into fixative after distention using a thin needle through the pleura. If fresh tissue is required for urgent frozen section (see below) or for immunohistochemistry, it must be cut in a category B cabinet.

Pleural biopsy may be obtained by thoracoscopy. Visceral or parietal lesions can be sampled. Blind needling with Abrams or Trucut needles less commonly gives diagnostic results.

The standard examination of such specimens is described by Gibbs and Seal.[11]

CULTURE, FROZEN SECTION, FIXATION, SECTION CUTTING, AND STAINING

Routine microbiologic and histopathologic techniques are used for biopsy specimens and resected and postmortem lung tissue. A few modifications may be added for pulmonary material.

It is important to submit any material obtained from immunocompromised patients fresh for microbiologic culture for a variety of microorganisms (see Chap. 4) rather than missing the opportunity by subjecting it all to fixation. In adults with acquired immunodeficiency syndrome (AIDS), culture and cytology of bronchoalveolar lavage material may be expected to detect 98 percent of *Pneumocystis carinii*, cytomegalovirus, mycobacterial, cryptococcal, and coccidioidomycosis infections.[12]

From any patient in whom there is a clinical suspicion of tuberculosis or other infection (particularly those who are immunocompromised or with sarcoidosis), fresh unfixed biopsy tissue *should* be submitted for culture. This tissue *should not* be submitted for frozen section because it puts the staff at risk and cryostats take 24 h to decontaminate. Nevertheless, for apical cavitating lesions, which may be malignant or inflammatory, the surgeon may still need a frozen section.

For routine histology and electron microscopy biopsy tissue has to be fixed. For histology 4% neutral buffered (pH 6–8) isotonic formaldehyde solution (i.e., concentrated, 40%, formaldehyde diluted 10 times, usually referred to as "10% formalin") is usually used. Adequate buffering to avoid a lower pH has been shown to be essential if morphometry of cells is to be undertaken.[13] Murray and Ewen[14] have described a combination of freeze-substitution and low-temperature plastic embedding, which avoids tissue fixation but gives good results with histochemical and immunohistochemical stains. For electron microscopy, small pieces of tissue are fixed in cold glutaraldehyde. Rowden and Lewis[15] have reported a rapid 3-h technique using half strength glutaraldehyde-paraformaldehyde fixative as the primary fixative.

Step-sectioning improves the yield of specific findings in some cases of transbronchial lung biopsies.[16]

Routine histologic, histochemical, and immunohistochemical stains are used for pulmonary material. The Ziehl-Neelson technique for acid- and alcohol-fast mycobacteria may be speeded up with a reduced fire hazard by the use of a microwave oven for the heating stage.[17]

HANDLING RESECTED SPECIMENS

In general, it is best for the pathologist to inspect, gently palpate, weigh, measure, and describe the specimen under safe conditions (see above) and decide whether it is necessary to perform any dissection or take any fresh tissue before the specimen is fixed in the usual fixative, namely, 10% formalin, or in any other preferred fixative.

The pathologist may decide to distend the specimen with fixative through the cut ends of any visible bronchi. Solid lesions may be incised to improve fixation. The whole specimen may be placed in fixative unopened if an infection risk exists. Before infusing fixative into any bronchus, a transverse ring of the resection margin should be sampled because the mucosa is easily damaged. This should be placed into a separately labeled container of fixative. Then a catheter or tube is used to infuse the fixative.

If the laboratory is unlikely to receive or deal with the specimen within a few hours, certainly overnight, it is best to put it whole into 10% formalin. Storage of the specimen in a sealed container in a refrigerator at 4°C is better than leaving it unfixed at room temperature. Freezing however, ruins the histology.

Careful distention with 10% formalin is carried out under safe conditions (see above) via the bronchi until the pleural surface becomes smooth. The resected end may be clipped with artery forceps. The specimen is immersed in fixative for 48 h to obtain good histology. Fixative may fail to penetrate solid areas and then a wide-bore needle can be used to pass fixative beyond the solid area. Theoretically the distention might displace airway contents and too large a distention pressure might tear the tissues; these do not present problems in practice. Nevertheless formalin fixation does cause shrinkage of the parenchymal tissue of up to 30 percent unless distention pressure is maintained. If necessary this can be monitored, as is done in some circumstances for postmortem studies (see below).

After fixation, the specimen is sliced and blocks are taken for histologic processing. The pathologist must decide in each case how best to slice the specimen so that any lesions may best be exposed and examined. In the case of recognizable tumor the aim is to locate its origin, to get a measure of its size, and to determine the extent of its spread. This is best done by first placing the specimen on its lateral surface and making a complete midsagittal slice cleanly through it. Study of both surfaces will show which lobes are involved, if the pleura is involved, and whether involved bronchi or vessels or lymph nodes are visible on the medial cut surface. A probe should be passed from the hilar bronchi to the medial cut surface and the bronchi opened with scissors or a knife to expose the most proximal extent of the tumor within the bronchus. Further sagittal slices may be made in the lateral or medial halves if necessary. Blocks for histology should be taken from the most proximal point, from the tumor itself, from all lymph nodes located, from the distal aspect of the tumor, from the pleura if involved, from any other lesions observed, and randomly from distal lung.

Transplant resections are handled in the same way as postmortem specimens (see below).

POSTMORTEM SPECIMENS

Full clinical information helps the pathologist to predict the techniques required. For some conditions considerable thought has been given to establishing a comprehensive protocol, for example, chronic nonspecific lung disease.[18]

Removing Lungs, Including Pleura and Diaphragm

If pulmonary tuberculosis is suspected, the prosector and others present should wear protective clothing and masks (see above). To avoid disseminating the organisms the trachea can be cannulated and 10% formalin infused into the respiratory tract before dissection is undertaken. Otherwise at postmortem, after removal of the sternum (which results in collapse of aerated lungs), it is best to remove the respiratory organs, including the visceral pleura and diaphragm, in the complete block of visceral organs from the tongue to the urethra, rectum, and vagina in female subjects.

In the thorax if there are adhesions between the visceral and parietal pleura, these should first be gently separated by blunt finger dissection. If this is impossible, a plane of cleavage should be made with the fingers between the parietal pleura and the chest wall, rather than tear the lung. With really dense fibrous, calcified, or malignant adhesions, it may be necessary to cut the lungs from the chest wall. The aorta and thoracic duct must be carefully separated posteriorly (rough pulling from above causes artefactual intimal aortic tears), and the diaphragm cut from its peripheral insertion. Some prosectors think it easier to divide the visceral block above the diaphragm into thoracic and abdominal blocks for removal, but this destroys the continuity of the transdiaphragmatic structures before they can be examined, thus breaching a fundamental principle of postmortem practice.

Inspection of the Thoracic Cavity

After the thoracic organs are removed, the parietal chest wall, including the ribs, is inspected, particularly for pleural plaques and tumors. Subpleural carbon deposits are often seen, particularly in the intercostal depressions on the parietal side and on the intercostal protuberances on the visceral side, although the mechanism for this is unknown.

Dissection

The respiratory system is best examined last—after the cardiovascular system and the esophagus. The aorta is best separated by blunt scissors from the underlying pulmonary artery so that the latter can be specifically incised and inspected, together with the site of the ductus arteriosus. If the base of the heart is then lifted up the pulmonary veins can be specifically identified and cut, and the whole pericardium carefully inspected. If the heart is just hacked out of the pericardium, as in ancient Peru, pulmonary venous pathology may be missed. For complete assessment of the pulmonary system it is important to obtain the weight of the right ventricle of the heart, which does not normally exceed 80 g in the adult, and the ratio of the weights of the right and left ventricles. This is done according to Fulton's method[19] in which the atria are removed by cutting around the atrioventricular ring and the right ventricle cut from the left ventricle along the margin of the interventricular septum, leaving the septum as part of the left ventricle. It makes no difference to the weights whether the heart is unfixed or fixed, but it is eas-

ier to do the cutting if the heart has been stuffed with cotton wool and fixed in 10% formalin overnight.

The lungs are best inspected lobe by lobe for externally visible lesions and palpated for nodules and differences in texture. The decision must now be made as to whether distention is required for either or both lungs. In general distention is good for anatomic studies of the parenchyma (especially emphysema) and preservation of the airways and lungs. Incision and slicing, followed by fixation is good for the histologic elucidation of consolidated areas and specific lesions. In either case the larynx and trachea are cut open longitudinally with scissors, usually down the membranous posterior part. They may be opened anteriorly from below the larynx if the membranous part is of particular interest or if the bronchoscopist wants to view the airways as in life.

Distention/Fixation

It is unwieldy to inflate both lungs together, so one lung is usually selected for distention under safe conditions (see above). It is best to distend the left lung because its main bronchus is longer, but the clinical history, radiographs, and postmortem findings may determine which one is selected. Distention can be simply done as above for the surgically resected specimen. An adult lung takes about 5 L of 10% formalin and can be cut to give excellent histology at the end of the postmortem examination. However, 48 h fixation before cutting gives better preservation. But for comparative quantitative studies of normal or abnormal parenchyma a more standardized technique is desirable.

Weibel[20] used formalin steam (which requires a specially ventilated room); Blumenthal and Boren,[21] Heard,[22] Wright and Slavin,[23] the author, and others have used various ways of maintaining constant or intermittent "respiration-mimicking" pressure of formalin to maintain inflation. Intermittent pressure is not necessary because formalin leaks rapidly from the lung by the veins and, as fixation progresses, through the pleura. The author ties a catheter into the bronchus and briefly inflates the lung with air to expel much of the blood (which inhibits fixation) from the veins. Then the catheter is connected to the 10% formalin supply of a closed tank and pump (Fig. 35-1) at room temperature. When fully distended the preshrinkage volume can be measured by water displacement. Then pumping can be resumed until the lung barely collapses when the pump is switched off, usually after 2 to 3 days. This gives a *wet* lung, whose

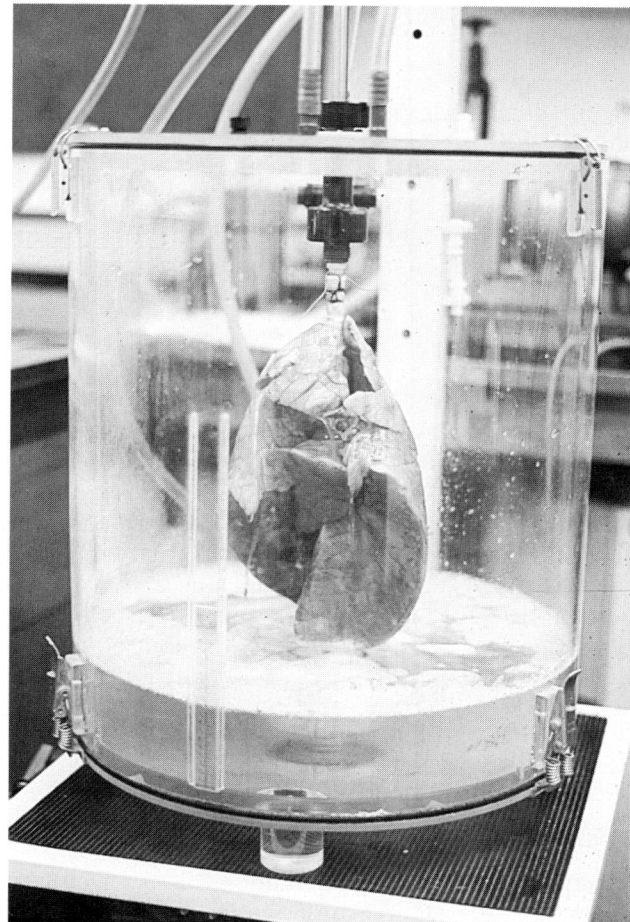

FIGURE 35-1
Lung distention machine (built for the author by Paul Richardson of the Biomedical Engineering Department, King's College School of Medicine).

volume can again be measured to assess shrinkage. The formalin can be drained off at this stage and the pump used to drive air through the lung to give the useful *semi-dry* format[24] in a further 2 or 3 days. A completely *dry* format, as shown by no further fall in weight, may be attained after 5 to 6 days.[25,26]

Lung Slicing/Description/Block Taking/Microscopy

Wet-fixed and semi-dry lungs can be sliced free hand with the longest sharpest knife available into two or three parasagittal slices. Five even slices are obtained with a slicing tray with sides 1.0 or 1.5 cm high. These slices may be examined on the tray or under water, which supports the parenchyma, with the naked eye, hand lens, or dissecting microscope. The slices may be further processed for whole lung paper sections (see below). Such examination of the slice restored to its anatomic configuration helps to bridge the conceptual leap from the macroscopic to the microscopic appearance. Slices may be enhanced,

especially for photography purposes, by impregnation with barium.[22] This highlights the areas of lung affected by carbon or other pigment.

Dry lungs may also be sliced free hand or into thinner slices (around 0.5 to 1.0 mm thick) on a bacon slicer. They are placed between perspex sheets and examined microscopically without staining under low power with transmitted or incident light. Blumenthal and Boren[21] produced excellent radiographs of such preparations. Small blocks may be placed on microscope slides for examination with shorter focal length objectives or with the confocal microscope (see below). Small blocks may also be floated through staining protocols such as hematoxylin and eosin or elastic-van Gieson. They are then viewed wet or mounted in a plastic mountant in a vacuum to remove air bubbles.

TECHNIQUES FOR SPECIAL PURPOSES

Macroscopic Methods

FIXATION BY ARTERIAL PERFUSION

Arterial perfusion[26] with fixative is sometimes advocated in preference to bronchial fixation on the grounds that the contents of the airways will remain for inspection as if in life. It is useful in asthmatic cases to bypass the bronchiolar obstruction but it does not restore the anatomic structure to normal.

RADIOGRAPHIC EXAMINATION OF THE VESSELS

Pulmonary or bronchial arterial or pulmonary venous perfusion can be carried out with a suspension of barium in water (e.g., Micropaque) with a teaspoonful of gelatin crystals. This should be warmed to body temperature to dissolve the gelatin before bronchial perfusion with 10% formalin. The latter fixes the gelatin in situ and gives excellent radiographs of the whole lung and its slices. If both the pulmonary arteries and veins are perfused, they may be distinguished by eye, but not radiographically, by adding different colored dyes to the two perfusates. (For radiology see reference 27). Histologically the barium is easily recognized, so that after arterial perfusion alone supernumerary arteries[28] may be distinguished from veins.

WHOLE LUNG PAPER SECTIONS

Paper-mounted sections of organs were invented by Gough and Wentworth.[29] The technique was subsequently simplified to a 48-h procedure[30] and has been revisited and modified several times since then for various purposes.[31,32] It is a useful technique for museum and record purposes and for the study of pulmonary emphysema.

MACROSCOPIC EMPHYSEMA SCORING

For scoring emphysema for epidemiologic purposes, Thurlbeck and colleagues[33,34] introduced the "eyeball" technique whereby a single paper section from each case was compared with a panel of paper sections to obtain a percentage figure that encompassed both the extent and severity of the emphysematous spaces. This is probably satisfactory for epidemiologic studies but is too subjective for other purposes.[32]

MACROSCOPIC POINT COUNTING FOR EMPHYSEMA

Turner and Whimster[35] used a simple point counting system to obtain the volume of emphysema in any given lung whereby a transparent perspex point counting sheet was laid over each slice and the points counted. This technique showed that the amount of emphysema in one slice could not be used to predict the amount of emphysema in that lung, nor could the amount of emphysema in one lung be used to predict the amount of emphysema in the other.

FETAL AND PERINATAL LUNGS

The techniques described above can be used to study fetal and perinatal lungs. It is best to clamp the main bronchus of the most consolidated lung and to distend the other with 10% formalin via the trachea and then to leave the whole specimen in 10% formalin overnight before slicing the lungs and taking blocks for microscopy. The uninflated microscope sections best represent the pathologic appearances, the inflated ones the anatomic relationships. With the smaller lungs whole lung slices can be hand processed and cut to give whole lung sections.

SPECIMENS FOR PNEUMOCONIOSIS ASSESSMENT

Assessment of postmortem lung specimens from patients known or suspected to have industrial lung disease continues to be required for clinicopathologic correlation and medicolegal purposes. Too often lungs are sliced fresh, a few random blocks of lung taken for histology, dumped in 10% formaldehyde, later retrieved, dried off, put into a plastic bag, and sent to a designated pulmonary pathologist. Such specimens are of limited value for assessment of the degree or distribution of fibrosis. Ideally, lungs should be distended as described above. The prevalence of "asbestos bodies" can be crudely assessed histologically in sections, stained with Perl's technique, and in unstained thick, 30-μm sections. Actual asbestos fibers can be extracted from lung samples and expressed in numbers of fibers per gram of dry weight of lung, for which therefore either inflated or uninflated lung tissue is appropriate. For further information see Chap. 15.

CORROSION CASTS

Tompsett[36] described how to make rigid resin casts of the airways and vessels in the respiratory tracts of man and other species. The tissue was then corroded away with acid. These give an excellent three-dimensional visualization of these tubes. Hollow casts can also be made for air flow studies. Horsfield[37] used casts to study the numbers and angles of airways branchings and also reported the use of a flexible plastic casting material. Pump[38] used casts to identify the relationships between alveoli and alveolar ducts.

Microscopic Methods

RANDOM BLOCKS

It may be necessary to take random blocks from lung slices.[20,39] It should not be thought that the parenchymal lung tissue is randomly distributed,[24] so such sampling may be inappropriate. It can be done by putting the slice onto a sheet of cork and placing a perspex sheet over the cut surface of the slice. This has a 2-cm grid with a hole drilled in each corner. Squares are chosen by random numbers and headless pins inserted through the relevant corner holes. The perspex sheet is lifted off and the pins used as guides to cut out the 2-cm blocks.[25] To get truly random blocks from what is a highly isotropic tissue, the principles of Gundersen may be used.[40]

BLOCK SURFACE STAINING

Block surface staining[41] is a technique whereby lead acetate is incorporated into the tissue during processing through to wax. The block is cut on a sledge microtome so that the cut surface is horizontal and sodium sulfide is pipetted onto the surface of the block, thus depositing insoluble lead sulfide differentially onto the tissue components within the block. These may be viewed in incident light with a dissecting microscope, levels are cut through the block and discarded, and the next surface stained and viewed again. This is a method of obtaining morphometric data quickly from surfaces such as transverse sections of bronchi.[42]

QUANTITATIVE TECHNIQUES IN THE PULMONARY FIELD

Quantitative pathology techniques, including morphometry, stereology, and three-dimensional reconstruction, have been particularly applied to the normal and abnormal respiratory system, including simple scoring techniques.[43]

Morphometry

Morphometry includes all the techniques of making simple measurements, for example, of tumor metastases on chest radiographs[44] to determine tumor growth rates. It includes making counts of, for example, tumor vessels,[45,46] mitotic figures,[13] nucleolar organizing regions,[47] and proliferation markers,[48] which can quite well be done on video images, to relate counts to prognosis.[49] Many software packages exist for making interactive computerized measurements derived from perimeters drawn around cells, nuclei, or other structures on microscopic or macroscopic images obtained with a video camera. These are also used as prognostic indicators. The relationships between cells may also be examined using the techniques of syntactic structure analysis, based on graph theory and including minimal spanning tree and Voronoi procedures, to assess prognosis.[50] (A computer program is available from Prodit, BMA, De Meern, Netherlands.) The new techniques of fractal analysis have also been applied to the respiratory system,[51-53] and a computer program is available to perform fractal analysis on microscopic images (PRISMview, supplied by Improvision*).

Stereology

Whereas morphometry consists of methods whereby elements in a tissue are *measured*, stereology is a set of methods whereby the tissue is sampled and the proportion of the tissue occupied by the element of interest is *estimated*. If the total area or volume of the tissue is known, the absolute number, area, or volume of the element of interest may be *calculated* from the estimate. The simplest method is point counting, which may be applied to lung tissue macroscopically or microscopically[54] and converted into an absolute area or volume.[35] Stereologic techniques have been particularly expounded by Weibel[20] to obtain, for example, alveolar number, by Hislop and Reid[55-57] to determine alveolar size in congenitally abnormal lobes, and by Gundersen,[40] who is particularly concerned with acquiring randomly oriented sections from isotropic tissues so that accurate estimations

can be made. Although computers cannot help with the acquisition of randomly oriented sections, a computer program is available to produce stereologic results relatively painlessly (Digital Stereology, supplied by Kinetic Imaging*).

In relation to the "internal surface area of the lung," Thurlbeck[39] laboriously applied the stereologic technique of linear intercepts in precomputer days. Gillooly and colleagues[58] and Javed and associates[24] have more recently applied the techniques of computerized image processing to measure the alveolar wall surface area per unit volume. In the normal lung this revealed that alveolar size is not randomly distributed, which has a bearing on random sampling (see above).

Three-Dimensional Reconstruction

Three-dimensional reconstruction of tissues, based on entering the data of stacks of two-dimensional slices of various types (macroscopic slices, serial microscope sections, serial electron microscope sections, optical sections, CT and magnetic resonance scans) into computerized systems to produce computerized three-dimensional models, has much to offer the respiratory system, a three-dimensional system par excellence, but has not yet been much used.[28,59,60]

ELECTRON MICROSCOPY

Transmission electron microscopy has helped to elucidate many features of pulmonary structure, for example, the epithelial nature of the alveolar lining,[61] and pathology, such as the common histogenesis of carcinoids and oat cell carcinomas from neuroendocrine cells.[62] It has, however, been somewhat overtaken by the techniques of immunochemistry and molecular biology in the identification and localization of tissue and cell components and microorganisms. Electron microscopy still has a place in the examination and identification of abnormal cilia. Scanning electron microscopy has also recently been used to good effect in the development of the human respiratory acinus[63,64] and the examination of the ciliarization of the developing fetal trachea.[65]

ELECTRON PROBE ANALYSIS

This technique is useful for the identification of particles seen histologically in the lung. If they are very scanty, the piece of the tissue section, together with the underlying glass of the slide, can be cut out. If

---

* Image Processing and Vision Company, Ltd., Coventry, United Kingdom.

* Kinetic Imaging, Liverpool, United Kingdom.

they are easily found the block can be recut. The specimen is mounted on a carbon stub and heated to 37°C to remove the wax. Irradiation with the electron beam causes x-rays to be emitted at different wavelengths, which are analyzed to give a picture of the elements contained in the particle, allowing it to be identified as, for example, inhaled talc. The x-ray diffraction pattern emitted from asbestos fibers can be used to identify the type of asbestos.

FUTURE TECHNICAL POSSIBILITIES

As shown above, much ingenuity has gone into devising techniques with which to examine normal and abnormal structure and function in the respiratory tracts of man and animals (for the latter see particularly reference 1). In the future new questions will bring new techniques for solving them, but three areas obviously require new techniques now. Firstly to discover how 300 million alveoli, 10 million alveolar ducts, and 10,000 terminal bronchi are organized, function, and interact in three dimensions in one pair of adult lungs remains to be elucidated beyond the point reached by Horsfield and Cumming[66] and ourselves.[59] This leads to the second area for technical innovation, namely, to elucidate the initial lesion in pulmonary emphysema and how it damages the three-dimensional structure to produce its loss of parenchyma (see Fig. 3 in reference 25). If elastases are at work techniques are needed to elucidate what they do and where they do it. Thirdly, similar or other techniques are required to show the function of supernumerary arteries.[28]

REFERENCES

1. Gil J (ed): *Models of Lung Disease. Microscopy and Structural Methods*, New York, Marcel Dekker, 1990.
2. Health Services Advisory Committee. *Safe working and the prevention of infection in clinical laboratories.* London, HMSO, 1991.
3. Health Services Advisory Committee. *Safe working and the prevention of infection in the mortuary and postmortem room.* London, HMSO, 1991.
4. Johnston WW, Elson CE: Respiratory tract. In Bibbo M (ed), *Comprehensive Cytopathology*. Philadelphia, WB Saunders, 1991, pp 320–398.
5. Williams AJ, Santiago S, Lehrman S, Popper R: Transcutaneous needle aspiration of solitary pulmonary masses: How many passes? *Am Rev Respir Dis* 136:452–454, 1987.
6. Rosenthal DL, Wallace JM: Fine needle aspiration of pulmonary lesions via fiberoptic bronchoscopy. *Acta Cytol* 28:203–210, 1984.
7. Wagner ED, Ramzy I, Greenberg SD, Gonzalez JM: Transbronchial fine-needle aspiration. Reliability and limitations. *Am J Clin Pathol* 92:36–41, 1989.
8. Rohwedder JJ, Handley JA, Kerr D: Rapid diagnosis of lung cancer from palpable metastases by needle thrust. *Chest* 98:1393–1396, 1990.
9. Mao L, Hruban RH, Boyle JO, Tockman M, Sidransky D: Detection of oncogene mutations precedes diagnosis of lung cancer. *Cancer Res* 54:1634–1637, 1994.
10. Mehta AC, Kavuru MS, Meeker DP, Gephardt GN, Nunez C: Transbronchial needle aspiration for histology specimens. *Chest* 96:1228–1232, 1989.
11. Gibbs AR, Seal RME: Examination of lung specimens. *J Clin Pathol* 43:68–72, 1990.
12. Weldon-Linne CM, Rhone DP, Bourassa R: Bronchoscopy specimens in adults with AIDS. Comparative yields of cytology, histology and culture for diagnosis of infectious agents. *Chest* 98:24–28, 1990.
13. Fleege JC, van Diest PJ, Baak JPA: Reliability of quantitative pathological assessments, standards, and quality control, in Baak JPA (ed), *Quantitative pathology in Cancer Diagnosis and Prognosis.* Berlin, Springer-Verlag, 1991, pp 151–181.
14. Murray GI, Ewen SWB: A novel method for optimum biopsy specimen preservation for histochemical and immunohistochemical analysis. *Am J Clin Pathol* 95:131–136, 1991.
15. Rowden G, Lewis MG: Experience with a three-hour electron microscopy biopsy service. *J Clin Pathol* 27:505–510, 1974.
16. Nagata N, Hirano H, Takayama K, Miyagawa Y, Shigematsu N: Step section preparation of transbronchial lung biopsy. *Chest* 100:959–962, 1991.
17. Hafiz S, Spencer RC, Lee M, Gooch H, Duerdon BI: Use of microwaves for acid and alcohol fast staining. *J Clin Pathol* 38:1073–1084, 1985.
18. Medical Research Council Committee on Pathology on Research into Chronic Bronchitis: Quantitative assessment of chronic non-specific lung disease at necropsy. *Thorax* 30:241–251, 1975.
19. Lamb D: Heart weight and assessment of ventricular hypertrophy, in Dyke SC (ed), *Recent Advances in Clinical Pathology*, Series 6. Edinburgh, Churchill Livingstone, 19, pp 133–148, 1973.
20. Weibel ER: *Morphometry of the Human Lung.* Berlin, Springer-Verlag, 1963; pp 41–43.
21. Blumenthal BJ, Boren HG: Lung structure in three dimensions after inflation and fume fixation. *Am Rev Tuberculosis* 79:764–772, 1959.
22. Heard BE: A pathological study of emphysema of the lungs with chronic bronchitis. *Thorax* 13:136–149, 1958.
23. Wright BM, Slavin G, Kreel L, Callan K, Saudin B: Postmortem inflation and fixation of human lungs. *Thorax* 29:189–194, 1974.
24. Javed AP, Whimster WF, Deverell MH, Cookson MJ: Distribution of alveolar wall per unit volume (AMUV) in the human lung. *Anal Cell Pathol* 6:129–136, 1994.
25. Whimster WF: The microanatomy of the alveolar duct system. *Thorax* 24:141–149, 1970.
26. Whimster WF, Lunkenheimer PP: A review of respiratory tract structure in relation to high frequency ventilation, with observations on collateral ventilation. *Br J Anaesth* 63:32S–37S, 1989.
27. Lallemand D, Duc TV: A technique for correlating the roentgenologic appearance of infant lung to its anatomy. *J Roentgenol Rad Ther Nucl Med* 118:58–65, 1973.
28. Elliott FM, Reid L: Some new facts about the pulmonary artery and its branching pattern. *Clin Radiol* 16:193–198, 1965.
29. Gough J, Wentworth JE: The use of thin sections of entire organs in morbid anatomical studies. *J Microsc Soc* 69:231–235, 1949.

30. Whimster WF: Rapid giant paper sections of lungs. *Thorax* 24: 737–741, 1969.

31. Gevenois PA, Koob M-C, Jacobovitz D, De Vuyst P, Yernault J-C, Struyven J: Whole lung sections for computed tomographic-pathologic correlation. *Invest Radiol* 28:242–246, 1993.

32. Gevenois PA, Zanen J, De Maertelaer V, De Vuyst P, Mumortier P, Yernault J-C: Macroscopic assessment of lung emphysema by image analysis. *J Clin Pathol* 48:318–322, 1995.

33. Thurlbeck WM, Anderson AE, Janis M, et al: A cooperative study of certain measurements of emphysema. *Am Rev Respir Dis* 98:217–228, 1968.

34. Thurlbeck WM, Dunnill MS, Hartung W, Heard BE, Heppleston AG, Ryder RC: A comparison of three methods of measuring emphysema. *Hum Pathol* 1:215–226, 1970.

35. Turner P, Whimster WF: Lung volumes in emphysema. *Thorax* 36:932–937, 1981.

36. Tompsett DH: *Casts from Lungs. Anatomical Techniques.* Edinburgh, E&S Livingstone, 1970, pp 123–137.

37. Horsfield K, Cumming G, Hicken P: A morphologic study of airways using bronchial casts. *Am Rev Respir Dis* 93:900–906, 1967.

38. Pump KK: Fenestrae in the alveolar membrane of the human lung. *Chest* 65:431–437, 1974.

39. Thurlbeck WM: Internal surface area and other measurements in emphysema. *Thorax* 22:483–496, 1967.

40. Gundersen HJG, Bendtsen TF, Korbo L, et al: Some new, simple and efficient stereological methods and their use in pathological research and diagnosis. *APMIS* 96:379–394, 1988.

41. Hegre ES, Brashear AD: The block-surface method of staining as applied to the study of embryology. *Anat Rec* 97:21–28, 1947.

42. Whimster WF: Distribution of submucous gland in the human trachea, main and segmental bronchi: Morphometric study. *Appl Pathol* 4: 15–23, 1986.

43. Ashcroft T, Simpson JM, Timbrell V: Simple method of estimating the severity of pulmonary fibrosis on a numerical scale. *J Clin Pathol* 41:467–470, 1988.

44. Geddes DM: The natural history of lung cancer: A review based on rates of tumour growth. *Br J Dis Chest* 73:1–17, 1979.

45. Macchiarini P, Fontanini G, Hardin MJ, Squartini F, Angeletti CA: Relation of neovascularisation to metastasis of non-small-cell lung cancer. *Lancet* 340:145–146, 1992.

46. Fontanini G, Bigini D, Vignati S, et al: Microvessel count predicts metastatic disease and survival in non-small cell lung cancer. *J Pathol* 175:000–000, 1995.

47. Soomro I, Patel N, Whimster WF: Distribution and estimation of nucleolar organizing regions in various human lung tumours. *Pathol Res Pract* 187:68–72, 1991.

48. Soomro I, Whimster WF: Growth fraction in lung tumours determined by K167 immunostaining and comparison with AgNOR scores. *J Pathol* 162:217–222, 1990.

49. Whimster WF: Quantitative nucleology: quantitative aspects of the study of nuclei. *Acta Stereologica* 11:25–34, 1992.

50. Kayser K, Stute H, Bubenzer J, Paul J: Combined morphometrical and syntactic structure analysis as tools for histomorphological insight into human lung carcinoma growth. *Anal Cell Pathol* 2:167–178, 1990.

51. Weibel ER: Design of biological organisms and fractal geometry, in Nonnenmacher TF, Losa GA, Weibel ER (eds), *Fractals in Biology and Medicine.* Basel, Birkhauser Verlag, 1993, pp 68–85.

52. Weibel ER: Fractal geometry: A design principle for living organisms. *Am J Physiol* 261:L361–369, 1991.

53. Kitaoka H, Takahashi T: Relationship between the branching pattern of airways and the spatial arrangement of pulmonary acini—a re-examination from a fractal point of view, in Nonnenmacher TF, Losa GA, Weibel ER (eds), *Fractals in Biology and Medicine.* Basel, Birkhauser Verlag, 1993, pp 116–131.

54. Aherne WA, Dunnill MS: *Morphometry.* London, Edward Arnold, 1982, pp 33–45.

55. Hislop A, Reid L: New pathological findings in emphysema of childhood: 1. Polyalveolar lobe with emphysema. *Thorax* 25:682–690, 1970.

56. Hislop A, Reid L: New pathological findings in emphysema of childhood: 2. Overinflation of a normal lobe. *Thorax* 26:190–194, 1971.

57. Hislop A, Reid L: New pathological findings in emphysema of childhood: 3. Unilateral congenital emphysema and hypoplasia and compensatory emphysema of the contralateral lung. *Thorax* 26:195–199, 1971.

58. Gillooly M, Lamb D, Farrow ASJ: New automated technique for the assessment of emphysema on histological sections. *J Clin Pathol* 44:1007–1011, 1991.

59. Cookson MJ, Davies C, Entwistle A, Whimster WF: The microanatomy of the alveolar duct of the human lung imaged by confocal microscopy and visualized with 3-D computer-based reconstruction. *Comput Med Imaging Graph* 17:201–210, 1993.

60. Whimster WF: Problems of the third dimension. *Pathol Res Pract* 185:594–597, 1989.

61. Low FN: The pulmonary alveolar epithelium of laboratory mammals and man. *Anat Rec* 117:241–263, 1953.

62. Bensch KG, Corrin B, Pariente R, Spencer H: Oat-cell carcinoma of the lung. Its origin and relationship to bronchial carcinoid. *Cancer* 22:1163–1172, 1968.

63. Dilly SA: Scanning electron microscope study of the development of the human respiratory acinus. *Thorax* 39:733–742, 1984.

64. Dilly SA: Microcorrosion casting of the human respiratory acinus. *Scanning Microsc* 3:1095–1101, 1986.

65. Moscoso GJ, Driver M, Codd J, Whimster WF: The morphology of ciliogenesis in the developing fetal human respiratory epithelium. *Pathol Res Pract* 183:403–411, 1988.

66. Horsfield K, Cumming G: Morphology of the bronchial tree in man. *J Appl Physiol* 24:373–383, 1968.

Chapter 36

Cytopathology

Jennifer A. Young

The inspection of sputum for evidence of blood or pus has been practiced by physicians since the dawn of medicine. The realization that cancer of the lung could be accurately diagnosed and typed by the microscopic study of expectorated cells is generally attributed to Dudgeon and Barrett[1] and the investigation of patients with suspected malignant disease remains the prime application of pulmonary cytopathology today. The technique of bronchoalveolar lavage (BAL), however, has widened the role of cytopathology to include the management of patients with benign diseases, most especially in the immunosuppressed individual, for whom it provides a safe and effective means of investigating pulmonary infiltrates of unknown etiology. BAL is also helpful in the management of patients with interstitial lung disease.

TECHNIQUES

Collection Techniques

Several different collection methods are available for obtaining specimens for cytopathology.[2] These are listed in Table 36-1.

TABLE 36-1
Specimen Collection Techniques

Sputum
  Spontaneous
  Induced
Bronchoscopic specimens for investigation of suspected tumor
  Bronchial brushing
  Bronchial washing
  Aspiration of bronchial secretions
  Transbronchial fine needle aspiration (TBFNA)
Percutaneous fine needle aspiration (PCFNA)
Bronchoalveolar lavage (BAL)

SPUTUM

Early morning "deep cough" sputum is preferable and the most efficient results are obtained if specimens are submitted on 3 consecutive days. Sputum produced during physiotherapy to the chest or after bronchoscopy is also suitable and is usually highly cellular, but the specimen collection method must be indicated because otherwise the hyperexfoliation may result in interpretative dilemmas. In patients who cannot expectorate spontaneously, *induced sputum* can be produced by inhalation of the vapors of a warmed (37°C) mixture of 15% sodium chloride and 20% propylene glycol for 20 min. Sputum should be collected into clean, dry containers with secure screw-on lids and transported to the laboratory in biohazard bags.

BRONCHOSCOPIC TECHNIQUES FOR INVESTIGATION OF SUSPECTED TUMOR

Bronchial Brushing
This is the method of choice for investigation of visible endobronchial lesions. If the surface of the suspected tumor appears necrotic at least two brush specimens should be collected because the first may only yield degenerate cell debris. Up to five brushings are advocated by some groups,[3] especially in instances where there may be risks associated with repeat bronchoscopy or more invasive procedures. Fluoroscopic radiographic guidance aids accurate collection particularly when no frank mass is visible endoscopically. Following withdrawal the brush should be firmly rolled onto several glass slides, each being placed in a container of 95% ethyl alcohol as it is prepared.

Bronchial Washing
Lavage is particularly useful when a sample is required from outside the area of direct bronchoscopic inspection. From 30 to 50 mL isotonic saline is instilled and reaspirated.

Aspiration of Bronchial Secretions
The material obtained often consists of only mucus and nondiagnostic debris. However, a few centers advocate this method.[4]

Transbronchial Fine Needle Aspiration
This method is used less commonly than brushing or washing but in selected cases may be the most appropriate means of diagnosis. It is of value for investigating external bronchial compression or submucosal lesions particularly if biopsy forceps are difficult to manipulate because the lumen of the bronchus is compressed. Hilar, subcarinal, and low right paratracheal lymph nodes can also be sampled. The Wang[5]

disposable device, which consists of a 120-cm long double-lumen retractable needle system, is the method most frequently used.

PERCUTANEOUS FINE NEEDLE ASPIRATION

This technique is also used almost exclusively for the investigation of suspected malignant disease. It is used particularly in patients with a lung mass, not amenable to diagnosis by endoscopic techniques, in whom a definitive answer is required without resort to open biopsy. In the patient with a solitary pulmonary nodule the nature of which remains uncertain after bronchoscopic investigation, if there are no contraindications to thoracotomy and the suspicion of malignancy is high, then open biopsy can be undertaken. However, if there is clinical suspicion of metastatic disease or the patient is unfit, then PCFNA is preferable. PCFNA is also occasionally useful for the investigation of a suspected localized infectious process, particularly in the immunosuppressed patient on whom sputum and BAL have not provided a diagnosis.

The procedure is not without complications. Pneumothorax is the most frequent and occurs in 8 percent[6] to 57 percent[7] of cases, but most episodes are minor. Coexisting chronic obstructive pulmonary disease, deep-seated lesions, and the use of the larger type of needle increase the risks.[8] Transient hemoptysis is common but is seldom of clinical consequence. Infectious complications are unlikely. Seeding of tumor cells in the needle track is always viewed as a potential hazard of fine needle aspiration (FNA). It is documented in the lung but, as elsewhere, is rare.[9]

The procedure is undertaken in the Department of Diagnostic Imaging. It is helpful, however, if a cytotechnologist attends to assist the radiologist in the preparation of the smears, so that optimum use is made of all available material. On occasion, if the lesion is a problem, it may aid the pathologist in arriving at a diagnosis if he views the image and discusses the case with the radiologist. A list of the items necessary for PCFNA is shown in Table 36-2.

Material aspirated from the center of a large tumor is often necrotic, and better preserved material is more often obtained from the periphery of the mass; this should be borne in mind when the needle route is planned. Special needles, for example, the Rotex screw needle, are available for sampling hard lesions such as mesenchymomas.[10] Plentiful air-dried and wet-fixed (95% ethyl alcohol) slides should be prepared, taking into account the possible need for special stains or research techniques.

TABLE 36-2
Equipment for Fine Needle Aspiration

Sterile gloves

Alcohol swabs for cleaning skin

Lignocaine 2% for local anesthesia

Small syringe for lignocaine

Syringes, 10 mL and 20 mL disposable, for aspiration

Cameco syringe holder

Needles, 21 to 25 gauge, disposable

Watch glass

Glass slides with frosted ends

Tray for spreading out slides

Pencil for labeling slides

Reagents for Diff-Quik staining (Merz Dade)

Hair dryer to dry slides

Jars or plastic containers filled with 95% ethanol for slide fixation

Rack for air-dried slides

Empty tubes for cyst fluids

Disposable centrifuge tubes containing aliquots of normal saline or Hank's solution, liquid fixative or cell culture medium

Tubes of 10% buffered formalin for tissue fragments

Tube of medium if microbiologic culture required

Tube of phosphate-buffered saline (pH 7.4) if immunocytochemistry required

Tube of 1% glutaraldehyde and 4% formaldehyde in cacodylate buffer if electron microscopy required

Paper tissues

Gauze swabs

Disposable plastic forceps

Adhesive tape

Ball point pen

Request forms

Biohazard bags and labels

Container for discarded needles, syringes, and unused slides

SOURCE: From Young,[11] with permission.

Further details on the handling of the material is provided elsewhere.[2,11]

BRONCHOALVEOLAR LAVAGE

Unlike the other sampling techniques, BAL is primarily performed for the investigation of benign lung disease.[2,12] It is a safe and efficient method for the investigation of pulmonary infiltrates in immunosuppressed patients and is also used for the assessment of patients with interstitial lung disease. It may be diagnostic in a few benign conditions.[2,13] A total of 100 to 300 mL istonic saline in 50-mL aliquots is instilled and aspirated. About 40 to 60 percent of the fluid is recovered.

Laboratory Processing

SPUTUM

Not all specimens submitted to the laboratory contain adequate lower tract cells, but the distinction between saliva, respiratory tract mucus, and lower tract material cannot be made by visual inspection and therefore all specimens must be processed for microscopic examination.

Within a category B1 safety cabinet, sputum is poured into a Petri dish and any blood-stained areas, thick streaks, or possible tissue fragments sampled with disposable plastic forceps. A small blob is placed on a slide, squashed down, and then spread with the edge of a second slide, each slide being immediately fixed with 95% ethyl alcohol as it is prepared. Unless special stains are anticipated four slides per specimen are adequate. Sputum can also be processed in the laboratory by the Saccomano technique[13-15] in which a pooled collection is emulsified in a blender or by formalin fixation of centrifuged cell blocks, which are then sectioned in the same manner as a biopsy.[16] There is a potential risk of aerosal infection with the former method and the latter is laborious. Sputum is stained by the traditional Papanicolaou method.

BRONCHIAL BRUSHINGS AND WASHING

Brushings reach the laboratory as prepared smears and are also best stained by the Papanicolaou technique. *Bronchial washings* are centrifuged in the laboratory and about six slides prepared from the deposit or this can be processed as cell blocks.[17] Again the Papanicolaou stain is the one of choice if malignancy is suspected.

FINE NEEDLE ASPIRATES

Fine needle aspirates, whether TBFNA or PCFNA, contain less mucoid material and debris than either sputum, brushings, or washings. The May-Grunwald-Giemsa (MGG) stain on air-dried slides is an excellent complement to the Papanicolaou technique. Any special stain can also be applied as required.

BRONCHOALVEOLAR LAVAGE

The BAL fluid should be processed without delay because cells suspended in saline start to deteriorate after about 30 min. Cytopreparation differs according to the clinical condition under investigation.

Interstitial Lung Disease
Determination of alveolar inflammatory cell populations in BAL fluid contributes to the management of patients with interstitial lung disease by monitoring disease activity and response to therapy. Total cell counts are now believed to be of little value.[18] A suitable technique for performing differential cell counts is the method described by Turner-Warwick and Haslam.[18] Separate the cells from the fluid by slow centrifugation and resuspend the deposit in buffered tissue culture medium to 2×10^6 cells/mL. Then using 100-μL aliquots of the suspension (2×10^5 cells/aliquot) prepare cytocentrifuged preparations by spinning the aliquots at 450 rpm for 4 minutes (Shandon Cytospin). At least eight slides should be prepared to allow for special techniques including immunocytochemistry. Stain two or three slides with MGG and differentially count 300 to 500 cells by random point counting with ×40 objective. Membrane filtration is an alternative method of processing BAL fluid.[19,20] The problems of sampling and technique are well discussed by Walters and Gardiner.[21] As well as cytomorphologic enumeration, cells can be characterized by their cytochemical and antigenic properties.

Pulmonary Infiltrates in Immunosuppressed Patients
The aliquots of BAL fluid can be pooled or processed separately. If the latter, diagnostic material, particularly with *Pneumocystis carinii* infection, may be more evident in the later aliquots because the early samples may recover only endobronchial material. The fluid is centrifuged and slides prepared from the deposit. In addition to Papanicolaou and MGG, stains for infectious agents (Gram, Ziehl-Nielsen, and Grocott), are necessary and if required these can be supplemented with fluoresence or immunocytochemical techniques.

Miscellaneous Applications
The role of BAL cytology in establishing a precise diagnosis in *benign disease* is limited. Eosinophilia (eosinophilic pneumonia) can be identified by examination of MGG slides. Application of special staining is occasionally helpful in a few uncommon conditions. These include Oil red O or Sudan (lipid pneumonia), Perls' (occult hemorrhage, asbestosis) and periodic acid-Schiff (PAS) (alveolar proteinosis). Special techniques including polarized light microscopy, immunofluorescence, immunocytochemistry, and electron microscopy are all possible.[12,18] For the latter (e.g., in suspected alveolar proteinosis, histiocytosis X) 25 mL of 5% glutaraldehyde in cacodylate buffer is added to 25 mL of the initial cell suspension.

The technique of BAL can also be used to investigate suspected malignant disease,[22] either possible diffuse infiltration or as a means of recovering cellular material from outside the range of more usual bronchoscopic sampling techniques in patients unsuitable for PCFNA or open biopsy. An overview of the potential applications of BAL for diagnosis and patient management is given in Table 36-3. Allied microbiologic studies extend the diagnosis of specific bacterial and viral pathogens, many of which are not identifiable by morphologic techniques.

INTERPRETATION

Findings in the Absence of Respiratory Disease

SPUTUM

Sputum consists of a mixture of oral cells, mucus, and respiratory tract cells. For a specimen to be considered "adequate" it must contain alveolar macrophages. Curschmann's spirals (bronchiolar casts) are also a marker of lower tract material. The presence of ciliated columnar cells in the absence of alveolar macrophages does not confirm lower tract sampling because the epithelial cells may originate from the nose. All sputum samples contain squamous cells and inflammatory disease in the mouth or larynx may produce quite marked morphologic alterations in these. Smokers may show squamous metaplasia (Fig. 36-1). The possibility of detecting early neoplastic disease of the oral cavity should not be overlooked. The finding of malignant squamous cells in a specimen that does not appear to contain alveolar macrophages should alert the pathologist to the possibility of carcinoma of the mouth or upper respiratory tract.

BRONCHIAL BRUSHINGS, WASHINGS, AND SECRETIONS

Bronchoscopic sampling is rarely carried out except as part of the investigation of suspected lung cancer.

Brushings from nonneoplastic areas of epithelium contain numerous ciliated columnar cells mixed with small numbers of mucus-secreting cells. Basal cells are not usually sampled. Alveolar macrophages are generally seen, but their presence or absence is immaterial.

Washings and Secretions

Material from the bronchial lumen is more abundant than with brushings. In washings, epithelial cells are quite plentiful (Fig. 36-2) and are mixed with mucus and alveolar macrophages. Aspirated secretions mainly consist of mucus mixed with debris and

TABLE 36-3

Contribution of Morphologic Examination of Bronchoalveolar Lavage Fluid to Diagnosis and Patient Management

Infectious Disease
  Diagnosis of
    Bacteria
      Gram-positive bacteria
      Mycobacteria
    Viruses
      Cytopathic effects due to
      Herpes simplex virus
      Cytomegalovirus
    Fungi
      Pneumocystis carinii
      Candida species
      Aspergillus species
      Cryptococcus neoformans
      Other common types
    Parasites
      Strongyloides Stercoralis
      Other uncommon types
Noninfectous Disease
  Diagnosis of
    Alveolar proteinosis
    Histocytosis X
    Diffuse malignant infiltration
  Contribution to diagnosis of
    Lipid pneumonia
    Eosinophilic pneumonia
    Pulmonary hemorrhage
    Hypersensitivity pneumonitis
    Sarcoidosis
    Asbestosis
    Silicosis
    Berylliosis
  Contribution to patient management
    Cryptogenic fibrosing alveolitis
    Sarcoidosis
    Pulmonary fibrosis associated with collagen vascular disease

macrophages. Epithelial cells are fewer and often poorly preserved.

FINE NEEDLE ASPIRATES

Regardless of whether TBFNA or PCFNA is used, aspiration will not be undertaken except from an area of abnormal lung and this will be reflected in the

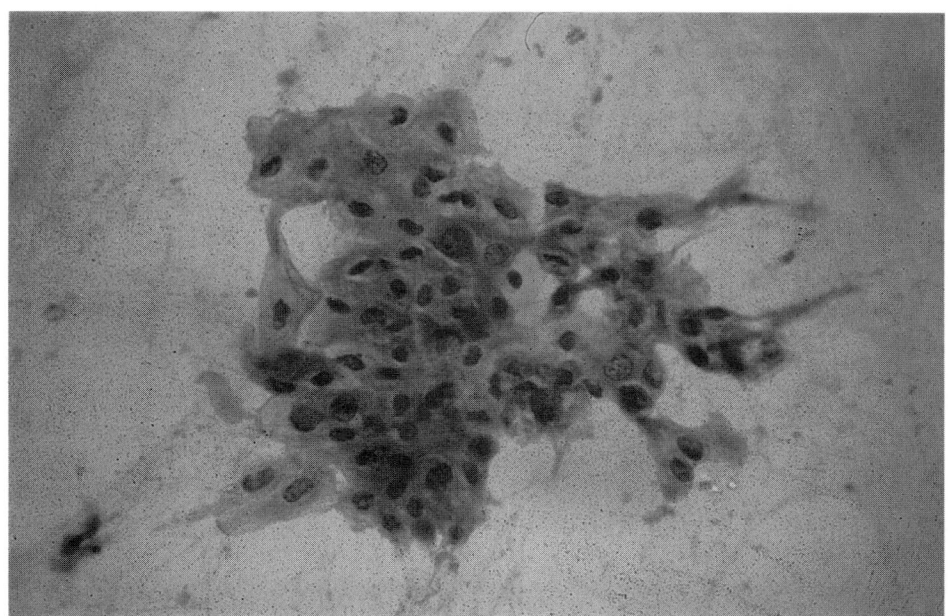

FIGURE 36-1
Squamous metaplasia of bronchial epithelium. Sputum from a smoker. (Papanicolaou, moderate magnification.) (*See also Color Plate 12.*)

contents of the specimens. However, unremarkable tissue may be sampled in addition. Alveolar macrophages are common, occur as single discrete cells, and most contain particles of carbon pigment although these may be inconspicuous in nonsmokers. Pneumocytes appear very similar apart from the absence of pigment. Bronchial epithelial cells are usually sparse and large numbers should raise the possibility of neoplasia. However, very occasionally in performing PCFNA the needle may run down the wall of a bronchus and this will result in aspiration of large groups of columnar epithelial cells. Fragments of pleura may also be sampled during the course of PCFNA.

BRONCHOALVEOLAR LAVAGE

Results of BAL from healthy controls or individuals with nonparenchymatous disease vary from center to center and according to the method of preparation, cytocentrifugation (CCF) or membrane filtration (MF). Taskinen and colleagues[20] found on MF preparations the proportions (mean + SD) of lymphocytes (30.6 ± 23.7), neutrophils (5.4 ± 11), and respiratory epithelial cells (2.4 ± 3.8) were higher than those on CCF slides, which were 21.0 ± 2.3, 3.6 ± 1.0, and 0.4 ± 1.5, respectively. Remaining cells are predominantly macrophages. Smoking increases the total cell count with raised percentages of macrophages, neutrophils, and eosinophils.

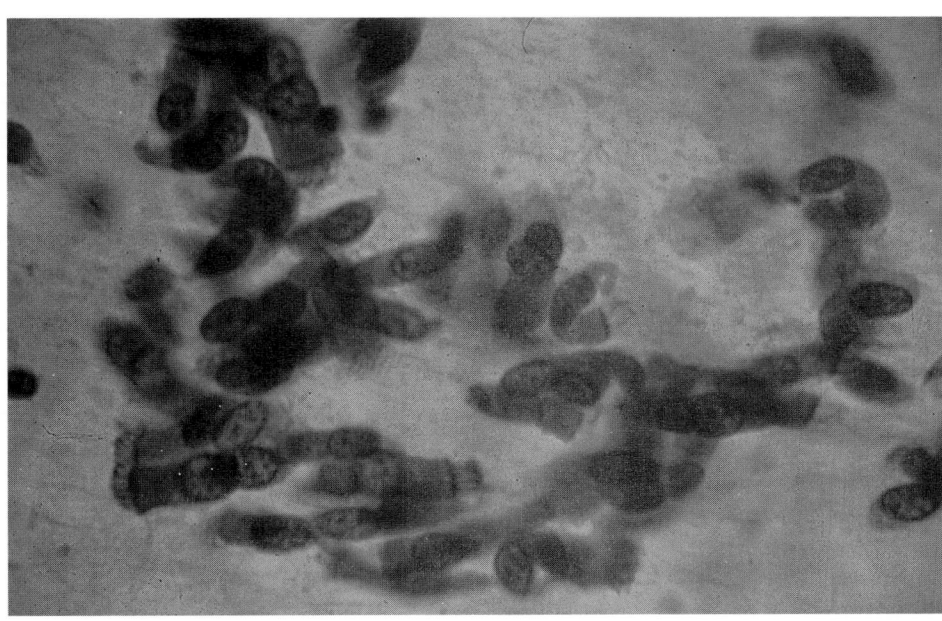

FIGURE 36-2
Normal bronchial epithelial cells. BAL from a heart/lung transplant recipient. (Papanicolaou, high magnification.) (*See also Color Plate 13.*)

Nonneoplastic Disease

Aside from the role of BAL in interstitial lung disease and the management of pulmonary infiltrates in the immunosuppressed patient, applications of cytopathology in the investigation of benign lung disease are limited. Several reviews are available[23,24] along with a discussion in general terms[25] and with particular reference to PCFNA.[26] In many instances material is sent for cytologic investigation because malignant disease is suspected or cannot be excluded by clinical or radiologic examination. Many nonneoplastic inflammatory, reactive, and degenerative conditions are capable of producing marked morphologic alterations at the cellular level. The appearances of the changes are frequently nonspecific, but it is essential that the pathologist be fully familiar with the range and extent of this atypia so that a false-positive diagnosis of malignancy is avoided.[25–29]

Cytology can aid the recognition of granulomatous disease including Wegener's granulomatous,[30,31] which clinically may raise suspicion of malignancy and has also been used to assess bronchopulmonary dysplasia in neonates.[32] The stigmata of allergic lung disease are commonly readily identifiable and include eosinophilia, hyperexfoliation of bronchial epithelial cells, generally in tightly rolled up clusters, numerous Curschmann's spirals, and showers of pointed, bright orange (with Papanicolaou stain) Charcot-Leydon crystals[25] derived from degranulated eosinophils. Additionally fungal hyphae will be present in allergic bronchopulmonary aspergillo-sis.[33] Clues to industrial lung disease, particularly the presence of asbestos bodies may be seen (Fig. 36-3). Leiman[34] found that asbestos bodies in PCFNA specimens from localized or dominant parenchymal lung masses were significant markers of underlying pathology whether or not cellular evidence of that pathology was observed in the aspirated material.

Certain infections are also amenable to cytologic diagnosis. Obviously sputum should be screened for acid-fast bacilli if clinically indicated and such examination is routine in patients infected with human immunodeficiency virus (HIV). In needle aspirates, tuberculosis should not be overlooked as the possible cause of a lung mass,[35] which may also be caused by *Actinomyces israelii*.[25] A good account of the cytology of mycotic infection is given by Johnston.[36] *P. carinii* infection, although more commonly investigated by BAL, can also be diagnosed by examination of induced sputum.[37,38] *Aspergillus* species and *Cryptococcus neoformans*[26] can be identified cytomorphologically and in endemic areas, *Blastomyces dermatitides,*[39] *Coccidioides immitis,* and *Histoplasma capsulatum* may be encountered. *Candida* species are common in sputum but generally arise from oral contamination. Some viruses produce distinctive cytopathic effects. *Herpes simplex* virus (HSV) tracheobronchitis results in diagnostic features in sputum. The changes are mainly seen in squamous cells, which contain characteristic multiple, molded, "ground glass" nuclei.[25] Adenovirus-induced effect can also be identified, particularly

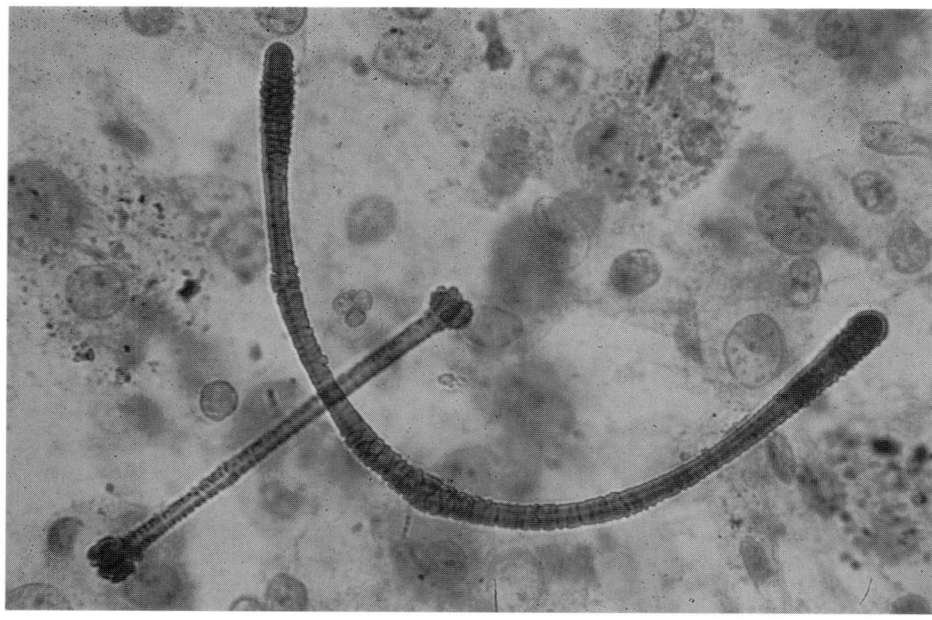

FIGURE 36-3
Asbestos bodies against a background of alveolar macrophages and a pink-staining squamous cell in BAL fluid. (Papanicolaou, high magnification.) (*See also Color Plate 14.*)

useful in the examination of tracheal aspirates from infants.[41] Rarely, evidence of parasitic disease is seen.[42–44] PCFNA of suspected hydatid disease is potentially dangerous because of the risk of anaphylactic shock. The main role of cytologic investigation in nonneoplastic disease is, however, examination of BAL fluid.

BRONCHOALVEOLAR LAVAGE

Interstitial Lung Disease

The cells collected by BAL are representative of the interstitial inflammatory infiltrate.[45] Two major cell patterns are found—neutrophilic and lymphocytic, the former in cryptogenic fibrosing alveolitis, (CFA), collagen vascular disease, and asbestosis and the latter in granulomatous conditions such as sarcoidosis (see Chap. 16), berylliosis, and extrinsive allergic alveolitis.[12,18] The interrelationship between different lymphocyte subpopulations is also important as is the presence or absence of eosinophils and mast cells. In CFA the percentages of neutrophils (9 ± 2) and eosinophils (5 ± 1) are raised in comparison to healthy volunteers (0.7 ± 0.3 percent and 0.1 ± 0.1 percent).[46] High eosinophil levels are associated with severe clinical impairment. However, some patients with CFA display lymphocytosis (17 ± 5 percent) compared to volunteers (7 ± 1 percent).[46] This appears to be a good prognostic sign and indicates that a clinical response to corticosteroids is likely.[46] In sarcoidosis (Fig. 36-4) the T-helper–to–T-suppressor cell ratio appears to be the most important factor, but the findings regarding the prognostic value of BAL lymphocytosis are less clear-cut than was originally thought. In some patients with chronic sarcoidosis, BAL lymphocyte counts may be normal but the neutrophils, generally without eosinophils, are raised.[18] These factors make the use of BAL for initiating and assessing therapy in sarcoidosis and CFA controversial.[18] In extrinsic allergic alveolitis the T-helper–to–T-suppressor cell ratio is low and mast cells are present. Cell counts vary from laboratory to laboratory depending on methodology and each center must establish its own range. A summary of published figures is included in the work by Stanley and colleagues.[12]

Pulmonary Infiltrates in Immunosuppressed Patients

A wide range of opportunistic infections and noninfectious disease processes can cause respiratory problems in patients with acquired immunodeficiency syndrome (AIDS) and other immunosuppressed individuals.[47–49] Etiologies to be considered are shown in Table 36-4. Although the great majority are nonneoplastic the possibility of malignant disease, particularly diffuse pulmonary infiltration, in patients with leukemia, lymphoma, and allied conditions most be included in the differential diagnosis. Several infections or other causes of pulmonary dysfunction may coexist.

Suspicion of opportunistic infection and in particular *P. carinii* pneumonia (PCP) is the most

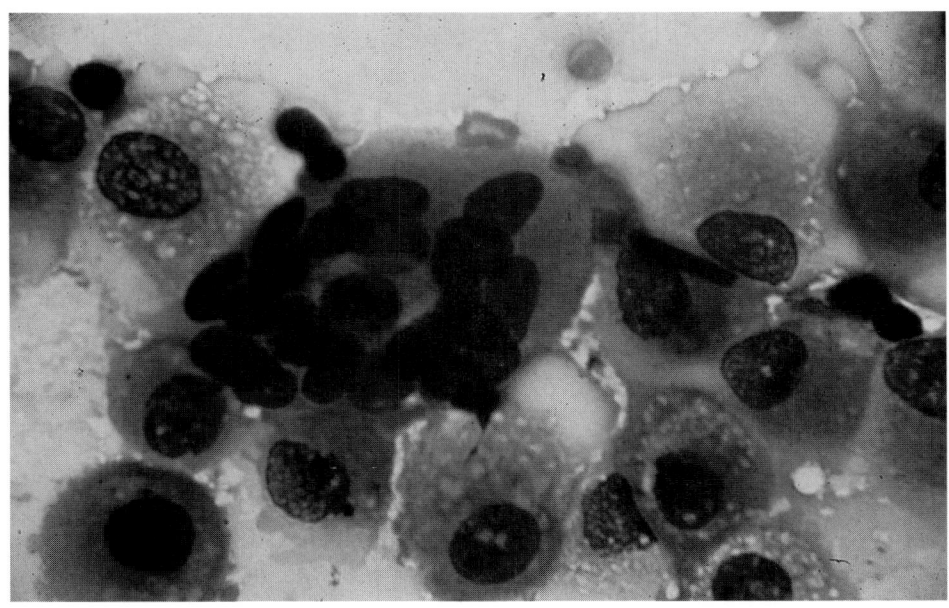

FIGURE 36-4
Normal mononuclear alveolar macrophages and a large multinucleated cell. BAL from case of quiescent sarcoidosis. (May-Grunwald-Giemsa, high magnification.) (*See also Color Plate 15.*)

TABLE 36-4

Principal Causes of Pulmonary Dysfunction in Immunosuppressed Patients

Infection
  Bacterial
    Staphylococcus aureus
    Streptococcus pneumoniae
    Haemophilus influenzae
    Pseudomonas aeruginosa
    Legionella pneumophila
    Mycobacterium tuberculosis
    Mycobacterium avium-intracellulare
  Viral
    Cytomegalovirus
    Herpes simplex
    Adenovirus
  Fungal
    Pneumocystis carinii
    Candida species
    Aspergillus species
    Cryptococcus neoformans
    Histoplasma capsulatum
    Coccidioides immitus
Neoplasia
  Lymphoma
  Leukemia
  Hodgkin's disease
  Myeloma
  Kaposi's sarcoma
  Bronchial carcinoma
  Secondary carcinoma
Iatrogenic Processes
  Irradiation
  Drug toxicity
  Oxygen toxicity
  Graft versus host disease
  Rejection of lung allograft
Idiopathic and Miscellaneous Processes
  Lipoproteinosis
  Alveolar hemorrhage
  Lymphocytic interstitial pneumonia
  Nonspecific interstitial pneumonia

SOURCE: From Young,[49] with permission.

common reason for performing BAL on the immunosuppressed patient. The incidence of PCP remains high in individuals with AIDS but in other immunosuppressed groups, especially transplant recipients, has been greatly reduced by prophylatic therapy. Examination of Papanicolaou-stained BAL fluid is rapid, sensitive, and specific. Characteristic foamy alveolar casts can be identified.[50,51] These contain small clear spaces due to the nonstaining, cystic forms of the organism. For confirmation these can be visualized by Grocott stain (Fig. 36-5). Sensitivity ranges from 78 percent to 95 percent.[51–54] Special techniques enhance detection. Fluoresence miscroscopy of the Papanicolaou-stained fluid shows up the cysts as 5-μm circular structures containing two reniform structures.[55] Wehle and coworkers[56] also noted fluorescent inclusions within macrophages, which they interpreted as cyst fragments. With this additional feature the sensitivity for the diagnosis of PCP was raised to 100 percent. Immunofluorescence using a murine monoclonal antibody (Northumbria Biological Ltd) also enhances sensitivity.[57] DNA amplification by polymerase chain reaction (PCR) also aids diagnosis.[57] Other opportunistic infections may be difficult to distinguish clinically from PCP and may coexist. A search for other pathogens including acid-fast bacilli, fungi, and viral inclusions should always be undertaken, even if a definitive diagnosis of PCP has been made.

Cytomegalovirus (CMV) pneumonitis is common in organ transplant recipients as well as AIDS patients and may coexist with PCP. Pathognomic "owl's eye" viral inclusions can be identified in pneumonocytes and alveolar macrophages in Papanicolaou-stained BAL fluid. Small, granular inclusions are sometimes also visible in the cytoplasm. Sensitivity of detection is enhanced by monoclonal antibody,[58] PCR,[59] or in situ hybridization studies.[60] In hematogenous-borne HSV pneumonia the characteristic cytopathic effects seen in squamous cells in tracheobronchitis may not be obvious. In situ hybridization can confirm the diagnosis.[61] *Candida* and *Aspergillus* species are not uncommon. Unusual fungi such as *C. neoformans* (Fig. 36-6), *H. capsulatum*, or *C. immitis* may also be found in BAL fluid as can evidence of parasitic disease including *Toxoplasma gondii*, *Strongyloides stercoralis*, or *Acanthamoeba*.[62]

Severe pneumonitis and various iatrogenically induced changes including radiation and drug toxicity and graft versus host disease can produce severe morphologic alterations in bronchial and alveolar cells.[63,64] The atypia may be sufficiently marked to raise the possibility of malignancy if details of the clinical history are not available. Very occasionally PCP is associated with lipoproteinosis when copious sudanophilic debris is present.

Miscellaneous Conditions
The organizing stage of acute adult respiratory distress syndrome (ARDS) is associated with reactive

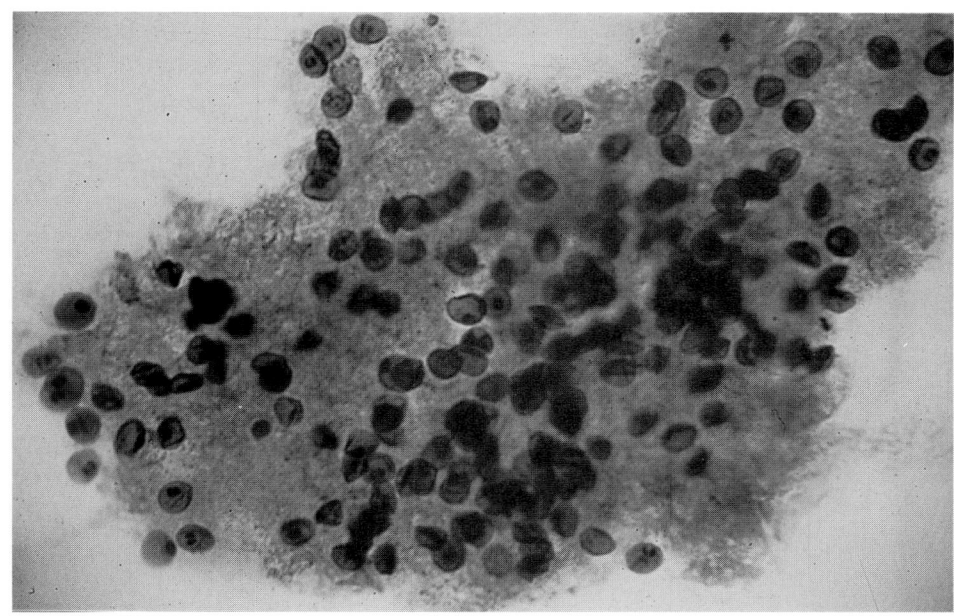

FIGURE 36-5
Alveolar cast containing cystic forms of *P. carinii* in BAL samples. (Grocott, high magnification.) (*See also Color Plate 16.*)

hyperplasia of type II pneumonocytes. Such cells may be present in large numbers in BAL fluid and display a spectrum of change from small cells to very large bizarre forms resembling the cells of adenocarcinoma (Fig. 36-7).[65] Careful clinical correlation is necessary to avoid a false-positive diagnosis of malignancy. The finding of numerous hemosiderin-laden (Perls' positive) macrophages is associated with the pulmonary hemorrhagic syndromes. Increased numbers are also found in many interstitial lung diseases, but the percentage of Perls' positive macrophages and the intensity of staining is greater in the hemorrhagic syndromes.[66] Large vacuolated macrophages containing sudanophilic material are found in lipid pneumonia. In industrial lung disease study of the macrophages may show evidence of silicosis (birefringent cystalline particles on polarization) or asbestos fibers or bodies,[67] and radiographic microanalysis aids identification of inorganic dust particles.[68] Special stains allied with electron microscopy, may be diagnostic in certain conditions. In alveolar proteinosis the BAL fluid is opaque and contains granular PAS-positive material consisting of osmiophilic bodies. In amiodarone toxicity the

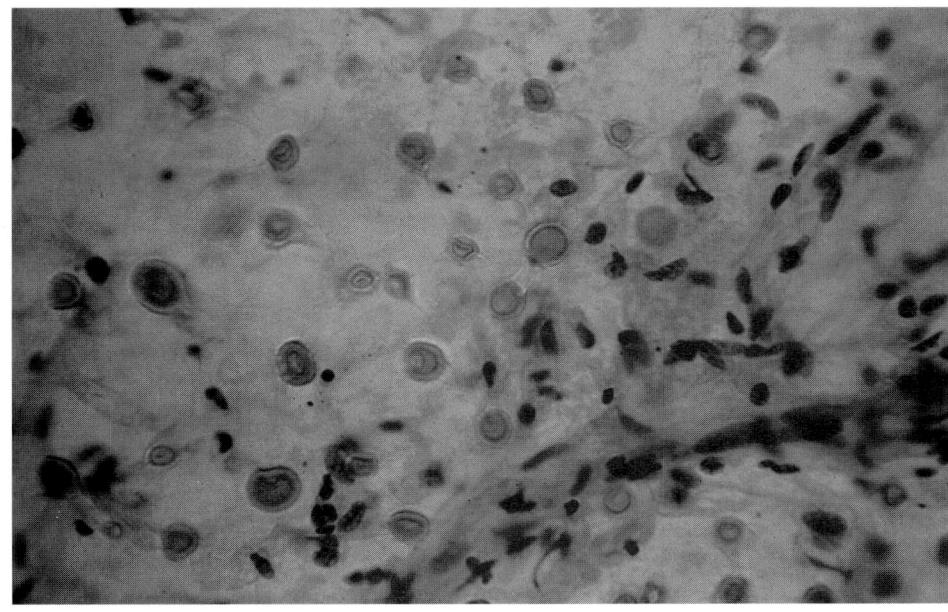

FIGURE 36-6
Numerous encapsulated yeast forms of *C. neoformans*, mixed with streaks of mucus and degenerate cells from FNA of lung. (Papanicolaou, moderate magnification.) (*See also Color Plate 17.*)

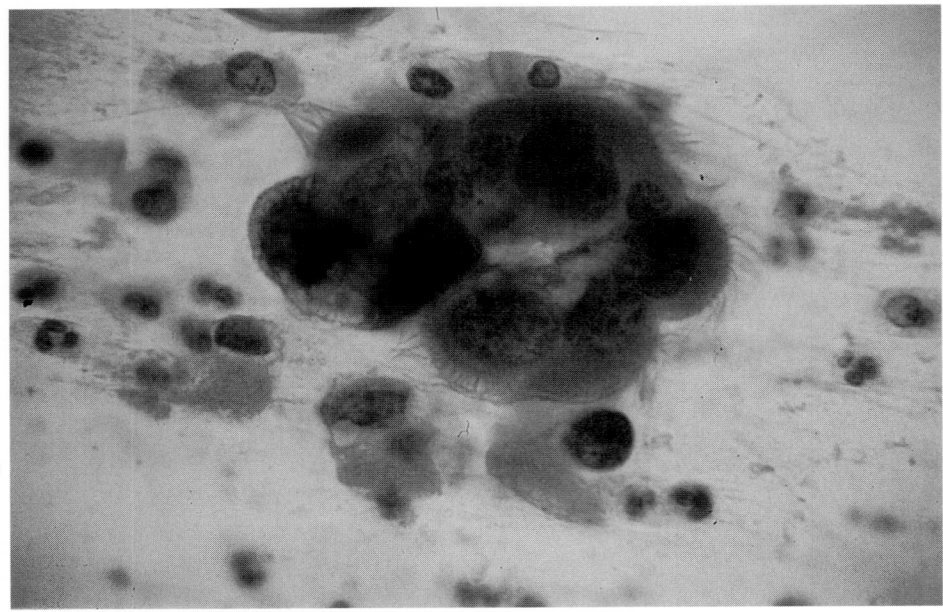

FIGURE 36-7
Abnormal bronchial epithelial cells with enlarged nuclei and well preserved cilia. BAL from case of respiratory distress following bone marrow transplantation. (Papanicolaou, high magnification.) (*See also Color Plate 18.*)

macrophages also contain lamellar bodies but the fluid is clear.[69] Excess S-100-positive Langerhans' cells occur in histiocytosis X. Diagnosis is confirmed by the demonstration of characteristic Birbeck granules.

Neoplastic Disease

BENIGN TUMORS AND TUMOR-LIKE CONDITIONS

A large number of benign tumors are described in the lungs.[70] Although many of these are rare, with the extended use of TBFNA and PCFNA several of the more common entities are likely to be seen in cytologic material.

Hamartomas

It has been suggested that these are benign tumors of connective tissue (mesenchymomas)[71] rather than congenital lesions. Hamartomas are amenable to diagnosis by PCFNA and represent one of the more frequent benign lesions investigated as suspected peripheral lung cancer. Hamartomas usually consist of a bronchial epithelium together with cartilage and fibromyxoid material (Fig. 36-8). Several descriptions of the cytologic findings have been pub-

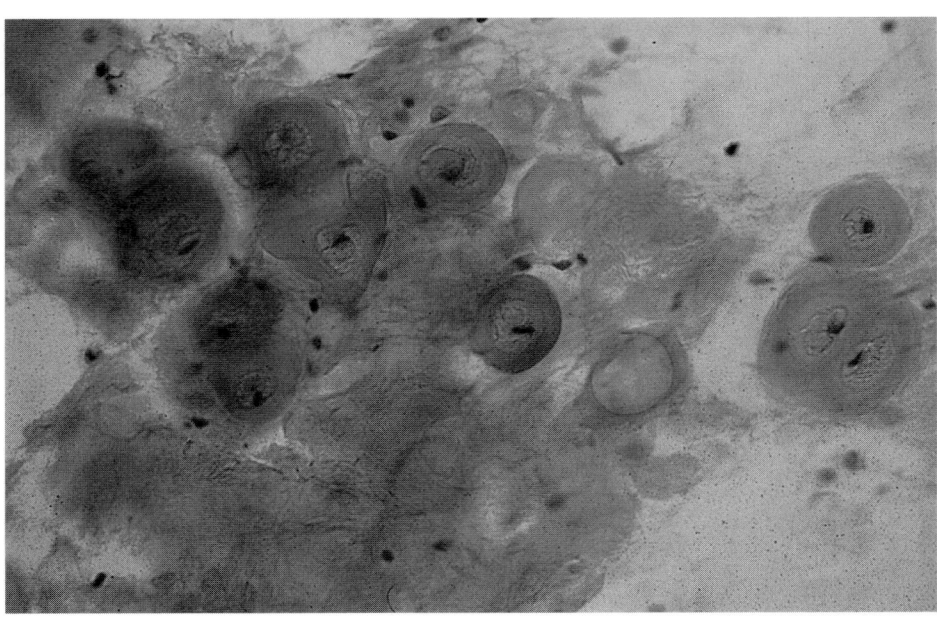

FIGURE 36-8
Hamartoma. Cartilage cells mixed with feathery mesenchymal material from FNA of lung. (Papanicolaou, low magnification.) (*See also Color Plate 19.*)

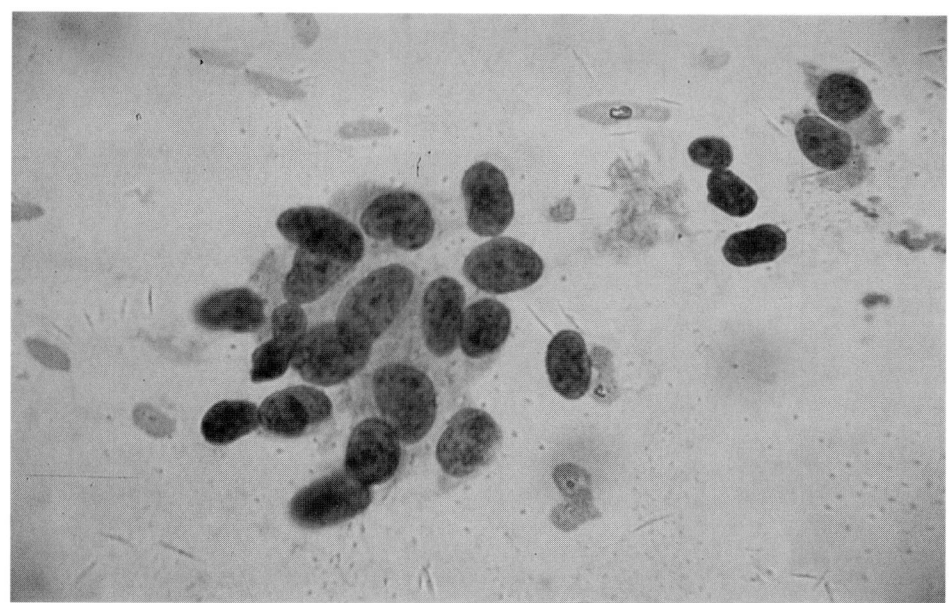

FIGURE 36-9
Carcinoid tumor. Cluster of small neoplastic cells in bronchial brushing sample. (Papanicolaou, moderate magnification.) (*See also Color Plate 20.*)

lished.[72–74] Aspirated material is often sparse, especially in chrondromatous lesions. The background is free of necrotic debris or inflammatory cells. The epithelial cells are unremarkable in appearance and diagnosis depends on recognition of the coexisting mesenchymal elements. Unless a screw-needle is used[10] cartilage cells are unusual and identification of the wispy, feathery fibromyxoid material provides the guide to diagnosis in most instances.[74]

Carcinoid Tumors

Classification of these tumors[75–77] is given in Chap. 29. Aspirates from *typical carcinoids*[26,78–80] generally contain plentiful epithelial cells against a clean background. The individual cells have scant cytoplasm and nuclei with granular chromatin and small prominent nucleoli (Fig. 36-9). Occasional macronucleoli may be seen. Although some single cells will be seen, cohesion is better than in small cell carcinoma. The cells in typical carcinoids have slightly more plentiful cytoplasm and nuclear molding, a distinctive feature of small cell carcinoma, is not obvious. In *atypical carcinoids*[81] the nuclei are more irregular in shape and size, the chromatin is more coarsely granular, and mitoses may be seen.[79] Some suggestion of molding is often evident but not to the same degree as seen in small cell carcinoma. The cytoplasm is more fragile than in typical carcinoids and cell cohesion less good, so that a small quantity of necrotic cellular debris may be present in the background. Atypical carcinoid tumors and small cell carcinoma may on occasion be difficult to separate from each other on purely cytologic grounds and therefore surgical resection of all stage 1 neoplasms with features of either tumor is advised.

Other Benign Tumors

Reports of the diagnostic cytology of endobronchial granular cell tumor[82] and sclerosing hemangioma[83] have been published. The wide range of other uncommon benign tumors mainly arise from the submucosal glands or mesenchymal tissue and do not differ in cytomorphology from salivary gland, soft tissue, or nerve sheath tumors seen in aspirates from other parts of the body.[84]

PRIMARY MALIGNANT TUMORS

Squamous Cell Carcinoma

Two subtypes, keratinizing (well differentiated) and nonkeratinizing (poorly differentiated), can be identified in cytologic material. The former is readily recognizable by the large bizarre-shaped keratinized cells, often with "pseudopodia" or the so-called tadpole (caudate) or 'spindle' forms (Fig. 36-10).[25,26] Papanicolaou is the stain of choice because the hyperkeratinized, glassy cytoplasm is highly orangeophilic. Well-differentiated carcinoma can usually be typed with a high degree of accuracy. However, cells of the poorly differentiated subtype are sometimes difficult to distinguish from those of

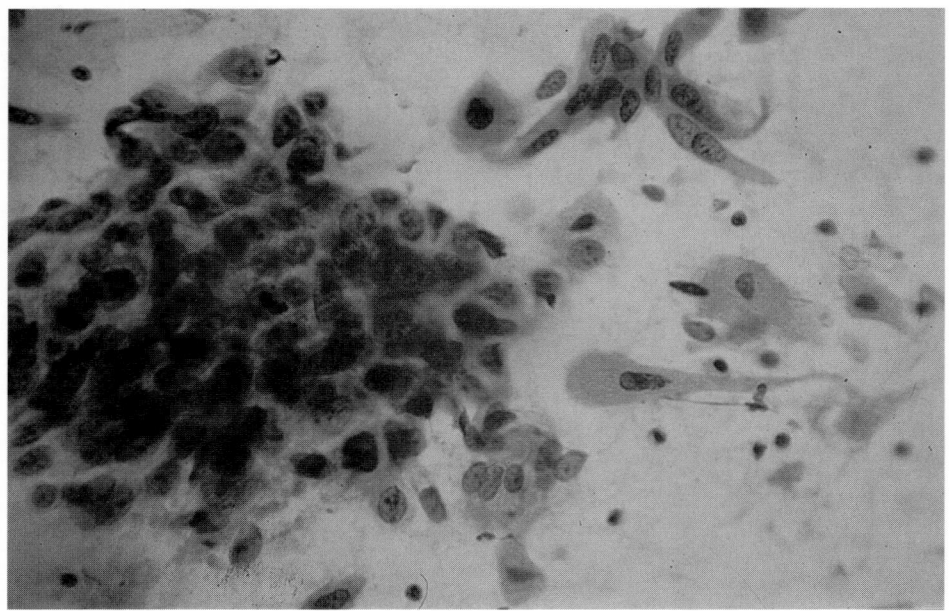

FIGURE 36-10
Squamous cell carcinoma. Well-differentiated orangeophilic "tadpole" cell next to group of poorly differentiated malignant cells in bronchial brushing samples. (Papanicolaou, low magnification.) (*See also Color Plate 21.*)

poorly differentiated adenocarcinoma or large cell carcinoma. This applies particularly to brushings and aspirates. The predominant cytologic features of squamous cell carcinoma are shown in Table 36-5.

If there is extensive necrosis in well differentiated carcinoma much anucleate hyperkeratinized debris is often present (Fig. 36-11), especially in

TABLE 36-5
Cytologic Features of Squamous Cell Carcinoma

Background	Inflammatory cells, necrotic debris.
Cell pattern	Single cells and flat sheets.
Cell morphology	Marked anisocytosis, especially in well differentiated subtype in which bizarre "tadpole" or "spindle" cells or forms with pseudopodia may be present.
Cytoplasm	Highly keratinized with distinctive orangiophilic (Papanicolaou) or bluish (MGG) dense appearance in well differentiated subtype. Less dense cytoplasm and absence of specific tinctorial properties in poorly differentiated subtype. Cytoplasmic lamination sometimes visible.
Nuclei	Frequently pyknotic in well-differentiated cells. Large with coarse chromatin in poorly differentiated cells.

sputum and PCFNA specimens. A definitive diagnosis of carcinoma should not be made on anucleate debris because this can be quite abundant in conditions such as chronic bronchitis, bronchiectasis, or lung abscess. At least a few malignant cells with viable nuclei must be identified within the debris for a secure diagnosis. The characteristic orangeophilic (Papanicolaou) or blue (MGG) cytoplasm (Fig. 36-12) is missing in the poorly differentiated subtype. In brushings in particular, the cytoplasm is sometimes quite fragile. Concentric cytoplasmic laminations are occasionally visible. A careful search through the specimen will often, however, identify a few better differentiated squamous carcinoma cells. If the guidelines above are followed, diagnosis is highly reliable. Some possible pitfalls for the inexperienced are given in Table 36-6. The cytomorphology of atypical squamous metaplasia has been discussed and illustrated by Young.[25]

Small Cell Carcinoma
The salient diagnostic features of small cell carcinoma are listed in Table 36-7 and possible mimics are shown in Table 36-8. In sputum the nuclei are generally pyknotic and therefore appear deeply hyperchromatic. Typically they occur in streaks against a background of very fine debris and are not admixed with other cell types. This is a helpful diagnostic point because streaks of lymphocytes are almost invariably intermingled with follicle center cells or granulocytes. In brushings and PCFNA, because the cells are sampled directly from the tumor, they are more likely to be viable and therefore appear larger.

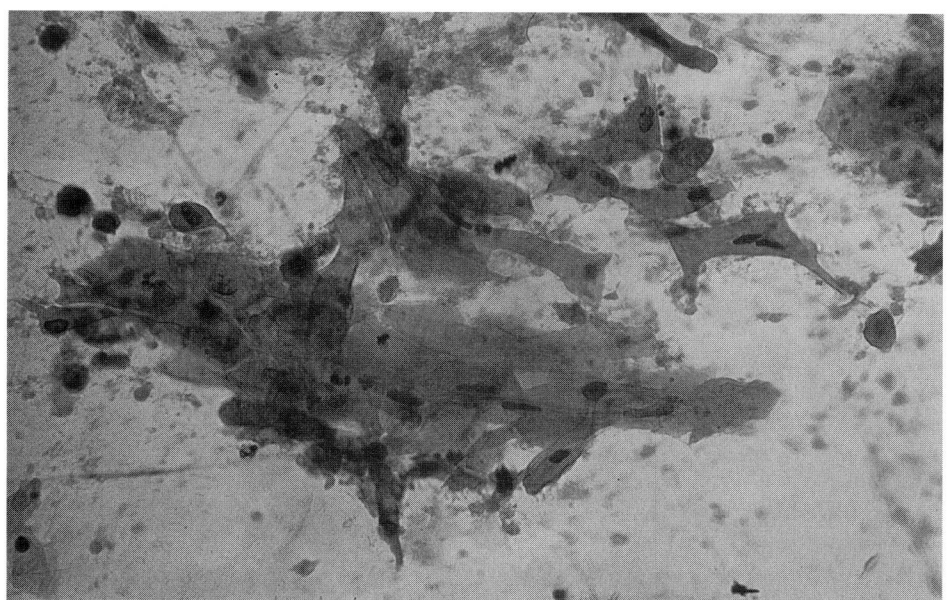

FIGURE 36-11
Squamous cell carcinoma. Copious, mainly anucleate, orangeophilic debris in sputum specimen. (Papanicolaou, low magnification.) (*See also Color Plate 22.*)

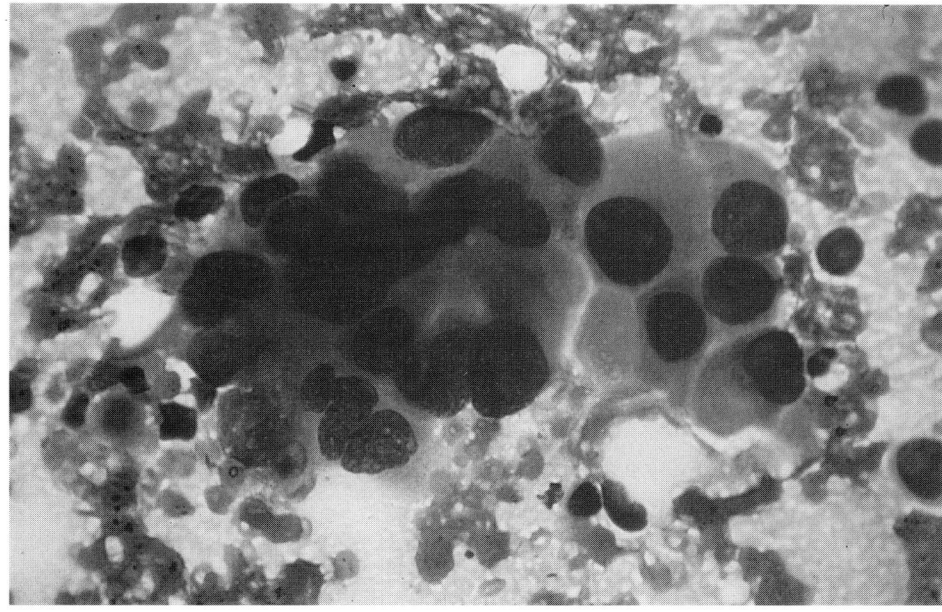

FIGURE 36-12
Squamous cell carcinoma. Malignant cells with dense flat cytoplasm and sharply articular borders from FNA sample. (May-Grunwald-Giemsa, moderate magnification.) (*See also Color Plate 23.*)

TABLE 36-6
Conditions That May Be Confused with Squamous Cell Carcinoma

Abnormal oral or laryngeal squamous cells*

Atypical squamous metaplasia and debris arising from same in cases of chronic lung disease

Radiation/chemotherapy-induced change

Viral infection

Charcot-Leyden crystals

Calcospherities

*Inflammation is the common cause but if changes persist the possibility of oral/laryngeal neoplasia should be excluded.

The characteristic feature of small cell carcinoma is nuclear molding within the groups (Figs. 36-13, 36-14, and 36-15). Lymphoma cells do not appear as compact groups nor display nuclear molding. With the intermediate subtype of small cell carcinoma the cells are larger than the classic oat cells. In the combined subtype evidence of squamous cell or adenocarcinoma may coexist. Limited experience suggests the cells of large cell neuroendocrine carcinomas are identifiably bigger than those of small cell anaplastic carcinoma.

1233

TABLE 36-7

Cytologic Features of Small Cell Carcinoma

Background	Fine particulate debris in sputum and brushings. More obvious and coarser debris in PCFNA.
Cell pattern	Streaks and loosely cohesive clusters in sputum. Single cells, loose clusters, and groups in bronchoscopic specimens and PCFNA.
Cell morphology	Uniform population of small cells although in some tumors occasional larger cells are present.
Cytoplasm	Minimal and fragile, barely discernible especially in sputum.
Nuclei	Small, hyperchromatic and pleomorphic with evidence of characteristic molding. Pyknotic in sputum. Better preserved in brushings and PCFNA and therefore appear somewhat larger. Fine granular chromatin pattern and indistinct nucleoli. Nucleoli more obvious in intermediate cell subtype.

TABLE 36-8

Conditions That May Be Confused with Small Cell Carcinoma

Lymphocytes, e.g., in follicular bronchitis

Basal cells, particularly if poorly preserved in brushings

Bare nuclei due to cytolysis

Carcinoid tumor

Large cell neuroendocrine carcinoma

Malignant lymphoma

Adenocarcinoma

The diagnosis of adenocarcinoma can be straightforward but of all types of lung cancer it is the variety that produces most problems for the cytopathologist, both of recognition and of typing. The main guidelines for diagnosis are given in Table 36-9. Single cells are less common in adenocarcinoma. Regardless of whether it is exfoliated material in sputum, endoscopic, or PCFNA sampling, cohesive cell groups (Fig. 36-16) with considerable depth of focus predominate. Sometimes an acinar (Fig. 36-17) or papillary pattern may be discernible[25,26] but this feature is not often seen. Similarly with vacuolation, if hard-edged vacuoles containing very pale pink (Papanicolaou) or light magenta (MGG)[26] material indicative of mucin are present then this is in keeping with adenocarcinoma. Such vacuoles are seldom numerous. Delicate or finely vacuolated cytoplasm alone is not diagnostic of adenocarcinoma because it may occur as a result of degeneration either in benign cells or in poorly differentiated malignant squamous cells.

Recognition of *bronchiolalveolar carcinoma* (alveolar carcinoma) is clinically important. Because of the copious sputum sometimes expectorated, this is commonly submitted for cytology. Numerous groups of cells are seen, often in tight rolled up formations with nuclei bulging from the periphery (Fig. 36-18). The cells are usually well differentiated and chromatin and nucleolar abnormalities seldom striking. However, if the sputum is carefully screened occasional, single, malignant cells with more obviously abnormal nuclei can often be found. The best

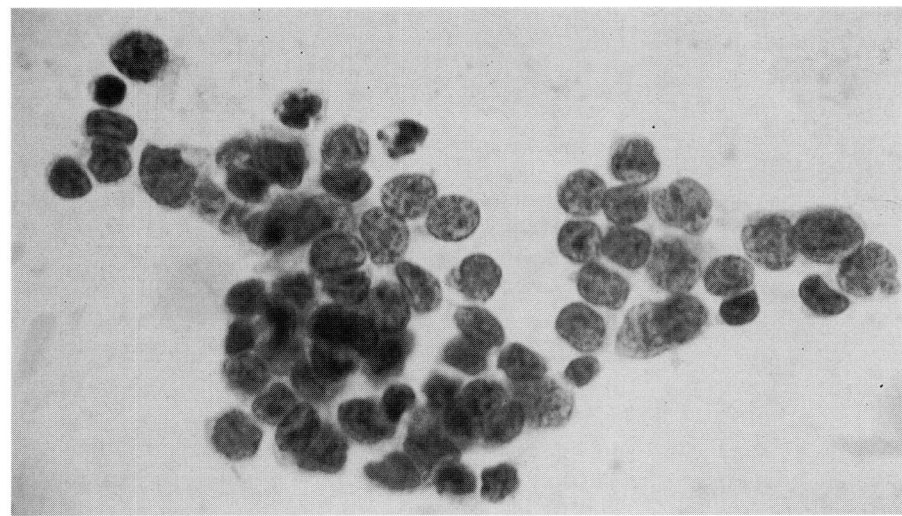

FIGURE 36-13

Oat cell carcinoma. Cluster of small malignant cells with minimal cytoplasm and granular nuclei in sputum specimens. (Papaniclaou, high magnification.) (*See also Color Plate 24.*)

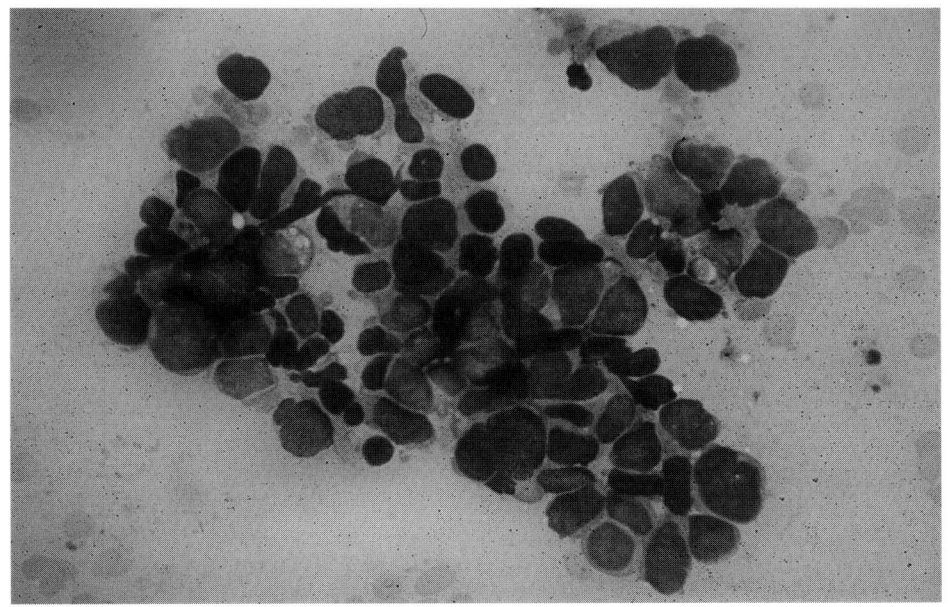

FIGURE 36-14
Oat cell carcinoma. Deeply hyperchromatic nuclei shows evidence of molding from FNA of lung. (May-Grunwald-Giemsa, moderate magnification.) (*See also Color Plate 25.*)

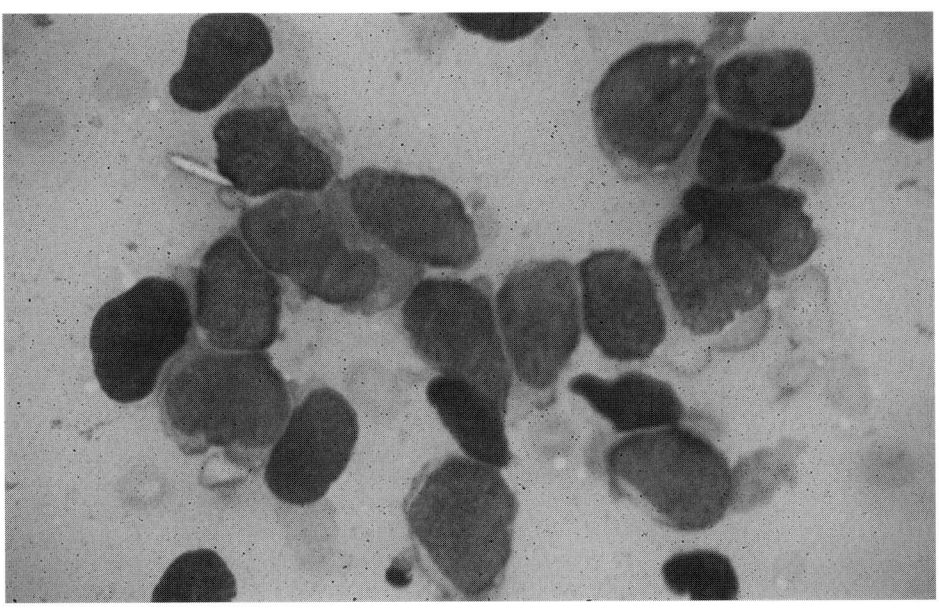

FIGURE 36-15
Oat cell carcinoma. Small pleomorphic nuclei with finely granular chromatin and inconspicuous nucleoli from FNA of lung. (May-Grunwald-Giemsa, high magnification.) (*See also Color Plate 26.*)

TABLE 36-9
Cytologic Features of Adenocarcinoma

Background	Sparse debris, often inflammatory cells.
Cell pattern	Cohesive, three-dimensional groups. Evidence of acinar or papillary pattern may be present
Cell morphology	Medium-sized or large cell with moderate anisocytosis.
Cytoplasm	Delicate. Vacuolation may be evident. Absent keratin.
Nuclei	Large, hyperchromatic, often eccentrically placed. Coarse chromatin with prominent nucleoli.

guidance for diagnosis is contained in the review by Spriggs and colleagues.[85] In PCFNA, epithelial cells are abundant with, in some cases, obvious cell balls or papillary clusters. Flatish sheets without the usual depth of focus associated with adenocarcinoma can also sometimes be seen. Often nuclei are bland and single cells may be difficult to distinguish from alveolar macrophages. Nuclear folds are common and both nuclear pseudoinclusions (invaginations) and psammona bodies have been described.[85] Silverman and coworkers[86] have the most extensive experience with PCFNA cases.

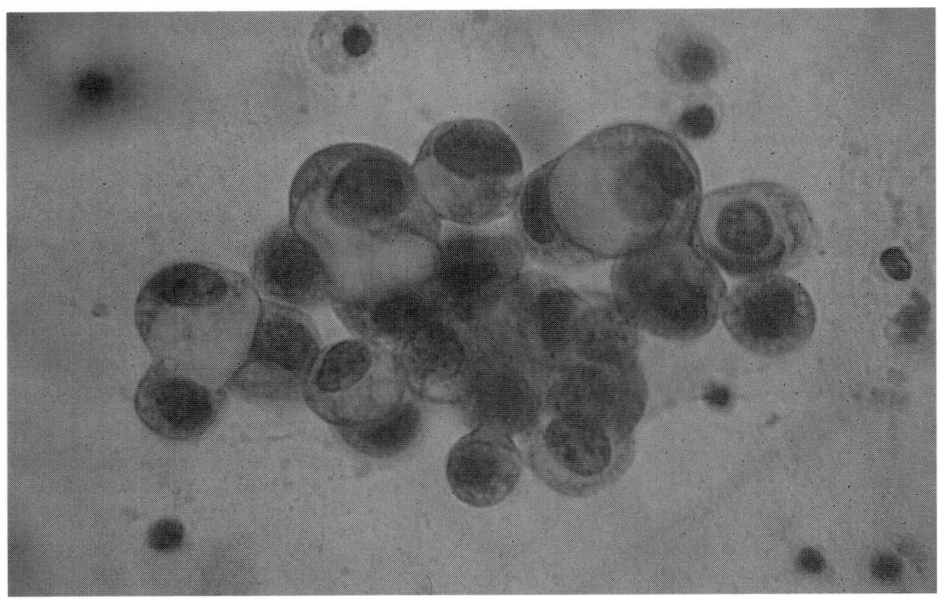

FIGURE 36-16
Adenocarcinoma. Three-dimensional group of malignant cells with eccentric nuclei and copious vacuolated cytoplasm in a sputum specimen. (Papanicolaou, high magnification.) (*See also Color Plate 27.*)

The possibility of making a false-positive diagnosis of well differentiated adenocarcinoma is a pitfall for the unwary. A selection of potential mimics is given in Table 36-10. Hyperexfoliation of groups of bronchial epithelial cells following instrumentation or as the result of infection or allergic bronchopulmonary disease may resemble the cytologic picture of alveolar carcinoma. However, malignant cell groups never bear cilia and if these can be identified it is reassuring evidence of a benign process. The converse of course it not a reliable diagnostic feature because benign cells may lose cilia due to degeneration. A more serious interpretation problem is the severe morphologic atypia that may be produced by chemotherapy, pneumonitis, or ARDS.

Bizarre penumocytes may closely resemble malignant cells even to the experienced, especially in BAL fluid.[64] Nuclei, however, almost invariably

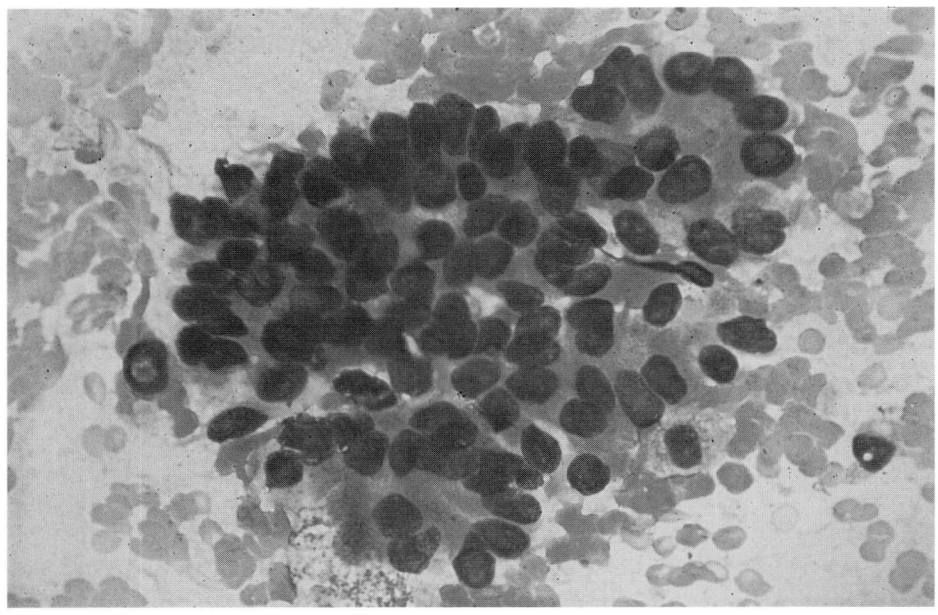

FIGURE 36-17
Adenocarcinoma. Large cluster of malignant cells with discernible acinar pattern in FNA of lung. (May-Grunwald-Giemsa, low magnification.) (*See also Color Plate 28.*)

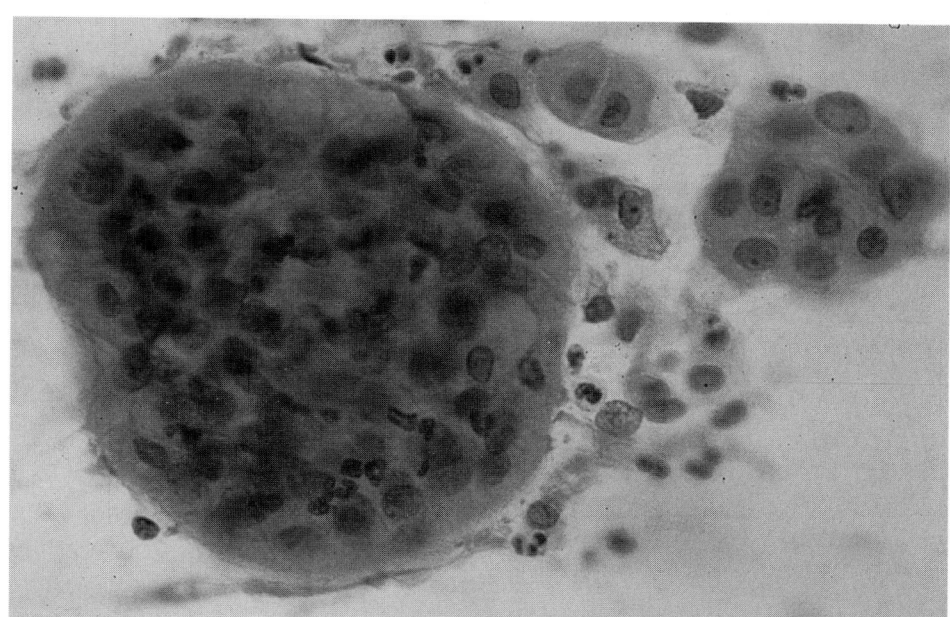

FIGURE 36-18
Bronchoalveolar carcinoma. Two spherical groups of well differentiated malignant cells in FNA of lung. (H&E, moderate magnification.) (*See also Color Plate 29.*)

display degenerative changes. Careful clinical correlation will avoid the possibility of a false-positive diagnosis in such cases.

Large Cell Carcinoma

Poorly differentiated squamous cell carcinoma, poorly differentiated adenocarcinoma, and large cell carcinoma are difficult to separate by cytologic examination and in most centers there is overdiagnosis of the last tumor because no distinctive evidence of keratinization or mucin production is evident. The cells of large cell carcinoma display florid stigmata of malignancy as described in Table 36-11. Providing

the lesion has been properly sampled there is little problem in establishing a diagnosis of carcinoma (Figs. 36-19 and 36-20). In the giant cell variety the cells are very large, some multinucleated, and active phagocytosis of neutrophils or small malignant cells may be evident (Fig. 36-21). Examples of cytologic material from large cell neuroendocrine carcinoma are few. A fine chromatin pattern may be identifiable in some cells. Apart from atypical pneumocyte hyperplasia and severe drug or radiation toxicity there is little that is not frankly malignant that can be confused with large cell carcinoma (Table 36-12).

TABLE 36-10

Conditions That May Be Confused with Adenocarcinoma

Postbronchoscopy sputum

Hyperexfoliation of bronchial epithelium due to inflammation or allergic bronchopulmonary disease.

Granulomatous disease

Pulmonary infarction

Chemotherapy

Atypical pneumocyte hyperplasia associated with acute lung injury/pneumonitis/adult respiratory distress syndrome

TABLE 36-11

Cytologic Features of Large Cell Carcinoma

Background	Coarse debris and inflammatory cells.
Cell pattern	Single cells and groups.
Cell morphology	Large cells with marked anisocytosis and pleomorphism.
Cytoplasm	Variable in quantity but often plentiful. Absence of keratinization or hard-edged vacuoles.
Nuclei	Large, hyperchromatic and pleomorphic with coarse chromatin and angular, often multiple nucleoli. Multinucleated malignant cells in giant cell subtype. Fine chromatin in large cell neuroendocrine carcinoma.

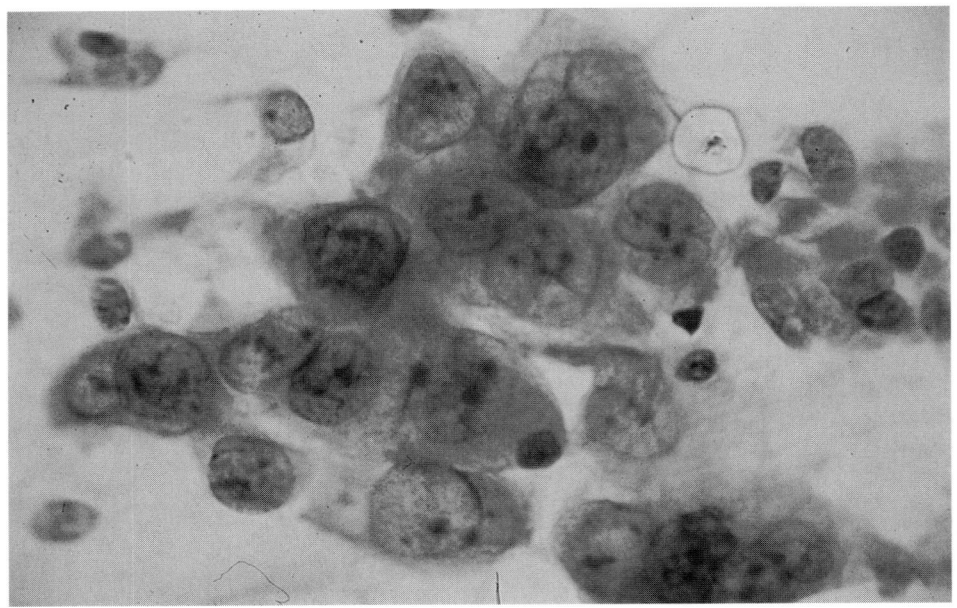

Bizarre vegetable cells from food debris are not uncommon in sputum but are so alien in appearance that the risk of error is minimal.

Other Primary Carcinomas

Of the various other types of primary carcinoma described in the lung[70] *adenosquamous carcinoma* is the entity most likely to be seen in cytologic material. Diagnosis depends on recognition of features of both adenocarcinoma and squamous cell carcinoma. In the survey by Zuman-Harach and associates [88] this dual differentiation was identifiable in PCFNA material from 58 percent of tumors in which the final diagnosis was adenosquamous carcinoma. Because of the potential for extended sampling, evidence of adenosquamous carcinoma is more likely to be seen in sputum[25] than in PCFNA. Several descriptions of the cytopathologic presentation of *submucous gland carcinomas* have been published including *adenoid cystic carcinoma* and *mucoepidermoid carcinoma.*[90]

Other Primary Malignant Tumors

The review of PCFNA in the diagnosis of *pulmonary sarcoma* by Crosby and coworkers[91] includes one

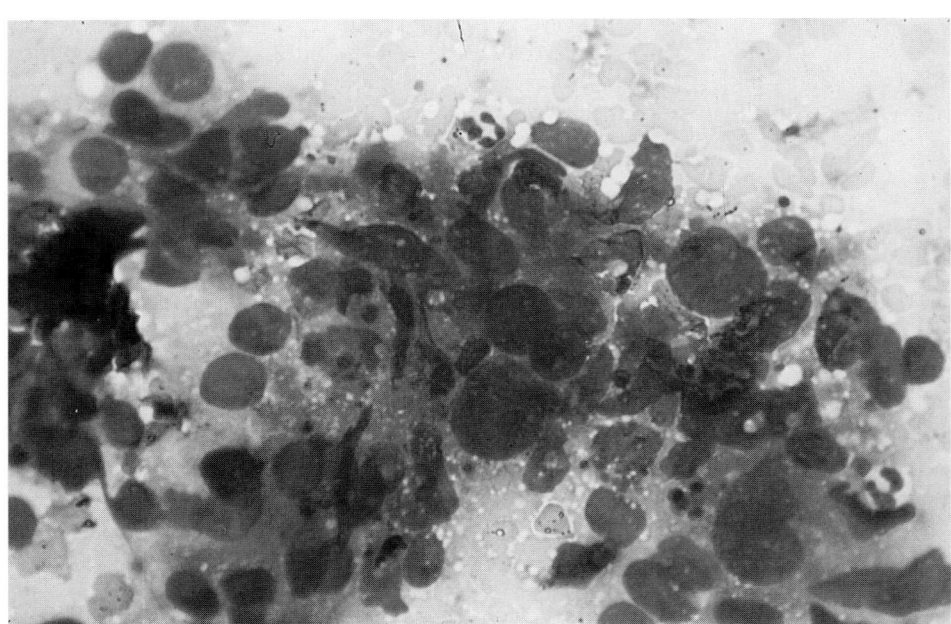

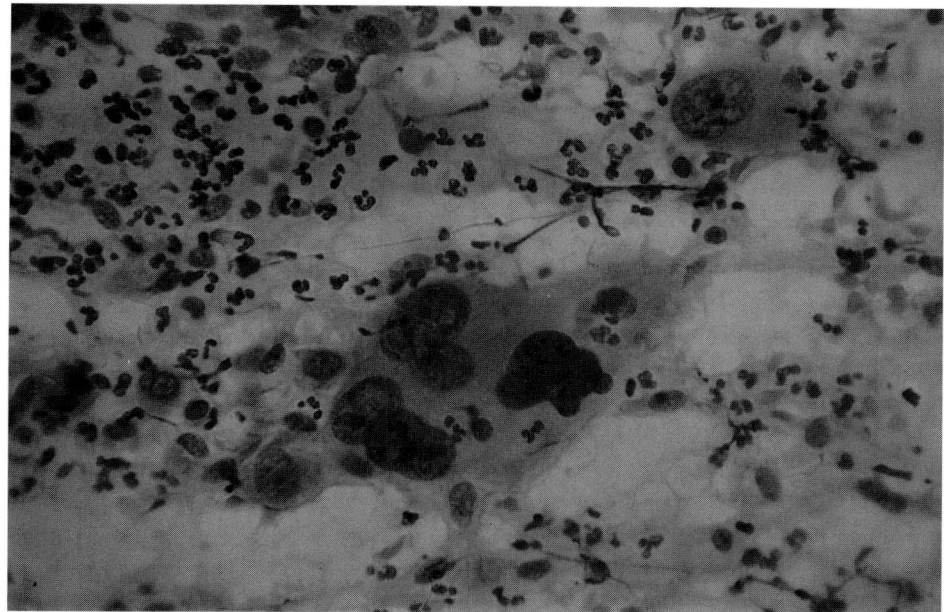

FIGURE 36-21
Giant cell carcinoma. One of numerous very large multinucleated malignant cells plus smaller mononuclear malignant cells against a background of inflammatory debris from a bronchial brushing. (Papanicolaou, moderate magnification.) (*See also Color Plate 32.*)

TABLE 36-12

Conditions That May Be Confused with Large Cell Carcinoma

Drug or radiation toxicity

Atypical hyperplastic pneumocytes associated with acute lung injury/pneumonitis/adult respiratory distress syndrome

Food debris

case of a primary tumor and brushings from another, confirmed by autopsy, are illustrated in Young.[25] Findings in *malignant fibrous histiocytoma*[92] and *carcinosarcoma*[93] have also been published and several case reports describe the cytopathology of *pulmonary blastoma*.[94–96] Primary *Hodgkin's disease*[97,98] is documented but is exceedingly rare. *Non-Hodgkin's lymphoma*[99,100] is also recognized but is very uncommon in comparison with secondary pulmonary infiltration.

SECONDARY MALIGNANT TUMORS

Before the practice of PCFNA of the lung was widely adopted, patients with suspected metastatic pulmonary disease were treated on the basis of radiologic findings alone or required thoracotomy for definitive diagnosis. Sputum cytology is seldom diagnostic and similarly unless there is visible endobronchial tumor brushing or forceps biopsy are of little value. Confirmation of malignancy in suspected secondary infiltration now constitutes about 15 percent of PCFNA of the lung in major centers.[101] *Metastatic carcinoma* from any primary site can occur but spread from the gastrointestinal tract, breast, kidney, female genital tract, and prostate are among the more common origins. Immunocytochemistry can be helpful, for example in prostatic cancer[102] in confirming the primary. However, distinction between metastatic adenocarcinoma or squamous cell carcinoma and a second primary may be impossible. *Metastatic melanoma* is not uncommon[25,26] and at least two papers detailing the findings in cases of *metastatic sarcoma* have been published.[91,102] Secondary infiltration by *Hodgkin's disease*, *non-Hodgkin's lymphoma*, *myeloma*, and *leukemia* quite often occur in patients with advanced disease. When this is a diffuse process BAL is a safe and appropriate means of investigation.[25,49,103]

DIAGNOSTIC EFFICIENCY

Investigative Protocol in Suspected Lung Cancer

The relative place of sputum, bronchoscopic sampling (cytopathology plus forceps biopsy), and PCFNA in the investigation of patients with suspected lung cancer is not entirely clear-cut. Many clinicians submit sputum, followed by bronchoscopy and cytopathology/biopsy, and only if these are nondiagnostic proceed to PCFNA. Others for reason of speed, accuracy, and cost effectiveness consider PCFNA as a suitable technique for the first investigation especially if the mass is peripherally situated within the lung. In an in-depth study of the changing use of cytopathology (sputum, brushing, washing, and FNA) versus histopathology, Fraire and colleagues[104] found an overall increase of cytopathology procedures as the sole means of diagnosis. There was a decrease in the utilization of sputum and an increase in FNA. There was also an increase in the sensitivity of cytologic diagnosis over the 20-year period studied. Sputum examination is noninvasive and causes no inconvenience to the patient. Definitive diagnosis and tumor typing are possible but localization is not. It is, however, labor intensive as a laboratory procedure both with regard to preparatory time and microscopy and many of the specimens received are unsatisfactory.[16] Sputum examination still remains a simple test for confirmation of malignancy in patients with clinically advanced disease and it also provides the only means of screening for severe dysplasia and possible early indications of neoplasia.[105,106] The various bronchoscopic sampling techniques are carried out during the course of endoscopic assessment with the advantages for staging and appropriate treatment that this confers. PCFNA is a highly accurate diagnostic method, but it is not without a risk and there are anatomatic limitations.

It is in my opinion prudent and the best use of available resources to adopt a flexible approach to investigation taking into consideration the clinical and radiologic findings, the assessment of risk, and the likelihood of surgical treatment in each individual patient.

Sensitivity and Specificity

The predictive value of *sputum cytology* was examined by Benbassat and coworkers[107] who reviewed the sensitivity and specificity of published studies and discussed the associated problems. Sensitivity varies from 28.6 percent[107] to 88.9 percent[108] and specificity from 89.7 percent[109] to 99.9 percent.[110] The wide range of sensitivity is influenced by the number of specimens examined, the size and location of the tumors, and whether unsatisfactory samples are included or excluded. For example Johnston and Bossen[109] found the first specimen was diagnostic in 34.3 percent of patients but that a conclusive diagnosis of cancer was possible in 63.9 percent of cases after examination of five satisfactory specimens.

The sensitivity of *bronchial brushing* ranges from 52 percent to 90 percent[111–118] but maximum diagnostic efficiency is obtained by using several bronchoscopic techniques and cytopathology and biopsy are complementary. In the series by Popp and colleagues[119] the sensitivity of brushing was 80.6 percent, biopsy histology 62.9 percent, and biopsy imprint cytology 84.9 percent, but when results of the three techniques were combined sensitivity rose to 97.3 percent. Multiple brushings also increase the diagnostic yield.[3]

Transbronchial FNA is used more selectively and in such series sensitivity varies from 46 percent[120] to 81 percent.[121] In the study by Harrow and associates[120] it was the only sampling method that provided the correct diagnosis in 17 percent of patients. Similarly Bhat and colleagues[122] found the overall sensitivity of bronchoscopic diagnosis rose from 54 to 72 percent by the addition of TBFNA. Sensitivity of *PCFNA* ranges from 62 percent[123] to 96.8 percent[124] with specificity of 100 percent in several series.[125,126] In a comparative study of all diagnostic techniques Johnston[127] found it the single most efficient test.

Tumor Typing

The intrinsic heterogenicity of lung tumors and the problem of their classification make consistent accurate typing difficult when only the limited material available in cytopathologic specimens or small biopsies is available for examination. Well-differentiated squamous cell carcinoma and small cell carcinoma are straightforward and well-differentiated adenocarcinoma is usually correctly typed. Poorly differentiated squamous cell and adenocarcinoma and large cell carcinoma are much more difficult to separate. In a careful study, with autopsy histology providing the final diagnosis, Bonito and coworkers[128] found 100 percent accuracy for small cell carcinoma, 90 percent for squamous cell, 80 percent for adenocarcinoma, and 50 percent for undifferentiated large cell.

Ancillary Studies

Any of the vast array of special stains and additional techniques can potentially be applied to cytopathologic material.[129] The value of ancillary studies in PCFNA of the lung was reviewed by O'Reilly and associates.[130] Mucin staining contributed to the correct typing of non-small cell carcinoma in 68 percent of tumors. Immunocytochemistry was helpful in 20 (40 percent) of cases in which it was attempted and electron microscopy provided significant additional information in 10 (67 percent) instances.

The role of immunocytochemistry in the diagnosis of small cell carcinoma was studied by Tome and coworkers[131] and the immunocytochemical characterization of large cell carcinomas considered by Schulte and associates.[132] Utilization of morphometry as an aid to the differential diagnosis of large cell carcinoma has been found to confer some help.[133] Assessment of nucleolar organizer regions is useful in discriminating small cell and non-small cell carcinomas.[134]

REFERENCES

1. Dudgeon LS, Barrett NR: The examination of fresh tissues by the wet-film method. *Br J Surg* 22:4–22, 1934.
2. Young JA: Techniques in pulmonary cytopathology. ACP Broadsheet 140. *J Clin Pathol* 46:589–595, 1993.
3. Popp W, Merkle M, Schreiber B, Rauscher H, Ritschka L, Zwick H: How much brushing is enough for the diagnosis of lung tumors? *Cancer* 70:2278–2280, 1992.
4. DiBonito L, Colautti, I, Patriarca S, Falconceri G, Barbozza R, Vieth P: Cytological typing of primary lung cancer: A study of 100 cases with autopsy confirmation. *Diag Cytopathol* 7:7–10, 1991.
5. Wang KP, Marsh BR, Summer WR, et al: Transbronchial needle aspiration in the diagnosis of lung cancer. *Chest* 80:48–50, 1981.
6. Zavaia DC, Schoeil JE: Ultrathin needle aspiration of the lung in infections and malignant disease. *Am Rev Respir Dis* 123:125–131, 1981.
7. Fontana R, Miller W, Beabout J, Payne W, Harrison E: Transthoracic aspiration of discrete pulmonary lesions. Experience in 100 cases. *Med Clin North Am* 54:961–970, 1970.
8. Poe RH, Kallay MC, Wicks CM, Odoroff CL: Predicting risk of pneumothorax in needle biopsy of the lung. *Chest* 85:232–235, 1984.
9. Mooloo Z, Finley RJ, Lefcoe MS, Turner-Smith L, Craig ID: Possible spread of bronchogenic carcinoma to the chest wall after a transthoracic fine needle aspiration biopsy: A case report. *Acta Cytol* 29:167–169, 1985.
10. Nordenstrom BEW: Technical aspects of obtaining cellular material from lesions deep in the lung. A radiologist's view and description of screw-needle sampling technique. *Acta Cytol* 28:233–242, 1984.
11. Young JA: Technique, in Young JA (ed), *Fine Needle Aspiration Cytopathology*. Oxford, Blackwell Scientific Publications, 1993, pp 6–12.
12. Stanley MW, Henry-Stanley MJ, Iber C: *Bronchoalveolar Lavage*. New York, Igaku-Shoin, 1991.
13. Risse EKJ, Von't Hof MA, Laurini RN, Vooijs PG: Sputum cytology by the Saccomanno method in diagnosing lung malignancy. *Diagn Cytopathol* 1:286–291, 1985.
14. Rizzo T, Schumann GB, Riding JM: Comparison of the pick-and-smear and Saccomanno methods for sputum cytologic analysis. *Acta Cytol* 34:875–880, 1990.
15. Tang C-S, Tang CMC, Lung ITM: Dithiothreitol homogenization of prefixed sputum for lung cancer detection. *Diagn Cytopathol* 10:76–81, 1994.
16. Bocking A, Biesterield S, Chatelain SU, Gien-Gerlach G, Esser E: Diagnosis of bronchial carcinoma on sections of paraffin-embedded sputum, *Acta Cytol* 36:37–45, 1992.
17. Flint A: Detection of pulmonary neoplasms by bronchial washing. Are cell blocks an aid? *Acta Cytol* 37:21–23, 1993.
18. Turner-Warwick DM, Haslam PL: Clinical applications of bronchoalveolar lavage. *Clin Chest Med* 8:15–26, 1971.
19. Thompson AB, Robbins RA, Chafouri MA, Linder J, Rennard

SI: Bronchoalveolar lavage fluid processing: Effect of membrane filtration preparation on neutrophil recovery. *Acta Cytol* 33:544–549, 1989.
20. Taskinen E, Tukiainen P, Renkonen R, Bronchoalveolar lavage: Influence of cytologic methods on cellular picture. *Acta Cytol* 36:680–686, 1992.
21. Walters EH, Gardiner PV: Bronchoalveolar lavage as a research tool. *Thorax* 46:613–618, 1991.
22. DeGracia J, Bravo C, Miravittles M, et al: Diagnostic value of bronchoalveolar lavage in peripheral lung cancer. *Am Rev Respir Dis* 147:649–652, 1993.
23. Bedrossian CWM, Acetta PA, Kelly LV: Cytopathology of non-neoplastic pulmonary disease. *Lab Med* 14:86–95, 1983.
24. Rosenthal DL: Cytology in the diagnosis of benign lung diseases. *Clin Chest Med* 8:147–159, 1987.
25. Young JA: *Colour Atlas of Pulmonary Cytology*. London, Harvey Miller and Oxford University Press, 1985.
26. Young JA: The lung, pleura and chest wall, in Young JA (ed), *Fine Needle Aspiration Cytopathology*, Oxford, Blackwell Scientific Publications, pp 97–121.
27. Machicao CN, Sorensenk, Abdul-Karim FW, Somrak TM: Transthoracic needle aspiration biopsy in inflammatory pseudotumors of the lung. *Diagn Cytopathol* 5:400–403, 1989.
28. Berman JJ, Murray RJ, Lopez-Plaza IM: Widespread post-tracheostomy atypia simulating squamous cell carcinoma: A case report. *Acta Cytol* 353:713–716, 1991.
29. Naryshkin S, Young NA: Respiratory cytology: A review of non-neoplastic mimics of malignancy. *Diagn Cytopathol* 9:89–97, 1993.
30. Fekete PS, Campbell WG Jr, Bernadino ME: Transthoracic needle aspiration biopsy in Wegener's granulomatosis: Morphologic findings in five cases. *Acta Cytol* 34:155–160, 1990.
31. Pitman MB, Szyfelbein WM, Nibs J, Fienberg R: Clinical utility of fine needle aspiration biopsy in the diagnosis of Wegener's granulomatosis: A report of two cases. *Acta Cytol* 36:222–229, 1992.
32. Rothberg AD, Miot A, Leiman G: Tracheal aspirate cytology and bronchopulmonary dysplasia. *Diagn Cytopathol* 2:212–216, 1986.
33. Chen KTK: Cytology of allergic bronchopulmonary aspergillosis. *Diagn Cytopathol* 9:82–85, 1993.
34. Leiman G: Asbestos bodies in fine needle aspirates of lung masses: Markers of underlying pathology. *Acta Cytol* 35:171–174, 1991.
35. Das DK, Bhambhani S, Pant JN, et al: Superficial and deep-sealed tuberculous lesions: Fine needle aspiration cytology diagnosis of 574 cases. *Diagn Cytopathol* 8:211–215, 1992.
36. Johnston WW: Cytopathology of mycotic infections. *Lab Med* 2:34–40, 1971.
37. Pitchenik AE, Ganjei P, Torres A, et al: Sputum examination for the diagnosis of Pneumocystis pneumonia in the acquired immunodeficiency syndrome. *Am Rev Respir Dis* 133:226–229, 1986.
38. Carmichael A, Bateman N, Nayagam M: Examination of induced sputum in the diagnosis of *Pneumocystis carinii* pneumonia. *Cytopathology* 2:61–66, 1991.
39. Johnston WW, Amatulli J: The role of cytology in the primary diagnosis of North American blastomycosis. *Acta Cytol* 25:103–107, 1970.
40. Guglietti LC, Rheingold IM: The detection of *Coccidioides immitis* in pulmonary cytology. *Acia Cytol* 12:332–334, 1968.

41. Bayon MN, Drut R: Cytologic diagnosis of adenovirus bronchopneumonia. *Acta Cytol* 35:181–182, 1991.

42. Chaudhuri B, Nanos S, Soco JN, McGrew EA: Disseminated *Strongyloides stercoralis* infestation detected by sputum cytology. *Acta Cytol* 24:360–362, 1980.

43. Rangdaeng S, Alperi LC, Khiyami A, Cottingham K, Ramzy I: Pulmonary paragonimiasis: Report of a case with diagnosis by fine needle aspiration cytology. *Acta Cytol* 36:31–36, 1992.

44. Ingram EA, Heikson MA: Echinococcosis (hydatid disease) in Missouri: Diagnosis by fine needle aspiration of a lung cyst. *Diagn Cytopathol* 7:527–531, 1991.

45. Weinberger SE, Kalman JA, Elson NA, et al: Bronchoalveolar lavage in interstitial lung disease. *Ann Intern Med* 89:456–466, 1975.

46. Walters LC, Schwarz MI, Cherniack RM et al: Idiopathic pulmonary fibrosis; pretreatment brochoalveolar lavage cellular constituents and their relationships with lung histopathology and clinical response to therapy. *Am Rev Respir Dis* 135:696–704, 1987.

47. Strigle SM, Gal AA: A review of pulmonary cytopathology in the acquired immunodeficiency syndrome. *Acta Cytol* 29:1047–1052, 1985.

48. Dawber JH, Paradis IL, Dummer JS: Infectious complications in pulmonary allograft recipients. *Clin Chest Med* 11:291–308, 1990.

49. Young JA: Cytological investigation of immune suppressed patients, in Gray W (ed), *Diagnostic Cytopathology*, London, Churchill Livingstone, (in press), pp. 543–554.

50. Young JA, Hopkin JM, Cuthbertson WP: Pulmonary infiltrates in immunocompromised patients: Diagnosis by cytological examination of bronchoalveolar lavage fluid. *J Clin Pathol* 37:390–397, 1984.

51. Young JA, Stone JW, McGonigle RJS, Adu D, Michael J: Diagnosing *Pneumocystis carinii* pneumonia by cytological examination of bronchoalveolar lavage fluid: Report of 15 cases. *J Clin Pathol* 39:945–949, 1986.

52. Broaddus C, Daka MD, Stulberg MS, et al: Bronchoalveolar lavage and transbronchial biopsy for the diagnosis of pulmonary infections in the acquired immunodeficiency syndrome. *Ann Intern Med* 102:747–752, 1985.

53. Golden JA, Holland H, Stulberg MS et al: Bronchoalveolar lavage as the exclusive diagnostic modality for *Pneumocystis carinii*. *Chest* 90:18–21, 1986.

54. Orenstein M, Webber CA, Cash M, et al: Value of bronchoalveolar lavage in the diagnosis of pulmonary infection in acquired immune deficiency syndrome. *Thorax* 41:345–349, 1986.

55. Ghali VS, Garcia RL, Skolom J: Fluorescence of *Pneumocystis carinii* in Papanicolaou smears. *Hum Pathol* 15:907–909, 1984.

56. Wehle K, Blanke M, Koenig G, Pfitzer P: The cytological diagnosis by fluorescence microscopy of Papanicolaou stained brochoalveolar lavage specimens. *Cytopathology* 2:113–120, 1991.

57. Leigh TR, Gazzard BG, Rowbottom A, Collins JV: Quantitative and qualitative comparison of DNA amplification by PCR with immunofluorescence staining for diagnosis of *Pneumocystis carinii* pneumonia. *J Clin Pathol* 46:140–144, 1993.

58. Martin WJ, Smith TF: Rapid detection of cytomegalovirus in bronchoalveolar lavage specimens by a monocional antibody method. *J Clin Microbiol* 23:1006–1008, 1986.

59. Olive DM, Simsek M, Al-Multi S: Polymerase chain reaction assay for detection of human cytomegalovirus. *J Clin Microbiol* 27:1238–1242, 1989.

60. Iwa N, Sasaki M, Yutani C, Wakasa K: Detection of cytomegalovirus DNA in pulmonary specimens: Confirmation by in situ hybridization in two cases. *Diagn Cytopathol* 8:357–360, 1992.

61. Grosby JH, Pantazis CG, Stigall B: In situ hybridization for confirmation of herpes simplex virus in bronchoalveolar lavage smears. *Acta Cytol* 35:248–250, 1991.

62. Newsome AL, Curtis FI, Cuthbertson CG, Allen SD: Identification of *Acanthamoeba* in bronchoalveolar lavage specimens. *Diagn Cytopathol* 8:231–234, 1992.

63. Huang M-S, Colby TV, Goellner JR, Martin WJ: Utility of bronchoalveolar lavage in the diagnosis of drug-induced pulmonary toxicity. *Acta Cytol* 33:533–538, 1989.

64. Walloch JL, Hong NY, Bibb LM: Effects of therapy on cytologic specimens, in Bibbo M (ed), *Comprehsive Cytopathology*. Philadelphia, WB Saunders, 1991, 860–877.

65. Stanley MW, Henry-Stanley MJ, Gayl-Peczalska KJ, Bitterman PB: Hyperplasia of type II pneumocytes in acute lung injury: Cytologic findings of sequential bronchoalveolar lavage. *Am J Clin Pathol* 97:669–677, 1992.

66. Perez-Arellano JL, Garcia J-EL, Macias MCG, Gomez FG, Lopez AJ, de Castro S: Haemosiderin-laden macrophages in bronchoalveolar lavage fluid. *Acta Cytol* 36:26–30, 1992.

67. De Vuyst P, Dumortier E, Moulin N, Yourassowsky M, Yernault JC: Diagnostic value of asbestos bodies in bronchoalveolar fluid. *Am Rev Respir Dis* 136:1219–1224, 1987.

68. Johnson N, Haslam PL, Dewar A, et al: Identification of inorganic dust particles in bronchoalveolar lavage macrophages by X-ray microanalysis. *Arch Environ Health* 41:133–144, 1986.

69. Mermolja M, Rott T, Debelzak A: Cytology of bronchoalveolar lavage in some rare pulmonary disorders: Pulmonary alveolar proteinosis and amiodarone pulmonary toxicity. *Cytopathology* 5:9–16, 1994.

70. Whimster WF: *Tumors of the Trachea, Bronchus, Lung and Pleura*. London, Pitman, 1983.

71. Tomashefski JF: Benign endobronchial mesenchymal tumors: Their relationship to parenchymal pulmonary hamartomas. *Am J Surg Pathol* 6:531–540, 1982.

72. Ramzy I: Pulmonary hamartomas: Cytologic appearances of fine needle aspiration biopsy. *Acta Cytol* 20:15–19, 1976.

73. Ludwig ME, Otis RD, Cole SR, et al: Fine needle aspiration cytology of pulmonary hamartomas. *Acta Cytol* 26:671–677, 1982.

74. Dunbar F, Leiman G: The aspiration cytology of pulmonary hamartomas. *Diagn Cytopathol* 5:174–180, 1989.

75. Gould VE, Linnoila I, Memoli VA, Warren WH: Neuroendocrine components of the bronchopulmonary tract: Hyperplasia, dysplasia and neoplasms. *Lab Invest* 19:519–537, 1983.

76. Paladugu RR, Benfield JR, Pak HY, Ross RK, Teplitz RL: Bronchopulmonary Kulchitsky cell carcinomas: A new classification scheme for typical and atypical carcinoids. *Cancer* 55:1303–1311, 1985.

77. Lamb D: Tumours: Classification and pathology, in Brewis RAL, Gibson GJ, Geddes DM (eds), *Respiratory Medicine*. London, Bailliere Tindall, 1990, pp 790–818.

78. Kim K, Mah C, Dominquez J: Carcinoid tumors of the lung: Cytologic differential diagnosis in fine needle aspirates. *Diagn Cytopathol* 2:343–346, 1986.

79. Szyfelbein WM, Ross JS: Carcinoids, atypical carcinoids and small-cell carcinomas of the lung: Differential diagnosis of

fine needle aspiration biopsy specimens. *Diagn Cytopathol* 4:1–8, 1988.

80. Anderson C, Ludwig ME, O'Donnel M, Garcia N: Fine needle aspiration cytology of pulmonary carcinoid tumors. *Acta Cytol* 34:505–510, 1990.

81. Frierson HF, Covell JL, Mills SE: Fine needle aspiration cytology of atypical carcinoid of the lung. *Acta Cytol* 31:471–475, 1987.

82. Mermolya M, Rott T: Cytology of endobronchial granular cell tumor. *Diagn Cytopathol* 7:524–526, 1991.

83. Chow L T-C, Chan S-K, Chow W-H, Tsui M-S: Pulmonary sclerosing haemangioma: Report of a case with diagnosis of fine needle aspiration. *Acta Cytol* 36:287–292, 1992.

84. Young JA: Salivary glands, in Young JA (ed), *Fine Needle Aspiration Cytopathology.* Oxford, Blackwell Scientific Publications, 1993, pp 48–67.

85. Spriggs AI, Cole M, Dunnill MS: Alveolar-cell carcinoma: A problem in sputum diagnosis. *J Clin Pathol* 35:1370–1379, 1982.

86. Silverman JF, Finley JL, Park HK, Norris HT, Strausbauch PH: Psammoma bodies and optically clear nuclei in broncholoalveolar cell carcinoma: Diagnosis by fine needle aspiration with histologic and ultrastructural confirmation. *Diagn Cytopathol* 1:205–215, 1985.

87. Silverman JF, Finley JL, Park HK, Strausbauch P, Unvergerth M, Carney M: Fine needle aspiration cytology of bronchioloalveolar-cell carcinoma of the lung. *Acta Cytol* 29:887–894, 1985.

88. Zusman-Harach SB, Harach HR, Gibbs AR: Cytological features of non-small cell carcinomas of the lung in fine needle aspirates. *J Clin Pathol* 44:997–1002, 1991.

89. Lozowski MS, Mishriki Y, Solitare GB: Cytopathologic features of adenoid cystic carcinoma: Case report and literature review. *Acta Cytol* 27:317–322, 1983.

90. Tao LC, Robertson DI: Cytologic diagnosis of bronchial mucoepidermoid carcinoma by fine needle aspiration biopsy. *Acta Cytol* 22:221–224, 1978.

91. Crosby JH, Hoeg K, Hager B: Transthoracic fine needle aspiration of primary and metastatic sarcomas. *Diagn Cytopathol* 1:221–227, 1985.

92. Kawahara EI, Nakanish I, Kuroda Y, Morishita T: Fine needle aspiration biopsy of primary malignant fibrous histiocytoma of the lung. *Acta Cytol* 32:226–230, 1986.

93. Finley KL, Silverman JF, Dabbs DJ: Fine-needle aspiration cytology of pulmonary carcinosarcoma with immunocytochemical and ultrastructural observations. *Diagn Cytopathol* 4:239–243, 1988.

94. Spahr J, Draffin RM, Johnston WW: Cytopathologic findings in pulmonary blastoma. *Acta Cytol* 23:454–459, 1979.

95. Cosgrove MM, Chandrasoma PT, Martin SE: Diagnosis of pulmonary blastoma by fine-needle aspiration biopsy. Cytologic and immunocytochemical findings. *Diagn Cytopathol* 7:83–87, 1991.

96. Yokoyama S, Hayashida Y, Nagaham J, et al: Pulmonary blastoma: A case report. *Acta Cytol* 36:293–298, 1992.

97. Kern WH, Crepean AG, Jones JC: Primary Hodgkin's disease of the lung. Report of 4 cases and review of the literature. *Cancer* 14:1151–1165, 1961.

98. Levin IS: A case of primary caviting Hodgkin's disease of the lungs. *Acta Cytol* 16:546–549, 1972.

99. Koss MN, Hochholzer L, Nicols PW, Wehunt WD, Lazarus AA: Primary non-Hodgkin's lymphoma and pseudolymphoma of the lung: A study of 161 patients. *Hum Pathol* 14:1024–1038, 1983.

100. Herbert A, Wright DH, Isaason PG, Smith JL: Primary malignant lymphoma of the lung: Histopathologic and immunologic evaluation of nine cases. *Hum Pathol* 15:415–422, 1984.

101. Johnston WW, Elson CE: Respiratory tract, in Bibbo M (ed), *Comprehensive Cytopathology.* Philadelphia, WB Saunders, 1991, pp 320–398.

102. Kim K, Naylor D, Han IH: Fine needle aspiration cytology of sarcomas metastatic to the lung. *Acta Cytol* 30: 688–794, 1986.

103. Wisecarver J, Ness MK, Rennard SI, Thompason AB, Armitage JO, Linder J: Bronchoalveolar lavage in the assessment of pulmonary Hodgkin's disease. *Acta Cytol* 33:527–532, 1989.

104. Fraire AE, McLarty KW, Greenberg SD: Changing utilization of cytopathology versus histopathology in the diagnosis of lung cancer. *Diagn Cytopathol* 7:359–362, 1991.

105. Frost JK, Ball WC, Levin ML, et al: Early lung cancer detection: Results of the initial (prevalence) radiologic and cytologic screening in the John Hopkins study. *Am Rev Respir Dis* 130:549–554, 1984.

106. Risse EKJ, Vooijs CD, Van't Hof MA: Diagnostic significance of severe dysplasia in sputum cytology. *Acta Cytol* 32:629–634, 1988.

107. Benbassat J, Regev A, Slater PE: Predictive value of sputum cytology. *Thorax* 42:165–172, 1987.

108. Clee MD, Sinclair DJM: Assessment of factors influencing the results of sputum cytology of bronchial carcinoma. *Thorax* 36:143–146, 1981.

109. Jay SJ, Wehr K, Nicholson DP, Smith AL: Diagnostic sensitivity and specificity of pulmonary cytology. *Acta Cytol* 24:304–312, 1980.

110. Johnston WW, Bossen EH: Ten years of respiratory cytopathology at Duke University Medical Center. I. The cytopathologic diagnosis during the years 1970–74. *Acta Cytol* 25:103–107, 1981.

111. Mak VHF, Johnston IDA, Hetzel MR, Grubb C: Value of washings and brushings at fibreoptic bronchoscopy in the diagnosis of lung cancer. *Thorax* 45:373–376, 1990.

112. Skitarelic K, Von Haam E: Bronchial brushings and washings: A diagnostically rewarding procedure? *Acta Cytol* 18:321–326, 1974.

113. Zavala DC: Diagnostic fiberoptic bronchoscopy: Techniques and results of biopsy in 600 patients. *Chest* 68:12–19, 1975.

114. Kvale PA, Bode FR, Kini S: Diagnostic accuracy in lung cancer. Comparison of techniques used in association with flexible fiberoptic bronchoscopy. *Chest* 69:752–757, 1976.

115. Chopra SK, Genovesi MG, Simmons DH, Gothe B: Fiberoptic bronchoscopy in the diagnosis of lung cancer. Comparison of pre- and post-bronchoscopy sputa, washings, brushings and biopsies. *Acta Cytol* 21:524–527, 1977.

116. Lyall KRW, Summers GD, O'Brien IMO, Bateman NT, Pike CP, Braimbridge OMV: Sequential brush biopsy and conventional biopsy: Direct comparison of diagnostic sensitivity in lung malignancy. *Thorax* 35:929–931, 1980.

117. Miers MF, Boddington MM, Cole M, Murphy D, Spriggs AI: Cytological sampling at fibreoptic bronchoscopy: Comparison of catheter aspirates and brush biopsies. *Thorax* 37:457–461, 1982.

118. Bronchial brushing and bronchial biopsy: Comparison of diagnostic accuracy and cell typing in lung cancer. *Thorax* 41:475–478, 1986.

119. Popp W, Rauscher H, Ritschka L, Redtenbacher S, Zwick H, Dutz W: Diagnostic sensitivity of different techniques in the

diagnosis of lung tumors with the flexible fiberoptic bronchoscope. *Cancer* 67:72–75, 1991.

120. Harrow EM, Oldenburg FA, Smith AM: Transbronchial needle aspiration in clinical practice. *Thorax* 40:756–759, 1985.

121. Horsley JR, Miller RE, Amy RW, King EC: Bronchial submucosal needle aspiration performed through the fiberoptic bronchoscope. *Acta Cytol* 28:211–217, 1984.

122. Bhat N, Bhaget P, Pearlman ES, et al: Transbronchial needle aspiration in the diagnosis of pulmonary neoplasm. *Diagn Cytopathol* 6:14–17, 1990.

123. Payne CR, Hadfield JW, Stovin PG, Barker V, Heard BE, Stark JE: Diagnostic accuracy of cytology and biopsy in primary bronchial carcinoma. *J Clin Pathol* 34:773–778, 1981.

124. Yazdi HM, MacDonald LL, Hickey HM: Thoracic fine needle aspiration biopsy versus fine needle cutting biopsy. *Acta Cytol* 32:635–640, 1988.

125. Young CP, Young I, Cowan DR, Blei RL: The reliability of fine needle aspiration biopsy in the diagnosis of deep lesions of the lung and mediastinum: Experience with 250 cases using a modified technique. *Diagn Cytopathol* 3:1–7, 1987.

126. Simpson RW, Johnson DA, Wold LE, Goellner JR: Transthoracic needle aspiration biopsy: Review of 233 cases. *Acta Cytol* 32:101–104, 1988.

127. Johnston WW: Fine needle aspiration biopsy versus sputum and bronchial material in the diagnosis of lung cancer. *Acta Cytol* 32:641–646, 1988.

128. DiBonito L, Colauti I, Patriavaca S, Falconier C, Barbazza R, Vielh P: Cytological typing of primary lung cancer: A study of 100 cases with autopsy confirmation. *Diagn Cytopathol* 7:7–10, 1991.

129. Buley ID: Update on special techniques in routine cytopathology. *J Clin Pathol* 46:881–885, 1993.

130. O'Reilly PE, Bruechkner J, Silverman JF: Value of ancillary studies in fine needle aspiration cytology of the lung. *Acta Cytol* 38:144–149, 1994.

131. Tome Y, Hirohashi S, Noguchi M, et al: Immunologic diagnosis of small cell lung cancer in imprint smears. *Acta Cytol* 35:485–490, 1991.

132. Schulte MA, Ramzy I, Greenberg SD: Immunocytochemical characterization of large-cell carcinomas of the lung: Role, limitations and technical considerations. *Acta Cytol* 35:175–180, 1991.

133. Burns TR, Teasdale TA, Greenberg SD: Use of morphometry as an aid in the differential diagnosis of large cell carcinoma of the lung. *Anal Quant Cytol Histol* 15:101–106, 1993.

134. Ascoli V, Barsott P, Facciolo F, Nardi F, Marinozzi V: Nucleolar organizer regions in fine needle aspirates of lung tumor. *Cytopathology* 1:277–286, 1990.

Index

Note: Page numbers in boldface type indicate principal discussions.

fibrinous, in tularemic pneumonia, 215
following liver transplantation, 759
in Legionnaire's disease, 219
in melioidosis, 240
necrotizing, herpes simplex, posttransplant, 742
pneumococcal, 201
in Sjögren's syndrome, 820–821
tuberculous, 229
Bronchopulmonary aspergillosis, allergic, 851
Bronchopulmonary dysplasia (BPD). *See* Dysplasia, bronchopulmonary
Bronchopulmonary foregut malformations, 74–75
lung cysts versus, 81
Bronchoscopy
in lung cancer, 1014
techniques for, 1221–1222
Bronchus(i)
Aspergillus colonization of, in asthma, 704
basement membrane of
anatomy of, 17
development of, 48
branching pattern of, 2
cells of
anatomy of, 7–17, 18–19
development of, 48
congenital malformations of, 64–67. *See also specific malformations*
glands of
anatomy of, 18
development of, 48
histiocytoma of. *See* Inflammatory pseudotumor (IP)
intrauterine development of, 46–47
lobar, 2
main, anatomy of, **6–19**
malignant mixed tumor of, **894–895**
mucous plugging of, in chronic obstructive pulmonary disease, 623
occlusion of, by lung cancer, 1012
papilloma of, 881
segmental, 1–2
structural anomalies of, 2
subepithelial tissues of, anatomy of, 17–18
Bronchus-associated lymphoid tissue (BALT), 19–22, 1111. *See also* BALT lymphoma
depletion of, following heart-lung/lung transplantation, 728, 737
Brown induration of the lung, 673
Brucella, **216–217**
Brucella abortis, 216
Brucella melitensis, 216
lung abscesses caused by, 361
Brucella suis, 216
Brugia buckleyi, vascular disease and, 679–680
Brugia malayi, 335
eosinophilia caused by, 456
Brush cells, 10
Bubonic plague, 212
Bullae, in emphysema, 613–614
Bunyaviridae, 171
Burns, adult respiratory distress syndrome and, 385

herpes simplex virus infections and, 390
oxygen toxicity and, 393
Busulfan, pulmonary toxicity of, 553, 555, 556
Byssinosis, 434, **500–501**
Byssinosis bodies, 501

Cadherins, on pulmonary epithelium, 32
Cadmium exposure, **487–488**
lung cancer and, 1011
Café coronary, aspiration pneumonia and, 357
Calcification
of bronchial cartilage, posttransplant, 749
of lung, **788–790**
clinical features of, 790
dystrophic, 789
following liver transplantation, 759–760
pathologic findings in, 790
of nodules, in varicella pneumonitis, 160
Calcitonin gene-related peptide (CGRP)
in argyrophil cells, 16–17
immunoreactive, 50
Calcium channel blocking agents, pulmonary toxicity of, 555, 571
Calcospherites, in pulmonary alveolar microlithiasis, 787–788
cAMP (cyclic adenosine monophosphate), lung maturation and, 52
Canalicular phase of lung development, 47
Candida, 257
in cystic fibrosis, 91
cytologic identification of, 1226
fungus balls and, 260
lung abscesses caused by, 361
perinatal infections caused by, 102
pneumonia caused by, 268–269
posttransplant infections caused by, 744, 759
pulmonary infiltrates and, 1228
Candida albicans, 267, 584
alveolar hemorrhage and, 871
pneumonia caused by, in polymyositis/dermatomyositis, 820
posttransplant infections caused by, 743
Candida (Torulopsis) glabrata, 267, 269
Candida guilliermondi, 267
Candida krusei, 267
Candida parapsilosis, 267
Candida pseudotropicalis, 267
Candida stellatoidea, 267
Candida tropicalis, 267
Candidiasis, **267–269**
clinical and pathologic features of, 268–269
bronchitis and, 268
fungemia and, 269
pneumonia and, 268–269
diagnosis of, 269
epidemiology and pathogenesis of, 267–268
intravenous drug abuse and, 584
mucocutaneous, in HIV-infected patients, 124
nosocomial, 268

Cantrell's pentad, 87
Capillaria, 340
Capillaria aerophila, 341
Capillariasis, **341**
Capillaries
alveolar
alveolar/capillary block and, **424–425**
dysplasia of, **82**
proliferation of, in adult respiratory distress syndrome, 378, 381
in chronic obstructive pulmonary disease, 623
development of, 49
engorgement of, venoocclusive disease and, 667
functions of, 32
intralipid embolism of, 99
in pulmonary venous hypertension, 674
in radiation injury, 545
Capillaritis
in pulmonary hemorrhage, 867–868
in Wegener's granulomatosis, 839–841
Capillary dysplasia, alveolar, **82**
Capillary endothelium, 709
Capillary hemangiomatosis, 682–684
diffuse pulmonary lymphangiomatosis versus, 947
Capillary thrombosis, in *Chlamydia psittaci* pneumonia, 181
Caplan's syndrome, 465, 468, 479–481
Carbamazepine, pulmonary toxicity of, 576
Carcinoembryonic antigen (CEA)
in adenocarcinoma, 1037
in malignant mesothelioma, 1193–1194
Carcinogens. *See also specific carcinogens*
metabolism of, in lung cancer, 988–989
Carcinoid tumor, **895–911**
acinar, 896
atypical
cytopathologic findings in, 1231
histology of, 903–904
classification of, 896
clinical aspects of, 895–896
cytopathologic findings in, 1231
differential diagnosis of, 910–911
glomus tumor versus, 920
hemangiopericytoma versus, 1082
histology of, 896–905
of atypical carcinoid tumors, 903–904
individual cell death and, 897, 900
of papillary carcinoid tumors, 900
of spindle cell carcinoid tumors, 900
stains for, 904–905
insular, 896
macroscopic pathology of, 896
oncocytic, 897
acinic cell carcinoma versus, 891
granular cell myoblastoma versus, 963
inflammatory pseudotumor versus, 971
mucous gland adenoma versus, 887
papillary, histology of, 900
paraganglioma versus, 914
ploidy in, 905, 908

ISBN 0-07-105448-0

90000>

EAN

9 780071 054485

HASELTON:US PATH LUNG